Advances in Cryogenic Engineering

VOLUME 45, PART A

Advances in Cryogenic Engineering

VOLUME 45, PART A

Chief Editor

Quan-Sheng Shu

AMAC International, Inc.
Newport News, Virginia

KLUWER ACADEMIC / PLENUM PUBLISHERS
New York, Boston, Dordrecht, London, Moscow

The Library of Congress cataloged the first volume of this title as follows:

Advances in cryogenic engineering, v. 1–
New York, Cryogenic Engineering Conference; distributed
by Plenum Press, 1960–
v. illus., diagrs. 26 cm.
Vols. 1– are reprints of the Proceedings of the Cryogenic Engineering
Conference, 1954–
Editor: 1960– K. D. Timmerhaus

1. Low temperature engineering—Congresses. I. Timmerhaus, K. D.,
ed. II. Cryogenic Engineering Conference

TP490.A3 660.29368 57-35598

Proceedings of the 1999 Cryogenic Engineering Conference,
held July 12–15, 1999, in Montreal, Quebec, Canada

ISBN 0-306-46443-8

233 Spring Street, New York, N.Y. 10013

http://www.wkap.nl

10 9 8 7 6 5 4 3 2 1

A C.I.P. record for this book is available from the Library of Congress

Printed in the United States of America

CONTENTS

PART A

PULSE TUBE CRYOCOOLERS (I)

PULSE TUBE CRYOCOOLERS (II)

G-M AND STIRLING CRYOCOOLERS

J-T AND OTHER CRYOCOOLERS

REGENERATORS: DESIGNS, MATERIALS AND ANALYSES

CRYOGENIC SPACE APPLICATIONS

OTHER CRYOCOOLER APPLICATIONS

SUPERCONDUCTING MAGNETS

SUPERCONDUCTING MAGNETS: MATERIALS

SUPERCONDUCTING CAVITIES (I)

SUPERCONDUCTING RF CAVITIES (II)

PART B

He II SYSTEMS

He II PHENOMENA

FLUID MECHANISM AND HEAT TRANSFER (I)

FLUID MECHANISM AND HEAT TRANSFER (II)

LARGE SCALE REFRIGERATORS AND LIQUEFIERS (I)

LARGE SCALE REFRIGERATORS AND LIQUEFIERS (II)

EXPANDERS, PUMPS, AND COMPRESSORS

SC POWER LEADS AND CABLES

CRYOGENIC ELECTRONICS

OTHER SUPERCONDUCTING DEVICES

CRYOSTAT AND THERMAL INSULATION

OTHER NOVEL DEVICES

CRYOGENIC INSTRUMENTS AND CONTROLS

FOREWORD

Scientists and engineers from all over the world arrived at the Palais des Congrès de Montréal Convention Center in Montreal, Québec, for the 1999 Cryogenic Engineering Conference and International Cryogenic Materials Conference (CEC/ICMC), the last conference of the millennium. The participants spanned all regions of the globe, eagerly looking forward to the advancements in cryogenics in the 21st century.

The editorial team is delighted to present to you an **Advances in Cryogenic Engineering** volume consisting of high quality articles representing much enthusiasm in the field of Cryogenics. Volume 45 is comprised of 248 individual papers presented at the 1999 Cryogenic Engineering Conference. Divided into two books, the contents reflect substantial interest in cryocoolers and their relevant space applications. Likewise, this volume presents exciting and novel approaches in superconducting magnets, SRF cavities, and their applications, thus providing a deeper understanding of superfluid phenomena and large-scale refrigeration than in any of the preceding years.

REVIEWING THE EDITORIAL PROCESS

For this volume qualified reviewers evaluated submitted papers on its scientific and practical relevance, as well as its academic presentation, specific to that individual reviewer's area of expertise. This specialized feedback enabled the editorial team to better evaluate each paper, offer more field-specific suggestions, and simultaneously increase the quality of each article. If satisfied with the standards set in the paper, reviewers then passed each paper to a technical sub-editor who decided if the paper would be accepted, rejected, or returned to the authors for further revision. Finally, each paper was sent to the Chief Editor who made a final decision on its suitability for inclusion in the published volume. Even at this stage, a paper could be returned for further editing before inclusion for publishing. However, the Chief Editor's main objective was to improve the quality of the paper and, consequently, the quality of the volume.

Though selection and review is an important part of the process, organizing for the publication is also an involved task. The database for assimilating all the information relating to the papers and the organization of the review system was created by Jin-Xing Yan and Hong-Yu Zhang. Ian King and Loren Loving worked closely with the Chief Editor to produce the final edition of this volume.

Finally, without the Reviewers and co-editors who generously volunteered their time, and the numerous contributions of Kluwer Academic/Plenum Publishers, **Advances in Cryogenic Engineering, *Volume 45*** would not have been published. A debt of gratitude is due to these groups of people.

Quan-Sheng Shu
AMAC International Inc.
Applied Research Center
Newport News, VA

SAMUEL C. COLLINS AWARD

In 1965 the Cryogenic Engineering Conference decided to establish an award in honor of the late Samuel C. Collins, Professor of Mechanical Engineering at Massachusetts Institute of Technology. One of Professor Collins' most notable works is his invention of the modern helium liquefier.

The CEC Awards Committee solicited 29 nominations for the 1999 award. The list was reduced to eight nominees, from which the Committee selected Professor Ralph G. Scurlock as the recipient of the Samuel C. Collins Award. Professor Scurlock is Emeritus Professor of Cryogenic Engineering and former Director of the Institute of Cryogenics and Engineering at the University of Southampton in the United Kingdom.

Professor Scurlock received his B.A. from St. John's College, Oxford University in 1954. Three years later, he received his D.Phil. from Clarendon Laboratory, Oxford University. From 1959–1966 he was Professor of Physics at Southampton University, where he began teaching cryogenics. In 1979 he became the Director of the Institute of Cryogenics, and in 1985 was appointed BOC Professor of Cryogenic Engineering, remaining in that prestigious position until he retired in 1996.

Making the presentation to Professor Scurlock, Professor Klaus Timmerhaus remarked that all have and would continue to benefit from the magnitude of Scurlock's research and teaching. Selecting from his profound research subjects and achievements, Scurlock designated the following as the most important:

1. The development of Cryogenic Fluid Mechanics through the study of convection and heat transfer in cryogenic liquids and vapors. Results of his research are now standard worldwide for the efficient design of containers, dewars, and tanks for the storage and use of cryogenic liquids.
2. Studies of the stable and unstable modes of evaporation of cryogenic liquids under large-scale storage, leading to safe industrial practices.
3. The development of cold electronics instrumentation with 16 bit precision (0.01%) over variable temperatures between 77K and 300K in multi-channel miniature systems.

In the near future, Professor Scurlock visions the most important area of cryogenic research and development over the next 5–10 years as the power application of high temperature superconductors using temperatures reduced to 50K.

Professor Scurlock is the first person outside the USA to have been awarded this prestigious prize, and it reflects the contribution he has made to cryogenic engineering. In accepting the award, Scurlock graciously thanked the CEC board for the great honor bestowed upon him, and he thanked his colleagues—student, academic, and technical—from the Southampton Institute of Cryogenics, United Kingdom.

RUSSELL B. SCOTT MEMORIAL AWARDS

The Russell B. Scott Memorial Awards honor the first head of the Cryogenic Engineering Laboratory of the Boulder Laboratories of the National Bureau of Standards, now the National Institute of Standards and Technology. Mr. Scott was the founder of the Cryogenic Engineering Conference, the first of which was held in 1954 in Boulder, Colorado. He was author of the book: **Cryogenic Engineering.** He retired in 1965 after 37 years at NBS and died in 1967.

The Scott Memorial Awards provide an incentive for the production and presentation of high quality papers at Cryogenic Engineering Conferences, and to provide recognition of authors who, in the judgement of the CEC board, presented the best papers at the preceding conference. The awards consist of a certificate and an honorarium.

The 1999 Scott Awards, for the best papers delivered at the 1997 CEC in Portland, Oregon, and published in **Advances in Cryogenic Engineering, Vol. 43,** were presented at the Awards Luncheon at the 1999 Montreal conference.

Best Research Paper

E. D. Marquardt, R. Radebaugh, and J. Dobak
for their paper
"A Cryogenic Catheter for Treating Heart Arrhythmia"

Best Applications Paper

P. J. Shirron and M. J. DiPirro
for their paper
"Suppression of Superfluid Film Flow in the XRS Helium Dewar"

HONORABLE MENTION PAPER AWARDS

Research Paper

O. Tsukamoto, M. Furuse, T. Takao, N. Tamada, S. Fuchino, I. Ishii, and N. Higuchi
for their paper
"Transient Heat Transfer Characteristics of Liquid Helium in a Centrifugal Acceleration Field"

Applications Paper

A. Lang, H.-V. Häfner and C. Heiden
for their paper
"Systematic Investigations of Regenerators for 4.2K Refrigerators"

STUDENT MERITORIOUS PAPER AWARDS

The CEC board, to encourage the participation of students in the conference, presents student awards and, to provide an incentive for students to write high quality papers, the awards are based on papers submitted prior to the conference. Twelve students were nominated by their research advisors for this award. Each paper was reviewed and edited copies were returned to the students to assist them in completing the final draft of their papers. The board presents student awards based on the merits of their research and written submission.

Student Meritorious Paper Awards

Mr. Zhang Peng

University of Shanghai
"Investigation of Noisy Film Boiling under Various Thermal Conditions in He II"
(P. Zhang, M. Murakami, R. Z. Wang, and Y. Takashina)

Honorable Mention Paper Awards

Mr. Hiroki Nagai

University of Tsukuba, Japan
"Study of Shock Waves and Phase Transition in He II by Using the Superfluid Shock Tube Facility"
(H. Nagai, N. Takuno, H. S. Yang, and M. Murakumi)

Travel Assistance Awards

After reviewing applications from twenty one students and references from their advisors, the CEC Awards Committee selected and gave Travel Assistance Awards to the following six students:

Maoqiong Gong	Cryo. Lab, Academia Sinica, Beijing, China
Khang Dug Le	UCLA, Los Angeles, California
Yan Liang Xi'an	Jiatong University, Xian, China
Hiroki Nagai	University of Tsukuba, Tsukuba, Japan
Laura Savoldi	Dept. of Energy, Politecnico, Torino, Italy
Ian Spearing	University of Victoria, Victoria, Canada

1999 CRYOGENIC ENGINEERING CONFERENCE BOARD

ACKNOWLEDGMENTS

The Cryogenic Engineering Conference Board wishes to thank the sponsors:

Iowa State University,

Institut de recherche d'Hydro-Quebec,

National High Magnetic Field Laboratory at Florida State University,

National Superconducting Cyclotron Laboratory at Michigan State University,

Thomas Jefferson National Accelerator Facility,

U.S. Department of Energy, Division of High Energy Physics,

and

U.S. Department of Energy, Superconductivity Program,

of the 1999 Cryogenic Engineering Conference.
These organizations greatly contributed to the success of the conference.

0.5 W CLASS TWO-STAGE 4 K PULSE TUBE CRYOREFRIGERATOR

C. Wang and P. E. Gifford
Cryomech, Inc.
Syracuse, NY 13211

ABSTRACT

Cryomech, Inc. has developed and commercialized a two-stage 4 K Pulse Tube Cryorefrigerator, the Model PT4-05. Because of the absence of moving parts and cold seals, the PT4-05 has extra lower vibrations, higher reliability and longer meantime between maintenance than the existing 4.2 K options. The PT4-05 is powered by the new Cryomech CP950 Compressor Package; consuming 4.7kW of input power, to provide 0.57 W at 4.2 K on the second stage with 18 W at 65 K on the first stage. The COP at 4.2 K is 1.2×10^{-4}. The PT4-05 reaches minimum temperature, 2.65 K, in 65 minutes. There has been no degradation in the performance of the PT4-05 presently under life test during more than 3,000 hours of operation and 90 cool-down and warm-up procedures. The meantime between maintenance is forecasted to be >20,000 hours.

INTRODUCTION

The unique characteristics of Pulse Tube Cryorefrigerator (PTR) will open new applications for cryorefrigerators due to the absence of displacer motion in the cold head. This characteristic promises greater reliability and lifetime while lowering the costs of operation and manufacturing when compared with GM and Stirling Cycle Cryocoolers.

To reach temperatures below 4 K in a multi-stage PTR , special consideration must be given to the configuration of the stages and to the composition or the magnetic regenerative materials. Matsubara and Gao[2] in Japan first obtained a temperature of 3.6 K in 1994 with a 3-stage pulse tube cryorefrigerator. The net cooling power of their refrigerator at 4.2 K of 30 mW would be too low for the most applications. More recently, prof. Heiden's group (which included the first author) at the University of Giessen developed a two-stage 4 K pulse tube, which obtained a lowest temperature of 2.2 K as well as several hundred milliwatts at 4.2 K in 1997. The group applied this PTR successfully to both liquefy ^{4}He with a rate of 3 liters/day[4] and to conductively cool a 2.8 Tesla superconducting magnet to below 4 K[5].

In this paper, we are reporting on the two-stage 4 K PTR that has been developed at Cryomech. The PT4-05 has achieved the highest COP,1.2×10^{-4} to date 4.2 K of any PTR.

DESIGN OF PT4-05

The two-stage 4 K pulse tube cryorefrigerator includes a pulse tube cold head and a helium compressor package. A new compressor, Cryomech Model CP950, was designed for the pulse tube with regard to a long meantime between maintenance (MTBM). The compressor package supplies the cold head with pressurized helium through flexible metal hoses. A rotary valve in the cold head, similar to that in GM cryorefrigerator, directs the helium gas in and out of the pulse tube system.

1. PT4-05 Cold head

Figure 1 shows a photograph of the PT4-05 cold head. There is a schematic of the pulse tube cryorefrigerator displayed in Figure 2. The two-stage pulse tube cryorefrigerator employs the double-inlet configuration. Two reservoirs, four orifices and a rotary valve control the helium flow in the PTR. They are integrated inside the motor mount assembly at the warm end, which also performs as the room temperature heat sink for the refrigerator. The 1st stage cold plate, which provides cooling power between 40 K to 75 K, thermally anchors the 1st stage cold heat exchanger, the cold end of the 1st stage regenerator and 2nd stage pulse tube. The 2nd stage heat exchanger can provide cooling power below 4 K.

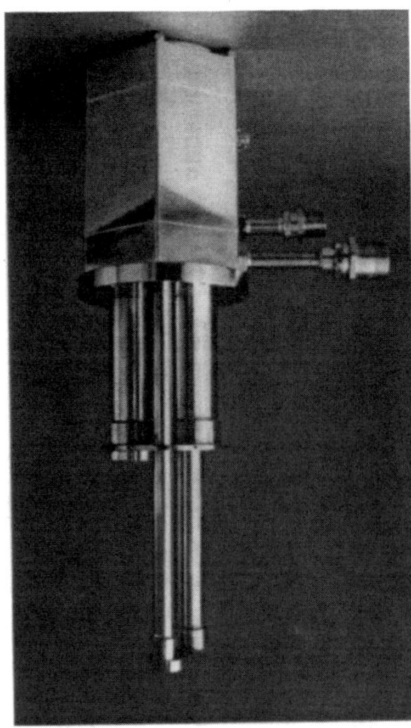

Figure 1. Photograph of the PT4-05 cold head

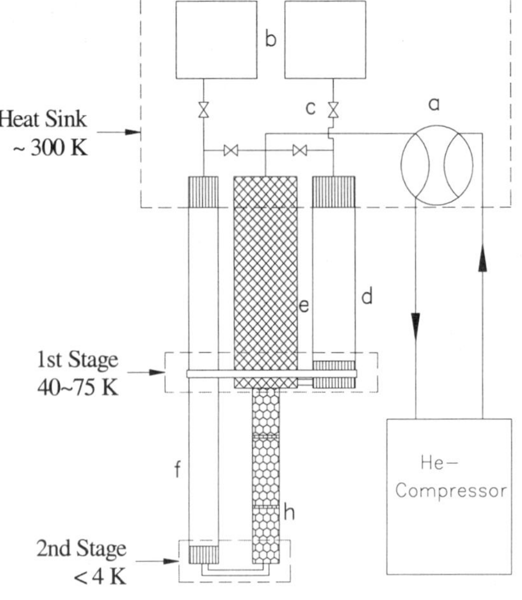

Figure 2. Schematic of PT4-05, a: rotary valve; b: 1st and 2nd stage reservoirs; c: orifices; d: 1st stage pulse tube; e: 1st stage regenerator; h: 2nd stage regenerator; f: 2nd stage pulse tube.

Table 1. Key Features & Benefits of PT4-05

Feature	Benefit
Cold head	
Pulse tube with double-inlet configuration	Extra low vibration, greater reliability and lifetime
Integrated rotary valve and pulse tube	Very compact and reliable, easy coupling
DC stepper motor driven valve	Very quiet operation and low electronic noise
Compressor package	
Helium scroll compressor	Highly reliable, efficient, and low in noise and vibration level
Highly efficient absorber design	MTBM > 20,000 hours

2. New Compressor Package, CP950

MTBM for a normal helium compressor packages is around 10,000 hours, which is too short when compared to that of the pulse tube cold head. Therefore, we developed a new compressor package, the CP950, which employs a helium scroll compressor and an highly efficient absorber with expected lifetime > 20,000 hours.

The key features and benefits of the PT4-05 are listed in Table 1. The DC stepper motor, which turns the rotary valve, is very quiet, and reduces electrical noise. Outside testing of vibration levels have not been able to quantify the level of vibration; because its vibration spectra is below that of the testing facilities environment; which was around 1.0×10^{-2} g. The meantime between maintenance of the PT4-05 is expected to be > 20,000 hours.

PERFORMANCE OF PT4-05

Figure 3 shows the cool-down characteristics of PT4-05. It takes 65 min for the 2nd stage to reach the minimum no load temperature of 2.65 K, and 120 min for the 1st stage to reach the no load temperature of 39 K. The temperature of the pulse tube at 4 K is extremely stable, fluctuating less than 0.2 K which is much smaller than the 1-0.5 K in a typical GM[6]. When the PT4-05 operates with no load, at the minimum temperatures, the input power of compressor is 4.3 kW.

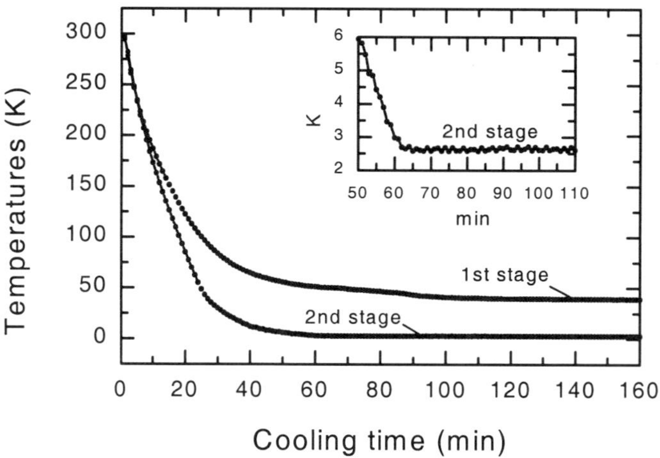

Figure 3. Cool-down performance of PT4-05

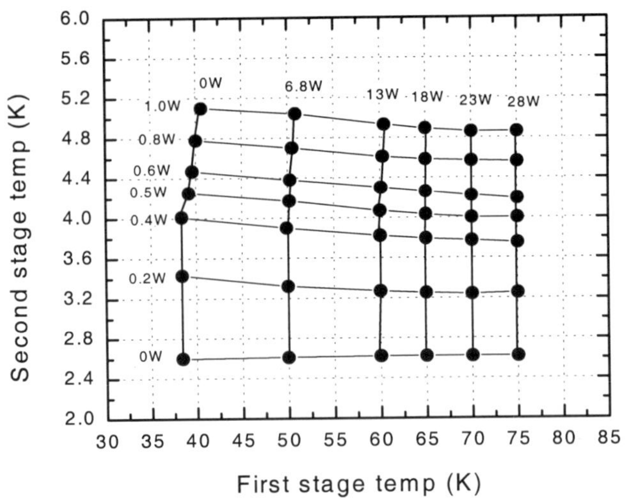

Figure 4. Typical cooling load map of PT4-05

A cooling load map of PT4-05 is given in figure 4. It provides 0.57 W at 4.2 K on the 2nd stage simultaneously with 18 W at 65 K on the 1st stage, while consuming 4.7 kW of input power. When the 1st stage provides 28 W at 75 K the 2nd stage can provide 0.61 W at 4.2 K. The increase in cooling power at 4.2 K upon increasing the 1st stage temperature (heat load) is caused by the increased pressure differential now available due to the higher temperature at the 1st stage. This also increases the compressor input power.

Figure 5 shows the cooling power curve of the 2nd stage, while the 1st stage has 18 W at 65 K. At higher temperatures the PT4-05 has a 2nd stage cooling power of 1.0 W at 4.8 K and 2.0 W at 7.0 K. In Table 2, the performance of PT4-05 is compared to the first two-stage 4 K pulse tube developed at University of Giessen[7]. The COP of the PT4-05 at 4.2 K is 1.2×10^{-4}, which has doubled.

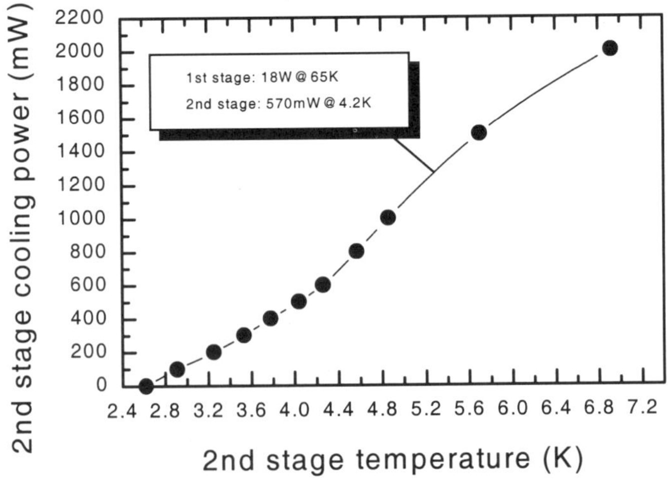

Figure 5. 2nd stage cooling power vs. temperature

Table 2. Performance comparison between two 4 K pulse tube cryorefrigerators

	2-stage PT at Uni. of Giessen[7]	*Cryomech PT4-05*
Input power	6.3 kW	4.7 kW
2nd-stage cooling power	0.42W @ 4.2K simultaneously 20W @ 67K	0.57W @ 4.2K simultaneously 18W @ 65K
COP @ 4.2K	0.6×10^{-4}	1.2×10^{-4}

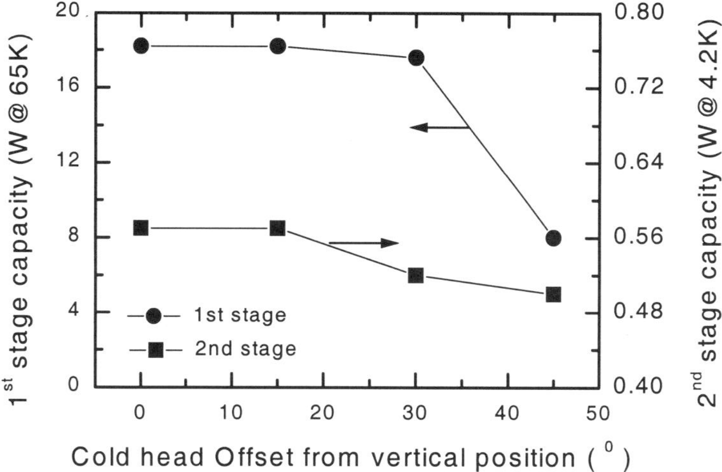

Figure 6. Orientation effects on the cooling performances

The PTR has been tested for orientation affects. It is well documented that tilting a PTR will degrade its performance. This is caused by the natural convection of the helium, and is more prevalent in low frequency pulse tubes. Figure 6 shows the effects of orientation on the performance of the PT4-05. The vertical axes show the cooling powers on the 1st temperature at 65 K and the 2nd stage at 4.2 K. There was no degradation in performance on both stages if the pulse tube cold was tilted 15^0 from the vertical position. For a tilt of 30^0, the cooling power on the 2nd stage drops down to 520 mW from 570mW. For a tilt of 45^0, the cooling power on the 2nd stage decreases to 500 mW, while the 1st stage decreases to 8 W from the 17 W at vertical. Therefore it is suggested that the pulse tube cold head operates as close to vertical as possible. But slight variations from vertical up to 30° will not degrade the performance greatly.

The pulse tube has been cooled down and warmed up more than 90 times, has operated for more than 3,000, and has continuously operated for more than 2,000 hours. Figure 7 shows the temperature stability on the 1st stage and 2nd stage during the continuously operation. There has been no performance degradation on the 2nd stage during the test. The 1st stage temperature slightly increased from 39 K to 40 K because of the variation of environment temperature. The life test of PT4-05 reported on in this paper, started in the past winter and ended this summer.

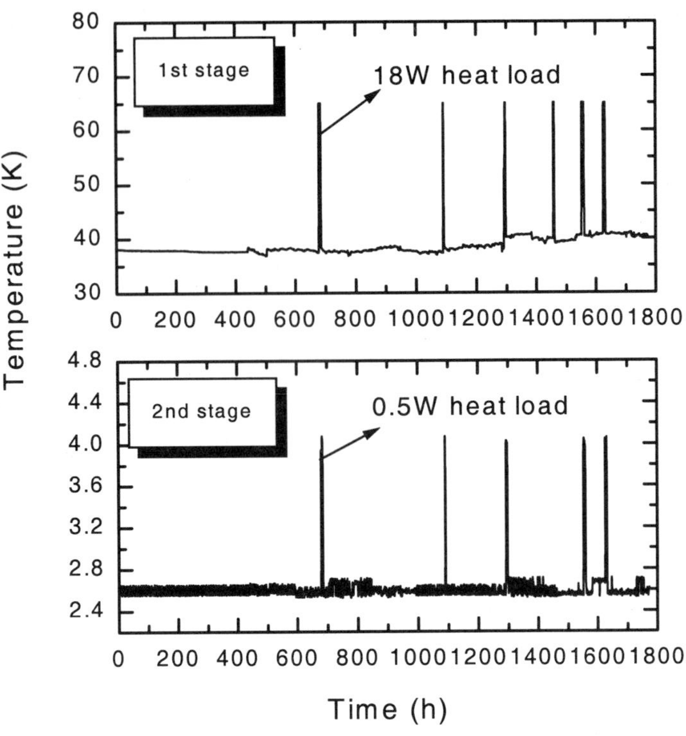

Figure 7. Temperature stability of PT4-05 during long term operation

CONCLUSION

A two-stage 4 K pulse tube cryorefrigerator has been was successfully developed and commercialized at Cryomech, Inc. The pulse tube can provide 0.57 W at 4.2 K on the 2nd stage and simultaneously 18 W at 65 K on the 1st stage, for 4.7 kW of the compressor input power.

The pulse tube cold head has the features of extra low vibration as well as 20,000 hours MTBM, and will open up attractive applications in cooling detectors, superconducting and cryoelectronic devices.

The two-stage pulse tube with cooling power of 1 W at 4.2 K is being developed at Cryomech now. A cooling power of 0.72 W at 4.2 K has already been obtained by using a larger compressor.

ACKNOWLEDGMENT

We would like to thank B. Zerkle and R. Dausman for useful discussions and help during the manufacture.

REFERENCES

1. W.E. Gifford and R.C. Longsworth, Pulse tube refrigeration, *Trans. ASME*, 86:264 (1964).
2. Y. Matsubara and J.L. Gao, Novel configuration of three-stage pulse tube refrigerator for temperatures

below 4 K, *Cryogenics*, 34:259(1994).

3. C. Wang, G. Thummes and C. Heiden, A two stage pulse tube cooler operating below 4 K, *Cryogenics,* 37:159(1997).
4. G. Thummes, C. Wang and C. Heiden, Small scale ^{4}He liquefaction using a two-stage pulse tube cooler, *Cryogenics*, 38:337(1998)
5. C. Wang, G. Thummes, K.-J. Best, B. Oswald and C. Heiden, Use of a two-stage pulse tube refrigerator for Cryogen Free Operation of a superconducting Nb_3Sn Magnet, Presented in ICEC17, Bournemouth, UK, July 14-17, 1998
6. R. Li, A. Onishi, T. Satoh and Y. Kanazawa, Temperature stabilization on cold stage of 4 K G-M Cryocooler, in:"Cryocooler 9", R.G. Ross, Jr., ed., Plenum Press, New York(1997), p.765.
7. C. Wang, G. Thummes and Heiden, C., Performance study on a two-stage 4 K pulse tube cooler, in: *Advances in Cryogenic Engineering*, Vol. 43B, Plenum Press, New York (1998), p. 2055.

1999 CEC Chairman Al Zeller, (Michigan State University), announces the opening of the Conference.

PULSE TUBE DEVELOPMENTS AT CEA/SBT

A. Ravex, I. Charles, L. Duband and J.M. Poncet

Département de Recherche Fondamentale sur la Matière Condensée
Service des Basses Températures
CEA-Grenoble
17, rue des Martyrs
38054 GRENOBLE Cédex 9, France

ABSTRACT

Pulse tube (PT) cryocoolers with no moving pieces in the cold part are potentially able to replace conventionnal Gifford Mac Mahon (GM) or Stirling cryocoolers for most applications. CEA/SBT has undertaken since several years basic and prototype developments of both GM or Stirling like PT. Systemactic studies have been carried out to caracterize d.c flow and inertance effects.

Based on a good comprehension of these phenomena, several prototypes have been built and optimized for various applications. These fundamental and applied research results are presented hereafter.

INTRODUCTION

It is now well known that the most attractive feature of a PT cryocooler is the absence of any moving component in the cold finger. This will allow for easier manufacturing and integration, lower production cost, lower level of induced vibrations. To insure a commercial success of PT cryocoolers by the replacement of traditionnal GM or Stirling cryocoolers in most of their applications, it is of major importance to optimise PT and to demonstrate performances and efficiencies as close as possible to those known for conventional coolers.

With this aim CEA/SBT has undertaken basic work on some of the specific aspects of PT operation such as d.c flow demonstration and cancellation or gas pressure versus velocity oscillations phase shift control by inertance effect.

Owing to the knowledge acquired, we have designed efficient PT prototypes operated either at high frequency (Stirling like) with inertance concept for spaceborne applications or at low frequency (GM like) in a two stage configuration with d.c. flow control.

BASIC DEVELOPMENTS

d.c. flow

A major improvement to PT performances has been introduced by Zhou[1] with the doublet inlet concept. It is generally stated that this configuration with a gas supply to the tube by both cold and warm end reduces the gas mass flow rate through the regenerator, thus increasing its performances and consequently those of the cooler. In fact the most important feature of the double inlet is the achievement of a better phase shift of pressure versus gas flow oscillations at the cold end of the tube.

But Gedeon[2], by theoretical calculations, has predicted that the double inlet configuration may allow for a d.c flow through the tube and regenerator, superimposed on the oscillating flow. Even if this d.c flow amplitude is limited to a small percentage of the oscillating flow amplitude, it can be shown that it results in a parasitic heat input of the order of magnitude of the PT gross cooling power. Duband[3] at CEA/SBT has observed this effect during Stirling like double inlet PT developments. Using an adjustable needle valve for the double inlet impedance control during prototype operation optimisation, he could observe a periodical (with needle valve angular position) performance degradation probably due to assymetries in the needle valve allowing for d.c flow establishment. An experimental evidence was also observed by Chen[4] during the development of a double stage 4 K GM type PT. He observed a large improvement of the performances of its prototype by the fine tuning of a so-called minor orifice (needle valve) which was set between the buffer and the compressor suction line. In fact a controlled d.c flow is created at room temperature by this connection which is used to cancel the parasitic d.c flow in the cold region. This arte-fact is now commonly used [5.6].

The evidence and capability to cancel d.c flow in GM type PT by temperature gradients measurements along both the regenerator and the tube was previously demonstrated and published by CEA/SBT. We have also performed a quantitative measurement of this d.c flow in a Stirling like PT configuration. The experimental arrangement is schematised on figure 1. A one-way valve is introduced on the pressure oscillator to PT connection line to fill a high pressure gas reservoir. The gas from this reservoir is introduced in the PT buffer volume through an adjustable needle valve and a mass flow meter to compensate and measure the parasitic d.c flow.

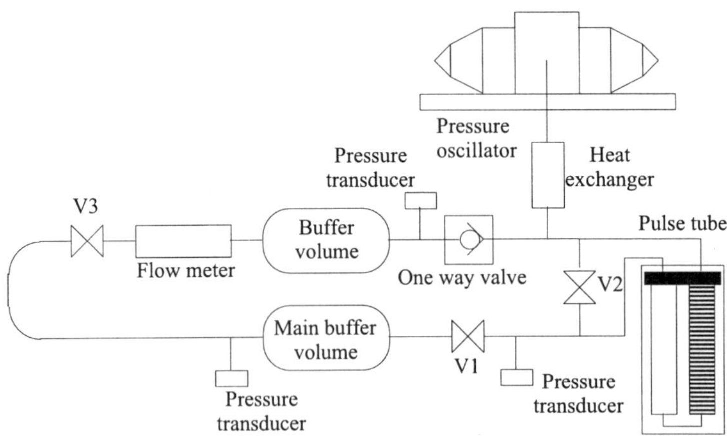

Figure 1 . DC controller set up

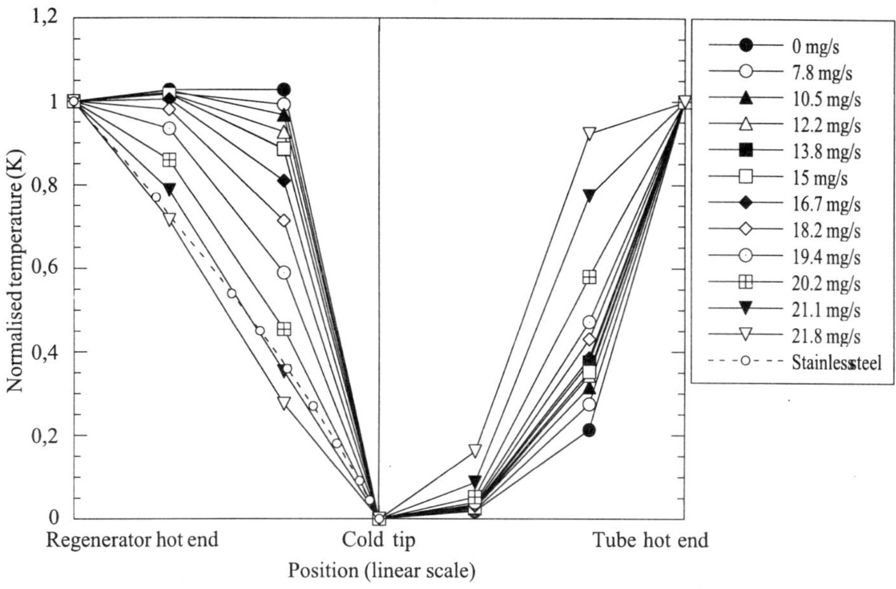

Figure 2 . Influence of the controlled DC flow on the temperature profiles

In figure 2, the temperature profile in both regenerator and tube is reported for various needle valve opening (associated mass flow rates are reported). The best performances of the PT are observed for a roughly linear temperature profile in the regenerator (corresponding to the stainless steel mesh natural temperature profile). The corresponding d.c flow is 21.1 mg/sec (to be compared to the 2 g/sec order of magnitude for the amplitude of oscillating flow at the cold end of the PT.

In conclusion, it can by stated that d.c flow can be clearly identified in double inlet configuration. For GM like low frequency PT, the use of a third orifice between the buffer volume and the compressor inlet or outlet lines (depending on the d.c flow orientation in the cold end) is a simple and efficient way to cancel this parasitic effect. Other methods or artefact can be used to cancel d.c flow such as non symetric opening time of the distribution valve for low and high pressure flow or control of the assymetry of the double inlet impedance.

For the Stirling type PT, d.c flow still remains an important problem to solve in double inlet configuration.

Inertance

As stated before, a beneficient effect of the double inlet is the gas pressure / flow oscillations phaseshift control at PT cold end. Several authors [5, 7, 8, 9] have been shown that this phase shift can be perfectly controlled in a single orifice configuration using an inertance. A simple electrical analogy of the PT cooler provides a straight forward demonstration of this effect.

In this analogy the pressure and mass flow rate correspond respectively to the voltage and current, and the impedance is defined as $Z = \Delta P / \dot{m}$. This analogy is valid if we assume the behaviour is laminar. As in an electrical circuit , the impedance Z can have three contributions : a resistance (pressure drop), a capacitance (dead volume) and a self inductance. In the acoustic community, a self inductance is called an inertance and is typically a pipe. For a capillary of diameter Ø and length l, a tube of internal volume V_0 at temperature T_0, it can be easily shown that for a laminar flow the equivalent self L and associated resistance R, and the capacitance C are respectively given by:

$$L = \frac{4l}{\pi Ø^2} \qquad R = \frac{128.l.\eta}{\pi.\rho.Ø^4} \qquad C = \frac{M.V_0}{\gamma.R.T_0}$$

where η, ρ and M are respectively the gas viscosity, density and molar mass, and γ is the ratio of isobaric to isochoric specific heats. In some cases (regenerator for example) the compression can be assumed to be isothermal.

An equivalent circuit for a pulse tube operated in orifice mode is shown in figure 3. The pressure oscillator is represented as a voltage source, the regenerator is represented as a resistance R_r (pressure drop in the regenerator matrix) in parallel with a capacitance (void volume) C_r. The pulse tube and buffer volume are pure capacitances C_t and C_b. "A" features a current amplifier, to take into account the temperature difference between the cold and warm end of the PT, and Z_1 is the impedance between the tube and reservoir (buffer). This circuit has been intentionally simplified for the purpose of the demonstration; the regenerator is subjected to a temperature gradient and thus it would be more accurately described by for instance a series of capacitances and resistances in parallel. In addition we are operating at high frequency and the buffer volume is fairly large so the term C_b will be neglected in the following. Using this simple approach we can evaluate the phasing between the cold mass flow rate and the oscillating pressure. From the above equations and the electrical junction rules and using complex impedances, the current and voltage, or mass flow rate and pressure, are described as :

$$\dot{m}_c = \alpha.\dot{m}_a + \dot{m}_t \text{ and } \Delta P = P - \bar{P} = Z_t.\dot{m}_t = \frac{1}{j.C_t.\omega}\dot{m}_t \text{ then } \dot{m}_c = \alpha.\dot{m}_a + j.C_t.\omega.\Delta P$$

where α is a coefficient which takes into account that the mass flow rates are at a different temperature and ω = 2πf, f = frequency of operation.

$$\dot{m}_a = \frac{\Delta P}{Z1} \text{ where } Z_1 = R_1 \text{ or } Z_1 = R_1 + j.L.\omega$$

for respectively a resistance or a resistance associated in series with an inertance.

Using the Fresnel geometrical construction and assuming P is on the x axis, we found the results displayed in figure 4. If the orifice is assumed to be a simple resistance R_1 (orifice, valve), we can see (figure 4a) that the phase shift at the cold end of the PT is never the optimal value (zero). This is due to the capacitance effect in the pulsation tube (C_t ω ΔP). This effect can significantly affect the performance. In order to counter balance the capacitance effect, it is necessary to introduce at the orifice level a self inductance L (inertance). As shown in figure 4b a "zero" phase shift between the cold mass flow rate and the pressure oscillation ΔP can be restored. One can easily show that if the capacitance of the buffer volume, C_b, is taken into account, the phasing due to the addition of the capacitances is further shifted and in this case the use of inertances is even more beneficial.

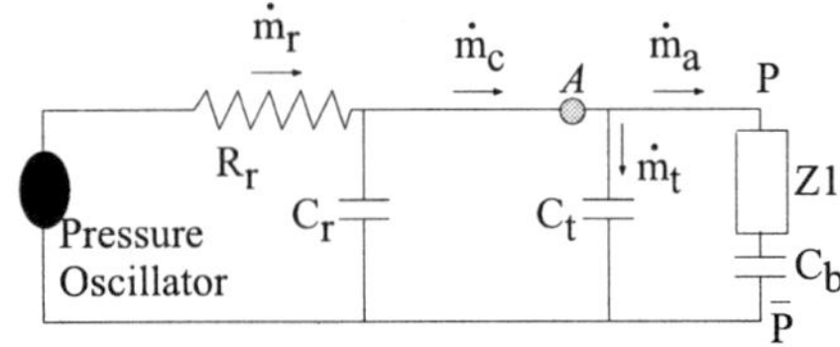

Figure 3 . PT orifice mode - equivalent electrical circuit

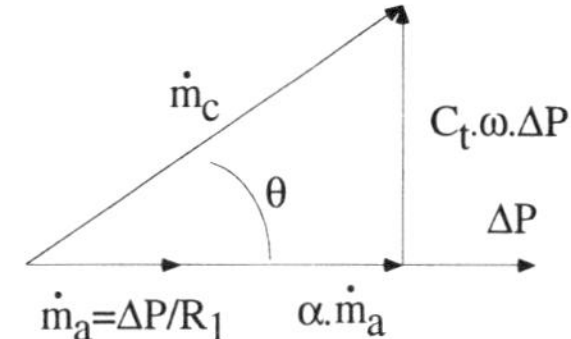

Figure 4 a . Z1 purely resistive

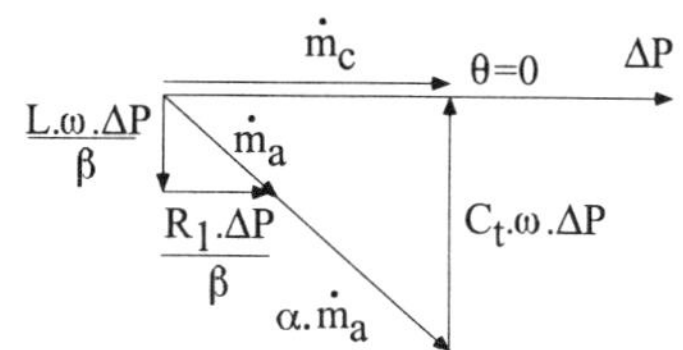

Figure 4 b . Z1 resistive + inertance

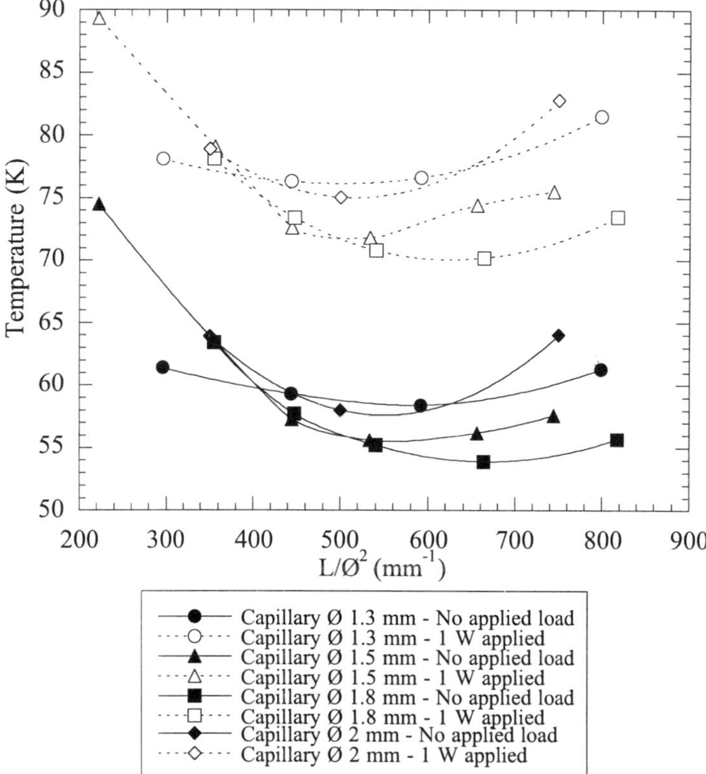

Figure 5 . Effect of the inertance term on the performance (orifice mode - 2 MPa, 45 Hz)

It is now evident that the impedance to control the flow between the pulse tube and the surge volume must include an inertance term.

In practice there is probably no equivalent to a pure self inductance; we use capillaries which are the association of a resistance (pressure drop) and a self inductance. In this case a compromise has to be found between the length l and the diameter Ø to fulfil simultaneously the inertance and resistance values required by respectively the phase shift control at the cold end and the mass flow rate in the surge volume.

All these aspects are clearly shown experimentaly. We have reported for illustration in figure 5 the influence of the inertance on the performance of a PT cooler (temperature for 1 W net cooling power and ultimate temperature at a given frequency for various diameter and length of the capillary used as inertance).

It is clear that the use of inertance avoids the etablishment of d.c flow and allows for an optimal adjusment of the gas pressure versus flow oscillations phase shift.

DEVELOPMENT OF PROTOTYPES

Stirling type PT cooler

In the framework of an European Space Agency (ESA) contract, in collaboration with Matra Marconi Space (MMS) and Rutherford Appleton Laboratory (RAL), a PT cooler for space purpose associated with a diaphragm springs oscillator (developed previously for a Stirling cooler : 2 x 3.4 cm^3 swept volume) has been developed.
This PT was initially designed to be operated in a double orifice mode. The d.c flow problems encountered during this program have lead us to operate it finally in an inertance mode. This prototype is shown on figure 6. It presents specific features (U shaped, prototype made of titanium, EDM channels for heat exchangers, electrical isolation of the cold finger using a specific insulating ceramic junction). Overall dimensions are ∅ : 72 mm and h : 200 mm for an overall mass of 800 grams. Inertance is incorporated in the buffer volume. The prototype has been successfully submitted to vibration tests, its lowest resonant frequency was found to be around 200 Hz. Exported vibration and thermal environment tests have also been performed. All these results are reported in an ESA report[10].

An ultimate temperature of 63 K and a temperature of 82 K with a 1 Watt net cooling power have been obtained in the following conditions :

Operation frequency : 45 Hz
Fill pressure : 2 MPa
Pressure oscillator strokes : 80% of nominal
Heat rejection temperature : 306 K

Recently, strongly improved performances have been obtained with a new in line design of the PT cooler optimised for inertance operation (53 K ultimate temperature, 1.7 Watt at 80 K). Figure 7 reports the evolution of the performances obtained during the ESA development program. To day a PV work of 23 Watt per Watt of cooling power is obtained at 80 K. More work is needed now to modify the resonance caracteristics of the MMS pressure oscillator which was previously developed for a Stirling cooler cold finger.

Figure 6 . Prototype pulse tube (partly assembled)

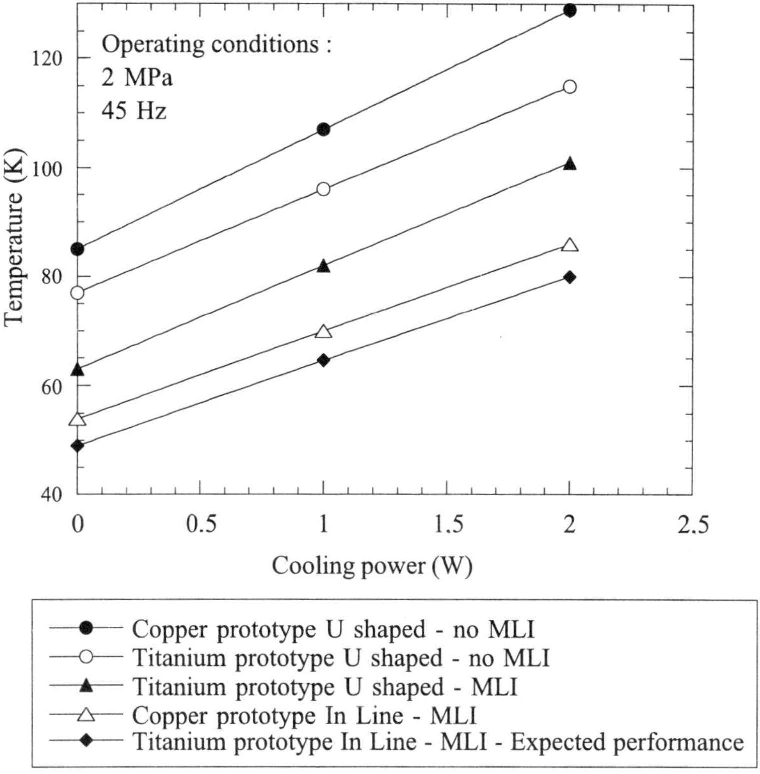

Figure 7 . Performance improvment of ESA prototypes

GM type PT coolers

At previous CEC[5] in Portland, we have reported results on a single stage GM type PT (ultimate temperature : 26 K, cooling power at 80 K : 100 W, operation frequency : 2 HZ, compressor : CTI 8500/5 Kw).

New prototypes have been extrapolated from this initial development the performances and applications of which are reported in table 1.

Among them, a single stage PT has been modified (addition of lead shots to stainless steel mesh in the regenerator) to achieve large cooling power in the 30 K/50 K temperature range for future high transition temperature superconducting materials cooling.

A double stage prototype (figure 9) has also been developed aimed to provide large cooling power at both the first (80 K) and second (20 K) heat stations.

The overall geometrical caracteristics of this prototype are as follow :

- overall length (300 K flange - second stage heat station) : 325 mm
- overall diameter : 95 mm

This prototype has been tested with 1 and 2 Leybold. Coolpak (6 Kw) compressors at an operation frequency of 2 Hz.

The performances are reported in figures 8a and 8b and are summarised for the usual 20 K / 80 K temperatures in table 2.

Table 1. Single tage GM type PT developed at CEA/SBT

Performances	Compressor	Comments
Tmin = 26 K 100 W/82 K 100 W/140 K	Leybold RW 6000 (6 Kw) CTI 9600 (5 kW)	. U-shape – not really optimised : compressor/tube volumetry . Detector cooling for telescope . Operated at 5 Hz in any orientation (ΔT < 5 K) . Valve/PT : 50 cm
20 W/45 K	CTI 9600 (5 Kw)	. Detector cooling . Valve/PT : 100 cm
25 W/40 K or 120 W/100 K	Leybold coolpak (6 Kw)	. Helium cryostat thermal shielding . Valve/PT : 50 cm
Tmin = 17 K 30 W/30 K 70 W/50 K	Leybold coolpak (6 Kw)	. Increased regenerator with lead shots

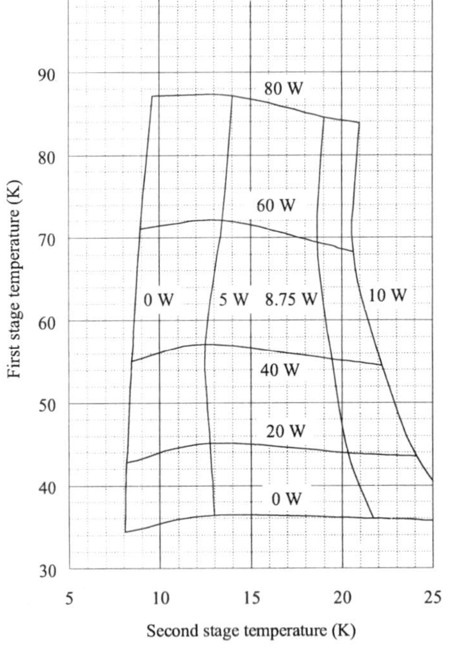

Fig 8a. Double stage prototype performances with 1 x 6 kW LEYBOLD compressor

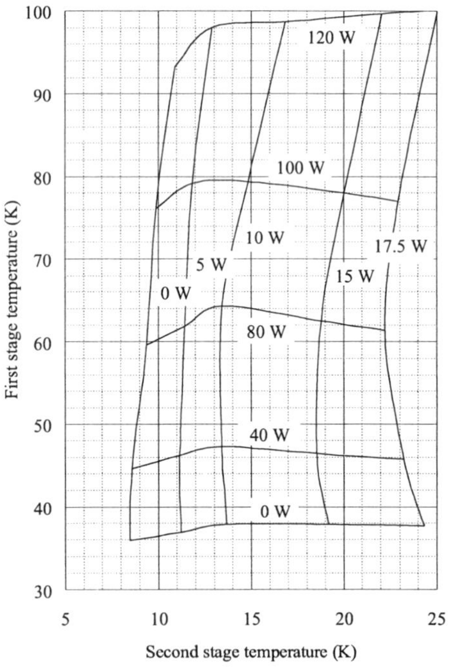

Fig 8b. Double stage prototype performances with 2 x 6 kW LEYBOLD compressors

Table 2. Double stage prototype performances

		1 x Leybold Coolpak (6 Kw)	2 x Leybold coolpak (2 x 6 Kw)
Ultimate	1st stage	34.5 K	36 K
Temperature	2d stage	8 K	8.5 K
Simultaneous	80 K	75 W	100 W
Cooling power	20 K	9 W	14.5 W

It can be observed that the performances obtained with a single compressor are excellent and equivalent to those of the corresponding GM coolers.

Incorporation of magnetic materials in the second stage regenerator are in progress and performances at 4 K will be soon available.

CONCLUSION

Specific behaviours of PT coolers such as d.c flow in double inlet configuration, inertance in single orifice configuration have been experimentaly studied.

Based on the comprehension of these phenomena, CEA/SBT has developed several prototypes either of Stirling or GM type. GM type prototypes of both single or double stage architecture have been built and optimised.

Experimental results are very encouraging. For several applications, prototypes have proven to be well suited.

A collaboration with industrial partners is now necessary to introduce commercially PT coolers on the traditionnal markets of GM or Stirling coolers.

Figure 9. Double stage GM like PT prototype on test bench

REFERENCES

1. S. Zhu, P. Wu and Z. Chen, Double inlet pulse tube refrigerators : an important improvement, *Cryogenics*, vol. 30 : 514 (1990).
2. D. Gedeon, DC gas flows in Stirling and pulse tube cryocoolers, in : "*Cryocoolers 9*", R.G. Ross Jr., ed., Plenum Press, New York (1997), p. 385.
3. L. Duband, I. Charles, A. Raex, L. Miquet and C. Jewell, Experimental results on inertance and permanent flow in pulse tube coolers, in : *Cryocoolers 10"*.
4. G.Chen, J.Zheng, L.Qiu, X.Bai, Z.Gan,P.Yan,J.Yu, T.Jin and Z.Huang, Modification test of staged pulse tube refrigerator for temperatures below 4K, *Cryogenics,*vol.37 :529-532 (1997)
5. A. Ravex, J.M. Poncet, I. Charles and P. Bleuzé, Development of low frequency pulse tube refrigerators, in : "*Advances in Cryogenics Engineering*", Plenum Press, New York, vol 43A :1957 (1997).
6. C.Wang, G.Thummes and C.Heiden,Control of DC gas flow in a single-stage double-inlet pulse tube cooler, *Cryogenics*, vol.38 : 843-847 (1998).
7. S.W. Zhu, S.L. Zhou, N. Yoshimara and Y. Matsubara, Phase shift effect of the long neck tube for the pulse tube refrigerator, *"Cryocoolers 9"*, Plenum Press, New York (1997), p. 269-278.
8. D.L. Gardner and G.W. Swift, Use of inertance in orifice pulse tube refrigerators, *Cryogenics*, vol. 37 : 117-121 (1997).
9. P. Roach and A. Kashani, Pulse tube coolers with an inertance tube : theory, modeling and practice, in *"Advances in Cryogenics Engineering"*, Plenum Press, New York, vol.43A :1895 (1997).
10. L. Duband, Development of a pulse tube 50-80 K cryocooler, ESA Contract # 11331/95/NL/FG. Final Report (1999).

CEC Awards Committee Chairman, Klaus Timmerhaus (University of Colorado), announces the award winners.

DEVELOPMENT OF LOW-COST PULSE TUBE CRYOCOOLER FOR HTS APPLICATIONS

S-Y Kim, K-Y Hong, S-T Kim, W-S Chung

LG Electronics Inc. Home Appliance Lab.
327-23, Gasan-Dong, KeumChun-Gu,
Seoul, 153-023, Korea

ABSTRACT

LG Electronics (LGE) has developed a pulse tube cryocooler (PTC) for HTS Applications such as cooling HTS RF filters in wireless communication systems. This paper reports on the development of the compressor and the pulse tube. The compressor in combination with an in-line pulse tube is based on a moving-magnet design combined with flexure supports that is both inexpensive and highly reliable. This approach aims for reducing production costs while maintaining the long life.

The PTC shows 4.9 W of cooling power at 65 K and 270 W of input power with a single acting compressor and better price-to-performance ratio. The key component of the PTC is a high efficiency of over 90% and low-cost linear motor. Our target cost is $1,000 for quantities of 10,000 units per year. This paper also presents predicted production costs for each of the parts in that volumes.

INTRODUCTION

LG Electronics is producing over 10 million compressors a year for air conditioners and refrigerators. The moving magnet type of linear motor has been recently developed for new type of compressor. Though the linear motor has a lot of benefits in efficiency and in the moving mechanism, the uses are very limited due to the relatively high production cost. The linear motor developed by LGE H/A Lab. has high efficiency and proper shape to be produced massively with low-cost. In addition, LGE H/A Lab. had tried to commercialize the Stirling Cycle for about 10 years before the PTC development.

A PTC has a lot of advantages over Stirling cryocoolers because it has no moving parts in the low-temperature region. This means much more reliable operation and much lower vibration in the cold region. Ideally, Stirling cryocoolers have better thermal efficiency than PTC. However, some recent studies [1,2,4] have suggested that PTC using 'inertance' tube can generate the phase shift needed to make PTC operate as good as Stirling cryocoolers. The inertance tube is a long, thin tube that allows the phase between the pressure and mass flow in the pulse tube to be adjusted to an extent that was not previously possible. The total heat load on the PTC developed by LGE when operating in 20°C ambient is 4.9 W at 65 K and the input power of the compressor is 270 W.

The design of the PTC was driven by the challenging low cost and reliability target to fulfill the requirements of the wireless company. M-CALC II report [5] suggested the goal of life and cost for the low-cost commercial cryocooler. The 40,000 hours continuous duty life limits the design options available and the cost goal of $1,000 for quantities of 10,000 a year for the complete cryocooler means that this cooler should achieve the goals that other cryocoolers have not yet met. To reduce the cost of PTC, we should decrease the cost of compressor because the cost of pulse tube is mainly due to the material cost. The compressor cost reported [3,4] is over 50% of the total system cost and the total cost of PTC is far from the cost goal. In our cost studies, the cost of compressor is 40% of the total cost of PTC for quantities of 10,000 units per year.

COMPRESSOR

The low-cost compressor consists of the main 3 parts: a clearance-sealed piston, spiral flexure linear bearings, and a moving magnet type of linear motor. These are enclosed in a common pressure vessel with an aftercooler interface to the pulse tube. The layout of the compressor in combination with in-line pulse tube is shown in figure 1. The layout is for the water-cooled prototype version. The production version is similar but uses an air-cooled aftercooler. A photograph of the PTC is shown in Figure 2. The pulse tube is enveloped with vacuum vessel. This unit has a motor diameter of 130 mm and a total prototype vessel dimension of 170 mm diameter including vessel flange and 215 mm long including aftercooler. The compressor mass is 10 kg and the total mass is 12 kg. Production vessels are intended to be welded shut, not bolted. This will save some mass.

Linear Motor

The moving magnet motor consists of an outer stator with coil, an inner stator, and a magnet assembly on which magnets attached. A magnet assembly is connected to the reciprocating shaft supported on the each end by flexures and connected to the piston. Moving magnet motors remove the need for problem-prone flexing leads, so it simplifies the structure greatly. The motor is designed for max. 400 W mechanical output and max. 20 mm of stroke at 220 VAC. The production version uses DC power with inverter electronics to convert DC to a sinusoidal waveform, so that it can be supported by battery backup system. In this project, the motor was tuned at 60 Hz by adjusting the moving mass and springs.

The high efficiency linear motor with radial-laminations stacked automatically can be produced as easily as a traditional motor in the mass production line and uses a very small Nd-Fe-B magnet.

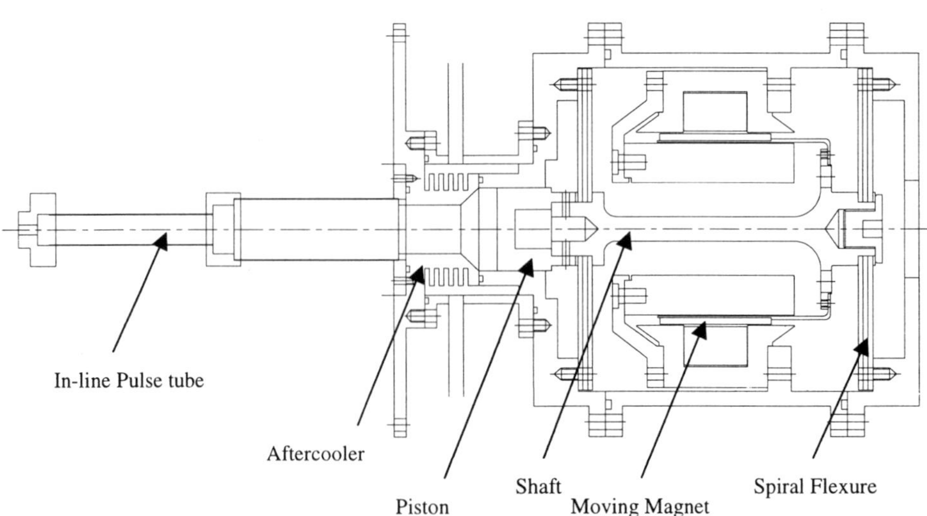

Figure 1. Layout of low-cost pulse tube cryocooler

Figure 2. Photograph of low-cost pulse tube cryocooler

Assembly and Alignment

The piston moves without rubbing contact in a cylinder to produce the pressure wave that drives the pulse tube. The clearance seal is maintained in a centered radial position by flexures and the radial clearance is under 25 μm. The clearance seal requires so tight tolerances that this high-precision alignment is too expensive for mass production. Therefore one of the primary concerns is to considerably reduce the cost of alignment. This is accomplished by reducing the high-precision machining process, devising a simplified mechanical architecture with fewer parts, and simplifying the alignment of piston supports. The front flexures combined with a piston and a shaft are bolted to cylinder assembly maintaining the piston in the center of the cylinder and the backside flexures bolted to a backside frame are combined with a shaft and a magnet assembly maintaining the pre-alignment. This robust and inexpensive design shows small amount of friction loss of delivered PV power.

PULSE TUBE

The LGE pulse tube has an in-line pulse tube with an inertance tube to provide the proper phase shift. Stacked brass 100 mesh screens are used in the cold end, warm end and aftercooler, and Stainless steel 400 mesh screens are used in the regenerator. The PTC uses 25 mm inner diameter thin stainless steel tube for regenerator, 12 mm for pulse tube and 3 mm long copper tube for inertance tube. The reservoir volume is 0.3 liter and the mean pressure in the compression space is 2.7 MPa with 0.5 MPa of pressure amplitude.

PERFORMANCE MEASUREMENTS

Figure 3 shows the heat load curve for the PTC with no multi-layer insulation and 10^{-4} Torr in the vacuum vessel. This test is conducted on the water-cooled version and cooling water temperature in this test was maintained at 20°C. The curve shows the 4.9 W at 65 K and no heat load temperature is 50 K. With no multi-layer insulation, there should be increased radiation heat loss, so we can expect somewhat better performance in better insulation conditions.

The compressor PV work is estimated by measuring the piston position and the pressure in the compression space. The PV work is 240 W in the 270 W of input power and the PV efficiency is 89%. Due to some losses such as gas leakage, friction and heat transfer, we can estimate the motor efficiency at over 90%.

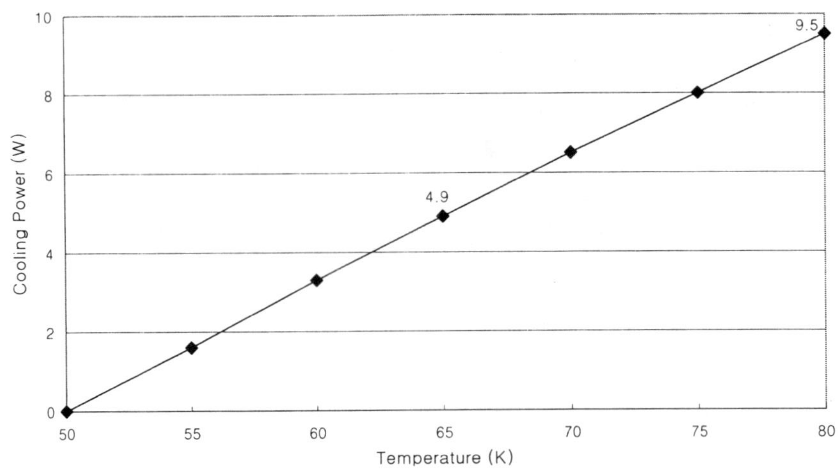

Figure 3. Cooling power at cold end

COST STUDIES

The cost studies conducted on this development were focused on the compressor, since its reported cost [3,4] is a major portion of the total system cost, and the reduction of parts and assembly cost is a key factor for success in the low-cost cryocooler manufacturing. Cost studies for the design were developed from complete engineering drawings and quotations of vendors.

Figure 4 shows our cost portion for each parts at these productions. The projected cost of the complete system is near $1,000 for quantities of 10,000 units per year, so we believe strongly the cost goal suggested [5] could be met. The cost of compressor is 40% of the total system cost and relatively low compared to other studies [3,4] due to low-cost linear motor and the robust and inexpensive assembly and alignment process. The cost of linear motor was reduced considerably by the design of LGE patent. In addition, the applications of the linear motor are so wide that its cost can be decreased with the increase of production quantities. The cost of pulse tube including an aftercooler and a reservoir is relatively high because the cost of materials like screens and tubes and tube brazing is very high in Korea. The electronics is an inverter to convert DC power to a sinusoidal wave form to drive the compressor. The labor & etc is usual labor fee in our factories and includes system test cost.

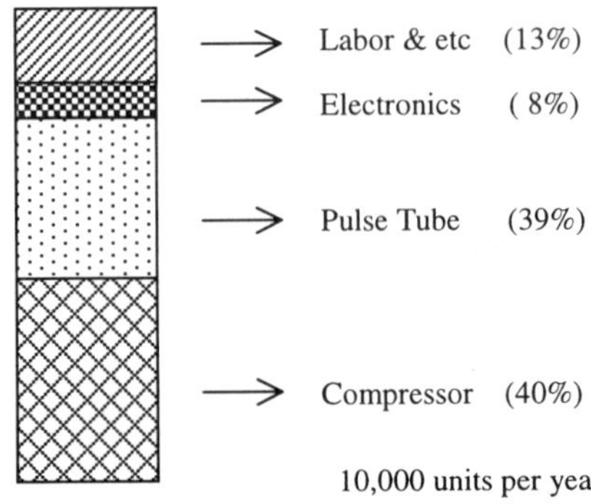

Figure 4. Production cost for the each parts of the PTC

SUMMARY AND CONCLUSIONS

The LGE aims for the commercialization of the low-cost PTC to be used widely for HTS applications, although the PTC has been developed specifically for the HTS RF filter in the wireless systems because it is an emerging market currently. The PTC developed has the potential for very low cost and high reliability. With performance measurements and cost studies, this PTC has better price-to-performance ratio than any other cryocoolers with comparable cooling power. In addition, these results indicate that the PTC will be able to meet the challenging cost goals suggested by M-CALC II [5], although some further work is needed to meet the long life time.

ACKNOWLEDGEMENTS

The authors would like to thank Mr. Seong-Je Park of Korea Institute of Machinery and Materials for the support and advice of his team.

REFERENCES

1. D.L. Gardner, and G.W. Swift, "Use of Inertance in Orifice Pulse Tube Refrigerator", *Cryogenics*, Vol. 37 (1997), pp. 117-121.
2. Pat R. Roach and Ali Kashani, "Pulse Tube Coolers with an Inertance Tube: Theory, Modeling and Practice", *Advances in Cryogenic Engineering*, Vol. 43, Plenum Press, New York (1998), pp. 1895-1902.
3. T. Nast, P. Champagne, and V. Kotsubo, "Development of a Low-Cost Unlimited-Life Pulse-Tube Cryocooler for Commercial Applications", *Advances in Cryogenic Engineering*, Vol. 43, Plenum Press, New York (1998), PP. 2047-2053.
4. J.L. Martin, J.A. Corey, and C.M. Martin, "A Pulse Tube Cryocooler for Telecommunications Applications", Cryocoolers 10, Plenum Press, New York (1999), pp. 181-189
5. M. Nisenoff, "Cryocoolers for Electronic Technologies", M-CALC II Workshop Report, San Diego, 1998.

CEC Board of Directors:
Front row left to right: A. Zeller (Michigan State University), R. Witt (University of Wisconsin), S. Breon (Goddard Space Flight Center), J. Pfotenhauer (University of Wisconsin), Q.S. Shu (AMAC International), C. Joshi (Energen Inc.).
Back row left to right: J. Hull (Argonne National Lab), P. Gifford (Cryomech), P. Kelley (Los Alamos National Lab), D. Coffey (Cryomagnetics), J. Theilacker (Fermilab), K. Gschneidner Jr. (Ames Lab).

A CONTAMINANT ICE VISUALIZATION EXPERIMENT IN A GLASS PULSE TUBE

J. L. Hall,[1] R. G. Ross, Jr.,[1] and A. K. Le[2]

[1]Jet Propulsion Laboratory
Pasadena, CA 91109
[2]Department of Chemical Engineering, UCLA
Los Angeles, CA 90095

ABSTRACT

Results are presented from pulse tube experiments designed to investigate the effect of 400 parts per million water vapor contamination of the helium working gas. The experiments were conducted in a glass pulse tube to enable visualization of ice formation on internal surfaces. Photographs of this ice formation were taken along with simultaneous coldtip temperature and compressor power measurements. Four types of regenerator elements were tested in various combinations: 200- and 400-mesh stainless steel screens, 1.6 mm diameter glass beads, and 1.6 mm thick perforated plastic plates. Internal spacers were also used to provide clear fields of view into the regenerator stack. Substantial water-ice formation was observed at the cold end of the regenerator and on the inside wall of pulse tube; it appeared to be highly porous, like snow, and was seen to accumulate only in a very localized region at the coldest end, despite changing the cold tip temperature across a range of 150 to 235 K. Ice formation degraded pulse tube thermal performance only in cases where screen regenerators were used at the regenerator cold end. It was concluded that flow blockage was the mechanism by which contaminants affected performance; coarse regenerator elements were largely immune over the tested time scale of a few days. Substantially reduced ice formation and minimal performance loss were also observed in repeated tests where the contaminated gas was reused after warming up and melting of the accumulated internal ice. Significant adsorption of the liquid water onto the regenerator was inferred, a process that depleted the gas phase concentration of water.

INTRODUCTION

Gas phase contaminants inside pulse tube cryocoolers will freeze. If enough frozen material accumulates, there will be a loss of overall thermal performance due to one or more possible mechanisms such as flow blockage, internal geometry change, or increased parasitic heat conduction. The source of these contaminants can be either impurities in the helium charge gas, or outgassing of internal cooler surfaces and materials. Both sources are typically present in actual systems with the relative importance a function of gas purity, cleaning procedures, materials of construction, and seal performance.[1] A recent investigation[2] focused on the charge gas impurity aspect of the problem and found modest perfor-

Advances in Cryogenic Engineering, Volume 45.
Edited by Shu *et al.*, Kluwer Academic / Plenum Publishers, 2000.

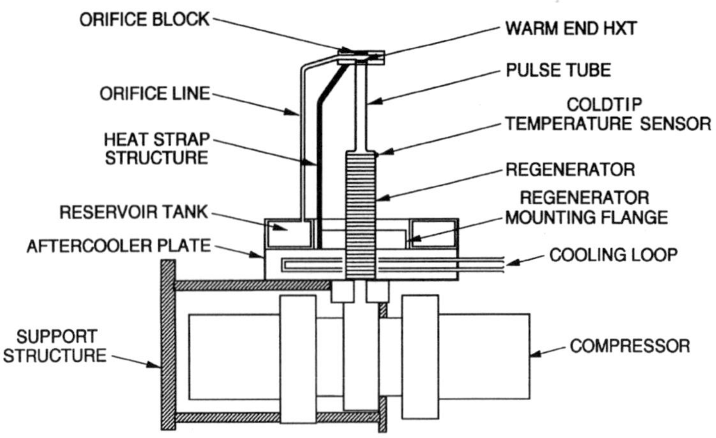

Figure 1. Schematic Diagram of Glass Pulse Tube.

mance losses due to water, carbon dioxide, argon and nitrogen contaminants at levels ranging from 10 to 100 ppm per component. The observed performance loss occurred over a time scale of approximately one week, a value that was estimated to be consistent with the gas diffusion time scale inside the pulse tube used for those experiments. Before and after measurements with a mass spectrometer did not reveal any appreciable accumulation of contaminants due to outgassing of internal surfaces over the one-week test period.

One of the limitations of the previous investigation was the lack of visual observations of contaminant ice formation inside the metal pulse tube. Thermal performance loss was determined from temperature and power consumption measurements, but it was not possible to deduce where the ice was forming or what specific physical mechanism was responsible for the performance loss. The present investigation was designed to address this problem by conducting experiments in a glass pulse tube that allowed direct observation of internal ice formation. The overriding intent was to enhance the visualization aspect of the experiment even at the cost of refrigeration performance. In practice, the use of a glass tube and the absence of multilayer insulation limited the coldtip temperatures to 150 K as opposed to the 60 K temperatures achieved in the earlier study.

EXPERIMENTAL APPARATUS

The experiments were performed with a simple orifice pulse tube mounted on a 5-cc Lockheed two-piston back-to-back linear compressor (Fig. 1). The pulse tube consisted of a quartz glass tube with stainless steel end connections. The glass to metal interfaces consisted of flared, grooved joints that were clamped together and sealed with rubber O-rings. The 1.6-mm wall thickness glass tube was comprised of two sections, a 19 mm O.D. lower part that housed the regenerator, and a 10 mm O.D. upper part that was the pulse tube itself. The two tubes were fused together with a smooth shoulder joint. An “L”-shaped copper busbar was used to thermally connect the orifice end of the pulse tube to the cooled reservoir plate. Indium shims placed between the copper bus bar and the pulse tube served the dual purposes of improving the junction heat transfer and increasing the pre-load on the stack to enhance the compression, and hence performance, of the O-rings.

The system was typically operated in the range of 36 to 40 Hz with no applied heat load. A fixed heat rejection temperature of 0°C was maintained by an external fluid-loop chiller attached to the pulse tube mounting plate. A gas pressure of 1.5 MPa (200 psig) was used in all tests. Two different gas compositions were employed: a “clean” gas comprised of five-nines pure helium, and a “contaminated” gas comprised of helium with 400 ppm of water vapor. The corresponding partial pressure of water at this concentration and pressure is 4 torr, which corresponds to a dew point of 275 K.[3]

Temperature data were provided by cryodiodes bonded onto the outside of the glass tube at three locations. The middle location, just below the shoulder joint, was defined to be the coldtip temperature for these experiments (see Fig. 1). Compressor input power was measured with a power analyzer and recorded by hand at irregular intervals.

Two different vacuum enclosures were used for these experiments. Initially, a small acrylic cylinder was bolted onto the pulse tube base plate and sealed with a Viton O-ring. This arrangement provided good optical access; however, outgassing of the acrylic limited the achievable vacuum levels to approximately 3×10^{-5} torr. This limitation became unacceptable once pulse tube coldtip temperatures of approximately 205 K or less were achieved because residual water vapor began to freeze on the outside of the pulse tube and obscured the internal view. The addition of a crude liquid nitrogen cryopump to the acrylic chamber succeeded in lowering the vacuum level to 1×10^{-6} torr, but this was still not sufficient to prevent external frost at the lowest coldtip temperatures. Therefore, some tests were conducted with the entire pulse tube placed in a 0.45 m diameter glass bell jar. A combination of a 1-stage Gifford-McMahon cryopump and mechanical roughing pumps succeeded in generating vacuum levels as low as 6×10^{-7} torr, sufficient to prevent all external icing. However, the glass bell jar was not constructed of optical quality glass and this degraded our ability to make high quality visual observations and photographs.

Visual inspections were enhanced by the use of a binocular microscope that provided an effective magnification of approximately 3x at standoff distances of up to 0.5 m. Photographs were taken with a hand-held 35 mm camera employing a 50 mm lens with extension tubes and using 200 ASA print film. Adequate depth of field was achieved by stopping the lens down to f/16 to f/22. One or two tungsten halogen lamps were used to provide camera illumination, with typical exposure times of 1/60 s or less.

Four different kinds of regenerator elements were tested in various combinations. These were: 1.6 mm glass beads; perforated 1.6 mm thick acrylic plates with typically 30% porosity and 1.6 mm holes; 200-mesh stainless steel screen; and 400-mesh stainless steel screen. The glass beads and perforated plates were primarily used to provide better visual access to the inside of the regenerator stack. Experiment duration with a particular regenerator and gas composition ranged from a few hours to a week. Every time the pulse tube was opened up to change the regenerator, the inside surface was cleaned with a propanol-soaked tissue. After reassembly, the pulse tube would be internally evacuated at room temperature with a turbopump for one to three days. Although the residual contamination was not measured, previous experience[2] suggests that a level of 20-30 ppm of water and nitrogen is achieved with this kind of cleaning.

RESULTS AND DISCUSSION

A total of 30 pulse tube tests was conducted in this study to explore contamination effects in the different regenerators across a coldtip temperature range of 150 K to 235 K. Several of these tests were done with clean helium gas to establish a baseline performance at various temperatures. All other experiments used 400 ppm water-contaminated helium gas. Over a hundred photographs were taken to document the extent of internal ice formation. Of these, four pairs of before-and-after photographs from four different experiments have been included to illustrate the key findings. Figure 2 shows a double-perforated-plate-over-glass-bead regenerator, both before and after one day of continuous operation at 235 K. Note the use of small glass columns to separate the plates and provide a clear view of the entire plate surface. Figure 3 shows the results, before and after two days of continuous operation, with a regenerator involving a perforated plate over a stack of 200-mesh stainless steel screens. Other configurations seen in later photographs included the use of an all-200-mesh stainless steel screen regenerator, and a regenerator with 50 layers of 400-mesh screens on top of a stack of 200-mesh screens.

A key result illustrated by the photographs is that every experiment with newly loaded 400 ppm water-contaminated helium gas produced enough internal frost to be visible to the

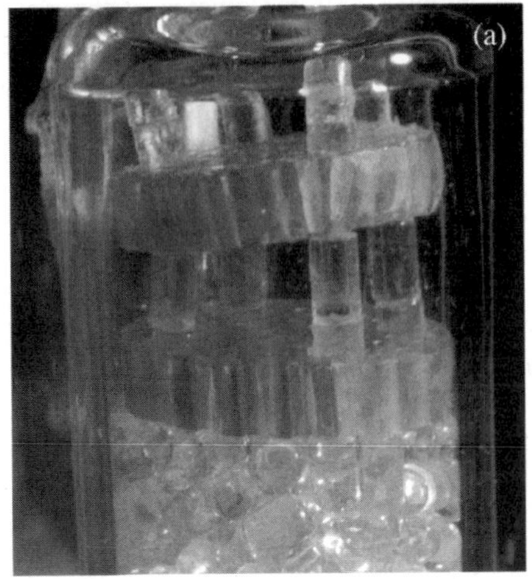

Figure 2. Double-plate-over-glass-bead regenerator, before and after icing (GPT 5).

Figure 3. Photos of plate-over-200-mesh-screen regenerator, before and after icing (GPT 9).

naked eye. In general, this frost became visible on the inside wall of the small diameter pulse tube section within half an hour of start-up. Figures 4a, 5a, and 5b show this wall frost after it had accumulated for one or more days. It took longer (2+ hours) to see frost accumulating on regenerator surfaces, presumably because of the poorer visibility rather than because of smaller accumulation rates. Interestingly, the frost always accumulated on the top surface of regenerator elements, and only on those elements at the very top of the stack. For example, both plates in Fig. 2b show frost on the top surface, while oblique views from below (not shown) revealed no frost underneath. A few of the glass beads right below the second plate also showed frost deposits, but no frost is seen lower down the stack. A similar pattern is seen in Fig. 3, where only the plate and the top one or two layers of screen show frost accumulation. In Fig. 4a and 4b we see frost only on the top screen and not in the gap that was located ten screen layers below the top. No counter-examples were seen to this pattern of frost accumulation occurring only at the very top of the regenerator stack despite achieving coldtip temperatures as low as 150 K in some tests.

The frost that collects on the regenerator is highly porous like snow. This is seen in Fig. 5a in which there is literally a pile of tiny ice shards stacked in the center of the screen.

Figure 4. Photos of screen regenerator with spacer below shoulder joint, after one day and after six days of continuous operation (GPT 10).

These ice shards were only seen in this experiment and its repeat. In every other case, microscope observations indicated that the frost tended to build up as a large number of thin tendrils side by side, very much like the frost that forms on an exposed liquid nitrogen pipe in air. Over time these tendrils merged to form more uniform patches like in Figs. 3b and 4a. The ice shard pile in Fig. 5a also disappeared in time and became the more uniform coating seen in Fig. 5b. Despite the fact that this porous "snow" accumulated on only the top of exposed surfaces, it was never observed to fall downwards. As far as our observations could determine, the frost structures only grew by accretion of gas phase material.

Figure 5. Photos of all-screen regenerator after one day and four days of operation (GPT 16).

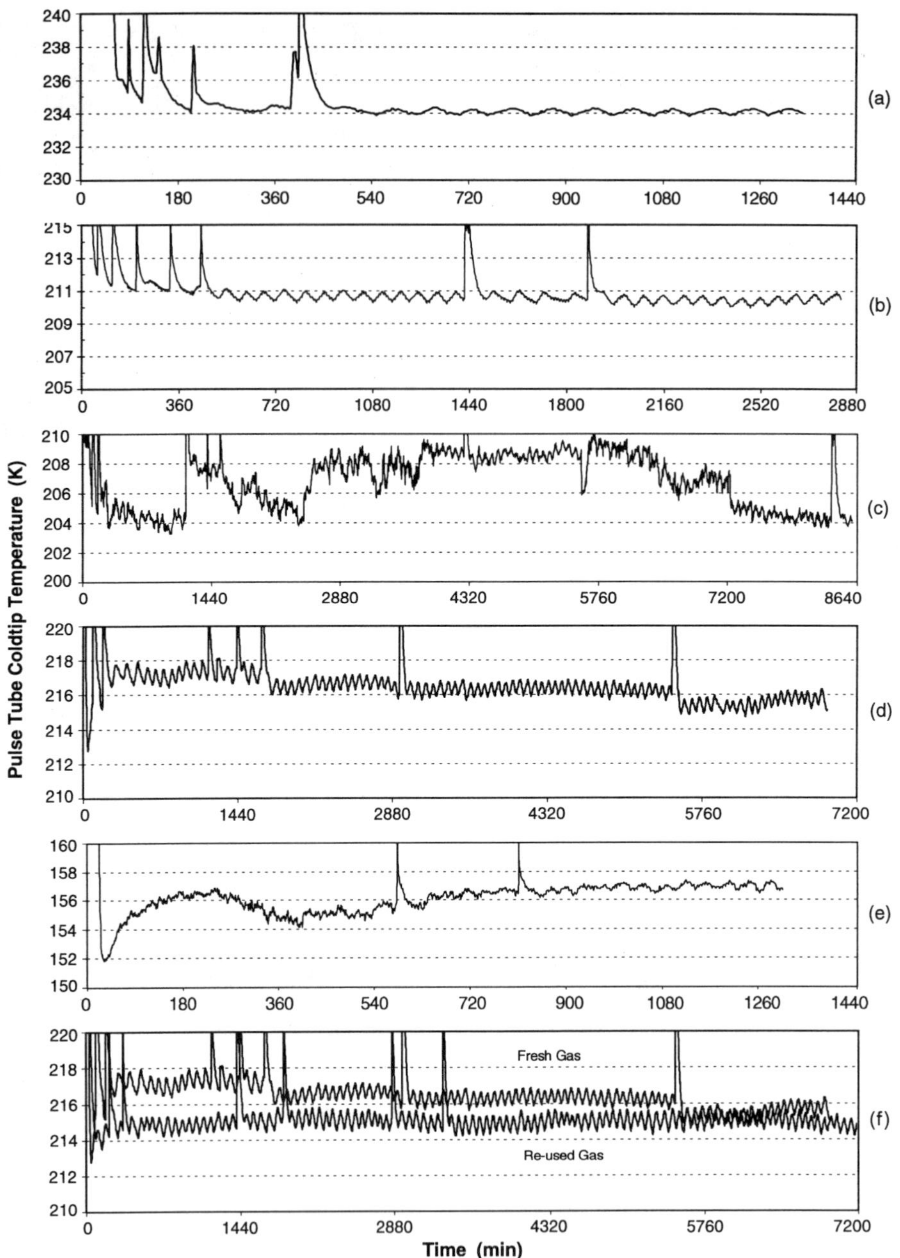

Figure 6. Coldtip temperature versus time traces for selected experiments.

Another result illustrated by the photographs is the tendency for internal frost to redistribute itself over time. For example, the wall frost in Fig. 4a has disappeared five days later in Fig. 4b. Also, the fairly uniform coating on the screen in Fig. 4a has developed some bare patches over the same time period. The disappearance of the ice shard pile from Fig. 5a to Fig. 5b is another example of this redistribution. However, it should be stressed that at lower temperatures, namely below 200 K, there generally seemed to be less redistribution of frost over the same few-day time scale.

Time traces of coldtip temperature from six different experiments are presented as Figs. 6a through 6f. The first four correspond in sequence to the tests whose photographs are shown in Figs. 2 through 5. Figure 6e shows data from the lowest temperature experiment in this study, while Fig. 6f compares the data from 6d with a repeat experiment at the same test conditions. All of the plots show temperature spikes at irregular intervals. This is due to the high power camera lamps that were used to illuminate the pulse tube for photography and microscopy. The radiative heat transfer from these lamps was clearly large compared to the nominal parasitic heat load on the uninsulated glass pulse tube. Another feature seen in Figs. 6a, b, d and f is a regular temperature oscillation with an amplitude of 1 K or less. The cause of this oscillation is speculated to be temperature fluctuations in the ethylene glycol and water chiller unit that removes heat from the pulse tube.

All of these time traces come from experiments where the compressor piston stroke and heat rejection temperature set point were held constant. Therefore, any change in the mean coldtip temperature can be attributed to internal contamination effects. The data indicate quite strongly that internal frost formation did not necessarily correspond with thermal performance loss. For example, the mean temperature in Figs. 6a and 6b is constant with time despite the large accumulations of frost as seen in the corresponding photographs in Figs. 2 and 3. This constancy was mirrored by compressor power measurements taken at the same time. Conversely, the mean temperature in Figs. 6c and 6d is not constant with time. In Fig. 6c, the mean temperature rises approximately 5 K by mid-experiment, only to decay back to the original value by the sixth day. A similar pattern is seen in Fig. 6d in which there is an initial temperature rise from 213 K to 217 within the first few hours, then a slow decay down to 215 K by the end of the experiment. These temperature movements generally coincided with the internal frost redistributions illustrated in the photographs, namely the formation of a bare spot on the screen in Fig. 4b and the disappearance of the ice shard pile in Fig. 5a. Experiments at lower temperatures, which featured much less internal frost redistribution, showed correspondingly less temperature recovery. For example, a 155 K all-screen regenerator result is shown in Fig. 6e. After a partial temperature recovery at the half-day mark, the temperature climbed back again to a value roughly 5 K higher than the start value.

Evaluation of all the temperature and frost data revealed that coldtip temperatures changed only in experiments using screen regenerators that were pushed up against the shoulder joint at the pulse tube/regenerator tube junction. All tests with spacers or plates at the shoulder joint showed no change of temperature with time. The conclusion we draw from this is that the mechanism by which frost degrades the overall thermal performance is flow blockage, and that only small pore regenerator elements (like screens) that have excellent circumferential contact are susceptible to this blockage. Large pore elements, like perforated plates or beads, and elements that do not form very small gaps at the glass wall junction were not sufficiently blocked by internal frost formation. The corollary to this conclusion is that poor contact at the outer edge of screens or other small pore elements will allow enough flow at the wall to mitigate the blockage of the pores themselves. This suggests that the use of small spacers at the top of any pulse tube regenerator can alleviate much of the blockage effect by allowing for the flow to slide around the circumference and away from the frosted center.

A handful of tests were conducted in which pairs of experiments were performed under identical conditions with the same 400 ppm water contaminated helium gas. The gas was allowed to heat back up to room temperature between tests, but was not replaced or otherwise processed. During the warming process, it was generally observed that all of the accumulated ice inside the pulse tube melted in the range of 275 to 280 K, forming visible water drops that typically evaporated over the course of a few hours. The striking results from the repeat experiments were significantly less frost formation and negligible coldtip temperature change. Fig. 6f shows time traces from one such repeated experiment in which it appears that the first experiment used contaminated gas and the second used clean gas, even though the gas was not replaced between experiments. We speculate that the water drops that form during the warm-up phase between experiments wet the surface of the regenerator screens to such an extent that a significant amount became physically adsorbed, thereby depleting the gas phase concentration of water. In addition, the time scale for desorption of this water must have been longer than the few day duration of the subsequent experiment in order for it not to reappear as visible frost. The large surface area of a screen regenerator coupled with a vast number of small crevices formed by wire to wire contact would seem to promote both physical adsorption and capillary condensation mechanisms; however, no detailed analysis has yet been performed to evaluate the plausibility of this adsorption theory for the apparent disappearance of gas phase water between experiments.

This dramatic effect of water adsorption onto regenerator screens suggested the possibility of adding specific adsorption elements to the regenerator to enhance the process. Two kinds of adsorbents were tested: a zeolite (molecular sieve) in 1.6 mm diameter bead form, and saran carbon in large pellet form. The zeolite was located near the high temperature end of the regenerator, while the saran carbon was placed near the cold end. In neither case, however, was frost reduction observed using fresh gas. Time traces of coldtip temperature likewise revealed no change due to blockage. One experiment was performed with a second saran carbon pellet placed at the top of the pulse tube on the possibility that significant water was reaching the coldtip from the downstream reservoir; however, this change also had no effect on the observed frost formation. A repeat experiment with the same gas after warm-up showed the same behavior as regenerators without saran carbon, namely apparent depletion of gas phase water and much reduced frost formation. The reasons for the zeolite and saran carbon null result are not yet clear. Two possibilities have been suggested: insufficient surface area of the zeolite and saran carbon to adsorb water at the necessary rate, and a limited vacuum bakeout temperature of only 100°C with this apparatus.

CONCLUSIONS

Glass pulse tube experiments were conducted with four different kinds of regenerator elements. Substantial internal ice formation was observed when 400 ppm water contaminated helium gas was tested across a coldtip temperature range of 150 to 235 K. The ice formation was limited to a highly localized region at the coldest end of the regenerator in all tests. The ice appeared to be highly porous like snow and was observed to redistribute itself over time, especially in the tests above 205 K. Internal ice formation only degraded pulse tube thermal performance in tests using screen elements at the regenerator cold end with good tube wall contact. Experiments with coarse regenerator elements or spacers at the end of the regenerator seemed immune to the blockage-induced performance degradation. Experiments in which water contaminated gas was reused after warm-up and melting of accumulated ice showed significantly reduced ice formation and negligible performance loss. It was concluded that perhaps chemical adsorption had occurred of the liquid water onto the regenerator surfaces, thus depleting the gas phase concentration of water.

ACKNOWLEDGMENTS

The work described in this paper was carried out at the Jet Propulsion Laboratory, California Institute of Technology and was jointly sponsored by the Caltech President's Fund and the NASA EOS IMAS TechDemo Project through an agreement with the National Aeronautics and Space Administration. The authors would like to acknowledge the laboratory assistance of Scott Leland at JPL.

REFERENCES

1. Getmanets, V.F. and Zhun, G.G., "Cryocooler Working Medium Influence on Outgassing Rate," *Cryocoolers 10*, Plenum Press, New York, 1999, pp. 733-742.

2. Hall, J. L. and Ross, R. G. Jr. "Gas Contamination Effects on Pulse Tube Performance," *Cryocoolers 10*, Plenum Press, New York, 1999, pp. 343-350.

3. O'Hanlon, J.F., *A User's Guide to Vacuum Technology*, John Wiley & Sons, NY, 1980, p. 364.

APPLICATION OF CO-AXIAL PULSE TUBE REFRIGERATION SYSTEM

L.W Yang[1], L.W Yan[1], J.T Liang[1], Y. Zhou[1], Y. S He[1], H. Li[1]
W.W. Xu[2], P. H. Wu[2], Q. Y. Ma[3], Edward S. Yang[3]

[1]Cryogenics Laboratory, Chinese Academy of Sciences
Beijing P.O.B.2711, 100080, P. R. China
[2] Department of Electrical Science and Engineering, Nanjing University,
Nanjing 210093, P. R. China
[3] The Jockey Club MRI Engineering Center, Hong Kong University
Pokfulam Road, Hong Kong

ABSTRACT

In recent years, high temperature superconductors (HTS) have been developing quickly, and their application needs the temperature range of 50-80 K. Pulse tube cooler may be the cheapest and most suitable cooler. In order to get competitive product, air-conditioning compressor is used in our laboratory to do the work. Through work of nearly two years, such a system has been completed. The compressor has a length of 500 mm, width of 250 mm and height of 500 mm; its weight is about 30 kg. All pulse tube coolers are designed to be co-axial. The temperature usually is lower than 40 K, cooling capacity >6 W/80 K, while the lowest temperature reached till now is 24.5 K. This performance is similar to U-type or linear structure pulse tube refrigerator. The input power of compressor is about 500 W-1000 W. In this paper, the latest experiments for cooperate HTS devices are presented. They are in our Laboratory and in Nanjing University. Experiments show pulse tube refrigerator can cool the HTS devices below 60 K. Similar results comparing to liquid nitrogen were acquired. But further work is still needed.

INTRODUCTION

Recently, pulse tube cryocoolers have developed quickly because they have no moving parts in cold head, which results in long life and low vibration compared with other types of cryocooler. This makes pulse tube refrigerator suitable for cooling HTS and other electronic devices [1]. Such applications usually require a few watts of cooling capacity at 80 K, high reliability and low cost. As to such an application, long life and low cost of compressors is becoming even more crucial. Pulse tube refrigerator driven by commercial air conditioning compressor may satisfy this requirement.

In 1997, we began to do this work under the support of Superconductivity Center of China[2, 3, 4].

In 1997, initial investigation acquired the lowest temperature of 51 K and 4 W/80 K cooling capacity with input power about 550 W [2].

Later, co-axial pulse tube refrigerator reached the lowest temperature 38.4 K and cooling power >6 W/80 K [3]. Several similar pulse tube coolers were fabricated and similar results were acquired. At the same time, stability problem was studied and long time operation was verified [4].

Recently, single-stage co-axial pulse tube refrigerator reaches the lowest temperature of 24.5 K. This makes the difference between co-axial structure and U-type structure or linear structure even small.

Our co-axial pulse tube refrigerator has been used to cool HTS in the last year. This paper introduces these works.

COUPLING OF REFRIGERATION SYSTEM WITH DEVICES

Structure of Refrigerator

As shown in Figure 1, the refrigeration system includes three main parts: compressor, rotary valve and pulse tube cooler.

A water jacket outside the surface of the compressor is used to remove compression heat and to condense the oil vapor in the compressor. A specially designed oil filter is used to further eliminate the oil vapor from the gas stream. The compressor and the oil filter are installed in a specially designed box with four wheels. The dimension of compressor is 500 mm long, 300 mm wide and 500 mm high; its weight is about 30 kg.

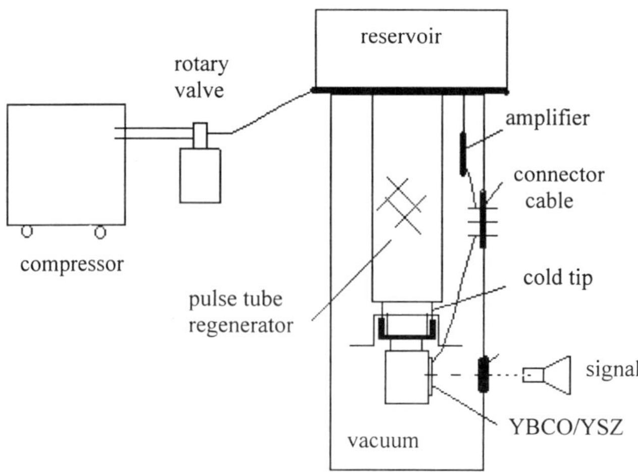

Figure 1. Pulse tube refrigeration system and its coupling with HTS mixer.

The rotary valve connected to the compressor is made by our lab. Its operation is quite stable. The operation speed of the motor could be adjusted at the beginning, but after the refrigeration effect is confirmed, the motor is kept unchanged at the best state. Generally, the operation frequency is 2-8 Hz. When a stable state is reached, the fluctuation of frequency is within 0.2 Hz.

The pulse tube is installed in the middle of the regenerator. The regenerator is filled with annular stainless steel screens. The cold tip is made of copper and the gas reservoir is made of aluminum alloy. The impedance for orifice, double-inlet and multi-bypass are all integrated within the pulse tube and regenerator. The cold head can be easily coupled with a vacuum jacket designed according to the specifications of the user.

Cooperating With HTS Mixer

The structure is as shown in Figure 1.

The HTS mixer is made by a 5×10 mm^2 YBCO/YSZ bicrystal Josephson devices, which is placed on the top of cold tip of the pulse tube refrigerator. In order to reduce heat leak, semi-rigid stainless steel cables are used for microwave connection. A special window, opposite to the Josephson device and vacuum sealed with quartz glass, is designed on one side of the vacuum cover of the refrigerator for transmitting microwaves from outside of the vacuum chamber to the Josephson device. The whole system can be cooled down to 60 K within one hour, which is considerably lower than the transition temperature of the bicrystal junction and provides very good working environment for the HTS mixer.

Cooperating With HTS Oscillator

Another work is cooperating with HTS low phase noise oscillator.

The whole system consists of a HTS cavity, a low noise conventional amplifier (LNA) and a pulse tube refrigerator, as shown schematically in Figure 2. The TE_{011} mode HTS cavity, comprising two 2 inch YBCO films and a sapphire cylinder, has a very high quality value ($>10^5$), leading to a very low phase noise of the oscillator (-120 dBC/Hz).

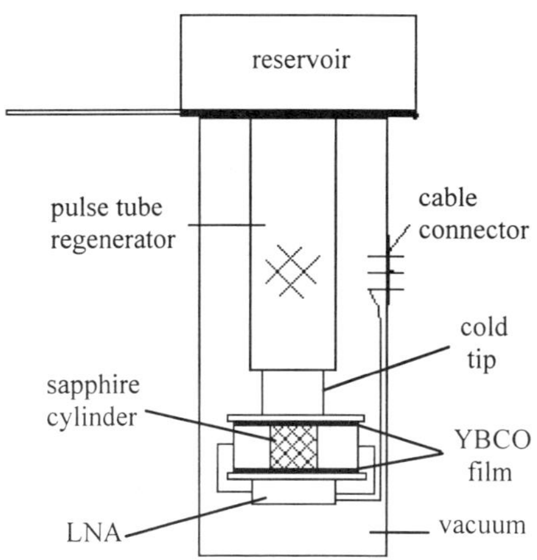

Figure 2. Structure of pulse tube cold tip cooperating with HTS oscillator

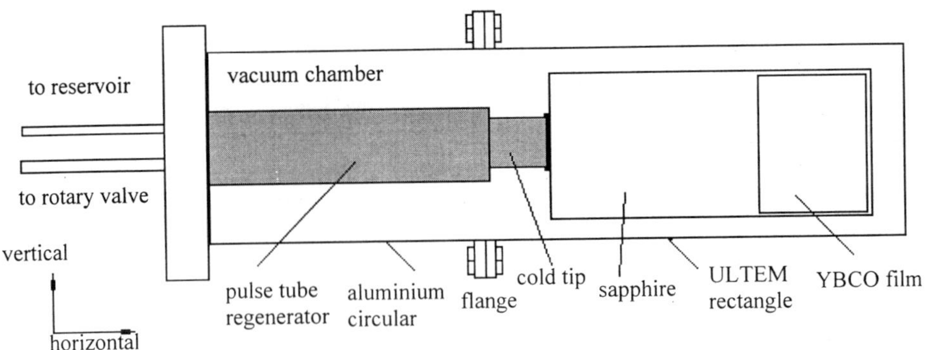

Figure 3. Structure of RF coils coupling with pulse tube cold tip

Cooperating with RF coil

The third work is cooperating with MRI RF coil.

It is believed that one can obtain a great gain in signal-to noise (SNR) of magnetic resonance image (MRI) by using superconducting receiver coils only at low field (0.3 T or below) or for small samples. This superconducting coil need low temperature of 50-60 K. The pulse tube refrigerator is designed to for this application.

The system is as shown in Figure 3. This system has some special requirements comparing with those shown in Figure 1, 2. Firstly, its vacuum chamber includes two parts, one is made of aluminium with a circular cross section, another part is made of ULTEM (a special material resemble rubber) with a rectangular cross section. Secondly, because pulse tube and regenerator are made of stainless steel and potential magnetic interfere may exist, a long sapphire piece has to be used to transfer cooling load to YBCO film. And thirdly, because total vacuum chamber is put under a strong magnetic field, the rotary valve should be put away from pulse tube and the distance between rotary valve and regenerator inlet should be at least 1.2 m.

REFRIGERATION PERFORMANCE

Performance on HTS Mixer

As shown in Figure 4, refrigerator cooperating with mixer reached a lowest temperature of 45 K while refrigerator itself has the lowest temperature of 37 K.

Some results have been acquired with the HTS Mixer. Measurements show that the odd and even order harmonic mixings of 8 mm wavelength signal are observed.

Performance on HTS Oscillator

As shown in Figure 5, refrigerator cooperating with HTS oscillator reaches the lowest temperature of 58 K while the pulse tube cooler alone has the lowest temperature of 36 K. Because the HTS oscillator has the weight of 0.45 kg, the cooling down curve is evidently slow.

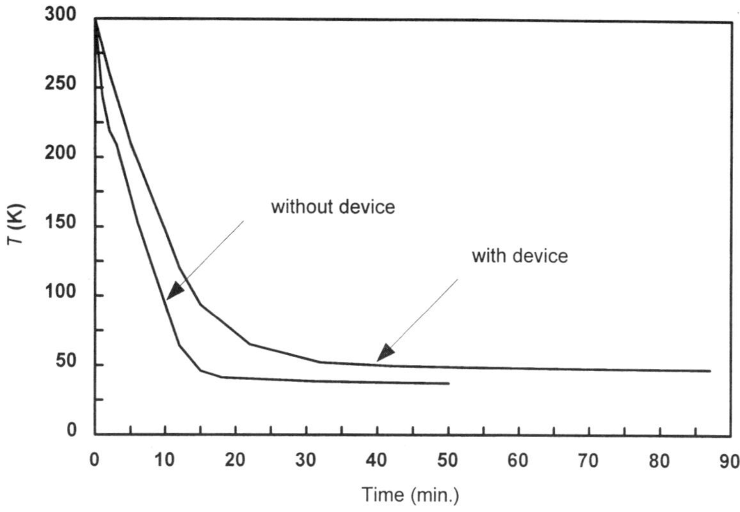

Figure 4. Performance of pulse tube refrigerator cooperating with HTS Mixer

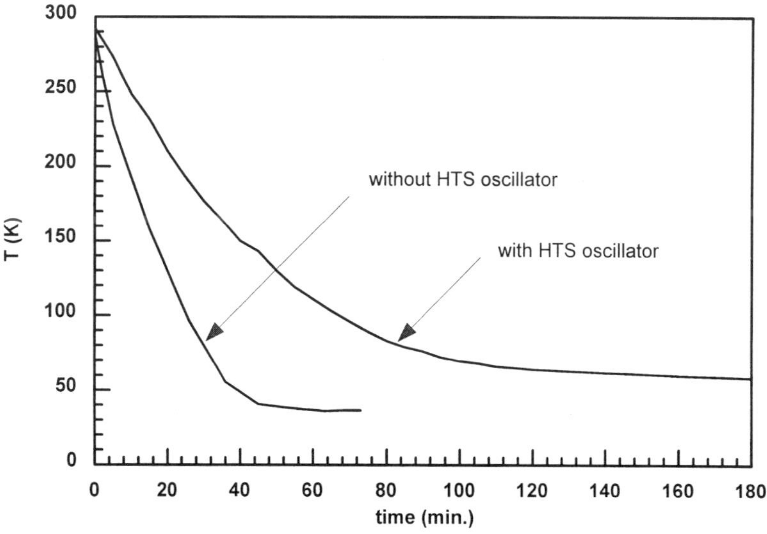

Figure 5. Performance of pulse tube cooperating with HTS oscillator

Refrigerator for RF Coil

The pulse tube refrigerator designed for Hongkong University has been fabricated. Main performance is as shown in Figure 6 and 7. Figure 6 gives two cooling down curves, one is when cold tip is placed downward and lowest temperature is 28 K. The other is when cold tip is placed horizontal and lowest temperature is 40K. In Figure 7, Cooling capacity is measured, which shows that >6 W/60 K for cold tip down and >4 W/60 K for cold tip horizontal. According to our design, such a performance is enough to cool the RF coil system to temperature below 60 K.

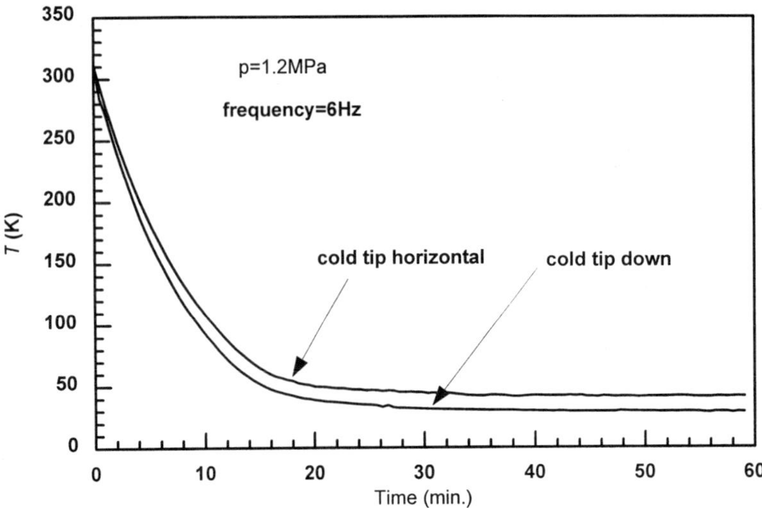

Figure 6. Cooling curves of pulse tube refrigerator used to cool RF coil. One curve is for cold tip down and lowest temperature is 28 K, while when cold tip is horizontal, refrigeration temperature is 40 K.

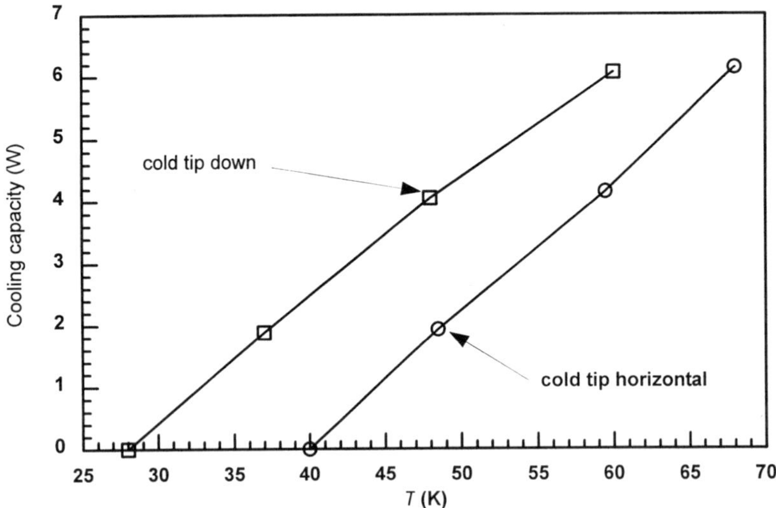

Figure 7. Cooling performance of pulse tube refrigerator

FURTHER WORK

Though pulse tube refrigerator cooperating with HTS can lower the temperature below liquid nitrogen temperature, there are still problems.

For pulse tube refrigerator, the problems are that compressor is large and weighty, that a vacuum could not keep for long time, and that the temperature of pulse tube refrigerator may oscillate with time.

For cooperating work, special problem exists for different applications.

For HTS Mixer, the problem is magnetic disturbance. Improvement is being made. The method is to eliminate magnet disturbance from vacuum pump, rotary valve and compressor. They have evident magnetic interface and they generally add side by side to signal source.

For HTS oscillator, we find that mechanical vibration from the gas in pulse tube produces additional noise to the spectrum.

CONCLUSION

In our laboratory, a co-axial pulse tube refrigerator driven by an air-conditioning compressor is used for the purpose of cooling HTS devices. In this paper, co-axial pulse tube refrigerator with lowest temperature of 40 K is used to cool HTS devices. Test shows that with pulse tube refrigerator, the devices can be lowered to a temperature below 60 K generally. At 60 K temperature, the devices show comparative quality to using liquid nitrogen. These tests are basis for further improvement of pulse tube refrigerator and further cooperating.

ACKNOWLEDGEMENT

This research is supported by the National Superconductor Center of China and National Natural Science Foundation of China. Thanks the help from Dr. Li Laifeng.

REFERENCES

1. G. Thummes, R. Landgraf, F. Giebeler, M. Much, C. Heiden, Pulse tube refrigerator for high-Tc SQUID operation, *Adv. Cryo.Eng.* , 41: 1463(1996)
2. L. W. Yang, J. T. Liang, Y. Zhou, P. S. Zhang, W. X. Zhu, Pulse tube refrigerator driven by commercial air-conditioning compressor, *Cryogenics and Superconductivity of China*, 25(3):1 (1997)
3. L. W. Yang, Y. Zhou, P. S. Zhang, J. T. Liang, W. X. Zhu, High efficiency coaxial pulse tube refrigeration system driven by air conditioning compressor, *Vacuum & cryogenics of China*, 4(3): 140(1998)
4. L.W. Yang, J.T. Liang, Y. Zhou, P. S. Zhang, W. X. Zhu, Stability study of co-axial pulse tube cooler driven by air conditioning compressor, *Cryocooler 10:* 337 (1999)

Left to right: Best Research Paper - E.D. Marquardt (National Institute of Science and Technology); Best Application Paper - P. Shirron (Goddard Space Flight Center).

A COMPACT LIQUID NITROGEN RECONDENSER WITH A PULSE TUBE CRYOCOOLER

R. Li,[1] A. Ishikawa,[1] T. Koyama,[2] T. Ogura,[2]
K. Aoki,[2] and T. Koizumi[2]

[1]Research and Development Center, Sumitomo Heavy Industries, Ltd.
Hiratsuka, Kanagawa, 254-0806, JAPAN

[2]Precision Products Division, Sumitomo Heavy Industries, Ltd.
Tanashi, Tokyo, 188-8585, JAPAN

ABSTRACT

A compact liquid nitrogen recondenser with an orifice pulse tube cryocooler has been developed. The recondensation capacity is about 170 cm^3/h. The orifice pulse tube cryocooler employed in the recondenser is a valved type, and is commercialized by Sumitomo Heavy Industries Limited. The cryocooler delivers a cooling power of more than 10 W at 77 K with a connecting tube of 4 m between the cold head of the cryocooler and a rotary valve unit. An air-cooled helium compressor unit with a rated input power of 1.6 kW is applied for the cryocooler. Excepting the compressor unit and the rotary valve unit , the recondenser is 113 mm in diameter, 567 mm in height, and 6 kg in weight. The recondenser and the cryocooler are handy and flexible for many scientific, medical and industrial applications. This paper describes the design and the test result of the recondenser, as well as the cooling performance of the pulse tube cryocooler.

INTRODUCTION

For a highly sensitive measurement and analysis, cryogenic environment is often required in a lot of scientific, medical and industrial equipment. Liquefied cryogenic fluids are very useful for keeping cryogenic environment and widely used in the past and the present. The reduction of liquefied cryogenic fluids owing to evaporation, however, is a seriously disadvantage of such cryogenic systems. A regular resupply of the liquefied fluids is extremely inconvenient for continuous operation of the equipment, as well as for a distant place.

To use a small cryocooler is a reasonable alternative choice in many cases. A dry cryogenic system is possible to build by using small cryocoolers instead of liquefied cryogenic fluids[1]. A recondenser of the liquefied fluids with small cryocoolers is also an effective means for a newly produced system or an existing system. Liquefied cryogenic fluids are still required when the recondenser is applied to a cryogenic system, but the recondenser makes possible zero boiloff or can greatly extend the regular resupply interval of the liquefied fluids[2].

Because of the reciprocating motion of the displacer in cold head, however, Stirling cycle cryocoolers and Gifford-McMahon (G-M) cycle cryocoolers generate considerable mechanical vibration, and therefore are hard to employ for a highly sensitive application. A pulse tube cryocooler has no moving parts in the cold head. A very low level of mechanical vibration compared with Stirling cryocoolers or G-M cryocoolers is a remarkable advantage and attraction of pulse tube cryocooler. Since the middle of the 1980's, great progress has been made in cooling power, efficiency, and reliability of pulse tube cryocoolers. The cryogenic system with a pulse tube cryocooler has been reported by a large quantity of papers in recent years.

In the present investigation, we developed a compact liquid nitrogen recondenser in which an orifice pulse tube cryocooler is applied. The design and the test results of the recondenser, as well as, the cooling performance of the pulse tube cryocooler are described in this paper.

DESIGN OF RECONDENSER

The liquid nitrogen recondenser we developed consists of a pulse tube cryocooler and the other components including a vacuum chamber, a recondensation chamber and a counterflow tube.

Pulse Tube Cryocooler

The pulse tube cryocooler employed in the recondenser is a valved orifice type. It is well known that an orifice pulse tube cryocooler[3] is inferior to a double inlet pulse tube cryocooler[4] in efficiency. Because of a simpler mechanism, the reliability of an orifice pulse tube cryocooler is much higher than that of a double inlet one. In the present development, reliability takes preference of efficiency.

The cryocooler (model SRP-2620) has been commercialized by Sumitomo Heavy Industries Limited. It is made up of a compressor unit, a rotary valve unit, a cold head and a reservoir. Figure 1 shows these assemblies. The main specifications and the design of the valved orifice pulse tube cryocooler are given in table 1.

Compressor Unit. The compressor unit (model CSA-11) generates a high pressure source and a low pressure source. The rated input power of the compressor unit is 1.6/1.9 kW for the power line cycle of 50/60 Hz. This air cooled compressor is able to be used for a G-M cryocooler.

Rotary Valve Unit. The rotary valve unit is linked with the compressor unit by two flexible hoses, a high pressure line and a low pressure one. The valve unit is driven by a 3-phase AC synchronous motor and alternately switches the cold head to connect with the high and low pressure sources of the compressor periodically. In the present system, there is no electric inverter used for changing the motor speed, so the cycle frequency of the cryocooler is fixed at 2/2.4 Hz for the power line cycle of 50/60 Hz.

Cold Head. The cold head is connected with the rotary valve unit by a copper connecting tube of 4 m. The periodic compression and expansion of working fluid in the cold head generates a cooling power below 77 K. The cold head mainly consists of a

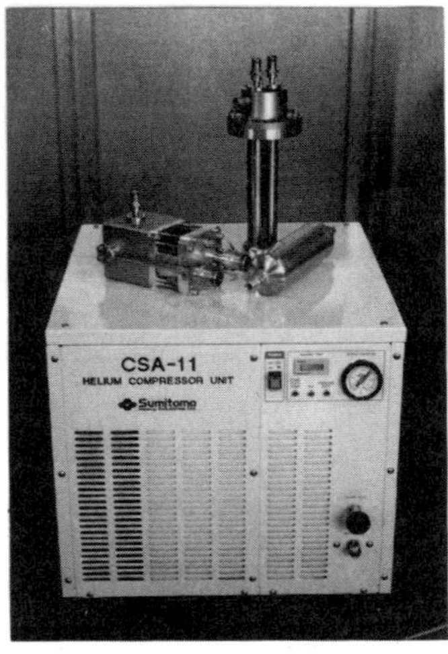

Figure 1. Rotary valve unit, cold head , reservoir and compressor unit of the orifice pulse tube cryocooler.

regenerator, a pulse tube, a cold stage and a hot end heat exchanger. The regenerator is 200 mm in length and 26 mm in diameter, and is stuffed with stainless steel mesh screens. The dimension of the pulse tube is 200 mm long and 20 mm in diameter. The orifice between the pulse tube and the reservoir is placed at the inside of the hot end heat exchanger. The opening of the orifice is optimized and fixed.

Reservoir. The volume of the reservoir is ~ 500 cm^3. The reservoir is connected with the cold head by using a short tube about 1 m. The reservoir and the orifice play an important role in phase shift between pressure and displacement of working fluid in the pulse tube, and strongly influence the cooling performance of the pulse tube cryocooler.

Other Components of Recondenser

The photograph and the schematic of the nitrogen recondenser are shown in figure 2

Table 1. Specifications and design of the orifice pulse tube cryocooler (model SRP-2620)

Component	Specification or design
Compressor unit (Model CSA-11)	
Electric power(50/60Hz)	200V×3Phase×1.6/1.9kW
Cooling	Air cooled
Dimension	W494×D494×H484(mm)
Weight	60kg
Rotary valve unit	
Motor	AC synchronous (200V×3Phase)
Dimension	W108×D214×H135(mm)
Weight	4kg
Cold head	
Configuration	U-shape
Regenerator	ϕ 26×L200(mm)
Pulse tube	ϕ 20×L200(mm)
Dimension	ϕ 115×H317(mm)
Weight	3kg
Connecting tube length	ϕ 6.35(mm)×4 m(Cu)
Reservoir	ϕ 207×64 (500 mm^3)

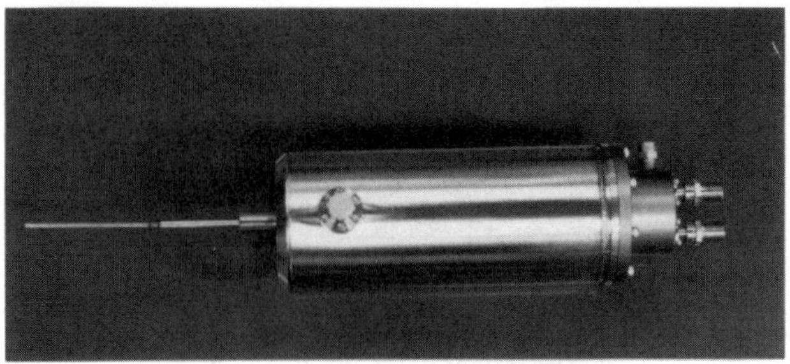

Figure 2. Photograph of the recondenser.

and 3. The recondenser is 113 mm in diameter, 567 mm in height and 6 kg in weight except the compressor unit, the rotary valve unit and the reservoir. The recondenser is compact and handy, and is easy to be directly held on the transfer port of nitrogen in a good number of cryogenic systems.

For keeping compactness and simplicity, a single counterflow tube of stainless steel is used for introducing nitrogen vapor into the recondensation chamber simultaneously with moving liquid nitrogen out from the chamber. In other words, the vapor flows up from a liquid nitrogen storing space to the recondensation chamber in the counterflow tube and the liquid drops down in the opposite direction in the same tube.

The copper recondensation chamber is directly united with the cold stage of the pulse tube cryocooler and is designed to have enough heat exchange surface for the recondensation of nitrogen vapor.

The regenerator, the pulse tube, the cold stage and the recondensation chamber are placed in a stainless steel vacuum chamber. The most of the surface of the counterflow tube is also surrounded by a vacuum space. The vacuum chamber is sealed off after baking out and vacuum pumping. The dimension of the vacuum chamber is 113 mm in diameter and 269 mm in height.

EXPERIMENTAL DETAILS

Three sets of the liquid recondenser were prepared in this investigation in order to

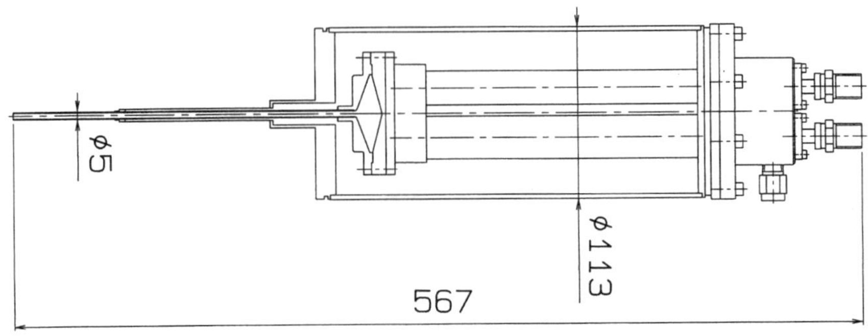

Figure 3. Schematic of the recondenser.

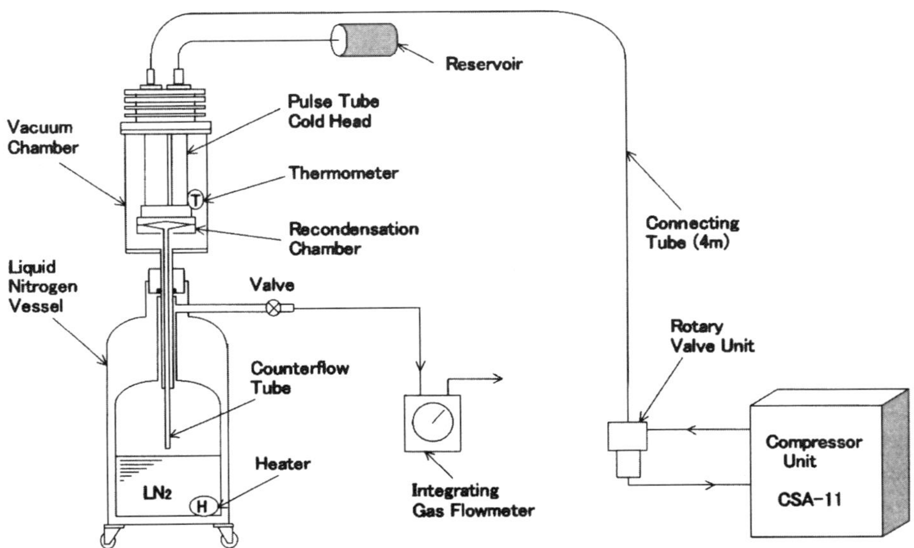

Figure 4. Setup of recondensation test.

evaluate the dispersion in recondensation capacity. The recondensation test for the three recondensers is performed, as well as the cooling power measurement for the three orifice pulse tube cryocoolers.

Measurement of Cooling Power

The cooling power of the pulse tube cryocoolers is measured before the cold head is assembled into the recondensers. In the measurement, the cold head is installed in a vacuum chamber, and is cooled down with the compressor unit of model CSA-11. The cooling power data is obtained by varying the heat load on the cold stage with an electric heater and by measuring the stable temperature of the cold stage. There is nothing installed on the cold stage except a Pt-Co resistance thermometer and an electric heater of manganin wire in this cooling power measurement.

Recondensation Test

Figure 4 illustrates the setup for the recondensation test. The recondenser is held on a liquid nitrogen vessel of 50 litter and the counterflow tube is inserted into the neck tube of the vessel. The vessel is opened to atmosphere via an integrating gas flowmeter at room temperature. Liquid nitrogen in the vessel is always kept to be 20 $\sim$ 30 litter during the test. On the cold stage, the Pt-Co resistance thermometer is still placed, but the electric heater is taken away and moved into liquid nitrogen in the vessel. The cold head is connected with the CSA-11 compressor unit via the rotary valve unit. The connecting tube length is 4 m between the cold head and the valve unit and there is 1 m between the cold head and the reservoir.

The method of recondensation test is as follows, (1) to measure the initial boiloff rate of the vessel by the integrating gas flowmeter, (2) to heat liquid nitrogen with the electric heater and to enhance the evaporation of the liquid, (3) to measure the enhanced boiloff rate, (4) to start cooling down and reducing the evaporation when the temperature of the cold stage is below 77 K, (5) to measure the reduced boiloff rate, (6) to calculate the recondensation capacity to be the enhanced boiloff rate minus the reduced boiloff rate.

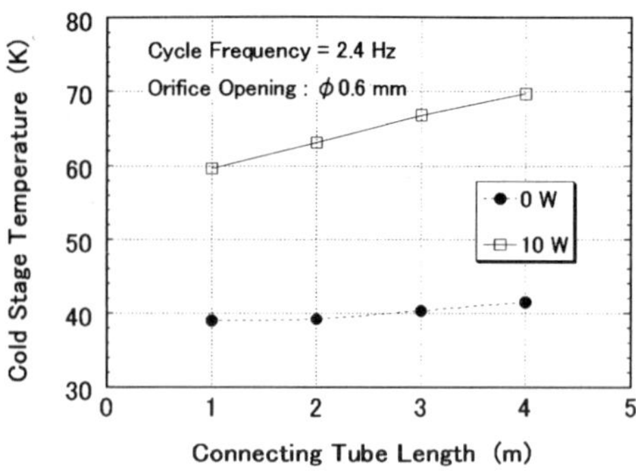

Figure 5. Connecting tube length dependence of cold stage temperature.

RESULTS AND DISCUSSIONS

Cooling Power of Pulse Tube Cryocooler

The cooling power of a valved pulse tube cryocooler is dependent on the length of connecting tube between the valve unit and the cold head. Figure 5 gives the connecting tube length dependence on cooling power for our orifice pulse tube cryocooler. Whether heat load is input or not, the cold stage temperature goes up with increasing length of the connecting tube. The increment of temperature is much larger when there is a heat load input to the cold stage. For the connecting tube of 1 m, the cryocooler reaches a no load temperature of 39 K. When a heat load of 10 W is input, the cold stage temperature is 59.7 and 69.7 K for the connecting tube length of 1 and 4 m respectively. From figure 5, the cooling power of the cryocooler is estimated to be 10 W at 77 K with a connecting tube of 6 m long.

Even though the cryocoolers have the same design and the same production process, the dispersion in cooling performance generally appears. Figure 6 shows the heat load versus cold stage temperature for the three pulse tube cryocoolers prepared in the present investigation. For the cooler A, B and C, the no load temperature is 43.8, 42.6 and 42.3 K respectively. When the heat load is 10 W, the cold stage temperature is 72.0, 71.6 and 71.8 K for the three cryocoolers. The dispersion is very small and almost negligible.

Compared with figure 5, the cold stage temperature is a little higher in figure 6. The reason is that the orifice opening is ϕ 0.5 mm for figure 6 but is ϕ 0.6 for figure 5. In our pulse tube cryocooler, the orifice opening of ϕ 0.6 mm is optimum for the cycle

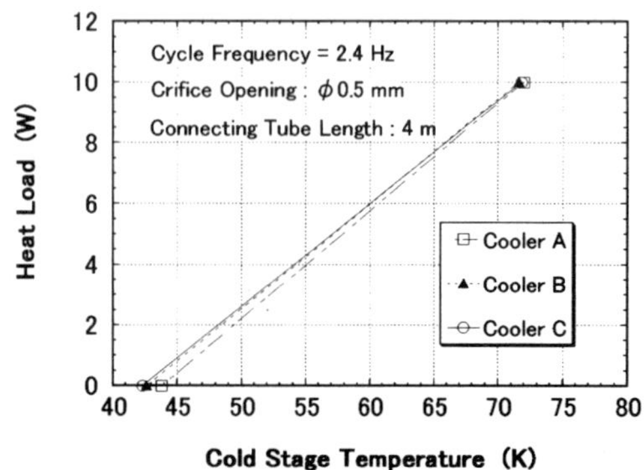

Figure 6. Heat load versus cold stage temperature for the three cold heads.

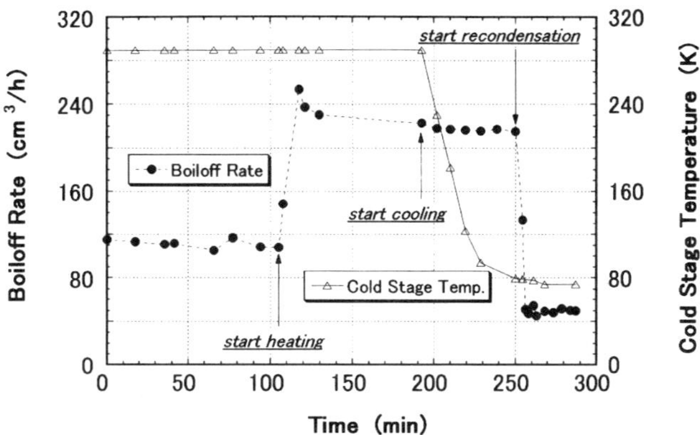

Figure 7. History of boiloff rate and cold stage temperature during recondensation test for recondenser C.

frequency of 2.4 Hz but not so good for 2 Hz. The orifice opening of ϕ 0.5 mm, however, is much better choice for both the cycle frequency of 2 and 2.4 Hz which correspond to the power line cycle of 50 and 60 Hz.

Recondensation Capacity

The typical result of recondensation test for the recondenser C according the test method mentioned above is given in figure 7. The initial boiloff rate of 108 cm^3/h from the vessel is obtained before the heating in liquid nitrogen and the cooling of the cold head. When the electric heater is turned on, a heat load of 6 W is input to the liquid for enhancing the evaporation. The boiloff rate increases rapidly, and becomes 223 cm^3/h as a stable value. The cold stage temperature goes down quickly after the cryocooler is turned on, the boiloff rate, however, does not change while the cold stage temperature is higher than 77 K. When the cold stage temperature becomes lower than 77 K, the recondensation starts, and the boiloff rate is reduced sharply down to 50 cm^3/h in a few minutes.

The main results of the recondensation test for the three recondensers are summarized in table 2. The cooling power at 77 K is estimated from the results of figure 6. The recondensation capacities (or recondensation rates) are calculated to be the enhanced boiloff rate minus the final boiloff rate, $Q_l=Q_e-Q_r$, and are 167.6, 168.4 and 172.9 cm^3/h for the recondenser A, B and C respectively. The three recondensers have the same design and the same production process as well as the three pulse tube cryocoolers. The

Table 2. Performance summary of pulse tube cryocoolers and recondensers

	Cryocooler A / Recondenser A	Cryocooler B / Recondenser B	Cryocooler C / Recondenser C
Cryocooler			
No load temperature (K)	43.8	42.6	42.3
Cold stage temp. for 10 W (K)	72.0	71.6	71.8
Estimated cooling power at 77K (W)	11.9	12.0	11.9
Recondenser			
Initial boiloff rate, Q_i (cm^3/h)	95.2	96.9	108.3
Enhanced boiloff rate, Q_e (cm^3/h)	208.8	209.9	222.8
Reduced boiloff rate, Q_r (cm^3/h)	41.2	41.5	49.9
Heat load input to liquid, Q_h (W)	5.97	6.05	6.04
Recondensation capacity, Q_l (cm^3/h)	167.6	168.4	172.9
Heat loss (W)	4.3	4.5	4.2

dispersion in recondensation capacity appears but is also very small.

The recondensation capacity can be introduced by another way, (1) to estimate boiloff rate due to the heat load input to liquid nitrogen, Q_h, (2) to sum up the estimated boiloff rate with the initial boiloff rate, Q_h+Q_i, (3) to calculate an equivalent recondensation capacity as that the summation minus the reduced boiloff rate, $Q_2=Q_h+Q_i-Q_r$. The equivalent recondensation capacity, Q_2, for each recondenser is almost 20 cm^3/h larger than the recondensation capacity, listed in table 2, Q_1. The difference is considered caused by the temperature distribution in the liquid nitrogen vessel. The temperature distribution in the vapor layer of the vessel depends on whether the heater and the cryocooler are turned on or not. A different temperature distribution usually makes difference in heat leak from room temperature to the inside of the vessel. In view of the temperature distribution, the recondensation capacity may increase or decrease in a real cryogenic system compared with the values given in table 2.

The cooling power for each cryocooler is estimated to be ~ 12 W at 77 K from figure 6, and the net cooling power used for recondensation is converted from the recondensation capacity, Q_1, as ~ 7.6 W. Therefore, the heat loss in the recondensers is ~ 4.3 W. The heat loss is mainly caused by (1) the thermal resistance between the cold stage and the recondensation chamber, (2) the thermal resistance on the recondensation surface, (3) the heat leak from room temperature to the counterflow tube. The heat leak to the counterflow tube is a little large in the recondensers, and may also be strongly influenced by the temperature distribution in the vessel.

SUMMARY

A compact liquid nitrogen recondenser with an orifice pulse tube cryocooler has been developed. The recondensation capacity of the recondenser is about 170 cm^3/h, and is considered that depends on the temperature distribution in the vapor layer of liquid nitrogen vessel. The recondenser has a potential of low mechanical vibration and high reliability because an orifice pulse tube cryocooler is employed. The recondenser is handy and flexible for many scientific, medical and industrial applications.

The cooling capacity of a valved orifice pulse tube cryocooler is strongly dependent on the length of connecting tube between the cold head and the valve unit. The orifice pulse tube cryocooler assembled in the recondenser is commercialized by Sumitomo Heavy Industries Limited. The cryocooler achieves a no load temperature ~ 40 K, and delivers a cooling power of more than 10 W at 77 K with a connecting tube of 4 m long.

In this investigation, there are three pulse tube cryocoolers prepared and assembled into the three recondenser respectively. The cryocoolers and the recondensers have the same design and the same production process. The dispersion in cooling performance and in recondensation capacity appears, but is very small.

REFERENCES

1. T. Hasebe, J. Sakuraba, K. Jikihara, K. Watazawa, H. Mitsubori, Y. Sugizaki, H.Okubo, Y. Yamada, S. Awaji, and K. Watanabe, Cryocooler cooled superconducting magnets and their applications. in: "Advances in Cryogenic Engineering", Vol.43, P. Kittel, ed., Plenum Press, New York (1998), p.291.
2. M. Nagao, T. Inaguchi, H. Yoshimura, S. Nakamura, T. Yamada, T. Matsumoto, S. Nakagawa, K. Moritsu, and T. Watanabe, 4 K three-stage Gifford-McMahon cycle refrigerator for MRI magnet. in: "Advances in Cryogenic Engineering", Vol.39, P. Kittel, ed., Plenum Press, New York (1994), p.1327.
3. E.I. Mikulin, A.A. Tasarov and M.P. Shkrebyonock, Low temperature expansion pulse tubes, in: "Advances in Cryogenic Engineering", Vol. 29, Plenum Press, New York (1984), p. 629.
4. S. Zhu, P. Wu, Z. Chen, W. Zhu and Y. Zhou, A single stage double inlet pulse tube refrigerator capable of reaching 42 K, Cryogenics **30**, ICEC 13 Supplement: 257 (1990).

LIQUID NITROGEN PRE-COOLED PULSE TUBE REFRIGERATOR WITH LOW TEMPERATURE SWITCH VALVE AND ITS SIMULATION

Wei Dai,* Weijia Sun, Jingtao Liang, Yuan Zhou,
and Ruzhu Wang*

Cryogenic Laboratory of Chinese Academy of Sciences
Beijing, China 100080
*Institute of Refrigeration and Cryogenic Engineering
Shanghai Jiaotong University, Shanghai, China 200030

ABSTRACT

In this article, a new kind of pulse tube refrigerator with a low temperature switch valve and a liquid nitrogen pre-cooler is introduced. The aim of this design is to eliminate the shortcomings of ordinary regenerators at temperature below 20K. Theoretical analysis shows that it's a promising alternative of ordinary pulse tube refrigerators. However, the experimental results are not good. Possible reasons are given.

INTRODUCTION

Pulse tube refrigerators (PTRs) have undergone great development in recent years. Liquid helium temperature has already been achieved on two-stage PTRs.[1] However, there are some problems occurring within this temperature range. Firstly, the specific heat of regenerator materials quickly decreases with the temperature, while that of helium gas increases to a level that may be greater than regenerator materials, which results in deteriorating performance of regenerators and reduction of refrigeration capacity. Although magnetic regenerative materials have been successfully used in liquid helium temperature regenerators, they still have some shortcomings, such as high price, large flow resistance and erosion, etc. The second problem is related to the void volume inside a regenerator, which generally plays a large part in the total volume of a refrigerator excluding the compressor. The gas inside this volume will be cyclically pressurized, increasing both the heat load of regenerator and the loss to refrigeration capacity. In addition, the pressure ratio will be greatly reduced if a compressor is fixed. To overcome these problems, a new type of pulse tube refrigerator with a low temperature switch valve has been proposed, whose details are given in another article[2] in this conference. Here we brief on its main differences from the ordinary pulse tube refrigerator: (1) The regenerator is replaced by a recuperative heat exchanger for the purpose of regeneration; (2) The rotary valve is moved from ambient surroundings to the pulse tube cold end. These design differences eliminate problems occurring in ordinary regenerators. In article,[2] some experimental results are given on this new design. The lowest temperature obtained is about 160K. However, the high efficiency of ordinary regenerators operating above liquid nitrogen temperature combined with their easy manufacturability makes it less meaningful to use the configuration within this temperature range. Thus, we decided to add a liquid nitrogen pre-cooler to obtain lower temperature and check effectiveness of this design at temperature below liquid nitrogen temperature.

EXPERIMENTAL SETUP AND RESULTS

Fig.1 is a schematic show of this new pulse tube refrigerator configuration. It mainly consists of six components: the high temperature recuperative heat exchanger (HEX), the liquid nitrogen pre-cooler, the low temperature recuperative heat exchanger(LEX), the low temperature switch valve, the pulse tube, the orifice and the reservoir. The pulse tube, orifice and reservoir used here are the same as those used in an ordinary pulse tube refrigerator. The low temperature switch valve is a kind of rotary valve driven through a long hollow thin–walled rod by an electromotor placed at the ambient temperature. LEX and HEX are simplified perforated plate heat exchangers. In them, two concentric thin-walled stainless steel tubes constitute an interior circular passage and exterior annular passage. Perforated copper disks are filled into these two passages alternatingly with stainless steel screens. The copper

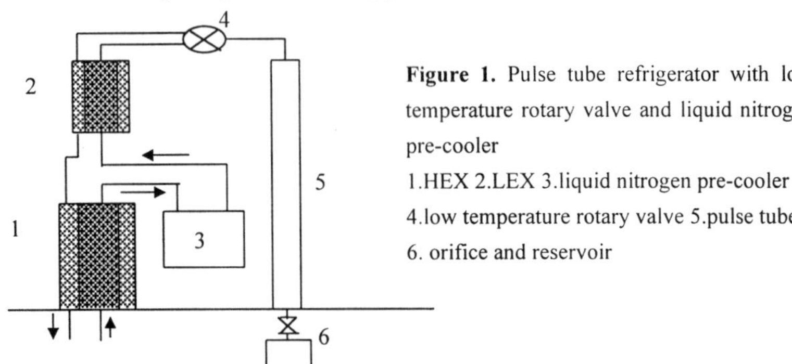

Figure 1. Pulse tube refrigerator with lo temperature rotary valve and liquid nitroge pre-cooler
1.HEX 2.LEX 3.liquid nitrogen pre-cooler 4.low temperature rotary valve 5.pulse tube 6. orifice and reservoir

disks are to conduct heat in the radial direction, while stainless steel screens are to reduce the axial conduction. A 2.2KW G-M compressor was used to provide high and low pressure helium gas.

In order to grasp the essence of working process, we neglect the influence of the void volume inside heat exchangers. Thus working process can be described as below. At first, the switch valve rotates to a position where the pulse tube and high pressure line connect. Then the high pressure gas begins to flow into pulse tube through interior passages of both HEX and LEX from compressor. It firstly goes through the interior passage of HEX from the compressor and cools down as a result of depositing heat to copper disks, which subsequently conduct the heat in radial direction through the passage-separating wall. Then the helium gas enters the liquid nitrogen pre-cooler and is cooled down to temperature near 77K. After leaving the pre-cooler, it moves into LEX and cools down further because of depositing heat to the coppers disks. As the switch valve rotates on, high pressure line is disconnected from the pulse tube. After some time, the valve rotates to connect the pulse tube with low pressure line and gas begins to expand out of the pulse tube and refrigeration effect occurs. The expanded gas has a lower temperature than incoming gas, so when it passes through exterior passage of the recuperative heat exchanger, it picks up the heat conducted across separating wall from interior passage to complete the heat regeneration.

The main dimensions of our experimental setup are: pulse tube $\Phi 20.4\times 0.2\times 260$mm; HEX's passages are formed by two stainless steel tubes $\Phi 32.4\times 0.2\times 120$mm and $\Phi 50.5\times 0.2\times 120$mm; LEX's passages are formed by stainless steel tubes $\Phi 18.4\times 0.2\times 120$mm and $\Phi 32.4\times 0.2\times 120$mm; copper disks filled in these two exchangers are similar with thickness 0.2mm, etched square holes $0.3\times 0.3\text{mm}^2$ with distance 0.6mm from each other; stainless steel screen are 100 mesh.

When performing experiments, we first filled the pre-cooler with liquid nitrogen, then turned on the rotary valve and compressor subsequently. Usually the lowest temperature would be stably obtained in less than an hour. Then we changed the orifice opening or other parameters to check their influence on the lowest temperature..

Far from our expectation, the lowest temperature at the cold end of pulse tube was even higher than liquid nitrogen temperature. At first, we placed the orifice and reservoir outside vacuum chamber. With operating frequency 5.2Hz, pressure difference 5bar, the lowest temperature was 135K. Then we used a long thin copper tube to replace the orifice and winded it around outer wall of the pre-cooler inside vacuum shield. The reservoir was also placed inside. The lowest temperature was 79K. Had we chosen suitable length of the copper tube, lowest temperature might have been lower. Anyway, the results are disappointing. To find the reasons for this anomalous phenomenon, we developed a simulation program introduced below.

THEORETICAL ANALYSIS

Before initiating the simulation, we made the following assumptions:

(1) Helium gas is ideal with constant specific heat;

(2) Pressure wave inside the pulse tube is sinusoidal;

(3) Gas flow inside the pulse tube is one-dimensional, and experiences adiabatic processes.

Usually, calculation of theoretical refrigeration capacity of a cryocooler includes calculation of both ideal refrigeration capacity and cold losses, which can be shown as:

$$Q_{theory} = Q_{ideal} - Q_{loss} \tag{1}$$

When Q_{theory} equals zero, the cold end temperature will be lowest. Below we will discuss it in detail.

Ideal Refrigeration Capacity

We use a modified theory of non-symmetry thermodynamic effect to calculate the ideal refrigeration capacity. In original theory,[3] gas entering the pulse tube at cold end is divided into a number of segments, whose movement and temperature changes are observed during a full cycle. Because of phase shifting mechanism of the orifice and reservoir, each element's temperature when it moves out of pulse tube is different from that when it enters pulse tube, which will finally results in refrigeration effect. This can be shown as:

$$Q_{ideal} = \sum_{i=1}^{n} m_i C_p \left(T_{i,in} - T_{i,out}\right) \tag{2}$$

where m_i is the *ith* gas element, T is gas element temperature, C_p the isobaric specific heat of helium. However, when it uses the following Eq.(3) to calculate mass flow at the cold end, the original theory neglected oscillating nature of gas temperature at both the cold end and the hot end, causing some errors in the value of mass flow rate, which is important to estimate the enthalpy loss inside the regenerator.

$$\tilde{m}_c = \frac{\tilde{T}_h}{\tilde{T}_c} \tilde{m}_h + \frac{V_{pt}}{kR\tilde{T}_c} \frac{d\tilde{p}_t}{dt} \tag{3}$$

where $\tilde{m}$ is mass flow rate, $\tilde{p}_t$ is pressure inside pulse tube, V_{pt} is pulse tube volumn. k is specific heat ratio and R is gas constant of helium. Subscript h and c mean hot and cold end, respectively. In modified theory, thermodynamic non-symmetry effect at the hot end of pulse tube is also considered by us to calculate both the refrigerating effect and oscillating gas temperature. Then mass flow rate at the cold end of pulse tube is corrected according to the results. By this, the calculation accuracy will be greatly improved. Another reason for calculating hot end no-symmetry effect is to observe movement of the hot end gas element to avoid its mixing with the gas element entering the pulse tube from cold end. In fact, this condition set one of the limits for the maximum orifice opening in our calculation.

Cold Losses

Cold losses are caused by several factors including axial heat conduction, irreversible losses inside the low temperature switch valve, inefficient heat regeneration inside the recuperative heat exchanger, etc. Because the heat flow carried by high pressure gas from HEX is intercepted by the liquid nitrogen pre-cooler, we neglect HEX's influence in calculation and assume that helium gas is at liquid nitrogen temperature when it leaves the pre-cooler.

Axial Conduction Loss Q_{cond}. As long as there is a temperature gradient in the axial direction, conduction loss is inevitable. Because of low conductivity of helium gas, we only consider the heat conducted by solid materials. Nevertheless, it's hard to estimate the heat conducted by the filling materials inside LEX. According to the reference,[4] we assume that the effective conductivity of the filling screens is one order smaller than the bulk materials. In fact, this loss only plays a very small part in total loss.

Losses Inside The Low Temperature Switch Valve Head Q_{valve}. Irreversible losses inside the valve head include loss due to sudden change of the flow path and friction loss, etc. In our switch valve, two solid faces, on which there are grooves and holes, are pressed together by difference between high pressure and pressure inside the pulse tube to provide a switch mechanism. Thus heat will be generated by friction at the pressed faces. Because the switch valve is located at the cold end, this loss is quite serious to the refrigerator's performance.

Inefficient Cold Regeneration Loss Q_{recu}. If heat regeneration inside the recuperative heat exchangers is imperfect, temperature of high pressure helium gas entering the pulse tube at cold end will be higher than cold end temperature, thus cause loss to refrigeration capacity. This loss can be expressed as:

$$Q_{recu} == (1-\eta)(\langle H_{i,hot}\rangle - \langle H_{e,cold}\rangle) \tag{4}$$

where $\langle H_{i,hot}\rangle$ and $\langle H_{e,cold}\rangle$ are time averaged enthalpy brought by high pressure gas at hot end of LEX and low pressure gas entering LEX at cold end, respectively. η denotes regeneration efficiency. In fact, the heat exchangers used by us is a kind of simplified perforated stack heat exchanger, whose heat transfer and friction characteristics are unknown. In addition, the flow pattern inside two separate passages is different from the continuous flow inside ordinary recuperative heat exchangers. When gas inside either passage is flowing, gas in the other passage is stagnant. This will deteriorate heat transfer rate. Because of difficulty of analysis, we simplify it by setting values of efficiency and check its influence on the performance. After all, highly efficient recuperative heat exchangers are commercially available, and our emphasis is on the ideal performance of this new design of pulse tube refrigerator.

SIMULATION RESULTS AND DISCUSSIONS

Based on the discussions above, a computer program was made to simulate both the cooler without the liquid nitrogen pre-cooler[2] and the one in this article. Fig 2 to 5 give some simulation results. Fig.2 and.3 show influences of orifice opening and recuperative heat exchanger efficiency on the lowest temperature obtainable on cryocooler introduced in ariticle[2] and in this article, respectively. Because the simulation focuses mainly on the influence of exchanger efficiency on the lowest temperature, the axial conduction loss and the valve loss are omitted in most cases, except when some comparisons are needed to show their influence.

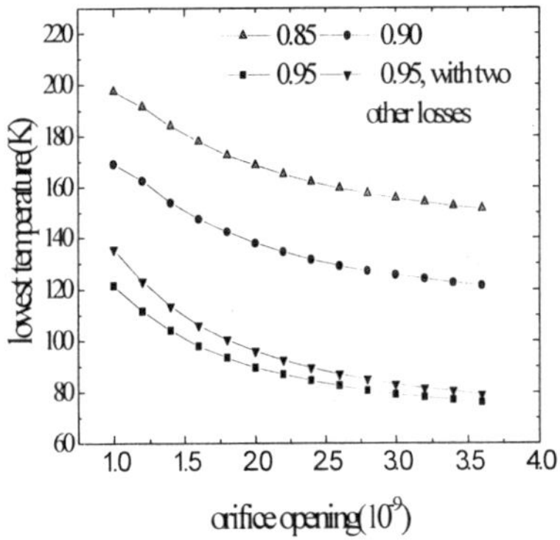

Figure 2. Lowest temperature vs. orifice opening with pressure amplitude 3 bar, frequency 5Hz on the crycooler without pre-cooler

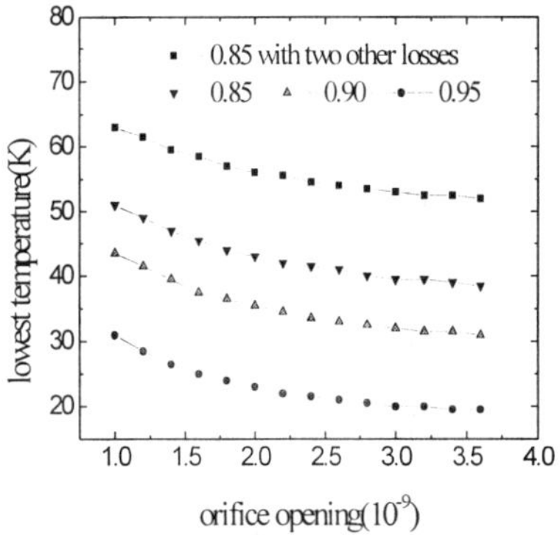

Figure 3. lowest temperature vs. orifice opening with pressure amplitude3 bar, frequency 5Hz on the cryocooler with pre-cooler

Fig.4 and Fig.5 show the influence of frequency and recuperative heat exchanger efficiency on the lowest temperature obtainable on the refrigerator without and with pre-cooler, respectively. Only the heat exchange loss has been considered. It can be noted that in these figures, the lowest temperature changes monotonously according to varied parameters. This is due to neglecting both the dependence of pressure amplitude on frequency when a compressor is determined and the dependence of exchanger efficiency on mass flow rate inside heat exchanger. From the simulation results given in Fig 2 and the experimental results given in the ariticle,[2] we can roughly estimate that the efficiency of the heat exchangers is higher than 0.85 since the ideal performance under this condition can only reach around 180K too. While according to Fig 3, with efficiency 0.85,

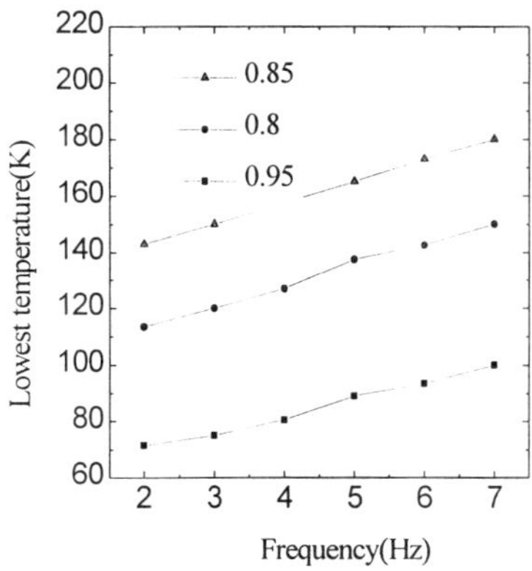

Figure 4. Lowest temperature vs. Frequency on cryo-cooler without pre-cooler, pressure amplitude 3bar, orfice opening 2e-9

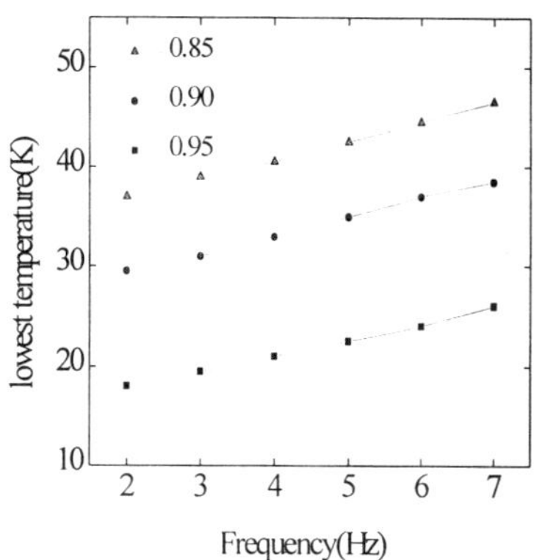

Figure 5. Lowest temperature vs. frequency on pre-cooled cryocooler with pressure amplitude 3bar, orifice opening 2e-9

which means rather poor performance as cryogenic recuperative heat exchangers, our liquid nitrogen pre-cooled cryocooler is still simulated to be able to reach temperature below 65K, which is much lower than the experimental results. There may be several reasons underlying this discrepancy. Firstly, the valve loss has been underestimated at lower temperature. Anyway, use of this kind of rotary valve at cold end is not a good idea. In further research work, we may use electro-magnetic valves, whose electro-magnetic parts are placed outside vacuum shield, to reduce this loss. Secondly, efficiency of the recuperative heat exchanger may decrease rapidly with the decrease of temperature. Our recuperative heat exchanger is less efficient than ordinary ones in two respects: degraded radial heat transfer because of separating wall at the cost of simplicity and the flow pattern inherent in our machine. The first problem can be easily solved because highly efficient recuperative heat exchangers are commercially available, and of course, with much higher prices. Another alternative is to use recuperative gap heat exchangers, as used in Boreas cryocooler. [5] However, the change of flow pattern inside our heat exchangers requires some fine design. In reference to Boreas cryocooler again, we can insert two gas reservoirs between the heat exchanger and the switch valve, whose volumes are large enough to ensure simultaneous continuous flow in both passages of the recuperative heat exchanger to enable higher heat transfer coefficients.

REFERENCES

1. C. Wang, G. Thummes and C. Heiden, A two stage pulse tube cooler operating below 4K, Cryogenics, 37:159(1997).

2. Jingtao Liang, Weijia Sun, Luwei Yang, Shiyao Bian and Yuan Zhou, Test of a recuperative pulse tube refrigerator with simpilified perforated plate heat exchanger, presented at 1999 CEC conference
3. Liang, J. Development and experimental verification of a theoretical model for pulse tube refrigeration, doctoral thesis, Institute of Engineering Thermodynamics, Chinese Academy of Sciences,1993
4. De.Waele A.T.A.M., Steijaert P.P and Gijzen J., Thermodynamic aspects of pulse tubes, Cryogenics, .37:.313(1997).
5. U.S patent, No. 5,345,769

DEVELOPMENT OF A 5W AT 80K STIRLING PULSE TUBE CRYOCOOLER

Y. Hiratsuka[1], Y. M. Kang[1], N. Fujiyama[2],T. Sotojima[2]
and Y. Matsubara[3]

[1]MEC Laboratory DAIKIN INDUSTRIES, LTD., 3 Miyukigaoka,
Tsukuba 305-0841, Japan

[2]Mechanical Engineering Laboratory, DAIKIN INDUSTRIES,LTD.,
1304 Kanaoka-cho, Sakai, Osaka 591-8511, Japan

[3]Atomic Energy Research Institute, Nihon university
7-24-1 Narashinodai, Funabashi 274-0063, Chiba, Japan

ABSTRACT

Pulse tube cryocoolers have recently received considerable attention due to their higher reliability and lower vibrations than other regenerative cryocoolers, e.g., GM and Stirling cryocoolers. The pulse tube cryocoolers used in this study are similar to Stirling cryocoolers except that they do not require any moving displacer in the cold head. This reduces mechanical complexity, increases reliability, reduces induced vibrations, and makes these cryocoolers suitable for electronics applications, especially for superconducting filter subsystems used in the base stations of mobile telecommunication systems.

We developed this Stirling type cryocooler with a design cooling capacity of 5W at 80K and we presented a progress report on the development at the 10th ICC.[1] Recently for a compressor input power of 200W, this prototype has achieved a cooling capacity of 5.51 W at 80K, with a minimum temperature of 47K and a maximum efficiency of 9.07% Carnot. Measured test parameters include cooling loads, cooling temperature, input power, influence of pulse tube inclination and the influence of expander shape, including in-line and U type expanders. Moreover, to aid the design of cryocoolers, we developed a numerical analysis technique, which we validated against our experimental results.

INTRODUCTION

The development of high temperature (HTc) superconducting devices has increased the need for small cryocoolers. Pulse tube cryocoolers, which have no moving parts in the cold section, are more attractive than other small cryocoolers because of their high reliability, simpler construction, and lower vibration levels. Pulse tube cryocoolers are cooled by adiabatic expansion of a gas piston, and the gas piston phase angle is controlled by the phase shifter. In 1992, E. Tward et al. developed a Stirling type pulse tube cryocooler with a cooling capacity of 1W at 80K,[2] similar to the power required for a conventional Stirling cryocooler. After that, many new pulse tube cryocoolers have been developed, mainly for military or for space use. In 1998, J. L. Martin et al. designed a low-cost prototype Stirling type pulse tube cryocooler for civilian electronics applications, especially for cooling superconducting filter subsystems in the base stations of mobile telecommunication systems.[3] We designed a prototype cryocooler to demonstrate the feasibility of making a long life, low noise cryocooler for cooling HTc superconducting devices. To optimize the cycle efficiency, we also developed a numerical model for evaluating pulse tube cryocooler cooling capacity and heat losses.

This paper describes the development of a prototype cryocooler with a target cooling capacity of 5W at 80K. We describe both measured and simulated cryocooler performance.

GENERAL DESIGN

A schematic of the proposed in-line type pulse tube cryocooler is shown in Figure 1 and its specifications are shown in Table 1. The cryocooler is a split Stirling type pulse tube with opposed pistons that are driven by a linear motor. The compressor has been developed for Stirling cryocoolers, with an outer diameter of 91mm and a length of 178.4mm, respectively. The compressor is connected to the expander by tubes that are about 100mm long. The piston position is monitored using a laser vibrometer. Pressure transducers are mounted near the compressor discharge head, near the hot side of the regenerator, and near the pulse tube, and used to determine the mass flow rate through the phase shifter. These measurements are used to calculate both the pressure-volume (P-V) work of the compressor and the equivalent P-V work of the expander.

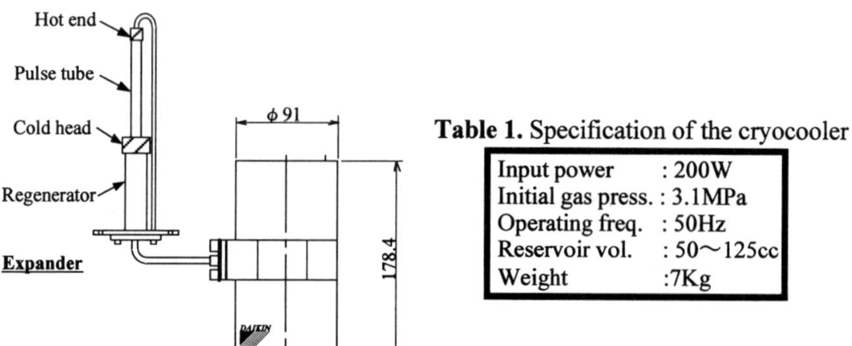

Table 1. Specification of the cryocooler

Input power	: 200W
Initial gas press.	: 3.1MPa
Operating freq.	: 50Hz
Reservoir vol.	: 50~125cc
Weight	:7Kg

Figure 1. Schematic of the prototype pulse tube cryocooler.

EQUIVALENT P-V METHOD

We measured and calculated the equivalent P-V work for various operating conditions, and used the P-V work in sensitivity studies to identify the key operating parameters that control the cooling performance (presented by Y.Matsubara[4]). The gas in the pulse tube can be divided into the three parts : I) the hot end of the pulse tube, II) the gas displacer, and III) the cold head of the pulse tube (Figure 2). At experimental value, We calculated the mass flow rate through the inertance tube and the regenerator from the measured pressure trace. The phase shifter of the pulse tube can be represented by the electric circuit shown in Figure 3, and it is possible to calculate the mass flow rate of the inertance tube and the regenerator as

$$m_{rg} = \alpha_{rg} (P_{co} - P_{pt}) \tag{1}$$

$$m_{it} = \alpha_{it} (P_{pt} - P_{re}) \tag{2}$$

where

$$\alpha_{it} = \frac{\mathrm{COS}(\omega t - \theta) \times \rho_m}{\sqrt{R_{it}^2 + \{\omega \times L_{it} - 1/(\omega \times C_{it})\}^2}} \tag{3}$$

with

$$\theta = \mathrm{TAN}^{-1}[\{\omega \times L_{it} - 1/(\omega \times C_{it})\} / R_{it}] \tag{4}$$

and α_{rg} and α_{it} are the coefficients of discharge of the regenerator and of the inertance tube, respectively. P_{co}, P_{pt} and P_{re} are the pressure of the compressor, of the pulse tube and of the reservoir, respectively. $C_{it} = KV_{it}$ is the capacitance, $L_{it} = \rho_m l / A_r$ is the inductance, $R_{it} = 8\pi \eta l / A_r^2$ is the resistance, and l, A_r, K, ρ_m, and η are the length, cross-section, compressibility of the fluid, density, and viscosity in the inertance tube, respectively. The gas displacer volume in the pulse tube can therefore be determined by solving Eqs. (1) and (2). The equivalent P-V work and the cooling capacity at the cold head can then be calculated as

$$W_{ch} = \oint P V_{III} \, dt \tag{5}$$

$$Q_{ch} = W_{ch} - \langle H_{rg} \rangle - \langle H_{pt} \rangle - \langle H_{st} \rangle \tag{6}$$

where H_{rg} is a regenerator heat loss. For an infinitely large matrix heat capacity, the regenerative effectiveness can be expressed as $\varepsilon = NTU / (NTU + 2)$. Then for packed screens, the friction factor and the Nusselt number are referred to as experimental counter-flow equations.[5] In Eq. (6) H_{st} is the combined conduction and radiation loss, and H_{pt} is the pulse tube enthalpy loss, which is calculated similar to the shuttle heat loss as

$$\langle H_{pt} \rangle = \pi \frac{d_{pt}}{2} k X_{pt}^2 \frac{(T_{ht} - T_{ch})}{l_{pt} h_{gap}} \tag{7}$$

where k is the thermal conductivities of helium gas in the pulse tube, X_{pt} is the gas-displacer stroke length, h_{gap} is the gap of between the pulse tube and the gas displacer, and d_{pt} and l_{pt} are the diameter and the length of the pulse tube, respectively.

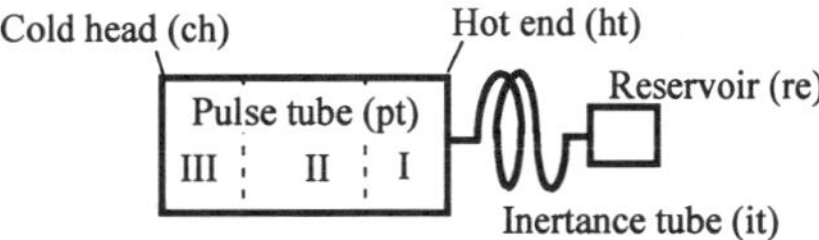

Figure 2. Equivalent P-V model of the cryocooler

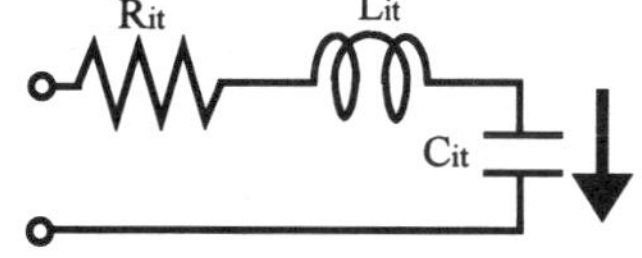

Figure 3. Equivalent electric circle

RESULT & DISCUSSION

Cryocooler Performance

Figure 4 shows the measured cooling performance of our prototype pulse tube cryocooler. The system was filled with helium gas up to 3.1MPa and the vacuum chamber was evacuated to about 0.13MPa (10^{-3} torr). The diameter and length of the regenerator, pulse, and inertance tubes were optimized to get the best cooling capacity. For 200W input power, the no load temperature was 47K, the cooling capacity was 5.51W at 80K and the P-V work was about 102W. The pressure ratio, P_{max}/P_{min}, was 1.19 in the compressor and 1.13 in the cold head. The efficiency of our compressor was 51%, but if a higher efficiency compressor is used the cooling capacity at a given temperature should increase. The cooldown time from the room temperature of 300K to 80K was about 16 minutes with about 100g weight cold head.

Figure 5 shows the efficiency and cooling capacity at 80K vs. compressor input power. The efficiency was 7.57% Carnot at 200W input power and the highest efficiency was 9.07 % Carnot at 80W input power.

Influence of Pulse Tube Inclination

The performance of pulse tube cryocoolers may be influenced by natural convection, which is affected by the pulse tube inclination. For pulse tube cryocoolers that are inclined during operation, therefore, the operating conditions must be determined to minimize the effect of natural convection on the cryocooler performance. In general, high operating frequency and small diameter pulse tubes reduce the convective heat loss. We measured the effect of the pulse tube inclination on the performance. The cooling capacity ratio at 80K and for vertical inclinations varying from 0° to 180° is shown in Figure 7. The cooling capacity ratio shown in Fig. 7 is the measured cooling capacity at each angle divided by the cooling capacity for the pulse tube cold head oriented downward ($\theta = 0°$). The data shown in Fig. 7 indicates that at 80K cooling temperature there is no influence of the pulse tube inclination.

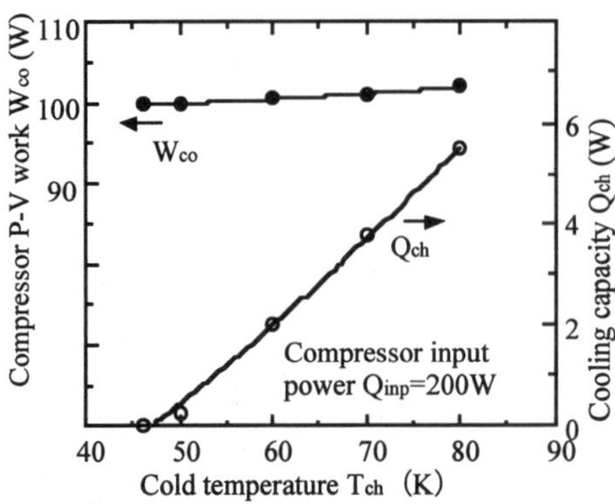

Figure 4. Measured cryocooler cooling capacity vs. cooling temperature

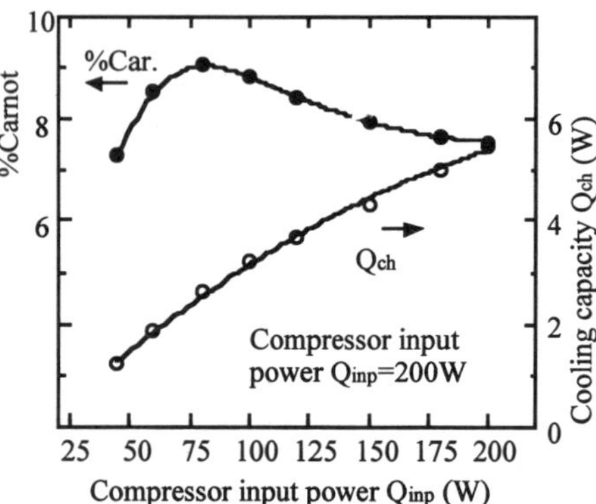

Figure 5. Measured cryocooler cooling capacity vs. compressor power

Difference of Expander type

Of the available cryocooler configurations, the U type is compact and therefore convenient for applications requiring a small unit. Although a compact form is desirable, the cooling performance should not be degraded. We therefore evaluated the effect of the expander configuration on the cooling performance of our prototype cryocooler. For in-line type pulse tube cryocoolers, the pulse tube is located on axis with the regenerator, and for U type cryocoolers, shown in Figure 8, the pulse tube and regenerator are parallel to each other and are the same size as that of the in-line type cryocooler. Figure 9 shows that the performance of our pulse tube U type cryocooler is 4.81W at 80K, with a minimum temperature of 48K. Compared with the cooling capacity of 5.41W at 80K for the in-line system, the cooling capacity of the U type cryocooler is about 11% lower. This indicates that the shape of the cold head strongly influences the cooling performance. Because we did not optimize the phase shifter specifically for a U type configuration, however, there is still the possibility of improving the cooling capacity of the U type configuration above the performance that we measured in this work.

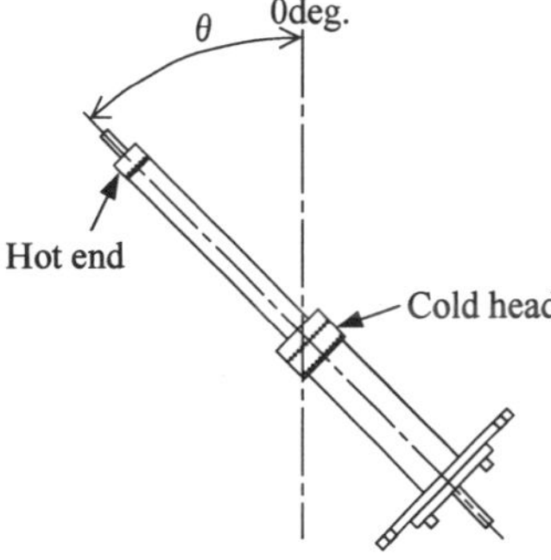

Figure 6. Cryocooler tilt angle

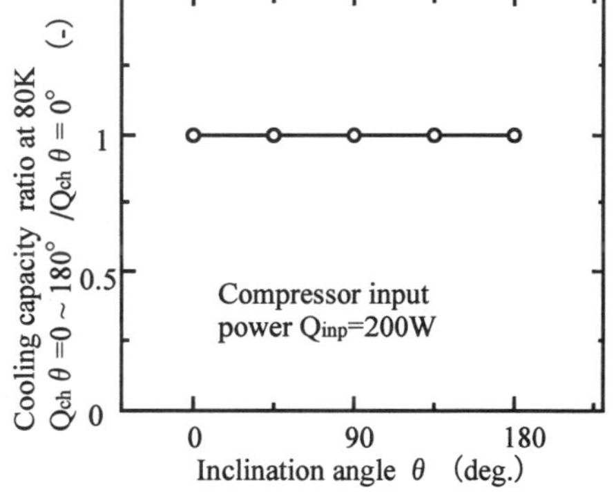

Figure 7. Measured cryocooler cooling capacity vs. tilt angle

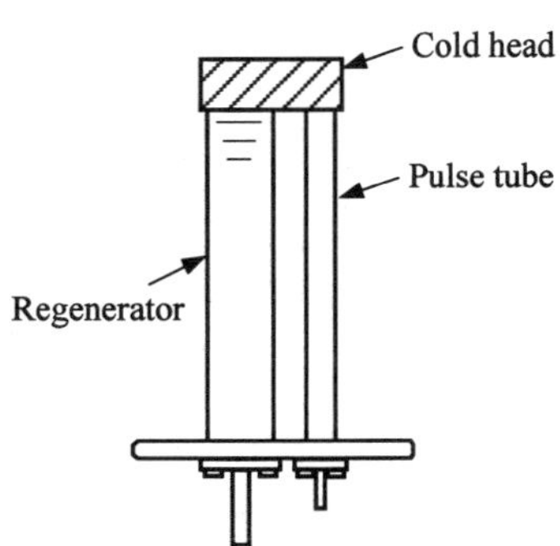

Figure 8. U type expander

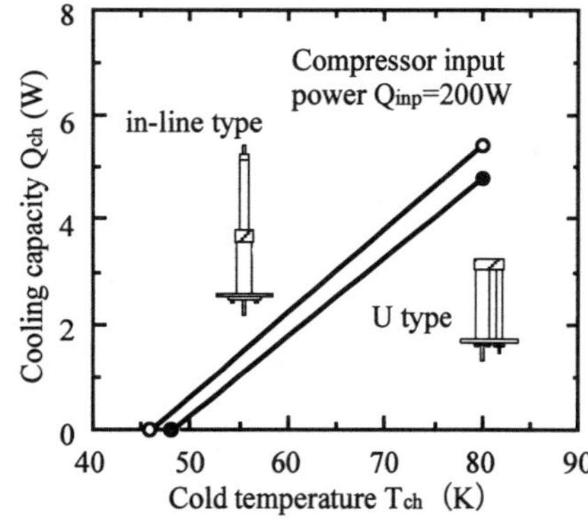

Figure 9. Measured cryocooler cooling capacity vs. cooling temperature for in-line and U type expanders

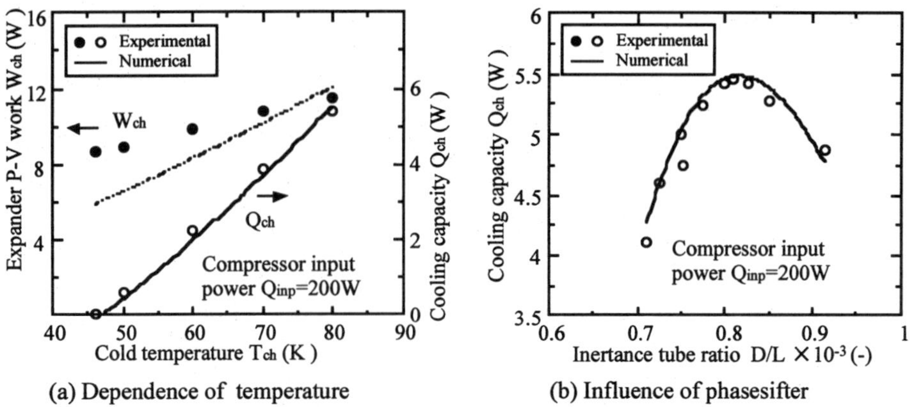

(a) Dependence of temperature (b) Influence of phasesifter

Figure 10. Measured and simulated cryocooler cooling capacity vs. (a) cooling temperature and (b) inertance tube diameter/length ratio

Table 2. Experimental and numerical heat loss analysis at 80K

	Experimental	**Equivalent P-V method**
Cooling capacity@80K	5.51W (Meas.)	5.58W
Expander P-V Work	12.46W (Cal.)	12.42W
Static heat loss	1.43W (Meas.)	1.43W
Estimated loss	5.52W	5.40W
Regenerator heat loss	2.72W (Cal.)	2.58W
Pulse tube heat loss	2.22W (Cal.)	2.82W
error	0.58W	—

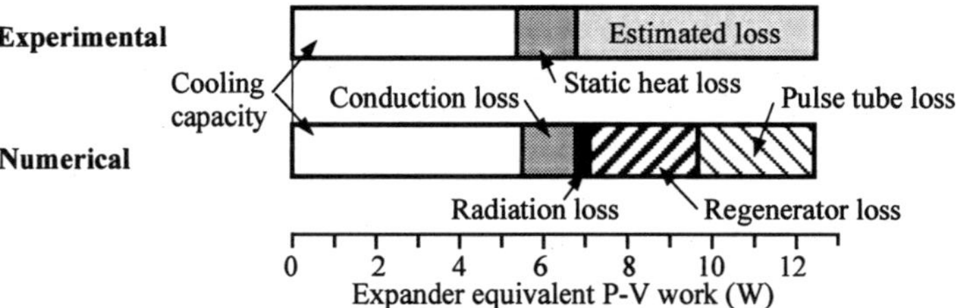

Figure 11. Heat loss analysis at 80K

Comparison of Experimental & Heat Loss Analysis

Figure 10 (a) shows measured and calculated an equivalent P-V work and cooling capacity for the expander vs. the cold head temperature, and Fig. 10 (b) shows measured and calculated cooling capacity vs. inertance tube diameter/length ratio. We note that the results shown in Figs. 10 (a) and (b) indicate that the equivalent P-V method roughly agrees with the measured values, indicating that the equivalent P-V method accurately represents the operating cycle of the cryocooler, and that this model can be used to design pulse tube cryocoolers. We are continuing to improve the model accuracy.

Figure 11 shows the heat loss of the expander due to various processes. The measured equivalent P-V work was 12.46W, composed of 5.51W cooling, 1.43W of combined conduction and radiation loss, and an estimated loss of 5.52W. This estimated loss may consist of regeneration loss of 2.72W and pulse tube loss of 2.22W. To maximize the cooling capacity, the origin of these heat losses must be identified and minimized.

CONCLUSION

We developed a split Stirling pulse tube cryocooler and analyzed its performance experimentally and numerically. The significant results of our research are:

1. For 200W input power, the no load temperature was 47K and the cooling capacity was 5.51W at 80K, for an efficiency of 7.57% Carnot. The highest efficiency was 9.07% Carnot at 80W input power.
2. There was no influence of pulse tube inclination on cryocooler performance.
3. The cooling capacity of the U type configuration decreased about 11% compared to that of the in-line type configuration.
4. The measured and simulated performance are in roughly agreement.
5. We successfully estimated the heat loss from the measured equivalent P-V work in the expander.

REFERENCES

1. Y. Hiratsuka al., : "Development of a 1 to 5W at 80K Stirling Pulse Tube Cryocooler" : ICC10 th`98
2. E. Tward et al., : "Miniature Pulse tube Cooler" : ICC 7th`92
3. J. L. Martin et al., : "Design Consideration for Industrial Cryocoolers" : ICC10 th`98
4. Y. Matsubara et al., : "Work-loss Distribution on GM-type Pulse Tube Coolers" : Cryogenic Engineering Vol. 33 No.4 `98 (in Japanese)
5. M. Tanaka et al., : "Flow and Heat Transfer Characteristics of the Stirling Engine Regenerator in an Oscillating Flow" : JSME Vol. 33`90

PERFORMANCE OF TRW NEW GENERATION PULSE TUBE COOLERS

C.K. Chan and Tanh Nguyen

TRW
One Space Park
Redondo Beach, CA 90278

ABSTRACT

Using the TRW next-generation cooler technology, two IMAS Engineering Models (S/N101 and 102) with different cold head configurations were designed, fabricated, and tested. This new generation of coolers was designed with much lighter weight and compact volume. They feature user friendly interface and use simpler electronics with lower EMI and ripple current than our first generation cooler. The refrigeration performance of the coolers was measured over a wide range of input power, cooling temperatures, and refrigeration loads, and at various operating strokes and frequencies and heat rejection temperatures. The design was verified for space application with a series of flight environmental tests and an ongoing life test at TRW.

INTRODUCTION

The Integrated Multispectral Atmospheric Sounder (IMAS) instrument is an advanced concept instrument being examined by JPL as a second-generation atmospheric sounder for making precision air temperature measurements from space[1]. Key to reducing the mass and power of the IMAS instrument is achieving a new generation of long-life cryocoolers (1) with significant mass and size reductions over the AIRS-class cryocooler[2], and (2) with significant cooling capacity increase over our miniature coolers[3]. Both the AIRS and the Miniature Pulse Tube belong to the first generation of pulse tube and Stirling Coolers, the mass of which is proportional to the maximum allowable compressor input power as shown in Figure 1. The mass of other Stirling space coolers[4] are included for comparison.

From the mass breakdown of the cooler system (shown in Table 1), the majority of the light-weighing potential for a pulse tube cooler lies in the compressor and its drive electronics. In the new generation design, the compressor size and mass are greatly reduced by using an entirely new approach to the basic layout of the motor design and coil suspension concept, which can produce larger mechanical PV work in a smaller package along with lower radiated DC and AC magnetic fields. In the past, control functions of DC

Advances in Cryogenic Engineering, Volume 45.
Edited by Shu *et al.*, Kluwer Academic / Plenum Publishers, 2000.

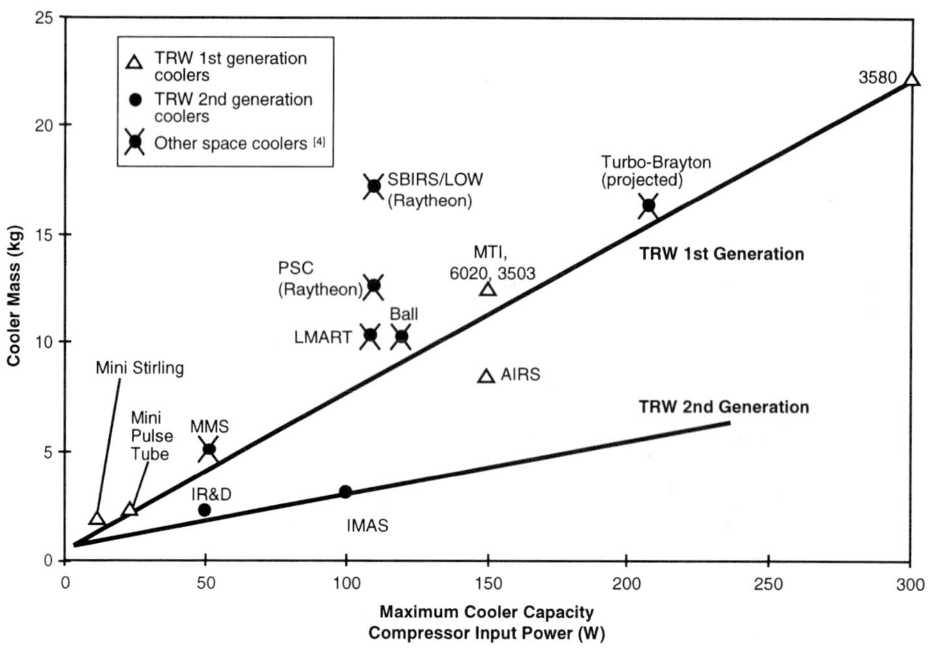

Figure 1. Mass comparison with first-generation coolers

Table 1. Mass budgets of first and second generation pulse tube coolers

	First Generation	Second Generation
Flight Instruments	AIRS, MTI, TES	IMAS, HEC
Cooler Mass	8 to 12 kg	3.6 kg
Control Electronics Mass	5 to 7 kg	2.5 kg (projected)
System Mass – Heat Spreader Vacuum Housing	2 to 8.7 kg	0.4 kg (S/N101)
Ripple Current Filter	8 kg (external)	Included in CCE mass
Total System Mass	23 to 35.7 kg	6.5 kg

offset and adaptive vibration have always complicated the control electronics and the control software. The new compressor was designed to help eliminate the needs of these two functions; the vibration force is low enough so that implementing adaptive vibration control is not necessary. Consequently, the electronics mass can be reduced (Table 1).

The system efficiency of the new generation cooler is also increased (1) by having a thermal/mechanical cooler interface that can be integrally mounted onto the radiator with a minimal temperature difference between the aftercooler on the center-plate and the ultimate heat rejection surface, and (2) by replacing the current AIRS external ripple current suppression box with active current suppression in the electronics. Since the completion of the IMAS program, this new ripple suppression was implemented in the Hyperion Flight electronics[3]. For the same cooling capacity, the new design offers a factor of two to three in cooler mass reduction (Figure 1) and a factor of four to five in system mass reduction (Table 1). Hence, our efforts are not limited to reducing mass and power of the new mechanical pulse tube cooler over current state-of-the art technology, but we also enhance total system efficiency by having advanced interface concepts and simpler system function.

This simplicity would lead to increasing reliability and manufacturability and ultimately cost reduction.

This paper describes the design, fabrication, and test of two IMAS flight-like coolers (EM coolers) with slightly different cold head configurations. Extensive measurements of the coolers have been performed. One cooler went through flight-level acceptance testing which includes launch vibration, thermal vacuum cycling, and 250 hour burn-in. At present, that cooler has been in life-testing. This paper reports the test data of these new generation coolers. The self-induced vibration force measurements which involve various electronics configurations is reported in a separate paper at this conference[5].

SECOND GENERATION COOLERS

Two flight-like coolers were designed and fabricated by combining the new compressor technology[1] with two different cold head configurations: one (S/N101) with a self-contained vacuum can over the cold head and the other (S/N102) with a more compact cold head. The envelope and photos of these two coolers are shown in Figure 2 and Figure 3. Both coolers have a linear integral Titanium cold head designed for 0.5W at 55 K. The component arrangement is very similar to the miniature pulse tube cold head configuration[3]. S/N101 has a Titanium surge tank surrounded by a vacuum can, which also serves as a contamination shield. Since the cold head is within a vacuum by itself, it is more convenient to use S/N101 for self-induced vibration and EMI measurements[5]. The S/N102 cold head is more compact; it is 3 inches shorter than cooler S/N101. Cooler S/N102 has an Aluminum surge tank which also serves as a support structure and a conduction path from the orifice block to the compressor center plate. Since S/N102 does not have its own vacuum can, thermal tests were conducted inside an 18-inch diameter vacuum bell jar.

During the development phase, the compressors were tested with LVDT position sensors that provide compressor stroke to input power correlation. The compressor input power was mapped as a function of fill pressure and compressor stroke. The LVDT position sensors were then removed, before the thermal characterization of the coolers.

In addition to the thermal characterization, cooler S/N102 also went through the environmental test of launch vibration level of 14.1 grms, thermal vacuum test between 220 K and 320 K, and a 250-hour burn-in test. As of June 14, 1999, cooler S/N102 accumulated 3,076 hours of operation.

Different features of S/N101 and S/N102 are listed in Table 2, which also includes the tests that have been performed with these two coolers.

COOLER PERFORMANCE

The coolers were tested for a broad range of cold block temperatures from 120 to 30 K at 54 Hz with different heat reject temperatures. The required operational frequency to synchronize with the IMAS instrument is 54 Hz. Tests with cooler S/N102 were performed in a thermal vacuum test facility while cooler S/N101, which has its own vacuum can, can be operated on a test bench. A schematic of the test facility is shown in Figure 4.

The heat rejection temperature on the baseplate was regulated by a cooling fluid loop and a heat exchanger bolted to the mechanical interface of the compressor (see Figure 3b and 3d). These temperatures which varied from –53 to 47°C were measured by silicon diodes.

The cold head was wrapped in multilayers of aluminized Kapton to minimize the parasitic radiation leakage while the vacuum level was kept less than 1 x 10^{-5} during the test. The cooler was driven by low distortion audio amplifier and sinusoidal voltage waveforms. Power to the cooler was measured by Vahalla true-RMS power meters.

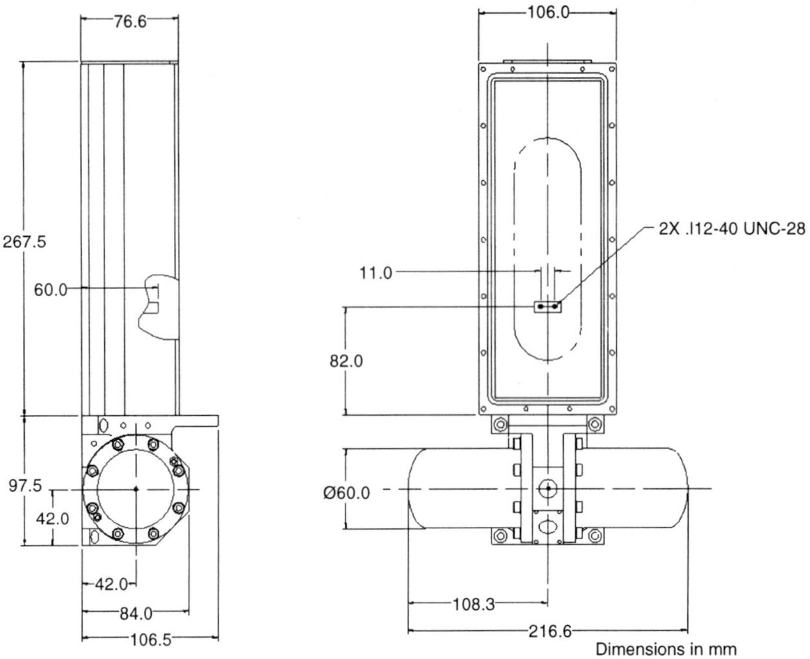

Figure 2(a). S/N101 cooler envelope

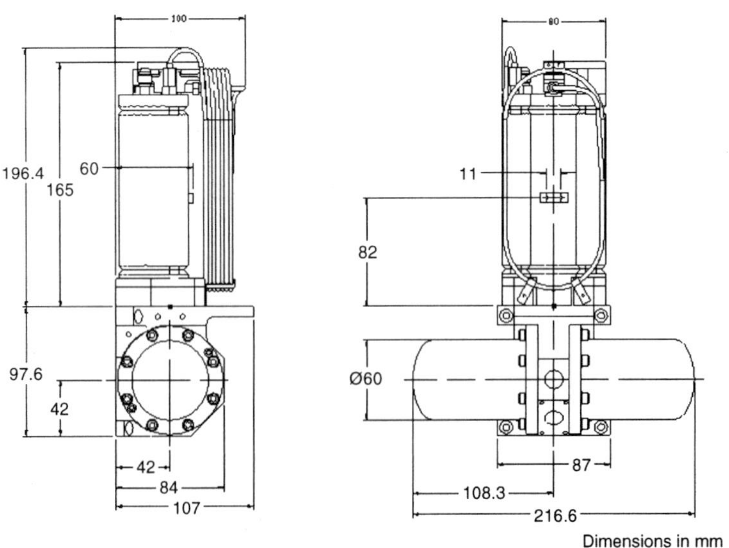

Figure 2(b). S/N102 cooler envelope

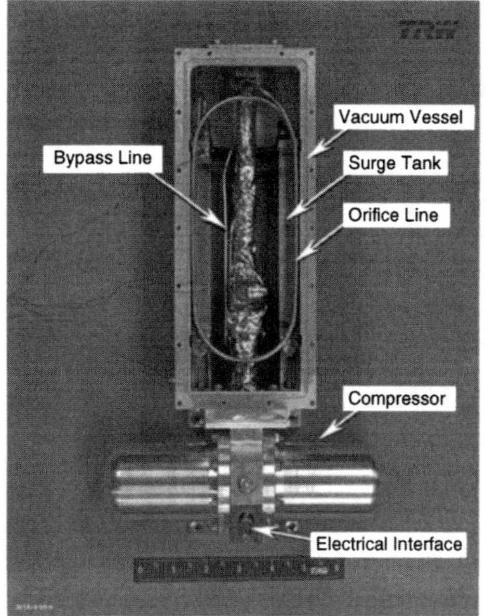

3(a). Front View of S/N101

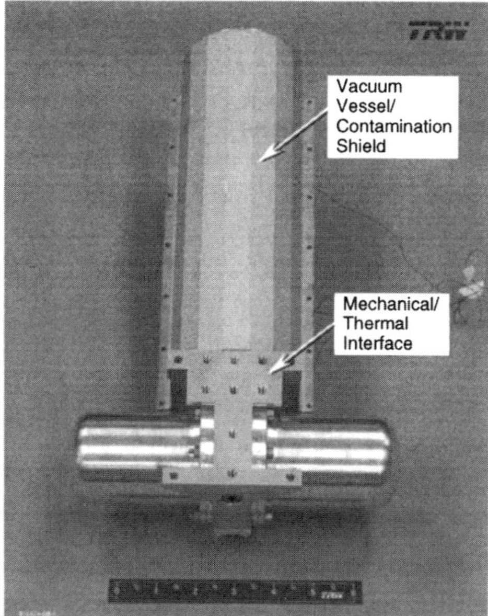

3(b). Back View of S/N101

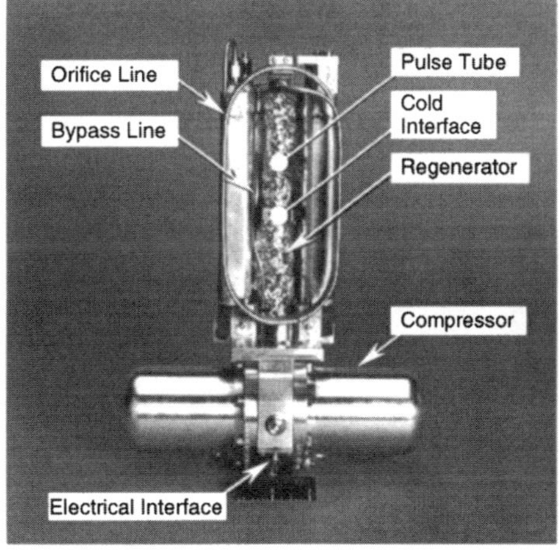

3(c). Front View of S/N102

3(d). Back View of S/N102

Figure 3. Photo of two IMAS coolers

The cold block was instrumented with a silicon dicode and an electrical heater to simulate the cooling load. For a given input power to the compressor, both the temperature and the applied load at the cold block were recorded at a steady equilibrium state. Figure 5 shows the load lines of constant input power levels. Despite of their configuration difference, the two coolers have similar performance as shown in the thermal performance map in Figure 6 which describes the thermal performance in terms of efficiency (or specific power) to the compressor input power, the cold block heat load, and the cold block temperature. Data were plotted for input power up to 85 W, cold block temperature from 50 to 100 K, and cold block heat load up to 3 W.

Table 2. Coolers S/N101 and S/N102 packaged differently for flexible use

	IMAS S/N101	**IMAS S/N102**
Interface/Dimension	Figure 2 (a)	More compact, Figure 2 (b)
Compressor	No position sensor	No position sensor
Cold Head	Linear integral pulse tube Ti surge tank Al vacuum can/contamination shield	Pulse tube length shortened 3 inches Al surge tank No vacuum tank/contamination shield
Test Performed	Cooler performance with power and heat reject temperature Self-induced vibration force measurement with and without waveform control No environmental test	Cooler performance with strokes and heat reject temperature No force measurement test Launch vibration Thermal vacuum Endurance test Life test
Mass		
Compressor	2.5 kg	2.5 kg
Cold Head	0.25 kg	0.24 kg
Ti Reservoir Tank	0.5 kg	N/A
Al Reservoir Tank	N/A	0.79 kg
Reservoir Line	0.1 kg	0.1 kg
Vacuum/Contamination Shield	0.68 kg	N/A
Total System Mass	4.03 kg (measured)	3.63 kg (measured)

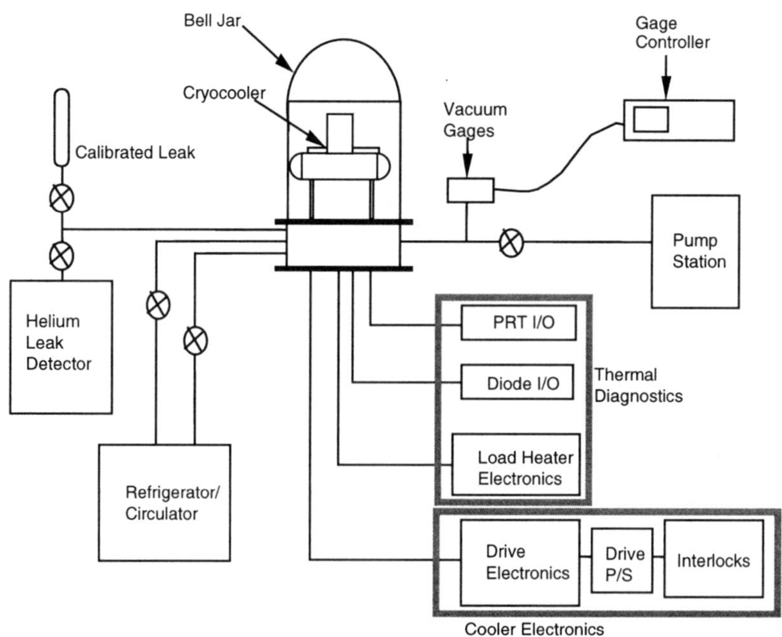

Figure 4. IMAS cryocooler test facility

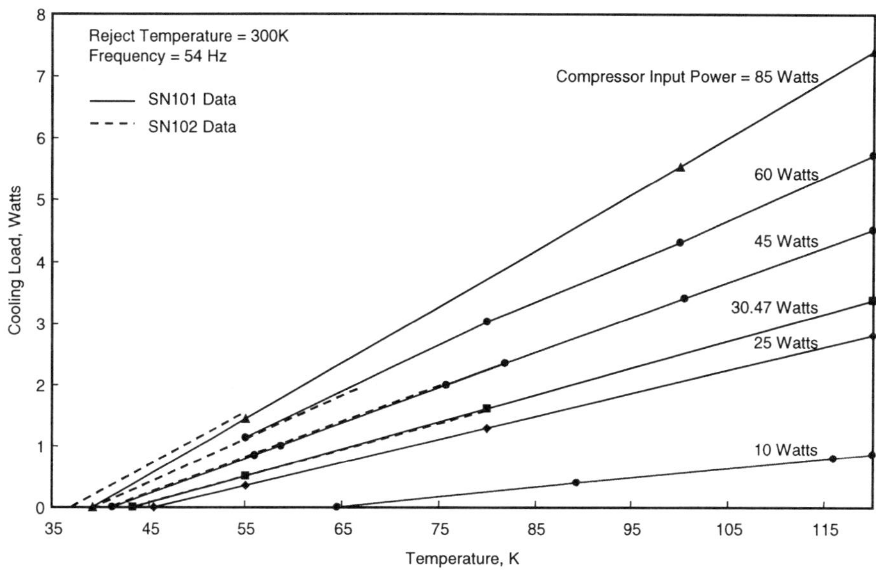

Figure 5. Load lines of constant input power

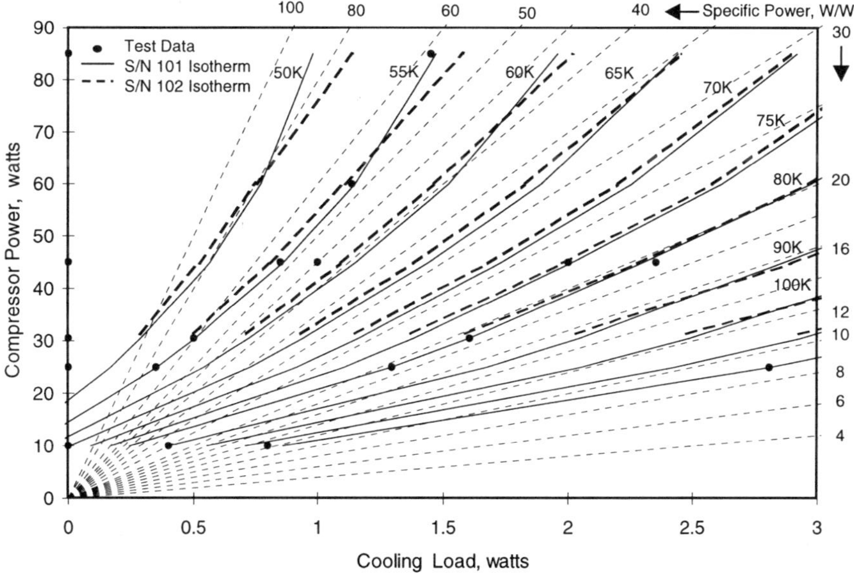

Figure 6. Thermal performance map

EFFECTS OF FREQUENCIES AND HEAT REJECTION TEMPERATURES

The sensitivity of S/N101 cooler performance as a function of the drive frequency is shown in Figure 7, where the cooler thermal efficiency peaks at a drive frequency of 53 Hz. However, the motor loss was not minimized at that frequency. Figure 8 shows the specific power of the cold head at different temperatures when the heat rejection temperature varied from 273 K to 320 K while the input power was maintained constant at 85 W. At 55 K, because of the larger Carnot factor, the reject temperature has more effect on the cooler efficiency.

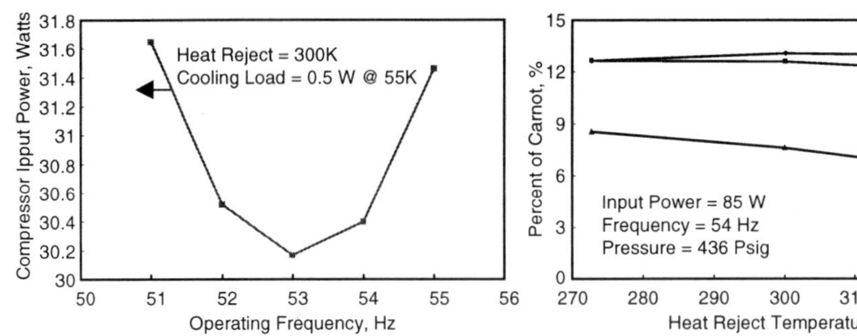

Figure 7. Operating frequency effects on input power

Figure 8. Heat rejection temperature effect on cooler efficiency

ENVIRONMENTAL TESTS AND COOLER ACCEPTANCE

The S/N102 cooler went through a series of environmental tests for space worthiness. Repeatable cooler performance after each environmental test is used as an acceptance criterion. Load lines and the helium leak rate are the measured parameters. Figure 9 displays load line data for the cooler taken prior to and after each of the cooler's environmental tests and shows the helium leak rate measurements after each test. The environmental tests include a launch random vibration of 14.1 grms and a thermal vacuum test from ambient temperature up to 320 K, and then lowered to 220 K. The temperature rise was maintained at 0.2 K/min when the cooler was operated at a fixed input power of 31 W. The burn-in test was performed 250 hours before the cooler was put to a life-test. No performance change of the load line was detected within experimental uncertainty. The measured helium leak rate was two orders of magnitude less than the 10-year life criterion which is 1.33 10^{-6} mbarl/sec.

Figure 10 shows the life test data from February 10, 1999 to June 14, 1999. In the first 1,000 hours, problems associated with the chiller system and the vacuum pump have caused the cold block temperature to rise. After fixing those problems the cooler has been

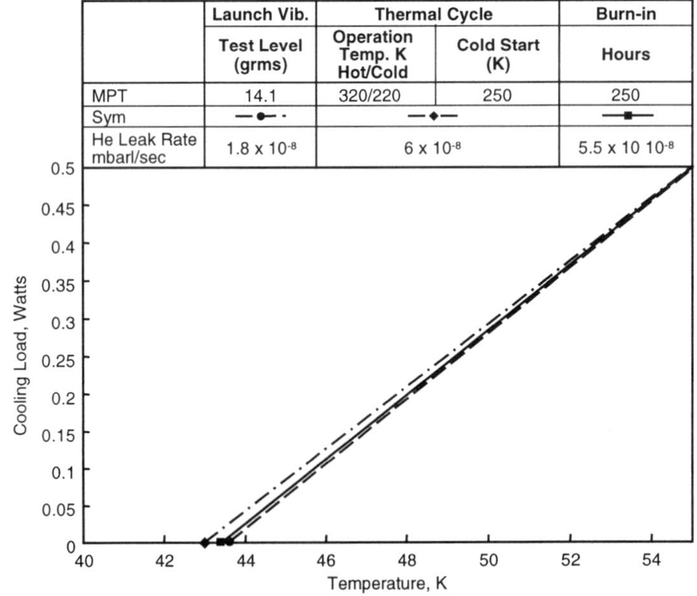

	Launch Vib.	Thermal Cycle		Burn-in
	Test Level (grms)	Operation Temp. K Hot/Cold	Cold Start (K)	Hours
MPT	14.1	320/220	250	250
Sym	—●- ·	—◆—		—■—
He Leak Rate mbarl/sec	1.8 x 10^{-8}	6 x 10^{-8}		5.5 x 10 10^{-8}

Figure 9. Repeatable cooler load lines used as acceptance criterion after each environmental test

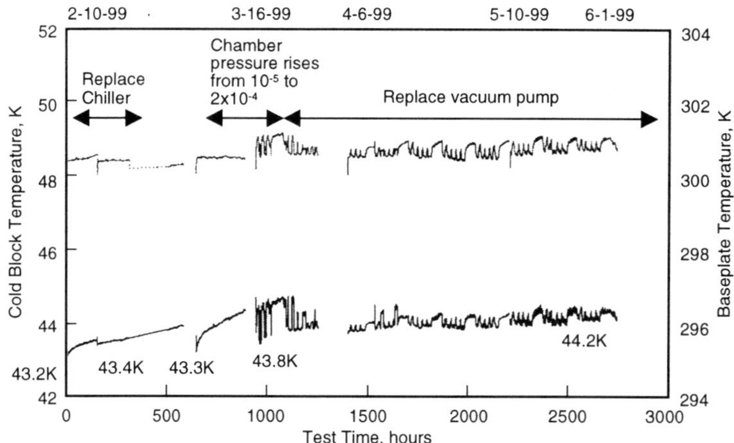

Figure 10. Life test of IMAS S/N102 cooler

running steadily now for more than 2,000 hours. The cold block temperature tracked the heat reject temperature variation as the room temperature varied over a 24-hour period. Both the chiller, which is air cooled, and the drive electronics are affected by the room temperature. When room temperature variation was factored into the result, the cold block temperature can be considered to be constant throughout this period.

CONCLUSION

Two flight-like coolers based on TRW second-generation technology were designed, built, and tested to IMAS specifications. These coolers have simple and efficient mechanical, thermal, and electronic interface. The cooler and system mass are improved by a factor of four over the first-generation cooler of equivalent cooling capacity. No DC-centering nor active vibration control is required, thus simplifying electronics hardware and software by a factor of four. The cooler efficiency has also improved over the first-generation cooler, as shown in Figure 11.

The design has been flight-qualified by launch vibration tested to 14.1 grms, by thermal vacuum functional test from 320 to 220 K and by life-test with no sign of degradation.

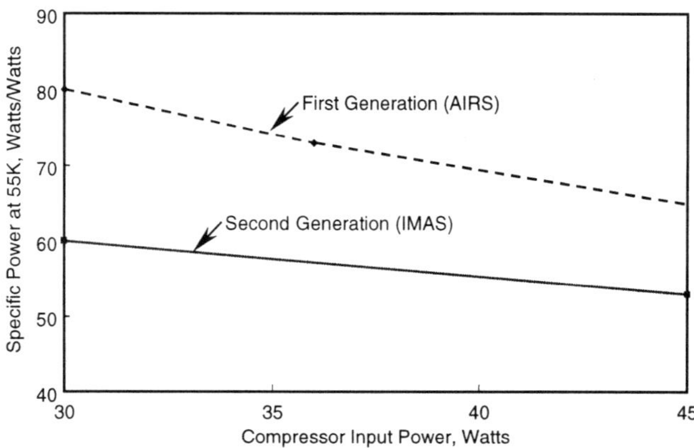

Figure 11. Coolers efficiency comparison

ACKNOWLEDGMENT

The work described in this paper was carried out by TRW. It was sponsored by the NASA – JPL IMAS project.

REFERENCES

1. C.K. Chan, T. Nguyen, R. Colbert, J. Raab, R.G. Ross and D. Johnson, "IMAS Pulse Tube Cooler Development", *Cryocoolers 10*, Plenum Press, New York (1999), pp.139-147.
2. C.K. Chan, J. Raab, R. Colbert, C. Carlson and R. Orsini, "Pulse Tube Cooler for NASA AIRS Flight Instrument" *ICEC 17 Proceedings* (1998), pp 77-87.
3. C.K. Chan, P. Clancy and J. Godden, " Pulse Tube Cooler for Flight Hyperspectral Imaging", to be published in *Proceedings of Space Cryogenic Workshop* (1999).
4. D. Glaister, M. Donabedian, D. Curran, "On Overview of the Performance and Maturity of Long Life Cryocoolers for Space Applications", *Cryocoolers 10*, Plenum Press, New York (1999), pp. 1-19.
5. C.K. Chan and C. Jaco, "Low Vibrational Performance of TRW Second Generation Pulse Tube Coolers", to be published in *Adv. In Cryogenic Engineering*, v. 45, Plenum Press, New York.

HYSTERETIC LOOP IN THERMOACOUSTIC OSCILLATIONS

G. B. Chen and T. Jin

Cryogenics Laboratory, Zhejiang University, Hangzhou 310027, P. R. China

ABSTRACT

A hysteretic loop, due to the difference between onset temperature and damping temperature, is recognized in thermoacoustic oscillation for the first time. It is verified by experiments on a thermoacoustic prime mover with nitrogen as the working fluid. The influence due to charging pressure has also been investigated.

INTRODUCTION

Onset is a fundamental topic on which many scientists have worked since the discovery of the thermoacoutic oscillation 200 years ago [1,2]. Onset oscillation becomes apparent from the start of experiments with thermoacoustical devices. Quantitative understanding of thermoacoustics begins with the work of N. Rott [1] and G. W. Swift [2,5]. The proposed "critical temperature gradient", temperature gradient between both ends of the stacks for the system begins to oscillate, is important for understanding thermoacoustic phenomena.

K. W. Taconis observed thermoacoustic oscillation in cryogenic systems for the first time in 1949 while he was making measurements of the vapor-liquid equilibrium of ^{3}He and ^{4}He below 2.19K. This kind of oscillation can be a severe nuisance in cryogenic apparatus. It raises a problem: how to avoid the oscillation and deal with damping of thermoacoustic oscillation.

Onset and damping are usually studied separately. A. Atchley et al.[9] attempted to combine the characteristics below and above onset, with a relation between reciprocal sound qualify and temperature difference. However, they assumed that a steady state exists, and the onset and damping points coincide. S. Zhou and Y. Matsubara [4] experimentally found a stop temperature (different from the onset one) in the oscillation when the heating power keeps a critical value, which is enough to excite while insufficient to maintain the oscillation.

Based on the characteristics of thermoacoustic phenomena, we propose a "hysteretic hoop" in thermoacoustic oscillation. It has been experimentally verified in a thermoacoustic engine with nitrogen as the working fluid.

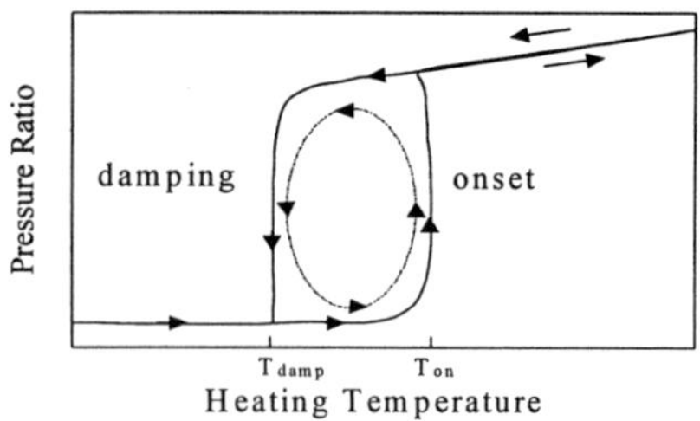

Fig.1. Predicted hysteretic loop in thermoacoustic oscillation

HYSTERETIC LOOP

Hysteresis can be widely found in nature, such as in light induced reaction [6], in boiling heat transfer [7], and in λ transition of liquid helium [10]. Summarizing all the hysteresis phenomena, we can find two common characteristics: 1) multiple steadies states and 2) catastrophe in the state transition.

The onset of the thermoacoustic oscillation is an abrupt process, rather than a gradual one. When the temperature difference of the stack reaches a critical value, oscillation arises abruptly, and the measured pressure ratio also suddenly jumps from a value near zero to a considerable one. The damping process reverses at some temperature, which is usually different from that for onset. So thermoacoustic oscillation also possesses the two characteristics mentioned above. Based on intuition, we think that it is reasonable to find the existence of the hysteretic loop in the thermoacoutic oscillation, as shown in Fig.1.

EXPERIMENTAL VERIFICATION

Experimental Apparatus

The thermoacoustic engine includes the following main parts: heater, stack, cooler, resonance tube and hot buffer [3], as shown in Fig.2. The stack is composed of the brass matrix of 6 mesh and 10 mesh alternatively with the ratio of 1:2. An inner electrical heater is used. And the cooler is a water cooler. The dimensions of the main parts are tabulated in Table 1.

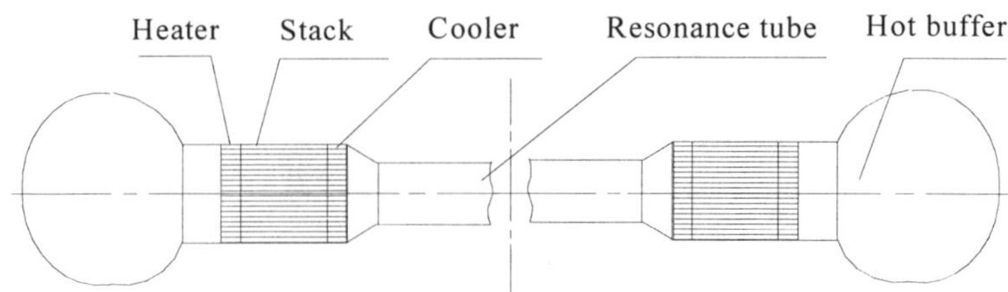

Fig. 2. Schematic of the experimental thermoacoustic engine

Table 1. Dimensions of main parts of the thermoacoustic engine

Items	Hot buffer	Heater	Stack	Cooler	Resonance tube
Diameter , mm	140	52	56	56	36
Length, mm	(1.2L)	50	288	50	4000

The parameters of the sound wave (frequency, amplitude and pressure ratio) can be measured by means of the pressure measuring system, which contains pressure sensor, strain gauge and optic oscillograph. The thermal couples are used to measure the temperatures, which can be read out from a digital recorder (display). Besides, electrical relay and adjuster are used to control and adjust the input power. The relay is connected to the temperature measuring system because its instruction is from the measured temperature.

Experiments

We have experimented on different pressures with symmetrical heating.

The main purpose is to observe the onset temperature and the damping temperature. In order to record the temperature data accurately, the slow heating is expected to lessen the measuring errors. The procedure for symmetrical heating is as follows:

1) Purging the system, and charging with nitrogen.

2) Heating before onset: A considerable temperature difference is required to excite the thermoacoustic oscillation. To speed up the experiment, a relatively high heating velocity (say 2°C/min) is applied, since the accuracy of the temperature measurement is not important for this stage.

3) Onset and heating above onset: When the temperature approaches to the onset temperature, the heating velocity should not be higher than about 0.5°C/min, until the onset point is reached. Meanwhile, it is necessary to adjust the input power of both ends to be balanced. After that, an increase of the corresponding input power will be required with the increase of transition from heat to sound. The sound wave should be recorded in a certain temperature interval until a satisfactory oscillation is reached.

4) Temperature falling and oscillation damping: When the temperature reaches a certain value, we begin on the test of temperature falling process by decreasing the input power. A special attention must be paid to record the damping point.

5) Repeating the same procedures for different charging pressures.

Results

A series of experiments have been carried out with nitrogen of different pressures.

With the input power of dozens of Watts, a constant cool-end temperature assumes due to the light heat load. The temperatures mentioned below refer to the hot-end heating temperatures.

For a typical instance, we focus first on the pressure of 2.4MPa. Fig.3 and Fig.4 present the relative amplitude ($=\Delta P/P_m$, where $\Delta P=(P_{max}-P_{min})/2$, $P_m=(P_{max}+P_{min})/2$) and pressure ratio ($=P_{max}/P_{min}$) with heating temperature, respectively. We can see evidently that when the temperature reaches 227°C, the amplitude abruptly jumps from 0.06% to 1.6%, and the pressure ratio from 1.001 to 1.033. This is the typical initial onset. And then, the amplitude and pressure increase gradually with the heating temperature. The system gradually becomes relatively steady. Some relations between the pressure ratio and temperature in higher heating temperature range are given in Ref [3].

As the temperature reaches 277°C, we begin temperature falling procedure by properly decrease input power. The temperature descends slowly, and the system remains oscillation

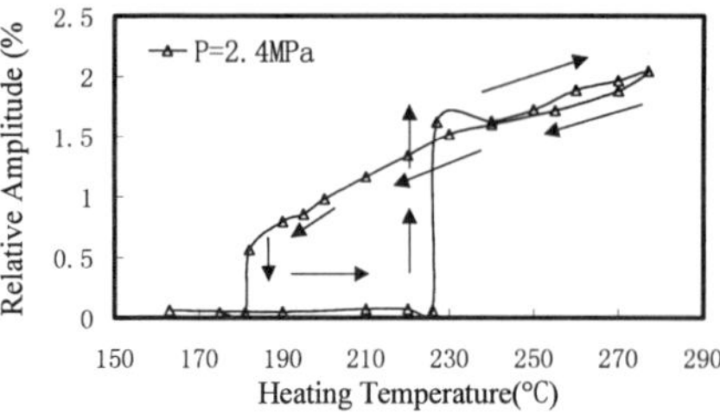

Fig.3. Relative amplitude versus heating temperature

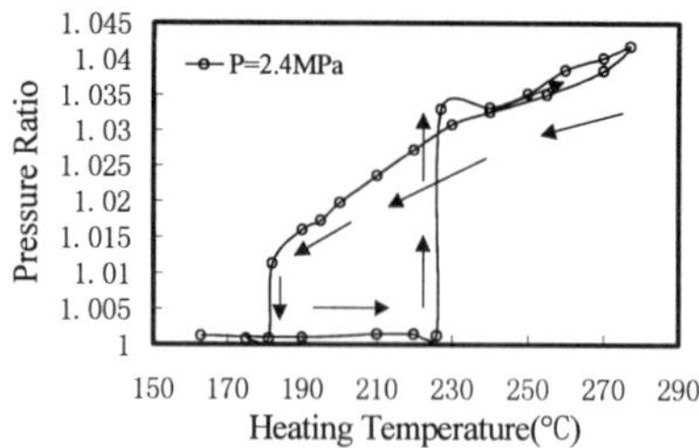

Fig.4. Pressure ratio versus heating temperature

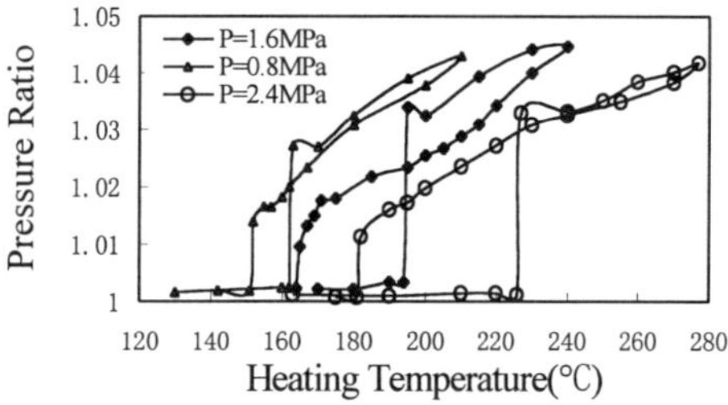

Fig.5. Pressure ratio versus heating temperature for different pressures

at the onset temperature 227°C and even lower. The oscillation disappears abruptly till the temperature falls to 181°C or so, when the relative amplitude and the pressure ratio is 50% of those at 227°C.

The onset and damping do not coincide, i.e., the damping lag the onset. As a result, the two vertical lines together with two approximately horizontal lines form a "hysteretic loop", which confirms our predicted loop as shown in Fig.1.

Fig.5 shows the pressure ratio-temperature curves for the cases of 0.8 MPa, 1.6 MPa and 2.4MPa. Apparently, both the onset temperature and damping temperature increase with the charging pressure, due to the increase of loss with the increasing density. Furthermore, the hysterisis also increases with the charging pressure. The lags for three pressures mentioned above are 12°C, 31°C and 46°C, respectively.

CONCLUSION

A hysteretic loop, due to the difference between the onset temperature and the damping temperature, is recognized in thermoacoustic oscillation for the first time. From experiments, we can see that the hysterisis increases with the increase of charging pressure.

We would like to emphasize that the results in the paper are experimental and intuitive. A theoretical discussion is needed for further understanding the phenomena.

ACKNOWLEDGMENT

The authors wish to express their appreciation for the financial support from the Zhejiang Provincial and National Natural Science Foundation, P. R. China. We would also like to thank Z.Q.Ying, Y.R.Xu and S.S.Feng for their assistance in the experiments.

REFERENCES

1. N.Rott, Thermoacoustics, Advances in Applied Mechanics 20:135 (1980).
2. G.W.Swift, Thermoacoustic engines, J. Acoust. Soc. Am. 84:1145 (1988).
3. G.B.Chen, T.Jin, et al. Experimental study on a thermoacoustic engine with brass screen stack matrix, Adv. Cryo. Eng. 43b:713 (1998).
4. S.L.Zhou and Y.Matsubara, Experimental research of thermoacoustic prime mover, Cryogenics 38:813 (1998).
5. G.W.Swift, Analysis and performance of a large thermoacoustic engine, J. Acoust. Soc. Am. 92:1551 (1992).
6. E.C.Zimmermann and John Ross, Light induced bistability in $S_2O_6F_2 \Leftrightarrow 2SO_3F$: Theory and experiment, J. Chem. Phys. 80:720 (1984).
7. B.A.Hands, Cryogenic engineering, Academic Press, London (1986).
7. A.Atchley, H.E.Bass, T.J.Hofler, and H.T.Lin, "Study of a thermoacoustic prime mover below onset of self-oscillation", J. Acoust. Soc. Am. 91:734 (1992).
8. A.A.Atchley, Analysis of the initial buildup of oscillation in a thermoacoustic prime mover, J. Acoust. Soc. Am. 95:1661 (1994).
9. S.Caspi, et al., Oscillations and hysterisis of helium during lamda transition above the thermodynamic critical pressure in the presence of heat flow, Adv. Cryo. Eng. 23:349 (1978).

R. Scurlock, winner of the Samuel C. Collins Award, delivers his speech at the Awards Luncheon.

EXPERIMENTAL INVESTIGATION OF LOSS MECHANISMS IN A 4 K PULSE TUBE

S.L. Zhou[1,2], G. Thummes[1,3], and Y. Matsubara[1]

[1]Atomic Energy Research Institute, Nihon University
Narashinodai 7-24-1, Funabashi, Chiba, 274-8501, Japan

[2]Cryogenic Division, Suzukishokan Co., Ltd.
Yoshinodai 2-8-52, Kawagoe, Saitama, 350, Japan

[3]Permanent address: Institute of Applied Physics,
University of Giessen, D-35392 Giessen, Germany

ABSTRACT

In case of multi-stage 4 K pulse tube coolers, the phase shifting assembly at the hot end of the 4 K pulse tube is generally located at room temperature. This arrangement facilitates the control of phase shift, but it increases the thermal losses through the pulse tube. In order to find out the thermal load to the cold end that is introduced through the pulse tube as compared to that resulting from the enthalpy flow through the regenerator, a 4K pulse tube cooler has been fabricated and tested. As a special feature, the set-up employs two heat exchangers which are located at the cold ends of pulse tube and regenerator and are connected in series by a narrow tube. During the cool down process and also in stationary operation, a temperature difference between the two heat exchangers was found which is ascribed to the imbalance between thermal losses through pulse tube and regenerator. The cooling performance increased under operating conditions with reduced temperature difference, e.g. by introducing a proper DC gas flow in the cooler. A moving plug was used to inhibit the DC flow from the double inlet and to measure the work flow at the hot end of pulse tube. Numerical analysis using input data from the experiments indicates that DC flow has a heat pumping effect at the cost of a higher enthalpy flow in the regenerator.

INTRODUCTION

The cooling performance of a cryocooler is determined by the gross cooling power and the thermal losses at the cold end. The losses in a pulse tube cooler are mainly coming from the regenerator and the pulse tube. A lot of work has been done to improve the performance of the regenerator. For instance, the thermal losses in the regenerator can be reduced by the double-inlet arrangement[1]. However, the double inlet can give rise to another problem originating from DC flow, i.e. a circulating gas flow due to asymmetric flow resistances or due to the pressure wave form in an oscillating flow system. DC flow will influence the cooling performance of the pulse tube cooler. Recent studies revealed that a proper DC flow can reduce the thermal losses from the pulse tube side and improve the cooling performance[2,3,4].

In the case of multi-stage 4 K pulse tube coolers, the phase shifting assembly at the warm end of the pulse tube is generally located at ambient temperature. This arrangement facilitates the control of phase shift, but it will increase the thermal losses through the 4 K pulse tube. In the present work, a 4 K pulse tube was fabricated and tested that employs two separate heat exchangers located at the cold ends of pulse tube and regenerator. This makes it possible to distinguish to some extent between the thermal load introduced through the pulse tube and that given by the enthalpy flow through the regenerator under different operating conditions, also including DC flow in the system.

EXPERIMENTAL APPARATUS

Figure 1 shows the schematics of the experimental apparatus. 1: 1st stage regenerator; 2: 1st stage heat exchanger; 3: 2nd stage regenerator; 4: 2nd stage heat exchanger; 5: 3rd stage regenerator; 6: 3rd stage heat exchanger at the regenerator side; 7: connecting tube; 8: cold heat exchanger at the pulse tube side; 9: pulse tube; 10: hot heat exchanger of pulse tube; 11: double inlet valve; 12: orifice valve; 13: reservoir; 14: DC flow valve; 15: rotary valve; 16: compressor; 17: Moving Plug; 18: mass flow meter.

The regenerator of the pulse tube cooler consists of three stages. The first stage

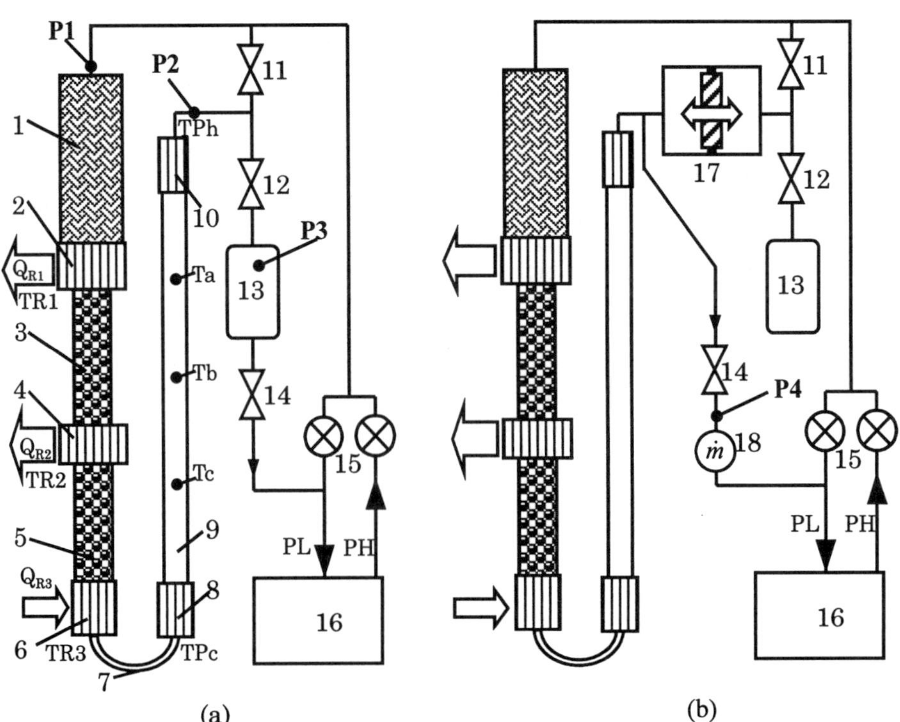

Figure 1. Schematics of the experimental apparatus.

regenerator has an inner diameter of 46 mm and a length of 120 mm, and is filled with 300 mesh stainless steel wire screens. The second stage regenerator filled with lead spheres has an inner diameter of 29 mm and a length of 108 mm. The third stage regenerator filled with Er_3Ni spheres has an inner diameter of 29 mm and a length of 106 mm. The pulse tube has an inner diameter of 19 mm, a wall thickness of 0.5 mm, and a length of 284 mm.

The first stage heat exchanger and second stage heat exchanger are cooled by the first and second stage cold ends of a GM-cooler, respectively. Under typical operating conditions temperatures were in the range 80-100 K (first stage) and 20-30 K (second stage).

The temperature at the hot end of the pulse tube (TPh) is measured by a Pt-thermometer, and the pulse tube wall temperatures at three different positions (Ta, Tb, Tc) are measured by three E-type thermocouples. The temperature at the cold end of the first and second stage regenerator (TR1 and TR2) are measured by a Pt-Co thermometer and a calibrated CGR thermometer, respectively.

There are two heat exchangers located at the cold ends of pulse tube and regenerator. The two heat exchangers are connected in series by a narrow tube. Calibrated CGR thermometers are provided for each heat exchanger and an electric heater is attached to the heat exchanger at the regenerator side.

In Figure 1(a), three needle valves (orifice valve, double inlet valve, DC flow valve) and a reservoir are used as the phase shifting unit for the pulse tube cooler. Here the DC flow valve is connected from the reservoir to the low pressure side of the compressor.

In Figure 1(b), an additional moving plug unit[5] is inserted between the phase shifting assembly and the hot end of the pulse tube. The position of the moving plug is measured by a displacement transducer. The DC flow valve in Figure 1(b) is connected from the hot end of the pulse tube directly to the low pressure side of the compressor, and the DC flow rate is measured by a mass flow meter.

Pressures (P1,P2,P3,P4) at the inlet of regenerator, the hot end of the pulse tube, the reservoir and the downstream side of the DC flow valve in Figure 1(b) are measured by means of pressure transducers. An oscilloscope records the pressure, mass flow rate and the displacement of the moving plug.

EXPERIMENTAL RESULTS

Temperature Difference and Cooling Performance

During the cool down process, the heat exchanger temperature at the cold end of the pulse tube (TPc) was found to be lower than that of the regenerator (TR3). The difference between them became gradually smaller and finally reversed sign. TR3 is expected to be mainly affected by the enthalpy flow through the regenerator, while TPc will be more affected by the thermal losses from the pulse tube side, such as the shuttle heat losses, conduction losses, etc.. The large heat capacity at regenerator side may cause the lag in the cool down speed of TR3, while near stationary state the temperature difference TR3-TPc is considered as a measure for the difference in losses from both sides.

A slightly higher temperature at the cold end of pulse tube indicates that the thermal losses from the pulse tube side are not small in comparison with the enthalpy loss from the regenerator. Pre-cooling of the pulse tube at an appropriate wall position to a suitable temperature by use of the upper stage might be a method to reduce the thermal losses. Changing the operating conditions, such as frequency, pressure ratio, or valve settings will be another effective way to balance the losses coming from the regenerator and the pulse tube in order to achieve minimum total losses.

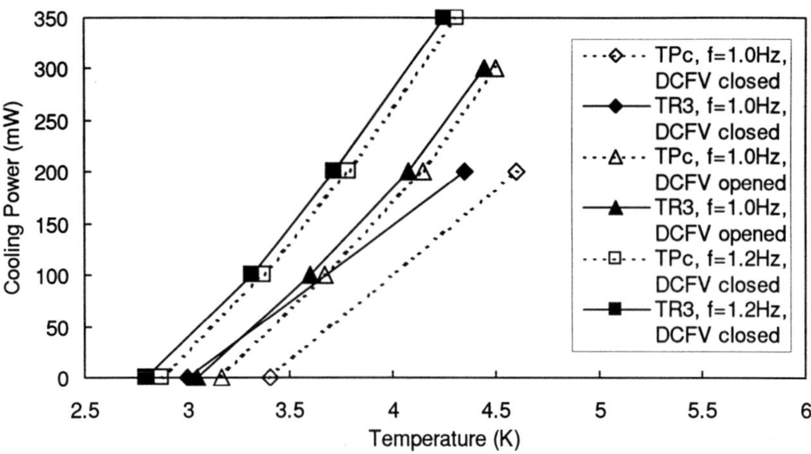

Figure 2. Cooling power versus temperature at the two heat exchangers located at the cold ends of regenerator and pulse tube.

Some results obtained with the set-up in Figure 1(a) at temperatures around 4 K with the latter method are shown in Figure 2. At an operating frequency of 1.0 Hz and with the DC flow valve fully closed, TPc was 0.4 K higher than TR3. This difference decreased to 0.12 K when the DC flow valve was opened. The data marked by square symbols in Figure 2 were obtained with the DC flow valve closed. But the temperature difference between TR3 and TPc is only 0.08 K, and the temperature rise per unit of heat input at the regenerator side is similar to that with DC flow valve opened (triangles). This finding indicates that with proper settings of orifice valve and double inlet valve, as well as choice of frequency and pressure ratio, one can achieve similar cooling performance as by opening the DC flow valve.

DC flow appears to be a simple and effective tool for balancing the thermal losses and diminish the temperature difference. In double-inlet mode of operation one can surmise whether DC flow exists or not inside the pulse tube from the temperature profile along the pulse tube, but can not clearly prove it. Therefore, several efforts have been made to experimentally clarify this point, and the results are shown in Figures 3 and 4.

DC Flow in Simple Orifice Mode

A DC flow valve is commonly used in double-inlet operation of the pulse tube[2-4]. Figure 3 shows the result when DC flow is applied to a simple orifice pulse tube cooler (i.e. set-up in Figure 1(a) with the double inlet valve 11 closed). A needle valve with a maximum flow coefficient of Cv = 0.004 at the opening of 10 turns was used as DC flow valve (DCFV). From Figure 3 it follows that TR3 = 14.4 K and TPc = 15.1 K when DCFV is closed. Upon opening DCFV by 3/5 turns, TR3 and TPc decreased to 11.14 K and 11.4 K, respectively. At an opening of 21.5/25 turns, TR3 and TPc decreased to near 10.5 K and only a very small temperature difference remained. When the DC flow valve was opened to one turn, TPc and TR3 both increased and TR3 was higher than TPc, which means that the increase of thermal losses from regenerator became larger than the decrease of the thermal losses from the pulse tube.

Comparing the cooling power curves in Figure 3, it is seen that they have almost the same slopes, which means that the gross cooling power at the cold end is the nearly same in all cases. So, even in a simple orifice pulse tube cooler, an appropriate DC flow can balance the thermal losses to a minimum value.

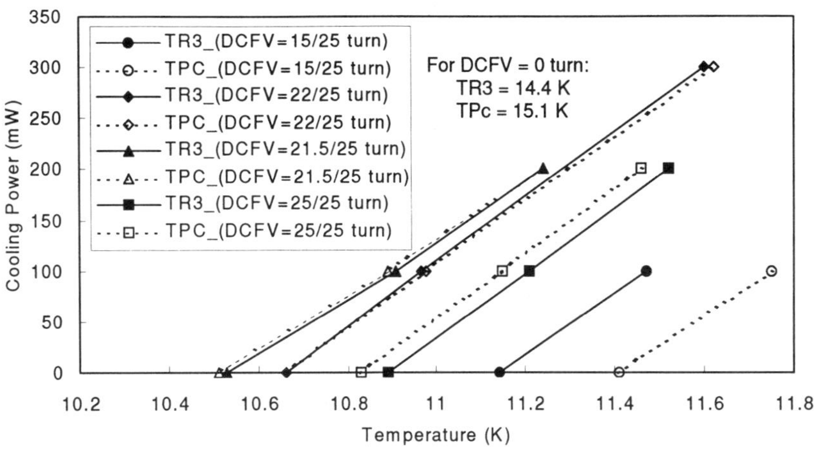

Figure 3. Cooling power vs. temperature of the two heat exchangers at different DC flow valve openings.

Separation and Measurement of DC Flow in Double Inlet Mode

It is difficult to measure a small DC flow component in an oscillating gas flow. To measure the DC flow rate through the pulse tube, it is necessary to inhibit the DC component from the double inlet valve. For this purpose, a moving plug[5] has been inserted at the hot end of the pulse tube, as shown in Figure 1(b). The orifice and double inlet valve still operate as phase shifters, but DC flow from the double-inlet route is now totally suppressed.

The friction between the seal of the moving plug and the cylinder wall will have some influence on the phase shifting characteristics. Under the same orifice opening and a slightly larger double inlet opening, a minimum temperature of 3.3 K has been attained which indicates that the moving plug works well in acting just as gas separator between the phase shifting unit and the pulse tube.

Figure 4 shows the cooling powers for different DC flow valve openings obtained with the moving plug. The orifice valve was fully closed to get temperatures comparable to those in Figure 3. The role of DC flow in reducing the losses through the pulse tube and increasing the losses through the regenerator is seen in the temperature difference between the two heat exchangers. For a general double-inlet type pulse tube cooler (the phase

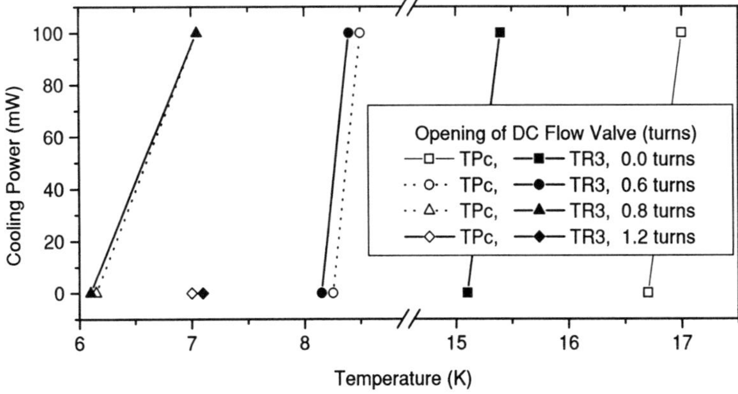

Figure 4. Cooling power vs. temperature at different DC flow valve openings in double-inlet mode operation.

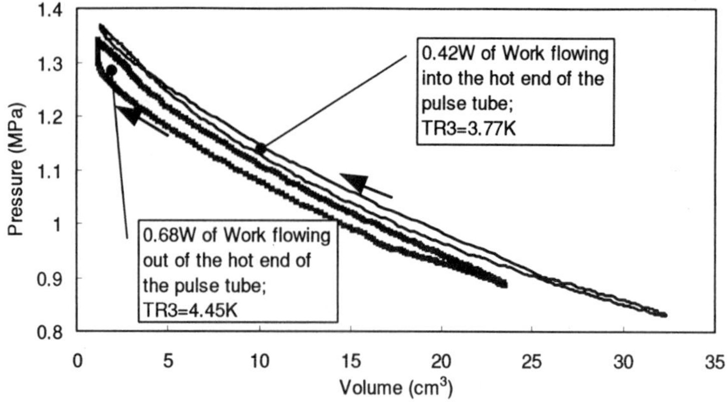

Figure 5. PV diagrams measured at the hot end of the pulse tube by use of the moving plug.

shifting unit is connected to the hot end of the pulse tube directly), DC flow from the hot end to the cold end, or from the cold end to the hot end of the pulse tube may exist. Adjusting the DC flow valve connected to the low pressure line or high pressure line of the compressor can balance the thermal losses from regenerator and pulse tube and improve the performance of the cooler.

Moving Plug Acting like a Compressor

In our experiment with the set-up shown in Figure 1(b), the pressure-volume (PV) diagram at the hot end of the pulse tube has been monitored. Usually the work flow from the pulse tube should be dissipated in the orifice valve. Consequently the direction of work flow measured at the hot end of the pulse tube should be from the pulse tube to the phase shifting unit.

Two PV diagrams are shown in Figure 5. With 0.68 W of work flowing out of the pulse tube, a temperature of 4.45 K was attained, while with 0.42 W of work flowing into the pulse tube, a lower temperature of 3.77 K was achieved. Because of the similar operating conditions as that in Figure 1(a) around 4 K, it is possible that the pulse tube in Figure 1(a) was also operated with work flowing into the hot end of pulse tube.

NUMERICAL ANALYSIS

Numerical calculations have been carried out to analyze the experimental results. The calculation model is shown in Figure 6. The position at the hot end of the pulse tube is defined as X = 0, and the positive direction is from hot to cold end of the pulse tube. A finite element method has been used to solve the continuity equation, the energy equation combined with the ideal gas state equation in the pulse tube section. Heat transfer between gas and the pulse tube wall has been considered. The heat transfer coefficient is treated as a

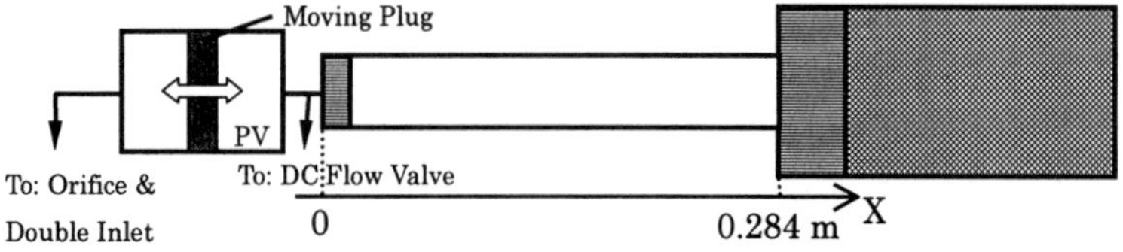

Figure 6. Physical model of the pulse tube cooler.

function of temperature. The input data for the calculations are: hot and cold end temperatures, heat capacity of the wall material, pressure in the pulse tube, and mass flow rate through the hot end of the pulse tube.

The initial temperature at the hot end of the pulse tube is fixed at 300 K, and the cold end temperature is the achieved temperature in our experiment. A linear temperature distribution along the pulse tube wall is assumed initially, and the heat capacity of the pulse tube wall material (SUS304) is calculated from the corresponding wall temperature.

From the PV diagram at the hot end of the pulse tube, the mass flow rate (AC component) at X = 0 is calculated assuming isothermal conditions. The DC flow rate is taken from the experiment using a mass flow meter (Figure 1(b)).

From the pressure in the pulse tube and the mass flow rate (AC+DC) through the hot end of the pulse tube, the mass flow rate at the cold end of the pulse tube and also the enthalpy flow $<\dot{H}>$, and the work flow $<\dot{W}>$ along the pulse tube can be calculated.

Some results concerning the role of DC flow and the explanation of the case with a work input to the hot end of the pulse tube are given in Figures 7 and 8. A positive value of work or enthalpy flow means that the work or enthalpy is flowing from the hot to the cold end of the pulse tube while a negative value means the opposite flow direction. The heat input to the cold end of the pulse tube is zero. If other losses are neglected, the enthalpy flow $<\dot{H}>$ in the pulse tube should equal the thermal loss caused by the enthalpy flow $<\dot{H}>_{reg}$ through the regenerator. It was confirmed by a calculation of $<\dot{H}>_{reg}$ with real gas properties that $<\dot{H}>$ and $<\dot{H}>_{reg}$ coincide with each other quantitatively.

Figure 7 shows the energy flow in the pulse tube for different DC flow rates. The calculated pressure-displacement (PVe) diagram at the cold end of the pulse tube is also shown. Both enthalpy and work are flowing from the cold end to the hot end. At the cold end (X = 0.284 m), the absolute value of enthalpy flow is greater than that of the work flow, which means that with DC flow, heat has been pumped to the hot end of the pulse tube instead of being pumped to the cold end, where the latter case is known as the shuttle heat loss. From case A to case B in Figure 7, the DC flow rate was increased and as a result the heat pumping effect was intensified. The cost for this is an increased enthalpy flow in regenerator.

Figure 8 shows the energy flow in the pulse tube when the work flow at the hot end of the pulse tube changes from negative in case C to positive in case D, as shown in Figure 5. From Figure 8, in case D the displacement of the gas at the cold end decreases, and a decrease in the enthalpy flow from the regenerator appears to be the reason for the observed temperature drop from 4.45 K to 3.77 K.

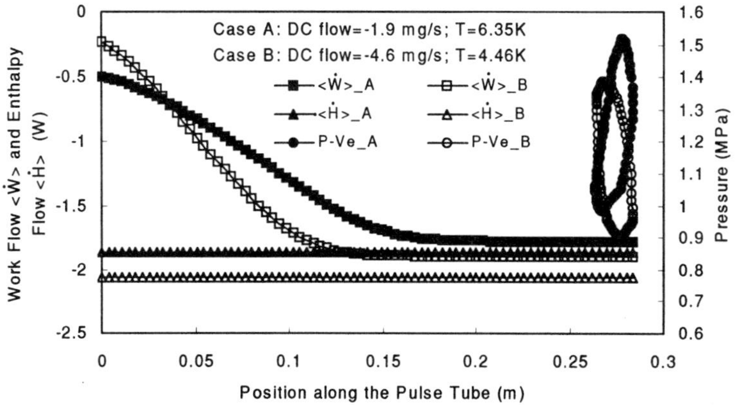

Figure 7. Work flow, enthalpy flow and PVe diagram for different DC flow rates.

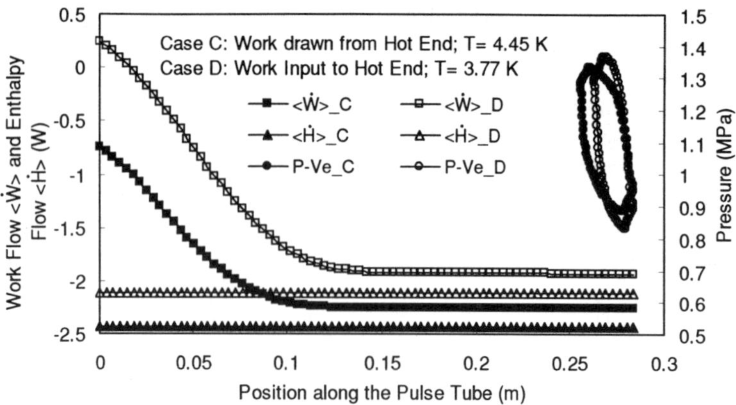

Figure 8. Work flow, enthalpy flow and PVe diagram for different work flows at the hot end of the pulse tube.

CONCLUSION

From experiments on a 4 K pulse tube cooler employing two separate heat exchangers at the cold ends of regenerator and pulse tube, it follows that generally there exists an imbalance between thermal losses through the pulse tube and the regenerator. This imbalance, seen as a temperature difference between the two heat exchangers, can be minimized and the cooling performance optimized by introducing a suitable DC flow. DC flow can pump heat from the cold to the hot end of the pulse tube, and reduce the thermal losses from the pulse tube at the cost of an increasing enthalpy flow from the regenerator.

With a moving plug inserted at the hot end of the pulse tube cooler, DC flow through the double inlet valve can be suppressed, thus making possible to measure the DC mass flow rate through the DC flow valve and also the PV diagram at the hot end. From a numerical calculation on the basis of the experimental data it was found that several watts of work dissipation exists in pulse tube and hot heat exchanger. Moreover it was shown that the cooler performance can be improved even with work flowing into the hot end of the pulse tube.

ACKNOWLEDGMENT

The authors wish to thank Dr. T. Kuriyama of Toshiba Corporation for providing magnetic regenerator materials.

REFERENCES

1. S. W. Zhu, P. Y. Wu, and Z. Q. Chen, Double inlet pulse tube refrigerator: an important improvement, *Cryogenics* 30:514 (1990).
2. G. Chen, L. Qiu, J. Zheng, P. Yan, Z. Gan, X. Bai and Z. Huang, Experimental study on a double-orifice two-stage pulse tube refrigerator, *Cryogenics* 37:271-273 (1997)
3. C. Wang, G. Thummes, and C. Heiden, Effects of DC gas flow on performance of two-stage 4 K pulse tube coolers, *Cryogenics* 38:689-695 (1998).
4. C. Wang, G. Thummes, and C. Heiden, Control of DC gas flow in a single-stage double-inlet pulse tube cooler, *Cryogenics* 38:843-847 (1998).
5. Y. Matsubara, A. Miyake, Alternative methods of the orifice pulse tube refrigerator, In: 5th Int. Cryocooler Conf. 127-135 (1988).

TESTBED FOR THE INVESTIGATION OF MINIATURE PULSE TUBE COLD HEADS

G. Kaiser and A. Binneberg

Institut für Luft- und Kältetechnik, Fachbereich Kryotechnik
Dresden, Saxony, D-01309, Germany

ABSTRACT

A testbed for the investigation and optimization of miniature pulse tube cold heads was developed at ILK Dresden. In order to drive the cold head a double piston compressor is used operating at a fixed frequency of 50 Hz. The swept volume of the compressor can be varied up to 6 cc by use of an adjustable power supply. A high-pressure helium gas supply unit is used for filling or venting of the system. This gas supply unit also enables a variation of the average pressure in the system during the operation up to 2.4 MPa. A vacuum pump is used for thermal insulation of the cold head as well as for evacuation of the inner system.

The temperature at the cold head can be measured and stabilized with a temperature controller. A line recorder is connected to the temperature controller to record transient scripts as well as for electronic data acquisition. In order to perform cooling power measurements heat can be applied to the cold head by means of a thermo-foil heater connected to a dc-current source. The testbed enables the operation of several types of miniature pulse tube cold heads (orifice, inertance tube, double inlet) in the cooling power range up to about 2 W at 80 K.

In this contribution we will present the details of our experimental set-up. First results of experimental investigations on miniature orifice pulse tube cold heads will be discussed.

INTRODUCTION

In order to prepare a research project on pulse tube cryocoolers for the cooling of highly sensitive cryoelectronic devices like HTS-SQUIDs or -bolometers a testbed for investigations on several kinds of pulse tube cold heads was developed. An important point of this development was the possibility to change many different operating parameters simultaneously and independently. After the description of the experimental set-up first results of investigations on orifice pulse tube cold heads are presented and compared with results derived from simulations.

EXPERIMENTAL SET-UP

Figure 1 shows block diagrams of the experimental set-up. The vacuum system is required to generate a thermally insulating vacuum around the cold head inside of the dewar. It consists of vacuum pumps and vacuum monitors. In order to prevent over pressure inside the vacuum system in the case that the cold head breaks a safety valve opening at a pressure difference of 115 kPa between vacuum system and environment is inserted in the connecting line between vacuum pumps and dewar.

The test bed is operated with Helium coming from the pressure gas system. This gas supply unit enables a variation of the average pressure in the system during the operation. At an internal pressure of 2.4 MPa a safety valve opens the pressure gas system to environment.

For operation of the pulse tube cold heads a linear motor driven double-piston compressor (SL200, AIM Heilbronn) is used. It is operated by use of an adjustable power supply at a fixed frequency of 50 Hz corresponding to the frequency of the power line. Thus the swept volume of the compressor can be varied up to 6 cm^3. The temperature at the cold head is measured by means of a temperature monitor (Model 208, Lake Shore). As temperature sensor a Silicon diode thermometer (DT470, Lake Shore) is used.

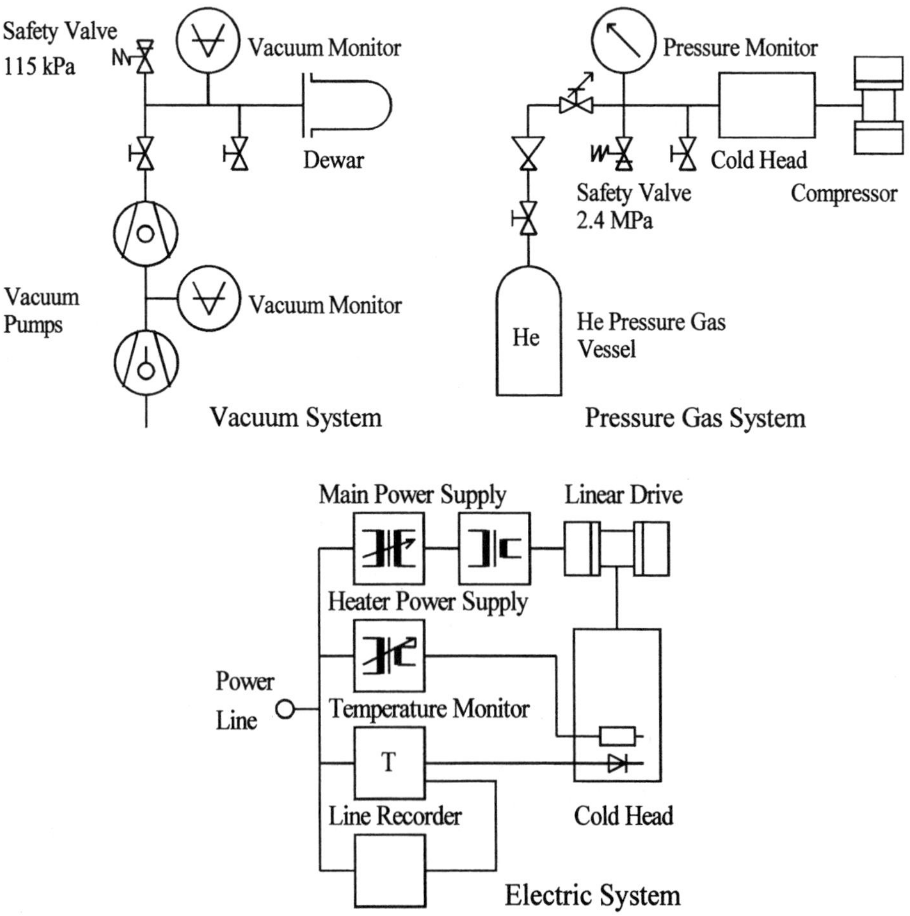

Figure 1. Block diagrams of the testbed for investigations on pulse tube cold heads.

In order to perform cooling power measurements heat can be applied to the cold head by use of thermo foil heater driven by an additionally power supply. The temperature behavior of the cold head can be recorded on strip charts as well as on disk for data acquisition.

First experimental investigations on orifice pulse tube cold heads were performed using this testbed. The design of the cold head is shown in Figure 2. Figure 3 shows the cold head mounted on the testbed and connected to the terminals.

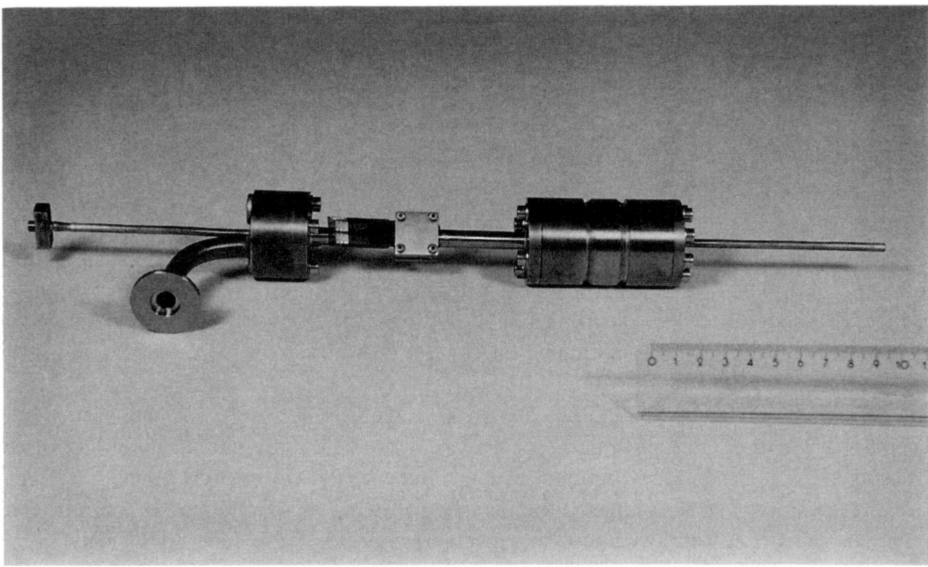

Figure 2. Orifice pulse tube cold head.

Figure 3. Orifice pulse tube cold head mounted on the test bed.

It has an in-line configuration of regenerator, cold heat exchanger, pulse tube, orifice and reservoir. The stainless steel tubes used for regenerator and pulse tube have a length of 50 mm, an outer diameter of 8 mm and a wall thickness of 0.15 mm. The regenerator tube was filled with stainless steel mesh screens (400 Mesh). As orifices screws with inner bores of 0.3...0.5 mm were used. In order to avoid gas jets coming from reservoir into pulse tube some stainless steel mesh screens (400 Mesh) were used as a flow straightener in part of the measurements. The cold heat exchanger has a length of 10 mm. It was filled with copper mesh screens (100 Mesh) which were soft soldered with the heat exchanger wall in order to improve the internal heat transfer resistance.

COLD HEAD SIMULATIONS

To get hints for a useful pulse tube cold head design simulations with ARCOPTR[1], available at http://arcoptr.arc.nasa.gov/arcoptr-options.html in the internet, were performed. In order to find the orifice settings for the best net cooling power the optimizer option of ARCOPTR was used. The parameter set used for the simulation is shown in Table 1. The length of the aftercooler was set to 1 mm because the program didn't accept a value of zero at this position and in the experiment no aftercooler was used. Therefore the length of regenerator was reduced by 1 mm. A hot heat exchanger length of 1 mm was entered in order to express the pressure resistance due to the flow straightener at the hot end of the pulse tube.

The results of the simulation are shown in Table 2. The results indicate that it is in general possible to operate this pulse tube cold head configuration with a net cooling power of around 1 W at 80 K. The required swept volume of the compressor is about 2.5 cm^3. Thus the maximum swept volume of the compressor is higher than required. The efficiency of the set-up is relatively low as to be expected for orifice pulse tube cryocoolers in this power range. It can be increased considerably by use of a double inlet configuration (simulation with ARCOPTR is not possible) or substitution of the orifice by an inertance tube.

Table 1. Parameter set used for ARCOPTR cold head simulation

Parameter	Value	Parameter	Value
Frequency	50 Hz	Mesh size of regenerator	400 in^{-1}
Average pressure	2.0 MPa	Length cold heat ex.	10 mm
Hot end temperature	300 K	ID of cold heat exchanger	7.7 mm
Cold end temperature	80 K	Wire dia. of cold heat ex.	0.1 mm
Dead volume	1 cm^3	Mesh size of cold heat ex.	100 in^{-1}
Length of aftercooler	1 mm	Length of pulse tube	50 mm
ID of aftercooler	7.7 mm	OD of pulse tube	8 mm
Wire diam. of afterc.	0.025 mm	Wall thickn. of pulse tube	0.15 mm
Mesh size of afterc.	400 in^{-1}	Length of hot heat ex.	1 mm
Length of regenerator	49 mm	ID of hot heat exchanger	7.7 mm
OD of regenerator	8 mm	Wire diam. of hot heat ex.	0.025 mm
Wall thickness of reg.	0.15 mm	Mesh size of hot heat ex.	400 in^{-1}
Wire diam. of regenerator	0.025 mm	Reservoir volume	30 cm^3

Table 2. Results of cold head simulation

Parameter	Value
Orifice setting	97.2 mg s^{-1}
Swept volume	2.53 cm^3
Cooling power	0.801 W
Driving power	82.6 W
COP	0.971 %

EXPERIMENTAL RESULTS

The experiments were performed with an average pressure of 2.0 MPa He in the system. The compressor was operated at its maximum swept volume. In the first run the orifice settings were changed from 0.3 mm up to 0.5 mm in steps of 0.1 mm. In the second run this procedure was repeated with an additionally flow straightener at the pulse tube hot end.

Figure 4 shows the cooling curves for both runs. The best result was obtained for an orifice setting of 0.3 mm in both runs. The minimum temperature reached without a flow straightener was 135 K. During operation with an additional flow straightener a minimum temperature of 109 K was reached. Because of the relatively low thermal mass of the cold head a large amount of the cooling temperature drop was reached within 10 minutes.

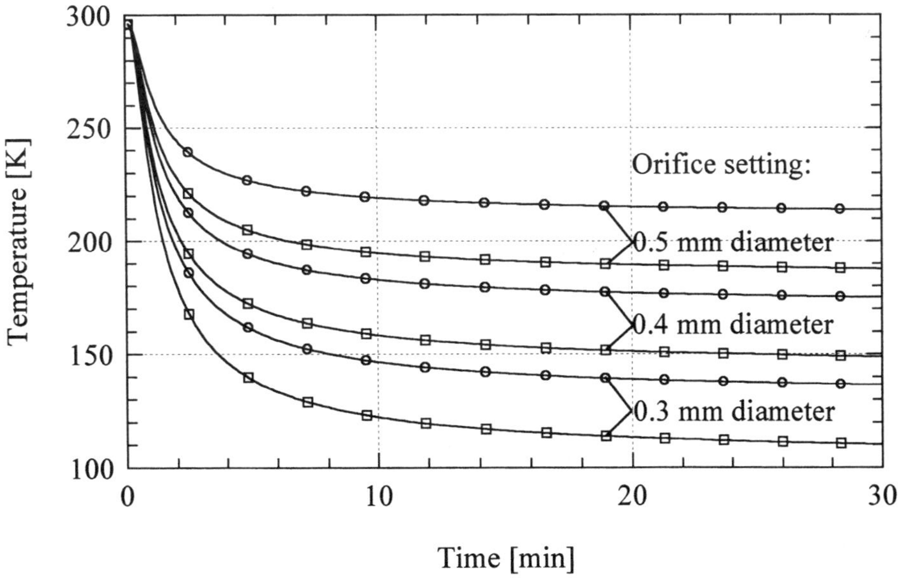

Figure 4. Cooling curves of the pulse tube cold head with flow straightener (boxes) and without flow straightener (circles) for different orifices settings.

CONCLUSION

A testbed for experimental investigations on several kinds of pulse tube cold heads was developed. First investigations on relatively simple orifices pulse tube cold heads were performed and the results were compared with simulation results obtained by use of ARCOPTR.

During the investigations of a pulse tube cold head with flow straightener and with a regenerator and pulse tube length of 50 mm and diameters of 8 mm a minimum temperature of 109 K at an orifice setting of 0.3 mm diameter could be reached. It could be shown that the use of flow straighteners for small scale pulse tube cryocoolers is very important. The minimum temperature could be reduced by 26 K in this way.

A difference between the results of simulation and experiment could be observed which is caused by differences between premises of simulation and the real situation in the experimental set-up.

ACKNOWLEDGMENT

The authors wish to express their thankfulness to Dr. Pat Roach and Dr. Jeff Lee from NASA Ames Research Center for their assistance in the use of ARCOPTR.

REFERENCES

1. P.R. Roach, A. Kashani, A simple modeling program for orifice pulse tube coolers, in: "Cryocoolers 9" R.G. Ross jr., ed., Plenum Press, New York (1997), p. 327.

RESEARCH OF SEPARATED VERSION TWO-STAGE PULSE TUBE REFRIGERATOR

L. W. Yang, Y. Zhou, J. T. Liang

Cryogenics Laboratory, Chinese Academy of Sciences
Beijing P. O. Box 2711, 100080, P. R. China

ABSTRACT

At the beginning of the article, three key aspects of pulse tube refrigerator, cooling capacity, pressure amplitude and DC flow, are analyzed and their characteristics are revealed. A special two-stage pulse tube refrigerator is designed and fabricated for the purpose of acquiring larger cooling capacity. This refrigerator has two separated pulse tube coolers, each has its own reservoir, regenerator and pulse tube. The two coolers are coupled together through a red copper plate between the cold head of one cooler and the regenerator middle part of the other. Three groups of experiments and results are introduced. The first group is for a single stage test result, which shows a 2.5K temperature increase per watt cooling capacity. The second group is for the initial two-stage cooler result, where a new second orifice is proposed to improve the performance. The third group is for a further improved two-stage cooler experiment, in which the pulse tube cooler reaches the lowest temperature of 11.7K and 3W cooling power at 20K.

INTRODUCTION

Cryocoolers within the 10-20K temperature range have been widely used in practice. The two-stage G-M refrigerator in cryopumps are a good example. Because of the low cost and longevity of pulse tube refrigerator, it has potential to replace other alternatives. And now in our laboratory, the focus in pulse tube cooler is to improve cooling capacity, because two-stage refrigeration temperature has been able to reach 11.5K since 1995.[1]

ANALYSIS

Cooling Capacity

According to the authors' analysis, pulse tube volume is an important parameter related to theory refrigeration. The largest theoretical cooling capacity $\dot{Q}_c$ derived is [2]

$$\dot{Q}_c = f\int_0^\tau RT_0 \ln(p_{pt}/p_r)\dot{m}\,dt \tag{1}$$

where f is frequency, τ is period, R is gas constant, T_0 is hot end temperature, p_{pt} is pressure in pulse tube, p_r is pressure in reservoir, and $\dot{m}$ is mass flow rate of orifice that connects reservoir and pulse tube hot end.

According to Eq. (1), large $\dot{m}$ will produce relatively large refrigeration. But, increasing of $\dot{m}$ accompanies increase of loss. When $\dot{m}$ is small, refrigeration increases faster than loss; and when $\dot{m}$ is too large, loss increases faster. Thus, there is a best point where a lowest temperature can be acquired, and there is another point that largest cooling capacity at given temperature can be acquired.

Determination of $\dot{m}$ makes analysis and design difficult. With some supposition, including sinusoidal wave, $\dot{m}\sim(p_{pt}-p_r)$, the following formula is derived [2]

$$\dot{Q}_c = \pi\varphi V_{pt} f p_r \cdot F(p_a/p_r) \tag{2}$$

where φ is reservoir coefficient, V_{pt} is volume of pulse tube, p_a is amplitude of p_{pt} and $F(p_a/p_r)$ is pressure determined parameter. φ is defined to be:

$$\varphi = V_m/V_{pt} \tag{3}$$

where V_m is the total gas volume that flows into and out of reservoir at pressure p_r. $F(p_a/p_r)$ is determined by:

$$F(p_a/p_r) = -0.009418 + 0.639888 p_a/p_r - 0.49267(p_a/p_r)^2 + 0.62077(p_a/p_r)^3 \tag{4}$$

From Eq. (2), we can find that φ makes analysis easier.

Neglecting change of φ, Eq. (2) shows that increasing of pulse tube volume, pressure amplitude, average pressure will produce more cooling capacity. Thus, large pulse tube volume is important to improve the cooling capacity, though increasing of V_{pt} may produce other problems, for example, refrigeration temperature may rise.

In practice, φ is different under different working condition, but generally φ is limited to 0.1~0.3. During starting-up process, both pressure amplitude and φ have a large value; when frequency changes, φ will have different value. High pressure ratio and low frequency correspond to large value of φ.

Pressure Amplitude

According to Eq. (2), increase of pressure amplitude is beneficial for refrigeration.

Based on some suppositions, the amplitude of pressure wave p_a is deduced.[2]

The main suppositions include: neglecting flow resistance through the regenerator, sinusoidal pressure wave, and orifice flow $m\sim(p_{pt}-p_r)$.

Based on mass conservation, the refrigeration system can be divided into five parts: compressor swept volume V_{a0}, heat-exchanger volume V_{he}, regenerator void volume V_r, pulse tube volume V_{pt}. Define

$$S = \frac{1}{2}V_{a0} + V_{he} + T_0\Sigma\frac{V_r}{T} + T_0\Sigma\frac{V_{pt}}{T} \tag{5}$$

An approximate value of pressure amplitude in pulse tube is deduced to be:

$$\frac{p_a}{p_{ave}} \approx \frac{V_{a0}}{2S}\sqrt{1-(\frac{\varphi V_{pt}}{V_{a0}})^2} \tag{6}$$

In normal condition, $\varphi \le 0.3$ and $V_{pt} \le V_{a0}$, this makes $\sqrt{1-(\varphi V_{pt}/V_{a0})^2} \ge 0.954$. Thus effect of reservoir coefficient on p_a is small. Mainly V_{a0} and S determine pressure amplitude. Then we have:

$$\frac{p_a}{p_{ave}} \approx \frac{V_{a0}}{2S} < 1 \tag{7}$$

This is the simplest method to predict pressure amplitude.

According to Eq. (7), when compressor is given and V_{a0} is predetermined, if we want to increase the pressure amplitude, decreasing S is a fundamental and most effective method. And according to Eq. (5), the most effective method to decrease S is small volume of regenerator.

From upper two analyses, large pulse tube and small regenerator are crucial to increase the cooling capacity. Thus, the second stage pulse tube volume in this paper is about $64\,cm^3$, while in our former experiment the volume is $18\,cm^3$, and the dimensions of regenerator change little.[1]

Direct Flow

The double-inlet method is widely used in pulse tube refrigerator, especially when operating frequency is not high. But the double-inlet method may result in direct flow, which will degrade the performance.[3]

Assuming pressure before the regenerator has the following form:

$$p_c(t) = p_{ave} + p_{a1}\cos(2\pi ft+\theta) + \beta p_{a1}\cos(4\pi ft+\theta) \tag{8}$$

and pressure form in the pulse tube is:

$$p_{pt}(t) = p_{ave} + p_{a2}\cos(2\pi ft) + \beta p_{a2}\cos(4\pi ft) \tag{9}$$

where β is the second-order coefficient of sinusoidal harmonic wave, θ is phase difference between two pressures.

The flow rate through the double-inlet valve can be expressed by:

$$m_d = \alpha \cdot \varepsilon \cdot A\sqrt{2\rho_H(p_H - p_L)} \tag{10}$$

where subscripts H, L corresponding high and low pressure.

Define the net flow amount through the double-inlet valve as:

$$\Delta_d = 2\oint m_d dt \Big/ \oint \left| m_d \right| dt \tag{11}$$

Calculation shows that generally Δ_d>0. This means that more gas flows into the pulse tube than that flows out. But with increasing of β, $\Delta_d < 0$ may occur, as shown in Figure 1, Δ_d decreases with β increasing. The amplitude of Δ_d is determined by pressure ratio, flow resistance and β at the same time.

Figure 1 shows that the direct flow is complex, its direction may change. For different direction, the method to eliminate it should be different, too.

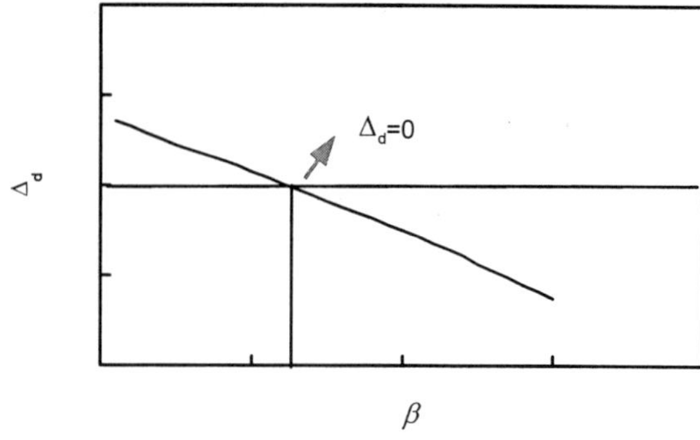

Figure 1. Effect of second-order sinusoidal wave on net mass flow

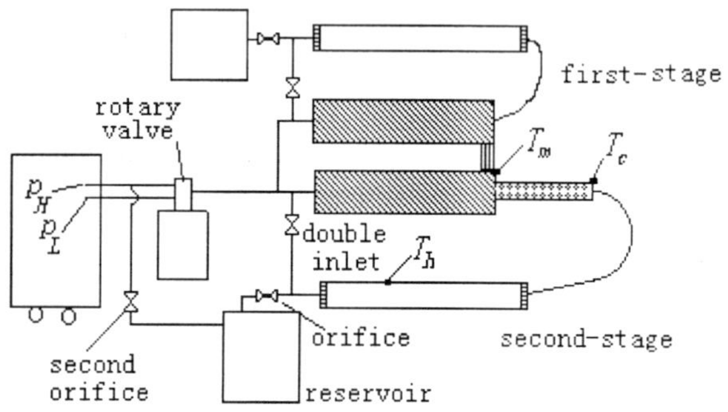

Figure 2. Structure of two-stage pulse tube refrigerator

STRUCTURE AND MEASUREMENT

Figure 2 shows the structure of the cooler.

The typical dimensions for the first stage cooler are: regenerator, 28mm outside diameter, 0.45mm thick wall, 120mm long stainless steel tube, filled with 250mesh stainless steel screen; pulse tube, 18.4mm outside diameter, 0.2mm thick wall, 200mm long stainless steel tube.

The second stage cooler has only one point different from the first stage, that is, the regenerator includes another part, which is 24.4mm outside diameter, 0.2mm thick wall, 80mm long stainless steel tube, filled with 0.2-0.3mm diameter Pb.

The two coolers are coupled together through a red copper plate between the cold head of the first stage cooler and the regenerator middle part of the second stage cooler.

In experiments, the main parameters measured include:

1. Cold tip temperature T_c, using Gold-iron thermocouple, calibrated by Cryogenics Temperature Calibration Station, Chinese Academy of Sciences. Another important temperature to be measured is T_h, which is 60mm from the hot end of the second-stage pulse tube, as shown in Figure 2.

2. Pressure wave amplitude, include average pressure, high pressure and low pressure, and pressure wave. The pressure was measured through pressure transducer. The pressure wave is shown using HP 54602B Oscilloscope.

3. Operation frequency, recorded through Oscilloscope.

4. Cooling capacity, using HP6634A SYSTEM DC POWER SUPPLY.

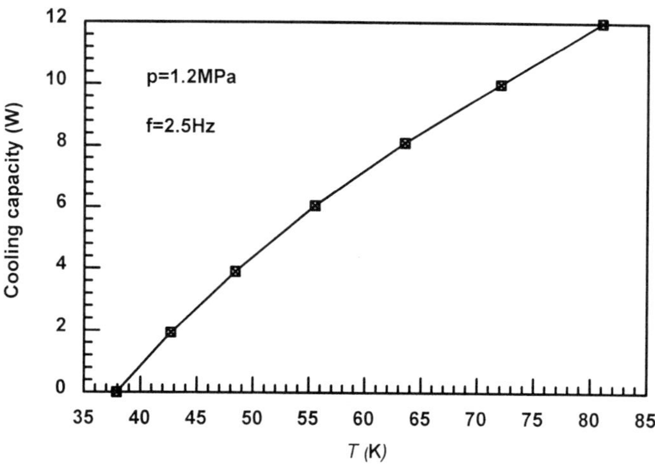

Figure 3. Cooling capacity of single-stage pulse tube cooler

5. Input power (including the compressor and the rotary valve), using AC power meter. The input power of rotary valve is rather low, which are generally below 20W and its value changes little.

6. Vacuum degree, Generally 1.5--2Pa.

RESULT AND ANALYSIS

Single-Stage Pulse Tube Cooler

First, the first-stage pulse tube cooler is fabricated and tested. Test shows this single-stage cooler has a good performance. As shown in Figure 3, the lowest temperature of 38K and >6W/60K cooling power is acquired. 2.5K temperature rising is corresponding to 1W cooling capacity. Such a cooling capacity makes it possible to acquire a large cooling when the temperature drops further.

New Double-orifice and Verification

After two single-stage coolers are coupled together, experiments are done.

The lowest temperature acquired is 22.5K. Test shows that 2K temperature rising per watt cooling capacity. At the same time, author finds that temperature distribution is abnormal. That is, when double-inlet is used, the temperature of hot end T_h becomes even lower than that with orifice opening only. Author concluded that this phenomenon is due to above-explained direct flow. Then a new version second-orifice different from former is arranged as shown in Figure 3:[4] connecting of the high-pressure pipe with the second stage reservoir with a small valve. The goal is to compensate the net flow and to raise T_h.

Further test shows that the new arrangement is effective. As shown in Figure 4, with second orifice closed and a larger double-inlet opened, the refrigerator is started. In 15 minutes, T_c and T_h reach 50K and 140K respectively. But in later 15minutes, both of them rise about 10K gradually. At about 31st minutes, the second orifice is opened at 13/50 turns: T_h rises to 350K quickly, then drops gradually to 300K; T_c drops quickly at the beginning then slowly, and stabilizes at 18.7K at last.

The cooling performance is shown in Figure 5. For the double inlet version, the temperature rises 4K for 2W, and for second orifice version, temperature rises 5.6K for 2W. The reason for this change needs further analysis. But, it is clear that the difference is due to the second orifice.

From upper experiments, author concludes:

(1) The second orifice is effective to lower the temperature if DC flow exists.
(2) The second orifice should be arranged in accordance with DC flow direction.
(3) The second orifice has a negative effect on the cooling capacity.
(4) The second orifice has limited function to improve the performance.

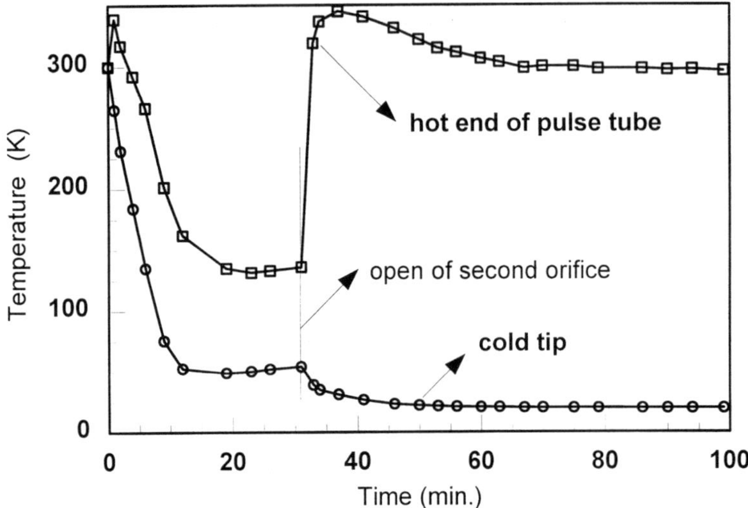

Figure 4. A typical cooling down curve with second orifice

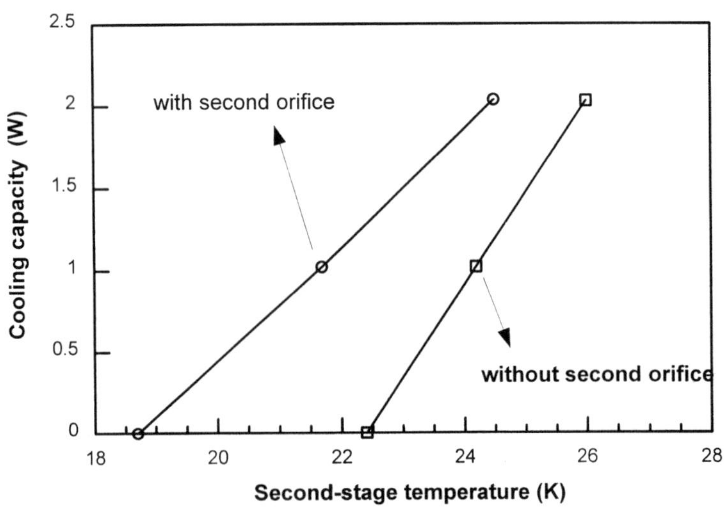

Figure 5. Cooling Capacity with and without second orifice

Improved Two-stage Cooler

The refrigeration temperature of upper two-stage cooler is still high even using the second orifice new scheme. As to this, further analysis is given and several improvements are done, which includes increasing pressure ratio, making second-stage regenerator second part longer and thinner, enhancing heat transfer between joint position of the first stage cooler cold tip and the second stage cooler middle.

Test with further improved pulse tube refrigerator shows that the improvement is effective even without second orifice. Figure 6 gives the cooling down curve, the lowest temperature of 11.7K for the second stage cold tip and 51K for the first stage cold tip are acquired with pressure ratio 1.89 and average pressure of 1.4 MPa. In Figure 7, the cooling capacity is given and 3W/20K cooling power is acquired. The actual input power of compressor is less than 3KW.

CONCLUSION

Three aspects, cooling capacity, pressure ratio and DC flow, are analyzed. Larger pulse tube volume and smaller regenerator volume is selected for actual design.

In experiments, possible change of the DC flow is found, and through a new second orifice arrangement, test shows that this DC flow can be suppressed and the cold tip temperature can be lowered further.

After further improvements, test acquired the lowest temperature of 11.7K and 3W/20K cooling capacity. This makes potential application possible.

ACKNOWLEDGEMENT

This research is supported by the National Superconductor Center of China and National Natural Science Foundation of China.

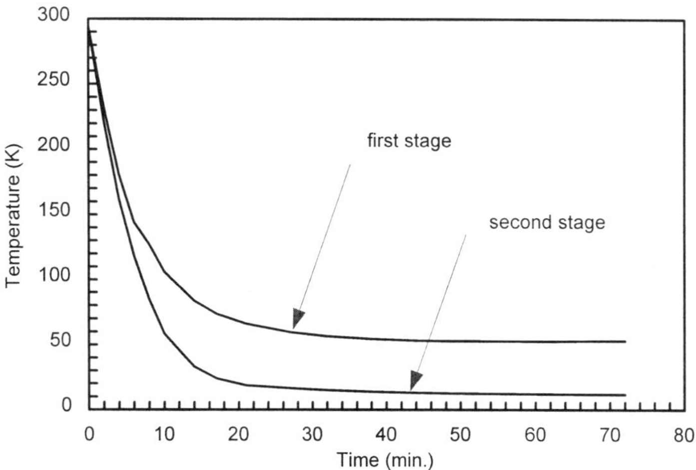

Figure 6. Cooling down curves of temperature

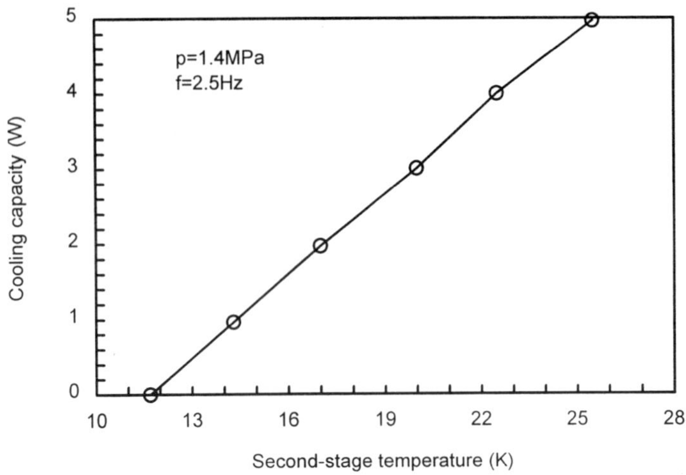

Figure 7. Cooling capacity of improved second-stage pulse tube cooler

REFERENCES

1. C. Wang, Y. L. Ju, Y. Zhou, Experimental investigation of a two-stage pulse tube refrigerator, Cryogenics, 36: 605 (1996)
2. L. W Yang, Research of pulse tube refrigeration mechanism and practical development, Post-doctor report, Chinese Academy of Sciences, 1998.10
3. Y. L. Ju, C. Wang, Y. Zhou, Dynamic experimental investigation of a multi-bypass pulse tube refrigerator, Cryogenics, 37: 357 (1997)
4. G. Chen, L. Qiu, J. Zheng, P. Yan, Z. Gan, X. Bai, Z. Huang, Experimental study on a double-orifice two-stage pulse tube refrigerator, Cryogenics, 37: 271(1997)

TEST OF A RECUPERATIVE PULSE TUBE REFRIGERATOR WITH SIMPLIFIED PERFORATED PLATE HEAT EXCHANGER

Jingtao Liang, Weijia Sun, Luwei Yang, Shiyao Bian, and Yuan Zhou

Cryogenic Laboratory, Chinese Academy of Sciences
P.O. Box 2711, Beijing, 100080, China

ABSTRACT

This paper discusses a new pulse tube refrigeration system that can be applied to gas-liquefaction, in which the regenerator is replaced with a recuperative heat exchanger. The rotary valve and the pulse tube work as a gas expander. Experimental apparatuses are set up to investigate the performance of the whole system. Using helium as the working substance, the pulse tube refrigerator reaches 163K at the cold end and achieves 1.0W/K as cooling power per unit temperature increases.

INTRODUCTION

Since W. E. Gifford and R. C. Longsworth invented the pulse tube refrigerator in 1963,[1] they observed that cyclic alternative pressurization and depressurization of a tube from one end of it, whilst the other end remained closed, could establish a considerable gradient temperature along the tube wall. They utilized this effect and installed the basic pulse tube refrigerator. In 1984, E. I. Mikulin published their innovative modification of the basic pulse tube refrigerator.[2] The hot end of the pulse tube is connected to a large volume gas reservoir via an orifice instead of being closed. The refrigerating performance had been improved significantly.

As a regenerative refrigerator, the cooling power of pulse tube refrigerator is low, usually below 200W at the liquefied nitrogen temperature. So, now the pulse tube refrigerator can only be used in the field of low power refrigeration. However, theoretically, the pulse tube refrigerator can be used not only in low power refrigeration, but also for large refrigeration power, such as gas-liquefaction.[3] We have installed a new recuperative pulse tube refrigeration system and analyzed it. The feasibility of this system has been proved by the experiment.

EXPERIMENTAL APPARATUS

Figure 1 showed the schematic of the experimental apparatus of the low-temperature rotary valve pulse tube refrigerator system. It includes a compressor, a pulse tube, a recuperative heat exchanger, a low-temperature rotary valve, a needle orifice valve, a reservoir and some connecting pipes. Helium has been used as the working substance in the experiments. The main components are stated in details as below.

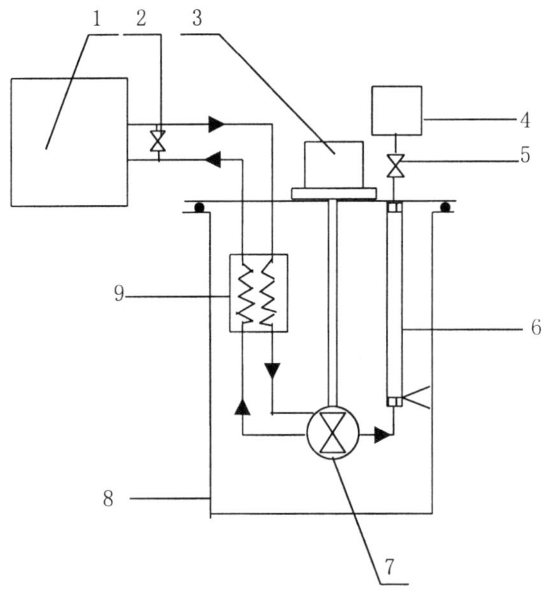

1. Compressor 2. Bypass Valve 3. Electromotor
4. Gas Reservoir 5. Needle Orifice Valve 6. Pulse Tube
7. Low-temperature Rotary Valve
8. Vacuum jacket
9. Recuperative Heat Exchanger

Figure 1. Low-temperature Rotary Valve Experiment Apparatus

In the experiment we use a C50W compressor. It is a Helium-compressor which have been used in the G-M regenerators. The input power is 2kW, the capacity is 40Nm^3/h and the highest discharge pressure is 2.0MPa.

The pulse tube has been made of thin stainless steel pipe. Its size is $\Phi 20.4 \times 260$mm. The thickness of it is 0.2mm.

The low-temperature rotary valve has been designed by ourselves. It was driven by a permanent magnetic DC electromotor of the 70LY54 model. The rating current is 1.26A and the rating voltage is 48V. The largest rotate speed is 900rpm and the lest torque is 0.637N•m. The frequency of the refrigerator has been regulated by adjusting the input voltage of the electromotor.

We utilize the recuperative heat exchanger as the heat exchanger of the system. It has some difference with the normal perforated-plate heat exchanger.[4] The high and low pressure parts of the recuperative heat exchanger are not separated by gaskets, but located in two coaxial stainless steel pipes. The high-pressure stream flows in the inside tube and the low-pressure part of the stream flows in the outside annular channel. It has been filled alternatively with a perforated-plate and a stainless steel screen. The perforated-plate has been made by the laser-etching technology.[5]

The mesh size of the stainless steel screen is 100. It can reduce the axial conduction loss of the heat exchanger. Compared with the perforated-plate heat exchanger, this kind of exchanger has no problem of leakiness. It also has the advantage of a larger heat transfer area per unit volume in contrast with normal heat exchangers. The heat exchanger is 120mm in length and 0.2mm in thickness. The diameters of inner and outer pipes are 32mm and 18mm, respectively. The recuperative heat exchanger is shown in Figure 2.

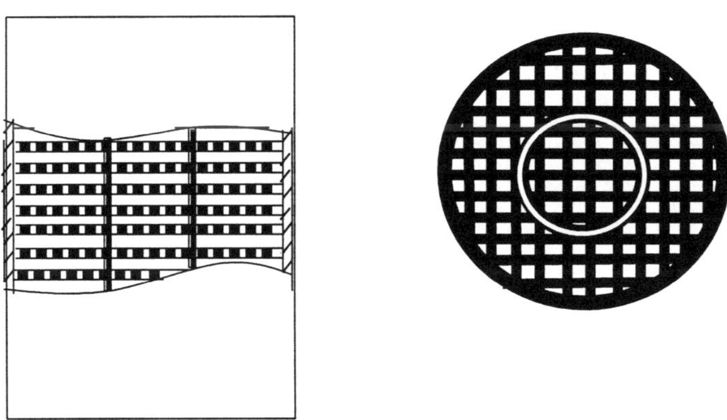

Figure 2. Schematic of the Recuperative Heat Exchanger.

The gas reservoir is 80mm in diameter and 1500cm^3 in volume.

RESULTS AND DISCUSSION

The cool down curve for the pulse tube has been shown in Figure 3. The pressure ratio is 2.5, the cycle frequency is 3Hz and the orifice opening is 1.8 turns. It took 50 minutes to get the lowest temperature.

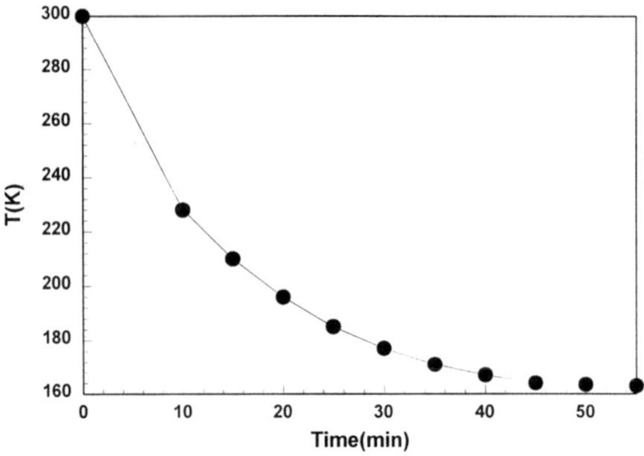

Figure 3. Cool down curve of the pulse tube refrigerator.

Figure 4(a) shows the relationship between temperature of cold end of the pulse tube and the cycle frequency under different pressure ratios. Figure 4(b) shows the dependence of old end temperature on the pressure ratio under different cycle frequencies. The opening of the orifice is 1.8 turns. From Figure 4(a), the temperatures decreased firstly and then increased with the rising frequency. While the opening of the needle valve is the same, reducing the frequency will increase the pressure ratio, which leads to lower temperatures. But too low frequency would also result in higher temperatures. The optimal frequency is 3Hz while the pressure ratio is 2. It can be seen from Figure 4(b) that the cold end temperature falls with the rising pressure ratio. So increasing pressure ratio improves refrigeration power and results in lower temperature at cold end of the pulse tube. But this trend is reversed when the pressure ratio is greater than 2.

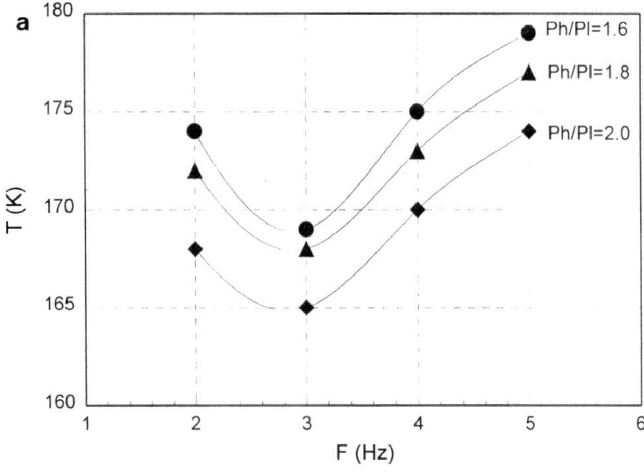

Figure 4. The relationship of the cold end temperature and the pressure ratio, the cycle frequency.

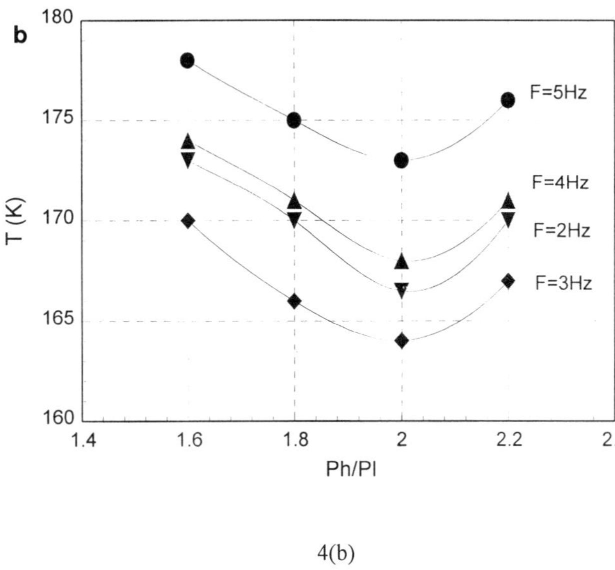

4(b)

Figure 4. The relationship of the cold end temperature and the pressure ratio, the cycle frequency.

Figure 5 shows the lowest cold end temperature of the pulse tube refrigerator achieved at different frequencies and opening turns of the orifice when the pressure ratio is 3.5. Because the mechanism of the refrigerator is the expansion of gas and the dissipation of power in the hot end, the throttling process of the orifice plays a very important role. So corresponding to the higher frequency, the optimal opening of orifice is larger. Because the flow of the gas is reduced along with the rising frequency, the opening of orifice should be increased to achieve the optimal temperature.

In Figure 6, the relation of the refrigerating power and the temperature with respect to different opening turns of orifice has been shown. The pressure ratio is 3 and the cycle frequency is 4Hz. We can find that the optimal orifice opening turn is about 2 to such pressure ratio and frequency. The refrigerating power is 1.0W/K under this condition.

The lowest temperature obtained by this refrigerator is 163K. It is higher than that of ordinary regenerative pulse tube refrigerators. There are two reasons for it. Firstly, the existence of low-temperature rotary valve has caused the losses such as the friction, the heat conduction, and the leakage from the high pressure to the low-pressure side and so on. In the experimental apparatus the mass flow rate of gas is quite small, so the loss is important in compared with the refrigerating power. The problem will be solved in practice because the proportion of the losses will decrease while the mass flow rate of the system increases. Secondly, the efficiency of the recuperative heat exchanger can only reach around 90%, but the efficiency of regenerators often can reach 99%. So improving the performance of our pulse tube refrigerator in practice should increase the heat exchange efficiency.

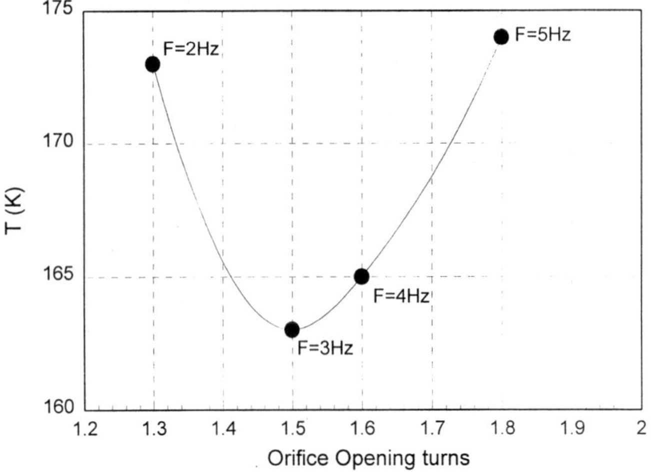

Figure 5. Variation of the cold end temperature under different frequencies and opening turns of orifice.

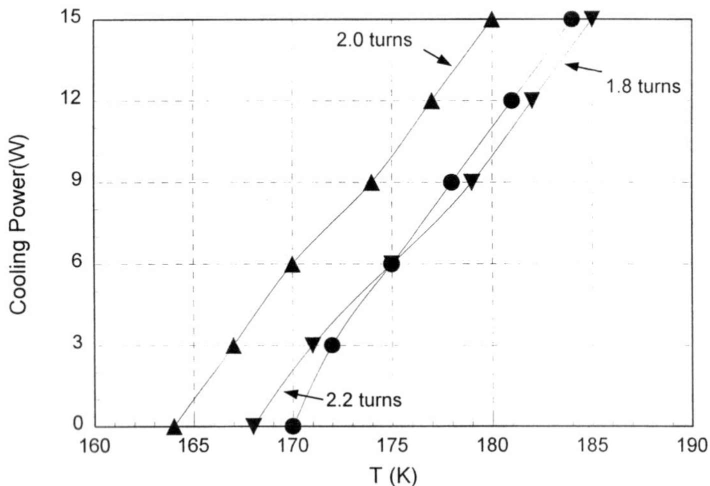

Figure 6. Variation of the cooling power with respect to the cold end temperature and the modified opening turns of orifice.

CONCLUSION

In this paper we have introduced a kind of recuperative pulse tube regenerator. It has an advantage of a simple structure and it can be used in situations such as the gas-liquefaction where large refrigeration power is needed. The relations between temperature of cold end of the pulse tube refrigerator and the cycle frequency, the pressure ratio, opening turns of the orifice have been obtained. The research on this system has set the basis for the application in gas-liquefaction. In the future, we will improve the structure of the rotary valve to reduce losses and enhance the efficiency of the recuperative heat exchanger to improve the performance of the system.

REFERENCES

1. W E Gifford, R C Longsworth. Pulse tube refrigeration. Trans ASME, J Eng Ind, 86:264 (1964).
2. Mikulin E I, Tarasov A A, Shkreb A,Shrebonock M P. Low temperature expansion pulse tubes. Adv Cryo Eng, 29:629 (1984).
3. J.Liang, Large power and low temperature switch-valve pulse tube refrigerator. Proceedings of Eighth conference of Cryocooler of China (1996)
4. Zhiyou Zhang, Bingsan Shi, The technologic principle and apparatus. Beijing: the Publishing Company of Mechanical Engineering (1987)
5. Yuhao Shen, The research of the compact perforated-plate heat exchanger, Cryogenics, 4: 95 (1981)

In recognition of their many years of faithful service on the CEC Directors Board, R. Witt (University of Wisconsin), P. Kelley (Los Alamos National Lab), P. Kittel (Ames Lab), and P. Gifford (Cryomech) received the sincere acclaim of their fellow CEC members.

EXPERIMENTAL STUDY OF HEAT TRANSFER PHENOMENA BETWEEN WORKING GAS AND TUBE WALL IN A PULSE TUBE REFRIGERATOR

K.Takamatsu,[1] M. Shiraishi,[2] M.Murakami,[1] and A.Nakano[2]

[1]University of Tsukuba
Tennodai 1-1, Tsukuba, Ibaraki, 305-8537 Japan
[2]Mechanical Engineering Laboratory, MITI
Namiki 1-2, Tsukuba, Ibaraki, 305-8564 Japan

ABSTRACT

The experiment is conducted to investigate how heat is exchanged between working gas and the tube wall over one cycle in a pulse tube refrigerator. The pulse tube has an inner diameter of 14.9 mm and a length of 306 mm. The radial velocity and temperature distributions were measured at six positions along the pulse tube by using a hot-wire anemometer and thermocouples, and the distributions over the pulse tube were obtained from the measurement data through the linear interpolation. From these distributions (Eulerian form), the change of radial temperature distribution through the pulse tube in the coordinate system fixed on the moving gas element (Lagrangian form) is obtained over one cycle. The thickness of boundary layer is determined to be approximately 1 mm or less by the experiment. The heat flux, which is transferred between the gas element and the wall, is calculated from the temperature distribution and the thickness of the boundary layer. It is found that each element releases heat to the wall when it displaces toward the hot end and absorbs heat when it goes toward the cold end. Moreover, it is also found that heat exchange phenomena can be explained qualitatively well based on the idea of surface heat pumping.

INTRODUCTION

Cryocoolers will be applied to various cryogenic engineering fields such as superconductor applications, space cryogenic applications and so on. The pulse tube refrigerator is one of the most common cryocoolers, but few practical systems are available now compared to Stirling and G-M cryocoolers. The main reason is that many aspects concerning pulse tube design and operation remain poorly understood so that a high reliability and efficiency similar to Stirling and G-M cryocoolers has not been proved in the practical field. One of the possible solutions is to understand the refrigeration mechanisms in more detail and to develop the realistic analytical model for the design of practical systems.

Several expositions the heat transfer mechanism in a pulse tube refrigerator have been presented by many researchers. The mechanism of surface heat pumping is widely accepted as an explanation of operation in a basic pulse tube.[1,2] On the other hand, for an orifice pulse tube, there are "enthalpy flow analysis", "non-symmetry effect", "pressure heat pumping", "thermodynamic model" and so on.[3,4,5,6,7] It is generally believed that there are several different operating mechanisms in basic and orifice pulse tube refrigerators.

Although strenuous efforts have been done to develop models based on these ideas, the experimental studies to prove these ideas have not been sufficiently conducted, especially from a view point of phenomenological thermo-fluid dynamics. For example, the surface heat pumping model is widely accepted, but it has not been clearly proved experimentally.[8] The most fundamental and important factors in refrigeration mechanisms are the work and heat flows in a refrigerator. Several experimental studies concerning the work of gas have been reported, but few experiments to investigate the heat transfer phenomena have been reported.[9,10]

The objective of this study is to investigate experimentally how heat is transported along a pulse tube refrigerator; especially, the difference in the heat transport phenomena between a basic and an orifice pulse tube refrigerators. As the first step, we introduce the procedure of the experiment and the typical results for the basic pulse tube refrigerator for an operating frequency of 10 Hz. This frequency is somewhat higher than the optimum one for the basic, but this frequency is selected as the optimum frequency for the orifice configuration. The details of the orifice pulse tube refrigerator and the comparison of them with the basic one will be presented in future.

EXPERIMENTAL APPARATUS AND PROCEDURE

A schematic of the experimental apparatus is shown in Figure 1. It is composed of three units, the compressor unit, the pulse tube refrigerator and the measuring instrument. The combination of a helium compressor unit originally designed for a Gifford-McMahon refrigerator and a rotary valve can supply periodical pressure oscillation to the pulse tube. The frequency of gas pressure oscillation can be varied from 0 to 20 Hz by varying the rotation speed of the rotary valve. The compression ratio, which is defined as the ratio of high and low pressures over one cycle, can be changed from 1 to 3.4 by adjusting the opening of the bypass valve connecting the high pressure exit to the low pressure inlet of the compressor. The pulse tube is

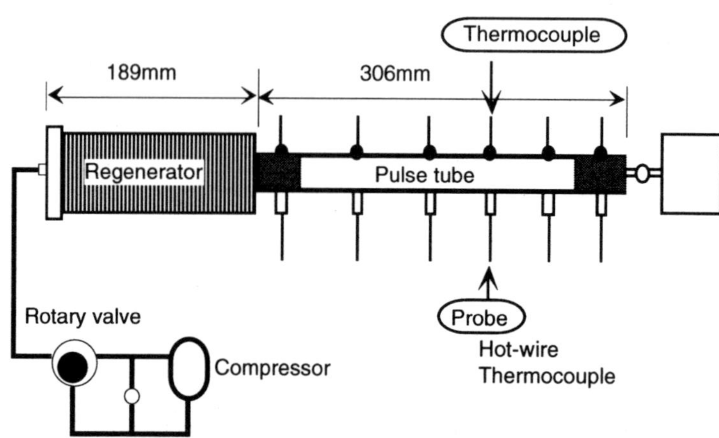

Figure 1. Schematic of experimental apparatus.

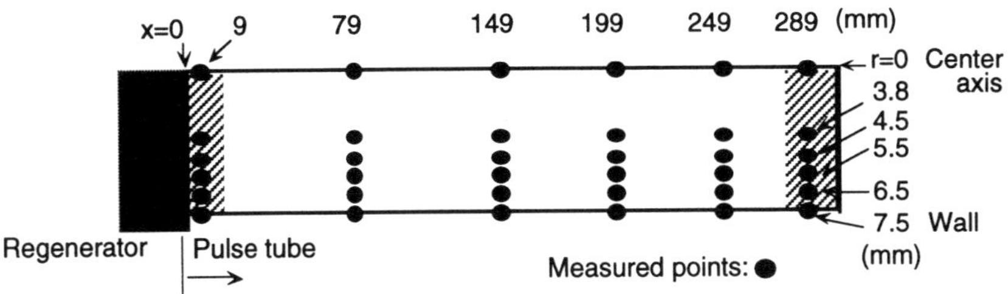

Figure 2. Measured points in the pulse tube.

made of stainless steel with an inner diameter of 14.9 mm, wall thickness of 2.4 mm and a length of 306 mm. The heat exchangers at both the ends are made of copper tube. The regenerator is composed of a stainless steel housing and a bakelite tube with an inner diameter of 35 mm and a length of 160 mm packed with 856 discs of 100 mesh stainless steel screen. The reservoir with a volume of 603 cm^3 is connected to the pulse tube via the orifice valve. Types of pulse tubes, which are a basic, an orifice and a double inlet types, are selected by using the orifice and bypass valves.

The gas pressure is measured with strain gauge type pressure transducers at the warm end of the regenerator, the pulse tube and the reservoir. The probes consisting of a hot-wire and a thermocouple are installed at six positions inside the pulse tube to measure the velocity and temperature of gas. The probe is fixed by a fitting with Teflon ring for the sake of sealing and easy traversing in the radial direction. Other thermocouples are installed on the pulse tube wall at same positions along the tube to measure the wall temperature.

The procedure of the experiment is as follows: After the whole refrigerator system is evacuated, helium gas is charged in it at the pressure level of 1.4 MP. Cooling water is circulated through the heat exchanger at the hot end, and then, pressure oscillation is introduced into the pulse tube. The pressure ratio is adjusted at about 1.4 with the aid of the bypass valve at the compressor. When the wall temperature reaches a steady state, the velocity and temperature of gas in the pulse tube and the wall temperature are measured at many positions as indicated in Figure 2 enough to estimate the detailed distributions of the temperature and the velocity over the pulse tube. The experiments are carried out in the range of the driving frequency between 1 and 15 Hz for three pulse tube configurations. In the experiment we investigated systematically the effects of the operating frequency, the opening of the orifice and double-inlet valves and the types of pulse tube configurations on the heat transfer phenomena in the pulse tube.

RESULTS AND DISCUSSION

As mentioned above, in what follows we introduce the typical results for the basic pulse tube refrigerator under the operating frequency of 10 Hz.

Variations of the velocity and temperature oscillations in the axial and radial directions

The typical measurement data of the velocity and temperature oscillations are shown in Figures 3 to 6. Figure 3 and 4 show the variations in the axial direction (x = 9, 149 and 249 mm) measured on the center axis of the pulse tube (r = 0 mm). Figures 5 and 6 show those in the radial direction (r = 0, 6.5 mm) at x = 149 mm. The origin of the time-axis is taken at the onset of the compression process. It is found from these figures that the velocity decreases with the distance from the cold end. On the other hand the amplitude of temperature oscillation slightly increases with the distance. The temperature level at x = 149 mm is higher than

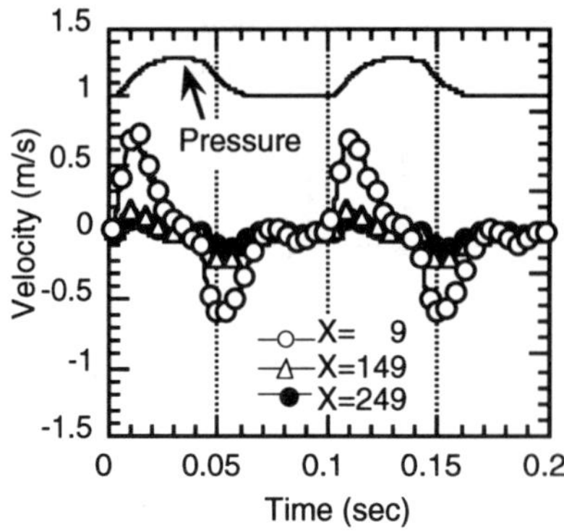

Figure 3. Change of velocity oscillation in the axial direction. $r = 0$ mm.

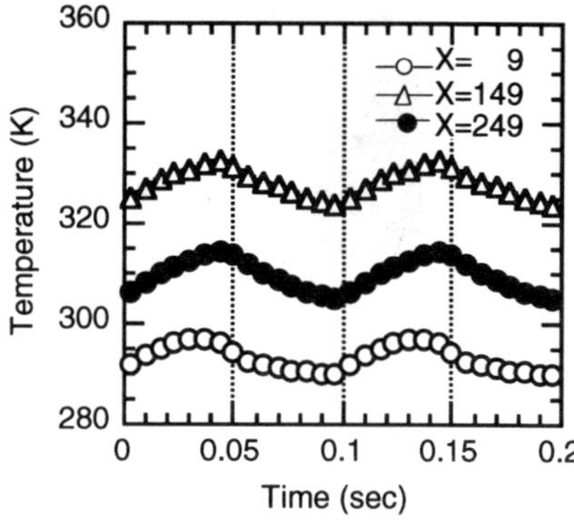

Figure 4. Change of temperature oscillation in the axial direction. $r = 0$ mm.

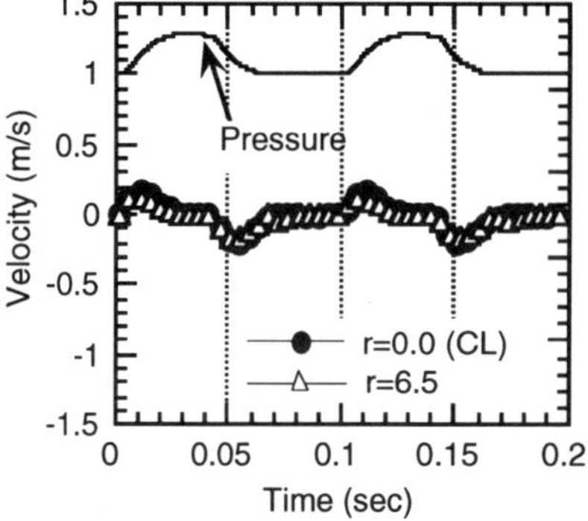

Figure 5. Change of velocity oscillation in the radial direction. x= 149 mm.

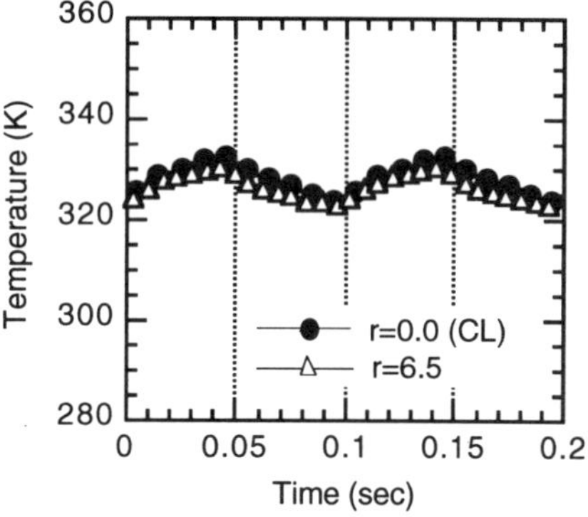

Figure 6. Change of temperature oscillation in the radial direction. x= 149 mm.

that at $x = 249$ mm, because in the case of 10 Hz the temperature distribution has a maximum between the hot end and the middle point of the pulse tube. The variation of the amplitudes in the radial direction in Figures 5 and 6 is negligibly small and other data also show a similar results so that the thickness of the boundary layer is thought to be approximately 1 mm or less.

Estimation on displacement of a gas element

The axial distribution of the velocity along the center axis can be constructed with the aid of the linear interpolation among the data measured at positions shown in Figure 2. The velocity u_i of element at x_i at the time t_i which is initially at x_0 is calculated from Equation (1) and the displacement is from Equation (2). Here, u_k and u_j denote the velocities measured at the position k and j in Eulerian coordinate system, and u_i is the velocity of element at x_i in

$$u_i(x_i,t_i) = \left[\frac{u_k(t_i) - u_j(t_i)}{l_{k-l}}\right](x_i - x_j) + u_k(t_i) \tag{1}$$

$$x_{i+1} = x_0 + \sum_{t=0}^{n} u_i \delta t \tag{2}$$

Lagrangian system. Here, the distance between the position k and j is l_{k-l}. The sampling period (= 0.4 msec) and the total data number during one period are denoted by δt and n. Figure 7 shows the derived movement of the element on the center axis initially at $x = 9$, 79, 149 and 199 mm over one cycle. The onset of the compression process corresponds to time of 0. It is found that the amplitude of the displacement decreases drastically with the increase of the distance from the regenerator for the basic configuration.

Radial temperature distribution near the wall

A gas element exchanges heat with the wall during one cycle. It is easy to know through the examination of the radial temperature distribution whether the gas element absorbs heat from or releases heat to the wall. The time variation of the temperature distribution in the whole region of the pulse tube over one cycle is obtained from the measured temperature data with the aid of the linear interpolation. From the results of the temperature distribution and the displacement, the radial distribution of the temperature in the coordinate system fixed on the moving gas element on the center axis is derived. Figure 8 shows the time variation of temperature distribution over one cycle. It is seen that the wall temperature varies with time due to the distribution of the temperature along the wall.

Heat flow at the wall

As a first step for the investigation of the heat exchange process between the gas element and the wall, our attention is focused on the understanding of the general tendency than on the quantitative explanation. Therefore, it is assumed that the heat flow to or from the wall in a sampling period of δt (= 0.4 msec) is determined by Fourier's law and that the thickness of thermal boundary layer is 1 mm as suggested from the above result. Based on these assumptions, it is seen that the heat flow at an instance during one cycle is calculated from the temperature gradient determined from the difference in the temperatures between at the wall and at $r = 6.5$ mm, and the thermal conductivity of gas evaluated at the average temperature at both positions.

Figure 9 shows the time variation of the heat flux over one cycle at each position in the

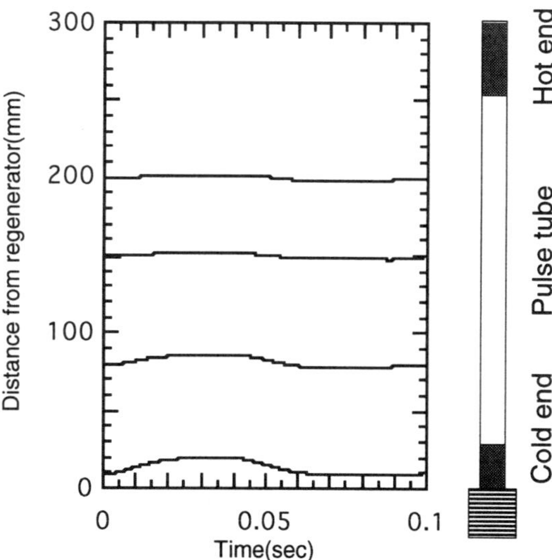

Figure 7. Change of displacement of the gas element at each position on the center axis.

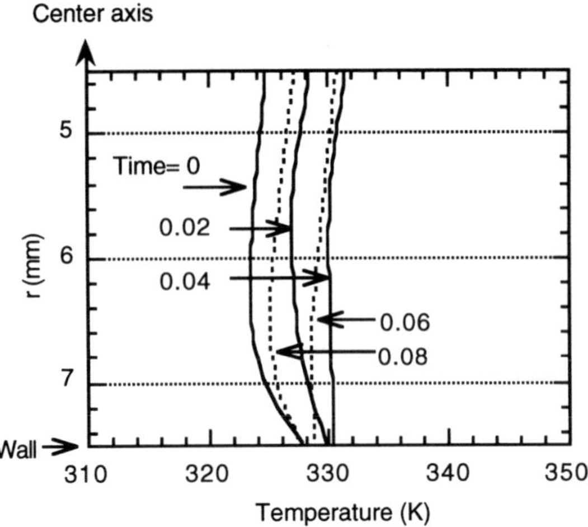

Figure 8. Change of radial temperature distribution near the wall in the coordinate system fixed on the moving gas element on the center axis (Lagrangian form). The gas element oscillates near the middle point of the pulse tube (x = 149 mm).

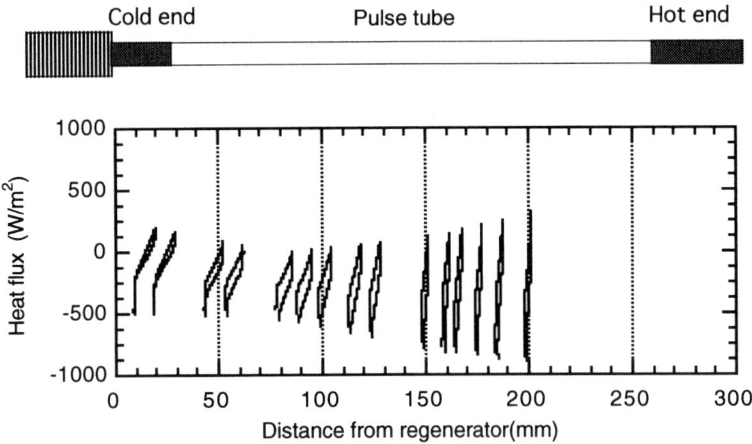

Figure 9. Variation of heat flux over one cycle as a function of displacement of each element at own position in the pulse tube. The positive heat flux corresponds to the heat transfer from the element to the wall and the negative to from the wall to the element.

pulse tube together with the displacement of each gas element. Here, the positive heat flux indicates the heat transfer from the gas element to the wall and the negative from the wall to the gas element. Each trajectory turns counterclockwise with time. The absolute value of the heat flux may not be satisfactorily corrects due to the accuracy of the experiment, but the heat transfer mechanism in the basic pulse tube can be explained well by this results. The top and bottom of each trajectory is inclined toward the hot and the cold end, respectively. This fact indicates that each gas element releases heat to the wall when it is displaced toward the hot end and absorbs heat when it is done toward the cold end. This phenomena can be explained qualitatively well by the idea of surface heat pumping as an explanation of the cooling mechanism in the basic pulse tube.

SUMMARY

The experiment is conducted to investigate how heat is exchanged between working gas and the tube wall over one cycle in the basic pulse tube refrigerator. The time variation of the heat flux in the coordinate system fixed on a moving gas element (Lagrangian form) over one cycle is clearly presented together with the displacement of each gas element.

The results indicate that each gas element releases heat to the wall when it is displaced toward the hot end and absorbs heat when it is done toward the cold end, and that the heat exchange phenomena can be qualitatively explained well by the idea of the surface heat pumping as an explanation of the cooling mechanism in the basic pulse tube refrigerator. Further study for the other configurations such as on orifice and double-inlet ones will be made in the future.

REFERENCES

1. W.E.Gifford and R.C.Longthsworth, Surface heat pumping, Adv Cryo Eng, 11:171 (1966).
2. R.N.Richardson, Pulse tube refrigerator-an alternative cryocooler?, Cryogenics, 26:331 (1986).
3. R.Radebaugh, A review of pulse tube refrigeration, Adv Cryog Eng 35:1191 (1990).
4. P.Kittel, A.Kashani,J.M.Lee, and P.R.Roach, General pulse tube theory, Cryogenics, 36:849 (1996).
5. J.Liang, A.Ravex and P.Rolland, Study on pulse tube refrigeration Parts1-3, Cryogenics, 36:87 (1996).
6. P.C.T. de Boer, Heat removal in the orifice pulse tube, Cryogenics, 38:343 (1998).
7. A.Tominaga, Thermodynamic aspects of thermoacoustic theory, Cryogenics, 35:427 (1995).
8. B.E.Evans and R.N.Richardson, An experimental investigation of the heat pumping mechanism in a pulse tube changes with frequency, "Cryocoolers 9", R.G.Ross,Jr. ed., Plenum Press, New York(1997), p.375.
9. M.Kasuya, M.Nakatsu,Q.Geng,J.Yuyama, and E.Goto, Work and heat flows in a pulse-tube refrigerator, Cryogenics, 31:786 (1991).
10. M.Shiraishi, N.Nakamura, K.Takamatsu, M.Murakami and A.Nakano, Visualization study on flow behavior and work of working gas in a pulse tube refrigerator, Proceedings of 17th ICEC, D.Dew-Hughes, et.al. ed., Institute of Physics Publishing, Bristtol(1998), p.73.

J. Pfotenhauer (University of Wisconsin), T. Peterson (Fermilab), C. Joshi (Energen Inc.) were elected as the new members of the CEC Board of Directors.

VISUALIZATION STUDY OF SECONDARY FLOW IN AN INCLINED PULSE TUBE REFRIGERATOR

M. Shiraishi,[1] K. Takamatsu,[2] M. Murakami,[2] and A. Nakano[1]

[1]Mechanical Engineering Laboratory, MITI
Namiki 1-2,Tsukuba, Ibaraki, 305-8564 Japan
[2]University of Tsukuba
Tennoudai 1-1, Tsukuba, Ibaraki, 305-0006 Japan

ABSTRACT

A secondary flow induced in an inclined orifice pulse tube refrigerator has been observed to investigate the reason for the effect of inclination angle on the performance. The pulse tube, which is 16 mm in diameter and 320 mm long, is made of a transparent plastic tube. Air is used as the working fluid and the pulse tube is operated under the frequency of 5 Hz. A smoke-wire flow visualization method is used to observe the flow behavior of secondary flow. The observation is done at the inclination angles of 0°, 60°, 90°, 120°, 150° and 180°. By comparing the smoke-lines, which are at the turn from compression to expansion processes at every cycle, the flow behavior of secondary flow is investigated and the nominal velocity is obtained.

Two types of secondary flow are observed: One is the acoustic streaming resulting from oscillatory phenomena and the other is the natural convection induced by the difference of gas temperature between at the cold and the hot ends. The flow patterns of secondary flow are classified into 3 patterns according to the inclination angle. The effects of the inclination angle on the performance are qualitatively well explained in terms of the convective heat losses produced by the secondary flow.

INTRODUCTION

Although Stirling and G-M cryocoolers are now widely used, these have such a moving part as a displacer in the cold region, which makes them mechanically complicated and demands periodic maintenance. A pulse tube refrigerator does not have a cold moving part.[1] This promises many advantages such as structural simplicity, high reliability and low cost compared to Stirling and G-M cryocoolers. On the other hand there is an inevitable disadvantage

Advances in Cryogenic Engineering, Volume 45.
Edited by Shu *et al.*, Kluwer Academic / Plenum Publishers, 2000.

such as the orientation dependence of the performance. The performance of a pulse tube changes drastically as it is inclined from the vertical to the inverted orientation.[2,3] Here, we defined the inclination angle as 0° is the vertical orientation with the cold end lower than the hot end and 180° is the inverted orientation (*see Figure 2*). According to the reference 2, the orientation dependence is as follows: the change in a minimum no-load temperature of the cold end is small until the angle reaches around 80°; further increase of angle leads to a steep rise of the temperature attaining a maximum value at about 120°, which is about 3 times higher than that at 0° ; the temperature decreases again with further increase of angle through to 180°. In this case, the pulse tube can not maintain a temperature level below the liquid nitrogen temperature in the region of 80° to 180°. This is a very serious problem for the practical cooling of high temperature superconductors. It is pointed out in the reference 2 that the cause of orientation dependence is the heat losses caused by gravity induced natural convection occurring for the angle larger than 0°. However, it is doubtful whether the natural convection occurs in the region of below 90° or not, because the cold end is lower than the hot end in the region. Moreover, it is of interest to note that the performance gets better again in the region over 120°. Although the details of secondary flow are needed for understanding of the orientation dependence, there is few experimental data on it.[4,5]

The objective of the present study is to observe what kind of secondary flow is induced in an inclined orifice pulse tube refrigerator and to investigate the cause of the orientation dependence.

EXPERIMENT

The experimental apparatus for visualization is shown schematically in Figure 1. The pulse tube, which is 16 mm in diameter and 320 mm long, is made of a transparent plastic tube. The regenerator is 18 mm in diameter, 170 mm long, and made of #100 stainless-steel

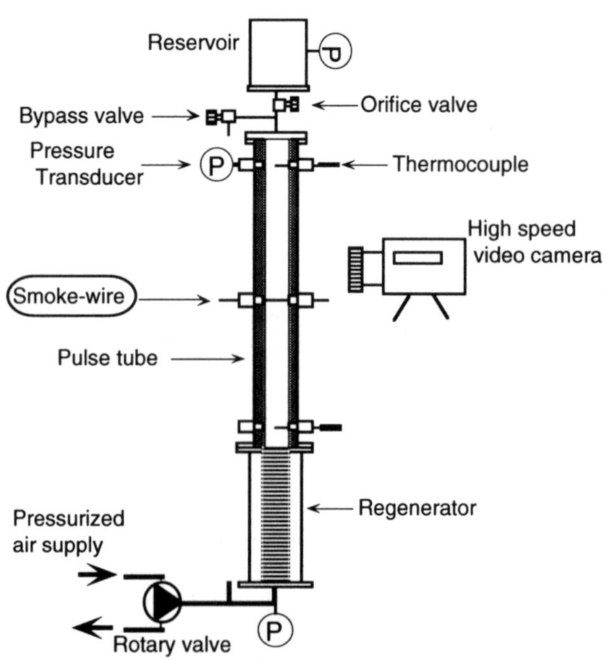

Figure 1. Schematic of the experimental apparatus.

screen with a wire diameter of 0.1 mm. The reservoir is composed of a plastic vessel with a volume of about 10 times as large as the pulse tube. The orifice is a needle valve. Air is used as the working gas. The pressure oscillation is generated by introducing pressurized air of about 0.2 MPa into the rotary valve and releasing it to the atmosphere. A smoke-wire is a tungsten wire with 0.1 mm diameter. Both ends of the wire are soldered to copper supports acting as electrodes and as a fixing agent to keep the wire taut. The wire is tightened through the tube fittings at the middle of the pulse tube. Pressure transducers of a strain gauge type are installed near the warm end of the regenerator, at the hot end of the pulse tube and in the reservoir. Mineral insulated thermocouples (CA) of 0.15 mm in diameter are inserted into the gas space at the cold and hot ends. The performance of the pulse tube is evaluated based on the difference of the gas temperatures at the both ends.

The pulse tube is fixed on the base with a rotatable horizontal axle. The pulse tube can be rotated from the vertical (0°) to the inverted orientation (180°). The experimental condition is determined as follows: The frequency is fixed at 5 Hz by taking the duration of smoke-line into consideration. The gas velocity increases with the increase of the amplitude of pressure wave and the observation becomes difficult so that we found the adequate amplitude to be 1.2, that is defined by the compression ratio between high and low pressures. The setting of the orifice valve is selected so as to maximize the gas temperatures difference between both ends of the pulse tube in the vertical orientation. The visualization is done at typical 6 orientations with angles of 0°, 60°, 90°, 120°, 150° and 180°.

After coating paraffin on the smoke-wire surface, it is installed in the tube and the lead wires are connected to the high voltage pulse generator. The pulse tube is operated under the prescribed conditions. After confirming a steady state is reached by monitoring the gas temperature, a smoke-line is emitted. The movement of smoke-line is viewed and recorded using a high speed video camera with a frame rate of 400 frames/sec over 5 cycles.

RESULTS AND DISCUSSION

Effect of inclination angle on the performance

The orientation dependence of the pulse tube performance is investigated in advance under same prescribed operating conditions, that is, the frequency is 5 Hz and the compression ratio is 1.2. Figure 2 shows the orientation dependence in the form of the variation of the gas temperature difference between the cold and hot ends. The temperature difference in the

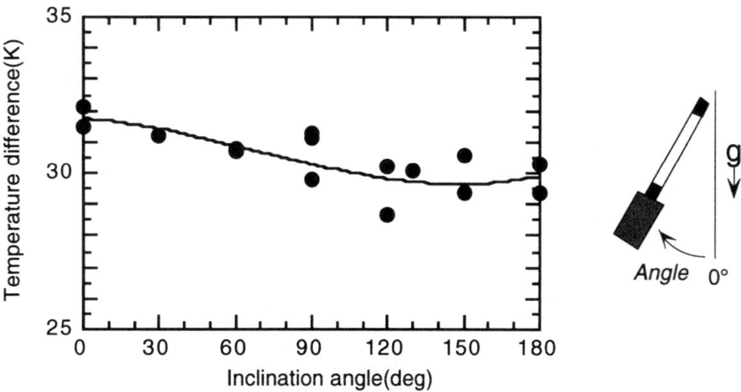

Figure 2. Variation of the difference in the gas temperature between at the hot and the cold ends as a function of inclination angle.

region of 120° is smaller than that in the region of small angle. Even though the change of temperature with the inclination angle in our experiment is extremely small compared to that in the reference 2, but we can also find the same orientation dependence of the performance on the inclination angle. This result confirms that the same flow phenomena occur in the pulse tube under these operating conditions.

Observation of oscillating flow

Figure 3 shows the oscillating motion of a smoke-line every 20 msec selected from the original video sequential pictures in the first two cycles after emitting the smoke-line at the middle point of the pulse tube with the inclination angle of 0°. The smoke-line is emitted at the onset of compression process so that it moves toward the hot end, and then, goes back in the expansion process. We can see that the shape of smoke-line is gradually deformed with time. The point to observe is that the distance between the tips of smoke-line near both walls (sides) and the vertex of parabolic smoke-line in the core increases with time. The slow drift of the smoke-line with time indicates the existence of a secondary flow.

To allow the quantitative evaluation of the secondary flow, the smoke-lines at the moment of a turn from the compression to expansion processes in each cycle are selected and compared. Figures 4 and 5 show the deformation of smoke-lines at every turn for 5 cycles at

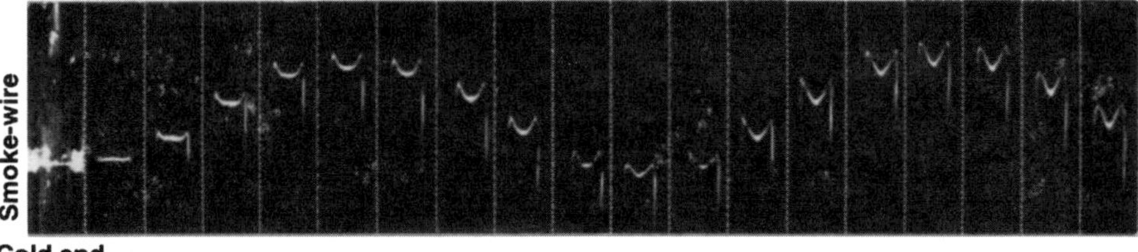

Figure 3. Sequential video pictures of a smoke-line in the pulse tube with the vertical orientation(0°).

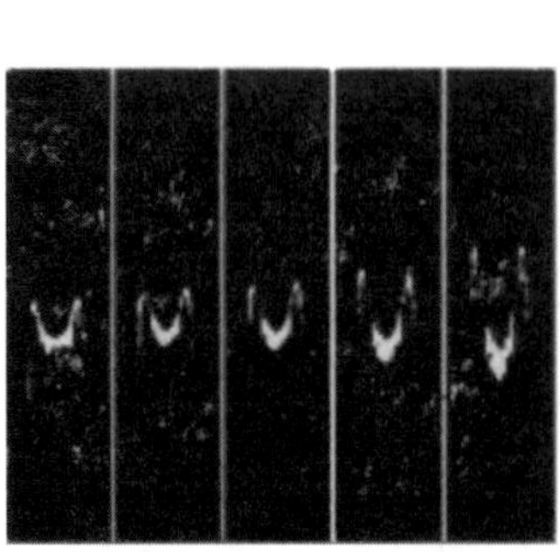

Figure 4. Deformation of the smoke-lines with time. Those are selected from the original video frame at a turn from compression to expansion processes in every cycle. The inclination angle is 0°.

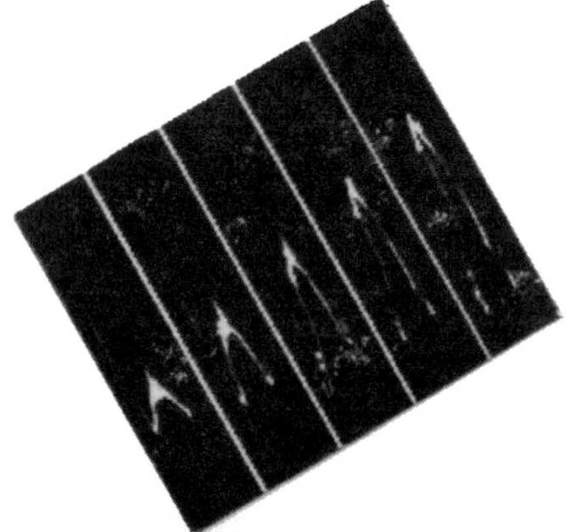

Figure 5. Deformation of the smoke-lines in every cycle. The inclination angle is 150°.

the inclination angles of 0° and 150°, respectively. In both figures it is seen that the tip of the smoke-line near the wall drifts toward the hot end and the vertex in the core region does toward the cold end with time. In the case of 150° the vertex shifts to the upper side wall and the distance between the tip and the vertex is longer than those in Figure 4. Even though the angle is 0°, the deformation of smoke-line is recognized. This fact indicates the existence of another type of secondary flow not originated from the gravity effect. The flow can be regarded as an acoustic streaming driven by oscillatory phenomena.[6]

Observation of secondary flow

The video frames at every turn as shown in Figures 4 and 5 are converted to digital images by using an image-processing system and computer-identified smoke-lines are obtained. Figure 6 shows the plots of the identified smoke-lines for all observed inclination angles. The plots for angles of 90° to 150° suggest the existence of natural convection that the fluid goes toward the cold end along the upper side wall and the associated return flow toward the hot end along the lower side wall. On the other hand, other plots indicate a different convection that the fluid goes toward the cold end in the core and the associated return flow does toward the hot end along both walls. The flow patterns of this kind of convection in the plots of 0° and 180° are the same, but the convective motion for 180° is much faster than that for 0°. On the other hand, little difference is recognized between the plots 0° and 60° so that

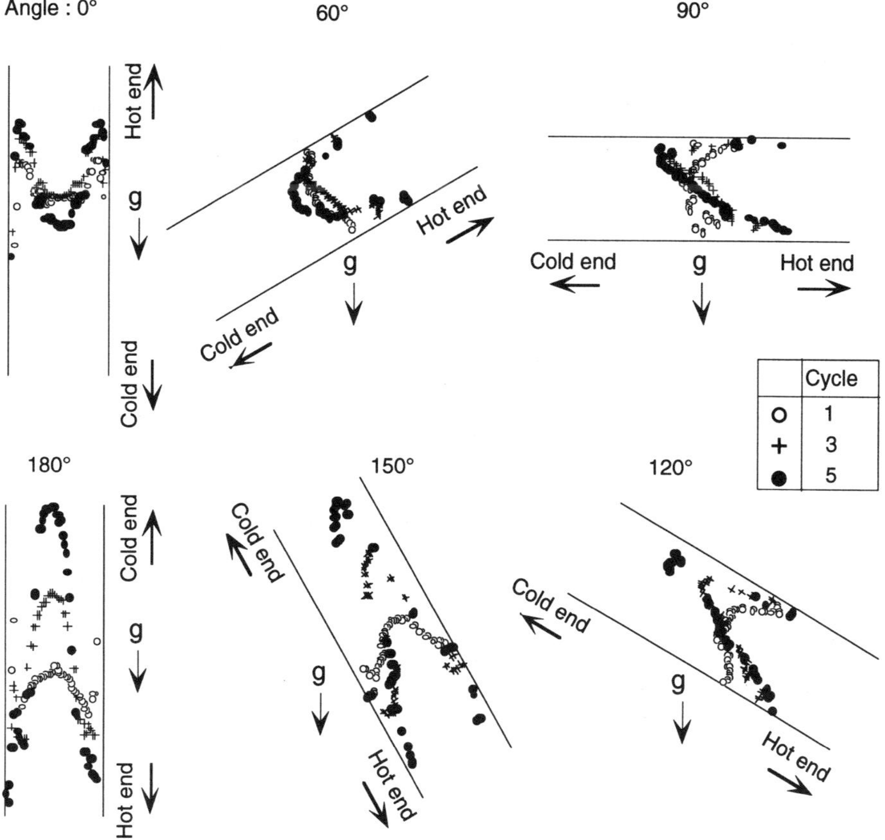

Figure 6. Plots of computer-identified smoke-lines based on the video sequence such as figures 4 and 5 for all observed orientations.

the inclination angle is considered not to have substantial effects in this region of the inclination angle.

Velocity of secondary flow

From the results of Figure 6, the axial velocities of the regions of the vertex and both the tips near the upper and lower sides are derived as typical velocities for the secondary flow from the change of displacement. Figure 7 shows the result. It can be concluded that the convection in the inclined pulse tube is classified into 3 flow patterns according to the angle: The first pattern in the region of angle less than about 90°, the second roughly 90° through 150°, and the third around 180°.

In the convection in the first region, the acoustic streaming resulting from the oscillatory phenomena dominates. On the other hand, in other regions the natural convection induced by the gravity effect superimposes on the acoustic streaming. Therefore, the flow patterns are characterized by the combination of the acoustic streaming and the natural convection. According to the knowledge of natural convection in an inclined confined channel, the convection in the second region is of a single loop circulation where the fluid goes upward along the upper side wall and the associated return flow does downward along the lower side wall: In the third region it is of an axisymmetric circulation where the fluid goes upward in the core and the associated return flow does downward along the peripheral wall.[7,8,9] In the second region the single loop circulation superimposes on the acoustic streaming. The flow near the upper side wall is canceled by each other and near the lower side wall it is intensified by each other so that the velocity near the upper side decreases and near the lower side it increases. In the third region the flow directions of both streamings, that is the acoustic streaming and the axisymmetric circulation of natural convection, coincide with each other so that the overall axisymmetric convection is produced. The velocity in the core increases and the difference in the velocity of both tips decreases.

According to the study of natural convection, the heat transfer rate in the second region is larger than that in the third region.[8,9] Although the velocity in the core is fastest at 180°, the heat transfer rate is supposed to be smaller compared with that in the second region. This is the explanation why the performance of an inclined pulse tube becomes good in the region of

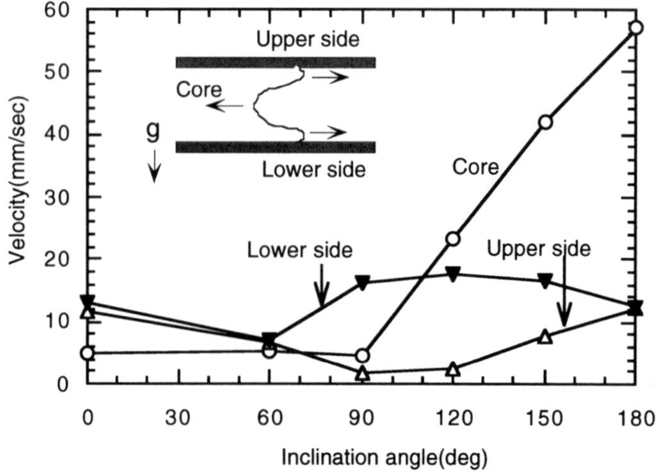

Figure 7. Velocity of the secondary flow in the core and at near both walls.

near 180°. There has been only a little amount of information about natural convection in an inclined confined channel like a pulse tube where the ratio of length to diameter is very large. More detailed study is needed for the quantitative evaluation of convective heat losses in practical inclined pulse tube refrigerators.

SUMMARY

We have observed a secondary flow induced in an inclined pulse tube refrigerator by using a smoke-wire flow visualization method. Some results are as follows:

1) Two types of secondary flows are observed: One is the acoustic streaming resulting from oscillatory phenomena and the other is the natural convection induced by the gravity effects due to the difference in the gas temperature between at the cold and the hot ends.

2) Natural convection is recognized in the region of inclination angle of 90° and above.

3) The flow patterns of overall convection can be classified into 3 patterns by the region of inclination angle.

4) The effects of inclination angle on the performance are qualitatively well explained by the convective heat losses produced by the natural convection.

REFERENCES

1. W.E.Gifford and R.C. Longthsworth, Pulse tube refrigeration, ASME paper, No.63-WA-290 (1963).
2. G.Thummes, M.Schreiber, R.Landgraf, and C.Heiden, Convective heat losses in pulse tube coolers: Effect of pulse tube inclination, “Cryocoolers 9”, R.G.Ross,Jr., ed., Plenum Press, New York(1997), p.393.
3. D.L.Johnson, S.A.Collins, M.K.Heun, and R.G.Ross,Jr., Performance characterization of the TRW 3503 and 6020 pulse tube coolers, “Cryocoolers 9”, R.G.Ross,Jr., ed., Plenum Press, New York(1997), p.183.
4. M.Shiraishi, N.Nakamura, K.Seo and M.Murakami, Visualization study of velocity profiles and displacements of working gas inside a pulse tube, “Cryocoolers 9”, R.G.Ross,Jr., ed., Plenum Press, New York(1997), p.355.
5. J.M.Lee, P.Kittel, K.D.Timmerhaus and R.Radebaugh, Steady secondary momentum and enthalpy streaming in the pulse tube refrigerator, “Cryocoolers 8”, R.G.Ross,Jr., ed., Plenum Press, New York(1995), p.359.
6. J.R.Olson and G.W.Swift, Acoustic streaming in pulse tube refrigerators: tapered pulse tubes, Cryogenics, 37:769 (1997).
7. D.Japikse, Advances in thermosyphon technology, “Advances in Heat Transfer”, T.F.Irvine,Jr. and J.P.Hartnett ed., Academic Press, New York(1973), Vol.9, p.2.
8. H.Ozoe and H.Sayama and S.W.Churchill, Natural convection in an inclined square channel, Int.J.Heat Mass Transfer, 17:401 (1974).
9. H.Ozoe and H.Sayama and S.W.Churchill, Natural convection in an inclined rectangular channel at various aspect ratios and angles-Experimental measurements, Int.J.Heat Mass Transfer, 18:1425 (1975).

1999 CEC-ICMC Awards Luncheon.

EXPERIMENTAL INVESTIGATION OF SOME PHASE SHIFTING TYPES ON TWO-STAGE GM PULSE TUBE CRYOCOOLER

S. Fujimoto,[1] T. Oodo,[1] Y.M. Kang,[1] T. Kanyama,[2] and Y. Matsubara[3]

[1]MEC Laboratory, DAIKIN INDUSTRIES, LTD.
3 Miyukigaoka, Tsukuba 305-0841, Japan

[2]Semiconductor Equipment Department, DAIKIN INDUSTRIES, LTD.
3-12 Chikkou-Sinmachi, Sakai 592-8331, Japan

[3]Atomic Energy Research Institute, Nihon University
7-24-1, Narashinodai, Funabashi, Chiba 274-0063, Japan

ABSTRACT

Two-stage pulse tube coolers are expected to be used for shield-cooling of MRI magnets, cryopumps, low Tc superconducting magnet systems for laboratory use, etc. For these applications they must be designed for a long-term, continuous operation, and they must be more efficient. We are studying the optimum design of the phase-shifting mechanism for two-stage GM pulse tube coolers.

We tested simple-orifice, double-inlet, and moving-plug phase-shifting mechanisms. We evaluated them experimentally by using a single-stage cooler to compare their cooling performance. We also used simplified numerical analysis based on the concept of equivalent PV diagram. Our experimental results indicate that the double-inlet configuration yields the highest cooling performance.

Our prototype two-stage pulse tube cooler uses a double-inlet configuration. The rated input power of the compressor unit is 3.3kW at 50Hz. At a cycle frequency of 2.0Hz, we achieved a minimum temperatures of 33K on the first stage and 5K on the second stage, and we achieved the maximum cooling capacity of 22.9W at 80K on the first stage and 5.9 W at 20K on the second stage. From preliminary measurements using the same compressor unit, the two-stage cooler was able to reach a minimum temperature of 2.70K when magnetic regenerative materials were used in the second stage regenerator. With a heating load of 7.3W at 60K on the first stage, the second stage can provide 190mW at 4.2K.

INTRODUCTION

Pulse tube cryocoolers have received considerable attention because of the absence of moving parts in the cold stages. The absence of moving parts increases their reliability and life time while reducing mechanical vibrations, electromagnetic noise, and cost. We previously developed a single-stage GM pulse tube cooler that uses a double-inlet configuration.[1] For a 1.5kW compressor unit, we achieved a minimum temperature of 25.7 K and a maximum cooling capacity of 18.0W at 80K. We also evaluated the dependency of the operating frequency on the inclination of the pulse tube. We measured the cooling performance by turning the cooler upside down, and found that the effect of the pulse tube inclination on the cooling performance at 80K was negligible when the operating frequency was higher than 6Hz.

In the future, two-stage pulse tube cryocoolers are expected to be used for shield-cooling of MRI magnets, cryopumps, low Tc superconducting magnet systems for laboratory use, superconducting devices such as SQUIDs, etc. For these applications cryocoolers must be suitable for continuous, long-term operation, and they must be more efficient.

The phase-shifting mechanism in pulse tube coolers strongly affects their reliability and cooling performance. Pulse tube coolers using a double-inlet configuration have performance superior to coolers using a simple-orifice configuration. However, the double-inlet configuration has an internal closed loop fluid path that allows a DC fluid current to flow either from ambient-temperature region to the cold end or from the cold end to the ambient-temperature region. In the two-stage pulse tube coolers, there are more internal closed loop fluid paths than in the single-stage coolers, therefore, the amount of the DC fluid current tends to vary. Compared with single-stage coolers, therefore, it is more difficult to achieve continuous, long-term stable operation with two-stage coolers.

To select a suitable phase-shifting mechanism for the two-stage GM pulse tube coolers, we evaluated the performance of simple-orifice, double-inlet, and moving-plug phase-shifting configurations by measuring the cooling capacity for a single-stage cooler. To evaluate the experimental results and to improve the cooler performance, we estimated the PV work in the cold-end volume of the pulse tube and the heat loss in the cooler by simplified numerical analysis based on the equivalent PV method.

EXPERIMENTAL APPARATUS WITH A SINGLE-STAGE COOLER

Figure 1 shows a schematic of the test apparatus for the measurement with a single-stage cooler. To test various types of valved pulse tube coolers, we used rotary type switching valve units. The on-off timing of these valves could be varied by stepper motors controlled by a personal computer. In some experiments of valved cryocoolers, solenoid valves have been used to control the on-off timing. The problems of using the solenoid valves, however, are that they are unreliable and they require additional valve design for applications to industrial processes. In this work, therefore, we used a maximum of three rotary valves for the measurements with a single-stage cooler. We are able to control both the on-off timing of each valve and the phase differences between the each valve.

To monitor the PV diagram in the hot end of the pulse tube, we inserted a moving plug in a cylinder connected to the hot end of the pulse tube. This plug was made of a thin plastic plate and a plastic seal ring, which could be moved smoothly by a relatively low pressure difference. We measured the position of the moving plug by using a laser-type displacement sensor. We calculated the mass of the gas flowing through the connection at the hot end of the pulse tube by the observed displacement. The compressor unit had a rated input power of 700W at 50Hz.

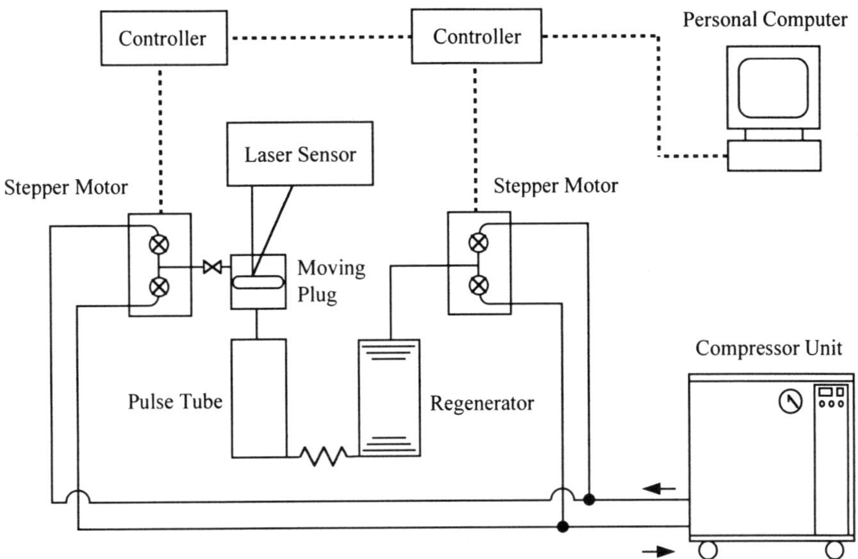

Figure 1. Schematic of test apparatus using a single-stage cooler.

The length of the pulse tube and regenerator cylinders was 200mm and 180mm, respectively. The inner diameter of the pulse tube and regenerator cylinders are 20mm and 25.4mm, respectively.

NUMERICAL ANALYSIS

To evaluate the experimental results and to improve the cryocooler performance, we estimated the PV work in the cold end volume and heat loss in the cooler by numerical analysis based on the equivalent PV method.[2] The pulse tube can be divided into three parts: I) the hot end of the pulse tube, II) the gas displacer, and III) the cold end of the pulse tube. The mass flow rate through the orifice and the double-inlet valves can be calculated as

$$m = \alpha \rho \sqrt{\frac{\Delta p}{T}} \tag{1}$$

where α is the coefficients of discharge of either the orifice or the double-inlet valves, and ΔP is the pressure difference either between the pulse tube and the reservoir tank or between the pulse tube and the regenerator hot end. For the coefficient of discharge, we used the values measured during steady-flow tests. For representing the flow through the regenerative matrix, the flow-friction characteristics for flow through cylindrical tubes was used. A correlation of friction factor and Reynolds number was defined during steady-flow tests.

We obtained the gas displacer volume in the pulse tube by solving the differential equations. The equivalent PV work at the cold end and the cooling capacity can be represented as

$$W_{ce} = \oint P_{pt} V_{ce} dt \tag{2}$$

$$Q_{ce} = W_{ce} - H_{rg} - H_{oth} \tag{3}$$

where H_{rg} is a regenerator loss. We used the Schalkwijk's equation for expressing the effectiveness of the regenerator.[3] To represent the heat transfer characteristics of the stacked screen matrix and of the randomly stacked sphere matrix, we used the experimental data by Kays and London.[4] To include the effect of the void volume in the regenerators, we calculated the inefficiency as a function of the operating pressure ratio, heat capacity ratio, and void volume ratio. We multiplied the inefficiency calculated by Schalkwijk's equation by a coefficient defined by the above results. H_{oth} represents other heat losses, such as conductive and radiative heat losses.

RESULTS OF A SINGLE-STAGE COOLER

To evaluate the efficiency of phase-shifting mechanisms, we measured the cooling performance of the single-stage cooler using either double-inlet, simple-orifice, or moving-plug phase-shifting mechanisms (Figure 2). The rated input power of the compressor unit used in this experiment was 700W. In the double-inlet configuration, the reservoir tank was connected to the suction line of the compressor unit through a needle valve. In the moving-plug configuration, the upper space of the plug was connected to a space containing the gas either being supplied to the compressor or being supplied from the compressor, which was switched by another rotary valve. The switching timing of the two rotary valves were controlled such that the pressure oscillations in the upper space of the plug had some phase lead to those in the regenerator. In this configuration, the hot end of the pulse tube was connected to the suction line of the compressor unit through a needle valve. We optimized the valve timing and the operating frequency for all experiments.

The measured cooling performance for each configuration is shown in figure 3. A cooling capacity of 8.5W at 80K was obtained for the double-inlet configuration, 6.1W at 8 0K for the simple-orifice configuration, and 6.3W at 80K for the moving-plug configuration. These results indicate that the double-inlet configuration has the best cooling capacity.

Simple numerical analysis indicates that either the four-valve configuration or the moving-plug configuration coolers are more efficient than the double-inlet configuration cooler.[5] For the moving-plug configuration, the reason for the cooler performance degradation may be related to the effect of the valve timing on the phase shift of the gas flow to the regenerator and the gas flow to the upper space of the moving plug. Figure 4 shows PV diagrams of the moving-plug configuration. The solid line shows the measured PV diagram in the hot end of the pulse tube, and the dashed line shows the calculated PV diagram in the cold end. The phase shift and the gas piston stroke in the cold end were not

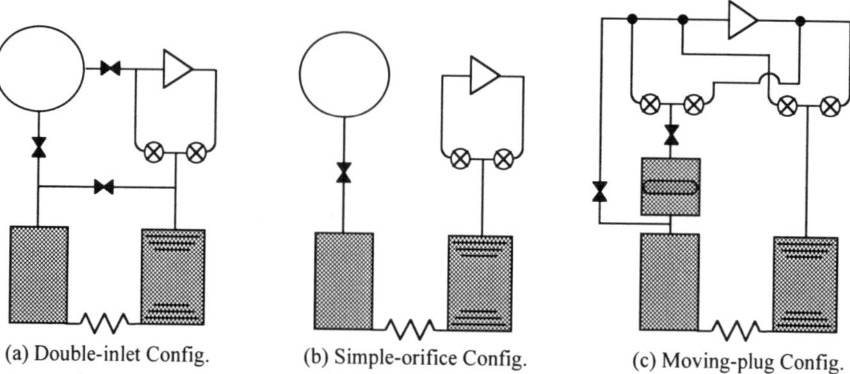

Figure 2. Schematics of each phase-shifting mechanism.

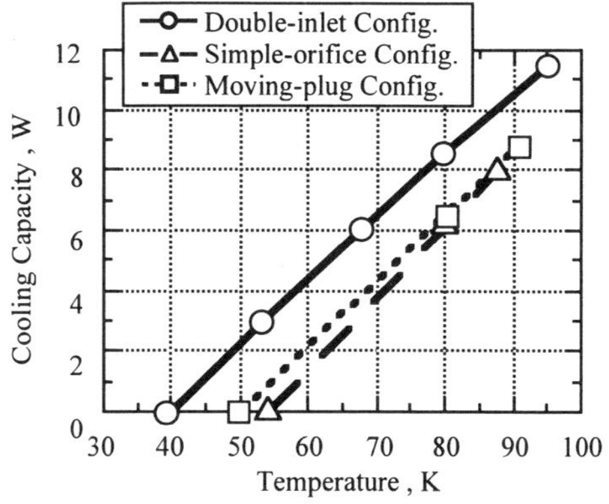

Figure 3. Effect of phase-shifting configuration on cooling capacity vs. cooling temperature characteristics.

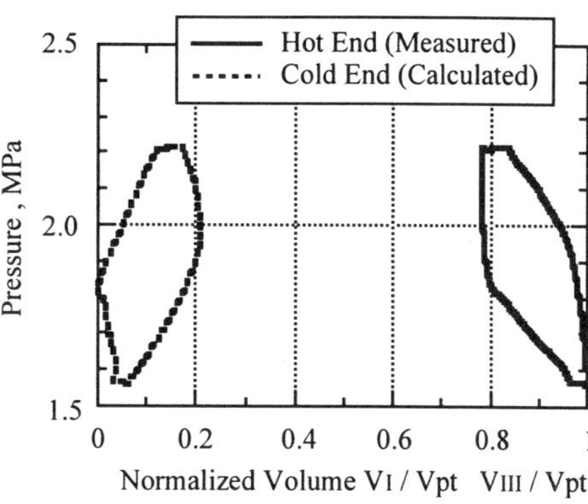

Figure 4. PV diagram for moving-plug configuration.

sufficient to accomplish the efficient PV diagram. From the measured results, when either the phase shift or the gas piston stroke was increased, the cooling performance degraded. We are continuing to optimize the PV characteristics to improve the cooler performance.

TWO-STAGE COOLER PERFORMANCE

Figure 5 shows a schematic of our first two-stage pulse tube cooler. The phase shifters for first and second stage were all maintained at ambient temperature. Two reservoir tanks, each with a volume of 500cm^3, were connected to the hot ends of the first and the second stage pulse tubes through needle valves, and connected to the suction line of compressor

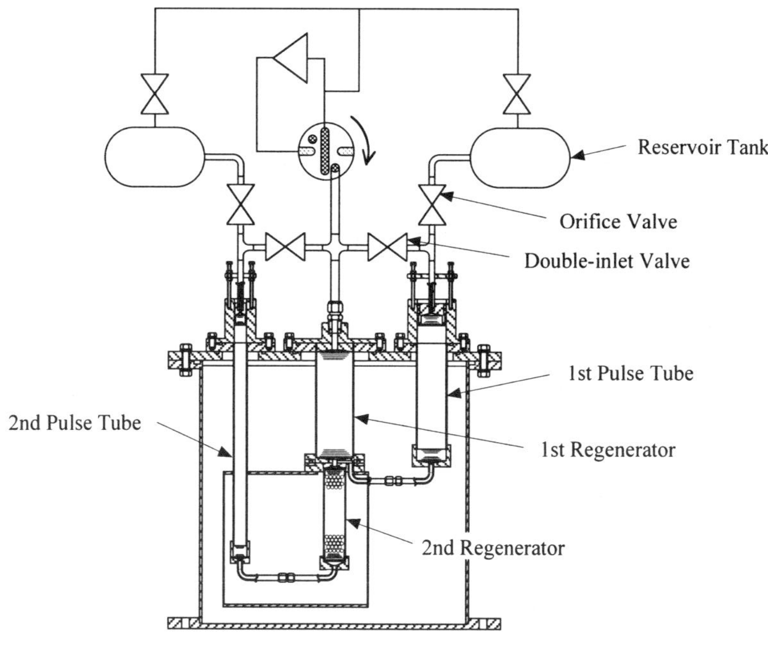

Figure 5. Schematic of two-stage cooler.

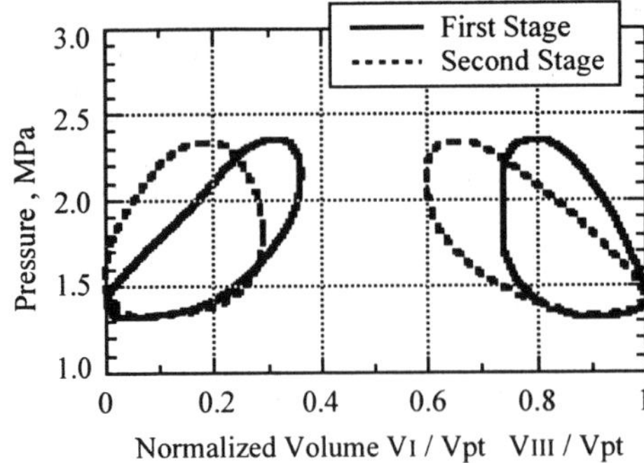

Figure 6. Calculated PV diagram of two-stage cooler.

Table 1. Loss analysis based on equivalent PV method.

	First stage (at 80K)	Second stage (at 20K)
PV Work	41.1W	18.1W
Conduction loss	3.2W	0.6W
Radiation loss	1.3W	0.0W
Regenerator loss	12.8W	8.3W
Shuttle loss	4.1W	3.6W
Cooling power	28.6W	5.5W

unit through other needle valves. The DC fluid current, flowing from the cold end of the pulse tube to the hot end, could be controlled by using these needle valves. This also permitted control of the cooling performance.[6] The rated input power of the compressor unit was 3.3kW at 50Hz.

The pulse tubes and regenerators were fabricated from stainless steel tubes. The first-stage regenerator was filled with about 1800 disks of 300-mesh stainless steel screen. The second-stage regenerator was filled with about 48cm^3 of lead spheres. The temperatures of the first and second stage cold heads were measured with calibrated Pt-Co resistance thermometers. A calibrated Germanium resistance thermometer was also attached to the second stage cold head to measure its temperature below 5K. The minimum temperatures of 33K at the first stage and 5K at the second stage were achieved with a cycle frequency of 2.0Hz and an initial pressure of 2.1MPa. The maximum cooling capacity of 22.9W at 80K on the first stage and 5.9W at 20K on the second stage was also obtained. For multi-stage coolers, the percent Carnot can be calculated by dividing the sum of the ideal work required to produce refrigeration of each cooling stage by the actual work input. The percent Carnot of our two-stage cooler was about 4.4%. By further optimization of the phase-shifting mechanism, we expect to further improve the two-stage cooler efficiency.

As a preliminary experiment to obtain the temperature below 4K, we used a magnetic regenerative material, a mixture of Er_3Ni and $HoCu_2$, in the lower half of the second-stage regenerator. The minimum temperatures of 2.70K at the second stage was achieved with cycle frequency of 2.0Hz. With heating load of 7.3W at 60K on the first stage, the second stage provided a cooling capacity of 190mW at 4.2K. If the cooler uses a second-stage regenerator containing three layers of $HoCu_2$ grains in the lower, Er_3Ni spheres in the middle, and lead spheres in the upper part, lower temperatures and higher cooling capacity at 4.2K can be attained.

For the experimental conditions corresponding to the maximum cooling capacity, we calculated PV diagrams in the cold and hot end of pulse tubes (Figure 6). For this calculation, we used some compensated coefficients of discharge at the orifice and double-inlet valves. We multiplied the coefficients of discharge, their values are measured during steady-flow tests, by the following coefficients; about 2% for the first double-inlet and orifice valves, less than 50% for the second double-inlet and orifice valves. These values of the compensated coefficient were defined by some experimental results we obtained by changing the cooler's specifications. Based on this equivalent PV diagrams, we calculated the cooling capacity of the first stage at 80K and the second stage at 20K (Table 1). Our numerical loss analysis indicates that the regenerator heat loss is dominant in both cooling stages.

DISCUSSION

If the calculated PV diagrams shown in Figure 6 represent those corresponding to the experimental conditions, then the PV diagram of the first stage is not efficient. The results of the loss analysis shown in Table 1 indicate that the ratio of the net cooling capacity to the PV work is not small. To increase the efficiency of two-stage cooler, it is necessary to realize the more efficient PV diagrams by optimizing the phase-shifting mechanism.

In the experiments of our first two-stage cooler, we had problems with stability. The cause of the problems may be a DC fluid current. We still have not been able to identify the cause yet. To develop an acceptable two-stage cooler that is stable for continuous, long-term operation, we are evaluating additional suitable phase-shifting mechanism.

CONCLUSIONS

We attempted to identify an appropriate phase-shifting mechanism for two-stage GM pulse tube cryocoolers, by evaluating the cooling capacity of single-stage GM pulse tube coolers equipped with simple-orifice, double-inlet, moving-plug phase-shifting mechanisms. We evaluated the cooling performance of a two-stage cooler equipped with a double-inlet phase-shifting mechanism. The major results are as follows.

1. The double-inlet configuration yielded the highest cooling capacity for the single-stage pulse tube cooler.
2. Our prototype two-stage pulse tube cooler has the following performance: a minimum temperatures of 33K at the first stage and 5K at the second stage, with cycle frequency of 2. 0Hz and a rated input power of 3.3kW. We achieved a maximum cooling capacity of 22.9 W at 80K on the first stage and 5.9W at 20K on the second stage. The percent Carnot was 4.4%.
3. As a preliminary experiment to obtain the temperature below 4K, we achieved a minimum temperature of 2.70K when magnetic regenerative material was used in the second-stage regenerator. With heating load of 7.3W at 60K on the first stage, the second stage provided cooling capacity of 190mW at 4.2K.

REFERENCES

1. S. Fujimoto, Y.M. Kang, and Y. Matsubara, "Development of a 5 to 20W at 80K GM Pulse Tube Cryocooler", CRYOCOOLERS 10, to be published
2. S. Zhu, P. Wu and Z. Chen, "Double inlet pulse tube refrigerators : an important improvement", Cryogenics, Vol. 30 (1990), p.514
3. W. F. Schalkwijk, "A Simplified Regenerator Theory", Trans. of the ASME (1959), pp.142
4. W. M. Kays and A.L. London, "Compact Heat Exchangers", McGraw-Hill Book Company (1976), p.140
5. Y. Matsubara, "Work-loss Distribution on GM-type Pulse Tube Coolers", Jap. Journ. of Cryog. Eng., Vol.3 3, No.4 (1998), p.207
6. G. Chen, L. Qiu, J. Zheng, P. Yan, Z. Gan, X. Bai, and Z. Huang, "Experimental study on a double-orifice two-stage pulse tube refrigerator", Cryogenics, Vol.37 (1997), p.271

REDUCTION OF CONVECTION HEAT LOSSES IN LOW FREQUENCY PULSE TUBE COOLERS WITH MESH INSERTS

S. Kasthurirengan*, G. Thummes, and C. Heiden

Institute of Applied Physics, Justus-Liebig-University,
D-35392 Giessen, Germany
*permanent address: Central Cryogenic Facility, Indian
Institute of Science, Bangalore 560012, India

ABSTRACT

Low frequency (f ~ 1 Hz) pulse tube coolers (PTC) usually exhibit a more or less pronounced dependence of their cooling performance on orientation due to gravity driven convection streaming in the pulse tubes. Following earlier work, several inserts into a pulse tube have been tested that were constructed in such a way as to effectively inhibit the convective secondary streaming without affecting the oscillatory gas flow too much. Inserts of metallic and nylon meshes in the form of a twisted ribbon were chosen for this purpose. In our case of a single stage double inlet PTC driven by a GM-type 2 kW-compressor, the no-load temperature as function of the tilt angle θ from the vertical varied between T_{min} = 32 K at $\theta = 0°$ and T = 99 K at θ = 120° for the empty pulse tube. When using metallic mesh inserts with an optimum mesh number of about 100, T(θ) did not exceed 1 K, at the expense of an increased T_{min} = 84 K. Under otherwise similar conditions, inserts made of Nylon showed a lower T_{min} compared to the metallic inserts.

INTRODUCTION

On of the attractive properties of pulse tube coolers (PTCs) is the inherent low-noise level of the PTC cold head. For low-noise cooling of SQUIDs and other highly-sensitive devices by use of a PTC, the cold head has to be spatially separated from the pressure wave generator. This is rather easily accomplished with low-frequency GM-type PTCs, where the cold head can be connected to the pressure-wave generating assembly by means of a flexible tube with a length of several meters. In this way the interference noise from helium compressor and rotary valve can be reduced to a negligible level, as sensed by a YBCO-SQUID gradiometer that was mounted on the cold head of a specially designed PTC operating at 2 Hz.[1]

Advances in Cryogenic Engineering, Volume 45.
Edited by Shu *et al.*, Kluwer Academic / Plenum Publishers, 2000.

The flexible connecting line makes it possible to easily change the orientation of the cold head as required in a number of applications. However, the cooling performance of such low-frequency PTCs was found to exhibit a pronounced degradation when the pulse tube axis was tilted with respect to the direction of gravity[2,3]. The effect of pulse tube orientation on cooling performance was systematically studied in Ref. 2, and was attributed to an enhanced heat transfer by natural convection of helium gas, which is superimposed on the oscillatory gas displacement in the pulse tube.

The orientation effect can be reduced by increasing the operating frequency of the PTC[2,4], since the oscillatory gas flow attenuates the natural convection. In miniature Stirling-type PTCs, operating at 40 to 60 Hz, the effect of orientation on cooling performance is absent[5]. However, in PTCs with a connecting tube of several meters the low-pass filter characteristic of that tube limits the working frequency to about 10 Hz, which is not high enough to suppress the orientational dependence of cooling performance[4].

Inserts of porous plugs within the pulse tube were employed[2] to hinder the development of the convective flow, but these plugs seriously impaired the cooling characteristic of the pulse tube. Alternatively, one can try to reduce the convection by modifying the pulse tube shape, such as increasing the length to diameter ratio or choosing an elliptical cross section, as has been suggested recently also by Scurlock and Beduz[6]. The results of experimental studies on elliptical pulse tubes will be presented elsewhere[7].

Following previous work[2], we have tested new types of pulse tube inserts made of metallic or nylon mesh in the form of twisted ribbons. The results of angular variation of the cold end temperature and cooling power with these inserts are reported in this paper.

BACKGROUND: CONVECTION IN CYLINDERS

The empty pulse tube is a thin-walled cylinder filled with He-gas, with the cold end at temperature T_c, and the warm end at ambient temperature T_h. Depending on the angle θ between pulse tube axis and vector of gravity $\vec{g}$, different convection patterns can occur in a cylinder (see Figure 1). In vertical orientation with the cold end down ($\theta = 0°$) there is no natural convection in the gas column, since there is no gravity-induced buoyant force. For $0° < \theta < 180°$ the radial component of $\vec{g}$ gives rise to the fundamental convection roll, with gas flowing upwards at the hot end and downwards at the cold end. For $90° < \theta \leq 180°$, at

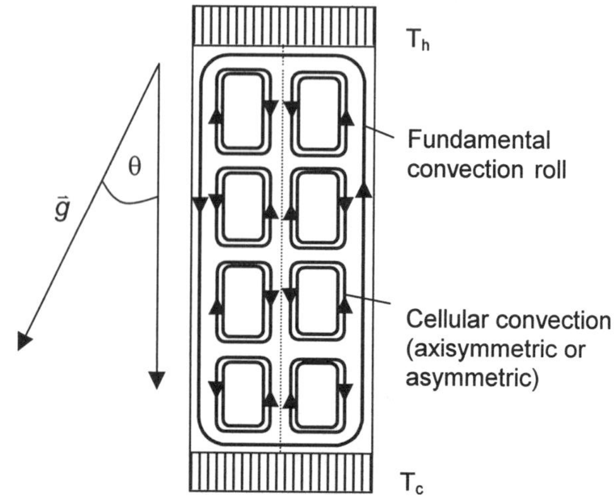

Figure 1. Qualitative picture of natural convection in a cylindrical pulse tube.

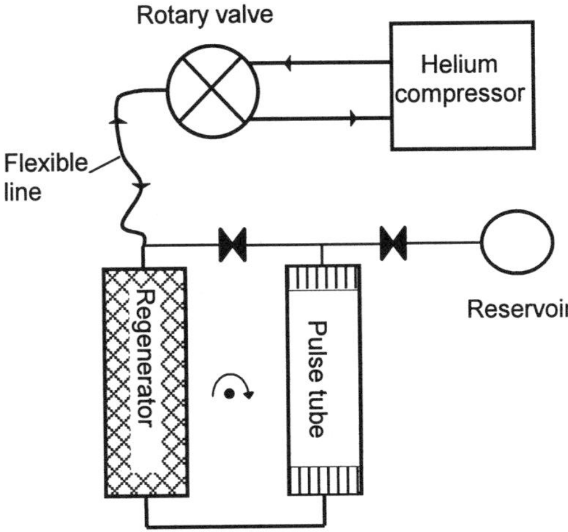

Figure 2. Schematic of the double-inlet pulse tube refrigerator.

high enough buoyant forces, i.e. for Rayleigh numbers Ra larger than a critical number Ra_c (as is valid for a PTC under typical operating conditions[2]), also cellular convection takes place that is driven by the axial component of $\vec{g}$. At $\theta = 180°$, only cellular convection is present because of the missing radial component of $\vec{g}$. A quantitative theory for heat transfer from free convection in long cylinders is only available for $\theta = 180°$.[8] In this case the measured convective heat transfer from warm to cold end of a pulse tube is in satisfactory agreement with the prediction.[2]

EXPERIMENTAL DETAILS

Description of experimental set-up. The schematic of the experimental set-up, which is similar to that in Ref. 2, is shown in Figure 2. The pulse tube and regenerator are made of stainless steel tubes with outer diameter, wall thickness and length of 14 mm × 1 mm × 250 mm and 19 mm × 0.3 mm × 210 mm, respectively. A standard U-shaped configuration is used. The regenerator matrix consists of a stack of bronze and stainless steel screens of mesh no. 247. The PTC is operated in double-inlet configuration with the help of two needle valves (Nupro, type M) and a reservoir with volume of 0.5 L

The pressure oscillation in the cooler is generated by means of a modified GM rotary valve in combination with a 2 kW He-compressor (Leybold RW2). The cold end temperature is monitored by means of a Pt100 sensor, and net cooling powers are measured using a calibrated resistive heater attached to the cold heat exchanger.

In order to facilitate the measurements of the cooler performance at different pulse tube inclinations, the refrigerator is connected to the rotary valve via a flexible polyamide tube with inner diameter of 4 mm and with a length of 1.2 m. The vacuum vessel that contains the PTC cold head is mechanically fixed to a horizontal axle, thus making possible to vary the tilt angle between $\theta = 0°$ and $\pm 180°$.

Fabrication of mesh inserts. Wire nets of copper, stainless steel, bronze, and nylon with mesh numbers from 20 to 295 (porosities: 0.62 – 0.68) were chosen for fabrication of the pulse tube inserts. A given mesh was cut to 14.0 mm width in the form of a long ribbon. The ribbon was fixed on a stainless steel rod of 0.5 mm thickness. Then twists were made on the ribbon such that the edges of the mesh follow a helical path. The number of twists could be

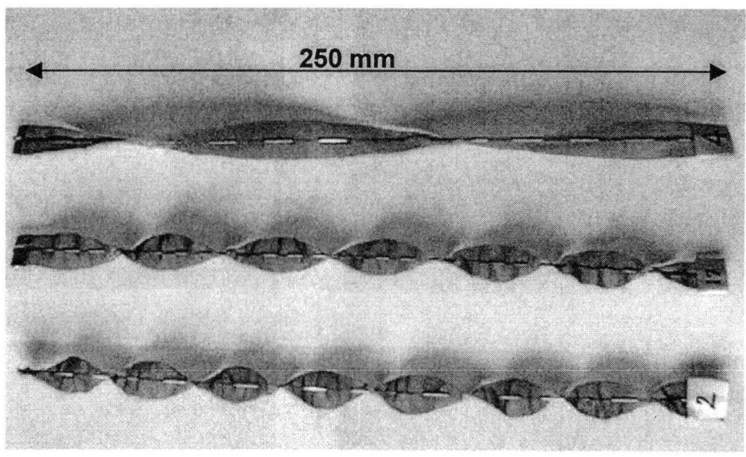

Figure 3. Photograph of typical mesh inserts with 2.5, 6, and 7.5 twists (from top).

varied up to about 8 twists over the pulse tube length without damaging the meshes used. The ends of the twisted mesh inserts were fixed by soldering them to the stainless steel rod. A photograph of typical inserts is shown in Figure 3. When such a mesh insert is mounted within the pulse tube, it lies along the pulse tube length such that the gas flow is not obstructed, but follows a helical path.

EXPERIMENTAL RESULTS AND DISCUSSION

In all experiments the PTC was operated at a frequency of 2 Hz, a peak to peak pressure variation of $\Delta p \cong 10$ bar at the regenerator inlet, and a charging pressure of 17 bar.

Cooling performance of the empty pulse tube. Figure 4 displays the variation of the no-load temperature as function of the tilt angle θ in stationary state. A lowest temperature of $T_{min} = 32$ K is reached at $\theta = 0$. The form of the $T(\theta)$-curve closely resembles that reported for other low-frequency PTCs in Refs. 2 - 4. In particular a maximum temperature is attained at $\theta \approx 120°$, which in the present case is $T_{max} = 99$ K.

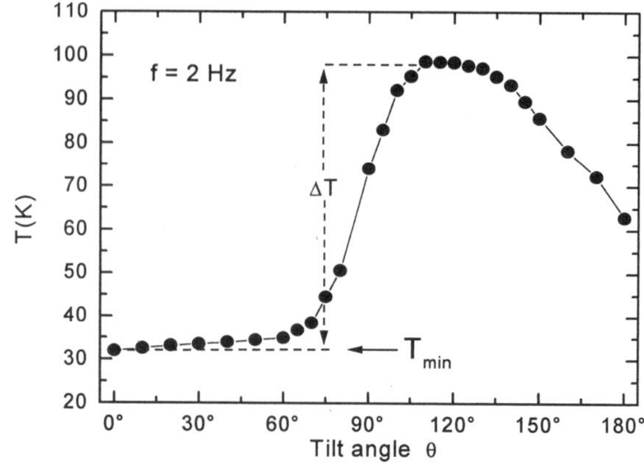

Figure 4. Variation of minimum no-load temperature with angle θ of inclination of the pulse tube.

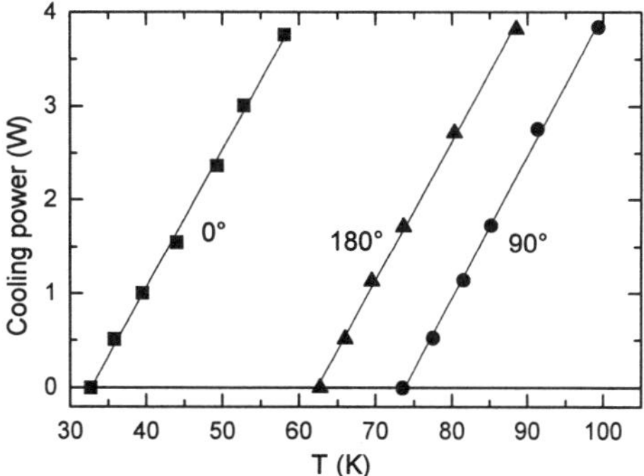

Figure 5. Net cooling power vs cold end temperature at distinct tilt angles of the empty pulse tube.

To our knowledge, there is no quantitative explanation available for the maximum in $T(\theta)$ from theory of convective heat transfer. Evidently, the maximum originates from the interplay between fundamental and asymmetric cellular convection. As discussed above, both mechanisms are present for $90° < \theta < 180°$, which should result in an enhanced gas mixing.

Figure 5 displays the net cooling power $\dot{Q}$ at distinct tilt angles. At $\theta = 0°$, $\dot{Q} = 2.6$ W is available at 50 K. Interestingly, the slope $d\dot{Q}/dT = 0.15$ W/K is nearly independent of tilt angle. This means that the gross cooling power, i.e. the enthalpy flow in the pulse tube, is not changed by the superimposed free convection.

Optimum number of twists of mesh inserts. In the following, an insert is characterized by its efficiency in reducing the orientation dependence of $T(\theta)$, represented by $\Delta T := T_{max} - T_{min}$ between $\theta = 0°$ and 180°, and by the achievable minimum temperature T_{min}.

Initially for a given material and mesh number, the effect of the number of twists on both T_{min} and ΔT were studied. Table 1 gives some results for inserts from copper mesh no. 98 with varying number of twists. As seen from Table 1, ΔT decreases rapidly with increasing number of twists, at the expense of an increase in T_{min}. A maximum number of about 7.5 twists is needed to reduce ΔT to as low as 1.0 K.

Orientational dependence with mesh inserts. Figure 6 displays $T(\theta)$ for some typical metallic mesh inserts with an optimum number of twists (7.5 – 8) but with different mesh sizes between 20 and 187. For comparison also $T(\theta)$ of the empty pulse tube is shown. For all inserts T_{min} is considerably increased, which is due to the regenerative effect of the insert material, as discussed below.

The variation of T_{min} and ΔT as function of the mesh number for metallic and also Nylon inserts with 7.5 to 8 twists is plotted in Figure 7. For a given class of materials, T_{min} decreases monotonically with increasing mesh size, while ΔT shows a minimum around mesh number 100. As seen from Figure 7a), for the same mesh number Nylon inserts exhibit a clearly lower T_{min} than metallic inserts. For example, T_{min} = 84 K and 72 K for stainless steel and Nylon inserts with mesh no. 78, respectively. The reduction of convective

Table 1. Effect of number of twists on T_{min} and ΔT for Cu mesh no. 98

No. of twists	ΔT (K)	T_{min} (K)
2	70.9	69.3
6	30.3	77.8
7.5	1.0	84.0

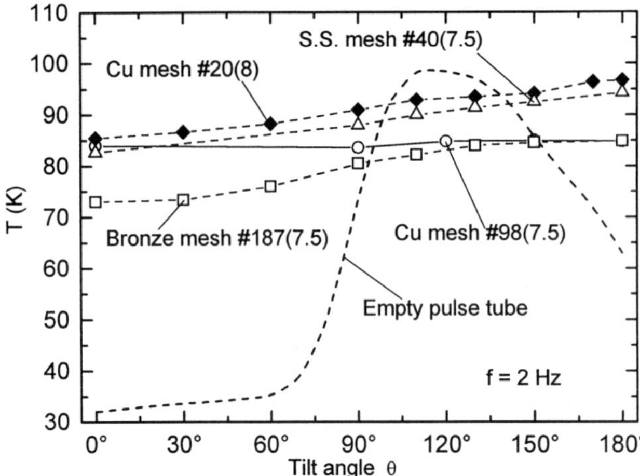

Figure 6. Variation of no-load temperature with tilt angle θ for some metallic mesh inserts with 7.5 to 8 twists and mesh numbers from 20 to 187, compared to T(θ) for the empty pulse tube.

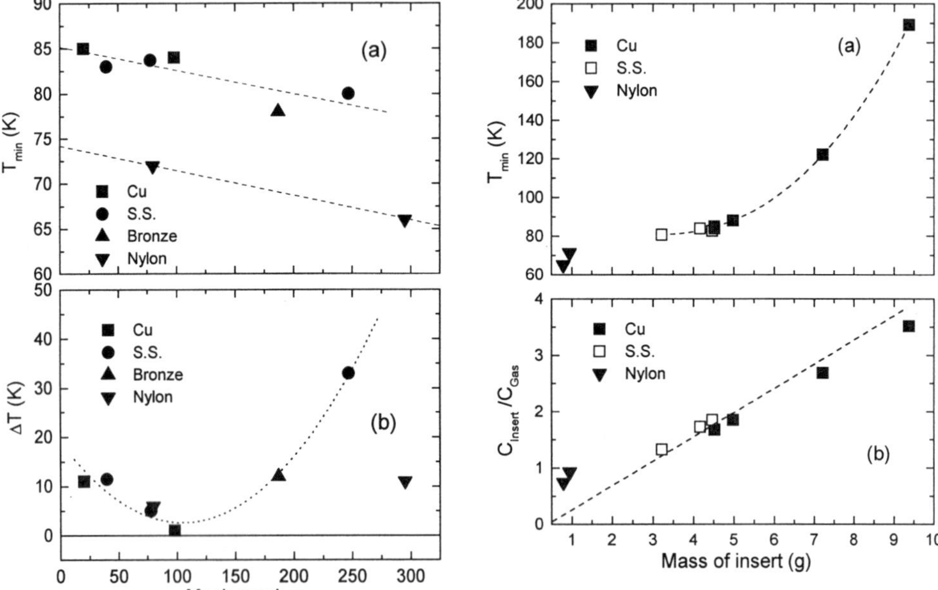

Figure 7. Variation of a) T_{min} and b) ΔT with mesh number for inserts with 7.5 to 8 twists.

Figure 8. Variation of a) T_{min} and b) heat capacity of insert normalized to that of the He-gas in the pulse tube versus mass of insert.

heat losses by a Nylon insert with mesh no. 78, as seen from ΔT in Figure 7b), is similar to that of a stainless steel insert with the same mesh size. A fine Nylon mesh of number 295 evidently gives a much lower ΔT than expected from the ΔT versus mesh number plot for the metallic inserts in Figure 7b).

The most effective insert with respect to reduction of convection in these experiments is that fabricated from Copper mesh no. 98 with ΔT = 1.0 K and T_{min} = 84 K. From the results in Figure 7 it is suggested that with a Nylon insert of mesh number ≈ 100 also a very small ΔT could be achieved, but at a considerably lower T_{min} near 70 K.

Effect of mass of insert. As seen from Figure 8a), the measured minimum temperature clearly increases with the mass of the insert. This finding is in coincidence with Figure 7a), since for a given material a lower mesh number means a higher mass per unit area.

This result can be interpreted in terms of the regenerative effect of the insert material. Heat transfer between the oscillating gas in the pulse tube and the insert reduces the magnitude of dynamic temperature variation, and thus the enthalpy flow in the pulse tube. The lower gross cooling power then gives rise to a higher minimum no-load temperature. The heat capacity of a given insert, C_{insert}, normalized to the heat capacity of the gas volume in the pulse tube, C_{gas}, can be taken as a measure for the regenerative effect. The ratio C_{insert}/C_{gas}, evaluated at an average pulse tube temperature of 160 K and an average pressure of 16 bar, is plotted in Figure 8b) as function of the mass of the insert. For copper and stainless steel with specific heats of 0.34 and 0.37 J/(gK), respectively, the points in Figure 8b) roughly follow a straight line. For Nylon with a specific heat of 1.25 J/(gK) at 160 K, C_{insert}/C_{gas} lies above this line. The relatively low regenerative effect of Nylon, as indicated by the low T_{min} in Figures 7a) and 8a), is ascribed to a reduced heat transfer originating from the low thermal diffusivity of Nylon which is more than one order of magnitude smaller than that of the metallic samples.

Cooling power with mesh inserts. Figure 9 shows the cooling power for some mesh inserts all fabricated with the optimum number of ≈ 7.5 twists. Here the measured slopes of cooling power are in the range $d\dot{Q}/dT$ = 0.09 – 0.12 W/K, as compared to $d\dot{Q}/dT$ = 0.15 W/K for the empty pulse tube (see Figure 5). This finding also indicates a reduced enthalpy flow caused by the regenerative effect of the inserts.

Effect of gap between insert and pulse tube wall. To obtain some information on the boundary layer thickness for convective heat transfer, some inserts from copper mesh no. 98 (7 twists) were prepared with a gap to the pulse tube wall. Figure 10 displays the measured T_{min} and ΔT obtained for zero gap and with gaps of 1 and 2.5 mm. As seen from Figure 10a), T_{min} decreases roughly linearly with increasing gap size, which can be related to the lower mass and heat capacity of the inserts with gap. In comparison, ΔT rises only slightly from 1.0 K at zero gap to 1.5 K at a gap of 1 mm, see Figure 10b). It follows that heat transfer by free convection mainly occurs in the core of the pulse tube at distances larger than 1 mm from the tube wall.

CONCLUSIONS

The above results show that convection heat losses in tilted pulse tubes can be greatly

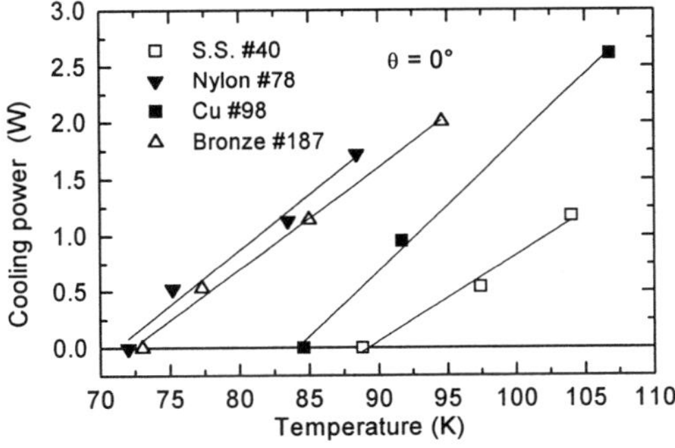

Figure 9. Cooling power vs cold end temperature at θ = 0° for some mesh inserts with 7.5 twists.

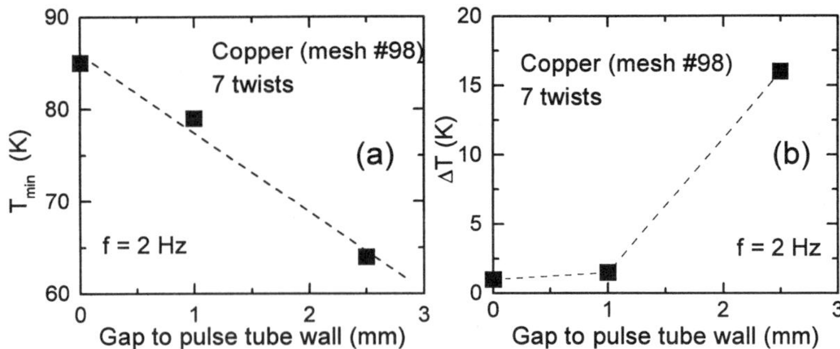

Figure 10. Effect of gap between insert and pulse tube wall on a) T_{min} and b) ΔT.

reduced by inserts of twisted mesh ribbons. There is an optimum number of twists of 7 to 8 over the pulse tube length that is probably related to the size of the convection cells, and an optimum mesh number of about 100 for which the restraining of convective flow is most effective. A gap of up to $\approx$ 1 mm between insert and pulse tube wall can be allowed without essential change in the orientational dependence of the cold end temperature.

For all tested inserts the gross cooling power is reduced to some extent because of the regenerative effect of the insert material that reduces the enthalpy flow in the pulse tube. In this sense Nylon mesh appears to be a good choice. With Nylon inserts the degradation of cooling performance is found to be considerably lower than with metallic inserts of the same size, which is related to the low thermal diffusivity of Nylon.

For some applications, such as HT-SQUID cooling for material inspection, pulse tube inserts could be a useful tool to achieve a nearly orientation independent operating temperature.

ACKNOWLEDGMENTS

One of the authors (SK) thanks the German DAAD for providing a scholarship. Useful discussions with R.G. Scurlock (University of Southampton) are gratefully acknowledged.

REFERENCES

1. G. Thummes, R. Landgraf, F. Giebeler, M. Mück, and C. Heiden, Pulse tube refrigerator for high-T_c SQUID operation, in: "Advances in Cryogenic Engineering", Vol. 41B, Plenum Press, New York (1996), p. 1463
2. G. Thummes, M. Schreiber, R. Landgraf, and C. Heiden, Convective heat losses in pulse tube coolers: effect of pulse tube inclination, in: "Cryocoolers 9", R.G. Ross, Jr., ed., Plenum Press, New York (1997), p. 393
3. A. Ravex, J.M. Poncet, I. Charles, and P. Bleuzé, Development of low frequency pulse tube refrigerators, in: "Advances in Cryogenic Engineering", Vol. 43B, Plenum Press, New York (1998), p. 1957
4. K. Klundt, C. Lienerth, G. Thummes, F. Steinmeyer, M. Vester, W. Renz, and C. Heiden, Use of a pulse tube cefrigerator for cooling a HTS-antenna for magnetic resonance imaging, in: "Advances in Cryogenic Engineering", Vol. 43B, Plenum Press, New York (1998), p. 2085
5. D.L. Johnson, S.A. Collins, M.K. Heun, and R.G. Ross, Jr., Performance characterization of the TRW 3503 and 6020 pulse tube coolers, in: "Cryocoolers 9", R.G. Ross, Jr., ed., Plenum Press, New York (1997), p. 183
6. R.G. Scurlock and C. Beduz, Reduction of the effects of asymmetrical convection in tilted pulse-tube refrigerators, *Cryogenics* 38: 359 (1998)
7. S.Kasthurirengan, G.Thummes, and C.Heiden, Experimental studies on elliptically shaped pulse tubes, unpublished
8. D.K. Edwards, and I. Catton, Prediction of heat transfer by natural convection in closed cylinders heated from below, *Int. J. Heat Mass Transfer* 12: 23 (1969)

EXPERIMENTAL INVESTIGATION OF HEAT PUMPING IN THERMOACOUSTIC REFRIGERATOR

J. V. Lubiez, F. Jebali Jerbi and M. X. François

LIMSI BP 133 91403 Orsay Cedex France

ABSTRACT

Heat pumping in regenerative heat engines is investigated by using a gas thermoacoustic refrigerator. The engine consists of a nitrogen-filled tube in which a stack of mylar parallel plates is inserted and placed between cold and hot heat exchangers. The total length of the tube $L = 2\ m$ is chosen so that the resonant frequency is approximately 52 Hz. A mechanical acoustic driver mounted at one end of the tube produces the gas pressure oscillations with frequencies varying from 5 to 80 Hz. Acoustic pressure and the stack temperature gradient are firstly measured as a function of frequency for fixed mean pressure and ambient temperature. The acoustic parameters (pressure and velocity) are then calculated using an analytical network model. In this calculation, any isothermal element of the engine such as heat exchangers is represented by a passive circuit of a series impedance per unit length, z, and a shunt admittance per unit length, y, when the regenerator (parallel plates stack), being submitted to a temperature gradient, is represented by an active circuit with a temperature gradient dependent current gain, $e(\partial T/\partial x)$, x being the longitudinal coordinate. The stack network is divided into a series of infinitesimal elements to take into account the local fluid thermodynamic properties. Thus, work and heat fluxes and the temperature profile along the stack are calculated in terms of oscillating pressure and velocity. The experimental results revealed clearly the role of the driving frequency in the heat pumping far and close to the resonance of the acoustic tube.

INTRODUCTION

Many researches in thermoacoustics have been driven by the design and construction of efficient prime mover or mechanically driven refrigerator and, more recently, by the design of so called thermoacoustic heat driven pulse tube refrigerator. The potential reliability and low cost manufacturing of these engines are indeed considered with a great interest. In particular, the possibility of using pure or mixture neutral gas in a structure with no moving parts is very attractive[1,2]. Some recent works have shown the interest of a network approach of such acoustic devices to get a good understanding of their behavior at least for small power that is for small acoustic pressure amplitude[3,4]. In this paper, we first recall briefly the background of the network representation of a thermoacoustic engine. We obtain then, for each part of the system (stack, heat exchanger, resonant tube, mechanical driver), its electrically equivalent circuit. In a second part, using an experimental set up of thermoacoustic refrigerator mechanically driven, we give some experimental results which show a good agreement with those got from the numerical simulation made with the previous network modeling. In particular, it is seen that, in spite of the inherent limitation of the linear modeling, it is nevertheless easy to take in account both the propagation of the pressure waves in the tube and the energy exchanges in the stack. Moreover, for this later, we may use, if necessary, its acoustic « image » obtained from impedance measurements[5]. As an example, we show the effect of the frequency on the performance of the refrigerator. Such calculation can be also easily made to show the influence of others parameters as spacing between stack's plates, length and position of the stack, or pressure amplitude generated by the driver.

LINEAR MODEL

According to the well known similarities between acoustic and electrical energy transmission lines, the transport of acoustic energy in a gas filled and isothermal pipe submitted to a pressure wave may be looked in the same way. Each element (j) of the pipe, of length δx, is characterized by a complex impedance $z(j)\delta x$ associated with a complex admittance $y(j)\delta x$ which represent the inertance and compliance of the element as well as its resistance to the energy transport. $p_1(x)$ and $q_1(x)$ are respectively the local acoustic pressure and the local volumetric flow rate and subscript **1** indicates the first order quantity. The transmission line becomes an association of many passive networks such as that associated to the element (j) and which is drawn with solid lines in figure 1.

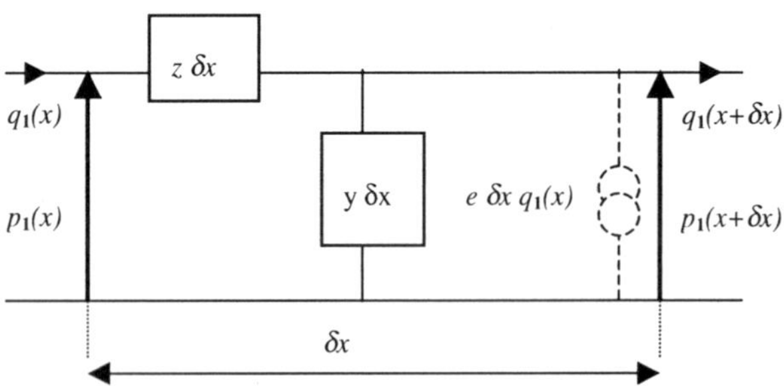

Figure 1. Network representation of an element (j)

Under the usual assumptions[6,7] of one dimensional problem, plane wave, the continuity and momentum equation integrated over the cross section of the tube, lead to the two following equations, which may be rewritten to give the transmission equations:

$$\frac{dp_1}{dx} = -\ z(x)\ q_1 \qquad (1)$$

$$\frac{dq_1}{dx} = -\ y(x)p_1 \qquad (2)$$

In a stack of parallel plates submitted to a temperature gradient, the heat exchange between the wall and the fluid in the thermal boundary layer (of thickness δ_k) modify strongly the previous result[6], and it has been shown that, in the same network approach, the thermoacoustic effect is taken in account by the addition of a source term $e(x)\delta x$ in the flow rate Eq. (2) (the source is represented by a dashed line in figure 1), that is:

$$\frac{dq_1}{dx} = -\ y(x)p_1 + e(x)\, q_1 \qquad (3)$$

Expressions of impedance, admittance and current gain constant could be found in the reference 6. The above equations lead to the transmission equations system within a circuit (j), and therefore, for the total line of length L and consisting of N circuits, one obtains:

$$\begin{bmatrix} p_1^{\ N} \\ q_1^{\ N} \end{bmatrix} = \coprod_{j=1}^{N} \begin{bmatrix} 1 & \dfrac{z_j \delta x}{1+e_j \delta x} \\ \dfrac{y_j \delta x}{1+e_j \delta x} & \dfrac{1}{1+e_j \delta x} \end{bmatrix} \begin{bmatrix} p_1^{\ 0} \\ q_1^{\ 0} \end{bmatrix} \qquad (4)$$

This matrix depends on the topography and the thermal state of each component of the system. For example, the resonator, as well as heat exchangers may be assumed to be isothermal. Thus, for these elements, the current gain constant $e(x)$ vanishes and they become passive. On the other hand, the stack and the heat exchangers are made of parallel plates arrangements. Their respective equivalent impedance and admittance are then obtained from those of circuits of parallel impedance and admittance. Knowing the boundary conditions of the system, i.e. the volumetric velocity at the piston and the closed end locations, one can calculate step by step the acoustic pressure and velocity along the whole line. Then, the total energy flow, which consists of the work flux, the energy flux due fluid oscillations and the heat conduction flux through the solid matrix (one notes that the heat conduction in fluid is much less than that in solid medium), is calculated as function of these parameters according to the following equation[8]:

$$\dot{E} = \langle p_1.u_1 \rangle_t\, A_f + \rho_0 T_0 \langle\langle s_1.u_1 \rangle_t \rangle_r A_f - \lambda_s \nabla T\, A_s \qquad (5)$$

where u_1 is the first order spatial average of the axial component velocity, A_f is the fluid cross section, ρ_0 and s_1 are respectively fluid density and oscillating entropy and T_0 is the mean temperature. Symbols $< . >_r$ and $< . >_t$ indicate respectively space average and time average of the quantity in the bracket. The subscript s indicates the solid medium. Applying locally the energy balance in the stack, and recurring the above operation, one can calculate step by step, the local temperature gradient and therefore the heat flux pumped at the cold end of the stack, Q_c.

EXPERIMENTAL SET-UP

The resonant tube refrigerator, shown in figure 2, consists of a gas nitrogen filled resonator of 36 mm diameter and 1995 mm length, comprising the stack and two heat exchangers, and related to an acoustic driver. The stack, made of mylar parallel plates, is of 15 cm length, 0.2 mm plates thickness and 0.7 mm plates spacing. The acoustic driver consists of a mobile piston of 100 mm diameter and ± 4 mm amplitude stroke. A clearance between the piston and the driver wall have been provided for in order to avoid seal problems at relatively high frequencies. Acoustic waves of 5 to 80 Hz frequencies and of few hundreds mbars pressure amplitude are generated with this system. The cold and hot heat exchangers are made of parallel copper plates of 0.2 mm thickness and of 0.8 mm spacing. Their temperatures are held constant by, respectively, a water circulation and an electrical heater.

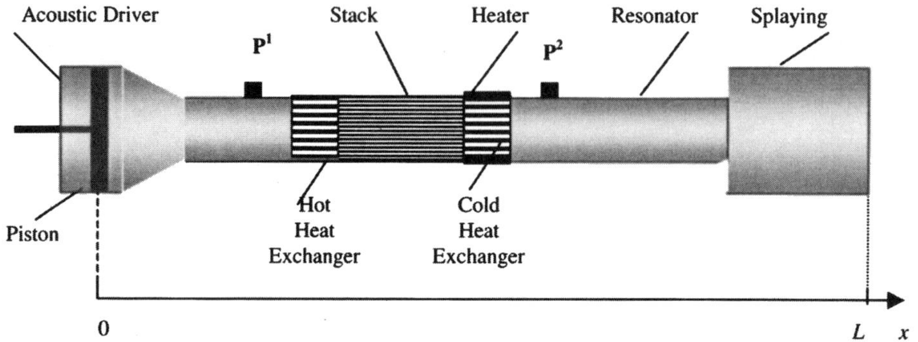

Figure 2. Experimental set-up

For acoustic pressures measurements, two piezoelectric transducers are mounted on upstream and downstream side of the stack. The signals $p^1(t)$ and $p^2(t)$ are recorded during 2s with a sampling frequency of 1250 Hz and then processed to obtain their corresponding FFT, respectively, $p^1(f)$ and $p^2(f)$.

The pumped heat at the cold source was measured as follows: the driver is started for a fixed frequency, f, and a fixed mean pressure, p_0, leading to a cooling at the cold heat exchanger. Then an electrical current is supplied in the heater to held a constant temperature gradient between the two heat exchangers, and the refrigeration power is measured taking into account the heat losses through the regenerator, the surrounding thermal insulation and the resonator walls, according to the following equation:

$$Q_c = Q_{elec} + Q_{loss} + Q_{leak} \qquad (6)$$

A second kind of experiments was made by measuring the temperature difference between the two heat exchangers, ΔT, for a given cold power, as a function of the working frequency.

RESULTS AND ANALYSIS

Acoustic Field. The first result concerns the pressure field inside the resonator for various mean pressures and at various locations in the tube. Although the clearance between the mobile piston and the wall is represented in the model by an additional network with an impedance $Z_{clear}(f)$, the value of this impedance and its frequency dependence may be got from the experimental data. On the figure 3, we have reported the pressure amplitude of $p^1(t)$ versus the frequency. The different curves correspond to the result given by the model for four cases: with no clearance, with the adjust clearance and with the adjust clearance ± 10%.

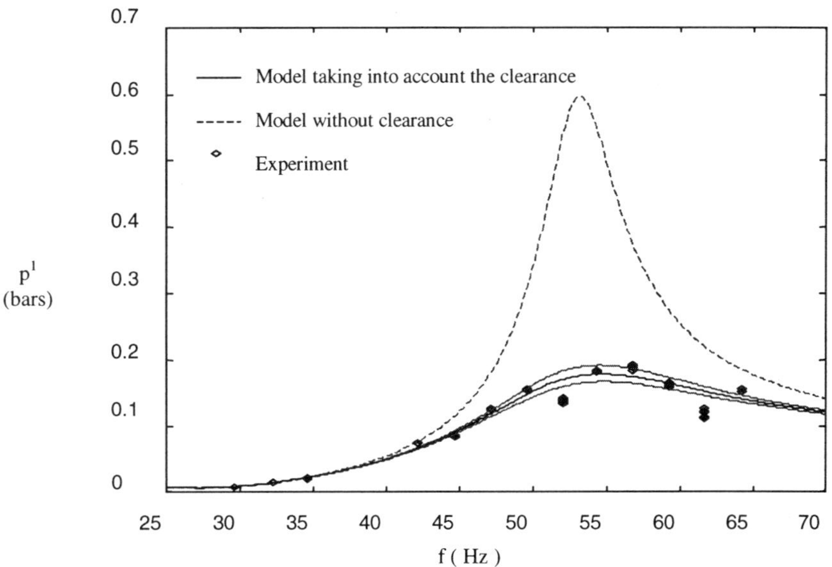

Figure 3. Acoustic pressure amplitude at the first transducer location as a function of the frequency

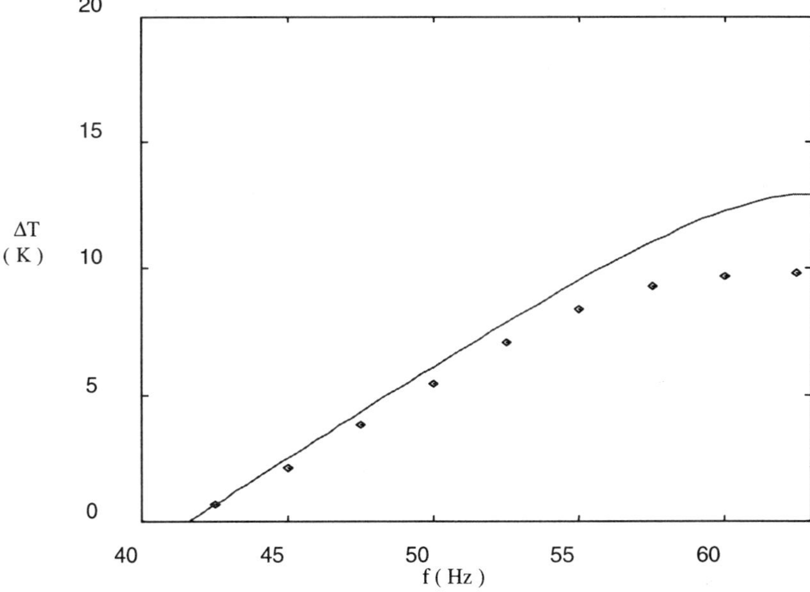

Figure 4. Temperature difference between the hot and the cold ends of the stack as a function of the frequency for Q_c = - 10 Watts

Temperature Span. As said above, in order to confirm the validity of the model, it is possible to measure either the temperature span at a given cold power for different frequency or the cold power at a given temperature difference imposed to the stack. The figure 4 shows an example of a temperature span for 10 Watts pumped. The agreement with the model is quite correct even if at high frequency a discrepancy starts to appear. This can be explain by a not accurate appreciation of $Z_{clear}(f)$ in this frequency range which is clearly a point easy to improve.

Refrigeration power. The refrigeration power is reported in figure 5. First we see again a good agreement between the model and the experimental results. Second, those obtained with a global temperature span of zero Kelvin (dashed line) allow to minimize the heat losses due to solid conduction or fluid convection. This is better seen if one compare the two results on figure 5. With a ΔT of 5 K the refrigeration power remains positive until the frequency reaches 35 Hz for which the pressure amplitude is high enough to drive a heat pumping just equal to the heat losses. This emphasizes the fact that in a thermoacoustic refrigerator or engine the heat losses due to heat conduction or convection and viscous dissipation may be prohibitive if the thermo-physical properties of materials as well as the porosity of the stack are not optimized.

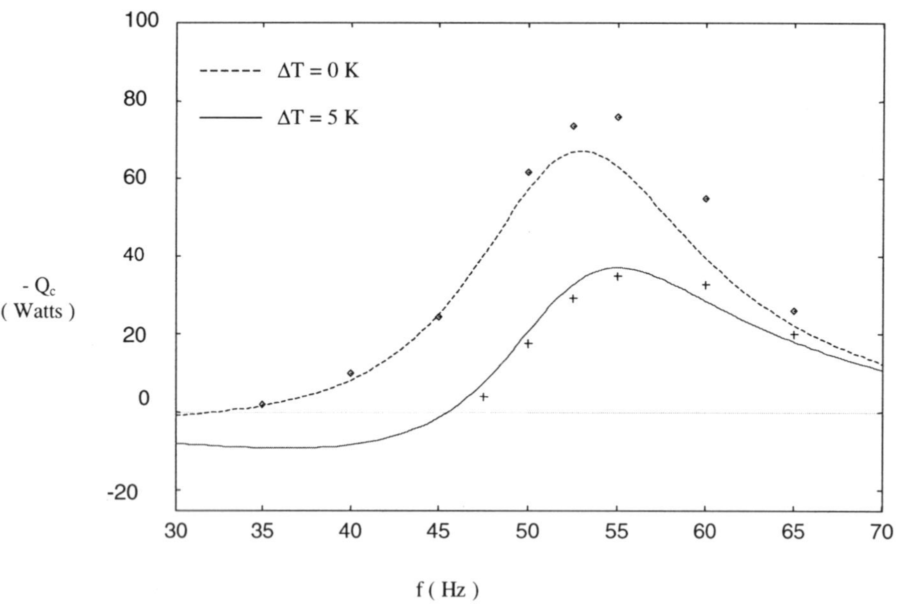

Figure 5. Refrigeration power as a function of the frequency

CONCLUSION

We have shown that a network modeling of a thermoacoustic system may give a very tool to analyze the behavior of such engine. The experimental results obtained from an experimental set up which is very far from the optimization and working with a very small acoustic power show nevertheless a good agreement and give confidence for a numerical and parametric experimentation of the system. The model may be thus used to see the behavior of the thermoacoustic refrigerator versus the various influencing parameters such as the geometry of the stack, the position of the stack in the acoustic field, the mean pressure and the frequency. It is also interesting to note that we get from the model a clear understanding of the contribution of the travelling and stationary parts of the waves. These results will be published elsewhere.

REFERENCES

1. G. W. Swift, "Thermoacoustic engines", *J. Acoust. Soc. Am.* October (1988), **84** (4)
2. J. R. Bechler et al, "Working gases in thermoacoustic engines", *J. Acout. Soc. Am.* May (1999), **105** (5), p. 2677-2684
3. F. Z. Guo et al., "Flow characteristics of a cyclic Flow regenerator", *Cryogenics* (1987), **27**, p. 152-155
4. J. H. Xiao and F. Z. Guo, "Analytical network model on the flow and thermal characteristics of cyclic flow cryogenic regenerators", *Cryogenics* (1988), **28**, p. 762-769
5. B. Zhao, F. Jebali and M. X. François, "Mesure des propriétés acoustiques d'un régénérateur utilisé dans une machine thermoacoustique", *4ème CFA*, Marseille (1997), p. 739-742
6. G. W. Swift, "Thermoacoustic engines and refrigerators", *Short Course on Thermoacoustics*, Berlin version (1999)
7. W. P. Arnott et al., "General formulation of thermoacoustics for stacks having arbitrarily shaped pore cross sections", *J. Acoust. Soc. Am.* December (1991), **90** (6)
8. A. Tominaga, "Thermodynamic aspects of thermoacoustic theory", *Cryogenics* (1995), **35** (7), p. 427-440

A NEW THERMALLY DRIVEN PULSE TUBE REFRIGERATOR OPERATING IN DOUBLE TRAVELLING WAVE MODES

LUO Ercang, LIU Haidong, LIANG Jingtao and ZHANG Liang

Cryogenic Laboratory, Chinese Academy of Sciences
Beijing 100080, P.R.China

ABSTRACT

A thermoacoustic engine is a new type of pressure-wave generator that possibly has a very long lifetime because of no mechanical moving parts. It should be more reliable and simpler in structure compared with mechanical compressors used in cryocoolers. Thus, a pulse tube refrigerator driven by such a thermoacoustic engine will be more reliable and simpler to manufacture. A thermoacoustic engine operating in standing wave mode is also called as a natural engine with intrinsically thermodynamic irreversibility, so it shows a poor thermodynamic performance from thermal energy to acoustic energy, with a low conversion efficiency around 10 percent of the Carnot engine. This paper analyzed the generation rate of acoustical workflow in the traveling-wave and standing wave modes, and proposed a new thermoacoustic engine that mainly operates in the traveling-wave mode. In addition, a new thermally driven pulse tube refrigerator operating in the double traveling-wave mode was suggested. Finally, the experimental result of the thermally driven pulse tube refrigerator was reported in this paper.

INTRODUCTION

A first thermally driven pulse tube refrigerator was introduced and developed by G.W.Swift and R.Radebaugh in the year of 1990[1]. In their prototype, the thermoacoustic wave generator was operating in the standing-wave mode with intrinsic thermodynamic irreversibility. Although P.Ceperley proposed a thermally driven thermoacoustic refrigerator operating in traveling-wave mode similar to the ideal Stirling cycle, he had never made a successful progress [2]. However, we have persisted in that the core of Ceperley's proposal was correct. Recently, T.Yazaki et al made an experimental progress in a looped tube, and they demonstrated that a self-excited thermoacoustic oscillation occurred under a lower onset temperature in the traveling-wave mode [3]. More recently, S.Backhaus et al made progress in the traveling-wave thermoacoustic engine, which achieved 42% of Carnot efficiency[4]. These two progresses encouraged us greatly.

In this paper, we proposed a thermally driven pulse tube refrigerator working in double traveling-wave modes. We plan to make both thermoacoustic engine and refrigerator operate in traveling-wave modes for obtaining a maximum efficiency.

THERMODYNAMIC CONSIDERATIONS

Onset Temperature of Traveling-Wave and Standing-Wave Modes

Based on linear thermoacoustic theory developed by N. Rott, G. Swift proposed the onset temperature gradient in a pure standing-wave field[5,6]. Later, Xiao gave the formula of onset temperature gradient under arbitrary acoustic fields[7]. In this section, we will analyze and discuss the formula of onset temperature gradient under a pure standing-wave and travelling-wave modes, respectively.

For arbitrary acoustic field, the onset temperature gradient can be described by the following equation:

$$\left(\frac{dT_0}{dx}\right)_{cr} = \frac{\gamma-1}{\beta_0 a_0}\omega\frac{g_{WK}+\frac{1}{\gamma-1}|Y_a|^2 g_{W\mu}}{\mathrm{Re}[Y_a f_{WT}]} \tag{1}$$

Where,

Ya is local acoustic conductivity, which is defined as $\frac{\rho_0 a_0 \tilde{U}}{A\tilde{P}}$. $\tilde{U}$ is the local volumetric velocity. $\tilde{P}$ is the local dynamic pressure. γ is the adiabatic index of fluids. $g_{w\mu}$ and g_{wk} are the work dissipation factors by the viscosity and finite thermal conductivity of fluids, respectively. f_{WT} is the work production factor. The definition of these factors can be found in Dr. Xiao's work.

Let $\theta = \frac{T}{T_0}$ and $\hat{x} = \frac{2\pi x}{\lambda}$, a formula about the dimensionless onset temperature gradient can be obtained:

$$\left(\frac{d\theta}{d\hat{x}}\right)_{cr} = \frac{(\gamma-1)}{\beta_0 T_0}\frac{g_{WK}+\frac{1}{\gamma-1}|Y_a|^2 g_{W\mu}}{\mathrm{Re}[Y_a f_{WT}]} \tag{2}$$

Where,

λ is the wavelength. $\hat{x}$ and θ are the dimensionless length and temperature, respectively.

According to Eq.(2), we calculate and analyze dimensionless the onset temperature gradients under the standing-wave and travelling-wave modes in the condition of different acoustic conductivities, respectively. The calculation results are shown in figure 1 to figure 4. The solid line is for the travelling-wave mode, while the dashed line is for the standing-wave mode. Based on the calculation result, the following conclusions can be drawn on:

1.In all calculation cases, the minimum onset temperature gradient under the travelling-wave mode is lower than that under the standing-wave mode;

2.As the Prandtl number decrease, the minimum onset temperature gradient under the travelling-wave mode is much lower than that under the standing-wave mode;

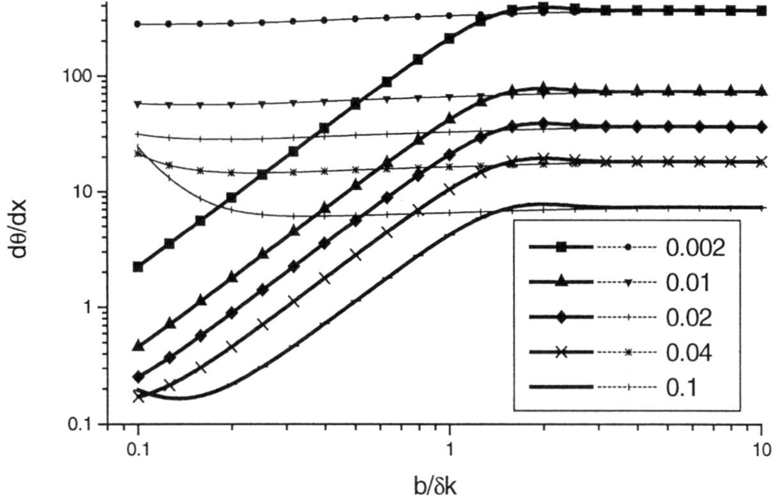

Figure 1. (Pr=0.01)

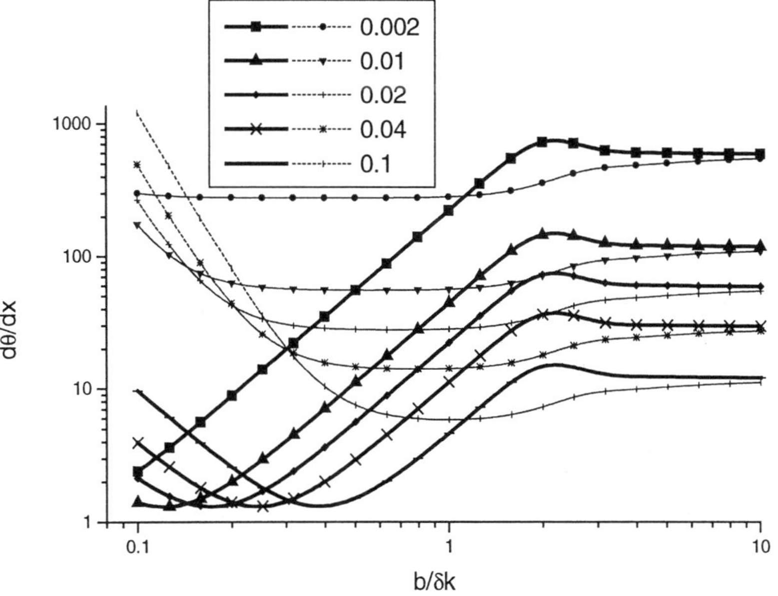

Figure 2. (Pr=0.64)

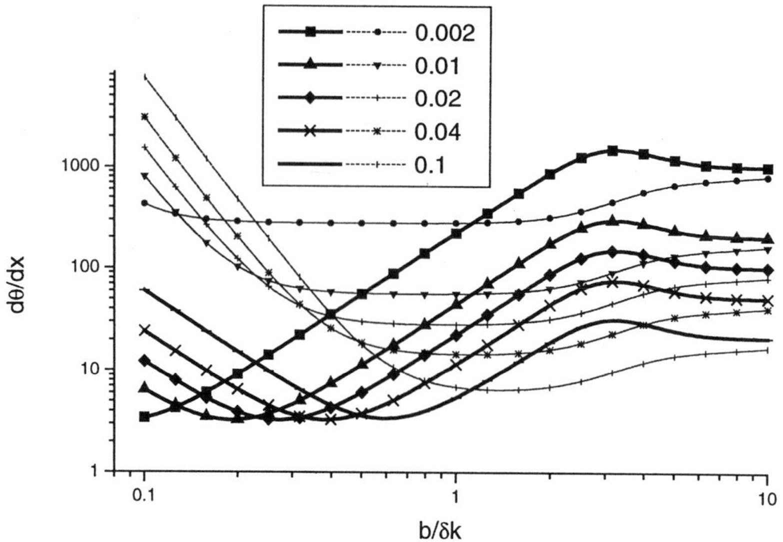

Figure 3. (Pr=4.0)

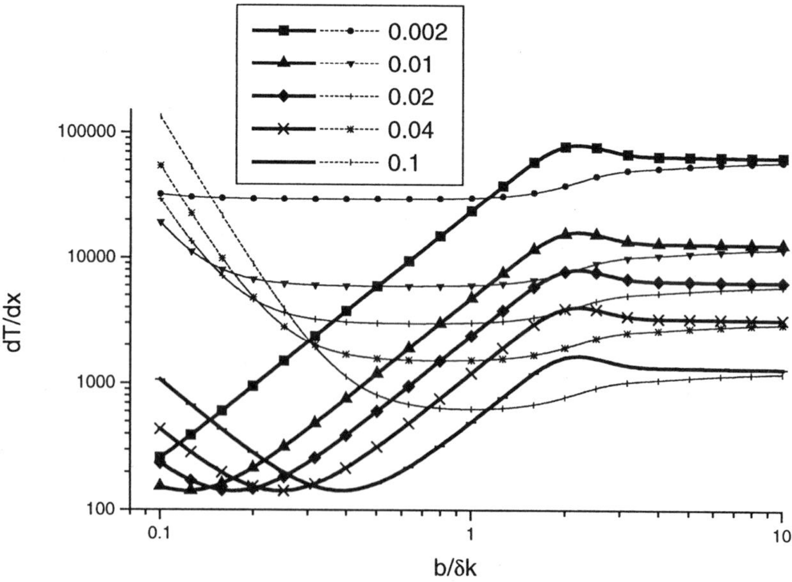

Figure 4. (Helium, P_0=1.0MPa,T_0=400K,f=50Hz)

3. For helium, the minimum onset dimensionless temperature gradient under the standing-wave mode is about 4 to 6 times as that under travelling-wave mode.

All in a word, a thermoacoustic engine operating in the travelling-wave mode can more easily generate a self-excited thermoacoustic oscillation, requiring a lower heat source temperature.

Formation of Double Traveling-wave Modes

Although Stirling engine and refrigerator operate with the pressure and velocity in phase, they are not real traveling-wave engine and refrigerator in acoustic sense. According to the author's opinion, they are different in nature. Now, we gradually realized and caught up with some features of thermoacoustic engine: (1) Although the pressure and velocity inside Stirling engine and refrigerator are almost in phase, the working zone is only a very short part of acoustical wavelength. On the contrary, the working zone of P.Ceperly's traveling-wave thermoacoustic engine and refrigerator is occupied with a whole wavelength; (2) The acoustic transmission rates, $\rho_0 a_0 \tilde{U} / A\tilde{p}$, inside P.Ceperley's design was 1, but typically about $10^{-2} - 10^{-3}$ inside the Stirling engine and refrigerator. Thus, a considerable loss was produced in Ceperley's design, and consequently the work dissipation overwhelmed the acoustic work or refrigeration production; (3) Because the volumetric velocity inside P.Ceprley's design was considerably large, a significant nonlinear effect such as acoustic streaming or DC flow would happen. Some theoretical and experimental results proved that this nonlinear effect would introduced a significant loss; (4) the intrinsic frequency of self-excited thermoacoustic oscillation system usually is higher than that of the Stirling engine and refrigerator. This also would increase the loss inside these self-excited thermoacoustic engine and refrigerator.

To avoid the shortcomings of H.Cepeley's travelling engines, a new thermally driven pulse tube refrigerator operating in double travelling-wave modes shown in figure 5 is suggested. It mainly includes three parts: a quarter wavelength resonator, a thermoacoustic prime mover and a thermoacoustic refrigerator. Because the feedback acoustic work flow required for the thermoacoustic prime mover is not equal to that of the thermoacoustic refrigerator, an additional feedback loop is employed in the figure 5. In this design, the quarter wavelength resonator predominantly decides the operating frequency of the whole system. Thus, the total effective length of the thermoacoustic prome mover and refrigerator is much smaller than that of the quarter wavelength resonator.

The operating principle of the thermally driven thermoacostic refrigerator is as follows:

First, the quarter wavelength resonator decides the operating frequency of the whole system. Secondly, the thermoacoustic prime mover generates acoustic work. One part of the acoustic work drives the thermoacoustic refrigrtaor, and the other part is feedbacked. Because the required feedback work of the thermoacoustic prime mover is different from that of the thermoacoustic refrigerator, an additional feedback loop for the thermoacoustic prime mover is introduced. Properly matching the acoustic system, one can make both the thermoacoustic prime mover and refrigerator operating in the Stirling engines, with the dynamic pressure and volumetric velocity in phase and a small amplitude of the volumetric velocity.

Electrical Analogy of Double Traveling-wave Modes

Assume that the room temperature, the refrigeration temperature and the heating temperature are T_0, T_C and T_H, respectively. Because the lengths of most thermodyanmic components except the quarter wavelength tube are much shorter than the wavelength, the

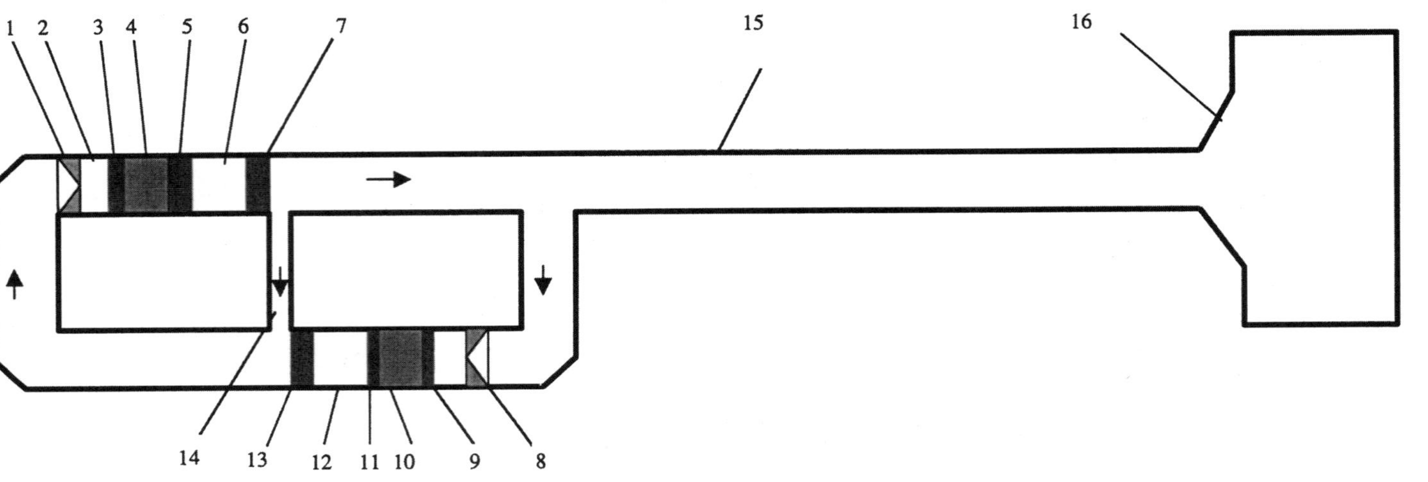

Figure 5. A thermally driven thermoacoustic refrigerator operating in double travelling wave modes

1. Asymmetric flowing channel 2. Laminar flow straighter 3. Aftercooler of thermoacoustic engine 4. Regenerator of thermoacoustic engine 5. Heater exchanger of thermoacoustic engine 6. Resonator tube of TE 7. Aftercooler of TE 8. Asymmetric channel of travelling feedback loop 9. Room temperature aftercooler 10. Regenerator of thermoacoustic refrigerator (TR) 11. Cold-end heat exchanger of TR 12. Resonator tube of TR 13. Room temperature aftercooler of TR 14. Feedback loop of TE 15. Quarter wavelength tube 16. Quarter wavelength resonator

centralized parameter method can accurately deescibe the features of these thermodyanmic components in the view of electrical network theory. Figure 6 is analogical electrical-circuit diagram of our suggested design.

In order to grasp the core of our design, it is convenient to adopt electrical analogy method to explain the operation of the double travelling-wave modes. In the view of acoustics, if the length of a pipe is much shorter than one wavelength, the centralized parameter method is feasible to describe the pipe. In our design, the length of all thermoacoustic elements, except the quarter wavelength tube, are deliberately designed shorter than one wavelength distance. Thus, the following analogy parameters can be defined:

Acoustic impedance

$$R = (2.5 \sim 4.0)\frac{8}{r_0^2}\mu_0 l \tag{3}$$

Acoustic inductance

$$L = \frac{4}{3}\frac{\rho_0 l}{A} \tag{4}$$

Acoustic capacitance

$$C = \frac{V}{\gamma p_0} \tag{5}$$

Where,

ρ_0 and μ_0 are the average density and viscosity of fluids, respectively. r_0, l and V are the hydraulic radius, length and volume of a thermodynamic element. p_0 is the average pressure. γ is the adiabatic index of fluids.

According to the above-mentioned method, the electrical analogy diagram is given in figure 6.

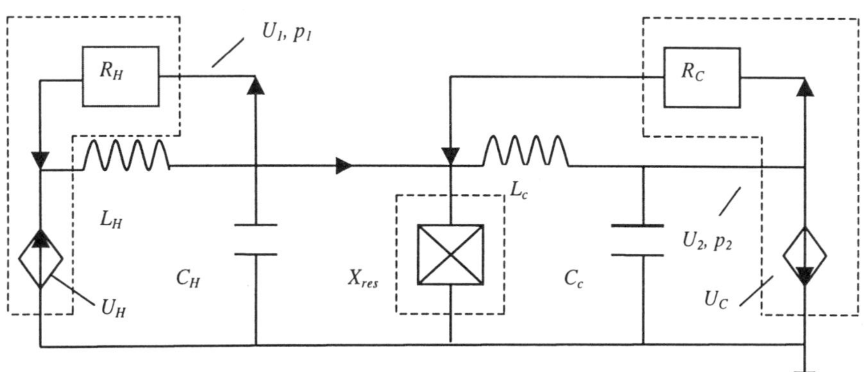

Figure 6. Electrical analogy diagram of TPTR operating in double traveling-wave modes

U_H and U_C are the equivalent acoustic constant sources for the thermoacoustic prime mover and refrigerator. R_H, L_H and C_H are the acoustic impedance, inductance and capacitance of the hot temperature section, while R_C, L_C and C_C are the acoustic impedance, inductance and capacitance of the cold temperature section. X_{res} is the complex acoustic resistance of the quarter wavelength resonator.

In the figure 6, the two regenerators are considered the acoustic constant-current resources:

$$U_H = (\frac{T_H}{T_0} - 1)U_1 \quad (6)$$

$$U_C = (1 - \frac{T_C}{T_0})U_2 \quad (7)$$

To eliminate the two shortcoming of H.Cepeley's engines, it is necessary to design the appropriate R_H, L_H and C_H of the hot temperature section, as well as R_C, L_C and C_C of the cold temperature section. One key point is to make p_1, U_1 and p_2, U_2 in phase. The other key point is to make $p_1 / U_1 >> \rho_0 c / A$ and $p_2 / U_2 >> \rho_0 c / A$. If these two conditions can be realized, the thermally driven thermoacoustic system will operate like the Stirling system.

Suppression of Acoustic Streaming/DC Flow

When the amplitude of thermoacoustic oscillation is big enough, there will exist an obvious acoustic streaming or DC flow. No matter how the direction of the acoustic streaming is, it must introduce the heat/cold energy loss. To eliminate this loss, it is possible to employ asymmetric hydraulic elements or an external DC flow. Thus, two asymmetric hydraulic channels are included in our suggested design.

CONCLUSIONS

In this paper, the onset temperature gradients of self-excited thermoacoustic oscillation under the travelling-wave and standing-wave modes are calculated and analyzed. Then, a new thermally driven thermoacoustic refrigerator is suggested. Finally, its operating principle and design methodology is described.

ACKNOWLEDGEMENT

This work is supported by the National Natural Sciences Foundation Of China under the contract number 5960002. The discussion with Professor Y.Zhou is greatly appreciated.

REFERENCES

1. R. Radebaugh, A review of pulse tube refrigerator, Adv. Cryo. Eng., 35:1191(1990)
2. H.Cepeley, Gain and efficiency of a short travelling wave heat engine. *J. Acoust. Sco. Am.*, 77:1239 (1985)
3. T.Yazaki, A.Iwata, T.Maekawa, and A.Tominaga, Traveling Wave Thermoacoustic Engine in a Looped Tube. *Phy. Rev.Lett.*, 81(15):3128(1998)
4. S.Backhaus, G.Swift, A thermoacoustic Stirling heat engine, *Nature*, 39:335(1999)
5. N. Rott, Thermoacoustics, *Adv. Appl. Mech*, 20:135(1980).
6. G.W.Swift, Thermoacoustic Engines, *J.Acoust.Soc. Am*, 84(4):1145(1988)
7. J.H.Xiao, Thermoacoustic theory for cyclic regenerator, Part 1: Fundamentals, *Cryogenics*, 32(10):895(1992)

PHASOR ANALYSIS FOR DOUBLE INLET PULSE TUBE CRYOCOOLER

M. D.Chokhawala,[1] K. P. Desai,[1] H. B. Naik,[1]K. G. Narayankhedkar[2]

[1]S.V.R. College of Engineering and Technology, Surat - 395 007, India.
[2]Indian Institute of Technology, Mumbai - 400 076, India

ABSTRACT

Concept of phasor analysis of pulse tube cryocooler was introduced by Radebaugh[12]. This concept was further explained by Kittel[13] and using Taylor series expansion an expression for refrigeration power was developed by them which does not involve phase angle between mass flow at the cold end and pressure vector for orifice pulse tube refrigerator(OPTR). In the present work the concept of phasor analysis is extended for double inlet pulse tube refrigerator (DIPTR). Phasor diagram for DIPTR is proposed using expression for mass flow rate across orifice, double inlet valve etc. An expression for refrigeration power for DIPTR is developed, using Taylor series expansion, which does not include the phase angle. The refrigeration power thus obtained was compared with that obtained using the expression given by Radebaugh[12]. Effect of pulse tube diameter and length, pressure difference, frequency and flow co-efficient for double inlet valve on phase angle and mass flow rate at cold end is studied for DIPTR and OPTR. It is found that pulse tube diameter and length, frequency and double inlet flow co-efficient have considerable influence on phase angle. There exists optimum value of pulse tube diameter and length, frequency and double inlet co-efficient to get minimum phase angle. Pressure difference is found to have no influence on phase angle.

INTRODUCTION

Pulse tube cryocooler are proved to have an edge over their counterparts in crucial applications like space SQUIDS, etc. due to low vibrations and relatively long maintenance free life. In last one and half decade many versions of PTR like orifice, double inlet, active buffer along with multi staging operations are developed to reach lower temperature and improve the thermodynamic efficiency. Theoretical analysis was also carried out to understand the underlying mechanism responsible for producing the refrigeration. Surface heat pumping theory[1] was the first effort in this direction. An isothermal model[2] was proposed based on the premise that a pulse tube refrigerator can

Advances in Cryogenic Engineering, Volume 45.
Edited by Shu *et al.*, Kluwer Academic / Plenum Publishers, 2000.

be considered to be a type of split Stirling refrigerator and the gas in the pulse can be divided in to 3 parts. Numerical analysis of PTR[3-5] is found to predict their performance quiet efficiently. Various thermodynamic models[6-10] are also proposed for qualitative understanding of the mechanism. Recently the thermodynamic non symmetry effect[11] was proposed to explain the mechanism which produces refrigeration in PTR system. This mechanism was quite effectively explained using the phasor diagram[12] for orifice pulse tube. The analysis was further extended[13] to analyze OPTR. In the present work the phasor diagram for DIPTR is proposed. The analysis was carried out to study the effect of few parameters on phase angle and refrigeration power.

PHASOR ANALYSIS

The phasor diagram for OPTR[12, 13] and that for DIPTR proposed in the present work are as shown in figure (1) and (2). It can be seen from this diagram that the presence of m_{di} vector shifts the m_c vector towards the pressure vector which reduces the phase angle ϕ thereby increasing the regeneration power.

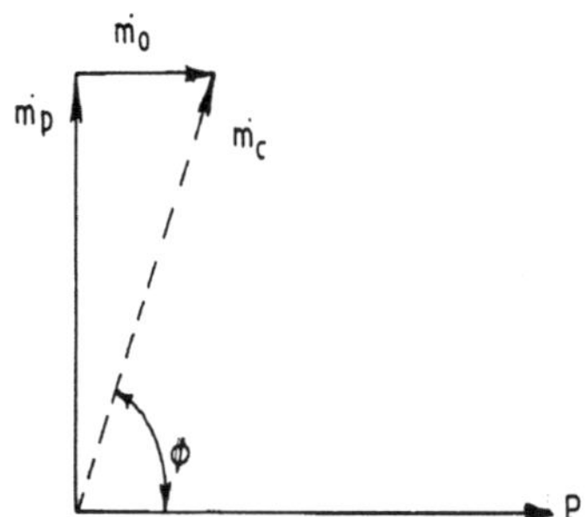

Figure 1 Phasor diagram for OPTR

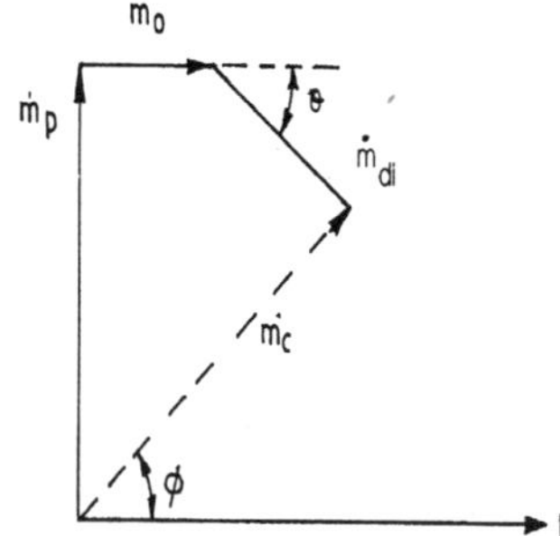

Figure 2 Phasor diagram for DIPTR

The simplifying assumptions like monoatomic ideal gas, ideal regenerator, adiabatic pulse tube, no axial heat conduction in pulse tube and regenerator etc., were made for the analysis. Pressure and temperature variations were assumed as[12-13].

$$P = P_0 + P_1 \ \mathrm{Cos}\,(\omega t) \tag{1}$$

$$T = T_0 + T_1 \ \mathrm{Cos}\,(\omega t + \phi_t) \tag{2}$$

The mass of gas in the pulse tube is found by integrating the density over the volume.

$$m_p = \int_0^L \rho \, dz \tag{3}$$

Assuming linear temperature variation in the pulse tube, the mass flow rate due to pressurization can be derived as :

$$\dot{m}_p = \frac{-3\,\omega}{5\,R} P_1 \ \sin\,(\omega t) \int_0^L To^{-1} \, dz \tag{4}$$

The mass flow rate across the orifice and double inlet transferred to the cold end

$$\dot{m}_o = C_1 \, [P_{pt} - P_r] \, \frac{T(L)}{T(Z)} \tag{5}$$

$$\dot{m}_{di} = C_2 \, [P_l - P_{pt}] \, \frac{T(L)}{T(Z)} \quad \text{for } 0 \le \tau \le \pi$$

$$= C_2 \, [P_h - P_{pt}] \, \frac{T(L)}{T(Z)} \quad \text{for } \pi \le \tau \le 2\pi \tag{6}$$

Net mass flow rate at the cold end ;

$$\dot{m}_c = \dot{m}_p + \dot{m}_o - \dot{m}_{di} \tag{7}$$

Which can be expressed with reference to figure (2) as ;

$$\dot{m}_c = \text{sqrt}\{ (\dot{m}_o + \dot{m}_{di} \cos\theta)^2 + (\dot{m}_p - \dot{m}_{di} \sin\theta)^2 \} \tag{8}$$

The expression for phase angle ϕ can be written as ;

$$\phi = \text{Cos}^{-1} \frac{\dot{m}_o + \dot{m}_{di} \, \text{Cos}\,\theta}{\dot{m}_c} \tag{9}$$

Here the angle θ is assumed to vary linearly over half cycle from 0 to $\pi/2$. Refrigeration power is equal to the enthalpy flow in the pulse tube[12-13] which is

$$\langle Q \rangle = \frac{1}{\tau} \int_0^{\tau} \dot{m}_c \, C_p \, T \, dt \tag{10}$$

Upon substituting equations for mass flow and temperature, performing the integration, using Taylor series expansion with higher order terms neglected ;

$$\langle Q \rangle = \frac{1}{5} \frac{C_p . T_h}{P_o} \left[C_1 P_1^2 - \frac{C_2}{2} (P_l P_1 - P_o P_1 - P_1^2) - \frac{C_2}{2} (P_h P_1 - P_o P_1 - P_1^2) \right] \tag{11}$$

It is to be noted that enthalpy flow from the cold end to hot end of the pulse tube takes places only in one half of the cycle. The similar expression for refrigeration power for OPTR, given by Kittel and others[13]

$$\langle Q \rangle = \frac{C_1 \, C_p \, T_o (L) \, P_1^2}{5 \, P_o} \tag{12}$$

and that given by Radebaugh[12]

$$<Q> = \frac{R \quad T_C}{2 \quad P_O} \dot{m}_c P_1 \cos\phi \tag{13}$$

can be applied to both OPTR and DIPTR.

RESULTS

The model discussed above was solved for studying the effect of various parameters like geometry of pulse tube, frequency and pressures on phase angle, mass flow rate at cold end and also refrigeration power obtained in case of OPTR and DIPTR. The set of common parameters used for investigations are shown in table 1. The results were plotted over the half cycle only, as the nature of variation is similar over other half of the cycle.

It can be seen from figure (3) that for DIPTR length has considerable effect on variation of phase angle ϕ over the half cycle. As length increases variation in phase angle ϕ decreases up to a limit after which increase in length increases the variation in phase angle ϕ. For the present set of parameters investigated, minimum variation of phase angle ϕ over the half cycle is observed for pulse tube length of 250 mm.

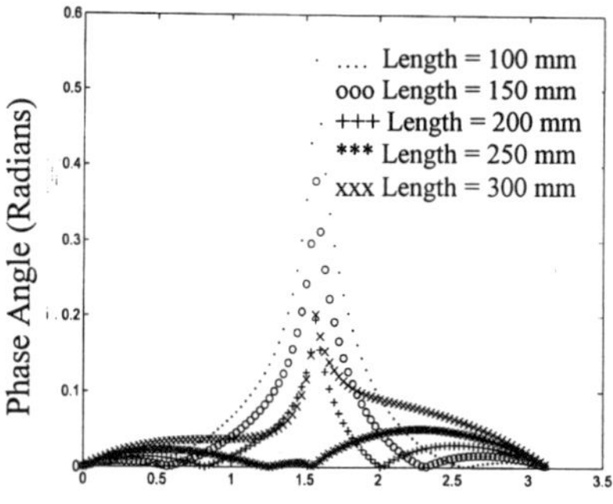

Figure 3 Effect of Length on Phasor Angle variation with time for DIPTR

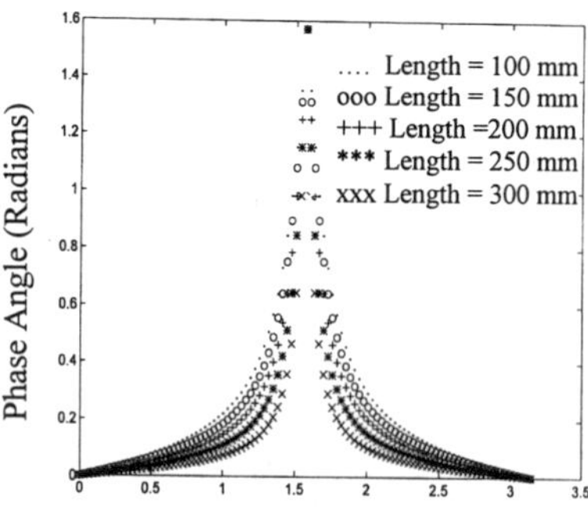

Figure 4 Effect of Length on Phasor Angle variation with time for OPTR

However as seen from figure (4) that for OPTR length does not have significant effect on phase angle ϕ. As length increases variation in phase angle ϕ increases. For the range of parameters investigated it is found that phase angle ϕ passes through the nearly same peak value over the half cycle.

Table 1. Common Parameters Used for Computation

Parameter	Value	Parameter	Value
Higher pressure P_h	= 17 bar	Flow coefficient of orifice valve C_1	= 6×10^{-10} kg/s/Pa
Lower pressure P_l	= 07 bar	Flow coefficient of DIPTR valve C_2	= 1×10^{-10} kg/s/Pa
Average pressure P_o	= 12 bar	Specific heat at constant pressure C_p	= 5198 J/kg-K
Frequency ω	= 5 Hz	Temperature at cold end T_c	= 80 K
Diameter of pulse tube	= 12 mm	Temperature at hot end T_h	= 285 K
Length of pulse tube	= 200 mm	Gas constant R	= 2078.75 J/kg-K

It is interesting to note here that in case of DIPTR the peak value of phase angle ϕ for a length of 100mm is approximately 0.6 radians that is 36° which is much lower than the corresponding value in case of OPTR for a length of 300mm which is approximately 1.55 radian that is nearly 90°. The smaller variation of phase angle ϕ over the cycle for DIPTR compared to OPTR clearly justifies the higher refrigeration power for the former.

It can be seen from figure (5) for DIPTR that diameter also has considerable effect on variation of phase angle ϕ over the half cycle. As diameter decreases variation in phase angle ϕ decreases up to a limit beyond which decreases in diameter increases the variation in phase angle ϕ observed. For the present set of parameters investigated there exits minimum variation of ϕ over the half cycle for a diameter of 12mm.

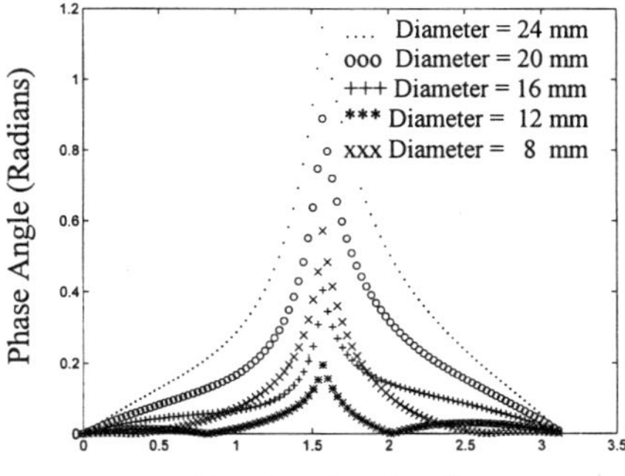

Figure 5 Effect of Diameter on Phasor Angle with time for DIPTR

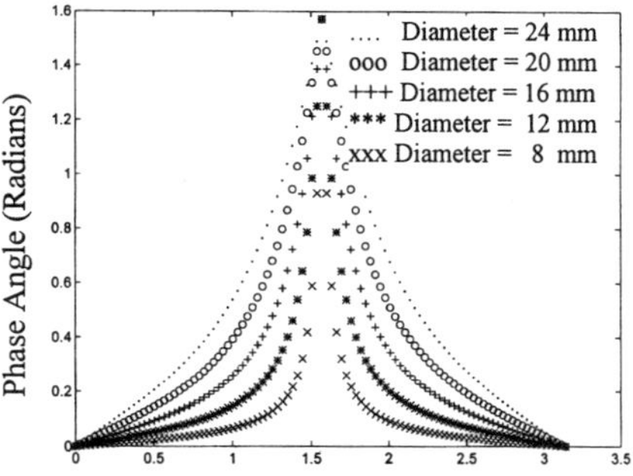

Figure 6 Effect of Diameter on Phasor Angle with time for OPTR

Figure (6) for OPTR shows that for all values of diameter investigated phase angle ϕ passes through a peak value, however for lower values of diameter variation in phase angle tends to be minimum for the major part of the half cycle. As diameter increases variation in ϕ increases.

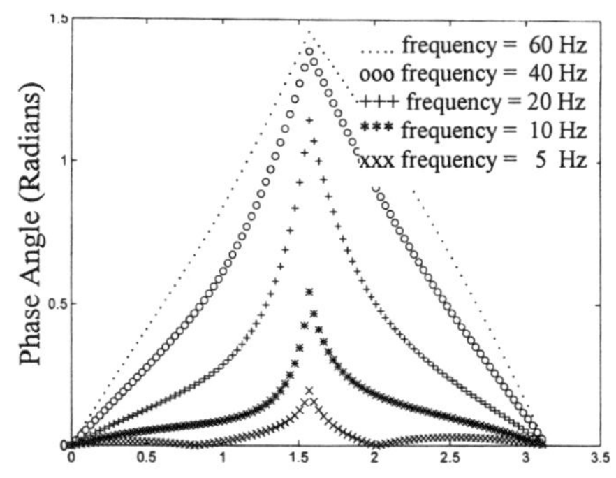

Figure 7 Effect of Frequency on Phasor Angle with time for DIPTR

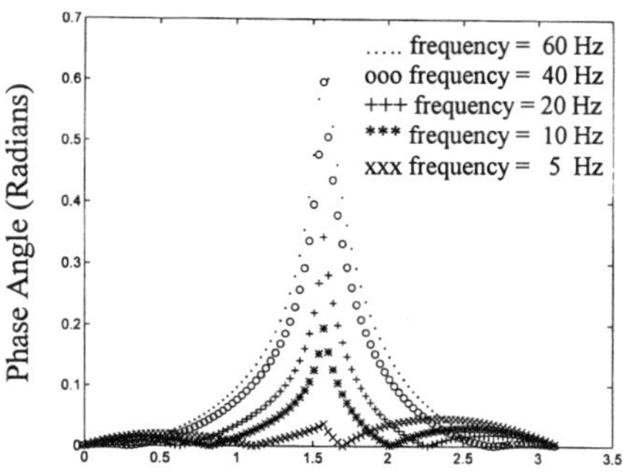

Figure 8 Effect of Diameter on Phasor Angle with time for DIPTR

It can be seen from figure (7) and (8) for DIPTR that as frequency is increased from 1 Hz onwards the variation in phase angle ϕ is seen to decrease however increasing frequency beyond 6 Hz onwards variation in phase angle over the cycle increases significantly. For the present set of parameters investigated it is observed that for a frequency of 6 Hz the variation in phase angle over the cycle is minimum and the peak value attained is approximately 2.5°.

From figure (9) for DIPTR it can be said that increasing the flow coefficient C_2 significantly increases the variation in phase angle ϕ over the cycle. For the present set of parameters investigated cyclic variation of phase angle ϕ is found to be minimum for a value of 1×10^{-10} kg/s/Pa.

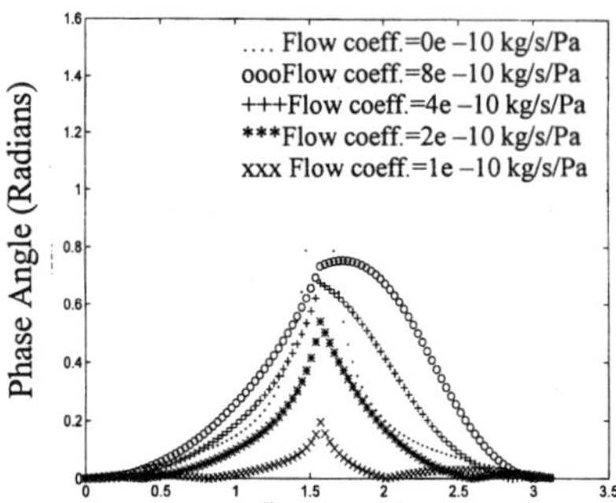

Figure 9 Effect of Flow coefficient of double inlet valve on Phasor Angle with time.

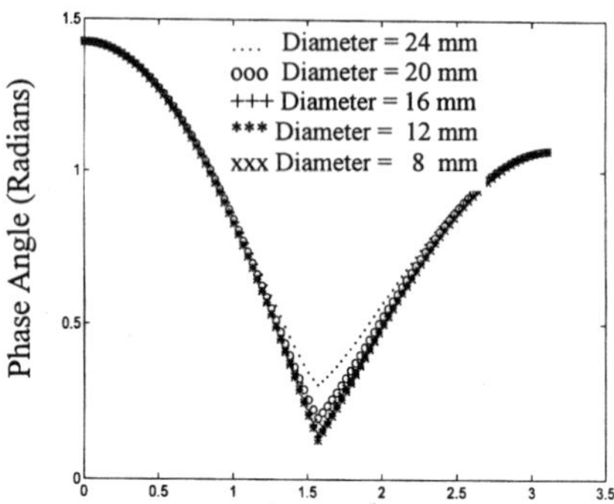

Figure 10 Effect of Diameter on Mass flow rate at cold end Vs Time For DIPTR

Figure (10) for DIPTR shows that diameter has negligible effect on mass flow rate at cold end whereas it has some effect in case of OPTR as seen from figure (11).

As seen from figure (12) that refrigeration power increases with increase in the pressure difference within the present range of investigation in case of both OPTR and DIPTR, however this effect reduces towards higher pressure difference. Also the refrigeration power for DIPTR is higher than that for OPTR through the range of pressure differences as expected. It is further observed that the refrigeration power between OPTR and DIPTR increases with the pressure difference.

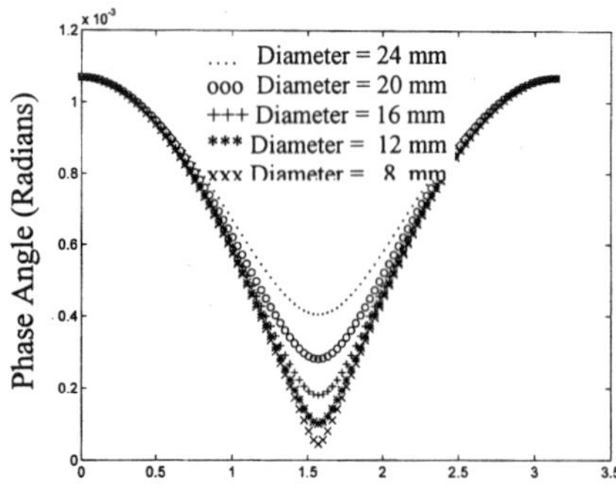

Figure 11 Effect of Diameter on Mass Flow rate at cold end Vs Time for OPTR

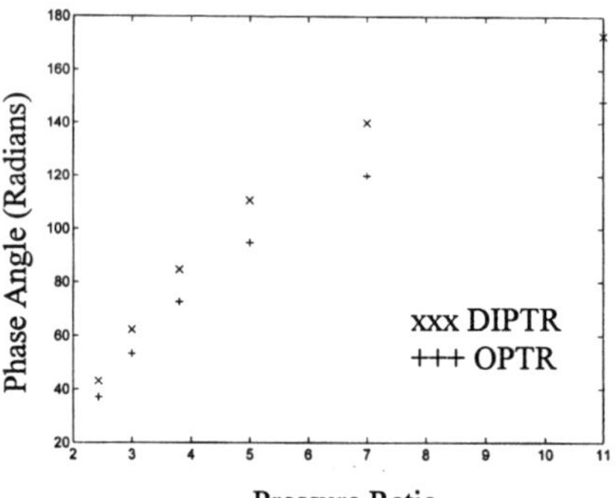

Figure 12 Refrigeration Power Vs Pressure ratio

CONCLUSION

The phasor diagram proposed for the DIPTR clearly explains the process of improvement of refrigeration power compared to the OPTR. The present analysis shows, under ideal conditions, how the phasor angle varies over the cycle. Length and diameter of pulse tube have more effect on phasor angle for DIPTR compared to the OPTR. Phasor angle is found to have lowest value and minimum variation over the cycle for certain value of frequency in case of DIPTR.

REFERENCES

1. Longsworth R. C. : "An experimental investigation of pulse tube refrigeration heat pumping rates", Advance in cryogenics, Vol. 12.
2. Zhu S. W. and Chen Z. Q. : "Isothermal model of pulse tube refrigerator", Cryogenics (1994), Vol. 34, Number 7, 591-595.
3. Chao Wang, Peiyi Wu and Zhonggi Chen : "Numerical modeling of an orifice PTR, Cryogenic (1992) Vol. 32, No. 9, 785 - 790.
4. Wang C., Wu P.Y. and Chen Z.Q. : "Numerical analysis of double inlet PTR", Cryogenics, (1993), Vol. 33, No. 5, 526 - 531.
5. Wang. C. : "Numerical analysis of 4K pulse tube coolers : Part I Numerical simulation", Cryogenics, (1997), Vol. 37, No. 4, 207 - 213.
6. Boer de P. C. T. ; "Analysis of basic pulse tube refrigerator", Cryogenics (1994), Vol. 34, No. 9, 699 - 711.
7. Boer de P.C.T., "Analysis of basic pulse tube refrigerator with regenerator", Cryogenics, (1995), Vol. 35, No. 9, 547 - 553.
8. Boer de P. C. T. : "Heat removal in the orifice pulse tube", Cryogenics (1998) Vol. 38, No. 3, 343 - 357.
9. deWaele A.T.A.M., Steijaert P. P. and Gizen J. : "Thermodynamical aspects of pulse tubes", Cryogenics, (1997), Vol. 37, No. 6, 313 - 324.
10. deWaele A.T.A.M., Steijaert P.P. and Koning. : "Thermodynamical aspects of pulse tubes II", Cryogenics, (1998), Vol. 38, No. 3, 329 - 335.
11. Liang J., Ravex, A. and Rolland P. : "Study on pulse tube refrigeration part I : Thermodynamic non symmetry effect part : 2 Theoretical modeling art 3 : Experimental verification", Cryogenics, (1996), Vol. 36, 87 - 106.
12. Radebaugh Ray : "A review of pulse tube refrigerator", Advances in cryogenic engineering, Vol. 35, Plenum press, New York, 1990.
13. Kittel. P., Kashani A., Lee J. N. and Roach P. R. : "General pulse tube theory", Cryogenics, (1996), Vol. 36, 849 - 857.

NOMENCLATURE

C_1 = Flow coefficient of orifice valve kg/Sec-Pa
C_2 = Flow coefficient of double inlet valve kg/sec-Pa
C_P = Sp. heat at Const. pressure, J/kg-k
L = Length of pulse tube, m
$\dot{m}_c$ = Mass flow rate at cold end, kg/sec.
$\dot{m}_{di}$ = Mass flow across doublet inlet valve, kg/sec.
$\dot{m}_o$ = Mass flow across orifice valve, kg/sec.
$\dot{m}_p$ = Mass flow rate due to pressurization, kg/sec.
M_p = Mass of gas inside the pulse tube, kg
P = Pressure kPa
P_o = Average pressure kPa
P_1 = Pressure amplitudes kPa
P_h = Reservoir pressure (High), kPa

t = time, Sec.
T = Temperature, K
T_o = Average temperature, K
T_1 = Temperature amplitude, K
T_h = Hot end temperature, K
z = Co-ordinate alongth the pulse tube from cold end, m
Z = Total distance from cold end, m
ω = Frequency, Hz
τ = Cycle time, Sec.
ρ = Density, Kg/m^3
ϕ = Phasor angle between the $\dot{m}_c$ and p vector

P_L = Reservoir pressure (Low), kPa
P_{pt} = Pressure inside the pulse tube, kPa
P_r = Orifice reservoir pressure, kPa

θ = Phasor angle between $\dot{m}_o$ and $\dot{m}_{di}$
$\langle Q \rangle$ = Refrigeration power, W
R = Gas constant, J/kg-K

NUMERICAL STUDY OF GAS DYNAMICS INSIDE OF A PULSE TUBE REFRIGERATOR

Yoshikazu Hozumi,[1] Masahide Murakami [2]

[1]Chiyoda Corporation, 2-12-1, Tsurumichuo, Tsurumi-ku, Yokohama, 230-8601, Japan
[2]Institute of Engineering Mechanics, University of Tsukuba, Tennodai 1-1-1, Tsukuba, 305-8573, Japan

ABSTRACT

Numerical simulation of the viscous compressible flow in a pulse tube is carried out to study the detail of the flow and fundamental mechanism of refrigeration. Axisymmetric two-dimensional Navier-Stokes equations are solved numerically by using finite volume method with implicit scheme. The phase difference in oscillating flow between the tube center and nearby the tube wall is observed by the simulation. The time variation of temperature profiles obtained by the simulation is in good agreement with experimental data. The particle path of working gas element and temperature difference between gas elements and tube side wall are evaluated in order to investigate the heat transfer from cold end to hot end. The secondary mass flux is observed by the particle path trace and the time-average mass flow. The simulation results suggest that the boundary layer on the tube wall might play an important role to the heat transfer.

INTRODUCTION

Pulse tube refrigerator is categorized in thermoacoustic refrigerators. The feature of thermoacoustic refrigerators is installation of a regenerator, which is an important element to determine the refrigeration performance. One approach to explain the refrigeration phenomenon in pulse tube refrigerator is that pulse tube refrigerator works with the same mechanism as Stirling cycle refrigerator. The displacer of Stirling refrigerator is replaced by a pulse tube in the pulse tube refrigerator. The pulse tube works as a gas piston similar to the displacer in a

Stirling refrigerator. The phase shift devices are installed at the hot end of pulse tube to improve the performance of refrigeration. An orifice and double inlet bypass connection are effective phase shift devices to improve the performance of pulse tube refrigerator. In order to complement the heat transfer in basic type pulse tube, "Surface heat pumping mechanism" was suggested by the investigators Gifford and Longsworth[1]. Although the gas elements do not move directly from closed end to hot end in the basic type pulse tube, the gas transfers heat from the open end to the closed end by these successive heat pumping.

The concept of heat transfer is suggested in 1960's, but the surface heat pumping mechanism is not testified by the experimental research. Because the mechanism is very local and short cyclic phenomenon that experimental probes, which measure temperature, velocity, and pressure, can not respond to the alternation of the working gas status. However, improvements in computer hardware, resulting in increased memory and efficiency, have made it possible to solve equations in fluid mechanics in practical way.

Thermoacoustic theory is available to describe the phenomenon of pulse tubes. Thermoacoustic theory is based on linearized and one-dimensional Navier-Stokes equation. Since the practical pulse tubes have a finite flow velocity field whole through the inside of pulse tube, the flow would be dominated by strong nonlinear advection phenomena. Thermoacoustic theory is good for small amplitude of pulse tube flow, but the theory could not be used for discussing the surface heat pumping mechanism.

The nonlinear approach is required to evaluate the heat transfer mechanism in pulse tube. The purpose of this research is to investigate the refrigeration mechanism in pulse tubes by using direct numerical simulation.

Simulation Model

Several approaches of numerical investigation of the flow inside pulse tubes have been made in recent years. The state of compressible viscous flow is expressed by the Navier-Stokes equations; the equation of mass conservation, momentum conservation, and energy conservation. A qualitative comparison among different approaches of modeling is made by Lee and Kittel, et al.[2]. Since the computation of discretized full three-dimensional Navier-Stokes equations is still heavy for current computer ability, several simplified modelings of governing equations have been investigated. Fortunately, the phenomena inside of a pulse tube can be treated as an axisymmetric flow.

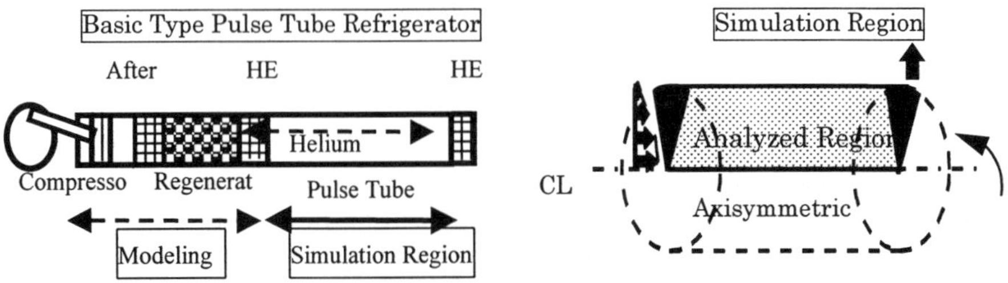

Figure 1. Schematic drawing of pulse tube refrigerator for simulation and axisynmetric two-dimensional model of pulse tube flow field

In this study, axisymmetric two-dimensional Navier-Stokes equations are directly solved for the pulse tube flow. Figure 1 shows the simulation model of the pulse tube. The dimension of simulated pulse tube is 288 mm x 15.6 mm (length x radius). In order to focus this study on how energy is transfered from the cold end to the hot end in the pulse tube refrigerator, only fluid motion inside the pulse tube is investigated. The regenerator is modeled as an energy buffer for the internal energy income and outgo through the pulse tube inlet. The heat exchanger is installed at the hot end. A certain amount of heat is transferred depending on the temperature difference between gas and pulse tube wall.

Mesh Geometry

50 x 21 (length x radius) grid points mesh is generated to carry out the calculation. Viscous boundary layer thickness is given as function of magnitude of internal flow velocity. For better resolution of the viscous shear stress on the wall surface to the simulation, the mesh points are concentrated toward the wall surface and approximately 8 mesh points are imposed within viscous boundary layer.

Simulation Method

Finite volume method is applied to the present simulation. The phenomena inside of the pulse tube can be treated as an axisymmetric flow. Primarily, simple two-dimensional Navier-Stokes equations are addressed to explain the procedure of differential formulation, and then the formulated governing equations are extended to axisymmetric two-dimensional by finite volume topology.

An implicit flux split difference approximation is applied to the Navier-Stokes equations. The flux split difference approximation has an advantage in describing the compressible flow. Gauss-Seidel line relaxation is used to solve the discretized Navier-Stokes equations. This numerical approach is popular method in the field of treating high compressible flow[3)].

The two dimensional Navier-Stokes equations are written at Cartesian coordinate as follows:

$$\frac{\partial U}{\partial t}+\frac{\partial F}{\partial x}+\frac{\partial G}{\partial y}+viscous\ terms=0 \qquad (1)$$

where
$$U=\begin{pmatrix}\rho\\ \rho u\\ \rho v\\ e\end{pmatrix},\quad F=\begin{pmatrix}\rho u\\ \rho u^2+P\\ \rho v u\\ (e+P)u\end{pmatrix},\quad G=\begin{pmatrix}\rho v\\ \rho u v\\ \rho v^2+P\\ (e+P)v\end{pmatrix} \qquad (2)$$

where $e = \varepsilon/\rho - 0.5(u^2 + v^2)$, ρ is density, P is pressure, u and v are velocity in x and y directions, $\gamma = Cp/Cv$ is specific heat.

In order to close the equations, the assumption of perfect gas is made. The perfect gas equation of state is $P = (\gamma - 1)\rho\varepsilon$, $\varepsilon = CvT$ is internal energy. The working gas is Helium in this study.

BOUNDARY CONDITION

A numerical simulation requires boundary conditions for solving the governing equations. Because of its great influence on the results of simulation, the practical velocity profile must be specified as the tube inlet boundary condition. The analytical solution of the Navier-Stokes equations for pulsating flow in an open end tube is obtained in terms of Bessel function [4].

The heat conduction is also important for heat pumping mechanism. The heat conduction wall model is applied. The wall temperature is determined by the heat conduction from the gas temperature.

SIMULATION RESULTS AND DISCUSSION

The first step of this numerical simulation is the verification of the presented simulation method and simulation model whether it is appropriate to describe the gas dynamical phenomena inside of a pulse tube. Throughout the whole pulse tube simulation, the time variation of velocity profiles in three locations, the cold end and middle of tube, and hot end, are shown in Figure 2.

The phase difference in oscillating flow between the tube center and nearby the tube wall is observed by the simulation. The time variation of temperature profile is also shown in Figure 2.

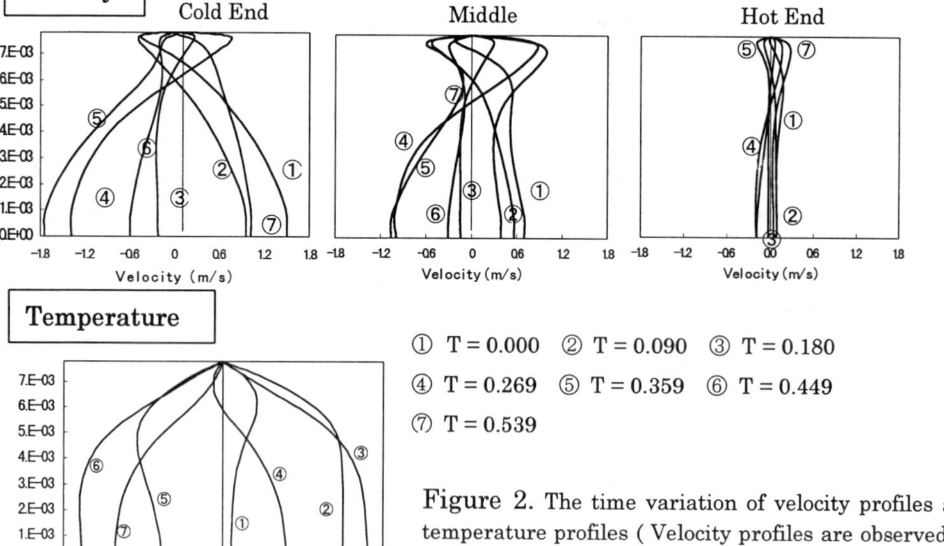

Figure 2. The time variation of velocity profiles and temperature profiles (Velocity profiles are observed on the cold end and middle of the pulse tube, and hot end)

The temperature profiles inside of a pulse tube with time increment are available by the experimental research, which had been carried out at Mechanical Engineering Laboratory MITI[5),6),7)]. The simulation results are compared with the experimental data to verify the simulation method and simulation model. The operational cycle is 2Hz which is the best performance operational frequency of this pulse tube. The comparison of simulation and experimental gas temperature profiles with time step is shown in Figure 3.

Very similar contour maps of gas temperature profile are observed at every instance. The comparison with the experimental results warrants the simulation method and the simulation model, because the results are qualitatively and quantitatively in a good agreement. The simulation well describes the gas dynamics inside of a pulse tube.

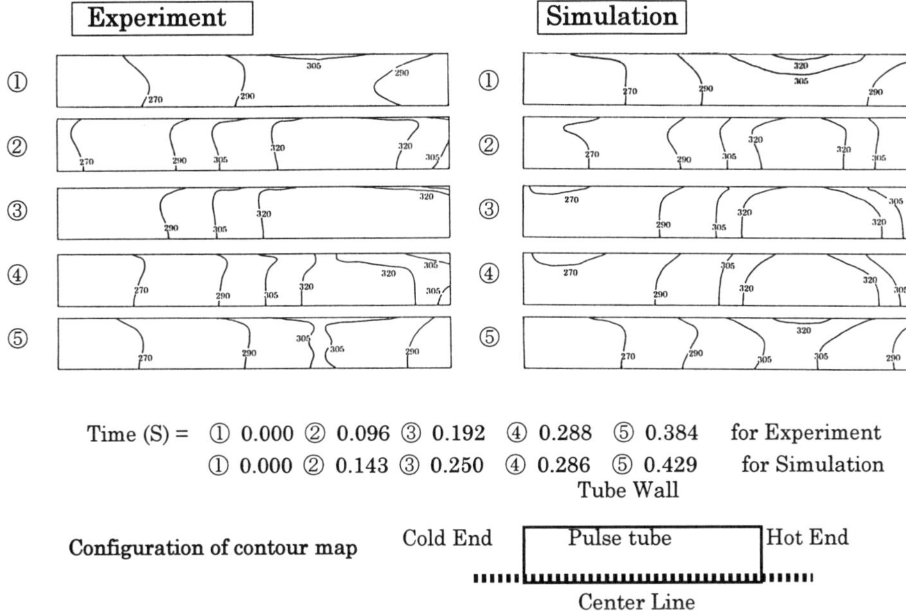

Figure 3. Time variation of temperature contour maps in the pulse tube (Operation Frequency = 2Hz)

The heat exchangers are installed at the both ends of the tube. The hot end heat exchanger releases gas heat out of the pulse tube and the temperature does not rise so much during compression process at the hot end. In the expansion process ($t = 0.384$ experiment, $t = 0.429$ simulation), some local parts of gas temperature fall quickly compared with the surrounding at closed region of the boundary layer. This result suggests that some thermodynamic phenomena happen in the wall boundary layer and the closed region of outside boundary layer. The temperature variation at cold end and hot end at start up is shown in Figure 4

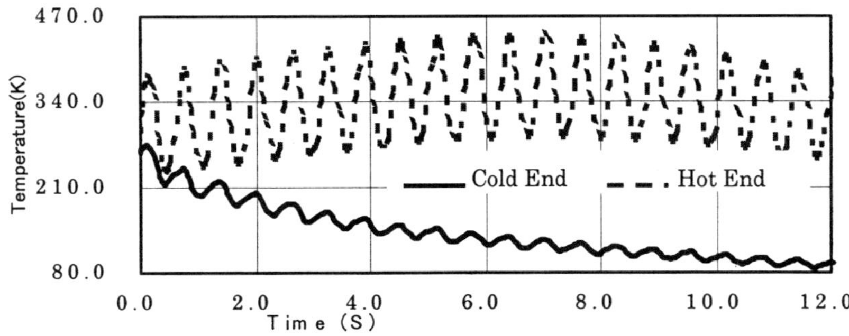

Figure 4. Temperature variation at cold end and hot end at start up process

SECONDARY MASS FLUX

In the basic type pulse tube, the heat transfer by the surface heat pumping is supposed to be an important refrigerating mechanism. On the contrary, the surface heat pumping mechanism is an obstacle to improve refrigeration performance in the orifice type and double inlet type pulse tube refrigerator. There is convective heat transfer within the pulse tube, which carries heat from the hot heat exchanger to the cold heat exchanger and thereby reduces the net cooling power. In the pulse tube, this convective driving can occur in the oscillatory boundary layer at the solid wall of the pulse tube. In the cyclic expansion and compression process, the gas elements close to the wall experience different viscous drag by the alternation of gas temperature and experience a net drift form the cold end to the hot end. This means that gas element around the boundary layer moves from cold end to hot end (called as the secondary mass flux). Since the net mass flux along the tube must be zero, the streaming is introduced in the center part of pulse tube to cancel out the mass flux.

The particle path of working gas element and temperature difference between gas elements and tube side wall are evaluated in order to investigate the heat transfer. The secondary mass flux and the streaming are observed by tracing particle path of gas elements in Figure 5.

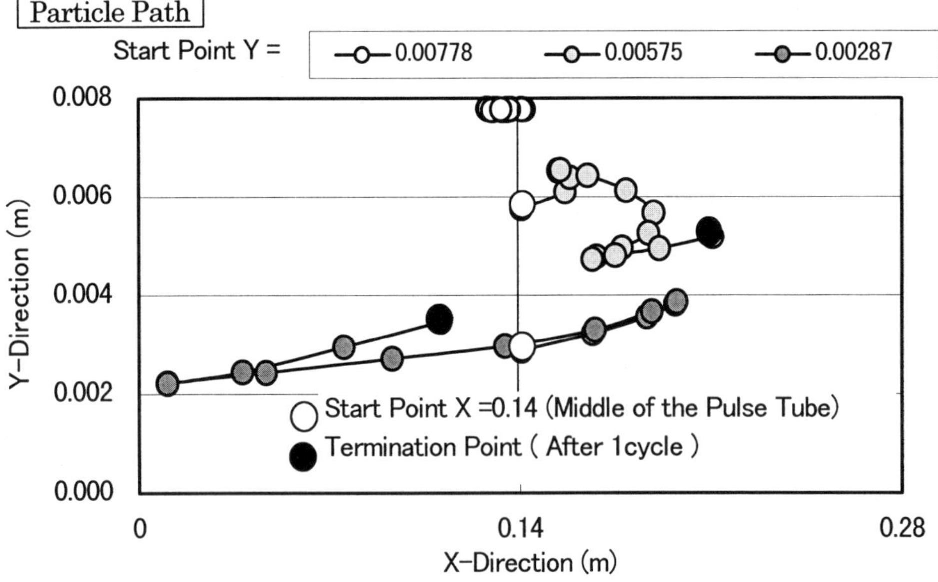

Figure 5. Particle path and Heat transfer

TIME AVERAGED COMPONENTS

The time averaged flow components are show in Figure 6 to Figure 8. The components are obtained by integrating these variables for the time of one compression – expansion process. The time averaged temperature profile of working gas inside of the pulse tube is shown in Figure 6. The closed end is heated up and the open end is cooled down. The temperature difference is obtained by taking the time averaged components.

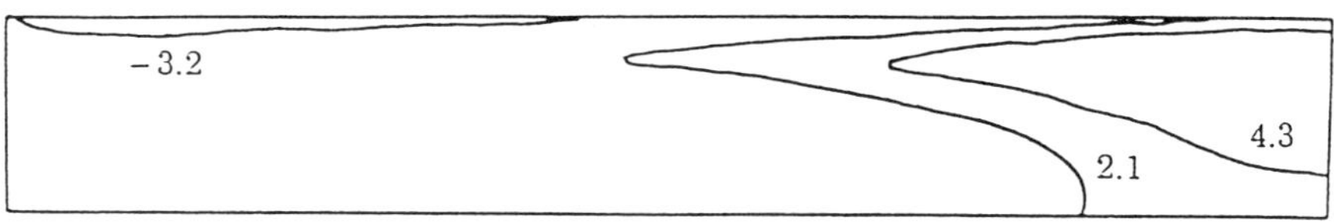

Figure 6. Time averaged of temperature profile in the pulse tube
(ΔT form ambient temperature. 1 cycle time average at 3 cycles after start up)

The time averaged mass flow, shown in Figure 7, passes trough the closed region of the outside of wall boundary layer and goes toward the hot end from the cold end. This results are in agreement with particle path tracing.

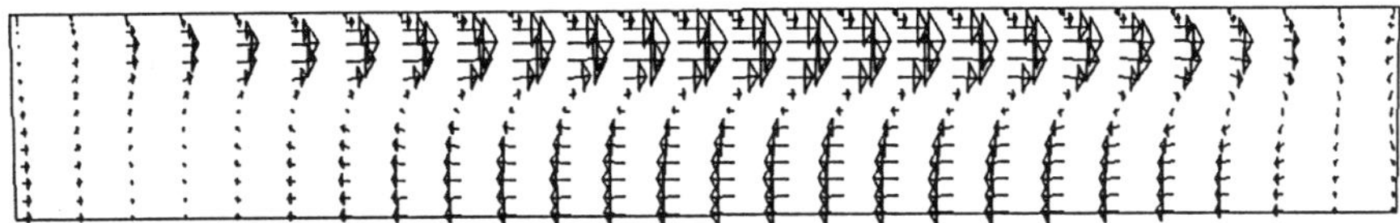

Figure 7. Secondary mass flux (Time averaged mass flow)

The energy flux vector distribution shows that the energy is transferred from the cold end to the hot end in Figure 8.

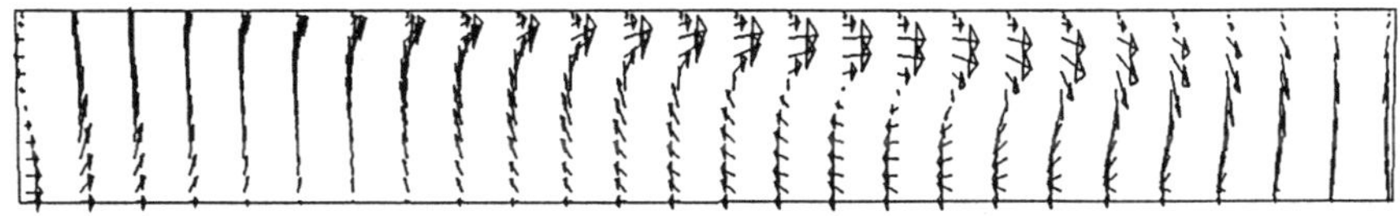

Figure 8. Time averaged energy flux

SURFACE HEAT PUMPING

Since the gas elements, inside of the boundary layer whose temperature is influenced by the heat conduction with the tube wall, move only very small distance along the wall, the conventional pumping mechanism could not be observed by the simulation. However, the energy flux vector distribution suggests that the enthalpy flows in the region close to the outside boundary layers from the cold end to the hot end. The energy vector is enlarged in this region. This phenomenon is also observed in the time averaged temperature profile. The high temperature region extends from the hot end to the cold end in this region. In the region closed to the outside boundary layers, the secondary mass flows toward hot end and the temperature is higher than that of surrounding region. Since the heat diffusion is not larger than the heat transfer by the convection, this temperature profile is not caused by the thermal diffusion but that is caused by the energy transfer toward the hot end from cold end. The simulation results suggest that the heat conducting buffer for the surface heat pumping would not be only the tube wall but the tube wall and boundary layer. The direction of heat conduction between the gas and wall on the particle path is shown in Figure 9.

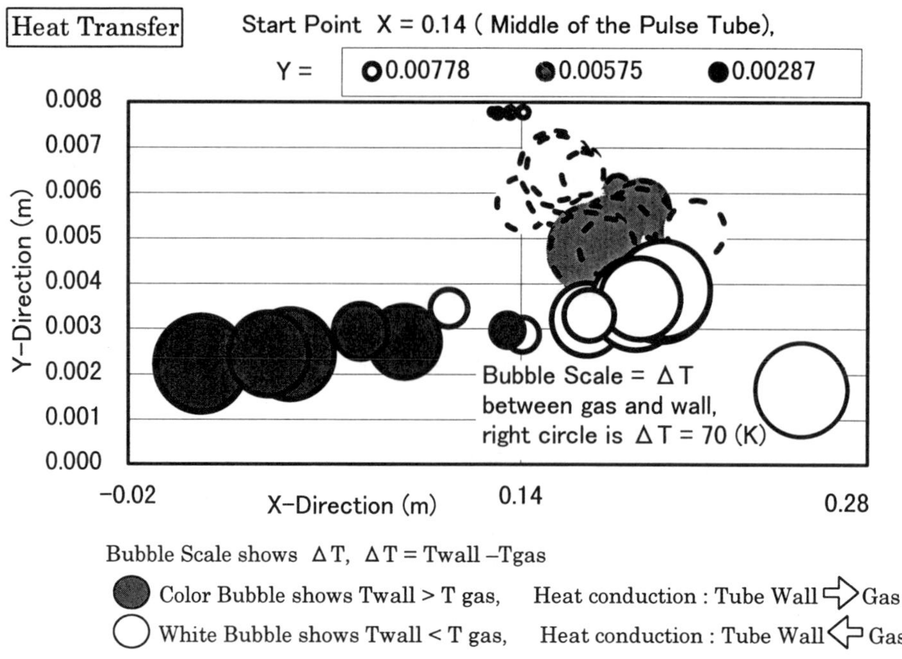

Figure 9. Heat transfer between working gas and wall

CONCLUDING REMARKS

The simulation program for the viscous compressible flow is developed to study the detail of fluid motion and gas dynamics in a pulse tube. The simulation results suggest that the boundary layer on the tube wall might play an important role as to the heat transfer in basic type pulse tube. The tube wall and boundary layer are thermal conducting buffer to transfer heat from the cold end to the hot end. Since the program is applicable for study of boundary layer flow in a regenerator, future work will incorporate for whole pulse tube refrigerator modeling.

REFERENCES

1) W.E.Gifford and R.C. Longsworth "Surface Heat Pumping" Advances in Cryogenic Engineering, Vol.11 (1965) 171

2) J.M.Lee and P.Kittel "Higher Order Pulse Tube Modeling" Cryocoolers, 9 (1997) 345

3) R.W. MacCormack and G.V.Candler, "The Solution of The Navier-Stokes Equations using Gauss-Seidel Line relaxation" Computer & Fluid Vol. 17 No.1 (1989) 135

4) A. F. D'Souza, and Oldenburger, Tran. ASME, D, 86-4 (1964-9) 589.

5) M.Shiraishi and N.Nakamura, and M.Murakami, "Visualization Study of Velocity Profiles and Displacements of Working Gas Inside a Pulse Tube Refrigerator" Cryocoolers, 9 (1997) 355

6) M.Shiraishi et al., "Visualization Study of Flow Fields in a Pulse Tube Refrigerator" Advances in Cryogenic Engineering, Vol.43 (1998)

7) M.Shiraishi et al., "Start-up Behavior of Pulse Tube Refrigerator" Advances in Cryogenic Engineering, Vol.41 (1996) 1455

THEORETICAL ANALYSIS OF REFRIGERATION AND LOSSES IN A PULSE TUBE

L.W Yang, Y. Zhou, J.T Liang

Cryogenics Laboratory, Chinese Academy of Sciences
Beijing P. O. Box 2711, 100080, P. R. China

ABSTRACT

Based on thermal analysis of hot end of pulse tube, a group of formula is deduced to calculate the theoretical refrigeration, which reveals that refrigeration is mainly determined by parameters of the pulse tube hot end. Then three important losses in the pulse tube that may reduce refrigeration are analyzed. They are jet flow loss due to gas out of and into the reservoir, surface heat pumping loss or shuttle loss due to heat transfer between tube wall and gas in tube, and gas mixing due to cyclic flow or density difference. The gas mixing loss always exists and is difficult to eliminate; but jet flow loss may be reduced. Jet flow loss is verified by experiment. Analysis of surface heat pumping shows that the ratio of high and low temperature is one key parameter. In the basic pulse tube refrigerator, the amplitude of surface heat pumping decreases till zero with temperature ratio increasing, while for the orifice version and low temperature, this heat pump effect is from high temperature to low temperature evidently.

This paper concludes: refrigeration is mainly determined by hot part of pulse tube, while the pulse tube is one key loss source though it is an irreplaceable part in the refrigerator.

INTRODUCTION

Pulse tube refrigerators have developed quickly during the last ten years. And as the term suggests, pulse tube is a unique part when compared with other refrigerators. The uniqueness of pulse tube cooler makes actual designing difficult.

The process in a pulse tube is complex. On one side, the flow in a pulse tube is cyclic, it is concerned with boundary layer, laminar and turbulent flow, secondary flow. On the other hand, thermodynamic effects and heat transfer are deeply involved in pulse tube due to refrigeration mechanism and large temperature difference. These problems have been recognized, theoretical analysis using numerical method and some special experiments have been done.

In this paper, some flow-heat-combined problem and phenomena are revealed. An important viewpoint is: once orifice and reservoir are used in pulse tube refrigerator, pulse tube is of little refrigeration function but becomes one main loss source [1].

Advances in Cryogenic Engineering, Volume 45.
Edited by Shu *et al.*, Kluwer Academic / Plenum Publishers, 2000.

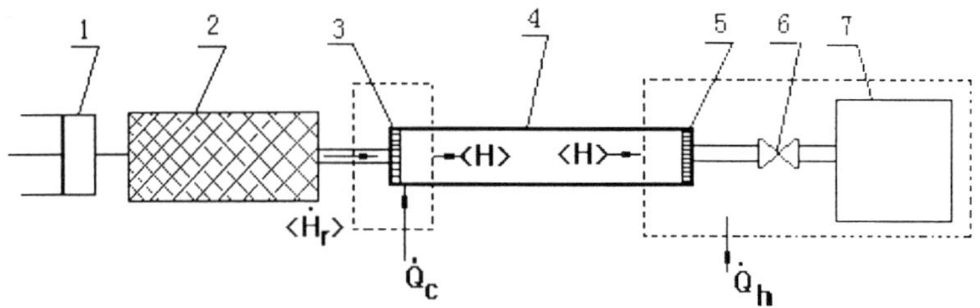

Figure 1. Structure of pulse tube refrigerator and enthalpy flow. Main parts are: 1. compressor, 2. regenerator, 3. cold tip, 4. pulse tube, 5. hot end, 6. orifice valve, 7. reservoir.

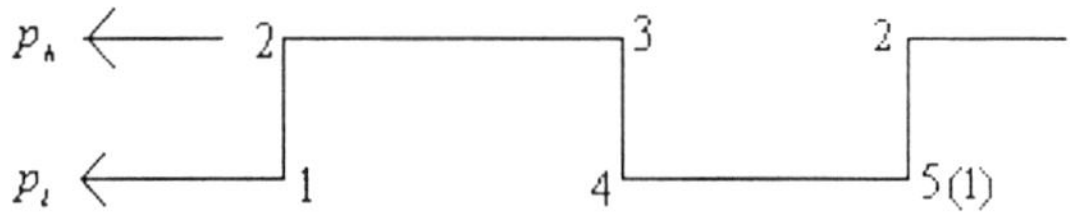

Figure 2. The typical square pressure wave in analysis.

REFRIGERATION MECHANISM

Enthalpy Flow And Refrigeration Power

The system to analyze refrigeration is shown in Figure 1 and it consists of working fluid in the tube. Due to the oscillating gas flow, enthalpy flows are time-averaged over one cycle. Applying the first law of thermodynamics to the cold end of the tube results in the following equation as for the heat absorbed by the system [2]:

$$\dot{Q}_c = <\dot{H}> - <\dot{H}_r> \tag{1}$$

where $<\dot{H}>$ is average enthalpy flow in the tube and $<\dot{H}_r>$ is average enthalpy flow from the regenerator.

According to Figure 1, since there is no heat transfer to outside along pulse tube, the first law of thermodynamics says that the heat rejected by hot end of pulse tube, orifice valve and reservoir is:

$$\dot{Q}_h = <\dot{H}> \tag{2}$$

Then:

$$\dot{Q}_c = \dot{Q}_h - <\dot{H}_r> \tag{3}$$

Eq. (2, 3) shows: (1) Refrigeration is mainly determined by hot part of pulse tube. (2) Acquiring $\dot{Q}_h$ through analyzing hot end thermal process is an effective method to get gross refrigeration.

Ideal Heat Rejection

For convenience, consider ideal square wave as shown in Figure 2, which includes four processes, compression process 1-2, isobaric process 2-3, expansion process 3-4, and isobaric 4-5. These four processes form a whole ideal cycle.

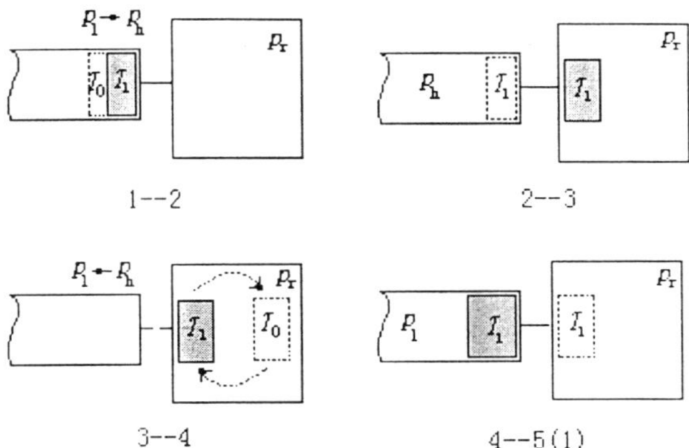

Figure 3. Adiabatic processes analysis with reservoir. If there is no heat transfer, energy is collected at hot end of pulse tube or reservoir, and its amplitude is the theoretical refrigeration according to Eq. (3).

The typical thermal processes are shown in Figure 3, assuming process 1-2 is adiabatic compression.

In compressing process 1-2, change of temperature at hot end is:

$$T_1 = T_0 \cdot (p_h / p_l)^{(\kappa-1)/\kappa} \tag{4}$$

T_0 is the initial temperature of gas in pulse tube hot end at pressure p_l, and T_1 is the temperature after compression at pressure p_h.

In process 2-3, some gas in hot end of pulse tube flows into reservoir and pressure $p_h \to p_r$, assuming this gas mass is m, its temperature stays T_1 due to throttling process.

In process 3-4, pressure in pulse tube drops to p_l, and in reservoir, m, T_1 keep unchanged and do not mix with other gas.

In process 4-5, gas m returns to pulse tube and T_1 keeps unchanged.

A whole cycle finishes. We find that in this ideal cycle, the gas temperature of pulse tube hot end m changes from T_0 to T_1.

After another such ideal process, T_1 will change to T_1':

$$T_1' = T_1 (p_h / p_l)^{\kappa-1/\kappa} = T_0 (p_h / p_l)^{2(\kappa-1)/\kappa} \tag{5}$$

In fact, the temperature of hot end gas m is not necessary and could not increase continuously. As shown in Figure 3 (3-4), supposing m, T_1 in reservoir is supposed to have a good heat transfer to make its state change to m, T_0. For example, other gas in reservoir with the state of m, T_0 may replace it and return to pulse tube hot end. Then, a whole cycle will not lead to a continuous increase of internal energy of m. In actual condition, mixing of T_1, T_0 will result in a temperature rise in reservoir: $T_0 + \Delta T$. This ΔT keeps a stable and effective heat transfer between reservoir and atmosphere.

When a steady state is reached, hot end gas temperature at low pressure and temperature in reservoir should be T_0, the atmosphere temperature should be $T_0 - \Delta T$. Heat rejected to atmosphere per cycle under ideal condition should be:

$$Q_h = h \cdot S \cdot \Delta T = m \cdot c_v (T_1 - T_0) \tag{6}$$

This heat equals to compression work of gas mass m:

$$w = m \cdot \frac{R}{\kappa - 1}(T_1 - T_0) = m \cdot \frac{RT_0}{\kappa - 1}[(p_h / p_l)^{\frac{\kappa-1}{\kappa}} - 1] \tag{7}$$

With above analysis, heat transferred to atmosphere in hot end per unit time is:

$$\dot{Q}_h = f \cdot c_v \cdot m \cdot (T_1 - T_0) \tag{8}$$

The above analysis shows that the reservoir and orifice accumulate heat at the hot end of the pulse tube. This is the essence of refrigeration when orifice and reservoir exist.

Eq. (7) resembles more to actual condition and is proposed by the authors for the first time. If isothermal compression assumption is adopted for hot end gas m, a different formula that is larger than Eq. (7) will be acquired [1, 3].

Theory Refrigeration

The considerations above give an ideal result for refrigeration. In practice, pressure wave may be a random form, then differentiating and integrating will get the result.

Assuming flow coefficient of orifice is c_1, flow rate of orifice is:

$$\dot{m}_1 = c_1 X(p, p_r) \tag{9}$$

We can divide hot end gas that flows in and out of reservoir into small parts $1, 2, \cdots, i, \cdots N$. Each portion of gas will flow into the reservoir at a certain time, and when all have flowed into the reservoir, they will return to the pulse tube hot end one by one. For each of them, assume that pressure is p_h^i, mass Δm_i and time Δt_i when it flows into the reservoir. On the other hand, when the Δm_i returns to the pulse tube hot end, the pressure at this hot end of pulse tube is p_l^i. Based on this supposition, according to Eq. (7):

$$dq_i = \Delta m_i \cdot \frac{RT_0}{\kappa - 1}[(p_h^i / p_l^i)^{\frac{\kappa-1}{\kappa}} - 1] \tag{10}$$

And for a whole cycle:

$$Q_h = \sum_{i=1}^{N} dq_i = \sum_{i=1}^{N} \cdot \frac{RT_0}{\kappa - 1}[(p_h^i / p_l^i)^{\frac{\kappa-1}{\kappa}} - 1]\Delta m_i \tag{11}$$

And for unit time:

$$\dot{Q}_h = f \sum_{i=1}^{N} \cdot \frac{RT_0}{\kappa - 1}[(p_h^i / p_l^i)^{\frac{\kappa-1}{\kappa}} - 1]\Delta m_i \tag{12}$$

$$= fc_1 \cdot \frac{RT_0}{\kappa - 1} \sum_{i=1}^{N} [(p_h^i / p_l^i)^{\frac{\kappa-1}{\kappa}} - 1] \cdot X(p_h^i, p_r)\Delta t_i \tag{12a}$$

Mass flows into reservoir and its volume are:

$$m = \sum_{i=1}^{N} \Delta \dot{m}_i = c_1 \sum_{i=1}^{N} X(p_h^i, p_r)\Delta t_i = c_1 \int_0^{\tau/2} X(p, p_r)dt \tag{13}$$

$$m = p_r V_m / RT_0 \tag{14}$$

$$c_1 = \frac{p_r V_m}{RT_0} \Big/ [\sum_{i=1}^{N} X(p_h^i, p_r)\Delta t_i] = \frac{p_r V_m}{RT_0} \Big/ \int_0^{\tau/2} X(p, p_r)dt \tag{15}$$

If c_1 in Eq. (12a) is replaced by V_m, then:

$$Q_h = f \cdot \frac{p_r V_m}{\kappa - 1} \sum_{i=1}^{N} [(p_h^i / p_l^i)^{\frac{\kappa-1}{\kappa}} - 1] \cdot X(p_h^i, p_r)\Delta t_i \Big/ [\sum_{i=1}^{N} X(p_h^i, p_r)\Delta t_i] \quad (16)$$

$$= f \cdot \frac{p_r \varphi V_{pt}}{\kappa - 1} \sum_{i=1}^{N} [(p_h^i / p_l^i)^{\frac{\kappa-1}{\kappa}} - 1] \cdot X(p_h^i, p_r)\Delta t_i \Big/ [\sum_{i=1}^{N} X(p_h^i, p_r)\Delta t_i] \quad (17)$$

where φ, here called reservoir coefficient, is defined to be:

$$\varphi = V_m / V_{pt} \quad (18)$$

and V_{pt} is volume of pulse tube.

Discussion

With orifice flow and pressure wave determined, gross refrigeration can be obtained easily through Eq. (16). This is the most important analysis result. On the other hand, if pressure wave is confirmed, for example, to be typical sinusoidal wave or trapeze wave, effect of pressure wave form and parameters can be acquired and expressed easily.

The meaning of φ is to show that flow amount through orifice is mainly limited by the structure. On the other hand, with φ determined, heat transfer along pulse tube produced by inconvenient flow will be easily analyzed. For example, surface heat pumping is evidently a loss; jet flow of gas in hot end will result in direct loss.

In fact, φ is an important parameter to calculate refrigeration, at the same time a parameter to evaluate the amplitude of losses.

LOSSES ANALYSIS

Jet Flow Loss

Just now, we calculate the hot end heat rejection amount. But this hot end heat may be reduced by unwanted gas flow. The first place considered is pulse tube hot end gas into and out of reservoir, where there is a streamline device and it generally includes tens of copper screens.

Let's consider the actual flow at hot end of pulse tube. According to fluid mechanics, Figure 4(A) gives the potential jet flow at hot end of pulse tube due to sudden change of flow area. The orifice has a very small diameter, the largest velocity through it may be very fast:

$$v_{max} = \sqrt{2(p_h - p_l)/\rho_h} = \sqrt{2 \times 200000/4} = 300\text{m/s} \quad (19)$$

For such a velocity, center of high velocity gas may go further into the pulse tube, which is just like jet flow in fluid mechanics. And apart from the center of orifice, vortex flow may occur. The final result will be gas flowing further into the pulse tube, and in the next cycle, other gas will flow into reservoir, though total amount is same. This means that actual gas volume or length in pulse tube hot end that flows into the reservoir may be larger than determined by φV_{pt}.

For a pulse tube with a large temperature difference, when reservoir gas flows further into pulse tube and gas mix happen, it results in a gas temperature T_1 in Eq. (4) that will be lowered.

Assuming the gas temperature T_1 flowing into the reservoir drops to $T_1 - \Delta T$, then the loss is:

$$\Delta Q = c_v \cdot M_{in} \cdot \Delta T \quad (20)$$

In order to verify jet flow loss, two special devices are designed and experiments are carried out, as shown in Figure 4(B), (C).

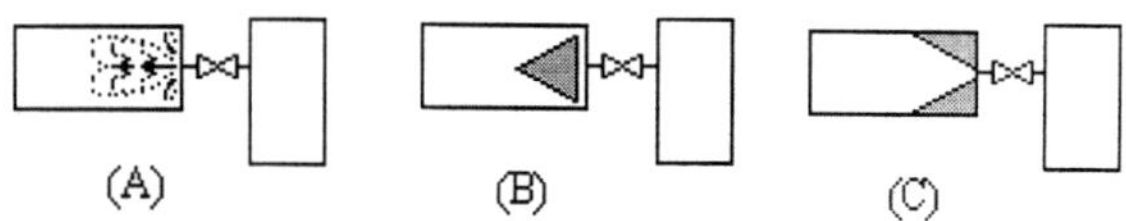

Figure 4. Several structures to verify jet flow loss effect. (A) is with no streamline device and jet flow is evident, (B) is a special flow channel designed to eliminate jet flow, (C) is a streamline device to eliminate hot end corner vortex while jet flow still exist.

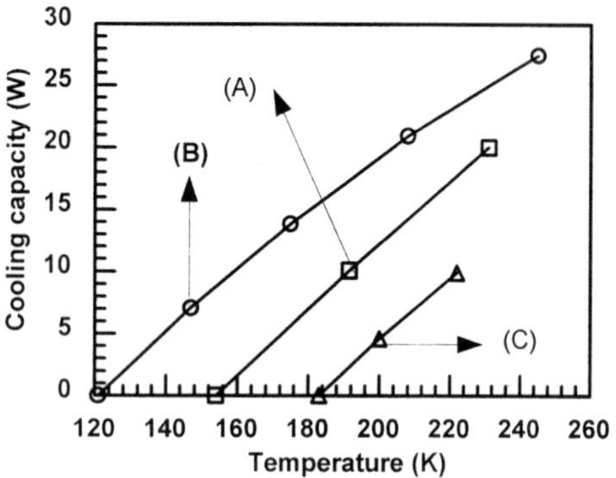

Figure 5. Experiments with 20mm diameter Pulse tube refrigerator

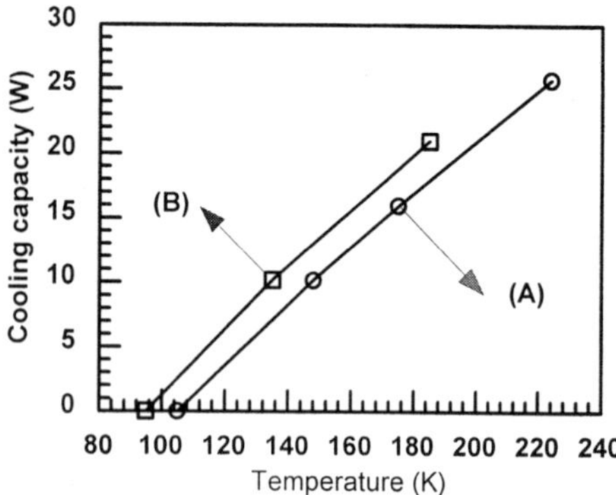

Figure 6. Experiments with 14mm diameter pulse tube refrigerator

In experiments, $55\,cm^3$ swept volume compressor and two pulse tube coolers are used. One pulse tube is of 20mm inside diameter and 150mm long, another is 14mm inside diameter and 150mm long, while regenerator is 18mm inside diameter and 150mm long, filled with 60 mesh stainless steel screen.

Figure 5 and 6 show the results of experiments. Evidently, structure (B) has lower temperature than structure (A), while structure (C) has the highest refrigeration temperature. The above results show that improvement of structure (B) is effective, and jet flow loss is large.

Surface Heat Pumping

In the pulse tube, gas oscillates and contacts with different part of the pulse tube wall. This effect is attributed to surface heat pumping, refrigeration mechanism of basic version pulse tube refrigerator. But for orifice-version, this process may result in a loss.

First let's consider temperature distribution in pulse tube. For convenience, assume temperature distribution in the pulse tube has exponent form:

$$T = T_h - (T_h - T_c)x^n / x_c^n \tag{21}$$

As shown in Figure 7, assuming $T_h = 300\,\mathrm{K}$ and $T_c = 30\,\mathrm{K}$, temperature change at different exponent n is given. One typical simulation result for basic version pulse tube is shown in Figure 8. In Figure 8, n=0.5 and temperature change with pressure for certain gas element at average position x=0, 0.2, 0.4, 0.7 is given.

When pressure increases, gas temperature will increase and gas will go forward to hot end of pulse tube. According to surface heat pumping mechanism, at high pressure, gas temperature is higher than wall temperature and gas transfer heat to wall. At low pressure, gas temperature is lower than wall temperature and gas absorbs heat from wall. Final result is pumping heat from low temperature to high temperature.

Evidently, for x=0, and x=0.2, gas temperature is lower than wall temperature before compression, and gas temperature is higher than wall temperature after compression. For such a condition, refrigeration due to surface heat pumping will occur.

But for position x=0.4, gas temperature and wall temperature are almost the same. At this point, there will be little heat transfer. At this position, $T \approx 130$K.

At position x=0.7, gas temperature is higher than wall temperature before compression, and gas temperature is lower than wall temperature after compression. For such a condition, heat will be transferred from high temperature to cold temperature. Refrigeration is impossible.

Figure 8 shows that temperature distribution is key to refrigeration in basic version pulse tube refrigerator. Further analysis shows that in order to reach 30K, n=0.074 is needed, but such a temperature distribution is impossible as shown in Figure 7.

From Figure 8, when temperature is lowered to a certain point, surface heat pumping to high temperature will terminate for a normal temperature distribution. This is key to basic version pulse tube and is proposed here for the first time. We know that single-stage basic version pulse tube refrigerator has the lowest temperature of 124K. Figure 8 explains the reason that a lower temperature than 124K is not easy or not possible to achieve.

Adding an orifice and reservoir will make an obvious difference. In Figure 9, an interesting result is shown, the difference to Figure 8 is that some gas in the pulse tube flows into the reservoir.

First, let's look at gas temperatures at any position for any part of gas. Obviously, every part of gas performs a circle, and one high and one corresponding low temperature at each position are acquired. This means refrigeration happens always without surface heat pumping.

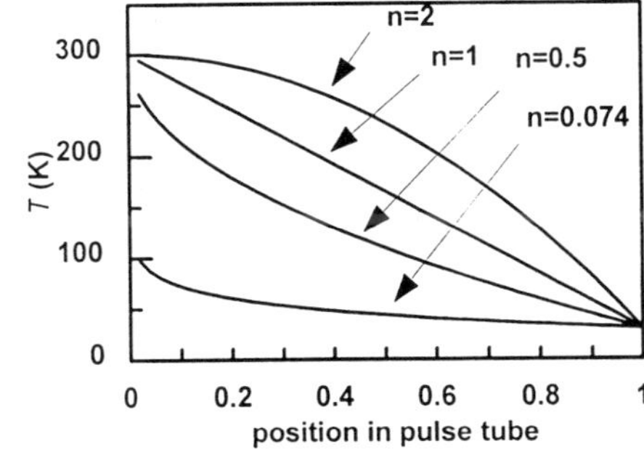

Figure 7. Gas temperature distribution in pulse tube

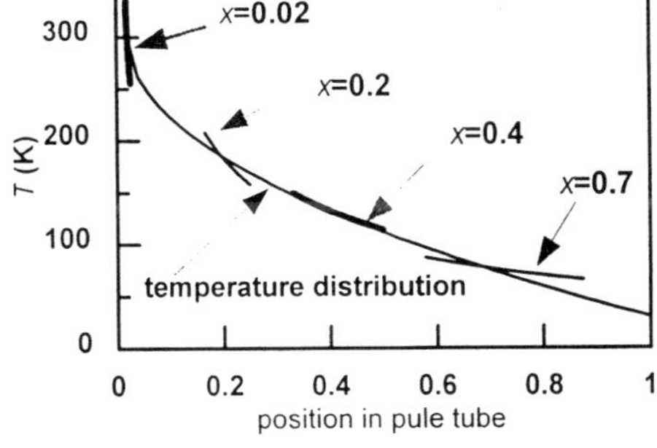

Figure 8. Surface heat pumping in basic version pulse tube refrigerator

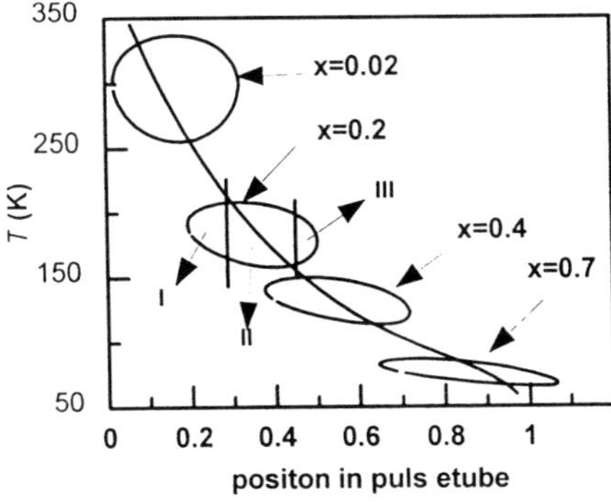

Figure 9. Surface heat pumping in orifice version pulse tube refrigerator

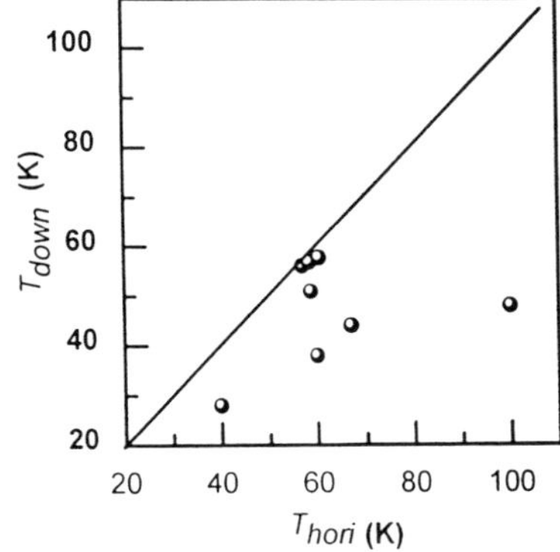

Figure 10. Effect of direction on performance of pulse tube refrigerator

Then let's look at the function of surface heat transfer. When comparing Figure 9 with Figure 8, we find that the gas moving some distance is different. In the basic version, the hot end gas moves a small distance and the cold tip gas moves a much larger distance, while for the orifice version, both hot end and cold end gas move over a rather larger distance.

Take x=0.2 as an example, the circle of gas path can be divided into three parts according to two joints of gas wall temperature and gas temperature. First part I is left to left joint. In this part, all gas temperature is lower than wall temperature and tube wall will transfer heat to gas. Second part II is between two joints. During one cycle, the gas temperature is higher than the wall at one time and lower than the wall at the other time. Gas may transfer heat to the wall and then absorb it, though the net heat may not be zero. Third part III is right to right joint. In this part, all gas temperature is higher than wall temperature and wall absorb heat. Addition of the three parts heat transfer effects shows that heat transfer between wall and gas results in a transfer of heat to low temperature continuously. Refrigeration mechanism in the basic version pulse tube refrigerator becomes an evident loss. When temperature difference is small, gas part I and III will become short and this loss becomes small.

Mechanism of surface gas pumping loss in pulse tube is explained for the first time.

Mixing Loss

This loss is due to gas characteristics. First, mass transfer and heat transfer between gas in pulse tube are unavoidable. Second, due to quick oscillating move, mass transfer will be enhanced. Third, due to density difference between gas temperature, convection heat transfer is unavoidable, too. And these losses are difficult to measure or to detect.

For example, due to density difference in pulse tube, the refrigeration temperature T_{hori} (pulse tube is placed horizontal) is evident higher than T_{down} (pulse tube cold tip is placed upright). In Figure 10, comparisons of results of several pulse tube coolers of our low frequency experiments are given. T_{down} is always smaller than T_{hori}, difference and uncertainty are large.

One point worth mentioning is that, all upper losses in pulse tube make temperature T_1 in equation (4) smaller and thus Q_h smaller, while Eq. (2, 3) is tenable always.

CONCLUSION

As to orifice version pulse tube refrigerator, a group of new formula about theory refrigeration and its amplitude are deduced. Then three losses in pulse tube are analyzed. Special mechanism of surface gas effect loss is explained for the first time.

ACKNOWLEDGEMENT

This research is supported by National Superconductor Center of China and National Natural Science Foundation of China.

REFERENCES

1. L.W Yang, *Research of pulse tube refrigeration mechanism and practical development*, Post-doctor report, Chinese Academy of Sciences (1998)
2. R. Radebaugh, J. Zimmerman, D. R. Smith, and B. Louie, *Adv. Cry. Eng.*, 31: 779 (1986)
3. Y. Matsubara, A. Miyake, *Cryocooler5:* 127 (1988)

EXPERIMENTAL INVESTIGATION ON PULSE TUBE REFRIGERATOR WITH BINARY MIXTURES

G. B. Chen, J. P. Yu, Z. H. Gan, T. Jin

Cryogenics Laboratory, Zhejiang University
Hangzhou, 310027, P.R.China

ABSTRACT

Pulse tube refrigerators enjoy advantages over most other types of refrigerators in reliability, lifetime, vibration and cost. It can be a candidate for cryogenic cooling in space cryogenics. Given the structural and operating parameters, we think the adoption of working fluids is a way to improve the refrigerating performance. We focus our interest in gas mixtures, which have been proved very useful in refrigeration, natural gas liquefaction and J-T cryocooler, due to the advantage of high COP.

An experimental apparatus has been constructed to probe into the possibility of using binary gas mixtures instead of helium to enhance the performance of pulse tube refrigerators in the 80K temperature region. Detailed experimental data for helium-hydrogen mixtures are given. Preliminary experimental results show a 10% gain of cooling capacity with an 80%He-20%H_2 mixture could be obtained, compared to that when using pure helium.

INTRODUCTION

To meet the requirement of various high-tech applications, cryocoolers should possess the following characteristics: (1) long-term non-maintenance operation and reliability; (2) minimum size and weight with the expected cooling capacity. The pulse tube refrigerator is an attractive candidate, for it has no moving component in the cold region, as do traditional coolers, and has few problems with cold head vibration and abrasion, etc., which provide it immense potential in space applications. The key point for the latter is to improve the efficiency of the cryocooler in the expected temperature region. Optimizing the structure and adopting novel regenerative materials are generally effective ways. On the other hand, it is worth studying the possibility of applying mixtures to a regenerative refrigerator for higher refrigeration capacity in the 80 K temperature region.

Advances in Cryogenic Engineering, Volume 45.
Edited by Shu *et al.*, Kluwer Academic / Plenum Publishers, 2000.

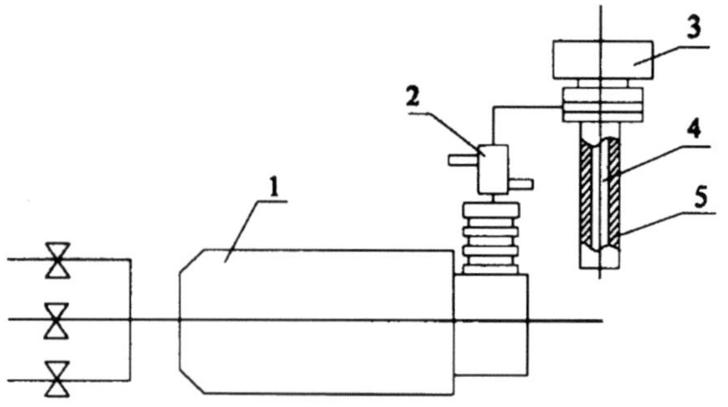

Figure 1. Schematic of the experimental set-up.
1. Compressor 2. After-cooler 3. Reservoir 4. Pulse Tube 5. Regenerator

Mixtures have been widely used in normal refrigeration, natural gas liquefaction, J-T cooler and so on. They provide the advantages of high COP, quick cooling down and large cooling capacity. Unfortunately, few experimental results have been reported of using gas mixtures to improve the refrigeration performance of regenerative cryocoolers such as for the pulse tube refrigerator.

Since helium has relatively higher thermal conductivity and lower viscosity compared with other working fluids, it is always regarded as the best choice of working medium for regenerative refrigerators. Based on our previous analysis[1-4], however, the regenerator performance of heat transfer and fluid flow of some gas mixtures such as helium-hydrogen mixture may be better than that of the pure helium in a regenerative cryocooler in the 80 K temperature region, which is meaningful for improving refrigeration capacity in this region. Furthermore, the cooling capacity of the refrigerator depends upon the adiabatic index or the specific heat ratio of the working medium to some extent. The specific heat of some gases changes with temperatures and pressures, which makes possible a higher specific heat ratio $k(=c_p/c_v)$ in some temperature regions. An experimental apparatus has been set up and helium-hydrogen mixtures of different composition were tested in a coaxial pulse tube refrigerator. Preliminary experimental results show that an improvement of refrigeration capacity in the 80 K region has been obtained.

EXPERIMENTAL APPARATUS

A schematic of the experimental apparatus is shown as Figure 1, including a pulse tube refrigerator, a compressor, a vacuum system and the measuring system. The pulse tube refrigerator used in our experiments is a coaxial one with a minimum refrigeration temperature of 53 K and a cooling capacity of 2 W or so at 80 K. The compressor is a valveless one with 33 cm^3 in swept volume and 40 mm in stroke.

Measuring System

The measuring system consists of temperature and refrigeration capacity measuring. A calibrated rhodium-iron resistance temperature sensor is used to measure the temperature on the cold head. A four-wire measuring method is adapted, the 1 mA measuring current is supplied by the Lakeshore constant current source of model 120CS, and the voltage is displayed on a digital voltmeter of Keithley 177. The temperature at the warm end of the pulse tube is measured with a copper-constantan thermocouple.

The cooling capacity is measured by a compensation method that uses an electricity-heater. Manganin wire of about 450 Ω resistance at ambient temperature is wrapped around the cold head of the pulse tube. The voltage in the heater is supplied by an adjustable constant voltage source. The voltage and the current of the heater determine the cooling capacity at a given working temperature.

Preparation of Gas Mixtures

Dalton's Law is used for preparing the helium-hydrogen mixture in our experiments, i.e., the volume proportion of the component in the mixture equals its proportion of partial pressure. For instance, the procedure of preparing 20%He+80%H_2 is as follows:

(1) After vacuum purging through repetitive replacement with hydrogen, the system is first charged with hydrogen to a pressure of 0.54 MPa (gauge pressure).

(2) Helium is then added until the total system pressure reaches 0.70 MPa (gauge pressure).

Thus, the total absolute pressure in the system is 0.80 MPa, the partial pressure of helium and hydrogen is 0.16 MPa and 0.64 MPa, respectively. Apparently, we obtain the mixture of 20%He and 80%H_2. Gas mixtures of any expected composition can be prepared in the same way.

Experiment

For comparison, experiments with pure helium should be first done. The charging pressure is kept as 0.8 MPa in all experiments. We regulate the opening of the orifice valve and double inlet valve to reach the lowest no-load temperature of the pulse tube refrigerator. Generally, about 35 min is required for the pulse tube refrigerator to reach its minimum temperature.

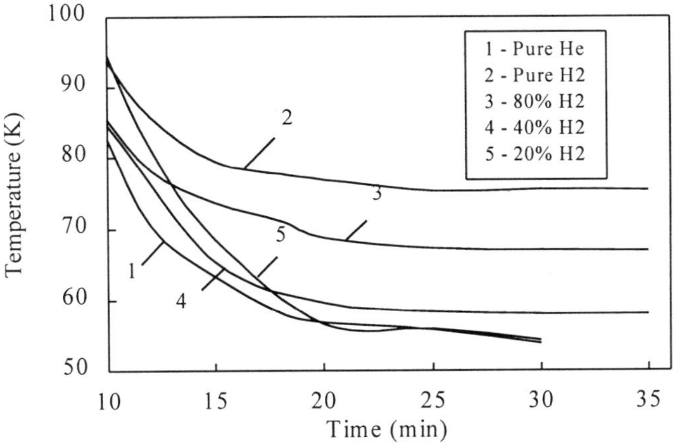

Figure 2. Cooling down process of the pulse tube refrigerator with different refrigerant.

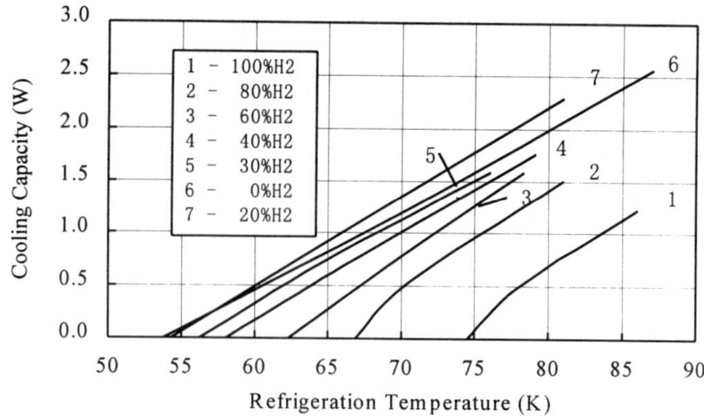

Figure 3. Refrigeration capacity versus temperature for different He-H_2 mixtures.

In our first step of experimental study, we want to compare the performance of the refrigerator with that of pure helium while keeping all the structural and operating parameters unchanged from that which is optimal for helium. We can carry out the experiments with mixtures of different compositions. First, the experiment is done for the machine filled with a mixture and without heat load to determine the minimum refrigeration temperature. Then, a certain value of heat load is added to the cold head at the initial condition at room temperature. When the temperature of the machine under this heat load is unable to decrease, we regard the heat load as the net refrigeration capacity at this certain temperature. Repeating the above procedures, a series of load minimum temperatures of the pulse tube with different loads can be obtained.

EXPERIMENTAL RESULTS

Seven groups of working fluids have been adopted in our experiments, including pure helium, pure hydrogen, and five sets of helium-hydrogen mixtures with different composition, i.e., He-H_2(20%+80%, 40%+60%, 60%+40%, 70%+30%, 80%+20%). The cooling down processes and the minimum temperatures reached are shown in Figure 2. The refrigerator can reach 53.8 K in 30 min with pure helium as the working fluid, and the minimum temperature is 74.5 K for pure hydrogen, while those of the mixtures decrease with the decrease of the hydrogen component in the mixture. When a hydrogen concentration of 20% adopted, the refrigerator reaches a minimum temperature of 54.3 K, which is almost the same level as for pure helium.

Figure 3 demonstrates the relationship of the refrigeration capacity and temperature for pure helium, pure hydrogen and helium-hydrogen mixtures with different compositions. The refrigeration capacity for pure helium is 2.0 W at 80 K, while those for the mixtures increase with the decrease of the hydrogen component in the mixture. Although the minimum temperature for the 80%helium+20%hydrogen is a little higher than that for pure helium, the refrigeration capacity of 2.21 W at 80 K is surprisingly 10% higher than that for helium.

Figure 4 presents the relation between the hydrogen component and the refrigeration capacity at 77 K. From the figure, we can see that the capacity of the mixture with 20% hydrogen is about 10% higher than that of pure helium in the liquid nitrogen region.

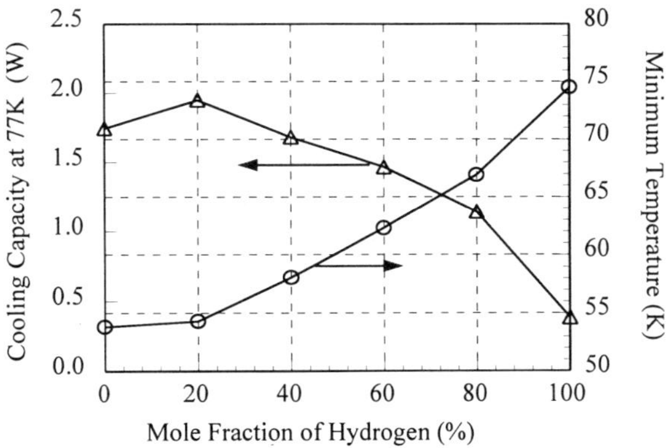

Figure 4. Refrigeration capacity versus hydrogen concentration in the mixture.

We tried to do some repeat tests. However, the compressor of our test cryocooler was out of order, and the performance of the machine became unstable. We think that further careful repeat tests must be done to confirm the above results.

CONCLUSION

The substitution of pure helium by helium-hydrogen mixtures is proposed to improve the refrigeration capacity of regenerative refrigerators in the temperature region of liquid nitrogen. Experiments have been done with a co-axial pulse tube refrigerator. Experimental results demonstrate a 10% gain of refrigeration capacity with 80%Helium–20%hydrogen mixture, while the other operating conditions are kept constant. This is a primary result, and it needs further verification in our future research.

ACKNOWLEDGEMENT

The project is financially supported by National Natural Science Foundation of China.

REFERENCES

1. J.P.Yu, G.B.Chen and Z.H.Gan, Investigation on the regenerator performance using gas mixtures, Proc. of ICCR'98, Hangzhou, 409-412(1998).
2. Z.H.Gan, J.P.Yu and G.B.Chen, Study on pulse tube refrigerator performance with two-component mixture, Proc. of ICCR'98, Hangzhou, 405-408(1998).
3. J.P.Yu, G.B.Chen et al., Discussion on regenerator performance improvement with binary gas mixtures, Proc. of ICEC-17, 117-122(1998).
4. G.B.Chen, Z.H.Gan and J.P.Yu, Study on two-component gas mixtures in rege-nerative refrigerators, Proc. of ICEC-17, 197-200 (1998).
5. G.Walker, Stirling-cycle cooling engine with two-phases, two-components working fluid, *Cryogenics,* 8:459(1974).
6. Patwardhan K.P. and Bapat S.L., Cyclic simulation of Stirling cycle cryogenerator using two component two phase working fluid combinations, Proceedings. of ICCR'98, Hangzhou, 397(1998).

FLOW-CONTROLLING DEVICES IN PULSE TUBES

A. T. A. M. de Waele[1], H.W.G. Hooijkaas[2]

[1]Department of Physics, Eindhoven University of Technology,
PO Box 513, NL-5600 MB Eindhoven, The Netherlands
[2]Signaal-USFA, PO Box 6034, 5600 HA Eindhoven, The Netherlands

ABSTRACT

This paper is a continuation of our work on the thermodynamical aspects of pulse tubes.[1] This time we discuss various flow- (or pressure) controlling devices, proposed to improve the performance of pulse tube refrigerators (PTR), such as orifice, expander, four-valve method, inertance tube, and first and second inlet. We have obtained analytical expressions for the various cases assuming a finite flow resistance in the regenerator, but with zero void volume, perfect heat exchange, and high heat capacity of the regenerator material.

GENERAL EQUATIONS

A schematic drawing of a double-inlet PTR is presented in Fig. 1. The flow control system is the subject of this paper. We assume that the fluid is an ideal gas and that the time dependence of the pressure in the tube is harmonic

$$p_{\rm t} = p_0 + p_1 \cos \omega t, \tag{1}$$

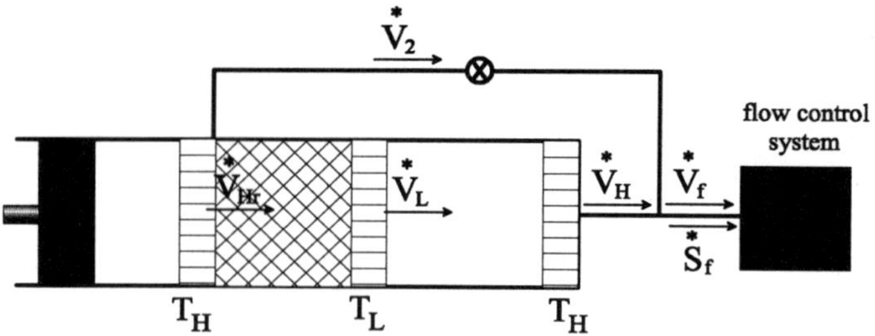

Figure 1: Schematic diagram of a Stirling-type double-inlet pulse tube. The flow-control system is represented as a black box.

with $p_0 \gg p_1$. The volume flow rate[1] towards the flow controlling system can be written as

$$\overset{*}{V}_{\mathrm{f}}=\overset{*}{V}_1 \, (\cos\omega t + \alpha_{\mathrm{f}} \sin\omega t). \tag{2}$$

The volume flow at the cold end of the tube, $\overset{*}{V}_{\mathrm{L}}$, is related to the volume flow at the hot end, $\overset{*}{V}_{\mathrm{H}}$, through the relation

$$\overset{*}{V}_{\mathrm{L}}=\overset{*}{V}_{\mathrm{H}} - \alpha \, \overset{*}{V}_1 \sin\omega t \tag{3}$$

with

$$\alpha = \frac{C_{\mathrm{V}}}{C_{\mathrm{p}}} \frac{V_{\mathrm{t}}\omega}{C_1 p_0}, \tag{4}$$

$C_1 = \overset{*}{V}_1 / p_1$, and $V_{\mathrm{t}} = A_{\mathrm{t}} L_{\mathrm{t}}$ is the volume of the tube. The volume flow rate through the second inlet is expressed as

$$\overset{*}{V}_2 = C_2(p_{\mathrm{c}} - p_{\mathrm{t}}) \tag{5}$$

where p_{c} is the pressure in the compressor. The volume flow rate at the hot end of the tube is

$$\overset{*}{V}_{\mathrm{H}}=\overset{*}{V}_{\mathrm{f}} - \overset{*}{V}_2 \, . \tag{6}$$

We assume that the void volume in the regenerator is very small. In this case

$$\overset{*}{V}_{\mathrm{L}} = C_{\mathrm{r}}(p_{\mathrm{c}} - p_{\mathrm{t}}). \tag{7}$$

[1]We adopt a sign convention in which flows going to the right (Fig.1) are counted positive.

For simplicity of notation we introduced $C_r = T_L C_{rH}/T_e$ with C_{rH} the flow conductance of the regenerator at room temperature and T_e an effective temperature[1] $T_e = \frac{1}{L_r}\int_0^{L_r}(T\eta/\eta_H)dl$. Here η is the viscosity of the fluid and η_H the viscosity at room temperature; L_r is the length of the regenerator. For simplicity of notation we also introduce $C_{2r} = C_2 + C_r$. Combining these relations gives

$$p_c - p_t = (\overset{*}{V}_1 / C_{2r})[\cos\omega t + (\alpha_f - \alpha)\sin\omega t] \tag{8}$$

and

$$\overset{*}{V}_L = \overset{*}{V}_1 (C_r/C_{2r})[\cos\omega t + (\alpha_f - \alpha)\sin\omega t]. \tag{9}$$

Using Eq.(3) and (9)

$$\overset{*}{V}_H = (\overset{*}{V}_1 / C_{2r})[C_r\cos\omega t + (C_r\alpha_f + C_2\alpha)\sin\omega t]. \tag{10}$$

The volume flow at the hot end of the regenerator is $\overset{*}{V}_{Hr} = (T_H/T_L)\overset{*}{V}_L$, so

$$\overset{*}{V}_{Hr} = \overset{*}{V}_1 \frac{T_H}{T_e}\frac{C_{rH}}{C_{2r}}[\cos\omega t + (\alpha_f - \alpha)\sin\omega t]. \tag{11}$$

THE COP

For the molar entropy holds

$$\delta S_m = \frac{C_p}{T}\delta T - \frac{V_m}{T}\delta p. \tag{12}$$

Defining $\mathcal{D} = \frac{1}{2}p_1 \overset{*}{V}_1$ the average entropy flow to the hot end of the tube is

$$T_H\overline{\overset{*}{S}_H} = \overline{\overset{*}{V}_H\,\delta p} = -(C_r/C_{2r})\mathcal{D}. \tag{13}$$

Consequently the average cooling power $\overline{\dot{Q}_L}$ is

$$\overline{\dot{Q}_L} = (C_r/C_{2r})\mathcal{D}. \tag{14}$$

The pressure in the compressor is, with Eqs.(1) and (8),

$$p_c = p_0 + p_1\cos\omega t + (\overset{*}{V}_1 / C_{2r})[\cos\omega t + (\alpha_f - \alpha)\sin\omega t]. \tag{15}$$

The rate of change of the volume behind the compressor is $\dot{V}_c = -(\overset{*}{V}_{Hr} + \overset{*}{V}_2)$. The power input to the compressor $P_c = -\overline{p_c \dot{V}_c}$, which gives

$$P_c = \frac{T_e C_2 + T_H C_{rH}}{T_e C_{2r}^2}[C_{2r} + (1 + (\alpha_f - \alpha)^2)C_1]\mathcal{D}. \tag{16}$$

The coefficient of performance (COP) is $\xi = \overline{\dot{Q}_L}/P_c$. Eqs.(16) and (14), and some algebra, give our most general result

$$\xi_{id}\xi^{-1} = \left[1 + \frac{T_e C_2}{T_H C_{rH}}\right]\left[1 + \frac{[1 + (\alpha_f - \alpha)^2]C_1}{C_2 + T_L C_{rH}/T_e}\right] \tag{17}$$

with $\xi_{id} = T_L/T_H$.

The maximum COP is obtained if $\alpha_f = \alpha$. If also $C_{2r} \gg C_1$ Eq.(17) reduces to

$$\xi_{id}\xi^{-1} \approx 1 + \frac{T_e C_2}{T_H C_{rH}} \quad \text{with} \quad \alpha_f = \alpha \text{ and } C_{2r} \gg C_1. \tag{18}$$

Based on Eq.(18) the second inlet should always be closed for the best performance ($C_2 = 0$). In order not to complicate the following calculations too much we will, from now on, consider the *single-orifice PTR* ($C_2 = 0$). In this case

$$\xi_{id}\xi_s^{-1} = 1 + [1 + (\alpha_f - \alpha)^2]\frac{T_e C_1}{T_L C_{rH}} \tag{19}$$

and $\overset{*}{V}_f = \overset{*}{V}_H$, so

$$\overset{*}{V}_L = \overset{*}{V}_1 [\cos\omega t + (\alpha_f - \alpha)\sin\omega t]. \tag{20}$$

FLOW CONTROLLERS

Some important options of flow-controlling devices are schematically represented in Fig. 2.

Orifice

If the flow-controlling device is a pure flow resistance, $\alpha_f = 0$, so, with Eq.(4),

$$\xi_{id}\xi_s^{-1} = 1 + \left[\frac{1}{\alpha} + \alpha\right]\frac{C_V}{C_p}\frac{V_t\omega}{p_0 C_{rH}}\frac{T_e}{T_L}. \tag{21}$$

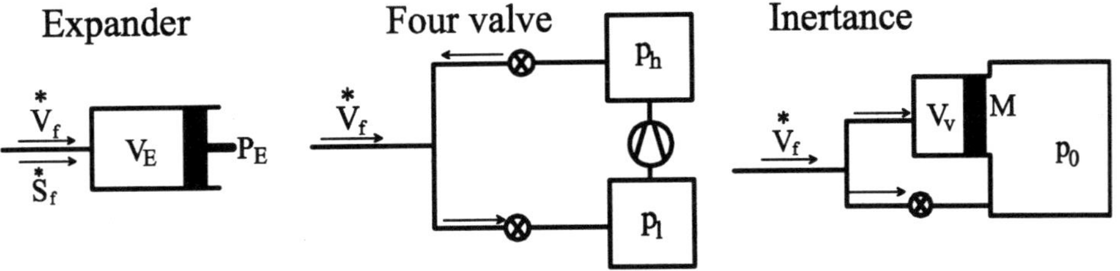

Figure 2. Schematic diagram of various flow controllers.

The optimum situation is found for $\alpha = 1$, so

$$\xi_{\mathrm{id}}\xi_{\mathrm{s0}}^{-1} = 1 + 2\frac{T_{\mathrm{e}}}{T_{\mathrm{L}}C_{\mathrm{rH}}}\frac{C_{\mathrm{V}}}{C_{\mathrm{p}}}\frac{V_{\mathrm{t}}\omega}{p_0}. \tag{22}$$

Expander

If the flow controlling device is an actively controlled piston (expander) the first and second law, applied to the expander, demand that the power generated by the expander

$$P_{\mathrm{E}} = \mathcal{D} - T_{\mathrm{H}}\overline{\dot{S}_{\mathrm{E}}}, \tag{23}$$

where $\overline{\dot{S}_{\mathrm{E}}}$ is the entropy production in the expander. In the ideal case it is zero. The energy comes from heat, extracted from the environment. The work done by the expander can be used to drive the compressor. In that case the overall efficiency of the system goes up since now the *net* power input is $P_{\mathrm{c}} - P_{\mathrm{E}}$. The COP of the PTR with an ideal active expander is

$$\xi_{\mathrm{E}}^{-1} = \xi_{\mathrm{s}}^{-1} - 1 = \frac{T_{\mathrm{H}} - T_{\mathrm{L}}}{T_{\mathrm{L}}} + [1 + (\alpha_{\mathrm{f}} - \alpha)^2]\frac{T_{\mathrm{e}}C_1}{T_{\mathrm{L}}C_{\mathrm{rH}}}\frac{T_{\mathrm{H}}}{T_{\mathrm{L}}}. \tag{24}$$

If $C_{\mathrm{rH}} \to \infty$ then ξ_{E} is equal to the Carnot COP. If $\alpha_{\mathrm{f}} = \alpha$ the COP is at its maximum, which means that the expander compensates exactly the volume-change due to compression in the tube (Eq.(20)).

Four-Valve Method

A third way to control the flow at the hot end of the pulse tube is to use two buffer volumes, one at a low pressure p_l and one at a high pressure p_h, the so-called four-valve method.[2,3] The valves are actively controlled in such a way that the desired flow-time dependance (Eq.(2)) is obtained. Typically $p_l = p_0 - p_1$ and $p_h = p_0 + p_1$. During half the cycle, gas flows from the tube to the low-pressure buffer, and during the other half, it flows from the high-pressure buffer to the tube. A compressor is needed to maintain the steady state. The average flow through this compressor is

$$\overline{\overset{*}{V}_{AB}} = \frac{1}{\pi}\sqrt{1+\alpha_f^2}\,\overset{*}{V}_1\,. \tag{25}$$

Assuming $p_h - p_l = 2p_1$, this requires an average power of

$$P_{AB} = 2p_1\overline{\overset{*}{V}_{AB}} = \frac{4}{\pi}\sqrt{1+\alpha_f^2}\mathcal{D}. \tag{26}$$

The efficiency of the cooler with an active buffer is given by

$$\xi_{AB} = \frac{\mathcal{D}}{P_c + P_{AB}}. \tag{27}$$

Substitution of Eq.(19) gives

$$\xi_{id}\xi_{AB}^{-1} = 1 + [1+(\alpha_f-\alpha)^2]\frac{C_1}{C_{rH}}\frac{T_e}{T_L} + \frac{4}{\pi}\sqrt{1+\alpha_f^2}\frac{T_L}{T_H}. \tag{28}$$

For best performance α_f should minimize this function.

Inertance

Now we will discuss the possibility of using an inertance system, as depicted in Fig. 2. It consists of an orifice O_1 and a buffer volume. The buffer has a volume V_v, cross sectional area A_v, length L_v, and contains n_v moles of gas at constant temperature T_H. It is connected to the tube at one end and closed at the other end by a mass M. This mass is elastically connected to an equilibrium position with spring constant k. The equilibrium value of the volume is V_{v0}. The mass moves under the influence of the pressure difference in the pulse tube p_t and the average pressure p_0. This can be a real mass, but it can also be the mass of a gas column in a tube.[4,5] In the latter case, Fig. 2 can be considered as a model system of the inertance tube. A little algebra shows that

$$\alpha_\mathrm{f} = \left(\frac{A_\mathrm{v}^2}{M\omega^2 - k} - \frac{V_\mathrm{v0}}{p_0}\right)\frac{\omega}{C_1}. \tag{29}$$

With Eqs.(19) and (4) we obtain

$$\xi_\mathrm{id}\xi_\mathrm{I}^{-1} = 1 + \frac{C_1}{C_\mathrm{rH}}\frac{T_\mathrm{e}}{T_\mathrm{L}} + \left(\frac{A_\mathrm{v}^2}{M\omega^2 - k} - \frac{V_\mathrm{v0}}{p_0} - \frac{C_\mathrm{V}}{C_\mathrm{p}}\frac{V_\mathrm{t}}{p_0}\right)^2 \frac{\omega^2}{C_1 C_\mathrm{rH}}\frac{T_\mathrm{e}}{T_\mathrm{L}}. \tag{30}$$

The optimal adjustment is obtained if

$$M\omega^2 = k + \frac{A_\mathrm{v}^2 p_0}{V_\mathrm{v0} + \frac{C_\mathrm{V}}{C_\mathrm{p}} V_\mathrm{t}} \tag{31}$$

in which case

$$\xi_\mathrm{id}\xi_\mathrm{I0}^{-1} = 1 + \frac{C_1}{C_\mathrm{rH}}\frac{T_\mathrm{e}}{T_\mathrm{L}}. \tag{32}$$

It is instructive to substitute some engineering values in Eq.(31). Let us take $k = 0$ (freely moving piston), $A_\mathrm{v} = 1\ \mathrm{cm}^2$, $p_0 = 10$ bar, and $V_\mathrm{v0} + (C_\mathrm{V}/C_\mathrm{p})V_\mathrm{t} = 100\ \mathrm{cm}^3$. Then $M\omega^2 = 100\ \mathrm{kg/s}^2$. For $\omega = 100\pi$ rad/s this gives $M = 1$ gram. For pulse tubes operating at low frequencies, e.g. 1 Hz, the mass should be as large as 2.5 kg. If the mass would consist of a freely moving gas column ($k = 0$) in the void volume itself, then $M = M_\mathrm{m} p_0 V_\mathrm{v0}/RT_\mathrm{H}$, and Eq.(31) can be written as

$$M_\mathrm{m} L_\mathrm{v0}\left(L_\mathrm{v0} + \frac{C_\mathrm{V}}{C_\mathrm{p}}\frac{V_\mathrm{t}}{A_\mathrm{v}}\right)\omega^2 = RT_\mathrm{H}. \tag{33}$$

The solution for the length of the tube is

$$L_\mathrm{v0} = -\frac{C_\mathrm{V}}{C_\mathrm{p}}\frac{V_\mathrm{t}}{2A_\mathrm{v}} + \sqrt{\left(\frac{C_\mathrm{V}}{C_\mathrm{p}}\frac{V_\mathrm{t}}{2A_\mathrm{v}}\right)^2 + \frac{RT_\mathrm{H}}{M_\mathrm{m}\omega^2}}. \tag{34}$$

With e.g. $V_\mathrm{t} = 100\ \mathrm{cm}^3$, $A_\mathrm{t} = 13\ \mathrm{mm}^2$, and $\omega = 100\pi$ rad/s gives a length $L_\mathrm{v0} = 1$ m.

CONCLUSION

In this paper we have obtained analytical expressions for various basic types of flow-controlling devices used to improve the performance of pulse tubes. They can be used for a trade off of various options and for optimizing the design parameters.

ACKNOWLEDGMENT

This work is partly supported by the Dutch Technology Foundation (STW) and by the ministry of Economic Affairs (BTS). We thank M. Xu, A.A.J. Benschop P. Bruins and for many stimulating discussions.

REFERENCES

1. A.T.A.M. de Waele, P.P. Steijaert, and J. Gijzen, *Cryogenics*, 37:313 (1997); A.T.A.M. de Waele, P.P. Steijaert, and J.J. Koning, *Cryogenics*, 38:329 (1998); A.T.A.M. de Waele, H.W.G. Hooijkaas, P.P. Steijaert, and A.A.J. Benschop, *Cryogenics* 38:995 (1998); A.T.A.M. de Waele, *Cryogenics* 39:13 (1999).

2. Y. Matsubara, J.L. Gao, K. Tanida, Y. Hiresaki, and M. Kaneko. Proc. of 7th Int. Cryocooler Conference (1993) P. 166.

3. K. Tanida, J.L. Gao, N. Yoshimura, and Y. Matsubara. "Advances in Cryogenic Engineering 41" (1996) P. 1503.

4. S.W. Zhu, S.L. Zhou, N. Yoshimura, and Y. Matsubara. Cryocoolers 9, edited by R.G. Ross, Jr. Plenum Press, New York (1997) P. 269.

5. D.L. Gardner and G.W. Swift. *Cryogenics*, 37:117 (1997).

ENTHALPY FLOW IN A PULSE TUBE REFRIGERATOR

G.Q. Lu, P. Cheng, and C.T. Hsu

Department of Mechanical Engineering
Hong Kong University of Science & Technology
Clear Water Bay, Kowloon, Hong Kong

ABSTRACT

In this paper, we analyze the performance of a double-inlet pulse-tube refrigerator by a cycle averaging of the transient one-dimensional governing equations. The cycle-averaged enthalpy flows at any cross section of the double-inlet pulse-tube refrigerator are obtained. The refrigeration capacity of a double-inlet pulse-tube refrigerator is expressed in an algebraic form in terms of the net heat transfer at the hot-end heat exchanger, the cycle-averaged enthalpy flows at the orifice valve and at the double-inlet valve, the cycle-averaged enthalpy flow at the hot end of the regenerator, the kinetic energy loss and thermal conduction loss in the regenerator. The effects of these parameters on the refrigeration capacity of a basic pulse-tube refrigerator, an orifice pulse-tube refrigerator and a double-inlet pulse-tube refrigerator are discussed. The importance of the optimal design of the hot-end heat exchanger on the refrigeration capacity is discussed in a Lagrange frame.

INTRODUCTION

In the literature, the heat transfer mechanism in a basic pulse tube refrigerator was analyzed based on the surface heat pumping mechanism by Gifford and Longsworth,[1] and the orifice pulse tube refrigerator was analyzed based on an enthalpy flow model by Radebaugh.[2] Liang proposed a compound model to explain the performance of a pulse-tube refrigerator.[3] He assumed that surface heat pumping prevailed in the domain inside of the thermal boundary layer and nonsymmetric effect existed outside of the thermal boundary layer. Rott,[4] Swift [5] and Xiao [6] proposed thermoacoustic theories to analyze the interaction between thermal energy and mechanical work.

The existing enthalpy flow model and the nonsymmetric model did not consider the performances of the hot-end heat exchanger and the regenerator. These components obviously would affect the system performance. In this paper, we shall first obtain the cycle-averaged enthalpy flow in any section of a double-inlet pulse-tube refrigerator (as shown in figure 1) by a cycle averaging of the governing transient one-dimensional equations as applied to the regenerator, the heat exchangers, and the pulse tube. The refrigeration capacity of the double-inlet pulse-tube refrigerator is obtained in terms of the net heat transfer at the hot-end heat exchanger, the cycle-averaged enthalpy flows at the orifice valve and at the double-inlet valve, the cycle average enthalpy flow at the hot end of the regenerator, the kinetic-energy loss and thermal-conduction loss in the regenerator. The effects of these parameters on the performance of a basic pulse-tube refrigerator, an orifice pulse-tube refrigerator and a double-inlet pulse-tube refrigerator are discussed.

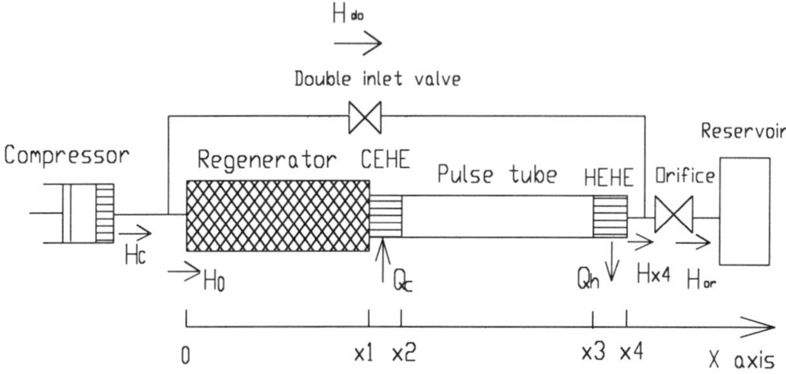

Figure 1. Schematic diagram of a double-inlet pulse-tube refrigerator.

CYCLE AVERAGE ENTHALPY FLOW

For simplicity, we analyze the performance of pulse-tube refrigerator by a transient one-dimensional model. The energy equation for the fluid is

$$C_p\left(\frac{\partial \rho T}{\partial t}+\frac{\partial \rho u T}{\partial x}\right)=\frac{\partial p}{\partial t}+u\frac{\partial p}{\partial x}+\rho\frac{fu^2|u|}{2d_h}+\frac{\partial}{\partial x}\left(k\frac{\partial T}{\partial x}\right)+\frac{h}{L_1}(T_s-T) \tag{1}$$

where ρ, T, p and u are the density, temperature, pressure and velocity of the fluid; h is the interfacial heat-transfer coefficient and C_p is the specific heat of fluid at constant pressure; Here the friction coefficient f is a positive quantity which is different in different components of the pulse-tube refrigerator and d_h is the equivalent hydraulic diameter. Substituting the state equation of the fluid $p=\rho RT$ in the above equation gives

$$\frac{C_v}{R}\frac{\partial p}{\partial t}+\frac{C_P}{R}\frac{\partial pu}{\partial x}=u\frac{\partial p}{\partial x}+\rho\frac{fu^2|u|}{2d_h}+\frac{\partial}{\partial x}\left(k\frac{\partial T}{\partial x}\right)+\frac{h}{L_1}(T_s-T) \tag{2}$$

where R is the gas constant, and k and C_v are the thermal conductivity and specific heat at constant volume of the fluid.

The energy equation for the solid phase is

$$\left(\rho C_p\right)_s \frac{\partial T_s}{\partial t} = \frac{\partial}{\partial x}\left(k_s \frac{\partial T_s}{\partial x}\right) + \frac{h}{L_2}\left(T - T_s\right) + q_s \tag{3}$$

where subscript s indicates the solid phase and q_s indicates the energy source term which is defined for different parts as follow:

$$\begin{aligned} & q_s = h_w \frac{1}{L_3}(T_w - T_s) && \text{in the hot end heat exchanger} \\ & q_s = 0 && \text{in the pulse tube and regenrator} \\ & q_s = \frac{Q_c}{V} && \text{in the cold end heat exchanger} \end{aligned} \tag{4}$$

with h_w being the heat-transfer coefficient between cooling water and the solid wall and Q_c the refrigeration capacity. L_1, L_2 and L_3 in equation (1)-(4) are the characteristic lengths, which are defined as the elementary representative volume divided by interfacial heat transfer area.

We now express equation (2), the energy equation of the fluid, in terms of the enthalpy flow which is defined as $H = mC_pT = \rho uAC_pT = \frac{AC_p}{R}pu$, where m is the mass flow rate and A is the cross-sectional area. Eliminating the interfacial heat-transfer term between equations (2) and (3) and express the resulting equation in terms of H, we have

$$\frac{C_v}{R}\frac{\partial p}{\partial t} + \frac{1}{A}\frac{\partial H}{\partial x} = u\frac{\partial p}{\partial x} + \rho\frac{fu^2|u|}{2d_h} + \frac{\partial}{\partial x}\left(k\frac{\partial T}{\partial x}\right) + \frac{L_2}{L_1}\left[\frac{\partial}{\partial x}\left(k_s\frac{\partial T_s}{\partial x}\right) - (\rho Cp)_s\frac{\partial T_s}{\partial t} + q_s\right] \tag{5}$$

Combining the momentum equation and the continuity equation of the fluid, it is easy to show that the first two terms at the right side of the above equation are equal to kinetic energy terms. The above equation becomes

$$\frac{C_v}{R}\frac{\partial p}{\partial t} + \frac{1}{A}\frac{\partial H}{\partial x} = -\left[\frac{\partial \frac{\rho u^2}{2}}{\partial t} + \frac{\partial\left(u\frac{\rho u^2}{2}\right)}{\partial x}\right] + \frac{\partial}{\partial x}\left(k\frac{\partial T}{\partial x}\right) + \frac{L_2}{L_1}\left[\frac{\partial}{\partial x}\left(k_s\frac{\partial T_s}{\partial x}\right) - (\rho Cp)_s\frac{\partial T_s}{\partial t} + q_s\right] \tag{6}$$

We now consider the situation when a pulse-tube refrigeration system has reached a cycle-steady state after a sufficiently long time. At such conditions, local cycle-average parameters such as pressure and temperature will not change with time; hence their local derivatives with respect to time are equal to zero. Therefore, integrating the above equation with respect to time, the resulting equation at cycle-steady state is

$$\frac{1}{A}\frac{\partial \bar{H}}{\partial x} = -\frac{\partial\left(\overline{u\frac{\rho u^2}{2}}\right)}{\partial x} + \frac{\partial}{\partial x}\left(\overline{k\frac{\partial T}{\partial x}}\right) + \frac{L_2}{L_1}\frac{\partial}{\partial x}\left(\overline{k_s\frac{\partial T_s}{\partial x}}\right) + \frac{L_2}{L_1}\overline{q_s} \tag{7}$$

where the bars denote cycle-average quantities.

REFRIGERATION CAPACITY

We now substitute equation (4) in equation (7), and integrate the resulting equation along the x-axis in different components of the pulse-tube refrigerator as shown in figure 1.

(i) In the regenerator where the energy source term q_s=0, we integrate equation (7) from 0 to x_1 to give

$$\overline{H}_{x1} = \overline{H}_0 + A_1\left[-\overline{u\frac{\rho u^2}{2}}\Big|_0^{x_1} + k\frac{\partial \overline{T}}{\partial x}\Big|_0^{x_1} + \frac{L_2}{L_1}k_s\frac{\partial \overline{T}_s}{\partial x}\Big|_0^{x_1}\right] \quad (8)$$

where A_1 is cross sectional area of the fluid channel in the regenerator and $\overline{H}_0$ is the cycle-averaged enthalpy flow at the hot-end heat exchanger. Note that in the regenerator,

$$\frac{L_2}{L_1} = \frac{1-\varphi}{\varphi} \quad and \quad A_1 = A_{reg}\varphi \quad (9)$$

where A_{reg} denotes cross-sectional area of the regenerator and φ denotes the porosity. Substituting equation (9) in equation (8) and noting that the cycle-averaged temperatures of the fluid and the solid have the same values in the regenerator (i.e., $\overline{T} = \overline{T}_s$), then

$$\overline{H}_{x1} = \overline{H}_0 + \left\{ -A_{reg}\varphi\overline{u\frac{\rho u^2}{2}}\Big|_0^{x_1} + A_{reg}\left[k\varphi + k_s(1-\varphi)\right]\frac{\partial \overline{T}}{\partial x}\Big|_0^{x_1}\right\} \quad (10)$$

Note that the last two terms in the bracket of the above equation represent the net kinetic-energy loss and the net heat-conduction loss in the regenerator.

(ii) In the cold-end heat exchanger, we can neglect the heat-conduction terms because temperature gradients in the cold-end heat exchanger are small. Also, we can neglect the kinetic-energy loss under the consideration that the works done against friction and pressure drop are small compared to Q_c. Thus, integrating equation (7) from x_1 to x_2 gives

$$Q_c = \overline{H}_{x2} - \overline{H}_{x1} \quad (11)$$

(iii) In the pulse tube, since friction and pressure drops are very small in the fluid, kinetic energy loss can be neglected. In addition, the solid wall of the pulse tube is thin compared to the diameter of the tube so that total heat conduction in the wall is very small, and can be neglected. Heat conduction by fluid is also neglected due to its small thermal conductivity compared to that of the solid. Since q_s=0 in the pulse tube, equation (7) gives

$$\overline{H}_{x2} = \overline{H}_{x3} \quad (12)$$

(iv) In the hot-end heat exchanger, heat-conduction and kinetic-energy loss terms can also be neglected for the same reasons as in the cold-end heat exchanger. Thus,

$$\frac{1}{A_4}\int_{x3}^{x_4}\frac{\partial \overline{H}}{\partial x}dx \approx \int_{x3}^{x_4}\overline{\frac{L_2}{L_1}h_w\frac{1}{L_3}(T_w - T_s)}dx \quad (13)$$

where A_4 is cross-sectional area of the fluid channel in the hot-end heat exchanger and h_w is the heat-transfer coefficient between cooling water and the solid wall. We define the total heat transfer over a cycle at the hot-end heat exchanger as Q_h, i.e.

$$Q_h = -A_4 \int_{x3}^{x4} \overline{\frac{L_2}{L_1} h_w \frac{1}{L_3} (T_w - T_s)} dx \tag{14}$$

The minus sign in the above equation indicates that heat is removed from the solid to the cooling water. Substituting equation (14) in equation (13) yields

$$\overline{H}_{x3} = \overline{H}_{x4} + Q_h \tag{15}$$

Substituting equations (10), (12) and (15) in equation (11), we obtain the following expression for the refrigeration capacity

$$Q_C = \left[Q_h + \overline{H}_{x4}\right] - \left\{ \overline{H}_0 - A_{reg} \varphi u \overline{\frac{\rho u^2}{2}} \Bigg|_0^{x_1} + A_{reg} \left[k\varphi + k_s (1-\varphi)\right] \frac{\partial \overline{T}}{\partial x} \Bigg|_0^{x_1} \right\} \tag{16}$$

From an energy balance consideration, $\overline{H}_{x4}$ is related to $\overline{H}_{or}$ and $\overline{H}_{do}$ by $\overline{H}_{or} = \overline{H}_{x4} + \overline{H}_{do}$, where $\overline{H}_{or}$ and $\overline{H}_{do}$ are respectively the cycle-averaged enthalpy flows at the orifice valve and at the double-inlet which can be expressed as $\overline{H}_{or} = C_p (\overline{\tilde{m}\tilde{T}})_{or}$ and $\overline{H}_{do} = C_p (\overline{m}\overline{T} + \overline{\tilde{m}\tilde{T}})_{do}$ with the " ~" indicating fluctuated quantities. The cycle-averaged enthalpy flow $\overline{H}_0$ can be obtained from an energy balance as $\overline{H}_0 = \overline{H}_C - \overline{H}_{do}$, where $\overline{H}_C$ is the cycle-averaged enthalpy flow at the exit of the compressor. Thus, equation (16) becomes

$$Q_C = \left[Q_h + \overline{H}_{or}\right] - \left\{ \overline{H}_C - A_{reg} \varphi u \overline{\frac{\rho u^2}{2}} \Bigg|_0^{x_1} + A_{reg} \left[k\varphi + k_s (1-\varphi)\right] \frac{\partial \overline{T}}{\partial x} \Bigg|_0^{x_1} \right\} \tag{17}$$

It is relevant to point out that equations (10), (11), (12), and (15) could have been obtained directly by applying the first law of thermodynamics to an open system (undergoing a cyclic process) for the regenerator, the cold-end heat exchanger, the pulse tube and the hot-end heat exchanger separately. The refrigeration capacity, given in the right-hand side of equation (17), consists of two terms: the square brackets term and the braces term. The square brackets term consists of the net heat transfer at the hot-end heat exchanger Q_h, and the cycle-averaged enthalpy flow at the orifice valve $\overline{H}_{or}$. Generally, Q_h and $\overline{H}_{or}$ are always positive quantities. The braces term consists of the cycle-averaged enthalpy flow at the exit of the compressor $\overline{H}_C$, kinetic-energy loss and heat-conduction loss in the regenerator. In the following, we shall consider the effects of the orifice valve and the double-inlet valve on the refrigeration capacity of a pulse-tube refrigerator for a given compressor (with prescribed value of $\overline{H}_C$) and a given regenerator (with given A_{reg}, x_1, and φ).

In a *basic pulse-tube refrigerator,* where both the orifice and the double-inlet valve are closed, we have, $\overline{H}_{or} = 0$ in equation (17). Thus, the square brackets term in equation (17) consists only of Q_h, which is a positive quantity as will be shown in the next section. However, the value of Q_h is small because only a small portion of fluid is cooled as a result of the closed end of the tube. This is the reason why a basic pulse-tube refrigerator has a low refrigeration capacity.

In an *orifice pulse-tube refrigerator* where the hot-end heat exchanger is connected to an orifice valve, the square brackets term in equation (17) consists of two terms, a net heat transfer Q_h which is a positive quantity as discussion in the next section, plus the cycle-averaged enthalpy flow at the orifice valve $\overline{H}_{or}$, which is also a positive quantity. The combination of these two terms is usually greater than the Q_h term in the basic pulse tube. Thus, the orifice pulse-tube refrigerator usually has a higher refrigeration capacity than the basic pulse-tube refrigerator.

In a *double-inlet pulse-tube refrigerator*, the square brackets term in equation (17) still consists of Q_h and $\overline{H}_{or}$. However, by the effects of the double-inlet valve, the kinetic energy loss in equation (17) will be decreased,[7] thereby increasing the refrigeration capacity.

In summary, equation (17) shows that the refrigeration capacity of a pulse-tube refrigerator can be improved by using a highly efficient hot-end heat exchanger, an optimal design of a regenerator (with minimal friction and minimal heat-conduction loss), and by adding an orifice valve and a double-inlet valve.

HEAT TRANSFER AT THE HOT-END HEAT EXCHANGER

Equation (17) shows the importance of the hot-end heat exchanger on the refrigeration capacity. For further insight, we will discuss the effect of the hot-end heat exchanger in a Lagrange frame during the period before the cycle-steady temperature profile in the pulse-tube refrigerator has established. In a basic pulse-tube refrigerator, assuming that a certain fluid is in state 1 at the beginning of a cycle (when the pressure wave angle is 0°) as shown in figure 2(a), where T_r indicates room temperature and P_a indicates average pressure. If there is no heat exchanger at the closed end, then the gas parcel will pass through 1-2-1 during a cycle; in that case, there is no net heat removed. When there is a hot-end heat exchanger that is cooled by cooling water, the fluid parcel will be cooled. Hence its trace will be 1-3 instead of 1-2 during the period of compression because of the cooling effect. At the beginning of the next half cycle (with the pressure wave angle from 180° to 360°) of expansion with temperature T_3, the temperature of the fluid parcel is still higher than the temperature of the cooling water at first, hence, the temperature will decrease sharply due to expansion with cooling by the trace 3-4 until the temperature of the fluid parcel equals that of the cooling water. Then, the temperature of the fluid parcel still decreases, but more slowly, due to further expansion, which absorbs heat from the cooling water by the trace 4-5 until its pressure reaches a minimum. Thus, the fluid parcel experiences the traces of 1-3-4-5 during a cycle, and there is a net temperature decrease during a cycle, which indicates that there is net heat removed at the hot-end heat exchanger, or $Q_h > 0$, and this net temperature decrease will not vanish until the cycle-steady temperature profile in the pulse-tube refrigerator has established. However, due to the closed end, only a small portion of the fluid passes through the heat exchanger, and the cycle-averaged heat removed by the cooling water is small.

Figure 2(b) shows the case of an orifice pulse-tube refrigerator, where T_r indicates room temperature and P_a indicates pressure in the reservoir (if we assume that the pressure fluctuation in the reservoir is small and can be neglected). Due to existence of the orifice valve and the reservoir, the flow of the fluid parcel depends not only on the pressure in the pulse tube but also on the pressure in the reservoir. Now we consider the pressure wave in the pulse tube from 0° to 180°, which is greater than the reservoir pressure; thus, the fluid parcel flows from the pulse tube to the reservoir. At first, the fluid parcel flows towards the reservoir by trace 1-2 with temperature increasing due to compression but cooled by cooling water until the pressure reaches a maximum value (pressure angle from 0° to 90°).

Then the fluid parcel still flows toward the reservoir but with the expansion period by trace 2-3-4-5 (pressure angle from 90° to 180°). The temperature of the fluid parcel decreases sharply due to expansion together with being cooled by cooling water during the trace 2-3, then expands adiabatically during the trace 3-4 after leaving the hot-end heat exchanger. When the fluid parcel flows into the reservoir, its temperature reaches room temperature due to dissipation in the reservoir. Beginning in the next half cycle (when the pressure angle is from 180° to 360°), the fluid parcel flows backward from the reservoir by trace 5-4-6-7-8. During the period 5-4-6, the gas parcel flows with adiabatic expansion, while during the period 6-7 the fluid parcel expands and absorbs heat from the cooling water until the pressure is a minimum (at pressure angle of 270°). Then the temperature of the fluid parcel increases sharply due to compression and being heated by cooling water by trace 7-8 (pressure angle is from 270° to 360°). Hence the fluid parcel experiences trace 1-2-3-4-5-4-6-7-8 during a cycle. After completing the cycle, the fluid parcel experiences a larger temperature decrease compared to that of a basic pulse-tube refrigerator. From Figure 2(b), we see that the decrease of fluid temperature after a complete cycle is caused by two mechanisms: one is the net heat removed by heat exchanger during 1-2-3 and 6-7-8; the other is heat dissipated in the reservoir during 3-4-5 and 5-4-6. The two mechanisms are represented by the two terms in the square bracket of equation (17).

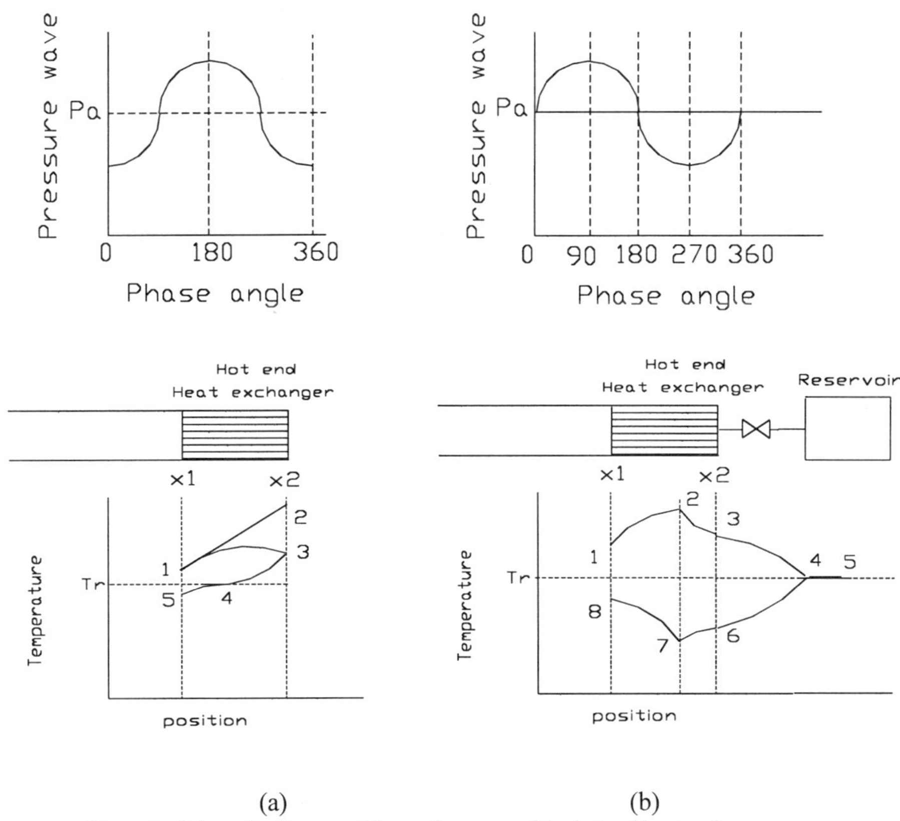

(a) (b)

Figure 2. Schematic diagram of the performance of the hot end heat exchanger.

From the above analysis, the heat-transfer area of a hot-end heat exchanger should have an optimal value for a given geometry of the pulse-tube refrigerator. For the basic pulse-tube refrigerator, as shown in figure 2(a), during the period 1-3 and 3-4, the fluid has a higher temperature than the cooling water and heat is removed by the cooling water, which has a positive effect on the refrigeration. However, during the period 4-5, the fluid temperature is lower than the cooling water, and fluid is therefore absorbing heat during the process 4-5, which has a negative effect on refrigeration capacity. Hence, there exists an optimum value of the heat-transfer area in the hot-end heat exchanger. The optimization of the heat-transfer area in the hot-end heat exchanger becomes more important for the orifice pulse-tube refrigerator. As shown in Figure 2(b), the fluid is absorbing heat during the process 4-6-7-8, which decreases the refrigeration capacity.

CONCLUSION

In this paper, we have applied the enthalpy-flow theory to different components in a pulse-tube refrigerator. The cycle-averaged enthalpy flows at different locations along the pulse tube are obtained by an integration of the governing transient one-dimensional equations. The refrigeration capacity of a double-inlet pulse-tube refrigerator is expressed in an algebraic form which shows clearly the effects of the regenerator, the hot-end heat exchanger, the orifice valve and the double-inlet valve. The reasons for the improvement of the refrigeration capacity of an orifice pulse-tube refrigerator over a basic pulse-tube refrigerator are explained. Analyses on the performance of the hot-end heat exchanger in a Lagrange frame show that the enthalpy flow in a basic pulse-tube refrigerator is caused by net heat transfer at the hot-end heat exchanger. An analysis following a fluid parcel (i.e., in a Langrange frame) suggests that the heat-transfer area of the hot-end heat exchanger in a pulse-tube refrigerator should have an optimal value for improving the system performance.

ACKNOWLEDGMENT

This work was supported by a RGC grant # HKUST 709/95E.

REFERENCES

1. W.E.Gifford and R.C. Longsworth. Surface heat pumping, *Adv. Cry. Eng.* 11:171 (1965).
2. R. Radebaugh. A review of pulse tube refrigeration, *Adv. Cry. Eng.* 35:1191 (1990).
3. J.Liang, A.Ravex. and P.Rolland. Study on pulse tube refrigeration. Part 1: Thermodynamic non-symmetry effect, *Cryogenics* 36:87 (1996).
4. N.Rott. Thermoacoustics, *Adv. App.Mech.* 20:135 (1980).
5. G.W.Swift. Thermoacoustic engines, *J. Acoust. Soc. Am.* 84:1145 (1988).
6. J.H.Xiao. Thermoacoustic heat transportation and energy transformations, Part 1:Formulation of the problem, *Cryogenics* 35:15 *(1995).*
7. S.Zhu, P.Wu and Z.Chen. Double inlet pulse tube refrigerator--An important improvement, *Cryogenics* 30:514 (1990).

ANALYSIS OF INERTANCE TUBE TO IMPROVE THE PERFORMANCE OF PULSE TUBE REFRIGERATOR

L.W Yang, Y. Zhou, J.T Liang

Cryogenics Laboratory, Chinese Academy of Sciences
Beijing P. O. B. 2711, 100080, P. R. China

ABSTRACT

An inertance tube (IT) replacing orifice is proposed to be a new and effective method to improve the performance of a pulse tube refrigerator. Related analysis has been mainly based on thermoacoustic theory or electrical analogy. In this paper, according to thermodynamic processes and flow characteristics, the authors reveal the underlying mechanism that an inertance tube evidently increases gross theory cooling capacity. But on the other hand, the volume of the inertance tube may affect the performance. Symmetry-nozzle has a similar affect on improving performance. According to this analysis, a double-inlet valve should be effective with either an inertance tube or an orifice.

INTRODUCTION

Since an orifice and a reservoir was used in a pulse tube refrigerator in 1984[1] and a double-inlet was proposed in 1991,[2] pulse tube refrigerators have been developing quickly. In 1995,[3] another method using an inertance tube was proposed to improve the performance of a pulse tube cooler. Gardner & Swift[4] explained this using thermoacoustic theory, and some researchers did experiments to verify this point.[5-7] This shows that, generally, inertance tubes used instead of orifices or needle valves will result in good performance. Because almost all the analysis has been based on electrical analogy or thermoacoustic theory, the authors of this paper hope to find the working mechanism from the viewpoint of fluid mechanics and thermodynamics. In fact, flow and heat transfer in an inertance tube is complicated.

At present the IT is used in a miniature pulse tube refrigerator with a working frequency of 30~60 Hz, which is related to its transient effect. For example, sound speed for helium gas is:

$$a = \sqrt{\kappa RT} = \sqrt{1.667 \times 2118 \times 300} = 1029 \text{ m/s} \tag{1}$$

When a tube is long enough, flow resistance, pressure phase change and mass flow will produce unique characteristics, including the transient effect.

There has been research on oscillation pipe flow.[8-11] In these research, different parameters besides traditional *Re* (define as the Reynolds number) are used to analyze the transition from laminar to turbulent flow. These parameters include:

Stokes-layer thickness	$\delta = \sqrt{v/(\pi f)}$
Reynolds number	$Re_{\delta} = u\delta / v$
Stokes parameter	$\eta = (\pi f / v)^{0.5} \cdot d/2$

Transition from laminar to turbulent flow is an important issue.[8, 9] Though transition *Re* is complex, there are some common features to the transition. One is that turbulent flow can recover to laminar flow when the frequency increases. Another is that laminar flow persists partially even though *Re* is high. These results show that laminar flow should always exist in the IT. This is helpful to the analysis in this paper. Gas exchange in laminar oscillatory flow has been experimentally investigated.[10] Experiments show gas exchange may be limited when frequency is high. Another analysis shows that heat conduction in fluids is enhanced due to sinusoidal oscillations.[11] But flow resistance, phase change in pressure and mass flow has been studied little, though they are important for pulse tube refrigerators. In this paper, most of the flow and corresponding equations are based on traditional thermodynamic theory and steady state or quasi-state fluid mechanics, while flow inerti is not considered.

EFFECT OF TUBE VOLUME

First, let us consider a gas flow including that in the hot end of pulse tube, inertance tube and reservoir. Assume the inertance tube has an inner diameter d and length L. The reservoir has a volume of V_r, as shown in Figure 1.

In an adiabatic process, we have:

$$pV^{\kappa} = \text{Constant} \tag{2}$$

Differentiate:

$$dp \cdot V^{\kappa} + \kappa p V^{\kappa-1} dV = 0 \tag{3}$$

Then

$$dV = -V dp / (\kappa p) \tag{4}$$

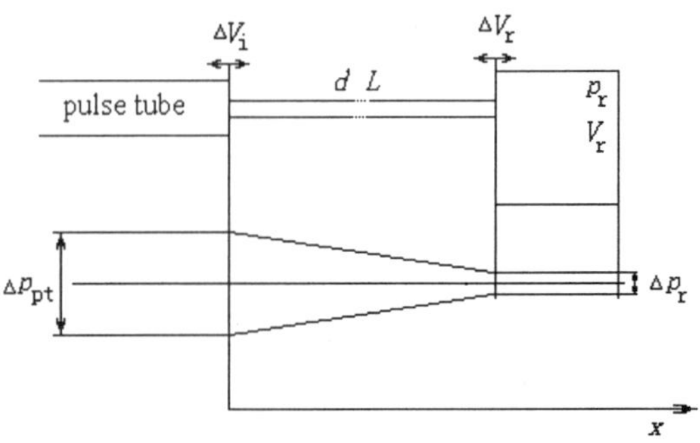

Figure 1. Structure of pulse tube hot end, including pulse tube, inertance tube and reservoir. Potential pressure differences are also depicted.

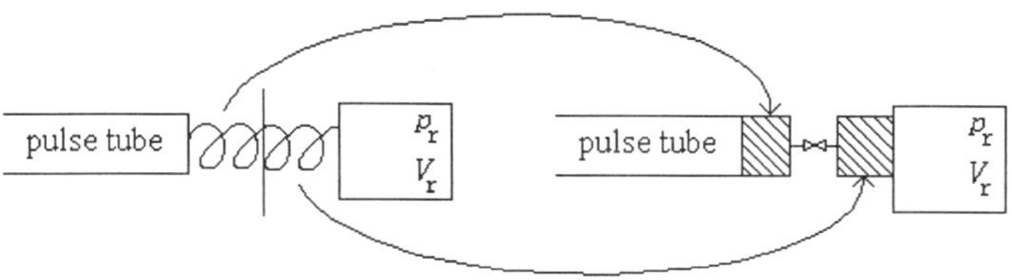

Figure 2. Analogy of the inertance tube. Inertance tube is separated into two parts, one forms part of pulse tube, while the other forms part of reservoir. This makes actual hot end of pulse tube prolonged into hypothetical pulse tube.

Focusing on the reservoir, if pressure change in the reservoir is Δp_r, then at the inlet port of the reservoir, the volume of gas that flows into or out of the reservoir is:

$$\Delta V_r = -V_r \Delta p_r /(\kappa p_r) \tag{5}$$

For the inertance tube, the volume is:

$$V_i = \pi d^2 L/4 \tag{6}$$

Assuming that the inertance tube has the same phase change and same pressure amplitude change, then the volume of gas that flows into or out of pulse tube hot end is:

$$\Delta V_i = -(V_r + V_i)\Delta p_r /(\kappa p_r) \tag{7}$$

Comparing Eq. (5) and Eq. (7), if $V_r >> V_i$ then:

$$\Delta V_i \approx \Delta V_r \tag{8}$$

Then let's consider flow resistance of the inertance tube. The pressure difference in the pulse tube is assumed to be Δp_{pt}. As shown in Figure 1, the inertance tube has a pressure drop from Δp_{pt} to Δp_r. Assuming a linear pressure drop distribution:

$$\Delta p_x = \Delta p_{pt} - (\Delta p_{pt} - \Delta p_r)\cdot x/L \tag{9}$$

V_i can be integrated to get:

$$\Delta V_i = -V_r \Delta p_r /(\kappa p_r) - \int_0^L \pi d^2/4 \cdot \Delta p_x \cdot dx/(\kappa p_r) \tag{10}$$

$$= -V_r \Delta p_r /(\kappa p_r) - V_i(\Delta p_r + \Delta p_{pt})/(2\kappa p_r) \tag{11}$$

$$= -(V_r + V_i/2)\Delta p_r /(\kappa p_r) - V_i/2 \cdot \Delta p_{pt}/(\kappa p_r) \tag{12}$$

From Eq. (12), the volume of the inertance tube is divided into two parts; one resembles the pulse tube volume, and the other part resembles the reservoir volume as depicted in Figure 2.

From the proceeding analysis, we could not see the advantage of inertance tube. However, the design becomes complex because the volume and the pressure drop are not easy to size.

FLOW CHARACTERISTICS OF TUBE

Flow includes two typical states: laminar and turbulent. For constant flow, the transition point from laminar to turbulent is *Re*=1600~ 2000. Here *Re* is defined as:

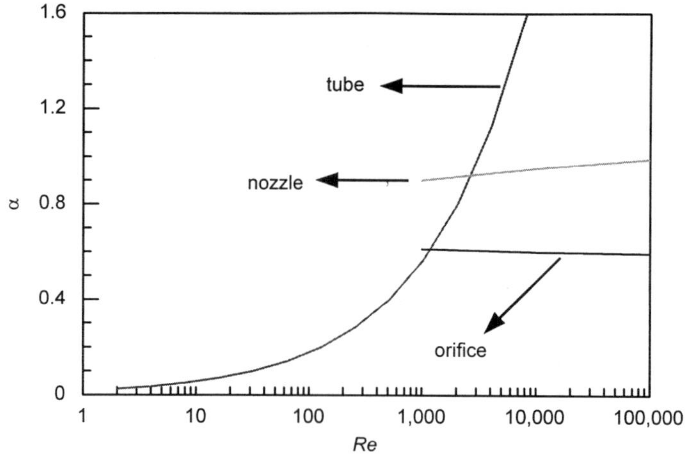

Figure 3. Flow coefficient changes with *Re*. Tube result is based on laminar assumption, while nozzle and orifice DATA are from reference where no result is given for *Re*<1000.

$$Re = ud / \nu \tag{13}$$

where u represents velocity, d tube diameter and ν viscosity.

For laminar flow, the flow resistance coefficient is:

$$\lambda = 64 / Re \tag{14}$$

and the pressure drop is:

$$\Delta p = \lambda \cdot L / d \cdot \rho u^2 / 2 \tag{15}$$

Considering the mass flow rate:

$$\dot{m} = \pi d^2 \rho u / 4 \tag{16}$$

Applying Eq. (13, 14, 15) to Eq. (16) one arrives at:

$$\dot{m} = \frac{\pi d^4 \Delta p}{128 \nu L} \tag{17}$$

Equation (17) means that the pressure drop is in proportion to the mass flow rate. But generally, the mass flow formula for an orifice or nozzle is expressed as follows[12]

$$\dot{m} = \alpha \cdot \pi d^2 / 4 \cdot \sqrt{2 \rho \Delta p} \tag{18}$$

where α is flow coefficient, which is generally constant when *Re* is large.

For comparison, Eq. (17) can be arranged into another form similar to Eq. (18):

$$\dot{m} = \sqrt{Re \cdot d / (32L)} \cdot \pi d^2 / 4 \cdot \sqrt{2 \rho \Delta p} \tag{19}$$

From Eq. (19), the flow coefficient of the tube with laminar flow can be expressed as:

$$\alpha_{tube} = \sqrt{Re \cdot d / (32L)} \tag{20}$$

In Figure 3, the change of the flow coefficient α_{tube} with *Re* is given assuming *L*/*d*=100. At the same time, the flow coefficient for an orifice and nozzle is also given according to the reference.[12] These curves are very different. With increasing *Re* number, the α_{tube} increases evidently. The α for a nozzle also increases, but α for an orifice slowly decreases. Similar results can be found in Reference 13.

REFRIGERATION PERFORMANCE

With similar initial conditions, the flow coefficient difference in Figure 3 will result in different refrigeration amplitude and different performance. Before this can be explained, refrigeration amplitude should be analyzed.

In 1988, based on implied assumptions, Y. Matsubara and A. Miyake reached the conclusion that theoretical refrigeration has the following form[14]

$$Q_c = m_0 R T_0 \ln(p_{pt} / p_r) \tag{21}$$

where m_0 is mass flow through orifice, p_{pt} is the pressure in pulse tube, and p_r is the pressure in reservoir. The time integration of Q_c through one cycle gives the ideal refrigeration.

A similar expression is acquired, after deduction, for an ideal condition.[15] The main assumptions are: (1) ideal square pressure wave in pulse tube; and (2) hot end gas or m_0 undergoes an isothermal compression process in pulse tube. The result is:

$$Q_c = m_0 R T_0 \ln(p_h / p_l) \tag{22}$$

$$= m_0 R T_0 \ln(p_h / p_r) + m_0 R T_0 \ln(p_r / p_l) \tag{23}$$

in which p_h, p_l represent the high and low pressure in the pulse tube. For an ideal gas isothermal compression process, process work equals to process heat. Eq. (22) represents this heat. This method to analyze refrigeration may be called "compression work dissipation".

According to Eq. (21), refrigeration in differential form should be:

$$dQ_c = R T_0 \ln(p_{pt} / p_r) \dot{m}_0 \, dt \tag{24}$$

For unit time with frequency f:

$$Q_c = f R T_0 \oint_\tau \ln(p_{pt} / p_r) \dot{m}_0 \, dt \tag{25}$$

The maximum mass m_0 that flows into reservoir is:

$$m_0 = \int_0^{\tau/2} \dot{m}_0 \, dt \tag{26}$$

where $\tau/2$ represents a half cycle.

When m_0 is unchanged and refrigeration is calculated according to Eq. (25), a different expression form of $\dot{m}_0$, such as Eq. (17, 18), will result in a different theoretical refrigeration amount.

This phenomenon is found in simulation first.[16] For example, when Eq. (17, 18) is used to calculate theoretical refrigeration, assuming pressure wave is sinusoidal and m_0 is unchanged, the difference may be 10% in theoretical amplitude. Eq. (17) will produce a larger figure.

Based on the upper analysis, a similar calculation is done. For this calculation, the pressure difference is the same and m_0 = constant, at the same time the inertance tube is treated as two parts just like that depicted in Figure 2, while flow rate is determined by Eq. (17). Figure 4 shows a typical result for Eq. (17, 18), where X-axis is for a non-dimensional length of pulse tube, and the Y-axis is the pressure in a pulse tube. Gas movements at both the hot end and cold tip are given. At the hot end, laminar tube flow produces an evidently larger circle than that of the orifice, though the maximum value of pressure and position are the same. At the cold tip, a larger circle is also acquired. The difference in the circle area represents the different refrigeration amplitudes.

In the analysis, the increase of theoretical refrigeration is about 8-10%. This makes net performance different. Generally for cryocoolers, COP at 80 K is about 10% of Carnot cycle efficiency. This means 90% of refrigeration capacity is consumed due to various losses. If theoretical capacity increases 10% while only 5% is consumed due to loss, COP at low temperature will increase 4~5%.

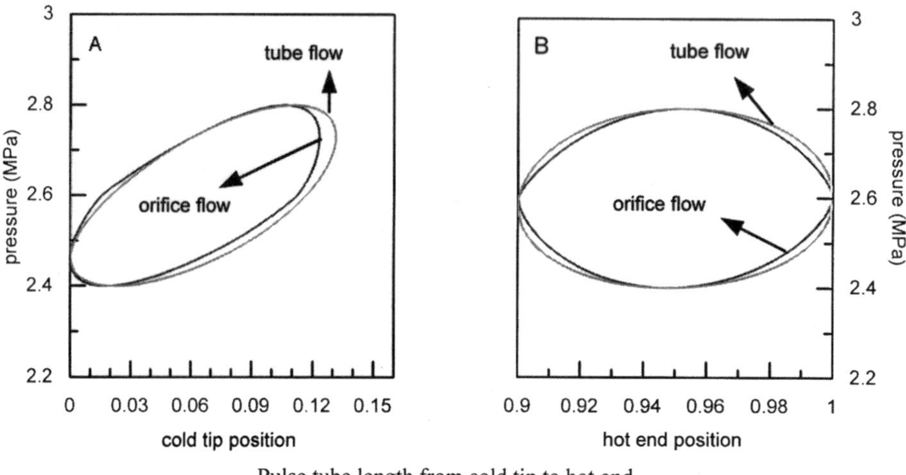

Figure 4. Gas piston movement in pulse tube hot end and cold tip

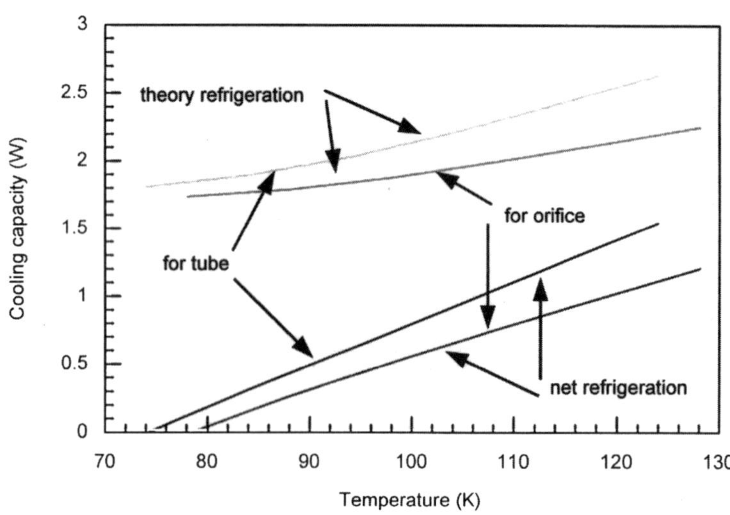

Figure 5. Simulating result about refrigeration amplitude and temperature. Inertance tube produces evident larger refrigeration amount than that of orifice, both for theoretical gross refrigeration and calculated net refrigeration. Lowest temperature with inertance tube is also lower than that with orifice.

Performance simulation based on one typical dimension of a miniature pulse tube cooler is calculated, and the results are shown in Figure 5. In Figure 5, the calculated theoretical cooling capacity and the net refrigeration are given. The difference between the theoretical refrigeration grows larger when the refrigeration temperature becomes higher. This is the same for the calculated net refrigeration. The lowest temperature with the inertance tube is about 4 K lower than that with the orifice.

COMPARISON OF SYMMETRY NOZZLE WITH INERTANCE TUBE

The previous analysis shows that the inertance tube flow instead of an orifice or check valve will improve performance. In reality, a symmetry nozzle or venturi tube has features similar to that of a tube, too, as shown in Figure 3. The nozzle flow coefficient increases with the *Re* number and is very similar to tube flow.

Assuming the mass flow rate through the tube or through a nozzle is the same, then

$$d_{tube}^2 u_{tube} = d_{nozzle}^2 u_{nozzle} \tag{27}$$

$$d_{nozzle} u_{nozzle} /(d_{tube} u_{tube}) = d_{tube} / d_{nozzle} \tag{28}$$

and

$$Re_{nozzle} / Re_{tube} = d_{tube} / d_{nozzle} \tag{29}$$

Tube flow has a smaller *Re* compared to a nozzle because tube flow has a larger diameter.

For a special condition,[17] the *Re* number for a nozzle reduces to:

$$u_{\max} = \sqrt{2(p_h - p_r)/\rho/\xi} = \sqrt{2 \times 2 \times 10^5 / 3.17/10} = 112 m/s \tag{30}$$

where ξ is the minor resistance coefficient and assumed to be 10.

Then:

$$Re = \frac{u_{\max} d\rho}{\mu} = \frac{112 m/s \times 0.0002 m \times 3.17 kg/m^3}{186 \times 10^{-7} N \cdot s/m^2} = 3818 \tag{31}$$

In a miniature pulse tube refrigerator, the gas flow *Re* in and out of the reservoir is small, even in nozzle flow. Considering past research where "laminar flow persists partially even when *Re* is high,"[8, 9] we can conclude that the nozzle will show similar features just like the inertance tube, and cooling capacity will increase and the coldest temperature will be lowered further.

In our laboratory, we find that when the symmetry-nozzle is used instead of a check valve, evident performance improvement is measured.[17]

Though both the inertance tube and the nozzle benefit from the increase in performance, there is a difference. Tube flow will be better than nozzle flow, however, a tube has a rather large volume when compared to pulse tube and this makes the design and experiments complex. The inertance tube volume may change actual pulse tube volume while a nozzle will not.

DOUBLE-INLET

According to the previous analysis, the double-inlet is still effective to drop temperature. The double-inlet is effective because gas from the hot end of a pulse tube makes the gas movement in the cold tip smaller. As seen in Figure 4, the gas movement in the cold tip is longer than that at the hot end, and adding the double-inlet will make this gas movement in the cold tip short. But because the typical feature of a miniature pulse tube is high average pressure and small pressure ratio, this makes the cold tip gas piston change small due to the fact that the compressibility and function of a double-inlet may be small.

In our experiments on a miniature pulse tube cooler, with a check valve or with a symmetry-nozzle, the double-inlet and multi-bypass will drop the lowest temperature further.[17]

CONCLUSION

Based on steady state or quasi-state assumptions, the difference between the tube flow and the orifice is revealed. The flow rate of the former is in proportion to pressure drop, while the latter is in proportion to the square root of the pressure drop. Because the flow of the orifice or inertance tube is directly related to refrigeration capacity in pulse tube refrigerators, the inertance tube will produce larger refrigeration capacity with the same pressure amplitude and mass flow amount. This can partly explain the reason that the inertance tube will produce better performance. The symmetry-nozzle has similar characteristics and has been verified by experiments. In reality, the symmetry–nozzle may be even more convenient and practical.

ACKNOWLEDGEMENT

This research is supported by the National Superconductor Center of China and National Natural Science Foundation of China.

REFERENCES

1. E. I. Mikulin, A. A. Tarasov, and M. P. Shkrebyonock, *Adv. Cry. Eng.*, 29: 629(1984).
2. S. Zhu, P. Wu. and Z. Chen, *Cryogenics*, 30: 514(1990).
3. S. Zhu, S. L. Zhou, N. Yoshimuru, and Y. Matsubara, *Cryocooler 9*: 269 (1997).
4. D.L. Gardner and G.W. Swift, *Cryogenics*, 37: 117(1997).
5. P. Roach and A. Kashani, *Adv. Cry. Eng.*, 43: 1895(1998).
6. L. Duband, L. Charles, A. Ravex, L. Miquet, C. Jewell, Experimental results on inertance tube and permanent flow in pulse tube coolers, *Cryocooler 10*(1999).
7. K.V. Ravikumar, Y. Matsubara, Experimental Results of pulse tube cooler with inertance tube as phase shifter, *Cryocooler10*(1999).
8. M. E. David, B.G.James, *J. Fluid Mech.*, 222: 320(1991).
9. Mikio Hino, Masaki Sawamoto, shuji Takasu, *J. Fluid Mech.*, 75: 193(1976).
10. C. H. Joshi, R. D. Kamm, J. M. Drazen, A. S. Slutsky, *J. Fluid Mech.*, 133: 245(1983).
11. U. H. Kurzweg, *J. of Heat Transfer*, 107: 459(1985).
12. W. F. Robert, Alan T. McDonald, *Introduction to fluid mechanics*, John Wiley & Sons, Inc (1978), p. 449.
13. Hongjun Jiang, *Flow Mechanics*, Higher Education Publication of China (1985), p. 308.
14. Y. Matsubara, A. Miyake, *Cryocooler 5*: 127(1988).
15. L.W.Yang, *Research of pulse tube refrigeration mechanism and practical development*, Post-doctor report, Chinese Academy of Sciences (1998).
16. L. W. Yang, S. Y. Bian, J. T. Liang, Y. Zhou, Code simulation and verification of pulse tube refrigerator, *Chinese J. of Cryogenics*, 99: 1 (1997).
17. Y. Zhou, Y. L. Ju, W. X. Zhu, J. T. Liang, E. C. Luo, High efficiency coaxial miniature pulse tube cryocooler with symmetry spray nozzle, *Chinese J. of Vacuum and Cryogenics*, 4: 150 (1998).

INFLUENCE OF A PHASE LEADING PRESSURE WAVE ON PULSE TUBE PERFORMANCE

H.W.G. Hooijkaas[1], A.A.J. Benschop[1] and S.C.M. Aerts[2]

[1]Signaal-USFA, Meerenakkerweg 1, PO-box 6034,
5600 HA, Eindhoven, The Netherlands
[2]Eindhoven University of Technology, PO-box 513,
5600 MB, Eindhoven, The Netherlands

ABSTRACT

The phase angle between the pressure wave and the gas flow in a pulse tube has an important effect on the pulse tube performance.[1, 2] Systems with a phase leading pressure wave at the warm end have in general a better efficiency and a higher cooling performance than the ordinary orifice pulse tube refrigerator.

In our recent study we analyzed the influence of this phase lead. With our simulation program, which is based on the harmonic approximation, we studied the influence of the phase angle on the various enthalpy flows in the pulse tube. These simulation studies have been verified with experimental pulse tube refrigerators. Besides these numerical simulations and verification tests, an experimental test system with a second linear compressor at the warm end was used to observe the overall effects of the phase angle. The use of the second compressor makes it possible to change both the phase with respect to the pressure wave and magnitude of the warm end volume flow.

It is shown that only the gas flow contributing to the compression process, i.e. the out phase part should be added at the warm end. The optimal situation for our system corresponds to a phase lead on the order of 54 degrees.

INTRODUCTION

Throughout the last two decades a lot of conceptual changes on the orifice pulse tube configuration have been introduced. All these systems tend to decrease the cold end phase angle between the pressure wave and the mass flow. This smaller phase angle results in a better cooling performance and a higher efficiency.

The most common way to decrease the cold end phase angle, is the introduction of an additional gas flow at the warm end. This additional gas flow contributes to the process of

Advances in Cryogenic Engineering, Volume 45.
Edited by Shu *et al.*, Kluwer Academic / Plenum Publishers, 2000.

compression and expansion in the pulse tube. In this paper the introduction of this flow at the warm end is shown beneficial for both hydrodynamic and thermodynamic aspects.

PULSE TUBE MODEL

The main cause for the phase shift between the pressure wave and gas flow is mass accumulation in the system. Mass conservation states:

$$\frac{\partial\rho}{\partial t} = -\frac{\partial\rho\mathbf{u}}{\partial x} \tag{1}$$

Applying the equation of state for an ideal gas to this equation yields:

$$\frac{\partial p}{\partial t} = -\alpha\frac{RT}{AM}\frac{\partial\rho\phi_v}{\partial x} - \beta\frac{R\rho}{AM}\frac{\partial T}{\partial x}\phi_v \tag{2}$$

where ϕ_v (=A**u**) denotes the volume flow. In the first term on the right hand side the parameter α depends on the rate of heat exchange between the gas in the volume and its surroundings. Two extreme values, which correspond to the two limiting thermodynamic regimes can be given. In the isothermal case the parameter α equals 1 and in an adiabatic environment the value of α is equal to γ (= c_p/c_v), the adiabatic expansion factor.

The second term on the right side of this equation takes into account the effect of heat transport by a flow of gas with an internal gradient. Parameter β depends on the thermodynamic situation. In the isothermal case it is zero, as heat transported by the flow is blocked by the ideal heat exchange. In the adiabatic case it equals γ.

In order to evaluate Eq. (2), we assume the actual pulse tube to be adiabatic and without pressure gradient. This leads to the much simpler equation:

$$\frac{\partial p}{\partial t} = -\frac{\gamma p}{A}\frac{\partial\phi_v}{\partial x} \tag{3}$$

From this relation it is seen that the rate of pressure build up in a particular volume is in first order proportional to the accumulation of gas ($-\Delta\phi_v$) in this volume. Consequently, the accumulation of gas will be leading the pressure wave. Hence, the downstream phase difference between the pressure wave and volume flow tends to shift towards a lag of 90 degrees.

The conservation of momentum states:

$$\frac{d\rho\mathbf{u}}{dt} = -\frac{\partial p}{\partial x} + Z\mathbf{u} \tag{4}$$

where Z is an effective friction factor, which accounts for the combination of drag and dissipation forces. For most parts of the cooler the change in momentum will be negligible. However, in one special case, the inertance tube, it is rather important.

In case of a dominating resistance the pressure drop is in phase with the flow, as can be seen from Eq. (4). This means that the phase difference decreases in sections with flow resistance. However, in adjacent void volumes the phase angle will increase. Therefore, the resulting phase difference at the compressor depends on the total geometry of the system.

The temperature oscillations and energy flows in the system are related by energy conservation, as stated in the first law of thermodynamics. In our analyses a system under steady operational conditions is considered. This restriction allows the use of time averaged quantities, which are denoted by means of brackets < >. Applying the first law in the enthalpy form, yields with time averaging:

$$\left\langle \frac{d\rho H_m}{dt} \right\rangle = \langle \dot{Q} \rangle + \left\langle \frac{dp}{dt} \right\rangle \tag{5}$$

In this equation all terms are normalized to the cross sectional flow area. The heat flow accounts both for heat exchange with the surroundings as well as conduction in the gas.

As the pressure wave is periodic, the term with the derivative (dp/dt) drops out. The enthalpy of an ideal gas is dependent on the temperature only. Using the isobaric specific heat and splitting the heat flow into a conduction and heat exchange term, yields:

$$\left\langle \frac{\partial \rho \mathbf{u} c_p T}{\partial x} \right\rangle = \left\langle \frac{\partial}{\partial x} \left\{ \kappa \frac{\partial T}{\partial x} \right\} \right\rangle + \langle \dot{Q}_{surr} \rangle \tag{6}$$

where the averaged contribution of the time dependent enthalpy fluctuations is omitted on the basis of periodicity, as the product of density and temperature is proportional to pressure.

Solving Eq. (6) requires the temperature oscillations to be known. These oscillations are caused by the compression effect, the gas flow and the heat exchange in the system. To calculate these oscillations the time dependent equation of energy conservation is used. Because the time dependent variations of the temperature gradient are small for relative small pressure oscillations, the effect of heat conduction is neglected. This yields:

$$\tilde{T} = G(s)\left(\frac{\partial T}{\partial p} \tilde{p} - \tilde{x} \frac{\partial T}{\partial x} \right) = G(s)\left(\frac{1}{\rho c_p} \tilde{p} - \tilde{x} \frac{\partial T}{\partial x} \right) \tag{7}$$

where G(s) is a transfer function in the Laplace domain. This transfer function accounts for the influence of heat exchange and heat storage in the matrix and tube walls. It is a complex valued function, and both the amplitude and phase of the fluctuations are affected.

Substitution of Eq. (7) in Eq. (6) yields the first law in terms of pressure wave, mass flow and time averaged temperatures. For harmonic signals the enthalpy and heat flows are found after integration with respect to the length coordinate x and multiplication with the cross sectional area, as given by Eqs. (8), (9), (10) and (11):

$$\langle \dot{H}_{comp} \rangle = G_{Re} A \tilde{p} \cdot \mathbf{u} = \tfrac{1}{2} G_{Re} \hat{p} \hat{\phi}_v \cos(\varphi_{p\phi}) \tag{8}$$

The enthalpy flow due to compression is proportional to the potential workflow (pdV). The proportionality factor (G_{Re}) is the real part of the defined transfer function, which is equal to 1 in an adiabatic tube. However, in reality there is always some heat exchange, and G_{Re} will be smaller. The heat exchange also enables two other contributions in the enthalpy flow to exist. The first one is due to a phase difference between displacement and pressure wave, which causes the gas to be colder due to expansion at one of its outer positions than at the opposite outer position. Consequently the gas absorbs and rejects heat at different positions, creating a heat flow. This effect, the so-called surface heat pumping, is the operating mechanism of the basic pulse tube refrigerator.

$$\langle \dot{H}_{sh} \rangle = G_{Im} A \tilde{p} \times \mathbf{u} = \tfrac{1}{2} G_{Im} \hat{p} \hat{\phi}_v \sin(\varphi_{p\phi}) \tag{9}$$

The downside of the capability to establish surface heat pumping due to heat exchange is the gradient effect. This effect rises from the temperature gradient of the surroundings, which causes the moving gas to meet different surrounding temperatures. The resulting variations in temperature difference give rise to heat absorption and rejection at different positions. As stated before, this causes a heat flow. In practice, the surface heat pumping and gradient effect will always occur at the same time and result in a total heat flow. However, for analysis it is more convenient to evaluate these effects separately.

$$\left\langle \dot{H}_{grad} \right\rangle = -G_{Im} A \frac{\rho c_p}{\omega} \frac{\partial T}{\partial x} \mathbf{u}^2 = -\tfrac{1}{2} G_{Im} \frac{\rho c_p}{\omega A} \frac{\partial T}{\partial x} \hat{\phi}_v^2 \tag{10}$$

According to Eq. (10), the enthalpy flow due to the gradient effect is always in the opposite direction of the gradient, i.e., the heat flow will be from the warm end to the cold end. So the gradient effect represents one of the irreversible losses in the system.

The last term in the time averaged energy balance is conduction. This is also an irreversible loss effect and according to the law of Fourier given by:

$$\left\langle Q_{cond} \right\rangle = -A\kappa \frac{\partial T}{\partial x} \tag{11}$$

The latter two equations only consider the gradient of the average temperature, as the fluctuations are relatively small and their contribution to the time average is even smaller.

MODEL ANALYSIS

In the previous section a set of equations on the pulse tube behavior was derived. The use of proper parameters for heat transfer and heat storage makes the relations applicable for both the regenerator and the actual pulse tube. In the present study we used the model to investigate the effect of various types of warm end configurations.

Considering Eqs. (8), (9), (10) and (11) shows that the compression and surface heat pumping in the pulse tube are the only two effects with a possible positive contribution to the cooling power. This suggests that those two effects should be optimized. However, as known from the analysis of basic pulse tubes, a critical gradient can be defined.[3] This is just that value of the temperature gradient at which the enthalpy flow from the surface heat pumping and the part of the gradient effect due to the out phase component of speed cancel each other. In practice, the orifice pulse tube or any of the improved concepts will have a gradient that is much larger than this imaginary, critical gradient. So consequently, it will not be beneficial to maximize the surface heat pumping.

Also the in phase speed, required for the compression effect, will contribute to the gradient effect. However, defining a similar critical gradient for these two contributions gives much larger values. This is caused by the fact that this critical gradient is proportional to the reciprocal value of the heat exchange, which will be rather small for the approximately adiabatic pulse tube. This implies that it is beneficial for the performance to establish a relatively large compression effect.

Based on this superficial analysis of the cooling performance, it appears that gas flow with a large out phase component is not favorable for this performance. On the contrary, both Eqs. (9) and (10) show that a relative large pressure oscillation is favorable for the cooling performance. This large pressure wave can only be established by means of high accumulation of gas, as seen in Eq. (3), the relation on pressure build up. Such a high level of accumulation requires the downstream out phase flow component to grow rapidly. But as shown before, these high out phase speeds degrade the performance.

However, the negative effects of the required high accumulation can be minimized by creating a negative warm end phase difference, i.e. a pressure wave phase lead. Such a phase leading pressure wave can be realized using a double inlet system, an inertance or some active phase controlling system. The effect of this particular phase shift is a decrease of the maximal amplitude of the out phase speed component. This implies a smaller average value for the squared speed, which appears in the losses due to the gradient effect, Eq. (10).

Table 1. Simulation results on mechanical input, enthalpy flows, cooling power and phase shift for our LPT-6/5 miniature pulse tube for different warm end configurations

	OPTR	IPTR	DIPTR*	DIIPTR*	DIPTR	DIIPTR	
P_{mech}	28.7	30.2	15.9	14.7	30.1	31.7	[W]
ΣH_{PT}	2.42	2.67	1.94	1.93	2.81	2.97	[W]
ΣH_{Reg}	1.96	2.00	1.13	1.05	1.77	1.80	[W]
Q_C	256	457	610	679	830	963	[mW]
θ_{WE}	4	-1	-26	-30	-15	-19	[°]

Although the surface heat pumping will decrease too (it will even become a loss effect at the warm end), the overall effect will be beneficial for the proper initial phase. Next to this direct influence of the speed on the enthalpy flows, another benefit is the improvement of regenerator performance. This results from the fact that less gas passes the regenerator, so the internal temperature fluctuations and the associated losses will reduce. The last positive effect of this phase lead rises from the fact that a part of the gas required for accumulation comes in via the warm end. For this part of the volume flow the freezing effect is omitted, consequently the volume flow at ambient temperature is smaller. This reduces the flow losses.

SIMULATIONS

We used our simulation program to analyze the impact of a double inlet, an inertance and the combination of both on a pulse tube refrigerator. The chosen input dimensions correspond to one of our miniature pulse tubes (LPT-6/5). This pulse tube is equipped with a combination of second inlet and inertance. It has a cooling power of 850 mW at 80 K for a mechanical input power of 26 W at a frequency of 40 Hz. The lowest temperature at this input is 56.6 K. This input requires 60 W electrical input. However, this is due to the poor match between cooler and compressor. We expected to need approximately 40 W after a redesign of the compressor. Simulations of this particular configuration predict a cooling power of 1141 mW at a mechanical input of 26 W. In these simulations the dewar losses are not taken into account.

In our simulations, we started with the single orifice type of pulse tube (OPTR) and chose the geometrical conditions, operating frequency and average pressure equal to those in the above experiments. The pressure wave amplitude in the pulse tube was chosen to be 0.24 MPa. This pressure wave in the pulse tube was kept constant in the simulations performed. The results of the simulations are given in Table 1.

The abbreviation DIIPTR is used to denote the combination of double inlet (DIPTR) and inertance (IPTR). In this table two columns are marked by an asterisk. For these simulations the first orifice respectively inertance was identical to those of the system without a second inlet. The two columns right of those marked with an asterisk contain data on the system with optimized first orifice and inertance in double inlet mode.

From the simulations it is seen that the efficiency is the highest for the biggest phase lead of the pressure wave (corresponding with a negative phase angle θ_{WE}). For the optimized settings in the latter two columns, the in phase flow is increased, which leads to a

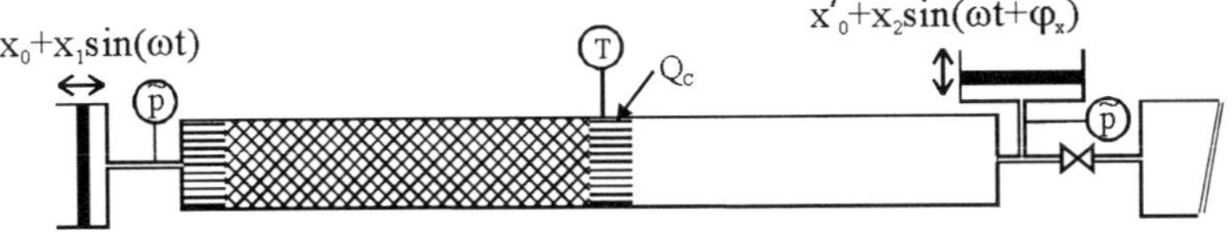

Figure 1. Experimental setup: Pulse tube with an additional warm end phase shifter.

smaller phase lead. The result of this optimization is a larger enthalpy flow in the pulse tube and also in the regenerator. This improves the cooling power, but decreases the efficiency.

EXPERIMENTS

For the experiments we used a special test mockup, which is depicted schematically in Figure 1. In order to adjust the warm end phase difference between pressure wave and gas flow, we added a compressor (expander) to the warm end. This linear compressor is supplied by a combination of amplifier and triggered tone generator. This provides the possibility to adjust the phase angle and stroke as required.

The pulse tube cooler used is of the u-shape type. The pulse tube is 70 mm long and has an inner diameter of 6.1 mm. The regenerator has the same length and a diameter of 8.77 mm. It is filled with 325 Mesh wire screen with a wire diameter 23 μm. The low temperature of the system in single orifice mode and a filling pressure of 2.4 MPa is 94 K at a mechanical input power of 35 W_{mech} and a driving frequency of 35 Hz. At the same input conditions and a filling pressure of 2.8 MPa, but with a second inlet and inertance the system reaches a low temperature of 79 K. The ratio between the mechanical and electrical input power is rather poor due to the unoptimized compressor geometry.

The phase between the pressure wave and the piston movement can be found from an evaluation of the work delivered by the compressor:

$$\langle W \rangle = -\frac{1}{T}\oint F dx = -\frac{A_p}{T}\oint_T p u dt = -\tfrac{1}{2}\hat{p}\hat{\phi}_v \cos(\varphi_{p\phi}) \tag{12}$$

where the volume flow at the surface is substituted for $A_p u$. This relation shows that only the in phase part of the piston speed contributes to the work. The out phase component affects the pressure wave, but has no influence on the work. In evaluating the phase angle at the warm end of the pulse tube, the influence of the void volume of the compressor has to be taken into account. This effect can be calculated using Eq. (2).

RESULTS

During all experiments, the filling pressure was 2.0 MPa and the frequency 50 Hz. First, some brief tests on the system with the second compressor passive were done. In these tests the orifice between the pulse tube and the buffer was closed. Operating the system with no load on the warm end expander, a low temperature of 281 K was reached at an input power of 50 W. This poor performance rises from the fact that the system has only small dissipation in the expander, which limits the enthalpy flow. Despite this small internal damping the phase lag of the piston speed to the pressure wave is 11 degrees. This relative large phase shift is because the ratio of driving frequency and resonance frequency of the expander is close to one. At this specific value the transition between 0 and 180 degrees phase difference for second order systems occurs. So even for a relative small damping factor, the phase shifts are noticeable. Applying a load of 4.7 Ω to the expander results in a temperature decrease to 162 K at 50 W input. Because of the much larger dissipation, the phase lag increased to 46 degrees.

In the next experiments, we installed an orifice with a diameter of 0.35 mm between the pulse tube and buffer. This is the optimal orifice for the experiments without an expander. The load on the expander was replaced by the amplifier. In the experiments the input power of the main compressor is kept at 30 W. The phase between pressure and piston movement is shifted from 0 to 360° by adjusting the phase of the piston movement.

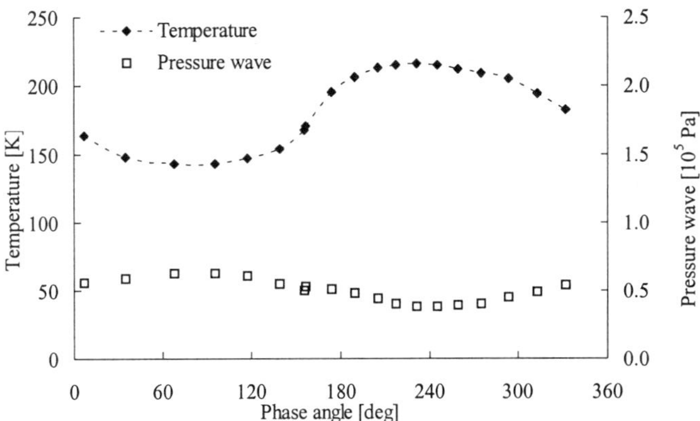

Figure 2. Low temperature and pulse tube pressure amplitude related to the phase angle between pressure wave and piston speed. Expander stroke in the order of 0.12 mm.

The results are depicted in Figure 2. It is clearly seen that an additional compression by the expander results in a better performance. The maximal pressure wave and minimum temperature are reached for a phase difference of 75 degrees between pressure and voltage. At this setting the phase difference between voltage and current is 90 degrees, corresponding to zero input power for the second compressor.

To verify the relation between the improved compression and the performance the expander stroke was raised from approximately 0.12 mm to 1.0 mm peak-peak. The results are depicted in Figure 3. It is seen that increasing the compression by the expander results in a better performance. However, the increase to the largest stroke shows a small rise in the lowest temperature. For the smaller amplitudes the input power is approximately zero for the optimal settings. However, for the larger strokes the phase angle for the lowest temperature is smaller than those for maximal pressure amplitude. This shift is also noticeable in the decrease of the phase angle between current and voltage (down to 69 respectively 64 degrees), which corresponds with a measured dissipation of power in the compressor. This means that the orifice is not optimal for the system under these operational conditions.

As we are mainly interested in the effect of additional out phase gas flow, we applied a series of tests for increasing stroke with zero input. This latter condition prevents additional dissipation in the expander. The input power was lowered to 20 W for these experiments. In Figure 4 the results for temperature and pressure amplitude at the driving compressor are depicted in relation to the pressure wave in the pulse tube. The increasing stroke causes a

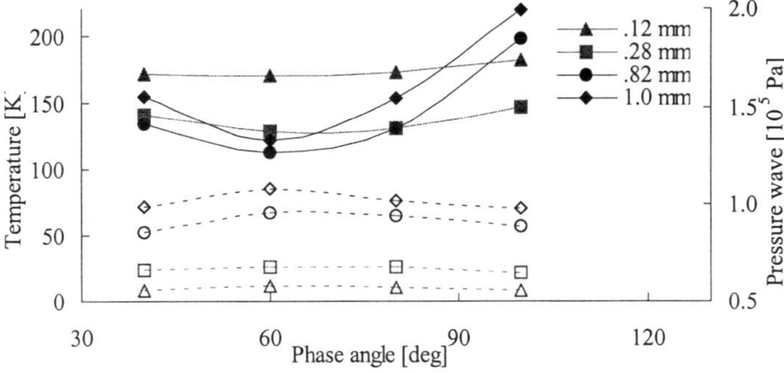

Figure 3. Low temperature and pulse tube pressure amplitude as function of the phase angle for different strokes. Solid lines denote temperatures, dashed lines pressure. Corresponding marker shapes are related.

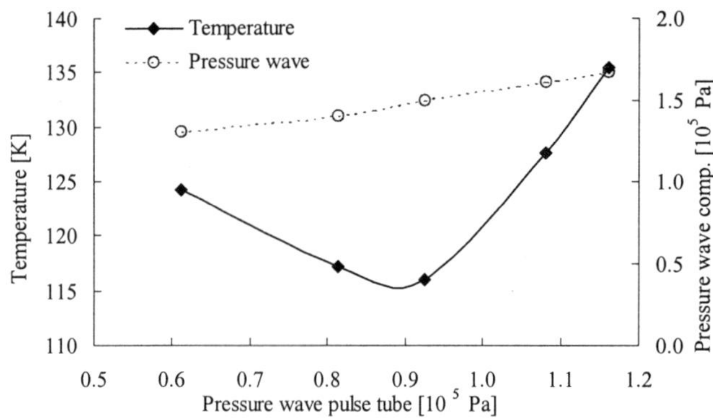

Figure 4. Low temperature and compressor pressure wave, for varying pressure wave in the pulse tube. The first compressor input power is constant at 20 W and the stroke of the second is varied for zero input power.

bigger pressure wave and a smaller gas flow through the regenerator, which is clearly seen from the increasing pressure amplitude in the compressor. The optimal performance is found for a pressure amplitude of 0.89×10^5 Pa in the pulse tube, with a corresponding stroke of 0.70 mm peak-peak. This value is quite similar to the best results of the experiments described before. The corresponding added flow is 28.2 mL/s and the in phase flow for this combination of orifice and buffer is 20.7 mL/s. Therefore, the optimal phase lead for the pressure wave, using a passive expander in this setup, is 54 degrees.

CONCLUSION

It is shown that the introduction of an out phase warm end flow improves the pulse tube performance. This flow can be generated without any additional input power. In case of large added flows, the out phase flow that contributes to the dissipation can even be slightly increased. The optimal pressure wave phase lead for the current system is 54 degrees.

ACKNOWLEDGEMENTS

This work is carried out as part of a research cooperation between Signaal-USFA and the Eindhoven University of Technology. Signaal-USFA and its affiliate Cryotechnologies, together form the SBU Cryogenics of Thomson-CSF.

REFERENCES

1. M. Kasuya, M. Nakatsu, Q. Geng, J. Yuyama and E. Goto, *Cryogenics 31*:786(1991).
2. D.L. Gardner and G.W. Swift, *Cryogenics 37*:117(1997).
3. M.J.A. Baks, Eenvoudig analytisch model voor een pulsbuis-koelmachine, Eindhoven University of Technology,(1990).

THERMODYNAMIC LOSSES IN PULSE TUBE COOLERS — THROTTLING IRREVERSIBILITY INCLUDING CONTAMINATION EFFECTS

L. Rohlin, A. K. Le and T. H. K. Frederking

Cryogenics Lab, Chemical Engineering Department
University of California Los Angeles 90095 CA

ABSTRACT

We have continued studies of the thermodynamic losses in pulse tube coolers. The topics addressed are loss functions for pressure drop in a single component, loss analog for irreversible pressure differences associated with irreversible temperature differences (Gifford-Longsworth model), Joule-Thomson [JT] throttling and orifice loss modeling. In the experiments we have determined thermodynamic loss effects including contamination influences during a long-term test of approximately 2000 hours.

INTRODUCTION

In preceding investigations,[1] thermodynamic loss functions have been determined for several orifice pulse tube [OPT] cooler components: the heat exchangers' irreversible temperature differences (Gifford-Longsworth [GL] model[2] including a modified model), the cold box irreversible temperature difference, the thermostatic Kittel model for the orifice, and the heat leak model for insulation heat leakage and other parasitic heat inputs. In the present studies, we extend OPT loss modeling. The components and subsystems considered are listed in Table 1. Emphasis is on flow power expenditure rates caused by irreversible pressure drops and JT throttling including a quasi-steady oscillatory flow model for the orifice. The results are consistent with several individual loss experiments, with specific cooler characteristics and with contamination effects. Contamination is caused by H_2O frost deposits carried into cold sections with the compressed air used in our experiments.

THERMODYNAMIC LOSS MODELS OF COMPONENTS

Our previous work on thermal irreversible effects[1] has been continued with emphasis on hydrodynamic loss effects. All functions are expressed in non-dimensional form using COP_C [= Carnot Coefficient of Performance] as argument. The loss function is expressed as

Advances in Cryogenic Engineering, Volume 45.
Edited by Shu *et al.*, Kluwer Academic / Plenum Publishers, 2000.

Table 1. Thermodynamic loss models for hydrodynamic components

System	Carnot Fraction [CF]
1. P-drop of single flow component	$CF = 1/[1 + \beta_{pl}\, COP_c]$
2. Dual P-drop , Gifford-Longsworth analog	$CF = COP_c^{-1}/[\{(1 + \beta)/(1 - \beta)\} - 1)]$
3. Joule-Thomson [JT] throttling	$CF = 1/[2 + COP_c]$
4. Oscillating flow through OPT orifice	$CF = CF[\, COP_c, f/f_{ref}]$

deficiency in Carnot fraction CF, i.e. $[1 - CF_j]$ for the jth component. COP_C is considered a non-dimensional low temperature.

Single-Pressure Drop Subsystem

The first entry in Table 1 is the CF-function associated with the pressure (P) drop in a single flow component, e.g. heat exchanger. $[1\text{-}CF_{pl}]$ is the related thermodynamic loss function. It increases from zero to large values as COP_C is raised, as the low temperature T_L is increased. A comparison with thermal losses of the modified GL model leads to the following result: the loss function has a high-T_L asymptote in common with the modified GL model if the loss coefficient $\beta_{pl} = 2\ \alpha_{ir}$. This result is a constraint of the original GL model assuming two equal irreversible T-differences at the cold end and near the environmental temperature T_E. In contrast, the single pressure drop is not consistent with the two T-differences of GL for the heat transfer-related ΔT_{ir} values at the cold box and at the environmental T_E.

Dual-Pressure Drop - Analog of GL Model

Because of the factor of 2 in the asymptote, we introduce two pressure drops in our analog of the modified GL model,[1] entry 2 in Table 1. In this manner, both loss contributions, the irreversible T-difference-related loss function and the irreversible pressure drop related loss function are consistent. The case $\alpha_{ir} = \beta$ is characterized by equal loss function values at a low T_L, i.e. at a specified COP_C-value.

Joule Thomson Throttling

JT throttling is to be avoided in JT coolers for ideal gas states. Since the ideal gas undergoes zero temperature change during P-reduction in a JT valve, no cooling can be achieved. In real compressed gases, the second virial coefficient is sufficiently large to produce a cooling effect, starting at $T_E = 300$ K, with the exception of the gases neon, hydrogen, helium including isotopes. The present JT-loss model constitutes an extreme case of unfavorable assumptions: we adopt ideal gas, equal mass flow rates at OPT entry and through the JT valve (in lieu of an optimum "orifice"), further, complete throttling away of the pressure difference achieved in the compressor. It is noted below that our 4th model has high loss function values in common with the JT model in a certain limit of the frequency ratio (discussed below).

Orifice Loss Model

The orifice-related loss model (entry 4 of Table 1) is based on the following assumptions: cylinder hole geometry (inset of Figure 1), one-dimensional oscillatory flow with quasi-steady shear forces on the lateral surfaces of the flow constriction, frequency [f]-dependent viscous penetration depth $[\eta/(\rho\, \pi\, f)]^{1/2}$, pressure drop $[\Delta P]$ resulting from shear stress on the lateral wall only (obtained from the momentum theorem for fluid-wall interaction). For a specified refrigeration load Q_L, the Carnot fraction is characterized by additional power input $[\Delta W]$ required to compensate for the orifice flow power loss.

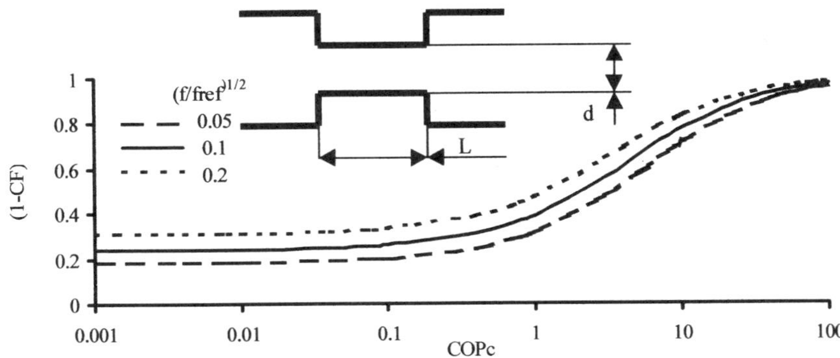

Figure 1. Orifice model loss function as a function of COP_C; Inset, orifice geometry

$$CF = 1/[1 + \Delta W/W_c] \qquad (1)$$

Carnot fraction is defined as COP/COP_c with Coefficient of Performance $COP = Q_L/W$; COP_C = Carnot value. For a specified Q_L - value, we have $CF = W_c/W$ where $W = W_c + \Delta W$. The resulting (non-dimensional) additional power input for the orifice loss model is expressed in Equation 1 as the ratio of ΔW to the Carnot power input W_c.

$$W_c = m\,(\Delta P_{eff}/\rho)/[COP_c + 1] \qquad (2)$$

Then the power ratio in Equation 1 is obtained as

$$\Delta W/W_c = [COP_c + 1]\ (f/f_{ref})^{1/2} \qquad (3)$$

The abbreviated result, Equation 3, contains the reference frequency

$$f_{ref} = (\Delta P_{eff})^2\ (\pi/256)\, d^4 / [m^2\,(L_o/d)^2\ (\eta\,\rho)] \qquad (4)$$

After insertion of Equation 3 into Equation 1 we obtain the thermodynamic loss function $(1 - CF_{of})$.

$$(1 - CF_{of}) = [COP_c + 1]\,(f/f_{ref})^{1/2} / [\,1 + (COP_c + 1)(f/f_{ref})^{1/2}] \qquad (5)$$

The function is shown in Figure 1 versus COP_c with frequency ratio as parameter. The loss function values increase as COP_c is increased and as frequency ratio is increased. For the special case of $f = f_{ref}$ we obtain the simple result $CF = 1/[2 + COP_c]$. This is the most unfavorable extreme case of ideal gas throttling (entry 3 of Table 1). Another pulsation-related loss model for fluid oscillations is the inertance tube model of Roach and Kashani.[3,4] For details, we refer to References 3 and 4.

Loss Function Superposition for Entire Cooler

As outlined in Reference 1, we add up pertinent thermodynamic loss functions in order to obtain the Carnot fraction of the entire cooler. There is the "insulation barrier" with $(1-CF_{ins})$ decreasing as COP_c is increased.[1] In contrast, all thermo-hydrodynamic loss functions considered so far show loss contributions increasing as COP_c, i.e. the low temperature of the cold box, is raised. An example is Figure 2 displaying a linear superposition of the insulation loss function and the heat exchanger loss function of the modified GL model. There is a clear minimum, at COP_c*, of the loss summation function. This minimum ought to be located at the specified cold box temperature of the cooler.

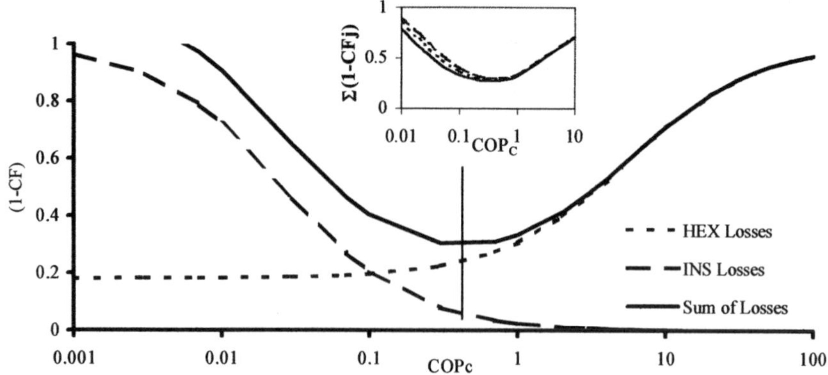

Figure 2. Loss function minimum as a result of optimization of the cold box;
Inset: loss function sum $\Sigma(1\text{-}CF_j)$ versus COP_c for superposition of insulation (INS) and heat exchanger (HEX) losses; insulation parameter (ref. 1) variation from 0.01 to 0.025.

Contamination Effects

Contamination effects are predicted by all thermo-hydrodynamic loss models of Reference 1 and Table 1. A condensate layer, e.g. of solid H_2O, CO_2, and of other deposits, alters unfavorably the cross section available to flow: the cross section is reduced. In addition, the layer deposited adds a thermal resistance thereby increasing the losses. Since regenerator heats and Q_L of the cold box have to pass through the condensate layer, the loss function value is enhanced accordingly. The only potentially positive effect is an enhanced heat capacity under certain conditions.

EXPERIMENTS

The experiments with a linear orifice pulse tube system[5] have been continued to include contamination effects in a long test run. We use compressed air as working substance from the internal supply in the building via a pressure regulator. The air is compressed by a Cooper centrifugal compressor in the UCLA-Ralph M Parsons cogeneration plant. The major contaminant for our OPT is water vapor converted into frost upon cool-down in the regenerator and cold box. There is a drier (desiccant) system downstream of the compressor with a specified humidity output given as wet bulb range from 266 K (20 °F) to 233K (-40 °F); (8 bar absolute; 100 psig).[6] Thus, there is an upper relative humidity (RH) value of approximately 15 % at 294 K (70 °F).

A five-phase stepping motor (Super VEXTRA) drives a rotary valve, which controls the pressure oscillation in the system. A function generator (Wavetek, San Diego, model 90) adjusts the speed of the stepping motor. In one rotation of the rotary valve,[4] the system is pressurized twice and exhaust gas is discharged twice. The exhaust is vented to the atmosphere. The regenerator is fabricated of 323 100 x 100 mesh 304 stainless steel screens (wire diameter 0.114 x 10^{-3} m) and 30 40 x 40 mesh copper screens (0.250 x 10^{-3} m). All screens are inserted in a 0.1016 m long, 0.0254 m inner diameter [schedule 40] PVC tube. The cold box section is a perforated copper plate.[7] A heater encloses the copper plate. The entire cold section is covered with fiberglass insulation. The pulse tube section has a length of 0.2478 m (inner diameter 0.0254 m). Phenolic flanges on each side of the system hold the system together with 4 tie rods. At the exit of the pulse tube section, a metering valve (Nupro "L" series) functions as an orifice. The metering valve is located before a reservoir with a volume of 0.664 liter. The purpose of the metering valve and reservoir is to offset the phase shift between the externally applied pressure wave, created in the upstream section, and the volumetric flow rate. Between the rotary valve and the regenerator a pressure transducer is inserted to measure the pressure oscillation in the pulse tube. A thermocouple is attached to the copper plate outside the pulse tube. A data acquisition system (National Instruments) collects the data from the thermocouple and pressure transducer. Labview 4.0 is the operating system employed for the data acquisition. Additional humidity data (Omega RH 70 and RH 70-ac) have been taken as well.

The frequency was set at 1.5 Hz and the inlet pressure at 3.75 bar [40 psig]. The orifice valve was set at 12 small units noting 25 small divisions per turn of the valve stem. The data acquisition system was set to collect 40 points with a 33 ms delay and an interval of 5 minutes.

DISCUSSION AND COMPARISON WITH THEORY

Observations and Data Discussion

The temperature curves for the cold box and the ambient temperature (T_E) (lab) are shown in Fig. 3. The cold box temperature reaches 256 K after 40 minutes. After a minimum is attained, the temperature slowly rises to 262 K. After 8.5 hours it reaches a steady state where it oscillates up and down for the next 50 hours. At the same time the ambient temperature (T_E) goes from 298 K to 299 K and then decreases to 296 K at 100 hours. The T_E thermometer does not observe the oscillation in the cold box temperature (T_L). Further, the overall trends in the cod box temperature are not observed in the T_E records. Instead, T_E of the laboratory air is the result of a combination of building ventilation, lab instrument dissipation rates and external weather conditions. The pressure behavior during the same time period is almost quasi-steady with no noticeable departures from mean frequency and mean amplitude. Figure 4 shows the results in the period where the temperature rises from its low value. A small decrease on the high-pressure side can be seen, and then P levels off on a steady state amplitude level. No correlation is noted between the pressure oscillation amplitudes and coldbox temperature oscillation records. The inset of Fig. 4 shows the pressure oscillation generated by the double acting rotary valve.

Figure 5 plots the cold box temperature $T_L(t)$ at four selected time intervals of the long-term test. The initial mean T_L-value is low in comparison to the later results. An apparent trend toward an increase in T_L for the first 500 hours is not reproduced subsequently. There is no statistically significant departure from the mean T_L-value in the next 1000 hours. These records are interpreted as support for quasi-steady cold box operation such that mean field temperature and mean field condensate thickness do not change significantly in the parameters range covered during the OPT runs.

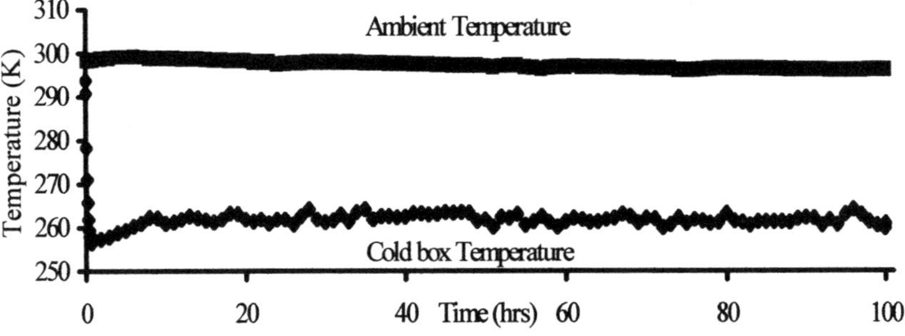

Figure 3. Temperature versus time during the first 100 hours of OPT long-duration test.

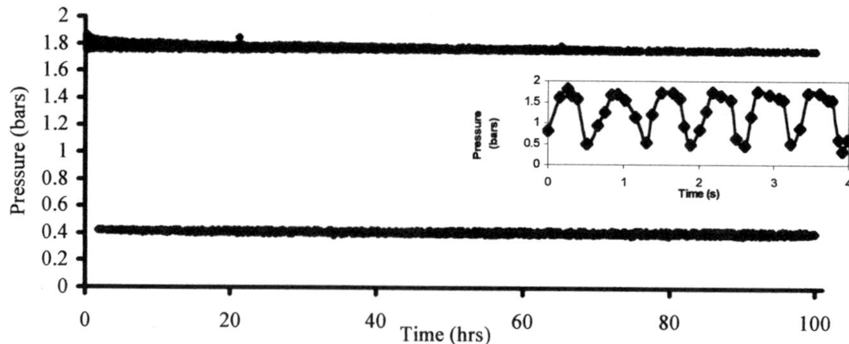

Figure 4. Pressure records versus time tracing $P_H(t)$ and $P_L(t)$; inset: P(t) oscillation pattern.

Figure 6 shows the temperature lift and cold box temperature as a function of time. The oscillation in the temperature lift is the mirror image of the oscillation in the cold box temperature. The temperature lift [T_E - cold box T] decreases when the cold box temperature increases.

Flow Visualization

In previous studies[4] we used a different setup with a plexiglass pulse tube section and regenerator. It was observed that ice or snow did build up at the cold end, and when the snow cover got too thick, the ice cover was broken off. Snowflakes were blown into the pulse tube section and settled 0.05 m away from the cold box. The time periods in which the temperature rises are much longer than the periods of temperature decrease. It is proposed after observing Fig. 3 that the snow builds up as the temperature steadily rises. When the snow breaks up, the temperature quickly decreases. The quick decrease is an indicator that the pulse tube is working properly and still retains its cooling effect.

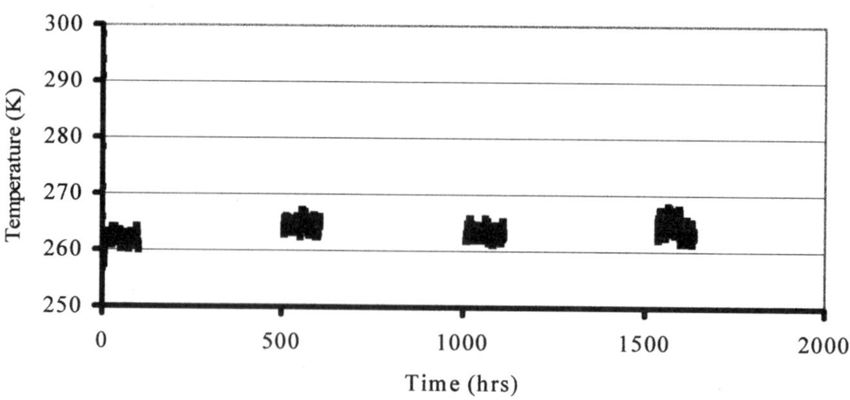

Figure 5. Cold box temperature versus time at discrete time intervals of the long-term test.

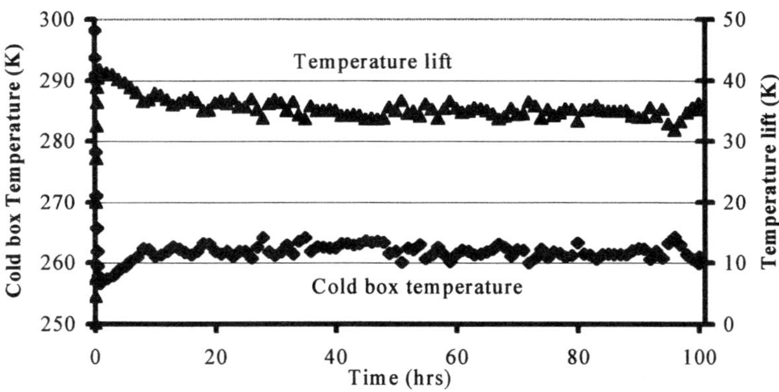

Figure 6. Cold box temperature and temperature lift versus time.

Long-Term Test

The 2000 hours of continuous OPT operation shows a robust behavior despite the continuous condensation of frost toward internals of the system. The lack of a significant disturbance of OPT operation by contamination with H_2O is interpreted as a quasi-steady input-output balance in mean field rates of H_2O inflow and outflow.

The prediction of theory is an enhanced cold box temperature and higher losses as the ice/frost layers increase in thickness. Simultaneously, the thermodynamic loss function is increased. For a nearly constant relative humidity, observed often during the long-term test, there is a significant diurnal variation in absolute humidity. The water vapor rate varies with the daily T-pattern. The mass fraction (water vapor-to-air) is 0.622 (RH) $P_{sat}(T)/P_{air}$, where the saturation pressure of H_2O vapor is a relatively strong function of temperature. Thus, the absolute mass of water vapor is high during the warm afternoon hours. This pattern is reflected in the cold box T-variations during the day.

CONCLUDING REMARK

The long term operation of more than 2000 hours shows a robust operational characteristics of the present OPT system despite considerable H_2O frost deposition inside the system. Condensation layer variations alter T_L without serious interruptions of quasi-steady transport rates in the range of parameters covered in the present investigations.

NOMENCLATURE

CF	Carnot fraction (= Carnot efficiency) $(1\text{-}CF_j)$ = thermodynamic loss function of jth component
COP	Coefficient of performance, Q_L/W
COP_c	Carnot coefficient of performance = $T_L/[T_H - T_L]$; COP_c* optimum value
d	orifice diameter
f	frequency, f_{ref} reference frequency, equation 4
L_o	orifice length
m	mass flow rate
P	pressure, ΔP_{eff}= effective P-difference; P_{sat} = saturation pressure of H_2O vapor
Q_L	refrigeration load at T_L
RH	relative humidity
t	time
T	temperature, T_L low temperature; T_H high T-value = environs T_E
W	power input into the cooler system;

ΔW	additional input power needed for loss compensation
α_{ir}	irreversible reduced temperature difference= $\Delta T_{ir}T$
β_{p1}	irreversible P-difference of a single component $\beta_{p1} = (\Delta P/P)\ \ln[P_H/P_L]$
β	irreversible reduced P-difference, $= \Delta P/P$
η	fluid shear viscosity
ρ	fluid density

Subscripts

c	Carnot value
H	high value ; e.g. $T_H = T_E$; P_H
ir	irreversible
L	low value , e.g. T_L , P_L
of	orifice model

ACKNOWLEDGMENTS

This work has been supported in part by NASA JPL / CIT President's Fund and NASA ARC. Additional information provided by Mr. David N. Johnson, UCLA, on the H_2O content of the compressed air system is appreciated. One of us, A. K. L. acknowledges support of his group project collaborators Andy Fong and Rebecca Mendoza [108B] on optimization studies. The senior author (thkf) has appreciated the hospitality of Prof. H. Quack, Chair of Refrigeration and Cryogenics (T U Dresden) and Prof. K. Andres, Director, Walther-Meissner Institut (Garching near Munich) during sabbatical leave. Frank Hung and Jane Tsai supported the contamination runs and studies. All of these inputs, and others not named, are acknowledged with thanks.

REFERENCES

1. Lars Rohlin, Comparison studies of thermodynamic losses in pulse tube cooler components, Proc. 17th Intern Cryog Eng Conf ICEC17, IoP Institute Physics Publ, Bristol and Philadelphia 1998, 89-92.
2. W. E. Gifford and R. C. Longsworth, Pulse Tube Refrigeration Progress, Int. Adv. Cryog. Eng. 10, (1965) 69-79.
3. P. R. Roach and A. Kashani, Pulse tube coolers with an inertance tube: theory, modeling and practice, adv. Cryog. Eng 41B (1998) 1895 - 1902.
4. Lars Rohlin, research results (unpublished) toward M S thesis, UCLA.
5. Lars Rohlin et al., Heat transfer phenomena during oscillations in an "air conditioning pulse tube", Refrig. Science and Technology, (1997), Intern. Institute of Refrigeration, Paris, 1997-5, pp 270-286.
6. David N. Johnson, Director, Energy Services, Ralph M. Parsons-UCLA cogeneration plant, private communication. 1999.
7. Henry Chen, Jeff Bennett et al., Pressure drop in pulse tube components, Cryocoolers 10, Proc. 10th Intern Cryocooler Conference, Monterey CA, May 1998. Ed. Ron Ross Jr., Plenum Press 1999, pp. 275-280.

IMPROVEMENT IN THE SECOND-STAGE COOLING POWER OF A TWO-STAGE GM REFRIGERATOR THROUGH MODIFICATION OF THE FIRST-STAGE REGENERATOR

T.W. Wysokinski, I.G. Spearing, P.G. Reedeker and J.A. Barclay

Cryofuel Systems Group, University of Victoria
Victoria, B.C., V8W 3P6, CANADA

ABSTRACT

Many research groups have made progress in improving the no-load minimum temperature and the 2nd-stage cooling power of a two-stage GM refrigerator by substituting rare-earth materials for lead in the 2nd-stage regenerator while leaving the 1st-stage regenerator unmodified. Because the 1st and 2nd-stages are a coupled system, modifications to the 1st-stage can impact the performance of the 2nd-stage. This work focuses on obtaining improved performance in the second stage solely by modifications to the 1st-stage regenerator. Simulation and experimental results for an improved 1st-stage regenerator that increases the cooling capacity of the 2nd-stage regenerator are presented. The results of the simulation were used to optimize the design of the 1st-stage regenerator. This new design is characterized by reduced void volume, improved flow impedance properties, and reduced turbulence in gas flow. Novel manufacturing techniques, which simplify the process of loading the regenerator with regenerative materials, and a custom-built GM system with a replaceable 2nd-stage cylinder used to compare modified and conventional regenerators are presented.

INTRODUCTION

Over the last several years, researchers have made steady progress in improving the no-load minimum temperature and the 2nd-stage cooling power of two-stage GM refrigerators. Design changes to date have focused on five main approaches to achieving this improvement. Nearly 25 years ago, Buschow *et al.*[1] pointed out the potential use of lanthanide materials as cryogenic regenerator materials. Replacement of the 2nd-stage regenerator matrix material with rare-earth/transition metal compounds such as Er_3Ni, $ErNi_{0.9}Co_{0.1}$, $HoCu_2$, and elemental Nd, which exhibit a heat capacity spike at low temperatures due to magnetic ordering, has yielded improved cooling performance[2,3]. In practice, many of these intermetallic compounds have proven brittle and researchers have

Advances in Cryogenic Engineering, Volume 45.
Edited by Shu *et al.*, Kluwer Academic / Plenum Publishers, 2000.

come up with novel means to prevent the materials from slowly pulverizing due to the helium gas pressure waves.[4] To increase the P-V work, researchers have optimized the valve timing[5] and increased the expansion volume[6]. These techniques serve to increase pressure swing and swept volume, thus enlarging the practical P-V work and bringing it closer in appearance to the theoretical cycle. Improvements in valve timing and regenerator materials increase the gross cooling power, but net cooling power depends on properly transferring that cooling power to a load. Ensuring sufficient heat transfer within the cold head has proven to be important through modeling and practical experiment by Zhang *et. al.*[7] and Inaguchi *et. al.*[8] Enhanced heat transfer and lower pressure drop play a significant role in the increased cooling capacity at lower operation frequencies. Researchers report increased cooling at lower frequencies in regenerative cycles despite the decrease in the number of cycles performed at lower frequency.[9] Practically, the regenerator seals in the cylinder housings have proven to be an important design consideration.

These techniques serve to address specific limitations of the regenerator for the 2nd-stage, and ultimately, the optimum GM will require their consideration. The 1st and 2nd-stages are a coupled system though, so modifications to the 1st-stage can impact the performance of the 2nd-stage. This work focuses on obtaining improved performance in the 2nd-stage solely by modifications to the 1st-stage regenerator.

CONVENTIONAL REGENERATOR DESIGN

Figure 1 shows a cross-section of a conventional 1st-stage regenerator. Typically, a regenerator core consists of multiple stacks of phosphor-bronze (or copper-bronze) and/or stainless steel wire screens. The screens are individually punched to give disks of uniform area and are hand-packed into the regenerator housing. Coarse screens are placed near the hot and cold ends of the regenerator to help provide uniform distribution of flow through the principal stack of fine screens making up the central core. The placement of coarse screens at the hot end of the regenerator also helps to reduce the pressure drop in this region where gas viscosity and velocity through the regenerator is relatively high.

A full three-dimensional (3-D) fluid flow simulation through a conventional regenerator using the commercially available software package CFX-TASCFlow[10] was carried out. This package allows solution of the Navier-Stokes equations in porous media in 3-D, with pressure drop through the media modeled using a user-defined momentum source term in the discretized momentum equations. The momentum source term was modeled to represent a local pressure drop corresponding to the relation

$$\Delta P = \frac{1}{2} f \cdot N \cdot \left(\frac{1-\alpha^2}{\alpha^2} \right) \cdot \rho \cdot V_b^2 \tag{1}$$

where ΔP is the pressure drop, f is the friction factor, N is the number of screens, α is the screen porosity, ρ is the gas density, and V_b is the gas bulk velocity. The friction factor is given by a non-linear curve fit to the graphical data presented by Kays and London.[11] A linear variation of temperature along the regenerator bed from 295 K to 30 K was used to determine local transport properties of the gas. Each control volume used local flow conditions to determine an appropriate momentum source term.

Typical results of computer simulation of the 3-D flow through the conventional regenerator for a 45° model section can be seen in Figure 2. This figure shows the gas flow streaklines for a fixed helium mass flow through the regenerator of 16 g/s from the hot side to the cold side. This flow roughly corresponds to the peak flow through the regenerator for a 4.6 kW input GM during the intake phase. Large re-circulation eddies can be seen in the hot-end coarse screen region; this suggests that simplified 1-D momentum and energy models will overestimate the regenerator effectiveness because they will inherently ignore the detrimental heat transport associated with the eddies.

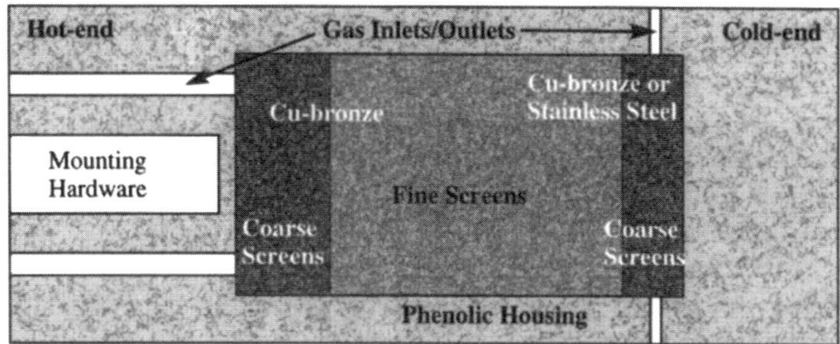

Figure 1. Schematic representation of a conventional 1st-stage regenerator.

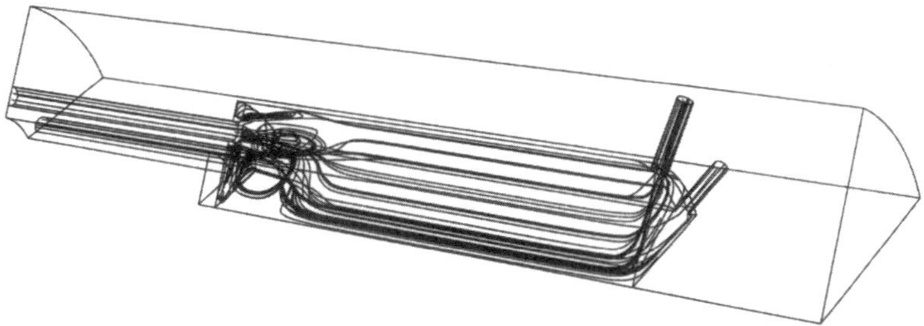

Figure 2. Streaklines for a conventional 1st-stage regenerator. Note the large re-circulation eddies in the hot-end coarse screen region that would compromise thermal performance.

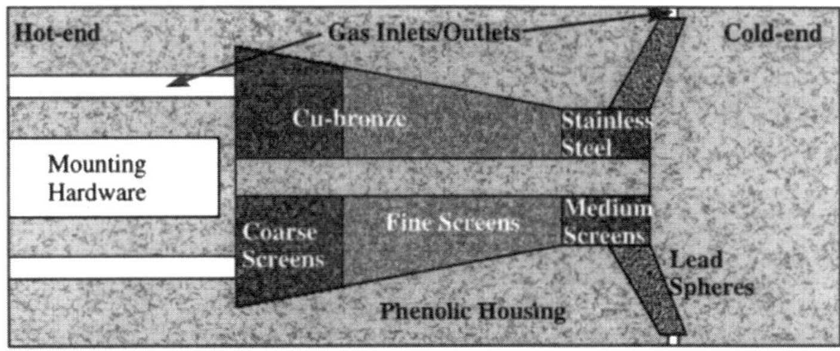

Figure 3. Schematic representation of a conical stacked-screen 1st-stage regenerator.

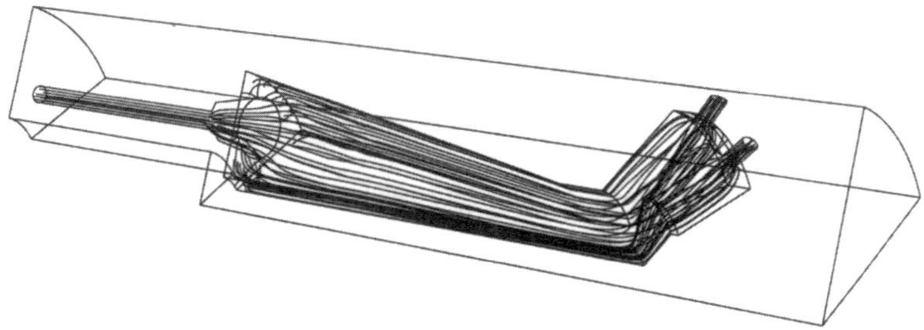

Figure 4. Streaklines for a conical stacked-screen 1st-stage regenerator. Diffuser regions and a streamlined tapered body ensure uniform flow without gas re-circulation regions.

MODIFIED REGENERATOR DESIGN

Figure 3 shows a schematic of a non-uniform cross-section regenerator. Fine and coarse annular-shaped screens make up the main body of the regenerator, arranged with coarse screens at the hot end, fine screens in the middle, and medium-coarse screens at the cold end. Lead spheres are packed in the annular space between the cold end screens and the cold end gas outlets/inlets.

Figure 4 shows streaklines on a 45° model section for the same inlet flow conditions as the uniform cross-section regenerator. The non-uniform cross-section forms a linear taper in screen outer diameter and serves to streamline the flow and eliminate re-circulation eddies. The tapered design compensates for the thermal dependence of the transport properties of the circulation gas, allowing a lower pressure drop in the screens due to a lower gas velocity at the hot end of the regenerator and a higher heat transfer coefficient due to higher gas velocity at the cold end. Along their length, the tapered screens allow a more uniform flow impedance and heat transfer coefficient, thereby better utilizing the maximum potential performance of each screen. Overall pressure drop and heat transfer for the implementation presented here also includes the annular packed bed region.

Figure 5 shows the partially assembled regenerator and regenerator housing constructed for this study. The tapered and straight stacks of screens were created using a simple, proprietary manufacturing process that leaves the screens without burrs, dimples, or rough edges. The screen materials selected for the regenerator were chosen mainly to verify the manufacturing process capability across a range of screen materials and mesh sizes, and generally to follow the thermo-fluid requirements of the GM cycle. Insertion of the tapered screen stack into the matching taper of the regenerator housing firmly locks each screen into place without any possible gas channels or scratched grooves along the walls. The assembled regenerator with its conical screen stack has been dubbed the "Cryocone" regenerator to distinguish this design from conventional, uniform-diameter stacked screen regenerators.

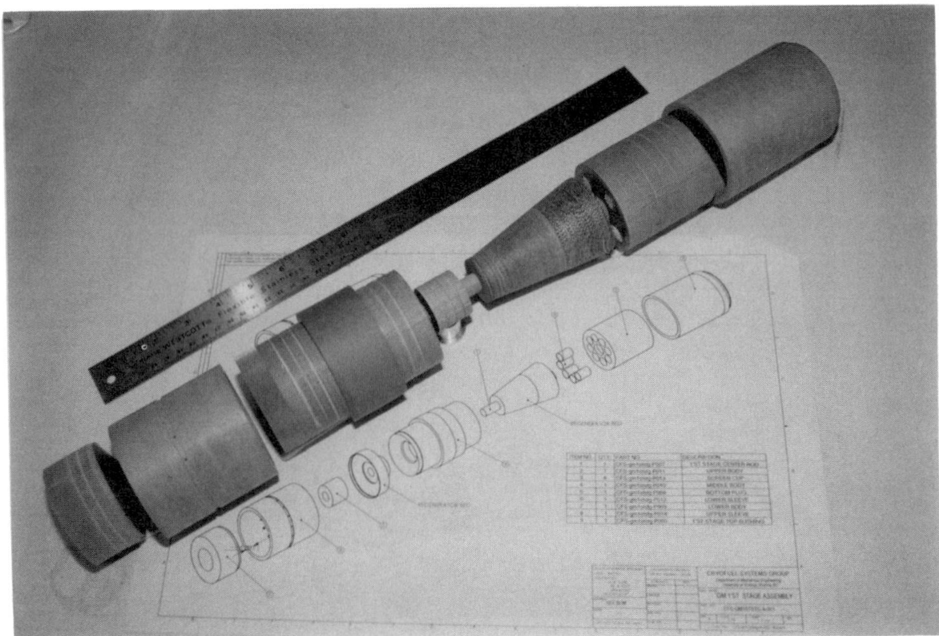

Figure 5. Partially assembled regenerator and housing showing the Cryocone design.

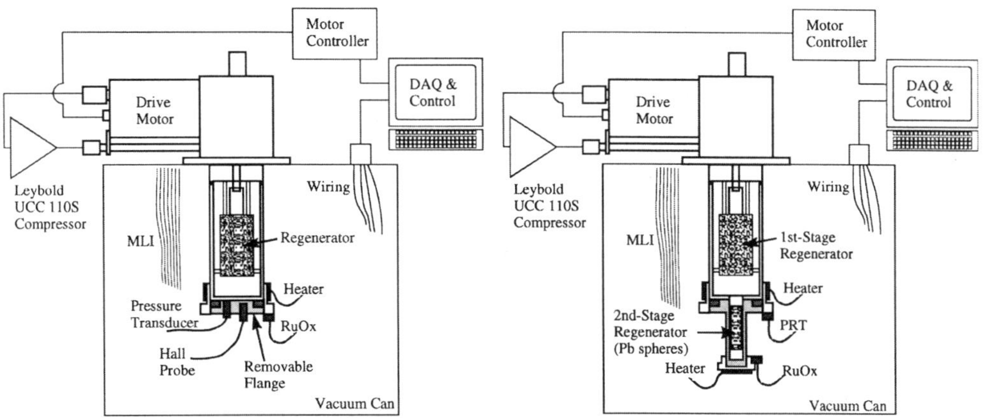

Figure 6a. Schematic of single-stage test apparatus. **Figure 6b.** Schematic of two-stage test apparatus.

EXPERIMENTAL APPARATUS AND PROCEDURE

A two-stage GM test apparatus was constructed to test the Cryocone regenerator and to compare its performance with a conventional regenerator. The coldhead cylinder assembly features a replaceable 2nd-stage cylinder, allowing the GM to be operated as a single-stage unit by substitution of a blank-off flange for the 2nd-stage cylinder. The 2nd-stage cylinder (or blank-off flange) bolts to the 1st-stage cylinder and uses an indium seal to prevent gas leaks to the vacuum space. The indium seal has proven reliable through multiple thermal cycles without need for replacement. Figures 6a and 6b give schematics of the test apparatus arranged for single-stage and two-stage testing.

For both the one-stage and two-stage tests, ruthenium oxide and platinum resistance thermometers were used to measure temperature, as indicated in the diagrams. For one-stage tests, two additional transducers were mounted in the blank-off flange. First, a Kistler[12] model 601B2 piezoelectric transducer was used to measure pressure variations in the expansion space. This transducer provided a linear relative signal but was not calibrated for these temperatures. Second, a Hall probe was used to sense a disc-shaped magnet mounted in the bottom of the regenerator housing. The output from the Hall probe was used as a timing signal to determine bottom dead center of the regenerator/displacer and thereby the volume of the expansion space determined from the known sinusoidal variation of regenerator position.

Using the Hall probe provided a convenient, non-intrusive, and inexpensive means to measure regenerator position compared to the use of an LVDT such as by Inaguchi *et. al.*[8] or laser displacement sensor such as by Kurihara *et. al.*[13] A photo of the one-stage test apparatus with a close-up view of the pressure transducer and Hall probe arrangement mounted in the blank-off flange is shown in Figure 7. Sample output from the Hall probe is given in Figure 8, showing maximum signal of the sensor corresponding to bottom dead center of the regenerator.

Tests using the one-stage apparatus consisted of measurement of the no-load minimum temperature for a range of operating frequencies and temperature under a modest load of 25 W at 2.4 Hz, for both the Cryocone and conventional regenerators. Pressure fluctuations were also recorded for future experiments on valve timing optimization; however, the sensor output was not calibrated *in situ* to give an absolute scale in units of pressure. Consequently, the P-V diagrams generated only allow relative comparison in this study.

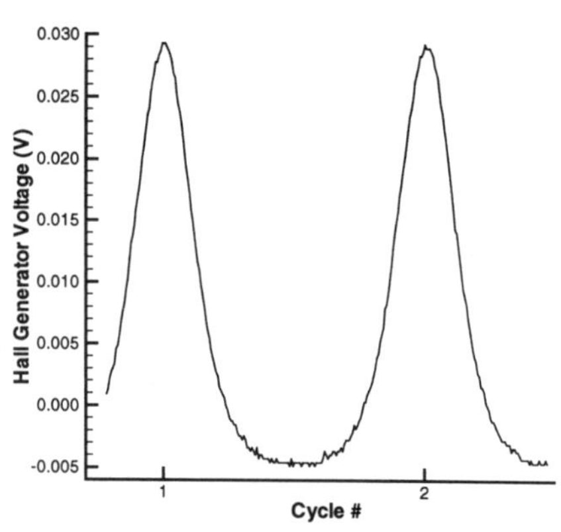

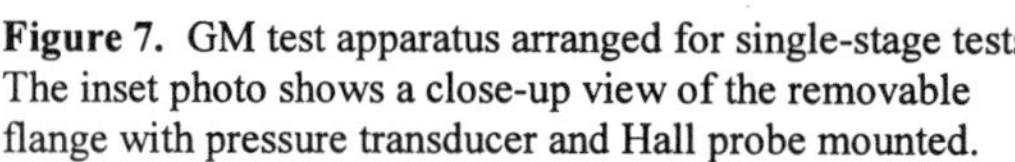
Figure 7. GM test apparatus arranged for single-stage tests. The inset photo shows a close-up view of the removable flange with pressure transducer and Hall probe mounted.

Figure 8. Sample Hall probe output used as a synchronization tool to determine displacer bottom dead center and thereby expansion volume from known displacer motion.

Tests on the two-stage apparatus consisted of measurement of the no-load minimum temperatures for the 1st and 2nd-stages, and 1st and 2nd-stage temperatures under a combination of loads to both stages using a 2nd-stage regenerator filled with lead spheres. No-load tests were conducted across a spectrum of operating frequencies, while load tests were performed solely at 1.5 Hz.

RESULTS

Figure 9 shows pressure transducer voltage vs. expansion volume for the Cryocone regenerator and for the conventional regenerator for an operation frequency of 2 Hz. These curves are identical to the P-V diagrams for this cycle except for scaling and offset for the y-axis. Minimum pressure is approximately 0.5 MPa and maximum pressure is approximately 2.2 MPa, as indicated at the gas compressor. The P-V diagrams for the two regenerators are almost identical, although the Cryocone regenerator has a slightly smaller enclosed area, and hence a slightly smaller gross P-V work. This is thought to be due to a slightly higher pressure drop in the Cryocone regenerator. The diagrams indicate that the as-built valve timing is far from the ideal for the GM cycle.[14]

The no-load tests on the single-stage apparatus revealed a frequency dependence of the 1st-stage temperature, as can be seen in Figure 10. The Cryocone regenerator has a stronger frequency-dependent no-load minimum temperature than does the conventional regenerator. Despite its consistently smaller gross P-V cooling power, the Cryocone regenerator has a lower no-load minimum temperature than does the conventional regenerator down to a frequency of approximately 1 Hz. For these frequencies, assuming equal external heat leaks, this implies that the Cryocone regenerator has a higher net cooling, and therefore lower internal losses in the regenerator.

The no-load tests on the two-stage arrangement revealed a frequency dependence of both the 1st and 2nd-stage temperatures. Results of these tests are shown in Figure 11. Other researchers have reported a similar optimum frequency of operation for attaining the lowest no-load temperatures.[9] At all frequencies, the Cryocone 1st-stage regenerator permitted a lower no-load minimum temperature on the 2nd-stage than did the conventional regenerator. The Cryocone regenerator also exhibited a lower 1st-stage temperature than the conventional regenerator at most frequencies tested.

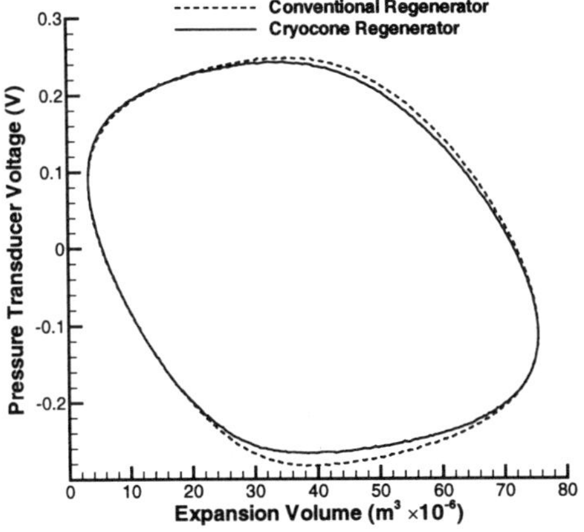

Figure 9. Comparison of pressure transducer voltage vs. expansion volume for single-stage GM operating at 2 Hz using conventional and Cryocone regenerators. Y-axis is analogous to pressure, except for scaling and offset.

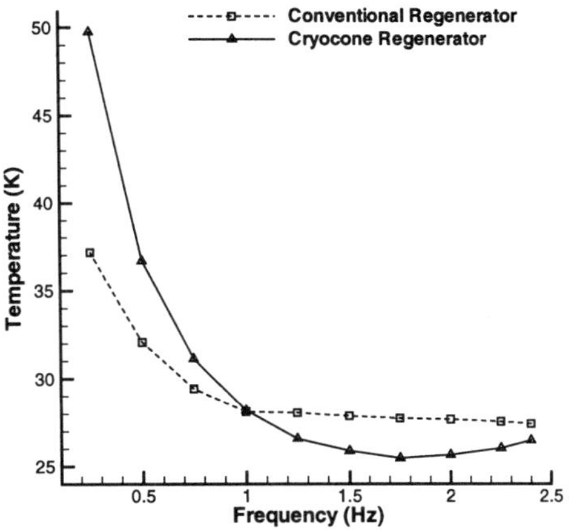

Figure 10. Comparison of the frequency dependent no-load minimum temperatures of the conventional and Cryocone regenerators in the single-stage GM.

Under load, the Cryocone regenerator gave mixed results compared to the conventional regenerator. Figure 12 displays the cooling performance curves using the Cryocone and conventional regenerators for the 1st-stage, with the conventional lead regenerator for the 2nd-stage. Applied heat loads were identical for both regenerators tested. For most loads, the Cryocone 1st-stage allowed a lower 2nd-stage temperature. Only at low 1st-stage loads did the Cryocone 1st-stage also consistently provide a lower 1st-stage temperature.

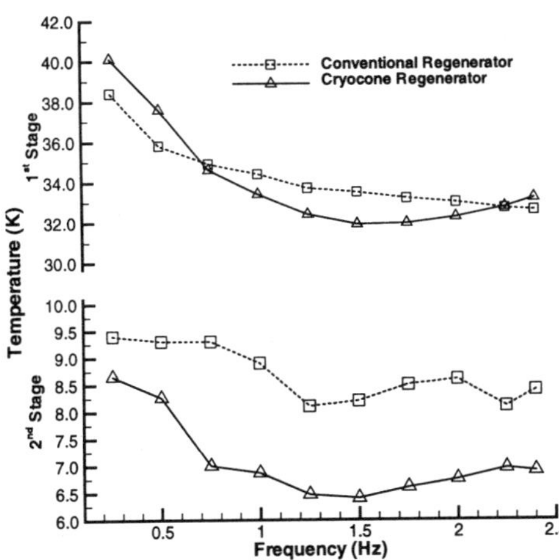

Figure 11. 1st and 2nd-stage no-load minimum temperatures as a function of GM operation frequency. Peak cooling performance is achieved at approximately 1.5 Hz for both the 1st and 2nd-stages for the Cryocone regenerator.

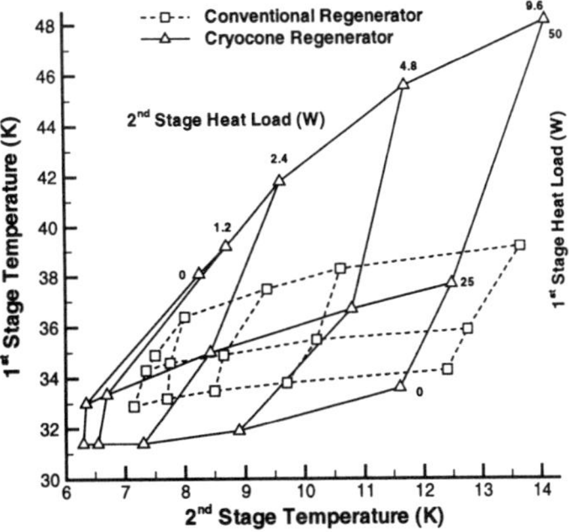

Figure 12. Comparison of cooling performance curves for the conventional and Cryocone regenerators. Loads for the conventional regenerator can be found on the corresponding points of the Cryocone curves.

SUMMARY AND CONCLUSIONS

Models show that current designs for the 1st-stage regenerator of a two-stage GM refrigerator can be improved to reduce turbulent and re-circulating gas flow by use of a non-uniform cross-section regenerator. Improved no-load minimum temperature and cooling performance in the 2nd-stage of the refrigerator can be achieved solely through modifications to the 1st-stage regenerator. For regenerator displacers with a known displacement function, a simple, non-intrusive Hall probe arrangement can be used as a synchronization signal to generate displaced volume information. Manufacture of uniform and non-uniform cross-section stacked screen regenerators is easily achieved across a range of mesh size and material compositions. Future work will focus on the further optimization of cross-section, screen meshes, and composition based on the thermo-fluid requirements of the GM cycle and the 1st-stage regenerator's relation to performance of the 2nd-stage.

ACKNOWLEDGEMENTS

The authors would like to acknowledge the Natural Sciences and Engineering Research Council of Canada (NSERC), Centra Gas/Westcoast Energy B.C., Natural Resources Canada (NRCan), and the University of Victoria for their support for this work and its affiliated projects.

REFERENCES

1. K.H.J. Buschow, J.F. Olijhoek, and A.R. Miedema, "Extremely Large heat Capacities between 4 and 10K", *Cryogenics*, 15, 261-264 (1975).
2. T. Inaguchi, M. Nagao, and H. Yoshimura, "Two-stage Gifford-McMahon cycle cryocooler operating at about 2K", Proceedings of the Sixth International Cryocoolers Conference, vol. II:25 (1990).
3. T. Kuriyama, R. Hakamada, H. Nakagome, Y. Tokai, M. Sahashi, R. Li, O. Yoshida, K. Matsumoto, and T. Hashimoto, High efficient two-stage GM refrigerator with magnetic material in the liquid helium temperature region, in: "Advances in Cryogenic Engineering," 35, Plenum Press (1990), pp. 1261-1269.
4. W.R. Merida and J.A. Barclay, Monolithic regenerator technology for low temperature (4 K) Gifford-McMahon cryocoolers, in: "Advances in Cryogenic Engineering," 43, Plenum Press, N.Y., (1998), pp. 1597-1604.
5. R. Li, A. Onishi, T. Satoh, and Y. Kanazawa, Influence of valve open timing and interval on performance of 4 K Gifford-McMahon cycle cryocooler, in: "Advances in Cryogenic Engineering," 41, Plenum Press, N.Y., (1996), pp. 1601-1607.
6. T. Satoh, A. Onishi, R. Li, H. Asami, and Y. Kanazawa, Development of 1.5W 4K GM cryocooler with magnetic regenerator material, in: "Advances in Cryogenic Engineering," 41, Plenum Press, N.Y., (1996), pp. 1631-1637.
7. L. Zhang, L.H. Gong, and X.D. Xu, Study of effect of heat transfer in the cold head to the performance of a 4.2 K G-M refrigerator, in: "Advances in Cryogenic Engineering," 41, Plenum Press, N.Y., (1996), pp. 1645-1651.
8. T. Inaguchi, M. Nagao, K. Naka, and H. Yoshimura, Effects of thermal conductance in the cooling stage of a 4K-GM refrigerator on refrigeration capacity, in: "Advances in Cryogenic Engineering," 43, Plenum Press, N.Y., (1998), pp. 1807-1814.
9. T. Kuriyama, "Technological Research on GM Refrigerators Employing Magnetic Regenerator Materials," (English translation), Doctoral Thesis, Tokyo Engineering University (1994).
10. CFX-TASCFlow, AEA Technology Engineering Software Ltd., 554 Parkside Drive, Waterloo, Ontario, Canada, N2L 5Z4.
11. W.M. Kays and A.L. London, Experimental correlations for simple geometries, in: "Compact Heat Exchangers," 3rd ed., McGraw-Hill Book Company, New York (1984), p. 149.
12. Kistler Instrument Corp., 75 John Glenn Drive, Amherst, NY, U.S.A., 14228-2171.
13. T. Kurihara, M. Okamoto, K. Sakitani, H. Torii, and H. Morishita, Numerical and experimental study of a 4 K Modified-Solvay cycle cryocooler, in: "Advances in Cryogenic Engineering," 43, Plenum Press, N.Y., (1998), pp. 1791-1798.
14. G. Walker, Gifford-McMahon, Solvay, and Postle cryocoolers, in: "Cryocoolers, Part 1: Fundamentals," Plenum Press, New York (1983), p. 253.

EXPERIMENTAL RESEARCH ON A PRECOOLED SINGLE-STAGE G-M REFRIGERATOR OPERATING AT THE LIQUID HELIUM TEMPERATURE

L. Zhang, L. Wang*, L. Fang, and X.D.Xu

Cryogenic Laboratory, Chinese Academy of Sciences
Beijing 100080, China

ABSTRACT

A single-stage G-M refrigerator operating directly between the room temperature and the liquid helium temperature is designed and constructed. Its main structure features are as follows: 1. the regenerator is put outside the cylinder; 2. a cold source for precooling, which could be a refrigerator or liquid nitrogen, is linked with the middle part of the refrigerator to offset the heat losses from ambient environment. The experimental system for testing the performance of the new structure refrigerator is established. The influences of some important parameters are experimentally studied.

INTRODUCTION

Recently, The 4.2K G-M refrigerator has been successfully applied to some cryogenic systems, such as SIS receivers for astronomical radio observation, helium recondensation in MRI systems and cryocooler-cooled superconducting magnets.

To improve the efficiency of the 4.2K refrigerator, its structure and operating parameters should be optimized because its optimum operational condition is quite different from that of the conventional G-M refrigerator[1]. In recent years, studies on many important parameters, such as valve timing[2], regenerator structure[3,4], stroke[5] and so on, have been done. Their investigations were mainly conducted on the two-stage G-M refrigerator. In this paper, a single-stage G-M refrigerator, which is precooled by a cold source, is designed and investigated.

Supported by the National Natural Science Foundation of China
*Present address: Brookhaven National Laboratory, Upton, NY 11973

DESCRIPTION OF THE NEW CONFIGURATION

Figure1 demonstrates the arrangement of the new G-M refrigerator. It is quite different from a conventional single-stage G-M refrigeration. It operates directly between the room temperature and the liquid helium temperature. The regenerator (8), whose hot end is linked with the hot chamber of the cylinder and the cold end with the cold chamber of the cylinder, is put outside the cylinder (5). The displacer (4), cylinder (5) and regenerator (8) of the refrigerator are all constituted by warm parts, middle parts and cold parts. The warm parts are over the middle parts and the cold parts are under them.

The warm and cold parts of the cylinder are both stainless steel pipes. The middle part of the cylinder is a copper circular, which is linked directly with the cold source (1) by a solid heat bridge (2). The warm and cold parts of the displacer are made of textolite. They are linked with a copper block, the middle part of the displacer , by epoxy adhensive. The copper circular in the cylinder and the copper block in the displcacer compose a gap heat exchanger (3), by which the heat losses in the displacer are transferred to the cold source. The middle part of the regenerator is a porous-medium heat exchanger (9), which is used to precool the helium gas and thus offsets the regenerator loss. The cold source (1) could be the cold head of a refrigerator or liquid nitrogen.

The main characteristics of the novel single-stage refrigerator are as follows:

First, the regenerator is not inside the displacer and so its structure can be adjusted easily. For example, the volume of the regenerator can be designed big enough to meet the requirement of V_R/V_C, where V_R represents the volume of the regenerator and V_C represents the cold chamber volume[1,6]. Moreover, the diameter of the regenerator can be enlarged to reduce the pressure drop across the matrix.

Second, the temperature and pressure profiles along the regenerator and the cylinder can be easily measured, which will provide some useful information about the principle of the refrigerator.

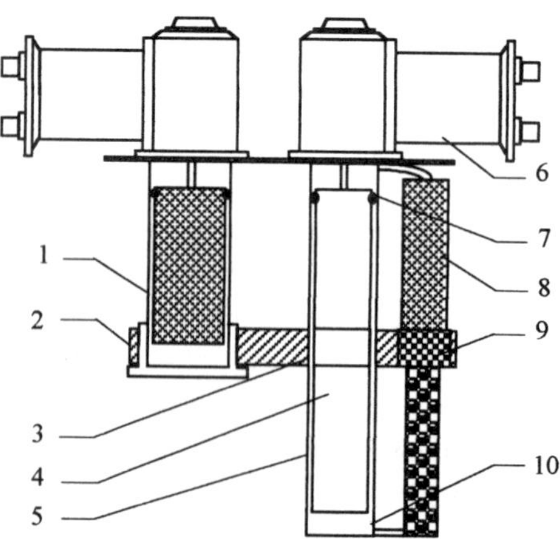

1. Cold source 2. Heat bridge 3. Gap heat exchanger 4. Displacer 5. Cylinder 6. Drive motor 7. Seal 8. Regenerator 9. Porous-medium heat exchanger 10. Cold head

Figure 1. Novel configuration of the single-stage G-M refrigerator

Third, the refrigerator is reliable because the seal operates at the room temperature.

EXPERIMENTAL APPARATUS

The experimental apparatus of the refrigerator is shown in Figure2, which consists of four units: refrigeration system, compressor, vacuum system and measuring system.

Table 1 shows the structure parameters of the refrigeration system. The warm part of the regenerator is 50mm in diameter and 150mm in length, filled with 250 mesh screens. Its cold part is a triple-layer regenerator, in which lead spheres, Er_3Ni and $ErNi_2$ grains are used. The cylinder and regenerator are both longer than that of a conventional single-stage refrigerator to reduce the conduction loss. The poprous-medium heat exchanger in the regenerator is 30mm in diameter and 20mm in length. It can be fabricated easily by filling a copper pipe with copper spheres and then sintering together. The basic features of this kind of heat exchanger are the large heat transfer area per unit volume and the compact structure[6,7]. The gap heat exchanger is 40mm in diameter and 50mm in length.

The heat bridge is crucial to the performance of the refrigerator. Its section area is determined by the following equation:

$$Q_{in} = Q_{cyl-loss} + Q_{reg-loss} = \bar{\lambda} A \Delta T / L \quad (1)$$

where, Q_{in} represents the cooling capacity of the cold source; $Q_{cyl-loss}$ represents the losses between the room temperature and the pre-cooling temperature in the warm parts of the cylinder and the regenerator--including the conduction loss of the wall, the shuttle loss and the pump loss; $Q_{reg-loss}$ represents the losses in warm part of the regenerator--including the regenerator loss and the conduction losses of the wall and the matrix;

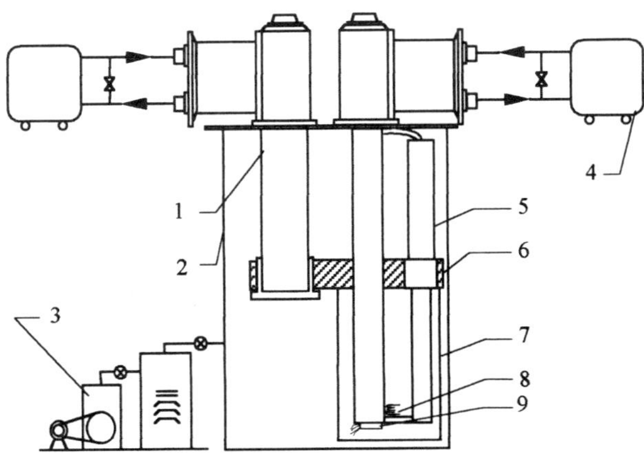

1.Cold source 2.Vacuum chamber 3.Vacuum pump 4. Compressor 5.Refrigerator 6.Heat bridge 7.Radiation shield 8.Heater 9.Temperature sensor

Figure 2. Schematic diagram of experimental apparatus

Table 1. Structure Parameters

	Warm part	Cold part	
Cylinder			
Diameter	40 mm	40 mm	
Length	180 mm	120 mm	
Thickness	0.7 mm	0.7 mm	
Stroke	25 mm	25 mm	
Regenerator			
Diameter	50 mm	45 mm	
Length	150 mm	100 mm	
Matrix			
Materials	Phosphor bronze	Pb	Er_3Ni+$ErNi_2$
Type	screen	Sphere	grain
Size	250 mesh	0.18-0.45 mm	
Mass	1500 sheets	1100 g	
Porosity	76%	39%	

ΔT means the temperature difference between the cold source and the heat exchangers; $\bar{\lambda}$, A and L represent the average thermal conductivity, section area and length of the heat bridge, respectively. When the thermal resistance of heat bridge increases, the temperature difference enlarges. Thus the irreversible heat transfer loss will also increase. In general, the better the thermal contact is, the better the performance of the refrigerator will be.

The input power of the compressor is about 6.5kW. The measuring system is controlled by a computer and can record the temperatures and pressures of the refrigerator immediately. The Rh-Fe resistance thermometers with 0.1K accuracy are used to measure the temperature at the cold head. Besides, a number of thermocouple thermometers are set on the wall of the cold part of the regenerator to obtain the temperature distributions.

TEST RESULTS AND DISCUSSIONS

To ensure that the whole system can operate smoothly, the refrigerator is tested without precooling at first. Figure3 shows the cooling down curve at the cycle speed of 60rpm and P_H=2.1MPa and P_L =0.7MPa. The cooling down time is within 100 minutes. Figure4 shows the cooling capacity of the refrigerator. We can see from Figure4 that its cooling capacity at 20K is about 4.4W when the cycle speed is 60rpm.

The refrigerator is tested with precooling and the results are presented in the following figures. In Figure5, curve 1 is the temperature of the cold source and curve 2 presents the temperature of the cold head. The lowest temperature of the new refrigerator reaches 4.6K when the running frequency is 24 rpm and P_H = 1.9MPa and P_L = 0.7MPa (see Figure7).

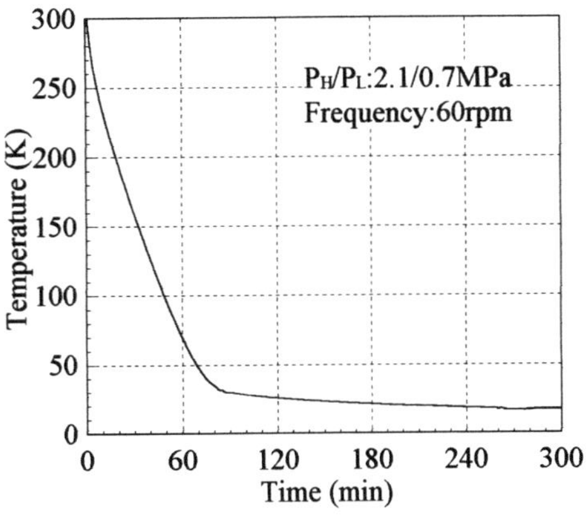

Figure 3. Curve of cooling down without precooling

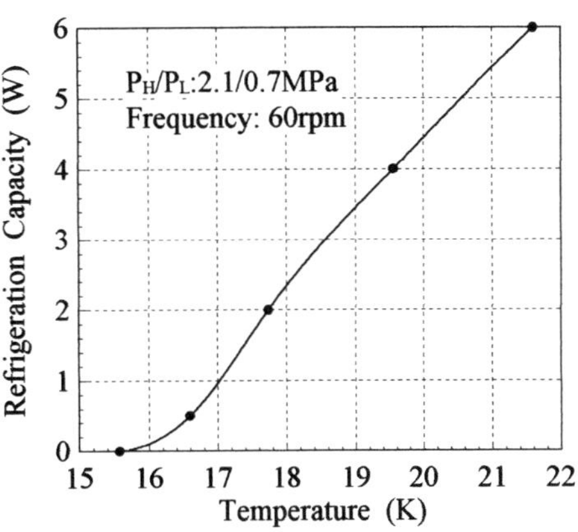

Figure 4. Refrigeration capacity without precooling

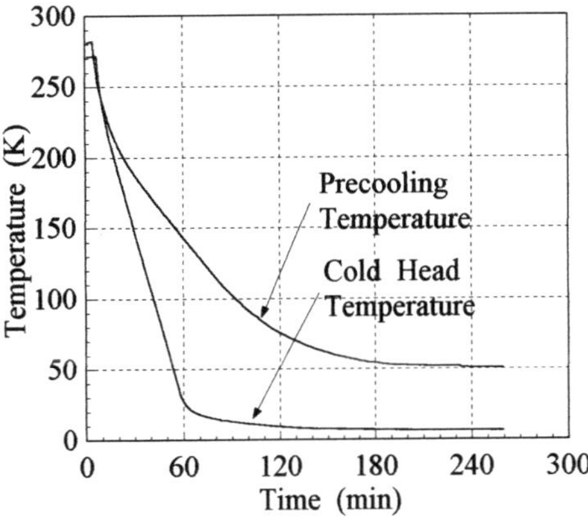

Figure 5. Curve of cooling down (precooled by a refrigerator)

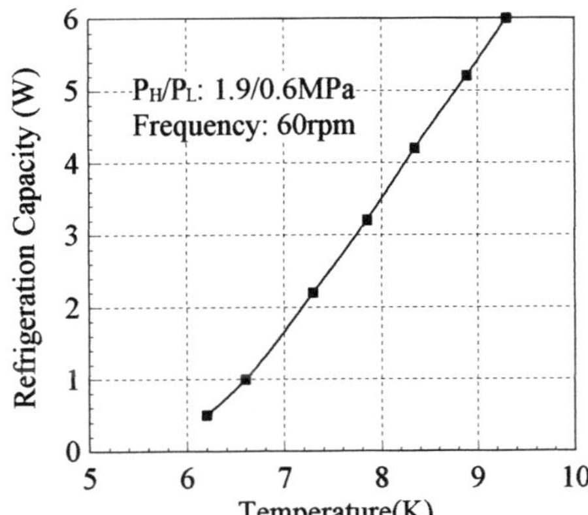

Figure 6. Refrigeration capacity (precooled by a refrigerator)

Figure6 indicates that the cooling capacity at 10K is above 6W. Figure7 shows the influence of the cycle speed on the lowest temperature of the refrigerator. When the cycle speed is decreased form 60rpm to 24 rpm, the lowest temperature is lowered from 5.8K to 4.6K.

The temperature distributions along the length of the regenerator are shown in Figure8 and Figure9. The temperature profiles are nearly linear when the temperature is above 20K. A temperature plateau appears in the cold end of the regenerator when the temperature is under 10K. But in the hot end of the regenerator, the temperature profiles are very steep. Approximately 75% of the temperature drop occurs here. Figure9 shows that the temperature distributions are almost not influenced by the frequencies.

The location of precooling parts and the temperature of the cold source should be optimized in order to improve the efficiency of the refrigerator. As shown in Figure10, the

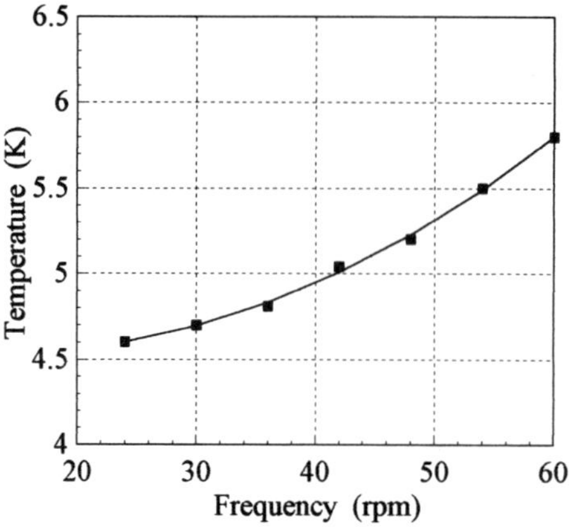

Figure 7. Lowest temperature vs frequency

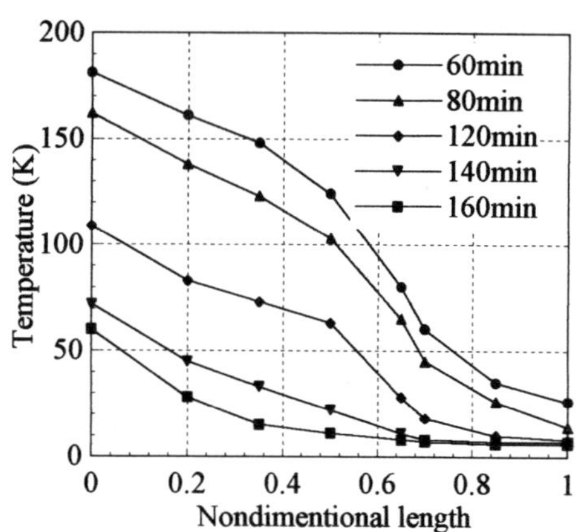

Figure 8. Temperature distributions along the cold part of the regenerator

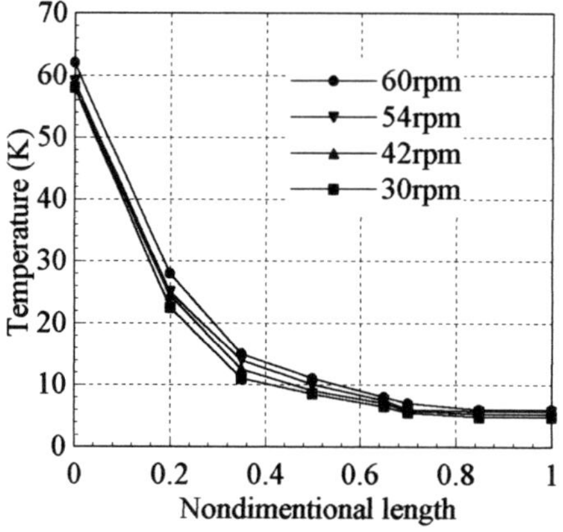

Figure 9. Temperature distributions under different cycle speeds

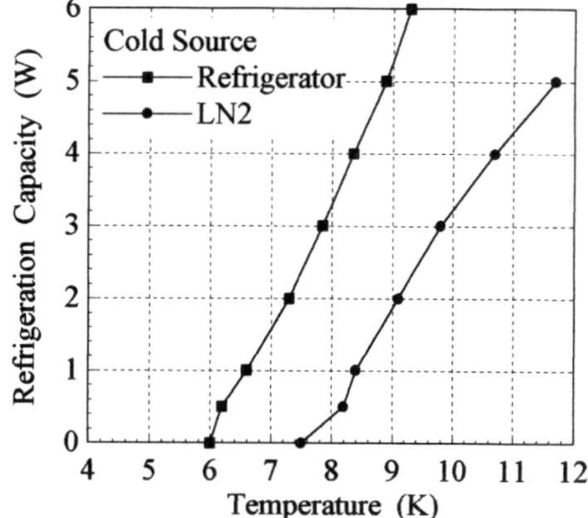

Figure 10. The refrigeration capacity with different cold sources

cooling capacity is enlarged when the pre-cooling temperature is lowered from 95K (pre-cooled by liquid nitrogen) to 60K (pre-cooled by a refrigerator). This is because some losses, such as the shuttle loss、the pump loss and the conduction loss, are all functions of the temperature difference between the precooling parts and the cold head. The losses increase when the temperature difference enlarges. The structure of the regenerator is dependent on the temperature of the cold source. The warm part should be big enough to meet the requirement of the heat exchanging of the helium from the room temperature to the precooling temperature. And the cold part of the regenerator should also have enough volume to meet the requirement of the heat exchanging of the helium from the precooling temperature to the liquid helium temperature.

CONCLUSIONS

1 A single-stage G-M refrigerator precooled by a cold source is established. The regenerator is put outside the cylinder so its structure can be optimized conveniently. In the preliminary experiment, the refrigerator reached the liquid helium temperature.

2 The performance of the refrigerator has been tested. A lowest temperature of 4.6K was achieved by this refrigerator. The influences of frequencies and temperatures of the cold sources have been studied. The temperature distributions along the axis of the regenerator are obtained.

3 To improve the performance of the refrigerator, further studies on the regenerator structure and location for precooling should be conducted.

ACKNOWLEDGMENT

This research project is supported by the National Natural Science Foundation of China.

REFERENCES

1. C.S.Hong, et al., On the unique features of the low temperature regenerator in a 4.2 K G-M refrigerator, JSJS4, (1993)
2. R.Li, et al., Influence of valve open time and interval on performance of 4K Gifford-McMahon cycle cryocooler, Adv. Cryo. Eng., Vol.41, (1996)
3. T.Tsukagoshi, et al., Optimum structure of multilayer regenerator with magnetic materials, Cryogenics, Vol.37, No.1, (1996)
4. L. Wang, et al., Performance analysis of the low temperature regenerator with multi-layered hybrid materials, Proceedings of ICEC17, (1998)
5. T. Kuriyama, et al., Optimization of operational parameters for a 4K-GM refrigerator, Adv. Cryo. Eng., Vol.41, (1996)
6. L.Zhang, et al., Study of effect of heat transfer in the cold head to the performance of a 4.2K G-M refrigerator, Adv. Cryo. Eng., Vol.41, (1996)
7. T.Inaguchi, et al., Effects of Thermal conductance in the cooling stage of a 4K-GM refrigerator on refrigeration capacity, Adv. Cryo. Eng., Vol.42, (1997)

THE EXPERIMENTAL STUDY OF A SINGLE-STAGE G-M REFRIGERATOR WITH THE REGENERTATOR SET OUTSIDE THE CYLINDER*

L. Wang,[1] L. Fang,[2] W. H. Lu,[2] and L. Zhang[2]

[1]Brookhaven National Laboratory
Upton, NY 11973
[2]Cryogenic Laboratory, Chinese Academy of Sciences
Beijing 100080, China

ABSTRACT

In this paper, a single-stage G-M refrigerator with the regenerator set outside the cylinder is presented. The experimental system for testing the performance of the cryocooler was constructed. The lowest temperature was 14K when the operating frequency was 0.6 Hz. The cooling capacity of 4.4W has been obtained at 20K. The effects of operating parameters of the refrigerator on cooling performance were also experimentally studied.

INTRODUCTION

The introduction of multi-layer structural regenerators with magnetic materials has greatly improved the performances of multistage G-M refrigerators at low temperature range[1][2]. GM cryocoolers are now widely applied to many cryogenic systems, such as cryopump, cryostat, SIS mixer cooling, helium recondensation in a superconducting MRI magnet and cooling of small superconducting magnets[3][4]. Up to date, a great deal of efforts have been put into the improvement of the cooling performances of 4K GM refrigerators by using various experimental apparatuses[5][6]. Some attention presently is given to adjust the structure of the GM cryocooler to meet the requirement of its special applications, as well as to improve its operation liability and stability. In the case of cooling cryoelectronics devices, in order to achieve the proper performance of the cryoelectronics devices, the disturbance of the magnet field of the magnetic regenerator materials must be reduced. Based on the above reasons, a new-type GM refrigerator operating at liquid helium region was designed and constructed[7]. Its major structure features are that the warm stage and the cold stage displacers were driven separately and the warm stage was used to precool the cold stage for removing heat losses from ambient

*Supported by the National Nature Science Foundation of China

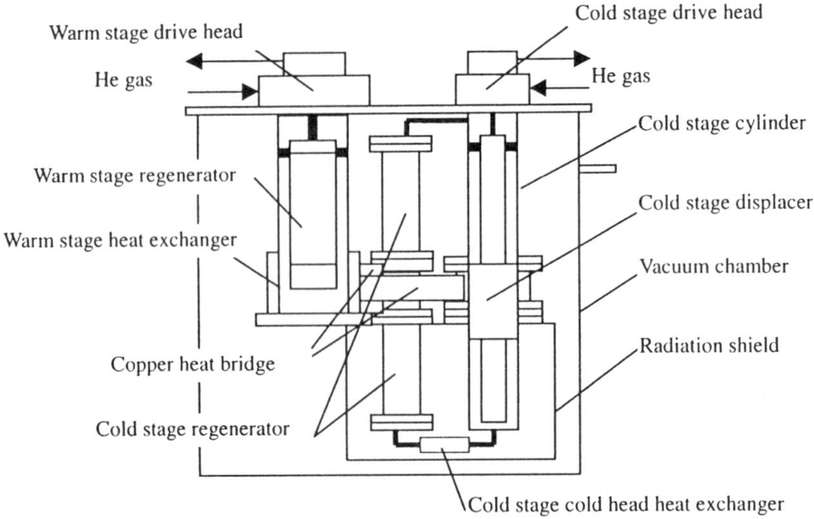

Figure 1. Schematic diagram of experimental apparatus for a new-type 4K GM cryocooler

environment, the cold stage regenerator was put outside the cylinder, and the cold head heat exchanger at the cold stage was somewhat different from the conventional one. The new-type refrigerator shown in Figure 1 is flexible in constructing experiments. It also can be used as a test station for the researches of GM coolers. In this paper, we preliminarily tested the performance of the cold stage GM refrigerator that can be used for a kind of single-stage cryocooler different from the conventional co-axial structure. Some results were obtained [7]. This paper also describes the principal design features of the single-stage GM refrigerator.

EXPERIMENTAL APPARATUS

Figure 2 shows the schematic of the single-stage GM cryocooler used as the experimental machine. The regenerator was placed outside the cylinder. The solid displacer reciprocated in the cylinder, driven by an AC induction motor. The cycle speed could be varied by changing the power supply frequency. The stroke of the displacer was 25mm. The displacer was divided into three parts. The middle part is a copper block. The rest of the displacer was made of textolite. They were connected by epoxy adhensive.

The regenerator outside the cylinder and the cylinder were also divided into three parts in order to precool the cold stage of the 4K GM cryocooler shown in Figure 1. The cylinder and regenerator are both longer than that of a conventional single-stage refrigerator to reduce the shuttle and conduction losses. The high temperature and low temperature parts of the cylinder and the regenerator are both stainless steel pipes. The middle part of the cylinder is a copper circular.

The phosphor bronze screens were stacked at the high temperature part of the regenerator. The copper spheres were sintered into a copper tube as its middle part. A combination of the triple-layer materials composed of lead spheres, Er_3Ni and $ErNi_2$ grains was stuffed in sequence from the hot to the cold end in the low temperature part of the regenerator. The cold head heat exchanger was located at the cold end of the regenerator, which was easily processed by sintering copper spheres into a copper tube to increase the

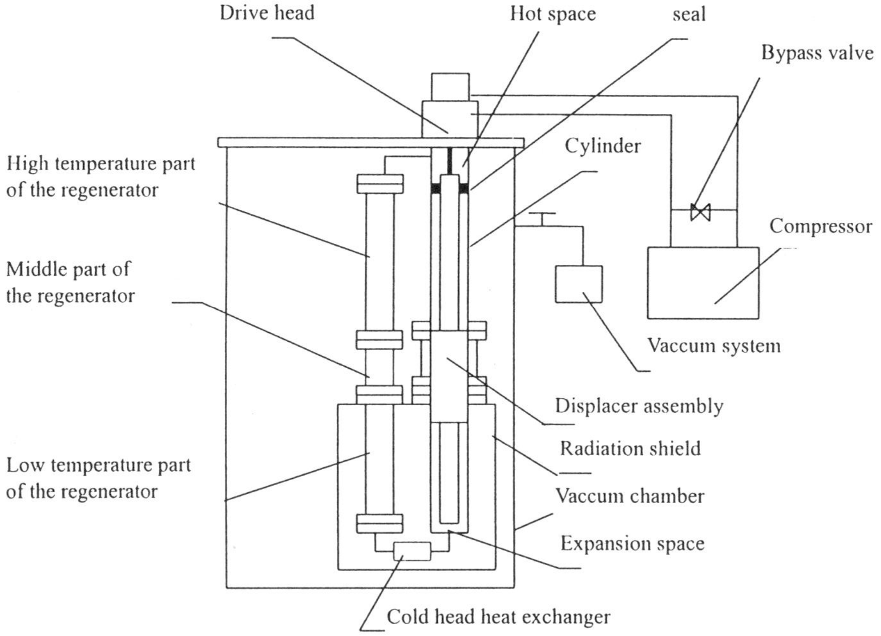

Figure 2. Schematic diagram of experimental apparatus for a single-stage GM cryocooler

heat transfer area per unit volume. A radiation shield was attached to the lower portion of the cryocooler to prevent the radiant heat from room temperature.

Two Rh-Fe resistance sensors with 0.1K accuracy were respectively installed at the middle of the regenerator and at the cold head heat exchanger for the cooling temperature measurement. The Manganin wire was wound around the cold head heat exchanger as electric heater. The main structural parameters are shown in Table 1.

EXPERIMENTAL RESULTS AND ANALYSIS

Cool Down Curves

Figure 3 shows the cool down curve of the single-stage GM refrigerator. The inlet and exhaust pressures of the compressor were respectively 2.08MPa and 0.70MPa. The cycle frequency was 1Hz. Tc is the temperature measured at the end of the cold head heat exchanger that was close to the expansion space. Treg is the temperature of the middle

Table 1. Main structural parameters

Cylinder	Diameter(mm)		40
Regenerator	High temperature part	Diameter(mm)	50
		Materials	Phosphor bronze meshes
	Middle part	Diameter(mm)	30
		Materials	Copper spheres
	Low temperature part	Diameter(mm)	45
		Materials	lead spheres, Er_3Ni and $ErNi_2$ grains
Stroke(mm)	25		

portion of the regenerator. It can be seen from Figure 3 that the cool down rate of Tc was much faster from the room temperature to 30K than that from 30K to the lowest temperature. Treg reached about 160K when Tc was about 16K. The cool down time is within 100 minutes. We concluded that the performance of the regenerator was better at the low temperature portion than that of the high temperature portion.

Effects of Cycle Frequency and Pressure Range on Cooling Performance

The effects of cycle frequency on no-load cooling temperature are indicated in Figure 4. The two curves were obtained under different operating pressure range. Here the operating pressures mean the inlet and the exhaust pressures of the compressor. One can observe that there exists the same optimum cycle frequency for the single-stage cryocooler under both kinds of operating condition. The optimum cycle frequency was 0.6Hz in the experiment.

It also can be seen that the cooling temperature varied with the operating pressure range. The ratio of the inlet pressure over the exhaust pressure of 1.86/0.50MPa is larger than that of 2.08/0.70MPa. The pressure differences are almost the same. Because the PV work diagram doesn't represent the cooling capacity when the cooling temperature is lower than 100K[8], we simply evaluated the ideal cooling capacity of the refrigerator and the heat loss under different operating pressures. It was found that the ideal cooling capacity was higher under 1.86/0.50MPa than that under 2.08/0.70MPa, and the major heat loss under the former was smaller than that under the latter. Therefore, the cooling performance was improved.

The cooling temperature was 14.5K under the inlet and the exhaust pressure of 1.86/0.50MPa. It was 15.6K under 2.08/0.70MPa when the cycle frequency was 1Hz. For the former the lowest temperature of 14K was obtained at the operating frequency of 0.6 Hz.

Cooling Capacity

Figure 5 gives the cooling capacity of the single-stage GM refrigerator. It indicates that the lowest cooling temperature was 15.6K and the cooling capacity was 4.4W at 20K, under the operating pressure of 2.08/0.70MPa and the cycle frequency of 1Hz.

In addition, the new-type G-M refrigerator shown in Figure 1 has successfully been operating at liquid helium region. The lowest temperature was 4.6K under the operating pressure of 1.90/0.70MPa when the operating frequency was 0.4 Hz. The cooling capacity of larger than 6 W has been obtained at 10K[5]. The further improvement has been planned.

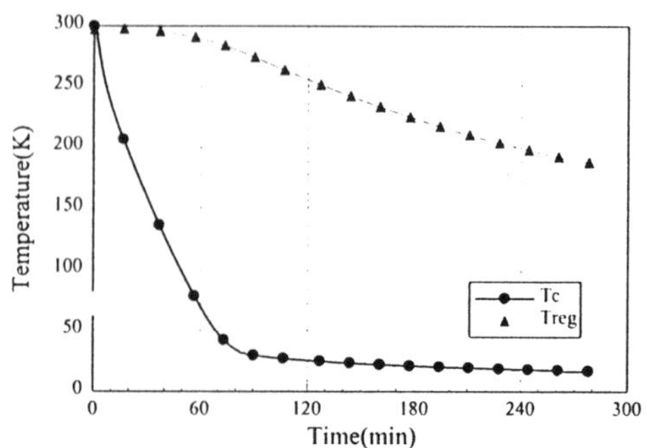

Figure 3. The cool down curve

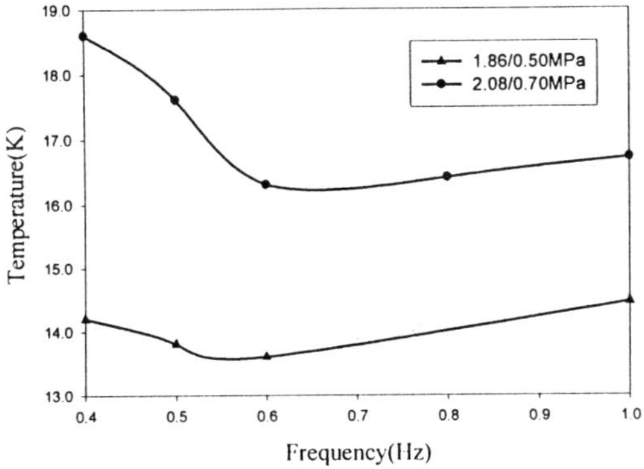

Figure 4. Effect of cycle frequency on no-load temperature under different operating pressure range

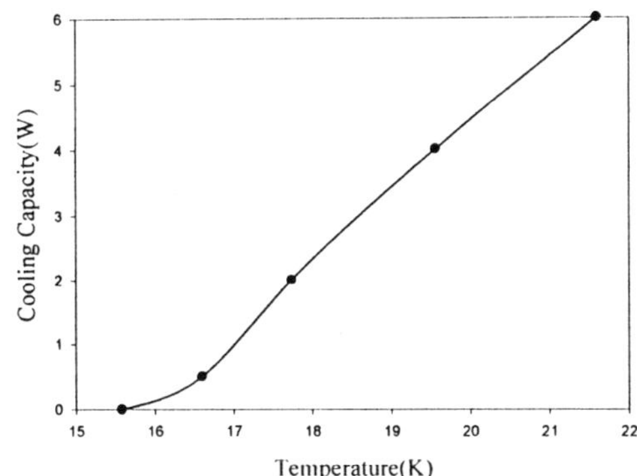

Figure 5. Cooling capacity of the GM refrigerator

CONCLUSIONS

1. A new-type single-stage GM refrigerator with the regenerator set outside the cylinder was fabricated and preliminarily tested. The lowest temperature was 14K when the operating frequency was 0.6 Hz. The cooling capacity of 4.4W has been obtained at 20K.
2. There exists the optimum cycle frequency for the cryocooler when the operating condition was changed. The operating pressure range has a large effect on the cooling performance.
3. In the new-type single-stage machine, because the regenerator was located outside the cylinder, there is no limitation on the regenerator structure. The larger regenerator diameter can be made than that of displacer to reduce the pressure drop across the regenerator. It is also convenient to change the geometry sizes and materials of the regenerator to investigate their effects on cooling performance and then to optimize the structure parameters. It becomes easier to measure the temperature and pressure profiles along with the regenerator and the cylinder to study the working mechanism of the cryocooler. The heat transfer efficiency of the cold head heat exchanger can be improved by putting it

between the regenerator cold end and expansion space where its heat transfer area per unit volume can be easily increased.
4. In the new-type GM cryocooler at liquid helium region, the regenerator matrix does not move along with the displacer, so the oscillation of the magnet field of the matrix does not exist. There is less magnetic noise because the magnetic matrix is static. It also becomes easy to meet the ratio of regenerator volume over cold chamber volume, which is important to a 4K GM cooler. Because of the parallel structure of the new-type machine, the operating parameters of the first-stage and second-stage can be optimized separately to reach the optimum cooling performance. Finally, the refrigerator is reliable because the seal operates at the room temperature.

ACKNOWLEDGMENT

This research was supported by the National Natural Science Foundation of China.

REFERENCES

1. T.Satoh, et al, Development of 1.5W 4K GM Cryocooler with Magnetic Refrigerator Material, *Advance in Cryogenic Engineering*, 41: 1631(1996).
2. T.Inaguchi, et al, Development of 2W Class 4K Gifford-McMahon Cycle Cryocooler, *ICEC16/ICMC Proceedings*, 335(1996).
3. M. Nagao, et al, *Advances in Cryogenic Engineering*, 39: 1327(1994).
4. K. Watanabe, et al, Liquid Helium-free Superconducting Magnets and their Applications, *Cryogenics*, 34: 639, ICEC Supplement (1994).
5. A. Onishi, et al, Development of 2W class 4K Gifford-McMahon cycle cryocooler, *ICEC16/ICMC Proceedings,* Part 1: 335(1996)
6. T. Tsukagoshi, et al, Optimum structure of multilayer regenerator with magnetic materials, *Cryogenics,* 37:11(1997)
7. L. Wang, Numerical Analysis of a GM Refrigerator Operating at Liquid Helium Temperature and Experimental Study of a New Type of GM Refrigerator, *Doctoral Thesis*(1998).
8. X. D. Xu, et al, Investigation of the Performance of a 4.2K G-M Refrigerator, *ICEC16/ICMC Proceedings*, 343(1996)

PERFORMANCE AND QUALIFICATION OF BEI'S 600 mW LINEAR MOTOR COOLER

S.W.K. Yuan, D.T.Kuo, A.S. Loc, and T.D. Lody

BEI Technologies, Cryocooler Group
Sylmar, CA 91342

ABSTRACT

BEI's B602 Linear Motor Cooler can deliver 600 mW of cooling at 78K with an input power of less than 30W. Cooldown time from room temperature to 78K is less than 3 minutes with a thermal mass of 250J. The cooler weighs less than 0.907 Kg and can operate at temperatures between 75°C and -54°C. In this paper, the performance of this cooler is discussed in detail. Environmental qualification of this cooler is also presented in this paper, in particular shock test, operational vibration, high and low temperature storage, interchangeability and vibration output, etc.

INTRODUCTION

Among a vast variety of linear motor coolers manufactured by BEI[1-8], the B602 model is widely used in Airborne GIMBAL and FSI SAFIRE systems. In this paper, performance and qualification of the B602 cooler are discussed. Weight and physical dimensions of the B602 cooler are listed in Table 1. Operating conditions of the cooler are listed in Table 2. The cooler has been tested extensively with charge pressure between 2.757 to 3.446 MPa. While the steady state performance of the cooler is more efficient at 2.757 MPa, the cool down time may be slightly worse.

Table 1. Physical Characteristics

Weight	0.789 Kg
Compressor Diameter	4.445 cm
Cold finger tip Diameter	0.660 cm
Cold finger Length	5.080 cm
Expander flange Diameter	2.794 cm
Expander flange Thickness	0.435 cm

Table 2. Operating Conditions

Charge Pressure	2.757-3.336 MPa
Frequency	60 Hertz
Voltage	11 Volts
Maximum Input Power	35 Watts
Ambient Temperature	-54°C to 75°C
Heat Load	600 mW

Advances in Cryogenic Engineering, Volume 45.
Edited by Shu *et al.*, Kluwer Academic / Plenum Publishers, 2000.

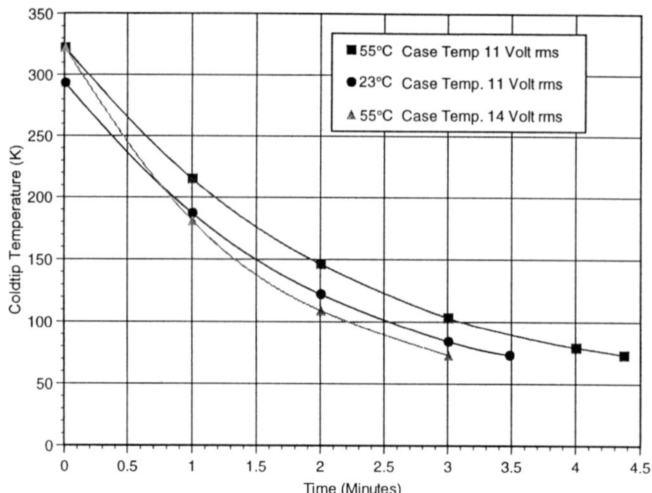

Figure 1. Cooldown time of the B602C cooler.

Table 3- Maximum voltage

71°C	17.5 V
23°C	16 V
-40°C	11.5 V

Performance

Figures 1 to 3 show the typical performance of the B602C cooler. The coolers discussed in this paper were charged to 3.336 MPa.

Cooldown Time. Figure 1 shows the typical cool down time of the B602 cooler at various case temperatures and input voltages. Typical cooldown time from room temperature to 78 K is about 3 minutes. At an elevated ambient temperature of 55°C, the cooldown time is approximately one minute longer. The maximum voltage for the B602C cooler is listed in Table 2. These are the maximum voltages that can be applied without having the compressor pistons hitting each other.

Input Power. Heat load curves of the B602C cooler at 73K are plotted in Figure 2. Average maximum refrigeration capacity was measure to be 550 mW at 71°C. The average input power was 35 W for 350 mW of heat load at 71°C. At room temperature, the B602C cooler can relieve 350 mW of heat load with an input power of less than 20 W.

Refrigeration. Figure 3 depicts the refrigeration as a function of cold tip temperature at 11 volt input power. Three ambient temperatures are included in the figure, namely –54°C, 23°C and 71°C. At 78K, the cooler can provide 0.9 W of cooling at 23°C ambient and 0.7 W of cooling at 71°C ambient. With no load, the cooler can reach a temperature of 33 K at 23°C ambient and 40 K at 71°C ambient.

Environmental Qualification

Three coolers with serial numbers 525, 526, and 527 were tested for environmental qualification.

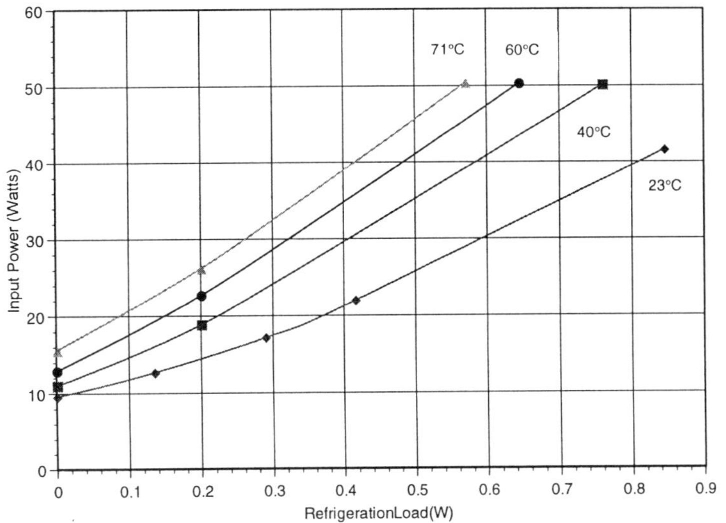

Figure 2. Input power versus refrigeration.

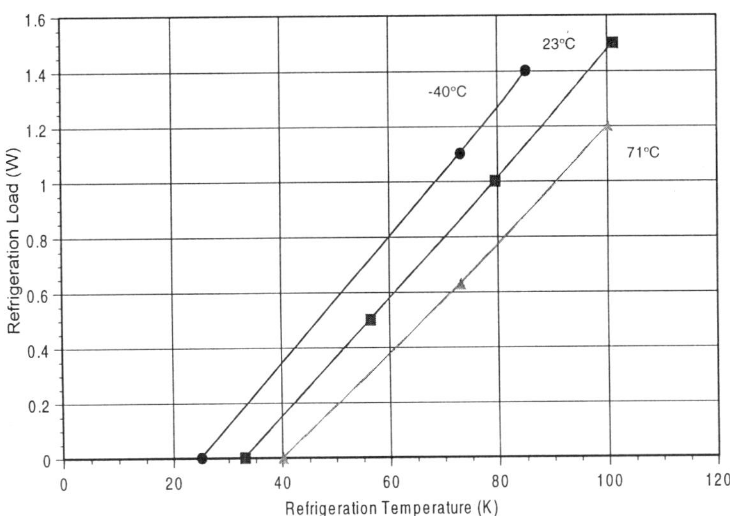

Figure 3. Refrigeration versus cold tip temperature.

Table 4. Specification for the B602C cooler

	33°C	75°C
Cooldown Time* w/ 250J Thermal Mass	2:30 min	3:30 min
Min. Refrigeration @11.5V	750 mW	500 mW
Max. Input Power @350 mW	25 W	35 W

*Cooldown starts at 23°C and 55°C respectively, instead of 33°C and 75°C.

High Temperature Soak. The coolers were soaked in an elevated temperature of 71°C for 48 hours in a BEMCO environmental chamber per MIL-STD-810C, Method 501.1, Procedure 1, and found to meet the specification (in Table 4) afterwards.

Low Temperature Soak. The coolers were stored at -57°C for 24 hours in a BEMCO environmental chamber per MIL-STD-810C, Method 501.2, Procedure 1, and found to meet the specification (in Table 4) afterwards.

Temperature Shock. The Thermal Shock Tests were performed according to MIL-STD-810C, Method 503.1, Procedure. A BEMCO environmental chamber was set to 71°C and used as the high ambient soak. A Tenney Jr. III environmental chamber was set to –57°C and used as the low ambient soak. The coolers withstood changes in ambient temperature between –57°C and 71°C at a rate of 15°C per minute, and were found to meet the specification (in Table 4) afterwards.

Operational Vibration Test. The Operational Vibration Tests were performed at the facility of Environment Associates, Inc. Chatsworth,. The coolers were subjected to a modified spectral density profiles (per MIL-STD-810D, Method 514.3, Category 6) as shown in Figure 4 for 60 minutes per axis. Acceptance tests were performed on the coolers at 71°C and they all passed the specification.

Shock Test. The Non-Operational Shock Tests were performed at the facility of Environment Associates, Inc. Chatsworth, California. The coolers were subjected to shock impulses (half sine wave) of 20g for 11ms in accordance with MIL-STD-810C, Method 516.2, Procedure 1 (Figure 5). Three shock impulses in each direction (+ and -) on each axis for a total of 18 shocks were applied. Acceptance tests were performed on the coolers at 71°C and they all passed the specification.

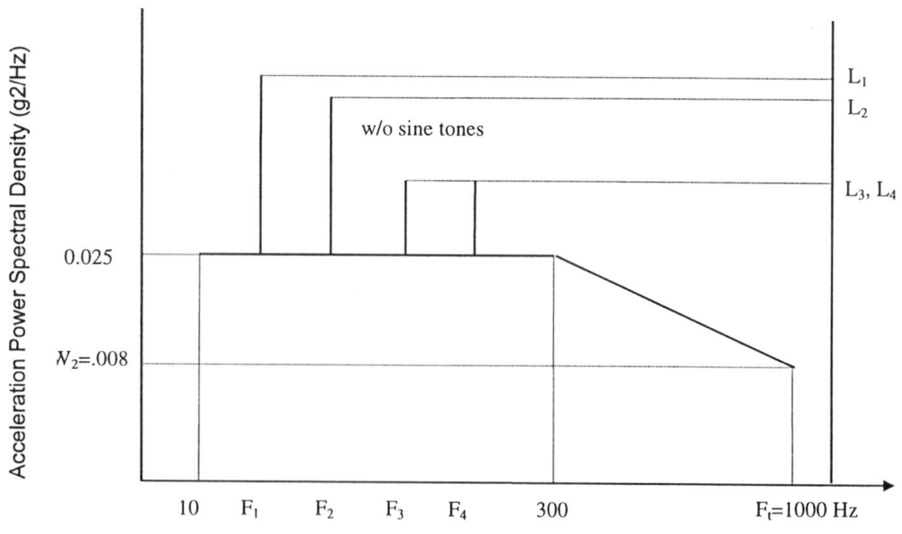

Figure 4. Spectral density profiles for operational vibration test.

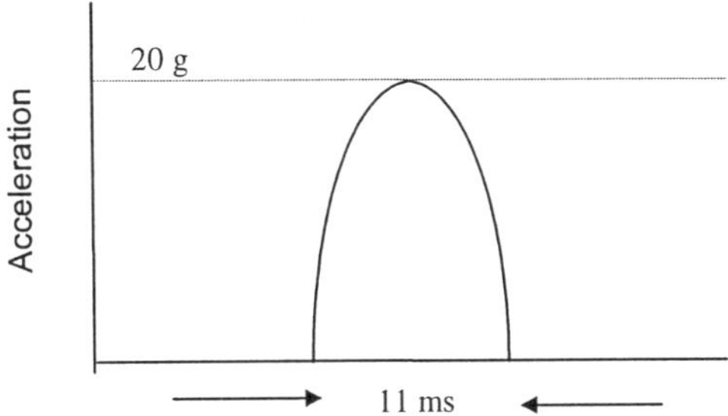

Figure 5. Shock Impulse.

Interchangeability. To satisfy the Interchangeability requirement, the expanders of the best performance cooler (Cooler Assembly S/N 525) and the worst performance cooler (Cooler assembly S/N 526) were swapped. Acceptance tests were performed on both coolers and both passed the specification.

Vibration Output Test. The Vibration Output Test was performed following the procedure supplied by Army's Night Vision Laboratory. In an open loop operation with an input voltage of 11.5 volts, vibrations were recorded along three axes for frequencies up to 20 KHz. All three coolers passed the vibration tests with less than 2.224 N in any of the three axes.

Life Test

Two B602C coolers were used in the life test. The coolers were run at 45°C case temperature with a heat load of 350 mW at 73K. Instead of running the cold tip in a dewar, the cold fingers were inserted in a vacuum chamber. At 1700 hours, a power failure took place in a section of BEI's laboratory, which stopped the vacuum pump with the coolers still running.

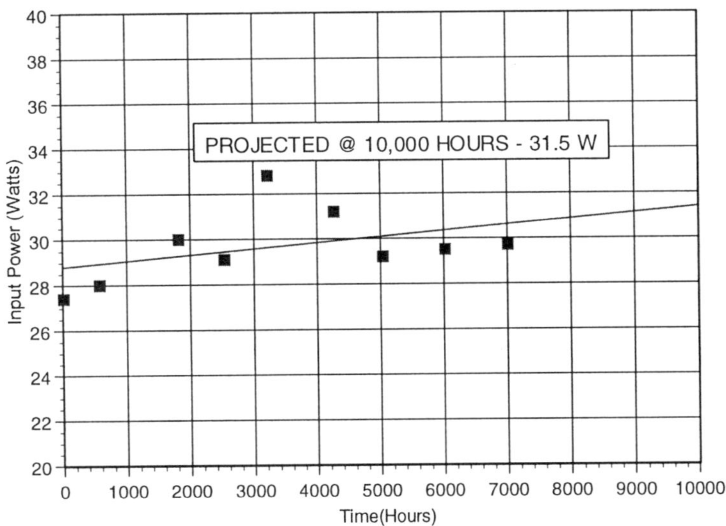

Figure 6. Input power vs. life for B602 cooler.

After losing vacuum, the coolers were overdriven by the drivers to meet the cold tip temperature, leading to premature wear out of one of the coolers. The result presented here belongs to the cooler which survived the catastrophe. Figure 6 is a plot of input power vs. life. Figure 7 is a plot of cooldown time vs. life and Figure 8 is a plot of refrigeration vs. life. The B602 cooler has demonstrated a life of over 7000 hours to date. Qualification of BEI's B512 cooler can be found in Reference 6.

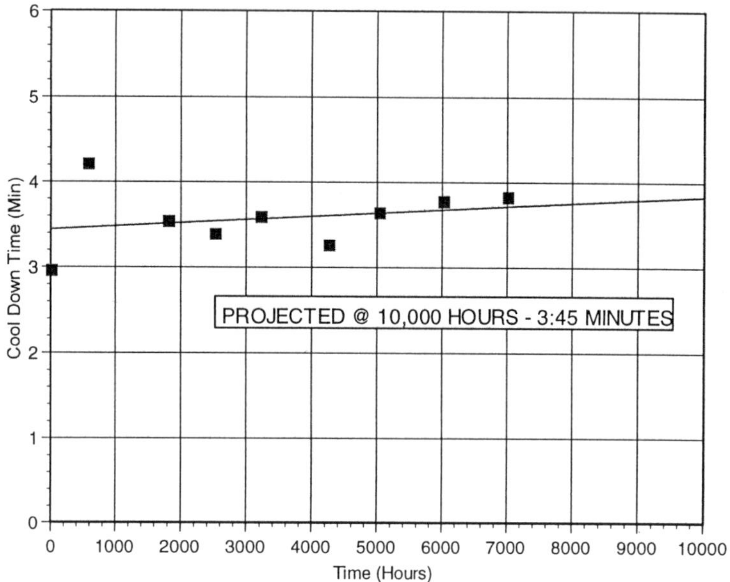

Figure 7. Cooldown time vs. life for B602 cooler.

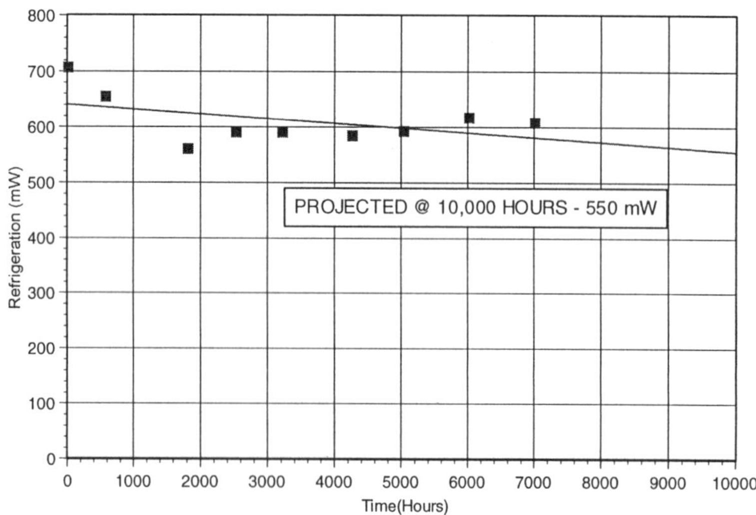

Figure 8. Refrigeration vs. life for B602 cooler.

CONCLUSIONS

The performance and qualification of the BEI B602C cooler have been reported in this paper together with life test data. The cooler takes about 3 minutes to cool down to 78K. At steady state, the cooler delivers 350 mW of cooling at 78K, with an input power of less than 20 W. The maximum refrigeration of this cooler is 0.7W at 71°C ambient and 0.9 W at room ambient (with a cold tip temperature of 78K). The B602C cooler passed all the environmental qualification tests and has a life of exceeding 7000 hours to date. The coolers discussed in this paper have a charge pressure of 3.336 MPa. These coolers were found to be more efficient when operated at a charge pressure of 2.757 MPa, (data to be reported in a separate paper).

REFERENCES

1. D.T. Kuo, A.S. Loc, and S.W.K. Yuan, Experimental and Predicted Performance of the BEI Mini-linear Cooler, in the Proc. of the 9th International Cryocooler Conference (1997) p.119.
2. D.T. Kuo, A.S. Loc, and S.W.K. Yuan, Design of a 0.5 Watt Dual Use Long-Life Low-Cost Pulse Tube Cooler, Advances in Cryogenic Engineering, 43:2039, 1997.
3. S.W.K.Yuan, D.T.Kuo, A.S.Loc, and T.D. Lody, Experimental Performance of the BEI One Watt Linear (OWL) Stirling Cooler, Advances in Cryogenic Engineering, 43:1855, 1997.
4. S.W.K. Yuan, D.T. Kuo, and A.S. Loc, Enhanced Performance of the BEI 0.5 Watt Mini-Linear Stirling Cooler, Advances in Cryogenic Engineering 43:1847, 1997.
5. S.W.K. Yuan, D.T. Kuo, and A.S. Loc, Design and Preliminary Testing of BEI's CryoPulse 1000, the Commercial One Watt Pulse Tube Cooler, to be published in the Proc. of the 10th International Cryocooler Conference, 1998.
6. D.T. Kuo, A.S. Loc, and S.W.K. Yuan, Qualification of the BEI B512 Cooler, Part 1 - Environmental Tests, to be published in the Proc. of the 10th International Cryocooler Conference, 1998.
7. S.W.K. Yuan, D.T.Kuo, and A.S. Loc, Cryocooler Contamination Study, parallel paper in the Cryogenic Engineering Conference, Montreal, 1999.
8. D.T. Kuo, A.S. Loc, T.D. Lody and S.W.K. Yuan, Cryocooler Life Estimation And its Correlation With Experimental Data, parallel paper in the Cryogenic Engineering Conference, Montreal, 1999.

DESIGN OF SINGLE STAGE AND TWO STAGE MINIATURE STIRLING CRYOCOOLERS USING CYCLIC SIMULATION

H. K. Agrawal and K. G. Narayankhedkar*

*Indian Institute of Technology, Bombay
Powai, Mumbai 400 076, India

ABSTRACT

Small capacity Stirling cryocoolers for space-borne applications generally use linear electromagnetic drives (also called linear motor) for long life, maintenance free operation and high reliability. Such motors are normally of moving coil type. The performance prediction of such small capacity (miniature) Stirling cryocoolers is very critical and is governed by principal design parameters. The performance of a cryocooler depends on various losses such as loss due to regenerator ineffectiveness, temperature swing loss, shuttle conduction loss, P-V loss etc. In this paper, performance prediction of single stage and two stage miniature split Stirling cycle cryocoolers has been carried out considering above losses using cyclic simulation reported by Atrey et al.[1] The systematic approach for performance prediction using cyclic simulation is explained in detail. The gross refrigerating capacity of a single stage cryocooler as predicted by the cyclic analysis is 5.4 W. The losses work out as 4.35 W. Thus the net refrigerating effect is 1.05 W. This net refrigerating effect is compared with the net refrigerating effect measured during experimental investigations. The comparison shows a good agreement. Thus, the cyclic simulation has been further utilized for design of a miniature two stage cryocooler to meet the specification of 2 W at 100 K (stage I) and 0.5 W at 50 K (stage II).

INTRODUCTION

The actual and the ideal Stirling cycles are distinctly different and performance prediction of the actual cycle has been a subject of interest. Cyclic analysis for single stage Stirling cycle cryogenerator developed by Atrey et al.[1] and modified for two stage Stirling cycle cryogenerator by Natu and Narayankhedkar[2], has been further extended for miniature cryocoolers. The cyclic simulation calculates ideal refrigerating effect and ideal power requirement. For finding actual power required, it also calculates losses due to pressure drop in the connecting tube, regenerator and expansion space. The ideal refrigerating capacity reduces due to the regenerator ineffectiveness, shuttle heat conduction, temperature swing, pumping action, instantaneous pressure drop and axial conduction through different parts. Thus the next step in the cyclic analysis calculates various losses as mentioned above and evaluates net refrigerating effects available.

ANALYSIS

The cyclic simulation developed by Atrey et al.[1] for Stirling cryogenerator is more realistic in approach. The major assumption is the adiabatic nature of compression process and the isothermal nature of expansion process. In the present analysis the cyclic simulation of miniature Stirling cryocooler is presented. The following assumptions are made:

Assumptions

1. the gas behaves as a perfect gas;
2. connecting tube volume is considered as part of the compression space clearance volume;
3. movement of piston and displacer are sinusoidal;
4. the pressure in the system remains constant at any instant;
5. the compression process is adiabatic and the expansion process is isothermal;
6. the regenerator mass flow rate is the average of the mass flow rates of the compression and expansion spaces.

Pressure - Volume Variation

The volume variations for the expansion and compression spaces are given by the following expressions

$$V_E = V_{EC} + V_{EM}\ (1 - \cos\ \phi) / 2 \tag{1}$$

$$V_C = V_{CC} + V_{EM}\ [1 + K + \cos\phi\ -\ K\cos(\phi\ - \alpha)\ ] / 2 \tag{2}$$

Where V_{CC} = Compression clearance volume
= Connecting tube volume + clearance in compressor + space below regenerator

The total mass of working fluid

$$M = M_C + M_{CT} + M_{RD} + M_E \tag{3}$$

Using perfect gas law

$$P = \frac{MR}{\left[\frac{V_C}{T_C} + \frac{V_{RD}}{T_{RD}} + \frac{V_E}{T_E}\right]} \tag{4}$$

The magnitude of pressure P_1 and volume V_C, V_E and temperature T_C vary with crank angle ϕ. The temperature T_E is assumed to remain constant. At the beginning, the regenerator temperature is assumed to be the logarithmic mean temperature of the two end temperatures.

Assuming the product of total mass in terms of moles of the gas, M, and the universal gas constants R, (MR) to be unity as the exact value of M is not known

$$P(1) = \frac{MR}{\left[\frac{V_C(1)}{T_C(1)} + \frac{V_{RD}}{T_{RD}} + \frac{V_E(1)}{T_E}\right]} \tag{5}$$

For adiabatic compression, assuming $\left(\frac{\gamma - 1}{\gamma}\right) = E$

$$T_C(2) = T_C(1) \times \left(\frac{P(2)}{P(1)}\right)^E \tag{6}$$

Substituting the value of $T_C(2)$, $V_C(2)$ and $V_E(2)$ in equation (4), the equation thus has only one variable, P(2). The value of P(2) could be obtained using the Newton – Raphson method. The value of P(2) obtained is thus used to calculate P(3) and so on. In the same way, all the values of pressure and temperature are calculated for the complete cycle, depending on the interval chosen ϕ. If ϕ is, for example, equal to 30, 13 points need to be calculated. The pressure and temperature at the first and thirteenth points should match within the tolerance limit, otherwise process of calculation restarts from the first point.

Mean pressure P_m, would therefore be equal to

$$P_m = P_{Total} / 12$$

The average existing pressure of the system, P_{avg} is known and the ratio of P_{avg} to P_m would give the correct value of MR. Previously, MR was assumed to be equal to unity. So, with new values of MR, all the pressure values calculated earlier need to be corrected so that P_m matches P_{avg}.

Calculation of mass flow rates, ideal power input, ideal refrigerating effect and loss analysis has been reported by Atrey et al.[1]

RESULTS AND DISCUSSIONS

The present analysis was applied to 1 W at 80 K miniature split Stirling cryocooler developed by Gaunekar et al.[3] The results are shown in Figures 1- 4. Tables 1 and 2 present respectively power requirement and refrigeration available for Gaunekar's unit using present analysis. Table 3 shows comparison of the overall performance of the cryocooler as obtained by the present analysis and experimental results reported by Gaunekar et al.[3] The efficiency of linear motor is assumed to be 75%.

The comparison shows that the prediction of performance by this analysis is in good agreement with the experimental results.

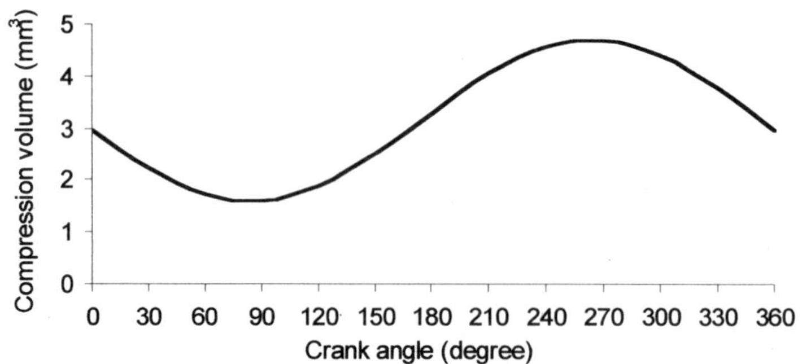

Figure 1 Compression volume variation

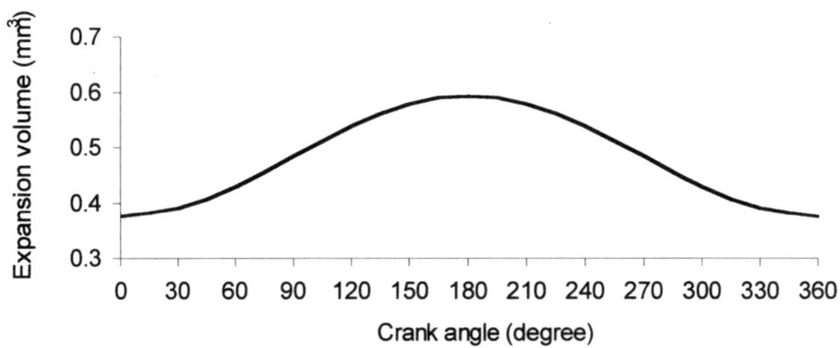

Figure 2 Expansion volume variation

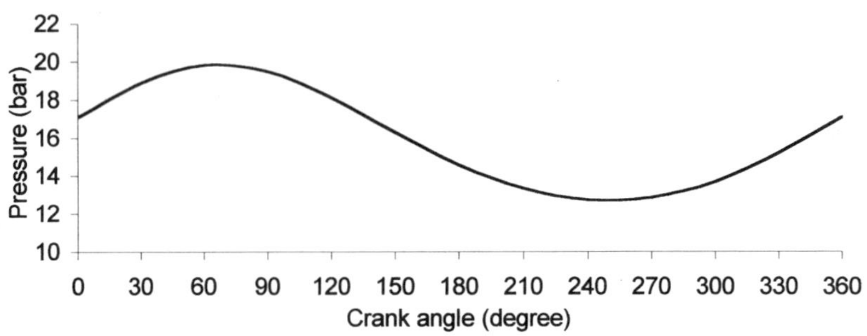

Figure 3 Pressure variation

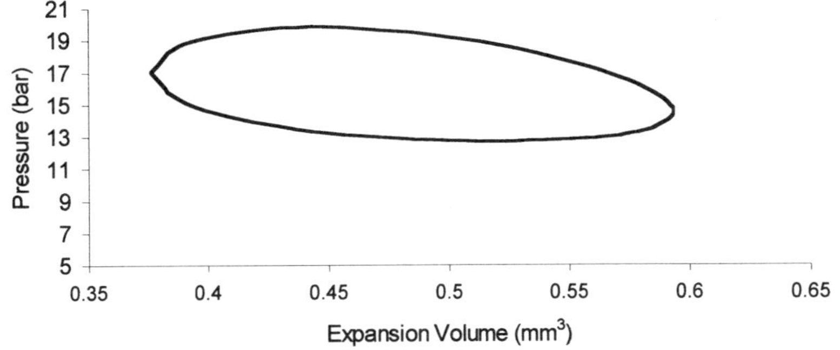

Figure 4a P -V diagram for expansion volume

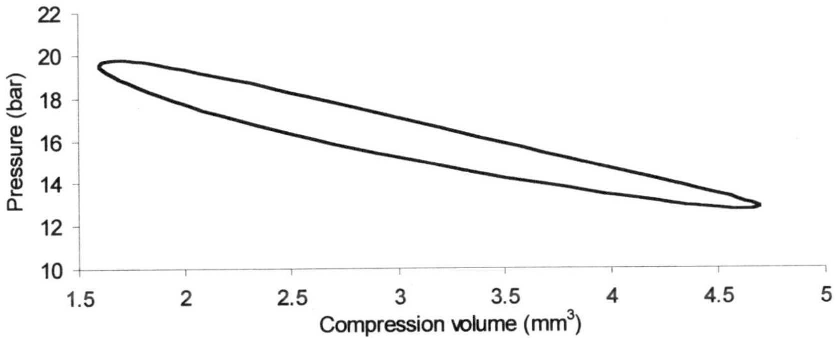

Figure 4b P -V diagram for compression volume

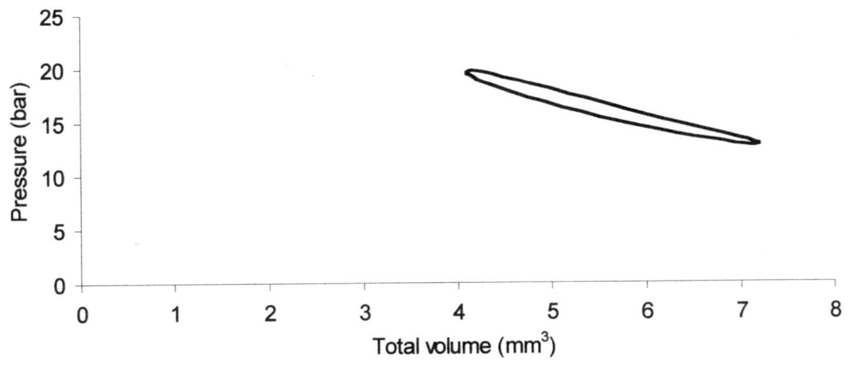

Figure 4c P - V diagram for total volume

Table 1. Power requirement and losses for Guanekar's unit[3] using present analysis

	Present results (W)
Basic Power**	16.25
Connecting tube flow loss	5.86
Regenerator flow loss	3.17
Mechanical loss*	8.43
Actual Power***	33.71

*Efficiency of linear motor assumed as 75 %
**Basic power is Ideal Power requirement of the system
*** Actual power is the summation of Ideal Power and Losses

Table 2. Refrigeration available and losses for Gaunekar's unit[3] using present analysis

	Pressure Results (W)
Basic refrigeration	5.40
Regenerator loss	1.61
Shuttle	1.57
Pumping	0.01
PV Loss duct to pressure drop	0.14
Conduction	1.02
Net refrigeration	1.05

Table 3 Comparison of overall performance predicted by present analysis with experimental data

	Gaunekar's Unit[3]	Present results
Power Input (W)	36.00	33.71
Net Refrigeration (W)	1.02	1.05
COP	0.0283	0.0311

ANALYSIS OF TWO STAGE CRYOCOOLER

After validating the analysis with the experimental results, the cyclic simulation is further extended for design of a miniature two stage cryocooler to meet the specification of 2 W at 100 K (stage I) and 0.5 W at 50 K (stage II).

Cyclic simulation starts with assumed dimensions such as diameter of piston, stroke, diameter of displacer (stage I), diameter of displacer (stage II) and other parameters such as matrix material of regenerator, connecting tube dimensions etc. By using computer programme following dimensions have been arrived at to meet required specifications.

Compressor (Opposed piston)		Displacer (Stage I)		Displacer (Stage II)	
Diameter	20 mm	Diameter	17 mm	Diameter	9 mm
Stroke	10 mm	Stroke	3 mm	Stroke	3 mm

Tables 4 and 5 present power requirement and refrigerating effect available respectively for the two stage cryocooler.

Table 4. Power requirement and losses for two stage cryocooler

	(W)
Basic Power	27.35
Connecting tube flow loss	5.64
Regenerator flow loss (stage I)	0.58
Regenerator flow loss (stage II)	0.06
Mechanical loss*	15.61
Actual Power	49.24

*Efficiency of linear motor assumed as 75 %

Table 5. Refrigeration available and losses for two stage cryocooler

	Stage I (W)	Stage II (W)
Basic refrigeration	7.120	2.000
Regenerator loss	0.560	0.346
Shuttle	2.340	0.620
Pumping	0.006	0.0006
PV Loss due to pressure drop	0.561	0.153
Conduction	1.540	0.269
Net capacity	2.113	0.606

CONCLUSIONS

Validation of the analysis with experimental results shows a good agreement. Consideration of the cyclic variation of mass flow rates and calculation of different losses for each interval of the cycle shows good agreement with actual values. Also the assumption of adiabatic compression and isothermal expansion makes the analysis more realistic. Experimental results of the two stage cryocooler could be used for validating the design based on cyclic analysis.

NOMENCLATURE

K Swept volume ratio of compression space and expansion space
M Mole of gas
P Pressure (bar)
R Universal gas constant (J g^{-1} mol^{-1} K^{-1})
T Temperature
V Volume (m^3)

Subscript

avg Average
C Compression space
CC Compression clearance
CM Maximum compression space
CT Connection tube space
E Expansion space
EC Expansion clearance
EM Maximum expansion space
m Mean

RD Regenerator dead space
Total Sum of all pressures 1 to 12

Greek symbols

γ Ratio of specific heat capacity
ϕ Crank angle

REFERENCES

1. Atrey, M. D. , Bapat, S. L. and Narayankhedkar, K. G., "Cyclic Simulation of Stirling Cryocoolers", Cryogenics, Vol. 30, 341-347 (1990).
2. Natu, P. V. and Narayankhedkar, K. G., "Cyclic Analysis of the Two Stage Stirling Cycle Cryogenerator", ICEC 16, 373-376 (1996).
3. Gaunekar, A. S., "Analysis, Design, Development and Experimental Testing of a Linear Motor Driven, Miniature Split Stirling Cryocooler using Flexure Bearings, for Satellite based Cooling Applications", Ph. D. Thesis, Department of Mechanical Engineering, IIT Mumbai (1996).

CRYOCOOLER LIFE ESTIMATION AND ITS CORRELATION WITH EXPERIMENTAL DATA

D.T. Kuo, A.S. Loc, T.D. Lody and S.W.K. Yuan

BEI Technologies, Cryocooler Group
Sylmar, CA 91342

ABSTRACT

While the life of a cryocooler is guaranteed under the manufacturer's specification, the manufacturer is constantly faced with the problem of trying to determine cooler life for customers' specific applications. An attempt has been made to estimate cooler life based on a Watt-Hour approach proposed by Scott Miskimins[1]. In the current paper, the total life of a cooler in Watt-Hour is expressed as a function of initial power and the failure criterion. BEI is currently conducting two life tests on the 0.5 W Stirling cooler (B512), with various heat loads at 78K (at room temperature ambient). Correlation of the experimental life test data with the theory will be discussed in detail, together with life estimate of operating the cooler at conditions other than that of the life tests, namely, high and low ambient temperatures, different heat loads and coldtip temperatures, etc.

INTRODUCTION

Miskimins[1] proposed to estimate cooler life using a watt-hour approach. For a given failure criterion, a cooler has certain life in terms of watt-hours, i.e., the product of input power and cooler life (in hours) is a constant. Therefore, if a cooler operates at low power it's life is extended and vice versa. The theory proposed in this paper differs from that of Reference 1, which assumes a linear correlation between cooler life (in Watt-Hour) and input power.

Figure 1 shows the life test data of BEI's B512C cooler with input power as a function of life. A least squares function is used to fit the data as represented by the solid line. The total life of the cooler in Watt-Hours is represented by the total area under the life test curve (Figure 1). It is BEI's experience in conducting life tests that coolers with same

Advances in Cryogenic Engineering, Volume 45.
Edited by Shu *et al.*, Kluwer Academic / Plenum Publishers, 2000.

configuration have almost the same slope, when input power is plotted as a function of life. Coolers with better alignment tend to have lower input power, but the slope remains almost the same (Figure 1).

Assuming a failure criterion of 14 W, one can integrate curves (e.g., a, b and c) of same slope from various initial input power to the failure criterion to get the cooler life curve as shown in Figure 2.

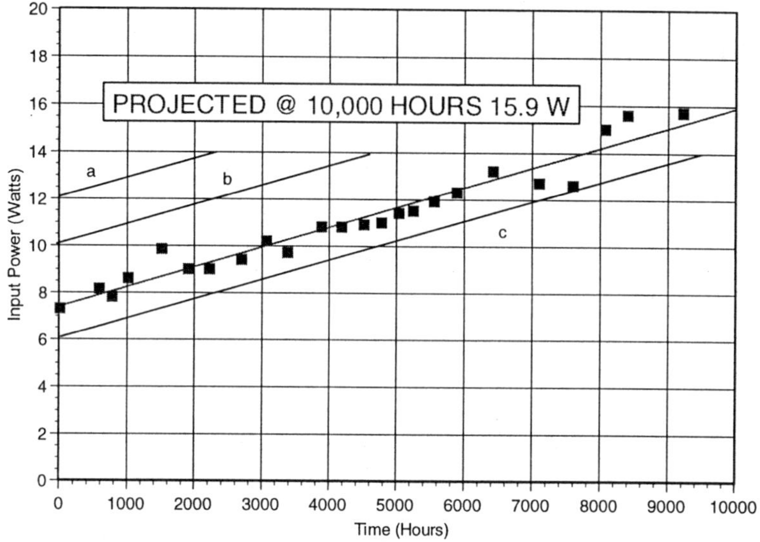

Figure 1. Life test data of BEI's B512C Cooler.

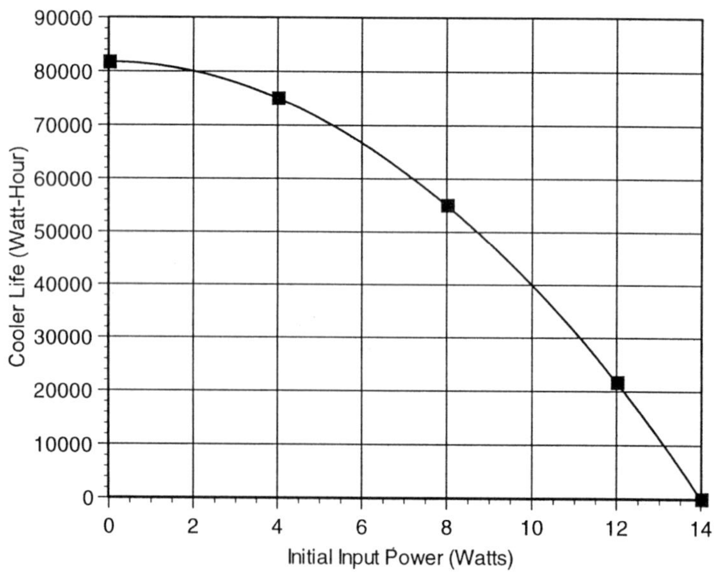

Figure 2. B512C cooler life curve (14 W as failure criterion).

Given Figure 2, one can proceed to estimate cooler life. For example, a cooler with an initial power of 10 W, has an estimated life of about 40000 Watt-Hours. The life of the cooler in time is then 40000W-Hr/10W = 4000 hours. If the initial power of the cooler is 14 W, the cooler has essentially no life.

Cooler life curves like that of Figure 2 can be easily constructed if one knows the slope of the life test data (W/Hour). The total area underneath the life test curve in Figure 1 with initial power (P_i) and failure criterion (P_f) can be written as

$$\text{Cooler Life (W-Hr)} = \text{Area Under Curve} = ((P_f+P_i)/2)*(P_f-P_i)/\text{slope}$$

$$= (P_f^2-P_i^2)/(2*\text{slope}) \qquad (1)$$

Note that Figure 2 is not a linear curve as suggested in Reference 1. Miskimins suggested that extrapolation of the linear curve towards low power might not be accurate. Figure 2, which is based on integration of the life test data should provide more accurate prediction of life at low power.

With Figure 2, one can also estimate cooler life at conditions other than that of the life test. The life test data in Figure 1 was taken with a heat load of 150 mW at room temperature ambient. Higher heat load or ambient temperature tends to shorten life and lower heat load or ambient temperature has the opposite effect. Moreover, the life test cooler has a transfer line of one inch in length. Longer transfer line tends to degrade life. If the initial input power of the cooler at conditions other than that of the life test are known, the cooler life at these conditions can be estimated from Figure 2. For example, a cooler with 150 mW load has an initial power of 6.5 W (Figure 3)[5]. From Figure 2, this cooler should have an estimated life of close to 9700 hours (63000W-Hr/6.5W). If the case temperature of this cooler is raised to 71°C, the input power becomes 12W (Figure 3) and the cooler life becomes ~1850 hours (22000W-Hr/12W).

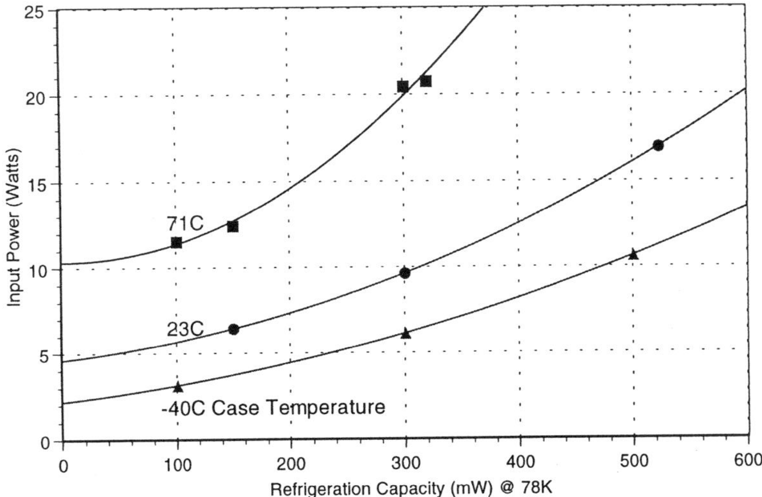

Figure 3. Typical performance of B512C coolers.

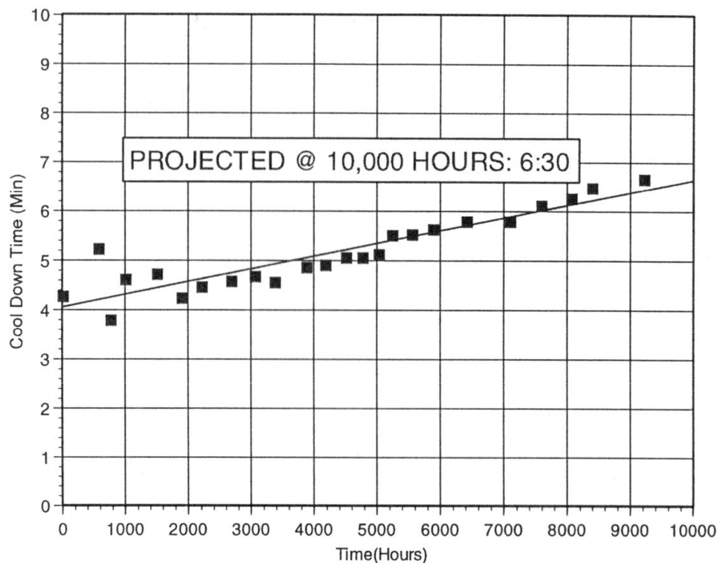

Figure 4. B512C cooler cooldown time vs. life.

By the same token, if the heat load is 300 mW (at 23°C case temperature), the life of the cooler is about 4400 hours (42000W-Hr/9.5W). Note that in the above calculations, a failure criterion of 14W input power was assumed.

Input power is not the only criterion for failure. Other criteria like cooldown time and minimum refrigeration are also important. Figure 4 shows the cool down time as a function of life. Although the life of the B512C cooler is close to 8000 hours (Figure 1), based on the criterion of input power, the life of the cooler is less than 4000 hours (Figure 4) if the criterion on cooldown time is ≤ 5 minutes.

Since both the B512C cooler and the B512B cooler use the same compressor (with different design of expander), it is conceivable that the life data collected for the B512C cooler can be applied to the B512B cooler. A separate life test was conducted on the B512B cooler with three coolers running at 212 mW (average value represented by dotted lines in Figures 5 to 7). These three coolers have an average initial input power of 8.5 W. From Figure 2, one can estimate the life of the B512B cooler to be around 6200 hours (53000W-Hr/8.5W). Unfortunately, the actual life test data fell short of the prediction. Further examination of the data showed that besides the difference in the expander design, the charge pressure of the B512B cooler (650 psig) was also much higher than that of the B512C cooler (450 psig). Due to the presence of larger dead volumes inside the B512B expander, a higher charge pressure is needed to prevent the compressor pistons from hitting. It is a known fact that higher charge pressure tends to enhance piston wear and thus shorten cooler life. From this one should note that the theory presented in this paper is only applicable to coolers of same configuration.

To meet the 4000-hour life requirement, BEI modified the material of the piston seal from a teflon to a teflon based material with glass. The life test data of coolers made of old and new seal materials are presented in Figures 5 to 7. Coolers with old piston seal material are represented by dotted lines (average of three coolers). Three coolers with the new piston seal material are also included in the figures. Out of the six life test coolers, five of them have heat loads of 212mW applied during the life test. Cooldown tests were performed with no heat load. The remaining unit with the new piston seal material has been tested to a much tougher specification, with 300mW load during life test and 120mW load during cooldown.

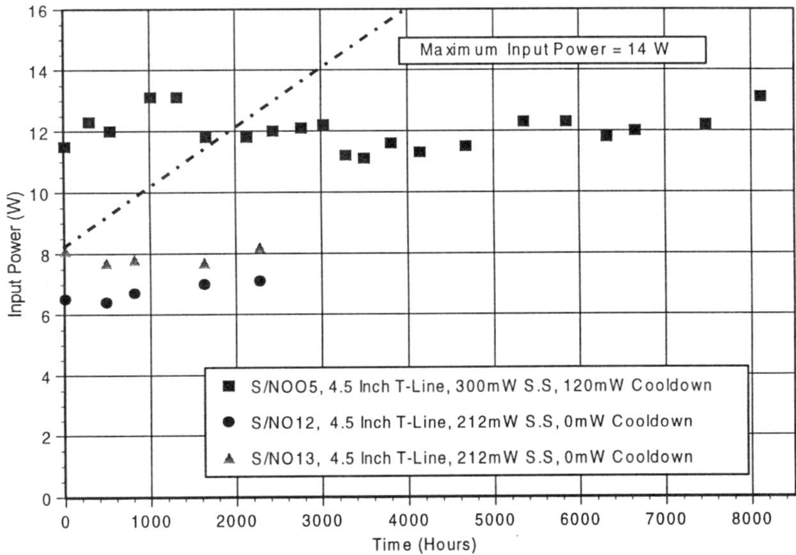

Figure 5. Input power vs. life for B512B cooler.

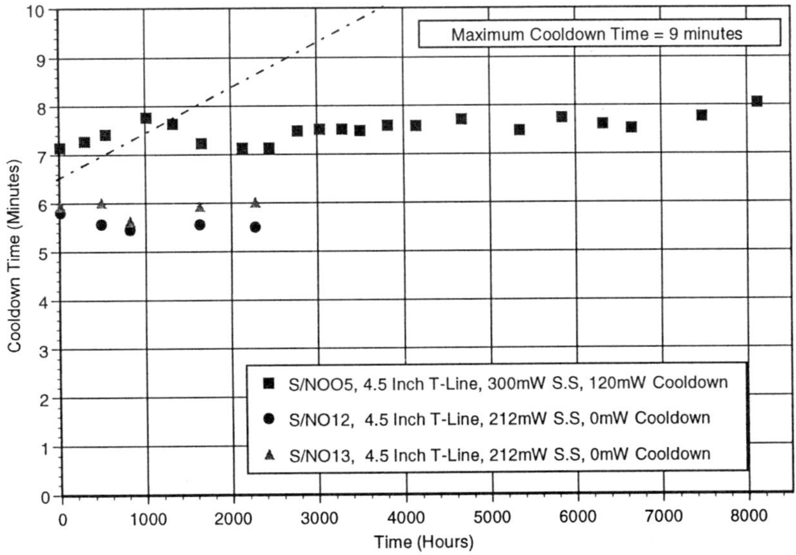

Figure 6. Cooldown time vs. life for B512B cooler.

Figure 5 is a plot of input power as a function of life. Figure 6 shows the cooldown time from ambient to 78K with 11 volt input power. Figure 7 is a plot of minimum refrigeration at 11 volt versus life. As one can see, the slope of the old piston seal material (dotted line) is far steeper than that of the new material, indicating a far shorter life. Due to this finding, BEI has changed the piston seal material in all coolers[2-9] across the board. To date, two of coolers with new piston seal have accumulated more than 2200 hours showing little or no sign of wear. The third unit which has been tested to a tougher specification (300mW) has been running for more than 8000 hours.

Life test data of B602 cooler. Life test data on the B602 cooler is presented in Reference 9. The B602 cooler is a 600 mW cooler. It can provide larger cooling capacity than the B512 cooler, but it also consumes more energy. Figure 8 shows the cooler life (in Watt-Hour) of the B602 cooler as a function of input power for failure criteria of 30W, 32W and 34W, as predicted by Equation 1. At this range, close to the failure criteria, the life curves appear to be a straight line. Currently, the B602 life test unit has more than 7000 hours accumulated, and it's input power is still below 30 W. The cooler's initial input power was 27.4W

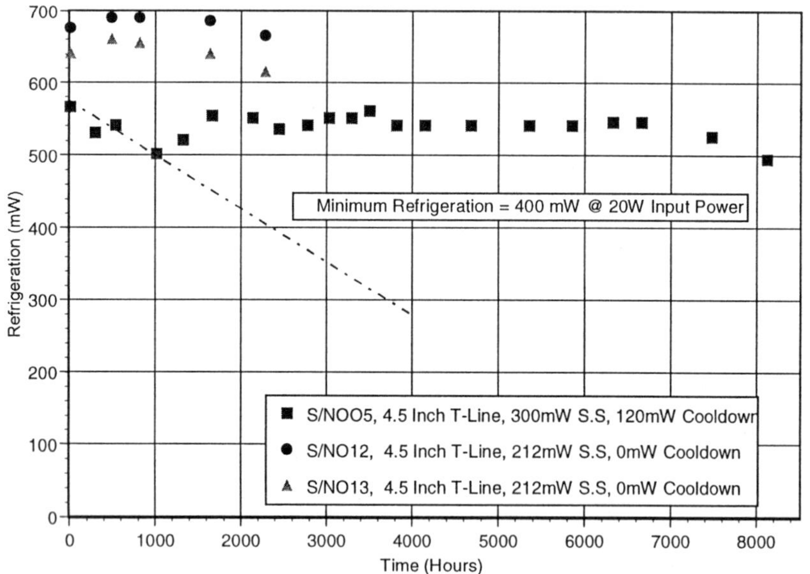

Figure 7. Refrigeration vs. life for B512B cooler.

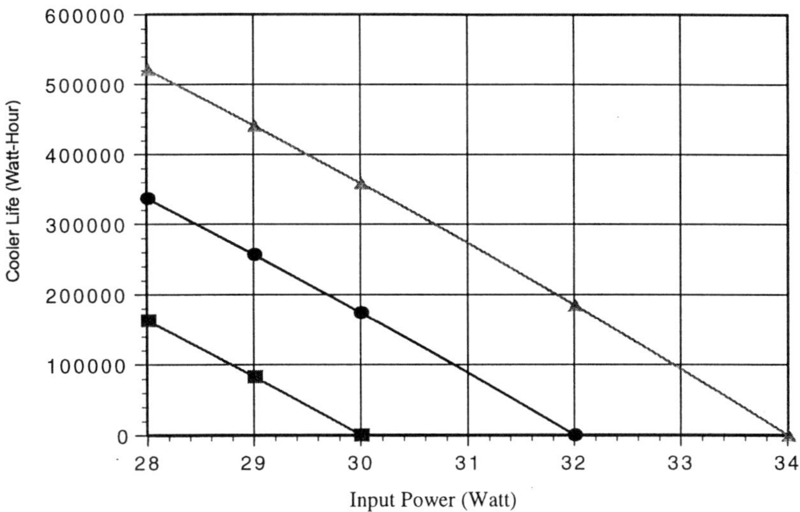

Figure 8. Cooler life curve for BEI's B602C Cooler.

CONCLUSIONS

A watt-hour approach based on integration of the life test data has been suggested to estimate life of cryocoolers. If the life of a cooler under a particular condition is know, the life of the same model of coolers under various conditions can be estimated. Life test data of various models of BEI Cryocoolers have also been presented.

REFERENCES

1. Miskimins, Scott, Estimating SADA II Cooler Life, private communication.
2. D.T. Kuo, A.S. Loc, and S.W.K. Yuan, Experimental and Predicted Performance of the BEI Mini-linear Cooler, in the Proc. of the 9th International Cryocooler Conference (1997) p.119.
3. D.T. Kuo, A.S. Loc, and S.W.K. Yuan, Design of a 0.5 Watt Dual Use Long-Life Low-Cost Pulse Tube Cooler, Advances in Cryogenic Engineering, 43:2039, 1997.
4. S.W.K.Yuan, D.T.Kuo, A.S.Loc, and T.D. Lody, Experimental Performance of the BEI One Watt Linear (OWL) Stirling Cooler, Advances in Cryogenic Engineering, 43:1855, 1997.
5. S.W.K. Yuan, D.T. Kuo, and A.S. Loc, Enhanced Performance of the BEI 0.5 Watt Mini-Linear Stirling Cooler, Advances in Cryogenic Engineering 43:1847, 1997.
6. S.W.K. Yuan, D.T. Kuo, and A.S. Loc, Design and Preliminary Testing of BEI's CryoPulse 1000, the Commercial One Watt Pulse Tube Cooler, to be published in the Proc. of the 10th International Cryocooler Conference, 1998.
7. D.T. Kuo, A.S. Loc, and S.W.K. Yuan, Qualification of the BEI B512 Cooler, Part 1 - Environmental Tests, to be published in the Proc. of the 10th International Cryocooler Conference, 1998.
8. S.W.K. Yuan, D.T.Kuo, and A.S. Loc, Cryocooler Contamination Study, parallel paper in the Cryogenic Engineering Conference, Montreal, 1999.
9. S.W.K. Yuan, D.T.Kuo, A.S. Loc, and T.D. Lody, Performance and Qualification of BEI'S 600 mW Linear Motor Cooler, parallel paper in the Cryogenic Engineering Conference, Montreal, 1999.

CRYOCOOLER CONTAMINATION STUDY

S.W.K. Yuan, D.T.Kuo, and A.S. Loc

BEI Technologies, Cryocooler Group
Sylmar, CA 91342

ABSTRACT

Eliminating or reduction of contamination is essential for the success of cryocoolers as far as performance and cooler life is concerned. Contaminants in a cooler can be defined as any foreign material that results in the rubbing of mechanical parts or blockage of flow, which in turn reduces cooler life and / or degrades performance. In this paper, we will limit the scope of our study to gaseous and liquid contaminants, which may freeze and are detrimental to cooler performance. It was found that most of the contaminants in BEI's coolers consist, carbon dioxide, water, cleaning solvents (acetone and alcohol), nitrogen and oxygen. Five 0.5 watt coolers (B512) were used for the study. These coolers were baked out continuously at 71°C for one week. Gas analysis was performed on the cooler working gas every day during the one-week period. Outgassing curves have been constructed on the contaminants listed above. Based on the analysis, optimum bake-out time of the cooler components can be derived. Optimum purge and fill procedure will also be discussed in this paper.

INTRODUCTION

Due to the presence of narrow passages in cryocoolers (clearance gaps and torturous paths in regenerators), any contaminant would tend to interfere with the gas dynamic or mechanism of the cooler operation, which in turn results in the degradation of performance or shortening of cooler life. In this paper, we will concentrate on the study of gaseous and liquid contaminants. Using a Varian 3800 Gas Chromatography (GC) Analyzer (which is connected to a Dell Workstation), the following contaminants were identified and tabulated in Table 1. The Varian GC Analyzer uses a Thermal Conductivity Detector (TCD) and a Flame Ionization Detector (FID) to analyze contaminants. TCD detects gases like nitrogen, oxygen, carbon dioxide, and water vapor. FID on the other hand detects hydrocarbons. Also included in Table 1 are the retention time of various contaminants.

Table 1. List of contaminants with retention time

Contaminants	Detector	Retention Time (min)
Nitrogen / Oxygen	TCD	0.48
Carbon Dioxide	TCD	1.78
Water	TCD	5.5
Methane	FID	0.51
Acetone	FID	0.83
Alcohol	FID	1.22

The Varian GC Analyzer was calibrated using known amount of nitrogen, oxygen, carbon dioxide and methane. Sample of water vapor, acetone and alcohol were prepared at BEI. Acetone and alcohol are used as cleaning solvents, and the rest of the above gases are readily available in the atmosphere. A Hayesep column (N80/100) and a DB wax capillary (15m x 0.53 mm) were used in the GC analyzer. The column temperature was ramped up from 35°C to 140°C in five minutes and held at 140°C for another five minutes. The retention times of nitrogen and oxygen are too close to be distinguished.

Outgassing Study

Five B512C coolers[1-3] were used for the contamination study for one week. The coolers were randomly selected from the production line. They were placed in an oven connected to the Varian GC Analyzer. The oven temperature was set at 71°C. Gas samples were injected into the GC Analyzer every 24 hours and the levels of contaminants recorded. Out of the five readings (from the five coolers) the maximum and minimum values were discarded and the average of the three medium values was recorded. Water, acetone, and alcohol tend to be absorbed onto the surfaces inside the cooler and outgas for an extended period of time showing a steady increase of concentration in the gas sample as shown in Figures 1 (water and alcohol) and 2 (acetone).

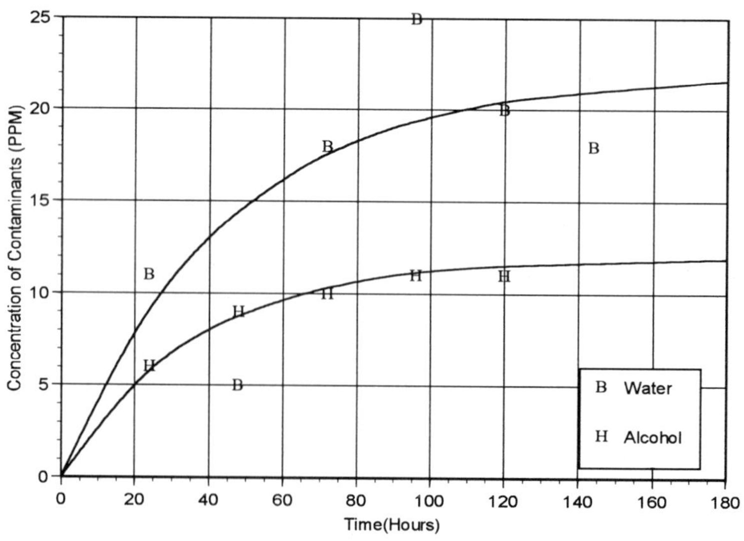

Figure 1. Concentration of water and alcohol as a function of time.

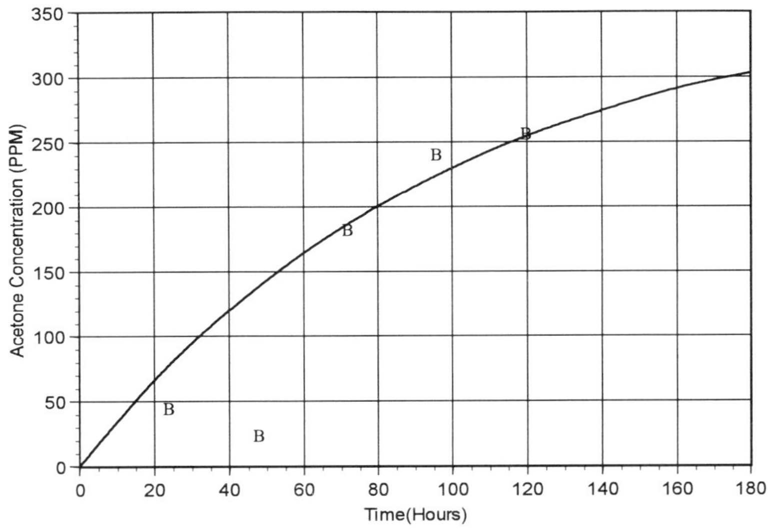

Figure 2. Acetone concentration as a function of time.

On the other hand, gases like nitrogen, oxygen and carbon dioxide do not get absorbed and outgassed, at least not substantially. The rate of increase in contaminant concentration (m) was found to follow the following equation.

$$m = B*T*(1 - e^{-t/A}) \tag{1}$$

Where A and B are constants, T is the ambient bake-out temperature in degrees Celsius and t is time in hours. Values of A and B for water, alcohol and acetone can be found in Table 2. The outgassing rate (dm/dt) can then be obtained by taking the derivative of Equation 1.

$$dm/dt = (B*T/A)*e^{-t/A} \tag{2}$$

Table 2. Outgassing constants

	A	B
Water	45	0.31
Alcohol	36.4	0.17
Acetone	100	5.13

Note that within the temperature range of this study, the temperature dependence of outgassing appears to be linear. Care should be taken in applying these equations at temperatures above 71°C. Figure 3 shows the outgassing rate of water and alcohol and Figure 4 shows the outgassing rate of acetone. Extensive bake out of cooler components and hot purge of coolers are essential to prevent contamination. Data like that of Figure 1 and 2 help to determine if the bake out and hot purge processes are adequate. Based on experience, the maximum amount of each contaminant allowable in a cooler was determined. This value must not be exceeded during the entire life of the cooler. Looking at Figure 1 and 2, one realizes that the rise in contaminant concentration tends to peak out after 200 hours or so, which can then be considered as the end-of-life value. This end-of-life value, therefore, must not exceed the maximum allowable contaminant concentration, obtained from experience.

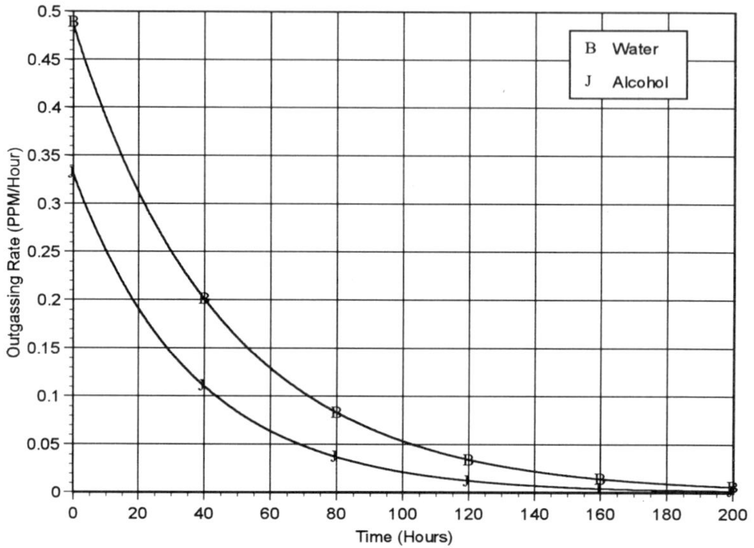

Figure 3. Outgassing rates of water and alcohol.

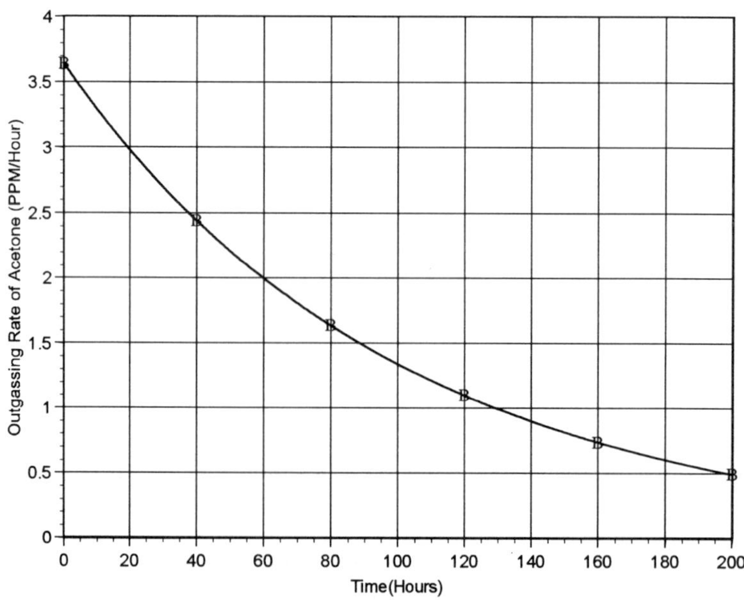

Figure 4. Outgassing rate of acetone.

However, testing each cooler for 200 hours is a time-consuming process. To shorten test time, one can test the cooler for a couple of hours and extrapolate the result (using Equation 1) to the end-of-life value at 200 hours. This end-of-life value can then be used to fine-tune the bake-out and hot-purge processes to optimize the process.

One interesting observation was the build-up of an unidentified contaminant as the cooler ages. This contaminant has a retention time of approximately 0.5 minutes and it shows up in the Flame Ionization Detector of the GC Analyzer. From the short retention time, one can speculate that the contaminant is a light hydrocarbon. By characterizing various light hydrogen carbons it was determined that the contaminant was methane. The presence of methane together with the disappearance of alcohol in other BEI coolers with long run-time, tends to suggest that alcohol in the coolers decomposes into methane. Figure 5 shows the accumulation of methane in the coolers under study, however the level is quite low (below 1 PPM) after one week. Although 90% of outgassing tends to be completed in 200 hours, the decomposition of alcohol into methane tends to take much longer, as supported by the trace amount of methane detected compared to the relatively high concentration of alcohol after 200 hours. Thus to limit the build up of methane, one must limit the initial concentration of the alcohol in coolers.

Outgassing is a function of surface area on which the contaminants are being absorbed. The findings in this paper can also be applied to other models of BEI coolers[1-8] via scaling if the internal surface area can be estimated.

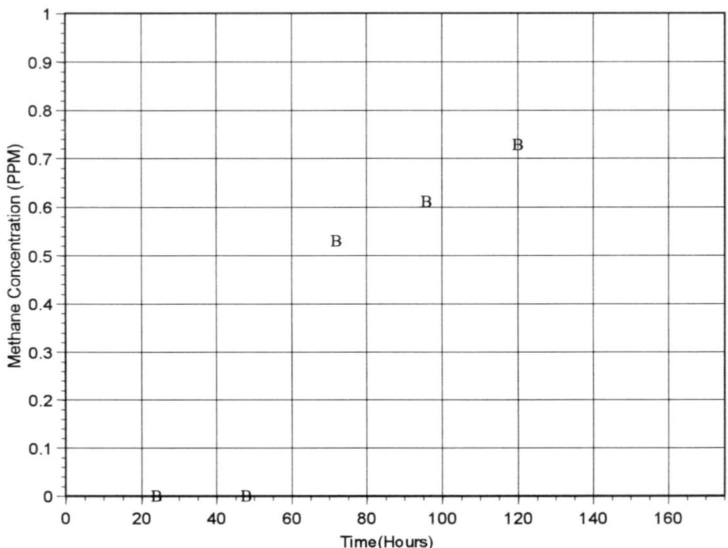

Figure 5. Detection of methane in coolers.

Room Temperature Purge and Fill

As mentioned earlier, gases like nitrogen, oxygen, carbon dioxide do not tend to outgas in the cooler. However, since these gases are readily available in the atmosphere, they must be carefully purged out of the coolers to prevent contamination. Thus, before the coolers are charged with helium gas, the coolers are generally purged with helium gas. The optimum number of purges must be determined so that the least number of purges or the shortest purge process would guarantee acceptable levels of these gases.

A separate study was conducted to determine this optimum value. A cooler was selected for this purpose. The expander was separated from the compressor and the cooler was allowed to come to equilibrium with the ambient air overnight. The cooler was then reintegrated and the appropriate number of purges applied. The cooler was then kept in an oven for two hours at 71°C before injecting the working gas into the GC Analyzer for evaluation. The process was then repeated with a different purge number. Since the retention times of nitrogen and oxygen are not distinguishable, we will group the gases together and call them "air".

Figure 6 is a plot of air concentration vs. number of purges at half-a-minute intervals. After about 15 purges, the decrease in air concentration levels off at a value (approx. 30 PPM) which is way below the acceptable level of 100 PPM (Ref. 9). When conducting the purge and fill test, it became quite evident that the higher the number of purges and the longer the duration of the purge process the cleaner the working gas became. The test was thus repeated with different combinations of purge duration and number of purges. It was found that when plotting the air concentration vs. product of purge number and total purge duration, the curve follows that of an exponential decay, as shown in Figure 7. Within the limits of this study (maximum number of purges = 20, minimum of purges = 4, maximum total purge time = 10 minutes, minimum total purge time = 1 minute) the data tend to fall on this exponential curve and are quite insensitive to the combination of purge number and total purge duration. This information is quite useful if one wants to shorten the purge process. In order to obtain the same result of a 10 minute process, with 20 purges at a 30 second interval (20x10min=200), one can adopt a 7 minute process with 28 purge at a 15 second interval (28x7min equals to 196).

The shortest purge interval used in this study was 15 second. It is not recommended to decrease the purge interval to below 15 seconds for the air inside the cooler (especially in the regenerator) will not have enough time to equilibrate with the purge gas.

CONCLUSIONS

The outgassing characteristics within the BEI B512C coolers have been studied. Outgassing equations for water, alcohol, and acetone have been suggested. The dependence of outgassing on ambient temperature should be further investigated. An optimized purge and fill process has also been proposed to eliminate air and moisture in coolers.

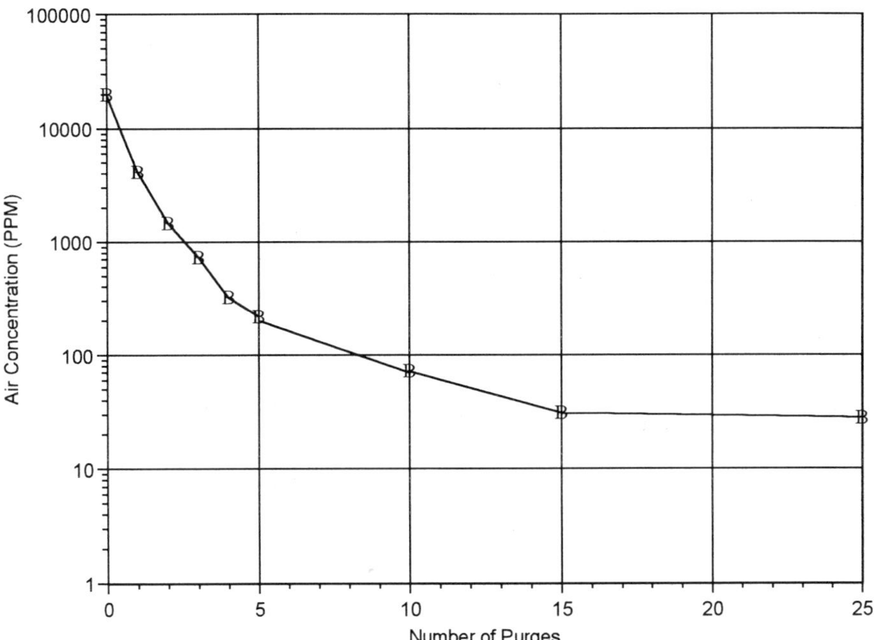

Figure 6. Air concentration vs. number of purges.

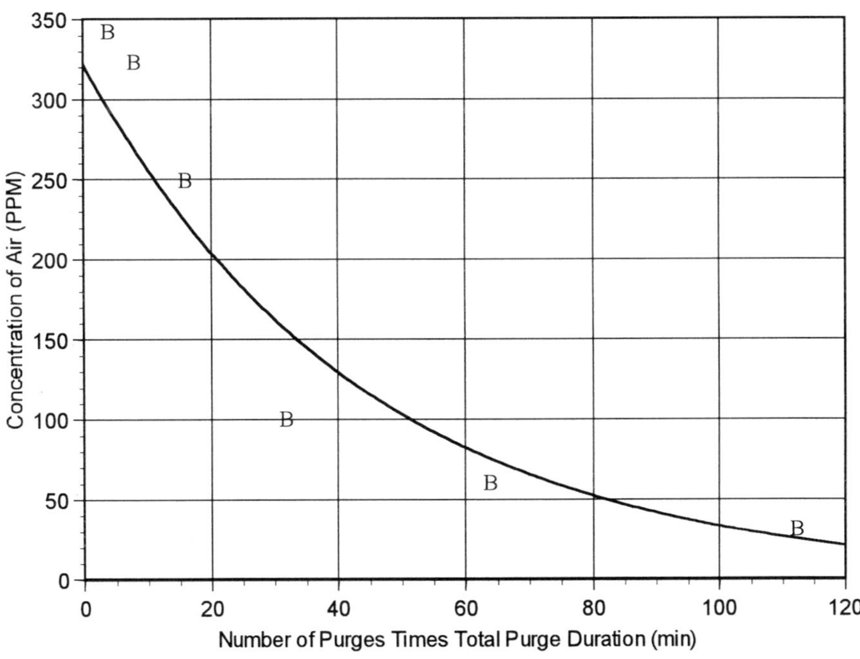

Figure 7. Study of purge duration and number of purges.

REFERENCES

1. D.T. Kuo, A.S. Loc, and S.W.K. Yuan, Experimental and Predicted Performance of the BEI Mini-linear Cooler, in the Proc. of the 9th International Cryocooler Conference (1997) p.119.
2. S.W.K. Yuan, D.T. Kuo, and A.S. Loc, Enhanced Performance of the BEI 0.5 Watt Mini-Linear Stirling Cooler, Advances in Cryogenic Engineering 43:1847, 1997.
3. D.T. Kuo, A.S. Loc, and S.W.K. Yuan, Qualification of the BEI B512 Cooler, Part 1 - Environmental Tests, to be published in the Proc. of the 10th International Cryocooler Conference, 1998.
4. D.T. Kuo, A.S. Loc, and S.W.K. Yuan, Design of a 0.5 Watt Dual Use Long-Life Low-Cost Pulse Tube Cooler, Advances in Cryogenic Engineering, 43:2039, 1997.
5. S.W.K.Yuan, D.T.Kuo, A.S.Loc, and T.D. Lody, Experimental Performance of the BEI One Watt Linear (OWL) Stirling Cooler, Advances in Cryogenic Engineering, 43:1855, 1997.
6. 5. S.W.K. Yuan, D.T. Kuo, and A.S. Loc, Design and Preliminary Testing of BEI's CryoPulse 1000, the Commercial One Watt Pulse Tube Cooler, to be published in the Proc. of the 10th International Cryocooler Conference, 1998.
7. S.W.K. Yuan, D.T.Kuo, A.S. Loc, and T.D. Lody, Performance and Qualification of BEI'S 600 mW Linear Motor Cooler, parallel paper in the Cryogenic Engineering Conference, Montreal, 1999.
8. D.T. Kuo, A.S. Loc, T.D. Lody and S.W.K. Yuan, Cryocooler Life Estimation And its Correlation With Experimental Data, parallel paper in the Cryogenic Engineering Conference, Montreal, 1999.
9. Based on BEI's experience with various models of coolers.

OPTIMUM COMPOSITION CALCULATION FOR MULTICOMPONENT CRYOGENIC MIXTURE USED IN JOULE-THOMSON REFRIGERATORS

M.Q. Gong, E.C. Luo, Y. Zhou, J.T. Liang, and L. Zhang

Cryogenic Laboratory, Chinese Academy of Sciences
Beijing 100080 China

ABSTRACT

An optimization calculation method is developed in this work to evaluate the optimum composition of the mixed gas used in the Joule-Thomson (J-T) refrigerator system. In this method, the ideal thermodynamic efficiency for the J-T refrigeration cycle is used as the objective function to be maximized. The influence of the temperature difference in the counterflow heat exchanger and the efficiency of the compressor are also considered. The Complex Optimization Algorithm (COA) is introduced in this paper. The calculation result can be useful in the experimental investigation of mixed-refrigerant J-T refrigerators. To understand the result better, the thermodynamic analysis of the optimal mixture obtained from the simulation is also presented.

INTRODUCTION

Extensive study shows that the usage of multicomponent mixtures of freons or light hydrocarbons with argon or nitrogen in Joule-Thomson (J-T) refrigeration cycles can greatly improve the thermodynamic performance of the J-T cryocooler [1,2,3]. The simplicity and reliability of using mixed refrigerants in a system with a single stage compressor and no mechanical expander provide opportunities to build relatively low cost and highly reliable cryocoolers for lower temperatures that broaden their commercial applications.

However, one must realize that not any mixture used in a J-T refrigerator can improve its thermodynamic performance. That is, there must be an optimum mixture or optimum composition of given components at certain conditions. The purpose of this paper is to discuss this problem of optimization of the mixed-refrigerant J-T refrigerator. An attempt is made in this work to develop a method to optimize the thermodynamic performance of the mixed refrigerants J-T refrigerator system by analyzing the composition of the mixture and adjusting the operating parameters.

CALCULATION METHOD

An attempt is made in this work to develop a mathematical method to optimize the composition of the mixture. The influences of the temperature difference in the counterflow heat exchanger and the efficiency of the compressor are also considered.

Mathematical Model

Generally, the optimization of a mixed gas J-T refrigerator is a systematical problem including the optimization of both the composition of the mixture and operating parameters. Usually in most applications, the required cold temperature and the ambient temperature are given; then the problem is essentially considered as an optimization of the composition of the mixture with the high and low operating pressures of the refrigerator. However, it is too complicated to simulate the optimization of the whole system. In this work, in order to describe the problem and solve it quickly, the optimizations for the composition of the mixture and the optimal operating pressures are considered, respectively.

For simplification we assumed the following:

- the mixture composition does not change in the cycle
- no pressure losses in the counterflow heat exchanger
- no heat load from the insulation or by heat conduction

The Inherent Optimum Composition of Mixture. When the operating parameters of the cycle are given, the problem is concerned with the optimization of the inherent optimum composition for the candidate mixture.

Mathematically, this optimization problem can be defined as follows:

To adjust the composition vector $\vec{X}$:

$$\vec{X} = X(x_1, x_2, \cdots, x_i, \cdots, x_n)^T \tag{1}$$

We can obtain:

$$\eta_{mix} = \frac{\min[\Delta H_T(x_i, T, P_H, P_L, \Delta T_{\min})]}{W_{T_0}} \times \left(\frac{T_0 - T_c}{T_c}\right) \tag{2}$$

with the constrained conditions of:

$$\sum_1^n x_i = 1 \qquad x_i > 0 \quad (i=1,\ 2,\ \ldots,\ n) \tag{3}$$

where:

η_{mix} - the Carnot efficiency of the mixture

$min\Delta H_T$ - the theoretical cooling power of the mixture

W_{T_0} - the theoretical input power of adiabatic compression

T_0, T_c - ambient and cold temperature, respectively

P_H, P_L - high and low pressure, respectively

ΔT_{min} - minimal temperature difference in the counterflow heat exchanger

x_i - mixture composition (molar %)

Equation (2) is the expression of the objective function for the ideal mixed-refrigerant J-T refrigeration cycle.

Influence of the Compressor. When the efficiency of the compressor is considered, the optimization objective function is changed as follows:

$$\eta = \eta_{mix}\left[\vec{X}, P_H, P_L, T_c, T_0, \Delta T_{min}\right] \cdot \eta_v \cdot \eta_e \cdot \eta_f \tag{4}$$

$$\eta_v = 0.94 - 0.085\left[\left(P_H / P_L\right)^{1/k} - 1\right] \tag{5}$$

where:

η - the Carnot efficiency of the refrigeration cycle

η_v - the volumetric efficiency of the air conditioning compressor, as obtained with Equation (5) [4]

k - the adiabatic index

η_e - the efficiency of the electric motor [4]

η_f - the frictional efficiency of the compressor [4]

With the above equations we can simulate the thermodynamic performance of the mixed-refrigerant JT refrigerator. By systematical variation of the high and low pressures, the optimum operating pressures can be achieved. The simulation pressure ranges are set within the limited operating pressure of the air-conditioning compressor, as follows:

$$1.5 \le P_H \le 2.5 \text{ MPa}, \;\; 0.1 \le P_L \le 0.3 \text{ MPa} \tag{6}$$

The Complex Optimization Algorithm (COA)[5]

The problem presented above is a classical non-linear constrained optimization problem. The complex optimization algorithm (COA) is introduced to optimize this problem.

In this method, there are *2n* vertexes that satisfy all the constraints specified to construct the complex in the *n* dimensional non-linear constrained space. All vertexes of the complex are compared one by another, with the least desirable point being replaced by a new point that can satisfy all constraints while improving the objective function. Step by step, the optimum point is achieved.

Detailed progress of this calculation can be presented as follows:

1) Generate *2n* vertexes of the complex, and construct the complex.

2) Evaluate the function at each vertex: $f\left(x^{(j)}\right)$, ($j=1,\ 2,\ \ldots,\ 2n$), to find the least desirable vertex $f\left(x^{(w)}\right)$ and the most desirable vertex $f\left(x^{(o)}\right)$; then turn to step 6.

3) Find the reflection point, and calculate the centroid $X(o)$ of all other vertexes without using the least desirable vertex $f\left(x^{(w)}\right)$. If the centroid $X(o)$ can satisfy all constraints, then the rejected point can be obtained with the following equation:

$$X^{(\alpha)} = X^{(o)} + \alpha\left(X^{(o)} - X^{(w)}\right) \tag{7}$$

where α is the reflection coefficient; in this paper $\alpha \ge 1$. Then turn to step 4. If the centroid $X(o)$ does not satisfy some constraints, the most desirable point is set as the optimum point.

4) If $X(\alpha)$ satisfies all constraints, turn to step 5, or if $\alpha = \alpha/2$ go back to step 3.

5) If this trial point is also the least desirable, it is moved halfway towards the centroid of the remaining points to provide a new trial point. The above procedure is then repeated until some constraint is violated.

6) If the complex has collapsed into the centroid, progress is ended, that is:

$$\left[\frac{1}{2}\sum_{j}^{2n}\left(f\left(x^{(o)}\right) - f\left(x^{(j)}\right)\right)^2\right]^{1/2} < eps \tag{8}$$

Otherwise, go back to step 3.

To calculate the thermodynamic properties of the mixture, the well-known two parameter cubic Peng-Robinson equation of state [6] is used.

OPTIMIZATION ILLUSTRATION

To better describe the optimization problem, several simulation examples are presented in this paper utilizing the method developed above.

Illustration 1. The Inherent Optimum Composition Optimization

The calculation conditions are described as:

1) The cooling temperature (T_C) is 173K, and the ambient temperature (T_H) is 300K
2) The high pressures are 2.0MPa and 2.5MPa, respectively; and the low pressures are 0.1 and 0.2 MPa
3) The components are R11, R12, R13, and R14

The optimization results are presented in Table 1. Figure 1 is the enthalpy vs. temperature (h-T) diagram of the No.4 mixture (mix.1) in Table 1, Figure 2 is the η_{mix} vs. temperature (η_{mix}-T) diagram of the No.4 mixture (mix.1) in Table 1.

Illustration 2. Influence of the Compressor and Temperature Difference in the Counterflow Heat Exchanger

The calculation condition is:

1) The cooling temperature (T_C) is 100K, and the ambient temperature (T_H) is 300K
2) $1.5 \le P_H \le 2.0$ MPa, $0.1 \le P_L \le 0.3$ MPa
3) The components are N_2, CH_4, C_2H_6, C_3H_8, and iC_4H_{10}
4) The refrigerator is driven by a single stage reciprocal air-conditioning compressor, its performance parameters such as η_e and η_f can be obtained from the reference[4]
5) The minimal temperature difference in the counterflow heat exchanger is set as 3K

Equation (4) is used to calculate the objective function. The results of the simulation are presented in Table 2. The h-T diagram and the $\eta_{mix} \times \eta_v$-T diagram are shown in Figure 3 and Figure 4 to provide a comparison of the No.3 and No.4 mixtures in Table 2.

Table 1. Optimization results of Illustration 1

No.	R11 %	R12 %	R13 %	R14 %	P_H MPa	P_L MPa	Δh_T kJ/mol	COP %	η_{mix} %
1	22.5	8.7	10.8	58.0	2.5	0.1	6.31	66.4	50.8
2	16.6	17.8	2.5	63.1	2.5	0.2	5.41	76.4	58.4
3	22.4	10.2	13.2	54.2	2.0	0.1	5.92	67.8	51.8
4	17.1	19.1	4.4	59.4	2.0	0.2	4.90	76.9	58.8

Table 2. Optimization results of Illustration 2

No.	N_2 %	CH_4 %	C_2H_6 %	C_3H_8 %	iC_4H_{10} %	P_H MPa	P_L MPa	ΔT_{min} K	ΔH_T J/mol	η_{mix} %	η_v %	η_e %	η_f %	η %
1	19.7	8.1	39.4	32.5	0.3	1.5	0.1	0	788	19.4	14.0	65	60	1.1
2	21.4	9.9	41.4	24.3	3.0	1.5	0.1	3	704	17.3	14.7	65	60	1.0
3*	31.5	28.3	5.2	13.4	21.6	2.0	0.3	0	1068	40.2	59.1	65	85	13.1
4*	35.3	20.4	10.7	13.7	19.9	2.0	0.3	3	840	31.5	59.1	65	85	10.3
5	28.3	19.9	7.4	26.7	17.7	1.8	0.3	0	897	36.5	62.2	65	87	12.8
6	31.2	15.9	19.7	14.9	18.3	1.8	0.3	3	710	28.6	62.6	65	87	9.9

* Optimal operating pressures in this simulation condition.

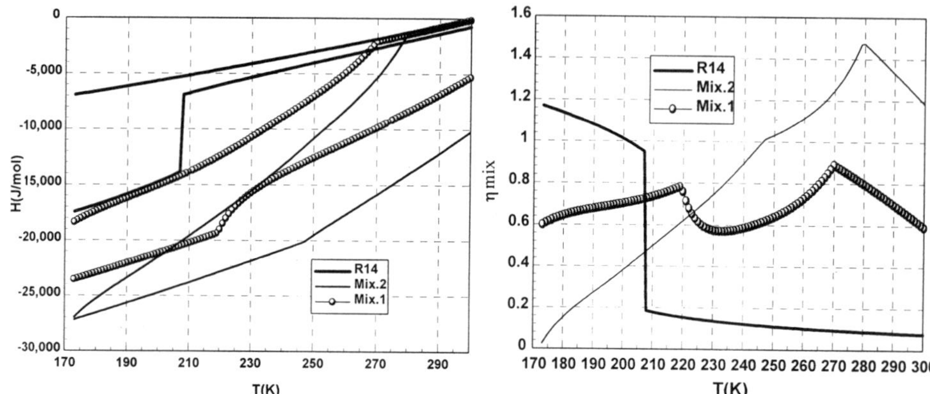

Fig.1. *h*-T diagram for three refrigerants (R14, mix.1 and mix.2 mixtures)

Fig.2. *η*-T diagram for three refrigerants (R14, mix.1 and mix.2 mixtures)

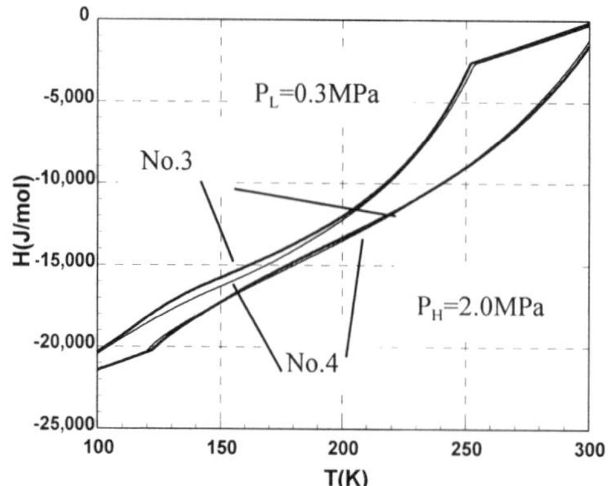

Fig.3. *h*-T diagram for the No.3 and No.4 mixtures in Illustration 2

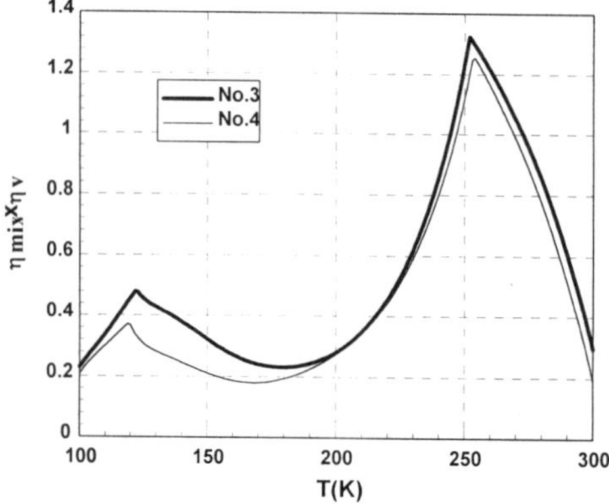

Fig.4. *η*-T diagram for the No.3 and No.4 mixtures in Illustration 2

THERMODYNAMIC CONSIDERATION

There are several simulation illustrations presented in the third part of this paper. From those optimization results, one notes that the ideal efficiency of a J-T refrigeration cycle with mixed gases can achieve an efficiency up to 58.8% when the cold temperature is 173K (-100°C). Are these really optimum mixtures in the given condition? Can every mixture improve the thermodynamic performance of the refrigerator? In order to answer these questions, a comparison is made using the same calculation conditions with mix.1 (the No.4 mixture in Table1) and mix.2 with the same components as mix.1 but with compositions of R11 (25%), R12 (25%), R13 (25%), and R14 (25%) and the pure gas R14. The *h*-T diagram of these three substances is shown in Fig.1, and Fig.2 is the η_{mix} vs. temperature (η_{mix}-T) diagram for these three refrigerants.

The exergy losses of the two mixtures and the R14 in different process stages of the J-T refrigeration cycle are presented in Table 3. From Table 3 and Figs.1 and 2, it is found that:

1) The exergy loss of the recuperative process in the heat exchanger is the major loss of the cycle for mix.1 and mix.2. However, the throttling loss of the pure gas is larger than that for any other process.
2) The exergy loss of mix.2 is larger than for R14; thus, not all mixtures can show an improvement in the thermodynamic performance of the refrigeration system.
3) The total exergy loss of mix.1 is the smallest.

From Illustration 2, it can be inferred that the efficiency of the compressor is a key factor when a mixed-refrigerant J-T refrigerator system optimization is considered. In fact, most exergy loss occurs during the compression process. It is important to obtain a new type of compressor with a high efficiency to drive the mixed-refrigerant J-T refrigeration system. In addition, the efficiency of the counterflow heat exchanger is also very important in the refrigeration system. From Table 2, we can find that when the minimum temperature difference in the counterflow heat exchanger is only 3K, the exergy loss will be up to 10 percent more than that without a temperature difference in the counterflow heat exchanger under ideal conditions. It is necessary to improve the efficiency of the counterflow heat exchanger based on the mixture used. However, it becomes a very complex process of flow and heat transfer in the counterflow heat exchanger for the multicomponent mixture with phase change. It is a task for further effort.

Table. 3 The exergy losses in the J-T refrigeration system

Refrigerant No.	Exergy losses in throttling process, %	Exergy losses in recuperative process, %	Total losses in the system, %
mix.1	13.9	27.3	41.2
mix.2	6.83	90.92	97.75
R14	71.35	20.08	92.43

SUMMARY

An optimization method has been developed in this paper. Using this method it is convenient to analyze a mixed refrigerant Joule-Thomson refrigeration cycle. By systematic variation of the high and low pressures, the optimal operating parameters can also be obtained. The thermodynamic performance of the entire refrigeration system is directly related to the mixture used, the heat exchanger, and the compressor. Influences attributed to the counterflow heat exchanger and the compressor for optimization calculation are considered in this method. The results can be helpful in the experimental investigation of the mixed refrigerant Joule-Thomson cycle refrigerator.

ACKNOWLEDGMENT

This work is financially supported by the National Natural Sciences Foundation of China under contract No. 59706002.

REFERENCES

1. M. J. Boiarski, V.M. Brodianski, and R.C. Longsworth, Retrospective of mixed-refrigerant technology and modern status of cryocoolers based on one-stage, oil-lubricated compressors, in: "Advances in Cryogenic Engineering", Vol.43, Plenum Press, New York, (1998), p.1701
2. E.C. Luo, Y. Zhou, and M.Q. Gong, Mixed-refrigerant Joule-Thomson cryocooler driven by R22/R12 compressor, in: "Advances in Cryogenic Engineering", Vol.43, Plenum Press, New York, (1998), p.1675
3. W. A. Little, Kleemenko cycle: low-cost refrigeration at cryogenic temperatures, *Proceedings of ICEC17*, Bournmouth, UK, (1998), p.1
4. S.S. Lee *et al,* The principle and equipment of refrigeration, Shanghai Science and Technology Publish House, Shanghai, (1980), p.33 (in Chinese)
5. M.J. Box, A new method of constrained optimization and a comparison with other methods, *Computer Journal*, 8(1): 42-52, (1965)
6. D. Y. Peng and D. B. Robinson, A new two-constant equation of state, *Ind. Eng. Chem. Fund.,* 15(1), (1976), p.59

ENHANCED REFRIGERATION PERFORMANCE OF THE THROTTLE-CYCLE COOLERS OPERATING WITH MIXED REFRIGERANTS

M. Boiarski, A. Khatri, and O. Podcherniaev

IGC-APD Cryogenics Inc.
Allentown, PA 18103 – 4783

ABSTRACT

Cryocoolers using single-stage compression and gas mixtures have proven to be efficient for many practical applications in a temperature range between 70 to 200 K. Computer models combined with limited experimental data allow for estimations of further improvements. Analytical and experimental results are summarized to describe the losses in refrigeration performance. Maximum values of the cooler efficiency compared to the Carnot cycle and expected refrigeration capacity for single-stage and two-stage compressors were estimated based on these computer models. Data on the refrigeration performance are presented for experimental coolers based on a single-stage compressor.

INTRODUCTION

Results achieved with the mixed refrigerant (MR) coolers have generated growing interest to use them in new fields of applications; some of them are related to a semi-cryogenic temperature range from 120 K up to 200 K.[1] The goal of this paper is to evaluate the refrigeration performance of such systems over a wide temperature range.

A cooler schematic may vary depending on the refrigeration temperature (T_R), refrigeration capacity (Q_R) and the field of application. A simple, low-cost cryocooler operating with MR has a system schematic similar to a traditional vapor-compression cycle providing a single-stage refrigeration.[2,3] Figure 1 shows a simplified schematic and temperature-enthalpy (T-h) diagram. Retrospective and evolution of the mixed refrigerant technology have been presented elsewhere.[2] It was shown that the MR technology may be efficient for both the constant temperature and the temperature distributed refrigeration process.[3] In the latter case, a heat load Q_{HL} (T) to the cooler should be distributed in a certain temperature range. For example, to cool the gas flow a system should support a temperature distributed heat load, but refrigerated electronic components need constant temperature refrigeration. It is more difficult to build a high-performance MR cooler that is cooling a load at a constant (uniform) temperature T_R because most of the mixtures simply

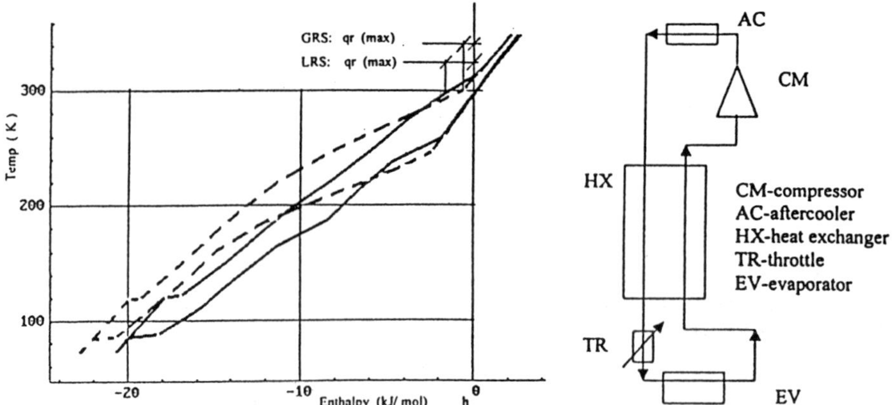

Figure1. Schematic of the single-stage cooler and the T-h diagram for a mixed refrigerant.

evaporate at variable temperatures. The "temperature glide" in the evaporator causes additional losses reducing the refrigeration efficiency.[3] Results presented in this paper are related to the coolers providing constant temperature refrigeration.

IDEALIZED CYCLE CHARACTERISTICS

The first step in the performance evaluation should be based on the cycle analysis.[4] Appropriate MR thermodynamic properties of acceptable accuracy can be obtained with the modern equations of state combined with limited experimental data.

For specified ambient temperature (T_A) and refrigeration temperature (T_R), the refrigeration capacity (Q_R) and compressor power consumption (PW_C) may be estimated with different assumptions on the equipment characteristics. They include a minimum temperature difference ΔT_{MIN} in the counter-flow heat exchanger (HX) and pressure-drop value (ΔP) in the heat exchanger and connection lines. The compressor performance should also be assumed for both volumetric efficiency (η_{CV}) and power efficiency -- either isothermal (η_{CT}) or adiabatic (η_{CS}).

The efficiency of the MR technology was initially estimated using a highly idealized cycle, which takes into account only intrinsic losses. We used the Carnot efficiency of the cycle, which is identical to the exergetic efficiency[3] (η_{EX}). A value of the exergetic efficiency was calculated as η_{EX} = COP (T_A / T_R – 1), where COP = Q_R / PW_C is a coefficient of performance related to a specified T_R and T_A.

Although η_{EX} is independent of the MR flow rate (G_{MR}), for practical applications it is of interest to evaluate Q_R and η_{EX} for a specific compressor using a computer model.[4]

Two different technologies could be considered to develop a cooler. The difference is related to the phase-state of the MR at the cryostat inlet. It is possible to use either a gas refrigerant supply (GRS) or a liquefied refrigerant supply (LRS) technology. The dew point temperature (T_D) of the MR designed for the GRS technology should be less than T_A at the discharge pressure; where as for the LRS: $T_D > T_A$. These approaches allow different opportunities in developing the MR coolers.

The simple design is a major advantage of the GRS technology.[5] The oil separator may be placed either before or behind the aftercooler. This technology avoids the issue of two-phase flow management at the cryostat inlet. Flexible lines (bellows type) may be used to make connections between the compressor unit and the cryostat. A charging procedure is simpler with a gas refrigerant. The amount of the MR is smaller in the system that allows the higher content of the flammable components. It leads to the simplest design and maintenance. However, the GRS technology restricts the maximum available capacity per unit of the refrigerant (q_R), which equals the enthalpy difference related to the ambient temperature (Fig.1): $q_R\ (max) = \Delta h\ (T_A)$.

The LRS technology can provide the higher capacity $q_R(max)$ but involves some resolvable problems on the two-phase flows and oil management.

Computer software was used to determine available T_R, $Q_R\ (T_R)$ and η_{EX} coefficient for the given MR composition.[4] The computation algorithm predicts vapor-liquid (V-L) and vapor-liquid-liquid equilibrium (V-L-L) as shown in Fig. 1. The optimization method[6] was used to determine the MR composition. It has been shown that for any mixed refrigerant $Q_R(T_R) = \Delta h\ (min)^* G_{MR}$, where $\Delta h(min)$ is the minimum enthalpy difference between the high (P_H) and low (P_L) pressure flows, in the temperature range of T_R to T_A.[6,7]

Idealized cycle characteristics were calculated for both GRS and LRS technologies with the following assumptions: $\Delta T_{MIN} = 0$, $\Delta P = 0$, $\eta_{CT} = 1$. A compressor volumetric efficiency was also assumed as $\eta_{CV} = 1$. Such a highly idealized cycle gives maximum available Q_R and η_{EX} values for a specific compressor for the cycle presented in Fig.1. The exergy losses are caused by the entropy growth in the throttle and in the counter-flow HX. They are related only to thermodynamic properties of the selected MR.

Performance data for a number of the idealized cycles are presented in Tables 1 and 2. They were obtained for $T_A = 300$ K, assuming the use of a single-stage compressor having a displacement volume of 28 l/min (1cfm). The compressor discharge pressure during the calculations was always less than $P_H \leq 20$. The tables also present a heat load of the counter-flow heat exchanger Q_{HX} and a pressure ratio $PR = P_H / P_L$ in the cycle. The constituents which were used to design the refrigerant and a number of components (N_C) in the MR are also listed. They include hydrocarbons (HC), fluorocarbons (FC) and hydro-fluorocarbons (HFC). Some MR compositions include neon (Ne) to provide a high Q_R.

Table 1. Characteristics of the idealized MR-cycles based on the GRS technology: $T_A = 300$ K, an ideal compressor displacement volume 28 l/min (1 cfm), $\Delta T_{MIN} = 0$ K, $\Delta P = 0$

No	MR	T_R,K	Q_R,W	η_{EX}	Q_{HX},W	PR	N_C
1	$Ne/N_2/HC$	82	26	0.33	805	10	5
2	N_2/HC	85	36	0.36	1015	8	5
3	N_2/HC	88	35	0.35	1000	8	4
4	N_2/HC	88	34	0.34	1010	8	5
5	$Ne/N_2/HC$	90	50	0.39	1610	5	5
6	N_2/HC	94	72	0.45	2425	3.3	4
7	N_2/HC	94	56	0.42	1655	5	4
8	$Ne/N_2/HC$	95	58	0.40	2400	3.3	5
9	Ar/HC	108	56	0.34	1555	5	4
10	Ar/HC	114	74	0.37	2310	3.3	4
11	Ar/HC	118	54	0.25	2050	3.3	4
12	Ar/HC	133	67	0.24	2100	3.3	4
13	FC/HC	150	65	0.20	2100	3.3	4
14	FC/HC	158	67	0.20	2210	3.0	4
15	FC/HC	158	69	0.21	2250	3.0	5

Table 2. Characteristics of the idealized MR-cycles based on the LRS technology: T_A =300 K, η_{CV} = 1, η_{CT} = 1, ΔT_{MIN} = 0.2 K, ΔP = 0

No	MR	T_R,K	Q_R,W	η_{EX}	Q_{HX},W	PR	N_C
1	$Ne/N_2/HC$	80	38	0.55	630	9	5
2	$Ne/N_2/HC$	82	36	0.53	650	8	5
3	$Ne/N_2/HC$	87	55	0.67	660	10	5
4	$Ne/N_2/HC$	91	66	0.60	1050	6.7	5
5	$N_2/Ar/HC$	92	81	0.71	960	6.7	5
6	$N_2/Ar/HC$	92	86	0.67	980	6.7	5
7	N_2/HC	95	84	0.75	1010	6.7	4
8	N_2/HC	97	81	0.70	1370	4	4
9	N_2/HC	98	69	0.74	1460	3	4
10	$N_2/Ar/HC$	108	116	0.68	1950	3	5
11	Ar/FC/HC	119	106	0.68	1420	4	4
12	Ar/FC/HC	125	136	0.72	2070	3	4
13	FC/HC	135	168	0.67	1950	3.3	4
14	FC/HC	151	181	0.70	1350	5	4
15	FC/HC	155	191	0.77	1070	7	4
16	HC	160	224	0.76	1920	3.3	3
17	FC/HC	170	258	0.76	1360	5	5

The exergetic efficiency of the idealized cycles is given in Fig. 2. The LRS technology provides η_{EX} = 0.7 and higher in a temperature range of T_R = 90 to 180 K which matches the data obtained earlier.[3] Efficiency is close to that which may be achieved with traditional vapor-compression cycles[3] using pure refrigerants at T_R > 240 K. Below 90 K the η_{EX} values lower to 0.6. For the GRS technology, maximum achieved η_{EX} values are related to T_R = 95 - 105 K. They are somewhat above 0.4. The refrigeration capacity $Q_R(T_R)$ is presented in Fig. 3. For the LRS technology it is almost a linear function of temperature. Refrigeration capacity increases from Q_R = 40 W at T_R = 80 K to Q_R > 200 W at T_R = 180 K. For the cycles based on the GRS technology, Q_R values increase linearly for T_R = 80 to 120 K. At higher temperatures, Q_R becomes approximately constant at 70 to 80 W due to the constraints on the GRS technology.

One of the potential ways to improve performance of the GRS cycles is to use a two-stage compression to increase the discharge pressure above 20 bar. Figure 4 presents typical data showing the influence of P_H on Q_R and η_{EX}. This data is related to the MR #5 and #6 selected from Table 1, which provide refrigeration at T_R = 90 and 94 K, respectively. Increasing P_H results in an increase of Q_R. However, at the same time the compressor power consumption also increases. This is why, for the idealized cycles, the exergetic efficiency remains almost identical for P_H = 20 to 30 bar.

PERFORMANCE OF REAL CYCLES

Real cycle characteristics can be estimated by assuming some finite ΔP and ΔT_{MIN} values and a compressor efficiency based on cooler design experience. The efficiency of such a cycle will be reduced due to the influence of the extrinsic factors that depend on the hardware design.

Table 3 presents a comparison of experimental and calculated data, which were obtained with different assumptions on the hardware characteristics. The values of ΔT_{MIN}, ΔP for the low-pressure line and the compressor efficiency are taken from the test. We can see that a computer model[4] describes the cooler performance with good accuracy. Using this model, the MR cooler [5] has been modified to improve the performance with both GRS and LRS technologies.

Figure 5 presents Q_R data for the high-performance cooler based on a single-stage compressor. Dashed lines show Q_R for the original cooler configuration. The optimized hardware allows for an increase in Q_R even for the GRS technology.

Performance data obtained in the LRS technology are given in Table 4. Both the connecting lines and compressor unit were modified to supply two-phase flow to the cryostat without carrying oil. The cryostat configuration remained the same. This table includes the exergy refrigeration capacity (E_R) defined as $E_R = Q_R (T_A/T_R - 1)$. It allows us to take into account the refrigeration temperature level while comparing the performance of different coolers and MR. For example, the experimental unit produced an E_R = 80 W similar to a traditional vapor-compression cycle providing Q_R = 400 W at T_R = 250 K. Values of η_{EX} are relatively high for such a small cooler and comparable with the performance of traditional refrigeration systems operating above 240 K.

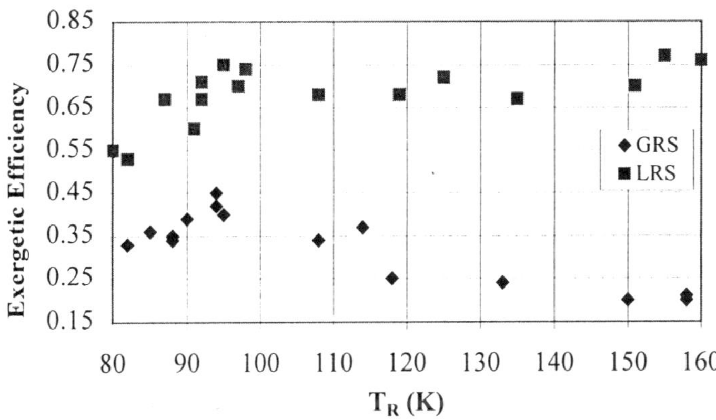

Figure 2. Exergetic efficiency of idealized MR coolers.

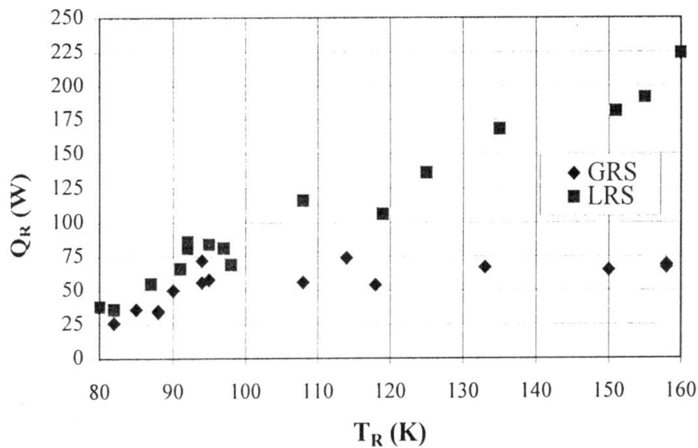

Figure 3. Refrigeration capacity of idealized MR coolers using 28 l/min displacement compressor.

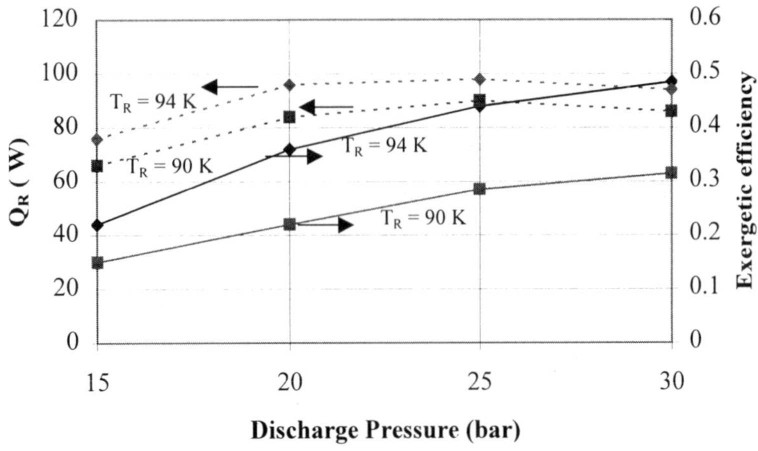

Figure 4. Influence of discharge pressure on refrigeration capacity and exergetic efficiency.

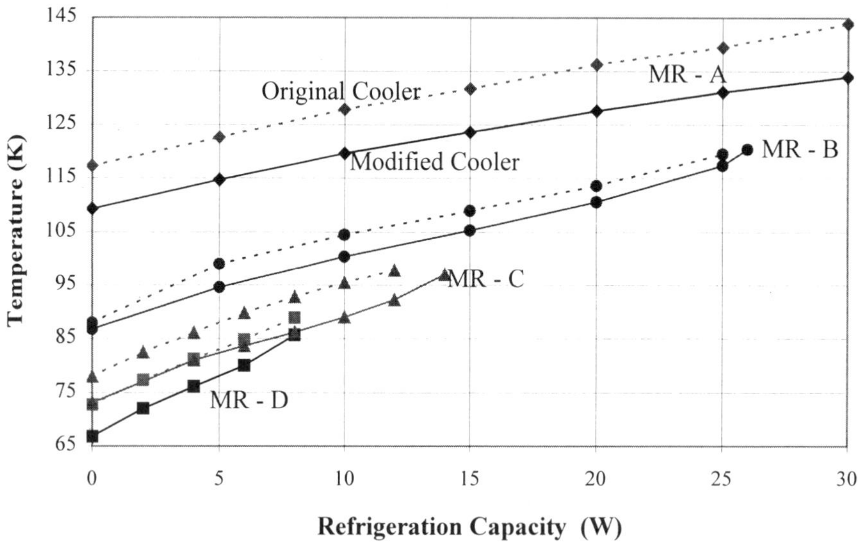

Figure 5. Refrigeration capacity of the high performance modified cooler.

Table 3. Comparison of experimental and calculated data based on different assumptions: T_R = 102 K; T_A =300K; compressor displacement volume 28 l/min (1 cfm)

No.	CYCLE	Q_R, W	PW_C, W	G_{MR}, mole/s	Q_{HX}, W	η_{EX}
1	Highly idealized	105	295	0.095	1600	0.69
2	$\Delta T_{MIN} = 0$, $\Delta P = 0$, $\eta_{CV} = 0.8$, $\eta_{CT} = 0.45$.	82	575	0.075	1250	0.28
3	$\Delta T_{MIN} = 0$, $\Delta P = 0.6$ bar $\eta_{CV} = 0.8$, $\eta_{CT} = 0.45$.	71	495	0.064	1080	0.28
4	$\Delta T_{MIN} = 8$ K, $\Delta P = 0.6$ bar, (model[8]) $\eta_{CV} = 0.8$, $\eta_{CT} = 0.45$.	37	490	0.063	1040	0.14

Table 4. Experimental data on the cooler performance using the LRS technology T_A = 300 K, compressor displacement volume 28 l/min (1 cfm).

MR	T_R, K	Q_R, W	PW_C, W*	η_{EX}	T_R (min), K	E_R, W
#1	78	8	480	0.05	67	22
#2	103	37	575	0.12	83	71
#3	163	100	550	0.15	148	84
#4	182	120	555	0.14	153	78

* PW_C includes 40 W consumed by the fans removing heat from the aftercooler.

Analysis shows that the LRS technology allows further improvements not only in the refrigeration performance but also in the cool-down time reduction, which directly depends on the refrigeration capacity.

SUMMARY

The mixed refrigerant technology provides highly-efficient, constant-temperature refrigeration at temperatures between 85 to 200 K. The Carnot or exergetic efficiency of the idealized MR cycles may be of the same level achieved for traditional vapor-compression cycles operating with pure refrigerants above 240 K. The high performance of the throttle-cycle coolers based on a single-stage, 28 l/min displacement compressor operating with different mixed refrigerants can produce more than 100 W capacity at T_R = 200 K. Minimum achievable temperature could be as low as 70 K.

ACKNOWLEDGMENT

S. Harold and H. Fasnacht of IGC-APD Cryogenics are thanked for their excellent help in testing our MR coolers. We greatly appreciate the support of J. Zoulcarneeva, V. Kovalenko and V. Mogorichny of the Moscow Power Engineering Institute for their contribution in modeling the properties of the refrigerants. We appreciate the participation by R. Longsworth in discussing these results.

REFERENCES

1. M. Ellsworth, E. Moser, A. Khatri, and M. Boiarski, Performance of a mixed-refrigerant system designed for computer cooling, presented at CEC/ ICMC meeting, July 12 – 16, 1999; Montreal, Canada.
2. M. Boiarski, V. Brodianski, and R. Longsworth, Retrospective of mixed-refrigerant technology and modern status of cryocoolers based on one-stage, oil-lubricated compressors, in: "Advances in Cryogenic Engineering," Plenum Press, New York; Vol. 43 (1998) pp. 1701 – 1708.
3. V. Brodianski, M. Boiarski, and A. Lunin, The exergy analysis of the throttle refrigerating systems on pure and mixed refrigerants, Proceedings of the EDSA, PD, Vol. 64 – 3; *Eng.Des.Sys.Anl.*, Vol. 3, ASME (1994).
4. M. Boiarski, A. Khatri, and V. Kovalenko, Design optimization of the throttle-cycle cooler with mixed refrigerant, in: "Cryocoolers 10," R.G. Ross, Jr., ed.; Plenum Press, New York (1999) pp. 457 – 465.
5. A. Khatri and M. Boiarski, A throttle-cycle cryocooler operating with mixed-gas refrigerants in 70 K to 120 K temperature range, in: "Cryocoolers 9," R.G. Ross, Jr. ed.; Plenum Press, New York (1997) pp. 515 – 520.
6. M. Boiarski, et al, Autonomic cryorefrigerators of small capacity, Energoatomizdat, V. Brodianski, ed., Moscow, Russia (1984).
7. A. Alexeev, H. Quack, and Ch. Haberstroh, Low-cost mixture Joule Thomson refrigerator, Proc. ICEC 16/ ICMC, T. Haruyama, ed.; Elsevier, Oxford, England (1996) pp. 395 – 398.

THE RESEARCH AND DEVELOPMENT OF CRYOGENIC MIXED-REFRIGERANT JOULE-THOMSON CRYOCOOLERS IN CL/CAS

E. C. Luo, M.Q. Gong, Y. Zhou, J. T. Liang and L.Zhang

Cryogenic Laboratory, Chinese Academy of Sciences
Beijing 100080, P.R.China

ABSTRACT

This paper summarizes the research and development of cryogenic mixed-refrigerant Joule-Thomson cryocoolers in the Cryogenic Laboratory of the Chinese Academy of Sciences (CL/CAS). In recent years, this laboratory has systematically studied several key thermodynamic problems of cryogenic mixed-refrigerant Joule-Thomson refrigeration cycles and successfully developed several prototypes for different applications.

INTRODUCTION

Because of high reliability and simplicity, Joule-Thomson cryocoolers have been widely used in the field of cryogenic engineering. However, for a long time the cryocoolers had used pure refrigerants, and consequently they had not operated efficiently in the cryogenic temperature range. Only recently, was it observed that the Joule-Thomson cryocoolers could be efficient by using selected cryogenic mixed refrigerants [1,2]. More recently, new progress has been achieved successfully by using highly reliable air-conditioning compressors[3], permitting wider applications for the Joule-Thomson cryocoolers.

In order to understand the internal working process of the mixed-refrigerant Joule-Thomson cryocoolers, the first step of the research was to develop accurate thermophysial properties of the cryogenic mixtures. The second step was to understand the essence of using mixtures to improve the thermodynamic performance of the Joule-Thomson

Advances in Cryogenic Engineering, Volume 45.
Edited by Shu *et al.*, Kluwer Academic / Plenum Publishers, 2000.

refrigerators and experimentally demonstrate the theoretical results to develop new products for practical applications.

THERMODYNAMIC PROBLEMS

Estimation of Thermophysical Properties

For estimating the thermodynamic properties of cryogenic mixtures, we successfully developed a program package based on several well-known cubic equations of state (EOS) such as the Peng-Robinson equation and the Soave-Redlich-Kwong equation. Figure 1 is the T-S diagram of an efficient cryogenic mixture developed by the authors over several years. This cryogenic mixture which includes nitrogen, methane, ethane, propane and iso-butane has a special feature with a coexistence state of vapor-liquid-liquid equilibrium. When in the vapor-liquid-liquid equilibrium, one liquid is substantially pure nitrogen, evaporating at a nearly constant temperature like pure nitrogen. Making use of this unique feature and optimizing the mixture, the authors successfully developed a mixed-refrigerant J-T cryocooler driven by a single-stage, air-conditioning compressor to efficiently provide a cryocooler operating at liquid nitrogen temperature. In addition, the software package provides assistance in developing other efficient mixtures for different applications.

Analysis of Phase Equilibrium

As mentioned above, some mixtures will form a special vapor-liquid-liquid equilibrium in which one liquid is substantially a pure substance.

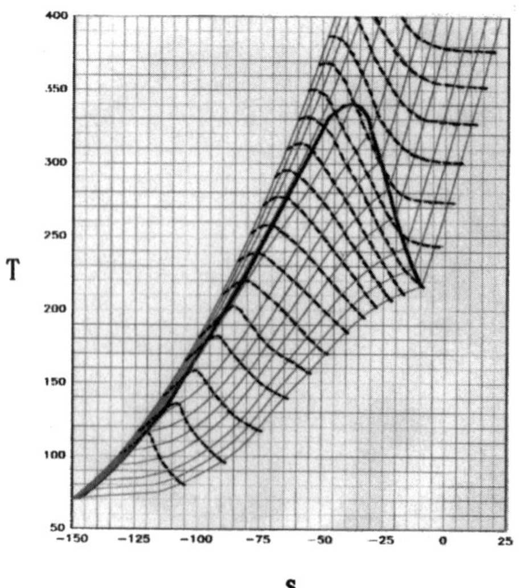

Figure 1. T-S diagram of a mixture.

According to the thermodynamics of vapor-liquid equilibrium, the equation of state (EOS) method is adopted to estimate the vapor-liquid-liquid equilibrium of a nitrogen mixture based on the Peng-Robinson equation of state. A comparison was made between experimental and predicted data of the vapor-liquid-liquid equilibrium of several ternary and quaternary nitrogen mixtures [5].

Very little previous work is available in the literature on the liquid-solid equilibrium of nitrogen mixtures. Table 1 gives the solubility parameters of the pure components: N_2,CH_4,C_2H_6, C_3H_8, i-C_4H_{10} and i-C_5H_{12}. From this table one notes that all hydrocarbons have higher freezing temperatures than that of pure nitrogen. To avoid the solidification of these components as they exit from the throttling valve and flow in the low temperature passage of the counter-flow heat exchanger, their contents in mixtures need to be maintained below the solubility limit. Based on the Scatchard-Hildebrand regular solution model[4], the liquid-solid equilibria of nitrogen mixtures have been analyzed[5]. As an example, Figure 2 shows the isothermal liquid-solid equilibrium diagram of a ternary nitrogen mixture.

Mixed-Refrigerant Refrigeration Cycle and Its Optimization

Figure 3 is a schematic of a typical mixed-refrigerant Joule-Thomson cryocooler driven by a single-stage, oil-lubricated compressor, described elsewhere.

For a closed-cycle mixed-refrigerant Joule-Thomson cryocooler, the actual thermodynamic performance of the complete refrigerating system depends not only on the refrigerant mixture but also on the heat exchanger and compressor used. The optimization of a mixed-refrigerant Joule-Thomson cryocooler is a complicated thermodynamic task. Consequently, we present a case study of the optimization principle of the mixed-refrigerant Joule-Thomson cryocooler driven by a single-stage compressor.

Assume the following conditions: (1) The refrigeration and environment temperatures, T_c, T_0 are 80K and 300K, respectively; (2) The discharge pressure of the compressor, p_H safely operates below 2.5MPa, while the suction pressure of the compressor, p_L, operates above 0.1MPa; (3) The mixture consists of Ne, N_2, CH_4, C_2H_6, C_3H_8 and iC_4H_{10}. The compressor with a reciprocal piston is hermetically sealed, thus the compression process inside the compressor is close to an adiabatic thermodynamic process.

For the compressor used in this thermodynamic simulation, the volumetric efficiency, is defined by[6]:

$$\eta_V = 0.94 - 0.085[(p_H / p_L)^{\frac{1}{k}} - 1] \tag{1}$$

Based on the analysis of the mixed-refrigerant Joule-Thomson refrigeration cycle, the exergy efficiency of the complete refrigeration system can be given by the following equation:

$$\eta = \eta_V \eta_{mix} = \eta_V \frac{\min[\Delta h(T)]}{W_{in}} \frac{T_0 - T_c}{T_c} \tag{2}$$

We take the exergy efficiency as the goal function of the optimization. Table 2 shows the optimization results.

Table 1. Thermodynamic parameters for liquid-solid equilibrium

Component	Fusion enthalpy $\Delta H(J/mol)$	Freezing point $T_{tp}(K)$	Liquid molar volume $\tilde{V}_l(cm^3/mol)$	Solubility parameter $\delta(J/cm^3)^{1/2}$
N_2	714	63.16	310.1	10.84
CH_4	941	90.7	52.0	11.62
C_2H_6	2859	90.0	68.0	12.38
C_3H_8	3524	85.5	84.0	13.09
iC_4H_{10}	4540	113.6	105.5	13.77
iC_5H_{12}	5153	113.3	117.4	14.36

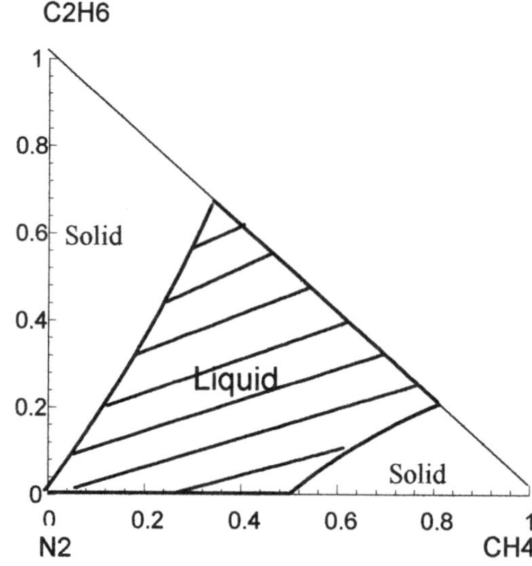

Figure 2. Liquid-solid diagram for a ternary nitrogen mixture at 80K.

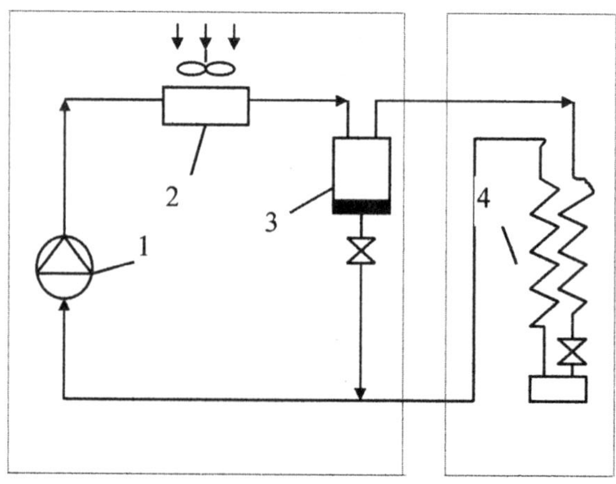

Figure 3. Schematic of a typical single-stage, oil-lubricated, mixed-refrigerant J-T cryocooler. (1. Compressor 2. Aftercooler 3. Oil separator 4. Joule-Thomson cooler unit)

Table 2. A case study for the optimization of mixed-refrigerant J-T cryocooler

Mixtures	P_H	P_L	ΔT_{min}	η
(Mol.)	(MPa)		(K)	(%)
Ne(0.181)-N_2(0.315)-C_1(0.189)-C_2(0.065)-C_3(0.001)-iC_4(0.249)	2.0	0.2	0	14.2
Ne(0.222)-N_2(0.290)-C_1(0.125)-C_2(0.100)-C_3(0.098)-iC_4(0.165)	2.0	0.2	3	8.6
Ne(0.226)-N_2(0.346)-C_1(0.112)-C_2(0.078)-C_3(0.025)-iC_4(0.213)	2.0	0.3	3	11.0

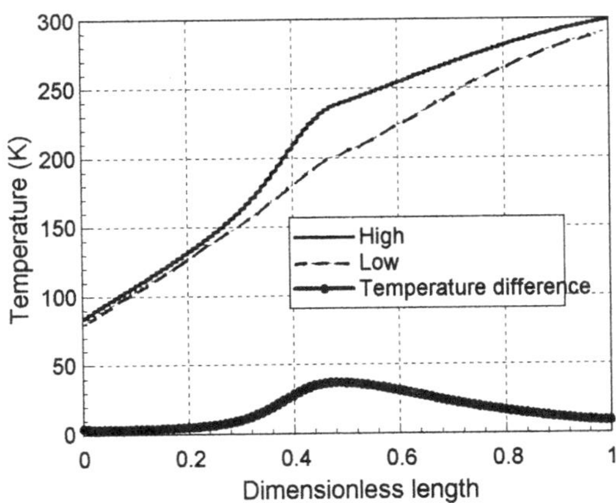

Figure 4. Temperature distribution in a counter-flow heat exchanger.

Design of Counterflow Heat Exchanger

To really make the mixed-refrigerant Joule-Thomson cryocoolers operate efficiently, a good understanding and design of the miniature counter-flow heat exchangers is critical. For the heat transfer between cryogenic multi-component mixture refrigerants used in the miniature counterflow heat exchangers, two complicated heat transfer processes of flowing condensation and flowing evaporation must to be solved. The Colburn-Hougen method[7] was used to estimate the forced convection heat transfer coefficient of flowing condensation, and the Chen correlation[8] combined with the Stefan method[9] was employed to predict the heat transfer coefficient of flowing evaporation of cryogenic mixtures. The general methodology for designing the heat transfer of miniature counterflow heat exchangers used in mixed-refrigerant J-T refrigerator was described by Luo[5]. Following this numerical simulation, the dimensionless temperature distribution of a miniature, counterflow heat exchanger used in our mixed-refrigerant J-T refrigerator is shown in Figure 4.

PROTOTYPES OF MIXED-REFRIGERANT J-T CRYOCOOLERS

Prototypes for High Tc Devices and Infrared Detectors (0.2W-5W)@80K

Because of high reliability and low cost, the mixed-refrigerant J-T cryocoolers have good potential for the cooling of high-Tc superconducting devices and infrared detectors. Generally, these devices require a refrigeration capacity from milliwatts to several watts around 80K. Thus, several prototypes have already been fabricated and demonstrated to operate in the temperature range from 65K to 90K. Figure 5 is a photograph of three cold fingers with about 0.5W, 2.5W and 4W at 80K. The counetrflow heat exchangers were fabricated of perforated plates with thin stainless steel tubes serving as their partitions. All of these prototypes have been subjected to lifetime testing.

Prototypes for Medical Knifes (10W-30W)@(120K-200K)

A microminiature J-T refrigerator also has been used for cryosurgery because of its easier miniaturization and faster cool-down. However, the use of high-pressure gas in earlier J-T refrigerators has provided a safety hazard. Moreover, the J-T refrigerator utilizing only a pure gas can not produce sufficient cooling capacity due to the low thermodynamic efficiency of the pure gas. Thus, using a mixture as the working fluid for the J-T refrigerator will have an obvious advantage in cryosurgery and other medical applications. Figure 6 is a prototype for cryosurgery, which can successfully operate over a temperature range from -140°C to -80°C with a refrigeration power from 5W to 30W. The refrigeration capacity and temperature range should be adequate for many cryosurgical applications. Different cold fingers for different medical needs are being developed and tested.

Prototype for High Vacuum Pumps (50W-100W)@(120K-150K)

By optimizing different mixtures and operating parameters, the mixed-refrigerant Joule-Thomson refrigerator can be quite efficient in the temperature range from liquid nitrogen temperature to room temperature, particularly for the temperature range from 80K to 200K. Below 200K, the classical, two-stage, cascade vapor-compressed throttle refrigerator will operate poorly with decreasing refrigeration temperature. Moreover, the classical cascade cycle is relatively complicated due to the necessity of two independent systems. Thus, the mixed-refrigerant Joule-Thomson refrigerator should find its application in this field. Figure 7 is a photograph of a prototype prepared for a turbopump. The thermodynamic performance of the refrigerator is under test.

CONCLUSIONS

In this study, several thermodynamic problems have theoretically been studied. At the same time, some prototypes of mixed-refrigerant J-T refrigerators for different applications have been developed.

A software program package for estimating the different thermophysical properties of mixed refrigerants of cryogenic interest has been made. With the assistance of this software package, the authors have developed specific mixtures for different applications. A general methodology is presented for the optimization of mixed-refrigerant J-T cryocoolers and the design of the heat transfer in counter-flow heat exchangers. Several prototypes of mixed-refrigerant J-T refrigerators for different applications have been fabricated and tested.

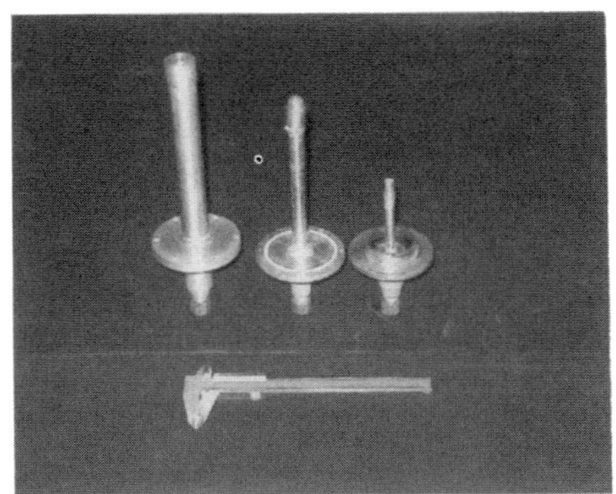

Figure 5. Photograph of three cold fingers.

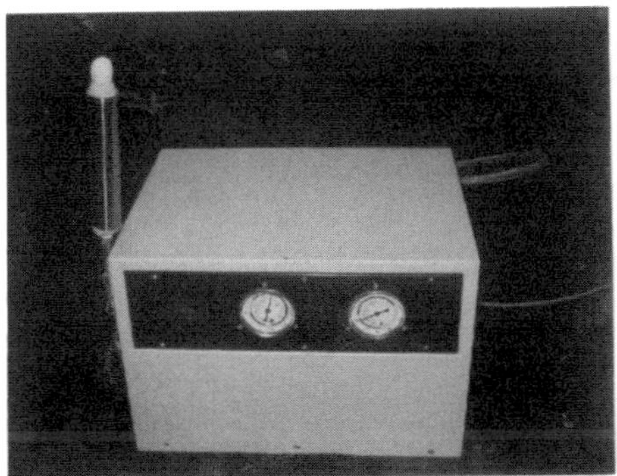

Figure 6. Photograph of a prototype for cryosurgery.

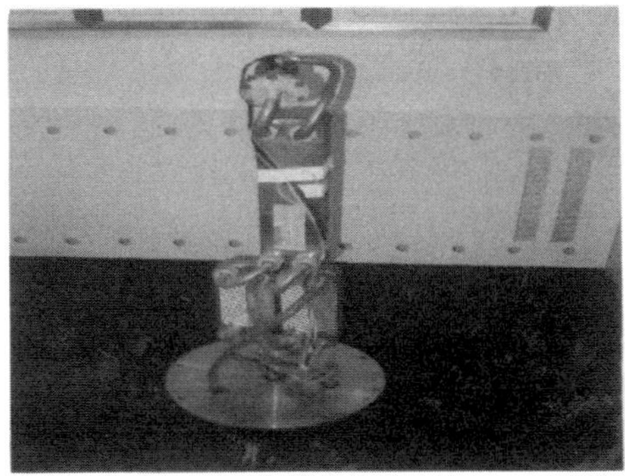

Figure 7. Photograph of a prototype for turbopump.

ACKNOWLEDGMENT

This work is financially supported by the Natural Sciences Foundation of China under Contract 59706002.

REFERENCES

1. V.M. Brodianski, et al., The use of mixtures as the working gas in throttle Joule-Thompson cryogenic refrigerator, in: "Proc. of 13th Intern. Congress of Refrigeration", Plenum Press, New York (1973), p.43.
2. W.A.Little, Recent developments in Joule-Thompson cooling: gases, coolers, and compressor, in: "Proc. of 5th Intern. Cryocooler Conference", Monterey, CA (1989), p.1.
3. R.C. Longsworth, et al., 80K closed cycle throttle refrigerator, in: "Cryocooler 8", R. Ross, ed., Plenum Press, New York (1995), p.537.
4. S. M.Walas, "Phase Equilibrium in Chemical Engineering", Butterworth, Boston (1985).
5. E.C.Luo, "The theoretical and experimental study on mixed-refrigerant Joule-Thomson crycoolers operating in the liquid nitrogen temperature range", Ph.D. Dissertation, Cryogenic Laboratory, Academia Sinica, Beijing(1997)
6. Q. Yan, "Refrigeration Technology for Air-conditioning Engineering", Architecture Industry Press, Beijing (1985).
7. A.P.Colburn and O.A.Hougen, *Ind.Eng.Chem.* 26:1178 (1934).
8. J.C.Chen, A correlation for boiling heat transfer to saturated fluid in convective flow, *ASME Paper 63-HT-34* (1962).
9. K Stefan and M Korner, Calculation of heat transfer in evaporating binary liquid mixtures, *Chem.Ing.Tech.* 41:409(1969).

STUDY OF BEHAVIOR IN THE HEAT EXCHANGER OF A MIXED GAS JOULE-THOMSON COOLER

A. Alexeev, A. Thiel, Ch. Haberstroh and H. Quack

Lehrstuhl für Kälte- und Kryotechnik
Technische Universität Dresden, Germany

ABSTRACT

The object of the investigation is a mixed gas Joule-Thomson (J-T) cooler. A computational model was developed, which makes it possible to investigate the steady-state behavior of the refrigerant in the heat exchanger of a mixed gas J-T system. The calculations show that the temperature distribution as well as the pressure distribution in the heat exchanger channels depends to a large degree on the pinch point locations in the process, which is strongly influenced by the mixture composition. This new understanding gives valuable information for the practical and optimal design of the heat exchanger as well as for the choice of the mixture composition.

INTRODUCTION

A mixed gas Joule-Thomson cooler has distinct advantages over other types of cryogenic refrigerators. The system has acceptable efficiency and the design is flexible for the interface with the cooled object. Based on the use of oil lubricated hermetic compressors this cooler combines such important benefits as reliability and "low cost". It can replace presently used systems (for example, Gifford-McMahon refrigerators) and be used for old and new applications in the temperature range between 70 and 150 K.

The thermodynamics associated with the mixed gas J-T cycle are discussed in the literature[1-3]. A more detailed description is given by Alexeev[4]. Figures 1 and 2 show the flow diagram and T-h diagrams of J-T cycles with a typical nitrogen-hydrocarbon mixture and with pure nitrogen for typical working pressures. The following details are worth noting: the heat exchanger duty in a mixed gas cycle is approximately three times larger than in the nitrogen cycle; the average temperature difference in a mixed gas cycle is smaller than in the nitrogen cycle and the NTU is more than 50. Because the heat transfer coefficient in a mixed gas exchanger is not higher than in a J-T system with pure nitrogen, a considerably larger heat exchange surface is necessary to realize a comparable mixed gas system.

The special degree of freedom of the mixed gas systems is that it is possible to influence the temperature profiles in the heat exchanger passages by a variation of the mixture composition. This is an important option, which is not available in systems with a single refrigerant.

In the development of an effective mixed gas system a solution for the following well-known contradiction has to be found: from a thermodynamic point of view an optimal mixture leads to minimal temperature differences in the heat exchanger. From a heat exchanger point of view, to transfer the heat at small temperature differences requires a larger surface and a higher pressure drop. This will lead to additional losses and to a degradation in cooler performance. Consequently, an optimization of cycle parameters is necessary to realize the maximum potential of the mixed gas J-T technology in a real system.

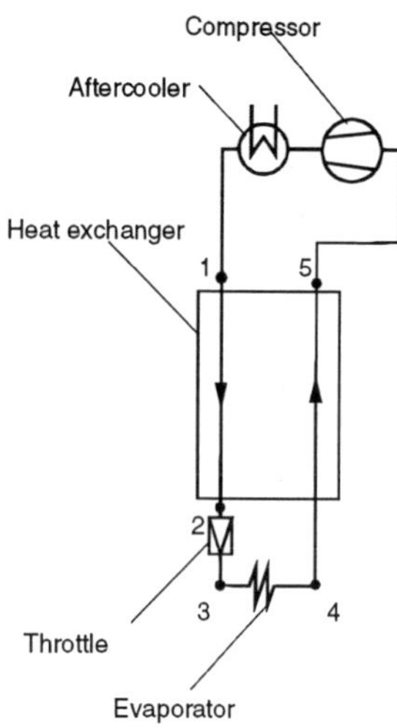

Figure 1. Simple mixed gas Joule-Thomson cycle.

A number of optimization methods are available in the literature[1-3]. These methods allow one to optimize the mixture composition and the working pressures by assuming minimal temperature differences and pressure drops in the heat exchanger and theoretical compressor volumetric and power efficiencies. The value of the calculated results depends mostly on the accuracy of the assumed input data. Because the compressor performances are normally known (or can be evaluated relatively simply), the main problem is a correct assumption of the heat exchanger parameters: minimal temperature difference and pressure drops as well as temperature and pressure distribution.

In principle if these values could be predicted, a numerical model of the heat exchanger could extend prior optimization methods. However, the problem is that generalized correlations for heat transfer by condensation and vaporization of the multi-component mixtures, as well as predictions of the pressure drop for two-phase flow of multi-component mixtures in narrow channels with the necessary accuracy, are not available at the present time.

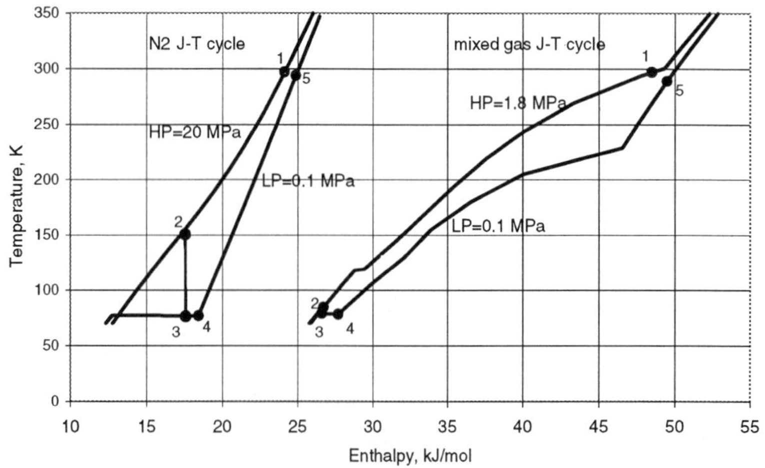

Figure 2. T-h diagrams of J-T cycles (T_{amb}= 297 K, ΔT_{min}=3K, T_o= 80 K, low pressure LP=0.1 MPa) with a mixture of N_2 (30 mol), CH_4 (23 mol), C_2H_6 (17 mol), C_3H_8 (10 mol), i-C_4H_{10} (20 mol), high pressure HP=1.8 MPa and with pure nitrogen, high pressure HP=20 MPa.

The goal of the present work was to obtain a deeper insight into the behavior of the mixed gas heat exchanger and to make further work in this field more efficient.

BACKGROUND

Very little information concerning the optimization of mixed gas JT-systems including simulation of the heat exchanger is available: Landa et al reported such a method[5] and another contribution was made by Boiarski et al[6].

In the paper by Landa an optimization of a mixed gas system based on an oil-free multi-stage compressor with a Hampson-type heat exchanger is described. The parameter optimized was the cool-down time. It was assumed that the heat transfer is limited by the heat transfer on the low-pressure side and the liquid fraction in the low-pressure stream is small; therefore, the main heat transfer mechanism is not vaporization, but convection. Consequently, Landa could use the known correlation for forced convection to calculate the heat transfer. This model did not include any correlations for pressure drop in the heat exchanger. It was not needed because of the special properties of the chosen system, namely the pressure drop on the low pressure side is minimal and the pressure after the compressor is relatively high, therefore the pressure drop on the high pressure side does not significantly affect the performance of the system.

In the paper by Boiarski the first attempt was made to develop a complete optimization method for small mixed gas J-T coolers based on a single stage compressor. Some results of this optimization were presented. For such systems the pressure drop on the low-pressure side as well as on the high-pressure side plays an essential role. Therefore the correlation for pressure drops in the heat exchanger is required. These equations, as well as equations for heat transfer used are semi-empirical correlations, based on the information obtained from the testing of real coolers. Because the test set up allowed the measurement of temperature and pressure values at the ends of the heat exchanger, only average data could be evaluated. The use of average data simplifies the calculation. But on the other hand, the use of average coefficients has some disadvantages; for example, it does not give any information concerning the temperature and pressure drop distribution. It is not possible to generalize these data and to use them for other kinds and sizes of heat exchangers. Therefore the authors of this publication tried to combine these average data with simple theoretical models (based on the assumption of a homogeneous structure of the flow) for predictions of local heat transfer coefficients and pressure drops. Although, in general, this methodology is questionable, because such simplified models cannot describe correctly the complexity of the two-phase heat exchange and fluid mechanics, this procedure can nevertheless be useful for some problems with limited complexity.

SIMULATION MODEL

Several types of the heat exchangers can be used in mixed gas Joule-Thomson coolers. In the present work the double pipe heat exchanger and the multi-tube heat exchanger are discussed. A double pipe heat exchanger consists of two concentric tubes connected by end closures. The high-pressure flow (warm side) is through the inner pipe, and the low-pressure flow (cold side) is through the annulus formed by the inner and outer pipe. The outer tube is designated as the shell. The multi-tube heat exchangers are similar in construction to the double pipe heat exchangers, except that the inner pipe is replaced with a bundle of tubes. Although these heat exchangers are not compact and have a relatively high-pressure drop on the shell side, they are often used in mixed gas systems, because their manufacture is simple. Further advantages are low longitudinal heat conduction and a uniform flow distribution on the shell side.

A model for the simulation of the behavior of a heat exchanger in a mixed gas J-T system has been developed. It is based on the following assumptions:

- Steady-state operation
- Pressure losses in the lines between the compressor and the J-T-stage are neglected
- Pressure drop in the evaporator is neglected
- Mixture composition does not change in the cycle
- Uniform flow distribution on the tube, as well as on the shell side of heat exchanger
- No heat leak from ambient
- Longitudinal heat conductivity is neglected

Some of these assumptions are actually incompatible. For example, for the thermodynamic calculation we assume that the mixture composition does not change in the cycle, although the calculation method for the pressure drop in two-phase flow assumes slip, and consequently a change of the local mixture composition in the heat exchanger. In the present work this effect is neglected. Later study of this effect is needed.

The input parameters are the following:

- Ambient temperature (corresponds to the temperature of the warm end of the heat exchanger, high pressure stream)
- Pressures before and after the compressor
- Data for compressor volumetric and power efficiency
- Mixture composition
- Geometric configuration of the heat exchanger, including shell inside diameter, tube outside diameter, tube inside diameter, number of tubes in the bundle and heat exchanger length

The output parameters are the following:

- Mixture flow rate
- Cooling capacity of the cooler
- Cooling temperature
- Minimal temperature difference and temperature distribution in the heat exchanger
- Pressure drops and pressure distribution in the heat exchanger
- Distribution of the heat transfer coefficients in the heat exchanger

For a prediction of the thermodynamic properties of mixtures a Peng-Robinson equation of state was used. We used the TRAPP-method[7] to predict viscosity and thermal conductivity for the mixtures. The surface tension for hydrocarbon liquids is estimated with the procedure 10 A3.1 described by API[8]. The Bell-Delaware method[9] was used to compute the pressure drops. The method was adjusted by introducing correction factors estimated from available experimental data. A modified method developed by Chen[10] was used to calculate the heat transfer coefficients. This method includes correlations for forced convection as well as for vaporization and condensation of mixtures. The software program HEXTRAN[11] was used for the calculation.

The calculation results were not compared with experimental data, except for the pressure drop at the shell side. The verification of calculations will be the next step in this investigation. However, the analysis of the calculated data does not show essential discrepancies in the behavior of the real systems and calculated values. This is the reason why it is plausible to use these data to study basic relationships between the heat exchanger parameters in a qualitative manner.

DISCUSSION OF RESULTS

A series of calculations for different mixtures were made. Each sequence included the calculation of the cooling capacity and optimal cooling temperature of the cooler for a set of fixed input parameters for different heat exchanger lengths. The combination of sequences allows one to compare the calculation results for different mixtures.

The simulation shows that the behavior in the heat exchanger initially depends on the mixture composition. For some mixtures this behavior is very similar, so one can use this fact to sort all mixtures into groups according to the kind of temperature distribution in the heat exchanger.

The following examples illustrate this consideration. They were evaluated for the following three nitrogen-hydrocarbon mixtures:

components	mixture # 1	mixture # 2	mixture # 3
nitrogen:	30 mol %	27.27 mol %	33 mol %
methane:	23 mol %	20.91 mol %	18 mol %
ethane:	17 mol %	15.45 mol %	12 mol %
propane:	10 mol %	13.64 mol %	12 mol %
i-butane:	20 mol %	22.73 mol %	25 mol %

for following parameters

- Ambient temperature 297 K
- Pressure after the compressor 1.5 MPa
- Pressure before the compressor 0.15 MPa
- Mixture flow rate 2.1 standard m^3/h
- Shell inside diameter 10 mm
- Tube outside diameter 3 mm
- Tube inside diameter 2 mm
- Number of tubes in bundle 7

Figure 3 shows the different temperature distribution curves for a system with a heat exchanger length of 6 m.

The temperature distribution in the heat exchanger depends on the location of the pinch points in the thermodynamic process. For example, mixture # 1 contains a reduced fraction of high-boiling components (propane and butane), the pinch point is located at the warm end of the heat exchanger, and the temperature curve is very flat at the warm end. Since mixture # 2 consists of an increased fraction of high-boiling components, the pinch point is located at the cold end of the heat exchanger, and the flat part of the curve also locates at the cold end. Mixture # 3 consists of an increased fraction of high-boiling as well as low-boiling components (nitrogen and methane), the pinch point is located in the middle, and results in a flatness of the curve in the middle of the heat exchanger.

From a thermodynamic point of view, this pinch point effect is positive, because the heat exchange at the pinch point is less irreversible. On the other hand, the heat transfer with the small temperature difference demands a large surface, and the heat exchanger surface is used ineffectively. This takes place, for example, in the area from 4.5 m to 6 m for mixture # 1, in the area from 0 to 2.5 m for mixture # 2 and in the area from 3.5 m to 4.5 m for mixture # 3.

The influence of the pinch points is thus very important for the understanding of the behavior of a mixed gas heat exchanger. One can subdivide all mixtures in the groups based on the "pinch-point" criterion: how many pinch points does a mixture have, and at which place in the heat exchanger do these pinch points occur. Such an understanding considerably simplifies further analysis.

The data for the pressure distribution on the shell side of the heat exchanger are presented in Figure 4. The pressure distribution depends on the location of the pinch-point in the heat exchanger similar to the temperature distribution. The pressure drop for mixture # 1 (with a pinch point on the warm end of the heat exchanger) is greater than that for mixture # 2 (with a pinch point at the cold end).

It is a well-known fact, that for a heat exchanger operating with high-density fluids, the pressure drop expenditure relative to the heat transfer rate is generally smaller than for low-

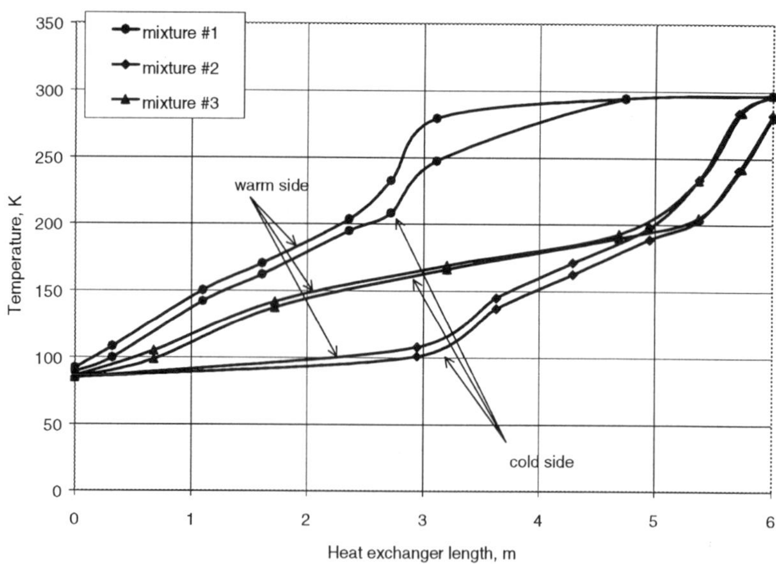

Figure 3. Typical temperature distribution in a heat exchanger.

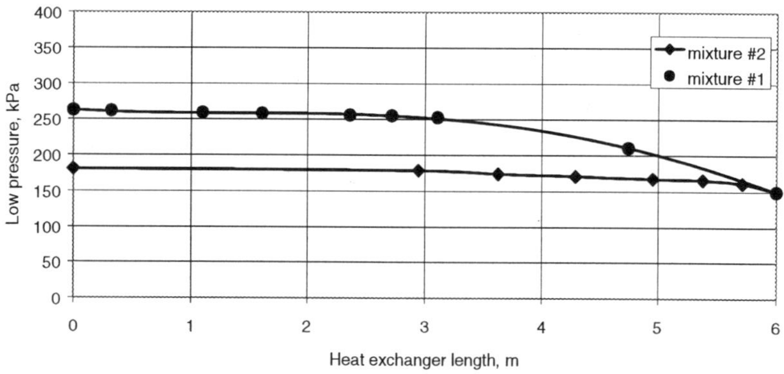

Figure 4. Pressure distribution in a heat exchanger.

density fluids. So the ratio of pressure drop to heat transfer coefficient for mixtures at lower temperatures (80-100 K) is smaller than that at higher temperatures (250-300 K).

Figure 4 illustrates this effect: in the heat exchanger working with mixture # 1 approximately 25 % of the surface operates at the warm temperature (pinch point area), therefore the heat exchange in this area is associated with a relatively high pressure drop (practically all losses are concentrated in the pinch area). On the other hand, in the heat exchanger working with mixture # 2, the fraction of the surface operating at the high temperature is relatively small. This results in a smaller pressure drop for the shell side and consequently a lower cooling temperature can be obtained.

Moreover, this understanding helps to explain the curves for the cooling capacities and cooling temperatures of mixed gas J-T coolers. Figure 5 shows these for mixtures # 1 and # 2 for different heat exchanger lengths. One can observe the following:

1. For both mixtures, the minimal length is approximately 2 m. If the heat exchanger is shorter than this value, a cooling temperature lower than 85 K can not be reached.

2. With a heat exchanger length of 3 m an acceptable cooling capacity (10 W @ 86 K for mixture #1 and 14 W @ 84 K for mixture # 2) can be achieved.
3. The maximum cooling capacity of 15-16 W can be produced with mixture # 1 at 88-90 K, if the heat exchanger is longer than 5 m.
4. Approximately the same heat exchanger length is required to realize the potential of mixture # 2; however, with mixture # 2 a higher cooling capacity (18 W) at a lower temperature (below 85 K) can be produced.
5. The cooling temperature increases with increased heat exchanger length. But for mixture # 1 the rise is higher than for mixture # 2.
6. The use of mixture # 2 is more favorable since a lower optimal cooling temperature and a larger cooling capacity of the cooler can be reached.

From this point of view, the use of mixtures with pinch points in the colder part of the heat exchanger is more favorable for a mixed gas J-T system.

From a thermodynamic point of view, each system has an optimal heat exchanger length. If the heat exchanger is shorter, the cooling capacity of the system decreases. If the heat exchanger is longer, it leads to a higher pressure drop and to an increased cooling temperature and increased cooling capacity. The optimal minimal temperature difference between the high and low pressure streams in the heat exchanger for the 80-90 K systems amounts to less than 0.5 K. However, the choice of the best heat exchanger length is more than a thermodynamic problem. In many cases it is reasonable to make the heat exchanger shorter than optimal. Although the cooling capacity decreases, it can bring some other advantages. For example, a smaller

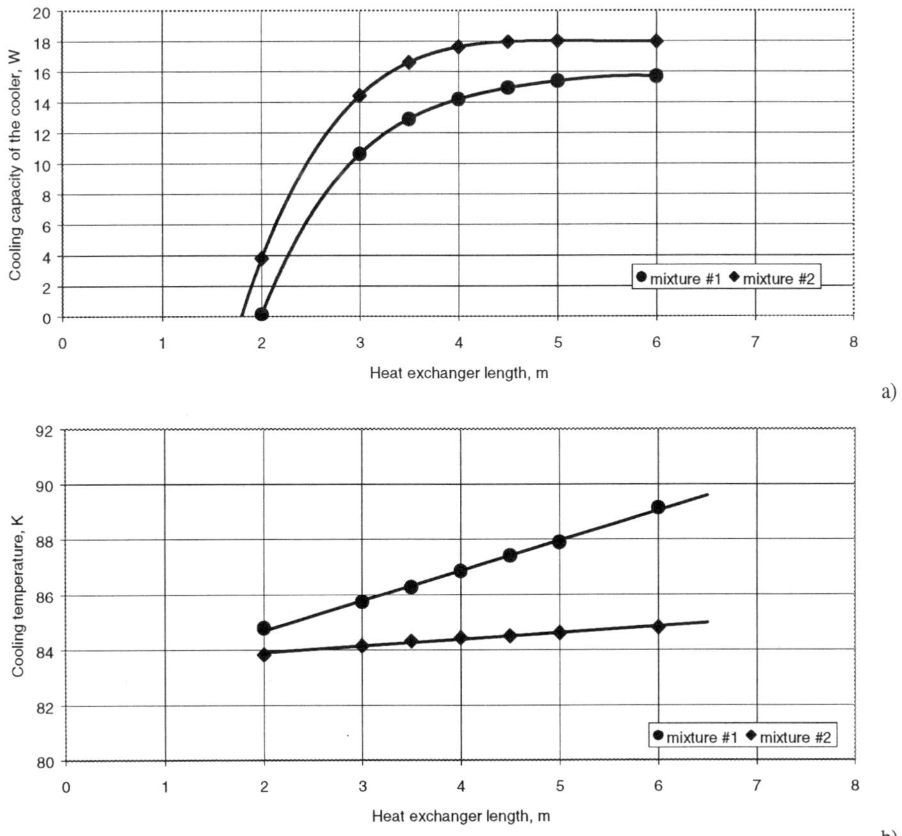

Figure 5. Cooling capacity and cooling temperature for different heat exchanger lengths

refrigerant hold-up results and this allows simpler compressor start-up, because of the lower stand-still pressure in the system. This is important for safety, particularly if flammable mixture components are used. Another advantage of a shorter heat exchanger is a lower mass and volume of the cold head, which can result in a faster cool down.

Analysis of the temperature distribution in real systems can give some useful information for adjusting the mixture composition, i.e., a flat part of the curve indicates the temperature range of a component, whose fraction in the mixture is smaller than optimal.

CONCLUSION

A model of a mixed gas J-T system was developed, which allows one to investigate the steady state behavior in the heat exchanger of a mixed gas J-T system.

The calculations show that the temperature distribution as well as the pressure distribution in the heat exchanger depends primarily on the pinch point locations in the heat exchanger. This understanding considerably simplifies the analysis. The study of the temperature and pressure profiles allows one to obtain some important data for the optimization of the mixture.

To use this model for the design of real systems, a check needs to be made of the computed results and adjustment made of its parameters, if necessary.

REFERENCES

1. W.A.Little, "Method for efficient countercurrent heat exchange using optimized mixtures", US Patent # 5,644,502 (1997)
2. A.Alexeev, H.Quack and Ch.Haberstroh, Further development of a mixed gas Joule-Thomson refrigerator, in: "Advances in Cryogenic Engineering", Vol. 43B, Plenum Press, New York (1998), pp. 1667-1674
3. G.G.Potapov, Optimization of the mixed gas throttle cycles and development of small-scale system based on such a cycle for producing of liquefied products, Ph.D. Thesis (in Russian), Moscow Power Engineering Institute (1994)
4. A.Alexeev, Untersuchung zur Weiterentwicklung von Gemisch-Joule-Thomson Kältemaschinen auf der Basis von ölgeschmierten Kompressoren, Ph.D. Thesis, TU Dresden (1999)
5. Yu.I.Landa and A.K.Gresin, Design method for J-T-microcooler heat-exchanger applying multicomponent refrigerant, in: "Proceedings 19th Intern. Cong. Refr.", Vol. IIIb (1995), pp. 1102-1106
6. M.Boiarski, A.Khatri and V.N.Kovalenko, Design optimization of the throttle-cycle cooler with mixed refrigerant, in: "Cryocoolers 10", R.G.Ross, Jr., ed., Plenum Press, New York (1999), pp. 457-465
7. J.F.Ely and H.J.M.Hanley, Prediction of transport properties. 1. Viscosity of fluids and mixtures. 2. Thermal conductivity of pure fluids and fluid mixtures, *IEC* 22: 90 (1981)
8. API Technical Data Book, 3rd Edition (1978), pp. 10-17
9. K.J.Bell, Final report of the cooperative research program on shell and tube heat exchangers, Bulletin No. 5, University of Delaware Engineering Experiment Station, Newark, Delaware (1963)
10. J.C.Chen, Correlation for boiling heat transfer to saturated fluids in convective flow, *IEC* 5: 322 (1966)
11. HEXTRAN, User Manual, Simulation Science Inc., 601 S.Valencia Ave., Brea, California 92621, USA

EXPERIMENTAL INVESTIGATION OF A MIXED-REFRIGERANT J-T CRYOCOOLER OPERATING FROM 30 TO 60K

E. Luo, M. Gong, Y. Zhou and L. Zhang

Cryogenic Laboratory, Chinese Academy of Sciences
Beijing 100080, P. R. China

ABSTRACT

A modified thermodynamic process for obtaining refrigeration over a temperature range from 30 to 60K was proposed based on the auto-cascade mixed-refrigerant (AMR) cycle. An experimental set-up was made to verify the proposed thermodynamic cycle. In the experiment, a no-load temperature of around 50K was achieved.

INTRODUCTION

Over the past few years, mixed-refrigerant Joule-Thomson cryocoolers have seen a rapid development in the temperature range around liquid nitrogen temperature [1,2] . Currently, several companies and laboratories have successfully developed some prototypes and even formal products [3,4]. By using specific mixtures, these cryocoolers can efficiently operate in the liquid-nitrogen temperature range, although they are driven by a single-stage, air-conditioning compressor. To eliminate the clogging problem in these cryocoolers, the auto-cascade method was introduced into micro-miniature Joule-Thomson cryoooclers[5,6]. Hitherto, the above-mentioned refrigeration cycle has been limited to liquid-nitrogen temperatures or higher. To obtain lower temperatures (e.g., around 50K), a modified thermodynamic process was proposed to achieve a temperature around 50K, based on the AMR cycle and a specific cryogenic mixture. This paper provides an analysis of the thermodynamic cycle, and describes an experimental set-up to verify the proposed idea.

REFRIGERATION CYCLE

The schematic of an AMR cycle with one separator is shown in Fig.1. The system includes a compressor, an aftercooler, a vapor-liquid separator, two counterflow heat exchangers, two throttling valves and an evaporator.

Advances in Cryogenic Engineering, Volume 45.
Edited by Shu *et al.*, Kluwer Academic / Plenum Publishers, 2000.

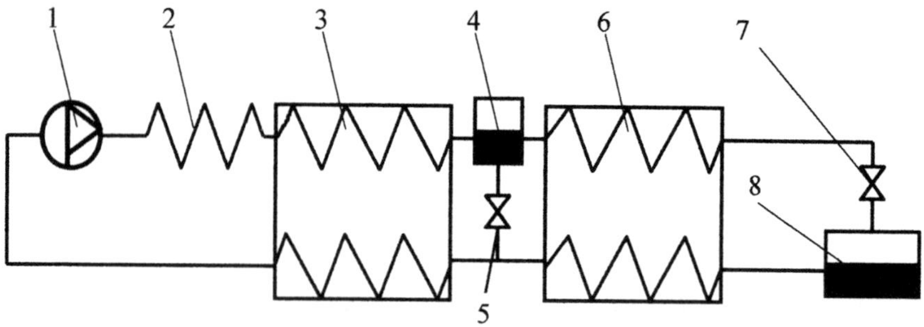

Figure 1. Schematic of 30-60K auto-cascade mixed-refrigerant refrigeration cycle.

(1.Compressor 2. After-cooler 3.Counterflow heat exchanger I 4. Vapor-liquid separator 5.Throttling valve I 6. Counterflow heat exchanger II7. Throttling valve II 8. Evaporator)

In this cycle, the specific mixture includes two groups of working substance: One is a very efficient nitrogen mixture, and the other is neon gas. Thermodynamically, the mixture initially goes through the first stage counterflow heat exchanger and is precooled to a temperature range between 70 to 90K by the auto-cascade throttle method. In this temperature range, all the components of the nitrogen mixture except neon condense as liquid and return to the low pressure side of the first stage counterflow heat exchanger after the first throttling valve. The cooled neon gas is separated in a vapor-liquid separation unit and passes through the second stage counterflow heat exchanger. The neon gas after being precooled in the second stage counterflow heat exchanger and expanded in the second throttling valve produces a cooling capacity at a temperature range between 30 to 50K. The concept for obtaining lower temperatures can be expanded by adding more separators and throttling valves and selecting specific mixtures.

THERMODYNAMIC ANALYSIS

Verification of the above concept involved the development of a thermodynamic design process with suitable operating parameters and utilizing a specific mixture. To analyze the thermodynamic performance of the refrigeration cycle shown in Fig.1, assume that one mole of a mixture is discharged from the compressor. All of the mixture will flow through the first counterflow heat exchanger, resulting in partial condensation of the mixture. Only a fraction of the total mixture is permitted to flow to the second counterflow heat exchanger after the vapor-liquid separator. The amount of the separated gas depends upon the mixture used, the first counterflow heat exchanger and the opening of the first throttling valve. Assuming a vapor-liquid separator temperature of T_m and vapor-liquid equilibrium to exist, we can calculate the vapor-liquid ratio and the equilibrium compositions of the vapor and liquid phases from the following relationships:

$$\bar{f}_i(T, p, \vec{\xi}^{(V)}) = \bar{f}_i(T, p, \vec{\xi}^{(L)}) \qquad (i=1,n) \qquad (1)$$

$$\sum_{i=1}^{n} \vec{\xi}_i^{(V)} = 1 \qquad (i=1,n) \qquad (2)$$

$$\sum_{i=1}^{n} \vec{\xi}_i^{(L)} = 1 \qquad (i=1,n) \qquad (3)$$

$$\sum_{i=1}^{n}\vec{\xi}_i^{(V)}\beta^{(V)}+\sum_{i=1}^{n}\vec{\xi}_i^{L}\beta^{(L)}=1 \qquad (i=1,n) \tag{4}$$

According to heat transfer theory, the refrigeration capacity of the evaporator at T_C is defined by Eq. (5), and the ideal thermodynamic efficiency can be given by Eq.(6):

$$q_c = \min[\Delta h(T)] \tag{5}$$

$$\eta_{mix}^{id} = \frac{\min[\Delta h(T)]}{W_{in}}\left(\frac{T_0 - T_C}{T_C}\right) \tag{6}$$

Although the auto-cascade method has a self-cleaning function, all components of the mixture except neon will probably solidify and clog the flow passages of the refrigeration system at temperatures below 70K. Moreover, the neon gas coming from the separator will still contain a small amount of the other components. Thus, it is still uncertain whether the residual components in the cooled neon gas will solidify to clog the flow channel of the second counterflow heat exchanger. To solve this problem, we used the Scatchard-Hildebrand regular solution model to estimate the liquid-solid equilibrium of the mixture used. According to Eq.(7), (8) and (9), we can predict the limitation solubility of each component at different cryogenic temperatures of interest:

$$\frac{x_{iS}\gamma_{iS}}{x_{iL}\gamma_{iL}} = \exp\left[\frac{\Delta h_{tp,i}}{R}\left(\frac{1}{T_{tp,i}} - \frac{1}{T}\right)\right] \tag{7}$$

$$RT\ln\gamma_{iL} = \tilde{V}_i^{L}\left(\delta_i - \sum_j \phi_j \delta_j\right)^2 \tag{8}$$

$$RT\ln\gamma_{iS} = \sum_{i}^{n-1}\sum_{j=i+1}^{n} A_{ij}x_{jS} \tag{9}$$

A mixture containing neon, nitrogen, oxygen and argon with molar concentrations of 0.4, 0.15, 0.35 and 0.1, respectively, was selected to examine this problem. Table 1 gives the predicted vapor-liquid equilibrium characteristics at different separation temperatures. From this table, we can observe the vapor-liquid ratio and the equilibrium compositions at the separator. The solubility limit of the components in the mixture is shown in Table 2 at different temperatures. If the residual amounts of the components with high freezing temperatures after the separator do not exceed their solubility limits, the mixture should normally not cause clogging. Analyzing the theoretical results in Tables 1 and 2, it appears that the process should operate normally by reasonably designing an appropriate separation temperature and mixture.

We can then predict the thermodynamic performance of the auto-cascade refrigeration with the given mixture. We assume that the high and low pressures of the compressor are 10.0MPa and 0.1MPa, respectively, and the ambient temperature, the separation temperature and the evaporation temperature are 300K, 80K and 30K, respectively. Based on the Peng-Robinson equation of state, the enthalpy-temperature diagrams of the mixture in the first and second heat exchangers are shown in Figs.2 and 3, respectively.

According to the results of Table 1, Figs. 2 and 3, the vapor fraction of the mixture at 80K is about 0.2302, and the minimum isothermal throttling effect in the first and second stages for one mole are about 413J/mol and 352J/mol, respectively. Thus, the refrigeration capacity at 30K will be equal to about 81J/mol.

Table 1. Vapor fraction and equilibrium compositions [Ne(40%)-N_2(15%)-O_2(35%)-Ar(10%)])

T (K)	β^V	Ne	N_2	O_2	Ar
	P=10.0MPa				
60	0.1812	0.9788	0.123e-1	0.506e-2	0.379e-2
70	0.2068	0.9724	0.147e-1	0.753e-2	0.530e-2
80	0.2302	0.9529	0.227e-1	0.146e-1	0.968e-2
90	0.2525	0.9167	0.362e-1	0.289e-1	0.180e-1
		P=7.0MPa			
60	0.2340	0.9904	0.600e-2	0.199e-2	0.159e-2
70	0.2655	0.9822	0.101e-1	0.442e-2	0.367e-2
80	0.2895	0.9620	0.193e-1	0.110e-1	0.756e-2
90	0.3142	0.9239	0.347e-1	0.252e-1	0.161e-1

Table 2. Solubility limit of components at different temperatures

T (K)	Ne-O_2-N_2-Ar			
	Ne	O_2	N_2	Ar
25	1	0.316	0.131	0.018
30	1	0.449	0.225	0.059
35	1	0.579	0.335	0.109
40	1	0.701	0.455	0.173
45	1	0.813	0.578	0.251
50	1	0.915	0.699	0.339
55	1	1	0.817	0.432
60	1	1	0.913	0.503

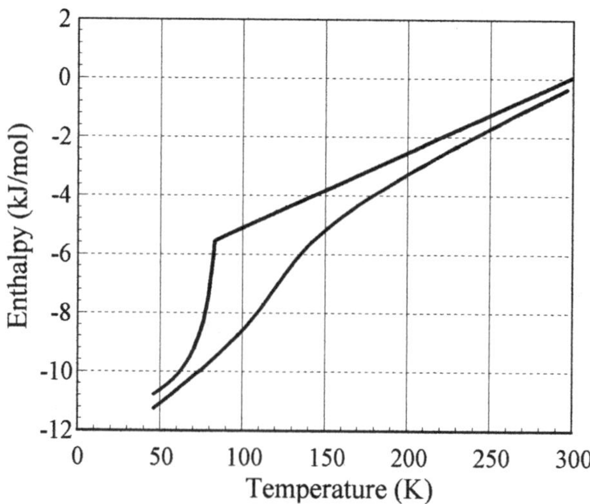

Figure 2. Enthalpy-temperature diagram of the mixture in the first heat exchanger.

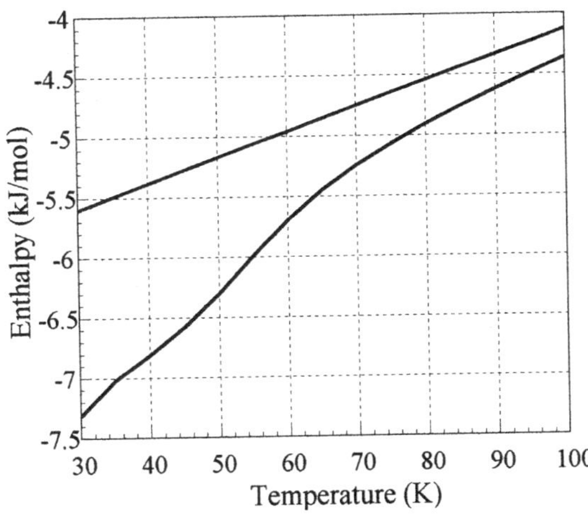

Figure 3. Enthalpy-temperature diagram of neon in the second heat exchanger.

If the compressor has a capacity of $4Nm^3/h$, the refrigeration system will produce a refrigeration capacity of about 4W at 30K and a thermodynamic efficiency of about 3.6%. In our laboratory, there is a three-stage oil-free compressor having a capacity of $4Nm^3/h$ at 10.0MPa with about 1.0kW electrical power input.

EXPERIMENTAL VERIFICATION

To verify the feasibility of the proposed thermodynamic cycle, the authors designed a cryogenic package shown schematically in Figure 4 and in a photograph in Figure 5. The cryogenic package consists of two counterflow heat exchangers fabricated of perforated plates, a vapor-liquid separation unit with the first throttling valve, an evaporator and the second throttling valve. The total length of the cryogenic package is about 220mm, and the outer diameter is about 14mm.

In the experiment, a mixture containing neon, nitrogen, oxygen and argon was tested in an open cycle mode. The working pressure was maintained between 9.5MPa and 7.5MPa. Figure 6 is the cool-down curve of the mixed-refrigerant Joule-Thomson cryocooler, which obtained a low temperature of about 51K. Although the unit did not achieve the temperature range of 30K, it did indicate that the new cryocooler could operate in the 50K temperature range. To achieve a design temperature range of 30K, the heat exchanger area needs to be increased and the openings of the two throttling valves need to be regulated.

CONCLUSIONS

In this paper, a modified auto-cascade mixed-refrigerant Joule-Thomson refrigeration cycle is presented for achieving temperatures below 60K. The theoretical analysis shows that the proposed thermodynamic process can operate in the temperature range of 30 to 60K. To avoid clogging of the high-temperature components, the authors analyzed the vapor-liquid and solid-liquid equilibrium characteristics of the proposed mixtures. An experimental set-up was made to verify the proposed refrigeration cycle. Although the final goal of 30K was not achieved, a temperature of around 50K for cooling high temperature superconducting devices was obtained in the experiment.

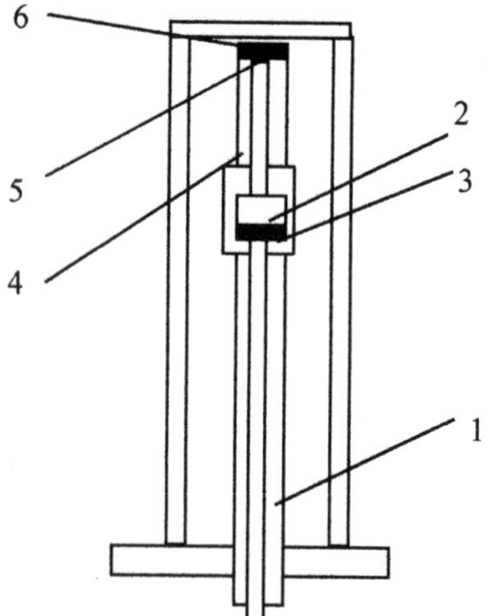

Figure 4. Schematic of the cryogenic package.
(1. Counterflow heat exchanger I 2. Vapor-liquid separator 3. Throttling orifice I 4. Counterflow heat exchanger II 5. Throttling orifice II 6. Evaporator)

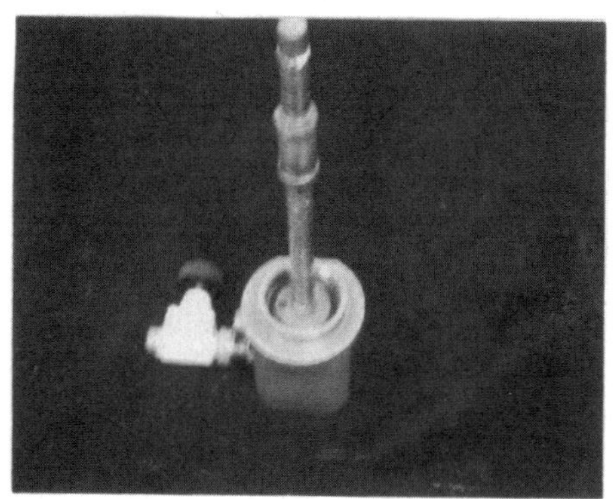

Figure 5. Photograph of 50K auto-cascade J-T cryocooler.

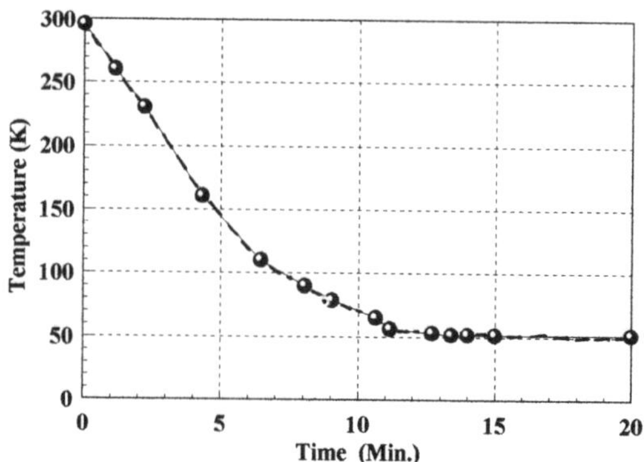

Figure 6. Cool-down curve.

To make the proposed refrigeration cycle operate in the 30K temperature range, we think that the heat exchange areas of the two counterflow heat exchangers and the openings of the throttling valves should be matched and optimized. Moreover, the mixture used needs to be optimized.

NOMENCLATURE

A	interaction coefficient in the solid phase	(-)	
f	fugacity	(Pa)	
Δh	isothermal throttling effect	(J/mol)	
Δh_{tp}	fusion enthalpy at the triple point	(J/mol)	
p	thermodynamic pressure	(Pa)	
q	refrigeration capacity at the evaporator	(J/mol)	
R	universal gas constant, 8.314	$(J/mol{\cdot}K)$	
T	thermodynamic temperature	(K)	
$\tilde{V}$	molar volume in liquid mixture	(m^3/mol)	
W	electrical power input of compressor	(J/mol)	
x	component concentration in liquid or solid phase	(-)	
β	phase splitting fraction	(-)	
ξ	equilibrium concentration in vapor or liquid	(-)	
γ	activity coefficient	(-)	
ϕ	volume fraction of component in liquid mixture	(-)	
δ	solubility parameter	$(J/m^3)^{0.5}$	

Subscripts

n	comp. no.
V	vapor
L	liquid
S	solid
H	hot
C	cold

ACKNOWLEDGMENT

This work is financially supported by the Natural Sciences Foundation of China under Contract 59706002.

REFERENCES

1. V.M.Brodianski, et al, The use of mixtures as the working gas in throttle Joule-Thompson cryogenic refrigerator, in:" Proc. of 13th Intern. Congress of Refrigeration", Plenum Press, New York (1973), p.43.
2. W.A Little, Recent developments in Joule-Thompson cooling: gases, coolers, and compressor, in: " Proc. of 5th Intern. Cryocooler Conference", Monterey, CA (1989), p.1.
3. R.C.Longsworth, et al, 80K closed cycle throttle refrigerator, in: "Cryocooler 8", R.Ross,ed., Plenum Press, New York (1995), p.537.
4. E.C.Luo, et al, Experimental investigation of an efficient closed-cycle, mixed-refrigerant J-T cooler operating in low pressures, *Vacuum and Cryogenics,* 1:215(1995).
5. A.Klimenko, One flow cascade system, in:"Proc. of Xth Intern. Congress of Refrigeration", Pergamon Press , London (1959), p.34.
6. W.A.Little and I.Sapozhnikov, Low cost cryocoolers for cryoelectronics, in:"Cryocooler 9", R.Ross, ed., Plenum Press, New York (1997), p.509.

FLOW-RATE PRESSURE-DEPENDENCE OF A FIXED-ORIFICE JOULE-THOMSON CRYOCOOLER

B-Z. Maytal

Rafael, Cryogenic Section
P.O.Box 2250(39) Haifa 31021, Israel

ABSTRACT

Experimental study shows that the free (non-recuperative) flow rates of a Joule-Thomson cryocooler are quite proportional with the inlet pressure. However, the recuperative flow rates may be non-proportional, though monotonically increase with pressure. Recuperative flow rate *per unit pressure* may increase but also significantly decrease at higher pressures: at 45 MPa this value was observed to be about half of the value at 15 MPa. Such information is essential for modeling and predicting the discharge pattern and period of a pressure vessel through a cryocooler or at evaluating the benefit of incorporating a pressure regulator. The proposed explanation is based on the reduction of the speed of sound in two-phase flow as the fraction of the denser (liquefied) phase is increased. The ratios of cold flow and free flow were measured with nitrogen and argon at pressures up to 68 MPa.

INTRODUCTION

Two types of flow rates are associated with a fixed orifice Joule-Thomson cryocooler. The "free flow", $\dot{m}_{FR}$, is obtained without recuperation as measured when the cooler is outside its encapsulation. When cooler operates inside an encapsulation (often a dewar) the recuperative heat exchanging lowers the temperature; thus, at a given pressure, P, the flow rate, $\dot{m}_{RE}$, increases.

The pattern of discharge of a pressure vessel through a fixed orifice Joule-Thomson cryocooler is directly related with the flow rate pressure dependence, $\dot{m}_{RE}(P)$. The character of a fixed orifice Joule-Thomson cryocooler in terms of this dependence should be predetermined before any attempt is made to simulate the discharge period. In some cases, for prolonging this period, a regulator as studied by Buller [1] reduces the pressure. Thus, a lower pressure drives the flow rate instead of the instantaneous pressure in the vessel. It is relevant to evaluate the extent of the degradation when the pressure regulator is deleted. Prior art applied a double fold assumption [2], which indeed did facilitate any analysis: (a) the ratio between the recuperative and free flows, $\dot{m}_{RE}/\dot{m}_{FR}$, is constant (equals 1.8 for nitrogen) and implicitly (b) the flow rates of an operating cryocooler are proportional with the applied pressure. Consequently, that leads to logarithmic pressure decay. The purpose of the present study was to challenge the validity of these assumptions by measuring $\dot{m}_{RE}(P)$ and $\dot{m}_{FR}(P)$. Another objective was to identify the mechanism or condition under which it is not valid.

THE EXPERIMENTS

Two fixed orifice cryocoolers (A and B) were used. Both have identical finned tube heat exchangers of 8.3 mm diameter, about 20 mm in length and were tested in a non-evacuated encapsulation. The size of the orifices of cryocooler A is characterized by free flow rates of 2.5 standard liters per minute (SLPM) of nitrogen under a pressure of 10 MPa and 3.5 MPa for B. The free and recuperative flow rates for both coolers were measured as a function of applied pressure (up to 68 MPa), once with nitrogen (Figure 1) and then with argon (Figure 2) at room temperature.

Two discharge experiments were performed with cryocooler A and a 4 L argon pressure vessel. The cooler initially was fed directly from the outlet of the vessel. In the second run, a pressure regulator between the vessel and the cooler ensured a constant pressure supply of 17 MPa. Figure 3 displays the discharge pattern of both experiments.

Another cryocooler (C) was fabricated with a heat exchanger of the same diameter but with twice the length (40 mm). Its fixed orifice delivered a free flow of about 6.6 SLPM nitrogen at a pressure of 10 MPa. This unit operated within an evacuated dewar and recuperated flow rates were measured at pressures up to 40 MPa. The orifice was then replaced with three different smaller orifices, and flow measurements repeated. Figure 4 summarizes $\dot{m}_{RE}(P)$ for the four orifices.

DISCUSSION

While the free flow rates (of cryocoolers A and B, with nitrogen and argon) seem to be quite proportional with pressure, the recuperative flow rates are not. Lines noted by L in Figures 1 and 2 draw the trend of recuperative flow being proportional with pressure for cooler B (which means constant value of $\dot{m}_{RE}/P$). The extent of deviation from proportionality for a given orifice may be obtained by the aid of the parameter $\dot{m}/P$. The reference case of proportionality (not just linearity) with pressure is characterized by a constant $\dot{m}/P$. For example, cryocooler A with argon at 14 and 41 MPa, the $\dot{m}_{RE}/P$ values for both pressures are 0.97 and 0.54 SLPM per each MPa, respectively. Put another way, the flow at 41 MPa is almost less than half that of what it should have been if it were proportional with pressure.

This result must have a reflection in the discharge pattern of a cooler. Figure 3 focuses on the difference between one discharge driven by the instantaneous pressure in the vessel and the other driven by a constant 17 MPa regulated pressure. The non-linearity of the latter curve (in spite of the constant discharge flow rate as a result of regulated pressure) is related to the pressure dependence associated with compressibility of the residual gas in the vessel. Starting at 41 MPa, discharge until 17

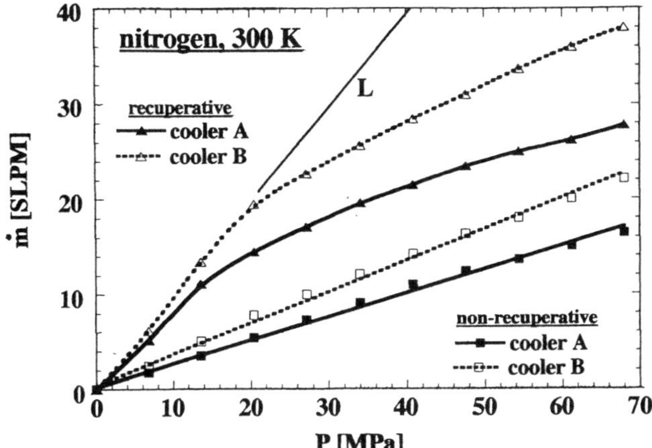

Figure 1. Recuperative and non-recuperative flow rate of nitrogen for cryocoolers A and B as a function of pressure, at room temperature.

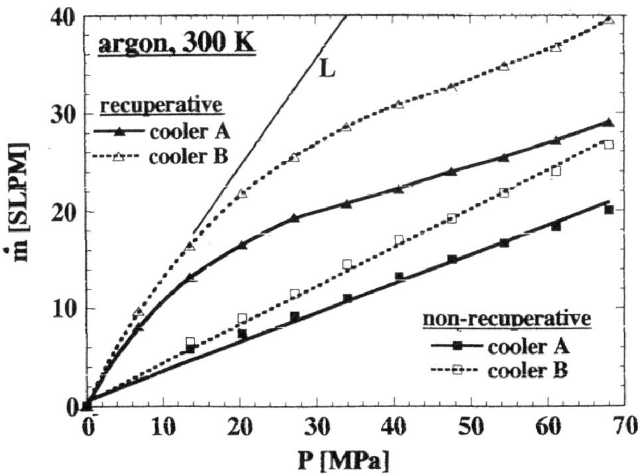

Figure 2. Recuperative and non-recuperative flow rate of argon for cryocoolers A and B as a function of pressure, at room temperature.

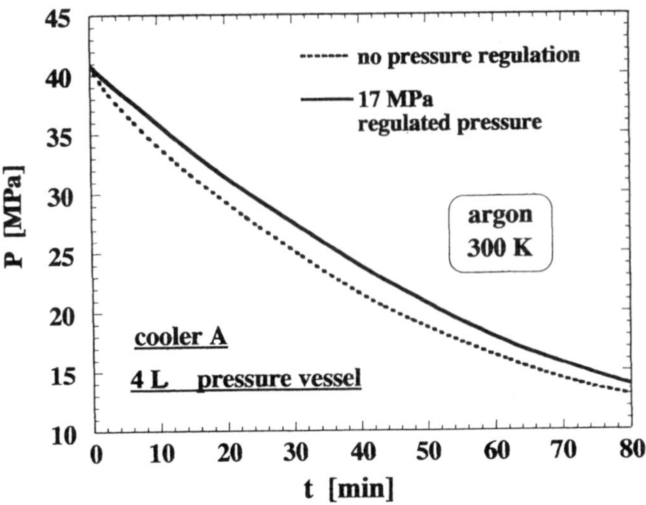

Figure 3. Discharge pattern of cryocooler A with and without a pressure regulator.

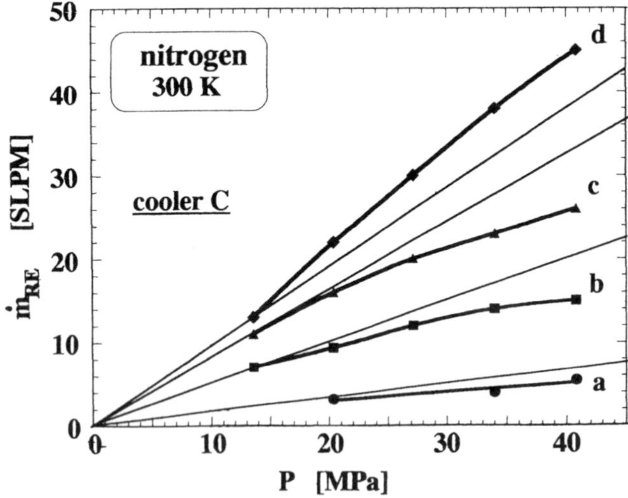

Figure 4. Recuperative flow rates as a function of pressures for cryocooler C at room temperature with four different orifices as a function of pressure.

MPa is prolonged only by about 14 percent (from 56 to 64 minutes). It has the nature of a self-regulating effect, where at higher pressures the "expected excess" flow rate is actually depressed. Thus, even such a sharp pressure reduction from 40 down to 17 MPa results in minor benefit. Extending the vessel by the same volume as the regulator could compensate for the extra few minutes of operation.

Figure 5 displays the ratio of $\dot{m}_{RE}/\dot{m}_{FR}$ for cryocoolers A and B operating with argon and nitrogen. This ratio is pressure dependent and ranges from about 1.5 at the higher pressures, to about 3.0 at the lower pressure. Obviously, if both, $\dot{m}_{FR}$ and $\dot{m}_{RE}$ would have been proportional with pressure, this ratio would have remained independent of pressure.

The fundamental reference case is the discharge through a nozzle under various inlet conditions and not associated with a cryocooler. Schmidt et al. (3) and Johnson (4) showed theoretically for air up to 40 MPa that the choked flow-rate per unit pressure at 300 K inlet, is relatively proportional with pressure. However, at lower temperatures such as 225 K, $\dot{m}/P$ increases with pressure. Hendricks et al (5) measured nitrogen choked flow rates through a nozzle down to 87 K but at pressures below 10 MPa. However, flow rates consumed by an operating cryocooler may exhibit decreasing $\dot{m}_{RE}/P$ values with pressure. A question still remains whether the special condition of a nozzle while incorporated in a cryocooler is responsible for this inconsistency with the above references.

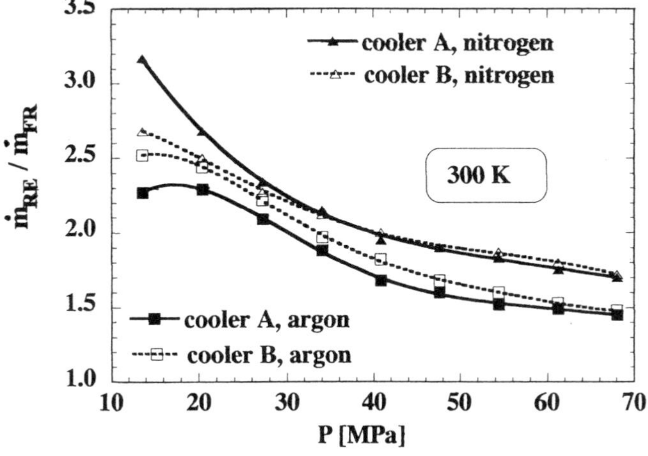

Figure 5. Ratios of recuperated and non-recuperated flow-rates of cryocoolers A and B as a function of pressure with nitrogen and argon at room temperature.

Elevated pressure in a cryocooler shows the following trends: (a) flow rate increases, (b) the integral Joule-Thomson effect, Δh_T, increases, (c) the yield of liquefaction increases as Δh_T increases, (d) the nozzle inlet temperature decreases slightly, and (e) the effectiveness of the heat exchanger decreases. As liquefaction occurs, the stream at the nozzle is a two-phase flow. The speed of sound sharply decreases as the fraction of the denser phase increases [6, 7], namely, the $\dot{m}/P$ decreases. The lower effectiveness of a heat exchanger on the one hand suppresses the flow, but on the other hand, it reduces the yield of liquefaction, thus elevating the speed of sound. The resulting pressure dependence of the recuperated flow is a consequence of these interconnected and conflicting, trends.

The experiment in Figure 4 supports this concept. The less effective the heat exchanger or the less efficient the cryocooler, both associated with the largest flow rates (nozzle d) exhibit $\dot{m}_{RE}/P$ values increasing with pressure, which is closer to the trends described in the literature.

REFERENCES

1. J.S. Buller, A miniature self-regulating rapid-cooling Joule-Thomson cryostat", *Adv. Cryo. Eng.*, Plenum Press, New York, (1971), **16**:205-213.
2. R.C. Longsworth, Joule-Thomson Cryocoolers: User's Manual, APD Cryogenics, Allentown, PA.
3. B. Schmidt, R. Martin and C. House, Behavior of high pressure air in critical-flow through nozzles, *Journal of Thermophysics,* **4**(1):37-41,(1990).
4. R.C. Johnson, Real-gas effects in critical flow-through nozzles and tabulated properties", NASA TN-D-2565, 1964.
5. R.C. Hendricks, R.J. Simoneau and R.C. Ehlers, Choked flow of fluid nitrogen with emphasis on the thermodynamic critical region, *Adv. Cryo. Eng.,* Plenum Press, New York, (1973), **18**:150-161.
6. R.V. Smith and K.R. Randall, Critical two-phase flow for cryogenic fluids, NIST (formerly NBS) Technical Note 633, January 1973.
7. E. Elias and G.S. Lellouche, Two phase critical flow, *Int. J. Two Phase Critical Flow,* Vol. 20 Suppl., pp. 91-168, (1994).

PERFORMANCE TESTING OF A 4 K ACTIVE MAGNETIC REGENERATIVE REFRIGERATOR

Stephen F. Kral, [1] and Carl B. Zimm[2]

[1]4133 SE Lincoln St.
Portland, OR 97214-5958
[2]Astronautics Corporation of America
Technology Center
5802 Cottage Grove Rd.
Madison, WI 53718-1387

ABSTRACT

This report describes performance tests of an active magnetic regenerative (AMR) refrigeration stage intended for operation between a conventional Gifford-McMahon (G-M) cryocooler sink cooling to below 10 K and a source at approximately 4 K. The AMR configuration used here achieves refrigeration through the magnetocaloric effect of ferromagnetic ErNi. The ErNi refrigerant in two porous packed beds reciprocates into and out of a magnetic field produced by a stationary, superconducting, Nb_3Sn magnet. Helium gas flowing through the beds between reciprocation effects heat transfer between source and sink.

The refrigerator performance target includes 1.5 W of cooling at a cold source temperature of 3.5 K. Testing explored refrigerator performance for fields of 1 T to 2.25 T, operating pressure up to 0.5 atm and periods between 16 s and 24 s. Evidence of helium liquefaction in the beds at the higher pressures and fields was observed. Loads up to 1 W at the cold heat exchanger were applied. Refrigerator performance was compared with previously published modeling results.

INTRODUCTION

Over the past decade, active magnetic regenerative (AMR) refrigerators with operation spanning temperature ranges from around 300 K, room temperature, to below 2.17 K, the superfluid transition temperature of He^4, have been described.[1-5] Performance testing of refrigerators spanning various temperature ranges within these bounds has been reported. Refrigeration extending to 4 K is of particular interest for scientific and commercial applications of low temperature superconducting magnet systems. We report here the first successful tests of an AMR refrigerator operating between a hot heat sink at approximately 10 K and a cold source temperature around 4 K.

Magnetic refrigeration is based on the magnetocaloric effect of refrigerant magnetic materials that undergo a temperature change when adiabatically magnetized and demagnetized. The AMR refrigerator combines the magnetocaloric effect of a magnetic material formed into a regenerator bed and heat transfer mediated by a heat transfer fluid to couple the hot sink, the cold source and the magnetic material bed in a thermodynamic refrigeration cycle. Four steps constitute the refrigeration cycle. The refrigerant material is warmed through magnetization. Heat transfer fluid is forced through the bed from the cold heat source to the hot heat sink, warming while traversing the bed and transferring heat to the hot sink after emerging from the bed. The refrigerant material is cooled through demagnetization. Heat transfer fluid is forced through the bed from the hot sink to the cold source, cooling while traversing the bed and absorbing heat from the cold source after emerging from the bed. The refrigerator design to achieve successful operation of this cycle include the choice of the magnetic refrigerant, the configuration of the regenerator bed, the method for bed magnetization and demagnetization, and specification of the heat transfer process including choice of the heat transfer fluid and heat transfer mass flow rates.

4 K AMR REFRIGERATOR DESIGN

A description of the design of this refrigerator has been published elsewhere.[6] The refrigerator is a laboratory test device and is capable of operating with independent variations of loads, magnetic fields, heat transfer gas pressures and operating frequencies. Information gleaned from testing of laboratory will be used in the design of practical AMR refrigerators in the 4 K range extracting loads from a few watts to kilowatts.

Essential to good AMR refrigerator performance are the thermal and magnetic characteristics of the magnetic material that forms the refrigerant, the characteristics of the regenerator bed, and the characteristics of the heat transfer fluid. For magnetic material selection, a proprietary finite reduced period AMR refrigerator model enabled comparison of the predicted COP of this refrigerator using a variety of recently characterized erbium compounds.[7] ErNi was selected as the refrigerant because modeling with it as the magnetic refrigerant and low pressure helium as the heat transfer fluid predicted a refrigerator COP 50% larger than the next best predicted performance with a different refrigerant.

The performance of the regenerator can be assessed through comparison of the heat capacity of the heat transfer fluid that flows through the regenerator bed during the heat transfer part of the refrigerator cycle to the heat capacity of the material in the regenerator bed. A figure-of-merit for such performance is obtained from the ratio R of the bed's reduced period, Π, and reduced length, Λ, defined as[7]:

$$R = \frac{\Pi}{\Lambda} = \frac{\dot{m} c_f \tau_b}{M_m c_m}. \tag{1}$$

Here m is the mass flow rate of the helium, M_m is the mass of the bed material, τ_b is the "blow period" during which the heat transfer fluid flows between the hot and cold reservoirs, c_f is the heat capacity of the heat transfer fluid and c_m is the heat capacity of the bed material. For a 4 K refrigerator, helium is the only possible heat exchange fluid. Modeling studies of refrigerator performance established that a practical refrigerator design results from choosing R ≅ 0.4.[7] With this choice, the refrigerator's cooling power varies linearly with R for large changes in R. R values obtained in testing were about this size and the refrigerator's R-dependent performance was consistent with predictions.

The isomagnetic volumetric specific heat of ErNi over the temperature range of this refrigerator is 5 to 10 times that of the volumetric isobaric specific heat of gaseous helium. However, when the helium gas condenses to liquid (eg at 3.4 K for 0.4 atm pressure), the volumetric heat capacity of the helium becomes larger than that of the ErNi solid, and the assumptions of the AMR model used to design this machine are not valid. Subsequent to the design of this cryogenic machine, a near room temperature AMR using liquid water for heat transfer has very successfully operated in a regime where the volumetric heat capacity of the fluid and the solid refrigerant are comparable.[4] The condition of two-phase flow that would occur when subcritical helium condenses would, however, be an uncommon range of operation for regenerators and may present difficulties for the operation of an AMR refrigerator. An alternate approach for the design of sub-4 K AMR refrigerators would be to use helium at supercritical pressure, although such pressures would not permit use of the present bellows design.[5]

Figure 1 shows the AMR refrigerator system. A stationary, conductively cooled, niobium-tin superconducting magnet that is specified for persistent mode operation up to 3 T magnetizes the beds. In order to compensate partially the large magnetic forces that act on the beds containing the magnetic material, the refrigerator uses two beds that are alternately magnetized and demagnetized. The beds reciprocate into the bore of the magnet, driven by an hydraulic actuator coupled to the beds through a bellows from room temperature by a low thermal conductivity shaft. Bed movement and heat transfer fluid flow through the beds is made possible through the use of flexible metal bellows. Stationary slide bearings guide the bellows motion. Heat transfer gas flows through these bellows to and from the beds. The use of bellows to facilitate bed motion limits the refrigerator's safe range of operating pressures to about 0.5 atm. In addition, heat generated at the slide bearings is thought to be major low temperature heat leak.

A metal bellows displacer drives the heat transfer gas. The two beds are in series. Fluid flow traces a path that begins at the hot end of the most recently demagnetized bed, passes through the bed and through the cold heat exchanger (CHEX) and finishes by traversing the magnetized bed and the hot heat exchanger (HHEX) to the displacer bellows. Bellows seal all moving elements internal to the refrigerator and between the refrigerator and the laboratory.

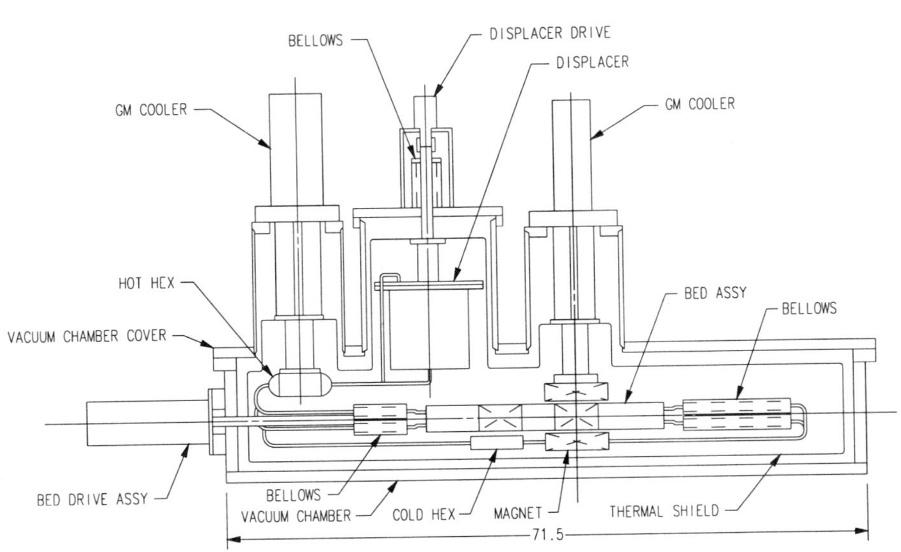

Figure 1. AMR refrigerator system.

The HHEX is coupled to a G-M cryocooler that operates between 9 K and 13 K depending on the operating mode of the AMR refrigerator. The magnet is cooled by a second cryocooler that operates at about 8 K.

INSTRUMENTATION AND TESTING

The test refrigerator configuration permitted control of the magnetic field, the system pressure, the operating frequency and the applied load. The operating frequency could be varied through varying the time for bed motion and displacer motion; these can be varied separately and independently.

Table 1 describes relevant system instrumentation. Care was taken to use sensors that are not sensitive to magnetic field. A computerized data acquisition system obtained the data. Sensors were read at a fixed time interval that varied from 6 seconds to two minutes.

In addition to the data-taking period, the data evince the different periodicities of the G-M refrigerator and the components of the AMR refrigerator cycle. Data analysis usually required averaging data over several cycles of these periodicities to extract equilibrium values.

Table 1. Performance testing instrumentation

Name	Type	Location	Comment
T4	Silicon Diode	Bed Bellows Bearing	Establishes slip bearing friction as a source of ambient heat leak
T5	Silicon Diode	G-M Cold Foot	AMR refrigerator sink temperature
T6	Silicon Diode	Displacer Bottom	Provides temperature of known volume of gas displaced during heat transfer fluid circulation
T14	Cernox2424	CHEX	AMR refrigerator source temperature
T13	Cernox2423	Bed, Cold Side	Gas temperature either immediately after it exits bed cold side & bellows or immediately after it passes through CHEX, depending on flow direction
T15	Cernox2425	Bed, Hot Side	Gas temperature after exiting bed & bellows or before entering bellows & bed.
P1	Psia Transducer	Displacer Bottom Exit	Operating pressure
P2	Psia Differential Transducer	Across Beds	Bed pressure drop

Limitations on the performance of the superconducting magnet's persistent mode switch also complicated testing. In general, we were unable to obtain data for fields greater than 2.25 T. When the magnet was in persistent mode, the field would decay by about 10% over an approximately 3-hour time interval. The magnet was recharged regularly to maintain consistent testing conditions.

RESULTS

Testing results are summarized in Figures 2,3 and 4. All load specifications refer to a load applied by a heater at the AMR refrigerator cold heat exchanger. Careful analysis of the zero load data suggests that the refrigerator also absorbs an additional load of order 1 W. A load of such magnitude exceeds that due to radiation and residual gas conduction and is thought to be due to friction in the moving bed system.

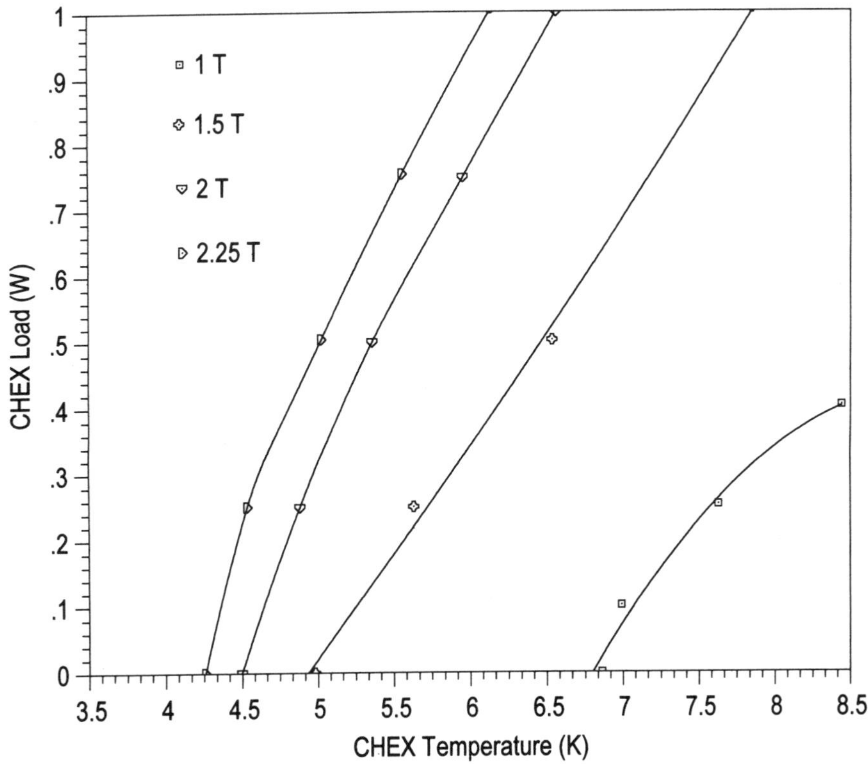

Figure 2. Load Curve. Cold heat exchanger temperature (CHEX) in response to load applied at CHEX for four different magnetic fields. Data are for 0.5 atm AMR refrigerator pressure and AMR period of 24 s.

Figure 2 shows the refrigerator's load curve for several different fields. Magnetic field dependence results from the AMR's magnetic material's adiabatic temperature change increase with field. The observed increase of cooling power with field is consistent with the increased adiabatic temperature change. In Figure 3, the zero load, minimum temperature exhibited by the helium heat exchange fluid as it exits the cold side of the beds is shown for varying refrigerator pressure and applied field. This minimum temperature decreases with applied field and operating pressure until conditions on the helium two-phase boundary are reached. Then, the helium follows the two-phase boundary with changes in field and pressure. We believe this is evidence of helium liquefaction in the AMR beds. Operation at the original design field of 3 T, and at a lower operating pressure of 0.25 atm, would have likely achieved the design temperature of 3.5 K.

Figure 4 shows the CHEX temperature variation with mass flow rate for the four field values used in the test with no applied load. Mass flow rate increases result in reduced CHEX temperature until flow rate of about 0.4 g/s to 0.5 g/s is reached. While at the higher fields and lower temperatures this behavior may be confounded with helium

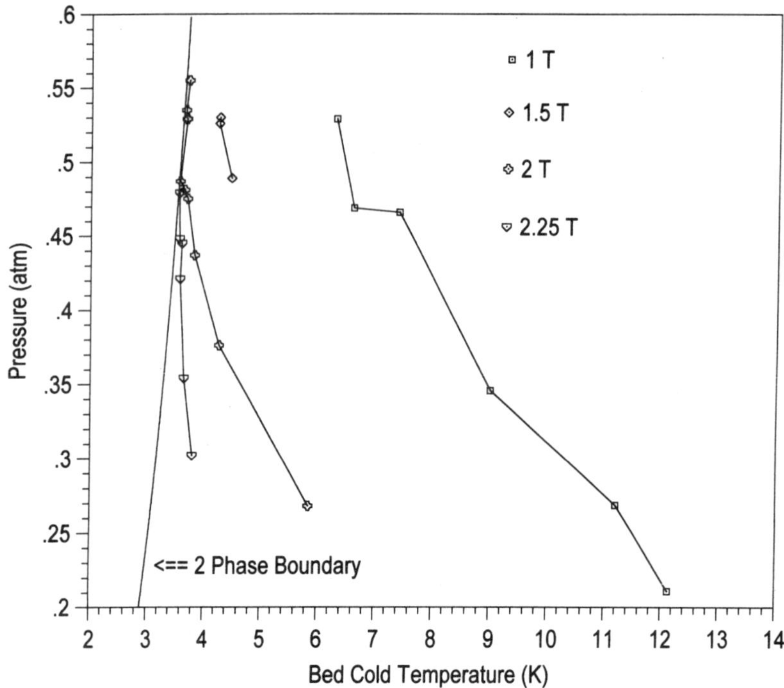

Figure 3. Minimum helium heat exchange fluid temperature for four different magnetic fields and varying refrigerator pressure. The helium two-phase boundary is shown and provides evidence of liquefaction in the AMR beds.

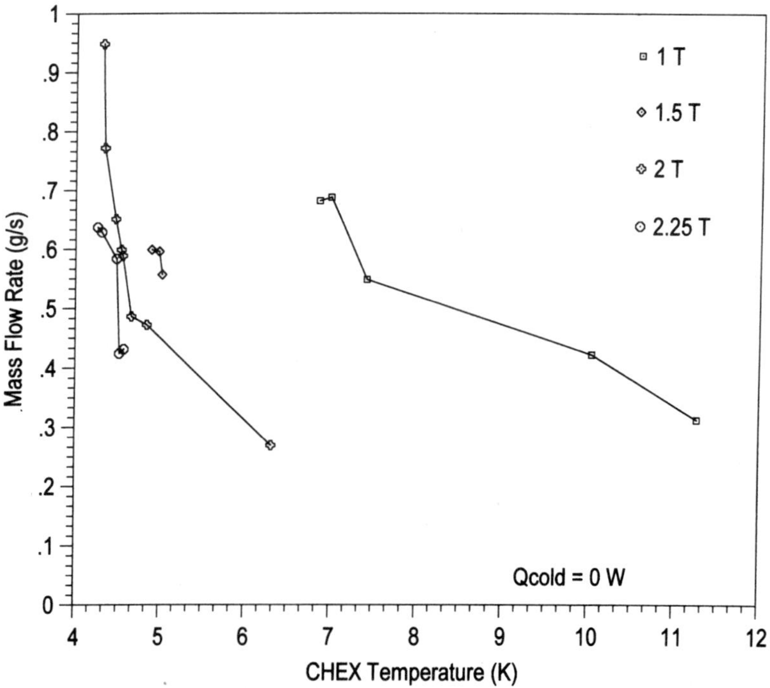

Figure 4. Cold heat exchanger temperature for zero load and varying mass flow rates. Mass flow rates vary through changes in refrigerator frequency and pressure.

liquefaction, the curve's "knee" occurs at temperatures well above the liquefaction temperature for lower fields.

We note that the differential pressure drop across the beds is relatively small, no greater than 5% of the system operating pressure. We also note that for periods between 16 and 24 s and fixed field, operating pressure and load, the cooling power exhibited a maximum at around 20 s.

These results can be considered from the perspective of the refrigerator modeling studies. Modeling predicts that as the refrigerator's cold end temperature increases, so also does the load that it can support. Figure 2 illustrates this feature of the AMR refrigerator. The model predicts that the cooling power varies almost linearly with mass flow rate at mass flow rates from 1 g/s to perhaps 3 g/s. Our studies do not achieve mass flow rates in this region. However, for the limited lesser mass flow rates (up to nearly 1 g/s) of these studies, the cooling power does exhibit a linear dependence on mass flow rate. The model also predicts that the refrigerator's cooling power is relatively insensitive to the hot sink temperature. The cooling power was relatively unaffected by hot sink temperatures changes between 9 K and 13 K. Above 13 K performance was affected significantly.

DISCUSSION

The successful refrigerator testing reported here provides operational data for the design of large, 4 K refrigeration systems for cooling superconducting magnet systems. This 4 K to 10 K refrigerator operates over the same temperature range as the lowest temperature stage of a four-stage AMR helium refrigerator that would operate between a heat sink at 80 K and a cold source, such as a 4 K superconducting magnet system. AMR refrigerators have operated over all the staging temperatures required and refrigerator test data using three of the four materials proposed for the four-stage device are available [2,8].

Modeling studies for a two stage AMR hydrogen liquefier system predicted an operating efficiency of 25 % of Carnot [9]. Related modeling studies for the four-stage helium refrigerator predict an efficiency that is lower than 25 % due to the extra stages. The previously described size and reliability advantages over conventional helium refrigerators remain.[5]

ACKNOWLEDGEMENTS

The testing reported here was supported jointly by the Astronautics Corporation of America, Milwaukee, Wisconsin and BWX Technologies, Inc., Lynchburg, Virginia. Design, manufacture and initial testing of this refrigerator received support from the U.S. Navy, Annapolis Detachment, Carderock Division, NSWC, Contract #N61331-93-C-0076.

REFERENCES

1. G. Green, J. Chafe, J. Stevens, and J. Humphrey, A gadolinium-terbium active regenerator, in: "Advances in Cryogenic Engineering", Vol 35, Plenum, New York (1990), p. 1165
2. A.A. Wang, J.W. Johnson, R.W. Niemi, A.A. Sternberg, and C.B. Zimm, Experimental results of an efficient active magnetic regenerator refrigerator, "Proceedings of the 8th International Cryocoolers Conference", Plenum, New York (1995), p.1165

3. C.B. Zimm, J.W. Johnson, and R.W. Murphy, Test results on a 50 K magnetic refrigerator, in: "Advances in Cryogenic Engineering", Vol 41, Plenum, New York, (1996), p. 1675
4. C.B. Zimm, A.G. Jastrab, A. Sternberg, V.K. Pecharsky, K.A. Gschneidner, Jr., et al, Description and Performance of a Near-Room Temperature Magnetic Refrigerator, in: "Advances in Cryogenic Engineering", Vol 43, Plenum, New York, (1998), p. 175
5. A.J. DeGregoria, J.A. Barclay, P.J. Claybaker, S.R. Jaeger, S.F. Kral et al, Preliminary design of a 100 W 1.8 K to 4.7 K regenerative magnetic refrigerator, in: "Advances in Cryogenic Engineering, Vol 35 (1990), p. 1125
6. C.B. Zimm, A.G. Jastrab, and J.W. Johnson, "Design of Active Magnetic Regenerative Stage Interfacing to a G-M Cryocooler", Proceedings of the 8th International Cryocooler Conference", Plenum, New York (1995), p. 65
7. J.W. Johnson and C.B. Zimm, Performance modeling of a 4 K active magnetic regenerative refrigerator, Journal of Applied Physics, 79:2171 (1996)
8. A.J. DeGregoria, L.J. Feuling, J.F. Laatsch, J.R. Rowe, J.R. Trueblood, et al, Test results of an active magnetic regenerative refrigerator, in "Advances in Cryogenic Engineering", Vol 37, Plenum, New York, (1992), p. 875
9. L. Zhang, A.J. DeGregoria, S.A. Sherif and T.N. Veziroglu, Optimization analysis on a two-stage AMR hydrogen liquefier, "Proceedings of the 7th International Cryocooler Conference", Phillips Laboratory, Kirtland AFB, (1992), p 586

MODELS FOR SCALABLE HELIUM-CARBON SORPTION CRYOCOOLERS

C. A. Lindensmith[1], M. Ahart[2], P. Bhandari[1], D. Crumb[3], C. Paine[1] and L. A. Wade[1]

[1]California Institute of Technology, Jet Propulsion Laboratory
Pasadena, CA 91109
[2]Princeton University
Princeton, NJ 08544
[3]Swales Aerospace
Pasadena, CA 91107

ABSTRACT

We have developed models for the rapid design of helium-based Joule-Thomson cryocoolers using charcoal-pumped sorption compressors. The models take as inputs the number of compressors, desired heat-lift, cold tip temperature, and available precooling temperature and provide design parameters, optimized for low power, as outputs. The output parameters are the compressor dimensions, required charcoal mass, cycle temperatures and pressures, heat exchanger dimensions and J-T orifice characteristics. We predict that a helium/carbon cryocooler with 18 K precooling can be as efficient as 60 W/W. These models, which run in MathCad and Microsoft Excel can be coupled to similar models for hydrogen sorption coolers to give designs for 2-stage vibration-free cryocoolers that provide cooling from ~50 K to 4 K. We present a preliminary design for a two-stage vibration-free cryocooler that is being proposed to support a mid-infrared camera on NASA's Next Generation Space Telescope.

INTRODUCTION

A number of space missions currently in development are expected to use detectors that require cooling to less than 10 K. These missions include NASA's Next Generation Space Telesope (NGST), Interstellar Probe, and Terrestrial Planet Finder (TPF), as well as ESA's FIRST/Planck and Darwin. At present there are only limited options for cooling the detectors on these missions. Stored cryogens are planned for use on the Far Infrared and Submillimeter Space Telescope (FIRST) and NASA's Space Infrared Telescope Facility (SIRTF), but provide limited lifetime and contribute a substantial mass, from both the cryogen and its storage system, that must be launched into space. The Planck mission uses a combination of hydrogen-sorption cryocoolers (to reach 18 to 20 K) and helium-based Joule-Thomson (JT) mechanical cryocooler to reach 5 K.

Although mechanical cryocoolers can achieve the desired low temperatures of these types of missions, they have several disadvantages. The mechanical compressors must be vibration-isolated from the instruments; JT coolers allow this isolation to be made on the warm side of the spacecraft, but most other types of cryocoolers must have the compressor located near the cold-head, making vibration isolation substantially more difficult. Mechanical coolers are also not very scalable—the mass and volume of a 10 mW, 4 K cooler are not very different from those of a 300 mW cooler[1]. Because many of these space missions will require cooling power on the order of ~10 mW, there is substantial inefficiency in using mechanical coolers.

We are developing sorption cryocoolers that are inherently vibration-free and whose mass and power scale with the required cooling power. This paper describes developments in the design models for 5-8 K low-power (10-100 mW) continuous coolers that use charcoal as the sorbent material and helium as the working gas. Periodic charcoal-helium based coolers have been used on the ground for many years[2,3], and have been demonstrated in space as well[4,5], but there has been limited development of continuous charcoal coolers[6]. These models will also form the basis for future designs of 1.5 K coolers. We will also describe a preliminary design for a charcoal-helium sorption cooler for a proposed mid-infrared camera for NGST.

BASIC PRINCIPLES

Charcoal-helium sorption coolers rely on the ability of charcoal to adsorb and retain substantial amounts of helium at temperatures above the liquefaction temperature. The helium-charcoal system obeys a Clapeyron-like saturated vapor-pressure relationship[7, 8]

$$\ln(P_s) = A + \frac{B}{T} \tag{1}$$

where A and B are constants and P_S is the effective saturated vapor pressure. At any given temperature and pressure, charcoal can adsorb some amount of helium onto its surface. The amount of helium adsorbed depends on the particular charcoal, and in the range that is useful for making sorption coolers can be described by the expression[9]

$$C = \frac{M\alpha}{b}\exp(-\gamma x) \tag{2}$$

where

$$x = \frac{RT}{\beta}\ln(\frac{P_s}{P})\,. \tag{3}$$

M is the molecular weight of helium, R is the gas constant, T is the temperature, P the total pressure, b specific volume of the adsorbed phase (taken to be the van der Waals volume), and α, β, and γ are experimentally determined parameters of the charcoal.

By changing the temperature, we can change the amount of helium stored in a compressor filled with charcoal. If we heat the charcoal, the helium is desorbed and can be driven through a JT expander to provide cooling. At the same time we must provide a charcoal-filled compressor depleted of helium and at a low temperature for the gas to adsorb onto after passing through the JT device.

DESIGN MODEL

We have developed a pair of complementary design models. The first model is a simplified cooler system, in which the precooling temperature (typically also the low compressor temperature) and the desired cooling power are the input parameters. The model then runs a program in Visual Basic and Microsoft Excel to determine the optimal high and low pressures in the compressors and the high temperature in the compressors by iteratively evaluating a simplified design. These parameters are then used in the more detailed MathCad-based model. We used MathCad for model development because it

Table 1. Power budget for a 6 K, 5 mW cooler with 18 K precooling

Item	Power (mW)
Charcoal heat capacity	25.4
Container heat capacity	81.8
Helium heat capacity	151.0
Heater heat capacity	7.0
Heat leak in gas-gap	12.2
Desorption	128.8
Void volume	22.2
Gas gap power	3.0
Total power	431.4

offers the general functions of a spreadsheet, including programmability, but displays mathematical relations in a relatively easy to read format, much like standard textbook mathematical style. This feature makes debugging and modification of the model easier, particularly as the model becomes more complex.

The ultimate performance of the model depends on the characteristics of the charcoal used in the compressors[6]. The sorption properties and the void fraction are the main charcoal parameters. The model presently uses values from Duband, as they comprise a set of data that can be reasonably described by Equations (1) and (2). Data from any other available charcoal can be readily substituted either as fitted equations, tabular values with a lookup procedure, or inserted by hand. The helium parameters (enthalpy, density, etc.) are obtained similarly from MathCad calls to Excel, which obtains the desired values from a GasPak dynamically linked library. The remainder of this section describes the subsections of the design model in the order in which they are evaluated.

Cooler Architecture

The basic design that we assume is that the cooler is composed of pairs of compressor elements, with the two elements connected via a bi-directional JT constriction and heat exchanger. A single pair of compressor elements would make up a periodic sorption cooler, in which the helium is transferred back and forth between the two compressors. In some circumstances the recycling time will be short compared to time constants at the cold head so that the cold head temperature will be sufficiently stable for the application and a single pair of compressors is all that is needed. Multiple pairs of compressors can be connected in parallel, with their heat exchangers and JT cold-heads tied together, and operated out of phase to smooth out the temperature fluctuations at the cold head. This arrangement also offers some redundancy in case of the failure of one of the compressor pairs or the plugging of one of the JT constrictions. Microfabricated valves could be used to simplify the plumbing of the system when they become available[10].

The compressor elements are located at the cold stage of the precooler, which should provide cooling to 12 to 20 K, and thermal conduction between them and the precooler is controlled by heat switches. The present design uses gas-gap heat switches, but mechanical suitable mechanical heat switches could be made from magnetostrictive materials. The gas from the hot, desorbing compressor is precooled back to the low compressor temperature before going to the cold counterflow heat exchanger. It may also be possible to improve the efficiency of the system by using the waste heat from the warm compressors to partially heat the cold compressors. Such a pairwise regeneration system could reduce the input power by 25 to 30%, which is less than the ~50% improvement possible in higher temperature systems[11] because the heat of desorption is comparable to the heat capacity of the cooler and is not recovered in the regeneration process.

Table 2. NGST Mid-IR cooler system properties for various 6 K cooling loads. The electronics mass is assumed to be 8 kg and consume 10 W for all cases

Heat lift at 6 K (W)	Charcoal input power (W) (at 18 K)	Charcoal sys. mass (kg)	Hydride input power (W) (at 270 K)	Hydride sys. mass (kg)	Total system power (W)	Total system mass (kg)	Passive cooling requirements (W) 35 K	270 K
0.005	0.43	0.56	66.4	12	76.8	20.6	0.44	66.4
0.007	0.58	0.71	72.9	12.7	83.5	21.4	0.59	72.9
0.010	0.81	0.95	82.6	13.7	93.4	22.6	0.82	82.6
0.014	1.12	1.27	95.6	15.1	106.7	24.4	1.13	95.6
0.015	1.20	1.34	98.9	15.4	110.1	24.7	1.21	98.9

Compressors

Given the enthalpy change, ΔH, removed by each unit mass of helium, determined by the compressor temperatures, cold-tip temperature and heat-exchanger efficiency, we can simply evaluate the required mass flow rate as the desired cooling power divided by ΔH. The cycle times for the compressors are determined by the user, and are typically chosen to be between 800 and 1200 s. Combining the cycle time and the mass flow rate gives the amount of helium that must be expelled from a single compressor during its desorption cycle. Additional gas is required to fill the dead volumes of the system and the void volume of the charcoal. Taking the total gas to be removed from the charcoal and relating it to the storage capacity given by Eq. (2), we can solve for the required charcoal mass.

To contain the charcoal and helium, we assume a cylindrical container with hemispherical endcaps and minimize the mass of constant-wall-thickness material to obtain the radius and length of the cylinder. The wall thickness of the container is then determined by the high pressure and safety margin required by the application. When determining the total volume to be enclosed we first solve the model with no heating element enclosed, and then repeat the calculation with the volume of a wire heater of appropriate power. It is important to keep the mass of container material as small as possible to reduce the heat capacity of the components that must be cycled between the minimum and maximum temperatures of the compressor elements. The power to drive the compressors is determined by the energy required to heat the compressors and the gas in them from the low, adsorption temperature to the high, desorption pressure and the heat of desorption of the helium-charcoal.

Gas-gap Heat Switch

Heat switches are required for coupling and uncoupling the compressor elements from the cold stage. It is desirable to have good heat switches to minimize the heat leak to the precooling system when the compressors are at their maximum temperature. Gas-gap heat switches have been demonstrated in space[5] and are ideal for this temperature range. Magnetostrictive materials may also provide suitable heat switches if the mass of the solenoid required to actuate the material can be made small enough.

In the design for the NGST coolers described below, the heat switches are integrated into the cold-head for the precooling stage. The cold-head is made from a block of high thermal conductivity material that has been bored out to provide individual cylindrical wells approximately 200 μm larger in diameter than the outer shells of the compressor elements. A fine capillary tube connects each gas-gap volume to its own charcoal pump. Because the volume of the gap is very small, the size of the charcoal pump for the heat switch is determined more by manufacturing limits than by the charcoal properties. We have assumed a switching ratio of 300 based on information in the literature[5].

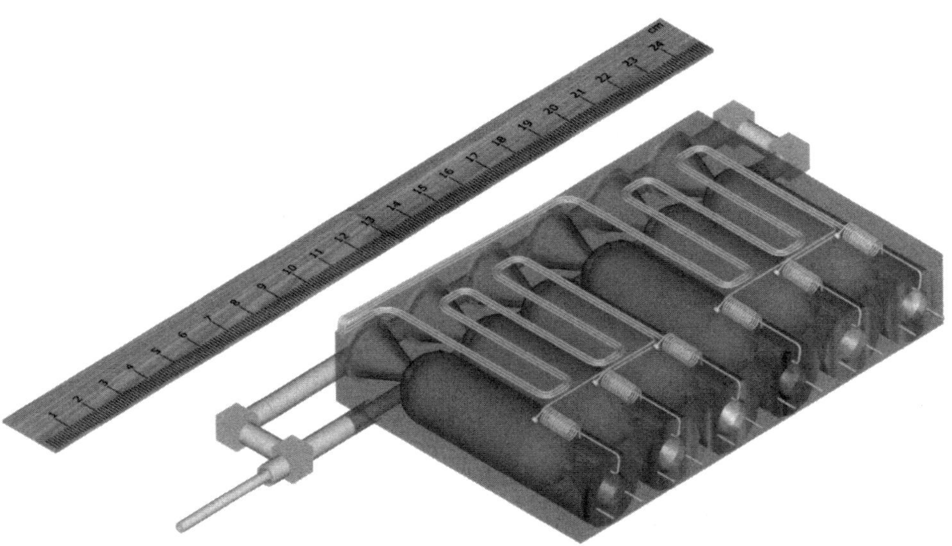

Figure 1. Concept rendering of charcoal compressors for an NGST mid-IR cryocooler located in the precooler cold head. The scale shows centimeters.

Counterflow Heat Exchanger

The tube-in-tube counterflow heat exchangers, located between the precooling stage and the low-temperature cold head, are all coupled together so that they provide a common heat exchanger for all of the gas lines. The mass of the lines is very small (less than 1 g per pair), so the mass of the heat exchanger is dominated by the structure used to couple the lines together. The mass of this structure is assumed to be 50 g. The diameters of the lines are selected[12] to keep the Reynolds number below 2000 and keep the pressure drop to a user determined value (typically ~1%). The wall thickness is then determined by the high pressure in the lines, the elastic modulus of the material and the required safety factor of the application. In many cases the required wall thickness is less than can be economically manufactured, so a nominal minimum wall thickness of 50.8 μm is set by the model. Capillary tubing of ~150 to 200 μm and ~50 to 75 μm wall thickness can be readily obtained in stainless steel and other materials, such as cupronickel alloy.

Summary of Expected Performance

As an example of the output from the pair of models, we assume a cooler with 5 mW of cooling power at 6 K and a precooling stage of 18 K, as provided by a hydrogen-sorption cooler. The high (desorption) temperature determined by the Excel model is 80 K, and the low and high pressures in the compressor elements are 0.23 MPa and 2.5 MPa, respectively. The input power required by the cooler is the sum of the power required to heat the charcoal, the helium, the charcoal and the heater, plus the power for desorption and the power for the gas-gap heat switches. The power to desorb gas into the lines is treated separately and added to the system power. All of these power requirements are summarized in Table 1, which gives a breakdown of the power input due to each component of the system, giving an efficiency of 86 W/W. The mass of this system will be 0.43 kg. The efficiency could be improved to 60 W/W by a pairwise regeneration scheme for recycling waste heat from the compressors.

COOLER FOR NGST MID INFRARED CAMERA

A cooler system designed to provide base temperatures of 6 to 8 K will consist of two stages – a precooler to reach ~18 K and the final low-temperature stage. The precooler will require precooling of the working fluid to less than ~60 K, which will be provided by passive radiation. For the NGST ISIM cooler, the passive precooling temperature is expected to be 35 to 40 K. In this section, we present a design for a two-stage continuous sorption cooler system capable of achieving 6 K. The precooling stage is assumed to be a hydrogen sorption cooler, similar to the Planck cooler described elsewhere in these proceedings. The coldest stage is a charcoal sorption cooler, as described above.

The heat load on the 6 K stage is expected to be less than 15 mW, depending on the number of detector arrays used in the MIR focal plane and their characteristics. Each array will dissipate approximately 1 mW, with parasitic heat leak from lead wires (20/array, plus 16 for temperature monitoring and control) contributing an additional 0.03 mW/array. Heat leak through the supporting structure will also be small, assumed to be 0.16 mW independent of the number of arrays. We expect that ten 1024x1024 pixel arrays would require ~13.5 mW of heat lift at 6 K (including parasitic heat loads). In order to provide 10 mW of heat lift, which would be sufficient for 7 detector arrays, the complete cooler system is expected to be 15 kg, requiring a total input power of 93.4 W, with a cooling efficiency of 9340 watts input power per watt of heat lift at 6 K. Table 2 shows the mass and power requirements of each of the stages, as well as the complete system, for a range of cooling powers at 6 K.

Figure 1 shows a semitransparent rendering of the precooler cold-head and charcoal sorption compressor system. The block for the cold head is 13.8 x 9.2 x 2.8 cm. The compressor elements are inserted in the block and isolated from it by the conical standoffs on the ends. The standoffs also serve to locate the compressor elements so that the walls of the compressor are concentric with and separated from the walls of the cold head by approximately 100 μm. Each compressor has a gas input/output line and pair of heater leads penetrating the gas-gap and the pressure cylinder. There is a small sorption pump (located outside the cold-head) for connection to each gas-gap. The large tube penetrating the cold-head along the longer side is the gas supply/return for the hydrogen JT cooler system.

CONCLUSIONS

We have developed detailed models for the design of charcoal-helium based vibration-free Joule-Thomson cryocoolers. The simpler Excel based model allows us to quickly choose design parameters for an efficient cooler given the design constraints, and the MathCad model provides detailed calculations to determine sizing of the various parts of the system, including the charcoal mass, the compressor housing, the counterflow heat exchanger, and the gas-gap heat switch. The calculations can be done quickly and the model can be readily modified to match design changes as they arise. As we begin hardware development we will compare the predictions to the observed results and modify the model as necessary. We have used these models to develop a crycooler concept for a mid-infrared camera instrument on the Next Generation Space Telescope.

ACKNOWLEDGMENT

The research described in this paper was carried out by the Jet Propulsion Laboratory, California Institute of Technology, under a contract with the National Aeronautics and Space Administration.

REFERENCES

1. D.S. Glaister, M. Donabedian, D.G.T. Curran, and T. Davis, An overview of the performance and maturity of long life cryocoolers for space applications, in "Cryocoolers 10", R. Ross, ed., Plenum Press, New York (1999) p.1.
2. G. Seidel and P.H. Keeson, *Rev. Sci. Inst.* **29**, 606 (1958).
3. G.K. White, "Experimental Techniques in Low Temperature Physics" Oxford University Press, Oxford, 1993).
4. J.J. Bock, L. Duband, M. Kawada, H. Matsuhara, T. Matsumoto, and A.E. Lange, ^{4}He refrigerator for space, *Cryogenics* **34**, 635 (1994).
5. L. Duband, L. Hui, and A. Lange, Space-borne ^{3}He refrigerator, *Cryogenics* **30**, 263 (1990).
6. L. Duband, Ph.D. Thesis "Etude et realisation d'une machine frigorifique a cycle de Joule-Thomson utilisant un compresseur thermique a adsorption." L'Universite scientifique, technologique, et medicale de Grenoble, Grenoble, France (1987).
7. L. Duband, A. Ravex, and J. Chaussy, Adsorption isotherms of helium on activated charcoal, *Cryogenics*, **27** 397 (1987).
8. L. Duband, J. Chaussy, and A. Ravex, High pressure isotherms of helium on activated charcoal, in "Advances in Cryogenic Engineering", Vol. 33, Plenum Press, New York (1987) p. 1023.
9. M.M. Dubinin, The potential theory of adsorption of gases and vapors for adsorbents with energetically nonuniform surfaces, *Chem. Rev.* **60**, 235 (1959).
10. J.F. Burger, M.C. van der Wekken, E. Berenschot, H.J. Holland, H.J.M. ter Brake, H. Rogalla, J.G.E. Gardeniers, and M. Elwenspoek, High pressure check valve for application in a miniature cryogenic sorption cooler, *preprint*, (1999).
11. S. Bard and J. Jones, Regenerative sorption compressors for cryogenic refrigeration, in "Advances in Cryogenic Engineering",Vol. 35, Plenum Press, New York (1989) p. 1357.
12. R.F. Barron, "Cryogenic Systems" Oxford University Press, New York, (1985).

DESIGN OF A NEW TYPE REGENERATOR

Y. Ishizaki,[1] E. Ishizaki,[1] H.R. Mueller,[1]
T. Ohtsuka,[1] and K. Hamaguchi,[2]

[1]ECTI
241 Yamanouchi, Kamakura 247-0062 Japan
[2]Dept. of Mechanical Engineering,
Meisei University, Hino,
Tokyo 191-8506 Japan

ABSTRACT

We report a textile type regenerator we have developed where three materials, ceramic fiber, stainless steel wire and fine thread are interwoven (textile type) and compared it with a stainless steel wire mesh screen regenerator (mesh type). A ceramic fiber about 0.03mmϕ with a very low thermal conductivity and high elasticity is used as threads running crosswise from room to low temperatures its length being about the same as the regenerator (80~200mm). The warp consists of very long stainless steel wires (0.02mm~0.03mmϕ) that act as the regenerator. The wires are rigidly fixed and insulated from each other by fine threads (e.g. Kevlar or Polyimide resin) of 0.01mm~0.03mmϕ.

A textile machine was used to interweave the three materials into a cloth. The cloth was wound around a Teflon stem or the pulse tube about the same length as the regenerator in a multi-spiral form (several hundred layers) to form the textile type regenerator.

In comparison to a 300 mesh type, the heat transfer area is large and from the geometry of the fluid path, the pressure loss is expected to be low. A textile type regenerator B which has a simpler structure and may be easier to construct will also be reported.

INTRODUCTION

The regenerator is the key component important in determining the efficiency of a regenerative cryocooler. Ideally[1] a regenerator which maintains the large temperature between room temperature and the cold end requires the following properties:

1) High heat transfer area. 2) Minimal pressure drop loss. 3) High heat capacity.
4) Minimal dead volume. 5) Low axial heat conduction. In practice, it is required that
6) the quality of mass produced products must not differ, 7) the matrix should be designed so that fluid vibration does not occur due to change in fluid velocity,

Advances in Cryogenic Engineering, Volume 45.
Edited by Shu *et al.*, Kluwer Academic / Plenum Publishers, 2000.

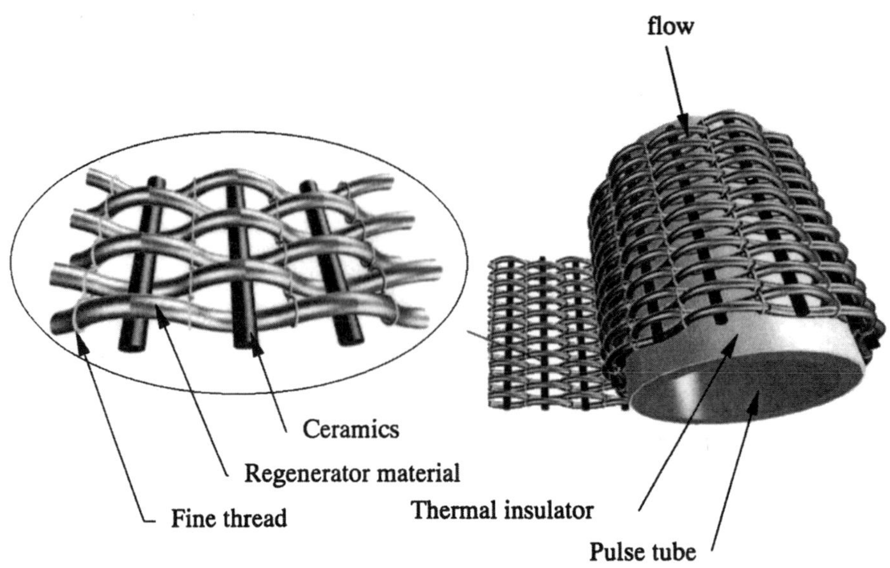

Fig.1 Outline of the structure A

8) minimal age deterioration, 9) reasonable price.
We have studied the structure of a regenerator which fulfills these requirements and have compared the properties with a wire mesh screen (mesh type) regenerator.

STRUCTURE

An ideal structure of a regenerator has been reported by R. Yaron et. al.[2]
Our approach differs in that we have examined a unique structure that may be produced at a low price by utilizing the thermal, mechanical and chemical properties of materials used widely in industry.

Fig.1 shows a sketch of a structure of the Textile Type Regenerator in which the pulse tube acts as a mandrel. The textile structure which is composed basically of ceramics, stainless steel wire and fine thread, is inserted in the cylinder to compose the regenerator. Zirconia or Tyranno fibers[3] which are elastic materials with very low thermal conductivity and high hardness is used as the ceramic filament which have about same length as the regenerator. These ceramics have been chosen as their thermal and mechanical properties may be controlled in their production process and, in particular, may be readily processed to have very low heat conductivity (about 1/25 of stainless steel). A very long stainless steel wire (304L or related materials) which acts as the regenerator material is entwined on the ceramics to compose a path for the flow of the working fluid helium. We have already produced 23K with a single stage pulse tube refrigerator[5] using several hundred discs of stainless steel mesh and lead spheres.

In this textile, however, the regenerator material is not restricted to stainless steel wire and any material such as bronze, lead plated copper wire which have large specific heat in the working temperature region may be used and we expect similar results with our new textile type regenerator using different wire materials. The fiber constitutes the weft and the wire the warp of the textile. Kevlar or fine thread of resin which do not deteriorate at low temperature are inserted as the weft of the textile to prevent contact between the multitude of wires, axial heat conduction and also the motion of the wires.

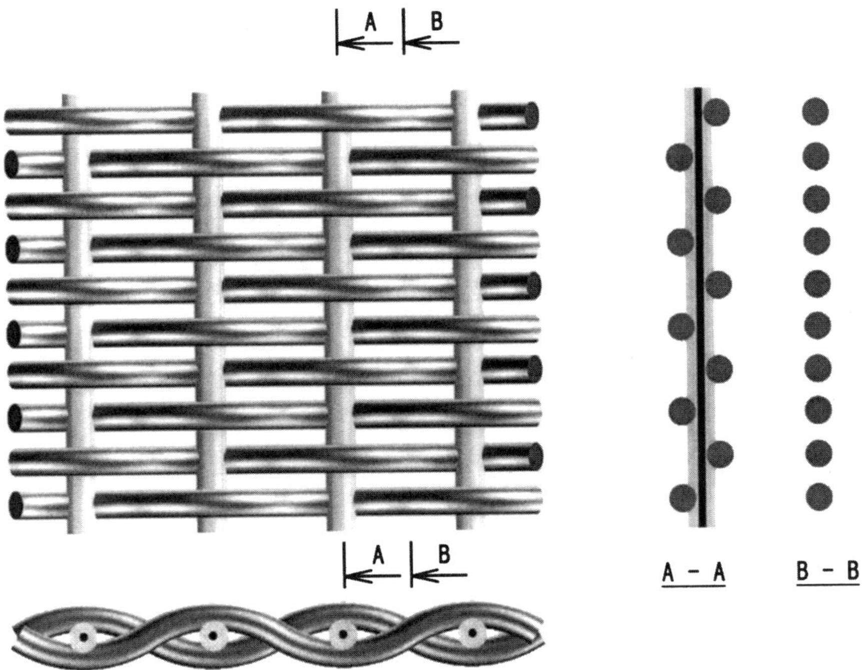

Fig.2 Outline of the structure B

We have also studied a textile type regenerator B shown in Fig.2 which has simpler structure and may be easier to construct. Hard ceramics with low heat conductivity together with other filaments constitute the central axis and instead of ceramic filaments shown in Fig.1, a two component composite 0.03~0.05mmϕ coated with Teflon, epoxy, polyethylene, polyimide, polypropylene, silicon or other resins are used. The resin will melt or soften when hot-pressed to the stainless steel wires during the weaving process (the hot-press process may also be performed after the structure is completed) and the wires will be rigidly retained and, as in the interwoven fine thread structure discussed before, there will be no contact between the stainless steel wires.

The textile type A and B shown in Figs.1 and 2 are wound around a Teflon mandrel slit into 4 sections in a multi-spiral form as shown in Fig.3 to form the regenerator.
The matrices which form the flow path of He are (a), (b) shown in Fig.4 or of intermediate structure.

DISCUSSION

Assuming the fluid path structure and dimensions shown in Fig.4, expressions for the void volume ratio φ, specific heat transfer surface normalized by the unit volume σ, and the open area ratio β, defined as the ratio of the minimum flow passage area within the area $(L_1 \times L_2)$ perpendicular to the flow direction, have been derived.
Preliminary numerical calculations based on these formulas indicate that the heat transfer area is larger and the pressure loss somewhat lower than a mesh type regenerator (300 mesh) between room temperature and 20K. The results will be reported after experimental confirmation.

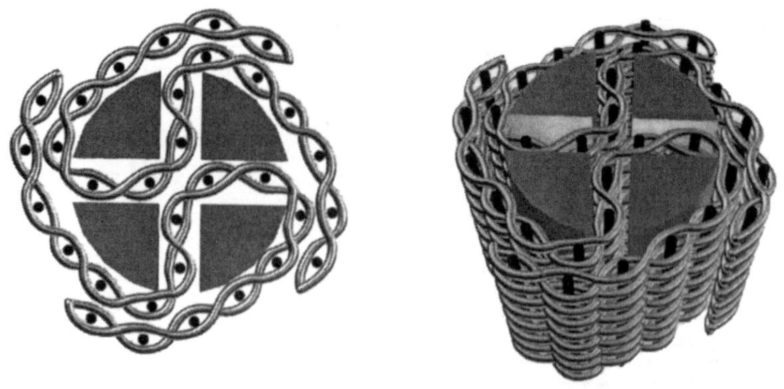

Fig.3 Structure of the regenerator

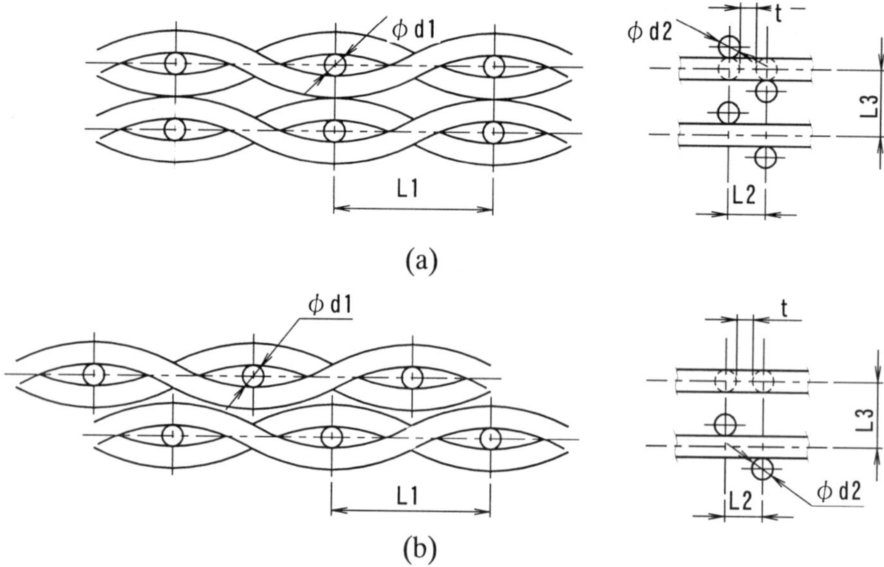

Fig.4 The fluid flows through the opening provided by the ceramic fiber inserted in the stainless steel windings

REFERENCES

1. Robert A. Ackermann. "Cryogenic Regenerative Heat Exchangers," Plenum Press, New York (1997)
2. R. Yaron, S. Shokralla, J. Yuan, P.E. Bradley and R. Radebaugh. "Etched Foil Regenerator," Advances in Cryogenic Engineering, Vol. 41B, Plenum Press, New York (1995)
3. Tyranno Fibers: (Si:50%,Ti:2%,C:30%,O:15%)
4. Professor S. Nishijima. ISIR, Osaka University, "Private Communication".
5. Y. Ishizaki & E. Ishizaki "Experimental Performance of Modified Pulse Tube Refrigerator below 80K down 23K," International Cryocooler Conference. Santa Fe (Nov. 17-19, 1992)

A REGENERATOR THAT WILL PERFORM AT MODERATELY HIGH FREQUENCY AND BELOW 10 KELVIN FOR USE IN A PULSE TUBE COOLER

J.M. Lee[1], A. Kashani[2], and B. Helvensteijn[2]

[1]NASA Ames Research Center
Moffett Field, CA 94035

[2]Atlas Scientific
Sunnyvale, CA 94086

ABSTRACT

We are in the process of demonstrating a 45 Hz pulse tube cooler that will achieve temperatures below 10 K. Of primary importance is the low temperature stage of the regenerator. Two difficulties arise when constructing a regenerator that will perform satisfactorily at moderately high frequency (40 Hz to 70 Hz) and below 10 Kelvin. The first is matching the thermal penetration between the working gas and the regenerator material. The second is avoiding large pressure drops. In this work we address both problems using Neodymium ribbon as the regenerator material. Neodymium has adequate heat capacity at low temperature and sufficient ductility to work into ribbon form. These properties can be used to address the problem of matched thermal penetration and large pressure drops. This paper discusses the construction of a Neodymium ribbon stacked regenerator and presents modeling results for its use in a 45 Hz pulse tube cooler.

INTRODUCTION

The purpose of the regenerator in a cryocooler is to enable the working gas to traverse large temperature gradients and to dampen local temperature oscillations. The ability of a regenerator to dampen temperature oscillations depends on efficient heat exchange between the regenerator material and the gas. The internal geometry of a regenerator is typically that of a porous matrix, such as a packed bed of spherical powder or stacked screens, which allows good thermal contact between the solid material and the gas. Parallel plates are another suitable matrix that has the advantage of low pressure drop. The regenerator should also be made from material with a high heat capacity throughout the expected temperature range so that the local gas temperatures remain nearly constant.

Features that are desirable in a cryocooler include compactness and high efficiency. In order to reduce overall system size, coolers are being developed to operate at high frequency (40 Hz to 70 Hz). High frequency operation results in small cooler systems with correspondingly small compressors and regenerators. The need for temperatures less than 10 K is also desirable. New regenerator materials with high heat capacity at these temperatures are being used in coolers running at low frequency (<3 Hz). Such coolers have even demonstrated continuous operation at liquid helium temperatures.

A significant obstacle for realizing 10 K coolers that operate at high frequency is the low effectiveness of regenerators. At high frequency and low temperature, the heat exchange between the oscillating gas and the regenerator material is limited due to the finite thermal diffusivities of the gas and the solid. Consequently, the gas and solid spacing perpendicular to the mass flow should be small. This increases the difficulty of regenerator construction and of achieving low flow resistance. This paper discusses our approach and our current progress in trying to solve this problem.

SIMPLE REGENERATOR DESIGN ANALYSIS

Assuming the oscillating flow in the regenerator matrix to be laminar, the heat transfer between the gas and the matrix takes place by thermal diffusion. Therefore, the diffusion terms in the differential energy equations can be used to describe the local thermal transport. By linearizing these equations the local characteristic length scales for the gas and the matrix may be determined. For the case of a parallel plate design, there are four relevant length scales: the half distance of the plate thickness, l; the half distance of the spacing between plates, d; the thermal penetration into the plate $(\alpha_r/\omega)^{1/2}$; the thermal penetration in the gas, $(\alpha_g/\omega)^{1/2}$, where α_r and α_g are the thermal diffusivities of the regenerator material and the gas, respectively, and ω is the angular frequency. The length scales are shown in Figure 1.

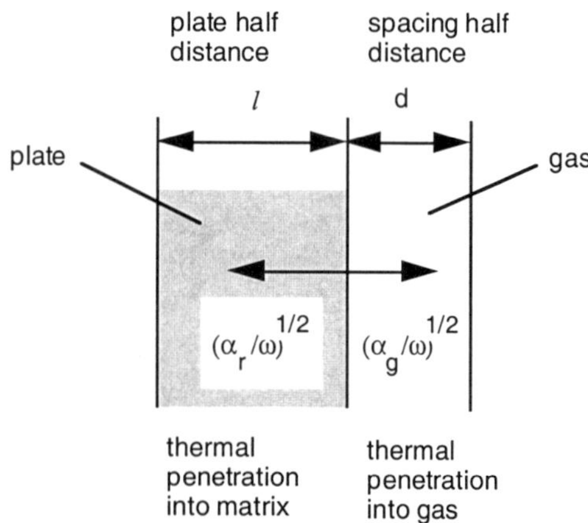

Figure 1. Length scales used for estimating local dimensions of a parallel plate regenerator matrix.

The system can also be characterized in terms of three time scales: the period of oscillation ($2\pi/\omega$); the time required by a thermal wave to propagate a distance l into the plate; and the time required by the damped thermal wave to penetrate a distance d into the working gas.

The aspects of time, length and thermal diffusivity can be collectively considered by using dimensionless numbers. The relevant dimensionless numbers are obtained from the governing energy equations. Assuming an ideal gas with constant thermal physical properties, the linearized energy equations applicable to the parallel plate geometry (two-dimensional rectangular coordinate system) are:

$$\frac{1}{\gamma}\frac{\partial p}{\partial t}+\frac{\partial(p\upsilon)}{\partial y}+\frac{\partial(pu)}{\partial z}=\frac{1}{\mathrm{Pr\,Va}}\frac{\partial^2 T}{\partial y^2} \tag{1}$$

$$\frac{\partial\theta}{\partial t}=\frac{1}{\mathrm{F}}\frac{\partial^2\theta}{\partial y^2} \tag{2}$$

where Eq. (1) applies to the gas and Eq. (2) applies to the plate. The two equations are coupled through the interface boundary conditions. In these equations t is the time, z is the lengthwise coordinate along the plate, y is the transverse coordinate, u is the lengthwise velocity, υ is the transverse velocity, p is the pressure, T is the gas temperature, θ is the plate temperature, and γ is the heat capacity ratio. For the resulting dimensionless numbers, Pr is the Prandtl number, Va is the Valensi number, and F is the inverse Fourier number. The Valensi number is an oscillating Reynolds number and comes from the linearized momentum equation,

$$\frac{\partial(\rho u)}{\partial t}=-\frac{\partial p}{\partial z}+\frac{1}{\mathrm{Va}}\frac{\partial^2 u}{\partial y^2} \tag{3}$$

where ρ is the density. The dimensionless numbers are defined as $\mathrm{PrVa}=d^2\omega/\alpha_g$, $\mathrm{F}=l^2\omega/\alpha_r$, and $\mathrm{Va}=d^2\omega/\nu$, where ν is the kinematic viscosity. The dimensionless numbers represent ratios of time scales or alternately, ratios of length scales. In practice one normally takes $\mathrm{PrVa}\leq 1$ and $\mathrm{F}\leq 1$. However, more detailed analysis (e.g. using a numerical modeling program) is needed to correlate temperature oscillations with mass flow, to assess losses associated with pressure gradients, and to determine thermal penetration depths. The ratio of F to PrVa is a useful parameter in determining the local thermal penetration time scale of the plate relative to the thermal penetration time scale of the gas,

$$\frac{\mathrm{F}}{\mathrm{Pr\,Va}}=\frac{l^2}{\mathrm{d}^2}\frac{\alpha_g}{\alpha_r} \tag{4}$$

As an initial estimate, this ratio is of order 1 so that the time scales for thermal penetration through the plate and the gas are about the same.

For coolers designed for relatively high frequency and low temperature, the small thermal diffusivity of helium requires a small spacing between plates. By comparison, the thermal diffusivities of a number of high heat capacity rare earth compounds (e.g. Er_3Ni), are quite large so that for Eq. (4) with F/(PrVa)=1, the l:d ratios may be as high as 5. The flat plate geometry, preferred for its low flow resistance, may be used to obtain such a high l:d ratio. However, trying to obtain this geometry is difficult with available rare earth compounds because they are oftentimes brittle and difficult to form into thin flat plates. When plates cannot be formed, a satisfactory alternative may be to employ spherical particles. A close packed spherical powder has an *l*:d ratio that is about 5. However, limited control over particle size tends to cause the powder to settle in a less orderly fashion than the close packed arrangement resulting on average in a smaller *l*:d ratio. This creates a potential for powder motility during operation of the cooler

REGENERATOR SPECIFICS

At present our approach is to employ Neodymium (Nd) ribbon as the regenerative matrix. Neodymium has adequate heat capacity at low temperature and sufficient ductility to work into ribbon form. The ribbon we consider is supplied by Concurrent Technologies Corporation[1] and is similar to that previously reported on by Chafe and Green[2]. Figure 2 plots the heat capacity of Nd and compares this to lead for the 4 K to 15 K range.

Figure 3 illustrates the temperature and frequency dependence of *l* for a Nd flat plate and d for helium at 10 atm. The plot has been constructed for PrVa=1 and F=1. At a frequency of 45 Hz and 12 K, Fig. 3 gives d~0.02 mm and *l*~0.18 mm. This is an *l*:d ratio of about 10:1. A comparable analysis for a flat plate of Er_3Ni whose thermal diffusivity is about 4 times less than Nd results in an *l*:d ratio of about 5:1. Unfortunately, Er_3Ni is not available in flat plate form; it is available in spherical powder form. Even though spherical powder packed beds do have an *l*:d ratio of about 5:1, the advantages of low flow resistance using parallel plates are not realized with Er_3Ni spheres.

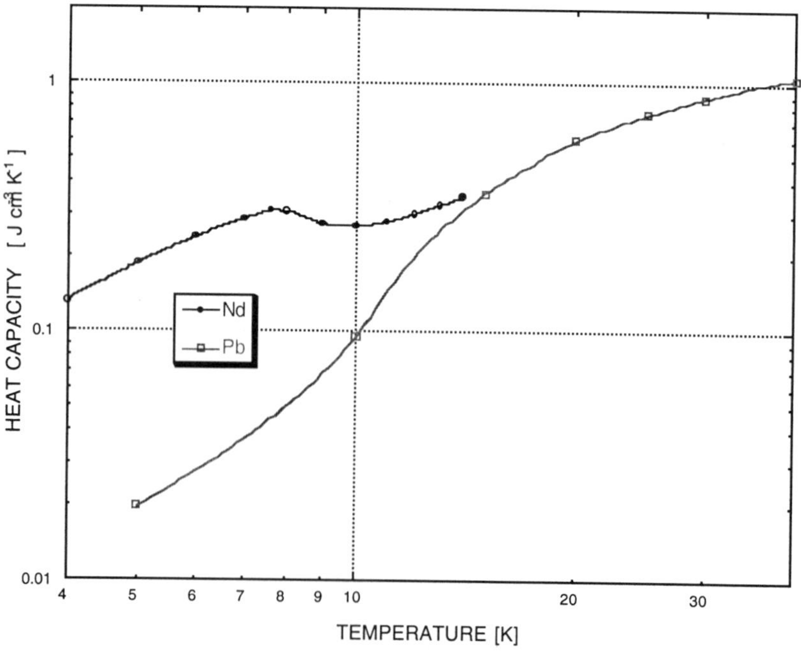

Figure 2. Heat capacities of Neodymium and Lead.

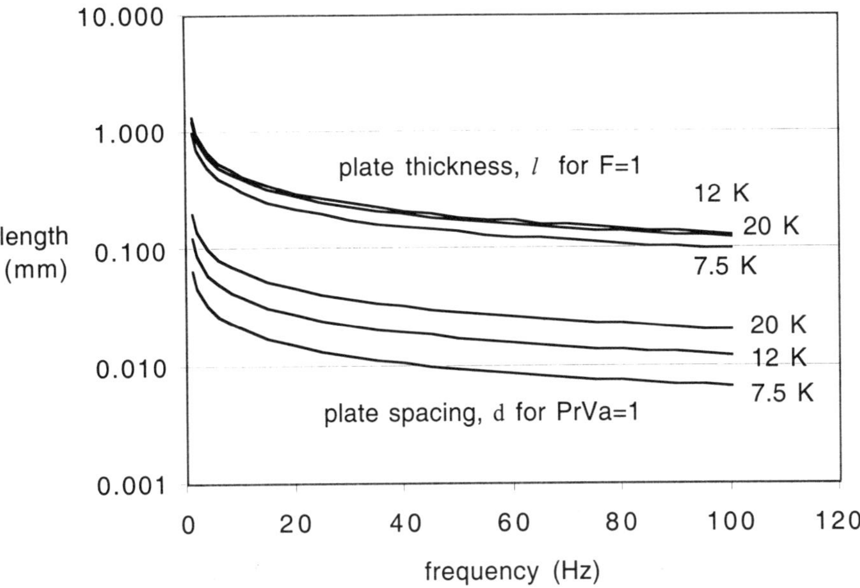

Figure 3. Characteristic lengths of plate thickness, d, and plate spacing, *l*, for helium gas and neodymium.

A sample of the Nd ribbon is shown in figure 4. The ribbon is composed of a 0.175 mm thick layer of Nd, wrapped in a 0.025 mm Niobium blanket, then clad in a 0.025 mm layer of Copper. The ribbon is 3.4 mm wide and has ridges 0.025 mm high, spaced 1.6 mm apart.

Figure 4. Neodymium ribbon. Dimensions: width, 3.4 mm; ridge height, 0.025 mm; ridge spacing 1.6 mm. The ribbon is composed of a 0.175 mm layer of Neodymium wrapped in a 0.025 mm Niobium blanket then clad in a 0.025 mm layer of copper.

To form an effective regenerator with a parallel plate matrix (parallel stacked-ribbon arrangement) it is necessary to maintain the plates with precise and uniform spacing. Attempts at winding the ribbon around a mandrel have proven less than satisfactory because of the difficulty in obtaining uniformity during winding, and because of the tendency of the coil to unwind after it is wound. In addition, the mandrel does not contribute to the regenerative effect and is a source of conduction losses.

We have taken a somewhat different approach in constructing a parallel plate arrangement with the ribbon. Our approach has been to stack the ribbon flat, layer-upon-layer, and then to cut circular plugs from the stack. Before the plugs are cut, the stack is bonded together with an adhesive, which is later removed. Plugs are cut from the stack using a water jet. Water jets produce little shear and so are ideally suited for cutting the bonded ribbon stack without de-lamination. An attempt to cut the stack with a diamond core drill bit proved unsatisfactory because of the shear stresses associated with the cutting and grinding. After the plugs are cut they are inserted into a tube and the adhesive is removed. At present, we are in the process of determining the correct water jet settings for a clean cut. Sample cuts are being conducted on an all copper ribbon test stack. Figure 4 shows a sample water jet cut. Each ribbon shown in Fig. 4 is 0.25 mm thick. The thickness of the plug is 4.8 mm and the plug diameter is 12.7 mm. Upon determining the required settings for a clean and precise cut we will proceed with the actual Nd ribbon stack.

MODELING

The modeling program SAGE was employed to assess the performance of the parallel-plate Nd regenerator in a pulse tube cooler. The objective was to optimize the efficiency of a single stage cooler operating from 20 K down to 7 K. In the modeled regenerator the Nd plates extend the length of the regenerator without any interruptions thus assuming the maximum possible conduction heat load to the cold end. An inertance tube is used to enhance the efficiency of the cooler. Real gas properties are used in all the cases modeled. The reservoir is maintained at 20 K, i.e., the temperature of gas entering the regenerator. Charge pressure, hot and cold end temperatures, frequency, and the plate

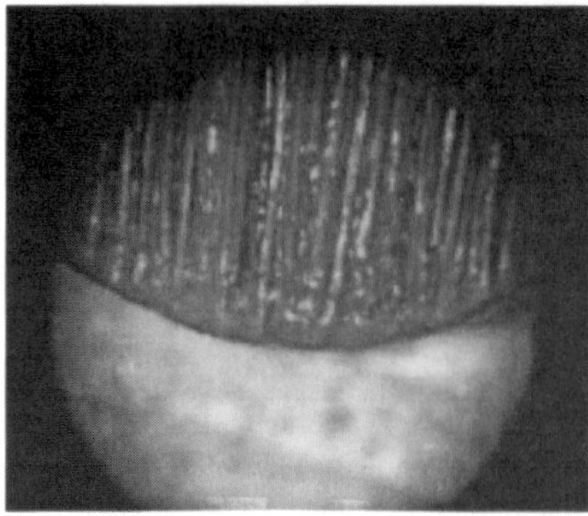
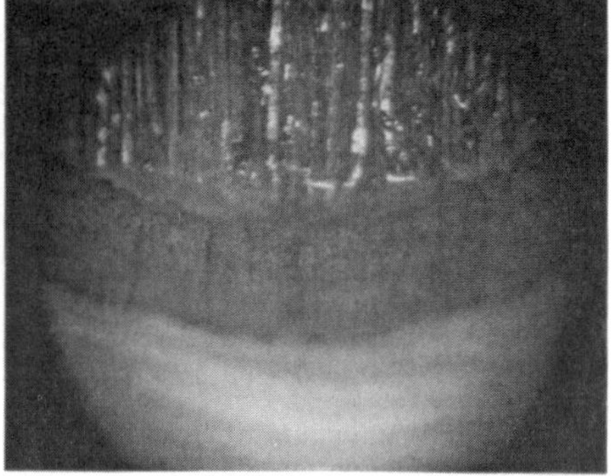

Figure 5. Copper stack cut with a water jet. Each ribbon is 0.25 mm thick. The thickness of the plug is 4.8 mm and the plug diameter is 12.7 mm.

thickness and spacing were constrained. The regenerator length and diameter, pulse tube length and diameter, inertance tube length and diameter, PV power, cooling power, and pressure drop were obtained from the optimization. In the first model run the plate thickness and spacing are taken to be those of the Nd ribbon fabricated as part of this study. The results are presented in Table 1 with a cooler efficiency of 1.2%.

A second model run was performed in which the plate thickness and spacing were optimized resulting in a cooler efficiency of 7.7%. The optimum plate thickness and spacing were 0.031 and 0.006 mm, respectively. Note that the *l*:d ratio for this case is nearly 5 and is of the same order as that of the simple scaling analysis. Fabricating a stacked Nd regenerator with such small dimensions is a difficult task, however. Increasing the plate spacing and thickness eases fabrication at the cost of reducing efficiency.

For comparison a regenerator packed with Nd spherical powder was also modeled. The optimum powder size was 0.029 mm yielding a cooler efficiency of 3.4%. Obtaining such a small Nd powder is difficult. Further, since Nd readily oxidizes and can not be coated in powder form it would have to be kept in an inert atmosphere at all times to prevent oxidation.

CONCLUSIONS

A technique has been devised to fabricate a stacked Nd ribbon regenerator for use in high frequency pulse tube coolers to achieve temperatures below 10 K. Neodymium ribbon is predicted to yield higher efficiencies than Nd powder, and can be clad in Copper to reduce oxidation. The modeling program SAGE was employed to validate the improvements that may be attained using Nd in parallel-plate geometries.

Table 1. Parameters for the 7 to 20 K pulse tube cooler

Parameter	Value
Frequency	45 Hz
Ambient pressure	20 bar
Hot end temperature	20 K
Cold end temperature	7 K
Regenerator length	10 cm
Regenerator diameter	4 cm
Foil thickness	0.178 mm
Foil spacing	0.025 mm
Pulse tube length	1.75 cm
Pulse tube diameter	4.3cm
Inertance tube length	101 cm
Inertance tube diameter	0.65 cm
PV power	8.19 W
Cooling power	0.096 W
Heat conduction through Nd plates	0.51 W
Pressure drop in regenerator	0.043 bar

REFERENCES

1. Concurrent Technologies Corporation, Johnstown, PA.
2. J.N. Chafe and G.F. Green, Neodymium-ribbon-regenerator cooling performance in a two-stage Gifford-McMahon refrigerator, Adv. Cryo. Eng., 43B:1589 (1998).

REPLACEMENT OF REGENERATOR WITH A STACK IN A STIRLING REFRIGERATOR

Shaowei Zhu[1] and Yoichi Matsubara[2]

[1]Second Development Department, AISIN SEIKI Co., Ltd.
2-1, Asahi-machi, Kariya, Aichi, 448-8650 Japan
[2]Atomic Energy Research Institute, Nihon University
7-24-1, Narashinodai, Funabashi, Chiba, 274-8501 Japan

ABSTRACT

The stack in an acoustic cooler or engine, and the stack as regenerator in Stirling refrigerator are discussed. The critical point of the stack is described by a critical void volume ratio. The COP or efficiency of the stack in the acoustic cooler or engine mainly depends on the void volume ratio. Near the critical void volume ratio, the COP is the maximum in acoustic cooler and in Stirling refrigerator. With limited heat transfer area, the efficiency of the stack can be over 80 % of the percent Carnot near the critical point in a Stirling refrigerator with limited input power when the temperature ratio is less than two.

INTRODUCTION

A stack is known as a component of a thermal acoustic engine or cooler, which convert the heat flow to the work flow or vice versa[1]. If we use a compressor piston to input the power at the hot end of the stack of an ordinary thermal acoustic cooler, and use an expander instead of the resonator, the thermal acoustic cooler becomes a Stirling refrigerator. The hydrodynamic diameter of the stack is larger than that of the regenerator. It allows higher operation frequency such as 50-500 Hz with smaller pressure drop and higher tolerance against the working gas impurity. This Stirling refrigerator is similar to the pulse tube refrigerator introduced by reference[2]. The enthalpy flow along the stack is negative in acoustic cooler. The enthalpy flow along the stack is possible negative or zero in this type Stirling refrigerator because the enthalpy flow direction does not suddenly change to positive if the input power at the hot end of the stack is gradually increased. Because the stack is the component of the acoustic cooler or engine, the stack for the acoustic cooler or engine is first analyzed by the nodal analysis method. Then, the work flow at the hot end of the stack is increased to analyze the stack working as the regenerator to know the regenerative performance. Finally, a Stirling refrigerator with a stack as regenerator is analyzed by a third order nodal analysis program[3] eliminated inertance term.

NUMERICAL METHOD

Figure 1 is a basic model of the Stirling refrigerator in which the regenerator is replaced by the stack. It includes a compressor, a hot heat exchanger, a stack, a cold heat exchanger, and an expander.

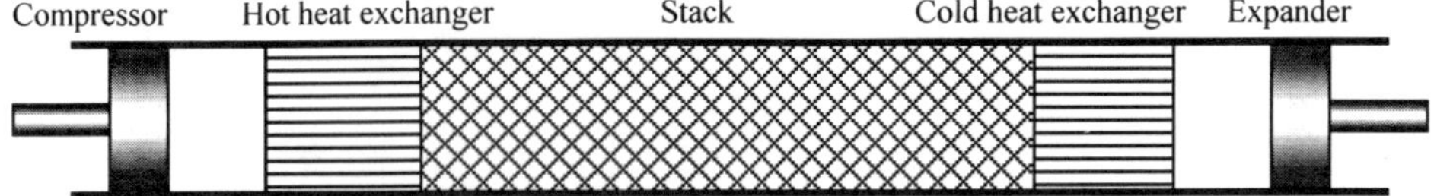

Figure 1 Stack as regenerator for Stirling refrigerator

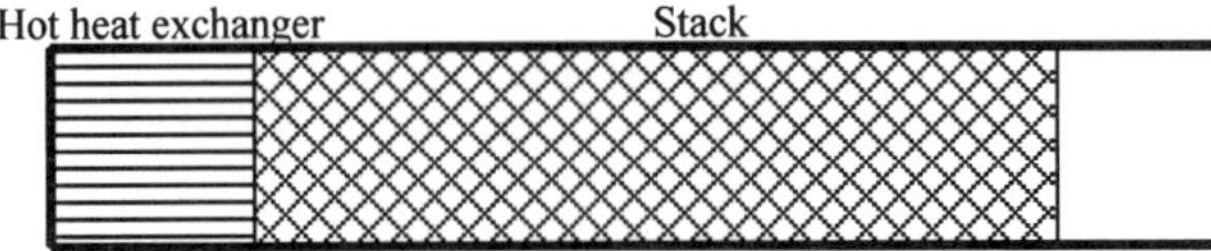

Figure 2 Calculation model

In order to simplify the discussion, we only discuss the stack. The calculation model is shown in figure 2. The void volume of the hot heat exchanger generates mass flow component which is in 90 degrees phase angle with pressure. The input power of the compressor is represented by a mass flow component which is in phase with pressure.

Three assumptions are made. 1. Ideal gas. 2. Temperatures of the gas in the compressor, the hot heat exchanger, the cold heat exchanger and the expander are constant. 3. The total length of the hot heat exchanger, the stack and the cold heat exchanger are very short compare to the sound wave length. The friction loss is neglected. The pressure in the hot heat exchanger, the stack and the cold heat exchanger are the same. It is sinusoidal.

The control equations of the stack are as follows.

Energy equation for the gas

$$AC_V \frac{\partial(\rho T)}{\partial t} + C_P \frac{\partial}{\partial x}(\dot{m}T) + hS(T - T_S) = 0 \tag{1}$$

A is the gas flow area, C_V is the constant volume specific heat of the gas, C_P is the constant pressure specific heat of the gas, ρ is the density of the gas, T is the temperature of the gas, t is the time, $\dot{m}$ is the mass flow rate, x is the coordinate along the stack, h is the heat transfer coefficient, S is the heat transfer area per unit length, T_S is the temperature of the matrix.

Energy equation for the matrix

$$A_S \rho_S C_S \frac{\partial T_S}{\partial t} + hS(T_S - T) = 0 \tag{2}$$

A_S is the matrix area along the regenerator, ρ_S is the density of the matrix, C_S is the specific heat of the matrix.

Continuity equation

$$A\frac{\partial \rho}{\partial t} + \frac{\partial \dot{m}}{\partial x} = 0 \tag{3}$$

Nodal analysis method is used to solve the above equations.

The pressure and the mass flow rate at the hot end of the stack are

$$P = P_0 + P_a \cos(\omega t) \tag{4}$$

$$\dot{m}_H = \dot{m}_{C0} \cos(\omega t) + V_H P_a \omega / (RT_H) \sin(\omega t) \tag{5}$$

Here, P_0 is average pressure, P_a is pressure amplitude. $\dot{m}_{C0}$ is the amplitude of mass flow component which is in phase with pressure wave, V_H is the void volume of hot heat exchanger, R is gas constant, ω is angle frequency, T_H is the temperature in the hot heat exchanger. The first term of Eq.(5) is generated by the compressor, which is called as progressive wave, and the second term is generated by the void volume of hot heat exchanger, which is called as standing wave.

The enthalpy flow rate through the stack is $H=\oint \dot{m}C_PTdt/\tau$, the work flow at the cold end of the stack is $W_C=\oint \dot{m}_CRT_C\ln Pdt/\tau$, which is the exergy flow under the assumption (2), here T_C is the temperature of cold heat exchanger, $\dot{m}_C$ is the mass flow rate at the cold end of the stack. The work flow at the hot end of the stack is $W_H=\oint \dot{m}_HRT_H\ln Pdt/\tau$. The heat flow at the cold end of the stack is $Q=H-W_C$. If the stack is in a cooler, heat flow is cooling capacity, its symbol is negative, its COP is $COP=(H-W_C)/(W_C-W_H)$, its Percent Carnot is $Carnot=100COP(T_H-T_C)/T_C$. If the stack is in an engine, the heat flow is discharged heat, and its efficiency *is* $\eta=(W_C-W_H)/H$.

There are six non-dimensional parameters which are pressure ratio $r=(1+a)/(1-a)$, temperature ratio $T_r=T_H/T_C$, void volume ratio $V_r=V_H/V_R$, heat transfer area $\Lambda=T_HhSL/(T_C \dot{m}_0C_P)$, heat capacity ratio $\Gamma=\dot{m}_0C_P\tau/(C_SA_SL\rho_S)$, and $\varepsilon=\dot{m}_{C0}/\dot{m}_0$. Here $a=P_a/P_0$, V_R is the void volume within the stack, L is the length of the stack, $\dot{m}_0=V_RP_0\omega/(RT_C)$ which is reference mass flow rate. It is convenient to use the normalized enthalpy flow $H/(C_P\dot{m}_0T_C)$, normalized work flow is $W_C/(C_P\dot{m}_0T_C)$, and normalized heat flow is $Q/(C_P\dot{m}_0T_C)$. Here, the void volume ratio is an important parameter. It represents the rate of the standing wave at the hot end of the stack, ε represents the work flow or progressive wave, or input power at the hot end of the stack. When ε changes from zero to some value, the acoustic cooler becomes a Stirling refrigerator. This is easy to see from the normalized mass flow rate at the hot end of the stack

$$\dot{m}_{HN}=\varepsilon\cos(\omega t)+V_ra/T_r\sin(\omega t) \quad (6)$$

STACK IN ACOUSTIC ENGINE OR COOLER

Stack for thermal acoustic engine or cooler is analyzed. In this condition, there are only standing wave at the hot end of the stack, ε=0 in Eq. (6). The work flow at the hot end of the stack W_H is zero. The work is added or taken out by the expander. The pressure is generated by the expander which is similar to the resonator in thermal acoustic engine. Helium gas is working medium.

Γ represents the heat capacity of the matrix. It has little effect if the heat capacity is enough. We choose it as 0.1 for the following calculation.

Figure 3 shows heat transfer area Λ effect. The temperature ratio T_r=1.2 and 2.0, pressure ratio r=1.2, the void volume ratio V_r=4, 3, and 1. When the heat transfer area is sufficiently large or sufficiently small, the enthalpy flow is small. Near Λ=1, the enthalpy flow value becomes maximum. At other temperature ratio, void volume ratio, and pressure ratio, the tendency is the same, and the peak enthalpy flow point is near Λ=1, too.

Figure 4 shows the enthalpy flow H, work flow at the cold end of the stack W_C, heat flow Q, and heat flux between the gas and matrix during the half cycle Qgs which is scaled by 0.1 to put in the figure. The temperature ratio T_r=1.2, the pressure ratio r=1.2, heat transfer area Λ=1. At point V_r=1.9, the normalized enthalpy flow, work flow, heat flow, and heat flux between gas and matrix become near zero. If the void volume ratio is larger than

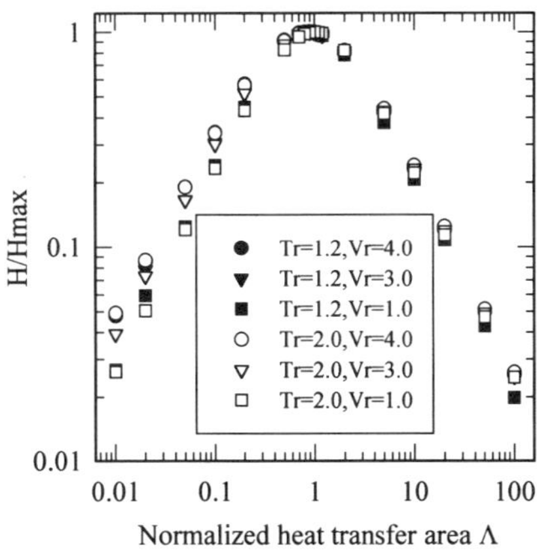

Figure 3 Maximum enthalpy flow position

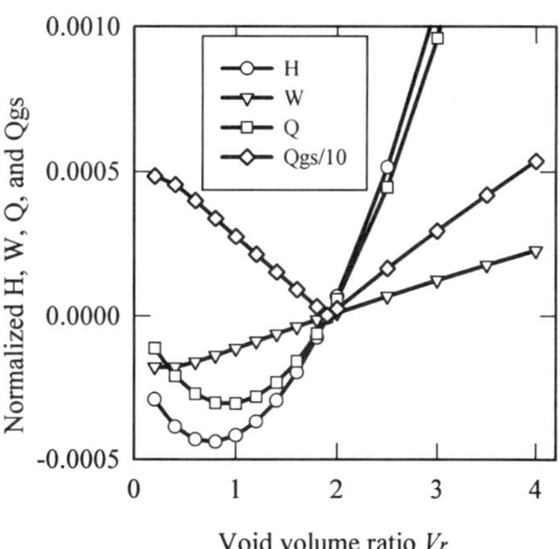

Figure 4 Void volume ratio effect

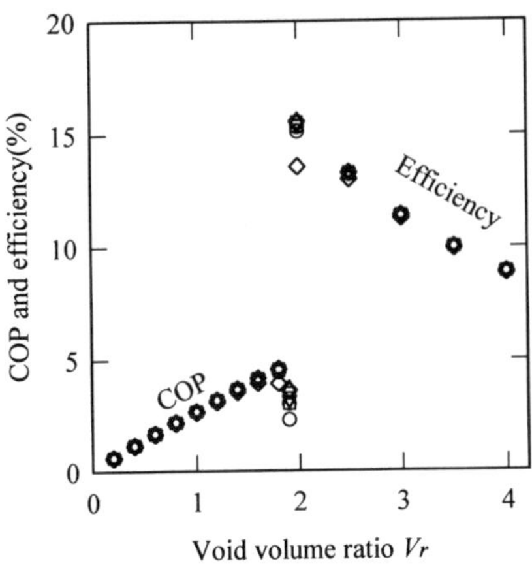

Figure 5 Efficiency and COP

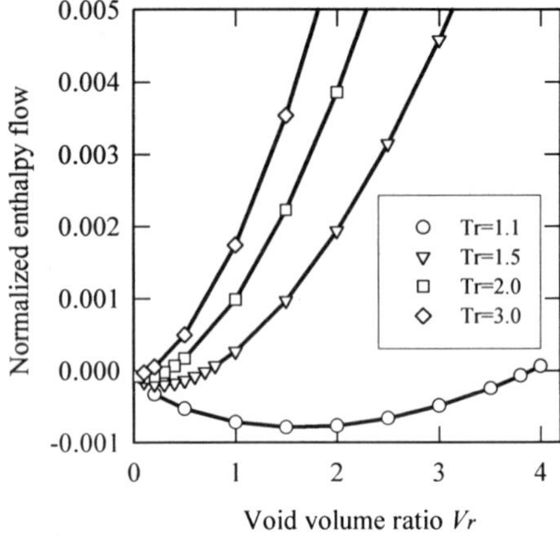

Figure 6 Temperature ratio effect

1.9, the enthalpy flow, work flow and heat flow is positive, and therefore the stack becomes an acoustic engine. If the void volume ratio is less then 1.9, the enthalpy flow, work flow and heat flow are negative, and the stack becomes an acoustic cooler. The relation between work flow and void volume ratio is near linear, and the relation between Qgs and void volume ratio is also near linear. When the void volume ratio is near 0.75, the enthalpy flow becomes to its minimum which means that the value of the negative enthalpy flow is maximum. When the void volume ratio is near 0.9, the cooling capacity which we use negative symbol is near maximum. Though the enthalpy flow, work flow and heat flow change with pressure ratio and heat transfer area, the tendency is same as figure 3, the zero enthalpy flow point, the maximum cooling power point, and minimum enthalpy flow point are still near V_r =1.9, 1, and 0.75, respectively.

Figure 5 shows the COP or efficiency changes with void volume ratio at Λ=0.5, 1.0, 1.5, 10.0 with r=1.2, and r=1.1,1.2,1.3,1.4,1.5 with Λ=1. Almost all the points are on one line, which means that the COP or efficiency mainly depends on the void volume ratio. Near V_r =1.9, the COP or efficiency becomes a maximum though the power density is near zero. Zero enthalpy point is a very important parameter. Here, we define that the critical void volume ratio V_{r0} is the void volume ratio when the enthalpy flow is zero at Λ=1 and r=1.2. At temperature ratio T_r=1.2, the critical void volume ratio V_{r0}=1.9.

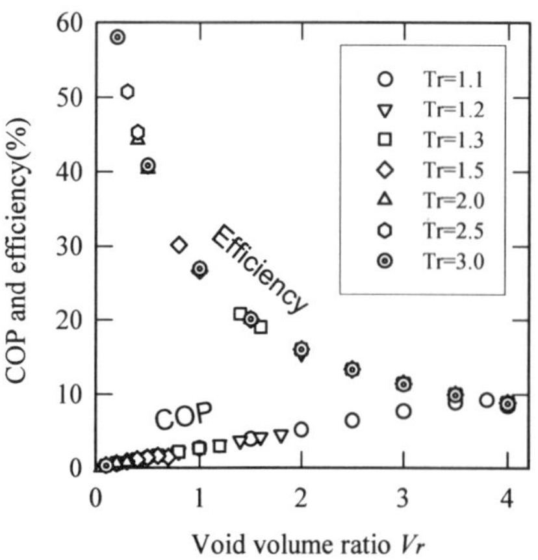

Figure 7 Efficiency and COP at different temperature ratio

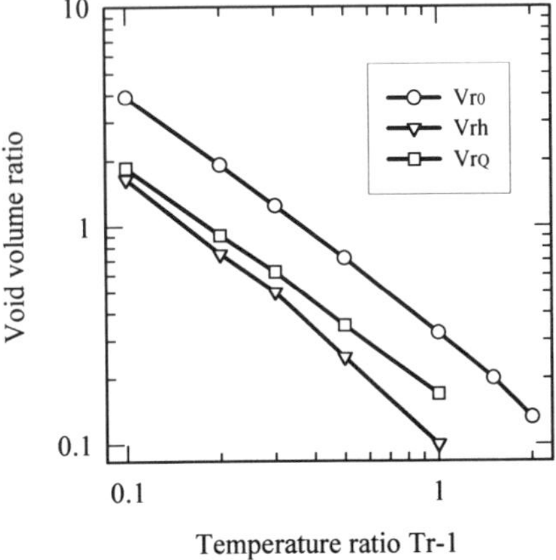

Figure 8 Relation between critical void volume ratio and temperature ratio.

Figure 6 shows enthalpy flow changes with void volume ratio at different temperature ratio T_r=1.1, 1.5, 2.0, and 3.0, respectively. The critical void volume ratio decreases with the increase of temperature ratio. The peak of the negative enthalpy flow quickly decrease with the increase of temperature ratio, the positive enthalpy flow rapidly increase with the increase of temperature ratio. The work flow, heat flow, and the heat flux between the gas and the matrix are similar with that in figure 4. Figure 7 shows that the COP and efficiency at different temperature ratio are on the same line. So the efficiency and COP mainly depends on the void volume ratio. The relation between COP, efficiency and void volume ratio are near $COP=0.1676+2.4V_r$ and $\eta=66.62(V_r+1)^{-1.276}$, respectively. This is very useful for the design of acoustic engine or cooler. Figure 8 shows the relation between temperature ratio and critical void volume ratio, it is near $V_{r0}=0.3119(T_r-1)^{-1.122}$. Another two lines V_{rh} and V_{rQ} are the minimum enthalpy flow and the maximum cooling capacity point. The left of the line V_{r0} is the acoustic cooler, the right of the line V_{r0} is the acoustic engine. On the line, the acoustic cooler gets highest COP with near zero power density.

STACK IN STIRLING REFRIGERATOR

An acoustic cooler can be considered as a special Stirling refrigerator with zero swept volume of compressor, which gives high COP and low cooling capacity near the critical void volume ratio. The possible high COP point for stack as regenerator is estimated near the critical void volume ratio because COP does not suddenly decrease when ε increase from zero. When ε increases from zero, the compressor size is increased from zero, progressive wave at the hot end of the stack increases from zero, and the acoustic cooler becomes a Stirling cooler. Helium gas is the working medium.

Figure 9-11 shows percent Carnot, cooling capacity, enthalpy flow rate, and heat flux between gas and matrix during the half cycle. Temperature ratio T_r=1.2, pressure ratio r=1.5, and heat transfer area Λ=2. When ε increase from 0.01 to 0.2, the maximum percent Carnot, minimum enthalpy flow and maximum cooling capacity are given near V_r=1.9, 0.75, and 1, respectively. The heat flux between the gas and matrix becomes a minimum at V_r=1.9. This is similar to that in acoustic cooler. Near V_r=0.75, the enthalpy flow gradually changes from negative to positive when ε increases. This is one of the interesting phenomena.

Fig.12 shows the Percent Carnot at heat transfer area Λ=20, pressure ratio r=1.5, temperature ratio T_r=1.2. Compare figure 9 and 12, the percent Carnot increases when the heat transfer area increases. Figure 13 shows percent Carnot at pressure ratio r=1.3, Λ=2, T_r =1.2. Compare figure 9 and 13, the percent Carnot increase when the pressure ratio increases near critical void volume. Though the COP changes with the heat transfer area and pressure ratio, the peak position is almost not changed. For a given temperature ratio, pressure ratio, and heat transfer area, the percent Carnot becomes very low if the input power is too large. Here we call the input power as upper limit input power when the percent Carnot is 80 %.

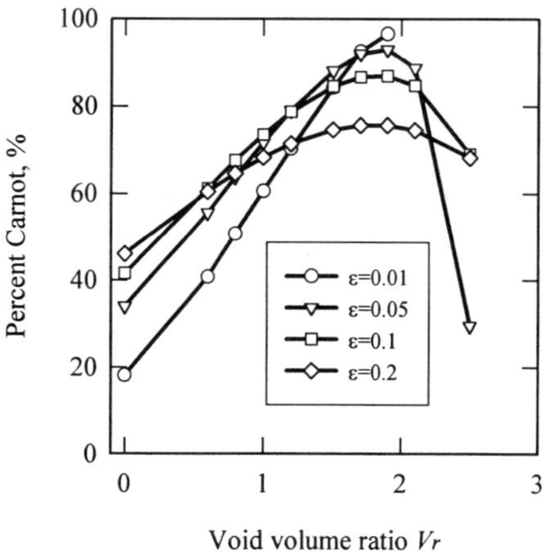

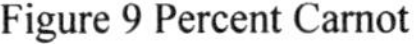
Figure 9 Percent Carnot

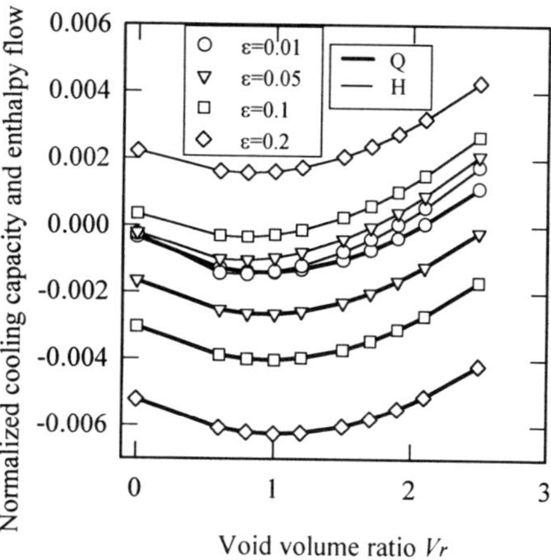

Figure 10 Enthalpy flow and cooling capacity

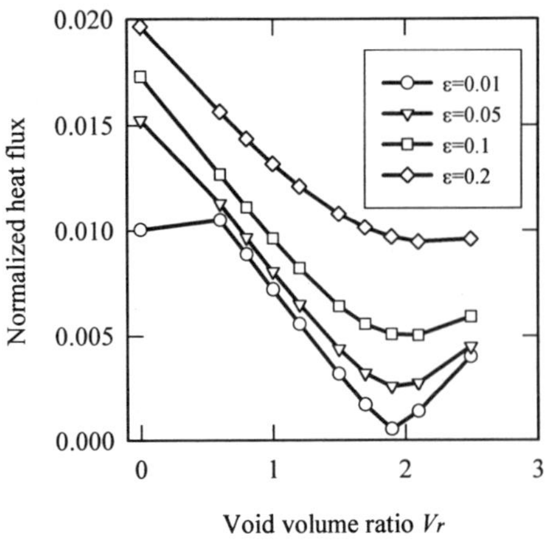

Figure 11 Heat flux between gas and matrix

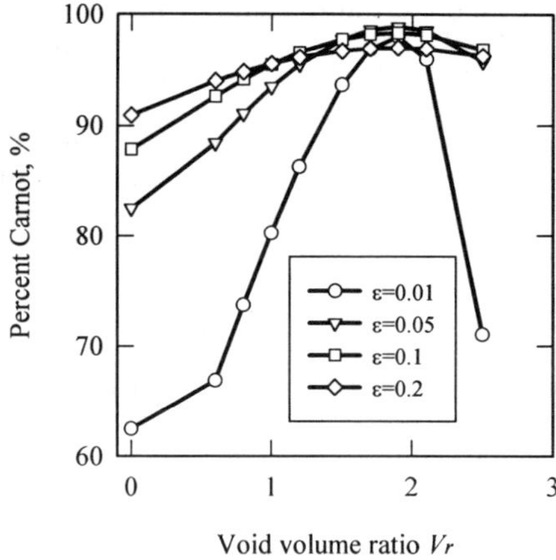

Figure 12 Percent Carnot at Λ=20

Figure 14 shows the percent Carnot at temperature ratio1.5 and 2, respectively. The pressure ratio and heat transfer area is r=1.5, Λ=2. In order to compare, the results of a regenerator with temperature ratio T_r=4 r=1.5 and Λ=100 are also shown in the figure. The peak point at temperature ratio T_r=1.5 and 2 is near void volume ratio 0.7 and 0.3, respectively, which are the critical void volume ratio at temperature ratio 1.5 and 2, respectively, as shown in figure 7. When Λ is the same, the upper limit input power at temperature ratio 2 is rather small compare to that at temperature ratio 1.2. In this figure, negative void volume ratio is used. The negative void volume ratio means the negative standing wave, which can be easily seen from Eq.(6). If void volume ratio is positive, the second term of Eq.(6) can be generated by a real heat exchanger. If it is negative, it can be generated by the compressor and expander with suitable phase angle. At temperature ratio 4, the peak percent Carnot is given at V_r=0. V_r=0 can be considered as a critical void volume ratio at temperature ratio 4. The minimum enthalpy flow and maximum cooling capacity point at temperature ratio 1.5 and 2 are also on the lines in figure 8. So the peak COP, minimum enthalpy flow, and maximum cooling capacity point is same in acoustic cooler and Stirling refrigerator.

In the above discussion, the volume of the compressor and expander can be gotten by the mass flow rate at both ends of the stack. It is depends on the temperature ratio, pressure ratio, and void volume ratio.

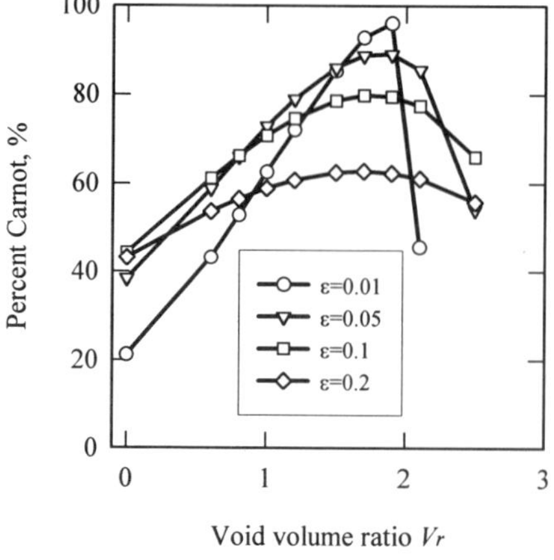

Figure 13 Percent Carnot at r=1.3

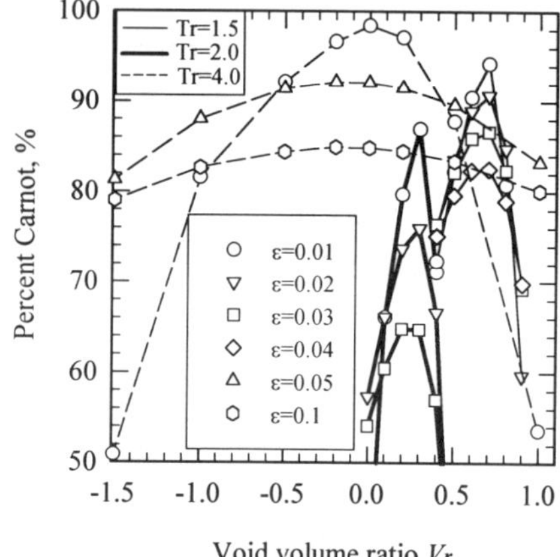

Figure 14 Percent Carnot at different temperature

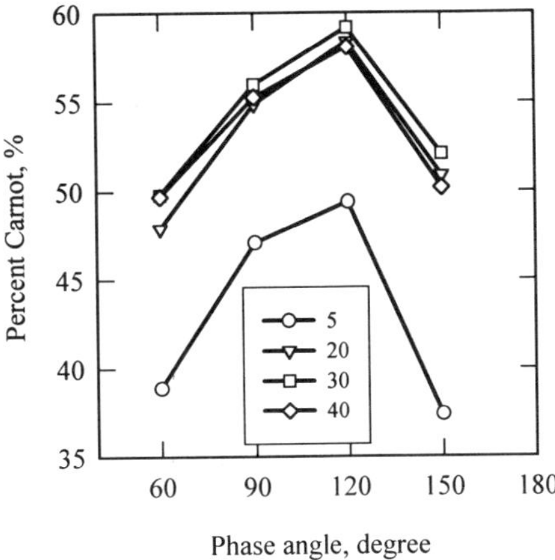

Figure 15 Percent Carnot of the Strirling refrigerator

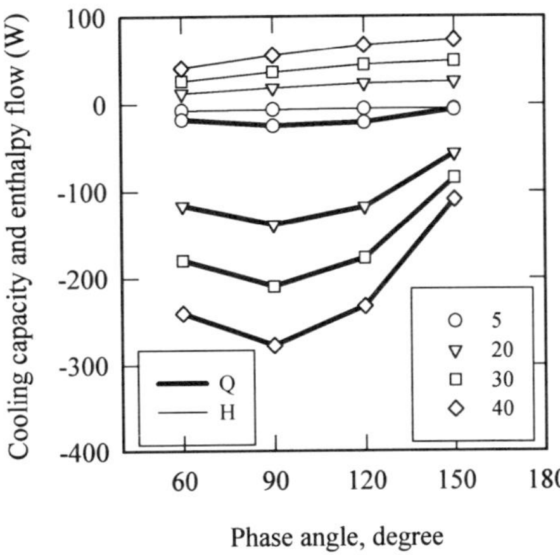

Figure 16 Enthalpy flow and cooling capacity of the Stirling refrigerator

Stirling refrigerator, large scale inertance tube pulse tube refrigerator and moving plug type pulse tube refrigerator are easily operated on the critical void volume ratio. It is very difficult for double inlet pulse tube refrigerator because large double inlet work loss is generated if bypass is largely opened. Hofler[5] cooler is also easy. But the expansion work is not cooling capacity because its resonator is in the vacuum chamber.

EXAMPLE STIRLING REFRIGERATOR

In the above discussion, we only consider the stack, here a Stirling refrigerator as shown in figure 1 is simulated by a numerical program which is similar to reference[3] without considering the inertance effect and kinetic energy of gas. Heat transfer coefficient and friction factor is from reference[4]. Stack diameter 50 mm, length 50 mm, mesh wire diameter 0.1 mm, porosity 0.832. The heat exchanger is gap type, length 10 mm, gap width 0.25 mm, height 20 mm, 100 gaps. T_H=300 K, T_C=150 K. Helium gas, average pressure 1 MPa, frequency 50 Hz. The phase angle between the compressor and expander is changed to get suitable standing wave. The expander swept volume is 40 cc, the compressor swept volume is changeable to change input power. The gas in compressor and expander are adiabatic.

Figure 15-16 shows percent Carnot, enthalpy flow and cooling capacity. The optimum phase angle is 120 degrees for percent Carnot, is 90 degrees for cooling capacity. The percent Carnot get highest at 120 degrees and 30 cc compressor. The pressure ratio is about 1.3 to 2. The enthalpy flow increases with the increases of compressor swept volume and phase angle. The enthalpy flow is negative at 5 cc compressor.

CONCLUSION

The void volume ratio of the hot heat exchanger over the stack is an important parameter. The stack efficiency or COP mainly depends on it. The critical point of the stack can be described by the critical void volume ratio which separates the stack from thermal acoustic engine to thermal acoustic cooler. Near the critical point, the stack efficiency or COP is maximum, the enthalpy flow, work flow, and heat flux between gas and matrix are near zero. The critical void volume ratio mainly depends on the temperature ratio of the hot heat exchanger over that of cold heat exchanger. When the stack is used as the regenerator for Stirling refrigerators, the COP also is maximum near the critical void volume ratio. This tendency almost does not change with heat transfer area, pressure ratio, and temperature ratio. If the temperature ratio is less than 2, it is possible to use stack as regenerator near the critical void volume ratio with high COP with limited input power. The results are also useful for the design of thermal acoustic engine, thermal acoustic cooler, and pulse tube refrigerator, too.

ACKNOWLEDGMENT

Thanks to Mr. Y. Momose, Dr. T. Inoue, and Dr. M. Nogawa in AISIN SEIKI CO. LTD. For useful discussion and support in writing of this paper.

REFERENCES

1. G. W. swift, Thermoacoustic engine, J. acoust. Soc, Am. 84, 1145, 1988
2. J. H. Xiao and Ray Radebaugh, A new approach for thermoacoustic refrigerator, advances in Cryogenic Engineering, Vol. 43, 1887, 1998
3. S. W. Zhu and Y. Matsubara, Theory and experimental study of thermal acoustic engine, Proceedings of 7th International Conference on Stirling Cycle Machines, 1995, p579
4. William M. Kays and A. L., London, Compact heat exchanger, McGraw-Hill Book Company, Third Edition.
5. T. J. Hofler, Concepts for thermoacoustic refrigerator and a practical device, Proceedings of Fifth International Cryocooler Conference, 1999, p93

WOUND PROFILE-WIRE REGENERATORS - FABRICATION AND TEST

I. Rühlich and H. Quack

TU Dresden, Lehrstuhl für Kälte- und Kryotechnik
01062 Dresden, Germany

ABSTRACT

Stacked screens and sphere beds today are the only regenerator configurations with importance for cryocoolers. On the one hand, many investigations over the past four decades have been performed to develop methods to identify optimal regenerator configurations involving hydraulic diameter, material, outer dimensions etc. On the other hand it is known that stacked screens and sphere beds have intrinsic disadvantages relative to the ratio of pressure drop to heat transfer.

Systematic investigations of fluid flow and heat transfer within matrices of regenerators have led to a concept of an optimal inner geometry. The ratio of pressure drop to heat transfer of the ideal geometry is almost as good as that of parallel plates. Opposed to the performance of parallel plate arrangements, the proposed geometry reduces losses due to longitudinal conduction within the matrix and due to maldistribution of the flow. A new fabrication technology and results of first tests for two prototypes are shown.

INTRODUCTION

The regenerator introduces two main losses within the cycle of the cryocooler. These are losses due to imperfect heat transfer and losses due to the pressure drop of the fluid. Heat transfer and pressure drop are closely connected. Therefore it is reasonable to define a ratio of pressure drop to heat transfer and to use this ratio as a figure of merit for different regenerator geometries. There are different dimensionless ratios which can be used. A ratio, which is advantageous due to a clear physical interpretation, is "Number of Pressure Heads pressure drop (NPH) per Number of Transfer Units NTU". NPH/NTU can be calculated as follows:

$$\frac{\mathrm{NPH}}{\mathrm{NTU}} = \frac{f}{j}\frac{\mathrm{Pr}^{2/3}}{4} \tag{1}$$

with

$$f = \frac{2 \cdot \Delta p \cdot D_h}{\rho \cdot U^2 \cdot L} \quad \text{and} \quad j = \frac{\mathrm{Nu}}{\mathrm{Re} \cdot \mathrm{Pr}^{1/3}} \tag{2}$$

Advances in Cryogenic Engineering, Volume 45.
Edited by Shu *et al.*, Kluwer Academic / Plenum Publishers, 2000.

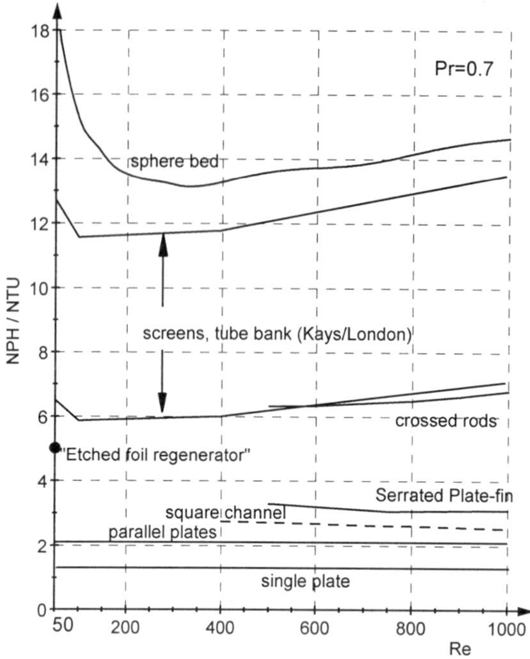

Figure 1. NPH/NTU vs. Re for typical heat transfer surfaces [6].

The graphic representation of this ratio is depicted in Figure 1 for some typical heat transfer surfaces.

The pressure drop in classic regenerators, sphere beds and stacked screen is about four times, or more higher than for parallel plates. In other words: The ratio is at least four times more than necessary for the heat transfer task.

According to Radebaugh [1] the pressure drop in the regenerator causes losses of the same magnitude compared with the net performance of the cooler, especially for pulse tube coolers with high mass flow rates through the regenerator. Using this fact one can estimate the potential of regenerators with an improved ratio of pressure drop to heat transfer. If a NPH/NTU ratio for a regenerator can attain a ratio of the same magnitude exhibited by parallel plates without additional losses, the net cooling performance of the cooler can be increased by about 75% with the same input power.

IDEAL INNER GEOMETRY

For many years it is known that screens and sphere beds don't have an ideal inner geometry, relative to the ratio of pressure drop to heat transfer.[2] Over the past decade parallel plates were regarded as an ideal inner geometry for regenerators. That led to the development of various prototypes of regenerators with parallel plate type structures.[3,4,5] All these prototypes did not attain the expected performance. Additional losses like maldistribution and losses due to longitudinal conduction may have accounted for the poor performance.

Recently we have presented results of systematic numerical simulations to identify what an ideal inner geometry of a regenerator should look like.[6] It turned out, that there are two main parameters of a matrix geometry that have substantial influence on the ratio of the pressure drop to the heat transfer. These two parameters are the aspect ratio (thickness to length) of the single element on the one hand and the ratio of the mean free flow area to the minimum free flow area within the matrix on the other hand. As a result of the CFD calculations, the ratio NPH/NTU is depicted for some typical geometries at two Reynolds numbers, Re = 500 and Re = 200, in Figure 2 and Figure 3.

From this quantified description of the dependency of NPH/NTU on these parameters it is easy obtain an ideal inner geometry of a regenerator. Slim streamlined elements in a staggered overlapping arrangement are required. The particular shape of the elements is of minor importance. There are almost no differences in the performance between elements with a sharp leading edge and those ones with a marked radius in flow direction. One example of an almost ideal geometry is shown in Figure 4. The corresponding performance of that geometry is marked in Figure 2. This arrangement of heat transfer elements is almost as effective as that of parallel plates.

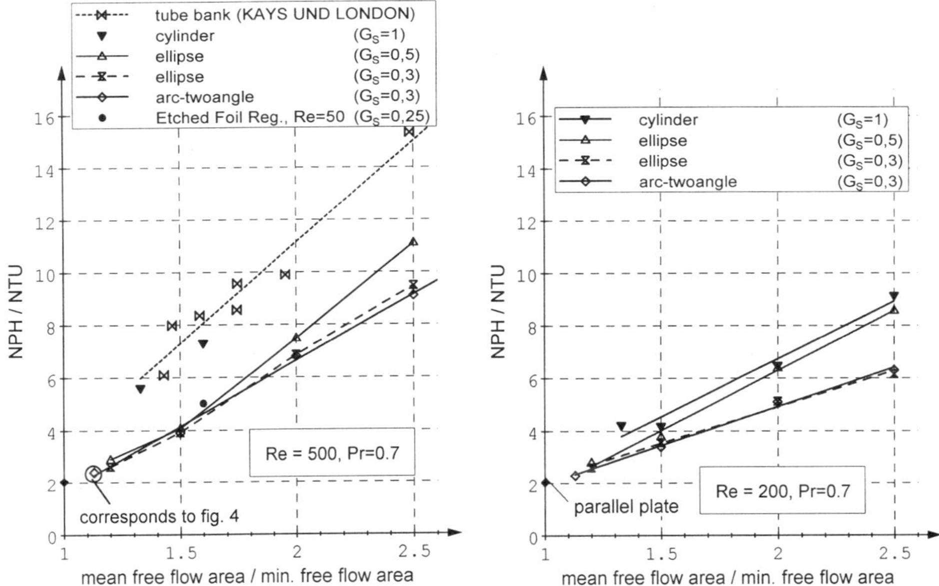

Figure 2. Ratio NPH/NTU vs. area ratio within the matrix for Re = 500 (All calculation results – except Kays/London).

Figure 3. Ratio NPH/NTU vs. area ratio within the matrix for Re = 200 (All calculation results).

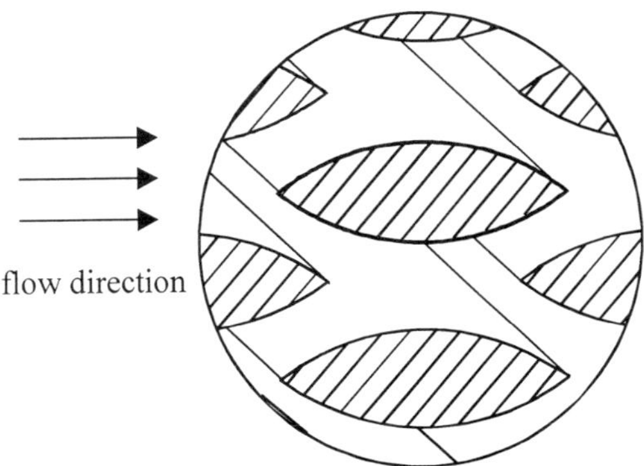

Figure 4. Example of an ideal geometry.

FABRICATION OF NEW REGENERATORS

After extensive investigations it turned out that no suitable fabrication technology for the requested geometry is available today. The main problem is given by the required dimensions. To achieve the necessary hydraulic diameters the smallest dimension of each matrix-element has to be about $0.7*10^{-3}$ m or less. Due to the lack of applicable technologies we decided to develop our own technology. Profile wires are wrapped, layer by layer, in a screw-wise manner around a core. After each layer spacers are inserted consisting of many parallel thin wires placed in the axial orientation. These spacing wires have two functions. They take care of the necessary spacing between the layers and also support the proper axial position of the profile wires. In Figure 5 the winding of a prototype is depicted.

Both, profile wire and spacing wire consist of phosphor bronze. The winding core is driven in a computer controlled winding-machine. The cross-section of the geometry achieved for prototype 1 is shown in Figure 6. The diameter of the spacing wire in this case is $0.06*10^{-3}$ m, for prototype 2 it is $0.05*10^{-3}$ m. Due to the reduced diameter of the spacing wire, the hydraulic diameter could be reduced by 15%. The dimensions of the complete concentric regenerator are as follows: 0.22 m in length, $16*10^{-3}$ m inner diameter, $32*10^{-3}$ m outer diameter. Details of the matrix can be seen in Table 1.

Figure 5. Winding of a profile wire regenerator.

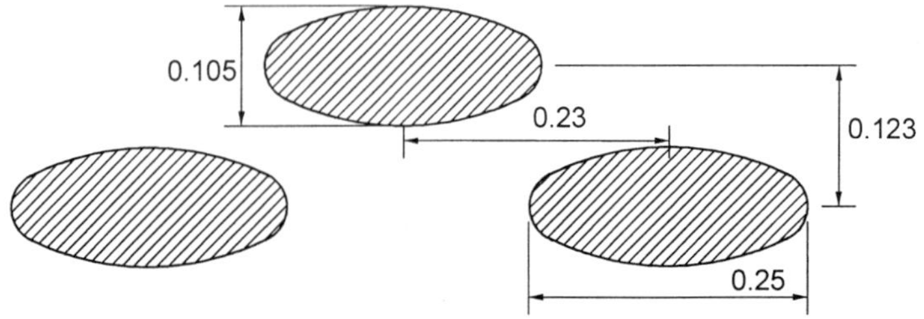

Figure 6. Matrix-arrangement for prototype 1, schematically.

TESTING OF TWO PROTOTYPES

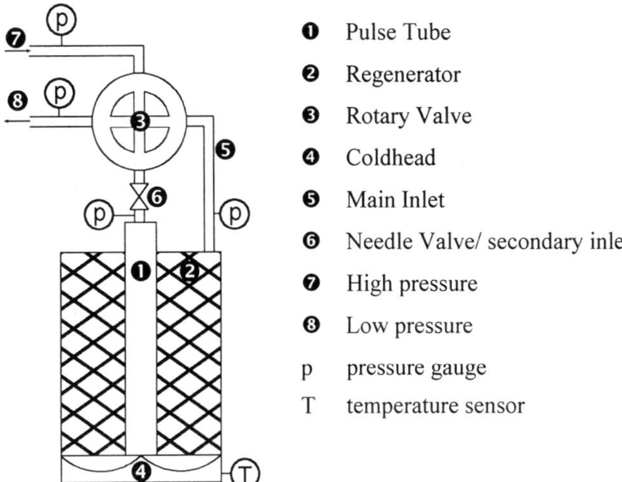

Figure 7. Schematic of Four-Valve-Pulse Tube.

The Low Temperature Physics working group of the Friedrich-Schiller-Universität Jena has developed several Four-valve pulse tubes, which because of a coaxial arrangement of pulse tube and regenerator [7], are particularly suited for the test of the prototypes. The assembly of the cooler is shown schematically in Figure 7.

From CFD calculation we obtained values for f and j, which were used to calculate the heat transfer and pressure drop for the prototypes. The results shown in Table 1 are calculated for steady-state flow at a mean pressure (12 bar) and mean temperature (170 K).

Table 1. Design values for different regenerators

		Screens, 150mesh	Prototype 1	Prototype 2
$\dot{m}$	kg/s	$3.4*10^{-3}$	$3.4*10^{-3}$	$3.4*10^{-3}$
Porosity	-	0.66	0.69	0.66
outer diameter	m	$32*10^{-3}$	$30.5*10^{-3}$	$31.8*10^{-3}$
surface S	m^2	2.6	1.2	1.32
D_h	m	$0.14*10^{-3}$	$0.27*10^{-3}$	$0.23*10^{-3}$
Re	-	85	177	160
NTU	-	890	360	454
Pressure drop	bar	0.95*	0.13	0.18
f	-	4.52	0.98	1.18
j	-	0.105	0.084	0.099
NPH / NTU	-	8.5	2.2	2.3

* measurement

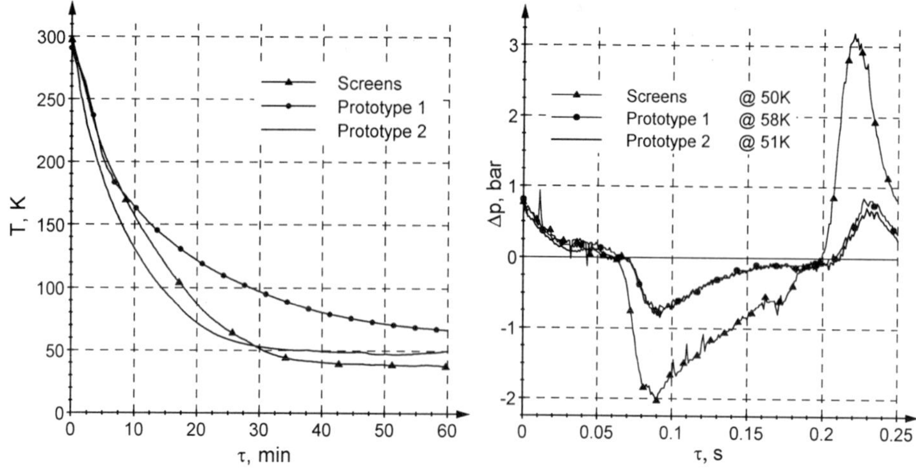

Figure 8. Cooldown performance with different regenerators.

Figure 9. Pressure drop for different regenerators.

In addition to the temperature of the cold head, the high and low pressure at the rotary valve were measured. The cooldown performance with different regenerators is depicted in Figure 8. The pressure drop vs. the cycle can be seen in Figure 9. With prototype 2 a more rapid cooldown was achieved than with the screen regenerator. The no load temperatures are 38 K for the screens (9.5W@80 K), 47 K for prototype 2 (7W@80 K) and 56K for prototype 1.

Both prototypes were tested with a slightly modified rotary valve. The working frequency and pressure level also were slightly modified.

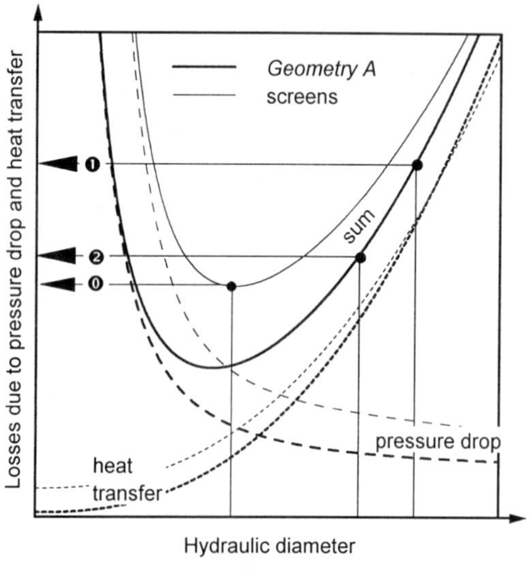

Figure 10. Losses (in quality) vs. hydraulic diameter [8].

Inspecting the results of the tests, one has to ask: Why did the new regenerators not result in an improvement of the cryocooler performance as predicted? From our point of view there are three main reasons for the unsatisfactory results:

1. The hydraulic diameter of both prototypes is too large (see Figure 10). The actual dimensions were limited by the fabrication technology. Because of this, both heat transfer surface and heat transfer coefficients are smaller compared than that of the screens. In Figure 10 losses due to pressure drop and heat transfer for the screen regenerator and the geometry A achieved in the prototypes are shown vs. the hydraulic diameter. Additionally, the sum of both losses is shown. The three different regenerators are indicated with reference to their hydraulic diameter. Although the minimum of the sum of the losses is lower for the new geometry, the prototype performance is still low due to the large hydraulic diameter.
2. The laminar flow regime with almost no flow separation in the new regenerator leads to an almost linear dependency of the mass flow rate on the pressure difference over the regenerator. This results in a strongly varying mass flow rate over the cycle and thus to a reduced average NTU.
3. The cooler is not optimized for the new regenerator. Normally the adaptation of all parameters of the pulse tube requires extensive experiments. This has presently not been done. One result of that non-optimized operation may be a DC-flow, which could contribute to the poor performance.

The very low pressure drop of the prototypes shows a big potential to further reduce the hydraulic diameter (Figure 9). Thus, it should be possible to reduce the losses due to imperfect heat transfer. This results in the reduction of the overall losses within the regenerator.

Pressure drop measurement under continuous flow conditions for prototype 2 revealed a friction factor that is about 36% higher than predicted for Re = 100...200 out of the CFD. These deviation can be explained by the existence of the spacing wires, asymmetries of the fabricated geometry in general and effects due to the glued and cut ends of the wound regenerator. The loss due to longitudinal conduction within the wires of the prototypes is less than 1 W.

CONCLUSIONS

With CFD simulation it was shown, that there are matrix geometries with almost the same ratio of pressure drop to heat transfer than the ones achieved for parallel plates. Matrix-geometries are almost insensitive to maldistribution and do not suffer from longitudinal conduction. Two prototypes with such a new inner geometry were built and tested. With one of these prototypes the cooldown rate was improved compared to the screen regenerator. But the no load temperature and the cooling performance of the screen regenerator could not be achieved. The main reason for the unsatisfactory results seems to be related to the large hydraulic diameter of the prototypes. The dimensions of the prototypes were limited by the fabrication technology. A slight decrease in the hydraulic diameter provide the potential to achieve better results than with the screen regenerator. A challenge to the future, from our point of view, is the development of an appropriate fabrication technology.

ACKNOWLEDGMENT

Financial support by a BMBF grant (13N6619/0) and the support from Mr. Thuerk and Mr. Gerster, FSU Jena, is gratefully acknowledged.

NOMENCLATURE

D_h	hydraulic diameter	m	NTU	number of transfer units	-
f	friction factor	-	Nu	Nusselt number	-
j	Colburn modulus	-	p	pressure	Pa
L	length in flow direction or plate length	m	Pr	Prandtl number	-
$\dot{m}$	mass flow rate	kg/s	Re	Reynolds number	-
NPH	number of pressure heads pressure drop	-	S	wall surface	m^2
			T	temperature	K
			U	velocity	m/s

Greek symbols

ρ density kg/m^3

REFERENCES

1. Radebaugh, R., Pulse tube refrigerators, Cryogenic Society of America, Short Course Symposium, Portland (1997).
2. Radebaugh, R., Louie, B., A first simple step to the optimization of regenerator geometry, Proc. of the 3rd Cryocooler Conference, NBS Special Publication 698, Washington DC (1985), p. 177.
3. Rawlins, W., Timmerhaus, K.D., Radebaugh, R., Measurement of the performance of a spiral wound polyimide regenerator in a pulse tube refrigerator, Adv. Cryo. Eng., Vol. 37B, Plenum Press, New York (1992), p. 947.
4. Yaron, R., Shokralla, S., Yuan, J., Bradley, P.E., Radebaugh, R., Etched foil regenerator, Adv. Cryo. Eng., Vol. 41B, Plenum Press, New York (1996), p. 1339.
5. Wild, St., Untersuchung ein- und zweistufiger Pulsrohrkühler, Fortschr.-Ber. VDI Reihe 19, Nr. 105, VDI-Verlag, Düsseldorf (1997).
6. Rühlich, I., Quack, H., Investigations on regenerative heat exchangers, 10th Int. Cryocooler Conference, Monterey (1998).
7. Blaurock, J., Hackenberger, R., Seidel, P., Thürk, M., Compact four-valve pulse tube refrigerator in coaxial configuration, Cryocoolers 8, Plenum Press, New York (1995), p. 395.
8. Rühlich, I., Strömungstechnische Optimierung von Regeneratoren für Gaskältemaschinen, Fortschr.-Ber. VDI Reihe 19, Nr. 117, VDI Verlag, Düsseldorf (1999).

REGENERATOR PERFORMANCE EVALUATION IN A PULSE TUBE CRYOCOOLER

J. P. Harvey[1], C. S. Kirkconnell[2], and P. V. Desai[1]

[1]Woodruff School of Mechanical Engineering
Georgia Institute of Technology
Atlanta, Georgia 30332-0405

[2]Raytheon Systems Company
El Segundo, California, 90245-0902

ABSTRACT

Pressure drop through the porous structure and temperature ineffectiveness in its thermal response contribute heavily towards a loss of performance in the regenerator of a Pulse Tube Cryocooler (PTC). A proper selection of matrix material and its arrangement within the regenerator thus become critical design issues. For low cost applications, an additional design trade may be sought with regenerator materials which provide a reduction in assembly cost. This work focuses on a comparative experimental evaluation of the performance of several regenerators, all with a homogeneous matrix, of varying structure. Such factors as the oscillatory and non-oscillatory pressure drop and ultimate cold temperature for a specific load are examined in evaluating the performance. Experimental results from this work are presented and compared with a view toward identifying the influence of matrix structure on the regenerator performance. A novel method for measuring mass flow rate and temperature in a quasi-steady flow is also presented.

INTRODUCTION

A pulse tube cryocooler has been tested with various regenerator materials, all composed of 300 series stainless steel. Table 1 describes the regenerators tested, and Figure 1 contains magnified photographs of the cross sections of the regenerator materials.

This research was carried out to establish a method for testing regenerators and to begin collecting information on homogeneous matrices. In future efforts, these test results will be used to improve modeling/design tools that are currently being used to design pulse

Table 1. Descriptive Summary of the Regenerators Tested

#	Name / Figure	Description
1	325 Mesh / Figure 1A	Wire mesh screens - 325 wires per inch Wire diameter: 27.9 μm Porosity: 0.696
2	400 Mesh / (Not shown)	Wire mesh screens - 400 wires per inch Wire diameter: 25.4 μm Porosity: 0.692
3	60 μm / Figure 1B Alabama Cryogenic Engineering[1]	Perforated disks Pore size: 60 μm Porosity: 0.717

Figure 1A. 325 Wire Mesh Screen (200X)

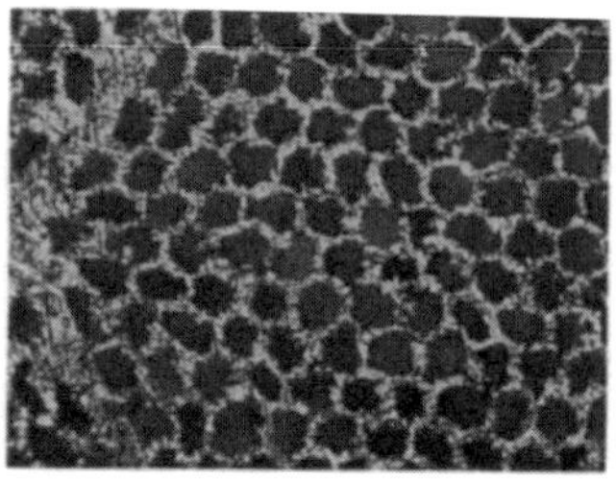

Figure 1B. 60 μm Perforated Disk (200X)

tube regenerators. The problem with many of the current design tools is that they are based on the experimental correlations published by Kays and London[2] which assume 1) steady, 2) incompressible, and 3) fully developed flow. The flow within the regenerator is less than adequately described using these three assumptions. In actuality, the regenerator flow violates all of these assumptions as follows:

1) The flow in the regenerator is unsteady, oscillating in a quasi-steady, periodic fashion.
2) Due to significant density variations in the presence of large temperature and pressure waves, the gas must be treated as compressible.
3) Due to flow reversals and end effects, the flow profile varies across the regenerator length and is constantly in a state of transition.

Each of these three assumptions contributes to the deviation in behavior for a set of operating conditions. Using in-house code developed at Raytheon Systems Company, the error in using the correlations from Kays and London has been estimated to be on the order of 30% with respect to pressure drop and heat transfer efficiency. This has motivated the present research.

EXPERIMENTAL APPARATUS

The apparatus used is of an Orifice Pulse Tube Cryocooler (OPTC) design and is illustrated in Figure 2. With the exception of the regenerator materials, the identical cooler assembly was used by Kirkconnell, et al. to characterize overall cryocooler performance as a function of pulse tube aspect ratio and volume[3]. Towards that end, the apparatus was designed to accommodate five interchangeable regenerator/cold heat exchanger/pulse tube assemblies, hereafter referred to as "expanders." Using the numbering scheme of the previous paper, expanders 1 through 3 are of identical volume, and expanders 1, 4, and 5 are of constant length. The performance of the constant volume pulse tubes (1-3) were found to be virtually indistinguishable, hence the use of these three is ideal for the present effort in which the unique impact of regenerator design on performance is sought. An unfortunate oversight resulted in the testing of one of the regenerators in expander 5, which

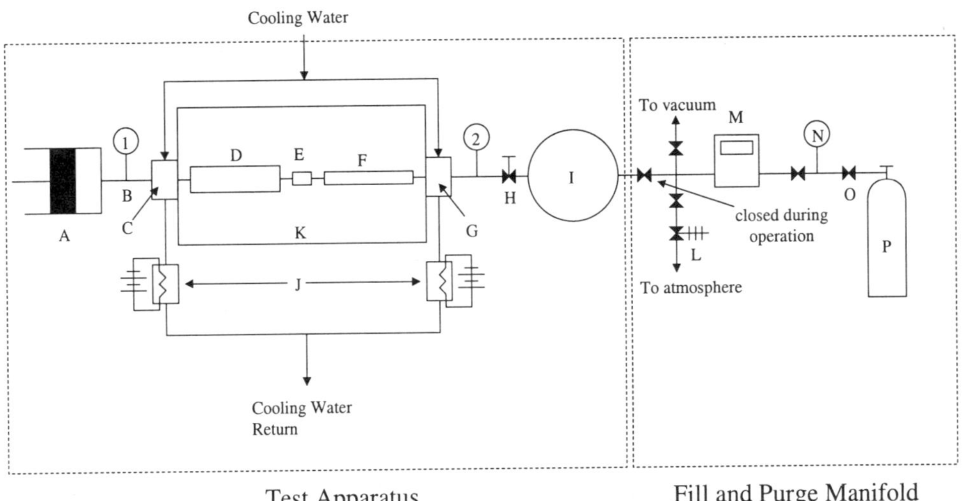

1=Inlet Instrumentation (P,T,m)
2=Outlet Instrumentation (P)
A=Compressor
B=Transfer Line
C=Inlet HX
D=Regenerator
E=Cold HX
F=Pulse Tube
G=Rejection HX
H=Orifice Valve
I=Surge Tank
J=Calorimetric HX
K=Vacuum Chamber
L=Relief Valve
M=Flowmeter
N=Test Gauge
O=Pressure Regulator
P=Helium Supply Bottle

Figure 2. Experimental setup, including pulse tube cooler and fill and purge manifold.

has a 17% larger pulse tube volume than the constant volume set and performed slightly better in the previous experiments (no load temperature of 76 K vs. 79 K). The differences between the performance of the various regenerators were much larger than this pulse tube volume effect, however, so the data from expander 5 are still considered relevant. The compressor used is a nominal 3 cc swept volume Hughes Condor on loan from Night Vision & Electronic Sensors Directorate (NVESD). The reader is referred to ref. 3 for additional details on the apparatus.

MEASUREMENTS

A full data acquisition and control system was designed and implemented for this research. The system contained a Hewlett Packard HP3852 Data Acquisition and Control Unit, which was programmed using a PC and HP VEE software. Among the abilities of this system were full data acquisition and logging, high-speed acquisition of waveforms, Fourier analysis of waveforms, and General Purpose Interface Bus (GPIB) control of instruments and processes. GPIB allows compatible laboratory instruments to be controlled with a desktop computer.

The compressor was driven by a power amplifier, which amplified a sine wave generated by a GPIB compatible function generator. The compressor power was then controlled by modulating the function generator signal amplitude. By measuring the compressor power, PID control was achieved which allowed automatic adjustment as the system proceeded towards steady state. Calculating compressor power involved sampling the current and voltage waveforms and performing Fourier analysis on the signals. Since the current waveform contained multiple harmonics, power factor was also calculated and displayed. This allowed "tuning" the frequency to obtain optimum power factor. Power factor represents the amount of distortion and phase shift between the voltage and current in a given circuit, which leads to less efficient electrical to mechanical conversion.

Other powers were controlled through the use of GPIB programmable DC power supplies. These power supplies powered the cold heat exchanger load heaters and two auxiliary calorimetric heaters installed on the two ambient heat exchangers (see below.)

Pressures were measured at two locations in the cooler according to Figure 2. P1 was measured in the transfer line between the compressor and the inlet heat exchanger. P2 was measured between the rejection heat exchanger and the orifice valve. These two pressures represent the pressures at the inlet and outlet of the expander which is composed of the regenerator, pulse tube, and the three heat exchangers. It would be desirable to measure the pressure in the cold heat exchanger, but it was deemed too difficult at this stage in the research with the existing hardware to install and calibrate a pressure transducer in a cryogenic region. This is a significant obstacle which requires a more complicated, fully compensated whetstone bridge strain gauge pressure transducer.

Pressure waveforms were decomposed into their Fourier components. This allowed noise rejection and accurate calculation of pressure wave amplitude. Important phase information was calculated as well such as the phase lag between P1 and P2.

Temperatures near ambient were measured using type K thermocouples which were installed in locations such as the heat exchangers and vacuum dewar. These instruments gave an indication of thermal equilibrium for the warm structure. Two Lakeshore DT470 silicon diodes were used to measure the cold tip temperature. These diodes are much more accurate than thermocouples at cryogenic temperatures (±0.25 K).

The inlet and rejection heat exchanger heat loads were measured experimentally using calorimetric heaters. These heaters were used to heat the water passing through the heat exchangers causing the temperature of the water to increase. The heat rate was controlled such that the two differential temperatures were identical. Neglecting any other heat transfer modes, the two heat rates would be matched. However, these heaters were in a convective environment, so the measurements could only be used as a comparison. Rawlins describes this measurement in detail[4]. He has used a thermocouple network to accurately measure the temperature differences.

A novel dynamic mass flow measurement method has been developed specifically for this research with the use of a thermal anemometer. The details of this technique can be found in Ref. 5. This technique involved operating a single Constant Temperature Anemometer (CTA) probe in a quasi-steady oscillatory flow field with oscillatory temperature variations. When measuring a mass flux in a flow with temperature variations, the signal from a CTA is a function of both the mass flux and the fluid temperature. If the fluid temperature is changing rapidly, a high frequency response thermocouple such as a Constant Current Anemometer (CCA) is typically used to resolve the fluid temperature accurately[4,6]. The CCA operates with a sufficiently small DC current to prevent self-heating. The voltage across the CCA can be related to the fluid temperature through a simple calibration procedure. The flow in a cryocooler at "steady" operation is periodic and quasi-steady, or identical from one cycle to the next. A single sensor is used to measure the temperature dependent mass flux over two whole cycles. The sensor is operated at different temperatures during each whole-cycle measurement. By operating the sensor at different temperatures, the temperature and mass flux of the fluid can be calculated based on the physics which govern the anemometer response. This method represents a significant improvement over the dual sensor technique. Leak and contamination paths as well as equipment are minimized. A single sensor disturbs the flow less and provides only one parasitic conduction path for measurements in a cryogenic region. Frequency response of the temperature signal is improved to nearly that of the CTA. This method has only recently been perfected, so complete data is not available. Future experiments will utilize this technique.

RESULTS AND DISCUSSION

Experimental results for three different regenerator designs are presented. For each configuration, oscillatory and non-oscillatory pressure drop and cooling performance are used to characterize their differences. The results of the experiments are presented in Figure 3-Figure 8. Note that these figures show the results at two slightly different compressor power settings, 55 W (325 mesh and 60 μm) and 50 W (400 mesh.) Except for the load curves which describe the overall performance over the temperature range tested, the data is plotted for the no-load (zero applied load to the cold tip) condition.

Efforts were made to assess the leak integrity and the cleanliness of the system. The system pressures were monitored over a significant time period when the cooler was essentially in thermal equilibrium to monitor the leak integrity. The cleanliness of the system was more difficult to assess but was indicated in the repeatability of steady cold tip temperature. Gradual changes in cold tip temperature over several hours indicated condensable contamination in the cold region where it was being condensed/frozen on the screens in the cold heat exchanger. Some contamination should be expected with an experimental apparatus as a result of instrument penetrations. Stable cold tip temperature with less than 0.2 K per hour drift was assumed to be acceptable[a].

Load Curves

Load curves, as shown in Figure 3, are the most widely used indicator of overall cooler performance. These curves indicate the net refrigeration versus the cold tip temperature, where net refrigeration is the difference between gross refrigeration and parasitic heat loads. These diagrams are usually constructed for constant compressor power, so first and second law efficiency can be derived from the curves. In addition, the slope of these curves suggests the incremental decrease in temperature that can be achieved with a given decrease in parasitic heat load. The 400 mesh regenerator performed the best with a no-load temperature of 76.1 K. In increasing order are the 325 mesh at 86.9 K and the 60 μm at 110.6 K no-load temperatures.

Porosity and Permeability

To correlate regenerator performance, detailed information must be known about the pressure drop and heat transfer characteristics of the porous material. Two parameters that have a strong dependence on pressure drop are porosity and permeability, which are shown for the regenerators in Figure 4. Porosity, φ, is the ratio of void volume to total volume, approaching unity for an open tube. Porosity in terms of the mass, m, volume, V, of a porous sample, and density of the base material, ρ, was calculated as

$$\varphi = 1 - \frac{m}{\rho V} \tag{1}$$

[a] During warm-up, the 325 mesh exhibited behavior indicative of condensable contamination (cold tip temperature plateaued at 220 K). The extent of the effect of the contamination on the cooler performance is suspected to be small since the temperature drift was <0.2 K/hr.

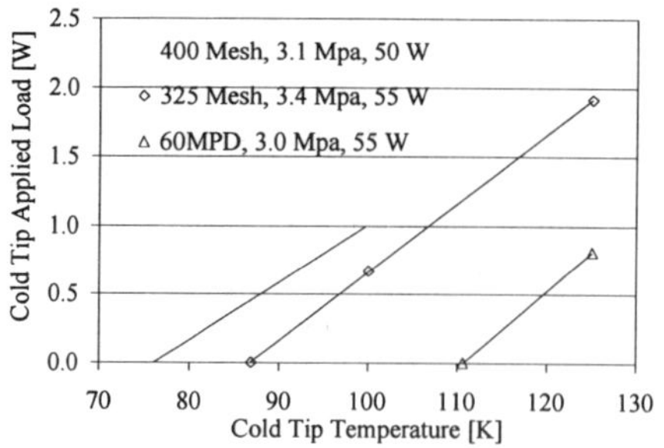

Figure 3. Load curves

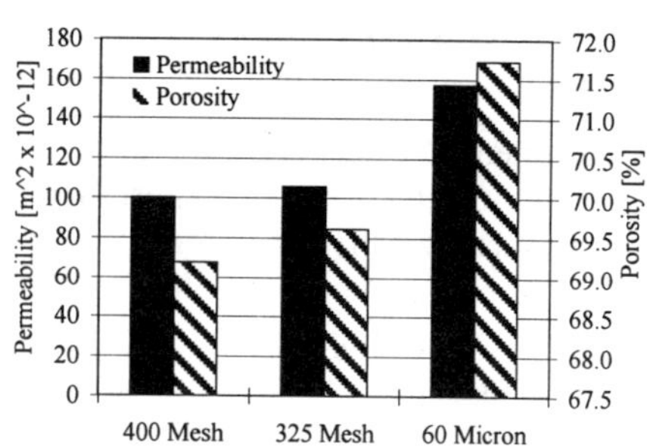

Figure 4. Permeability and porosity

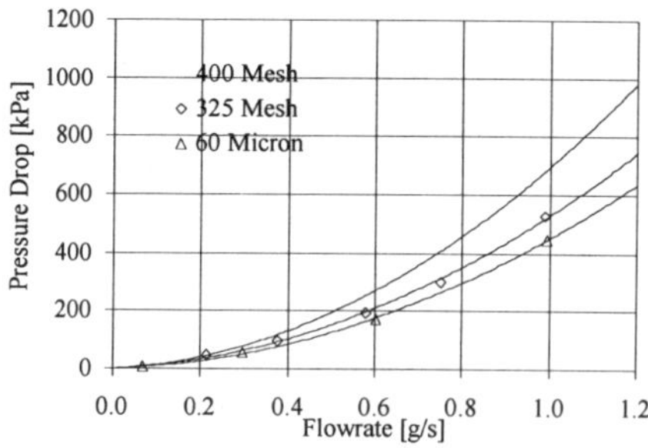

Figure 5. Non-oscillatory (steady flow) pressure drop

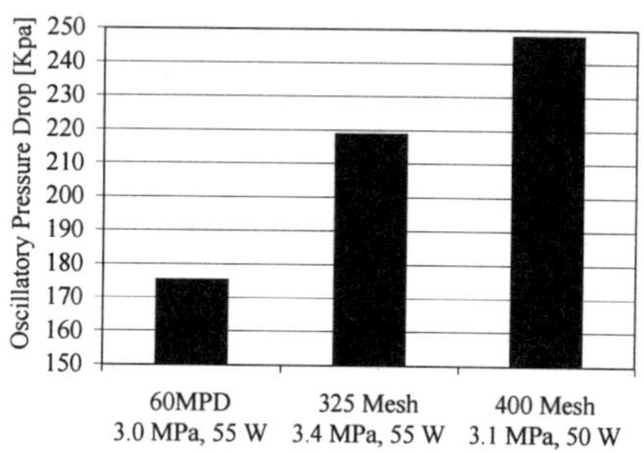

Figure 6. Oscillatory pressure drop

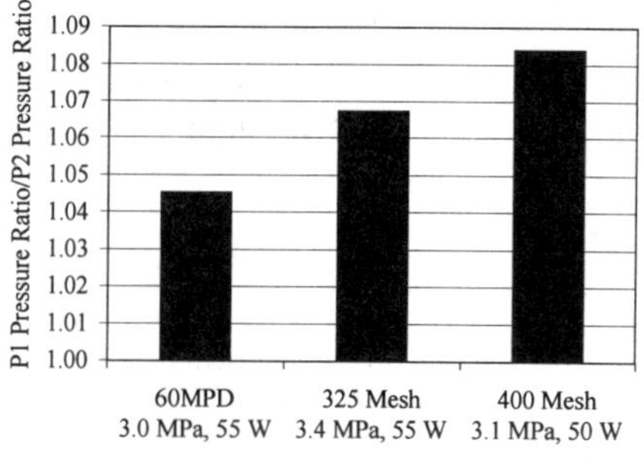

Figure 7. Pressure ratio attenuation

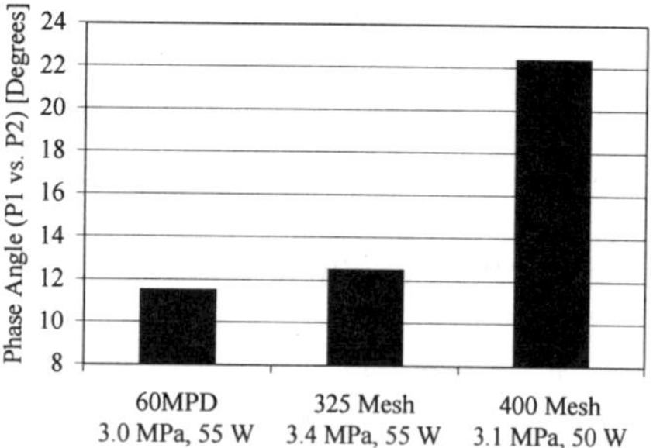

Figure 8. Pressure wave phase angle (P1 vs. P2)

The Darcy law defines permeability, K, as follows

$$u = \frac{K}{\mu}\left(-\frac{dP}{dX}\right) \Rightarrow K = \frac{u\mu}{\left(-\frac{dP}{dX}\right)} \qquad (2)$$

where u is the pore velocity, μ is the viscosity of the fluid, and dP/dX is the pressure gradient in the regenerator. The Darcy Law assumes that the velocity, u, is a linear function of pressure gradient, dP/dX. This model is valid only when inertial effects are negligible (small u). Experimentally measured pressure drop (Figure) over the range of applicability for the regenerator exhibits quadratic behavior indicating significant inertial effects. The pressure drop data was fitted to a quadratic equation utilizing two coefficients which were determined by least squares regression. The zeroth order coefficient was assumed to be zero such that that the pressure gradient would be identically zero for zero flow rates. The model for pressure drop as a function of mass flow rate was then given in terms of the two least squares coefficients, α and β, as

$$\left(-\frac{dP}{dX}\right) = \alpha\dot{m} + \beta\dot{m}^2;\ \dot{m} = \rho u A_{flow} \tag{3}$$

For oscillatory flow, the inertial effects become even more important leading to a frequency-dependent model. Permeability is still a meaningful parameter representing a length scale squared. With the quadratic model, permeability was calculated as

$$K = \frac{\mu}{\alpha\rho\varphi A_{cs}};\ A_{flow} = \varphi A_{cs} \tag{4}$$

which is a direct manipulation of Eq. 3 assuming $\dot{m} >> \dot{m}^2$. Note that the permeability of the 60 μm disks is significantly larger. Together with the load curves in Figure 3, the data seem to indicate that the permeability for the mesh screens is closer to the optimum for this cryocooler.

While permeability indicates the magnitude of the pressure drop, porosity partially describes the heat transfer interaction through surface area dependence. However, the porosity does not take into account the amount of surface area that is actively involves in heat transfer. In general, a low porosity, high permeability matrix should perform best.

Oscillatory Pressure Drop

Pressure Drop and Pressure Ratios. Pressure drop, both non-oscillatory and oscillatory, trends inversely with permeability and porosity. Figure 6 shows the oscillatory pressure drop of the three expanders under conditions corresponding to the no-load conditions in Figure 3. Figure 7 shows the pressure ratio attenuation for the same conditions, where pressure ratios were calculated as the ratio of maximum to minimum pressure over a cycle. This ratio decreases across the expander due to pressure drop (mean pressure remains essentially constant.)

Pressure Wave Phase. The pressure waves propagate with finite speed through the expander resulting in a phase lag between P1 and P2. This phase angle is driven by the RC time constant of the orifice and surge volume. Interestingly in Figure 8, the phase angle increase with the more restrictive 400 mesh is much more dramatic than the observed pressure drop increase. Measurement error may be partly responsible. Graphical oscilloscope measurements were made for the 325 mesh and 60 micron disks while FFT measurements were used for the 400 mesh.

Frequency

Frequency of operation is also an important parameter. The frequencies chosen for these experiments such that compressor efficiency was maximized were 34 Hz ± 2 Hz. For the system tested, frequency has a large impact on overall system performance mainly due to electrical efficiency. Because of the effect of the regenerator design on the pressure wave and resulting pressure forces on the compressor pistons, the optimum frequency has some functional dependence on regenerator design. To experimentally measure this behavior, it is necessary to mathematically isolate the compressor performance from that of the rest of the cooler. The best method to achieve this is by measuring the actual PdV power delivered by the compressor. Meaningful correlations can then be developed using constant hydraulic power and not the electrical power consumed by the compressor. In future experiments, measurements of the hydraulic power delivered to the regenerator will be made using the aforementioned anemometry techniques.

CONCLUSIONS

A pulse tube cryocooler test station and accompanying data acquisition and reduction system has been developed for the characterization of various regenerator designs. Preliminary experiments have been performed on a limited number of homogeneous regenerator matrices. Of the regenerators tested, the best was the 400 mesh based on overall cooler performance. However, the wire mesh regenerators required more time to pack than the 60 μm perforated disk. A high efficiency, low-assembly time regenerator design is ultimately the desired choice.

The performance of the 400 mesh can be attributed to a favorable trade-off between pressure drop and effective heat transfer. Additional matrices will be tested in future experiments. These include two additional perforated disks to attempt to identify matrices which yield an even better balance. The 60 μm pore size was not expected to be optimal for this expander due to the comparatively large permeability of the matrix structure; the 45 and 30 μm sizes should perform better. These disks should provide a porosity less than the 400 mesh with less pressure drop. These trades will hopefully yield the optimum balance between regenerator performance and cost/ease of assembly. Future experiments will also include carefully controlled warm-ups to assist an approximation of the regenerator conductance.

In addition to the experiments mentioned above, future work will include more in-depth data analysis utilizing existing pulse tube system level models to help isolate the regenerator effects.

ACKNOWLEDGEMENT

The authors would like to recognize the support of Raytheon Systems Company and the Georgia Institute of Technology. The testing was performed in the Fluid Mechanics Research Laboratory of the Department of Mechanical Engineering at Georgia Tech under financial support from Raytheon Systems Company. Howard Dunmire of NVESD graciously provided the compressor used to drive the cooler.

REFERENCES

1. J. B. Hendrix, A new method for producing perforated plate recuperators, *Advances in Cryogenic Engineering* 41:1329 (1996).
2. Kays and London. "Compact Heat Exchangers," McGraw-Hill, Inc, New York (1964).
3. C. S. Kirkconnell, S. Soloski, and K. Price, Experiments on the effects of pulse tube geometry on PTR performance, *Cryocoolers* 9:285 (1997).
4. W. C. Rawlins, The Measurement and Modeling of Regenerator Performance in an Orifice Pulse Tube Refrigerator, Ph.D. thesis, University of Colorado, Boulder (1992).

5. J. P. Harvey, Parametric Study on Cryocooler Regenerator Performance, Master's thesis, Georgia Institute of Technology, Atlanta (1999).
6. W. C. Rawlins, R. Radebaugh, P. E. Bradley, and K. D. Timmerhaus, Energy flows in an orifice pulse tube refrigerator, *Advances in Cryogenic Engineering* 39:1449 (1994).

DYNAMIC BEHAVIOR IN OSCILLATING FLOW REGENERATORS WITH THERMAL DIFFUSION EFFECTS

Y. L. Ju,[1] Y. Zhou,[2] and A. T. A. M. de Waele[1]

[1] Department of Physics, Eindhoven University of Technology
P. O. Box 513, NL-5600 MB Eindhoven, the Netherlands
[2] Cryogenic Laboratory, Chinese Academy of Sciences
P. O. Box 2711, Beijing, 100080, China

ABSTRACT

The understanding of the dynamic flow behavior and heat transfer characteristics in the oscillating flow regenerator is of crucial importance in the optimal design of pulse tubes and other coolers. In this paper we present an analytical model incorporating both hydrodynamics and thermodynamics to describe the dynamic flow behavior with thermal diffusion effects of the regenerator in Stirling-type pulse tube coolers. The model is based on mass, momentum and energy conservation. The energy equations for the regenerator, including the gas flow and the matrix, are derived and solved coupled with the dynamic flow behavior. A closed set of expressions is obtained by using the perturbation treatment approach. It is shown that the mass flow and pressure fluctuations are affected by thermal diffusion effects, which can not be neglected.

INTRODUCTION

Regenerators used in pulse tubes and Stirling coolers are operated under cyclic- or oscillating flow conditions. As a necessary component, the regenerator is essentially a cylindrical tube filled with a porous medium of metallic wire screen mesh or powder, and plays an important role in the performance of coolers. The right condition for cooling to occur requires that the regenerator does not only have small pores and fine wires to give a large surface area for heat exchange, but little resistance for the gas flow as well. There are many differences between steady and oscillating flow. The significant influence of the dynamic flow characteristics of the oscillating flow regenerator on the performance of cryocoolers has been recognized recently by many researchers[1,2]. The understanding of the momentum and energy diffusion mechanism of the oscillating flow regenerator is of crucial importance in the optimum design of the pulse tube and other coolers.

Advances in Cryogenic Engineering, Volume 45.
Edited by Shu *et al.*, Kluwer Academic / Plenum Publishers, 2000.

The working process in the oscillating flow regenerator is complex due to the unsteady, compressible, porous media in addition to the coupled flow and thermal diffusion effects. The dynamic flow and the heat transfer associated with these conditions have a strong effect on the performance of the regenerator, but are not well understood. Many previous analyses[3-5] of the regenerator are based on the method of quasi-steady approximation in order to simplify the analysis. However, the quantitative accuracy of the analytical results was not satisfactory. An understanding of the mechanisms associated with regenerator requires a full solution of the Navier-Stokes equations. However, the complexities of these equations make an analytical solution for the oscillating flow regenerator essentially impossible. Therefore, numerical simulation[6-8] is commonly used and has been extensively developed in recent years to investigate the performance of regenerators. However, owing to the complexity of partial differential equations for the mass, momentum, energy conservation, and equation of state, evaluation of these compressible equations for the velocity pressure and temperature is difficult and needs much more attention. These models are not completely satisfactory.

It is well known that the oscillating compressible fluid can be considered as acoustic oscillations. Many researchers[9-12] consider that the working mechanisms of the oscillating flow regenerator is based on time-averaged energy effects caused by the thermal interaction of the oscillating gas and the porous medium. However, for the moment the thermoacoustic theory may not be adequate for the practical design of regenerators. Although it introduced some new concept which is helpful to the understanding of the time-averaged thermoacoustic heat transportation and energy transformation caused by the thermal interaction between the oscillatory fluid and the solid.

The viewpoints of system dynamics and thermoacoustics have in common that the linear flow network analysis[13-16] can be applied to describe the dynamic behavior of regenerators by using the governing equations in conjunction with a linearization approach and some approximations. This approach was developed for the system analysis of the regenerator by considering the pressure as the electric voltage and the mass flow as the electric current. Huang, et al.[15-16] investigated the dynamic response of isothermal and cyclic flow in a regenerator by using a basic transfer-function model. A linear dynamic model of Darcy's form for a regenerator was first derived[16].

However, most of the above research was limited to study independently the flow dynamic behavior and the energy transformation of the regenerator under oscillating flow conditions. Only a few papers discussed the coupled mechanism between the dynamic flow behavior and thermal diffusion in the regenerator. In fact, the effect of thermal diffusion on regenerator behavior in cryocoolers is very important. For example, de Waele, et al.[17] derived a set of expressions describing regenerator dynamics in harmonic approximation. They showed that the temperature profiles were determined by thermal conduction and gave the amplitudes of the harmonic components.

We attempt, in this paper, to deal with the problem by considering the mass, momentum and energy conservation laws simultaneously. A closed set of expressions is derived based on perturbation treatment approach. The energy equations for the regenerator including the gas flow and the matrix are also derived and solved coupled with the dynamic flow behavior.

MATHEMATICAL MODEL

A regenerator is essentially an energy-storage element in which the working fluid is heated or cooled by the regenerator matrix in a thermodynamic cycle. We use the porous medium approach, in which the regenerator is treated as a uniformly distributed flow resistance unit and a one-dimensional model to reduce the CPU time and to avoid any difficulty involved with the two- or three-dimensional models. In a one-dimensional model, two problems need to be considered: (1) fluid parameters are averaged over the cross section normal to the principal flow direction and the fluid dynamic equations are expressed in terms of these mean parameters, and (2) the point of needing heat transfer and pressure drop correlation to insert into the one-dimensional model. Thus, the regenerators are treated as one-dimensional steady flow problems commonly reported in the iterature[18,19] in terms of cross section-averaged flow parameters correlating heat transfer and pressure drop. The basic assumptions are as follows:

(1) One-dimensional ideal gas flow;
(2) The matrix is homogeneous, stationary, and has uniform porosity;
(3) The axial heat conduction is negligible.

The mass continuity equation is

$$\frac{\partial \rho}{\partial t}+\frac{\partial(\rho U)}{\partial x}=0 \tag{1}$$

The momentum equation of a regenerator is the revised Darcy's law [15,16]:

$$\frac{\partial(\rho U)}{\partial t}+\frac{\partial(\rho U|U|)}{\partial x}+\frac{\partial P}{\partial x}+\alpha\mu(\varepsilon U)+\beta\rho(\varepsilon^2 U|U|)=0 \tag{2}$$

The energy equation for the gas flow is

$$\varepsilon\rho C_P\frac{\partial T}{\partial t}+\varepsilon\rho C_P U\frac{\partial T}{\partial x}+h\frac{A_r}{V_r}[T-T_r]-\frac{\partial P}{\partial t}=0 \tag{3}$$

The energy equation for the matrix is

$$(1-\varepsilon)\rho_r C_r\frac{\partial T_r}{\partial t}+h\frac{A_r}{V_r}[T_r-T]=0 \tag{4}$$

Where T, U, P are the gas temperature, velocity and pressure, respectively. T_r is the regenerator matrix temperature, ρ the gas density, μ the gas viscosity, C_P the gas heat capacity, ρ_r the bulk matrix density, C_r the matrix heat capacity, ε the porosity of the regenerator, h the convection heat transfer coefficient, V_r the regenerator volume, A_r the heat exchange surface area of the regenerator, α and β are the first viscous coefficient and the second viscous coefficient of the gas flow, respectively.

For convenience, the gas velocity terms U in Eqs. (1) ~ (4) can be rewritten in terms of the gas mass flow rate $\dot{m}$, that is $\dot{m} = \rho A_f U$, where A_f is the free-flow cross-sectional area of the regenerator. Therefore, Eqs. (1) ~(4) become

$$\frac{\partial P}{\partial t}+\frac{RT_0}{A_f}\frac{\partial \dot{m}}{\partial x}=0 \tag{5}$$

$$\frac{\partial \dot{m}}{\partial t}+\frac{1}{A_f}\frac{\partial(\dot{m}|\dot{m}|/\rho)}{\partial x}+A_f\frac{\partial P}{\partial x}+\alpha\frac{\varepsilon\mu}{\rho}\dot{m}+\beta\frac{\varepsilon^2}{\rho A_f}\dot{m}|\dot{m}|=0 \tag{6}$$

$$\varepsilon\rho C_P V_r \frac{\partial T}{\partial t} + C_P \frac{\partial \dot{m} T}{\partial x} dx + hA_r (T - T_r) - V_r \frac{\partial P}{\partial t} = 0 \tag{7}$$

$$(1-\varepsilon)\rho_r C_r V_r \frac{\partial T_r}{\partial t} + hA_r (T_r - T) = 0 \tag{8}$$

The first term in Eq. (6) represents the unsteady term, the second represents the inertia term, the third represents the pressure drop term, the fourth represents viscous term, and the last represents the inertia dissipation. Since the gas flow velocity in the regenerator is not very high in practice, the inertia term (second term) can be neglected. Furthermore, the last term can be ignored and the associated effect included in a modified parameter α'. Thus, Eq. (6) can be reduced to

$$\frac{1}{A_f}\frac{\partial \dot{m}}{\partial t} + \frac{\partial P}{\partial x} + \frac{\alpha' \varepsilon \mu}{\rho A_f}\dot{m} = 0 \tag{9}$$

$$\alpha' = \alpha + \beta \mathrm{Re}_h \tag{10}$$

$$\alpha = \frac{A}{2D_h^2}; \beta = \frac{B}{2D_h^2} \tag{11}$$

with $A = 175$ and $B = 1.60$, according to Ref. [15]. D_h is the hydraulic diameter of the regenerator, which is defined based on the wire diameter D_W and the porosity of the regenerator:

$$D_h = \frac{\varepsilon D_W}{(1-\varepsilon)} \tag{12}$$

The Reynolds number is defined based on the regenerator hydraulic diameter and the maximum velocity of gas inside the regenerator $U_{\max}$

$$\mathrm{Re}_h = \rho U_{\max} D_h / \mu \tag{13}$$

THEORETICAL ANALYSIS

Under oscillating flow conditions, these equations can be solved with perturbation methods[20,21] assuming the magnitude of the oscillating gas pressure is small (a small amplitude of piston motion). Therefore, the fluctuations of the gas mass flow, temperature and pressure are small. Considering a pulse tube cooler with angular frequency $\omega = 2\pi f$, the general parameters q $(=T, T_r, \dot{m}, P)$ can be expressed as a sum of mean value q_0 plus higher-order harmonic perturbation terms q_i $(i = 1,2 \ldots N)$:

$$q = q_0 + q_1 + q_2 + \cdots = q_0 + \tilde{q}_1 e^{iwt} + \tilde{q}_2 e^{i2wt} + \cdots \tag{14}$$

where $\tilde{q}_i$ is the amplitude of the high-order parameters q_i. Substituting Eq. (14) into Eqs. (5)~(8) and keeping only the first-order terms (terms of higher order are neglected) we can obtain the following first-order equations (first harmonic perturbations):

$$\frac{\partial P_1}{\partial t} + \frac{1}{C_F}\frac{\partial \dot{m}_1}{\partial x} = 0 \tag{15}$$

$$L_F \frac{\partial \dot{m}_1}{\partial t} + \frac{\partial P_1}{\partial x} + R_F \dot{m}_1 = 0 \tag{16}$$

$$\frac{\partial T_1}{\partial t} + \frac{1}{C_{PF}}\frac{\partial P_1}{\partial t} + \frac{1}{C_{TF}}(T_1 - T_{r1}) + \frac{\partial T_0}{\partial x} = 0 \tag{17}$$

$$\frac{\partial T_{r1}}{\partial t} + \frac{1}{C_{TR}}(T_{r1} - T_1) = 0 \tag{18}$$

Eqs. (15) and (16) are similar to an RLC electric circuit. R_F, C_F, L_F represent the flow resistance, the flow capacitance and the flow inductance per unit length of the regenerator, respectively. By analogy with a RLC electric circuit, these terms are defined as

$$R_F \equiv \frac{\overline{\alpha}\,\varepsilon\mu}{\rho A_f}; C_F \equiv \frac{A_f}{RT_0}; L_F \equiv \frac{1}{A_f} \tag{19}$$

where $\overline{\alpha}$ represents the parameter α' evaluated at the peak value of the mass flow rate $\dot{m}_1$. C_{TF}, C_{TR}, C_{PF} represent the gas time constant, the matrix time constant, and the thermal capacitance, respectively.

$$C_{TF} = \frac{\varepsilon\rho C_P V_r}{hA_r};\ C_{TR} = \frac{(1-\varepsilon)\rho_r C_r V_r}{hA_r};\ C_{PF} = \varepsilon\rho C_P \tag{20}$$

Eqs. (15)~(18) can be solved by Laplace transform with the boundary conditions

$$\dot{m}_1(x,s) = \frac{\dot{m}_1(L_r,s)}{\sinh(\lambda_s L_r)}\sinh(\lambda_s x) + \frac{\dot{m}_1(0,s)}{\sinh(\lambda_s L_r)}\sinh[\lambda_s(L_r - x)] \tag{21}$$

$$P_1(x,s) = \frac{P_1(L_r,s)}{\sinh(\lambda_s L_r)}\sinh(\lambda_s x) + \frac{P_1(0,s)}{\sinh(\lambda_s L_r)}\sinh[\lambda_s(L_r - x)] \tag{22}$$

$$T_1(x,s) = \frac{C_{TF}}{1+\xi}\cdot\left(\frac{1}{C_{PF}}P_1(x,s) - \frac{dT_0}{dx}\right) \tag{23}$$

$$T_{r1}(x,s) = \frac{1}{1+\xi}\cdot\left(\frac{1}{C_{PF}}P_1(x,s) - \frac{dT_0}{dx}\right) \tag{24}$$

where s is the Laplace transform variable and λ_s is the characteristic value which is defined as follows

$$\lambda_s^2 = s(R_F C_F + sL_F C_F) \tag{25}$$

ξ is the ratio of the heat capacity of the matrix materials and the gas per unit regenerator volume:

$$\xi = \frac{(1-\varepsilon)\rho_r C_r}{\varepsilon\rho C_P} \tag{26}$$

The value of dT_0/dx is determined self-consistently using energy conservation at no-load temperature at the cold end of the regenerator[17]. Eqs. (21) and (22) represent a distributed-parameter dynamic model accounting for the distributions of the gas mass flow rate and pressure as functions of the position inside the regenerator. Apart from the energy Eqs. (23) and (24), these two expressions were already derived by Huang and Lu[15]. By using the form of Eqs. (21) and (22), the amplitudes of gas mass flow $m_1(x,s)$ and pressure $P_1(x,s)$ can be solved analytically. The gas temperature amplitude $T_1(x,s)$ and the matrix temperature $T_{r1}(x,s)$ can be obtained with Eq. (23) and (24) by using the solutions of $P_1(x,s)$ and $m_1(x,s)$. With Eqs. (23) and (24) we obtain that the temperature difference between the gas and the matrix in the regenerator is determined by the gas time constant C_{TF}, which means the gas heat capacity for a given regenerator volume per heat exchange parameter.

RESULTS

The dynamic behavior of regenerators is determined by four average parameters and four first-order harmonic perturbations. In this section we will apply our analytical model to a practical case. The parameters of the regenerator we take as an example are as follows: diameter $D = 20mm$ and length $L_r = 200mm$; system mean pressure $P_{av} = 1.5MPa$; $T_H = 300K$; $T_L = 70K$. 250-mesh stainless steel screen is adopted in the analysis, its physical properties are: wire diameter $D_W = 0.041mm$; porosity $\varepsilon = 0.70$; hydraulic diameter $D_h = 0.0956mm$. In this case: $\alpha = 0.957 \times 10^{10} m^2$ and $\beta = 0.875 \times 10^8 m^2$. The working medium is helium, the operating frequency is 2Hz.

Figure 1 gives the amplitude of the gas mass flow rate $\dot{m}_1$. Figure 2 gives pressure amplitudes P_1. Figure 3 gives the amplitude of the gas temperature oscillations T_1 in regenerator. The time variations of the mass flow rate $\dot{m}_1(L,s)$ and pressure $P_1(L,s)$ of the gas in this regenerator is shown in Figure 4.

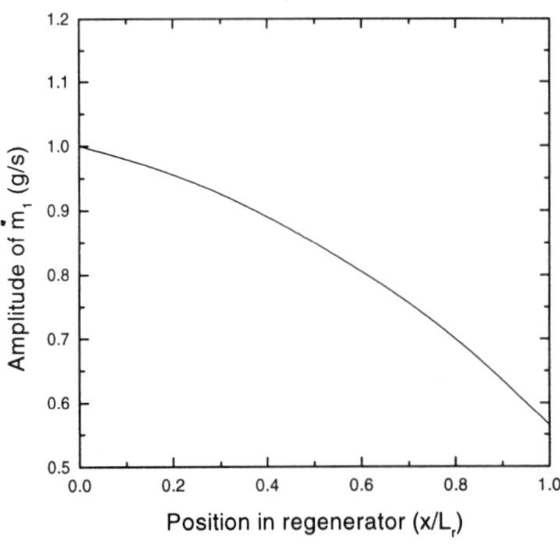

Figure 1. Amplitude of the gas mass flow rate $\dot{m}_1$.

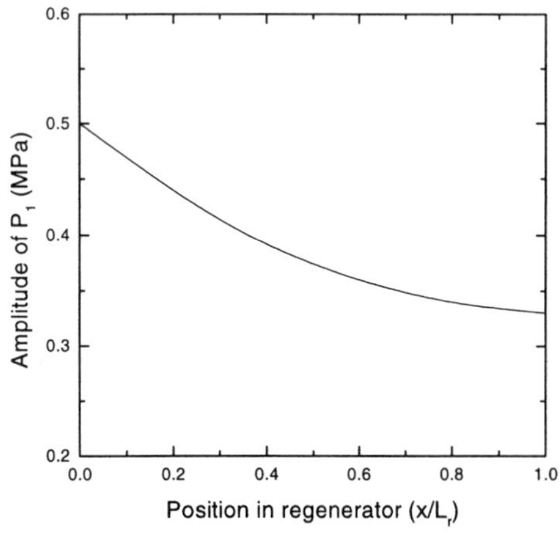

Figure 2. Amplitude of the gas pressure P_1.

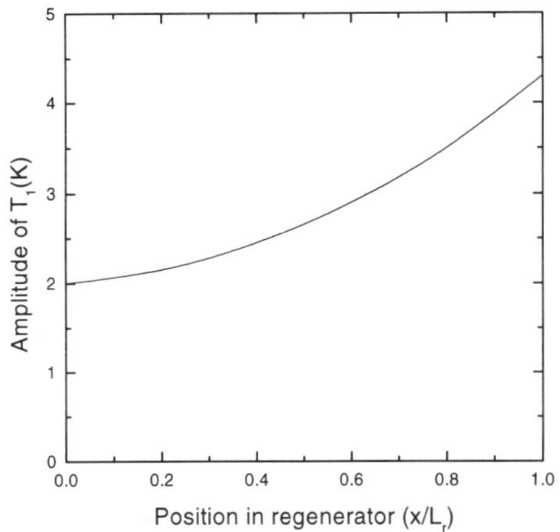

Figure 3. Amplitude of the gas temperature T_1

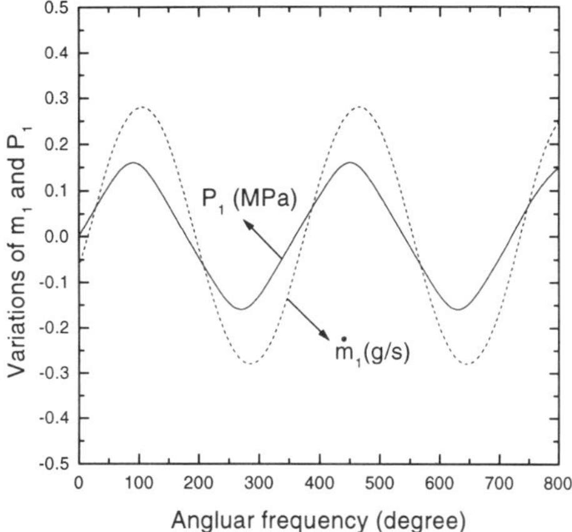

Figure 4. Variations of $\dot{m}_1(L,s)$ and $P_1(L,s)$

DISCUSSION

Most previous linear flow network analyses only considered the dynamic flow characteristics of the oscillating flow regenerator. Therefore, the governing equations of the flow network analysis were derived solely from the continuity and momentum equations. However, the solutions for the network model cannot be obtained without the temperature solution since the effect of temperature variation is involved in the governing equations of the flow network of the regenerator used in pulse tubes or other coolers. Therefore, the thermal diffusion effects cannot be evaluated since the gas temperature solution was not known. That is why previous thermoacoustic theory and linear flow network analysis can only discuss the fluid motion in a homogeneous temperature field. In this paper we present an analytical model incorporating both hydrodynamics and thermodynamics to describe the dynamic flow behavior with thermal diffusion effects of the regenerator in Stirling-type pulse tube coolers. The model is based on mass, momentum and energy conservation. The energy equations for the regenerator, including the gas flow and the matrix, are derived and solved coupled with the dynamic flow behavior. A closed set of expressions is obtained by using the perturbation treatment approach.

The dynamic expressions (21) and (22) form the basic transfer-function model of a regenerator which accounts for the dynamic responses of the regenerator between the two boundaries. They show the dynamic flow behavior, and the effects of the pressure drop of the gas in the regenerator. Eqs. (23) and (24) give a general understanding of the physical mechanism of heat generation and heat transfer characteristics of the gas and matrix in the regenerator. These analytical solutions are helpful for understanding the physical mechanism of the dynamic behavior and the thermal transport processes of the oscillating gas flow in the regenerator. Besides having theoretical significance, the present model can be used to simplify and develop numerical models of regenerators used in pulse tubes or other coolers.

REFERENCES

1. Swift, G. W. and Ward, W. C. Simple harmonic analysis of regenerator, *J. Thermophysics and Heat Transfer.* 10: 652 (1996).
2. Zhao, T. S. and Cheng, P. Oscillatory pressure drops through a woven-screen packed column subjected to a cycle flow, *Cryogenics.* 36: 333 (1996).

3. Walker, G. and Vasishta, V. Heat transfer and friction characteristics of wire-screen Stirling engine regenerator, *Adv. Cry. Eng*. 16: 324 (1970).
4. Modest, M. F and Tien, C. L. Analysis of real-gas and matrix-conduction effects in cyclic cryogenic regenerators, *ASME J. Heat Transfer.* 15: 199 (1972).
5. Modest, M. F and Tien, C. L. Thermal analysis of cyclic cryogenic regenerators, *Int. J. Heat Mass Transfer.* 17: 37 (1974).
6. Kohler, J. W. L. Stevens, P. F. de Jonge, A. K. and Beuzekom, D. C. Computation of regenerators used in regenerative refrigerators, *Cryogenics.* 15: 69 (1975).
7. Gary, J. Danry, D. E. and Radebaugh, R. A computational model for a regenerator, 3th. Cryocooler Conference, Boston, USA 199 (1985)
8. Minas, C. Dynamic modeling of stirling cryorefrigerator, *Cryogenics.* 34: 37 (1994)
9. Rott, N. Thermoacoustics, *Adv. Appl. Mech.* 20: 135 (1980)
10. Wheatley, J. C., Hofler, T., Swift, G. W. and Migliori, A. An intrinsically irreversible thermoacoustic heat engine, *J. Acoust. Soc. Am.* 74: 153 (1983)
11. Swift, G. W. Thermoacoustic engines, *J. Acoust. Soc. Am.* 84: 1145 (1988)
12. Xiao, J. H. Thermoacoustic theory for cyclic flow regenerators. Part 1: Fundamentals, *Cryogenics.* 32: 895 (1992).
13. Guo, F. Z. Chou, Y. M. Lee, S. Z. et al. Flow characteristics of a cyclic flow regenerator, *Cryogenics.* 27: 152 (1987).
14. Guo, F. Z. Network model of a cyclic flow regenerator, *Cryogenics.* 30: 199 (1990)
15. Huang, B. J. and Lu, C. W. Dynamic response of regenerators in cyclic flow system, *Cryogenics.* 33: 1046 (1993).
16. Huang, B. J. and Lu, C. W. Linear network analysis of regenerator in a cycle-flow system, *Cryogenics.* 35: 203 (1995).
17. de Waele, A. T. A. M., Hooijkaas, H. W. G., Steijaert, P. P. and Benschop, A. A. J., Regenerator dynamics in the harmonic approximation, *Cryogenics.* 38: 995 (1998).
18. Kays, W. M. and London, A. L. Compact heat exchanger, 2nd edition, McGraw -Hill. New York, USA (1964).
19. Tong, L. S. and London, A. L. Heat transfer and flow friction characteristics of woven-screen and cross-rod matrices, *Tran. ASME* 1558 (1957).
20. Milton, V. D. Perturbation methods in fluid mechanics, The Parabolic Press, Standford, USA (1975).
21. Schlichting, H. Boundary Layer Theory, McGraw-Hill, New York, USA (1979).

STUDY ON THE VOLUMETRIC RATIO OF THE HYBRID GM REGENERATOR OVER THE COLD CHAMBER WORKING AT 4K

L. Wang,[1] X. D. Xu,[2] L. Fang,[2] L. Zhang [2]

[1]Brookhaven National Laboratory
Upton, NY 11973
[2]Cryogenic Laboratory, Chinese Academy of Sciences
Beijing 100080, China

ABSTRACT

This paper presents a computational analysis on the hybrid GM regenerator at liquid helium region. The effects of the ratio of the regenerator volume over the cold chamber volume on cooling performance were simulated numerically. The results show that there exits a minimum ratio of the regenerator volume over the cold chamber volume for a G-M cryocooler at 4K. The cooling capacity of the refrigerator drops sharply with smaller regenerator, and rises slowly with larger regenerator. The effects of the material combinations of the hybrid regenerator on the volumetric ratio were also studied. The design principle of the geometry size of the regenerator working at 4K was discussed.

NOMENCLATURE

A	Heat transfer area
C	Volumetric specific heat
h	Specific enthalpy of fluid
m	Mass of control volume
$\dot{m}$	Mass flow
P	Pressure
Q	Heat exchange
T	Temperature
t	Time
V	Volume

Greek letters

α	Heat transfer coefficient

Subscripts

i	The i-th control volume
w	Matrix of control volume
f	Boundary of control volume, or gas flow

Supported by the National Nature Science Foundation of China

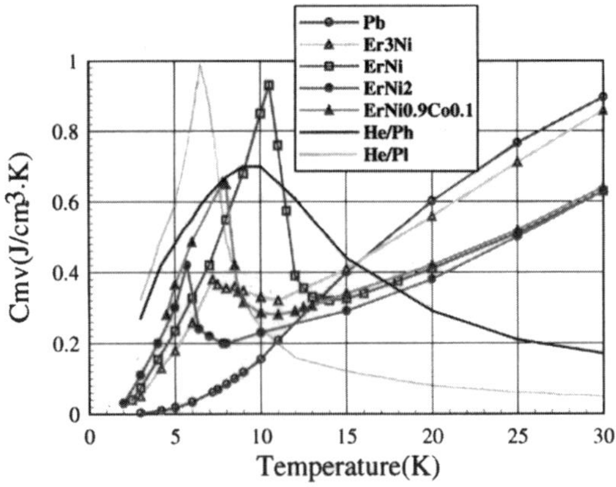

Figure 1. Volumetric specific heat of the regenerator materials[1]

INTRODUCTION

The heat of compression of helium gas causes the major loss on cooling capacity of the 4K GM cryocooler, because of the non-ideal physical properties of helium and the low specific heat of the regenerator materials near 4K region. The specific heat of matrix at the cold end of the regenerator is the key factor for further improving refrigerator performance[2][3][4]. At present, to obtain larger cooling capacity and increase regenerator efficiency, hybrid regenerators are greatly developed[5][6]to compensate the narrow peak of the specific heat of a single magnetic material, as shown in Figure 1. It has been realized through early numerical analyses[7] and experimental investigations[8][9] that the combinations for currently used materials and the ratio for the materials packed in the low temperature regenerator have larger effects on the cooling performance.

In this paper, we mainly studied the effects of the ratio of the volume of the regenerator with multi-layered materials over the cold chamber volume on cooling performance by numerical method. The effects of different combinations of materials on the volumetric ratio are also analyzed in detail. The design principle of the geometry size of the regenerator working at 4K was discussed.

NUMERICAL SIMULATION

The physical model is shown in Figure 2, including the regenerator with multi-layered hybrid materials, the cold end heat exchanger and cold chamber. The right hand side is the surface of the imaginary piston. The regenerator was assumed stationary. At the left hand side, pressure oscillation[2] is given in Figure 3. The basic assumptions in the model are: one-dimensional flow of gas, negligible axial heat conduction and axial pressure drop, and constant wall temperature of the cold chamber. The fundamental equations are as follows:

The mass equation for gas flow:

$$\frac{\partial m_i}{\partial t} = \dot{m}_{f_{i+1}} - \dot{m}_{f_i} \tag{1}$$

The energy equation for gas:

$$\frac{\partial(mh - PV)_i}{\partial t} = (\dot{m}h)_{f_i} - (\dot{m}h)_{f_{i+1}} + \frac{\delta Q}{dt} - P_i \frac{dV_i}{dt} \tag{2}$$

Where $\delta Q = \alpha_i \cdot A_i \cdot (T_{wi} - T_i) \cdot dt$ (3)

The energy equation for the matrix in regenerator:

$$C_{mi} \cdot V_{mi} \cdot \frac{\partial T_{wi}}{\partial t} = \alpha_i \cdot A_i \cdot (T_i - T_{wi}) \tag{4}$$

It should be noted that the mass flow rate is defined as positive if the gas flows from the hot end to the cold one of the regenerator and as negative for the opposite direction. The real gas properties of helium were obtained from the NIST TN-1334. The heat transfer coefficient was from the reference[10]. A Finite difference method was used to solve the above equations[3][7]. The volume change of the cold chamber with the crank angle is also shown in Figure 3.

COMPUTATIONAL RESULTS AND ANALYSES

Main Calculation Parameters

The main calculation parameters of regenerators are shown in Table 1. The

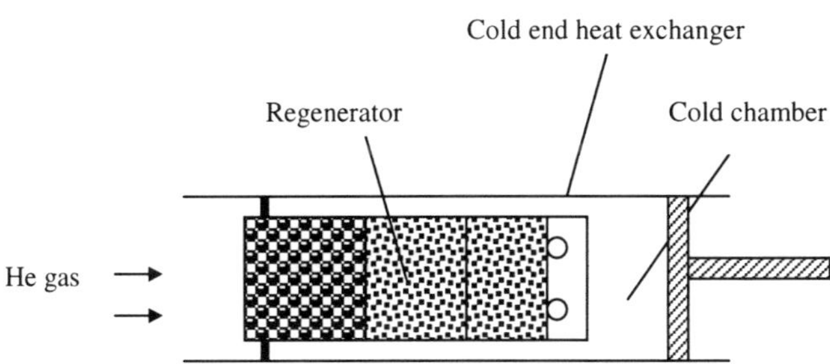

Figure 2. Physical model of the multi-layered hybrid regenerator

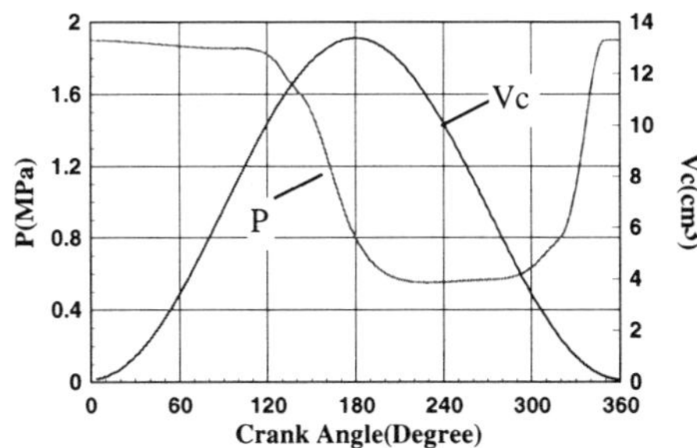

Figure 3. Pressure oscillation at the hot end of the regenerator

temperatures of the inlet gas and that at the wall of the cold chamber are 30K and 4.2K, respectively. The intake/exhaust opening and closing angles are 320°/145 ° and 130 ° /305 °. The volumetric ratios of three kinds of materials from the hot end to the cold end in regenerators are 40%, 30%, and 30% in sequence[6]. The volumetric ratios of two kinds of materials from the hot end to the cold end in the regenerators are both 50%. Lead sphere, ErNi, Er_3Ni, $ErNi_2$ and $ErNi_{0.9}Co_{0.1}$ grains were used for the regenerator materials.

Computational Results and Analyses

The calculation results are shown in Figure 4-6. T_f is the temperature of helium gas in the regenerator. Q_c is the cooling capacity. V_r/V_c is the ratio of the regenerator volume over the cold chamber one. In this paper, the cold chamber volume and the regenerator diameter constant, the variation of V_r/V_c means the change of the length size of the regenerator.

Figure 5 shows the effect of the ratio V_r/V_c on the cooling capacity Q_c and the effect of different combinations of materials in regenerators on the cooling capacity under different V_r/V_c. From the figure one can observe that with V_r/V_c increasing, Q_c increases. For the multi-layered regenerator composed of lead spheres, Er_3Ni, and $ErNi_2$ grains, when V_r/V_c is larger than 4.5, the cooling capacity rises slowly. For example, Q_c rises 12% with V_r/V_c from 2.86 to 4.30. However, it's only 1% with V_r/V_c from 5.44 to 6.3.

Figure 4 gives the temperature distribution of gas flow in regenerators under different V_r/V_c. By comparison, the gas temperature distribution in the regenerator with V_r/V_c =2.86 is higher than that in the regenerator with V_r/V_c=5.44 at the end of gas admission process of the refrigerator. There exits larger temperature range near 4.2K at the regenerator cold end for the latter than that for the former. For the former, the high temperature portion moves further towards the cold chamber. The gas temperature of the former entering the

Table 1. Main calculation parameters of regenerators

Regenerator: diameter (mm)	22
Length (mm)	190
Materials	Pb, Er_3Ni, $ErNi_2$, ErNi, $ErNi_{0.9}Co_{0.1}$ grains
Porosity	0.41
Stroke (mm)	25

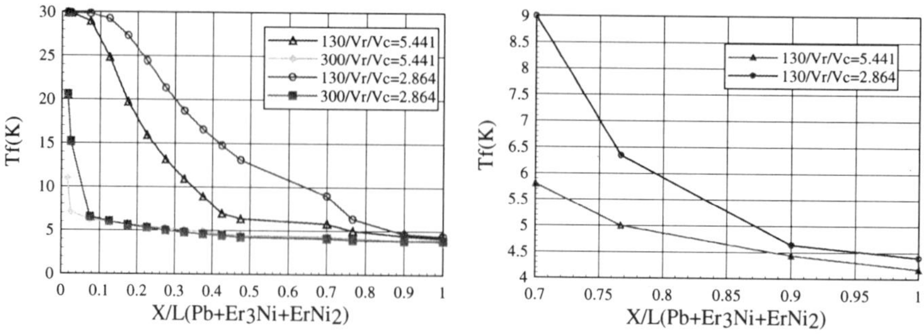

Figure 4. Temperature distribution of gas flow in the regenerators under different V_r/V_c

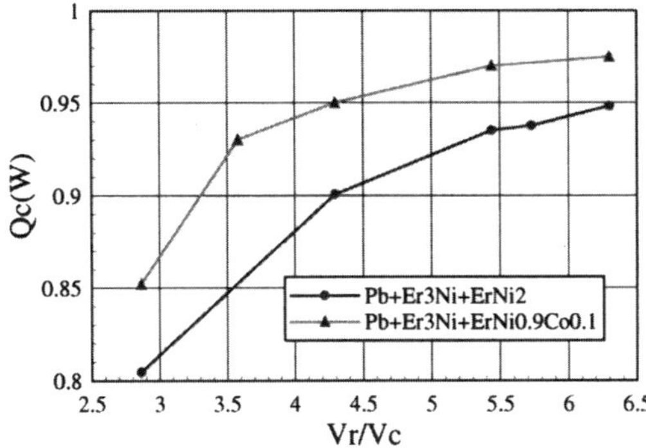

Figure 5. Effect of the ratio of the regenerator volume over the cold chamber one on the cooling performance and effect of different combinations of materials in regenerators on the cooling capacity under different V_r/V_c

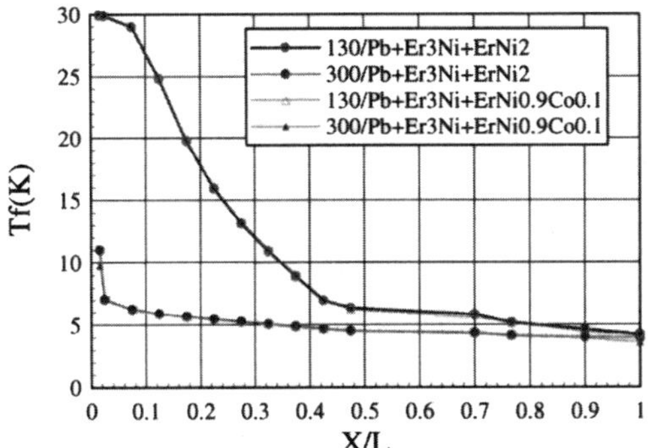

Figure 6. Temperature distribution of gas flow in regenerators with different combinations of materials

cold chamber becomes higher, and Qc is greatly reduced. With V_r/V_c increasing, the temperature region near 4.2K becomes larger, and then the gas temperature into the cold chamber lowers down, so Qc is enhanced. On the other hand, it is known that[1] the dominant heat loss for a 4K GM cryocooler is the compression heat of gas in the voids at the cold end of the low temperature regenerator. During the cycle the compression heat is partially stored in the regenerator matrix. The rest of the heat is carried into the cold chamber by the compressed gas, causing the reduction of Qc. The stored heat is determined by the ratio of the volumetric heat capacities of the matrix and of the helium in the regenerator voids at the temperatures of compression. When the matrix at liquid helium region is chosen, the compression heat stored in the regenerator matrix is almost constant. With the further increasing of V_r/V_c, Qc increases slightly. As a result, there is the minimum ratio V_r/V_c for a 4K GM refrigerator. In order to enhance the cooling capacity greatly, the matrix with high specific heat should be used.

Figure 5 also gives the comparison between the regenerator composed of lead spheres, Er_3Ni, and $ErNi_2$ grains(named the former) and that composed of lead spheres, Er_3Ni, and $ErNi_{0.9}Co_{0.1}$ grains(named the latter). The two regenerators were made up of the same materials at the hot end and different matrix at the cold end. We can find that the minimum ratio V_r/V_c of the latter is smaller than the former. For the former, V_r/V_c=4.2 may be the minimum size. For the latter, it is 3.6.

According to Figure 1 and Figure 6, the volumetric specific heat of $ErNi_{0.9}Co_{0.1}$ is a little higher than that of $ErNi_2$ near 4K, and the temperature of gas at the cold end of the former flowing into the cold chamber is slightly higher than that in the latter. The compression heat ratio stored in the regenerator of the former is approximately 67% and the latter is about 90% at 4K. Therefore, with the same size of the regenerator, the cooling capacity of the latter is larger than that of the former. To obtain the same cooling capacity as that of the latter, the large size for the former may be used. However, because the specific heat of the matrix at 4K region is the critical factor for the enhancement of Qc, to simply increase the size of the regenerator can not completely make up for low specific heat at the regenerator cold end. In addition, by comparing the two regenerators with the same materials at the cold end, we found that the small geometry size can be used for the regenerator that is composed of the materials with high specific heat at the hot end.

CONCLUSIONS

1. There exits a minimum ratio of the low temperature regenerator volume over the cold chamber volume for a G-M cryocooler at 4K. The cooling capacity of the refrigerator drops sharply with smaller regenerator, and rises slowly with larger regenerator.
2. The minimum volumetric ratio should ensure that there is the gas temperature region near 4K at the cold end of the hybrid regenerator.
3. To simply increase the size of the regenerator can not completely make up for low specific heat at liquid helium region because the specific heat of the matrix at 4K region is the critical factor for the enhancement of Qc.
4. Low specific heat at the regenerator hot end can be compensated by properly increasing the geometry size of the regenerator.

ACKNOWLEDGMENT

This research was supported by the National Natural Science Foundation of China.

REFERENCES

1. G. Ke, et al., Improvement of two-stage GM Refrigerator Performance using a hybrid regenerator, *Advance in Cryogenic Engineering*, 40A: 639(1994)
2. C. S. Hong and X. D. Xu, On the Thermodynamic Cycle of the Low Temperature G-M Refrigerator, *Cryogenics,* 34: 183, ICEC Supplement (1994)
3. L. Wang et al, A Numerical Simulation Method of a Two-stage G-M Refrigerator and Comparison with Experiment, *Advance in Cryogenic Engineering*, 43: 1799(1998)
4. X.D.Xu, L. Wang et al, Analysis on the Colder Regenerator of 4.2K G-M Refrigerator, *Advance in Cryogenic Engineering*, 43: 1605(1998)
5. A. Onishi et al, Development of 2W Class 4K Gifford-McMahon Cycle Cryocooler, *ICEC16/ICMC Proceedings,* Part 1: 335(1996)
6. T. Tsukagoshi et al, Optimum Structure of Multilayer Regenerator with Magnetic Materials, *Cryogenics,* 37:11(1997)
7. L. Wang et al, Performance Analysis of the Low Temperature Regenerator with Multi-Layered Hybrid Materials, *Proceedings of ICEC17*, 315(1998)
8. T. Tsukagoshi et al, Refrigeration Capacity of a GM Refrigerator with Magnetic Regenerator Materials, *Advance in Cryogenic Engineering*, 41: 1623(1996)
9. T. Hashimoto et al, Excellent Character of Multi-layer Type Magnetic Regenerator near 4.2K, Cryocoolers 8, 677(1995)
10. Kays, W.M. and London, A.L., *Compact Heat Exchangers* 2nd Edn(1964)

REFRIGERATION CAPACITY OF A GM REFRIGERATOR UTILIZING HoSb AS ITS REGENERATIVE MATERIAL

H. Nakane[1], T. Hashimoto[1], T. Numazawa[2], M. Okamura[3], T. Kuriyama[4], and Y. Ohtani[4]

[1]Kogakuin University
Shinjuku-ku, Tokyo, 163-8677, Japan
[2]National Research Institute for Metals
Tsukuba-shi, Ibaragi, 305-0003, Japan
[3]Toshiba Corporation, Material & Components Div.
Yokohama-shi, Kanagawa, 235-8522, Japan
[4]Toshiba Corporation.
Kawasaki-shi, Kanagawa, 210-0862, Japan

ABSTRACT

In order to obtain better regenerative effectiveness at very low temperatures, the heat capacity of the regenerator materials must be larger than that of He as the working gas. The specific heat of pressurized He gas has a large peak below 15 K. Only magnetic materials, whose phase-transition temperatures exist in the low temperature region, can be effective as regenerative materials and only multi-layer regenerators with magnetic materials have practical use. Two magnetic compounds were found to have remarkably large peak values. One such compound is HoSb with the peak value of 2.7 J/K·cm^3 at 5 K, and the other is DySb with 2.0 J/K·cm^3 at 9 K. Materials with a very large peak in heat capacity were packed into a regenerator and the structure of the materials was analyzed by computer simulation. The combination of multi-layer structure with synthetic materials are considered best in the structure of a regenerator. HoSb and $HoCu_2$, which has a broad peak below 10 K, were packed into the second stage of a GM refrigerator as the multi-layer materials and the refrigeration capacities of the materials were compared. It seems that HoSb is effective for use in the refrigerator below 6 K.

INTRODUCTION

A large enough cooling capacity for a 4 K GM refrigerator can not be obtained yet. In a regenerative refrigerator, the refrigeration capacity depends on the response speed of the heat exchange based on the heat emission and absorption between the working gas and the materials packed into the regenerator. In order to obtain higher effectiveness in the regenerator, the heat capacity of the regenerator materials must be larger than that of He used as the working gas. The specific heat of Pb, which is conventionally used as a regenerative material, is small below 10 K. Only magnetic materials with a large specific heat based on magnetic phase transition is effective below 10 K. However, as shown in Fig. 1, the specific heat of He gas of 8 and 20 atm is very large below 15 K. Since the heat-exchange region of He gas is wide, a single magnetic material such as Er_3Ni can not cover the specific heat of He gas. A two-layer regenerator is constructed for the second cylinder of a GM refrigerator. A refrigeration capacity of about 1 W at 4 K is commer-

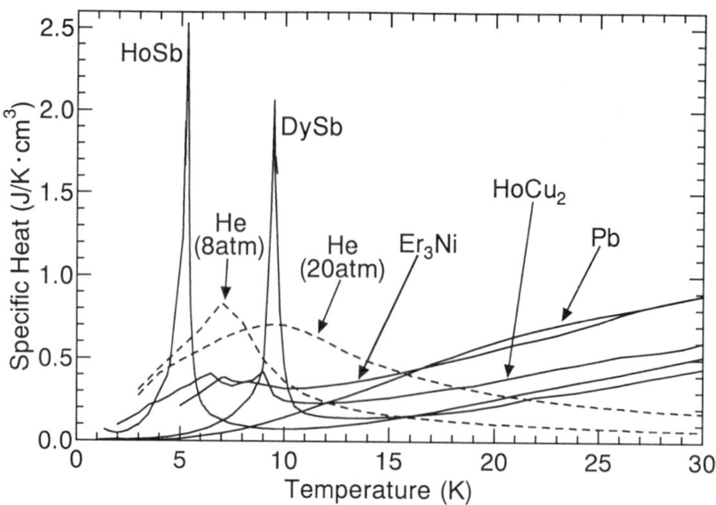

Figure 1. Specific heat of regenerative materials.

cially available by packing Er_3Ni and $ErNi_{0.9}Co_{0.1}$ into a comparatively large second stage regenerator of a GM refrigerator. However, as shown in Fig. 1, the authors noticed that HoSb and DySb, when used as the regenerative materials, have very large peaks around 5 K and 9 K, respectively. The regenerator effectiveness of magnetic materials having such very large peak was analyzed by computer simulation and the results are reported[1]. Obviously, the relation between the refrigeration capacity and such materials must be studied. The authors are trying to make a simulation program to analyze the relation among the refrigeration capacity, the property of the regenerative materials and the structure of the regenerator, which is near completion. Investigation to obtain the relation between the refrigeration capacity and the property of regenerative materials will be carried out after the program is completed.

In this paper, the problems that occur when materials with large specific heat peak are packed into a multi-layer regenerator are analyzed. In order to ascertain the analytical results, an experiment using a GM refrigerator was carried out, The measured refrigeration capacity are reported.

MATERIAL COMPOSITION IN THE REGENERATOR

When a high temperature and high pressured gas (20 atm in Fig. 1) passes through the regenerator, the regenerative materials absorb the heat of the He gas. Conversely, when a low temperature and low pressured gas (8 atm in Fig. 1) passes through the regenerator, the regenerative materials add heat to the He gas. By repeating the above process, the cold end of the regenerator is cooled. The temperatures in the high and low pressure periods in the second stage of a GM refrigerator were calculated at each position in the regenerator by using an equation[2], and the results is shown in Fig. 2. The temperature characteristics of the materials and He gas at each place in the regenerator are shown by a solid line (for materials) and a broken line (for gas) in Fig. 2, when Pb (58 %), Er_3Ni (17 %), $HoCu_2$ (13 %) and HoSb (12 %) are packed into the regenerator from high toward low temperatures. From this figure, it can be seen that a considerably large difference in the temperatures between high and low pressure periods exists at each place in the regenerator. A method to effectively use the regenerative materials which have a very large specific heat peak must be considered in such a case.

When a number of regenerative materials with very large specific heat peak are packed into a multi-layer regenerator, and the number of layers in the regenerator with a constant volume are increased, the volume of partition used to separate the layers are not negligible[3]. Hence, the two structures in the regenerator, as shown in Fig. 3 were compared to see if these regenerative materials having a very sharp specific heat peak enables the heat in the He gas be absorbed and emitted. Fig. 3 (a) shows the conventional

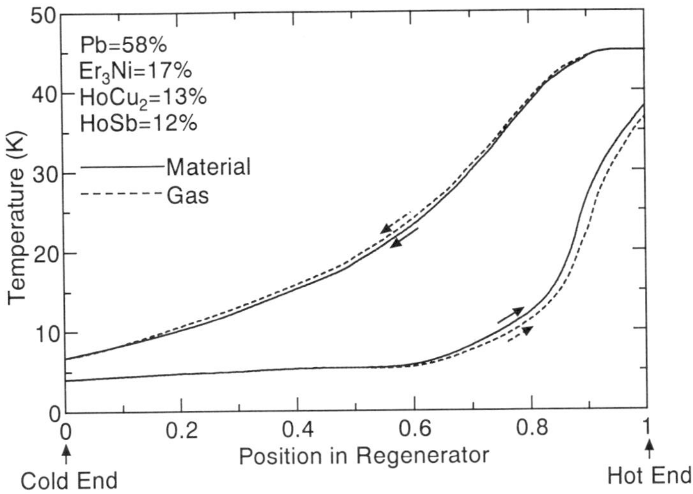

Figure 2. Temperature distribution in regenerator.

way the regenerative materials have been arranged, while Fig. 3 (b) shows a combination of multi-layer structure with synthetic materials in which the number of layers are limited and various materials with different peak temperatures are packed in one layer. A large synthetic specific heat peak was obtained in the latter case.

The regenerative effectiveness, η was calculated by using the enthalpy[4] (= Δ*Hreal* / Δ*Hideal*) for the two cases in Fig. 3, and the results are shown in Fig. 4. Δ*Hreal* and Δ*Hideal* are the actual and ideal enthalpy changes in the gas, respectively. In an ideal regenerator, η is 100 %. The specific heat distribution for the regenerator operation is considered best when η is close to 100 %. As regards the difference in the valves of η for Fig. 3 (a) and (b) was confirmed to be negligibly small. Thus the most practical structure of a regenerator would be a combination of a multi-layer structure with synthetic materials as shown in Fig. 3 (b).

ENTHALPY PROPERTY OF THE SYNTHETIC REGENERATIVE MATERIAL

The enthalpy properties for the cases of regenerative materials having a very sharp specific heat peak and a broad specific heat peak was investigated. The measured specific heats of HoSb and DySb are shown in Fig. 5. In this figure, three specific heats assumed from $Ho_xDy_{1-x}Sb$ (compound of HoSb and DySb) are also shown. These hypothetical specific heats are named A, B and C from the low temperature side. The same quantities

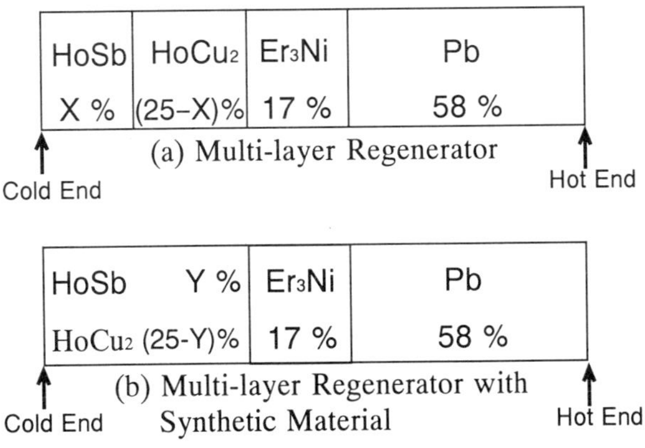

Figure 3. Structure of regenerator.

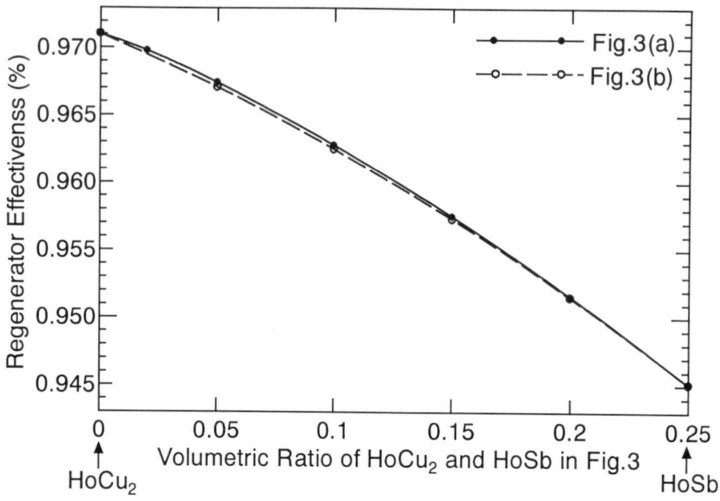

Figure 4. Regenerative effectiveness for regenerator in Fig. 3.

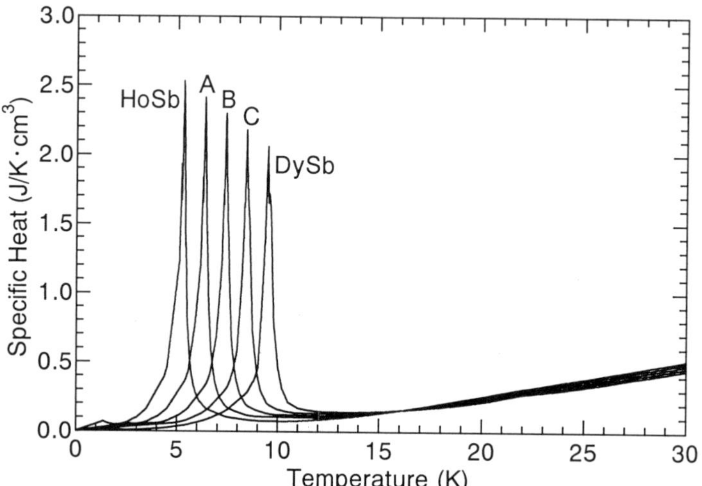

Figure 5. Hypothetical specific heat of $Ho_xDy_{1-x}Sb$.

of HoSb, DySb and B, and HoSb, DySb, A, B and C are mixed. Enthalpies calculated from the synthetic specific heat are shown in Fig. 6. The enthalpies of HoSb, DySb and $HoCu_2$ are also shown.

EXPERIMENT IN THE REFRIGERATION CAPACITY

Experiments to measure the refrigeration capacity were carried out with a two-stage GM refrigerator in order to compare their regenerative effectiveness using HoSb with a very large specific heat peak and $HoCu_2$ with a broad specific heat peak as regenerative materials.

The diameter of the spherical $HoCu_2$ was between 0.18 and 0.35 mm. An HoSb ingot was made by arc welding at the melting point of 2433 K (2160℃). The ingot was crushed and sifted through a sieve having a mesh of 0.149 to 0.355 mm. The selected pieces were not spherical.

Since the optimum conditions for HoSb were unknown, the best conditions for $HoCu_2$ were adopted in this experiment. The materials were packed into the second stage of the GM refrigerator (with compressor of 5 kW) having the inner radius of 16 mm and 20

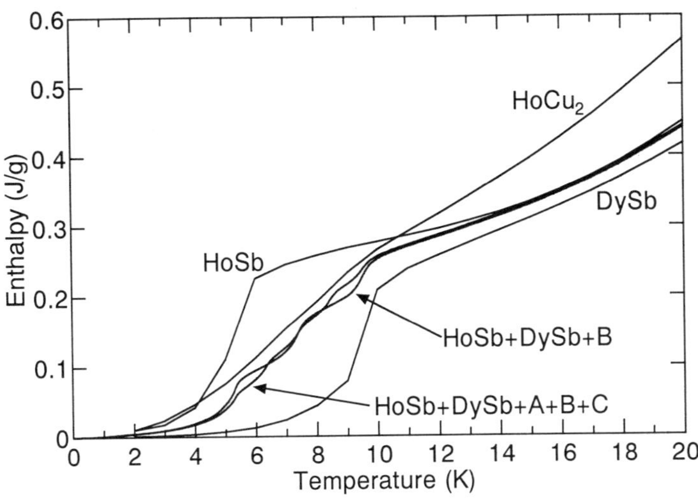

Figure 6. Enthalpy of compound and synthetic materials in same volume.

mm in length as shown in Fig. 3 (a). Spherical Pb with a diameter of 0.2 to 0.3 mm (58 %), spherical Er_3Ni with a diameter of 0.18 to 0.35 mm (17 %) were packed in the regenerator from the high temperature end to the low. The remaining 25 % was packed with HoSb (109.2 g) and $HoCu_2$ (130 g).

When the temperature at the cold end of the first stage regenerator was fixed at 45 K with the driving cycle of the displacer set at 74 rpm up to the refrigeration capacity of 10 W, the relation between the refrigeration capacity and the temperature at the cold end (the low temperature side) of the second stage regenerator is shown in Fig. 7. Superior property was obtained when a single material of 25 % $HoCu_2$ was used. When the property of 25 % HoSb packed into the low temperature side was compared with the property of 13 % $HoCu_2$ at the high temperature side and 12 % HoSb at the low temperature side, the difference between their properties under 6 K was negligible and at above 6 K, the latter was better than the former. When the properties of the above two cases and that of a single material of $HoCu_2$ were compared, the former were inferior to that of the latter except at around 6K. However, taking into account that because the HoSb was not spherical, the packed volume decreased by 10 %, and that the best driving conditions for $HoCu_2$ were adopted, the authors think that the property for HoSb below 6 K are equal to that for $HoCu_2$. Based on the enthalpy in Fig. 6, the authors think that the

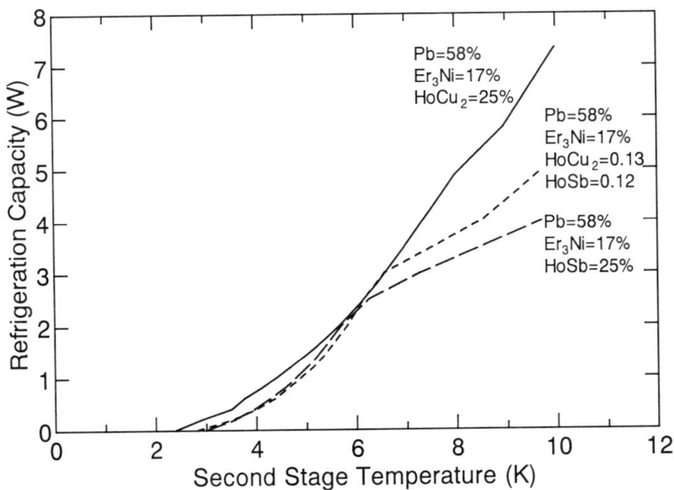

Figure 7. Refrigeration capacity used HoSb and $HoCu_2$ as regenerative materials.

refrigeration capacity of HoSb is superior to that of $HoCu_2$ in some region below 10 K.

CONCLUSION

A study to determine effective use of regenerative materials having a very large specific heat peak under 10 K was carried out. Conventional studies propose that these materials be packed in the multi-layer regenerator. However, our study indicates that a fairly large difference in the temperatures between the high and the low pressure periods exists, and the volume of the partition separating the layers are not negligible when the number of layers was increased. The authors propose in this paper that the number of layers in the multi-layer structure be limited and that different synthetic materials having sharp peaks or materials with sharp peak and broad peak be mixed in the one layer. It was confirmed by computer simulation that this combination is effective.

An experiment was carried out to compare the refrigeration capacities when $HoCu_2$, which has a broad specific heat peak, and HoSb, which has a very sharp specific heat peak at 5 K were packed into a 4 K GM regenerator. It was confirmed that HoSb has the potential to provide effective regeneration below 6 K. What remains to be done is that precise refrigeration capacity be obtained by computer simulation and experiments when a synthetic regenerative material is made by mixing a small amount of various different regenerative materials having a sharp specific heat peak packed into the low temperature side of the multi-layer regenerator.

REFERENCES

1. H.Nakane, T.Hashimoto, M.Okamura, H.Nakagome and T.Miyata, "Multi-layer Magnetic Regenerators with an Optimum Structure around 4.2K," Cryocoolers 10, Plenum Press, New York, (1999).
2. T. Hashimoto, M.Yabuki, T.Eda T.Kuriyama and H.Nakagome, "Effect of High Entropy Magnetic Regenerator Materials on Power of the GM refrigerator," *Adv. Cryog. Eng.* 40 :655 (1994).
3. T.Hashimoto, H.Nakane, T.Tsukagoshi and H.Nakagome, "Recent Progress in the Application of Magnetic Regenerator Materials," *Adv. Cryog. Eng.* 43:1541 (1998).
4. H.Seshake, T.Eda, K.Matsumoto and T.Hashimoto, "Analysis of Rare Earth Compound Regenerators Oprating at 4K," *Adv. Cryog. Eng.* 37B, Plenum Press, New York, p.995 (1992).

MEASUREMENT OF HEAT TRANSFER COEFFICIENTS IN HIGH NTU REGENERATIVE HEAT EXCHANGERS

J. A. Ramirez and K. D. Timmerhaus

University of Colorado, Boulder, CO 80309

ABSTRACT

An experimental study is summarized that incorporates the simple periodic temperature variation technique under controlled conditions for evaluating the heat transfer characteristics of high *NTU* regenerative heat exchangers. Details of the experimental apparatus and testing procedure are outlined. Application of the technique with actual matrices used in cryogenic systems shows that the heat transfer coefficients obtained correlate quite well with literature values obtained with the more complex ineffectiveness measurement technique.

INTRODUCTION

The prediction of heat transfer performance of regenerative heat exchangers was first performed independently by Anzelius [1], Hausen [2], Schumann [3], and Glaser [4]. However, the development of regenerative heat exchangers with high heat transfer area to solid core volume in the early 1960s for use in space and defense related applications called for still better heat transfer correlations.

Traditionally, the heat transfer characteristics of regenerative heat exchangers have been measured either under steady-state or transient conditions. The former approach was first proposed by Kays and London [5] in 1950 and was originally applied to compact recuperative heat exchangers designed to operate at low *NTU* (number of transfer units) ranges, specifically $0.2 < NTU < 3.0$. A variation of this approach applicable to regenerative heat exchangers fabricated of stacked wire screens and utilized in many cryocooler units was developed by Kim and Qvale [6] in 1973. However, the latter variation was recommended for use only at low *NTU* ranges.

The transient approaches presently are the most widely applied techniques and utilize many variations. However, all use a steady flow of the test fluid but differ in the nature of the temperature forcing function. These techniques are generally classified as (1) the step function single blow, (2) the exponential function single blow, and (3) the periodic or cyclic temperature method. The first technique is based on the mathematical solution to the problem of incompressible steady flow through a porous solid which is subjected to a sudden step change in the inlet fluid temperature. Many variations of the step function forcing method have been proposed; however, the principal deficiency with the technique has been the failure to experimentally reproduce a perfect step function. Liang and Yang [7] solved the problem analytically for an exponential forcing function of known steady-state gain and time constant. This more accurately represents the actual conditions of a test as the step becomes an exponential rise or decay due to the finite thermal inertia of the heating element. However, these two techniques experience significant inaccuracies at high *NTU* ranges.

An alternative to the step or exponential function based single blow methods is the use of a periodic or cyclical forcing function to heat the gas at the inlet of the regenerator. If this system is described by a set of linear equations, a periodic wave of the same frequency as at the inlet will appear downstream from the regenerator. Depending on the heat transfer characteristics of the matrix and the capacity of the matrix to store heat, the outlet temperature wave will exhibit an amplitude attenuation and a phase shift with respect to the original inlet signal. This may be used to determine the heat transfer coefficient between the matrix and the fluid. An important feature of this method as noted by Stang and Bush [8] is that there is no necessity to have a perfect sinusoidal input, since the system is linear and will require only the first harmonic of both temperature signals.

Researchers working with cryocoolers have taken a somewhat different approach to evaluating the heat transfer characteristics of regenerative heat exchangers, generally by measuring the ineffectiveness of the regenerator. The ineffectiveness measurement approaches are based on the technique proposed by Gifford [9] in 1969. Several variations have been developed, particularly by researchers evaluating Stirling engines; however, these variations are all based on Gifford's original concept. Gifford's test rig consisted of two identical regenerative heat exchangers through which the flow was alternated via a rotary valve. The heat losses were measured by submerging the connecting line between both regenerators in liquid nitrogen and measuring the boil-off rate.

The difficulty with the ineffectiveness approach is in the complexity (relative to the single blow experimental arrangement) of the experimental test rig because of the need for an alternating flow pattern. The approach also requires the necessity of operating at conditions which deviate significantly from ambient in order to be able to measure the heat loss term accurately.

In order for any method to be effective in measuring the heat transfer characteristics of high *NTU* regenerative heat exchangers one must require that it have a low sensitivity index with respect to the uncertainties in the measured parameters and variables. Based on existing literature, the periodic or cyclical technique appears to provide the best opportunity to obtain reliable heat transfer coefficients at high *NTU* values. The most important source of error in this method is an uncertainty in the measurement of the amplitude or phase shift, which result from uncertainties in the temperature measurements. The development of a mathematical model appropriately defining the periodic method along with a comprehensive sensitivity analysis of this technique has been detailed by Ramirez. [10]

APPARATUS AND TEST PROCEDURE

Test Apparatus

To obtain the heat transfer characteristics of certain regenerator matrices using the cyclic or periodic technique, it is necessary to measure the temperature of the test fluid at both the inlet and outlet of the regenerator matrix. It is also necessary to measure the mass flow rate of the test fluid passing through the matrix at any given time. The use of these two measurements along with known values of a modified dimensionless frequency Ω^* and an estimation of a dimensionless frequency λ_f (that relates thermal properties of the fluid) are sufficient for the extraction of an *NTU* value for any given experimental run. It is of key importance to reproduce the conditions that validate the assumptions under which the mathematical model was developed.

Figure 1 provides a schematic of the test apparatus. The system consists of two stages. The first stage is where the pressure and mass flow rate of the gaseous nitrogen serving as the test fluid are controlled. The second stage includes the heater and the test regenerator with associated instrumentation.

The heater is located inside the flared tee. The heater wire is supported in the flow stream with an annular shaped piece of glass fiber. The annular piece is epoxied to a stainless steel pipe stub which, in turn, is soldered to the centerpiece of a flared cap. The piece is fabricated in such a way that when the cap is tightly screwed in place, the axis of the annular heater support coincides with the centerline of the pipe.

The sinusoidal temperature wave is obtained with the aid of a function generator. A square root sine wave voltage is applied to the heater terminals to provide a sinusoidal power, and thus a fairly precise sinusoidal temperature perturbation. The instantaneous voltage is monitored by a digital multimeter and transmitted to a data acquisition program which calculates and records the corresponding instantaneous power.

The insertion of a flanged connection after the heater assembly permits changing of pipe diameters to accommodate the different regenerators tested in this study. The connecting flow system between the heater and the regenerator was designed in such a manner to ensure near plug flow conditions at the entrance to the regenerator. A schematic of the assembly is provided in Figure 2. The long, thin walled tube inside the larger diameter pipe stub provides a critical length for the velocity and temperature profiles to develop and smooth out. Twenty 200 mesh screens further assure a near plug flow regime immediately before the entrance into the regenerator.

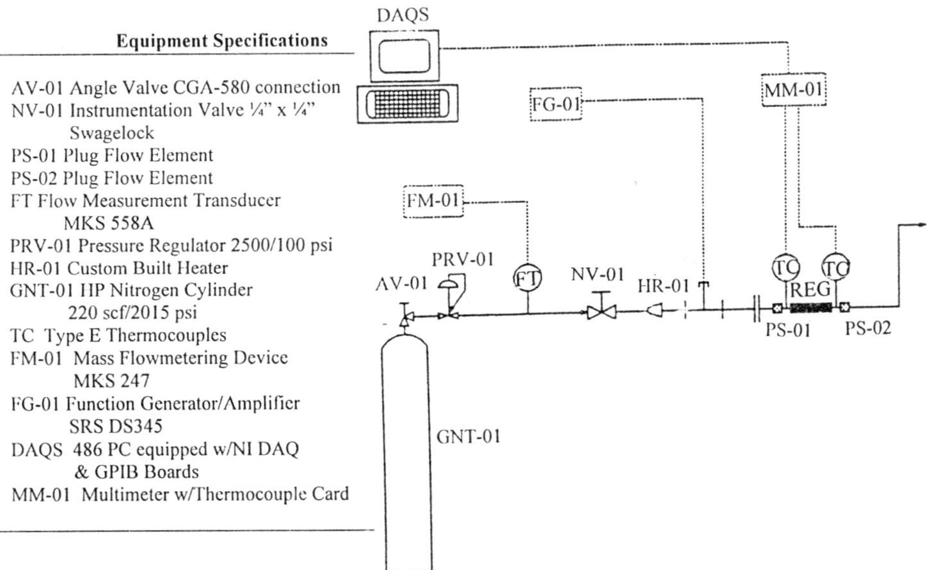

Figure 1. Test apparatus and instrumentation diagram.

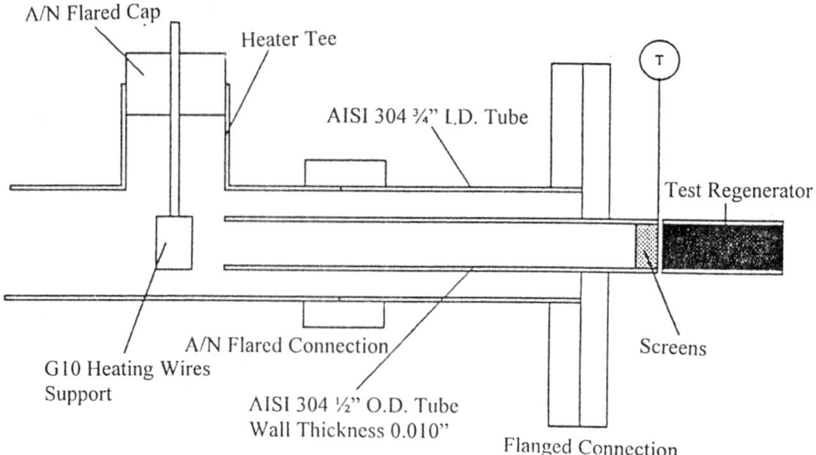

Figure 2. Schematic of the flow system prior to the entrance to the test regenerator core.

Five chromel-constantan (type E) thermocouples of equal length were evenly spaced across the cross-sectional area of flow at both the entrance and exit to the test regenerator. The connection in parallel results in an average voltage measurement and the resulting temperature will approximate an average measurement of the actual temperature distribution across the tube cross section. Care must be exercised that the thermocouples are located in immediate acess to the ends of the regenerator; otherwise, the measured amplitude ratio and phase shift will contain the effects of heat transfer and subsequent storage of energy in the enclosing tube wall. It is also important to assure a plug flow regime upstream and downstream at the locations where the temperature measurements are made. This validates the boundary conditions used in developing the model.

Test Procedure

Although the test apparatus is quite simple and straight-forward to operate, other factors pertaining to the methodology have to be considered for successful application of the technique. This involves not only selecting the optimum excitation frequency for the test but also determining the best combination of test core length and diameter.

From the mathematical model developed, it is desirable to operate the testing program at the lowest values of the product $\Lambda_m NTU$ and NTU, where Λ_m is a dimensionless quantity associated with the matrix material. From the mathematical analysis, it is evident that for a given mass flow rate a small core diameter for the regenerator configuration is desirable to minimize the product $\Lambda_m NTU$. The length of the regenerator core does not have an effect on this product; however, a long test core will result in a high NTU since it will increase the value of the matrix heat transfer area A_m. A short test core length would, in theory, appear to have a positive effect since this maintains the NTU at a low value. However, a test matrix that is too short will also have a low value of heat storage capacity because of the small mass contained within the core. It is apparent that when the product $M_m Cp_m$ is too small, the excitation will be very small (or the excitation frequency will be very high), and perhaps be unattainable in practice. Thus, the selection of the final dimensions of the test core implies an interactive process in which the actual slope of the NTU versus amplitude ratio curves is calculated for the corresponding A_m and the anticipated NTU for a desired Reynolds number. When a satisfactory low value of this slope is identified, the search process is terminated and the test core dimensions are finalized.

Following the above approach, a 150 x 150 mesh AISI 304 stainless steel screen Matrix contained in a casing with an inner diameter of 1.2×10^{-2}m and a length of 3.4×10^{-2}m was adequate for testing over a Reynolds number range from 30 to 250. Similarly, it was determined that for a packed bed of lead shot 3×10^{-4}m in diameter, dimensions of 1.2×10^{-2}m for the casing inner diameter and 1.5×10^{-2}m for the length were adequate. For coarser stainless steel screens and for larger diameter spheres in packed beds, greater lengths are advisable in order to maintain the excitation frequency at a reasonable value. On the other hand, for finer meshes and smaller spheres, smaller casing diameters are required while retaining the regenerator length to control the mass of matrix material and thus the value of the excitation frequency.

In order to select the optimal excitation period at which to test a given regenerator one must anticipate the *NTU* value for a given flow condition.* This is generally achieved by using a preexisting correlation if available or by extending the range of an existing correlation for a similar geometry. Solution of the key relationships developed in this study[10] over a desired range of modified dimensionless frequencies provides a series of curves which exhibit different slopes. Since the slope actually represents the sensitivity index $\partial NTU/\partial A$, the curve with the lowest value in the slope is of prime interest. This lengthy selection process may be performed on a plot that portrays the solution to these relationships for the desired flow conditions or may be solved numerically with the aid of a computer. Since the latter approach was more convenient, a computer program was developed to obtain the curve with the lowest slope. This same program was also used in the selection of an acceptable combination of test matrix casing diameter and length configurations to determine whether the given dimensions would result in an acceptable uncertainty in the *NTU* evaluation.

After the temperature and flow rate data are obtained for a test run, the amplitude ratio and phase shifts between the inlet and outlet signals are calculated. For this purpose a code based on the Fast Fourier Transform was written to incorporate the stored temperature data, the sampling time used and the number of data points recorded. This program provides the amplitude and phase shift spectrum of both in graphical and numerical form. Amplitude and phase data are presented for every frequency that is an integer multiple of f_s/N where f_s is the sampling frequency and N is the number of data points recorded. It should be noted that Shannon's sampling theorem should not be violated, i.e., the sampling frequency should always be higher than the highest expected signal frequency. However, when working with a sampling frequency that is near the optimal signal frequency it is necessary to gather data over several periods of the wave. Furthermore, the closer the sampling frequency is to the Nyquist frequency, the more periods must be included in a given sample. This is necessary in order to preserve the information contained in the original measured signal.

Once the amplitude and phase values at the fundamental frequency are obtained, the amplitude ratio and phase shift are calculated from

$$A = A_f / A_o \quad \text{and} \quad \Delta = \Phi_f - \Phi_o \tag{1}$$

where A_o represents the amplitude of the inlet temperature wave at the fundamental frequency, A_f the amplitude of the outlet temperature wave at the fundamental frequency, and Φ_f and Φ_o the outlet and inlet temperature wave phase angles in radians. The amplitude ratio value is used in the solution of the above relationships to evaluate the number of transfer units.

* This is a major deficiency of the periodic or cyclic technique under high *NTU* conditions.

APPLICATION OF THE PERIODIC TECHNIQUE

Two typical regenerator matrices were tested in this study to evaluate the practical application of the periodic technique and compare the heat transfer results obtained under high *NTU* conditions with existing correlations developed for lower *NTU* ranges.

Mesh Screen Regenerator

The geometrical dimensions of the 120 mesh AISI 304 stainless steel mesh regenerator include a wire diameter of 66μm, openings of 104.1μm, hydraulic diameter of 149.6μm, mass of 8.36×10^{-3}kg, porosity of 0.75, and a heat transfer area of 0.0786 m^2. The regenerator dimensions of diameter (1.2×10^{-2}m) and length (3.4×10^{-2}m) were obtained using the iterative procedure outlined earlier. This combination of regenerator diameter and length provided values of $\Lambda_m NTU$ and NTU that resulted in acceptable values for the sensitivity index $\partial NTU/\partial A$ over the desired range of Reynolds numbers. These dimensions also provided adequate heat capacity in the matrix so that operation at the optimal dimensionless frequency would not result in unattainable excitation periods.

The measured amplitude ratios and phase shifts as well as the test frequencies are presented in Table 1. The dimensionless Λ_m parameters which are mass flow rate dependent are also shown. The dimensionless frequencies were obtained with the aid of a computer program and the preliminary *NTU* estimation (required when using the periodic technique) was obtained from the empirical correlation of Kays and London [11]. The resulting *NTU* value is then found by numerically solving the relationships developed in the mathematical analysis. An alternate method is to plot the solution curves for each experimental case and from these determine the intersection of the measured amplitude ratio with the solution curve.

Table 1. Measured amplitude ratio and phase shift for the fundamental harmonic of the inlet and outlet temperature waves for stainless steel mesh screen regenerator

Run no.	m_f, L/s †	A_o/A_i	Phase shift, rad.	Λ_m	Ω^*	Freq., Hz
1	0.428	0.26	-8.73	0.0072	9.05	0.20
2	0.515	0.28	-8.79	0.0060	9.04	0.24
3	0.687	0.27	-9.29	0.0045	9.60	0.34
4	0.772	0.24	-9.61	0.0040	10.05	0.40
5	0.857	0.28	-9.43	0.0036	9.51	0.42
6	1.250	0.42	-7.62	0.0025	7.76	0.50

† At STP conditions

Table 2 summarizes the heat transfer results. The measured heat transfer coefficient is obtained from

$$h = NTU_{exp} \, (mCp)_f / A_m \tag{2}$$

Table 2. Measured heat transfer values for the stainless steel mesh screen regenerator

Run no.	Re	Nu	NTU	h, W/m^2K
1	50	3.47	98	696
2	60	4.04	95	810
3	80	5.33	93	1069
4	90	5.78	90	1159
5	100	6.39	89	1281
6	146	8.40	81	1684

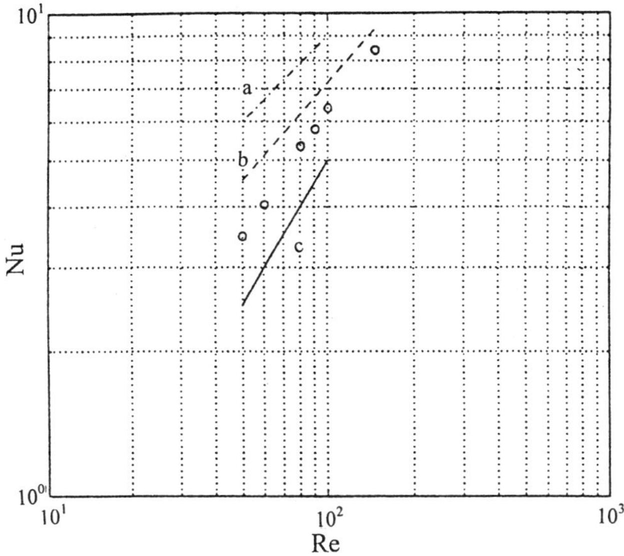

Figure 3. Comparison of measured data with correlations from (a) London, (b) Tanaka et al, and (c) Rice for the same type of stainless steel mesh regenerator.

The measured data points are shown as a function of the Reynolds and Nusselt numbers in Figure 3 and compared with the well-known correlations of Kays and London [11], Tanaka et al [12], and Rice [13] for the same type of mesh screens. For the Reynolds number range considered, the measured data agree reasonably well with the predicted values.

Lead Sphere Regenerator

The experimental studies involving lead spheres as the matrix material covered a Reynolds number range between 500 and 1000. The geometrical characteristics of the lead shot are sphere diameter 3.05×10^{-4}m, mass 0.0142 kg, porosity 0.31, and heat transfer area $0.0242m^2$. The casing diameter and appropriate regenerator length were determined using the same methodology as outlined for the mesh screen matrix. The initial estimate of the *NTU* was obtained from the literature[14]. The selection of the optimal dimensionless frequency Ω* was again determined with the aid of the computer program noted above. The heat transfer coefficient is calculated as before with the Nusselt number based on the sphere diameter. The heat transfer results are summarized in Table 3. A comparison with established correlations [14, 15] shows that the measured data agree reasonably well with such values.

Table 3. Measured heat transfer values for the lead sphere matrix regenerator

Run no.	Re	Nu	NTU	h, W/m^{2}K
1	500	24.4	51	2401
2	700	26.9	40	2635
3	800	27.7	36	2711
4	1000	28.9	20	2835

CONCLUSIONS

The periodic technique used for determining heat transfer characteristics has only been applied in the evaluation of reasonably low *NTU* matrices. In this study the technique has been extended to geometries possessing considerably higher heat transfer areas where the traditional methods of steady state and single blow step testing are not applicable. It has been demonstrated that the experimental apparatus and procedure are rather simple in comparison to the other viable method of ineffectiveness measurement utilizing double regenerators. A typical test run requires less than an hour, including the recording of four data samples. Data analysis is equally straightforward, requiring approximately the same amount of time.

ACKNOWLEDGMENT

The authors express their thanks to the Fundacion Gran Mariscal de Ayacucho for providing financial support for this study and to Ray Radebaugh, Peter Bradley, Mike Lewis and Eric Marquardt of the National Institute of Standards and Technology (NIST) for their technical guidance in the experimental phases of the research performed.

REFERENCES

1. A Anzelius, *Z. Angew. Math.* 6:291 (1926).
2. H. Hausen, *Z. Angew. Math.* 9:173 (1929).
3. T.E. Schumann, *J. Franklin Inst.* 208:405 (1929).
4. H. Glaser, *Z. VDI, Beih. Verfahrenstech.* 4:112 (1938).
5. W.M. Kays and A.L. London, *Trans. ASME.* 72:1075 (1950).
6. J.C. Kim and E.B. Qvale. *Adv. Cryogenic Eng.* 16:302 (1971).
7. C.J. Liang and W.J. Yang, *J. Heat Transfer.* 97:16 (1975).
8. J.H. Stang and J.E. Bush, *J. Eng. Power.* 96A(2):87 (1974).
9. W.E. Gifford, A. Acharya, and R.A. Ackerman, *Adv. Cryogenic Eng.* 34:353 (1969).
10. J.A. Ramirez, "The Measurement of Heat Transfer Coefficients in High NTU Regenerator Heat Exchangers", University of Colorado MS Thesis (1995).
11. W.M. Kays and A.L. London, "Compact Heat Exchangers", McGraw-Hill, New York (1984).
12. M. Tanaka, I. Yamashita, and F. Chisaka, *JSME Int. J. Serles II* 33:283 (1990).
13. G. Rice, M. Thonger, and M.W. Dadd, *Proc. 20th ICEC* 859 (1985).
14. K.D. Timmerhaus and T.M. Flynn, "Cryogenic Process Engineering", Plenum Press, New York (1989).
15. A. Jeschar, *Arch. Eisenhüttenw.* 35 (1964).

DESIGNING METHODS OF CRYOGENIC REGENERATORS FOR GAS HIGH PURIFICATION

Anisia Bornea[1], I.Cristescu[1], M.Peculea[2]

[1] Institute of Cryogenics & Isotope Separations
P.O. BOX 10,1000 Rm.Valcea, Valcea, Romania
[2] Romanian Academy
Calea Victoriei 125, 71102 Bucharest, Romania

ABSTRACT

Many cryogenic chemical processes require using high purity gases. For high purification of gases, from several methods, we chose the retention of the impurity on a cryogenic regenerator. The retention of gas impurities is achieved by condensation on the cooled package of the regenerator.

In order to design a regenerative system for purifying the gases at low temperatures, it is necessary to experimentally determine the specific coefficients for the thermal and mass exchange processes. These coefficients are experimentally determined as functions of, the gas nature and the impurity's nature, the hydrodynamic parameters and the regenerator type.

A mathematical model is developed that describes the thermal and mass exchange in regenerator together with the functions of thermal and mass exchange coefficients versus hydrodynamic parameters allowed the designing of the regenerative system for gas purification.

As an example we chose an experimental model for designing a regenerative system for hydrogen purification at cryogenic temperatures with carbon dioxide impurities, using metallic package regenerators.

1. INTRODUCTION

In cryogenic processes regenerative heat exchangers are used as heat exchangers and/or mass exchangers when they have a prevailing role for gas purification. For designing cryogenic regenerators it is necessary to determine experimentally the average thermal and mass exchange coefficients of the impurities condensation. These coefficients are specific for the type of regenerator used and for the hydrodynamic parameters.

We analyzed the thermal exchange and experimentally determinated the values of the mean thermal coefficients as a function of the hydrodynamic parameters. After that we

Advances in Cryogenic Engineering, Volume 45.
Edited by Shu *et al.*, Kluwer Academic / Plenum Publishers, 2000.

analyzed the successive thermal and mass exchange by retaining the impurities by condensation on the cooled regenerator package.

The regenerative system is composed of two regenerators which alternately functioning of the gas purification process.

One of the regenerators works on retaining impurities by condensation on the cooled regenerator package. At the same time, the second regenerator works on the cleaning process, flowing hot gas over the regenerator package. In this way the impurities are sublimated from the regenerator package.

Then is begun the cooling process of the regenerator package in order to prepare it for the gas purification process. That is realized by flowing cool gas over the regenerator package until it has obtained a constant temperature along the regenerator which has a value smaller then the value of the temperature corresponding to the sublimation of the impurities.

This paper presents an experimental model for designing a regenerative system for hydrogen purification in a continuous process.

The designing model consists of regenerator operating parameters determined at the cleaning process and at the cooling process. The model is based on the fact that the sum of the functioning times corresponding to the cleaning and cooling for the first regenerator must be equal to the functioning time for gas purification in the second regenerator.

2.EXPERIMENTAL SET-UP

The experimental set-up is presented in figure 1.

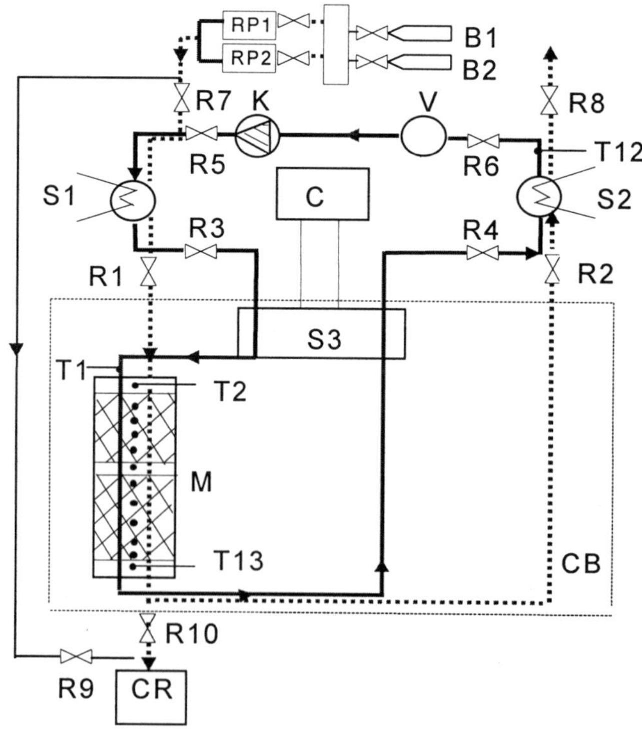

Figure 1. M - regenerator; S - heat exchanger; K - compressor; C - cryogenerator; V - buffer; R - valve; RP- pressure regulator; CB - cold box; T - thermocouple; B - bottle; CR – chromatograph.

We chose a cylindrical regenerator, with a metallic package made from lead balls, in order to have a greater thermal capacity of the regenerator.

The regenerator functioning is alternative: cooling, purification and cleaning. The first step of the experiment is cooling the regenerator to liquid nitrogen temperature (~80K) with a pure hydrogen gas. After that, at a controlled temperature equal with the ambient, a gas-hydrogen impurified with carbon dioxide flow over the cooled package. We analyzed the retention of impurities' particles - carbon dioxide from hydrogen, by condensation on the cooled package, following the time variation along the package of these parameters: gas temperature, package temperature, impurity concentration of gas and the pressure drop. The temperatures were measured with Fe-constantan thermocouples. The thermocouple's signal is introduced in an analogue multiplexor that selects it for a milivoltmeter and then is stored by an interface in a PC. We analyzed the influence of hydrodynamic parameters and impurity concentration in thermal and mass exchange.

3.REGENERATOR MODELLING

3.1. Thermal regenerator modeling

In the thermal regenerator the thermal exchange is realized between the gas and the package. In order to describe the thermal process, which occurs along the regenerator, we made the following assumptions:

- the gas temperature at the inlet of the regenerator package is constant;
- the package temperature at the zero time is knows along the regenerator;
- the mass flow rate and the specific heat are constants along the regenerator and in time;
- mass density and the specific heat of the package are constants along the regenerator and in time;
- the thermal conductivity of the package along the regenerator is zero;
- the mean thermal coefficient is constant along the regenerator and in time;
- the thermal losses of the regenerator are zero;

The thermal balance[1,2] refered to a surface and to an infinitesimal time is:

$$dQ = \overline{\alpha}(T - T_m)df\,dt = -C\left(\frac{\partial T}{\partial f}\right)_t df\,dt \qquad (1)$$

The stored heat in the package is:

$$dQ = dC_s\left(\frac{\partial T_m}{\partial t}\right)_f dt \qquad (2)$$

The equations (1) and (2) and the boundary conditions allowed the determination of the time temperature variation of the gas and package along the regenerator.

We introduced the reduced length and reduced time by the following equations:

$$\xi = \frac{\overline{\alpha}}{C} f$$
$$\eta = \frac{2n\overline{\alpha}}{\rho_m c_m \delta} t \qquad (3)$$

and we obtained the following system of equations:

$$\frac{\partial^2 T}{\partial\xi\partial\eta}+\left(\frac{\partial T}{\partial\xi}\right)_\eta+\left(\frac{\partial T}{\partial\eta}\right)_\xi=0$$
$$\frac{\partial^2 T_m}{\partial\xi\partial\eta}+\left(\frac{\partial T_m}{\partial\xi}\right)_\eta+\left(\frac{\partial T_m}{\partial\eta}\right)_\xi=0 \tag{4}$$

We solved the equations (4) using finite difference substitutions:

$$\frac{\partial T_m}{\partial\xi}=\frac{1}{2}\cdot\left(\frac{T_{m2}-T_{m1}}{\Delta\xi}+\frac{T_{m4}-T_{m3}}{\Delta\xi}\right)$$

$$\frac{\partial T_m}{\partial\eta}=\frac{1}{2}\cdot\left(\frac{T_{m3}-T_{m1}}{\Delta\eta}+\frac{T_{m4}-T_{m2}}{\Delta\eta}\right) \tag{5}$$

$$\frac{\partial^2 T_m}{\partial\xi\cdot\partial\eta}=\frac{1}{\Delta\eta}\cdot\left(\frac{T_{m4}-T_{m3}}{\Delta\xi}-\frac{T_{m2}-T_{m1}}{\Delta\xi}\right)$$

The gas temperature T is simultaneously determined with package temperature T_m using the similar equations and we obtained:

$$T_{m4}=\frac{1}{(2+\Delta\xi+\Delta\eta)}\cdot((2-\Delta\xi+\Delta\eta)\cdot T_{m3}+(2+\Delta\xi-\Delta\eta)\cdot T_{m2}+(-2+\Delta\xi+\Delta\eta)\cdot T_{m1})$$

$$T_4=\frac{1}{(1+2\cdot\Delta\xi)}\cdot(2\cdot\Delta\xi\cdot T_{m4}+T_1-T_2+T_3) \tag{6}$$

3.2. Thermal and mass regenerator modeling

In a transversal section of the regenerator, on an infinitesimal exchange surface df, with a thermal capacity dC_s, into an infinitesimal time dt, the heat transfer from package regenerator to gas is consumed by impurity condensation and decreasing gas temperature[1,2,3] :

$$dC_s\left(\frac{\partial T_m}{\partial t}\right)_f dt+r_s\left(\frac{\partial m_w}{\partial t}\right)_f df\cdot dt+C\cdot dt\left(\frac{\partial T}{\partial f}\right)_f df=0 \tag{7}$$

The thermal exchange between the gas and the package regenerator is described by:

$$C\cdot dt\left(\frac{\partial T}{\partial t}\right)_t df=\alpha\cdot df\cdot(T_m-T)dt \tag{8}$$

and the mass exchange is described by:

$$\left(\frac{\partial m_w}{\partial t}\right)_f df\cdot dt=\sigma\cdot df(x''-x)dt \tag{9}$$

The equations (7) – (10) are solved with the definition of the reduced impurity in gas φ and of the Lewis number[1] ε and using the boundary conditions:

$$\begin{cases} T = T_0 \\ T_m = T_{m0} \quad for\ \eta = 0 \\ x = x_0 \end{cases} \qquad \begin{cases} T = T_i \\ T_m = T_{mi} \quad for\ \xi = 0 \\ x = x_i \end{cases} \tag{10}$$

We obtained the variation in time of the gas temperature and regenerator temperature, along the regenerator:

$$\varepsilon \frac{\partial^2 T_m}{\partial \xi \cdot \partial \eta} + \left(\frac{\partial T_m}{\partial \eta} \right)_\xi + \left(1 + \frac{d\varphi"}{dT_m} \right) \left(\frac{\partial T_m}{\theta \xi} \right)_\eta = 0 \tag{11}$$

This equation has a single variable T_m, because $d\varphi''/dT_m$ can be calculate in function of T_m, using the condensation pressure equations for the impurity.

We obtained:

$$T_{m4} = T_{m1} + \frac{2\varepsilon \left(T_{m3} + T_{m2} - 2T_{m1} \right) + \left[\Delta\xi - \left(1 + \frac{d\varphi"}{dT_m} \right) \Delta\eta \right] \left(T_{m2} - T_{m3} \right)}{2\varepsilon + \Delta\xi + \Delta\eta \left(1 + \frac{d\varphi"}{dT_m} \right)} \tag{12}$$

$$T_4 = \frac{1}{1 + 2\Delta\xi} \left(2\Delta\xi T_{m4} + T_1 - T_2 + T_3 \right) \tag{13}$$

$$x_4 = \frac{1}{\varepsilon + 2\Delta\xi} \left[2\,\Delta\xi\, x_4^{"} - \left(x_2 - x_1 - x_3 \right) \varepsilon \right] \tag{14}$$

Because the condensation enthalpy of carbon dioxide has a small variation with temperature, it was considered constant.

4.RESULTS

We investigated hydrogen purification from carbon dioxide. For various functioning regimes of the regenerator we analyzed the retaining phenomena of the carbon dioxide on the cooled package. We determined Nusselt criteria for thermal exchange: $Nu=0.825 \cdot Re^{0.48} \cdot Pr^{1/3}$. For the mass exchange we determined the sublimation coefficient: $\sigma=2.3994074 \cdot 10^{-4}+1.5988529 \cdot 10^{-6} \cdot Re - 7.6885779 \cdot 10^{-10} \cdot Re^2$ [$kg \cdot m^{-2} \cdot s^{-1}$] for carbon dioxide.

The flow rate was varied in range 0.6 - 4 $Nm^3 \cdot h^{-1}$, at a gas pressure value of 2 bar and for a concentration of carbon dioxide in hydrogen gas $\approx$ 1 - 10 % [m^3imp$\cdot m^{-3}$ gas], measured with a chromatograph. For every experiment, it was analyzed the variation of the gas temperature and regenerator temperature, along the regenerator and the variation of carbon dioxide concentration at the outlet of the regenerator. For example we considered the case of gas purification at the flow rate 1.3 $Nm^3 \cdot h^{-1}$, pressure value 2 bar and carbon dioxide in hydrogen -concentration $\approx$9 % [m^3 $CO_2 \cdot m^{-3}$ gas]. The experimental data about carbon dioxide concentrations in gas allowed to determine the maximum effective time of the regenerator for purification. This time is approximately equal with 8800 s and it corresponds to the maximum admissible impurity concentration 100 ppm.

In figure 2 are presented the experimental data of gas temperature along the regenerator at different times.

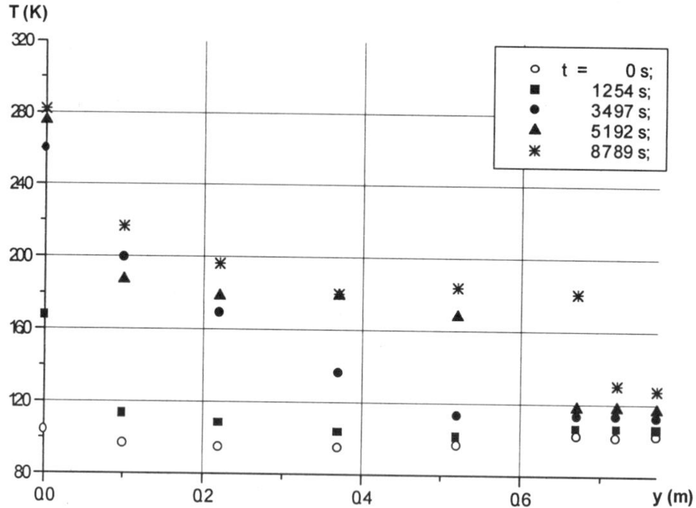

Figure 2. The experimental data of gas temperature along the regenerator at different times for the purification process.

The partial pressure of carbon dioxide for this experiment was ≈ 0.074 bar. At this pressure value, in the T-S diagram of carbon dioxide it is observed that the solidification phenomena occurrs at a temperature lower than 170 K. Thus, the thermal analysis of the purification process is very important. It points out in every moment of time the part of the regenerator where the retaining of the impurity by solidification is taking place.

We tried to determine the hydrodynamic parameters at the continue functioning of the regenerator, in a technological flow.

The experimental analysis of the regenerator during the retaining process of the impurities, allowed to obtain the necessary values for designing the cleaning process and cooling process of the regenerator.

Consequently the sum of the functioning times of the regenerator during the cleaning process and during the cooling process not be greater then the maximum functioning time of the regenerator on the purification process, that is ≈ 8800 s.

Also, the temperature profile obtained from the experimental data on the package, at the end purification process, represents the initial temperature profile for computation of the cleaning process.

For the cleaning process computation we used the pre-determinated Nusselt criteria. It is important to notice that, this equation of the Nusselt criteria is specific to the regenerator type and it is specific for the hydrodynamic conditions.

The cleaning process of the regenerator occurs until it has attained the value of the package temperature at the outlet of the regenerator, approximately equal with 200 K.

We chose this value of the temperature, greater then 170 K that represents the maximum temperature for carbon dioxide impurity condensation.

The time corresponding for attaining that value of temperature is the maximum functioning time of the cleaning process.

The temperature profile obtained corresponding to this time represents the initial profile for the cooled process of the regenerator.

The cooling process of the regenerator occurs until it is attained the value of the package temperature along the regenerator equal with 80 K.

We chose this value of temperature, smaller then the value 170 K corresponding to carbon dioxide condensation temperature.

In order to have the same value of gas velocity both for cleaning and cooling processes, there were computed the following parameters: the flow rate 7.73 $Nm^3 \cdot h^{-1}$, the gas pressure value 2 bar. These parameters were computed considering the basic assumption of equal times for cleaning and cooling process in regenerator 1 and purifying process in regenerator 2.

The maximum functioning time for cleaning process is 1600 s.

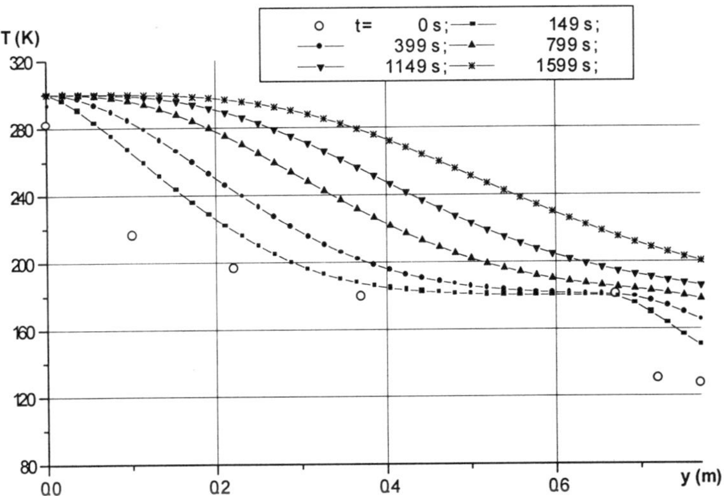

Figure 3. The variation of the gas temperature along the regenerator at different times for the cleaning process.

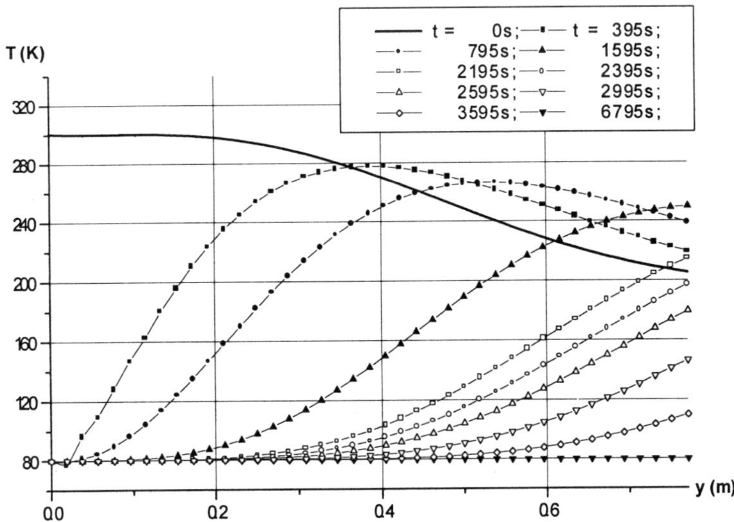

Figure 4. The variation of the gas temperature along the regenerator at different times for the cooling process.

In figure 3 is represented the variation of the gas temperature along the regenerator at different times for the cleaning process. The maximum functioning time for cleaning process is 6800 s.

In figure 4 is represented the variation of the gas temperature along the regenerator at different times for the cooling process.

The experimental results represent the background for designing a hydrogen purification plant from impurities (carbon dioxide or water) at low temperature in a regenerative system with alternative function.

NOMENCLATURE

Roman symbols: C - thermal capacity of the fluid ($W \cdot K^{-1}$); C_s - thermal capacity of the regenerator ($J \cdot K^{-1}$); c_m – specific heat ($J \cdot kg^{-1} \cdot K^{-1}$); f - exchange surface (m^2); m_w - mass of condensate of the impurity, respectively ($kg \cdot m^{-3}$); Nu - Nusselt number; Pr - Prandtl number; Re - Reynolds number; r_s – sublimation enthalpy of the impurity ($kJ \cdot kg^{-1}$); t- time (s); T - gas temperature (K); T_m - regenerator temperature (K); x - impurity fraction of gas (kg imp$\cdot kg^{-1}$ gas); x" - impurity fraction of saturate gas at the temperature T_m (kg imp$\cdot kg^{-1}$ gas); **Greek symbols:** α - mean heat transfer coefficient ($W \cdot m^{-2} \cdot K^{-1}$); δ – diameter of lead balls [m]; ε - Lewis number; η - reduced time; φ - reduced fraction of the impurity in gas; φ'' - reduced fraction of the impurity in saturate gas at the temperature T_m; ρ_m - package density ($kg \cdot m^{-3}$); σ - medium sublimation coefficient ($kg \cdot m^{-2} \cdot s^{-1}$); ξ - reduced length;

REEFERENCES

[1] H. Hausen - Wärmeübertragung im Gegenstrom, Gleichstrom und Kreuzstrom, Springer - Verlag Berlin Heidelberg, New York 1976;

[2] R. Gregorig - Wärmeaustausch und Wärmeaustauscher, Verlag Sauerländer Aaran und Fraukfurt am Main 1973;

[3] N. Afgan, E.U. Schunder - Heat exchangers: Design and Theory Sourcebook, Mc.Graw - Hill Book Company 1974;

[4] M. Peculea, A.Mosteanu – "Transfert de chaleur dans un regenerateur thermique a basse temperature", Revue Generale de Thermique, 1996, N.35, pp. 208-213, Elsevier, Paris;

A SIMPLIFIED LINEARIZED THERMOACOUSTIC THEORY FOR REGENERATIVE CRYOCOOLERS

LUO Ercang, LIU Haidong, XIAO Jiahua, ZHOU Yuan, ZHANG Zhong

Cryogenic Laboratory, Chinese Academy of Sciences
Beijing 100080, P. R. China

ABSTRACT

A simplified linearized thermoacoustic model for regenerative refrigerators is presented to give an analytical correlation of thermodynamic performance depending on the structure and operation parameters. With the guidance of the model, it is feasible to fast predict the thermodynamic performance of the regenerative refrigerators.

INTRODUCTION

Linearized thermoacoustic theory is a powerful new tool for understanding the internal working mechanism of regenerative cryocoolers such as Stirling cryocooler, pulse tube cryocooler and thermoacoustic refrigerator, but it is a locally analytical one. Without a further numerical simulation, it is unrealistic to obtain their dynamic and thermodynamic performance depending on operating parameters and structure parameters. Thus it is difficult to quickly estimate the performance of the cryocoolers. Based on the linearized thermoacoustic theory founded by N.Rott (1980) [1] and later improved by G. Swift (1988) [2] and J.H. Xiao (1991) [3], a simplified linear thermoacoustic theory [4] is presented to obtain the internal dynamic parameters and the thermodynamic performance of the above-mentioned cryocoolers. In the simplified theory, the simplified models of the different components such as regenerator, pulse tube, compressor chamber, connecting pipe and orifice are first given. Then, a straightforward correlation of the thermodynamic performance related to the operating parameters and structure parameters of the Stirling and pulse tube refrigerators are obtained.

SIMPLIFIED LINEARIZED THERMOACOUSTIC THERORY

The linearized thermoacoustic theory gives the equations describing the dynamic parameters and the time-average energy flux and temperature of any thermodynamic elements (Xiao 1992,1995) [5]:

$$\frac{d\tilde{p}}{dx} = -Z_f \tilde{U} \tag{1}$$

$$\frac{d\tilde{U}}{dx} = -\frac{1}{Z_c}\tilde{p} + f_{wt}\beta_0 \frac{dT_0}{dx}\tilde{U} \tag{2}$$

$$\frac{dT_0}{dx} = \frac{\frac{1}{2}\mathrm{Re}[\tilde{U}\tilde{p}^*(1 - T_0\beta_0 f_{qx})] - E_x}{A_f K_{eff}} \tag{3}$$

$$\frac{dE_x}{dx} = 0 \tag{4}$$

Eq.(1) to eq.(4) are constant differential equations, but thermodynamic variables, $\tilde{p}, \tilde{U}$, E_x and T_0 are interdependent. To obtain these thermodynamic parameters usually needs to make a further numerical simulation. However, by making reasonable simplifications, it is possible to obtain a straightforward correlation about thermodynamic parameters.

Introducing two auxiliary functions $\Theta(x), \tilde{V}(x)$ and letting $\tilde{U}(x) = \Theta(x)\tilde{V}(x)$ can yield the following equations:

$$\frac{d\tilde{p}}{dx} = -Z_f \Theta \tilde{V} \tag{5}$$

$$\frac{d\tilde{V}}{dx} = -\frac{1}{Z_c \Theta}\tilde{p} + \tilde{V}(f_{wt}\beta_0 \frac{dT_0}{dx} - \frac{1}{\Theta}\frac{d\Theta}{dx}) \tag{6}$$

Letting $f_{wt}\beta_0 \frac{dT_0}{dx} - \frac{1}{\Theta}\frac{d\Theta}{dx} = 0$, we can obtain the auxiliary function $\Theta(x)$ as the following:

$$\Theta(x) = e^{\int f_{wt}\beta_0 dT_0} \tag{7}$$

Bring $\Theta(x)$ into eq.(5) and eq.(6) and integrating them from the inlet to outlet of a thermodynamic element will produce the following equations:

$$\tilde{p}_{out} - \tilde{p}_{in} = \int(-Z_f \Theta \tilde{V})dx \tag{8}$$

$$\tilde{V}_{out} - \tilde{V}_{in} = \int(-\frac{1}{Z_c \Theta}\tilde{p})dx \tag{9}$$

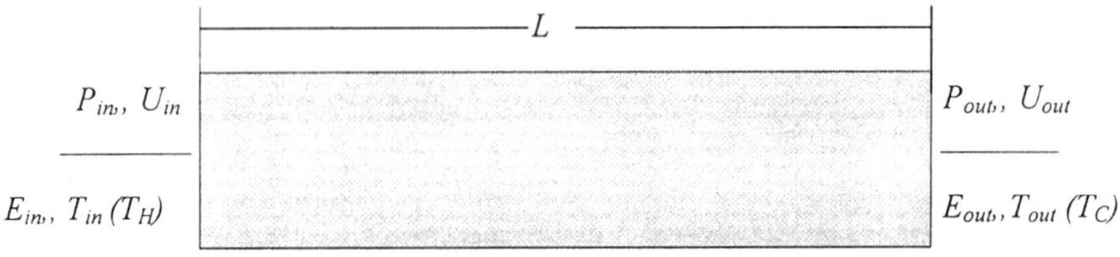

Figure 1. The physical and analytical model of any thermodynamic element

Replacing $\widetilde{V}$ with $0.5(\widetilde{V}_{in}+\widetilde{V}_{out})$, $\widetilde{p}$ with $0.5(\widetilde{p}_{in}+\widetilde{p}_{out})$, and $\overline{Z}_f$ at the average temperature of $0.5(T_{in}+T_{out})$, we can derive the transmission function matrix of the thermodynamic element:

$$\begin{pmatrix}\widetilde{p}_{out}\\ \widetilde{U}_{out}\end{pmatrix}=\begin{pmatrix}Z_{11} & Z_{12}\\ Z_{21} & Z_{22}\end{pmatrix}\begin{pmatrix}\widetilde{p}_{in}\\ \widetilde{U}_{in}\end{pmatrix}=Ł\begin{pmatrix}\widetilde{p}_{in}\\ \widetilde{U}_{in}\end{pmatrix} \tag{10}$$

For different thermodynamic elements, they have different characteristic size and undergo the different thermodynamic processes, so they have different simplified models. Fig.1 is the physical and analytical model of any thermodynamic element.

For regenerator, $Z_f, Z_c, f_{wt}, f_{qx}, K_{eff}$ are defined as follows:

$$Z_f=\frac{j\omega\,\rho_0}{A_f}(1+\beta_t)+\frac{\alpha\,\mu_0}{A_f},\quad Z_c=\frac{j\omega\,A_f}{\rho_0 a_0^2}[1+(\gamma-1)f_h],$$

$$f_{wt}=f_h=\frac{(1-\phi\,)\Omega_H}{j\omega\,\phi\,(1-\phi\,)+\Omega_H}$$

$$K_{eff}=K_0+\frac{A_s}{A_f}K_{s0}+\frac{1}{2}\rho_{\,0}C_{p0}\frac{|\widetilde{U}|^2}{A_f^2\omega}g_{qx},\quad f_{qx}=f_h^{*},\quad g_{qx}=\mathrm{Re}[jf_h]$$

Because the metallic materials of regenerator usually being screen-packed have much more bigger heat capacity than that of working gas (helium) above 40K. Moreover, the angle frequency of characteristic heat exchange Ω_h is much bigger than the operating frequency of cryocoolers, ω. Bearing these features of the regenerators in mind, we can obtain the simplified expressions about f_{qx}:

$$f_{qx}\approx(1-\phi\,)[1+j\frac{\phi\,\omega\,(1-\phi\,)}{\Omega_h}] \tag{11}$$

The factor f_{qx} defined by eq.(10) will be used in later energy analysis. To derive the transmission function of regenerator, we make a further simplification about f_{wt}, letting $f_{wt}\approx 1$. In addition, the thermal expansion coefficient β_0 of idea gas is equal to $1/T_0$. Eventually, we can obtain the transmission function matrix of regenerator:

$$Ł=\begin{pmatrix}1 & -\frac{1}{2}\frac{\alpha\,\overline{\mu}_{\,0}L}{A_f}\\ -j\frac{1}{2}\frac{T_c}{\overline{T}_{\ln}}\frac{\omega\,V_f}{\gamma\,p_0} & \frac{T_c}{T_H}\end{pmatrix}\begin{pmatrix}1 & \frac{1}{2}\frac{\alpha\,\overline{\mu}_{\,0}L}{A_f}\\ j\frac{1}{2}\frac{T_c}{\overline{T}_{\ln}}\frac{\omega\,V_f}{\gamma\,p_0} & 1\end{pmatrix}^{-1} \tag{12}$$

For screen-packed heat Exchanger, the energy flux factor f_{qx} and the transmission function matrix are obtained as follows:

$$f_{qx}\approx(1-\phi\,)[1+j\frac{\phi\,\omega\,(1-\phi\,)}{\Omega_h}] \tag{13}$$

$$\text{Ł} = \begin{pmatrix} 1 & -\dfrac{\alpha\,\bar{\mu}_0 L}{A_f} \\ -j\dfrac{\omega\,V_f}{p_0} & 1 \end{pmatrix} \tag{14}$$

For circular pulse tube, Z_f, Z_c, f_{wt}, f_{qx} and g_{qx} are as follows:

$$Z_f = \frac{j\omega\,\rho_0}{A_f(1-f_\mu)},\quad Z_c = \frac{j\omega\,A_f}{\rho_0 a_0^2}\left[1+(\gamma-1)\frac{f_K}{1+\varepsilon_s}\right]$$

$$f_{wt} = \frac{f_K - f_\mu}{(1-\sigma^2)(1+\varepsilon_s)(1-f_\mu)},\quad f_{qx} = \frac{f_K^* - f_\mu}{(1-\sigma^2)(1+\varepsilon_s^*)(1-f_\mu)}$$

$$g_{qx} = \frac{1}{1-\sigma^2}\,\mathrm{Im}\left[\frac{(f_\mu^* - f_K)(1+\varepsilon_s f_\mu / f_K)}{(1+\sigma^2)(1+\varepsilon_s)} - f_\mu^*\right] / |1-f_\mu|^2$$

$$f_\mu = \frac{2J_1(Z)}{ZJ_0(Z)},\ f_K = \frac{2J_1(\sigma Z)}{\sigma ZJ_0(\sigma Z)} \qquad (Z = (j-1)b/\delta_\mu)$$

Generally, the viscous penetration depth δ_μ of working gas inside pulse tube is much smaller than its transversal flow size, *b*, leading to $J_1(Z)/J_0(Z) \approx -j$. Thus, the transmission function matrix of pulse tube and heat flux factor can be simplified as the follows:

$$\text{Ł}_p = \begin{pmatrix} 1 & 0 \\ -j\dfrac{\omega\,V_p}{\gamma\,p_0} & 1 \end{pmatrix} \tag{15}$$

$$f_{qx} = \frac{1+j}{1+\sigma}\frac{\delta_K}{b_p} \tag{16}$$

For compressor, displacer, connecting pipe and reservoir, the following simplified models are derived:

$$\text{Ł}_c = \begin{pmatrix} 1 & 0 \\ -j\dfrac{\omega\,V}{\gamma\,p_0} & 1 \end{pmatrix} \qquad (V = 0.5V_{com}, 0.5V_{dis}, V_{con}, V_{res}) \tag{17}$$

$$f_{qx} \approx 0 \tag{18}$$

Usually, the channel of the orifice or valve used in pulse tube refrigerator is very short and small, so it has no chamber to save working gas. Thus, its transmission function matrix can be simplified as follows:

$$\text{Ł} = \begin{pmatrix} 1 & -R_o \\ 0 & 1 \end{pmatrix} \tag{19}$$

$$f_{qx} \approx 0 \tag{20}$$

APPLICATION OF SLTT TO REGENERATIVE CRYOCOOLERS

Having obtained the simplified models of any thermodynamic elements of regenerative refrigerators, we can apply them to the analysis of regenerative refrigerator systems such as Stirling refrigerator and pulse tube refrigerator as our case study.

Stirling Refrigerator

The schematic and analytical model of integral Stirling refrigerator is shown in fig.2. According to the derived model above-mentioned and the linearized thermoacoustic theory, we have the following equations:

$$\begin{pmatrix} \tilde{p}_d \\ \tilde{U}_d \end{pmatrix} = Ł_{dis} \cdot Ł_{cold} \cdot Ł_{reg} \cdot Ł_{hot} \cdot Ł_{com} \cdot Ł_{com} \begin{pmatrix} \tilde{p}_c \\ \tilde{U}_c \end{pmatrix} \tag{21}$$

Refrigeration power Q_c:

$$\begin{aligned} Q_c = E_2 - E_1 = \frac{1}{8}\frac{\gamma\, p_0}{V_{eff}}\omega\, V_{com} V_{dis} \frac{T_c}{T_H}[(1-\phi)(\sin\theta - Z_{eff} \cdot \frac{\omega\, V_{eff}}{\gamma\, p_0} \cdot \frac{T_H}{T_C}\cos\theta) \\ -\frac{\phi\omega}{\Omega_H}(\cos\theta + Z_{eff} \cdot \frac{\omega\, V_{eff}}{\gamma\, p_0}\frac{T_H}{T_C}\sin\theta)] - A_{reg} K_{eff} \cdot \frac{(T_H - T_C)}{L_{reg}} \end{aligned} \tag{22}$$

The net input power ΔW:

$$\Delta W = W_{com} - W_{dis} = \frac{1}{8}\frac{\gamma\, p_0}{V_{eff}}\omega\, V_{cold} V_{dis}[(1 - \frac{T_c}{T_H})\sin\theta + Z_{eff} \cdot \frac{\omega\, V_{eff}}{\gamma\, p_0}\cos\theta] \tag{23}$$

Where,

$$V_{eff} = 0.5 V_{dis} + \gamma\, V_{cold} + \frac{T_c}{\overline{T}_{\ln}}\gamma\, V_{reg} + \frac{T_c}{T_H}(\gamma\, V_{hot} + V_{con} + 0.5 V_{com})$$

$$Z_{eff} = \frac{\alpha_{hot}\mu_{T_H} L_{hot}}{A_{f,hot}} + \frac{\alpha_{reg}\overline{\mu}\, L_{reg}}{A_{f,reg}} + \frac{T_c}{T_H}\frac{\alpha_{cold}\mu_{T_C} L_{cold}}{A_{f,cold}}$$

$$K_{eff} \approx K_0 + \frac{A_{s,reg}}{A_{f,reg}} K_{s0} + \frac{1}{8}\rho_0 C_{p0}\omega \frac{V_{dis}^2}{A_{f,reg}^2}[(1-\phi)^2 \phi\omega / \Omega_H]$$

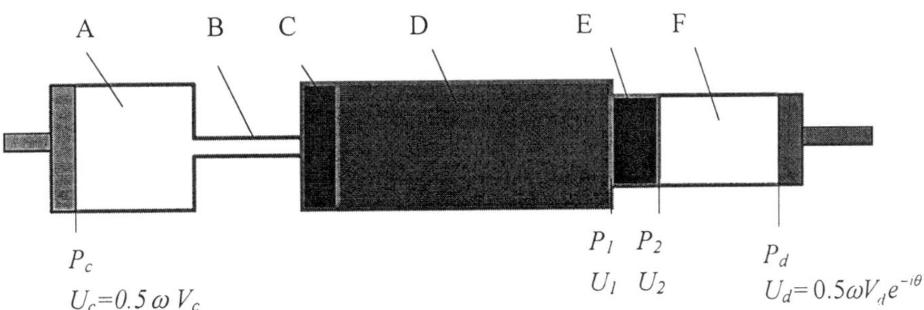

Figure 2. The physical and analytical model of Stirling refrigerator
(A-Compressor; B-Connecting pipe; C-Hot-end heat exchanger; D-Regenerator; E-Cold-end heat exchanger; F-Displacer). θ is the lag phase of the displacer to the compressor.

Eq.(22) and eq.(23) clearly describe the performance of the Stirling refrigerator. Some important real losses such as the friction loss and regenerative loss are included. Giving the structure and operating parameters, one can quickly evaluate the thermodynamic performance of the Stirling refrigerator. The predicted accuracy of the simplified model can be comparable with that of the numerical simulation model developed in our research group. Due to the limitation of the paper pages, the comparison is not shown in the paper. The entire comparisons of the simplified model with other numerical simulation and experimental data will be in the future paper.

Pulse Tube Refrigerator

A pulse tube refrigerator driven by a valveless compressor is the Stirling-like refrigerator. The only difference between them is in that a pulse tube and some auxiliary adjustable elements such as Helmholtz resonator, double-inlet and multi-inlet valves, inertial tube and other elements replace the displacer of Striling refrigerator. Fig.3 is the physical and analytical model of the pulse tube refrigerators. According to the derived model above-mentioned and the linearized thermoacoustic theory, we have the following equations:

$$\begin{pmatrix} \tilde{p}_d \\ \tilde{U}_d \end{pmatrix} = Ł_{dis} \cdot Ł_{cold} \cdot Ł_{reg} \cdot Ł_{hot} \cdot Ł_{con} \cdot Ł_{com} \begin{pmatrix} \tilde{p}_c \\ \tilde{U}_c \end{pmatrix} \tag{24}$$

$$\frac{\tilde{p}_0}{\tilde{U}_0} = R_o - j\frac{\gamma\, p_0}{\omega\, V_{res}} \tag{25}$$

The refrigeration capacity Q_c:

$$Q_c \approx \frac{1}{8}\frac{T_c}{T_H}\left\{(1-\phi-\frac{1}{1+\sigma}\frac{\delta_k}{b_p})\left[\frac{R_o\frac{T_c}{T_H}+Z_{eff}(1+\frac{V_{eff}}{V_{res}})}{(1+\frac{V_{eff}}{V_{res}})^2+(R_o\frac{\omega V_{eff}}{\gamma p_0})^2}-Z_{eff}\right]-(\frac{\phi\,\omega}{\Omega_H}-\frac{1}{1+\sigma}\frac{\delta_k}{b_p}) \right.$$
$$\left.\left[\frac{(R_o\frac{T_c}{T_H}+Z_{eff})R_o\frac{\omega V_{eff}}{\gamma p_0}+(1+\frac{V_{eff}}{V_{res}})\frac{T_c}{T_H}\frac{\gamma p_0}{\omega V_{eff}}}{(1+\frac{V_{eff}}{V_{res}})^2+(R_o\frac{\omega V_{eff}}{\gamma p_0})^2}\right]\right\}(\omega V_{com})^2 - A_{f,reg}K_{eff}\frac{T_H-T_c}{L_{reg}} \tag{26}$$

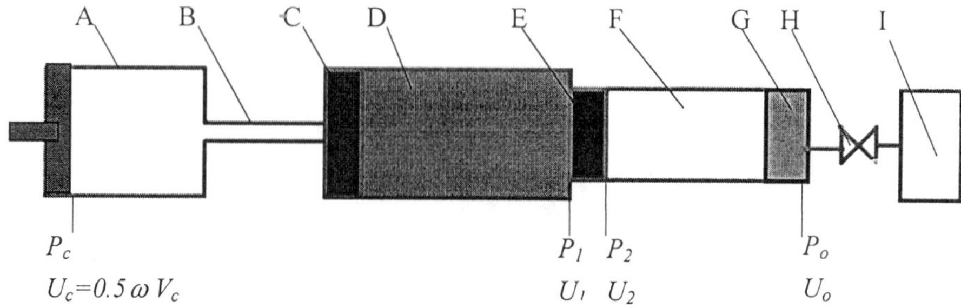

Figure 3. The physical and analytical model of orifice pulse tube refrigerator (A-Compressor; B-Connecting pipe; C-Hot-end heat exchanger I; D-Regenerator; E-Cold-end heat exchanger; F-Pulse tube; G-Hot-end heat exchanger II;H-Orifice valve; I-Reservoir). R_0 is the acoustic impedance of the orifice.

The input p-V power of the compressor:

$$W_{in}=\frac{1}{2}\mathrm{Re}[\tilde{U}_c\tilde{p}_c^{\bullet}]=\frac{1}{8}\frac{R_o\dfrac{T_c}{T_H}+Z_{eff}(1+\dfrac{V_{eff}}{V_{res}})}{(1+\dfrac{V_{eff}}{V_{res}})^2+(R_o\dfrac{\omega\, V_{eff}}{\gamma\, p_0})^2}(\omega\, V_{com})^2 \tag{27}$$

Like the Stirling refrigerator, if the structure and operating parameters are given, one can easily and quickly estimate the refrigeration capacity and input p-V power of the pulse tube refrigerator according to eq.(26) and eq.(27). In fact, the simplified theory can be easily extended to the other types of the pulse tube refrigerator. This work has already been done. Also, due to the limitation of the paper pages, the work is not included in the paper. We will publish them in the future.

DISCUSSIONS

In this paper, we firstly found the simplified thermoacoustic models of the thermodynamic elements of regenerative refrigerators. Then as case study, we derive the system models of the Stirling, pulse tube refrigerators based on the simplified theory, and obtain thermodynamic performance correlation which are directly dependent of their operating conditions and structure parameters. These analytical correlations are straightforward, so one can quickly predict the thermodynamic performance of these regenerative refrigerators without a complicated numerical calculation.

With the help of the simplified thermoacoustic theory, it is easy to obtain the dynamic parameters of the interface of any thermodynamic elements. In addition, it is easy to expand the simplified theory to the analysis of thermally driven thermoacoustic refrigerator and other types of pulse tube refrigerators such as double-inlet and multi-inlet pulse tube refrigerators.

NOMENCLATURE

			Subscript	
A	cross-sectional area	(m^2)	*0*	time-averaged
a	sound velocity	(m/s)	*1*	inlet of cold exchanger
b	hydraulic radius	(m)	*2*	outlet of cold exchanger
E	total energy flux	(W)	*C*	cold or of compressor
f	factor of heat flux or work flux	(-)	*cold*	of cold exchanger
g	dynamic factor of entropy flux	(-)	*com*	of compressor
J	Bessel function	(-)	*con*	of connecting pipe
K	thermal conductivity	(W/m. K)	*D*	of displacer
L	length	(m)	*dis*	of displacer
Ł	transmission function matrix	(/)	*eff*	effective
p	dynamic pressure	(Pa)	*f*	of fluid or flowing
Q	refrigeration capacity	(W)	*H*	hot or of heat exchange
R	acoustical impedance	(Pa.s/m^3)	*j*	imaginary
T	temperature	(K)	*k*	thermal
U	volumetric velocity	(m^3/s)	*k*	thermal
V	volume or auxiliary function	(m^3)	μ	viscous
W	work flux	(W)	*o*	of orifice

x	coordinate in flowing direction	(m)	p	of pulse tube
	Greek letter		reg	of regenerator
α	flowing friction coefficient	($1/m^4$)		
β	thermal expansion coefficient	(1/K)		
δ	penetration depth	(m)		
ϕ	heat capacity ration of fluid to solid	(-)		
γ	adiabatic exponent	(-)		
μ	dynamic viscosity	(Pa)		
θ	phase lag of displacer to compressor	(-)		
ρ	density	(kg/m^3)		
σ	square root of Prantl number	(-)		
ω	angle frequency	(1/s)		
Θ	auxiliary function	(-)		
Ω	characteristic angle frequency of heat exchange	(1/s)		

ACKNOWLEDGMENT

The work is financially supported by the Natural Sciences Foundation of China under the contract No.5960002. Also, the discussion with Dr. J.H.Xiao and Professor Y.Zhou is greatly appreciated.

REFRENCES

1. N.Rott, Thermoacoustics, *Adv. Appl. Mech,*20:135(1980).
2. G.W. Swift, Thermoacoustic Engines, *J.Acoustics Soc. Am,*84:1145(1988).
3. J.H.Xiao,.*Thermoacoustic effect and thermoacoustic theory for regenerative engine/refrigerator*, Ph.D. Dissertation, Physic Institute, Academia Sinica, China, 104 p.(in Chinese,1990)
4. E.C. Luo, *The simplified thermoacoustic theory and experimental study of pulse tube refrigerators*, Master Thesis, Cryogenic Laboratory, Academia Sinica, China, 120 p.(in Chinese,1993)
5. J.H.Xiao, Thermoacoustic theory for cyclic regenerator, Part 1: Fundamentals, *Cryogenics*, 32:895

MULTIPLE STAGE VS. SINGLE STAGE EFFICIENCY CALCULATIONS FOR STIRLING CYCLE SPACECRAFT CRYOCOOLERS

Benny Joe Tomlinson, Jr.

Air Force Research Laboratory
Kirtland AFB, NM 87117-5776

ABSTRACT

In recent years, spacecraft developers have been provided Stirling mechanical cryocooler efficiencies in the form of percent Carnot efficiency or specific power (cryocooler input power divided by the useful heat lift measured at a constant stroke, rejection temperature, and cold end temperature). However, recently there have been extensive advancements in multi-stage Stirling cycle cryocoolers with the capability of lifting heat loads at different stages. This presents a problem with the traditional analysis of efficiency (single load Carnot analysis or specific power) used in trade studies for these coolers due to the fact that the heat lift is occurring at different cold end temperatures. Some trade studies have calculated the efficiency of multi-stage Stirling coolers by translating the heat loads from one stage to another by means of a "Carnot correction factor" to obtain an efficiency value that could be compared to single stage coolers. However, this factor can create large errors in the overall efficiency calculation. In order to provide space system integrators with a more accurate overall cryocooler efficiency to compare with single stage units, this paper will compare the efficiency calculations for multi-stage Stirling cycle cryocoolers where the expansion of the gas is provided by a displacer or expander (pulse tube cryocoolers are not examined) by examining traditional Stirling analysis and stage-by-stage efficiency.

INTRODUCTION

The use of long-life, active cryocoolers provides a significant improvement to DoD space surveillance and missile tracking missions. Cooling of infrared sensors in space at temperatures of 80K and below has mainly been accomplished using stored cryogens. These expendable cryogenic systems require the launch of heavy and complex dewars, which at best have a one or two year life. Cooled detectors allow vast improvements in identification and discrimination capability with a minimum of sensor aperture growth.

Other space missions such as communications, remote sensing, and weather monitoring can benefit from subsystems using cryogenic technology including super conducting electronics, high data rate signal processors, and high speed/low power analog to digital converters.

The Air Force Research Laboratory cryocooler technology effort is developing and demonstrating space qualifiable cryogenic technologies required to meet future requirements for Air Force and Department of Defense (DoD) missions. In order to meet customer requirements, cryocooler development has explored different thermodynamic cycles and variations of thermodynamic cycles in order to develop a base of cryocoolers for specific and general applications. Recent programs have focused on the development of multiple stage, multiple load Stirling cryocoolers for various space applications.

Air Force and DoD system developers have had the daunting task of analyzing and comparing various cryocooler technologies for their respective payload applications. One key component of technology trades is system efficiency, where the system developer can compare the advantages or disadvantages of a particular technology on the basis of a power or converted mass-to-power efficiency figures of merit. However, multistage cryocoolers do not easily lend themselves to the traditional coefficient of performance analysis used for a single stage device and erroneous calculations of efficiency may unfairly bias the selection of one cryocooler technology over another.

COEFFICIENT OF PERFORMANCE CALCULATION FOR A MULTISTAGE REFRIGERATOR

The calculation for the coefficient of performance (Γ) for a multistage refrigerator is not as straightforward as a single stage machine. In addition to having intermediate temperature stages between the coldest reservoir and the hot reservoir at the rejection temperature, a further complication arises due to the fact that heat is being lifted at these intermediate temperatures. Figure 1 and 2 represent the refrigerator systems for a single stage cooler and a three-stage cooler.

It can be seen that for a single stage cooler the coefficient of performance, Γ, is equal to the heat lift, Q, divided by the input power, W. However for the multistage cooler, the input power is divided up between the different stages and there are additional heat loads at the intermediate stages. In order to calculate the coefficient of performance for this system, a model of the multistage cooler must be created. Figure 3 shows another model of the three-stage cryocooler, where $T_{C3} < T_{C2} < T_{C1} < T_R$, meaning T_{C1} is the warmest cold stage and T_{C3} is the coldest cold stage.

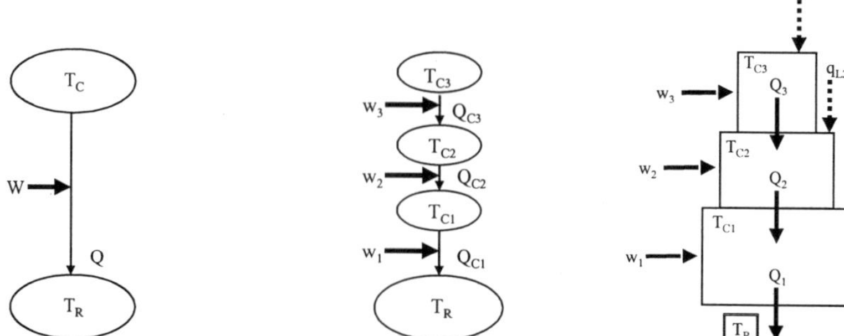

Figure 1. Single Stage Cooler **Figure 2**. 3-Stage Cooler **Figure 3**. 3-Stage Cooler Model

The model used here is similar to a model presented in a paper from Colgate and Petschek. However, the model presented here has the added heat lift at an intermediate stage, but is essentially the same as presented in the paper from Colgate and Petschek. In this model the coefficient of performance for each stage is defined as,

$$\Gamma_3 = \frac{Q_3}{w_3} = \frac{q_{L3}}{w_1},\ \Gamma_2 = \frac{Q_2}{w_2} = \frac{(Q_3 + q_{L2} + w_3)}{w_2},\ \text{and}\ \Gamma_1 = \frac{Q_1}{w_1} = \frac{(Q_2 + w_2)}{w_1} \quad (1, 2, 3)$$

Where w_1, w_2, and w_3 and the input powers to each stage (division of total input power). Combining these efficiencies to obtain the total effective efficiency,

$$\Gamma_{3-1} = \frac{(\Gamma_1\Gamma_2\Gamma_3)}{(1 + \Gamma_1 + \Gamma_2 + \Gamma_3 + \Gamma_1\Gamma_2 + \Gamma_1\Gamma_3 + \Gamma_2\Gamma_3)} \quad (4)$$

To obtain the Carnot efficiency for the multistage cooler, a similar calculation is made for each stage (i = 1 to 3 corresponding to each stage; $T_{E(0)}$ is the rejection temperature T_R) such that,

$$\Gamma_{Ci} = \frac{T_{Ei}}{(T_{E(i-1)} - T_{Ei})} \quad (5)$$

And the combined Carnot efficiency from the coldest stage to the rejection temperature is,

$$\Gamma_{C,3-1} = \frac{(\Gamma_{C1}\Gamma_{C2}\Gamma_{C3})}{(1 + \Gamma_{C1} + \Gamma_{C2} + \Gamma_{C3} + \Gamma_{C1}\Gamma_{C2} + \Gamma_{C1}\Gamma_{C3} + \Gamma_{C2}\Gamma_{C3})} \quad (6)$$

The calculation for the Carnot efficiency may be made without any other knowledge of the cooler except the cold end temperatures T_{E3}, T_{E2}, and T_{E1} and the rejection temperature T_R. However, the coefficient of performance calculation cannot be performed without knowing how much of the input power goes into each stage.

The first attempt to define the input power division was a simple ratio of the cold end expansion volumes, V_E. This simply amounted to using the cold end expansion volume ratios to determine the input power for that stage. For this paper this method will be called the Multistage Volume Analysis, or MSVA.

$$w_1 = w_{total}\, V_{E1} / V_{E,\,total}\ ,\ w_2 = w_{total}\, V_{E2} / V_{E,\,total}\ ,\ \text{and}\ w_3 = w_{total}\, V_{E3} / V_{E,\,total} \quad (7,\ 8,\ 9)$$

As a second approach, called the Multistage Stirling Analysis (MSSA), an expression for the input power to each stage was gained as an expression for the heat transferred in the expansion spaces from a ideal Stirling cycle analysis for a multiple expansion machine and is defined as (i=1 to 3 for each stage),

$$\sum W = [(\tau_1 - 1)V_{E1} + (\tau_2 - 1)V_{E2} + (\tau_3 - 1)V_{E3}],\ \text{where}\ \tau_i = \frac{T_R}{T_{Ei}}, \quad (10)$$

Cold end volume has been substituted into Equation 10 in place of the heat lift, Q. Within the ideal Stirling analysis, the equation for the heat lift can be represented as follows,

$$Q_{E(i)} = V_{E(i)}\zeta \text{, where } \zeta = \pi\overline{p}\left[\frac{\delta}{1+\left(1-\delta^2\right)^{1/2}}\right]\sin\theta \tag{11}$$

The variables represented in ζ are properties that relate to the Stirling cycle including pressure, phase angle, temperature ratios, etc., but are not important to this analysis due to the fact that ζ may be treated as a constant and divided out in the ratio (Equation 12).

Using equation 10, a ratio for the total input power for each stage is obtained,

For each stage (i = 1 to 3),

$$\frac{(\tau_i - 1)V_{Ei}}{(\tau_1 - 1)V_{E1} + (\tau_2 - 1)V_{E2} + (\tau_3 - 1)V_{E3}} \tag{12}$$

This ratio now gives us another means of dividing the input power up between the expansion stages with the only additional knowledge of the expansion volumes.

There is an alternate method within the Stirling analysis where the ratio of the input power can be obtained using the distribution of mass in the expansion volumes. This relationship for mass is reduced to the expansion volume divided by the volume temperature. For this paper, this is called the multistage mass analysis, or MSMA.

COMPARISON OF MULTISTAGE ANALYSIS METHOD AND THE SINGLE STAGE CARNOT CORRELATION METHOD

In a paper presented at the 9th International Cryocooler Conference, a calculation of the system efficiency for the multistage cryocooler was attempted using a *Single Stage Carnot Correlation* (called SSCC, in this paper). This correlation took the form

$$Q_{reference} = \frac{T^{cold}_{reference}\left(T^{reject}_{data} - T^{cold}_{data}\right)Q_{data}}{T^{cold}_{data}\left(T^{rejection}_{reference} - T^{cold}_{reference}\right)} \tag{13}$$

Basically, the load from intermediate stages could be translated to the coldest stage and then the cooler could be treated as a single stage cooler acting between the hot reservoir at the rejection temperature and the coldest stage.

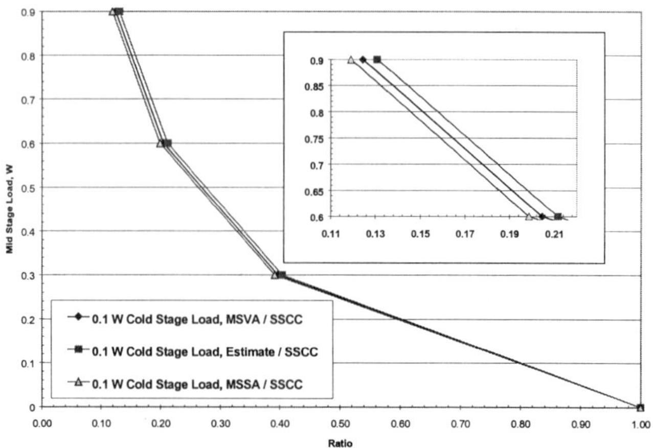

Figure 4. MSVA, MSSA, and Estimate ratios vs SSCC

Using the SSCC method and the MSSA method, several comparisons were made of the real and Carnot coefficient of performance calculations with actual data from the Ball Aerospace 3-stage 35/60K cryocooler. The comparisons used the simple ratio of the cold end expansion volumes (MSVA), the modified cycle analysis (MSSA), and an arbitrary percentage estimate of the input power distribution.

The Carnot calculation using the SSCC, MSVA, or MSSA method between the coldest temperature and the rejection temperature produces the exact same results. That is, when the heat load is transferred to the coldest stage using the SSCC method and then the Carnot efficiency for the cooler is calculated, it is identical to the multistage Carnot efficiency calculation using Equation 6. This includes situations where there is no load on the mid stage and the cooler is acting like a single stage cooler. This is to be expected though; no matter how many intermediate steps, the best that can be done is the Carnot efficiency between the coldest and hottest reservoirs.

There is a discrepancy in the calculation of the COP for each stage using the different multistage analysis methods. The MSVA shows the first stage as having slightly greater than Carnot efficiency and the MSSA and MSMA methods show a much larger than Carnot efficiency in addition to a larger value for the second stage COP (Table 1). As a check, I selected an arbitrary distribution of the input power for each stage until the COP for each stage was reasonable. These numbers, called 'Estimate' in Table 1, turned out to be numerically close to the percentage distribution of the input power values calculated using the MSVA method. Although the abnormal COP values were included, the MSVA, MSSA, and estimate methods showed very similar results for the combined COP (as can be seen in Table 1 and in ratios vs the SSCC method in Figure 4) and only vary up to +/-0.0002 in combined COP values. It is possible that for a narrow band, the division of input power and the aberrant 1st and 2nd stage COP is muted or compensated for in the combination of the individual stage COP values. The MSVA and MSSA methods probably are not accurate enough (they are based on an "ideal" Stirling cycle), but they do provide an alternate

Table 1. Sample Calculation of COP

0.3 W @ 30 K (3rd stage), 0.6 W @ 60.9 K (2nd stage), and 0.0 W @ 180 K (1st stage), rejection temperature of 300 K, and input power of 61.8 W							
Method	**Division of Input Power (%)**		**COP, 1st Stage**	**COP, 2nd Stage**	**COP, 3rd Stage**	**Combined COP**	**% Carnot**
SSCC	This method "translates" the load from the midstage to the coldest stage.		N/A	N/A	N/A	0.00996	8.96%
MSVA	1ST	35.6%	1.853	0.3793	0.0291	0.0051	4.59%
	2ND	47.8%					
	3RD	16.7%					
MSSA	1ST	6.6%	14.46	0.8278	0.0117	0.0049	4.41%
	2ND	51.9%					
	3RD	41.5%					
MSMA	1ST	12.8%	6.897	0.7368	0.0134	0.0049	4.41%
	2ND	51%					
	3RD	36.1%					
Estimate	1ST	45%	1.255	0.2546	0.0485	0.0053	4.78%
	2ND	45%					
	3RD	10%					
Carnot	N/A		1.5	0.5113	0.9709	0.1111	100%

method of calculating the efficiency in a relatively simple calculation for spacecraft developers.

Within the Stirling analysis, the amount of input power required for each stage can also be analogous to the mass present in each expansion volume. In the ideal analysis, this is proportional to the ratio of the expansion volume to the volume temperature. Table 1 has this method tabulated as the MSMA method and has similar results to the MSSA method for overall system efficiency.

The three-stage analysis cannot be used if there is no external load applied to the last stage. This has the effect of making $\Gamma_3 = 0$ and reducing Equation 4 to zero. In this case, you must pretend the cooler has three stages for purposes of division of input power, but the COP calculation has to be for a two-stage cooler, effectively throwing away the input power on the last stage.

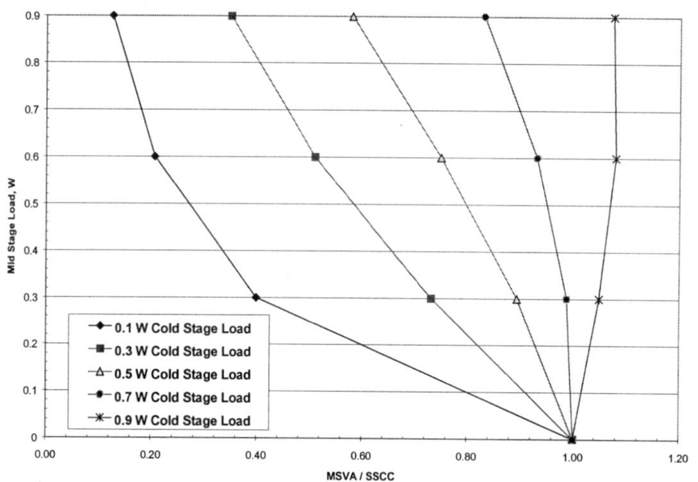

Figure 5. % Carnot MSVA / SSCC Comparison

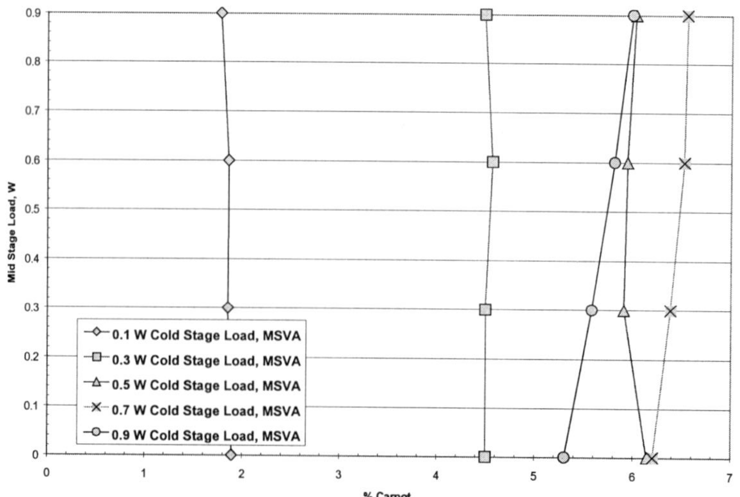

Figure 6. % Carnot MSVA

As expected, the COP calculations from the multistage methods and the SSCC show a large discrepancy. This is very evident when the coldest stage load is low and the midstage load is high (Figure 5). The multistage efficiency calculation for this condition has a very low value for this data (the midstage has the largest expansion volume), around 1.5% Carnot, where the SSCC predicts from 1.5% to 11% Carnot (Figure 7). Figure 6 plots the efficiencies calculated from the MSVA method.

It can be seen in Figure 6 that the efficiency for the multistage cooler is fairly flat for the incremental change in the midstage load for a given cold stage load. Figure 7 shows that the SSCC method calculates a larger change in efficiency for each mid stage load as the cold stage load decreases. The deviation of the MSVA analysis from the SSCC analysis is less pronounced as the cold stage load increases.

The specific power ($1/\Gamma$) plots are a bit clearer. Figure 8 shows the specific power using the MSVA calculation. The plot clearly shows that the dominant factor in the specific power term is the cold stage load and is only slightly influenced by the mid stage load. This is expected if one considers that the heat being lifted at the midstage includes the cold stage load and the work input to the cold stage, which dominates over the change in applied heat load at the midstage.

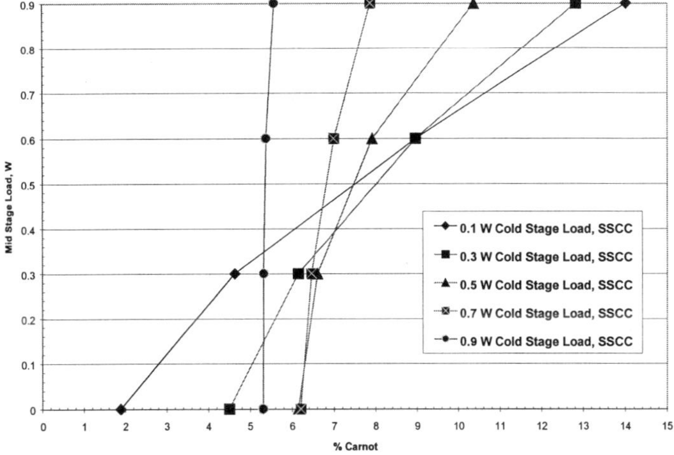

Figure 7. % Carnot SSCC

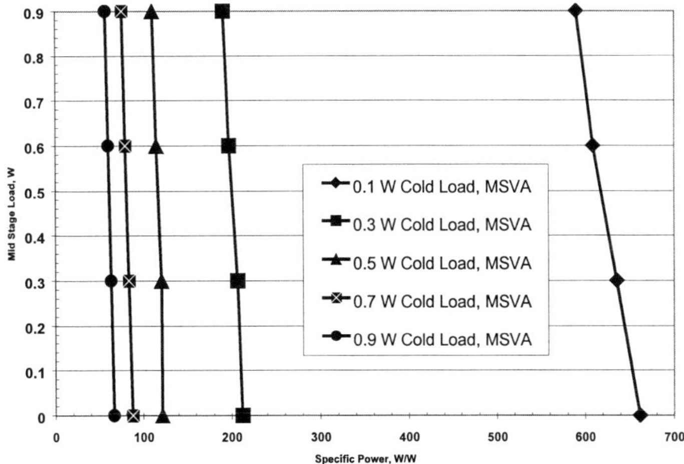

Figure 8. Specific Power ($1/\Gamma$), MSVA Method

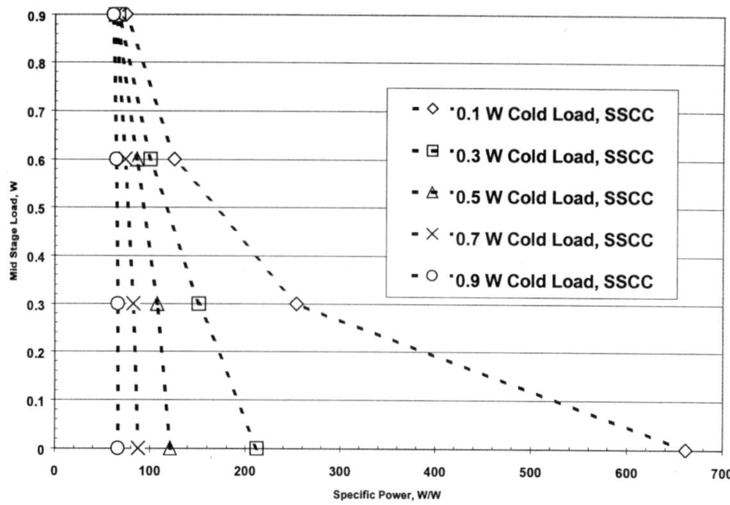

Figure 9. Specific Power ($1/\Gamma$), SSCC Method

Figure 9 shows the specific power using the SSCC calculation. Due to the way that the efficiency is calculated, the midstage load and the cold stage load both have a significant influence over the efficiency of the cooler. Again, the 'no load' mid stage calculations are identical for the SSCC and multistage methods.

SUMMARY

This model shows promise as a more accurate tool for spacecraft developers to compare the system level efficiencies of multiple stage and single stage Stirling cryocoolers on even terms with a overall system efficiency calculation for the thermodynamic cryocooler. A more accurate calculation of the stage-by-stage efficiency will require additional knowledge, beyond the expansion volumes, of the cryocooler internal process to allow for a closed form model that includes pressure drop, dead volume, and other losses to more accurately calculate the division of pV work among the expansion volumes.

REFERENCES

1. Colgate, S. A., and Petscheck, A. G., "Regenerator optimization for Stirling Cycle Refrigeration", Advances in Cryogenic Engineering, Vol. 39, 1993 proceedings of the Cryogenics Engineering Conference / International Cryogenic Materials Conference, Albuquerque, NM, July 12-16, 1993.
2. Walker, G., "Cycle Analysis for Stirling Refrigerator with Multiple Expansion Stages, Perfect Regeneration and Isothermal Processes", International Journal of Refrigeration, v. 13(#1), pp. 13-19, 1990.
3. "*Final Report for Phase 4 of the 30 K Multistage, Long Life, Low Vibration Cryocooler Program, 35 K / 60K Dual Temperature Supplement,*" Final Report, Ball Aerospace and Technologies Corp., Boulder, CO, May 1998.
4. Correspondence with Dr. W. Gully, Ball Aerospace, August 1998 to June 1999.

PARAMETRIC THERMAL AND CRYOGENIC COOLING SYSTEM MODELS FOR SPACE BASED INFRARED INSTRUMENTS

D. S. Glaister[1], M. Donabedian[2], D. G. T. Curran[2], S. D. Miller[2]

[1]Ball Aerospace and Technologies Corporation
Boulder, CO 80306
[2]The Aerospace Corporation
El Segundo, CA 90009

ABSTRACT

This paper describes the development at The Aerospace Corporation of two spreadsheet based models for the parametric analysis of entire cryogenic cooling systems involving infrared instruments for space applications. The models were developed for and have application to the conceptual and preliminary design of infrared instruments and payloads. The capability is included for the analysis of up to four sensors including long to very long, medium, and short wavelength detectors. The two models are linked to each other and were developed to be linked with similar models for other spacecraft subsystems.

The first model denoted THERMAL predicts the temperatures and heat loads internal to the infrared instrument. The model encompasses detectors/cold cavities, cold stops, benches, thermal shields, aft optics, fore optics, and payload mount. The calculated heat loads include detector electrical dissipation, absorbed scene loads, electrical lead parasitics, support conduction, and radiation parasitics. THERMAL calculates temperatures at the multiple instrument-to-cooling system interfaces which are input into the second model denoted CRYOSIZE. CRYOSIZE then analyzes the associated cryogenic cooling system. CRYOSIZE allows the user to parametrically choose cooling system components including mechanical refrigerators (or cryocoolers), cryogenic radiators, expendable cryostats, hybrid cryostats with cryocoolers, and thermal storage units. CRYOSIZE includes algorithms to estimate the input power, mass, volume, and area of the cooling system components as a function of technology freeze date, temperature, operating duty cycle, and heat load.

INTRODUCTION AND BACKGROUND

A predecessor to the THERMAL model was developed by D. Glaister for the Air Force Research Laboratory at Kirtland AFB in New Mexico. That model was used to analyze potential future cryogenic applications to determine cooling requirements to drive cryocooler development programs. In particular, the model was used to predict cooling requirements for a 10 K cryocooler (see Reference 1) for application to space based missile tracking infrared sensors as well as long wave hyperspectral instruments. This model was based upon and correlated to the design of more detailed Raytheon models for the SBIRS (Space Based Infrared System) Low FDS (Flight Demonstration System) Track Sensor.

Then, the authors revised the THERMAL predecessor and added the CRYOSIZE model in support of the Aerospace Corporation Concept Design Center (CDC). The CDC is employed by customers to develop conceptual designs and perform quick turnaround trade studies of future satellites and missions. The CDC essentially consists of about a dozen system specialists running spreadsheet based models which are linked in real time to each other and system engineering models. The primary inputs for the payload thermal model come from the sensor models while the outputs are primarily sent to system engineering. Spacecraft design parameters can be parametrically varied and their impact cascaded through each of the critical payload and bus systems to produce an overall system mass, power, area, and cost. The THERMAL revisions eliminated most of the previous close associations to SBIRS Low to produce a broader and more parametric analysis tool.

The models have proven to be very useful for concept level studies of future complex infrared payloads. Their initial use in the CDC was to analyze relatively complex hyperspectral systems with long wave infrared (LWIR), medium wave (MWIR), short wave (SWIR), and visible and very near infrared (VNIR) sensor combinations. However, the models were also generically developed to support future analyses of other electro-optical systems. In general, these top level spreadsheet models are very useful for parametric system analyses which involve a large number of configurations and runs. They can be used to narrow the trade space to focus support of more detailed analyses .

THERMAL MODEL DESCRIPTION

Overview

THERMAL is essentially a steady state, thermal math model with about 20 nodes. It has a central worksheet with about 500 lines which is linked to 10 supporting worksheets. The central worksheet calculates temperatures and heat loads while the supporting worksheets calculate parameters such as material properties and orbit environmental fluxes. The central worksheet has sections for external inputs, internal inputs, internal outputs, and external outputs. When linked into the CDC network, an input/output worksheet is added which contains every external output from all of the CDC models.

The basic payload configuration for THERMAL is shown in Figure 1 and includes 4 detectors: LWIR, MWIR, SWIR, and VNIR/Visible. This configuration can be changed significantly using the input parameters. For instance, any of the detectors can be removed by setting the number of pixels to zero. For significant deviations from the basic configuration (such as separate aft optics), changes can also be easily made to the code.

The basic configuration has the sensors sharing a common fore optics. The LWIR and MWIR detectors also share an aft optics. There are a substantial number of shielding and support options. The standard configuration has the detector surrounded by the aft optics cavity or bench. To minimize parasitics on the coldest LWIR sensor, a cold stop can be used around the detector and a bench shield around the optical bench. The shields can be eliminated by changing their temperature to be the same as the area which encompasses the shield. Low conductivity structural supports are provided for each component.

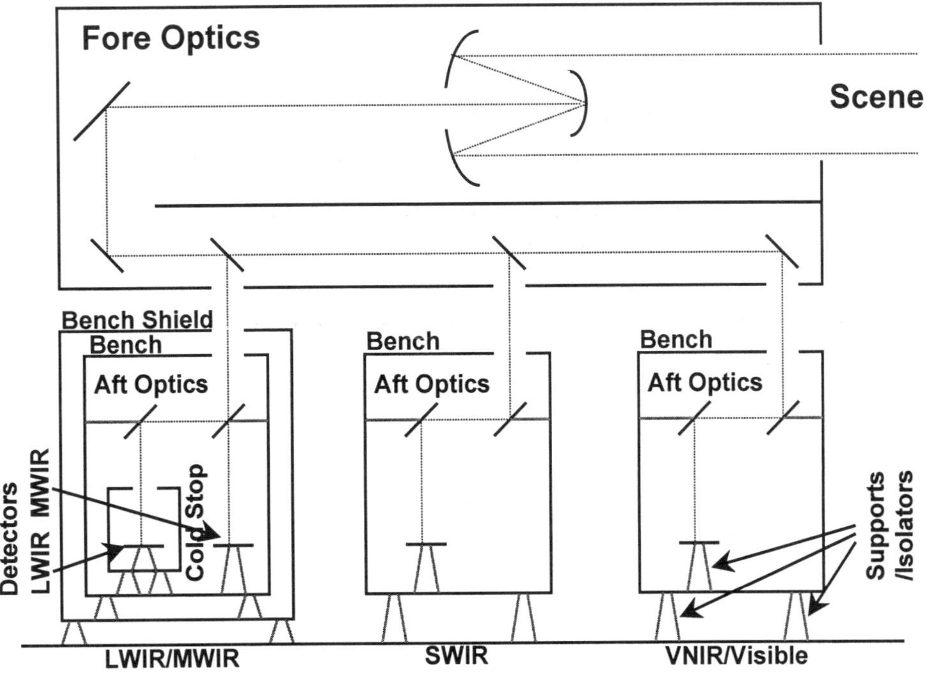

Figure 1. Basic Infrared Payload Configuration

Model Description

The sensor detectors or cold cavity models include the following heat loads: detector electrical dissipation, scene heat loads from the aperture reflected through the optics, conduction through the power and signal leads (which are heat sunk to the cold stop or bench), conduction through the supports, and radiation from the cold stop or bench. If applicable, the cold stop heat loads include conduction from the bench through the supports and the leads and radiation from the bench. The aft optics or optical bench includes absorbed scene loads, conduction through supports and leads, and radiation from the bench shield or ambient. If used, the bench shield includes conduction through supports and leads and radiation from the ambient enclosure. For the fore optics, the scene load is calculated based the aperture environmental flux loads which are determined by simplified flat plate solar, earth infrared, and earth albedo worksheets. When the fore optics are cooled, the heat loads also include conduction and radiation from the ambient enclosure or telescope.

External Inputs

The model inputs can be categorized into those which will be externally changed and those that are internal. The internal inputs can also be changed, but are likely to be kept constant. A sample set of external inputs is shown in Figure 2. The external inputs include orbital parameters which are used in the scene or aperture loading calculations. The sensor duty cycle (fraction that the sensor will be powered) can be specified to determine the minimum, average, and maximum detector heat loads. For each sensor's cold cavity, the following can be specified: the maximum detector interface temperature, the number of detector pixels, the pixel dissipation and size, and the number of signal and power leads.

External Inputs	Value
General	
Technology Freeze Date	1998
Mission Life (yrs)	2
LWIR Cold Cavity	
LWIR Cold Cavity Temp (K)	38
LWIR No. Pixels	131072
LWIR Total Heat Load (W)	0.17
LWIR Heat Load Per Pixel (uW)	1.27
LWIR Pixel Size (um)	40
LWIR Scene Heat Load (mW)	0.0005
No. DC Signal Leads	14
No. AC Signal Leads	14
No. Power Leads	4
Altitude Perigee (km)	325
Altitude Apogee (km)	645
Inclination (deg)	63
Orbit Period (min)	94.3
LWIR FP Operating Time per Orbit (min)	0.9
LWIR FP Duty Cycle (%)	1.0
LWIR Cold Stop/Shield	
LWIR Cold Stop Temperature (K)	65
LWIR Cold Stop Surface Area (cm2)	2270
MWIR Cold Cavity	
MWIR Cold Cavity Temp (K)	80
MWIR No. Pixels	65536
MWIR Total Heat Load (W)	0.08
MWIR Heat Load Per Pixel (uW)	1.27
MWIR Pixel Size (um)	40
MWIR Scene Heat Load (mW)	0.0000
No. DC Signal Leads	14
No. AC Signal Leads	14
No. Power Leads	4
MWIR FP Operating Time per Orbit (min)	10.0
MWIR FP Duty Cycle (%)	10.6

LWIR/MWIR Bench/Aft Optics	
Bench/Aft Temp (K)	88
Bench Surface Area (cm2)	3770
LWIR/MWIR Bench Shield	
LWIR Bench Shield Temperature (K)	220
SWIR Cold Cavity	
SWIR Cold Cavity Temp (K)	220
SWIR No. Pixels	65536
SWIR Total Heat Load (W)	0.11
SWIR Heat Load Per Pixel (uW)	1.69
SWIR Pixel Size (um)	16
SWIR Scene Heat Load (mW)	0.0000
No. DC Signal Leads	10
No. AC Signal Leads	10
No. Power Leads	4
SWIR FP Duty Cycle (%)	1.0
SWIR Bench/Aft Optics	
Bench/Aft Temp (K)	293
Bench Surface Area (cm2)	3770
Fore Optics	
Fore Temp (K)	293
Altitude (km)	325
Altitude (nm)	175
Aperature (m)	1.000
Telescope DC (%)	100
Optics Surface Area (cm2)	62832

Figure 2. Sample THERMAL External Inputs

Internal Inputs and Assumptions

The general internal inputs include the effective emissivity of multi-layer insulation (MLI) blankets (0.02 default). A total Area to Length (A/L) ratio is assumed for each cold cavity, bench, and shield support and checked by the external payload structures models. Based on the supported mass and area, the structures models assume a geometry such as a box and compute dynamic modes. The baseline design minimum A/Ls were 0.75, 1.50, 0.75, 4.50, 0.75, and 4.50 cm for the LWIR cavity, the LWIR cold stop, the MWIR cavity, the LWIR/MWIR bench, the SWIR cavity, and the SWIR bench, respectively.

For each detector/cold cavity, internal inputs also include detector effective infrared emittance (default is 0.70), radiation surface area ratio, and the material, cross section, and length of the signal and power leads to the next shield. For each shield, the length of leads to the next warmer element can be specified. For the benches/aft optics, total effective infrared and solar absorptances (defaults are 0.04 for each) are specified. These absorptances are used to calculate the absorbed aperture loads. They are effective in nature as they include the cumulative absorptance of all of the mirrors and optical surfaces.

To calculate the aperture heat loads, the environmental flux constants are specified. A single fore optics surface emissivity is specified to calculate radiation. Similar to the aft optics, effective infrared and solar absorbtances are specified to calculate the total aperture loads absorbed on the optical components.

Outputs

The THERMAL outputs can also be categorized into internal and external parameters. The internal outputs are those which are typically not utilized outside the model. These include itemized heat loads, material thermal conductivities, and lead and support total conductances. The external outputs are a subset of the total outputs. The primary external outputs are minimum, average, and maximum heat loads for the system components. Along with the temperatures, these outputs are provided to CRYOSIZE.

CRYOSIZE DESCRIPTION

Overview

CRYOSIZE analyzes the associated cryogenic cooling system and the penalties involved to support the cooling loads generated in THERMAL. It is made up of about 400 lines of a primary spreadsheet plus 12 additional worksheets which include databases and sub-models. CRYOSIZE allows the user to select various types of cooling systems in combination including cryocoolers, expendable cryostats (dewars), hybrid systems (cryocooler plus cryostat), cryoradiators and includes the option to include thermal storage units (TSU). CRYOSIZE includes the algorithms, databases and various matrices to estimate the mass, power, volume and area (for cryoradiators) for the system components as a function of technology freeze date, temperatures, operating duty cycles and heat loads.

Model Description

Cryocooler Sizing. Two of the secondary CRYOSIZE worksheets contain a matrix of information on cryocoolers. The source for the cryocooler data is the database described in Reference 1. Based on the temperature and technology freeze date (one worksheet is for cryocoolers with technology freeze dates prior to 2001 and the other is for after 2001), the main program interpolates these matrices to determine maximum capacity, mass and power input required for the specified load. The model interpolates based on the cold tip temperature and computes the total motor power using the load and the specific power. Electronics and total input power are then calculated based on a fixed tare and an efficiency. If the load is more than the stated capacity, multiple coolers are assigned.

Thermal Storage Units (TSU) for 10 K Systems. The TSU model contains the algorithms and properties to size the storage system using compressed helium gas, solid lead, or a rare earth material. The TSU allows the use of the average load instead of the maximum to size the cryocoolers. Its use is very valuable for 10 K detectors because of the limited capacity of available cryocoolers and the high power penalty. Pressurized helium gas is generally the lightest TSU option near 10 K since the heat capacity of solid materials are relatively low and, therefore, have large mass penalties.

Cryostats (or Dewars) using Stored Expendable Solid Cryogens. The cryostat model actually contains 3 separate designs covering the range of 8.5 K to 90 K. A single stage solid hydrogen cryostat is used for temperatures between 8.5 K and 14 K. Between 14.1 K and 60.9 K, two-stage neon-ammonia configurations are used, whereas between 61 K and 90 K, two-stage methane-ammonia configurations are utilized. Inputs include the load, temperature, and mission life. The output provides the mass and the volume.

The solid hydrogen (SH2) cryostat model uses the results of prior parametric analyses made for solid hydrogen cryostats using the Aerospace CRYO4 computer program (see Reference 2) The parametric analysis assumed a vacuum shell temperature of 250 K. The CRYO4 model was correlated with the MSX Spirit III design. Data from parametric CRYO4 analyses were fitted with third degree polynomials and utilized in the model.

A model was developed to estimate the significant amount of cooling at intermediate temperatures from the exhaust vapor of solid hydrogen cryostats Using the Spirit III design as a basis, two vapor cooled shields (VCSs) were assumed; one at 65 K and one at 220 K. The hydrogen flow rate is based on the load from the cryostat model and the enthalpy of the vapor is computed based on outlets of 65 K and 220 K at the VCS and inlets at 10 K and 65 K. The net cooling capacities for the two VCSs can then credited against the loads calculated for the LWIR cold stop and the LWIR/MWIR bench shield.

The basis for the two stage cryostat models are analyses using an old Aerospace program called "cooler". For the neon unit (used from 14.5 K to 60.8 K), the neon triple point of 24.5 K would normally be the limit. But, in order to have a continuous range, it was assumed that a thermal resistance link could be used between 24.5 K and 60.8 K.

Hybrid Designs (SH2 cryostat plus cryocooler). The SH2 hybrid model computes the mass of a solid hydrogen cryostat which has an intermediate shield controlled to 65 K with a cryocooler. This is a special case which has been previously analyzed by Aerospace using the CRYO4 thermal model off line. A cryocooler is sized to handle the 65 K shield load as well as additional loads at the same temperature such as the aft optics/bench. In this application, only one vapor cooled shield is assumed and is based on a 220 K temperature. The cooling capacity for the VCS can be credited towards the bench shield load.

Cryoradiator Sizing Model. A cryoradiator can be used to cool the bench shields or higher temperature detectors such as the SWIR. The cryoradiator model is a relatively simple set of equations for the calculation of the area and mass of a flat plate radiator in low earth orbit using a deployable earth shield. The model assumes no direct sun on the cold face of the radiator and uses a black painted open cell honeycomb radiator surface. To use this configuration, the spacecraft must be either a) in a sun-synchronous orbit or b) must utilize continuous yaw maneuvering to prevent direct solar impingement. The coefficients in the equations were correlated against the Raytheon SBIRS-Low FDS design.

External Inputs

Figure 3 is a sample of the external inputs for the CRYOSIZE model. Most of the inputs are linked outputs from the THERMAL model. There are user friendly drop down menus for the selection of cryogenic cooling sources for the LWIR cold cavity (cryocooler, dewar, or hybrid choices), MWIR cold cavity (currently only cryocooler), LWIR/MWIR bench/aft optics (currently only cryocooler), LWIR/MWIR bench shield (cryocooler or

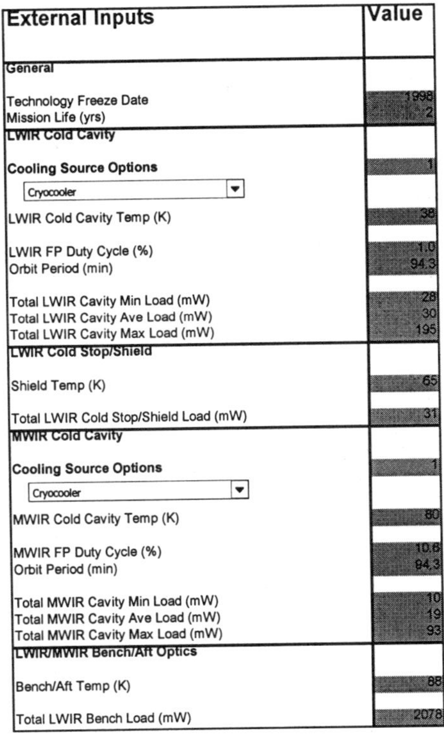

External Inputs	Value
General	
Technology Freeze Date	1998
Mission Life (yrs)	2
LWIR Cold Cavity	
Cooling Source Options	1
Cryocooler	
LWIR Cold Cavity Temp (K)	38
LWIR FP Duty Cycle (%)	1.0
Orbit Period (min)	94.3
Total LWIR Cavity Min Load (mW)	28
Total LWIR Cavity Ave Load (mW)	30
Total LWIR Cavity Max Load (mW)	195
LWIR Cold Stop/Shield	
Shield Temp (K)	65
Total LWIR Cold Stop/Shield Load (mW)	31
MWIR Cold Cavity	
Cooling Source Options	1
Cryocooler	
MWIR Cold Cavity Temp (K)	80
MWIR FP Duty Cycle (%)	10.6
Orbit Period (min)	94.3
Total MWIR Cavity Min Load (mW)	10
Total MWIR Cavity Ave Load (mW)	19
Total MWIR Cavity Max Load (mW)	93
LWIR/MWIR Bench/Aft Optics	
Bench/Aft Temp (K)	88
Total LWIR Bench Load (mW)	2078

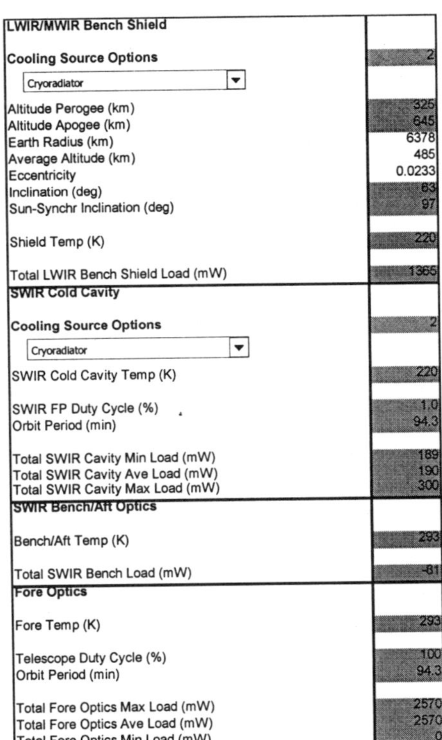

LWIR/MWIR Bench Shield	
Cooling Source Options	2
Cryoradiator	
Altitude Perogee (km)	325
Altitude Apogee (km)	645
Earth Radius (km)	6378
Average Altitude (km)	485
Eccentricity	0.0233
Inclination (deg)	63
Sun-Synchr Inclination (deg)	97
Shield Temp (K)	220
Total LWIR Bench Shield Load (mW)	1365
SWIR Cold Cavity	
Cooling Source Options	2
Cryoradiator	
SWIR Cold Cavity Temp (K)	220
SWIR FP Duty Cycle (%)	1.0
Orbit Period (min)	94.3
Total SWIR Cavity Min Load (mW)	189
Total SWIR Cavity Ave Load (mW)	190
Total SWIR Cavity Max Load (mW)	300
SWIR Bench/Aft Optics	
Bench/Aft Temp (K)	293
Total SWIR Bench Load (mW)	-81
Fore Optics	
Fore Temp (K)	293
Telescope Duty Cycle (%)	100
Orbit Period (min)	94.3
Total Fore Optics Max Load (mW)	2570
Total Fore Optics Ave Load (mW)	2570
Total Fore Optics Min Load (mW)	0

Figure 3. Sample CRYOSIZE External Inputs

cryoradiator), and SWIR cold cavity (cryocooler or cryoradiator). The baseline assumes that the fore optics and SWIR bench/aft optics do not require cryogenic cooling.

Internal Inputs and Assumptions

The internal inputs include the cryogenic integration penalty (0.075 kg of spacecraft penalty per W of cooling system input power which covers the mass of the cryogenic heat transport, shields and cables, but doesn't include vehicle power system and ambient heat transport/radiators which are sized by the spacecraft models) and the margins on electrical and thermal heat loads (default is 20% on electrical and 50% on thermal). Additional information on the cryogenic integration penalty is contained in Reference 3. Other internal inputs include cryocooler redundancy factor, allowable detector temperature rise, and the thermal resistances from the element to the cooling source. The allowable temperature rise (default is 0.2 K) is used to size a sensible heat thermal storage device. The nominal thermal resistances are 3 K/W for a cold cavity (including resistances within the detector elements and a thermal strap or link to the cooler) and 2 K/W for benches and shields.

Outputs

A partial listing (outputs for the LWIR cold cavity and the thermal system summary) of the CRYOSIZE external outputs is shown in Figure 4. The outputs summarize the key element parameters such as the loads, temperatures, the number of cryocoolers, maximum cryocooler capacity, the cryocooler power and mass (including electronics), dewar volume

LW Cryocooler External Outputs	Value
LWIR Cold Cavity	
Assumes load leveling TSU for <30 K	
Cavity Ave Load w/o Margin (mW)	30
Addtl. Load (i.e.-No Shield) Adjustment (mW)	0
Ave Load without Margin (mW)	30
Min Load with Margin (mW)	42
Ave Load with Margin (mW)	44
Max Load with Margin (mW)	242
Cryocooler Load with Margin (mW)	242
Cryocooler Cold Tip Temp (K)	37.9
Cryocooler Max Capacity (mW)	1268
Cryocooler Motor Specific Power (W/W)	78
Cryocooler Total Specific Power (W/W)	101
Cryocooler Unit Mass (kg)	13
Electronics Mass (kg)	6
Total Cooler Mass (kg)	19
Number Operating Units	1
Total Number of Units	1
Cryocooler Motor Power (W)	19
Electronics Power (W)	5
Total Cryocooler Power (W)	24
Cryocooler Mass (kg)	13
Electronics Mass (kg)	6
Total Cooler Mass (kg)	19
TSU Storage (J)	0
TSU Volume (cm3)	0
TSU Mass (kg)	0.00

Thermal System Summary	
Total Number of Operating Units	2
Total Number of Units	2
Total Cryocooler TMU Power (W)	85
Total Electronics Power (W)	24
Total Cryocooler Power (W)	110
Total Cryocooler TMU Mass (kg)	25
Total Electronics Mass (kg)	12
Total Cryocooler Mass (kg)	37
Cryogenic Integration Penalty (kg)	8
LW TSU Storage (J)	0
LW TSU Volume (cm3)	0
LW TSU Mass (kg)	0.0
LW/MW Bench Shield Cryoradiator Summary	
Radiator Area (m2)	0.026
Radiator Mass (kg)	1.5
SWIR Cryoradiator Summary	
Radiator Area (m2)	0.000
Radiator Mass (kg)	0.0
Overall Cryoradiator Summary	
Radiator Area (m2)	0.026
Radiator Mass (kg)	1.5
Overall Summary	
Overall Thermal Power (W)	110
Overall Thermal Mass (kg)	47
Power Penalty (kg)	33
Total Mass Penalty (kg)	80

Figure 4. Partial Sample of CRYOSIZE External Outputs Including System Summary

and mass, cryoradiator area and mass, and TSU storage capacity, mass and volume.

One of the most important parameters is called "Additional load' which allows the user to combine loads of elements and, thus, minimize the number of individual coolers or radiators. This adjustment option is a very powerful tool for simplifying a complex system with multiple detectors, optics, and shields. The user can set a shield or bench temperature of one detector (i.e. - LWIR/MWIR bench shield at 220 K) to coincide with the cold cavity of another (i.e. - SWIR cold cavity). In this way, a single cooling source can be provided for 2 or more loads. One may also wish to combine two loads at different temperatures into one load at the lower temperature. Also, the use of a 2-stage cryoradiators or cryocoolers (for example the Ball 35K/65K) can be accommodated by using this adjustment factor.

The thermal summary output shows the number of coolers, the total cooler mass and power, the cryoradiator area and mass, dewar volume and mass, integration penalties, and the total mass and power. This information would normally be linked to the master payload spreadsheet. It also displays a number for the power penalty and the total equivalent mass penalty which is not linked (as they are calculated in the spacecraft model), but shown for the benefit of the thermal operator for evaluating the optimum configuration.

REFERENCES

1. Glaister, D. S., Donabedian, M., Curran, D. G. T., "An Overview of the Performance and Maturity of Long Life Cryocoolers for Space Applications," 10th International Cryocooler Conference, Monterey, CA, May 26, 1998.
2. Donabedian, M., "Revised Computer Program (CRYO4) for Solid Hydrogen Cooler Analysis," Aerospace ATM Report No. 89(4033-02)-6, 1989.
3. Glaister, D. S., and Curran, D. G. T., "Spacecraft Cryocooler System Integration Trades and Optimization," Cryocoolers 9, R. Ross Jr., Ed., Plenum Press, New York, NY, 1997.
4. Reilly, J., Glaister, D. S., and Levenduski, R., "Development of a 10 K Space Cryocooler," to be released at 1999 Cryogenic Engineering Conference.

PROPELLANT PRESERVATION FOR MARS MISSIONS

P. Kittel[1] and D. W. Plachta[2]

[1]NASA, Ames Research Center
Moffett Field, CA 94035

[2]NASA, Glenn Research Center
Cleveland, OH 44135

ABSTRACT

There has been extensive technology planning in the last few years for human missions to Mars. The human missions will make extensive use of cryogenic propellants. Some of these propellants will be transported to Mars. Additional propellants will be manufactured, liquefied, and stored on Mars. The missions that use these propellants could start as early as 2011. While many of the plans are still evolving, it has been possible to derive a set of cooler requirements. Recent estimates of these requirements are given here along with a discussion of whether the requirements can be met with existing coolers and coolers currently being developed.

INTRODUCTION

NASA is in the early stages of planning missions to take humans to Mars in the next millennium. These missions will make extensive use of cryogenic propellants, some of which will be manufactured on Mars. Preceding the human missions, several robotic missions may use cryogenic propellants to return samples from Mars and will test propellant manufacturing schemes on Mars. A thorough description of the earlier scenario is given elsewhere.[1] That design was based on using a nuclear thermal reactor (NTR) for the Trans-Mars Injection (TMI) stage. Other stages used chemical propulsion. The cryogenic requirements have been given previously.[2,3] Since those references were prepared, the human mission has been extensively redesigned. The scenarios for these missions are continuing to evolve with time. This paper is based on the updated version, version 4, of the Design Reference Mission.[4,5] This version has an updated bimodal NTR concept and an alternate, non-nuclear Solar Electric Propulsion (SEP) concept in combination with chemical propulsion stages.

Advances in Cryogenic Engineering, Volume 45.
Edited by Shu *et al.*, Kluwer Academic / Plenum Publishers, 2000.

In the earlier designs, each of the many tanks was custom built for the particular requirements of its application. The original design also assumed that the tanks were not cooled. Thus, the tanks were large enough to allow for evaporation during the missions.[1,2] Consequently, some of the tanks had to be quite large to accommodate storage times of up to 1605 days before the propellants were used. Previous analysis showed that using cryocoolers to eliminated boil off could reduce the mass launched from Earth (Injected Mass to Low Earth Orbit, IMLEO).[2] Reducing the IMLEO has a significant effect on cost.

The SEP baseline concept characteristics are

1) A SEP tug boosts the TMI stage to a highly elliptical orbit. This phase requires 400 days of propellant storage for the TMI piloted stage and 250 days for the TMI cargo stages.
2) The ascent and descent stages require about 580 days of propellant storage for the piloted mission and 550 for the cargo case. These stages use O_2 and CH_4 for propellants.
3) The TEI (Trans Earth Injection) stage requires a storage duration of 1200 days. This stage uses H_2 not CH_4 as the propellant.
4) All tanks are cooled by cryocoolers to eliminate boil off.
5) The tank design is standardized as much as possible.

The last of these changes results in having two or three standard designs. Fixing the volume results in many of the tanks not being full at launch. The effects of these requirements on coolers for the SEP mission are discussed in this paper.

There are also a number of changes to the bimodal NTR baseline concept, which significantly effect the cryogenic systems. These changes and the coolers for the NTR mission will not be discussed here, but will be presented in a later paper.

The current selection of tanks is given in Table 1. The propellants are stored at 2 atmospheres (0.2 MPa) under saturated conditions. For comparison, a reference H_2 tank is also analyzed. This tank is a standard and is held for 37 days in Low Earth Orbit (LEO) before the fuel is used.

THERMAL MODEL

The thermal model used to estimate the cooler requirements has been revised. The basis of the model has been described previously.[1,2] This model could estimate the tank size and mass for MLI (multi-layer insulation) insulated tanks with and without coolers. The cooling power and the mass of the power source and radiators were included for Zero Boil-Off (ZBO) storage. In the first case, the volume of the tank was variable. The volume was adjusted until it was large enough to accommodate the boil off during the mission and still preserve the required propellant until it was needed. In the latter case, the volume was fixed. The earlier model did not include cooled shields. An optional cooled shield has been added to the model for hydrogen tanks. Several of the empirical relations have been changed to better reflect test data and progress in cooler development. The penetration heat load is now based on a study of tank supports for flight systems that looked at the effect of varying the external temperature, resonant frequency, and of tank size for two types of supports.[6] For S-glass epoxy struts, the following correlations are good fits to the results of that study:

For O_2 tanks:

$$\dot{Q} = 1.28\text{x}10^{-4}\,(f/35)^2\,(T/250)^{2.3}\,m \qquad (1)$$

and for H_2 tanks:

$$\dot{Q} = 2.70\text{x}10^{-4}\,(f/35)^2\,(T/250)^{1.6}\,m \qquad (2)$$

Table 1. Tank size and propellant selection

Stage §	Liquid propellant	Number of tanks	Propellant mass (kg)	Dimensions † (m)	Percent full (%)
TMI cargo	H_2	3	2607	3.29	71
	O_2	1	15640	3.29	78
TMI piloted	H_2	4	4393	3.29	89
	O_2	2	26355	3.29	66
reference	H_2	1	1110	3.29	90
TEI	H_2	4	3720	3.29	76
	O_2	2	22319	3.29	56
Ascent	CH_4	2	8778	3.29	60
	O_2	2	30722	3.29	78
Descent – opt 2	CH_4	1	2195	3.29	31
	O_2	1	7683	3.29	39
Descent – opt 1	CH_4	2	2623	2.05	73
	O_2	2	9182	2.05	94
ISPP seed	H_2	3	4107	3.29 x 3.60	97

§ TMI: Trans Mars Injection – the Earth to Mars stage
TEI: Trans Earth Injection – the Mars orbit to Earth stage
Ascent: the Mars surface to Mars orbit stage
Descent: Martian lander
ISPP seed: hydrogen required for In-Situ Propellant Production - used to make CH_4 and H_2O from CO_2.and H_2.

† Single dimension: inner diameter of spherical tank
two dimensions: inner diameter x inner overall length of cylindrical tank with hemispherical ends.

where $\dot{Q}$ is the heat load, f is the resonant frequency, T is the outside temperature, and m is the supported mass. Since the storage temperature of CH_4 and O_2 are nearly the same, Eq. (1) was also used for CH_4 tanks. Because the supports account for only a small part of the total heat load on these tanks, this approximation does not result in significant errors. When a cooled shield is included, the shield intercepts some of these heat loads. One half of the mass of the cooled shield is included in the supported mass. The mass of the shield is assumed to be 2 kg/m^2 for actively cooled shields and 2.5 kg/m^2 for vapor cooled shields (VCS).[7] The latter mass includes the additional plumbing required. The mass used for the shields is conservative; a density of 1 kg/m^2 is thought possible.[6] The MLI is modeled using the relation developed by Nast and co-workers using 15 layers/cm and 0.018 kg/m^2/layer.[8] The mass of solar cells, including additional system mass to support the cells and 30% contingency, is 0.026 kg/W.[9] The solar cell mass was not included in the TMI stages because those stages have an excess of available power. The radiator mass is taken to be the middle of the range of masses given by Glaister and Curran with the addition of 30% contingency:[10]

$$m = (1.3)(10.5)\ \dot{Q}\ [3.08\ (T/100)^4 - 8.93]^{-1} \tag{3}$$

where $\dot{Q}$ is the power rejected by a radiator operating at a temperature, T. The cooler efficiency is the efficiency found by Strobridge increased by a factor to account for advances in the last 25 years:[11]

$$\dot{Q}_{ip} = \dot{Q}_c\ (T_h\text{-}T_c)\ (T_c\ \eta\ 10^{\Sigma})^{-1} \tag{4a}$$

where

$$\Sigma = -\ 1.73586 + 0.59998 \log(\dot{Q}_c) - 0.1474 \log(\dot{Q}_c)^2 + 0.021323 \log(\dot{Q}_c)^3 - 0.00125 \log(\dot{Q}_c)^4, \tag{4b}$$

$\dot{Q}_c$ is the cooling power, $\dot{Q}_{ip}$ is the electrical input power, T_h is the heat rejection temperature, T_c is the propellant temperature, and η is the improvement factor. The values used for η were 2.5 for CH_4 and O_2 tanks and 2.0 for H_2 tanks. These factors are estimates based on current and projected near term capabilities.[12,13] Similarly, the mass of cryocoolers is based on Strobridge's correlations adjusted by a factor of 0.2 to account for recent advances:

$$m = 0.2\ \dot{Q}_c^{0.7}\ [(T_h - T_c) / T_c]^{1.45} \tag{5}$$

The coolers on the liquid hydrogen tanks will most probably be 2-stage coolers. Strobridge gives no guidance on how to estimate the effect of the cooling power of the two stages on the overall input power or mass. One approach is to assume that each stage was an independent cooler. This approach overestimates the power and mass of coolers for H_2 tanks and places an upper estimate on the mass. An alternate approach is to assume that the effect of the first stage is already accounted for in the second stage estimates. This may be true if the first stage is only used to reduce internal losses in the cooler. However, applying an external load must have some effect on the overall power requirement and, thus, on the system mass. This approach gives a lower estimate on the mass. In the cases looked at here, using the first approach results in the 2 stages contributing about equally to the overall mass. Using the second approach shifted the optimum system to a tank with less MLI (less insulation mass) but with a more powerful cooler. For either approach the minimum was broad and the change in total mass was not significant when compared to likely errors in the model. In all cases, The coolers and associated masses were considerable smaller than the mass of the tank and propellant. For the results reported here, the more conservative approach was used with only the upper estimate reported.

The CH_4 and O_2 tanks were modeled without cooled shields. The H_2 tanks were modeled with and without shields. For comparison, we also modeled some of the tanks without coolers. These tanks were allowed to vary in size as described above. The un-cooled H_2 tanks were also modeled with and without shields. Here the shields were cooled by the effluent vapor. To make use of the enthalpy change during conversion of para-hydrogen to ortho-hydrogen, a catalytic converter was inserted in the vent line at the shield. The effectiveness of both the catalyst and the heat exchanger was assumed to be 0.9.

The model for all tanks minimized mass with the restriction that the total number of MLI layers did not exceed 100 layers. This restriction is somewhat arbitrary. It reflects the limited benefit of additional layers in thick blankets.[14] The parameters that were varied to minimize the system mass for the different models are given in Table 2. The models are steady state models. They do not account for the transients that occur at the beginning of the mission nor for the environmental changes that occur during the mission.

Table 2. Parameters varied in order to minimize mass

Model	Variables
fixed volume – 1-stage cooler	layers of MLI
fixed volume – 2-stage cooler	total layers of MLI location of cooled shield within MLI temperature of cooled shield
variable volume – without VCS	layers of MLI
variable volume – with VCS	total layers of MLI location of cooled shield within MLI

MODEL RESULTS

The model was run for all the configurations listed in Table 1. Even though the tank sizes are mostly standardized, the model did not result in identical cooler requirements for tanks operating at the same temperature. This is a result of the model adjusting the supports to the differing masses of the tanks. Thus, the thermal load on the tank increases with increasing initial propellant load. Since the purpose of this design exercise is to develop standardizes parts, the supports and coolers will be standardized as much as possible. For each tank size and propellant, it is assumed that a single support size, MLI thickness, shield location (if used), and cooler will be used. Thus, only a single set of cooler specifications is reported for each tank size and propellant. These represent the largest cooler for the application. The resulting requirements are shown in Table 3. The coolers reported there are those for the fullest, heaviest tanks. The H_2 cooler is from the Seed tank; the O_2 and CH_4 coolers are from the Ascent tanks.

DISCUSSION

The motivation to minimize the number of different coolers that need to be developed is one of cost minimization. Space flight qualified cryocoolers are expensive to develop and have been expensive to produce. One of the causes of the high cost is the low production run of these coolers. Typically, only a few copies of each cooler have been made.[12] The human missions to Mars could require more than ten H_2 coolers and more than ten O_2/CH_4 coolers per mission. Minimizing the number of different coolers that are developed and produced, could result in considerable cost savings over several missions. A minimum set of coolers that meets the requirements is listed in Table 4. This list gives the nominal design

Table 3. Estimates of cooler specifications for minimum mass assuming heat rejection at 250 K

Propellant	Tank cooler (W) @ (K)	MLI (layers)	Shield cooler (W) @ (K)	MLI (layers)	Cooler mass (kg)	Cooler input Power (W)
H_2 ¶	7.8 @ 22.8	60			23.7	784
H_2 ‡	1.2 @ 22.8	25	18.6 @ 85	20	10.3	507*
O_2 §	4.4 @ 97.2	30			1.1	70
O_2 †	11.8 @ 97.2	30			2.2	128
CH_4 §	5.6 @ 121	20			0.7	54
CH_4 †	14.2 @ 121	20			1.4	98

¶ single stage cooler without cooled shield
‡ 2-stage cooler with cooled shield
§ 2.05 m diameter tank
† 3.29 m diameter tank
* overestimate – see discussion in text

Table 4. Minimum set of coolers assuming 250 K heat rejection temperature

Cooler	Tank cooler Power (W)	Temperature (K)	Shield cooler Power (W)	Temperature (K)	Input power (W)
Single stage	11.8	97.2			128
2-stage	1.2	22.8	18.6	85	507*
Optional	5.2	97.2			77

* overestimate – see discussion in text

point for the coolers. The CH_4 cooling requirement can be met by the same cooler designed for the O_2 tanks but operated at reduced input power. The optional cooler is for the Descent option 1 configuration with smaller tanks. Although, the larger cooler could also be used, with reduced input power, in this application.

A comparison of the various options for hydrogen tanks is shown in Figures 1-4. There are two passive options: to use insulation only or to insert a VCS with a catalytic converter into the insulation. Both of these options require carrying extra hydrogen and sizing the tanks to be 97% full at launch. There are two ZBO options: to use a single stage cooler without a cooled shield or to use a 2-stage cooler with a cooled shield. The storage time is an important factor in selecting the best option. Figures 1-4 show comparisons passive versus cooled tanks with increasing storage time. Figure 1 shows the comparison for the reference H_2 tank that has a storage life of 37 days. Figure 2 shows the comparison for the TMI cargo stage H_2 tank with a 250-day hold time. Figure 3 shows the comparison for the Seed H_2 tank with a 430-day hold time and Figure 4 shows the comparison for the TEI H_2 tank with a 1200-day storage time. In all of these examples, the two options using coolers are about the same. The slight advantage of using a single-stage cooler is due to the conservative estimates used for the mass of the shield and the mass of a 2-stage cooler.

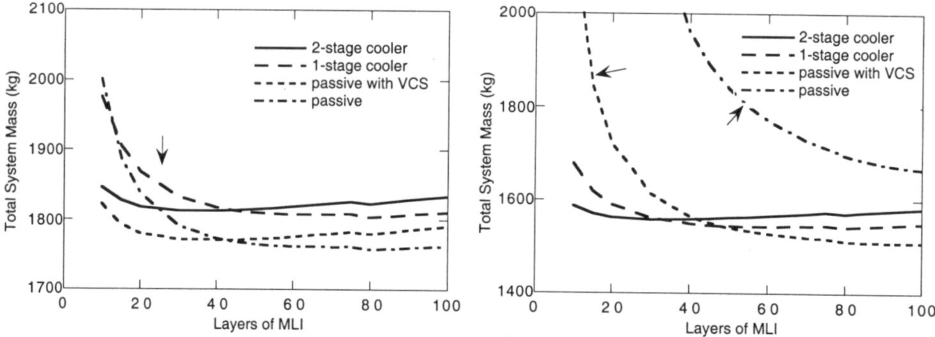

Figure 1. The system mass with and without coolers for the reference H_2 tank. The arrows indicate where the passive tank have the same volume as the cooled tanks.

Figure 2. The system mass with and without coolers for the TMI Cargo H_2 tank. The arrows indicate where the passive tanks have the same volume as the cooled tanks.

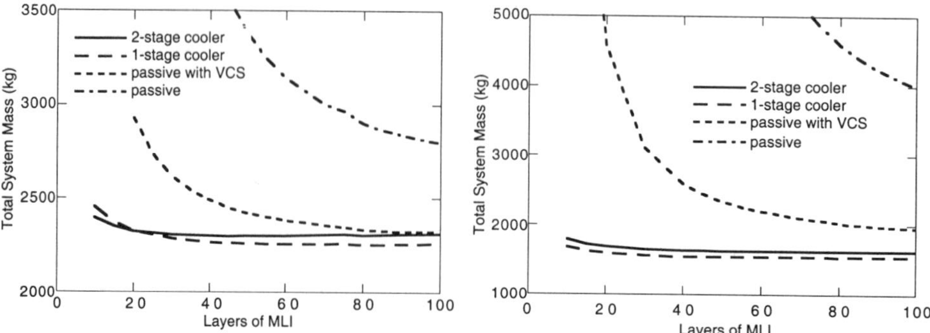

Figure 3. The system mass with and without coolers for the Seed H_2 tank. The passive tanks are larger than the cooled ones.

Figure 4. The system mass with and without coolers for the TEI H_2 tank. The passive tanks are larger than the cooled ones.

This difference is probably within the accuracy of the model. These figures show a general trend for H_2 storage:

1) Passive storage without a VCS results in a low mass system only for very short (≤30 days) duration missions.
2) Passive storage with a VCS and catalytic converter has a mass similar to cooled systems for all but the longest storage times. For small tanks (Fig. 2) the passive system may even be lighter and can fit within the volume constraints. This is a result of the cooled tank not being full.
3) Only for very long storage times, does the actively cooled systems have an apparent advantage over passive storage with a VCS and catalytic converter.
4) If it were not for the constraint of standardizing the tank size, the actively cooled systems would have a smaller volume than the passive systems. Active cooling presents an advantage in a volume-limited payload.

The situation is quite different for the storage of oxygen and methane. This is illustrated in Figures 5 and 6. For the oxygen tanks on the short-lived TMI cargo stage, Figure 5, a cooler saves about 3% of the mass. On the longer lived Ascent stage, Figure 6, the savings is about 21 % and 17% for both the oxygen and methane tanks, respectively.

SUMMARY

A comparison of various methods of propellant storage for human missions to Mars has been made. In the scenarios studied, eliminating the boil off of O_2 and CH_4 tanks by using a cooler always resulted in the lowest mass tank. This comparison included the added system mass due to the solar cells and radiators needed by the cooler. For H_2 tanks, using a cooler to eliminate boil off did not always result in the lowest mass system. For short duration tanks, allowing boil off and carrying extra H_2 resulted in the lowest mass. For long duration tanks, the mass of the cooled systems were comparable with a passive system that included a vapor-cooled shield and catalytic converter.

If the cooled option were chosen, only 2 or 3 different coolers would have to be developed. Suitable O_2 and CH_4 coolers are already being developed for cooling IR instruments and other applications.[13] There has been less development of coolers for H_2 tanks. Although there are some recent developments in regenerator materials that promise

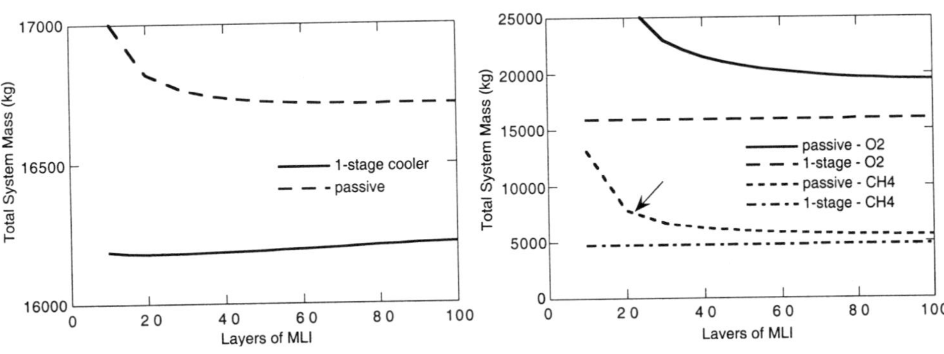

Figure 5. The system mass with and without coolers for the TMI Cargo O_2 tank. The passive system has more liquid at launch, but uses a smaller tank.

Figure 6. The system mass with and without coolers for the Ascent stage O_2 and CH_4 tanks. The arrow indicates where the passive CH_4 tank is the same size as the cooled tank. The O_2 tank is always larger.

to significantly improve 20 K pulse tube and Stirling coolers.[15]

The estimates given here are only intended to give preliminary estimates of the performance of coolers that will be required for human missions to Mars. By their very nature, these results, based on hypothetical coolers, can only approximate the mass of a final system. To improve the accuracy of the model, the performance characteristics of real coolers must be incorporated. But first, such coolers need to be developed.

REFERENCES

1. S.J. Hoffman and D.I. Kaplan, "Human Exploration of Mars: The Reference Mission of the NASA Mars Exploration Study Team," NASA SP 6107 (1997) and http://exploration.jsc.nasa.gov/expoHome.html
2. L.J. Salerno and P. Kittel, Cryogenics and the Human Exploration of Mars, Cryogenics 39 (1999), to be published
3. P. Kittel, L.J. Salerno, and D.W. Plachta, Cryocoolers for Human and Robotic Missions to Mars, Cryocoolers 10, Plenum Press New York (1999) 815
4. S.K. Borowski, L.A. Dudzinski, and M.L. McGuire, "Artificial Gravity Vehicle Design Option for NASA's Human Mars Mission Using Bimodal NTR Propulsion," AIAA-99-2545 (1999)
5. R. Alexander and L. Kos, private communication, MSFC (1999)
6. R.T. Parmley, W.C. Henninger, S.A. Katz, and I. Spradley, "Test and Evaluate Passive Orbital Disconnect Struts," NASA CR 177368 (1985).
7. R. T. Giellis, et. al., "Long Term Cryogenic Storage Study," US Air Force, AFRPL TR-82-071 (1982)
8. T. Nast, Multilayer Insulation Systems, "Handbook of Cryogenic Engineering," J.G. Weisend II, ed., Taylor & Francis, Philadelphia, (1998) pp. 186-202.
9. C.A. Withrow, and N. Morales, "Solid Electro-Chemical Power System for a Mars Mission," NASA TM 106606 (1994)
10. D.S Glaister and D.G.T. Curran, Spacecraft Cryocooler System Integration Trades and Optimization, Cryocoolers 9:873 (New York, 1997).
11. T.R. Strobridge, "Cryogenic Refrigeration - An Updated Survey," NBS Tech Note 655 (1974)
12. D.S. Glaister, M. Donabedian, D.G.T. Curran, and T. Davis, "An Overview of the Performance and Maturity of Long Life Cryocoolers for Space Applications," The Aerospace Corp, El Segundo, Report No. TOR-98(1057)-3 (1998)
13. E. Tward, T. Nast, W. Swift, S. Castles, and T. Davis, private communications.
14. R.J. Stochl, "Basic Performance of a Multilayer Insulation System Containing 20 to 60 Layers," NASA TN D-7659 (1974)
15. B.J. Korte, V.K. Pecharsky, and K.A Gschneidner, The Influence of Multiple Magnetic Ordering on the Magnetocaloric Effect in RNiAl Alloys, Adv. Cryo. Engin., 43:1737 (1998).

A LIQUEFIER FOR MARS SURFACE PROPELLANT PRODUCTION

L.J. Salerno[1], B. P. M. Helvensteijn[2], and P. Kittel[1]

[1]NASA Ames Research Center
Moffett Field, CA 94035-1000

[2]Atlas Scientific
Sunnyvale, CA 94086

ABSTRACT

NASA's planned Mars exploration missions will require that cryogenic propellants be manufactured on the surface. The present scenario calls for oxygen and methane gases to be produced using the carbon dioxide atmosphere plus seed hydrogen brought from Earth. Gases will require liquefaction for both storage on the Martian surface and for use in the ascent vehicle. The planned liquefaction rates range from 12.6 g/hr of oxygen for the 2003 robotic mission to 2500 g/hr for the later human missions.

This paper presents the results of a nitrogen liquefaction demonstration using a commercially available cryocooler. The experiment was set up to liquefy nitrogen gas instead of oxygen to limit laboratory safety concerns. A nitrogen gas condenser, attached to the cooler's cold tip, was sized to liquefy up to 42 gN_2/hr at the intended storage pressure (0.2 MPa). The experiment was conducted inside an atmospheric, air-filled, refrigerated chamber simulating the average Martian daytime temperature (240 K). In this demonstration a liquefaction rate of 9.1 gN_2/hr was realized, which is equivalent to 13 gO_2/hr.

INTRODUCTION

NASA is planning an extensive set of exploration missions. Prominent in this set of missions is a series of human missions to Mars. These missions plan to make extensive use of cryogenic propellants, some of which will be manufactured on Mars[1]. This In-Situ-Consumable-Production (ISCP) is part of the scheme to reduce the mass launched from Earth through In-Situ-Resource-Utilization (ISRU). Maintaining propellants on Mars requires liquefaction, storage, and transfer of cryogens on Mars. Some of the proposed chemical processes for ISCP require hydrogen as feed stock. The hydrogen will be transported from Earth as a liquid. The oxygen is to be manufactured on location by ISCP.

One of the goals of the planned 2003 mission to Mars is to demonstrate that oxygen gas produced on its surface can be liquefied at a rate of 12.6 g/hr and stored at 2 atm pressure (0.2 MPa). The liquefier will run for 60 days, roughly 8-12 hours each day, while being exposed to a typical daytime Mars surface temperature of 240 K. The present paper demonstrates that the planned 2003 liquefier requirement can be met with existing, off-the-shelf hardware. The following discusses our test apparatus, the dimensioning of the condenser used in liquefaction and the experimental results.

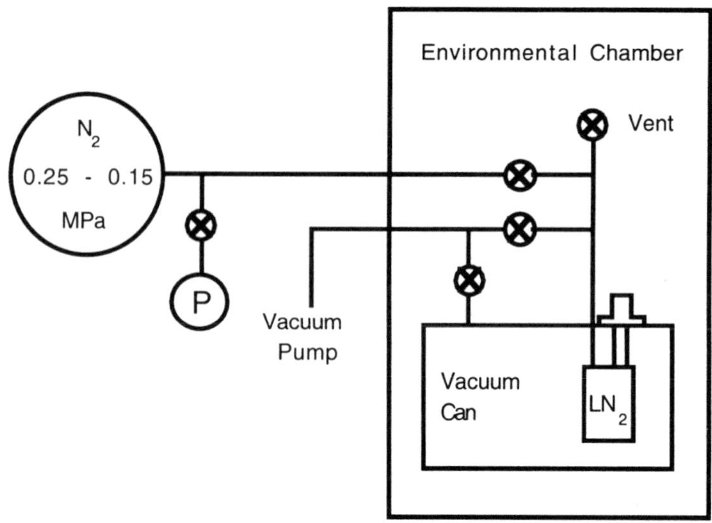

Figure 1. Schematic of the test setup.

EXPERIMENTAL SETUP AND PROCEDURE

A schematic of the experimental setup is shown in Figure 1. The liquefier consists of a cryocooler and a condenser (copper) that is epoxied directly to the cryocooler cold tip. The cooler selected is an off-the-shelf BEI 1000E cooler. The liquefier is mounted on a small dewar which is merely used as a vacuum can. The assembly is placed inside an environmental chamber. The chamber contains refrigerated air (245 K) at atmospheric pressure that is circulated by means of a fan. The nitrogen gas that is to be condensed is stored outside the chamber in a 36.7 L sphere at 0.2 MPa pressure and 293 K. A gas handling system allows evacuation of the apparatus and pressurization of the condenser as required. The condenser is pressurized through a 4.8 mm (3/16"; inner diameter) stainless steel fill tube. The assembly, as placed in the environmental chamber, is shown in Figure 2. The liquefier, as mounted on the dewar top flange, is shown in Figure 3.

Figure 2. Environmental chamber with test apparatus.

Figure 3. The liquefier (shown upside down).

The liquefaction procedure is as follows. The storage sphere is pumped, purged and pressurized with dry nitrogen gas and then is valved off. The Dewar and condenser are pumped out before and during cooldown of the environmental chamber. Once the chamber and the apparatus have cooled down to near Martian temperature (245 K) the cryocooler is turned on. A silicon diode thermometer attached to the condenser is used to monitor its temperature. When the condenser reaches the liquefaction temperature (85 K) valves are opened allowing gas to flow from the sphere to the condenser as liquefaction occurs. Measuring the pressure drop in the sphere versus time, the mass transfer rate, i.e. the liquefaction rate, is derived.

Minor issues concerning the test procedure are the following. As for selecting to liquefy nitrogen rather than oxygen in this demonstration, nitrogen has the obvious advantage in safety. Further, demonstrating that the liquefier is suitable for nitrogen provides only greater assurance that it is appropriate for oxygen liquefaction since oxygen has a higher boiling point. In terms of quantities liquefied at the same heat load to the cooler, roughly 1.3 gLO_2 corresponds to 1 gLN_2. Lastly, the environmental chamber has been run at 245 K rather than the average Mars temperature of 240 K in order to stay within the operating margin of the various rubber vacuum seals. Replacement of all vacuum seals by other materials has not been pursued considering the benefit gained for this test would be minor.

CONDENSER

In order to achieve the maximum liquefaction rate the condenser must have sufficient internal surface area for condensation. Insufficient surface area will result in under-utilization of the available cooling power. We will assume that the cooler provides cooling only at the cold tip, i.e. all the enthalpy associated with gas cooldown and liquefaction is rejected at the condenser attached to the cryocooler cold tip. Thus, the effective heat load is:

$$\dot{Q}_l = \left[h_{fg} + \Delta h_g\right]\dot{m} \tag{1}$$

(symbols are defined in the nomenclature at the end of this paper).

In terms of the average conductance at the wall Eqn. (1) becomes:

$$\dot{Q}_1 = h\,A\,\delta T \tag{2}$$

In our tests the condenser consists of a cylindrical copper can. For this case, laminar film condensation on a vertical plate applies. The conductance at the condenser wall is a function of the mass flow rate, latent heat and enthalpy of the gas, the condensation surface area and the temperature difference (subcooling) between the exchanger and the liquid. The applicable correlation is[2]:

$$h = 0.943\left[g\,\rho_p\left(\rho_p - \rho_v\right)k^3 h_{fg} / L\,\mu_p\,\delta T\right]^{1/4} \tag{3}$$

Although the available refrigeration capacity of the cooler allows only a liquefaction rate of roughly 10 g/hr, the condenser has been sized according to the equations stated above for a maximum liquefaction rate of 42 gN_2/hr. The condenser is constructed of a copper can of 41.4 cm^3 total volume and 79.3 cm^2 inside surface area, containing a tube heat exchanger of 26.6 cm^2 surface area suspended from the can's top. The total condenser surface area is 106 cm^2.

RESULTS AND DISCUSSION

Preceding both liquefaction and load testing the cooler has been run establishing the no-load cold tip temperature for the conditions in the environmental chamber (245 K). A no-load temperature of 46±1 K has been measured. After the no-load bottom temperature is reached and before liquefaction, the cooler is turned off temporarily until the condenser temperature rises above 70 K. By this measure solidification of nitrogen in the fill line is avoided. Then when the condenser is pressurized, the storage vessel pressure begins to drop and the condenser temperature is observed to rise and to stabilize at roughly 85 K. The fall in pressure is monitored over time. Converting the pressure data to mass flow and liquefaction rate, results in an average liquefaction rate of approximately 9.1 gN_2/hr over a 3.55 hour period. The liquefaction rate versus time is shown in graphical form in Figure 4.

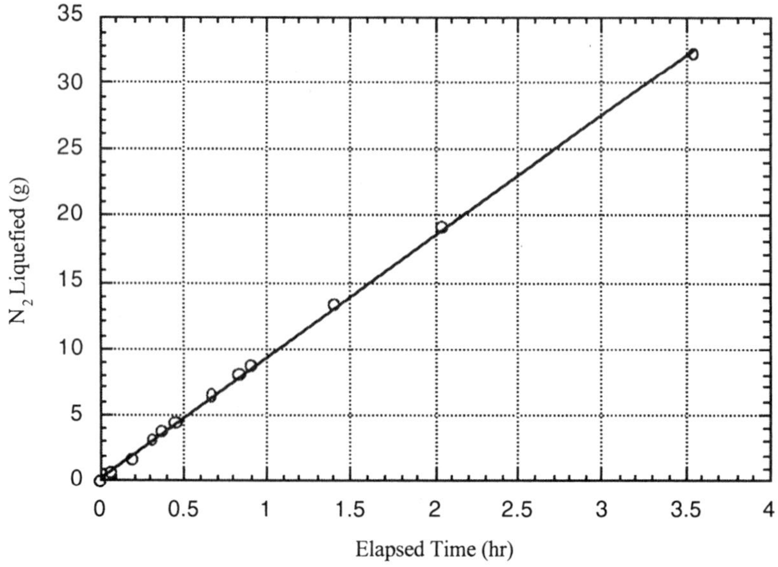

Figure 4. N_2 liquefaction versus time.

The oxygen liquefaction rate equivalent to the nitrogen liquefaction rate is derived in the following manner. The nitrogen liquefaction rate of 9.1 g/hr was achieved at 85 K. The cooler manufacturer's data show a 33% increase in refrigeration capacity at 97 K over that at 85 K (97 K is the 0.2 MPa liquefaction temperature of oxygen). Also to be taken into account is the ratio between the total enthalpy changes of oxygen and nitrogen when cooled from 293 K and liquefied at their respective liquefaction temperatures (400 J/gO_2 and 426 J/gN_2). It follows that liquefying nitrogen at a rate of 9.1 gN_2/hr corresponds to an oxygen liquefaction rate of 12.9 gO_2/hr. This exceeds the planned 2003 Mars mission goal of 12.6 gO_2/hr by 2 %.

Two load tests on the cooler, with the condenser pumped out, have been performed. At 0.1 W the condenser temperature stabilized at 48.5 K; at 1.18 W the condenser temperature stabilized at 85 K. The cooling power measurement at 85 K was made in order to verify the cooling power exerted by the cryocooler when it is liquefying nitrogen gas at 85 K. The measured cooling power should agree with the heat load to the cooler as expressed in Eqn. (1). From the enthalpy change (426 J/gN_2) and the measured mass flow the heat load is calculated to have been 1.08 W during liquefaction. Considering the close agreement of the measured cooler power during the load test and the heat load required for liquefaction, it should be noted that the limiting factor for the liquefaction rate was the available cooling power and not the size of the condenser. As previously stated, the condenser was sized for up to 42 g/hr liquefaction.

An analysis was performed to calculate the parasitic heat input to the cooler due to both conduction and radiation from the 245 K environment to the condenser at 85 K. The radiation heat input through three layers of aluminized mylar was estimated to be on the order of 0.10 W. The conduction along the stainless steel nitrogen fill tube was calculated as 0.070 W. The conduction along the instrumentation wires was calculated to be 0.019 W. All this adds up to a total parasitic heat input of about 0.19 W.

CONCLUSIONS

In summary, it has been demonstrated that oxygen liquefaction rates of the planned 2003 Mars mission can be met with existing off-the-shelf hardware. Future work will focus on the more formidable challenge of demonstrating that the 2500 g/hr requirement for the later human missions can be met with an economically feasible package.

REFERENCES

1. Salerno, L. J., and Kittel, P., Cryogenics and the Human Exploration of Mars, *Cryogenics* (in process)
2. Kreith, F.; "Principles of Heat Transfer", 3rd Ed., Intext Press, NY (1973), p. 527

NOMENCLATURE

A	surface area	Q_l	liquefier load
g	gravitational acceleration	T_c	cold temperature
h	conductance	T_h	hot temperature
h_{fg}	latent heat of condensation	μ	dynamic viscosity
h_g	enthalpy of gas	δT	subcooling of condenser
k	thermal conductivity of liquid	ρ_v	density of gas
L	fin length of liquefier	ρ_p	density of propellant

PULSE TUBE OXYGEN LIQUEFIER*

E.D. Marquardt and Ray Radebaugh

National Institute of Standards and Technology
Boulder, Colorado 80303

ABSTRACT

The performance of a single-stage coaxial pulse tube refrigerator used as an oxygen liquefier is discussed. The liquefier is a half-size laboratory unit for NASA/JSC to demonstrate the technology of liquefying and storing oxygen on Mars. The liquefier would be part of a plant that extracts oxygen from the Martian carbon dioxide atmosphere and stores the oxygen in liquid form. The liquid oxygen could then be used either for the oxidizer of a fuel or for breathing in a manned mission. The pulse tube and regenerator, which are arranged coaxially, provide 18.8 W of cooling at 90 K with 222 W of PV input power. The resulting coefficient of performance (COP) is 20.0% of the Carnot value based on PV input power, among the highest ever achieved. Liquid nitrogen was produced at a rate of 1.75 g/min (2.17 cm^3/min) with 215 W of PV input power. Oxygen should be produced at about 2.7 g/min (2.37 cm^3/min). While this cooler is not a flight version, it is a flight-like design similar to what would be used on a mission. This paper presents details of the liquefier geometry and results of thermal performance tests.

INTRODUCTION

Manned missions to Mars will require in situ resource production[1] (ISRP), the use of resources found on Mars. The robotic missions planned over the next decade will be used to prove some of the technology required for ISRP. One primary requirement is oxygen for both breathing and as an oxidizer for fuel. NASA is currently developing the technology to produce oxygen from the carbon dioxide atmosphere of Mars. Another requirement is the long-term storage of the oxygen in liquid form. NASA/JSC is building an environmental chamber to simulate the Martian environment to study the complete system of oxygen production and storage. This liquefier will be used within that chamber.

Research partially funded by NASA/JSC and NASA/Ames.

LIQUEFIER DESIGN

The pulse tube liquefier is made of several components: the compressor, connecting lines, aftercooler, regenerator, cold end, cold end heat exchanger, pulse tube, warm end heat exchanger, orifice, inertance tube, reservoir volume, and condensing fins. Table 1 lists the performance goals. The design of the cold head is summarized in Table 2.

In order to be used as a liquefier, the pulse tube cold head must completely fit within the neck of a dewar. Clearly, an in-line pulse tube would not be appropriate and a U-tube geometry would require a larger dewar neck. A coaxial pulse tube makes the integration with the dewar easy. Performance will suffer compared to an in-line design by about 5-10%. Figure 1 shows a photo of the complete system. Figure 2 shows a schematic diagram.

Compressor

The compressor, designed with a swept volume of 20 cm^3 and a resonant frequency of 40 Hz, is capable of delivering 300 W of PV power. It uses a clearance seal and linear-arm flexure bearings for long life. In this system, it is being used with a swept volume of 15.5 cm^3 at 45 Hz. The compressor is not very efficient, partially because it is operating off the resonant frequency. Thus, we will report PV power in this paper. Under the liquefier operating conditions, the compressor converts AC electrical power to PV power at 62% efficiency. The most advanced linear compressors have an efficiency of 80% to 85%.

Connecting Lines

The connecting line between the cold head and compressor has a diameter of 4.62 mm and a length of 111.3 mm. Next, it splits into two lines, each with a diameter of 3.80 mm

Table 1. Liquefier design goals

Parameter	Goal
Operating temperature	90 K
Cooling load	15 W
Heat rejection temperature	300 K
Oxygen Liquefaction rate	2.2 g/min 1.93 cm^3/min
Assumed compressor efficiency (PV Power/Electrical)	85%
PV Input Power	260 W
Specific Power (load/PV Power)	17.3 W/W
% Carnot (based on PV power)	13.5 %

Table 2. Cold head design

Parameter	Value
Frequency	45 Hz
Average pressure	2.0 MPa
PV Input Power	~220 W
Compressor swept volume	~15.5 cm^3
Regenerator	28 mm OD x 12.75 mm ID (annular) x 50 mm length 400 mesh SS screen
Pulse tube	Ø12.45 mm ID x 70 mm length, 8.52 cm^3 volume
Inertance tube	Ø2.57 mm ID x 2.20 m length
Secondary orifice	Ø0.25 mm x 1.25 mm length

Figure 1. Pulse tube liquefier system.

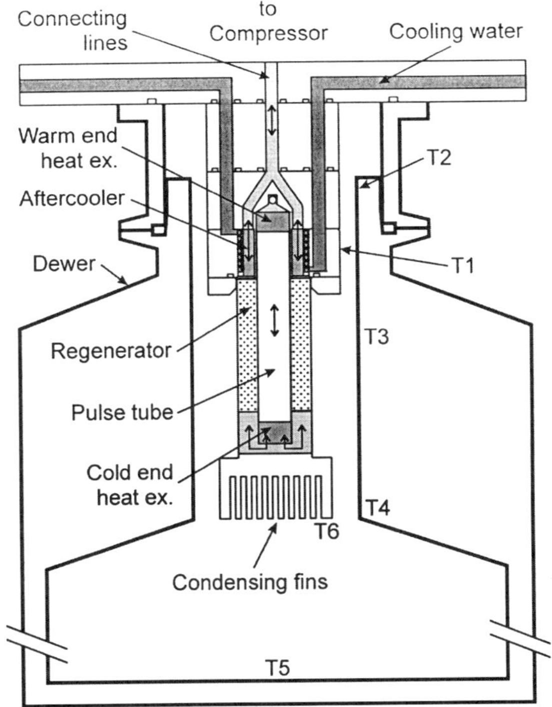

Figure 2. Schematic diagram of the liquefier. T1 through T6 show the temperature sensor placement referred to in Figure 5.

and a length of 22.5 mm. The split lines end in an annular space, for re-mixing, which has a 22.86 mm OD, 15.24 mm ID, and a 1.78 mm depth. The compressor pressure is measured here. Also connected to the annular space is a line 2.53 mm in diameter and 14.5 mm long, connected to one side of the secondary bypass orifice.

Aftercooler

The aftercooler is a radial parallel plate design. It is composed of 104 radial gas flow passages, each with a gap 150 μm by 4.11 mm, and 17.40 mm long. A radial slot heat exchanger is a very good match for a coaxial pulse tube. The pulse tube fits into the large central hole of the annular slots. Uniform slots are cut from the large central hole using a wire electro-discharge machine (EDM). This results in a heat exchanger with a low pressure drop, low void volume, and high heat transfer area.

Regenerator

The regenerator is placed in the annular space surrounding the pulse tube. The outer diameter of the regenerator tube is 28.57 mm, the wall thickness is 0.28 mm, and the length is 50.0 mm. The outer diameter of the pulse tube, which forms the inner wall of the regenerator, is 12.75 mm. The annular space is filled with 400 mesh stainless steel screen with a wire diameter of 25 μm and a porosity of 68.7%. The regenerator tube is made of 718 Inconel. Testing a coaxial pulse tube makes it difficult to independently change the size of the regenerator and pulse tube. A possible way to change the regenerator length is to insert a spacer in place of some of the screen at the warm end. The spacer must be designed to have low pressure drop, void volume, and heat transfer. We have made two spacers, 5 and 10 mm long. They were made of stainless steel with the same inner and outer diameters as the regenerator but with 16 radial slots cut from the inner diameter. Each gap was 0.44 mm by 6.35 mm.

Cold End

The cold end reverses the flow between the regenerator and the pulse tube. This is an important part of a coaxial pulse tube design as it is easy to add extra undesirable void volume; we must balance pressure drop and void volume. We have used 16 slots each with a gap 200 μm by 7.00 mm, and length of 15.50 mm. After the slots, the flow enters a small volume below the cold end heat exchanger with a volume of 18.9 mm^3. The slots also act as part of the cold end heat exchanger, providing about 25% of the required heat transfer.

Cold End Heat Exchanger

The cold end heat exchanger is made of 100 mesh, 114 μm diameter wire, copper screen. It is 12.75 mm in diameter and 8.08 mm long. This provides the bulk of the heat transfer and provides flow straightening for the pulse tube.

Pulse Tube

The pulse tube is made of stainless steel tubing with an outer diameter of 12.75 mm and a wall thickness of 0.15 mm. The length is 70 mm and the volume is 8.52 cm^3. In a coaxial design, the pulse tube volume may be changed by using a thin walled insert; G-10 composite with its low thermal conductivity, is a good choice of material. To prevent turbulence, the insert must extend the entire length of the pulse tube. It is also possible to taper the insert, allowing for the cancellation of any flow streaming[2] within the pulse tube. We did not experiment with any inserts.

Warm End Heat Exchanger

The warm end heat exchanger is made of 100 mesh, 114 μm diameter wire, copper screen. The diameter is 12.75 mm and the length is 7.32 mm. There is a volume of 0.19 cm^3 after the heat exchanger where the pulse tube pressure is measured. Between this volume and the orifice is a connecting line with a diameter of 2.53 mm and length of 18.7 mm.

Orifice

The pulse tube can be run with either a single orifice, double orifice, and/or with an inertance tube. Each of the two orifices, shown in Figure 3, is a plug into which a channel may be cut. The orifices fit together in a recess on the side of the aftercooler into which three lines enter: one each to the compressor, pulse tube warm end, and the reservoir volume. A cover plate allows access. Two orifices are required, although the secondary orifice is sometimes filled with a blank plug. When the inertance tube is used, the primary orifice is a channel 2.5 mm wide, 1.25 mm high, and 2.5 mm long. This provides only a small pressure drop and is considered 'wide open'. The secondary orifice is more difficult to make. A symmetric pressure drop element cannot be used since it will (and did) cause a very large DC flow[3] through the pulse tube and regenerator. With a symmetric pressure drop element, such as the one used for the primary orifice, the system achieved only temperatures of around 270 K. Figure 3b and 3c show an asymmetric pressure drop element, which can eliminate the DC flow as indicated by the temperature gradient along the regenerator tube. The flow rate is controlled by the hole size connecting the top and bottom sides, typically with a diameter ranging up to 0.6 mm. Following the primary orifice

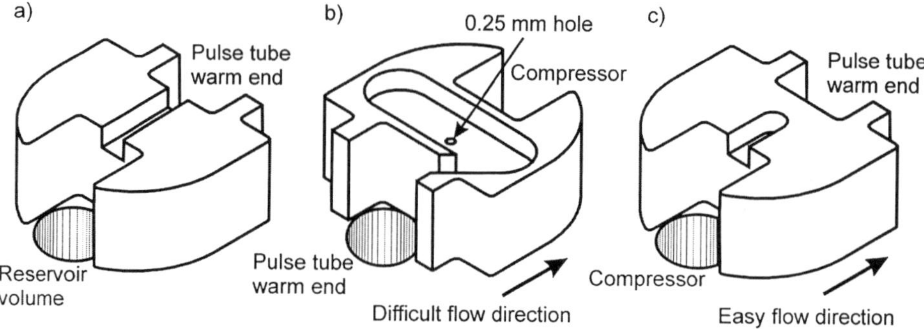

Figure 3. a) Primary orifice b) Secondary orifice top c) Secondary orifice bottom

is a connecting line with a diameter of 3.18 mm and length of 100 mm. Next follows another connecting line of 2.62 mm diameter and 52 mm length. Then comes either the inertance tube or the reservoir volume.

Inertance Tube

The best performance was achieved with an inertance tube[4,5] in place but the system was also run without the inertance tube. The best performing single diameter inertance tube had an inner diameter of 2.57 mm and a length of 2.20 m. The best overall results were obtained using a double inertance tube made with two different diameters with a smooth transition between them. These two stainless steel tubes were: 2.57 mm ID, 1.20 m long and 4.23 mm ID, 3.10 m long, with the larger tube adjacent to the reservoir volume.

Reservoir Volume

Most tests were performed with an external reservoir volume of 500 cm^3. The final configuration uses a 500 cm^3 volume internal to the compressor but with a separate pressure boundary from the backside of the compressor.

Condensing Fins

The condensing fins were of gold plated copper with a diameter of 42.5 mm and height of 18.9 mm. In this block, 9 slots 2.0 mm wide and 15.7 mm deep were cut. This allows a large surface area to be exposed to the fluid while not allowing a meniscus to form in any slot, even in the 0.38 g gravity on Mars.

EXPERIMENTAL RESULTS AND DISCUSSION

Background Heat Leak

The background heat leak was measured by observing the warm up rate of the cold end with the compressor off and with different applied heat loads. By assuming the thermal degradation factor[6] for the regenerator screens to be 0.1, we can separate the radiation and conduction terms. This method assumes that all the heat capacity is at the cold end. We have about three times as much mass in the cold head as in the regenerator. Table 3 summarizes the thermal background losses measured with a cold end temperature of 90 K.

Cryocooler Refrigeration Performance

All experiments use the same pulse tube and regenerator. We have not experimentally optimized the pulse tube or regenerator volumes. The volumes used were those optimized from our modeling, not from experimental tests. For this pulse tube and regenerator, we have found the optimized inertance tube length for a calculated diameter and for a double inertance tube. We also experimentally determined the optimized secondary orifice size.

Table 4 summarizes the results of four experiments. The first run was done with a single diameter inertance tube. Run two was done with a double inertance tube; both lengths were optimized. Run three optimized the

Table 3. Thermal losses at 90 K

Parameter	with MLI	No MLI
Regenerator tube	0.95 W	0.95 W
Regenerator screen	0.76 W	0.76 W
Pulse tube	0.32 W	0.32 W
Helium	0.04 W	0.04 W
Radiation	0.17 W	0.90 W
Total measured	2.24 W	2.97 W

Table 4. Experimental results at 90 K

Parameter	Run 1	Run 2	Run 3	Run 4
Inertance tube (ID x length)	Ø2.57 mm x 2.20 m	Ø2.57 mm x 1.20 m/ Ø4.23 mm x 3.10 m	Ø2.57 mm x 1.20 m/ Ø4.23 mm x 3.10 m	Ø2.57 mm x 1.20 m/ Ø4.23 mm x 3.10 m
Secondary orifice	closed	closed	Ø0.25 mm x 1.25 mm	Ø0.25 mm x 1.25 mm
MLI	none	none	none	aluminized mylar
No load temp.	45.35 K	42.49 K	42.61 K	40.38 K
Cooling power	16.18 W	17.51 W	18.07 W	18.78 W
Rejection temp.	302 K	302 K	302 K	302 K
PV input power	221.5 W	217.3 W	221.7 W	221.6 W
Specific power	13.7 W/W	12.4 W/W	12.3 W/W	11.8 W/W
% Carnot	17.2	19.0	19.2	20.0
V_{co}	15.49 cm^3	15.50 cm^3	15.47 cm^3	15.57 cm^3
P_{co}	0.303 MPa	0.317 MPa	0.320 MPa	0.320 MPa
P_{pt}	0.221 MPa	0.244 MPa	0.244 MPa	0.247 MPa
$\dot{m}_{res}$	1.184 g/s	1.741 g/s	1.710 g/s	1.712 g/s
$\dot{m}_{pt}$	1.08 g/s	1.14 g/s	1.12 g/s	1.12 g/s
$< \dot{W}_{pt} >$	36.3 W	37.3 W	30.2 W	30.2 W
Regenerator loss	9.2 W	7.0 W	5.7 W	5.7 W
FOM_{pt}	0.78	0.74	0.885	0.885
PA $\dot{m}_{res}$	-24.8 degrees	-56.2 degrees	-56.3 degrees	-56.2 degrees
PA $\dot{m}_{pt}$	-7.8 degrees	-33.9 degrees	-43.0 degrees	-43.0 degrees
PA V_{co}	148.9 degrees	148.6 degrees	148.0 degrees	148.2 degrees
PA P_{co}	13.1 degrees	12.2 degrees	12.4 degrees	12.3 degrees

V_{co}: Compressor swept volume
P_{co}: Pressure amplitude measured at the compressor
P_{pt}: Pressure amplitude measured at the pulse tube
$\dot{m}_{res}$: Mass flow rate amplitude measured at the reservoir volume
$\dot{m}_{pt}$: Mass flow rate amplitude calculated at the pulse tube warm end
PA $\dot{m}_{res}$: Phase angle by which the reservoir volume mass flow leads the pulse tube pressure
PA $\dot{m}_{pt}$: Phase angle by which the warm end pulse tube mass flow leads the pulse tube pressure
PA V_{co}: Phase angle by which the compressor swept volume leads the pulse tube pressure
PA P_{co}: Phase angle by which the compressor pressure leads the pulse tube pressure

secondary orifice with the double inertance tube. Run four repeated run three except the cold head was wrapped in about 10 layers of aluminized mylar. Figure 4 shows a performance plot of the cryocooler with the double inertance tube, a secondary orifice, and aluminized mylar (run 4). The Carnot efficiency of 20% relative to PV power is among the highest ever achieved with a pulse tube refrigerator or any other small cryocooler.

We can define a figure of merit for the pulse tube, FOM_{pt}, as the ratio of the time-averaged enthalpy flow, $< \dot{H}_{pt} >$, to the hydrodynamic work flow, $< \dot{W}_{pt} >$, through the pulse tube. The hydrodynamic work flow is the time average of the instantaneous product of the volumetric flow and pressure. The best way to measure the enthalpy flow along the pulse tube is to measure the heat rejected at the warm end heat exchanger. In our coaxial pulse tube, this is not possible since the warm end heat exchanger is built inside the aftercooler. We can also measure the enthalpy flow by adding the thermal loss terms to the net cooling power, $\dot{Q}_c$. We have measured the conduction, $\dot{Q}_{cond}$, plus radiation term, $\dot{Q}_{rad}$, but we must calculate the regenerator loss, $\dot{Q}_{reg}$, from a numeric model. We then have

$$FOM_{pt} = \frac{< \dot{H}_{pt} >}{< \dot{W}_{pt} >} = \frac{\dot{Q}_c + (\dot{Q}_{cond} + \dot{Q}_{rad}) + \dot{Q}_{reg}}{\frac{1}{\tau}\int_0^{\tau} P_{pt}\dot{V}_{pt}dt}. \quad (1)$$

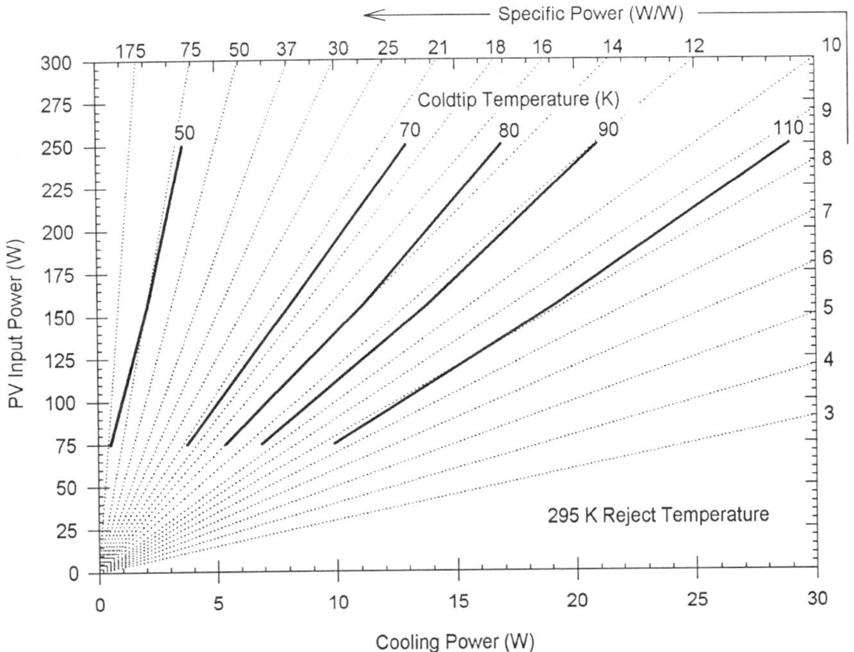

Figure 4. Measured cooler performance from run 4.

In general FOM_{pt} for a cryogenic pulse tube will be 1.0 if all losses in the pulse tube are zero. We cannot eliminate the boundary layer or non-adiabatic losses, but other pulse tube losses could be due to turbulence or acoustic flow streaming, which we can reduce. Rawlins et. al.[7] have shown that typical small pulse tubes have a FOM_{pt} of 0.60 to 0.85. Swift et. al.[8] described a large tapered pulse tube (1.5 kW cooling power) with a FOM_{pt} of 0.96. Figures of merit between 0.74 and 0.885 have been measured here.

Liquefier Performance

The condensing fins were added to the cold head and the dewar was filled with nitrogen gas at atmospheric pressure. Nitrogen has a specific enthalpy difference similar to that of oxygen from room temperature to the liquid state at atmospheric pressure, but has a lower liquefaction temperature. For safety reasons, we chose to liquefy nitrogen.

Figure 5 shows the experimental results for a period of about 27 hours. Thermocouples were placed at three locations along the neck of the dewar as well as on the cold head and at the bottom of the dewar (placement is shown in Figure 2). First, the dewar had to be cooled to the liquefaction temperature by convection. This process took about 12 hours. Liquid began to condense and drip to the bottom of the dewar after about 9 hours. After about 13 hours, the liquefaction process reached its equilibrium and liquid was produced at a rate of 1.75 g/min (2.17 cm^3/min) with 215 W of PV input power. The dewar had a measured heat leak of 0.15 g/min or 1.1 W. The liquefier should produce about 2.7 g/min (2.37 cm^3/min) of liquid oxygen.

There was a 1.4 K temperature difference between the cold head and the liquid. This shows that the condenser surface area was acceptable. The liquefaction rate decreased slightly with time. The initial reduction was probably due to a reduction in average pressure of the cryocooler caused by a small helium leak. After about 23 hours, the temperature difference between the cold head and liquid increased to 1.5 K and the liquefaction rate decreased somewhat more, due probably to contamination on the condenser.

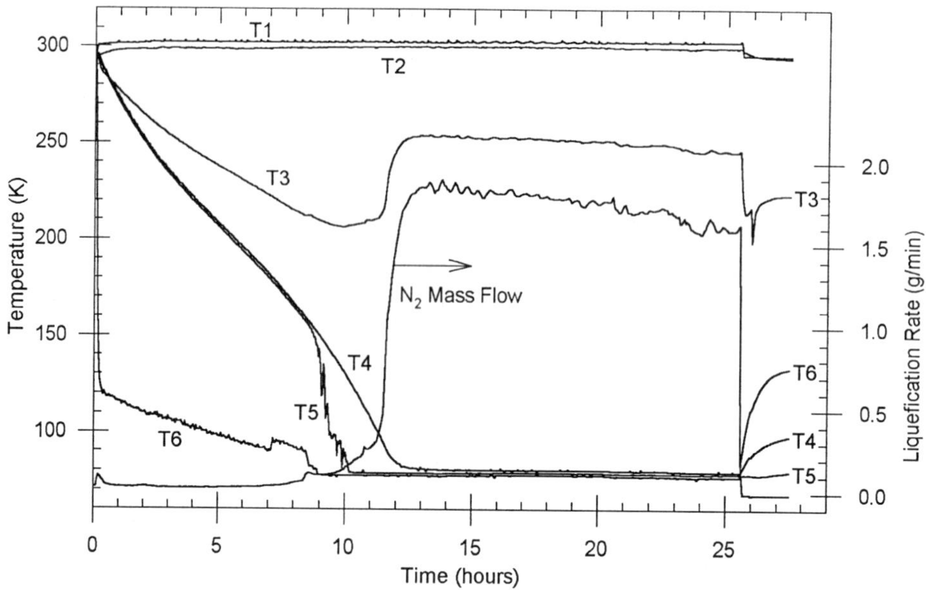

Figure 5. Liquefier performance using nitrogen. The locations of the temperature sensors T1-T6 are shown in Figure 2.

The cooler was switched off after about 25.5 hours; this would be normal operation for the liquefier running on solar energy, where it would be switched off at night. We were interested in finding the equilibrium temperature of the cooler to determine the background heat leak at night. The boil-off caused by this heat leak would have to be recondensed in the morning before the full liquefaction rate could be resumed. The cold fins remained at about 134 K with the cooler off, resulting in an additional heat leak of about 2.5 W from the cold head. This means that for every hour the cooler is off, it must run for 11.7 minutes to re-condense the boil-off from the cold head. While this cooler was designed for optimum performance, it may be possible to optimize the cold head for less off-state thermal conduction for a system design where the liquefier will run for only part of a day. A longer, smaller diameter regenerator may not perform as well but it will not have to re-liquefy as much boil-off and the liquefier may use less energy over the entire day/night cycle.

REFERENCES

1. L. J. Salerno, P. Kittel, "Cryogenics and the Human Exploration of Mars", *Cryogenics* **39**, pp. 381-388, Elsevier Science (1999).
2. J. R. Olson, G. W. Swift, "Acoustic Streaming in Pulse Tube Refrigerators: Tapered Pulse Tubes", *Cryogenics* **37**, pp. 769-776, Elsevier Science (1997).
3. D. Gedeon, "DC Gas Flows in Stirling and Pulse Tube Cryocoolers," Cryocoolers Vol. 9, pp. 385-392, Plenum Press (1996).
4. K. M. Godshalk, J. C. Kwong, Y. K. Hershberg, G. W. Swift, R. Radebaugh, "Characterization of 350 Hz Thermoacoustic Driven Orifice Pulse Tube Refrigerator: Measurements of the Phase of the Mass Flow and Pressure," Advances in Cryogenic Engineering Vol. 41, pp. 1411-1418, Plenum Press (1996).
5. R. Radebaugh, "Advances in Cryocoolers,"16th International Cryogenic Engineering Conference, pp. 33-44, Elsevier Science (1996).
6. M. Lewis, T. Kuriyama, F. Kuriyama, R. Radebaugh, "Measurement of Heat Conduction Through Stacked Screens," Advances in Cryogenic Engineering Vol. 43, pp. 1611-1618, Plenum Press (1997).
7. W. Rawlins, R. Radebaugh, K. D. Timmerhaus, "Energy Flows in an Orifice Pulse Tube Refrigerator," Advances in Cryogenic Engineering Vol. 39, pp. 1449-1456, Plenum Press (1993).
8. G. W. Swift, M. S. Allen, J. J. Wollan, "Performance of a Tapered Pulse Tube," Cryocoolers Vol. 10, pp. 315-320, Plenum Press (1998).

HYBRID THERMAL CONTROL TESTING OF A CRYOGENIC PROPELLANT TANK

D. W. Plachta
National Aeronautics and Space Administration
Glenn Research Center
Cleveland, Ohio 44135

ABSTRACT

This report presents the experimental results of a hybrid thermal control system, one that integrates a passive system (multi-layer insulation) with an active system (a mechanical cyrocooler) applied to cryogenic propellant storage. These experiments were performed on a 1.39 m diameter spherical propellant tank filled with LH2 while installed in an evacuated chamber. The tank heat transfer to the cryocooler was accomplished with a condenser installed in the ullage of the tank and mated to the second stage of the cooler, and by conduction, through copper leaves mated to the first stage of the cooler. The first hybrid system test was performed with both the condenser and the leaves, a configuration that had excess capacity to remove the heat entering the tank; the second test was performed with only the condenser, with a capacity closely matched to the tank heating rate. In both of these tests, the goal of zero boil-off was achieved.

INTRODUCTION

A fact of cryogens is that they will boil-off because of environmental heating. Planning for these losses, larger than necessary tanks and associated storage systems are used. In the case of long term missions in-space, the added mass due to the cryogen boil-off and the oversized tanks and storage systems can make the use of cryogenic propellants prohibitive.

An alternate storage option is to eliminate the boil-off and associated larger-than-necessary storage system by integrating cryocoolers into the tanks. This option has been looked at in previous studies[1,2] but has never been incorporated. Recent advances in cryocooler technology[3,4,5] however, have greatly enhanced this concept. Analysis indicates that added mass for new generation cryocoolers and associated power and heat rejection systems are less than the mass of the boil-off and larger-than-necessary storage tanks if the mission storage time is greater than roughly 45 days.[6] As the storage duration increases, the mass savings increases. The limiting factor is the cryocooler life, which the current maximum is ~10 years.

While this zero boil-off (ZBO) approach looked promising based on the analysis performed, the concept had not been tested on a cryogenic storage tank that holds propellant of the quantity approximating that of an upper stage of a launch vehicle. A test on such a tank would help determine the feasibility of the concept.

EXPERIMENTATION

The design goal of the experimentation was a simple and inexpensive integration of a cryocooler with an existing insulated propellant tank.

Facility

NASA Glenn Research Center's Supplemental Multi-layer Insulation Research Facility (SMIRF), documented in Ref. 7, is a multi-purpose test bed designed to evaluate the performance of thermal protection systems. One modification to the vacuum chamber was that the cold GN_2 shroud, used to provide a controlled thermal environment to the test article, was removed in order to accommodate the test tank

Tank

The liquid hydrogen test tank used is briefly described below and in detail in Ref. 8. This 2219 aluminum tank is spherical, 1.39 meters diameter and 1.42 cubic meters in volume. This tank has a 0.3 m diameter opening and cover to allow access to the tank interior. The tank was suspended from a tubular, stainless-steel support ring by six stainless-steel wire support struts. The tubular support ring was, in turn, suspended from the lid of the vacuum chamber by three support cables. The tank (insulated) is shown in Fig. 1.

Insulation

The insulation installed on the test tank was two multi-layer insulation (MLI) blankets to cover and thermally protect the entire tank surface. This concept is also discussed in Ref. 8. Each blanket consisted of 15 double aluminized mylar (DAM) radiation shields alternatively spaced with double silk net spacers. This results in a total of 17 layers of mylar radiation shields, or MLI, for each blanket, for a total of 34 layers of MLI. The blankets are held together with Nylon fasteners and reinforced Mylar cover sheets. Nylon button-pin studs epoxied to the tank wall supported the blankets. Approximately 2/3rds of the tank was covered with this insulation, which was fabricated and tested in 1977 and was in storage for the last 15 years at Plum Brook Station. The missing third of the insulation, which consisted of one of the six gore sections and the top and bottom dome sections that mate to the tank, were constructed at GRC using similar materials and were fitted to the tank. Also, constructed were fitted MLI and cryoglass blankets that insulated the tank supply, vent, and drain lines as well as the cryocooler. Aluminized mylar tape was used to attach the new MLI on the tank and also to attach the MLI/cryoglass blankets around the lines and the cryocooler. The insulated tank, attached to the vacuum tank lid is shown in Fig. 1.

Cryocooler

The cryocooler was borrowed from our colleagues at NIST. It was a two-stage Gifford/McMahon cycle, commercial cryocooler. The cryocooler capacity rating of the second stage was 17.5 W at 18K while, simultaneously, providing 20 W of cooling at 35K at the first stage. The cryocooler is shown in Fig. 2. The cryocooler also included a separate helium scroll compressor (not shown), which was located outside the vacuum chamber.

Heat Exchangers

The cryocooler second stage coldhead was mated to a condenser heat exchanger, shown in Fig. 3. Indium foil was used to increase the thermal contact area between the mating surfaces. Copper No. 101 was used for the flange/rod because of its very high thermal conductivity at cryogenic temperatures. Brazed to the rod was a stainless steel sleeve, which was welded to a stainless steel flange that was bolted to the tank access cover. Stainless steel was used because of its

Figure 1. MLI insulated tank supported by tubular support ring and hung from vacuum chamber lid. Cryocooler is located at top of tank. Flexible helium lines, which go to the scroll compressor located outside the vacuum chamber are shown.

low thermal conductivity, less than 1/1000th that of copper, and effectively stopped the heat transfer from the access cover to the cryocooler. Bolted to the bottom end of the rod was our 248 cm^2 Copper 101 condenser, which extended 12.7 cm below the top of the tank, at the 97 percent full level. Note that none of the tests had liquid levels over 95.5 percent full. This condenser was designed to remove the predicted tank environmental heating by condensing the hydrogen vapor in the ullage of the tank. Natural fluid convections of the LH2 moved the heat from the tank wall to the vapor.

In addition to the condenser, other heat exchangers were used in one of the tests to take advantage of the heat removal capacity of the cryocooler's first stage. These heat exchangers were made from 0.16 cm thick Copper No. 101 sheet. Attached to the cryocooler first stage was a hexagon plate with 19.2 cm wide sides. Attached to this plate were 2.5 cm wide straps used to

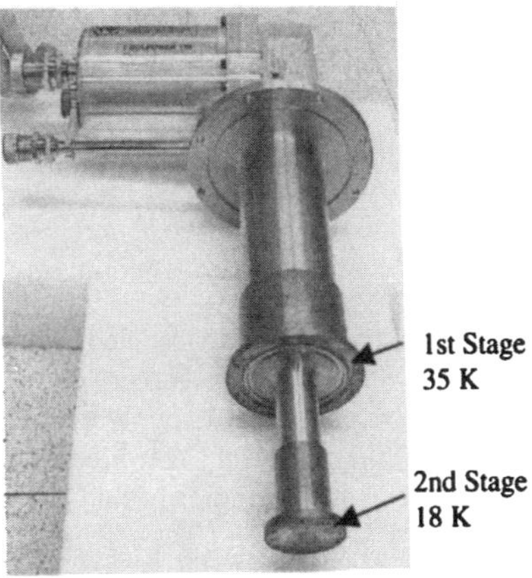

Figure 2. Two stage Gifford-McHahon cycle cryocooler used in testing. Cryocooler capacity was 17 W at 18 K and 20 W at 35 K.

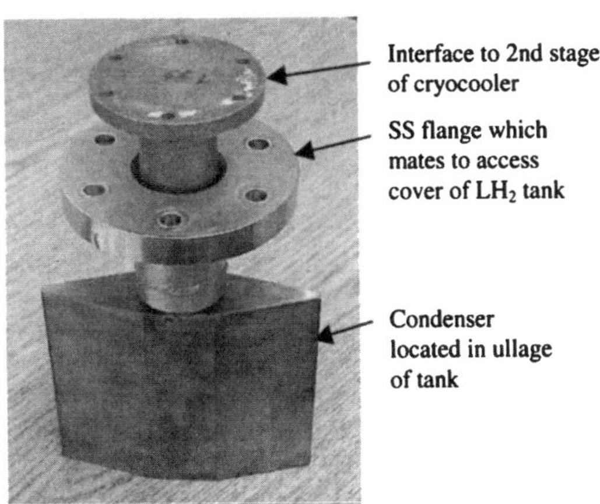

Figure 3. Second stage condenser heat exchanger.

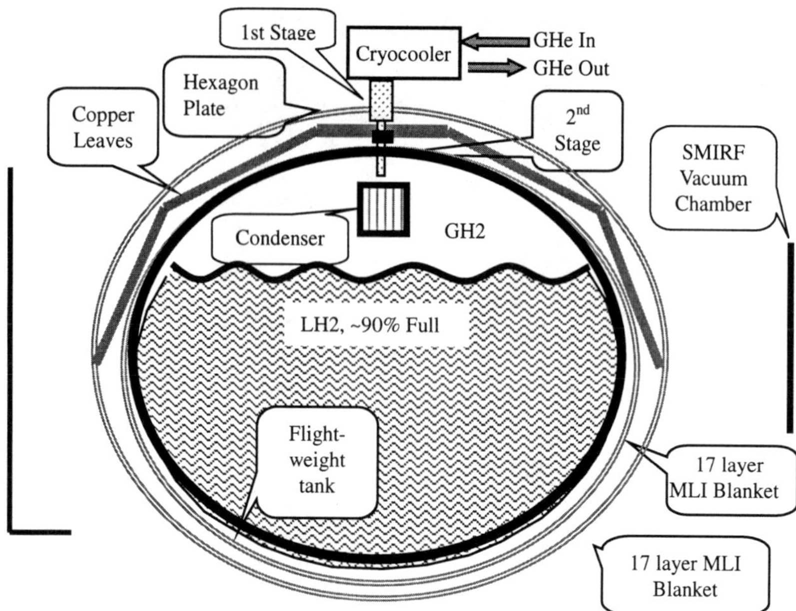

Figure 4. Schematic showing the tank, cryocooler, heat exchangers, and insulation installed for the Cooler On test.

conduct heat from the vent and fill lines as well as six quadrilateral plates or leaves that were inserted between the two 17 layer MLI blankets covering the tank. These leaves were trimmed to fit under the six gore sections of the insulation blankets. The purpose of the leaves was to intercept the heat entering the tank and conduct that heat to the cryocooler's first stage. The leaves were 1 m long and extended near the mid-section of the tank. The leaves plus the hexagon plate covered 2.5 m^2, or ~40 percent of the tank surface. Figure 4, shown later in the paper, is a simplified schematic that shows how the leaves fit in with cryocooler, insulation and tank.

Instrumentation

Temperatures on the tank wall, the cryocooler, and the condenser were taken with silicon diodes. The temperature sensors on the copper leaves, tank piping, MLI, and vacuum chamber were chromel-constantan thermocouples. Liquid level in the tank was determined with silicon diode point sensors and a delta pressure transducer located at the tank drain.

Capacitance-type pressure transducers were used to measure tank and facility pressures. Vacuum instrumentation with a range between atmospheric pressure and 10^{-5} torr was used. Tank boil-off was measured by one of the four calibrated mass flowmeters. These meters were calibrated with air and had a full scale of 0 to 3200, 0-400, 0-8, and 0-1 standard ft^3/hr.

Limitations

Despite the fact that the test was performed in a vacuum chamber, space environment conditions were not simulated. The cryocooler used was not a flight-type cryocooler, thus the compressor was not integral to the cooler. Because the compressor would add heat, our test heating rates are lower than predicted for a flight cryocooler. In addition, the heat exchanger used in the tank was located in the ullage and liquid condensed off of it. This configuration would not be possible in-space.

A few minor problems were noticed during testing and afterwards. There was a small hydrogen leak in the test tank access cover which permeated the MLI, reducing its performance and causing a higher than predicted heat leak. To minimize this leak, we conducted the test at lower tank pressures than preferred. Saturation temperatures at these reduced pressures are lower, correspondingly, the cryocooler had to operate at lower temperatures where it has a lower performance. Post test, the vacuum chamber mass spectrometer measured 3.6×10^{-4} torr partial pressure of hydrogen gas. In addition, the heat entering the cryocooler second stage was not accurately measured and is not reported.

RESULTS

Baseline Conditions

The tank liquid level for all tests was ~90 percent full. The vacuum chamber pressure fluctuated for each test, with the hardest vacuum being 3×10^{-5} torr and minimum vacuum 2×10^{-4} torr. The vacuum wall temperature was not controlled and was typically about 295K.

No Cooler Test

This was a baseline test to characterize steady state heating rate of the insulated tank in the vacuum chamber. The cryocooler and associated hardware were not installed. The measured steady-state boiloff rate was ~0.12 kg/hr, equating to a heating rate of ~14.5 W. The vacuum chamber pressure was 2.8×10^{-5} torr.

Cooler On Test

This test incorporated both the first and second cryocooler stages with the test tank. The heat transfer to the first stage was via solid conduction through the copper leaves. Also, straps from this stage were used to remove heat from the tank lines. The second stage heat transfer was by condensation of the hydrogen vapor in the ullage. A schematic of this test configuration is shown in Fig. 4.

After the tank was filled with liquid hydrogen, the cryocooler was turned on. During the first few hours, the chamber vacuum gage measured a relatively soft vacuum of ~1.5×10^{-4} torr causing the boil-off rate to be about 0.3 kg/hr (35 W). Even so, the cryocooler and heat exchanger temperatures as well as the boil-off rate quickly decreased. Within 8 hr, the boil-off rate decreased to zero and the vent valve was then closed. The first stage temperature, which had a relatively low load, was at 35K and would remain at about that temperature throughout the test. The temperature at the end of the warmest copper leaf was at 43K and also remained close to that temperature through the test. At these temperatures, an estimate of the copper thermal conductivity was made and the heat transfer through the leaves was calculated to be 9 W. Besides that heat removal, the copper straps to the plumbing removed ~60 percent of the heat that had entered the tank during the No Cooler test. The second stage of the cryocooler was also effective. After the vent valve was closed and over the next 54 hr, the condenser temperature decreased at a steady rate of 0.02 K/hr. In addition, the average LH2 tank temperature steadily decreased by 0.017 K/hr and the tank pressure dropped at a steady rate of 0.55 kPa/hr. The propellant liquid level did not change noticeably during the test. The testing was terminated when the tank pressure dropped to 103 kPa, which was 4.4 kPa above atmospheric pressure. This was to prevent air from being accidentally drawn into the tank through the vent or fill lines. The temperature and pressure data are shown in Fig. 5.

Cooler Off Test

Upon completion of the Cooler On test, the cryocooler was turned off to simulate a failed cryocooler configuration. The goal of this test was to reach steady state and compare the measured heating rate with the results from the No Cooler test. For the first 2.5 hr the boil-off rate was zero. After that, the boil-off rate quickly climbed up to 0.12 kg/hr (14.5 W), as it was for the No Cooler test, and then slightly increased for the duration of the test, ending at 0.14 kg/hr (17 W). Note that this flowmeter data was considered accurate, as the tank pressure and chamber vacuum level were consistent during this test. The condenser temperatures also increased, apparently due to the heat

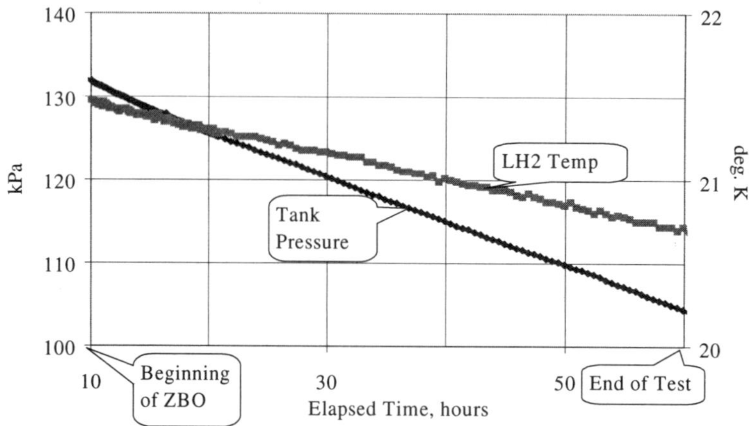

Figure 5. Cooler On test plot of tank temperature and pressure decay during testing.

from the copper leaves conducting through the cooler and into the condenser. After 77 hr of testing, the test was stopped. Steady state was never reached.

Second Stage Only Test

After the Cooler Off test was completed, the tank was removed from the vacuum chamber. The copper leaves and tank line straps were removed so that the cryocooler's first stage would have zero load. The tank would thus be cooled by only the cryocooler's second stage. As stated earlier, this stage capacity was 17.5 W at 18K, a much closer match to the insulated tank's heating rate of 14.5 W determined in the No Cooler test. This test configuration resembled a one-stage cryocooler integration. Single stage cryocoolers would likely be used to cool oxygen and methane propellants in space.

During the first 55 hr of testing, the boiloff varied as did the vacuum pressure, which was not nearly as hard for this test. The average boil-off rate was 0.005 kg/hr (0.6 W), which meant that the cryocooler removed 96 percent of the heating rate experienced in the No Cooler test. At the end of 55 hr of testing, the chamber vacuum pressure finally improved somewhat, from 1×10^{-4} torr to $\sim8.5\times10^{-5}$ torr, we achieved zero boiloff (0.00 kg/hr) and closed the vent valve. At this time, the tank pressure started decreasing very slightly. The pressure decrease rate was ~0.35 kPa/hr, while the condenser temperature remained steady. That data is shown in Fig. 6. As in the Cooler On test, testing was terminated to prevent the tank pressure from dropping below atmospheric pressure. In this case, the test was stopped when the tank was 1 kPa above atmospheric pressure.

DISCUSSION OF RESULTS

The condenser heat exchanger, attached to the second stage of the cryocooler and installed in the ullage of the tank, worked effectively. When combined with the copper leaves in the Cooler On test, it overmatched the heat entering the tank and caused the pressure and temperatures to steadily drop. In the Second Stage Only test, it closely matched the heat entering the tank and, at the end of the test, removed the heat and caused the pressure to decay, although at a lower rate (0.35 kPa/hr vs. 0.55 kPa/hr). Also, integration of the first stage was effectively accomplished with solid conduction. Despite the fact that the cooper leaves were 1 m long, the ends were only 8K warmer than the first stage cryocooler temperature. Also, the heat removed by the leaves was more than half the heat that entered the tank during the No Cooler test. Simple 2.5 cm wide straps were affixed to the plumbing and removed a majority of that heat.

Cryocooler operation, while installed on a simply supported tank, did not cause noticeable tank movement or fluid sloshing. The cryocooler with its first stage heat exchangers damped out

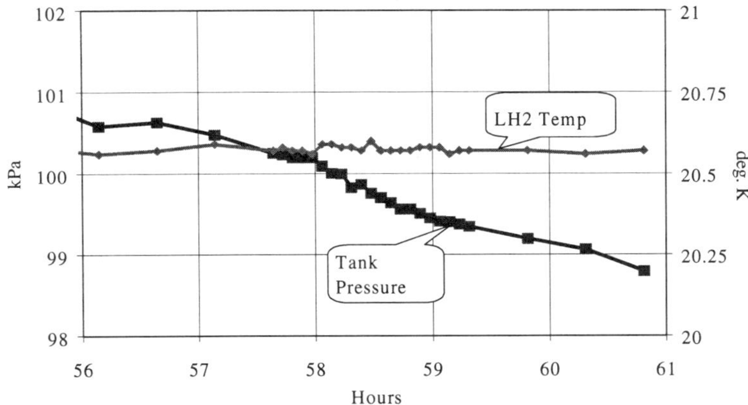

Figure 6. Second Stage Only test data showing tank pressure and LH_2 temperature versus time. This portion of the data was at the end of the test, when the boil-off rate was zero.

transient facility vacuum fluctuations and reduced the operating vacuum. With the cryocooler off, slightly higher heating rates occurred and the heat exchanger temperatures steadily increased.

In the Cooler On and Second Stage Only tests, the condenser lowered the pressure because it removed the heat that entered through the tank wall and had, by natural convections, made its way through the liquid hydrogen and into the ullage. Without these convections, which are absent in low-g, the fluid behavior and heat transfer will be different. Also, in low-g, the location of the ullage would be unknown. Thus, these factors require a different heat exchanger design that would either have much greater surface area or would include a propellant mixing device, creating forced convective heat transfer.

While the first stage heat exchangers removed a majority of the tank heating during the Cooler On test, the benefit if applied to a cryogenic propellant tank in-space is undetermined. This is because the designs for a flight-type first stage heat exchanger has not been done. Lightweight, rigid designs that include fluid convection in addition to solid conduction need to be investigated.

Also needing study are two thermal issues of in-space operations. The first issue is the heat conduction from the "off" cryocooler to the tank propellants. If the cryocooler operation in-space necessitates an on-off duty cycle due to power supply availability or low pressure limitations on the storage tankage, then a thermal shunt or diode incorporated into the cryocooler-heat exchanger assembly could be necessary to prevent this undesirable heat flux. Alternatively, this problem might be mitigated if the cryocooler is sized for a greater capacity than the heat that enters the tank. This overcapacity, the second thermal issue needing study, was shown in this test to reduce the tank pressure and temperatures.

CONCLUSIONS

This zero boil-off demonstration of cryogenic propellant storage has provided an early proof of concept that, combined with advances in cryocooler technology, offer exciting possibilities for long-term cryogenic fluid storage applications in-space. The potential mass reductions for zero boil-off cryogenic storage applied to long-term space missions are substantial. The possibility to treat cryogens like space storable propellants reduces safety concerns for in-space operations, such as rendezvous and docking. Delays in those operations or other mission delays could now be tolerated because cryogen boil-off margins would not be exceeded. In addition, there is the potential for the cryocooler to damp out unexpected transient heating. Because of these benefits, zero boil-off cryogen storage offers more opportunities to utilize high-performing cryogenic propellants for long duration space missions.

While the demonstration was successful and the benefits are substantial, this effort only provides the foundation for remaining work. Analytical modeling and testing with cryocoolers and heat exchanger configurations designed for a low-g space environment are recommended. In addition, thermal effects from in-space power supply and heat rejection systems need to be included.

REFERENCES

1. R.T. Giellis, et. al., "Long Term Cryogenic Storage Study," US Air Force, AFRPL TR-82-071 (1982).
2. "Investigation of External Refrigeration Systems for Long Term Cryogenic Storage," Lockheed Missiles & Space Report A981632 (1971).
3. Glaister, Donabedian, Curran, and Davis, "An Overview of the Performance and Maturity of Long Life Cryocoolers for Space Applications," The Aerospace Corp, El Segundo, Report No. TOR-98(1057)-3 (1998).
4. E. Tward, T. Nast, W. Swift, S. Castles, and T. Davis, private communications.
5. A. Kashani, private communication.
6. P. Kittel, L.J. Salerno, and D.W. Plachta, "Cryocoolers for Human and Robotic Missions to Mars," Cryocoolers 10, Plenum Press (New York, 1999) to be published.
7. P.J. Dempsey, R.J. Stochl, "Supplemental Multi-layer Insulation Research Facility," NASA TM–106991 (1993).
8. I.E. Sumner, "Thermal Performance of Gaseous He Purged Tank-Mounted Multilayer Insulation System During Ground-Hold and Space-Hold Thermal Cycling and Exposure to Water Vapor," NASA TP–1114 (1978).

TESTING OF DENSIFIED LIQUID HYDROGEN STRATIFICATION IN A SCALE MODEL PROPELLANT TANK

John M. Jurns[1], Thomas M. Tomsik[2], and William D. Greene[3]

[1]Dynacs Engineering Company Inc.
Cleveland, Ohio

[2]National Aeronautics and Space Administration
Glenn Research Center
Cleveland, Ohio

[3]Lockheed Martin Michoud Space Systems
Huntsville Operations
Huntsville, Alabama

ABSTRACT

This paper describes a test program that was conducted at NASA to demonstrate the ability to load densified LH_2 into a subscale propellant tank. This work was done through a collaborative effort between NASA Glenn Research Center and the Lockheed Martin Michoud Space Systems (LMMSS). The Multi-lobe tank, which was made from composite materials similar to that to be used on X-33, was formed from two lobes with a center septum. Test results are shown for data that was collected on filling the sub-scale tank with densified liquid hydrogen (DLH_2) propellant that was produced at the NASA Plum Brook Station. Data is compared to analytical predictions. Data collected for this test series agrees well with analytical predictions of the environmental heat leak into the tank and the thermal stratification characteristics of the hydrogen propellant in the tank as it was filled with DLH_2.

INTRODUCTION

NASA has identified propellant densification as a critical technology in the development of the single stage to orbit (SSTO) reusable launch vehicle (RLV) designated by Lockheed Martin as the VentureStar™. The densification of cryogenic propellant through subcooling allows 8 to 10 percent more propellant to be stored in a given volume. This allows for higher propellant mass fractions than would otherwise be possible with conventional, normal boiling point cryogenic fluids.

To date, several aspects of densification technology have been investigated. Previous tests at NASA have been conducted with a subscale liquid hydrogen densifier.[1] This test hardware was built and tested to prove the ability to produce DLH_2 at 0.9 Kg/sec. The next step after production of densified LH_2 was to investigate the particulars of performing the sequential process necessary to load a propellant tank with densified propellants. This

Advances in Cryogenic Engineering, Volume 45.
Edited by Shu *et al.*, Kluwer Academic / Plenum Publishers, 2000.

report details the tests that were conducted at NASA Plum Brook Station in Sandusky, Ohio to demonstrate the ability to load DLH_2 into a subscale propellant tank.

Lockheed Martin Michoud Space Systems (LMMSS) personnel have developed an analytical tool that models the loading of a tank of specific geometry with densified liquid hydrogen. For a propellant tank filled with normal boiling point (NBP) liquid hydrogen, and a specified external heat flux, the tool models the total time required to replace the NBP liquid hydrogen with densified liquid hydrogen and the final fluid conditions.

The test program described here was conducted to demonstrate the ability to load a subscale model propellant tank with densified liquid hydrogen and to validate the LMMSS analytical model. The plan was to produce, in a batch process, densified liquid hydrogen and then transfer it to the propellant liquid hydrogen tank. This test program was conducted jointly through a Space Act Agreement between NASA Glenn Research Center in Cleveland, Ohio, and Lockheed Martin Michoud Space Systems in New Orleans, Louisiana.

TEST OBJECTIVES

The purpose of the tests reported here was to perform the sequential propellant tank loading and simulate the recirculation process using DLH_2 propellant with a tank configuration that is traceable to RLV. The tests were designed to characterize and demonstrate tank thermal stratification, a necessary requirement for optimum loading, densifier design and production operations in the RLV environment. Another important objective was to evaluate the thermal conditions at which LH_2 could be maintained in its subcooled state during a simulated recirculation of DLH_2. Furthermore, a tank loading demonstration program could provide the data and information necessary to firmly ground the analytical tank models developed by LMMSS. The specific test objectives were as follows:

1. Characterize the LMMSS Multi-lobe tank environmental heat leak.
2. Produce batch quantities of 15 K LH_2 working fluid in a facility dewar to simulate tank recirculation of densified propellant.
3. Demonstrate the ability of LH_2 to flow upwards out of the tank via a vertical siphon.
4. Evaluate "load-and-go" tank filling with DLH_2 on an ambient temperature and pre-chilled Multi-lobe tank.
5. Demonstrate and verify tank thermal stratification characteristics over a range of inlet LH_2 recirculation flow rates.
6. Obtain sufficient LH_2 thermodynamic data of tanking operations to allow validation of mathematical-analytical tank models developed.

FACILITY AND TEST HARDWARE

Test Facility

The K-Site test facility at NASA Glenn Research Center Plum Brook Station is used for large-scale tests that utilize liquid hydrogen. Utilities were provided for filling, pressurizing, draining, and inerting the Multi-lobe tank.

The Multi-lobe tank was situated within a structural steel frame, which had a roof to protect the tank, as well as scaffolding to provide access to the tank and piping. The tank sat on a platform ~1.2 m above grade.

Liquid hydrogen was provided at the site with two liquid hydrogen trailers. The capacity of each trailer was 50,150 L. Working pressure for each trailer was 793 kPa. One trailer (designated H24) was used to produce densified liquid hydrogen by pulling vacuum on the ullage using vacuum pump VP-5. VP-5 was a mechanical vacuum pump with a capacity of 24,070 L/m. H24 was modified to allow the ullage to see negative pressure. The second trailer (designated H25) was used as a catch tank during the recirculation simulation.

Liquid hydrogen was transferred from the trailers to the test tank through 51-mm vacuum-jacketed piping. Piping was also provided for venting the tank ullage, and pressurizing the tank with either gaseous hydrogen or gaseous helium. A schematic of the test facility is shown in Fig. 1.

Instrumentation was provided to monitor test and facility parameters. Temperatures, pressures, liquid hydrogen flow rates, and tank strains were all monitored. Data was recorded using a facility data acquisition system. Data was updated once per second, and could be recorded at the rate of 1 scan/sec, 1 scan/10 sec, or 1 scan/min. The total number of channels monitored was 364.

Test Article

The Multi-Lobe tank is fabricated from composite material, and consists of two lobes. Each lobe measures 152.4 cm in diameter, is 518.2 cm long, is joined at a 10-degree angle, and contains a barrel section, a web and two domes. The lobe to lobe joint contains a continuous 6.35-mm K-type Raco seal. The joint is assembled with 176 (15.9-mm diam) bolts. The assembled tank is covered with foam insulation to prevent ice formation and to maintain liquid hydrogen temperatures.

Tank empty weight is ~680 kg. Total tank volume including ullage is ~13,677 L. Surface area of the tank is estimated to be ~37.2 m^2.

There are two inlet ports, one in the bottom of each lobe of the tank. These inlet lines are 38-mm pipe size flanged connections. There are two outlet ports, one in the top of each lobe of the tank. The outlet port in lobe one is a vapor vent line. The outlet port in lobe two is a siphon line to drain liquid from the top of the tank. The end of the siphon line is located ~91-cm below the top of the tank.

Instrumentation

The Multi-lobe tank was instrumented with a number of silicon diode temperature sensors. These sensors were located inside the tank. Several silicon diodes were mounted on the wall of the tank, several were mounted on the web connecting the two halves of the tank. The balance of the silicon diodes were mounted on rakes inside the tank, and were used to monitor the temperature profile of liquid inside the tank. There were two rakes, one mounted in each lobe. A schematic indicating the location of these silicon diodes is shown in Fig. 2. A total of 54 silicon diodes were monitored during testing to determine the LH_2 temperature profile in the tank.

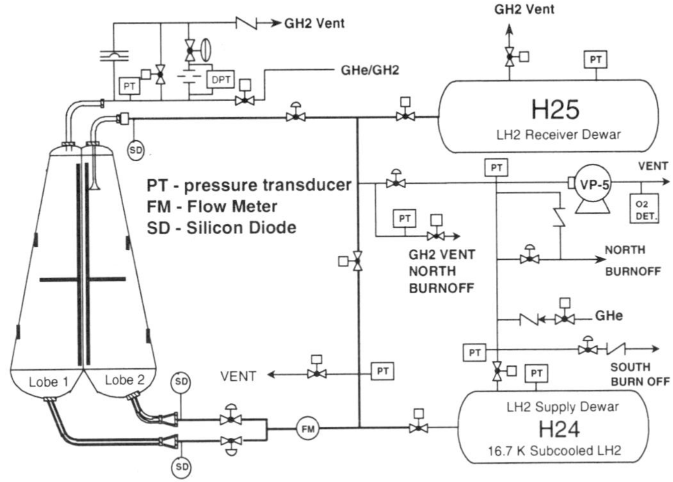

Figure 1. Multi-Lobe tank Thermal Stratification Simplified Schematic

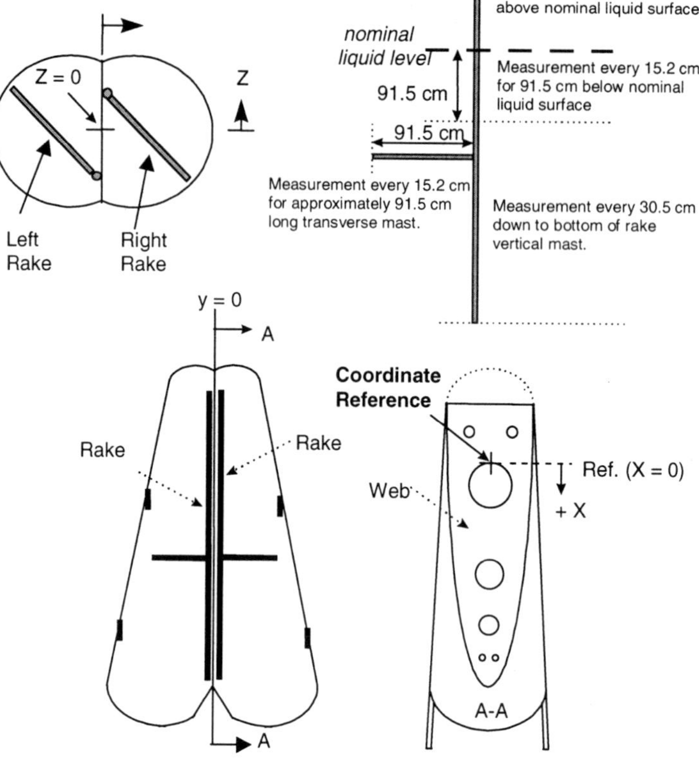

Figure 2. Silicon diode rake subassembly installation for test tank.

TEST MATRIX

For the tests reported here, densified LH_2 was generated inside supply dewar H24 by gradually reducing pressure over the fluid with the vacuum pumping system. The dewar pump-down time from 103.4 to 10.3 kPa took between 12 and 16 hr. Once DLH_2 at the target temperature of 15 K was produced, it was pressure transferred from dewar H24 into the bottom of the Multi-lobe tank while normal boiling point fluid was siphoned off the top. This flow scheme (filling DLH_2 from the bottom while siphoning warmer LH_2 from the top) provides the best method for filling the tank with densified propellant.[3] The liquid siphoned off the top of the tank flowed to the second dewar H25. Table 1 is a summary of the tests.

Table 1. Test Matrix

Test No.	Test Description	Initial Tank Conditions	Number of Fill Lines Used	DLH_2 Flow Rate (Kg/sec)	Run Time (min)	Tank Pressure (KPa)
1	Boil off	Filled w/NBP LH_2	n/a	0.0	125	103
2	Saturated Siphon	Filled w/sat'd LH_2 at 30 psia	2	0.27 – 0.68	25	207
3	Straight Load-and-go	Empty, Ambient	2	0.68	155	124-138
4	Pre-chilled Load-and-go	Empty, Cold	1	0.68	250	124-138
5	Recirculation Simulation	Filled w/NBP LH_2	1	0.23	120	207
6	Recirculation Simulation	Filled w/NBP LH_2	1	0.45	60	207
7	Recirculation Simulation	Filled w/NBP LH_2	1	0.68	40	207
8	Recirculation Simulation	Filled w/NBP LH_2	2	0.23	90	207
9	Recirculation Simulation	Filled w/NBP LH_2	1	0.23	125	207

ANALYTICAL MODEL

The mathematical simulation of the densification process developed by LMMSS is based upon a multiple-node, Lagrangian approach in which the liquid is subdivided into a number of moving small bundles of fluid. The model is called the Multi-Lobe, Multi-Layer Densification Model (ML2DM). The diagram in Fig. 3 shows the general schematic of the model for a number of different multiple lobe tank configurations.

The fluid is subdivided into the separate lobes of the tank and then further subdivided into layers. During the recirculation cycle, subcooled fluid is flowed into the bottom of the tank and warmer fluid is siphoned from the upper portion of the tank. Within the simulation, as the fluid enters, nodes are created at the bottom of the tank. The conditions of each node are then tracked as it is pushed upward by the recirculation process. At the top of the tank where fluid is being removed, the nodes shrink and eventually collapse.

In addition to the slow bulk movement of the liquid caused by the recirculation flow there are a handful of other significant effects that control the effectiveness and efficiency of the densification process. One of these is the formation of buoyancy driven boundary layer flows and another is the lobe-to-lobe interaction of the fluid. Both of these phenomena are discussed in somewhat more detail.

Buoyancy driven boundary layer flow is significant in the densification process because it causes a redistribution of the energy within the tank. The continuous heat leak into a nonvacuum jacketed propellant tank can be quite significant. This influx of energy tends to work against the densification process which is, at its core, essentially a bulk enthalpy lowering exercise. However, due to the existence of buoyancy driven boundary layer flow, the detrimental effects of the continuous heat leak is significantly minimized since it tends to draw the fluid warmed along the walls upward towards the top of the fluid where it can be siphoned off. Thus, this boundary layer flow amplifies the thermal stratification within the tank and thereby enhances the densification process.

There are no exact analytical solutions to the governing differential equations for motion and temperature distribution of the fluid within such a thermal convection boundary layer of fluid along the sidewall of the tank. The fluid boundary layer profiles used in this analysis come from a unique approximate solution in which the temperature profile follows a third order polynomial and the velocity profile follows a fifth order polynomial. All of the boundary conditions imposed at the wall and at the boundary layer to bulk fluid interface are satisfied by this system of equations though conservation of energy is not satisfied along the full breadth of the profile. This fact is compensated by the assumption that the boundary layer thickness is proportional to the Grashoff number to the $1/4^{th}$ power in the laminar region and to the $1/3^{rd}$ power in the turbulent region. The empirical nature of these last correlations, largely verified by this testing, far outweighs the slight errors inherent within the approximate solution to the boundary layer equations. The result is a straightforward algorithm, which correlates well with the test results presented in the next section.

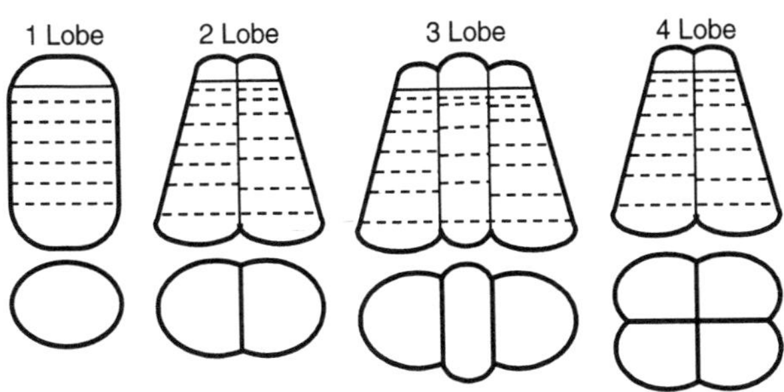

Figure 3. General Schematic of ML2DM

The second significant feature of the analytical model and the second significant phenomenon occurring within the tank during the densification process is the lobe-to-lobe interaction of the fluid. This is especially significant to the densification process when the inflow, outflow, and/or tank configuration has a large asymmetry as was the case for these tests where inflow and outflow was located in different lobes. The Multi-Lobe tank described within this paper is a two-lobe configuration with a single separating septum. This septum is not solid, but it does offer some restriction in the lobe-to-lobe cross flow. The simulation handles the calculation of this cross flow by first reducing the number of lobe-to-lobe, node-to-node interactions into eleven possibilities. These interaction possibilities for each combination of adjoining, moving nodes is evaluated at a given time step and is then superimposed against the fixed geometry of the septum. The relative fluid flow resistance and fluid acceleration damping used within the simulation was based on standard orifice flow calculations. The model to test data comparisons presented in the next section confirmed the validity of using such values.

Other features of the model include a single node, multiple gas ullage simulation, bulk boiling and surface condensation algorithms, as well as an approximate algorithm to simulate convection motion of the bulk fluid.

TEST RESULTS

The results from the testing can be broken into three categories. First there was the saturated siphon test which was an operations demonstration. Second, there were two "load and go" tests, which were attempts to demonstrate potential alternative tank loading methods with densified propellant. Finally, there were five recirculation tests that were intended both to fully demonstrate the baseline densification process and to provide data for the validation of the analytical model.

The saturated siphon test was intended to demonstrate the fact that saturated cryogenic fluid could be flowed upwards and out of the tank via a vertical siphon. The fluid in the tank was fully saturated at ~207 KPa and then flow was initiated. Because the fluid was static and saturated within the tank initially, the rise to a higher elevation and the acquisition of velocity should have cavitated the fluid slightly. The question going into the test was whether this low level cavitation would be detrimental to the siphoning process.

Figure 4 shows two plots. The first is the inflow of liquid hydrogen to the tank while liquid hydrogen flowed out the siphon. Because the liquid level within the tank was held constant and because there was little or no thermal stratification within the tank, the outflow up the siphon is equated to this inflow value. Reasonably steady flow was obtained at both a lower value (~0.32 kg/s) and at a higher value (~0.73 kg/s). The second plot in Fig. 4 shows the measured temperatures both in the vicinity of the siphon inlet within the tank and

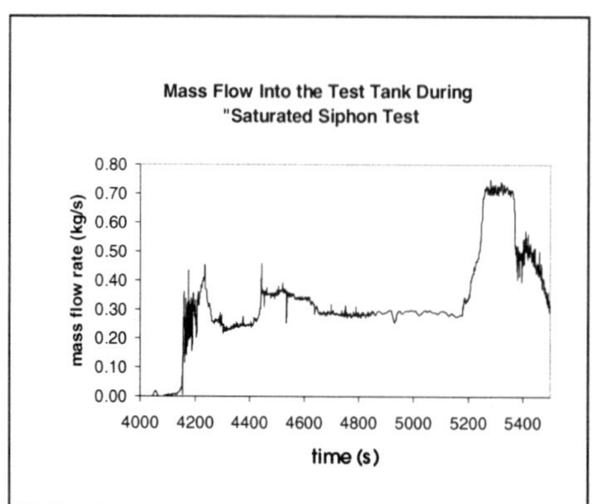

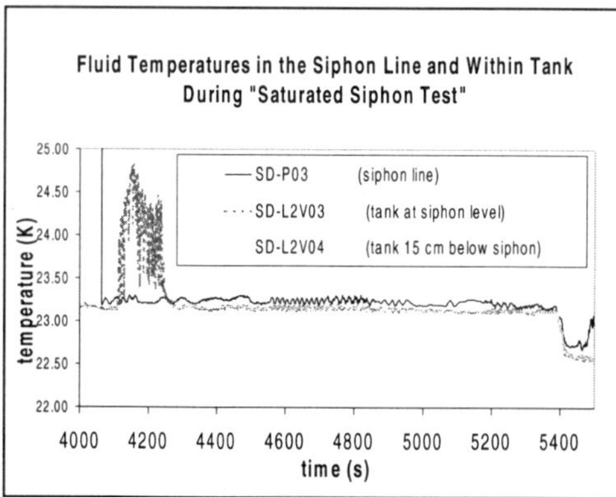

Figure 4. Saturated Siphon Test

in the siphon line itself. This confirms, based upon the tank pressure, that the fluid was fully saturated.

The second category of tests undertaken was two "load and go" tests. The purpose of these tests was to load densified propellants directly into an empty tank with no recirculation. If this could be done on a launch vehicle such that an acceptable degree of overall bulk subcooling was achieved, then there would be no need to have a recirculation system in place. The first "load and go" test, called the "straight load and go", began with a warm empty tank. For the second "load and go" test, called the "prechilled load and go", the tank was first prechilled by filling with normal boiling point fluid, then drained, and then refilled with the densified propellant. As the tank was filled from the bottom, the diodes recorded cryogenic temperatures. The diodes at the bottom of the tank near the in-flowing subcooled liquid became and remained the coldest. However, as the fluid rose it picked up heat so that by the time that the uppermost diode saw liquid it was nearly saturated fluid. The prechilling of the tank appeared to alleviate this effect somewhat as the temperatures were generally lower through the tank than during the "straight load and go" test. However, neither loading scenario achieved a complete top-to-bottom subcooling of the liquid. Also, in both of these cases, as soon as the liquid inflow was stopped the overall bulk temperature rose quickly. If this tank had been within a flight vehicle on the launch pad experiencing a prelaunch hold it is likely that the undensifying, expanding liquid would have been expelled out the vent.

The final category of tests undertaken within this test plan were five recirculation and densification tests run at various inflow rates and configurations. These tests were intended both to fully demonstrate propellant densification as it would be applied to a launch vehicle and to validate the mathematical simulation of the in-tank densification process. There were five tests, listed as Test 5 to 9 in Table 1. Two plots for Test 7 are shown in Fig. 5, showing comparisons between simulation reconstructions of the test conditions and the actual test data itself after completion of loading with DLH_2. The test data points presented are the silicon diode temperature measurements up the vertical rakes, one in each lobe, and the lines of simulation output are the predicted thermal stratification profiles. Such data to simulation comparisons were made at many points along the densification timeline but due to space restrictions only the representative plots are presented.

For nearly all the tests, the agreement between the measured test data and the simulation was quite good. Table 2 summarizes the accuracy of the simulation based upon estimated total mass loaded within the tank. Total mass was estimated from final fluid density and the tank geometry. With the exception of one test, Test 5 the first recirculation test conducted, all of the error values are within the accuracy range resulting from the accuracy of the diodes. The attempt to come to a better understanding of the relatively large error on Test 5 is ongoing.

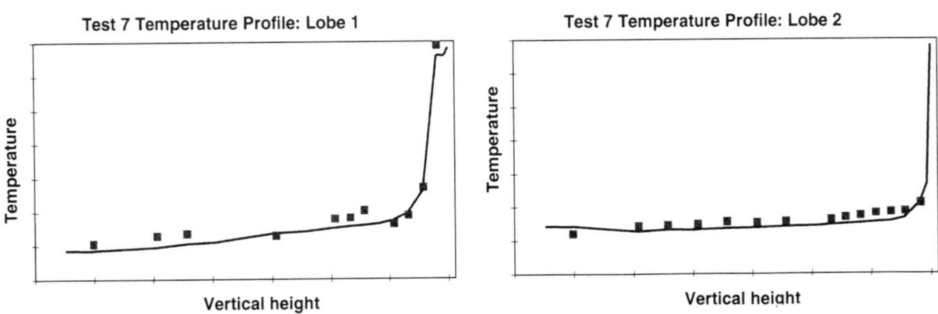

Figure 5. Test 7 Thermal Stratification Profiles at end of DLH_2 fill.

Table 2. Approximate Error in Loaded Mass

Test	Description	Approximate Error in Loaded Mass
5	≈0.23 Kg/s - 1 inlet	0.53%
6	≈0.45 Kg/s - 1 inlet	0.33%
7	≈0.68 Kg/s - 1 inlet	0.16%
8	≈0.23 Kg/s - 2 inlets	0.06%
9	≈0.23 Kg/s 1 inlet	0.36%

CONCLUSIONS

Data collected from this test program leads to the following observations and conclusions:

1. Environmental heat leak into the Multi-lobe tank was consistent with predictions.
2. Saturated liquid hydrogen can be flowed upwards and out of the tank via a vertical siphon, proving that siphoning saturated liquid hydrogen from the top of the tank will not adversely affect recirculation.
3. Neither "load and go" loading scenario (either with a warm tank or a pre-chilled tank) achieved a satisfactory top-to-bottom sub-cooling of the liquid.
4. For recirculation simulation tests, agreement between measured test data and analytical predictions was good, which validates the model.
5. Tank thermal stratification characteristics over a range of inlet LH_2 recirculation flow rates was demonstrated and verified.

REFERENCES

1. Tomsik, T., "Performance Tests of a Liquid Hydrogen Propellant Densification Ground System for the X33/RLV" (NASA TM–107469, AIAA-97–2976, 1997).
2. Bailey, T., Vandekoppel, R., Skardtveldt, G., and Jefferson, T., "Cryogenic Propellant Stratification Analysis and Test Data Correlation" AIAA Journal, Vol. 1, No. 7, pp. 1657–1659, 1963.
3. Fazah, M.M., "STS Propellant Densification Feasibility Study" (NASA TM–108467, 1994).
4. Greene, W. and Vaughan, D., "Simulation and Testing of In-Tank Propellant Densification for Launch Vehicles" (AIAA-98–3688, 1998).
5. Incropera, F. and DeWitt, D., Fundamentals of Heat and Mass Transfer (John Wiley & Sons, 1996).
6. Lak T., Lozano, M., and Tomsik, T., "Advancement in Cryogenic Propulsion System Performance through Propellant Densification" (AIAA-96–3123, 1996).
7. Anthony, M. and Green, W., "Analytical Model of an Existing Propellant Densification Unit Heat Exchanger" (AIAA-98–3689, 1998).
8. White, F.M., Fluid Dynamics (McGraw-Hill, 1979).

FLIGHT TEST RESULTS FOR THE NICMOS CRYOCOOLER

F. X. Dolan,[1] J. A. McCormick,[1] G. F. Nellis,[1] H. Sixsmith,[1] W. L. Swift,[1] and J. A. Gibbon[2]

[1]Creare Incorporated
Hanover, NH, 03255

[2]NASA Goddard Space Flight Center
Greenbelt, MD, 20771

ABSTRACT

In October 1998 a mechanical cryocooler and cryogenic circulator loop were flown on NASA's STS-95 as part of the HST Orbital System Test (HOST). The system will be installed on the Hubble Space Telescope (HST) during Service Mission #3 in 2000 and will provide cooling to the Near Infrared Camera and Multi-Object Spectrometer (NICMOS). It will extend the useful life of that instrument by 5 to 10 years. This was the first successful space demonstration of a Turbo Brayton cryocooler. The cooler is a single stage reverse Brayton type, using low-vibration high-speed miniature turbomachines for the compression and expansion functions. A miniature centrifugal cryogenic circulator is used to deliver refrigerated neon to the instrument. During the mission, the cooler operated without anomalies for approximately 185 hours over a range of conditions to verify its mechanical, thermodynamic and control functions. The cryocooler satisfied all mission objectives including maximum cooldown to near-design operating conditions, warm and cold starts and stops, operation at near-design temperatures, and demonstration of long-term temperature stability. This paper presents a description of the cooler and its operation during the HOST flight.

INTRODUCTION

The Near Infrared Camera and Multi-Object Spectrometer (NICMOS) is a cryogenically cooled instrument on the Hubble Space Telescope (HST) that provides the HST with near-infrared and limited spectrographic capabilities in the 800 nm to 2.5 μm wavelength range.[1] Cooling for the detectors is provided by the sublimation of solid N2. The solid nitrogen is enclosed in a dewar in such a way that the detector is cooled by the nitrogen through an aluminum conductive matrix. This conductive matrix also incorporates cooling tubes that were used during pre-launch operations to maintain the nitrogen in a solid condition. Bayonet connections are available at a valved interface manifold on the cryostat to permit the connection of helium cooling lines during pre-launch cooling.

The cryogen in NICMOS was prematurely depleted in late 1999. In order to extend the life of this instrument, NASA embarked on an effort to install a mechanical cryogenic cryocooler on the HST that will provide cooling to the NICMOS instrument. The cryocooler is a closed loop reverse Brayton system using turbomachines for compression and expansion. The system also incorporates a circulation loop that will interface to the cryostat at the existing bayonet connections. In October 1998 the cryocooler and associated system was flown on STS-95 on the Hubble Orbital System Test (HOST). This mission served to qualify the cooler and provide important information about the operation of the system relevant to its integration on HST.

The NICMOS Cooling System (NCS) is designed to meet the specific constraints imposed by the HST, the NICMOS instrument and the flight demonstration. These constraints include thermal loads, packaging, mechanical support and heat rejection limitations. The NCS is designed as a package that can be installed by astronauts in the aft shroud of the HST. Because of the existing HST configuration, the NCS package will be separated from the bayonet coupling interface on the NICMOS dewar by the two flexible tubes, each of which is about 1 m long. In total, the circulating loop includes about 5 m of tubing in the flexible and fixed portions of the loop. The heat load at the NICMOS instrument is about 400 mW. However, the parasitics at the cold end of the system are substantial. At least 7 W of cooling must be provided by the cryocooler to overcome conductive losses from the bayonets at the bayonet coupling interface, radiation from the flexible lines, and conductive losses through the mechanical support structure for the cooler in its frame.

One of the objectives of the HOST flight was to demonstrate the performance of the entire system under conditions simulating those to be encountered after integration with NICMOS. The NICMOS Cooling Loop Simulator (NCLS) embodied the necessary thermal and physical features of the NICMOS dewar to serve this purpose during the HOST flight. It was attached to the circulator loop at the bayonet connections that would interface with NICMOS.

NICMOS COOLING SYSTEM (NCS)

The NICMOS Cooling System includes three fluid loops that transport heat from the NCLS to the radiator.[2] A schematic of the system is shown in Figure 1. A cryogenic circulator pumps neon through the circulator loop between the NCLS and the cryocooler. Heat from the NCLS, external parasitics, and the circulator is passed into the cryocooler loop through the Cold Load Interface (CLI) at the cold end of the cryocooler. The cryocooler is a closed loop reverse Brayton cycle that conveys heat from the CLI to the Heat Rejection Interface (HRI) at the warm end of the system. The main components in the cycle are a centrifugal compressor, a counterflow recuperator and a cryogenic turboalternator. Heat is removed from the system by means of an ammonia filled Capillary Pump Loop (CPL).[3] The accumulator and purge and fill valves in the cryocooler loop provide for gas management and system decontamination.

Circulator Loop

The circulator loop consists of a centrifugal circulator that pumps a constant flow of neon between the cryocooler and the NCLS. The centrifugal circulator uses a 7.5 mm

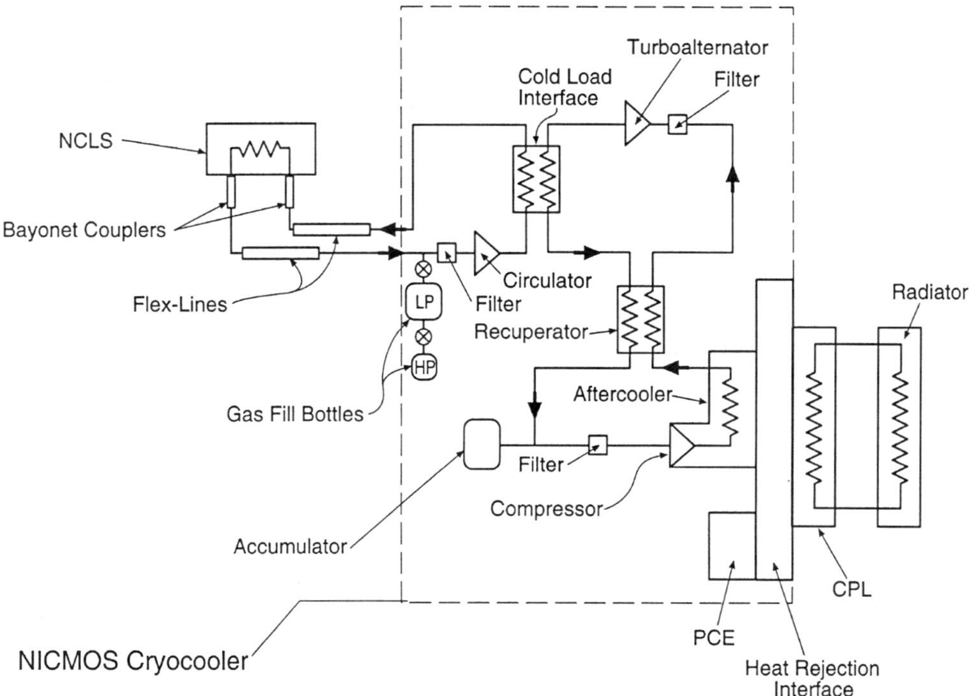

Figure 1. NICMOS Cooling System configured for HOST

diameter impeller on a shaft supported by gas bearings at a nominal speed of 1500 rev/s. At 70 K, the circulator provides a pressure rise of about 6 kPa at a neon flow rate of about 0.4 g/s. The tubing system is made up of flexible stainless steel tubes that interface between the cryocooler and the NCLS, bayonet couplings at this interface, and the cooling coil within the NCLS that simulates the tubing in the NICMOS dewar. The loop contains high pressure and low pressure fill tanks. These tanks will be used after interfacing the NCS to NICMOS to charge the system. They were inactive during the HOST flight. The valves and bayonet couplings at the NCLS interface are attached to a surface at nearly ambient temperature. The parasitic heat leak from these devices to the circulator loop account for a major portion of the heat load handled by the cryocooler.

The CLI is a stainless steel counterflow fin-type heat exchanger that transfers heat from the circulating loop to the cryocooler loop with low pressure loss. At design point, the pressure loss in the CLI on the circulator side is about 90 Pa. The thermal effectiveness of this component is 0.963. The CLI serves as the support structure for the circulator, the turboalternator and the cold end mechanical assembly. Kevlar straps attached between the CLI and the frame support the cold end of the cryocooler against launch loads.

NICMOS Cryocooler (NCC)

The NICMOS cryocooler is a single stage reverse Brayton (SSRB) cooler that uses small high-speed turbomachines for the compression and expansion of the refrigerant. The machines contain low-mass shafts that rotate at high speed with negligible vibration. The lack of vibration in the system is critical to meeting the requirements of the HST. The cooler is derived from an earlier design[4,5] that was designed for 5 W of cooling at 65 K. An engineering model of this system has been undergoing life tests at the Air Force Research Laboratory at Kirtland AFB since April 1995. It has accumulated over 28,000 hours of operation in these tests. Table 1 summarizes the important operating parameters for the NCC at design point. Figure 2 shows views of the cryocooler assembly and its configuration when packaged for the HOST mission.

The Power Conversion Electronics (PCE) provides the main conversion from unregulated DC input to variable frequency AC output to drive the circulator and compressor. It also contains a variable electrical resistance load for the turboalternator and provides power to the instrumentation.[6] The power electronics are contained in a compact package with a 165 mm x 190 mm footprint on the HRI. The input power to the PCE is nominally up to 450 W. Up to 67 watts of heat are dissipated in the PCE and rejected to the HRI.

The compressor is a centrifugal type machine with a 15.2 mm diameter impeller driven by a three phase induction motor. The shaft is 6.35 mm diameter and is supported on self acting gas bearings. The maximum operating speed for the compressor is 7,500

Table 1. Nominal operating conditions for the NICMOS CryoCooler

Cryogen	Neon
Cooling load	7 W
Load temperature	67 K
Compressor inlet pressure	1.45 atm
Pressure ratio	1.62
Flow rate	1.5 g/s
Rejection temperature	305 K
Electric power to the cooler	280 W

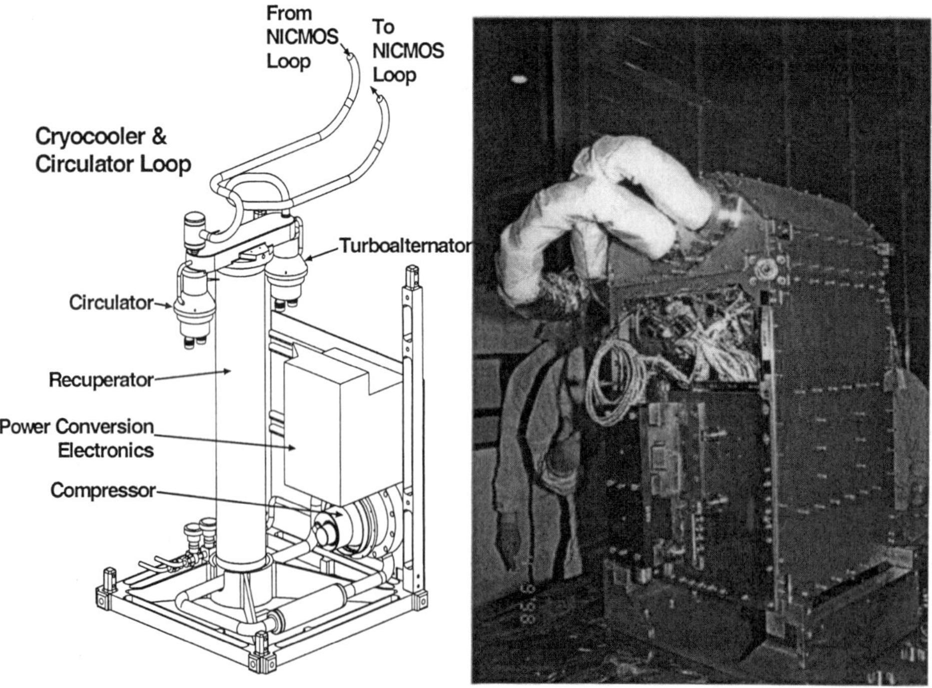

Figure 2. The NICMOS Cryocooler system. View on left shows the cryocooler components and portions of the circulator loop attached to the assembly. View on right is the system packaged for the HOST mission.

rev/s. The compressor housing contains an integral aftercooler in the base plate where it attaches to the HRI.

The recuperator provides for efficient heat transfer between the high pressure and low pressure streams of gas. The recuperator consists of 300 slotted copper disks that are equally spaced axially along the length of a stainless steel outer shell. The high and low pressure gas streams are separated by stainless spacer rings between adjacent pairs of disks. The thermal effectiveness of the heat exchanger is 0.992 and the total pressure loss in both streams is about 1 % of the system pressure. The warm end of the recuperator is mounted directly to the frame/support structure. The cold end is supported through the CLI by three Kevlar straps that are attached to the frame and tensioned to meet launch requirements with low heat leak.

At the cold end of the cryocooler, a turboalternator provides the necessary refrigeration. Work is extracted by expansion of the gas through a 7.6 mm diameter rotor. The resulting shaft work is then converted to electric power by a synchronous three phase alternator and conveyed to a resistive load in the PCE assembly where it is dissipated as heat and rejected to the HRI. The turboalternator has a nominal output of 10 W at 70 K at a speed of 3500 rev/s. Like the circulator, the low-mass shaft is supported on gas bearings with no vibration.

MISSION SUMMARY

The NCC was launched aboard Discovery on October 29,1998 at 13:19 EST as part of the HOST payload of instruments intended for installation on the HST in 2000. The cryocooler was started approximately 20 hours after the launch and operated under test until it was deactivated approximately 9 hours before Discovery landed on November 7, 1998. During the test sequence, the cryocooler cooled to a nominal operating temperature and was then exercised through various transient and steady operating conditions to demonstrate performance and functional behavior of the cooler and support systems. These transients included a cold shutdown and restart of all machines, a partial power down to idle conditions and re-cooling to a stable temperature to demonstrate system thermal control stability. Table 2 lists the primary events during the test sequence that lasted approximately 185 hours. Figure 3 shows some of the key cryocooler parameters during the mission.

Table 2. NCC operation timeline

Mission Elapsed Time (da:hr:min)	Event
00:19:54	NCC startup
00:19:54 to 02:23:41	Maximum power cooldown to first operating point
03:00:01 to 05:01:28	Set-point operation at cold plate temperatures from 76 K to 78 K
05:01:28 to 05:12:03	Reduced power idle operation
05:20:09 to 06:14:53	Compressor off during CPL deprime tests; circulator operating
08:04:28	Minimum cold plate temperature of 72.65 K
08:07:19 to 08:08:04	Cold stop/start tests
08:12:50	Circulator disabled at end of warm-up
08:13:33	System power off

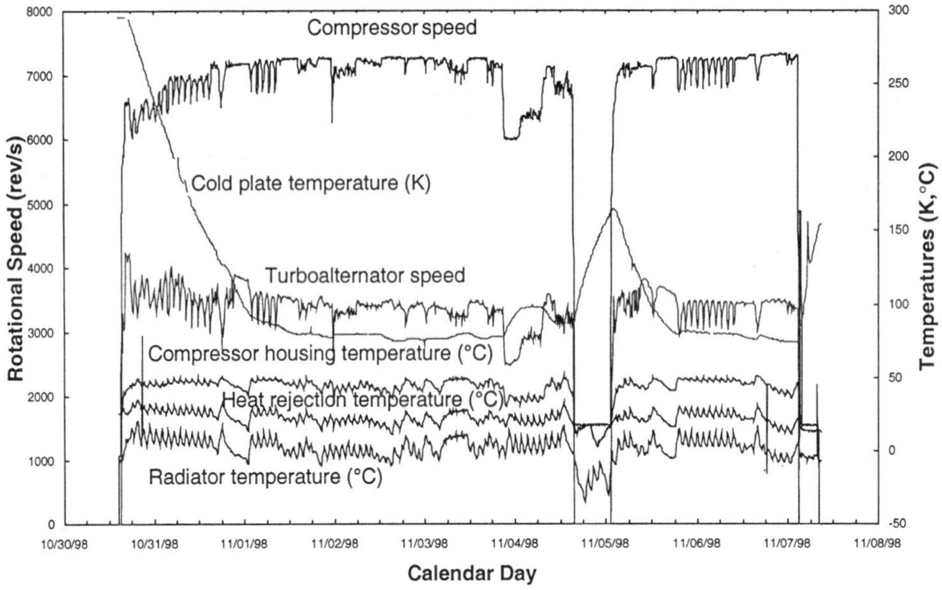

Figure 3. NCC operating profile during the HOST mission.

Following the start of the cryocooler, the system was set for maximum cooldown rate. During this interval, the compressor speed was set to a maximum value consistent with limits on the compressor housing temperature and the turboalternator speed. At warm conditions, the characteristics of the turboalternator and compressor are such that the turbine would operate at excessive speed if the compressor operated at full speed. The compressor speed must be restricted until the temperature at the turbine approaches the steady design condition near 70 K. Compressor speed is gradually increased during cooldown to keep the turbine operating at maximum output.

The duration of the cooldown interval is governed by the thermal mass of the cryocooler and the NCLS, the thermal environment and the input power to the cooler. The initial cooldown interval lasted approximately 52 hours. During this time, the net refrigeration from the turbine varied from about 16 W at the start of cooldown to about 12 W as the cold end approached 80 K. During the cooldown (and during most of the mission), the HOST payload and radiator for the NCC was exposed to a thermal environment that varied regularly during the 90 minute orbit. The radiator temperature changed by up to 15 °C during each orbit. These temperature variations produced corresponding variations through the heat rejection system. The effects are shown in the lower three traces in Figure 3. As temperatures in the radiator varied, the capacity of the radiator and CPL also varied causing the temperature of the compressor housing to vary. The controller continually adjusted compressor speed (and hence power) to maintain the housing below the upper allowable value of about 50 °C.

Following the initial cooldown, a series of six tests were performed at steady operating conditions. The tests were performed at constant loads and load temperatures in the NCLS consistent with the thermal environment that was available from the orbiter orientation. The performance of the cryocooler was measured under steady operating conditions at load temperatures between 76 K and 78 K and loads from 0 - 0.4 W at the NCLS. The system was then powered back to an idle condition. The speed of the compressor was reduced, and the temperature of the NCLS was permitted to rise to about 90 K. The compressor was then returned to full power and the NCLS temperature decreased to about 86 K.

The compressor was shut off in preparation for de-prime tests with the CPL. The circulator continued to operate during this period. The compressor was restarted approximately 10 hours later. During the interval that the cryocooler was off, the temperature at the NCLS rose from about 86 K to 165 K. The cryocooler remained under maximum power for the duration of the mission in order to attain the lowest temperature possible. During the last 10 hours of operation, the orbiter orientation was adjusted to give the coldest thermal environment available during the mission. The NCLS cold plate reached a low temperature of 72.65 K during this period. Then the cold plate temperature was set to 72.9 K. The PID controller maintained this temperature within the +/- 0.1 K stability target for more than an hour. During this period, the radiator temperature varied by up to 9 °C.

At the conclusion of the final steady state test point, the compressor was shut off. The circulator was then shut off and restarted about 17 minutes later. The compressor was restarted approximately one hour from the time it had been shut off. These shutdowns and restarts occurred without anomalous behavior. The figures below show the behavior for each of the three machines during these cold starts. Circulator motor current, speed and differential pressure are plotted in Figure 4. Compressor speed, turboalternator speed and the pressures at the inlet and outlet to the compressor are shown in Figure 5. At the time of the restarts, the circulator housing temperature was 81 K and the turboalternator temperature was 72 K. The cryocooler was then again shutdown to prepare for landing. The circulator continued to operate for several hours in order to assist in warming the cold end of the system prior to entry into the earth's atmosphere.

CONCLUSIONS

The operation of the NCC during the HOST mission demonstrated that the system would meet the requirements for NICMOS. Valuable test data and operating experience were gathered during the flight. This information is already being used to implement modifications to the system to reduce power and improve performance. The flight served to qualify the technology for space applications. All mission objectives for the cooler were met.

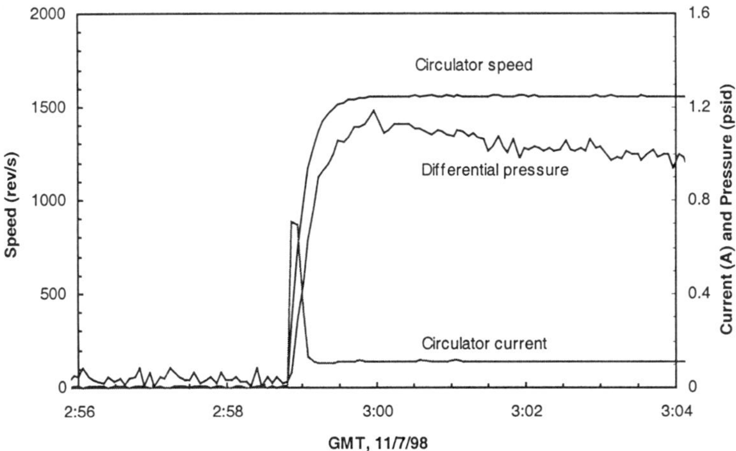

Figure 4. Circulator performance during HOST cold restart.

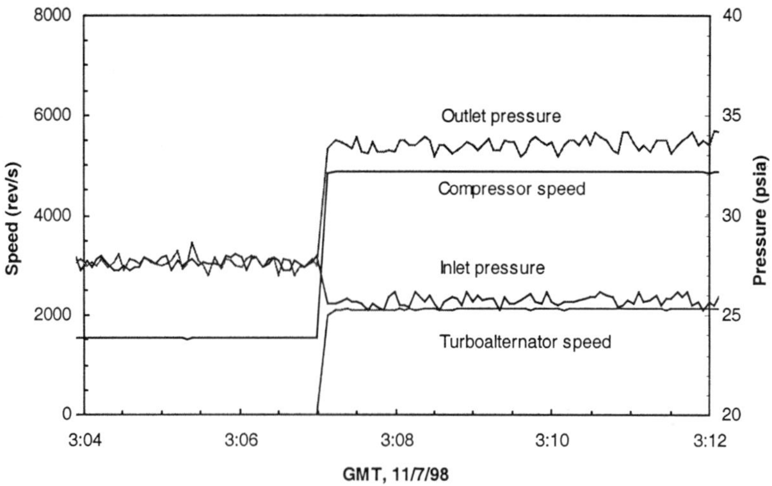

Figure 5. Compressor and turboalternator characteristics during HOST cold restart.

ACKNOWLEDGMENT

Support for the development of the Turbo Brayton cooler has been provided by NASA/GSFC, BMDO, Air Force Research Laboratory, and the Hubble Space Telescope program. The authors gratefully acknowledge this support and the advances that have been made possible.

REFERENCES

1. Cheng, E. S., et al, "The Near Infrared Camera and Multi-Object Spectrometer Cooling System," Presented at the SPIE 1998 Symposium on Astronomical Telescopes and Instrumentation, Kona, HI, March 20-28, 1998.
2. Nellis, G., et al, "Reverse Brayton Cryocooler for NICMOS", *Cryocoolers 10*, Plenum Press, New York, 1999, pp.431-438.
3. McIntosh, R., et al, "A Capillary Pump Loop Cooling System for the NICMOS Instrument", presented at the 28th International Conference on Environmental Systems, Danvers, MA, July 13-16, 1998.
4. Dolan, F. X., et al, "A Single Stage Reverse Brayton Cryocooler: Performance and Endurance Tests on the Engineering Model", *Cryocoolers 9*, Plenum Press, New York, 1997, pp. 465-474.
5. Swift, W. L., "Single Stage Reverse Brayton Cryocooler: Performance of the Engineering Model", *Cryocoolers 8*, Plenum Press, New York, 1995, pp. 499-506.
6. Konkel, C., et al, "Design and Qualification of Flight Electronics for the HST NICMOS Reverse Brayton Cryocooler", *Cryocoolers 10*, Plenum Press, New York, 1999, pp. 439-448.

INTEGRAL SPECTROMETER CRYOSTAT DESIGN AND STM PERFORMANCES

R. Briet

CNES - Thermal Control Department
TOULOUSE, 31401, France

ABSTRACT

The INTERnational Gamma-Ray Astrophysics Laboratory (INTEGRAL) is an ESA satellite which uses a recurrent platform from XMM and will be launched in September 2001 by PROTON. The payload module consists of four instruments, one of them being dedicated to gamma rays energy resolution. This spectrometer (around 1300 kg, 2.5 m high, 1.1 m diameter) is developed by an international consortium under management of CNES where this instrument is assembled and tested before delivery to ESA.

The detection plane contains 19 Germanium detectors and weights 25 kg which need to be cooled at 85 K during observation and periodically annealed up to 483 K (110°C). The operational temperature is achieved by the mean of four 50-80 K MMS Stirling coolers while a passive radiator provides, through a heat pipe connection, an intermediate stage at 205 K around the detectors and thus reduces the cryocoolers heat loads.

This paper describes the design and technologies used on this three-stage cryostat, points out the major difficulties encountered during development, and presents performance predictions as well as those achieved on the Cryostat Structural and Thermal Model.

INTRODUCTION

Gamma emissions coming from celestial objects is the trace of violent phenomena which occur in environment of particular objects such as supernovae, neutron stars, black holes, active galaxy nuclei, quasars... Detailed observation of γ emission allows understanding of these phenomena and nucleosynthesis in the universe.

The 2nd mid-term mission scientific plan of European Space Agency (ESA) includes the INTEGRAL observatory which will carry four complementary instruments.

The SPectrometer for INTEGRAL (SPI) has a very high energy resolution of Gamma rays in the range of 20 KeV - 8 MeV. An Imager allows accurate geometrical location of the emitting sources while XRM and OMC complete the observation in different wavelength (respectively Xrays and visible range).

INTEGRAL satellite (shown on figure 1) will be nominally launched in September 2001 by Proton . Final orbit parameters are the following : 10000 - 153000 km with a 51.6° inclination and a 72 hours period.

Professor Gilbert Vedrenne of CESR (Centre d'Etudes Spatial des Rayonnements) succeeded in gathering the scientific community studying gamma rays in a highly international consortium. Indeed German, Spanish, Belgium, Italian, English, American and French scientists participate in the SPI project. Laboratories involved have to find funds in their own country and are responsible for the development of a sub-assembly of the Spectrometer.

Overall management, development plan, technical specifications and consistency, interface with ESA, integration and validation tests are under CNES responsibility. CNES also takes part in this instrument by providing, from conception to manufacturing and tests, some sub-assemblies such as the Lower Structure Assembly (LSA) which supports all the instrument, the on-board science and technological data management software, as well as the Cryostat which provides the required temperature for detectors.

This paper describes the design and technologies used on this three-stage cryostat, points out the major difficulties encountered during development, and finally presents the performance achieved on the Cryostat Structural and Thermal Model.

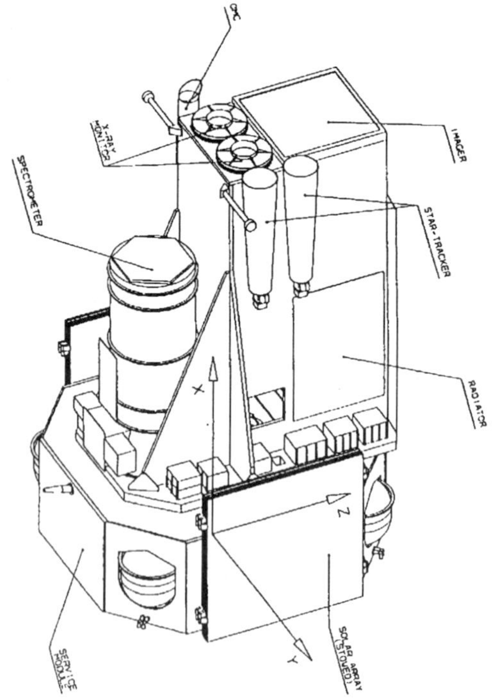

Figure 1. Overview of the INTEGRAL satellite and instruments

INSTRUMENT DESCRIPTION AND OPERATION

The incident radiation, coming from sources located in the field of view, is intercepted by a passive MASK composed of pixels which either stop γ rays (tungsten blocks) or allow them to pass through. Once the photon beam has been coded by the mask, it reaches a pixelized 500 cm^2 detection plane and projects on it a coded mask shadowgram characteristic of the source location in the sky.

The energy range and the energy resolution impose the choice of Germanium crystals detectors. Photons react with the detectors by creating a current proportional to the energy deposited by incoming photons. Relevant sensitivity requires the detectors to be operate lower than 110 K. The optimised 85 K working temperature is achieved by two twin cryocoolers (ACC). The detection plane is enclosed in a Beryllium structure (CBX) controlled around 205 K through a passive cooling device (PAC) to reduce the heat loads.

In order to shield the Germanium detector against background, the cryostat with its Ge detector array is surrounded by an anticoincidence sub-assembly (UVS, LVS and VCU), composed of crystal scintillators and photomultipliers, keeping free the field of view. A plastic scintillator (PSAC) located under the mask also reduces background.

All the signal output data produced from both detection plane and active shielding are then processed by several electronic modules (PSD, AFEE1 and 2, DFEE).

Mechanical loads are supported by the lower structure (LSA) which also acts as a single mechanical interface with the platform.

Figure 2 presents a split view of the instrument, before final assembly in CNES leads to the Spectrometer as shown on figure 3.

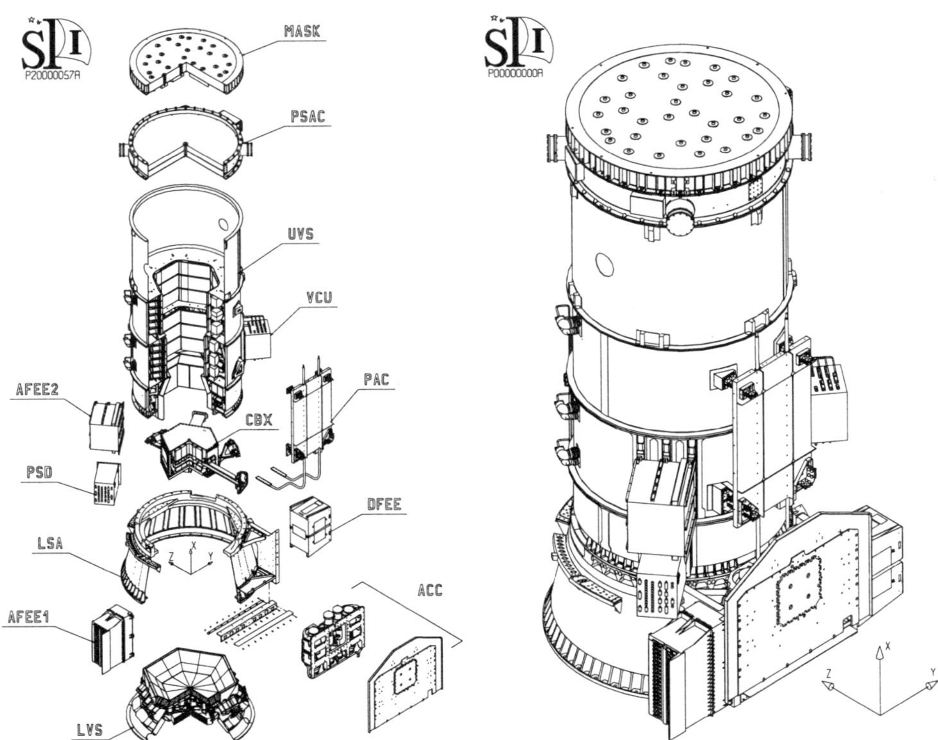

Figure 2. Equipment of the Spectrometer

Figure 3. General spectrometer design

The instrument main features and performances are :

- Mass : 1300 kg	- Field of view : 16°
- Height : 2.5 m high	- Energy range : 20 keV to 8 MeV
- Diameter : 1.1 m	- Resolution : 3 keV at 1.33 MeV
- Power : 340 W	- Telemetry rate : 22 kb/s

THREE STAGE CRYOSTAT DESCRIPTION

The detection plane is composed of 19 Germanium detectors mounted on a Beryllium plate, the total weighting 25 kg. Its overall volume can be described as a cylinder 100 mm high with a 305 mm diameter. In order to gain a high sensitivity, Ge detectors must be maintained between 78 K and 85 K during observation, whereas they have to be heated up once a year to 483 K for annealing purpose. In addition, two main detector specifications had to be taken into account for the thermal design : a drift $\Delta T/\Delta t$ lower than $\pm$ 0.5K in observation, and a gradient ΔT lower than 1K between detectors.

To achieve these requirements, CNES conceived a cryostat composed of the following sub-assemblies : ACtive Cooling (ACC), PAssive Cooling, and the Cold BoX (CBX).

Active cooling (ACC)

The operating temperature of the detectors is achieved by the use of four 50-80 K MMS Stirling coolers fixed on an outside radiator to reduce background noise on detectors. These coolers are controlled by independant command electronics located on the platform.

Cold box (CBX)

The heat lift of the cold stage is transferred to the cooler's cold tips through a cold rod and a complex flexible link assembly. The detection array and associated 19 preamplifiers are located inside a chamber controlled at about 205 K. To limit the background noise, the detector plate, the chamber structure and the cold bus bar are made of beryllium. The enclosure composed of this chamber and a tube surrounding the cold rod is vacuum tight to allow simple and low cost ground testing. A "pumping tube" is thus fitted to the bottom plate of the chamber for connection to an external pump or space. These equipment as well as the detectors to chamber, and chamber to main structure (insulating) supports constitute the CBX.

Passive cooling (PAC)

The location of the Spectrometer on the anti-solar face of the satellite was requested early since it permits the use of an intermediate stage. Indeed the structural envelop of the cold box is cooled at about 205 K by a sub assembly called PAssive Cooling. It consists of two ammonia heat pipes linked to an external radiator fixed on the instrument by means of insulating supports. This radiator rejects to space the intermediate stage heat leaks as well as preamplifiers' dissipation and the antifreeze safety system heating power.

Indeed, to optimize the intermediate stage efficiency, heat pipes are used near the ammonia freezing point (196 K). Notice that a thaw system is available to start the heat pipes after coast phase or after survival mode.

The cryostat developed for the spectrometer is thus shared in three stages :
- A first "85 K stage" for the Ge detectors (and part of preamplifiers). It is cooled through a beryllium cold rod and five flexible copper braids (2 stages) for conductive coupling and mechanical decoupling with regards to the displacers of the mechanical coolers;
- A second "200K stage" enclosing and supporting the detection plane and the other part of preamplifiers. It is composed of a beryllium cold box, a stainless steel tube around the cold rod, two ammonia heat pipes and associated radiator;
- A third "300K stage" consisting of four "MMS 50-80 K Stirling coolers", their common structure and radiator (control electronics are located directly on the platform).

Figure 4 presents the cryostat concept and main heat fluxes as explained further whereas a cryostat overview (without MLI) is shown figure 5.

The mechanical supports for the detection plane on the cold box (85 K - 200 K stages) use glass fibre plate technology. These three couples of plates support about 32 kg, have an overall 300 mm² section, sized for 15 g and a first eigenfrequency > 120 Hz. The passive radiator (200 K stage) is also fixed on the anticoincidence system (300 K stage) by six glass fibre plates. The cold box is fixed on the LSA by means of three titanium supports. These plates support about 60 kg, have a 40 cm² global section sized for 15 g and F > 80 Hz.

This concept results from a mechanical and thermal compromise.

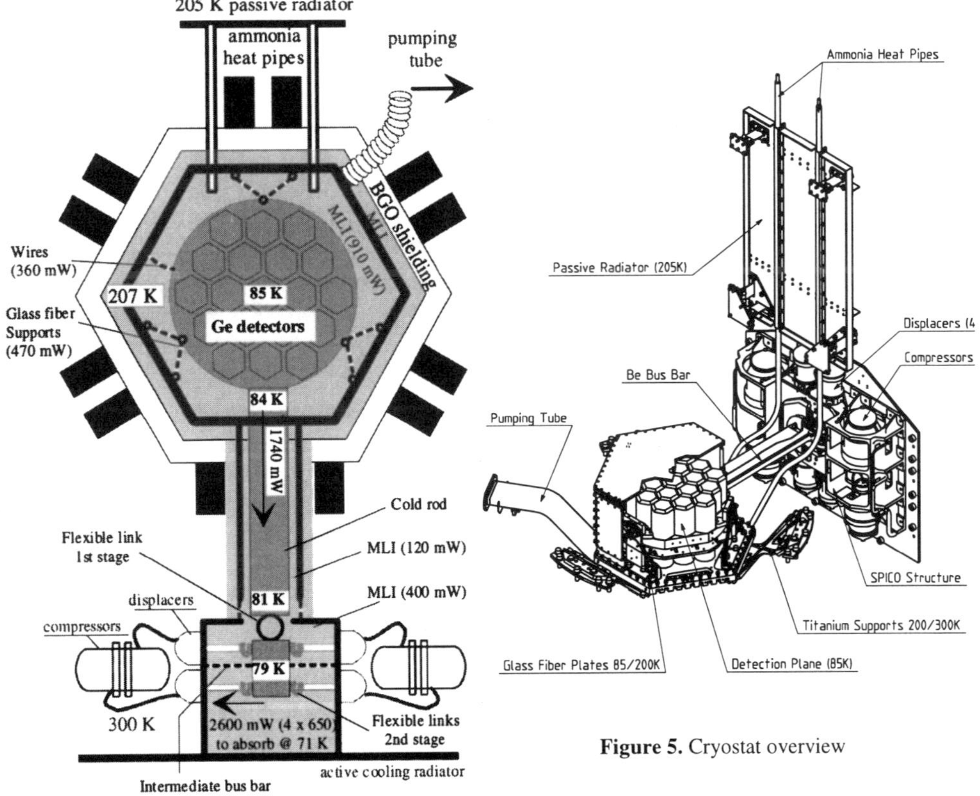

Figure 4. General cryostat concept

Figure 5. Cryostat overview

The radiative insulation between 85 K and 200 K stages uses 50 insulative layers. The 200 K stage is insulated from the instrument by 10 to 20 layers.

Inside the cold box, both scientific wires (19 signal + 19 High Voltage), housekeeping hardware (10 sensors), and annealing heaters cables use low conductive material such as manganin, or constantan. Specific integration techniques lead to a 5 cm thermal gradient length through the MLI blankets. An electrical to thermal compromise has been achieved to reduce the heat leaks through cables dedicated to redundent annealing heaters (2 x 30 W).

The same kind of design and integration techniques have been applied to reduce the heat leaks on the intermediate stage due to the 133 scientific wires exiting from the cold box and the 20 housekeeping (sensors + heaters) coming from the PAC subassembly.

The cold box and displacers enclosure is vacuum-tight for ground tests purpose. This constraint applies for MLI efficiency which requires a pressure lower than 10^{-4} hPa, and permits the detectors to be supplied in high voltage (typ. 5000 v) without dangerous discharges. The airtight tube around the cold rod is fixed on one side on the cold box structure, and one the other side on the coolers supporting structure.

On the cooler side, this tube ends with a bellow system in order to disconnect the mechanical link before launch vibration. This can be achieved through a trap door located on the coolers radiator. This trap door also allows the connection to a cryogenic ground support equipment (to decrease cooling down duration) and gives access for integration of the flight cooling system to the end of the beryllium bus bar and its first stage flexible link.

MAJOR DIFFICULTIES ENCOUNTERED

As a general and common rule for cryogenic hardware conception and development, a very close collaboration and good cooperation of thermal, electrical, mechanical engineers and scientists is required to achieve the best compromise.

The development of the cryostat was subjected to other standard problems such as accommodation and validation of the concept. Indeed the dimensions of the cryostat, especially the inner and outer volume of the cold box, had to be minimized since half the mass of the Spectrometer comes from the anticoïncidence system using BGO crystals. A small increase of the "radius" of the cold box would have a big impact on the total mass of the instrument (1mm led to several tenth of kilograms).

Predictions of the performance of a cryostat can be considered quite complex since everything cannot be computed easily. Indeed MLI blankets efficiency, as well as the thermal gradient length and materials properties as a function of the temperature are driven parameters for the cold stage thermal balance. Many characterisation tests have been performed early and a mock-up has been built to validate assumptions in terms of leaks through supports and wires, and mainly MLI efficiency. This can only be appreciated with a representative integration taking into account the slots for wires.

In this field all wires have been specifically developed in the frame of a respect of functional performances while limiting thermal leaks, either conductive by the material itself, or radiative by the slots done in the MLI blankets. Definition of the blankets and routing of the cables have also been optimised to present a maximum thermal gradient length through a complex MLI. Thermal insulation has also been fitted to ensure the maximum thermal gradient length along the mechanical supports.

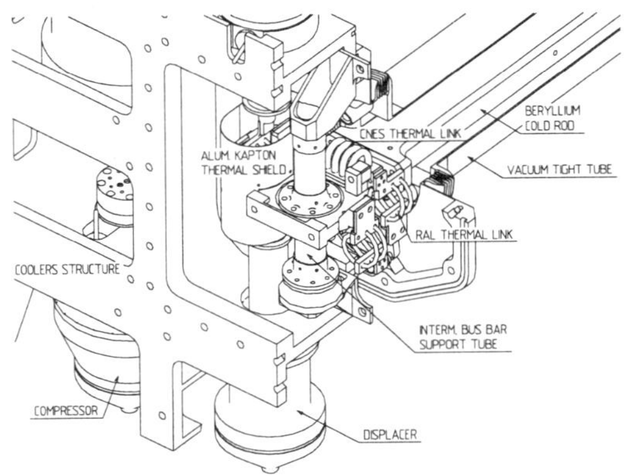

Figure 6 . Intermediate bus bar and flexible links

Ignoring alignment issues, manufacturing issues (especially as far as beryllium parts and thermal flexible links are concerned), vacuum tight enclosure requirement, we will just point out problems in connection with the mechanical cooler system : At a first step, four flexible links thermally coupled directly the end of the beryllium bus bar to the cold tips of the four Stirling displacers. Prototypes of these braids have been manufactured and tested (dynamic and static stiffness as well as thermal conductivity).

The maximum acceptable transverse force on operating displacers is only 0.4 N. Taking into account the stiffness of the braids, integration and manufacturing tolerances, thermal contractions of the 420 mm long beryllium cold rod ... this leads to stiction inside displacers. The solution adopted is a double-stage thermal link assembly.

A "1st stage" flexible link connects the end of the cold rod to an intermediate bus bar which is maintained rigidly on the cooler structure by a GFRP tube. Thus large deflections of the cold rod result in small bendings of the intermediate bus bar. Four "2nd stage" flexible links developed by RAL then filter the force transmitted to the displacers.

Drawbacks of this late concept are an additional conductive path through the GFRP tube between 70 K and 300 K, and a higher thermal gradient on the cold heat flux path through flexible links and contacts at additional interfaces.

Figure 6 shows the intermediate bus bar area with the two stages of flexible links.

CRYOSTAT THERMAL PERFORMANCES PREDICTIONS

A Thermal Mathematical Model of the cryostat has been built to predict performances at detectors level and compute heating power required for annealing purpose, for the antifreeze safety system, for heat pipes thaw and for cryocoolers thermal control.

Margins philosophy, transient computations performed will not be discussed here.

Both breadboard models results, characterisation tests and computations allowed us to be quite confident in the 85 K stage thermal balance. Thermal interface between CNES and MMS parts is located at the first stage flexible link / intermediate bus bar interface, further named cold interface.

Table 1 presents the various heat leaks contributions, the heat lift at this cold interface as well as the coolers heat loads assuming worst hot case conditions as follows :
cold box (CBX) at 211 K and preamplifiers (PA) at 218 K.

Table 1. "85 K stage" thermal balance

	Heat leaks towards detectors (DET) at 85 K
Dissipations	0
MLI DET-CBX	400 mW
MLI DET-PA	510 mW
Mechanical Supports	470 mW
LV wires (DET-PA)	150 mW
HV wires (DET-CBX)	150 mW
Temperature sensors wires	10 mW
Annealing heaters	50 mW
MLI Cold rod -Tube	120 mW
Heat lift at cold I/F (79 K)	**1860 mW**
Intermediate bus bar support	300 mW
MLI around coolers ends	400 mW
Inactive displacers	0
Coolers heat lift (71 K)	**≈ 2560 mW**

Table 2. "200 K stage" thermal balance

	Worst cold case	Worst hot case
PA Dissipations	4.4 W	4.4 W
Vacuum tight tube	0.4 W	0.7 W
MLI around CBX	1.1 W	2.1 W
Pumping tube	0.7 W	1.4 W
CBX supports	2.9 W	5.9 W
CBX wires	1.9 W	3.4 W
Leaks => 85 K stage	-1.6 W	-1.9 W
Heat pipes heat lift	**9.8 W**	**16.0 W**
MLI radiator	0.4 W	0.7 W
Mechanical supports	0.5 W	1.0 W
PAC heaters cables	0.7 W	1.3 W
PAC sensors cables	0.2 W	0.3 W
Antifreeze power	6.5 W	0
Radiator rejection	**18.1 W**	**19.3 W**
Radiation to space	-17.3 W	-18.4 W
Radiation to PF MLI	-0.8 W	-0.9 W

It shall be noticed that the very first design of this cryostat was based on the use of pulse tube split coolers. The pulse tube head was planned to be fitted on the cold box itself and directly coupled to the detection plane through a flexible braid, so that the coolers heat load would have been only 1740 mW at about 82 K. Problems encountered during pulse tube developments led to elimination of this option.

Finally, after issues due to low acceptable forces on Stirling displacers, the coolers heat lift is about 2560 mW at 71 K.

This impressive difference in the cold tips temperature mainly comes from the thermal gradient in the 2nd stage flexible links and especially thermal resistances at contacts.

The "200 K stage radiator" has been sized with the objective of avoiding any consumption of the antifreeze safety system in the hottest environment, this is an ammonia temperature target of 208 K. Table 2 presents the "200 K stage" thermal balance in the two following conditions : cold case (cold box at 208 K, ammonia at 205 K, interface at -13°C)
hot case (cold box at 213 K, ammonia at 208 K, interface at 40°C).

Both annealing (473 K / 483 K) and antifreeze (205 K / 207 K) safety regulations use electronical thermostats while mechanical coolers thermal control just requires mechanical thermo-switches. All these power lines are redunded.

The antifreeze safety system uses a specific electronical thermostat supplied through two transistor switches controlled by telecommand. This allows two different maximum consumptions (either 12 W or 18 W) depending if one relay or both relays are closed.

This design has been accepted since heating power needs vary according to the instrument functional modes : only one line is activated when preamplifiers are supplied.

Through telemetry available whatever the SPI functional status is, one can know if ammonia contained in the heat pipes is frozen. When needed a telecommand can be thus sent to the spacecraft to supply a 45 W heaters stuck on the "200 K stage radiative plate". The effective white painted area of the PAC radiator is 0.21 m².

Notice that both annealing and heat pipes thaw heaters are also used for outgassing.

Table 3. Cryostat performances predictions (mathematical model)

Environment temperature / coolers configuration	Detectors temperature	Coolers (ON) temperature	Heat load at cold I/F	Cold I/F temperature	Ammonia temperature
40°C / 4 x 25 W	82.4 K	24.5 °C	1850 mW	74.5 K	208.3 K
-10°C / 4 x 25 W	78.4 K	12.2 °C	1790 mW	70.7 K	205 K
40°C / 2 x 44 W	105.6 K	26.5 °C	1620 mW	98.9 K	208.6 K
-10°C / 2 x 44 W	100.8 K	13.5 °C	1570 mW	94.3 K	205 K

The allocated mean input power to the mechanical coolers is 100 W (150 W including drive electronics). In eclipse mode the total budget for the coolers and electronics is 100 W.

The "300 K stage radiator" has been optimised to present a heat rejection temperature as low as possible to increase the coolers efficiency, but with the constraint of non requiring thermal control heating power after a maximum 1.8 hour long eclipse, this is a minimum -12°C eclipse outlet temperature deduced from thermostats tolerances and stiction tests.

SPI cryostat nominally uses two pairs of MMS 50-80 K Stirling coolers controlled by two drive electronics for standard back to back operations. Even if a three running coolers configuration is available and acceptable with regards to microvibrations, the degraded configuration studied just assumes the availability of a pair of coolers.

The four coolers have thus been fitted on a single piece bracket structure to distribute dissipations in degraded mode. This bracket is mounted on a radiator which is fixed on the Lower Structure Assembly. Indeed, high susceptibility of the detectors imposed to fit this active cooling subsystem outside the shielding as shown previously on figures 2 and 3. The effective white painted area of the ACC radiator is 0.3 m^2.

In nominal configuration all four coolers operate with typical 6.5 mm compressors and 3.4 mm displacers strokes leading to a 100 W mechanical coolers mean input power. A specific design of the electronics has been developed to achieve a 90° phase angle between the pairs of coolers in order to reduce peak currents demands to the platform bus.

In degraded mode two coolers are operated at about 8.5 mm compressors strokes with a mean input power lower than 88 W (without electronics).

A single heater power line has been fitted to be able to switch-on the coolers. The minimum coolers switch-on and operating temperatures are equal, this is -20°C. This leads to 80 W heating power (peak value computed in worst survival conditions).

Table 3 above presents performances predictions as a function of the environment interface temperatures as well as the compressors plus displacers input power. Notice that performances given with only two coolers suppose the use of the worst pair (flight spare).

STM CRYOSTAT MAIN RESULTS

In addition to many mock-ups and characterisation units, manufacturing and testing of an Electrical Model (EM) and a Structural and Thermal Model (STM) was foreseen in the Spectrometer development plan. Thermal vacuum tests have been performed on the STM cryostat. Its configuration was thermally fully flight representative since it used flight materials, a flight model (FM1) and the flight spare (FS) pairs of coolers but controlled by laboratory electronics. All wires contributing to heat leaks have been nominally integrated and routed through MLI blankets.

Figure 7. STM Cryostat integration in the TV tests chamber (before fitting external MLI)

As shown on figure 7, the STM cryostat mounted on the STM LSA has been fitted inside a vacuum chamber whose internal shrouds were maintained at 106 K during tests. The cryostat radiative and conductive thermal interfaces with the Spectrometer have been simulated and controlled in temperature to achieve representative test conditions.

Temperatures achieved at detectors level were quite satisfactory as shown in table 4. As discussed before, specifications require detectors colder than 85 K, knowing that up to 110 K this does not affect the resolution but just means more frequent annealing processes (in degraded mode).

Moreover annealing and antifreeze safety regulations have been successfully qualified.

CONCLUSIONS

After completion of satisfactory mechanical environment tests, the STM cryostat has been submitted to quite satisfactory thermal vacuum tests.

Even if the major aim of these tests was to validate and characterise the cryogenic performances, it also gave useful information on antifreeze, annealing device as well as "standard" techniques used to thaw the heat pipes or control the coolers minimum temperature. These tests shall be repeated soon at Spectrometer level which will thus be fully qualified as far as mechanical and thermal aspects are concerned.

The current status of the Spectrometer instrument and Cryostat is phase D. While not yet available, flight cryostat units are under manufacturing to start a five months period of integration and validation tests process.

Table 4. Cryostat performances measured on STM : *(resp. **) with FM (FS) coolers

Environment temperature / coolers configuration	Detectors temperature	Coolers (ON) temperature	Ammonia temperature
40°C / 4 x 25 W	79 K	18 °C	209 K
40°C / 2 x 44 W *	96 K	18 °C	211 K
40°C / 2 x 44 W **	101.5 K	18.5 °C	211 K
-10°C / 2 x 44 W **	95 K	8.5 °C	208.5 K

HYDROGEN SORPTION CRYOCOOLERS FOR THE PLANCK MISSION

L. A. Wade[1], P. Bhandari[1], R. C. Bowman, Jr.[1], C. Paine[1], G. Morgante[2], C. A. Lindensmith[1], D. Crumb[3], M. Prina[4], R. Sugimura[1], D. Rapp[1]

[1]California Institute of Technology, Jet Propulsion Laboratory
Pasadena, CA 91109
[2]CNR-Te.S.R.E.
Bologna, Italy
[3]Swales Aerospace
Pasadena, CA 91107
[4]Politecnico di Milano
Milano, Italy

ABSTRACT

Two continuous operation 18 K/20 K sorption coolers are being developed by the Jet Propulsion Laboratory (JPL) as a NASA contribution to the European Space Agency (ESA) Planck mission that is currently planned for a 2007 launch. The individual sorption coolers will each be capable of providing a total of 230 mW of cooling at 18 K and 1.45 W at 20 K given passive radiative precooling at 50 K. The hydrogen sorption coolers will directly cool the Low Frequency Instrument HEMT amplifiers to approximately 20 K and will also serve to intercept parasitics and precool a RAL 4.5 K closed-cycle helium J-T cooler to 18 K for the separate High Frequency Instrument. The operating conditions and mission requirements for the Planck sorption cooler are presented. The concept design of the 20 K coolers is described along with the predicted performance. $La_{1.01}Ni_{4.78}Sn_{0.22}$ hydride sorbent beds are currently in fabrication and initial test data on their performance are presented.

INTRODUCTION TO PLANCK

Planck, the third Medium-sized ESA mission M3, will be launched in 2007 in combination with the Far InfraRed and Sub-millimetre Telescope (FIRST). Planck will carry two instruments: the High Frequency Instrument (HFI) and the Low Frequency Instrument (LFI). Together they will observe and image the full sky in nine spectral bands between 30 and 857 GHz. The LFI utilizes an InP HEMT amplifier radiometer cooled to 20 K through a combination of passive cooling to <50 K and the hydrogen sorption coolers. The HFI observes using bolometers cooled to 100 mK through a combination of passive coolers, the 18 K/20 K sorption cooler, a 4.5 K RAL Mechanical J-T cooler and an Benoit style open cycle dilution cooler[1].

Planck is the third generation space mission (following COBE and MAP) to be designed for observation of the Cosmic Microwave Background (CMB) anisotropies. Planck will observe the full sky and produce maps with an accuracy limited only by cosmic variance and astrophysical foregrounds at all angular scales larger than 10' for the LFI and 6' for the HFI. In addition, both instruments will be capable of measuring the polarization in the CMB. The unprecedented angular resolution and sensitivity ($\Delta T/T \sim 2 \times 10^{-6}$) will allow the primary cosmological parameters (Hubble constant, deceleration parameter, curvature of space, baryon density, amplitude and spectral index of the primordial scalar density perturbations, and the gravity wave content of the Universe) to be determined to an accuracy of a few percent. In doing so a number of truly fundamental questions will be answered about our Universe: How fast is the Universe expanding? What is the ultimate fate of the Universe? What are its material constituents (baryons, dark matter,...)? Where did the initial perturbations come from? When did the first structures form in the Universe? What is the nature of particle physics at energies $\simeq 10^{16}$ GeV?

PLANCK PAYLOAD DESIGN OVERVIEW

As with most fundamental physics experiments, the data to be taken by Planck are more coupled to the specifics of the experimental apparatus and environment than to the science of interest. Uncertainties or oscillations in pointing, supply voltages, thermal fluctuations, straylight (much of this is thermal emission from different surfaces) all impact the data collected by the instrument. As an example, the spacecraft emits approximately 10^3 W. The HFI is sensitive to energy levels on the order of 10^{-18}W. As a consequence, even extraordinarily minute spacecraft thermal emission fluctuations that reach the HFI detector could be confused with sky emission.

Changes over some timescales, especially at about the 1 minute spin period, are particularly critical, as they are difficult to distinguish from the observed sky. Even worse, this does not mean that these changes necessarily have a frequency of 0.01667 Hz; but only that they have a non-zero Fourier component at this frequency. The result is that the mission configuration and design, along with the instrument designs, operational and data analysis strategies, must all be driven to minimize systematic effects on the final science data products by reducing the level of these effects and ensuring that they are well understood so that the residual effects can be confidently removed in software without adversely affecting the science results. Ensuring the successful accomplishment of the science goals is the major factor affecting the mission design, and therefore the 18/20 K coolers design.

Planck will fly in a Lissajous orbit around the Earth-Sun L_2 point, and spins at about 1 rpm about an axis that points towards the sun. The 1.5 m aperture telescope looks 80° to the spin axis and scans circles in the sky. Figure 1 shows a crossection view of a likely configuration for the Planck spacecraft. As the Architect Study is still underway, some changes from this figure will occur.

All warm components of the instruments are mounted on the service module. This includes not only the instrument electronics but the warm compressors of the sorption coolers and the RAL 4.5 K mechanical J-T cooler, the gas storage tanks for the dilution refrigerator and all their associated electronics. The optical bench and telescope are thermally isolated from the warm spacecraft bus by low conductance struts and v-groove shields. V-groove shields are a set of two or more angled low-emissivity specular surfaces typically 2 to 5 degrees out of parallel. The radiative heat transfer between two facing surfaces of the V-groove shield with emissivity ε is proportional to ε^2, while that between infinite parallel planes is $\varepsilon/2$. The extra energy in the V-groove case is radiated to space. The V-groove shield therefore can provide extremely good radiative isolation between objects at different temperatures even with surfaces of only moderately low emissivity. In addition, it can be highly efficient at intercepting conductive thermal loads and radiating them to space. The reduction in heat transfer in this arrangement is significantly superior to that typically achieved by conventional Multi-Layer Insulation (MLI) between two parallel plates. The v-groove design concept was originally invented by Ray Garcia at

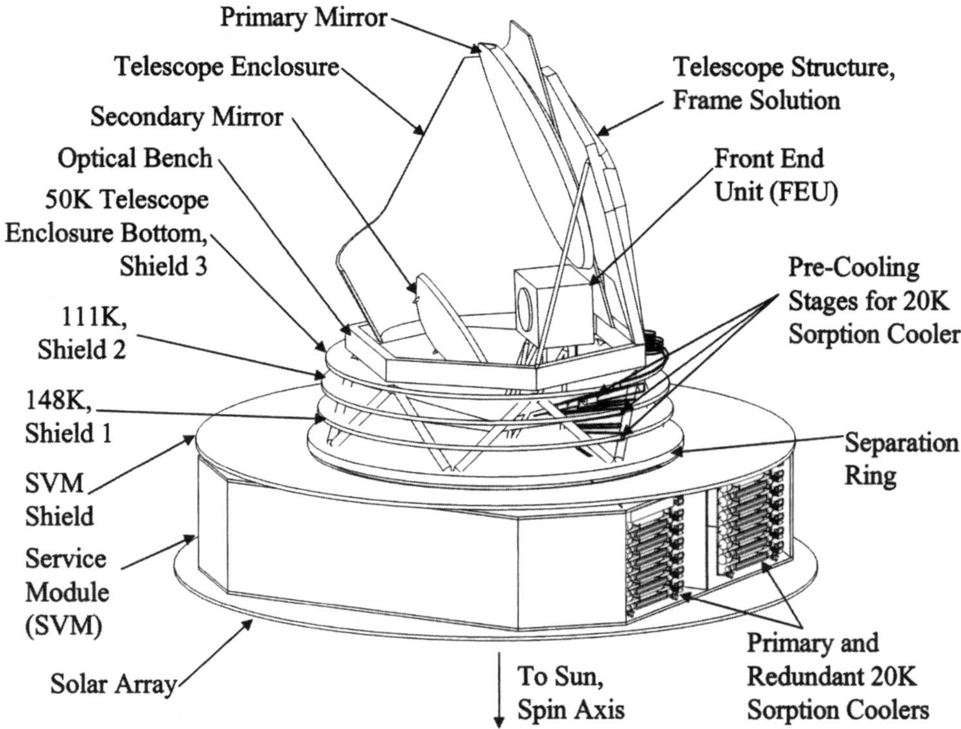

Figure 1. The Planck configuration optimizes passive radiative cooling and minimizes straylight and other systematic effects on the final science data.

JPL. Since then this new technology has been validated in thermal vacuum and vibration tests[2,3]. Similar 'bounce-view-of-space' tricks are commonly employed in high performance radiators (e.g. the NIMS radiative cooler for the Galileo orbiter[4]) and therefore substantial flight heritage date exists for evaluating the long term behavior of such surfaces.

The current estimates for shield temperatures are 148 K for the first thermal shield, 111 K for the second and 50 K for the telescope enclosure[5]. The Front End Unit (FEU) is located within the telescope enclosure. This contains the frontend radiometer of the LFI as well as the HFI bolometer array.

PLANCK 18 K/20 K COOLERS OVERVIEW

Two sorption coolers will be flying on the Planck mission. One will provide the primary cooling of up to 230 mW at 18 K for HFI parasitic interception and RAL 4.5 K J-T cooler precooling, and up to 1.45 W of cooling at 20 K for the LFI. The second will be turned off and used as a backup unit should anything happen to the primary cooler. Each cooler is sized to achieve a two year operating life. The cooler input power is 520 W at end-of-life plus an additional 30 W estimated for the cooler electronics. Figure 2 shows a schematic of the sorption cooler that will be used on the Planck spacecraft.

The inherently split design of continuous operation sorption coolers both enables, and maximally benefits from, the passively cooled design needed to minimize systematic impacts on the data gathered but which requires the spacecraft to be located over a meter from the 20 K Front End Unit. The very long operating cycle of the sorption cooler enables easy thermal fluctuation characterization and removal from the science data.

The compressor assembly shown in figure 3 is composed of six identical sorption compressor elements, each filled with metal hydride and provided with independent

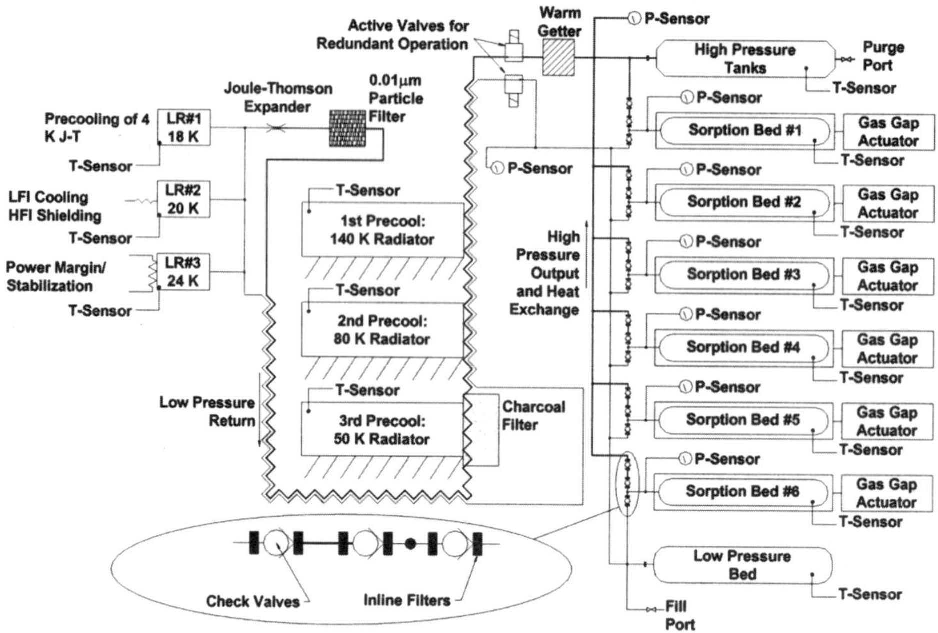

Figure 2. Planck sorption cooler schematic.

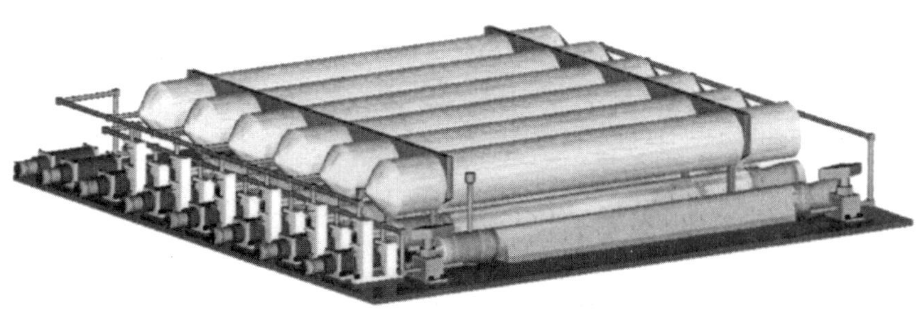

Figure 3. Planck sorption cooler compressor assembly

heating and cooling which will be described later. Each compressor element is connected to both the high pressure and low pressure sides of the plumbing system through check valves, which allow gas flow in a single direction only. The check valves are indicated on the schematic as single arrows, which indicate the direction of gas flow through them. In addition to the compressors, there are five one-liter high-pressure stabilization tanks connected to the high pressure side of the system to damp out oscillations of the high pressure gas, and a low pressure stabilization sorbent bed to damp out pressure fluctuations of the low pressure gas.

Refrigerant travels from the compressors through a series of heat exchangers and radiators, which provide precooling to approximately 50 K, through the J-T expander at the FEU. The cold end assembly shown in figure 4 includes three liquid reservoirs in the system: a first reservoir providing 18 K cooling of the 4.5 K RAL mechanical J-T and the HFI thermal shielding, a 20 K reservoir providing cooling to the LFI, and an overflow reservoir for unused cooling capacity. Each of the reservoirs is filled with a wicking material in order to retain the liquid in the reservoirs without gravity. The third reservoir is maintained above the hydrogen equilibrium temperature, to wick and then sublimate any liquid which reaches it, providing an even gas flow back to the sorbent bed.

All regulation of the system is done by simple heating and cooling, with no active control of valves being necessary. Normal operation of the compressors can be done with only a minimum of active feedback: The heaters for the compressors are controlled by a simple timed on-off heater system. In addition to the on-off heater power, each heater will have up to 30 W of additional heating supplied by a proportional controller to compensate for any degradation of hydride or gas-gap properties that might occur.

Compressor Assembly

The compressor assembly is comprised of the six compressor elements, high-pressure stabilization tanks, the low pressure stabilization bed, check valves, and manifolding. The switch time is 667 s leading to an overall cycle time of 4000s. The compressor assembly mounts directly onto the heat rejection radiator. This radiator is sized to reject the cooler input power at 270 K +10 K/-20 K. The low heat rejection temperature was selected to ensure precooling of the 4.5 K RAL cooler at less than 19 K. The compressor assembly mass is 40 kg. It's volume is 0.25 m x 0.8 m x 0.8 m.

A single compressor element is comprised of two concentric cylinders closed with end caps. The inner of these tubes contains the $La_{1.01}Ni_{4.78}Sn_{0.22}$ hydride material and the outer forms a vacuum jacket around the inner cylinder. An exploded view of an compressor element is shown in figure 5. This vacuum jacket is used as a gas-gap heat switch. A recently assembled compressor element which is being used for gas-gap thermal switch characterization is shown in figure 6. The hydrogen gas for the gas-gap heat switch is supplied through a tube penetrating the side of the outer cylinder.

The heater passes through the hydride material and is designed to uniformly distribute heat to ensure a high degree of temperature uniformity when at the maximum temperature of 465 K. Heat transfer to the hydride is provided by aluminum foam that fills the inner cylinder and makes tight contact to the heater. The foam is 89% empty, and is cut to allow penetration by the various other components, which are located in the inner cylinder. A vent tube passes through the center of the hydride material. This vent tube is fabricated from sintered 316 stainless steel, and has a sub-micron porosity, which excludes the powdered hydride from the hydrogen gas flow.

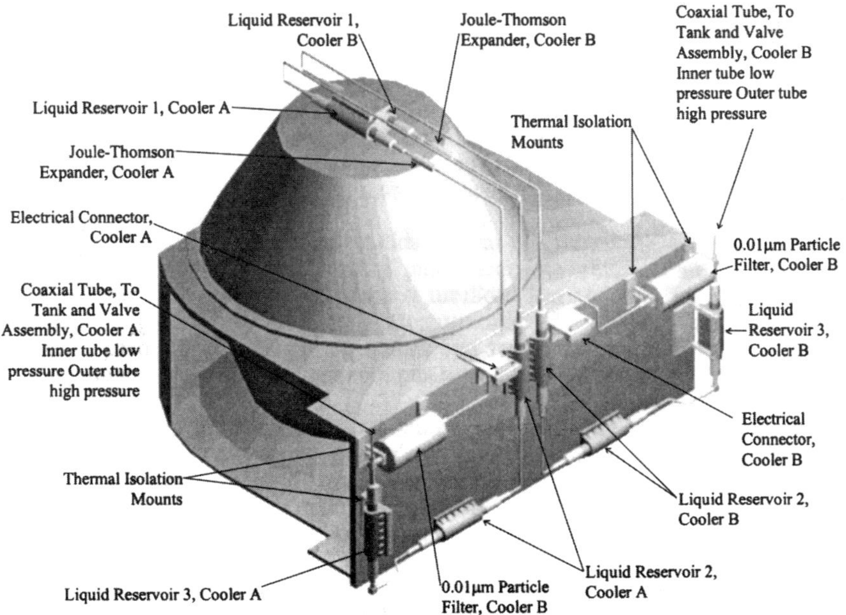

Figure 4. Planck sorption cooler cold end assembly

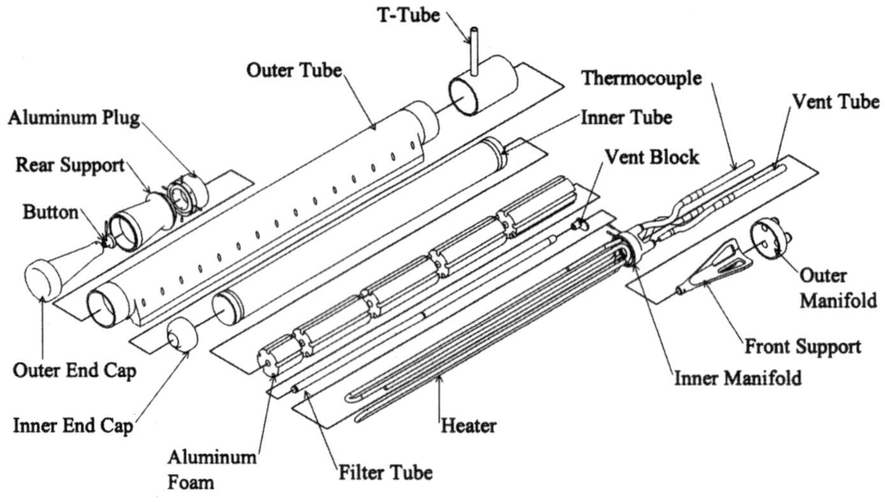

Figure 5. Exploded view of compressor element piece parts.

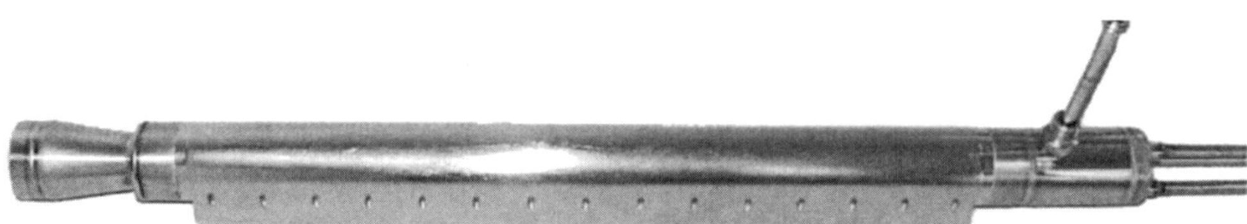

Figure 6. Compressor element used for gas-gap characterization tests.

A heater, thermocouple, and vent-tube lead run from the end cap of the inner cylinder to that of the outer cylinder. Each has a small bend for stress reduction, which defeats the potential for any of these elements to carry load. Therefore the only elements taking significant load are those designed to be structural, load carrying elements. There are two of these structural end supports: one at either end of the tube assembly. At one end provision is made for thermal compliance, without strain, between the inner and outer tubes. The outer tube assembly is primarily fabricated of 6061-T6 aluminum. This outer tube also provides the primary structural attachment point for the single compressor bed.

Nearly all parts of the compressors and the gas-handling system which come in contact with hydrogen are made of 316L vacuum arc remelt (VAR) stainless steel which has been electropolished on the surfaces exposed to the hydrogen. This choice of material also serves to prevent the degradation of the compressors structural parts by reaction with the hydrogen. There are two components which are made of other materials. The aluminum foam, which provides heat conduction to the hydride in the compressor, and the seals of the check valves, which are made of Viton. Stringent cleaning and assembly methods are used during construction.

The space between the inner compressor vessel and the outer tube assembly is used as a gas-gap thermal switch. It is preferred that this switch be operated in a closed cycle using a hydride to pump the gap to approximately 0.01 Torr when 'off' and to approximately 10 Torr when 'on.' A test program is currently in place to characterize candidate materials and to determine the operational pressure requirements for closed cycle operation[6]. Closed cycle operation will require approximately 8 W per switch which is 'on.'

In figure 7 conductance test data at approximately 300 K is presented for the compressor assembly shown in figure 6. The sum of conductive and radiative parasitics at the highest temperature difference (475 K inner vessel and 289 K outer vessel), which occurs in operation with the gas-gap volume evacuated, is 31 mW/K. This is slightly better than the predicted performance of this assembly. The conductance across the gas-

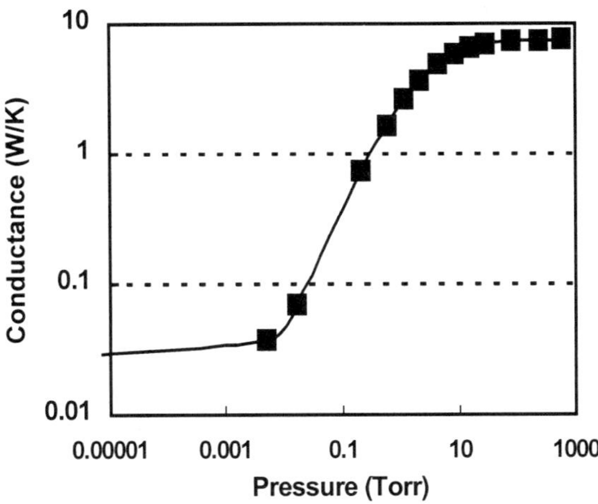

Figure 7. Gas-gap conductance from rarified gas regime through continuum flow at approximately 300 K.

gap matches theoretical predictions based on a model with one adjustable parameter, the accommodation coefficient, which has a value of 0.4. At high pressure (above approximately 30 Torr) the gas-gap switch conductance saturates at approximately 7.8 W/K, in good agreement with expectations.

J-T Cryostat Assembly

The J-T cryostat assembly includes a warm contaminate filter, active valves to allow redundant operation of the coolers, four heat exchangers, an approximately 50 K charcoal filter, an approximately 30 K 0.01 micron particulate filter, a porous plug J-T expander, and three liquid reservoirs. The four tube-in-tube heat exchangers are all 6.35 mm outer diameter tubing with a 3.18 mm outer diameter inner tube. The high pressure gas is held in the annulus. The heat exchangers are heat sunk at each thermal shield.

Contaminate trapping is done at room temperature with either a hydride getter or a resin bed (to be determined), with activated charcoal and a particulate filter at 50 K and a 0.01 micron filter at ~30 K. All of the components at the cold end were shown in figure 4. The J-T expansion is done through a porous plug device nearly identical to those described previously[7]. The first two liquid reservoirs are designed to separate the liquid refrigerant from the two-phase fluid leaving the J-T expander. This fluid is wicked to the wall of the reservoirs. The fluid in each reservoir is at essentially the same pressure and temperature. However the heat flux into each is very different. The first liquid reservoir interface temperature is approximately 18 K when providing < 230 mW of refrigeration to the HFI and the 4.5 K RAL cooler. The second reservoir has an approximately 20 K interface temperature due to the higher heat flux from the LFI. The third reservoir is designed to vaporize any excess liquid refrigerant to ensure a stable mass flow and pressure at the cryostat. This helps significantly to remove temperature fluctuations in the range of 1 Hz.

Figure 8 presents the predicted performance of the Planck sorption coolers as a function of temperature as compared with the predicted combined instruments cooling requirement. The difference between these predictions is the predicted margin. At a precooling and telescope enclosure temperature of 50 K, a 77 % margin is predicted. With an environment of 60 K the margin is 0%.

The flight cooler electronics and software will be supplied by the Institut d'Astrophysique Spatiale in Orsay France and an industrial partner. The electronics consist of a CPU, relay boards and several temperature and pressure sensor boards. The details of these electronics will be presented in a future paper.

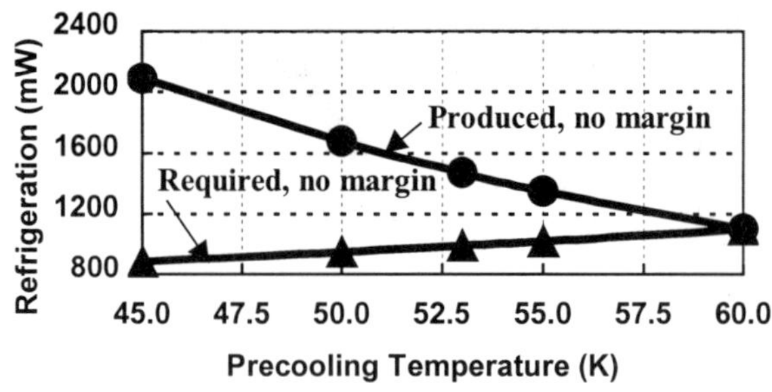

Figure 8. Predicted performance of the Planck sorption coolers and predicted combined Planck instruments cooling requirement as a function of precooling/telescope enclosure temperature.

DEVELOPMENT PLAN SUMMARY

1999 and 2000 will be primarily spent completing the technology development for this program. In 1999 an accelerated lifetest of the hydride material will be performed, several compressor elements will be built and begin parametric performance tests, gas-gap and characterization and actuation materials testing will be completed, and a demonstration of a three liquid reservoir cryostat will be made. In 2000, this development will continue leading up to initiation of testing of a flight-like 'elegant breadboard cooler' in December. The qualification model cooler is scheduled for delivery in January 2003. After qualification, this cooler will be refurbished and fly as the redundant cooler. The flight model cooler will be delivered in January 2004.

SUMMARY

A pair of sorption coolers are being developed for the ESA Planck mission. These coolers will be combined with passive cooling, a 4.5 K RAL cooler and a Benoit style dilution cooler to provide refrigeration. The resulting cryogen-free mission design will likely prove a pathfinder for many other future astrophysics missions.

ACKNOWLEDGMENT

The research described in this paper was carried out by the Jet Propulsion Laboratory, California Institute of Technology, under a contract with the National Aeronautics and Space Administration.

REFERENCES

[1] B. Collaudin and T. Passvogel, *Cryogenics* 39:157 (1999).
[2] S. Bard, *J. Spacecraft and Rockets*, 21:150 (1984)
[3] S. W. Petrick, and S. Bard, presented at AIAA 26th Aerospace Sciences Meeting, Reno, NV, 11/14/88.
[4] T. T. Cafferty, in: "Spacecraft Radiative Transfer and Temperature Control," T. E. Horton, Editor, 1982.
[5] Alcatel Planck Payload Architect Team, Plank Payload Module Architect Technical Assistance Phase 1 Presentation, 29/04/99 ESTEC, private communication.
[6] M. Prina, P. Bhandari, R. C. Bowman Jr., C. G. Paine, and L. A. Wade, "Advances in Cryogenic Engineering," Vol. 45 (In Press).
[7] A. R. Levy and L. A. Wade, in: "Cryocoolers 10," R. G. Ross, Jr., Ed., Plenum Press, New York, (1999) p. 545.

THERMAL PERFORMANCE OF THE XRS HELIUM INSERT

Susan R. Breon, Michael J. DiPirro, James G. Tuttle, Peter J. Shirron, Brent A. Warner, Robert F. Boyle, and Edgar R. Canavan

Cryogenics and Fluids Branch
Code 552
NASA/Goddard Space Flight Center
Greenbelt, MD 20771

ABSTRACT

The X-Ray Spectrometer (XRS) is an instrument on the Japanese Astro-E satellite, scheduled for launch early in the year 2000. The XRS Helium Insert comprises a superfluid helium cryostat, an Adiabatic Demagnetization Refrigerator (ADR), and the XRS calorimeters with their cold electronics. The calorimeters are capable of detecting X-rays over the energy range 0.1 to 10 keV with a resolution of 12 eV. The Helium Insert completed its performance and verification testing at Goddard in January 1999. It was shipped to Japan, where it has been integrated with the neon dewar built by Sumitomo Heavy Industries. The Helium Insert was given a challenging lifetime requirement of 2.0 years with a goal of 2.5 years. Based on the results of the thermal performance tests, the predicted on-orbit lifetime is 2.6 years with a margin of 30%. This is the result of higher efficiency both in the ADR cycle and in the low temperature top-off, more than compensating for an increase in the parasitic heat load. This paper presents a summary of the key design features and the results of the thermal testing of the XRS Helium Insert.

INTRODUCTION

The Astro-E/XRS instrument has been described in previous papers.[1-6] To summarize briefly, a three-stage cryogenic system is used to cool the X-ray detectors to their operating temperature of 60 mK. The detectors are calorimeters capable of detecting single X-ray photons. When an X-ray strikes the detector, its energy is absorbed and the resulting temperature rise is characteristic of the energy of the X-ray. By operating at 60 mK, the XRS calorimeters are able to measure the energy of the X-rays to an unprecedented resolution of 12 eV over the energy range 0.1 to 10 keV. The cryogenic system includes a solid neon dewar which provides a 17 K environment for the superfluid helium cryostat. The helium bath at 1.3 K is the heat sink for an Adiabatic Demagnetization Refrigerator (ADR), which is capable of operating down to 55 mK, although the nominal operating temperature is 60 mK.

The helium insert, shown in Fig. 1, is made of the helium cryostat, the ADR, and the detector package designated as the Front End Assembly (FEA). The helium insert has completed its performance and launch qualification testing and has been integrated into the

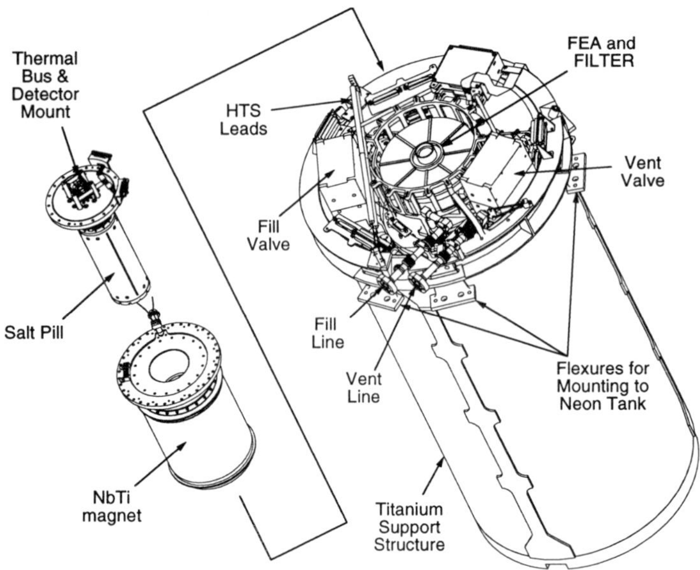

Figure 1. Partially exploded view of the XRS Helium Insert. The ADR salt pill is mounted inside the bore of the magnet; the magnet and ADR in turn are mounted below the FEA.

neon dewar. Based on the results of these tests and verification of the thermal model, the predicted lifetime on-orbit is 2.6 years with a 30% margin. Interestingly, the predicted heat flow into the helium tank is 784 μW—exactly the same as it was when we presented the baseline thermal design at the 1995 Space Cryogenics Workshop.[2] At the time of this writing, the cool-down and initial performance checks of the entire cryogenic system are beginning. The launch of Astro-E is scheduled for February 2000.

HELIUM INSERT THERMAL MODEL

The helium insert thermal model is an Excel spreadsheet finite difference model with 32 nodes and 64 conductors. It is primarily a model of the helium cryostat. The ADR and FEA heat loads on the helium tank are inputs to the model, although the model does include conductive paths from the FEA JFETs to the titanium shield and to the neon tank. The ADR and FEA inputs are based on the results of various tests performed separately on each subsystem and on the integrated helium insert. These tests are described in Reference 6, although this paper presents updated ADR performance data.

The helium cryostat consists of a 32.3 L helium tank supported from a titanium shield by 12 graphite/epoxy straps. A system of copper wire buses thermally couples the 6 forward straps to warm and cold heat exchangers on the vent line. The 6 aft straps are not vapor-cooled. The fill line is mechanically and thermally anchored to the vent line at the warm and cold heat exchangers. Twisted-pair stainless steel wiring is provided for thermometers, heaters, valve microswitches and liquid level detectors in the cryostat and ADR as well as thermometers and detector bias leads in the FEA. The wires are bundled and routed to various destinations within the helium insert, constrained by fine Kevlar strings on an aluminum structure referred to as the "halo." Currents of 0.5 A and greater are carried on high-temperature superconducting (HTS) assemblies. The assemblies are thermally connected to the cold heat exchanger at the transition point between the HTS and low-temperature superconducting (LTS) sections of the assembly.

Table 1 shows a comparison of the thermal model prediction and the measured performance of the cryostat with no FEA or ADR present. As expected, the temperature predictions vary as much as 1 K from the actual temperatures: the model is relatively

Table 1. Comparison between thermal model predictions and cryostat performance

Parameter	Thermal Model	Measured Performance
Neon tank simulator temperature (K) *	17.0	17.0
Helium temperature (K) *	1.34	1.34
Titanium support structure temperature (K)	16.94	16.69
Warm strap temperature (K)	12.09	11.08
Warm heat exchanger temperature (K)	12.03	10.89
Cold strap temperature (K)	5.65	5.06
Cold heat exchanger temperature (K)	5.42	4.82
Heat load on helium tank (μW)	591	629

*Thermal model input parameters

simplistic compared to the actual hardware. The heat load prediction is quite good—within 6% of the measured value. We have not tweaked the model parameters to force agreement with the measured performance.

Comparison to 1995 Baseline Model

In 1995 the conceptual design of the helium insert was complete. The thermal model reflected a best estimate of the detailed design that had been started, but was not yet finalized. Inputs to the model were based on numerous calculations using material properties reported in the literature. The present model has been updated to reflect the as-built hardware, with more detail added to better define thermal paths. Much thermal testing has been completed at the component, subsystem, and helium insert levels and the results of those tests are incorporated in the model. The following sections describe the present model and compare it to the model presented in 1995. Thermal maps from the 1995 baseline[2] and the present model are given in Fig. 2. We found it instructive to review key areas of the thermal model where the initial assumptions were modified, usually because the design in 1995 did not have all the detail necessary to give an accurate prediction. It is fortuitous that increases in some heat loads were counterbalanced by decreases in other heat loads: in 1995 we did not expect to still have a 30% margin available when the testing was complete.

Plumbing and Plumbing Supports. The most significant change since the 1995 thermal model that resulted in an increase in the predicted heat load is in the plumbing and plumbing supports. The helium fill and vent lines are welded assemblies of thin-walled tubing and bellows. The bellows are the dominant thermal restriction in the plumbing. The bellows are also very flexible and the plumbing as a result is not self-supporting. A combination of clamps and Kevlar fibers are used to support the plumbing. The conflat flanges at the warm end of the plumbing are clamped to the structure of the neon shroud. The plumbing is also clamped to the titanium shield. A bellows is installed between the titanium shield and the neon shroud to add compliance in the interface. Copper radiation baffles in the fill and vent lines near the conflat interface intercept radiation coming down the neon tank plumbing. The vent line is supported from the titanium shield by Kevlar fibers attached to the warm heat exchanger. Another set of Kevlar fibers, attached to the vent valve support on the helium tank, supports the vent line at the cold heat exchanger. The fill line is anchored to the vent line at the warm and cold heat exchangers.

The addition of the Kevlar supports to the model adds a small amount of heat to the warm heat exchanger and to the helium tank. More importantly, the as-built geometry has a significantly higher conductance between the titanium shield and the warm heat exchanger than was assumed in the 1995 baseline. The warm heat exchanger temperature has increased from 8.5 K to 11.2 K, and is a dominant factor in increasing the heat load from the fore straps from 22 μW in 1995 to 98 μW today. It is also partially responsible for the increase of the cold heat exchanger temperature from 3.4 K to 4.3 K, which in turn affects the fore strap heat load on the helium tank. The as-built plumbing has a lower conductance between the cold heat exchanger and the helium tank, counteracting the effect of the higher cold heat exchanger temperature on the plumbing heat load.

Straps. Although the heat load down the fore straps has increased significantly, only minor changes have been made to the fore and aft straps themselves. We performed a

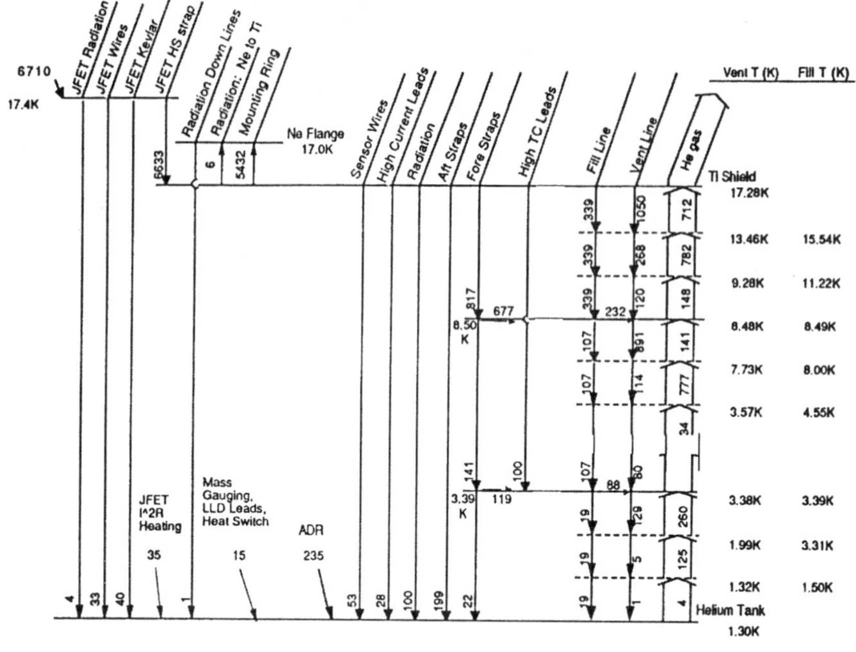

(a) 1995 baseline[2] thermal model

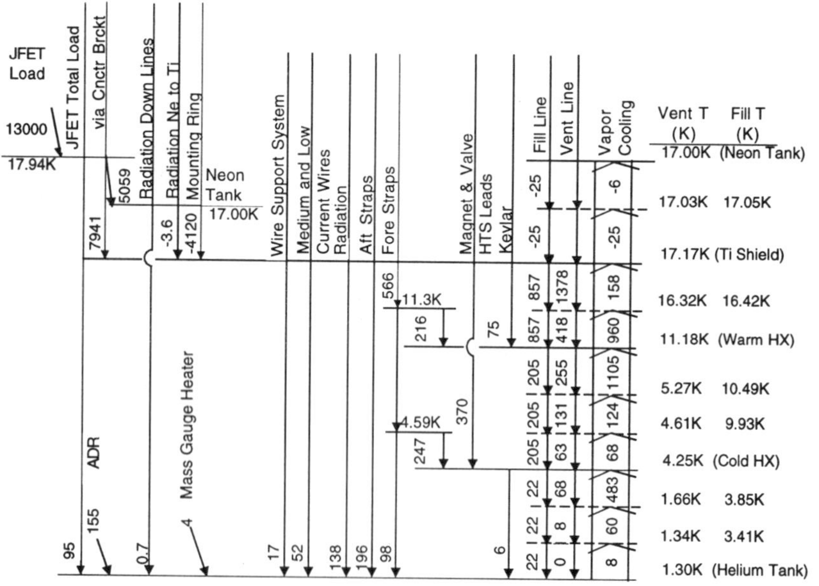

All heat flows in microwatts.

Heat flow into He tank: 784μW
Mass flow: 3.60E-05 g/s

(b) 1999 thermal model

Figure 2. Comparison of 1995 baseline thermal model (a) and present model based on as-built helium insert (b). It is coincidental that the heat flow into the tank is identical in both models.

thermal optimization study of the straps which resulted in the use of slightly longer aft straps than the 1995 baseline. The fore straps did not change dimensions, but the size and placement of the heat sinks were optimized. A system of copper wires are routed from the fore strap heat sinks to the warm and cold heat exchangers on the vent line. The thermal resistance of the copper bus has been added to the thermal model to more accurately reflect the temperature distribution in the fore straps. Another minor change to the model was the use of measured thermal conductivity for the strap material used in the helium insert. The conductivity was very close to that for M50J graphite/epoxy used in the 1995 baseline.

HTS Lead Assemblies. The HTS lead assemblies[7] were found to have a significantly higher thermal conductance than had been estimated in the 1995 baseline. There are two redundant HTS lead assemblies, each with 12 YBaCuO filaments epoxied to a fiberglass tube. The fill and vent valves require 5 leads each to drive the 4-phase stepper motors. The magnet current is carried on the remaining two leads. The assemblies are supported from the titanium shield by Kevlar strings. Heavy copper wires are used to transition from the HTS leads to the neon dewar wiring. The cold end of each of the HTS lead assemblies is heat sunk to the cold heat exchanger by a copper braid. Copper-clad NbTi wires are soldered to the cold ends of the HTS leads to connect to the leads coming from the valves and magnet. A 2.5 cm length of the copper cladding near the NbTi/YBaCuO transition has been stripped away to provide thermal isolation from the helium tank. In 1995, we calculated a heat flow through the HTS lead assemblies of 100 μW, assuming that the YBaCuO leads were the main thermal path. The measured value was 185 μW per assembly, for a total of 370 μW. We attribute the difference to the thermal conductivity of the fiberglass/epoxy tube that supports the YBaCuO leads. This increased heat load is a significant factor in raising the cold heat exchanger temperature from 3.4 to 4.3 K.

One surprising lesson we learned with the HTS lead assemblies was the vulnerability of the NbTi leads to radiative heating. Through careful heat sinking and operational constraints we ensured that the NbTi leads would remain superconducting under nominal operating conditions. By stripping some of the copper cladding, however, the leads were susceptible to off-nominal radiative heating. In one instance, the common lead from the fill valve burned out after a series of valve operations with both valves. It was determined that the leads to the microswitches were probably close to room temperature and had a direct view of the unclad section of the NbTi leads. This problem was solved by reducing the current to the microswitches and by limiting the length of time that the microswitches are powered. Another time, a magnet lead burned out when a probe was inserted to measure the magnetic field. The probe was not properly heat sunk and, again, the NbTi leads had a direct view to high temperature radiation. These events reemphasized the need to ensure that no high temperature radiation sources are visible to the helium insert at any time. Subsequent tests showed that the NbTi leads in the HTS lead assemblies remained superconducting and capable of carrying 2.5 A in a radiation environment of 45 K: the warmest radiation environment the leads are expected to see in normal operations is 30 K.

Wiring and Wiring Supports. The 1995 baseline model represented the wiring in several different ways. In addition to the HTS lead assemblies, there were low current "sensor wires" and "high current leads" as conductors in the model. The effect of Joule heating in the wires was added in with the time-averaged mass gauge heater input. For the present model, we have constructed a wiring submodel to account for the combined effects of conduction, Joule heating, and, radiation exchange between the wires and the environment. The submodel includes actual cross-sectional areas and lengths of the wires. The output of the wiring submodel is represented in the helium insert model by the "medium and low current wires" conductor. All wiring that runs from the titanium shield to colder temperatures is stainless steel twisted pair wire. The smallest wire is 42 AWG (64 μm diameter). Larger wire is used for the liquid level detectors, ADR heaters, and mass gauge heaters to accommodate the higher currents to these devices.

The present thermal model includes a term for the wiring supports. Wire bundles are tied to Kevlar strings to route the wires throughout the helium insert. The Kevlar strings are attached to an aluminum structure that is mounted on the helium tank. The actual geometry of the wires and Kevlar strings is too complex to model completely. Since the Kevlar is a very good thermal isolator, we have used a simple assumption that half of the Kevlar strings are held at the temperature of the titanium shield at their midpoint, and that heat is conducted to the helium tank. In practice, we attempted to separate "warm" wires from "cold" wires by choosing separate strings or tying warm wires near the center of the strings and cold wires closer to the ends, which are thermally anchored to the helium tank.

The heat load from the mass gauge heater remains a direct input to the model. The resistance of the mass gauge heater has been increased from 1000 Ω to 3694 Ω, with a corresponding decrease in the current to keep the total energy input to the tank the same as the 1995 baseline, 10 J. The mass gauge is assumed to be operated monthly.

Radiation. Radiation remains one of the dominant heat loads on the helium tank. In both the 1995 baseline and the present model, the neon tank, neon shroud, titanium shield, and most of the helium tank are assumed to have an emissivity of 0.05. The top of the helium tank—where the valves, wiring, plumbing, ADR, and FEA are mounted—is assumed to have an emissivity of 0.08. To achieve low emissivities, the neon tank, neon shroud, and cylindrical sections of the helium tank are polished aluminum; the titanium cylinder is gold-plated; and larger surfaces on the top of the tank are gold-plated (FEA, ADR flange, halo) or covered with aluminized mylar (fill and vent valves). It was difficult to quantify the relative radiative versus conductive heat loads, even by changing the simulated neon tank interface temperature in the test dewar. However, the thermal model predictions matched the test results well enough to demonstrate that the radiative term in the model is reasonable.

Film Suppression and Getter. Two important design changes were implemented subsequent to the 1995 baseline model that resulted in no changes to the thermal model. In both cases, the design changes were needed to mitigate potential heat loads that had been overlooked in the original design. The most critical change was the suppression of film flow in the vent line. Initially, the helium insert used a 2.54 cm diameter sintered stainless steel porous plug. Calculations showed that conditions downstream of the porous plug would be such that some of the vapor would recondense inside the vent line. The resulting heat of condensation and conductance through the film would add a heat load on the helium tank in excess of 1 mW—more than the total thermal budget. In fact, the measured heat load through a film in the fill and vent lines with the valves open is 1.42 mW.[6] The solution to this problem is described in detail in Reference 8. To summarize, the porous plug diameter has been reduced to 7.6 mm and a heat exchanger has been added downstream of the porous plug. The larger pressure drop in the porous plug and the rise in temperature at the heat exchanger change the thermodynamic conditions such that the helium does not recondense. In case the modification were not 100% effective, a knife-edge film suppressor and an orifice were added downstream of the heat exchanger to prevent flow of film to warmer sections of the plumbing. Tests at the component level and helium insert level indicate that the film has been totally eliminated.

The other design change was to add a carbon/carbon composite getter to the helium tank. A helium gas pressure as low as 10^{-8} torr in the vacuum space would result in a heat leak of 25 μW to the cryostat.[9] The getter is sized to adsorb 1.5 standard liters of helium gas. The getter is oversized for its primary task, which is to adsorb the helium gas permeating through the Vespel shell of the ADR heat switch at room temperature. Other sources of helium gas include permeation through rubber o-rings during ground servicing and small leaks in the system that are undetectable at room temperature. The entire cryogenic system has been leak-checked at room temperature and found to have no detectable leaks at 10^{-10} standard cubic centimeters per second.

FEA. The cold electronics for the detectors include JFET amplifiers which operate at 124K. The heat input required to maintain the JFETs at 124 K is 13 mW, nearly double the budget in the 1995 model. Despite this increase, the JFET heat load on the helium tank has decreased slightly, from 112 μW to 95 μW. To reduce the heat load on the helium tank, wires that connect the JFETs to the dewar wiring are no longer heat sunk at the helium bath. In the thermal model, the heat load from the JFETs to the helium tank is extrapolated from measurements of the cryostat parasitic heat loads with and without the FEA. The extrapolation is necessary to account for the increased vapor cooling. The model has also been modified to reflect the connection of the JFET thermal strap and wires to a connector bracket mounted on the titanium shield and the presence of an additional thermal strap from the bracket to the neon tank.

ADR. The counterbalance to the various increases in the heat load on the helium tank comes from the improved efficiency of the ADR. The ADR heat load on the helium tank is a composite of three different energy sources, time-averaged over the ADR cycle. The largest amount of energy comes from the salt pill itself. When operating, the salt pill intercepts heat from the heat switch, Kevlar supports, and wiring to maintain the detectors at 60 mK. On-orbit, the salt pill will also be subject to a small amount of heating from cosmic rays. Heat input from the detectors is negligible. When the salt pill can no longer absorb energy and maintain a constant temperature, the magnet is charged and the heat switch closed. The energy that has been absorbed by the salt pill is deposited in the helium bath. Once the magnet is fully charged and the salt pill is within 50 mK of the bath temperature, the switch is opened and the magnet discharged until the salt pill reaches its operating temperature.

In 1995, it was estimated that the salt pill would release 10 J into the bath on each cycle with the detectors operating at 65 mK. The heat switch was calculated to produce 6.3 J, and hysteresis in the magnet was estimated[10] to be 3.4 J for a total of 19.7 J. In various performance tests we have measured the magnet hysteresis energy, 6.8 J, and the total energy added to the helium bath during an ADR cycle, 20.4 ± 1 J with the ADR operating at 60 mK. Energy from the heat switch is easily calculated from the current profile and heater resistance to be 2.7 J. The remaining 10.9 J are attributed to the salt pill. The improved efficiency results from a significant increase in the hold time of the ADR, estimated to be 35.5 hours on-orbit compared to 24.6 hours in 1995. The time to cycle the ADR remains 1 hour.

The differences in performance since the establishment of the baseline can be attributed mostly to changes in the ADR heat switch. During the initial tests of the helium insert, the ADR performance showed significant degradation each time it was cycled to room temperature and cooled back down. The problem was traced to the formation of a ^{4}He film in pockets that developed between the Vespel shell and an outer layer of titanium foil.[11] A two-pronged approach was followed to solve this problem. The first step was to replace the ^{4}He with ^{3}He, which will not condense in the pockets. In retrospect, ^{3}He is a better choice overall because it has a higher thermal conductivity when the switch is on and desorbs from the getter at a lower temperature. As a result, less heat is required to operate the switch and the switch is on for a shorter period of time. The second approach was to improve the technique for wrapping the foil on the Vespel and to use a different epoxy, Scotchweld 2219, which has a lower thermal conductivity than Armstrong A-12. Improving the wrapping of the foil reduced the off-conductance of the heat switch by approximately 15%, which increased the salt pill hold time. We have observed an anomalous behavior in the heat switch when it is first operated after being cooled down, as described in more detail in Reference 11. Once the behavior is eliminated, the heat switch performs properly and the anomaly is not expected to affect on-orbit behavior.

Some of the ADR performance improvements can also be traced to the salt pill. The salt pill cooling power was calculated very conservatively assuming a 70% fill fraction for the salt pill. All salt pills fabricated have had fill fractions in excess of 95%. The flight salt pill has an estimated fill fraction of 99% as well as a greater salt pill volume due to changes in the container design. The result is a longer hold time.

The magnet design is unchanged, so the discrepancy between the calculated and measured heat load is unexplained. The measurement was performed with no ADR or FEA present using the same calorimetric technique as the mass gauge[12] and the magnet ramp rates were varied to rule out contributions from eddy current heating.

In the 1995 heat load calculation, a 10% margin was added to the ADR contribution to account for interruptions in the ADR cycle before the cycle was complete. Shortening the cycle might be done to meet command and telemetry requirements or to take advantage of occultation periods in the Astro-E orbit. This margin has been eliminated in the present estimates. Since the salt pill is the dominant source of heat, cutting the cycle short by 10% increases the time-averaged ADR heat load by less than 5%, from 155 μW to 162 μW, which has a negligible impact on the helium lifetime.

One important assumption in estimating the ADR heat load is that the bath temperature is 1.3 K. The on-orbit bath temperature could be even lower, between 1.25 and 1.28 K. To achieve this low temperature, however, it is important to pump the helium tank to as low a temperature as possible prior to launch and to have the bath temperature rise as heat is introduced from the ADR and FEA operations. If the bath is relatively warm at launch and must cool down to its operating temperature, the porous plug will not be fully wet and the helium will cool slowly. If the ADR and FEA are operated while the porous plug is still in the hysteretic range, the bath may not ever reach 1.3 K.

Lifetime Calculation

Even though the heat flow into the helium tank in the present model is identical to that in 1995, the predicted helium lifetime is different—2.46 years in 1995 versus 2.60 years in 1999. The difference is not in the size of the helium tank, which has remained unchanged. Instead, practice with the engineering model dewar and cryostat demonstrated that a higher fill fraction can be achieved during the low temperature top-off than originally predicted. In 1995, it was assumed that the ullage on-orbit at the start of the mission was 15%. We revised that in the 1999 model to 10%. Actual operations on the engineering model hardware have demonstrated the possibility of reaching as low as 6.5% ullage.

CONCLUSION

Historically we have maintained a 30% margin on the lifetime calculations for the XRS helium insert. The 2.60-year prediction includes this margin, as did the 2.46-year prediction in 1995. The purpose of the margin was to provide some flexibility in accommodating errors in the helium insert design as well as uncertainties in the neon dewar interfaces and on-orbit conditions. The 30% margin is equivalent to approximately 350 to 400 μW, so there is still not much room for error. With the completion of the testing of the helium insert, and the successful operation of the ADR in the flight dewar, many of the uncertainties have been eliminated. The thermal interfaces between the dewar and the helium insert still remain to be measured, but the performance of the engineering model dewar gives us confidence that the thermal interfaces have been conservatively modelled. It is predicted that the helium insert will meet its goal of 2.5 years on-orbit, unless the neon runs out first.

REFERENCES

1. S.R. Breon, H.D. Branch, G.J. Blount, M.L. Jackson, R.F. Boyle, J.G. Tuttle, Design of the XRS Helium Insert, *Adv in Cryo Engr* 41:1129 (1996).
2. S.R. Breon, J.A. Gibbon, R.F. Boyle, M.J. DiPirro, B.A. Warner, and J.G. Tuttle, Thermal design of the XRS helium cryostat, *Cryogenics* 36:773 (1996).
3. S.M. Volz, K. Mitsuda, H. Inoue, Y. Ogawara, M. Kyoya, M. Hirabayashi, The X-Ray Spectrometer (XRS)—A multi-stage cryogenic instrument for the Astro-E x-ray astrophysics mission, *Cryogenics* 36:763 (1996).
4. A.T. Serlemitsos, M. SanSebastian, and E. Kunes, Final design of the Astro-E/XRS ADR, *Adv in Cryo Engr* 43:957 (1998).
5. A.T. Serlemitsos, M. SanSebastian, and E. Kunes, Design and performance of a spaceworthy ADR, Accepted for publication in *Cryogenics*.
6. S. Breon, P. Shirron, R. Boyle, E. Canavan, M. DiPirro, A. Serlemitsos, J. Tuttle, and P. Whitehouse, The XRS low temperature cryogenic system: ground performance test results, Accepted for publication in *Cryogenics*.
7. J.G. Tuttle, T.P. Hait, R.F. Boyle, and S.R. Breon, A high Tc superconducting current lead assembly for the XDS helium cryostat, *Adv in Cryo Engr* 43:965 (1998).
8. P.J. Shirron and M.J. DiPirro, Suppression of superfluid film flow in the XRS helium dewar, *Adv in Cryo Engr* 43:949 (1998).
9. R.J. Corruccini, Calculation of gaseous heat conduction in dewars, *Adv in Cryo Engr* 3:353 (1960).
10. Y.M. Eyssa and S.W. Van Sciver, Analysis of AC losses in a superconducting magnet for adiabatic demagnetization refrigeration, Final Project Report for NASA/Goddard Space Flight Center Grant NAG5-2879.
11. E.R. Canavan and J.G. Tuttle, Performance of the XRS ADR heat switch, To be presented at the 1999 Cryogenic Engineering Conference
12. M.J. DiPirro, P.J. Shirron, and J.G. Tuttle, Mass gauging and thermometry on the Superfluid Helium On-Orbit Transfer flight demonstration, *Adv in Cryo Engr* 39:129 (1994).

DEVELOPMENT OF THE FACET CRYOSTAT

Alfred Nash[1], Peter Shields[2], Munir Jirmanus[2], and Zuyu Zhao[2]

[1]Jet Propulsion Laboratory, California Institute of Technology
Pasadena, CA 91109

[2]Janis Research Company
Wilmington, MA 01887

ABSTRACT

A "proof of concept" prototype cryostat has been developed to demonstrate the ability to accommodate low temperature science investigations within the constraints of the Hitchhiker siderail carrier on the Space Shuttle. The Fast Alternative Cryogenic Experiment Testbed (FACET) hybrid Solid Neon - Superfluid Helium cryostat has been designed to accommodate instruments of 16.5 cm diameter and 30 cm length. In this paper the design requirements, the implementation experiences and test results will be discussed.

INTRODUCTION

The start of the build era for the International Space Station (ISS) has resulted in the end of regularly scheduled microgravity science opportunities such as the United States Microgravity Payload (USMP) missions on which the Lambda Point Experiment (LPE) and Confined Helium Experiment (CHeX) were performed. In addition, the ISS is not scheduled to be completed enough for the planned Low Temperature Microgravity Physics Facility (LTMPF) to conduct experiments until 2003 at the earliest. This situation creates a backlog of selected low temperature flight definition experiments. It also compromises the ability to conduct any incremental tests of scientific or technological concepts in microgravity until after the start of the Space Station era.

To address this gap in manifest opportunities, several approaches were investigated, alternate carriers for the existing Low Temperature Platform (LTP) cryostat[1], an early flight of the LTMPF, and FACET. The mass and volume of the LTP and LTMPF constrain these platforms to cross bay carriers that are not in the baseline shuttle manifest. In the case of an ISS schedule slip, the carriers most likely to be manifested are pressurized modules, such as SPACELAB, which are not compatible with crossbay carriers. Also, an earlier flight of the planned LTMPF requires an accelerated funding schedule for early completion of that facility.

The Fast Alternative Cryogenic Experiment Testbed (FACET) project was a one year proof of concept study to demonstrate, through the design, construction, and test of a prototype, the feasibility of flying cryogenic payloads aboard the Space Transportation System (STS) (a.k.a. the space shuttle) during the ISS build era. This paper describes the development of the cryostat. The remainder of the payload and its development is described elsewhere[2]. This paper will describe the objectives, requirements, and constraints for the cryostat and its development, the design approach, and the results of the prototype hardware development.

Table 1 Cryostat Performance Requirements

Parameter	Requirement
On Orbit Lifetime†	$\geq$ 6 days
Bath Temperature	< 2.17 K
Instrument Volume	> 4.7 Liters

† Predicted flight system performance after first launch opportunity

OBJECTIVES AND REQUIREMENTS

The ultimate objective of the FACET project is to produce a simple, low cost, facility providing frequent flight opportunities before the availability of the Low Temperature Microgravity Physics Facility (LTMPF) for existing flight definition Principal Investigators. The prototype was to demonstrate, within tight schedule and cost constraints, the feasibility of a flight system by the test of ground hardware of which the technical approach could be used to develop low cost flight hardware. It was desired that the flight system should be compatible with multiple reflights, each capable of supporting a different investigation.

In this development, cost and schedule were the driving constraints, with technical scope and performance secondary. The performance requirements, shown in Table 1, were derived from minimum mission requirements negotiated with the backlogged science investigators or their representatives. Among the features considered, but not included in the prototype due to the development constraints, were porous plug phase separators, and motor driven cold valves.

Early in the development, a carrier trade study identified the Hitchhiker Siderail (HH-S) carrier as having the optimal mass, volume, telemetry, ease of integration and manifesting opportunities consistent with the needs of the low temperature microgravity fundamental physics program. Hitchhiker siderail carrier payloads have flown an average of four times a year. They have flown with a multitude of other payloads, including a TDRSS satellite deployment, MIR servicing missions, SPACELAB missions and the USMP missions. During the station build era, the hitchhiker office has an agreement to fly on a "mass available" basis. In fact, several Hitchhiker Siderail payloads flew during the mission which deployed the first US module of the ISS, the Unity module, and the first servicing mission.

Baselining the Hitchhiker Siderail (HH-S) as the carrier placed challenging constraints on the design[3]. The body of the cryostat needed to fit within an envelope 0.64 X 0.64 X 0.99 m. The total mass, including siderail mounting hardware, had to be less than 880 kg. The fundamental structural frequency of the cryostat had to exceed 35 Hz.

Baselining the Hitchhiker Siderail (HH-S) as the carrier also placed operational constraints on the design. Probably the most design driving constraint is the ability of the cryostat to safely operate unattended for at least 65 hours, but as long as 161 hours (for a 96 hour launch window), before launch. In addition, the Hitchhiker Siderail (HH-S) carrier does not have "T0" power during the aforementioned 65 - 161 hour period, precluding the use of a vacuum pump to keep the helium within the cryostat superfluid as was done with the LTP. The orientation of the cryostat during servicing on the launch pad was also dictated by the choice of carrier.

DESIGN APPROACH

In early feasibility analyses, it was decided that, given the volume constraints, and the conducted heat load from the electrical leads and plumbing associated with a typical low temperature microgravity fundamental physics instrument, it was unlikely that a cryostat that relied on liquid helium alone would meet the lifetime requirements. Therefore a hybrid approach was chosen.

Although cryocoolers were initially considered for interception of the conducted heat loads to the helium due to the instrument, there were eventually abandoned due to concerns over induced vibration, and the size of the radiators required to reject the waste heat.

Instead, solid Neon was chosen as a "guard" cryogen for a variety of reasons. The temperature of solid Neon ($\leq$ 24K) is much lower than that of the more commonly used cryogen Nitrogen (65 K). The radiation heat load on the Helium reservoir is thus nearly 2 orders of magnitude smaller. Solid neon has an appreciable heat of sublimation (~ 105 Joule/gm). Solid neon is very dense (1.444gm/cc) and will therefore not occupy too much space. The stresses created during solidification from the liquid state do not deform typical metal tanks. In addition, neon is much safer than hydrogen.

Another alternative considered, but ultimately abandoned, was the concept of designing a cryostat to operate *within* a Get Away Special (GAS) can. Ultimately, the volume and mass constraints proved too restrictive for the performance objectives. Instead, the design is to make the vacuum shell identical to the GAS can. This allows direct mounting of the cryostat to the siderail using existing, flight qualified, hardware. In turn, we then imposed the same requirements on the interior elements of the cryostat that are levied on GAS payloads[3].

Throughout the design phase, projections of performance were accomplished with the aid of a self consistent spreadsheet model[4]. The model solves for the temperatures of 6 nodes assuming values for the outer shell, neon reservoir, and helium reservoir temperatures. The model also included vapor cooling, when appropriate, using the product of the vapor specific heat and the temperatures of adjacent shields. The model is iterated until the node temperatures "relax" consistent with the temperature dependent thermal links.

The thermal model results indicated that without sufficient helium boiloff, it would be impossible to keep the neon from partially melting between last servicing and launch, even if the neon was cooled to below 10K at the last servicing.

Moreover, the thermal analysis also indicated, that due principally to the small thermal mass of helium within the cryostat, it would be impractical to rely on thermal stratification to provide sufficient boiloff while maintaining some helium in a superfluid state between the last servicing and launch[5].

However, thermal modelling did indicate that even with the nominally 40% loss of helium volume pumping down from 4.2 K to below 2 K, the on orbit lifetime requirements could be met if the heat load to the helium during on orbit operations was significantly less than during the period between last servicing and launch. Therefore the baseline operating scenario chosen was, during the time between last servicing and launch, to have the helium vent at its normal boiling point, and the neon to be held, subatmospheric, below its triple point. Once on orbit, the neon would begin sublimating to the vacuum of space, and the helium would be pumped superfluid via a phase separator of the type developed for SHOOT[6], which would be fed by Liquid Acquisition Devices (LADs)[7], before switching over to steady state operation with phase separation via a "standard" sintered stainless steel porous plug. The helium boiloff rate during the time between last servicing and launch would be maintained with a small battery powered heater, which once on orbit, would be turned off, decreasing the heat load to the helium.

The thermal model results also aided in the optimization of on orbit lifetime via the relative sizing of the helium and neon reservoir volumes. Estimates of the helium boiloff necessary to keep the neon frozen during the time between last servicing and launch, estimates of the efficiency of pump down from normal to superfluid, and estimates of the heat loads during steady state on orbit operation eventually lead to a choice of a helium reservoir nearly half again as voluminous as the neon reservoir.

CRYOSTAT DESIGN

A cross section of the FACET cryostat is shown in shown in Figure 1. The cryostat is a hybrid solid neon - liquid helium with "folded tube" G-10 supports. There exist field joints in all vapor cooled shields at the same axial location as the cold flange joint to aid in instrument integration. The shield closure plates and warm flange (collar) interface are modular to accommodate a wide variety of instrument input/output. The instrument cavity has an interior diameter of 16.51 cm and a depth of 30.48 cm. The instrument cavity has a vacuum independent of the cryostat vacuum when sealed with the instrument cold flange (and pumped through an instrument provided instrument guard vacuum vent). The instrument cavity is surrounded by an annular 15 liter (+ 2 liter ullage volume) liquid helium reservoir. Heat transfer from the instrument to the helium is accomplished via conduction through the (annular) reservoir's walls (i.e., the instrument is in a "dry well"). Axially displaced from the helium reservoir is the 9.5 liter (+ 1.5 liter ullage volume) solid neon reservoir. Anchored within the solid neon reservoir is an aluminum foam[8] to aid in heat transport within the neon.

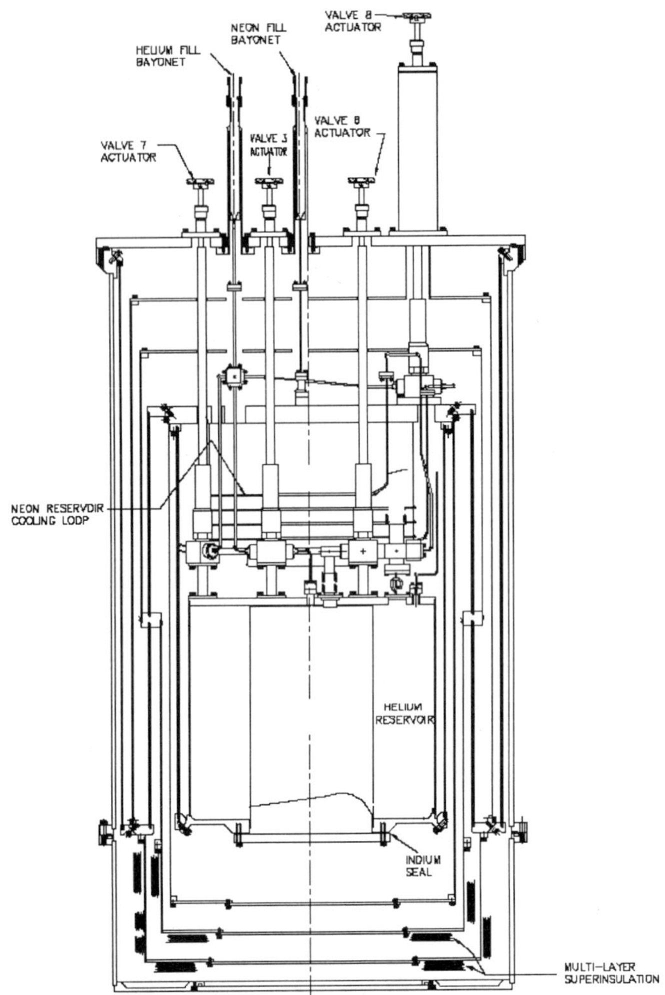

Figure 1 Cross section of FACET cryostat

The cryogen tanks, as well as the rest of the cryostat internal elements, are structurally, as well as thermally separated by "folded tube" G-10 supports. There are two vapor cooled shields, in addition to the shield attached to the neon reservoir. There exist field joints in all vapor cooled shields at the same axial location as the instrument cavity cold flange joint to aid in integration. The shield closure plates and warm flange (collar) interface (not shown) are modular to accommodate a wide variety of instrument input/output. The prototype was built with two vent lines from the helium reservoir, one for each of the type of phase separators that would be necessary for a flight version of the cryostat. The vent lines from the solid neon reservoir, as well as from the helium reservoir, were anchored to the vapor cooled shields.

The valving of the interior cryostat manifold allows for the precooling of the lines and vapor cooled shields before topping off the helium, thus avoiding undesired increases in instrument temperature. A bypass in the helium cryogen manifold creates a cooling loop for solidifying the neon. For the prototype, the neon was solidified after filling the reservoir with liquid at its normal boiling point. Due to tight schedule and cost constraints, the cryogenic valves used for the helium manifold in the prototype cryostat were manually actuated. For a flight cryostat, stepper-motor actuated valves similar to the ones used on

SHOOT[9] would be used. Determining the feasibility of the flight design to meet on orbit lifetime objectives from prototype's demonstrated lifetime thus involves taking into account the difference in heat load due to the cryovalves for the prototype and for the flight design. In the prototype, the majority of the heat leak due to the cryovalves is from the radiated heat leak around the valve actuators. In the flight design, the majority of the heat leak due to the cryovalves is from the conducted lead resistance, and is much less than the radiated heat leak in the prototype. The thermal model was used to select the number of layers of multi-layer insulation[10] so that radiation was comparable in heat leak to other sources. The cryostat has Germanium Resistance Thermometers and Silicon Diode thermometers throughout for the monitoring of housekeeping data. The pressure within the helium and neon reservoirs is also monitored. A heater for mass gauging, as well as a commercial superconducting transition level gauges were installed in the helium reservoir. The as built prototype has a dry weight (excluding vacuum shell) of 82 kg. This is 6 kg below the requirement, including the allocated mass for cryogens (8.6 kg), instrument (15kg) and external manifold (1.8 kg). The prototype design provides for 5.6 liters of instrument volume, exceeding the requirement by 19%. A finite element analysis of the design using NASTRAN was performed. The lowest natural frequency (lateral mode) of the internal cryostat structure, at 48 Hz, is greater than the 35 Hz requirement, and close to the goal of 50 Hz.

TEST AND ANALYSIS

Projections for the performance of a flight cryostat are achieved using a numerical model that reproduces the performance of the ground prototype. All radiative and conductive heat flow path resistances are temperature dependent. Heat flow through MLI was modelled using the empirical formula[10] and a conservative loft of 24 layers/cm . The effects of MLI penetrations were included by adding a 1/4 inch perimeter of blackbody coupling to adjacent shields around each penetration and field (assembly) joint. The emissivity on the LHe & SNe cooled surfaces (a single layer of aluminized mylar) was taken at a conservative 0.03.

In this conservative model, the radiated heat loads due to nearly unavoidable imperfections at the penetrations in the MLI are nearly an order of magnitude larger than the heat loads through the rest of the blanket on both the OVCS & IVCS. We note that in the thermal model developed during the CHeX project for a conventional helium dewar[11], the dominant heat load (by nearly two orders of magnitude) to the helium reservoir was conduction along the supports. By the incorporation of a solid neon reservoir, the dominant heat load to the helium reservoir becomes "stray radiation". Although controlling radiation leaks to the helium reservoir will ultimately extend the total lifetime of the cryostat, it may become necessary to incorporate the activation of a heater during the launch hold to ensure adequate vapor cooling.

Measurements of ground performance during simulated on orbit operation agreed well with numerical modelling of the cryostat. The model predicted temperatures and heat flow data inside the prototype cryostat during simulated on orbit operation are shown in Figure 2. The numbers in parenthesis are the actual data from simulated operation. Note that the heat flow into the helium tank is dominated by radiation of 36 mW (from leaks around the cold valve actuators). On orbit, vapor cooling from both the solid neon and the liquid helium can be used. With the addition of Neon vapor cooling, the on orbit heat load to the neon is less than half of heat load to the neon during the launch hold.

In addition, the model has been used to estimate the "heater power" necessary to keep the neon solid during the launch hold (It's ~49 mW for a total of ~ 58.5 mW). We have also determined the efficiency of the conversion from normal to superfluid in the test to be ~57%. From these numbers and the temperature dependent latent heat of helium we calculated the estimated lifetimes.

To predict the on orbit performance of a flight cryostat, first the decreased temperature of the outer shell needs to be taken into consideration. In the cargo bay of the space shuttle, the vacuum shell of the cryostat drops to temperatures (which vary depending on orbiter attitude) around the freezing point of water (based on data from the LPE and CHeX missions). Secondly, the proposed flight cryostat would utilize the commercial off the shelf (COTS), flight qualified, stepper motor driven cryovalves for the internal cryostat manifold. These valves were first used on SHOOT and have subsequently been successfully used in many other space qualified cryostats. Without the 36 mW of "stray" radiation to the

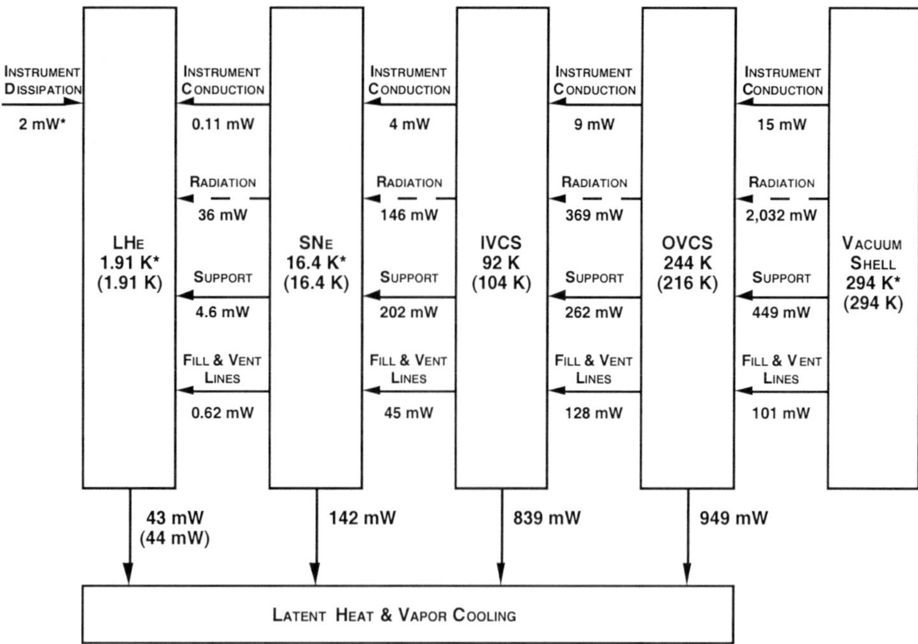

Figure 2 Steady State Model Predictions. Prototype test data are given in parenthesis. Model inputs are marked with an asterisk (*).

helium reservoir, the total heat load to the neon increases, but the helium lifetime increases dramatically. Without the stray radiation, the principal heat loads to the helium reservoir consist of two sources: conduction through the structure from the neon reservoir, and internal dissipation due to the operation of the instrument (see Figure 3). In this scenario the cryostat lifetime is limited by the neon reservoir. Since the heat load to the neon reservoir in this scenario is more than double the heat load in the ground simulation, the neon lifetime would "decrease" to 15 days. With an order of magnitude less heat load to the helium than in the prototype cryostat, orbit helium lifetime could have been as long as 21 days.

The result is that the predicted helium lifetime, for the proposed flight cryostat, depending on whether the launch is at first (65 hours) or last (161 hours) opportunity, is 29.5 or 6.5 days (respectively).

We therefore conclude that it would seem more than feasible, with mechanisms in place to assure adequate helium vapor cooling on the launch pad, and to minimize radiation heat loads to the helium reservoir, to construct a flight cryostat that could provide liquid helium cooling to a science instrument for the full duration of even the longest shuttle missions (16 days).

SUMMARY

The development of the FACET prototype cryostat has proven the feasibility of a multi-use, simple, low cost, facility to accommodate low temperature cryogenic payloads within the constraints of the Hitchhiker siderail carrier on the Space Shuttle. Such a facility could provide frequent flight opportunities during the build era for the International Space Station.

All requirements (performance, hitchhiker payload and development constraints) were met or exceeded. Test data from the prototype indicate that flight cryostat that could provide superfluid liquid helium cooling to an instrument for the full duration of even the longest shuttle missions (16 days).

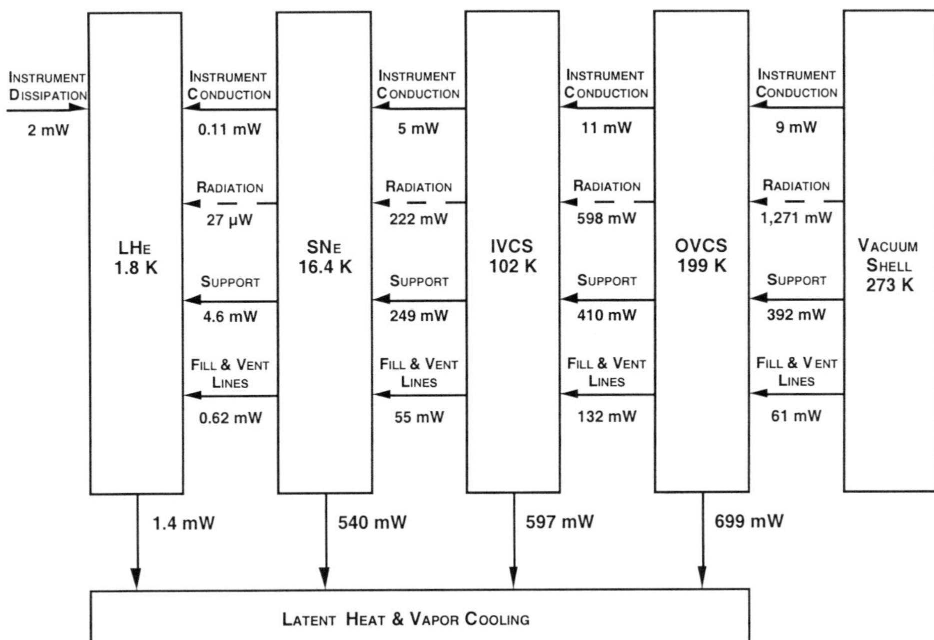

Figure 3 Predicted steady state heat flows for a flight FACET cryostat

ACKNOWLEDGMENT

This work was carried out by the Jet Propulsion Laboratory, California Institute of Technology under contract to the National Aeronautics and Space Administration. The work was funded by NASA Microgravity Research Division.

REFERENCES

[1] D. Petrac, U. Israelsson, and T. Luchik, The Lambda Point Experiment: Helium Cryostat, Cryoservicing, Functions and Performance", *Advances in Cryogenic Engineering*, 38, (1993).
[2] A. Nash, , P. Sheilds, M. Jirmanus, Z. Zhao, R. Abbott, and W. Holmes, The Fast Alternative Cryogenic Experiment Testbed, *To be presented at the 1999 Space Cryogenics Workshop*.
[3] HHG-730-1503-07, Hitchhiker Customer Accomodations & Requirements Specifications, 1994.
[4] A. Nash, Simple Thermal Spreadsheet Models for Cryogenic Applications, *Advances in Cryogenic Engineering*, 41, (1996).
[5] P.J. Shirron and M.J. DiPirro, Low Gravity Thermal Stratification of Liquid Helium on SHOOT, *Cryogenics*, 32, (1992). and J. Tuttle, M.J. DiPirro and P.J. Shirron, Thermal Stratification of Liquid Helium in the SHOOT Dewars, *Cryogenics*, 34, (1994).
[6] P.J. Shirron, J.L. Zahniser, and M.J. DiPirro, A Liquid/Gas Phase Separator for He-I and He-II, *Advances in Cryogenic Engineering*, 37, (1992).
[7] P.J. Shirron, M.J. DiPirro and J. Tuttle, Flight Performance of the SHOOT Liquid Acquisition Devices, *Cryogenics*, 34, (1994).
[8] 5083 Aluminum foam from ERG Materials Aerospace Corporation.
[9] R. H. Haycock, A cryogenic Vacuum Valve for Use in Space, Report CSE 87-007, *National Technical Information Service*, (1987).
[10] T.C. Nast, A review of Multilayer Insulation, Theory, calorimetry measurements, and applications, *Recent Advaces in Cryogenic Engineering*, 267, (1993).
[11] T.S. Luchik, U.E. Israelsson, D. Petrac and S. Elliott, Performance Improvement of the CHeX Flight Cryostat, *Advances in Cryogenic Engineering*, 41, (1996).

A PULSE-TUBE CRYOCOOLER FOR THE SKA RADIO TELESCOPE

X. Deng, P. Mayzus, L. Bauwens, O. R. Fauvel and G.T. Reader

Department of Mechanical and Manufacturing Engineering
University of Calgary
Calgary, AB, T2N 1N4
Canada

ABSTRACT

A pulse-tube cryocooler prototype is being built as a proof-of-principle for the Square Kilometer Array radio telescope project. In this paper, the cryocooler design code is described. The pulse-tube cryocooler model formulation, including the machine description with orifice and reservoir, and the approximations made to the conservation equations are discussed. A one-dimensional model is adopted making use of empirical models for viscous losses and convective heat transfer with the heat exchanger walls and the regenerator matrix. The numerical solution is based upon a Godunov-like algorithm, consisting of a Lagrangian half step followed by reconstruction on an Eulerian grid. The need for such an algorithm suitable for rapid changes results from the presence of sudden changes in cross section in the machine and a discretisation including relatively large volumes and also much smaller ones. Some simulation results are presented.

BACKGROUND

In August 1994, NRC's (National Research Council of Canada) Herzberg Institute of Astrophysics (HIA) sponsored an international workshop to take the first step in identifying future trends in radio astronomy. The Square Kilometer Array (SKA) was recommended as the next generation radio telescope for which Canada will provide national facility support within an international collaboration [1]. The SKA telescope will have a collecting area of one square kilometer, making it one hundred times more sensitive than the Very Large Array (VLA), the benchmark instrument at those wavelengths of decimeter and centimeter.

The SKA will consist of an array of radio antennas with a total aperture of $10^6 m^2$. Implementation of the SKA will require considerable technological advances compared with current state-of-the-art. The telescope sensitivity requirement is one of the most important specifications and can be only compromised to a limited extent. The most important noise sources are the input amplifier noise, losses within the antenna and the path to the amplifier,

Advances in Cryogenic Engineering, Volume 45.
Edited by Shu *et al.*, Kluwer Academic / Plenum Publishers, 2000.

noise from the sky, and blackbody radiation from the ground. The low background noise means that at microwave frequency receivers should be designed with a minimum amplifier noise, losses, and spillover. In practice, this can be obtained by implementing cryogenic cooling for crucial amplifier stages [1].

Requirements for the receiver cooling system are [1]: 1) to cool to less than 80 K, 2) a small size, 3) closed cycle system, 4) low vibration, 5) low maintenance, 6) lifetime more than 3 years. Pulse-tube refrigeration is the technology that appears to be the most suitable. The SKA receiver system will require an inexpensive distributed cooling system. This is due to the distributed nature of phased arrays and the need for low electromagnetic noise which forces the low-noise amplifiers to be placed as close as possible to the antenna elements. However, the requirement for cooled amplifiers in a phased array appears to be difficult to meet because of the large physical size of such an array. For the SKA telescope, there are about 400 separate amplifiers to be cooled. After careful consideration and comparison on techniques and costs, the following cooling system arrangement is recommended: all the receivers are cooled by a distributed liquid nitrogen, while the liquid nitrogen is generated by a few pulse-tube cryocoolers with sufficient cooling capacity.

The objective of the ongoing research is then to establish the feasibility of pulse-tube technology as the approach of choice for cooling the receivers of the SKA radio telescope. This entails designing, building and testing a proof-of-principle prototype of an efficient, lightweight and reliable orifice-type pulse-tube refrigerator that can be incorporated seamlessly in the feed system of the radio telescope. Since the feed system will be located above the ground on a balloon with serious payload limitation, weight is crucial, and because of the extra weight required by the power supply, so is efficiency. This paper presents one of the key tools being used in the design of the pulse-tube cryocooler.

MODEL FORMULATION

Although conceptually the pulse-tube cryocooler is a very simple device, numerical simulation of the thermodynamic processes turns out to be challenging [2, 3, 4]. Different parcels of working substance in different parts of the cryocooler undergo different thermodynamic cycles, and as a result, the convective processes, including transport of mass, energy and momentum by the fluid in motion, play a dominant role. These processes are non-linear and hyperbolic in nature; they are numerically difficult and as a result, the simulation is computation-intensive. Furthermore, the device includes a combination of small and large volumes and the interaction between them needs to be dealt with appropriately. In the large volumes such as the compressor and the reservoir, significant mixing but little heat transfer occurs; however one cannot reasonably expect, or even need to resolve the instantaneous turbulent flow details. These are inherently three-dimensional, transient and characterized by very small scale, but they only play an indirect role and ultimately, the resulting processes do not depart too much from the stratified adiabatic model [5]. But in the small volumes such as the heat exchangers and regenerator, accurate performance prediction requires that the convective interaction between flow motion and heat transfer with the wall or matrix must be well resolved. For the small flow passages a one-dimensional formally high-order accurate model is in all likelihood feasible under suitable assumptions. Technically, a high order accurate model of some one-dimensional representation of the large volumes might be feasible also, but its meaningfulness would be illusory and in any event, estimates of the overall error remain based upon the coarse mesh used in the large volumes. In other words, no matter what detailed model is used for the large volumes, flows supplied by the large volumes to smaller ones cannot really be predicted qualitatively better than first-order accuracy. Thus predicted process outcome must be looked upon as no better than a discrete, discontinuous, jump at every time step.

There is no rationale for looking at them instead as second-order accurate piecewise linear functions constructed by straight connections between values, which would lend the information a higher quality than the model intrinsically contains. Under these conditions, there is no value in implementing a second order model at any location in the machine.

One should note that so far, this limitation originates in the physical model, even before considering algorithms and numerical implementation. In this respect, while resolving molecular diffusion in the longitudinal diffusion would be unreasonable, heat exchange with walls or matrix yields predominantly diffusive physics [6]. As a result, one would expect that any front-like profiles delivered by the large volumes would be rapidly smoothed by diffusion. Unfortunately, there are flow passages such as the tube, in which there is little heat transfer hence insignificant diffusion, and in which spatial and temporal gradients need to be properly resolved. The limitation expressed by the Godunov theorem [7] applies: no algorithm that is at the same time conservative, monotone and more than first order accurate exists. In these spaces, a conservative, strictly second-order accurate solution will thus result in oscillations that may ultimately compromise global stability, and would render the solution useless. Workarounds based upon nonlinear schemes exist, effectively letting accuracy deteriorate to first-order only whenever and wherever gradients are so sharp that the strict second order model would induce oscillations. But since the physical model imposes serious limitations upon accuracy anyway, the benefits would be illusory, and as a result, one might as well stick to a simple, robust first order-accurate solution. This is the limitation that was already encountered in simulation of Stirling refrigerators. The algorithm used and its implementation are effectively identical as in the MS*2 Stirling Cycle code [8, 14], adapted for the orifice pulse-tube configuration.

As already mentioned above, the pulse-tube geometry is actually three-dimensional. The geometry is especially complex in the regenerator. Actual time scales range from the travel time of acoustic waves to the time for the machine to reach a periodic behavior, with the same processes more or less repeating themselves until reaching the limit cycle. But as a practical matter, cost- and time-effectiveness mandates a one-dimensional model in which time scales shorter than the convective time, including acoustic phenomena, are neglected. While implicit time integration would yield a stable algorithm formally incorporating acoustics even if the CFL condition that time steps must be smaller enough than acoustics do not fully cross any cell over one step is not met, accurate resolution of the acoustics would still require a time step of the same order required to meet the CFL condition, which again would require unreasonably long computations. Neglecting acoustics is not a big deal, but the one-dimensional model is a major compromise; viscous stresses and convective heat transfer are inherently multidimensional. A one-dimensional model cannot be based strictly on Newton's and Fourier's laws. Instead, empirical models for viscous and convective heat transfer are unavoidable and for the latter, empirical models are demonstrably inconsistent with the true physics whenever pressure fluctuations play a significant role in the energy exchanges [9]. Still, in the pulse-tube, the pressure amplitudes are typically relatively small, and one must hope that the resulting error remains acceptable.

Recognizing that the cylinder volumes are bulky spaces, with no clear distinction between length and cross-section the model of which can only remain rather rudimentary, and focusing instead on the flow in the heat exchangers and regenerator, formally, the periodic boundary value problem being solved can be obtained in a two-step process: first, a formal magnitude analysis based upon neglecting terms in the equations that scale either with the Mach number or d/L, both of which are small, followed by replacing transverse diffusion by phenomenological models. In the order of magnitude analysis, the objective is to determine the importance of the physical mechanism in the real machine, but keeping the real physics of conduction and viscosity.

Using coordinates shown on Figure 1, the dimensionless form of the conservation laws were given in Ref. [8] After the order of magnitude analysis, the equations were

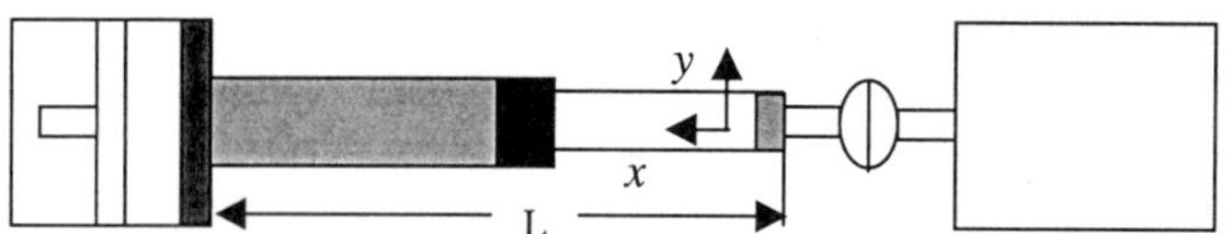

Figure 1. Coordinates for orifice Pulse-tube cryocooler

reduced to a two-dimensional model. However the two-dimensional model is still too complicated and expensive, especially since at best, the detailed regenerator geometry would be known in a probabilistic sense. The next step then is to replace the true diffusion processes by empirical models, in an equivalent one-dimensional representation:

$$\frac{\partial p}{\partial x} = \frac{-\gamma M^2}{\text{Re}} \frac{L}{d} f(\rho u \text{Re}) \frac{u^2 \rho}{2} \tag{1a}$$

$$\frac{L}{U\tau} \frac{\partial \rho}{\partial t} + \frac{\partial (\rho u)}{\partial x} = 0 \tag{1b}$$

$$\rho \left[\frac{L}{U\tau} \frac{\partial T}{\partial t} + u \frac{\partial T}{\partial x} \right] - \frac{\gamma - 1}{\gamma} \frac{dp}{dt} = \frac{1}{\text{Pr}\,\text{Re}} \frac{L}{d} Nu(\rho U \text{Re})(T_w - T) \tag{1c}$$

$$p(x,t) = T\rho \tag{1d}$$

where M is the Mach number, *Re* is the Reynolds number, *Pr* is the Prandtl number, *D/Dt* is the material derivative, γ is specific heat ratio, d is the characteristic lateral dimension of fluid passage, L is the characteristic length of the system (also see Ref. [8]).

The boundary conditions are provided by the machine geometry. For heat exchangers Reynolds numbers and Nusselt numbers (*Nu*) are based on the tube diameter; for regenerator they are evaluated in the same way as Ref. [11]. A periodic solution to the boundary value problem is obtained starting from some initial conditions and solving it marching in time, until the solution becomes periodic.

To complete the model, empirical correlation formulae are needed for the friction factor and the Nusselt Number based on actual local instantaneous Reynolds number. For heat exchanger tubes, the exact steady solution for laminar flow is used:

$$f = 64/\text{Re} \tag{2}$$

For higher Reynolds number flow (>1514), the following correlation has been used [8]:

$$\log_{10}(f/4) = -1.34 - 0.2\log_{10}\text{Re} \tag{3}$$

For the Nusselt number in the heat exchanger tubes, the following correlation is used:

$$Nu = 0.021\,\text{Re}_d^{0.8} \tag{4}$$

For screen mesh, the following data [8, 11] are used.

$$f = 4(0.337 + 3.35/\text{Re}_d) \tag{5}$$

$$Nu_d = 0.42\,\text{Re}_d^{0.56} \tag{6}$$

The orifice and the reservoir are the parts that differ in the pulse-tube, in relation to Stirling devices. In a slender orifice, longitudinal diffusion is negligible. If the length of the orifice is short compared to the sweep by the flow, then the flow becomes quasi-steady, and all time derivatives are negligible. In particular, mass conservation then results in a spatially uniform mass flow rate and the inertia terms are negligible in the momentum equation. Finally, for velocities smaller than the speed of sound, acceleration terms are negligible. So yields the Poiseuille velocity profile:

$$u(r,t) = -\frac{\partial p}{\partial x}\frac{R_0^2}{4\mu}\left(1-\frac{r^2}{R_0^2}\right) \quad 7)$$

where R_0 is the radius of the orifice. The thermodynamic process in the reservoir is considered as adiabatic and pressure is spatially uniform.

One severe limitation currently implemented refers to the pressure gradients, which are being dealt with in a decoupled manner. In other words, the problem of mass and energy conservation, and heat transfer with the wall and matrix, are dealt with assuming pressure p to be time-dependent but spatially uniform. A perturbation p' evaluated from equation (1a) is then calculated separately based upon the velocities resulting from the mass and energy conservation problem. This approach can be justified when pressure gradients are small compared with the amplitude of the temporal fluctuation, but it becomes increasingly inaccurate as the gradients become larger, which is typically the case in pulse-tubes. A better approach would be to solve the fully coupled problem, but this would require a completely different algorithm, requiring an upwind scheme, in which the solution is computed in the direction of the flow. This is difficult because the direction of the flow changes over the period, with moments during which flow originates in the compressor, at the orifice, at both or neither. These various situations would have to be identified and different loops built for each.

NUMERICAL SCHEME AND RESULTS

A variety of numerical techniques for the above initial boundary value problem are available. However, considering the stability and consistency of numerical solution resulting from the discontinuity, the sudden changes of cross-section in the machine, a conservative, monotone, semi-Lagrangian approach, Godunov-like algorithm, is implemented in the code. The conservative formulation means that the integral form of the conservation laws is satisfied even in the presence of discontinuity, so that the numerical error depends only on the step sizes, but not on the difference in the values of the field variables between cells. An operator-splitting technique is used, with a sequence of three operations in the code carried on at each time step. First, heat transfer and piston motion is allowed to occur assuming no flow between cells. The resulting pressures are not uniform. Next, the flow is dealt with in two half steps. The first one is the Lagrangian half step. For an ideal gas with constant specific heat, global conservation of energy implies that $\int_V c_v \rho T dV$ remains unchanged under the flow process. Using * to indicate state variable obtained after dealing with heat transfer, but before fluid motion, then $\int_V c_v \rho^* T^* dV = \int_V c_v \rho T dV$. For ideal gas, $p=\rho RT$, so that $\int_V p^* dV = \int_V p dV = pV$, where V is the total volume of the machine system. The final pressure is spatially uniform. The mass of gas within the volume of each space step relaxes then isentropically to the final uniform pressure. As a result, each mass moves as a front into one of the neighboring volumes. Finally, the new average densities are computed in each volume and the decoupled pressure gradients are evaluated. The flow in the orifice is computed based upon the difference between the pressure in the reservoir and the corrected pressure, including the effect of the gradients.

Table 1. The dimensions and the operating conditions of a orifice Pulse-tube cryocooler

Frequency	40Hz	Heat transfer area of aftercooler	$264cm^2$
Mean pressure	25bar	Length of freezer	15mm
Diameter of regenerator	9mm	Heat transfer area	$33cm^2$
Length of regenerator	68mm	Length of hot heat exchanger	10mm
Fill factor	0.3	Heat transfer area of hot heat exchanger	$26cm^2$
Diameter of wire of matrix	0.044mm	Length of tube	20mm
Diameter of pulse-tube	7mm	Diameter of tube	6mm
Length of pulse-tube	50mm	Temperature of freezer	80K
Volume of reservoir	$160cm^3$	Temperature of aftercooler	300K
Resistance of orifice	5000 /ms	Temperature of hot heat exchanger	300K
Length of aftercooler	40mm		

As an example, the following orifice Pulse-tube cryocooler shown in Table 1 is simulated with the above numerical scheme. A relatively small volume of reservoir is chosen to obtain a periodic solution in a shorter CPU time. Also since most orifice pulse-tubes use a needle valve as orifice, a resistance is chosen to represent the character of a needle valve.

Figure 2 shows the pressure amplitude distributed along the system, with a relatively large pressure drop in the regenerator. Figure 3 shows the metal temperature along the system including regenerator matrix and pulse-tube. Figure 4 shows the helium gas temperature variation, with the two curves showing a large difference between maximum and minimum gas temperature in the tube where heat transfer with the wall is little but a small difference in the regenerator. Also, the large temperature variation in the aftercooler and the hot heat exchanger may imply that these heat exchangers are not sufficient enough. Figure 5 shows the mass flow rate along the system (in all figures, the position 0 represents the orifice).

For the design of orifice Pulse-tube cryocooler, optimization of the dimensions of the system is crucial, especially for the dimensions that play important role in the cryocooler system, such as those dimensions of the regenerator and pulse-tube. However, the optimization for the cryocooler system is proven to be very difficult because there are more than two dimensions that determine the efficiency of the system; the effects of these dimensions on the efficiency are not linear, and the effects of these dimensions interact. One

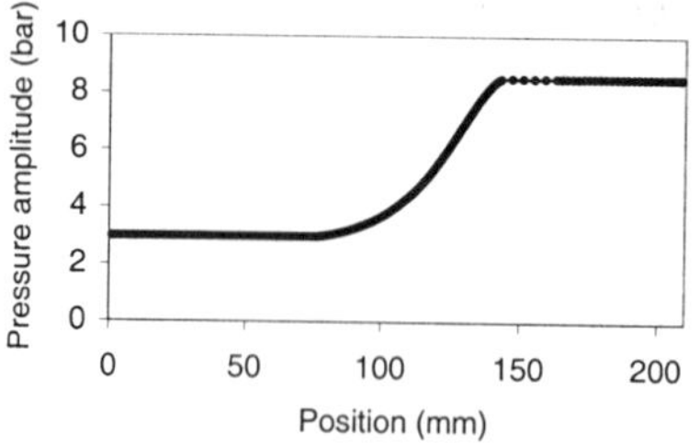

Figure 2 Pressure amplitude spatial distribution

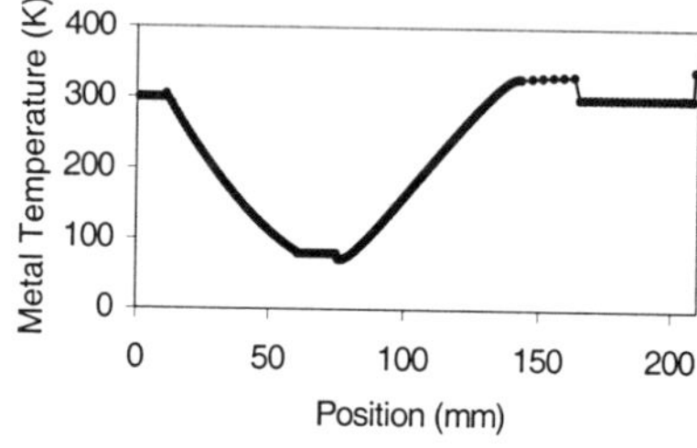

Figure 3. Metal temperature spatial distribution

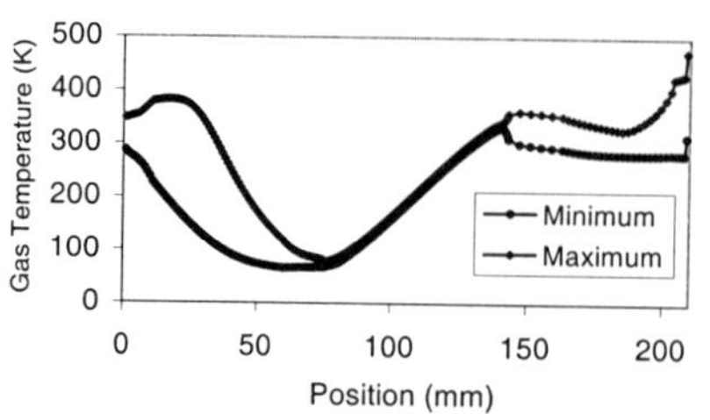

Figure 4. Spatial distribution of minimum and maximum gas temperature

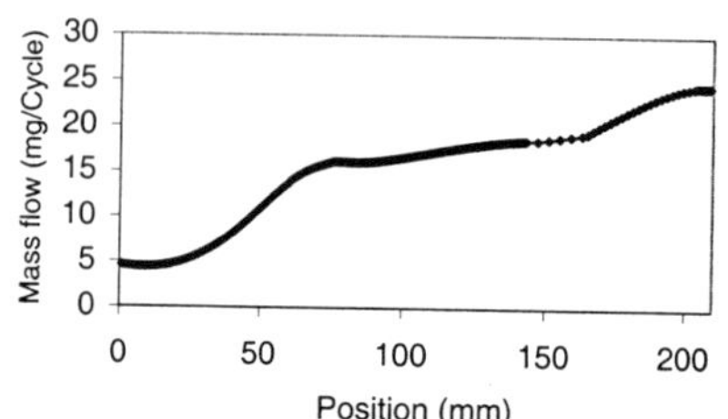

Figure 5. Mass flow rate spatial distribution

method for optimization is to find the trend of the effect of one parameter while keeping the others fixed, then keep the optimized former parameter value fixed and see the effect of a later parameters. When the optimization of the latter parameter is found, again check the trend of the former parameter. This should be carried over and over again. Obviously, the procedure is time consuming, and it is being done on our design for the SKA Pulse-tube cryocooler proof-of-principle prototype with the help of some analytical work [12, 13]. Figures 6-9 shows some results of the effect on the system efficiency by varying the dimension of regenerator and pulse-tube respectively. Figure 10 shows the effect of the resistance of orifice on the system efficiency. Actually the resistance value represents the opening of the needle valve. Figure 11 show the effect of frequency on the refrigeration of the system. These results present guides to the operation of the prototype.

CONCLUSIONS

A one-dimensional numerical pulse-tube simulation code has been implemented for the SKA project. The model makes use of empirical correlation of viscous and heat transfer. The code is conservative and valid across the discontinuity because of its Godunov-like algorithm. The code can be used to optimize the design of Pulse-tube cryocooler and to present guides for operating conditions although it is a time consuming procedure. Some results have been presented.

ACKNOWLEDGMENTS

This work is supported by the Natural Science and Engineering Research Council of Canada.

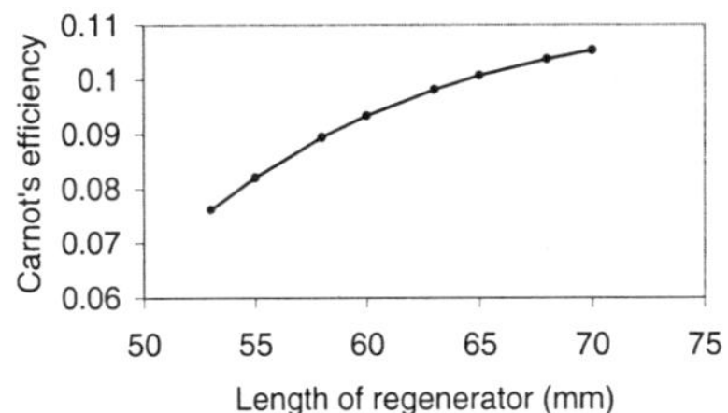

Figure 6. Efficiency versus length of regenerator

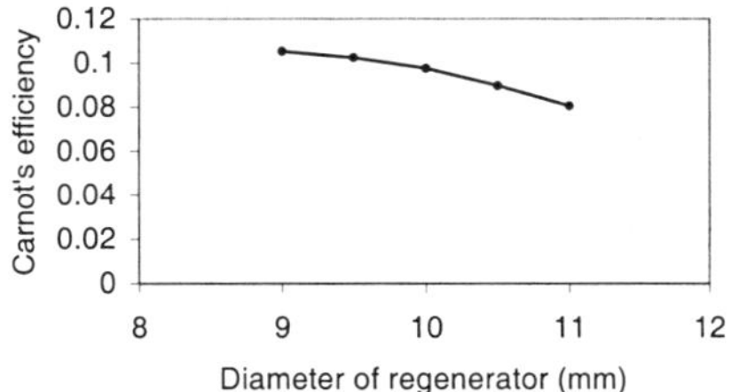

Figure 7. Efficiency versus diameter of regenerator

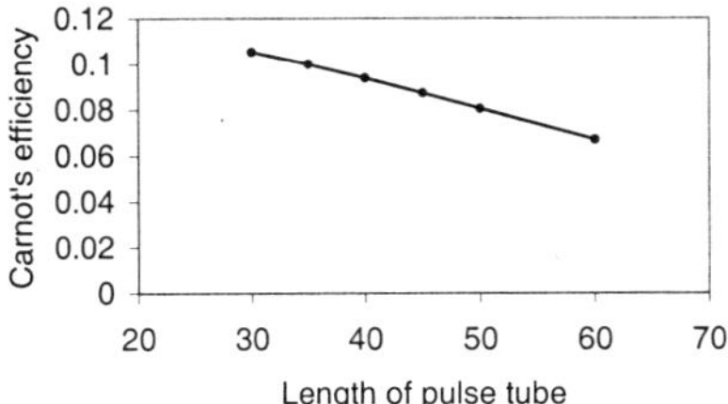

Figure 8. Efficiency versus length of pulse-tube

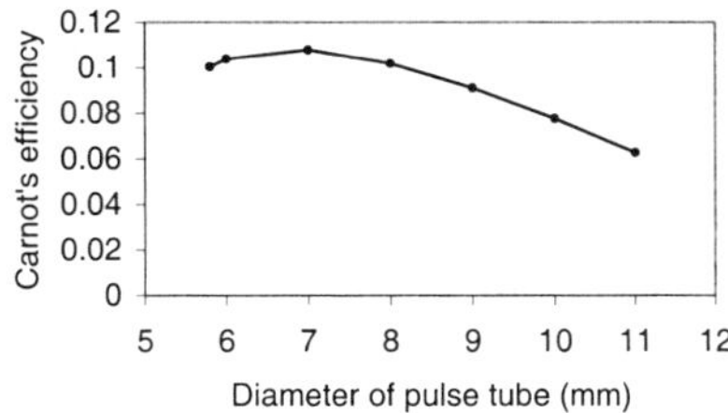

Figure 9. Efficiency versus diameter of pulse-tube

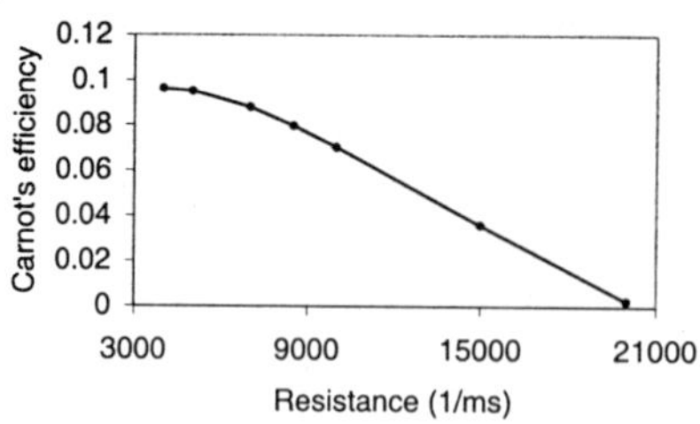

Figure 10. Efficiency versus resistance of orifice

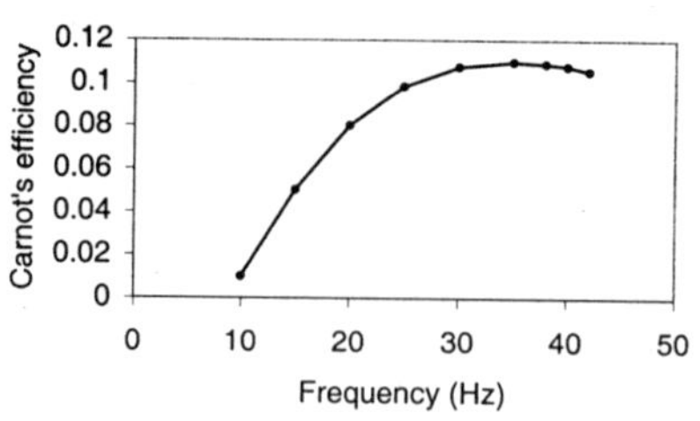

Figure 11 Efficiency versus frequency

REFERENCES

1) P. Dewdney & B. Veidt, "The Development of the Large Adaptive Reflector for the Square Kilometer Array", National Research Council of Canada, Herzberg Institute for Astrophysics, 1997.
2) A. Kashani and P. R. Roach, "An optimization program for modeling Pulse-tube coolers", Advances in Cryogenic Engineering, Vol. 43, pp. 1903 (1998).
3) M. Lewis, T. Kuriyama, J. H. Xiao and R. Radebaugh, "Effect of regenerator geometry on Pulse-tube refrigerator performance", Advances in Cryogenic Engineering, Vol. 43, pp. 1999 (1998).
4) C. Wang, et al, "Numerical modeling of an orifice Pulse-tube refrigerator", Cryogenics, Vol. 32, pp. 785 (1992).
5) L. Bauwens, "A stratified flow model for adiabatic losses in regenerative thermal devices", Journal of Energy Resources Technology, Vol. 117 (1995).
6) L. Bauwens, "Thermoacoustics: Transient regimes and singular temperature profiles", Physics of Fluids, Vol. 10, No. 4 (1998).
7) S. K. Godunov, Mat Sb. 47, 271 (1959), translation at Cornell Aeronautical Lab.
8) L. Bauwens. "Consistency, stability, convergence of Stirling engines model", Proc. of 25th IECEC, pp352 (1989).
9) L. Bauwens, "Near-isothermal regenerator: complete thermal characterization", Journal of Thermophysics and Heat Tansfer, Vol. 35, No. 3 (1998).
10) G. T. Reader and D. R. Taylor, "Comments on the computer-aided design of Stirling engines", Proc. of 18th IECEC, pp. 737 (1983).
11) H. Miyabe, et al., "An approach to the design of Stirling engine regenerator matrix using packs of wire gauze", Proc. of IECEC, pp.1839 (1982).
12) L. Bauwens, "Interface loss in the small amplitude orifice Pulse-tube model", Advances in Cryogenic Engineering, Vol. 43, pp. 1933 (1998).
13) L. Bauwens, "Entropy balance and performance characterization of the narrow basic Pulse-tube refrigerator", Journal of Thermophysics and Heat Transfer, Vol. 10, No .4, pp. 663 (1996)
14) L. Bauwens, "Stirling engine Modeling: the MS*2 code", Proc. of 6th ISEC, Eindhoven, The Netherland, pp.371 (1993)

DEMONSTRATION OF A LOW COST CRYOCOOLER ON A LONG DURATION BALLOON MISSION

E.F. James [1], I.S. Banks [1], R.B. Ray [2], S.H. Castles [3]

[1]Orbital Sciences Corporation
Greenbelt, Maryland

[2]NASA / Wallops Flight Facility
Wallops Island, Virginia

[3]NASA / Goddard Space Flight Center
Greenbelt, Maryland

ABSTRACT

NASA/GSFC has been evaluating the use of low cost Stirling cycle cryocoolers for aerospace applications since 1994. These include the M77B and M77C cryocoolers built by Sunpower Corporation. To date NASA has tested eight M77B and two M77C cryocoolers, with 8 additional M77C units now under construction. The intent of this work is to determine the flight worthiness of these cryocoolers. The Sunpower M77 coolers are candidate for use on the Ultra Long Duration Balloons presently under development by NASA. The flight on the Long Duration Balloon (LDB) in July 1998 represented an opportunity to test the cryocooler in the high altitude balloon environment in order to gain experience in preparation for possible opportunities on the Ultra Long Duration Balloon (ULDB) missions. The Long Duration Balloon is typically a 10 to 15 day mission. Typical ULDB missions might be as long as 100 days or more, and it is this duration which now forces many science groups to consider the use of cryocoolers in place of stored cryogens. This paper presents the basic design of the cryocooler experiment, and data acquired during the flight. The paper also includes a general perspective on the use of cryocoolers on future ULDB flights.

INTRODUCTION

NASA Goddard Space Flight Center has been involved in the development of long life space-based cryogenic coolers for several decades. In recent years Goddard has been cooperating with aerospace and commercial companies to facilitate the commercialization

Advances in Cryogenic Engineering, Volume 45.
Edited by Shu *et al.*, Kluwer Academic / Plenum Publishers, 2000.

of cryocoolers that can also be used in space. In addition to providing a service to U.S. industry, Goddard's goal is to develop coolers that can meet the requirements of both space-based missions and ground-based applications. This dual use provides advantages to NASA. Specifically, the coolers will have lower production costs because of the increased production rate and will be readily available. Also, once production rates have increased, the fabrication processes will become more uniform from cooler to cooler. Over time, this uniformity will improve reliability since variations from cooler to cooler can result in subtle defects that may limit the cooler lifetime.

The development of "dual use" cryogenic coolers is an example of NASA's drive to find ways to carry out scientific missions in a faster, better and cheaper manner. The use of balloon-borne instruments instead of space-based instruments is another example of NASA's push to reduce costs. Some scientific investigations can be carried out on high altitude balloons that are above most of the atmosphere. Examples include cosmic ray studies and certain X-ray studies. To facilitate these investigations, NASA has embarked on a program to extend the duration of high altitude balloon experiments. These programs include the Long Duration Balloon Program (LDB) and the Ultra-Long Balloon Program (ULDB).

NASA SCIENTIFIC BALLOON PROGRAM

The NASA ULDB Program at Wallops Flight Facility is developing a new super pressure balloon capable of sustaining a 1000 kg. science payload at an altitude of 35,000 meters for 100 days duration. The first full scale demonstration of this balloon will be an engineering test flight from Alice Springs, Australia in December 2000. The first experiment to fly on this vehicle is scheduled for December 2001. Although this experiment does not require cryogenics for their detector, future ULDB flights will. Four candidates for early ULDB flights require long duration cryogenic cooling. The Sunpower Stirling cycle cryocooler is a candidate to fill this anticipated requirement in future flights.

A Long Duration Balloon "flight of opportunity" was performed from Fairbanks, Alaska in 1998. This flight provided an opportunity to expose the Sunpower Stirling cycle cryocooler to a long duration balloon environment. This was a combined effort between the Balloon Program at Wallops, the Cryogenics Branch at NASA / Goddard Space Flight Center, and the National Scientific Balloon Facility in Palestine, Texas.

DESIGN OF CRYOCOOLER EXPERIMENT ON LDB MISSION

Objective

The primary objectives for this mission were to determine the effectiveness of the passive radiator at removing heat from the cryocooler, to characterize the cryocooler's thermodynamic performance, and to confirm mechanical and electromagnetic compatibility with other systems on the balloon.

Radiator Design

The Sunpower M77 Stirling cycle cryocooler produces approximately 100 watts of heat when running at maximum power. To dispel this heat, a passive radiator consisting of a flat aluminum plate, pointed at cold space, was used. The radiator was oriented away from the sun by rotating the gondola using a sun tracking motorized rotator in the flight train. In order to reject the 100 watts of heat generated by the cryogenic cooler while maintaining it under its maximum operating temperature limit, the passive radiator thermal design maximized conduction from its heat source to the radiator surface and radiation from the radiator surface to cold radiation sinks such as deep space (-273 deg C), earth (-18.7 deg C) and the balloon (variable). Also, the radiator design minimized the solar flux on the radiator surface and the view factors to other relatively warm balloon/payload system components. To accomplish this, the design of the radiator surface called for it to be inclined 55 degrees from the horizontal, located on the anti-sun edge of the payload with the radiator surface pointed in the anti-sun direction. The radiator surface was coated with Sherwin Williams A6W40 Super White paint (alpha=0.236, epsilon=0.895) to maximize the infrared radiation and minimize the solar absorption. The computed view factors, calculated with a Monte Carlo ray tracing technique, from the radiator surface to other sources/sinks for this configuration were: 0.667 to space, 0.136 to earth, 0.099 to the balloon, 0.085 to the other science components, 0.007 to the rotator assembly, cables and rigging, and 0.005 to the photo-voltaic panels. Under hot-case conditions, this radiator design resulted in a maximum predicted radiator temperature of +60 deg C. Under cold-case conditions, the radiator temperature was predicted to be +25 deg C.

Cryocooler Interface Issues

The cryocooler cold finger was enclosed in a vacuum bonnet and instrumented with a LakeShore DT-400 series temperature sensor and a resistive heating element to simulate the detector load.

Ground testing prior to flight included numerous tests to ensure that the cryocooler did not interfere with other scientific payloads on the balloon gondola. Of particular concern was a liquid nitrogen cooled germanium X-ray detector. Given the disparity between the cooler's input power and the energy detected by the detector, there are many potential sources for noise coupling. Among them are electromagnetic interference and microphonics. Microphonic noise consists of electronic signals induced by mechanical vibrations stemming from the operation is the of the cooler. One goal of the experiment was to evaluate the signal degradation at the detector caused by EMI and vibration from the cryocooler.

The germanium radiation detector was located about half a meter from the mechanical cooler. The test consisted of an evaluation of the noise performance as the cryocooler power level was stepped up. At no time was there any indication that the cooler was interfering with the detector. This demonstrates the feasibility of using a mechanically cooled detector on a long duration balloon mission.

Control Electronics

The electronics developed for the cryocooler experiment had to be kept very small and have minimal complexity to meet the schedule dictated by this rapid deployment mission. The circuit design was based on electronics developed for use with Sunpower cryocoolers in the NASA/GSFC laboratory.

The basic electrical system diagram is shown in Figure 1. The power was supplied from solar panels, charge converter and batteries provided by the balloon gondola electrical system. The power system provided 28 V dc power to the cryocooler experiment for the duration of the flight. The power input to the cryocooler system was fused at 10 amps.

The input commands for control of the cryocooler were performed through 4 latching relays. Power on / off, cooler start / stop, hi / low power, and hi / low heat load. The hi power level was set to about 87 watts, and the low power was set to about 78 watts. The Hi heat load was set to 4 watts, and low heat load was set to 2 watts.

The cryocooler data file, which was included in the telemetry stream every 30 seconds, provided radiator temperature, cooler heat sink temperatures, cold tip temperature, heat load voltage (indicating which load was on) and the cryocooler system bus current.

The cryocooler compressor motor was driven using a 28 V PWM servo amplifier, Model 4122Z made by Copley Controls. These amplifiers can deliver 10 amps continuous and 20 amps peak current. The balancer motor was also driven from the same amplifier but through a simple phase shift network which set up the proper amplitude and phase relative to the compressor such that the vibration was reduced. The proper relationship of phase and amplitude is primarily a function of cold tip temperature and was obtained from previous charcterizations of this cryocooler. The vibration being generated during the flight was not quantified, but was determined to be at a level low enough to not interfere with other systems on the balloon gondola.

The function of the gain control network was to vary the amplitude of the sinewave drive signal. This closed loop circuit read the cold tip temperature and adjusted the signal level accordingly. The compressor power level would only increase if the cooler cold tip was getting colder. The purpose of this is to ensure that the free displacer piston is not overdriven. When the helium is warm there is insufficient damping of the displacer piston to allow full power to be commanded.

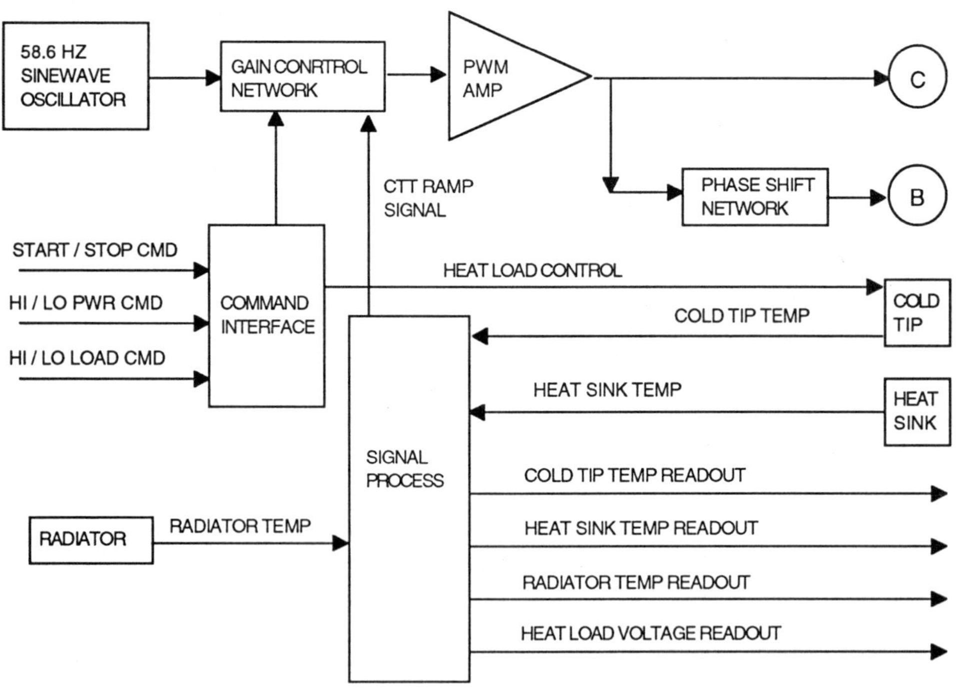

Figure 1. Block diagram of cryocooler control electronics

FLIGHT DATA

The first flight of the balloon was aborted after about 7 hours into the flight. The balloon only achieved about 100,000 ft. altitude, rather than the 120,000 ft. required. The balloon was recovered, and launched again some 10 days later. Although the first flight was short, sufficient data was acquired to evaluate cooler performance at one power level and one heat load. Figure 2 illustrates the performance of the radiator and heat sink for the cryocooler while on the runway, with the cooler running, during the ascent phase, and at altitude before the abort was commanded. The cryocooler was in low power mode for all of flight 1. The cold tip temperature as a function of compressor power for the first flight is presented in Figure 3.

A problem with the electronics, which was not realized prior to flight, was that the gain control network was temperature sensitive. As the heat sink temperature increased, the compressor power increased. Figure 3 includes data collected on the runway through the abort. A power level change of about 10 watts per 40 degrees Celsius change in heat sink temperature was observed.

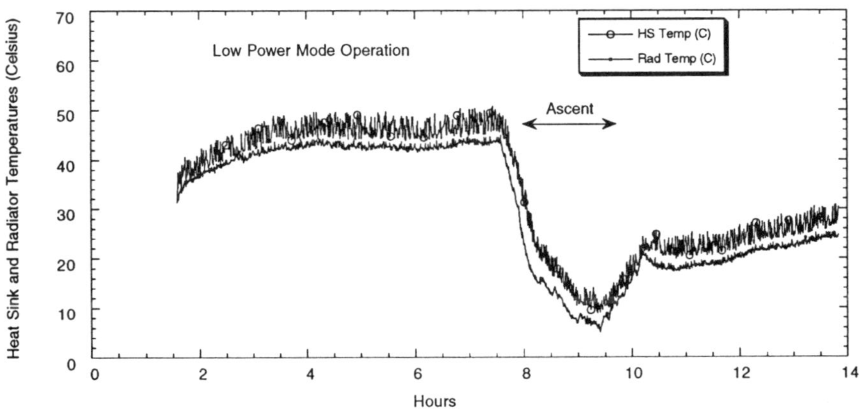

Figure 2. Heat sink and Radiator temperatures during the first flight

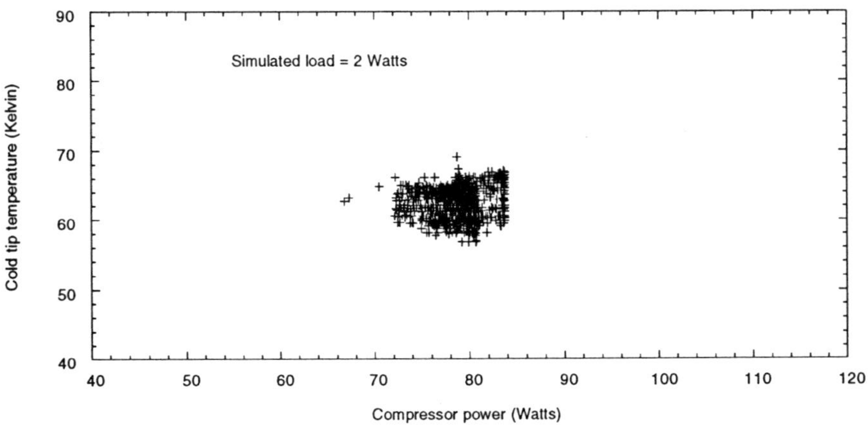

Figure 3. Cold tip temperature performance for the first flight

Flight 2 lasted 13 days. Data was collected with the cooler at low power and with low load, high power and with low load, and with the cooler off, with a 45 Watt heat load on the radiator. Unfortunately, data was not obtained for high power and high load, because the cryocooler compressor fuse blew sometime after the cooler was commanded to high power. This was due to the electronic gain control problem mentioned earlier, which resulted in excessive power to the compressor.

A communication blackout with TDRSS occurred 47 hours into the flight. The cryocooler was commanded to high power mode about 6 hours prior to the blackout. The compressor motor fuse blew sometime during the 10-hour blackout. When communications returned, the cryocooler was not running. Data analysis revealed that the cooler was still in low power mode. At that point the system was commanded to cooler stop mode, which automatically turned on the main 45 watt heater on the radiator. This permitted continued data collection of heat sink and radiator temperatures for the remainder of the flight with a fixed 45 watt heat load on the heat sink. The data link, however, continued to periodically blackout for the remainder of the flight.

Figure 4 shows the time base performance of cold tip and heat sink temperatures of the cryocooler while on the runway, through the ascent phase, and at altitude up to the point at which all data communications were lost. Notice the periodic nature of the heat sink temperature. Figure 5 illustrates the radiator and heat sink temperatures for the three power levels for the entire 13 day flight. Constant power operation of a cryocooler will result in a cold tip temperature which will vary as the heat sink temperature varies. On this flight, since the compressor power level was increasing as the heat sink temperature increased, the cold tip temperature remained relatively constant.

The cyclic nature of the radiator and heat sink temperature is primarily due to the day / night cycle, and was compounded by the power fluctuation problem. Constant cold tip temperature operation will most likely be the requirement for a cryocooler cooling a detector on a balloon flight. On future missions, either a robust controller will be required to maintain a constant cold tip temperature under such conditions, or some technique for keeping the radiator temperature more constant will be required.

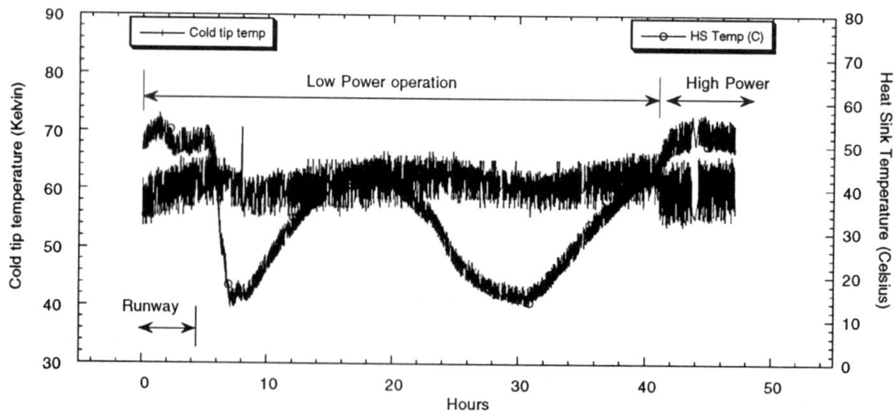

Figure 4. Heat sink and cold tip temperatures . Cold tip with a 2 watt simulated load

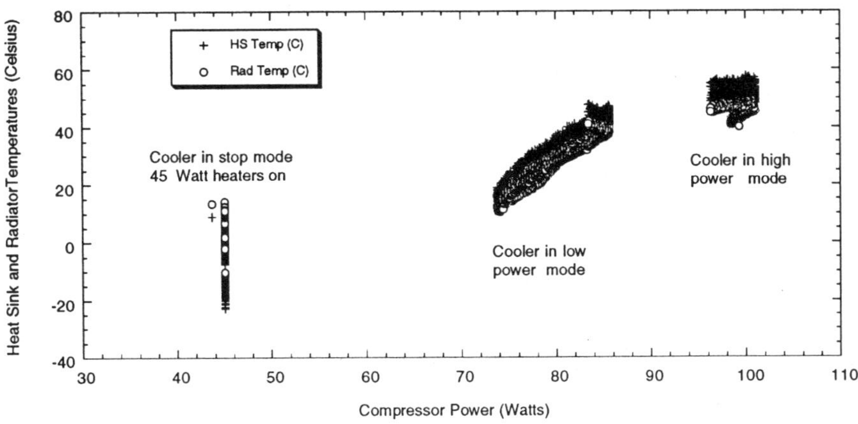

Figure 5. Heat sink and Radiator performance at low power, high power, and with 45 watt load.

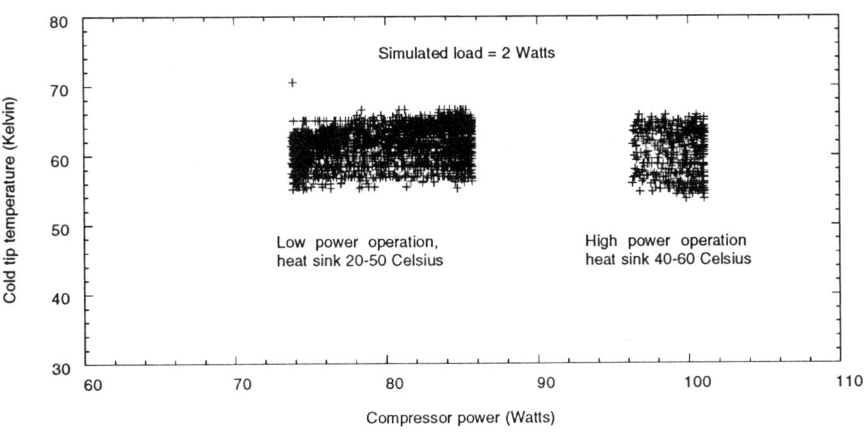

Figure 6. Cold tip temperature measurement for both high power and low power modes

Figure 5 includes only data acquired at altitude. The three levels are low power mode, high power mode, and a constant 45 watt load simulating the compressor.

The cryocooler cold tip temperature during the second flight is presented in Figure 6. Note that the heat sink temperature was considerably higher in the high power mode than in the low power mode. The cold tip temperature performance shown in Figure 6 is very close to what was measured in the NASA cryogenics branch thermal vacuum chamber during tests performed on another M77B Sunpower cryocooler. Two thermal vacuum chamber test results were with (1) a 60 Kelvin cold tip temperature, with 100 watts compressor power, 30 degree celsius heat sink, and (2) a 70 Kelvin cold tip temperature with 60 watts compressor power, and 30 degree celsius heat sink. Both tests were conducted with a 2 watt simulated load on the cold tip.

CONCLUSIONS

The sequence of commands established for the cryocooler during the flight were to run first at low power and low load, then high power and low load, then high power and high load, and finally low power and high load. Data was collected for low power and low load, and high power and low load only. While most of the desired thermodynamic data was obtained, a complete thermodynamic load line was not possible since the cryocooler stopped running prior to the command to high load.

A fluctuation of the gain control electronics with heat sink temperature was discovered and has been subsequently investigated. An alternative design has been implemented for use on possible future balloon missions that eliminates the gain control temperature sensitivity.

Sufficient data was collected to characterize the radiator effectiveness at removing heat from the cooler. The cryocooler was mounted off center on the radiator, which probably resulted in some loss of cooling effectiveness. Other methods to improve radiator effectivness are being investigated.

The flight revealed a need to enclose the electronics unit and make it accessible without removing the cooler and radiator from the gondola.

No electromagnetic incompatibility with other payloads was seen while the cryocooler was running.

Prior to launch, all systems were running on a generator which ensured that the batteries were not drained. Moments before launch the 28 V dc supplied to cryocooler was hot switched over to the batteries, and a significant voltage transient was measured which could cause damage to the electronics. Design modifications have been investigated to protect the cooler electronics from this voltage transient.

ACKNOWLEDGEMENTS

The authors wish to thank Vincent Arillo of Raytheon for electrical technician services as well as Scott Cannon of the Physical Science Laboratory, New Mexico State University for the thermodynamic analysis of the radiator.

PERFORMANCE COMPARISON OF M77 STIRLING CRYOCOOLER AND PROPOSED PULSE TUBE CRYOCOOLER

R.Z. Unger and J.G. Wood

Sunpower, Inc
Athens, OH 45710

ABSTRACT

This paper compares the performance of an existing Sunpower M77 free-piston Stirling cryocooler with the simulated performance of a proposed pulse-tube cryocooler that uses the same linear compressor. Sunpower has built and delivered numerous M77 free-piston Stirling cryocoolers. These cryocoolers lift 5 W at 77K with 100 W of electrical input power. These coolers are state of the art both in thermodynamic performance and in mechanical design. The mechanical design includes a high efficiency permanent magnet linear motor; piston and displacer supported by gas bearings; mechanical compliance to reduce manufacturing costs; and effective folded fin heat exchangers. While the M77 is a successful design that is reliable and able to be manufactured cost effectively, it is not as mechanically simple as a pulse tube cryocooler might be. A pulse tube design would eliminate the moving displacer and replace it with a tuned volume connected to the expansion space. However, the expansion space work would be lost, unlike in the moving displacer design. To assist in understanding which arrangement would have the best cost and performance trade-offs, a preliminary pulse tube cryocooler design, using the linear motor/piston combination (linear compressor) of the M77, is presented. Simulated performance is compared with test results for the M77.

INTRODUCTION

Sunpower's M77 free-piston Stirling cryocoolers are noted for their high performance and low cost. The low cost is largely the result of the compressor that is designed for mass production and the use of a patented compliant suspension system for the piston and displacer. Recently Sunpower completed work supported by NASA SBIR Phase I funds to study the projected performance of inertance tube PTRs (pulse tube refrigerators) driven by Sunpower linear compressors. Part of this study included a direct pulse tube replacement for the M77, the primary results of which are reported here.

SUNPOWER M77 CRYOCOOLER

Since 1994, Sunpower has delivered numerous Model M77 free-piston Stirling cryocoolers to commercial licensees and to various university and Government laboratories. Sunpower's Model M77 cryocoolers operate with stable thermal performance and incorporate a patented low-cost, compliant suspension mechanism with gas bearings to eliminate mechanical contact, friction, and wear on internal moving parts (USP 5,525,845, 1996).

At the heart of the M77 cryocooler is a linear compressor. A linear compressor is a positive-displacement type compressor in which the piston is driven directly by a linear motor, rather than by a rotary motor coupled to a conversion mechanism as in a conventional crankshaft driven reciprocating compressor. Linear motors are simple devices in which axial forces are generated by currents in a magnetic field. Because all the driving forces in a linear compressor act along the line of motion, there is no radial thrust on the piston, substantially reducing bearing loads. By resonating the mass of the piston with planar springs linked to the piston by a compliant rod, radial forces due to tolerance stackups and misalignments during assembly are reduced to levels that can be overcome by gas bearings. This virtually eliminates mechanical contact, friction and wear and greatly reduces manufacturing cost. The typical power consumption of the gas bearing system is 1 to 2% of the input power.

The patented linear motor is a moving magnet type characterized by constant electromechanical coupling as long as the magnets remain within the boundaries of the motor lamination poles. (U.S. Patent 4,602,174, 1986). Assembly of the linear motor magnets becomes cost effective with the aid of a magnet ring support system (U.S. Patent 5,642,088, 1997), which is a structurally sound, low loss construction for holding magnets in the air gap of a linear motor.

PULSE-TUBE BACKGROUND

The common element of all PTRs is the *pulse tube* itself, which is essentially just a gas-filled tube that transmits pressure and volume displacements to the cold expander via an insulating gas column. The role of the pulse tube is similar to the role of the displacer in a Stirling-cycle refrigerator and the objective is to duplicate the thermodynamics of the Stirling cycle without using any moving parts, besides those in the compressor.

The optimal performance for both Stirling and pulse-tube cryocoolers hinges on the pressure and velocity fluctuations being close to in phase within the regenerator. With Stirling technology this is easy because of the control provided by the independently moving displacer. In pulse-tube technology, the recent advancement that has made this possible is the introduction of pressure-velocity phase-shifting by use of an *inertance* tube (a gas filled tube long enough that inertial pressure drops are significant) placed at the warm end of the pulse tube.

There are actually two families of PTRs. One employs modified air-conditioning compressors driving low-frequency (< 4 Hz) expanders, similar to Gifford-McMahon coolers. The other employs high-frequency (30 – 100 Hz) valveless compressors similar to those used in Stirling coolers. Iwatani International Corp. is currently selling a line of PTRs of the low-frequency type. TRW and Lockheed-Martin have developed PTRs of the high-frequency linear-compressor type but they are not sold commercially. Our PTR designs are of the high-frequency linear-compressor type and are distinguished from the competition by the use of a linear compressor configuration that is based on Sunpower's free-piston linear machine technology designed for reliability and low-cost production.

PULSE TUBE REFRIGERATOR DESIGN

Table 1. Specifications for single-stage 77 K PTR

Electrical input	90 W, 115V single phase
PV power	75 W
Cooling capacity @ design point	7 W @ 77 K
No-load minimum temperature	40 K
Heat rejection	fan air @ 25 °C
Operating frequency	60 Hz
Charge pressure	25 bar

Our principal thermodynamic design tool for the PTR was the Sage computer program. Our first task was to validate the Sage software against actual test data for the Sunpower M77 Stirling cryocooler. We found Sage to predict the dynamics and performance of an instrumented M77 very accurately. The prediction was within the accuracy of our test instruments, on the order of a few percent.

We next used Sage with confidence to optimize all the important design variables such as pressure, frequency and heat exchanger dimensions, while simultaneously maximizing an objective function (usually maximum cooling capacity) and maintaining several constraints (such as constant power input and dimensional stack-up constraints). Optimization turn-around time was on the order of one or two hours each, enabling us to run a large number of optimizations to progressively refine the choice of optimized variables and constraints, based on feedback from layout drawings and the optimizations themselves.

The objective for this design was a PTR that would fit on essentially the same compressor as currently used for the Sunpower model M77 Stirling cryocooler. *Essentially* meaning the same frequency (60 Hz), piston amplitude (5.5 mm) and PV power capacity (75 W). We allowed charge pressure, piston diameter and reciprocating mass to vary within reasonable limits. The design-point temperature is the same as the M77 cryocooler but the cooling capacity turned out to be a bit more.

Table 1 gives the overall specifications for the preliminary design of our single-stage 77K PTR.

Figure 1 shows layout drawings that compare the size of the pulse tube design to that of the existing M77. Figure 2 presents a plot of the projected cooling capacity versus cold end temperature of the PTR. The Sage computer optimization sequence that led to this design started with the M77 compressor, however we found some benefit to increasing the charge pressure, reducing the piston diameter, and re-tuning the planar spring stiffness. For the regenerator matrix, we eventually settled on stainless-steel screens after investigating random-fiber felt and other materials such as polyester. We reduced the reservoir volume until just before the point of noticeable performance degradation.

There is also the possibility of arranging the PTR in a split-cycle arrangement where the cold head would be isolated from the compressor by a flexible tube. This would reduce vibration at the cold head and might be the preferred packaging arrangement for many applications. Figure 3 shows a split-cycle layout.

Regenerator Study

The regenerator of the final PTR design contains stacked stainless-steel screens with 75% porosity (void / total volume) and wire diameter 19.8 microns. Both porosity and wire diameter were optimized but constrained to a maximum of 75% and a minimum of 19.8 microns, respectively. The porosity constraint was based on a rough upper limit for available woven screens and the wire diameter constraint on the smallest available wire diameter. Sage's choice of 75% porosity and 19.8 micron wire diameter for the chosen (25

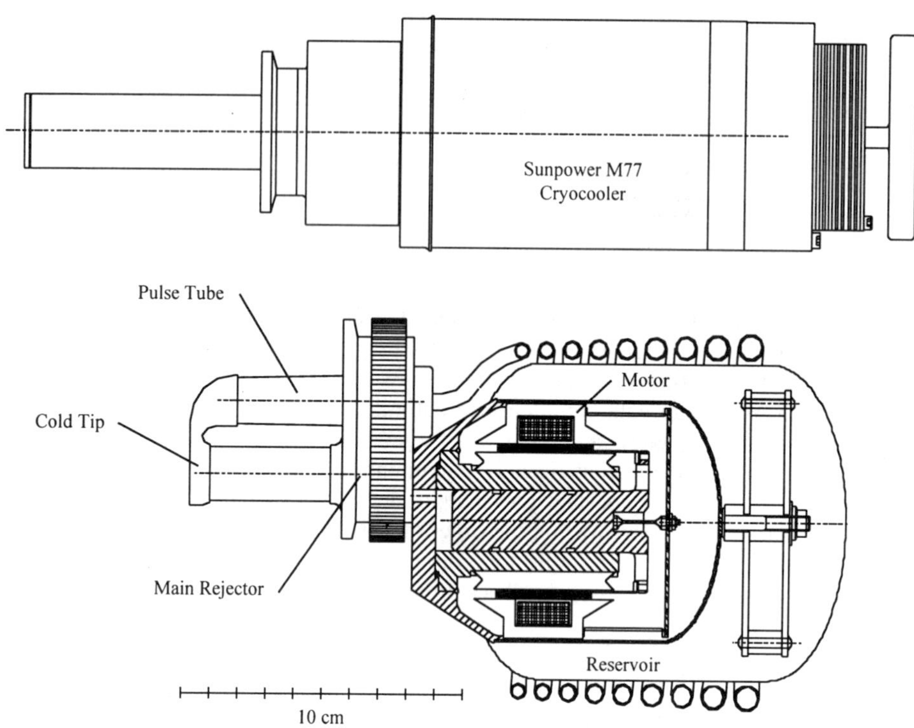

Figure 1: Single-stage 77K PTR layout (bottom) compared to existing M77 Stirling cryocooler (top).

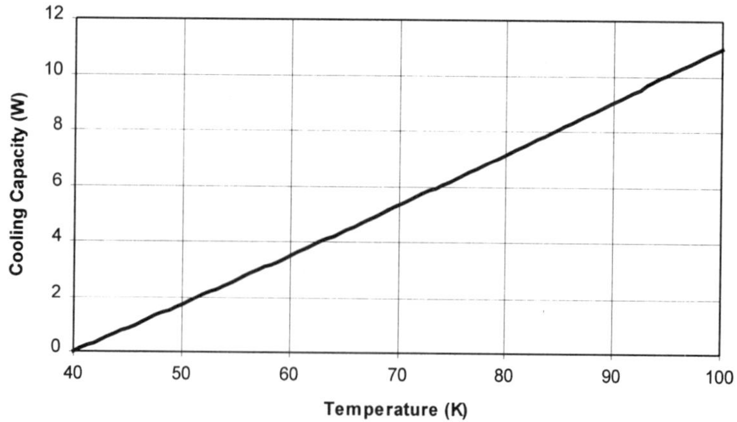

Figure 2. Cooling capacity as a function of cold-end temperature for the single-stage PTR.

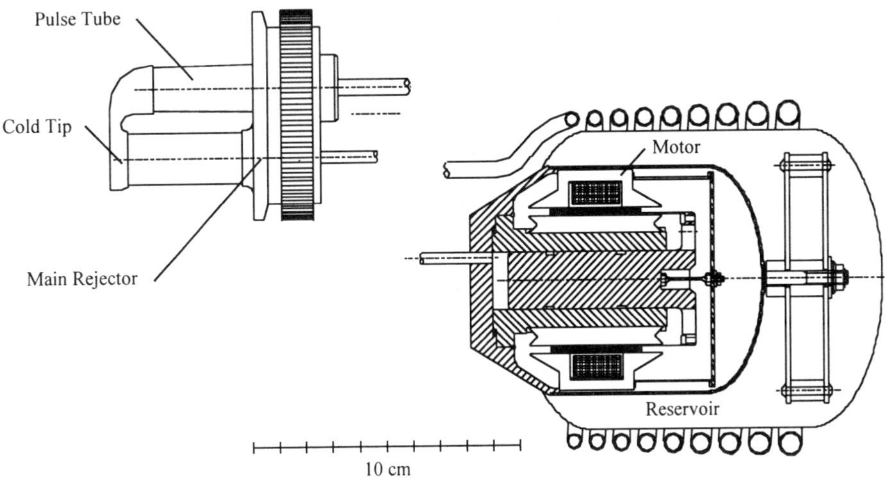

Figure 3. Split-cycle arrangement for the PTR.

bar) design may not actually correspond to any commercially available screen. We will have to check further into this during detailed design. There is a 500 mesh (per inch) screen available with 70% porosity and 19.8 micron wire diameter, which may be close enough. We plan to evaluate several potential screens near the Sage optimum before making a final decision.

Prior to selecting stainless-steel screens, we also evaluated stainless-steel random fibers, polyester screens, polyester random fibers, and the proprietary regenerator of the M77. Figure 4 compares the performance for the five different types of regenerator material as a function of charge pressure. Each point in the figure corresponds to a complete optimization of all key dimensions of a single-stage PTR at fixed charge pressure, cold-end temperature and compressor input power. For comparison purposes, the figure also shows the performance of the Sunpower M77 cryocooler. The conclusions drawn from Figure 4 are that stainless steel screens offer the best performance and that gains in heat-lift with charge pressure diminish at pressures above the 20 to 25 bar range. The falloff is believed to result from constraints placed on the regenerator wire diameter and porosity (wire diameter > 19.8 micron, and porosity < 0.75). On the basis of this study we selected stainless screens and a charge pressure of 25 bar for our single-stage PTR preliminary design.

DISCUSSION

The pulse tube design presented here is projected to lift 7 W at 77K compared to 5 W for the Sunpower M77 with the same input power. However this does not mean that PTRs have higher thermodynamic performance than displacer-type cryocoolers. The pulse tube presented in this report performs better only because of the use of a much more costly regenerator. An examination of Figure 4 comparing the M77 cryocooler to optimized PTRs with the same regenerator (lower most curve), shows that the pulse tube has significantly lower performance than the M77 when the two are compared on equal terms.

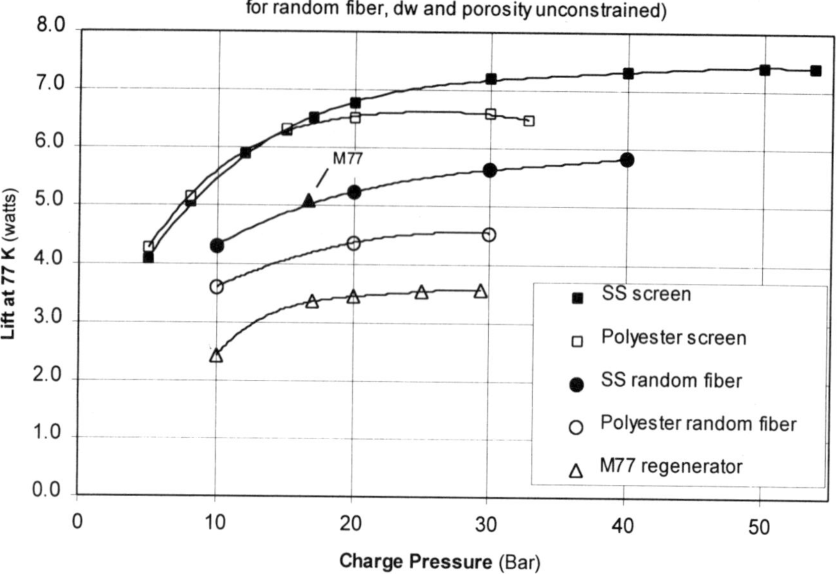

Figure 4. Performance versus regenerator material and charge pressure for optimized single-stage PTRs, with the performance of the Sunpower M77 Stirling cryocooler shown for comparison.

Pulse tube refrigerators are expected to be lower in performance than a displacer type cryocooler. The primary reason for this is that the expansion space work is not recovered in a PTR. An additional reason is that a PTR will usually have higher conduction losses than a displacer type cryocooler where the regenerator is internal to the displacer. The PTR in this case has additional conduction losses because of the additional pressure containing wall (wall of the pulse tube itself), as well as through the gas contained in the pulse tube.

ACKNOWLEDGEMENT

This study was supported in large part by NASA Goddard Space Flight Center, SBIR Contract NAS5-99052. We were aided in the effort reported here by David Gedeon of Gedeon Associates, Athens, Ohio, who is the author of Sage and is a respected consultant in the pulse tube field. We wish to acknowledge Mr. Gedeon's contribution to this study.

PERFORMANCE OF THE XRS ADR HEAT SWITCH

E. R. Canavan, J. G. Tuttle, P. J. Shirron, and M. J. DiPirro

NASA-Goddard Space Flight Center, Code 552
Greenbelt, MD 20771

ABSTRACT

The X-Ray Spectrometer (XRS) aboard the Astro-E satellite uses an adiabatic demagnetization refrigerator (ADR) to cool its detectors to the operating temperature of 60 mK. A ^{3}He-filled gas-gap heat switch thermally connects the ADR's salt pill to a pumped liquid helium bath while the pill is magnetized. The switch thermally opens to isolate the pill during the demagnetization phase. At 1.3 Kelvin the XRS heat switch has open and closed-state conductances of about 4 μWatts/Kelvin and 78 mWatts/Kelvin respectively. The flight model heat switch showed a performance anomaly in which it was latched in the on state after a normal cooldown. Repeatedly cycling the getter on and off returned the switch to normal operation, but the number of cycles required to do this increased each time the switch was thermally cycled to room temperature. It is believed that this anomaly was due to water contamination of the getter. A procedural work-around allowed this switch to be used despite the anomaly. We present the results of the switch's performance testing, and we describe in detail the anomaly and the work-around.

INTRODUCTION

Bolometers and superconducting tunneling junctions operating at temperatures below 0.1 K offer unprecedented sensitivity for photon detection, and promise vast improvements in the capabilities of astronomy over broad ranges of the electromagnetic spectrum. The first x-ray satellite to make use of such a low temperature imaging spectrometer will be the XRS instrument on the Astro-E satellite, to be launched in January 2000.

The XRS project and other proposed missions will use adiabatic demagnetization refrigerators (ADRs) to provide the final stage of cooling. An ADR consists of a mass of paramagnetic salt connected through a thermal bus to a heat switch, which is attached to the next refrigeration stage, typically, a bath of liquid helium. The salt is suspended in the bore of a large magnet. Initially, the magnet is at full field and the salt is at the bath temperature. The heat switch is put in its low conductance state and the field is ramped down, rapidly at first to bring the thermal bus down to the operating temperature, then slowly to counteract the parasitic heat leak and the detector dissipation. When the field reaches zero, the magnet

Advances in Cryogenic Engineering, Volume 45.
Edited by Shu *et al.*, Kluwer Academic / Plenum Publishers, 2000.

is ramped up. As the salt temperature exceeds the bath temperature, the heat switch is put in its high conductance state, dumping the heat of magnetization to the bath and eventually pulling the salt back down to bath temperature.

The parameters of the heat switch in large measure determine the ADR's operating characteristics . The off-state conductance sets the contribution of the switch to the heat load on the thermal bus. The heat load, the mass and composition of the salt pill, and the maximum field determine the hold time, the time the system spends at the operating temperature. The on-state conductance of the switch, along with the time to turn it on and off, determines the duration the system must spend at the bath temperature, which for the detector is "dead time."

A type of heat switch with many attractive features for use in orbit is the getter-pumped gas-gap heat switch. In the off-state, the cool getter insures that there is a hard vacuum inside the body of the switch, so the only significant heat leak between the warm and cold ends is due to conduction through the outer shell. Applying power to the heater raises the getter's temperature, causing it to release helium gas. As the pressure rises, the conduction through the gas across the gap between the warm and cold ends becomes large. If the device's geometry is arranged so that there is a narrow gap with a large surface area, but the shell's conductance is small, the switch can have a large on/off conductance ratio.

The gas-gap heat switch has several advantages over its principle contender, the mechanical clamp heat switch: it has no moving parts to cause vibration or stick, it requires very little power to operate, it can be compact and lightweight, and it can have a very high on-state conductance at relatively low temperatures (~1K). In principle, it should be very reliable, because there are no parts that can fatigue or wear out. It has the disadvantage that it must have a shell impermeable to helium that encloses the region between the warm and cold ends. The conductance of the shell determines the switch's off state conductance.

XRS HEAT SWITCH DESIGN

The requirements on the XRS heat switch were tightly coupled with the requirements on the ADR. The ADR must absorb 0.3 μW from the detector assembly and have a hold time of at least 19 hours. With the flight salt pill this requires the open switch to conduct less than 5.5 μW from 1.3 Kelvin to 65 mK. The ADR must be able to cycle, that is to magnetize, dump its heat to the bath, and demagnetize to the operating temperature, within 60 minutes. This means that the closed switch must conduct at least 20 mW from 1.8 Kelvin to 1.3 Kelvin. To keep the system compact, the heat switch is located in the bore of the magnet, mounted directly to the salt pill and to a flange at the bath temperature. It is thus part of the salt pill suspension, providing stiffness in the axial direction. Kevlar lines between the salt pill and a frame surrounding it provide stiffness in the other degrees of freedom.

Figure 1 shows a cross section of the XRS heat switch. The cold (salt pill) end of the switch consists of a flange with a coaxial rod and cylinder. The warm (bath) end consists of a flange with two coaxial cylinders. These parts, both of gold plated copper, interleave so that they are separated by a 0.5 mm gap with a total surface area of approximately 72 cm^2. A cylindrical Vespel shell encloses these fins. Vespel was chosen because of its extremely high strength to thermal conductivity ratio in this temperature range. The Vespel thickness, 0.5 mm, is determined by the maximum load on the switch. Because Vespel is permeable to helium, the shell is covered by a 12 μm thick foil of Ti15V3Cr3Sn3Al bonded with a ~10 μm thick layer of Scotchweld 2216 epoxy.

The getter is Varian Vacusorb Molecular Sieve Charge, which is sold for charging low temperature sorption pumps. Chemically, it is zeolite 5A bound with 20% "inert clay binder" to form pellets. Fifty five pellets, 1.5 mm in diameter and 5.4 mm in length, are epoxied into an aluminum cup with Epo-Tek H20-E silver filled epoxy. Silver filled epoxy

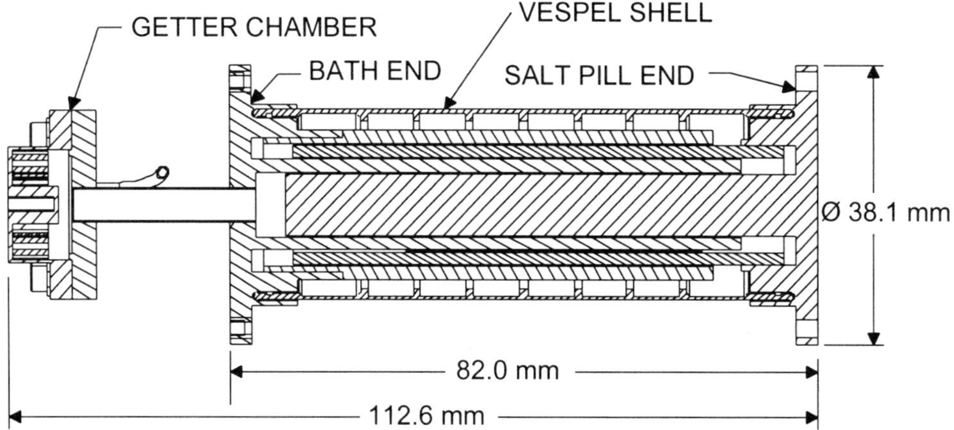

Figure 1. Cross section of the XRS heat switch.

is used to improve the thermal conductance of the getter to the cup. The cup is attached to a flange to form a chamber which is connected to the warm end of the switch through a 0.125 mm wall stainless steel tube 4.5 mm in diameter and 18.6 mm long. A thin gold plating on the tube increases its thermal conductance to the desired value. The conductance between the getter chamber and the bath, along with the getter chamber's heat capacity, determine the time constant for the heat switch to change states. This conductance is chosen to keep the time to turn the heat switch off of the same order as other components of the cycle time, such as the magnet ramp time. Making the conductance substantially larger increases the heat load on the bath without significantly reducing the cycle time.

Three space-flight heat switches were made according to this design. The first one, known as switch F, is currently part of the flight cryostat. The second one, switch G, was kept in the lab for extensive testing. The third switch, H, is in the early stages of testing. The performance results for these switches are presented later in this paper.

TEST RESULTS: EARLIER DESIGNS

Engineering model heat switches were made in 1996 using Vespel wrapped with the Ti foil bonded with Armstrong A-12 epoxy. No problems were observed with the switches degrading over time or with thermal cycles, and few problems with debonding of the Ti. Typical off-state heat conductances (from 1.3K to .065K) were 3.2 to 3.8 μW.

A second series of switches, considered to be the space flight series at the time, was made in 1997 and 1998. The design of these switches was the same as the final flight design except for the foil bonding technique and some structural details, including a smaller gap (0.25 mm) between the warm and cold cylinders.

By this time the Armstrong A-12 epoxy was being sold by a new manufacturer. It evidently had a new formulation, as it no longer completely cured at room temperature in a reasonable amount of time. When Ti foil was bonded to the Vespel body of a switch with this new epoxy, the foil began to peel off after a few thermal cycles. This increased the rate at which the helium leaked out of the switch, reducing the switch's useful life. In addition, some helium gas which permeated through the Vespel at room temperature became trapped in voids between the Vespel and Ti. At low temperatures, however, the gas in the voids could not permeate back through the Vespel into the switch body. At temperatures below about 1.6 Kelvin this gas condensed into a superfluid film which thermally shorted part of the switch. The voids resulting from this debonding grew and increased in number with thermal cycles, leading to an observed degradation in off state conductance.

Various new bonding techniques were tried in order to solve this problem, including crimping the overlapping ends of the foil together, wrapping the foil coating with Dacron strings soaked in A-12 epoxy, and curing the epoxy at a slightly elevated temperature after the initial room temperature cure. While these techniques solved the debonding problem in one switch, the results were not easily reproducible in all switches.

One of the switches with gaps under its foil was pumped out and refilled with ^{3}He to eliminate superfluid films under the foil. After it was sealed, the switch was tested as an individual component and then as part of an ADR. The parasitic heat load between 1.3 Kelvin and 65 mK was 3.9 μW. Because of its superior qualities, ^{3}He was used in all subsequent heat switches. However, bonding the Ti foil on the switches with Armstrong A-12 epoxy was not sufficiently reliable for a space-flight application.

TEST RESULTS: FLIGHT DESIGN

Heat switch F was filled with ^{3}He and tested as a component in July, 1998. At 1.37, 1.56 and 4.27 Kelvin its on conductance was found to be 0.088, 0.097 and 0.176 W/K respectively. The off conductances at 1.25, 1.32, 1.50, 1.94 and 4.30 Kelvin were 3.20, 3.53, 4.59, 7.23, and 14.3 μW/K respectively. The integrated switch conductance from 1.3K to 65 mK plus the other known parasitic heat loads were calculated to be 2.3 μW total.

Following this test the switch was integrated with a salt pill, and the resulting ADR was performance tested. The measured total parasitic heat load was 2.6 μW. After a few days at room temperature the ADR was retested with similar results. It was then warmed up and integrated with the XRS flight cryostat. After about a month at room temperature it was cooled back down and held at 4.2 Kelvin for 4 days and 1.3 Kelvin for 2 days before being retested. This time the ADR required most of its cooling power to reach its target temperature on the initial demagnetization. Analysis indicated that the cold plate temperature dipped during demagnetization, implying that the heat switch had not opened properly. However, by the fifth ADR cycle, the ADR was operating normally.

The XRS cryostat was warmed to room temperature for about a month two more times, and the ADR was tested after each of these warm periods. In both cases the heat switch did not open during the initial cooldown and required a number of getter cycles before it performed normally. Figure 2 shows the number of cycles required as a function of the number of days at room temperature since the switch was filled with ^{3}He.

After anomalous behavior was first observed in the operation of the flight ADR, an identical heat switch, designated HS-G, was tested as an individual component. The initial off-state conductance was approximately 4 μW/K at 1.3 K. The switch was turned on by applying nominal operating power to the getter heater (1.9 mW) for 20 min, the typical on time during ADR operation. The getter was then allowed to cool back down to the bath temperature. It was cycled ten times in this manner. The only unusual behavior noticeable in the data was that the getter temperature at which the switch conductance crossed a certain threshold increased with each cycle. On the next cooldown, the conductance with the getter at the bath temperature was 78 mW/K, which is the on state conductance. Thus, the switch was latched in the on state. After warming up to inspect for physical damage and cooling back down, the switch had a conductance of 74 mW/K with the getter at the bath temperature. However, after cycling the getter heat power (20 minutes on, 40 minutes off) 6 times, the conductance began to drop noticeably. The threshold for detecting the drop was approximately 20 mW/K. With additional cycles, the minimum conductance continued to drop, and it stabilized after approximately 25 cycles. Minimum conductance occurred at the end of each cycle, just before the getter heater turned on.

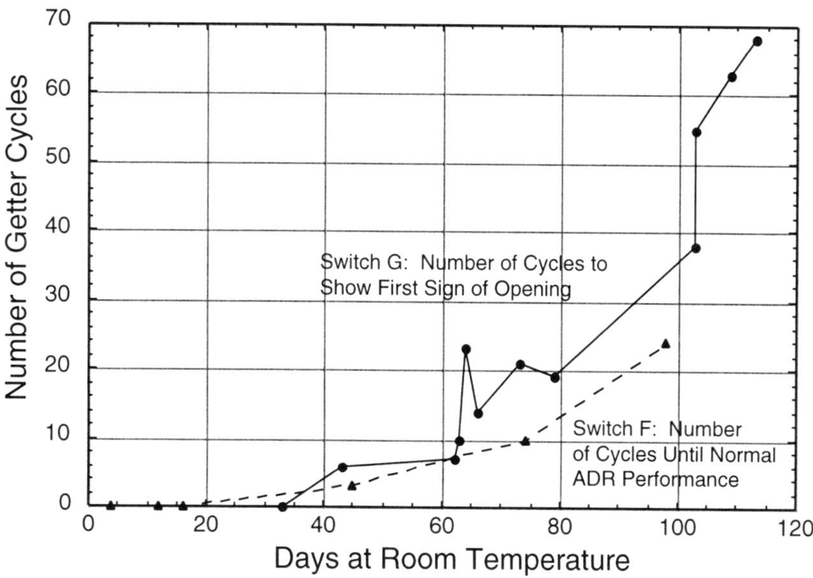

Figure 2. Performance degradation of heat switches HS-F and HS-G as a function of time at room temperature.

The number of heat switch cycles before the conductance dropped below 20 mW/K was set as a criterion. Using this criterion, the recovery of the heat switch on many additional cool-downs was monitored. Figure 2 shows the number of cycles to start of recovery versus the time spent at room temperature. Clearly, the trend is for the number of cycles to increase with time, although the data are not perfectly monotonic. Fitting number of cycles verses total time elapsed, time spent at room temperature, or number of thermal cycles between room temperature and helium temperature results in a similar dependence. Figure 3 show the minimum conductance vs. the number of cycles for the most recent test of switch G. The 20 mW/K criterion was met after 68 cycles.

The rate of the cool down does not appear to affect the trend. In one test, power was applied to the getter heater to maintain the getter at room temperature as the switch body cooled down, and then to maintain a large temperature difference between the getter and the switch body. Once the switch body reached 4.2 K, the getter was held at 240 K for two hours before the power was removed. If the anomalous behavior were due to contamination of the getter by highly volatile materials that are not strongly bound to the getter at room temperature, such as N_2 or O_2, these would have condensed onto the walls of the switch body, leaving the getter in a cleaner state. However, the number of cycles to recover increased after this treatment.

It was found that raising the getter to 30 K for 90 minutes reset it to the same latched on state that it was in following cool-down. That is, the same number of getter cycles was required to recover proper operation of the switch. This was convenient, because other techniques of recovering the heat switch could be probed without warming up and cooling down the dewar. Testing new methods became important as the number of one hour cycles to completely recover the heat switch started to approach one hundred. One technique that was found to be faster was to hold the getter at its nominal on temperature, 13.8 K, for several hours, then slowly step down its temperature in approximately 0.5 K intervals, holding it at each temperature three hours. Once the temperature was below 8.5 K, the heater was turned off completely, and the switch opened thermally as the getter cooled to the bath temperature.

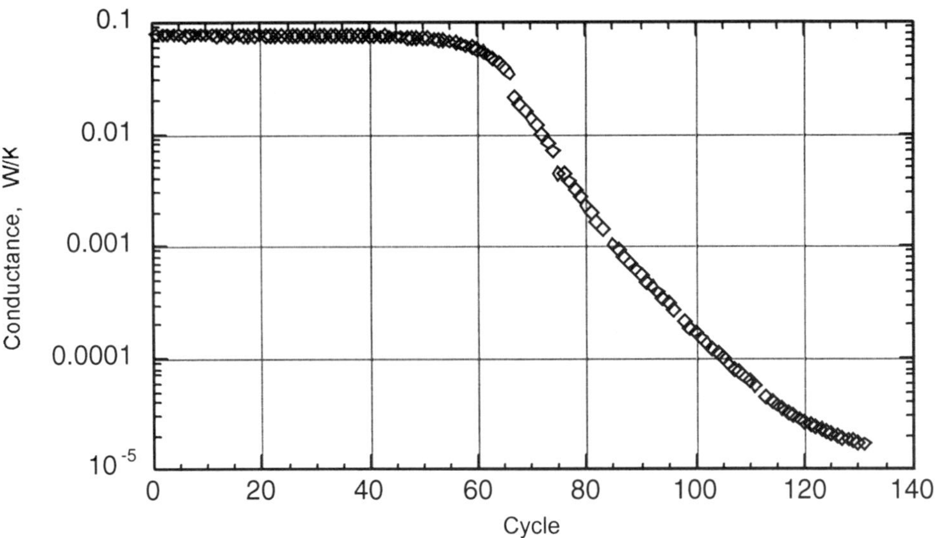

Figure 3. Heat switch conductance at the end of each cycle. The "off" criterion was met after 68 cycles.

Switch H was made in the spring of 1999. Extreme care was taken to dry out the Vespel shell and to keep it dry during the switch assembly. When it was first cooled down for testing, switch H turned completely off as the getter cooled to the bath temperature.

The thermal conductances of switches G, E and H were measured with the switch bodies at 1.30 Kelvin as a function of the getter temperature. The gas gap conductance, obtained by subtracting the off-state conductance from the total switch conductance, is shown in Figure 4. The data for switch E, one of the pre-flight switch series containing ^3He, were obtained in July, 1998. The switch G data were measured at three different times, as indicated in the figure. The October and January data were taken by stepping the getter temperature up after the getter had been warmed to 13.8 Kelvin and cooled to 1.3 Kelvin. The March data were taken during the step-down process described above. The switch H data were taken in July, 1999 by stepping the getter temperature up.

The curve labeled "theory" is calculated using the relation for heat transport in the free molecular flow regime. The pressure is estimated from data of Daunt and Rosen[1] for adsorption of ^3He on zeolite 13X, assuming a constant isosteric heat of adsorption. Data are not available for the heat of adsorption of ^3He onto zeolite 5A, but measurements by Flapper and Gijsman[2] at 20 and 77 K indicate that the adsorption isotherms for He on these two zeolites are similar.

It is clear that switch G's getter did not perform as well as those of switches E and H below about 13 Kelvin. Interestingly, the March data for switch G show better gettering than the earlier results, while the number of getter cycles to open the switch after cooldown was much higher in March. It is as though there were two different gettering sites in switch G: one which gettered ^3He quickly but had limited capacity, and another which had high capacity and gettered slowly at high temperatures but stopped gettering below about 8 Kelvin.

EXPLANATION OF DEGRADATION: H_2O CONTAMINATION

It seems likely that the getters of switches F and G were contaminated. Zeolite 5A contains approximately 2.9×10^{-4} moles of alpha cages. Since we filled each switch with 5.2×10^{-4} moles of helium, the average occupancy of an off-state getter should have been

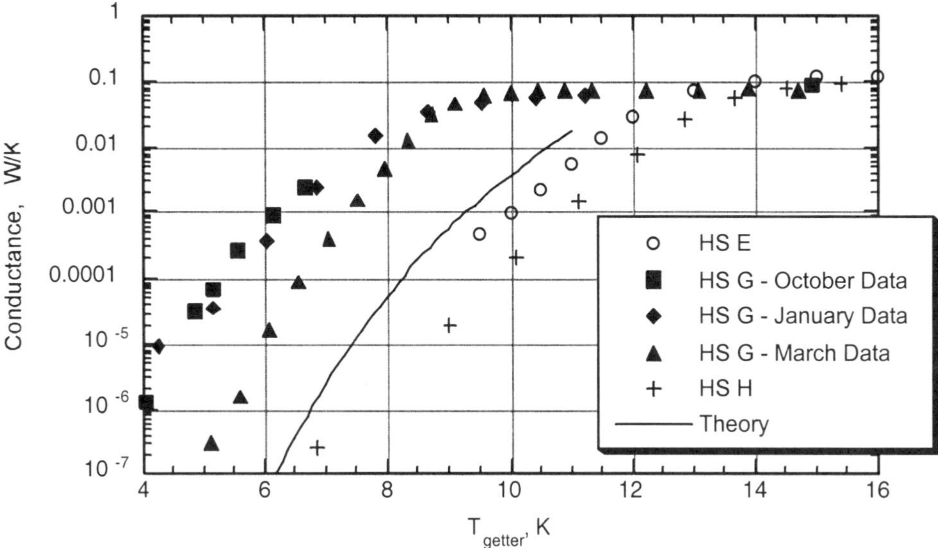

Figure 4. Gas gap thermal conductance data for switches E, G and H with switch body at 1.30 Kelvin. Here the measured off-state conductance has been subtracted from the total switch conductance data.

1.8 helium atoms per cage. The maximum helium occupancy is about 12 atoms per cage[5]. The Vespel in the heat switch bodies is known to adsorb water. Vespel shells have been machined and processed exactly like those in the switches and then degassed at elevated temperature on a microbalance. The total amount of water in the shells is as high as 3.6 millimoles, or 12.4 molecules per zeolite cage. The time constant for removing this water from the Vespel in a vacuum at room temperature is approximately 125 hours; much longer than the length of time any one of the switches before H was pumped before installing its getter.

It seems that the zeolite in a switch's getter should adsorb a significant amount of water from the Vespel in the days or weeks following the getter installation. It is possible that an excessive amount of water in the zeolite slows the adsorption of helium in the getter and even stops the adsorption altogether or limits the total amount that can be adsorbed at low enough temperatures (at 1.3 Kelvin, for example). If this were the case, the zeolite might not remove enough helium from the switch during the part of the cooldown when the getter was above 10 to 14 Kelvin, and the switch would not open. Subsequent cycling of the getter to 14 Kelvin or performing the step-down process might allow more of the helium to be adsorbed, and the small amount of remaining gas could be gettered away even at 1.3 Kelvin.

This qualitative model is consistent with many aspects of the observed behavior of switches F and G. However, it does not explain why switches D and E always opened on cooldown or why switches F and G continued to deteriorate months after they were sealed up. It's possible that some yet-to-be-determined difference in the processing of the new switches led to a higher water content in the Vespel. The second problem could be explained if the Vespel and zeolite each reach some equilibrium water content, while maintaining a constant water partial pressure in the switch. Perhaps thermally cycling the switch to room temperature temporarily redistributes the water in the zeolite, allowing it to getter more water at the given partial pressure. If this were true, the performance might be expected to deteriorate with each thermal cycle, which is roughly what has been observed.

This model also fails to explain another observation. In an attempt to remove water from switch G's getter, the getter was heated to 48 Celsius while holding the switch body at –90 Celsius (where the vapor pressure of water is 7×10^{-5} torr) for 5.5 hours. A 50 Celsius sample of zeolite 5A in a water vapor pressure of 10^{-4} torr has an equilibrium water content of less than 1% of the zeolite mass[4]. A getter hydrated by extended storage at ambient conditions was degassed in vacuum at 50 Celsius to determine its desorption rate, and it lost 75% of the water in 5 hours. Clearly, switch G's getter should have lost a significant quantity of water while at 48 Celsius. After this process the switch body was cooled quickly to 77 Kelvin, and the getter was allowed to cool slowly to the same temperature. The next day the switch was cooled to 1 Kelvin, and the getter was then cycled to 13.8 Kelvin until it began to open. Surprisingly, the number of cycles to begin opening the switch rose to 55 from 38 on the previous cooldown. The switch body never warmed above 230 Kelvin during this entire process. One possible explanation for these results is that the water distribution in the zeolite may be more important than the actual amount of water. Measurements of water transport in zeolite beds indicate that the time scale for transport at 50 C is more than an order of magnitude lower than that at 23 Celsius[5]. The large temperature gradient set up by quickly cooling the getter to might maintain a large concentration of water near the outer surface of the getter. This might degrade the getter performance more than a larger quantity of water evenly distributed through the getter.

CONCLUSION

The XRS cryostat, including the ADR and heat switch F, was shipped to Japan in January of 1999 to be integrated with the XRS flight dewar. The results of testing switch G from January to June were used to develop a plan for getting switch F into a working mode after its cooldown. After five months of integration, the cryostat was cooled down in June of 1999. The plan, including the getter temperature step-down process, was carried out, and the switch opened completely during the first cycle of the ADR. The cryostat should remain cold until its launch in early 2000. It is assumed that the ADR heat switch will continue to perform adequately as long as it stays cold.

ACKNOWLEDGEMENTS

The authors wish to acknowledge Armando Morell and his team for building the flight heat switch bodies. This work is supported by NASA's Office of Space Science.

REFERENCES

1. J. G. Daunt and C. Z. Rosen, Multilayer adsorption of ^{3}He and ^{4}He on zeolite from 4 K to 80 K, *J. Low Temp. Phys.* **3**:89 (1970).
2. G.L.B.F. Flapper H.M. Gijsman, Gravimetric measurements of gas adsorption on zeolites at low temperatures, *Cryogenics* **14**:150 (1974).
3. H.J. Halama and J.R. Aggus, Measurements of adsorption isotherms and pumping speed of helium on molecular sieve in the 10^{-11} – 10^{-7} Torr range at 4.2 K, *J. Vac. Sci. Tech.* **11**:333 (1974).
4. R.R. Bannock, Molecular Sieve Pumping, *Vacuum.* **12**:101 (1962).
5. P.D.M. Hughes *et. al.*, Water diffusion in zeolite 4A beds measured by broad-line magnetic resonance imaging, *Phys. Rev. B*, **51**:11334 (1995).

DEVELOPMENT OF GAS GAP HEAT SWITCH ACTUATOR FOR THE PLANCK SORPTION CRYOCOOLER

M. Prina,[1] P. Bhandari,[2] R. C. Bowman Jr.,[2] C. G. Paine,[2] L. A. Wade [2]

[1] Politecnico di Milano
Milano, Italy
[2] Jet Propulsion Laboratory
Pasadena, CA, 91109

ABSTRACT

A heat switch has been designed and characterized for the thermal control of the compressor beds in ESA Planck Surveyor 20 K sorption cryocooler. The heat switch has been designed to minimize the overall cryocooler power and to maximize the stability of the cold tip temperature, both of which are strongly affected by the performance of the gas-gap heat switch actuator. The gas gap thermal switching is obtained by varying the pressure of hydrogen in a closed volume defined by the two surfaces to be thermally connected or isolated. Hydrogen pressure within this volume is controlled by thermally varying the equilibrium hydrogen content of a hydriding metal. Switching characteristics are determined by the mechanical and thermal properties of the switch actuator containing the hydriding material to give a heat conduction ratio of over 1000 between the ON and OFF states. We present the general principles of gas gap heat switches that impact their performances for the Planck cryocooler, develop two figures of merit relevant to this application, and present initial characterization results of switches using uranium or ZrNi as the hydriding material.

INTRODUCTION

An 18 K continuous-cycle sorption cryocooler is being developed[1,2] for the ESA Planck mission[3], which will map the anisotropy of the cosmic microwave background. This cooler will produce liquid hydrogen by a closed-cycle Joule-Thomson expansion of gas compressed and circulated using metal hydride sorbent beds in the compressor assembly. The compressor assembly will be located on the spacecraft bus and is attached to radiators for passive cooling to 280 K in order to provide a heat sink for cooling the sorbent beds to their absorption temperature and rejecting the heat of absorption. Each compressor bed will be provided with a gas gap heat switch for thermal coupling or decoupling of the sorbent material from the 280 K radiator. While gas-gap switches have been used previously on cryocoolers[4,5], the operating conditions and power limitations for the Planck sorption coolers require modified design and performance. In this report we concentrate on the development and behavior of the gas-gap actuators for the Planck mission; the overall behavior of the sorbent compressor elements will be reported elsewhere.

HEAT SWITCH DESCRIPTION

The gas gap heat switch is used to thermally connect or disconnect two objects, by filling the gap between the surfaces of the objects with gas at a pressure P_{ON}, or evacuating the gap to a pressure P_{OFF}. The pressure in the gap determines the heat transfer, as indicated in Figure 1. The pressure P_{ON} will be close to P_2 , and P_{OFF} will be lower than P_1.

Optimal heat switching requires a high gas thermal conductivity and an actuator to reversibly insert and extract the gas. Hydrogen and hydriding metals have been selected as the approach for this gas-gap switch actuator.

OPERATION CONCEPT OF THE HEAT SWITCH

The hypotheses assumed in the heat switch design are:

- Hydrogen with a maximum pressure below 20 Torr and minimum temperature above 280 K can be considered an ideal gas
- Switch assembly is isothermal. The thermal gradient is only in the tube connecting the actuator to the compressor bed.
- The tube is considered only as a conductance and not as an additional thermal mass.
- Infinite kinetic speeds of the H_2 absorption and desorption reactions.
- The heater power supplied during the ON state is constant and equivalent to the switch ON state equilibrium power (see Eq.(2)).
- The switching operations are synchronous with the compressor bed operations
- The heat of re-absorption is negligible.
- The heater has a thermal mass equal to the shell mass (hydrides, shell and filter)

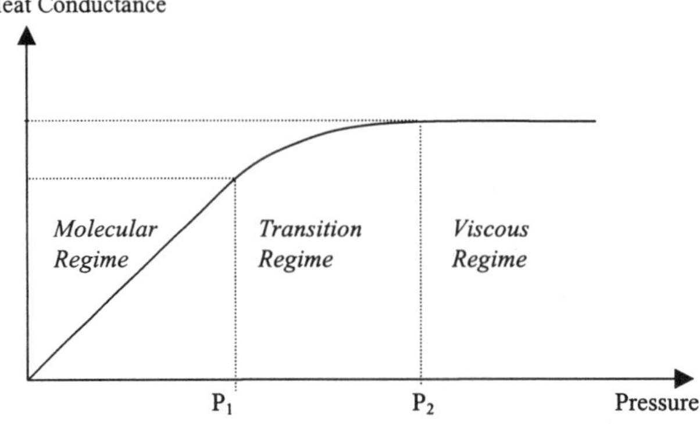

Figure 1. Heat conductance versus pressure.

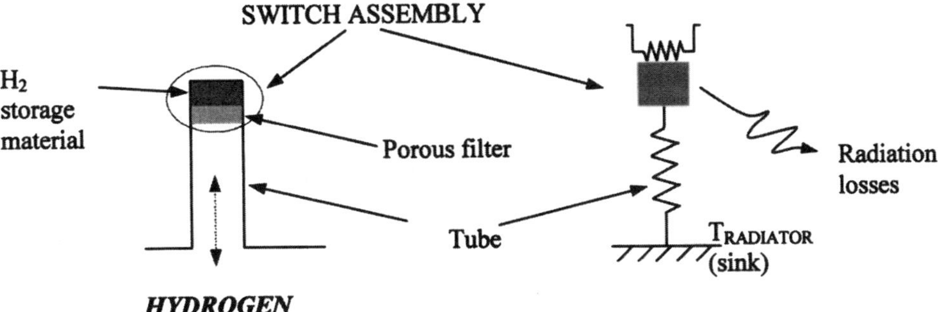

Figure 2. Schematic and equivalent thermal circuit of the gas gap heat switch.

The general configuration of a gas gap heat switch is shown in Figure 2.

The relations describing the switch actuator thermal behavior according to the previous hypothesis are:

- thermal switching of the assembly

$$(mc_p)_{SwitchAss.} \frac{dT_{SwitchAss.}}{dt} = Power - G(T_{SwitchAss} - T_{Sink}) - \varepsilon\sigma_0 A_{SwitchAss}(T^4_{SwitchAss} - T^4_{Sink}) \qquad [1]$$

where, G is the heat conductance of the stainless steel tube (length L and diameter D), $(mc_p)_{Switch\ Ass.}$ is the thermal mass of the Switch assembly,
$A_{Switch\ Ass.}$ is the area of the Switch assembly
ϵ is the emissivity of the switch assembly,
σ_0 is the Stefan-Boltzmann constant and T is the temperature [K].
The previous expression, solved for steady state condition, provides the value of the $Power_{Switch}$ required to maintain the switch in the ON state:

$$Power_{Switch} = G(T_{ONstate} - T_{Sink}) + \varepsilon\sigma_0 A_{SwitchAss.}(T^4_{ONstate} - T^4_{Sink}) \qquad [2]$$

- Van't Hoff equation for the hydride hydrogen pressure.

$$\log P_{Assembly} = B - \frac{A}{T_{Assembly}} \qquad [3]$$

The constants A and B depend on the H_2 storage material and define the pressure $P_{Assembly}$ produced within the gas gap assembly.

- Flow resistance from porous filter

Pressure drop across the porous filter is considered using the equations:

$$\dot{n} = \frac{V_{Gap}}{RT_{Gap}}\left(\frac{dP_{Gap}}{dt}\right) \qquad [4]$$

And

$$\left(P_{Gap} - P_{Assembly}\right) = \frac{4\dot{n}M_{H_2}RT_{Assembly}}{\pi D^2\sqrt{\frac{T_{Assembly}}{294}}\,\overline{C}\,\frac{t_{Filter}}{t_{sample}}} \qquad [5]$$

where P_{gap} is the hydrogen pressure in the gap, V_{gap} is the volume between the two surfaces, R is the universal gas constant, D is the diameter of the porous filter (the same of the tube) and t_{sample} is the thickness of a the test filter that at 294 K gave the flow conductance C, while t_{filter} is the thickness of the used filter. n is the hydrogen mass flow [moles/s] and M_{H2} is the hydrogen molecular mass.

- Thermal conductance as a function of pressure[6]

$$G = Area \frac{K_{H_2}}{Gap + (9\gamma - 5)\frac{2-\alpha_{tot}}{\alpha_{tot}} L_{mfp}} \qquad [6]$$

Where:

K_{H2} is the hydrogen conductivity at the temperature of the gas.

γ is the ratio of specific heat at constant pressure to that at constant volume for hydrogen at the temperature of the gas.

Gap is the distance between the two surfaces.

L_{mfp} is the mean free path of hydrogen at a pressure P_{Gap} and at a temperature T_{Gap}.

Area is the area of the surfaces on the gap.

α_{tot} is the global accommodation coefficient defined as

$$\alpha_{tot} = \frac{\alpha_{comp.unit} \cdot \alpha_{radiator}}{\alpha_{comp.unit} + \alpha_{radiator} - \alpha_{comp.unit} \cdot \alpha_{radiator}} \qquad [7]$$

and the individual α are accommodation coefficient for the two surfaces, a measure of the efficiency of thermal transport between the surface and the gas: $0<\alpha<1$.

SYSTEM REQUIREMENTS

Gas-gap switches are incorporated in the Planck sorption compressors to allow the sorbent material (a $LaNi_{4.8}Sn_{0.2}$ hydride) to be isolated from its surroundings while heated to approximately 480 K for desorption of stored hydrogen, then to cool the sorbent material to approximately 280 K for re-absorption of the hydrogen refrigerant fluid into the rare earth alloy material. The sorbent portion of each compressor element is cycled through this temperature range once in approximately 4000 seconds. Constraints on available power and radiator size impose performance requirements on the gas-gap switch actuator. The actuator must

1) supply gas at sufficient pressure that the ON-state conductance is adequate for cooling of a sorption bed to the temperature at which re-absorption occurs, and rejection of the heat of absorption at that temperature
2) extract gas to low enough pressure that OFF-state residual gas conduction is tolerable while the sorbent material is at high temperature
3) perform both transitions between ON- and OFF-states quickly enough that there is adequate time to cool a bed for re-absorption, and time to heat a bed for desorption.

Consideration of the Planck operational conditions indicates that switching OFF-to-ON time and the OFF-state steady-state conductance at high temperature are the most stringent performance requirements for this application. From this, and from constraints on total system electrical power, we find it useful to express the behavior of the switch actuators in terms of two figures of merit, as discussed below.

Figure Of Merit: OFF-ON Switching Margin

As is seen in Figure 1, there is a limiting conductance for a gas gap; even if the gas-gap became conductive instantaneously, the time to transfer heat across the switch is finite. For a real actuator, which requires some time to change the conductance within the gas-gap, the time required to transfer heat is related to the gas-gap pressure as a function of time. In the evaluation of designs presented below, the Figure of Merit "Margin" is the fractional enthalpy which could be transferred across the switch within the allowed time interval, in excess of the requirements for the Planck system design:

$$\text{Margin} = \frac{(mc_p) - (mc_p)_{Comp.unit}}{(mc_p)_{Comp.unit}}$$

where (mc_p) is the maximum enthalpy transferable by a given switch design, and $(mc_p)_{Comp\ unit}$ is the requirement for the PLANCK system design.

Figure Of Merit: Total Power

The gas-gap switch actuators are controlled via heaters, which dissipate electrical power. The primary sorption compressor storage material also employs electrical heaters for cycling the working fluid; gas remaining in the gas-gap switch during the actuator OFF-state can increase this latter power requirement. The total system electrical power requirement is dominated by the sum of these two contributions. It is this electrical power which we wish to minimize through design of the actuator.

It is clear that these two figures of merit give rise to conflicting design requirements; minimizing the switching time is easily done by increasing thermal conduction between the actuator storage material and its thermal sink, but this increases power dissipation required to heat the actuator. Similarly, a choice of hydrogen storage material that yields lowest pressure at low temperature (OFF-state) will require higher heater power to maintain adequate pressure at high temperature (in the ON-state). We have examined the effects of configuration and material selection on the overall system performance, relative to the above figures of merit; these considerations and results are presented next.

GAS-GAP ACTUATOR DESIGN

We have selected two materials, ZrNi and Uranium, for evaluation as hydrogen-storage materials for gas-gap switch actuators. We have parameterized the response of these actuators in terms of the material properties, thermal characteristics of the actuator design, and gas-gap switch behavior. We have constructed switch actuators utilizing these materials based upon the results of the parameterization; these actuators are now undergoing characterization.

The physical design of the switch actuator is as indicated in Figure 2. A tube of 12.7 mm OD, 0.15 mm wall thickness, 316L stainless steel tubing provides thermal isolation between a thermal sink (which is part of the gas-gap switch) and the switch actuator containing the hydriding material; 0.2-0.50 grams of hydriding material was encapsulated in the actuator, which included a stainless steel filter and an external heating element. With these parameters fixed, the thermal mass of the actuator calculated, and the behavior of the gas-gap switch for varying gas pressure calculated, we examined the effect upon gas-gap performance due to variation in four parameters associated with the actuator design and gas-gap characteristics:

- the temperature to which the actuator must be raised to produce the gas-gap ON condition at a pressure $P_{ON\ state} > 2$ kPa (15 Torr) (i.e. T ~ 470 K for ZrNi, T ~ 560 K for Uranium);

- the length of the thermal isolation tube, treated as a free parameter;
- the diameter of the isolation tube (including changes in gas flow related to the change in restriction due to changing diameter of the filter);
- the accommodation coefficient of the walls of the gas-gap (which is influenced by other system constraints that will not be addressed here).

The results of variation about a design point, relative to the Figures of Merit detailed above, are indicated in Figure 3a and 3b, for the two materials under consideration. Variation of the design parameters in this manner has allowed us to perform tradeoffs amongst the competing figures of merit while retaining reasonable physical and thermal designs for the gas-gap actuator.

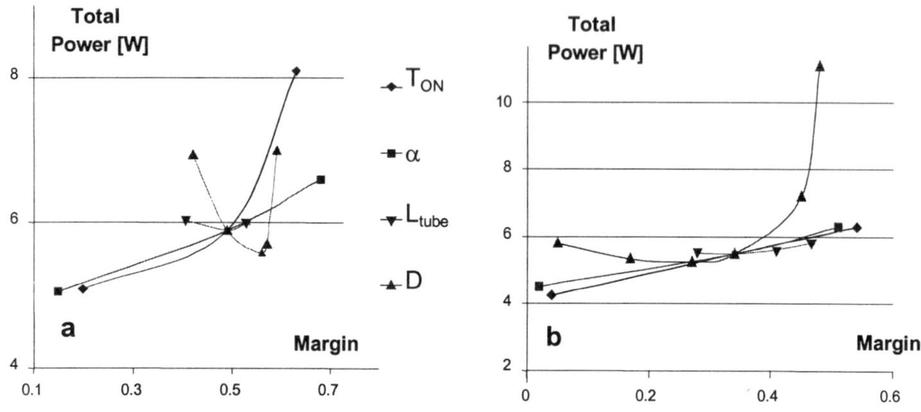

Figure 3. Variations in the gas gap switch performance about the design points for the ZrNi (a) and U (b) actuator hydride.

TEMPERATURE CYCLING TESTS

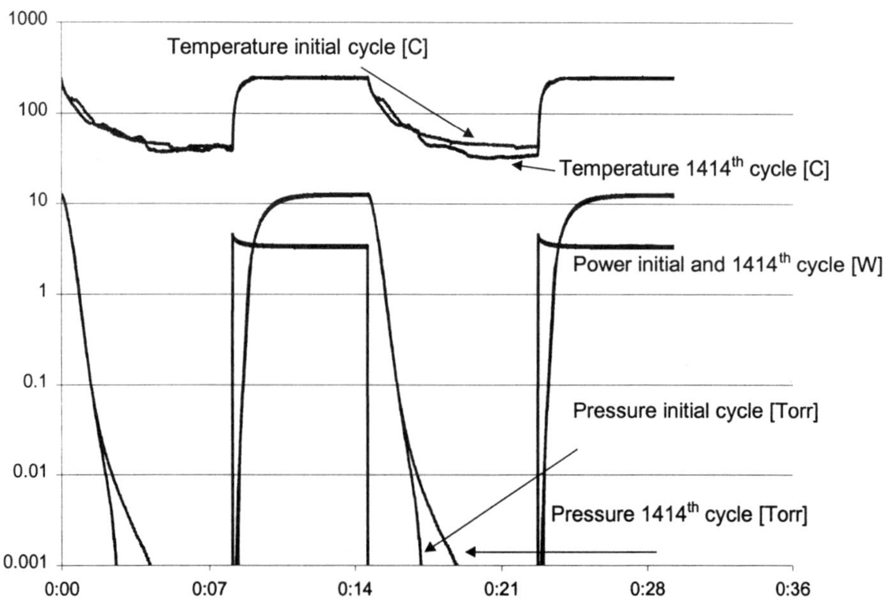

Figure 4. Pressure, Power, and Temperature profiles of gas gap switch containing ZrNi hydride observed during cycles numbers 1 and 1414.

Prototype versions of the gas gap switch configuration from Figure 2 have been fabricated to permit testing in the laboratory of the operational parameters and performance of the hydride materials during extended thermal cycling.

Heat switches have been built containing ZrNi and uranium metal with the hydride encapsulated in the 126 mm^3 volume between the cap and porous filter. The hydride sorbent was inserted as a bulk metal into the volume that was sealed by e-beam welding the cap, the filter and the tube together. After assembly of the heat switches and activation of the sorbent, the hydride was charged with the appropriate quantity of hydrogen before starting the cycling experiments. The response for temperature and pressure during the initial cycles agreed closely with the behavior predicted by our analyses.

The effect of continued cycling on a ZrNi actuator is shown in Figure 4, which compares the initial and 1414th cycles. No changes in either the input power or the temperature profile are seen. The ON state pressure values have remained constant, but rate of pressure decrease below 1.3 Pa (10 mTorr) is somewhat slower during the 1414th cycle. The source of this latter behavior is currently unknown and will be subject of further study as the cycling is continued.

During the initial cycles with the uranium gas gap switch, the actuator was heated to 558 K to produce an ON state pressure of ~ 2 kPa (~ 15 Torr). However, there were problems with attachment of the heater after a few cycles. After several modifications in heater attachment method and the maximum temperature, a stable configuration was found where a maximum temperature of 513 K was used to produce a 0.56 kPa (4.2 Torr) ON state pressure. No further problems have been detected during the next several hundred cycles. As an example, the parameters obtained during the cycles 430 and 970 are compared in Figure 5.

During these cycles, the power and temperature profiles are identical with only a small decrease in the pressure.

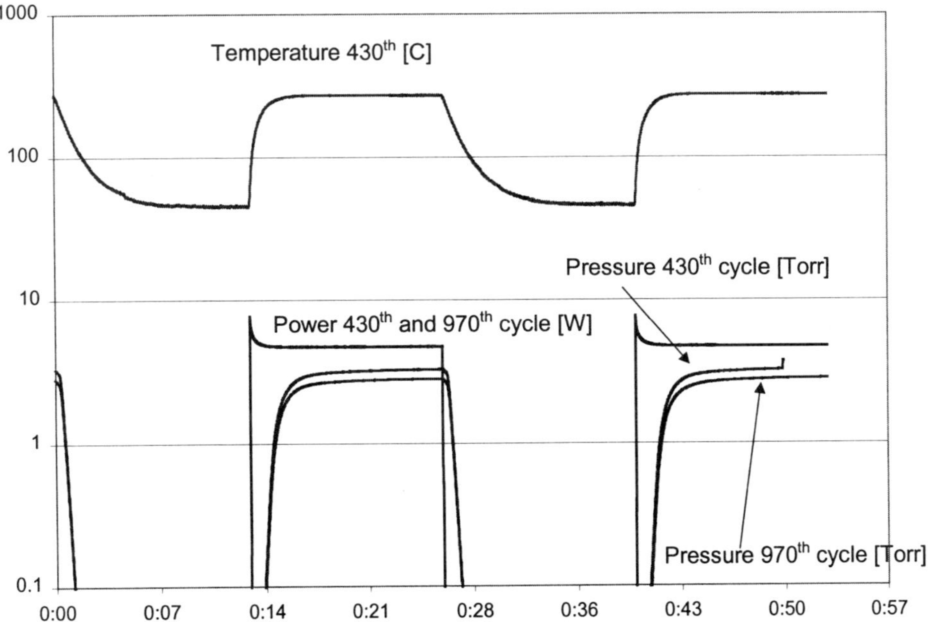

Figure 5. Pressure, Power, and Temperature profiles of gas gap switch containing uranium hydride observed for cycles numbers 430 and 970.

CONCLUSIONS

Operational requirements have been derived for the gas gap heat switch according to the present operational behavior of the Planck sorption cooler. Realistic mechanical designs for the candidate actuator materials ZrNi and U have been analyzed in terms of variations in design options. The designed configurations have been built and tested. Initial results validate the model predictions with respect to input power requirements and time constants for turning the switches on and off. Little degradation in the behavior of the actuator hydrides have been seen after ~1000 cycles. Future work involves long term cycling of the actuators as well as assessment of the heat transfer characteristics of the Planck compressor sorbent bed. These results will be used to refine the design of the gas gap switch and selection of the optimal actuator hydride.

ACKNOWLEDGEMENT

The research described in this paper was carried out by the Jet Propulsion Laboratory, California Institute of Technology, under contract with the National Aeronautics and Space Administration.

REFERENCES

1. P. Bhandari, R.C. Bowman, R.G.Chave, C.A. Lindensmith, G. Morgante, C. Paine, M. Prina, and L.A.Wade, Astrophysical Letters and Communications, (in press).
2. L. A. Wade, P. Bhandari, R. C. Bowman, JR., C. A. Lindensmith, C. Paine, G. Morgante, M. Prina presented at Cryogenic Engineering Conference, Montreal, July, 1999
3. B. Collaudin and T. Passvogel, *Cryogenics* 39:157 (1999)
4. S. Bard, J. Wu, P. Karlmann, P. Cowgill, C. Mirate, and J. Rodriguez, in: "Cryocoolers 8", R. G. Ross, Jr., ed., Plenum, New York (1995), p. 609.
5. D. L. Johnson and J. Wu, in: "Cryocoolers 9," R. G. Ross, Jr., ed., Plenum, New York (1997) p. 795.
6. S. Dushman, Scientific Foundations of Vacuum Technique, John Wiley & Sons, NY 1963, p. 43.

DEVELOPMENT AND SPACE FLIGHT VERIFICATION OF CRYOGENIC FLEXIBLE DIODE HEAT PIPES

D. S. Glaister[1], P. J. Brennan[2], M. Buchko[2], and M. Stoyanof[3]

[1]Ball Aerospace and Technologies Corporation
Boulder, CO 80306
[2]Swales Aerospace, Inc.
Beltsville, MD 20705
[3]The Air Force Research Laboratory
Albuquerque, NM 87119

ABSTRACT

The first ever cryogenic flexible diode heat pipes were developed and verified under micro-gravity conditions on the Space Shuttle on STS-94 (July 1997) and the previous, minimum mission STS-83. The heat pipe working fluids were oxygen (with an operating range of about 60 to 145 K) and methane (about 95 to 175 K). The space flights represented one of the few times that cryogenic heat pipes have been operated in 0-g, only the third time that an oxygen heat pipe has flown, and the first time that flexible cryogenic heat pipes have been operated in space. Overall, the micro-gravity verification proved that both heat pipes met or exceeded their requirements for transport capacity, diode reversal, and thermal resistance. Because these requirements were based on several potential applications, the heat pipes should meet the integration needs of the users.

INTRODUCTION

The CRYOFD (Cryogenic Flexible Diode) heat pipes were verified as part of the CRYOFD experiment which was a reflown Get Away Special (GAS) cryogenic test bed. CRYOFD was managed and developed by the Air Force Research Laboratory with NASA GSFC co-sponsoring the experiment. The Air Force Space Test Program and NASA GSFC Hitchhiker provided the Shuttle integration and support, and Swales under a subcontract to J&T Engineering designed and manufactured the heat pipes and experiment test bed. Mission support, data analysis, and reduction were performed by The Aerospace Corporation, AFRL, and Nichols Research. CRYOFD was a SBIR (Small Business

Innovative Research) program initiated in 1994 and culminating in 2 micro-gravity verification flights aboard the Space Shuttle in 1997.

Due to Shuttle fuel cell problems, the first flight aboard STS-83 was aborted after only 4 days of intermittent operation from April 4 to 8, 1997. During this flight, the startup and priming of both heat pipes were verified. However, no operating or capacity data were obtained. Because of the aborted flight, the entire Shuttle mission was successfully reflown aboard STS-94 from July 1 to 17, 1997. The CRYOFD experiment operated for nearly the entire duration of this flight and over 250 hours of micro-gravity characterization were performed on the heat pipes. The space flights represented one of the few times that cryogenic heat pipes have been verified in 0-g, only the third time[1,2] that an oxygen heat pipe has flown, and the first time that flexible cryogenic heat pipes have been tested in space.

TECHNOLOGY APPLICATION

The primary application for the heat pipes is to facilitate integration of cryogenic refrigerators (or cryocoolers) to space based infrared sensors. The use of a thermal strap or conduction bar to interface one or more cryocoolers to an instrument is acceptable only when spacing permits the close placement of the coolers (cold tip less than 0.15-0.25 m from instrument interface). Often, the cryocooler must be a significant distance away from the instrument interface. This particular situation could be due to heat rejection requirements, a desire to minimize or isolate the instrument from the cryocooler motor vibrations, or the need to cross a gimbaled joint. Over any significant distance, the temperature drop associated with solid conduction quickly becomes excessive. This temperature drop drives down the cryocooler cold tip temperature and quickly increases the system power consumption. The heat pipes provide over 100 times the heat transport efficiency of equivalent solid conductors and, thus, allow minimal temperature drops for transport across significant distances. The flexibility of the heat pipes allows for improved integration, assembly, and, if required, on orbit deployment and/or pointing.

The diode capability originally was intended to enable the use of a higher temperature cryogenic radiator as an alternate cooling source to the primary cryocooler[3]. For a larger set of applications, the diode capability also provides for the minimization of redundant or off-cryocooler parasitic heat loads. Cryocooler redundancy can significantly increase the cooling loads by adding parasitics from the non-operating or off cryocooler (which is attached at one end to ambient and at the other end to the cooled instrument). For typical Stirling or pulse tube cryocoolers, this additional heat load is about 0.3 to 0.8 W at 60 K. If cooling margins are tight and/or the cryocooler has a small capacity relative to this off-cryocooler load, cryogenic diode heat pipes can be very beneficial. Their diode action has the capability to thermally isolate the off cooler from the instrument and reduce the off cooler parasitic heat loads by at least a factor of 3 to 5.

HEAT PIPE DESCRIPTION AND REQUIREMENTS

The CFDHPs (Cryogenic Flexible Diode Heat Pipes) are illustrated in Figures 1 and 2. The oxygen and methane heat pipes are 0.60 and 0.75 m, long, respectively, with flexible bellow sections of 0.27 and 0.4 m, transport sections of 0.38 and 0.53, and both with 0.09 m condensers and 0.14 m evaporators. Thus, the effective transport lengths of the OFD and MFD are 0.50 and 0.60 m, respectively. The composite wick (with a fine screen mesh

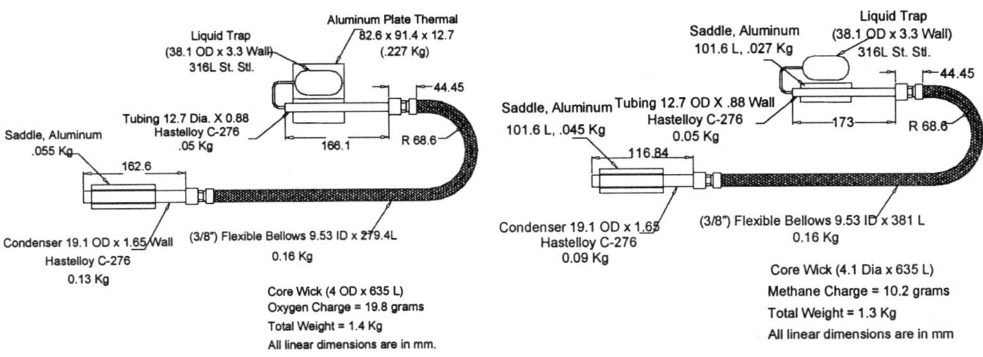

Figure 1. OFD Design **Figure 2.** MFD Design

surrounding a coarse wick) provides over 5 times the transport capacity over a homogeneous wick once the pipe is primed. The MFD (Methane Flexible Diode) was designed to operate over the range of 100 to 170 K. The OFD (Oxygen Flexible Diode) was designed to operate from about 60 to 150 K.

A SS wire cloth (200 mesh) wrapped cylindrical core is installed inside the reservoir to "trap-out" liquid and prevent reverse heat piping. The liquid traps are sized to accommodate all of the working fluid with 20% excess. In an actual application, the diode heat pipes could be used to connect 2 cryocoolers or cooling sources to one instrument. One of these cryocoolers would be operating while the other was turned off and isolated by the shut down diode heat pipe. The heat pipe would shut down when the temperature of the condenser (attached to the off cryocooler) exceeded the temperature of the evaporator and trap (attached to the still cooled instrument). At this time, the liquid within the heat pipe would condense and accumulate within the trap and "dry out" the pipe. For the experiment, only one cooling source was available. When that source was turned off, a thermal storage unit was used to keep the evaporator (and trap) end of the heat pipe cold in order to simulate the actual application's redundant cooling source. The Brilliant Eyes Thermal Storage Unit (BETSU) phase change material (PCM) canister that was previously flown in CRYOTP[4] is thermally coupled to the MFD's evaporator and liquid trap. When the heat pipe shuts down during diode reversal, the melting of the PCM (2-methyl pentane) at 120 K permits the TSU to store the heat pipe's transient shutdown energy and parasitic heat flows due to conduction or radiation. An aluminum thermal mass is used with the OFD to provide energy storage.

Special features that are incorporated into each of the CFDHPs are a composite wick structure for increased transport, a large diameter condenser section to facilitate startup, and offset fittings that interface the condenser and evaporator section with the flexible section. This last feature was used to center the core wick within the flexible bellows while permitting it to rest on the bottom of the evaporator and condenser tubes to facilitate startup in 1-g. The larger diameter condenser was used to ensure that all of the liquid could be condensed within the condenser section to insure startup. To also guarantee startup, reduction of the pressure to subcritical was insured by the additional containment volume of the liquid trap and the flexible bellows, and the larger volume of the condenser.

Once primed, the pumping derived from the effective pore size of the fine mesh outer wrap increases the capillary limit of the wick. As a result the primed or composite transport capability is approximately 3.5 times greater than that of the homogeneous core. This allows the wick area to be reduced for a given transport requirement which in turn reduces the fluid inventory and the associated pressure containment required and the amount of energy back flow that is required for diode shutdown.

The requirements for the heat pipes are given in Table 1. These requirements were derived from assessing the potential user or system needs as well as accounting for the projected technology capabilities.

CRYOFD EXPERIMENT DESCRIPTION

The CRYOFD was the third (and fourth) flight of the Cryogenic Test Bed[1,4] (CTB). The CTB has been developed to provide a platform for quick turnaround of cryogenic flight demonstrations. The cooling of the flight test articles was provided by five tactical Hughes 7044 cryogenic refrigerators (three on the OFD side and two on the MFD). The heat pipes were thermally connected to the cryocoolers using a copper braided interface coupling. The heat pipes were isolated from the support structure using Kevlar straps and G-10 brackets. Note that once integrated, each of the heat pipe configurations are three dimensional. Hence, only reflux operation was verified in 1-g after installation into the test bed.

GROUND AND FLIGHT VERIFICATION

Ground and Flight Tests and Data Analysis

Due to an extremely tight schedule minimal ground characterization was performed on the heat pipes. Before integration into the experiment, some limited startup and heat transport capacity tests were performed on the heat pipes at the component level in a co-planar configuration with a 12.7 mm adverse tilt. Once integrated into the experiment, only reflux mode measurements were possible as the pipes were not mounted in a single plane.

Table 1. Heat Pipe Requirements and Performance.

Parameter	Oxygen Heat Pipe		Methane Heat Pipe	
	Requirement	Performance	Requirement	Performance
Min Temperature Performance (W-in)	>50 @ 58K	~200 @ 58K from 85K 0-g	>400 @ 100K	~700 @ 100K from 105K 1-g
Nominal Temp. Perform.				
Rqmt - W-in	>250 @ 63K	~300 @ 63K	>600 @ 115K	~750 @ 115K
- W	>7.5 @ 63K	~15 @ 63K	>20 @ 115K	~29 @ 115K
Goal - W-in	>325 @ 63K	from 85K 0-g	>750 @ 115K	from 105K
- W	>10 @ 63K	data	>25 @ 115K	1-g data
Max Temp Perform (W-in)	>70 @ 100K	420 @ 100K from 0-g data	>100 @ 150K	450 @ 150K from 0-g data
Thermal Resistance (K/W)	<0.2	~0.05 from 85-120K 0-g data	<0.1	~0.1 from 103K 1-g data
Diode Performance				
Thermal Resist. (K/W)	>1000		>1000	
Shutdown Energy (W-hr)	<2 rqmt <1 goal	1.1 from 0-g data	<1.25 rqmt <1 goal	1.2 from 0-g data
Mass	<1 kg	1.4	<1 kg	1.3
Flexibility				
Bend Radius	<5.5 in	-	<5.5 in	-
Bend Angle	>180°	-	>180°	-
Bend Force	<5 lbf @ 135° <30 lbf@180°	-	<5 lbf @ 135° <30 lbf@180°	-
Bend Cycles	>1000	-	>1000	-

The majority of the effort during the flights was directed toward characterizing the transient startup, the transport capacity, and the diode operation of the heat pipes. Steady state heat transport operation was difficult to obtain due to the limited net cooling capacity at the heat pipes. This was the result of the combination of intermittent problems with one of the OFD cryocoolers, limited experiment heat rejection due to Shuttle mission orientations, cryocooler degradation during continuous operation, and high parasitic losses (calculated at about 7 W on each pipe). For the OFD with three cryocoolers operating, the calculated net cooling was less than 5 W at 100 K with almost nothing available below 70 K. Thus, the nominal design point performance at 63 K could not be evaluated in 0-g. For the MFD with only 2 cryocoolers, the net cooling was about 3.5 W at 115 K and less than 2 W at 100 K. Because of the limited net cooling capacity, the majority of heat pipe dryouts were performed at transient conditions.

Diode shutdown with each unit was tested by turning off the cryocooler and in some cases applying heater power at the condenser. Restart from a shutdown condition was also demonstrated. Faster restart was observed by applying heat to the liquid trap to drive the fluid back into the heat pipe. Freeze/thaw characteristics and temperature control with the TSU were evaluated as part of the MFD forward mode and diode shutdown operations. The diode characterization was negatively impacted by the large heat parasitics which warmed up both the condenser and evaporator after the cryocoolers were turned off.

During the post-flight data reduction, heat transport characterization tests were divided into four categories: (1) dryout (defined as when the condenser temperature separated from the evaporator and actually decreased), (2) partial dryout (where the condenser temperature separated, but no decrease was observed and the conditions were not maintained long enough to show whether the pipe may have recovered), (3) steady state hold (when the pipe held a heat load, as indicated by minimal temperature differences across the pipe, at steady state temperatures), and (4) transient hold (when the pipe held a heat load, but the temperatures were not steady). Unfortunately, during the flight, the experiment operators often terminated a dryout test after observing a separation of the condenser and evaporator temperatures, but before the condenser temperature decreased which would indicate a complete separation per our data reduction dryout definition.

During the flight, multiple steady state calibration points were run where the condenser heater power was adjusted to maintain a constant heat pipe temperature with no applied evaporator heater power. Following the flight, the cryocoolers were characterized by the AFRL cryocooler laboratory. The parasitic heat loads as a function of temperature were then estimated as the difference between the predicted cooling capacity and the condenser heater power. Based on analyses of the parasitics on each pipe, the amount of parasitic heat load that would have to be transported by the heat pipes was estimated as about 70% for the OFD and 78% for the MFD. This additional load was added to the evaporator heater loads to estimate the total heat transport of the pipes for each flight test.

The intermittent anomalous operation (with significant cooling losses) observed during the flight in one of the OFD cryocoolers also occurred during the ground characterization. Because of the inconsistent behavior, there was no accurate method for estimating what the instantaneous cooling capacity of the cooler was during the flight. Thus, for the parasitic analysis, it was assumed that the cooler operated nominally. This assumption leads to some errors in the analysis which result in increased scatter in the heat pipe transport capacity data as well as some random over estimation of the cooling capacity by several watts.

The estimate of the heat pipe operating thermal resistance required temperature measurements of the evaporator and condenser walls. Unfortunately, due to several factors, the flight temperature measurements were probably not accurate to more than 1-2 K on an absolute scale. Also, due to physical constraints, the PRTs often could not be directly

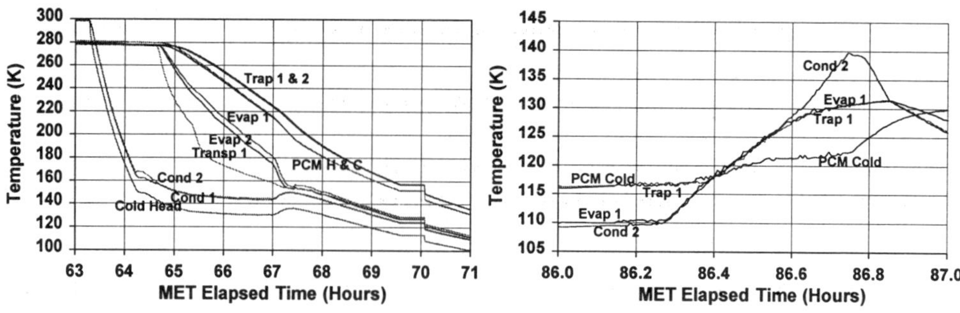

Figure 3. MFD Cooldown/Startup on July 3, 1997. **Figure 4.** MFD Diode Reversal/Recovery on July 4.

mounted on the desired heat pipe location. Overall, these temperature measurement errors led to significant scatter in the resistance calculations, especially for the lower heater power tests where the heat pipe temperature gradient was only a few degrees.

MFD Performance

Figure 3 shows an on-orbit cooldown and priming of the methane heat pipe. Isothermalization of the heat pipe occurred at 155 K approximately 4 hours into the cooldown. The rapid cooldown and high temperature for isothermalization indicates that the system pressure became subcritical early in the cooldown phase with condensation beginning very near the critical temperature. The net effect of the parasitics on the MFD is to increase the heat load that must be transported by approximately 4.7 watts at 155 K. This is the amount that must be transported by the non-composite wick structure prior to the pipe being fully primed. The theoretical transport capability of the non-composite core is 3.49 W-m at 155 K. With an effective length of 0.648 m, the maximum theoretical heat load that can be transported is 5.4 watts, which is greater than the estimated parasitics. Therefore, isothermal startup and full priming at 155 K could be achieved as demonstrated. Higher parasitics would have resulted in isothermalization at a lower temperature.

Figure 4 shows an MFD diode reversal at 110 K with 3.5 watts applied to the condenser at MET (Mission Elapsed Time) 86.3 hours, and 5.5 watts at 86.4 hours. The shutdown time is about 9 minutes and the corresponding shutdown energy is estimated to be 1.2 W-hr, which is 85% of the MFD's methane inventory. It should be noted that during this cycle an average of approximately 3.5 watts were being transferred to the TSU as it was melting. This corresponds to 12 minutes of melt time for the 2500 joules of stored energy. As shown in the figure, the heat pipe recovered and reprimed in less than 1 hour.

Figure 5 shows the predicted heat pipe capacity and program requirements versus the flight and ground data. Because of the limited cooler capacity, only a few complete dryouts were obtained and only at the upper end of the operating temperature range of the heat pipe. The measured test results tend to be 10-20% below theoretical in the flight tests. The fact that steady-state partial dryouts occur is indicative that dryout is occurring in the circumferential threads, since a partial dryout cannot exist in the composite core except as a short transient without any possibility of recovery. This is the probable explanation for the difference between actual and theoretical. The transport model is based on the maximum capability of the composite wick and does not take into account the circumferential liquid flow, which appears to be defining the transport limit. Increased vapor losses due to the bends or increased liquid losses in the wick due to stretching of the wick in the bends are also possibilities since the transport model assumes the heat pipe is straight.

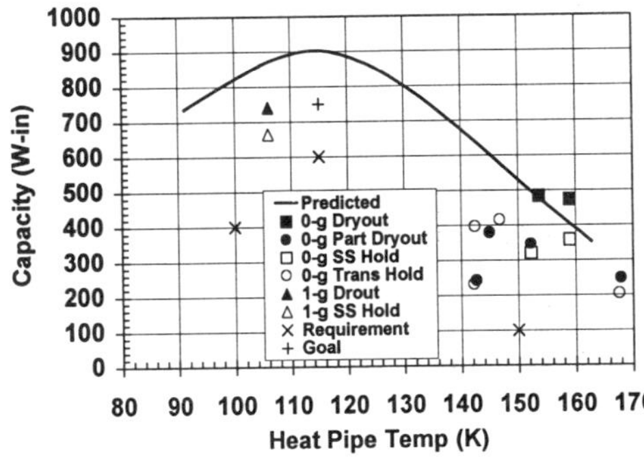

Figure 5. MFD Transport Capacity in Watt-inches.

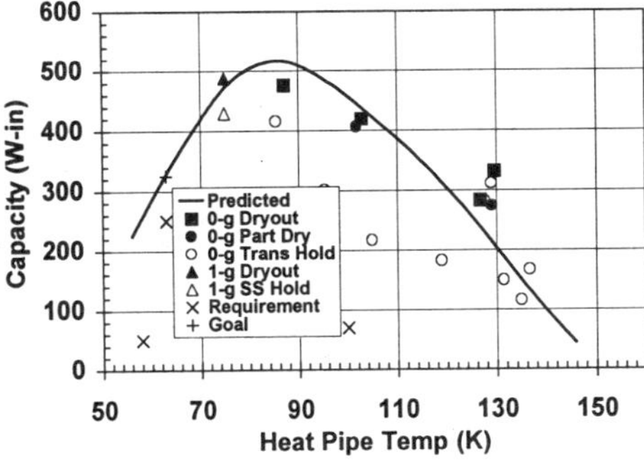

Figure 6. OFD Transport Capacity in Watt-inches.

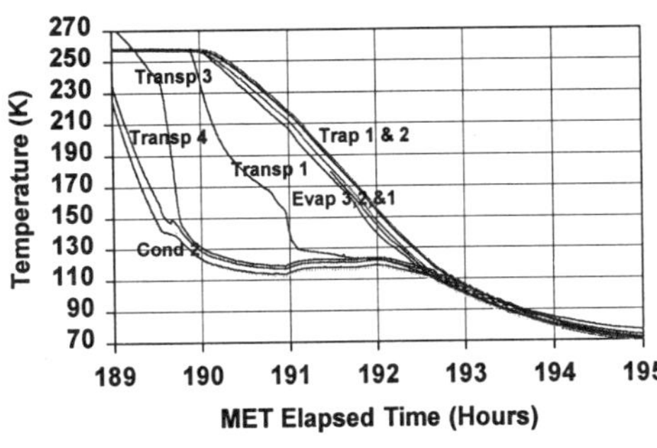

Figure 7. OFD Cooldown/Startup on July 8, 1997.

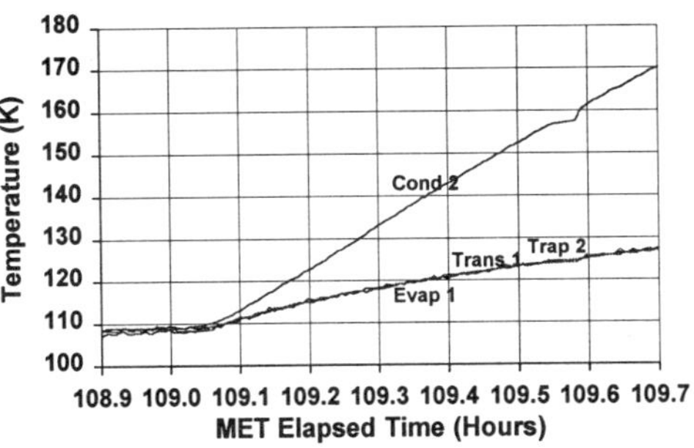

Figure 8. OFD Diode Reversal on July 5, 1997.

OFD Performance

Figure 7 shows a cooldown and priming of the oxygen heat pipe. Isothermalization occurs at 112 K at approximately four hours from the start of cooldown. Cooldown of the liquid trap lags slightly (1-2 K) due to its high thermal mass and the parasitic heat input. Isothermalization with the rest of the pipe is not required for forward mode operation. Figure 8 shows an OFD diode reversal at 110 K with 3.5 watts applied at the condenser. Complete shutdown occurs in about 8 minutes. The corresponding shutdown energy was estimated to be 1.1 watt-hr which is equivalent to 100% of the oxygen inventory.

Figure 6 shows the predicted and required heat pipe transport capacity compared with the flight and ground test data. For the oxygen heat pipe (which had 3 attached cryocoolers), we were unable to obtain dryout conditions at temperatures below 85 K. Thus, we were unable to verify the nominal design point requirements at 63 K. The OFD data shows very good agreement with the predictions and, in general, there is better correlation than with the MFD data. This would imply, along with the ground and flight data for the MFD, that the pumping radius may not be as small with the methane composite wick. It is possible that the methane wick could have been damaged during insertion into the heat pipe. The "transient-hold" data for the OFD is provided as reference and only indicates that there was no dryout at these transient heat loads.

Heat Pipe Performance versus Requirements

Table 1 includes a comparison of the technology program requirements versus their actual performance. Because of the limited experiment net cooling capacity, micro-gravity verification of the heat pipe performances at their minimum and nominal temperatures was not possible. However, through a correlation of the flight data to the predictions, the probable performance at the lower temperatures can be extrapolated with a high degree of confidence (especially for the OFD). Thus, it was estimated that the OFD exceeded the minimum 58 K transport capacity and the nominal 63 K W-in. With minimal extrapolation, the flight data confirmed an OFD transport capacity at 100 K which was six times greater than the requirement. For the MFD, it was estimated that the heat pipe far exceeded the minimum 100 K transport capacity and met the nominal 115 K W-in and heat capacity goals. The 0-g data confirmed that at 150 K, the MFD transport capacity was over 4 times the requirement.

Due to the previously described temperature measurement inaccuracies, the flight data for the operating thermal resistance of both pipes had a significant amount of scatter. For the OFD, over the typical operating range of 80 to 120 K, the flight data indicate a resistance of about 0.05 K/W, which is significantly better than the requirement. For the MFD, the flight data was only taken over the upper operating regime of the heat pipe where the apparent resistance exceeded the 0.1 K/W requirement. Based on ground test data in the nominal MFD operating range, the estimated resistance was near the 0.1 K/W requirement.

Both heat pipes met the diode performance requirements for off resistance and shutdown energy. Due to significant design changes during the program, both heat pipes exceeded their mass requirements. However, the heat pipes would still be significantly lighter than other integration alternatives such as conduction bars. Due to programmatic constraints, the flexibility requirements of the heat pipes were never verified. But, tests on several ambient heat pipes made by Swales with similar flexible bellows sections indicate that the CRYOFD heat pipes would probably have met their flexibility requirements.

Overall, although the micro-gravity data was limited, it can be stated with good confidence that both heat pipes easily met their program requirements. And, because these requirements were based on several potential applications, the heat pipes should meet the integration needs of the users.

ACKNOWLEDGMENTS

The authors would like to thank Lt. Mike Rich and Lt. B. J. Tomlinson of AFRL, Charlotte Gerhart of Nichols, Ted Swanson of the Goddard Space Flight Center, Chuck Stouffer and Robert Hagood of Swales Aerospace, Pete Thomas of Aerospace, and John Vermillion and Mike Lewis of Jackson and Tull for their dedicated efforts on this program.

REFERENCES

1. Brennan, P., Stouffer, C., Thienel, L., and Morgan, M., "Performance of the Cryogenic Heat Pipe Flight Experiment" SAE Technical Paper Series number 921408, July 1992.
2. Final Report, Cryo System Experiment, Hughes Aircraft Company, Electro-Optical Systems, February, 1996.
3. Glaister, D. S., Curran, D. G. T., Mahajan, V. N., Stoyanof, M., "Application of Cryogenic Thermal Switch Technology to Dual Focal Plane Concept for Brilliant Eyes Sensor Payload," 1996 IEEE Aerospace Applications Conference, Snowmass, CO, February 3, 1996.
4. Glaister, D. S., Bell, K. D., Bello, M., Edelstein, F., "Development and Verification of a Cryogenic Brilliant Eyes Thermal Storage Unit," AIAA Journal of Spacecraft and Rockets, Vol. 33, No. 4, 1996.

DEVELOPMENT OF A THERMAL ISOLATION STRUCTURE FOR AEROSPACE CRYOGENIC INSTRUMENTS

A. E. Nash and L. S. Robeck

Jet Propulsion Laboratory, California Institute of Technology
Pasadena, CA 91109

ABSTRACT

A "proof of concept" structure has been developed to demonstrate the capabilities of a lightweight, rigid, compact, thermally isolating structure for aerospace low temperature instruments. The modular design not only provides thermal isolation of the instrument from its surroundings, but also between the components within the instrument, to avoid bias resulting from thermal "crosstalk". In this paper the design requirements as well as the implementation experiences will be described.

INTRODUCTION

The Critical Dynamics in Microgravity Experiment (DYNAMX), required a "proof of concept" prototype sensor package in their development of an experiment for a space shuttle flight in early 2002. This paper describes the development of the tension/compression thermal isolation structure for this prototype sensor package, as well as the structural, thermal, electrical and hydrodynamic links between components. The design of some of the components is described elsewhere[1]. This paper will describe the objectives, requirements and constraints for the thermal isolation structure and its development, the design approach, the as-built design, as well as the analyses and tests conducted to verify the feasibility of the design approach.

DYNAMX will examine the critical behavior of a second order phase transition driven far from equilibrium by measuring the thermal conductivity of liquid helium, as a function of applied heat flux, at temperatures around the normal/superfluid transition temperature in ^{4}He. To achieve the scientific objectives of the experiment requires that the apparatus have sub-nanoKelvin temperature control and resolution, sub-nanoWatt heat control and resolution, as well as control and readout resolution of the helium superfluid/normal fluid interface within the cell at the micrometer level. In addition to the science requirements, the apparatus must also meet environmental requirements, (e.g., shuttle launch loads), as well as all safety verification requirements that accompany a shuttle-borne experiment. These levels of requirements, however, have not previously been simultaneously met by any cryogenic apparatus, and the current challenge was to prove the feasibility of meeting all requirements within the inherited experiment configuration .

OBJECTIVES AND REQUIREMENTS

The prototype development requirements were derived from the scientific objective of measuring the non-linear thermal transport properties of ^4He in the region very near the lambda transition in the limit of vanishingly small heat flux to better than 30%. The two principal aspects of this are: (1) to measure the thermal profile and the thermal conductivity as a function of applied heat flux, and (2) to measure the depression of the transition temperature as a function of applied heat flux. This is depicted in Figure 1.

The measurement technique for the DYNAMX experiment is to apply heat to one end of a right cylindrical sample cell of helium and control its extraction from the other end by controlling the temperature of a highly stable thermal reservoir. Over the range of interest, normal and superfluid phases of helium can simultaneously exist in the sample cell, separated by an interface region. By varying the heat flow in and out of the cell, the interface can be quasi-statically moved back and forth between the hot and cold ends of the cell, allowing the temperature with respect to interface position to be measured by thermal probes embedded in the walls of the sample cell. The heat flow into the cell is merely the heat applied by an external heater. The heat removed is determined by the difference between the temperature of superfluid helium and the temperature of the thermal reservoir, divided by the resistance of the thermal link between them. This is the thermal equivalent of a "current bias" measurement, in contrast to the more commonly used "voltage bias" technique where the temperatures of the cell endcaps are the control parameters. The parameter range of interest is temperatures within ± 220 nK of the heat flux dependent transition temperature, $T_\lambda(Q)$, for applied heat fluxes of 5 to 70 nW/cm^2.

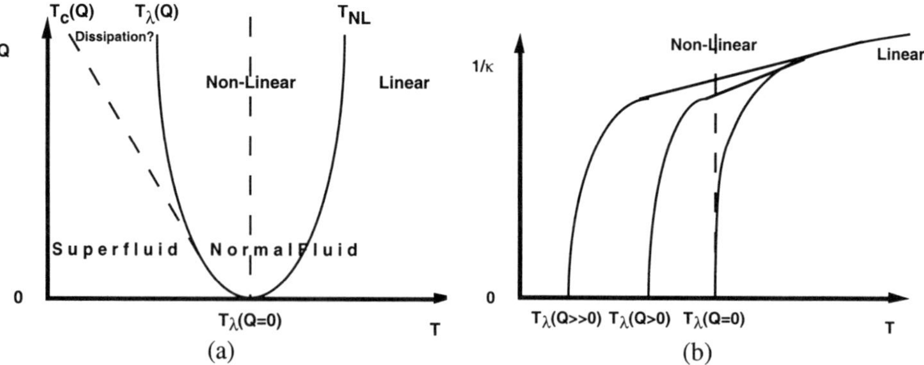

Figure 1 Schematic depiction of the phase diagram (a), and conductivity (b) for non-linear heat transport near the superfluid transition in 4He. In (a), T_{NL} is the heat current (Q) dependent transition temperature from linear to non-linear thermal transport, $T_\lambda(Q)$ is the heat current (Q) dependent superfluid - normal fluid transition temperature, and $T_C(Q)$ is the heat current (Q) dependent temperature of dissipation below the superfluid transition[2].

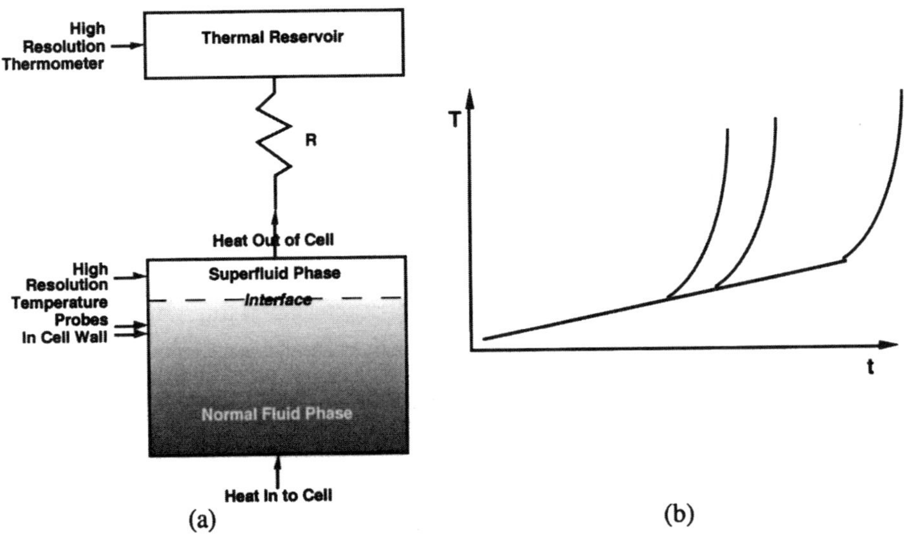

Figure 2 Schematic depictions of the DYNAMX measurement technique (a), and sidewall thermometer readout time series in 1g (b).

As the interface traverses the length of the cell, the time series of temperatures is measured by the thermal probes embedded in the walls of the sample cell, $T_i(t)$. This is depicted in Figure 2. This allows for a reconstruction of the thermal profile, T(z), from the measured interface velocity, $v_{I/F}$. From the heat flow equation, the known values of heat flow, Q, and thermal profile gradient, ΔT(Z)/ ΔZ, the nonlinear thermal conductivity can be extracted.

$$Q = -\kappa\left(Q, T - T_{\lambda}(Q,P)\right)\frac{\Delta T(Z)}{\Delta Z} \qquad (1)$$

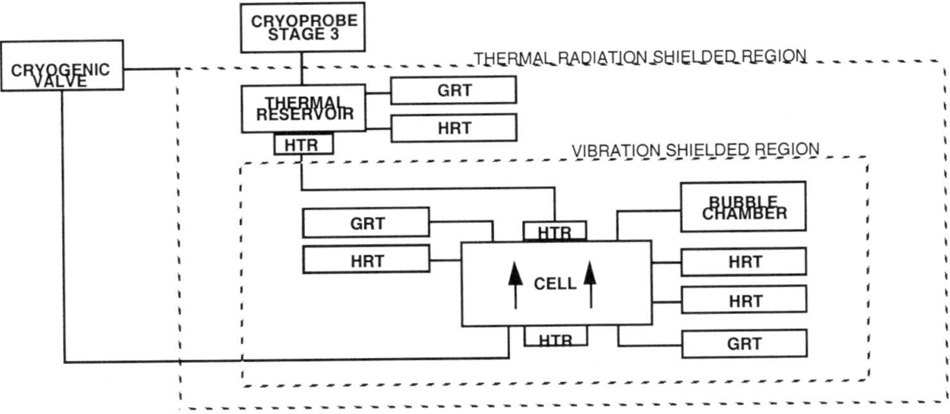

Figure 3 Thermal schematic block diagram of DYNAMX sensor package components.

A block diagram of the principal thermal connections of the DYNAMX sensor package components is shown in Figure 3. The components are: a sample cell with three sidewall temperature probes, a nano Kelvin stable thermal reservoir, a superfluid leak-tight cryogenic valve, and a pressure stabilization device (a saturated vapor pressure bubble chamber). The temperature of these elements is monitored and controlled using Germanium Resistance Thermometers (GRTs), High Resolution Thermometers (HRTs) and heaters (HTRs) made from metal thin film resistors.

The sensor package was designed to connect to a thermal control structure developed for the Lambda Point Experiment (LPE)[3]. The connection was to be via a cryogenic vibration isolation system to reduce the milli-g level g-jitter present on the shuttle, particularly at the 17 Hz dither frequency of the shuttle's Ku - Band antenna. This vibration isolation system was to be co-axial with the sensor package, attaching at a ring around its center of mass.

Another design driver was the need to accomodate two different kinds of HRTs: one (larger) type based on the previously flight proven design[4], but modified internally, as well as a new miniaturized version in development.

The science objectives led to stringent top level and derived performance requirements. Table 1 summarizes the key performance requirements starting with the requirements on temperature, heat flux, and interface position as well as key derived requirements (shown by indenture).

Of particular concern in the development of this prototype was mitigating, to the extent possible, the 0.6 - 2.1 pW/g heating that comes with the cosmic ray environment of the microgravity low earth orbit of the space shuttle.

At the time of this development, DYNAMX was baselined to be the fourth in a series of experiments conducted in the cryogenic environment provided by the Jet Propulsion Laboratory (JPL) Low Temperature Platform (LTP) Facility in the Space Transportation System (STS)[5]. This placed several stringent constraints on the system. Primary among these was the limited number of servo control channels (4 with High Resolution Thermometers and 4 with Germanium Resistance Thermometers) and the allowable envelope (just over 6 cm in diameter and 30 cm in length).

In this development, technical scope, performance and schedule were the driving constraints, with cost secondary.

DESIGN APPROACH

The design drivers were related to the derived requirements to control heat flow between the sensor package components, which to complete the science objectives, needed to change temperature, albeit only on the order of 100 nK. Table 2 lists the performance requirements that were the design drivers, as well as the related derived requirements on maximum allowable heat flow.

Table 1 Sensor Package Performance Requirements

Parameter	Requirement
Temperature relative to transition	≤ 0.3 nK
Variation in heat flux through sidewall probes	< 3 pW
RMS heat flux through sidewall probes	< 13 pW
Pressure Variation	< 0.8 ppm
Bubble Chamber Temperature Variation	< 245 nK
Heat Flux	< 1%
Interface Position Knowledge	< 12 μm
Probe Spacing Knowledge	< 1%
Interface Velocity	
Accuracy in Q_{IN} - Q_{OUT}	< 0.5 % Q_{IN}
Variation in Q_{IN} - Q_{OUT}	< 0.03% Q_{IN}
Reservoir Temperature Variation	< 32 nK

Table 2 Performance Requirement Design Drivers

Performance Requirement	Derived Heat Flow Requirement
< 0.3 nK variation in sidewall probe temperature	< 3 pW
< 0.03% Q_{IN} variation in Q_{IN} - Q_{OUT}	< 6.3 - 87.3 pW
< 0.33 nW/cm^2 parasitic through sidewall probes	< 13 pW
< 0.5% accuracy in Q_{IN}	< 105 - 1460 pW

From numerical models that predict the helium thermal profiles in the non-linear region of interest, a multi-node thermal model that included the interface position dependent thermal resistance throughout the sample cell for various applied heat currents was constructed. From this model, the interface position dependent temperature of the sidewall probes, as well as the cell endcaps and thermal reservoir could be calculated. These temperatures, combined with the primary and derived requirements, set lower limits on the allowable resistances for interconnections within the sensor package components and the supporting structure. Limits arising from quasi-static requirements could be calculated from $R_{MIN} = \Delta T/\Delta Q_{MAX}$. Limits from noise requirements in the temperature of item (1) due to connections to item (2) were calculated from

$$R_{MIN} = \frac{1}{2\pi f C_1}\left[\left(\frac{\sigma T_2}{\sigma T_1}\right)^2 - 1\right]^{1/2} \tag{2}$$

where R_{MIN} is the minimum resistance, C_1 is the heat capacity of item (1), and σT_i is the variation in the temperature of item (i), evaluated at the measurement bandwidth frequency of 1 Hz.

DESIGN

The design for the sensor package structure is a tension/compression structure made up of a cylindrical "exoskeleton" composed of identical rings, separated axially by posts, which mount to the rings at two different clock angles, and provide cavities for the internal suspension under tensile preload of the individual components. The exoskeleton was made from 6061 aluminum. Suspension was achieved using Kevlar cords that were tied off at tabs on the components, and at adjustable stainless steel "L" brackets mounted on the rings. The brackets compensated for the difference in coefficient of thermal expansion between Kevlar (which is negative) and Aluminum (which is positive). A thermal cycle test confirmed the design was able to retain the necessary tensile preload. Each component is held at six points, with a cord angle of nominally 30 degrees to the cylinder axis (for maximizing location control).

A detachable, thin (127 µm), aluminum skin provided a baffle for radiative heat transfer, while greatly increasing the stiffness of the exoskeleton. The "kit of parts" fabricated allowed the flexibility to assemble a structure to accommodate the thermometers chosen late in the development process. The modularity of the design also facilitated a change in valve implementation that occurred late in integration and test. Figure 4 shows a photograph of the sensor package structure skeleton and a component undergoing test fit. Figure 5 contains a photograph as well as an annotated rendering of the fully populated sensor package.

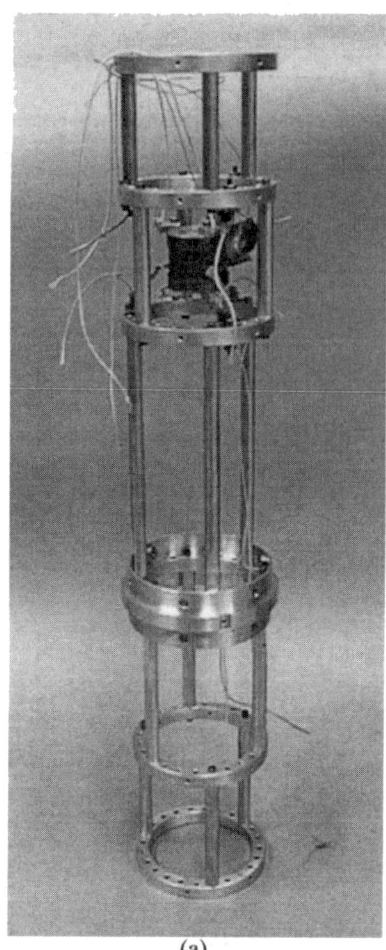

(a)

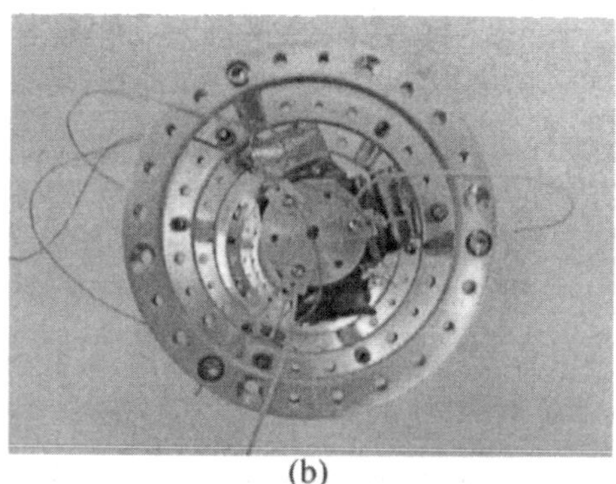

(b)

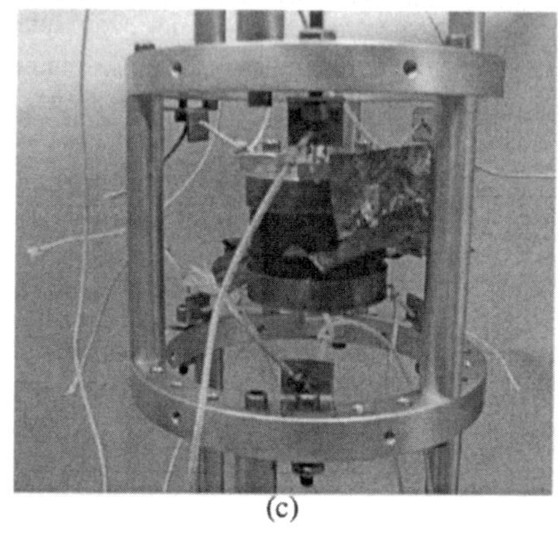

(c)

Figure 4 Photograph of sensor package structure skeleton (a, b) and suspension detail (c) showing a test cell.

Because the cross-sectional area of the cords is small, and the conductivity of the cord material is low[6], the cords provide a high level of thermal isolation. For the 445 N test Kevlar 29 braided cord used in this design, the thermal resistance is 5 X 10^8 K/W•m at 2 K, providing ~ 10^7 K/W of thermal isolation. This exceeded all performance requirements. The high level of thermal isolation, combined with the independent suspension of each component, minimizes the inter-component thermal couplings, minimizing the bias resulting from thermal "crosstalk".

Heat leak along the electrical leads was minimized by using Ø 76 μm Nb-Ti wiring. All fluid capillaries were constructed from 30 gauge 304 stainless steel tubing (305 μm O.D, 152 μm I.D.) with Ø 127 μm wires inside.

This design mitigates the effects of cosmic ray heating in two ways. The choice of aluminum for the exoskeleton structure reduces the absorbing mass, and the high thermal conductivity improves its recovery from South Atlantic Anomaly (SAA) passages. The low density and small size of the cord segments also limits the deposition of heat from cosmic rays. The tension/compression design also permitted a compact package, meeting the stringent envelope requirements.

The tension/compression structure offers enormous advantages in rigidity over cantilevered designs. The lowest natural frequency of our structure, determined by analysis,

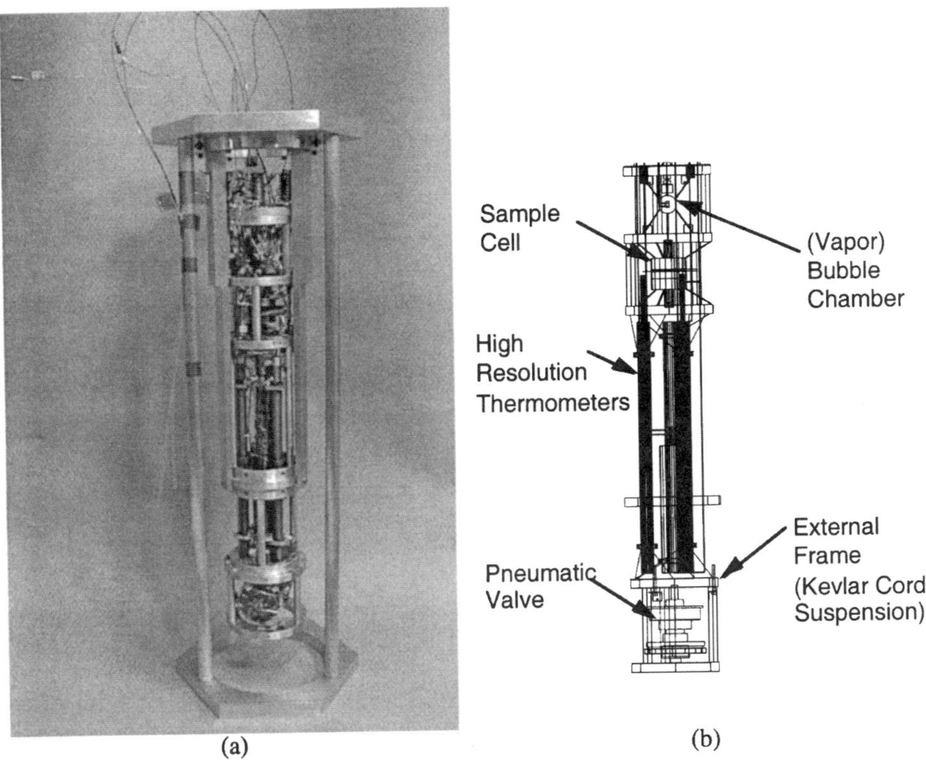

Figure 5 Photograph (a) and annotated rendering (b) of the fully populated sensor package.

is 1.3 kHz. High structural rigidity is necessary not only to withstand loads encountered during launch, but also to prevent amplification of low frequency vibrations from the shuttle that can generate heat in the sensor package, biasing the results of the experiment.

SUMMARY

A "proof of concept" structure has been developed that demonstrates the capabilities of a lightweight, rigid, compact, thermally isolating structure for aerospace low temperature instruments. The modular design not only provides thermal isolation of the instrument from its surroundings, but also between the components within the instrument, to avoid bias resulting from thermal "crosstalk", while at the same time allowing a variety of different components to be accomodated in the package. The development described here was accomplished within 6 months.

The design approach for this structure is also adaptable to scientific experimentations in normal Earth gravity cases in which there are requirements for lightweight, rigid, thermally isolating structures in confining geometries.

ACKNOWLEDGMENT

This work was carried out by the Jet Propulsion Laboratory, California Institute of Technology under contract to the National Aeronautics and Space Administration. The work was funded by NASA Microgravity Research Division.

REFERENCES

[1] A. Nash, Y. M. Mukharsky, K.M. Aaron and C. G. Paine, Development of Structural and Thermal Elements for Critical Dynamics in Microgravity Experiment, *Proceedings of the 35th AIAA Aerospace Sciences Meeting & Exhibit*, Reno, NV (1997).

[2] F. C. Liu and G. Ahlers, New Dissipative Region below the Superfluid Transition of ^{4}He in a Heat Current, Physical Review Letters, 76:(8) 1300 - 1303 (1996).

[3] J. A. Lipa, T. C. P. Chui, J.A. Nissen, and D.R. Swanson. A low temperature system for advanced thermal control. *Temperature its measurement and control in Science and Industry* Vol. 6; No. 2 : 949-952 1992.

[4] T.C.P. Chui, P. Day, I. Hahn, A.E. Nash, D.R. Swanson, J.A. Nissen, P.R. Williamson, and J.A. Lipa, High Resolution Thermometers for Ground and Space Applications. *Cryogenics*, 34:417 (1994).

[5] JPL D-11781. Low Temperature Platform Briefing Document. 1994.

[6] L. Duband, L. Hui, and A. Lange, Thermal isolation of large loads at Low-Temperature Using Kevlar Rope. *Cryogenics* 33:(6) 643-647 (1993).

LOW VIBRATIONAL PERFORMANCE OF TRW SECOND-GENERATION PULSE TUBE COOLER

C.K. Chan and C. Jaco

TRW
Redondo Beach, CA 90278

ABSTRACT

The next-generation TRW pulse tube cooler has been designed to minimize self-induced vibration, with an ultimate goal to mechanically balance the system to such an extent that no active vibration control-algorithm is required. Elimination of the active vibration control greatly simplified the drive electronics and the control software. Force measurements were performed with the newly developed IMAS S/N101 cooler on the Kistler Table using simple drive electronics with set-and-forget vibrational control when the cooler provided 1.5 W cooling at 55 K. The fundamental and other 15 harmonics were below 10 mN for the drive axis and below 200 mN for the other two off-axes. A typical (e.g., AIRS) sensor vibration requirement is 220 mN for the first two harmonics and 440 mN for the others. Effects of environmental factors on vibrational force such as changing cooler heat reject temperature and structural interaction under gravity were also examined.

INTRODUCTION

The vibratory force produced by the mechanical cooler onto its supports is the reaction force to moving masses within the cooler that undergo accelerations during various phases of the cooler's operational cycle. The accelerations come from 1) controlled motion such as the reciprocating motion of the piston(s), and 2) natural vibratory resonances of the compressor elastic structural elements. The force can also come from the "oil canning effect" of the cold finger, which is caused by a combination of pressure-driven elongation at the drive frequency or lateral cantilever vibration in response to higher frequency cooler harmonics. Although the level of vibratory force that is acceptable is a strong function of the specific application, a value on the order of 0.2 N (0.05 lb) has gained acceptance as a reasonable design goal. The AIRS sensor vibration requirement is 0.22 N for the first two harmonics and 0.44 N for the others [1].

Pulse tube coolers, which have no moving cold displacer or turbine and produce less vibration than Stirling or Reversed Brayton coolers, still require some form of control algorithm to reduce vibration to the acceptable level. In designing our first-generation coolers, a counter-balancer configuration is used in the miniature cooler [2] while a back-to-back dual-acting compressor module configuration is used in the mid-size (e.g., AIRS) cooler [1]. In both configurations the signal from an axial accelerometer is used as the error signal to closed-loop control the cooler's drive signal and waveform. This adaptive control waveform algorithm reduces the harmonics, but its effectiveness is limited to the following:

- The control can only cancel the cooler's axial vibration and is ineffective for vibration in the other two off-axes
- The waveform can only be distorted to a certain level before efficiency becomes an issue
- The algorithm is sensitive to the natural frequencies of the mounting structure when the harmonics of the drive frequency are close to the structural frequencies.

The adaptive vibration control requires complicated electronics and software, the complexity of which would lower total system reliability. As we moved into the second-generation cooler design, one of our goals was to mechanically balance the cooler to such an extent that no active vibration control is required. Since non-linearity and non-symmetry are fundamental causes of the force, both the first and the second-generation cooler enhanced linearity and symmetry by:

- Having a linear motor
- Having non-rubbing piston/cylinder assembly with clearance seal
- Operating the piston within the linear regime of the supporting flexure springs
- Having two nominally identical compressor modules mounted back-to-back.

In addition to the above features, our second-generation compressor was designed with a more rigid internal structure to eliminate the natural resonances of structural elements. Two flight-like coolers based on second-generation technology have been built and tested for IMAS application. [3] [4] One of these two coolers (S/N101), which had its vacuum can over the cold head, as shown in Figure 1, was used in this self-induced vibrational force investigation. The objective is to understand:

- Drive electronics configuration and control algorithm
- Effect on vibratory force due to environmental conditions such as gravity and compressor temperature variation

Results of the vibration measurements are summarized for a variety of operational conditions including different drive electronic configurations, control algorithms, harmonic nullings, operational strokes (or power), gravity orientations of the cold head, and compressor temperatures.

TEST APPARATUS

Measurements of the cooler vibration are conducted in TRW's dynamometer facility. The cooler was mounted on the Kistler Dynamometer 9257B, usually with the pulse tube in a horizontal position, as shown in Figure 2a. During the gravity effect study phase, the pulse tube was mounted in a vertical position as shown in Figure 2b. The Kistler dynamometer can measure the three components of the vibratory force from 8 mN to 7 kN full scale. The dynamometer was rigidly mounted on a 2-inch thick steel plate, which was placed on top of a 70-ton seismic mass built into the floor. All these assure a stable and rigid base for calibrated measurements. During cooler operation at steady state, the forces (Fx, Fy, and Fz) are simultaneously recorded in real time using a high-speed signal analyzer. All tests were performed with 85 W input power producing 1.45 W cooling at 55 K. Unless otherwise noted, the baseline case is 300 K heat reject, electronic configuration #1, and pulse tube in the horizontal position.

SENSITIVITY TO DRIVE ELECTRONICS CONFIGURATIONS

Two configurations of amplifier were studied:

- Single amplifier to drive the back-to-back compressor module (A) and module (B) in series or in parallel referring to configurations 3 and 4 in Figure 3
- Two amplifiers (1 and 2) to drive the compressor modules (A) and (B) separately. In configuration 1, amplifier 2 was fed with a waveform with fixed phase shift and magnitude difference in comparison to the amplifier 1 waveform; while in configuration 2, amplifier 2 was fed with a frozen waveform that could provide multi-harmonic nulling.

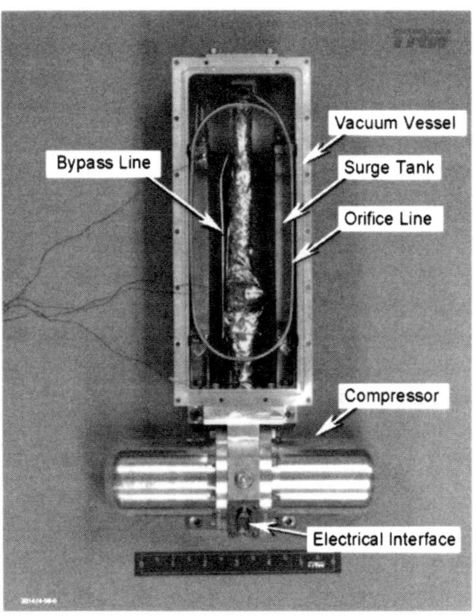

Figure 1. IMAS S/N101 cooler

(a) Horizontal

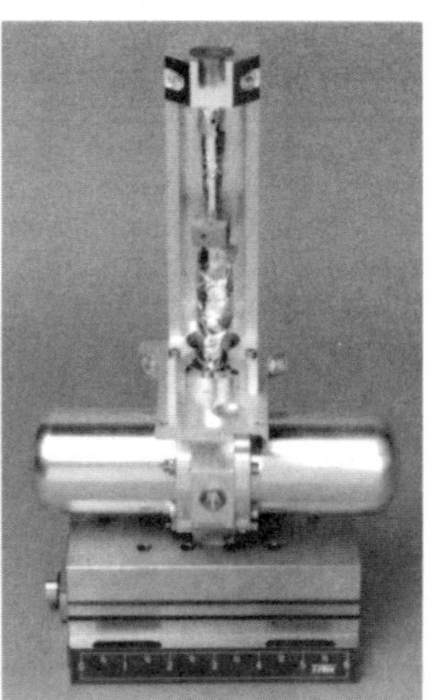

(b) Vertical

Figure 2. Orientations of pulse tube cold head for gravity effect study (fixed coordinates Y-drive axis, Z-gravity axis)

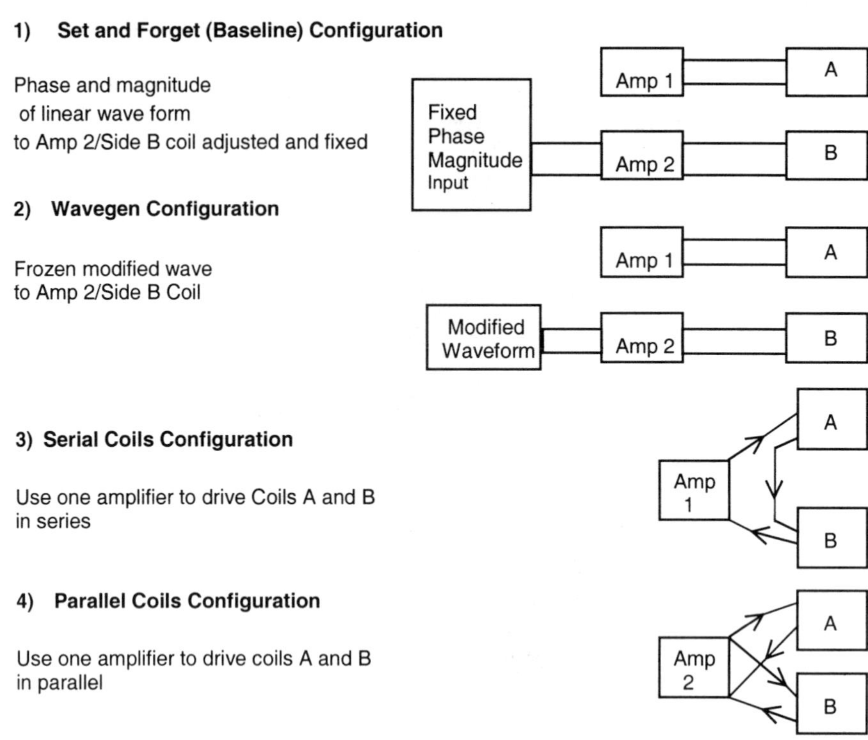

Figure 3. Drive electronic configurations

Figure 4 displays the measured peak forces of the cooler over 16 harmonics for the four configurations at the same input power of 85 W which provides cooling of 1.45 W at 55 K when the pulse tube is in the horizontal position, and the compressor temperature is maintained at 300 K, operating at 54 Hz. The single amplifier configuration is not adequate to provide vibration cancellation for the on-axis vibration. The dual amplifier with frozen waveform reduces the vibration to the noise level (10 mN). Improvement of the second-generation cooler over the first-generation cooler can be seen by comparing the off-axes results with the AIRS data [2] that shows higher off-axes vibration for similar power levels. The relatively large fundamental and the 15th harmonics in the pulse tube axis are due to the "oil canning" effect and the cantilever effects, respectively. Details will be discussed in the gravity effect section.

SENSITIVITY TO STROKE OR INPUT POWER

By comparing the vibration spectra of the input power for the 30 W to the 80 W case, it can be concluded that the vibration level was fairly well scaled by the input power.

SENSITIVITY TO GRAVITY EFFECT

The study of the gravity effect (Figure 5) revealed that the fundamentals of the off-axes are dominated by the oil canning effects of the pulse tube because the fundamentals are fairly constant in the pulse tube axis (X or Z axis), independent of the gravity orientation. The higher harmonics discrepancy (especially the 15th harmonics) is due to the gravity-induced cantilever effects.

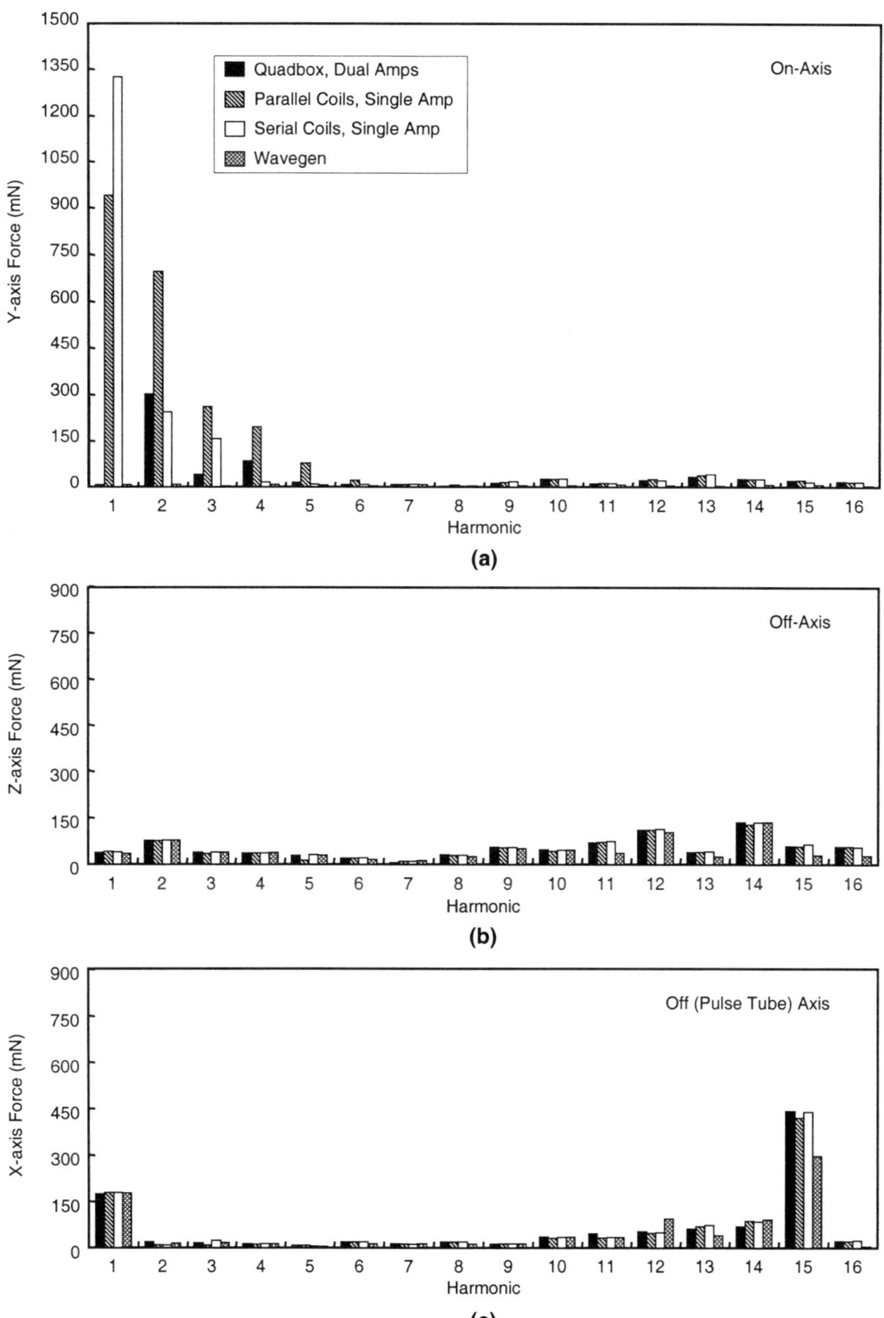

Figure 4. Vibration spectrums of different drive electronics and control algorithm configurations

SENSITIVITY TO COMPRESSOR TEMPERATURES

The phase shift and the magnitude difference between the two amplifiers were set when the cooler was tuned at 300 K. Hence, it is not a surprise that the 300 K case has the minimum fundamental in comparison to the other two temperatures of 290 K and 320 K (Figure 6a). In general, the harmonics decrease with higher temperature as shown by the harmonics in Figures 6(a), (b), and (c), but the variation is quite small.

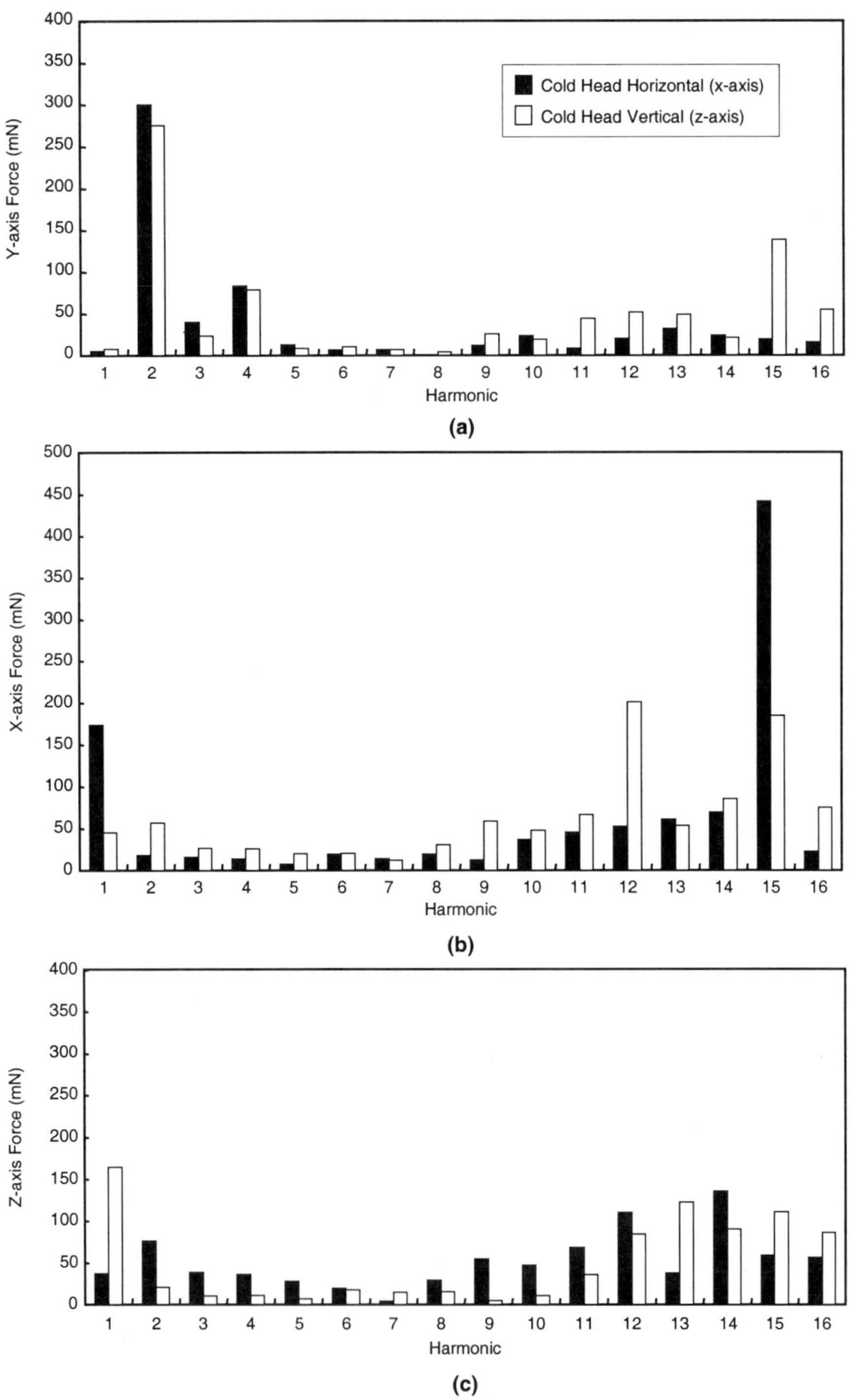

Figure 5. Gravity effect on force spectrum

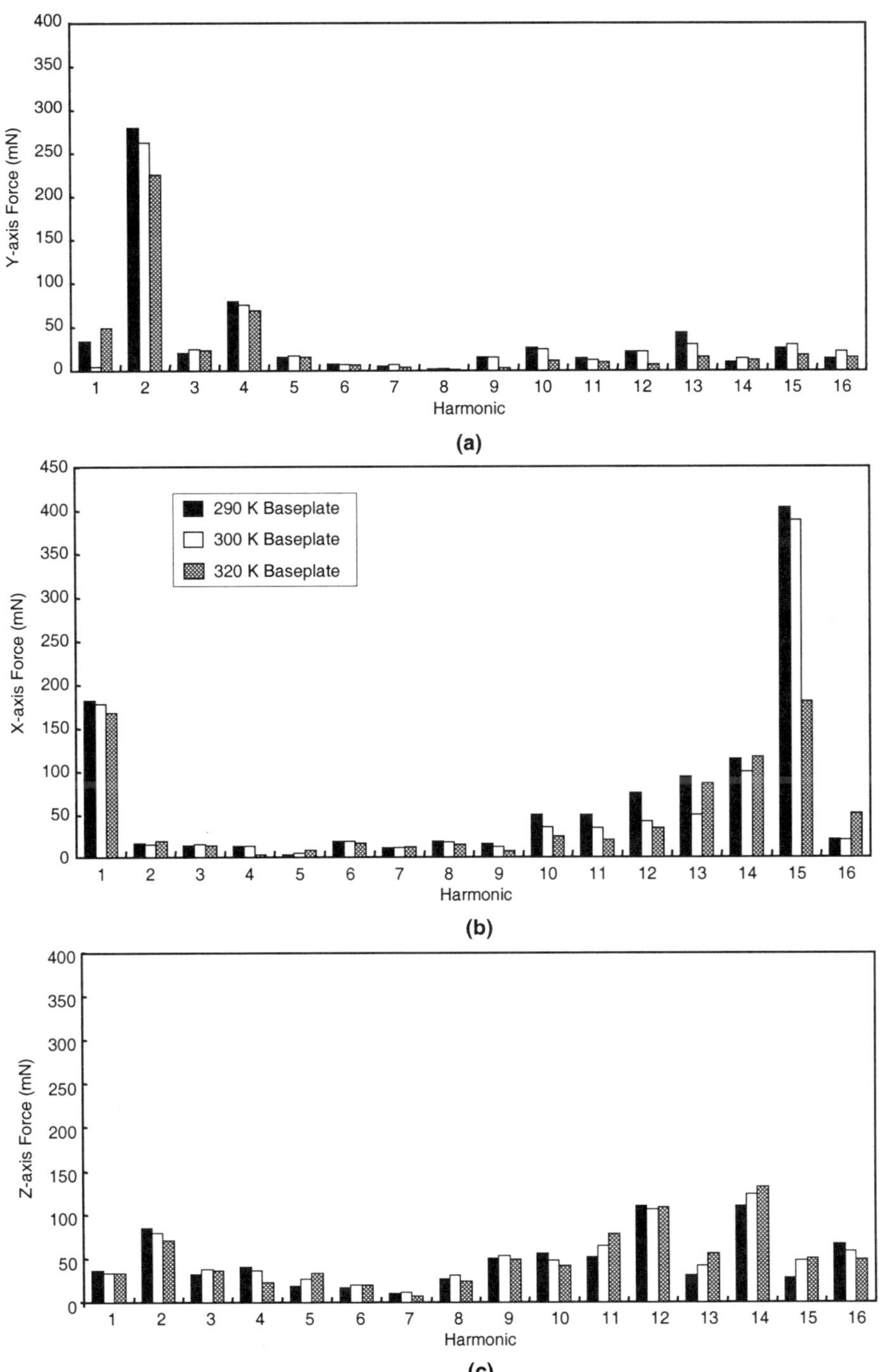

Figure 6. Base plate temperature sensitivity on force spectrum

CONCLUSION

The self-induced vibration was measured on the IMAS S/N101 cooler at 30 and 80 W input power to the compressor for the two cases of simple vibration control and two cases of no vibration control (simple amplifier sinusoidal drive only). The cooler with simple electronics drive and no adaptive vibration control already exceeds the general vibration requirement of less than 0.2 N at AIRS-type cooling capacity. Self-induced vibration is proportional to input power. The 'oil canning' effect of the pulse tube and cantilever structural mode contributed to the off-axis vibration. A frozen waveform tuned at the fixed compressor temperature is adequate in reducing vibration below the requirement, even when the compressor temperature varied over a 30 K range.

ACKNOWLEDGMENT

The work described in this paper was carried out by TRW. It was sponsored by the NASA JPL IMAS Project.

REFERENCES

1. C.K. Chan, J. Raab, R. Colbert, C. Carlson, and R. Orsini, Pulse tube cooler for NASA AIRS flight instrument, in: "ICEC 17 Proceedings" (1998), pp. 77-87.
2. C.K. Chan, P. Clancy and J. Godden, Pulse tube cooler for flight hyperspectral imaging, to be published in: "Proceedings of Space Cryogenic Workshop" (1999).
3. C.K. Chan, T. Nguyen, R. Colbert, J. Raab, R.G. Ross, and D. Johnson, IMAS pulse tube cooler development, in: "Cryocoolers 10", Plenum Press, New York (1999), pp.139-147.
4. C.K. Chan and T. Nguyen, Performance of TRW second generation pulse tube coolers, to be published in: "Adv. in Cryogenic Engineering", v. 45, Plenum Press, New York.

CRYOCOOLER STATE OF THE ART FOR SPACE-BORNE APPLICATIONS

D. G. T. Curran,[1] M. Donabedian,[1] D. S. Glaister,[2] T. Davis [3]

[1] The Aerospace Corporation
El Segundo, CA, 90245
[2] Ball Aerospace and Technologies Corporation
Boulder, CO, 80301
[3] Air Force Research Laboratory
Kirtland Air Force Base, NM, 87119

ABSTRACT

A survey is presented which identifies more than 30 cryocoolers for space-borne, long-life applications. The cryocoolers cover a variety of thermodynamic cycles and configurations with single and multi-stage temperature cooling levels and range in cooling capacity from milliwatts at 10 K to over 10 W at 120 K. The survey's primary objectives are to provide hardware summaries and performance data to assist potential users in the design of cryogenic cooling systems and to identify areas for future development. Funding of the cryocoolers listed in this summary includes the Department of Defense (DOD), NASA, various government agencies in the United States and overseas as well as worldwide company research and development. Several existing flight qualified cryocoolers and those to be qualified by the turn of the century from at least 12 programs are identified. An overview and status of the maturity of the various cryocoolers is made including performance comparisons at 35 K, and 60 K.

INTRODUCTION

The authors have provided technical support to numerous United States DOD spacecraft programs involving cryogenic cooling systems, as well as to the Air Force Research Laboratory (AFRL) at Kirtland Air Force Base, New Mexico, which is the center for space cryocooler and cryogenic technology development. They have compiled information on a variety of long-life, mechanical cryogenic refrigerators (cryocoolers) for space that are available or under development to provide cooling of infrared sensors and spectrometers, optical elements, low noise amplifiers, superconductivity devices, zero boil-off or reduced volume refrigerant storage and for atmospheric monitoring and astronomy. The intent of this paper is to present a brief overview of an evolving cryocooler data package[1] compiled for the AFRL to assess the state of the art and provide guidelines for development of space cryocoolers in the next millennium. The intent of the AFRL is to continue to make this cryocooler data package available as a hard copy or CD to any user or vendor who requests it.

SPACE-BORNE CRYOCOOLER STATUS AND OVERVIEW

Since the large number of coolers represent several different thermodynamic cycles, different hybrid combinations and many different vendors, space does not permit including all coolers. Therefore, at the authors' discretion, the described coolers represent a sample of the current inventory and are grouped by the nominal operating temperature range near their design point. Since these coolers provide off-design performance over a broad range of temperatures, several are represented in more than one range. The four temperature ranges are: 1) 4-12 K, 2) 18-45 K, 3) 50-100 K, and 4) over 100 K.

It should be noted that on-gimbal optics applications at ~100 K and lower, demand weight reductions and higher efficiencies for cooling due to mass and motor torque limits restricting on-gimbal heat rejection capabilities. Also, high specific powers at ~10 K have restricted the use of coolers in the lowest temperature range.

Since clearance seal concepts with flexure and gas bearings show promise for long-life, attention needs to be focused on improving integration capabilities and performance by reducing cooler weight and volume while reducing irreversible losses. For these reasons, user friendly interfaces for heat transport and structural support, more efficient motor designs, and optimization techniques will need to be introduced to minimize vibration as well as temperature and pressure drop in the cooler components. The latter involve critical cold-end and hot-end working fluid heat exchangers, regenerators and recuperators. Cooler and payload/spacecraft system designers should optimize the cooling capabilities and load requirements by utilizing temperature staging for overall system efficiency. Vendors should have the capability of scaling their designs over a sufficient performance range to meet user needs without development efforts or the need to re-qualify and life-test the scaled unit. The latter is a challenge but its realization will serve the user's need to make cooler change-outs with minimum schedule impact due to modified cooling requirements resulting from late test verification of the payload.

4-12 K Operating Range. Traditionally, this range has involved the use of super-fluid helium dewars such as the Infrared Astronomical Satellite (IRAS) and the Cosmic Background Explorer (COBE) at temperatures near 2 K or solid hydrogen cryostats such as the Space Infrared Imaging Telescope (SPIRIT-III) flown on the Mid-Course Space Experiment (MSX) spacecraft for cooling near 10 K. However, there is an incentive to develop cryocoolers for operation in this range because of the desire for longer life with reduced mass and volume relative to dewars. Mechanical coolers are emerging with capacity for continuous cooling in the 50-250 mW range. This range has the least mature cooler technology.

Matra Marconi Space (MMS) 4 K Hybrid Joule-Thomson (J-T)/Stirling. This cooler is the culmination of a development activity started at Rutherford Appleton Laboratory (RAL) in the mid 1980's. Under European Space Agency (ESA) funding, British Aerospace Space Systems (BAe now MMS) developed a 4 K cooler that has been qualified for space flight. It consists of a two-stage Stirling pre-cooler integrated with a

J-T cryostat to achieve liquid helium cooling. The unit produces 5-10 mW of cooling for 200 W of input power. The total mass is 50 kg and is a prime candidate for several astronomy missions including the Far Infrared Space Telescope (FIRST).

MMS 10 K Two-Stage Stirling. This program was initiated by AFRL/BMDO in early 1997 with the objective to quickly develop a protoflight quality cryocooler with at least 45 mW of cooling at 10 K. To demonstrate a fast turn-around, the program leverages off of the MMS 20 K two-stage expander with an improved cold-head using a double set of MMS 50-80 K back-to-back compressors. Through regenerator improvements with rare earth materials, the program now has the potential to achieve up to 75-100 mW of cooling with a motor power of 2000 W/W or less. The program completion date for a flight quality thermo-mechanical unit is January 2000. The intent of this program is not only to produce a 10 K flight cooler for customers who may need a cryocooler in the next three years, but also to serve as a pathfinder for future, more optimized Stirling and Pulse Tube 10 K programs.

Jet Propulsion Laboratory (JPL) 10 K Hybrid Sorption Cooler. The Brilliant Eyes Ten Kelvin Sorption Cooler Experiment (BETSCE) was developed under USAF/BMDO/AFRL funding to demonstrate hydride periodic sorption cryocooler technology for operation near 10 K. The flight demonstration flew on STS-77 in May of 1996 supplemented with a 70 K Stirling pre-cooler. The flight unit provided 100 mW of cooling at 10 K (15-minute per 24-hour) with a power input of about 200 W at 290 K.

Ball 10 K Hybrid J-T/Stirling. The AFRL initiated this program with the state of the art goal of developing a very efficient (<1000 W/W motor input power), lightweight, 10 K cryocooler for space applications. The design will utilize an enhanced 35/60 K Stirling three-stage unit (with regenerator changes and upper stage cooling added) to provide precooling at 15-18 K to a J-T cooler which in-turn provides at least 250 mW of cooling at 10 K. The J-T components, based on a Redstone concept, include two oil-lubricated compressors using a rotary-vane design with heritage from commercial applications, but which requires development for long-life higher-pressure application with a dry helium working fluid. Additional J-T loop hardware includes recuperators, filters, and a thermal storage device to level-out cold-end heat spikes. This hybrid design takes advantage of the Stirling efficiency in cooling to cryogenic temperatures from ambient and the efficiency of the recuperative J-T cycle near 10 K.

18-45 K Operating Range. There are several coolers designed primarily for operation near the lower end of this temperature range and several that are best suited for the 35-45 K range but operate efficiently in the 50-80 K range. Engineering model coolers have been available for several years and flight-qualified coolers are now available. Qualified control electronics are being developed and should be available in the next year.

MMS 20 K Two-Stage Stirling. The MMS 20 K cooler is based primarily on the two-stage technology achieved at RAL under ESA funding. The program was initiated in 1994 to provide 20 K cooling capability for the FIRST/Planck instrument. The back-to-back compressors are derived from the 50-80 K technology combined with a two-stage displacer producing 120 mW at 20 K plus 400 mW at 30 K for a total power of 122 W. Extensive flight qualification was completed in mid 1998.

Ball 30 K Two-Stage Stirling. NASA Goddard Space Flight Center (GSFC) developed this cooler from earlier coolers using RAL/Oxford technology. The cooler provides 0.3 W at 30 K using integral back-to-back compressors and a split expander with balancer. The expander incorporates a fixed regenerator design allowing for a lightweight displacer/piston and improved cooling efficiency. The cooler has undergone successful environmental and performance testing conducted at NASA GSFC with flight-type electronics. It has accumulated approximately 10,000 hours of operation but was recently shut down for a cooling anomaly investigation.

Ball 35/60 K Three-Stage Stirling Cooler. This cooler was delivered to the AFRL as a candidate for cooling sensor infrared focal plane assemblies (IRFPA's) such as Space Based Infrared System-Low (SBIRS-L). The upper stage is used as a shield for the lower stages that provide 0.4 W at 35 K and 0.6 W at 60 K. The design was derived from the two-stage NASA GSFC 30 K program that also used the upper stage as a shield. The cooler provides very efficient cooling with a specific power estimated to be 30 W/W at 60 K and 80 W/W at 35 K.

JPL 20 K Hybrid Continuous Sorption/Radiator. JPL is currently developing a 20 K continuous sorption cooler for the Planck Surveyor program sponsored jointly by NASA and ESA. The system utilizes hydride compressors connected to a J-T cryostat assembly to produce liquid hydrogen operating near 20 K. A passive radiator operating in deep space at about 60 K allows the system to produce about 1.2 W of cooling at 20 K with around 400 W of input power. The schedule calls for an engineering development model (EDM) delivery by the year 2000 with two flight units delivered to ESA in 2003.

Creare 35 K Single-Stage Reverse Brayton (SSRB). This AFRL cooler is a candidate for cooling sensor IRFPA's such as those used for the SBIRS-L. NASA GSFC has also supported this technology and is currently using components of the SSRB for the Near Infrared Camera and Multi Object Spectrometer (NICMOS) flight cooler and circulator. Component improvements include higher efficiency at lower cooling loads and temperatures. Program requirements were changed from 0.4 W at 35 K and 0.6 W at 60 K to a single-stage providing 1.0 W at 35 K for 100 W of power. The miniature components involve using permanent magnets for both the compressor motor and cryogenic turboalternator and the use of a more compact recuperator design for the heat exchanger. The turboalternator replaces the SSRB turboexpander reducing parasitics and eliminating brake flow/cooling. This cooler will be delivered to the AFRL for performance and endurance testing with laboratory electronics.

Lockheed Martin Missiles and Space (LMMS) 35 K Single-Stage Stirling. A modification of the L1710C based on previous Lucas-built coolers was developed for the Low Altitude Demonstration System (LADS). The cooler utilizes Oxford-type technology with back-to-back compressors. The modifications have included improved compression space porting and change in position sensor design. Two L1710C units have been built but have not been life tested. These units have accumulated a few thousand hours in various laboratory tests and one Lucas-built cooler has accumulated over 16,000 hours. Control electronics including provision for cancellation of axial residual vibration and temperature stability of the cold-tip have been flight qualified on a predecessor program. An LMMS compressor has also demonstrated low lateral vibrations using tangential-arm flexures in limited testing. In February 1999, prior to undergoing test verification of the predicted 0.5 W at 35 K for a specific power of 138 W/W, the LADS flight program was canceled.

TRW 35 K Single-Stage Pulse Tube Cooler (PTC) Models PTC-010A-035-I and PTC-020C-035-I. These coolers were developed under separate AFRL programs and used previously developed technology under NASA/DOD contracts. The back-to-back compressors utilize flexure-bearing supports for clearance seals and linear motor compressor drive as do their Stirling counterparts and are derived from Oxford technology. The PTC's have been sized to deliver 0.3 and 0.85 W at 35 K. These coolers are at the AFRL for performance and endurance testing and were earlier performance characterized at JPL. The larger capacity unit has accumulated about 14,000 hours while the smaller unit has about 12,000 hours. The latter unit's ground test heat rejection loop failed in November 1998 and the cooler was seriously over-heated. The unit was able to be subsequently operated but with a 2 K increase in cold-end temperature. In April 1999, one of the back-to-back compressor motors tripped-off. The unit will undergo investigation at TRW to determine the nature of the over-heat damage. Control electronics including residual vibration and temperature control have been flight qualified.

Raytheon 35 K Single-Stage Stirling Protoflight Spacecraft Cooler (PSC)/Space and Missile Tracking System (SMTS). These coolers have heritage to the standard spacecraft cooler (SSC) designed under an AFRL development program to provide 2 W at 60 K. An improved SSC (ISSC) was developed under IR&D and used for SBIRS-L life testing. Two units accumulated 46,000 hours at 60 K and another unit flew on a shuttle mission. The PSC and SMTS Flight Demonstration System (FDS) programs were initiated in late 1992 and 1996, respectively, but the latter was canceled in February 1999. The PSC unit incorporates tangential-arm flexure springs for clearance seal support in the back-to-back compressor. The unit demonstrated reduction in residual vibrations in both axial and lateral axes at JPL. The SMTS units incorporate a JPL heat-intercept concept. The PSC was delivered to the AFRL for performance and endurance testing and has accumulated 1500 hours. An SMTS life test unit for the FDS program accumulated 9000 hours under contract and is at 13,000 hours under IR&D funding. The SMTS electronics are radiation-hard and were flight qualified prior to program cancellation but were not stand-alone and were integrated with other FDS functions.

50-100 K Operating Range. The largest group of coolers representing the more mature technology is in this range with the majority designed to operate between 55-75 K. Over ten coolers including Oxford-type Stirling and pulse tubes have been flown since the early 1990's primarily to support science experiments or technology demonstrations. The first units utilized simple flight electronics without vibration control but the recent pulse tube flight included fully qualified electronics with temperature and vibration control.

Ball 58 K Single-Stage Stirling. Ball began development of Oxford-type Stirling coolers in 1990 under a RAL license. Subsequently, in 1992, Ball developed two improved IR&D designs for operation at 60 K. The initial design achieved 1.5 W of cooling at 55 K and is currently on life test at Ball with >23,000 hours. The second version of this unit (SA160) provides 2.5 W of cooling at 60 K for 116 W to the compressors. In parallel Ball developed a single-stage version of its SB230 cryocooler using a fixed regenerator design capable of lifting 1.6 W at 60 K for 53 W to the compressor. This cooler is scheduled to fly on the High Resolution Dynamic Sounder (HIRDLS) instrument on the NASA Earth Observation System (EOS) Chem Platform in 2002.

Ball 65/120 K Two-Stage J-T. Ball has developed a cryogenic on-orbit long-life active refrigeration (COOLLAR) two-stage J-T that was successfully verified in micro-gravity and characterized aboard the Space Shuttle in August of 1997. The design uses an oil-lubricated compressor to provide the high-pressure ratio for the J-T cold head (with thermo-electric cooler pre-cooling) producing 5 W at 120 K and 1.25 W at 65 K. This J-T cooler produces and stores liquid nitrogen at the cold interfaces, allowing a variable load (peak of 3.5 W) to be averaged at the 65 K stage. The cooler provides long, flexible lines leading to the J-T cold head for remote mounting of the compressor.

Creare 65 K Single-Stage Reverse Brayton. This cooler was designed for long-life using high speed, small turbomachines with gas bearings allowing vibration and wear-free operation to provide 5 W of cooling at 65 K. The cooler consists of small-size precision devices. The cycle operates with continuous flow of neon gas circulated by a compressor through a recuperative heat exchanger and turbine expander. Additional components include a high efficiency inverter/controller, after-cooler and load heat exchanger. Both NASA and the AFRL supported the technology. The cooler achieved its highest efficiency with a system (cooler and electronics) specific power of 37 W/W for 7.5 W of cooling at 65 K and a heat rejection temperature of 280 K. The cooler accumulated 29,000 hours in life test until December 1998 when the test was aborted due to a chiller pump failure. However, the turboexpander failed to re-start and a disassembly investigation is planned.

Creare 65 K Single-Stage Stirling. In the late 1980's, both the USAF and Ballistic Missile Defense Operations (BMDO) recognized the need for a long-life cryocooler capable of handling small loads in the 50-80 K temperature range. As part of the AFRL SSC program, Creare was funded to develop an alternative technology to the flexure spring cryocoolers of other vendors. The Creare diaphragm development resulted in an engineering model of a cooler designed to provide 2 W of cooling at 65 K. The EDM was delivered to the AFRL in early 1994 and its performance was characterized before accumulating over 10,000 hours of endurance testing at program end.

Creare 70 K SSRB. This cooler will be used for the Near-Infrared Camera and Multi-Object Spectrometer (NICMOS). NICMOS is a second generation Hubble Space Telescope (HST) science instrument whose detectors are cooled to about 58 K by a solid nitrogen cryostat. An anomaly that was discovered in late 1997 caused the operating life of the dewar to be reduced from a predicted 48 months to about 1.7 years. As a result, NASA GSFC decided to attempt to retrofit the solid cryostat with an improved version of the Creare 5 W SSRB. The new cryocooler is designed to provide approximately 7.7 W at 70 K for a total input power of around 315 W. The new system, along with a newly developed flight electronics package, was successfully flight tested aboard the STS in November 1998. Final integration into the HST is planned for 2000.

LMMS 60 K 1710-C/Space Cryogenic Refrigeration Systems (SCRS) Single-Stage Stirling. Since 1987, LMMS has been developing advanced cryocooler systems based on Oxford-type technology under a teaming arrangement (since terminated) with Lucas Aerospace. Several different models emerged from this activity including the 1710-C and SCRS discussed in this paragraph and the LADS unit previously discussed. The 1710-C consists of back-to-back compressors connected to a single displacer, providing about 2.0 W at 60 K with a total power input of 130 W including a flight qualified control electronics package. Utilizing similar hardware with a slightly larger compressor piston diameter, an integrated system was developed and denoted as the SCRS. This unit was funded jointly by several USAF organizations as part of the SPAS-III flight experiment. Two compressors and two displacers which have their cold tip attached via flex couplings in a common vacuum housing provided about 1.2 W of cooling at 59 K. This assembly was tested and delivered to Utah State University for overall system testing.

Oxford and RAL 80 K Single-Stage Stirling. The original design, now an industry standard, for linear motor, spiral-flexure support and clearance seals is based on single-stage Stirling coolers developed by Oxford University beginning in 1978. The technology was based on Oxford pressure modulator technology that was flown successfully for up to seven years in the 1970's before the instruments (Pioneer Venus, Nimbus 6 and 7) were retired. Two were flown as part of the Improved Stratospheric and Mesospheric Sounder (ISAMS) instrument aboard the Upper Atmosphere Research Satellite (UARS) in September 1991. These coolers accumulated well in excess of 25,000 hours in orbit. Several ground test units have accumulated >50,000 hours (8.1 years maximum). RAL has built two similar single-stage coolers. The first unit funded by ESA flew as part of the Along Track Scanning Radiometer (ATSR) instrument aboard the Earth Resources Satellite (ERS-1) in July 1991. The second unit flew on ATSR-2 replacing ATSR-1. Together both RAL coolers have exceeded 60,000 hours.

Raytheon 60 K PSC/SMTS/ISSC/SSC Single-Stage Stirling. These coolers were designed for cooling 2 W at 60/65 K. Both the PSC and SMTS (SBIRS-L) coolers provide 35 and 60 K cooling. The SMTS life test cooler has been operated at 60 K and at 35 K as part of 13,000 hours of operation. One ISSC has accumulated nearly 24,000 hours while the other life test unit accumulated 22,000 hours under a prior SBIRS-L program. The respective specific powers for the ISSC and SSC units at 65 K are 45 W/W for 1.75 and 1.2 W. The respective specific powers for the PSC and SMTS (with heat intercept) are 27.5 and 22.4 W/W at 60 K (27.5 W/W without heat intercept). ISSC #4 was modified to improve motor efficiency and thermal interfaces.

TRW 60 K Single-Stage PTC. This cooler was developed for both NASA and Air Force programs; namely, the Atmospheric Infrared Sounder (AIRS), the Tropospheric Emission Spectrometer (TES), the AFRL 6020 version, and the Integrated Multi-spectral Atmospheric Sounder (IMAS) in both integral (I) and split (S) configurations. It has common heritage to the above mentioned TRW 35 K coolers. Two flight models of the PTC-010B-055-S (55 K) have been delivered to the NASA AIRS program and two flight units of the PTC-010C-057-I (57 K) will be delivered to JPL in 1999 for TES. PTC-010A-060-I (60 K) was delivered as a flight unit to MTI in 1998. An EDM of this unit was delivered to the AFRL for endurance and performance testing after being performance characterized at JPL. PTC-004A-055-I was delivered to JPL in June 1998 for the recently canceled IMAS program. The IMAS unit had accumulated 1500 life test hours at TRW.

TRW 60-150 K Single-Stage Integral Miniature Stirling Cooler (MSC), SC-0-6A-65-I and Integral Miniature Pulse Tube (MPT), PTC-001A-115-I. These coolers were developed for the DOD requirements and have been delivered for several applications including life testing and optics cooling for SBIRS-L. Two mini-Stirling coolers were life tested for 15,000 hours each to provide 0.16 W at 60 K for an initial SBIRS-L IRFPA requirement. The corresponding specific power at 60 K is 88 W/W. At the same time,

TRW ran life tests on two IR&D coolers one of which has accumulated 40,000 hours to date. The Stirling cooler is part of the HTSSE-II payload launched on the ARGOS spacecraft in February 1999. Two miniature PTC's are currently operating on the CX payload and have accumulated one-year of operation.

MMS 50-80 K Single-Stage Stirling. This cooler is a modification of the single-stage BAe 80 K. The 80 K cooler was built and qualified for an ESA contract using Oxford University technology. Over 20 units of this design were built with one unit launched on ARGOS mentioned above. The basic 50-80 K unit is also flight qualified and produced in batches of 10 and as of this writing about 20 units have been manufactured. A number of programs sponsored by both NASA and ESA are scheduled to fly this unit over the next several years including the Michelson Interferometer for Passive Atmosphere Sounding (MIPAS) and the Measurement for Pollution in the Tropopause (MOPITT) instruments. Nominal performance of this unit is quoted as 1.7 W at 80 K, but it is used over a wide range of temperatures from 50 K-90 K. Several life tests on these units have accumulated >100,000 hours with 6.5 years on an 80K unit.

Above 100 K. The TRW MPT and MSC units have been characterized for cooling optics or shields. The MPT specific power for 1.5 W at 115 K is 13.4 W/W, for 2.5 W at 150 K is 8.2 W/W, and for 1.25 W at 175 K is 5.8 W/W. The MSC specific power for 0.5 W at 100 K is 15.5 W/W, for 0.8 W at 120 K is 9.7 W/W, and for 1.3 W at 150 K is 6.4 W/W. The larger capacity Raytheon PSC/ISSC, TRW PTC's, LMMS, Ball and Creare coolers have been considered for operation above 100 K. For SBIRS-L gimbaled optics with limited heat rejection capabilities, projected requirements are cooling loads of 6 to 10 W at 100 K with specific powers as low as 8 to 10 W/W and mass of 3 to 5 kg. Typical performance of Stirling and PTC's are 12 to 14 W/W at this temperature with masses of 7 (PTC)-12kg. To meet the 100 K cooling and mass requirements, the AFRL has initiated the High Efficiency Cooler (HEC) program.

SPACE-BORNE CRYOCOOLER PERFORMANCE COMPARISON

The cryocooler capacity and power efficiency data were normalized to a reference (ref) cooling load temperature and a 300 K rejection temperature. It was based on the following Carnot correction equation where Q is the cooling capacity and T is the temperature either at the cold tip (ct) or heat rejection (rej) interface:

$$Q_{ref} = \frac{T_{ref}^{ct} * (T_{data}^{rej} - T_{data}^{ct}) * Q_{data}}{T_{data}^{ct} * (T_{ref}^{rej} - T_{ref}^{ct})} \tag{1}$$

The ratio assumes that the Carnot efficiency (or refrigerator coefficient of performance) is the same for both the data and reference conditions for cooling load temperature differences <5 K and rejection temperature differences <20 K. Also, where flight quality electronics controllers are not available, the input power has been estimated to be 85% efficiency and 10 W of quiescent power, respectively.

Figures 1 and 2 are plots of cryocooler motor (input divided by cooling capacity) and total (motor and electronics) specific power, respectively, for cooling at 60 K as a function of cooling capacity. The figures show the expected effect of efficiency increasing with capacity and indicate the general trend at 60 K of the higher efficiency at low to medium

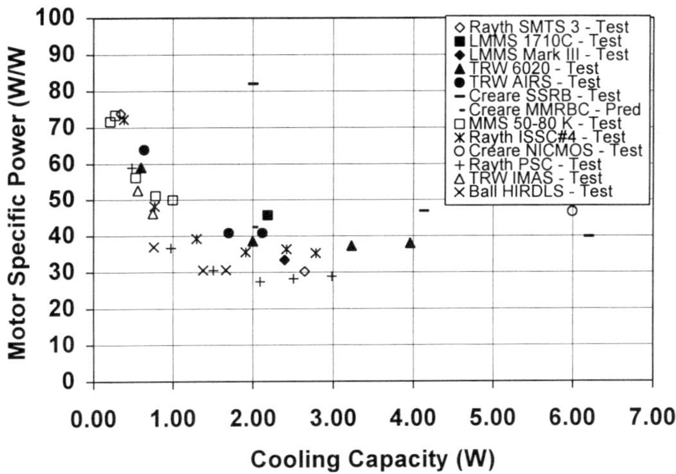

Figure 1. Cryocooler motor specific power interpolated to 60 K cooling and 300 K reject.

loads of Stirling (especially the Raytheon PSC) and Pulse Tube cryocoolers compared with the reverse Brayton. Figures 3 and 4 are similar plots of the specific power as a function of cooling capacity at 35 K. At 100 K the AFRL HEC program goals of 8-10 W/W of motor power for 10W of cooling will advance the state of the art for this temperature level.

SUMMARY AND CONCLUSIONS

The status and performance for a wide variety of long-life cryocoolers being developed for space applications ranging in temperature from 10 to 175 K. This survey identifies more than 30 coolers covering a variety of thermodynamic cycles and cooler types. Capacities vary from a few milliwatts to over 10 W for single and multi-stage designs. The survey indicates that several coolers have been flown and others are near flight model status and are undergoing flight qualification. The latter should be available by the turn of the century. Performance comparisons were made for specific power versus cooling capacity at 35 K and 60 K. Comparisons show Stirling coolers generally have the highest relative efficiencies at 35K with pulse tube efficiencies comparable at 60 K.

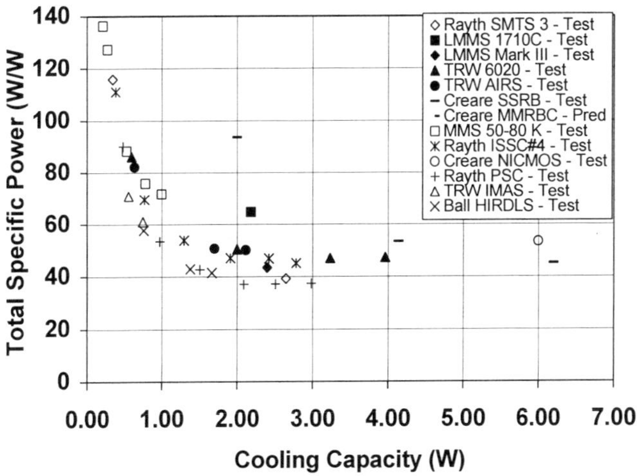

Figure 2. Cryocooler total specific power interpolated to 60 K cooling and 300 K reject.

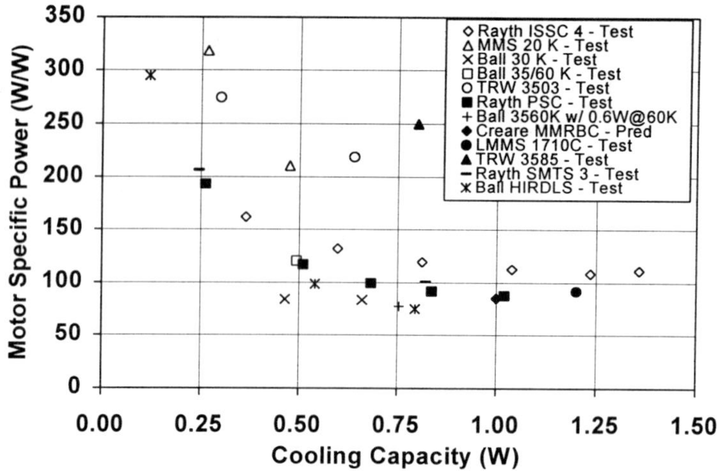

Figure 3. Cryocooler motor specific power interpolated to 35 K cooling and 300 K reject.

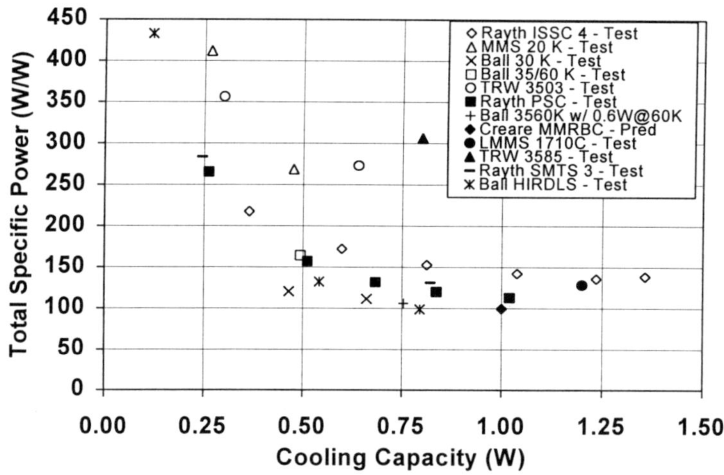

Figure 4. Cryocooler total specific power interpolated to 35 K cooling and 300 K reject.

ACKNOWLEDGMENT

The authors acknowledge the support of the SBIRS-L Program, AFRL and BMDO personnel. Special thanks are given to J. P. Reilly and B. J. Tomlinson of AFRL, S. Bard, B. Bowman, R. Ross, Jr., and L. Wade of the Jet Propulsion Laboratory, R. Fernandez, W. Gully, W. Kiehl, R. Levenduski, and R. Reinker of Ball, W. Swift of Creare, D. Gilman and K. Price of Raytheon, T. Nast and I. Spradley of Lockheed Martin, B. G. Jones formerly of MMS, and C. K. Chan and M. Tward of TRW for providing a large portion of the cryocooler data presented here.

REFERENCES

1. D. S. Glaister, M. Donabedian, D. G. T. Curran, "An Overview of the Performance and Maturity of Long Life Cryocoolers for Space Applications," The Aerospace Corporation TOR-98 (1057)-3, 8/1998.

AIR FORCE RESEARCH LABORATORY CRYOCOOLER CHARACTERIZATION STATUS AND LESSONS LEARNED

B. J. Tomlinson, Jr.[1], C. Yoneshige[2], and N. Abhyankar[3]

[1,2] Air Force Research Laboratory
Kirtland AFB, NM 87117-5776

[3] Dynacs Engineering Co., Inc.
Albuquerque, NM 87123-2710

ABSTRACT

The Air Force Research Laboratory's (AFRL) Cryocooler Characterization Laboratory (CCL) is an independent government laboratory established by the Ballistic Missile Defense Organization (BMDO) and the US Air Force for the unique purpose of establishing cryocooler performance and increasing endurance hours to demonstrate the long life characteristics of strategic cryocoolers. In order to meet these objectives, AFRL has implemented standardized testing procedures, testing methods, empirical and analytic analysis of performance data, and data feedback mechanisms to industry partners. AFRL is striving to demonstrate cryocooler technology that is capable of achieving a ten year on-orbit life, improving the accuracy of cryocooler modeling and simulation tools, identifying life limiting mechanisms and performance degradation, and studying and characterizing the issues surrounding the prediction, demonstration, and accelerated testing of cryocoolers. The data gathered in the laboratory has been standardized for cataloging and ease of analysis. In order to provide standardization, the laboratory has utilized modular approach to data acquisition software, instrumentation, and measurement equipment use, common to each test stand. In order to present the characterization data, performance prediction models for the cryocoolers are derived from empirical data. These models aid space system designers and cryocooler developers in evaluating the entire performance envelope of specific cryocoolers and modeling the effects on cooler performance due to changes in sensor duty cycle, heat load, input power, and rejection temperature change. The cryocoolers under characterization and endurance testing vary in design point, thermodynamic cycle, and technical maturity. The current characterization and endurance activities in the laboratory and specific lessons learned are presented.

Advances in Cryogenic Engineering, Volume 45.
Edited by Shu *et al.*, Kluwer Academic / Plenum Publishers, 2000.

ACRONYM LIST

SSC	Standard Spacecraft Cryocooler	CSSC	Creare Standard Spacecraft Cryocooler
PSC	Protoflight Spacecraft Cryocooler	AFRL	Air Force Research Laboratory
MMS	Matra Marconi Space	CCL	Cryocooler Characterization Laboratory
PT	Pulse Tube	BMDO	Ballistic Missile Defence Organization
SSRB	Singe Stage Reverse Brayton	NASA	National Aeronautics and Space Administration

INTRODUCTION

AFRL has been actively engaged in cryocooler development for over 10 years and characterization for over 8 years. The Cryocooler Characterization Laboratory (CCL), sponsored by the Ballistic Missile Defense Organization, has evolved into a state-of-the-art facility with the capability of performing all aspects of cryocooler characterization and long term endurance evaluation, except for electromagnetic interference and electromagnetic compatibility measurements. The CCL has evolved automated data acquisition, parametric performance models, and state-of-the-art thermal-vacuum environment chambers.

AFRL objectives for cryocooler characterization have emerged to meet the following cryocooler technology needs for AF and DoD:

1. Performance characterization, demonstration, and validation of emerging cryocooler and cryogenic technologies to meet Air Force, BMDO, and DoD technology requirements.
2. Long life endurance evaluation of cryocooler technologies with the focus on the characterization of long term performance degradation, reliability, and the development of proven methods for accelerated testing of long life space cryocoolers.
3. Provide data and lessons learned feedback to cryocooler and cryogenic technology developers to aid in follow-on developments of emerging technology.

The AFRL CCL is a 5,800 ft^2 laboratory and a 1,400 ft^2 high bay and model shop. The laboratory has the capability to support three tabletop endurance evaluation rooms, each with the capability to support three coolers under test. The CCL has one 18" slide bell, eight 24", two 36", thermal-vacuum chambers. Each chamber has high vacuum, multiple feedthroughs, fluid feedthroughs for chiller fluid, and computerized environmental control in addition to the control electronics for the cryocooler and the measurement electronics / computer for data acquisition.

PERFORMANCE CHARACTERIZATION STATUS

At present AFRL has three cryocoolers in characterization. The Raytheon 2 W @ 60 K Prototype Spacecraft Cryocooler (PSC), the Ball Aerospace 0.4 W @ 35 K, 0.6 W @ 60 K three stage dual load cryocooler, and the Raytheon 2 W @ 60 K Standard Spacecraft Cryocooler II (SSC II) are all progressing at different stages of their respective characterization plans.

Raytheon's PSC is a split stirling cooler designed for 2W @ 60K or 1W @ 35K. It has been successfully integrated in a 36" vacuum chamber for characterization and its initial baseline acceptance evaluation has been completed. It meets the design specifications and runs at specific power below 30W/W. The optimization evaluation has also been completed and is in the analysis stage. It will be followed by detailed load lines at different reject temperatures and thermal transient trials before it is put under endurance evaluation.

Table 1. AFRL Endurance Evaluation Status Summary

Cryocooler	Run Time Hours	Status
		Characterization
Raytheon SSC II	500	This cryocooler is prepared for re-characterization after re-assembly for a misaligned component in the cold finger and gas cleaning. The characterization activities will bound the performance of the cooler and allow a projected endurance evaluation start date of Oct 99.
Raytheon PSC	1,000	This cryocooler is currently under characterization, having completed baseline characterization. Projected endurance evaluation start date is Nov 99.
MMS 10K	N/A	This cooler is scheduled for delivery in Sep 99. The projected start date for characterization is Nov 99 and endurance evaluation start date of May 00.
TRW mini PT	200	The flight-quality unit has been completed by TRW and is currently at Utah State University as a flight spare for the NASA SABER payload. Projected delivery to AFRL is Jan 00 and an endurance evaluation start date of Jun 00.
Ball Aerospace 35/60K	300	This cooler is currently undergoing a baseline characterization. The 35/60 will then be transferred to Ball Aerospace to repair a design flaw in the cold head. The cooler will then be returned to AFRL for a projected Jan 00 characterization and an endurance evaluation start date of Jun 00.
		Endurance Evaluation
6020 PT	16,872	This cooler is currently in Endurance Evaluation. Data to date shows the cooler operating within design margins and instrument error. Resolution of apparent discrepancies in the characterization data and the data correlation with JPL have not been completely quantified.
3585 PT	12,386	This cooler is currently in Endurance Evaluation. Data to date shows the cooler operating within design margins and instrument error.
3503 PT	12,093	The 3503 has completed endurance evaluation. After undergoing a significant overheat condition, the cooler continued to operate at degraded performance until April 99 when the cooler tripped and could not be restarted. The cooler internal gas was sampled and then taken to TRW for disassembly inspection. Discussion is underway with TRW to determine the future of the hardware.
SSRB	29,063	The SSRB shutdown in Dec 98 due to an alarm condition on the compressor. Subsequent attempts by AFRL and Creare personnel to restart the cooler were unsuccessful. They couldn't get the turboexpander to lift up and start spinning. Creare and AFRL personnel disassembled the turboexpander for inspection and analysis. Discussion is underway with Creare to determine the future of the hardware.
CSSC	13,358	The CSSC shut down in December 98. Subsequent attempts by AFRL personnel to purge, re-fill, and re-start the cooler were not successful. At the present time, the CSSC endurance evaluation is complete and discussion is underway with Creare to determine the future of the hardware.

The Ball Aerospace 35/60 K cryocooler, another split Stirling design, has been integrated into a 36" chamber for characterization activities. Due to the need for a retrofit of the expander cold head, this cooler will remain at AFRL until around October 1999 when it will be de-integrated and returned to Ball for the retrofit. Once the retrofit is complete the cooler will return to AFRL for the continuation of characterization and endurance evaluation.

The SSC II is also a split Stirling cooler, and was built by Raytheon (then Hughes Aircraft Corporation). It was designed for 2W at 65K with a 30W/W specification. During the initial baseline characterization, however, the specific power was found to be above 30W/W mark and did not meet design specifications. It also developed hardware problems and was returned to Raytheon for refurbishing. It was repaired and sent back to AFRL. It will be placed under limited re-characterization for design verification and endurance. Its design served as a stepping stone for Raytheon's PSC design.

The Matra Marconi Space 10 K cryocooler is scheduled for delivery to AFRL in late August 1999. This cryocooler will require the use of the 36" vacuum chamber to accommodate the four compressors, two-stage expander, and structural support plate. Development of the characterization plan for this cryocooler is underway and AFRL personnel are scheduled to support on-site contractor acceptance testing as well as contractor personnel visiting AFRL for initial setup and demonstration after delivery.

AFRL CCL has designed and is procuring the necessary components to build up an off state conduction experiment stand. This experiment stand will be integrated with one of the existing 36" thermal-vacuum chambers in the laboratory and will provide AFRL with the capability to make more precise measurements of the parasitic load that an "off" cooler would penalize a cryogenic system.

AFRL CCL engineers have, out of necessity designed an automated load line program to aid in characterization. Some cryocoolers have very long settling times to reach steady state operation. This LabVIEW™ sub-program allows the operators to engage an automated sequence of heater loads that will proceed through the user defined list once the cold end temperatures have reached predetermined steady state criteria. It will also allow utilization of the nighttime hours to complete load lines and cut the overall characterization time significantly.

ENDURANCE EVALUATION STATUS

AFRL currently has two cryocoolers in endurance evaluation and two coolers in post-endurance tear down and inspection. AFRL engineers are currently supporting the on-going endurance evaluation and supporting the tear down and inspection; both on-site and at the contractor facilities.

The coolers in endurance at this time are the TRW 2 W @ 60 K (6020) Pulse Tube Cryocooler and the TRW 0.85 W @ 35 K (3585) Pulse Tube Cryocooler. The 3585 cooler is a 20-cc compressor unit that was the engineering design model pathfinder under this contract. The cooler has accumulated over 12,000 hours of operation time, both in a thermal-vacuum chamber for characterization and in a tabletop configuration for endurance evaluation. The 6020 cooler is a follow-on design from the 3585 and utilizes an optimized 10-cc compressor. This cryocooler has several brothers/cousins that are currently flying or are scheduled for flight within 1999. This cooler has remained in a thermal-vacuum chamber for characterization and endurance evaluation and has accumulated over 16,000 hours of operation.

The TRW 0.3 W @ 35 K (3503) Pulse Tube Cryocooler is currently in post-endurance tear down and inspection. This cooler suffered an environmental failure that exposed it to very high rejection temperatures for an extended period of time. After the overheat condition, the cryocooler showed typical contamination performance degradation, but was allowed to continue endurance evaluation. In April 1999, the 3503 cooler tripped, and was unable to be restarted. Subsequent investigation isolated the thermomechanical cooler as the culprit and AFRL and TRW undertook a joint tear down and inspection of the device. As it turns out, the source of the tripping was a loose retainer cap and capacitance sensor target that was not staked during the original build. This is a testament to the robustness of the cryocooler that the environmental overheat was not the cause of the tripping. The cryocooler is currently at AFRL undergoing further stiction measurements and plans are in the works to evaluate the need for a rebuild, refill, and re-enter the cooler into characterization.

The Creare 5 W @ 65 K Single Stage Reverse Brayton (SSRB) cryocooler is also in post-endurance tear down and inspection. This cryocooler tripped and shut down in December 1998, and was unable to spin up the turboexpander to re-start. All attempts by AFRL personnel and Creare personnel on site did not achieve restart. During June 1999, Creare and AFRL personnel removed the turboexpander for disassembly and inspection. Preliminary results include the detection of tiny foreign particulates in the components. At this time, the investigation will continue with final analysis and recommendations from Creare on the cause and potential repair of the failure.

LESSONS LEARNED

At present, AFRL is in the process of characterizing several generations of cryocooler technology. The cryocoolers in test represent a large cross section of available long life cryocooler technology, including pulse tubes, Stirling, and reverse Brayton cryocoolers. Over recent months, some critical lessons learned have cropped up. The following bullets represent some of these critical lessons learned.

Disciplined Approach to Cryocooler Characterization

AFRL has re-directed the characterization approach to optimize the cryocooler characterization method for maximum return while maintaining the rigorous detailed laboratory examination of the technology and allowing for experiment flexibility for characterization of sponsor/user specific technology objectives. AFRL has incorporated government wide expertise and feedback on cryocooler characterization planning by utilizing the Interagency Cryocooler Test Working Group (IACTWG), NASA, academia, and industry partners as guidance and sounding boards for short and long term planning.

Extensive Documentation of the Developed Cryocooler Technology and AFRL Characterization of the Technology

The need for extensive documentation of the cryocooler technology cannot be stressed enough. As long life space cryocooler technology emerges from one-of-a-kind engineering development units, the government development and characterization team needs to examine every detail of the electronic and thermomechanical aspects of the technology. AFRL has implemented a process for the structured planning, execution, documentation, and reporting of characterization activities. This process has been documented, in draft form, in the "*Air Force Research Laboratory Master Guidelines for Cryocooler Characterization and Endurance Evaluation*." This document, after completion of peer feedback and approval by Air Force Research Laboratory management, will serve as a tool for AF program managers and laboratory engineers to ensure proper and thorough characterization of cryocooler technology.

Cryocoolers Are Not Always Used at Their Optimum Design Point Operation

It has become apparent that over the development time for cryocooler technology, from concept to characterization, the user community often specifically requests performance characterization that is significantly outside the original "design point" of the cryocooler technology. Interest has ranged from quantifying the ultimate heat lift capacity of the cooler at various off nominal conditions, and operation of the cryocooler at elevated or significantly lower rejection temperatures, to time dependent temperature control of step changes in the cryogenic heat load. Anticipating and integrating these potential data points of interest to the user community into a comprehensive cryocooler characterization plan has proved to be challenging but necessary in order to provide users and spacecraft developers unbiased understanding of the technology.

Fail-Safe and Alarm Conditions

Unfortunately, 24 hour automated characterization and endurance evaluation will conspire to find every weakness in the laboratory fail-safe protection systems. As a specific example, a test engineer alarmed the system chiller but not the rejection temperature of the cryocooler. This was based on the belief that the only way the cooler could be deprived of chilling fluid is if the chiller failed. However, an analog-to-digital converter in the environmental control electronics failed and commanded the valves in the fluid lines to the chamber to close. The cooler experienced a significant overheat condition for over 20 hours and did not trip (avoiding another alarm condition). The lesson learned here is that the laboratory did not foresee the need to alarm the rejection temperature. As a result, AFRL has conducted an extensive system analysis of all the cryocooler experiments in the CCL to ensure alarm sensors are placed on all possible failure modes.

Use of Industry and Government Standard Practices and Military Standards as Guidelines

In the past, AFRL relied on haphazard interpretations of military standards as guidelines for cryocooler characterization (mainly for thermal, vibration, and EMI / EMC acceptance, proto-qualification, and flight qualification of developmental hardware). AFRL has now refined the characterization process to include industry and government (Air Force, DoD, and NASA) standard practices and a unified interpretation of applicable military standards for the characterization of space cryocooler hardware.

Contamination of the Cryocooler Internal Working Fluid

One of the critical factors affecting the long life characteristics of cryocooler technology is the initial level and time dependant evolution rate of gaseous working fluid contamination and physical particulate contamination of the working spaces within the cooler. AFRL is tracking the long-term performance of cryocooler technology and quantifying the short and long term performance degradation due to contamination in the laboratory. In addition to the laboratory work, AFRL CCL engineers are working with the AF program managers and industry to identify the sources, effects, and mitigation techniques of contamination. Specifically, during a post-endurance evaluation tear-down, a cryocooler in the laboratory was found to have foreign particles present in the cryocooler.

Integration of the Technology Development Process and the Characterization / Endurance Evaluation Activities

AFRL has developed a team approach to the development and characterization of cryocooler technology. This team incorporates AF program managers, laboratory engineers and technicians, sponsors, and industry partners to ensure that developed technology is seamlessly carried from program concept to post-endurance tear down.

Data Transfer to Technology Users and Technology Developers

The largest concern with the transfer of lessons learned and the characterization data is *timeliness*. AFRL has relied on a slow and incomplete method of piecemeal data dissemination and substantial one to two year delays in publication of reports and papers. AFRL is currently working to develop a "fast-track" multi-path method of disseminating data to the user and developer community. Current efforts include the "*An Overview of the Performance and Maturity of Long Life Cryocoolers for Space Applications*", worked jointly with the Aerospace Corp., a more extensive World Wide Web page, and a CD-ROM database of cryocooler data, all in addition to the traditional conference paper / presentation of data.

SUMMARY

AFRL is striving to meet the development and characterization objectives for emerging cryocooler technology through disciplined planning and teaming with government and industry partners. Significant strides have been made to forge a team of government and contractor personnel that are leading the development and characterization of state-of-the-art long-life cryocoolers to meet AF, BMDO, and DoD technology needs. In addition, AFRL has incorporated the significant lessons it has learned to improve the laboratory characterization process and ensure user satisfaction and confidence in AFRL developed technologies.

REFERENCES

1. Tomlinson, B. J., Gilbert, A., and Bruning, J., "Endurance Evaluation of Long-Life Space Cryocoolers at AFRL-An Update", Cryocoolers 10, Plenum Press, New York (1999).
2. Davis, T., Tomlinson, B. J., and Reilly, J., "Air Force Research Laboratory Cryocooler Technology Development", Cryocoolers 10, Plenum Press, New York (1999).
3. Glaister, D. S., Donabedian, M., Curran, D. G. T., and Davis, T., "An Overview of the Performance and Maturity of Long Life Cryocoolers for Space Applications", prepared for the Air Force Research Laboratory, Aerospace Report No. TOR-98(1057)-3, August 1998.
4. *Air Force Research Laboratory Master Guidelines for Cryocooler Characterization and Endurance Evaluation*, Draft Reference Document, Air Force Research Laboratory, Kirtland AFB, NM, May 1999.
5. Air Force Instruction 99-113, *Space Systems Test and Evaluation Process Direction and Methodology for Space System Testing*, 1 May 1996.

HTS MAGNETS FOR ADVANCED MAGNETOPLASMA SPACE PROPULSION APPLICATIONS*

S. W. Schwenterly[1], M.D. Carter[1], F.R. Chang-Díaz[2], and J.P. Squire[2]

[1] Oak Ridge National Laboratory**
Oak Ridge, TN 37831
[2] Lyndon B. Johnson Space Center
Houston, TX 77058

ABSTRACT

Plasma rockets are being considered for both Earth-orbit and interplanetary missions because their extremely high exhaust velocity and ability to modulate thrust allow very efficient use of propellant mass. In such rockets, a hydrogen or helium plasma is RF-heated and confined by axial magnetic fields produced by coils around the plasma chamber. HTS coils cooled by the propellant are desirable to increase the energy efficiency of the system. We describe a set of prototype high-temperature superconducting (HTS) coils that are being considered for the VASIMR (Variable Specific Impulse Magnetoplasma Rocket) thruster proposed for testing on the Radiation Technology Demonstration (RTD) satellite. Since this satellite will be launched by the Space Shuttle, for safety reasons liquid helium will be used as propellant and coolant. The coils must be designed to operate in the space environment at field levels of 1 T. This generates a unique set of requirements. Details of the overall winding geometry and current density, as well as the challenging thermal control aspects associated with a compact, minimum weight design will be discussed.

INTRODUCTION

Conventional chemical rocket engines for space applications have limited ability to modulate thrust. Consequently, the spacecraft trajectory is controlled by alternately firing the engine for a short time and coasting for long periods. For interplanetary missions, this can lead to very long transit times that would be impractical for human space exploration. If spacecraft speed is increased to shorten the mission, the propellant is used less efficiently, leading to much greater initial launch weight and mission cost.

The thrust in a rocket engine is the product of propellant mass flow and relative exhaust velocity. If these two parameters can be independently controlled, they can be modulated during the mission to produce a much shorter transit time with a given amount of propellant than with a conventional engine. If the engine is operated at constant power, it can be shown[1] that increasing the exhaust velocity and decreasing the propellant mass flow

*Research supported by the National Aeronautics and Space Administration under Interagency Agreements, D.O.E. #2044-AA18-Y1 and NASA #T-7139V.
**Managed by Lockheed Martin Energy Research Corp. for the U.S. Department of Energy under contract number DE-AC05-96OR22464.

in the middle of the trajectory leads to minimum propellant usage. The VASIMR thruster[1] is one way of accomplishing this exhaust modulation. In this concept, an RF-heated plasma is constrained to stream out of a magnetic nozzle at very high exhaust velocities. The thrust and exhaust velocity can be varied by controlling the inlet mass flow, plasma parameters, and magnetic field. A collaboration between NASA–Johnson Space Center, Oak Ridge National Laboratory (ORNL), University of Texas-Austin, University of Maryland, and MSE Technology Applications is underway to develop such a thruster for the Radiation Technology Demonstration (RTD) satellite, which may be launched as early as 2002. This satellite will be launched into low earth orbit by the Space Shuttle, and will then use electric propulsion by either the VASIMR or a Hall thruster to spiral out to a distance of 5 earth radii. Along the trajectory, it will release several small micro-satellites that will monitor radiation levels in the Van Allen Belts. ORNL is performing theoretical and experimental plasma studies in support of the VASIMR project, as well as studies on the development of its superconducting magnetic nozzle and confinement systems that are the subjects of this paper.

MAGNET CONFIGURATION

Figure 1 shows an overall schematic of the VASIMR thruster. Since the RTD satellite will be launched by the Space Shuttle, for safety reasons liquid helium will be used as both propellant and magnet coolant. In principle this would allow conventional Nb-Ti coils for the magnet, but since the ultimate goal is the use of hydrogen propellant for interplanetary missions, NASA wants an early demonstration of an HTS magnet in the spacecraft.

The thruster assembly is supported off the aft end of the liquid helium (LHe) dewar by low-conductivity composite tubes. The magnet coils are built up from a set of nested cylindrical solenoids. The coils are well isolated from the hot plasma chamber by multi-layer insulation. Cold He vapor enters the aluminum magnet case at a chill ring located midway along its axis over the highest-field coil. If needed, a cryocooler can be tied to this ring to provide additional refrigeration. Cold vapor will exit at roughly 30 K. It will then cool a set of HTS current leads and an intermediate shield to less than 80 K. Finally, it will cool a set of vapor-cooled copper leads before entering the plasma chamber at roughly room temperature. The coil set has four main sections. The helium inlet end of the plasma chamber contains a RF antenna to ionize the helium and form the plasma. The low-field ion source coils surround this area. A choke coil that produces a maximum magnetic field of 0.8 T on axis is located downstream of the ion source section. The next section surrounds another RF antenna that further heats the plasma ions to 100 eV by ion cyclotron resonance heating (ICRH). The last section is the nozzle area, where the hot plasma exhaust exits the thruster. All four sections will be connected in series. The expected thrust is about 0.2 N.

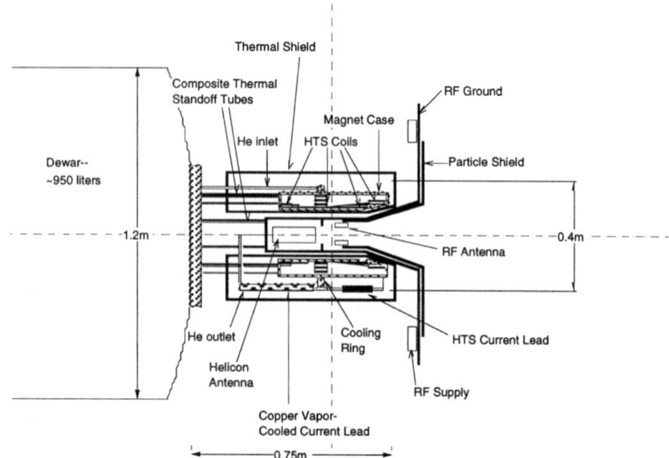

Figure 1. Schematic of VASIMR thruster.

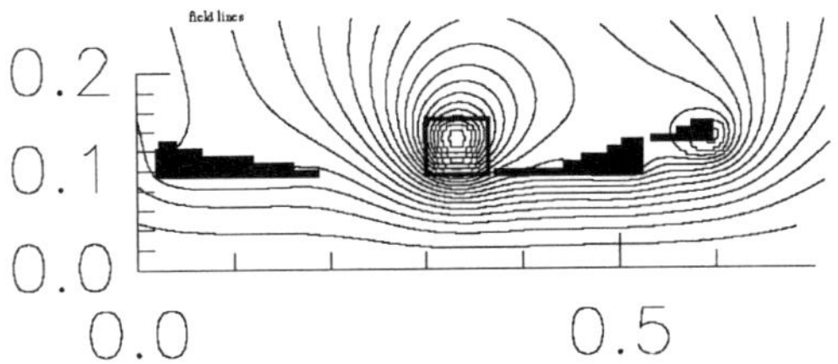

Figure 2. VASIMR thruster coils and magnetic flux lines. Axis units are meters.

General Requirements

The magnet must be able to withstand the Space Shuttle's 3-g launch load, as well as vibration which peaks in the 125-300 Hz range. It must be resistant to the radiation levels that occur in the Van Allen Belts over the mission duration of a few months. This issue is probably more important for the magnet insulation than the HTS material. During the times that the RTD satellite will be in the Earth's shadow, solar power for the magnet power supply will not be available. A persistent switch is therefore desirable. The magnet windings must be accurately constructed to give a field error of less than 1%. The target mass is 20 kg.

DESIGN CALCULATIONS

Magnetic Field Calculations

Figure 2 shows a detail of the coils along with calculated flux lines. The coils contain about 200,000 Amp-turns, with a nominal operating current of 60 A. The overall winding current density is 2.5 kA/cm^2 in the low-field ion source section and 5 kA/cm^{2}in the remaining sections. These values were chosen so as to be within the limits achievable by existing commercial Bi-2212 coated conductors or Bi-2223 PIT conductors operating in the neighborhood of 30 K.

Detailed field calculations were made using the ORNL code SOLE. This code uses current density distributions rather than current filaments, and is accurate within the windings as well as in free space. Table 1 shows the maximum radial and axial field components in each section of the magnet. The axial fields determine the maximum hoop stress in the conductor. The radial fields (which are perpendicular to the HTSC tape and limit its current density) are used along with manufacturer conductor specifications to verify that the conductor has the specified current density.

Mechanical Loads on Winding

The axial attractive forces between the coils are reacted by the coil mandrel. These forces were calculated with the code LF2COAX, by P.L. Walstrom[2]. These forces are as follows:

Table 1. Maximum magnetic fields in winding sections

Magnet Section	Max. Radial Field (T)	Max. Axial Field (T)
Ion source	0.24	0.29
Choke	0.79	1.19
ICRH	0.31	0.69
Nozzle	0.56	0.65

1. Between the ion source section and the choke section— 578 N (130 lb)
2. Between the choke section and the ICRH section— 4467 N (1003 lb)
3. Between the ICRH section and the nozzle section— 3646 N (818 lb)

The mandrel section where the maximum axial force of 4467 N is exerted is about 25 cm long and 19 cm in diameter. For a short column such as this, failure occurs by compressive stress in the material. By this criterion, even a 1-mm mandrel wall thickness would easily handle the compressive load, given a material with at least 34 MPa (5000 psi) working compressive strength. The ultimate compressive strength of G-10 fiberglass edgewise to the laminations, for example, is 320 MPa (47,000 psi)[3]. However, if the mandrel is the inside wall of the magnet case, it is under external pressure. Standard ASME Pressure Vessel Code calculations indicate that an aluminum mandrel would need to be about 1.5 mm thick to withstand 1 atm in the vacuum of space.

The winding hoop stress at radius R is given by

$$\sigma = R\, J_{cond}\, B_z\ , \tag{1}$$

where J_{cond} is the conductor current density and B_z is the local axial field. At the point of maximum axial field in the choke coil, B_z = 1.19 T and R = 0.095 m. If the conductor packing fraction in the winding is 70%, then J_{cond} is 71.4 MA/m^2. The maximum hoop stress is then 8.1 MPa (1171 psi). This is well within strain limits for commercial HTS conductors, even without considering any support from fiberglass and epoxy impregnating the windings.

Heat Loads

The refrigeration available for cooling the magnet is stringently limited by the 3.3-mg/sec flow of helium gas needed to operate the thruster. Since the LHe dewar alone would normally boil off much more than this, it is expected that the dewar will have an actively cryocooled shield. About 140 J/gm of sensible heat is available for cooling the magnet with a He gas inlet temperature of 4.4 K and an outlet temperature of 30 K. This allows a total heat load of 0.46 W on the magnet. The major heat loads are:

1. Joule heating from residual resistance and joints in the conductor,
2. Conduction through the magnet supports,
3. Heat flow through the superinsulation,
4. Heat flow through the current leads.

Design calculations have been performed to allocate about 0.1 W to each of these heat loads.

Winding Resistance. The total length of conductor in the winding is over 3000 m. This would generate over 18 W at 60 A with operation near the critical current at 1 μV/cm. The conductor voltage is proportional to I^n. An n-value of at least 10 is necessary to allow operation at 60% of critical current. This would limit resistive heating in the conductor to 0.1 W. Joints in the winding will need to be minimized and be of very low resistance.

Support Conduction. The magnet would be supported from a baseplate on the LHe dewar by composite tubes to minimize heat leak. To limit the support loads to about 0.1 W, three G-10 tubes of diameter 2.54 cm, length 20 cm, and thickness 1.7 mm could be used. A heat sink at 77 K would be needed about halfway along the tubes. Support tubes with larger diameter or thickness would need to be made of advanced low-conductivity carbon composites.

Superinsulation. The area of the inner surface of the magnet case that faces the warm plasma chamber is about 0.38 m^2. Recent measurements[4] on evacuated multilayer insulation (MLI) indicate that a practical effective conductivity is about 0.2 mW/m-K. An intermediate shield cooled by the helium vapor is assumed between the magnet case and the plasma chamber. For simplicity, assume that this shield has a linear temperature gradient along its axis from 30 K to some intermediate temperature T_i. Then the average temperature differentials can be used to calculate the heat conducted from the 300-K surface to the shield and from the shield to the magnet. The difference between these heat loads is the heat absorbed by the He vapor as it passes through the shield, warming from 30 K to T_i. If the MLI on each side of the shield is 1 cm thick, T_i comes out to 72 K and the heat load on the magnet is about 0.16 W. However, it must be considered that the outer circumference of the magnet may also be facing a warm surface, which could more than double this heat load. This could be alleviated by installing a much thicker MLI blanket around the outer circumference of the magnet, where more room is available.

Current Leads. To minimize heat loads, commercial HTS leads will be used in the temperature range between the shield and magnet. The vapor from the shield will be used to cool copper leads that extend to the room-temperature power supply buss. Commercial 60-A HTS leads are available with heat loads below 0.03 W per pair. The HTS leads will need to be mounted in a low-field area, and their ability to handle the Shuttle launch and vibration loads needs to be verified.

The copper leads have not been analyzed in detail. However, following Iwasa's[5] demonstration, in the "high" current limit where the thermal gradient along the lead is determined mainly by Joule heating, the heat transferred out of the cold end at temperature T is

$$Q(T) = \frac{\kappa(T)\rho(T)I^2}{\dot{m}C_p(T)} \tag{2}$$

where the thermal conductivity κ, resistivity ρ, and specific heat C_p are evaluated at the bottom end temperature of 72 K. With a mass flow of 3.3 mg/sec, this relation gives 0.25 W per lead. This 0.5 W is in addition to the MLI heat load on the shield. At a mass flow of 3.3 mg/sec, it could raise the shield temperature by another 29 K, too high for operation of the HTS lead.

Mass

The winding mass will depend on the conductor chosen. The winding contains about 3040 m of conductor. On the basis of current commercial conductors known to the authors, a conductor mass of about 15 kg is estimated. This leaves 5 kg for the remaining insulation and structure. Clearly, lightweight structural materials such as advanced composites will be required.

CONCLUSION

The preliminary calculations summarized above suggest that a HTS coil would be feasible for an advanced magnetoplasma thruster. However, the desired He mass flow appears to be marginal for operation at 30 K. Hopefully, continuing advances in HTS conductor performance will allow operation at higher temperatures. In the future, a more detailed and self-consistent thermal model will be generated to take into account the interactions between the magnet, lead, and shield heat loads. Further investigations will explore alternative plasma configurations and use of liquid hydrogen instead of He.

REFERENCES

1. F.R. Chang-Díaz et al, "Rapid Mars Transits With Exhaust-Modulated Plasma Propulsion," NASA Technical Paper 3539, March 1995.
2. P.L. Walstrom, "AXISYS: A System of Computer Programs for Calculation of Magnetic Fields, Forces, Inductances, and Flux Lines for Collections of Coaxial Solenoid Elements," Univ. of Wisconsin Fusion Technology Institute, Report #UWFDM-751, January 1988.
3. J.R. Benziger, "Properties of Cryogenic Grade Laminates," 1979 Electrical/Electronics Insulation Conference, October 1979.
4. J.E. Fesmire, NASA, Kennedy Space Center, private communication.
5. Y. Iwasa, "Case Studies in Superconducting Magnets," Plenum, New York (1994), p. 125.

AIR FORCE RESEARCH LABORATORY SPACECRAFT CRYOCOOLER ENDURANCE EVALUATION UPDATE: FY98-99

B. J. Tomlinson, Jr.,[1] C. Yoneshige,[1] and N. Abhyankar[2]

[1]Air Force Research Laboratory
Kirtland AFB, New Mexico 87117
[2]Dynacs Engineering Co. Inc.
Albuquerque, New Mexico 87123

ABSTRACT

The need for long term endurance evaluation data on space cryocoolers has long been an issue due to the 10-year plus design life of this technology and the absence of any accepted accelerated testing methodology. The Air Force Research Laboratory (AFRL), under sponsorship from the Ballistic Missile Defense Organization (BMDO), has been evaluating the long term performance of space cryocooler technology for the past five years to provide feedback to system designers and cryocooler developers to help mitigate or correct identified deficiencies. Cryocooler technology under evaluation at AFRL includes units from TRW, Creare, and Raytheon utilizing a variety of refrigeration cycles including Stirling, Pulse Tube (Stirling variant), Reverse Brayton cycles, and technological maturity. Cryocoolers are instrumented in state-of-the-art experiment stands utilizing software driven data acquisition systems that collect temperature, power, current, and voltage measurements. This allows critical cryocooler operating parameters to be closely tracked in order to quantify the technology's ability to meet the 5 to 10 year lifetime. It also allows AFRL to observe and document any long-term changes in cryocooler performance. Endurance data has been collected on 5 cryocoolers, some of which have been in operation for nearly 5 years. This paper includes experiment philosophy, lessons learned, and conclusions from the endurance evaluation program at AFRL for FY 98-99.

INTRODUCTION

Air Force, BMDO, and Department of Defense (DoD) investment in the development of space cryogenic coolers rests primarily on the critical characteristics of *lifetime and reliability*. This distinguishes the capabilities of long-life space cryocoolers from short-lived tactical units. Assessing cryocooler lifetime, reliability, and thermodynamic performance is the primary focus of AFRL's cryocooler characterization and endurance evaluation activities.

Cryocooler technology developers can underestimate the factors that contribute to cooler unreliability. Some of the potential contributors to cryocooler unreliability include wear, drift, fatigue, material creep, gaseous contamination, particulate/compound contamination and clogging, material and workmanship defects, inadequate machining process development, magnetic circuit degradation, assembly errors, material thermal expansion mismatches, and long-term alignment or mechanical instability.

All potential failure modes and life limiting mechanisms have not been identified, and those that have are not completely understood or quantified. Also, various cryocooler concepts can have vastly different inherent failure modes and effects, as well as "graceful degradation" characteristics. Because of the importance of validating cryocoolers as long-life spacecraft components, a very high priority is placed on gathering cooler-specific life, reliability, and long-term performance data. However, for an endurance evaluation to have meaning and correlate to the long life and reliability characteristics of the cryocooler technology, care must be taken when examining the data from the cryocooler experiments. Experimental setups that include make-up gas (no hermeticity), gas cleaning, periodic rebuilding, operation at levels significantly below the capacity of the cooler, and inadequate or overly favorable environments will bias the endurance data gained from the experiment.

AFRL objectives for long term cryocooler endurance evaluation have evolved to meet the requirements for the demonstration of the lifetime and reliability of cryocooler technology for the Air Force, BMDO, and the DoD:

1. The focus of long life endurance evaluation of cryocooler technologies is on the characterization of long term performance degradation, system reliability, reliability contributors / detractors, the lifetime of the technology, and the development of proven methods for accelerated testing of long life space cryocoolers.
2. AFRL provides feedback in the form of technical reports and conference presentations of data and lessons learned to cryocooler and cryogenic technology developers, users, and spacecraft developers to aid in follow-on developments of emerging technology.

The AFRL Cryocooler Characterization Laboratory's (CCL) experiment facilities, procedures, equipment and instrumentation all serve to enable long-duration endurance evaluation data collection in environmental conditions closely simulating the space environment for cryocoolers with significant heritage to flight systems. Some cryocoolers are in experimental setups that have a vacuum dewar provided for the cold tip assembly and have ambient (>280 K) rejection temperature cold plates. Once underway, endurance evaluation is normally run until the cooler meets pre-defined failure criteria. If the test article continues to exhibit adequate performance past its design life, it will be allowed to run until it does meet the predetermined failure criteria. On the other hand, an endurance evaluation that requires 5 years of continuous data to support lifetime and reliability predictions as well as program requirements, will not meet the technical insertion freeze dates for critical Air Force and DoD programs. Although endurance data is valuable for understanding the life and reliability characteristics of a cryocooler, AFRL is working toward the development of proven accelerated testing methods to increase the understanding of the reliability of the design and reduce the technology insertion cycle time for emerging technologies. Information gathered during endurance evaluation is then made available to sponsors, technology users, cooler developers, and system integrators to help them refine system trades for both the coolers and their intended use on space platforms.

ENDURANCE EVALUATION OVERVIEW

Five cryocoolers are currently undergoing endurance evaluation or post-endurance disassembly and inspection at AFRL. Table 1 lists these coolers and their nominal operating conditions. Shaded areas are coolers in post endurance disassembly and inspection. These operating conditions are maintained at constant levels during the entire endurance trial phase except for periodic heat load/cold end temperature comparisons to previously established benchmark conditions. Cold end load / temperature checks against the baseline performance are usually conducted at 1 to 2 month intervals and are used to quantify time-dependent performance drift.

It should be noted that all of the coolers' heat rejection temperatures (T_R) are cycled above and below the nominal condition listed in Table 1, except for the SSRB. Cycling rejection temperatures during the experiment somewhat emulates the transient thermal effects experienced under normal space environmental usage and allows AFRL engineers to conduct baseline checks at different rejection temperatures. For comparison purposes, 300 K is defined as the nominal rejection temperature for all coolers. The rejection temperature cyclic range is varied depending on intended orbital transient profiles, cooler design sensitivity to coefficient of thermal expansion effects, and thermodynamic performance limitations usually experienced at higher reject temperatures.

Endurance evaluation experiments can be run either in a thermal vacuum chamber or on in an experimental tabletop setup, where a vacuum dewar is provided for the cold tip and the ambient rejection temperature (>280 K) is provided by a fluid loop conductively mounted to the cooler in laboratory air. The thermal vacuum chamber provides the most realistic vacuum and conductive thermal environment for the cryocooler. These chambers are reserved for cryocooler technology that is most representative of cryocooler family lines leading to flight technology. The tabletop configuration is used for cryocoolers that meet the criteria for long life endurance evaluation and have some heritage to the flight systems. These cryocoolers are reasonably hermetic and contribute to the heritage of emerging flight technology or technology alternatives in that they contribute to the cryocooler lifetime and reliability database, but may not lead directly to flight technology.

During FY 99 three coolers entered the post-endurance evaluation disassembly and inspection phase. The Creare SSRB, the TRW 3503, and the Creare SSC completed their endurance evaluation at AFRL and are in the process of final inspection and analysis; by AFRL and by the cryocoolers' respective contractors. These coolers represent an opportunity to examine the long-term effects on the internal mechanisms and working fluids of the cryocooler. Post-endurance inspection of coolers that have stopped functioning normally is not all bad. The contributing factors and overarching circumstances in these instances need to be examined to ensure the transition of lessons learned to the cryocooler developers for inclusion in follow-on designs. In addition, there

Table 1. Cryocooler Nominal Operating Conditions

Croycooler	T_C	Q_L	P_{in}	T_R	Endurance Phase
6020 PT	60 K	2.0 W	76 W	300+/-10 K	Endurance
3585 PT	35 K	0.85 W	200 W	300+/-10 K	Endurance
3503 PT	35 K	0.3 W	110 W	300 +/-10 K	Post-Endurance
SSRB	65 K	5 W	<240 W	300 K	Post-Endurance
CSSC	65 K	0.8 W	55 W	300+/-10 K	Post-Endurance

is the possibility for further evaluation of this technology after the post-endurance inspection. Further discussion on the initial observations from these experiments can be found in the following sections.

FY 98-99 ENDURANCE EVALUATION STATUS

Fiscal years 1998 and 1999 have been busy years for the AFRL CCL. In addition to normally busy characterization activities, three cryocoolers under endurance evaluation stopped operating for various reasons. As discussed in the introduction, there are a myriad of different mechanisms by which a cryocooler fails to continue operation or falls below some threshold of acceptable thermodynamic performance. Some of these mechanisms were clear in the initial observations during post-endurance evaluation disassembly and inspection. Table 2 summarizes the current endurance evaluation activities at AFRL and the projected start dates of endurance evaluation of new coolers in the CCL. Subsequent discussion elaborates on the current status of each current experiment.

The TRW 6020 (2 W @ 60 K) Pulse Tube Cryocooler is currently under nominal conditions in endurance evaluation and has accumulated over 16,000 hours. The 6020 was integrated and instrumented in a 24" thermal vacuum chamber with multiple vacuum feedthroughs for instrumentation and power. Its rejection temperature is set by a copper block interface to a computer controlled chilled fluid loop. The cryocooler is controlled by a set of contractor supplied laboratory electronics that allows test engineers to control and explore a large range of thermodynamic performance for this machine. In addition, data acquisition is provided by computer software and plots sensor readings in real time and writes data to file for later retrieval and reduction. This DAQ setup is part of a standardized system for the characterization of cryocooler technology at AFRL.

Apparent discrepancies in the characterization data have led AFRL to initiate a detailed review of all data on this cryocooler, the TRW 3503, and the TRW 3585. The first indication that there were data discrepancies was a distinct difference in thermodynamic performance compared to Jet Propulsion Laboratory data. Further data indicates that there was a possible shift in the parasitic load on the cold block of the cryocooler. Data reduction is continuing and should culminate in several technical reports that will be distributed to the appropriate users and developers. Additionally, the data will be summarized and reported in a conference forum for publication.

The TRW 3585 (0.85 W @ 35 K) Pulse Tube Cryocooler is still under endurance evaluation and has accumulated over 12,000 hours. This cryocooler is in a tabletop configuration with a vacuum dewar provided for the cold tip / pulse tube assembly and a chiller fluid loop for ambient rejection temperature (>280 K). The tabletop setup does not provide the optimal endurance evaluation environmental conditions (the heat rejection now includes convective components as well as conductive), but it is being used since this particular cooler has limited follow-on applications. The data acquisition system for this cryocooler is similar to the setup for the TRW 6020 and is a tailored derivative of the standard setup for all cryocoolers under characterization in the AFRL CCL. The TRW 3583 is being evaluated because it has significantly contributed to the family heritage of the TRW / Oxford Stirling compressor technology.

The TRW 3503 (0.3 W @ 35 K) Pulse Tube Cryocooler is currently in post-endurnace disassembly and inspection. The cryocooler was integrated into a 24" thermal vacuum chamber and was identical to the control and data acquisition systems that were provided for the TRW 6020 cryocooler. This cooler suffered an environmental failure that exposed it to rejection temperatures up to 421 K for over 20 hours (Figure 1). After the overheat condition, the cryocooler showed typical performance degradation due to internal contamination, but it was allowed to continue its endurance evaluation. In April 1999, the 3503 tripped and was unable to restart. Subsequent investigation isolated the thermomechanical cooler as the culprit. AFRL and TRW performed a joint disassembly and inspection of the device. Prior to the disassembly inspection, AFRL and Pernicka Corp. conducted a sampling of the internal gas of the cryocooler. As expected, the analysis showed trace contamination gases. However, the source of the tripping was not a cooler damaged by the high temperature transient, but it was a loose locking nut on the capacitance sensor target that was not staked during the original build. This is a testament to the extreme robustness of the cryocooler design to be able to endure a huge environmental transient and still function to a reasonable level of performance.

Table 2. AFRL Endurance Evaluation Status Summary

Cryocooler	Hours	Status
TRW 6020 Pulse Tube	16872	This cooler is currently in Endurance Evaluation. Data to date shows the cooler operating within design margins and instrument error. Resolution of apparent discrepancies in the characterization data and the data correlation with JPL have not been completely quantified.
TRW 3585 Pulse Tube	12386	This cooler is currently in Endurance Evaluation. Data to date shows the cooler operating within design margins and instrument error.
TRW 3503 Pulse Tube	12093	The 3503 has completed endurance evaluation. After undergoing a significant overheat condition, the cooler continued to operate at degraded performance until April 99 when the cooler tripped and could not be restarted. The cooler internal gas was sampled and then taken to TRW for disassembly inspection. Discussion is underway with TRW to determine the future of the hardware.
Creare Single Stage Reverse Bryaton	29,063	The SSRB shutdown in Dec 98 due to an alarm condition on the compressor. Subsequent attempts by AFRL and Creare personnel to restart the cooler were unsuccessful. They couldn't get the turboexpander to lift up and start spinning. Creare and AFRL personnel disassembled the turboexpander for inspection and analysis. Discussion is underway with Creare to determine the future of the hardware.
Creare Standard Spacecraft Cryocooler	13358	The CSSC shut down in December 98. Subsequent attempts by AFRL personnel to purge, re-fill, and re-start the cooler were not successful. At the present time, the CSSC endurance evaluation is complete and discussion is underway with Creare to determine the future of the hardware.
Projected Endurance Evaluation Experiments (hours are at current levels)		
Raytheon Standard Spacecraft Cryocooler II	500	Projected endurance evaluation start date of Oct 99.
Raytheon Protoflight Spacecraft Cryocooler	1000	Projected endurance evaluation start date of Nov 99.
Matra Marconi Space 10K	N/A	Projected endurance evaluation start date of May 00.
TRW mini Pulse Tube	200	Projected endurance evaluation start date of Mar 00.
Ball Aerospace 35/60K	300	Projected endurance evaluation start date of Jun 00.

At the time of the loss of operational capability, the TRW 3503 had over 12,000 hours of operation. The cryocooler is currently at AFRL undergoing further stiction measurements on the compressor assembly. Plans are in the works to evaluate the need to rebuild, refill, and return the cooler to AFRL for characterization and endurance evaluation.

The Creare 5 W @ 65 K Single Stage Reverse Brayton (SSRB) cryocooler is also in post-endurance disassembly and inspection. The SSRB is a reverse Brayton device that employs fast spinning, micro-machined turbines to produce compression and expansion of the working fluid (in this case, neon). This cryocooler engineering design model was delivered to AFRL (then Phillips Laboratory) with known gaseous contamination and atmosphere and water permeable o-ring seals. With these known flaws and characteristics, the decision was made to carry on with characterization and then long life endurance evaluation.

This cryocooler tripped and shut down in December 1998 and was unable to spin up the turboexpander to re-start. All attempts by AFRL and Creare personnel on site did not achieve restart. During June 1999, Creare and AFRL personnel removed the turboexpander for disassembly and inspection. Preliminary results include the detection of tiny foreign particulates in and near the moving components. At this time, the investigation is still underway. The final failure analysis will be completed jointly by Creare and AFRL, where Creare will outline the recommendations for the potential repair of the cryocooler. The total tally for operational hours on this unit was over 29,000 hours.

The Creare 65K SSC Diaphragm Standard Spacecraft Stirling Cryocooler (CSSC) was configured in a tabletop experiment format, similar to the TRW 3585, for portions of characterization and all of endurance evaluation. Due to a delivered flaw, the cryocooler was not hermetic and required a make-up helium gas bottle and periodic purge and fill gas cleaning cycles performed in roughly 6 month intervals. Although the configuration of this endurance evaluation included a make-up gas system to compensate for the gas loss in the cooler, some useful data was gathered on the lifetime of the mechanical components of this

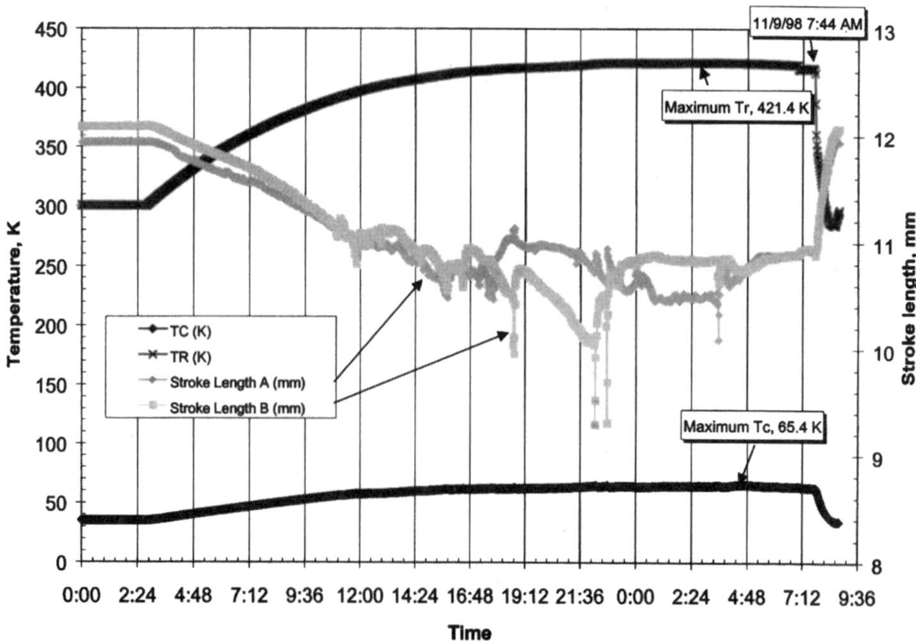

Figure 1. TRW 3503 Overheat Condition

diaphragm compressor design that is a technology alternative to the linear flexure bearing compressors that are in general use now.

The CSSC tripped and shut down in December 1998. Subsequent attempts by AFRL personnel to purge, re-fill, and re-start the cooler were not successful. At the time of the shutdown, the CSSC had over 13,000 hours of operation. At the present time, the CSSC endurance evaluation is complete and discussion is underway with Creare to determine the future of the hardware.

FUTURE ENDURANCE EVALUATION AT AFRL

Endurance evaluation of cryocooler technology is a time consuming and painstaking process. The lack of proven accelerated methods for cryocooler life time demonstration has forced cryocooler developers and users to seek precious lifetime data in real time demonstrations of cryocooler life in endurance evaluations. Significant thought has been given to the validity of the experimental setup for endurance evaluations, but additional input is being sought to ensure the best, most technically acceptable methods to demonstrate cryocooler lifetime. Reliability issues associated with cryocoolers are intertwined into the fabrication, assembly, bakeout, and gas fill of the cryocoolers. Performance tracking over long periods of time point out the robustness of the technology to maintain the desired level of performance expected over the entire mission duration.

AFRL recognizes the importance of the lifetime data that the laboratory is producing and will ensure this data is available for users and developers over the coming years. In the coming fiscal year five additional cryocoolers will enter endurance evaluation. The Raytheon PSC, Matra Marconi Space 10K, Ball Aerospace 25/60K, and TRW mini Pulse Tube are cryocoolers with significant pedigree in flight cryocooler technology. These coolers represent some of the most advanced cryocooler technology and, alongside the current endurance evaluation experiments, will provide developers and users with significant data on the long-term characteristics of cryocoolers.

ACKNOWLEDGEMENTS

Personnel from the Air Force Research Laboratory and the Dynacs Engineering Corporation, Inc. carried out the work described in this paper. Mr. Erwin Myrick of the Ballistic Missile Defense Organization (BMDO/TOS) has played a crucial role in supporting both the requirements generation, laboratory vision, and necessary funding for spacecraft cryogenic cooler characterization and endurance evaluation. Ms. Karen Basany of the SIBRS Low Program Office has provided advocacy to insure technology objectives for the SBIRS program and the Air Force are available to meet schedule need dates.

SUMMARY AND CONCLUSIONS

Endurance evaluation continues to add to the growing database of lifetime and reliability information on cryocooler technology. Cryocoolers in endurance evaluation at AFRL are producing volumes of long-term performance data. AFRL is actively engaged in improving the experimental test setup and remaining in close contact with government and industry partners to facilitate performance data feedback and share lessons learned.

Even cryocooler endurance "failures" are providing cryocooler developers and users with valuable lessons on manufacturing, hermeticity and working fluid cleanliness that will aid in the development of follow-on technology. It is certain that continuing endurance evaluation at AFRL will provide ever more meaningful lifetime and reliability data for mission enabling cryocooler applications in the future.

REFERENCES

1. Tomlinson, B. J., Gilbert, A., and Bruning, J., "Endurance Evaluation of Long-Life Space Cryocoolers at AFRL-An Update", Cryocoolers 10, Plenum Press, New York (1999).
2. Davis, T., Tomlinson, B. J., and Reilly, J., "Air Force Research Laboratory Cryocooler Technology Development", Cryocoolers 10, Plenum Press, New York (1999).
3. Bruning, J., Pilson, T., "Phillips Laboratory Space Cryocooler Development and Test Program", *1997 ICEC Conference Proceedings*, Plenum Press, New York (1998).

THE BALLISTIC MISSILE DEFENSE ORGANIZATION CRYOGENIC COOLING TECHNOLOGY DEVELOPMENT FOR SPACE-BASED INFRARED SYSTEM LOW

T. M. Davis,[1] B. J. Tomlinson, Jr.,[1] and E. Myrick[2]

[1]Air Force Research Laboratory
Kirtland AFB, New Mexico 87117
[2]Ballistic Missile Defense Organization
Washington, DC

ABSTRACT

The Air Force Research Laboratory (AFRL) and its predecessors, Phillips Laboratory and the Air Force Space Technology Center, has been the primary agent of the Ballistic Missile Defense Organization (BMDO) for the development of low capacity cryogenic refrigerators and integration technologies for space applications since the mid 1980s. These cryocooler development programs concentrated on addressing the negative impacts of mechanical refrigerators on optical space systems: induced line of sight vibration, longevity, power consumption, and mass.

The initial focus of development efforts was on relatively large capacity machines to support cooling requirements for the Space Surveillance and Tracking System (SSTS). The protoflight Cryocooler program produced two three-stage 10K cryocoolers (Contractors: Air Research and Arthur D. Little) for cooling of the long wave silicon focal plane arrays. Both of these programs were terminated in the final stages of assembly as a result of the cancellation of SSTS in 1989. An additional program aimed at 10K performance was jointly funded with the Brilliant Eyes program office (successor to SSTS) and focused on sorption technology. Primarily developed by NASA's Jet Propulsion Laboratory and Aerojet, this program culminated with the BESTCE Shuttle flight experiment in 1995. While overall performance was adequate, the sorption cryocooler was considered too large for current small satellite payloads.

The Standard Spacecraft Cryocooler program (SSC) initiated in 1990 marked a change in emphasis from relatively large machines to more compact and efficient cryocoolers aimed at meeting cooling needs in the range from 60K to 150K for mid-wave infrared (MWIR) applications. Using Oxford Stirling cycle technology developed primarily in the United Kingdom, these machines utilized linear drive motors and tight clearance seal non-contacting piston shafts. The pulse tube cryocooler, a variation of this technology, replaces the actively moving expander piston with a non-moving regenerator and pulse tube.

AFRL has also pursued alternate cryocooler concepts including reverse Brayton cycle designs. The principal technological challenge of these simple Brayton machines is accurately micro-machining the components (impellers, shafts, motor magnets, etc). For extremely low temperature cooling (~10K), variants using Joule-Thomson combinations and improved Stirling and pulse tube designs are being considered.

CURRENT DEVELOPMENT PROGRAMS

The primary focus of the current BMDO funded cryocooler development efforts is to support the Engineering Manufacturing Development (EMD) requirements for the Space Based Infrared System Low (SBIRS Low) satellite program. Several machines developed by AFRL were basedlined for the SBIRS Low Flight Demonstration System (recently terminated) and Cobra Brass flight experiment. The Cobra Brass experiment and later the Brilliant Eyes program office developed requirements for the TRW 150K Protoflight Spacecraft Cryocooler (PSC). This "mini pulse tube" machine has a large capacity (>2W) and high efficiency for 150K cooling, and more limited capability at colder temperatures (down to about 65K). Two TRW mini pulse tubes were flown on the unsuccessful NASA SSTI satellite launched in 1997 and two other units are currently flying on the Cobra Brass payload. The NASA/Langley SABER flight experiment scheduled for late 1999 launch is also using a mini pulse tube supplied by AFRL.

Figure 1. TRW 150K Protoflight Spacecraft "Mini Pulse Tube" Cryocooler

Figure 2. Ball Three-Stage Cryocooler

Although recently cancelled, the TRW/Raytheon Flight Demonstration System (FDS) cryocooler needs were accommodated by the use of both the TRW miniature pulse tube and the Raytheon Improved Standard Spacecraft Cryocooler (ISSC). The ISSC was used to cool the FDS tracking sensor and is based on designs of the 60K SSC and PSC, both funded by BMDO. It is a Stirling cycle cryocooler optimized to perform at 60K, but retains cooling margin down to 25K (0.7W@35K). Two mini pulse tubes were designed to cool the FDS track sensor fore-optics at 115K (2.45W).

Table 1. Current AFRL Cryogenic Technology Programs

PROGRAM	DESIGN OBJECTIVES	APPLICATION	STATUS
TRW Mini Pulse Tubes	2W @ 150K	FDS Optics Cooler	Program complete; life-testing in progress
Raytheon 60K PSC	2 W @ 60K	EMD Tracking Sensor	Delivered to Air Force Dec 97; in characterization/life test at AFRL
Ball Aerospace 35/60K	0.4W @ 35K, 0.6W @ 60K	EMD Tracking Sensor	Delivered to AFRL in Sep 98; in characterization/life test at AFRL
Creare MRBC	1W @ 35K	EMD Tracking Sensor	CDR in Sep 99; delivery to AFRL in Jun 00 for test
MMS Stirling	0.045W @ 10K	VLWIR Missile Detection and Space Surveillance	CDR in Jul 98; delivery to AFRL in Aug 99
Ball Aerospace Joule-Thomson Stirling EDM	0.25W @ 10K	VLWIR Missile Detection and Space Surveillance	PDR in Jan 98; delivery of Engineering Development Model Dec 00
TRW High Efficiency	10W @ 95K	SBIRS Low EMD Optics cooling and NASA Mars mission	Contract award Sep 1998; PDR in Jan 99; delivery to AFRL Oct 00
Raytheon High Efficiency	10W @ 95K	SBIRS Low EMD Optics cooling	Contract award Jun 99; delivery to AFRL Jun 00
Swales CRYOBUS	Integration Components	Redundant Coolers for EMD optics and tracker	Thermal switch flight demo on STS-95 Oct 98

ENGINEERING MANUFACTURING DEVELOPMENT (EMD) NEEDS

The SBIRS Low operational system (EMD) is currently scheduled for deployment in 2006. The major drivers for EMD cryocooler requirements are increased duty cycle (60-100%), higher cooling loads for the tracking sensors (0.5W at 35K and 0.7W at 60K), increased cooling loads for the fore optics (up to 12 W at 100K), 10 year design life with complete mechanical and electronic redundancy, and higher tolerance to radiation environments. The cryocooler must also have a low system mass penalty and improved efficiency. While the final EMD design is still evolving, current concepts require two focal planes in the track sensor imager to stare at a target simultaneously in both the MWIR and LWIR bands. Dual temperature cryocoolers (lifting heat loads at 35K *and* 60K) offer attractive system benefits over single stage cryocoolers if redundant cryocoolers are mandated.

System mass penalty is defined as the sum of the cryocooler mass, the mass of the electrical power system necessary to drive it, control electronics mass, and the radiator area mass needed to reject the waste heat. It is used as a measure of cryocooler impact on spacecraft design.

The Air Force Research Laboratory is addressing the issue of system mass penalty in several ways. Thermal Storage Units (TSU) have been developed to absorb the wide thermal load variations during peak duty cycle and allow system designers to size the cryocooler for the average load instead of the peak heat load. An Air Force funded 60K TSU was successfully flight demonstrated in October 1998 aboard the STS-95 Shuttle mission. Under the Swales Aerospace CRYOBUS program, several cryogenic integration technologies are being designed and fabricated for potential ground demonstrations. This program is developing cryogenic Thermal Switches with high thermal resistance in the "OFF" state and low resistance in the "ON" state to make redundant cryocooler designs feasible. Additionally, the program is addressing various integration approaches for the optics cooling and heat rejection problems. A cryogenic capillary pumped loop, flexible cryogenic link, ambient loop heat pipe, and cryogenic loop heat pipe are being developed to support EMD needs.

No multiple temperature cryocoolers are space qualified at this time, however, AFRL recently completed the development of a protoflight multi-stage Stirling cryocooler with Ball Aerospace. The objective of this effort is to simultaneously provide 0.4 Watts of cooling at 35K and 0.6 Watts of cooling at 60K. The cooler is a technology option for the SBIRS Low EMD tracking sensor. As a result, the cryocooler must be able to achieve <0.1 Nrms induced vibration requirement and greater than 10 years lifetime.

The Ball effort was a BMDO/SBIRS Low funded program managed by NASA/GSFC with technical oversight provided by AFRL. The Air Force program leveraged technology previously developed for a NASA 30K two-stage cryocooler. Ball modified the NASA compressor design in order to help double the cooling capacity. To satisfy requirements, the displacer was redesigned from a two stage to a three-stage design. The cold finger features a fixed regenerator, which improves life, efficiency, and reduces induced vibration. Additional performance improvement is realized by the incorporation of precision piston alignment techniques that eliminate piston/cylinder contact in the cryocooler. A Protoflight cryocooler with associated flight electronics was delivered to the Air Force Research Laboratory in September 1998 and is currently in acceptance/characterization testing and preparing for endurance evaluation.

Another potential candidate for tracking sensor cooling is the 35 K Creare Miniature Reverse Brayton Cryocooler (MRBC), which is a next generation of the 65K Single Stage Reverse Brayton (SSRB) cryocooler developed with BMDO and Air Force funding. The NICMOS cryocooler, a joint effort with the NASA Goddard Space Flight Center, leverages the engineering unit heat exchanger and develops a miniaturized turboexpander and compressor. Miniaturization of these components results in a technical challenge to reduce component size without significantly decreasing efficiency because parasitic losses are a larger percentage of the overall system when miniaturized. The cryocooler was flight demonstrated on STS-95 in October 1998. Subsequently, the cooler will be installed on the Hubble Space Telescope during a year 2000 servicing mission.

Figure 3. NICMOS Cryocooler System

The Creare MRBC has the objective to develop dual load cooling, but has been realigned to meet the requirement of 2 Watts at 35K. Total input power for the cooling requirement is not to exceed 125 Watts and must demonstrate an operating life in excess of 10 years. Vibration of this unit is negligible due to the operation of the turbine at such high speeds that results in high vibration frequencies that have negligible effects on the vibration of the sensor.

In order to achieve these requirements, Creare plans to implement a cryogenic turboalternator that will reduce parasitic losses and result in increased efficiency of the unit. Current reverse Brayton cycle machines rely on the working fluid and external turboexpander braking loads to control the speed of the turbine. Addition of the turboalternator will allow the turbine speed to be continuously optimized by the alternator. Advanced materials will be used in the design of the unit, which should result in additional motor efficiency improvements. The turboalternator is projected to have 50% less parasitic heat loads and achieve 70% more efficiency compared to the turboexpander design (used in the SSRB unit). A radial flow heat exchanger is also being incorporated into the design will reduce volume by 75% and be 70% lighter than the existing SSRB heat exchanger. Final design of the turboalternator and heat exchanger is currently being completed. Because of technical development issues and program schedule conflicts, the program has been restructured and plans now call for the Critical Design Review to be held in September 1999, with the delivery of Engineering Development level hardware to AFRL in June 2000.

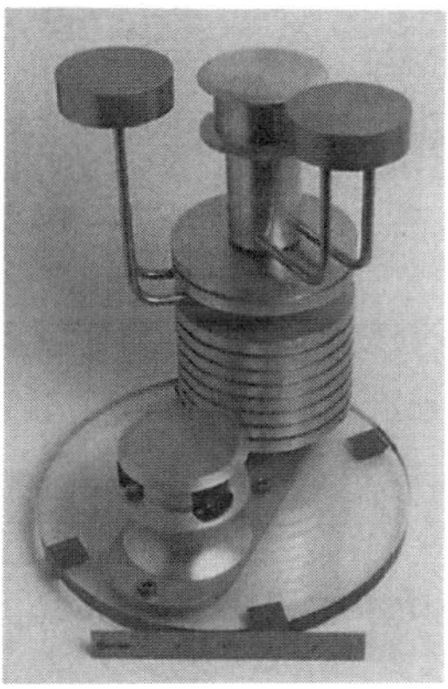

Figure 4. Miniature Reverse Brayton Cryocooler

Figure 5. Raytheon 60K Protoflight Spacecraft Cryocooler.

An alternative for cooling the tracking sensor is the Raytheon developed 60K Protoflight Spacecraft Cryocooler (PSC). This cooler has the potential to meet the requirements for the MWIR and LWIR tracking sensor needs for the SBIRS Low EMD system. This design is the most mature in a series of cryocoolers developed by government and company sponsored internal research and development resources (SSC, ISSC, and PSC). The specific objectives of this program are to develop a unit requiring less than 100 Watts of input power with 2 Watts of cooling at 60 K, life in excess of 10 years, and a total system mass less than 33 kg. Design features include incorporation of a linear tangential flexure into the compressor. This innovation allows for improvement in radial stiffness and is expected to result in improved cryocooler reliability, performance, and significantly reduced off drive axis vibration. Titanium has been utilized within the housing and piston assembly to reduce system mass. Raytheon also demonstrated the adequacy of piston alignment techniques necessary for ensuring adequate sub-mil clearance of moving parts. The protoflight unit was delivered in December 1997, and subsequently subjected to acceptance tests at JPL and performance characterization and endurance evaluation at the Air Force Research Laboratory.

An additional requirement for SBIRS Low EMD is a cryocooler to support cooling of the increased heat loads on the fore optics. Two AFRL programs, jointly funded by BMDO and SBIRS Low, have the objective to design and develop an advanced high efficiency cryocooler to meet on-gimbal optics cooling requirements of 10 Watts @ 95K. The 24-month programs will emphasize ease of integration, low EMI signatures, low EMC susceptibility, and the ability to survive the launch loads of any existing Air Force launch vehicle.

A contract was awarded to TRW in September 1998 for a cooler producing 10 Watts of cooling at 95K with the added objectives of minimal cryocooler mass and input power. TRW's design objectives are for a pulse tube cryocooler with mechanical mass of ~3.6 kilograms and a specific power of less than 10 W/W (goal of 8 W/W). The AF, BMDO, and NASA are jointly funding the program. NASA is considering the cooler as a candidate for an oxygen liquifier on the Mars Exploration Mission. A second contract was awarded to Raytheon in June 1999. This parallel effort uses a unique two-stage hybrid Stirling expander / pulse tube cold head design to achieve the desired cooling loads.

Figure 6. Second Generation TRW 95K High Efficiency Cryocooler and the First Generation TES Pulse Tube Cryocooler

While not specifically required for SBIRS Low EMD, improved discrimination utilizing high performance multicolor and multi-spectral focal planes would provide a significant improvement in operational capability for surveillance and missile tracking and detection.

Focal Plane operation beyond 14 microns requires a cryocooler capable of cooling at 10K. A BMDO funded near term effort with Matra Marconi Space (MMS) and Rutherford Appleton Laboratory is developing a 3 stage Stirling cryocooler with a projected capacity of between 0.045 and 0.1 Watts at 10 K. MMS has completed all development and fabrication activities and plans to deliver the cryocooler to AFRL in August 1999 for performance characterization and endurance evaluation.

An Air Force funded program with design goals of 0.1-0.3W @ 10K and with a specific power of <1000W/W was initiated in May 1998 with Ball Aerospace Corporation in Boulder, Colorado. This cryocooler uses an existing Stirling pre-cooler combined with a Joule-Tompson based Redstone Interface to achieve cooling at 10K.[3] The current program schedule calls for delivery of an Engineering Development Model to AFRL in December 2000.

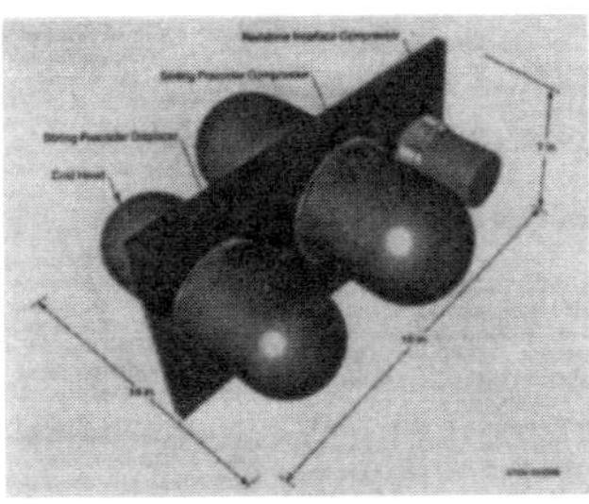

Figure 7. Ball 10K Cryocooler

Figure 8. MMS 10K Cryocooler Compressors

CRYOCOOLER RELIABILITY INITIATIVES

A major concern for the SBIRS Low EMD in applying cryocooler technology is their unproven reliability and lifetime. AFRL has several initiatives to investigate issues to meet the program requirements for high reliability for > 10 years life. The AFRL Cryocooler Characterization Laboratory (CCL) and Aerospace Corporation are also playing a critical role in improving reliability confidence of cryocoolers. Leveraging existing capabilities at JPL, GSFC, other government laboratories, and private industry, the AFRL CCL supports performance characterization / verification and reliability evaluation of developed hardware in a thermal vacuum environment that closely simulates operational space conditions. The principal focus of the CCL is to improve reliability confidence in cryocooler hardware and provide a better understanding of life-limiting and performance degradation factors. Cryocooler characterization and endurance evaluation is being accomplished on the Creare SSRB, three TRW pulse tubes, and Raytheon (Hughes) SSC units. Both the Raytheon 60K PSC and Ball Aerospace 35/60K began a three-year endurance evaluation in thermal vacuum chambers in early 1999.

AFRL CCL cryocooler performance and endurance evaluation data feedback is being improved to allow access for program offices and next-generation cryocooler designs. A near-term solution recently implemented was to incorporate current "load line" data into the Aerospace Corporation developed Cryogenic System Integration Model (CSIM).[6] Users of this model will now have near real time information from AFRL performance evaluations. A handbook of cryocooler life performance data has been compiled and is available through the Aerospace Corporation.[7] Making cryocooler test data and other relevant information available on the AFRL Web page is also being evaluated. Characterization test plans have expanded to include a broader range of potential operating conditions during performance evaluation. The results will provide additional reliability confidence to the SBIRS Low program office and development contractors. The end result of this process is to influence improvements in designs, incorporation of heritage lessons learned, and improved reliability of future cryogenic systems.

AFRL recently established a working group to more formally address the range of cryocooler reliability issues. The group's charter is to identify and implement actions to improve reliability and confidence in spacecraft coolers. A collaborative effort with the Ukrainian Institute of Low Temperature Physics is assessing the possibility of applying accelerated lifetime testing methodology to selected AFRL cryocoolers.

An AFRL sponsored Reliability Workshop was conducted by Georgia Institute of Technology in September 1998. Participation included cryocooler users and developers from industry, government, and academia. The recommendations from the workshop will be implemented in future AFRL programs. A follow-on workshop is planned for August 23-24, 1999 in Los Angeles. The areas of interest will be expanded to include discussions on cryocooler electronics and software related issues.

The Aerospace Corporation has also made significant progress in critical component reliability through the development of tangential linear flexure bearings to satisfy low induced vibration requirements in cryocooler designs. The Aerospace developed tangential spring has been basedlined in the Raytheon 60K PSC and Lockheed Martin Missiles and Space Coaxial Pulse Tube (CAPT) cryocooler. To date, life testing has shown a reliability of 10^8 cycles without failure.

SUMMARY

The BMDO and AFRL consolidated approach to cryocooler development is providing high confidence that the schedule for SBIRS Low will not be impacted by cryocooler design and development concerns. Cryocooler mass has been steadily reduced and cryocooler induced line of sight vibration has fallen below the value allocated for other sources in the flight demonstrations and experiments. Promising data continues to be accumulated on these machines at the AFRL CCL. Power efficiency for some designs during 35K operations has already exceeded the target of 80 W/W. A flight qualifiable dual temperature 35/60K cryocooler was completed and demonstrated by early spring 1999. Additional dual temperature and optics coolers are being developed for cost and risk reduction. Incremental improvements in common components and the high capacity optics cryocoolers will provide design margin and risk reduction.

ACKNOWLEDGEMENTS

Personnel from the Air Force Research Laboratory and the Aerospace Corporation carried out the work described in this paper. Mr. Erwin Myrick of the Ballistic Missile Defense Organization (BMDO/TOS) has played a crucial role in supporting both the requirement and necessary funding for spacecraft cryogenic cooler development. Ms. Karen Basany of the SIBRS Low Program Office has provided technical requirements, advocacy, and funding to insure technology objectives for the SBIRS program are available to meet schedule need dates.

REFERENCES

1. Swift, W.L., "Single Stage Reverse Brayton Cryocooler: Performance of the Engineering Model", Cryocoolers 8, Plenum Press, New York (1995), pp. 499-506.
2. Ross, R.G., "JPL Cryocooler Development and Test Program Overview", Cryocoolers 8, Plenum Press, New York (1995), pp. 173-184.
3. Levenduski, R., Gully, W., and Lester, J., "Hybrid 10 K Cryocooler for Space Applications", Cryocoolers 10, Plenum Press, New York (1999), pp. 505-511.
4. Burt, W.W, and Chan, C.K, "Demonstration of a High Performance 35 K Pulse Tube Cryocooler", Cryocoolers 8, Plenum Press, New York (1995), pp. 313-319.

5. Sparr, L., et al, "NASA/GSFC Cryocooler Test Program Results for FY94", Cryocoolers 8, Plenum Press, New York (1995), pp. 221-232.
6. Donabedian, M., et al, "Cryogenic Systems Integration Model (CSIM)", Cryocoolers 8, Plenum Press, New York (1995), pp. 695-707.
7. Glaister, D. S., Donabedian, M., Curran, D. G. T., and Davis, T., "An Overview of the Performance and Maturity of Long Life Cryocoolers for Space Applications", prepared for the Air Force Research Laboratory, Aerospace Report No. TOR-98(1057)-3, August 1998.
8. Kawecki, T., "High Temperature Superconducting Space Experiment II (HTSSE II) overview and Preliminary Cryocooler Integration Experience", Cryocoolers 8, Plenum Press, New York (1995), pp. 893-900.

APPLIED CRYOCOOLING OF A RESISTIVE HTS - FAULT CURRENT LIMITER

B. Gromoll[1], H.-P. Kraemer[1], J. Niewisch[1] and H.-W. Neumueller[1]
R.R. Volkmar[2], S. Fischer[2]

[1]Siemens AG, Corporate Technology, 91050 Erlangen, Germany
[2]Siemens AG, Power Transmission and Distribution, 13599 Berlin, Germany

ABSTRACT

According to the modular concept of the Siemens resistive superconducting fault current limiter (FCL) switching modules for the next functional model of the 1MVA class are being assembled and tested. The modules consists of 4 YBCO switching elements. The modules are bath cooled in a liquid nitrogen cryostat. During normal operation at nominal current, there are heat losses caused by the current leads, the connections between the switching elements and the cryostat. The evaporating nitrogen is recondensed at a coldhead of a refrigerator. Decisive for the heat losses are the current leads. During a fault a switching action time of 1 ms is followed by a limitation time of 50 ms. During this time the normal conductive elements heat up to about room temperature; then the limiter is separated from the network by a circuit breaker. The elements are recooled by evaporation of nitrogen until they return to the superconducting state again. The switching of the elements is monitored by a quench detector and the operational recovery time is determined by a bubble detector. In this paper we report on the switching behavior of the modules, the recovery time, the pressure rise after limitation, recooling time of the nitrogen bath and the operational monitoring.

INTRODUCTION

In an electric power system the task of Fault Current Limiters (FCL) is the limitation of mechanical and thermal loads on busbars, insulators and circuit breakers during a short circuit. Several types of FCL concepts based on superconducting materials have been proposed so far, an overview is given in [1,2]. A 1.2 MVA prototype of the inductive "shielded iron core" type has been built and tested by ABB[3]. The so-called resistive FCL relies on the switch-like onset of resistivity during a quenching process caused by exceeding the superconductor's critical current. This results in a self-triggered limitation of a fault-current passed through the device and the FCL therefore works as a fast and self-

Advances in Cryogenic Engineering, Volume 45.
Edited by Shu *et al.*, Kluwer Academic / Plenum Publishers, 2000.

restoring fuse with passive triggering. Applications range from new systems to upgrading existing networks with increased fault currents, or it may be used as an integrated component of superconductive systems like cables, transformers and SMES or in the feed line of power generators[4].

Based on the results of our 100 kVA limiter[5] this paper reports the implementation and successful testing of modules for the 1 MVA-limiter at Siemens Erlangen.

DESIGN OF THE 1 MVA FUNCTIONAL MODEL

The next step within the Siemens limter program is a 1 MVA-functional model. This model will be equipped with all components which are required for larger limiters later on. Fig. 1 shows the design of the three phase 1 MVA limiter. In our modular setup of the functional model, 3 x 7 switching modules have to be assembled to achieve the overall nominal switching power of $P_{nom,}$ = 1 MVA, 340 kVA per phase. The modules are bath cooled in liquid nitrogen. The cryostat has 6 conduction cooled current leads and a leadthrough for the glass fiber sensors for quench detection and bubble detection. The modules are fixed on a support which is mounted below the cover of the cryostat. The modules are interconnected by copper leads which are soldered and screw-connected. For recooling of the heat losses caused by the cryostat itself, the current leads and the normal conducting connections a coldhead is integrated in the cover of the cryostat for recondensation of the evaporated nitrogen. The cryostat is a closed system regarding the nitrogen. The coldhead is connected to an external compressor cooled with water.

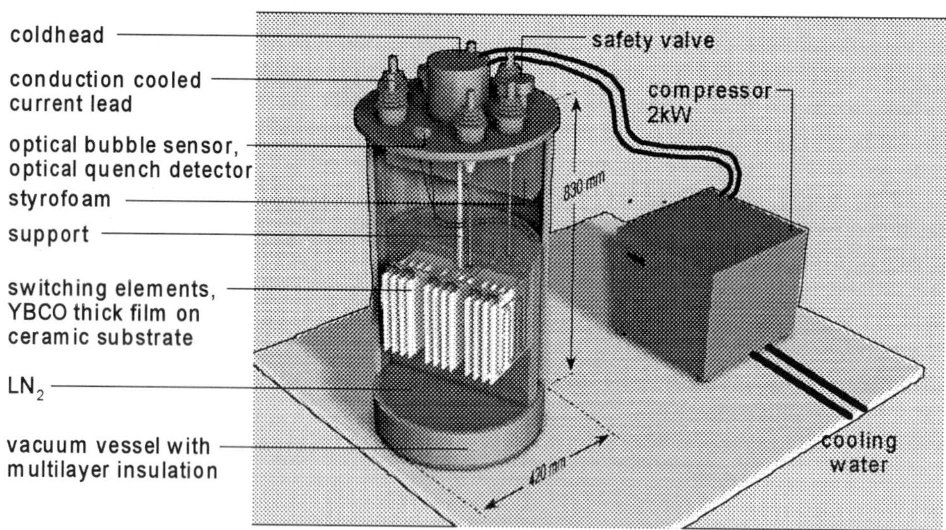

Figure 1. Design of the 3 phase 1 MVA model (6kV, 95A)

SET UP OF THE SWITCHING MODULE

Each module consists of 4 switching elements. The switching element is a YBaCuO film with a thickness of 250 nm deposited on 4" sapphire wafers by means of thermal coevaporation (TU Munich[6,7]). The resistively measured critical currents I_c of the switching elements vary between 33 A and 36 A. Inductive scans of the critical current density show that the spatial variations within one element are below 20%. Since these inhomogeneities can cause the formation of hot spots during switching the YBaCuO film is covered with an additional 100 nm gold shunt layer. A spiral shape with a length of 80 cm and a path width of 7 mm was etched into the layer structure.

At first the switching elements were tested individually. They were bath-cooled in a cryostat filled with liquid nitrogen and exposed to a short circuit simulated in the laboratory: After carrying the nominal current $I_{nom} = I_c/\sqrt{2}$ for some AC-cycles, they were facing a short-circuit starting in the middle of a half cycle and persisting another four AC half cycles, i.e. $\Delta t = 45$ ms in total. The resulting peak let-through current I_{peak} and the limited current $I_{lim,5}$ in the first respectively fifth half cycle were taken to characterize the switching behavior of the single element. The resistance normalized to its room temperature value RR is used as an indication of the sample temperature. In the measurement with a nominal power of 20 kVA per switching element the temperature reaches a maximum near 200 K.

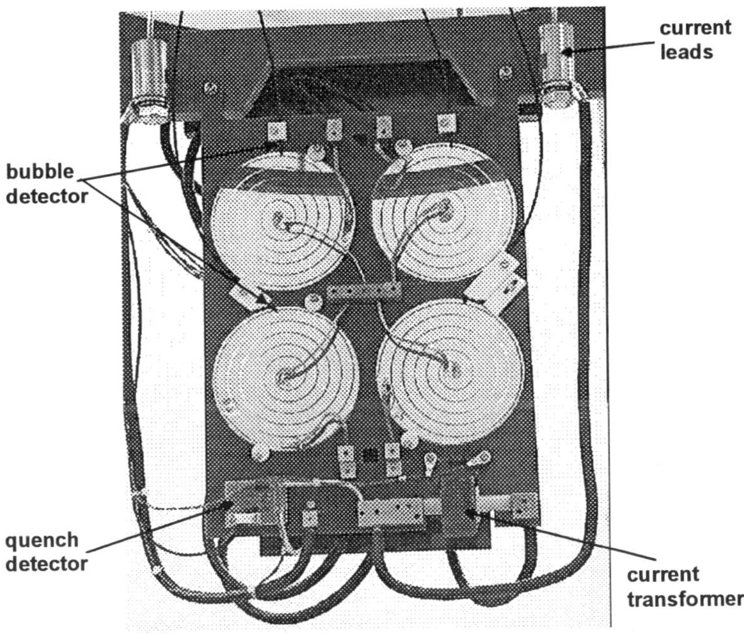

Figure 2. Switching module of the 1MVA limiter

In order to achieve P_{nom} = 340 kVA/phase with a nominal current I_{nom} = 95A and a maximum voltage of U_{nom} = 3,5 kV the elements have to be arranged in parallel and in series. To achieve the nominal current 4 plates have to be switched in parallel thus forming one module. For the nominal voltage, 7 modules are arranged in series. Fig. 2 shows the set up of one module. The plates are mounted in one plane in a sample holder and connected in parallel. The modules are equipped with bubble and quench detectors.

LN_2 bubble detector[8]

Each plate is equipped with a bubble detector. The detector is a pointed glass fiber. The tip is mounted in a holder on the top edge of the plate. The principle of the bubble detection with a pointed glass fiber is shown in Fig. 3. The deposition of energy in the conductor in case of a quench results in an increase of temperature which in turn produces bubbles of nitrogen. So these bubbles are an indicator for the conductor's temperature. This is important for the determination of the moment when the device is superconducting again and can be switched back to the network. This happens about 3s after the fault current is switched off.

Principle of operation. For discriminating N_2 bubbles from LN_2 one uses the difference in the refractive index. In our case it is n=1 for bubbles and n=1.28 for the liquid. To minimize light losses the end of the POF is shaped parabolically. The high light level allows the use of cheap LEDs and simple optical receivers with low gain and thus higher bandwidth.

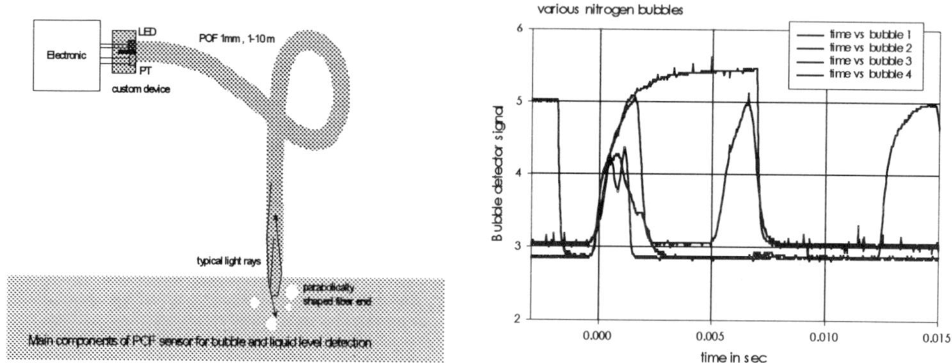

Figure 3. Principle setup of the bubble detector and its typical time response to various bubble sizes

Quench detector[8]

Each module has a quench detector for the 4 plates switched in parallel. The schematic set up of the Polymer Optical Fiber Quench Detector (POFQD) is presented in Fig. 4. Under superconducting condition the voltage between the current leads is zero. In case of a quench, caused by a current higher than the critical one, a fast voltage rise is generated at the leads (see Fig.4). The task of the quench detector is to detect the voltage rise in a time short enough to possibly switch off the current before destruction in the line or the limiter

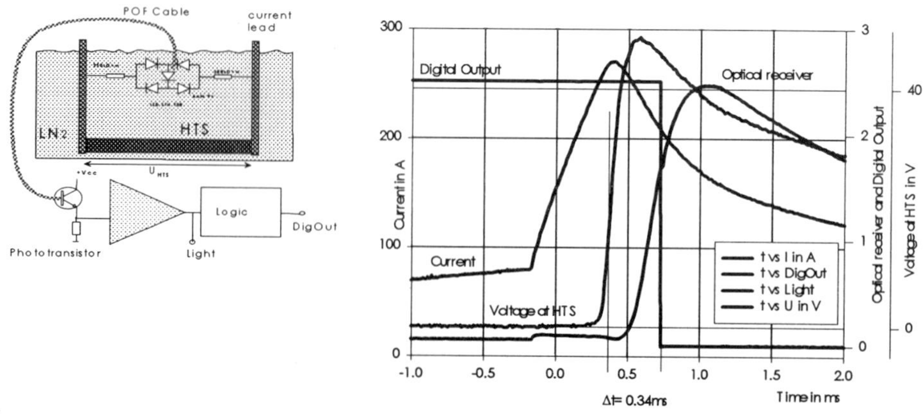

Figure 4. Principle setup of quench detector and its time response during switching

occurs. The necessary electromechanical elements react in about 1ms. The current limiter has to be designed thermally in a way that it can bear the energy loss for this time interval. The indicator for the voltage rise is the light generated in the LED by the current determined by the resistors. Tens of milliamperes are sufficient for easy noise free detection. Because of the AC current, two LEDs in antiparallel configuration are used to detect both half waves with the advantage that the two LEDs protect themselves from over voltage. The light is detected at the end of some meters of PMMA fiber with photo diodes or photo transistors. An electronic interface is provided between the optical receiver and the electromechanical circuit breaker. Figure 3 shows the current, the voltage and the responses of the detector during switching. One is the light intensity and the other curve the digital switching signal. The reaction time is about 340 μs.

TEST OF THE MODULE LIMITING CAPABILITY

To test the modules and the sensors, the modules are mounted radially on a centered support fixed below the cover of the cryostat. The modules are connected to the current leads which can be seen in Fig.2. The support can carry up to 24 modules.

Figure 5. Cover of the cryostat with module support and 1 module fixed

For the selection of the elements of the modules, the following criteria were applied:

- Obviously, the sums of critical currents I_c in different modules should be equal, because the lower value determines the nominal current of the device.
- Since I_{peak} reflects the speed of the resistive onset in an individual element and homogenous switching is essential, the sums of the peak currents should be the same as well.
- Finally, the limited currents $I_{lim,5}$ should also be equal to ensure that each module is exposed to approximately the same voltage during the current limitation.

The modules are being tested in the same way like the single plates. In the test result of one module in Fig. 6 the peak let through current I_{peak} of the module is 480 A, the limited current $I_{lim,5}$ = 100A and $I_{nom.}$= 130A. Due to the high j_c of the YBaCuO-films the nominal current of the tested module is higher than the design value of 95 A. The nominal power of one module is 83kVA.

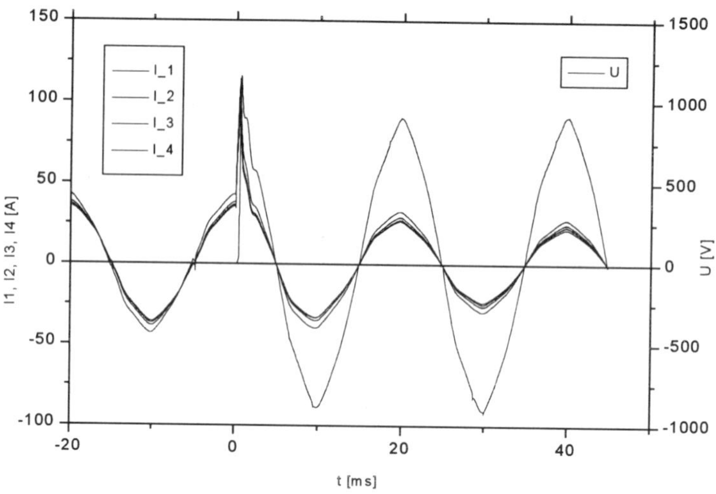

Figure 6. Test of one module, U_{nom} = 640V, I_{nom} = 4 x 33A, P_{nom} = 83kVA, I_{peak} = 4 x 120A, I_{lim} =4 x 25A

SENSOR SIGNALS AND OPERATIONAL RECOVERY TIME

Through a window of the cryostat the module can be observed during switching. Fig. 7 shows the evaporation of the liquid nitrogen recorded with a video camera at the beginning and near the end of the recovery phase. Due to the heat input of 4.5 kJ and a heat flux of 2.5 W/cm² there is film boiling at the beginning of the recovery phase which changes to bubble boiling at the end. In fig. 8 the signals of the optical sensors during switching and recovery phases are shown. The quench detector signal is roughly proportional to the rectified voltage. The signals of the bubble detectors in fig. 8 show the end of the recovery time after which the limiter can be switched on again, here after 3,5 s.

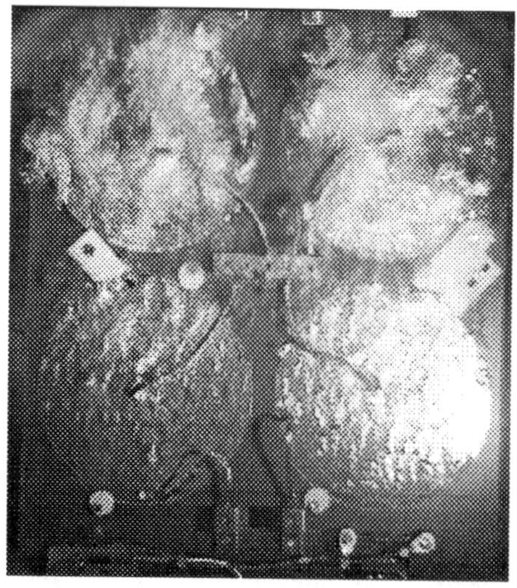

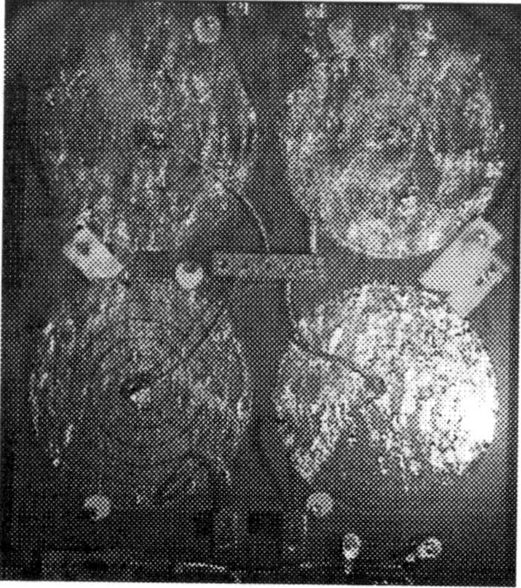

Figure 7. Pictures at the beginning and near the end of the operational recovery phase

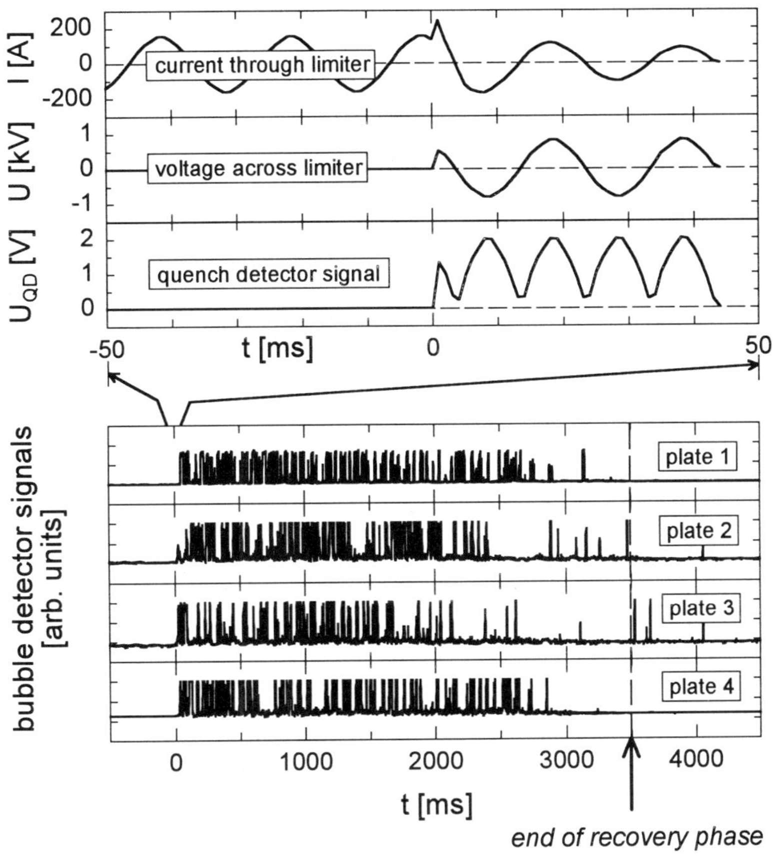

Figure 8. Signals of quench and bubble detectors during switching and recovery phase

REFRIGERATION

The cryostat for the 1 MVA limiter has an inner diameter of 600 mm, an outer diameter of 800 mm and a height of 930 mm. The cryostat is filled with 200 l liquid nitrogen. The measured heat loss of the cryostat is 40 W. The conduction cooled current leads with thermally optimized cross sections contribute 5 W each, adding up to 30 W. The connections between the plates and the modules have heat losses of approximately 6 W. The overall heat loss is 76 W. To keep the operating temperature of 77 K and to have enough back up cooling power to recondensate the evaporated nitrogen after switching within a short time a coldhead with a cooling power of 120 W @ 77 K and a compressor power of 6 kW is employed. During the quench of the plates of all three phases 50 kJ of heat are transferred to the 200 liter nitrogen bath. In an experiment the heat input during the quench of the limiter was simulated with an electric heater in the nitrogen bath with a power of 1 kW and a heating time of 50 seconds. The bath is heated up thereby for 0.25 K. With the extra back up power of 44 W it takes 19 minutes to cool the whole bath down to 77 K again. The pressure rise of 4.8 kPa is lower than expected. The reason is, that already during heating, recondensation of the nitrogen vapor takes place within the bath. Only part of the generated vapor leads to a pressure rise in the vapor room of the cryostat above the bath.

The recondensation is dependent on the heating time and will be tested with a quench off all modules installed in the bath.

CONCLUSION

It has been demonstrated that with 4 high j_c YBCO films on sapphire substrates it is possible to realize a very efficient switching module with a nominal power of 83kVA. This module will be the basis module to build a limiter of the 1MVA class. With detectors made of fiberoptical sensors the quench of the plates can be monitored and the end of the operational recovery time can be determined. The heat losses during normal operation in a closed cryostat system are compensated by recondensation of the evaporated nitrogen on a coldhead. The pressure rise in the above set up with a simulated quench is 4.8 kPa and the recooling time for the nitrogen bath with a back up cooling power of 44 W is 19 minutes. The back up cooling power is needed to reach the operational temperature after the heat input caused by switching. At present, more plates are being produced to build more modules for the limiter of the 1 MVA class.

ACKNOWLEDGMENT

This work has been supported by the German Federal Ministry for Education, Science, Research and Technology BMBF under contract no. 13N6842A.

Part of the work is done within the joint collaboration between Hydro Quebec, Canada and Siemens, Germany.

REFERENCES

1. R.F. Giese and M. Runde, Fault Current Limiters, Argonne National Laboratory, November 1, 1991
2. R.F. Giese and M. Runde, FaultCurrent Limiters-A Second Look, Argonne National Laboratory, March 16, 1995
3. W. Paul, M. Lakner, J. Rhyner, P. Unternaehrer, Th. Baumann et al., Test of 1.2 MVA High-T_c Superconducting Fault Current Limiter, *Applied Superconductivity 1997,* vol. 2, p. 1173-1178, IoP, 1997
4. S.Fischer, D, Povh, H. Schmitt Siemens AG, Erlangen, Superconducting fault current limiters – getting a grip to short circuit currents, *Proceedings CIRED 6/1999*
5. B. Gromoll, G. Ries, W. Schmidt, H.-P. Kraemer et al, Resistive Fault Current Limiters with YBCO Films - 100 kVA Functional Model, unpublished, submitted to ASC '98
6. H. Kinder, W. Prusseit, R. Semerad, B. Utz et al., YBCO Deposition on Large Substrates up to 9" in Diameter, *Adv. in Superconductivity IX (ISS '96)*, vol. 2, p. 1011-1016, Springer-Verlag, Tokyo, Japan, 1997
7. A. Heinrich, H. Kinder, H. Mosebach, Fault Current Limiting Properties of YBCO-Films on Sapphire Substrates, unpublished, submitted to ASC '98
8. J. Niewisch, POF Sensors for high temperature superconducting fault current limiters, *Proceedings of POF Conference 97'* Kauai, Hawaii, Sept. 22-25, 1997, page 130-131

TESTS OF A GM CRYOCOOLER AND HIGH TC LEADS FOR USE ON THE ALS SUPERBEND MAGNETS

J. Zbasnik[1], J. Y. Chen[2], M. A. Green[1], E. H. Hoyer[1], C. E. Taylor[1] and S. T. Wang[2]

1. Lawrence Berkeley National Laboratory
University of California
Berkeley, CA 94720

2. Wang NMR Incorporated
Livermore CA 94550

ABSTRACT

A 1.5 W (at the second stage) Gifford McMahon (GM) cryocooler was selected for cooling the superconducting SuperBend dipoles for the Advanced Light Source (ALS) at Berkeley. A GM cryocooler is a reasonable choice if conduction cooled leads are used to provide current to the superconducting magnet. The expected parasitic heat leaks are expected to range from 0.1 to 0.5 W at 4.2 K depending on the temperature of the shield and the cold mass support intercepts. Heat flow to 4 K down the SuperBend 350 A high Tc superconducting leads is expected to vary from 0.11 to 0.35 W depending on the intercept temperature and the current in the leads. The high Tc leads are designed to carry 350 A without significant resistive heating when the upper end of the lead is at 80 K. The 1.5 W cryocooler is expected to provide 45 to 50 W of refrigeration at the first stage at 50 K. The parasitic heat load into the first stage of the cryocooler will be about 8 W. The heat flow from 300 K down the upper copper leads is expected to be around 30 W. The cryocooler and high Tc lead test will measure the performance of the cryocooler and the high Tc leads. The heat leak down the cryocooler, when it is not operating, is also of interest.

INTRODUCTION

The Advanced Light Source (ALS) at the Lawrence Berkeley National Laboratory (LBNL) is a national user facility that produces high brightness vacuum ultraviolet (energies from 6 eV to 6 keV) and soft x ray (energies >6 keV) synchrotron radiation. The 200 meter diameter, 1.9 GeV, ALS electron storage ring consists of twelve cells each cell having three combined function C shaped 1.3 T gradient bending magnets (with the return flux leg pointing to the inside of the ring) about one meter long. Each dipole generates photons with a critical energy of 3.1 keV when the storage ring electron beam energy is 1.9 GeV. These photons can be delivered to users through forty-eight ports to users outside of the synchrotron shielding. Some years ago, a study was commissioned on ways to increase the photon energy from the ALS bending magnets[1]. The selected approach is to increase the maximum photon energy from the ring by increasing the bending field in some of the ring dipoles. (The critical energy of photons from the storage ring dipole is proportional to the dipole induction times the electron energy squared.) To maximize the utility of the ALS, three ring dipoles could be replaced with 5 T superconducting dipoles. This has minimum disturbance to the ring lattice and user experiment currently underway at the ALS.

A number of experimental SuperBend dipole coils were built and tested. The fourth set of SuperBend coils were successful in that they did not train and had acceptable field quality. They could be charged to the their full design current in less than 100 seconds. Once a successful SuperBend test magnet with a suitable magnetic field was made, the SuperBend project could move forward[2,3]. At the photon extraction points the induction is 5 T, which will generate photon beams with a critical energy of about 12 keV.

A study of possible cooling systems[4] for the SuperBend dipoles yielded some interesting results. The least attractive cooling option appeared to be a central refrigerator or liquefier with transfer lines to the three SuperBend magnets. Cooling by using liquid helium and liquid nitrogen did not fare much better. Small pure Gifford McMahon (GM) cryocoolers, without J-T circuits, appear to be the most cost effective and reliable cooling system for the three SuperBend dipoles for the ALS. Pure GM cryocoolers can provide refrigeration at 50 K and 4 K simultaneously. The use of high Tc superconducting (HTS) current leads between 50 K and 4 K permits the SuperBend magnets to be powered continuously while operating on a pure GM cryocooler. It was found that the size of the GM cryocooler is dictated by the cooling requirements at 50 K, not the 4 K cooling requirements. In an emergency, the SuperBend magnets can be kept cold using stored liquid helium and liquid nitrogen. The HTS current leads must carry up to 350 A with their upper end at 80 K, since the cooling system is a GM cryocooler with back-up cooling, in an emergency, supplied by liquid helium and liquid nitrogen.

Reliability is a key issue for the ALS. In order to insure that we understand the operating characteristics of cryocoolers and HTS current leads, the Lawrence Berkeley National Laboratory (LBNL) had contracted with Wang NMR to build a test stand for a GM cryocooler and 350 A HTS current leads. This report describes the LBNL test stand and presents the results of the cryocooler and HTS current lead tests.

THE GM CRYOCOOLER AND THE HTS LEADS

After some study of the available options, LBNL selected the Sumitomo SRDK-415D pure GM cryocooler that develops up to 1.5 W at 4 K while developing up to 60 W of cooling at a temperature between 50 and 60 K. The cryocooler selection process was driven by the following factors: 1) The 4 K heat load for SuperBend is expected to be about 0.5 W while the projected heat load at 50 K is about 40 watts. The size of the cryocooler was driven by the heat load on the first stage. 2) LBNL selected a pure Gifford McMahon cryocooler, because we felt that it would be more reliable. 3) LBNL selected a cryocooler company that already had a large number of 4 K cryocoolers operating in the field. The pure GM cryocooler business is changing rapidly. In the near future, we expect that there will several more pure GM and pulse tube cryocoolers on the market.

A set of typical refrigeration capacity curves for the 1.5 W Sumitomo SRDK-415D pure GM cryocooler is shown in Figure 1[5]. Figure 1 shows that the temperature on the first stage (50 K stage) is strongly dependent on the first stage heat load Q1. There is a weak first stage temperature dependence on the second stage (4K stage) heat load Q2. The refrigeration capacity curves show that the second stage temperature is dependent primarily on the second stage heat load Q2. Figure 1 suggests that the second stage temperature will go down as the heat load on the first stage Q1 is increased. The minimum second stage temperature is about 2.8 K, which means the superconducting magnet temperature margin could be larger during normal operation with the cryocooler.

LBNL specified that the HTS current leads should carry 350 A with a high end temperature of 82 K with a low end of heat leak of 0.3 W. The two types of HTS leads tested were: 1) The American Superconductor Corporation (ASC) HTS leads were made with BSCCO silver tape that appears to be encapsulated in a NEMA G-10 rod. 2) The Intermagnetics General Corporation (IGC) HTS leads were made from melt textured BSCCO inside a stainless steel tube. In both cases, the leads were sold to LBNL as commercially available leads . We do not know the details of the design of either of these HTS leads.

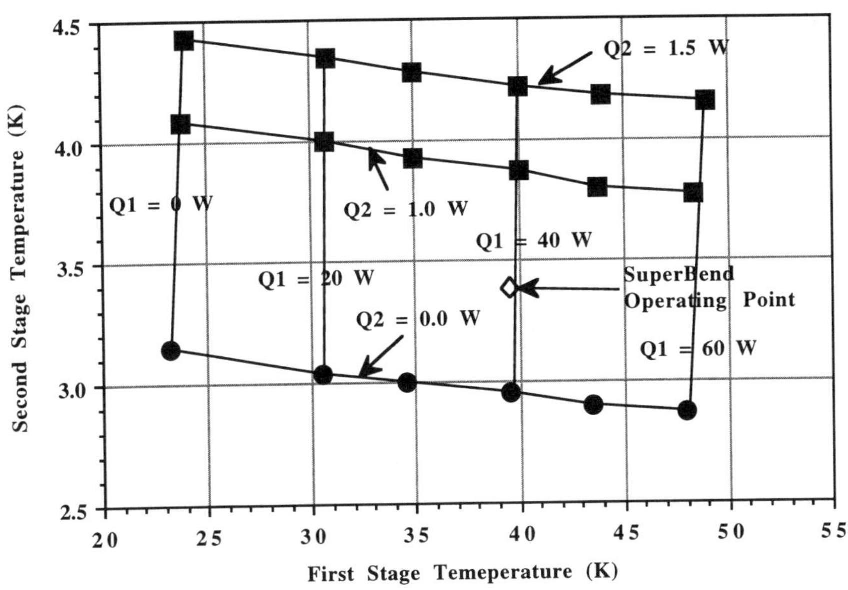

Figure 1 Typical Refrigeration Capacity Curves for the 1.5 W Sumitomo SRDK-415D Cryocooler

Table 1 Estimated Heat Leaks into the Cryocooler Experiment and the Mass of Various Components in the Experiment

Parasitic Heat Leaks at 50 to 80 K	
Static Heat Leak MLI and Cold Mass Support	5.8 W
Heat Leak Down the Solid Copper Leads, Current = 0 A	19.0 W
Heat Leak Down the Solid Copper Leads, Current = 290 A	26.5 W
Heat Leak Down First Cryocooler Stage when Off (Est.)	>50. W
Estimated 50 K Heat Leak, Cooler On, Current = 0 A	23.8 W
Estimated 50 K Heat Leak, Cooler On, Current = 290 A	32.3 W
Estimated 80 K Heat Leak, Cooler Off, Current = 0 A	>73.8 W
Estimated 80 K Heat Leak, Cooler Off, Current = 290 A	>82.8 W
Parasitic Heat Leaks at 4 K	
Static Heat Leak through MLI and Cold Mass Support	0.22 W
Heat Leak Down Leads, Upper End = 50 K	0.17 W
Heat Leak Down Leads, Upper End = 80 K	0.30 W
ASC Lead Resistive Heating at 290 A, Upper End T = 50 K	0.11 W
ASC Lead Resistive Heating at 290 A, Upper End T = 80 K	0.22 W
Heat Leak Down Cryocooler Second Stage when Off (Meas.)	~1.96 W
Estimated 4 K Heat Load with No HTS Leads Installed	0.22 W
Estimated 4 K Heat Load with Cooler On, Current = 0 A	0.39 W
Estimated 4 K Heat Load with Cooler On, Current = 290 A	0.50 W
Estimated 4 K Heat Load with Cooler Off, Current = 0 A	~2.48 W
Estimated 4 K Heat Load with Cooler Off, Current = 290 A	~2.70 W
Experiment Component Masses	
Estimated Cryostat Mass at 50 to 80 K	20.0 kg
Estimated Stage 1 Cold Head Mass (50 to 80 K End)	~2.0 kg
Estimated Mass at 50 to 80 K	22 kg
Estimated Cryostat Mass at 4 K	25.0 kg
Estimated Stage 2 Cold Head Mass (4 K End)	~0.5 kg
Estimated Mass at 4 K	25.5 kg

THE CRYOCOOLER AND HTS CURRENT LEAD TEST STAND

During the first quarter of 1999, a test stand was built for testing the cryocooler and the HTS leads in a reasonable simulation of the SuperBend configuration. Table 1 on the previous page shows an estimate for the heat leaks at 50 to 80 K and at 4 K for the test apparatus shown in Figure 2 below. The test stand was 495 mm in diameter and 760 mm high. The HTS leads are located on the outside of the helium tank so that they can be changed when the vacuum vessel and super insulated shield have been removed. Connected to the test stand is a pair of twenty meter long flexible metal hoses that connect a helium compressor to the cryocooler. The hoses also serve as a surge volume for pressure pulses. Cables from a 600 A power supply were also connected to the test stand.

The test stand shown in Figure 2 below includes the 1.5 W Sumitomo SRDK-415D cryocooler, a 30 liter storage tank for liquid helium, a 14 liter storage tank for liquid nitrogen, conduction cooled copper leads from 300 K, HTS current leads that are accessible when the experiment is disassembled, a copper based Nb-Ti superconducting interconnect between the HTS leads, a 50 to 80 K shield, twelve silicon diode temperature sensors, eight voltage taps and four resistive heaters to provide additional heat to various parts of the test stand. Figure 3 at the top of the next page shows a schematic representation of the test stand that shows the location of the temperature sensors, voltage taps and the resistive heaters.

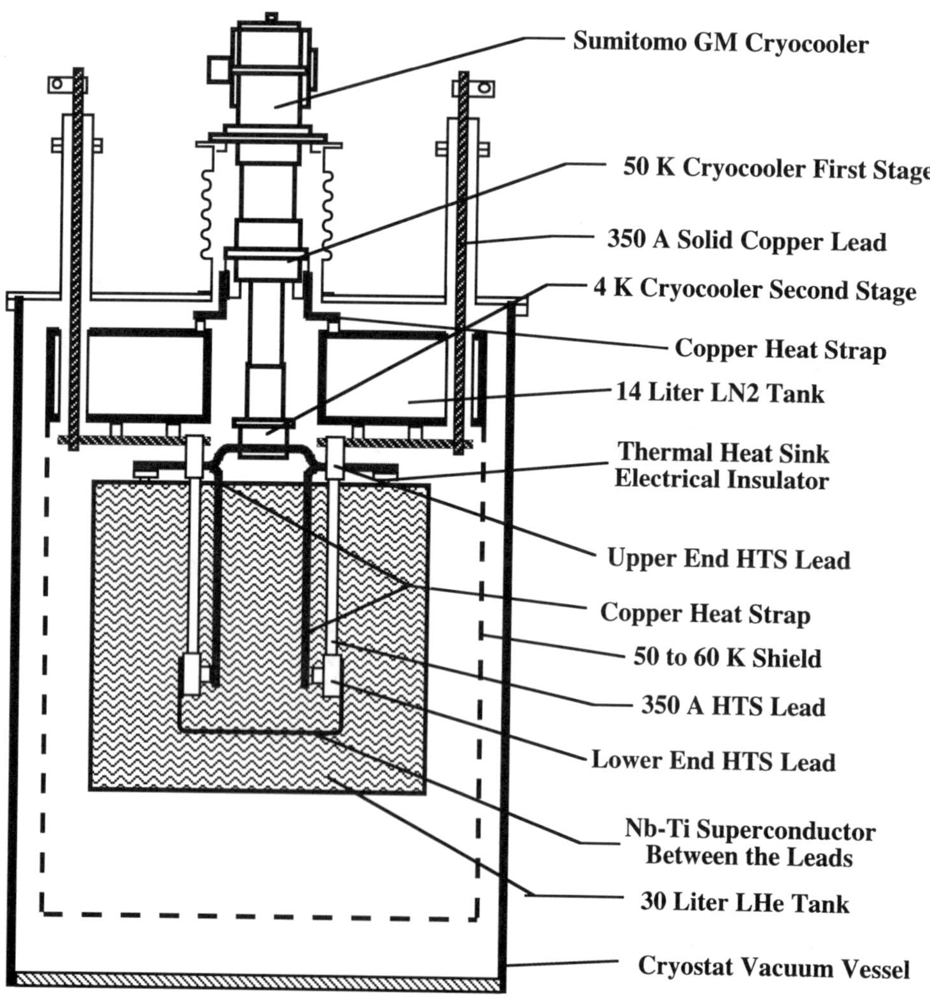

Figure 2 A Cross-Section View of the SuperBend Cryocooler and HTS Lead Test Stand

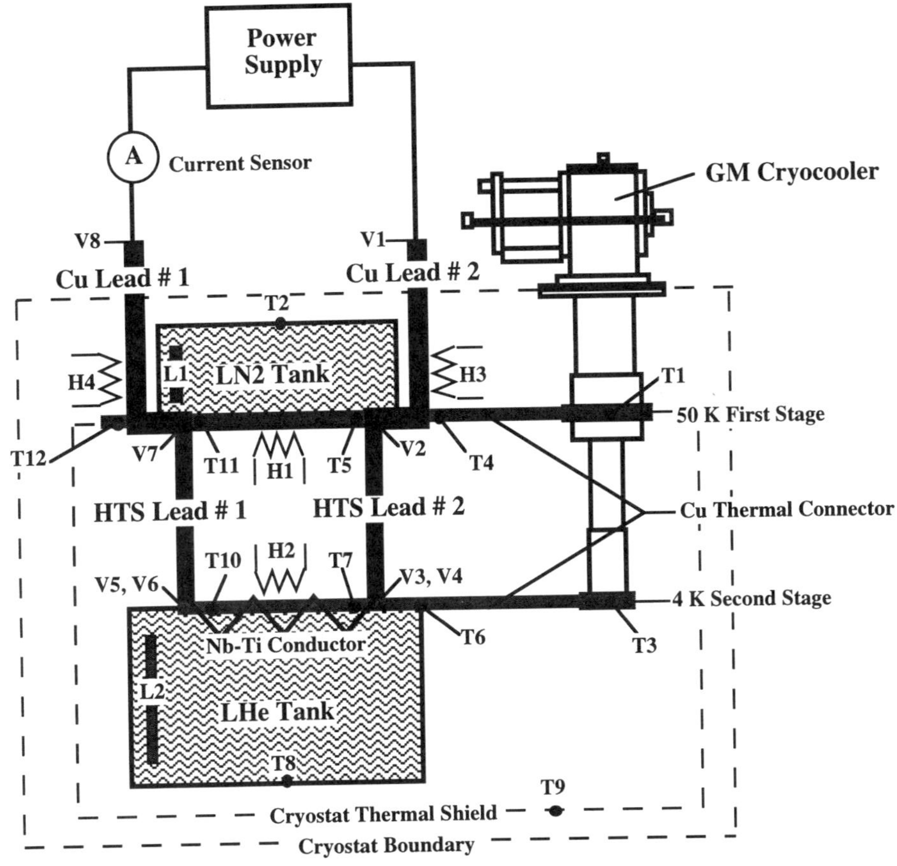

Figure 3 A Schematic Representation of the SuperBend Cryogenic Test Stand for the 1.5 W Sumitomo Cryocooler and HTS Leads

Figure 3 above shows the location of eight voltage taps labeled V1 through V8, twelve silicon diode temperature sensors labeled T1 through T12, and four resistive heaters labeled H1 through H4. Voltage drops can be measured across the HTS leads and the copper leads using the voltage taps. The temperature sensors are located as follows: T1 measures the temperature of the cryocooler first stage (the 50 K stage). T2 measures the temperature of the nitrogen tank. T3 measures the temperature of the cryocooler second stage (the 4 K stage). T4 and T12 measure the temperature at the cold end of the conduction cooled current leads. T5, and T11 measure the temperature of the upper ends of the HTS leads. T6 measures the temperature of the end of the copper strap connecting the second stage of the cryocooler with the helium tank. T7 and T10 measure temperature at the lower ends of the HTS leads. T8 measures the temperature of the helium tank. T9 measures the temperature of the shield. The diodes T3, T6, T7, T8, and T10 are accurate to about 0.1 K, with proper calibration. The diodes T1, T2, T4, T5, T9, T11 and T12 are accurate to about 0.3 K

L1 and L2 are liquid cryogen level sensors for nitrogen and helium. The heaters H1 through H4 serve the following functions: H1 heats the nitrogen tank. Heat from H1 must pass through the copper strap to the first stage of the cryocooler. H2 heats the helium tank. Heat from H2 must pass through the copper strap that carries heat to the second stage of the cryocooler. H3 and H4 heat the lower end of the copper current leads and the upper end of the HTS current leads. The current entering the experiment was measured and controlled using the current sensor shown in Figure 3. The apparatus shown in Figure 3 allows one to vary the heat load (and temperature) on both stages of the cryocooler. System control and data acquisition were performed using a Lab View based program written by LBNL and run on a PC. A quench detection circuit independent of the PC was built by LBNL to turn off the power supply, in the event of an HTS lead quench.

EXPERIMENTAL RESULTS

The test stand was cooled down using the cryocooler on a number of different occasions. The 25 kg of cooled mass connected to the second stage of the cryocooler took about 13 hours to cool from 300 K to 4 K. The 22 kg of mass connected to the first stage of the cryocooler took somewhat longer to cool down, about 15 hours. If the masses are scaled to the projected cold masses for SuperBend, the expected cool down time for the magnet would be from 550 to 850 hours depending on how the masses are counted in the scaling process. At 150 K, the cooling from the first stage is estimated to be about 80 W while the cooling rate for the second stage is about 45 W. It is clear that liquid cryogens must be used to cool down SuperBend rapidly.

Table 2 shows test results once the experiment has been cooled down to its normal operating temperatures. Cases A through N are cases where the upper copper leads were powered, but there were no HTS leads connecting the upper and lower temperature stages of the experiment. Cases O, P and Q are tests with the ASC HTS current leads between the upper and lower temperature stages. Cases R, S and T are tests with the IGC HTS current leads between the upper and lower temperature stages. The table shows cases where various amounts of heat were put into the experiment using heaters H1 and H2. Q1 is our best estimate for the heat entering stage 1 of the cryocooler. Q2 is our best estimate for the heat entering stage 2 of the cryocooler. The temperatures on sensors T1 and T3 were chosen so that the performance given in Table 2 can be compared directly with the typical Sumitomo cryocooler performance curves given in Figure 1.

Table 2 and Figure 4 show cases where the estimated heat entering the first stage of the cryocooler varies from 24 W to 62 W. The estimated heat entering stage 2 of the cryocooler varies from 0.22 W to 1.22 W in Table 2 and Figure 5. The range of first stage temperatures was from 29 K to 48.6 K. The second stage temperatures varied from 2.9 K to 4.1 K. Figures 4 and 5 compare our measured data with the standard refrigeration curves for the Sumitomo SRDK-415D Cryocooler.

Table 2 Temperatures versus Heat Loads for Various Refrigeration Cases

CASE	Current (A)	H1 (W)	Q1* (W)	H2 (W)	Q2* (W)	T1 (K)	T3 (K)
No HTS Leads are Installed; Tested with Upper Cu Leads Only							
A	0	0.0	23.8	0.0	0.22	31.1	3.2
B	290	10.0	42.3	0.0	0.22	37.3	3.1
C	290	10.0	42.3	0.2	0.42	37.3	3.3
D	290	10.0	42.3	0.37	0.59	37.3	3.4
E	290	20.3	52.6	0.2	0.42	42.5	3.4
F	290	20.4	52.7	0.37	0.59	43.2	3.6
G	290	20.4	52.7	0.62	0.84	43.2	3.7
H	290	20.4	52.7	0.75	0.97	43.2	3.9
I	290	20.4	52.7	0.87	1.09	43.5	4.0
J	290	20.4	52.7	1.00	1.22	44.0	4.1
K	0	0.0	23.8	0.25	0.47	29.0	3.3
L	0	0.0	23.8	0.50	0.72	30.0	3.5
M	290	0.0	32.3	0.50	0.72	33.7	3.6
N	290	0.0	32.3	0.75	0.97	34.2	3.8
American Superconductor Corp. (ASC) Leads Installed and Tested, Upper Leads are Cu							
O**	290	10.0	42.3	0.20	0.64	38.1	3.5
P**	290	10.0	42.3	0.74	1.18	38.2	3.9
Q**	290	30.0	62.3	0.75	1.19	45.9	4.0
Intermagnetic General Corp. (IGC) Leads Installed and Tested, Upper Leads are Cu							
R	290	10.0	42.3	0.20	0.64	37.8	3.5
S**	290	20.0	52.3	0.37	0.81	37.3	3.4
T**	290	30.0	62.3	0.74	1.18	48.6	4.0

* Q1 is the estimated heat flow into Stage 1 of the cryocooler. Q2 is the estimated heat flow into stage 2 of the cryocooler.

** Liquid Helium and Solid Nitrogen are in the reservoirs.

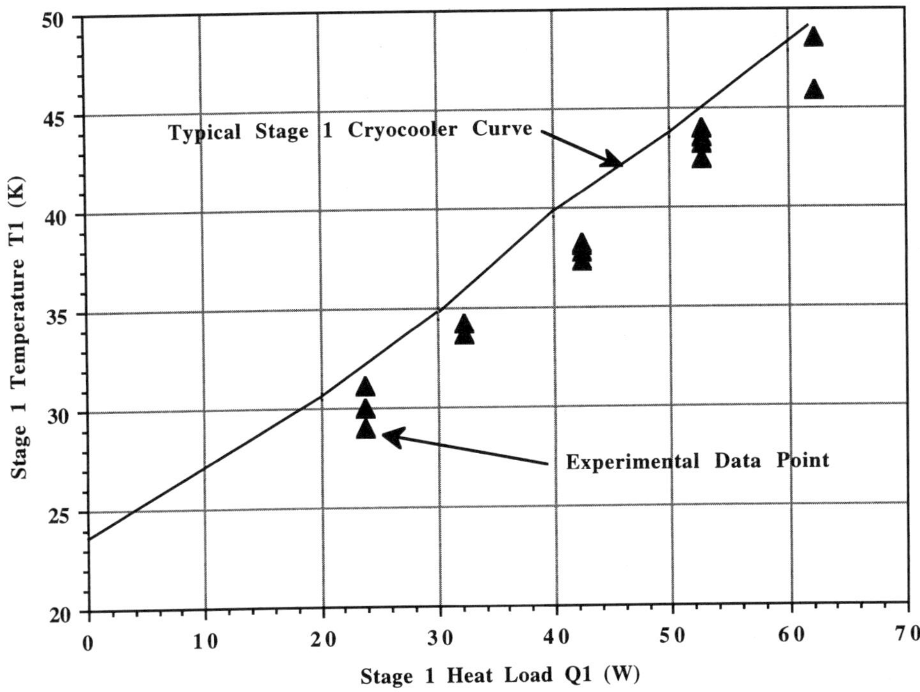

Figure 4 First Stage Temperature T1 as a Function of First Stage Heat Load Q1
A comparison of Experimental Data with Sumotomo Data

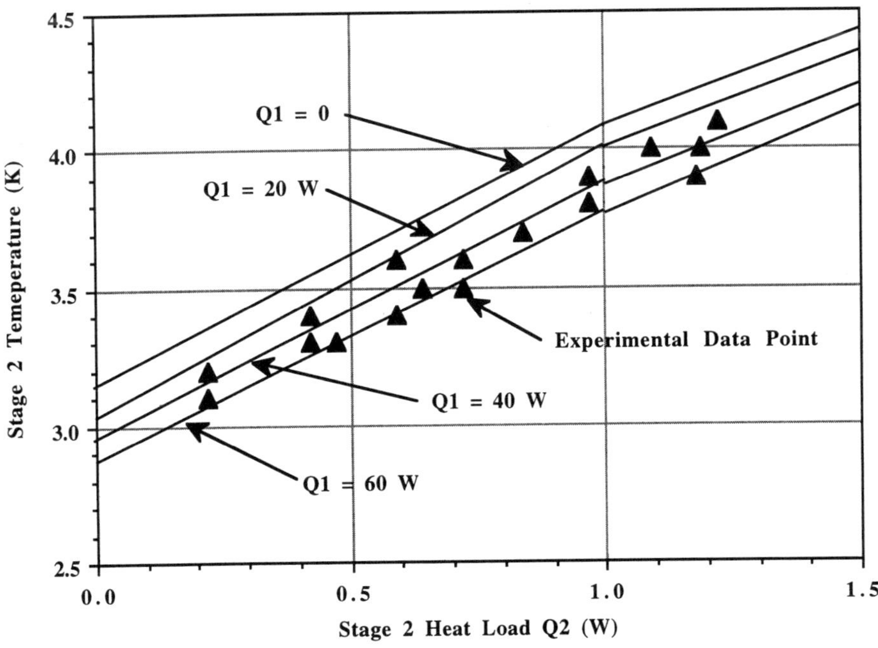

Figure 5 Second Stage Temperature as a Function of Second Stage Heat Load Q2 and Q1
A Comparison of Experimental Data with Sumitomo Data

The data shown in Figure 4 suggests that the value of Q1 was over estimated by a few watts. The data shown in Table 2 and Figure 5 do not show the stage 2 temperature going down with heat load on stage 1. There was little change in the stage 2 temperature seen by T3 as the heat load Q1 in stage 1 increased. One explanation for this is that there is an increased heat flow through the cold mass supports between the nitrogen tank and the helium tank as the temperature at T1 goes up. Data was taken from the other temperature sensors besides T1 and T3. At the beginning of the experiment, there was a difference between the temperature measured by sensors T6, T7, T8 and T10 from that measured by sensor T3. This temperature difference increased with the heat from H1. A change in the copper strap between the cryocooler second stage and the helium vessel eliminated this thermal drop.

Both the ASC and the IGC leads performed well on the test stand. There appears to be little difference in the heat leak to the low temperature end of the experiment between two brands of leads. The ASC leads were tested using heater H3 and H4 to put heat into the upper end of the leads. The ASC leads operated well at 333 A even when the temperature at the upper end was as high as 88K. The ASC leads, we tested, appeared to be slightly resistive. The resistance of the leads was dependent on the upper end temperature. One explanation for this is that LBNL measured the resistance of the joint between the copper and the superconductor at the top of the lead and not the resistance of the superconductor in the lead. The resistance experiment has not been repeated using the IGC leads.

The GM cryocooler was shut off and the experiment was operated at full current while the cryocooler was shut off. The cold end heat leak down the cryocooler was about 2 W while stage one of the cryocooler was kept at 80 K by the liquid nitrogen in the tank attached to stage one. The heat leak down stage one from 300 K with the cryocooler off was not measured because of a faulty level gauge in the liquid nitrogen tank.

CONCLUSIONS

The test of the 1.5 W Sumitomo cryocooler was successful. The cryocooler operated over a range of heat loads at both 40 K and 4 K simultaneously. The performance of the cryocooler was satisfactory over the range of operating conditions tested. Our experiments indicate that the thermal bond between the cryocooler heads and the rest of the cold mass is very important. The HTS leads from ASC and IGC appeared to perform in a satisfactory way under the conditions of the LBNL experiment. More experimental work is needed on both the ASC and the IGC leads. From the experimental data taken to date, it can be concluded that small GM cryocoolers are an attractive and cost effective way to cool the three superconducting SuperBend dipoles for the ALS.

ACKNOWLEDGMENTS

The authors gratefully acknowledge the efforts of the J. Hinkson, M Chin, M. Fahmie, A. Geyer, and F. Ottens of the ALS Electrical Engineering group in preparing the electrical equipment and the computer software needed to carry out the measurements. This work was supported by the Director of the Office of Science, United States Department of Energy under contract number DE-AC03-76SF00098.

REFERENCES

1. C. E. Taylor and S. Caspi, "A. 6.3 T Bend Magnet for the Advanced Light Source," IEEE Trans. Magnetics **32**, No 4, (1996)
2. C. E. Taylor, et al, "Test of a High-Field Bend Magnet for the ALS," to be published in IEEE Transactions Applied Superconductivity **9**, No. 2 (1999)
3. A. Lietzke, "SuperBend 4 Magnetic Measurements," LBNL Superconducting Magnet Group Internal Report SC-MAG-633, September 1998
4. M. A. Green, et al, et al, "Refrigeration Options for the Advanced Light Source SuperBend Dipole Magnets," presented at the 1999 Cryogenic Engineering Conference, Montreal Quebec, Canada, 13 to 16 July 1999, this Proceedings Advances in Cryogenic Engineering **45**, Plenum Press, New York (1999)
5. The Sumitomo Data sheet for the SRDK-415D cryocooler from Janis Research Inc, Wilmington MA 01887 (Janis is the US seller of Sumitomo cryocoolers

A COMPACT SWEEPER MAGNET FOR NUCLEAR PHYSICS

A. F. Zeller[1], D. Bazin[1], M. Bird[2], J. C. DeKamp[1], Y. Eyssa[2], K. W. Kemper[3], L. Morris[1], S. Prestemon[2], B. S. Sherrill[1], M. Thoennessen[1], and S. W. Van Sciver[2]

[1]National Superconducting Cyclotron Lab
Michigan State University
E. Lansing, MI 48824

[2]National High Magnetic Field Lab
Florida State University
Tallahassee, FL 32310

[3]Department of Physics
Florida State University
Tallahassee, FL 32313

ABSTRACT

A superconducting dipole, designed for use as a sweeper magnet, is being constructed jointly at the NSCL and NHMFL. The completed magnet will be installed at the NSCL and used in nuclear physics experiments. The magnet operates at a peak field of 4 T in the 140 mm gap. Since it is necessary that the area that corresponds to the outgoing beam direction be open to allow the exit of the neutrons, the magnet is built as a modified "C"-magnet. The actual coil shape is in the form of a "D", which keeps the magnet compact and removes the need for a negative curvature side. Resistive trim coils on the side of the iron yoke reduce the external fringe field to levels which allow operation of photo-tubes as close as 300 mm from the iron.

INTRODUCTION

The National Superconducting Cyclotron Lab (NSCL) is in the process of upgrading[1] the capabilities of the lab for the production of secondary beams and higher energy heavy ions. At

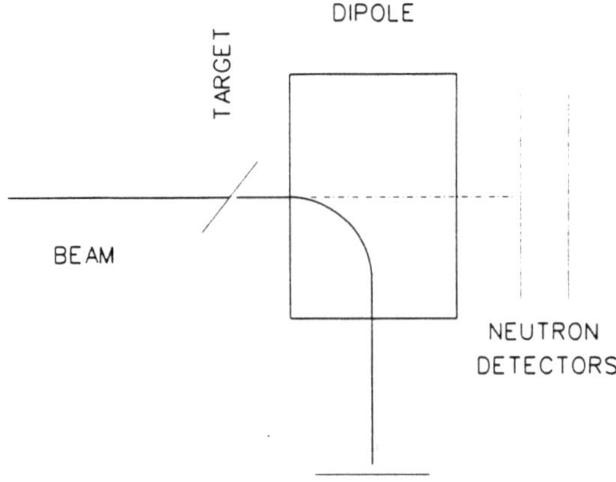

Figure 1. Schematic representation of a sweeper magnet experiment.

the same time, the National Science Foundation has provided funding to build a system that will allow simultaneous detection of neutrons and charged particles. The main instrument for detection of charged particles will be the S800 Spectrograph[2], although the magnet will be capable of stand alone operation. The principle of the operation is shown in Figure 1. The beam produces the reactions of interest in the target. Neutrons produced in short-lived species continue in the beam direction and are detected at some distance from the magnet. Charged particles produced in the reaction are deflected by the sweeper magnet into the S800 Spectrograph, as shown in Figure 2. Other experiments detect gamma rays in the beam direction with an array of detectors. Finally, the sweeper may be set up in a stand alone mode, as shown in Figure 3. To accommodate this set of different requirements the magnet has to be:

- compact
- have a large gap
- be portable

The large gap is needed to allow the neutrons to reach a large detector (2000 mm by 2000 mm) placed sufficiently far away to provide adequate angular resolution. The geometry of the S800 is

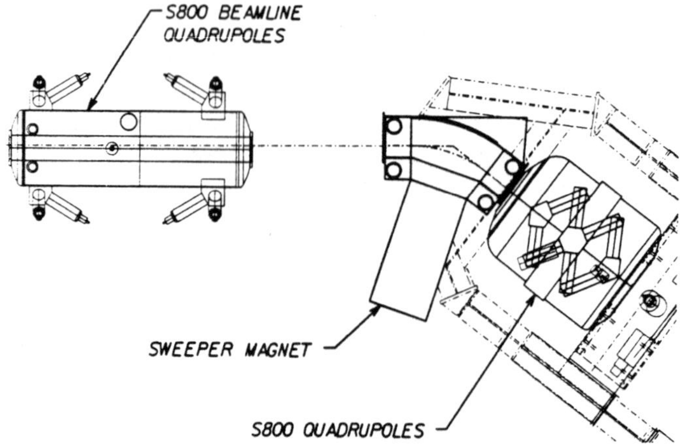

Figure 2. The sweeper magnet in position with the S800 Spectrograph.

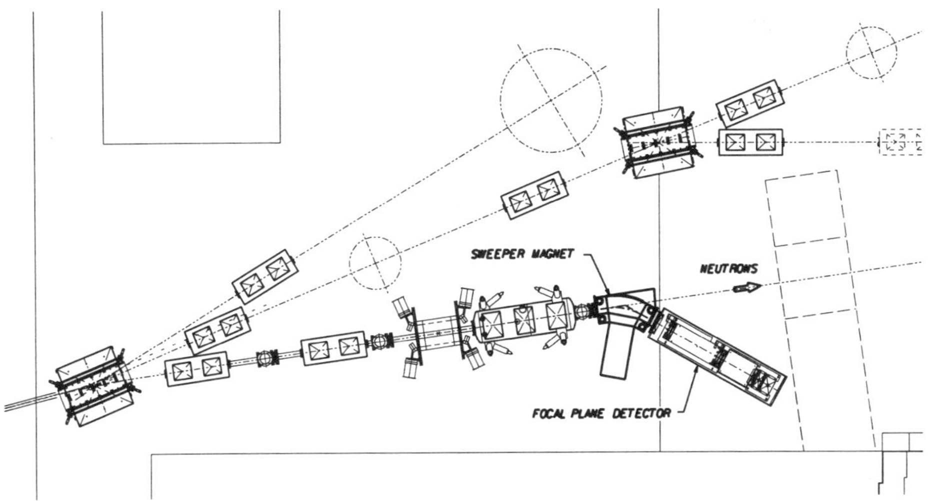

Figure 3. The sweeper magnet in position for independent operation.

fixed, so the magnet has to fit into place and have a high enough field (4 T) to bend the fragments 40^0 into the first quadrupole of the S800. The pieces of the magnet have be small enough to be moved around with a small crane in a restricted space and have a maximum weight of 2000 kg.

MAGNETIC FIELD REQUIREMENTS

The large gap requirement and the beam rigidity dictate the superconducting requirements for the magnet. A bend angle of approximately 40 degrees and the relatively small radius of one meter requires a central field of 4 T. Because the saturation magnetization of low carbon iron is approximately 2.2 T and the required gap is 140 mm, most of the field must be generated by the coil. The iron then serves mainly as a clamp to reduce stray field. The large gap and compact

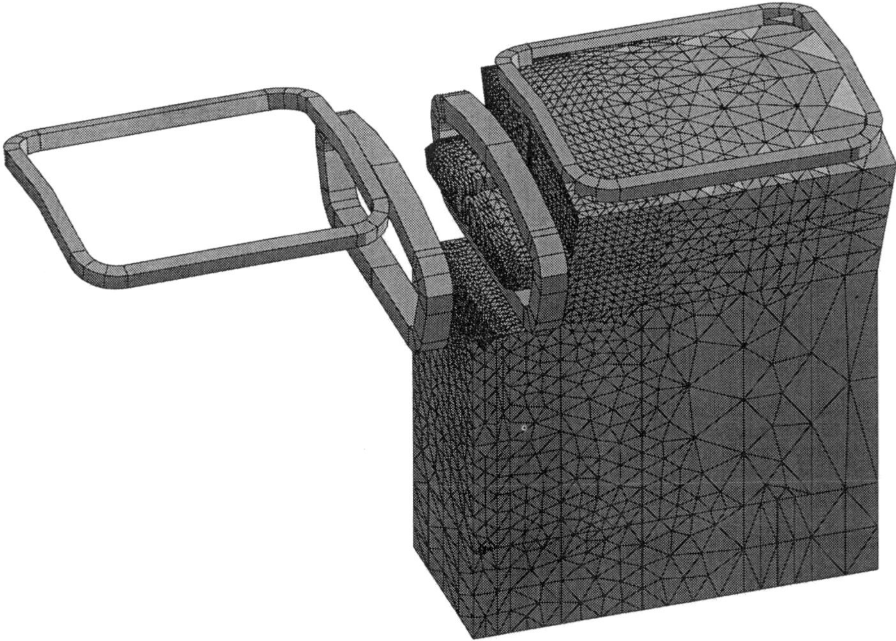

Figure 4. Coil and iron configuration for the sweeper magnet.

shape means the coils must be relatively small and the current density high.

Another problem with high field operation is the high fringe field levels. Since one of the operational modes is has gamma detectors close to the beam, the fringe fields must be kept below 10 mT, otherwise the photo-tube pre-amps will not work. Moving the detectors farther away from the magnet increases the cost because more tubes would be needed to cover the same solid angle. The mass of the iron was chosen to help reduce the external field and an external, resistive coil is used to buck the fringe field further.

The requirement for open space in the beam direction results in having to build a C-magnet. The requirement for compactness and minimum mass would suggest a sector magnet, with a negative curvature side. To avoid this major complication, the negative curvature side is replaced with a straight section, resulting in a "D" shaped coil. Similar magnets have been built at the NSCL previously[3]. A cross section of the upper half of the magnet is shown in Figure 4. The trim coil is also shown. The upper half of the trim coil is actually just the return leg of the lower coil. The cryostat is also complicated because the neutrons and gammas should traverse as little mass as possible before reaching the detectors, so the coils should not be supported across the median plane on the curved side of the magnet.

Preliminary magnetic field calculations were done with a two-dimensional code, but because the magnet is very is only mid-plane symmetric, all final calculations were done with TOSCA[4]. Constraints on reducing the iron mass were: keeping the fringe field below the desired 10 mT requirement, keeping the current density to approximately 20 kA/cm^2 (sufficiently low coil stress so the supports do not become complicated, and giving a magnetic field profile with sufficiently low harmonics (sextupole content less than 5%) that the necessary resolution could be achieved. The final iron configuration is also shown in Figure 4. The final angular extent of the iron, 37^0, is determined by the effective bend angle. Because the bend angle changes with field level the iron length is an average of low field and high field effective lengths. The uniformity perpendicular to the beam path is sufficient that the resolution is preserved for any normal trajectory. The magnet and iron parameters are given in Table 1.

The trim coils are wound from normal, resistive wire. To keep the area across the median plane free from material, there are actually two trim coils. The section of the coils near the median plane is of opposite polarity from the main coil to provide for field cancellation. The return leg

Table 1. Parameters of the sweeper magnet

Parameter	
Nominal bend angle	40 degrees
Gap	140 mm
Maximum gap field	4 T
Radius	1000 mm
Iron mass	14 850 kg
Total height	2400 mm
Iron length	1570 mm
Stored energy	790 kJ

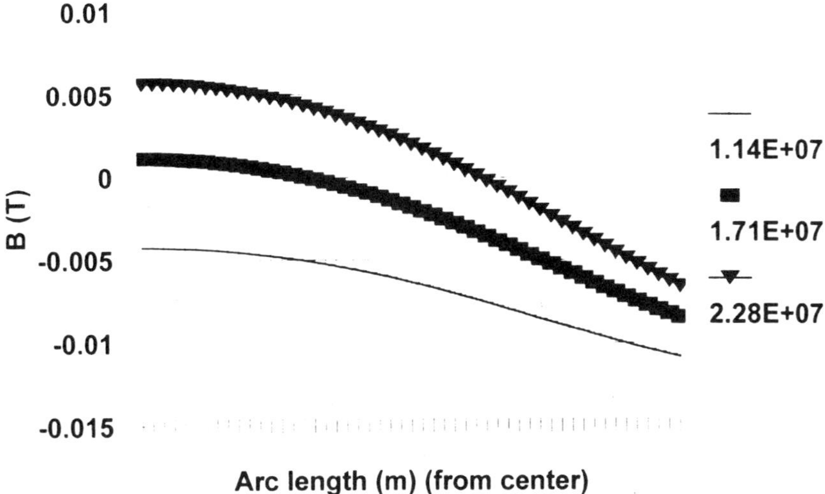

Figure 5. Stray field outside the magnet where the phototubes will be placed. Current density in A/m^2.

for each coil is at the top of the iron. The effects of the trim coils in moving the 10 mT line closer to the iron are shown in Figure 5 for several trim coil currents. The line parallels the outer, curved section of the iron yoke, where the photo-tubes for the gamma detectors will be located.

COIL AND CRYOSTAT DESIGN

The rather small space available for the coils means the current density must be about 20 kA/cm^2 at a peak conductor field of 5.5 T. To achieve the maximum packing factor an ordered winding with rectangular conductor is used. Because of a desire to keep the liquid helium

Table 2. Parameters of the sweeper magnet coil.

Parameter	
Bare wire size	1.898 X 0.898 mm^2
Cu:SC	1.8:1
Operating current	387 A
Critical current at 5.5 T and 4.2 K	1100
Number of turns per layer	35
Number of layers	62
Current density	18 kA/cm^2

Figure 6. The lower bobbin showing the cold link between the outer and inner sections of the coil. The connection to the upper bobbin is also shown. Note the upper and lower sections are only connected in the inner leg to keep the beam direction open.

consumption low, the desired maximum current is less than 500 A, consistent with safety during a quench (low induced voltages). Since the designed coils are potted, no active quench protection is used so the coils will be self-protected. The wire and coil parameters are listed in Table 2. The calculated quench voltages and hot spot temperatures are low and acceptable.

The forces on the coil and the coil case are significant, and act both in the radial and axial directions. The radial forces act outward from the dipole center. Due to the large gap, the mutual attraction of the two coils is smaller than the attraction between the coils and the iron. The net axial force is thus upward. The radial loads are transmitted between the outer (curved) leg and the inner (straight) leg through a direct cold link slicing through the iron pole piece, as shown in Figure 6, thereby avoiding large load bearing cold-to-warm links. The result is a very stiff structure, which minimizes strains within the coil structure, and in particular within the epoxy. The axial loads on the inner (straight) leg are accommodated by connecting the encased coils

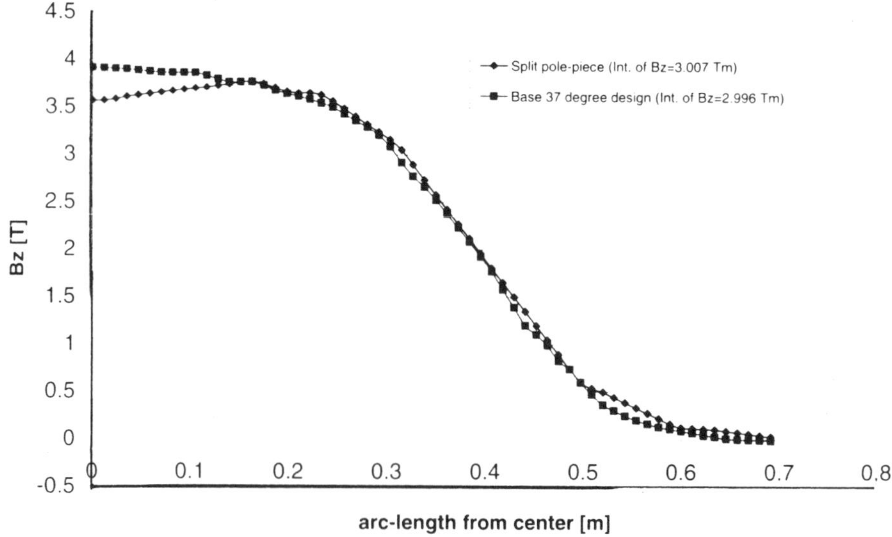

Figure 7. The change in magnetic length due to the cold link. This is the field integral in T-m.

directly through cold link members. In order to keep the median plane free on the outer leg, a similar structure was not used. The cold link between the inner and outer legs is capable of supporting most of the load generated by the axial forces on the outer legs. A G10 link connecting the outer leg of the (cold) case to the (warm) vacuum vessel is used to further reduce the strain levels. The reduction in magnetic field due to the cold link is small, as shown in Figure 7. Note the effective length is essentially unchanged. Loads due to the magnet weight and net in-plane forces (directed towards the iron mass) are supported by G10 links between the bobbin and the vacuum vessel, which is anchored to the iron.

CALCULATED HEAT LOADS

The device will be operated attached to a closed refrigeration system. It will be operated in a continuous fill mode, which minimizes the helium reservoir requirements. This allows a heat load budget of as much as 10 W. The thermal radiation load is currently estimated to be about 0.4 W. Numerous small cold-to-warm links are required to position the case within the vacuum vessel, and yield an estimated 1.5 W conduction load. The links are necessarily short due to the very restrictive space constraints.

SCHEDULE

The majority of the magnet system will be constructed by the NHMFL. The magnet will be assembled and tested there, before being shipped to the NSCL for installation. The NSCL is in the midst of a two year shut down for upgrading the facility. At the end of that time, the magnet will be installed and operated attached to the S800 Spectrograph.

ACKNOWLEDGMENT

This work is supported in part by a grant from the US National Science Foundation and from Michigan State and Florida State Universities.

REFERENCES

1. R. C. York, et al, "Proc. Of the Fourth EPAC", London, 1994, p. 554 and R. C. York, et al "Proc. Of the 1995 PAC", IEEE Press, Dallas, TX, p. 345.
2. J. Caggiano, Ph.D. Thesis, Michigan State University, 1999 (unpublished).
3. A. F. Zeller, et al, Adv in Cryo Eng 43A, 1998, p.245.
4. TOSCA ver. 6.3, Vector Fields, Inc. Aurora, IL

DESIGN OF AN APPARATUS FOR A 5T CONDUCTION COOLED NBTI SOLENOID WITH A 203 MM ROOM TEMPERATURE BORE

A. Rowe[1], J. A. Barclay[1] and S. Dost[2]

[1]Cryofuel Systems Group, University of Victoria
Victoria, B.C., V8W 3P6, CANADA
[2]Crystal Growth Research Group, University of Victoria
Victoria, B.C., V8W 3P6, CANADA

ABSTRACT

The design, fabrication, and operation of an apparatus to generate magnetic fields in a 203 mm diameter, room temperature, open bore are discussed. The device is used in experiments pertaining to the growth of crystals in high magnetic fields, examining the behaviour of novel magnetic materials, and as a test bed for various projects related to magnetic refrigeration. The field generator consists of an 840 mm x 540 mm high cylindrical chamber with a 203 mm diameter clear bore and is designed around a converted immersion cooled solenoid. The magnet is wound with NbTi superconductor and is conduction cooled by a single two-stage GM cryocooler. The design operating current is 362 Amps for a 5 T field.

INTRODUCTION

Until recently, the generation of high magnetic fields has traditionally been achieved through the use of immersion-cooled superconducting magnets. This type of apparatus is made up of a complex arrangement of vacuum housing, dewars, liquid nitrogen and liquid helium transfer lines, radiation shielding, structural supports and insulation to maintain the temperature of a magnet in the superconducting state with minimum boil-off. In the case of a typical immersion cooled solenoid made with NbTi wire, the transition temperature is up to 9.5 K and the magnet is designed with the expectation that it will be operating in a liquid helium environment of approximately 4.2 K. With the development of commercially available High Temperature Superconducting (HTS) leads and closed cycle cryocoolers that achieve substantial cooling powers at temperatures near 4 K, it is now possible to design superconducting magnet devices that are cooled conductively.[1-5]

Conduction cooling is achieved by thermally linking the magnet to the cold finger of a cryocooler via a solid medium and no longer requires liquid cryogens or as much associated ancillary equipment. Magnet systems designed to be conduction cooled have several features that are different from immersion cooled magnets as described in the recent literature. However, there are few reports of immersion cooled magnets that have been successfully conduction cooled. This paper describes our successful conversion of an existing 5 T, NbTi solenoid from being immersion cooled to conductively cooled.

BACKGROUND

The primary characteristic that motivated the field generator design was to provide up to 5 Tesla in a cylindrical region of space freely accessible from the laboratory environment. The bore has a diameter of approximately 203 mm (8 in.) and is 540 mm (21 in.) high. The cold-box fits through a standard door. A cylindrical furnace is inserted into the bore of the field generator and, within the furnace, ternary semiconductor crystals are grown in an 1133 K environment. The magnetic field and the furnace are vertical in direction. The entire apparatus is supported on a platform that allows access to the bottom and top of the furnace. By growing the crystals in a magnetic field the effects of gravity, as seen by convection currents in the melt, can be reduced.[6-8] It is expected that larger and more uniform growth will result. Future uses of the apparatus entail magnetic material characterization, magnetic refrigerator testing and related magnetic experiments.

APPARATUS DESIGN

The magnetic field is generated by a 250 mm I.D. pancake wound superconducting solenoid designed and built by Darien Magnetics. This magnet was salvaged from a liquid helium cooled device that had the clear access room temperature bore oriented in a horizontal direction. The immersion-cooled apparatus was disassembled and the solenoid removed. A picture of the bare solenoid is shown in Figure 1. The spool of the magnet is made of 304 SS and a layer of machined G-10 separates the spool flanges from the windings. The G-10 is approximately 6 mm thick with radial grooves machined to allow

Figure 1. Recovered solenoid from immersion cooled apparatus (left). Solenoid with addition of OFHC copper conduction plates and clamping (right).

Table 1. Solenoid specifications

Type	Pancake Wound Solenoid
Clear Bore	253 mm
Hmax/H(0,0)	1.43
Current at 5T	362.3 Amps
KG/Amp	0.138
I/Ic (B(0,0)=5T, 4.2K)	0.5
Inductance	5.5 H
Mass	136 Kg
Superconductor	Nb-48%Ti
Total Winding Length	4570 m
Number of filaments	2000
Cu/SC ratio	1.8:1
Twist Pitch	1 per 50 mm
Cu matrix RRR	180

for the circulation of liquid helium around the windings. The circumference of the winding is wrapped with aluminum wire. Specifications of the solenoid are listed in Table 1.

To convert to conduction cooling, the magnet was sandwiched between two 6 mm thick OFHC copper plates that act as thermal conductors. A thin layer of thermally conductive grease was spread between the spool flanges and the plates to ensure the contact area was as large as possible. The plates are held together by four aluminum clamps on the outer diameter, and the plate centres were located with stainless steel rings bolted to the magnet spool. The magnet was positioned by eight G-10 suspenders that attach to the clamps on the magnet and to tabs on an aluminum plate termed the strongback. The strongback was connected to the first stage of a 1W/4.2 K GM Sumitomo cryocooler. A copper radiation shield attaches to the strongback and surrounds the inside and outside of the magnet. An OFHC copper busbar is bolted to the second stage and acts as a thermal anchor point for leads and connections from the first stage.

Conventional current leads optimized for minimum heat leak with a current of 250 Amps connect room temperature feed-throughs to the warm end of the HTS leads. The aspect ratio was optimized for a lower current rating than that needed for rated field operation in order to facilitate testing, and also because the initial application of the field

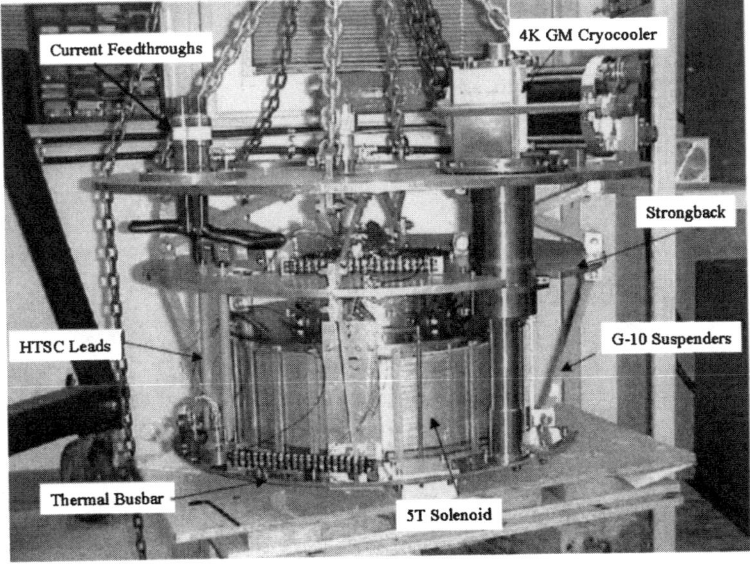

Figure 2. Partial magnet assembly.

Table 2. Calculated heat leaks assuming a charging rate of 0.08 Amps/sec

Heat Source	First Stage	Second Stage
Conduction		
Suspenders	0.3 W	20 mW
Leads	18 W	130 mW
Radiation	2.5 W	300 mW
Steady-State Load	20.8 W	450 mW
Joule Heating	30 W	50 mW
Eddy Currents (structure only)	negligible	280 mW
Transient Load	50.8 W	780 mW

generator is for a field of 2 Tesla. Two HTS leads rated at 500 Amps lead from the strongback to the busbar where the magnet leads are connected.

Thermal links from the second stage to the copper plates on the magnet were manufactured with flexible 0.4 mm thick OFHC copper foil. Ten layers of foil were soldered together to produce a connection with large cross-sectional area and minimum length. One connector attaches to each copper plate on the magnet. Figure 2 shows the magnet support assembly. The entire assembly is placed in a 304 SS vacuum chamber. Multilayer insulating blankets of aluminized mylar surround the magnet inside the radiation shield as well as outside the shield. Table 2 shows the calculated heat loads expected for steady-state operation, and while charging at a rate of 0.08 Amps/s. A diode is connected in parallel with the magnet to clamp the voltage in the event of a quench. The complete apparatus is shown in Figure 3.

TESTING

To date the magnet has been cooled down and tested without a persistent current switch which is the worst situation for cooling load. The maximum operating current has been limited to 185 Amps (2.5 Tesla) due to the effects of stray field, and also because the current leads are optimized for only 250 Amps. The apparatus is cooled without the aid of a thermal accelerator as is used on other conductively cooled devices.[1,2] The cool-down curve for the magnet is shown in Figure 4 and is typical of cooling curves presented by other authors for similar magnets.[1,2] Initially, the temperature decrease of the magnet is

Figure 3. Field Generator with furnace in the bore.

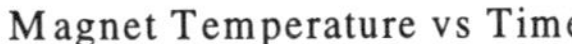

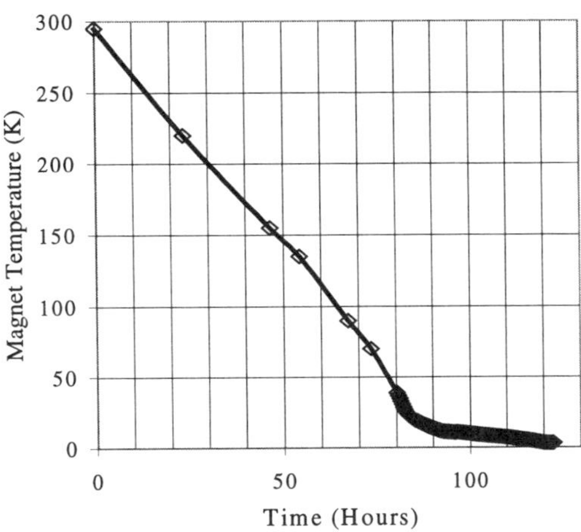

Figure 4. Magnet temperature during cool-down. No thermal switch is used.

approximately linear and then begins to accelerate at about 75 K. Below 20 K the rate of temperature decrease is slow. This curve may be explained in terms of changing thermal diffusivity and decreasing cooling capacity of the second stage.

Figures 5 and 6 show the response of the magnet when charged to 20 Amps at 0.02 Amps/s. Figure 5 are the results when held at 20 Amps and allowed to reach steady-state, while Figure 6 shows the magnet temperature for a ramp up and down without a pause. The magnet temperature is measured by a Cernox™ sensor in one of the cooling channels in the G-10 end plates and is in contact with the end of the windings. Although the degree of thermal contact between the windings and the sensor is difficult to verify, test results suggest that while the sensor may not accurately reflect winding temperature it does give useful data reflecting winding thermal response. In all figures the temperatures have been non-dimensionalized according to Eq. (1) where T_{min} is 3.69 K and T_{max} is 6.2 K,

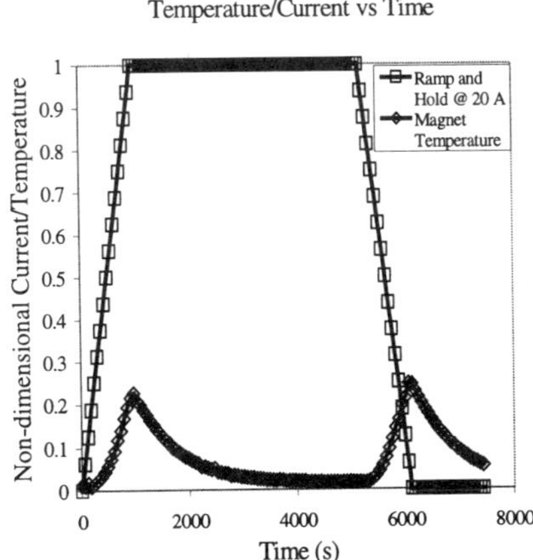

Figure 5. Magnet response when ramped to 20 Amps at 0.02 Amps/s and allowed to come to equilibrium.

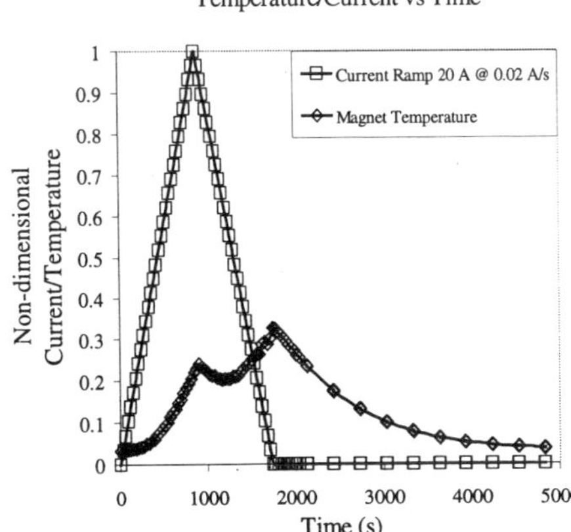

Figure 6. Magnet response when ramped up and down at 0.02 Amps/s to a peak of 20 Amps.

$$T^* \equiv \frac{T - T_{\min}}{T_{\max} - T_{\min}}. \tag{1}$$

(3.69 K and 6.2 K correspond to the minimum and maximum magnet temperatures measured during the experiments.) The current has been non-dimensionalized by the peak current that is most useful for comparing different runs (i.e. 20 Amps in Figures 5 and 6: 145 Amps in Figures 7-9). Figure 6 shows that although the joule heating effect decreases during the ramp down, the a.c. effects are significant enough to cause the magnet temperature to continue to rise even with a modest ramp rate of 0.02 Amps/s.

Figures 7 and 8 further highlight the effects of ramp duration (i.e. peak current) on time-varying losses and joule heating effects. These tests show the magnet and HTS cold-end temperatures for three cycles to 20 and 50 Amps respectively at 0.07 Amps/s. These figures clearly show the effect of ramp duration on hysteresis and eddy current heat generation. Once the cycling is complete, the magnet temperature decays slowly which is in accordance with what one would expect given the design of the solenoid. During testing it was noted that the copper plates on the magnet showed little change in temperature for currents less than 100 Amps, indicating that the windings of the magnet were well isolated thermally.

Charging characteristics for a ramp rate of 0.07 Amps/s and a peak current of 145 Amps (2 T) are shown in Figure 9 where the magnet was ramped up and down twice at a constant rate. The magnitude of joule heating has increased significantly. After 14 hours of operation at 145 Amps the magnet non-dimensional temperature decayed to 0.36 (4.59 K) while the HTS steady-state temperature rose slightly to 0.59 (5.17 K). It is interesting to note from Figures 7-9 that while there is a large increase in magnet temperature when the peak current is changed from 20 to 50 Amps, the magnet temperature for two cycles to 145 Amps is only slightly higher than the ramps to 50 Amps. Although increasing temperature may enhance the rate of heat transfer, this reduced temperature rise is better explained by the rapidly increasing heat capacity of the magnet. For a typical epoxy impregnated NbTi winding, the volumetric enthalpy increases by nearly an order of magnitude when the temperature is increased from 4 to 6 K.[9]

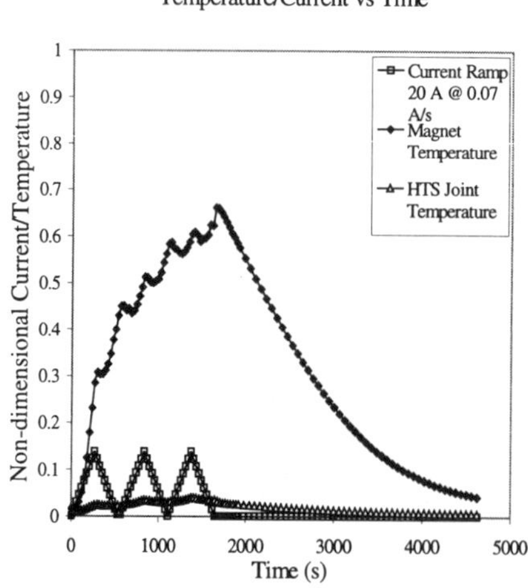

Figure 7. Magnet response to 3 cycles of ramping to 20 Amps at 0.07 Amps/s. (Current of 1 = 145 Amps.)

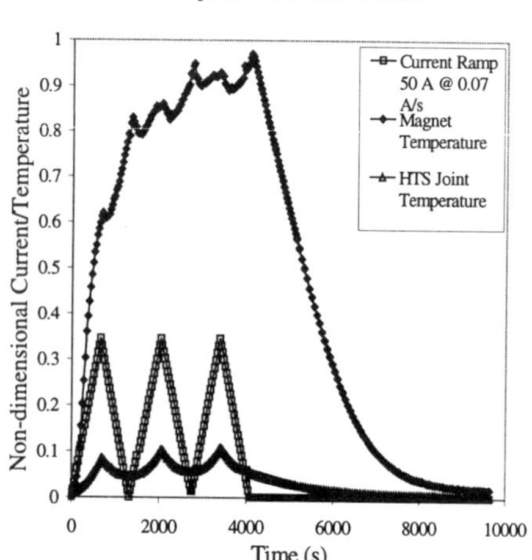

Figure 8. Magnet response to 3 cycles of ramping to 50 Amps at 0.07 Amps/s. (Current of 1=145 Amps.)

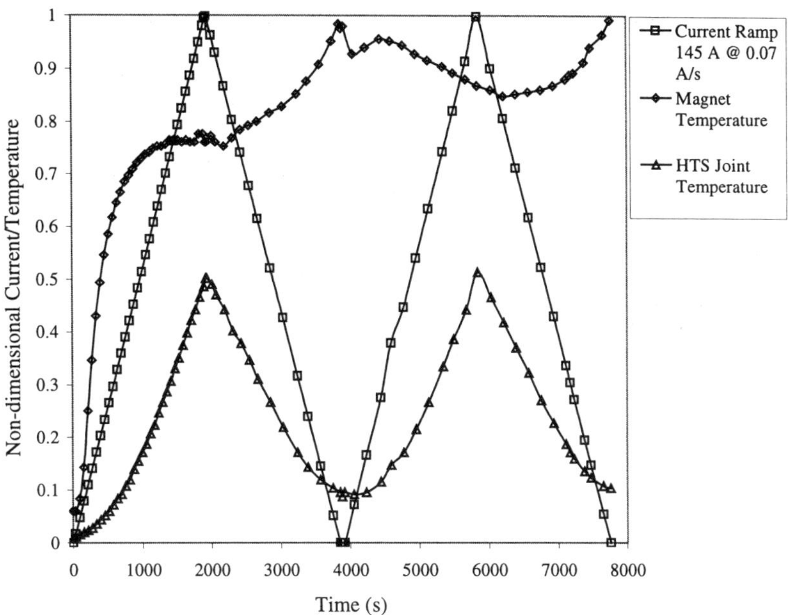

Figure 9. Magnet response when cycled twice to 145 Amps at 0.07 Amps/s.

CONCLUSION

An apparatus to produce magnetic fields of up to 5 Tesla in a 203 mm room-temperature open bore has been designed and tested. The device uses a NbTi superconducting magnet cooled by a 4.2 K/ 1 Watt GM cryocooler. The solenoid was previously part of an immersion-cooled device. The design operating current is 362 Amps for 5 T; however, to date testing of the magnet has used a maximum of 185 Amps producing a field of 2.5 T. Test results show that the effects of hysteresis and eddy-current heat generation significantly influence the magnet temperature even with moderate charging rates. As one would expect, given the magnet construction, the thermal diffusivity of the magnet is low thereby restricting the charging rate and ramp shape. Results indicate that it is possible to make effective use of solenoids previously designed for immersion cooling in a conductively cooled mode.

ACKNOWLEDGEMENT

The authors are grateful for the support provided by the Natural Sciences and Engineering Research Council (NSERC) of Canada.

REFERENCES

1. T. Hasebe, J. Sakuraba, K. Jikihara, K. Watazawa, H. Mitsubori, Y. Sugizaki, H. Okubo, Y. Yamada, S. Awaji, and K. Watanabe, Cryocooler cooled superconducting magnets and their applications, in: "Advances in Cryogenic Engineering," Plenum Press, New York (1998), 43:291.
2. K. Shibutani, S. Itoh, O. Ozaki, T. Takagi, T. Miyazaki, R. Hirose, S. Hayashi, M. Shimada, R. Ogawa, Y. Kawate, K. Matsumoto, N. Kimura, and K. Takabatake, Development of two types of cryogen free superconducting magnet, in: "Advances in Cryogenic Engineering," Plenum Press, New York (1998), 43:299.

3. K Watanabe, S. Awaji, J. Sakuraba, K. Jikihara, K. Watazawa, T. Hasebe, F. Hata, C.K. Chong, Y. Yamada, and M. Ishihara, Cryogen-free split-pair superconducting magnet with a 50 mm x 10 mm room temperature gap, in: "Advances in Cryogenic Engineering," Plenum Press, New York (1996), 41:319.
4. T. Kuriyama, M. Urata, T. Yazawa, K. Yamamoto, Y. Ohtani, K. Koyanagi, T. Masegi, Y. Yamada, S. Nomura, H. Maeda, H. Nakagome and O. Horigami, Cryocooler directly cooled 6T NbTi superconducting magnet system with 180 mm room temperature bore, *Cryogenics*, 34:643 (1995).
5. M. Takahahi, R. Hakamada, K. Yamamoto, T. Kuriyama, H. Nakagome, S. Masuyama, H. Yamamoto, S. Tanaka and Y. Matsubara, A 7.7 T NbTi superconducting magnet system cooled by a 4 K GM refrigerator, in: "Advances in Cryogenic Engineering," Plenum Press, New York (1994), 39:343.
6. H. Sheibani, S. Dost, LPEE growth of semiconductors under external magnetic field, to appear in Proc. of CANCAM'99, p.405 (1999).
7. S. Dost, Numerical simulation of liquid phase electroepitaxial growth of GaInAs under magnetic field, ARI the Bulletin of ITU (in press).
8. S. Dost and H. Sheibani, LPEE growth of ternary GaInAs crystals under external magnetic field, Proc. ASME Mech. and Materials Conference, p.152 (1999).
9. M. Wilson. "Superconducting Magnets," Oxford University Press, New York (1983).

BENT SOLENOID SIMULATIONS FOR THE MUON COOLING EXPERIMENT

M. A. Green[1], Y. M. Eyssa[2], S. Kenney[2], J. R. Miller[2] and S. Prestemon[2]

1. E. O. Lawrence Berkeley National Laboratory
Berkeley, CA 94720

2. National High Magnetic Field Laboratory
Tallahassee FL 32310

ABSTRACT

The muon collider captures pions using solenoidal fields. The pion are converted to muons as they are bunched in an RF phase rotation system. Solenoids are used to focus the muons as their emitance is reduced during cooling. Bent solenoids are used to sort muons by momentum. This report describes a bent solenoid system that is part of a proposed muon cooling experiment. The superconducting solenoid described in this report consists of a straight solenoid that is 1.8 m long, a bent solenoid that is 1.0 m to 1.3 m long and a second straight solenoid that is 2.6 m long. The bent solenoid bends the muons over an angle of 57.3 degrees (1 radian). The bent solenoid has a minor coil radius (to the center of the coil) that is 0.24 m and a major radius (of the solenoid axis) of 1.0 m. The central induction along the axis is 3.0 T There is a dipole that generates an induction of 0.51 T, perpendicular to the plane of the bend, when the induction on the bent solenoid axis is 3.0 T.

INTRODUCTION

The proposed muon cooling experiment[1,2] consists of a pair of S shaped solenoid bends each of which has two bent solenoids, and straight solenoids for four TPC detectors and an 805 MHz RF cavity[3]. Between the two S shaped bend sections is a straight muon cooling section. The energy and momentum of the muon beam are analyzed by the S shaped bent solenoid systems before and after the straight solenoidal muon cooling section. A schematic representation the muon cooling experiment is shown in Figure 1.

After the beam goes through first S shaped solenoid for energy and momentum analysis, the muon beam enters a muon cooling section that is about 15 meter long. The cooling section consists of ten alternating solenoids with a peak induction of 15 T. The muons are cooled by a liquid hydrogen section that is about 400 mm long and 70 mm in diameter. After the muon momentum has been reduced by about 25 MeV/c, the muon are re-accelerated by a 900 mm long 805 MHz RF cavity section back to their original energy of about 180 MeV. The second S shaped solenoid section analyzes the energy and momentum of the muons after they have gone through cooling.

The bent solenoid separates muons by momentum spatially across the bore of the solenoid following the bent solenoid. A coil that generates a dipole field in the bent solenoid moves the momentum separated muons so that their momentum separation is distributed around the center of the solenoid following the bent solenoid.

Advances in Cryogenic Engineering, Volume 45.
Edited by Shu *et al.*, Kluwer Academic / Plenum Publishers, 2000.

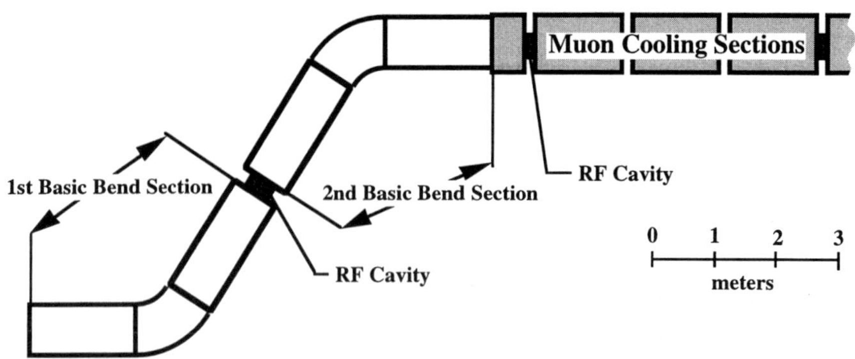

Fig. 1 The S Bend Bent Solenoid System at the Start of the Muon Cooling Channel

THE BENT SOLENOID SYSTEM DESIGN

The following design assumptions were used for the preliminary design of the superconducting bent solenoid system: 1) Magnetic flux is conserved in the solenoids. Magnetic flux conservation means that all of the solenoid coils have the same average current radius. This minimizes the leakage flux into the TPC detectors. 2) The warm bore diameter for the for the two TPC solenoids and the bent solenoid need only be about 320 mm. The solenoid around the RF cavity must have a warm bore diameter of about 440 mm. 3) The solenoid bend angle used for this study is 57.3 degrees (1 radian). The exact bend angle required for the muon cooling experiment will depend on the average momentum of the muons cooled and the muon momentum spread that will be measured. 4) The average bend radius for the bent solenoid axis will be 1000 mm. Simulations of particle motion within the solenoid suggest that the bend radius should be variable with a longer radius at the beginning and end of the bend and a shorter radius for the center of the bend. The variable bend radius may improve the efficiency of the transfer of the muons through the solenoid. 5) The bent solenoid induction on axis will be 3.0 T. 6) In order to have momentum separation at the center of the solenoid, a bent solenoid dipole is of 0.51 T is needed on the axis. The on axis dipole field can be provided using separate dipole winding or by tilting the bent solenoid coils. 7) Standard MRI superconductor has been assumed for the solenoid windings. The solenoids are hooked together electrically in series. 8) In the TPC, the integral of the r component of field over the integral of the z component of field is less than 0.0002 within the TPC active volume. The TPC active volume is assumed to be 160 mm in diameter by 500 mm long near the center of the TPC solenoid. 9) The TPC cables are fed out of the solenoid through room temperature slots between coils that are located at the ends of the TPCs away from the bent solenoid. 10) The two TPC solenoids, the bent solenoid and the RF solenoid share a common cryostat vacuum vessel. The two straight solenoids and the bent solenoid are assumed to have separate cold mass support systems. The cryostat ends at the center of the RF cavity so that the RF wave guide can enter the cavity between the solenoids. 11) The superconducting coils are assumed to be cooled by conduction to a system of pipes attached to the outside of the magnet, which carry two-phase helium.

The first basic bend section shown in Figure 1 is shown in Figure 2. This section goes from the first TPC solenoid to the solenoid that covers half of the RF cavity. The average coil diameter for the magnets shown in Figure 2 is 480 mm. The smallest coil diameter in the TPC magnet section is about 450 mm. The coils around the RF coil clear the warm bore of the cryostat by about 10 mm. The coils around the RF cavities are located inside of an aluminum support structure that has the helium cooling tube attached to it. The bent solenoid shown in Figure 2 is a constant bending radius solenoid. The dipole coils for the bent solenoid are not shown in Fig. 2. The solenoidal coils parameters are shown in Table 1. All of the solenoidal coil shown in Fig. 2 and described in Table 1 are designed to be powered by a single power supply. The total stored energy in the string shown in Fig. 2 is 3.06 MJ. The high current density magnet coils (163.3 A per square mm) would require that the coils be subdivided so that they can be protected by cold diodes and resistors.

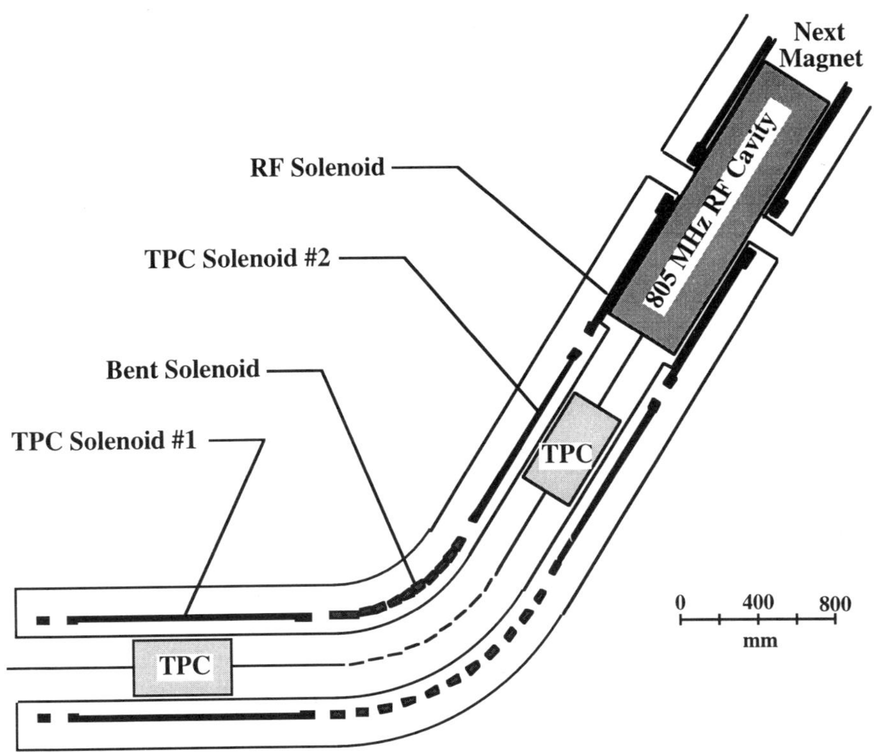

Fig. 2 Cross-section of One Bend Magnet Section for the muon Cooling Experiment Showing the Locations of the TPC Detectors and the 805 MHz RF Cavity

Table 1 Parameters for the Solenoids in the Bent Solenoid Assembly Shown in Figure 2

	TPC # 1	Bent Sol.	TPC #2	RF Sol.
Warm Bore Diameter (mm)	320	320	320	440
Cryostat Outer Diameter (mm)	800	800	800	800
Cryostat Section Length on axis (m)	1.6	1.2	1.1	0.8
Number of Coils	2	11	1	1
Coil Average Diameter (mm)	480	480	480	480
Coil Thickness (center) (mm)	17.9	25.7	17.9	17.9
Coil Thickness (ends) (mm)	34.8	NA	34.8	34.8
Center Coil Length (mm)	1100	50	870	550
End Coil Length (mm)	50	NA	50	50 & 100
Number of layers (center)	16	31	16	16
Number of Layers (ends)	32	NA	32	32
Number of Turns per Layer Center	686	30	544	344
Number of Turns per Layer Ends	31	NA	31	31 & 62
Central Induction on axis (T)	3.0	3.0	3.0	3.0
Coil Design Current (A)	257.3	257.3	257.3	257.3
Coil Section Stored Energy (MJ)	1.04	0.78	0.72	0.52
Coil Section Inductance (H)	31.4	23.7	21.7	15.7
Conductor Current Density (A mm^{-2})	163.3	163.3	163.3	163.3
EJ (10^{22} J A^2 m^{-4})*	2.8	2.1	1.9	1.4

* Quench protection must be supplied to each magnet. Cold diode and resistor quench protection is attractive for these magnets. A separate power supply can be used for each string of magnets. HTS current leads could be used.

THE BENT SOLENOID THE ASSOCIATED DIPOLE

The bent solenoid is used to separate charged particles by momenta. A bent solenoid will cause the charged particles to be separated along a line away from the solenoid axis. It is desirable to have the separated particles centered on the solenoid axis. A dipole field that perpendicular to the bend plane is used to center the separated charged particles in the solenoid at the end of the bend. The magnitude of this dipole field is proportional to the axial field at the center of the bent solenoid, the particle charge, and the mean charged particle momentum. An expression for the required dipole induction on the bend solenoid axis, perpendicular to the plane of the bend can be given as follows:

$$B_d = G(p_c,q)\frac{B_s}{r_s} \tag{1}$$

where B_d is the dipole induction required to center the separated charged particles on the solenoid axis; B_s is the on axis solenoidal induction; r_s is the local bend radius (given in SI units) for the bent solenoid; and $G(p_c,q)$ is the fitting parameter that is a function of the mean particle momentum pc and the particle charge q. For a muon with a charge $q = 1$ or $q = -1$ and a mean momentum p_c of 180 MeV/c, the value of $G(p_c,q)$ is about 0.17.

A dipole field can be generated using a cosine theta distribution of currents flowing in a direction parallel to the solenoid axis, on the surface of the solenoid. The highest current (where cosine theta = 1) must be on the bend plane that includes the bent solenoid axis. This distribution of current will produce a perfect dipole superimposed on the solenoidal field if a cosine theta distribution of current is put around a straight solenoid. When the solenoid is bent, a cosine theta current distribution will no longer produce a perfect dipole. The magnitude of the dipole will vary as r over r_s. Near the bent solenoid axis one will have a dipole plus a quadrupole field in the direction perpendicular to the plane of the solenoid bend. Studies indicate that one can alter the cosine theta distribution of the current only a few degrees to achieve a pure dipole in a bent solenoid.

Particle tracking studies were done with a dipole superimposed on a bent solenoidal field. The greatest efficiency of particle transport through the bent solenoid occurred when the impressed dipole had a small amount of quadrupole added to the dipole field in the bent solenoid. The amount of this quadrupole term is between the quadrupole produced when a cosine theta distribution of currents is used to generate the dipole and no quadrupole at all (the perfect bent dipole case). This suggests that the bent solenoid dipole winding may be more complicated than originally thought.

The original bent solenoid studies were done using bent solenoids with a constant bend radius. Particle tracking studies show that the efficiency of particle transmission through the solenoid can be improved considerably by varying the radius of curvature in the solenoid so that the kick in the particles in the horizontal direction increases adiabatically as the particle enters the bent solenoid and decreases adiabatically as the particles leave the bent solenoid. This also means that the impressed dipole B_d must vary along the length of the bent solenoid. This further complicates the design of the bend solenoid and the dipole needed to center the particles spread by momentum.

One way to produce the dipole in a bent solenoid with a variable bend is to tilt the coils in the bent solenoid. Tilting the coils a small angle will produce a dipole field that will vary as r over rs. A small angle tilt will cause the on axis solenoidal field to be reduced as the cosine of the tilt angle Ω. This reduction can be compensated for by increasing the current in the coils that make up the bent solenoid. The tilt angle Ω can be estimated using the following expression:

$$\Omega = \sin^{-1}\left[\frac{B_d}{B_s}\right] \tag{2}$$

where Bd and Bs have been previously defined. The only extra windings needed are quadrupole windings used to tune the bent solenoid for maximum through put of charged particles. Figure 3 shows a cross-section of a bent solenoid with a variable bend radius. The variable bend sections are coils 1 through 5 and 11 through 15. The first and fifteenth coils have the same axis as the adjacent solenoids. The first five coils and last five coils each produce a total bend of 11.46 degrees. The middle five coils bend of 34.38 degrees.

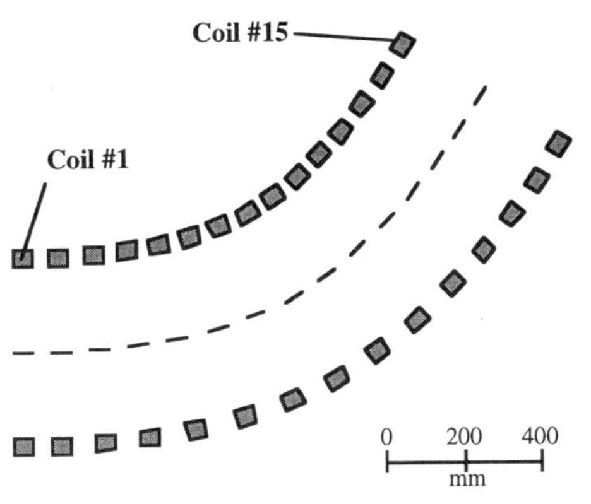

Number of Coils = 15
Total Bend Angle = 57.3 degrees
Coil Average Radius = 240 mm
Coil Radial Thickness = 32.7 mm
Coil Physical Length = 56 mm
Inside Coil Spacing = 17.5 mm
Outside Coil Spacing = 70.5 mm
Number of Coil Layers = 31
Number of Turns per Layer = 30
Number of Turns per Coil = 930
Design Current = 256.7 A
Design on axis Induction = 3.0 T
Bend Length on Axis = 1300 mm
Maximum Bend Radius = 1000 mm

Fig. 3 A Cross-section View of a 57.3 Degree Bend Magnet Section with Variable Bend Radii
Coils 1 through 5 and 11 through 15 Variable, Coils 6 through 10 Bend R = 1000 mm

The cross-section shown in Figure 3 is in the plane of the bend of the bent solenoid. Each of the 15 coils shown in Figure 3 has the same cross-section and the same number of turns. The table next to Figure 3 is a parameter list for the coils and the variable bend radius bent solenoid. Each coil is assumed to be wound using a 0.955 x 1.65 mm MRI superconductor with 155 filaments about 51 microns in diameter in a copper matrix. The copper to superconductor ratio is 4. This conductor is designed to carry a minimum current of 760 A at 5 T and 4.22 K. The Jc at 5 T and 4.22 K is at least 2400 A mm^{-2}. There is enough margin in the coil to permit the central induction of the bent solenoid to be increased to over 4 T and still have a current margin of 85 percent of critical current along the load line

The bent solenoids shown in Figure 2 and 3 can be built using identical coil packages. The bend and tilt for the solenoid coils is achieved by machined spacers between the standard coil packages. These spacers can be welded to the coil winding forms. The coils can be wound on aluminum winding forms and have aluminum spacers between the coils. Once the coils are assembled into a bent solenoid and welded, the coils can be connected together in series. The spacers allow one to vary the bend radius and tilt angle of the coils to produce the needed dipole field as well as the solenoidal field. Table 2 below shows the parameters of the variable bend radius solenoid shown in figure 3. The dipole is produced by tilting the superconducting coil packages. Table 2 shows the dipole needed in various parts of the bent solenoid shown in Figure 3 along with the coil tilt angle needed to produce that dipole field. At a current of 256.7 A, the solenoid produces and integrated solenoid induction of 3.859 Tm with an average solenoidal induction of 2.969 T. An increase of the bent solenoid current by 1.06 percent will bring the average solenoidal field up to 3.0 T.

Table 2 Parameters for the Variable Bend Angle Bent Solenoid Shown in Figure 3

	Bend R (mm)	Coil Bend Angle (Degrees)	Dipole B (T)	Coil Tilt Angle (Degrees)
Coils 1 and 15*	infinite	0.00	0.000	0.00
Coils 2 and 14	5000	1.15	0.102	1.95
Coils 3 and 13	2500	2.29	0.204	3.90
Coils 4 and 12	1667	3.44	0.306	5.85
Coils 5 and 11	1250	4.58	0.408	7.82
Coils 6 through 10	1000	5.73	0.510	9.79

* Coils 1 and 15 are part of the straight solenoids before and after the bent solenoid.

BENT SOLENOID FIELD UNIFORMITY STUDIES

A study was done to see what effect the bent solenoid would have on the field uniformity within the straight solenoids where the TPC are to be installed. The integral of the r component of field over the length of the TPC over the length of the axial component of field over the length of the TPC should be less than a few parts in 10000. This field uniformity should be occur within the 150 mm in diameter by 400 mm long active volume of the TPC. The following can contribute to the field errors in the two TPC detectors: 1) the holes in the coil were the TPC electronic cables pass out of the cryostat, 2) the field error due to the field from the bent solenoid and 3) the field non uniformity due to the dipole in the bent solenoid. The field uniformity along the axis of the bent solenoid is nominally set to be better than one part in 1000. The off axis solenoidal field is nominally proportional to 1 over R from the center of the solenoid bend. At 75 mm from the axis, on the inside of the bend, the induction is 3.243 T. On the outside of the bend, 75 mm from the solenoid axis, the induction is 2.791 T. In the non bent sections of the solenoid, the solenoid field is supposed to be uniform across the solenoid.

A computer model of the TPC magnets and the bent solenoid was made. The model has the following characteristics: 1) TPC solenoid and the bent solenoid have an average coil diameter of 400 mm. The coils were assumed to be 20 mm thick sheets. 2) The bent solenoid had a constant radius bend of 1000 mm, and it consists of ten straight coils that are 80 mm long. On the inner part of the bend the ten straight coils touch each other and on the outer part of the bend (in the plane of the bend) the coils are separated by a gap of 40 mm. 3) The dipole is located outside of the bent solenoid. The current distribution in the dipole coils is a skewed cosine theta current distribution that produces a perfect dipole at the center of the bent solenoid. The point of zero current is moved a few degrees toward the outside of the solenoid bend. The ends of the dipole make a simple up and over curve over the ends of the bent solenoid, so that the ends of the bent dipole correspond to the ends of the bent solenoid. Figure 4 is a plot of field error defined as 1-BR/BoRo on axis and on curves ± 75 mm from the axis on the plane of the bend. R is the radius of the line from the bending axis and Ro is the radius of curvature of the bend. B is the calculated induction; and Bo = 3.00 T. Within the straight solenoid sections, R over Ro is assumed to be one.

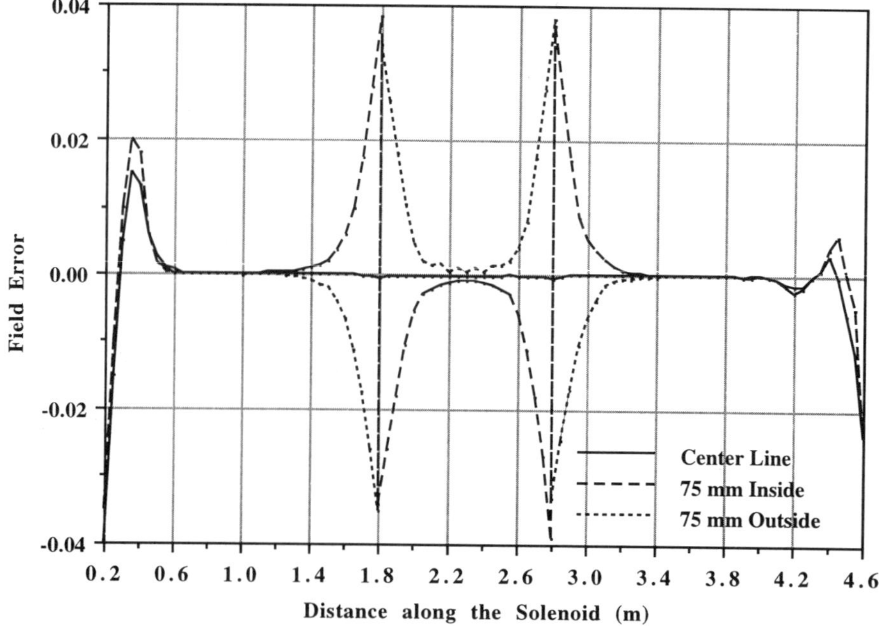

Fig. 4 Field Uniformity Versus Distance along the Solenoid and Distance from the Solenoid Axis in the Bend plane (Note: The field uniformity is corrected for radius in the bent solenoid.)

The two regions of near zero field error in the solenoid between z = 0.7 and z = 1.2 m and between z = 3.3 m and z = 3.9 m are the regions where the TPC detectors would be located. Within these regions, the maximum field error is ±0.0002 (2 parts in 10000), which is good enough for TPC detectors. The integrated field error within both TPC detectors is less than ±0.0001. The field error excursions at z < 0.6 m and z > 4.1 m are due to field fall off at the physical ends of the straight solenoids.

The bent solenoid extends from z = 1.8 m to z = 2.8 m in the model studied. At the ends of the bent solenoid, the field error approaches ±4 percent. The large excursions of the field error curves shown in Figure 4 are caused by the round ends of the bend dipole, that is mounted on the outside of the bent solenoid. The ends of the dipole cause the largest field errors observed in the study. In the central region of the bent solenoid, one can see that the field error curve for 75 mm inside of the solenoid axis is rather smooth whereas the field error curve for 75 mm outside of the axis has a number of oscillations with a period of about 0.1 m. The amplitude of the field error oscillations is less than 0.001 peak to peak. The oscillations are due to the 40 mm gap between the coils on the outside of the bent solenoid. The field error curve inside the axis of the solenoid does not exhibit these oscillations because there are no gaps between the coils.

Moving the average current radius from 200 mm to 240 mm from the solenoid axis will improve the field uniformity (maybe a factor of two) both in the TPC detectors and in the bent solenoid. Using a variable bend radius bent solenoid, with tilted coils to produce the dipole, should also improve the field uniformity in the bent solenoid. The quadrupole windings needed to correct the dipole generated by the tilted coils should produce a far smaller field error than the separate dipole windings used in the study. The end effect due to the quadrupole windings will be shorter. One should be able to reduce the field errors in the bent solenoid by over an order of magnitude by using a combination of a larger diameter solenoid, a variable bend radius, and tilted coils to produce the dipole.

CONCLUSION

A combination of two straight solenoids on the ends of a bent solenoid was studied. The bent solenoid can be fabricated from a number of identical straight superconducting coils. The bent solenoid can be fabricated so that its bend radius can be variable. A bent solenoid must also have a dipole to move the center of the particle momentum spread to the center of the solenoid following the bend. This dipole can be provided by a separate winding, or it can be provided by tilting the bent solenoid coils. If this dipole is produced by tilting the solenoid coils, separate quadrupole windings must be used. Field errors studies indicate that the field in a straight solenoid near a bent solenoid can be made good to a few parts in 10000, which is suitable for a TPC detector. The magnitude of the field error due to individual bent solenoid coils can be made to be less than one part in 1000 in a region that is 150 mm in diameter within a 400 mm diameter bent solenoid. It appears that the field uniformity can be improved by using larger solenoid coils, a variable bend radius in the bent solenoid and tilted bent solenoid coils to produce the dipole.

ACKNOWLEDGMENTS

This work was supported by the Director of the Office of Basic Energy Science, High Energy Physics Division, United States Department of Energy under contract number DE-AC03-76SF00098.

REFERENCES

1. The Muon Collider Collaboration, "Status of Muon Collider Research, Development and Future Plans," BNL-65-623, Fermeilab-PUB-98/179 and LBNL-41935 (1998)
2. K. T. McDonald, "Muon Colliders: Status of R&D and Future Plans," (This Proceedings) the Proceedings of the 1999 Particle Accelerator Conference, New York, NY 29 March to 2 April 1999, IEEE Publications (1999)
3. S. A. Kahn, H. Guler, C. Lu, K. T. McDonald, E. Prebys and S. E. Vahsen, "The Instrumentation Channel for the MUCOOL Experiment," (This Proceedings) the Proceedings of the 1999 Particle Accelerator Conference, New York, NY 29 March to 2 April 1999, IEEE Publications (1999)

A CRYOCOOLER DIRECTLY COOLED 5 TESLA NIOBIUM TITANIUM SUPERCONDUCTING MAGNET SYSTEM

N. H. Song[1], L. Z. Lin[1], L. Zhang[2], X .D. Xu[2], T. C. Wang[3], Y. R. Cai[3], B. Q. Fu[3], Y. Feng[3]

[1]Institute of Electrical Engineering
Chinese Academy of Sciences, Beijing 100080, China

[2]Cryogenic Laboratory
Chinese Academy of Sciences, Beijing 100080, China

[3]Northwest Institute for Nonferrous Metal Research, Xi'an 710016, China

ABSTRACT

A 5 Tesla Niobium Titanium magnet system directly cooled by a 4K Gifford-McMahon cryocooler was made. The 4K cryocooler with a capacity of 0.5W at 4.2K is used with Bi2223 high temperature superconductor current leads. The Niobium Titanium multifilamentary wire was fabricated in China. The magnet, with an inner diameter of 60 mm, outer diameter of 100mm, height of 172mm, achieves a central field of 5 T at a current of 98A. Powdered Aluminum Nitride was mixed in the impregnation epoxy to increase the epoxy thermal conductivity. The current leads were thermally anchored at the first and second stage through Aluminum Nitride spacers to obtain high thermal conductivity and electrical insulation. Liquid nitrogen was used to shorten the cool down time for the magnet system, to reduce the total operation time of the cryocooler and to compensate the lack of cooling capacity of the first stage of the cryocooler. The details of the system and test results are presented.

INTRODUCTION

Recently, GM cryocooler-cooled superconducting magnet systems are being developed quite rapidly in some countries.[1,2,3,4,5,6] The qualities of easy operation and compactness are important factors for the final choice of the end users. This kind of superconducting magnet system is expected to be accepted by industrial users as well as researchers, especially in

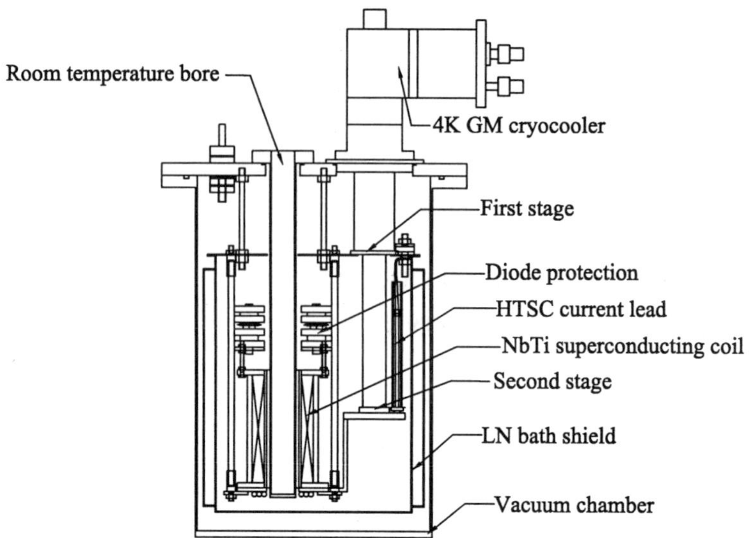

Figure 1. Cross section of the magnet system.

China where the problems in cryogenic engineering associated with liquid helium impedes the utilization of superconducting magnet systems. This kind of superconducting magnet system will be an aid to the technological development of China.

In this paper we describe the design and the test results of a cryocooler cooled 5 T superconducting magnet system.

DESIGN AND CONSTRUCTION

Components of system

Our magnet system consists of a 5 T NbTi superconducting coil, protection diodes for the coil, Bi2223 HTSC current leads, 4K GM Cryocooler, thermal radiation shield cooled by liquid nitrogen, vacuum chamber, power source and measurement equipment. The parts in the vacuum chamber (components at low temperature) are shown in Figure 1.

NbTi Coil Specifications

The coil, 60 mm inner diameter, 100 mm outer diameter and height of 172 mm, is designed to generate 5 T center field at the normal operating current of 98.15 A (corresponding an overall current density of 220 A/mm^2). The calculated field uniformity of 10 mm diameter sphere at the center is better than 0.1%.

Multifilamentary NbTi superconducting wire is used in the coil winding. The wire parameters are 0.65 mm diameter, 18 μm filament diameter, 1.5:1 Cu/NbTi volume ratio and a composition of Nb-50.0wt%Ti. A weight of 5 kg of NbTi wire is used in the coil. Total inductance is 1.52 H.

The field dependence of the critical currents at various temperatures are obtained by using the reduced critical state method.[7] Figure 2 shows critical current (I_c) versus magnetic field (B) curves estimated at every 1.0 K step from 4.0 to 7.0 K including the measured data at 4.22 K. The load line of the coil is also shown in Figure 2. The critical temperature of the NbTi coil is estimated at 5.8K for 5 T central field.

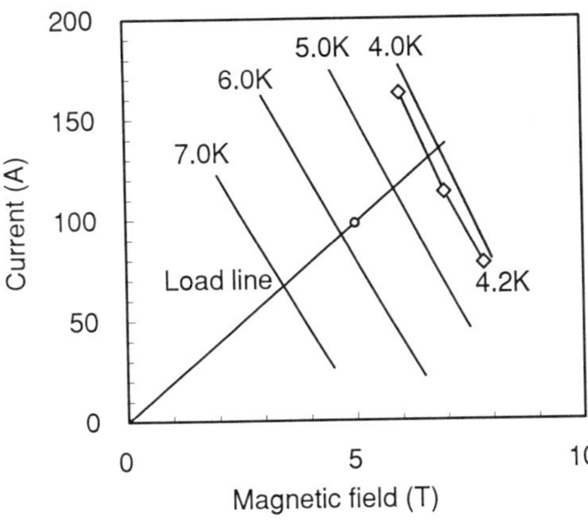

Figure 2. Critical current versus magnetic field at various temperatures for NbTi wire and the load line.

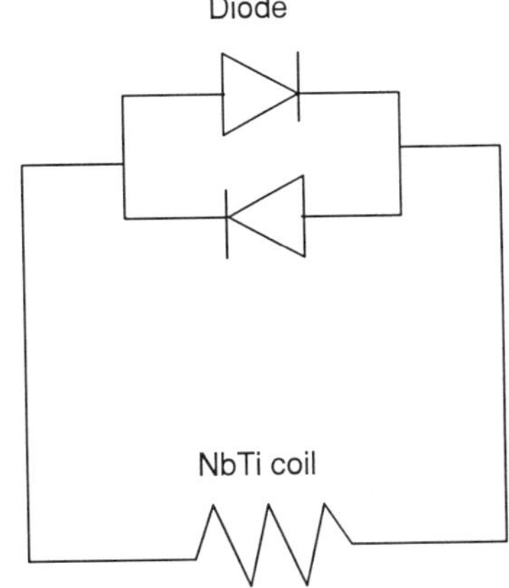

Figure 3. The protection circuit of the coil.

Coil Fabrication and Cooling Structure

The coil was wound on a bronze mandrel and flanges with the wet method using epoxy. The epoxy was mixed with aluminum nitride (AlN) powders to increase the epoxy thermal conductivity. The coil is covered with a pair of split cylindrical copper plates and radially compressed by the outer stainless steel binding wire to keep the thermal contact to the coil winding. The coil is cooled by thermal conduction through the copper outer cylinders.[8,9]

At the bottom flange of the magnet a copper tube heat exchanger was installed in order to cool the magnet from room temperature to liquid nitrogen temperature during the initial cool down.

The coil was tested in a liquid helium bath for training. The highest quench current 132 A (6.75 T) reached 100% of the critical current of the NbTi wire measured from the short sample test.

Coil protection

A diode shunt circuit is mounted at the 4 K stage to protect the coil shown in the Figure 3. The stored energy of the coil is 7.5 kJ at 5 T, which is not large. When a coil quench occurs, the stored energy is absorbed adiabatically by the magnet mass and the protection diodes only. The magnet mass, including the NbTi coil, bobbin, heat transfer copper plates, is about 16 kg. The estimated average magnet temperature is about 36 K after quenched at 5 T if the stored energy is entirely absorbed by the magnet. The maximum temperature in the coil during the current decay following a quench is about 64 K using the analysis method in the reference of Martin N. Wilson.[10] A diode shunt circuit without resistor will protect the coil and the temperature rise is quite low after quenches. In order to protect the coil should the high temperature superconducting current leads burn out, we choose to install the diodes at the 4 K second stage instead of the first stage at which the cooling power is larger. This is a correct decision because we experienced HTS current lead burn out once during the experiments.

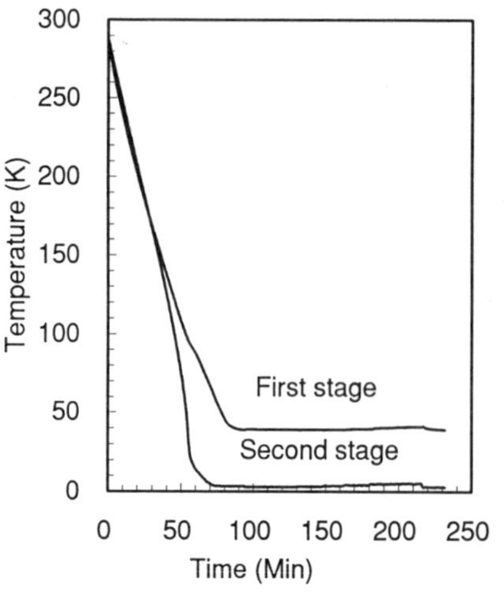

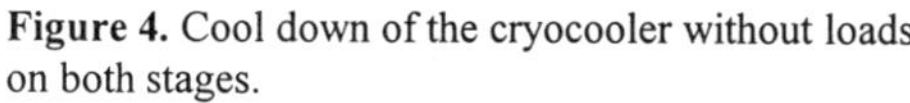

Figure 4. Cool down of the cryocooler without loads on both stages.

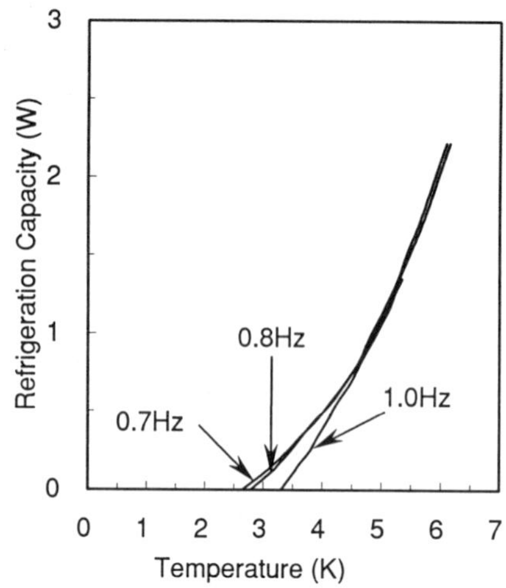

Figure 5. Refrigeration capacity at various frequencies.

4K-GM Cryocooler

The cryocooler is a double stage 4K Gifford McMahon cryocooler. The hybrid regenerators of magnetic material of Er_3Ni and Pb spheres are used as the second stage regenerator material. Copper meshes are used for the first stage. The refrigeration power for the second stage is about 0.58 W at 4.2 K under almost no heat load of the first stage at 40 K. Figure 4 shows the cool down of the cryocooler without loads on both stages. Figure 5 is the refrigeration capacity at different operation frequencies. The first stage temperature is expected to be about 70 K with a heat load of 15 W at the first stage. For this reason, liquid nitrogen was used to cool the shield to a bath temperature of 77 K. Liquid nitrogen was also used for the initial cool down of the coil to reduce the total operation time of the cryocooler.

HTSC current leads

$Bi_2Sr_2Ca_2Cu_3O_{10}$ oxide superconducting current leads are used for reducing the heat leakage into the 4 K stage. These bulk rods of 10 mm diameter, 120 mm length, were fabricated by the multiple cold isostatic process (MCIP) method. The process was that the Bi2223 powder was formed into rods or tubes by cold isostatic pressing method, and the formed materials was sintered under certain conditions. Then the bulk materials were cold pressed and sintered again to obtain dense and textured structure. By using MCIP, high quality Bi2223 bulk rods could be obtained with quite good shape, high density and smooth surface. In addition, the joint between superconductor and current wires was fabricated by Ag-Mg alloy foils and Sn-Pb alloy during the preparation of Bi2223 bulk rods. Then, Bi2223 bulk rods could be linked directly with current wires to form current leads unit which was strengthened with an epoxy tube. As a result, the joint resistance was measured as $10^{-6}\ \Omega$ at 77 K and $10^{-8}\ \Omega$ at 5 K.

Figure 6 shows the field dependence of the critical current of the leads at liquid nitrogen temperature. It is worth noting that the Bi2223 current leads have transport current capacities of 190A in fields up to 0.1T at 77K. The warm and the cold ends of the current leads experience a leakage field of less than 0.08 T when the central field of 5 T is generated. Figure 7 shows distribution of stray magnet field at Bi2223 current leads. The current leads maintain the superconductivity at the operating current of 100A and a temperature of 60K.

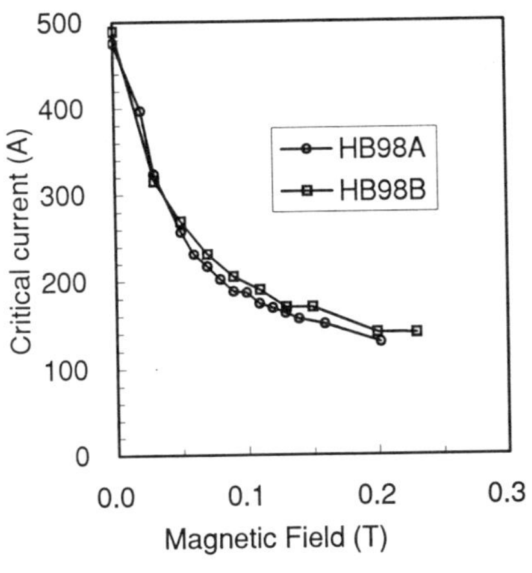

Figure 6. Field dependence of the critical current of the leads at liquid nitrogen temperature.

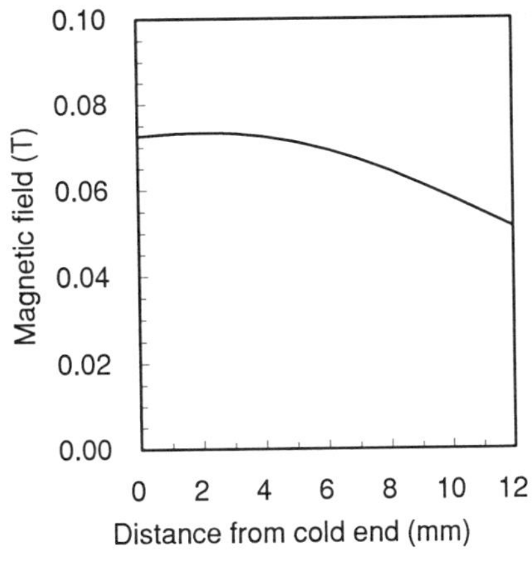

Figure 7. Distribution of stray magnetic field at HTS current leads with a central field 5 T.

The estimated heat loads from 60K to 5K for a pair of leads is about 0.182W.

All current leads are thermally anchored through aluminum nitride insulation spacers at each cooling stage. AlN is a good electrical insulator, meanwhile it presents a high thermal conductivity larger than that for copper.

EXPERIMENTAL RESULTS

Initial Cool Down

In the first test, liquid nitrogen was used to cool the thermal radiation shield all the time and to cool down the magnet to about 80 K to shorten the cool down time and,

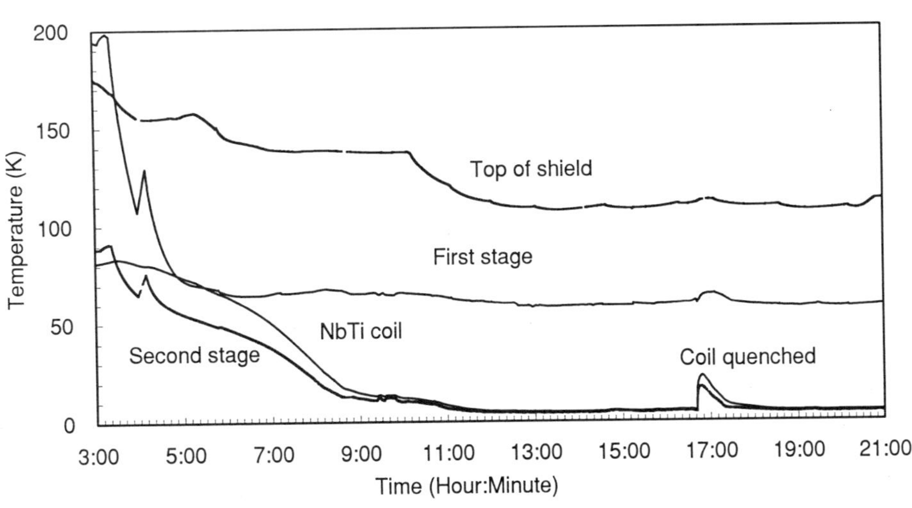

Figure 8. Temperatures of the system cool down after cryocooler started and temperatures during and after quench.

especially, to reduce the operation time of the cryocooler. The 4K GM cryocooler was started after the temperature of magnet reached about 80 K. Figure 8 shows the cool down characteristics of the first and second stages, the NbTi coil and the top of the shield after the cryocooler started. It took 12 hours to cool down the magnet to 80 K and took 10 hours for the magnet to be cooled from 80 K to the stable temperature of 5.0K at zero current. The stable temperature of the second stage is 4.5 K, the first stage 60 K.

Magnet Performance

The magnet was excited twice to quench. The magnetic field reached 4.7 T and 4.6 T at quench currents of 92.5 A and 90.6 A, respectively. The coil temperatures rose up to 5.2 K and 5.5 K before quenches, respectively. The temperature rise is caused by A.C. losses of the NbTi conductor and is decreased by reducing the ramp rate. The increased temperature of the coil could decrease somewhat by maintaining the current at constant level during the excitation.

After quench the coil temperature rose up to 26 K. The coil was cooled down to stable temperature of 5.0 K again after about 2.5 hours.

Test using Sumitomo 4 K cryocooler

Recently, the system was reassembled and tested again using a new 4K GM cryocooler bought from Sumitomo Heavy Industries, Ltd.. The cooling capacity of the first stage of the Sumitomo cryocooler is 1 W at 4.2 K and the second stage 31 W at 40 K. In the new system, no liquid nitrogen was used and the system becomes cryogen free. The coil was cooled down to 3.6 K in about 21 hours. The stable temperature of the second stage is 3.1 K, the first stage 33 K. Figure 9 shows the temperature of the coil, second stage and ramping current of one of the excitations. The current ramp rate is 1.5 A/min. Figure 10 shows the details of the temperature rise and the ramping current before quench at the same excitation. The temperature of the top flange of the coil (far from the heat sink) rose to 3.65 K at the

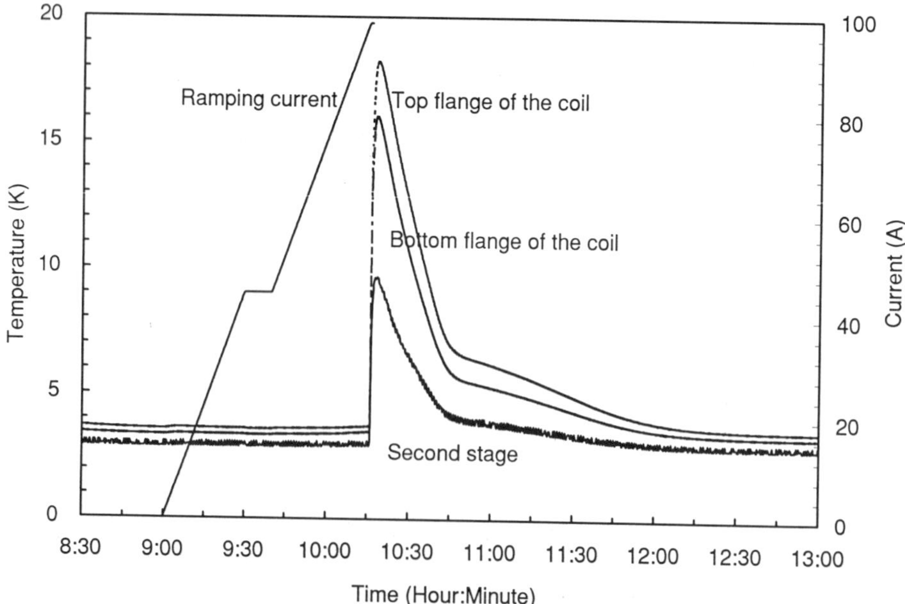

Figure 9. The temperatures of the magnet, second stage and ramping current of one of the excitations.

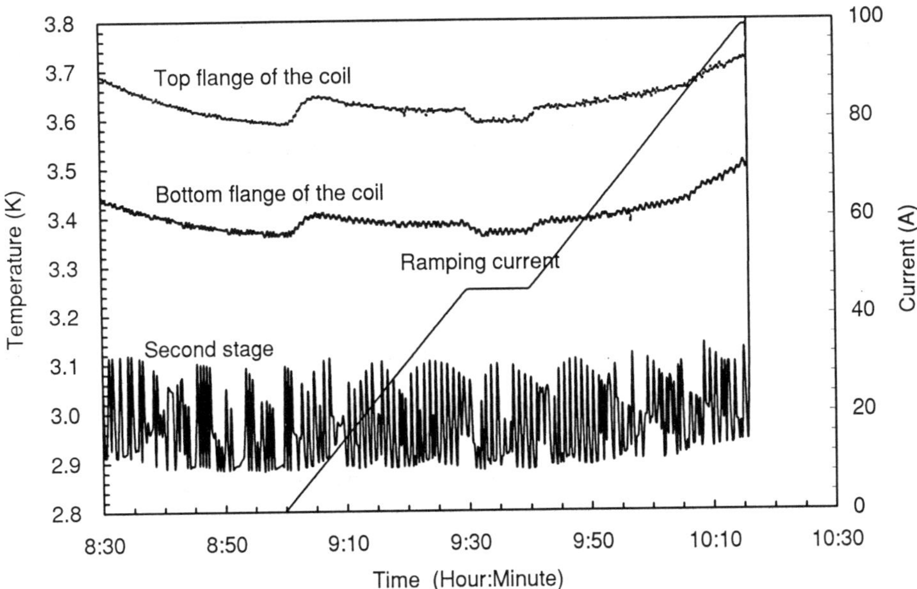

Figure 10. Details of the temperature rise and the ramping current before quench.

ramping current of 9 A (in the low field of 0.45 T) and gradually decreased to 3.61 K at 45 A (field of 2.3 T). Then the temperatures of the coil decreased by maintaining the current at constant level. The coil quenched after maintaining at a current of 98.78 A (field of 5.03 T) for about 40 seconds. The coil temperature rose up to 19 K after quench and was cooled down to stable temperature of 3.6 K again after about 2 hours.

CONCLUSION

A cryocooler cooled NbTi superconducting magnet was designed and constructed. The 4K cryocooler, Bi2223 high temperature superconductor current leads, the NbTi wire were fabricated in China. The magnet system was cooled down to 5.0 K for about 20 hours. A central field of 4.7 T was obtained at the first test. Liquid nitrogen was used to compensate the cooling capacity lack of the first stage of the cryocooler.

A field of 5.0 T was achieved at a recent test using Sumitomo 4K cryocooler. This reassembled system was a cryogen free superconducting magnet system.

ACKNOWLEDGMENT

This work was supported by the National Center for R&D on Superconductivity of China.

REFERENCES

1. T. Kuriyama, Y. Ohtani, M. Takahashi, H. Nakagome, A 4 K G-M refrigerator for direct cooling of a 6 T NbTi superconducting magnet system, Cryocoolers 8, p.785 (1995).
2. K. Watazawa, J. Sakuraba, F. Hata, T. Hasebe, C.K. Chong and Y. Yamada, K. Watanabe, S. Awaji and T. Fukase, A cryocooler cooled 6 T NbTi superconducting magnet with room temperature bore of 220 mm, *IEEE Trans. Magn.* 32:2594 (1996).

3. K. Shibutani, S. Itoh, T. Takagi, O. Ozaki, R. Horise, S. Hayashi, M. Shimada, R. Ogawa, Y. Kawate, Y. Inoue, K. Matsumoto and N. Kimura, Design and fabrication of cryogen free superconducting magnet, *Proc. ICEC/ICMC16*, 1129 (1996).
4. K. Koyanagi, M. Urata, Y. Ohtani, T .Kuriyama, K. Yamamoto, S. Nakayama, T. Yazawa, K. Tasaki, S. Nomura, Y. Yamada, H. Nakagome, S. Murase, H. Maeda, O. Horigami, A cryocooler-cooled 10 T superconducting magnet with 100 mm room temperature bore, *IEEE Trans. Magn.* 32:2558 (1996).
5. K. Watanabe, S. Awaji, J. Saburaba, K. Watazawa, T. Hasebe, K. Jikihara, Y. Yamada and M. Ishihara, 11 T liquid helium-free superconducting magnet, *Cryogenics*, 36:1019 (1996).
6. Y. Ohtani, H. Hatakeyama, H. Nakagome, K. Koyanagi, T. Yazawa, and S. Nomura, Development of a 11.5 T liquid helium-free superconducting magnet system, *Proc. ICEC/ICMC16*, 1113 (1996).
7. Michael A. Green, Calculating the J_c, B, T surface for niobium titanium using a reduced-state model, *IEEE Trans. Magn.* 25:2119 (1989).
8. M. Urata, T. Kuriyama, T. Yazawa, K. Koyanagi, K. Yamamoto, S. Nomura, Y. Yamada, H. Nakagome, S. Murase, H. Maeda, and O. Horigami, A 6 T refrigerator-cooled NbTi superconducting magnet with 180 mm room temperature bore, *IEEE Trans. Appl. Supercond.* 5:169 (1995).
9. T. Yazawa, K. Koyanagi, M. Urata, T. Kuriyama, Y. Ohtani, S. Nomura and H. Maeda, Cooling Structure for 6 T NbTi Superconducting Magnet Directly Cooled by Cryocooler, *IEEE Trans. Appl. Supercond.* 5:181 (1995).
10. M. N. Wilson, "Superconducting Magnets", Oxford University Press, New York (1983).

DESIGN AND PERFORMANCE OF COOLING SYSTEM OF HELICAL COILS FOR THE LHD

S. Imagawa, T. Mito, K.Y. Watanabe, H. Tamura, N. Yanagi, H. Sekiguchi, R. Maekawa, T. Satow, S. Satoh and O. Motojima

National Institute for Fusion Science
Toki, GIFU, 509-5292 JAPAN

ABSTRACT

Helical coils for the Large Helical Device are large scale pool-cooled superconducting coils. The coils are designed to produce a central toroidal magnetic field of 3 T at 4.4 K. The composite conductors are packed into thick cases which are fixed from a thick shell structure to sustain large electromagnetic forces. The total weight of the coils and the cryogenic supporting structures exceed 800 ton including three pairs of poloidal coils. The cryogen for the helical coils is supplied from the ten bottoms of the coil cases, and the gas generated in the coils is vented to a header tank from the ten tops of the coils. There are 20 parallel paths in the coils. The supporting structures and 80 K thermal shields are cooled by conduction from the cooling pipes attached by metal cleats and epoxy resin of high thermal conductive grade. Because of huge area of the structures, the total length of the cooling pipes exceeds 7.8 km, and the pipes also form a large number of parallel paths. In the first and second campaign, the whole structures were successfully cooled down within four weeks.

INTRODUCTION

The Large Helical Device (LHD) is a toroidal fusion experimental facility with a pair of superconducting helical coils and three pairs of superconducting poloidal coils for producing strong magnetic fields to confine the high temperature plasma of several 10 keVs. [1] In order to attain high accuracy of coil winding and high current density, composite conductor of pool-cooled type was selected for the helical coils. The design current of the helical coil is 13.0 kA at 4.4 K and 17.3 kA at 1.8 K, and the standard operation current at 4.4 K is 12.5 kA which produces 3.0 T of toroidal magnetic field at a major radius of 3.75 m.

The construction was completed at the end of 1997, and the first cool-down was performed by 4 weeks from the middle of February. [2] The first excitation tests up to 1.5 T were carried out successfully in March 1998. After two months' plasma experiments at 1.5 T, the coils were warmed up for upgrade of the plasma heating devices and diagnostics. The second cool-down was performed from the middle of August. In this campaign, we tried the excitation up to standard 3.0 T. The helical coil was designed to satisfy 'cold-end' stability at 13.0 kA. However, a wide propagation of normal zone occurred when reaching 11.4 kA. The quench detection system acted, and all the coils were discharged by protection resistors. Even though large amounts of AC losses were produced from the helical coils and the surrounding structures, the cooling system worked precisely and could recover within 1 day. This paper intends to summarize the design and performance of the cooling system for the helical coils and the surrounding structures.

DESIGN OF COOLING HELICAL COILS AND STRUCTURES

The cryogenic system for the LHD consists of the helium refrigerator/liquefier [3], a main cryostat, current-leads cryostat, superconducting bus-lines, and two valve-boxes. Three different cooling schemes are utilized for each cooling objects; a pool-cooling for the helical coils, a forced flow of supercritical helium for the poloidal coils, a forced flow of two-phase helium for the supporting structures and the superconducting bus-lines. This system has cooling capacities of 5.65 kW at 4.4 K, 650 liter/hr liquefaction and 20.6 kW at 80 K. These powers were determined for the steady heat loads of all the components. Liquid nitrogen is used to produce low temperature helium necessary for the initial cool-down to 100 K around.

The helical sub-system of the LHD cryogenic system consists of the helical coils, the shell-arms, the supporting structures, the cryogenic supporting posts, and thermal shields, as shown in Fig. 1. The cryogen is distributed by the helical coil valve box (HC.VB), that equips 49 control valves and 4 m^3 buffer tank with a sub-cooler. The weight and heat loads are summarized in Table 1. During cool-down, the pressurized gas helium is distributed in proportion to each weight, and the temperature of helium gas was controlled by mixing cold and warm gases in order to keep the temperature differences in the coils and the structures less than 50 K. In steady state, cryogen for the helical coils and the structures is switched to two-phase helium from the reservoir. The cryogen is cooled again in the sub-cooler in HC.VB, and the flow rates are controlled to keep the liquid level.

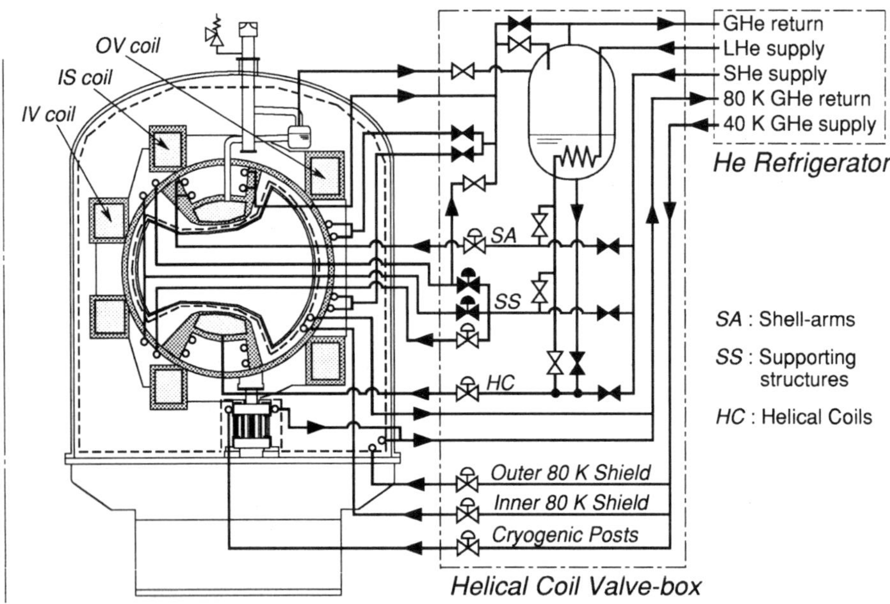

Figure 1. Schematic flow diagram of the helical sub-system of the helium cooling system for LHD.

Table 1. Weights and steady heat loads of cryogenic components of LHD

item	weight (ton)	surface area (m^2)	steady heat load (W) design	steady heat load (W) measured	flow rate (g/s) initial cool-down	flow rate (g/s) steady
Helical coils	155	159	60	87	65	–
Shell-arm	85	106	17	–	20	–
Supporting structures	400	677 [both sides]	201	–	190	–
Poloidal coils	182	–	210	217	130	250
(Sum of 4K)	*(822)*	*(942)*	*(510)*	*(–)*	*(405)*	*(250)*
Inner 80 K shield	10	422	5.54×10^3	3.23×10^3	23	33
Outer 80 K shield	13	587	3.06×10^3	3.83×10^3	16	27
Cryogenic post	11	–	0.82×10^3	0.80×10^3	3	6
(Sum of 80K)	*(34)*	*(1009)*	*(9.42×10^3)*	*(7.86×10^3)*	*(42)*	*(66)*

Helical Coils and Coil Leads

The helical coils are pool-cooled, and the conductors are packed into thick cases to endure large electromagnetic forces. The cryogen is supplied from the ten bottoms of the coil cases through two inlet headers, as shown in Fig. 2. The gas generated in the coils is vented to a header tank from the ten tops of the coils. There are 20 parallel paths in the coils. The inlet headers and the header tank are connected to a large buffer tank in the HC.VB. Longitudinal cooling channels inside the helical coils are placed at both sides of each layer of conductors and both top corners, the areas of which are 25 $mm^2 \times 2 \times 20$ layers and 306 $mm^2 \times 2$, respectively. These areas were required to vent helium bubbles generated by steady heat loads. Electrical insulators between conductors are 2.0 and 3.5 mm thick, and they are positioned at intervals to create transverse cooling channels. Against the cryogen inventory of 2.2 m^3 in the two helical coils, the volume of the header tank in the main cryostat and buffer tank in the HC.VB were determined to be 2.3 and 4.0 m^3, respectively, to stabilize the liquid level control and to delay the pressure increase during an emergency. The amounts of liquid helium in both the tanks are controlled to be 1 m^3. The total inventory of cryogen reaches almost 15 m^3 including the structures and the thermal shields, as shown in Table 2.

The coil leads are derived from the outer mid-plane and connected electrically to the superconducting power cables, which are called SC Bus-lines, with S-shaped flexible leads at the bottom of the cryostat. Since the coil leads move inside by 20 mm during cool-down from 300 K to 4.4 K, the flexible leads are necessary to keep the counter force to the terminal of the SC Bus-line under 200 kg. Terminals of both the coil lead and the SC Bus-line are copper plates with superconductors inside. The flexible lead consists of two copper plates, compacted superconducting strands buried in them, and copper pipes welded to both sides. A set of the coil leads are contained in a pipe and cooled by liquid helium from the header tank through each branching pipe. The cryogen is bounded at the terminal of the leads by ceramic breaks. The flexible leads are cooled by two-phase helium from the SC Bus-lines. Since the number of strands are twice the conductor of the helical coils, the cryostability will be sufficient at less than 8.5 K. The cross-section of the copper plate was designed to be 8 mm thick and 110 mm wide in order to avoid a thermal runaway. The joint resistance between the flexible leads and the terminals are estimated in the order of nΩ.

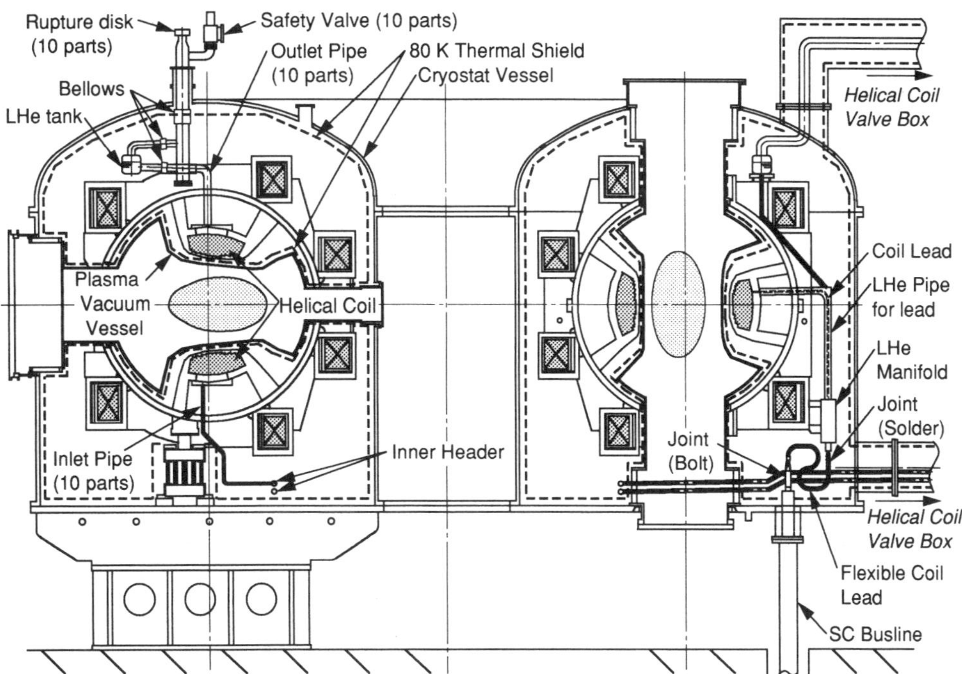

Figure 2. Cross-sectional view of helical coils and surrounding structures of LHD.

Table 2. Helium inventory of helical sub-system

Parts	Inventory (m^3)	LHe contents (m^3)
Helical coils and inner headers / HC leads (6 lines)	2.57 / 0.36	2.57 / 0.36
A header tank / Connecting pipes	2.36 / 0.65	1.0 / 0.20
Cooling pipes for Shell-arm / Supporting structures	0.27 / 1.14	0.27 / 1.14
Cooling pipes for Outer 80 K shield / Inner 80 K shield	0.35 / 0.52	0 / 0
Cooling pipes of cryogenic posts / R.T. return pipe	0.01 / 1.33	0 / 0
Buffer tank / Inner pipes in HC.VB	4.34 / 0.89	1.0 / 0.28
Sum	14.80	6.82

The heat loads of the helical coils are radiation and molecular conduction from 80 K shields, thermal conduction through measuring cables and cooling pipes, joule losses in the joint, and AC losses. The steady heat loads by radiation and molecular conduction were estimated to be 40 and 2 W, respectively, with setting 0.12 as the emissivity of the surface and 1.33×10^{-4} Pa as the vacuum pressure in the cryostat. The steady heat loads by thermal conduction were estimated to be 18 W. By reason of helical coil winding, the maximum length of a conductor was limited to be less than 1200 m, and the conductors were jointed between layers except between the first and second layers. The copper sheath of the conductor was removed in the jointing part for 400 mm, and both strands were soldered with sandwiching superconducting strands pieces. The other joints in the coil leads are also soldered. The total joule losses in the 32 internal joints are estimated to be less than 3.6 W at 13.0 kA. The joule losses in the joints in the six sets of coil leads are estimated to be 4.8 W. The total joule losses are estimated to be less than 8.4 W.

Supporting Structures and 80 K Thermal Shields

The supporting structures are heavy and wide, and 80 K thermal shields are wide, as shown in Table 1. Since the temperature difference between parallel cooling paths should be enhanced with the procedure of cool-down, we discussed the shape of cooling channels in the design phase. One leading plan was rectangular channels with wide cross-section enough to realize a single cooling path. However, this idea was given up, because it will take much time to make the channels on the surface of the complicate supporting structures and because the possibility of helium leakage should be enlarged. At the same time, we estimated the temperature instability during cool-down. It was revealed that the final temperature is determined uniquely by the heat loads in the case of gaseous helium cooling. Then, we selected seamless pipes for the cooling channels, and distance of the pipe was determined to keep the temperature increase from the pipe within 1 K against steady heat loads. The pipes were attached on the objects by metal cleats mechanically and by epoxy resin of STYCAST 2580FT thermally. The total length of pipes was determined from the area and the distance of pipes. The size of pipe was limited by the working efficiency. The number of parallel paths was determined to gain a rated mass flow from the beginning of cool-down. The details and the pressure drops are shown in Table 3. The pressure drops of the pipes ΔP_{pipe} were calculated by Blausius' equation as

$$\Delta P_{pipe} = 0.3164\left(\frac{\rho \cdot d \cdot v}{\mu}\right)^{-0.25} \cdot \frac{\ell}{d} \cdot \frac{\rho \cdot v^2}{2} = 0.2414 \cdot \frac{\mu^{0.25} \cdot q^{1.75} \cdot \ell}{\rho \cdot d^{4.75}} \qquad (1)$$

where ρ, v and μ are a density, a velocity and a viscosity of helium. d and ℓ are diameter and length of the cooling pipe. q is the mass flow. The pressure drop of a 90 degree elbow ΔP_{elbow} and that of a 180 degree bend ΔP_{bend} were calculated by the following equation

$$\Delta P_{elbow} = 1.265 \cdot \frac{\rho \cdot v^2}{2} = 1.025 \cdot \frac{q^2}{\rho \cdot d^4} \qquad (2)$$

$$\Delta P_{bend} = 0.15 \cdot \frac{\rho \cdot v^2}{2} = 0.122 \cdot \frac{q^2}{\rho \cdot d^4}. \qquad (3)$$

Table 3. Design of cooling pipes for the supporting structures and 80 K shields

item	number of paths	length per path (m)	diameter (mm)	Pressure drop (kPa) design	measured [outlet temp.]
Helical coils	20	4	15.5, other	0.34	<2.0 at 66.2 g/s [280 K]
Shell-arm	4	110	23	36.7	32 at 20.2 g/s [275 K]
Supporting structures	80 (*1)	28.8	17.5	8.8	13 at 186 g/s [274 K]
Inner 80 K shield	10+10	112.6, 118.8	8.1	250	247 at 23.1 g/s [275 K]
Outer 80 K shield	10	280	10.5	296	250 at 15.4 g/s [284 K]
Cryogenic post	2	96	8.1	316	378 at 2.8 g/s [297 K]

(*1) The number is changed to 20 in the two-phase helium cooling.

The distances of pipes on the supporting structures are less than 400 mm, and the total length exceeds 2.3 km. The number of parallel paths for the supporting structures was enlarged from 20 to 80 by considering the increase of capacity of the helium refrigerator to shorten a cool-down period. Heat exchange between the helium and objects almost finish within several meters because of the high heat transfer of helium. The supporting shell was determined to be cooled from the inside in order to reduce the temperature difference among the parallel paths by good heat exchange between them around the inside. The actual position of the inlet is between IV and IS coils by the reason of layout of cooling pipes and headers. In the two-phase helium cooling, two lower and two upper paths come in a series by closing the return from two outlet-headers and one of three inlet-headers in order to avoid the effect of gravity. The pipes on the shell-arms are located in four corners, and they are in a series.

The thermal shields are grouped into Inner 80 K shields and Outer 80 K shields which are attached to the plasma vacuum vessel and the vacuum chamber of the cryostat with a distances of 50 and 100 mm, respectively. Multilayer insulation (MLI) made of aluminum coated polyster films are inserted between them to reduce heat loads. The thickness of the thermal shield is 3 mm, and the distances of pipes are 200 mm around. The Inner 80 K shield is divided into three groups which are a "bottom" confronting the half pitch of helical coil cases, a "side" confronting the half pitch of shell-arms, and a "torus" confronting the upper or lower half of 1/10 of the supporting shell. Two "bottoms" and four "sides" are in a series, and two "tori" are in a series. The shields surrounding the ports are included in the "torus". One path for the Outer 80 K shield covers 1/10 in toroidal. The cryogen goes through 1/20 part of the base plate, inner cylinder, the rest of the base plate, 1/20 part of the outer cylinder, top roof, and the rest of the outer cylinder. The shields of the both cylinders are suspended by long rods made of stainless steel. The other shields are fixed by grass-fiber-reinforced plastics (GFRP) bolts with sliding mechanism to avoid thermal stress.

RESULTS OF COOL-DOWN

The first cool-down of the LHD started on February 23 in 1998, and finished on March 22. [3] The helical coils and the poloidal coils became superconducting on March 17, when the supply gas temperature was 8.5 K. The cool-down was successfully completed in 28 days. The second cool-down started on August 18, and the condition for coil excitation was satisfied on September 11, as shown in Fig. 3. Since the temperature difference in the coil should be small due to high thermal conductivity of the conductor, the average temperature is considered to be almost equal to the outlet temperature. The temperature difference of the two helical coils was well controlled within 2 K during cool-down to 100 K, but it increased up to 7 K at 50 K around. This may be caused by a small error of the orifice flowmeter in a low temperature, because the each apparent flow rate is always controlled by a needle-valve. The temperature difference among the 80 outlet pipes of the supporting structures was kept within 30 K in spite of the large number of parallel paths without control valves. The increase of difference should be suppressed by thermal conduction in the thick supporting shell itself and by heat exchanges with the coils. The temperature difference among the ten sectors of the Inner and Outer 80 K shield became less than 2 and 10 K, respectively, at steady-state. Since each piece of the thermal shields is separated electrically, heat exchanges between the parallel paths are small except on the shield of the inner cylinder. Such small temperature difference proves the quality of construction.

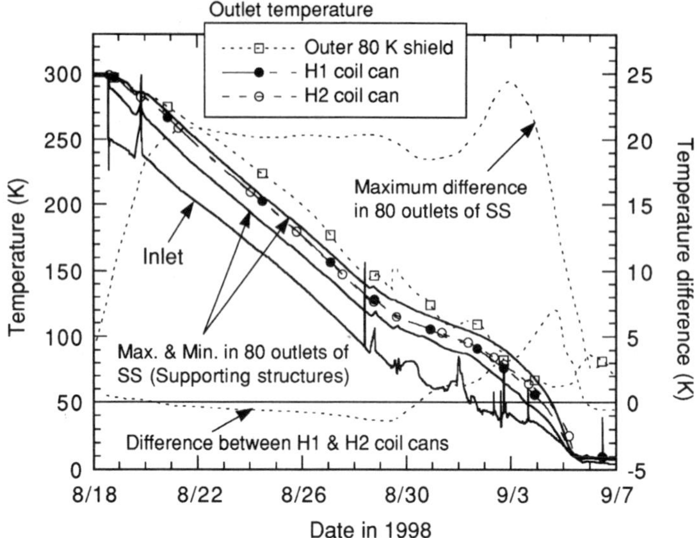

Figure 3. The temperature change of the supporting structures, the helical coil cases and the 80 K shields in the 2nd cool-down.

The pressure drops in the cooling pipes change widely according to the temperature and the pressure of the cryogen, and these become the largest in beginning of cool-down. The pressure drops measured on the August 20 in 1998 are summarized in Table 3. These are in good agreements with the design values based on circular pipes and bends in turbulent flow. These are also evidence of the quality of the construction.

The steady heat loads of the thermal shields can be measured from the temperature increase of the cryogen and the mass flow. The design and measured heat loads of the thermal shields are listed in Table 4, and the temperature changes during the 95 °C baking of the plasma vacuum vessel are shown in Fig. 4. The measured heat loads of the Inner 80 K shields are moderately less than the design. On the contrary, the heat loads of the Outer 80 K shields are slightly larger than the design. The former reasons are considered to be over-estimation of conduction through the support and heat flux of MLI. The latter reason is considered to be underestimation of radiation. Since the Inner 80 K shield is closed to the room temperature surface, 5.7 W/m^2 was used for the average heat flux of MLI between 300 and 80 K with considering contingencies. On the other hand, 3.3 W/m^2 was used for the Outer 80 K shield. The actual value is considered to be 4.5 W/m^2 in both the shields from the measured heat loads. Furthermore, the contact thermal resistance is not negligible especially in the sliding mechanism. The conduction in the Inner 80 K shield may become almost half of the design values.

Table 4. Design and measured heat Loads of 80K thermal shields

item	area (m^2)	number of support	heat load at 300 [368] K of V/V (*1) (kW)			
			radiation	conduction	design total	measured total
Inner 80 K shield						
bottom	97.6	1440	0.55 [1.32]	1.33 [1.71]	1.89 [3.03]	0.86 [1.34]
side	99.6	1680	0.56 [1.35]	0.71 [0.91]	1.28 [2.26]	0.88 [1.39]
torus	222	2244	1.27 [3.04]	1.11 [1.42]	2.38 [4.46]	1.49 [2.50]
Outer 80 K shield						
base plate and inner cylinder	208	580	0.69 [1.05]	0.87 [0.87]	1.55 [1.92]	1.88 [2.26]
outer cylinder and top roof	367	545	1.21 [1.46]	0.24 [0.24]	1.45 [1.71]	1.71 [1.94]

(*1) V/V means the plasma vacuum vessel. The radiation was calculated by using 5.7 and 3.3 W/m^2 of the heat flux of MLI between 300 and 80 K for the Inner and Outer 80 K shield, respectively.

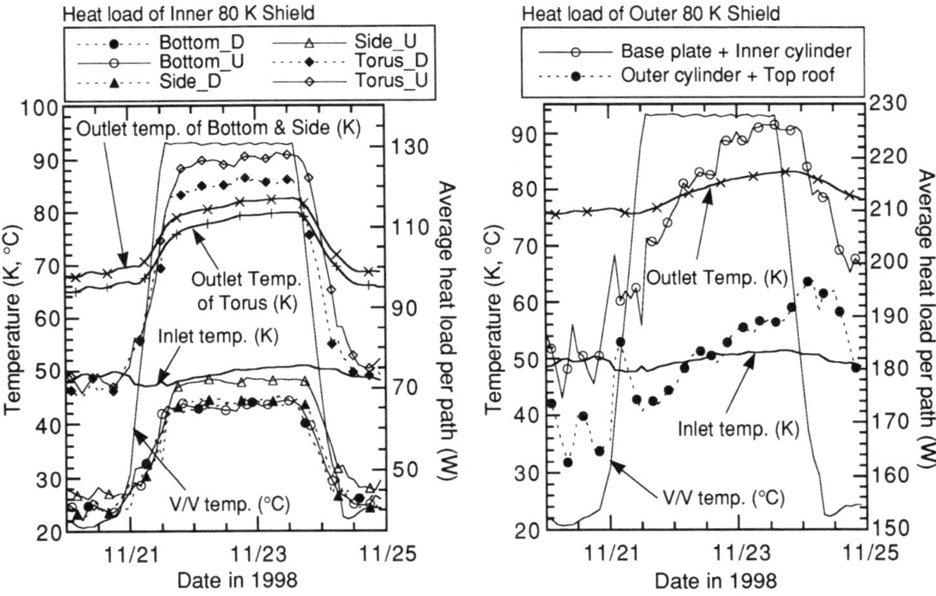

Figure 4. Temperatures and heat loads of the Inner and Outer 80 K shield when baking the plasma vacuum vessel up to 368 K. Each heat load was calculated by utilizing average temperature rise of ten parallel paths and one tenth or twentieth of the total mass flow. U and D mean upper and lower half, respectively.

AC LOSSES AND SAFETY DEVICES

Since the coil case is a bath of helium, all the joints are welded, and there are no electric breaks. The supporting structures were also assembled by welding in order to attain high reliability for strength and rigidity. Then, large amounts of AC losses are induced by various discharges. The estimated values are listed in Table 5 for the normal discharge of 30 minute and two types of resistor dumps in emergencies. The resistor dumps are named "1Q" and "1M" in which the discharge time constants are 20 and 300 s, respectively. 1Q is only selected when a coil quench is detected. The AC losses of the conductors were derived from the measured overall time constant of eddy current of the conductor. [4] AC losses of the other objects were estimated by being approximated by the equivalent current circuits for the coil. Since the AC losses are very large, the header tank contains ten safety valves, ten rupture disks and two kinds of venting lines. When the pressure increases, the controlled release-valves act first to exhaust the helium in the tank to the outside of the building. In the 1Q dump from high field, the helical unit was disconnected from the refrigerator, and the pressure in the coil case increases rapidly. In the actual 1Q dump from 2.75 T, the liquid helium in the coil cases was almost vaporized in several minutes, and 900 m^3 of helium gas was successfully exhausted through the release-valves to atmosphere. Since the cryogenic system worked precisely, it could recover within a day.

The AC losses and steady heat loads of the coils and the cases were measured by the decrease of liquid helium level during the 1M dump from 1.5 T, as shown in Fig. 5. These AC losses are in good agreement with the design values. The steady heat loads are slightly larger than the design value. The reason may be the conduction from the shell-arms.

Table 5. Design values of AC losses during discharges from 3 T [MJ]

	1Q (20 s)	1M (300 s)	Normal
Conductors	4.51	0.353	0.119
Coil case	6.49	0.433	0.144
Header tank	0.12	0.008	0.003
Shell-arms	2.61	0.174	0.058
Supporting structures	10.83	0.722	0.241

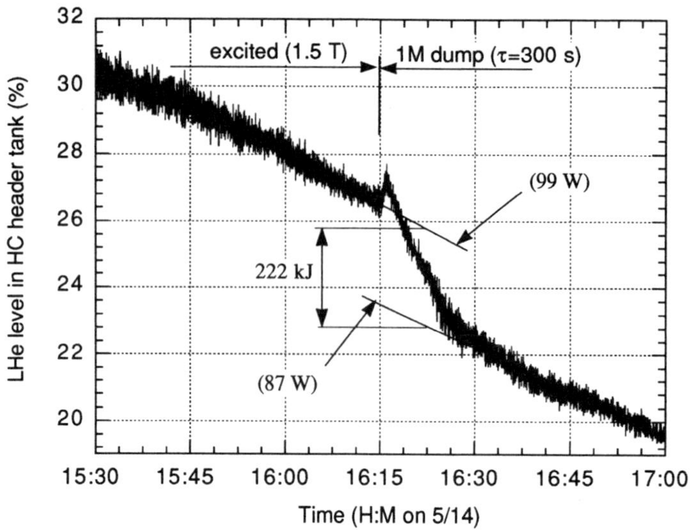

Figure 5. Liquid helium level in the the header tank of the helical coils during 1M dump from 1.5 T. The supply of liquid helium was stopped from 15:15 to 18:00.

CONCLUSION

The helical sub-system of the LHD cryogenic system consists of the helical coils, the shell-arms and the supporting structures, the thermal shield for the cryostat and the plasma vacuum vessel, and cryogenic posts. Since the weight and surface area are heavy and wide, large number of parallel paths is indispensable to obtain high flow rates from the beginning of cool-down. In spite of large numbers of parallel paths without control valves, the cool-down was successfully performed in four weeks while keeping the temperature differences in the coils and structures within 50 K. The pressure drops of cooling pipes were in good agreements with the design values based on circular pipes in turbulent flow. The measured heat loads of the thermal shields of the plasma vacuum vessel were about 60% of the design values. The reasons are considered to be contact thermal resistance of GFRP supports and better performance of the multilayer insulators. The steady heat load and AC losses during a fast dump of helical coils were measured, and the values are confirmed to be adequate. These results prove the validity of the design and construction of the cryogenic system.

ACKNOWLEDGMENT

The authors are indebted to Dr. A. Iiyoshi, the previous Director General of National Institute for Fusion Science, and Dr. M. Fujiwara, the Director General of National Institute for Fusion Science, for their continuous encouragement. We are also grateful to many staff of Hitachi Ltd. who joined this work.

REFERENCES

1. A. Iiyoshi, S. Imagawa and the LHD Group, Design, construction and the first operation of LHD, the 20th Symposium on Fusion Technology, Marseille, France, Sep. 16-20 (1998) invited talk.
2. T. Mito, R. Maekawa, S. Yamada, A. Nishimura, et al., First cool-down performance of the LHD, ASC'98 Palm Desert, CA, Sep 13-18 (1998) LIC-04.
3. S. Satoh, T. Mito, S. Yamada, J. Yamamoto, O. Motojima and LHD group, Construction and commissioning tests of a 10 kW-class helium refrigerator for the Large Helical Device, *Cryogenics*, Vol. 34 ICEC supplement (1994) 95-98.
4. N. Yanagi, S. Takacs, T. Mito, K. Takahata, A. Iwamoto and J. Yamamoto, Measurement of time constants for superconducting cables with Hall probes, *Cryogenics* 37:783 (1997).

ANALYSIS AND COUNTERMEASURE FOR A PROBLEM OF ABNORMAL VOLTAGE GENERATION IN A SMES MAGNET

H.Hayashi,[1] K.Tsutsumi,[1] F.Irie,[1] T.Teranishi,[2] S.Hanai,[2] L.Kushida[2]

[1] Kyushu Electric Power Co., Inc.
2-1-47, Shiobaru, Minami-ku, Fukuoka, 815-8520, Japan
[2] Toshiba Corporation
2-1, Ukishima, Kawasaki-ku, Kawasaki, 210-0862, Japan

ABSTRACT

We have experienced a problem due to the abnormally high voltage that damaged one of six coils in a 1kWh/1MW SMES (Superconducting Magnetic Energy Storage). The damaged coil has already been replaced and the whole system is now in operation without any problem.

We will report here the analysis and discussion of the problem of high voltage generation based upon the recorded current wave form. It was estimated that the problem had originated by an accidental grounding of a thin potential lead wire connected to a point of interconnection between two coils, which was then followed by a disconnection of the short-circuit by wire melting. This disconnection of the wire bearing current caused a generation of very high voltage and the resulting damage of the coil. Theoretical considerations with simulated calculations are made to confirm the above estimation quantitatively. A countermeasure for this sort of problem is also presented.

INTRODUCTION

We have experienced a trouble due to an abnormal high voltage generated by an accidental grounding of a thin potential lead connected to a point of a coil in a 1kWh/1MW two module-type SMES[1,2](Superconducting Magnetic Energy Storage) during its preliminary test operation.

We postulated the following mechanism for the trouble. Large currents flowed in one of potential lead wires by its accidental grounding, because the trouble happened at the time of discharging the magnet. The lead wire then melted and disconnected. A very high voltage was generated by an inrush current due to the disconnection of the lead wire. The waveform of this voltage becomes oscillatory due to stray capacities. Discharge and damage happened at the position of the highest voltage to ground of a coil.

The above experience of problem is thought to be very important for developing industrial use of large scale superconducting magnet. However there is no detailed report on analysis of this problem. Therefore the detailed analysis on the problem is made with the aid of numerical calculations for an equivalent circuit. A countermeasure to remove the problem is proposed. The damaged coil has already been replaced with the new one, and the total SMES system with that countermeasure is now in operation at the experimental site without any problem.

STATE OF THE PROBLEM

The trouble happened during the preliminary exciting test of the SMES system. The circuit of one module is shown in Fig.1. The current in the coil was measured by the shunt resistor, which was recorded during the trouble (Fig.2).

State of the Trouble

The trouble happened in a module having three coils, where the SMES has two modules. It happened in a coil at the stage of current holding at 1042A, where suddenly the power source had tripped, and the protection system operated automatically. The voltage drop in the protection resistor at that time was ~1000V. There was heard sparking noise, when the terminal assembly of lead wires for measurements burned out, and smoke came out from the valve of the cryostat. By the later investigation, it was found that an end of the coil had burned out in the extent of 15cm length, 3 cm width and 5 cm depth.

Estimation for the Problem

The insulation of the coil might have had broken down by an induced abnormal high voltage connected with the lead wire burning. Finally the current decreased steadily with the time constant of 2.7 second after a period of unstable discharge. This time constant corresponds to the reduction of to the inductance in a half of that of the coil due to that of short-circuiting. The shunt current decreased to 448A with the time constant of 0.5 second. This was thought to be due to an accidental grounding of the lead wire. The maximum current of the lead wire was then estimated to be 594A. After a half second, the lead wire melted and opened. This disconnection of the short-circuiting path should induce an AC wave with a high frequency voltage which have a large amplitude as the result of a resonance in the coil having stray capacities.

ANALYSIS AND COUNTERMEASURE FOR THE ABNORMAL HIGH VOLTAGE GENERATION

Circuit model for numerical simulation

The structure of a coil in a module is shown in Fig.3. The coil is clamped by stainless steel (SUS304) plates. The clamp metal is floated from the earth potential. The capacitance between flat edges of the coil and the clamp plate facing to them is 900pF, the capacitance

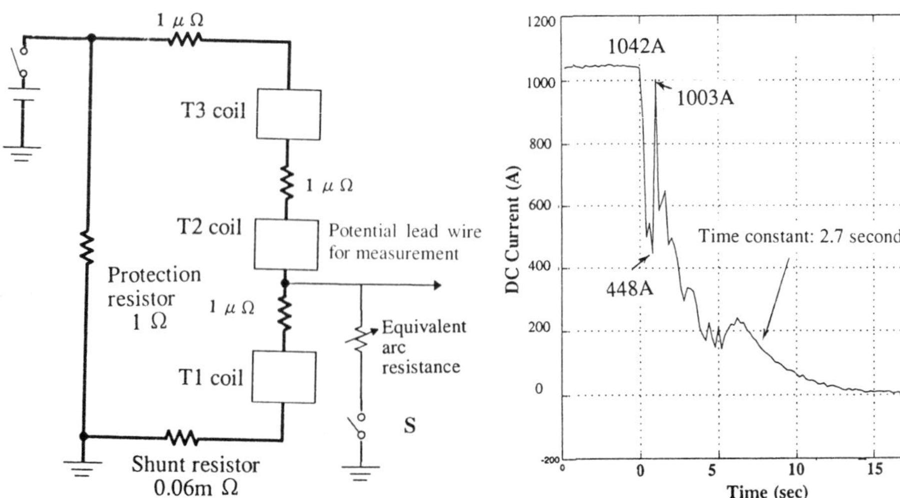

Fig.1. Test circuit of the troubled module. The switch "S" simulates for an accidental break down of a thin potential lead wire for measurement.

Fig.2. Recorded current of the shunt resistor during the trouble of module.

between side edges of the coil and the side parts of the clamp plate is 300pF. One coil consists of 20 pancake coils i.e.10 double pancakes. The arrangement, values and parameters of it are shown in Fig.4(a). Other coils, T2 and T3, have the same configuration. The equivalent circuit model for the numerical calculation and the corresponding pancake arrangement are shown in Fig.4(b). Each pancake coil has mutual inductance with each other. The total inductance of one coil is 1.0H. Stray capacitances in each point of pancake coil are also shown in Fig. 4(b).

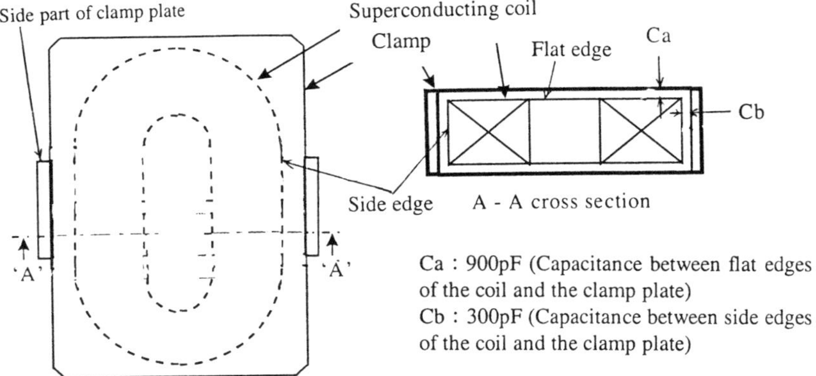

Fig.3. Calculation model of capacitances of one coil

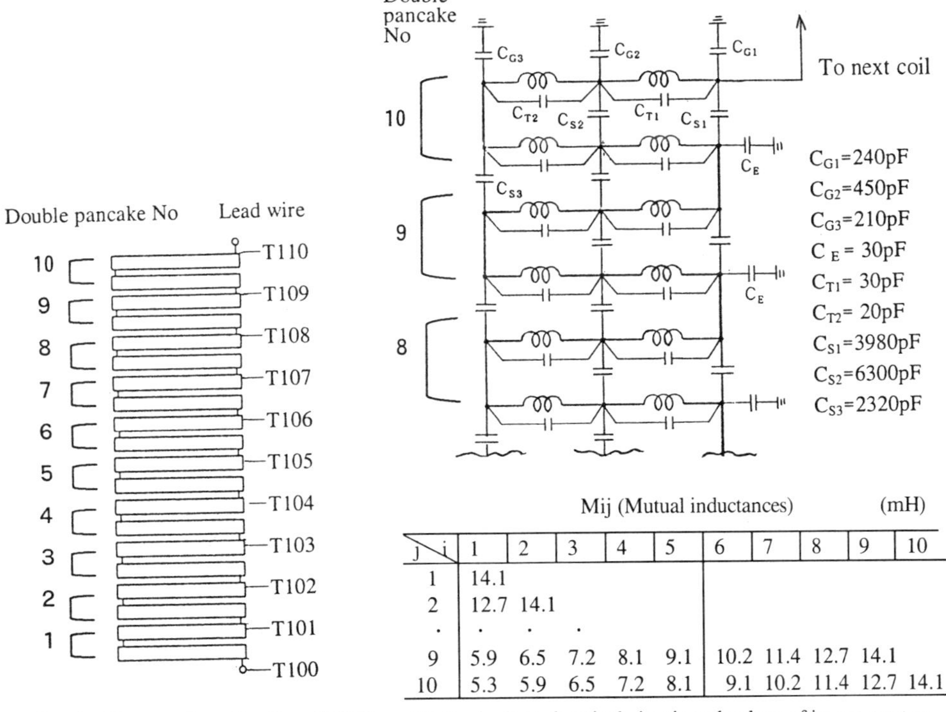

j \ i	1	2	3	4	5	6	7	8	9	10
1	14.1									
2	12.7	14.1								
.	.	.	.							
9	5.9	6.5	7.2	8.1	9.1	10.2	11.4	12.7	14.1	
10	5.3	5.9	6.5	7.2	8.1	9.1	10.2	11.4	12.7	14.1

(a) Ten double pancake coils of a coil (T1).

(b) Equivalent electrical circuit and values of its parameters, where "i" and "j" denote pancake number.

Fig.4. Full equivalent electrical circuit of a coil (T1) for the numerical calculation. Other coils (T2,T3) have the same equivalent circuit.

Conditions for the Calculation

At the trouble high frequency AC voltage should have been superposed on DC voltage which should be much smaller than the AC one. Therefore in this calculation only the induced AC voltage is dealt with. We assumed no loss in coils, which should be enough for our purpose to obtain the highest voltage. From the fault condition, the current of the lead wire at the time of opening is assumed to be 10A considering the characteristics of arc resistance in the break down. EMTP(Electro Magnetic Transient Program) was used for this numerical calculation.

Results of the Simulation

Calculations were carried out for five cases of the point of lead wire connections. The point of connection is T110 (T2 side of T1 coil) for Case1, T105 (center of T1 coil) for Case2, T205 (center of T2 coil) for Case3, T210 (T3 side of T2 coil) for Case4 and T305 (center of T3 coil) for Case5. The AC voltage change in time in these case depend on the above mentioned cases are shown in Fig.5~Fig.7.

In Case1, the highest value of the maximum voltage is 190kV which is induced at T110. Its waveform is almost sinusoidal with a frequency of 4.5kHz as shown in Fig.5.

In Case2, waveform in time is not sinusoidal which is due to the existence of harmonic oscillation mode. The obtained maximum value of the voltage at the point of connection to lead wire at T105 is 70kV , and at T110 the maximum voltage is 97kV which is the highest one of this case as shown in Fig.6.

In Case 3, waveform in time is not sinusoidal again. The maximum value of the voltage is 170kV at the point of T205. At T200 and T210, both ends of T2 coil, the maximum voltage is 145kV which is highest one of this case as shown in Fig.7.

In Fig.8, the maximum voltages of pancake coils in five cases are shown. The highest value among those maximum voltages of around 200kV is seen at T110 and T210. These points seem to be a candidate for the breaking point. The event of the break down was really happened at T110, which coincide with the above result of calculation.

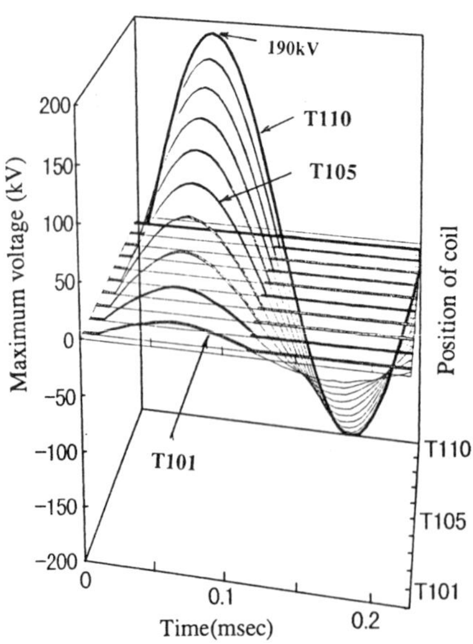

Fig.5. Calculated results of voltage waveforms in each point of T1 coil for case1, the terminal connected to the lead wire being T110 (thick line).The highest maximum voltage of this case is obtained at T110 which is 190kV.

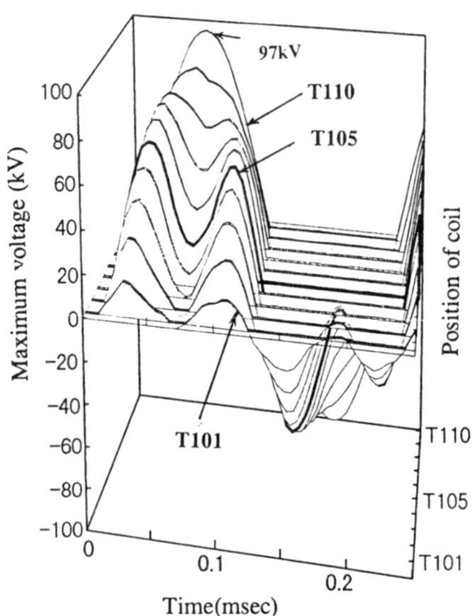

Fig.6. Calculated results of voltage waveforms in each point of T1 coil for case2, the terminal connected to the lead wire being T105 (thick line). The highest maximum voltage of this case is obtained at T110 which is 97kV.

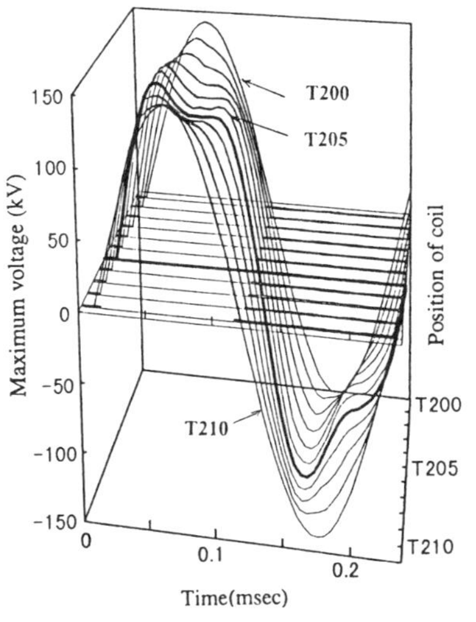

(a) First waveforms in each point

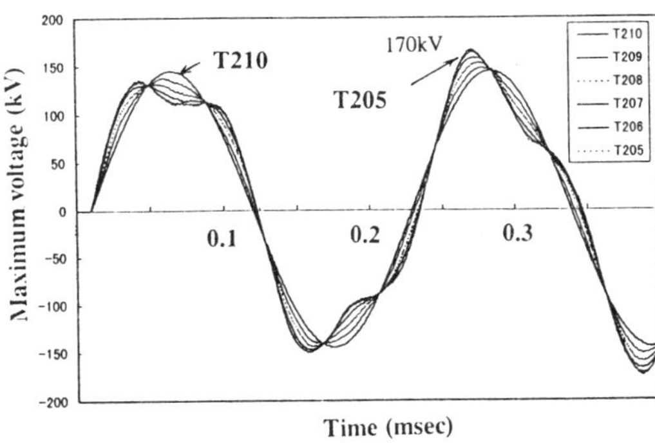

(b) First and second waveforms in each point

Fig.7. Calculated results of voltage waveforms in each point of T2 coil for case3, the terminal connected to the lead wire for measurement being T205 (thick line). The highest maximum voltage of this case is obtained at T205 which is 170kV.

It has been shown in the above that the disconnection of the current path induces very high AC voltage and finally it leads to a serious damage even when the current is rather small compared with that in coils. This suggest that limiting of the grounding current should be effective for weakening the induced voltage.

Countermeasure for the High Voltage Generation

The effect of a resistor inserted in the potential lead as shown in Fig.9 is examined. The break down to the ground and the disconnection afterwards is simulated by the switch in the figure. The effect to limit the break down current is calculated by this circuit. The reduction of the maximum current with the increase of the resistance can be seen in Fig.10.

It is seen in the figure that the maximum current become around 1A for 500 Ω as the resistor. This means that the break down through the air can not be developed in our case become the current is much less then 10A at which break down current should be disconnected in our case due to the increase of arc resistance. This result, 500 Ω for the current limiting resistors, is already used as a countermeasure against damages for a 1kWh/1MW module-type SMES.

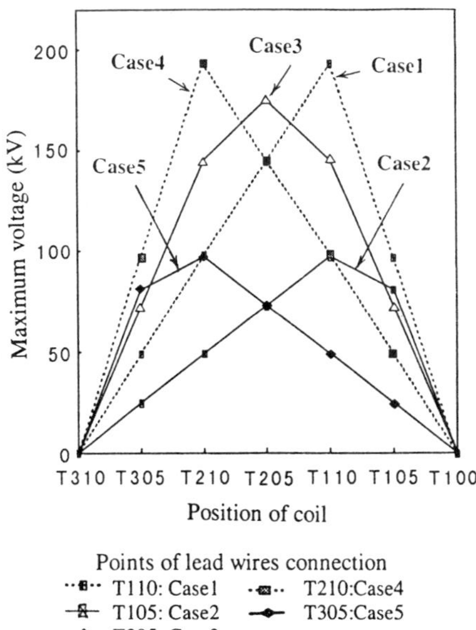

Fig.8. Calculated results of the maximum voltages for 5 cases of lead wires connection.

CONCLUSION

We analyzed the problem of a break down in our SMES magnet. We modeled the events

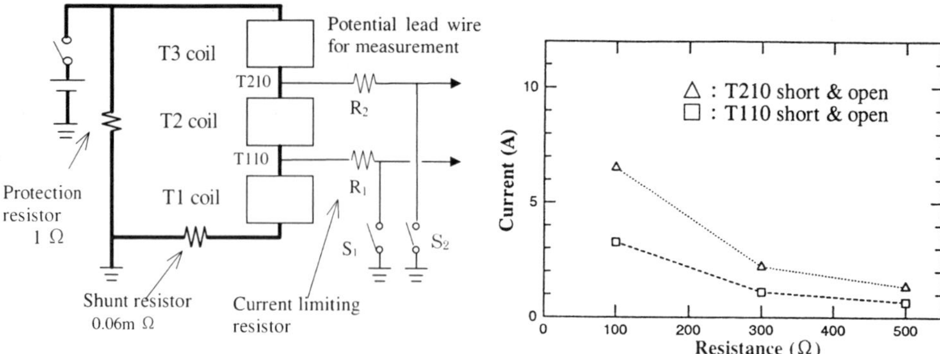

Fig.9. The circuit of numerical calculation for the case with current-limiting resistors (R_1,R_2) as a countermeasure. The switch "S_1"and "S_2" simulate for accidental break downs. "R_1"and "R_2" are current limiting resistors of 100 Ω, 300 Ω and 500 Ω.

Fig.10. Calculated results of maximum currents in each line with current-limiting resistors of 100 Ω, 300 Ω and 500 Ω.

caused by a lead wire grounding and disconnection by melting, on the basis of recorded time dependence of the current fluctuation. Numerical calculations were made for the oscillatory voltage generation by the introduction of rush current at the point of the lead wire.

The results are found to explain the possible generation of high voltage that could lead to the trouble. From the results of the above analysis, a method of current-limiting resistors are proposed as the countermeasure for this problem. For the industrial applications of large scale superconducting magnets, the protection from damage shall be very important issue. The above results might increase the robustness of a superconducting magnet safety design.

REFERENCES

1. H.Hayashi, T.Imayoshi, H.Kanetaka, K.Tsutsumi, F.Irie et al., "Test Results of Elementary coils for Toroidal SMES", Advances in Cryogenics Engineering, 43B:1085 (1998).
2. H.Hayashi, H.Kanetaka, T.Imayoshi, K.Tsutsumi, F.Irie et al., "Results of Tests on Components and the System of 1kWh/1MW Module-type SMES", IEEE Trans. on Appl. Supercon.,1.9(2):313 (1999).

SAFE AND FAST QUENCH RECOVERY OF LARGE SUPERCONDUCTING SOLENOIDS COOLED BY FORCED TWO-PHASE HELIUM FLOW

L. X. Jia,[1] M. A. Green [2]

[1]Brookhaven National Laboratory
Upton, NY 11973
[2]Lawrence National Laboratory
Berkeley, CA 94720

ABSTRACT

The cryogenic characteristics in energy extraction of the four fifteen-meter-diameter superconducting solenoids of the g-2 magnet are reported in this paper. The energy extraction tests at full-current and half-current of its operating value were deliberately carried out for the quench analyses and evaluation of the cryogenic system. The temperature profiles of each coil mandrel and pressure profiles in its helium cooling tube during the energy extraction are discussed. The low peak temperature and pressure as well as the short recovery time indicated the desirable characteristics of the cryogenic system.

INTRODUCTION

The g-2 muon storage ring began operation at BNL in April 1997. Successful experience in the operation of the large superconducting magnet reflects well on many aspects of the cryogenic design. The g-2 magnet design was reported in 1994[1]. The cryogenic system design was reported in 1994[2] and 1996[4,6]. Early system tests were reported in 1996[3,5,8]. Design parameters for the g-2 power supply and quench protection system were evaluated by M. Green in 1993[7].

The g-2 magnet ring consists of four superconducting solenoids 15 meters in diameter. The total cold mass at 4.5 K is 6.2 tons which is indirectly cooled by forced two-phase helium flow in tubes 200 meters in length. The magnet is operated at 5200 A and the central field is 1.45 T. Total stored energy is 6.1 MJ. The refrigerator plant is capable of delivering 625 W at 4.5 K.

This paper discusses the cryogenic characteristics of its safe and fast recovery after the system quenches. The energy extraction tests at full-current (5175 A) and half-current (2450 A) were deliberately carried out for the quench analyses. The temperature and pressure profiles in the coil mandrels and cooling tubes during the energy extraction were measured. A maximum temperature of 38 K and a maximum pressure of 0.68 MPa were observed when the magnet quenched at full current. The four large solenoids were re-cooled down to superconducting temperature within 30 minutes after the full-current quench.

THREE SOLENOIDS AND COOLING SYSTEM

A cross section of the g-2 muon storage ring is shown in Figure 1. The four coil assemblies were built into three cryostats. The two OUTER solenoids supported by one mandrel share one common cryostat. The cold mass of the superconductor and aluminum mandrel in OUTER cryostat is 4010 kg. The thermal heat load in the OUTER cryostat is 60 W. The two cooling tubes of 16.2 mm hydraulic diameter and total length of 96 meter are arranged in a parallel flow path in opposite directions in OUTER mandrel. The inlet and outlet of helium to the cooling tubes are located at the same interconnect position. The two INNER solenoids are built in separate cryostats. The cold mass of superconductor and aluminum mandrel in each INNER cryostat is 1040 kg. The thermal heat load in the INNER cryostat is 61 W. One cooling tube of 16.2 mm hydraulic diameter is attached to the coil mandrel in each inner cryostat. The total length of cooling tube in each INNER cryostat is 42 meters. The inlet and outlet of helium to the cooling tubes are located at the same interconnect position.

There are three major parallel flow circuits in the cryogenic system shown in Figure 2: the main solenoid cooling circuit, the gas cooled leads circuit, and the cooling circuit for the Inflector solenoid and its leads. The low quality two-phase helium flow is delivered from the sub-cooler so called control dewar.

The three cryostats are linked together through a cryogenic interconnect chamber. In order to reduce the number of cryogenic control valves and the heat load, the main solenoid cooling circuit was designed so that the three cryostats can be cooled individually, or any two of the three, or three together. When the three rings are cooled together, they can be cooled either in series or in parallel flow schemes. The cooling circuit for the liquid nitrogen shield was constructed in the same pattern as the liquid helium flow circuit. These flow patterns are shown in Figure 3.

Using the combination of the four valves one can distribute the mass flow rates to each of the three rings and get the desired cooling performance according to cooling power requirements. It is important because the pressure drops in the OUTER and INNER cooling tubes are different even though the parallel-cooling pattern is employed. Because each ring can be isolated from others, the cooling circuit has great advantages in various tests during engineering runs. The pressure relief valves located at the inlet and outlet of the cooling flow are also directly connected to the interconnect between the OUTER cryostat and each of the two INNER cryostats. This reduces the tube length between the relief valves so

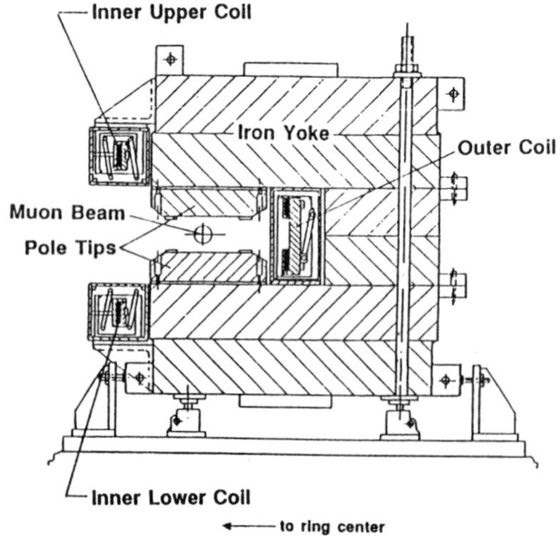

Figure 1. Cross section of the g-2 storage ring

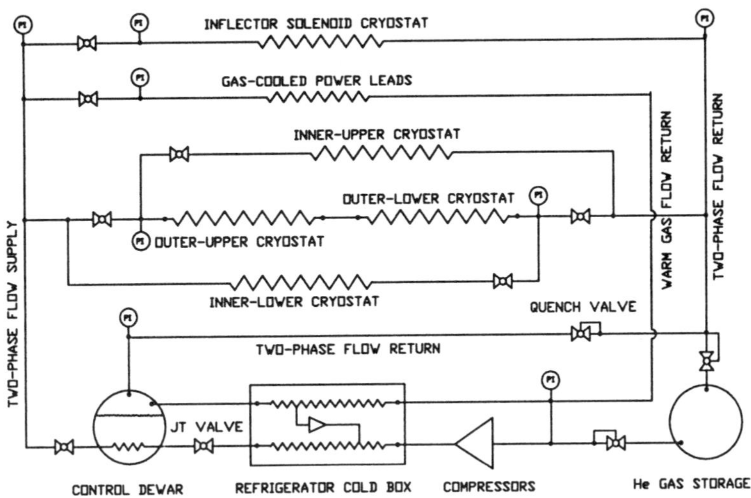

Figure 2. Simplified flow diagram of the g-2 cryogenic system

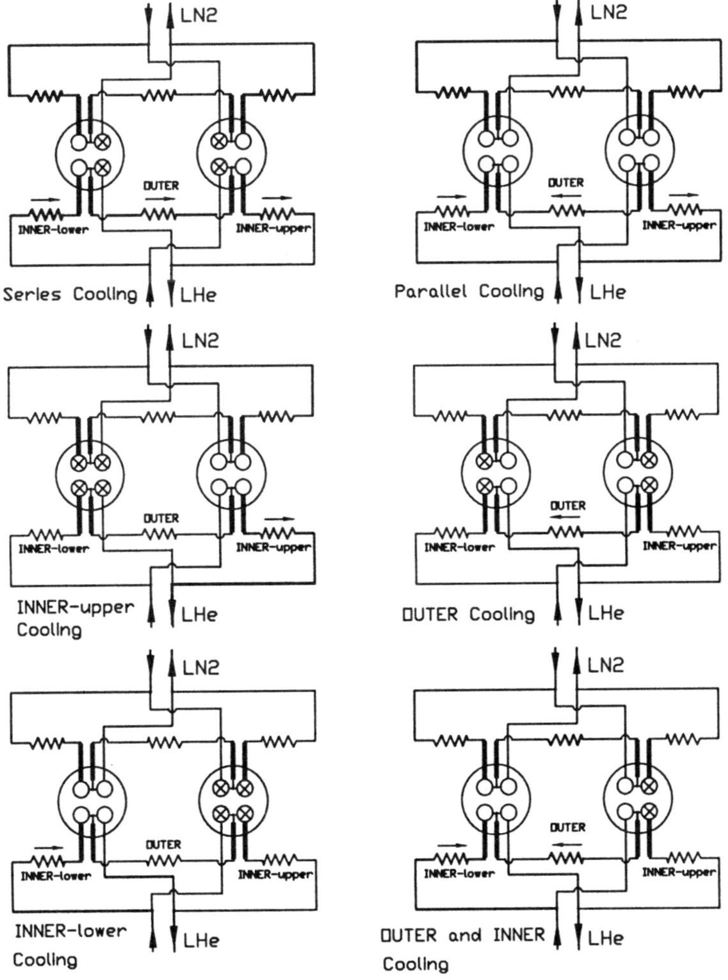

Figure 3. Interconnections of three cryostats for different cooling purposes

as to reduce the maximum pressure when the magnet quenches.

ENERGY EXTRACTION

To study the quench behaviors of the four solenoids and to test its cooling system, a series of energy extraction tests were carried out at low current, half-current and full-current comparatively with the magnet operating value. The magnet discharge was initiated by introducing a dump resistor in quench protection system or switching off the power supply of the magnet. Since the solenoids are all surrounded by the aluminum mandrels, when the dump resistor is put across the electrical power leads eddy currents are induced in the aluminum mandrel. The heat generated by the eddy current in the mandrel drives the superconducting solenoid above the critical temperature, which causes them to become normal. This mechanism also applies to the situation when part of the superconducting solenoid quenches. The phenomenon is called the "quench-back". The quench-back is beneficial if a spontaneous quench has occurred, as the energy gets distributed uniformly across the solenoid mandrel assembly. As a result, the hot spot temperature at the point where the magnet quench occurred is reduced. The quench-back can also occur in between each solenoid as the case of the g-2 magnets. One quenched solenoid may drive the other solenoid packages to quench, which will also avoid the entire stored magnetic energy to dissipate in a single solenoid. In this paper, the temperature and pressure histories of solenoid mandrels and their cooling flows as the g-2 magnet quenches are presented and discussed.

Two groups of temperature and pressure profiles of solenoid mandrels undergoing energy extraction are given in Figure 4 for half-current and full-current quench respectively. In each group there are four plots, three temperature plots for each of the three solenoids and one pressure plot for the cooling fluid. The four temperature profiles given in each temperature plot contain four measurements at evenly positioned points around each ring, which is also illustrated in Figure 5. The flow direction in the cooling tube of each solenoid mandrel is also shown in each figure. It should be noted that the cooling flows in the three cryostats were not isolated from each other during the tests. This allows the high pressure and high temperature gas in one quenched solenoid to be quickly diverted to the others and dissipate the stored energy. The cryogenic characteristics of this process are discussed in the following sections.

HALF-CURRENT ENERGY EXTRACTION

The OUTER solenoid quenched 7.6 seconds after the magnet started to discharge. The temperature of the mandrel was 7.8 K (critical temperature) at this time. It took about 90 seconds to reach the average peak temperature of 20 K. The entire mandrel was cooled back to 5 K in 12 minutes. The helium flow direction in the cooling tube close to these four temperature sensors was clockwise. The temperature sensor at 3 o'clock was the one upstream and close to the inlet of cooling flow. This portion of the mandrel started to drop the temperature at 30 seconds. The temperature profiles showed a clear gradient along the ring because of continuous cooling flow through the tube. The initial rate of temperature rise was 0.27 K/sec. The rate of re-cooling of the mandrel was 0.2 K/sec.

Neither of the INNER solenoids quenched at half-current energy extraction. The eddy current heating induced by quick discharging of the magnet reached its first peak in 2 seconds and raised the temperatures about one degree. This early dissipation of heat continued for 7 seconds. The temperatures increased again when the OUTER solenoid quenched. One can observe that the entire ring was not heated up uniformly along the

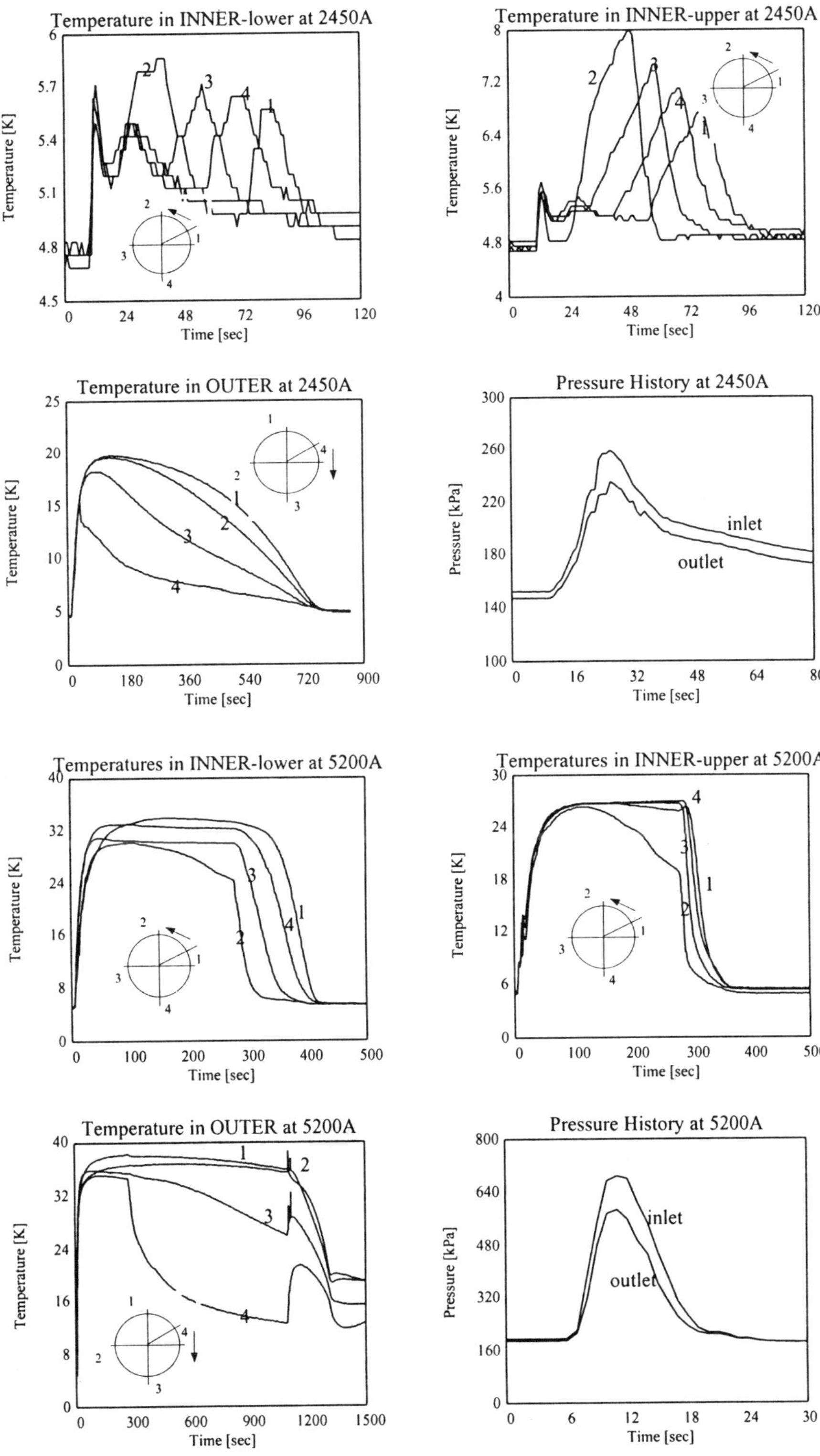

Figure 4. Temperature and pressure profiles during energy extraction

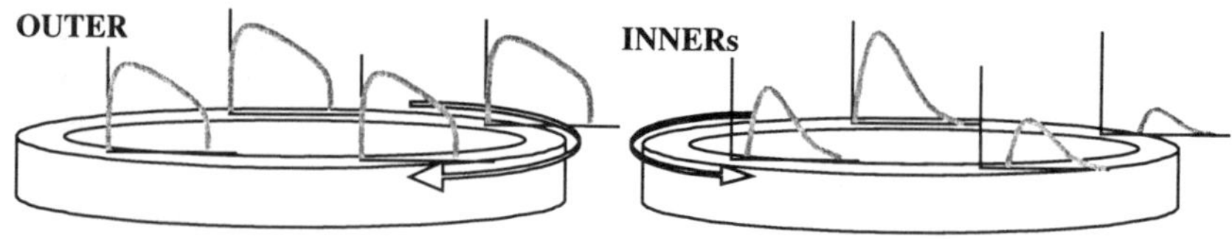

Figure 5. Schematic of temperature profiles along the solenoid mandrels

INNER mandrel ring. Every portion of the ring experienced an "A" shape of temperature up-and-downs in about 40 seconds in sequence. There is also a decay pattern in the peak temperature along the ring in the counter clockwise flow direction. The front wave of peak temperature travels at a velocity of 1.5 m/sec. It indicated that not only did cooling fluid continuously carry the heat in the INNER mandrels away, but also the heat was carried in by warm fluid in the quenched OUTER mandrel tube. This phenomenon suggested that not only did the quench-back distribute the stored energy from one solenoid to others, but also the back-flow diffused the energy through each other ring. Like the quench-back, the back-flow also helped to reduce the peak temperature in the cooling tubes. The two INNER solenoids were cooled back to 5 K in 90 seconds.

FULL-CURRENT ENERGY EXTRACTION

For the 5200 A energy extraction, all four solenoids quenched and quench-back occured in 2 seconds. About 60 percent of total stored energy was deposited into three solenoids and mandrels. The maximum temperature of the INNER-lower mandrel was 34 K and that of the INNER-upper was 27 K. The maximum temperature of the OUTER mandrel reached 38 K. The recovery time for the INNER-lower mandrel was 7 minutes and that for INNER-upper mandrel was 6 minutes. The recovery time for the OUTER mandrel was about 30 minutes. The temperature histories are also given in Figure 4.

The peak pressure in the cooling tube increased with mandrel temperature and the rate of thermal energy transfer to the helium in the cooling tube. The pressure relief valves were installed at the upstream and downstream ends of each cryostat. Once the magnet quenched, the pressure in the flow circuit was regulated by these valves, which allowed the warm gas to by-pass the control dewar and to exhaust directly into the gas buffer tank. Because of these control valves, the pressure in the control dewar as well as that in the mandrel cooling circuits is maintained at a desired low level during the magnet quenches. Meanwhile, the sub-cooled helium in the control dewar can continue to flow into the cryostats as long as the quench pressure is relieved, which carried away thermal energy to reduce their maximum temperature.

The highest measured pressure in the cooling tube for a quench at full-current was 0.68 MPa. The pressure reached the maximum in 5 seconds and decayed back in another 10 seconds. Within this 15 seconds, about 4 kilograms of helium stored in the cooling tubes were expelled through the pressure control valves. Once the pressure dropped, the cooling fluid started to flow in and to carry the heat away from the warm mandrel and to drop the temperature of mandrels. The pressure transducers were located in the inlets and outlets of three cryostats. The points in the cooling circuit some distance from the pressure transducer may have been at a pressure up to 40 percent higher than the pressure measured at the highest pressure transducer.

CONCLUSION

The cryogenic characteristics of the g-2 magnet quenches was investigated by discharging the magnet using the dump resistor and power supply switch. The phenomena of the quench-back by eddy current heating and the back-flow of warm gas from quenched solenoid have been observed. The low peak temperature and pressure in long cooling tube was obtained at low and high level current quenches. The operating temperature of superconducting solenoids was recovered within a half-hour. The cryogenic system may be ready for re-powering these large superconducting solenoids within an hour. Safe and fast quench recovery is obtained in operating the g-2 magnet. This was achieved not by using large capacity of refrigeration but by rational system design. There are some advantages shown in the cryogenic design of the g-2 superconducting magnet, which includes the use of a large buffer volume, sub-cooled control dewar, by-pass quench gas circuit, versatile interconnect of three cryostats, and automatic cryogenic control.

ACKNOWLEDGMENT

This project was supported by Office of Energy and Nuclear Physics, High Energy Physics Division, United States Department of Energy.

REFERENCES

1. The g-2 Collaboration, The g-2 storage ring superconducting magnet system, *IEEE Trans. on MAG-30*, 4:2423 (1994)
2. L. X. Jia, et al., Cryogenics for the muon g-2 superconducting magnet system, *Cryogenics*, 34:87 (1994)
3. G. Bunce, et al., Progress report on the g-2 storage ring magnet system, *IEEE Trans. on MAG-32*, 4:3057 (1996)
4. L. X. Jia, et al., Cryogenic system for the muon g-2 superconducting magnet, *Proc. of 16th ICEC/ICME*, 871 (1996)
5. L. X. Jia, et al., Cryogenic tests of the g-2 superconducting solenoid magnet system, *Advance in Cryogenic Engineering*, 41:383 (1996)
6. L. X. Jia, et al., Forced two-phase helium cooling of four 15-meter-diameter superconducting solenoids, *Proc. of 16th ICEC/ICME*, 875 (1996)
7. M. Green, Design parameters for the g-2 solenoid power supply and quench protection system, *Muon g-2 Note 159*, (1993)
8. L. X. Jia, et al., Cryogenic transient heat transfer in the g-2 cryostats at thermal insulation vacuum loss, *Advance in Cryogenic Engineering*, 43:621 (1998)

SIMULATION OF THERMAL-HYDRAULIC TRANSIENTS IN TWO-CHANNEL CICC WITH SELF-CONSISTENT BOUNDARY CONDITIONS

L. Savoldi,* L. Bottura,+ and R. Zanino*

* Politecnico, Dipartimento di Energetica
Torino, I-10129, Italy
+ CERN, Division LHC
Geneva, CH-1211, Switzerland

ABSTRACT

The use of boundary conditions at the conductor ends, taken from the experiment, has recently allowed an accurate thermal-hydraulic simulation of both quench [1] and heat slug [2,3] transients in the two channel cable-in-conduit conductors (CICC), using the 2-fluid MITHRANDIR code [4]. However, in order to be used as a design tool, i.e., to achieve a predictive capability, the code should be independent as much as possible of input from the experiment. Therefore it is necessary to couple MITHRANDIR to a hydraulic network simulator such as FLOWER [5], providing a self-consistent description of thermal-hydraulic transients in a cryogenic plant. We show here how the coupling is achieved and demonstrate the reliability of the coupled codes against quench and heat slug propagation runs from the QUELL experiment [6] in the SULTAN facility at Villigen PSI, Switzerland. The results show good agreement with experimental data and with simulations performed using experimental boundary conditions. Different levels of detail in the modeling of the hydraulic network are investigated for different types of thermal-hydraulic transient.

INTRODUCTION

The 2-fluid MITHRANDIR code [4] was developed specifically to simulate thermal-hydraulic transients in super-conducting cables with a two-channel topology, and was validated against quench [1] and heat slug propagation [2,3] in the QUELL experiment [6], assuming a given (experimental) pressure at the inlet and outlet of the conductor sample. Two-channel cable-in-conduit conductors have now been chosen for the Toroidal Field Model Coil (TFMC) and the Central Solenoid Model Coil (CSMC), in the frame of the International Thermonuclear Experimental Reactor (ITER) project, and the MITHRANDIR code will be used in the assessment of the test program of the TFMC. Therefore, it becomes

Advances in Cryogenic Engineering, Volume 45.
Edited by Shu *et al.*, Kluwer Academic / Plenum Publishers, 2000.

critical that the code can be used in a *predictive* mode, as opposed to interpretative simulations. This condition can be satisfied by numerically modeling the *entire* hydraulic circuit, provided its parameters are known, in order to compute self-consistent pressure p_{in} and temperature T_{in} at the sample inlet, and outlet pressure p_{out} – a typical set of boundary conditions.

Here we shall model the super-conducting cable (i.e., the sample) using the MITHRANDIR code, and the rest of the cryogenic circuit using a specific hydraulic network solver, i.e., FLOWER [5].

FLOWER has been developed specifically to supply self-consistent boundary conditions to the 1-fluid GANDALF code. The coupling of the two modules was recently validated against quench initiation and propagation in the QUELL experiment [7].

Here we couple FLOWER to MITHRANDIR, and compare the simulations with experimental results of quench initiation and propagation, and of heat slug propagation, in QUELL.

MITHRANDIR-FLOWER COUPLING

A cryogenic system for a super-conducting coil cooled by forced-flow of supercritical helium can be simulated by FLOWER after identifying the principal components of the circuit. These components are divided into two main categories: junctions (i.e., pipes, pumps or compressors, heat exchangers, valves) and volumes (i.e., reservoirs or manifolds). They are modeled in the code by means of a restricted set of parameters, e.g., cross section, length, hydraulic diameter and friction factor, for the pipes, or volume V, temperature T and pressure p, for the reservoirs, see Bottura and Rosso [5], Marinucci and Bottura [7] and references therein for details. All the different elements must then be connected in a loop, closed by the super-conducting sample.

The coupling between MITHRANDIR and FLOWER is achieved through an explicit staggered time integration of the hydraulic network and the cable. At each time step MITHRANDIR provides inlet and outlet mass and energy flux, as input to FLOWER. The latter uses the flux values to compute pressure and temperature in all the junctions and volumes. It thus can feed back MITHRANDIR with p_{in}, T_{in}, p_{out} in a self-consistent way. No iteration is performed because MITHRANDIR uses the boundary conditions computed by FLOWER in the previous time step.

THERMAL-HYDRAULIC TRANSIENTS IN QUELL

Description of the Cryogenic Circuit

In Fig.1 a simplified sketch is shown of the cryogenic circuit of the SULTAN facility supplying helium to the QUELL sample. Pressurized helium flows from the cold-box (a two-phase component) through a valve-box that provides plant regulation, and then through the cryostat, i.e., a liquid bath heat exchanger. The supercritical helium flow from the cryostat outlet is split between sample and sample terminals plus current leads. Note that a significant fraction of the helium flow is needed in the refrigeration of the terminals and current leads (~50 % in most cases). A control valve CV and a heater are used to manage the flow and inlet temperature in the sample. The helium flows from the sample and the parallel path back to the cryostat, through two Joule-Thomson valves (JTV), and from there to the cold-box. If a strong increase of the inlet and outlet pressure takes place (e.g., in the case of quench of the sample), the fluid can vent into a reservoir through two relief valves (RV). Unless otherwise mentioned, all circuit data are obtained from Bruzzone and Marinucci [8].

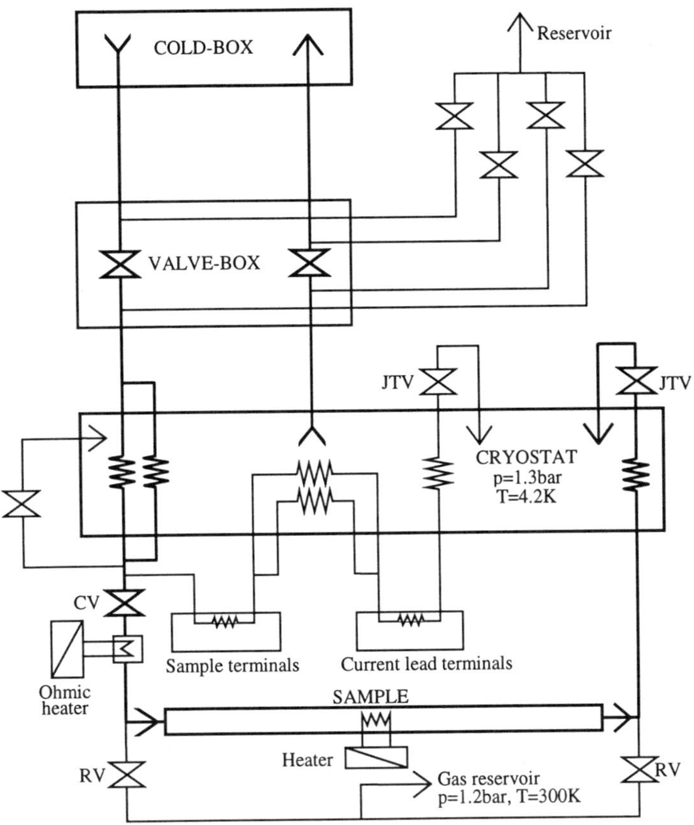

Figure 1. Simplified sketch of the cryogenic circuit of the SULTAN facility for the QUELL experiment.

Model of the Cryogenic Circuit for Quench Studies

In order to decide which components of the circuit are to be considered in quench simulations, we can observe that in quench studies p_{in} and p_{out} are expected to be driven mostly by quench evolution inside the sample (because of the strong heating induced flow), while the cryogenic circuit is less important. Furthermore, the time scale O(1-10s) of the circuit response to an external perturbation is comparable or longer than the time scale O(1s) of quench propagation. Therefore, even a relatively simple cryogenic circuit model can lead to acceptable results in quench simulations.

For this reason we use here a very simplified circuit model, shown in Fig.2a. The choice of components is the same as presented by Marinucci et al. [7], but their quantitative characterization is different. Two manifolds M_{in} and M_{out} are located at the inlet and outlet of the sample, respectively. M_{in} and M_{out} are considered here to account for the physical helium volume contained in the pipes connecting the sample to the cryostat. The two manifolds are linked by a *fictitious* compressor that gives the needed pressure head. The compressor emulates the cold-box and the pipes linking the latter with the cryostat. The cryostat is assumed to provide only a localized pressure drop, and it is "hidden" in that part of the circuit modeled as a compressor. The parallel path through the sample terminals and current lead terminals is neglected in this model by assuming that the compressor operates with ~50% of the helium volume actually flowing through the cold-box. The relief valves are assumed to open at an absolute pressure of 10.5bar.

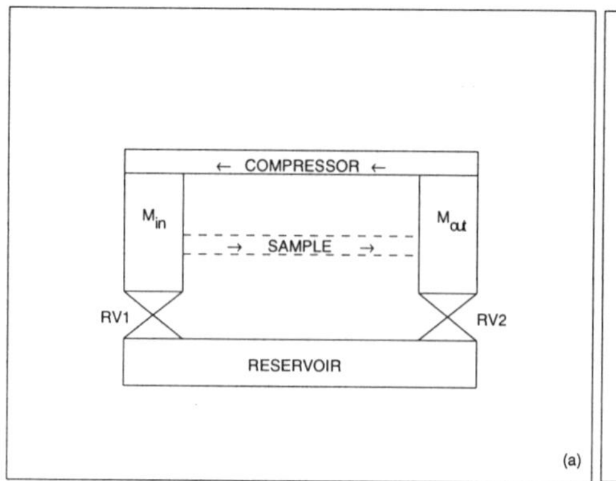

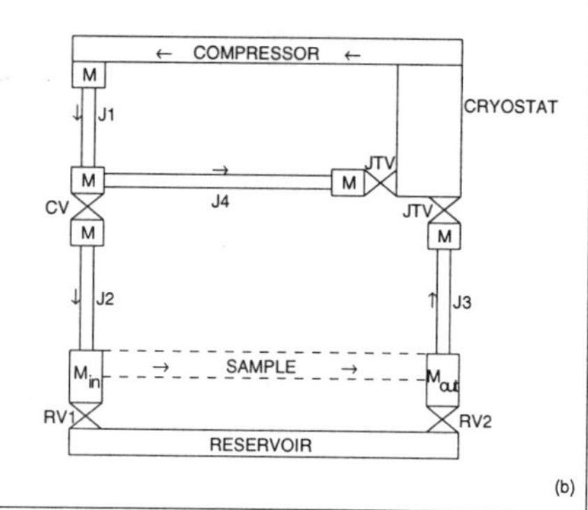

Figure 2. Two models of increasing complexity for the circuit in Fig.1. (a) The active component (compressor) acts directly on the sample. (b) Tubes (J1-J3), JTV's, "cryostat" and CV are interposed between the sample and the compressor; a simple model (J4) of the parallel path is also included. Fictitious manifolds M are included to link different junctions.

The compressor is defined by its characteristic in the (G,Δp) plane, where G is the mass flow rate and Δp is the pressure head, which is determined by two parameters. Since we have only one (G,Δp) experimental point, i.e., the operation condition at the beginning of the quench experiment, the second parameter has been chosen to give a qualitatively good global agreement with the experimental results. Notice that this is not an "ad hoc" recipe because, on the quench time scale, the helium pressure in M_{in} and M_{out}, influenced somehow by the compressor characteristic, is much more strongly influenced by the correct definition of the helium volume contained in the two manifolds.

Notice finally that in an earlier publication [7] only the part of the circuit up to the cryostat was considered, with smaller M_{in} and M_{out}, and less fluid flowing in the compressor. This results in a significant underestimation (more than a factor of 20) for the helium volume inside the circuit, compared to the present case.

Analysis of a Quench Run

We consider a "standard" quench (Run #2) [1] of QUELL.

The QUELL input for the MITHRANDIR code (both conductor geometry and material properties) is the revised one used in earlier studies [2,3]. The external heating is supplied by a resistive heater wound directly around the jacket.

The results of the simulation are presented in Fig.3. Figure 3a shows a very good agreement between the experimental and the computed total voltage drop. In Fig.3b the experimental quench front propagation, as deduced by the switching on of the voltage signal at the different voltage taps, is also very well reproduced by the simulations both with FLOWER and with experimental boundary conditions. The pressures p_{in} and p_{out} (Fig.3c and 3d, respectively) are somewhat underestimated by FLOWER (relative standard deviation σ of p_{in}, $p_{out} \approx 8\%$). Still, since the experimental pressure drop along the sample is well reproduced by the simulation (not shown), the computed quench evolution is in good agreement with the experiment. The pressure underestimation is partly due to an overestimation of the volume of helium available for the sample refrigeration. Indeed, if only a part of the cryogenic circuit is taken into account [7], p_{in} and p_{out} increase too quickly (not shown), leading to $\sigma \approx 12\%$, because the volume of helium to be pressurized in the cryogenic circuit is too small. The sensitivity of the results to the calibration of the compressor characteristic is small.

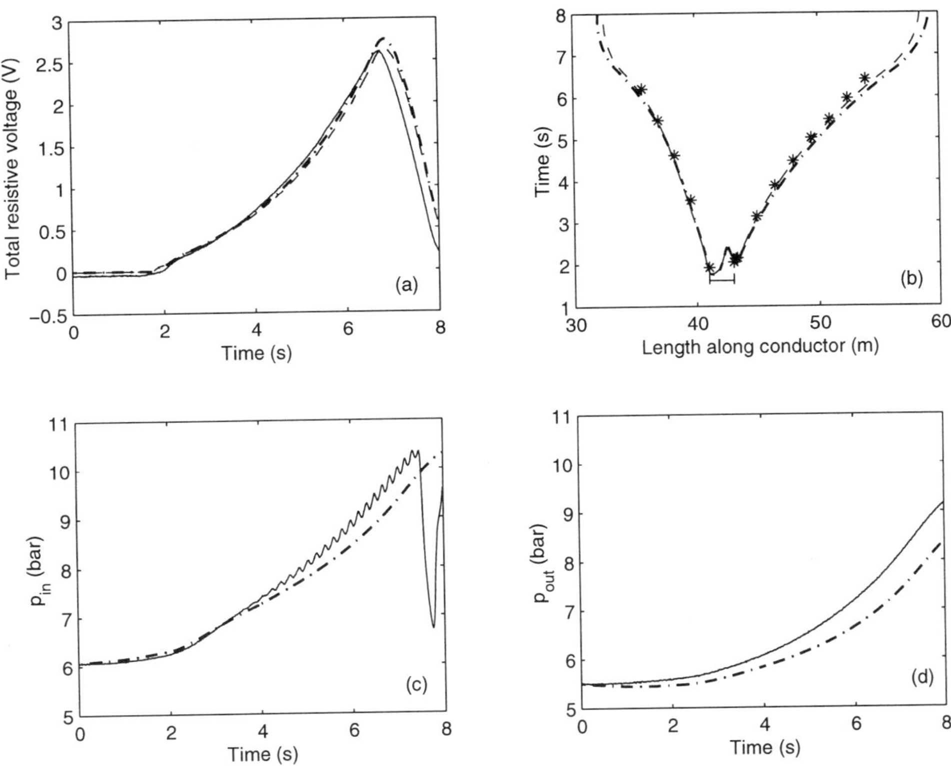

Figure 3. QUELL Quench #2 [1]: experimental data (solid,*), results computed with FLOWER boundary conditions using the circuit model in Fig.2a (dot dashed), results computed with experimental boundary conditions (dashed). (a) Total resistive voltage as a function of time. (b) Quench front propagation along the conductor as a function of time. (c,d) Pressure evolution at the sample inlet and outlet, respectively.

In the experiment the opening of the relief valve leads to a collapse of the inlet pressure This is not observed in the simulation because of the slower increase in the computed p_{in}.

Analysis of an Inductive Heat Slug Run

If we perform simulations of heat slug propagation (zero current and field) with the circuit model in Fig.2a, we obtain different degrees of accuracy in the results depending on the amount of energy deposited and on the external heating duration.

In an inductively heated case (Run #008 [2,3], linear input power Q0=63125W/m supplied on a length L_H=0.12m), a reasonable agreement is found between computed results and experimental data, see Fig.4. The computed jacket temperature evolution at different sensors along the conductor (Fig.4a) is shown to be as accurate as that with experimental boundary conditions. The characteristic of the compressor in the (G,Δp) plane has been changed with respect to the quench run, emulating the operation of CV and JTV in the real circuit, in order to provide the correct G_{in} at steady state (see Fig.4b). This explains the smaller phase shift at the temperature sensors (Fig.4a) with respect to the results obtained with experimental boundary conditions, which give an overestimation of G_{in}. The computed p_{in} and p_{out} (Fig.4c and 4d, respectively) show a qualitatively different behavior with respect to the experiment. In the simulation, the pressure tends to stabilize around a value higher than the initial one because the circuit, that can be thought here as a 0-D object, reacts to the energy supply, pressurizing. In the experiment, the cryostat possibly absorbs the energy input as latent heat of condensation/evaporation, and the pressure, after a few oscillations, tends to return to its initial value. The accuracy in the results, notwithstanding the relatively simple circuit model, is mainly due to the fact that the inductive heater acts on a short time scale (10ms). Indeed on this time scale the perturbation of the hydraulic circuit conditions is very small (i.e., the variation of p_{in} and p_{out} is < 0.5bar, see Fig.4c and 4d).

Model of the Cryogenic Circuit for Resistive Heat Slug Studies

In a resistively heated case (Run #012 [2,3], Q0=2306W/m, L_H=2.3m), the circuit model in Fig.2a gives a significant disagreement with the experiment (not shown). The external heating, which effectively acts on a much longer time scale (~10s) than the inductive one, leads here to significant changes in pressure and mass flow (see below). Furthermore, in the second phase of the transient the circuit response to the perturbation becomes the driving force of the transient evolution. A more accurate circuit model is thus needed in order to get a realistic response of the cryogenic circuit to the perturbation caused by the sample. Notice that the heat slug conditions in QUELL were rather different from those expected in the TFMC [2,3].

A more detailed circuit model is shown in Fig.2b. The cryostat is implemented as a one-phase reservoir (V=O(m^3), i.e., much larger than the real two-phase volume). Provided V is sufficiently large, the helium temperature and pressure in the reservoir will not change significantly from the initial values, behaving as in the cryostat. Differently from the previous model, the parallel path is included and pipes, CV and JTV are present here.

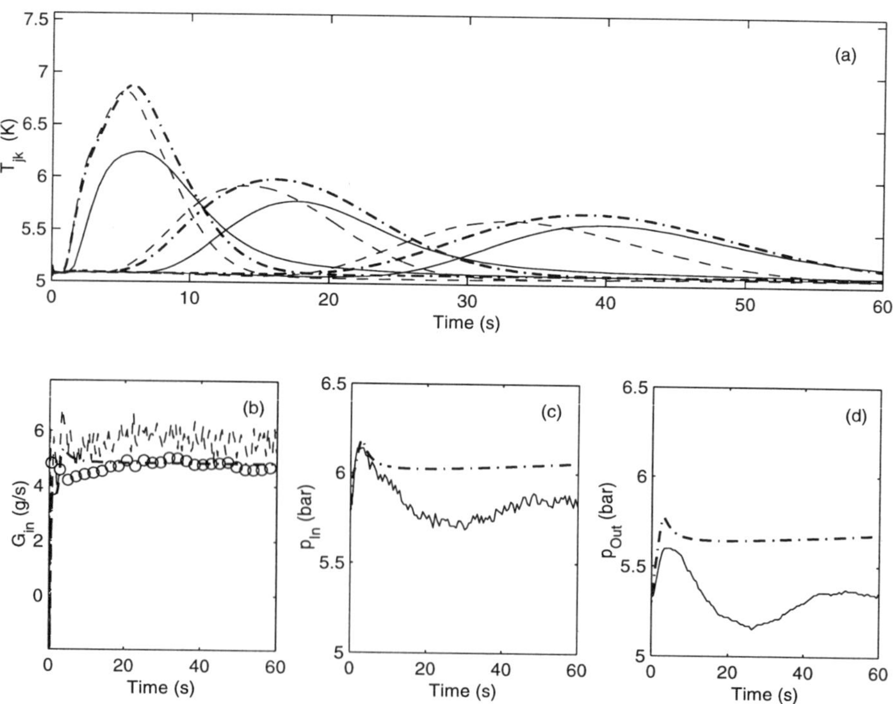

Figure 4. QUELL inductive heat slug #008 [2,3]: experimental data (solid, open circles), results computed with FLOWER boundary conditions using the circuit model in Fig.2a (dot dashed), results computed with experimental boundary conditions (dashed). (a) Jacket temperature at the sensors TA5, TA6, TA8, respectively. (b) Sample inlet mass flow rate as a function of time. (c,d) Pressure evolution at the sample inlet and outlet, respectively.

The compressor now takes into account *all* the helium flowing through the cold-box. The JTV's (implemented as valves with a high localized pressure drop coefficient) regulate the helium access to the cryostat. The CV allows, together with the calibration of the compressor characteristic, a fine regulation of G_{in}. For the parallel path J4 no heat exchange (see Fig.1) is taken into account, but the higher hydraulic resistance, due to the helium heating and to the path tortuousness, is emulated by a high friction factor. Since J4 is approximately a factor of five longer than the other tubes, and has a higher pressure drop, it is worthwhile to simulate it with compressible helium flow, because its hydraulic evolution is expected to strongly affect the sample. Pipes J1, J2 and J3 are defined by their physical parameters [8].

The compressor calibration is now expected to have an even smaller influence on the transient evolution, at least because of the presence of the parallel path J4 in the circuit model.

Analysis of a Resistive Heat Slug Run

The results obtained for Run #012 with the circuit model in Fig.2b are shown in Fig.5. The difference between the peak jacket temperature computed with FLOWER, and the experimental ones, is < 0.5K, while the phase shift is < 4s (Fig.5a). The computed G_{in} (Fig.5b) decreases faster than the experimental one in the first part of the transient. This effect can be partly attributed to some kind of delay in the response of the two-phase components (not implemented in FLOWER at present), due to condensation/evaporation of

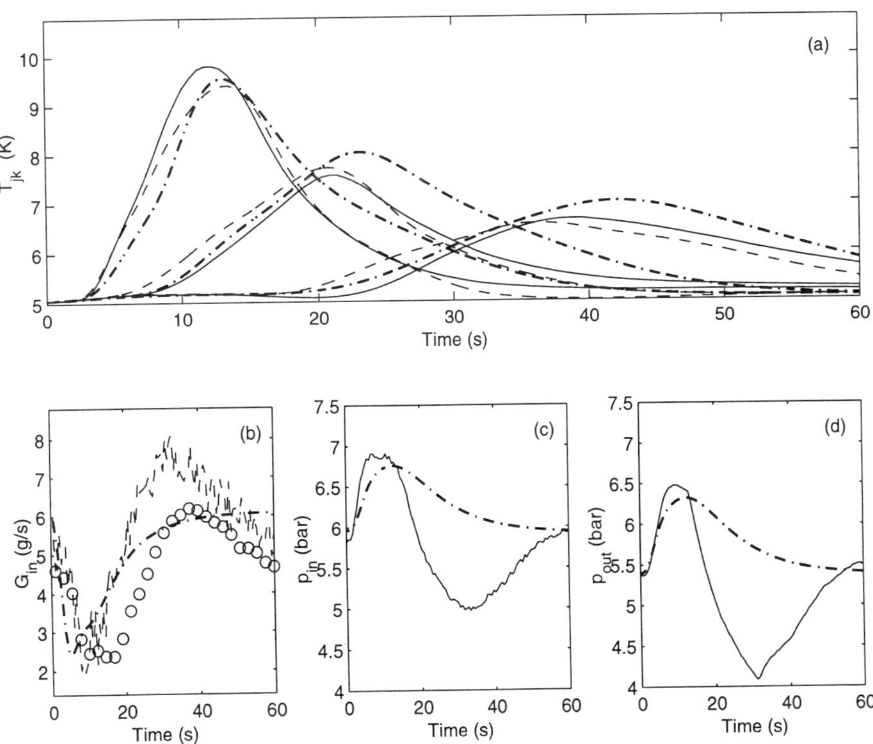

Figure 5. QUELL resistive heat slug #012 [2,3]: experimental data (solid, open circles), results computed with FLOWER boundary conditions using the circuit model in Fig.2b (dot dashed), results computed with experimental boundary conditions (dashed). (a) Jacket temperature at the sensors TA5, TA6, TA8, respectively. (b) Sample inlet mass flow rate as a function of time. (c,d) Pressure evolution at the sample inlet and outlet, respectively.

helium. After the first decrease, the computed G_{in} monotonically increases till the end of the transient, while both the experimental signal, and the evolution computed with experimental boundary conditions give a non-monotonic behavior, linked to the experimental pressure drop evolution (not shown). The boundary pressures provided in this case by FLOWER are again only qualitatively in agreement with the experimental ones (Fig.5c and 5d), which behave similarly to the inductive run. After a first increase, p_{in} and p_{out} decrease to their initial value because the supplied energy can now be absorbed by other components of the circuit (as opposed to what occurs with the circuit in Fig.2a, where p_{in} and p_{out} asymptotically increase, not shown). However, the decrease is slower than in the experiment, possibly due again to the absence of two-phase components in the model.

The circuit model in Fig.2b has also been used (not shown) for the analysis of the quench run considered in the previous section, giving comparably good results.

CONCLUSIONS

A hydraulic network solver, FLOWER, has been coupled to the two-fluid MITHRANDIR code. The cryogenic circuit of the QUELL experiment has been modeled with different levels of detail in order to simulate different kinds of transient. For fast heating transients (i.e., quench or inductive slugs), even a simple model of the circuit leads to reliable results in good agreement with the experiment, albeit for different reasons, provided a realistic estimation of the total helium volume in the circuit is given. For slow heating transients (resistive slugs), where the effective heating time scale is comparable to the circuit response time scale, a more accurate model of the circuit is needed (larger number of components, and suitable fluid equations, i.e., compressible flow) to obtain acceptable results.

ACKNOWLEDGEMENTS

The work at POLITO has been supported by the EURATOM Fusion Technology Program, under contract to R.Z.. The authors would like to thank C.Marinucci and P.Bruzzone of CRPP, Villigen, for providing information and data on the SULTAN facility.

REFERENCES

1. R.Zanino, L.Bottura and C.Marinucci, Computer simulation of quench propagation in QUELL, *Adv. Cryo. Eng.* 43:181 (1998).
2. R.Zanino and C.Marinucci, Heat slug propagation in QUELL. Part I: experimental setup and 1-fluid GANDALF analysis, to appear in *Cryogenics* (1999).
3. R.Zanino and C.Marinucci, Heat slug propagation in QUELL. Part II: 2-fluid MITHRANDIR analysis, to appear in *Cryogenics* (1999).
4. R.Zanino, S.DePalo and L.Bottura, A two-fluid code for the thermohydraulic transient analysis of CICC superconducting magnets, *J.Fus. Energy* 14:25 (1995).
5. L.Bottura and C.Rosso, Hydraulic network simulator model, Internal Cryosoft Note, CRYO/97/004 (1997).
6. A.Anghelo, et al., The QUench Experiment on Long Length QUELL - Final Report, EPFL - CRPP, JAERI, MIT-PFC and SINTEZ-NIIEFA Report (1997).
7. C.Marinucci and L.Bottura, The hydraulic solver Flower and its validation against the QUELL experiment in SULTAN, to appear in *IEEE Trans. Appl. Supercond.* (1999).
8. P.Bruzzone, C.Marinucci, private communication (1999).

RESEARCH AND DEVELOPMENT OF SUPERCONDUCTING MAGNETIC SYSTEMS FOR HIGH POWER PULSED MHD-GENERATORS

E.P.Polulyakh,[1] A.V.Spiridonov,[1] V.A. Afanas'ev,[1] M.I. Kharinov,[1]
A. A. Yakushev[1] and E. Yu. Klimenco[2]

[1]SRC Troitsk Institute for Innovation and Fusion Research (TRINITI)
Troitsk, Moscow reg., 142092, Russia

[2]RRC Kurchatov Institute
Moscow, 123182, Russia

ABSTRACT

Superconducting Magnetic Systems (SMS) of various configurations aimed at generation of magnetic fields up to 5 T induction in the operating volume of pulsed propellant MHD generators with up to 10 MW power and higher have been developed and built in TRINITI for gaining experience in design, optimization and technology of manufacturing. Possibilities of their weight - dimension characteristic optimization have been also presented. The main efforts were made to look for optimal decisions at the choice of a winding shape, operation current density and mechanical structure design. The conducted experimental investigations of SMS as a part of MHD setups have confirmed their reliability and agreement between their calculated and operation parameters. Further improvement of SMS aimed at upgrading their reliability and simplifying their maintenance is being considered. Light 20 - 50 MW MHD generators with superconducting magnetic systems are the most promising in the aerospace industry. The applicablity of these SMS for other purposes have been studied.

INTRODUCTION

Pulsed MHD generators with superconducting magnets gain absolutely new qualities and a higher efficiency. They become continuously available for service and work in a repetitive pulse regime. These unique qualities allow their use for nuclear power plant supply in difficult and severe situations such as accidents. They also use as power, transportable source of primary electromagnetic field for electric prospecting in geology, geophysics and earthquake forecasting. In the MHD-program, 5.0 T SMS for pulsed MHD generators at 10 MW have been designed, manufactured and tested.

MAGNET SYSTEMS DESCRIPTION AND TEST RESULTS

The cryostats construction promotes free access for placing and maintaining MHD channels in the operating area of the magnet system. A superconducting Nb-Ti alloy in a copper stabilizing matrix was used as a current-carrying material for windings. The SMS windings are distinguished both by shape (round, race-track and saddle) and stabilizing means (cryostatic, combined) (see Table 1).

As a first step in the development of a technique for design and manufacture of a SMS of an immersed type with round coils has been chosen as it seems to be the most reliable and simplest. The general view of the winding and a separate section is given in Figure 1 and 2. Among the SMS IM - 05 structural features a free access to a operating volume in three mutually perpendicucular directions should be noted, which presents a wide range of opportunities not only for experimental studies of MHD channels, but for applications of this magnet when conducting other physical, clinical - biological and technological researches. The R & D results obtained at creating of this SMS have served as basic for developing and building a large -scale SMS IM - 114 with race- track coils industrially (see Figure 3).

Table 1. Specifications of SMS for pulsed MHD generators TRINITI

Main characteristics	IM-05	IM – 114	IM – 06
Winding shape	Ring pare	Race-track pare	Sadlle
Operating volume, m × m × m	0.11×0.3×1.0	0.27 × 0.6×2.0	∅ 0.42 ×1.5
Outer diameter, m	1.1	2.0	0.72
Effective length,m	0.5	1.0	0.9
Magnetic field, T	4.5	2.8	3.0
Operating current density, A/m^2	5.7×10^7	5.6×10^7	1.7×10^8
Stored energy, MJ	3.7	5.0	1.5
Mass, kg	2.500	8.000	3.500
Conductor:			
cross-section, mm^2	10 ×1	10×1	2 × 3.5
material	NbTi / Cu	Nb Ti / Cu	Nb Ti / Cu
fraction of Nb Ti , %	5 - 18	10 - 20	35

Fig.1. The section of the ring winding.

Fig. 2. The coil of the magnet IM – 05 with the support structure.

The superconducting saddle-shaped magnet IM-06 (Fig.4), designed for the experimental pulse-repetition rate MHD generators has been created and tested in the operating cryostat. The cryostat has a horizontal "warm" operating zone of 400 mm diameter, with a uniform field length 1000 mm and the field value of 3.0 T at a 1150A feeding current. The main feature of this superconducting magnet is an incompletely-stabilized superconducting braid used as a current-carrying element in the monolithic (without cooling channels) saddle-shaped winding. The cross-section of the braid of NT50 alloy in copper matrix is 2×3.5mm NbTi/Cu ratio is 2:1. The braid has an insulating coating of lavsan yarn which was impregnated with an epoxy compound and boron nitride while being mounted and each layer of the winding was bandaged with glass band impregnated with the same compound. Then the winding was placed into a ring vessel formed by two concentric shells and two end connecting flanges.

Fig. 3. View of the industrial magnet IM – 114.

Fig. 4. The coils of the saddle magnet IM – 06 during winding.

The ring vessel was filled in with the said compound under pressure which resulted in a completely monolithic winding. All inner connections of the winding were brought outside and mounted on the fiberglass plate. The connections, lap-soldered with tin, are in good contact with liquid helium. In the experimental cryostat (1400 mm diameter) the magnet winding was tested three times: in the outer removable duralumin bandage, with no outer bandage and in an outer permanent bandage of stainless steel. It has been revealed that stiffness of the outer bandage defines the extent of degradation of the current-carrying capacity of the magnet system.In the temporary duralumin bandage a prolonged training of the magnet took place. After 20 quenches, the critical current reached 1160A. In the experiments with the removed temporary bandage during the first run the winding quenched at 900 A. The maximum diametral deformation (compression) of the inner support tube, measured at its limiting cross-section, was 10 mm, and residual deformation was 2 mm (Fig.5). Before mounting of the permanent bandage, the inner and outer jackets of the ring vessel containing the superconducting winding were turned on a carousel machine to attain a round cross-section. Then the permanent bandage was mounted. 15 I-rings of stainless steel were put on the outer cylindrical surface after its treatment and packed out (Figure 6). Each ring mass was 7 kg. The mass of the winding was about 1000 kg, the mass of the cryostat - about 2500 kg, the total mass of the magnet system - about 3500 kg. The winding in the permanent bandage was first tested in the experimental cryostat of 1400 mm diameter. At the first run 1164 A critical current in the winding was obtained with the value being obtained earlier in the temporary bandage after a prolonged training. Spontaneous quench of the winding in the permanent bandage occurred 6 times (Fig. 7). After the first two quenches the winding warmed up to room temperature. The critical current of 1222 A attained at the second run fell to 1190 A after warming up and then after three runs monotonically increased to 1266 A. At the maximum current diametral deformations in the middle cross-section of the inner tube were about 10 mm.

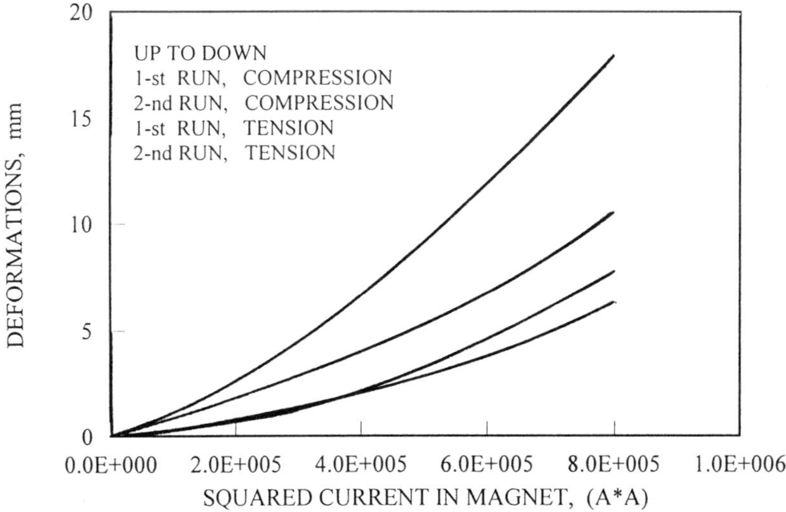

Fig. 5. Results of the experimental investigation of the SMS IM – 06 winding deformation.

Figure 6. Supporting system

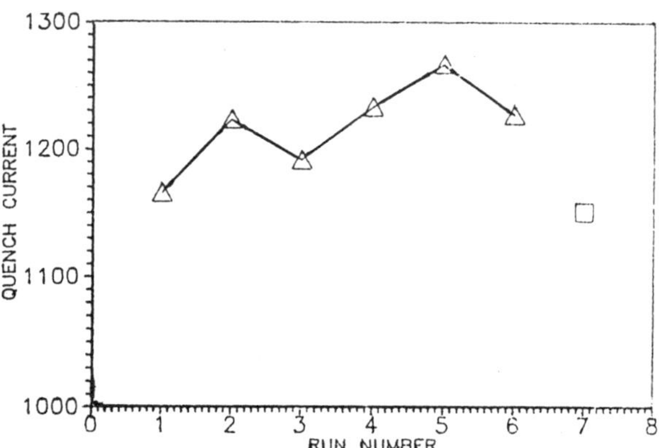

Figure 7. Behavior of SMS during the " training"

The final stage was testing the winding in the operating cryostat (Fig.8). Fiberglass chains with low heat conductivity were used in the fixing system of the helium vessel. Heat losses by each of four chains do not exceed 0.6 W. Evaporation of liquid helium was 7-8 liters per hour. At the current input the heat losses increased resulting in an additional evaporation of 2.5-3 liters of liquid helium per hour at 1000-1150 A. The liquid helium store in the cryostat allows to maintain the magnet in the operating state for 5 hours. In these experiments 12 runs took place. At five runs (N1,3,4,5 and 6) current did not exceed 625A. At the rest of runs (N2,7-11) the magnet successfully operated at a current level of 1000-1150 A and at current rates of 20-60 A/min; the "shelf" duration of the maximum current reached 30 min (at 1150 A corresponding to a field of 3.0 T in the operating zone).
The fragment of the record in Fig.9 shows a change of performances of the winding and cryostat during the tests.

Fig. 8. The general view of the MHD facility "Sever-M"with SMS IM-06.

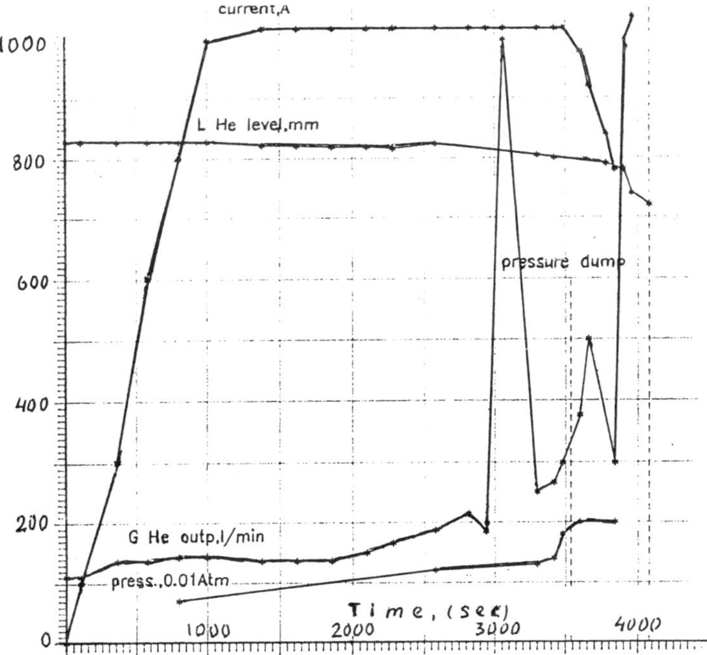

Fig. 9. The change of some performances of the cryostat during the tests.

The data in the left part of the figure corresponds to the values characteristic of the magnet. A certain rise in pressure and rate of gas helium flow observed in the experiment was due to warming up of one of the vapor-cooled current leads. Two sharp peaks of the rate of gas helium flow are concerned with attempts to damp the pressure in the cryostat by dumping to theatmosphere through the valve. After the beginning of current output from the magnet at 617 A, a defective current lead burnt out and an electrical break-down occurred, which resulted in burning the cryostat shell. Because of coming to a violent boil of liquid helium the cryostat opened. The damage to the cryostat has been eliminated, and the defective current lead has been modernized. After that this magnet successfully worked as a part of the MHD-generator(1).

The IM-06 magnet tested is more light-weight compared to a similar superconducting IM-114 magnet, where a cryostatic stabilization of the winding was implemented (see table 1). At present the magnet is used both in an autonomous regime and as a part of a pulsed MHD generator.

SUMMARY

Pilot superconducting magnet systems with a large aperture for pulsed 10 MW MHD generators have been developed and successfully tested. With these systems the MHD generators gain a higher efficiency and become continuously available for service and work in a repetitive pulse regime. The opportunities of these magnet systems applications for other purposes have been studied: high-gradient magnetic purification of liquids with paramagnetic impurities; a clinical and biological research; generation of additional magnetic field for electromagnetic launchers (2).

REFERENCES

1. Ye.P. Velikhov, V.P. Panchenco, E.P. Polulyakh et al., ''Experimental MHD facility with superconducting magnet ''Sever-M'', Proc. 11 th Int. Conf. MHD Electr. Power Generation, Beijing,China, 1992, p.519-523
2. E.P. Polulyakh, A.V. Spiridonov and V.Ye. Cherkovets, ''Superconducting magnet system application for clinical and biological research'', Soviet-American Symposium on Research,Technology and Trade, San Francisco, USA, 1991

SMALL CURRENT TRANSFORMER USING OXIDE SUPERCONDUCTOR FOR TRANSPORT AC LOSS MEASUREMENT

E. S. Otabe,[1] Y. Morizane,[1] T. Matsushita,[1,2]
J. Fujikami,[3] K. Ohmatsu[3]

[1]Department of Computer Science and Electronics
Kyushu Institute of Technology
680–4 Kawazu, Iizuka 820-8502, Japan
[2]Graduate School of Information Science and Electrical Engineering
Kyushu University
6–10–1 Hakozaki, Higashi-ku, Fukuoka 812–8581, Japan
[3]Electric Power System Technology Research Laboratories
Sumitomo Electric Industries, Ltd.
1–1–3 Shimaya, Konohana-ku, Osaka 554–8511, Japan

ABSTRACT

A Bi-2223 tape was prepared by wind and react method for fabrication of a compact superconducting transformer which was designed for an AC current source for a measurement of the transport energy loss density. The tape was used for the secondary winding and wound 2 turns on a bobbin of diameter 28 mm. The critical current of the tape was 37.5 A at 77.3 K and zero external magnetic field. Four tapes were connected in parallel to attain a sufficient secondary current. An iron core was used to get a good coupling between the primary and secondary windings. The size of the transformer was 55 mm × 60 mm × 250 mm and the weight was 0.4 kg. The peak secondary current reached 372 A at 100 Hz when the primary current was 2.27 A in liquid nitrogen. This secondary current is about 2.5 times larger than the total critical current. Any harmonics were not detected in the secondary current in the frequency range of 50–2000 Hz.

INTRODUCTION

Application of oxide superconductors to AC equipment has been studied as well as that to DC equipment.[1] For such application, the energy loss density due to the AC transport current is a key parameter to be measured.[2] However, the critical current of specimens such as Y-123 bulk superconductor or Bi-2223 conductor for current leads exceeds sometimes several hundred A, and the measurement of AC loss up to a sufficient transport current is not easy. A superconducting transformer is useful for a current source in such measurements. In our previous study, 500 A class small AC transformer was designed and fabricated using a Bi-2223 superconducting tape for an AC current source of laboratory-scale for this purpose.[3] For the primary winding, a copper wire of 0.2 mmϕ was wound by 300 turns on a bobbin of 54 mm diameter. A silver-sheathed Bi-2223 multifilamentary tape with 61 filaments was used for the secondary winding and wound on a bobbin of 60 mm diameter by 2 turns. The size of the transformer is limited by the minimum bending radius without degradation of the critical current of the tape, since a reacted tape was used. In addition, the leakage of the magnetic flux is significant due to the shape of the secondary winding, resulting in a low coupling coefficient, $k = 0.72$.

To realize a more compact transformer with high coupling coefficient, the wind-and-react (W&R) method is promising for the superconducting tape used in the secondary winding. In this study, a compact superconducting transformer is designed and fabricated using this method. The diameter of the secondary coil is 28 mm and the designed maximum AC current is 150 A. Discussion is given for the parameters of the transformer.

DESIGN AND FABRICATION

Specifications of the primary and secondary coils are listed in Table 1. A copper wire was used for the primary winding, and a Bi-2223 multifilamentary tape prepared by wind-and-react method on a stainless bobbin of 28 mm diameter was used for the secondary winding. Specifications of the Bi-2223 tape are shown in Table 2. The critical current of the tape was 37.5 A at 77.3 K and zero external magnetic field. This value is a little lower than the tape used in previous transformer of 45 A. Four tapes were connected in parallel to attain the maximum secondary current of 150 A, and wound for 2 turns on a bakelite bobbin. Since the total length of the secondary winding is short, a transposition was not introduced. The secondary winding was not fixed by resin on the bobbin. One edge of the tape was soldered on a copper plate of length 50 mm which worked as a current lead to a superconducting specimen for AC loss measurement. The total length of the superconducting tape used in the transformer was about 1.2 m. The whole size of the transformer was 55 mm × 60 mm × 250 mm and the weight was 0.4 kg. A schematic figure of the transformer is shown in Fig. 1.

Table 1. Specifications of coils used in superconducting transformer.

	primary	secondary
number of turns N_1, N_2	575	2
wire	Cu(0.2mmϕ)	Bi-2223 tape (4 tapes parallel)
diameter of bobbin [mm]	23	28
height of bobbin l_1, l_2 [mm]	26	23

Table 2. Specifications of Bi-2223 multifilamentary tape used for superconducting transformer.

number of filaments	61
twist pitch	∞
I_c(77.3 K, 0 T) [A]	37.5
T_c [K]	110
cross section [mm^2]	3.0×0.22
superconducting volume fraction [%]	24

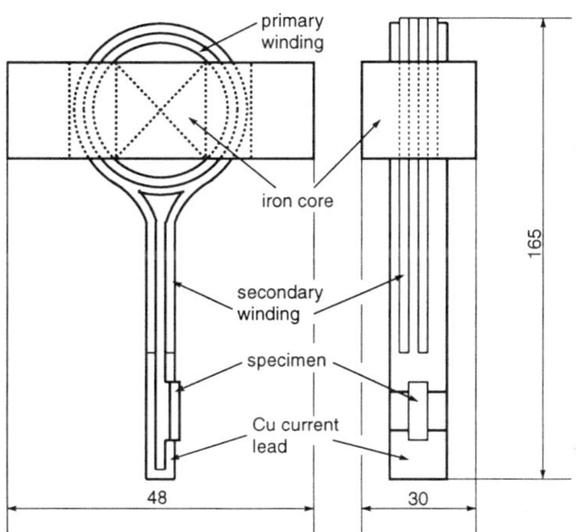

Figure 1. Schematic figure of superconducting transformer operating in L. N_2.

RESULTS AND DISCUSSION

Parameters of Transformer

The measured parameters of the superconducting transformer at 77.3 K are given in Table 3. The coupling coefficient $k = M/\sqrt{L_1 L_2}$ is 0.79. This relatively low value is caused by the leakage of magnetic flux. However, it is better than 0.72 attained in the previous transformer. Here we shall theoretically estimate these parameters.

If the primary winding is approximated by a simple long solenoid, the self induc-

Table 3. Measured parameters of superconducting transformer at 77.3 K.

primary coil self inductance L_1 [H]	5.2×10^{-1}
secondary coil self inductance L_2 [H]	9.0×10^{-6}
mutual inductance M [H]	1.7×10^{-3}
coupling coefficient k	0.79

tance, L_1, is given as

$$L_1 = \frac{N_1^2}{l_1}[\mu S_c + \mu_0(S_1 - S_c)], \tag{1}$$

where N_1 is the number of turns, l_1 is the height of coil, μ is the permeability of the iron core, S_c is the cross-sectional area of the iron core and S_1 is the cross-sectional area surrounded by the primary coil. The self inductance of the secondary winding, L_2, is also given as

$$L_2 = \frac{N_2^2}{l_2}[\mu S_c + \mu_0(S_2 - S_c)], \tag{2}$$

where the quantities with subscript 2 represent the parameters for the secondary coil. The mutual inductance is given as

$$M = \frac{N_1 N_2}{l_1}[\mu S_c + \mu_0(S_1 - S_c)]. \tag{3}$$

Here $S_c = 2.25 \times 10^{-4}$ m^2 and if μ is assumed as $140\mu_0$ for attaining good agreement between theoretical and experimental results, the parameters are calculated as $L_1 = 5.1 \times 10^{-1}$ H, $L_2 = 7.0 \times 10^{-6}$ H and $M = 1.8 \times 10^{-3}$ H. Although good agreement is obtained between theoretical and experimental results for L_1 and M, the deviation in L_2 is relatively large. The measured, larger L_2 is considered to be due to an extra inductance of the region of straight tapes including current leads.

The resulting coupling coefficient, 0.79, is smaller than the theoretical estimation of $k = 0.93$ because of the deviation of L_2. To achieve high coupling coefficient, it is necessary to increase the inductance of the secondary winding itself. For this purpose, N_2 should be increased. If the current ratio is kept constant, N_1 should also be increased. The other option is to make S_c large. In this case the size of the transformer becomes large. Therefore, an optimized design is necessary.

Performance of Transformer

Fig. 2 shows the relationship between the primary and secondary currents at liquid nitrogen temperature (77.3 K) in the frequency range of 50–2000 Hz under the condition where the secondary winding is short-circuited with a copper plate. The secondary current was detected by a Rogowski coil. The transformer ratio was about 150 in the entire range of primary current in the measurement. The peak output current reached 372 A when the peak input current was 2.27 A. It should be noted that a current about 2.5 times the critical current flowed stably in the superconducting tape. This is quite different from cases of conventional metallic superconductors. This is ascribed to both

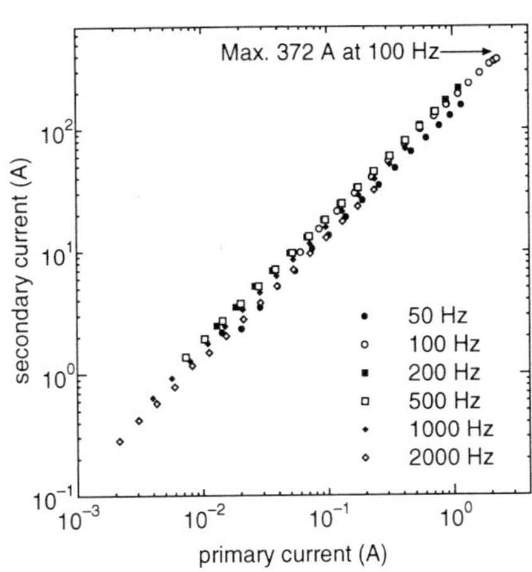

Figure 2. Primary current vs secondary current at various frequencies. Secondary current reaches 372 A in peak at 100 Hz.

the relatively slow increase in the electric field with increasing current due to a low n-value and the large specific heat at high temperature. A FFT analyzer was used to observe harmonics in the secondary current. Typical harmonics were less than 1 % (-40 dB) of the fundamental component in the entire frequency range. Although the heat cycle between room temperature and 77.3 K was repeated more than 10 times, no degradation was observed in the performance of the superconducting transformer.

AC Loss Measurement by Transformer

The AC transport current loss was measured for a Bi-2223 superconducting multi-filamentary tape specimen to confirm the performance. The specification of this specimen is similar to the present tape listed in Table 2, except that the critical current was 45 A. Fig. 3 shows the AC energy loss density vs AC peak current at various frequencies measured with the present and previous superconducting transformers. The same results were obtained for the two transformers in a wide range of AC transport current.

Loss of Transformer

The copper loss in the primary winding is considered the main loss of the present superconducting transformer. In estimating the loss, we assume the condition of the primary current of 0.92 A at 100 Hz, i.e., the condition at which the secondary current reaches the critical current. The resistance of the primary winding is $R_1 = 2.6\ \Omega$ and the loss power is estimated as 2.2 W. The corresponding secondary current is 150 A and the energy loss density of the secondary winding is estimated to be of the order of 10^4 J/m^3 from an extrapolation of the result in Fig. 3, since the critical current density of the present tape used in the transformer is smaller than that of the measured specimen. In this case, the loss power is approximately estimated as 0.2 W. On the other hand,

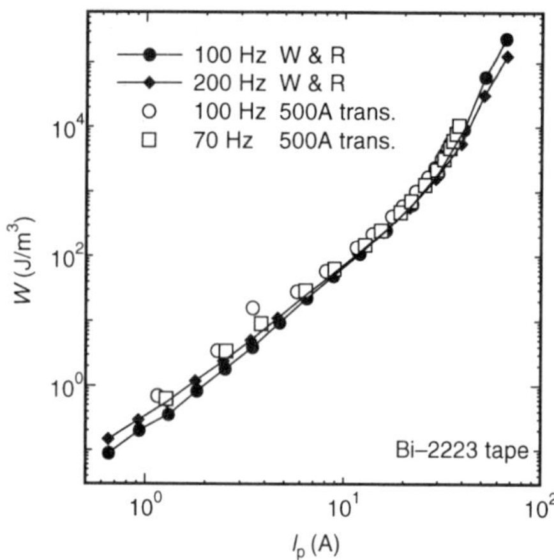

Figure 3. Energy loss density vs AC peak current for a Bi-2223 multifilamentary tape specimen.

the core loss is considered to be significantly smaller. The magnetic field amplitude is estimated as 8.8×10^{-3} T at core region under the same condition. Since the estimated energy loss density from an area of magnetic hysteresis loop of the iron core is 3.5 J/m^3 at 77.3 K, the corresponding loss power is estimated as 2.0×10^{-3} W. The eddy current loss in the iron core is considered to be of the same order of magnitude as the iron loss at these frequencies. Therefore, the main loss originates from the copper loss in the present transformer.

SUMMARY

In this study, a compact superconducting current transformer was designed and fabricated with superconducting Bi-2223 multifilamentary tape prepared by a wind and react method for the secondary winding.

1. The peak secondary current of 372 A was achieved at 77.3 K under a short-circuited condition. This current is about 2.5 times as large as the total critical current of the secondary winding. This stable operation is attributed to both the relatively slow increase in the electric field with increasing current due to a low n-value and the large specific heat at high temperature.

2. Higher harmonics were not observed in the secondary current in the frequency range of 50–2000 Hz.

3. The coupling coefficient is better than that in the previous transformer, but still small (0.79). This is due to the small inductance of the secondary winding. There is still room for improvement of the coupling coefficient.

4. The energy loss density of a Bi-2223 multifilamentary wire was successfully measured using the present transformer. And good agreement with the measurement

by the previous transformer was obtained. This proves that this type of superconducting transformer can be generally used as a current source.

5. The main loss of the transformer was the copper loss in the primary winding. It was estimated as 2.2 W at the primary current of 0.92 A at 100 Hz.

REFERENCES

1. K. Funaki, M. Iwakuma, M. Takeo, K. Yamafuji, J. Suehiro, M. Hara, M. Konno, Y. Kasagawa, K. Okubo, Y. Yasukawa, S. Nose, M. Ueyama, K. Hayashi, K. Sato, "Design and Construction of a 500 kVA-Class Oxide Superconducting Power Transformer Cooled by Liquid Nitrogen", Proc. of ICEC 16/ICMC, Kitakyushu: 1009 (1996)
2. H. Ishii, S. Hirano, T. Hara, J. Fujikami, K. Sato, "The A.C. Losses in $(Bi,Pb)_2Sr_2$ $Ca_2Cu_3O_x$ Silver-Sheathed Superconducting Wires", Cryogenics **36**: 697 (1996)
3. E.S. Otabe, Y. Morizane, H. Matsuoka, M. Izawa, T. Matsushita, J. Fujikami, K. Ohmatsu, "500 A Class Small AC Transformer Using Oxide Superconductor Operating at Liquid Nitrogen Temperature", to be published in Adv. in Supercond. XI (Proc. of ISS'98)

DEVELOPMENT OF TOROIDAL SUPERCONDUCTING MAGNETIC ENERGY STORAGES (SMES) FOR HIGH - CURRENT PULSED POWER SUPPLIES

E.P.Polulyakh,[1] L.A.Plotnikova,[1] V.A.Afanas'ev,[1] M. I. Kharinov,[1] A. K. Kondratenco,[1] E.Yu. Klimenco[2] and V.I. Novicov[2]

[1]Troitsk Institute for Innovation and Fusion Research (TRINITI)
142092,TRINITI, Troitsk, Moscow reg., Russia
[2]RRC Kurchatov Institute, Moscow, Russia

ABSTRACT

A brief description of superconducting magnetic energy storage (SMES) construction, performance and experimental results are presented. Possibilities of their weight-dimension characteristic optimization are shown. Opportunities for "Dual Use" of such superconducting magnetic systems are considered.

INTRODUCTION

The TRINITI has been developing superconducting magnets for pulsed power supplies since 1969, including magnets for energy storage. Main efforts were directed to designing, manufacturing and testing large superconducting magnet systems (SMS) and associated cryogenic systems. A test facility for superconducting magnets allows for a long-term testing of large bore magnets, especially SMES. Two combined helium refrigerators/liquefiers with a capacity of 500 W and a 20 kA, 12 V power supply are available. Magnets and current leads with currents up to 20 kA and with charge/discharge rates of up to nearly 1000 A/s can be tested.

The use of superconducting magnetic energy storage (SMES) is very vital now, because of high density of the stored energy (up to $10^8 J/m^3$), unlimited time of storing and high efficiency of the energy output. The SMES can be used both to improve the power quality of utility supplied power and provide higher quality utility power, as well as increase the transmission capability of existing transmission systems. SMES are very helpful for load leveling during unequal energy consumption, for protection of critical loads in distribution substations, and pulsed power amplifiers. The use of SMES looks very attractive for delivering large amounts of power for short periods of time, as required for fusion research and electromagnetic launch technology.

Advances in Cryogenic Engineering, Volume 45.
Edited by Shu *et al.*, Kluwer Academic / Plenum Publishers, 2000.

A great experience was gained by scientists and engineers of the TRINITI in the development and construction of SMES for use in facility power supply systems for energy accumulation and as pulsed power amplifiers in conditions of a restricted space [1]. In this case toroidal coils were chosen which produce practically no stray field outside the installation, minimizing the influence on personnel health and electronic equipment and control systems. The scheme of ''current multiplication'' is used for getting pulsed currents up to 1 MA and power up to 10 GW.

SMES COMPONENT DESCRIPTION AND TEST RESULTS

The superconducting magnetic energy storage consists of (see Fig. 1): a superconducting toroidal winding, with its sections linked successively through superconductive multishot switches; a helium cryostat maintaining the temperature of the winding and switches at 4.5 K; a low voltage power supply, and a control system. In the charging and energy storage regimes the switches have zero resistance and the current flows through the superconducting toroidal winding, with the energy stored in the toroidal magnetic field (Fig. 2). During energy output on the load, which is set in parallel to the winding sections, the state of switches is changed by the control system to normal with a finite resistance. The voltage drop on them increases sharply and the current of every section is switched to the load, achieving a maximum value equal to the sum of initial currents in the winding sections. With the aim of providing safety for processes of energy output and reducing liquid helium consumption, the liquid helium surrounding the winding and switches is substituted for gaseous helium before connecting the storage to the load. This is accomplished by displacing the liquid into an independent reservoir located in the storage cryostat.

The change from energy storage in a "frozen current" mode to its release is promoted by foil-type superconducting switches actuated by magnetic field. The critical current of the foil drops sharply at low magnetic fields (below 1T). Basic performance of the switches are presented in Table 1 and shown in Fig. 3. The experimental investigations of superconducting switches were aimed at improvements in their design and perfection of the technology for batch production. A design of the module of powerful high current superconducting switches based on Nb-Ti foil has been developed. The current-carrying characteristics of this foil (80mm wide by 20μm thick) are very close to those of the short sample (I=7-8kA). The design of superconducting multi-module switches with operating currents over 20kA has also been developed. (Fig.4).

Superconducting inductive storage of 3 and 5 MJ energy have been created at TRINITI (Table 2). Great attention was paid to critical current degradation, stability of a current carrying element and its fixing in the winding. Cross-sections of the winding were oval and D type, the latter shown in Figure 5. The conductor in the winding was prevented from displacement by making separate sections monolithic with the help of an epoxy compound for the SMES-3 and sticking it to special plates of the support structure for the SMES-5. The current-carrying capability of the toroidal windings was preliminarily tested in the series of experiments.

Table 1. Basic performance of superconducting switches for SMES-5

Material	foil NbTi-50
Foil thickness (μm)	20
Critical current of switch (kA)	28
Resistance at 10K (Ω)	1.25
Overall dimensions (mm)	125×100×130

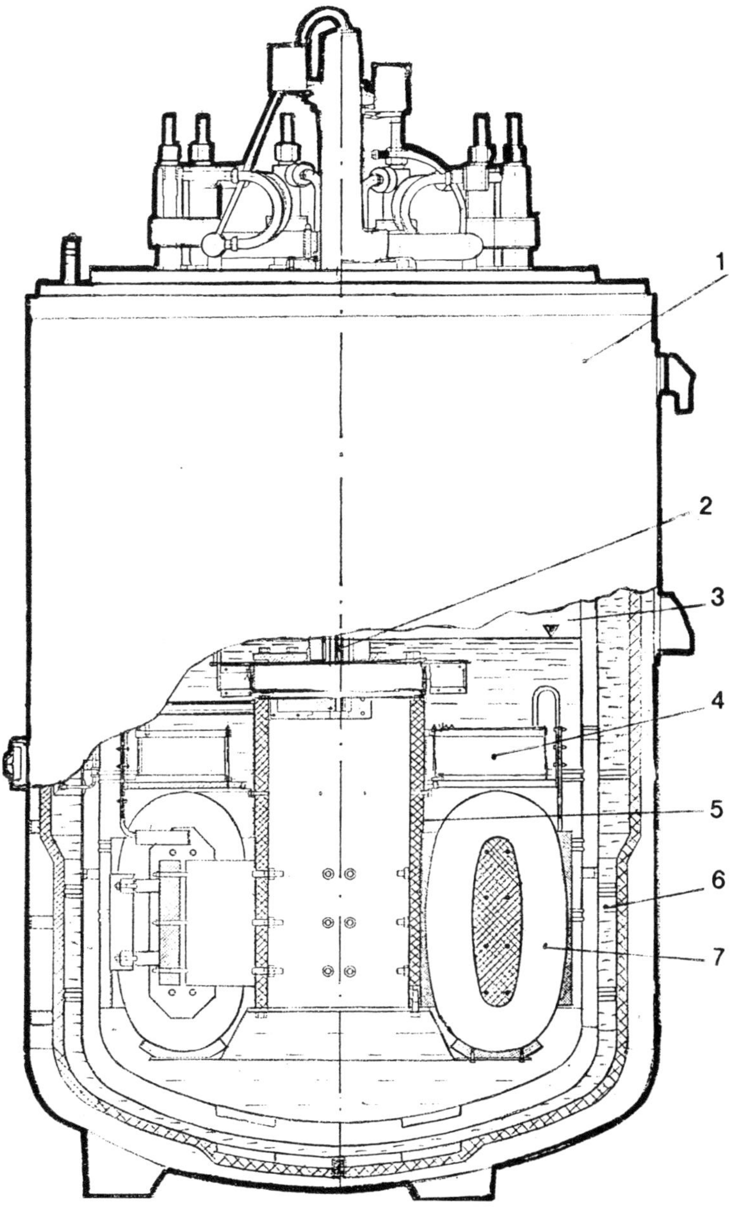

Figure 1. Schematic drawing of the SMES.
1- cryostat; 2- pulsed current lead; 3- level of liquid helium; 4- superconducting switch; 5- support structure; 6- liquid nitrogen; 7- section of superconducting winding.

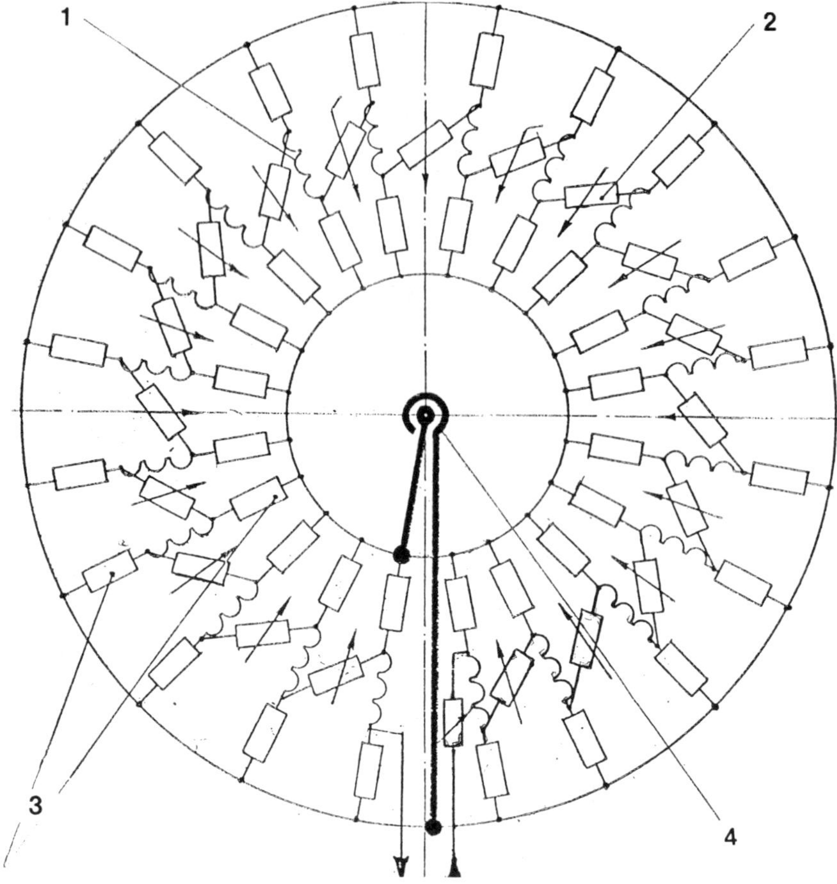

Figure 2. Electrical scheme of the SMES.
1- section of winding; 2- superconducting switch; 3- resistors; 4- pulsed current lead.

Figure 3. Superconducting switch during SMES assembly.

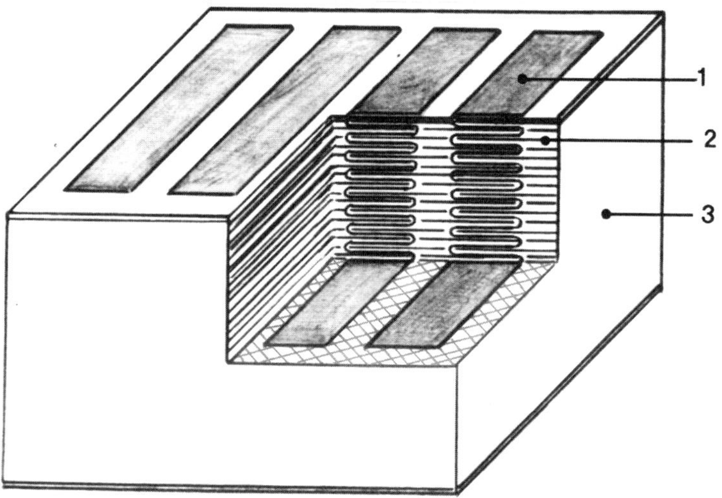

Figure 4. Cross section of the multi-module superconducting switch.
1- superconducting foil; 2- isolating spacer; 3- support structure .

Figure 5. The D-type section of the SMES-5.

The test results show that in the SMES-3 winding a critical current equal to 0.65 of the short specimen critical current has been obtained. The maximum operating current of the SMES-5 winding without quenching was 20.5 kA (fairly close to the value of the short specimen) and limited by properties of the power source.

The progress was stipulated by both the optimal coil shape (D type), reducing stress on conductors as well as bending moment in coils, and a more developed technique of fixing conductors and junctions. Some special devices have been developed aimed at providing mechanical stresses inherent for large coils at medium size model tests.

Table 2. A comparison of main parameters of the SMES.

SMES parameters	SMES-3	SMES-5
Storage inductance (H)	0.068	0.0163
Critical current (kA)	6.4	25
Stored energy (MJ)	1.4	5.1
Number of sections	20	24
Overall dimensions of section (m)	0.75×0.4	0.5×0.6
Inner diameter of torus (m)	0.5	0.95
Outer diameter of torus (m)	1.3	1.9
Storage weight without cryostat (kg)	1670	2600
Current carrying element:		
Cross section (mm)	∅11.5	9.5×19
Superconductor	NT-50	NT-50
Cross-section of conductors (mm)	∅0.85	2×3.5
Number of conductors	6×8	17

Fig. 6. SMES - 5 in the process of its assembling

Tests of the SMES-5 in the energy output regime with the current multiplication scheme were carried out at operating current up to 7.2 kA. Each of 24 sections of the SMES was equipped with a superconducting switch (Fig.6). The current-carrying characteristics of these switches were different than those described in Table 1 and it resulted in low operating current for the SMES in these experiments. An energy output was carried out to an equivalent load of 0.011 Ω. The period of the current output was about 1 ms in accordance with the calculated value. The magnitude of the peak current of the pulse was 115 kA. The rate of the magnetic field change was about 10^3 T/s, and the winding of the SMES remained superconducting.

At present the SMES -5 is being accommodated to a real load and possibilties of the use of this installation as a power source for rail electrodynamic accelerators are being studied.

SUMMARY

The 5 MJ toroidal SMES with superconducting switchees for 20 KA have been designed and constructed to generate high power pulses (maximum current up to 600 kA). An optimized winding shape results in a superconductor with maximum efficiency and eliminates a fringe magnetic field. The reduction of the space required for this toroidal design might make it competitive with solenoidal magnets for SMES with a stored energy of more than 3.000 MJ.

REFERENCES

1. S. A. Egorov, V. V. Andrianov, E. P. Polulyakh et al., "Magnet energy storage", IEEE Transection on Magnetic Vol 28 (1992) 398- 401

TEST RESULTS FOR THE HIGH FIELD CONDUCTOR OF THE ITER CENTRAL SOLENOID MODEL COIL

P. Bruzzone,[1] A. M. Fuchs,[1] G. Vecsey,[1] E. Zapretilina [2]

[1]CRPP – Technologie de la Fusion
5232 Villigen-PSI, Switzerland
[2]ITER Joint Central Team
801-1 Mukoyama, Naka-machi, Naka-gun, Ibaraki 311-01, Japan

ABSTRACT

Two Nb_3Sn cable-in-conduit conductor sections, left over from the winding of the ITER Central Solenoid Model Coil (CSMC), have been assembled to a sample for the SULTAN facility. The test program included dc and ac performance. The quench current, I_q, of the conductor is measured up to 11 T background field, 100 kA and a broad range of operating temperature. The test is repeated for different current distribution in the conductor, obtained by altering the geometry of the bottom joint. The conductor dc performance is compared to the predictions from the strand data, deduced for homogeneous and unbalanced current distribution. The ac losses under sinusoidal field sweep are measured by gas flow calorimetry. The impact of the background field and transport current on the ac losses is explored for a number of typical operating conditions. The results indicate that the coupling loss is very small.

INTRODUCTION

In the scope of the R&D large tasks for the ITER fusion project, a Model Coil of the central solenoid (CSMC) has been wound[1] and will be tested by the end of 1999. Two sections of the high field conductor (CS1)[2], left over after the winding of the innermost layers, have been assembled at CRPP into a standard short sample for testing in the SULTAN facility.[3] The test results reported below are the only ones available for the high field conductor of the CSMC and represent a unique link between the Nb_3Sn strand database[4] and the future coil performance assessment.

The sample for SULTAN consists of two paired conductor sections, about 3.5 m long, joined at one end and connected at the other end to the secondary circuit of a superconducting transformer[5] which acts as a current supply with an operating range up to 100 kA. The sample is lowered in the gap of the SULTAN split coil system; the homogeneous high field section, up to 11 T at the sample location, is perpendicular to the sample axis and about 450 mm long. The field along the conductor drops quickly toward the joint with an average gradient around 15 T/m. The distance between the edge of the joint and the beginning of the high field section is ≈ 350 mm, see Fig. 1.

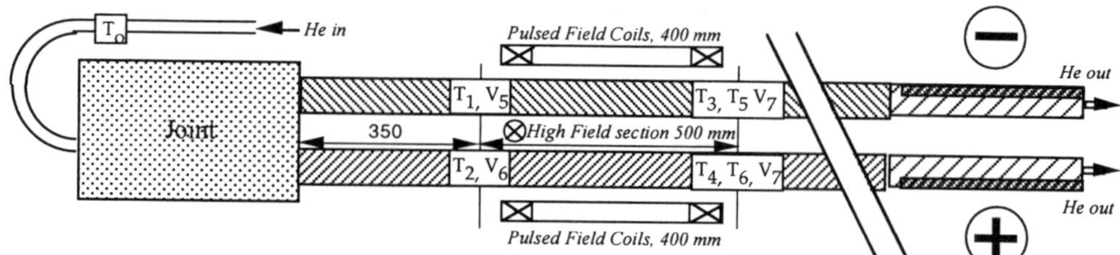

Figure 1 Schematic layout of the two joined conductor sections with location of the applied magnetic field and the crucial sensors for temperature (T) and Voltage (V)

A set of race-track coils, wound with a thick Cu wire, are placed in the free bore of the SULTAN coils and generate a pulsed field perpendicular to both the sample axis and the background field. The copper windings are conduction cooled and operate for a short time with adequate interval for re-cooling. In the ac loss tests reported below, a bipolar power supply (± 500 A, ± 100 V) is used to generate a sinusoidal field of variable frequency with 0.3 T peak-to-peak amplitude over the conductor cross section. The duration of the pulsed field sweep is 40 s and the re-cool time about 2 hours. For stability experiments (not reported here), the amplitude can be up to 2 T for a short duration of 110 ms, without need of a long re-cooling interval.

The main conductor data are gathered in Table 1. More detail on conductor layout and manufacture can be found in [2, 6]. The interstrand resistance on a 0.4 m section is reported in[7]. The two conductor sections ("legs") have been heat treated before joining. After heat treatment, the "+" leg has been bent and re-straightened up to ≈ 0.25% bending strain at the location of the high field section, to simulate the manufacturing loads applied during the react&transfer coil winding process.

The sample is cooled by supercritical Helium at 9.5 bar. The common inlet is at the bottom joint. The mass flow rate can be adjusted up to 11 g/s individually in the two conductor sections by cryogenic throttle valves, placed at the He outlet. The inlet temperature can be varied from 4.5 up to ≈ 9 K using the facility heaters. However, acoustic oscillations are observed at T > 7.5 K. Small heaters attached to the sample can unbalance the temperature in the two legs and are also used for calibration of the gas flow calorimetry.

The sample has been tested in three test campaign, with the length of the hairpin joint reduced from initially 400 mm down to 320 mm and 240 mm [8]. The actual cable pitch at the joint is ≈ 500 mm: a large current unbalance occurs for short joint length, as only few cable elements face each other and transfer the current through low resistance paths. The dc test program, including I_c vs. T at 9, 10 and 11 T as well as T_{cs} measurements at constant current, has been repeated in all the three test campaign.

Table 1. Main conductor data of CS1.4

Cr plated strand diameter	0.81 mm
Nb_3Sn Strand manufacturer	VAC
Manufacturing technique	Bronze
Number of strands	1152
Cu : non-Cu ratio	1.5
Cabling pattern	3x4x4x4x6
Nominal pitches, mm	45/90/130/160/410
Cable space diameter, mm	38.5
Nominal void fraction	36%
Jacket material	Incoloy 908
Outer size, mm	51 x 51
Recommended heat treatment	570° C/220 hrs + 650° C/175 hrs

Central spiral, ø =12 x 1 mm, 33% opening

Subcable wraps, 0.1 mm, 90% coverage, Inconel

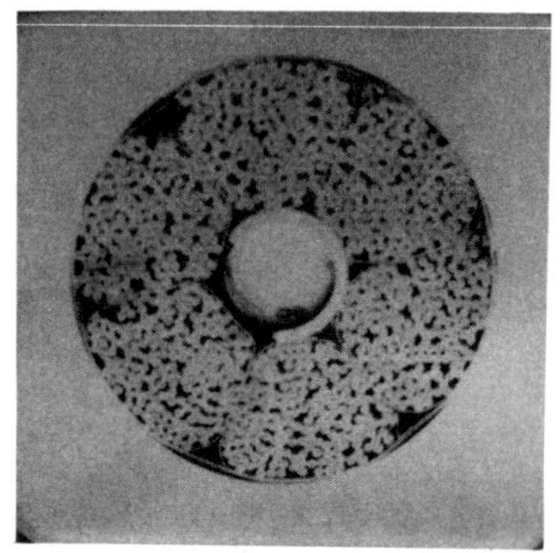

DC PERFORMANCE

The conductor dc performance is measured for the three joint lengths. The I_c test turned actually into a quench current test, I_q, as early voltage builds up gradually far away from the irreversible take-off. For a given current rate, 100 A/s, the I_q results are reproducible. Quench temperature tests, T_q, are carried out holding a constant current and increasing the operating temperature till a quench occurs. The T_q results, mostly with the chopped joint, depend on the heating rate, not fully reproducible in SULTAN. The "+" leg is observed to be systematically "weaker" than its twin, with T_q, about 0.2 K lower than in "-" leg.

The dc tests are carried out in a background field of 9, 10 and 11 T, in a temperature range from 5 to 9 K. The discussion on the strand to conductor comparison, as well as the impact of the joint chopping, is reported for the 10 T background field, but can be applied in the same way to 9 and 11 T.

Expected Performance with initial uniform Current Distribution

The formulae in the ITER design and the scaling law parameters [4] (B_{c20m} = 32.2 T and T_{c0m} = 17.4 K) are used to scale the strand performance to the SULTAN test conditions. The assumed operating strain is the intrinsic strain ε_0 = -0.25%. From the strand I_c data (verified by witness specimens attached to the sample during the heat treatment), the average C_0 is $8.03 \cdot 10^9$ A/m^{2}T$^{0.5}$ (J_c @12T, 4.2 K = 566-675 A/mm^2). The non-Cu strand cross section is 238 mm^2. Assuming that all the strands are identical to the "average strand" and the initial current distribution is uniform, the scaled performance is the dotted line in Fig. 2.

The effect of the self-field of the paired conductor sample cannot be neglected. The average self-field, ΔB_{av}, over the cross section is not 0 due to the current in the neighboring conductor. The peak self field, ΔB_{max}, occurs at the inner edge of the cable space.

$$\Delta B_{max} = 1.66 \cdot 10^{-5} I \qquad \Delta B_{av} = 3.85 \cdot 10^{-6} I$$

If the interstrand resistance in the cable is very low, the current distribution re-adjusts with minimum voltage at high field to the local field distribution. In this case, all the strands quench simultaneously and only the average self-field must be added to the background field. In case of insulated strands, i.e. no interstrand current sharing, the peak self field must be added: the most loaded strand quenches first and drives the quench of the whole cable. In

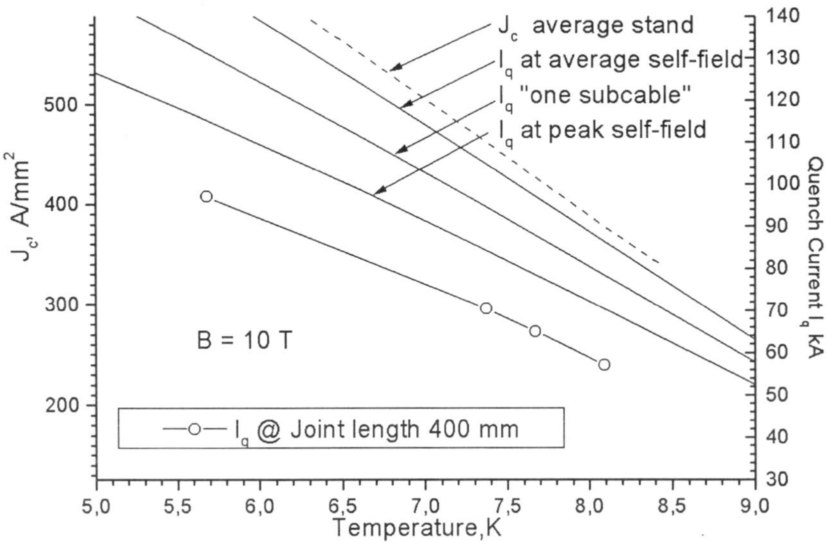

Figure 2 Quench current predictions with homogeneous current distribution from average strand data (dashed line), with different assumptions on interstrand resistance and self-field (solid lines) and test results

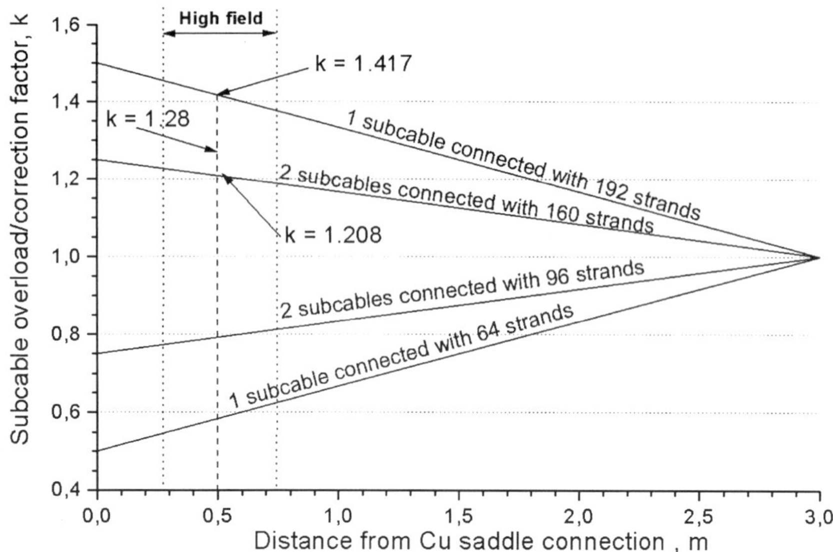

Figure 3 Current distribution in the subcables, for joint length 320 mm. The current overload factor for the subcables depends on the number of strands connected at the joint and on the distance from the joint (in SULTAN, 0.5 m between the center of the high field section and the joint)

a more realistic way, we may assume that the interstrand resistance is very low within the individual subcables, and very high between the subcables wrapped with high resistivity Inconel foils. Each subcable carries the same overall current; the current re-adjusts freely within one subcable, but cannot be shared among subcables. The three predictions, named "maximum self-field", "average self-field" and "one subcable" are plotted in Fig. 2, together with the I_q results with joint length 400 mm.

Correction for non-uniform Joint Resistance

The above predictions assume a balanced current distribution among subcables. However, the joint is shorter than the cable pitch and some subcables have either poor or no contact to the copper saddle. A correction factor must be applied to the above "one subcable" prediction, to account for the non-uniform current distribution in the subcables, which is a function of the joint length and of the distance from the joint to high field.

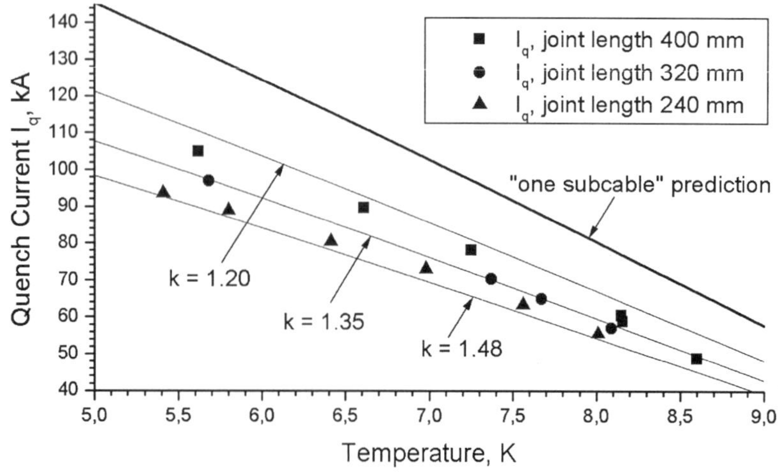

Figure 4 Quench current results at 10 T for different joint lengths (symbols) and the "one subcable" prediction, with applied correction factors

The transverse conductance between subcables [7] is 10^5 $(\Omega\cdot m)^{-1}$; in the lower cable stages, it ranges from 10^7 (triplet) and $5\cdot10^5$ (4th cable stage). In the 400 m long joint, one out of six subcables is assumed to be non-connected, with an overload/correction factor k =1.2 for the other subcables. With joint length 320 mm, see Fig.3, only one subcable is "fully connected", with 192 strands and k = 1.417, half meter away from the connection; two other subcables are moderately good connected with 160 strands and k = 1.208, the remaining three subcables are poorly connected. Assuming that, when voltage builds up at the high field, current re-distribution occurs at the joint (a weak increase of the joint voltage is actually observed when current sharing occurs at high field) among the best connected subcables, k = (1.417 + 2·1.208)/3= 1.28 at the quench. The proper correction factor for joint length 320 mm ranges between 1.417 (no redistribution at the joint) and 1.28.

Fig. 4 summarizes the I_q results for the three joint lengths, with the "one subcable" prediction corrected respectively by a factor 1.2, 1.35 and 1.48. The model to obtain the correction factors contains many parameters (transverse conductance, number of connected strands and resistance of the connections, additional current re-distribution at the joint in current sharing mode) which can be used and abused to fit the results. However, the general trend of the predictions confirms that the observed I_q at joint lengths shorter than a cable pitch is an issue of local current distribution, rather than a conductor degradation. According to the above discussion, the full, non-corrected, prediction should be achieved either with a balanced resistance distribution at the joint, or at a larger distance from the joint.

AC LOSSES

Gas flow calorimetry in cable-in-conduit conductors

To measure the ac loss in the sample, a sinusoidal field is applied as long as steady state temperature is obtained at the downstream sensors, see Fig.1. The time for thermal equilibrium is in the range of 20 to 30 s, depending on the speed of the coolant. During the field sweep, the pressure and mass flow rate are constant. The power loss is the product of the enthalpy change ΔH (from the temperature rise, ΔT) and the mass flow rate, $\dot{m}$.

In the first test campaign, the accuracy of the result was poor, possibly due to the non isothermal cross section: the temperature is measured at the outer radius of the cable space, but cooler Helium flows at higher speed in the central channel. To improve the accuracy for the second test campaign, the central channel was blocked by a sealed rubber pipe, a new heater was attached on one side of the jacket and additional sensors (T5 and T6) were placed opposite to T3 and T4, see Fig. 5.

The response of the sensors T4 and T6 to the heater power is shown in Fig.5. The coolant flows in spirals following the cable pitch with marginal radial heat exchange, due to the subcable wraps. Half cable pitch after the heat source, thermal gradients of the order of 0.5 K are observed between the different spiral streams in the cable space.

Using heaters wrapped around the conductor (not at one side as in Fig.5), the response of the two facing sensors is more balanced. However, the temperature rise (e.g. T3 and T5) is not identical, proving that, even for isotropic heat loads, the non-homogeneous coolant

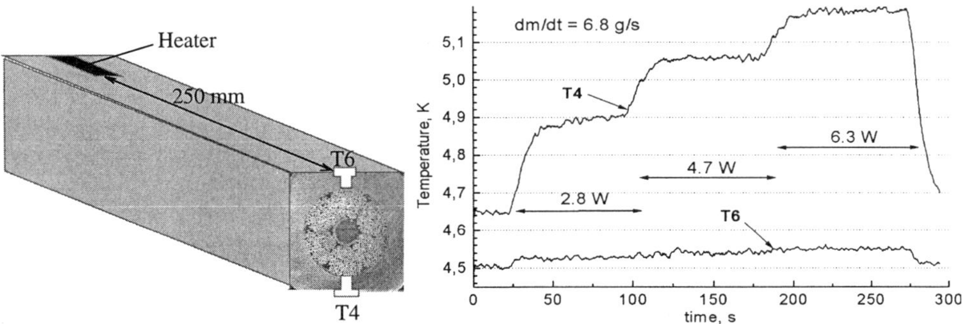

Figure 5 The response of the temperature sensors T4 and T6 to the heat load (W) of a heater placed on one side of the conduit, half cable pitch upstream of the sensors

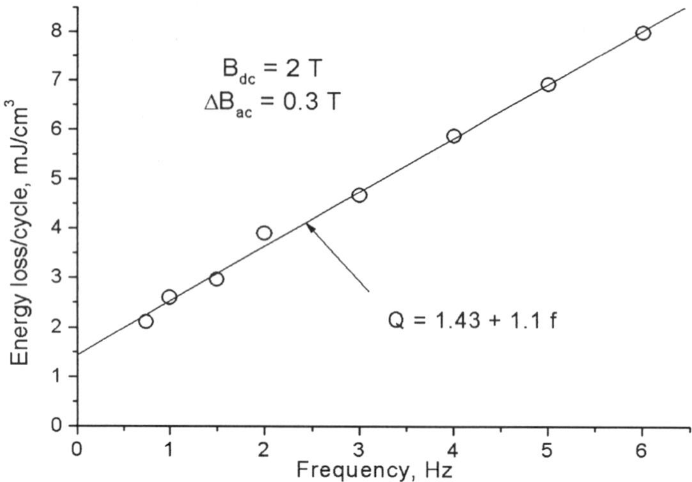

Figure 6 Loss curve for sinusoidal applied field. The reported dots are the average of + and - leg

speed in the cable space (e.g. higher speed the free "corners" created by the subcable wraps) is responsible for non-isothermal conditions.

The accuracy of the ac loss results is eventually driven by the non-identical temperature rise, ΔT, in the downstream sensors. The average of ΔT4 and ΔT6 (respectively ΔT3 and ΔT5) is assumed for the loss assessment and error bar estimate. The loss is normalized to the strand volume (200 cm^3), according to the effective conductor length (338 mm) exposed to the ac field.

The smallest frequency for the loss curve using the gas flow calorimetry is driven by the minimum power loss. The temperature rise must be $\Delta T \geq 0.05$ K to maintain an acceptable accuracy and the mass flow rate should be $\dot{m} \geq 2$ g/s to quickly achieve steady state temperature. The minimum detectable power is in the range of 0.3 – 0.5 W.

Coupling current loss without transport current

The loss curve is measured with a background field of 2 T, to reduce the weight of the hysteresis loss on the overall loss. The peak-to-peak amplitude of the sinusoidal field is ΔB = 0.3 T and the frequency ranges from 0.75 Hz to 6 Hz. From the slope of the loss curve, see Fig. 6, the coupling currents loss constant, nτ, is calculated as

$$n\tau = 1.1 \cdot 10^{-3} \frac{2\mu_o}{\pi^2 \Delta B^2} = 3.1 \pm 0.3 \ ms$$

The loss increase, compared to the interfilamentary loss quoted in [4] for the VAC strand (nτ = 0.62 ms), is due to marginal interstrand coupling loss and proves that the strands are not insulated and interstrand current sharing may occur.

The loss at f = 2 Hz and ΔB = 0.3 T is measured as a function of the background field, see Fig. 7; the solid line is an interpolation of the average results for both legs. The loss decreases with increasing magnetic field because of the smaller hysteresis loss, due to the decrease of J_c, and the smaller coupling loss, due to the copper magnetoresistance. The dotted line in Fig. 7 is the loss estimate after subtraction of the hysteresis loss, suggesting that nτ above 10 T is half the value observed at 2 T, i.e. nτ (10 T) ≈ 1.5 ms.

Removing the subcable wraps, the coupling loss scale roughly with the square of the ratio of the cable pitch (410 mm) to the subcable pitch (160 mm), leading eventually to

$$n\tau_{no\ wraps} = \frac{\ell_{t\ cable}^2}{\ell_{t\ subcable}^2} n\tau_{with\ wraps} \leq 10 \ ms$$

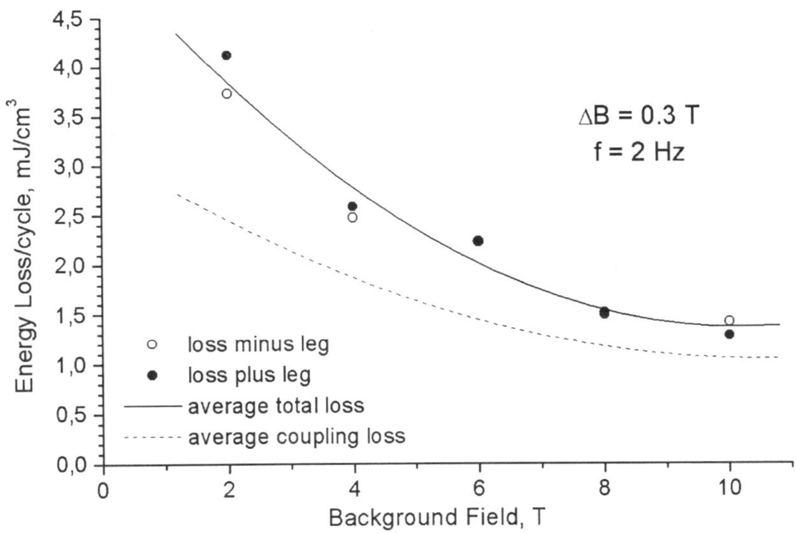

Figure 7 AC loss without transport current as a function of the background field.

AC losses with transport current

The ac losses in a CIC of Cr plated Nb_3Sn strands increase when a transport current is applied with a background dc field [9]. The loss rise may be due either to the transverse load reducing the interstrand resistance and increasing the coupling loss or to the saturation loss (overlapping of transport and coupling currents, generating current sharing).

The same field sweep (f = 2Hz, ΔB = 0.3 T) was applied in SULTAN for different combinations of B_{dc} and I, keeping constant the transverse load B_{dc} x I (600 and 300 kN/m). The loss results, after subtracting the hysteresis loss which is a function of B_{dc}, are plotted in Fig. 8 as a function of the transport current: the lines interpolate the points measured at constant transverse load. The slope of the interpolation lines vs. current is an evidence that the loss increase is mostly due to saturation loss rather than to a decrease of the interstrand resistance: otherwise the loss should be strictly constant for constant load. The strongly

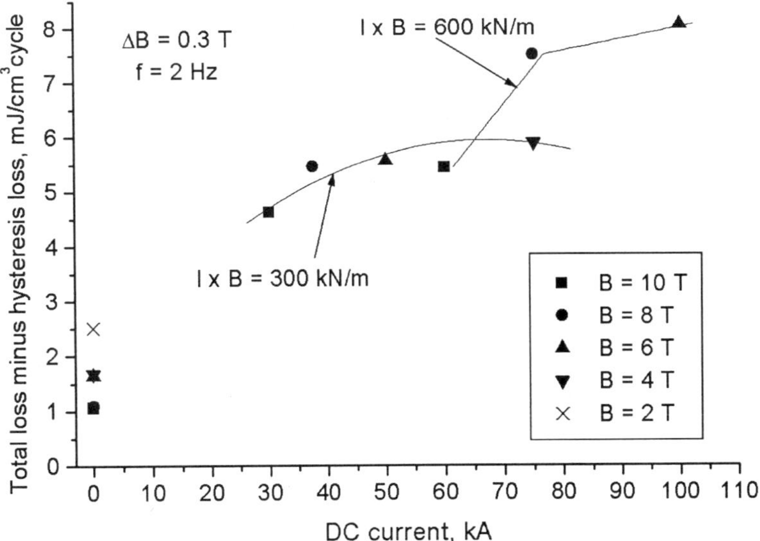

Figure 8 AC loss as a function of the transport current for two constant transverse loads I x B

unbalanced current distribution (the ac losses have been measured in the second test campaign, with the joint length reduced to 320 mm) is likely responsible for the local occurrence of saturation loss, even at moderate values of I/I_q and dB/dt.

The power dissipated by the saturation loss is supplied by the current source, not by the applied ac field. The saturation loss is not reduced by high resistivity barriers, which hamper the interstrand current sharing. At the test conditions, the overall loss with transport current is up to a factor of 5 larger than at 0 current. However it is not correct to assume $n\tau \approx 10$ ms, as the saturation loss scales differently compared to the coupling currents loss.

CONCLUSION

The conductor performance can be predicted from the strand data, through a modeling which includes the self-field and the interstrand resistance for current sharing. In the SULTAN sample, with the joint very close to the high field section, the non-homogeneous joint resistance must be also accounted as an additional boundary condition for current unbalance. Correction factors, specific to the joint and SULTAN geometry, are estimated. The test results, including the degradation of the quench current at shorter joint lengths, agree with the trend of the predictions which account for the current unbalance at the joint.

A realistic extrapolation to the I_q in the CS Model Coil, with a much longer distance between joint and high field, can be done applying the "one subcable" prediction with a correction factor between 1.2 (worst case = SULTAN sample) and 1 (best case, balanced subcables). The removal of the subcable wraps would allow a better current re-distribution among subcables and a more optimistic prediction of the quench current.

The coupling loss of the Cr plated Nb_3Sn ITER conductor after several operating cycles at high field is small, $n\tau \approx 3$ ms @ $B_{dc} = 2$ T. Compared to the interfilamentary coupling loss, $n\tau = 0.6$ ms, the loss increase is marginal. At higher background field, the coupling loss drops further to $n\tau \approx 1.$ ms. Removing the subcable wraps, $n\tau$ is expected to increase not beyond 10 ms.

The loss increase with transport current is not constant for the same transverse load and cannot be explained with a drop of the interstrand resistance. Saturation loss, occurring locally in the conductor sections already carrying I_c, is the likely mechanism to understand the higher loss for increasing transport current.

The high resistivity foils wrapping the subcables of the Cr plated Nb_3Sn cable-in-conduit conductors should be removed. They restrict the radial mass exchange of coolant, they hamper the interstrand current sharing and jeopardize the conductor stability for local disturbances. The coupling loss remains small even without wraps.

ACKNOWLEDGMENT

The authors are indebted with J. Minervini, PFC-MIT, for carrying out the heat treatment of the conductor sections and with A. Nijhuis, Univ. of Twente (NL), for testing the witness strands. The friendly assistance of M. Böhler and P. Erismann during the test in SULTAN is greatly appreciated.

REFERENCES

1. R.J. Thome et al., *Fusion Technology 1998, 803 (1998)*
2. N. Mitchell et al., *Proceeding of MT-15, 347,* Beijing, September 1997, Science Press 1998
3. B. Blau et al., *IEEE Trans. Appl. Supercond. 3, 361 (1993)*
4. H.G. Knoopers et al., *Applied Superconductivity 1997, Inst. of Physics Conference, 1271 (1997)*
5. G. Pasztor et al., *Proceeding of MT-15, 839, ,* Beijing, September 1997, Science Press 1998
6. P. Bruzzone et al., *IEEE Trans. Mag. 32, 2300 (1996)*
7. A. Nijhuis et al., *Proceeding of ASC 98,* Palm Spring September 1998
8. P. Bruzzone, Paper CCC-2 presented at this Conference
9. A. Nijhuis et al., *IEEE Trans. Appl. Supercond. 7, 262 (1997)*

MANUFACTURE AND PERFORMANCE RESULTS OF AN IMPROVED JOINT FOR THE ITER CONDUCTOR

P. Bruzzone

CRPP – Technologie de la Fusion
5232 Villigen-PSI, Switzerland

ABSTRACT

An improved, simplified layout is proposed for the joints and coil termination of the ITER Nb_3Sn cable-in-conduit conductors. The main features include the preparation of the cable before heat treatment, soldering of the superconducting strands and all metal joining carried out after heat treatment. The manufacture and assembly procedure for both joint and coil termination is realistically demonstrated by a full size sample consisting of two short sections of ITER CS1 conductor. The test in the SULTAN facility includes dc resistance over a broad range of operating conditions (B up to 9 T, I up to 100 kA, T up to 9 K), ac losses, quench temperature and stability under pulsed transverse field. After the first test campaign, the hairpin joint has been chopped twice, reducing the active length from the original 400 mm down to 320 and 240 mm. The dc and ac tests have been repeated as a function of the joint length.

INTRODUCTION

The Nb_3Sn cable-in-conduit conductor sections of the ITER fusion magnets need to be frequently joined to connect layers or pancakes in a winding. The performance specification and the design guidelines [1, 2] for joints aim for a resistance in the range of 1-2 nΩ and low ac loss for stable operation during the field transient at plasma disruption.

The development work in the scope of the ITER Model Coils led to different joint layouts and manufacturing procedures. From the lessons learned in the R&D, an improved joint has been proposed at CRPP [3]. The main features include:

- solder the strand-to-copper contacts after heat treatment for a low and reliable resistance
- maintain 30% void fraction in the cable for stability and ac loss optimization
- reduce the overall copper volume and segment it for low eddy currents loss
- avoid critical metal joining to go through the heat treatment
- allow generous fitting tolerances for the final joint assembly

The sample assembled at CRPP for testing in SULTAN consists of two sections of Central Solenoid Model Coil conductor [4] (CS1.4), joined at one end. The other ends of the sections are finished like a conductor termination in a coil.

THE MANUFACTURING PROCEDURE

The preparation of the conductor ends is identical for both coil termination and joints inside the winding, see Fig. 1: the jacket and the outer cable wrap are removed over 400 mm. The cable is compacted into two shaped steel shells, by a unidirectional load up to 200 t/m. The steel shells are longitudinally welded under load. After compacting, the cable has a slightly oval cross section, with the major axis identical to the original diameter. The steel spiral in the central channel withstands the compaction with minimum deformation. A 40 mm long tapering of the groove in the steel shell provides a smooth transition between the round and oval cable cross section. The void fraction is reduced from 36% down to 30%. Except for the vacuum tightness, no structural requirement applies to the welds, as they will be removed after the heat treatment.

Neither spring back nor distortion is observed at the heat treated cable after removing the welded shells. The Inconel subcable wraps are cut from the cable side to be contacted, using a grinding tool, Fig. 1c. The Cr plating is removed from the exposed strand surface by a steel brush. An Al profile with heaters is fitted to the cable on the side with subcable wraps, Fig. 1d. When the exposed strands are at 200°, they are tinned by Sn60Pb40 alloy (no flux penetrates the heated cable and no solder is sucked in by the Cr plated strands).

Hairpin joint assembly

An eye glass shaped steel piece is fitted to the two conductors ends and welded to the jacket from the side where the cables are protruding, see Fig. 2a. The alloy Inco 82 is used to weld the stainless steel eye-glass to Incoloy 908 jacket. After leak test of the weld, a stack of 20 copper saddle pieces, each 20 mm long and separated by a 0.2 mm Teflon foil is slid with minimum gap between the two tinned cables. In case of distortion or misalignment, the copper saddles can be machined to match the actual cable separation. The saddle pieces are prepared from two OF copper plates, vacuum brazed to 0.5 mm stainless steel foil to reduce the eddy currents (the copper is subdivided in cells of 20 x 18 mm). Despite the brazing, the RRR of the saddles is only 107.

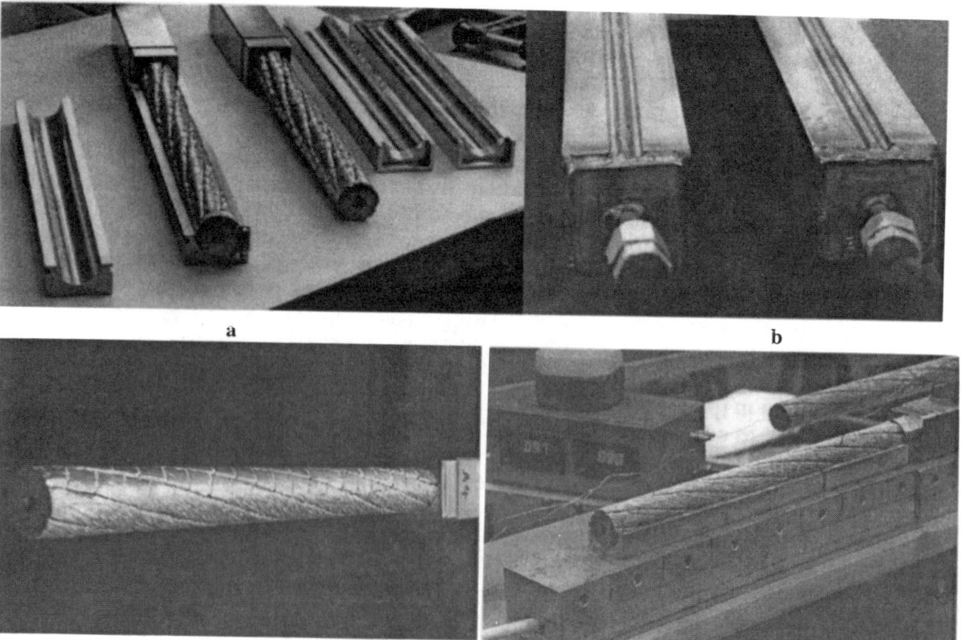

Figure 1 Preparation of the conductor end: a) the cable end is fitted in the steel shells after dismantling the jacket, b) the welded shells ready for the heat treatment, c) the subcable wraps are partly removed from the heat treated cable, d) the strand facing the joint are tinned after removing the Cr plating

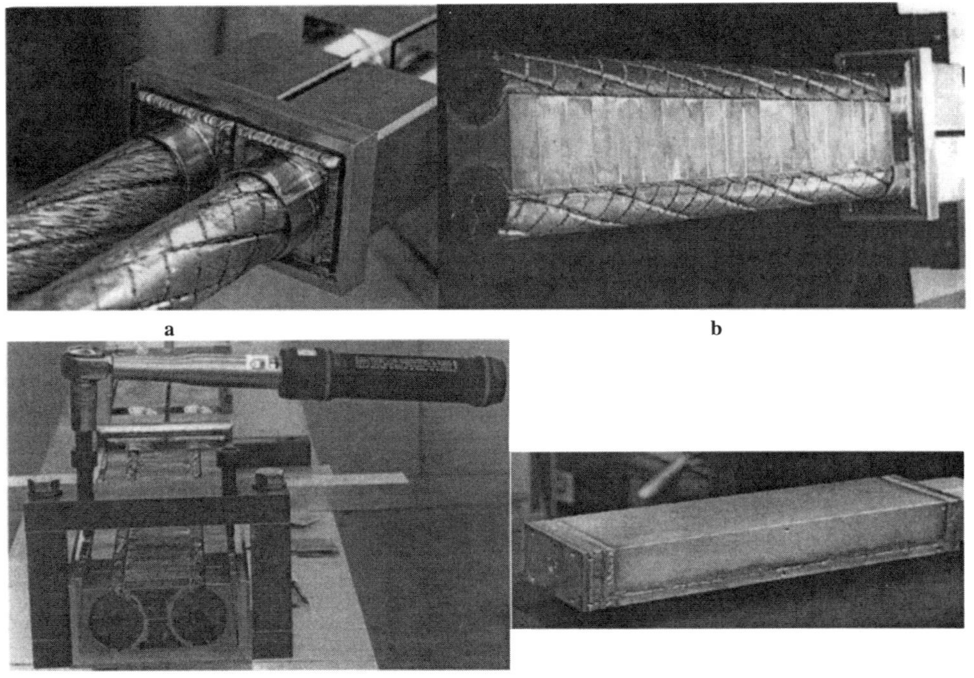

Figure 2 Assembly of the hairpin joint: a) the steel eye-glass welded to the machined jackets, b) the segmented copper saddles soldered to the tinned cables, c) the Ti wedges pre-loaded into the joint box, d) the fully welded joint box with the Helium inlet at the bottom

The surface of the copper saddles is pre-tinned with Sn60Pb40 alloy and a 0.15 mm thick foil of Sn40Pb60 alloy is attached to the saddles before sliding them between the cables. The joint assembly is heated up to 230° by 1 kW heater: for a better bonding, the melting point of the solder foil is ≈10° higher than the solder applied to the cable and saddle surfaces. Only ≈ 20% of the solder mass drops out of the joint assembly, Fig. 2b.

A U-shaped steel profile is welded to the eye-glass and a number of Ti shells and wedges are fitted between the cables and the steel walls (the Ti offsets during cool-down the differential contraction of the copper shells and steel box). The wedges are loaded, see Fig. 2c, to pre-compress the joint to ≈ 8 kN/wedge (≈200 kN/m), just above the expected operating load. The joint box is closed by a steel cover. A single He pipe provides the inlet to the joint box, with the coolant stream dividing into the two conductor branches, Fig. 2d.

The actual cable pitch of the cable ends is about 500 mm, instead of the nominal 410 mm: the original joint length, 400 mm, is ≈ 0.8 pitch length, i.e. one subcable is only partly soldered to the saddles. After a first test campaign, the joint has been chopped with a saw. The active joint length has been reduced to 320 mm (≈ 0.64 pitch length) and later to 240 mm (≈ 0.48 pitch length), to explore the performance changes and to induce a strongly unbalanced current distribution in the conductor [5].

Sample termination assembly

In some cases, the assembly procedure described above may be not applied to connect the coil termination to the current leads, e.g. when in situ welding or local soldering is not allowed and a bolted connection is required. A conventional termination can be assembled starting from the same cable end preparation shown in Fig. 1.

Two fitting shells, similar to the ones used for the heat treatment, are machined to fit the pre-tinned cable end. One shell is made of steel and the other is machined from a vacuum brazed steel-copper composite. The inner copper surface is tinned and the cable end, with an interleaved solder foil, is encased between the two shells. A load of 200 kN/m is applied to

Figure 3 Assembly of the termination: a) the shell machined from a vacuum brazed copper-steel composite, b) the final termination, with the cable encased in the welded shells

pre-compress the heat treated cable. The two shells are welded together and to the conductor jacket. Eventually the copper plate is heated to melt the solder foil and bond the cable to the inner face of the copper plate. A Helium in/outlet is machined in the steel shell, see Fig. 3.

TEST RESULTS AND DISCUSSION

The quench temperature of the joint at 9 T, see Table 1, is initially higher than the plain conductor (8.7 K at 9 T, 62.5 kA), due to the larger amount of copper. After reducing the length to 320 and 240 mm, T_q decreases, as the soldered subcables are saturated and the current transfer to the non-soldered subcables is hampered by the high resistivity wraps.

Resistance

The joint voltage is sensed at constant current by voltage taps attached to the conductor jacket at different distance from the joint. The voltage profile, see Fig. 4, is similar for

Table 1. Summary of joint dc performance before and after chopping

	Joint length 400 mm	*Joint length 320 mm*	*Joint length 240 mm*
Quench temperature, T_q @ 9 T, 50 kA	9.3 K	9.2 K	8.5 K
Quench temperature, T_q @ 9 T, 62.5 kA	9.0 K	8.6 K	7.7 K
Joint Resistance @ 0 T, 75 kA	0.76 nΩ	1.06 nΩ	1.84 nΩ
Joint Resistance vs. B @ 50 kA	$0.713 + 0.125 \cdot B$, nΩ	$0.994 + 0.175 \cdot B$, nΩ	$1.705 + 0.288 \cdot B$, nΩ

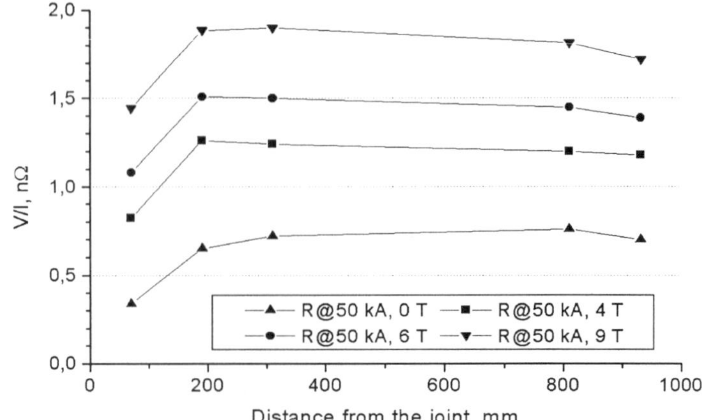

Figure 4 Joint voltage at different location of the voltage taps

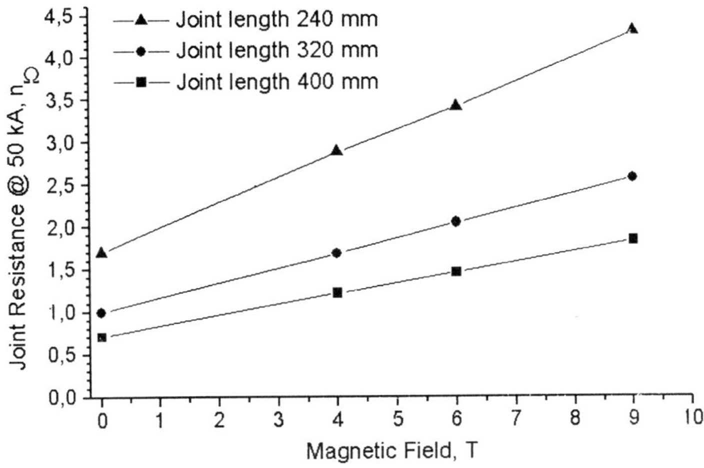

Figure 5 Joint resistance vs. background field

different background field. The first voltage taps, about 80 mm from the joint edge, underestimate the voltage drop at the joint. For the assessment of the joint resistance, the average of the 2nd, 3rd and 4th voltage pair is divided by the operating current. The joint power is also measured by gas flow calorimetry. The results from the voltage taps and calorimetry, at 9 T, 50 kA, agree within 1%. The accuracy of the joint voltage (average of three readouts) is mostly better than ± 3%.

The sample termination is connected to the transformer plates by pressure contact with an interleaved In foil. The resistance of the connections is in the range of 1 nΩ (- leg) and 2 nΩ (+ leg). The difference is due to the poor flatness of one of the transformer plates.

The joint resistance does not change significantly as a function of the operating current up to 100 kA (at 0 T a weak dependence is observed). Over a broad range of temperature, the joint resistance is constant: an increase by 10% is observed when T approaches T_q.

The resistance behavior vs. magnetic field is plotted in Fig. 5. The magnetoresistance ratio at 4.5 K, R(B)/R(B=0), is 30% smaller for the joint compared to the copper saddle and does not change as a function of the joint length.

For a typical operating condition, 4 T and 50 kA, the joint resistance is plotted in Fig. 6 vs. the reduced joint length. The bulk resistance of the copper saddle contributes, at 400 mm

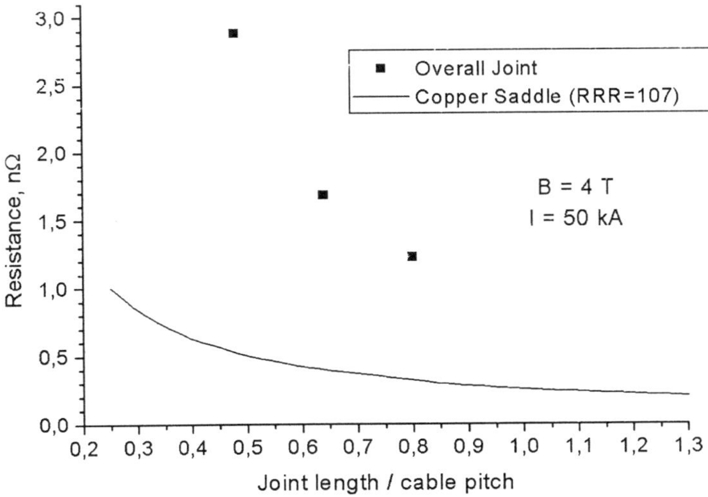

Figure 6 Joint resistance vs. joint length, normalized to the actual cable pitch (≈ 500 mm)

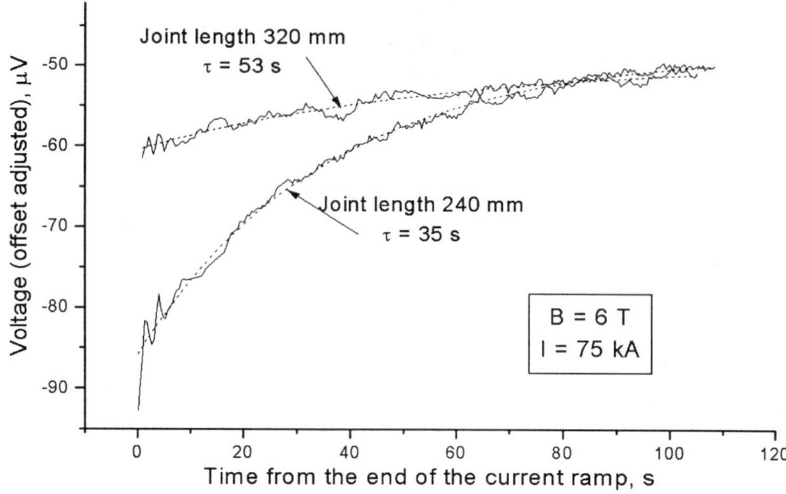

Figure 7 Joint voltage decay after a current ramp, as the current re-adjusts for the lowest resistance path

length, less than 30 % to the overall resistance. If really necessary, the joint resistance could be further reduced extending the length up to 1-1.2 cable pitch, using copper with higher RRR and reducing the minimum distance of the cables from 17 mm down to ≈ 10 mm. The minimum resistance achievable with this layout is expected to be around $0.4 + 0.1 \cdot B$ nΩ.

In many applications, e.g. in the pancake joints of the ITER Central Solenoid, severe space constraints call for a very compact joint size. The results in Fig. 5 and 6 show that the resistance increase at short lengths may still be acceptable. However, a strong current unbalance may affect the conductor performance near the joint [5]. An evidence of the current re-distribution in the joint is shown in Fig. 7: during a current ramp-up, the inductive voltage ($L\dot{I}$) forces a balanced current distribution at the cost of a higher joint voltage. At the current flat top, the current re-adjusts for the minimum voltage (i.e. lowest joint resistance) at the cost of a current unbalance. The shorter the joint, the larger the drop of joint voltage. The time constant for current re-distribution is in the range of 30-50 s.

AC Losses

The joint ac losses are measured by gas flow calorimetry with a transverse field in the plane of the joint (as it is expected in the ITER CS Model Coil). The peak-to-peak amplitude

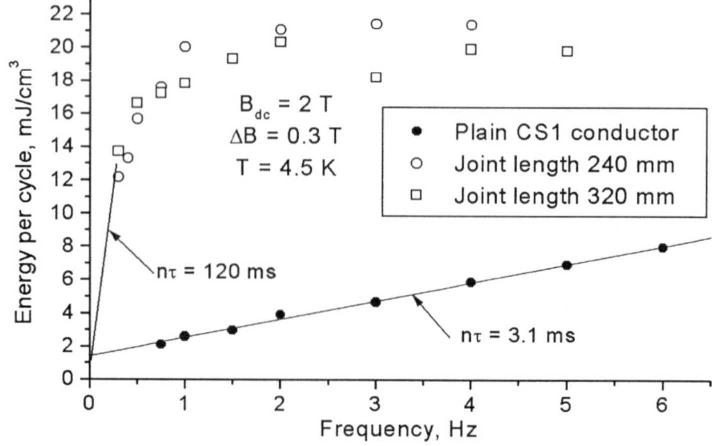

Figure 8 AC loss curve for joint and plain conductor

of the sinusoidal field is 0.3 T. The frequency range is 0.3 to 5 Hz and the sweep duration to achieve steady state power generation is 40 s. The energy loss is normalized to the strand plus copper saddle volume.

The loss curve of the joint, at 320 and 240 mm length, is compared in Fig. 8 to the plain conductor loss ($n\tau$ = 3.1 ms [5]). The joint loss is not a function of the joint length, i.e. the coupling loss in the last cable stage does not play a significant role. The loss peak is at about 2-3 Hz, suggesting a major screening time constant about 65 ms. The initial slope of the loss curve cannot be accurately assessed: a reasonable guess is $n\tau = 120 \pm 20$ ms.

The dominant contribution to the joint ac losses comes from the eddy current loss in the copper saddle, whose constant is estimated about 150 ms, from the size of the largest cell and the RRR. A further reduction of the joint ac losses, say below 100 ms, could be obtained, if required, by a further segmentation of the copper saddle, e.g. 15 instead of 20 mm and two instead of one brazed steel foils.

Stability in transient field

The stability under transverse field is measured applying a single sinus field cycle at 9 Hz, variable amplitude. The detailed test procedure, as well as the results for the plain conductor test, is reported in [6].

A comparison of the joint and conductor stability is shown in Fig. 9 as a function of the temperature margin, defined as the difference between quench temperature and operating temperature. The joint length in Fig. 9 is 400 mm. It may sound surprising that both, conductor and joint data, can be fitted approximately by the same line. However, at 9 Hz, the joint ac loss is well in the screening range, while the conductor is full penetrated, see Fig. 8; the loss ratio at 9 Hz is much smaller than at low frequency.

Further on, the ac loss energy is one order of magnitude smaller compared to the He enthalpy from the operating temperature up to T_q. The quench threshold occurs well before the nominal energy margin (calculated for uniform current distribution) is exploited. It is suggested [6] that a fraction of strands, already carrying I_c and with temperature margin close to 0, exceed locally the quench temperature and are not able to transfer the excess of current because of the high resistivity subcable wraps, driving eventually all the joint/conductor into an irreversible temperature runaway.

After chopping the joint, the stability is measured again at the same operating temperature. This actually means a smaller temperature margin, as the quench temperature decreases after chopping the joint, see Table 1. The results are gathered in Fig. 10: the stability remains constant from 400 to 320 mm length and drops eventually after the second chopping, due to the strong current unbalance.

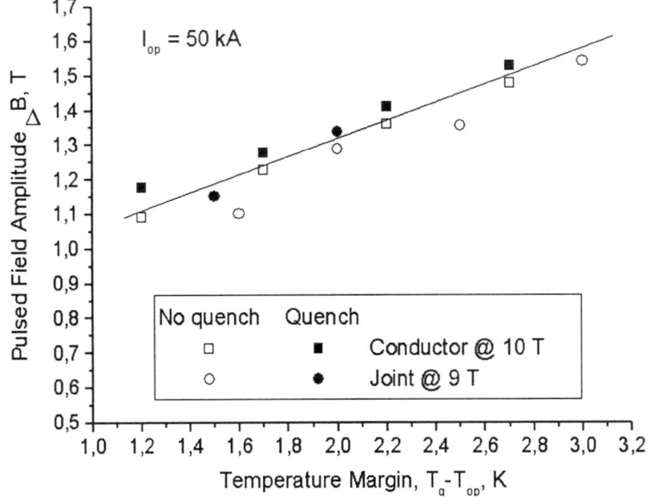

Figure 9 Comparison of joint and plain conductor stability under transverse pulsed field

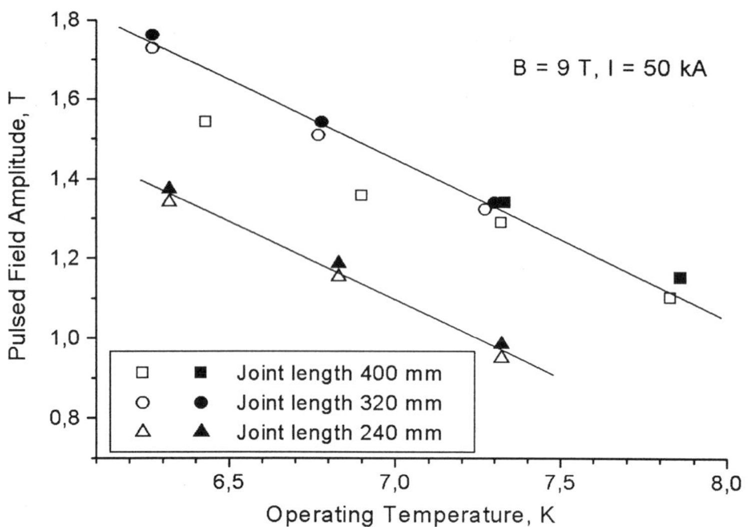

Figure 10 Joint stability under transverse pulsed field, before and after chopping the joint: open symbols are "no quench", full symbols are "quench"

CONCLUSION

The proposed joint layout has been manufactured without complicated tools, using technologies readily available in any workshop, at low cost and low risk.

The dc performance is satisfactory, with quench temperature in excess of that of the plain conductor. Even at joint length smaller than one cable pitch, with one or more non-connected subcables, the joint resistance is as low as 1 nΩ.

The coupling and eddy currents constant is only ≈ 120 ms, thank to the moderately high void fraction in the joint (30%) and the segmentation of the copper elements.

The ability to withstand transverse field pulse with moderate temperature margin exceeds by far the ITER requirement for plasma initiation and plasma disruption. The stability performance is comparable to the plain conductor.

After reducing the joint to a very compact size, the dc and stability performance is worse, but still acceptable, because of the strong current unbalance. In this case, the conductor performance close to the joint needs to be carefully watched.

ACKNOWLEDGMENT

The skill of E. Binder and F. Roth in the joint assembly, as well as the assistance of M. Böhler and P. Erismann during the test in SULTAN, is greatly appreciated.

REFERENCES

1. P. Bruzzone, F. Iida, N. Mitchell, "Layer-to-layer joints for the ITER superconducting coils" *Fusion Technology 1994, 897 (1995)*
2. P. Bruzzone et al., "Design and R&D Results of the Joints for the ITER Conductor" *IEEE Trans. Appl. Supercond. 7, 461 (1997)*
3. P. Bruzzone, "Strand resistance distribution in an improved full size joint for the ITER conductors" To be published in *IEEE Trans. Appl. Supercond. 9 (1999)*
4. N. Mitchell et al., "Conductor development for the ITER magnets" *Proceeding of MT-15, 347,* Beijing September 1997, Science Press 1998
5. P. Bruzzone et al., "Test results of the high field conductor of the ITER central solenoid Model Coil", Paper CCC-1 presented at this Conference
6. P. Bruzzone, "Stability under transverse Field Pulse of the Nb_3Sn ITER cable-in-conduit Conductor" To be presented at MT-16 Conference

TEMPERATURE RISE AND STRAIN BEHAVIOR OF LARGE HELICAL DEVICE DURING COIL EXCITATION

A. Nishimura, H. Tamura, S. Imagawa, T. Mito, S. Yamada, H. Chikaraishi, K. Takahata, N. Yanagi, R. Maekawa, A. Iwamoto, T. Satow, S. Satoh and O. Motojima

National Institute for Fusion Science
Oroshi, Toki, Gifu, 509-5292, Japan

ABSTRACT

The Large Helical Device has a large superconducting coil system to generate a strong magnetic field and a large cryogenic support structure to sustain the huge electromagnetic force. The magnetic field of 2.75 T at magnetic axis was achieved under the standard plasma position and the temperature and strain measurement of the support structure was carried out during the coil excitation. As a result, it is recognized that the temperature rise occurs by an eddy current on the ramp up and down processes of the magnetic field. Also, it is clarified that the measured stress on the equator of the support structure is in good agreement with the stress estimated by FEM analysis. The principal stress direction on the inner equator rotates depending on the coil excitation mode. These results show that the temperature and strain measurement are performed successfully and that the support structure deforms elastically by the electromagnetic force and works well.

INTRODUCTION

The Large Helical Device (LHD) has a pair of helical coils and three pairs of poloidal coils.[1] All coils were made of NbTi superconductors and sustained by a cryogenic support structure made of SUS 316 steel.[2] To cool down the coils and the structure, of which cold mass is over 800 tons, a large refrigeration system of about 9 kW at 4.5 K was installed.[3] The design of the support structure was performed under the magnetic field of 4 T at magnetic axis using a finite element method (FEM).[4] Since a standard mode (designated as #1-o, a plasma major radius (R) is 3.75 m), a vertical elongation mode (#1-a, R = 3.75 m), a horizontal elongation mode (#1-b, R = 3.75 m), an outward shift mode (#1-c, R = 3.9 m) and an inward shift mode (#1-d, R = 3.6 m) have been planned for the LHD experimental program, these operation modes were also taken account in the design.

The support structure of the superconducting coils in the LHD is the largest cryogenic welded structure in the world for a plasma experimental device. Therefore, special attention was paid to the design and the construction.[5] In parallel with the construction, research and development for a strain measurement system in high vacuum, under high magnetic field and at cryogenic temperature, was planned and carried out, to compare actual stresses with the ones estimated by FEM analysis, and to evaluate the soundness of the support structure by precise measurement. During this period, a new strain gage was developed and a new lead cable for the temperature and strain measurement was fabricated.

Advances in Cryogenic Engineering, Volume 45.
Edited by Shu *et al.*, Kluwer Academic / Plenum Publishers, 2000.

Based on these activities, the temperature and strain measurement system was finally designed and installed in the LHD.

The LHD operation started in February 1998 and the first plasma burned on March 31. During the Japanese fiscal year of 1998, two series of the LHD operation were performed without any significant trouble. On the second operation of the LHD, 2.75 T at magnetic axis was achieved successfully under #1-o mode and the mode change operation was performed in succession. When coils were excited, the temperature and strain measurement was carried out continuously with a sampling rate of 0.1 Hz or faster.

In this report, an outline of the strain measurement system and an apparent strain of the strain gage developed are described. Then, the results of the temperature and the strain measured on the equators of the structure are presented, and the deformation behavior of the structure and the relation between the measured stress and the one estimated by FEM are discussed. Also, change in the maximum principal stress direction on the equators during the successive mode change operation is shown and the reason for the rotation is discussed in connection with the electromagnetic force distribution of the helical coil.

STRAIN MEASUREMENT

The strain gage used in the LHD has a gage length of 5 mm, an electric resistance of 350 Ω, a base diameter of 20 mm and a gage factor of about 2.0. The gage was newly developed to make the apparent strain at 4.2 K small and to have polyimid coated gage lead wires. Before applying the gage to the LHD, over 150 cold thermal cycles between 6 K and 300 K were performed and the stability for cooling and warming was confirmed.[6]

The apparent strain measured using a 316 stainless steel plate is shown in Figure 1. So called Kondo effect is observed below 26 K. It is recognized that the strain output will shift to the negative when the active gage temperature rises. This shift is indeed the apparent strain by temperature rise of the active gage.

To reduce the temperature effect on the apparent strain and to cancel the apparent strain induced by fluctuating magnetic field, a Wheatstone bridge circuit consisting of active and dummy gages was constructed using a dummy stage shown in Figure 2.[7] The stage was made of a copper clad 316 stainless steel, and tri-axial strain gages were attached as dummy gages on the backside. Then the dummy stage was welded near the active gage and the Wheatstone bridge was formed.

An adhesive of a polyurethane system which solidifies at room temperature was used for the active gage because the heat treatment was very hard. For the dummy gage, a phenol system was applied and the dummy stage was heat-treated at 353-423 K.

A cable for the strain measurement was newly designed to reduce electric or magnetic noise and heat load to the cryogenic structure. Two pairs of twin leads for a constant input current and an output voltage were twisted, insulated electrically and covered with a fluoroethylene polymer. The cable has a final diameter of about 1.8 mm, an electric resistance of below 541 Ω/km, a break voltage of over 300 Vrms-min. and an insulation

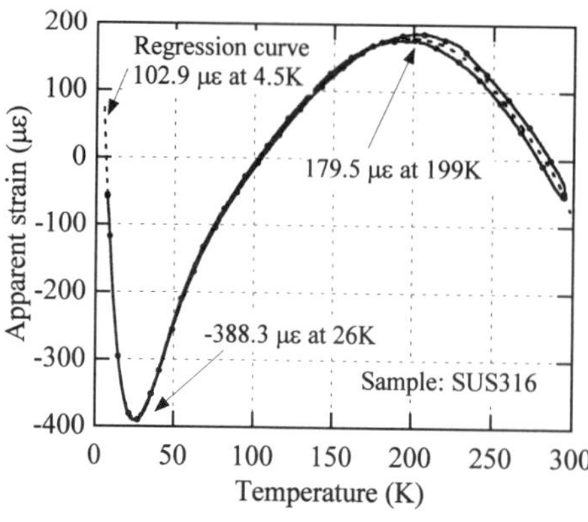

Figure 1. Change in apparent strain of strain gage against temperature.

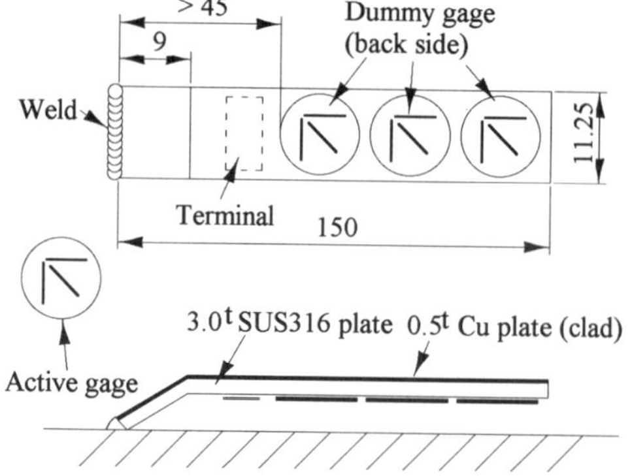

Figure 2. Dummy stage for strain measurement. (Units in mm)

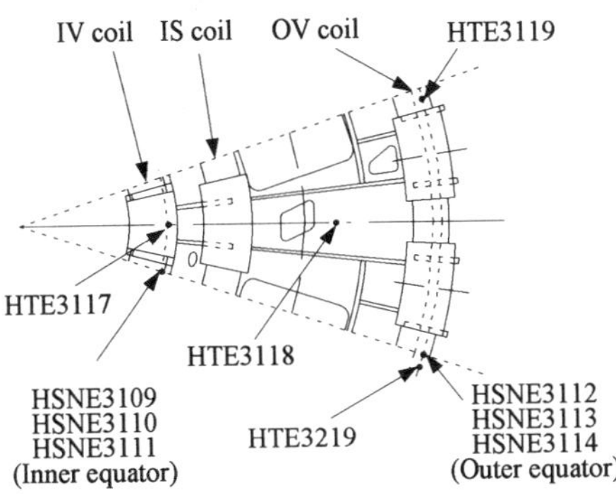

Figure 3. Top view of Sector 1.

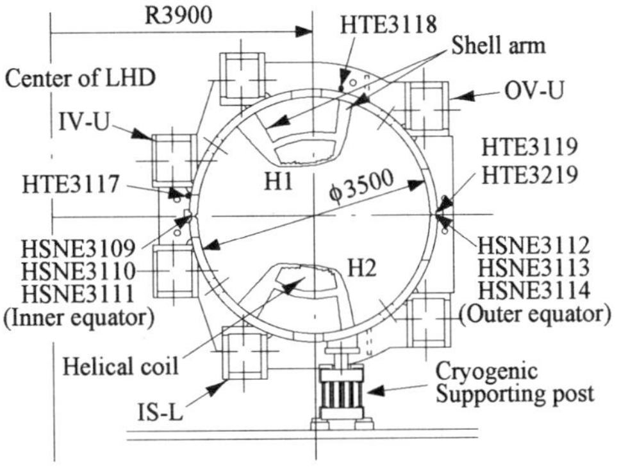

Figure 4. Cross sectional view of Sector 1.

resistivity of over 1500 MΩ-km. The total cross sectional area of copper is 0.318 mm^2 including the shield wires.

The LHD support structure consists of 10 sectors and one sector covers 36 deg. The top and the cross sectional views of Sector 1 are shown in Figure 3 and 4. On this section, (a toroidal angle is 18 deg.) one helical coil designated as H1 runs at the topside and the other (H2) is located at the bottom. The helical coils are fixed to the support structure by two shell arms, and the strain gages were attached at the center of these shell arms on the inner and the outer equators of the supporting shell. HSNE 3109 and 3112 show the strains in the toroidal direction, 3111 and 3114 are those in the poloidal direction, and 3110 and 3113 are those in 45 deg. direction to the torus. HTE 3117 and 3219 are thermo-sensors (Cernox) located under IV-U coil frame and on the outer equator. Each sector has four cooling channels of two-phase helium and the cooling was carried out continuously during the coil excitation.

RESULTS AND DISCUSSION

Temperature Rise and Strain Behavior

The change in temperature during the ramp up near IV-U coil frame and on the outer equator is shown in Figures 5 and 6. It is clear that the higher ramp up rate generates larger temperature rise and the temperature rise inside the torus becomes larger than that on the outer equator. Since no electric break is installed in the torus, an eddy current causes this temperature rise. The change rate of the temperature rise becomes smaller as the

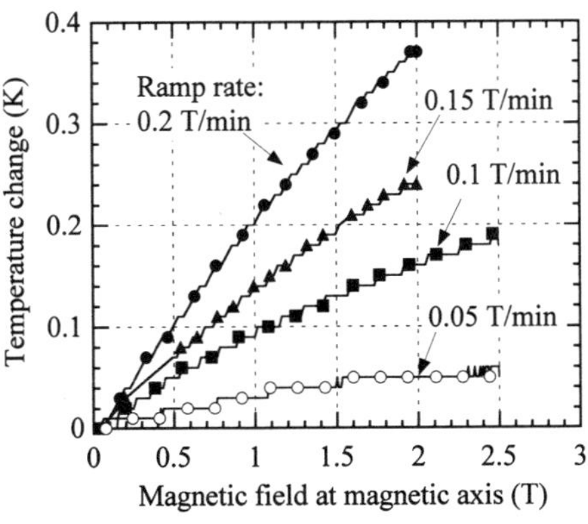

Figure 5. Temperature change against magnetic field under IV-U frame.

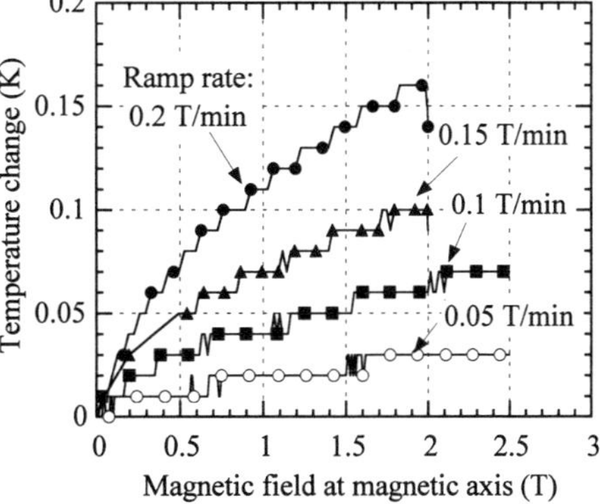

Figure 6. Temperature change against magnetic field on the outer equator.

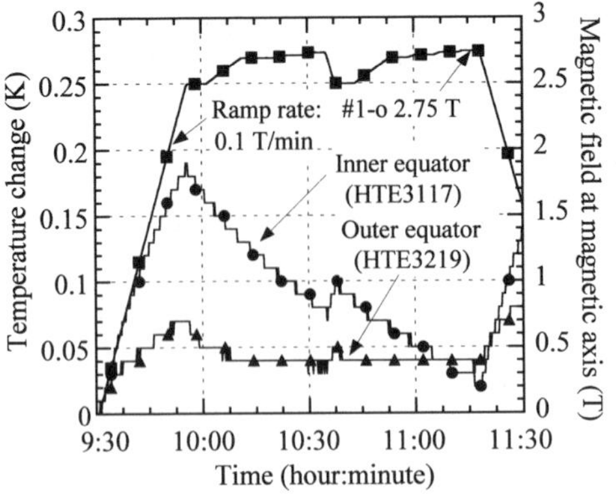

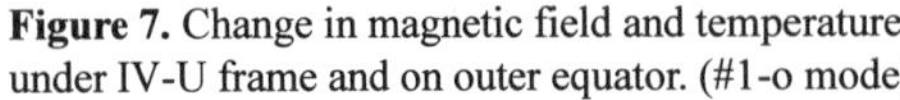

Figure 7. Change in magnetic field and temperature under IV-U frame and on outer equator. (#1-o mode)

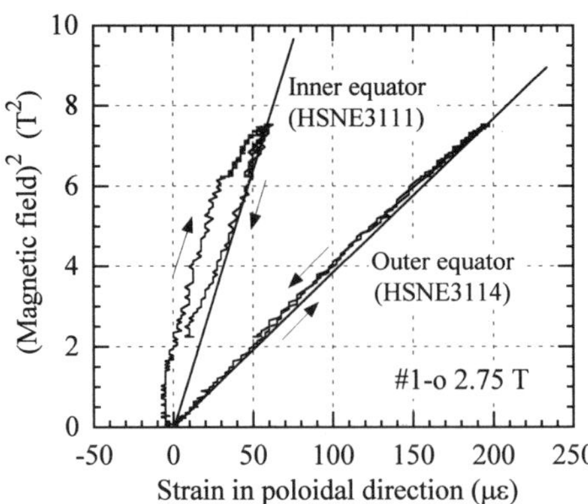

Figure 8. Relation between the square of magnetic field and strain in poloidal direction. (#1-o mode)

temperature difference between the structure and the coolant becomes larger, because the cooling with two-phase helium was carried out continuously. However, a linear relation between the temperature rise and the ramp rate could be found below around 1 T.

The variation of the magnetic field at magnetic axis under #1-o mode is shown in Figure 7 together with the temperature change of HTE 3117 and 3219. In this run, the magnetic field was increased up to 2.5 T at the rate of 0.1 T/min, then a step excitation was performed and 2.75 T at magnetic axis was achieved. This magnetic field is the maximum field under #1-o mode which the LHD experienced up to now. The temperature rose initially and then decreased during the step-up excitation. The change of the strain in the poloidal direction during this run is shown in Figure 8. The unit of the strain is 10^{-6} and described as με in the figure. The vertical axis shows the square of the magnetic field corresponding to the electromagnetic force. The strain on the inner equator shifted to negative direction on the ramp up process and was plotted on a certain line during the step excitation. This shift is caused by the apparent strain generated by the temperature rise of the active gage. On the other hand, the strain varied linearly against the square of the magnetic field on the outer equator.

Figure 9 shows another excitation result. In this case, the magnetic field was ramped up under #1-b mode at the rate of 0.05 T/min up to 1.5 T, and then the ramp rate was decreased to 0.02 T/min. At 2.47 T, a fast discharge named 1-M happened, caused by a trouble of the power supply system, and the coil currents were dumped at the time constant of 300 seconds. On the ramp up process to 1.5 T, small temperature rise occurred on the inner equator, but the temperature recovered under the ramp rate of 0.02 T/min. When 1-M motion happened, the large eddy current ran and heated up the structure remarkably.

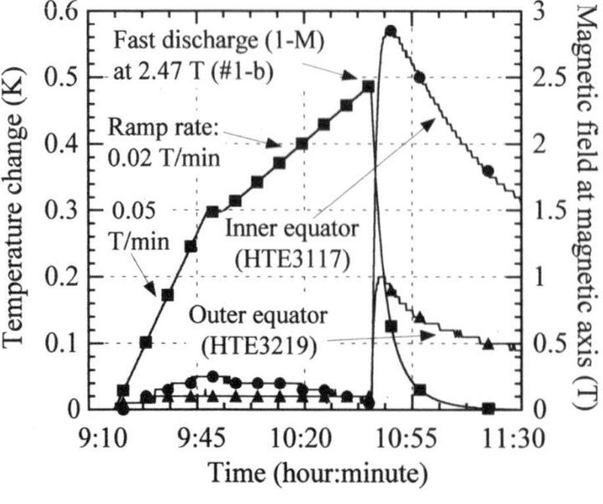

Figure 9. Change in magnetic field and temperature under IV-U frame and on outer equator. (#1-b mode)

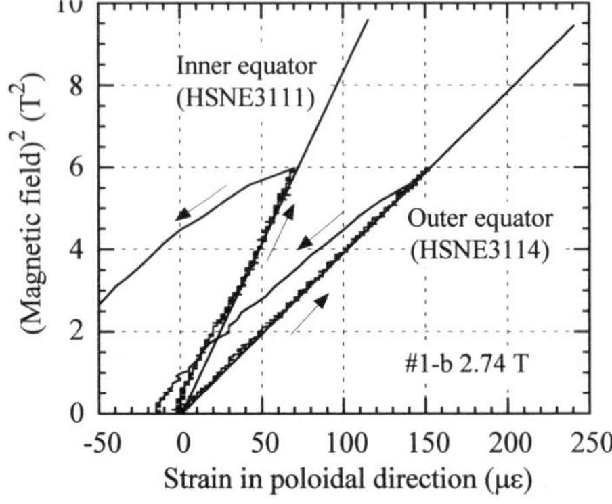

Figure 10. Relation between a square of magnetic field and strain in poloidal direction. (#1-b mode)

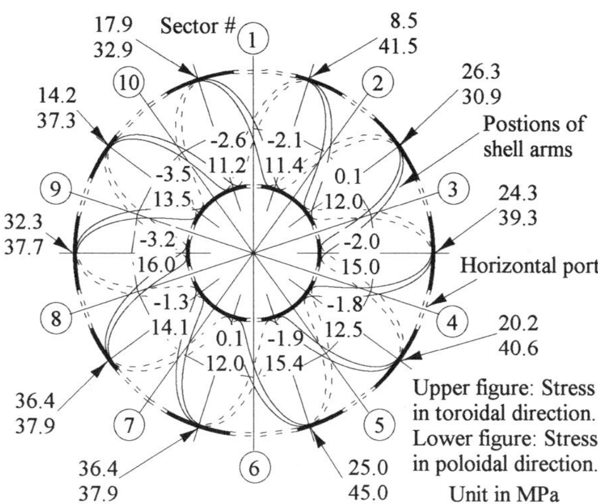

Figure 11. Summary of measured stress on equators. (#1-o mode, at 2.75 T)

The relation between the square of the magnetic field and the strain in the poloidal direction is shown in Figure 10. Although the slope of the straight line is different from that in Figure 8 because of the different operation mode, the data especially on the inner equator moves up on the line clearly comparing with the case in Figure 8. The big jump to the negative at around 6 in the vertical axis was caused by 1-M motion.

From these results, it is recognized that the measured strain shifts by the temperature rise due to the eddy current and that the cryogenic support structure deforms elastically and sustains the large electromagnetic force well.

Stresses on Inner and Outer Equators

Stresses on the inner and the outer equators at 2.75 T at magnetic axis under #1-o mode were calculated using the measured strain. Young's modulus of 200 GPa and Poisson's ratio of 0.33 were used in the calculation. Figure 11 describes the mid-plane of the support structure and evaluated stresses. The circles inside and outside correspond to the inner and the outer equators, and the dotted lines and a pair of curved lines indicate the horizontal ports and the shell arms of the helical coils, respectively. A pair of figures show the stresses at each position; the upper figure indicates the stress in the toroidal direction and the lower is that in the poloidal direction. The difference of the stress in the same direction among the sectors is considered to come from the difference of the size and shape of the port and the rib. From these results, it is found that a small compressive stress exists in the toroidal direction and a tensile stress of about 11 – 16 MPa works in the poloidal direction on the inner equator, while both the toroidal and poloidal direction stresses are tensile on the outer equator. This means that the inner equator is pushed to the center of the device slightly and pulled to up and down directions and that the outer equator is pushed to the outside of the major radius and pulled upward and downward. Since the upper and the lower poloidal coils would move upward and downward separately and the helical coils would expand to both the major and the miner radius directions, the evaluated stresses are considered to be reasonable in a qualitative sense.

One sector was analyzed by FEM using the cyclical symmetry condition and a shell element of 100 mm thickness under 4 T at magnetic axis.[4] However, the FEM model is not the same as the real support structure. The big differences which should be pointed out are as follows:

(1) The upper and the lower structures are welded on the equators by a partial welding, and the thickness of SUS 316 stainless steel is reduced to 50 mm on the weld joint.
(2) Stress concentration exists around the joint because of the reduction of thickness.
(3) The size and the shape of the horizontal and vertical ports and ribs are not the same among ten sectors.

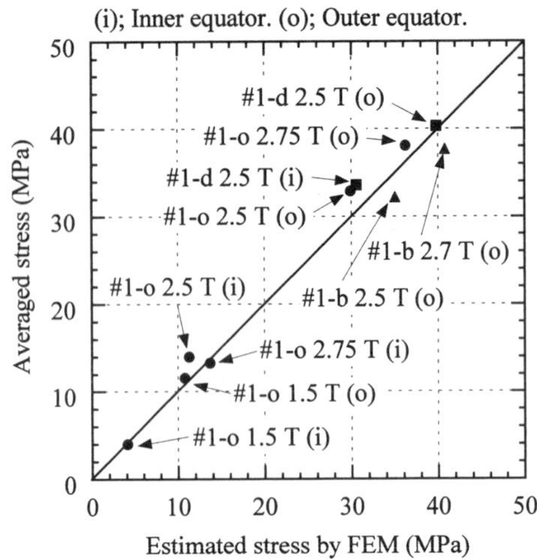

Figure 12. Relation between averaged stress measured in the experiment and estimated stress by FEM. (#1-o mode, at 2.75 T)

(4) The magnetic field at magnetic axis is 2.75 T under #1-o mode in the experiment, while the calculation is carried out at 4 T.

Therefore, the analytical results were modified by the following equation.

$$\sigma_E = \sigma_Z \times (100/50) \times K_t \times (MF/MF_0)^2$$

where, σ_E is an estimated stress in the poloidal direction, σ_Z is the calculated stress by FEM in the poloidal direction, (100/50) is a term for thickness correction on the equator, K_t is a stress concentration factor, and MF and MF_0 (= 4 T) are the magnetic fields at magnetic axis in the experiment and the FEM analysis, respectively.

On the other hand, the stresses in the poloidal direction obtained in the experiment were averaged to make a comparison with the analytical results.

Figure 12 shows the relation between the averaged stress in the experiment and the estimated stress by FEM. All data are plotted near the one to one correspondence line and the averaged stress is in good agreement with the estimated one. This means that the support structure was fabricated well following the design and the strain measurement was performed successfully.

From these results, it is expected that the stresses will become $(4/2.75)^2$ times larger when the 4 T operation will be performed under #1-o mode, and the average stress on the outer equator will reach about 85 MPa. This stress is small enough in comparison with the yield stress at 4.2 K of over 600 MPa.

Rotation of Principal Stress Direction during Mode Change Operation

To confirm the mode change operation goes on well, the operation changing the mode from #1-b, #1-c, #1-d to #1-o was performed successively under the 2.5 T at magnetic axis by controlling the coil currents.

The change in the magnetic field at magnetic axis is shown in Figure 13 together with the measured strain in the poloidal direction. After achieving 2.7 T at magnetic axis under #1-b mode, the magnetic field was reduced to 2.5 T, then the mode change operation was carried out. The poloidal direction strain on both equators changed remarkably from #1-c to #1-d mode, and the strain on the inner equator decreased under #1-d mode, while the strain on the outer equator increased. This was caused by the change in the electromagnetic force. Under #1-d mode, the electromagnetic force outside of the plasma increases to push the plasma inward, and the force inside decreases to pull the plasma toward the center of the device. In #1-o mode, both strains on the inner and outer equators become almost the average values of those in #1-c and #1-d mode.

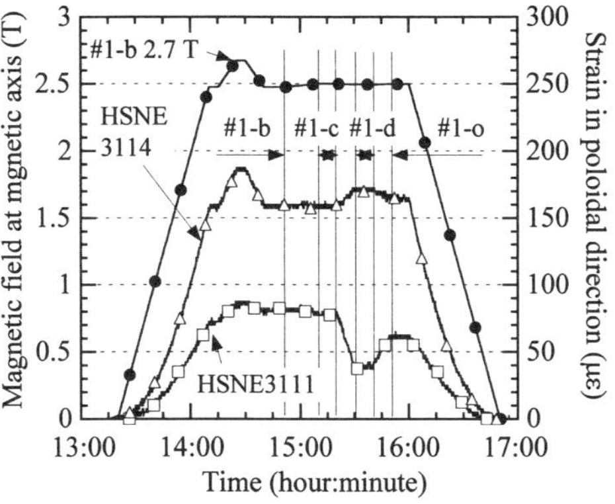

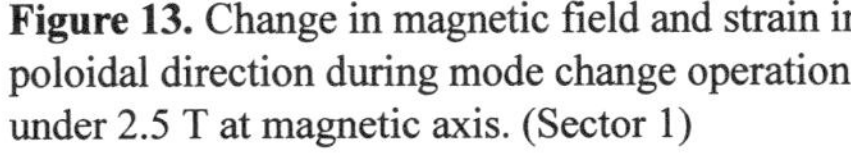

Figure 13. Change in magnetic field and strain in poloidal direction during mode change operation under 2.5 T at magnetic axis. (Sector 1)

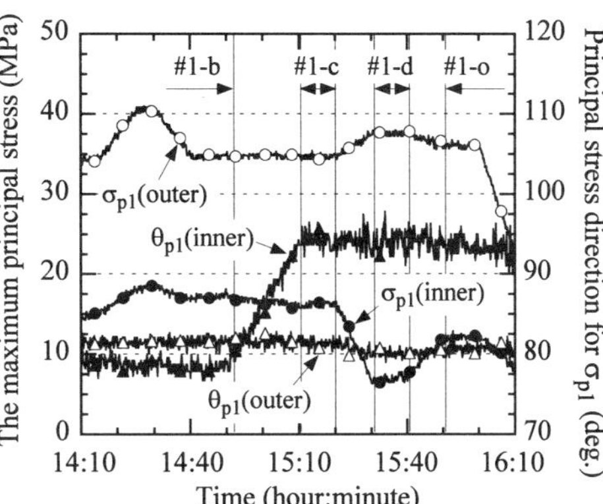

Figure 14. Change in the maximum principal stress and the direction on inner and outer equators during mode change operation. (Sector 1)

Since the strains in three directions could be measured on each sector during the mode change operation, the maximum principal stress and the direction were calculated by the measured strains. The results are shown in Figure 14. The maximum principal stress changed as seen in Figure 13 when the operation mode was varied from #1-c, #1-d to #1-o. On the other hand, the maximum principal stress direction rotated obviously to anti-clockwise on the inner equator, while the direction on the outer equator did not show such a remarkable change. This rotation is very clear, even though the results involve some error caused by the temperature rise. Figure 15 shows the electromagnetic force distribution of the helical coil against the toroidal angle under each operation mode. The calculation was performed under the condition of 4 T at magnetic axis.[4] The hoop force of the helical coil under each mode shows almost the same tendency, but the over turning force of the helical coil changes from the negative to the positive under #1-b mode, while the force under other modes varies from the positive to the negative. From these results, it is considered that the load balance between two shell arms would change by the operation mode change from #1-b to #1-c, and this results in the rotation of the maximum principal stress direction on the inner equator.

This fact means the strain measurement was successful during the coil excitation.

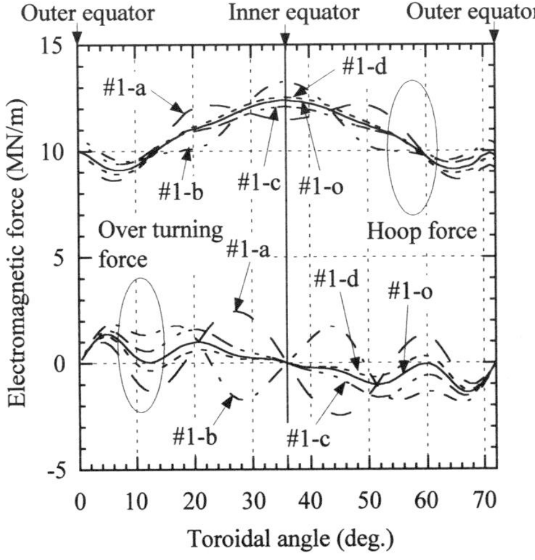

Figure 15. Distribution of electromagnetic force on helical coil under 4 T at magnetic axis. One helical coil runs on outer equator in case toroidal angle is zero or 72 deg. and on inner equator in case toroidal angle is 36 deg.

SUMMARY

The operation of the LHD started and the plasma experiments began on April in 1998. The maximum magnetic field of 2.75 T was achieved successfully under #1-o mode on the second operation. To measure the stress of the cryogenic support structure and evaluate the soundness of the structure, a temperature and strain measurement system was developed and installed in the LHD.

This report describes the results of the temperature rise and the change of the strain during the coil excitation of the LHD. The main results are summarized as follows:

(1) The strain measurement could be performed successfully for a long time at cryogenic temperature, in high vacuum and under high magnetic field.
(2) The temperature rise of the support structure occurs during the ramp up and down. Consequently, apparent strain is generated by the temperature difference between the active and the dummy gages, and this results in producing the negative shift on the diagram of the electromagnetic force against the strain. Sufficient cooling recovers the strain measurement, and the apparent strain disappears when the fluctuation of the magnetic field becomes small or zero.
(3) The relation between the square of the magnetic field and the measured strain shows the elastic deformation of the support structure. The structure works well to sustain the large electromagnetic force.
(4) The averaged stress in the poloidal direction on the equators is in good agreement with the one estimated by FEM analysis.
(5) During the successive mode change operation from #1-b, #1-c, #1-d to #1-o under 2.5 T at magnetic axis, the stress on the outer equator becomes larger under #1-d mode, while the stress on the inner equator becomes smaller. Also, the direction of the maximum principal stress changes remarkably when the mode changes from #1-b to #1-c. These facts could be explained by the change of the electromagnetic force distribution.
(6) From these results, it is clarified that the strain measurement is successful and useful to evaluate the soundness of the cryogenic support structure.

ACKNOWLEDGMENT

The authors wish to express our thanks to Dr. A. Iiyoshi, the former Director General of National Institute for Fusion Science (NIFS), for his continuous encouragement. Also, we would like to thank the members of Engineering Division in NIFS, Kyowa Electric Instruments Co., Ltd., and Hitachi Ltd., for their joint work.

REFERENCES

1. M. Fujiwara, et al., "Large Helical Device (LHD) Program," J. Fusion Energy, 15 (1996) pp. 7-154.
2. T. Satow, et al., "Assembly of the Superconducting Coils and the Cryostat for the Large Helical Device," Proc. MT-15, (1998) pp. 377-380.
3. T. Mito, et al., "Development of a Cryogenic System for the Large Helical Device," Adv. Cryo. Eng., 43 (1998) pp. 589-596.
4. S. Imagawa, et al., "Structural Analysis of the Large Helical Device," Adv. Cryo. Eng., 39 (1994) pp. 309-316.
5. H. Tamura, et al., "Design and Construction of Coil Supporting Structure and Cryostat Vessel for LHD," Adv. Cryo. Eng., 43 (1998) pp. 253-260
6. A. Nishimura, et al., "Strain Measurement in Fluctuating Magnetic Field and during Cold Thermal Cycles," Proc. MT-15, (1998) pp. 1275-1278.
7. A. Nishimura, et al., "Mechanical Deformation of the Large Helical Device during Cool-down and Excitation Tests," Proc. SOFT-20, (1998) pp. 873-876.

DEFORMATION BEHAVIOR OF CRYOGENIC COMPONENTS IN LHD DURING COOLING DOWN

H. Tamura, A. Nishimura, S. Imagawa, T. Mito, R. Maekawa, N. Yanagi, K. Takahata, T. Satow, S. Satoh and O. Motojima

National Institute for Fusion Science
Oroshi-cho, Toki-shi, Gifu 509-5292, Japan

ABSTRACT

Cryogenic components in LHD are cooled down to the liquid helium temperature. Since the coolant flows through long plumbing paths from the inlet to the outlet, the temperature gradient in a component induces a mechanical stress distribution during the cooling process. Although the strain measurement system was not stable under a condition of temperature changing, stress distribution was estimated by using FEM with an assumption of temperature gradient. The total weight of 822 ton is supported by ten support posts that are made of CFRP plate and stainless steel. The post was designed to prevent a heat leak from the base plate of the cryostat vessel (room temperature) to the coil support structure (4K). It also must be flexible against a displacement of 13 mm which is caused by the contraction of the support structure. The thermal anchor in each post has a different temperature because the coolant flows from one post to another. Mechanical behavior among the posts was studied considering the temperature difference and contact condition between the CFRP and stainless steel.

INTRODUCTION

LHD, the Large Helical Device, is a plasma experimental device with superconducting magnets[1] and has experienced two cycles of operation so far. Cryogenic components such as coil vessels, a coil supporting structure, and the superconducting coils have to be cooled down to liquid helium temperature. The components are mainly made of 100 mm thick 316 stainless steel. Total cold mass and the outside diameter of the cryogenic components are 822 tons and 13 m, respectively.

A thermal contraction of the 316 stainless steel between room temperature and liquid helium temperature is about 0.3 %. The components can not be cooled uniformly so that spatial temperature distribution will be induced. If the temperature difference is too large, the components could be damaged by the thermal stress. It is important to estimate the stress distribution under all conditions of cooling.

Generally, there must be a connection between a cryogenic structure and a component at room temperature, usually a supporting post. Deformation of the cryogenic structure links directly to the support post. It is better to maximize the length of the post and to use low thermal conductive material such as fiber reinforced plastic (FRP) and ceramics to minimize heat leak. On the other hand, the post has to be rigid enough against a load from the cryogenic structure such as weight, deformation during cooling down or a coil excitation etc. LHD uses ten cryogenic support posts that are set on the base plate of the outer cryostat vessel. The support posts are made of carbon fiber reinforced plastic (CFRP) plates and stainless steel plates. CFRP plate is used to reduce the heat leak from the base plate to the support structure. It was designed to be flexible against thermal contraction of the cryogenic components. The stainless steel plates were used for a thermal anchor around 80 K. In this paper, mechanical behavior of the coil support structure and the cryogenic support post is investigated to confirm if the structure worked as designed.

COIL SUPPORTING STRUCTURE

The coil supporting structure in LHD has a torus shape with major radius of 3.9 m and minor radius of 1.8 m (after completion of cooling). It primarily consists of two parts, the upper and lower hemispheres. The only difference between them is that the lower part has the base for the cryogenic support posts. They fasten the helical coil between them through the shell arms which were connected to the coil vessels. Each part has the symmetric shape of 10 for its toroidal direction so that the support structure consists of 20 parts of identical shape. Figure 1 shows a schematic of one sector of the support structure. Not only the shape of each sector but also the layout of cooling pipes attached to the surface is identical. It is made of 100 mm thick 316 stainless steel and all parts were assembled by welding.

There are 8 X 10 parallel paths for cooling of the support structure; 8 paths are distributed symmetrically on the upper and lower hemisphere of one sector. The small gap between the pipe and the surface of the structure was filled up with high thermal conductive resin. Temperature of the coolant was controlled not to exceed 50K between the inlet and the highest outlet temperature among the cooling paths[3]. Since the coolant flows through long plumbing paths from the inlet to the outlet, spatial temperature distribution is generated and it induces a mechanical stress generation.

Strain Measurement

We adopted a temperature compensation system for a strain measurement by setting dummy gauges on a stress free plate welded on its edge near an active gauge. The strain is

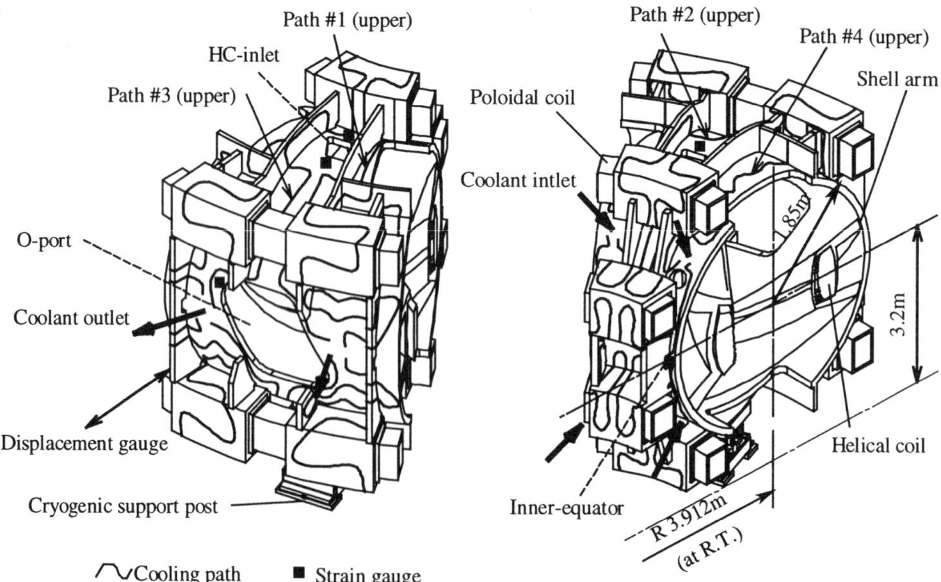

Figure 1. Schematic of the coil support structure. 1/10 part is shown with the coolant pipe and measurement location.

obtained from the bridge circuit consisting of one active gauge and three sets of dummy gauges. Temperatures of those four strain gauges are expected to be the same under stable cooling condition at room temperature or after completion of cooling down, for example. However, if the temperatures among gauges are different, the output from this gauge system might include a temperature dependent apparent strain[2]. In fact, data obtained from this system showed such behavior so that the original value of the strain included an apparent strain. Output strain data can be expressed by the following equation:

$$\varepsilon_{output} = \varepsilon_0 + \left(\varepsilon_{ap}(T_{active}) - \varepsilon_{ap}(T_{dummy})\right) \tag{1}$$

where ε_{output} is strain output, ε_0 is a true mechanical strain, $\varepsilon_{ap}(T)$ is an apparent strain at temperature T. We defined ε_d as a strain difference between toroidal and poloidal directions. A bi-axial gauge, with two gauges printed on the same base film, was used to measure those strains. Since the temperature in a bi-axial gauge was uniform, the apparent strain terms in Eq. (1) must be canceled by the subtraction. The strain difference was compared with the analytic result, and exact strain/stress distribution was estimated.

Analytic Model

We have previously proposed an analytic method for this kind of cryogenic component according to the first cycle operation[4]. In this paper, a calculation was done by using the same method. However, parameters needed in the analysis were reconsidered for detailed investigation. Furthermore, a cooling condition in the second cycle operation was chosen since the temperature difference between the inlet and outlet at low temperature was larger than during the first cycle operation.

To obtain a temperature distribution on a surface of the supporting structure, a temperature gradient along a cooling path should be known. We assumed that the temperature gradient as,

$$T = T_{inlet} + \left(T_{outlet} - T_{inlet}\right)\frac{1-\exp(-\beta\frac{L}{L_0})}{1-\exp(-\beta)} \tag{2}$$

where T is temperature at length L, L is a distance from the inlet, L_0 is total length and β is a parameter. Eq. (2) has a similar form of solution for governing energy equation under the condition of perfect heat transfer. Although the parameter β has to be decided by considering the change of thermal property such as thermal conductivity and specific heat, it was determined according to a temperature measurement in the actual structure. Figure 2 shows an example of measurement of the inlet, outlet and intermediate temperature along coolant

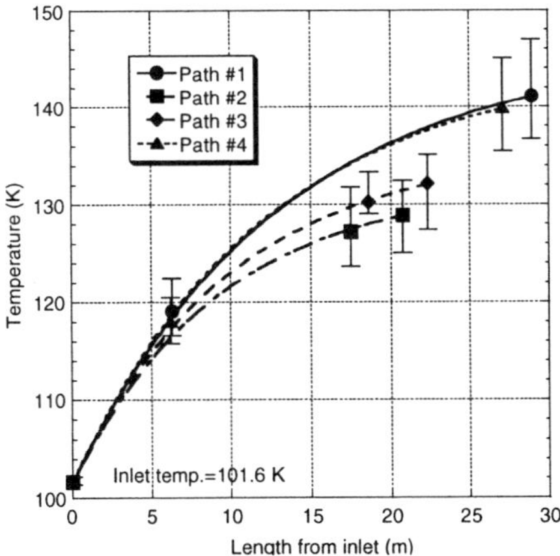

Figure 2. Temperature gradient along coolant paths. Plots are an average of 10 sectors. Error-bars describe the maximum and the minimum temperature among the sectors. Curves are estimated temperature gradient calculated from Eq. (2) with β=2.2.

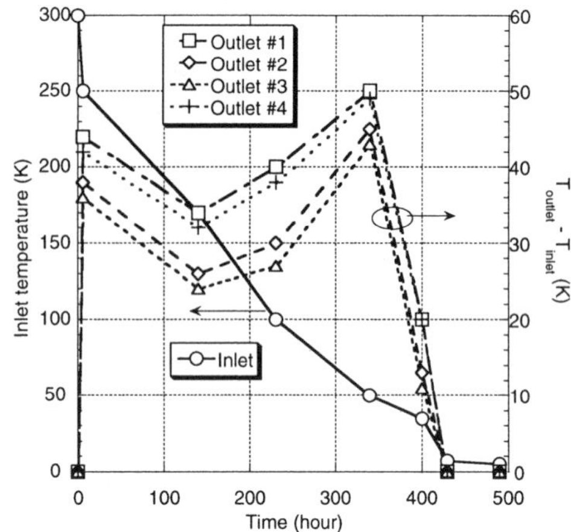

Figure 3. Inlet temperature and temperature difference between the outlet and inlet assumed in the analysis.

paths. The temperature change seemed to be decreased as the distance from the inlet was increased. The tendency of the temperature changing should be expressed by the parameter β in Eq. (2). The estimated β which was fit to the measurement value did not change much after 100 hours when the inlet temperature became 200 K. The parameter β was decided as 3.0 for T_{inlet} >= 170 K, 2.2 for T_{inlet} < 170 K. The ramp rate of inlet and outlet temperature of the cooling paths was assumed as shown in Figure 3, which also simulated the second cycle operation.

Physical properties needed in the analysis were defined as temperature dependent properties[5]. Spatial temperature distribution was calculated at first to decide transient temperature for every node point, and then the result was given to structural calculation as a body load. Temperatures of paths on the shell arms were assumed as an average of the inlet and outlet temperature of the supporting structure. Radiation heat leak to the components was omitted. FEM model was prepared and was solved by using ANSYS®5.5.

Result

Figure 4 shows the strain difference ε_d in the analysis and the measurement. The measurement data used in Figure 4 were obtained from the point HC-inlet (see Figure 1) in the sector #4 of the supporting structure. In the actual operation, coolant supply was stopped

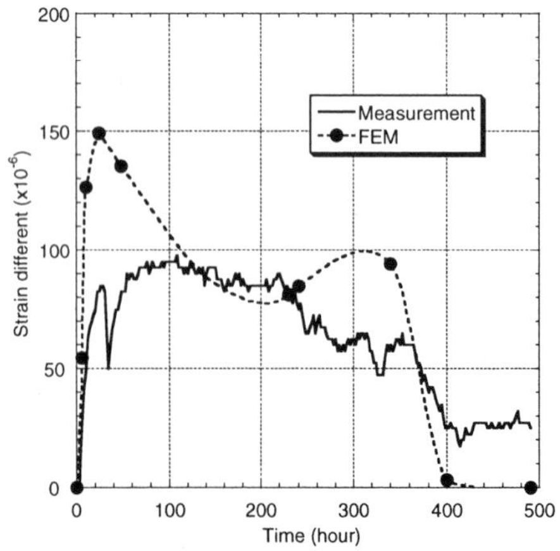

Figure 4. Strain different between poloidal and toroidal direction at HC-inlet area.

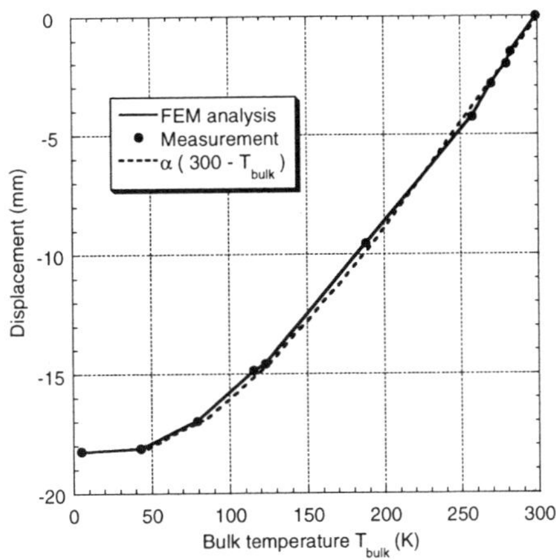

Figure 5. Displacement of outer radius against the bulk temperature.

accidentally at the time around 25 hours and also sometimes in the period of 240 to 320 hours. The times when the result of FEM and the measurement were different coincide with those times. The assumed temperature in the analysis omitted the suspended supply. Excepting those times, they agreed well.

The maximum equivalent stress of 270 MPa appeared at the root of the support rib for the outer poloidal coil. The area had been notched to release a stress concentration. Stress in other regions did not exceed 100 MPa. The tensile strain in *inner-equator* was the largest among them. The inner area of the component was cooled first but the other area, especially the outer area, was still warm and did not deform. Since the volume of warm area was much larger than inner area, the inner area was subjected to large tensile stresses.

A bulk temperature T_{bulk} that describes an average temperature through the length of the coolant pipe has been defined:

$$T_{bulk} = \frac{1}{l_0}\int_0^{l_0} T dl = T_{inlet} + (T_{outlet} - T_{inlet})\left(\frac{1}{1-\exp(-\beta)} - \frac{1}{\beta}\right) \tag{3}$$

Figure 5 shows the displacement of outer diameter area showed in Figure 1. The results of the measurement, FEM analysis and the calculation result simply multiplied thermal contraction coefficient by the temperature difference between 300 K and T_{bulk} are plotted. They showed good agreement with high accuracy.

The mechanical behavior of the coil support structure can be analyzed by the method described here and approximation of displacement can be calculated from Eq. (3) with thermal contraction coefficient if the inlet, outlet and certain intermediate temperatures are known.

CRYOGENIC SUPPORT POST

The weight of the supporting structure and the coils are sustained by the cryogenic support posts which were set on the base plate of the cryostat vessel. Thermal insulation between room temperature and liquid helium temperature is given by these posts. They also absorb a thermal contraction of the coil support structure during the cooling down. The post was also designed to stand against earthquake of 0.3 G level[6].

Figure 6 shows a schematic of the cryogenic support post. It was attached to the coil support structure at the upper stage. The base of the support post was set at -6 mm inside from the neutral position at first. Once the cryogenic components are started cooling down, the upper stage of the support post moves to the inner direction by the thermal contraction. The last position of the support structure must be –13 mm inside from the initial radius so that the upper stage would be at –7 mm inside from the neutral position.

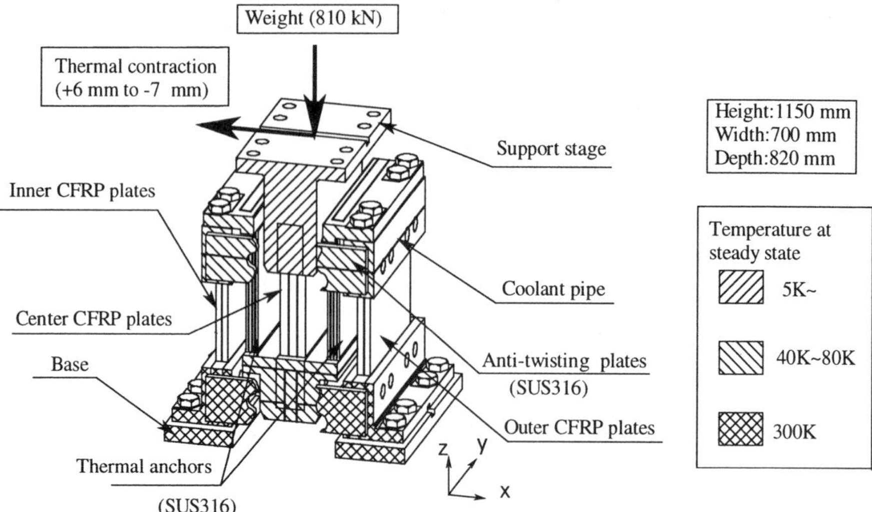

Figure 6. Schematic of the cryogenic support post.

CFRP was chosen for thermal insulation since the thermal conductivity below 50 K is lower than GFRP and it has high strength for buckling. It also has higher elastic modulus. To keep a factor of safety for buckling, CFRP plates were located at the center and inner/outer sides through the thermal anchor made of stainless steel.

The thermal anchor of the support post is cooled to 40 to 80 K at the steady state condition. Coolant for the thermal anchor was supplied from two cooling paths and each path cools five posts from one post to another. Therefore, there are temperature differences among five posts' thermal anchor. The temperature differences affect a change of free height of the post because thermal contraction of the thermal anchor plates among them would not be the same. Since the thermal contraction of CFRP material is much less than that of stainless steel, the colder the thermal anchor is, the taller the post becomes. This will cause unequal load distribution among the posts since the support structure is rigid enough against elasticity of the post.

Strain Measurement

The strain in the support post was measured at three locations. They were at the thermal anchor plate, the outer/inner CFRP plates, the anti-twisting plate and the support stage on the

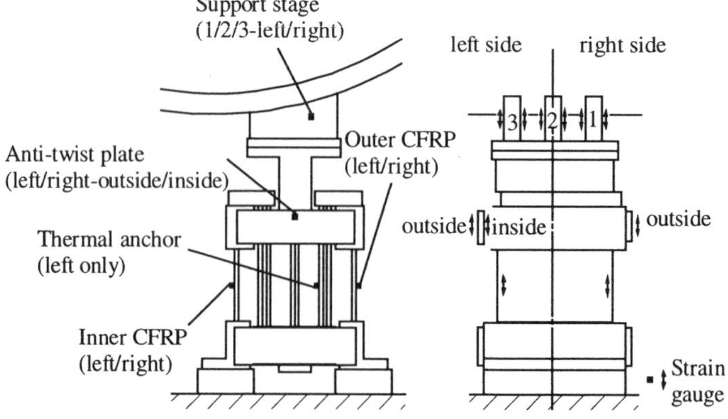

Figure 7. Strain measurement in the support post.

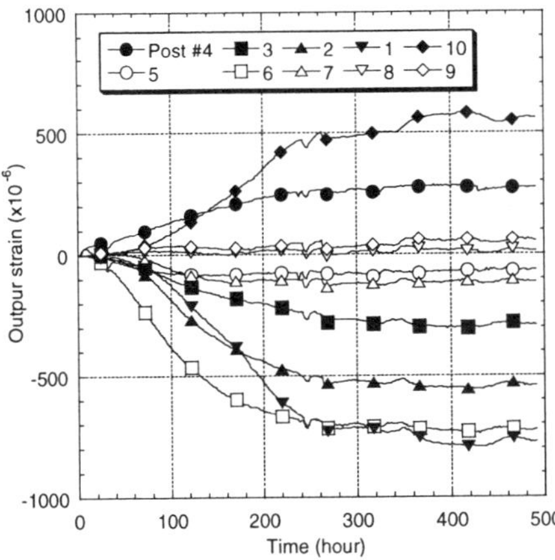

Figure 8. Output strain at the thermal anchor plates. Close; posts at coolant path 1 (#4-#3-#2-#1-#10), open; at path 2 (#5-#6-#7-#8-#9).

Table 1. Output strain and estimation of load for the support stage in post #3.

		Support stage			
		1	2	3	
Output strain	left side	13	-53	16	
	right side	-17	42	-85	$\times10^{-6}$
Axial strain		-2	-5.5	-34.5	
Bending strain		15	7.5	50.5	
Estimated load change from room temperature		43.2	21.6	145.4	(kN)

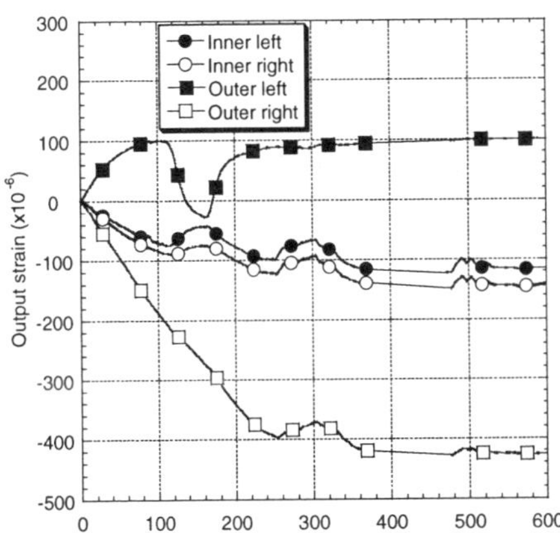

Figure 9. Output strain at the outer and inner CFRP plates in post #3.

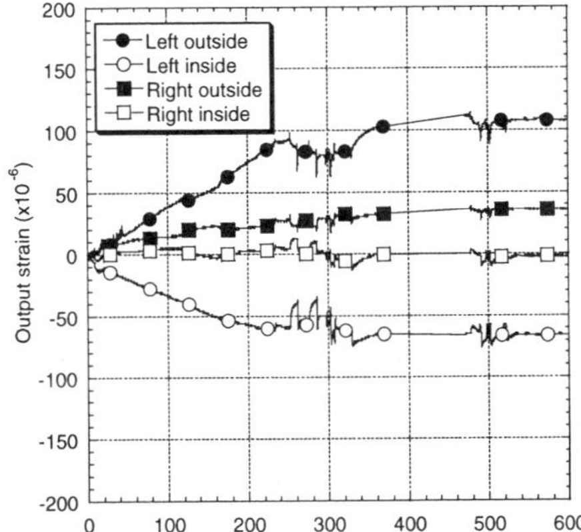

Figure 10. Output strain at the anti-twist plates in post #3.

top of the post as shown in Figure 7. The strain of the support stage was obtained by the same system as the coil support structure, i.e. one active gauge (put to coincide with the height direction) and three dummy gauges set on a stress free dummy stage. Therefore, the output strain after completion of cooling would be valid. Other gauges were combinations of 4 active gauges (2 for height direction and others for perpendicular to the height). In this case, temperature dependency is compensated and $2(1+\nu)$ (ν; Poisson's ratio) times as much as the original axial strain could be obtained when a load is applied from the axial direction to it. Strains in the thermal anchor were measured from all posts and others were from the support post #3.

Results and Discussion

All the measured output strain was initialized to the one at room temperature. Figure 8 shows the output strain at the thermal anchor plate. There was an observed dispersion of strains among 10 posts. One reason for the dispersion could be the load balance among the posts caused by the changing in free height of the post. For example, from the measurement of the strain at the support stage in the post #3 as shown in Table 1, the total load to the post was estimated to be increased by 210.2 kN after cool down. Therefore, the load was increased in some posts and decreased in other posts. However, the order of the strain value did not coincide with the temperature of each post's thermal anchor. The load among support stage 1 to 3 was different so that the load to the thermal anchor plate and to the CFRP plate must have been unequal through the right side to the left side. Furthermore, the strain was obtained from one side of the plate.

Figure 9 and 10 show strain output from the outer/inner CFRP plate and the anti-twisting plate, respectively. There was an unequal strain in both edges of these plates. The average strain of the inside and outside surface represents an axial strain. The axial strain of the anti-twist plate at the left side and the right side were almost the same. On the other hand, a bending strain of the plate at the left side, which is calculated from the axial and the original strain, was twice as much as that of the plate at the right side. It means that a small non-horizontal load had been applied to the post and a twist deformation was generated in the post. Furthermore, the CFRP plates were fitted to the square slot in the stainless steel block and they were fixed by trough keys. There was possibly unequal contact between the CFRP plate and the block caused by difference of thermal contraction at low temperature. As a result, the load distribution through the width of the CFRP might not have been uniform. It is considered that this is another reason of dispersion not only in a post but also among the posts. However, the stress level estimated by the measurement was small enough compared with a permissible stress of 200MPa for the thermal anchor and the CFRP plate so that the post maintain a sound condition during cooling down.

CONCLUSIONS

The deformation behavior of the coil supporting structure and the cryogenic support post was investigated. FEM analysis was prepared to support estimation for a stress distribution of the coil supporting structure. Although the exact strain values were unknown, the subtracted value between certain axis and its perpendicular axis was known so that the subtracted strain was compared with analytic results and the stress distribution could be estimated. Displacement during cooling down could be estimated by the bulk temperature in a coolant pipe, which was calculated from a temperature gradient in the coolant pipe.

Behavior of the cryogenic support post was understood as a combination of change in free height and a contact problem between the CFRP plates and the stainless steel block. We need to analyze this by considering those phenomena for a detailed investigation, but the stress generated in the post is thought to be within a safe level from these measurement.

For both the coil supporting structure and the cryogenic support posts, it was confirmed that the stress distribution and the displacement were within acceptable designed values.

REFERENCES

1. M. Fujiwara, et al., Large helical device program, *J. Fusion Energy* 15:7 (1996).
2. A. Nishimura, et al., Thermal and mechanical strain measurement of support structure of the Large helical device, in: "ICEC17 Bournemouth" D. Dew-Hughes, R. G. Scurlock, J. H. P. Watson, Institute of Physics Publishing, London (1998), p.847.
3. T. Mito, et al., First cool-down performance of the LHD, presented at Applied Superconductivity Conference, 1998, Palm Desert, CA, USA (1998).
4. H. Tamura, et al., Mechanical behavior of cryogenic components in LHD, in: "ICEC17 Bournemouth" D. Dew-Hughes, R. G. Scurlock, J. H. P. Watson, Institute of physics Publishing, London (1998), p871.
5. National Bureau of Standards, "LNG Materials and Fluids –User's Manual", USA (1977).
6. H. Tamura, et al., Structural and mechanical design of cryogenic support system for LHD, in: "Fusion Technology 1996", C. Varandas and F. Serra, Elsevier Science B.V., Amsterdam (1996), p.1019.
7. G. Hartwig, Reinforced polymers at low temperature, "Advances in Cryogenic Engineering 28", (1985), p.179.

THERMAL AND STRESS ANALYSIS OF A 0.8 m DIAMETER HTS MAGNETIC SEPARATOR

C.M. Rey, W.C. Hoffman, and J.G. Sloan

DuPont Superconductivity
Wilmington, DE, 19880-0304

ABSTRACT

The thermal and stress/strain analysis of a 0.8 m diameter cryocooled High Temperature Superconducting (HTS) magnet is presented. The intended use of the magnet is in the separation of mineral contaminants from kaolin clay and titanium dioxide. The design central field of the magnet is ~ 2 T. The HTS magnet will be cryocooled with a target operating temperature of 20 K. A non-linear stress/strain analysis was performed to determine the stress/strain values on the HTS conductor and its corresponding support structure at four critical stages of manufacture and operation: 1) tension winding, 2) axial pre-compression, 3) cool down to 20 K, and 4) magnet energization. Results indicate the HTS conductor and its support structure remain below acceptable stress/strain levels (< 0.2 % mechanical strain) during the fabrication and operation of this device. In addition, a steady state thermal analysis was also performed in order to insure adequate heat removal within the HTS conductor windings. Results indicate a maximum temperature rise of < 3 K for a 12 W input, which represents the worst case operational scenario.

INTRODUCTION

DuPont Superconductivity is investigating the commercial viability and technical feasibility of designing and building a large High Temperature Superconducting (HTS) magnetic separator.[1] The development project is divided into four separate phases: 1) conceptual design, 2) final design, 3) fabrication, and 4) testing. Each phase will last approximately one year.

The intended use of the magnet is in pilot scale research of the separation of mineral contaminants from kaolin clay and titanium dioxide (TiO_2). Both kaolin clay and TiO_2 are white pigments with a wide variety of uses and are used extensively in the paint, paper, and plastics industries.

Advances in Cryogenic Engineering, Volume 45.
Edited by Shu *et al.*, Kluwer Academic / Plenum Publishers, 2000.

REQUIREMENTS

HTS coil performance specifications were developed for the magnetic separator. Performance specifications were based upon existing commercial LTS magnetic separators presently used by the kaolin industry. Major requirements for the HTS coil are given in Table 1.

A central magnetic field of 2 T was chosen because it represents what is commercially available in batch type separators.[2] Dimensional requirements for the HTS coil were determined by those presently used in the kaolin and TiO_2 industries. Other requirements in terms of useful life, magnetic and thermal cycles, ramp time, etc., are representative of those used in commercial LTS magnetic separators, with the exception of the refrigeration system. Presently, the LTS coils used in commercial magnetic separators of this diameter have been limited to using liquid helium as a refrigerant.

Table 1. Major HTS Coil Performance Requirements

Parameter	Requirements
Magnetic Field	2 T
Coil inner diameter	0.81 m
Coil outer diameter	0.838 m
Coil height	50 cm
Refrigeration	Cryocooled
Thermal cycles	25
Magnetic cycles	200,000
Cost	equivalent LTS unit
Ramp time	< 2 minutes
Useful life	5 years

DESIGN

Based upon the HTS coil performance requirements, an HTS coil conceptual design was generated. Key elements of the design are summarized in Table 2.

The HTS conductor for the coil will consist of several strands of silver sheathed $Bi_2Sr_2Ca_2Cu_3O_x$ (Bi-2223) powder-in-tube multi-filamentary tape. Each Bi-2223 tape will be approximately 4 mm wide and 0.25 mm thick and contain over 61 HTS filaments.[3] The HTS tapes will have a normal metal to superconductor ratio of ~ 2.5 to 1. The conductor will be tension wound into double pancake coils on a pre-formed winding mandrel. A stainless steel support strip will be used to support the hoop tension in the coil. Each double pancake coil will be stacked and spliced together to form a single continuous HTS coil. The stacked HTS coil assembly will be axially pre-compressed via tension rods to minimize conductor movement during cool-down and energization. Kapton® and G-10CR will be used for the turn-to-turn and pancake to pancake insulation, respectively. The current leads will consist of a binary–type HTS current lead with an intercept temperature of about 60 K. The lower stage (HTS portion) of the current leads will consist of silver - alloyed Bi-2223 powder-in-tube tape. A 6061-T6 aluminum shield operating at about 60 K will be used to intercept the room temperature

Table 2. HTS Coil Conceptual Design

Parameter	Design
Operating current	400 A
HTS conductor	Bi-2223 (PIT)
Winding type	Double pancake
No. of pancakes	102
No. of turns per pancake	28
Intercept temperature	60 K
Operating temperature	20 K
Iron yoke	Yes
Refrigerator	Two 2 stage G-M, ~ 25 W @ 20 K
Current leads	HTS – binary type Ag – alloyed PIT

radiation heat load. The HTS coil assembly will be mounted in a donut shaped vacuum cryostat. The cryostat will be mounted inside an iron yoke structure. The purpose of the iron yoke is to: a) reduce the stray field at the magnet perimeter and b) lower the required number of ampere-turns for the HTS coil. The entire magnet assembly (including the iron yoke) is being designed for transportability among potential on-site customer test facilities.

ANALYSIS

Stress/Strain Analysis

A non-linear finite element analysis (FEA) was used to determine the stress/strain with the HTS conductor and its corresponding support structure at four critical states of HTS coil fabrication through operation: 1) tension winding, 2) axial pre-compression at room temperature, 3) cool-down to 20 K, and 4) magnet excitation. For the magnet excitation stress/strain analysis, the magnetic forces (including the iron yoke) were determined by two dimensional (2D - MAGNETO)[4] and three dimensional (3D - AMPERES)[4] electromagnetic analysis using a commercial Boundary Element Method code.

ABAQUS Version 5.8 FEA code[5] was used to complete the structural analysis. Figure 1 shows the 2D axisymmetric analysis model of ½ of a double pancake cross section with the various material parts shaded. The analysis elements included the insulation/metallic/insulation layers along with half of each pancake coil above and below the layers. By using axisymmetric elements, the model approximates the spiral wound coil as a series of concentric rings. Because of the stress gradient (i.e. magnetic field gradient) from the inner diameter to the outer diameter within the winding pack, the first 3 inner diameter (ID) and last 3 outer diameter (OD) coil plies included separate HTS and stainless steel (SS) support plies to better simulate the actual conditions of these extremes. The central plies were simulated with average properties to model the basic behavior without capturing the details. Contact surfaces were defined between each part of the model.

The lower cut plane through the pancake coil winding is constrained against axial movement. The upper cut plane is constrained to move together axially. These boundary conditions simulate the central portion of the full magnet.

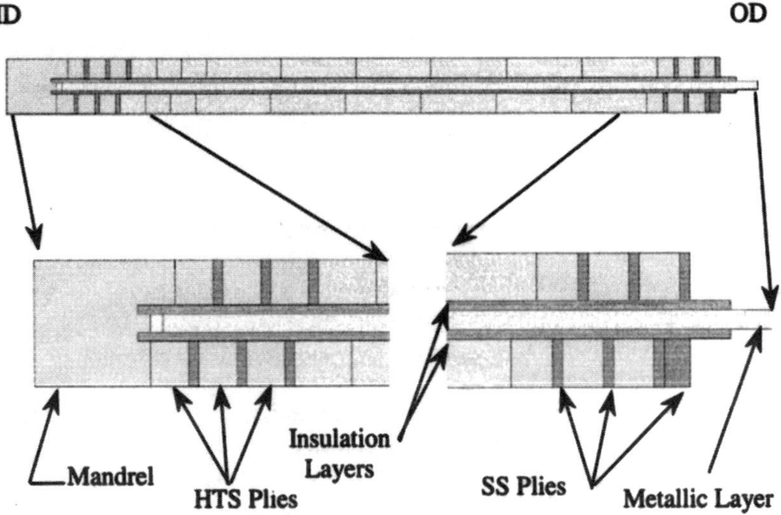

Figure 1. ABAQUS FEA 2D Axisymmetric Model – ½ cross section of a Double Pancake.

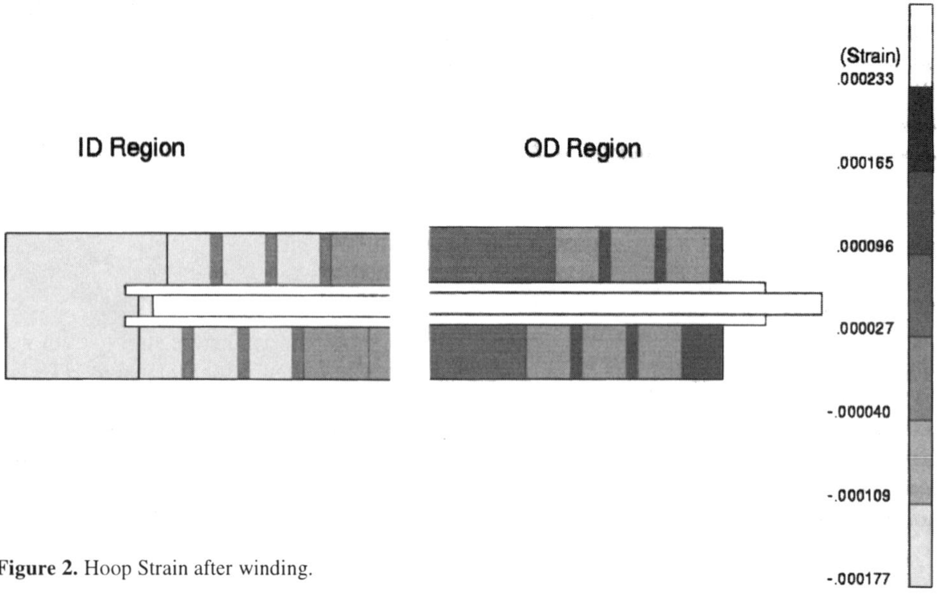

Figure 2. Hoop Strain after winding.

The simulation of the pancake winding process involved defining an initial hoop stress in each HTS and SS ply pair. These stressed ply pairs were released one pair at a time to mimic the actual winding process. The initial stress was adjusted such that the hoop stress in the HTS and SS plies, after release, matched the winding tension. As each new pair is released, the previously released ID plies are compressed and eventually lose their initial tension.

Figure 2 shows the resulting hoop strain in the wound pancake simulation for only the enlarged ID and OD regions that have separate HTS and SS plies. The mandrel and ID HTS plies are all in compression. The ID SS plies are still in tension although considerably less than the initial winding tension. The OD SS plies are all in tension at about the winding tension level. Friction is not included between the contact surfaces during the winding simulation.

Applying an axial load to the model simulated the clamped assembly of the stacked pancakes. The resulting axial compression loads to the wound pancake model are not large compared to the energized loads, so the results are not shown here.

The next simulation involved cooling the magnet to 20 K. For this state, friction between the contact surfaces was included in the analysis. Due to the difference in coefficient of thermal expansion, the HTS, SS, insulator and metallic materials all shrink differing amounts. This tends to unload the compressive stress holding the HTS/SS plies tightly against the mandrel. The challenge of the modeling task is to determine the initial level of wind tension necessary to cause the ID HTS plies to maintain contact with the mandrel without overstraining the HTS material. If insufficient winding tension is used, the ID HTS plies will lose contact with adjacent plies and jeopardize heat transfer capability.

Figures 3 shows the radial strain in the ID and OD regions for the 20 K cooled condition for the magnet. The intervening plies fall between these bounding stress states. The important result is that the HTS plies are all in radial compression with the adjacent SS support plies. The only tensile radial strains are due to friction between the insulation layers and OD SS plies.

The final state simulated the excitation loads. This was accomplished by applying a radial acceleration to the HTS material in the model. The acceleration varied linearly in a 2:1 ratio from ID to OD of the coil. The acceleration magnitude was set to approximate the

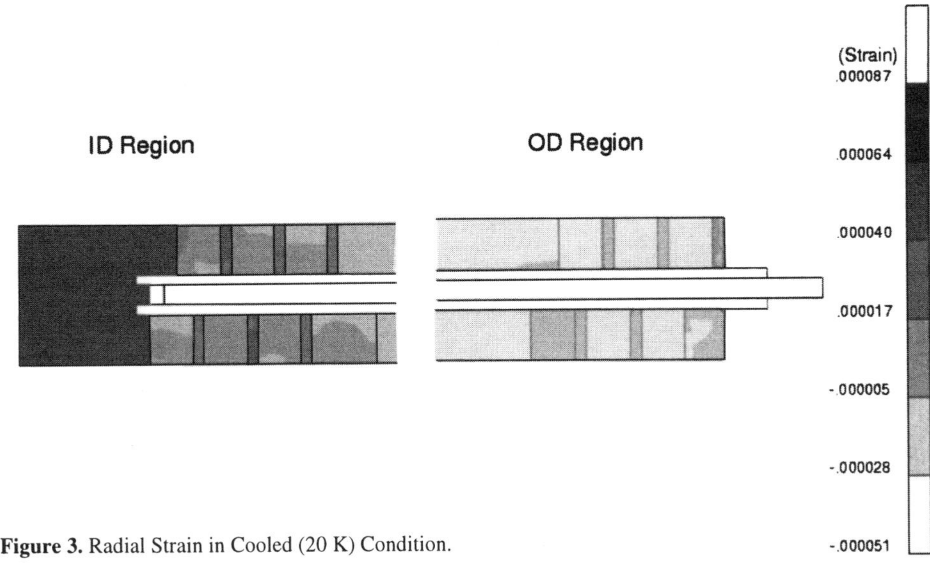

Figure 3. Radial Strain in Cooled (20 K) Condition.

total radial force generated in the energized magnet, which included the influence of the iron yoke in magnitude of the magnetic field but not direction. In addition, an axial compression load from the magnetic analysis was applied to the top HTS/SS pancake cut plane.

Figure 4 shows the HTS hoop strains in the ID and OD regions of the energized coil. As shown, the strain levels are well below the 0.2 % capability of the HTS material.[3]

The HTS axial strains are shown in Figure 5. These strains are compressive due to the large axial energized load. Since the SS support plies carry a large portion of the axial force, the load carried by the HTS is low.

Table 3 details the strains in the HTS and SS support plies for the first and last plies at the three highest load states. Strains are within the capability of the HTS material.[3]

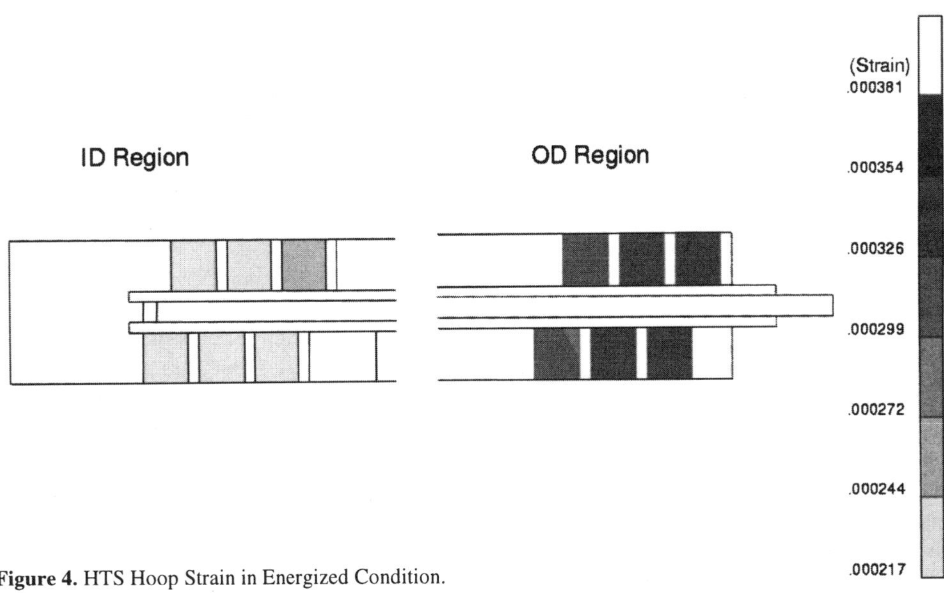

Figure 4. HTS Hoop Strain in Energized Condition.

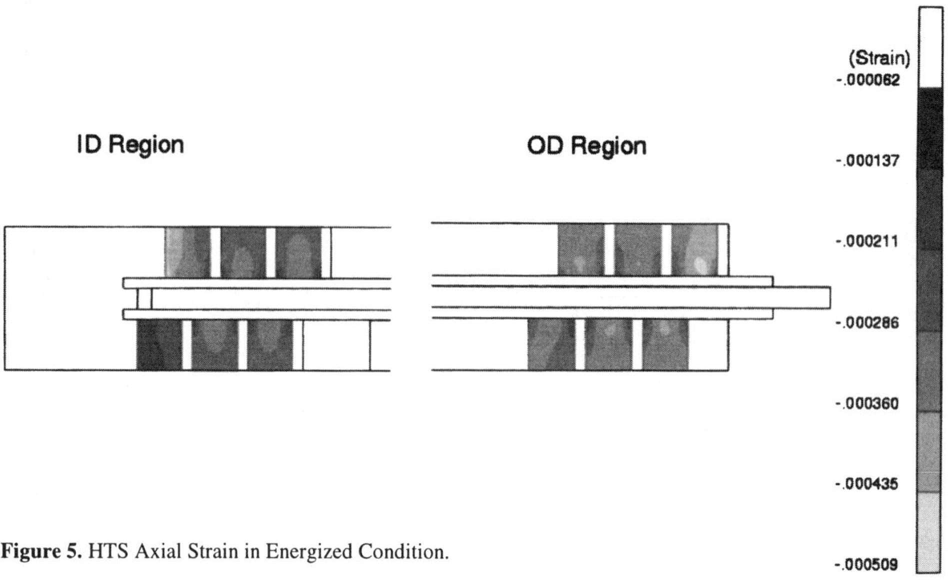

Figure 5. HTS Axial Strain in Energized Condition.

Table 3. Strains in Pancake Model

State	Wound			Cooled to 20 K			Energized		
Direction R – Radial A – Axial H – Hoop	R (%)	A (%)	H (%)	R (%)	A (%)	H (%)	R (%)	A (%)	H (%)
Mandrel	.0049	.0051	-.018	.0087	.0099	-.031	.0014	.0014	-.0049
1st HTS Ply	.0057	.0045	-.015	-.003	-.003	-.0038	.0137	-.051	.022
1st SS Support Ply	-.002	-.005	.007	.006	-.010	-.006	.013	-.049	.019
Last HTS Ply	.0013	-.003	.0003	-.0051	-.005	.015	-.013	-.046	.038
Last SS Support Ply	-.006	-.010	.023	-.0034	-.020	.015	-.006	-.059	.037

Thermal Analysis

A static linear FEA was used to determine the maximum coil temperature under normal operating conditions. The AC loss (eddy current, hysteresis, etc.)[6] and DC static heat loads (radiation, index loss, splice, etc.) used as inputs to the thermal analysis (~ 12 W at 20 K) were calculated in a separate analysis. The AC loss portion of the heat load was ~ 8 W and consists primarily of hysteresis loss. A temperature intercept of 60 K was assumed for the radiation shield and the HTS current leads.

Pro/ENGINEER® and Pro/MECHANICA® THERMAL Release 19.0 software[7] were used to create the models and perform analysis, respectively. Two types of steady-state thermal conductive analysis were performed: A 2D Axisymmetric analysis of a ½ cross-section through the magnet and a 3D analysis. Thermal model inputs are summarized in Table 4.

For the 2D Axisymmetric model, both the number of coil pancakes and turns per coil (see Table 2) were consolidated ~ 10:1 while maintaining overall physical dimensions of the coil. This consolidation yielded an equivalent 10 pancake structure, each pancake consisting of a detailed 3 turn structure (see Figure 6). A thin outer copper shell simulated the actual axial cooling feature. The 12 W heat load was distributed to the HTS elements with a decreasing linear radial gradient. Figure 6 is the 2D analysis thermal fringe plot for

Table 4. Thermal Model Inputs

Parameter	2D Axisymmetric	3D Solid
Heat Load & Distribution	12 W	12 W
Radial (ID → OD)	Linear Decrease	Linear Decrease
Axial	Uniform	Uniform
Operating Temperature	20 K	20K
Material Properties	Isotropic	Orthotropic & Isotropic
Thermal Contact Resistance	0	0
Eddy Current Breaks	None	Included
Splice features	None	Included (per pancake)
Current Lead features	None	Included
Cryocooler Connection	Distributed at each end	Discrete at each end
Outer Conductive Shell	Uniform	Discrete

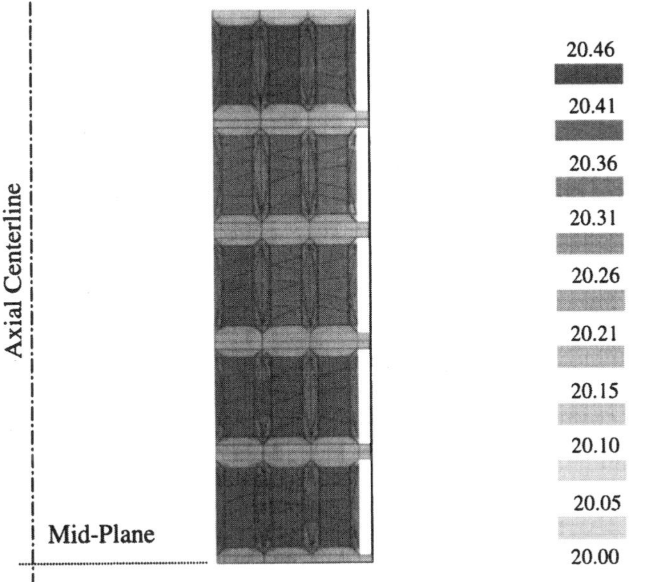

Figure 6. 2D Axisymmetric thermal fringe plot of ½ HTS coil cross-section. The maximum ΔT is 0.46 K for a 12 W heat load.

the ½ consolidated structure cross section. For a starting operating temperature of 20 K, the 2D analysis predicts a maximum temperature of 20.46 K within the HTS material.

For the 3D model, the pancake coil structure was lumped together to give an equivalent single pancake of 3 turns. Orthotropic thermal properties, calculated and verified with the 2D model, were assigned to the lumped structure. The 3D model includes the asymmetric features of eddy current breaks, pancake splices, current lead connections and the discrete mounting of the two cryocoolers (see Table 4).

Figure 7 is the 3D thermal analysis fringe plot of the lumped coil structure. For a starting operating temperature of 20 K, the 3D analysis predicts a maximum temperature of 22.19 K within the HTS coil. The addition of a 0.50 K calculated temperature difference across the flexible cryocooler connection (not shown) yields a maximum HTS coil temperature to 22.69 K.

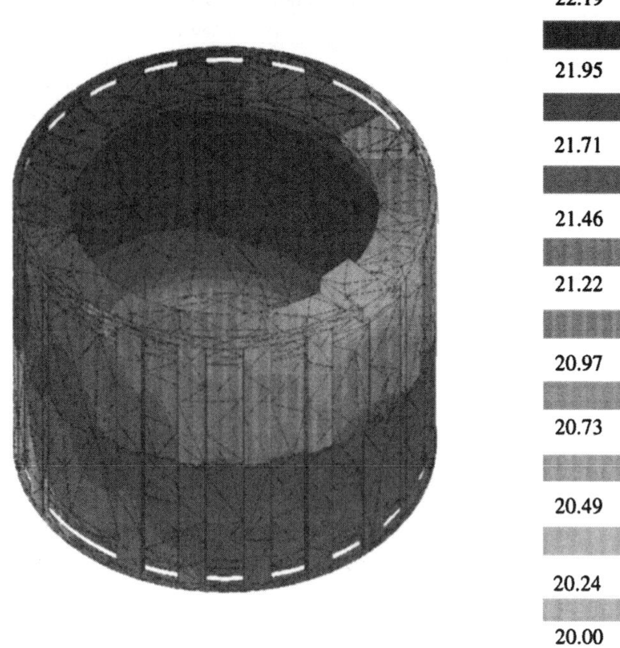

Figure 7. 3D thermal fringe plot of the HTS coil model – isometric view. The maximum ΔT is 2.19 K for a 12 W heat load.

CONCLUSION

DuPont has begun the conceptual design of a 0.8 m diameter HTS magnetic separator. The separator will be used for the removal of mineral contaminants from kaolin clay and TiO_2. The central field of the magnet is designed to be ~ 2 T. The design operating current of the magnet is 400 A at 20 K.

A 2D axisymmetric stress/strain analysis was performed to simulate the four states of magnet assembly through operation. Results of the model indicate that the mechanical stress was less than 10 MPA at room temperature and 32.6 MPA at 20 K and 400 A. These values are well within the operating limits of the HTS material.

A 2D axisymmetric and 3D thermal analysis was performed to determine the maximum temperatures of the coil under steady state operating conditions. For a 12 W heat load, the 2D and 3D model results show a maximum ΔT of ~ 0.5 K and 3K, respectively. These values are well within the safe operating envelope of the HTS magnet system.

REFERENCES

1. J. Iannicelli et al., "Magnetic Separation of Kaolin Clay Using a High Temperature Superconducting Magnet System," *I-EEE Transactions on Applied Superconductivity*, **7**:1061 (1997)
2. Products catalog of Eriez Magnetics, Erie, PA (1997)
3. K. Hayashi, N. Saga, K. Ohkura, K. Sato, "Development of Ag-Sheathed Bi2223 Superconducting Wires and their Applications", *Advances in Cryogenic Engineering Materials*, **44**:799 (1998)
4. Integrated Engineering Software, Winnipeg, Canada, 1997.
5. Hibbitt, Karlsson & Sorensen, Inc., Pawtucket, Rhode Island (1998)
6. M. Wilson, Superconducting Magnets, Oxford University Press, New York, **8**:159 (1986)
7. Parametric Technology Corporation, Waltham, Massachusetts (1998)

DEVELOPMENT AND TESTING OF A 3 T BI-2212 INSERT MAGNET

H.W. Weijers,[1] Q.Y.Hu,[1] Y. Viouchkov,[1] E. Celik,[1] Y.S. Hascicek,[1] K. Marken,[2] J.Parrell,[2] and J. Schwartz[1]

[1] National High Magnetic Field Laboratory
Tallahassee, Florida

[2] Oxford Superconducting Technology
Carteret, New Jersey

ABSTRACT

This paper describes the development and testing of a 3 T Class Bi-2212 insert magnet. The magnet consists of three sections, each built by stacking double pancakes using Powder-In-Tube conductor and the Wind&React approach. Conductor with a pure Ag matrix was used for the inner section and conductor with a mixed Ag and AgMg matrix was used for the outer two sections. Elements of the design, conductor properties, construction, and testing are presented. The successful generation of 3 T in a 19 T background magnetic field and some of the development issues are discussed.

INTRODUCTION

One of the goals of the National High Magnetic Field Laboratory (NHMFL) is to build a 1.066 GHz NMR system, requiring a 25 T magnet. Metallic superconductors like NbTi and Nb_3Sn will generate about 20 T. We focus on Bi-Sr-CaCu-O superconductors as the material for the inner coils to generate the additional 5 T. To develop magnet technology, we set an intermediate goal to build a 3 T class insert magnet, to be tested in a background magnetic field of 19 T as provided by the Large Bore Resistive Magnet (LBRM) at the NHMFL. The design and some initial results have been published previously[1]. This paper focuses on the construction of the magnet, in-field testing and a discussion of the results. The Bi-2212 coil generated 3.0 T in a 19 T background magnetic field. This 3 T magnet project was a collaboration between the NHMFL and Oxford Superconducting Technology (OST).

DESIGN

The design of the 3 T class insert coil is based on 3 stacks of double pancakes of Bi-2212 Powder-In-Tube tape conductor produced by Oxford Superconducting Technology

Advances in Cryogenic Engineering, Volume 45.
Edited by Shu *et al.*, Kluwer Academic / Plenum Publishers, 2000.

(OST) and using the Wind and React approach. This choice was motivated by the success of a 1.2 T single-stack insert coil[2]. It expands the concept by increasing the number of concentric sections from one to three and increasing the number of double pancake units to achieve the desired central field. Projected current densities are based on results previously obtained with OST conductor. Conductors with a pure silver matrix (labeled Ag) and conductor with a mixed matrix of silver and silver alloy (labeled AgMg) were used. The Ag conductor, which typically has a higher current density then AgMg conductor, was used in the inner section, where the strain levels are acceptable for this type of conductor. The geometric parameters of the magnet as built are given in Table 1. All critical currents are determined using the $1*10^{-4}$ V/m criterion, equivalent to 1 μV/cm. Using a $5*10^{-13}$ Ω.m criterion would have resulted in comparable critical currents

Table 1. Geometric parameters of the 3 T insert magnet.

Property	Unit	Section A	Section B	Section C
Conductor width	mm	3.0	3.1	3.1
Conductor thickness	mm	0.2	0.2	0.2
Free bore	mm	12.7	54.2	94.8
Winding ID	mm	16.0	56.6	97.2
Winding OD	mm	46.0	86.6	127.2
Total OD	mm	50.2	90.8	131.4
# turns/pancake	-	67.2	65.4	66.3
# double pancakes	-	14	13	10
Winding height	mm	96.3	98	75
Field contribution	mT/A	23.3	17.4	11.5
Total conductor length	m	184	383	467.4
Packing factor	-	78%	72%	73%

CONDUCTOR PROPERTIES

All conductors used in this project were supplied by OST. The conductor in the 3 T magnet came from two batches of Ag-sheathed conductor and eight batches of AgMg sheathed conductor. Batch lengths ranged from 150 to 300 m. Typical short sample I_c and mechanical properties are listed in Table 2. The measured Young's modulus at cryogenic temperatures is 67±2 GPa[3]. The heat treatments used are discussed in the next section.

Table 2. Conductor properties at 4. 2 K.

Property	unit	Ag	AgMg
Routinely obtained I_c	A	300-360	220-300
Estimated 4.2 K stress limit	MPa	60	160
Estimated 4.2 K strain limit	%	0.18	0.4

To characterize the $I_c(B)$ relations, short samples were measured in background magnetic fields, with B either parallel (B//) or perpendicular (B⊥) to the flat side of the tape and perpendicular to the current. The initial critical currents varied from 328 to 348 A for the Ag conductor and from 216 to 278 A for the AgMg conductor. Typical data normalized to the I_c value at zero background field are presented in Figure 1. The relation for Ag conductor is comparable to that of AgMg. The $I_c(B_\perp)$ curve for Ag presented in [1] is a-typical.

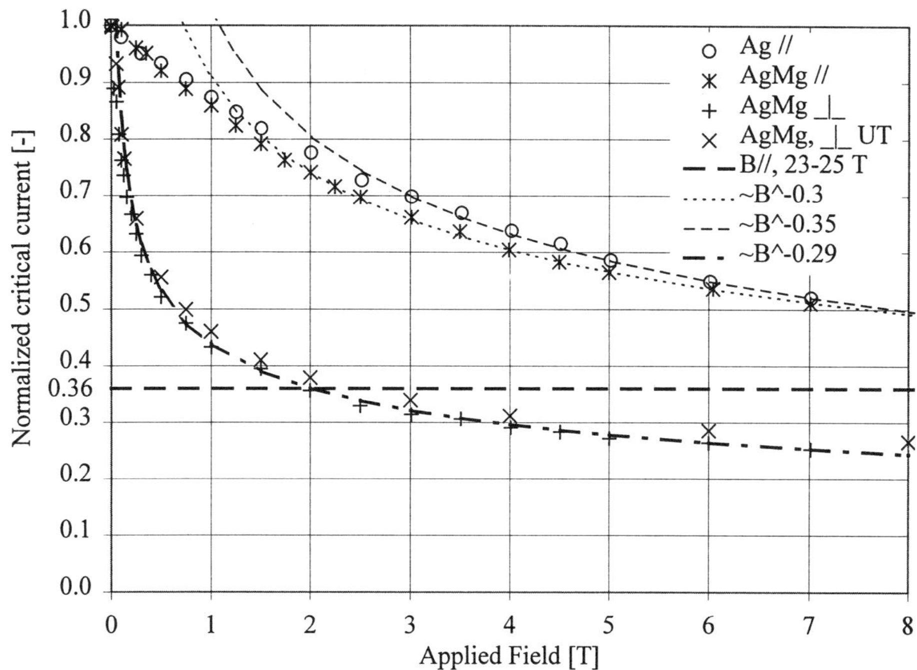

Figure 1. The reduction of the normalized I_c in Ag and AgMg samples for applied field parallel and perpendicular to the flat side of the tape and perpendicular to the current, corresponding to axial and radial field components in a magnet. Markers indicate measured data; fit relations are indicated with dashed lines. Fit relations apply above 0.05 T for the perpendicular orientation and above 2-3 T for the parallel orientation. The dashed horizontal line represents the reduction for both Ag and AgMg conductors at 23-25 T.

DOUBLE PANCAKE PRODUCTION

The conductor was insulated at the NHMFL before winding, using a sol-gel process that deposits a uniform layer of 7 μm ZrO_2 [4]. Part of the conductor batches were insulated with MgO doped ZrO_2 to reduce the processing time and improve cohesion of the insulation [5].

Each double pancake is wound from a single piece of conductor on a thin-walled Macor tube until a winding thickness of 15 mm is reached. This thickness was selected because of prior success with the 1.2 T insert coil[2]. Thicker pancakes could easily be wound but strain management then becomes a more significant issue. A sheet of alumina paper separates the pancakes. Winding a silver wire around each pancake prevents unwinding during heat treatment, after which the wire is removed. Some coils had a layer of Al_2O_3 paper around the bore tube that allowed the bore tube to be removed after heat treatment, before impregnation. This allows a higher packing factor if used on all double pancakes. Each coil was vacuum-impregnated with Stycast 1266 after heat-treatment and tested at 4.2 K.

Heat treatment

Some of the coils were heat-treated at OST using their proprietary procedures. At the NHMFL, coils were heat-treated in a trickle-flow of oxygen, according to one of two schedules as listed in Table 3. Schedule A was used for Ag sheathed conductors (A-units), and some of the AgMg conductors (B, C-units). Schedule B was used at the NHMFL only for AgMg conductors. The coils were heat-treated in tube-furnaces with both the coil and the furnace axis horizontal. Bore tube support was required to maintain the shape of the coil.

The peak temperature was optimized for each conductor batch using short samples. Figure 2 shows the relations between peak temperature as measured and the critical currents of both short samples and A-coils from one batch. The critical currents of A-coils from another conductor batch, but made using the same ceramic starting powder, are also included. The following observations can be made: 1) The lowest peak temperature that gives high critical currents is higher for coils than short samples; 2) The highest critical currents in the coils are obtained using the same peak temperature for both batches; 3) The I_c's for both coils and short samples from batch 2 are slightly higher that for batch 1; 4) The critical current of both coils and short samples are reduced by 50% when the peak temperature is 10° C above the value that results in the highest I_c. Figure 3 shows the I_c-distribution of the double pancakes used in the 3 T magnet.

The peak temperature that gives the highest I_c in short samples is not the optimal peak temperature for a coil. Since one or more iterative steps were required to find the optimal peak temperature for coils, and because a batch of conductor would yield no more then 3 to 4 B and/or C coils, a significant fraction of a batch might be heat treated at non-optimal peak temperatures. This can be alleviated with longer conductor batches.

Occasionally AgMg conductors show a slight thickness increase during heat treatment, resulting in residual stresses in the windings at room temperature. There is no apparent relation between such occurrences and either the heat treatment schedule, the conductor deformation, the ceramic starting powder or other factors. The residual stress causes a pancake to bend out-of-plane. When it is possible to remove the bore tube, the stresses are released and the coil is easily returned to a flat shape. The swelling may also occur in Ag sheathed conductor, but the extremely low yield stress of annealed pure silver may prevent the build –up of residual stress. Winding a sheet of alumina paper around the bore tube before winding makes removal of the bore tube after heat treatment easy. This procedure may also result in higher critical currents. The two coils with the highest I_c, indicated with a "<" symbol in Figure 2, had the bore tube removed after heat treatment. However, there is not enough data available to make a conclusive statement. See reference 6 for a further discussion on the heat treatments.

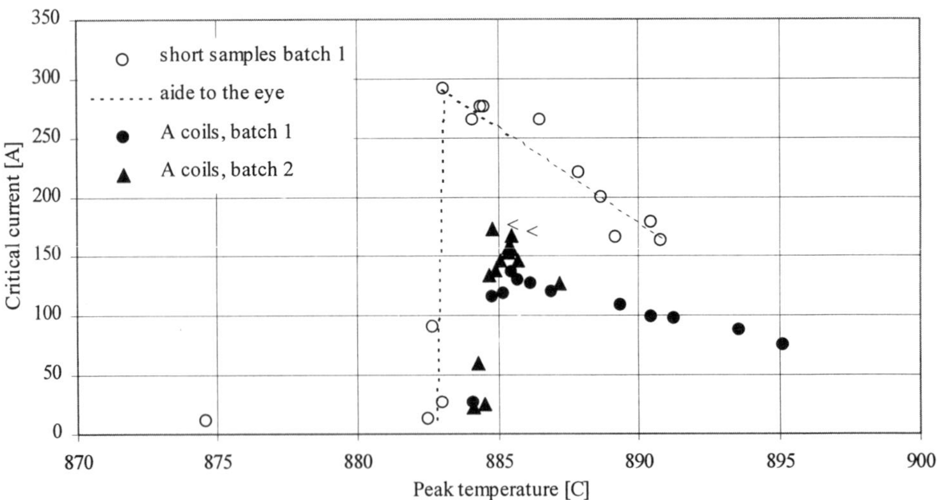

Figure 2. Obtained critical currents in short samples and coils from conductor batch 1 as a function of the peak temperature during the heat treatment. The coil critical currents of batch 2,with the same ceramic starting powder but processed independently, are also shown. The peak critical current for batch 2 was 360 A, which explains the higher I_c values. The two coils with the highest I_c, indicated with a "<" symbol, had the bore tube removed after heat treatment.

Table 3. Heat treatment schedules

Schedule A			Schedule B		
Rate to level [°C/hour]	Temperature [°C]	Time at level [hours]	Rate to level [°C/hour]	Temperature [°C]	Time at level [hours]
400	650	0	260	800	24
50	820	24	85	peak	0.2
50	peak	0.2-0.4	3-3.3	860	0
10	825	10	1-2.8	837	0
furnace cool	ambient	-	furnace cool	ambient	-

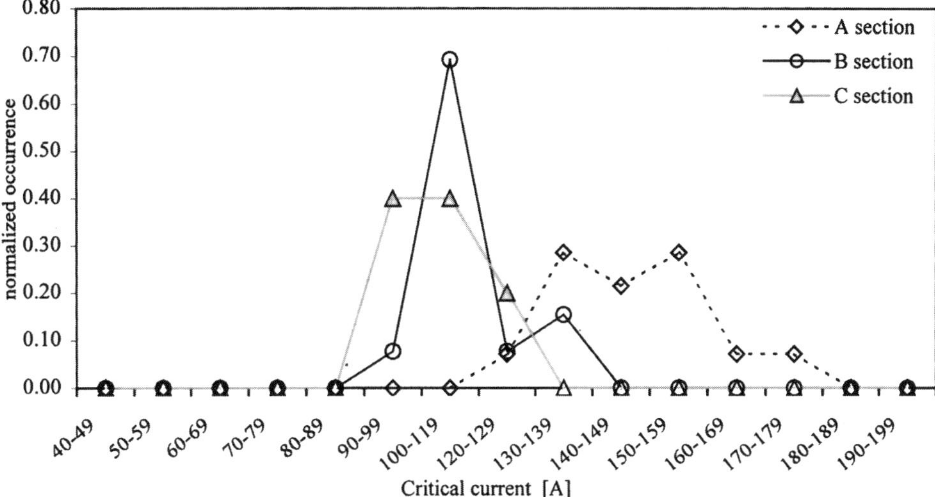

Figure 3. Distribution of the self-field critical currents in the units selected for stacking.

Selecting the units for stacking

Double pancake units with the higher self-field I_c were placed at the top and bottom of the stack, such that the entire stack would be axial field limited by the pancakes on the mid-plane in high background fields. The point at which to stop adding to the stack is when the projected I_c of the 3 sections in a 19 T background decrease more than the field-to-current ratio of the stack would increase by adding the next-best coil on the mid-plane. This represents an optimization for maximum field generation based on a given distribution of double pancake I_c.

Stacking

A thin-walled G-10 tube served to stack selected A-units. Soldering short sections of Ag/Bi2212 superconductors joined adjacent double pancakes. Nylon flanges were mounted on the top and bottom of the stack before placing the assembly in a brass can for a second vacuum impregnation. The brass can extends about 10 mm beyond the flanges and centers the stack on a G-10 ring attached to the stack flanges. This positions the section accurately and allows the windings to thermally contract, and to expand under Lorentz-forces, independent of the G-10 flanges and rings. The brass is separated from the windings by about 1.5 mm epoxy and therefore does not act as re-enforcement.

Figure 4. The 3 T insert magnet mounted on its support structure. Visible are the outer brass tube, the G-10 stack flanges and 12 threaded stainless steel rods that position the stack flanges. The cylindrical bottom end of the support structure and an intermediate mounting flange are visible on the right.

A similar procedure was used for the B and C-section, using a Nylon stack tube that was removed after the vacuum impregnation. The top and bottom-flange thickness for each section was chosen such that their heights are equal. The stack flanges were held together by stainless steel rods that also connect to the sample support structure in the dewar. The stainless steel rods were dimensioned to take up the maximum possible axial forces in case of a worst-case failure of the Large Bore Resistive Magnet and do not provide axial pressure during normal operation. Radial gaps between the sections and a series of small holes in the stack-flanges provide for helium cooling of the surface of each section.

Test facility

The Large Bore Resistive Magnet (LBRM) of the NHMFL was used for in-field testing of the 3 T insert magnet. The insert dewar provided a 170 mm cold bore. The sample support structure, equipped with three pairs of vapor cooled leads, is capable of supporting the mechanical loads in worst possible, albeit unlikely, case of resistive magnet failure. In another worst-case scenario, a trip of the resistive magnet power supplies at full field, induced voltages in the order of 500 to 600 V would be present across the B and C-sections. To protect the 3 T power supplies against the surge current in that event, diodes were mounted across the leads inside the terminal. The combination of surge currents and background magnetic field would result in large mechanical stresses in the 3 T magnet, at or above the levels that correspond to the onset of significant critical current degradation in short samples. A protection system using dump resistors and circuit breakers is under development for future insert magnet testing

Measurement

The instrumentation consisted of a voltage tap just inside the current leads on each section and a Hall sensor located at center field. Each section was powered by a separate power supply. The following procedure was employed to determine the critical current of each stack:

The current in all three sections was increased simultaneously in 1 A steps until the voltage across one section exceeded the critical current criterion. The current in the other sections was further increased, while making minor adjustments to the current in the section that reached I_c, maintaining its voltage close to the critical voltage. This process continued until all three sections were close to I_c. Typically, two to three current adjustments per section were required to get within 0.1 to 0.2 A of the critical current for all three sections simultaneously.

IN-FIELD MEASUREMENTS

The 3 T insert magnet was tested at self-field, and in background fields of 1, 5 , 10, 15, 19, 10 and 0 T, in the LBRM. Tables 4 and 5 list properties of the three sections at self-field in a 19 T background field. At self-field, the 3 T magnet generated 4.6 T. With the 3 T insert magnet at I_c, in a 19.0 T background field, the hall sensor registered a central field of 21.95 T. This corresponds to a Δ B of 2.95 T created by the 3 T insert magnet. A calculation of Δ B, using the measured I_c and calculated B/I ratios of each section resulted in 2.94 T. N-values range from 5 to 7. The stress and strain distributions within each section on the mid-plane were calculated using a generalized plane strain model[7].

Table 4. Properties of 3 T sections at self-field.

Property	Unit	Section A	Section B	Section C
I_c	A	111	79	77
$J_{engineering}$	A/mm^2	185	128	124
$J_{average\ in\ windings}$	A/mm^2	144	91	90
Packing factor	%	78	72	73
$B_{axial,\ max}$	T	4.94	2.40	0.98
$B_{radial,\ max}$	T	1.12	1.11	0.98

Table 5. Properties of the 3 T magnet in a 19 T background.

Property	unit	Section A	Section B	Section C
I_c	A	66	49.2	43.6
$J_{engineering}$	A/mm^2	110	79	70
$J_{average\ in\ windings}$	A/mm^2	86	57	51
$B_{axial,\ max}$	T	21.96	20.50	19.71
$B_{radial,\ max}$	T	0.81	0.94	0.87
$B_{radial,\ max\ from\ LBRM}$	T	0.14	0.26	0.30
Added central field	T	1.56	0.87	0.52
Estimated peak stress	MPa	63	77	96
Estimated peak strain	%	0.11	0.13	0.16

Note that the maximum radial field component in the 3 T insert magnet at self-field was higher than in a 19 T background. The radial self-field at I_c decreased more than the radial component of the background field increased, with increasing background field.

The relation between I_c and maximum axial field in each section is given in Figure 4. When the background magnetic field was returned to zero, the critical currents recovered to their initial values, except for the A-section that showed a marginal 1% degradation. The I_c(B) curves for the B and C-section are parallel, indicating similar behavior. The curve for the A-section shows a steeper slope for low field values. This can be explained based on short-sample properties.

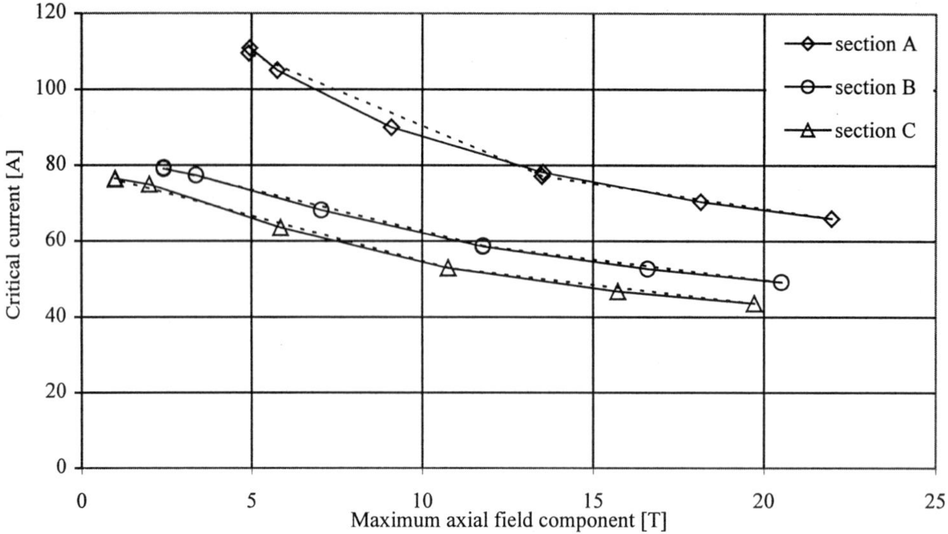

Figure 5. Measured critical currents in the 3 T insert magnet as a function of the maximum axial field component in each section. The lines connecting the datapoints are a guide to the eye only. Dashed lines connect datapoints taken while reducing the background field.

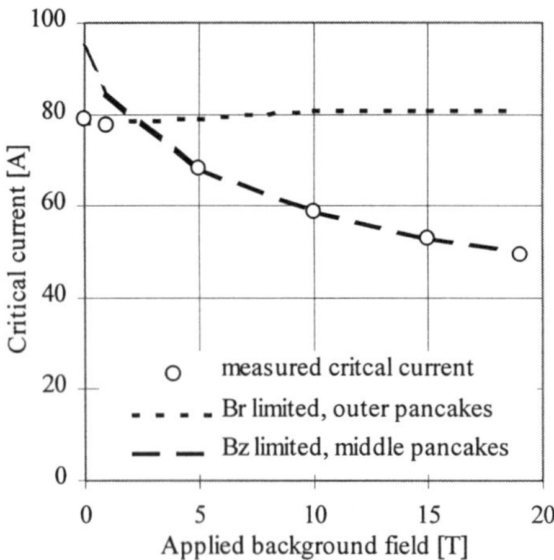

Figure 6. The measured critical current of the B-stack as a function of the background magnetic field, the calculated critical current of the outer units (I_c=137 A) as a function of local radial field component and the middle units (I_c=100 A) as a function of the axial field components at the mid-plane. This suggests that the B-stack is axial field limited in applied magnetic fields above about 2 T, and limited by radial field components at the outer units in applied field up to about 2 T.

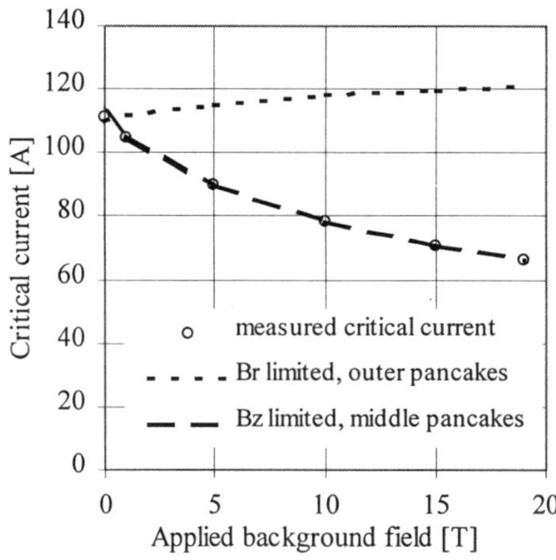

Figure 7. The measured critical current of the A-stack as a function of the background magnetic field, the calculated critical current of the outer units (I_c=170 A) as a function of local radial field component and the middle units (I_c=127 A) as a function of the axial field components at the mid-plane. This suggests that the A-stack is axial field limited in applied magnetic fields, and equally limited by axial fields at the middle units and radial field at the outer units at self-field.

Assume that, in a 19 T background field, the critical current of the B stack is completely determined by the unit in the middle of the stack, which has the lowest I_c, and the local maximum axial field component. By normalizing the short sample B// data to 49.2 A at 20.5 T, we can construct an I_c(B//) curve for the middle unit and find a zero-field value of

183 A. With a measured self-field I_c of 137 A, this corresponds to a self-field reduction to 75 % of the zero-field value for the conductor in the unit.

If the B-stack were radial field limited, the properties of the top and bottom units would determine the in-field performance since the radial field-components are strongly localized at the outer units. Assuming a similar self-field reduction ratio for the outer units, the expected reduction of I_c with radial field components, $I_c(B\perp)$, can also be calculated.

One would expect the measured $I_c(B)$ data for the stack to follow the lowest of the $I_c(B//)$ and $I_c(B\perp)$ curves. The results of these calculations are given in Figure 6. The points and curves are plotted against the applied magnetic field for simplicity. The curves are calculated with the maximum axial and parallel field-components present in each section at I_c at the given applied magnetic field. The lower of the calculated curves corresponds to the measured data. These results suggest that the B-stack is axial-field limited in applied magnetic fields above about 2 T. Radial field components at the outer units are the limiting factor at self field and in applied fields up to about 2 T.

To verify the calculated ratio between zero-field I_c for the conductor in the coil and self-field I_c of the coil, part of both outer turns were cut from one of the B-units after heat treatment. These sections were measured separately at 4.2 K, with the results summarized in Table 6. The I_c of the short sample from the end of pancake 1 corresponds to the expected zero–field value, indicating that the assumption of a self-field reduction to 75% of the zero-field I_c is reasonable. The second sample has a lower critical current, possibly related to the presence of a pinhole in that section.

Table 6. Properties of a B-coil and short samples cut from the coil.

Coil self-field I_c	99 A	I_c of sample from end of pancake 1	134 A
Projected zero field I_c	132A	I_c of sample from end of pancake 2	112 A

Repeating the calculations as for the A-section results in $I_c(B\perp)$ and $I_c(B//)$ relations as presented in Figure 7. The results suggest that the A stack is axial-field limited by the units at the mid-plane. At self-field, the radial field components at the outer units are equally limiting.

DISCUSSION

The 3 T magnet carried an average current density in the windings of 86, 57 and 51 A/mm^2 respectively for the A, B and C-section. This approaches design values of 70 –100 A/mm^2 for high field NMR applications[8,9], although these specifications are usually based on the 10^{-13} Ω.m critical current criterion.

The 3 T magnet operated with a maximum of about 0.13 % and 0.16% strain in the AgMg conductor of the outer sections. The strain limit, as determined using short samples, is 0.4 %. This suggests that the conductor is capable of operating at significantly higher current density and winding radius, without reinforcement. The Ag sheathed conductor would require reinforcement, if it were used in the inner sections of a NMR magnet, given the higher inner radius of the windings.

The 3 T magnet is axial field limited above 2-3 T. In future NMR magnets, operating at higher current densities, radial field component will be a more significant factor. Axial fields will reduce the critical current to about 36% of the zero-field I_c of the conductor. As shown in Figure 1, the same reduction occurs with a radial field component of 2 T. This suggest that, in a magnet with AgMg conductor with a constant critical current density, the maximum radial field component in the windings should be less then 2 T, to ensure that the critical current is limited by axial field components. Radial self-field components are highest at the top and bottom surfaces of a magnet section, and can be reduced by limiting the thickness of a section and varying the height of adjacent sections.

CONCLUSIONS

- A Bi-2212 Wind-and-React insert magnet consisting of three concentric sections was built and successfully generated 3.0 T in a 19.0 T background magnetic field without irreversible critical current degradation.
- The average current density in the section windings was 51, 57, 86 A/mm^2 in axial fields between 19.7 and 22 T.
- The measured $I_c(B)$ relations of the three sections can be modeled based on short sample properties and the self-field critical currents of the double pancakes before stacking.
- In axial fields of around 23 T, the AgMg conductor will be axial field limited with radial field components up to 2 T.
- Longer conductor batches with more homogeneous starting powder would improve the coil manufacturing success rate and reduce production time.
- The AgMg conductor was operated at a 0.16% strain level in the 3 T magnet without degradation. Short sample data suggest it can be operated at up to 0.4 % strain without reinforcement.

ACKNOWLEDGMENT

The authors like to thank I. Cubukcu of the NHMFL for measuring the many short samples and H.J.N. van Eck of the University of Twente (UT) for performing in-field measurements of AgMg samples at the UT, allowing an independent comparison.

REFERENCES

1 H. W. Weijers, Q.Y. Hu, Y.S. Hascicek, A. Godeke, Y. Viouchkov, E. Celik, J. Schwartz, K. Marken, W. Dai, J. Parrell, Development of 3 T class Bi-2212 insert coils for high field NMR, to be published in IEEE Transactions on Applied Superconductivity, Proceedings of the '98 ASC conference.

2 H.W. Weijers, D. Hazelton, L. Cowey, Y.S. Hascicek, I.H. Mutlu, U.P. Trociewitz, and S.W Van Sciver, Bi-2212 Coils for 1 T class insert coils, Advances in Cryogenic Engineering, Vol. 34, 1998.

3 Y. Viouchkov, H.W. Weijers, M. Meinesz, Q.Y. Hu and J. Schwartz, Methods of stress-strain-Ic characterization of Bi-2212 tapes, proceedings of the '99 ISS Conference, Fukuoka, Japan

4 I.H. Mutlu, and Y.S. Hascicek, Insulation for wind and react high temperature superconducting coils, *Advances in Cryogenic Engineering*, Vol. 44, 1998

5 E. Celik, E. Avci, and Y. S. Hascicek, MgO-ZrO_2 Insulation coatings on AgMg sheathed Bi-2212 superconducting tapes by sol-gel technique, *ibid.*

6 Q. Hu et al., Heat treatment of AgMg and Ag sheathed $Bi_2Sr_2CaCu_2O_x$ tapes for a 3 T insert magnet, *ibid.*

7 W. D. Markiewicz et al, Generalized plane strain analysis of solenoidal magnets, *IEEE Trans. on Mag.*, vol. 30, no 4, p.2233, 1994.

8 W. D. Markiewicz, I.R. Dixon, J. Schwartz, C.A. Swenson, S. W. Van Sciver, and H.J. Schneider-Muntau, 25 T High resolution NMR Magnet Program and Technology, IEEE Transactions on Magnetics 32:4:2586 1996

9 T. Kiyoshi, K. Inoue, M. Kosuge, H. Kitaguchi, H. Kumakura, H. Wada and H. Maeda, *NRIM R&D* Program on HTS coils for 1 GHz NMR Spectro*meter,* Proceedings of the 16th ICEC/ICMC conference, Part 2, p. 1099-1102, 1997.

DESIGN AND PERFORMANCE OF A 5 kJ HTS μ-SMES

J. Paasi,[1] R. Mikkonen,[1] T. Kalliohaka,[1] J. Lehtonen,[1]
J. Ollila,[1] L. Söderlund,[1] B. Connor,[2] and S.S. Kalsi[2]

[1] Tampere University of Technology
P.O. Box 692, FIN-33101 Tampere, Finland
[2] American Superconductor Corporation
Two Technology Dr., Westborough, MA 01581

ABSTRACT

A Bi-2223 magnet has been constructed by American Superconductor for a 5 kJ micro-SMES system at Tampere University of Technology. The nominal energy of the magnet was obtained by 160 A magnet current. The magnet was cooled to the operating temperature of 20 K with a two stage Gifford-McMahon type cryocooler with a cooling power of 8 W at 20 K. The HTS micro-SMES was tested in an uninterruptible power supply application, where the SMES supplies power for a load during a break in mains. In this paper the design of the magnet and the cryogenic system as well as test results are presented and discussed together with problems related to the cryointegration of the magnet.

INTRODUCTION

Superconducting magnetic energy storage (SMES) for improving power quality in electric networks is a potential large-scale application of high-temperature superconductors (HTS). HTS coils can operate at much higher temperatures than NbTi or Nb_3Sn coils and, therefore, they can accommodate larger heat loads due to AC currents or current ramps than Nb-based coils because of the increase of specific heat with increasing temperature and a larger margin between operation and critical temperatures. This gives rise to use LHe-free closed cycle cooling systems for the refrigeration. Several cryocooler refrigerated HTS magnets operating at around 20 K have been already constructed or are under construction.[1-6] There is also interest in LN_2 cooled HTS μ-SMES systems.[7] This paper addresses design and test results of a 5 kJ μ-SMES system at Tampere University of Technology. The Bi-2223 magnet of the μ-SMES was manufactured by American Superconductor Corporation. The nominal operation temperature of the magnet, 20 K, was obtained by a two stage Gifford-McMahon type cryocooler.

CRYOGENIC DESIGN

The cryostat system of the HTS μ-SMES is presented in Figs. 1-2. The HTS solenoid is cooled with a two stage Gifford-McMahon cryocooler. The first stage is cooled down to 40 K and it is thermally anchored to the copper radiation shield and the lower end of the copper current leads. The second stage having a cooling power of 8 W at 20 K includes four hydrogen heat pipes, which are attached to the thermal interface at the top of the magnet. The HTS coil is installed inside the radiation shield and attached to the thermal interface made of four copper sections. Thin sheets of indium are used between the magnet and the thermal interface in order to decrease the thermal contact impedance. The cold ends of the CryoSaver^{TM} HTS current leads are also connected to the thermal interface via thermal impedances. Bolt connections are used between the current leads and the magnet terminals. Magnet and thermal interface temperatures are monitored by Cernox temperature sensors on the bottom of the magnet and on the interface. In the cryogenic design special attention was paid to minimize heat loads due to eddy currents induced during the ramping of the magnet. By using the hydrogen heat pipes we could place the heat exchanger of the second stage of the cryocooler far away from the magnet. The thermal interface is slotted into four segments, each of which has one hydrogen pipe. Similarly, the radiation shield around the coil has a segmented configuration. Finally, the HTS magnet and the heat pipes were wrapped with superinsulation foils to minimize the radiation heat load.

The HTS magnet has inside and outside diameters of 252 mm and 317 mm, respectively, and axial length of 66 mm. It was prepared by stacking 11 double pancake coils, joining the pancakes with low resistivity joints, and vacuum impregnating the magnet with epoxy. The nominal design current of the magnet was 160 A at 20 K. The magnet carries a total of 160 kA-turns to store the 5 kJ of magnetic energy. The design current of the coil was obtained by using a conductor bundle consisting of three multifilamentary Bi-2223/Ag tapes. The 160 A operating current corresponds to an overall coil current density of 8500 A/cm^2, which was a state-of-the-art value in 1997 when the coil was manufactured. The effective magnet inductance is 0.4 H. The coil is of low field type with the center field of 0.8 T at the 160 A current in order to reduce AC losses during charging/discharging of the coil. In the magnet winding the highest flux density of 1.7 T (at 160 A) is at the mid-point of the inner surface of the coil. For the radial field the highest values of 1.3 T are found at the mid-point of the top and bottom surfaces of the coil. The influence of the radial field on the magnet performance was reduced by designing the end sections of the magnet to operate at lower current ratio, I_{op}/I_c, than other parts of the magnets. This was realised by using a tape with higher J_c in the end section pancakes. The design of the magnet as well as the cryogenic system have been described in more detail elsewhere.[8-9]

Before the cooldown of the system the cryostat is evacuated with a pump to a pressure of about 4×10^{-4} mbar. After starting the cryocooler it will take about 60 hours before the magnet reaches a stable temperature of about 20 K. However, this requires a second stage temperature of 17 K. Thus there is a temperature difference of 3 K between the copper interface and bottom of the magnet indicating a high thermal impedance at the interface and in the magnet. Although the thermal conductivity inside the epoxy impregnated magnet is low, we believe that the high thermal impedance is mainly due to the fact that the effective surface area for the cooling is essentially lower than the surface area of the magnet because of non-ideal top surface of the magnet (quasi-flat surface containing fine epoxy powder particles formed during thermal cycling of the magnet due to different thermal contraction in components).

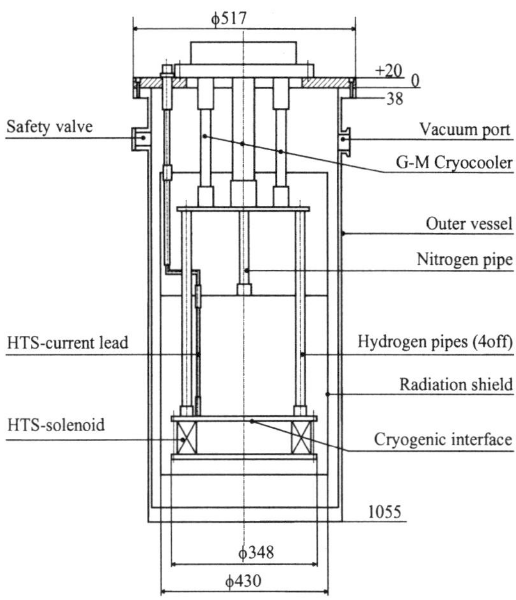

Fig. 1 The cryostat system of the HTS μ-SMES.

Fig. 2 HTS solenoid, current leads and cryocooler assembly before the installation of the radiation shield.

POWER CONVERTER

The HTS μ-SMES was aimed at an uninterruptible power supply (UPS) application, where the SMES magnet supplies power for a load during a break in mains. The system constructed at Tampere University of Technology is an on-line UPS with superconducting backup in the intermediate circuit. The block diagram of the UPS system is presented in Fig. 3. The main power path of the on-line UPS consists of a conventional rectifier and a voltage source inverter. The HTS storage magnet is connected to the intermediate circuit via step down and step up DC/DC converters. The system is controlled by a microcontroller. The activities are distributed to three printed circuit boards: a microcontroller board, a board for hardware protection, measurement and interface logic, and finally a board for the power processor. Special attention was paid in the design to minimize the on-state losses of the free wheeling circuit. Therefore, the bi-directional freewheeling switches for rapid charging/discharging of the magnet were realized by a parallel connection of 24 low-resistivity MOSFET transistors. Similar MOSFETs were used also elsewhere in the power processor. The total losses of the power processor are less than 50 W at the maximum design current of the freewheeling path 200 A. An example of measured SMES magnet current during a break in mains is given in Fig. 4 for a case where the UPS is connected to a PC computer load.

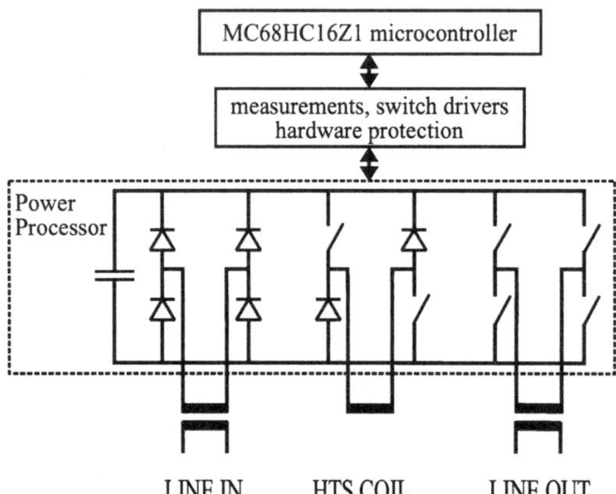

Fig. 3 The block diagram of the UPS system including a HTS magnet.

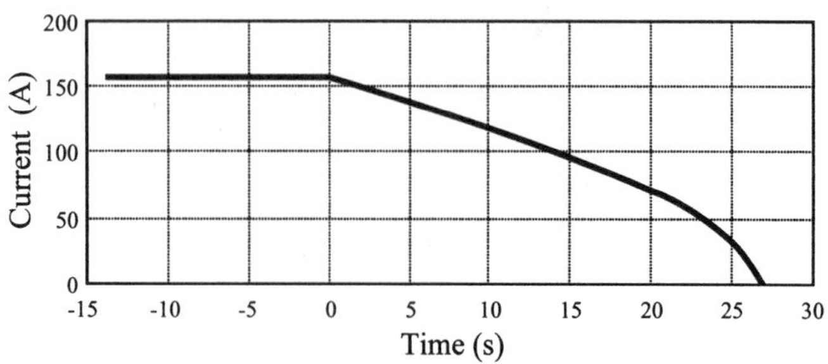

Fig. 4 Magnet current during a break in mains ($t \geq 0$) when the UPS is connected to a PC load.

TEST RESULTS

Current-voltage -curve of the HTS magnet at 20 K is given in Fig. 5 after the compensation of the inductive voltage component due to dI/dt=0.7 A/s. The potential taps were positioned on the current terminals of the magnet. Thus the measured voltage includes both the response of the Bi-2223/Ag conductor as well as the resistive joints inside the magnet and at the current terminals. From the $U(I)$-curve we can see that there is a relatively high linear voltage component in the magnet, corresponding to a resistance of 0.165 mΩ, indicating one or more poor joints between the terminals. At the nominal current of 160 A this means a 4 W additional heat load for the system. The heat load due to losses in the Bi-2223/Ag conductor is only 0.6 W at 160 A and 20 K. Monitoring of the temperatures during the operation suggests that the poor joint is in the current terminal/bottom pancake interface.

The stability of magnet operation was tested by first ramping the magnet current from zero to a constant value, I_{op}, with a ramp rate of 2 A/s and then keeping the current at I_{op}. Measured temperatures at the bottom of the magnet as a function of time are presented in Fig. 6 (a) and (b) for I_{op} = 160 A and 110 A, respectively. At the 160 A nominal design current of the magnet the operation was unstable. At first the magnet temperature increased rapidly due to ac losses generated during the ramping. Then dT/dt become slower until at about t=20 min it started to increase again. Finally at t≈25 min the magnet quenched. The behaviour of the $T(t)$ when 5 min < t < 20 min suggests that the increase of the magnet temperature was mainly due to heat generated in normal conducting parts of the magnet (i.e. joints). Only after t > 20 min when the magnet temperature has increased to about 30 K the losses in the superconductor started to be the main heat source in the magnet. This can be seen from the increase of dT/dt after T≈30 K. The second case of I_{op} = 110 A, Fig. 6(b), corresponds to the maximum operation current allowing a stable long-term operation. Also here the magnet temperature at first increases during the magnet current ramping due to ac losses, but after establishing the constant operation current the temperature slowly stabilizes to about 21.4 K. In both cases also the temperature of the second stage of the cryocooler is

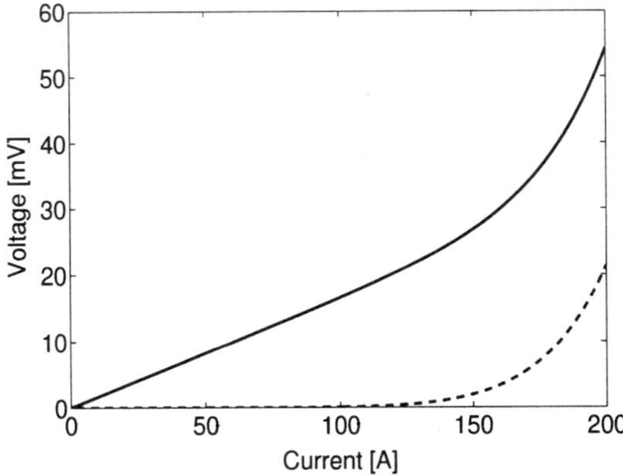

Fig. 5 Current voltage -curve of the HTS magnet at 20 K: a real $U(I)$-curve measured between the magnet current terminals (solid line) and a $U(I)$-curve after subtracting a linear voltage component due to resistive joints inside the magnet and at the current terminals (dashed line). The inductive voltage due to dI/dt has been numerically compensated out of the curves.

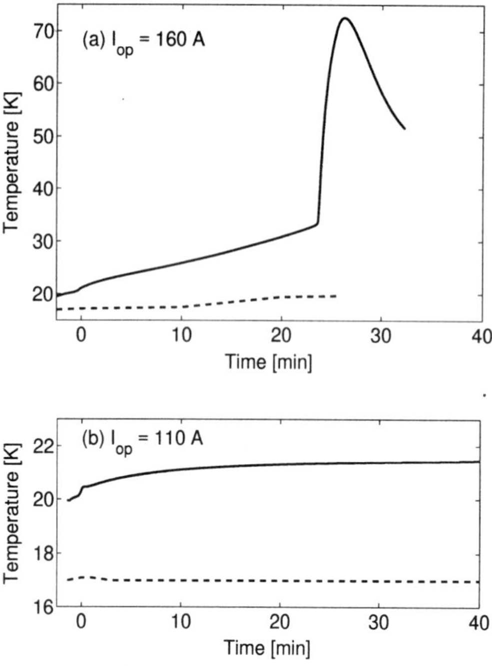

Fig. 6 Magnet temperature (solid line) and the temperature of the second stage heat exchanger of the cryocooler (dashed line) as a function of time when the magnet current is at first ramped from zero to I_{op} ($t<0$) and then kept as constant ($t\geq0$): (a) I_{op} = 160 A, (b) I_{op} = 110 A.

shown in the figures as a function of time. We can see that for I_{op} = 110 A the temperature remains constant, but for I_{op} = 160 A it slowly increases indicating that the heat load is higher than the cooling power of the cryocooler.

Although the realized maximum stable operation current, 110 A, is much lower than the design value, it does not mean that the magnet cannot be used for energy storing purposes for higher currents than the 110 A. As we could see from Fig. 6 (a) the magnet can be momentarily overloaded. For I_{op} = 160 A the time for safe overload operation is 20 min. For I_{op} = 180 A the corresponding time is about 10 min. The maximum value of the operation current where the magnet was tested as a SMES was 194 A corresponding to energy of 7.5 kJ. Then the HTS μ-SMES - UPS system could supply power for a 60 W bulb load for 85 s.

DISCUSSION

Cooling is a paramount issue in applied superconductivity. The use of closed cycle cryocoolers instead of liquid coolants is a very user-friendly solution but very challenging for the design and manufacturing of a magnet system. The main problem in the use of cryocoolers is not the small cooling power of the devices but how to effectively transfer the generated heat out of the magnet into the cold head of the cryocooler. Two problems must be solved to assure a good heat transfer, i.e. good cooling of the magnet. At first, the thermal impedance between the magnet and the copper interface of the cryocooler (as well as between the HTS current leads and the copper interface) must be low. In a high vacuum environment a good bolt connection between two different materials is a challenging task when temperatures are varying a lot. In the present epoxy impregnated magnet there was surface roughness of the order ±0.1 mm and, furthermore, fine epoxy powder particles were formed during the thermal cycling of the magnet due to different thermal contraction of copper and epoxy. Although thin sheets of indium were used between the magnet and the thermal interface, the thermal impedance at the interface gives rise to a significant temperature drop over the interface. Secondly, thermal conductivity inside the magnet should be increased. In a magnet consisting of a stack of (double) pancake coils the heat conduction to the important axial direction is poor. In the present magnet with small cross-sectional dimensions the heat conduction inside the magnet is not a big problem, but the larger the magnet the bigger the problem. In order to improve the axial heat conduction inside the magnet, especially in magnets for AC use, solenoid magnet structures with electrically insulating but thermally conducting heat drains inside the winding would be preferable. Both oxidized expanded metal Ni layers (i.e. NiO-layers)[6,10] and AlN plates[11] have been successfully applied as heat drains in Bi-based HTS magnets.

CONCLUSIONS

A cryocooler refrigerated HTS μ-SMES system was constructed and tested as an uninterruptible power supply application, where the SMES supplies power for a load during a break in mains. A long term stable operation of the Bi-2223 SMES coil at the nominal operation temperature of 20 K was obtained at 110 A magnet current. For a short period energy storing (a couple of minutes), however, the magnet was successfully overloaded with 194 A operation current which corresponds to a stored energy of 7.5 kJ.

ACKNOWLEDGMENTS

The authors acknowledge the assistance of H. Nieminen, M. Masti and J. Vuorinen at Tampere University of Technology and J. Kellers at American Superconductor Europe GmbH.

REFERENCES

1. T. Hase, K. Shibutani, S. Hayashi, M. Shimada, R. Ogawa, and Y. Kawate, Generation of 1 T, 0.5 Hz alternating magnetic field in room temperature bore of cryocooler-cooled Bi-2212 superconducting magnet, *Cryogenics* 36:971 (1996).
2. S.S. Kalsi, D. Aized, B. Connor, G. Snitchler, J. Campbell, R.E. Schwall, J. Kellers, Th. Stephanblome, A. Tromm, and P. Winn, HTS SMES magnet design and test results, *IEEE Trans. Appl. Supercond.* 7:971 (1997).
3. T. Kato, K. Ohkura, M. Ueyama, K. Ohmatsu, K. Hayashi, and K. Sato, Development of high-Tc superconducting magnet using Ag-sheathed Bi2223 tapes, Proc. MT-15 Conference, October 20-24, 1997, Beijing, China, p. 793.
4. H. Kumakura, H. Kitaguchi, K. Togano, H. Wada, K. Ohkura, M. Ueyama, K. Hayashi, and K. Sato, Performance tests of Bi-2223 pancake magnet, *Cryogenics* 38:639 (1998).
5. G. Snitchler, S.S. Kalsi, M. Manlief, R.E. Schwall, A. Sidi-Yekhlef, S. Ige, R. Medeiros, T.L. Francavilla, and D.U. Gubser, High-field warm-bore HTS conduction cooled magnet, *IEEE Trans. Appl. Supercond.* 9:553 (1999).

6. J.F. Picard, M. Zouiti, C. Levillain, M. Wilson, D. Ryan, K. Marken, P.F. Herrmann, E. Béghin, T. Verhaege, Y. Parasie, J. Bock, M. Baecker, J.A.A.J. Perenboom, and J. Paasi, Technologies for high field HTS magnets, *IEEE Trans. Appl. Supercond.* 9:535 (1999).
7. A. Friedman, N. Shaked, E. Perel, M. Sinvani, Y. Wolfus, and Y. Yeshurun, Superconducting magnetic energy storage device operating at liquid nitrogen temperatures, *Cryogenics* 39:53 (1999).
8. R. Mikkonen, M. Lahtinen, J. Lehtonen, J. Paasi, B. Connor, and S.S. Kalsi, Design considerations of a HTS μ-SMES, *Inst. Phys. Conf. Ser.* 158:1483 (1997).
9. J. Paasi, M. Lahtinen, J. Lehtonen, R. Mikkonen, L. Söderlund, B. Connor, and S.S. Kalsi, Design of HTS magnets for μ-SMES applications: differences to classical principles, Proc. MT-15 Conference, October 20-24, 1997, Beijing, China, p. 781.
10. P.F. Herrmann, E. Béghin, G. Duperray, F. Grivon, D. Legat, A. Leriche, J.P. Tavargnier, P. Marlin, Y. Parasie, and J. Bock, Pre-industrial PIT conductor and coil development at Alcatel, ASC'98 Conference, September 13-18, 1998, Palm Desert, CA, USA.
11. A. Tomioka, T. Bohno, S. Nose, M. Konno, M. Iwakuma, K. Funaki, K. Kajikawa, H. Kanetaka, H. Hayashi and K. Tsutsumi, Experimental results of the model coil for cooling design of a 1 T cryocooler-cooled pulse coil for SMES, *IEEE Trans. Appl. Supercond.* 9:932 (1999).

CRYOSTATS FOR THE KEKB IR SUPERCONDUCTING MAGNETS

Norihito Ohuchi,[1] Toru Ogitsu,[1] Kiyosumi Tsuchiya,[1] and Shin Nakamura [2]

[1] KEK, High Energy Accelerator Research Organization
1-1 Oho Tsukuba Ibaraki 305-0801 Japan
[2] Hitachi, Ltd. Hitachi Works
1-1 Saiwai Hitachi Ibaraki 317 Japan

ABSTRACT

Two magnet cryostats for the interaction region of the KEKB collider were constructed and cooldown tests were performed. During the initial test, the temperatures of the front end plates of the helium vessels were higher than 10 K and the magnets did not reach their nominal currents. Temperature measurements and calculations showed that the higher temperature was due to a thermal balance between the cooling condition of subcooled liquid helium and a heat leak through the support rods of the helium vessel. After improving the cooling circuits, we succeeded to lower the temperature of the front end plates and operate the magnets at their nominal current. The heat load of the cryostats, which was measured by the enthalpy increase of subcooled liquid helium between the inlet and outlet of the cryostat, was 22.3 W, slightly smaller than the design.

INTRODUCTION

KEKB is an asymmetric-energy, two-ring electron-positron collider for B-meson physics at KEK[1]. Commissioning has been ongoing since construction was completed in November 1998. In the interaction region (IR), a pair of compensation solenoids (S-R and S-L) and a pair of final focusing quadrupoles (QCS-R and QCS-L) with three corrector coils inside their bores[2] were installed. The two cryostats for these superconducting coils are placed close to the interaction point and inside the experimental detector, as shown in Fig. 1. The size of the cryostats is severely constrained by the detector. Large magnetic forces, which are created through the interactions with the magnetic fields of the coils and the detector solenoid, act on the cryostats. The magnets are cooled by subcooled liquid helium (4.5 K and 0.15 MPa). Subcooled liquid helium was selected for cooling the magnets since there is no vapor in the liquid helium flow[3]. The liquid helium is supplied through cryogenic multi-transfer lines from a subcooler-cold-box unit to the cryostats.

The fabrication of the magnet cryostats was completed in 1997. During the first cool-down test, the temperature of the end plate of the helium vessel for the solenoids was found to be higher than 10 K. The high temperature limited the operation of the S-L and the S-R to currents of about 200A and 450 A, respectively, while the nominal currents are 474 A and 603 A. After improving the cooling circuit in the helium vessel, the temperature of the end plate was decreased, and we successfully operated the solenoids at the design points. In this paper, we report these improvements and the performance of the cryostats.

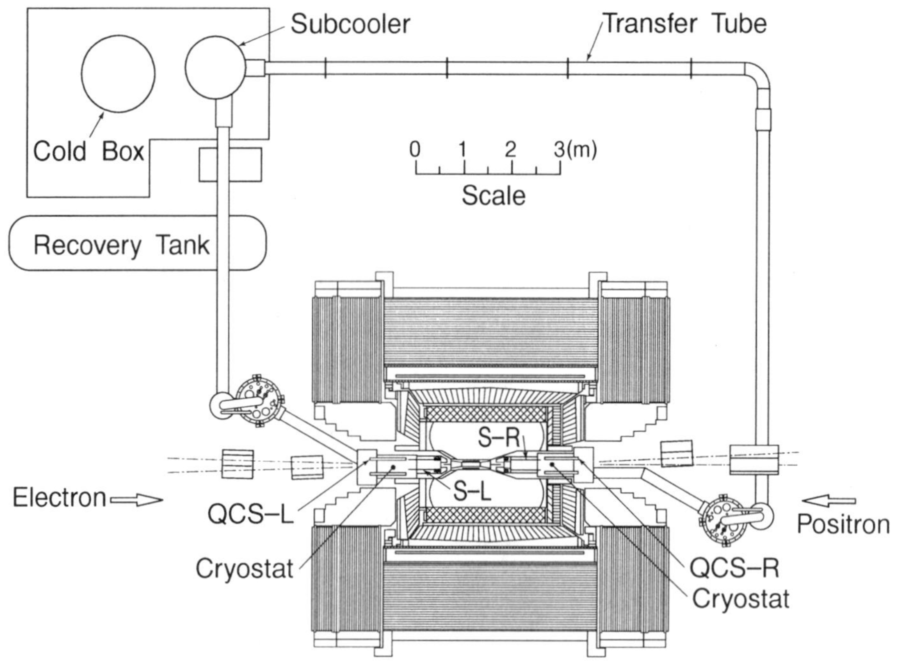

Figure 1. Schematic view of the KEKB interaction region.

OVERVIEW OF THE CRYOSTAT

The schematic view of the cryostat in the left side of the interaction point is shown in Fig. 2. The cryostat in the right side has alomst the same geometry as the left one. The cryostat has a service dewer for a connection between the cryostat and the cryogenic transfer tube from the refrigerator. The length of the cryostat is 1720 mm. The main components of the cryostat are a helium vessel, thermal shield, suspension system and a vacuum vessel. Figs. 3 and 4 show the details of the cryostat and the transverse cross section of the magnet cryostat. The solenoids are installed in the front part of the cryostats, and the QCSs and correctors are in the back part.

The helium vessel is made of 316L stainless steel. The material is selected not to disturb the magnetic field in the magnet bore. The outer diameters of the vessel in the front and the back are 258 mm and 352 mm, and the inner diameters of the bore are 174 mm and 230 mm, respectively. The inner cylinder works not only as a pressure vessel but also as a mechanical support of the magnets. The thickness is 7 mm in the front and 8 mm in the back. The axis of the cylinder for the S-L is shifted horizontally by 35.1 mm with respect to the axis of the QCS-L while those for the S-R and the QCS-R are on the same line.

As shown in Figs. 3 and 4, thermal shields, which are cooled by liquid nitrogen, shield the helium vessel from room temperature. The outer shield is a cylindrical 304 stainless steel shell with a cooling pipe of 10 mm diameter. The inner shield is a 316L stainless steel cylinder with two helical grooves for liquid nitrogen. The thickness of the cylinder is 7 mm, and the groove is 25 mm wide and 2.5 mm deep. The distance between the inner shield and the helium vessel is 7 mm. Between the vacuum vessel and the shields, multi-layer insulation of 20 layers for the outer shield and 10 layers for the inner shield are installed, respectively.

In Fig. 3, the suspension system of the helium vessel is shown. It consists of eight titanium alloy (Ti-6Al-4V ELI) rods. The rods are thermal-anchored to the liquid nitrogen pipe. The back rods are 140 mm long and 10 mm in diameter. The front rods are 9 mm in diameter, and the length is 67 mm long for the S-R, and 75 mm long for the S-L.

The vacuum vessel encloses the low temperature equipment. The outer cylinders, made of 304 stainless steel, function as the mechanical frame of the suspension system. Therefore, the thickness of the front and back cylinders is 10 mm and 12 mm, respectively. The material of the inner cylinder is 316L stainless steel.

The service dewer is 800 mm in diameter and 880 mm high. On the top plate of the dewer there are six 50 A current leads for the corrector coils, two 650 A current leads for the solenoid, two control valves and three bayonet joints. The superconducting bus lines for the QCSs are connected with solder in the dewer.

MECHANICAL DESIGN

Because the S-L and the S-R are operated in a magnetic field of 1.5 T generated by the detector solenoid, axial repulsive forces act on the solenoids. The calculated forces on the S-L and the S-R are 22 kN and 2.8 kN, respectively. The support rods must withstand these forces. The maximum stress in the rods for the left cryostat was calculated. The conditions of the calculation were that the weight of the helium vessel was 400 kg including the magnets, and the magnetic force in the axial direction was 25 kN. The stress on the rods was estimated to be 294 N/mm^2. The stress is sufficiently smaller than the yield strength of the titanium alloy

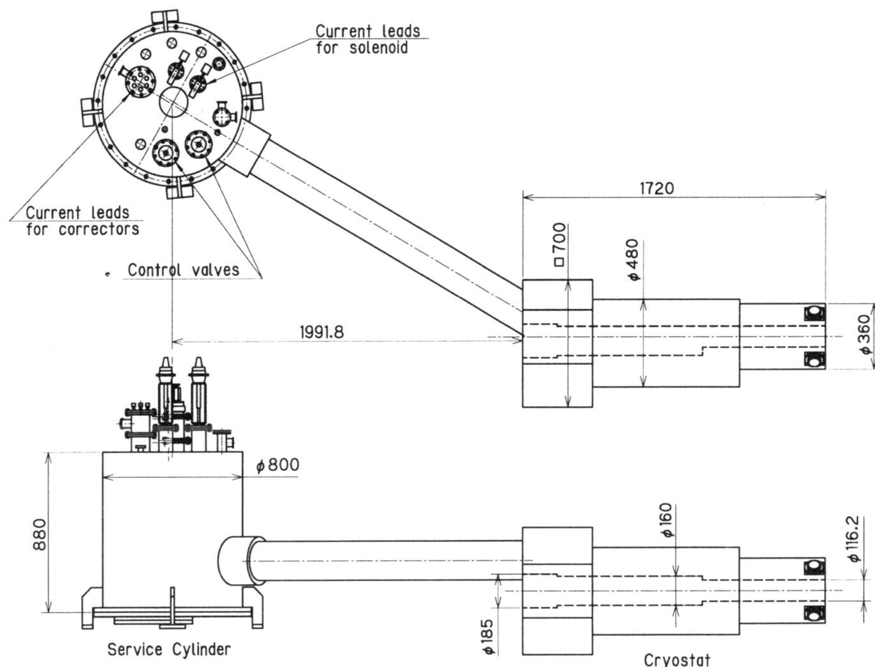

Figure 2. Schematic view of the cryostat in the left side with respect to the interaction point.

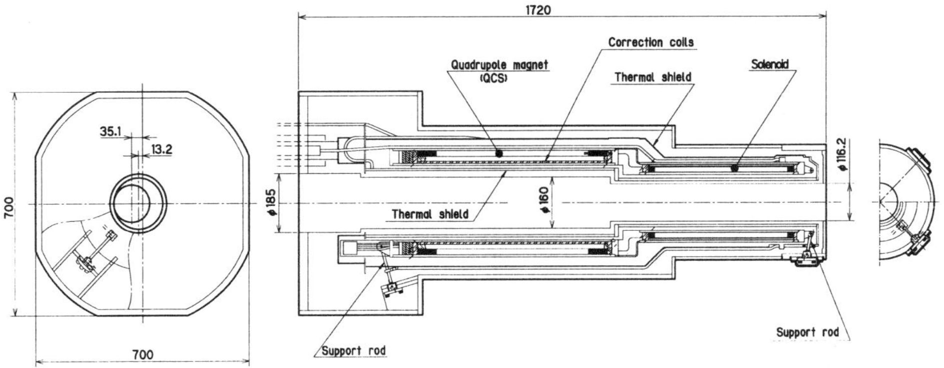

Figure 3. Details of the cryostat in the left side.

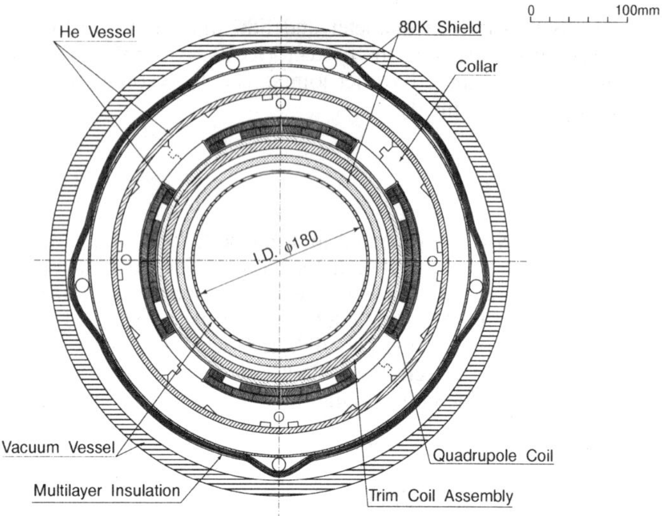

Figure 4. Transverse cross section of the magnet cryostat.

at the room temperature (792 N/ mm^2).

The tilt angles of the rods are designed so that the magnet axes do not change due to thermal contraction after the cool-down of the magnets. The tilt angles of the front and back rods at the room temperature are 10.33 degree and 14.22 degree, respectively. After the cool-down, the angles change to 8.42 degree and 13.20 degree, but the magnet axis is kept in the same position.

THERMAL DESIGN

The thermal design is based on the condition that the liquid helium should be in subcooled condition at the outlet of the cryostat. The helium temperature at the inlet and the maximum mass flow rate per one cryostat from the capacity of the cooling system are estimated to be 4.4 K and 13 g/s. Therefore, the allowable thermal load for one cryostat is calculated to be 21.3 W. The thermal budgets of the cryostats are summarized in Table 1. In the calculation, the temperatures of the thermal shields and the thermal anchors are assumed to be 100 K. The maximum thermal load is conduction through the support rods. They are larger than 3 W for each cryostat. In the service dewer, the thermal load amounts to 8.58 W because there is tubing for current leads and signal wires. The total thermal loads for the right side and the left side are 14.23 W and 13.78 W, respectively. The mass flow rate of subcooled liquid helium is designed to be 12 g/s for each cryostat. The calculated temperature increases of the liquid helium, which is 4.4 K and 0.15 MPa at the inlet of the cryostats, are 0.22 K and 0.21 K. The liquid helium at the outlet is still in the subcooled condition.

Table 1. The calculated thermal loads of the cryostats

		right side (W)	left side (W)
cryostat	radiation	0.94	0.87
	conduction through support rods	3.38	3.18
transfer line	radiation	0.56	0.38
	conduction through FRP supports	0.77	0.77
service dewer	radiation	0.33	0.33
	conduction through bayonet joints	2.12	2.12
	conduction through tubing	2.81	2.81
	conduction through valves	0.54	0.54
	conduction through signal wires	2.78	2.78
total		14.23	13.78

INSTRUMENTATION

Two carbon-glass resistor thermometers (CGR) and two platinum-cobalt alloy thermometers (PtCo) are installed on the magnets to measure magnet temperatures precisely below 15 K and to control the cooling process from room temperature to 4.5 K. In the service dewer, two CGRs are inserted in the helium lines to measure the heat load by way of the temperature increase. On the end plates of the helium vessel, eight CGRs are installed as a microcalorimeter[4] to measure beam loss. Thermo couples are used to monitor the temperatures of the support rods and the thermal shields. The stresses in the support rods are measured by strain gages. Pick-up coils are installed on the inner cylinder of the vacuum vessel to locate the position where a normal transition happens in the coil.

COOLDOWN TESTS

In the initial cool-down test of the magnets, the S-L and the S-R did not reach their nominal currents while the QCSs reached the current with no quench. The temperatures in the helium vessels were measured by the CGR thermometers including the microcalorimeters. The temperature profiles are shown in Figs. 5 and 6. In the figures, F-A to F-D and B-A to B-D correspond to the microcalorimeters on the front and back end plates, respectively. The positions on the front end plate are shown in Fig. 7 (a). The QCS and S correspond to the CGR thermometers embedded in the FRP spacers of the magnets. The helium temperature at the inlet and outlet in the service dewers are shown in the same figures. During the initial cooldown test, the temperatures of the FRP spacers of the solenoids and the front end plates of the helium vessel were higher than 5 K and 10 K, respectively. The average temperatures of the front microcalorimeters for the right and left sides were 9.3 K and 10.9 K while the averages in the back end plates were 6.1K and 6.8 K, respectively. The higher temperatures limited solenoid operation. Figure 7 (b) shows the cooling circuit at the initial cooldown. The vent line of the liquid helium was located 70 mm from the edge of the vessel and the distance

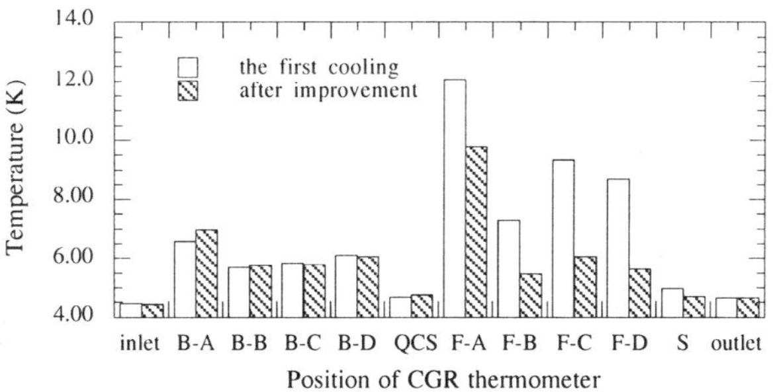

Figure 5. Temperature changes of the helium vessel and the magnet in the right side. B-A to F-D are the temperatures of micro-calorimeters. QCS and S are the temperatures of the CGRs on the magnets.

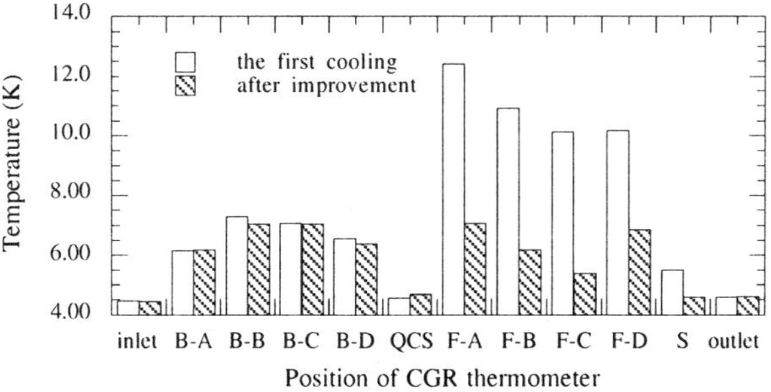

Figure 6. Temperature changes of the helium vessel and the magnet in the left side. B-A to F-D are the temperatures of micro-calorimeters. QCS and S are the temperatures of the CGRs on the magnets.

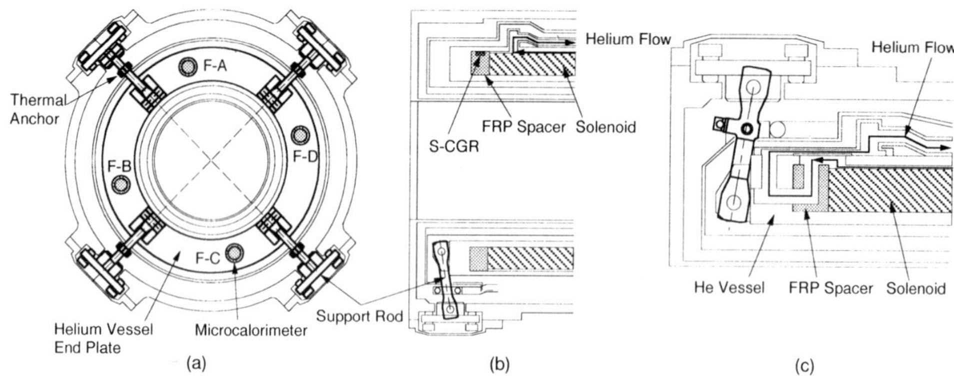

Figure 7. Positions of the support rods and the microcalorimeters on the front end plate of the helium vessel (a), the helium flow in the solenoid before the improvement (b) and the helium flow after the improvement (c).

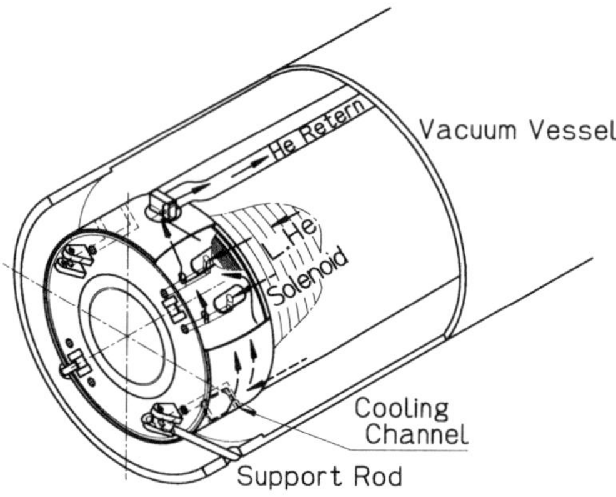

Figure 8. Helium cooling channels in the front of the helium vessel after the improvement.

between the helium vessel and the solenoid was 2.9 mm. Therefore, in the front section of the solenoid, the helium was stagnant and in poor cooling condition. In order to remove the heat inleak through the support rods and to get good cooling performance of the subcooled liquid helium in the front of the solenoid, the cooling circuit was improved, as shown in Fig. 7 (c). Eight cooling paths at the both sides of the connections of the support rods were made in the end plate of the vessel, as shown in Fig. 8. Figures 5 and 6 show the temperature profiles after the improvement. The average temperatures of the front end plates for the right and left sides decreased to 6.7 K and 6.4 K, and the solenoids were cooled down 4.7 K. From these improvement, the solenoids successfully reached their operating currents.

THERMAL ANALYSIS OF THE FRONT END OF THE HELIUM VESSEL

To understand the temperature profile of the cryostat, 3-dimensional calculations were performed with a finite element model. The calculated model is shown in Fig. 9. Because the cryostat has thermally a quadrupole symmetry, the calculated region was from 45 degree to 90 degree in the azimuthal direction. The temperatures of the outer surfaces of the FRP plate and the solenoid were constrained at 4.6 K. The temperature of the end of the support rod was constrained at 300 K. The calculations were carried out in two cases with the temperatures at the end of the thermal anchor of 100 K and 200 K. The calculated results are shown in Figs. 10 and 11. The open circles, closed circles and diamonds show the surface temperature distributions of the end plate at radii of 87 mm, 105 mm and 123mm. The microcalorimeters are located at the position of r=105 mm and θ= 60 degree. From Fig. 10, the temperatures at this position are 10.7 K and 8.7 K for the 200 K and 100 K anchors, respectively. The calculated maximum temperatures in the solenoid are 6.9 K and 7.9 K.

The average temperatures measured by the microcalorimeters were 9.3 K and 10.9 K for the S-R and the S-L, respectively. These values are close to the calculation for the 100 K and 200 K anchors. The solenoid temperatures estimated from the currents at the normal transitions were 6.8 K and 8.2 K for the S-R and the S-L. These results are good agreement with calculations.

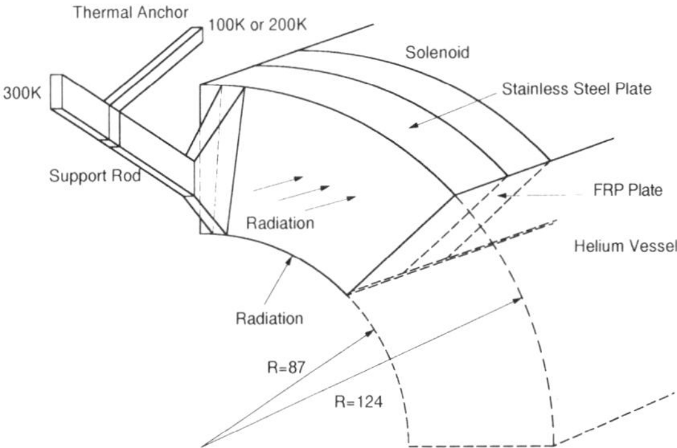

Figure 9. 3-dimensional calculation model of the front part of the cryostat.

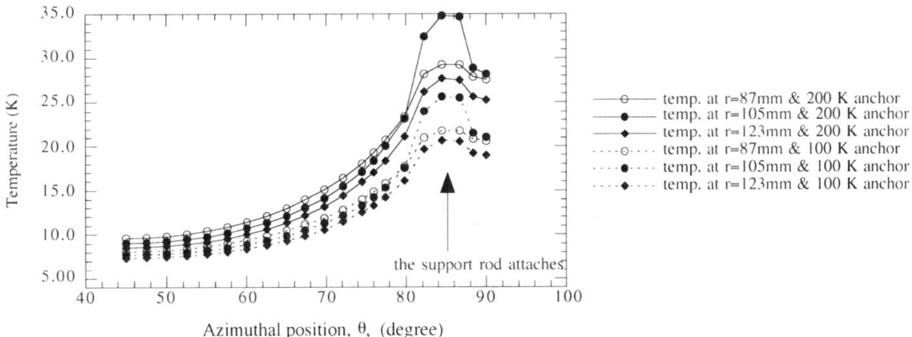

Figure 10. Calculated temperature profile on the end plate of the helium vessel before changing the cooling circuit.

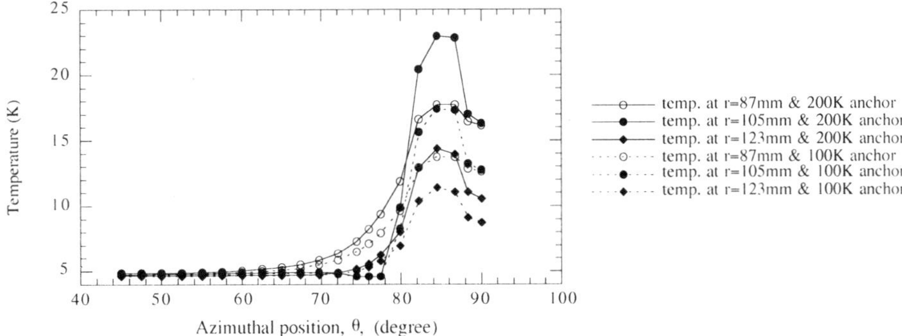

Figure 11. Calculated temperature profile on the end plate of the helium vessel after changing the cooling circuit.

Figure 11 shows the calculated temperature profile on the end plate after improving the cooling circuit. The temperature at the microcalorimeter and the maximum temperature in the solenoid decrease to 4.8 K and 5.3 K, respectively. The average temperatures measured by the microcalorimeters were 6.7 K and 6.4 K for the S-R and the S-L.

HEAT LOAD MEASUREMENTS

The thermal load of the cryostats after the improvement was estimated by the increase of the enthalpy of the subcooled liquid helium between the inlet and the outlet at the service dewer. The total mass flow rate of the liquid helium to the two cryostats was 22 g/s. The temperatures of helium at the inlets were 4.45 K for the right side and 4.44 K for the left side, and the temperature increases at the outlets were 0.20 K and 0.16 K, respectively. The heat loads of the cryostats were calculated to be 12.1 W and 10.2 W, assuming that the mass flow rates to the cryostats were same. Before the improvement, the heat loads for the right side and the left side were 12.4 W and 9.0 W, respectively. The change of the thermal load by improving the cooling circuit was small. The measured thermal load for each cryostat was smaller than the design, and the liquid helium at the outlet of the cryostat was subcooled.

CONCLUSION

(1) Two cryostats for the superconducting magnets in the KEKB interaction region were constructed. The cryostat shapes were very complicated since their size was severely constrained by the experimental detector and they enclosed 5 superconducting coils: the quadrupole, the solenoid and three corrector coils.

(2) At the initial cooldown test, the temperatures of the front end plates were higher than our expectation. However, after improving the cooling circuits, the end plates were cooled down and the system was operated successfully.

(3) The measured thermal loads of the two cryostats were 22.3 W in total, slightly smaller than the design.

ACKNOWLEDGMENT

We are very grateful to Professor S. Kurokawa and our colleagues of the KEKB magnet group for their continuous support. We are also grateful to Dr. Ph. Lebrun for showing important information on the microcalorimeter. Our thanks also go to the engineers and technicians of Hitachi Ltd. for their responsible and skillful work.

REFERENCES

1. S. Kurokawa, Present status of KEKB Project, presented at the 6th European Particle Accelerator Conf. (EPAC98), Stockholm, June 22-26, 1998.
2. K. Tsuchiya, T. Ogitsu, N. Ohuchi, T. Ozaki and N. Toge, Superconducting magnets for the interaction region of KEKB, presented at 1998 Applied Superconductivity Conference, California, Sept. 13-18, 1998.
3. K. Tsuchiya et al, Helium cryogenic systems for the superconducting insertion quadrupole magnets of the TRISTAN storage ring, in: *Advances in Cryogenic Engineering*, Vol. 37, Part A, Plenum Press, New York (1992), p667.
4. Ph. Lebrun, H. Blessing, T. Taylor and L. Walckiers, Cryogenic microcalorimeters for the measurement of energy disposition by beam losses in superconducting accelerator magnets, Proceedings of 15th International Conference on High Energy Accelerators (HRACC'92), p592 (1993).

THERMAL PERFORMANCE OF THE SUPPORTING SYSTEM FOR THE LARGE HADRON COLLIDER (LHC) SUPERCONDUCTING MAGNETS

M.Castoldi,[1] M.Pangallo,[2] V.Parma,[1] and G.Vandoni[1]

[1]LHC Division-CERN, European Organization For Nuclear Research
1211 Geneva 23, Switzerland
[2]SGG EEIG/ AMSE - Technoparc Gessien
01630 St.Genis-Pouilly, France

ABSTRACT

The LHC collider will be composed of approximately 1700 main ring super-conducting magnets cooled to 1.9 K in pressurised super-fluid helium and supported within their cryostats[1,2] on low heat in-leak column-type supports[3]. The precise positioning of the heavy magnets and the stringent thermal budgets imposed by the machine cryogenic system, require a sound thermo-mechanical design of the support system. Each support is composed of a main tubular thin-walled structure in glass-fibre reinforced epoxy resin, with its top part interfaced to the magnet at 1.9 K and its bottom part mounted onto the cryostat vacuum vessel at 293 K.

In order to reduce the conduction heat in-leak at 1.9 K, each support mounts two heat intercepts at intermediate locations on the column, both actively cooled by cryogenic lines carrying helium gas at 4.5-10 K and 50-65 K.

The need to assess the thermal performance of the supports has led to setting up a dedicated test set-up[4,5] for precision heat load measurements on prototype supports.

This paper presents the thermal design of the support system of the LHC arc magnets. The results of the thermal tests of a prototype support made in industry are illustrated and discussed. A mathematical model has been set up and refined by the comparison with test results, with the scope of extrapolating the observed thermal performance to different geometrical and material parameters. Finally, the calculated estimate of the heat load budgets of the support system and their contribution to the total cryogenic budget for an LHC arc are presented.

INTRODUCTION

Accurate supporting and stable positioning of the LHC super-conducting magnets within their cryostats (Figure 1) is essential for the machine alignment which is carried out through surveying and aligning of cryostat-mounted fiducials. High stiffness (mainly flexural) of the supports is therefore the basic mechanical requirement to guarantee that magnets are stable

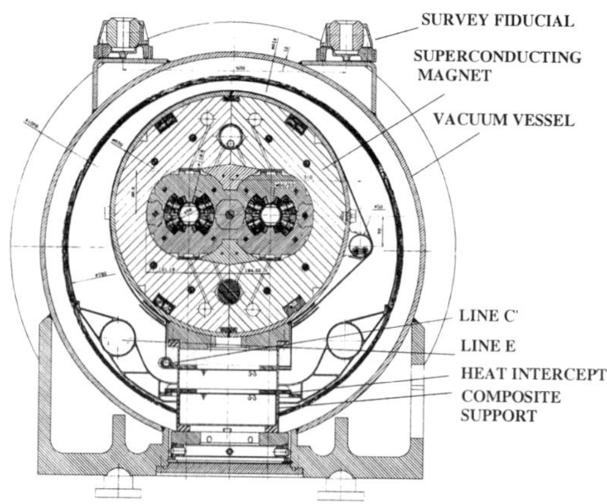

Figure 1. Cross-section of an LHC super-conducting magnet within its cryostat

within 0.3 mm under variation of external forces (magnet interconnection forces, weight components due to cryo-magnet inclination, etc.). A three-point support configuration is adopted for the 1300 15 m-long, 25,000 kg dipole magnets, whereas two points of support are sufficient for the 400 quadrupoles which are shorter (6 m-long) and lighter (6,000 kg).

To limit conduction heat in-leaks from the cryostat vacuum vessel at room temperature to within specified maximum values per support (<0.05 W to the magnets at 1.9 K, <0.5 W to the cryogenic line C' at 4.5-10 K, <5 W to the cryogenic line E at 50-65 K), materials featuring a low thermal conductivity-to-stiffness modulus ratio are required.

CHOICE OF THE MATERIAL

A wide variety of low thermal conductivity plastic materials, ranging from charged thermo-plastics to carbon or glass fiber reinforced epoxies, were tested at CERN during several years of R&D[6]. For tubular thin-walled structures of a given diameter, as for LHC supports, both conduction and flexural stiffness are proportional to the wall thickness therefore the merit figure can be expressed in terms of the conductivity-to-flexural modulus ratio. Figure 2 compares this ratio, in the 300-4 K range, of two among the best performing candidate materials, ULTEM 2300® (a short glass-fiber reinforced PEI resin) and G-10 (a long glass-fiber reinforced epoxy) with a more conventional material such as stainless steel.

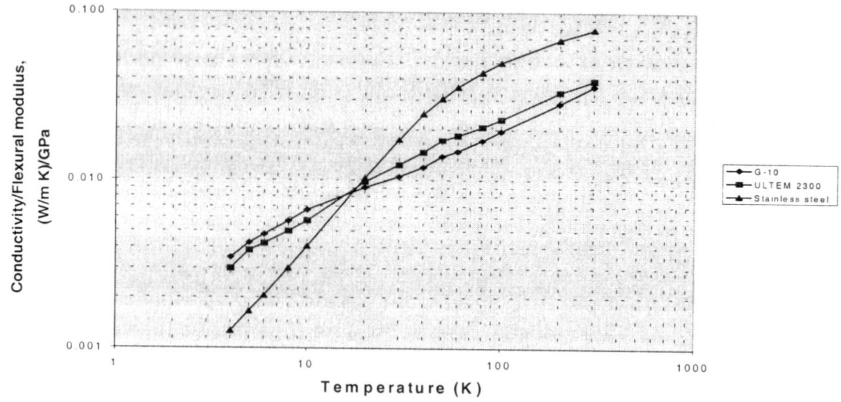

Figure 2. Thermal conductivity/flexural modulus versus temperature

Although stainless steel exhibits the best performance at very low temperatures, high thermal conductivity above 20 K limits its range of application for LHC supports. ULTEM 2300®, featuring very low thermal conductivity across the full temperature range, is however penalised by a low stiffness; when compared with glass-fiber reinforced epoxy composites, whose stiffness is more than twice as high, the resulting conductivity-to-stiffness ratios are of the same order across the temperature range of interest. However, the higher sensitivity of thermo-plastics to creep, unlike the more stable thermo-sets, limits the allowable stress for ULTEM supports, leading to higher thickness of the columns thus thermally less performing.

Therefore, glass-fiber epoxy composite is the material retained for the LHC supports. The wide availability of manufacturers in European industry, should assure a competitive market for the mass production of supports. Among all manufacturing processes, industrialisation experience suggests the use of hand lay-up of pre-preg material processed in autoclave or Resin Transfer Moulding (RTM). The latter presents the advantage of being highly automated reducing production costs, despite a large initial investment in tooling, and improving the quality control over the production of some 5000 supports required for LHC. Testing on prototypes at CERN confirmed that long glass-fiber in epoxy, with a 50-60% volume fraction of fiber, allows the required stiffness properties to be achieved.

However, thermal conductivity in composites is characterised by a wide dispersion depending on the properties of its constituents (fiber and resin) and on the composition of the material (lay-up, volume fraction of constituents, fibre orientation, etc.). Figure 3 reports data for glass-fiber epoxy composites available in literature and the estimated thermal conductivity for an LHC prototype support, reconstructed from test measurements as explained in this paper. The stringent cryogenic budgets demand a precise characterisation of the thermal performance of the supports, which requires qualification testing of prototypes at CERN.

THERMAL DESIGN

The maximum available vertical space between magnet and vacuum vessel, together with compatibility with integration and assembly constraints, limits the support height to 213 mm, whereas

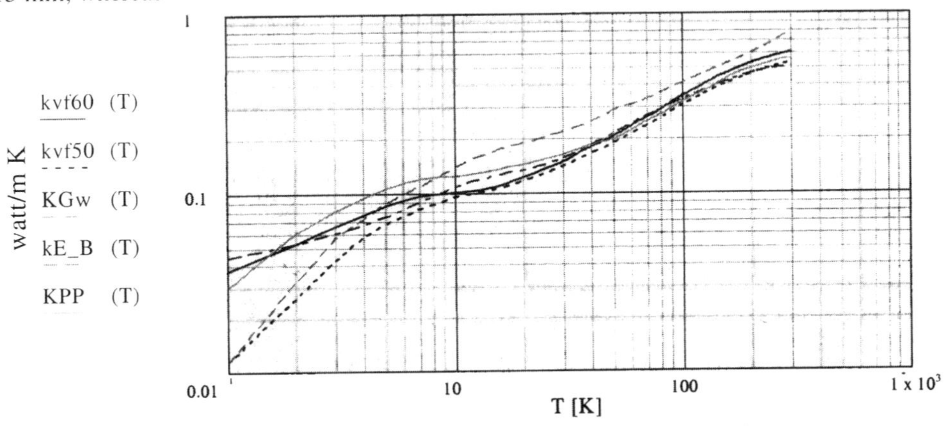

Figure 3. Thermal conductivity for various epoxy based composites versus temperature:

Gw:	G-10
E_B:	Glass fiber, E glass and B resin
vf50:	Glass fiber Vf = 50%
vf60:	Glass fiber Vf = 60%
PP:	Extrapolated conductivity for prototype dipole support

the diameter and thickness of 240 mm and 4 mm respectively, are defined by the stiffness constraints already discussed.

Cryogenic lines C' and E, run through each LHC cryogenic sector of 3.3 km length providing cooling capacity to the cryostat and the supports by helium gas flow. Line C' provides cooling between 4.5 K and 10 K whereas line E operates between 50 K and 65 K, depending on the location considered along the sector.

A reduction of conduction heat to the magnet, to meet the cryogenic requirement at 1.9 K, is obtained by extracting heat at two intermediate sections of the support, by so called *heat intercepts*, cooled by the lines C' and E. For the thermal design of the supports, the C' and E lines are considered at average temperatures of 7.5 K and 57.5 K, respectively.

The optimal spacing of the heat intercepts along the support is chosen to minimise the cryogenic cost function expressed by equation 1, which uses weight factors, suitably chosen for LHC, expressing the power required for extracting heat at 1.9 K, 5-10 K and 50-65 K:

$$Q_{tot} = (Q_{1.8K} \cdot 1000 + Q_{5\text{-}10K} \cdot 125 + Q_{50\text{-}65K} \cdot 15) \text{ [W]} \tag{1}$$

The heat intercepts are 10-mm thick aluminium plates externally glued to the composite column, and connected to the cryogenic lines by an all-welded solution for the aluminium line E, and a low-thermal-impedance aluminium-to-stainless steel transition for line C'. The heat intercepts shrink fit onto the composite column during cool down assuring an improved heat exchange due to contact pressure. However, an unavoidable thermal impedance exists at this interface, resulting in a not fully efficient heat extraction. This effect was measured during testing and mathematically modelled as explained in the following sub-section.

Aluminised Mylar foils, between the line E heat intercept and the vacuum vessel, are used to insulate the composite material from radiation in-leaks from the vacuum vessel at room temperature. With the same purpose, aluminium foil disks are also mounted inside the support column at the level of the heat intercepts.

Calculation Model and Results

With the aim of supporting and checking the measurements campaign on a prototype dipole support, a mathematical model has been set up[7]. This is a lumped-parameter model for a simplified geometry of the supports, which takes into account non linear solid conduction through the composite column and its aluminium heat intercepts, and radiation heat transfer at temperatures above 50 K. An analytical and non linear formulation for the steady state heat transfer was used, leading to a system of equations of an equivalent electrical network, as shown in Figure 4. An iterative finite difference algorithm applied to a step-wise linearized system of equations, leads to the simultaneous solution of the heat loads at the nodes.

The model also allows, by simple parameter changes, to estimate heat loads for supports having different geometry (as for the quadrupole magnets) or having different boundary condition temperatures. This last feature permits the extrapolation of the heat loads for a support in any point of a cryogenic sector, or the heat loads induced by a faulty heat intercept operating at a higher temperature. Moreover, the model also yielded a useful tool for tuning of the thermal design of the supports.

Due to the lack of data on the thermal conductivity of the composite material used in manufacturing the prototype support, the conductivity curve (see Figure 3) was reconstructed and fine-tuned to meet reliable measurement results obtained from the test set-up. 10 measurement points, covering most of the operating temperature range of a support in a cryogenic sector, were used with the aim of minimising the error between calculated and measured heat loads. The values for some of these working points are displayed in Table 1.

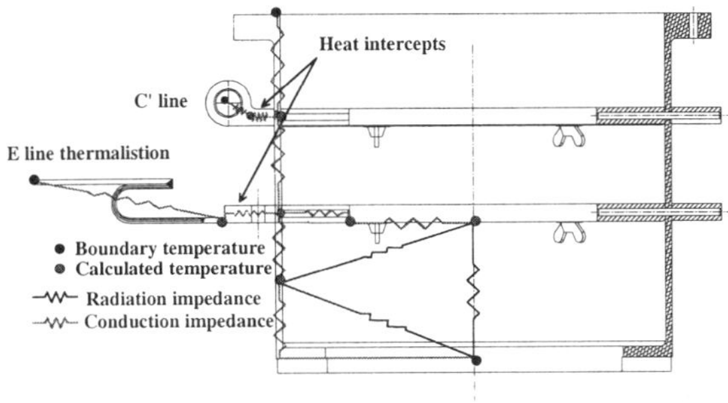

Figure 4. Equivalent electrical network of the calculation model

Table 1. Measured and calculated heat loads for the dipole support post

		Measurements				Calculation			
		Magnet	C' line	E line	Vacuum Vessel	Magnet	C' line	E line	Vacuum Vessel
1	T [K]	1.63	6.8	76	288	1.63	6.8	76	288
	HL[W]	**0.051**	**0.716**	**6.69**		**0.056**	**0.639**	**6.90**	
2	T [K]	2.19	19.6	94	289	2.19	19.6	94	289
	HL[W]	**0.171**	-	**7.09**		**0.159**	**0.688**	**6.61**	
3	T [K]	1.67	6.55	53.3	295	1.67	6.55	53.3	295
	HL[W]	**0.036**	**0.379**	**7.82**		**0.046**	**0.433**	**7.75**	
4	T [K]	1.87	17.43	52.0	296	1.87	16*	52.0	296
	HL[W]	**0.112**	-	**8.01**		**0.116**	**0.312**	**8.04**	
5	T [K]	1.83	7.33	95.3	299	1.83	7.33	95.3	299
	HL[W]	**0.070**	**0.880**	**7.14**		**0.068**	**0.876**	**7.29**	

It appears that the model remains affected by a residual error which, for the majority of the measurements, is below 10% and for only few cases, when the magnet temperature in the test set-up is below 1.9 K, is up to 25%, due to the uncertainty of the conductivity curve in that temperature range.

The measurements demonstrate the imperfection of the heat intercepts. Working points 1, 3 and 5 show similar temperatures on the magnet and C' line, however, the heat load to the magnet are quite different, and corresponds to the scaling between the temperatures on the E line, indicating that the C' line intercept is not fully effective in intercepting the heat.

Figure 5 shows the temperature profiles along the column with or without heat intercepts, whereas Table 2 presents the corresponding calculated heat loads breakdown and the total required cryogenic power. It can be concluded that the double heat intercept solution

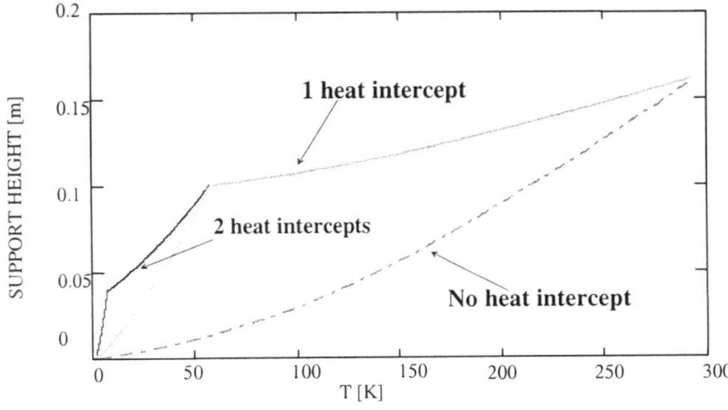

Figure 5. Temperature profiles along the support

Table 2. Calculated heat loads for various heat intercept configurations

	Q_{magnet} [W]	$Q_{c'line}$ [W]	Q_{Eline} [W]	Q_{tot} (exergetic) [W]
No heat intercept	2.06	-	-	2060
1 heat intercept	0.27	-	7.1	378
2 heat intercepts	0.049	0.45	7.1	212

provides a cost reduction by a factor 1.8 and 10 as respectively compared to the simpler alternatives of having only one heat intercept at the C' line, or no heat intercepts at all.

Figure 6 shows the calculated heat loads to the magnet over the full temperature range of the cryogenic lines, underlining the importance of having an efficient C' line thermalisation, to limit these heat loads. The somewhat weak correlation of heat loads with the temperature of line E, for a given C' line temperature, visible in the diagrams, is due to the partial efficiency observed for the heat intercept. Finally, calculation indicates that the effect of radiation at temperatures higher than 50 K up to the ambient temperature of the vacuum vessel, is responsible for approximately 50% of the total heat leak.

Cryogenic Test Set-up

In order to simulate the thermal behaviour of the support posts, the test set-up should reproduce as closely as possible their working environment in the dipole cryostat. Unavoidably, the measuring device perturbs this working environment, hence a compromise has to be made between measurement precision and perturbation induced on the working conditions. In the present set-up, illustrated schematically in Figure 7, heat-flow measuring devices, so called heat-meters, are mounted between the heat intercepts of the support and the heat sinks, representative of cryogenic lines E and C'. A heat-meter is a calibrated thermal impedance which measures a temperature difference induced by the heat flow. This additional thermal impedance inserted between the intercepts and the heat sinks slightly shifts the working temperature of the support intercepts, so a correction of this effect needs to be taken into account.

The support is mounted into a vertical cryostat consisting of a central cylindrical vessel which is surrounded by two annular concentric vessels; the internal one is the saturated HeII vessel, the other two are filled with HeI and with LN2 or with gaseous helium. Two radiation screens are suspended from the bottom of the two annular vessels. They can be cooled by conduction, or actively by a gaseous helium flux taken from the HeI vessel.

Each heat intercept of the support post is connected to the shields through two heat-meters. The bottom flange of the support post is bolted to an aluminium plate, which is maintained at 293 K by a resistive heater. Adequately tightened bolts and use of vacuum grease reduce thermal contact impedance. The comparison between the power applied to the

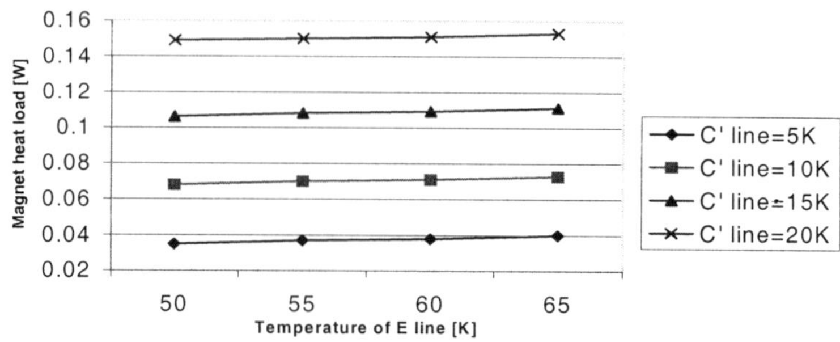

Figure 6. Magnet heat loads as a function of C' and E line temperatures

heater and the sum of the heat flows measured on the heat-meters provides a crosscheck of the the measurement precision, helping in detecting parasitic thermal leaks. The temperature sensors are mounted on the shields at the connection of the aluminium straps, or on the plate closing the support towards the 1.9 K heat sink. Tests were made with temperatures ranging from 7 K to 20 K, and from 50 K to 95 K at the C' and E line shields respectively, while the magnet and vacuum vessel interfaces were kept at around 2 K and 293 K, respectively.

Some problems were encountered during the tests. The heat-meters at the C' line level are equipped with carbon resistor sensors, whose sensitivity at 20 K is very low ($dR/dT = 5.5\ \Omega/K$). The error on their reading at this temperature is therefore very high, and their calibration not reliable. Also, the time required by the support posts to reach thermal equilibrium is very long (~10 h) and comparable to the operating time of the cryostat itself, raising stability problems. Nevertheless, data points show a very good reproducibility (1%). The heat-meter error varies between 2 and 10%, depending on the temperature range. Parasitic thermal leaks could be kept below 1%, and the error arising from non-equilibrium conditions is also smaller than 1%. Globally, the relative error on one data point can fluctuate between 4 and 15%.

HEAT LOAD BUDGETS FOR AN LHC ARC

The heat load budget, for the supports of an LHC arc, are calculated using the model presented, calibrated on the tested prototype support, and on the basis of average temperatures of the C' and E lines. Their contribution over the total budget for an LHC arc is shown in Table 3.

Some design improvements were introduced to the dipole supports since the measurements were made, namely an increase by 3 mm of the composite column height and a change in the position of the E line heat intercept. The values presented include these changes.

It can be observed that the heat loads at 1.9 K represent only 4% of the total at this temperature, which is however compensated by the heat intercepted at the cryogenic lines where cooling cost is lower. Globally the corresponding cost of total heat extraction is around 60 kW per arc.

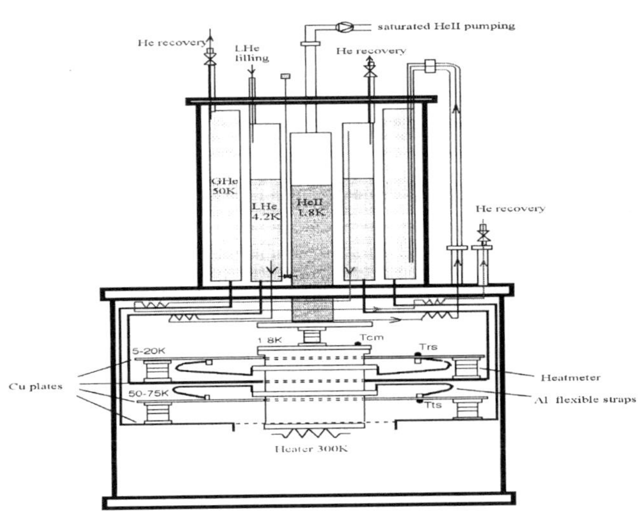

Figure 7. Schematic of a support in the cryogenic test set-up

Table 3. Calculated heat loads for an LHC arc

	No of units	Heat Load to magnet (1.9 K) [W]	Heat Load to C' line (4.5-10 K) [W]	Heat Load to E line (50-65 K) [W]
Heat Load per Dipole support	462	0.049 ±10%	0.45 ±10%	7.1 ±5%
Heat Load per Quadrupole support	108	0.047 ±10%	0.42 ±10%	7.2 ±5%
Totals	570	27.7 ±10%	253 ±10%	4057 ±5%
% of total arc static heat loads		4%	35%	31%

CONCLUSIONS

The LHC magnet supports are designed to fulfil minimum stiffness requirements to guarantee stability of the magnets within their cryostats. The achievement of the stringent cryogenic requirements imposes the use of low thermal conductivity-to-stiffness ratio materials for the tubular thin-walled supports. An intense R&D phase has led to the selection of glass-fiber epoxy resin composite. The thermal performance of each support is improved by two optimally spaced heat intercepts, obtained by aluminium plates shrink fitted onto the composite column, extracting conduction heat to cryogenic lines at intermediate temperatures.

Precision measurements on a dedicated test bench have been carried out on a prototype dipole support, allowing the characterisation of its thermal properties. A calculation model, tuned on test results from a prototype support, is used to estimate the heat loads of the supports across the full range of operating temperatures in an LHC arc, allowing the estimate of heat load budgets. Over the total heat loads of an arc, the support system contributes by only 4% at 1.9 K, but 35% and 31% of the total loads at 4.5-10 K and 50-65 K respectively, is the cost of heat extraction through the intercepts.

REFERENCES

1. W.Cameron, Ph.Dambre, T.Kurtyka, V.Parma, T.Renaglia, J.M.Rifflet, P.Rohmig, B.Skoczen, T.Tortschanoff, Ph.Trilhe, P.Vedrine, D.Vincent, "The new superfluid helium cryostats for the Short Straight Sections of the CERN Large Hadron Collider (LHC)", Advances in Cryogenic Engineering, Vol. 43-A (411-417).
2. J.C.Brunet, V.Parma, G.Peon, A.Poncet, P.Rohmig, B.Skoczen, L.R.Williams, "Design of the second series 15 m LHC prototype dipole magnet cryostats", Advances in Cryogenic Engineering, Vol. 43-A (427-434).
3. M. Mathieu, V. Parma, T. Renaglia, P. Rohmig and L.R. Williams, "Supporting systems from 293 K to 1.9 K for the Large Hadron Collider (LHC) cryo-magnets", Advances in Cryogenic Engineering, Vol. 43-A (427-434).
4. H. Danielson, P. Lebrun, J.m. Rieubland, "Precision heat in-leak measurement on cryogenic component at 80 K, 4,2' K and 1,8 K", Cryogenics Vol. 32, ICEC supplement, p. 215 (1882).
5. M.Blin, G.Ferlin, P.Gauss, C.Policella, J.-M. Rieubland, G.Vandoni, "Cryogenic R&D at the CERN Central Cryogenic Laboratory", 17th International Cryogenic Engineering Conference ICEC-17, Bournemouth UK, (14-17 July 1998).
6. M.Mathieu, Th.Renaglia, Ch.Disdier, "Supportage de la masse froide du LHC", Technical Note EST-ESM/96-01, CERN - Geneva, unpublished.
7. M.Castoldi, V.Parma, "Thermal performance calculation model for the LHC cold supports", CERN/LHC-CRI Technical Note 98-18, unpublished.

COOLING SCHEME FOR BNL-BUILT LHC MAGNETS*

K. C. Wu,[1] S. R. Plate,[1] E. H. Willen,[1]
R. van Weelderen[2] and R. Ostojic[2]

[1]Brookhaven National Laboratory
Upton, N.Y. 11973
[2]LHC Division, CERN
CH-1211 Geneva 23, Switzerland

ABSTRACT

Brookhaven National Laboratory (BNL) will provide four types of magnets, identified as D1, D2, D3 and D4, for the Insertion Regions of the Large Hadron Collider (LHC) as part of an international collaboration. These magnets utilize the dipole coil design of the Relativistic Heavy Ion Collider (RHIC) at BNL, for performance, reliability and cost reasons. The magnet cold mass and cryostat have been designed to ensure that these magnets meet all performance requirements in the LHC sloped tunnel using its cryogenic distribution system. D1 is a RHIC arc dipole magnet[1]. D2 and D4[3] are 2-in-1 magnets, two coils in one cold mass, in a cryostat. D3 is a 1-in-1 magnet, one coil in one cold mass, with two cold masses side by side in a cryostat. D1 and D4 will be cooled by helium II at 1.9 K using a bayonet heat exchanger similar to the main cooling system of LHC. D2 and D3 will be cooled by liquid helium at 4.5 K using a Two-Feed scheme. A detailed description of the cooling scheme for these magnets, their cryostats, special features and interfaces with the LHC distribution system is given.

INTRODUCTION

In 1996, BNL began to work on the D3 and D4 magnets for LHC. In 1998, the scope of work was extended to include D1 and D2. During this time, the cooling schemes for these magnets have gone through several iterations in a search for optimum ways to integrate with the LHC cryogenic distribution system. The operating temperature, type of cooling, the layout of cryogenic modules, the effect of dynamic heat load and system operation have been studied extensively. Requirements on the beam screen, superconducting bus and easy replacement of spare magnets were addressed. Cryostats were designed to include pipes for all modes of operation. This paper presents the results of this work.

MAGNETS TO BE BUILT BY BNL

There are eight insertion regions, identified as IR 1 through 8, in LHC[2]. Presently, BNL will provide magnets D1, D2, D3 and D4 for IR 1, 2, 4, 5 and 8 as shown in Table 1. Each D3 and D4 consists of two magnets a and b as shown in Table 1. D3a and D3b are

***Work performed under the auspices of the US Department of Energy**

Table 1. Magnets to be built by BNL

Insertion Region	Magnet
IR 1	D2
IR 2	D1, D2
IR 4	D3a, D3b, D4a, D4b
IR 5	D2
IR 8	D1, D2

the same magnet except for a slight difference in the spacing between the two cold masses. Similarly D4a and D4b are the same except in the beam tube separation. The length of all these magnets is about 10 meters. In each region, one set of magnets is needed for the left side of the insertion point and one set for the right side as viewed from inside the ring. Twenty magnets will be installed in LHC. In addition five spare magnets will be produced.

SLOPE AND ELEVATION CHANGE IN IR REGION

The LHC tunnel floor is sloped[2] and installed magnets experience a change of elevation from one end to the other. The elevation change is in proportion to the number of magnets cryogenically connected. It affects the natural flow direction and therefore the choice of feed points for both single and two phase cooling. Slope and change of elevation for typical IR cryogenic modules consisting of one or two magnets are given in Table 2.

LHC CRYOGENIC DISTRIBUTION SYSTEM

The LHC machine will be cooled using its cryogenic distribution system[4]. It contains two supply headers (C and E) and three return headers (B, D and F) to provide the necessary cooling at different temperature levels. Header C is designed for supplying 4.6 K helium at 3 bar. Header B is the low pressure pumping return line for superfluid helium heat exchangers and is to be operated at 0.016 bar. Header D is the 1.3 bar pressure return line for 20 K helium. Headers E and F are used for the 50 - 75 K heat shields. For 1.9 K cooling, Headers B, C and D are used. For 4.5 K cooling, Headers C and D are used. The cooling schemes for the D1, D2, D3 and D4 magnets were developed utilizing this cryogenic distribution line.

1.9 K COOLING FOR D1 AND D4

Both D1 and D4 can be operated at 4.5 K to meet the performance requirements. However, D1 is to be operated in the LHC IR region where, when LHC is operated with ultimate beam current intensity, a beam induced heating of about one to two watts per meter is expected. In addition, helium circulation (cooling channel between the beam tube and the coil) is restricted in D1 because of the required large diameter beam tube. After performing a heat transfer study of the magnet, we concluded that D1 must be cooled by superfluid helium II (1.9 K) rather than helium I (at 4.5 K) since D1 requires the large heat transfer capability of superfluid helium. 1.9 K is selected for D4 because its neighboring LHC dispersion suppressor magnets are operated at 1.9 K and their superconducting buses pass through the D4 and Q7 cold mass. Isolation of a 4.5 K thermal system from a 1.9 K system in the D4-Q7 region is considered impractical.

Table 2. Slope and elevation change for typical IR cryogenic module

Insertion Region	Slope	Change of elevation per 10 meter	20 meter
IR 1	+ 1.23 %	12.3 cm	24.6 cm
IR 2	+ 1.39 %	13.9 cm	27.8 cm
IR 4	- 0.36 %	- 3.6 cm	- 7.2 cm
IR 5	- 1.24 %	-12.4 cm	-24.8 cm
IR 8	+ 0.36 %	3.6 cm	7.2 cm

In the LHC 1.9 K cooling scheme, a heat exchanger tube[2,4,5] is used to provide cooling to the magnet coil via a bath of helium II at 1 bar. The exchanger tube is connected to the B Header and is kept at 0.016 bar. Two phase helium inside the exchanger tube must flow downward for flow stability.[5]

D4 is a two-in-one magnet: two coils in one cold mass. Its cold mass has an elliptical cross section. The minor diameter is 550 mm, which is about the same as the diameter of the LHC arc magnet. The major diameter is 630 mm. We designed a 60 mm opening at the same position as that for the helium II cooling tube of an LHC arc magnet. The cryostat is essentially identical to that of the LHC arc dipole magnet[6,7] and is compatible with the LHC cooling system. It has a 914 mm diameter vacuum vessel and is supported on three fiberglass posts. We performed heat load and pressure drop calculations and found that D4 and Q7 can be combined into the cooling module of the neighboring dispersion suppressor (DS) magnets. The exception being the beam screen cooling circuits that have to be implemented separately for the DS magnets and for D4. The reason is the imbalance in flow impedance generated by adding about 23m in length with respect to its mirror circuit in the DS.

D1 is a RHIC dipole magnet, one coil in one cold mass. The diameters are 267 mm and 610 mm for its cold mass and its vacuum vessel respectively. It is undesirable to incorporate a 60 mm opening in the D1 cold mass. Instead, we use 28.6 mm copper tubes in the 30 mm diameter by-pass hole in the iron yoke lamination. The cooling capacity of a heat exchanger tube is calculated from pressure drop and liquid phase velocity according to reference 6. A cooling capacity of 1.4 W/m is obtained for each such tube. Since there are two by-pass holes available, one or two cooling tubes can be incorporated depending on the dynamic heat load. The RHIC dipole cryostat will be used for D1 with minor piping modifications for all modes of operations.

4.5 K COOLING FOR D2 AND D3

In the LHC cryogenic distribution line, Headers C and D can be used for a 4.5 K cooling system. Header B is avoided to save compressor power. The operating pressure in Header D is 1.3 bar, which corresponds to the 4.5 K saturation temperature of liquid helium. As shown in Figure 1a, Headers C and D can be used to provide a bath of liquid helium at 4.5 K for cooling a superconducting magnet. A level gauge is used to control the amount of helium from Header C and to maintain the proper liquid level in the pot.

Application of pool boiling cooling for the D2 and D3 magnets requires special considerations on the cooldown requirement, the natural flow direction of vapor, change of elevation, and minimization of vapor in the system. Figure 1b shows the Two-Feed piping layout chosen for the 4.5 K pool boiling cooling of D2 and D3. From Header C, liquid helium is fed to the low elevation end during cooldown but to the high elevation for steady state operation. Feeding helium from the high elevation end eliminates unnecessary vapor flow through the magnet cold mass. Helium vapor always returns from the high elevation end to Header D.

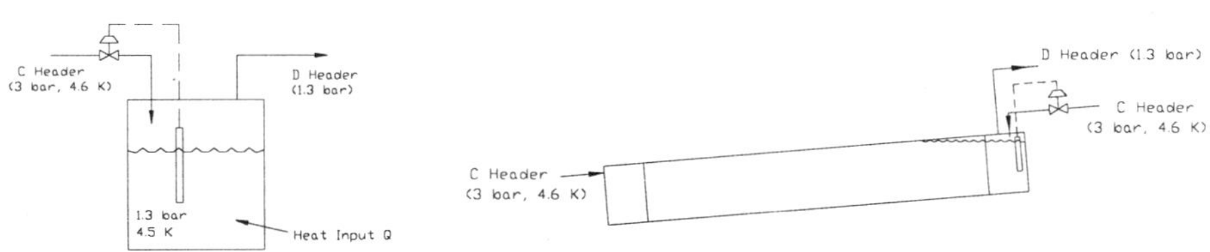

Figure 1. a) Pool boiling scheme using Headers C and D
b) Two-feed pool boiling cooling of D2 and D3

HEAT LOAD AND BEAM SCREEN

In LHC, the heat loads consist of two types. The static heat load comes from conduction through the supports, and radiation past the thermal insulation of the cryostat. The dynamic heat load comes from synchrotron radiation, image currents, beam scattering and beam-induced electron clouds.

The dynamic heat load of LHC is very complicated. It depends on the operating conditions and the location and the type of magnet in the ring. In some of the IR regions under consideration, the dynamic heat load around the beam pipe is large compared to that of the static heat load. This is particularly true for the (single aperture) D1's, in which preliminary dynamic heat load estimates are of the order of one to two watts per meter.

The Beam Screen[2] is a device installed inside the beam tube to shield some of the dynamic heat load from the tube. Its temperature is kept between 4.5 and 20 K by flowing helium inside its two small tubes. Removal of the majority of the dynamic heat load is expected. Beam screens will be installed in all BNL-built magnets, with the exception of the D1's at IP8, by CERN. There, a different solution is needed because of the need for a maximum physical aperture.

MAGNET CROSS SECTION AND CRYOSTATS

The cryostats for D1, D2, D3 and D4 are designed to match the requirements of the cooling system. Pipes are incorporated in the cryostat and will be connected to the LHC distribution system and/or relevant magnets for all modes of operation. The cross-sectional views of these magnets in their cryostats are given in Figure 2a, 2b, 3a and 3b. Nozzles in the end volume of the magnet, for liquid feed and vapor return are not shown. The beam screens installed inside the beam tube are not shown.

The D1 cold mass is supported by three Ultem plastic posts and is enclosed in an aluminum heat shield. A copper tube of 28.6 mm diameter in the upper left by-pass hole serves as the 1.9 K cooling tube. There are six cold pipes inside the cryostat for the left side of the IR. Only four will be used for the right side of the IR. The heat shield is cooled by 50 K helium using the lower two lines as identified in Fig. 2a. A 4.5 K line is used as the cooldown supply or quench relief. A small 4.5 K line is used as the beam screen supply. For the left side of IR 2 and IR 8, a small tube attached to the surface of the cold mass is used as the 1.9 K supply and a cold line is used as a 1.9 K vapor return.

The cross sectional view of D2 in its cryostat is given in Fig. 2b. The size difference between D1 and D2 are to the scale. There are five pipes inside the D2 cryostat. One 4.5 K line is used as the cooldown supply. A small 4.5 K line is used as feed for the beam screen. Two lines are for the 50 K heat shield.

D3 is a 1-in-1 magnet, one coil in one cold mass. Two cold masses are packaged side-by-side in a cryostat as shown in Fig. 3a. This configuration is selected since the distance between the two beam tubes is large compared to the coil and the magnet would become excessively heavy if a single cold mass were used. This layout utilize the available space and provides cost saving. Inside the cryostat, there are two pipes integral to the 50 K heat shield, two for the 4.5 K supply and one for the beam screen supply. The cooling system is designed so that each side can be controlled independently. However one cold mass cannot be warm while the other is cold since they are supported on a common cradle.

The cross-sectional view of D4 is given in Fig. 3b. The main difference between D2 and D4 comes from different operating temperature and cooling requirements. D4 is designed for 1.9 K cooling and needs a helium II heat exchanger tube. D2 is designed for 4.5 K pool boiling cooling and needs a liquid level gauge in its end volume.

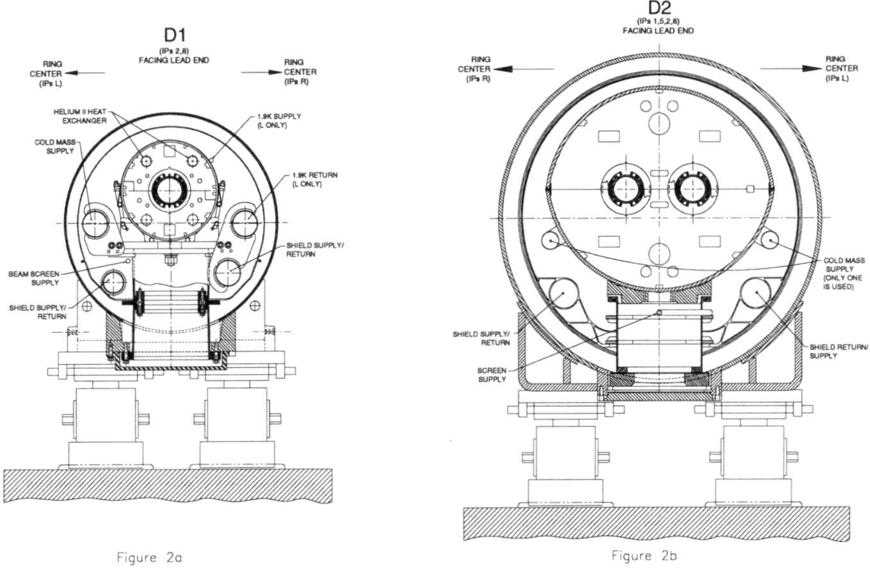

Figure 2. Sectional view of the D1 (Fig. 2a) and D2 (Fig. 2b) cryostats

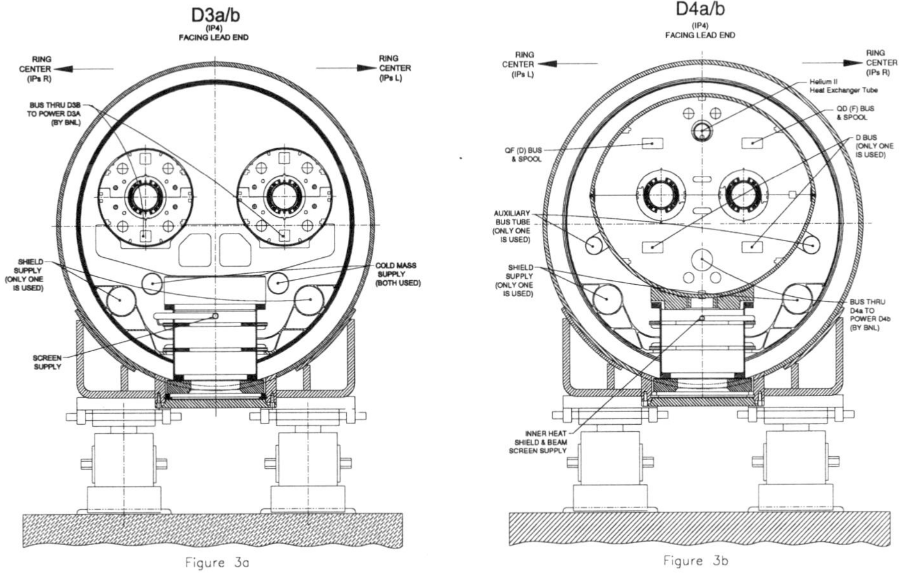

Figure 3. Sectional view of the D3 (Fig. 3a) and D4 (Fig. 3b) cryostats

CRYO MODULE

In the LHC arc region, one cell of eight magnets are combined into a cryogenic module for cooldown, steady state, quench venting, beam screen cooling, etc. In the present design, D1 is in a cryogenic module connected to the Inner Triplet Feedbox (DFBX) and the Insertion Triplet magnets. D2 and its neighboring quadrupole Q4 are in one module. D3a and D3b are in one module. D4a, D4b and quadrupole magnet Q7 are combined into the cryogenic module of their neighboring LHC dispersion suppressor cell.

COOLING SCHEMES

The cooling scheme for D1, D2, D3 and D4 in the left side of the IR are given in Figures 4, 5, 6 and 7. Those in the right side of the IR are based on the same idea but with a small difference in piping layout and are not shown. The machine optics is a mirror image between the left and the right side of the IR but not the cryogenic piping. All magnets are interchangeable between the left and the right sides with minimum piping alterations. The dashed box indicates the responsibility of BNL.

In Figure 4, the D1 cryogenic module is shown connected through the DFBX to the inner Triplet magnets and to the LHC distribution line. The magnet leads face the DFBX. Inside the magnet cold mass, X is the 1.9 K helium II heat exchanger tube and Y is the 1.9 K supply. CL is the cooldown line, C' is the beam screen supply and CY is the 1.9 K liquid fill. The CL, C' and CY lines are connected to the C Header. XB is the 1.9 K vapor return and is connected to the B Header. E1 and E2 are for the 50 K heat shield and are connected to the E and F Headers. CFV, JT and TCV are control valves for cooldown, steady state and shield flow respectively. The SRV is a pressure relief valve.

In Figure 5, the D2 and Q4 are connected in a cryogenic module of about 20 meters length. The magnet leads for D2 always face Q4. The cooldown supply is fed from the low elevation end of Q4. The vapor returns and helium supply during normal operation are from the high elevation end of D2. A pipe connects the top of the D2 and Q4 cold masses, allowing vapor flow from one magnet to the other without upsetting the coil and bus. CL1 is the cooldown line and CL2 is the liquid fill. CL1 and CL2 are connected to the C Header. LD is the vapor return line and is connected directly to the D Header, so no SRV is needed. E1 and E2 are for the 50 K heat shield and are connected to the E and F Headers. CFV, LCV1 and TCV are control valves for cooldown, steady state and shield flow respectively.

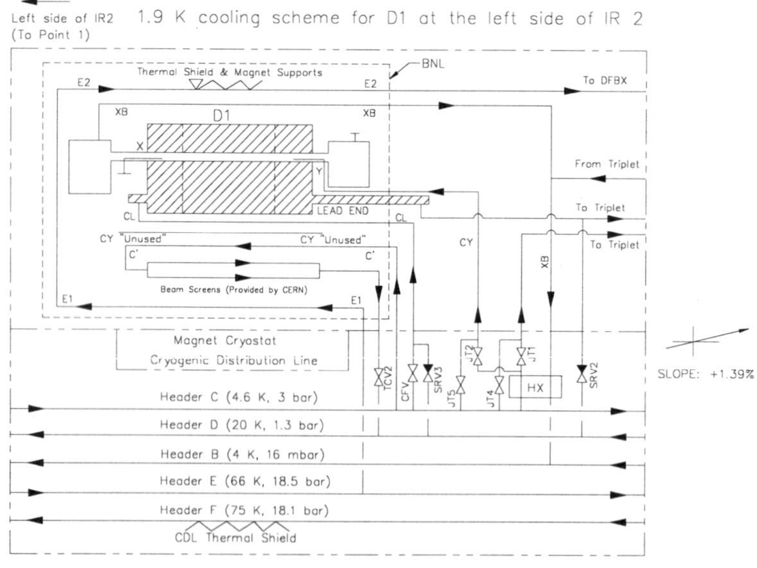

Figure 4. 1.9 K cooling scheme for D1 in the left side of IR 2

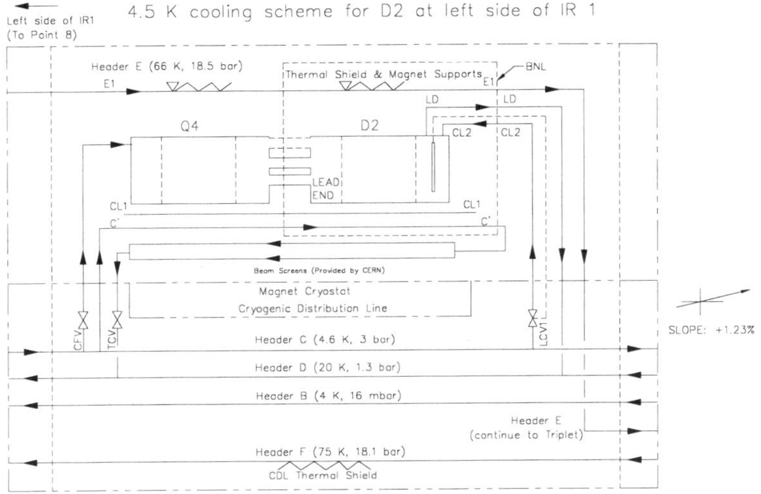

Figure 5. 4.5 K cooling scheme for D2 in the left side of IR 1

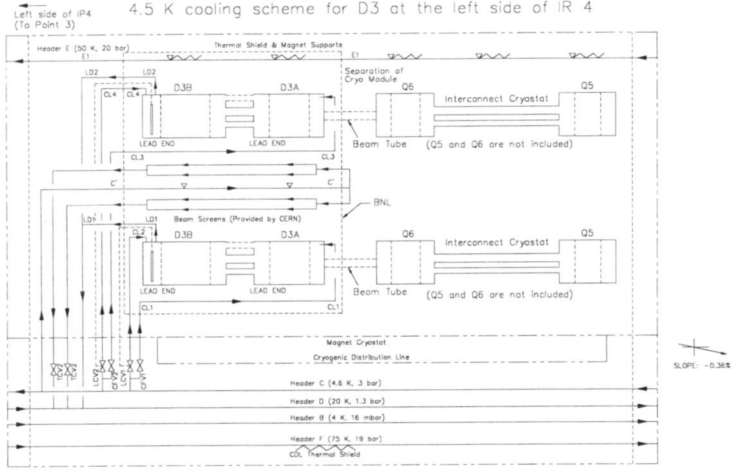

Figure 6. 4.5 K cooling scheme for D3 in the left side of IR 4

The cooling scheme for D3 is similar to that of D2 except that there are two parallel cooling passages. The cryogenic module consisting of D3a and D3b is about 20 meters long. The cooling schematic for D3 in the left side of IR 4 is given in Fig. 6. Since D3 contains side-by-side, single aperture cold masses, additional piping has been incorporated in the D3 cryostat so the cooling of each side can be controlled independently. In Fig. 6, D3b is on the high elevation end and the magnet leads for D3a and D3b face the left side. To comply with the Two-Feed scheme, the left side uses the end volume of D3b for phase separation and that of D3a for cooldown supply. The top piping connection between D3a and D3b prevents a vapor trap. CL1 and CL3 are the cooldown lines. CL2 and CL4 are for liquid fill. These four lines are connected to the C Header. LD1 and LD2 are vapor return lines and are connected directly to the D Header, so no SRV is needed. E1 is for the 50 K heat shield and is connected to the E Header. CFV1, CFV2, LCV1 and LCV2 are control valves for cooldown and steady state operation. The C' line is for beam screen cooling and is connected to the C Header. The two TCV2 valves are used to control flow through the two beam screens.

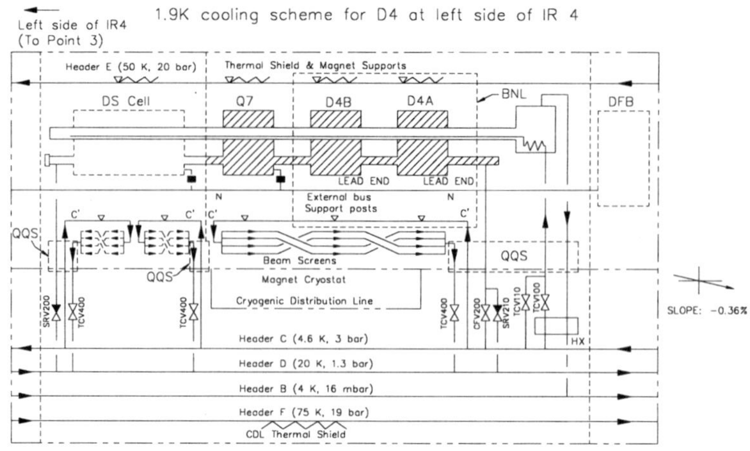

Figure 7. 1.9 K cooling scheme for D4 in the left side of IR 4

The flow schematic for D4 and Q7 in the left side of IR4 is given in Fig. 7. As one can see, D4a, D4b and Q7 are combined into the cryogenic module of the neighboring LHC dispersion suppressor magnets. A separate cooling line is provided for the beam screen for D4 and Q7 from the pressure drops consideration. The magnet leads for D4a and D4b face the DFB box. C' line is for cooling the support post at 4.5 K and is connected to the C Header. The N line is for the superconducting bus of the LHC arc magnets and is connected to the DFB box and Q7. Header E is used for 50 K heat shield. The cryogenic module is connected to the distribution headers. A counter flow heat exchanger is used to cool the supply helium from the C Header. Helium vapor, at 0.016 bar, returns to the B Header. Safety relief valves, SRV200 and 210, are connected to the D Header. Temperature in the beam screen is controlled by TCV400. Both the heat exchanger and the valves are located in the jumper between the cryogenic distribution line and the magnets.

CONCLUSIONS

The cooling schemes and the cryostats for the BNL-built IR magnets D1, D2, D3 and D4 have been designed. The magnets and cryostats will meet the performance requirement in a sloped tunnel environment using the LHC cryogenic distribution system.

ACKNOWLEDGMENT

The authors would like to thank L. Williams of CERN, T. Peterson, P. Pfund and J. Strait of FNAL, and J. Zbasnik of LBNL for valuable discussions.

REFERENCES

1. RHIC Design Manual, Brookhaven National Laboratory, Upton, N.Y.
2. The Large Hadron Collider, Conceptual Design, The LHC Study Group, CERN/AC/95-05 (LHC)
3. R. Gupta, et al., Coldmass for LHC Dipole Insertion Magnets, Beijing, 1997.
4. M. Chorowski, et al., A Simplified Cryogenic Distribution Scheme for the Large Hadron Collider, "Advances in Cryogenic Engineering", Vol. 43A, Plenum Press, NY, 1997.
5. A. Bezaguet, et al., Cryogenic Operation and Testing of the Extended LHC Prototype Magnet String, Proceedings of 16th International Cryogenic Engineering Conference, Kitakyushu, Japan, 1996.
6. Ph. Lebrun, et al., Cooling Strings of Superconducting Devices below 2 K: The Helium II Bayonet Heat Exchanger, "Advances in Cryogenic Engineering", Vol. 43A, Plenum Press, NY, 1997.
7. V. Benda, et al., Measurement and Analysis of Thermal Performance of LHC Prototype Cryostats, "Advances in Cryogenic Engineering", Vol. 41, Plenum Press, NY 1995.

THE CRYOGENIC SUPPLY FOR THE NEW S.C. MAGNETS AROUND THE INTERACTION REGIONS USED FOR THE HERA LUMINOSITY UPGRADE

H. Lierl

Deutsches Elektronen-Synchrotron, DESY
22603 Hamburg, Germany

ABSTRACT

A method to increase the instantaneous luminosity of the HERA electron-proton collider at the DESY laboratory in Hamburg, Germany, by a factor of five has been proposed. The foreseen accelerator redesign incorporates new types of septum magnets to decrease the proton β functions and new superconducting machine magnets near the interaction point inside the existing colliding-beam detectors. The s.c. magnet cryostats have diameters of 168 mm and 182-192 mm and lengths of 3.75 m and 1.80 m respectively. They will be positioned at either side of the two interaction points. The magnets consist of a one layer dipole, a three layer quadrupole coil of up to 13 T/m and a correction dipole and correction quadrupole within one layer. The cooling is performed with supercritical single phase Helium between 4 and 4.6 K supplied by thermally shielded, concentric, flexible transfer lines coming from a common precooler box located near to the experimental detector and supplied from the existing HERA cryogenic system. The 40 K to 80 K shield flow of the transfer lines is used inside the magnet cryostats to cool the beampipe. The lack of space forced a magnet design without actively cooled outer thermal shield and with gas-cooled current leads located separately from the magnet cryostat.

The s.c. magnets and the solution for their cryogenic supply especially w.r.t. their tight geometric and spacial constraints and as an extension of the existing HERA cryogenic system are described.

INTRODUCTION

HERA is a 6.3 km circumference colliding beam storage ring at the DESY laboratory in Hamburg, Germany. At two interaction points (I.P.) electrons or positrons with energies up to 27.5 GeV are being head on collided with protons of 920 GeV circulating in a s.c. magnet ring. The large detector facilities H1 and ZEUS are measuring at these interaction points

Advances in Cryogenic Engineering, Volume 45.
Edited by Shu *et al.*, Kluwer Academic / Plenum Publishers, 2000.

since 1992. The luminosity which is a measure of the attainable collision rate has been increased from year to year and has reached its design value now. For a further increase (up to a factor 5) a redesign of the magnet lattice around the interaction regions was proposed[1]. Stronger focussing "mini β" quadrupoles will be positioned as near as possible to the interaction points to increase the particle density. The necessary early beam separation will be done by combined function superconducting magnets[2] which bend and focus the electron beam at the same time. They are located at ±1700 mm at both sides of the I.P.'s, which is inside the existing detectors and inside their solenoidal fields (fig. 1). This gives very severe constraints to the radial dimensions (free diameter 180 mm), forces and supports and provokes challenging solutions for the magnet design[3] and for their cryogenic supply. Both will be described in this paper.

S.C. MAGNETS

Figure 2 shows the radial layout of the 3.75 m long upstream magnet denoted "GO". A different design was necessary for the 1.8 m long tapered downstream s.c. magnet "GG", since it is differing in space, beam deflection and aperture. Because of the strong bending and focussing in front of the I.P. a horizontally emitted synchrotron radiation fan of about 28 kW has to be passed through the detector within an oval shaped beam pipe. Only 39 mm (GO) and 26 mm (GG) radial space outside the beam pipe is available for the cryostat containing the coil layers which are wound onto a circular support tube and an annular space around the coil for the cooling with helium. The gap between the beam pipe and the coil is used for helium supply and return tubes. The radiative thermal shielding is achieved by 18 layers of superinsulation. No space is left for an actively cooled outer thermal shield. The beam pipe has to be cooled with helium between 40 K and 80 K to absorb about 100 W/m originating from higher order mode losses in the beam pipe. The helium vessel will be supported at three positions with four radial spacers made from stainless steel and G10. An endcan in front of the magnets houses a connection box for the s.c. cables, ports for the helium lines and a pumping port for the insulating vacuum (figure 3). The endcan for the GO magnet is designed to be completely demountable in order to fit into the H1 detector. The current leads have to be placed at a distance some 0.5-2.3 m in front of the endcans, differently for the H1 and ZEUS

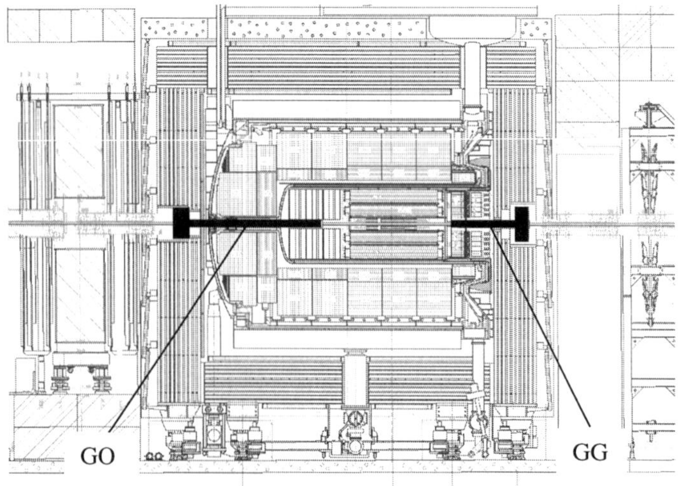

Figure 1 The s.c. magnets GO and GG inside the existing H1 detector

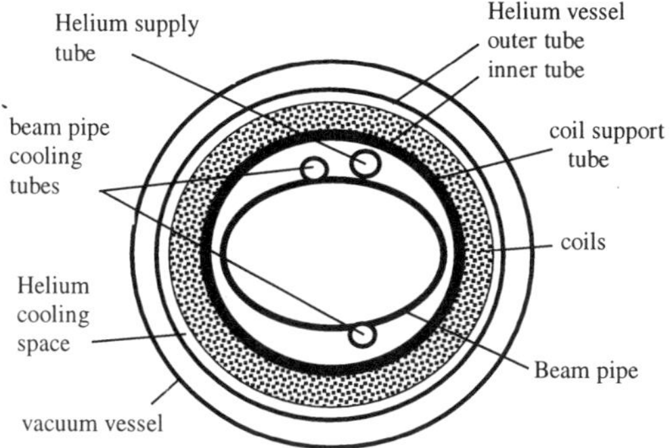

Figure 2 Cross section of the s.c. magnets

Table 1. Parameters of the super-conducting magnets GO and GG

	GO	GG
shape	cylindrical	conical
total length	3750 mm	1800 mm
magnetic length	3200 mm	1300 mm
beam pipe inner diameter	60 x 90 mm	60x(110–120 mm)
outer diameter	168 mm	182 -192 mm
tilt	4 mrad	0
shift w.r.t. I.P.	2 mm	20 mm

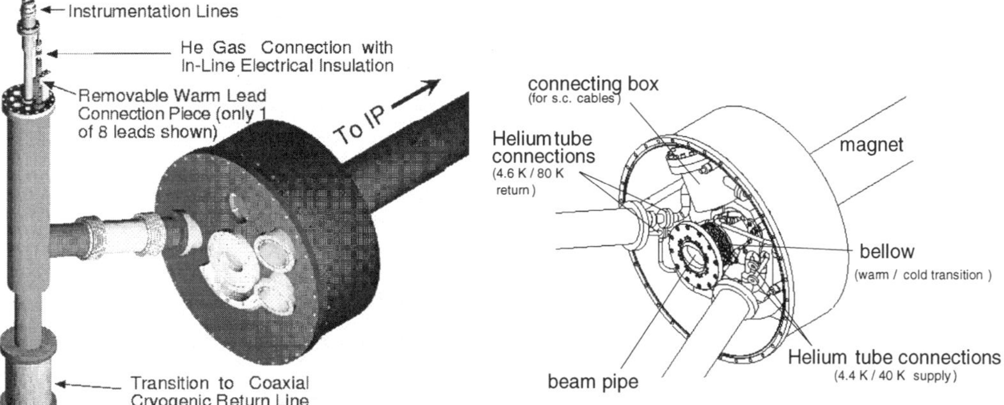

Figure 3 Layout of the GO magnet with its helium endcan

detector. A cryogenic transferline with s.c. cables inside the helium tube connects the endcan with the current lead tower. The magnets are designed and will be produced by BNL[2,3]. Table 1 summarizes the magnet parameters. The coil packages with a radial size of not more than 10 mm consist of up to 5 layers. The innermost layer is a dipole coil, layer 2 to 4 (“GO“) and layer 2 to 3 (“GG“) form a quadrupole coil and within the last layer two coils are wound over half the length to form a correction dipole and a correction quadrupole. Around the last layer an annular space of 2.9 mm is used for the helium cooling. The currents in the correction coils will be 150 A and about 500 A in the main quadrupoles giving gradients of 13 T/m (“GO“) and 7 T/m (“GG“), respectively.

CRYOGENIC SUPPLY

The superconducting coils will be cooled by supercritical, single phase helium between 4.0 K and 4.6 K. The beam pipe will be operated at a temperature between 40 K and 80 K. Both circuits can be supplied from the existing HERA helium distribution system[5] by establishing a branch from the helium supply box at H1 or from the existing helium transfer line to the ZEUS experiment, respectively (figure 4). At one side of the I.P. there will be a subcooler box from where two pairs of concentric, flexible transferlines run to each side of the I.P. to supply the magnets GO and GG.

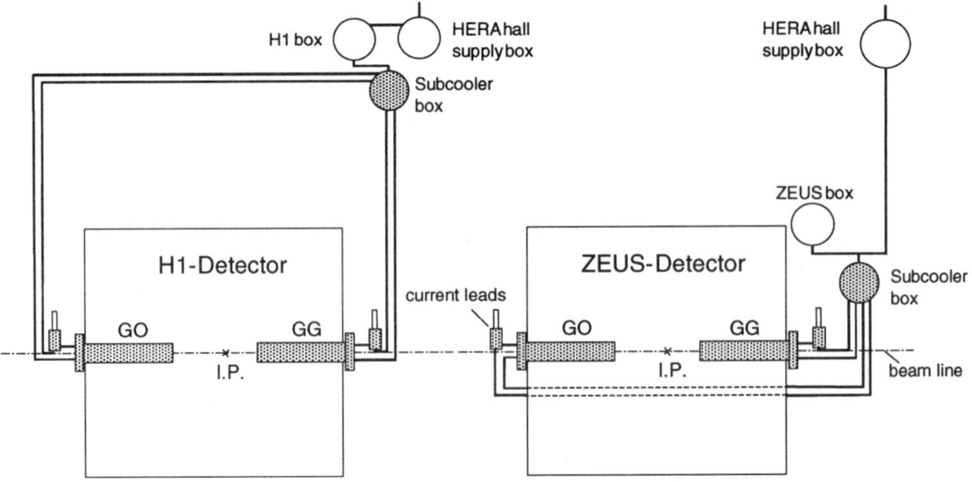

Figure 4 Cryogenic supply of the s.c. magnets GO and GG within the detectors H1 and ZEUS

Transfer Lines

Helium at temperatures of 4.0 K to 4.4 K and 40 K is supplied from the subcooler box to the magnets by pairs of flexible, transfer lines[4] made from four concentric, corrugated stainlees steel tubes. The lengths are 32 m and 27 m for the supply of the GO magnets at H1 and ZEUS, respectively and 15 m and 7 m for the GG magnets. The lines have an outer diameter of 91 mm and a minimum bending radius of 1.2 m (0.50 m for a single bend done only once). The lines will be prefabricated, delivered with insulation vacuum, will be leak checked and can easily be installed within the small gaps between the detector components. One line houses the 4.0 K supply tube shielded by the 40 K supply line (defined by two concentric tubes). In the second transfer line the 4.6 K return tube is surrounded by the 80 K return line. At both ends the lines have vacuum barriers and the concentric arrangement is changed to parallel lines. Adsorbers inside the insulation vacuum establish a "permanent" vacuum. Pumping ports at the endpieces allow in addition for an active vacuum pumping. The guarantied maximum heat loads are 0.03 W/m on the 4 K line and 1.5 W/m on the shield line.

Cooling Circuit

Figure 5 shows the proposed cooling circuit. Supercritical, single phase helium of about 4.9 K at a pressure of ca. 3 bar coming from the existing HERA helium distribution system will be subcooled by means of two helium bath coolers down to temperatures of 4.4 K to 4.0 K defined by the equilibrium vapour pressure in the subcooler vessel (suction pressure). Since the HERA p-ring magnets are now operated at 4.0 K [6] it is foreseen to cool also the magnets GO and GG down to 4.0 K by lowering the pressure in the bath cooler vessels to subatmospheric values. The subcooled single phase helium is then transfered to the magnets where it heats up to 4.6 K. The coils are cooled from the rear end (nearest to the I.P.) so that the the return flow is passing the vapour cooled current leads and no heat inleak from the current leads is transfered into the magnet coils. The beam pipe is cooled from the shield circuit (13 - 18 bar) at temperatures between 40 K and 80 K. These operating temperatures will be controlled with the mass flow rates. Heat loads of 23 W in the coils and up to100 W/m

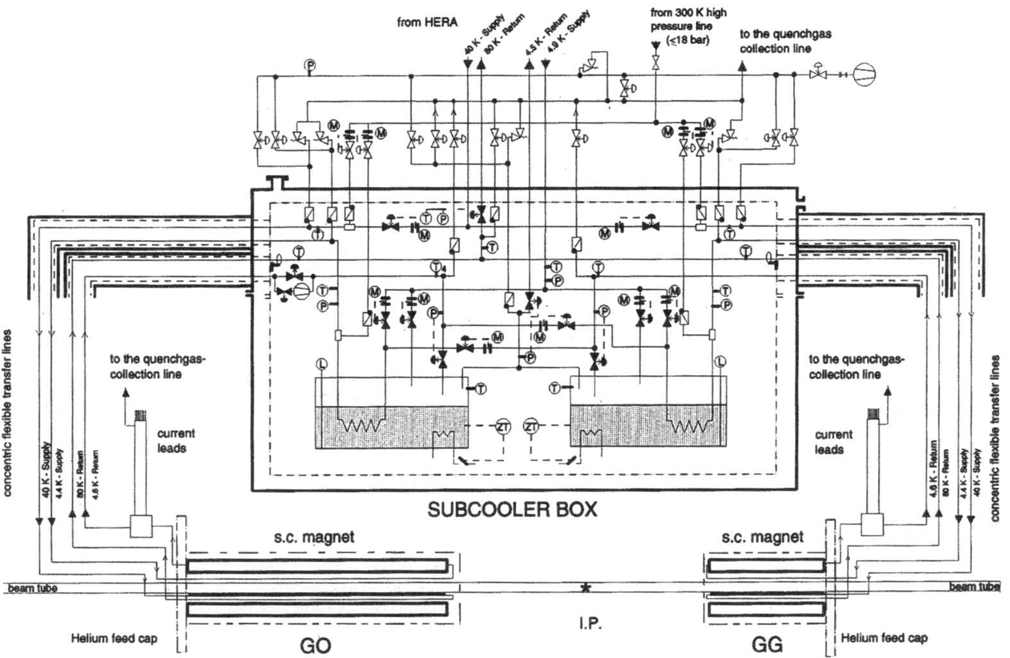

Figure 5 Cooling circuits for the s.c. magnets GO and GG

on the beam pipe have to be cooled. Returning to the precooler box the single phase helium can be expanded in a J.T. valve to be reliquified into the precooler vessel. Since there is a big excess of liquid being produced which has to be evapourated by applied heater power, the subcooler box will allow in addition for a cooling of both magnets in series. The returning single phase helium is then not expanded but passed to the second bath cooler to be recooled before it is transfered to the second magnet. Whereas the parallel cooling allows for a more independent handling of the two magnets the series cooling is the more economic one.

Additional warm up lines from the 300 K high pressure line (<18 bar) are foreseen for both cooling circuits. Warm gas returning during cool down operations or coming from the gas cooled current leads is collected and returned to the HERA quench gas collection line normally operating at 1.03 bar. The maximum operating pressure of the helium system is 20 bar. For a smooth cool down and warm up with limited temperature gradients (< 40 K) for both circuits mixing chambers are foreseen to mix cold and warm helium to the right temperature. In order to avoid a wrong mixing temperature the helium flow can be first directed and purged into the HERA quench gas collector before it is transfered to the magnets.

The stored energy in the magnets is low enough not to increase the pressure in the lines above 20 bar in case of a quench. Safety valves are foreseen to blow into the quench gas collection line. Together with a seperate pump and purge system the magnets can be operated, warmed up and cooled down without perturbation of the stationary cooling of the HERA ring.

The cryogenic system will be instrumented with level sensors and heaters in the precoolers, mass flow, pressure and temperature sensors and with the necessary check valves. The read out and control will be performed with the extended but existing HERA cryo-controlsystem.

REALISATION

The luminosity upgrade program will be realized within the HERA shut down scheduled starting May 2000 and lasting up to the end of January 2001. The boxes and transfer lines will be installed in June 2000 and connected to the magnets in November 2000. End of 2000 the first cool down will be performed. Already from January up to June 2000 the magnets will be tested cold in a separate test stand at DESY and the field quality will be measured.

ACKNOWLEDGEMENT

Thanks to the big effort of all people working in the luminosity upgrade group this difficult task up to now could only be managed by the splendid collaboration between the HERA accelerator group, the crews of the experiments H1 and ZEUS and the collaborating institutes. Special thanks go to the s.c. magnet team at BNL, especially to B. Parker and to H. Brück from DESY for providing me with pictures and data of the s.c. magnets.

REFERENCES

1. U. Schneekloth (ed.),"The HERA Luminosity Upgrade", DESY-HERA 98-05, Hamburg (July 1998).
2. B. Parker et al., "Superconducting Magnets for use inside the HERA e p Interaction Regions", Proceedings EPAC 98, Stockholm, p. 308 - 310.
3. Magnet Design and Production (3 of each type) by Brookhaven National Laboratory, BNL, Upton, NY. Magnet pictures and parameters given by courtesy of BNL (B. Parker)
4. H. Blessing et al., "Very low-loss liquid helium transfer lines with long flexible cryogenic lines", CERN LEP-MA 89-38, Geneva, 1989. Lines produced by: ALCATEL Kabelmetal, Hannover, Germany.
5. H. Lierl, S. Wolff, "Superconducting magnet and cryogenic system of HERA", review article in Japanese Cryogenic Engineering (Teion Kogaku), Vol. 31, No. 7, June 1996, p. 360 – 372, (ISSN 0389-2441).
6. H. Lierl, H. Herzog, "HERA at lower temperatures – Operational test of the HERA cryogenic system at subatmospheric pressure", proceedings ICEC16/ICMC, p. 147-150, Kitakyushu, Japan, 1996

THERMAL HYDRAULIC PERFORMANCE FOR A SEXTANT OF RHIC MAGNETS*

K. C. Wu, J. Sondericker, M. Iarocci, Y. Farah, C. Lac, A. Morgillo, A. Nicoletti, E. Quimby, J. Rank, M. Rehak and D. Zantopp

Brookhaven National Laboratory
Upton, N.Y. 11973

ABSTRACT

During the RHIC sextant test performed in spring of 1997, six recooler heat exchangers and a circulating compressor were used to determine the thermal hydraulic performance for a 600 meter string of magnets. Heat loads were measured from the temperature rise through the magnets and from recooler boil off. Cumulative heat loads were measured by turning off the cooling supply to some of these recoolers. The breakdown of the distribution of the sextant heat load is discussed and it is shown that the total heat load for a RHIC sextant is slightly higher than the design value. Pressure drop of the cooling loop was evaluated from the head rise generated by the circulating compressor. Total pressure drop is shown to be less than the RHIC design values. The cooling system also has a margin to manage unexpected heat loads. As a result, RHIC magnet temperatures in the sextant were able to be kept below 4.7 K without using the five recoolers in the sextant. Extrapolating from this experience, successful operation for the entire cryogenic system of RHIC is expected. Detailed description of the tests and results are given.

INTRODUCTION

The Relativistic Heavy Ion Collider, RHIC, is composed of two rings, identified as Blue and Yellow, of superconducting magnets arranged to intersect with each other at six experimental areas. The collider is 3.8 kilometers in circumference and each of the six sextants of magnets is approximately 600 meters long. Twelve Valve Boxes, six per ring, are installed for controlling the cryogenic operation of the collider. The first sextant test is carried out on Sextant 5 of the Yellow ring together with its two adjacent valve boxes at the 4 and 6 o'clock locations of the ring. The sextant cooldown was successful and the integrity of the system were demonstrated.[1,2,3] This paper presents the thermal hydraulic characteristics of the sextant including pressure drop, heat load, flexibility of cooling the sextant without liquid helium in some recoolers, effect of high heat load, and response to magnet quenches.

SYSTEM DESCRIPTION

The RHIC cryogenic system utilizes circulation of supercritical helium through distributed recooler heat exchangers to maintain its ring temperature below 4.6 K. There

*** Work performed under the auspices of the US Department of Energy**

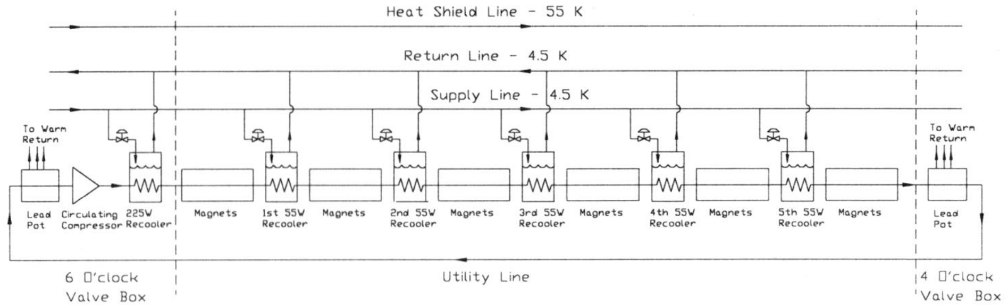

Figure 1. Flow diagram for RHIC 1st sextant test

are two cooling loops each consisting of one ring with six sextants of magnets in series. A total of seventy-two recoolers, thirty-six per ring, are used to achieve the desired temperature uniformity. Each ring has a circulating compressor installed in the 6 o'clock Valve Box. In addition to the magnet cooling line, there are four cold pipes in each ring. These four pipes are the 4.5 K supply line for the recooler feed, the 4.5 K vapor return line from the recooler, a multiple purpose utility line and a 55 K heat shield line.

The flow schematic for the sextant test is given in Figure 1. The circulating compressor is used to provide 100 grams per second of helium from the 6 o'clock Yellow Valve Box. The helium flows through the Magnet line to the 4 o'clock Valve Box where it returns to the 6 o'clock Valve Box through the Utility line. Although this is not a normal RHIC operation, however the cooling helium through the magnets are at RHIC design conditions.

There are six recooler heat exchangers in this loop, one 225-watt recooler in the 6 o'clock Valve Box and five 55-watt recoolers distributed along the sextant. The recooler is a liquid helium vessel that contains a heat exchanger[4]. It is used to cool helium in the Magnet line after the helium flows through a series of magnets. The names for these recoolers are derived from the design heat load. The heat exchanger has the capacity to handle at least 150% of the design heat loads. Major components in the sextant are the dipole and quadrupole magnets, recoolers, vacuum jacketed transfer lines (VJR)[5] and dummy magnets. A list of components between recoolers in the sextant is given in Table 1.

Table 1. Major components between recoolers of RHIC Sextant 5

Locations between	Major components**
6 o'clock Valve Box and 1st 55 W Recooler	One 20 m VJR and one 25 m VJR 1 D + 2 Q + 2 DU + 1 Triplet magnet
1st and 2nd 55 W Recoolers	One 15 m VJR and one 10 m VJR 7D + 8 Q + 1 DU
2nd and 3rd 55 W Recoolers	7 D + 7 Q
3rd and 4th 55 W Recoolers	7 D + 7 Q
4th and 5th 55 W Recoolers	7 D + 9 Q + 3 DU
5th 55W Recooler and 4 o'clock Valve Box	One 40 m VJR and one 70 m VJR 1 D + 2 Q + 1 DU + 1 Triplet magnet

**D: dipole magnet of about 9.7 m in length
Q: quadruploe magnet of about 3.5 m in length
DU: dummy magnet varies in length
A dummy magnet is a vacuum vessel contains cold pipes but no magnet. It can be considered as a long interconnecting piece and it will be replaced by actual magnets in the future.

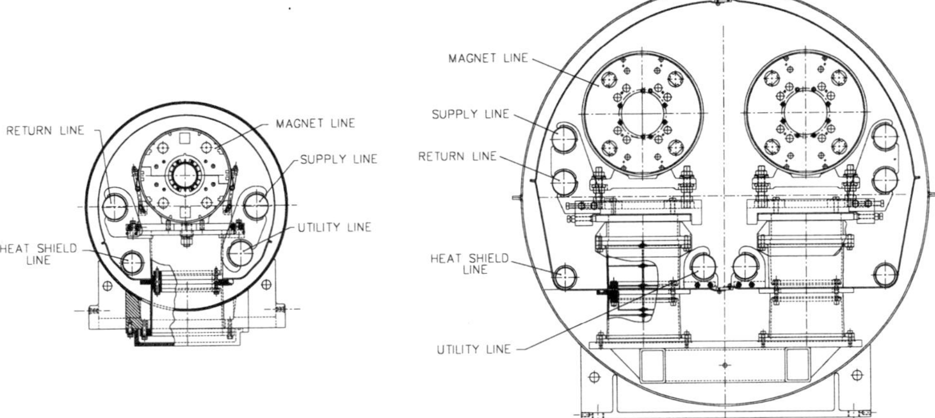

Figure 2. Cross-section of RHIC dipole

Figure 3. Cross-section of the Triplet magnet

Table 2. Flow area for the magnet lines

	Description	Flow Area
Magnet (majority)	Four 3-cm diameter by-pass holes	28 cm^2
55 W recooler	Ten 0.5" O.D. Tubes	9.3 cm^2
Interconnects between Magnets	One 2.5 inch pipe and one 3 inch pipe, each contains a 1" x 1" bus	57 cm^2
Magnet Line in VJR	One 4 inch pipe with two 2.5 cm bus flex line	64 cm^2
Magnet Line in Dummy Magnet	Two parallel 3" O.D. tubes each contains a 1" x 1" bus	61 cm^2

The cross section of a RHIC arc dipole is given in Figure 2. The four cold pipes in the RHIC distribution system are integrated in the 24-inch diameter cryostat. The superinsulation is not shown for clarity purposes. Other types of RHIC magnets differ in cross section but all have the five-pipe configuration. Near the intersection regions at the ends of each sextant, magnets in the Yellow and the Blue rings are closely located. There is a Triplet magnet consisting of two sets of four cold masses supported side-by-side in a 48-inch diameter cryostat[6] as shown in Figure 3. There are two sets of cold pipes in this cryostat. During normal RHIC operation, both rings are at 4.5 K. There are no thermal problems associated with the Triplet. In the Sextant Test, only the Yellow side is cooled to 4.5 K while the Blue side remains warm (275 to 250 K). This creates a large heat load in the Triplet.

The pressure drop of the sextant depends mainly on the flow areas in the magnet line. The flow areas of the major components are given in Table 2. As seen, the flow area in the interconnect region, the vacuum jacket lines, and the dummy magnets are all greater than that of a RHIC dipole magnet. The smallest flow area occurs in the 55-watt recooler heat exchanger due to the limited space available in the corrector-quadrupole-sextupole (CQS) cryostat as shown in Figure 4. The diameter of the recooler vessel is 8 inches and the heat exchanger consists of ten 0.5" O.D. tubes in parallel. The flow area of the 55-watt recooler is about one third that of RHIC dipole magnet.

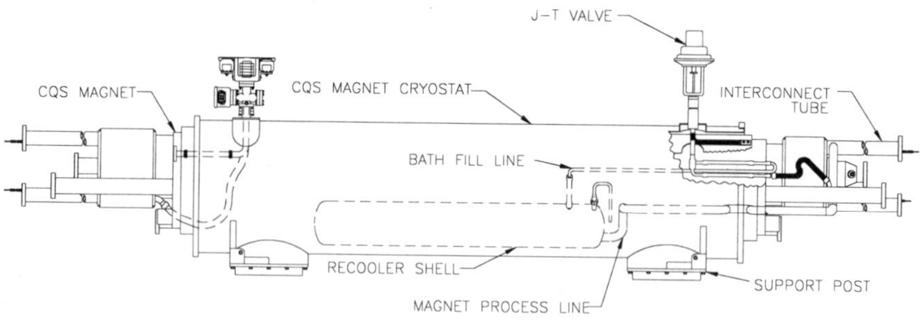

Figure 4. Side view of a CQS magnet with a 55-watt recooler

Table 3. Calculated pressure drop through the M line in a sextant at 100 g/s

Components	Velocity Head Loss - atm	Friction Loss – atm	Sum -atm	Percentage
Magnets	0.014	0.019	0.033	52 %
Recoolers	0.020	0.005	0.025	39 %
Others			0.006	~ 10 %
Sextant total			0.064	100 %

PRESSURE DROPS THROUGH THE MAGNET LINE

The pressure drops were calculated from the RHIC geometry, including all entrance and exit losses from each component. The results for 100 g/s of helium, at 5 atm and 4.5 K, through the sextant is given in Table 3. As seen, pressure drops of the five 55 watt recoolers and the magnets are more than 90% that of the sextant. Total velocity loss is slightly greater than that of friction loss. In Table 3, the friction factor through the iron lamination is assumed to be twice that of a smooth tube according to an earlier experiment performed at BNL.

Since the sextant is 600 meters long, it is impractical to use a differential pressure gauge with two connecting tubes. At 4.5 K, the pressure drop is small and accurate results cannot be obtained by subtracting two pressure readings from the ends. The pressure drops were obtained by first determining the pressure drop ratio of the Magnet line to that of the Utility line at 300 K. Total pressure drop of the Magnet and the Utility lines at 4.5 K were measured using a differential pressure transducer across the circulating compressor. The pressure drop of the Magnet line at 4.5 K was then calculated from the pressure drop ratio obtained at 300 K.

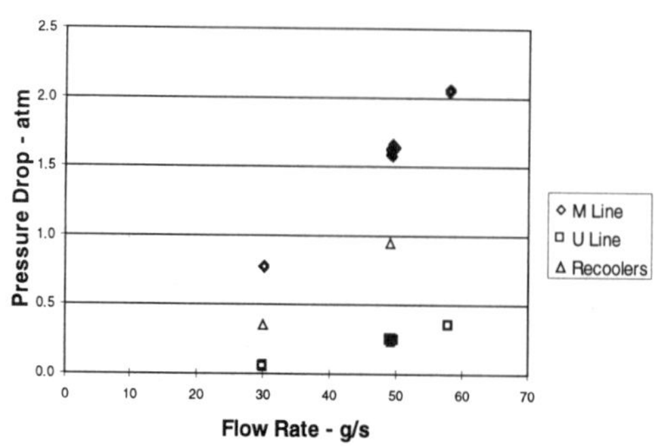

Figure 5. Pressure drop for the Utility and the Magnet Lines in Sextant 5 at 300 K

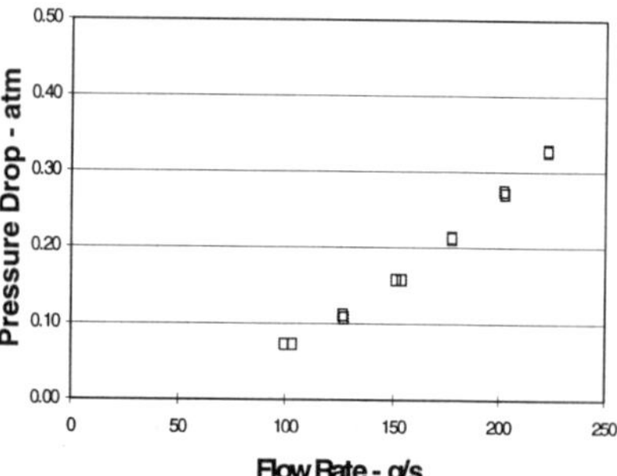

Figure 6. Pressure drop of the Magnet and Utility Lines at 4.5 K in Sextant 5

The pressure drops through Utility and the Magnet lines at 300 K for flow rates at 30, 50 and 60 g/s are given in Figure 5. The pressure drop across the Utility line is approximately 20 % that of the Magnet line. The contributions from five 55-watt recoolers at 30 and 60 g/s according to the measurement from a differential pressure gauge across a CQS unit are also shown. The pressure drops from these recoolers are approximately 50% of the Magnet line.

Total pressure drop of the Magnet and the Utility lines for helium at 4.5 K, 5 atm, and flow rates between 100 to 225 grams per second are given in Figure 6. Note that Fig. 6 includes a small pressure drop from the Valve Boxes. The pressure drop of the Magnet line is calculated from the ratio established at 300K. For 100 g/s flow at 4.5 K, the pressure drop through one sextant is found to be 0.06 atm, corresponding to 0.36 atm for the entire ring. The RHIC baseline design value is 0.42 atm.

4 K HEAT LOADS

The 4 K heat loads for the circulating loop were obtained from the temperature measurements across the magnet and from the boil off rate in the recoolers. A venturi flow meter is used in the heat load calculation. The venturi flow meter, located at the discharge of the circulating compressor, was used successfully for evaluating pump curves of the circulator. Temperature sensors were installed along the magnet line at both the entrance and the exit of each recooler, the Triplet and the VJR connecting the Valve Box. The heat load between two temperature sensors in the flow path was calculated from the enthalpy increases of the helium flow. The following results were taken when the temperature and flow were stable over extended periods of time.

The shield temperature was between 50 and 60 K. Total flow for cooling the main magnet leads in the 6 and 4 o'clock lead pots were 3 to 4 g/s. Although the installed temperature sensors were not accurate enough to determine heat loads less than approximately 10 watts, the heat load pattern was identified. In one incident, a thermal acoustic oscillation induced heat load, approximately 50 watts in magnitude, in a relief line was detected from the temperature rise across a nearby recooler. The temperature rise returned to the normal value as soon as the oscillation was eliminated by adding a 7-liter volume in the warm end of the line. Large temperature rise associated with a large heat load can also be confirmed from the larger percentage opening of the recooler feed valve.

In this test, a substantial amount of the 4.5 K heat load came from the lead pots and the valve boxes. Within the sextant, the heat loads in the ends are larger than that in the middle section. The Triplet magnet with the blue side at room temperature presents a large heat load, 50 to 80 watts. The temperature in the blue side was found to decrease by 1.5 K/day. The cooling system, however, is capable of keeping the Triplet and its neighboring magnets below 4.7 K. The VJR and the Warm to Cold transition pieces, between the 6 o'clock Valve Box and the 1st 55-watt recooler, appear to contribute to higher heat load as well. Overall, the 1st and the 2nd recoolers have the largest and second largest heat load respectively. In the region associated with the 3rd, 4th and 5th recoolers, the magnets are connected through standard RHIC interconnects and the heat loads are much less.

The daily average for February of total heat load from the six recoolers is given in Figure 7. The 225-watt recooler and the 1st 55-watt recooler (identified as Q6 in the figure) are given in the same figure. The largest heat load occurs around February 17. In those few days, all five 55-watt recoolers were shut off. Only the 225-watt recooler removed the heat. The high heat loads are due to a larger compressor work from circulating 200 g/s flow. During the reaming time, the compressor flows are mostly between 100 and 150 g/s.

Total heat load is approximately 850 watts with 600 watts in the 225-watt recooler. Since the total heat load includes the 4 o'clock lead pot, the Utility line, a major portion of the 6 o'clock Valve Box, the 6 o'clock lead pot and the work from circulating compressor, extrapolation of heat load for RHIC requires adjustments.

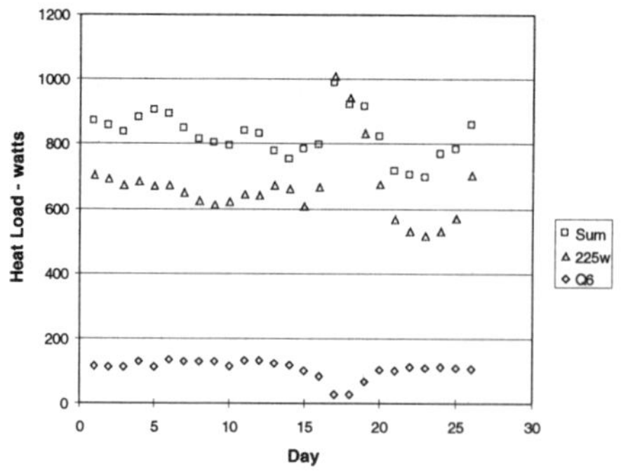

Figure 7. Recooler heat loads from temperature measurement

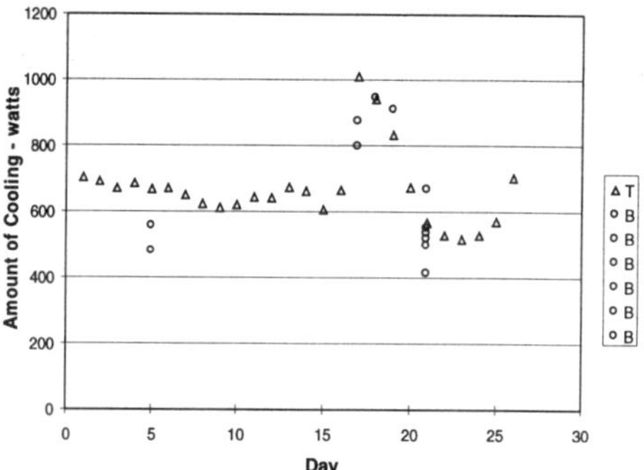

Figure 8. Comparison of heat load for the 225 watt recooler

RECOOLER BOIL-OFF RATES

Measuring the heat removing capacity in the recooler from liquid helium boil-off is a special feature in RHIC. Heat load can be evaluated from the change of liquid level after the feed valve was shut off.[4] While the boil-off measurement requires shutting off the feed valve and only measures heat load between recoolers, it provides a valuable alternative to the temperature measurement.

Several tests have been conducted on recoolers under typical operating condition. A few tests were carried out by purposely keeping some recoolers empty for the cumulative heat load. Larger temperature rises improve the accuracy of the measurements. A comparison of the results between the temperature measurement and boil off for the 225 watts recooler is given in Figure 8. For heat loads between 400 and 1000 watts, the agreements between the two methods are mostly within 20% of each other. For small loads, uncertainties associated with the temperature measurements have a large impact on the results. Table 4 presents additional results for comparison. For heat load greater than 50 watts, the two methods agree within 30 % of each other.

Table 4. Additional heat load measurements in Sextant 5

225 W Recooler	1st 55 W Recooler	2nd 55 W Recooler	3rd 55 W Recooler	4th 55 W Recooler	5th 55 W Recooler
679 (555)	132 (151)	76 (85)	8 (30)	16 (26)	51* (72)*
	131 (96)				
	92 (97)				
			12 (19)		
				24** (27)**	
		60 (42)			
					74*** (69)***
	104 (80)				
		37 (42)			

Number without parenthesis – from temperature and flow measurements.
Number with ( ) parenthesis – from boil off rates.
* Thermal Acoustic Oscillation occurs in relief line near the Q9 magnet
** The 3rd recooler is empty.
*** The 2nd, 3rd and 4th recoolers are empty.

COMPARISON WITH RHIC DESIGN HEAT LOAD

Total heat load in this test includes that from Sextant 5, the 4 o'clock Valve Box, the Utility line, a major portion of the 6 o'clock Valve Box and the pump work. The two Triplet magnets in Sextant 5 were operated with one side warm and the heat load was greater than its design value. Adjustments of heat load are necessary for comparison with the RHIC design value. Total heat load in the loop is about 1000 watts for the tests with heat removed only by the 225-watt recooler. The heat load is about 150 watts for the 4 o'clock Valve Box. Heat load resulting from the warm side of the Triplet magnet is assumed 50 watts. The pump work at 200 g/s is about 100 watts more than that at 100 g/s. The 4 K heat load for one sextant and one Valve Box is estimated at 700 watts, which is about 25 % more than the RHIC design heat load. Detailed values will be investigated further in the on-going RHIC commissioning.[8]

CONDITIONS WITH ALL FIVE 55 W RECOOLERS OFF

The recoolers provide opportunities for cumulative heat load measurements and for operational flexibility. In one test, the feed valves of all 55-watt recoolers were shut off and there was no liquid helium in these recoolers. The sextant is cooled only by the 225-watt recooler in the 6 o'clock Valve Box. The circulating flow is operated at 200 g/s to keep the sextant temperature between 4.5 and 4.7 K, permitting the RHIC magnets to be powered at 5,000 ampere.

It is not expected to have RHIC running solely on the 225-watt recoolers. The RHIC cryogenic system certainly does not need to rely on all 55-watt recoolers for normal operation. It is also not expected to have a flow of 200 g/s for RHIC. One can conclude that the circulator is capable of providing more than 150 g/s flow through the ring.

The travelling time between recoolers is about 30 minutes at 100 g/s flow. Increasing the flow can reduce this time. The RHIC magnets are very robust and it is not easy to trigger a magnet quench. Should a quench occur, the recovering time is directly proportional to the travelling time between recoolers with liquid. Therefore it is advantageous to use all 55-watt recoolers for normal operation. If there is a need to keep the temperature in the ring more uniform or to cool a hot spot quickly, one can always increase the circulating flow.

55 K PRESSURE DROP AND HEAT LOAD

The magnets in RHIC use a heat shield at 55 K to minimize the 4.5 K heat load. In this test, the shield flow is introduced from the Yellow 6 o'clock Valve Box through the Yellow Sextant, to the Yellow 4 o'clock Valve Box where a temporary jumper is connected to the 4 o'clock Blue Valve Box. The shield flow returns to 6 o'clock Valve Box through the Blue ring cryostat in which the magnet is at 300 K.

The average shield heat load for the Yellow Sextant 5 is about 1 kW. There are some uncertainties associated with the shield heat load in the 6 o'clock Valve Box. The best estimate of the total shield heat load for the 6 o'clock Yellow Valve Box and Sextant 5 is 2.4 kW. The expected shield load for RHIC is approximately twelve times the above stated value of 2.4 kW and equals 30 kW. The baseline design shield load for RHIC is 37 kW.

The pressure drop in the heat shield between the 6 o'clock and 4 o'clock Valve Box was 0.8 atm at 210 g/s. For RHIC, the designed shield flow is 150 g/s per ring. The expected pressure drop in a RHIC ring will be about 2.4 atm as extrapolated from the flow rate and the length.

Recent experience on the RHIC ring[8] suggests both the heat load and the pressure drop in the shield are lower than that given in this paper. Heat load values for RHIC will be revised and reported in the near future.

RESPONSE TO MAGNET QUENCH

The maximum pressure and the rate of pressure rise in the helium cooling system after a magnet quench[7] is of great interest and importance to the operation of the particle accelerator. For this purpose, electric heat was introduced to a magnet near the 3rd recooler to initiate a quench. A total of four quenches were initiated when the magnets were powered to about 5,500 ampere. Each time, a magnetic stored energy of approximately 400 kilo-joules was released into the cooling helium. In all cases, the loop pressure increases by less than one atmosphere. The small pressure increase is due to the large helium inventory in the loop and ample flow area throughout the magnet cooling loop.

The maximum temperature after a magnet quench, as observed at the recooler heat exchanger, is about 7 K. In all cases the heat release vaporized all liquid helium in the recooler immediately downstream of the quenched magnet. The heat also vaporized most liquid in the subsequent recooler, and in one case the third recooler, downstream of the quenched magnet. The recovery time after a magnet quench is approximately one hour which is the duration for helium to travel a distance of two recoolers.

SUMMARY

The pressure drop and heat load in Sextant 5 were reported. The pressure drops are dominated by the 55-watt recoolers and the magnets. Most heat loads are dominated by end effects and regions with penetrations, and not from magnets with standard interconnects. The pressure drops are smaller than the RHIC design allowance, and proper circulation of helium flow in the ring is expected. The heat load is slightly greater than the design allowance, but the RHIC refrigeration system can comfortably handle it. The system is versatile and can be operated under unexpected situations such as the Triplet with one side at room temperature or the occurrence of thermal acoustic oscillation in the relief line. The sextant can be operated without liquid in some of the 55-watt recoolers. Quench of a few magnets in RHIC will have negligible effects on the cooling helium, and the system can be recovered promptly.

ACKNOWLEDGEMENT

The authors would like to thank G. Ganetis, R. Grandinetti, H. C. Hseuh, E. Killian, G. McIntyre and all personnel in the RHIC cryogenic Section for the help they provide.

REFERENCES

1. J. Sondericker, et al., Performance and Operating Experience of the RHIC First Sextant Test, "Advances in Cryogenic Engineering", Vol. 43, p237, Plenum Press, N.Y. 1997.
2. K. C. Wu, et al., Status, Operation and Performance of the RHIC Helium Refrigerator, "Advances in Cryogenic Engineering", Vol. 43, p. 483, Plenum Press, N.Y. 1997.
3. M. Iarocci, et al., RHIC 25 kW Refrigerator and Distribution System, Construction, Testing and Initial Operating Experience, "Advances in Cryogenic Engineering", Vol. 43, p. 499, Plenum Press, N.Y. 1997.
4. A. Nicoletti, et al., Performance of RHIC 50 Watt Recoolers, "Advances in Cryogenic Engineering", Vol. 43, p. 517, Plenum Press, N.Y. 1997.
5. E. C. Qumiby, et al., VJR/VJRR Design, Construction, Installation and Performance, "Advances in Cryogenic Engineering ", Vol. 43, p. 531, Plenum Press, N.Y. 1997.
6. R. Grandinetti, et al., Design Considerations and Experience of the RHIC Dual Magnet Cryostat Installations, "Advances in Cryogenic Engineering", Vol. 43, p. 365, Plenum Press, N.Y. 1997.
7. K. C. Wu, Thermal Characteristics of the MAGCOOL Cryogenic System After Quenches of RHIC Dipoles, "Advances in Cryogenic Engineering", Vol. 39, p356, Plenum Press, N.Y. 1993.
8. M. Iarocci, et al., RHIC Accelerator Commissioning, Cryogenic Tests and Initial Operating Experience of the 25 kW Refrigerator and Distribution System, paper presented in the 1997 CEC.

THE TESLA TEST FACILITY (TTF) CRYOMODULE: A SUMMARY OF WORK TO DATE

J. G. Weisend II,[1] C. Pagani,[2] R. Bandelmann,[1] D. Barni,[2] A. Bosotti,[2] G. Grygiel,[1] R. Lange,[1] P. Pierini,[2] B. Petersen,[1] D. Sellmann,[1] and S. Wolff[1]

[1]Deutsches Elektronen-Sychrotron, DESY
Notkestrasse 85, 22607 Hamburg, Germany

[2]INFN Milano- LASA,
Via F.lli Cervi 201, I-20090 Segrate (MI), Italy

ABSTRACT

The proposed TeV Superconducting Linear Accelerator (TESLA) is a 30 km long electron/positron collider. The cryomodule is one of the principal building blocks of TESLA. The cryomodule contains the superconducting RF cavities that accelerate the beam, the superconducting focusing quadrupoles as well as the associated cryogenic piping and thermal shielding. The cryomodules must meet strict requirements for alignment, heat leak vibration and cost. Approximately 2500 of these cryomodules are required for TESLA 500. Since 1997, three prototype cryomodules (of two different designs) have been built and tested in the TESLA Test Facility Linac. This paper sums up the design, construction and operating experience with these prototypes. Measurements of alignment, heat leak, and vibration of both designs are reported. The installation and performance of the new style thermal shields on the second prototype design are compared to the shields on the first prototype, as well as to computer model predictions. Future cryomodule designs, allowing for fixed power couplers, are also discussed.

INTRODUCTION

An important aspect of the ongoing TESLA[1] project is the development of the cryomodules. Since 1997, a total of three cryomodules representing two different designs have been built and tested in the TESLA Test Facility Linac. The purpose of this paper is to summarize the experience with these prototypes and indicate the future direction of TESLA cryomodule design.

Role of the Cryomodules in TESLA

The proposed TeV Superconducting Linear Accelerator (TESLA) is an electron/ positron collider roughly 30 km long. The accelerating portion of the machine is built up of cryomodules. Each module contains 8 superconducting niobium cavities cooled to 2 K. Many modules also contain a superconducting magnet package consisting of quadrupoles and steering dipoles. The modules provide support, alignment and thermal shielding for the cavities and magnets as well as feed throughs for the RF power and instrumentation. The cryomodule is the fundamental building block for the TESLA machine. Approximately 2500 cryomodules are required for TESLA. These modules must meet strict requirements for alignment, heat leak, vibration and cost.

Cryomodule Requirements

Alignment. In order for TESLA to function properly as both a collider and a FEL driver, the cavities and magnets must be aligned to within certain tolerances. These tolerances must be maintained throughout transport, vacuum pumping and thermal cycling. For reasons of cost and complexity, there is no plan to allow adjustment of individual cavities once the module is assembled. Table 1 lists the alignment tolerances for the TESLA cryomodules. The axial tolerance is parallel to the accelerated beam while the transverse tolerance is in the plane perpendicular to the beam.

Static Heat Leak. As a superconducting RF device, the majority of the cryogenic load in TESLA comes from the RF power. However, the size of TESLA dictates that the static heat leak into the cryomodules also be kept as small as reasonable. Table 2 gives the expected static heat leaks at the 2 K, 4.5 K and 70 K levels.

Movable Couplers. The fixed point of the cryomodule is at the center. Thus the ends of the 12 m long module move 15 mm towards the center during cool down. This leaves the designer with a choice. Either fix the power couplers (and thus the cavities) with respect to the 300 K vacuum vessel and let them move relative to the rest of the cold mass, or fix them to the cold mass and design the coupler to move relative to the 300 K vacuum vessel. Early on in the TESLA project it was decided to take the second option (that of movable couplers). Couplers[2] have been designed to meet these requirements. It may however be preferable to use fixed couplers and the third generation cryomodule design in principle permits the use of both fixed and movable couplers.

Vibration. Excessive vibration can affect the cavity tuning and resulting beam performance. The cryomodule should be designed to reduce the vibration to the cavities and magnets. The resonant frequency of the cryomodule should be far away from the 10 Hz repetition rate of the accelerator.

Table 1. Alignment tolerances for TESLA cryomodule

Component	Transverse tolerance	Axial tolerance	Rotational tolerance
Cavity	+/- 0.5 mm	+/- 5 mm	-
Quadrupole	+/- 0.1 mm	+/- 5 mm	0.1 mrad

Table 2. Predicted TESLA cryomodule static heat leaks

Temperature Level	Predicted Static Heat Leak (W)
70 K	76.8
4.5 K	13.9
2 K	2.8

Cost. The size of TESLA dictates that the cost of each cryomodule must be kept as low as possible. While there is not a specific goal, reducing the cost of the cryomodule has been a consideration from the beginning. This has led to the design of long cryomodules to minimize the number of expensive interconnects. Cost was the principal reason for changing the thermal shield design between cryomodule #1 and cryomodule #2.

CRYOMODULE DESIGN

First Prototype

The cryomodule is 12 m long and contains 8 superconducting RF cavities. All cryomodules built so far also contain a superconducting quadrupole and steering dipole package. The cavities are bath cooled by saturated He II at 2 K. The cavity baths are supplied by a parallel two-phase He II line. The heat deposited in the He II evaporates vapor at the surface of the two-phase line and the resulting helium gas is returned to the refrigerator via a large diameter (300 mm) gas return pipe. The gas return pipe and two-phase line are connected together at the end of each module. The superconducting magnets are cooled by a separate 4.5 K flow. The same flow also cools the 4.5 K thermal shields. A second set of shields is cooled by a separate 70 K flow. The cryomodule also contains multilayer superinsulation (MLI) blankets on the 4.5 K and 70 K thermal shields and a separate warm up/cool down pipe parallel to the cavity string. All process lines are contained within the cryomodule.

Figure 1 shows a cross section of the first prototype cryomodule. The outer diameter of the vacuum vessel is approximately 1.2 m. The cavities and quadrupole package are directly attached to the 300 mm gas return pipe. This pipe is itself attached to the 3 composite support posts arranged axially along the length of the module. The middle post is fixed. The 2 outer posts along with the 300 mm pipe, cavities, magnets and shields move towards the center of the cryomodule as a result of thermal contraction during cool down.

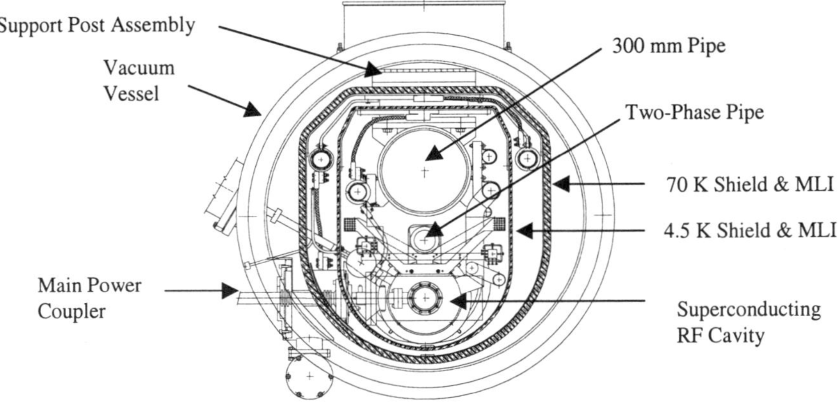

Figure 1. Cross section of 1st prototype cryomodule.

The gas return pipe is the structural backbone of the cryomodule and is key to the alignment of the cavities and quadrupole. The gas return pipe is aligned relative to the ideal beam axis via adjusting screws on the support posts. The cavities and quadrupole are then each aligned relative to the ideal beam axis via adjustment screws attaching them to the 300 mm pipe. Once this is accomplished, the cavity alignment is determined by the pipe alignment. As long as the gas return pipe is in the proper position relative to the beam axis, the cavities will be as well. This design only works properly if the cavities don't move relative to the 300 mm pipe once aligned, and the 300 mm pipe doesn't move in an unexpected manner once aligned relative to the beam axis. Upon cooling, thermal contraction will cause the 300 mm pipe, cavities and quadrupole to move vertically upwards by 1.8 mm relative to their warm position. This effect is allowed for in the alignment process.

In the first prototype cryomodule, the thermal radiation shields at 4.5 K and 70 K are constructed from aluminum and the cooling pipes are stainless steel. Large copper braids fastened between the shields and the cooling pipes cool the shields. Each thermal shield is divided into 18 pieces connected together by thread fasteners.

One example (cryomodule #1) of this first prototype design has been built.

Second Prototype

The second prototype design is virtually identical to the first prototype with the exception of the thermal radiation shields. There were two problems with the initial shield design. First, the copper braids used to cool the shields were bulky, expensive and not completely reliable. Second, the use of threaded fasteners to connect the shield pieces together was very expensive in both material and manpower. To solve these problems, it was decided to cool the shields with aluminum cooling pipes that were directly welded to the shields. Welding was also used instead of fasteners to connect the shield pieces together. In order to prevent excessive stress and deformation of the welded joints during thermal cycling a finger welding technique was used. Extensive FEA modeling was performed[3] to predict the behavior of the new shield design. The use of welded shields does place limits on the cryomodule cool down rate, but these limits are consistent with those imposed to prevent excess deformation of the gas return pipe during cool down.

Based on the experience of cryomodule #1, the mechanical tolerances of the second prototype design were adjusted. Those tolerances that were tighter than necessary were relaxed and the reference points of all the tolerances were adjusted to better match the manufacturing and assembly process.

The changes to the thermal shield design and the tolerances resulted in a cost savings of almost 50% for the cold mass (that is everything except the cavities, quadrupole and their related components) between the first and second prototype design. Two cryomodules (cryomodules #2 and #3) were built to this second prototype design.

CRYOMODULE ASSEMBLY

So far, 3 cryomodules have been assembled at DESY. The assembly process begins when the assembled string of eight cavities and one quadrupole is removed from the clean room. The two-phase He line connecting the cavities is then welded and leak checked. Next MLI, magnetic shields and temperature sensors are added to the cavities. The cold mass (provided by INFN) is then placed on the assembly tooling above the cavity string. The three support posts of the cold mass are adjusted so that the 300 mm gas return pipe is level straight and aligned with the reference beam axis. The cold mass is then carefully lowered onto the cavity string and attached to it. The resulting assembly is then lifted off the string support carts. A precision screw gear linkage system in the assembly tooling permits this to be done without damaging the cavities. Next the cavity tuner linkages and motors, remaining sensors, cables and magnetic shielding are installed. The system is now ready for final alignment.

The alignment is one of the most critical steps in the assembly. Each cavity and the quadrupole are aligned to the reference beam axis using adjustment screws that connect them to the cold mass. Experience showed that tightening the supports after the alignment in some cases altered the alignment. Thus, it is necessary to measure the alignments after the supports are tightened and realign the components as needed. In cryomodules #1 and #2 the alignment was not done to within the required tolerances to avoid damaging the cavities. However, in assembling cryomodule #3, all the components met the final alignment tolerances. The alignment takes 3 days.

After the alignment is finished, the beam tube vacuum is leak checked and the cryomodule is moved to another assembly position. Here the 4.5 K thermal shields, 4.5 K MLI, the 70 K thermal shields and 70 K MLI are installed. Temperature sensors are installed as needed on the thermal shields. Next, the 300 K vacuum vessel is slid over the cold mass and attached to it via the support posts. The support posts are then realigned to bring the gas return pipe (and thus the cavities and quadrupole) back into alignment with the ideal beam axis. Lastly, the warm parts of the main couplers, the quadrupole current leads and all the instrumentation feed throughs are installed. The cryomodule is now ready for installation in the linac.

The assembly of the welded thermal shields used in cryomodules #2 and #3 went very well. One team of two welders was able to install all the shields of a module in less than 2 days. No damage was done to any of the cryomodule components during the welding.

Eight weeks were needed to assemble cryomodule #1. Cryomodule #2 and #3 each required 6 weeks. Additional time savings should be possible in the future with elimination of the sensors now needed for research and some automation of the assembly process.

OPERATING EXPERIENCE

Introduction

Two cryomodules have been cooled down and operated with RF power and beam. Cryomodule #1 has been in operation since June 1997 and has undergone 5 thermal cycles. Cryomodule #2 was first cooled down in November of 1998 and has undergone 2 thermal cycles. The performance of the cryomodules was measured during these experiments.

Heat leak

The static heat leak to the 70 K and 4.5 K levels was found by measuring inlet and outlet temperatures and pressures, calculating the change in enthalpy and multiplying by the measured mass flow rate. Measurements with test heaters wrapped around the 70 K and 4.5 K cooling lines indicate that the error in these measurements is less than a few percent. The 2 K result was calculated by multiplying the latent heat of helium at 2 K times the measured vapor mass flow rate at the vacuum pumps; after subtracting out the amount of vapor generated during the J-T expansion at the inlet to the two-phase line.

Table 3 compares the predicted and measured static heat for the two cryomodules. Note that there are two sets of results given for cryomodule #1. When cryomodule #1 is operated alone, the end of the cryomodule is connected to a special cryostat known as the

Table 3. Static heat leak results

Temperature Level	Predicted Heat Leak (W)	Measured Heat Leak (W) Cryomodule #1 (alone)	Measured Heat Leak (W) Cryomodule #1 (with #2)	Measured Heat Leak (W) Cryomodule #2
70 K	76.8	90	81.5	77.9
4.5 K	13.9	23	15.9	13
2 K	2.8	6	5	4

Endcap. When cryomodules #1 and #2 are operated together the end of cryomodule #1 is connected to cryomodule #2 via 12 m long bypass transfer line. This transfer line is simpler in function than the original endcap and causes less heat leak to cryomodule #1. In particular, the old end cap contained optical windows which penetrated to the 2 K space and added heat to both the 4.5 K and 2 K levels. The absence of these windows is seen in the data. The heat leak in cryomodule #2 is smaller than that of cryomodule #1 and closer to the predicted values. This results from the smaller number of sensors in cryomodule #2 and a corresponding reduction in heat leak due to instrumentation cables.

Performance of new shield design

Figure 2 shows the measured temperature distributions on the 4.5 K and 20 K shields in cryomodule #2. The shields performed quite well. The temperature at each level is quite uniform and the temperature of the shield is within a Kelvin or two of the desired level. No damage was done to the shields during the controlled cool down and warm up of the cryomodule. The results here are consistent with the FEA predictions.[3] In particular, the temperatures on either side of the predicted thermal neutral line (180° from the cooling pipe) are equal as expected. The 8 K and 9 K temperatures at the end of the 4.5 K shield are most likely a result of the heat sinking of wires to the shield in that region.

The performance of the new shields in cryomodule #2 shows that the simpler, less costly design will meet the cryomodule requirements.

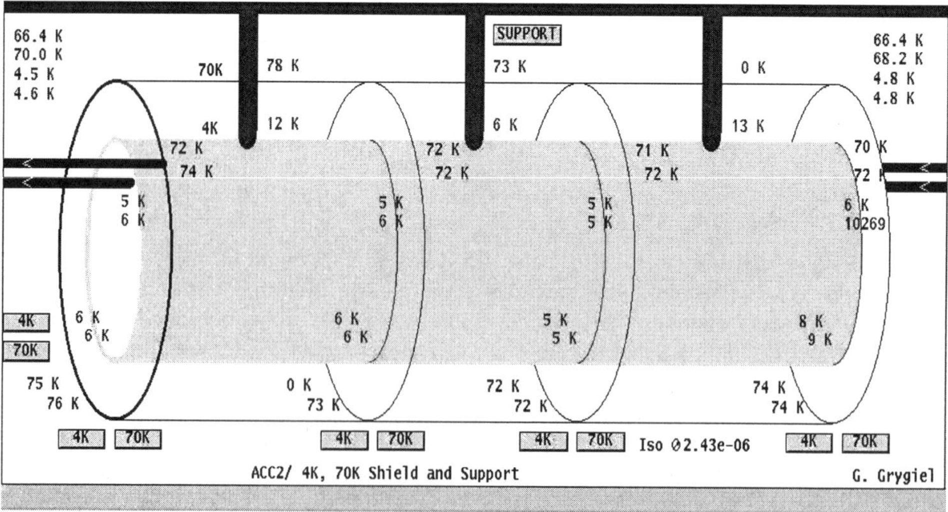

Figure 2. Measured temperature distributions on the 4.5 K & 70 K shields for cryomodule #2.

Alignment

The change in the alignment of the cavities and quadrupole is measured by a wire position monitor system[4] produced by INFN. This system permits the real time measurement of position of the components with a resolution of 50 microns. Figures 3 and 4 show the change in horizontal position for the cavities and quadrupole in cryomodules #1 and #2. Similar results are seen for the change in the vertical system. In these plots, the horizontal axis represents the position along the cavity string with the quadrupole as the rightmost data point. The remaining points represent cavity positions. Notice that the greatest deflection comes at either end of the cryomodules. This results from unbalanced forces generated on the 300 mm gas return pipe during vacuum pumping and cool down. These forces deflect the pipe and thus the attached cavities. Unfortunately, the quadrupole, which has the tightest alignment tolerance (+/- 0.2 mm), is located at the end of the cryomodule. Thus, it is always out of tolerance. The cavities, by contrast, are almost always within their tolerance of +/- 0.5 mm. Note also, that the 300 mm pipe does not return to its original position after being warmed up. Some residual forces remain. While the changes in the quadrupole position do not effect the performance of the TTF linac, they will be too large for proper performance of the TESLA 500 machine. One of the changes in the 3rd generation cryomodule is designed to solve the alignment problems.

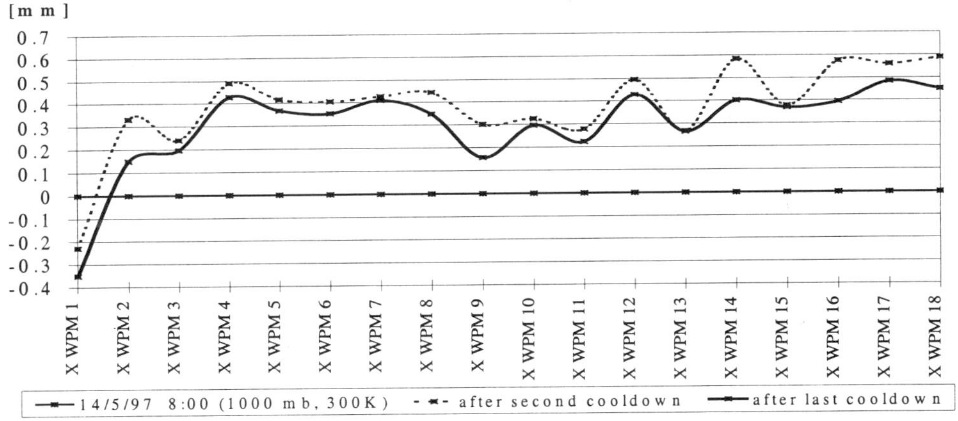

Figure 3. Horizontal displacements of the cavity string for Cryomodule #1 as measured by the WPM system.

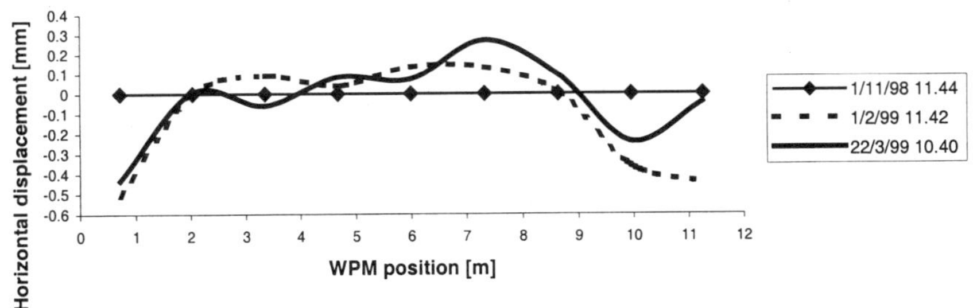

Figure 4. Horizontal displacements of the cavity string for Cryomodule #2 during cooldown (1/2/99) and warmup (22/3/99) compared to the original position (1/11/99).

Vibration

The best test of the cryomodule's resistance to vibration is operational. Excessive vibration will result in a detuning of the cavities off their resonant frequencies or perhaps in beam jitter and defocusing. None of these effects have been seen during several years of operation despite the cryomodules operating over a range of helium flow rates and heat loads. The cryomodules thus meets the vibration requirements.

THIRD GENERATION DESIGN

Based on the results from the first two generations of cryomodule design, a third generation design has been developed. This new design:

- Reduces the outer diameter of the cryomodule from 1.2 m to 0.98 m. This was accomplished by changing the connection of the cavities to the cold mass and by slightly moving the two-phase pipe. The smaller size frees up room in the proposed TESLA tunnel.
- Stiffens the 300 mm tube near the quadrupole by moving the support posts . Analysis shows that this should permit the quadrupole to stay within its alignment tolerance with unbalanced forces of up to 1000 N.
- Allows the use of rigid or semi-rigid main couplers by connecting the cavities to the cold mass by a series of roller bearings and a parallel Invar rod. This may result in significant cost savings for the main coupler as well as an easier and less expensive cryomodule assembly procedure.
- Uses the improved thermal shield design tested in cryomodules #2 and #3.

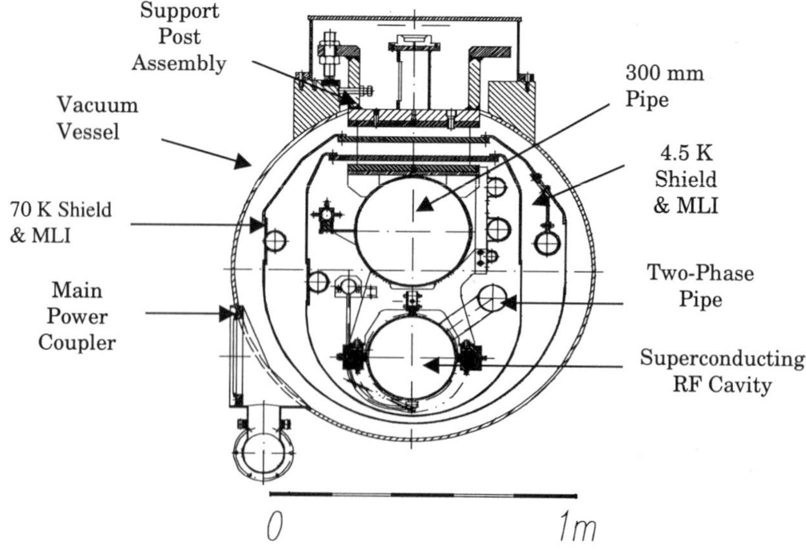

Figure 5. Cross section of the 3rd generation cryomodule design

Figure 5 shows a cross sectional view of the new design. The first of these new cryomodules should be built at the end of 1999. More details about the design are presented elsewhere in this conference.[5]

SUMMARY

In the past 5 years, significant progress has been made in the development of cryomodules for the TESLA project. 3 cryomodules representing 2 different designs have been constructed. By changing the thermal shield design and adjusting the mechanical tolerances, the second generation cold mass is significantly less expensive than the first. Two of these cryomodules (one of each design) have been tested in the TTF linac. The results show that the design meets or comes close to meeting the requirements for vibration and heat leak. The alignment of the quadrupole tends to exceed the tolerances. In order to solve this problem as well as to reduce the outer diameter of the cryomodule and allow for the possibility of fixed power couplers, a third generation design has been created.

REFERENCES

1. R. Brinkman, et al., eds. Conceptual design of a 500 GeV e+e- linear collider with integrated X ray laser facility, DESY Report, 1997–048 (1997).
2. K. Koepke, "Design of Power and HOM Couplers for TESLA", Adv. Cryo. Engr., Vol. 41a, P. Kittel, ed., Plenum Press, NY(1996), p. 877.
3. C. Pagani, D. Barni, M. Bonezzi, P. Pierini, J. G. Weisend II, "Design of the Thermal Shields for the New Improved Version of the TESLA Test Facility (TTF) Cryostat", Adv. Cryo. Engr., Vol. 43a, P. Kittel, ed., Plenum Press, NY (1998), p. 307.
4. D. Giove, et al., "A wire position monitor (WPM) system to control the cold mass movements inside the TTF cryomodule", in: "Proceedings of the 1997 Particle Accelerator Conference", M. Comyn et al. eds., IEEE, Piscantaway (1998). p. 3657.
5. C. Pagani, D. Barni, M. Bonezzi, J. G. Weisend II, "Further Improvements of the TTF Cryostat in View of the TESLA Collider", These proceedings.

THE LHC SUPERCONDUCTING RF SYSTEM

D. Boussard and T. Linnecar

CERN, Geneva, Switzerland

ABSTRACT

The European Laboratory for Particle Physics (CERN), the largest high energy physics laboratory worldwide, is constructing the Large Hadron Collider (LHC) in the existing 27 km circumference LEP (Large Electron Positron) collider tunnel. For the LHC, superconducting cavities, operating at 4.5 K, will provide the required acceleration field for ramping the beam energy up to 7 TeV and for keeping the colliding proton beams tightly bunched. Superconducting cavities were chosen, not only because of their high acceleration field leading to a small contribution to the machine impedance, but also because of their high stored energy which minimises the effects of periodic transient beam loading associated with the high beam intensity (0.5 A). There will be eight single-cell cavities per beam, each delivering 2 MV (5.3 MV/m) at 400 MHz. The cavities themselves are now being manufactured by industrial firms, using niobium on copper technology which gives full satisfaction at LEP. A complete cavity prototype assembly including cryostat, tuner and couplers is now being tested at CERN. In addition to a description of the LHC RF superconducting system, results on the prototype cavity assembly will be reported.

INTRODUCTION

The Large Hadron Collider (LHC), approved in December 1994, is now under construction at CERN, the European Laboratory for Particle Physics, near Geneva, Switzerland. This large-diameter circular particle accelerator will bring into collision intense beams of protons at high energy and luminosity, as well as heavy (Pb) ions at more modest luminosity.1 The two counter-rotating beams will be guided and focused by high-field superconducting magnets operating in pressurised superfluid helium, installed in the 26.7 km circumference tunnel of the existing LEP collider. The main parameters of the machine are given in Table 1. Two high-luminosity insertions are located at diametrically opposite straight sections, Point 1 (ATLAS) and Point 5 (CMS). A third experiment, optimised for heavy-ion collisions (ALICE) will be located at Point 2. A fourth experiment (LHCb) has now been approved and will be located at Point 8. The beams cross only at these four locations. Points 2 and 8 also contain the injection systems for the 450 GeV/c beams provided by the SPS. The other four long straight sections do not have beam crossings. Points 3 and 7 are practically identical and are used for collimation of the beam halo and Point 6 contains the beam abort system. Point 4 contains the RF systems which are independent for the two beams, the beam separation being increased from 194 mm in the regular arcs to 420 mm in order to provide the transverse space needed.

Table 1. Main parameters of the LHC

Parameter	Unit	Value
Collision energy	(TeV)	7.0
Dipole field	(T)	8.3
Distance between apertures	(mm)	194
Luminosity	($cm^{-2}s^{-1}$)	10^{34}
Beam-beam parameter		0.0032
Injection energy	(GeV)	450
Circulating current/beam	(A)	0.530
Bunch spacing	(ns)	24.95
Particles per bunch		1.1×10^{11}
Stored beam energy	(MJ)	332
Normalised transverse emittance	(μm)	3.75
R.m.s. bunch length	(m)	0.075
Beam lifetime	(h)	22
Luminosity lifetime	(h)	10
Energy loss per turn	(keV)	6.9
Total radiated power per beam	(kW)	3.7

In the CERN tradition and for the sake of economy, the LHC will re-use the chain of existing, older accelerators as injectors (PS machine as pre-injector to 26 GeV, SPS as main injector to 450 GeV). This imposes the choice of the RF frequency of the LHC (400.8 MHz), which must be a multiple of the SPS RF frequency (200.4 MHz) to permit fast transfer of long bunch trains.

Intrabeam scattering (or multiple Coulomb scattering) between particles during beam storage at 7 TeV results in an increase in transverse emittance that can rapidly degrade the luminosity, unless the six-dimensional phase space density is artificially diluted by increasing the longitudinal emittance. In the LHC, the latter will be increased from its injection value of 1 eVs to 2.5 eVs at collision energy. This determines the maximum RF voltage needed to ensure a bunch length much shorter than the region of minimum transverse beam size (low-β) at the interaction points. The design value of the RF voltage is 16 MV per beam, to provide an r.m.s. bunch length of 7.5 cm. As will be seen in the following, operating RF voltages significantly higher than the design value can be anticipated giving more freedom in the choice of parameters (longitudinal emittance or bunch length); this is especially attractive in the case of heavy ion collisions where the luminosity lifetime due to intrabeam scattering gets shorter.

The beam structure along the orbit is composed of trains of bunches (bunch spacing: 24.95 ns, i.e. ten RF periods) spaced by empty gaps of various lengths necessary to accommodate the rise times of the various injection kickers (200 ns, 1 μs) and the risetime of the beam dump kicker (3 μs). The effect of the beam gaps on the RF system is to generate a strong periodic transient beam loading which manifests itself as periodic modulations (mostly in phase) on the beam and on the RF voltage. For a single gap of length τ and no acceleration, the maximum phase modulation $\Delta\phi$ is given by:[2]

$$\Delta\phi = \frac{1}{2}\frac{R}{Q}\omega_o\frac{I_b}{V}\tau$$

where R/Q is the geometric cavity parameter, $\omega_0/2\pi$ the RF frequency, I_b the RF component of the beam current and V the cavity voltage. Phase modulation in turn entails a displacement of the collision points, which depends upon the position of the colliding bunches with respect to the beam gaps. This effect can be reduced to a negligible value in the LHC by choosing **superconducting (SC) cavities**, for which the critical parameter $(R/Q \times 1/V)$can be made one order of magnitude smaller than for normal-conducting copper cavities. Phase modulation could also be suppressed by brute force with any type of cavity, but at the expense of very high continuous RF power. Superconducting cavities offer instead the possibility of running the LHC with much less RF power (higher reliability, better robustness against power generator failures) and fewer units (small contribution to the overall machine impedance).

During the injection process at 450 GeV, each ring is filled by 12 consecutive injections and there is only partial filling of the ring; i.e. the beam gap is very large, as would be the phase modulation. Although the displacement of the crossing points is an irrelevant problem at injection, it was decided to compensate the phase modulation using the RF generator power in order to greatly simplify the injection procedure. As the RF voltage is low in this case (8 MV maximum as compared to 16 MV during storage), the necessary RF power ($VI_b/8$ = 1 MW per beam in the case of SC cavities) remains acceptable. The landing phase of a newly injected bunch is now fixed instead of being dependent on the previous injection history and beam intensity, as was originally envisaged, and much better injection tolerances can be anticipated.

Having different modes of operation for the cavities at injection (full beam compensation) and storage (minimum power), a **variable RF coupler** was necessary to optimise each mode. Moreover the coupling factor must be changed under power, during beam acceleration.

The nominal injection scheme, as described in the conceptual design, assumes small emittance bunches (0.63 eVs) from the SPS directly injected into 400 MHz buckets produced by an 8 MV per beam RF voltage. If the bunches leaving the SPS turn out to have a higher emittance (e.g. 1 eVs) this scheme no longer works and an alternative scenario is considered, in which the bunches would be captured by a 200 MHz copper cavity system (3 MV/beam), their injection phase oscillations damped and the whole beam transferred adiabatically to the 400 MHz RF system at the end of the injection process. In this scenario the 400 MHz SC RF system would be used as a harmonic of the 200 MHz RF system, to linearize the waveform and facilitate the damping of injection oscillations. Another operating mode of the SC cavities would then be to run them at very low voltage and opposite phase to the main 200 MHz voltage. In such a high beam-loading situation, full beam compensation is mandatory, leading again to the choice of a variable RF coupler for the SC cavities.

In all modes of operation the net energy delivered to the beam by the RF system is fairly small: at top energy the total synchrotron radiation power is only 3.7 kW per beam, and during the slow (20 min) ramping each cavity provides 32 kW to increase the beam energy. Contrary to that of lepton machines and high-current proton linacs, the installed RF power is not determined by the beam power, but by the need to maintain the specified RF voltage irrespective of the beam current.

ACCELERATING CAVITIES

Single-cell superconducting cavities with their individual RF power couplers are preferred to multicell cavities. This is to minimise the RF power requirements on each RF window and to simplify the design of the RF feedback circuit. Starting from the usual quasi-spherical shape which reduces the risk of multipacting, one can considerably decrease the R/Q of the fundamental mode by increasing the radius of the beam tube. In doing so the maximum voltage capability of the cavity at constant peak electric and magnetic fields is only slightly reduced. We have selected a beam tube diameter of 300 mm which gives $R/Q = 44\ \Omega$ and a nominal voltage of 2 MV/cell for 11.8 MV/m peak electric field and 27.3 mT peak magnetic field on the surface. Referred to a cavity length of $\lambda/2$ (as used for multicell cavities), this corresponds to an accelerating field of 5.3 MV/m. There will be eight single-cell cavities per beam in order to produce the nominal voltage of 16 MV during storage. The two groups of cavities for the two beams will be symmetrically arranged in straight section 4. The beam separation of 420 mm is enough to accommodate the vacuum tank radius of 360 mm, but not to bring the other beam outside the cryostat. Consequently the second beam tube is also cold.

The frequency of the first higher-order mode (HOM) (deflecting mode H_{111}) approaches that of the fundamental when the beam tube diameter is increased.[3] This is undesirable and can be corrected by reducing the length of the cavity cell (320 mm).

The eight single-cell cavities of a beam are arranged in two identical cryomodules. The four cavities of a cryomodule have a cell-to-cell distance of $3\lambda/2$ at 400 MHz (1122 mm) and are connected by the large diameter ($\varnothing = 300$ mm) beam tubes. The coupling between adjacent cells is negligible at the fundamental frequency, weak for the two lowest higher-order modes, but strong above the cut-off frequency of the 300 mm diameter tube. Above 700 MHz the location and magnitude of the HOMs differ markedly in the case of the coupled cavities as compared to that of a single cell with identical conical tapers. Moreover the peak and average values of the corresponding R/Qs are significantly more favourable for the coupled cavities than for a set of four individual cavities with conical tapers.

There is a so-called "trapped" mode at 1240 MHz which couples very weakly to the beam tube modes and which can be potentially dangerous.[4] Its Q_{ext} is critically dependent on the cell length ($Q_{ext} > 10^5$ for a cell length of 332 mm; $Q_{ext} = 300$ in the LHC case of a cell length of 320 mm).

The cavity technology is similar to that used successfully on a large scale for LEP2;[5] it is based on niobium film on copper cavities operating at 4.5 K. Bare cavities are produced by spinning and electron-beam welding and are coated with a thin (1 to 2 μm thickness) film of niobium by magnetron sputtering. The series production of 21 bare cavities is now being carried out by industry; 17 cavities have already been accepted at CERN. Their typical performance is displayed in Figure 1 together with the acceptance curve. It shows a large safety margin in the voltage capability of the cavities. The copper wall thickness results from a compromise between tuning force and mechanical stability against buckling. With a thickness of 2.8 to 3 mm, the cavity axial spring constant is about 20 kN/mm and the tuning sensitivity 240 kHz/mm.

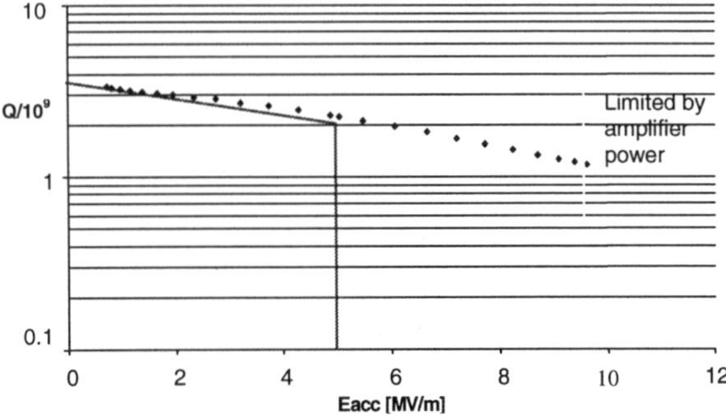

Figure 1. Typical cavity performance and acceptance curve.

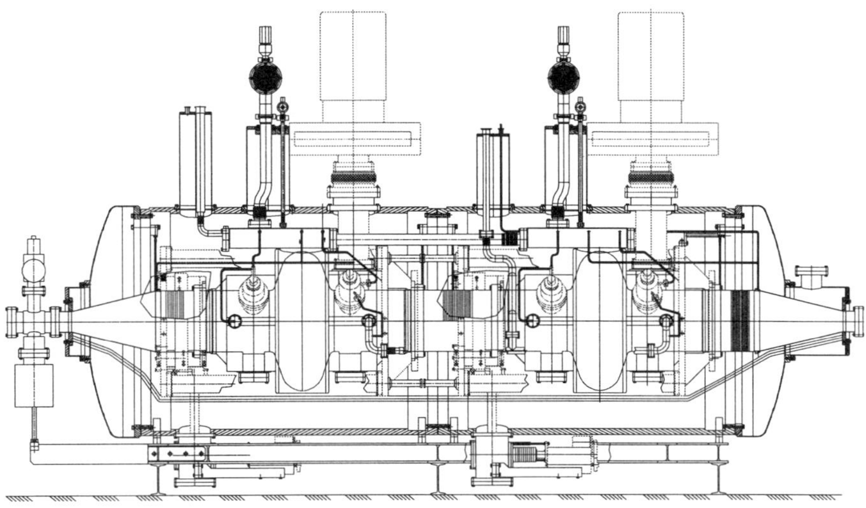

Figure 2. The prototype cryomodule with two cavities (a series cryomodule will have four cavities).

CRYOMODULES

Each cryomodule contains four single-cell cavities, each having its own helium tank. A prototype version having only two cavities was constructed (Figure 2) and tested.

A modular construction was adopted for the vacuum tanks of the cryomodule. Each tank is a stainless steel cylinder, without any welds, with four large lateral openings to permit easy access to the cavity. These openings are sealed by aluminium panels with long rubber rings. Each tank is joined to its neighbours or to the end flanges with Helicoflex® metallic joints (combined with rubber rings to allow vacuum testing before cavity assembly).

The four cavities are connected together with the wide bellows in a clean room; this assembly is then rolled inside the complete vacuum tank. The main couplers are mounted last, again in a clean room. It is also possible to disassemble and reinstall a single cavity in the middle of a cryomodule, without disassembling its neighbours.

The helium tank of each cavity is made of 2 mm thick stainless steel. Its cross-section is cylindrical around the cavity cell and octagonal at the location of the ports. The four helium tanks within a cryomodule are interconnected at the liquid and gas levels in such a way that a common helium feed and a common gas return are sufficient (Figure 2). Individual safety exhaust pipes with rupture disks are, however, provided for each cavity.

The helium supply at 4.5 K is provided from the main LHC cryogenic distribution line (QRL).[6] The line is moved transversely in the LHC tunnel to accommodate the SC cavities' cross-section (**Figure 3**).

As in LEP, each cavity cradle is suspended inside the cryostat to allow for contraction during cooldown. The longitudinal fixed point corresponds to the main coupler position to avoid stresses on the double walled tube of the coupler. Neither a magnetic shield nor a heat shield is necessary. The vacuum tubes for the second beam are attached to the side of each cavity cradle and connected together with standard shielded bellows. The measured static losses of the prototype cryomodule (having only two cavities, no couplers and no second beam tube) amount to 25 W.

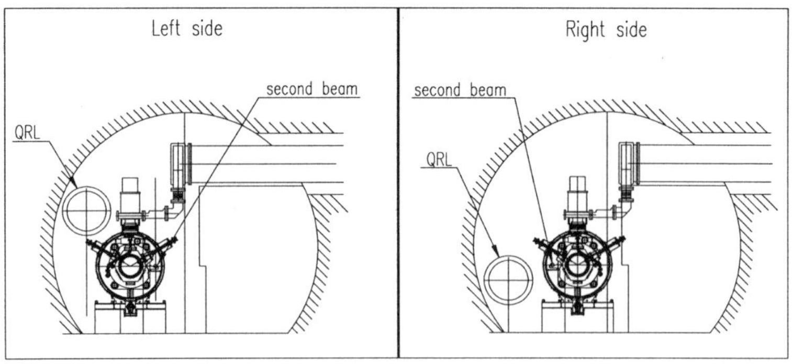

Figure 3. Tunnel cross-sections with cryoline and cavity.

TUNER

A purely mechanical tuner was chosen to provide the large tuning range at full speed required to compensate beam loading and thus minimise power requirements at injection. The large spring constant of the cavity (20 kN/mm) imposes a very rigid structure surrounding the cavity to take the return forces with little deformation. The stainless-steel (type 304) cavity cradle, with its two thick end plates joined by four columns, forms a structure free of harmful resonances and very rigid (< 0.08 mm axial shrinkage at a force of 20 kN). The cavity is always under tension, its end plate and the cradle end plate being pulled together via thin (1 mm thickness, 200 mm high) aluminium foils which act also as torsion shafts (Figure 4). A high-performance aluminium alloy (2219 T851) was chosen for these critical elements because of its excellent fatigue properties and elastic limit at low temperature (better than stainless steel). The maximum constraint for a 1 mm displacement of the cavity (170 MPa) is far from the elastic limit (500 MPa). The two torsion shafts (foils and shaft are made of a single piece machined by electro-erosion) are driven by long lever arms which provide a lever action (ratio 14:1) without sliding parts or backlash.

The axial force and movement at the extremities of the two lever arms inside the cold cradle are transferred to the outside of the cryomodule via two thin-walled stainless-steel cylinders acting as counter-rotating torsion shafts. The latter are driven by stainless-steel cables (∅ = 3 mm) providing again a transmission without friction or backlash.

Furthermore, this system allows displacement of the cradle during cooldown and provides a low heat conductance. A slightly different version of this tuner was successfully tested on the prototype cryomodule; the achieved tuning range and speed (limited by the stepping motor) were 180 kHz and 9 kHz/s respectively. The resolution is too small to be measurable.

In a hadron collider RF phase noise at the synchrotron frequency (f_s) is of great importance, as it may limit the beam lifetime. The contribution of the cavity microphonics, including the tuner, to the overall RF phase noise at f_s ($f_s = 20$ Hz in the LHC) must be evaluated. Preliminary measurements on the LHC prototype cryomodule (without RF couplers) indicate a microphonics phase noise density $S_\phi \cong 2 \cdot 10^{-8} \mathrm{rad}^2/\mathrm{Hz}$ which is adequate for the LHC. Note that the tuner in itself is too slow to compensate microphonics at f_s.

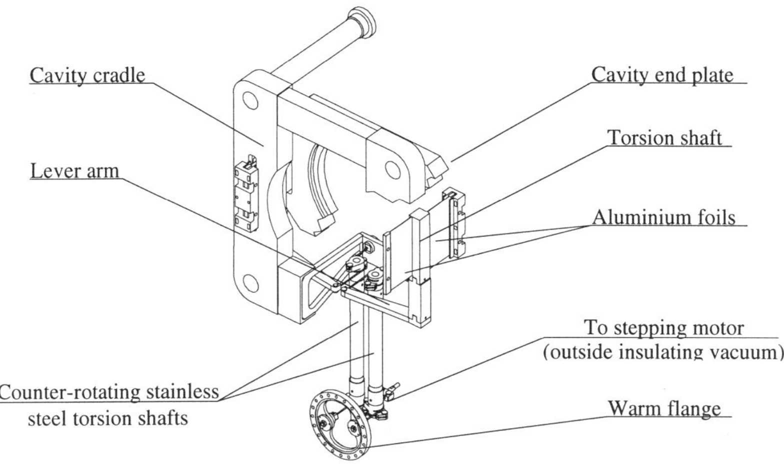

Figure 4. Tuner mechanism

VARIABLE POWER COUPLER

The LHC variable coupler (Figure 5) is an upgraded version of the LEP2 fixed coupler. The general layout and improvements of the latter have already been described in detail.[7,8] An open-ended 75 Ω coaxial line provides coupling to the cavity. The outer conductor (not represented in Figure 5) is made of copper-plated stainless-steel (double-walled) and cooled with 4.5 K helium gas, while the inner conductor (antenna) is a copper tube cooled by forced air. A cylindrical ceramic window, with solid copper rings brazed on its edges, is placed in the waveguide-to-coaxial transformer. A reduced height waveguide directly provides the matching to the coaxial line, avoiding the usual "doorknob". In order to suppress multipactor during operation, a d.c. bias of 3 kV is applied to the antenna, isolated from ground with a coaxial capacitor mounted in the waveguide. Air cooling is provided on the window and other critical elements of the coupler. A vacuum gauge and an electron pick-up antenna are located close to the window and are used for coupler conditioning and interlocks.

The antenna can be moved (60 mm stroke) by making use of bellows about $\lambda/4$ long. This changes the Q_{ext} of the cavity by a factor 20. A low impedance (7Ω) $\lambda/4$ line transformer brings the current in the bellows to low enough values, which do not then require copper plating of this stainless steel part. The displacement of the antenna is guided by a (motor driven) high precision device.

Two prototype LHC couplers have been manufactured, assembled and vacuum-tested. They are now mounted on the prototype cryomodule. Technical problems occurred during electron-beam welding of the ceramic window to the copper body and during titanium coating of the vacuum side of the ceramic, resulting in the breaking of some ceramic windows. Solutions to avoid these failures have been found. We also suffered from the bad quality of the OFE copper material used for the copper body, which developed vacuum leaks after baking out at 200°C for 24 h. In the future forged OFE copper will be used.

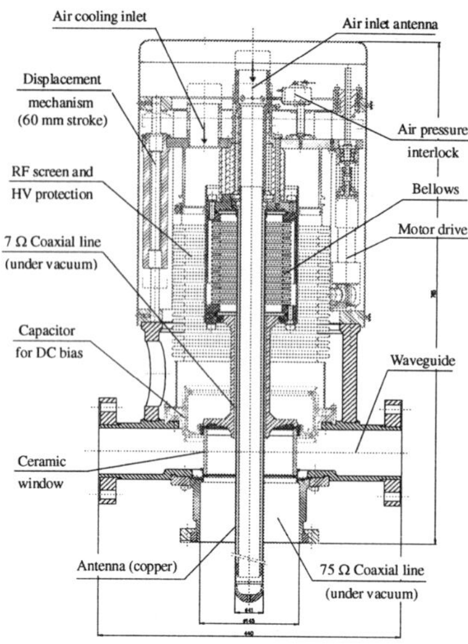

Figure 5. The LHC variable coupler

High-power RF tests at room temperature were done with two couplers, mounted horizontally on a 400 MHz copper test cavity. One coupler is connected to a 500 kW 400 MHz klystron,[9] via a circulator, the second to either a 1 MW load or to a mobile short circuit. In travelling-wave mode, and with a pressure limit of 2×10^{-7} mbar, the RF power could be ramped between 15 and 500 kW, crossing several multipactor levels, for which d.c. bias was effective. Below 15 kW multipactor occurred, with no electrons picked up on the antenna and no influence of d.c. bias. It is suspected that multipactor occurred inside the 7Ω λ/4 line, and therefore an additional bias on this is being studied. After conditioning of the low-power multipacting levels, the maximum power could be sustained for long periods (400 kW for 150 hours, 500 kW for 50 hours) with no sign of damage inside the coupler.

At full reflection, for any phase and any coupling the coupler sustains a 500 kW forward power (2 MW travelling-wave equivalent power) provided it is pulsed (50 ms on, duty cycle 10%) to avoid local overheating. After these tests the couplers were disassembled and thoroughly examined. No traces of damage were found on the surfaces exposed to RF. Before remounting the couplers on the cryomodule, all their components were properly cleaned (high-pressure water and alcohol rinsing), reassembled and conditioned again on the RF test bench.

HIGHER-ORDER-MODE COUPLERS

A special problem appears when the beam tube is enlarged to reduce the R/Q of the fundamental mode: the first dipole mode (TE_{111}) gets closer to the fundamental and its R/Q increases, making damping of this mode more difficult.[10] The solution adopted in high current e^+e^- machines is to let this mode propagate towards beam tube ferrite loads using fluted or widened beam tubes. In the LHC design with four cavities in a cryomodule warm ferrite loads are ruled out and a more conventional approach with two types of HOM coupler is used. This is suggested by the HOM spectrum which shows the first two dipole modes near 500 MHz which do not propagate in the 300 mm diameter beam tube and a cluster of monopole modes above 750 MHz.

The first type of HOM coupler damps the first two dipole modes. Coupling is with a loop perpendicular to the cavity axis (Figure 6a) (the modes have a strong longitudinal magnetic field component at the vacuum tube wall). Rejection of the fundamental mode is achieved in a classical way with an LC notch filter. By a careful design of the coupler elements, the equivalent circuit of the coupler (two coupled resonators) exhibits two resonances (Figure 7) located at the two HOM frequencies to be damped. Optimum damping is thus achieved for a given loop size. The Qs achieved with a single coupler per cavity are 200 for both the TE_{111} mode (500 MHz) and the TM_{110} mode (534 MHz).

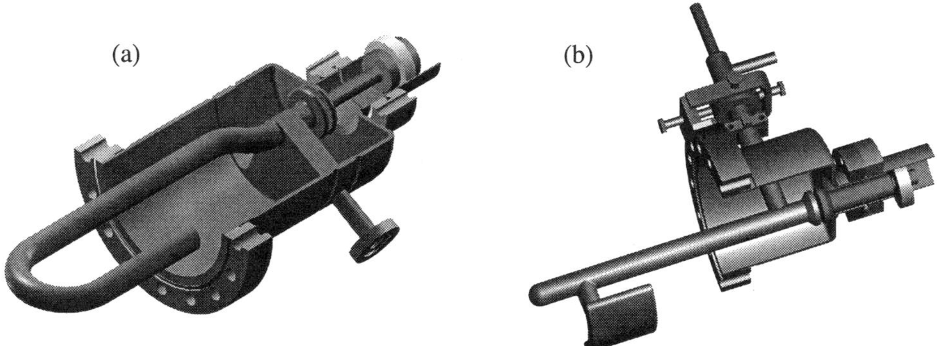

Figure 6. (a) Narrow band dipole mode HOM coupler; (b) Broad band HOM coupler

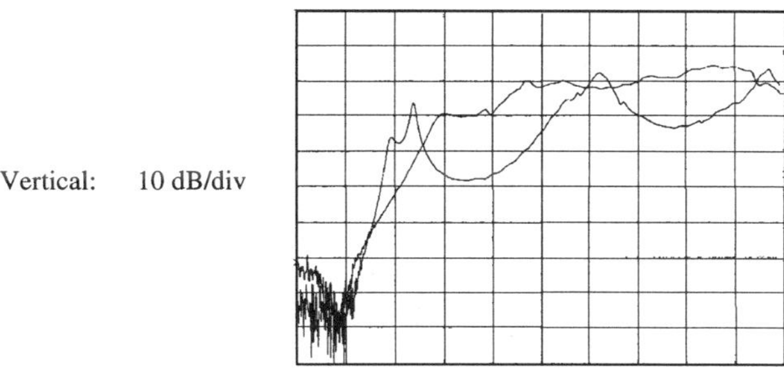

Figure 7. Frequency response of HOM couplers

There are also two broad-band HOM couplers per cavity cell. The coupling to these HOMs is predominantly electric with an open-ended antenna of about $\lambda/4$ length at the mid-band frequency (Figure 6b). The coupler includes a notch filter at the fundamental frequency and exhibits three resonances (Figure 7) to achieve the required bandwidth. The response peaks near the frequencies of the high R/Q mode TM_{011} (770 MHz) and the trapped mode TM_{012} (1240 MHz). On a model cavity external Qs of 500 have been measured for these modes, ensuring that even if mode excitation by the LHC beam is resonant the RF power coupled out will not exceed 500 W.

Both types of HOM coupler are made out of solid niobium. They are cooled by independent liquid-helium circuits. The RF connection from the cold HOM connector to the outside is made, as in LEP, with a thin-walled stainless-steel (copper-plated) 25 Ω coaxial line. The inner and outer conductors of the line are fitted with spring RF contacts at each end to allow for displacement during cooldown. This arrangement has been tested up to 800 W HOM power in the lab.

HIGH-POWER RF SYSTEM

Contrary to lepton machines, and high-current proton linacs, the RF power necessary to operate LHC is not determined by the power delivered to the beam, but rather by the need to control the RF voltage precisely under all transient beam loading conditions. From the analysis of the various situations (injection with or without a 200 MHz additional RF system, ramping, storage) it appears that a useful power of about 200 kW per cavity is adequate. With one klystron per cavity, the specified saturation power of the klystron should be in the range 250 to 300 kW, taking into account linear operation of the klystron, and circulator and waveguide losses. The choice of one klystron per cavity is justified technically: better control of microphonic noise, minimum perturbation in the case of a klystron trip. Klystrons of this power at a slightly different frequency are commercially available.

The klystrons will be installed vertically (4 m height available) in the klystron gallery running parallel to the machine tunnel at a distance of 10 m. They will be powered by the existing LEP HV supplies via HV feed boxes. The circulators are those used in LEP modified to operate at 400 MHz instead of 352 MHz. They are also located in the klystron gallery. Half-height waveguides connect the circulator output to the RF coupler via 900 mm diameter holes to be drilled between the two tunnels. Two waveguides for two cavities run in parallel in a common hole.

ACKNOWLEDGEMENTS

The authors wish to thank the many persons who contributed to the design and construction of the LHC SRF system.

REFERENCES

1. The LHC Study Group, The Large Hadron Collider, conceptual design, CERN/AC/95-05 (1995).
2. D. Boussard, RF Power Requirements for a High Intensity Proton Collider, PAC, San Francisco 1991.
3. E. Haebel, V. Rödel, The effect of the beam tube radius on higher-order modes in a pillbox RF cavity, CERN/SL/Note 93-17, 1993.
4. Z.T. Zhao, V. Rödel, E. Haebel, On trapped higher-order modes in a single-cell RF cavity, CERN SL-Note 96-80, 1996.
5 D. Boussard, Performance of the LEP2 SRF system, PAC, San Francisco 1991.
6 R. Trant, The cryogenic distribution line for LHC, this conference.
7. H.P. Kindermann et al., Status of RF power couplers for SC cavities at CERN, EPAC, Sitges (Spain), 1996.
8. H.P. Kindermann, M. Stirbet, RF power tests of LEP2 main couplers on a single-cell SC cavity, 8th Workshop on RF Superconductivity, Padova (Italy), 1997 and CERN/SL/97-64 RF.
9. H. Frischholz, W.R. Fowkes, C. Pearson, Design and construction of a 500 kW CW, 400 MHz klystron to be used as RF power source for LHC/RF component tests, PAC, Vancouver, 1997 and CERN/SL/97-39 RF.
10. E. Haebel et al., The higher-order mode dampers of the 400 MHz SC LHC cavities, 8th Workshop on RF superconductivity, Padova (Italy), 1997, and CERN SL/98-008 RF.

HIGH POWER CW SUPERCONDUCTING LINACS FOR NUCLEAR WASTE PROCESSING

Carlo Pagani and Paolo Pierini

INFN Milano-LASA and University of Milano
Via Fratelli Cervi, 201, I-20090 Segrate (MI), Italy

ABSTRACT

The interest of developing high intensity, CW, proton accelerators, with energies in excess of 1 GeV, to solve the problem of nuclear waste accumulation is growing worldwide. The large flux of fast spallation neutrons, produced through the proton beam and further multiplied in a subcritical reactor, could open the possibility of closing the fuel cycle in nuclear energy production, burning or transmuting the long lived radioactive nuclei. This new accelerator application asks for high overall plug efficiency and reliability, together with very low particle losses for hands-on maintenance. The superconducting RF technology seems to be the best solution, above 100-200 MeV, in order to design a cost effective machine, in terms of both capital and operational costs. In Europe a few R&D programs are being recently funded and the effort to unify them in a unique European project is under way, the aim being the construction of a large scale demonstration plant. In this paper we discuss the accelerator requirements as emerged so far, the design options which are being considered and the rationale behind them.

ACCELERATOR REQUIREMENTS FOR AN ADS SYSTEM

A critical part of any Accelerator Driven System (ADS) for waste transmutation[1] is the high power proton accelerator. The accelerator design call for a performance which is far from what has been currently achieved in existing machines. While the final choice of the beam parameters (final energy and average current) will come as a result of a complex optimization of the whole ADS system - in terms of performances, capital and operational costs - a few guidelines can be used for the optimization process.

The neutron yield per proton delivered by the beam depends mainly on the proton energy and from the target design. The total neutron flux needed by the ADS system is determined by the k_{eff} of the subcritical reactor at the core of the transmutation process and by the choice of the reactor thermal power. Moreover, the accelerator should be capable of the necessary beam adjustments required to compensate for the variations of the k_{eff} factor during the reactor operation, in order to keep constant the thermal reactor power.

While the number of neutrons per proton increases monotonically with the beam energy, the total neutron flux per Watt of incident beam reaches a nearly constant value above the energy of 1 GeV, as shown by the two curves displayed in Figure 1.

An upper limit on the beam energy is certainly given by the beam losses and the required shielding, while a lower limit is set by the efficiency of the spallation process and by the power lost in the beam window.

In addition to the needed performances in terms of beam power, the major challenges for the ADS driver are the required efficiency and its reliability and availability. In order to minimize the stresses in the subcritical system an availability close to 100% is needed, with additional requirements in terms of number and duration of beam interruptions (trips).

The high proton beam power - a minimum beam power in the range from 20 to 40 MW is foreseen for a demonstrator plant - sets very stringent requirements on the acceptable beam losses in the machine. The need to guarantee "hands-on" maintenance on the accelerator, i.e. the capability of accessing the accelerator without having to wait for the decay time of activated components, will significantly improve the system availability.

Another issue to be considered in the design of such an accelerator is its high capital costs and its efficiency — in terms of the ratio of the proton beam power to the electrical plug power needed by its operation.

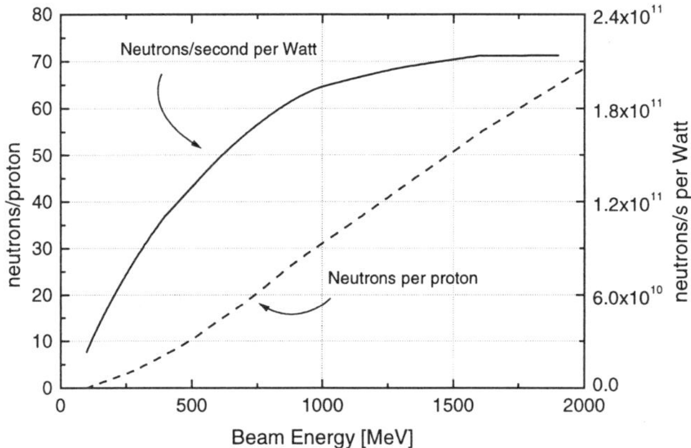

Figure 1. Neutron yield from the spallation process, as a function of the proton beam energy. The two curves are the total number of spallation neutrons per incident proton (left axis) and the total neutron flux per Watt of beam power (right axis).

THE EUROPEAN SCENARIO

For a few years, R&D activities have been carried on by several European groups for the investigation of a proton linac driver for an ADS system. The IPHI and ASH Projects[2-3] at CEA-CNRS in France and the TRASCO Program[4] in Italy are investigating a superconducting linac scenario for the ADS driver and have an experimental program intended to test the technological components of a similar machine. In the context of the 5th Framework Proposal Program to the EU, these activities are in the process of merging into a common group for the design of the accelerator subsystem.

The general considerations presented in this paper are based on this European scenario, although similar projects exists both in the USA and in Japan.

The conceptual design of the ADS accelerator will need to meet the specifications on beam current and energy arising from the design of the spallation target and subcritical core of the transmutation system, and address all the related reliability and availability issues.

The superconducting option

The use of the superconducting cavity technology seems to be most promising in terms of the needed accelerator plug power efficiency. For a CW high power accelerator, RF superconductivity is the most efficient technology for the minimization of the operational costs. Very low RF power is dissipated in the cavity walls and — even taking into account the fact that the power is deposited in the cryogenic bath — the acceleration gradient may be increased with respect to a normal-conducting machine.

The superconducting option has also a dramatic impact on the capital cost, by reducing the total length of the accelerator. The very low losses allow the operation at a much higher accelerating gradient than a normal conducting machine, where a limit to 1-2 MeV/m is rapidly imposed by the cooling requirements. Practical accelerating gradients for the superconducting option are in the 10 MV/m range.

The choice of parameters for the superconducting cavities that have been assumed so far are certainly conservative with respect to the outstanding performances of superconducting cavities (accelerating fields well above 25 MV/m) reached in various laboratories around the world (DESY, CEA, KEK). This margin on the cavity parameters adds an intrinsic advantage to the reliability of such a machine.

An additional merit of the superconducting cavities is the large bore, that highly reduces the risk of beam losses and the accelerator activation by halos of the high intensity proton beam.

THE MAIN ACCELERATOR COMPONENTS

All the accelerator components for an ADS driver rely on well-tested technologies that are currently used in existing accelerator complexes around the world. However, it is worth noticing that these machines are mainly motivated by high-energy physics experiments, where, generally, only the total integrated luminosity is of great concern. The demanding requirements on the reliability and availability of an ADS driver, and the constraints on beam trips frequency and duration, impose to verify the suitability of these technologies in this context.

The superconducting accelerator can be schematically divided in three sections:

1. **A proton source and intermediate energy section**, consisting of a 75-95 keV proton source, followed by a Radio-Frequency Quadrupole (RFQ) and either a conventional - warm - Drift Tube Linac (DTL) or a SC linac based on single gap reentrant cavities, up to approximately 80-100 MeV.
2. **The high energy (superconducting) linac**, a linac with elliptically shaped, reduced beta, superconducting cavities, divided in few (2 or 3) families on the basis of the desired machine energy.
3. **The beam transport system and beam expander**, needed to transport the high energy beam to the beam window and expand its transverse dimensions to provide a uniform density profile on the target.

A schematic layout of the SC linac is presented in Figure 2.

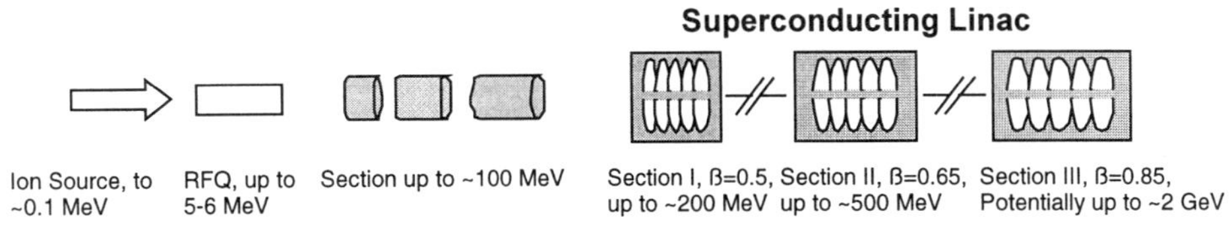

Figure 2. Schematic layout of a superconducting linac as a driver for an ADS transmutation system.

Proton Source and Linac Front End

The SILHI 2.45 GHz ECR source[5] built at CEA/CNRS as part of the IPHI Project[2] and operating in Saclay has successfully produced a 95 keV proton beam at its nominal current of 100 mA. The source has operated in test runs with an availability of 98%. One of the objectives of the TRASCO program is to build and test at INFN/LNS TRIPS[6] an ECR source similar to SILHI, designed for an 80 keV proton beam with a current greater than 33 mA. The main difference with respect to the SILHI design is the geometry of the extraction channel, modified in order to reduce the occurrence of sparks and increase the source availability. The commissioning of the TRIPS source is expected in mid 2000. These activities, and the existing sources operated at Chalk River and Los Alamos, demonstrate that the needed beam characteristics (in terms of current and beam quality) for the ADS driver are within the range of the present technologies. Further studies should address the availability and reliability issues imposed by the subcritical reactor design.

The proton beam out of the source is a continuous beam that needs to be bunched at (an harmonic of) the frequency used for the intermediate and high energy sections of the accelerator. This task, and an initial acceleration boost up to a few (5-6) MeV, is performed by a Radio-Frequency Quadrupole (RFQ). A 350 MHz prototype RFQ has been designed, built and commissioned for the LEDA accelerator in Los Alamos[7]. The LEDA RFQ aims at high transmission (~ 95%) at the nominal current of 100 mA. The IPHI Project at CEA has designed a high current (100 mA), high transmission (~ 98%), 5 MeV RFQ, that will proceed to the fabrication stage in the near future. As part of the TRASCO Program, INFN/LNL is designing a moderate current (30 mA) 352 MHz RFQ, aiming at 95% beam transmission, and a cold aluminum model[8] has been built to check tuning and matching criteria. The fabrication of the 3 sections of the 6.6 m RFQ is foreseen by mid 2002, to be tested with the beam provided by the TRIPS source. This moderate current design has looser constraints on fabrication tolerances, asking for a lower transmission, and is investigated as a possible simpler alternative to the very complex high current (100 mA) designs, because of the reduced beam currents envisaged for the ADS demonstrator plants.

While there is a wide international consensus on the choice of the combination ECR source-RFQ for energies up to 5-6 MeV, different ideas are being pursued for the intermediate energy section, up to approximately 100 MeV.

The APT design in Los Alamos[9] chose a combination of a Coupled Cavity Drift Tube Linac (CCDTL), followed by a Coupled Cavity Linac (CCL) up to 211 MeV; the IPHI/ASH design uses a classical Drift Tube Linac (DTL) structure, while the TRASCO design is following both the classical DTL approach and is investigating the possibility of extending the superconducting part of the accelerator down to low energies with an Independently phased Superconducting Linac (ISCL), using reentrant cavities[10]. All these three possible choices present technical challenges and an accurate comparison to assess the relative merits and demerits of the DTL - ISCL solutions will be carried out by the ongoing European collaboration.

The Superconducting Linac

Elliptically shaped superconducting cavities in the frequency range from 350 MHz to 1.3 GHz are successfully used in big electron accelerator complexes (in LEP200 at CERN, in CEBAF at TJNAF and in the TTF at DESY, to quote a few examples), either as bulk niobium or copper structures sputtered with a thin niobium film. In the last few years a strong effort has been aimed to push the cavity fabrication technology and chemical processing in order to reach high gradients. The TESLA Test Facility (TTF) is now routinely producing bulk niobium cavities with different European industries that, after the chemical treatments in DESY, operate at a gradient greater than 25 MV/m, with peak magnetic fields in excess of 100 mT[11].

The cheaper sputtering technology used for the 352 MHz LEP cavities was used for the industrial production of cavities operating reliably at 6 to 8 MV/m. In this case the gradients are limited by the lower peak magnetic fields allowed by the thin sputtered niobium film (of the order of 30 mT). The cavity sputtering technique is strongly dependent on the cavity geometry, and it has been verified to work well for the highest β=0.85 section, but cannot be applied "as is" for cavities with smaller β. Due to the very high incident angles of the sputtered niobium on the cavity walls a bad film quality has been achieved from the standard sputtering procedure, leading to a cavity performance well below that of the LEP cavities[12].

The choice of cavity fabrication technology (i.e. sputtering vs. bulk niobium) is also connected to the choice of the linac operating frequency. The low 352 MHz RF used for LEP200 would benefit from the existence of large infrastructures at CERN and from the existence of a great amount of ancillary components, like klystrons, couplers, tuners, RF control and cryostats. The reliability of these components has been widely proven during LEP operation. The drawback of this choice is first the limited accelerating gradient and second the fact that it is very expensive to duplicate the necessary infrastructures for the cavity treatment and measurement, that presently exist only at CERN, due to their very large size. As an additional note, the mechanical stability of the lowest β cavities at 352 MHz is poor and stiffening needs to be provided to keep the cavity from imploding under vacuum.

One of the advantages of 704 MHz (twice the LEP frequency) is that the cavities are 8 times smaller in volume with respect to the CERN cavities, and several infrastructures for cavity fabrication, treatment and measure exist or are being set up in Europe. The 704 MHz cavities show also a better mechanical stability. As a drawback, all the ancillary components listed above need to be developed and their reliability proven. The choice of the higher frequency (and higher fields due to the use of bulk niobium cavities) allows a reduction by a factor of two the superconducting linac length, thus greatly reducing the civil engineering costs for such a machine.

We may conclude the discussion on the frequency choice stating that the 352 MHz RF could be considered if the energy required for the beam is high enough to take advantages of the cavity sputtering technique (that is, to say, above 1 GeV) and a demonstrator machine needs to be built in a short time frame, in close collaboration with CERN and making use of its huge infrastructures. Conversely, the 704 MHz choice leaves more space for making use of the technological improvements obtained in the past years by the TTF/TESLA linear collider project, as it has been successfully proven by the single cell measurement at CEA/Saclay of a β=0.64 single cell cavity.

As a result of these considerations, the TRASCO/IPHI/ASH collaboration is focussing on the design and fabrication of several 704 MHz prototypes, both in Saclay and Milano, and aims at the definition of a common design for a prototypical machine based on the choice of 704 MHz. The activities of the collaboration will also investigate the possibility of reducing the cost of the bulk niobium cavities by fabricating them with thin niobium sheets and providing the necessary mechanical stability depositing a thick copper layer by means of a spraying technique (VPS, APS, VHOF or similar)[13].

It is necessary to note that, while a high acceleration gradient is desired in order to reduce the linac length, the superconducting portion of the ADS driver surely will not aim to reach the high gradients of TTF[11]. Longitudinal beam dynamic issues, together with the need of limiting either the power transferred by the coupler to the high current (20 to 40 mA) beam and the CW RF cryogenic losses, are suggesting a lower gradient.

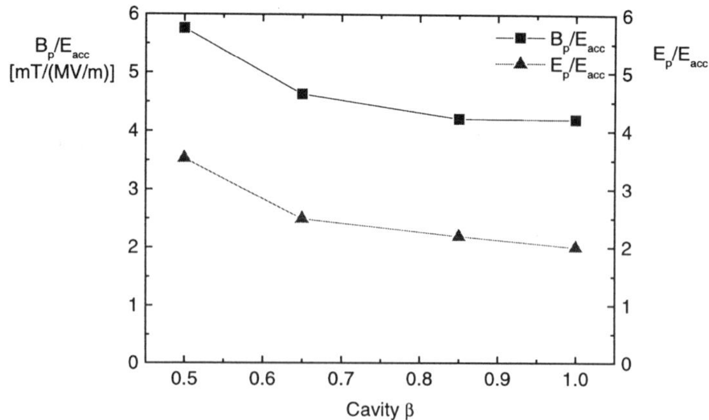

Figure 3. Plot of the ratio of the peak magnetic field on the surface with respect to the accelerating field (squares) and of the ration of the peak electric field on the surface with respect to the accelerating field (triangles), as a function of the cavity β value. The TRASCO and ASH cavities data are plotted for β<1. The points at β=1 refer to the TTF cavities.

A further constraint to the choice of the maximum cavity gradient comes from the necessity to adapt the multi-cell electron cavity geometries in order to satisfy the resonance condition at the reduced longitudinal velocity of the proton beam. The lower the proton velocity, the shorter needs to be the cell length ($L=\lambda_{rf}\beta/2$), and the worse is the ratio of the peak surface fields (both electric and magnetic) with respect to the accelerating field. This behavior is illustrated by Fig. 3, where we plot the ratio of the peak surface (electric and magnetic) fields with respect to the cavity accelerating field as a function of the cavity synchronous β value, for the TRASCO/ASH cavities. The points at β=1 are for the TESLA/TTF (electron) cavity and are included for reference.

On the basis of these considerations, and keeping a safety margin on the operating peak surface magnetic field by choosing a maximal design value in the range of 50-75 mT, the bulk niobium superconducting cavities can be run at a gradient around 10-15 MV/m.

The number of linac sections, and consequently the values of the synchronous velocity for the cavities depend on a series of factors. First, the multi-cell cavities have a velocity acceptance which decreases with the number of cells. Higher cell numbers are preferred in order to maximize the active length of the cavity with respect to its physical length. A wider velocity acceptance is preferred in order to minimize the number of different cavity types in the accelerator. Considering 5 cells per cavity and an energy range between 100 MeV to ~ 500 MeV, only two types of cavities are needed, loosing no more than 15% of acceleration performance due to the lack of proton synchronicity at the section edges.

In the case the subcritical reactor design will ask for an higher energy, an additional high β (~0.85) section has to be added in order to cover the rest of the linac with the same acceleration efficiency. This design option extends the maximum beam energy up to ~2 GeV and the cavity performance in the last section is close to that of an electron cavity.

Prototypical work on low β superconducting cavities is being carried out in different places in Europe. CEA/Saclay has exceeded the ASH cavity requirements[3] with 704 MHz, β=0.65 bulk niobium single cells, while the INFN Milano/Genova group, in collaboration with CERN, reached the TRASCO cavity specifications with a 5 cell niobium sputtered 352 MHz cavity. Both the APT Project at Los Alamos[9] and the JAERI transmutation project in Japan[14] are fabricating and testing single cell and multi-cell bulk niobium cavities up to this project's specifications.

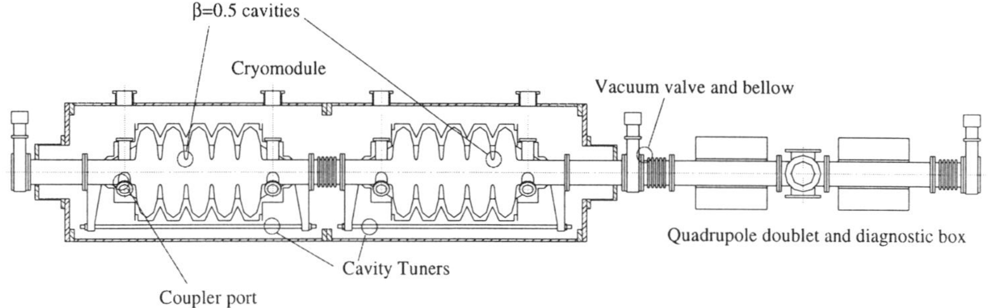

Figure 4. A sketch of the lattice period of the low β=0.5 section. The cryomodule contains two cavities, and the warm quadrupole doublet provides the transverse beam focussing. Space for beam diagnostics is in the drift between the two quadrupoles.

One other important component that influences the linac design is the main power coupler. The coupler should be able to provide from 200 to 300 kW to the cavities, with high reliability. CEA/Saclay, in collaboration with Los Alamos, is working on a prototype 300 kW, 700 MHz, coupler design for the ASH Project.

The superconducting cavities and the coupler will be assembled in cryomodules, and the necessary transverse focussing will be provided either by warm quadrupole doublets or by superconducting quadrupole doublets placed in the same cryomodule. A sketch of the resulting lattice period for the low β=0.5 section, with warm quadrupoles, is shown in Fig. 4. On the basis of the successful design of the TTF/TESLA cryomodules[15], the design of a cryomodule prototype for the TRASCO/ASH Project has begun.

BEAM DYNAMICS IN THE LINAC

The reliability and availability of the superconducting linac is deeply impacted by the possible beam losses on the beamline. In order to avoid the linac activation and allow for "hands-on" maintenance, a tight control of the particle loss should be achieved. It is therefore very important to base the linac on safe design criteria[16] in order to minimize any possible source of beam halo, and to allow for a large aperture in the high energy section. In this last respect the superconducting option allows to use much larger cavity apertures than the values allowed by a normal conducting linac design.

Beam mismatches can induce, via the non linear space charge forces, particle losses[17] that populate a "halo" region around the beam envelope. The particles in the halo may, in principle, be driven to very large amplitudes and eventually be lost at the linac aperture. This mechanism has been extensively studied both analytically and with detailed numerical simulation codes[18-21].

In all analyses, the halo growth induced by parametric resonance between the particle and beam envelope oscillations is found to be self-limiting[16], and no evidence of particles drifting to arbitrarily large extensions have been found. A safe linac design imposes the requirement of a continuous average focussing both in the longitudinal and transverse planes across the accelerator transitions (from the low energy to the high energy section and between the different β sections of the SC linac) and adiabatic variations in all accelerator sections. In addition to that, a large ratio of the beam aperture with respect to the rms beam size (from 20 to 50 in most designs) is required to allow the containment of possible mismatch-induced halo growth.

Other possible sources of beam halo formation have been investigated, such as the intrabeam scattering[22], but no beam losses are expected from the models.

CONCLUSIONS

We reviewed here the ongoing European R&D activities aiming at the design of a high intensity superconducting proton accelerator for nuclear waste transmutation, placing them in the context of similar projects in the USA and Japan.

The basis for the technical feasibility of such a machine exist and various alternatives are possible for most of the components. The final choice for the machine parameters and components will be made meeting the requirements, both in terms of performances and reliability, coming from the design of the subcritical reactor at the heart of the transmutation system, which will be deeply studied in the context of an Action of the 5th Framework Program of the European Union.

REFERENCES

1. C. Rubbia, J. Rubio, A Tentative Program Towards a Full Scale Energy Amplifier, CERN/LHC/96-11 (EET).
2. J.M. Lagniel et al., IPHI, the Saclay High-Intensity Proton Injector Source, Proceedings of the Particle Accelerator Conference 1997, Vancouver, Canada, p. 1120.
3. H. Safa et al., A Superconducting Proton Linear Accelerator for Waste Transmutation, Proceedings of the Acceleration Driven Transmutation Technology Conference 1999, Prague.
4. C Pagani, Activities on the High Current SC Proton Linac for the TRASCO Project, OECD/NEA Workshop on Utilization and Reliability of High Power Accelerators, Japan, 1998.
5. R. Gobin et al., New performances of the CW High-Intensity light-Ion Source SILHI, Proceedings of the European Particle Accelerator Conference 1998, Stockholm, Sweden.
6. G. Ciavola, S. Gammino, A Microwave Source for High Current of Protons, INFN/LNS Report, 21/10/96.
7. K.F. Johnson et al., Commissioning of the Low-Energy Demonstration Accelerator (LEDA) Radio-Frequency Quadrupole (RFQ), Proceedings of the Particle Accelerator Conference 1999, New York, USA.
8. G.V. Lamanna, A. Lombardi, A. Pisent, Test of a Radio-Frequency Quadrupole Cold Model, Proceedings of the Particle Accelerator Conference 1999, New York, USA.
9. J. Tooker, G. Lawrence, Overview of the APT Accelerator Design, Proceedings of the Particle Accelerator Conference 1999, New York, USA.
10. M. Comunian, A. Facco, A. Pisent, A 100 MeV Superconducting Proton Linac: Beam Dynamics Issues, Proceedings of the Particle Accelerator Conference 1999, New York, USA.
11. C. Pagani, Status and Perspectives of the SC Cavities for TESLA, in these Proceedings.
12. C Benvenuti et al., Production and Test of 352 MHz Niobium Sputtered Reduced Beta Cavities, Proceedings of the Eight Workshop on RF Superconductivity, Abano Terme, PD, Italy 1997, p. 1038.
13. S. Bousson et al., A New Fabrication and Stiffening Method of SRF Cavities, Proceedings of the European Particle Accelerator Conference 1998, Stockholm, Sweden.
14. N. Ouchi et al., Proton Linac Activities in JAERI, Proceedings of the Eighth Workshop on RF Superconductivity, Abano Terme, PD, Italy 1997.
15. C. Pagani, D. Barni, M. Bonezzi and J. G. Weisend II, Further Improvements of the TESLA TEST FACILITY (TTF) Cryostat in View of the TESLA Collider, in these Proceedings.
16. T.P. Wangler et al., Basis for Low Beam Loss in the High-Current APT Linac, Proceedings of the XIX International Linear Accelerator Conference 1998, Chicago, USA, p. 657.
17. A. Cucchetti et al., Simulation Studies of Emittance Growth in RMS Mismatched Beams, Proceedings of the Particle Accelerator Conference 1991, San Francisco, USA.
18. H. Takeda, J. Billen, Recent Developments in the Accelerator Design Code PARMILA, Proceedings of the XIX International Linear Accelerator Conference 1998, Chicago, USA.
19. R. Ryne et al., The U.S. DOE Grand Challenge in Computational Accelerator Physics, Proceedings of the XIX International Linear Accelerator Conference 1998, Chicago, USA.
20. M. Pabst et al., Progress on Intense Proton Beam Dynamics and Halo Formation, Proceedings of the European Particle Accelerator Conference 1998, Stockholm, Sweden.
21. P. Pierini et al., A Multigrid A Multigrid-Based Beam Dynamics Code for High Current Proton Linacs, Proceedings of the Particle Accelerator Conference 1999, New York, USA.
22. N. Pichoff, Intrabeam Scattering on Halo Formation, Proceedings of the Particle Accelerator Conference 1999, New York, USA.

DEVELOPMENT OF THE SUPERCONDUCTING CRAB CAVITY FOR KEKB

H. Nakai,[1] K. Hara,[1] K. Hosoyama,[1] A. Kabe,[1] Y. Kojima,[1] Y. Morita,[1] K. Okubo,[2] H. Hattori,[2] and M. Inoue[2]

[1]High Energy Accelerator Research Organization (KEK)
1-1 Oho, Tsukuba-shi, Ibaraki-ken, 305-0801 Japan

[2]Mitsubishi Heavy Industries, Ltd., Kobe Shipyard & Machinery Works
1-1 Wadasaki-cho 1-chome, Hyogo-ku, Kobe-shi, Hyogo-ken,
652-8585 Japan

ABSTRACT

The superconducting crab cavity is under R&D phase to be installed in KEKB. The test stand consisting of a cryogenic system and an RF measurement system has been constructed to measure the crab cavity performance. Recent experimental results of the crab cavity measured performance at 4.2 K and at 2.8 K are presented in this paper. It is revealed that the surface electric peak field could exceed its design value of 21 MV/m and reach up to 32 and 40 MV/m at respective temperature. It is found that dust on the cavity surface causes performance degradation of crab cavity, and this degradation can be recovered with high pressure ultra pure water rinsing. A simplified coaxial coupler was made to study the effect of the coaxial coupler on the cavity performance, and it is found that the heat loss generated in this coupler may degrade the total cavity performance. It is also described in this paper a fabrication method established to obtain high performance crab cavities.

INTRODUCTION

The crab crossing scheme will be adopted in KEKB to eliminate luminosity reduction due to geometrical effect and beam-beam instability caused by synchrotron-betatron coupling resonance in the finite crossing angle collision. Bunches of electrons and positrons moving toward the interaction point (IP) are tilted by time-dependent (508 MHz) transverse kicks in RF deflectors, i.e., crab cavities, and bunches can make head-on collision at IP. The R&D program for the KEKB superconducting crab cavity was started at KEK in 1994. As a basic design of the KEKB crab cavity, we have employed the squashed-cell cavity which was designed and studied at Cornell University in 1991 and 1992 for CESR-B under KEK-Cornell collaboration.[1] After the R&D of 1/3-scale 1.5 GHz squashed-cell superconducting Nb crab cavities, a full-scale 508 MHz squashed-cell superconducting Nb crab cavity has been designed and fabricated at KEK.

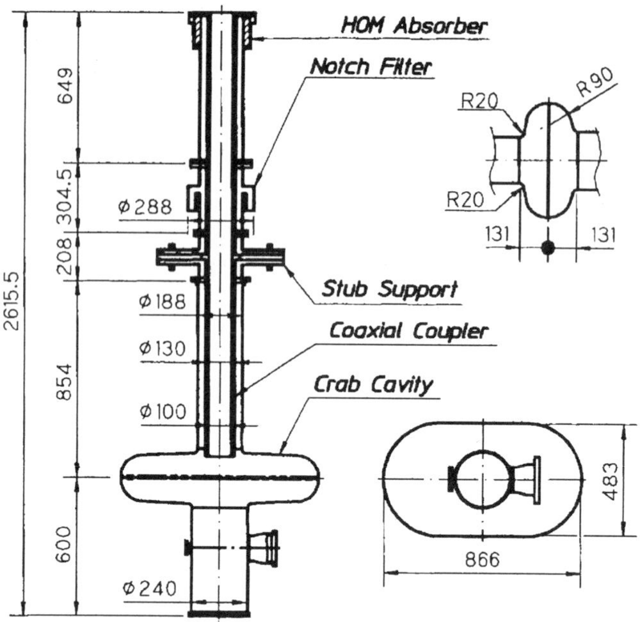

Figure 1. 508 MHz squashed-cell superconducting Nb crab cavity.

CRAB CAVITY

Design Parameters

The final design of the crab cavity has a coaxial coupler described below in detail, a stub support, a notch filter and a higher order mode (HOM) absorber, as shown in Figure 1. The size of the cavity is determined to make the frequency of the dominant crabbing mode in the cavity coincide with the KEKB frequency of 508 MHz. The crab cavity has a squashed-cell to make the resonant frequency of the other polarization mode higher than the cut-off frequency of the coaxial coupler. The stub support is required to fix the long coaxial coupler inside the beam pipe, while the notch filter prevents the crabbing mode from being absorbed by HOM.

Fabrication Method

A block diagram of fabrication procedure of crab cavities is shown in Figure 2. Half cells of the cavity are made of pure niobium (RRR≈200) sheet of 5 mm thickness with the hydro-forming method. These halves are welded together with beam pipes by electron beam. The inner surface of the cavity is polished with barrel polishers after grinding of the equator of the welded cell part. During these steps, we observe the inner surface of the cavity and beam pipes with a small CCD camera suspended inside the cavity. The surface skin of the cavity is removed by electro-polishing up to 100 μm. Chemicals used in electro-polishing are washed out with the high pressure ultra pure water rinsing (HPR) at the water pressure of 7.8 MPa (80 kgf/cm^2), and then annealing at 700 °C is applied to the cavity for 3 hours typically, depending on the out-gas rate. We carry out HPR again after annealing as the final surface preparation.

The cavity is then installed into the measurement setup with indium seals and evacuated with a turbo-molecular pump and an ion pump. After baking for a few days, the cavity is placed inside the cryostat and its performance at low temperature is measured.

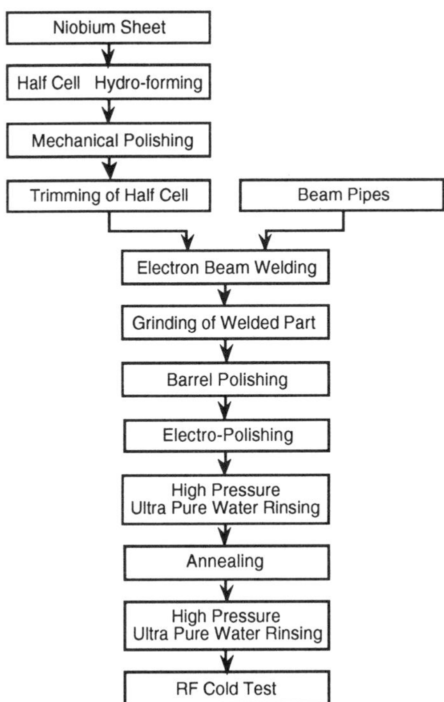

Figure 2. Block diagram of fabrication method and surface preparation.

Coaxial Coupler

A unique component of the crab cavity is a coaxial coupler inside the cavity as shown in Figure 1. This coupler is adopted to withdraw unnecessary modes from the cavity and to conduct them to the HOM absorber located at the end of the beam pipe. Since the effect of this coupler on the cavity performance has not been clear so far, we have constructed a simplified coaxial coupler made of pure niobium to study the effect, as shown in Figure 3. The surface of this coupler was prepared with electro-polishing and HPR. The coupler has a mechanism which can make the coupler be tilted inside the cavity to adjust the tip position of the coupler.

TEST STAND

Cryogenic System

A cryostat for the crab cavity, schematically shown in Figure 4, is directly connected to the 6.5 kW helium refrigerator system, and liquid helium is supplied to the cryostat from this refrigerator system. This refrigerator system was constructed for the TRISTAN superconducting accelerating cavities,[2 - 4] and is now adopted for the KEKB superconducting accelerating cavities.[5] The helium vessel of the cryostat is 3.5 m in height and 1 m in inner diameter. Inside the helium vessel, a magnetic shield made of Permalloy is installed to avoid the effect of the external magnetic field on the cavity performance. A liquid nitrogen shield is attached around the helium vessel to reduce the external heat load.

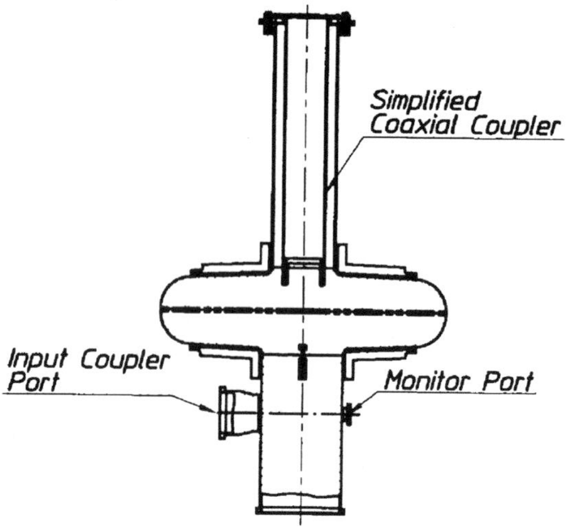

Figure 3. Simplified coaxial coupler inside crab cavity.

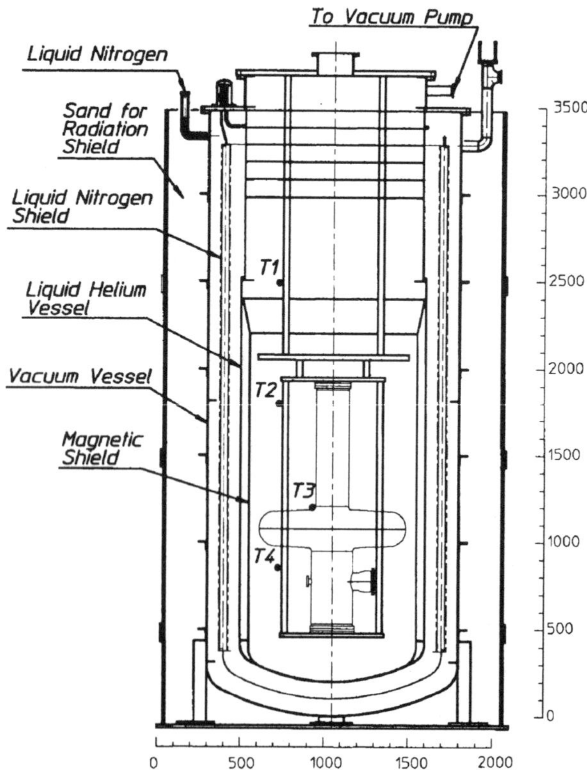

Figure 4. Cryostat for crab cavity.

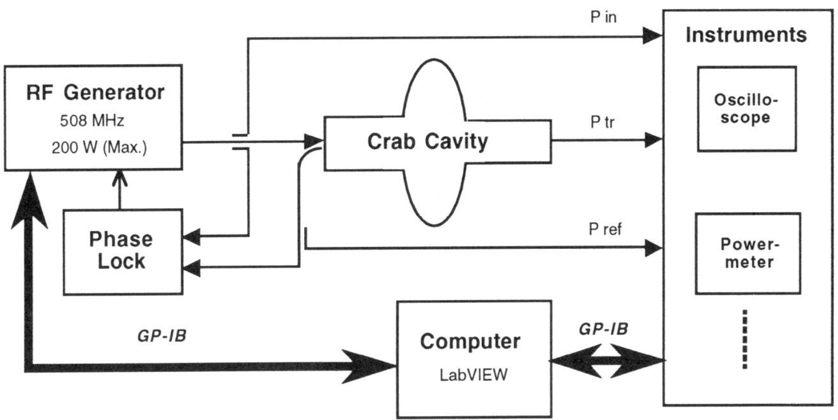

Figure 5. Conceptual block diagram of RF measurement system.

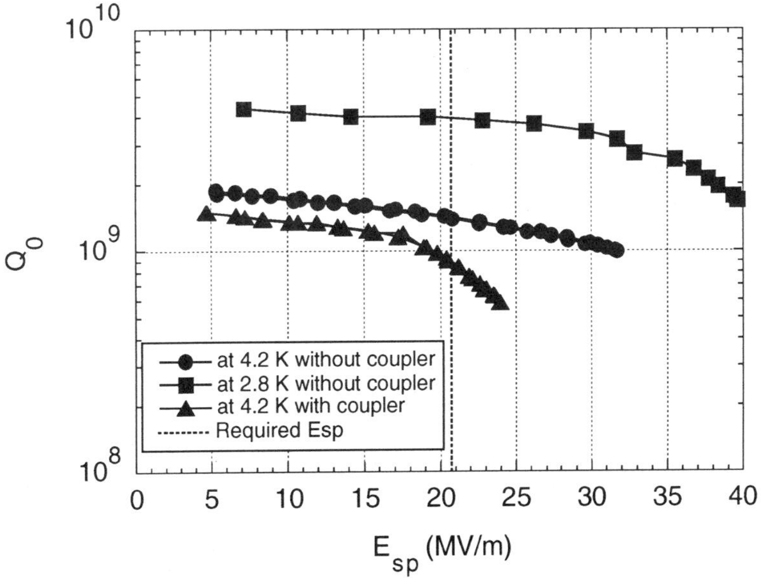

Figure 6. Cavity performance: ●; cavity without coaxial coupler at 4.2 K, ■; cavity without coaxial coupler at 2.8 K, and ▲; cavity with coaxial coupler at 4.2 K.

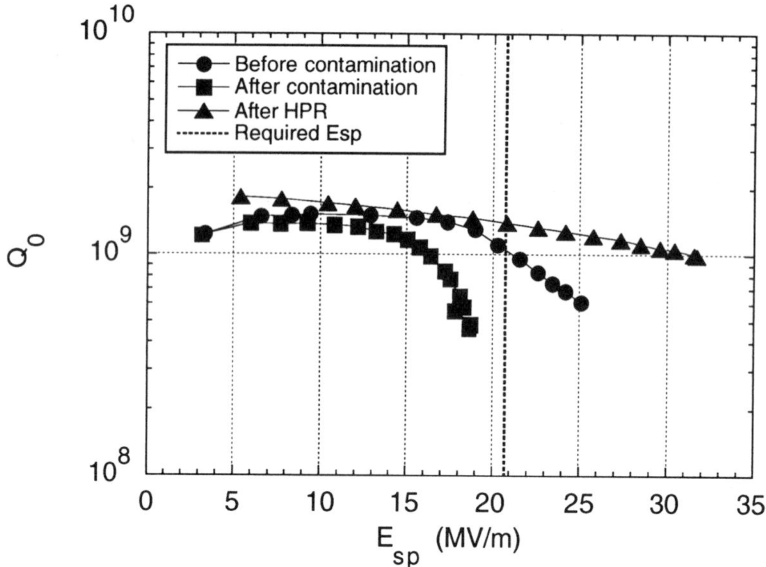

Figure 7. Degradation and recovery of cavity performance with high pressure ultra pure water rinsing: ●; cavity performance before contamination, ■; degraded cavity performance by contamination, and ▲; recovered and improved cavity performance with HPR.

RF Measurement System

An RF measurement system for the crab cavity has been assembled, and Figure 5 shows the conceptual block diagram of the system. The system is consisted of two major parts: a phase lock loop (PLL) and an RF power measurement. The RF power measurement part is controlled by a personal computer and the results are displayed on a CRT and stored in a hard disk of the computer. Once we measure a resonant frequency of the cavity, f_0, the decay time of the stored energy, τ, the incident power, P_{in}, the reflected power, P_{ref}, and the transmitted power, P_{tr}, we can calculate the quality factor, Q_0, and the surface electric peak field, E_{sp}, by measuring only P_{in}, P_{ref} and P_{tr} at various P_{in} to get the performance curves such as in Figure 6.

CAVITY PERFORMANCE

Cavity Performance at 4.2 K

To establish the fabrication method of the cavity and to check the cryogenic and the RF measurement systems, we have measured the cavity without any other parts such as a coaxial coupler at 4.2K, several times. The cavity performance has been getting better by the helium processing and the RF processing. The best performance of the cavity so far shows the highest E_{sp} of 32 MV/m as shown with solid circles in Figure 6. The cavity performance was greatly degraded when air was introduced into the cavity without any filters. However, the performance was again recovered with HPR for 1 hour and significantly better than that before contamination, as shown in Figure 7. From these results it is concluded that the principal cause of the performance degradation is dust on the cavity surface, hence HPR has the key role for the cavity performance.

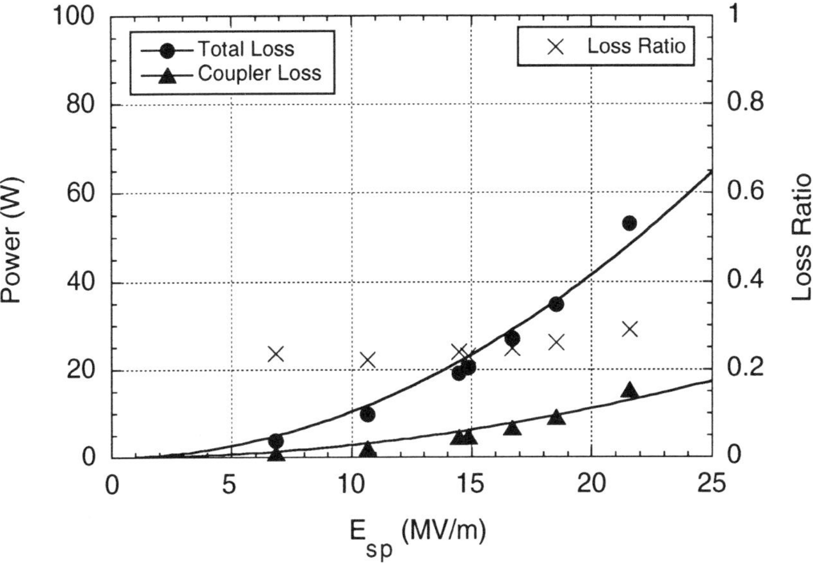

Figure 8. Variations of total power loss (●), coupler loss (▲) and their ratio (×) with surface electric peak field, E_{sp}.

Cavity Performance at Lower Temperature

It is easily supposed the lower is the ambient temperature of the cavity, the better is the cavity performance. Hence we have decreased the liquid helium temperature from the saturation temperature (4.2 K) to 2.8 K by reducing the helium pressure in the cryostat. It can be seen in Figure 6 (solid squares) that the cavity performance is greatly improved by the temperature reduction and the highest E_{sp} increased up to 40 MV/m.

Cavity Performance with Simplified Coaxial Coupler

The major issue of this crab cavity is the requirement of a coaxial coupler to withdraw unnecessary modes. An example of the cavity performance with the coupler is presented with solid triangles in Figure 6. The total performance is worse than that of the cavity only. This degradation may arise from the heat loss at the coupler, as shown in Figure 8. This heat loss ("coupler loss" in the figure) was calculated from the boil-off rate of the liquid helium inside the coaxial coupler. The heat generated in the coupler increases as a function of the square of E_{sp}, as shown with solid curves in the figure. In this case, the coupler loss is 20 to 30 % of the total loss in the cavity at various E_{sp}. This result means that if the coupler surface can be prepared more properly, the total performance would be better. Figure 9 shows the effect of the tip position of the coupler inside the cavity on the resonant frequency and quality factor of the cavity. In this case, the neutral position of the coupler inside the cavity happened to be the best position for the cavity performance.

CONCLUSION

The crab cavity R&D program has been progressing continuously by solving problems step by step. Our next step is to manufacture the coaxial coupler with the final design and to design the cryostat. A clean room for the crab cavity is now under construction, and an ultra pure water facility is installed at the clean room area.

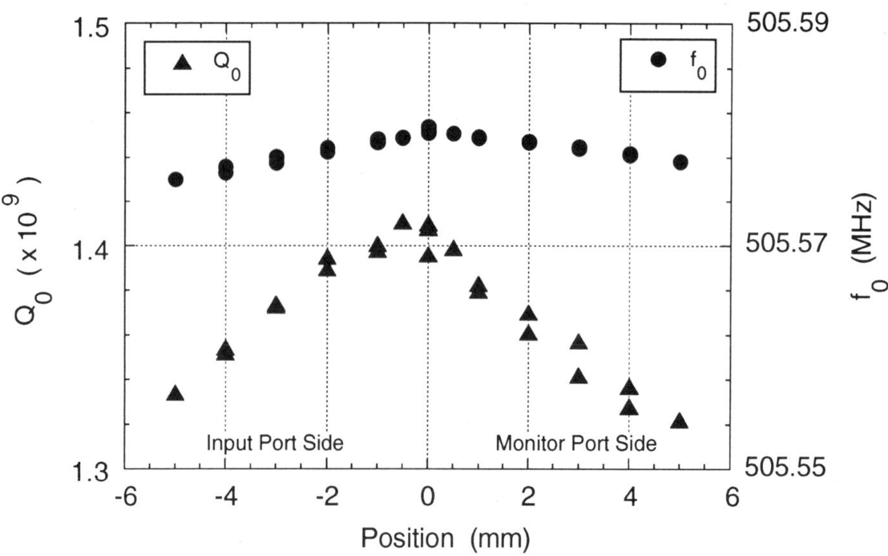

Figure 9. Variations of the resonant frequency, f_0, (●) and Q factor, Q_0, (▲) with the tip position of the simplified coaxial coupler measured from its neutral position inside the cavity.

REFERENCES

1. K. Akai, J. Kirchgessner, D. Moffat, H. Padamsee, J. Sears, T. Stowe, and M. Tigner, Crab Cavity for the B-Factories, *Int. J. Mod. Phys. A* 2B:757 (1993).
2. K. Hara, K. Hosoyama, Y. Kimura, Yuuji Kojima, Yuzo Kojima, S. Mitsunobu, M. Morimoto, H. Nakai, S. Noguchi, T. Ogitsu, Y. Sakamoto, T. Nakazato, S. Kawamura, K. Matsumoto, and S. Saito, Cryogenic system for TRISTAN superconducting RF cavity, in: "Advances in Cryogenic Engineering 33," Plenum Press, New York (1988), p. 615.
3. K. Hosoyama, K. Hara, A. Kabe, S. Kurokawa, Y. Kimura, Y. Kojima, Y. Kojima, S. Mitsunobu, M. Morimoto, H. Nakai, S. Noguchi, T. Ogitsu, Y. Sakamoto, S. Kawamura, K. Matsumoto and S. Saito, Cryogenic system for the TRISTAN superconducting RF cavities: performance test and present status, in: "Advances in Cryogenic Engineering 35," Plenum Press, New York (1990), p. 939.
4. K. Hosoyama, K. Hara, A. Kabe, Y. Kojima, T. Ogitsu, Y. Sakamoto, S. Kawamura, and K. Matsumoto, Cryogenic system for TRISTAN superconducting RF cavities: upgrading and present status, in: "Advances in Cryogenic Engineering 37," Plenum Press, New York (1992), p. 683.
5. K. Hosoyama, S. Mitsunobu, and T. Furuya, Design and performance of KEKB superconducting cavities and its cryogenic system, in: "Advances in Cryogenic Engineering 43," Plenum Press, New York (1988), p. 123.

SUPERCONDUCTING CAVITIES FOR THE KEK B FACTORY

S. Mitsunobu, K.Akai, E.Ezura, T.Furuya, K. Hara, K.Hosoyama,
A. Kabe, Y. Kojima, Y. Morita, H. Nakai and T.Tajima

KEK, High Energy Accelerator Research Organization
1-1 Oho, Tukuba, Ibaraki, Japan

ABSTRACT

KEKB is an asymmetric electron-positron collider at 8GeVx3.5GeV, which aims to provide electron-positron collisions at a center of mass energy of 10.58 GeV. For the high energy electron ring, 4 superconducting cavities were used to accelerate a high current. All 4 superconducting cavities reached more than 10 MV/m field gradient in the B factory tunnel. And 4 more superconducting cavities will be added and the current will be increased up to 1.1 A. The cavity is single cell and has large diameter beam pipe to damp beam induced higher order mode power. The superconducting cavities have been operated in the KEKB tunnel up to a high current of 0.51 A and high beam power of 380 kW/cavity. The current is not limited by the superconducting cavities.
KEKB uses finite angle crossing and may need SC crab cavities to increase the luminosity. A full scale Nb model crab cavity has been tested and exceeded required field.

INTRODUCTION

The KEKB, an asymmetric energy double-ring electron-positron collider for B-physics,

Advances in Cryogenic Engineering, Volume 45.
Edited by Shu *et al.*, Kluwer Academic / Plenum Publishers, 2000.

has been commissioned in December 1998. Two types of heavily-damped cavities are used in KEKB. The normal conducting cavity system (ARES) used in the low energy ring (LER) and the high energy ring (HER) and 4 superconducting cavities used in the HER. Four additional superconducting cavities will be install in HER next year. Fig.1 schematically shows the KEKB double ring accelerator systems. KEKB adapted a finite-angle crossing scheme of 2x11 mrad at the interaction point. The place of the superconducting crab cavities is also shown in Fig. 1. The machine parameters are shown in Table.1. The design current of HER is 1.1 A. The superconducting accelerating cavities will be operated at high current and high power condition.

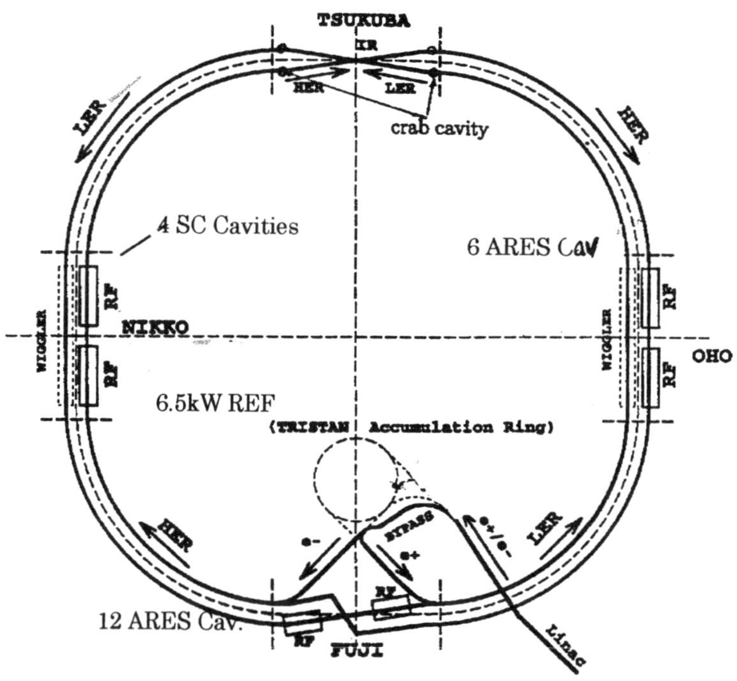

Figure 1. RF cavity layout of KEKB. 4 superconducting cavities have been installed in Nikko area. 4 superconducting crab cavities will be set at Tsukuba area. 6 ARES cavities have been set at Oho area. And 12 ARES cavities have been set at Fuji area. 8 more ARES cavity will be installed this summer. 4 more superconducting cavities will be installed in Nikko area.

Table 1. Parameters of KEKB

	LER	HER	
Beam Energy (GeV)	3.5	8	
Beam Current (A)	2.6	1.1	
Bunch Length (mm)	4	4	
Total RF Voltage (MV)	10	18	
Beam Power (MW)	4.5	4.0	
RF Frequency (MHz)	508.887	508.887	
Harmonic Number	5120	5120	
Cavity Type	ARES	SCC	ARES
Number of Cavities	20(12)	8(4)	12(6)
R/Q (Ω/cav)	14.8	93	14.3
Voltage (MV/cav)	0.5	1.5	0.5
Beam Power (kW/cav)	221	240	173
Wall loss (kW/cav)	154	-	154

SUPERCONDUCTING CAVITY

Based on the long term operation of 32 superconducting cavities in TRISTAN and the successful beam test of the prototype cavity at the TRISTAN Accumulation Ring (AR) reaching a maximum current of 0.57 A in 1996[1-3], we started construction of 4 superconducting cavity modules for KEKB HER. Figure 2 shows the whole module. The superconducting cavity has a single cell structure with large aperture beam pipes on both ends so that Higher Order Modes (HOM) can propagate out of the cavity and damped by ferrite absorbers bonded on the inner surface of beam pipe.

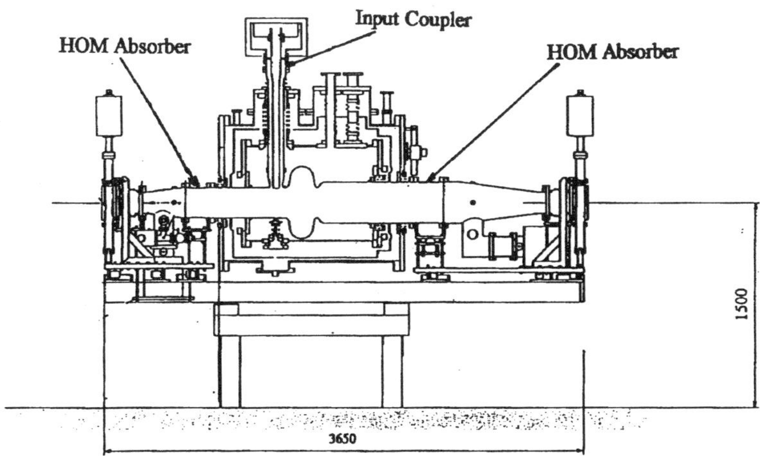

Figure 2. Superconducting cavity module for KEKB.

Before cooling down the cavities, we conditioned the input coupler up to 300 kW with full reflection condition, and up to 200 kW with DC bias voltage applied to the inner conductor up to $\pm$ 2kV. This conditioning decreased the secondary electron emission coefficients of the inner and outer conductors and the ceramic window to less than or nearly equal to one. During this room temperature conditioning, the desorbed gas was evacuated to the vacuum ion-pumps. The multipactor around the input coupler, which is known to induce break down of the superconducting cavity during high power beam operation, was strongly reduced. By doing this the cavity could be operated at high power without beam processing of the input coupler. Figure 3 shows the vacuum activity at a given power that resulted from at bias processing.

When we applied DC bias voltages to the inner conductor up to $\pm$ 2 kV, on the starting at a bias voltage of +100 V , we observed light emission near the window on the test stand.[1] As the bias was increased throughout the range, many out gassing events were observed, as shown in figure 3.

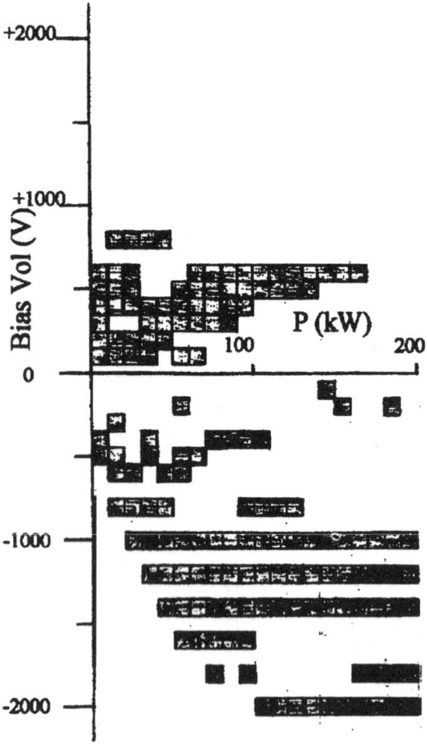

Figure 3 RF processing with bias voltage. Filled boxes show increasing vacuum pressure, which indicates multipacting is occurring.

The bias-type door-knob transitions were tested 450 kW transmission condition and 300 kW full reflection condition.

The bias voltage will be also used to suppress multipactoring in the input coupler after long term operation when multipactoring is known to reappear due to condensed gases on the outer conductor. The couplers handled RF power up to 380 kW to the beam with a beam current of 0.5 A. This is the highest power record for continuous operation in the world to date.

The 4 superconducting cavities installed in KEKB tunnel are shown in figure 4. Figure 5 shows the adjustable liquid helium transfer line with liquid nitrogen shield[4] and the electrically driven valve to vent helium gas if a cavity quench occurs and RF power couples to the quenched cavity. The cavities have a loaded Q value of 7×10^4 . A large amount of power might be dumped to the liquid helium bath, if the RF power is not turn down in the event of a cavity quench.

After the liquid helium vessel is filled to about 90 %, we try to raise the coupler power to 300 kW in an off-resonance condition, and raise the cavity voltage , Vc, up to 3 MV which corresponds to 12 MV/m or till breakdown(quench) occurs. Then we shift the phase of cavity up to $\pm 30^\circ$ so that the field profile in the coupler changes in order to check and condition the less-conditioned parts of the coupler. Usually, no or slight out gassing is observed during this phase change conditioning. Conditioning at low temperature might be a source of gas condensation at the input coupler and superconducting cavity, so for the stable operation of coupler, the room-temperature bias conditioning is important. A more detailed study for this problem is being done.

Figure 4. Four superconducting cavities installed in KEKB tunnel

Figure 5. Adjustable liquid helium transfer line between valve box (old TRISTAN main line) and superconducting cavity.

When we see some degradation on the attainable cavity field, we apply a partially-modulated pulse power conditioning. It takes a few hours at most to recover. This takes about one day.

Finally, we measure the Q_0 at a certain voltage, e.g. at 2.0 MV and 2.5 MV, with liquid helium consumption rate. This also takes about one day. Then we start beam operation.

The superconducting cavities were stably operated at 6 MV/m for more than 4 months. The maximum current was reached at 0.51 A. The maximum power transferred to beam is 380 kW with no difficulty. Each cavity experienced a few quenches (breakdown) but no dependence on beam intensities was observed. The vacuum level around the superconducting cavity is 10^{-7} Pa with beam. This good vacuum and room temperature RF coupler processing with bias voltage may contribute to the stable operation.

Figure 6 shows a Q-E curves of the cavities before and after beam operation, the cavity performances did not change.

The performance of the cavities measured after beam operation are shown in Table 2.

Table 2. Performance of the SC module after beam operation

Cavity	RA	RB	RC	RD
Q_0 at 2MV (x10^9)	2.1	2.0	1.3	1.4
Eacc,max(MV/m)	12.3	13.0	10.7	11.3
Vc/cavity(MV)	2.9	3.2	2.6	2.8

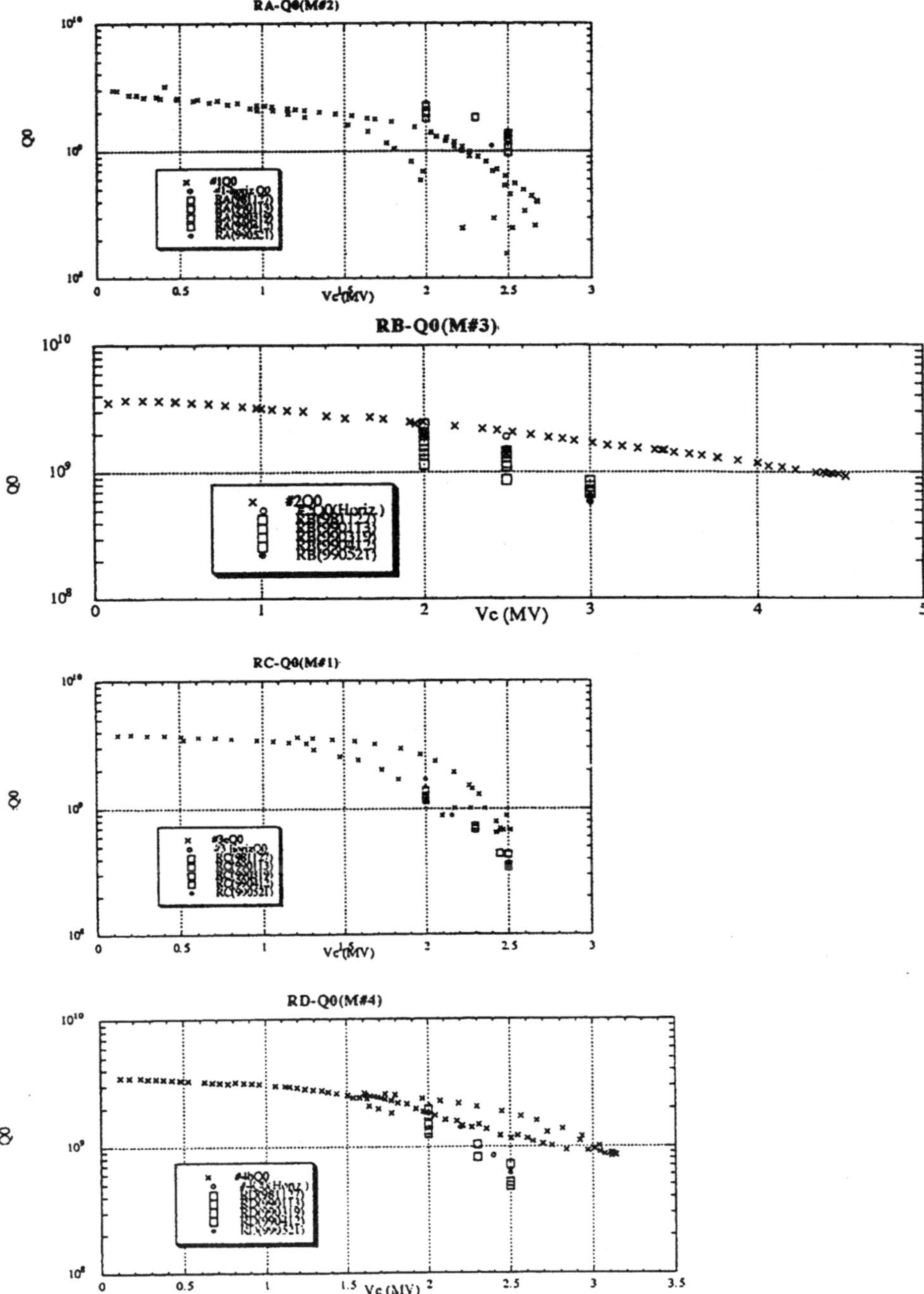

Figure 6. Q-E curves of the four superconducting cavities for KEKB.

During a shutdown period for the installation of the BELLE detector, the RB cavity module was taken out of the tunnel to repair the insulation vacuum leakage. For the repairing, the cavity was exposed to clean air for about 3 days. This clean air exposure required cavity processing of 6 hours with pulse power to reach the same field level as before.

CRAB CAVITY

KEKB adopted a finite angle crossing scheme of 2x11 mrad at the interaction point to reduce the background rates and to simplify beam optics at the interaction region.(Figure 7) By this scheme the luminosity reduction due to geometrical effects and the possibility of beam-beam instability. The crab crossing scheme was proposed to eliminate these effects. In this scheme electron and positron bunches to the interaction point are tilted by a time-dependent transverse kick in RF deflectors (crab cavities) and head-on collide. After the bunches are kicked back to the original orientations by another crab cavities. The selected parameters for KEKB crab cavity system are listed in Table 3.[5]

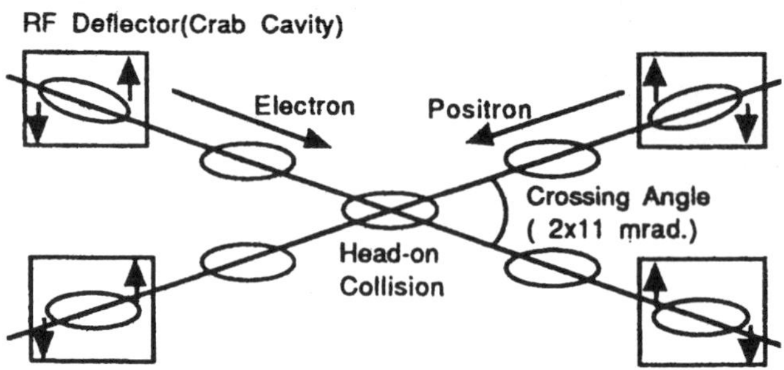

Figure 7. Crab crossing scheme for KEKB

Table 3. Parameters for the Crab Cavity in KEKB

Ring	LER	HER
Beam Energy (GeV)	3.5	8.0
RF frequency (MHz)	508.887	
Crossing Angle (mrad)	+- 11	
β_x^* (m)	0.33	o.33
β_{crab} (m)	20	100
Required kick (MV)	1.41	1.44
Esp (MV/m)	20	21

Fig 8 shows the dimensions of the squashed cell type crab cavity for KEKB. This type of cavity was originally studied at Cornell under KEK-Cornell collaboration. A full scale Nb model crab cavity has been fabricated and measured in a large vertical cryostat with 1.1 m in diameter and 3.5 m in height.

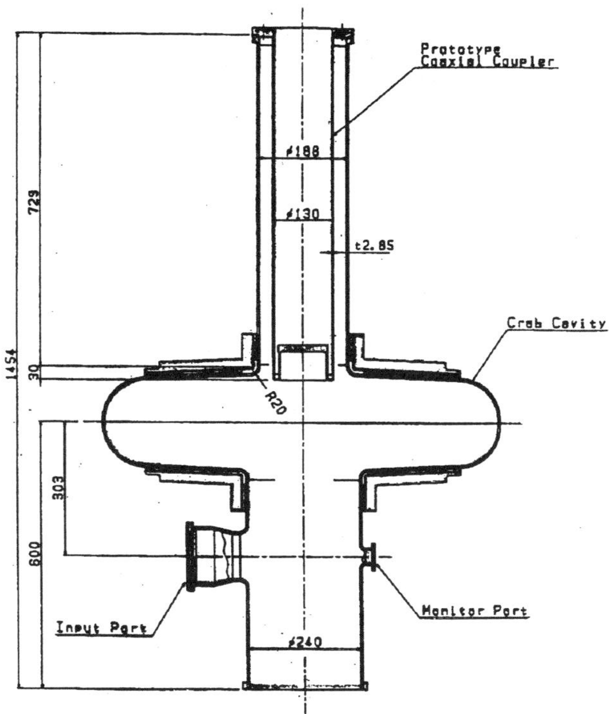

Figure 8. Cross section of the Crab Cavity with the Prototype Coaxial Coupler

The half cells of the Nb cavity were hydro-formed out of 5 mm thick Nb sheet with RRR=190 supplied from the Tokyo Denkai. The inner surface of the half cell were buff-polished to remove scars on the surface. The cell and beam pipes were assembled into a cavity by electron beam welding . The electron beam welded equator weld was ground by a specially designed grinding machine. The inner surface of the cavity was barrel polished where 100μm was removed and then electropolishing to remove another 100μm. After rinsing by ultra pure water the cavity was annealed in vacuum furnace at 700° C for about 3 hours. Before the cavity was assembled into the vacuum system for cold test, the inner surface of the cavity was rinsed by 8 MPa high pressure ultra pure water for about 45 minutes.

The maximum surface peak field Esp exceeded the design value of 21 MV/m and reached about 34 MV/m at 4.2 K and 40MV/m at 2.8 K. Now the full system with the coaxial coupler has been measured . The maximum Esp exceed the design value.

SUMMARY

Four superconducting accelerating cavities have been successfully operated in the KEKB asymmetric collider ring. The cavities were operated with 0.51 A of beam and of 380 kW of RF power. This result opens the possibility of using SC cavities in high current and high power accelerator application.

The development of the superconducting crab cavity has been successfully accomplished. A full scale Nb model crab cavity has been tested and exceeded the field specification.

ACKNOWLEGMENT

The authors would like to express their gratitude to professor S. Kurokawa for his strong support. They are deeply indebted to the vacuum group for their help for good vacuum beside the superconducting cavities. They also would like to thank the members of RF group for construction of the RF system.

REFERENCES

1. S.Mitsunobu et al.; Proc. PAC97, Vancouver, B.C.. Canada, May 12-16,1997. P.2908
2. T.Furuya et al.; ibid. p.3087
3. T.Tajima et al.; ibid. p.3092
4. K.Hosoyama et al. ; Proc. CEC/ICMC-97, Portland, Oregon, July 27-August 1,1997
5. K.Hosoyama et al. ; Proc. of 8th Workshop on RF Superconductivity (1997)

SUPERCONDUCTING RF CAVITIES AND CRYOGENICS FOR THE CESR III UPGRADE

E. Chojnacki and J. Sears

Cornell University, Laboratory of Nuclear Studies
Ithaca, NY 14853

ABSTRACT

The Cornell high energy electron/positron storage ring (CESR) is undergoing a phased upgrade to replace its normal-conducting copper RF cavities with superconducting niobium cavities. The accelerator continues to operate at ever increasing record luminosity part way through the upgrade (2 SRF cavities and 2 NRF cavities), with completion scheduled for Fall 1999. Details of the SRF cryomodule, its fabrication, refrigeration, and accelerator performance will be presented along with lessons learned regarding design and operation.

INTRODUCTION

The Cornell high energy electron/positron storage ring is nearing completion of a luminosity upgrade, CESR III, one facet of which replaces the four normal-conducting multi-cell copper RF cavities with four single-cell superconducting (SRF) niobium cavities.[1] The principal reason for the SRF upgrade is to achieve ever higher stored beam current, shorter bunch length, and thus higher collision luminosity in CESR for the CLEO high energy physics detector as well as beam for the CHESS synchrotron light source.

SRF cavities are well suited for high beam intensity by their inherently high fundamental mode shunt impedance R and consequent high accelerating gradient, high unloaded Q_o, low beam impedance loss factor $\propto R/Q_o$, and ease of thorough higher order mode (HOM) damping.[2] There is a compound benefit by which the high gradient allows fewer cavities to achieve the same accelerating voltage and each cavity has a lower loss factor.

In a storage ring, the loss factors and Q's presented by the RF cavities in the HOM's as well as fundamental mode determine the threshold in beam current at which various beam instabilities arise. Multi-bunch beam instabilities had indeed been encroached in CESR as the current was pushed to record levels of several hundred mA, arising from high-loss normal conducting RF (NRF) cavities which had been in place until recently.[1,3] The CESR run in Spring of 1999, however, was the first with only SRF cavities in place (three of four) and all NRF cavities removed. Thresholds for multi-bunch instabilities during the SRF-only run were greatly advanced,[4] thus they are expected to present no limitation to the upgrade goal of 1 Amp total beam current and luminosity 1.7×10^{33} cm^{-2}s^{-1}.

Technologies other than SRF cavities are available to combat storage ring beam instabilities, such as RF feedback in conjunction with exacting NRF cavity design. Indeed, modest RF feedback is utilized even in SRF cavity systems. But the NRF approach tends to be power hungry with both feedback and fundamental mode RF. SRF cavities and

Advances in Cryogenic Engineering, Volume 45.
Edited by Shu *et al.*, Kluwer Academic / Plenum Publishers, 2000.

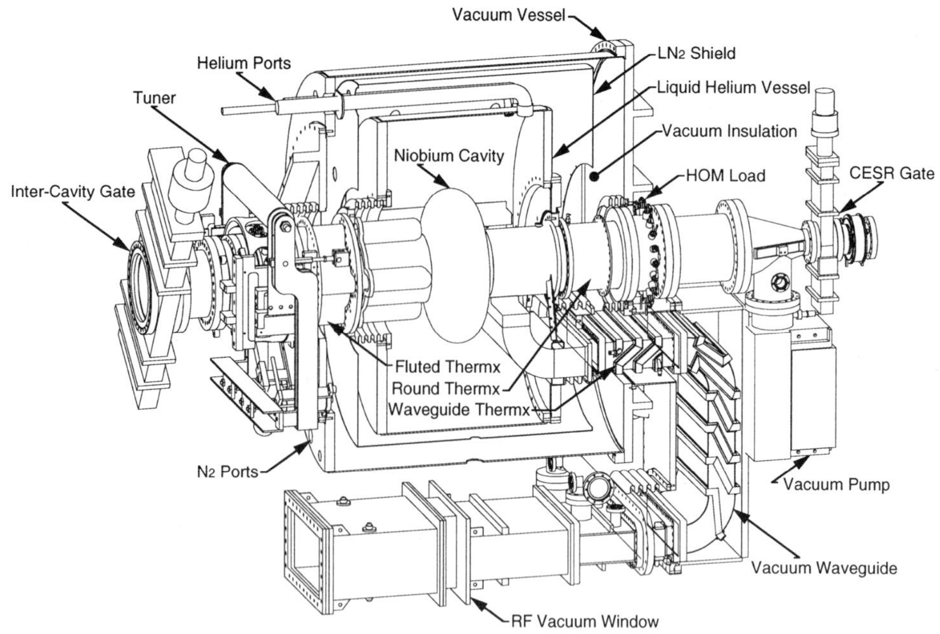

Figure 1. Illustration of the Cornell Mark II SRF cryomodule.

cryostats are admittedly technically complex, but great insight and streamlining of SRF installations is occurring world-wide at numerous accelerator facilities. This CESR III SRF upgrade represents a forward-looking approach to maintaining extremely high luminosity in a storage ring, taking a step closer to the ideals of textbook treatments of particle acceleration. Presented below are descriptions of salient features of the Cornell Mark II SRF cryomodule, design and fabrication details, a brief description of the refrigeration system, a performance update, and suggested improvements throughout.

CRYOMODULE LAYOUT

Shown in Figure 1 is an illustration of the Cornell Mark II SRF cryomodule, the fourth and final of which will be installed Fall 1999. The 500 MHz single-cell superconducting niobium cavity has a 24 cm aperture and resides inside a liquid helium (LHe) vessel, external to which is vacuum insulation, a thermal radiation shield maintained at liquid nitrogen (LN_2) temperature, layers of magnetic field shielding μ-metal, layers of mylar/fiberglass superinsulation, followed by the room temperature vacuum vessel enclosure. The He vessel is gravitationally suspended from the vacuum vessel by four $^1/_2$" diameter Invar rods, Invar chosen for its low thermal expansion and high tensile strength. Conduits for LHe, He gas, electronic wiring, and safety venting connect the He vessel to ports mounted near the top of one of the vacuum vessel end plates. Ports for LN_2, N_2 gas, electronic wiring, vacuum pumping, and safety venting are similarly mounted near the bottom of the same vacuum vessel end plate.

There are three main structural/thermal transitions between the cavity and external environment: a "fluted" beampipe, a round beampipe, and the rectangular waveguide RF feed. The fluted beampipe is required to allow two low frequency dipole HOM's to escape the cavity. Outboard of each beampipe thermal transition is a room temperature beamline HOM load. Following the round beampipe HOM load is a taper and small gate valve leading to oval CESR beampipe. The fluted beampipe HOM load has an integral fixture ring around which the cavity tuner clamps. The tuner mechanically adjusts cavity frequency around 500 MHz via longitudinal elastic deformation allowed by bellows joints

Table 1. Parameters of the Cornell B-cell SRF cavity

Frequency	500 MHz
Aperture	24 cm
Effective gap	30 cm
Gradient	> 6 MV/m
Unloaded Q_o at 6 MV/m gradient	$> 10^9$
Q_{ext}	2×10^5
Shunt Impedance R	89 GΩ
Delivered power at 1 Amp beam current	325 kW
HOM power at 1 Amp beam current	10 kW
Number of cavities in CESR	4

between the beampipe and He/vacuum vessels, with a large beamline sliding joint outboard of the HOM load. The cavity round beampipe is anchored to both the He and vacuum vessels.

Following the large sliding joint is a large inter-cavity gate valve (24 cm aperture) which maintains large-aperture beampipe to an adjacent mirror-image cryomodule. The beampipe between cryomodules is maintained large aperture since a significant portion of the SRF cavity's beam impedance is contributed by the tapers up and down to oval CESR beampipe, which is nominally 9 cm × 5 cm.

The rectangular waveguide RF feed immediately exterior to the He vessel is cooled for 30 cm by cold He gas boil-off from the He vessel flowing through tracing welded to the waveguide walls. Next is a waveguide double-E bend similarly cooled by LN_2 flowing through tracing. Following this is a short thermal transition from LN_2 to room temperature, a waveguide vacuum pumping section, and finally the waveguide vacuum window. This slightly convoluted waveguide path was arrived at by tight space considerations in the CESR tunnel and the need to mate to existing waveguide feeds.

DESIGN AND FABRICATION

The design of the Cornell Mark II cryomodule evolved disparately over several years, handed off a number of times at Cornell and volleyed between Cornell and the vendor Meyer Tool & Mfg. A coherent vision of the entire cryomodule assembly process would have made fabrication and assembly faster and efficient. Happily, a workable process developed and has proceeded quite well to meet all scheduled installations.

All components that share the CESR ultra-high vacuum, except the cavity and RF window, as well as numerous ancillary components were fabricated and assembled at Cornell. The bulk of the vacuum and He vessels were fabricated and partially assembled at Meyer Tool & Mfg., Chicago. The niobium cavity was fabricated by ACCEL Instruments, Germany. The waveguide vacuum window was fabricated by Thomson Components and Tubes Corp., France.

The SRF cavity known as the Cornell B-cell is a single cell elliptical type with 4" × 17" rectangular waveguide coupling.[2] It has design parameters as listed in Table 1. CESR operation at 1 Amp beam conditions will have the cavity matched at about 9 MV/m gradient, where the cavity tuning angle ψ will be about equal to the synchronous phase $\varphi = 80°$, requiring 325 kW forward power delivered by each of the four cavity couplers. Operation at lower cavity voltage will require slightly mismatched tuning, slightly higher forward power, and some reflected power. Computations of cavity modes have indicated total HOM power of up to 10 kW for 1 Amp beam current.[5] Design and fabrication details for the HOM loads[6] and RF window[7] have been presented elsewhere.

Prior to assembly, all major cryomodule components are subjected to acceptance tests. The cavity must achieve >6 MV/m gradient with unloaded $Q_o > 10^9$ in a vertical test. Each HOM load must dissipate >10 kW RF power at the test frequency of 2.45 GHz.[6] The RF window must transmit >400 kW CW at 500 MHz traveling wave and experience >125 kW CW standing wave with the electric field maximum at the window ceramic.[7] The assembled He and vacuum vessels (with a dummy copper cavity insert) must remain vacuum tight, demonstrate mechanical integrity, and have <40 W static heat leak upon cooling to LN_2 temperature.

Since thermal impedance is a central theme throughout the cryomodule, nearly all of the beamline vacuum components are made from thin-walled sheet metal welded into heavier duty flanges. The majority of the SRF cavity is made from 3 mm thick niobium sheet so the interior surface has low thermal impedance to the LHe bath. The beamline thermal transitions are made from 1.25 mm thick stainless steel sheet to have high thermal impedance along their length. The interior of the beamline transitions are electroless plated with 3.8 μm of copper to reduce wall-current heating from beam image currents and HOM's propagating to the HOM loads. The vacuum waveguide within the cryostat is made from 1.60 mm thick stainless steel sheet with the interior electroplated with 25 μm of copper. The thicker waveguide plating is required to ensure good electrical conductance, since as much as 400 kW CW of 500 MHz RF propagates through it.

Dimensional tolerances on beamline mating surfaces and bolt patterns were specified as ±0.002", required due to long lever arms and numerous joints on the beamline assembly that must eventually mate to opposite ends of the vacuum vessel, CESR beampipe, and the waveguide feed. This tolerance was easily achieved on individual pieces, but upon welding of, e.g., flanges to thin-walled beam tube, distortions frequently degraded tolerance to ±0.010" in linear dimensions and flatness. The specified tolerance on finished assemblies could be easily achieved by the common practice of first rough machining parts, then welding, then finish machining with inclusion of precision dowel-pin holes for alignment during assembly. This is the technique by which ACCEL Instruments fabricated the electron-beam welded niobium cavity to meet tight tolerances on all mating flanges. Though tolerances on the vacuum and helium vessels are somewhat relaxed from that of the beamline, there again, post-machining weld distortions along with the omission of fixturing during welding allowed tolerances to stray to nuisance levels. Achieving the specified dimensional tolerances on all cryomodule components and inclusion of alignment pins would have significantly expedited cryomodule assembly.

The volume of the entire He vessel is 520 L, typically maintained at 470 L to have the LHe level a few cm above the cavity equator top. This large volume is due in part to the coupler E-plane elbow oriented below the cavity, necessitating a significant LHe volume just to reach the equator bottom. Further, the He vessel is not contoured to the cavity nor weld sealed. Removable He vessel end plates with indium tongue and groove seals were implemented to facilitate potentially repetitive assembly/disassembly.

REFRIGERATION

Liquid He is provided by two 600 W refrigerators for a total of 1200 W supplying a 2000 L storage dewer. LN_2 is periodically delivered by a vendor to a 56000 L storage tank. Due to the refrigerator's suction pressure being about 3 psi above atmosphere and an additional 2 psi differential between the cryomodule and refrigerator, the cryomodule LHe temperature is nominally 4.5 K. Rigid transfer lines transport LHe, cold gaseous He, and LN_2 between the refrigerator and LHe dewer to a centrally located main valve box. From the main valve box, rigid transfer lines lead to satellite valve boxes supplying:

1) a pair of SRF cryomodules in CESR's East RF station,
2) a pair of SRF cryomodules in CESR's West RF station,
3) an SRF cryomodule in CESR's RF processing area,
4) the CLEO detector superconducting solenoid,
5) a pair of interaction region superconducting quadrupoles to be installed as part of the Phase III upgrade.

The main valve box also has one spare transfer line port.

The rigid transfer lines have a heat leak to LHe of < 0.5 W/m, contributing about 12 W per cavity feed. The largest heat leak is in the valving and flexible lines, contributing about 50 W per cavity feed. Thus, delivering LHe to four SRF cryomodules consumes about 250 W of refrigeration power, which does not include the cryomodule heat load discussed in the next section.

A key control for stable refrigerator operation was found to be maintaining a steady cold He gas return flow rate from the various LHe heat loads. Since the SRF cavity heat deposition varies with RF amplitude, an analog electronic circuit was implemented to heat a resistive load in the LHe vessel at the difference between a chosen "level load" and the dynamic RF load. The leveled load is nominally set at 60 W.

CRYOMODULE HEAT LEAK

There are seven main sources of heat loading LHe within the cryomodule itself:

1) Round beampipe thermal transition,
2) Fluted beampipe thermal transition,
3) Waveguide thermal transition,
4) Radiation from LN_2 shield to LHe vessel,
5) LHe boil-off gas cooling the waveguide thermal transition,
6) RF wall dissipation in the SRF cavity,
7) Infra-red radiation impinging from beamline apertures and waveguide duct.

Calculations of heat leak for the beampipe and waveguide thermal transitions were performed with a computer code that takes into account thermal and electrical conductivity variation with temperature. These results are presented below. Heat leak due to impinging IR radiation has not been thoroughly addressed, but briefly discussed below.

Round and Fluted Beampipe Thermal Transitions

The 24 cm diameter round beampipe wall is comprised of 1.25 mm thick stainless steel with 3.81 μm of copper plated on the interior. The end attached to the LHe vessel is at 4.5 K and the end attached to the vacuum vessel is at 300 K. A copper ring indirectly cooled by LN_2 is welded to the beampipe two thirds of the way from the LHe to vacuum vessel. Thermometers on the beampipe have shown the copper ring to be close to 100 K. The *static* heat leak for the round beampipe is then calculated to be 4.31 W. The additional heat due to image current of a 1 Amp beam is 0.10 W. The additional heat due to 5 kW of HOM RF power assumed to be at 1 GHz in the TM_{01} mode is 0.16 W. The total calculated *operational* heat leak of the round beampipe thermal transition with beam is then 4.57 W.

The fluted beampipe thermal transition is very similar to the round, with the flutes approximately doubling the cross-sectional area. Thus, simply multiplying the round beampipe calculations by two, the *static* heat leak for the fluted beampipe is calculated to be 8.62 W. Including image current from a 1.0 Amp beam and 5 kW of HOM power, the total calculated *operational* heat leak for the fluted beampipe thermal transition with beam is then 9.14 W.

Waveguide Thermal Transition

The waveguide wall is comprised of 1.60 mm thick stainless steel with 25.4 μm of copper plated on the interior. The end attached to the LHe vessel is at 4.5 K and the end attached to the LN_2 cooled waveguide double-E bend is taken to be 100 K. The waveguide thermal transition is cooled for 30 cm by cold He gas boil-off from the He vessel flowing through tracing welded to the waveguide walls. Commissioning of three cryomodules between 1997-1999 showed a mass flow of 134 mg/s yielded the most stable cavity operation.

With no cold He gas flowing through the waveguide tracing and no RF power, the *static* heat leak for the waveguide thermal transition is calculated to be 8.81 W. With 134 mg/s cold He gas flow and no RF power the heat leak drops to 1.99 W. With 134 mg/s cold He gas flow and 350 kW of 500 MHz RF propagating through the waveguide the *operational* heat leak for the waveguide thermal transition is calculated to be 5.30 W.

Radiation from LN_2 Shield to LHe Vessel

The surface area of the LHe vessel exterior is approximately 4.1 m^2, including beampipe apertures. The radiative heat transfer from the LN_2 shield to the LHe vessel is

Table 2. Liquid He heat loads within the Cornell Mark II SRF cryomodule

Heat source	Operation [W]	Static [W]	Avg measured [W]
Round beampipe	4.57	4.31	—
Fluted beampipe	9.14	8.62	—
Waveguide	5.30	8.81	—
LN_2 shield radiation	8.2	8.2	—
Waveguide gas flow	14.0	—	—
RF cavity wall dissipation	72.8	—	—
Total	**114.01**	**29.94**	**34.2**

then approximately $\sigma\,(T_{LN}^4 - T_{LHe}^4) \times Area$, where $\sigma = 5.67 \times 10^{-8}$ W/m^{2}K^4 is Boltzman's constant. This yields a heat leak of 8.2 W.

LHe Boil-off Gas Cooling the Waveguide Thermal Transition

The LHe boil-off gas cooling the waveguide thermal transition is warmed to room temperature and returned to the refrigerator. This He gas heat load on the Cornell refrigerator/liquifier with 134 mg/s mass flow consumes 2.8 W for refrigeration and 11.2 W for liquefaction, or a total of 14 W.

RF Wall Dissipation in the SRF Cavity

The CESR B-cell has a design $R/Q_o = V^2 / Q_oP_{wall} = 89$. Operating at $Q_o = 10^9$ and a gradient of 6 MV/m with an effective gap of 0.3 m gives 1.8 MV acceleration and $P_{wall} = 36.4$ W. However, two of three CESR B-cells to date have achieved only $Q_o = 5 \times 10^8$ which raises $P_{wall} = 72.8$ W.

Infra-red Radiation Impinging from Beamline Apertures and Waveguide Duct

It is expected that nearly all IR radiation along the beampipe simply passes through the cavity apertures unreflected. The few divergent IR rays that impinge on the cell are most likely completely reflected unattenuated by the highly IR reflective niobium cavity.

IR radiation impinging from the waveguide duct is also most likely completely reflected unattenuated by the highly IR reflective niobium waveguide coupler. The worse-case IR radiation impinging the coupler is launched by the exposed cross section of the ceramic window (251 cm^2) and the remaining cross section of 4" × 17" room-temperature waveguide (188 cm^2). The hottest possible ceramic window has a quadratic radial temperature profile with 353 K at the center and 293 K at the edge, yielding an average ceramic temperature of 333 K. This gives a radiation power launched from the ceramic of 17.5 W. The remaining cross section of 4" × 17" room temperature (293 K) waveguide launches 7.8 W. The maximum IR radiation power impinging the coupler via the waveguide duct is then 25.3 W. Most of this IR power is attenuated in the waveguide double-E bend if it has even a slightly IR dull surface.[8] And as mentioned previously, the small IR power making its way to the waveguide coupler will most likely be completely reflected unattenuated by the niobium surface.

Heat Leak Measurements

Measurements of heat load in the Cornell Mark II cryomodule have been performed in several instances, the most reliable of which was during a recent warm-up of three cryomodules installed in CESR in which the decreasing LHe levels were monitored with time. In this static condition there was no waveguide thermal transition gas flow and of course no beam or RF. The measurements yielded an average heat load of 34.2 W.

Table 2 summarizes the above calculated heat loads for operational and static conditions, with measurements to be compared to static conditions. There is good agreement between calculated and measured static conditions. Operational conditions were taken as: 134 mg/s waveguide thermal transition mass flow, 350 kW waveguide RF at 500 MHz, 1 Amp beam, total of 10 kW HOM RF, cavity $Q_o = 5 \times 10^8$, and cavity

$V = 1.8$ MV. Obviously, operational conditions are strongly dependent on cavity Q_o. Despite the low Q_o achieved by two of three CESR B-cells to date, the Cornell refrigerator has sufficient capacity to supply the full complement of CESR III upgrade LHe needs.

PERFORMANCE

The three Cornell SRF cryomodules installed is CESR to date have performed quite satisfactorily. All cavities have exceeded the 6 MV/m minimal gradient, Q's of all HOM's are <100, and cryogenic heat loads have been within tolerance.

This SRF upgrade has been phasing in over the last few years, wherein during bi-yearly standard-maintenance downs a single-cell SRF cavity/cryostat replaces a five-cell NRF cavity. Thus, until Spring 1999, CESR operated with a mix of SRF and NRF cavities. As mentioned in the Introduction, multi-bunch beam instabilities arising from the NRF cavities occurred during this time as the beam current was pushed to record levels. Even with the instability, a maximum beam current of 550 mA and luminosity of 8×10^{32} $cm^{-2}s^{-1}$ were achieved. A short run in Spring 1999 with three SRF and no NRF cavities in CESR showed great advance in the instability threshold current.[4] A new feedback system and these high thresholds for beam instabilities should allow achievement of the upgrade goal of 1 Amp beam current and luminosity 1.7×10^{33} $cm^{-2}s^{-1}$.

As is becoming apparent with SRF installations world-wide, the most difficult aspect of commissioning the Cornell SRF cryomodules has been RF-based discharges tripping vacuum thresholds in the input power coupler.[7,9] Signals from photo-multiplier arc detectors directly viewing the Cornell RF window ceramic rarely accompany such vacuum trips, though they do so prodigiously during unattached window processing.[7] It is then very likely that condensed gases in the cold regions of the waveguide coupler enhance the metal's secondary electron emission coefficient, making familiar multipactor barriers more virulent. Further, the high coupling coefficient characteristic of SRF cavities, e.g., $\beta = 5000$ for the Cornell B-cell, dictates that negligible traveling wave power is present without beam loading. As such, processing without beam can at best condition narrow regions of the waveguide at the standing wave's electric field crests. Indeed, during conditioning, vacuum trips in the Cornell B-cell coupler occur at strictly repeatable traveling wave power thresholds as beam loading increases.[9] Fortunately, considerable progress in traveling wave power has been made, employing a number of commissioning strategies since the first barrier at 90 kW made itself apparent in the first cryomodule installed in Fall 1997.[9] To date, over 220 kW has been coupled to the beam in a single cryomodule and the pattern is expected to continue by which the barriers become milder with increasing power threshold.

To make matters more challenging in the power coupler, about a meter of the interior of the vacuum waveguide double-E bend is corrugated with a sharp $^1/_{16}$" sawtooth pattern to ostensibly deter multipacting and attenuate IR radiation. This array of electric field singularities very likely generously seeds many of the multipactor resonances encountered. The following modifications were thus implemented for the input coupler waveguide in all Mark II cryomodules succeeding the first:

1) The corrugated waveguide was aggressively acid etched prior to copper electroplating to dull the sharp ridges,
2) Sharp corners were removed elsewhere in the waveguide and vacuum pumping holes were moved from the waveguide centerline to near the sidewalls where the electric field is much lower,
3) All components of the vacuum waveguide, except as part of the niobium cavity, were vacuum baked prior to final assembly,
4) Vacuum pumping speed was modestly increased by larger waveguide pumping ports,
5) Cold He gas mass flow through the waveguide thermal transition was increased to stabilize and lower its operating temperature,
6) The thickness of copper plating on the waveguide was increased to ensure minimal RF heating, especially on the hydrogen-condensing LHe-LN_2 thermal transition.

These improvements and clever in situ processing strategies[9] have enabled every cryomodule to RF power process faster and further than its predecessor. More advanced

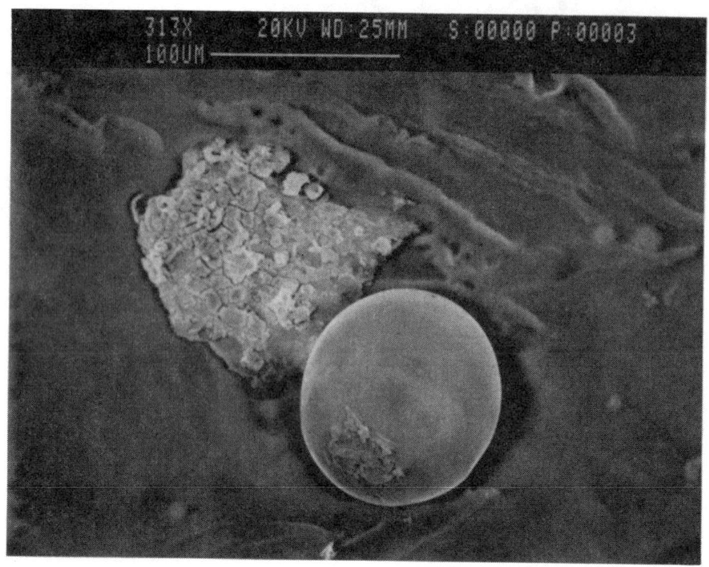

Figure 2. SEM photo of a partially melted stainless steel sliver wiped from the equator bottom of a B-cell.

modifications which could implement a potentially multipactor-free coupler, such as using elliptical or wedge-shaped waveguide, will have to await next-generation cryomodule R&D.

In an interesting and amusing note, when the second cryomodule was tested at high power in a processing area, the cavity quenched at a gradient around 5 MV/m, below the 6 MV/m acceptance threshold. A thermometer on the equator bottom, used to verify LHe level, showed a >50 K rise during quench, indicating a contaminant found its way into the cavity and settled in the gravitational well. Since tear-down and re-assembly of the cryomodule to re-etch the cavity would require many months turn-around, there was nothing to lose in performing the risky operation of using a long stick to wipe the cavity equator bottom in situ. Thus, a 2 m × 15 cm L-shaped, Teflon sheathed, ultra-clean stick with a methanol-soaked clean-room cloth tied to the end was inserted through the HOM load and fluted beampipe, then lowered and rotated inside the cavity to wipe the equator bottom. Shown in Figure 2 is an SEM photo of the largest contaminant picked up by the cloth. X-ray analysis showed the spherical object to be nearly pure Fe and the attached flake to contain Fe, Cr, and Ni, indicating partially melted stainless steel. Apparently, a stainless sliver from a sheared tongue and groove seal migrated to the cavity bottom and partially melted when subjected to high RF wall current during testing. Fortunately, the melted stainless did not wet well to the niobium, and after removal, cavity performance improved considerably, exceeding the 6 MV/m gradient acceptance threshold. The same thermometer indicates that quenches at the higher gradient still originate at the cavity equator bottom. This cryomodule now serves as CESR's E1 RF station.

CONCLUSION

Having already achieved a beam current of 550 mA and luminosity of 8×10^{32} cm^{-2}s^{-1} with the phased SRF installations, there is great anticipation of continued forging of frontiers in accelerator science, high energy physics, and synchrotron light sources as we near completion of the CESR III upgrade. This gives encouragement for continued competitiveness on the world stage of electron/positron storage rings by achieving our upgrade goal of 1 Amp beam current and 1.7×10^{33} cm^{-2}s^{-1} luminosity.

ACKNOWLEDGMENT

The success of this SRF upgrade to CESR was made possible by the National Science Foundation and many talented members of the Cornell University Laboratory of Nuclear Studies, especially H. Padamsee, S. Belomestnykh, and R. Ehrlich.

REFERENCES

1. D.L. Rubin, Results of the CESR upgrade, *Proc. 6th European Part. Accel. Conf.*, Stockholm, (1998).
2. H. Padamsee, *et. al.*, Design challenges for high current storage rings, *Part. Accel.* 40:17 (1992).
3. M. Billing, Observation of a longitudinal coupled bunch instability with trains of bunches in CESR, *Proc. 1997 Part. Accel. Conf.*, Vancouver, BC, 2317 (1997).
4. M. Billing, private communication.
5. S. Belomestnykh and W. Hartung, Calculations of the loss factor of the BB1 superconducting cavity assemblies, *Cornell LNS Report SRF960202-01* (1996).
6. E. Chojnacki and W.J. Alton, Beamline RF load development at Cornell, *Proc. 1999 Part. Accel. Conf.*, New York, NY, (1999).
7. E. Chojnacki, *et. al.*, Tests and designs of high-power waveguide vacuum windows at Cornell, *Part. Accel.* 61:[309]45 (1998).
8. N. Jacobsen and E. Chojnacki, Infra-red propagation through various waveguide inner surface geometries, *Cornell LNS Report SRF990301-01* (1999).
9. S. Belomestnykh, *et. al.*, Commissioning of the superconducting RF cavities for the CESR luminosity upgrade, *Proc. 1999 Part. Accel. Conf.*, New York, NY, (1999).

STATUS AND PERSPECTIVES OF THE SC CAVITIES FOR TESLA

Carlo Pagani for the TESLA Collaboration

INFN Milano-LASA and University of Milano
Via Fratelli Cervi, 201, I-20090 Segrate (MI), Italy

ABSTRACT

To fulfill the requirements of TESLA, the superconducting option for future linear electron-positron collider, a TESLA Test Facility (TTF) is under construction and commissioning at DESY by an International Collaboration. In this framework an extensive R&D program started in 1992 at DESY, and worldwide in the different laboratories participating in the Collaboration, to establish a reliable and cost efficient technology to industrially produce 9-cell pulsed cavities, at 1.3 GHz, with accelerating field exceeding 25 MV/m. In this paper the outstanding results obtained so far, as demonstrated by vertical tests and operation in TTF at DESY, are shown. In addition the perspectives of further improvements emerging from new cavity fabrication and processing techniques under development by the Collaboration are presented.

INTRODUCTION

A high energy e^+e^- linear collider with a center of mass energy of 500 GeV and higher is considered an essential tool to search for fundamental constituents of matter and their interactions and to address the problem of mass generation in the Standard Model. Worldwide a number of groups are pursuing different design efforts towards a next generation TeV Linear Collider. The different accelerator designs can be divided in two main categories: the high frequency, room temperature approach (NLC, JLC, VLEPP & CLIC) and the low frequency, superconducting one (TESLA). The combination of high conversion efficiency from mains to beam power (17-23%), together with small emittance dilution in the low-frequency (1.3 GHz) superconducting linac, makes the last choice ideal for an optimum performance in terms of the achievable luminosity [1].

From the work started in 1990 [2], a concept of a 500 GeV centre-of-mass energy superconducting linear collider emerged, based on 9-cell cavities operating at 1.3 GHz. An accelerating field of 25 MV/m @ $Q=5\cdot10^9$ was envisaged, the expected luminosity being $5\cdot10^{33}$ cm^{-2} sec^{-1}. In 1991 several institutions decided to join the efforts in the TESLA Collaboration – formally established in 1994 – to set up at DESY the necessary infrastructure, while building a new generation superconducting linac, that is the TESLA Test Facility (TTF) [3]. At present, more than 30 institutes from Armenia, P.R. China,

Advances in Cryogenic Engineering, Volume 45.
Edited by Shu *et al.*, Kluwer Academic / Plenum Publishers, 2000.

Finland, France, Germany, Italy, Poland, Russia and USA participate in the TESLA Collaboration and contribute to TTF.

The conceptual design report (CDR) of the TESLA 500 GeV collider was published in May 1997 [4] giving a complete description of the machine, including all the subsystems. Due to the low RF frequency, which implies better emittance preservation, and the long macro-pulses, a superconducting linac based on the foreseen TESLA technology lends itself as the best choice for the driver of a X-ray Free Electron laser (FEL), working in the Self-Amplified Spontaneous Emission (SASE) regime. So that, a SASE FEL user facility has been included as an integral part of both TTF and TESLA projects [3,4].

With respect to the existing large-scale installations of superconducting cavities (e.g. LEP and CEBAF), the major challenge for the feasibility of a superconducting linear collider was to reduce the cost per unit energy gain (MeV) by more than an order of magnitude. This means both to reduce the cost per unit length and to increase the accelerating gradient by about a factor of five, to 25 MV/m and more.

THE TESLA TEST FACILITY (TTF)

The TESLA Test Facility (TTF) is under construction at DESY, with major components flowing in from the members of the TESLA Collaboration. Figure 1 shows the general layout of TTF Phase I, in Building 28 (Hall 3). It comprises the complete infrastructure for the treatment, assembly and test of the 9-cell superconducting cavities, and a superconducting linac, composed by 3 cryomodules with 8 cavities, for an integral test of the machine critical components. An average cavity gradient of 15 MV/m was the initial goal, as a first step to reach the design gradient of 25 MV/m required by TESLA.

TTF Phase I is now close to completion and Phase II is under construction. It will include 5 more cryomodules, of the final TESLA design [5], placed in a building extension simulating the TESLA tunnel. A 25 m long undulator will be installed ad the end of the linac by fall 2001, to feed a FEL user facility with photon energies up to 200 eV.

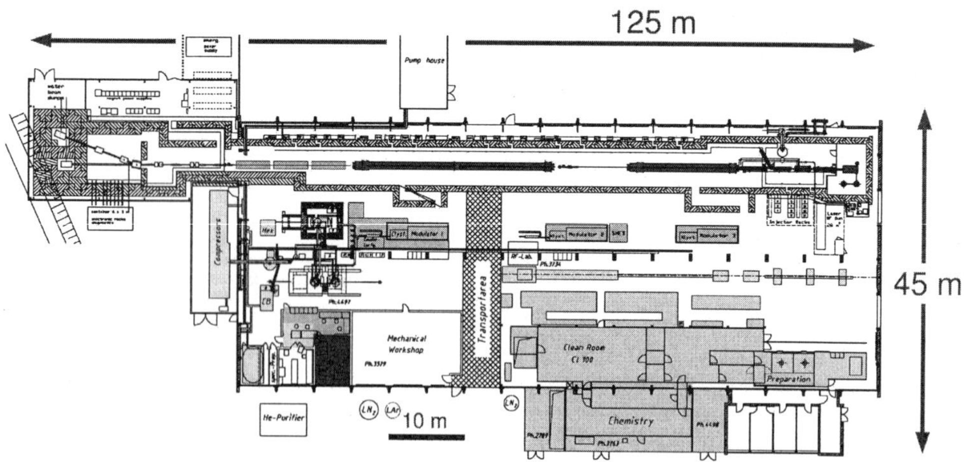

Figure 1. TESLA Test Facility linac (Phase I) and infrastructures at DESY. In the upper part, from right to left: injector, capture cavity, bunch compressor, first cryomodule, second bunch compressor, string of cryomodules #2 and #3, three undulator sections, beam analysis and dump. On the lower right: the cavity preparation infrastructures and the cryomodule assembly station. On the lower left: cryogenics and cavity cold test area.

The TTF linac

A 4 MeV laser-driven RF photoinjector [6] is producing the TESLA beam, that is a one millisecond long bunch train with an average current of 8 mA and a bunch charge of 8 nC. The electrons are then captured by a single 9-cell cavity to be injected in the first cryomodule at an energy of the order of 20 MeV. In 1997 the first cryomodule containing eight 9-cell cavities with an average gradient of 15 MV/m has been successfully commissioned while the second module was installed in summer 1998. An important step toward TESLA specs was made by achieving a minimum cavity gradient of 20 MV/m. Four cavities actually reached the TESLA goal (25 MV/m). The cavities of the third module, which has been recently assembled and will be cooled down by end of July, are expected to operate at an average gradient of 25 MV/m. This module is being installed in place of module # 1. The old module # 1, renewed in cryogenics and equipped with high performance cavities, will complete TTF Phase I by the end of the year. By August two modules together with photoinjector, capture cavity, bunch compressor and a 15 m long three section undulator will be operated for the SASE FEL proof-of-principle experiment. A beam energy up to 380 MeV will be available, the same as that originally foreseen for TTF Phase I with 3 cryomodules.

As stated above, we expect to conclude the installation of Phase II by fall 2001. The new generation cryomodules [5], which are smaller in diameter and allow for semi-rigid coupler and superstructures[9] are now in fabrication. The delivery of the first two is expected this year while a new batch of 24 cavities will be delivered starting January 2000.

The TTF infrastructures

The TTF infrastructure for cavity preparation and test[6], completely operational since the end of 1995, consists of a complex of clean rooms (from class 10000 to class 10), a chemical etching facility and an ultra-clean water supply. A UHV furnace is used to improve the niobium thermal conductivity via heat treatment at 1400 °C in the presence of Ti gettering. The last step of cavity preparation consists of a high pressure (100 bar) rinsing with ultra pure water.

The cavities are first tested in a vertical bath cryostat in superfluid helium at 2 K, the TESLA operation temperature. High peak power processing [7] as well as temperature mapping [8] and local RRR measurements through eddy currents can be applied. The 9 cell structures, once they have passed the vertical test, are welded into the helium tank. The fully assembled cavity can be tested in a horizontal cryostat in the TTF pulsed power mode (500 μs rise time, 800 μs flat-top and 10 Hz repetition rate). The performance of the main coupler, the high-order-mode (HOM) absorbers and the cold tuning mechanism is thus checked before the cavity is installed into the cryomodule.

Cavity performance can be measured either in the horizontal cryostat or after the installation into the cryomodule. In both cases the quality factor Q is obtained by measuring the RF heat load as the difference between total cryogenic losses at 2 K and the static ones. The TTF cryoplant permits precise heat load measurements at 2 K with a resolution of 50 mW (as a reference, with the TTF time structure, a cavity operated at 25 MV/m and a Q of $1 \cdot 10^{10}$ gives a heat dissipation of 0.9 W).

CAVITY FABRICATION AND PREPARATION

The cavities are fabricated from RRR 300 niobium sheets by electron beam welding (EBW). Up to now 55 cavities have been ordered to 4 European companies: a first series of 28 in 1994 and a second series of 27, with Nb-Ti flanges, in 1997.

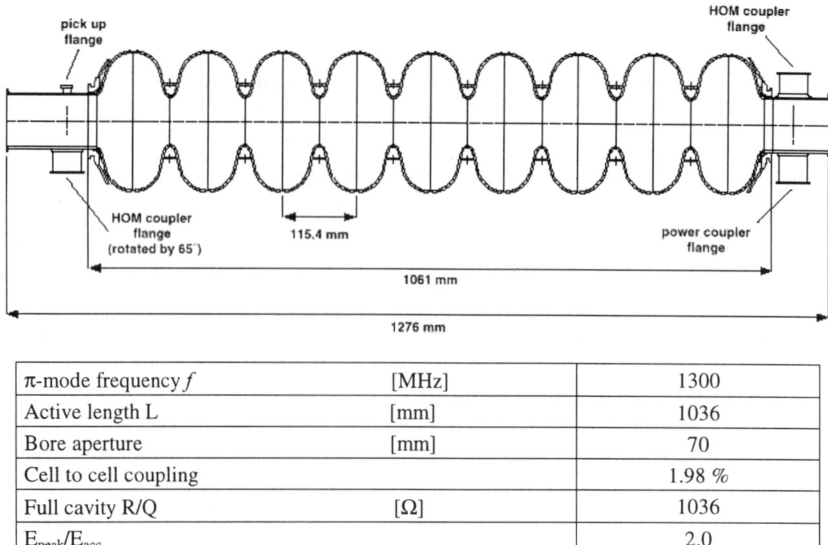

π-mode frequency f	[MHz]	1300
Active length L	[mm]	1036
Bore aperture	[mm]	70
Cell to cell coupling		1.98 %
Full cavity R/Q	[Ω]	1036
E_{peak}/E_{acc}		2.0
B_{peak}/E_{acc}	[mT/(MV/m)]	4.2
Tuning sensitivity $\Delta f/\Delta L$	[kHz/mm]	315
Loaded cavity bandwidth ($Q_{ext}=3\cdot10^6$)	[Hz]	433

Figure 2. Cross section and major design parameters of the TTF cavity. Stiffening rings are shown.

A new order for 24 cavities has being signed and the delivery is expected by January 2000. In parallel 6 special cavities (7-cell) will be fabricated to assemble a superstructure[9]. A cross section of the TTF 9-cell cavity is shown in Fig. 2, together with the main design parameters.

At present the cavity fabrication procedure includes the three following basic steps, the last two driven by the presence of the stiffening rings:

- Complete scanning of the 2.8 mm, RRR 300, Nb sheets by an eddy current apparatus [10], to discard all sheets with detected, 100 μm size, inclusions or marks.
- EB welding of the deep-drawn half cells to obtain the dumb-bells, including the stiffening ring. The iris weld is "finished" from inside.
- Completion of the cavity by equatorial EB welds performed from the outside in two subsequent passes with a fast wiggling beam. This technique[11] was found to be very powerful in making the welding parameters much less critical.

Once checked for surface status, dimensions and vacuum tightness, the cavity is delivered to DESY for acceptance tests, treatments, RF measurements and assembly. Among the large number of steps required[12] to prepare the cavity for vertical test, a partial, although significant, list is given in the following:

- Removal of 80 μm from the inner cavity surface by Buffered Chemical Polishing (BCP) and rinsing with ultrapure water, until the output resistivity is higher than 18 MΩ·cm
- Removal of 30 μm from the outer cavity surface by BCP.
- Hydrogen degassing and recrystallisation at 800 °C for two hours;

- High temperature heat treatment at 1400 °C (4 hours with Ti gettering) to improve the niobium thermal conductivity, rising the RRR from 300 to 500-700;
- 80 μm internal and 30 μm external BCP for titanium removal;
- Tuning and field profile adjustment;
- Final 20 μm BCP on the inner surface;
- High pressure (100 bar) ultrapure water rinsing (HPR);
- Drying by laminar flow in class 10 clean room, flange assembly and leak check;
- 2 additional HPR, drying as above and assembling of the input coupler.

VERTICAL TEST RESULTS

RF vertical tests include the excitation of the other 8 modes of the fundamental pass band, to determine the performance of individual cells. If required to cure field emission high power processing is applied, while T-mapping and local eddy current measurements are used to determine defect locations.

By the end of June 1999, 45 9-cell cavities have been tested in the vertical cryostat of the TESLA Test Facility. The majority of the cavities exceeded the TTF design goal of 15 MV/m and gradients up to 29 MV/m have been reached. Field values up to 35 MV/m (E_{acc}) have been achieved in individual cells by mode excitation.

Figure 3 shows the vertical test results of the 16 cavities exceeding the TESLA collider goals. Note that the cavities showing a Q_0 at low field exceeding $2 \cdot 10^{10}$ (the BCS value at 2 K) have been measured at a temperature below 2 K. All these cavities have shown a residual surface resistance of the order of a few nΩ. The majority of them reached their excellent performance already in the first vertical test. Only occasionally it was necessary to repeat the last part of the preparation procedure.

The average accelerating field of all 45 cavities measured up to now exceeds 20 MV/m and that of cavities measured in the last two years is close to 25 MV/m. An overview of the time dependence of the vertical test results (the best for each cavity) is presented in Fig. 4. Since 1997, the major fabrication errors have been corrected and all the niobium sheets used for the cavity production were eddy current scanned.

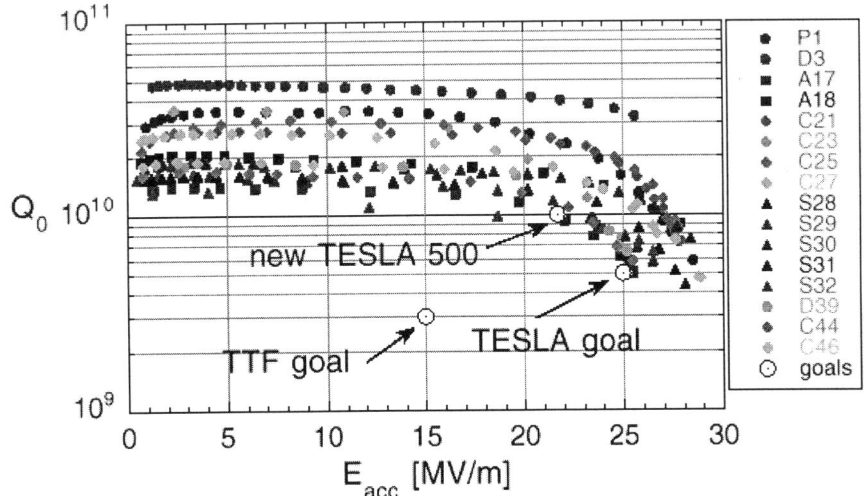

Figure 3. Vertical test results of the 16 best cavities out of the 45 measured up to now that exceed the TESLA 500 requirements. They are: E_{acc}=25 MV/m @ $Q_0=5 \cdot 10^9$ for standard 9-cell cavities and E_{acc}=21.4 MV/m @ $Q_0=1 \cdot 10^{10}$ for superstructures [9].

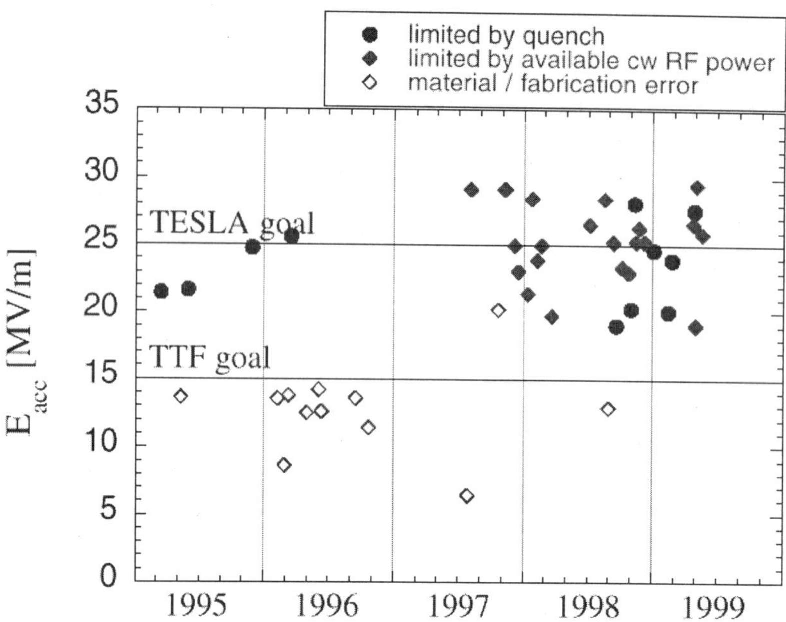

Figure 4. Overview of the time dependence of the vertical test results (the best for each cavity) for all 45 TTF 9-cell cavities measured by June 1999.

As said above, four European companies shared the production of the first two sets of cavities, for a total of 55 plus 2 prototypes. It is important to note that, once the fabrication errors were corrected, all companies have been able to produce cavities with performances exceeding the TESLA goals. At the beginning of the production the most crucial step of the fabrication procedure has been the equatorial weld[12]. Each company had to independently develop proper tooling and electron beam welding parameters. As a consequence we now have different possible procedures, all qualified for high field. In particular the final cavity assembly, by the eight equatorial welds, can be done either welding together one dumb-bell after the other, or mounting all components in a proper tool and performing the welds in a single step. Even the machining of the dumb-bell equatorial edges was based on two different philosophies and both solutions adopted, step like and flat, have proven to be adequate. At present, if no mistakes are done during the weld preparation or execution, the weld seam is no longer the region where the possible high field quenches occur. One very interesting result has been recently obtained with a cavity that had a repair performed on one of the equatorial welds. This cavity, which was supposed to be limited like the previous ones with the same defect, reached 27.9 MV/m at $Q_0=9.7\cdot10^9$. In this case the limitation was the available RF power and no quench has been detected.

HORIZONTAL TESTS AND LINAC OPERATION

After being welded into the Titanium He-tank, equipped with a motorized tuning system [13], and after the RF power coupler and the HOM couplers have been assembled, a last test prior to installation into the cryomodule is usually performed in a horizontal test stand, CHECHIA, designed and built at Saclay. The cavity performance in the horizontal cryostat in the pulsed mode is generally comparable with the results of the vertical tests [14], and with the cavity performance in the module. The small differences revealed in some cases, in both directions, can be interpreted as due to dust particles or to the further conditioning of the field emission limit respectively.

The average gradient obtained in the 18 cavities tested so far in the horizontal cryostat is 22.5 MV/m, which does not differ substantially from the average value of 22.3 MV/m obtained in the vertical tests. One particular case is cavity C23 that reached 33 MV/m in the horizontal test with a quality factor $Q=4\cdot 10^9$. In the vertical test, the cavity was limited to 25 MV/m by available RF power [15]. Most of the good cavities are at present limited by RF breakdown in the main power coupler.

The eight cavities installed in module #1 have been operated with an average gradient of 15 MV/m and the three best cavities performed close to the vertical test results [14].

The eight cavities installed in module #2 all exceeded 20 MV/m in the vertical test, but one that was limited at 19.7 MV/m by field emission. Given that no individual cavity test has been performed in the linac, it is worthwhile noting that, after proper coupler conditioning, all cavities reached 20 MV/m at full TTF pulse length and repetition rate. No further gradient increase was possible because of coupler limitations. In particular, after in-situ high peak power processing (HPP) on one cavity that was showing a very high field emission, a global RF heat load of 6.5 W was cryogenically measured at 2 K, with all cavities operating at 20 MV/m. By detuning the cavity that was still limiting the global performance, the measured RF cryogenic load given by the remaining 7 cavities was reduced to 2.9 W, which corresponds to an average Q_0 at 20 MV/m of $1.3\cdot 10^{10}$, a little higher than the result from the vertical tests.

The third module has been assembled with cavities ranging from 23 to 28 MV/m, the average value being close to 25 MV/m. This module, that has been installed in place of module #1, will be cooled down and tested by the end of July 1999.

R&D ACTIVITIES AND PERSPECTIVES

The R&D activity on superconducting cavity development has grown worldwide in the recent years. This activity, that is mainly concentrated on the TESLA cavity parameters, is being done in different laboratories which are either part of the International Collaboration, or linked to it through specific Memoranda of Understanding.

Two lines can be identified according to the main tasks that are at the basis of the TESLA activity: field enhancement and/or cost reduction. In the following I just quote some promising example that can be taken as a reference. For simplicity they are divided in two subchapters: field emission and Q drop, and alternative fabrication techniques.

Field Emission and Q Drop

Since mid 1997 the average accelerating gradient of the tested TTF cavities exceeds the TESLA goal of 25 MV/m. Applying the present fabrication and processing technique developed at TTF we are now routinely obtaining cavities which exceed 20 MV/m, at $Q=10^{10}$, without field emission. Above this field level, Q starts to drop and x-ray, associated with electron emission, are usually observed. The improvement of the EB welding technique, associated with 1400 °C Ti-gettering and eddy current scanning of the niobium sheets, has raised the quench limit above the present Q drop and field emission limits.

Efforts are undertaken to further reduce field emission by improving of the HPR system and the in-situ dust particle control during assembly in the clean room. Nevertheless in the few cavities not limited by field emission the performance is limited by a different Q drop, with no electrons, of the type first observed in single cell cavities at Saclay [16]. This phenomenon has different interpretations but seems to be intrinsically related to the actual processing procedure, based on the buffered chemical polishing (BCP).

Using commercial high purity Niobium, by different suppliers, outstanding results have been obtained at KEK with a different processing technique based on electropolishing. A number of single cell cavities reached accelerating field up to 40 MV/m with a moderate Q drop starting above 30 MV/m [17]. Since this technique could have good potentiality for TESLA, both for field enhancement and cost reduction, an R&D activity in this direction on single and multi-cell cavities was started in collaboration with KEK, CERN, Saclay and industry. Recent results on multi-cell cavities at TJNL and KEK have shown that the extrapolation of the technique developed for single cell to multi-cell cavities needs further investigation. The cavity prototype P1 (Fig. 3) has been sent to KEK for electropolishing. Up to now, after two steps of moderate electropolishing, the cavity is still limited by quench at 22 MV/m. Figure 5 compares the recent KEK results with those obtained at DESY.

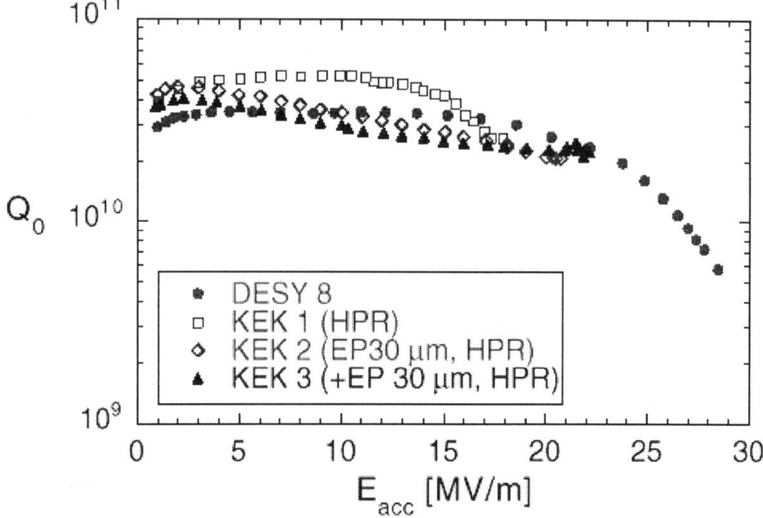

Figure 5. Preliminary results obtained at KEK with a 9 cell cavity prototype from DESY.

Alternative Fabrication Techniques

The aim of cutting cavity costs in the large scale production foreseen for TESLA drove an important R&D activity to find new cheaper techniques for cavity fabrication and stiffening. Two lines are being pursued:

- seamless Nb cavities, by spinning or hydroforming, to eliminate the equatorial welding and to slightly reduce the required niobium inventory [18];
- external thick coating of a thinner niobium cavity, to reduce the niobium inventory and mainly to perform the cavity stiffening, required to counteract the Lorenz forces in pulsed operation, at a lower cost[19].

The first line is pursued at DESY, INFN LNL, and Saclay. The results obtained so far are now approaching those of welded cavities and the application to multi-cell cavities is well advanced. At present the main limitation is still the need, for high performance, of a very deep internal surface removal (> 500 μm) to eliminate the damaged layer. A mono-cell cavity spun at LNL, while treated and measured at TJNAF, reached 33 MV/m, after a total removal of about 700 μm from the internal surface by alternating grinding and BCP [20].

The second line, proposed by IPN Orsay, using standard plasma jet spray of copper, gave initial very promising results and is now pursued in collaboration with DESY, INFN Milano and a few other partners. Different coating techniques and materials are being considered and extensive tests are under way for thermal and mechanical properties.

SUPERSTRUCTURES

Following an idea of Jacek Sekutowicz, since 1997 a growing effort is dedicated to the development and test of superstructures, which could be a challenging alternative to the present 9-cell cavities for the TESLA collider [9].

In the TESLA design [4], for a cavity accelerating gradient of 25 MV/m, the average energy gain produced by a cavity string is limited to 17.8 MeV/m because of the relatively poor cavity filling factor, 75%, defined as the percentage of the cavity active length with respect to actual cavity length. This value comes from the fact that the number of $\lambda/2$ cells per cavity is 9 and the cavity separation is $3\lambda/2$.

The proposed superstructure scheme is based on sub-units, called superstructures, made of four 7-cell standing wave cavities, coupled together through an enlarged $\lambda/2$ long beam pipe. The superstructure is fed by a single main coupler, while each 7-cell sub-structure needs an independent tuner. Five HOM couplers are envisaged for the superstructure and the risk for trapped modes is reduced. Applying the same criteria for the definition of the filling factor, with the new scheme it will grow from the present 75% to about 83%. This important saving may be spent either to increase the final energy for the same cavity gradient or to reduce the required gradient for the same energy. In addition to the increased cavity filling factor, some other advantages are expected, among then I wish to quote:

- the number of main couplers is reduced by a factor of more than 3 and the power distribution is simplified;
- the unflatness of the accelerating field, for the same geometrical errors, is reduced because it scales as the square of the number of cells (7 instead of 9).

The experimental results obtained with a warm model of the superstructure have been very encouraging and the expected RF behavior in terms of filling time, field distribution and sensitivity has been confirmed. Figure 6 shows the simplified design of the actual superstructure being ordered, to be made according to our standard fabrication and processing technique. In particular, to be compatible with the actual infrastructures at DESY, a flanged solution has been preferred, with respect to the final welded one, to connect together the four 7-cell sub-structures that constitute the superstructure prototype. We expect to start the cold RF measurements with beam, in a modified TTF cryomodule, by the end of 2000.

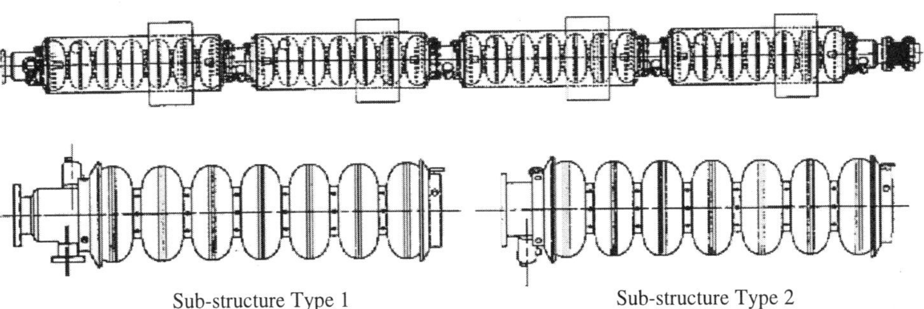

Figure 6. Simplified mechanical drawing of the niobium superstructure whose design has been completed and is being ordered in July 1999, to be cold tested by the end of 2000. An enlarged view of the two types of 7-cell sub-structures is shown in the lower part of the figure. While in the eventual final design the four 7-cell sub-structures should be welded together, we preferred the flanged solution for the prototype.

CONCLUSIONS

The results obtained so far in the framework of TTF are very encouraging and the technical possibility to build TESLA is becoming a reality. In particular the cavities are now routinely reaching the TESLA requirements and no degradation of cavity performance is found between vertical test, horizontal test and linac operation. Recent results and ideas show that even higher fields can be probably reached in the near future.

ACKNOWLEDGEMENTS

Thanks are due to all the members of the TESLA Collaboration for the excellent work done so far, and a special one to Matthias Liepe for figures and transparences.

REFERENCES

1. R. Brinkmann, Low Frequency Linear Colliders, Proc. EPAC'94, London, England.
2. H. Padamsee Ed., Proc. 1st Int. TESLA Workshop, Cornell, USA, 1990, Cornell Report, CLNS90-1029.
3. TESLA Test Facility linac - Design Report, DESY Report, TESLA 95-01, 1995.
4. Conceptual Design of a 500 GeV e+e- Linear Collider with Integrated X Ray Laser Facility, R. Brinkmann et al. Eds, DESY Report, 1997-048, 1997.
5. C. Pagani, D. Barni, M. Bonezzi, J.G. Weisend II, Further Improvements of the TESLA Test Facility (TTF) Cryostat in View of the TESLA Collider., Paper presented at this Conference.
6. S. Wolff, The Infrastructure for the TESLA Test Facility, Proc. PAC'95, Dallas, Texas.
7. C. Crawford et al., High Gradients in Linear Collider Superconducting Accelerator Cavities by High Pulsed Power to Suppress Field Emission, Part. Acc. 49, pp. 1-13, 1995.
8. Q. S. Shu et al., An Advanced Rotating T-R Mapping & its Diagnoses of the TESLA 9-cell Superconducting Cavities, Proc. PAC'95, Dallas, Texas.
9. J. Sekutowicz, et al., Superconducting Super-structures for the TESLA Collider, Proc. of EPAC'98, Stockholm, Sweden, June 1998 and DESY Report, TESLA 98-08.
10. W. Singer et al., Diagnostics of Defects in High Purity Niobium, Proc. of the 8th Workshop on RF Superconductivity, Abano Terme, Italy, Oct. 1997.
11. J. Brawley et al., Electron Beam Weld Parameter Set Development and Cavity Cost, Proc. of the 8th Workshop on RF Superconductivity, Abano Terme, Italy, Oct. 1997.
12. D. Proch, Status of Cavity Development for TESLA, Proc. of HEACC'98, Dubna, Russia, 1998.
13. Ph. Leconte et al., TESLA Test Facility Cold Tuning System, DESY Report, TESLA 93-09.
14 A. Gössel et al., Vertical and Horizontal Cavity Test Results in Comparison with Performance in the TTF Linac, Proc. of the 8th Workshop on RF Superconductivity, Abano Terme, Italy, Oct 1997.
15. C. Pagani, Developments and Achievements at the TESLA Test Facility (TTF), Proc. Applied Superconductivity Conference, Palm Desert, California, September 1998.
16. M. Pekeler, Experience with Superconducting Cavity Operation in the TESLA Test Facility, Proc. of PAC'99, New York, NY, March 1999.
17. M. Ono et al., Achievements of 40 MV/m in L-Band SC Cavity at KEK, and K. Saito, Superiority of Electropolishing over Chemical Polishing on High Gradients, Proc. of the 8th Workshop on RF Superconductivity, Abano Terme, Italy, Oct 6-10, 1997.
18. V. Palmieri, Seamless Superconducting RF Cavities, Proc. of PAC'99, New York, NY, March 1999.
19. S. Bousson et al., An Alternative Scheme for Stiffening SRF Cavities by Plasma Spraying, Proc. of PAC'99, New York, NY, March 1999.
20. P. Kneisel and V. Palmieri, Development of Seamless Niobium Cavities for Accelerator Application, Proc. of PAC'99, New York, NY, March 1999.

DESIGN OF SUPERCONDUCTING RADIO-FREQUENCY CAVITIES FOR A CONTINUOUS-WAVE PROTON LINAC

R. Gentzlinger,[1] B. Campbell,[1] K.C. Chan,[1] G. Ellis,[1] J.P. Kelley,[1]
F. Krawczyk,[1] J. Kuzminski,[2] M. Manzo,[1] R. Mitchell,[1] D. Montoya,[1]
B. Rusnak,[3] H. Safa,[4] D. Schrage,[1] and B. Smith[1]

[1]Los Alamos National Laboratory
Los Alamos, NM 87545

[2]General Atomics
San Diego, CA 92186

[3]Lawrence Livermore National Laboratory
Livermore, CA 94550

[4]C.E.A.Saclay
Gif-sur-Yvette, 91191, France

ABSTRACT

To date, all high-energy proton linear accelerators have been room temperature (or higher) machines. To produce a high-energy beam over a reasonable accelerator length (high accelerating gradient), these machines require large amounts of radio-frequency (rf) power to maintain accelerating fields. A substantial amount of this power, up to 50% in some cases, is dissipated as heat in the cavity walls. Using a superconducting accelerator minimizes this wasted power. In addition to the power savings, a superconducting cavity has a large velocity acceptance. This velocity acceptance minimizes the number of cavity designs necessary for a given accelerator. Also, superconducting cavities permit a large bore diameter, which reduces the interaction with the beam halo. These types of cavities have been successfully applied in electron accelerators. At the Los Alamos National Laboratory, a prototype design of proton superconducting cavities has been developed for the Accelerator Production of Tritium (APT) project. These cavities will operate at 2.15 K in a liquid-helium bath contained in an unalloyed, Grade 2 titanium vessel. The cryomodule, which contains the cavities and helium vessels, uses stainless-steel cryogenic piping for which special fittings were developed to transition from the titanium vessel to the stainless-steel pipes. This paper discusses the design and fabrication of the cavity and helium vessel and the experience gained during fabrication of the several cavities.

Advances in Cryogenic Engineering, Volume 45.
Edited by Shu *et al.*, Kluwer Academic / Plenum Publishers, 2000.

INTRODUCTION

The superconducting radio-frequency (SCRF) portion of the APT accelerator[1] only requires two cavity designs. The cavity design is driven by the velocity of the particles passing through it, as the accelerator gap of the cell is $\beta\lambda/2$, where λ is the free-space wavelength of the rf power resonating cavity and β is the ratio of the particle velocity relative to the speed of light. Only two cavities are required for APT ($\beta = 0.64$ and $\beta = 0.82$) because SCRF structures have very large velocity acceptance. Having only two cavity shapes minimizes design, fabrication, and prototyping costs.

Elliptical cavities having a $\beta \approx 1$ are used throughout the world for accelerating electrons, but a cavity of $\beta < 1$ used to accelerate protons had two major unknowns: (1) could the flatter shape of the reduced-beta cavity reach the same accelerator gradients as the $\beta \approx 1$ cavities without multipacting, and (2) what effect would a continual dose of protons on the superconducting cavity walls over the expected 40-year life of the machine have on the performance of the niobium. An engineering, development, and demonstration (ED&D) program was undertaken to study these issues. Four single-cell cavities (two each of $\beta = 0.48$ and $\beta = 0.64$) were fabricated and tested to see if they would multipactor at lower-than-expected field levels. Two 3-GHz elliptical cavities were irradiated with a 6×10^{16} protons to evaluate what effect the dose had on the superconducting performance of the cavity. In both cases the results were very encouraging.[2,3]

The other emphasis of the ED&D program is to build prototypes of multi-cell cavities and test their performance. The tests will include measuring the Quality factor (Q) vs. the accelerating gradient (E_{acc}) for the cavities in vertical and horizontal orientations, measuring the actual thermal loads to the 2 K bath, comparing these loads to the predicted ones, and evaluating cleaning and assembly procedures.

APT ACCELERATOR PHYSICS DESIGN

The APT proton linac design has already been described in detail.[1,4] This article will highlight the features of the present design.

The present APT linac design is based on a modular configuration. The proton linac that operates at 100-mA continuous wave (cw) will accelerate protons to a beam energy of 1030 MeV. It can be upgraded to 1700 MeV by extending the superconducting portion of the high-energy linac. The APT linac is designed as a two-stage machine, using both normal-conducting and superconducting accelerating-cavity technologies.

The superconducting high-energy linac consists of cryomodules containing two, three, or four 5-cell, 700-MHz, niobium superconducting accelerating cavities. Focusing is provided by normal-conducting quadrupoles in a doublet lattice located in the warm regions between cryomodules. There are two kinds of superconducting cavity shapes, each designed for efficient acceleration in a different velocity range. The cavity shape is optimized in the medium-beta section (211.4–471.4 MeV) for $\beta = 0.64$ and in the high-beta section (471.4–1700 MeV) for $\beta = 0.82$.

APT ACCELERATOR ENGINEERING DESIGN AND ANALYSIS

The APT SCRF cavities went through many design iterations before the configuration described herein was adopted.[5] The design incorporates a 5-cell, 700-MHz cavity designed for betas of 0.64 and 0.82. The titanium helium vessel is made up of inner and outer vessels. The inner vessel provides the load-bearing capability for reacting the active tuner loads applied to the cavity. It also could provide transverse support to the cavity if microphonics are problematic. The outer vessel provides liquid helium storage and ullage for cavity operation. The outer vessel for the $\beta = 0.64$ cavity will store approximately 120 liters of liquid helium at a pressure of 0.031 atm.[6]

Titanium was chosen as the helium vessel material for the following reasons. (1) Residual magnetic fields in stainless steel could adversely affect performance of the superconducting cavity. Shields are generally installed over the helium vessel to reduce the earth's magnetic field to less than 10 mGauss in the vicinity of the cavity. However, the proximity of the helium vessel relative to the cavity would not allow any shielding to be installed. (2) Thermal expansion for niobium and titanium is almost identical. Thus, the stresses caused by going to cryogenic temperatures and by the detuning resulting from differential thermal expansion between the vessel and the cavity would be minimized. (3) Titanium can easily be electron-beam-welded to niobium. This allows a good helium-tight joint to be made between the cavity and the helium vessel. If a stainless-steel helium vessel were adopted instead, a special interface, leak-tight to super-fluid helium, would have to be developed.

The structural and thermal analyses were performed using the ABAQUS® finite element code. The operating pressures of 2.2 atm and 3.0 atm were applied to the model at room and operating temperatures, respectively. The 2.2 atm pressure was the design case for determining sizes of the components. The yield and ultimate strengths of niobium and titanium go up by an order of magnitude at cryogenic temperatures. The analyses were hampered, however, because the fracture toughness for these two materials at cryogenic temperatures is unknown. Determination of the maximum flaw size could not be made without knowing this material property. A test program to address this difficulty was undertaken and will be discussed later.

CAVITY AND HELIUM VESSEL FABRICATION

Cavity and helium vessel fabrication is broken into three main sections. Fabrication and installation of the cavity, inner vessel, and outer vessel are described here. Component assembly is a major consideration in combining the three subsystems.

Cavity Fabrication

Six prototype five-cell cavities (Figure 1a) were manufactured, five by industry and one by Los Alamos National Laboratory (LANL). The cavity shown in Figure 1b, one of four fabricated by CERCA in Romans, France, is made from two types of niobium. Similar cavities (omitting the helium vessel, however) were fabricated by LANL and Advanced Energy Systems (AES) of New York. The CERCA cavity ½-cells and coupler beamtube are fabricated of high (>250) residual resistivity ratio (RRR) niobium material. The balance of cavity components (except for the liquid helium vessel heads) is fabricated from low RRR material.

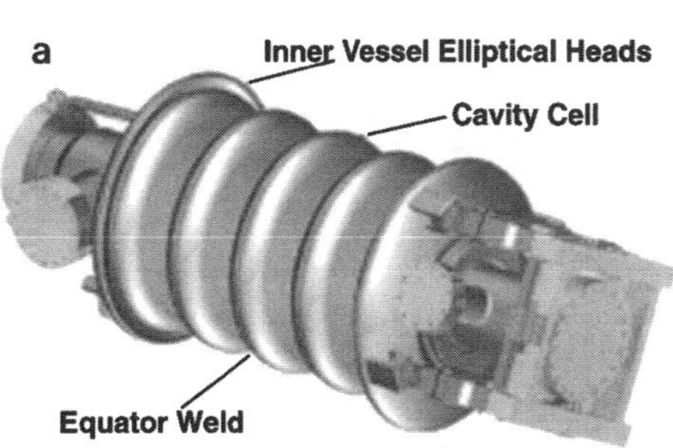

Figures 1a and 1b. APT $\beta = 0.64$ ED&D cavity.

Of the four cavities being fabricated by CERCA, the second set will have high RRR coupler ports. The higher thermal conduction has a negligible effect on the heat load to the 2 K helium.[7]

The cavity ½-cells are made from 4-mm-thick (0.157-in.) niobium and are formed by a spinning process into their final shape. Spinning the cavity ½-cells allows the minimum inner radius-of-curvature on the ellipse to have an R/t ratio of 2.5. Afterward, the equator and iris weld preps are machined. After machining, the ½-cells are electron-beam-welded together (at a pressure greater than 5×10^{-5} torr) at the iris to make a *dog-bone*. The full-penetration weld is performed from the inside. Prior to equatorial welding, the inside surface of the ½-cells are carefully inspected for scratches, inclusions, and holes. Imperfections must be removed. The dog-bones are then electron-beam-welded together at the equators, from the outside, using beam parameters that yield a smooth cosmetic-under-bead joint. The welding tooling must also incorporate adequate conductance to keep the pressure at 5×10^{-5} torr everywhere within the cavity. The additional gas load resulting from gas evolution due to the higher temperatures of the weld region will contaminate the heat effect zones adjacent to the welds if pumping conductance is too small.

The ports are fabricated from low RRR niobium (ASTM B391, Grade 1) tubing and stainless-steel flanges. A rough-machined piece of stainless steel with a closely sized center hole (tapered or cylindrical) is closely fitted to the niobium tube. Provisions for braze material are machined into the stainless steel. A gold alloy or copper can be used as the braze alloy. The prototype cavities fabricated at Los Alamos featured Nioro® (82% gold, 18% nickel) as the braze alloy. Nioro was used to minimize the amount of nickel that pure gold would leach out of the stainless steel during the brazing cycle. Initially, Conflat® flanges were purchased and brazed to the tubes, but the tight alignment tolerances resulted in many of the ports being rejected. Machining the flange afterward allowed the alignment tolerances imposed during brazing to be relaxed. An expansion ring was also incorporated inside the niobium tube in the region of the flange where it uniformly kept the tube in contact with the flange. This greatly reduced the cocking of the tube.

Inner Vessel Fabrication and Installation

The inner vessel, shown in Figure 2a, is fabricated from unalloyed Grade 2 titanium. It contains five components, two elliptical heads, two edge-welded bellows, and a cylindrical shell.

The elliptical heads are fabricated from 6.4-mm-thick material by Titanium Fabricators of Fairfield, New Jersey. A deep-drawing process forms the re-entrant portion of the head. After the re-entrant portion is formed, the 63.5- × 228.6-mm semi-minor by semi-major ellip-

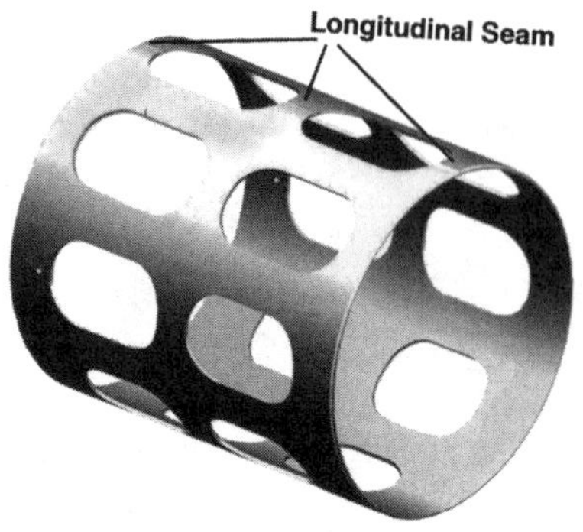

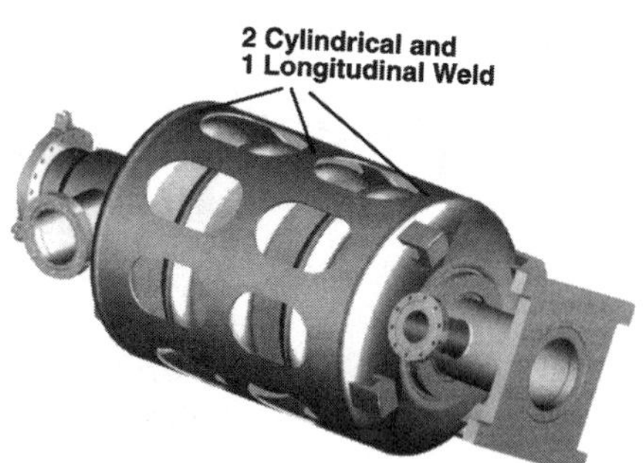

Figures 2a and 2b. The inner vessel shown and installed.

tical head is formed by *hot-spinning*. The surfaces are then ground to remove any deep scratches. A pickling process to remove the oxidized layer follows the grinding. As a result, the surface will have a 32 root-mean-square (rms) micro-inch (μ-inch) finish, which will be easy to clean in the cleanroom. Afterward, the tuner interface and strut support brackets are tungsten inert gas (TIG) welded to the heads in an open room using specially developed argon gas shields. Final machining is then done on the outer diameter weld prep, the tuner interface brackets, and the bellows interface. Afterward the edge-welded bellows is welded to the assembly. A leak check of 1×10^{-10} torr-liters/sec helium is conducted on the assembly to ensure it is leak-tight. The assemblies are shipped to Los Alamos, where the heads are leak-checked again. After the second leak-check, the heads are repackaged and shipped to the cavity manufacturer for integration into the cavity assembly.

Senior Flexonics of Sharon, Massachusetts, fabricated the titanium edge-welded bellows. A cryogenic lifetime-repetition count of 16,000 cycles was determined by calculating the number of days the cavities would have to be re-tuned over the expected 40-year life of the accelerator. This established the fatigue life of the bellows. The limited experience of this, or any, bellows manufacturer in delivering a product meeting these cryogenic design requirements made product reliability questionable. A test program was undertaken at Jefferson National Accelerator Test Facility (JLab) to cycle the bellows by ± 1.5 mm at 77 K until the bellows failed. The temperature of 77 K was chosen over 4 K because titanium does not undergo a ductile-to-brittle transition at this temperature, most of the thermal contraction has occurred by 77 K, and liquid nitrogen is much cheaper than liquid helium. The bellows were helium leak-tested every 5,000 cycles. The test frequency was decreased as the bellow's cycles increased without failure. The test was halted after the bellows was cycled 270,000 times without failure.

The cylindrical shell will be fabricated from 4.7-mm-thick titanium. The large racetrack holes (see Figure 2a), which allow the liquid helium to reach the cavity, will be cut using a high-pressure water-jet before the shell is rolled. The shell will be pickled before being installed on the cavity. Special care must be taken to ensure that any resulting weld spatter does not contaminate the outer surface of the cavity during the installation of the inner vessel shell over the cavity. Any weld-spatter shielding can be removed through the racetrack holes. Figure 2b shows the inner helium vessel installed on the cavity.

Outer Vessel Fabrication and Installation

The outer vessel, shown in Figure 3a, is also fabricated from unalloyed Grade 2 titanium. The 6-mm-thick × 101.6-mm × 355.6-mm (semi-minor by semi-major, respectively)

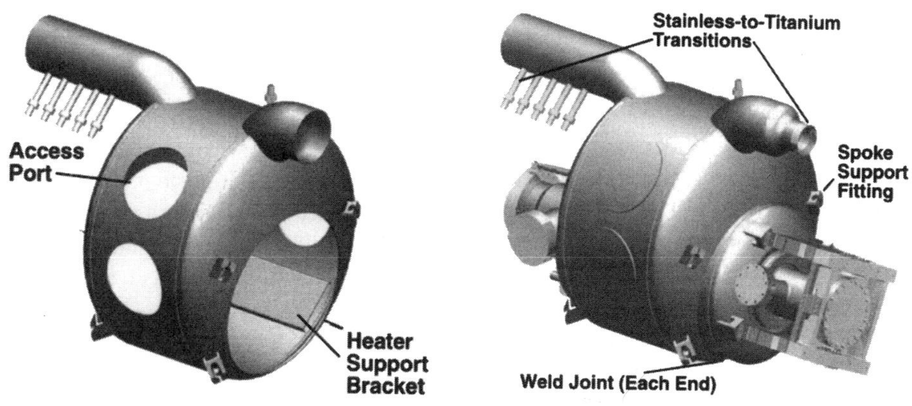

Figures 3a and 3b. Outer helium vessel shown and installed.

elliptical heads will be hot-spun as the inner vessel heads were. The shell of the vessel is also made from 4.8-mm-thick titanium. The identical weld preps and procedures are used to attach the heads to the cylindrical shell.

The ED&D outer vessel will have four access ports around the cylindrical shell to provide access to the inner vessel, cavity, and instrumentation installed inside. The design allows the covers to be welded closed, but they can be opened by grinding the weld off without compromising the integrity of the cylindrical shell or contaminating the inside of the vessel with chips. The weld design also allows the cover to be welded on without contaminating the cavity with weld spatter.

Machined parts for mounting the cryomodule spokes are welded to the elliptical heads at each end of the helium vessel. At operating temperature the load in each spoke exceeds 9000 newtons.[8] The design of the fittings must allow the spokes to be cleaned easily when the vessel is in the clean room. All welds are designed to have complete joint penetration to eliminate the possibility of virtual leaks.

One large disadvantage of a titanium vessel is interfacing to standard off-the-shelf components. For example, the cryomodule plumbing is planned to be 304 stainless-steel, which cannot be TIG-welded to titanium. Thus, a development program was undertaken to fabricate transitions between stainless steel and titanium using inertia welding.[9] The transition piece must be leak-tight and show sufficient fracture toughness at the cryogenic operating temperature. Transitioning to stainless steel will allow inexpensive common components to be attached to the vessel.

The outer vessel is installed over the inner vessel by machining two 457-mm holes on each end in the elliptical heads. The interface on the head is beveled before welding to ensure a complete joint penetration is achieved. A filet weld is then added to give additional strength to the joint. Figure 3b shows the inner vessel assembly installed in the outer vessel.

Instrumentation

The following list of instrumentation is for the helium vessel only. Some planned instrumentation has been omitted due to lack of availability. Mounting provisions and an assembly plan have been integrated into the helium vessel assembly. The access holes in the helium vessel can be used if any instrumentation fails to operate during testing of the cryomodule.

No modifications to the vessel were required to install a cavity positioning system. A wire-position monitoring system will be installed on the cavity flanges to measure cavity vibrations. Accelerometers will be used during modal testing of the cavity before its installation in the cryomodule. Modal testing will be discussed later.

A liquid-level probe will be installed on each cavity. A leveling tube welded at the bottom of each helium vessel keeps the level of helium between the cavities constant. Having a sensor on each cavity allows a redundancy in the measurement. A carbon resistor will be fastened to the bottom of the liquid level probes to determine when liquid helium is starting to collect at the bottom of the vessel.

Heaters will be incorporated into the helium vessel design. Two 200-watt heaters are planned for each helium vessel, allowing the load on the cryogenic system to remain constant. If a cavity has to be detuned and taken off-line, the wattage in the heater will be turned up to equal the thermal load the cavity had. The heaters will be installed in the outer vessel prior to installation of the inner vessel. It would be possible to install heaters from the access ports, but the task would be very difficult.

Four diodes will be installed on each end of the inner helium vessel (top and bottom) to measure incoming and outgoing helium temperatures. A tab was previously added to the inner vessel for mounting diodes.

A small quench detection system will be installed on each cavity. Four carbon resistors will be mounted on each cavity cell (total of 20) for determining where a quench may have originated.

Ceramic wire feed-throughs will be used to transfer power and signals through the vessel wall. The ports are located on the 168-mm-diameter vent tube on top of the outer vessel.

SUPPORTING ENGINEERING TEST

Material Testing

A test program is underway at the High Magnetic Laboratory at Florida State University (FSU) to determine at liquid helium temperature the yield and ultimate strengths, as well as the fracture toughness, of both the welded and parent materials used in the cavity-helium vessel assembly. Data will be incorporated into the finite element models to determine safety factors and allowable maximum flaw sizes in the fabricated assemblies. Manufacturers of the various cavities and helium vessels are also fabricating coupons for testing at FSU to qualify their manufacturing processes.

Testing will also be performed on inertial-welded coupons for property evaluation. Edison Welding Institute and Los Alamos National Laboratory will develop the welding parameters for the 88.9-mm (3.5-inch) and 19-mm (.75-inch) diameter joints, respectively. Los Alamos National Laboratory has found inertially welded joints are very sensitive to the parameters used for fabrication. Thus, it is essential that these joints are well-characterized and the parameters documented to control the fabrication process. The high sensitivity of the joints to small variations in the parameters does not allow them to be scaled reliably.

Microphonic Testing

The cavities will also be tested at LANL to evaluate their sensitivity to microphonics, or how the resonant radio-frequency of the cavities varies due to small displacements caused by ambient noise. Additional rf power is needed to maintain the correct accelerating field when the resonant frequency of the cavity changes. For example, a shift of 77 Hz in resonant frequency will require 1% more power. The resonant radio-frequency of the cavity will change by 319 Hz per micron of axial displacement of the irises toward or away from each other. The control margin available from the rf system to the cavity is 5% of the total cavity power. Of that, 20% of the control margin or 1% of the total cavity power is allocated for correcting for microphonics. If the resonant radio-frequency varies by 77 Hz, the microphonics control margin will be depleted. Therefore, if ambient noise causes an axial displacement in the cavity of greater than 0.22 micron, the additional power available to maintain the correct accelerating fields will not be available. Calculations for determining the displacements caused by ambient noise are made by using a finite element code. Small errors in the model could have a significant effect on the results. If the calculated mechanical resonance is near a spike in the facility power spectral density (PSD) curve, the impact might be determined to be negligible. If the resonance is actually on the spike, the impact could be considerable.

A multipart testing program was established to corroborate the finite element model with actual cavity results. The first phase involved comparing measured and expected resonant frequencies and mode shapes of a prototype cavity with free-free boundary conditions. This work has been concluded, and the results showed good agreement between the model and the measurements. Phase II will measure and predict resonant frequencies and mode shapes for an ED&D cavity having the same boundary conditions. An ED&D cavity uses thicker niobium for the cavity cells, is equipped with helium elliptical heads, and has the struts and tuners installed. Phase III will do the same thing, except the inner vessel will also be installed. In Phase IV, the outer helium vessel will be installed on the cavity. Having the outer vessel installed, the cavity and helium vessel assembly can be suspended by using spokes attached to a test frame that simulates the cryomodule. This will provide the same boundary conditions to the cavity as when it is installed in the cryomodule. If the predicted and actual resonant frequencies and mode shapes are the same, a high degree of confidence in the model would result for determining the cavity's response when it is installed in the cryomodule. With a high degree of confidence in the Finite Telement Analysis model, accurate cell displacements could be determined for a particular PSD. A code for determining the expected frequency shift caused by the predicted displacements needs to be developed. The frequency shift caused by a pure axial displacement in the cavity can be calculated using Slater's Perturbation Theorem, but determining the shift caused by transverse modes will not be as straightforward. One proposal is to force the cavity into the predicted mode shapes and greatly exaggerate the displacements so the shift in frequency can be measured. This test would be performed on the tuning bench where all equipment to do the measurement is available.

ACKNOWLEDGMENT

The authors acknowledge CERCA of Romans, France; Senior Flexonics of Sharon, Massachusetts; and Titanium Fabricators of Fairfield, New Jersey, for their contributions. This work was supported by the U.S. Department of Energy under contract No. DE-AC04-96AL89607.

REFERENCES

1. G. Lawrence, "High-power Proton Linac for APT Status of Design and Development," Proc of the XIX International Conference 1998, Chicago, IL (ANL-98/28).
2. W. Haynes *et al.*, "Medium-Beta Superconducting Cavity Tests at Los Alamos National Laboratory for High-Current, Proton Accelerators," 8th Conference on RF Superconductivity, Abano Terme, Italy (LANL Memo: LA-UR-98-682).
3. B. Rusnak *et al.*, "*In Situ* Irradiation and Measurement of Superconducting RF Cavities Under Cryogenic Conditions," PAC '97 Conference, Vancouver, B.C., Canada (LANL Memo: LA-UR-97-1663).
4. K.C. Dominic Chan, "Application of RF Superconductivity to a High-Current Linac," 1998 Applied Superconductivity Conference, Palm Desert Conference (LANL Memo: LA-UR-98-4669).
5. R. Gentzlinger *et al.*, "Design, Analysis, and Fabrication of the APT Cavities," 1999 Particle Accelerator Conference, New York, NY (LANL Memo: LA-UR-99-1462).
6. B. Campbell *et al.*, "Design Status of the Cryomodules for the APT Linac," 1999 Particle Accelerator Conference, New York, NY (LANL Memo: LA-UR-99-1462).
7. J. Waynert, Private Communication, July 2, 1999.
8. M. Fagan *et al.*, "Engineering Analysis of the APT Cryomodules," 1999 Particle Accelerator Conference, New York, NY (LANL Memo: LA-UR-99-1485).
9. M. Cola *et al.*, "Dissimilar Metal Joints for the APT Superconducting Cavity's Cryogenic Plumbing System," 1999 Particle Accelerator Conference, New York, NY (LANL Memo: LA-UR-99-1512).

EXPERIMENTAL INVESTIGATION OF THE THERMAL RESISTANCE IN NIOBIUM SAMPLES FOR SUPERCONDUCTING RF CAVITIES

M. R. Smith[1], T. Zhang[1], Y. Xiang[1], S. W. Van Sciver[1,2]
J. G. Weisend II[3], P. Schmueser[3], and M. Fouaidy[4]

[1]National High Magnetic Field Laboratory
Tallahassee, Florida 32301
[2]Department of Mechanical Engineering
FAMU-FSU College of Engineerng
[3]Deutsches Elektronen-Synchrotron
Hamburg, Germany
[4]Institut de Physique Nucléaire d'Orsay
91406, Orsay Cedex, France

ABSTRACT

While Kapitza conductance is fairly well understood in terms of phonon mismatch between a given surface and the adjacent liquid helium, accurate values depend strongly upon the condition and other microscopic characteristics of the surface, and must often be determined experimentally. The niobium cavities designed for the TESLA superconducting linear accelerator undergo a series of chemical and heat treatments, making it very difficult to predict their thermal conductivity or Kapitza conductance. We have constructed an apparatus to accurately characterize the total heat transfer, including Kapitza conductance, across typical TESLA niobium samples, at temperatures spanning the range from 1.7 to 2.1 K, with heat fluxes as high as 1000 W/m^2. Results for pure niobium, Titanium coated niobium and prototype niobium-copper composite samples are presented.

INTRODUCTION

Superfluid helium has excellent cooling properties which render it especially useful in applications involving large scale superconducting magnets and RF cavities for particle accelerators (CEBAF, LHC, TESLA). Due to the large surface areas involved, and the detuning effect of any local warming, the process of transporting heat from a solid body to a surrounding bath of He II is especially important to RF cavity design. The associated thermal conductance may be expressed as shown in Eq. (1).

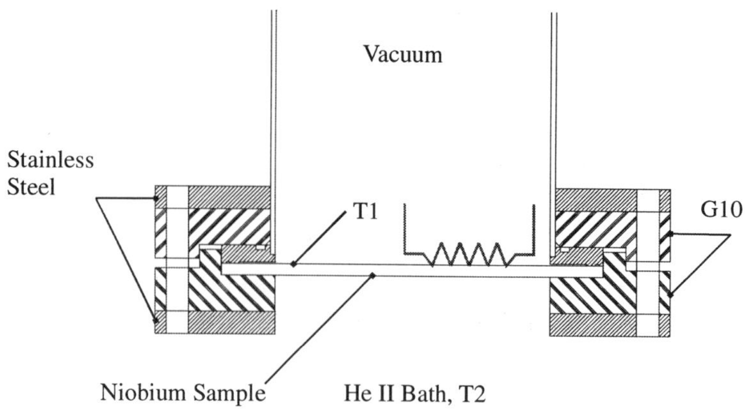

Figure 1. Schematic layout of thermal resistance experiment. The resistive heater covers the entire vacuum side of the sample. The different metallic coatings were applied to the helium II side only. G-10 flanges and mechanical contacts were designed to minimize heat transfer off the sample axis.

$$h = \lim_{\Delta T \to 0} \frac{q}{\Delta T} \tag{1}$$

Here q is the heat flow per unit area across the interface, and ΔT is the temperature difference between the inner surface, and the He II bath. The coefficient, h, consists of the thermal conductance of the bulk niobium sample (k/t, where t is the thickness) in series with the Kapitza surface conductance. Kapitza conductance is of great technical interest because it often results in the largest temperature difference in problems of low temperature heat transfer. A heat flux of 1000 W/m^2 for example, can lead to a temperature difference of about 100 mK at the interface due to Kapitza resistance. It would require about 1000 m of straight channel to produce the same temperature difference for turbulent He II flow with the same heat flux.

There have been many theoretical and experimental investigations of Kapitza conductance for various materials. Variations by up to an order of magnitude are seen for the same material, arising from variations in the surface condition, such as chemical treatment, cleanliness, oxidation and coating. Previous studies on niobium[1,2] have not dealt extensively with the effect of metallic coating on the overall heat transfer. Our experiment examines the impact of a copper or titanium layer on the overall heat transfer through a niobium sample to a He II bath. Results for Nb specimens with copper and thin titanium coatings are presented and compared with measurements on niobium samples without any coating.

EXPERIMENTAL SETUP AND PROCEDURE

The experimental setup is shown schematically in Figure 1. The test samples are circular plates, approximately 85 mm in diameter, and 3 mm thick. Samples were machined from pure niobium, and subsequently heat treated at 1400 C. This process produced relatively soft specimens, which required careful handling to avoid deformation when pumping on the vacuum side. As part of the process, samples were coated with a thin layer of titanium. One such sample was later chemically etched to remove this layer, resulting in the uncoated niobium sample. A copper layer was applied to a pure (uncoated) niobium sample at IPN by means of a plasma jet, sprayed on at approximately 200 C.

The overall thickness was approximately 6 mm, with the copper making up nearly half of that. Copper sprayed on in this manner exhibits a porosity of 3-6% by volume, making Kapitza conductance particularly difficult to anticipate. Samples are mounted to a vacuum can via indium seals. Two 10 mm thick G-10 flanges clamp the sample in place, and reduce the heat flux in the radial direction. Two 5 mm thick stainless steel flanges provide mechanical support for the bolts which hold everything in place. On the vacuum side, a 60 mm diameter resistive heater is mounted to the sample to supply heat flux. Resistance for these heaters is typically 164 Ohms (room temperature), and the heater covers nearly the entire upper surface. These factors, together with the thickness being much less than the diameter, allow us to treat the heat flux as approximately one-dimensional along the axial direction.

Temperatures are measured with Cernox thin film resistors, made by Lakeshore Cryogenics Inc. These were calibrated prior to use. One sensor is in the He II bath, and the other affixed to the top surface of the sample, using GE varnish. The thermometer leads leave the cryostat via a copper heat sink to reduce heat transfer from room temperature. Resistance values are determined using a standard four wire technique. The temperature of the He II bath is controlled by saturated vapor pressure, regulated by a control valve and MKS controller. Accuracy of the temperature measurement is ±1mK below 10 K. A superconducting level sensor is used to monitor the liquid level. A four wire technique is also used to measure the heat flux.

In operation, the vacuum on the heated side is always maintained below 10^{-4} Torr. Once the pressure set point was attained, and the bath temperature stabilized at the desired temperature (1.6, 1.8, and 2.0 K, as dictated by DESY operational concerns) the heat flux is increased incrementally from zero to 1000 W/m^2. The result is a series of ΔT versus heat flux plots, characteristic of each sample.

RESULTS AND DISCUSSION

Figure 2 shows Q versus ΔT for the niobium sample with an uncoated surface. The response is nearly linear over the entire range of heat flux. When the experiment is carried

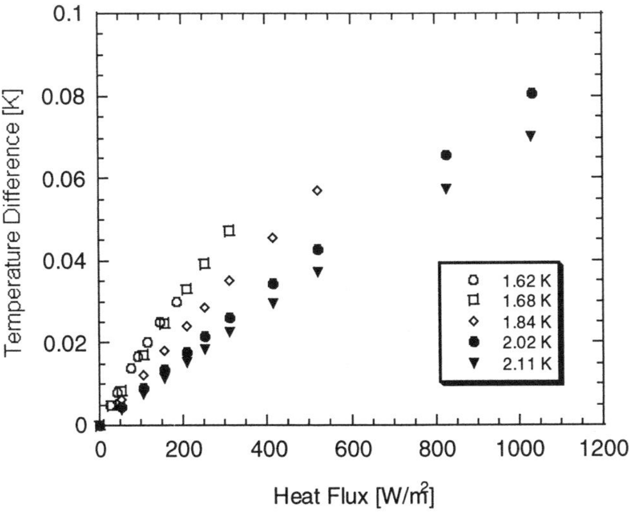

Figure 2. Temperature difference versus heat flux for uncoated niobium sample.

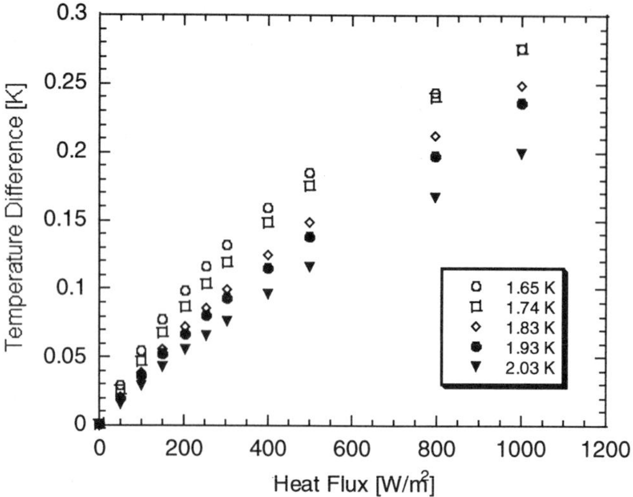

Figure 3. Temperature difference versus heat flux for niobium sample with titanium coating on exterior surface

out at higher temperatures, the increased thermal conductivity results in a lower temperature difference across the sample. Neither the Kapitza conductance, nor the bulk thermal conductivity dominate the heat transfer, since both are of approximately the same order of magnitude[2]. It is worth noting however, that niobium properties are strongly dependent upon heat treatment and surface conditions[1,2].

The effect of a thin titanium coating is apparent in Figure 3. Although we anticipate that the Kapitza conductance at the liquid-solid interface is now an order of magnitude larger than the bulk thermal conductivity of the niobium sample[2,3], the overall conductance appears to drop by nearly a factor of three. This could be the result of contact resistance at the bimetallic interface, or the formation of an alloy of titanium and niobium at the interface, with lower thermal conductivity than either of the two constituents. Note that maximum temperature difference across the sample at 1000 W/m^2 goes from approximately 80 mK for the niobium surface up to nearly 280 mK for the sample with the titanium coating.

The most interesting data however, are those corresponding to the sample with the copper surface, shown in Figure 4. As previously mentioned, the copper was applied via a plasma sputtering technique. The copper layer was applied to a thinner niobium substrate, and built up so that the total thickness was again approximately 3 mm, with the copper making up approximately half of this. The goal in this process was to enhance the effective surface area by means of the porosity of the copper layer, taking advantage of the coppers lower cost and higher mechanical strength in the construction of future RF cavities. Additionally, copper's bulk thermal conductivity is well over two orders of magnitude larger than that of niobium[2,3]. The resultant layer however, was highly non-uniform, and heavily oxidized. Since the layer is actually comprised of small bits of copper in point contact with one another, with 3-6% void fraction, any oxidation at those contact points would greatly reduce the overall thermal conductivity. While the present technique certainly resulted in a somewhat higher effective surface area (keeping in mind that the hydraulic path between the helium in the voids and that in the bath is far from perfect), contact resistance and surface oxidation appear to make the copper layer behave more like a blanket than an enhancement to overall heat transfer. The strong departure from linear

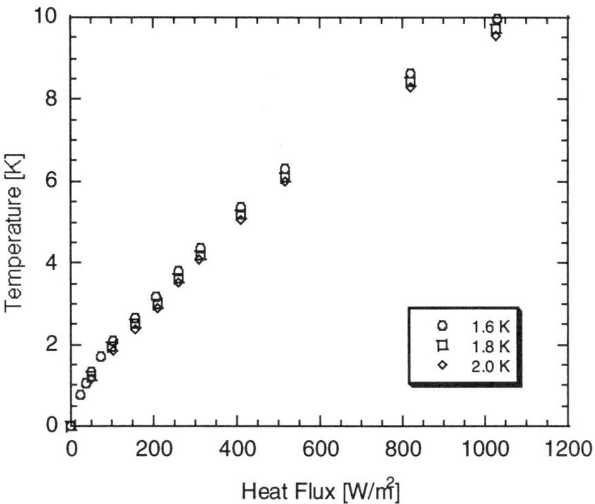

Figure 4. Temperature difference versus heat flux for niobium sample with copper coating on exterior surface

behavior in Figure 4 results from the large temperature difference across the sample, and the corresponding temperature dependence of bulk thermal conductivity in the copper and niobium.

The overall thermal conductance is shown in Figure 5 for all three samples. Note that the effective conductivity for the copper sample is over two orders of magnitude smaller than that of pure niobium. The titanium coating results in almost a factor of 3 difference inthe effective conductivity. The lines represent power law fits to the data of the form,

$$k_{eff} = \alpha \cdot T^{\beta} \tag{2}$$

Where k_{eff} is the overall conductance of the sample, and α and β are summarized in table 1.

CONCLUSION

Overall thermal conductivity measurements of niobium with and without copper and titanium coatings suggest that the pure, untreated niobium surface offers the best heat transfer to a surrounding helium bath. Although the bulk thermal conductivity for copper is higher than that of niobium, and the plasma spraying technique results in a higher effective surface area exposed to the helium bath, heavy oxidation and unknown contact resistance within the copper layer result in an overall thermal conductance which is over two orders of magnitude smaller than that of the plain niobium sample. Results were fit and the fit parameters presented for easy use in the future.

ACKNOWLEDGEMENTS

Work was supported by the US Department of Energy, Division of High Energy Physics under grant DE-FG02-96ER-40952. We also wish to acknowledge partial support from DESY, and the NHMFL.

Table 1. Fit parameters for overall thermal conductance

Exposed Surface	α [kW/m^{2}K$^{1+\beta}$]	β
Niobium	0.955	3.54
Titanium	0.363	2.99
Copper	0.0404	0.447

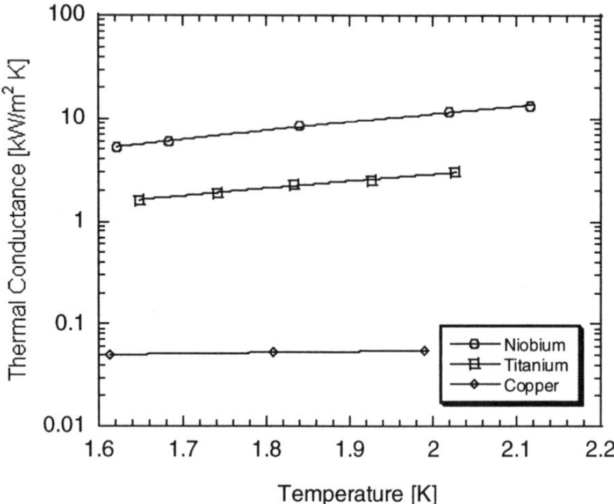

Figure 5. Overall thermal conductivity for niobium samples under three surface conditions. Lines represent fits to the data. Fit parameters are summarized in table 1.

REFERENCES

1. M.X. Francois, Heat transfer between niobium and helium superfluid: recent results, in "Proceedings of ICCR '98," Hangzhou, China, April 1998.

2. A. Boucheffa, M.X. Francois and F. Koechlin, Kapitza resistance and thermal conductivity for niobium, *Cryogenics* 34:297 (1994).

3. S.W. Van Sciver, "Helium Cryogenics," Plenum Press, New York (1986).

FRICTION MEASUREMENTS FOR SC CAVITY SLIDING FIXTURES IN LONG CRYOSTATS

D. Barni, M. Castelnuovo, M. Fusetti, C. Pagani and G. Varisco

INFN Milano-LASA, Via Fratelli Cervi 201, I-20090 Segrate (MI), Italy

ABSTRACT

Filling factor and static cryogenic loss considerations suggest the use of long multi-cavity cryomodules for high energy superconducting linacs. In the case of TTF/TESLA a further gain has been obtained by including the Helium Gas Return Pipe (HeGRP) in the cryostat vacuum vessel and using it as the reference supporting beam for the active elements. One limit of this solution has been the need of developing expensive flexible couplers to feed RF power into the cavities. For cost and reliability considerations in view of TESLA, a new spring loaded sliding fixture, to connect the cavities to the HeGRP, has been developed and qualified at the different operating conditions, from room to cryogenic temperature. Static and quasi-static friction has been measured in a dedicated apparatus for different material coupling and surface finishing. The icing effect in case of a vacuum break has also been studied. The results of this analysis are presented in this paper.

INTRODUCTION

The third[1] cryomodule generation designed for the TESLA Test Facility (TTF)[2] Super Conducting accelerator opens the possibility to use semi-rigid power couplers and superstructures[3]. This result has been obtained by designing a new fixture scheme that decouples the longitudinal position of the cavity string from the structural supporting beam, that is the Helium Gas Return Pipe (HeGRP). During the cooldown, the thermal contractions of the cavity string and of the HeGRP are independent, but the cavity alignment guaranteed by the HeGRP is preserved. Each cavity is now connected to the HeGRP by four low friction sliding fixtures. The qualification tests of this new fixture design are presented in this paper.

SLIDING SUPPORT

As in the previous generation cryostats, in the present design the SC cavities and the quadrupole package are anchored, through the Helium vessels, to the HeGRP which acts as

Advances in Cryogenic Engineering, Volume 45.
Edited by Shu *et al.*, Kluwer Academic / Plenum Publishers, 2000.

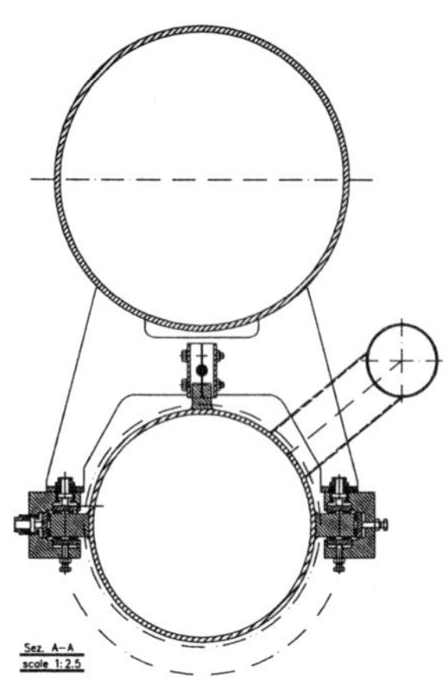

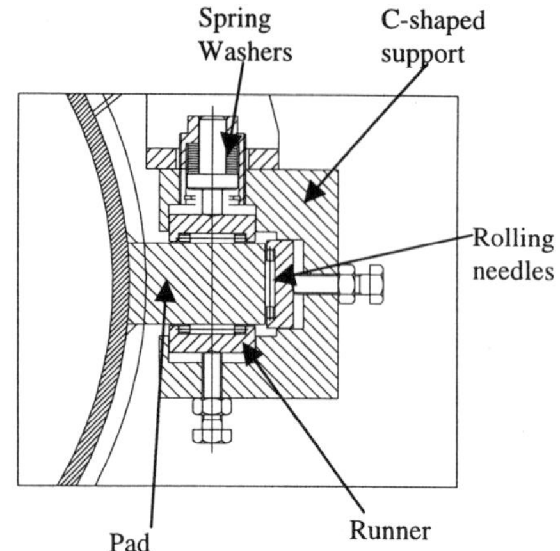

Figure 1. The cavity support scheme. Each cavity is attached to the HeGRP by means of four fixtures. Each fixture includes rollers, runners and reference screws to define the vertical and lateral position. Spring washers are used to give a constant 80 kg load. Rolling needles keep the cavity longitudinal position free.

a structural beam for the cold mass and preserves its alignment from room temperature to cryogenic operation.

In the third generation of TTF Cryomodules, a new fixture design has been developed in order to adjust and keep fixed the transverse position of the active elements, while leaving the cavity longitudinal position independent from the HeGRP displacements due to the thermal contraction/elongation during cooldown/warmup cycles. This freedom is obtained through a set of low friction rolling needles. The drawing of a cavity anchored by the sliding fixtures is presented in Fig. 1. Each cavity is connected to the HeGRP with four sliding fixtures, each clamping a titanium pad welded on the cavity helium vessel. Each pad fixture consists of a C-shaped stainless steel element that defines the transverse position (with respect to the beam axis) through the sequence of rolling needles, runners and reference screws. In each direction one screw is used for adjustment, while the other is transferring a 80 kg_f force via spring washers. Except for the low friction, the longitudinal displacement is independent from that of the HeGRP. The cavity longitudinal position, that could be determined by the use of a rigid coupler solution, is presently fixed via a connection to an independent Invar rod that runs over the cavities and has a fixed point at the cryomodule center. This produces a total longitudinal motion of less than 3 mm, which should be compatible with a semi-rigid coupler design.

MEASUREMENT APPARATUS

To qualify the new fixtures a mock-up has been built and a measuring apparatus has been set up. The scheme of the system is shown in Fig. 2. The sliding fixture prototype is assembled in a vacuum chamber that is part of a properly designed insert for a test cryostat. A motorized computer-controlled linear actuator moves the sliding element with a bellow and a "load cell". The system is designed to test the fixture either under vacuum or at atmospheric pressure. The test temperature, measured by an adequate thermometer system, can range from room temperature to that of liquid Helium. Measurements at 2 K can be taken pumping on the He bath.

The sliding fixture prototype consists of a remotely actuated titanium pad (a copy of that welded on the cavity helium vessel) moving between two rolling needles arrays. Both stainless steel and titanium runners have been tested, the load being controlled through a spring washer set. Measurements have been performed with different surface roughness and the results compared and discussed.

The test apparatus is assembled in a stainless steel vacuum chamber which is part of the insert of a vertical cryostat that can be filled by liquid nitrogen or liquid helium. The vacuum system is designed to allow a controlled venting of the vacuum test chamber with different gasses and pressures. Ice effect on the sliding surfaces has been produced and its influence on the system performance has been measured.

To read the applied force, a commercial low cost load cell (from Philips) has been chosen and its measured characteristic curve is presented in Fig. 2. Its linear range reaches more than 10 kg_f, with a few gram accuracy and an elastic constant of 72.5 kg_f/mm. In our friction measuring system we chose a working point close to the center of the load curve, by means of applying an external weight to the actuation rod.

The motorized linear actuator uses a slide guide and is controlled by a programmable PID. The position is read with micrometer precision, while the velocity and acceleration are set via computer. In order to compensate pre-load effect on the reading of the load cell, a calibration has been performed on the basis of the data from a set of four different movement patterns and kinetic parameters. The settings used are summarized in Table 1. The movement shape can be easily deduced also from the time domain plots as shown in Fig. 4.

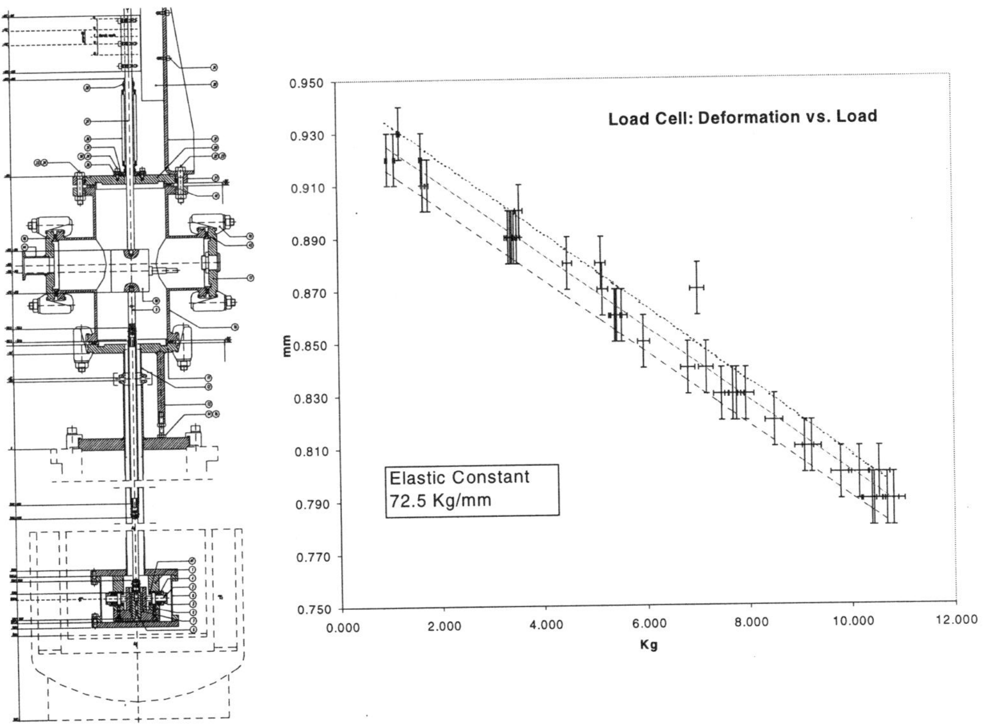

Figure 2. On Left side the schematic drawing of the system developed for the sliding fixture measurements. On the right the isplacement vs. force characterization of the load cell. The device shows a good linearity and reproducibility, with an elastic constant of 72.5 kg_f/mm.

Table 1. Actuator settings for the four different movement patterns used for calibration

Pattern type	Velocity [mm/s]	Acceleration [mm/s^2]	Wait time [s]	Movement Pattern
1	0.25	0.25	5	I: Up 5 mm Wait Down 5 mm
2	0.25	0.25	2	II: Up 2.5 mm Wait Up 2.5 Wait Down 5
3	0.5	0.5	10	I: Up 5 mm Wait Down 5 mm
4	0.5	0.5	10	II: Up 2.5 mm Wait Up 2.5 Wait Down 5

DATA ANALYSIS AND INTERPRETATION

The system composed by the sliding pad, the runner, the rolling needles and the load cell is driven by the laws of static and dynamic friction. To describe and simulate this system we have developed a dynamic non linear numerical simulation. By exploring the parameter ranges of the model we have tested different movements, and chosen the movement patterns and parameters listed in Table 1, in order to allow a good sensibility to the static and dynamic friction effects. Fig. 3 summarizes some simulation results showing the difference in movement patterns of type I and II. This analysis predicts that the maximum extension of the signal from the load cell is equal two times the static friction, while the oscillation amplitude is equal to two times the difference between the static and dynamic friction. Unfortunately experimental results are not as clear as the simplified model used for the simulations, because surfaces roughness, the acquisition system and the instrumentation introduce noise that needs to be filtered by repeating several times the

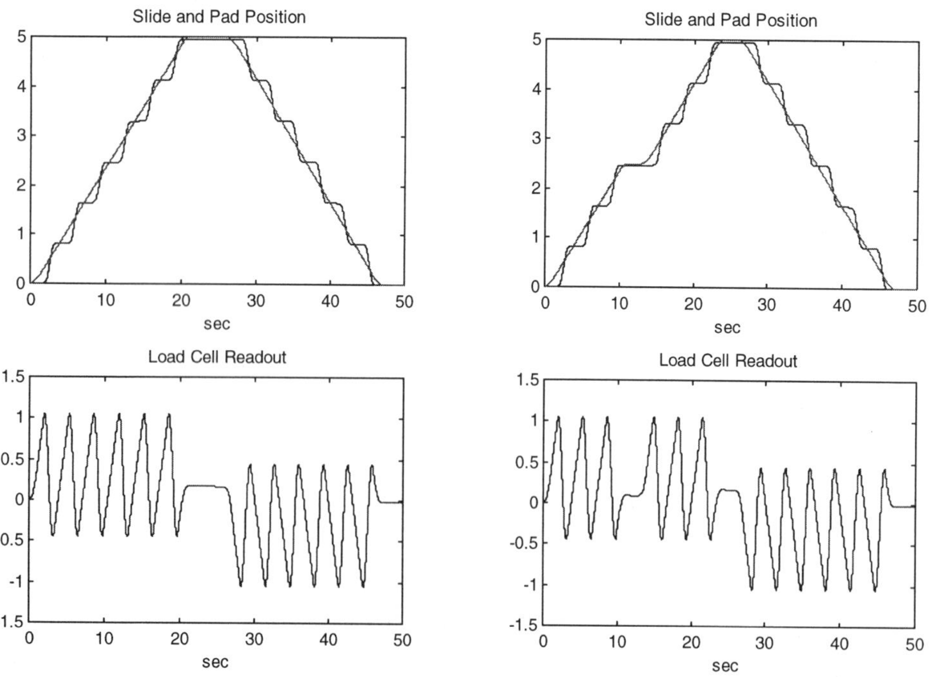

Figure 3. Simulated behavior of the readout system for movement patterns of type I (left) and type II (right).

measurements. To discriminate the amplitude of the oscillations, that depends on the value of static and dynamic friction, we had to fit result including a proper error estimate.

MEASUREMENTS RESULTS

In order to qualify the cavity support system, stainless steel and titanium slides have been built.and tested. The slides have been worked to a range of surface roughness, from lapped to extremely rough, in order to test the starting life, end life and the case of extremely damaged components. Tests have been performed under different conditions of environment pressure, vacuum, and temperature. A failure in the vacuum system and the consequent surface icing has been also simulated. Each situation has been measured with the four movement patterns, each pattern has been repeat five to ten times, in the same slide position, with a very good repeatability to filter quantization noise. The same pattern has been then tested at different slide positions. An automatic LabView[4] procedure operated the movements and the data acquisition, while a Matlab[4] routine analyzed data and compute the measured static and dynamic friction, producing a report of the measured run (Fig 4 and Fig 5). All the four movement patterns (two sets for each pattern) are reported and plotted with the calculated friction value, and a summary comparing the eight measurements is generated. Using the eight results (of friction values and corresponding errors) a mean static and dynamic friction value is computed. The previous procedure has been repeated for all samples under the conditions previously listed. Table 1 summarizes the results acquired by

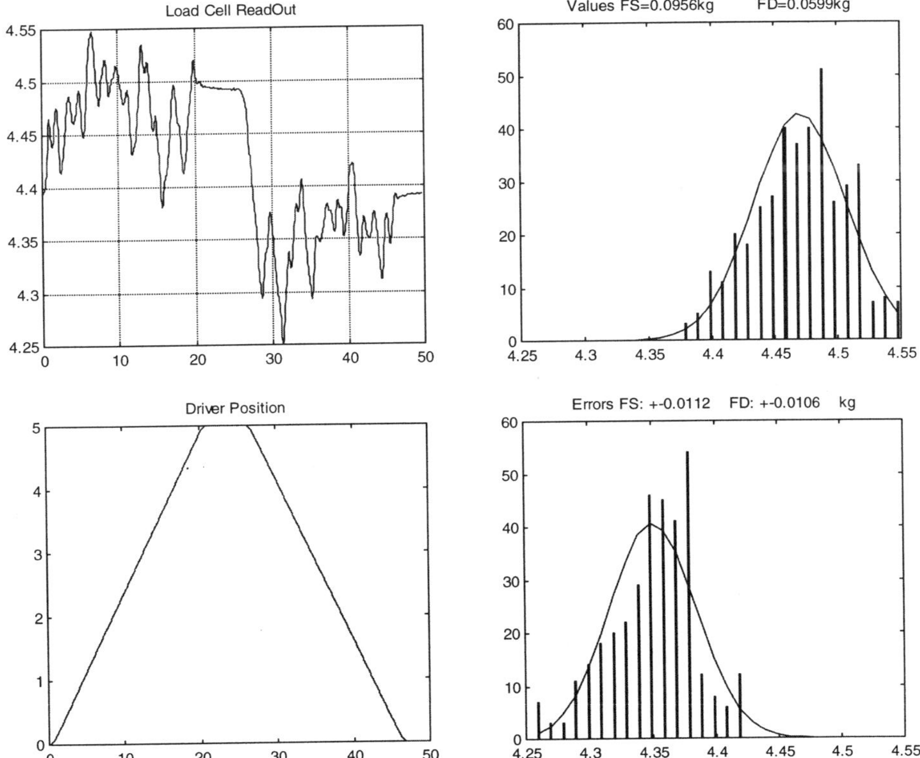

Figure 4. Data as output by the "Data Analysis Tool" developed to analyze the friction system results, for the case of the movement pattern of type I (Slow motion and single step). Studying the Load Cell tracks and checking them with the driving motor data, the oscillations predicted by friction theory are reconstructed, and static and dynamic friction are evaluated, with their errors.

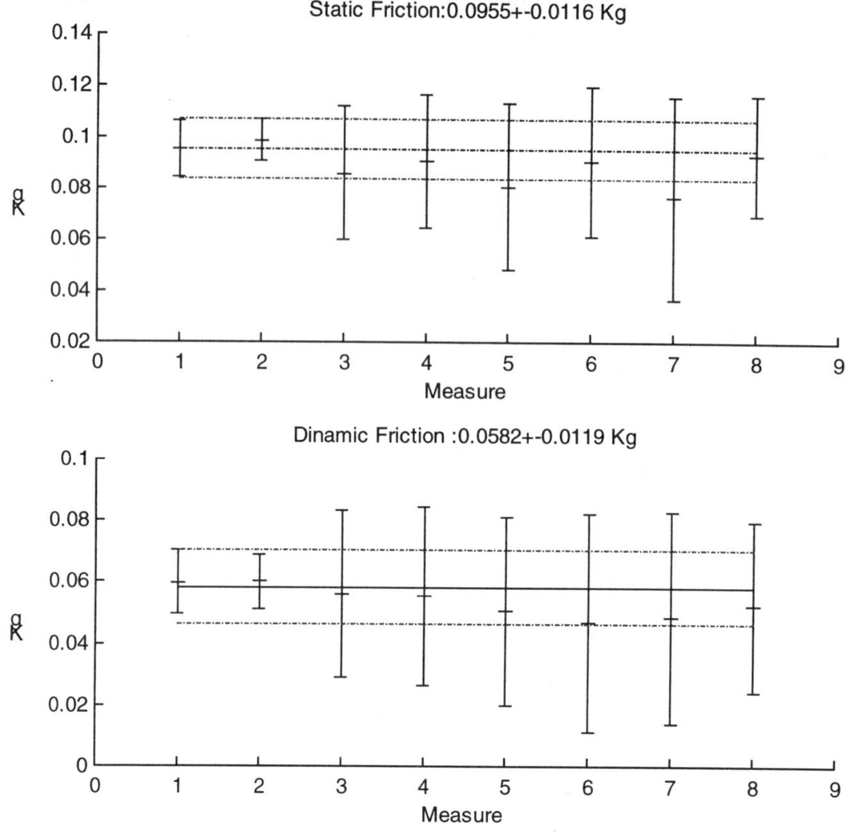

Figure 5. Results summary prepared by the "Data Analysis Tool". Results from the eight measures (four movement patterns, two run for each trip) are compared to achieve a final result value (and error) for dynamic and static friction.

the test system. The stainless steel slides have been used and tested also with a double force load (corresponding to about 80 kg), in order to check for the linear dependence of the measured friction values. The average friction force results in about 0.1 kg for stainless steel and about 0.8 kg for the titanium slide. The behavior of the slide does not depend strongly on the cryogenic temperature and no differences have been seen from the measurements. Icing effects have been "simulated" manually producing a vacuum break in the chamber at the liquid nitrogen temperature, raising the pressure to about 0.5 bar for 10 minutes (enough to condense water and other gases), but the rolling system performance was not impaired. The extremely damaged slides are characterized by an average friction value of about 0.8 kg, but some samples reached values up to 1.2 kg. This value is however compatible with the actual design of the cavities fixing feature[3] and should not be able to produce either deformation in the cavity string or damage in the supporting structure.

Table 2. Summary of the measurement results.

Sample[a]	Condition	Static Friction [kg_f]	Error [kg_f]	Dynamic Friction [kg_f]	Error [kg_f]
SS18	Liq. Nitrogen	0.0942	0.0266	0.0491	0.0203
SS18	Atm	0.0643	0.0078	0.0385	0.0061
SS18	Vacuum	0.0569	0.0135	0.0346	0.0118
SS36	Liq. Nitrogen	0.0901	0.0121	0.0449	0.0108
SS36	Ice	0.0715	0.0314	0.0362	0.0279
SS36	Atm	0.0979	0.0036	0.0564	0.0037
SS36	Vacuum	0.0955	0.0116	0.0582	0.0119
SS36Dam[b]	Liq. Nitrogen	0.5631	0.0761	0.2807	0.0751
SS36Dam[b]	Ice	1.0259	0.0381	0.6943	0.0460
SS36Dam[b]	Atm	0.4302	0.0387	0.2501	0.0375
SS36Dam[b]	Vacuum	0.4527	0.0330	0.2568	0.0346
Ti18	Liq. Nitrogen	0.1738	0.0433	0.0892	0.0352
Ti18	Atm	0.0756	0.0206	0.0458	0.0200
Ti18	Vacuum	0.0835	0.0210	0.0537	0.0189
SS18	Helium	0.0427	0.0153	0.0315	0.0214

[a] The "sample" notation is 'SS' for stainless steel and 'Ti' for titanium. The number is the spring load over the rolling needles, where 36 correspond to about 80 kg.

[b] The samples with the 'Dam' suffix have been heavily damaged before the tests.

CONCLUSION

In this work the sliding support for SC cavities in long cryostats, proposed for the third generation of the TTF cryomodule has been qualified in its working condition. A test system has been built and used to simulate the cryogenic operation conditions of the support. Accurate data analysis has been necessary for the interpretation of the measurement results and to obtain static and dynamic friction in stainless steel and titanium sliding supports. A mechanical and vacuum failure has been simulated using severely damaged slides and icing the support during the test. Static friction measurements results in 0.1 kg reaction force with about 80 kg applied.

REFERENCES

1. TESLA Test Facility linac - Design Report, DESY Report, March 1995, TESLA 95-01.
2. Conceptual design of a 500 GeV e+e- linear collider with integrated X ray laser facility, R. Brinkmann et al., editors, DESY Report, 1997-048, 1997.
3. Pagani, C., Barni, D., Bonezzi, M., Weisend II, J. G Further Improvements of the TTF cryostat in view of the TESLA collider, presented at this conference.
4. LABVIEW and MATLAB are trademarks.

DISSIPATED POWER MEASUREMENTS IN THE A0 SRF CAVITY SYSTEM

J. D. Fuerst and W. H. Hartung

Fermi National Accelerator Laboratory
Batavia, IL, 60510

ABSTRACT

Fermilab operates a single TESLA 9-cell superconducting RF cavity in support of a photoelectron R&D beam line. Power going into the 1.8K cryogenic system via static heat leak and RF dissipation is measured from the rate of rise of the pressure in the helium bath. This paper describes the techniques used to determine the cryostat heat load and the RF performance of the cavity.

INTRODUCTION

A photo-injector has been constructed at Fermilab to produce a low-energy (14-18 MeV) electron beam with high charge per bunch (8 nC), short bunch length (1 mm RMS), and small transverse emittance (<20 mm mrad)[1]. The facility was used to commission a photo-cathode RF gun for the TESLA Test Facility (TTF) Linac at DESY[2]. At present, the Fermilab machine is being used for R&D in bunch length compression and fast beam diagnostics; experiments in plasma wake field acceleration and channelling acceleration are in preparation.

Beyond the 4 MeV delivered by the RF gun, acceleration is done by a 9-cell superconducting cavity. The 1.3 GHz cavity also provides a position-momentum correlation for subsequent bunch length compression. The cavity was built by industry for TTF, tested at DESY, and then sent to Fermilab where it accelerated its first beam in October 1998. The maximum beam energy so far is 18.5 MeV, measured with a spectrometer[3]. The cavity is one of several with low quench field, attributed to contamination in the equator welds. Measurements of the cryogenic performance of the cavity and cryostat were done between periods of beam operation, as will be discussed herein.

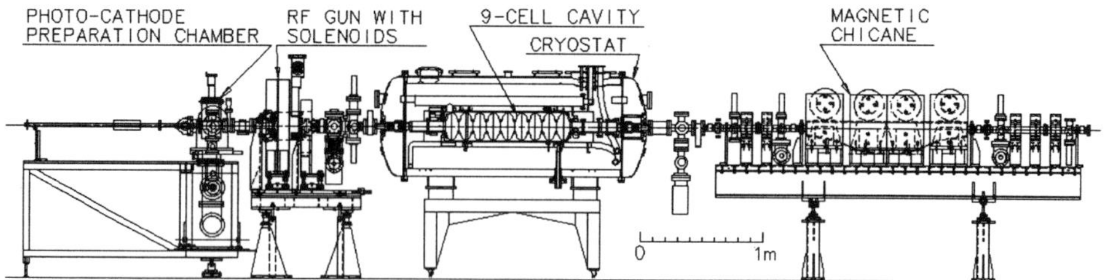

Figure 1. Elevation view of the Fermilab A0 Photo-injector beam line.

SYSTEM DESCRIPTION

The photo-injector beam line is shown in Figure 1. The cryostat is a copy of that used for the TTF capture cavity[4] and operates in a saturated bath of 1.8K helium II. Liquid helium is supplied from a manifold to which as many as four 500 liter helium dewars may be connected. The helium tank is radiation shielded and heat intercepted at 5K and 80K. Helium boil-off gas exhausts to the atmosphere through several vacuum pumps which maintain the cavity bath at 1.6 kPa (12 torr). Figure 2 shows a schematic of the cryogenic system.

The cavity and RF system were designed for pulsed operation at 10 Hz repetition rate with a 500 μs fill time and an 800 μs "flat top" of constant field. The input coupler is optimized for the heavy beam loading provided by a train of 800 bunches of 8 nC each, spaced 1 μs apart. Thus, without the beam, the cavity is highly over-coupled (the external Q can be adjusted between 10^6 and 10^7), and the RF power dissipation cannot be determined with any accuracy from RF measurements. Cryogenic loss measurements must therefore be used instead. The amplitude and phase of the cavity field are regulated via a digital control system developed for TTF[5].

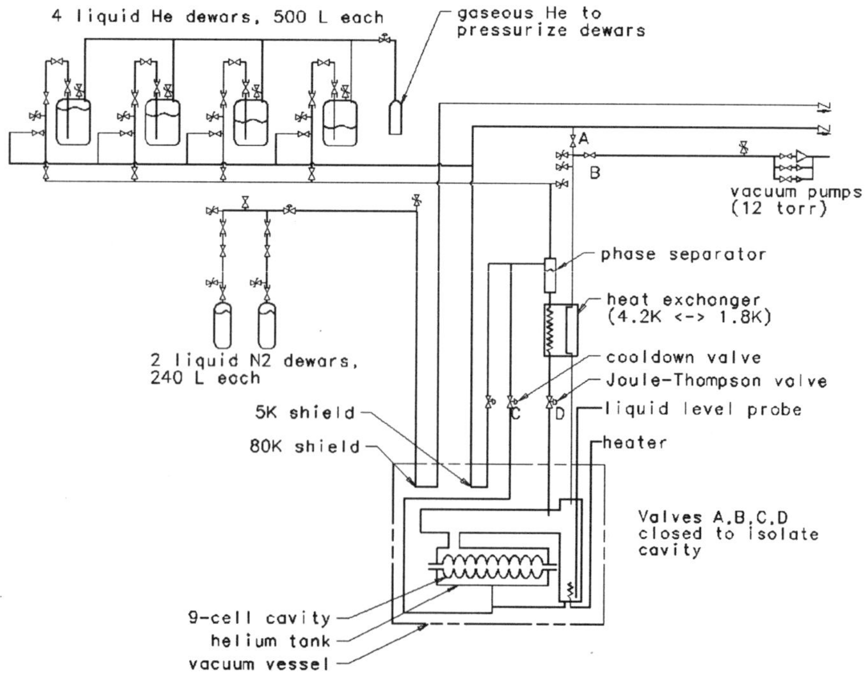

Figure 2. Schematic of the cryogenic system for the 9-cell superconducting RF cavity.

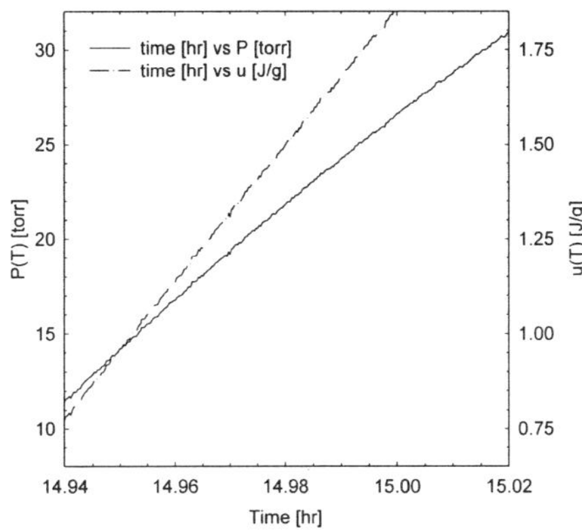

Figure 3. Rise in helium bath pressure and saturated liquid internal energy with supply and return valves closed.

MEASUREMENT TECHNIQUES

Several techniques, varying in accuracy and simplicity, exist for measuring the RF power dissipation to the cryogenic system.[2] We chose a slight modification of a method used at TJNAF for the CEBAF cryomodule commissioning[6] whereby the cryogenic system supply and return valves are shut while three distinct heat loads are applied to the cavity cryostat. First, RF power is applied over some pressure interval. Next, the RF is switched off and the pressure is allowed to rise further under the influence of the static heat load. Finally a known power level is applied to the bath through a resistance heater as the pressure rises still further. For small Δp, the rise in pressure over time is almost linear and the power dissipated by the RF can be inferred directly. In contrast to the CEBAF cryomodules, for which the pressure rise was on the order of a few millitorr over several minutes, the pressure rise for our system was several torr for each data point. This is due to the large volume difference between the systems (about 20 liters of liquid for our system vs. several hundred liters at CEBAF) for comparable static heat loads.

Plots of heater power against $\Delta p/\Delta t$ show inconsistencies in slope, particularly among data gathered on different days. For example, the isolation valves (particularly the LHe supply valve "D" in Figure 2) don't appear to seal completely when closed, causing error. The "3-point" measurement technique reduces error associated with valves as well as error caused by slow changes in the system. The calibration data (known heater power and static heat load alone) for each RF data point is taken before the isolation valves are opened again.

As the observed $\Delta p/\Delta t$ is not constant, we evaluate the system's change in internal energy instead. Our cryostat contains 21.5 liters saturated liquid, 6 liters saturated vapor, and 11.5 liters superheated gas. The superheated gas temperature is about 4K. By adding up the internal energy of the helium in these three regions at 9 torr (1.2 kPa) and at 21 torr (2.8 kPa), we find that 2700 Joules of energy are added to the helium between 9 and 21 torr. Under static conditions, it takes about 225 s for the pressure to rise between these two values, implying a static heat load of 12 W. Evaluating the saturated liquid component alone, we find an energy change of 2590 Joules, giving an 11.5W static load. Figure 3 shows the rise in system pressure as well as the rise in saturated liquid internal energy over time with the helium supply and return valves closed. Several "3-point" test measurements were done in which a known heater power was substituted for the RF dissipation. The heater power was inferred from the measured pressure rise.

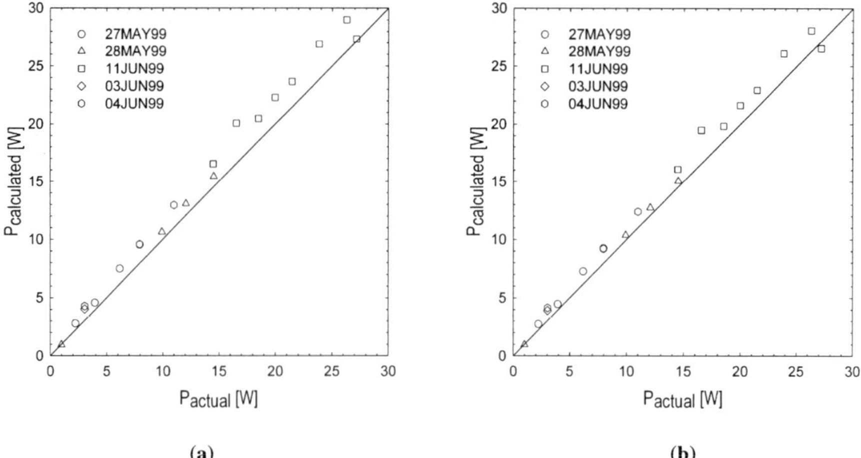

Figure 4. Calculated vs. actual power from $\Delta p/\Delta t$ data (a) and $\Delta u/\Delta t$ data (b). The solid line corresponds to the case in which the calculated power is equal to the actual heater power.

Figure 4a shows a plot of actual vs. calculated power using $\Delta P/\Delta t$ data while Figure 4b shows the same plot using $\Delta u/\Delta t$ data. If we plot heater power vs. $\Delta u/\Delta t$ (Figure 5) for the data of Figure 4b, we see the cumulative effects of valve position error, slow changes in system equilibrium, etc. This scatter is reduced using the "3-point" analysis technique. Nevertheless, our analysis technique systematically overpredicts the dissipated power by less than one watt at low power and about three watts at high power, with maybe one watt scatter in the data. We have not corrected the data for this effect.

The cavity was typically tuned to be on resonance at 11 torr bath pressure. RF data were collected between 9 and 13 torr. The RF control system was used to maintain a constant field level in the cavity, even as the resonant frequency shifted with the changing bath pressure. Static heat load data was taken from 13 to 17 torr, and the control heater data was taken from 17 to 21 torr. Figure 6 shows a single set of three measurements for a control data point (that is, using the heater instead of RF).

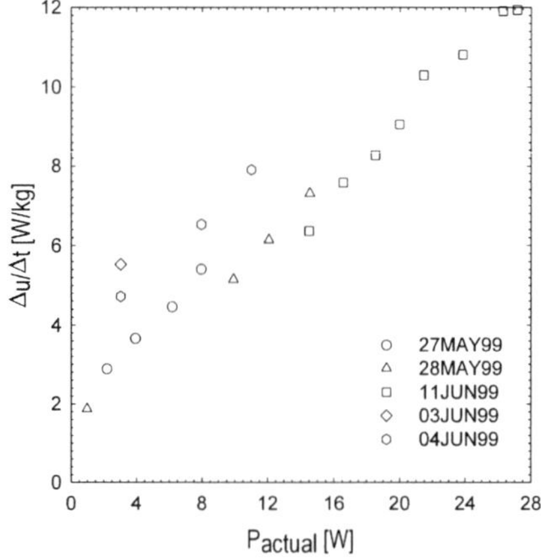

Figure 5. Heater power vs. $\Delta u/\Delta t$ for the data of Figure 4b, showing scatter between data sets.

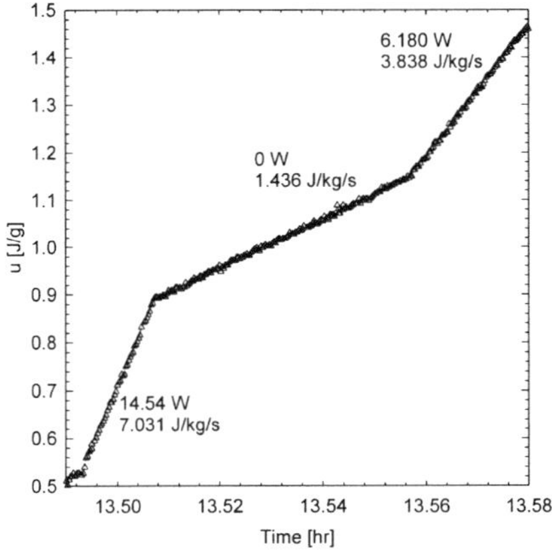

Figure 6. Data collected with heater settings of 14.54W and 6.18W, separated by static load.

MEASUREMENT RESULTS

Figure 7 shows the power dissipated into the helium as a function of cavity field level for several different RF input parameters. Initial tests of this cavity in a vertical dewar at DESY[7] indicated that the CW quench field was about 13 MeV/m. Dissipation into the cryogenic system begins to become noticeable as the accelerating gradient E_a exceeds 10 MeV/m. The attainable gradient was limited by trips in the RF interlock system. Dissipation rates above 20W caused fairly rapid pressurization of the helium tank making data difficult to collect. In any case, these rates approach the sustainable capacity of the helium vacuum pumps and are beyond the expected operating envelope. As can be seen in Figure 8, the scaling with repetition rate is as expected: for a given flat top duration, the energy loss per RF pulse is approximately the same, whether the repetition rate is 1 Hz, 5 Hz, or 10 Hz. Figure 8 also compares the cavity performance at Fermilab with earlier performance in the CHECHIA horizontal cryostat at DESY. Note the degradation in performance with time, perhaps caused by particulate contamination during the removal at DESY, handling, and subsequent installation at Fermilab. High peak power pulsed processing may allow us to reverse this degradation.

The data are most easily collected if the isolation valves seal completely and reliably. We also believe it is best if the volume trapped between the isolation valves is kept to a minimum and kept cold, to maximize the fraction of energy residing in the liquid.

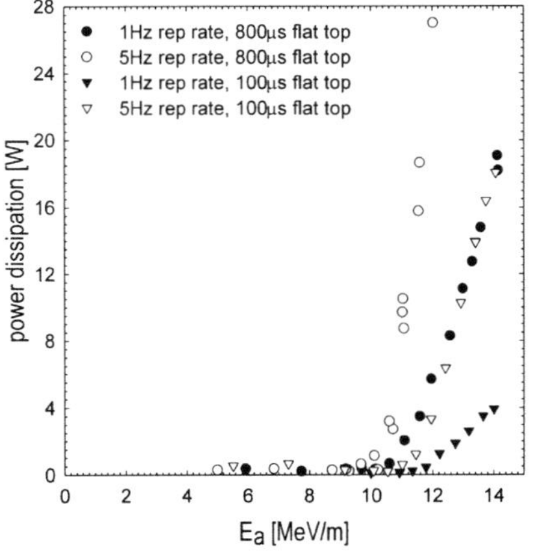

Figure 7. Average power dissipated in the cavity cryostat as a function of cavity accelerating gradient for several sets of RF parameters.

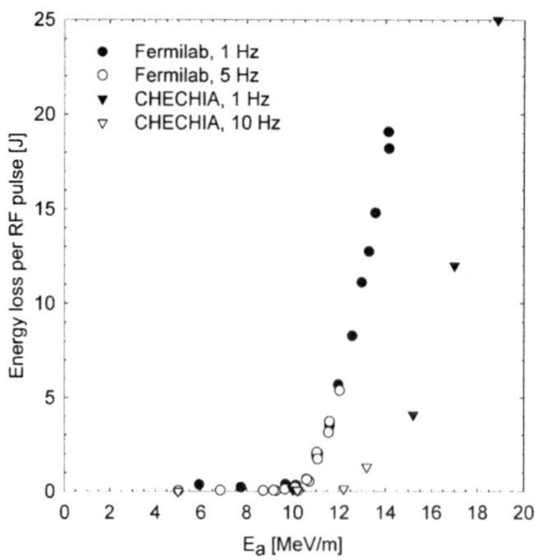

Figure 8. Comparison of cavity performance at Fermilab (June 1999) and DESY (April 1997). The CHECHIA data was obtained via measurement of the helium gas flow rate. The flat top duration is 800 μs in all cases. Dissipated power is divided by the repetition rate to give energy dissipation per RF pulse.

Finally, we learned early on that the response from our cavity pressure transducer was overly sluggish, caused by a long thin sensing line. Relocation of the room temperature transducer closer to the tap allowed us to see the small but rapid changes in bath pressure caused by changing heat loads.

CONCLUSIONS

A TTF 9-cell cavity is being used at Fermilab for a high-brightness electron photo-injector. The dissipation as a function of gradient was measured for several combinations of repetition rate and flat top duration. The dissipated power was measured by closing isolating valves on the supply and return sides of the helium cryostat and allowing the system to pressurize under the influence of the heat load. Following TJNAF, we collected calibration data with each RF data point ("3-point" measurements) in order to reduce the measurement error. This technique can be adapted to our system, in spite of the rate of pressure rise being about 100 times larger. The results indicate that the high-field losses in the cavity are presently higher than in initial measurements at DESY. However, accelerating gradients up to 10 MeV/m are possible without noticeable power dissipation in the cryogenic system.

ACKNOWLEDGMENT

Work supported by the U.S. Dept. of Energy under contract No. DE-AC02-76CH03000.

REFERENCES

1. J. P. Carneiro et al., *Proceedings of the 1999 Particle Accelerator Conference*, p. 2027.
2. D. A. Edwards, ed., "TESLA test facility linac - design report," TESLA 95-01, DESY (March 1995).
3. W. Hartung et al., *Proceedings of the 1999 Particle Accelerator Conference*, p. 992.
4. S. Bühler et al., "Advances in Cryogenic Engineering," Vol. 41, Plenum Press, New York (1996), p. 863.
5. S. N. Simrock et al., *Proceedings of the 5th European Particle Accelerator Conference*, p. 349.
6. M. Drury et al., *Proceedings of the 1993 Particle Accelerator Conference*, p. 841.
7. W.- D. Möller and M. Pekeler, "Advances in Cryogenic Engineering," Vol. 43, Plenum Press, New York (1998), p. 97.

CRYOSTAT TO TEST THE SUPERCONDUCTING RF GUN FOR THE DROSSEL PROJECT

Y.P.FILIPPOV and A.M.KOVRIZHNYKH

Joint Institute for Nuclear Research, Laboratory of Particle Physics, Dubna, Moscow Region, Russia

ABSTRACT

The aim of this work was to find a way which allows one to locate the 1.3 GHz RF-gun within the existing cryostat of LPP, JINR. This cryostat was designed, fabricated and tested for a beam test with the L-band cavity at 1.8 K and presented at ICEC16. For this purpose the initial version of the superconducting cavity was redesigned to be prepared for mounting within the cryostat as a separate node with blank flanges, pipes for liquid nitrogen flows to cool the photocathode stem, and potential leads. The feature of the redesigned cryostat is that its part with the beam pipe and corresponding flanges and horizontal part with cryogenic input/output nodes and couplers have remained unchanged. In its turn the flanges of the opposite side (near the preparation chamber for photocathode) were redesigned. The technical solution is presented, and assembly procedures are described.

INTRODUCTION

Photocathode RF guns are the most advanced type of electron injectors for high brightness accelerators. Based on the experience of the University of Wuppertal [1], the Drossel Collaboration was established to lay the foundations for the build-up of a superconducting RF gun at the Research Center Rossendorf, Dresden, Germany.[2] One of the steps is the test of a half-cell SC-cavity in order to prove the feasibility of the design concepts. It is necessary to have a horizontal cryostat for such tests. Several years ago a suitable cryogenic and control equipment was built at the Laboratory of Particle Physics of JINR to carry out a series of experiments with S-band and L-band superconducting RF-cavities.[3] The aim of this work [4] was to improve the existing cryostat taking into account the features of the superconducting RF gun designed and fabricated by Budker Institute of Nuclear Physics, Novosibirsk, for the Drossel Collaboration.[2]

INITIAL RF GUN AND EXISTING CRYOSTAT

A draft of the cavity is shown in Figure 1 borrowed from Reference 2. The half-cell cavity is based on the TESLA endcup 1 design. It is shortened at the equator and connected by an elliptical arc to the flat back wall of the cavity. The frequency is 1.3 GHz and the peak electric field is reached at the cathode center. A choke flange filter is attached directly at the cavity back wall. Its purpose is to avoid RF power transmission through the coaxial line, which is formed by the cathode stem and the cavity. The cavity and filter will be produced of high RRR niobium and cooled with superfluid helium.[2]

The feature of the design is that the cathode stem is made of OFHC copper. It has to be located inside a special holder cooled with liquid nitrogen. In other words, it is a minicryostat inside the vessel at $T < 2$ K, which should be supplied with its own coolant circuit at comparatively high temperature, $T \cong 70$ K. For the series of experiments with this cavity we proposed to insert it within the existing cryostat whose side view is shown in Figure 2.[3]

This cryostat includes the helium vessel for the cavity, He-gas and LN2-cooled shields, two nodes for RF couplers, and a device for mechanical stretching/squeezing of the cavity. The outer dimensions of the cryostat are : diameter - 680 mm and length between flanges - 920 mm. To seal the flanges at cryogenic temperatures, we have used indium and annealed copper gaskets allowing one to replace or remove the cavity if necessary. The assembly procedures are described in Reference 3. Cooling down to 1.5 K is performed with saturated helium pumped by a 500 L/s unit. The tests have shown [3] that the heat leak into the helium vessel from ambient space is about 3.5 W when the temperature of the liquid helium is 4.25 K. At the additional heat loading of 5 W, the 500 L/s pump easily maintains the temperature of 1.8 K in the helium vessel.

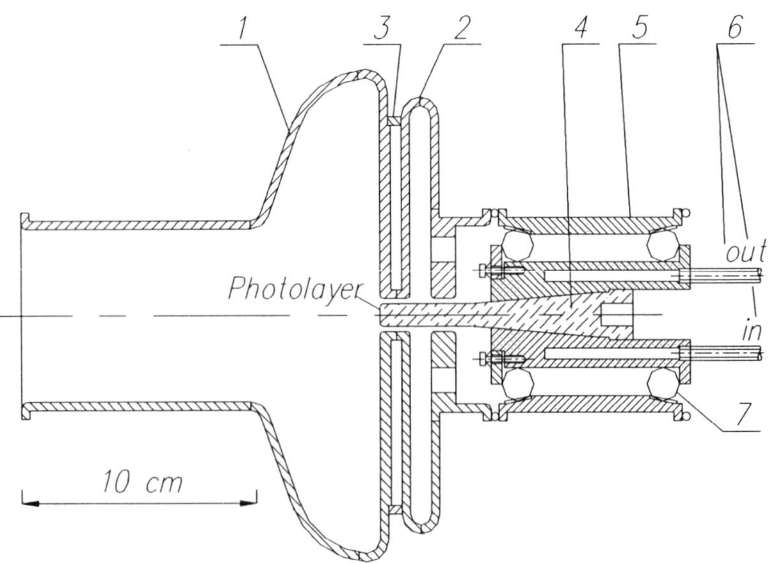

Figure 1. Draft of the superconducting test cavity [2]. 1 - TESLA endcup 1, 2 - choke flange filter, 3 - stiffening segment, 4 - copper cathode stem, 5 - cooled insert, 6 - tubes for liquid nitrogen, 7 - insulator insert.

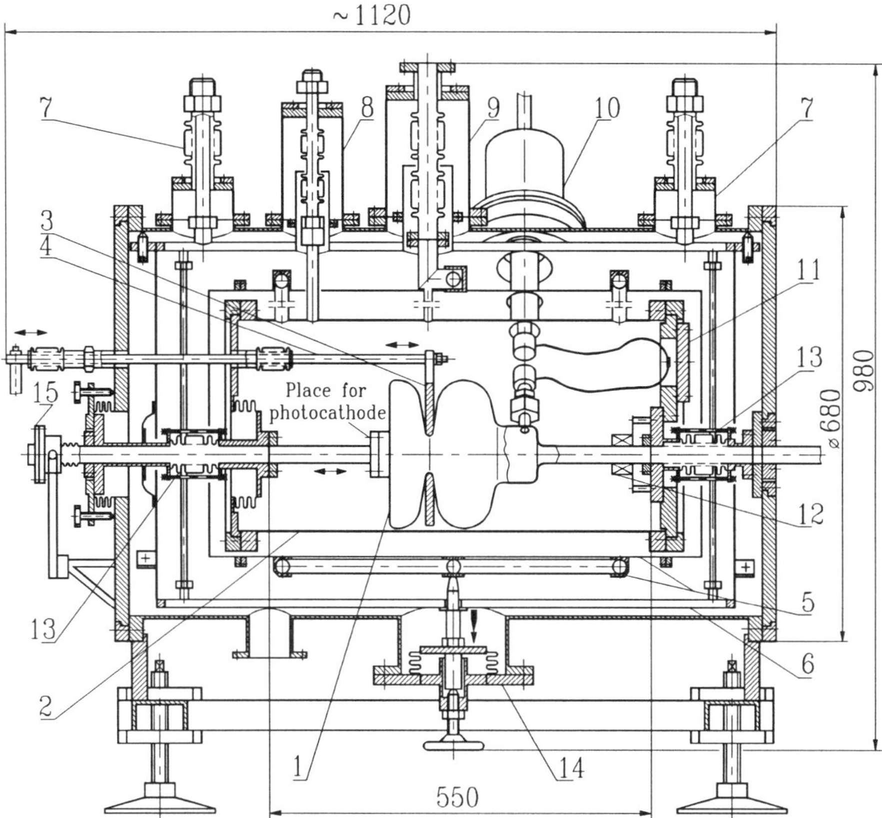

Figure 2. Side view of the existing cryostat for the 1.3 GHz cavity. 1 - cavity, 2 - helium vessel, 3 - yoke, 4 - two rods to tune the cavity, 5 - helium shield, 6 - nitrogen shield, 7 - input/output of nitrogen, 8 - input of LHe, 9 - output of He, 10 - two couplers, 11 - window to connect the RF-cables and wires, 12 - focusing solenoid, 13 - squeezed bellows, 14 - support, 15 - tuner.

REDESIGNED CAVITY AND CRYOSTAT

A preliminary analysis has shown that it is very difficult or perhaps impossible to mount the prepared RF gun shown in Figure 3 within any cryostat analogous to the one shown in Figure 2 without changing the backward support tube of the cavity. It is caused by the following reasons: the necessity to feed and return the nitrogen flow via the backward support tube within the space of the cryostat, to meet the requirement of cleanliness within the cavity during all the assembly procedures, and to balance the forces due to the pressure difference and temperature deformations. The analysis has shown that to satisfy these requirements, it is necessary to add an intermediate node to provide the mounting of the SC-cavity with the assembled photocathode within the cavity, the input and output tubes for nitrogen, and blank flanges, that is illustrated in Figure 4.

The cryostat for the assembled cavity shown in Figure 4 is presented in Figure 5. Briefly the main assembly procedures may be described as follows. The cavity with the blank flanges and intermediate node (1) is delivered into the clean room and connected to the two bellows nodes (2) and (3). To decrease the heat leak into the helium vessel (6), both bellows nodes are made of annealed stainless steel with a big number of corrugations in a relatively short total length due to squeezing of the primary bellows. Besides, the relatively

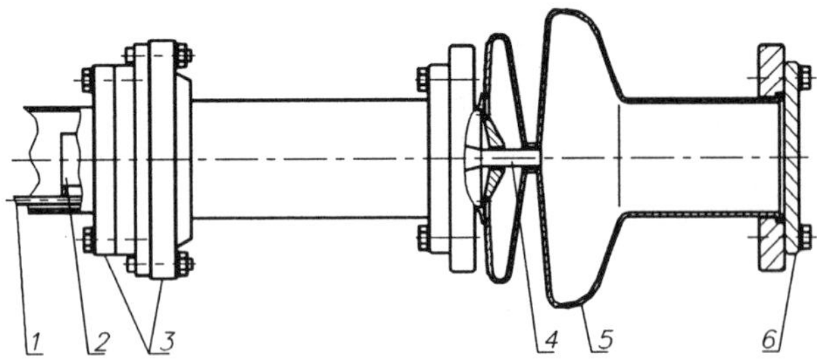

Figure 3. The assembled cavity before mounting within the cryostat: an initial version. 1 - input/output tube for liquid nitrogen, 2 - part of the cooled insert, 3 - flanges, 4 - copper cathode stem, 5 - TESLA endcup, 6 - blank flange.

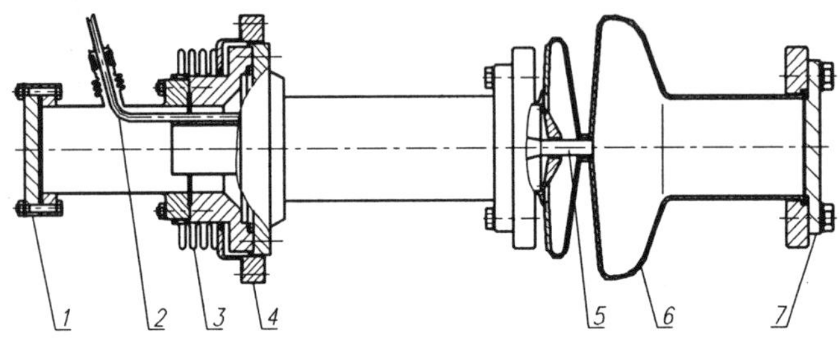

Figure 4. A new version of the cavity before mounting within the cryostat. 1- blank flange, 2 - input/output tube for liquid nitrogen, 3 - bellows, 4 - intermediate flange, 5 - copper cathode stem, 6 - TESLA endcup, 7 - blank flange.

warm ends are at nitrogen temperature due to the copper tubes (20 and 21) which connect the fixing bolts with the vessel (10) for liquid nitrogen. The fixing bolts are made of glass-fibre-plastic armoured with stainless steel. After mounting the blank flanges from both sides, the assembled unit is checked for vacuum tightness. Then it is removed from the clean room for further assembly. At first the cavity is connected with the cap (5) in the vertical position. Afterwards the thermometers and yoke to tune the cavity with two rods are mounted (they are not shown in Figure 5, one can see them in Figure 2), and the cavity is covered by the helium vessel (6) which is fixed with cap (5) via a copper gasket. Before this procedure the helium shield, cooled with the pumped vapor, is mounted around the helium vessel rigidly, and the mounting of the levelmeter and other devices are carried out at the inner wall of the helium vessel. Then the cap (7) is mounted near the right hand bellows node, sealed via the copper gaskets, and vacuum checking is performed.

The next step is to deliver the assembled unit into the clean room where the flange (8) should be installed near the bellows node (3) instead of the blank flange, and checking for vacuum tightness should be done.

Now it is necessary to insert the nitrogen shield (10) into the vacuum shell (9) that should be performed in the horizontal position. The screws (11) are used to regulate and fix the necessary position. The cap (13) is installed preliminarily (without the seal procedure) and the vessel is rotated into the vertical position. After that the helium vessel with the cavity inside and the helium shield outside is put into the nitrogen shield (10) up to the contact with flange (8). The adjusting screw of the supporting node (14) is used to regulate

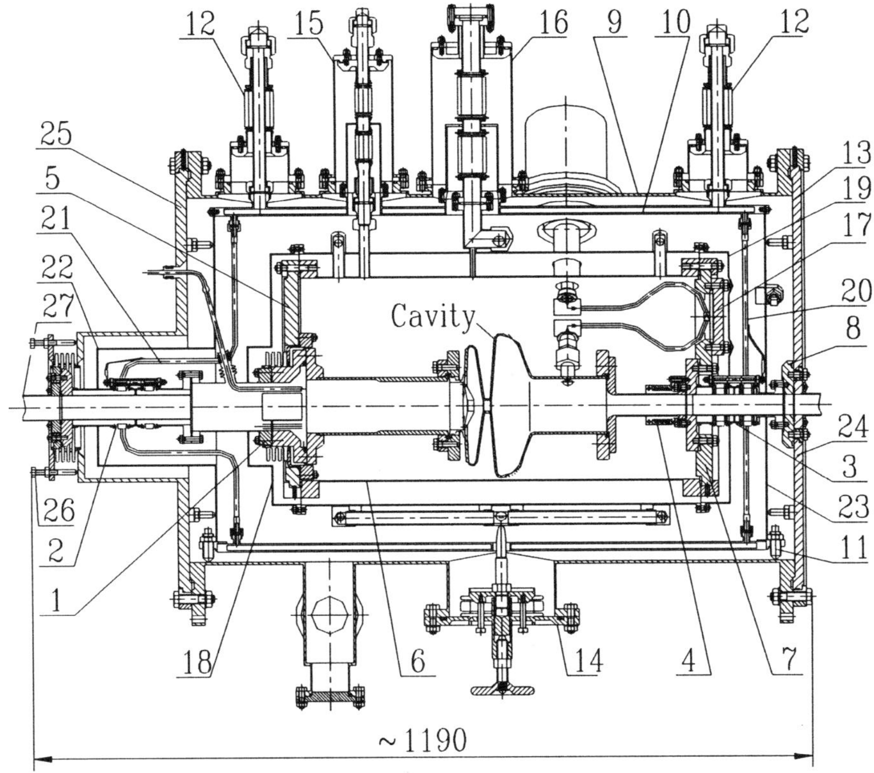

Figure 5. The side view of the redesigned cryostat for the cavity.

the necessary position of the cavity, and the helium vessel is fixed temporarily.

The following step is to place the cryostat into the horizontal position and mount the nodes of input/output (12) of liquid nitrogen, feeding line (15) of LHe and line (16) to pump the helium vapor, and couplers (two units located under angle of 25 degree). Then it is necessary to remove the right hand cap (13) and to connect the parts of the couplers by the RF-cables as well as the wires of the focusing lens(4) using the window (17) in cap (7). Such a design allows one to avoid the lateral forces due to pressure difference between the helium vessel and the ambient space. After vacuum checking all the technological supports and fixers are removed, and the helium shields (18 and 19) with the magnetic shields and superinsulation are mounted. Then the nitrogen tubes (20 and 21) are connected.

Afterwards, the caps (22 and 23) of the nitrogen shield are mounted. Then the right hand cap (24) is bolted, and the left hand cap (25) with the rods tuning the cavity and liquid nitrogen lines for cooling the photocathode, are sealed. At last, the cryostat is delivered into the clean room where tube (27) (with a vacuum shutter) should be connected with the preparation chamber for photocathode.

One can notice that the bellows at the tuner and intermediate node (1) contacting with liquid helium, are made of the same size to balance the corresponding forces caused by the pressure differences when the pressure in the helium vessel is equal to atmospheric pressure. The necessary compensation of the changed force due to pumping the helium vessel up to the given value may be performed by means of the tuning device (26).

CONCLUSIONS

The cavity and existing cryostat for experiments with L-band cavities were redesigned to meet the technical requirements of the Drossel project. The found technical solutions allow to perform a series of experiments with a superconducting RF-gun to demonstrate that the gradient goal can be achieved.

REFERENCES

1. A.Michalke, Photocathodes inside Superconducting Cavities, PhD thesis, University of Wuppertal, WUB DIS 92-5
2. D.Janssen, P.vom Stein, M.Karliner, S.Konstantinov, V.Petrov, S.Popov, I.Sedlyarov, A.Tribendis, V.Volkov, On the Way to a Superconducting RF Gun: Status of the Drossel Collaboration, Internal preprint of Research Center Rossendorf, Dresden, Germany, February 13, 1997.
3. Yu.P.Filippov, A.M.Kovrizhnykh, V.I.Batin and S.V.Uchaikin, Tests of the Cryostat for 1.3 GHz Superconducting Cavity at T<1.8 K, «Proceedings of the 16th International Cryogenic Engineering Conference», Part I, Kitakyushu, Japan, May 1996, Elsevier Science, p. 439-442.
4. Agreement on Collaboration between Joint Institute for Nuclear Research and Research Center Rossendorf «R&D on Superconducting RF Gun» of 18.12 1997.

AN RF-CAVITY FINE-TUNING ACTUATOR

J. P. Voccio, C. M. Gervais, and N. Kotsifakis

American Superconductor Corporation
Westboro, MA 01581

ABSTRACT

A prototype actuator for fine-tuning rf-cavities was designed, fabricated and tested. The actuator consists of a driver rod, fabricated from the large-strain, single crystal cryogenic magnetostrictive material, terbium-zinc (TbZn). The driver is energized by a high-temperature superconducting (HTS) coil. The device was designed to reside in the vacuum space of a helium cryostat, cooled through a braided copper strap, which could be attached to a cold radiation shield. The actuator was tested via a cryocooler at various temperatures, ranging from 50 K to 77 K, and under various preload conditions, including a maximum design load of 22.4 kN (5000 lbs). The results show that the actuator is capable of achieving at least twice the specified displacement of 12.7 μm, while working against the specified preload of 22.4 kN. The next stage in the development of this device would be to demonstrate its performance in controlling an rf-cavity. Expected challenges would be those concerning the natural hysteresis of the magnetostrictive material.

ACTUATOR CONCEPT

The objective of this program was to fabricate and test a "state-of-the-art" high power HTS magnetostrictive actuator which can be used for fine tuning of a superconducting rf-cavity. The operational requirements for the actuator are listed in Table 1.

Advances in Cryogenic Engineering, Volume 45.
Edited by Shu *et al.*, Kluwer Academic / Plenum Publishers, 2000.

Table 1. Actuator specifications.

Frequency Range:	± 1000 Hz*
Displacement:	± 6.4 μm (or 12.7 μm total travel)
Maximum Applied Load:	22.2 kN (5000 lbs)
Envelope:	45.6 cm (18 in.) long 7.5 cm (3 in.) diameter
Cooling Method:	Actuator located in vacuum space

Many particle accelerators, such as the one at Thomas Jefferson National Accelerator Facility (TJNAF, formerly CEBAF), use rf-cavities to accelerate the electron beam. These cavities must be precisely tuned in order to properly function. Tuning is performed by adjusting the length of the cavity. Typically, there are two types of adjustment: (1) fine-tuning, to compensate for the ponderomotive forces, and (2) coarse tuning, to compensate for the differential thermal contraction which occurs upon cooldown to cryogenic temperatures. The fine tuning stroke requirements are on the order of 25 μm (or 0.001 in.); whereas, the coarse tuning requires a much greater stroke, as much as 2 mm (or 0.080 in.).

Presently, the coarse tuning is performed with a mechanical link to the room temperature, or ambient, environment. The fine tuning is performed using piezoelectric actuators which are coupled into the coarse tuner. TJNAF would like to eliminate the link to ambient by using actuators which can reside in the vacuum space of the cryostat.

The prototype fine-tuning actuator described in this paper uses HTS coils and TbZn magnetostrictive material. This actuator is designed to sit in the cryostat vacuum space being cooled only via a thermal strap to the 50 K radiation shield. An iron magnetic return path provides a lower reluctance path for the magnetic flux which is used to energize the magnetostrictive rod. . A photograph of this prototype actuator is shown in Figure 1.

Figure 1. Photograph of prototype fine-tuning actuator.

* The present coarse adjustment worm gear drive has a frequency range of ± 200 kHz, which corresponds to ± 1.3 mm (± 0.050 in.). The cryogenic actuator is used for fine tuning, and ± 1000 Hz is the desired frequency range; this corresponds to a displacement of ± 6.4 μm (± 250 μ-in.), or 12.7 μm (500 μ-in.) total travel.

ANALYSIS AND DESIGN

A solid model rendering of the actuator and test assembly is shown in Figure 2. The braided copper thermal strap is connected to a copper ring which in turn is thermally connected to, but electrically insulated from, the HTS coil through the iron enclosure. The opposite end of the copper strap is connected to the coldhead of the cryocooler servicing the 50 K radiation shield.

HTS Coil Design

An ANSYS Magnetics finite element model was used to design the magnetic circuit of this test actuator. This model consisted of the TbZn rod, an HTS coil and a soft iron magnetic return path. The return path consists of a 3-mm thick soft iron tube which surrounds the HTS coil. The HTS coil was designed to produce a peak flux density of 1.8 tesla in the TbZn rod with an overall current density of 1000 amps/cm^2 in the HTS coil; this corresponds to an operating condition of 4100 amp-turns (or 37 amps for the resulting 110 turn coil). This coil was fabricated with 20 meters of fully-reacted HTS wire with a nominal cross-section of 4 mm x 0.2 mm using the double pancake approach. The coil contains 110 turns—55 turns per pancake. The power dissipation is shown as a function of current in Figure 3.

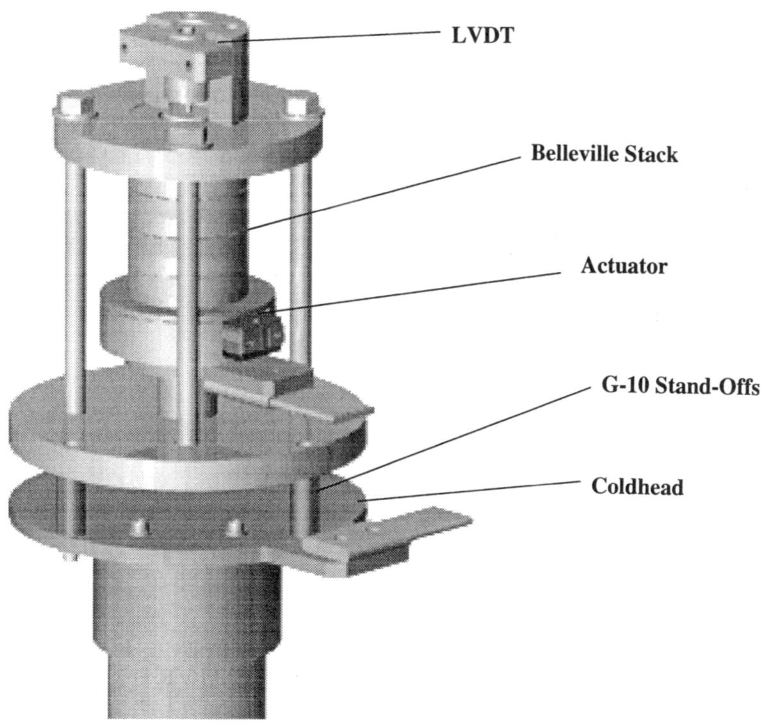

Figure 2. Solid model rendering of actuator and test assembly.

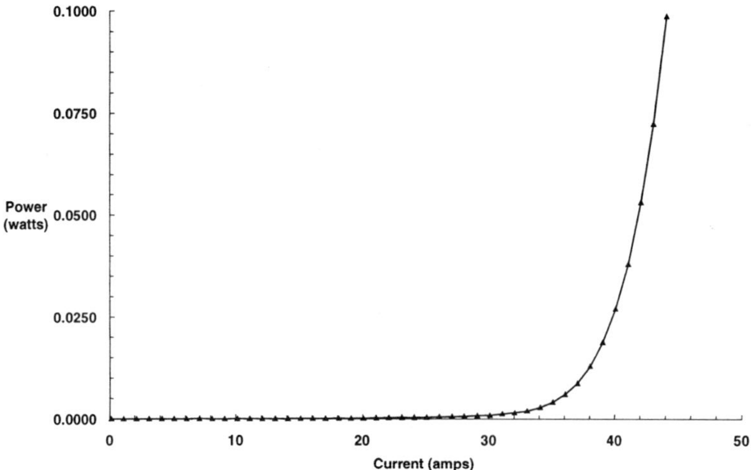

Figure 3. Power dissipation curve for actuator coil (measured at 77 K).

Magnetostrictor Design

The actuator design assumes a 12.7-mm long*, 25-mm diameter TbZn driver; the actual driver which was fabricated has a 21.8-mm diameter and is 16.5 mm long. This driver was assembled from two disks which were machined out of two of the TbZn single crystal ingots made by Ames Labs. The geometry of the ingot and the machined disks is depicted in Figure 4.

Under a previous DoD/Navy program, several large single crystal ingots of TbZn were fabricated and tested at 77 K under various preload conditions. Results for the highest measured preload condition of 24 MPa is shown in Figure 5. These data show that considerable magnetostriction (approximately 3000 ppm) is achievable with no preload and that higher magnetostrictions (greater than 4000 ppm) are achievable at higher preloads. The present actuator was designed at higher preload of 60 MPa, as a nominal diameter of 1 inch is the largest currently available for single crystal TbZn.

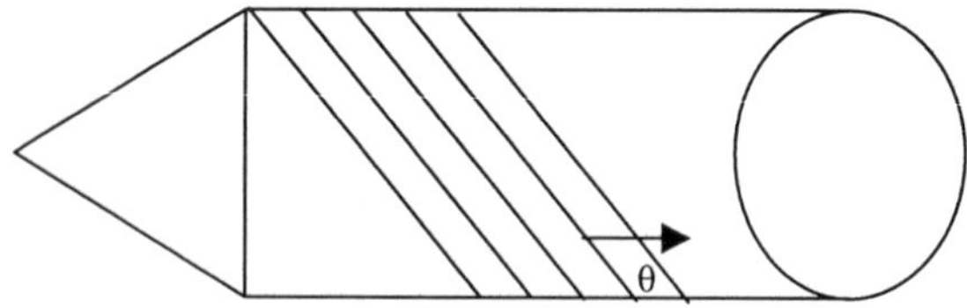

Figure 4. Geometry for machining TbZn disks from a single crystal ingot.

* Assuming a typical strain of 4000 ppm (or 0.40%), the 12.7-mm length can provide a displacement of 50 μm, which is considerably more than the required amount of 12.7 μm. This was intentionally overdesigned to provide margin in this prototype actuator.

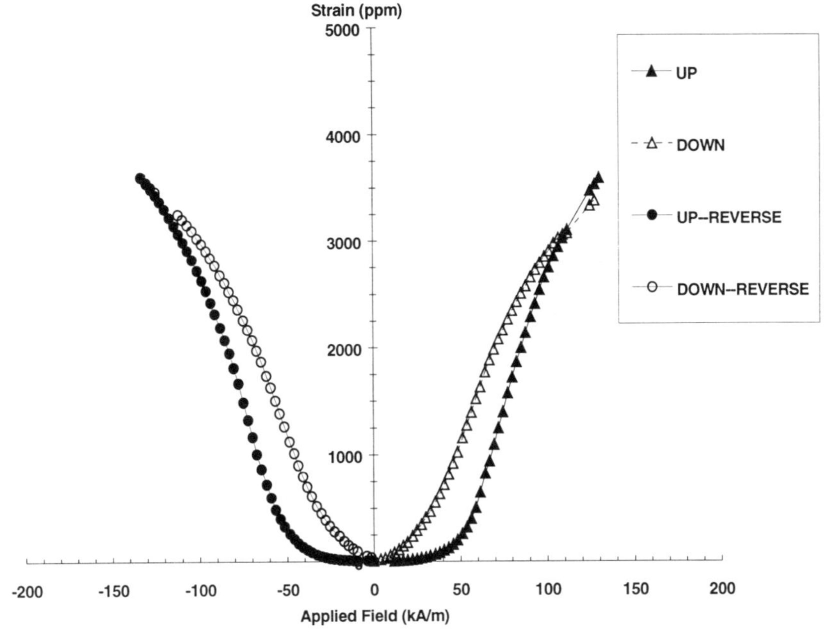

Figure 5. TbZn magnetostriction with 24 MPa preload (77 K).

Test Assembly Design

In addition to the actuator, a test assembly was designed to test the actuator performance. A set of Belleville washers along with some stainless steel bolts are used to provide precompression to the magnetostrictive rod. The nominal stiffness of the Belleville stack is 10.5 MN/m (60,000 lb/in); this is the same stiffness as the rf-cavities used at TJNAF. The actuator displacement is measured by a linear voltage differential transformer (or LVDT), which is mounted to the top plate of the test assembly; the smaller LVDT bobbin is rigidly connected to the top of the actuator.

The entire test assembly is mechanically grounded to a stainless steel plate which is mounted to the lower copper plate via some G-10 standoffs. The purpose of the G-10 standoffs is to provide a poor thermal path so that the cooling is limited to the thermal strap, as will be the case in the actual operating environment.

TESTING

A block diagram of the instrumentation used for these tests is shown in Figure 6. A 100-amp power supply is used to energize the HTS coil. The HTS coil activates the TbZn driver and causes it to elongate; this elongation is measured through an LVDT sensor. The outputs of the current shunt (amps), HTS Coil (voltage) and LVDT amplifier (displacement) are monitored with voltmeters.

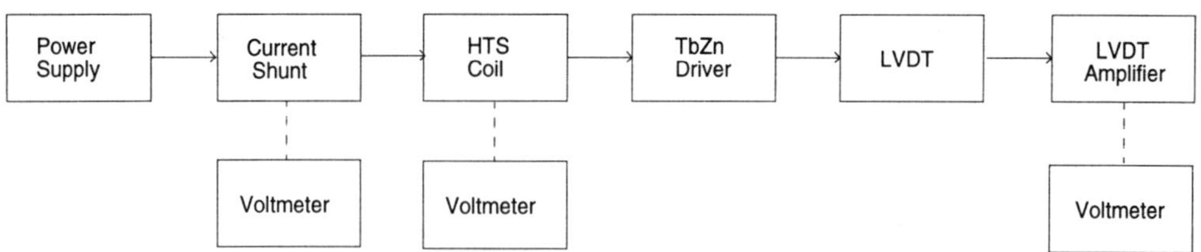

Figure 6. Instrumentation block diagram.

Typically, before testing, the LVDT is adjusted to read 4 volts as this represented the steepest section of the calibration curve; this adjustment is performed by manually changing the position of the LVDT bobbin. The output sensitivity at this point in the curve is approximately 0.45 μm/mV.

Once the device was assembled, it was first "checked-out" in liquid nitrogen to make sure it was functioning properly electrically. Next, the device was mounted on the cryocooler as shown in Figure 7. The device was tested under three sets of preload conditions: (1) no preload, (2) 13.3 kN (3000 lbs) preload, and (3) 22.2 kN (5000 lbs) preload.

For each load condition, the device was initially cooled down to the baseline condition; in general, it took approximately one day for the device to cool down. Next, a heater on the second-stage coldhead was activated to raise the coldhead temperature, and consequently the actuator temperature as well. In general, there was a temperature difference of 1 K between the actuator and the cryocooler coldhead. During testing, even at high currents (i.e., > 50 amps), there was negligible heating and temperature rise in the actuator.

SUMMARY

A summary of the performance test cases is provided in Figure 8. These data show that the actuator successfully met the performance specification of 12.7 μm (500 μ-in.) total travel, even with the maximum preload of 22.2 kN (5000 lbs). The data in Figure 9 demonstrates the hysteresis of the actuator when ramping the current up and down, even for small excursions in current; this hysteresis would be compensated in actual accelerator applications by using a control feedback loop.

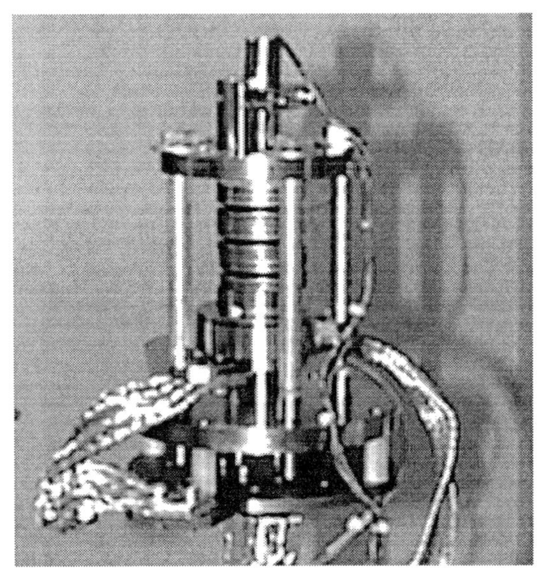

Figure 7. Actuator assembled on top of cryocooler coldhead.

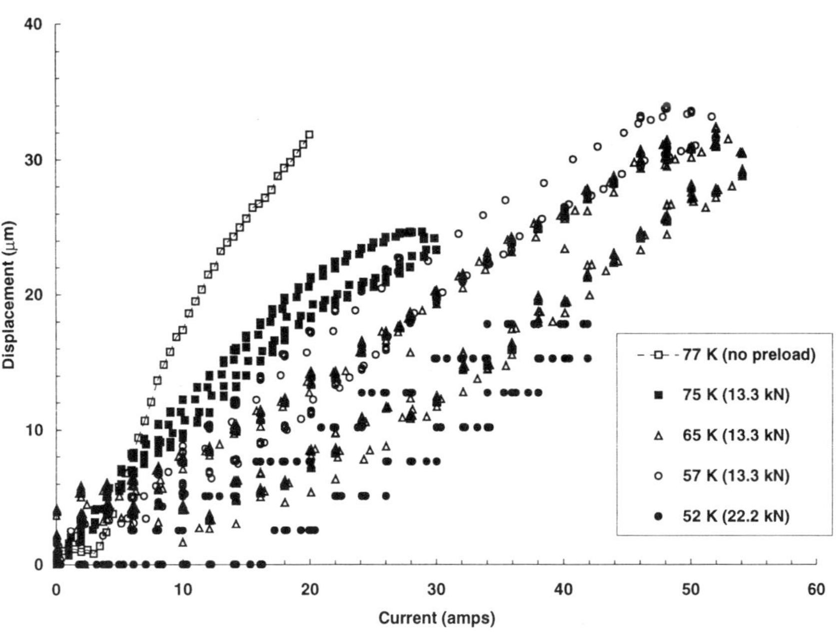

Figure 8. Summary of load cases.

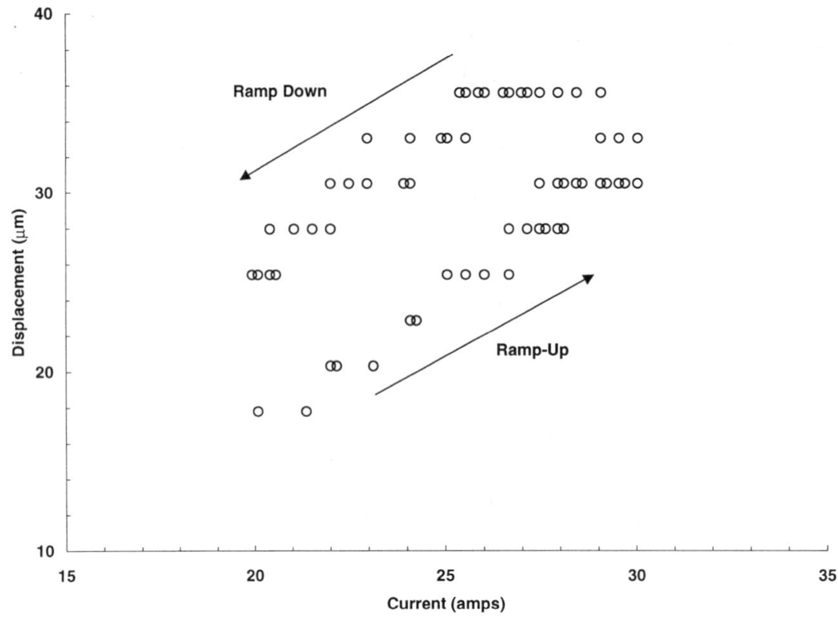

Figure 9. Hysteresis behavior of actuator (50 K).

ACKNOWLEDGEMENT

The authors would like to acknowledge the following people for contributing to this work: Tom Lograsso of Ames Labs of Iowa State University, who developed and provided the single crystal material for this program; Kristl Hathaway of the Office of Naval Research (ONR) in Arlington, VA, who served as technical advisor for the DoD/Navy Phase II STTR, which allowed us to fabricate and test large single crystal TbZn; Arthur Clark of Naval Surface Warfare Center (NSWC) in Carderock, MA, who provided expertise on the behavior of the TbZn material; Joe Prebel and Viet Nguyen of Thomas Jefferson National Accelerator Facility (TJNAF, formerly CEBAF) in Newport News, VA, for providing specifiations and design information for the rf-cavity actuator; Dawood Aized of ASC, for his help in setting up and performing the crycooler testing. Regarding the authors, Chris Gervais performed the solid modelling and design work; Nick Kotsifakis assembled the actuator and performed the cryocooler testing; and John Voccio designed the actuator and served as program manager. This work was sponsored by DoE Phase I SBIR (contract no. DE-FG02-98ER82531) and also in part by DoD Phase II STTR (contract no. N00014-96-C-4018).

VACUUM BREAK-DOWN IN THE BEAM TUBE OF LARGE ACCELERATORS WITH SUPERCONDUCTING CAVITIES

Ch. Haberstroh, H. Winkler and H. Quack

Technische Universität Dresden
Lehrstuhl für Kälte- und Kryotechnik
D-01062 Dresden, Germany

Abstract

For the TESLA accelerator superconducting cavities have been chosen as accelerating components. One aspect, important also for most of the existing accelerator installations and nearly unregarded up to now, is the behavior of the system in case of an unintentional vacuum break-down in the beam tube. Both massive inleakage from outside in case of a large average and small mass flows due to minor vacuum leaks must be taken into account.

In a first experimental approach measurements have been performed with a small scale experimental set-up at a temperature level of 4.2 K. Effects of air and dry nitrogen condensation had been compared. Both the solidification process inside the "beam tube" and pressure rise in the helium container had been quantified.

In this contribution the experimental set-up is presented and experimental results are shown.

Introduction

The beam tubes of large accelerator facilities are often kept at helium temperatures for a large extent. For the TESLA accelerator continuous cryogenic strings of about 2.5 km length are foreseen. The working temperature will be 2 K for the superconducting cavities and 4 K for magnet sections respectively.[1]

A widely unknown scenario is the behavior of these systems in case of an accidental admission of ambient air into the beam tube vacuum. This spreads from minor leakage in case of an imperfect seal to massive gas breakthrough caused by rupture of the beam tube at the warm end.

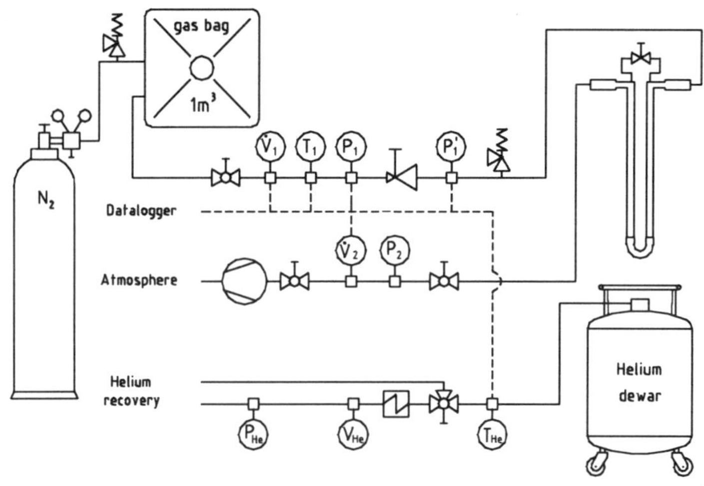

Figure 1. Experimental setup

Experimental Setup

For an experimental approach a test rig was assembled as indicated in the flow diagram Figure 1.

The beam line is simulated by a u-shaped tube immersed into liquid helium. A gas bag allows to provide constant feed gas parameters. Vacuum conditions are simulated by means of a high capacity (100 m³/h) rotary vane vacuum pump. Gas admission is adjusted with a suitable throttling valve. Pressure, temperature and gas flow are monitored at all interesting points. Both dry N_2 and ambient air were applied.

Tube parameters:	Material	X 5 CrNi 1810
	Inner diameter	11.4 mm
	Wall thickness	0.3 mm
	Immersed length	max. 1000 mm
Feed gas	Dry nitrogen; air	
	Pressure	0.1 MPa
	Temperature	295 K
	Volumetric flow:	0.007 .... 1.7 m³/h (at standard conditions)
Helium bath	liquid volume	max. 100 l
	p;T	0.1 .... 0.15 MPa; 4.2 .... 4.6 K

Results

An important aspect for the cryostat layout is the maximum heat impact on the helium system. Using the following assumptions:

- complete condensation and cool down to 4.2 K of the entering gas
- evaporation enthalpy of LHe (4.2 K) operative only

an upper limit for the resulting He boil-off rate versus a given gas flow was calculated.

More usefully this volumetric gas flow into the vacuum system can be converted to an equivalent opening diameter, i.e., leak size. Some representative values for air admission are listed in Table 1.

The Helium boil-off rates determined with the test rig described above (Figure 1) were below this limit at least by a factor of two. This is evidently due to the configuration of the test rig allowing some heat transfer from the evacuated u-tube to the boil-off helium before leaving the cryostat.

Test runs were performed with both dry nitrogen and ambient air. The volumetric flow rate (at standard conditions, 288 K; 0.1 MPa) spread from $2 \cdot 10^{-6}$ m³/s up to $470 \cdot 10^{-6}$ m³/s. The upper limit was given here by the cryostat performance. In each case a complete occlusion was observed. The duration from start of the gas admittance until complete blockage of the tube is shown in Figure 2. For small leak rates the condensate accumulation is dominated by the availability of admitted gas. For higher gas streams a widely constant value for the time until complete obstruction is observed. A minor decrease in time to blockage at the highest flow rate ($\dot{V}_1 = 1685\,\mathrm{L/h}$) can be understood by the improved heat transfer related to turbulent gas flow inside the u-tube.

A number of representative experimental data are given in Table 1. The total mass of the trapped gas inside the u-tube was determined during the warm up.

Table 1. Experimental data (condensation of ambient air)

Admitted gas flow (max.)	Equivalent leak diameter	Time until blockage	Total condensate mass	Helium boil-off (@ 288 K, 1 bar)
[L/h]	[mm]	[s]	[g]	[m³]
7.2	0.11	8660	13.0	0.11
35.6	0.25	3510	15.5	0.162
64.3	0.34	1250	22.7	0.336
92.7	0.41	1200	33.3	0.503
447.0	0.89	1180	142.9	5.757
1685.0	1.73	810	233.0	11

A typical course of the occlusion is shown in Figure 3 for small leak rates and in Figure 4 for a larger leak rate. In both cases a smooth convergence to complete obstruction is visible.

Similar investigations have been performed earlier for water freezing at the inner walls of a cooled tube.[2] With appropriate parameters for this system a very complex behavior was reported with partly re-thawing due to the increasing flow velocity in the narrowing channel and flipping between laminar to turbulent flow regime. As a result a permanent, wave-like dislocation of the condensate without complete obstruction of the channel takes place.

In the present work no evidence for such phenomena were observed. All experiments indicated a straight forward progress of the occlusion. There are hints that the occlusion is located close to the entrance of the cold section.

As an exception occasionally an unsteady behavior as shown in Figure 5 was observed only. This can be interpreted as fracture of the brittle condensate agglomeration.

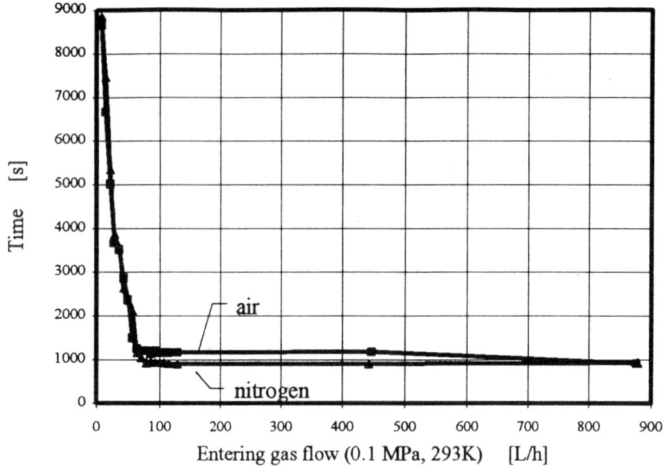

Figure 2. Time period of gas flow until tube blockage

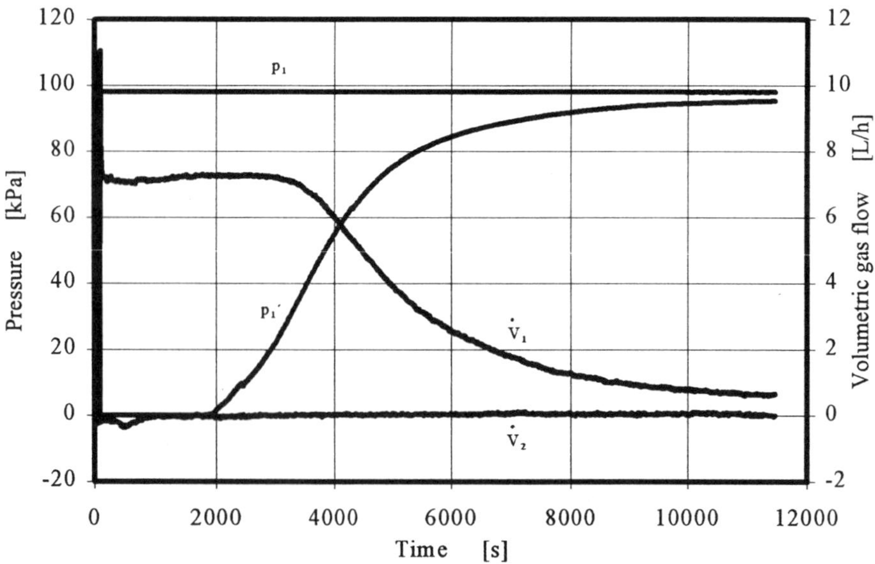

Figure 3. Experimental data for air, $\dot{V}_1 = 7.2$ L/h initially ($\dot{V}_1$, $\dot{V}_2$, p_1, p_1' according to Figure 1)

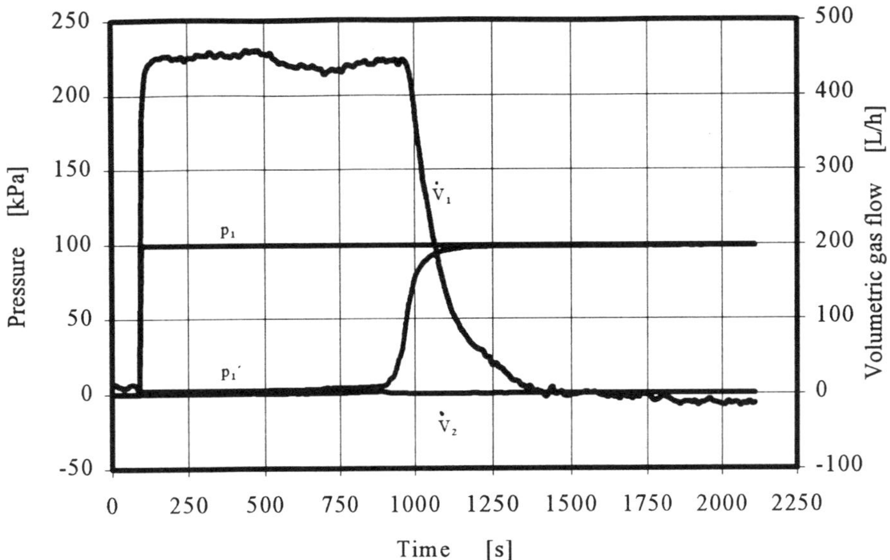

Figure 4. Experimental data for air, $\dot{V}_1 = 447$ L/h initially ($\dot{V}_1$, $\dot{V}_2$, p_1 , p_1' according to Figure 1)

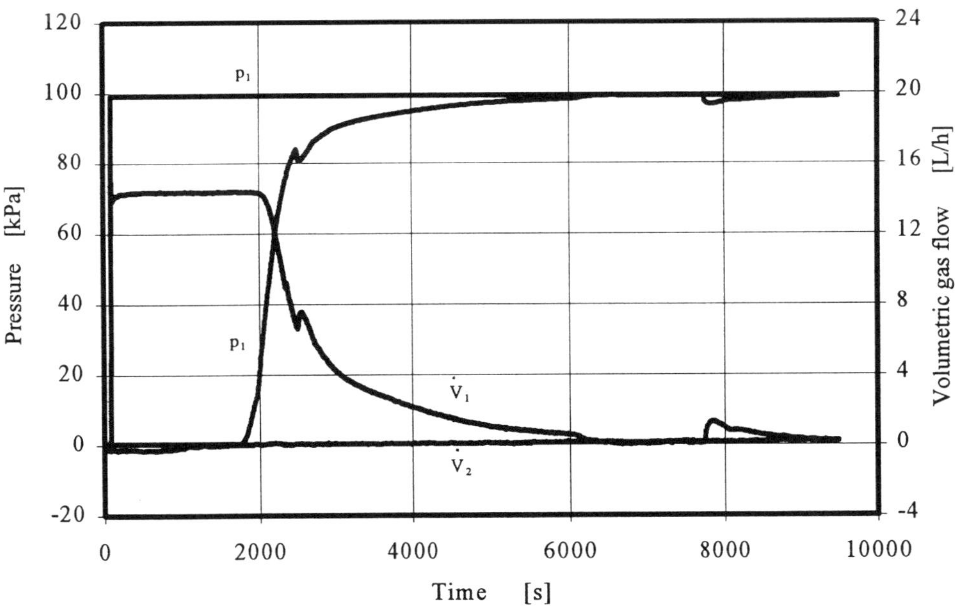

Figure 5. Unsteady progress of the obstruction ($\dot{V}_1$, $\dot{V}_2$, p_1 , p_1' according to Figure 1)

Summary

Experimentally a swift and complete occlusion close to the cryostat entrance could be demonstrated. Hence the consequences of a gas breakthrough into an evacuated beam line could be limited on the first cold sections in case of an appropriate configuration. Further work should focus on scaling up these investigations on larger tube diameters.

Acknowledgment

Financial support by DESY Hamburg is gratefully acknowledged.

References

1. R. Brinkmann et al. (eds.). "Conceptual Design of a 500 GeV e^+e^- Linear Collider with Integrated X-ray Laser Facility“, DESY Report 1997-048, Hamburg (1997)
2. Weigand, R., Erstarrungsvorgänge in einer Flüssigkeit in einem ebenen, geraden Kanal, PhD thesis, Darmstadt Germany (1992)

FURTHER IMPROVEMENTS OF THE TESLA TEST FACILITY (TTF) CRYOSTAT IN VIEW OF THE TESLA COLLIDER

C. Pagani,[1] D. Barni,[1] M. Bonezzi,[1] and J. G. Weisend II[2]

1 INFN Milano-LASA
Via F.lli Cervi 201, 20090, Segrate (MI), Italy
2 DESY,
Notkestrasse 85, D-22607 Hamburg, Germany

ABSTRACT

The experience gained in the commissioning and operation of the first and second generation of the TTF cryostat lead to a new and improved design which should fit the requirements of the TESLA collider in terms of cost and performance. The redistribution of the components in the cryostat cross-section allows us a reduction of 15% the vacuum vessel diameter and the use of a standard 38'' pipe. The thermal shields have been adapted to fit the new vacuum vessel, while the finger-welding technique has become a standard. A more stable quadrupole package position (in spite of the possible asymmetrical forces acting on the Helium Gas return Pipe (HeGRP) edges during pumping and cooldown) has been obtained by moving the three post positions. To allow the possible use of rigid couplers and superstructures, a sliding support scheme has been developed for the cavities. In connection with a reference Invar bar it lets the cavities stay fixed and aligned while the HeGRP slides over them during cooldown and warmup.

INTRODUCTION

The development of the TESLA[1] collider cryomodule has arrived at the third generation, exploiting the experience gained from the commissioning and operation of the three cryomodules already assembled at DESY. The cryomodule contains 8 cavities and a superconducting quadrupole package. It must provide cavity support, alignment, cooling, thermal insulation as well as feed-through for RF power, instrumentation, and connection to adjacent cryomodules or to the cryogenic supply lines.

The third generation cryomodule needs to meet the requirements (costs, mechanical requirements and cryogenic performances) for the TESLA collider to prepare the mass production cryomodule design.

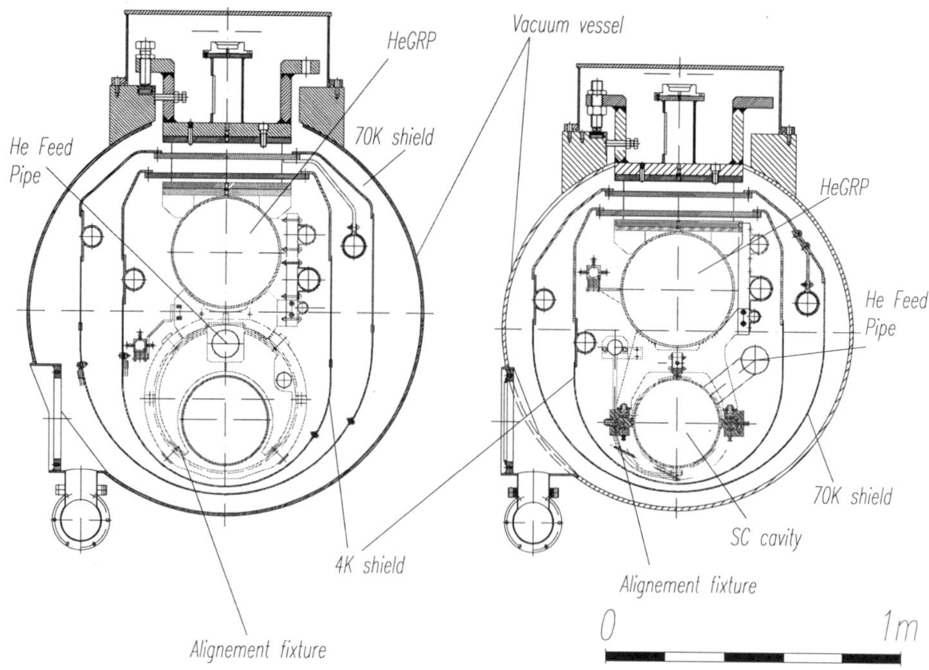

Figure 1. Comparison between second (left) and third (right) generation TESLA cryomodules. The components have been redistributed to fit the smaller vacuum vessel (38'' standard pipeline tube).

GENERAL LAYOUT AND CROSS SECTION

In order to improve the design of the third generation cryomodule, an evaluation of each component and its influence in the general layout lead to the decision to use a standard pipeline tube for the vacuum vessel. Checking all possible solutions a 38 inch (0.98 m) standard pipe (3/8''thick) was chosen.

To arrange all components in the smaller vacuum chamber, the thermal shields were redesigned, filling the spaces and keeping margin to follow the tolerances on the standard pipe. The major improvement in the reduction of the cross section has the decision to use an off-axis helium feed pipe. The requirements of superfluid helium distribution required it to stay above the helium tank level and below the gas return pipe. In this way, the "off-axis solution" allowed the cavity axis to move closer to the HeGRP.

The new cavity support described in the following also reduces the volume occupation in the lower region of the cross section, with respect to the previous design.

THERMAL SHIELDS

The cryogenic optimization of the TESLA module[1] string requires two thermal shield at 4.2K and at 70K. The shields are manufactured from aluminum (1050) and are both made of a roof (divided in two sections, each about 6 meters long) that supports eight panels. The panel length is 1.4 m, except for the first and the last. The shield panel length has been chosen in order to adjust the aluminum-stainless steel differential contraction and to allow the coupler holes to follow the cooling cones.

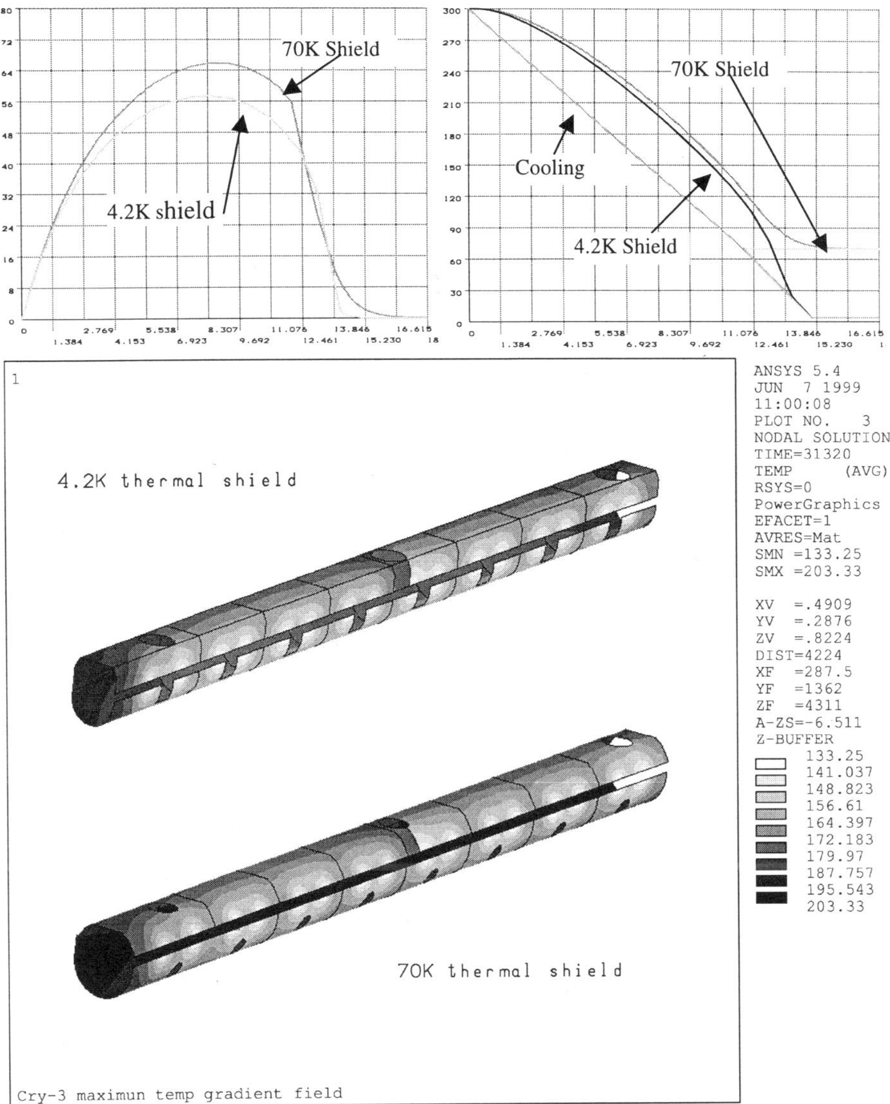

Figure 2. Cooldown simulation of the 4.2 K and 70 K aluminum thermal shields (upper right graph). We used a simultaneous 12 hour linear cooldown. The maximal thermal gradient on the shields (upper left graph) is below 60 K, a safe value. The temperature fields plotted in the lower figure show that the gradient is concentrated in the welding region, where the fingers unload the structure.

The shield cross-section have been adapted to the smaller vessel, with a rounder shape. The cooling pipe, as in second generation cryomodule, is an aluminum pipe, finger-welded[2] to the shield roof and to the panels. A bimetallic junction makes the transition to stainless steel in the connection region between cryomodules. The shield geometry has been checked by a finite element code to test for cooldown deformations and thermal inertia.

The results show that a 12 hour linear cooldown produces thermal gradients of about 60 K (Fig. 2) that induce deformations of ~10 mm (Fig. 4), which are compatible with the geometric free space in the section. In order to ease the fabrication and to decrease the cost, the 6 meter shield roofs have been divided in two sections which are rigidly connected by screwed plates. This heat transfer discontinuity has been included in the model, in order to check its influence.

Using the temperature field calculated during the cooldown, a thermo-mechanical analysis has been performed in order to obtain the deformation and stress distributions in the aluminum shields. The results show a lateral bending of the shields, due to the asymmetric cooling ("banana" effect[3]) of about 10 mm, while the maximal stresses are within 30Mpa (the results are shown in Fig. 4). The shield panels and cooling pipes are used to cool other parts of the cryodomule. The post plates, which support the cold mass and the shield roofs, need to be kept at different temperature. In particular, the lateral post plates are connected to the shields with sliding supports which do not assure a good heat exchange. To achieve the post cooling, short and very flexible copper braids have been used. The same solution has been used for the coupler cones. The cones has been redesigned (Fig. 3) to ease the fabrication and the assembling. This solution has been, in part, already tested during the cooldown of cryomodule number two and sensors showed it worked correctly.

CAVITIES SUPPORTS

The experience in the assembly of cryomodules #1, #2 and #3[4] showed evidence that the cavity support system could be improved in terms of maintaining the cavity alignment. Due to the geometry of the system, it was too complicate to keep cavities and quadrupole positions within the alignments requirements. The necessity to develop a system that could, in principle, be compatible with superstructure[5] or semi-fixed couplers lead to the decision to complete redesign the cavity supports. The solution that has been proposed and studied needs to keep the cavity transverse position fixed, while leaving the cavity longitudinal position independent from the HeGRP thermal contraction and extension during the cooldown-warmup cycles. These objectives have been reached with a set of low friction sliding supports. A support, whose section is presented in Fig. 5, consists of a C-shaped stainless steel element that clamps to a titanium pad welded on the cavity helium tank. The connection is mediated by a sequence of rolling needles, runners and reference screws. In each constrained direction (vertical and lateral) a reference screw defines the cold position of the cavity axis, and a spring washer package loaded at about 80 kg_f keeps the pad in contact. The friction between the cavity and the support has been measured and results in about 0.6 kg_f[6] This low friction value results in a decoupling of the cavity longitudinal position with respect to the supporting HeGRP. In order to be compatible with semi-rigid couplers and superstructures, an Invar rod fixture determines the longitudinal position of the coupler ports. This solution results in a maximum total longitudinal motion of the coupler port of about 3 mm (Table 1 reports the computed hot and cold positions of each coupler port with respect to the external position).

Table 1. Reference positions of the coupler port calculated for a semi rigid coupler. The positions are relative to the fixed point. Hot to cold positions have been computed by the Invar stainless steel and titanium relative contractions. The maximum displacement is less than 3 mm.

Coupler N	1	2	3	4	5	6	7	8
External (mm)	-5113	-3733	-2353	-973	+407	+1787	+3167	+4547
Hot Position (mm)	-5113.5	-3733.5	-2353	-973	+407.5	+1768	+3168	+4548.5
Cold Position (mm)	-5122.3	-3732.8	-2352.9	-973.4	+406.5	+1786.5	+3165.9	+4545.9
Displacement (mm)	-0.5+0.7	-0.5+0.2	-0+0.1	-0+0.4	+0.5-0.5	+1.0-0.5	+1.0-1.1	+1.5-1.1

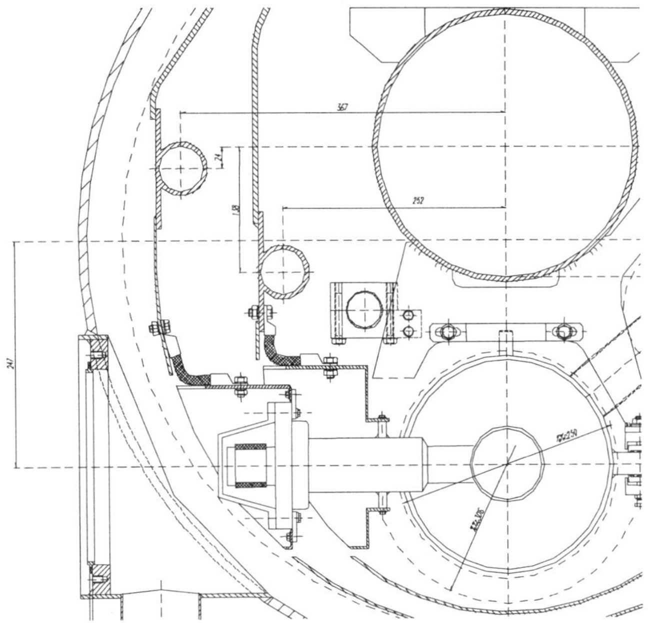

Figure 3. Coupler cones have been redesigned for an easiest fabrication and assembling. The cooling is guaranteed by copper braids connected to the shield panels.

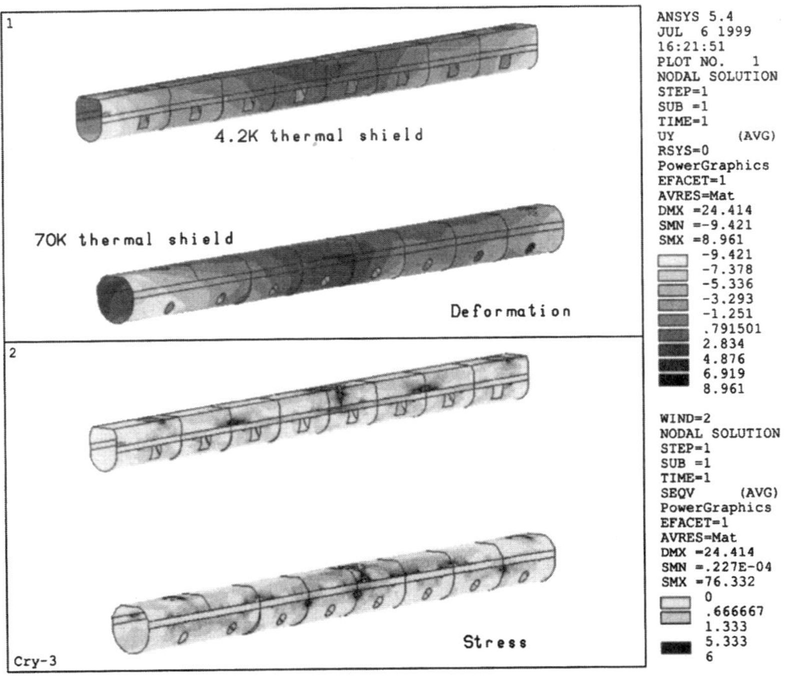

Figure 4. Thermo-mechanical analysis of the shield panels. Applying the computed temperature field the deformations and stress distribution can be easily computed. The stresses (lower plot) are within 30 Mpa, while the deformation due to asymmetric cooling is below 10 mm (upper plot).

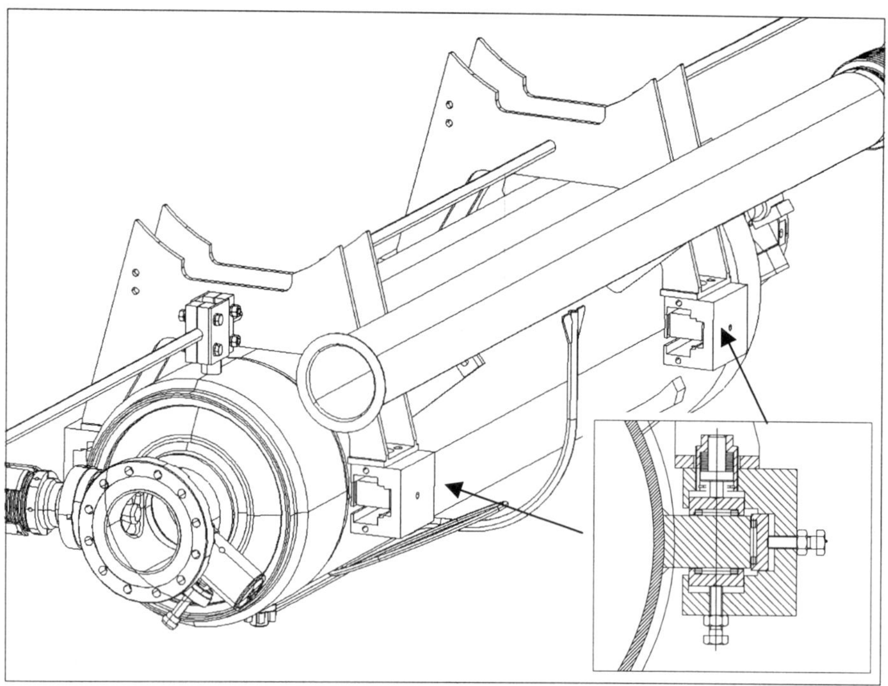

Figure 5. The cavity support system. Four C-Shaped stainless steel elements clamp a titanium pad welded to the helium tank by using rolling needles that reduce drastically the longitudinal friction, leaving cavities independent from the elongation and contraction of the HeGRP. Lateral and vertical position are defined by reference screws.

COLD MASS SUPPORT AND COLD ALIGNEMENTS REQUIREMENTS

Analyzing the motion of the active elements during the cooldown of cryomodules #1 and #2, as measured by the installed Wire Position Monitor[7] system, we have determined the presence of asymmetric forces in the feed-cap and end-cap sections of the module. These forces probably are generated by the misalignment of the HeGRP of the two consecutive modules, deforming the HeGRP, and moving the active elements out of alignment tolerances. To reduce this effect, two major improvements have been included in the present design.

First of all, we changed the longitudinal positions of the support post. The most critical component, from an alignment point of view, is the quadrupole, so a support post has been placed over the fixture of the quadrupole package. The other two posts have been placed, compatibly with the assembling procedures, at positions that reduce the static deformation. To check this new solution, a combination of 1 kN forces in all directions have been applied to the extremes of the HeGRP, and the induced deformations have been computed. The displacement in the cavity region is within 0.2 mm in the worse situation, while at the quadrupole it is estimated to be less than 0.1 mm. In order to simplify further the design, the support of the vacuum chamber has been modified. In the new design the same supports are used during assembling and in the linac operation. The positions have been chosen using the same philosophy as the HeGRP support, in order to minimize the reaction to unknown forces during operation. Figure 6 shows a comparison of the support positions in the second and third generation cryomodules, the support is located very close to the quadrupole to minimize its possible movement. The position of the other has been chosen as a compromise between good stability and space requirements during the assembly. Due to these changes, all the assembling tools have been redesigned to fit the new geometry.

One other major improvement follows from the decision to include the bellows that links the HeGRP of two different modules in the fabrication of the HeGRP itself. This has increased the free space in the interconnection of the modules, and has extended to the bellow flange the fabrication tolerances. The fabrication tolerances of the HeGRP are achieved with a 12 m milling machine that references the flanges and support element defining the axis of the tube. This operation can now be extended to the bellows flange. As a consequence, the bellows are referred to the cryomodule axis, and the connection between two consecutive HeGRP is more precise, and consequently less stressed. The result of this choice in the fabrication philosophy should be a sensible reduction of the external forces during pressurizing and cooldown.

CONCLUSIONS

We have presented here the improvements introduced in the third generation of the TESLA cryomodule. The changes have been done in the direction of achieving the industrial mass-production requirements for the TESLA 500 Collider Project. The vacuum vessel has been reduced to a standard pipe (38''); the thermal shields have been redesigned to fit the space and to reduce the number of components. To insure a full compatibility with superstructures and semi-rigid couplers, a new cavity support has been designed. A sliding fixture decouples the longitudinal motion of the cavities from the supporting HeGRP. In order to reduce the sensitivity to misalignments and to external forces the post and the vacuum vessel support positions have been changed. In this way the TESLA tolerances, especially in the quadrupole package region, should be achived.

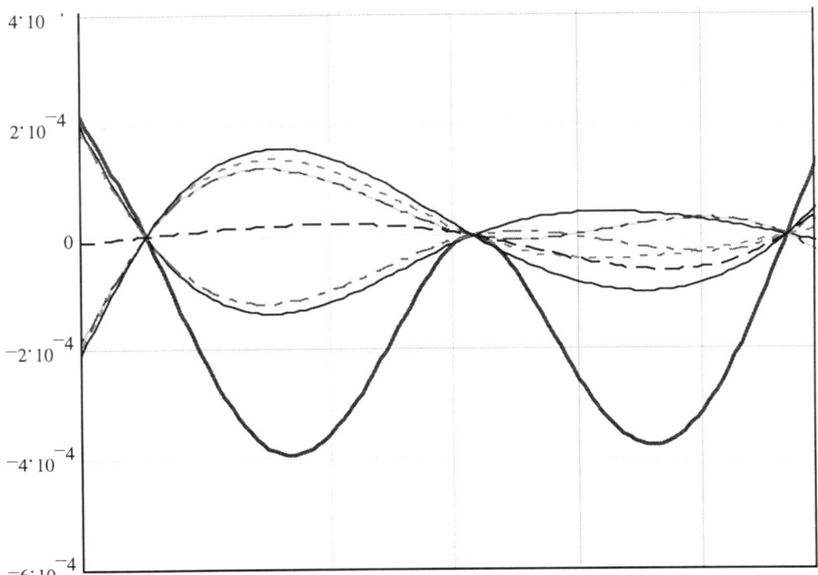

Figure 6. Comparison of second (upper) and third generation post positions (lower). The new solution improves the stability with respect to external unknown and asymmetric forces acting on the HeGRP and on the vacuum vessel during cooldown and under vacuum.

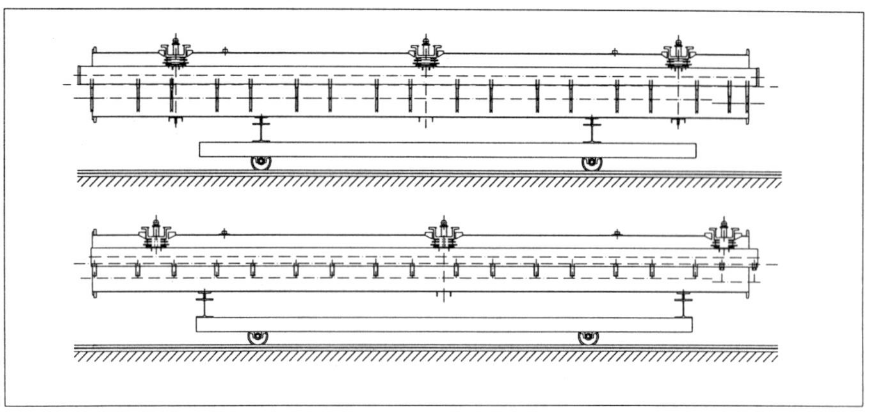

Figure 7. Computed deformation of the HeGRP when stressed by 1 KN asymmetric forces in the end cap and feed-cap position. The overall deformation is less than 0.2 mm in the cavities region and less than 0.1 mm for the quadrupole package. The solid line is the static absolute deformation with respect to which the cavities and quadrupole are aligned, while other lines are the displacements around the equilibrium position.

The combination of the new support positions and fabrication process should achieve the cold alignment requirements for the cavities and the quadrupole package needed by the TESLA Collider Project.

REFERENCES

1. TESLA Test Facility linac - Design Report, DESY Report, March 1995, TESLA 95-01. Conceptual design of a 500 GeV e+e- linear collider with integrated X ray laser facility, R. Brinkmann et al., editors, DESY Report, 1997-048, 1997.
2. C. Pagani, D. Barni, M. Bonezzi, J. G. Weisend II, Design of the thermal shields for new improved TESLA Test Facility, Adv. Cryo. Eng. Vol 43a, 307 (1998).
3. C. Pagani, D. Barni, M.Bonezzi, P.Pierini,J.G.Weisend II, Cooldown simulations for the TESLA Test Facility (TTF) cryostat, Adv. Cryo. Eng. Vol 43a, 315 (1998)
4. J. G. Weisend II, C. Pagani, R. Bandelmann, D. Barni, A. Bosotti, G. Grygiel, R. Lange, P. Pierini, B. Petersen, D. Sellmann, S. Wolff, The TESLA Test Facility (TTF) Cryomodule: A Summary of Work to Date, Paper #CCB1, 1999 CMC/ICMC Montreal.
5. J. Sekutowicz, et al., Superconducting Super-structures for the TESLA Collider, Proc. of EPAC'98, Stockholm, Sweden, June 1998 and Desy Report, TESLA 98-08.
6. D. Barni, M. Castelnuovo , M. Fusetti, C. Pagani and G. Varisco, Friction Measurements for SC Cavity sliding Fixture in Long Cryostat, Paper #CCA3 , 1999 CMC/ICMC, Montreal.
7. D. Giove, A. Bosotti, C. Pagani, G. Varisco, A Wire Position Monitor (WPM) system to control the cold mass movements inside the TTF cryomodule, PAC97, Vancouver.

THERMAL OPTIMIZATION OF THE APT CRYOMODULE WITH REGARD TO THE ASSOCIATED CRYOSYSTEM

J.A. Waynert, F.C. Prenger, and J.P. Kelley

Los Alamos National Laboratory
Los Alamos, NM 87545

ABSTRACT

The Accelerator Production of Tritium (APT) linac uses superconducting radio frequency (RF) cavities to accelerate a 100 mA proton beam from 211 MeV to 1700 MeV. The niobium cavities are cooled by superfluid helium at 2.1 K. Each cavity is energized by two 700 MHz co-axial RF power couplers (PCs). In the high-energy portion of the linac, there are four cavities within a cryomodule, and each of the PCs powering the cavities is continuously supplied 210 kW. Intermediate temperature intercepts and a counterflow heat exchanger on the outer conductor, have been proposed as methods to reduce the low temperature heat load of the PC. This paper presents a trade study that was performed to optimize the cooling approach to the PC. The study includes the issues of integration of PC cooling with cooling of the thermal shield, beam tube, and structural supports. The trade-study also considers the impact of the coolant stream(s) temperature, pressure, and flow rates on the cryo-plant design and operating cost, and the associated distribution system.

INTRODUCTION

The APT utilizes a linear accelerator (linac) to produce energetic protons, which strike a metal target of tungsten and lead, producing neutrons through spallation. The neutrons are captured in 3He, producing tritium. After protons are accelerated through a normal-conducting portion of the accelerator, they enter the superconducting (SC) section. The SC portion of the linac has cryomodules containing SC RF accelerating cavities separated by resistive focusing quadrupole magnets. The RF cavities are cooled by saturated superfluid helium at 2.1 K. Although there are three variations of cryomodules, this study focuses on the β=0.82 (β is the ratio of speed of the particle to the speed of light) cryomodule, because it has the highest heat loads and also represents the greatest number of modules in the linac.

The PC transports RF power that energizes the SC cavities and provides a thermal link to room temperature. Figure 1 shows a cut-away view of a portion of a cryomodule; the

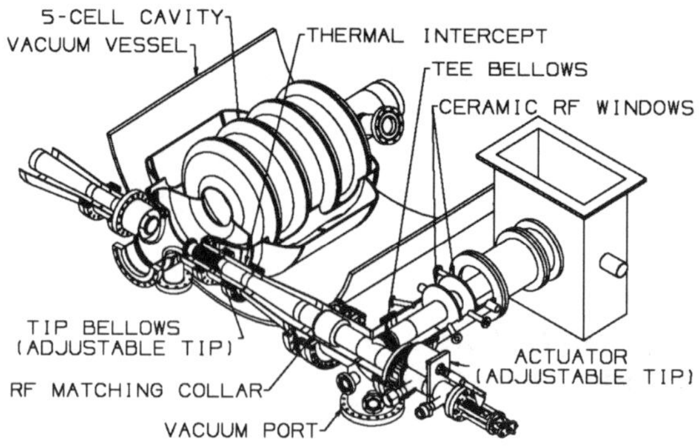

Figure 1. A cut-away view of the power coupler, vacuum vessel, and cavity within the helium vessel.

main components are labeled. Several cooling approaches have been proposed for the 210 kW PC. This paper presents the results of a trade study aimed at optimizing the impact of the PC cooling with the cooling of other loads in the cryomodule, the cryomodule design, and, the cryo-plant and its distribution system.

CRYO-PLANT AND DISTRIBUTION SYSTEM CONSTRAINTS

In discussions with the cryo-plant and distribution system designers, a number of constraints were developed: cryomodule return coolant temperatures should be below about 55 K to be compatible with the refrigeration cycle and transfer line shielding; the thermal shield coolant pressure should be around 450 kPa or 1900 kPa with typically a 50 kPa pressure drop to be compatible with the refrigeration cycle; and helium coolants at temperatures above about 4.5 K will be returned to the cold box at about 120 kPa.

BASELINE HEAT LOADS

Because of the high availability required of the APT linac, optimizations and trade-study comparisons are done for the cryomodule with the RF on. The main contributions to the heat loads of the cryomodule are from the cavities (96.4 W), the power couplers (40 W), higher order mode RF losses (6.8 W), joule heating in vacuum flange connections in the outer conductor (referred to as RF joint losses, 4 W), thermal shield radiative load (2.2 W), beam tube (2 W), instrumentation leads, structural supports (2 W), female portion of bayonets, and relief pipes (0.5 W), where the preliminary estimates of the heat loads to 2.1 K are given in parentheses. The first requirement of any cryomodule cooling scheme is that it must provide a primary coolant stream for the cavities. There may also be one or two intermediate temperature coolant streams for the secondary heat loads, essentially thermal intercepts. In addition, there must be a stream for cooling the inner conductor of the power coupler. Figure 2 shows an example of a simplified flow diagram for the β=0.82 cryomodule with a single secondary cooling loop.

In order to evaluate different cooling configurations, the total room temperature refrigeration input power required to meet the calculated heat loads was selected as a figure of merit. A survey of large-scale refrigeration systems[1] was performed to obtain the actual machine efficiencies relative to Carnot. Table 1 shows the results. In addition to the figure of merit, a thermal model is needed to determine the heat loads to the refrigeration system from each of the flow streams. The thermal model includes the dependence of the 2.1 K

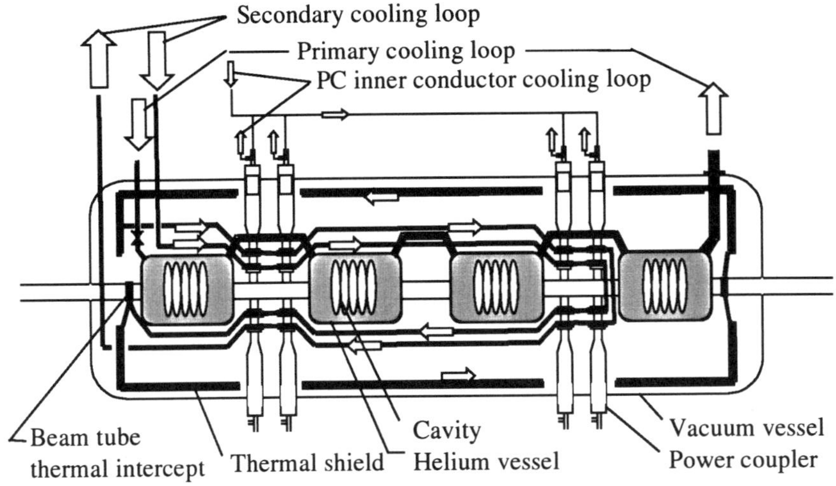

Figure 2. Schematic of β=0.82 cryomodule showing main components, and primary, secondary, and PC inner conductor cooling flow.

Table 1. Refrigeration efficiencies, as a percent of Carnot, for large scale helium plants.[1]

Cycle	% efficiency
2.1 K liquefaction	13
2.1 K refrigeration	18
4.5 K liquefaction	25
4.5 K refrigeration	30
15 - 50 K refrigeration	27
60 - 80 K refrigeration	30
LN_2 liquefaction	35

and intermediate temperature loads on the intermediate coolant fluid temperature. Generally, heat loads at lower temperatures are of more importance to the cryo-plant design and operation, because it is more difficult and costly in terms of room temperature refrigeration input power to produce the low temperature. For example, it requires about 800 W of room temperature, assumed to be 310 K, input power to provide 1 W of refrigeration at 2.1 K, but only 22 W of room temperature input power to provide 1 W of refrigeration at 45 K.

Cavity Heat Loads

The cavity heat loads were considered constant in this study because they are dependent on the achievable value of Q_o which has been established for the baseline for the APT linac.

Power Coupler Heat Loads

The power coupler heat loads depend on the method of cooling the outer conductor and the temperature of the inner conductor. A detailed thermal model of the power coupler was developed and is presented elsewhere.[2] Three different cooling schemes were considered for the PC outer conductor: a single localized intermediate temperature intercept; two localized thermal intercepts (referred to as double point); and a counterflow (distributed) cooling scheme. The outer conductor of the PC has a niobium nipple connecting the PC to the beam tube. (See Fig. 1). The heat loads to 2.1 K dramatically increase if a portion of the Nb nipple is resistive rather than superconducting. For cryomodule cooling schemes in which the secondary stream temperature is above the superconducting transition temperature of Nb, $T_c = 9.2$ K, a small copper strap was added to the upper temperature end of the Nb connecting it to the 2.1 K helium vessel. The strap was sized to ensure the Nb remained superconducting.

The efficiency of the various optimized PC outer conductor cooling schemes is shown in Table 2. The schemes shown are optimized in the sense that the thermal intercept(s) are located at a position that minimizes the refrigeration input power. The variation in heat loads and intercept temperatures for configurations in which the PCs are plumbed in series are accounted for through use of a fitting function. The fitting function is a second order polynomial that relates the heat loads to the respective intercept temperature and higher temperature intercept location. In integrating the PC into a cryomodule, it is assumed that the low temperature intercept is at the SS/Nb boundary and the upper temperature intercept is located in the same place for each PC, but such that the room temperature refrigeration input power was a minimum for the cryomodule.

Because of the coaxial arrangement of the inner and outer conductors, the inner conductor generates a high infrared (IR) heat load on the low temperature portion of the outer conductor, in particular, at the Nb nipple. This IR load can be reduced by lowering the temperature of the inner conductor. The optimum temperature for the inner conductor is about 150 K, and reduces the PC total room temperature input power nearly 35% for any of the outer conductor cooling schemes. However, in the cryomodule, the heat loads are dominated by the cavities, such that the refrigeration input power required to handle the total cryomodule loads is only reduced 4 to 9%. Given that our PC design will not presently accommodate such a significant lowering of the inner conductor temperature, it is not discussed further in this paper.

Temperature Dependent Heat Loads

The heat loads for the beam tubes, which connect the cavities to the vacuum vessel, were calculated from a parametric 1-D finite difference model. The model includes the thermal radiation down the beam pipe and the effect of a bellows which accommodates the thermal contraction of the beam tube relative to the vacuum vessel on cooldown. Two fitting functions were then developed; one to provide the heat loads at 2.1 K, and another, at the intercept as functions of the intercept temperature assuming the intercept was at the location of minimum refrigeration input power.

The heat loads on the thermal shield were considered independent of the shield temperature. This is not strictly correct, but the heat load variation in combination with the Carnot factor and refrigerator efficiency for shield temperatures in the range of 15 K to 100 K make very small changes to the total refrigeration input power. However, the shield temperature does have an impact on the 2.1 K load. This effect was included. The local

Table 2. Efficiency of various optimized PC outer conductor cooling schemes. SS = stainless steel; Nb = niobium; Ptot = total room temperature refrigeration input power

Single point thermal intercept	
4.6 K at SS/Nb interface	Ptot = 5.5 kW
15 K 4.5 cm from SS/Nb interface	Ptot = 7.6 kW, requires copper strap
Double point thermal intercepts	
7.8 K and 56 K	Ptot = 3.6 kW
15 K and 56 K	Ptot = 4.4 kW, requires copper strap
Counter-flow heat exchange	
2.6 K inlet, 0.06 g/s flow rate	Ptot = 3.1 kW
4.6 K inlet, 0.07 g/s flow rate	Ptot = 3.4 kW

vacuum was assumed to be sufficiently good that the effect of residual gas conduction was negligible. A baseline heat flux of 0.2 W/m^2 from the 45 K thermal shield to 2.1 K was used.[3] Using formulas developed by Ted Nast[4], a simple quadratic function was developed to describe the contribution to the 2.1 K heat load as a function of the thermal shield temperature.

The HOM cables and instrumentation lead heat loads were treated as conductive loads down a titanium rod whose area-to-length ratio was determined by matching the 2.1 K and 45 K heat loads to those in the APT baseline.[5] Titanium was chosen because it gives 2.1 K and 45 K heat loads closest to the APT baseline compared to copper or stainless steel. Clearly, the HOM and instrumentation leads will not use titanium, but whether the instrumentation will use low thermal conductivity materials like manganin or phosphor bronze or large gauge copper wire is unknown. The HOM cable is likely to be coaxial with copper or silver coated stainless steel conductors. Also, the HOM and instrumentation are a relatively small contribution to the total loads, especially at 2.1 K, so the error in the approximation of using titanium properties is not large. Again, temperature dependent fitting functions were developed to yield the 2.1 K and intermediate temperature heat loads as functions of the intermediate intercept temperature (the location of the intercept being such that the refrigeration input power was minimized).

Temperature Independent Heat Loads

The heat load from the RF joint in the outer conductor at the interface of the SS and Nb is estimated from joule heating. It is assumed constant in this study. The HOM losses are based on electromagnetic code simulations and are considered constant in this study.

The heat loads due to the thermal intercepts on the structural supports, and the cavity RF tuner levers represent less than 5% of the heat load at 2.15 K and were a similarly small contribution to the intermediate temperature heat load. Therefore, these heat loads were treated as constant, independent of the thermal intercept temperature in this study.

CRYOMODULE SYSTEM ANALYSIS

Being armed with the temperature dependence of the appropriate heat loads, and adding the ones that are being treated as constant, it is possible to configure various cooling schemes and calculate the resulting total refrigeration input power for each. Four different configurations were analyzed that utilize double point thermal intercepts on the PC. All the configurations provided a 2.6 K, 3 bar primary stream for the cavities. The configurations considered included: 1) a single secondary coolant stream that cools both PC thermal intercepts and the thermal shield, 2) changing the order of the elements cooled by the secondary stream, for ex. cool the lower PC intercept, high PC intercept, then thermal shield, or cool the lower PC intercept, thermal shield, then the high PC intercept, 3) two secondary coolant streams, one for the lower PC intercept and one for the high temperature PC intercept and thermal shield circuit, and 4) two secondary coolant streams, one for the PC low and high temperature intercepts, and the other for the thermal shield circuit.

Similarly, three different cryomodule configurations utilizing counterflow cooling were considered. One considered tapping off the primary coolant loop to provide the relatively small flow rate required for the PC counterflow cooling, and with a secondary loop for the thermal shield circuit. The second considered two secondary coolants, one for the PC counterflow circuit, and the other for the thermal shield circuit. The third approach considered mixing the output of the PC counterflow circuit with the thermal shield return flow.

The single point thermal intercept was considered too inefficient and was not utilized. For the same reason, configurations using the copper strap in the PC cooling were not considered.

To our initial surprise, for all the cases analyzed, the total refrigeration input power was within ±5% of the mean (121 kW) for all configurations. This is because, for APT, the cavity dominates the 2.1 K heat load and typically accounts for 75% of the total refrigeration input power. In addition, the ±5% variation in refrigeration input power is probably not statistically significant given the simplifying assumptions made in the analysis. This means the cryomodule cooling system trade-off can be made largely on other factors such as the impact of the coolant stream choices on the cryomodule design (number of vacuum vessel penetrations, number of control valves, ease of assembly), and cryo-plant and distribution system supply and return temperatures.

Based on this assessment, the double point intercept cooling with only two inlets and two outlets and the power couplers, thermal shield and other thermal intercepts plumbed in the same series circuit was the selected configuration for the APT cryomodules. Compared to other options: 1) it has fewer vacuum vessel penetrations; 2) it appears easier to assemble; 3) it permits easier access in the tunnel with only four bayonets; 4) it integrates the outlet temperatures and pressures well with the cryo-plant and the number of required lines in the distribution system; 5) it has one series coolant line requiring only one control valve which makes it cheaper; and 6) it has only one control valve, hence less hardware and software complexity, and increased reliability.

SUMMARY

A 'systems' approach for optimizing the cooling configuration in a cryomodule is presented. The approach considers not only the variations in heat loads as a function of coolant stream temperatures, but also the impact on the cryomodule design, the cryo-plant and its associated distribution system. For the cryomodule considered, all the cooling configurations required essentially the same room temperature refrigeration input power, thus permitting a decision based on other factors. The double point intercept cooling with only two inlets and two outlets, and the power couplers, thermal shield and other thermal intercepts plumbed in the same series circuit was the selected configuration for the APT cryomodules.

ACKNOWLEDGMENT

We thank Thomas Jefferson National Accelerator Facility and Greg Laughon of General Atomics, in particular, for their detailed review of the this work, and their assistance in the specifying the cryo-plant requirements.

REFERENCES

1. Private communication with C.H. Rode and R. Ganni of TJNAF.
2. J.A. Waynert, F.C. Prenger, A Thermal Analysis and Optimization of the APT 210 kW Power Coupler, *Proceedings of the XIX International Linac Conference* 2:950 (1998).
3. See for example, G.E. McIntosh, Layer by Layer MLI Calculation Using a Separated Mode Equation, *Adv. Cry. Eng.* 39B:1683 (1993).
4. T. Nast, Radiation Heat Transfer, in: "Handbook of Cryogenic Engineering," J.G. Weisend II ed., Taylor and Francis, Philadelphia, PA (1998).
5. Private communication with B. Campbell of LANL.

PLASMA SPRAY COATING OF NIOBIUM SUPERCONDUCTING RF CAVITIES

T. Junquera, S. Bousson, A. Caruette, M. Fouaidy, H. Gassot, and J. Lesrel

Institut de Physique Nucleaire
CNRS-IN2P3, Orsay, France

ABSTRACT

The high gradient niobium superconducting RF cavities proposed for particle accelerators (i.e. electron-positron colliders, high intensity proton accelerators,...) need to exhibit a near perfect mechanical stability. Perturbations generated by pressure variations of the cryogenic fluids and mechanical vibrations are typically encountered, but the main difficulty comes from the detuning Lorentz forces acting on the cavity walls submitted to very high electromagnetic fields. This problem has critical consequences in pulsed accelerators (e.g., TESLA collider project) and it leads to complicated mechanical stiffening solutions associated with RF feedback systems. In this paper, a new stiffening method is proposed. It is based on metal coating of the niobium cavity walls using plasma spray techniques. Several prototype cavities have been successfully tested, and the characterization of thermal and mechanical parameters were performed on samples. Numerical modeling of the expected behaviour of multicell cavities is also presented. All the tests and calculations have confirmed the interest and potential application of this technique to the construction of large scale accelerators.

INTRODUCTION

The high Q value of Superconducting Radio Frequency (SRF) cavities makes them very sensitive to geometry modifications which produce important resonant frequency shifts. This effect leads to amplitude and phase errors which reduce the stability and increase the energy spread of the accelerated beam. In a typical accelerator environment, mechanical vibrations are generated by different sources: machinery, ground motion, cryogenic fluids,... The frequency spectrum of these perturbations is rather large, and some vibrational modes of the cavities could be excited.

Another source of mechanical deformation is the Lorentz force created by the electromagnetic fields and acting directly on the cavity walls. These forces apply a pressure P on the walls proportional to the surface cavity fields E,H (Figure 1 a). The small

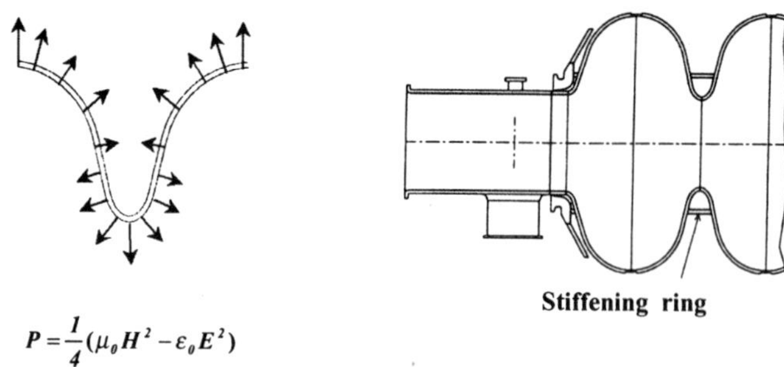

$$P = \frac{1}{4}(\mu_0 H^2 - \varepsilon_0 E^2)$$

Figure 1. **(a)** Pressure on cavity walls induced by Lorentz forces. **(b)** Stiffening method of TESLA cavities.

modification of the cavity volume induces a detuning, Δf ,which is proportional to the square of the surface fields. This effect can be quantified by a detuning factor K relating the frequency shift to the square of the accelerating gradient E_{acc} ($\Delta f = K \cdot E_{acc}^2$). This phenomenon has to be strictly controlled in order to keep the amplitude and phase perturbations within very narrow limits.

TESLA cavities

The TESLA 500 GeV e^+e^- collider project is based on high performance 1.3 GHz SRF cavities working at E_{acc} = 25 MV/m, in pulsed mode (1 ms pulses at a repetition rate of 5 Hz). An optional energy upgrade keeping the same accelerator length needs higher gradients (E_{acc} = 34 MV/m). High stability of field amplitude and phase are required to meet the final luminosity performance ($L > 10^{34}$ $cm^{-2}s^{-1}$). In order to reduce the Lorentz force effect, a mechanical stiffening of the cavity was obtained by adding welded rings between the cells (Figure 1.b). This system can reduce the frequency detuning of an unstiffened cavity by a factor of 2.

An optimized fabrication method has given very good cavity performances [1] in the first series of cavities fabricated by different companies. Sixteen cavities assembled in two cryomodules have been recently tested with beam, and the accelerating fields of the modules were 16 and 20 MV/m respectively. Precise measurements of the frequency detuning of stiffened cavities confirmed the influence of the Lorentz forces, giving a detuning factor K ranging from 0.8 to 1 Hz/(MV/m)^2. A fast feedback and feedforwad control of the cavity's amplitude and phase was tested with beam. The stability results ($\sigma_A/A \leq 10^{-3}$, $\sigma_\Phi \leq 0.05°$) confirm the ability of the RF control system to overcome the Lorentz forces effects. [2]

The time dependence of the accelerating field and the corresponding detuning of a TESLA cavity is presented in Figure 2. An initial detuning of the cavity is necessary in order to reach the resonant condition $\Delta f = 0$ in the middle of the beam pulse. This contributes to reduce the RF peak power needed to obtain a constant field during the beam pulse. Due to the mechanical time constant τ_m of the cavity ($\approx$ 300 μs), the resonant frequency keeps on shifting during the flat top of the RF pulse. Dynamical measurements of detuning were performed with the help of the digital RF control system: Δf = 100 Hz during the beam pulse for an accelerating gradient of 15 MV/m. For higher operating gradients, an improved stiffening system will be necessary to keep the RF peak forward power injected into the cavity within reasonable values. It will be also interesting to consider alternative stiffening methods offering simpler fabrication processes in order to lower the accelerator cost.

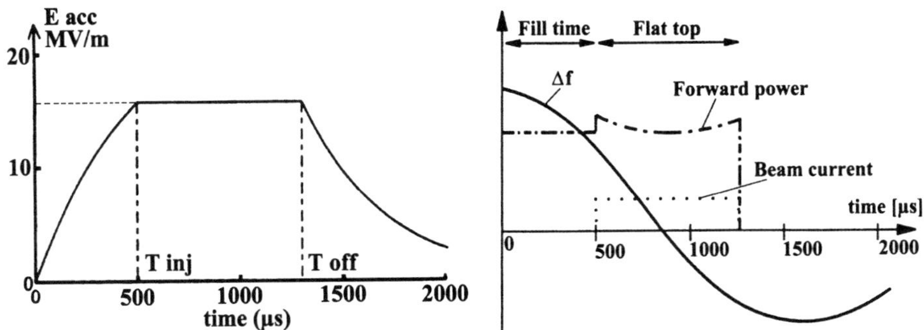

Figure 2. Accelerating field and cavity detuning in a TESLA cavity.

Low Frequency Cavities

The high current proton linac accelerators and the cavities for storage rings operate at lower frequencies, in the range 350 - 700 MHz. Most of these cavities run in a CW mode but for some applications, i.e. pulsed neutron spallation sources, they can be operated in pulse mode, and the Lorentz forces can induce fastly varying detuning. Nevertheless, in the CW mode two problems have to be considered: a) the structural rigidity of cavities of large dimensions with unfavourable shapes to support a vacuum load (mainly in the case of low velocity protons where the lateral walls are close to the vertical plane with angles ranging from 6° to 9°), b) the sensitivity to microphonics which is related to the mechanical vibrational modes of the cavities. Some kind of stiffening methods are needed, either by increasing the cavity wall thickness, or by adding special metal pieces welded to the cavity structure.

ALTERNATIVE STIFFENING SCHEME

In this paper, we propose an alternative stiffening method based on Plasma Spray (or more generally Thermal Spray) metal coating of a thin Niobium wall. This method could become a technical solution provided that the following requirements would be fulfilled: 1) Low cavity fabrication costs (reduction of niobium material, elimination of welded stiffening pieces). 2) High mechanical stability to fight Lorentz forces and microphonics. 3) High heat transfer between the inner niobium wall and the helium bath.

Thermal spray can be used with a large variety of coatings: pure metals or alloys, and several deposition methods which can be compared and optimized. One difficulty is the lack of information on the low-temperature heat transfer characteristics of these coatings. The mechanical behavior is better known but strongly depends on the deposition procedure. Both considerations lead us to initiate a research program to measure the main properties of these coatings on samples, and, in parallel, to assess the global behavior of the layer by measuring the performances of cavities monocell prototypes.

FEM model calculations have supplied an initial evaluation of the potential performances of this stiffening method. The detuning of a 9-cell TESLA cavity as a function of the square of the accelerating field is presented in Figure 3. The result using a classical stiffening welded ring is compared to two different configurations: a) homogeneous copper layer, and b) variable thickness copper layer. All the calculations where performed assuming a perfect bonding between the copper layer and the niobium wall (thickness 2.5 mm), and the mechanical properties of the copper layer were those of the bulk material.

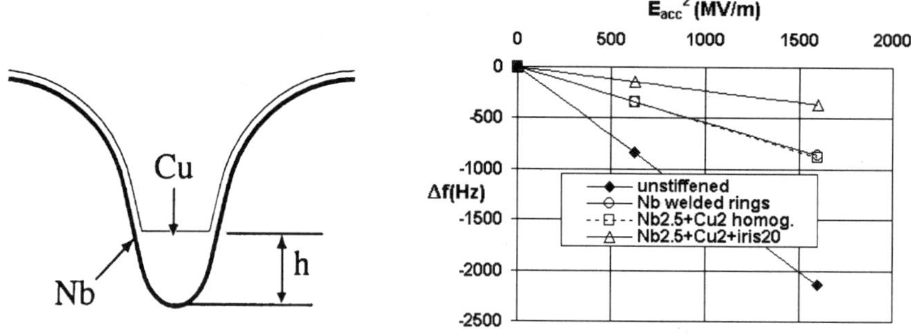

Figure 3. Model calculations of the Lorentz forces detuning of a 9-cell TESLA cavity with a Thermal Sprayed Copper layer.

The best result is obtained with a layer presenting a uniform thickness of 2 mm and a reinforcement at the iris region (height h = 20 mm). A uniform layer of 2 mm gives the same result as the actual stiffening system using welded rings. All these results describe the static behavior of the cavity; only the steady state value of the detuning is considered. The transient results obtained with the TESLA cavities in pulsed condition, presented in a previous section, confirm the importance of the dynamic characteristics. Modelling of the transient response needs a more complete knowledge of the mechanical properties of the coatings.

THERMAL SPRAY COATINGS

During the last 15 years, thermal spray has achieved outstanding technological progress together with an impressive improvement of the coating qualities. The principle (Figure 4) [3] is based on the formation of a layer from the small molten particles arriving at the surface of the substrate with high velocity and temperature. The coating quality depends on substrate preparation, environmental conditions, precise control of the temperatures (particles and substrate) and particles velocity. The main goals in the actual development are a good bonding strength, and mechanical and thermal characteristics compatible with the SRF cavities requirements.

In this study we have considered two thermal spray methods: *Plasma Spray* (PS) and *High Velocity Oxygen Fuel* (HVOF) Spray. Both methods cover a large number of applications in the aerospace and automotive industry. In the PS method the energy comes from a plasma created by an electric arc (T_{max}≈12000 K), the metal powder (particle size between 10 and 50 μm) is injected with the help of a carrier gas, and the particles are melted in the plasma jet. The molten particles travel towards the substrate at speeds of several hundred of meters per second. During their flight, the particles deccelerate, cool down and may undergo oxidation, evaporation or phase transition. Finally the particles hit the substrate; upon contact they flatten, cooldown at a high rate (> 10^7 K/s) and solidify a few microseconds after impact. The spray under low pressure conditions (Vacuum Plasma Spray VPS) gives high quality layers from the point of view of the bond strength (> 100 Mpa) and lowest porosity (<1%). Using a more simple Atmospheric pressure Plasma Spray (APS), porosities ranging from 5% to 10%, and bond strength higher than 50 Mpa can be easily achieved. Another important aspect is the in-flight oxidation of the molten particles which can reduces the thermal and mechanical qualities of the layer. From this point of view, the VPS is superior to the APS method.

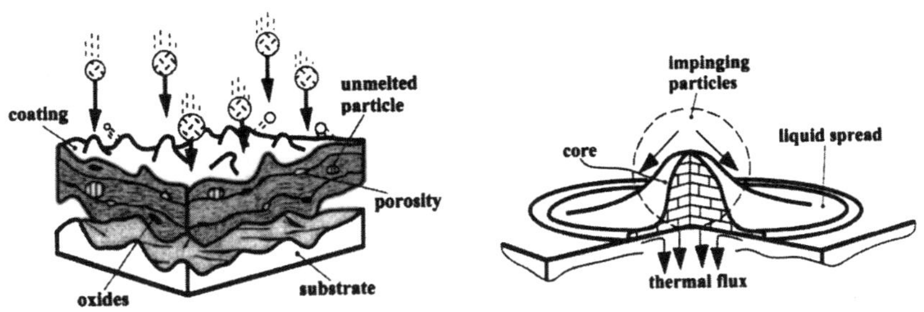

Figure 4. **(a)** Thermal spray coating build-up. **(b)** Molten particle at the substrate surface.

Keeping the substrate at high temperature contributes also to obtaining high quality recrystallized layers, but some drawbacks can be encountered, such as residual stress at the interface, which can originate cracks and failures of the layer or even limit the growth of thicker layers. In some cases, the substrate is kept at a lower temperature using a cryogenic gas or fluids during the deposition process.

A more recent technique is the HVOF Spray. In this method, the energy input is supplied by the combustion of high pressure gases, i.e. propylene and oxygen. A high velocity exhaust jet (2000 m/s) is generated where the powder is injected using a carrier gas. The temperature in the combustion chamber is lower than with the Plasma Spray method (2000 - 3000 K), but the velocity of particles is higher, reaching 1000 m/s. Two interesting results are obtained: lower in-flight oxidation due to a reduced particle temperature, and higher kinetic energy which increases the compressive stress needed to compensate the tensile stress generated by the fast cooling of particles. High density coatings with porosities between 1% and 2%, as well as small residual stress can be obtained with this method. HVOF spray, usage of which is rapidly expanding now, offers also lower operating and investment costs together with higher flexibility.

Mechanical and Thermal Requirements of Coating Layers in SRF Cavities

The two major goals to be achieved with the use of thermal spray coating of SRF cavities are: 1) good structural characteristics: elastic, bonding and cohesion properties, 2) good thermal characteristics: thermal conductivity of the layer and thermal interface conductance with respect to the HeII bath. Concerning the first point, the ideal result will be to approach the bulk material characteristics of a metal. Unfortunately, the thermal sprayed layers exhibits a strong dependence of the elastic properties on the porosity. Theoretical models and experimental results confirm that high values of elastic coefficient can be eventually obtained with copper deposited by VPS having porosities < 1%. But even in this case the Young modulus can be reduced by a factor of 2 compared to the bulk material value. Some preliminary tests using this technique show that the condition to reach such high density layers was to operate at high substrate temperature (T > 800°C). The result on niobium samples was a significant reduction of the RRR ($RRR_{initial}$ = 130, RRR_{final} = 15) which can be explained by a strong contamination of niobium by oxygen in the residual vacuum of the VPS chamber.

A quantitative estimation of the thermal requirements for the sprayed layers applied to SRF cavities was made by model calculations of a defect free cavity wall submitted on one side to an RF surface magnetic field, and cooled on the other side by an HeII bath at 2 K.

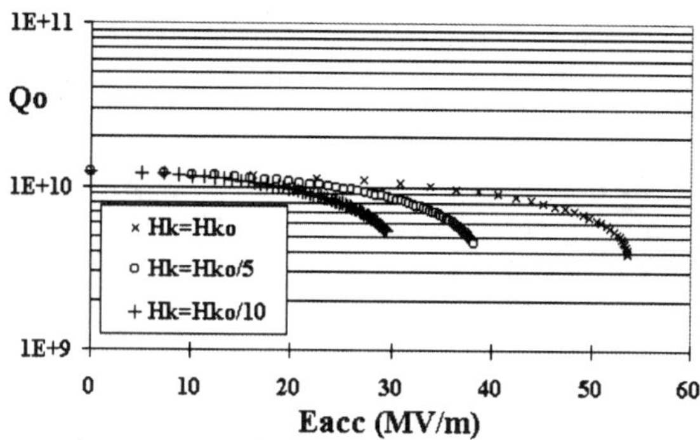

Figure 5. Simulation of a defect free cavity (frequency 1.3 GHz, T=2 K, Nb wall 2.5 mm, RRR=200).

The thermal characteristics of the wall were represented by a thermal conductivity term corresponding to the Nb sheet and a Kapitza conductance term at the interface between the cavity wall and the Helium bath. The supplementary thermal resistance added by the coating layer, was introduced into the calculations in a simple way, varying the value of the Kapitza conductance term (H_K). The results of this simulation are presented in Figure 5. The quality factor of the cavity (Q_0), inversely proportional to the losses, is plotted as a function of the accelerating gradient (E_{acc}). The maximum attainable gradient E_{acc}^{max} depends on the cooling conditions of the composite wall, which are represented by the value of H_K . If we assume a typical value of the Kapitza conductance of a Nb surface at 2 K, H_{K0}= 7000 W/m^2·K, to achieve E_{acc} > 30 MV/m, with low losses ($Q_0 \approx 5 \cdot 10^9$), we can tolerate a reduction of the conductance by a factor of 9, $H_K = H_{K0}/9 \approx 780$ W/m^2·K. It corresponds to an overall thermal resistance of the wall: R_{total} = 1.7.10^{-3} K·m^2/W , with a thermal resistance contribution of the niobium wall (2.5 mm) of 3.8·10^{-4} K·m^2/W, and a contribution of the coating including both the conductivity term and the Kapitza resistance term, of 1.3·10^{-3} K·m^2/W.

The presence of defects on the internal cavity surface limits the maximum available gradient due to the strong heating around the bad spot, which produces quenches at lower field values. To fight against these defects, an high thermal conductivity of the niobium wall (high RRR) is necessary to operate at high gradients. In this case the coating layer does not play an important role in the temperature distribution around the quench location. Only a good thermal conductivity in the close proximity (< 1 mm) will inhibit the quench.

EXPERIMENTAL RESULTS

The experimental work started with an initial assessment experiment in order to check the feasibility and the general aspects of this method. We use two types of Nb monocell cavities: 1.3 GHz cavities (wall thickness 2.5 mm) and 3 GHz cavities (wall thickness 0.5 mm). The first coatings were made using a simple industrial APS process, depositing a copper layer of 2.5 mm. The porosity obtained with this method was rather high, varying over the surface between 20 and 30 %. The cavities were prepared using specially adapted chemical polishing of the internal cavity surface, and high pressure water rinsing was systematically applied for each cavity before test.

The 1.3 GHz cavity (RRR=200) gives near identical maximum field (limited by quench) before and after coating: E_{acc}^{before} = 28.8 MV/m , E_{acc}^{after} = 27.6 MV/m. In this experiment we have checked the stiffening improvement of the cavity, measuring the

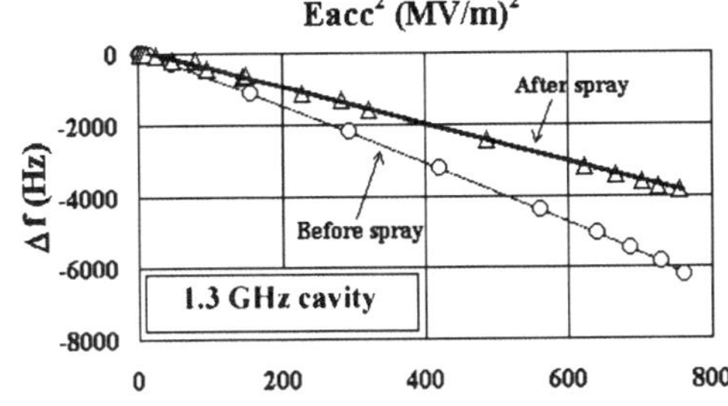

Cavity #	E_{acc}^{max} before coating	E_{acc}^{max} after coating
1	not tested	10 MV/m
2	16.5 MV/m	16.5 MV/m
3	14.5 MV/m	13.5 MV/m

3 GHz monocell Cavities

Figure 6. a) Performance of a 1.3 GHz monocell cavity, before and after APS copper coating of 2 mm.
b) Results with 3 GHz monocell cavities, before and after APS copper coating of 2 mm.

detuning as a function of the accelerating gradient (Figure 6.a). From an initial detuning factor K_{before} = 8.2 Hz/(MV/m)^2, we obtain after coating, a factor K_{after} = 5.2 Hz/(MV/m)^2.

Five 3 GHz cavities were fabricated from low quality Nb sheets (thickness 0.5 mm, *RRR*=40). Three of them were tested and the results are presented in the table (Figure 6.b). The cavities performances are very slightly modified. Quenches were localized using special surface thermometers and the location do not change between the two tests.

A more recent test was performed using a monocell 1.3 GHz cavity coated using a different method. This time a copper layer of 3 mm was deposited by HVOF spray. The porosity was considerably reduced, measured values of $\approx$ 2% were obtained on samples treated with the same deposition parameters. The resulting stiffening improvement was clearly observed: K_{before} = 9.2 Hz/(MV/m)^2, K_{after} = 2.2 Hz/(MV/m)^2. This result confirms the influence of the porosity on coating mechanical properties.

Thermal and Mechanical Measurements on samples

Two experimental apparatus were developed to measure the thermal conductivity and the Kapitza conductance. In the Figure 7.a, the measurements at HeII temperatures of the global thermal resistance of a composite wall (Niobium + Thermal Spray Layer) are presented. Two different metals were deposited using two different spray methods: titanium by APS (porosity > 20%) and copper by HVOF (porosity ~2%). The titanium layer thickness was 1.95 mm, and the copper's one was 3 mm. The result seems to be contrary to expectation: the titanium layer exhibits a lower thermal resistance than the copper layer. An initial hypothesis may bring a first explanation, the penetration of HeII in the porosities could greatly enhance the cooling and partially overcome the bad thermal conductivity of the layer. Comparing these results with the model calculations of TESLA cavities at 2 K presented in a previous paragraph, we observe that the additional resistance of the layer could be a limitation of the thermal performances. At 2 K, the global thermal resistance with a copper layer (HVOF), R_{total} = 1.6.10^{-3} K.m^2/W, is close to the estimated limit for E_{acc} = 30 MV/m and $Q_0 \approx$ 5.10^9. A careful optimization of the coating method is needed in order to reduce the thermal resistance while keeping low values of porosity necessary for a good mechanical behaviour.

Concerning the mechanical properties of the coatings, three different measurement campaigns have been initiated. First, the elastic properties can be roughly estimated from the comparison between model calculations and measurements of the steady state and transient detuning of cavities. A second possibility is to measure the response of monocell cavities submitted to tensile or compressive stress, comparing the results to those obtained by model calculations. Finally, the elastic properties can be directly measured on specially

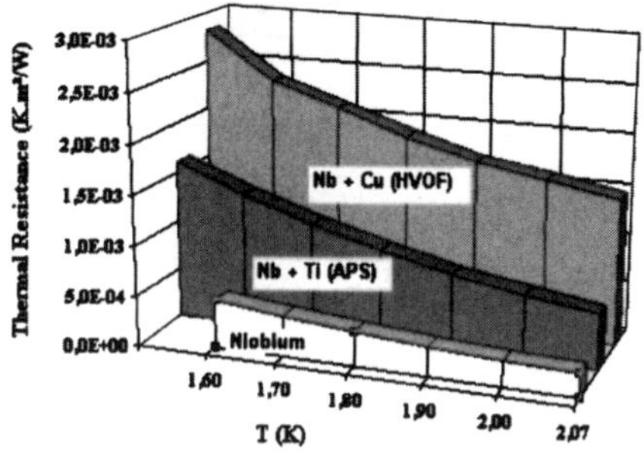

	bulk	VPS coating
Young's modulus	120 GPa	30 - 40 GPa
Shear modulus	45 GPa	30 - 35 GPa
Elongation	45 %	10 - 30 %
UTS	230 MPa	140 - 190 MPa
Tensile Properties of VPS Copper coatings		

Figure 7. **a)** Thermal measurements on Nb samples (RRR 160 , thickness 1 mm) coated with Ti (APS) and Cu (HVOF).
b) Tensile properties: comparison between bulk Copper and VPS sprayed Copper.

prepared samples using different techniques: tensile or compressive stress, two or three point bending tests, and X-ray difraction. In Figure 7.b, the results of a study on tensile properties of VPS copper deposits [4], are compared to those of bulk material. The dispersion on the VPS coatings measurements corresponds to different porosities values, ranging from 0.6% to 4%. Several tests performed with complete cavities or with specimens fabricated with the same deposition parameters as the cavities confirm the strong dependance on porosity. For example, copper deposits by APS or HVOF gives Young modulus values ranging between 25 to 50 GPa. The elongation was also measured on APS and HVOF specimens, giving lower values ($e < 1\%$) than those obtained with VPS deposits. This result could be related to the oxidation of copper particles that leads to a lower ductility of the coatings. The deposition parameters and the environmental conditions have to be optimized to improve the elongation value. Fatigue and thermal cycling tests will be necessary to confirm the ability of the coatings to handle both the cool down thermal stress and the cavity length modifications needed to tune the cavity frequency to the desired value.

ACKNOWLEDGEMENTS

This work was performed within the frame of a collaboration between the IPN-LAL (CNRS-IN2P3,Orsay) and the DAPNIA/SEA (CEA Saclay) for the study and development of SRF cavities. We would like to express our gratitude to Prof. C. Coddet and Dr. C. Verdy from the LERMPS (Institut Polytechnique de Sevenans, Belfort), and, Dr. M. Jeandin and Dr. V. Guipont from Centre des Materiaux (Ecole de Mines de Paris), for their interest and advice on thermal spray techniques. The fabrication of the samples and the first deposits on cavities, were made in these laboratories, in the frame of a fruitful collaboration.

REFERENCES

1. M. Pekeler. "Experience with Superconducting Cavity Operation in the TESLA Test Facility" Particle Accelerator Conference. New York USA (1999)
2. A. Gamp, *et al.* "Experience with the Control of the Vector Sum at the TESLA Test Facility". European Particle Accelerator Conference. Stockholm, Sweeden, (1998)
3. C. Verdy. Thesis U.T.B.M. - L.E.R.M.P.S. Belfort (France), 1998.
4. G. Montavon, *et al.* "Characterization of the Tensile Properties of Vacuum Plasma Spray Copper Deposits". Thermal Spray: Practical Solutions for Engineering Problems. Published by ASM International, Materials Park, Ohio USA (1996).

Advances in Cryogenic Engineering

VOLUME 45, PART B

Advances in Cryogenic Engineering

VOLUME 45, PART B

Chief Editor
Quan-Sheng Shu
AMAC International, Inc.
Newport News, Virginia

CEC Editorial Committee

William Burt
TRW, Inc.
Redondo Beach, California

Michael DiPirro
NASA/Goddard Space Flight Center
Greenbelt, Maryland

David Glaister
Ball Aerospace & Technologies Corp.
Boulder, Colorado

John Hull
Argonne National Laboratory
Argonne, Illinois

Patrick Kelley
Los Alamos National Laboratory
Los Alamos, New Mexico

Peter Kittel
NASA/Ames Research Center
Moffett Field, California

Vitalij Pecharsky
Ames National Laboratory
Ames, Iowa

Ray Radebaugh
National Institute of Standards and Technology
Boulder, Colorado

Charles Reece
Thomas Jefferson National Accelerator Facility
Newport News, Virginia

Jay Theilacker
Fermi National Accelerator Laboratory
Batavia, Illinois

Klaus Timmerhaus
University of Colorado
Boulder, Colorado

Steven Van Sciver
National High Magnetic Field Laboratory
Tallahassee, Florida

John Zbasnik
Lawrence Berkeley National Laboratory
Berkeley, California

Albert Zeller
Michigan State University
East Lansing, Michigan

KLUWER ACADEMIC / PLENUM PUBLISHERS
New York, Boston, Dordrecht, London, Moscow

The Library of Congress cataloged the first volume of this title as follows:

Advances in cryogenic engineering, v. 1–
New York, Cryogenic Engineering Conference; distributed by Plenum Press, 1960–
v. illus., diagrs. 26 cm.
Vols. 1– are reprints of the Proceedings of the Cryogenic Engineering Conference, 1954–
Editor: 1960– K. D. Timmerhaus

1. Low temperature engineering—Congresses. I. Timmerhaus, K. D., ed. II. Cryogenic Engineering Conference

TP490.A3 660.29368 57-35598

Proceedings of the 1999 Cryogenic Engineering Conference,
held July 12–15, 1999, in Montreal, Quebec, Canada

ISBN 0-306-46443-8

233 Spring Street, New York, N.Y. 10013

http://www.wkap.nl

10 9 8 7 6 5 4 3 2 1

A C.I.P. record for this book is available from the Library of Congress

Printed in the United States of America

EXPERIMENTAL INVESTIGATION OF THERMAL SHOCK WAVE INDUCED BY GAS DYNAMIC SHOCK WAVE IMPINGEMENT

H.S. Yang, H. Nagai, N. Takano and M. Murakami

Institute of Engineering Mechanics and Systems, University of Tsukuba
Tennoudai 1-1-1, Tsukuba, Ibaraki, 305-8573, Japan

ABSTRACT

The thermal shock wave in He II is investigated by measuring the temperature variation with a superconductive temperature sensor. It is generated by the impingement of a gas dynamic shock wave onto a He II free surface in the newly developed superfluid shock tube facility. The profile of thermal shock is found to be of a single triangular waveform with a limited shock strength. It is suggested from the experimental result that a thermal shock wave is generated not at the moment of the impingement nor propagates from the original free surface, but it is generated at slightly lower than the original free surface and a little later than the impingement. This means that a thermal shock wave is generated at the lower end of a thermal boundary layer after it is established. In the thermal boundary layer with a thickness of several mm, the thermodynamic state changes from supercritical at the original free surface to compressed He II via compressed He I. It is sometimes found in the experiments near lambda temperatures, that no thermal shock wave is detected in shock compressed He II. It can be understood that shock compression makes He II convert to He I where no thermal shock wave is excited.

INTRODUCTION

He II is an excellent coolant for the cooling of superconducting magnets and infrared detectors in space. In spite of the wide spread use, there are still unsolved open problems concerning highly transient heat transport in it. It is one of such problems that a large amplitude thermal shock wave is reduced to a single peak wave with a limited shock strength in spite of increasing the heat input[1,2]. It is very important to understand this process for practical uses under highly transient high heat flux conditions. For one instance, in case that superconducting magnet quenching occurs, a strong heat pulse transmits through He II. Furthermore, as pressurized He II has been frequently used for better cooling performance, it is indispensable to investigate the heat transport process in pressurized He II.

The wave form deformation of thermal shock waves with a limited shock strength is one of the main research subjects in the present work, through many research papers have been published in the past three decades. Turner[2] interpreted this phenomenon as follows. Large amplitude thermal shocks initiate a breakdown in He II, which is caused by large relative velocity beyond a critical value. This kind of limiting profile was also observed in the case of large heat input over 30W/cm^2 and long heating duration time beyond 0.2msec by Shimazaki et al.[1]. The author reasoned that high density quantized vortices which developed as a result of large heating caused strong interference with the thermal shock wave. However, these interpretations still leave several open questions. It would be one of the reasoning for this that a thermal boundary layer formed between a heater surface and He II is as tiny as inaccessible to direct measurement. To obtain clearer understanding to the heat transport process in the thermal boundary layer, we have chosen a different experimental method that can generate a shock wave within quite a short time, enables very large heat input into He II and furthermore the thermal boundary layer is accessible directly.

For these purposes the newly developed superfluid shock tube facility is used, in which a gas dynamic shock wave impinges onto a He II free surface to generate a thermal shock wave as well as a transmitted compression shock wave in He II. Liepmann et.al[3] and Cummings[4] have originally adopted this method to obtain large Mach number and large Reynolds number flows in cryogenic helium vapor.

EXPERIMENTAL APPARATUS

The superfluid shock tube facility

The schematic overview of the superfluid shock tube facility and the cross-sectional view of the test section of cryogenic shock tube are illustrated in figures 1- (a) and (b) respectively.

The shock tube facility consists of two major parts, a diaphragm-free shock tube with an M-O quick opening value and a cryostat for maintaining He II. The shock tube consists of two parts, the high pressure chamber and the low pressure tube. In the former high pressure helium gas is charged as the driver gas. The latter has a rectangular cross section (35 x 25 mm) and is equipped with two pairs of optical windows with a diameter of 25 mm in the test section. It is charged with saturated vapor as the driven gas.

The test section, where shock waves are measured, located in the lower portion of the low pressure tube is filled with He II. The measurement of the pressure is made by three piezo-electric pressure transducers located in He II and in helium vapor. The propagation speed of gas dynamic shock wave is measured by using upper two pressure transducers of the three. For the measurement of the temperature variation, we used superconductive temperature sensors. Detailed description is found in our previous paper[5].

Measurement of the Temperature Variation

The temperature variation is measured through the variation in the electric resistance of the superconductive sensing element with a length of 2 mm. The thin film sensing element consists of tin film with a thickness of 1000 Å of tin and 230 Å of gold films both of which are vacuum-deposited on the surface of a glass fiber with a diameter of 40 μ m.

This sensor is calibrated by referring to He II temperature regulated through the saturated vapor pressure controler. Accurate measurement of the temperature up to 1mK is possible with this sensor. Owing to the high sensitivity and quick response of the sensor, the measurement of the temperature variation induced by very fast compression shock waves is possible.

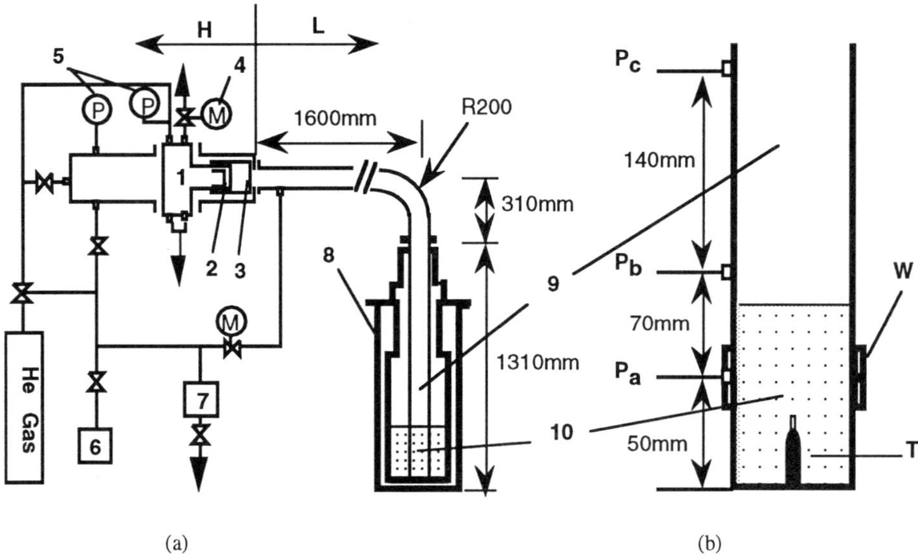

Figure 1. Superfluid shock tube facility
(a) Schematic overview of the superfluid shock tube facility (b) Enlarged cross-section of test section
H. High pressure chamber L. low pressure tube 1. M-O main valve 2. Sub-piston 3. Main-piston
4. Electro-magnetic pilot value 5. Pressure gauges 6. Rotary vacuum pump 7. Buffer tank
8. Cryostat 9. 10. Test section filled with helium vapor and He II respectively. W. Optical window
T. Superconductive temperature sensor Pa, Pb, Pc. Pressure transducers.

GENERAL SHOCK WAVE PERFORMANCE IN THE FACILITY

Theoretical predition of thermal shock wave

We introduce the approximation made by Khalatnikov[6], which have been frequently compared with experimental data. It is a weak shock wave approximation for both the thermal and compression shock waves. It is assumed that the jumps in all quantities across a shock wave are very small and the jumps in the temperature, the pressure and the connterflow velocity chosen as the independent variables are taken into account only up to second-order. The results for the temperature jumps are given for the thermal and the compression shock waves, respectively by

$$\Delta T = 2(M-1)\left[\frac{\partial}{\partial T}\ln\left(\frac{a_2^3 C_p}{T}\right)\right]^{-1} \qquad (1)$$

$$\Delta T = 0 \qquad (2)$$

The equation (2) indicates that no temperature jump is induced in a compression shock wave in He II in the level of the approximation. The denominator of Eg. (1) can be either positive or negative depending on the temperature. It is positive for T <1.88 K under the

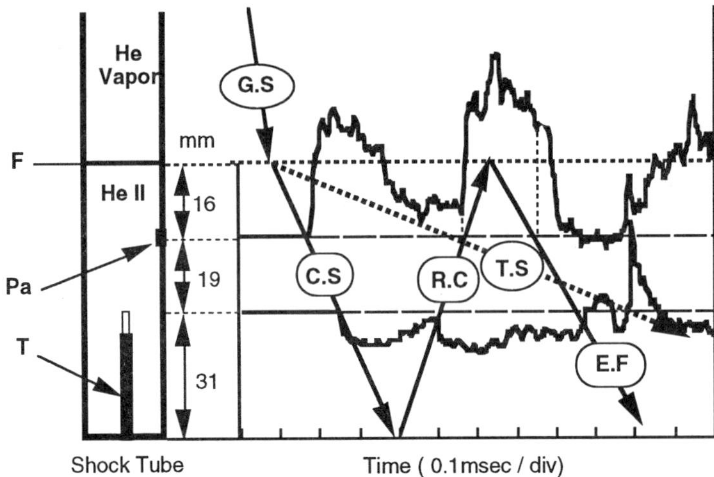

Figure 2. Typical waveforms detected by a pressure transducer (Pa) and a superconductive temperature sensor (T) after the impingement of a gas dynamic shock wave onto a He II free surface (F).

The geometrical arrangement of the free surface, the pressure and the temperature sensors and the bottom of the test section is also drawn in this figure. Driver gas; helium gas (700KPa, 300K), driven gas; saturated vapor of He II at 2.7KPa, initial He II temperature: 1.95K. The speed of the gas dynamic shock wave (G.S): 354.3m/sec, The speed of the transmitted compression shock (C.S): 245.2m/sec, The speed of the compression shock reflected from the end wall (R.C): 276.5m/sec. E.F: Expansion fan transformed from R.C. reflected at the free surface. The velocity of the thermal shock wave (T.S): 20m/sec.

saturated vapor pressure condition, which means that the thermal shock becomes a front shock wave as ordinary shock waves. And for T>1.88 K, it has negative sign, a back shock wave is formed and the temperature drops across back side of a thermal pulse.

Typical Profiles Induced by a Gas Dynamic Shock Impingement onto He II Free Surface

A typical waveform detected by the pressure transducer and the superconductive temperature sensor in He II is shown in figure 2. After a gas dynamic shock impinges into He II, a transmitted compression shock (C.S) propagates through He II and is detected firstly by the pressure transducer (Pa) and then by the superconductive temperature sensor (T). It is evident that the temperature jump in a compression shock is negative and the Khalatnikov approximation does not hold in the level of the present experiments. The negative temperature variation is a natural consequence of compressed of He II because it has negative thermal expansion coefficient. The second temperature and pressure jumps resulted from the reflected shock (R.C). The thermal shock wave signal (T.S) with a positive temperature jump can be recognized only by the temperature sensor after the detection of the expansion fan (E.F), because the speed of the second sound is much slower than that of the first sound. The triangular profile of the thermal shock suggests that intense heat is imposed after the impingement to generate high density quantum vortices which causes strong wave form deformation as a result of interaction.

RESULT and DISCUSSION

Occurrence of thermal shock wave

Figure 3 shows the temperature variations induced by a gas dynamic shock wave impingement measured at three different distances from the free surface. The profile (A) is different from these of (b) and (c) in the point that no thermal shock wave is detected though the thermodynamic state is still He II behind the compression shock. The fact that the liquid helium is He II is confirmed by the experimental facts that the temperature drops in the compression shock wave and it rises in the expansion fan. It is seen from the result in figure 3 (A) that the thermal shock wave is not generated at the moment that the gas dynamic shock impinges onto the He II free surface. It is of interesting to note that the compressed free surface region in the present case where the temperature rises to 30K accroding to the calculation from Rankine-Hougoniot equation in helium vapor is in the super critical state. Of course, the free surface that initially divides helium vapor from He II disappears in the case shown in this figure.

Figures 4 are a series of photographs of the free surface region taken at 0.02, 0.2, 0.3 msec after the impingement of the gas dynamic shock wave. Since the propagation of the transmitted compression shock wave is very fast, it has already passed out of view. As the lapse of time, the discontinuity surface, the original He II free surface, is getting thicker.

It is thought that violent evaporation occurres through sudden heating by the compressed gas dynamic shock wave in the very initial phase. After this initial phase, the thermodynamic state of the interface region turns to supercritical state because of quite large temperature and pressure rises due to shock compression. The dark zone in the free surface region must result from the critical opalescence phenomenon in the pseudo critical state. It is considered that the region with very large temperature gradient in which the critical opalescence region is included develops between the free surface and bulk He II, which may be called as the thermal boundary layer.

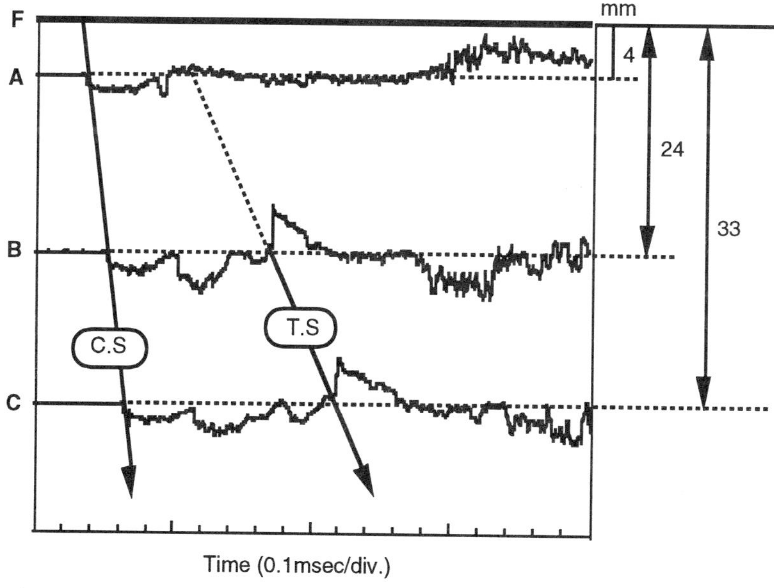

Figure 3. The temperature variations detected by the superconductive temperature sensor at three different positions measured from the free surface. No thermal shock wave is observed at the location A. F: free surface A, B and C are the measuring locations 4mm, 24mm and 37mm below the free surface. It should be noted that the pressure rises of the transmitted compression shock of cases A and C are 0.55MPa and 0.65MPa for the case of B. T.S: Thermal shock wave

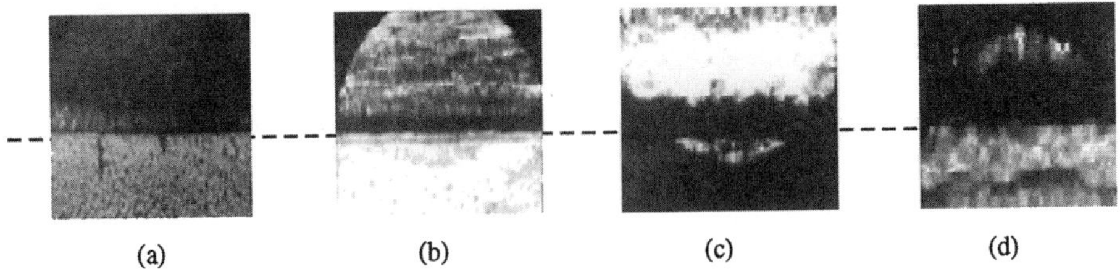

Figure 4. Schlieren visualization photos of the free surface region. (a) before impingement (b) 0.02, (c) 0.2 and (d) 0.3 msec after the impingement, respectively.

Thermal boundary layer

Figure 5 illustrates the physical model of the thermal boundary layer that exists between the free surface and bulk He II, where the temperature distribution is depicted.

As seen from the temperature variation in figure 3 and the visualization photos in figure 4, the thermal boundary layer extends to several millimeters in thickness within a very short time in the case of heating by the compressed gas dynamic shock. It consists of three different thermodynamic states of helium, compressed vapor to compressed He II via compressed He I. It is considered that in the thermal boundary layer thermal conduction is a dominant heat transport process. However, such other heat transport mechanism as the piston effect may be effective to supply heat through the supercritical state towards He II to generate the thermal shock wave. It is understood concerning the mechanism of the wave deformation with a limiting shock stregth that high density quantized vortices which develops as a result of large heating cause strong interference with the thermal shock wave.

The generation of the quantized vortices is responsible for thermal conduction in the thermal boundary layer. As the results, the amplitude of thermal shock is limited.

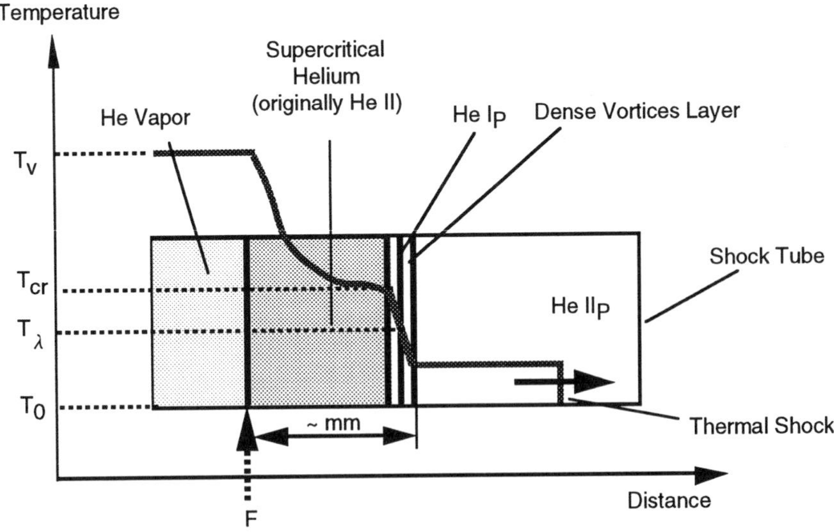

Figure 5. Physical model of the thermal boundary layer induced by the impingement of a gasdynamic shock wave. The temperature is indicated by marking with several symbol (0: initial state, λ : lambda point, cr: critical point, v: compressed vapor). F. free surface

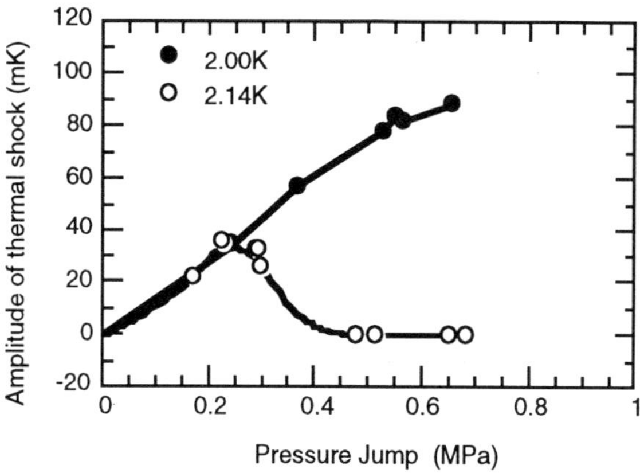

Figure 6. Amplitude of thermal shock wave as a function of pressure jump of the transmitted compression shock

Thermal shock wave generation near lambda temperature

Figure 6 shows the temperature amplitude of the thermal shock which may be regarded as the shock strength of the thermal shock wave plotted against the pressure jump appeared in corresponding transmitted compression shock for two initial temperatures 2.00 K and 2.14 K. It is seen that the strengths increase almost linearly with the pressure jump in the case of 2.00 K. However, near the lambda temperature at 2.14 K, the temperature amplitude reaches the maximum value at the pressure jump around 0.2 Mpa, and it vanishes above 0.5 MPa. That is to say, the thermal shock disappears above 0.5 MPa.

The temperature jump data of the transmitted compression shock is plotted against the pressure jump in figure 7. This quantity is always negative in the case of 2.00 K.

However, at the temperature of 2.14 K, though the tendency is the same as the case of 2.00 K up to about 0.5 MPa. The temperature jump turns to positive above 0.7Mpa.

This can be well explained in terms of the sign of the thermal expansion coefficient of liquid helium. It is negative for He II (over 1.2K) with the negative peak value near the lambda temperature, while for He I it changes its sign from negative to positive near the

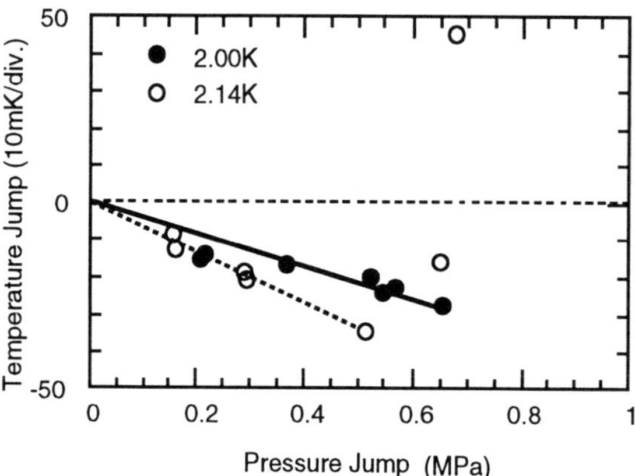

Figure 7. Temperature jump of transmitted compression shock as a function of pressure jump of the transmitted compression shock

lambda temperature as the temperature rises. It is understood from this discussion that thermal shock is not exited at 2.14K in the strongly compressed region behind the compression shock as the result of conversion of He II to He I.

SUMMARY

Thermal shock waves induced by a gas dynamic shock impingement are investigated by measuring the temperature variation with the superconductive temperature sensor. The temperature profile is found to be a single triangular waveform with a limited shock strength almost irrespectively of the impinging shock strength. It is found that the thermal shock wave is generated not at the moment of the impingement nor propagates from the free surface, but it is formed at slightly lower than the original free surface. There is a thermal boundary layer with a thickness of several mm between the impinging supercritical helium vapor portion and the compressed He II. In the experiments near lambda temperature, the phase transition from He II to He I is induced by shock compression, where no thermal shock is induced.

REFERENCES

1. T. Shimazaki, M. Murakami and T. Iida, Second sound wave heat transfer, thermal boundary layer of formation and boiling: highly transient heat transport phenomena in He II, Cryogenics. 35-10:645-651 (1995).
2. T. N. Turner, Using second-sound shock waves to probe the intrinsic critical velocity of liquid helium II, Phys. Fluid. 26-11:3227-3241 (1983).
3. H.W. Liepmann, J.C. Cummings and Viviane C. Rupert, Cryogenic shock tube, Phys. Fluid. 16-2:332-333 (1973).
4. J.C. Cummings, Development of a high-performance cryogenic shock tube, J. Fluid Mech. 66:177-187 (1974).
5. H. Nagai, H.S. Yang, N. Takano and M. Murakami and S. Teraoka, Development of superfluid shock tube facility. Adv. Cryog. Eng. 43:1393-1400 (1998).
6. I.M. Khalatnikov, "An Introduction to the Theory of Superfluidity," Academy of Science, Moscow (1965).

APPROXIMATE SOLUTION FOR TRANSIENT HEAT TRANSFER IN STATIC TURBULENT HE II

B. Baudouy

CEA/Saclay, DSM/DAPNIA/STCM
91191 Gif-sur-Yvette Cedex, France

ABSTRACT

Analytical solution in one dimension of the heat diffusion equation in static turbulent superfluid helium (He II) is proposed by mean of integral method. Although this is an approximate method, it has proven that it gives solutions with fairly good accuracy in non-linear fluid dynamics and heat transfer, especially in boundary layer theory. This analytical method is adequate for this class of equations because of its capability of solving non-linear problems and it proposes also a simpler alternative method to numerical calculation. To present the method and compare its accuracy, a simple case solution is compared with the exact solution and experimental data. A more general solution, taking account of the temperature dependence of the thermodynamic properties is also proposed.

INTRODUCTION

Analytical treatment of transient heat transfer in He II has received not enough attention, considering the substantial interest as it relates to the cooling and stability of magnet systems. Dresner using similarity solutions method has developed three analytical solutions[1,2]. These cases deal with linear boundary conditions and temperature independent properties for semi-infinite media. Several solutions are still of interest of designer to investigate the cooling performance and stability of magnet systems, such as solutions in a finite media and temperature dependent properties. An adequate method in the solution of heat diffusion problems is the integral method because of its capability of solving non-linear problems where the non-linearity can be found either in the differential equation itself or in the boundary conditions. This method is analogous to the method employed to solve thermal and momentum boundary layer in fluid mechanics[3]. With exact method, the resulting solution satisfies locally the system over the entire range of space and time. Such solutions are rather difficult to obtain when the differential equation is non-linear or if the boundary conditions involved are non-linear. Integral method, in the solution of time-dependent boundary-value problems, gives solutions, which satisfy the differential system only on the average over the region considered rather than considering a local solution. It is often sufficient for engineering calculations in which many more approximations are used to model complex cryogenics systems.

SOLUTION IN A SEMI-INFINITE MEDIA WITH CONSTANT PROPERTIES

Case for an Clamped Heat Flux

In this paper we examine the solution of a system in which the differential equation is non-linear but not the boundary conditions. The simplest case that can be studied is where the thermodynamic properties are temperature independent and the media is considered semi-infinite. We detail, to present the method, the case of a heat flux step where at t=0 a constant heat flux q_0 is applied at the boundary x=0. It has been already proven that the diffusion equation is able to model He II transient heat transfer[1,2]. The main reason is because for sufficient heat flux and length (~1 m) the heat transfer is dominated by the enthalpy variation of the He II. For a fully developed turbulent state, the heat flux is given by the Gorter-Mellink law[4], neglecting the dissipation effects in He II, the partial differential equation modeling our system for one space dimension is,

$$\frac{\partial T}{\partial t}=\frac{f^{1/3}}{\rho C_p}\frac{\partial}{\partial x}\left(\frac{\partial T}{\partial x}\right)^{1/3} \text{ in } 0\leqslant x\leqslant\infty \text{ and for } t>0, \quad (1)$$

where ρ is the density, C_p the specific heat at constant pressure and f the He II turbulent thermal conductivity function. The first boundary condition is

$$-\left(f\frac{\partial T}{\partial x}\right)^{1/3}=q_0 \text{ at } x=0 \text{ and for } t>0, \quad (2)$$

where q_0 is the heat flux at x=0. At the initial time, the entire media is at constant temperature, so the initial condition is

$$T=T_b \text{ in } 0\leqslant x\leqslant\infty \text{ and at } t=0. \quad (3)$$

As it is a semi-infinite media, the necessary second boundary conditions is a constant temperature when $x\rightarrow\infty$ or practically for large x, i.e. the temperature field is not disturbed for large x. This conditions is expressed by

$$T=T_b \text{ for } x\rightarrow\infty \text{ and for } t>0. \quad (4)$$

We are only interested by the solution of the disturbed temperature field which is limited by a distance $\delta(t)$, called the thermal layer, after which the temperature field is not disturbed. For a semi-infinite media, the thermal layer is defined as being always inferior to the length of the system. From this definition we can modify the boundary condition Eq. (4),

$$T=T_b \text{ at } x=\delta(t) \text{ and for } t>0. \quad (5)$$

Introducing a set of non-dimensional variables as

$$\theta=\frac{T-T_b}{T_\lambda-T_b},\ \chi=\frac{x}{L} \text{ and } \tau=\frac{f^{1/3}}{\rho C_p L^{4/3}(T_\lambda-T_b)^{2/3}}t, \quad (6)$$

where T_λ is the temperature corresponding to the lambda transition, L the length of the domain which is supposed to be thermally semi-infinite in this case, θ, χ and τ respectively the non-dimensional temperature, space dimension and time, the system is transformed into

$$\frac{\partial\theta}{\partial\tau}=\frac{\partial}{\partial\chi}\left(\frac{\partial\theta}{\partial\chi}\right)^{1/3} \text{ in } 0\leqslant\chi\leqslant\infty \text{ and for } \tau>0, \tag{7-a}$$

$$-\left(\frac{\partial\theta}{\partial\chi}\right)^{1/3}=\phi \text{ at } \chi=0 \text{ and for } \tau>0, \tag{7-b}$$

$$\theta=0 \text{ at } \chi=\Delta \text{ and for } \tau>0, \tag{7-c}$$

$$\theta=0 \text{ in } 0\leq\chi\leq\infty \text{ and at } \tau=0, \tag{7-d}$$

where we define a non-dimensional heat flux ϕ and thermal layer Δ as

$$\phi^3=\frac{q_0^3 L}{f(T_\lambda-T_b)} \text{ and } \Delta=\frac{\delta}{L}. \tag{8}$$

If Eq. (7-a) is integrated with respect to space over the thermal layer the resulting equation is called the Heat-Integral Equation. With this integration, terms in space gradient can be removed from the energy equation. Following these directions, the energy equation is then transformed into

$$\int_0^\Delta\frac{\partial\theta}{\partial\tau}d\chi=\int_0^\Delta\frac{\partial}{\partial\chi}\left(\frac{\partial\theta}{\partial\chi}\right)^{1/3}d\chi=\left(\left.\frac{\partial\theta}{\partial\chi}\right|_\Delta\right)^{1/3}-\left(\left.\frac{\partial\theta}{\partial\chi}\right|_0\right)^{1/3}. \tag{9}$$

With the use of the boundary conditions (7-b) and noticing that in our system $\left.\frac{\partial\theta}{\partial\chi}\right|_\Delta$ is null because of the definition of the thermal boundary Δ, Eq. (9) is reduced to

$$\int_0^\Delta\frac{\partial\theta}{\partial\tau}d\chi=-\left(\left.\frac{\partial\theta}{\partial\chi}\right|_0\right)^{1/3}. \tag{10}$$

When the rule of differentiation is used on Eq. (10), the integral on the left hand-side is transformed into

$$\frac{d}{d\tau}\int_0^\Delta\theta d\chi-\left[\theta\frac{d\chi}{d\tau}\right]_0^\Delta=-\left(\left.\frac{\partial\theta}{\partial\chi}\right|_0\right)^{1/3}. \tag{11}$$

One can notice that due to the boundary conditions the second term of the left hand-side of Eq. (11) is null which reduces it to a simpler formulation,

$$\frac{d}{d\tau}\int_0^\Delta\theta d\chi=-\left(\left.\frac{\partial\theta}{\partial\chi}\right|_0\right)^{1/3}=\phi. \tag{12}$$

Eq. (12) is the Heat-Integral Equation for the clamped heat flux problem; it could be used to treat non-linear boundary condition too. Let assume that the temperature has a polynomial form as $\theta=a+b\chi+c\chi^2+d\chi^3$ where the coefficients a, b, c and d are function of the thermal layer Δ. Obviously, θ is an approximate solution of the system and to find the different coefficients, we need to use different boundary conditions: the natural conditions, which ensues from the problem, and derived conditions, which are constructed from either the differential equation or the natural boundary conditions. For this expression of the solution we need two extra boundary conditions. The first one we choose is straightforward and comes from the definition of the thermal layer,

$$\frac{\partial\theta}{\partial\chi} = 0 \text{ at } \chi=\Delta \text{ for } \tau>0. \tag{13}$$

One can notice that this condition has been already used to construct the Heat-Integral Equation. The second one comes from the differential equation at $\chi=\Delta$ where derivative of the temperature with respect to space is null because of condition Eq. (7-c). We have, what it is called a derived condition,

$$\frac{\partial^2\theta}{\partial\chi^2} = 0 \text{ at } \chi=\Delta \text{ for } \tau>0. \tag{14}$$

By the use of the natural boundary conditions Eq. (7-b), (7-c) and (13) and the derived one, Eq. (14), we can formulate a solution of θ as a function of Δ,

$$\theta = \frac{\phi^3\Delta}{3}\left(1-\frac{\chi}{\Delta}\right)^3. \tag{15}$$

By substituting Eq. (15) into the Heat-Integral Equation, Eq. (12), we obtain a first order ordinary differential equation for the thermal layer thickness Δ,

$$\frac{d}{d\tau}\Delta^2 = \frac{12}{\phi^2}. \tag{16}$$

The solution of Eq. (16) subjected to the initial condition Eq. (7-d) gives

$$\Delta = \frac{2\sqrt{3}}{\phi}\sqrt{\tau}. \tag{17}$$

The negative solution of Eq. (17), which has no physical meaning in our problem, has been eliminated. The general solution is composed of Eq. (15) and Eq. (17). The time constant of the system is $\phi^2/12\times t$, when transformed into dimensional variables it is consistent with a dimensional analysis of Eq. (1), giving a time constant of $\rho C_p L^2 q_0^2/12f$.

Comparison with Existing Solution and Experimental Data

If we compare the solution of Eq. (15) at $\chi=0$ which is given by

$$\theta_0 = \frac{2\sqrt{3}}{3}\phi^2\sqrt{\tau}, \tag{18}$$

and can be expressed as a function of the dimensional variables as

$$\frac{T_0 - T_b}{T_\lambda - T_b} = \frac{2\sqrt{3}}{3}\frac{q_0^2}{\sqrt{\rho C_p f(T_\lambda - T_b)}}\sqrt{t}, \tag{19}$$

one can notice that this formulation is similar to Dresner's model with the exception of the coefficient $2\sqrt{3}/3$ (1.15), which is 0.83 in his solution[1]. We can formulate also the time where the temperature of the helium reaches the lambda temperature at the boundary. According to this, $\theta_0 = 1$ and the time t_λ is defined by

$$t_\lambda = \frac{3}{4}\frac{f\rho C_p(T_\lambda - T_b)^2}{q_0^4}. \tag{20}$$

This formulation is also similar to Dresner's formulation with the exception of the coefficient 3/4 which is 1.43 in his model but it agrees on the quartic dependence on the heat flux with experimental results reported by Van Sciver[5]. Dresner's coefficients are found by identification with experimental results reported by Van Sciver which means that these coefficients are only valid for the thermodynamic conditions of Van Sciver's experiment. A comparison with experimental data is encouraging, when we look at the proportional function between the time t_λ and q_0^4. The experimental work of Van Sciver give a value of 110 W^4scm^{-8} for bath temperature 1.802 K whereas Eq. (20) gives a value comprised between 52 and 141 W^4scm^{-8} for temperature between 1.8 K and 2.0 K.

The approximate solution Eq. (15) is plotted on Figure 1 with the data obtained by Van Sciver. As we are assuming that the thermodynamic properties are constant, we need to define an average temperature to evaluate ρ, C_p and f. For this plot, the best match have been found to be for an average temperature of 1.99 K. There is a good agreement for small temperature variation and small x whatever the time but for large x the solution reaches a null temperature variation too soon. It is the limitation of the model and in fact the null temperature variation space location is a function of time and corresponds to the thermal layer. The model underestimates the length of the thermal layer which comes from the profile of the approximate solution and also the associated boundary conditions taken to calculate the solution. Other profile and boundary conditions have been investigated in the following paragraph.

Other Solutions

For heat conduction problems, it has been shown that taking a polynomial form with a degree higher than three does not improve necessarily the accuracy of the solution. The reason is that for each of the polynomial coefficient, which are time dependent i.e. function of the thermal layer, a boundary condition has to be provided. For four coefficients, we have to provide two supplementary conditions, one natural and another derived, whereas for a fourth degree polynomial, another derived boundary condition has to be used for the fifth coefficients and the choice of the extra condition can reduce the accuracy.

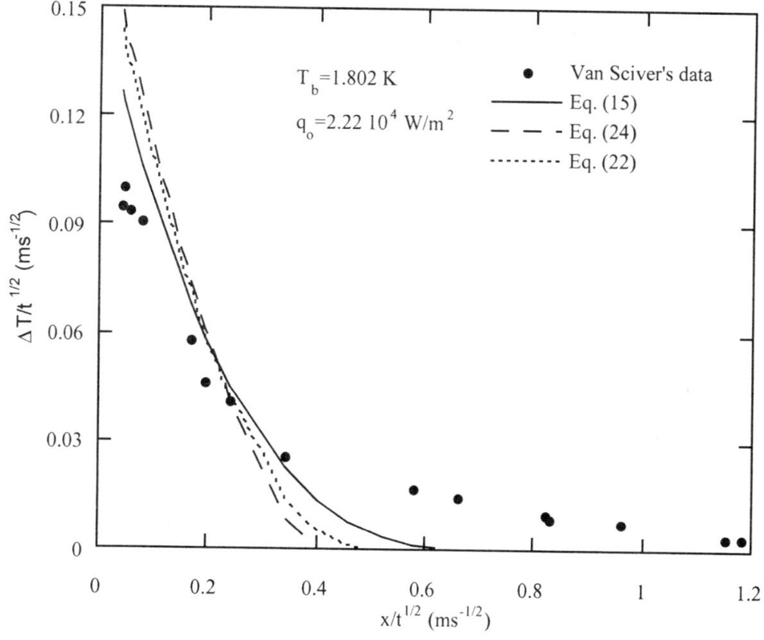

Figure 1. Comparison between different solutions and Van Sciver's experiment[5].

We solved the same problem with a quadric polynomial form. In order to do so, we have to use an extra boundary condition which is,

$$\frac{\partial^2\theta}{\partial\chi^2} = 0 \ \text{at } \chi=0 \text{ for } \tau>0, \tag{21}$$

which comes from the derivation of the boundary condition at $\chi=0$ Eq. (7-b). The solution for that case is

$$\theta = \frac{\phi^3\Delta}{2}\left(1 - 2\frac{\chi}{\Delta} + 2\left(\frac{\chi}{\Delta}\right)^3 - \left(\frac{\chi}{\Delta}\right)^4\right) \text{ with } \Delta = 2\sqrt{\frac{5}{3}}\frac{\sqrt{\tau}}{\phi}. \tag{22}$$

Another solution can be calculated with a cubic polynomial form to illustrate the effect of the boundary condition on the solution. We use a different boundary condition then Eq. (14) derived from the boundary condition Eq. (7-b), that is to say,

$$\frac{\partial^2\theta}{\partial\chi^2} = 0 \ \text{at } \chi=0 \text{ for } \tau>0. \tag{23}$$

The solution found is

$$\theta = \frac{2}{3}\phi^3\Delta\left(1 - \frac{3}{2}\frac{\chi}{\Delta} + \frac{1}{2}\left(\frac{\chi}{\Delta}\right)^3\right) \text{ with } \Delta = 2\frac{\sqrt{\tau}}{\phi}. \tag{24}$$

These two solutions are also plotted on Figure 1 and it is interesting to note that the solutions do not differ by a lot even if we can note that the quartic polynomial form is less accurate that the others. Solutions given by Eq. (22) has a lower accuracy than solution given by Eq. (15). Accuracy of these solutions depends on the boundary conditions and the profile of the approximate solution and is hard to predict unless by comparison with the exact solutions.

SOLUTION IN A SEMI-INFINITE MEDIA WITH TEMPERATURE DEPENDENT PROPERTIES

The system to solve is similar to the one defined by the system of Eq. (1), Eq. (2), Eq. (3) and Eq. (5), such as the differential equation,

$$\rho C_p \frac{\partial T}{\partial t} = \frac{\partial}{\partial x}\left(f\frac{\partial T}{\partial x}\right)^{1/3} \text{ in } 0\leq x\leq\infty \text{ and for } t>0, \tag{25}$$

where in this case we consider the thermodynamic properties ρ, C_p and f temperature dependent. The boundary conditions are identical to the previous case. By applying the Kirchhoff transformation

$$\Theta = \int_{T_b}^{T} f(T)dT, \tag{26}$$

the system Eq. (25), Eq. (2), Eq. (3) and Eq. (5) is transformed into,

$$\frac{1}{\alpha}\frac{\partial\Theta}{\partial t} = \frac{\partial}{\partial x}\left(\frac{\partial\Theta}{\partial x}\right)^{1/3} \text{ in } 0\leq x\leq\infty \text{ and for } t>0, \tag{27-a}$$

$$-\left(\frac{\partial\Theta}{\partial x}\right)^{1/3} = q_0 \text{ at } x=0 \text{ and for } t>0, \tag{27-b}$$

$$\Theta=0 \text{ in } 0\leqslant x\leqslant\infty \text{ and for } t=0. \tag{27-c}$$

$$\Theta=0 \text{ at } x=\delta(t) \text{ and for } t>0, \tag{27-d}$$

where $\alpha=f/\rho C_p$. One can remark that $\alpha=\alpha(\Theta)$. The Heat-Integral Equation is now written

$$\frac{d}{dt}\int_0^\delta \Theta dx = \alpha q_0, \tag{28}$$

where in a first approximation we consider $\alpha(\Theta)$ constant. If we use the same polynomial form $\Theta=a+bx+cx^2+dx^3$ for the expression of the temperature, the solution is expressed as

$$\Theta = q_0^3 \frac{\delta}{3}\left(1-\frac{x}{\delta}\right)^3, \tag{29}$$

when the following natural and derived boundary conditions used are,

$$\left.\frac{\partial\Theta}{\partial x}\right|_\delta = 0 \text{ and } \left.\frac{\partial^2\Theta}{\partial x^2}\right|_\delta = 0. \tag{30}$$

One can observe that the solution is again a function of the boundary condition at x=0, which is defined as $\Theta_0=q_0^3\delta/3$. The expression of δ is obtained by inserting Eq. (29) in Eq. (28) and using Eq. (27-d) is

$$\delta = \frac{2\sqrt{3}}{q_0}\sqrt{\alpha t}. \tag{31}$$

Since the boundary surface temperature Θ_0 is not yet known, Eq. (31) cannot be directly used to evaluate δ but we can eliminate the thermal layer in the expression of Θ_0 and have a transcendental equation for Θ_0 when α_0 is given as a function of Θ_0,

$$\Theta_0\sqrt{\alpha(\Theta_0)} = \frac{2\sqrt{3}}{3} q_0^2 \sqrt{t}. \tag{32}$$

It is possible to have the expression of $\alpha_0=f_0/\rho_0 C_{p0}$ as a function of Θ_0 by using analytical expressions of the He II turbulent thermal conductivity function and the specific heat as a function of the temperature[6],

$$f = \frac{\rho^2 s_\lambda^4 T_\lambda^4}{A_\lambda}\left[\left(\frac{T}{T_\lambda}\right)^{5.7}\left(1-\frac{T}{T_\lambda}\right)^{5.7}\right]^3 \text{ and } C_p=KT^{5.6}. \tag{33),(34}$$

where $A_\lambda\approx145$ ms/kg and $K\approx117$ J/kgK$^{6.6}$. The solution is easy to find when the boundary temperature difference (at x=0) is set, we can evaluate α_0 and Θ_0 and calculate the time needed to reach this temperature difference. As the solution is defined, by evaluating the thermal layer defined from the initial temperature, it is also easy to find the value of the temperature for location different than the boundary, i.e. $x\neq0$ with

$$\Theta = \Theta_0\left(1-q_0^3\frac{x}{3\Theta_0}\right)^3 \tag{35}$$

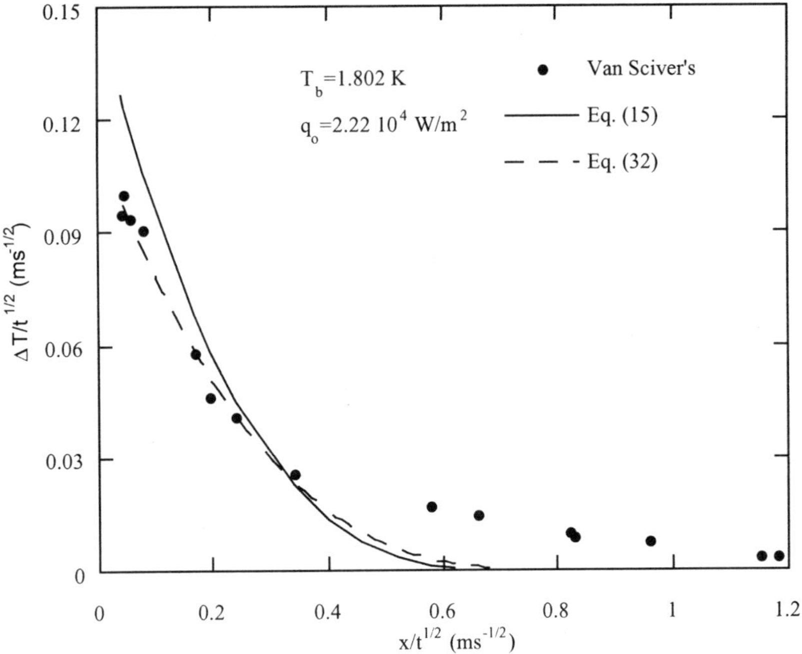

Figure 2. Comparison between Eq. (15), Eq. (32) and Van Sciver's experiment[5].

The solution is plotted on Figure 2 and compared with Eq. (15) and data obtained by Van Sciver. Not only the solution is more accurate than the solution given by Eq. (15), this solution gives a direct result, even with the constant α approximation, without the need to evaluate an average temperature to find the best fit of the experimental data. For the same reason than the other solutions, this one predicts with less accuracy the evolution of the temperature for large x. It comes from the profile of the approximate solution which gives a thermal layer shorter, as a function of time, than the experimental data's one.

CONCLUSION

Integral method are suited to solved the non-linear heat diffusion equation for superfluid helium with acceptable accuracy, but further work is needed to improve it, especially in the choice of the approximate solution profile and the associated boundary conditions. Taking account of the temperature dependence of the thermodynamics properties gives better accuracy and a direct result without the need of the evaluation of an average temperature to define the thermodynamic properties. Further work should involve other boundary conditions such as clamped temperature or pulsed-source problem.

REFERENCES

1. L. Dresner, Transient heat transfer in superfluid helium, *Adv. Cryo. Eng.* 27:411-419 (1982).
2. L. Dresner, Transient heat transfer in superfluid helium. Part II, *Adv. Cryo. Eng.* 29:323-333 (1984).
3. H. Schlichting. "Boundary-Layer Theory", 7th Ed. McGraw-Hill Inc., New-York (1976).
4. C. J. Gorter and J. H. Mellink, On the irreversible process in liquid helium II, *Physica.* XV:2851 (1949).
5. S. W. Van Sciver, Transient heat transport in He II, *Cryogenics.* 19:385-392 (1979).
6. S. W. Van Sciver. "Helium Cryogenics", Ed. Plenum Press, New York (1986).

A HE II HEAT EXCHANGER TEST UNIT DESIGNED FOR THE LHC INTERACTION REGION MAGNETS

Ch. Darve, Y. Huang, J. Kerby, T. Nicol, and T. Peterson

Fermi National Accelerator Laboratory
P O Box 500
Batavia, IL, 60510

ABSTRACT

The LHC interaction region (IR) inner triplets are cooled with stagnant pressurized He II at 1.9 K. All heat loads deposited in the pressurized He II bath will be carried away by saturated He II via a He II heat exchanger. The heat exchanger, made from a corrugated copper tube inside a stainless steel pipe, will be placed outside of and parallel to the cold mass in the cryostat. The current paper details the design work of a full scale He II heat exchanger test unit for the verification of LHC inner triplet magnet cooling system.

INTRODUCTION

The LHC interaction region (IR) inner triplet quadrupoles are cooled with stagnant pressurized He II at 1.9 K. A detailed description of the LHC inner triplet cryogenic system has been previously published.[1-4] As compared to the LHC arc magnet cryogenic system, the inner triplets are subject to larger dynamic loads at 1.9 K. This, in addition to system integration considerations, lead to the proposed placement of the He II heat exchanger to a position external to the cold mass. All heat loads absorbed in the pressurized He II bath at 1.9 K will be removed by the saturated He II flowing inside the heat exchanger pipe. The heat exchanger, made from a corrugated copper tube inside a stainless steel pipe, is about 30 m long and will be placed alongside the cold mass in the cryostat. The comparison between the heat exchanger arranged inside and outside of the cold mass has been made in a previous report.[2] The thermal design requirement for the external heat exchanger is to remove up to 180 W at 1.9K with a temperature drop limited to 50 mK.

A full-scale heat exchanger test unit has been designed at Fermilab and will be tested at CERN to provide experimental verification for the LHC inner triplet heat exchanger system design.

THERMAL DESIGN

This He II heat exchanger is a key component for the LHC inner triplet system. The heat exchanger test unit is designed for measurement of the temperature profile within the heat exchanger at different heat loads. The experimental data from the heat exchanger test unit will serve to guide us in the heat exchanger design of the LHC inner triplet. The updated heat loads to the LHC inner triplet magnet system at various temperatures are listed in Table 1.

The temperature drop in each section can be estimated for a given heat load and channel dimensions. The total temperature drop consists of three parts: that in the pressurized He II from the dummy magnet pipe center to the heat exchanger, across the HeII heat exchanger wall to the saturated He II, and that caused by the vapor pressure drop over the length of inner triplets.

The one dimensional turbulent heat transport in a channel containing He II is expressed by the Gorter-Mellink equation

$$\frac{dT}{dx} = -f(T)q^3 \tag{1}$$

where q is the heat flux in W/cm^2 and the quantity 1/f(T) is the effective thermal conductivity of He II and can be considered as a constant, 1200 W^3/cm^{5}K, at 1.9K for pressurized He II at atmospheric pressure. The temperature drop in pressurized He II from the magnet simulator pipe center to the heat exchanger, ΔT_1, is obtained by the following expression

$$\Delta T_1 = \frac{q^3 L}{1200} \tag{2}$$

where L is the heat transfer length in which a temperature difference is needed to carry the heat flux q. The total temperature drop can be calculated by applying (2) on each section along the heat transport length.

The total thermal resistance across the heat exchanger wall consists of three parts: Kapitza thermal resistance on each side of the wall and the thermal resistance across the heat exchanger pipe wall thickness. The total temperature drop is estimated by assuming a constant heat conductance between the pressurized He II and saturated He II.

$$\Delta T_2 = q\left(\frac{1}{h_{k1}} + \frac{1}{h_{k2}} + \frac{t}{k}\right) \tag{3}$$

where h_{k1} and h_{k2} are Kapitza thermal conductance on each side of the pipe between He II and a solid wall, and t and k are the wall thickness and thermal conductivity of the heat exchanger pipe, respectively.

Table 1. Inner triplet heat loads

Temperature levels	50 to 75 K	4.5 K	1.9 K
Static heat loads (W)	210	9	18.0
Dynamic heat loads nominal (W)	0	20	162
Total heat loads (W)	210	29	180

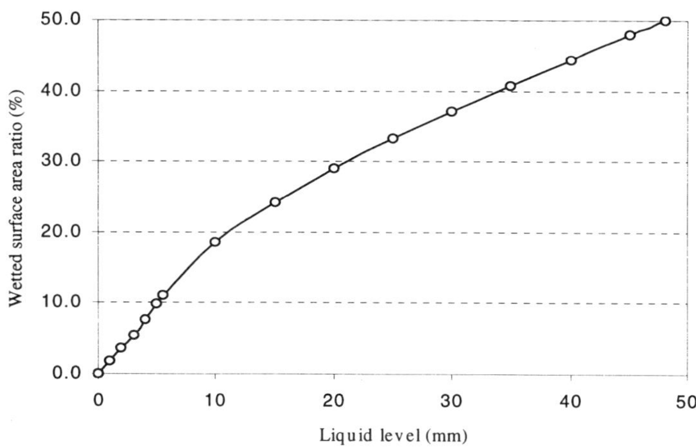

Figure 1. The wetted surface area of corrugated pipe as a function of liquid level

The temperature drop across the heat exchanger pipe wall can be obtained from (3). There are two unknowns in (3): one is the available heat transfer surface area from which the averaged heat flux q can be estimated and the other is Kapitza thermal conductance on both sides of the heat exchanger. The Kapitza conductance, h_k, can be estimated using the following suggested forms[5]

$$h_k = 0.9T^3 \text{ kW/m}^2\text{-K} \quad \text{for clean surface} \tag{4}$$

$$h_k = 0.4T^3 \text{ kW/m}^2\text{-K} \quad \text{for dirty surface} \tag{5}$$

The estimated Kapitza conductance on each side of the heat exchanger pipe is 2650 W/m^{2}K at an average temperature of 1.88 K for a fabricated surface with no special treatment of the surface.

The heat exchanger pipe inner surface is only wetted partially by the flowing saturated He II. The wetted surface area is a function of the liquid level inside the corrugated pipe. The percentage of the wetted surface area as a function of the liquid helium level is shown in Figure 1. The total inner surface area of the heat exchanger pipe, 30 m long, is about 12.54 m^2.

The pressure drop inside the corrugated pipe is calculated using the correlation given in the literature.[6] The temperature drop ΔT_3 is due to the vapor pressure drop in the 30 m long corrugated pipe. Table 2 summarizes each temperature drop along the heat transport path from magnet simulator pipe center to the exit of the corrugated pipe. The temperature drop across the wall is estimated at a heat load of 180 W. The wetted surface is assumed to be 30% of the inner surface area and the solid wall thermal resistance is negligible.

Table 2. Temperature drop calculation results

Location of temperature drop	Temperature drop (mK)
ΔT_1	11.2
ΔT_2 (wetted surface area is 30%)	36.1
ΔT_3	3.7
ΔT total	51.0

MASS FLOW AND YIELD

The mass flow rate required for the heat exchanger test unit depends on the heat load to the system, latent heat, and liquid yield after the JT valve

$$\dot{m} = \frac{Q}{y\lambda} \tag{6}$$

where Q is the heat load to the system in W and λ is the latent heat of saturated He II in J/g. The yield y can be obtained by equating the enthalpies before and after expansion[5]

$$y = \frac{h_v - h_1}{h_v - h_l} \tag{7}$$

where h_l is the enthalpy before expansion, h_v and h_l are enthalpies of saturated vapor and liquid respectively. For the helium at 1.2 bar and 2.6 K before the JT valve, the yield will be 82% after expansion. The required mass flow is 10 g/s for a heat load of 180 W and 15 g/s for a heat load 270 W.

HEAT EXCHANGER TEST UNIT

The heat exchanger test unit consists of a feedbox, four identical modules and one turnaround end. The four identical modules are connected in series and one end is interfaced with the turnaround end and the other is connected to the feedbox. Each one of the modules is 7.5 meters long and 508 mm in diameter. A simplified flow schematic for the He II heat exchanger test unit is shown in Figure 2. The saturated 1.8 K helium vapor is pumped through a JT heat exchanger to cool the incoming helium supply for the corrugated tube heat exchanger in the cryostats to around 2.6 K. A phase separator is connected to the helium supply of the test stand and helium will be separated into liquid and vapor phases. The liquid helium is used to cool the heat exchanger test unit down to liquid helium temperature via a control valve. The JT valve is opened to cool the system down to 1.8 K once the system is filled with liquid helium. The temperature drops to 1.8 K past the JT valve and vaporized helium is pumped by a cold compressor. The thermal shield is cooled by cold helium vapor from the phase separator to intercept the static heat load from room temperature. A stainless steel pipe, 143 mm in diameter, serving as a dummy magnet, is equipped with resistive heaters, capable of up to 250 W to simulate beam heating, and thermometers to measure the temperatures.

The corrugated copper tube serves as the heat exchanger pipe and is inserted inside of an outer diameter of 168 mm stainless steel pipe. The corrugated pipe is separated from the outer shell pipe by four equally spaced spiders. Another stainless steel pipe with an outer diameter of 143 mm, serving as a dummy magnet pipe mounted with heater and thermometers, is placed below the heat exchanger. The two pipes are connected to each other at the end through an 89 mm outer diameter reducer and tee. The reason to choose the smaller size pipe at the heat exchanger end is because the space at the interface of the inner triplet interface is restricted. On the other hand, the reduced pipe size at the inner triplet interface will not cause significant increase of the temperature drop in that region. Table 3 lists all relevant parameters of the corrugated copper tube.

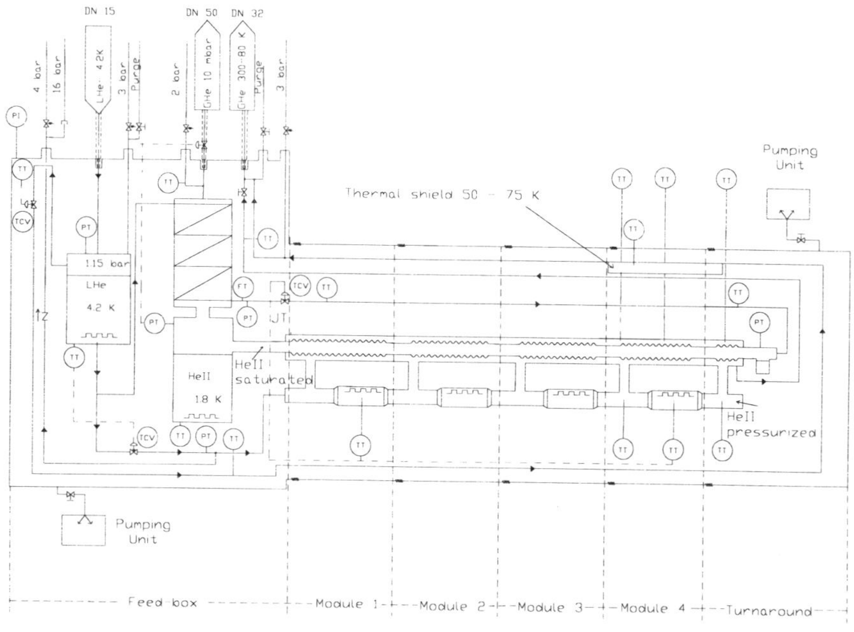

Figure 2. A simplified flow schematic of the He II heat exchanger test unit

Figure 3 shows the side view of one of the four modules. Besides the heat exchanger pipe and dummy magnet pipe mentioned above, two 38 mm tubes located on opposite sides are used to cool the thermal shield with cold helium vapor. A 19 mm stainless steel tube, which is connected to the outlet of the JT valve, is used to supply the saturated helium to the inside of the corrugated copper pipe at the turnaround end. During the system cool down from room temperature to liquid helium temperature, a 38 mm tube, which is welded to the shell-side of the heat exchanger outside of the corrugated pipe, serves as a cooldown return.

32 layers of MLI are used to reduce the thermal radiation from the room temperature to the thermal shield. Another 32 layers of MLI are wrapped on the 1.9 K test section to further reduce the heat leak from the thermal shield to the 1.9 K environment.

All thermometers and heaters are mounted in the pressurized helium side as marked in the flow schematic in Figure 2. The wires are routed out via a 6.3 mm OD tube to the room temperature. All joints are welded except the instrumentation pin connectors, where a rubber O-ring is used to seal at room temperature.

Table 3. The parameters of the corrugated tube

Corrugated tube material	Copper
Outer diameter (mm)	97
Wall thickness (mm)	0.7
Corrugation pitch (mm)	12.7
Corrugation depth (mm)	6.0

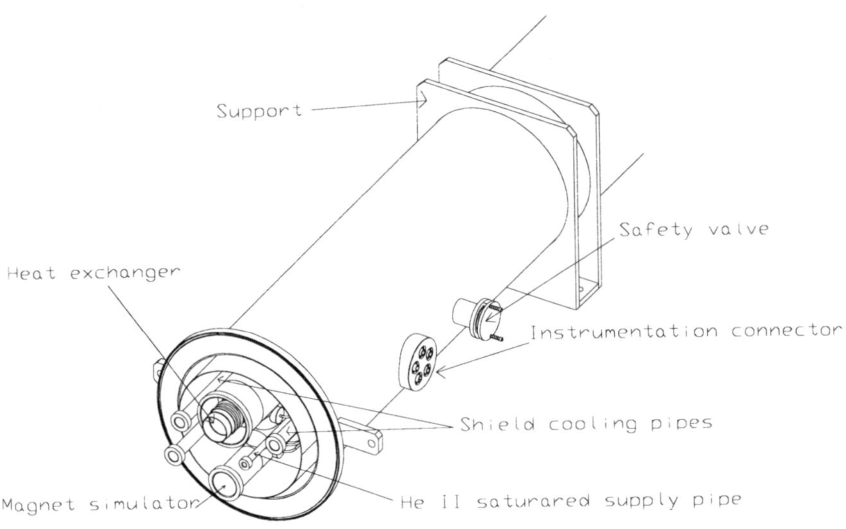

Figure 3. Side view of one of the four identical modules

FEEDBOX

The feedbox is between the CERN test stand and the heat exchanger test unit. The main function of the feedbox is to regulate and control the flow during different stages of the test run. The feedbox consists mainly of the valves, phase separator, counter flow heat exchanger (also called JT heat exchanger), liquid helium accumulator, instrumentation sensors, and safety valves. An overview of the feedbox without the thermal shield and vacuum vessel is shown in Figure 4.

All control valves and safety valves are mounted on the top flange. The interface connection between the CERN test stand and feedbox is located on the top flange as well. The other end of the feedbox is connected to one of the four identical modules, using the standard interconnect. The instrumentation connectors are located on the side of the feedbox. The instrumentation sensors include a turbine flow meter, heater, level indicator, pressure transducer and thermometer; all used to control and fine-tune the parameters of the system during the experiment. The flow meter is mounted just before the JT valve to measure the total flow rate of liquid helium into the heat exchanger. A heater is located in the helium accumulator to vaporize excess liquid helium to keep the counterflow heat exchanger from overflowing. Thermometers mounted inside the counterflow heat exchanger are used to monitor the thermal performance of the heat exchanger. The pressure sensor inside of the helium accumulator can be used to measure the saturated He II temperature and regulate it if necessary.

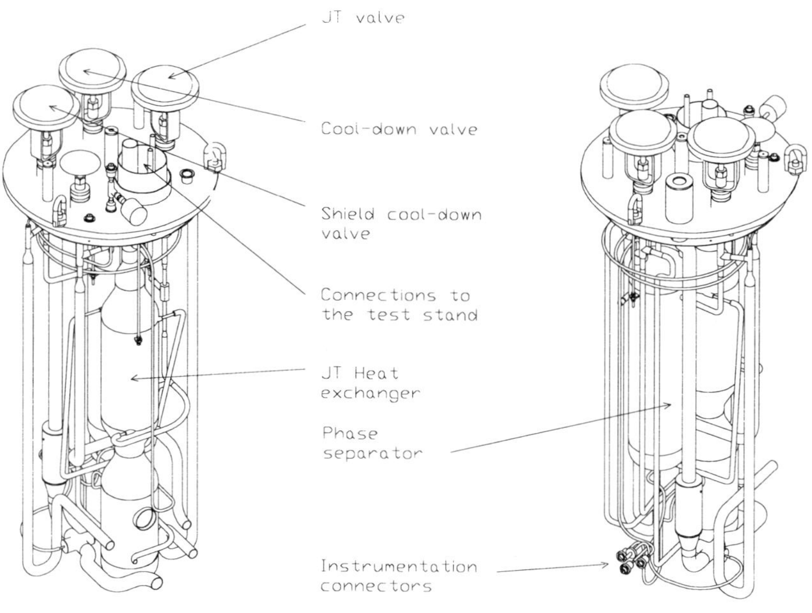

Figure 4. The feedbox without the thermal shield and vacuum vessel

The interface between the feedbox and the CERN test stand is simply three pipes approximately 3 meters above the ground. One pipe supplies saturated helium at about 1.2 bar. The second pipe is the low-pressure return and is used for the initial cooldown and thermal shield cooling return. The third pipe is connected to the cold compressor and serves as the very low-pressure return. The other parts of the test stand include the control valves and instrumentation sensors but are beyond the scope of this report.

SUMMARY

The full size He II heat exchanger test unit for verifying the LHC inner triplet cryogenic system has been designed. The four modules and turnaround end are complete. The feedbox is now in the manufacturing process and should be delivered to Fermilab in later October this year. The experimental data from the test unit will give us a verification of the LHC inner triplet cryostat design. The heat exchanger test unit is expected to remove a heat load up to 180 W from pressurized He II at 1.9 K with a temperature drop around 50 mK. The current schedule is to ship the heat exchanger test unit from Fermilab to CERN by the end of 1999; in order to foresee the measurement campaign by mid 2000.

REFERENCES

1. R. Byrns, Y. Huang, J. Kerby, Ph. Lebrun, L. Morrison, T. Nicol, T. Peterson, R. Trant, R. van Weelderen, and J. Zbasnik, The Cryogenics of the LHC Interaction Region Final Focus Superconducting Magnets, ICEC, England, 1998.
2. Y. Huang, J. Kerby, T. Nicol, and T. Peterson, Cryogenic System and Cryostat Design for the LHC IR Quadrupole Magnets, Advances in Cryogenic Engineering, 43A:403 (1998).
3. Ph. Lebrun, Superfluid Helium Cryogenics for the Large Hadron Collider Project at CERN, Cryogenics 34, ICEC Supplement (1994) p. 1.
4. A. Cyvoct, Ph. Lebrun, M. Marquet, L. Tavian and R. van Weelderen, Heated Two-Phase Flow of Saturated Helium II over a Length of 24 m, CERN-AT/91-28 (CR), LHC Note 169, (1991)
5. S.W. Van Sciver. "Helium Cryogenics" Plenum Press, New York, (1986).
6. J.G. Weisend and S.W. Van Sciver, Pressure drop from flow of cryogens in corrugated bellows, Cryogenics, Vol. 30, (1990), p.935.

A FACILITY FOR ACCURATE HEAT LOAD AND MASS LEAK MEASUREMENTS ON SUPERFLUID HELIUM VALVES

A. Bézaguet, L. Dufay, G. Ferlin, R. Losserand-Madoux, A. Perin,
G. Vandoni, R. van Weelderen

LHC Division
CERN, European Organisation For Particle Physics
1211 Geneva 23, Switzerland

ABSTRACT

The superconducting magnets of the Large Hadron Collider (LHC) will be protected by safety relief valves operating at 1.9 K in superfluid helium (HeII). A test facility was developed to precisely determine the heat load and the mass leakage of cryogenic valves with HeII at their inlet. The temperature of the valve inlet can be varied from 1.8 K to 2 K for pressures up to 3.5 bar. The valve outlet pipe temperature can be regulated between 5 K and 20 K. The heat flow is measured with high precision using a Kapitza-resistance heatmeter and is also crosschecked by a vaporization measurement. After calibration, a precision of 10 mW for heat flows up to 1.1 W has been achieved. The helium leak can be measured up to 15 mg/s with an accuracy of 0.2 mg/s. We present a detailed description of the test facility and the measurements showing its performances.

INTRODUCTION

CERN, the European Laboratory for Particle physics is presently building a new particle collider (LHC) where the high magnetic fields required for particle guiding will be produced by superconducting magnets operating at 1.9 K in a static bath of pressurized superfluid helium. To protect the magnets, pressure relief valves will be mounted between the cold masses and a separate cryogenic distribution line[1]. As there will be approximately 400 of these valves on the LHC, their contribution to the total heat load must be kept below 0.3W per valve. It is of crucial importance for the proper operation of the LHC cryogenic system to ensure that the contribution of theses valves to the thermal load remains below the specified value. In addition, as all mass leakage directly translates into heat load, it is also essential to limit the mass leakage of helium across the valve. A dedicated test bench

Advances in Cryogenic Engineering, Volume 45.
Edited by Shu *et al.*, Kluwer Academic / Plenum Publishers, 2000.

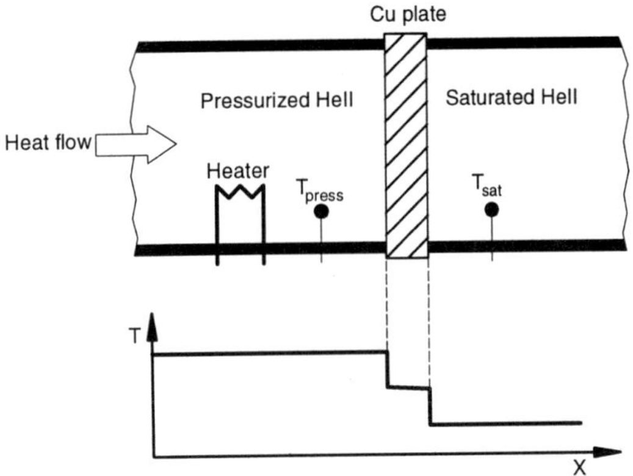

Figure 1. Schematic view of the Kapitza resistance heatmeter and temperature profile

has been developed to simulate the operating conditions of the LHC and to precisely determine the heat loads and mass leakage of such valves.

FACILITY CONFIGURATION

Kapitza resistance heatmeter

The LHC magnets will operate in superfluid helium at 1.9 K and 1.4 bar. The heat flow measurement is performed by measuring a temperature gradient across a calibrated thermal impedance[2]. As the safety valves operate in superfluid helium, a Kapitza resistance was used as thermal impedance[3].

The Kapitza resistance heatmeter of our test bench consists of two volumes of superfluid helium separated by a copper plate. The measuring principle of such a heatmeter is described in figure 1. When heat flows across such a system a sharp temperature gradient appears at the interface between the copper and the superfluid helium. Thanks to the high conductivity of copper, the temperature gradient in the copper plate is very small compared to the Kapitza resistance effects and it will be neglected in the following.

The Kapitza conductance h_k is strongly dependent on temperature, for copper an empirical relation is given by[4] Eq. (1)

$$h_k = \frac{q}{\Delta T} \cong C_k \cdot T^3 \text{ W/m}^2 \cdot \text{K} \qquad (1)$$

where q is the heat flux, T the temperature of the superfluid helium and C_k varies from about 900 for a clean surface to about 400 for a dirty surface. For the configuration described in figure 1 the characteristic temperature profile is shown in the graph and the relation between the heat flow and the temperature difference between the two superfluid helium temperatures T_{press} and T_{sat} is given by Eq. (2):

$$Q = S \cdot C_k \cdot \frac{T_{sat}^n \cdot T_{press}^n}{T_{sat}^n + T_{press}^n} \cdot (T_{press} - T_{sat}) \qquad (2)$$

Where Q is the heat flow, S the area of the copper plate in contact with the superfluid helium and $n=3$.

General configuration and dimensions

A schematic description of the test bench developed for measuring the heat loads and mass leak of the safety relief valves for LHC is shown in figure 2.

Kapitza cell. As the temperature difference must be kept small enough to keep helium superfluid on both sides of the heatmeter, the dimension of the Kapitza heatmeter is determined by the maximal heat flow that shall be measured on the test bench. A Kapitza heatmeter with a diameter of 40 mm was chosen which should allow the measurement of heat flows up to 1.4 W.

As can be seen on figure 2, the saturated helium is supplied to the Kapitza cell trough a flexible corrugated pipe. A flexible pipe was chosen to allow an ample vertical mobility of the test cell, for the valves do not have all the same length. A similar flexibility has been provided for all the piping connected to the Kapitza cell. The pressurized helium is generated by the condensation of gaseous helium supplied from a gas bottle via a pressure regulator that feeds the helium at pressures between 1 and 3.5 bar. Before reaching the

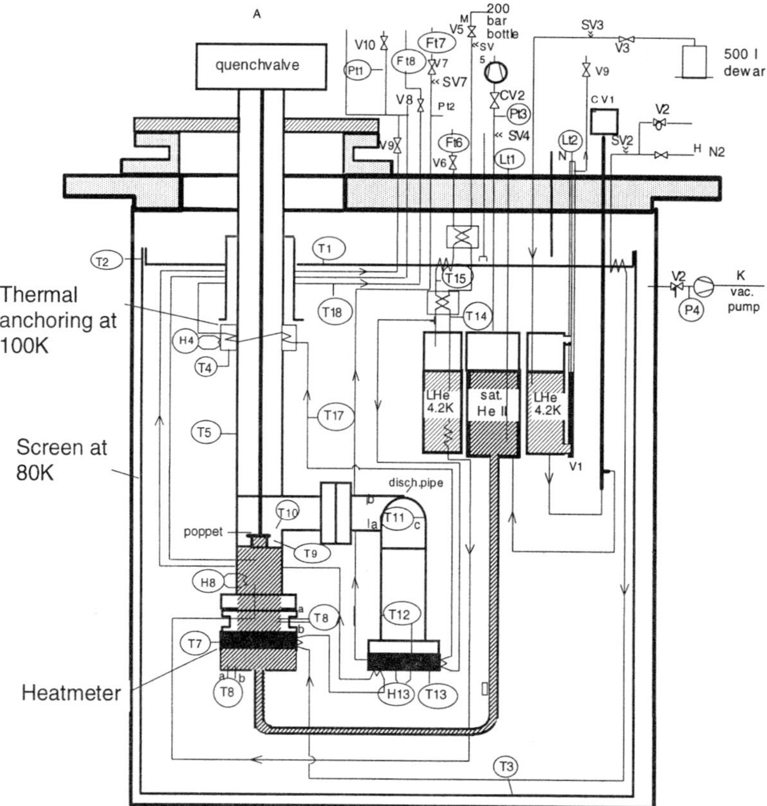

Figure 2. flow scheme of the test bench, Txx, : thermometers; Hxx: heaters; Vxx: valves; Ptx , DPtx : pressure sensors; Ftx : flow controllers/meters

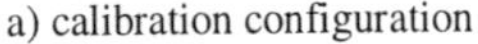

a) calibration configuration

b) valve measurement

Figure 3. Pictures of the Kapitza cell . 1. pressurised HeII ; 2. Kapitza heatmeter ; 3. saturated HeII ; 4. discharge pipe; 5. valve outlet

volume under the valve inlet, the helium is first cooled by two cross-flow gas-gas heat exchangers and then by a heat exchanger immersed in the main 4.2 K helium container.

Discharge pipe and thermal anchoring. The temperature of the discharge pipe is controlled between 5 K and 20 K by a heater coupled with a heat exchanger using the cold gas from the boil-off of the main 4.2 K reservoir, the same gas is also used to keep the heat sink connection at 100 K. The flow of helium leaking from the valve is measured by a mass flowmeter at room temperature.

Ancillary systems. The test bench also includes the ancillary systems that permit a fast cooldown with LN2 and a gas cooled thermal screen at 80K.

Figure 3 shows two pictures of the Kapitza cell. In figure 3.a) the valve is replaced by a stainless steel cap for a calibration run and in figure 3.b) the Kapitza cell is seen connected to a safety relief valve under test.

Instrumentation, data acquisition and control

Temperatures above 30K are measured by platinum resistors. Below 30K, calibrated Allen-Bradley carbon resistors and Lake Shore Cernox® resistors are employed. All measurements are performed by a 4 wires and offset suppression measuring method.

To achieve the desired precision, all temperatures on the heatmeter and on the valve are measured with sensitive digital voltmeters equipped with scanners.

All data is recorded on a computer trough a GPIB interface and a multifunction data acquisition card. A specially developed software allows the control of the data acquisition system, the preliminary treatment of data and the remote monitoring of the bench parameters.

The regulated temperatures are handled by temperature controllers either with heaters (for the discharge pipe and the thermal anchoring at 100 K) or with gas flow regulators (for the thermal screen).

Table 1. Main functional parameters of the test bench

Characteristic	value
T_{press} stability	± 1 mK for >1000 s
(T_{press} – T_{sat}) stability	± 2 mK for >1000 s
Heat flow precision	± 10 mW at 2 K, ± 8 mW at 1.9 K
Maximal heat flow	1.1 W at Tpress = 2K
Discharge pipe temperature	5±0.2 K to 30±0.2 K
Thermal anchoring temperature	70±0.2 K to 120±0.2 K
Thermal screen temperature	80±0.5 K
He leak mass flow	10^{-4} g/s to 10^{-2} g/s

Pressurized helium temperature regulation. As heat conductivity of superfluid helium and the Kapitza resistance are strongly dependent on temperature, the same, stable, temperature T_{press} must be kept in the pressurized helium for the calibration and for the valve measurements. The control of T_{press} is performed by varying the temperature of the saturated helium with a control valve placed on the pumping line. For this purpose a special control algorithm has been developed, which ensures a very good stability of the temperature while allowing a fast response when sharp variations of the heat flow occur. A temperature stability better than ±1 mK could be obtained for periods lasting several hours.

RESULTS AND DISCUSSION

The main performances of the test bench are summarized in table 1.

Heatmeter

For calibrating the heatmeter, the valve was replaced by a stainless steel cap as can be seen in figure 2a). The heat flow was generated by a resistance heater immersed in the pressurized superfluid helium, the power of which was measured by the four wire method.

The calibration procedure consists of measuring the temperature difference T_{press}-T_{sat} while varying the applied power. The measurements were averaged on periods of several thousand seconds. Figure 4 presents the temperature difference during the calibration of the Kapitza heatmeter for applied powers of 0 and 295 mW. The results are raw, unfiltered data, showing a stability of temperature difference better than 1 mK for a period of 1200s.

Calibrations have been performed for T_{press}=1.9 K and T_{press}=2 K. The calibration curves can be seen for both temperatures in figure 5. The experimental data are presented as points, the continuous lines are fitted theoretical curves of the form described by Eq. (2). The graph shows a very good agreement of the data with the theoretical behavior of a Kapitza heatmeter, the parameter C_k, with a value of approximately 980 being close to those found in literature[4]. As can be seen on the extrapolated curve on the graph, the zero-power heat load on the Kapitza cell is approximately 140 – 150 mW.

With a precision of ±2 mK for the temperature difference, the heatmeter precision can be estimated to be approximately ±10 mW at T_{press}=2 K and ±8 mW at T_{press}=1.9 K

Measurements on safety relief valves

Five prototypes of safety relief valves for the LHC cold masses have been mounted and extensively characterized using the test bench[5]. The Kapitza cell has been checked for every measurement by measuring the offset calibration curves at 1.9 K and 2 K for each

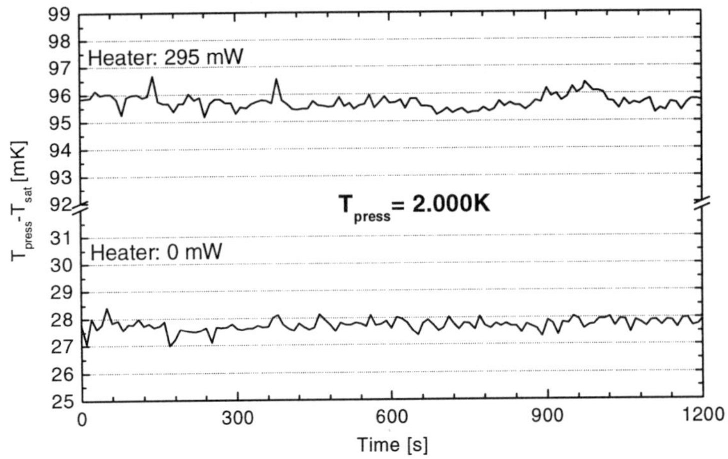

Figure 4. Temperature difference T_{press}-T_{sat} versus time during calibration for two values of the applied power (raw unfiltered data).

valve. All heat load values have been cross-checked with vaporization measurements. The results have shown that most valve manufacturers have underestimated the conduction heat flow. In addition to the determination of the thermal flow to the superfluid helium, mass leakage across the seat, heat load to the 100 K thermal anchoring and the effect on convection in the discharge pipe were measured.

The convection heat transfer effects in the discharge pipe were of particular interest, for they confirmed the importance of the geometry of the connection to the 20 K recovery line. Such a measurement is presented in figure 6, where it can be seen that the convection totally offsets the conduction heat load for one of the valves while for the other one the effect is much less pronounced. The different behavior is due to the different temperature levels in the outlet pipe of the two valves.

Reproducibility

The reproducibility of the calibration curves is of particular importance for the precision of the measurements. In particular, the influence of the time evolution of the

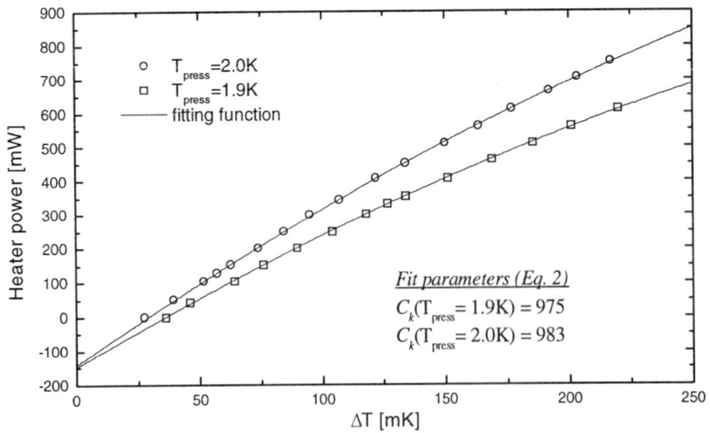

Figure 5. Calibration curves for the Kapitza resistance heatmeter

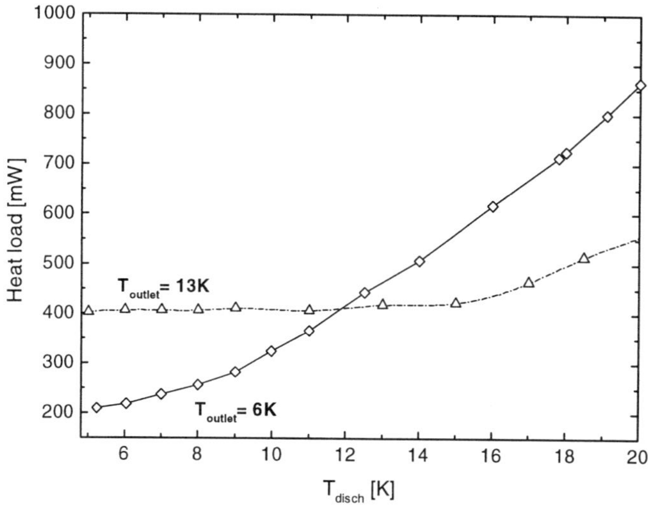

Figure 6. Heat flow to the pressurized HeII as a function of the temperature of the end of the discharge pipe (T_{disch}) (all values with zero leakage)

surface quality was investigated, for the Kapitza resistance is essentially a liquid-solid interface effect. Two calibration runs, separated by several weeks, were initially performed. During this period the Kapitza heatmeter was left in contact with air. The second calibration showed a drift of several mK in the heatmeter characteristic, probably due to a slight surface oxidization. In the successive measurements and re-calibrations the heatmeter was always kept under either vacuum or helium atmosphere, subsequently no significant variation could be observed between calibrations separated in time by several months.

CONCLUSIONS

A dedicated test bench, based on a Kapitza resistance heatmeter, has been developed for the determination of the thermal properties of superfluid helium cryogenic valves. The Kapitza resistance heatmeter has been calibrated and has shown a very good agreement with the calculated design parameters. A precision of ±10 mW could be achieved thanks to an accurate stabilization of the operating temperatures and by keeping the exposure of the Kapitza heatmeter surfaces to air/oxygen to the absolute minimum. The measurements performed during the calibration procedure showed the importance of the surface quality on the Kapitza heatmeter and the necessity of strict handling procedures to avoid any modification of the surface.

After calibration, the test bench has been successfully employed for the characterization of the properties of superfluid helium safety relief valves.

ACKNOWLEDGEMENTS

The authors gratefully thank the personnel of the workshop of the LHC/ACR/SR section for their contribution to the development of the test facility.

REFERENCES

1. W. Erdt, G. Riddone, R. Trant, The Cryogenic Distribution Line for LHC: functional specification and conceptual design, paper presented at this conference.
2. M. Kuchnir, J.D. Gonczy, and J.L. Teague, in "Advances in Cryogenic Engineering, vol. 31", Plenum Press, New York (1985), 1285.
3. H. Danielsson, G. Ferlin, B. Jenninger, C. Luguet, S.-E. Millner, J.-M. Rieubland, Peformance of a superfluid helium safety relief valve for the LHC superconducting magnets, in "Advances in Cryogenic Engineering, vol. 41", Plenum Press, New York (1995), 805.
4. S.W. Van Sciver. "Helium cryogenics", *Plenum Press* (1986) 178.
5. L. Dufay, A. Perin, and R. van Weeldeeren, Characterization of prototype superfluid helium safety relief valves for the LHC magnets, paper presented at this conference.

FLOW IN HORIZONTAL TWO PHASE HE II /VAPOR

S.W. Van Sciver, J.S. Panek and D. Celik

National High Magnetic Field Laboratory
Florida State University
Tallahassee, Florida 32310

ABSTRACT

Analytic and experimental studies of forced flow two phase He II and vapor in a horizontal channel are reported. A simple analytic model is presented that describes the steady state flow phenomena for both isothermal and non-isothermal conditions. Results are compared to measurements performed on a 2 m long test section with cross section 3 mm wide by 65 mm high. Reasonable agreement is obtained. Time dependent phenomena are also discussed.

INTRODUCTION

We consider the conditions present with simultaneous flow of He II and vapor in a partially filled horizontal channel. Such a system may represent a segment of a transfer line for distribution of He II in a large accelerator system. The liquid flow is driven by the addition of He II upstream and simultaneous extraction of He II downstream, while the coexisting vapor flows to remove the evaporated fluid due to the transfer of heat, see Fig.1. The boundary is sloping to compensate for the head loss. Also, the finite temperature difference will result in a slope of the liquid due to the variation of the vapor pressure along the channel. We assume that this is a steady state problem and total mass flow rate is conserved.

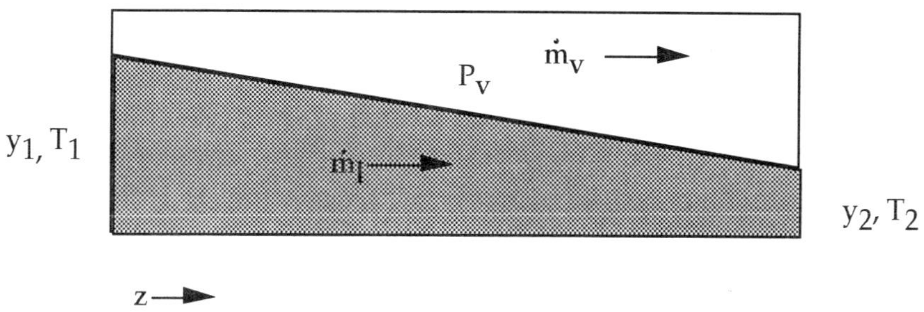

Figure 1: Schematic of horizontal He II/vapor two-phase flow system

Isothermal flow conditions

The first case to consider is that of isothermal flow conditions, $dT/dz = 0$, where z is the distance measured along the channel in the direction of liquid flow. The vapor pressure, p_v, is constant along the top surface and the pressure anywhere within the liquid is given by the sum of the vapor pressure and the hydrostatic head, y. The sloping boundary is assumed to be gradual such that $|dy/dz| << 1$. In this case, dy/dz is always negative for flow in the positive z direction.

The solution of this problem can be obtained by use of the differential form for the one-dimensional energy equation[1],

$$\frac{dy}{dz} = -\frac{u_l}{g}\frac{du_l}{dz} + \frac{dh_l}{dz} \tag{1}$$

where the first term on the right hand side is due to fluid acceleration. dh_l/dz is the head loss due to viscous drag along the channel and for isothermal flow and p is replaced by the hydrostatic head, y, since p_v is constant. For an ordinary channel, dh_l/dz may be described by the classical pipe flow expression,

$$\frac{dh_l}{dz} = -2C_{fl}\frac{u_l^2}{gD_h} \tag{2}$$

where the friction factor, C_{fl}, is determined using classical correlations for laminar or turbulent flow condition and D_h is the hydraulic diameter. For the He II flow regime of interest, the friction factor is always turbulent and is a weak function of the Reynolds number.

For a given flow rate in a rectangular channel, the velocity gradient may be written,

$$\frac{du_l}{dz} = -\frac{\dot{m}_l}{\rho_l y^2 t}\frac{dy}{dz} = -\frac{u_l}{y}\frac{dy}{dz} \tag{3}$$

where t is the channel width. Substituting into (1), we get,

$$\left(1 - \frac{u_l^2}{gy}\right)\frac{dy}{dz} = \frac{dh_l}{dz} \tag{4}$$

which becomes,

$$\frac{dy}{dz} = \frac{dh_l/dz}{\left(1 - Fr^2\right)} \tag{5}$$

where Froude number, $Fr = u_l/(gy)^{1/2}$ is a ratio of the fluid velocity to the velocity of a surface wave. The Froude number is a relative measure of the effect of disturbance to a flow. Small Fr means that the fluid is smooth and gradually varying. Large Fr implies a lot of disturbance and wave action. For most test conditions of interest, $Fr < 0.1$, so that the profiles should be smoothly varying and well behaved. For $Fr << 1$, Eq. (5) reduces to a simple statement that the interface slope is determined almost entirely by the viscous drag along the channel.

Using Eq. (2) to define dh_l/dz, substituting $u_l = \dot{m}_l / \rho_l t y$ for the fluid velocity, and integrating over the length of the channel and the level change, $(y_1 - y_2)$, one obtains an expression for the average friction factor,

$$C_{fl} = \frac{\left(y_1^3 - y_2^3\right) g t^2 D_h \rho_l^2}{6 \dot{m}_l^2 L} \tag{6}$$

which can be compared to experiment. For small Δy, Eq. (6) predicts a parabolic dependence between Δy and $\dot{m}$.

Non-Isothermal flow conditions

The more general case to consider is for a finite temperature difference, ΔT, across the channel. In this case, the vapor pressure, p_v, varies along the top surface and thus needs to be included in the energy equation. Substituting for the pressure, we obtain a modified form,

$$\frac{dy}{dz} + \frac{1}{\rho_l g} \frac{dp_v}{dz} = \frac{dh_l}{dz} \tag{7}$$

where as before, we neglect the acceleration term since the slope is small and $Fr << 1$. Since the liquid is in equilibrium with the saturated vapor, the vapor pressure gradient may be related to the temperature gradient,

$$\frac{dp_v}{dz} = \left(\frac{dp}{dT}\right)_{svp} \frac{dT}{dz} \tag{8}$$

such that simultaneous knowledge of the level and temperature difference across the channel allows one to predict the head loss using Eq. (7).

Substituting for the head loss, Eq. (7) becomes,

$$\frac{dy}{dz} + \frac{1}{\rho_l g} \left.\frac{dp}{dT}\right)_{svp} \frac{dT}{dz} = -\frac{2 C_{fl} \dot{m}_l^2}{g D_h t^2 \rho_l^2 y^2} \tag{9}$$

The fluid slope will decrease when the heat flux and mass flux are co-current in the positive z-direction and increase when they are counter-current. Thus, the fluid slope should change as $\dot{m}^2$ although there will be a constant offset associated with the temperature gradient.

Alternatively, since the vapor pressure gradient is related to the vapor velocity, u_v, through the ordinary turbulent flow condition,

$$\frac{dp_v}{dz} = \mp \frac{2 C_{fv} \rho_v u_v^2}{D_h} \tag{10}$$

where the -/+ sign refers to the heat flux being in the positive or negative direction. One can combine Eqs. (2) and (10) into Eq. (7) to obtain a simple relationship for the level change in a channel with simultaneous vapor and liquid flow,

$$\frac{dy}{dz} = -\frac{2}{\rho_l g D_h}\left(C_{fl}\rho_l u_l^2 \mp C_{fv}\rho_v u_v^2\right) \tag{11}$$

Note that Eq. (11) predicts either positive or negative sloping interface dependent on the relative values of the kinetic energy density of the liquid and vapor. If heat and mass flow are in parallel, $u_l > 0$, $u_v > 0$, the slope will become more positive relative to the isothermal case. On the other hand, if heat and mass flow are anti-parallel, the slope will become more negative. For small liquid mass flow rate $(\rho_l u_l^2 << \rho_v u_v^2)$, Eq. (11) may be approximated by the static fluid case.

In the complete solution to the combined heat and mass transfer one needs to integrate Eq. (9). This requires knowledge of the relationship between *dy/dx* and *dT/dx*, which can be obtained by simultaneous solution of the mass, momentum and energy equations. An example of a full numerical approach is presented in Reference 2.

EXPERIMENTAL APPARATUS

We have constructed an experiment to study two phase He II flow phenomena, a schematic of which is shown in Figure 2. Detailed description of this equipment is available in References 3 and 4. Two vertical end stacks contain He II reservoirs, which are used to fill the bellows pumps and the experiment end reservoirs. Suspended below each end stack is a welded bellows pump with a total displacement of about one liter. A computer-controlled room temperature linear stepper motor moves a coaxial inner tube welded to the inside of the bottom pump flange. Volumetric flow rates of up to 15 cm^3/second are achievable with this system.

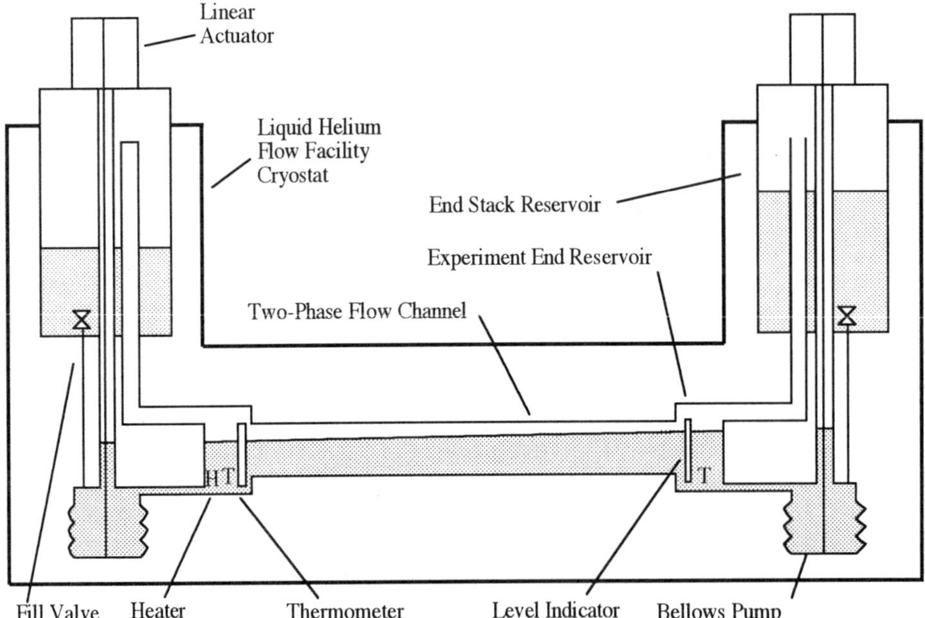

Figure 2. Schematic of experimental apparatus

The two-phase flow channel is constructed of stainless steel and has inner dimensions of height 65 mm, width 3 mm and length 2 m. The channel is welded between the two end reservoirs, cylindrical cans of diameter 135 mm and length 146 mm. Each experiment end reservoir contains a thermometer, heater, superconducting wire liquid level probe and capacitive liquid level gauge.

Temperatures in the cylindrical cans are measured with doped germanium and Cernox^{TM} resistance thermometers which have resistances in the range of 1 kΩ to 1 MΩ for the temperature ranges studied. The 4-wire method of resistance measurement was used. The direction of the excitation current was reversed between readings and an average thermometer resistance is used to avoid thermal emfs. The final temperature resolution, after analog to digital conversion in the data acquisition system, is ±0.5 mK.

The level in each of the cylindrical cans was measured by two methods: a commercial superconducting level gauge and a capacitive level gauge[5]. The capacitive level gauges were used to measure relative changes in the liquid level during the operation of the experiment, the primary advantage being the very high resolution (μm range). However, a slow zero drift of the output required periodic comparison to the absolute level as determined by the superconducting level gauge. Liquid level in the channel itself is calculated from the fact that the capacitive probe is centered to within 1 mm vertically in the reservoir with respect to the channel. Resolution of this system is remarkable, being ±20 μm during quiescent conditions, but variation of the *in situ* calibration limited the sensor absolute accuracy to ±2 mm.

RESULTS

The fluid dynamics experiment has been performed at three bath temperatures, 1.4 K, 1.6 K, and 2.0 K. The procedure involves establishing an initial liquid level in the channel, somewhat above half full, at a given bath temperature. The temperature difference across the channel is then set by regulating the heater and cooling power. Once steady state is achieved, the liquid can be forced through the channel by simultaneously and at the same rate, expanding and contracting the bellows pumps. A flow rate in the oposite direction is achieved by reversing this procedure. Measurements then consist of a dynamic record of the end temperatures and levels while the fluid is flowing. Isothermal ($\Delta T = 0$) data were acquired at 1.4 K and 2 K. Non-isothermal data were acquired at 1.4 K, $\Delta T = 10$ mK, 1.6 K, $\Delta T = 4$ and 7 mK and 2 K, $\Delta T = 1.5$ mK.

Isothermal Data

The magnitude of the liquid level differences, measured under isothermal conditions for mass flow rates up to +/- 2 g/s, are displayed in Fig. 3. For both temperatures, the initial liquid level in the channel is about 40 mm and $\Delta y = 0$. During the forced flow condition, the level on the upstream side increases and on the downstream side decreases; however, the average level remains approximately unchanged.

The results displayed in Fig. 3 are fit to Eq. (6) to yield a friction factor for liquid flow in the channel, C_{fl} = 0.0057+/- 0.0006. This value corresponds to the friction factor calculated for the Blausius correlation, $C_f = 0.079/\mathrm{Re}^{1/4}$, for the range of Reynolds number accessed by the experiment ($10^4 < \mathrm{Re} < 10^5$).

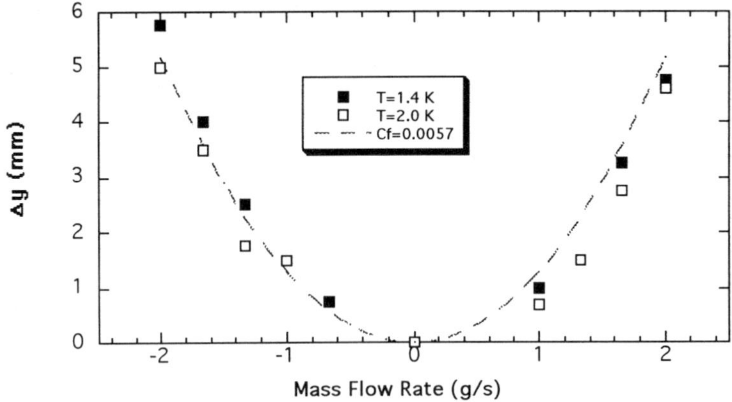

Figure 3. Level height versus mass flow rate in an isothermal flow experiment.

Non-Isothermal Data

Figure 4 displays the magnitude of the level change versus mass flow rate for different conditions with finite ΔT across the channel. In this case, for the initial condition with static fluid, the liquid interface is sloped upward from warm to cold due to the vapor pressure gradient, Eq. (7). However, for the data displayed in Fig. 4, this initial Δy is subtracted from the measured level change for flowing He II. Thus, the data in Fig. 4 should be due only to the hydraulic resistance of the channel and be proportional to $\dot{m}^2$. Also plotted in Fig. 4 is the predicted level increase based on the isothermal flow measurements. Clearly, the level increase for the non-isothermal flow is significantly larger than expected. Such an increase in level difference could result from a small increase in the ΔT across the channel, however, no measurable shift was observed in the data. Alternatively, the shift could be caused by an increase in the friction factor, but there is no mechanism to justify such a shift. We currently have no explanation for this observation.

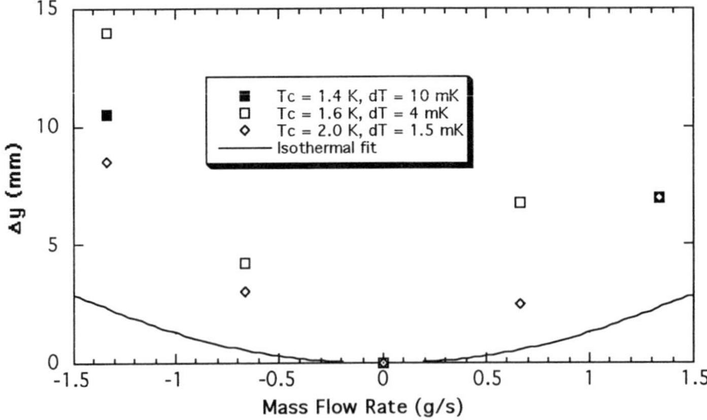

Figure 4. Level height versus mass flow rate in an isothermal flow experiment.

Transient Effects

The horizontal two phase He II/vapor channel shows interesting transient behavior, which can be qualitatively understood based on our understanding of the system hydrodynamics. These effects are best described by reference to a typical set of data, Fig. 5. Displayed is the time evolution of the temperatures and levels in the two end vessels. For these data, a 2 W heat flux is transported by the channel, which produces a $\Delta T = 4$ mK and $(y_2-y_1) = 8$ mm, while the liquid in the channel is static. In this case, the forced flow of the liquid in the positive direction begins at 150 s. During the time the bellows are in motion, the level difference gradually shifts from 8 mm to -12 mm. Simultaneously, the liquid temperature in the end cans increases at a nearly constant rate ($dT/dt \approx 0.03$ mK/s in this case) until motion stops at approximately 270 s. Overall, results indicate that the slope of this increase is roughly proportional to the mass flow rate. After recovery to the static condition, the bellows are exercised in the opposite direction and the level difference increases to about 22 mm. Coincident with this process, the temperatures in the end cans decrease at close to the same rate as previously. We would like to understand both the temperature changes and the time constant for the stabilization of the liquid level.

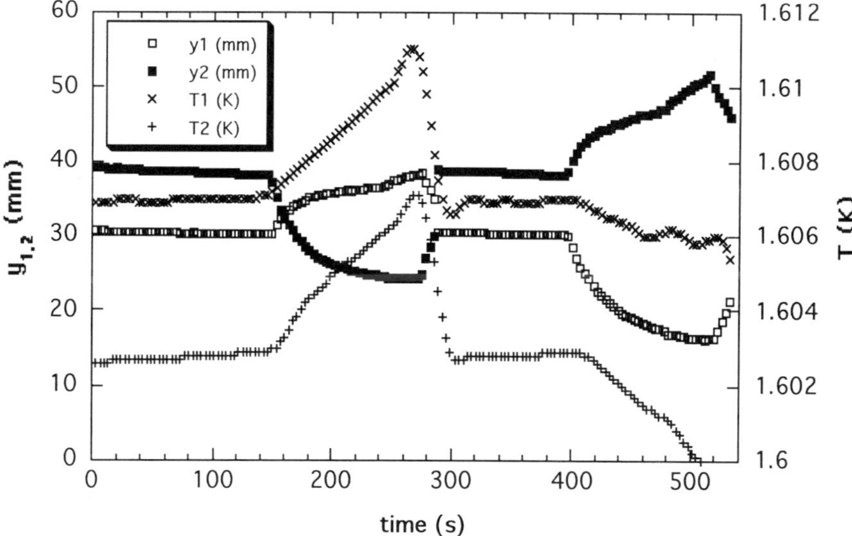

Figure 5. Temperature and level changes with time for $\dot{m} = 1.32$ g/s

To understand the temperature increase and decrease, we approximate the channel, end cans and bellows pumps as a closed system. As we pump warm liquid from the bellows into the channel, warm liquid simultaneously pushed toward the cold side. This process in turn raises the cold end temperature of the channel. However, since the overall conductance of the channel remains about the same, the (T_2-T_1) does not change and the therefore the temperature on the warm side will increase in correspondence to that of the cold side. This effect should continue as long as liquid is being pumped from the warm to cold side and should be proportional to the rate of transfer from the warm to cold side. If we then reverse the flow, the temperatures on both sides will decrease as cold liquid from the pump is circulated through the system. Therefore, the most of the measured increase in temperature is not entropy generation, but rather a redistribution the thermal capacity of the system. To check this argument, we balance the rate of enthalpy transfer to the rate of increase in the enthalpy stored in the end can. To first order, this balance may be written,

$$\left(T_1 - T_2\right)\frac{dm_2}{dt} = m_2\frac{dT_2}{dt} \tag{12}$$

For the case displayed in Fig. 5, Eq. (12) predicts a dT/dt ≈ 0.03 mK/s, in close correspondence to the experimental result.

The other transient effect worth noting in Fig. 5 is the rate of change of the two liquid levels brought on by the initiation of forced flow. There is an effective time constant for the establishment of steady flow conditions, which appears to be of order 25 s for all our experiments. This appears to be a hydraulic effect associated with establishing steady state between the pressure gradient or level shift and the head loss due to friction. This relatively long time constant limits the mass flow rate for the present experiments to less than 2 g/s, above which steady level shifts are not achieved during the limited time stroke of the bellows pump.

CONCLUSION

Analysis of forced flow He II/vapor in a rectangular channel has been performed and the results compared to experiment. In general, it is observed that He II stratified flow can be modeled using a simple analysis based on open channel flow. Thus, the vapor-liquid interaction does not play a major role in these systems. However, for the non-isothermal flow case, there is a significant discrepancy between experiment and our model. Transient flow phenomena are understood in terms of enthalpy conservation and hydraulic resistance of the experimental system.

ACKNOWLEDGMENT

Work supported in part by the US Department of Energy-Division of High Energy Physics under grant: DE-FG02-96ER-40952 and by the National Science Foundation, Thermal Transport Systems under grant: CTS-9806725. Current address for JSP is Jet Propulsion Laboratory, Pasadena, CA.

REFERENCES

1. B.R. Munson, D.F. Young, and T.H. Okiishi, Fundamentals of Fluid Mechanics, 2nd Ed. John Wiley, & Sons, NY, p. 651
2. Y. Xiang, S. W. Van Sciver, J.G. Weisend II and S. Wolf, Numerical Study of Two-Phase He II Stratified Channel Flow, paper CCA-4, 1999 Cryogenic Engineering Conference.
3. J. Panek, Heat Transfer in Horizontal Two Phase He II and Vapor, PhD Thesis, Florida State University (1998).
4. J. Panek and S. W. Van Sciver, Heat Transfer in a Horizontal Channel containing Two-Phase He II, Cryogenics (submitted).
5. D. K. Hilton, J.S. Panek, M.R. Smith and S.W. Van Sciver, A Capacitive Liquid Helium Level Sensor Instrument, Cryogenics (accepted)

NUMERICAL STUDY OF TWO-PHASE HELIUM II STRATIFIED CHANNEL FLOW

Y. Xiang,[1,2] N. N. Filina,[3] S. W. Van Sciver,[2] J. G. Weisend II,[1] and S. Wolff[1]

[1]Deutsches Elektronen — Synchrotron, DESY
Notkestrasse 85, 22607 Hamburg, Germany
[2]National High Magnetic Field Laboratory (NHMFL)
Florida State University
1800 E. Paul Dirac Dr.
Tallahassee, Florida 32310
[3]Moscow State Technical University
Moscow, Russia

ABSTRACT

A key feature of the cryogenic system for the proposed TeV Superconducting Linear Accelerator (TESLA) is a He II two-phase flow line that provides cooling for the superconducting RF cavities. Understanding the behavior of this line is vital to the proper functioning of the accelerator. A numerical model has been developed that allows predictions of the pressure, temperature and mass flow rates of the components in the two-phase line and in the connected gas return line. The model also predicts the helium level in the two-phase line. This is a rigorous model using the conservation laws of mass, momentum and energy for each phase and allowing for interactions between the phases and the walls as well as between each other. The model also takes into account the specific geometry of the TESLA cooling system. This paper describes the model and presents its predictions for two alternative cooling schemes.

INTRODUCTION

The TeV Superconducting Linear Accelerator (TESLA) is a proposed 30 km long e^+/e^- collider. The acceleration is accomplished using superconducting RF cavities cooled by saturated He II baths to 2 K. These baths are connected in parallel by a two-phase line which provides liquid He II to the baths. Heat deposited into the baths is transferred via internal convection to the liquid-vapor interface in the two-phase line where it evaporates vapor. This vapor is then transferred via the two-phase line to gas return line. The two-

Advances in Cryogenic Engineering, Volume 45.
Edited by Shu *et al.*, Kluwer Academic / Plenum Publishers, 2000.

phase and gas return line are connected together every 12.2 m. The basic cooling subunit of TESLA is a string consisting of 12 cryomodules connected together and supplied by the same J-T valve. Additional details of the TESLA cryogenic system have been published previously. [1]

Being able to predict the behavior of the He II two-phase flow is very important to ensure the proper operation of the cryogenic system. A knowledge of the temperature, pressure, flow rate and liquid level distribution along the length of the line is particularly important. It is also very useful for the cryogenic system designers to investigate the impact of changing the length, diameter and interconnections of the two-phase line on the flow parameters. Given the complexity of two-phase flow, such predictions require numerical modeling. An earlier model was developed [2] to help size the refrigeration plants for TESLA. While the model was sufficient for this purpose, it neglected important effects such as the friction between liquids and vapor phases. A rigorous model was developed [3,4,5] to investigate problems in counter-current He II two-phase flow but didn't consider the interconnection of the two-phase line and the gas return line. The work described here is the start of an effort to develop a rigorous, geometrically correct model of the TESLA two-phase flow system for use as a design tool. [6]

SIMULATION MODEL

Basic Model for Two-Phase He II Flow

To understand He II and vapor flow in the two-phase tube of one string, mass, linear momentum and energy conservation of the components should be considered. As a start, we make the following assumptions about the He II two-phase flow:

- One dimensional steady state flow for both phases;
- Saturation state at any cross section of channel;
- Stratified flow with a smooth interface between liquid and vapor phases.

Based on these assumptions, the conservation equations of He II and He vapor in the two-phase tube are given by:

Mass Continuity Equation for He II Flow

$$\frac{d}{dx}(\rho_l \cdot A_l \cdot V_l) = -\dot{m}_{gl} \tag{1}$$

Where. ρ_l is the He II density, A_l is the cross sectional area for He II, V_l is the He II velocity, and $\dot{m}_{gl}$ is mass evaporation rate on the interface.

Linear Momentum Equation for He II Flow

$$\frac{d}{dx}(\rho_l A_l V_l^2) = -\dot{m}_{gl}(V_g - V_l) - \tau_l S_l + \tau_i S_i + \rho_l A_l g \sin\beta - \frac{d}{dx}(A_l p_l) + p_{il}\frac{dA_l}{dx} \tag{2}$$

Where. τ_l, S_l are the shear stress and the wetted perimeter of the He II with the wall, τ_i, S_i are the interface shear stress and the interface perimeter between the phases, β is channel inclination, g is gravitational acceleration, p_l is pressure in He II, and p_{il} is the interface pressure on the He II side.

Energy Equation for He II Flow [7]

$$\frac{d}{dx}(\rho_l V_l A_l e_l)+\frac{dQ}{dx}+\tau_l S_l V_l-\tau_i S_i(V_g-V_l)=\frac{Q_{External}}{L}-\dot{m}_{gl}\cdot h_{fg} \tag{3}$$

Where. $e_l=I_l+\frac{V_l^2}{2}+g\cdot x\cdot\sin\beta$ is energy transfered by the convection of unit mass of He II, I_l is specific enthalpy of He II, $\frac{Q_{External}}{L}$ is heat load per unit length, h_{fg} is the latent heat of vaporization and the heat flow in He II is $Q=A_l\cdot q=A_l\left(-f_{Sat}^{-1}\right)^{1/3}\left(\frac{dT}{dx}\right)^{1/3}$.

Mass Continuity Equation for He Vapor Flow

$$\frac{d}{dx}(\rho_g\cdot A_g\cdot V_g)=\dot{m}_{gl} \tag{4}$$

Where. ρ_g is the vapor density, A_g is the cross sectional area for the vapor, and V_g is the vapor velocity.

Momentum Equation for He Vapor Flow

$$\frac{d}{dx}(\rho_g A_g V_g^2)=\dot{m}_{gl}(V_g-V_l)-\tau_g S_g-\tau_i S_i+\rho_g A_g g\sin\beta-\frac{d}{dx}(p_g A_g)+p_{ig}\frac{dA_g}{dx} \tag{5}$$

Where. τ_g, S_g are the shear stress and the wetted perimeter of the vapor with the wall, p_g is vapor pressure, and p_{ig} is the interface pressure on the vapor side. In fact, $p_{il}-p_{ig}=-\sigma\frac{\partial^2 h}{\partial x^2}$, σ is surface tension of HeII, and h is HeII level.

Energy Equation for He Vapor Flow

$$\frac{d}{dx}(\rho_g V_g A_g e_g)+\tau_g S_g V_g+\tau_i S_i(V_g-V_l)=\dot{m}_{gl}\cdot h_{fg} \tag{6}$$

Where. $e_g=I_g+\frac{V_g^2}{2}+g\cdot x\cdot\sin\beta$ is energy transfered by the convection of unit mass of vapor, I_g is specific enthalpy of He vapor.

Model for Horizontal He Flow in Two Arrangments of the TESLA String

Two schemes of He II and vapor flow arrangement (shown in Fig. 1a, and 1b) have been proposed in TESLA 500 machine design. [1] The main difference between these two

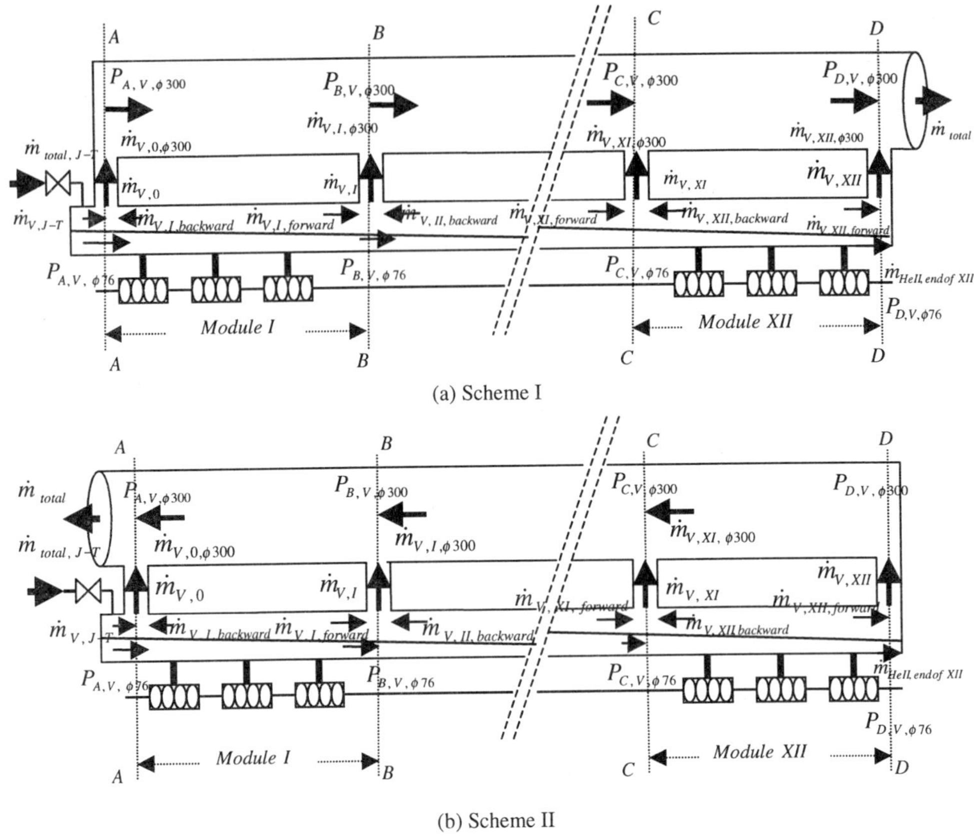

Figure 1. One horizontal cooling string in TESLA 500 machine, (a) scheme I, (b) scheme II

schemes is the relative flow direction of He II in two-phase tube and He vapor in gas return line, co-current in scheme I (Fig.1a) and countercurrent in scheme II (Fig.1b). Flow conditions representing coupling between the two-phase tube and the vapor return line at each position of vapor return interconnection tube change slightly upon the scheme type. They should be satisfied in the simulation module by module. From Figure 1 it can be seen:

- The total mass flow rate out of the string equals the flow through the J-T valve.
- The He II flow rate at the end of module XII equals zero.
- At any connection point between the gas return pipe and the two-phase pipe, the pressure in the gas return pipe equals that of the two-phase pipe plus the pressure drop in the connecting tube.

For scheme I, there are:

- $\dot{m}_{total} = \dot{m}_{V,XII} + \dot{m}_{V,XII,\phi 300}$, $\dot{m}_{V,XII,\phi 300} = \dot{m}_{V,XI} + \dot{m}_{V,XI,\phi 300}$, ..., $\dot{m}_{V,0,\phi 300} = \dot{m}_{V,0}$ (7)
- $\dot{m}_{V,XII} = \dot{m}_{V,XII,forward}$ (8)

As a result of the pressure balance between the gas return pipe and the two-phase line, the vapor in the two-phase line travels in both directions requiring a point of zero vapor mass flow in each two-phase line. Correspondingly, mass flow condition in two-phase tube should be:

- $\dot{m}_{V,0} = \dot{m}_{V,J-T} + \dot{m}_{V,I,backward}$, ... , $\dot{m}_{V,XI} = \dot{m}_{V,XI,forward} + \dot{m}_{V,XII,backward}$ (9)

Similarly, for scheme II, there are:

- $\dot{m}_{total} = \dot{m}_{V,0} + \dot{m}_{V,0,\phi 300}$, ... , $\dot{m}_{V,X,\phi 300} = \dot{m}_{V,XI} + \dot{m}_{V,XI,\phi 300}$ (10)

- $\dot{m}_{V,XI,\phi 300} = \dot{m}_{V,XII} = \dot{m}_{V,XII,forward}$ (11)

- $\dot{m}_{V,0} = \dot{m}_{V,J-T} + \dot{m}_{V,I,backward}$, ... , $\dot{m}_{V,XI} = \dot{m}_{V,XI,forward} + \dot{m}_{V,XII,backward}$ (12)

Numerical Solution Techniques

Ordinary differential equations (1) to (6) reveal an initial value problem which can always be reduced to the study of sets of first-order differential equations. Thus, the classical fourth-order Runge-Kutta formula could be used as a practical numerical method for such a problem. To establish the accurate coupling of energy relationship with the hydrodynamic relations in He II two-phase flow, an iteration technique has also been employed. In simulation, the mass flow rate $\dot{m}_{V,forward}$ before each interconnection of cryomodules is the only adjustment to keep the pressure balance conditions between two-phase tube and gas return line satisfied.

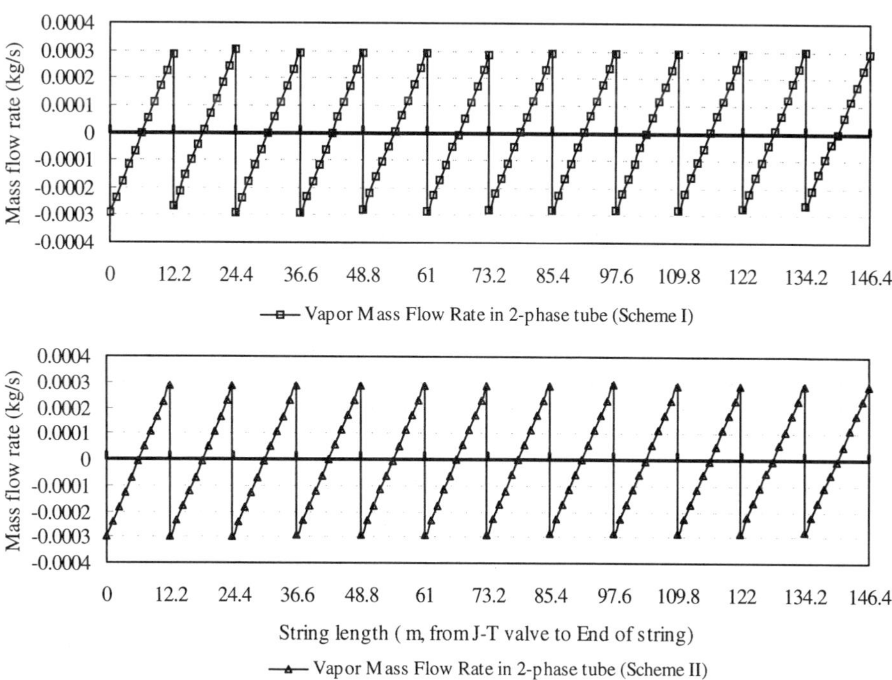

Figure 2. He vapor mass flow rate in two-phase tube of one string (+ : flow forward, - : backward)

NUMERICAL RESULTS

Using a total design heat load of 16.5W/module at 2K, simulations have been made for two schemes. Figure. 2 shows He vapor mass flow rate distribution in two-phase tube of one string. It is seen that the He vapor flows in two directions, forward and backward, in each cryomodule. Thus, a certain section of no net vapor flow exists in two-phase tube which is the requirement of pressure balance condition between two-phase and gas return tubes. Such a section is close to the middle section of each module, with only slight deviation due to the variation of pressure drop in the gas return line. Each sudden drop on the curve corresponds to an interconnection between two-phase and gas return tubes, which is also one end of module. Obviously, He vapor comes from two neighboring modules, joins together and flows into gas return line. It seems that scheme I and II have almost the same He vapor flow behavior in the two-phase tube.

Figure. 3 shows net mass flow rate in two-phase tube and gas return line as well as He II mass flow rate. The net mass flow rate in two-phase and gas return tubes of any module changes upstairs or downstairs because all the He vapor evaporated by heat load in each module is pulled into the gas return line through interconnection tubes of its own. In scheme I, the decrease of net mass flow rate in two-phase tube equals the increase of vapor mass flow rate in gas return line module by module. In scheme II, however, the net mass flow rates in two-phase tube and in gas return line step down at the same magnitude from J-T valve to the end of string. The mass flow rate of He II, by contrast, decreases smoothly to zero in both schemes since He II flows with evaporation from J-T valve to the end of string continuously. The calculated qualities of vapor after J-T valve for two schemes are in a very good agreement with the design value.

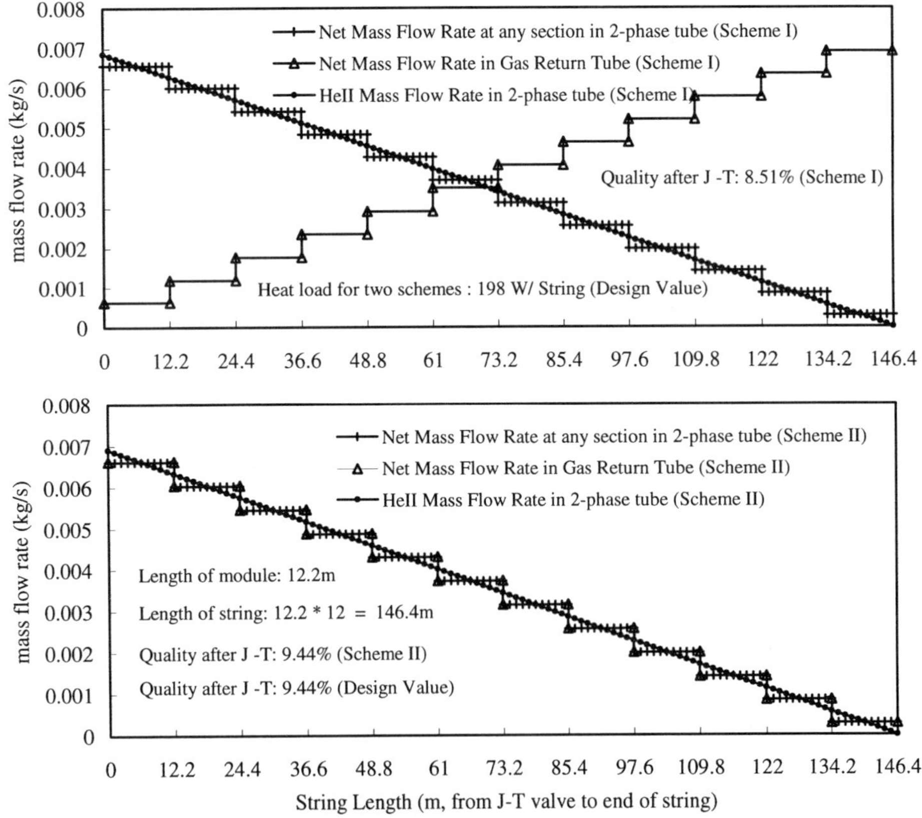

Figure 3. Net He II and vapor mass flow rate in two-phase tube and gas return tube of one string

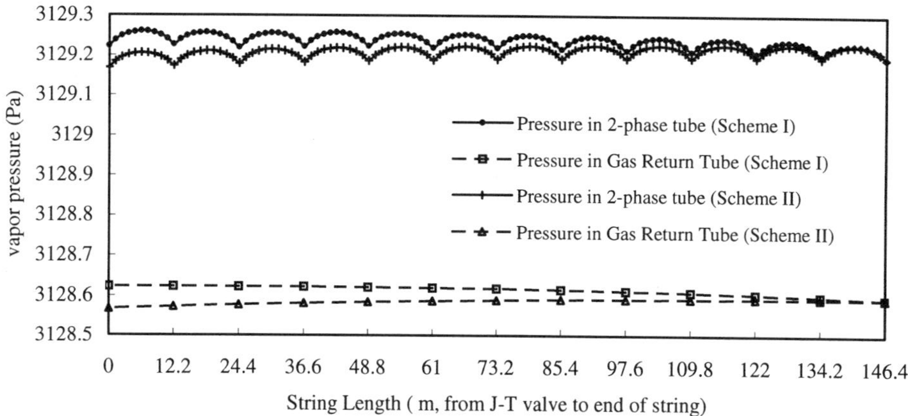

Figure 4. He vapor pressure in two-phase tube and gas return tube of one string

Vapor pressure in two-phase tube and in gas return tube are shown in Fig. 4. It can be seen that the pressure profile in two-phase tube in every module looks like an arc with upward bulge. This is due to the fact that vapor flows out of the module in two opposite direction from a certain section of higher pressure inside to the outside of lower pressure. The vapor pressure in gas return line, by contrast, increases slowly and smoothly from the end of string to J-T valve in scheme I, and decreases in scheme II. The vapor pressure in two-phase tube are always higher than that in gas return line because vapor flows from the former into the latter. Note that the absolute difference in pressure at each position of interconnections is contributed by the vapor pressure change due to height difference of two tubes and the pressure drop through the interconnection tube.

Figures. 5 and 6 are the He II level and temperature distribution in one string. It is interesting that the He II level seems change bumpily although the magnitude is minor through the string. This is caused by the longitudinal vapor pressure differences in two-phase tune and shear stress on He II / vapor interface. In simulation, the He II temperature is determined by the saturation vapor pressure so the profile of He II temperature in Figure. 6 is the same as that of vapor pressure in two-phase tube in Figure. 4. In fact, the magnitude of the change in He II temperature through one string is far less than it looks like in Figure. 6 as the pressure drop through the string is very small.

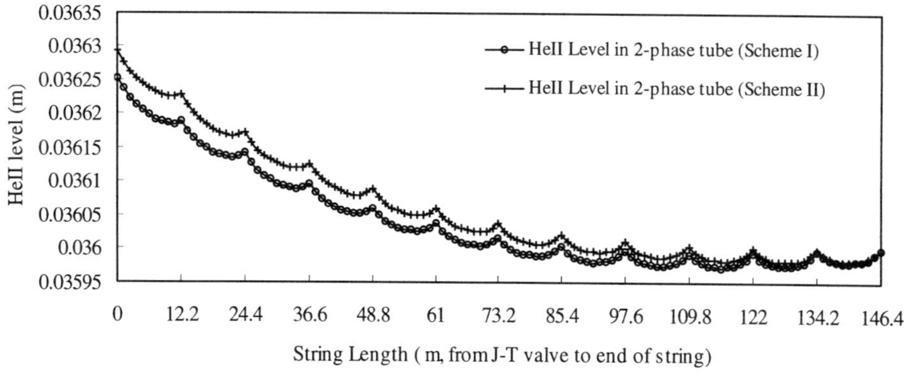

Figure 5. He II level distribution in two-phase tube of one string

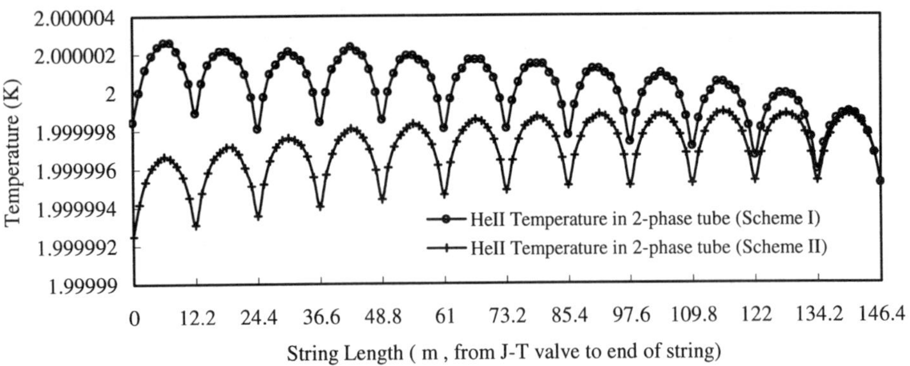

Figure 6. He II temperature distribution in two-phase tube of one string

CONCLUSIONS AND REMARKS

Simulation of He II and vapor flow in one horizontal cooling string of two schemes has been carried out for TESLA 500 machine design. The results show that two schemes have almost the same flow behavior when their boundary conditions at one end of string are the same. The predicted temperatures, pressures and He levels are well within those required for proper operation of the TESLA machine. While these results provide more detail they also agree in general terms with the earlier model. [2]

He II and vapor flow simulation in one refrigerator sector, consisting of 18 strings, with or without a small slope is underway. Furthermore, since the behavior of He II and vapor flow in one string is significantly dependent upon the pressure drop in gas return tube and interconnection tubes, calculation of such structure dependent pressure drop with enough precision becomes a key issue, which could be solved by experimental measurement and improvement of the friction correlation of He vapor.

ACKNOWLEDGEMENTS

Thanks to Dr. H. Lierl, Dr. B. Petersen, Dr. D. Sellmann and Mr. A. Zolotov of DESY for their support and advice.

REFERENCES

1. S. Wolff, H. Lierl, B. Petersen, and J. G. Weisend II, The cryogenic system of TESLA 500 – an update, in: " Proceedings of the ICEC 17 ", D. Dew Hughes et. al. ed., Institute of Physics Publishing, Bristol (1998)
2. G. Horlitz, " A Study of Pressures, Temperatures and Liquid Levels in the 2 K Refrigeration Circuit of a 1830 m TESLA Subunit under different Conditions and System Configurations. ", TESLA 94 – 17 (1995)
3. B. Rousset, A. Gauthier, B. Jager, R. Van Weelderen, J. G. Weisend II, Hydraulic behaviour of He II in stratified counter-current two-phase flow, in " Proceedings of the ICEC 17 ", D. Dew Hughes et. al. ed., Institute of Physics Publishing, Bristol (1998)
4. B. Rousset, A. Gauthier, L. Grimaud, Stratified Two-Phase Superfluid Helium Flow – Part I, *Cryogenics* 37:733 (1997)
5. B. Rousset, A. Gauthier, L. Grimaud and J. M. Delhaye, Stratified two-phase superfluid helium flow – Part II, *Cryogenics* 37:739 (1997)
6. N. N. Filina and J. G. Weisend II, "Cryogenic Two-Phase Flow : Applications to Large Scale Systems " Cambridge University Press, New York (1996)
7. S. W. Van Sciver, "Helium Cryogenics ", Plenum Press, New York (1986).

HE II TWO PHASE FLOW IN AN INCLINABLE 22 m LONG LINE

B. Rousset[1], A. Gauthier[1], B. Jager[1], R. van Weelderen[2]

[1]C.E.A. Grenoble / Département de Recherche Fondamentale sur la Matière Condensée/SBT, 17 rue des Martyrs, 38054 Grenoble Cédex 9, France

[2]CERN, European Organization for Nuclear Research, AT/CR, CH-1211 Geneva, 23, Switzerland

ABSTRACT

In the line of previous work done at CEA Grenoble, large size experiments were performed with the support of CERN for the validation of the LHC two phase superfluid helium cooling scheme. In order to be as close as possible to the real configuration, a straight, inclinable 22 m long line of 40 mm I.D. was built. Very accurate measurements of temperatures and pressures obtained after *in situ* re-calibration and verified by independent sensors allowed us to validate our two-phase flow model. Although we focus on pressure losses and heat exchange results in relation to power injected, additional measurements such as quality, void fraction, and total mass flow rate enable a complete description of the two-phase flow. Experiments were carried out to cover the whole range of the future LHC He II two-phase flow heat exchanger pipe: slope between 0 and 2.8 %, temperature between 1.8 and 2 K, total mass flow rate up to 7.5 g/s. Results confirm the validity of choice for the LHC cooling scheme.

INTRODUCTION

LHC[1] will use He II two-phase flow as the cold source. In order to maintain the magnet temperature as low as possible, two constraints have to be taken into consideration: the two-phase flow pressure losses (which means temperature difference between inlet and outlet of He II saturated flow, i.e. longitudinal ΔT) and the heat exchange between the saturated flow and the static He II pressurized bath in which the magnets are immersed (which means temperature difference between saturated flow and pressurized helium in a magnet cross section, i.e. transversal ΔT).

In the framework of LHC cooling scheme studies, previous experiments were performed[2,3,4] using a 40 mm I.D., 1.4 % slope helical pipe and a He II two-phase flow model was constructed. Obtained results gave us confidence in this co-current two-phase flow model and justified further experiments to fully validate the LHC cooling scheme. In particular, the influence of the slope (ranging from 0 to 1.4 % in case of LHC) had to be checked. Furthermore, the excess thermal heat exchange behaviour observed at high Vgs (Vgs represents the superficial vapour velocity, i.e. the vapour velocity if the vapour would occupied the whole cross section) had to be confirmed for a long straight line to avoid entrance and exit effects. In this paper, results on an inclinable straight pipe are presented.

DESCRIPTION OF TEST LOOP AND EXPERIMENT

The "Cryoloop" (Fig. 1) mainly consists of an inlet box used to create the inlet quality, the 22 m inclinable straight line, and an outlet saturated bath where excess liquid is evaporated.

The main measurements consist of the total mass flow rate (taken at room temperature), the saturated temperature and differential pressure between the extremities of the line (both used to calculate pressure losses), all powers injected (in order to calculate the quality), temperature increase in the pressurized chamber when power is on (which gives access to the thermal heat exchange) and visualization of the flow.

RESULTS

Hydraulic Results

First, the tube friction factor was determined in Reynolds similitude using nitrogen flow at room temperature. Best fit is smooth tube correlation (Fig. 2), which was adopted for wall friction factor in our two-phase flow model. The interfacial friction factor introduced in the model follows the Hanratty correlation as previously reported[3,4].

In situ calibration of the thermometers located along the two-phase flow enabled temperature measurements with 10^{-3} K accuracy.

On each figure, mass flow rate represents the sum of liquid and vapour mass flow rates.

Figure 3 compares pressure loss measurements derived from various sensors. Square symbols represent pressure losses between inlet and outlet boxes, respectively, measured with a differential pressure transducer and by means of immersed thermometers (see Fig. 1). The very good agreement confirms thermodynamic equilibrium in each box and good sensor accuracy. Line and circle illustrate comparison of measured and predicted pressure losses over the 22 m long line only (thereby excluding entrance and exit effects). The agreement is excellent and justifies the code assumptions[3]. Difference between square symbols (showing results given by sensors connected to the inlet box where liquid and vapour are at the rest) and open circle (showing results given by sensors connected to the line) are mainly due to entrance effects, i.e. acceleration of the two-phase flow at the inlet of the line.

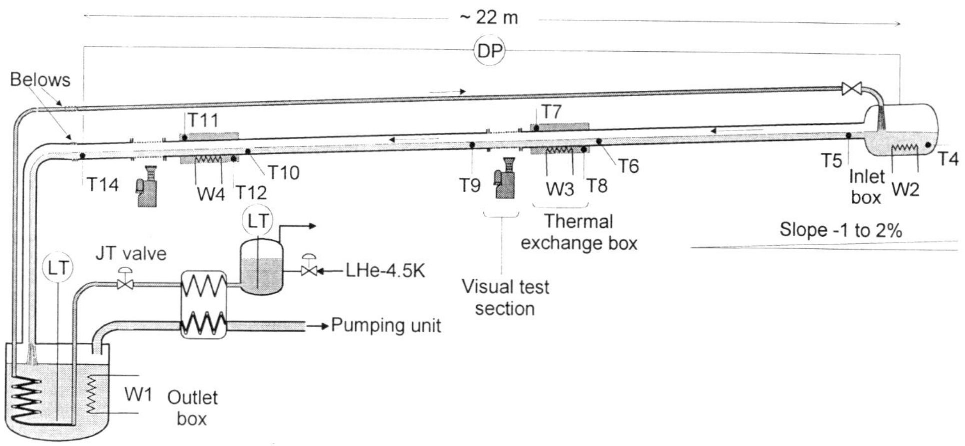

Figure 1. Test facility scheme

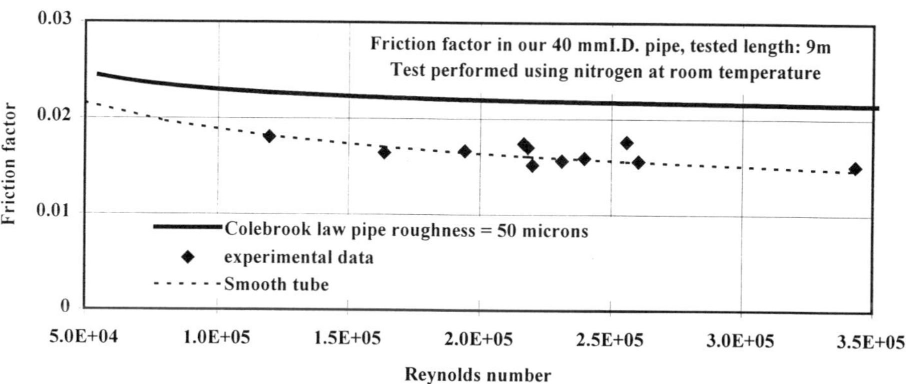

Figure 2. Friction factor calibration

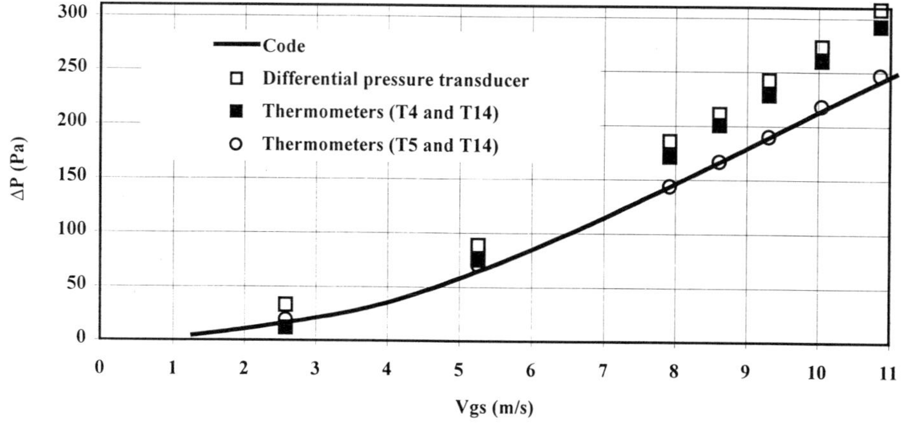

Figure 3. Various pressure losse estimations (slope 0%, total mass flow 7.2 g/s, temperature 1.77 K)

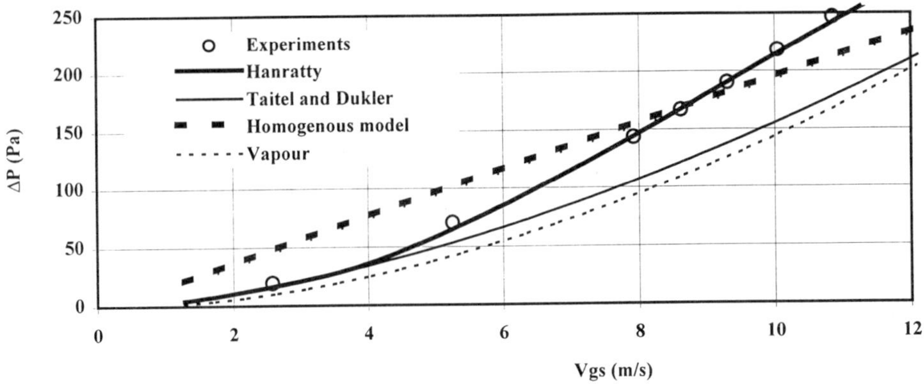

Figure 4. Experiment and model comparisons (slope 0%, total mass flow 7.2 g/s, temperature 1.77 K)

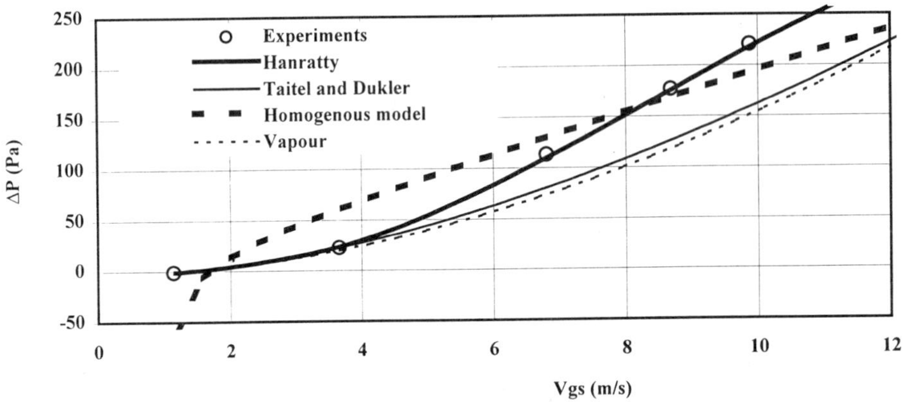

Figure 5. Experiment and model comparisons (slope 2.8 %, total mass flow 7.3 g/s, temperature 1.77 K)

Figures 4 and 5 compare the experiments with various models. As flow is stratified, the homogenous model gives worst results. Pressure losses are mainly due to the friction of vapour flow which explains the good agreement of vapour, Taitler and Dukler[5] and Hanratty[6] models at low Vgs, especially for large slopes where liquid occupies a small fraction of the cross section. Increasing Vgs results in a wavy interface with droplets dragging and Hanratty correlation has to be employed to represent flow behaviour.

On Figure 6, the curve obtained for 1.77 K is below the 1.91 K one. This result is different if injected power replaces Vgs in the abscissa axe. Due to vapour density changes with temperature, the same power injected results in different Vgs, e.g. 120 watt injected correspond to a Vgs of 11 m/s at 1.77 K and a Vgs of 7.6 m/s at 1.91 K, and it is clear on Fig. 6 that resulting pressure loss is higher at low temperature.

Finally, the influence of liquid mass flow rate can be deduced from Figure 7. This parameter has negligible effect, which confirms that pressure losses are mainly due to vapour flow.

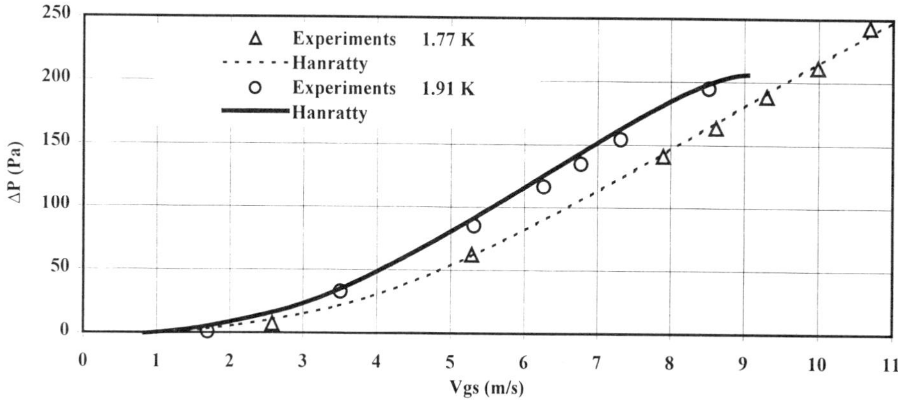

Figure 6. Temperature influence (slope 1.4 %, total mass flow 7.2 g/s)

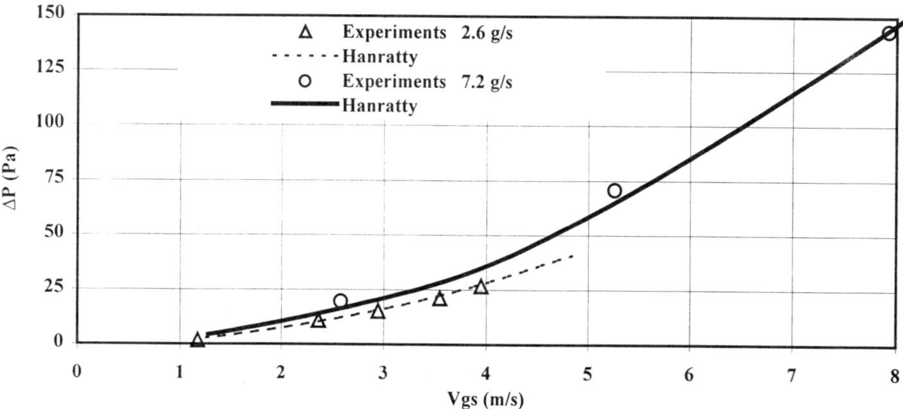

Figure 7. Total mass flow influence (slope 0 %, temperature 1.77 K)

Thermal Results

One other goal for this new test circuit was the confirmation of excess heat exchange with respect to stratified flow observed on the helicoidal line (*i.e.* heat exchange increases at high Vgs while liquid mass flow decreases) and the potential influence of using a straight line. As shown on figure 1, two heat exchange boxes instrument the Cryoloop. One is located 10 m downstream of the inlet, and the other 20 m downstream of the inlet. Results obtained are identical for the two boxes, and only data acquired on the second box are presented here.

Figure 8 compares the helicoidal line with the present Cryoloop. In both cases, heat flux to be transferred from pressurized bath to saturated flow is equal to 22 W/m^2 (value calculated using the whole pipe perimeter). Excess wetting is always present but appears at higher Vgs in the Cryoloop. This excess wetting consists probably of a tiny liquid layer due to interception of droplets by the pipe wall.

To verify this assumption, heat flux to be transferred was increased in order to dry out the wall liquid layer. The result is that this excess wetting can be partially evaporated when heat flux is increased (figure 9). Furthermore, the wetting dependence with heat flux exhibited a critical vapour velocity which is consistent with the apparition of excess wetting.

Low Vgs. On Figures 10, 11 and 12, numerical and experimental curves are parallel at low Vgs. The under-estimation of the wetting prediction could be due to wavy flow. Gravity plays a major role and wetting varies considerably with slope. As Vgs decreases, two-phase flow tends to an open channel liquid flow. Experimental slope uncertainties can also explain difference between measured and calculated values, especially for the horizontal case.

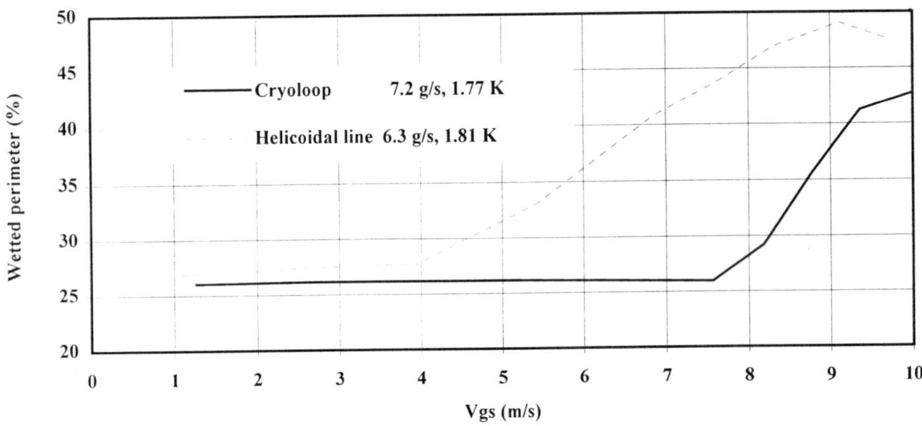

Figure 8. Comparison of thermal heat exchange for helicoidal line and cryoloop (slope 1.4%, heat flux 22 W/m²)

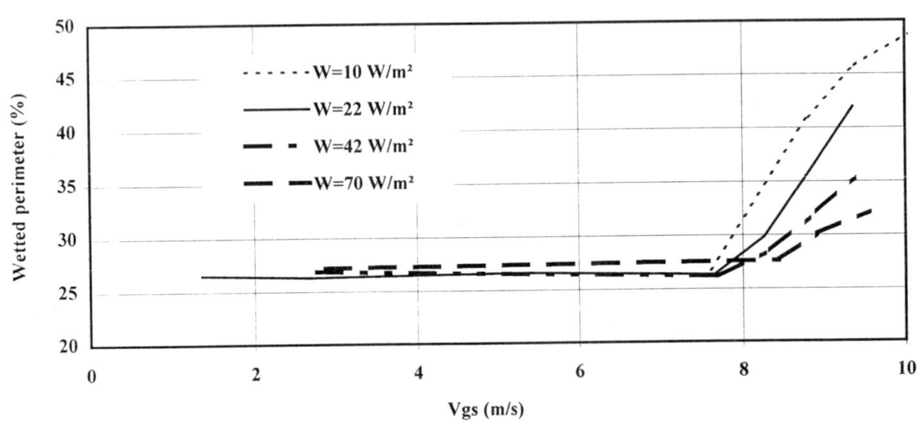

Figure 9. Influence of heat flux on apparent wetted perimeter (slope 1.4%, total mass flow 7.2 g/s, temperature 1.77 K)

High Vgs. Gravity has no influence, even on over-wetting. Unfortunately, it was not possible to access the mass flow or temperature influence, because in our experiments critical Vgs was not obtained for all curves. Over wetting is probably due to liquid droplet deposition on wall. The model, which use separate vapour and liquid flow can't take into account such phenomena.

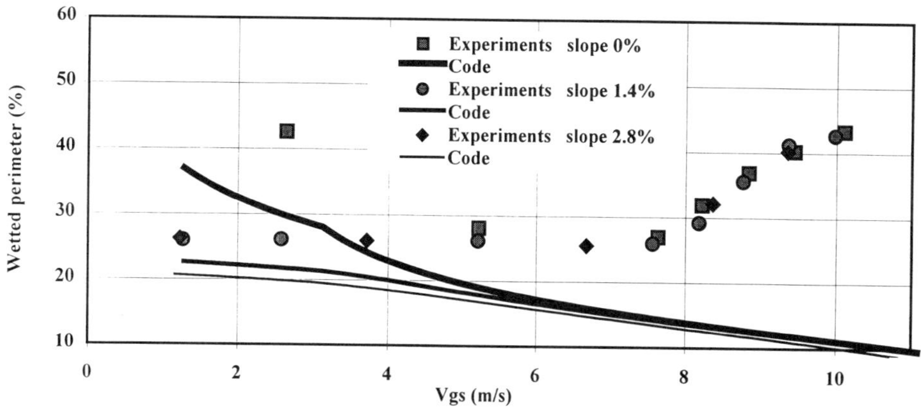

Figure 10. Slope influence (total mass flow 7.2 g/s, temperature between 1.77 and 1.8 K, heat flux 22 W/m^2)

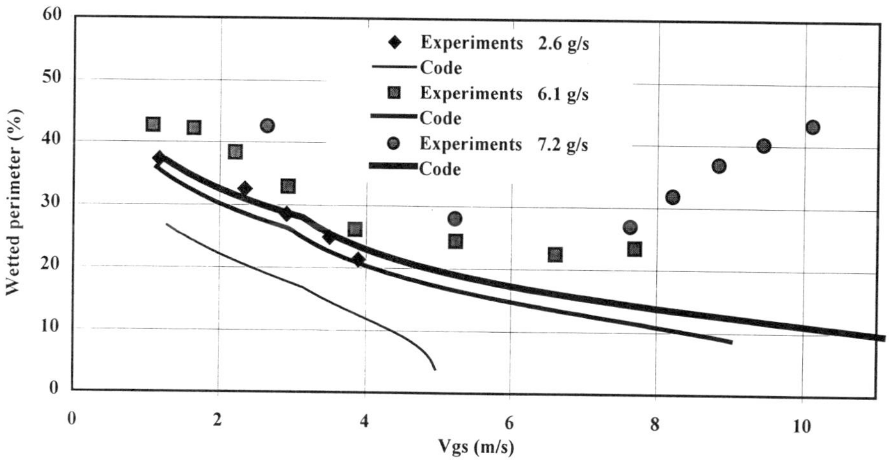

Figure 11. Total mass flow influence (slope 0%, temperature between 1.77 and 1.8 K, heat flux 22 W/m^2)

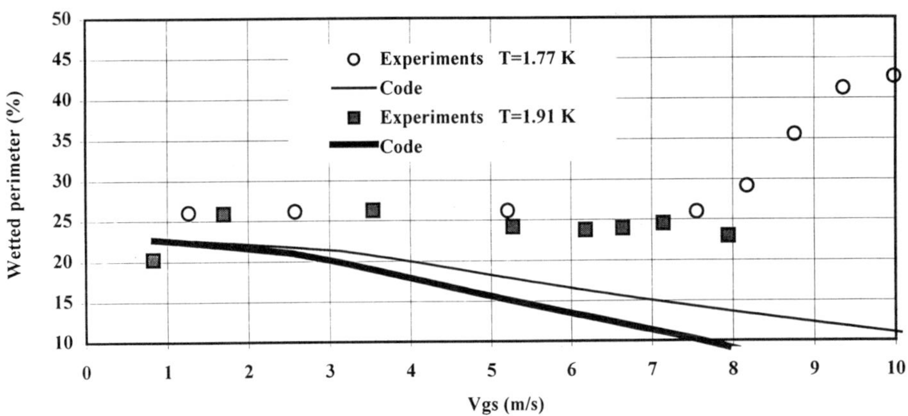

Figure 12. Temperature influence (slope 1.4%, total mass flow 7.2 g/s, heat flux 22 W/m^2)

CONCLUSION

We have investigated the thermohydraulic behaviour of He II two-phase flow. The large amount of data acquired in this experiments (only the most representative are presented here) confirms the excellent pressure loss prediction given by the Hanratty model.

Concerning heat transfer, apparent excess wetting appearing at high Vgs might possibly be due to an onset of change in flow pattern. Further measurements are needed to verify the presence of liquid droplets in the vapour flow. The code gives a conservative (under estimated) value for heat transfer.

REFERENCES

1. Lebrun P., "Superfluid Helium Cryogenics for the Large Hadron Collider Project at CERN" *Cryogenics 34* (1994)
2. Rousset B., Gauthier A. & Grimaud L., "Stratified Two-phase Superfluid Helium Flow: I" *Cryogenics 37* (1997)
3. Grimaud L., Gauthier A., Rousset B. & Delhaye J.M. "Stratified Two-phase Superfluid Helium Flow-Part: II" *Cryogenics 37* (1997)
4. Rousset B., Gauthier A. , Grimaud L. & van Weelderen R., "Latest Developments on HeII co-current Two-phase Flow Studies" *Advances in Cryogenics Engineering 43* (1998)
5. Taitel Y. & Dukler A.E., "A Model for Predicting Flow Regime Transitions in Horizontal and Near Horizontal Gas-liquid Flow", *AIChE Journal* (1976), Vol. 22, 47-55.
6. Andritsos, N. & Hanratty, T.J., "Influence of Interfacial Waves in Stratified Gaz-liquid Flows", *AIChE Journal* (1987), Vol. 33, 444-454.

INVESTIGATION OF NOISY FILM BOILING UNDER VARIOUS THERMAL CONDITIONS IN HE II

P. Zhang,[1] M. Murakami,[2] R.Z. Wang,[1] and Y. Takashima[2]

[1]Institute of Refrigeration and Cryogenics
Shanghai Jiao Tong University, Shanghai, 200030, China
[2]Institute of Engineering Mechanics and Systems
University of Tsukuba, Tsukuba-city, 305-8573, Japan

ABSTRACT

In the present study, some experimental results on noisy film boiling in He II are presented. Based on the study of the pressure oscillation of noisy film boiling under various thermal conditions, e.g. different bath temperature, immersion depth, heat flux, geometrical configuration and size of the heater, it is found that the bath temperature has a vital effect on the boiling state when it is close to the λ temperature. The critical immersion depth dividing different boiling states varies as bath temperature changes. The dominant frequency of noisy film boiling state shows little dependence on heat flux, while the geometrical configuration and size of the heater affects it drastically. Simultaneous measurement results from pressure sensor and superconductor temperature sensor show that a strong correlation exists between the pressure oscillation and the temperature oscillation during noisy film boiling.

INTRODUCTION

Superfluid liquid helium, in particular subcooled He II, has very high cooling capacity. In the applications of He II, the cooling performance is reduced by various mechanical or thermal disturbances which may be caused during noisy film boiling, and therefore it is necessary to understand its mechanism.

It was experimentally proved that there are two basic boiling modes in saturated He II: silent film boiling and noisy film boiling. The latter is accompanied by a loud acoustic noise and a violent mechanical vibration. A loud noise is generated by a vapor bubble of a comparable size with a heater collapsing repeatedly on the planar surface[1]. The critical hydrostatic pressure in He II for the occurrence of noisy film boiling decreases as the bath temperature drops[2]. It is also well known that the heat transfer coefficient in noisy film boiling is lower than that in silent film boiling, which may be detrimental to practical cooling applications of He II. Many experiments have been conducted using a thin wire heater[2,3,4], and a few experiments have also been conducted using a planar heater[1]. Some visualization methods, such as Schileren and shadowgraph methods were introduced into the studies of noisy film boiling[1,4].

A boundary map for noisy film boiling in the case of a thin wire heater has been presented by Leonard[2], but it was only approximately drawn. In particular, it seems unreasonable in the region of bath temperatures close to the λ-temperature. It may be concluded from the previous studies that there are many factors, such as the bath temperature, heat flux, geometrical configuration and size of the heater and other, which can affect the boiling state of He II, but they are not systematically studied. Thus, in the present study, we concentrated on the effect these factors have on noisy film boiling. Simultaneously, the pressure oscillations and the temperature oscillations were measured to investigate the mechanism of noisy film boiling.

EXPERIMENTAL SETUP AND PROCEDURES

Experimental Setup

A specially designed glass cryostat was used in our experiments, as shown on the left of Figure 1. During each experiment, a video camera to monitor the boiling process through the unsilvered strip of the cryostat was located in front of the cryostat. An evacuation system was used for regulating the bath temperature, the primary component of which was a rotary vacuum pump with a speed of 500 L/min and the automatic vapor pressure control system[1].

A planar heater, as shown on the right of Figure 1, was used to cause boiling. A Ni-Cr thin film (about 60nm), as heating element, and Cu thin film (about 600nm) as electrode were vacuum deposited on the surface of a Pyrex glass substrate. The heater used in the experiments was mounted horizontally in the test section, and was about 2.5 cm $\times$2.5 cm in size in most experiments. It was excited by a dc stepwise electric current.

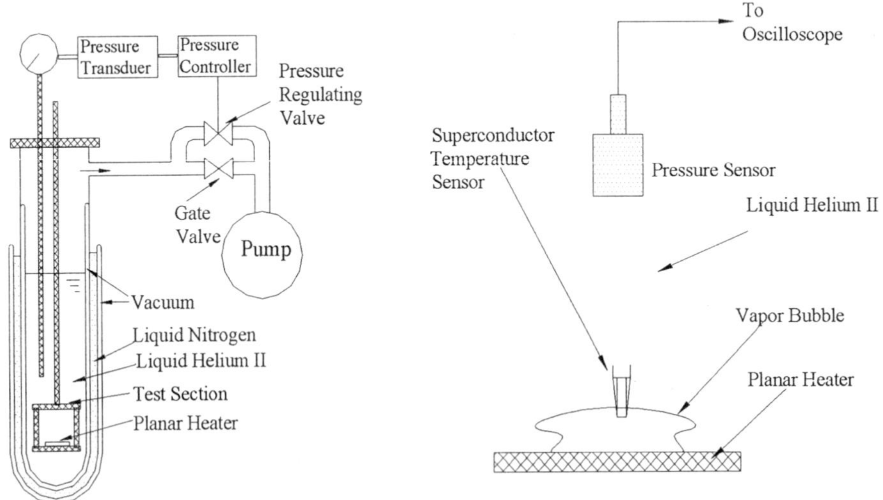

Figure 1 Schematic illustration of the experimental setup

A superconductor temperature sensor was mounted at a distance d_T above the heater, as shown in Figure 1. This type of superconductor temperature sensor was originally developed at the Max-Planck Institute[5], and was further modified at the University of Tsukuba[6]. The sensing element consists of gold (about 20nm) and tin (about 100nm) thin films which were vacuum deposited on outer surface of a fine quartz fiber 40μm in diameter and 1.5mm in length. Constant electrical bias current was applied to the sensor. The transition temperature of the sensor can be shifted to around superfluid temperature by trimming the thickness ratio of two films and by adjusting the bias electric current. Typical calibration result of the superconductor temperature sensor is shown in Figure 2. Its sensitivity is very high and the response time is very short[6]. We define the dimensionless sensitivity as

$$S_d = d(\ln V)/d(\ln T) \tag{1}$$

It can be seen from Figure 2 that the dimensionless sensitivity is higher than some popular cryogenic temperature sensors[7]. The superconductor temperature sensor was calibrated before and after experiment to ensure its stability.

A piezoelectic-type pressure sensor was used to measure the pressure oscillations. The distance d_P between the pressure sensor and the planar heater was adjustable, as shown in Figure 1. The pressure sensor could not be calibrated at the liquid helium temperature because it is an ac-type pressure transducer.

Experimental Procedure

The experiments were conducted under nearly saturated vapor pressure and under subcooled (pressurized) state conditions with the hydrostatic head produced from an immersion depth, h, up to 60cm. The bath temperature was controlled by regulating the vapor pressure in the liquid helium bath, and thus the vapor pressure was precisely controlled by the pressure regulating system. To cause boiling in He II, stepwise heat was generated from the planar heater. The pressure oscillations were measured by the pressure sensor. The signal of the temperature oscillations measured by the superconductor temperature sensor was amplified by a low noise pre-amplifier. All data were recorded by a storage oscilloscope and then stored in a computer for further data analysis.

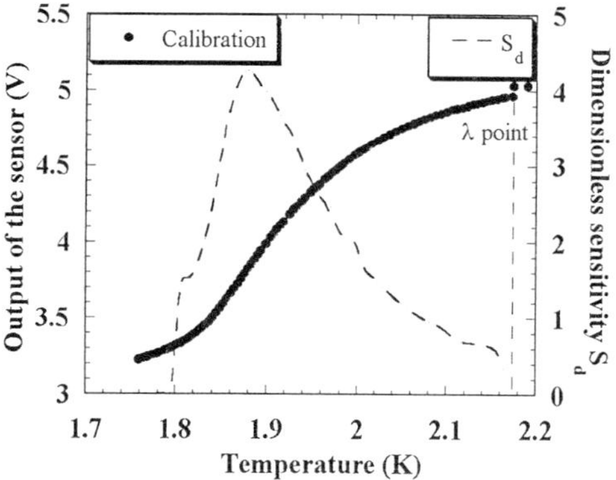

Figure 2 Calibration curve and dimensionless sensitivity of the superconductor temperature sensor

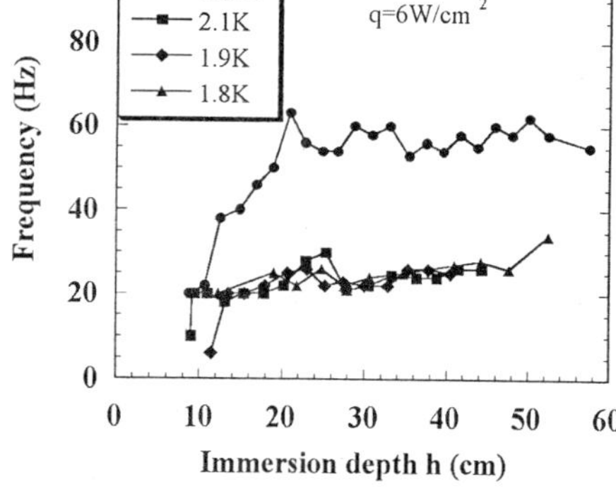

Figure 3 Variation of the dominant frequency of Pressure oscillations with the immersion depth at different bath temperatures

RESULTS AND DISCUSSION

Effect of Immersion Depth, h, on Noisy Film Boiling

When a large vapor bubble is expanding and crushing on the planar heater, it is accompanied by a loud noise and a violent mechanical vibration during noisy film boiling. Noisy film boiling generally occurs at low bath temperature and high immersion depth, h. As the immersion depth, h, decreases, boiling states may change from noisy film boiling to silent film boiling or transition boiling in which noisy film boiling and silent film boiling occurs intermittently making the dominant frequency lower. The frequency of expanding and crushing of the vapor bubble is defined as the dominant frequency of noisy film boiling. At different bath temperatures, the variation of the dominant frequency with the immersion depth, h, is shown in Figure 3. It is found from the figure that the dominant frequency is almost independent of the immersion depth, h, when it is larger than about 15cm. The dominant frequency decreases considerably as the immersion depth, h, further drops, which indicates that current boiling state changes to another boiling state.

Effect of Heat Flux on Noisy Film Boiling

Large pressure spikes generate loud noise through direct contact of He II with overheated heater surface when a vapor bubble crushes. The frequency of bubble expansion and crush determines the dominant frequency of noisy film boiling. From previous studies it can be concluded that the dominant frequency in the case of lower heat flux is slightly higher because lower heat flux results in a smaller vapor bubble.

The boiling state does not change to noisy film boiling state when the heat flux exceeds the upper critical heat flux above which noisy film boiling disappears in spite of large immersion depth. (We designate the critical heat flux required for the appearance of noisy film boiling as the lower critical heat flux. The critical heat flux leading to the disappearance of noisy film boiling is considered the upper critical heat flux)[8,9]. This indicates that the critical immersion depth varies also with the heat flux. In order to examine this, the variations of boundary map for boiling states with different heat fluxes are drawn at different bath temperatures. One example is shown in Figure 4.

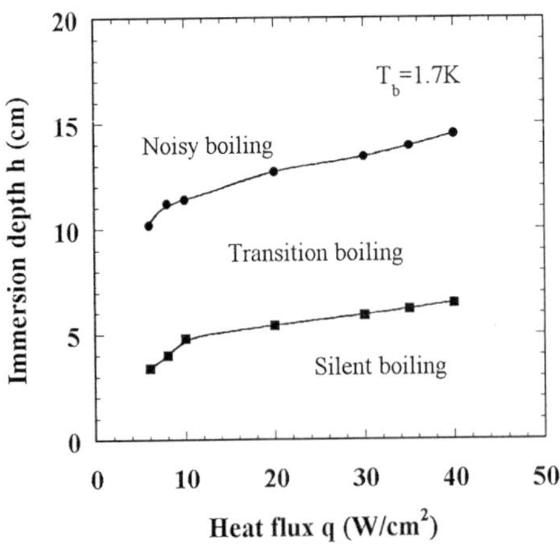

Figure 4 Effect of heat flux on film boiling state boundaries

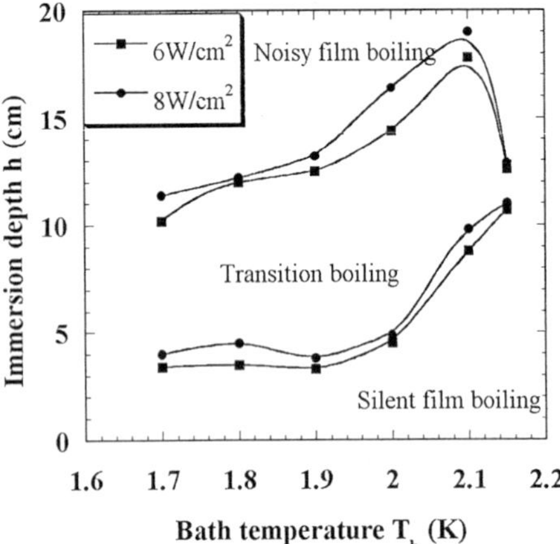

Figure 5 Boundary map for noisy and silent film boilings at two values of heat fluxes

Effect of The Bath Temperature on Noisy Film Boiling

Bath temperature is a vital factor which affects film boiling state in He II. In the present experimental range, weakly subcooled state of liquid helium II near the heater surface can be achieved when the bath temperature is close to the λ-temperature (i.e. 2.15K), while it is always saturated state when the bath temperature is below 2.1K. Different boiling phenomena can be found under two conditions discussed above, see reference[10,11] for details. It is seen form Figure 3 that the effect of the bath temperature is notable only at the conditions very close to the λ-temperature. The dominant frequency resulting from noisy film boiling is about 20~30 Hz when the bath temperature is below 2.1K. This frequency is less than half of that at the bath temperatures close to the λ-temperature (i.e. 2.15K). The higher frequency at the bath temperature close to the λ-temperature, provided that He II is in weakly subcooled state, may be due to the appearance of He I phase adjacent to the heater during boiling.

A boundary map of the critical immersion depth as a function of bath temperatures is shown in Figure 5. Some discrepancies are found between this result and the diagram presented by Leonard[2]. The possible reasons may be: first, the heater geometry is different, that is a thin wire was used to cause boiling in his case; second, in his case, the boundaries among boiling states were only approximately drawn, and the critical immersion depth for the boundaries were not definitely obtained and, furthermore, the heat flux effect was not considered. Another distinction is the variation of the two boundaries near the λ point. In the present study, the two branches, the noisy branch and the silent branch are close to one another when the bath temperature approaches the λ-temperature. This fact has also been confirmed qualitatively in another study[1].

Effect of Geometrical Configuration Size of The Heater on Noisy Film Boiling

The geometrical configuration and size of the heater has a remarkable effect on vapor bubble size and shape and therefore on the dominant frequency of the pressure oscillation. For a thin wire heater, it can be seen from the picture[9] that many small bubbles are generated along the thin wire heater during noisy film boiling. In the case of a rectangular planar heater, the heater aspect ratio has considerable effect. Single bubble comparable to the heater size on a square heater may transform into several connected small bubbles on heater[8] with a larger aspect ratio.

In order to investigate size effect, different size heaters were tested. It is obvious that the dominant oscillation frequencies of the smaller heaters are higher than that of the larger ones, and the amplitude of the pressure oscillation waves of the larger heater is larger than that of the smaller ones.

The experimental result is shown in Figure 6 and Figure 7, where the data for the heater smaller than 0.1 cm are taken from the reference[12], with thin wire heaters. Here, a new parameter to elucidate the data of both the thin wire and the planar heaters is defined as follows

$$D = \frac{4S}{P} \tag{2}$$

where, D is the parameter indicating the heater size, S is the heater surface area, that is a^2 for planar square heaters, and the cross sectional area πr^2 for thin wires, and P is the perimeter, 4a for planar square heaters, $2\pi r$ for thin wires. It can be seen from Figure 6 and Figure 7 that the dominant frequency has strong dependence on geometrical configuration and size of the heater. It increases as the size factor D becomes smaller. This is due to the smaller vapor bubble oscillating on the smaller heater, which results in the higher frequency. It is seen from the above two figures that the dominant frequency of noisy film boiling can be approximately fitted by logarithmic functions which are shown as dashed lines.

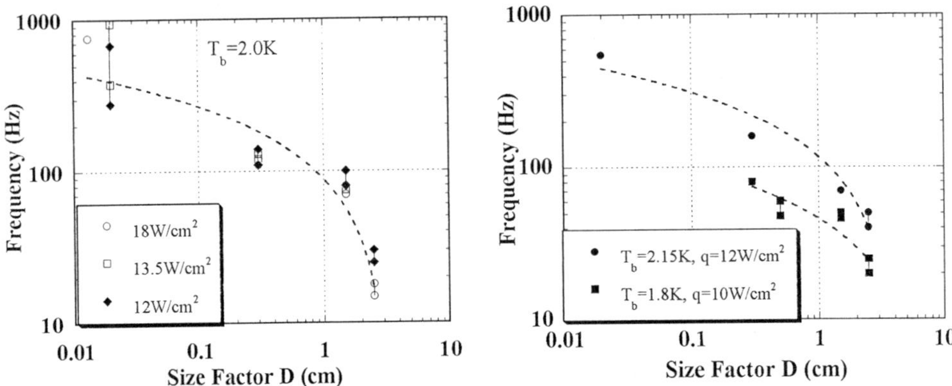

Figure 6 Heater size effect on noisy film boiling at 2.0K

Figure 7 Heater size effect on noisy film boiling at 1.8K and 2.15K

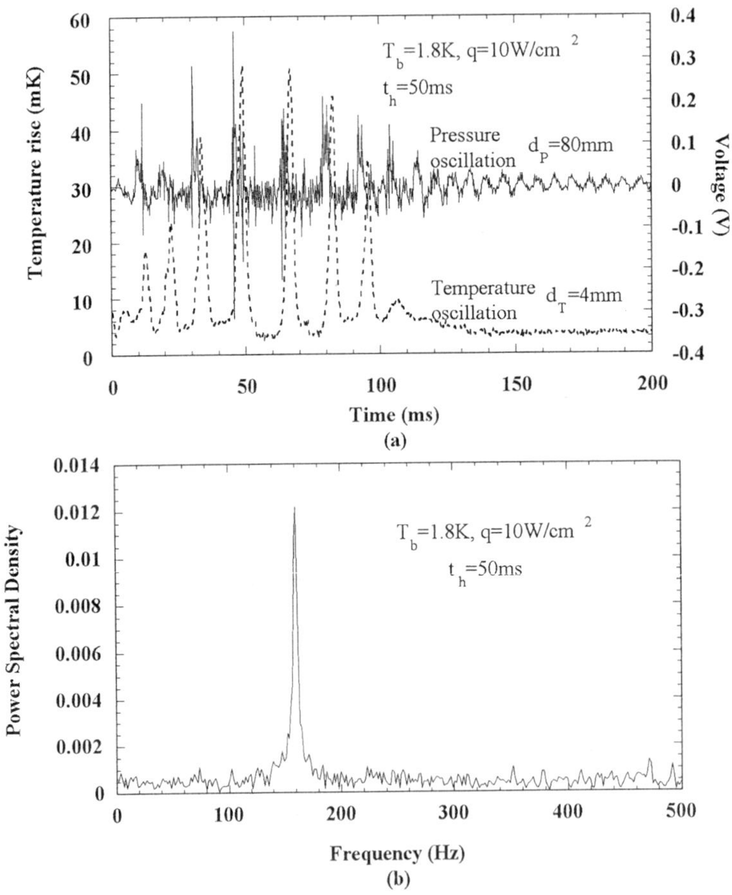

Figure 8 Correlation between the pressure and temperature oscillations and the FFT analysis of the latter stage of the pressure oscillation wave form

Correlation between The Pressure and Temperature Oscillations during Noisy Film Boiling

Pressure spikes caused by expanding and crushing of a vapor bubble are measured by the pressure sensor and the temperature oscillations are measured by the superconductor temperature sensor. It can be seen that there is a strong correlation between the pressure and temperature oscillations (Figure 8(a)). The Fast Fourier Transform (FFT) analysis shows that the dominant frequencies of the temperature and pressure oscillations are the same.

As is also seen in Figure 8, the latter part of the pressure oscillation wave is different from the early stage. FFT analysis shows that the frequency of the latter part of the pressure oscillation wave in the figure is about 160Hz, as shown in Figure 8(b), which is in good agreement with the frequency of the liquid column oscillation[11] for corresponding immersion depth. The frequency of the latter stage which is induced by liquid column oscillation is much larger than that of the early part. The pressure signal is regarded as propagating at the first sound speed about $200 m/s$, while the speed of temperature propagation is far lower than the speed of a second sound of $20 m/s$, which indicates the temperature rise is not caused by second sound but by the oscillatory movement of a thermal boundary layer.

CONCLUSIONS

1. Various factors affecting noisy film boiling are studied in details using the pressure oscillations measurements under various thermal conditions. It is found that the occurrence of noisy film boiling depends on immersion depth, h. The low heat flux causes smaller vapor bubble which results in the higher dominant frequency. Bath temperature effect is notable only when the bath temperature is close to the λ-temperature and a boundary map for different boiling state is established. Heater geometry and size have a remarkable effect on vapor bubble size and shape, and as a result, on the dominant frequency of the pressure oscillations.
2. It is found that during noisy film boiling there is a strong correlation between the pressure oscillations and the temperature oscillations. The pressure signal is propagating at a speed of first sound, while the temperature oscillation is a result of the oscillatory movement of a thermal boundary layer.

ACKNOWLEGEMENT

This study is partly supported by Chinese National Nature Science Foundation under contract No. 59876018. One of the authors, P. Zhang, would like to thank AIEJ (Association of International Education, Japan) for financial support during his stay in Japan.

REFERENCES

1. M. Yamaguchi, and M. Murakami, Study of pressure oscillation during noisy film boiling in He II, *Cryogenics,* 37:523 (1997).
2. A. C. Leonard, Helium-2 noisy film boiling and silent film boiling heat transfer coefficient values, *Proc. ICEC3,* 109 (1970).
3. P. Bussieres, and A. C. Leonard, Noise associated with heat transfer to liquid helium II, *Bull. IIR. Annexe,* 5:61 (1966).

4. D. M. Coulter, A. C. Leonard, and J. G. Pike, Heat transport visualization in helium II using focused shadowgraph and schlieren techniques, *Adv. Cryog. Eng.*, 13:640 (1968).
5. M. v. Schwerdtner, W. Poppe, and D. W. Schmidt, Distortion of temperature signals in He II due to probe geometry, and a new improved probe, *Cryogenics*, 29:132 (1989).
6. T. Shimazaki, M. Murakami, and T. Iida, Second sound wave heat transfer, thermal boundary layer formation and boiling: highly transient heat transport phenomena in He II, *Cryogenics*, 35:645 (1995).
7. LakeShore Company, Temperature measurement and control, (1995).
8. M. Murakami, Y. Katsuki, and M. Yamaguchi, Thermo-fluiddynamic aspect of noisy film boiling phenomena in He II, *Adv. Cryog. Eng.*, 41:257 (1996).
9. Y. Katsuki. Visualization study of transient boiling phenomena in He II, M. S. thesis, University of Tsukuba, (1995).
10. P. Zhang, M. Murakami, R. Z. Wang, and H. Inaba, Influence of bath temperature on noisy film boiling in He II, *Proc. ICCR'98*, 341 (1998).
11. M. Murakami, M. Yamaguchi, N. Yanase, and H. Inaba, Various film boiling states in He II at hydrostatic pressure from staturated vapor pressure to 1 atm, *Adv. Cryog. Eng.*, 43:1425 (1998).
12. F. L. Ebright, and R. K. Irey, High-speed motion-picture studies of film boiling in liquid helium II, *Adv. Cryog. Eng.*, 16:386 (1971).

THIN-FILM THERMOMETER AND HEATER DESIGN FOR THE DETECTION AND GENERATION OF SECOND SOUND SHOCK PULSES

D. K. Hilton[1,3], M. R. Smith[3], and S. W. Van Sciver[2,3]

[1]Department of Physics
Florida State University
[2]Department of Mechanical Engineering
FAMU-FSU College of Engineering
[3]National High Magnetic Field Laboratory
Tallahassee, Florida

ABSTRACT

Using a known heater design, and a common type of pencil lead graphite in a novel thermometer design, second sound shock (SSS) pulses comparable to those of previous researchers were successfully transmitted and received in a 1.7 K He II bath. Two sample traces of SSS pulses are presented. Extensive design requirements are outlined for high-speed thin-film thermometers and heaters used to transmit and receive SSS pulses. The heater and thermometer can be described by lumped parameter first order ordinary differential equations for the thermal case because their Biot numbers are less than 0.1 in the presented design. Also, their thermal time constants are minimized. However, the heater and thermometer can be described by lumped parameter first order ordinary differential equations also for the electrical case because they are embedded in an electronic instrumentation system. The time constants in both cases must be considered with respect to the required minimum pulse duration to arrive at a correct design for the heater and thermometer.

INTRODUCTION

Research making use of second sound shock (SSS) pulses continues to have applications to transient heat transfer problems in He II, such as the cooling and stability of superconductors. A thorough understanding of SSS pulse propagation is required for such applications. SSS pulses in He II of sufficient amplitude (0.1-100 W/cm^2) and short duration (0.01-10 ms) generate wakes of quantum vortex tangles, quantum turbulence, and are distorted by the quantum turbulence they induce. A review and classification of the existing experimental data was presented by Nemirovskii and Tsoi.[1] The parameter space defined by pulse initial power flux density and duration, along with bath temperature, reveals distinct regimes of qualitatively different transient heat transfer behavior in He II. The region centered near 10 W/cm^2 and 1 ms has not been explored, and is perhaps the most interesting in that it is the intersection between developing quantum vorticity, fully developed vorticity, and film boiling. The device described in this article will be one of three such devices to explore that region.

SSS pulses manifest themselves as rapid temperature jumps, consequently requiring fast, compact thermometers and heaters. The present design is capable of detecting and generating SSS pulses with durations between 10 μs and 6 ms, and with initial power flux densities up to 45 W/cm^2, which allows a tentative exploration of the region described above. However, a survey of journal articles relevant to second sound experimental techniques reveals that they do not emphasize details requisite to an effective design of a SSS pulse research device. Specifically, the thermometer and heater are embedded in an electronic instrumentation system. How the thermometer and heater interact with that system as well as how they behave in the He II bath are both important to an operating SSS device. The present article outlines extensive design requirements for high-speed thin-film thermometers and heaters used to transmit and receive SSS pulses. Furthermore, experimental SSS pulses resulting from such a design are also presented, and discussed with respect to the design requirements.

DESIGN

A thermometer and a heater for SSS pulses are considered properly designed when they are sufficiently thin in the direction of pulse propagation that their thermal time constants are minimized. However, more than this must be finalized in their design. In a correct design, the thermal time constant is comparable to the electrical time constant for the heater, whereas the electrical time constant is dominant for the thermometer. Furthermore, the heater must have a resistance comparable to that of its power pulse source for maximum power transfer, and an active area small enough to give the required initial power flux density. The thermometer current must be as high as possible without introducing significant self-heating for the thermometer to deliver the maximum possible voltage signal. The electrical as well as the thermal time constants for the heater and thermometer must be considered, to arrive at the correct design.

A flow chart for the SSS experiment as shown in Fig. 3 is mentioned here as a motivation for the following. The equivalent circuit for the SSS thermometer, shown in Fig. 1, is described by a first order ordinary differential equation with a characteristic time constant. This electrical time constant for the best SSS thermometer designs is on the order of 10 μs, whereas the thermal time constant is on the order of 10 ns. The electrical and thermal time constants for the thermometer are given by the following.

$$\frac{1}{\tau_E} = \left(\frac{1}{r_S} + \frac{1}{R_\infty} + \frac{1}{r_M} \right) \frac{1}{C_C} \tag{1}$$

$$\frac{1}{\tau_\theta} = \frac{h_K A - \dfrac{i_0^2 R_\infty s}{T_\infty}}{\rho V c} \tag{2}$$

where

τ_E = thermometer electrical time constant [s]
r_S = current source internal resistance [Ω]
R_∞ = thermometer mean operating resistance [Ω]
r_M = voltage measurement input resistance [Ω]
C_C = coaxial cable capacitance [F]
τ_θ = thermometer thermal time constant [s]
h_K = Kapitza conductance [W/m^2K]
A = thermometer active area [m^2]
i_0 = current source output [A]
s = thermometer operating sensitivity [-]
T_∞ = bath operating temperature [K]
ρ = thermometer mass density [kg/m^3]
V = thermometer active volume [m^3]
c = thermometer specific heat [J/kgK]

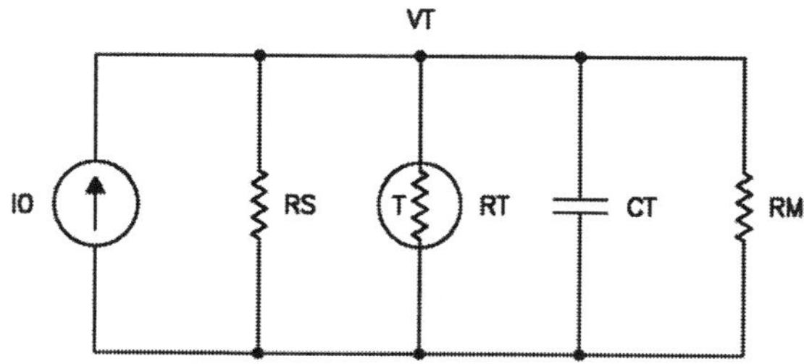

Fig. 1. Second sound shock (SSS) thermometer equivalent circuit.

The thermometer mean operating resistance is measured at the He II bath operating temperature. The coaxial cable capacitance includes that of both the current source cable as well as the voltage signal cable. The Kapitza conductance is estimated at about 1000 W/m^2K because reliable Kapitza conductance data for the thermometer (and heater) materials used in the present designs, such as graphite and nichrome, are scarce.[2] The current source output is the current delivered with the source output shorted. The following is the relationship assumed between the bath temperature and mean resistance of the thermometer, and describes the meaning of thermometer operating sensitivity, with α a proportionality constant [Ω/K^s].

$$R_\infty = \alpha T_\infty{}^s \tag{3}$$

The form of the thermal time constant for the thermometer, given by Eq. 2 expressed above, is slightly different than classically derived in that it takes into account the effect of self-heating. The Biot numbers for the thermometers (and heaters) were calculated, and found to be less than 0.1 for all designs, permitting the use of lumped parameter first order ordinary differential equations for the thermal case as in the electrical case.

The equivalent circuit for the SSS heater, not shown but similar to that for the thermometer, is also described by a first order ODE with a characteristic time constant. This electrical time constant for the best SSS heater designs is on the order of 10 ns, comparable to the thermal time constant. In the equivalent circuit for the SSS heater however, the power pulse source is represented as a Thevenin voltage source rather than a Thevenin current source. The power transfer from the pulse source to the heater is maximized by making the heater resistance comparable to the internal resistance of the source.

The following expression is for the self-heating of the SSS thermometer above the bath operating temperature. This self-heating is due to the thermometer excitation current.

$$T_0 = \frac{1}{\frac{h_K A}{i_0^2 R_\infty} - \frac{1}{T_\infty}} + T_\infty \qquad (4)$$

where T_0 is the thermometer mean operating temperature [K]. The implication to thermometer design is that a thin thermometer with a larger active area will suffer less self-heating than one with a smaller active area, because the self-heating is dependent on the active area whereas the thermal time constant is primarily dependent on a characteristic length, the active thickness (V/A).

APPARATUS AND EXPERIMENTS

The SSS thermometer, shown in Fig. 2, consists of a G-10 sheet, 0.23 mm thick, 2.11 mm x 6.02 mm, with one face, the active face abraded in every direction of the surface with medium grit (240 grit) sand-paper to insure retention of the graphite thin film. The abrasion cuts micro-channels in the surface, forming parallel, temperature sensitive graphite resistors in the direction of the abrasion, when the graphite is later applied. Thus, the direction of abrasion in part determines the final mean resistance of the thermometer, which in turn determines the electrical time constant. The source of the graphite is a particular brand of pencil lead (Pentel, 0.5 mm, hardness B). The final mean resistance and sensitivity of the thermometer is strongly dependent on the graphite source and hardness, graphite being a semi-metal whose resistivity is strongly dependent on its doping and impurity. After surface treatment of the active face of the G-10 sheet, the graphite thin film is formed also by abrasion. As shown in Fig. 2, the sheet is then impaled for mechanical support and electrical contact on two gold-plated size-10 needles (0.46 mm dia., 25.4 mm long) a fixed distance apart (3.43 mm center-to-center). Electrical contact is secured between the needle tips and the opposite ends of the graphite thin film using a two component, electrically conductive, silver filled epoxy, Epo-Tek H20E, produced by Epoxy Technology, Inc. of Billerica, Massachusetts. Enameled, 30 A.W.G. solid copper wires are connected to the mid-points of the needles using standard solder (60% Sn / 40% Pb).

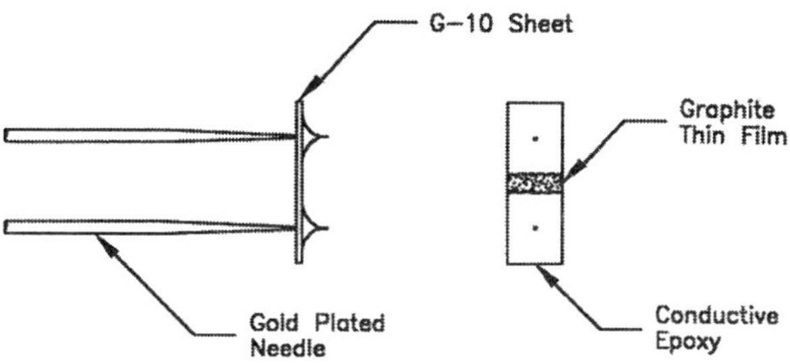

Fig. 2. Second sound shock (SSS) thermometer consisting of a graphite thin film deposited onto a G-10 sheet by abrasion.

The SSS heater, not shown, consists of a fused silica (amorphous SiO_2) substrate, 31.75 mm in diameter and 0.83 mm thick, with both faces ground and the upper active face polished. The ground and polished substrate is produced by Superconductive Components, Inc. of Columbus, Ohio. Upon the upper face is deposited by physical vapor deposition (PVD) a 34.7 nm thin film of nichrome (80% Ni / 20% Cr by weight), 29.92 mm in diameter. Nichrome has a nearly constant resistivity of 1.05×10^{-4} Ω-cm at $T < 4.2$ K, and that resistivity does not vary significantly with temperature up to $T =$ 298 K. The thinness of the film in large part determines the thermal time constant of the heater. Electrical power pulse connections are made to the heater at opposite edges through 200 nm thin film strips of gold deposited by PVD, each 4.76 mm wide, the 99.9999% pure gold having a resistivity of 2.44×10^{-9} Ω-cm at $T < 4.2$ K. Tinned, 24 A.W.G. stranded copper wires are connected to the gold thin film strips, again using silver filled epoxy. From calculations of the electric potential across and the current density through the heater thin film, the power density was shown to not vary significantly throughout the film, insuring uniform heating across the circular disk. Geometrically, the resistance of the circular disk is inversely proportional to its thickness only, and is not dependent on its diameter.

The thermometer and heater are arranged with their active faces directed toward each other within a SSS tube assembly submerged in a He II bath. The SSS tube consists of a G-10 tube 203.2 mm long, with a 31.75 mm O.D. and a 25.4 mm I.D. The smoothness of the inner surface is as received. The heater closes off the bottom end of the tube with an O-ring, although the seal is most likely not superfluid leak-tight. The gold-plated needles of the thermometer are secured to the end of a hollow G-10 rod not shown, which is secured to the end of a hollow stainless steel rod, both rods being 9.53 mm in outer diameter, and concentric with the SSS tube. The connected rods are movable along the longitudinal axis of the SSS tube. Thus, the thermometer can be positioned at any distance from the heater to characterize the propagation of SSS pulses. Otherwise, the top end of the tube is open. The maximum duration pulse that can be investigated is determined by the tube length. Reviewing selections by previous researchers, the propagation of SSS pulses without significant artifacts being introduced does not seem to be strongly dependent on the material choice or inner surface smoothness of the tube.[3,4,5]

The electronic instrumentation that facilitates SSS pulse experiments is shown in Fig. 3. The experiments are initiated, monitored, and documented by a Macintosh IIfx computer running LabVIEW 4.0 software. The computer initiates a square voltage pulse from the pulse generator (HP 8116 A), which is immediately converted to a power pulse by the high-speed power amplifier (NF 4015). This power is transferred by RG-58/U coax cable and 24 A.W.G. stranded copper wire twisted pair to the SSS heater. The resulting voltage signal from the SSS thermometer is transferred by Lake Shore SC and

RG-178 B/U coax cable to a high-speed, high-gain preamplifier. The preamplifier is comprised of three high-precision operational amplifier stages (PMI OP 27 EP), each set to a gain of about 10, and each with a gain-bandwidth product of about 5 MHz. This gain-bandwidth was chosen to allow the signal to pass without significant distortion as well as without excessive noise. The digitizing oscilloscope (TDS 744 A) displays the signal, and stores it for later retrieval by the computer. The 1 μA thermometer excitation current is supplied by the constant current source.

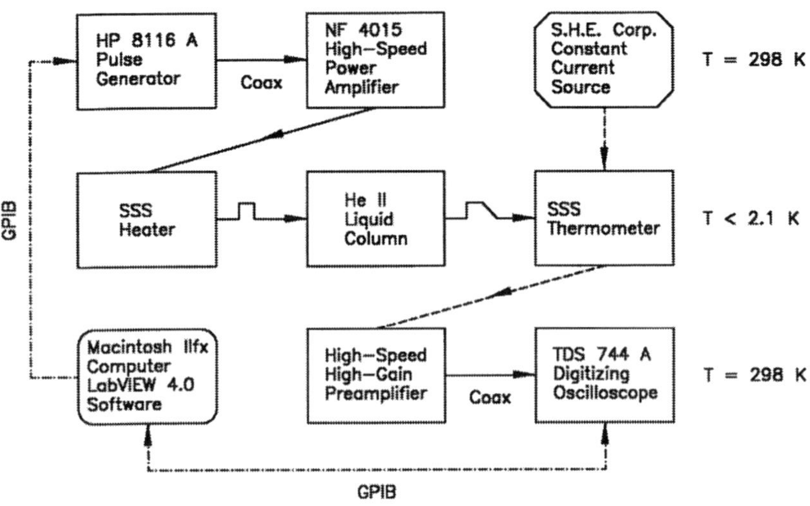

Fig. 3. Second sound shock (SSS) experiment flow chart.

RESULTS AND DISCUSSION

Shown in Fig. 4 are two samples of received SSS pulses. Each was generated by an initial square pulse of 20 W/cm^2 power flux density, 150 μs duration, initiated at zero time into a He II bath at 1.7 K. Each trace is an average of 5 repeated measurements. The Δz values noted next to the pulses are the distances from the heater to the thermometer. The vertical relative displacement of the two traces corresponds to the difference in bath operating temperatures at the respective times the traces were recorded. The trace upward deviations preceding the pulses are measurement artifacts resulting from electrical pick-up by the thermometer from the heater and its power connections. The noise that is present is also electrical in origin, and suggests that the thermometer excitation current may not be optimized. The pulse speed is 20.7 m/s for the upper trace, as measured by the leading edge displacement. The speed of second sound is about 20 m/s at 1.7 K operating temperature.[2] Although the least significant digit is uncertain, the slightly larger experimental value may be due to nonlinear wave propagation.[1] The change in shape with propagation of the pulse from trapezoidal to triangular that is apparent in the traces also suggests a nonlinear, dispersive mechanism.[5]

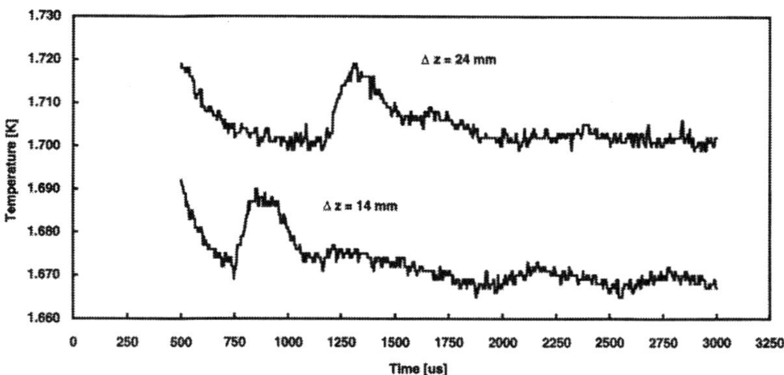

Fig. 4. Second sound shock pulses from an initial square pulse of 20 W/cm^2 power flux density and 150 μs duration.

The thermal and electrical performance of different resistive filament or film thermometer (RFT) types tested as SSS thermometers is summarized in Table 1. The first, CFT 4.2-F, is a carbon filament on the order of 5-7 μm in diameter, such as the kind used in composites. The second, CCT 1.1, is a Lake Shore CX-1080-BG Cernox® thermometer bare chip. The third, PFT 2.4-G, is an S-glass filament on the order of 5-7 μm in diameter with a polypyrrole thin film about 0.1 μm thick . The fourth, GFT 1.1-G10, is the G-10 sheet with the graphite thin film described in the apparatus section above, and that received the SSS pulses shown in Fig. 4.

Table 1. Resistive Filament / Film Thermometer (RFT)
Thermal and Electrical Performance Summary at $T_\infty = 1.7$ K

Thermometer	s [–]	i_0 [μA]	τ_E [μs]*
CFT 4.2-F	-0.50	1.0	3.48
CCT 1.1	-0.60	1.0	538
PFT 2.4-G	-3.68	-0.001	285000
GFT 1.1-G10	-0.41	1.0	2.05

*C_c = 651 pF

As can be deduced from the table, CCT 1.1 was not tested as a SSS thermometer because of its high electrical time constant. However, a parallel multifilament form of PFT 2.4-G was tested despite its very high time constant. A low-temperature current source and source follower circuit was built in an attempt to reduce the thermometer mean operating resistance by impedance matching. Although the circuit operated at 4.2 K, it failed to operate at 1.7 K. This is unfortunate because the thermometer operating sensitivity is exceptionally high compared to the others. CFT 4.2-F shows promise as a SSS thermometer since its electrical time constant is comparable to that of the GFT 1.1-G10. However, when this thermometer was tested, the heater then installed was too thick, with a thermal time constant of about 10 μs, and the high-speed, high-gain preamplifier was not present. A parallel multifilament form of CFT 4.2-F will be tested again in the next set of experiments.

CONCLUSIONS

Second sound shock pulses comparable to those of previous researchers were successfully transmitted and received in a He II bath at 1.7 K, using a known heater design, and a common type of pencil lead graphite in a novel thermometer design. The heater and thermometer can be described by lumped parameter first order ordinary differential equations for the electrical as well as the thermal case. The time constants in both cases must be considered with respect to the required minimum pulse duration to arrive at a correct design for the heater and thermometer. The internal resistance of the power pulse source determines the SSS heater resistance, and thus, the heater thickness. The required initial power flux density then determines the heater active area, and thus, the SSS tube cross-sectional area. The required maximum pulse duration determines the tube length. The thermal time constant and Biot number of the thermometer or the heater are determined by their respective thickness. The electrical time constant is determined by the mean operating resistance of each, the thermometer and heater. The self-heating of the SSS thermometer is determined by its volume.

ACKNOWLEDGMENTS

This research is supported by U. S. Department of Energy Grant No. DE-FG02-96ER40751. The authors thank Keith Bartholomew, Powell Barber, Andrew Powell, Soren Prestemon, Robert Goddard, Deborah Hilton, and Dr. James Brooks for their technical support, and Dr. Robert Gammon of the University of Maryland for discussions concerning graphite thin films.

REFERENCES

1. S. K. Nemirovskii, A. N. Tsoi, Transient thermal and hydrodynamic processes in superfluid helium, *Cryogenics*, Vol. 29 (1989), pp. 985-994.
2. S. W. Van Sciver, "Cryogenics," Plenum Press, New York (1986).
3. T. N. Turner, Using second-sound shock waves to probe the intrinsic critical velocity of liquid helium II, *Phys. Fluids*, Vol. 26, No. 11 (1983), pp. 3228-3241.
4. J. R. Pellum, Investigations of pulsed second sound in liquid helium II, *Phys. Rev.*, Vol. 75, No. 8 (1949), pp. 1183-1194.
5. T. Shimazaki, T. Iida, M. Murakami, Experimental study of thermal shock wave deformation and decay due to tangled mass of quantized vortices in He II, in: *Advances in Cryogenic Engineering, Vol. 39*, Plenum Press, New York (1994), pp. 1859-1864.

STEADY STATE HEAT TRANSFER CHARACTERISTICS IN HE I AND HE II WITH A COPPER SURFACE

A. Iwamoto, R. Maekawa, T. Mito and S. Satoh

National Institute for Fusion Science
322-6 Oroshi, Toki, Gifu 509-5292, Japan

ABSTRACT

Steady-state heat transfer in He I and saturated He II has been measured. The bath temperatures are 4.20 K, 1.90 K and 1.78 K in saturated pressure. The sample is an oxygen free copper cylinder 20mm in diameter. It is mounted in a vacuum chamber for thermal insulation with an exposed surface facing upward into the He coolant. The heat transfer surface is polished by 0.5 μm alumina powder or oxidized by the chemicals. A thermofoil heater is attached on the opposite side of the heat transfer surface. The surface temperature is estimated by a temperature gradient in the cylinder. The temperatures are measured by calibrated germanium resistance temperature sensors. Dependence of the heat transfer characteristics from the copper surfaces to He I and He II on the surface treatment are discussed.

INTRODUCTION

Heat transfer in the non-boiling region from a metal to He II depends on Kapitza conductance. Many studies have been conducted for various surfaces.[1-4] Generally, for large sized superconductors, the conductor surface is copper or aluminum because of the stabilizer. But, in some cases, some surface treatment is applied to improve the heat transfer. It has been already reported that surface treatment influences the Kapitza conductance.[1,4] There are still surface treatments which have not been studied on the Kapitza conductance. The study of the dependence of the Kapitza conductance on surface treatment is technically important for the stability estimation of superconductors.

The oxidation was applied on the superconductor surface of the helical coil for the large helical device (LHD) in National Institute for Fusion Science (NIFS).[5] The chemicals are used for the oxidation. The helical coil would be cooled by He II in the future.[6]

In the present work, we are concerned with the chemically oxidized surface. The Kapitza conductance of the chemically oxidized surface is measured compared with that of a polished copper surface. At the same surface preparation, heat transfer at 4.2 K is also measured. Dependence of the heat transfer characteristics of both He I and He II on the surface treatments, namely the polished and the oxidized surface, is studied.

EXPERIMENTAL PROCEDURE

A diagram of the present sample for heat transfer measurement is shown in Fig. 1. The sample, which is an oxygen free copper cylinder (RRR = 500) with 70 mm in length and 20 mm in diameter, is mounted in a vacuum can. The can is made of stainless steel. The exposed surface is facing upward to the bath of He I or saturated He II. To minimize the heat leak to the bath via the can, the cylinder is attached to an FRP flange with low thermal conductivity, and then the flange is installed to a stainless steel flange which is a part of the can. Indium was used for the seal between the flanges. The gap between the cylinder and the FRP flange was filled up by vacuum grease. A schematic illustration of the experimental apparatus is shown in Fig. 2. The can is connected to the high vacuum system at room temperature. It was evacuated through a pipe. During measurements, a vacuum of 8×10^{-4} Pa was achieved.

Two cylinders were machined from a copper block. Two different surface preparations, polish and oxidation, were applied for each cylinder as the followings.

1) The surface was polished by 2000-grade emery paper and further finished by alumina powder of 0.5 μm with water, and then wiped by alcohol.
2) To oxidize a surface, the chemicals whose principal ingredient is NaOH was applied. At first, the surface was polished by 600-grade emery paper. For the chemical reaction, the chemicals in a beaker was heated around 100 °C on a hotplate, then the surface was soaked in it for 8 minutes. After the reaction, the chemicals was rinsed away by water. Just before putting into the cryostat, the surface was wiped by alcohol.

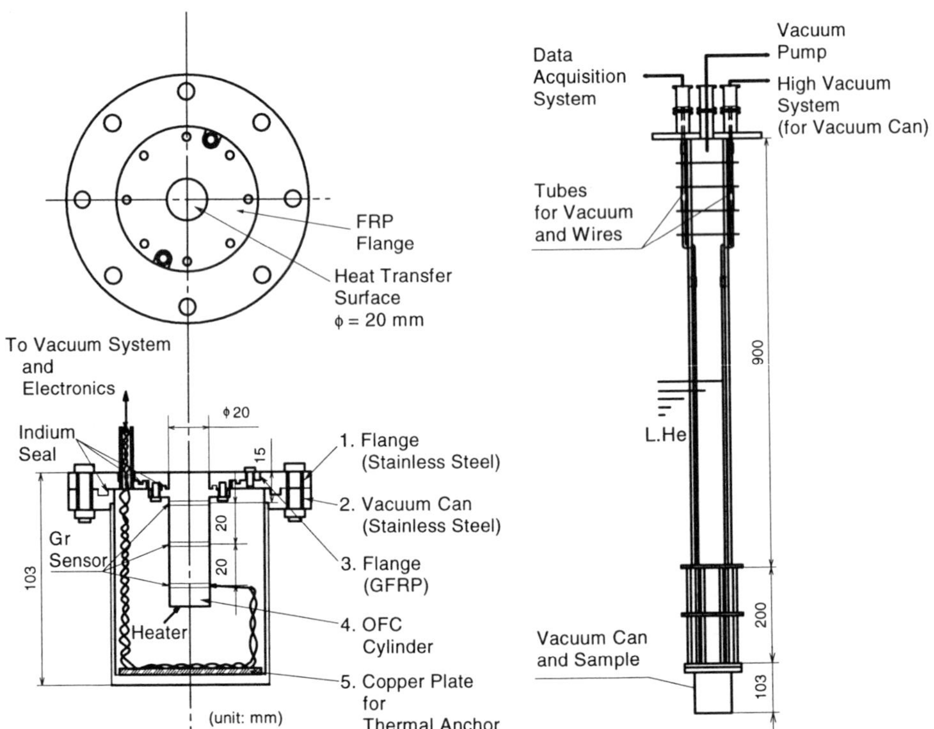

Figure 1. Diagram of sample.

Figure 2. Schematic illustration of experimental apparatus.

Within 4 hours after every surface treatment, the assembled sample was put into the cryostat, and then the cryostat was evacuated to a vacuum. After that, helium gas was filled in the cryostat.

The bath temperatures are 4.20 K, 1.90 K and 1.78 K. At first, the measurement at 4.20 K was carried out, and then liquid helium was transferred and cooled down to 1.78 K (for the polished and the oxidized surface) or 1.90 K (only for the polished surface). The vacuum system with a throttle valve kept the He II bath temperature constant. The bath temperature was verified by two calibrated germanium resistance thermometers. The hydrostatic head of the saturated He II was 350 – 200 mm during the measurements.

The sample with each surface treatment was cooled down a few times as the followings.

1) For the polished surface, the measurements at 1.78 K were carried out three times, and the sample was warmed up to room temperature between cool-downs. Before the second cool-down, the surface was finished by alumina powder of 0.5 μm and wiped by alcohol again. For each cool-down, the surface preparations were apparently identical. After the measurements at 1.78 K, the sample was warmed up to room temperature and cooled down at 1.90 K.
2) The measurements for the oxidized surface were carried out twice. The sample was warmed up to room temperature and exposed in air between the cool-downs.

The surface temperature is calculated from the temperature gradient in the cylinder. Three calibrated germanium thermometers were used for the temperature measurement. The thermometers, which are placed in holes drilled in the cylinder with Cry-Con Thermal Conductive Grease for good contact, are put at 15mm, 35 mm and 55 mm from the heat transfer surface. To prevent heat leak from room temperature through wires for electronics, a thermal anchor is put on the bottom of the vacuum can.

The sample was heated by a thermofoil heater on the opposite side of the heat transfer surface. The heater was bonded to the cylinder by a kind of epoxy resin. The resistance of the heater is 74.5 ohm at room temperature. Heat flux is calculated from the voltage and current through the heater. The heat flux was inputted up to 1.2 W/cm^2 for He I measurement and 5 W/cm^2 for He II measurement.

RESULTS AND DISCUSSIONS

Observation of Treated Surfaces

The treated surfaces were observed by a scanning electron microscope. The roughness of the polished surface is less than a few μm. On the other hand, the thickness of the oxidation layer is a few μm. The oxidation layer is formed by oxidized copper crystals and seems like velvet. The chemically oxidized surface has different appearance from the surface oxidized by exposing or heating in air. It is expected that the Kapitza conductance is different between two kinds of the oxidation treatments.

Surface Temperature Calculation

A typical experimental result of the temperatures and the pressure in the cryostat is shown in Fig.3. The sample is the polished copper surface. The surface heat flux is 148 mW/cm^2. The bath temperature is 1.78 K. At zero heat flux, the temperatures in the cylinder are nearly equal to the bath temperature. According to the temperature measurement, the thermal anchor is effective to prevent the heat leak from the room temperature. Figure 4 shows the temperature distribution in the cylinder and the calculated surface temperature for the heat flux of 148 mW/cm^2. The temperature gradient in the cylinder is a linear function, so that thermal conductivity in the cylinder can be assumed constant at the surface temperature calculation. The surface temperature is calculated by extrapolating the three measured values.

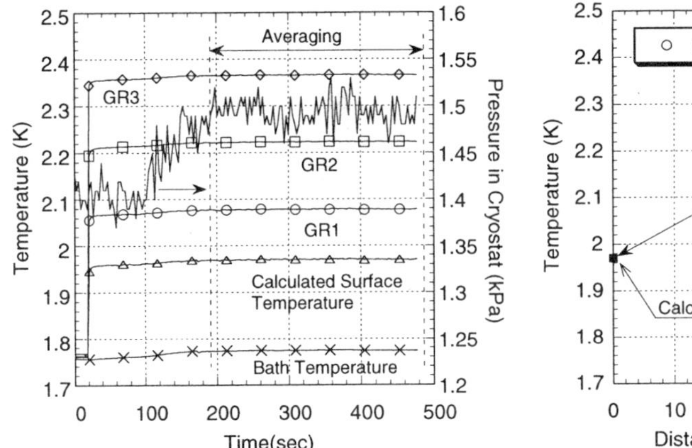

Figure 3. A typical measurement of temperatures.

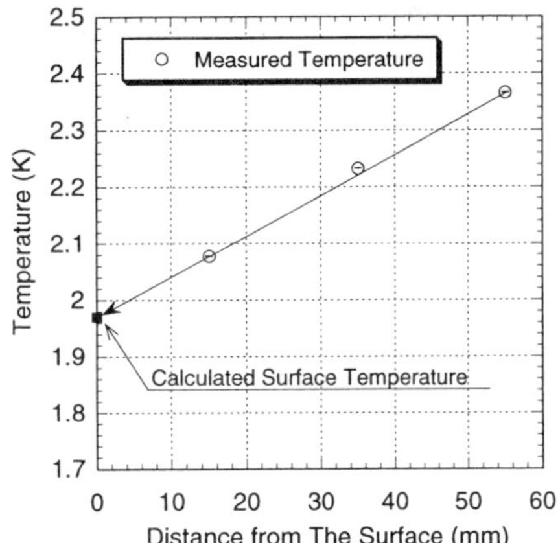

Figure 4. Surface temperature calculation.

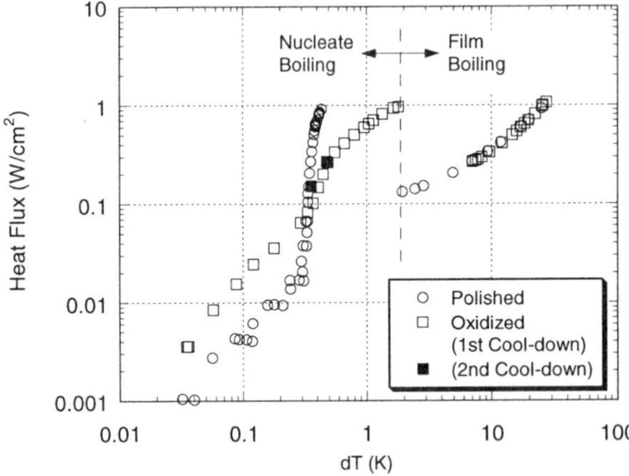

Figure 5. Heat transfer features for the polished and the oxidized surface at 4.20 K of bath temperature.

He I Heat Transfer

Figure 5 shows the experimental results of the heat transfer with the polished and the oxidized surface. In the nucleate boiling region, the oxidized surface has larger temperature difference between the surface and the bath than the polished surface. Especially at high heat flux, the difference becomes large. At the critical heat flux (CHF), the temperature differences with the polished and the oxidized surface are 0.434 K and 1.80 K, respectively. This large temperature difference is caused by not the thermal resistance in the oxidation film but boiling condition on the surface because, at low heat flux, the oxidized surface has lower temperature difference than the polished surface. The CHF with the oxidized surface is 0.97 W/cm^2. The polished surface has less CHF of 0.92 W/cm^2 than the oxidized surface. On the other hand, in the film boiling region, heat transfer coefficients is independent of surface treatment. The minimum film boiling heat fluxes (MFBHF) with the polished and the oxidized surface are 0.14 W/cm^2 and 0.27 W/cm^2. It is thought that the surface tension of the oxidation coating is different from that of the polished surface, so that the CHF and MFBHF depend on the surface treatments.

He II Heat Transfer

Kapitza Conductance of a Polished Copper Surface. The experimental results of Kapitza conductance of the polished copper surface are shown in Figs. 6(a) and (b). The bath temperatures are 1.78 K and 1.90 K. The experimental results are consistent with the other studies.[1,3] The variation of the surface temperatures among the cool-downs at 1.78 K is negligibly small. Judging from the measurements, the surface preparations by 0.5 μm alumina powder polish were practically the same for every cool-down.

Dependence of Kapitza Conductance on Oxidation. The experimental results of the Kapitza conductance with the oxidized surface at 1.78 K is shown in Fig.7, compared to that of the polished surface. The appearance of the oxidized surface was apparently identical for two cool-downs. Nevertheless, the experimental results for the runs are different. At the heat flux of 1.69 W/cm^2, the surface temperatures are 4.86 K for the first cool-down and 5.29 K for the second cool-down. As the heat flux increases, the difference between the surface temperatures becomes large. It is thought that the surface condition was changed because of exposing in air and wiping by alcohol between the cool-downs.

Generally, heat flux in the Kapitza regime is written:

$$q = a\left(T_s^n - T_b^n\right) \tag{1}$$

where a and n are fitting parameters, and T_s and T_b are the surface and the bath temperature, respectively. The fitting parameters of the Kapitza conductance for the polished and the oxidized surface are given in Table 1.

The oxidized surface has lower Kapitza conductance than the polished surface. The

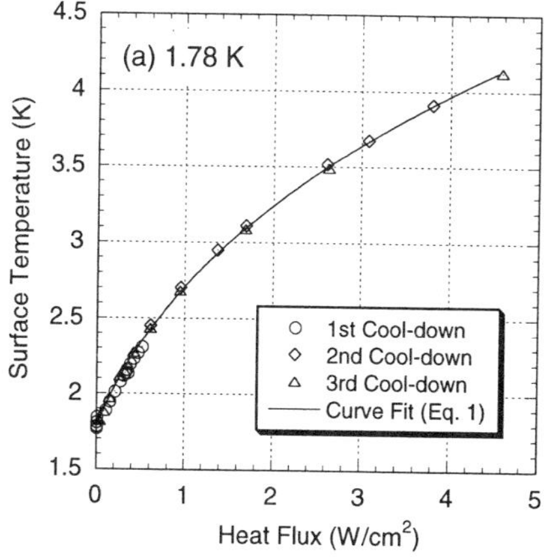

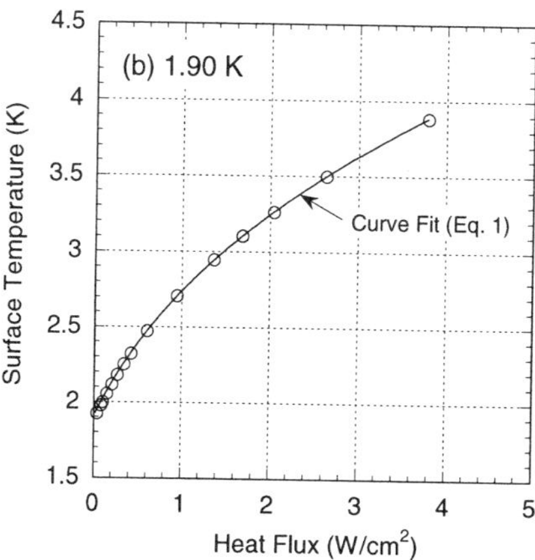

Figure 6. Surface temperature versus heat flux for the polished surface at 1.78 K and 1.90 K of bath temperature.

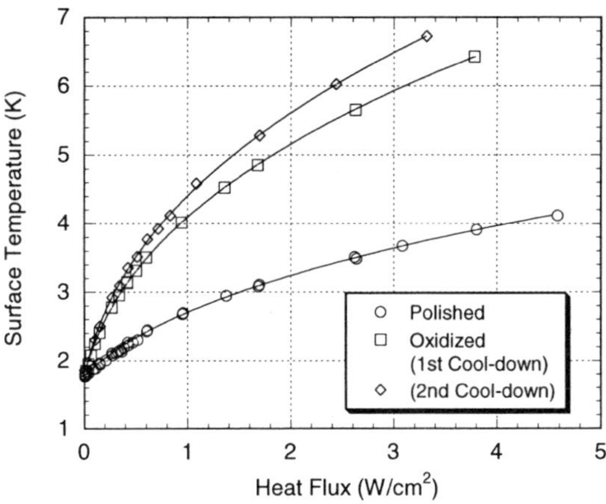

Figure 7. Surface temperature variation for the polished and the oxidized surface at 1.78 K of bath temperature.

Table 1. Fitting parameters of the Kapitza conductance at 1.78 K and 1.90 K

Sample	*a*		*n*	
	at 1.78 K	at 1.90 K	at 1.78 K	at 1.90 K
Polished by 0.5 μm alumina powder	0.07016		3.001	
		0.07346		2.996
Oxidized (the first cool-down)	0.02170		2.789	
Oxidized (the second cool-down)	0.02039		2.685	

Kapitza conductance of the oxidized surface is close to that of the surface coated by varnish.[1] By the way, it was already reported that the surface oxidized in air has higher Kapitza conductance than a polished surface.[1,3] Our experimental result contradicts the other studies. The resistance of the oxidized surface is a sum of three resistances: the copper–oxidation layer interface resistance, the thermal resistance in the oxidation layer and the Kapitza resistance. The resistance of the copper-oxidation layer interface may be independent of the oxidation method. The oxidation layer is thin enough to ignore the thermal resistance compared to other terms. It is thought that the chemically oxidized surface has higher Kapitza resistance of the oxidation layer-He II interface than the surface oxidized by exposing or heating in air.

CONCLUSIONS

The heat transfer from the polished and the oxidized surface to He I and saturated He II was measured. Dependence of the heat transfer characteristics on surface treatments was discussed.

For He I heat transfer, the critical heat flux and the minimum film boiling heat flux depend on surface treatments because of different surface tensions between the two surface treatments. In nucleate boiling region, the oxidized surface has larger temperature difference than the polished surface. This large temperature difference is caused by not the thermal resistance in the oxidation film but boiling condition on the surface.

The variation of the Kapitza conductance measurements on the polished surface among the cool-downs is negligibly small. The Kapitza conductance of the 0.5 μm alumina powder polished surface almost agree with the results obtained by other studies.

For the oxidized surface, the surface appearance for two cool-downs is apparently identical. Nevertheless, the Kapitza conductances are slightly different between two runs.

The surface oxidized by the chemicals has different appearance from the surface oxidized by exposing or heating in air. The chemically oxidized surface may have higher Kapitza resistance of the oxidation layer-He II interface than the surface oxidized by exposing or heating in air, so that the Kapitza conductance of the chemically oxidized surface is lower than that of the polished surface.

ACKNOWLEDGMENT

The authors wish to thank K. Kikuchi and S. Inaba from Hitachi-Cable, Ltd. for offering the chemicals for the surface coating of the oxidation.

REFERENCES

1. A. Kashani and S. W. Van Sciver, High heat flux Kapitza conductance of technical copper with several different surface preparation, *Cryogenics* 25:238 (1985).
2. S. W. Van Sciver, Kapitza conductance of aluminum and heat transport through subcooled He II, *Cryogenics* 18:521 (1978).
3. G. Claudet and P. Seyfert, Bath cooling with subcooled superfluid helium, in: "Advances in Cryogenic Engineering Vol.27," Plenum Press, New York (1981), p.441
4. E. A. Jones and J. C. van der Sluijs, Some experiments on the influence of surface treatment on the Kapitza conductance between copper and He^4 at temperatures from 1.2 K to 2.0 K, *Cryogenics* 13:535 (1973)
5. N. Yanagi, et al., Development, fabrication, testing and joints of aluminum stabilized superconductor for the helical coils of LHD, in: "Proceedings of ICEC16/ICMC," T. Haruyama, T. Mitsui and K. Yamafuji, ed., Elsevier Science, Tokyo (1997), p.751.
6. O. Motojima, et al., Physics and engineering design studies on the Large Helical Device, *Fusion Engineering and Design* 20:3 (1993)

ICMC Chairman, T. Hartwig (Texas A & M University), addresses the CEC-ICMC Awards Luncheon.

Seated from left to right: K. Timmerhaus (Universtiy of Colorado), S. Breon (Goddard Space Flight Center), Q.S. Shu (AMAC International), R. Witt (University of Wisconsin), V. Bardos (Synchrony), A. Zeller (Michigan State University), T. Hartwig (Texas A & M University), and P. Kelley (Los Alamos National Lab).

FILM BOILING ON A HORIZONTAL CYLINDER IN SATURATED AND SUBCOOLED HELIUM II

K. Hama[1], K. Hata[1] and M. Shiotsu[2]

[1]Institute of Advanced Energy, Kyoto Univ.
Uji, Kyoto 611-0011, Japan

[2]Dept. of Energy Science & Technology, Kyoto Univ.
Uji, Kyoto 611-0011, Japan

ABSTRACT

Film boiling heat transfer coefficients on horizontal cylinders with the diameters of 0.08 and 0.5 mm were measured under saturated conditions for the liquid temperatures of 1.8, 1.9, 2.0 and 2.1 K (He IIs), and under subcooled conditions for the same liquid temperatures at atmospheric pressure (He IIp). At the maximum heat flux, q_{max}, of Kapitza conductance regime in He IIs, the heat transfer state changes from Kapitza conductance regime to film boiling regime with a jump of the heater surface temperature. On the other hand, the heat transfer changes to film boiling with little jump of the heater surface temperature at q_{max} in He IIp. The film boiling heat transfer coefficients under saturated conditions are dependent on the pressure at the level of the test cylinder, surface superheat and bulk liquid temperature.

A model for film boiling heat transfer from a horizontal cylinder in He II based on latent heat transport, non-equilibrium kinetic effects at the vapor-liquid interface and the heat transport in the superfluid is presented. An equation of film boiling heat transfer was obtained based on the numerical solutions of the model and experimental data. The equation can predict the experimental data of film boiling heat transfer coefficient under saturated and subcooled conditions within ±20 % difference. The fact that the surface temperature jump occurs at q_{max} in He IIs, but does not occur at q_{max} in He IIp can be explained by the model.

INTRODUCTION

The knowledge of film boiling phenomena in saturated and subcooled He II is important for the analysis of instability for He-II-cooled superconducting magnets caused by local thermal disturbances in the windings.

Many workers (Haenssler and Rinderer[1], Frederking[2], Lemieux and Leonard[3], Leonard[4], Steed and Irey[5], and Shiotsu et al.[6]) experimentally investigated film boiling heat transfer from a horizontal cylinder in He IIs. However, there have been few experimental data of film boiling heat transfer in He IIp until recently. Shiotsu et al.[7] measured the film boiling heat transfer on a 0.08-mm-diameter horizontal cylinder in He II at atmospheric pressure. They observed that the heat transfer state changed from Kapitza conductance regime to film boiling with little jump of the heater surface temperature when the heat flux reached the maximum heat flux, q_{max}, of Kapitza conductance regime.

Advances in Cryogenic Engineering, Volume 45.
Edited by Shu *et al.*, Kluwer Academic / Plenum Publishers, 2000.

On the other hand, some workers presented analytical models of film boiling heat transfer in He II. Rivers and McFadden[8] analyzed the problem by using standard boundary-layer-type model. However, they could not predict the film boiling heat transfer due to the lack of knowledge on the heat flux at the vapor-liquid interface. Labuntzov and Ametistov[9] presented a model based on non-equilibrium molecular kinetic considerations. However, their correlation was applicable to the liquid temperatures below 1.95 K, because they could not analytically evaluate the liquid temperature at the vapor-liquid interface and it was evaluated empirically. Shiotsu et al.[6] revised the model given by Labuntzov and Ametistov to evaluate the liquid temperature at the vapor-liquid interface using their heat transfer equation from a cylinder in He II based on Gorter-Mellink equations. They reported that the model can describe their and other workers' experimental data for He IIs on horizontal cylinders with the diameters ranging from 80 μm to 0.254 mm in a surface superheat range below 250 K for the bulk liquid temperatures from 1.4 to 2.15 K. However, their model neglecting the convection heat transfer in the vapor phase would not be applicable for higher surface superheat. There have been no theoretical model for film boiling heat transfer in He IIp as far as the authors are concerned.

The purposes of the present study are fourfold. First is to obtain the experimental data of film boiling heat transfer from 0.08 and 0.5 mm-diameter horizontal cylinders in He IIs and He IIp for the liquid temperatures from 1.8 to 2.1 K. Second is to present a film boiling model for saturated and subcooled He II by extending the model by Shiotsu et al.[6] to consider the convection heat transfer in the vapor phase. Third is to present a film boiling heat transfer equation based on the numerical solutions for the model and the experimental data. Fourth is to make clear the availability of the equation by comparing the predicted values with the experimental data.

EXPERIMENTAL APPARATUS AND METHOD

The test heaters used are 0.08 and 0.5 mm in diameter and about 60 mm in length, made of an Au-Mn(0.25 Weight %) alloy. The gradient of the electrical resistivity versus temperature curve for this alloy is positive for the entire experimental range from around 1.8 K to several hundred Kelvin. The test cylinders were annealed in an inert atmosphere and their electrical resistance versus temperature relations were calibrated in a temperature-controlled vessel by immersion in liquid helium and in liquid nitrogen.

The heating current to the test heater was supplied by a power amplifier which can supply a direct current up to 300 A at a power level of 6 kW. The input signal of the power amplifier was controlled by an analog computer so that the heat generation rate in the test heater agrees with a desired heat input regardless of the variation in the electrical resistance of the test heater due to its temperature change.

The average temperature of the test heater was measured by resistance thermometry. A double bridge circuit including the test heater as a branch of the bridge was used to measure the electrical resistance of the test heater during its heating. The output voltages of the bridge circuit were amplified and passed to a digital memory system. The average temperature of the test heater was calculated from its electrical resistance with the aid of the previously calibrated temperature-resistance relation.

The surface temperature of the test heater was obtained from the measured average temperature and heat generation rate by solving the conduction equation for the test heater with the heat flux set as a boundary condition at the surface. The thermal conductivity of the test heater that becomes lower with the decrease in temperature was approximately taken to be that for the average temperature in calculating the surface temperature. Experimental error is estimated to be ± 0.1 K for the surface temperature of the cylinder and ± 2 percents for the heat flux.

EXPERIMENTAL RESULTS

Heat Transfer Characteristics in He IIs and He IIp

Figures 1(a) and 1(b) show the typical examples of the heat transfer characteristics for He IIs and He IIp, respectively.

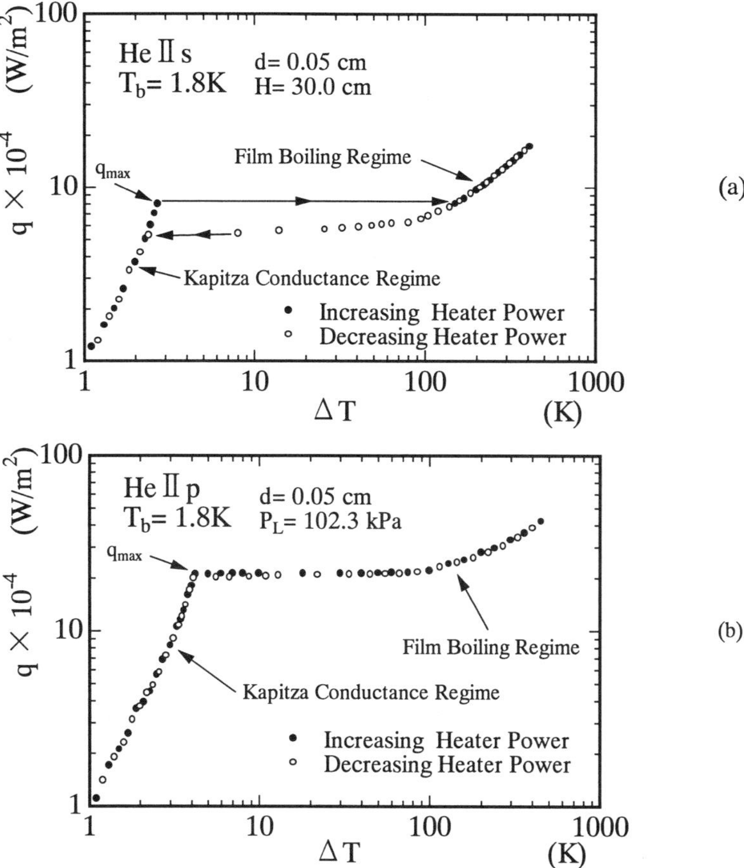

Fig. 1 Steady-state heat transfer from a 0.5 mm-diameter horizontal cylinder to (a) He IIs and (b) He IIp.

In the saturated condition, the heat flux gradually increases up to a value called the maximum heat flux, q_{max}, where the heater surface temperature rapidly jumps from the non-boiling heat transfer regime to the film boiling regime as shown in Fig. 1(a). Film boiling curve in the figure was obtained by quasi-steadily increasing the heat input afterwards and then decreasing it down to the minimum film boiling point at which the transition from the film boiling to Kapitza conductance regime occurs.

Under atmospheric pressure, the heat transfer state changes at q_{max} from Kapitza conductance regime to film boiling regime without a jump of the heater surface temperature as shown in Fig. 1(b). No hysteresis was observed at q_{max} for increasing and decreasing the heat flux.

Film Boiling Heat Transfer Coefficients

The average film boiling heat transfer coefficients, h, were measured for bulk liquid temperatures, T_b, from 1.8 to 2.1 K under saturated and subcooled conditions. Under saturated condition, liquid head above the test heater was varied from around 10 to 30 cm.

These data for 0.08 and 0.5 mm-diameter cylinders are shown versus surface superheat, $\Delta T_{sat}(P_L)(=T_w - T_{sat}(P_L))$, for bulk liquid temperatures of 1.8, 1.9, 2.0 and 2.1 K in Figs. 2 to 5, respectively. Solid symbols show the data for 0.5 mm-diameter cylinder and open symbols show those for 0.08 mm-diameter cylinder. Here P_L is the pressure at the level of the test heater axis ($P_L=P+\rho gH$), T_w is the surface temperature of the test heater and $T_{sat}(P_L)$ is the saturation temperature corresponding to P_L. It can be seen from these figures that the heat transfer coefficients are higher for smaller diameter cylinder, lower value of T_b and higher value of P_L. The heat transfer coefficients for a fixed value of P_L on a cylinder

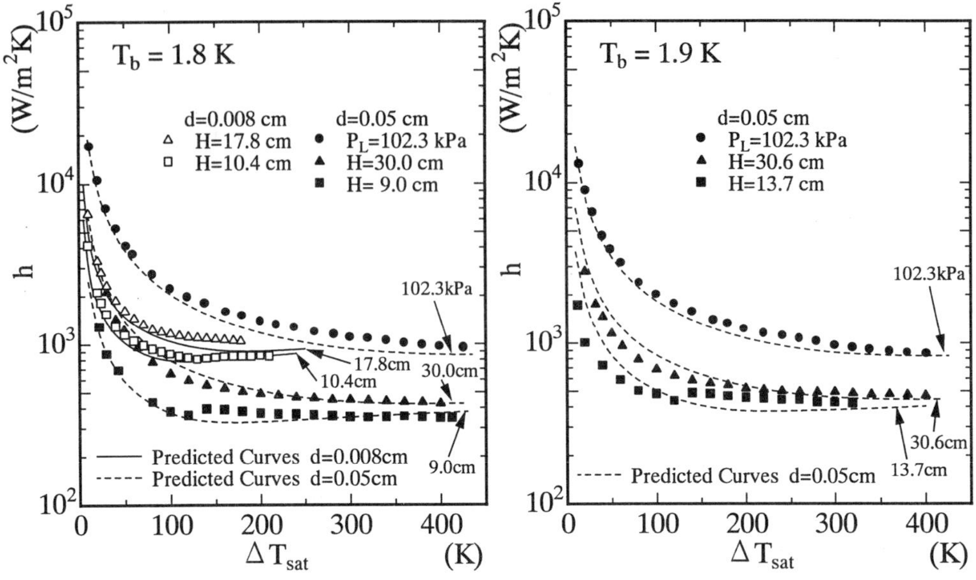

Fig. 2 Average heat transfer coefficients versus heater surface superheat for T_b=1.8 K.

Fig. 3 Average heat transfer coefficients versus heater surface superheat for T_b=1.9 K.

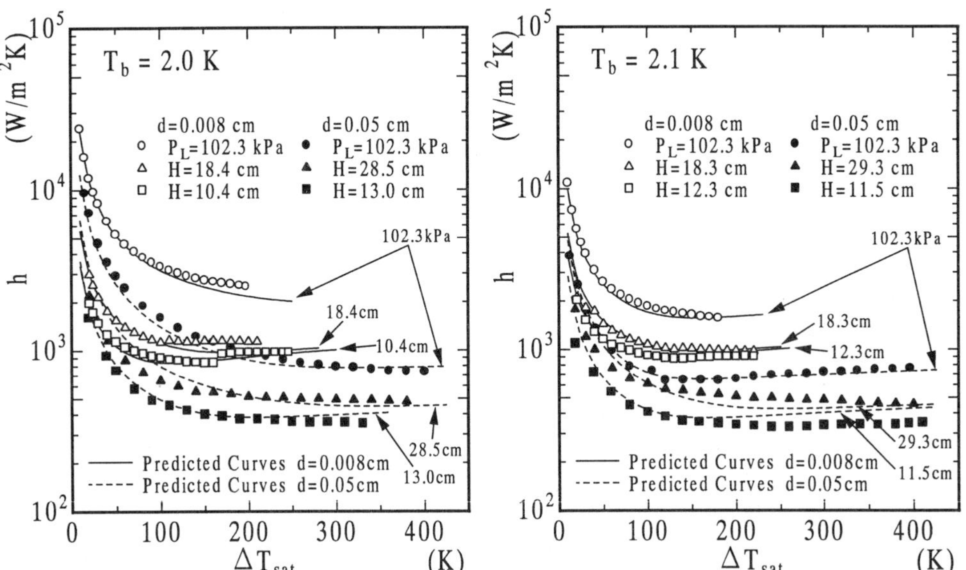

Fig. 4 Average heat transfer coefficients versus heater surface superheat for T_b=2.0 K.

Fig. 5 Average heat transfer coefficients versus heater surface superheat for T_b=2.1 K.

significantly decrease at first and then become almost constant with the increase in the surface superheat from that near the minimum film boiling point. The first decreasing range becomes wider to higher ΔT_{sat} with the increase in P_L.

A FILM BOILING MODEL

A model for film boiling on a horizontal cylinder in He II is considered based on the following assumptions:

1) Surface temperature of the cylinder, T_w, is uniform and the thickness of the vapor film around the cylinder is far thinner than the radius of the cylinder.
2) In case of saturated film boiling ($T_{sat}(P_L)\le T_\lambda$), a temperature gap exists at the vapor liquid interface due to the non-equilibrium vaporization and condensation as suggested by Labuntzov and Ametistov[9].
3) In case of film boiling in pressurized He II ($T_{sat}(P_L)>T_\lambda$), thickness of liquid He I layer outside vapor layer is negligible.
4) The convection, which occurs within the film is steady, is two-dimensional and boundary layer type flow.

According to these assumptions, the model for He II film boiling on a horizontal cylinder is schematically illustrated in Fig. 6. The laws of conservation of mass, momentum and energy are expressed as follows in the form of boundary layer equations.

$$\frac{\partial u}{\partial x}+\frac{\partial v}{\partial y}=0 \tag{1}$$

$$u\frac{\partial u}{\partial x}+v\frac{\partial u}{\partial y}=\frac{1}{\rho_v}g(\rho_b-\rho_v)\sin\frac{x}{r}+\frac{\mu_v}{\rho_v}\frac{\partial^2 u}{\partial y^2} \tag{2}$$

$$u\frac{\partial T}{\partial x}+v\frac{\partial T}{\partial y}=\frac{k_v}{\rho_v c_{pv}}\frac{\partial^2 T}{\partial y^2} \tag{3}$$

The boundary conditions are expressed in the following form: At the cylinder surface (y=0),

$$u=v=0 \tag{4}$$

$$T=T_w \tag{5}$$

At the vapor-liquid interface $(y=\delta)$,

$$u=u_\delta \tag{6}$$

$$v=v_\delta \tag{7}$$

$$T=T_\delta \tag{8}$$

$$\rho_b v_b=\rho_v(v_\delta-u_\delta\frac{d\delta}{dx}) \tag{9}$$

$$\mu_v\left(\frac{\partial u}{\partial y}\right)_\delta=\rho_b v_b u_\delta \tag{10}$$

$$k_v\left(\frac{\partial T}{\partial y}\right)_\delta=\rho_b v_b L-q_b \tag{11}$$

Boundary layer equations (Eqs. (1) to (3)) were transformed into the following integrodifferential form by first integrating the partial differential equations from 0 to δ and then substituting the interface boundary conditions into the result.

$$\rho_v\frac{d}{dx}\int_0^\delta u\,dy=-\rho_b v_b \tag{12}$$

$$\rho_v\frac{d}{dx}\int_0^\delta u^2dy=g(\rho_b-\rho_v)\int_0^\delta \sin\frac{x}{r}dy-\mu_v\left(\frac{\partial u}{dy}\right)_w \tag{13}$$

$$\rho_v c_{pv}\frac{d}{dx}\int_0^\delta u(T-T_\delta)dy=\rho_b v_b L-q_b-k_v\left(\frac{\partial T}{\partial y}\right)_w \tag{14}$$

The velocity and temperature profiles are approximately given by the following fourth degree polynomials.

$$u(x,y)=a_1+a_2y+a_3y^2+a_4y^3+a_5y^4 \tag{15}$$

$$T(x,y)=b_1+b_2y+b_3y^2+b_4y^3+b_5y^4 \tag{16}$$

Following boundary conditions were introduced to solve the problem as done by Rivers and McFadden[8].

At y=0,

$$\mu_v\left(\frac{\partial^2 u}{\partial y^2}\right)_w = -g(\rho_b - \rho_v)\sin\frac{x}{r} \tag{17}$$

$$k_v\left(\frac{\partial^2 T}{\partial y^2}\right)_w = 0 \tag{18}$$

At $y=\delta$,

$$\mu_v^2\left(\frac{\partial^2 u}{\partial y^2}\right)_\delta = \rho_b^2 v_b^2 u_\delta - \mu_v g(\rho_b - \rho_v)\sin\frac{x}{r} \tag{19}$$

$$k_v^2\left(\frac{\partial^2 T}{\partial y^2}\right)_\delta = c_{pv}\rho_b v_b\left(\rho_b v_b L - q_b\right) \tag{20}$$

Expressions for the coefficients a_i and b_i are obtained as functions of x by applying the boundary conditions of u and T in the y direction.

The heat flux removed from the vapor-liquid interface to the liquid, q_b, is given dependent on the conditions of a) and b) as follows.

a) Saturated film boiling, where $T_{sat}(P_L)\leq T_\lambda$

In this case, the liquid adjacent to the vapor-liquid interface is He II. Labuntzov and Ametistov[9] suggested that film boiling in He II is a non-equilibrium process involving vaporization and condensation at the vapor-liquid interface. Namely, temperature gap exists between the vapor and liquid at the interface. Let the interface temperature of liquid is T_i, the heat flux removed from the interface to He II, q_b, is given by

$$q_b = 2.27\left[P_L - P_{sat}(T_i)\right]\sqrt{2RT_i} \tag{21}$$

where $P_{sat}(T_i)$ is the saturation pressure corresponding to the liquid temperature at the interface, and R is the gas constant for helium (R=2079 J/kg K). On the other hand, Shiotsu et al.[7] and Sakurai et al.[10] reported that the maximum heat fluxes, q_{max}, on horizontal cylinders in saturated and subcooled He II are well described by the following equation.

$$q_{max} = 0.58\left(\frac{2}{r}\int_{T_b}^{T^*} f(T)^{-1}dT\right)^{\frac{1}{3}} \tag{22}$$

where $T^* = T_{sat}(P_L)$ for $T_{sat}(P_L)\leq T_\lambda$, $T^* = T_\lambda$ for $T_{sat}(P_L)>T_\lambda$

$$f(T)^{-1} = g(T_\lambda)\left[T_R^{6.8}\left(1-T_R^{6.8}\right)\right]^3, \quad g(T_\lambda) = \rho^2 S_\lambda^4 T_\lambda^3 / A_\lambda$$

$T_R=T/T_\lambda$, S_λ=1559 J/(kg K), A_λ=1150 ms/kg

It is assumed from this result that the heat flux, q_b, from the interface whose liquid temperature is T_i will be given by the following equation, when the thickness of vapor layer is neglected based on the initial assumption 1):

$$q_b = 0.58\left(\frac{2}{r}\int_{T_b}^{T_i} f(T)^{-1}dT\right)^{\frac{1}{3}} \tag{23}$$

It is possible to calculate q_b by finding the value of T_i which satisfies Eqs. (21) and (23) simultaneously.

b) Pressurized film boiling, where $T_{sat}(P_L)>T_\lambda$

In this case, the vapor layer would be surrounded by a thin He I layer and outside which He II layer would exist. Although the temperature gap may exist at the interface of vapor and He I layers, there is no temperature gap at the interface of He I and He II. The heat flux, q_b, from the vapor-liquid interface is given by the following equation, when the thickness of vapor and He I layers is neglected based on the initial assumptions of 1) and 3):

$$q_b = 0.58\left(\frac{2}{r}\int_{T_b}^{T_\lambda} f(T)^{-1}dT\right)^{\frac{1}{3}} \tag{24}$$

The integral equations, the profile expressions and the boundary conditions described above were non-dimensionalized. The characteristic of the phenomena is dependent on the

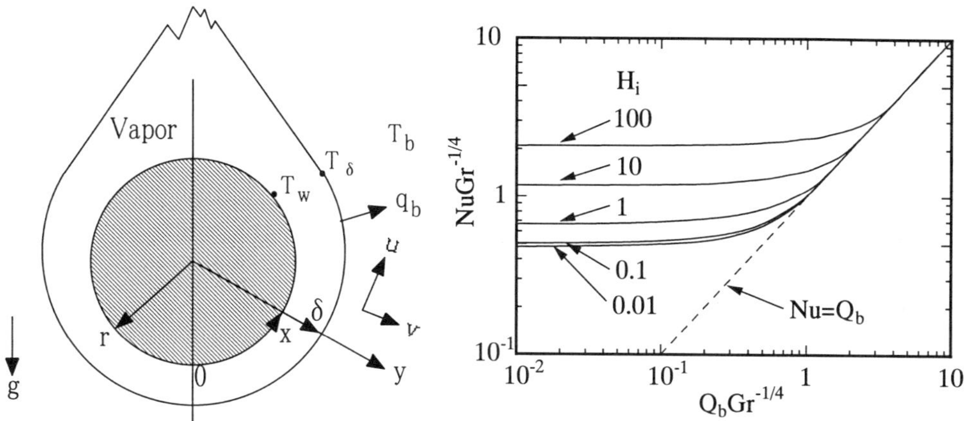

Fig. 6 Physical model and coordinate system.

Fig. 7 Calculated results of average heat transfer Coefficients ($Nu=hd/k_v$).

following five dimensionless parameters: the Prandtl number, $Pr=\mu_v c_{pv}/k_v$, Nusselt number, $Nu=hd/k_v$, Grashof number, $Gr=gd^3\rho_v(\rho_b-\rho_v)/\mu_v^2$, the "interface entahalpy" parameter, $H_i=L/(c_{pv}\Delta T_{sat})$, and the "interface heat flux" parameter, $Q_b=dq_b/(k_v\Delta T_{sat})$. The Q_b and Gr always appear together in the form of $Q_bGr^{-1/4}$. Using the profile expressions with the formulated coefficients a_i and b_i, the integral equations were reduced to three first-order ordinary differential equations. Numerical solutions for the differential equations were obtained for the $Q_bGr^{-1/4}$ ranging from 0.01 to 10 by using Runge-Kutta method. Calculated results are shown in Fig. 7 on $NuGr^{-1/4}$ versus $Q_bGr^{-1/4}$ graph with H_i as a parameter. It seems that the results for a fixed value of H_i are very weakly dependent on Q_b for $Q_bGr^{-1/4}$ lower than 0.1, and they approach a line of $Nu=Q_b$ with increasing $Q_bGr^{-1/4}$. The former is the convection dominant region, and the latter is the conduction dominant region where q_b is large and film thickness is small. The results for the H_i ranging from 0.01 to 100 obtained here were expressed within ±5 % difference by the following equation.

$$NuGr^{-\frac{1}{4}} = \left[\left(Q_bGr^{-\frac{1}{4}}\right)^7 + \left[0.728\{(H_i+0.25)Pr\}^{\frac{1}{4}} + 0.24\left(Q_bGr^{-\frac{1}{4}}\right)\right]^7\right]^{\frac{1}{7}} \tag{25}$$

It was found by comparing the experimental data and the corresponding values given by Eq. (25) that this equation predicts well the experimental data for lower ΔT_{sat} range but under-predicts the data with the increase in ΔT_{sat}. This equation was based on the numerical solutions for the boundary layer film-boiling model, which neglected the radial dissipation of heat flow based on the assumption of negligible vapor film thickness in comparison with the radius of the cylinder. With increasing the ΔT_{sat}, vapor film becomes thicker and the ratio of vapor film thickness to the cylinder radius becomes higher. The assumption of negligible vapor film thickness may become inappropriate for the higher ΔT_{sat} regime and that would be the cause of the under-prediction. The Eq. (25) was modified as follows to account for the increase of vapor film thickness.

$$\frac{Nu}{1+\dfrac{2}{Nu}} = Gr^{\frac{1}{4}}\left[\left(Q_bGr^{-\frac{1}{4}}\right)^7 + \left[0.728\{(H_i+0.25)Pr\}^{\frac{1}{4}} + 0.24\left(Q_bGr^{-\frac{1}{4}}\right)\right]^7\right]^{\frac{1}{7}} \tag{26}$$

The compensation factor introduced in the denominator of the left-hand side of Eq. (26) would be approximately expressing the ratio of the vapor-liquid interfacial area to the surface area of the cylinder. This is because, if the temperature distribution in the vapor layer of thickness, δ, is linear, $1+2/Nu=(d+2\delta)/d$.

COMPARISON OF THE EXPERIMENTAL DATA WITH THE PREDICTED VALUES

Experimental data of film boiling heat transfer coefficients for the liquid temperatures of 1.8, 1.9, 2.0 and 2.1 K previously shown in Figs. 2, 3, 4 and 5 are compared with the values predicted from Eq. (26). It can be seen from these figures that all the experimental data on different diameter cylinders in saturated and subcooled He II for the surface superheats from slightly above the minimum film boiling value to the highest one obtained here are in agreement with the predicted values within ±20 %.

It was already shown in Fig. 1 that the surface-temperature-jump occurs at q_{max} in He IIs, but does not occur at q_{max} in He IIp. This is explained by the film boiling model presented here as follows. In the former case, liquid temperature close to the heater surface that is $T_{sat}(P_L)$ at q_{max} suddenly decreases by the value of the temperature gap suggested by Labuntzov and Ametistov[9] with the generation of vapor film. This causes the temperature jump due to the reduction of heat removal by the He II. In the latter case, however, the temperature gap exists at the vapor-He I interface, but does not exist at the He I-He II interface. The highest value in the temperature distribution of He II is still T_λ after the generation of vapor film.

CONCLUSIONS

Film boiling heat transfer coefficients were measured on 0.08 mm- and 0.5 mm-diameter horizontal cylinders in He IIs and He IIp for bulk liquid temperatures ranging from 1.8 to 2.1 K.

The heat transfer coefficients are higher for smaller diameter cylinder, lower value of T_b and higher value of P_L. The heat transfer coefficients for a fixed value of P_L on a cylinder significantly decrease at first and then become almost constant with the increase in the surface superheat from that near the minimum film boiling point. The first decreasing range becomes wider to higher ΔT_{sat} with the increase in P_L.

A model for film boiling on a horizontal cylinder in saturated and pressurized He II based on latent heat transport, non-equilibrium kinetic effects at the vapor-liquid interface and the heat transport in the superfluid was presented. The numerical solutions of the model predict well the experimental data for lower ΔT_{sat} range but under-predict the data with the increase in ΔT_{sat}.

An equation of film boiling heat transfer was presented based on the numerical solutions of the model and experimental data. It was confirmed that this equation can describe the experimental data within ±20 % difference.

REFERENCES

1. F. Haenssler, and L. Rinderer, "Heat Transfer in Superfluid Helium," Helvetica Physica Acta, Vol.32, (1959), p.322.
2. T.H.K. Frederking, "Warmeubergang bei der Verdampfung der Verflussigten Gase Helium und Stickstoff," Forschung, Vol.27 (1961), p.17.
3. G.P. Lemieux, and A.C. Leonard, "Film-Boiling Heat-Transfer Properties of Liquid Helium 2 for a 76.2-mu-dia Horizontal Wire at Depths of Immersion up to 70 cm," ASME paper 67-WA/HT-37 (1967).
4. A.C. Leonard, "Helium-2 Noisy Film Boiling and Silent Film Boiling Heat Transfer Coefficient Values," Proc. of the 3rd International Cryogenic Engineering Conf., West Berlin (1970), p.109.
5. R.S. Steed and R.K. Irey, Correlation of the depth effect on film boiling heat transfer in liquid helium II, in: "Advances in Cryogenic Engineering," Plenum Pub. Corp. New York, Vol.15 (1970), p.299.
6. M. Shiotsu, K. Hata, and A. Sakurai, Film boiling heat transfer from a horizontal wire in He II, in: "Heat Transfer in Phase Change," I.S. Habib et al., ed., ASME HTD.Vol.205 (1992), p.63.
7. M. Shiotsu, K. Hata, and A. Sakurai, Transient heat transfer for large stepwise heat inputs to a horizontal wire in subcooled He II, in: "Advances in Cryogenic Engineering," Plenum Pub. Corp. New York, Vol.37A, (1992), p.37.
8. W.J. Rivers and P.W. McFadden, "Film Free Convection in He II," ASME J. Heat Transfer, Vol.88 (1966), p.343.
9. D.A. Labuntzov, and Ye.V. Ametistov, "Analysis of Helium II Film Boiling," Cryogenics, Vol.19, (1979), p.401.
10. A. Sakurai, M. Shiotsu, and K. Hata, Transient heat transfer for large stepwise heat inputs to a horizontal wire in saturated He II, in: "Advances in Cryogenic Engineering," Plenum Pub. Corp. New York, Vol.37A, (1992), p.25.

STUDY OF SHOCK WAVES AND PHASE TRANSITION IN HE II BY USING A SUPERFLUID SHOCK TUBE FACILITY

H. Nagai, N. Takano, H.S. Yang and M. Murakami

Institute of Engineering Mechanics and Systems,
University of Tsukuba, Tsukuba 305-8573, Japan

ABSTRACT

In the experimental study, we focus on the thermo-fluid dynamics of shock waves and the lambda phase transition, He II to He I, induced by shock compression utilizing the newly developed superfluid shock tube facility. In the experimental facility, a compression shock wave is generated in He II by the impingement of a gas dynamic shock wave propagating through the helium vapor onto a He II free surface. He II can be shock-compressed to convert to He I across the lambda line in the case of He II initially at temperatures rather close to the lambda temperature. The highly transient lambda phase transition is detected very rapidly (a few μsec) by a superconductive temperature sensor, where the temperature variation is positive, in striking contrast to the temperature drop that occurs during He II shock compression without the lambda transition. Photographs of these phenomena are taken by the Schlieren visualization method.

INTRODUCTION

It is well known that superfluid helium (He II) exhibits several unique properties, such as an extremely high apparent thermal conductivity and the ability to flow through even extremely fine channels without any appreciable pressure drop. He II is expected to be an excellent coolant for superconductive magnets and space devices, which need low-temperature environment. A deep understanding not only of steady He II heat transfer but also transient heat transfer is needed for further practical application of He II to cooling. The quench mechanism of a He II cooled superconductive magnet involving a strong pressure oscillation and boiling heat transfer is not fully understood. It is very desirable to understand the highly transient heat transfer characteristics of He II that can lead to the He II-He I lambda transition. For these reasons the experiment utilizing the newly developed superfluid shock tube facility would be of great value.

In our research, we focus on the two modes of shock wave propagation in He II, which are a compression and a thermal shock wave, and on shock induced lambda phase transition. The lambda phase transition, a second order phase transition, can be induced by shock compression in the shock tube if a shock wave is generated in He II at a temperature near the lambda point. When He II converts to He I crossing the lambda transition line, the stable cooling due to super-thermal-conduction of He II would be lost. Accordingly, it is very important to understand highly transient heat transfer accompanied by the lambda phase transition.

Many studies of lambda phase transition have been made, mostly by heating, cooling or a static pressurization of He II[1] We cause He II to He I lambda phase transition by a sudden pressure rise as a result of shock wave compression, which doesn't require heating. In the experiment, a gas dynamic shock wave propagating through saturated helium vapor impinges on a He II free surface, and consequently generates a transmitted compression shock wave (originating from a first sound mode) and a thermal shock wave (from a second sound mode) in He II. The process realized in the shock tube is an extremely high-speed one with a characteristic time of the order of a few μsec and consequently we must treat this as a non-equilibrium phase transition. Accordingly, some advanced experimental technology is necessary to conduct research for this kind of phase transition. We recently developed the superfluid shock tube facility[2] to investigate the two modes of shock wave phenomena in He II. We have succeeded in the detection of a compression shock wave and a thermal shock wave[3]. The high-speed measurement of the temperature depended critically on the development of superconductive temperature sensors. Visualization of both shock waves is also successfully carried out by using the Schlieren visualization method. It is one of the objectives of the present research to make successful transfer of experimental techniques developed in conventional thermo-fluid dynamics to cryogenic research.

EXPERIMENTAL APPARATUS

Superfluid shock tube facility and measurement

Details of the experimental facility were reported at the last CEC[2]. The superfluid shock tube facility consists of two major parts, a diaphragm-free shock tube and a He II cryostat with optical windows. In this paper, the schematic illustration of the He II cryostat section of the facility, which is the most important section, is only shown in Fig.1. The cryogenic shock tube with optical windows for visualization studies is inserted in the cryostat and is maintained at around 2 K at the bottom portion (immersed in He II) and at about 80 K at the top. A LN_2 tank is installed at the top to minimize the conduction heat leak from the normal temperature portion of the cryostat to the 2 K portion of the shock tube. The test section of the cryogenic shock tube is filled with He II and the rest of the low-pressure tube is filled with saturated helium vapor. The cryogenic shock tube has a rectangular cross section (35 mm$\times$25 mm) and is equipped with two pairs of visualization windows with a diameter of 25 mm in the test section. Three piezo-electric pressure transducers are fixed on the wall of the shock tube. A superconductive temperature sensor[2] is located in the He II. The temperature variation resulting both from a compression and a thermal shock waves are measured by a superconductive temperature sensor, which is made of gold-tin thin film vacuum deposited on the side surface of a fine quartz rod with a diameter of 40μm. A characteristics feature of this sensor is its very high sensitivity and rapid time response within a few micro seconds. Any temperature variation as small as 1 mK results in detectable resistance change. Some experimental detail is described in the accompanying paper.[4]

In this experiment, a gas dynamic shock wave is generated in the helium vapor by using a diaphragm–free shock tube and impinges on a free surface of He II. Some portion of the shock wave penetrates into the He II and the remainder is reflected back into the vapor. The schematic illustration of a number of propagating shock waves in the cryogenic shock tube is shown in Fig.2.

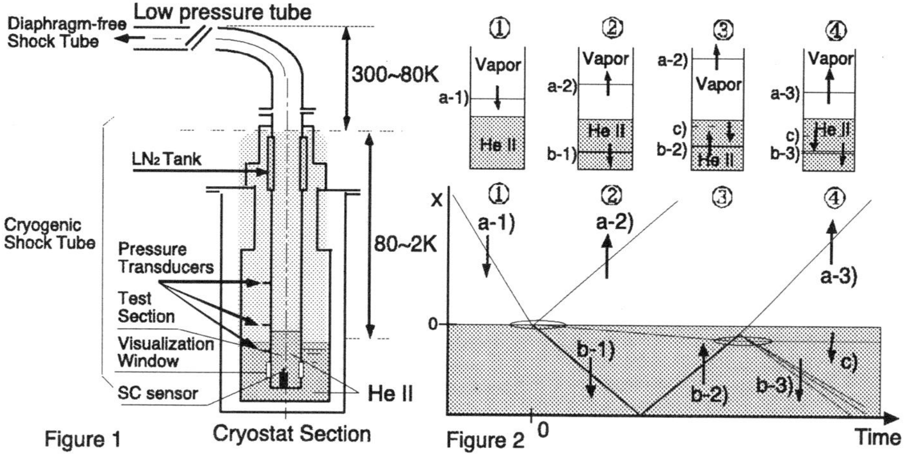

Figure 1. Cryostat section of the superfluid shock tube facility

Figure 2. Schematic x-t diagram illustration of shock waves propagating through the helium vapor and He II. a-1) Gas dynamic shock wave, a-2) Reflected gas dynamic shock wave, a-3) Transmitted gas dynamic shock wave, b-1) Transmitted compression shock wave in He II, b-2) Reflected compression shock wave in He II, b-3) Expansion fan, c) Thermal shock wave.

RESULTS AND DISCUSSION

Shock wave performance of superfluid shock tube facility

Figures 3(a), (b), (c) and (d) show the fundamental shock wave performance of the superfluid shock tube facility. All these results are shown as a function of the initial pressure ratio P_{41}, which is defined as the ratio of the pressure in the driver section to that in the driven section, $P_{41}=P_4/P_1$. The impinging shock Mach number M_{SV} is plotted against P_{41} in Fig.3(a), the pressure jump of impinging shock wave $P_{21}=P_2/P_1$ in Fig.3(b), the shock Mach number in He II in Fig.3(c) and the pressure jump of transmitted compression shock wave ΔP in Fig.3(d). The magnitudes of M_{SV} and M_{SL} are considered to be the indices of the intensity of the shock wave. In these figures, the solid lines in (a) and (b) are the theoretical curve calculated at 2 K on the basis of the simple shock tube theory, Rankine-Hugoniot theory, and the solid lines in Figs.3(c) and (d) are just guide lines for each temperature.

It is noted in Figs.3(a) and (b) that the variations of M_{SV} and P_{21} with P_{41} hardly agree with the Rankine-Hugoniot (R-H) curve. They are much smaller than the theoretical predictions, approximately half to one third of the theoretical value. This fact seems to suggest that the shock wave still has not been fully developed at the moment it reaches the test section. It should also be taken into account that the opening time of a diaphragm-free shock tube is generally longer than that of a diaphragm-type shock tube and the effect of the very large temperature gradient in the test section of the tube. Accordingly, the impinging shock Mach number M_{SV} and pressure jump P_{21} are less than the theoretical prediction based on P_{41}. It is, however, required in practical application of the facility to predict the quantities of M_{SV} and P_{21}. On the other hand, it is noted in Fig.4 that the relation between P_{21} and M_{SV} is in good agreement with the theoretical curve. This implies that the shock waves generated in this shock tube satisfy the gas dynamic shock wave condition. This means every jump quantity can be estimated with the aid of the shock condition provided that P_2 is measured. It is seen in Figs.3(c) and (d) that the transmitted shock Mach number M_{SL} and the pressure jump ΔP in the He II almost linearly increase with P_{41}.

The transmitted compression shock waves propagate at a Mach number M_s about1.05~1.15, which indicates they are rather strong shock waves in liquids. The

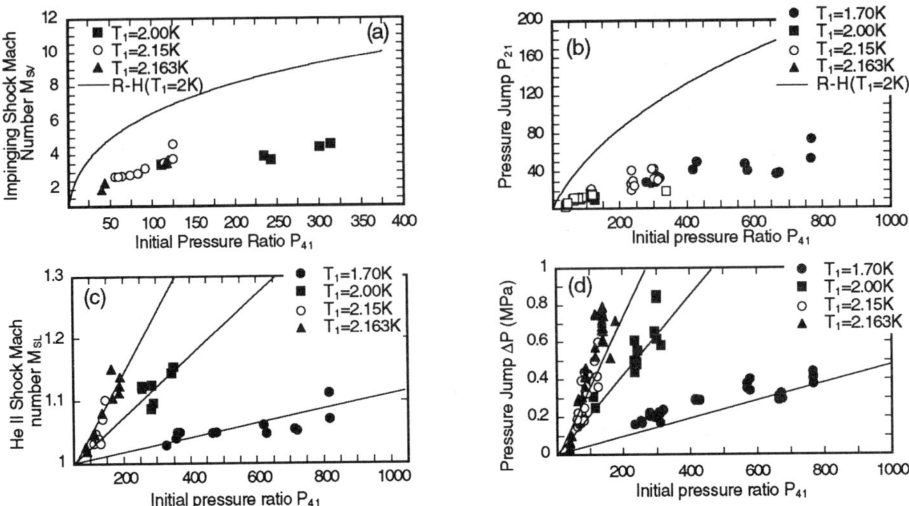

Figure 3. Shock wave performance of superfluid shock tube facility.
(a) P_{41} vs. M_{SV}; M_{SV}= Impinging shock Mach number in the vapor. $P_{41} = P_4/P_1$, P_4=Pressure of high-pressure helium in the driver section. P_1 = Initial pressure of saturated helium vapor in the driven section.
(b) P_{41} vs. pressure jump P_{21}; $P_{21}=P_2/P_1$, P_2 = Wave front pressure of impinging shock wave in the vapor.
(c) P_{41} vs. M_{SL}; M_{SL}= Shock Mach number of transmitted compression shock wave in He II.
(d) P_{41} vs. ΔP in He II ; ΔP = Pressure jump of transmitted compression shock wave in He II.

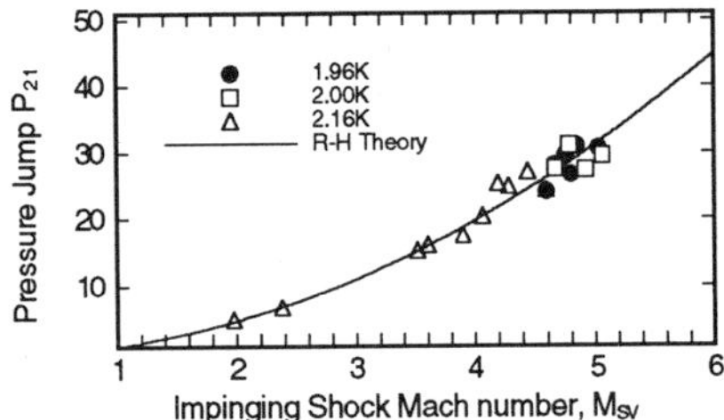

Figure 4. Impinging shock wave condition in the helium vapor. Solid line is the Rankine-Hugoniot relation.

dependence on the bath temperature is clearly evident within the measurement accuracy of this experiment.

Propagating shock waves in cryogenic shock tube

The transient variations of the pressure of a propagating shock wave in the cryogenic shock tube are shown in Fig.5, which are measured by the pressure transducers installed on the wall surface in the vapor phase (a, b, c) and in the He II phase (d, e), respectively. These are not representative of a single test but of several individual tests, because of satisfactory reproducibility. We focus on the pressure profile measured in the vicinity of the He II free surface. The traces (a) and (b) are detected by transducers located in the helium vapor and the trace (c) is measured in the vapor phase in the vicinity of the He II free surface. The traces (d) and (e) are the pressure histories in He II. The first and second stepwise pressure rises of (a) result from the arrivals of an impinging shock and a reflected one from the He II free surface. The reflection is similar to that obtained from the solid boundaries. It is, however, noted in trace (b) that the second large pressure jump which results from the shock reflection from the He II free surface is highly unstable. In the data trace (c), which is

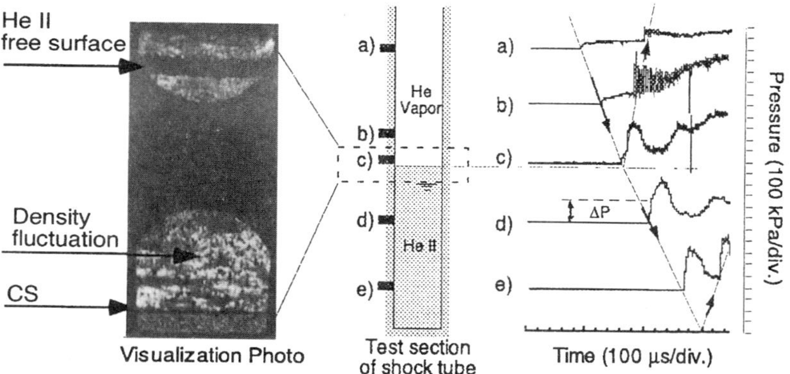

Figure 5. Pressure variations associated with the propagating shock waves.
a), b), c) measured in helium vapor and d), e) measured in He II. P_1=18.1 torr, T_1=2.00K, U_{SV}=327.7m/s, M_{SL}=1.09, U_{SL}=249.1m/s. CS: Transmitted compression shock wave.

measured closer to the free surface, the reflected shock wave shows a very large amplitude, which is considerably different from reflected shocks from the solid boundaries. In traces (d) and (e), the pressure jumps are the transmitted compression shock wave and the reflected shock wave from the bottom of the test section, which is the second jump in (e). The pressure jump ΔP in He II is also very large as observed in the case of (c). From the visualization photograph, one observes a transmitted compression shock wave, accompanied with a fluctuating density state behind it and a thickened free surface region.

Transmitted compression shock wave in He II

The pressure jump data of the transmitted compression shock wave, $\Delta P/P_1$, is plotted against the shock Mach number M_{SL} in Fig.6. Here P_1 is the initial (upstream) pressure of He II, equal to the saturated vapor pressure at the temperature T_1, and ΔP is the pressure jump across a transmitted compression shock wave. The lines in Fig.6 are the Khalatnikov prediction[5] for the compression shock in He II, which is a weak shock wave approximation derived by neglecting terms higher than the second order of smallness relative to the

$$\Delta P = 2(M_{SL}-1)\left(\frac{\partial}{\partial P}\ln(\rho a_1)\right)^{-1}$$

$$\Delta T = 0$$

counterflow velocity w as given below;
Here, ρ is the density of He II and a_1 is the speed of the first sound. It is seen from Fig. 6 that $\Delta P/P_1$ increases almost linearly with M_{SL} and the experimental results are in good agreement with the Khalatnikov prediction. However, in the case of the initial temperature at 2.15 K, the data points deviate slightly from the Khalatnikov prediction for M_{SL} larger than 1.05. It is assumed that this data deviation is because the magnitude of the thermal expansion coefficient becomes very large near the lambda point but this effect is neglected in the approximation.

The visualization photographs are shown in Figs.7 (a) through (e). These are a series of propagating compression shock waves in He II with a propagation speed of 230 m/s. These photos are not from a single shot but from several independent shots. In a Schlieren photograph, a compression shock wave appears as a single line. It is of interest to note that in the region behind the shock wave the density is highly unstable. The reason for the disturbance has not been established at this time.

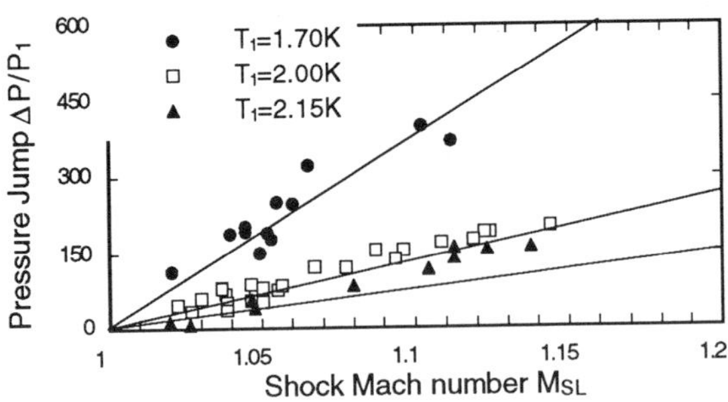

Figure 6. Pressure jump result at the wave front of the transmitted shock wave in He II as a function of shock Mach number M_{SL}. Solid lines are Khalatnikov prediction.

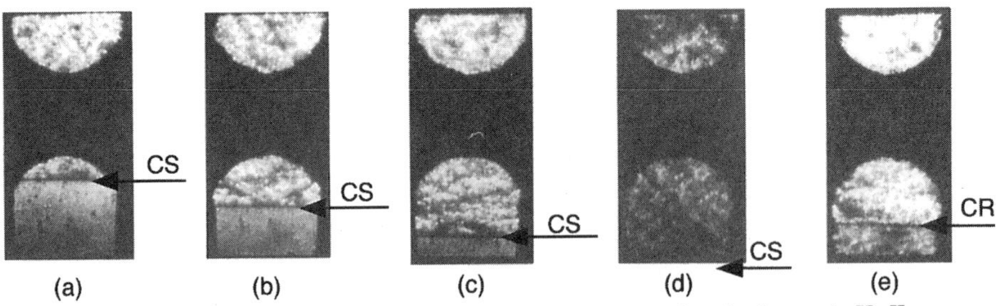

Figure 7. Visualization photographs of propagating transmitted compression shock wave in He II.
td : Time delay of photographing from signal detection by pressure transducer C.
CS: Compression shock wave in He II. CR: Reflected compression shock wave in He II.
(a) td=28μsec, (b) td=46μsec, (c) td=68μsec, (d) td=116μsec, (e) td=340μsec.

Shock induced lambda phase transition

Figure 8 shows two data records of temperature variations detected by the superconductive temperature sensor. The data (a) is the He II compression result without lambda phase transition and (b) the lambda. In trace (a), there are two sharp temperature drops induced by a transmitted and a reflected compression shock. These temperature drops caused by shock compression result from the negative thermal expansion coefficient that is characteristic of He II. This temperature variation indicates that the He II still remains He II after shock compression. The thermal shock wave is observed in the latter half of the data record in Fig.8(a). On the other hand, the trace in Fig.8(b) is the record of the lambda transition from He II to He I induced by shock compression, where the initial temperature T_1 is 2.15K which is rather close to the lambda temperature. In fact, this is the evidence that the He II converts to He I which has positive thermal expansion coefficient. It may be concluded that the He II converts to He I by crossing the lambda line due to shock compression without any appreciable time delay. Since a positive temperature jump appears from the beginning of the shock compression, it is found that the lambda transition occurs in a relatively short time even under this highly transient compression induced by the shock wave. Considering the fact that the lambda transition is a second order phase transition which does not involve latent heat, the actual occurrence of this highly transient phase transition may be quite reasonable. However, strictly speaking, the thermal expansion coefficient is still negative in the He I in a narrow limited region at slightly higher temperature than the lambda point as shown in Fig.9. Accordingly, one needs to confirm the non-existence of the second sound wave in order to verify the He II to He I lambda phase transition. The visualization photographs are shown in Figs.10(a)-(d) for the case of initial temperatures of T_1=2.10K and 2.15K, respectively. In Figs.10(a)-(c), a thermal shock wave is observed near the He II free surface. The density variation associated with the thermal shock wave is very small compared with that caused by a compression shock. On the other hand, it is observed in these figures that the He II free surface zone becomes thick as time proceeds. Strong density fluctuations resulting from dynamic instability of the free surface and from anomalous behavior around the critical point may cause thickening of the free surface zone. In Fig.10(d), no thermal shock is seen in the case of the initial temperature T_1=2.15K. It is believed that a thermal shock wave is not excited due to the phase transition to He I in this case.

The pressure and the temperature variations induced by transmitted compression shock waves in He II are examined referring to the p-T diagram given in Fig.11. The experimental data are plotted together with curves for the isentropic process applied to liquid helium and at four initial temperatures. Since the compression by a shock wave is not really an sentropic process, these curves should be considered as approximate ones, substituting for the shock adiabatic process. Roughly speaking, compression causes a temperatu re drop in the He II. It is found from this result that the temperature still drops as a result of shock

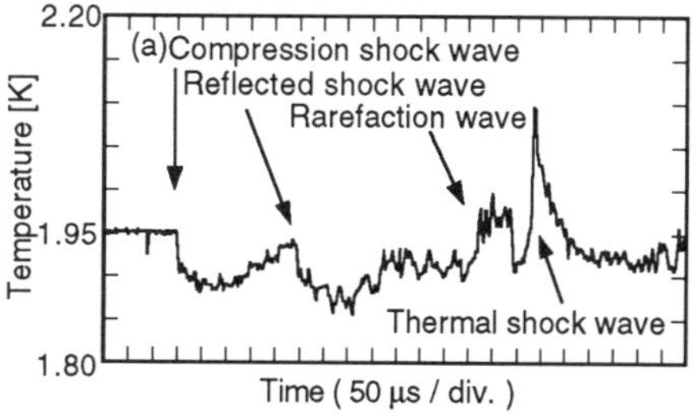

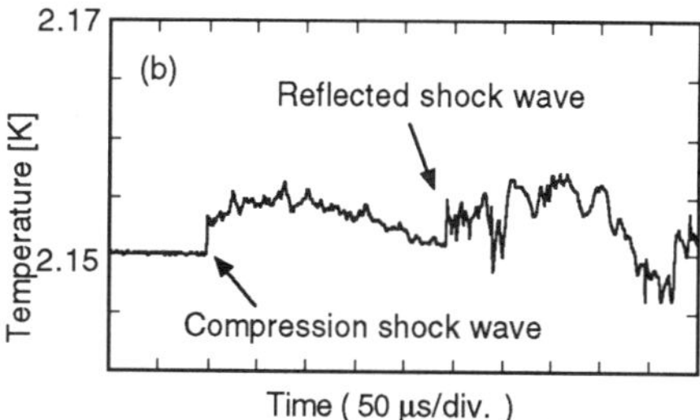

Figure 8. Transient records of temperature variation detected by the superconductive temperature sensor in He II for comparison among; (a) Simple shock compression without lambda phase transition. T_1=1.95K. (b) Lambda-transition, from He II to He I, caused by shock compression by transmitted shock wave. T_1=2.15K.

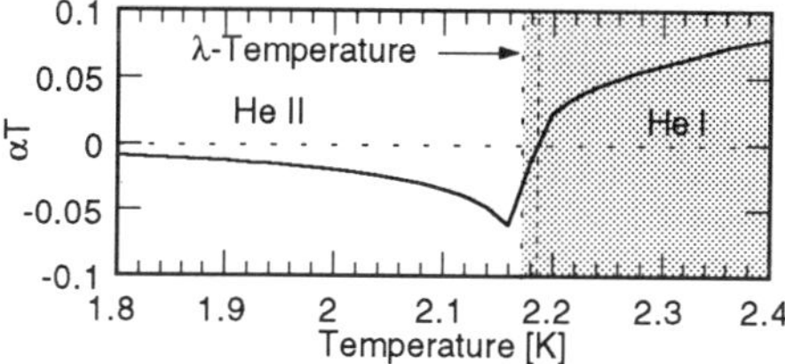

Figure 9. Thermal expansion coefficient of liquid helium at saturated vapor pressure. α: Thermal expansion coefficient.

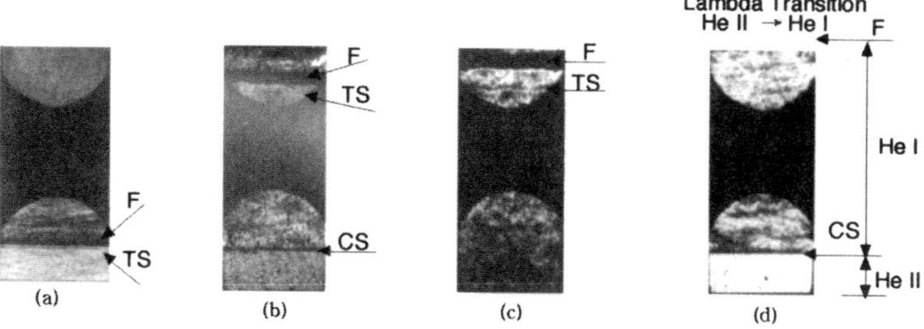

Figure 10. Visualization photographs of propagating thermal shock wave and He II-He I lambda-transition. F: He II free surface, CS: Transmitted compression shock wave, TS: Thermal shock wave. Thermal shock waves are located; (a) 0.7 mm below He II free surface, (b) 3.0 mm, (c) 6.8 mm, (d) He II-He I lambda-transition. Thermal shock is not excited.

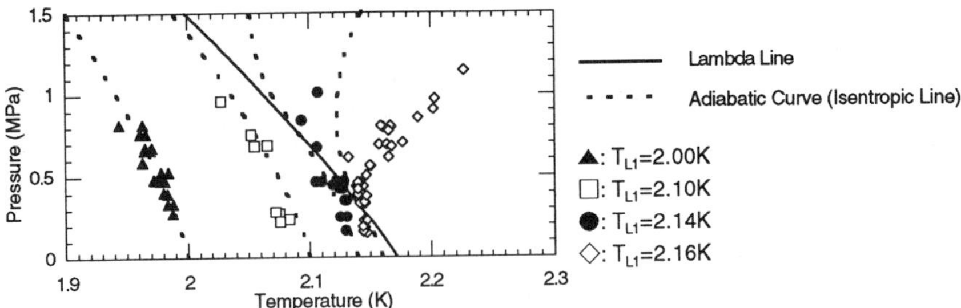

Figure 11. The temperature and pressure at the wave front of compression shock wave in He II, plotted for four initial temperatures of He II.

is still negative in the narrow region near the lambda line in He I as shown in Fig.9. It is consequently seen that the lambda-transition from He II to He I as the result of shock compression cannot be confirmed just by a positive temperature jump. Th ere is a small possibility that the lambda-transition from He II to He I leads to a negative temperature jump as mentioned above. Furthermore, the experimental data deviates considerably from the isentropic curve beyond the lambda-line in the He I region. In the shock compression process crossing the lambda line, the entropy increases and thus the temperature rises appreciably. The reason why such a large deviation appears in this case is not yet wholly clear, though such aspects can be related to the physical anomaly around the lambda-line. There is a small possibility that the calibration of the superconductive temperature sensor may not be accurate because of the difficulty in the calibration procedure in the trans-lambda-line region. The heat transfer mechanism in He I is different from that in He II. In He II the effective thermal conductivity is extremely large due to internal convection mechanism, and thus the temperature of the sensor element can be regarded to be essentially that of the surrounding He II. However, the thermal conductivity of He I is much smaller than and consequently, the temperature of the sensor element is, in general, higher than that of the surrounding He I.

ACKNOWLEGEMENT

This experiment is financially supported by Grant-in-Aid for Scientific Research (A) (11305068) and JSPS Research Fellowship.

REFERENCES

1. R. Wang , “Special physical problem related to heat transfer in bath of superfluid helium. ” , *Advances in Cryogenic. Engineering*, Vol. 41A (1996) pp233-240
2. H. Nagai, H.S. Yang, N.Takano and M. Murakami, S. Teraoka, “Development of Superfluid Shock Tube Facility”, *Advances in Cryogenic. Engineering*, Vol. 43, pp.1393-1400 (1998)
3. H. Nagai, H.S. Yang, N.Takano and M.Murakami, “Investigation of two modes of shock wave in He II by using newly developed superfluid shock tube facility”, *ICEC-17*, Bournemouth, UK : 843-846 (1998)
4. H.S. Yang, H. Nagai, N. Takano and M. Murakami, “Experimental investigation of thermal shock wave induced by gas dynamic shock wave impingement” *CEC'99*, Montreal, Canada (1999)
5. I.M. Khalatnikov. “Introduction to the Theory of Superfluidity.” Benjamin, New York (1965)
6. J.C. Cummings, “Experimental investigation of shock waves in liquid helium I and II.” *J. Fluid Mech..*, vol. 75, part 2, pp 373-383 (1976)
7. D.M.Moody and B.Sturtevant, “Shock waves in superfluid helium” *Phys. Fluids 27(5)* pp1125-1137 (1984)
8. H.W. Liepmann and G.A. Laguna, "Non-linear interactions in the fluid mechanics of helium II." *Ann.Rev.Fluid Mech.* Vol.16 pp139-177 (1984)

EXPERIMENTAL STUDY ON THE TRANSITION OF HE II BOILING MODES WITH THE HYDROSTATIC PRESSURE AROUND THE LAMBDA TEMPERATURE

Y.Takashima, M. Murakami and T.Toyoshima

Institute of Engineering Mechanics and Systems, University of Tsukuba
Tennodai 1-1-1, Tsukuba, Ibaraki, 305-8573, Japan

ABSTRACT

The transition of film boiling modes in He II with the change of the hydrostatic pressure is studied experimentally. Special attention is paid to the pressure dependence of boiling modes around the λ temperature. For this purpose, detailed investigation is carried out with the aid of optical visualization and the pressure oscillation and the temperature oscillation measurements. All the experiments are conducted by using a single cryostat over the entire pressure range from saturated vapor pressure to subcooled states because the excited pressure oscillation is highly affected by the geometrical configuration of the test space in a cryostat. In the case of subcooled boiling, flickering of bulk vapor film is seen on the planar heater instead of repeated growth and crush of a vapor bubble as observed in the noisy film boiling. It is an interesting finding that a regular pressure oscillation is excited at slightly lower temperature than the λ-temperature for a moderate value of the hydrostatic pressure, which is obviously different from rather irregular ones of noisy and subcooled film boiling modes. With the decrease in the temperature and the increase in the heat flux in the temperature range close to the λ-temperature, low frequency spiky oscillations are superposed on the regular oscillation. At still lower temperatures than λ-temperature, the regular oscillation disappears and boiling turns to noisy film boiling.

INTRODUCTION

The development of high field superconducting magnets is needed for a number of technical applications to MRI diagnosis systems and accelerators. He II has been often utilized as an excellent cryogen for superconducting magnet cooling. In particular, the cooling performance of subcooled superfluid helium (He II_p) is superior to that of saturated superfluid helium (He II_s) for superconducting magnet cooling. This means that in the of He II_p cooling, the violent boiling may be avoided because of high limiting heat flux. It is known that two film boiling modes appear in He II_s; noisy and silent film boiling modes.[1,2] Noisy film boiling produce very loud acoustic noise and a large-scale unstable motion of vapor bubbles. Silent film boiling is accompanied by very weak sound generation and small-scale fluctuating motion of vapor layer. The occurrence of each film boiling depends on the magnitude of the hydrostatic pressure and the bath temperature. In He II_p at higher pressure than the lambda pressure P_λ, it is recognized that two kinds of the phase transition phenomena take place, that is He II to He I and He I to vapor transitions. In other words, a triple-phase co-existing state may appear. The subcooled film boiling looks like silent film boiling without any violent unstable vapor layer motion, but is characterized by a higher frequency of the pressure oscillation than silent boiling in He II_s. The studies of fundamental mechanism of subcooled boiling and its control is essential for various practical equipment in which He II is utilized as cryogen. Studies of heat transfer in He II_p have been reported frequently.[3,4] However, there are few studies in which a visualization technique was applied. We have attempted to measure the pressure and temperature oscillations during film boiling in He II_p with a pressure sensor and a small temperature sensor, in addition to visual observation. The mechanism of subcooled film boiling in He II_p is investigated by comparing with the noisy film boiling in the present study. It is another objective to investigate the newly found regular pressure oscillation excited at slightly lower temperature than the λ-temperature for moderate values of the hydrostatic pressure.

EXPERIMENTAL APPARATUS

Cryostat with Optical Windows

The cryostat illustrated in Fig.1 was originally designed for the experiments on subcooled He II (He II_P). It is of a Claudet-type cryostat with the Joule Thomson (J-T)

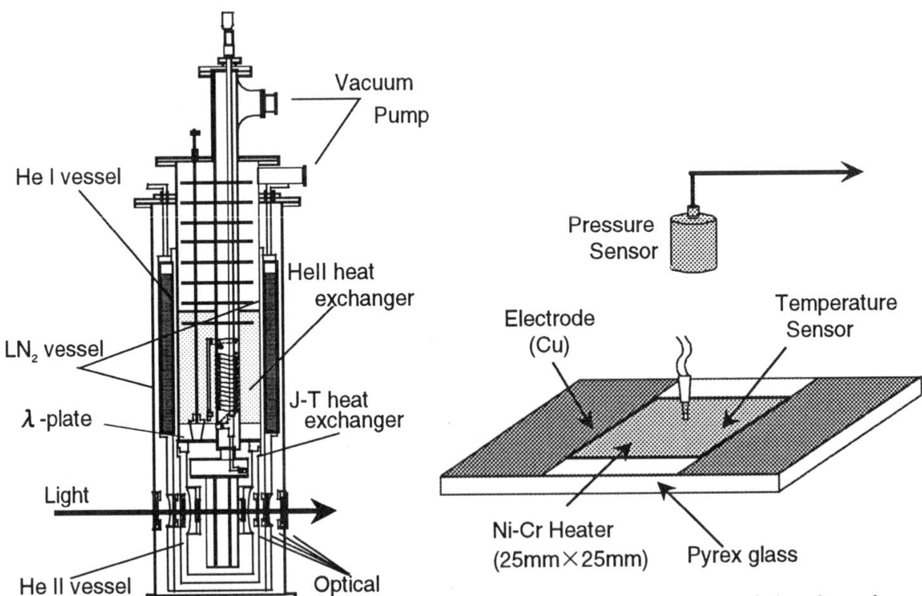

Fig.1 Schematic illustration of the experimental He II_p cryostat with the optical windows.

Fig.2 Schematic illustration of the planar heater and the pressure and temperature sensors in the test section

cooler. There is the He II vessel under the He I vessel, and the λ-plate between each vessel. In the λ-plate with the narrow slit, the He II vessel links to the He I vessel through the pressure, but not through the temperature. The pressure in the He II vessel is kept at the pressure of the saturated vapor pressure in the He I vessel and thus He II vessel is in the subcooled state. He II_P in the He II vessel is cooled by the J-T cooler through the heat exchanger in the He II vessel. The cryostat is equipped with observation windows with an aperture of 50mm in diameter. These windows are used to observe the thermo-fluid dynamic phenomena in the test section. The outermost window is vacuum tight and the innermost window of the H II vessel is super-leak tight. The optical glass plates of intermediate windows, which are coated with infrared absorbing thin film, are cooled to liquid nitrogen temperature to absorb the thermal radiation heat leak from the room temperature environment.

Test Section

The heater and the pressure sensor are drawn in Fig. 2. The square planar heater to cause boiling in He II_p is mounted horizontally in the test section. It consists of vacuum-deposited Ni-Cr thin film on a Pyrex substrate for heating part and a Cu thin film as an electrode. Above the planar heater, the pressure and the temperature sensors are controlled. The heater is heated by step-wise electrical current. The boiling state in the test field is visualized through some optically and the image is recorded by a still camera and a high speed video. The pressure oscillation induced by boiling is measured with the pressure sensor in order to investigate the detailed mechanism of boiling in He II_s and He II_P. The temperature oscillation is also measured by a small temperature sensor at the same time. The pressure and the temperature sensors are located at 40mm and 5mm above the heater, respectively. Since the pressure sensor is supplied in uncalibrated state the output signal is directly compared. The temperature sensor is in-house calibrated in liquid helium, but it is not calibrated in helium gas.

RESULTS AND DISCUSSION

Variation of Appearance of Boiling Mode With the Pressure

In order to compare the difference in the appearance of boiling modes between noisy film boiling in He II_s and subcooled film boiling in He II_p, the visualization pictures and the phase diagram of helium are shown in Figs.3 (a), (b), (c) and (d), respectively. Boiling in each case is caused by the heat flux q of $10W/cm^2$ or $20W/cm^2$ from the planar thin film heater with dimensions of 25mm×25mm. These pictures were taken at 500 msec after the onset of heating. The only difference in the experimental condition between the noisy (a) and the subcooled states (b) and (c) is the hydrostatic pressure, P.

In the case of noisy film boiling in He II_s, a large single hemispherical vapor-bubble is periodically generated as seen in Fig.3(a). It is known that pressure spikes are periodically generated, synchronizing with the vapor-bubble formation and crush and result in loud audible noise generation. It is found that these pressure spikes are generated at the moment when rapid and violent burn-out like evaporation occurs after the surrounding He II comes into direct contact with the superheated heater surface subsequent to the vapor-bubble crush. It should be noted that in the case of smaller hydrostatic pressure in He II_s, silent film boiling appears, which is completely different from noisy film boiling[3].

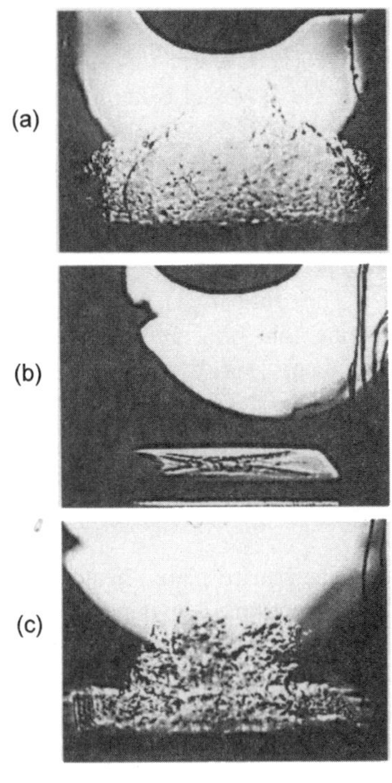

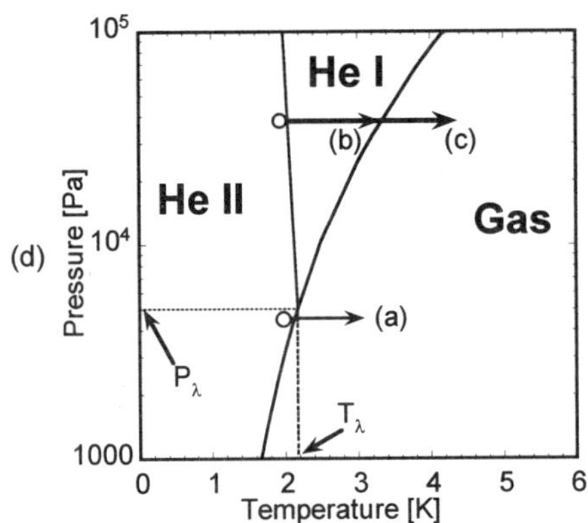

Fig.3 Visualizations of two boiling modes in He II, (a) through (c), and corresponding helium phase diagram (d).
(a) Saturated He II. Noisy film boiling, T=2.10K, q=10W/cm^2
(b) Highly subcooled He II film boiling, T=2.10K, P=39.9kPa, q=10W/cm^2
(c) Highly subcooled He II film boiling (high heat flux), T=2.10K, P=39.9kPa, q=20W/cm^2
(d) Phase diagram indicating boiling states

On the other hand, in the case of subcooled film boiling in He II_p at the pressure of 39.9kPa for the heat flux q of 10W/cm^2, it is seen from the picture (b) that violent boiling does not occur. This is partly because the critical heat flux for film boiling is larger in the subcooled state than in the saturated state. Even in the case of large heat flux of 20W/cm^2, as seen from Fig.3-(c), no large single vapor bubble is generated, but instead, a vapor-layer with irregular thickness, which is rather stable as a whole, is excited at the heater. It is emphasized that for the case of the Fig.3-(c) the vapor-layer develops in the center portion of the heater, but it is very stable it the edge region. An important result is that the selective appearance of film boiling mode strongly depends on the hydrostatic pressure.

Comparison of Temperature and Pressure Oscillations between Noisy Film Boiling and Subcooled Film Boiling

The simultaneous measurement data of the pressure and temperature oscillations measured in film boiling states caused by stepwise heating are compared between the saturated pressure state as seen in Fig.4, and the subcooled state in Fig.5.

It is seen in Fig.4 that there is a strong correlation between the pressure and the temperature oscillations in the case of noisy film boiling. The detailed examination of the ΔT and ΔP data as given in Fig.4 and the visual observation record (a)-(e) taken by the high-speed video camera revealed some phenomenological aspects of noisy film boiling. At first, a vapor-layer with a heater-size grows up over the planar heater as seen in Figs.4 (a) through (c). As soon as the large bubble leaves the vapor-layer as seen in Fig.4 (d), it begins to shrink as seen in Fig.4 (e). When the babble vanishes above the thin vapor-layer, a strong pressure spike is emitted. This spiky-pressure wave seems to cause the collapse of a portion of the vapor-layer, and thus, He II makes direct contact with the heater at the portion. The contact of He II and the super-heated heater causes explosive vaporization on the heater surface to induce another pressure peak forming a double-peak shaped pressure spike as seen in each pressure spike in Fig.4 (f). By comparing the visualization pictures

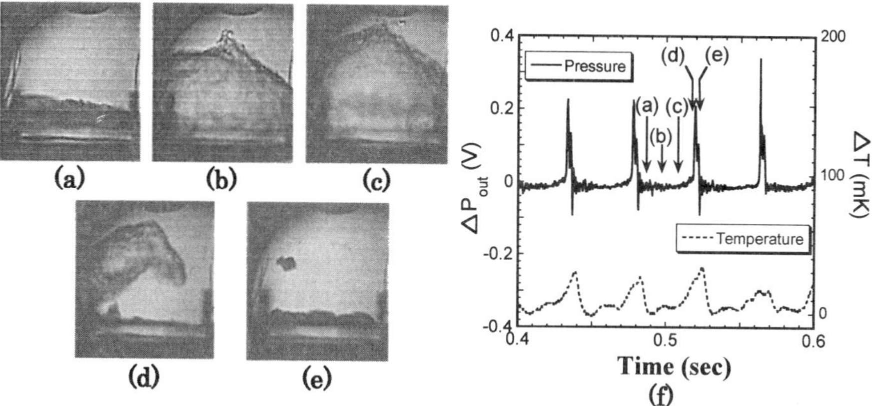

Fig.4 Photographs (a) to (e), and the simultaneous measurement data of the pressure and the temperature oscillations (f) in the case of noisy film boiling. The pressure and the temperature sensors are located at 40mm and 5mm above the heater, respectively.T=2.10K, q=10W/cm^2. The photos are taken at (a)485ms, (b)497, (c)509ms, (d)517ms, (e)521ms, after the onset of stepwise heating as indicated in graph (f).

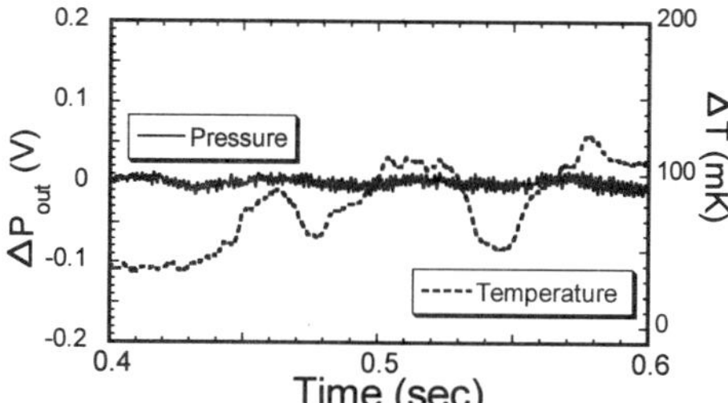

Fig.5 Simultaneous measurement data of the pressure and the temperature oscillations in the case of subcooled film boiling. The pressure and the temperature sensors are located at 40mm and 5mm above the heater, respectively.
T=2.10K, P=13.3kPa, q=10W/cm^2

with the temperature variation data given in Fig.4 (f) in the case of the noisy film boiling, it is found that the temperature hardly changes while the thermometer is included in the vapor-layer as seen in Figs.4 (a) to (c). When the vapor-bubble grows as large as the size of the planar heater as seen in Fig.4 (c), the growth of the vapor-bubble stops, and the temperature begins to rise. This can be understood as follows: in the growing phase of vapor-layer, the temperature in the vapor-layer does not rise because of evaporation at constant pressure. After the vapor-layer fully develops, the heater surface gradually tends to dry-out, and then the heating results in the vapor temperature rise. Finally, since the vapor-layer collapse after the bubble takes off and shrinks quickly, the temperature drops rapidly.

In the case of the subcooled film boiling, there is no strong correlation between the temperature and the pressure oscillations, nor the spiky pressure oscillation as seen in Fig.5. In the visualization pictures, only small scale flickering of vapor-liquid interface is observed. There is no breaking down of the vapor-layer. The vapor-layer is very stable without a bubble breaking off the surface of the heater because of the strong pressurization, and thus the heater surface is always covered with a vapor-layer. It is found in Fig.5 that the average temperature above the heater surface, which rises with the time, is higher than that of the noisy film boiling as seen in Fig.4. There are two reasons for this: one is that the saturated vapor temperature in the subcooled state is higher than that in the saturated state. The other is that it is easy for the temperature above the heater surface to rise because of the thermal absorption of the stable vapor-layer.

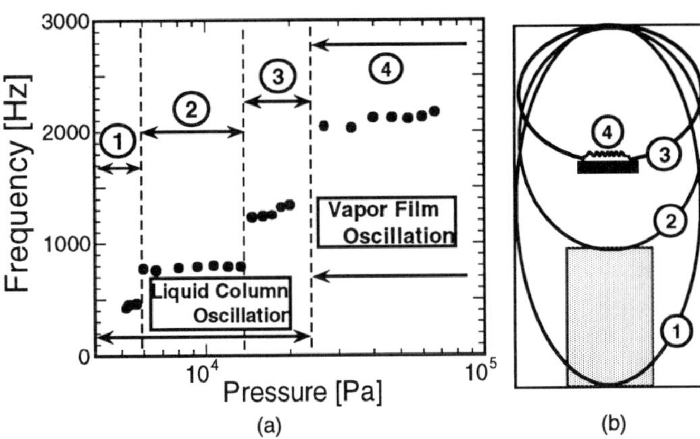

Fig.6 (a) Fundamental frequency for the pressure oscillation data is plotted against the pressure. (b) Illustration of the pressure oscillation modes in the test section. ① through ③ indicate each elastic liquid column oscillation with the nodes at the bottom surface of the lambda plate and at the bottom of the test section ①, at the top of the bakelite block ② and at at the heater surface ③, respectively. T=2.10K, q=10W/cm^2

Dependence of Pressure Oscillation in Subcooled Film Boiling on Hydrostatic Pressure

In order to investigate the detail of the mode transition among film boiling states with the hydrostatic pressure, FFT analysis results are summarized for comparison in Fig.6. The illustration of the pressure oscillation modes is also added in Fig.6(b), where each solid line indicates each fundamental mode of liquid column oscillation generating the primary pressure oscillation with the largest FFT analysis coefficient. It is found that liquid column oscillations are a kind of elastic liquid oscillation in the first sound mode. In the figure, the peak frequency of the power spectral density diagram for the pressure oscillation data is plotted against the pressure. It is seen from these two figures that there appear four pressure oscillation modes in the pressure range between 5.3kPa and 66.6kPa above P_λ. Each mode corresponds to the liquid column oscillation with a liquid column height h of 25 (①), 15(②) and 10cm(③). The pressure peak of each primary pressure oscillation gradually becomes small as the increase of the pressure. It seems that the pressure oscillation with a frequency of several hundreds Hz in subcooled film boiling is not so harmful to technical applications, since the pressure peak in subcooled film boiling does not become so large as the one in noisy film boiling.

However, in the high pressure case, over 26.6kPa, the frequency is found to reach a nearly constant value, roughly 2000Hz, irrespectively of the pressure, as seen in the region ④ in Fig.6(a), though it is predicted theoretically that no liquid column oscillation modes are excited at such high frequency as 2000Hz under the present experimental conditions. When film boiling is caused, some pressure and temperature oscillations are excited in a certain frequency range as the result of irregular fluctuation of the vapor-liquid phase boundary and the formation and crush of a vapor bubble. A particular column oscillation mode for which the fundamental frequency is within this frequency range of vapor oscillation, is mainly excited as the fundamental mode, and furthermore the n-th harmonics may also be excited. It may be supposed that as the pressure increases, the frequency range resulting from film boiling may shift towards higher frequency. The main reason for the shift of the frequency is that the geometrical scale of the oscillation of vapor-layer and vapor bubble becomes small as the increase of the pressure resulting in pressure oscillation with higher frequency.

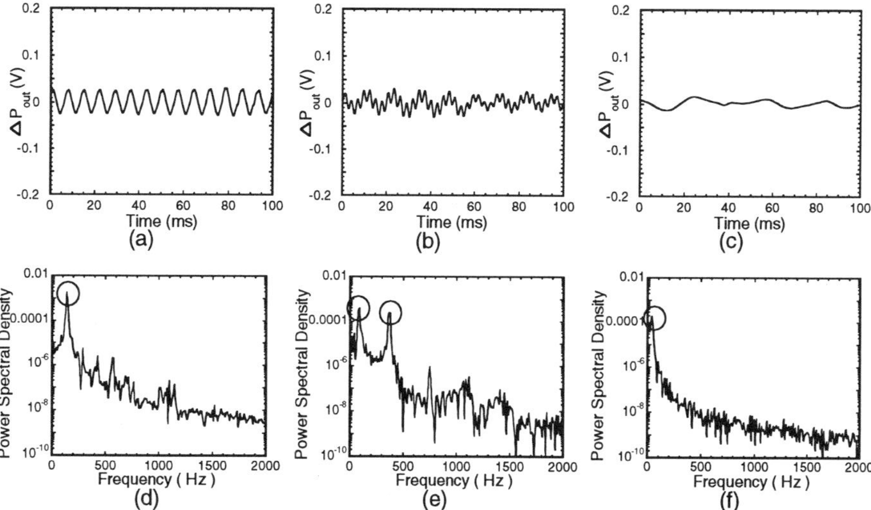

Fig.7 Transition of the pressure oscillation and the FFT analysis result with the heat flux. The hydrostatic depth between the He II free surface and the planar heater, h is 30cm, T=2.170K. (a)(d):q=0.6W/cm^2, (b)(e): q=5W/cm^2, (c)(f): q=10W/cm^2

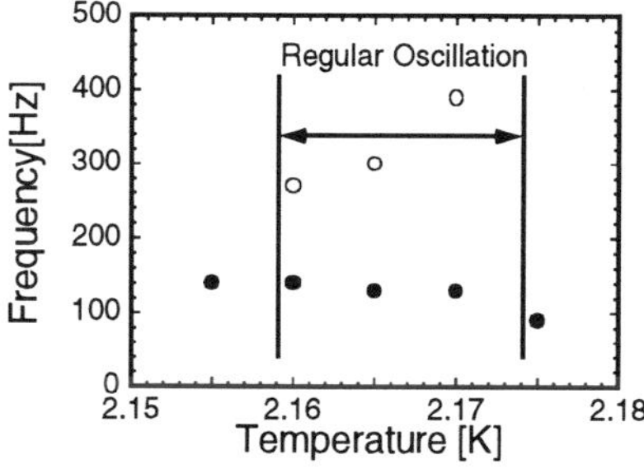

Fig.8 Variation of the frequencies of the pressure oscillation with the temperature. ●:fundamental mode, ○:secondary mode, q=1W/cm^2, h=30cm.

Fig. 9 Map of the regular pressure oscillation and no regular pressure oscillation with the change of the hydrostatic pressure and the temperature.
○: regular oscillation mode, ●: without regular oscillation mode.

Pressure Oscillation around T_λ

Special attention is paid to the quite regular pressure oscillation found in the case of the temperatures around T_λ. It is generated under a certain limited condition such as small heat flux and the temperature very close to T_λ. The transition of the regular pressure oscillation with the heat flux is shown in Figs.7(a), (b) and (c), and the corresponding FFT analysis result is in Figs.7(d), (e) and (f). In the case of small q, there appears the regular oscillation with the dominant frequency of 140Hz, as shown in Figs.7(a) and (d). As the heat flux increases, in the case of q=5W/cm^2 there appears not only the dominant oscillation mode with a frequency of 90Hz, but also the second oscillation mode with a frequency of 380Hz as seen in Figs.7(b) and (e). At still larger q, 10W/cm^2, the regular pressure oscillation is not excited, and the pressure oscillation becomes irregularly as seen in Fig.7(c). It follows from the experiments on the transition of the pressure oscillation that as the heat flux increases, the calm film boiling with a thin vapor-layer is changed into the fully developed film boiling, and furthermore He I boiling is excited. In short, the transition of boiling state starts with the appearance of the regular oscillation. The appearance of the secondary mode is the turning point into the irregular state. In order to examine the influence of the temperature on the regular pressure oscillation, the variation of the frequencies of the fundamental and the secondary modes are plotted against the temperature in Fig.8. It is found in Fig.8 that the frequency of the fundamental mode hardly changes with the temperature, but the frequency of the secondary mode drastically decreases as the temperature drops. From the result of the visualization, it seems reasonable to suppose that as the bath temperature increases, the vapor layer becomes so thick that the vapor layer can vibrate flexibly, and then the secondary mode with higher frequency is excited. To investigate the influence of the hydrostatic pressure, the appearance of the regular pressure oscillation is tested by changing both the temperature and the hydrostatic pressure. The result is plotted on the phase diagram in Fig.9, where the open circles indicate the regular pressure oscillation occurs, and the closed circles indicate it does not occur. The pressure corresponds to the sum of the saturated vapor pressure and the hydrostatic pressure between He II free surface and the heater surface. It can be seen in Fig.9 that the regular pressure oscillation is induced under the condition that the hydrostatic pressure is higher than P_λ. It is supposed that in the case of the higher pressure than P_λ He I phase exists between He II and the vapor phase caused by boiling. The He I phase plays a part in the thermal barrier to suppress rapid diffusion of heat and vaporization. The He I nucleate boiling and-or the He I phase vibration are detected as the regular pressure oscillation.

CONCLUSION

A series of film boiling experiments around the T_λ is conducted, and the following conclusions are drawn. Boiling phenomena is visualized under the noisy film boiling in He II_s and the subcooled film boiling in He II_p. It is observed that the selective appearance of two boiling modes strongly depends on the hydrostatic pressure. In case of the noisy film boiling, it is found from the visual observation and detailed measurement that the spiky pressure oscillation is closely related with the shrinkage of a bubble, and then there is a strong correlation between the pressure and the temperature oscillation. In the case of the subcooled film boiling, the pressure oscillation with the high frequency depends on the geometrical structure of the experimental space, and there is no strong correlation between the temperature and the pressure oscillation, nor the spiky pressure oscillation. It is an interesting that a regular pressure oscillation excited at slightly lower temperature than T_λ and higher hydrostatic pressure than P_λ is obviously different from rather irregular ones of noisy film boiling modes.

REFERENCES

1. A. C. Leonard, Helium-2 noisy film boiling and silent film boiling heat transfer coefficient values, *Proc. ICEC3*, 109 (1970).
2. M. Yamaguchi, and M. Murakami, Study of pressure oscillation during noisy film boiling in He II, *Cryogenics*, 37:523 (1997).
3. Van Sciver S.W., Transient heat transport in He II, Cryogenics, 19:385 (1979).
4. Kobayashi, H., Fujimura, Y., Murata, T. and Sakata, M., Heat transfer through subcooled He I layer from distributed heat source in a pressurized He II channel, *Cryogenics*, 37:851 (1997).
5. M. Murakami, M. Yamaguchi, N. Yanase, and H. Inaba, Various film boiling states in He II at hydrostatic pressure from staturated vapor pressure to 1 atm, *Adv. Cryog. Eng.*, 43:1425 (1998).

TRANSIENT HEAT TRANSFER PRODUCED BY A STEPWISE HEAT INPUT TO A FLAT PLATE ON ONE END OF A RECTANGULAR DUCT CONTAINING PRESSURIZED HELIUM II

M. Shiotsu[1], K. Hata[2], K. Hama[2] and Y. Shirai[1]

[1]Dept. of Energy Science & Technology, Kyoto Univ., Uji, Kyoto, 611-0011, Japan

[2]Institute of Advanced Energy, Kyoto Univ., Uji, Kyoto, 611-0011, Japan

ABSTRACT

Transient heat transfer on a flat plate in subcooled He II produced by stepwise heat inputs was measured for the bulk liquid temperatures ranging from 1.8 to 2.1 K at atmospheric pressure. Four test plates with the same dimensions were used. To see the effect of expansion on heat flow, one of them is supported in a pool of subcooled He II and other three are located on one end of rectangular ducts with various cross sections. Other ends of the ducts are opened to a pool of pressurized He II.

The transient heat transfer is such that the quasi-steady state exists with a certain lifetime, t_L, and rapidly changes to film boiling regime after the depletion of lifetime. The lifetime is shorter for higher value of the step height.

The values of t_L on the flat plates in the ducts with the ratio of cross sectional duct area to the heater area, A_d / A_h, higher than unity agree well with the values for the duct with $A_d / A_h = 1$ for q_s higher than a certain value. With the decrease of q_s from the value, they become longer than the values for $A_d / A_h = 1$. This trend is more significant for the ducts with higher values of A_d / A_h. Correlation of lifetime based on the two fluid model is given.

INTRODUCTION

The knowledge of transient heat transfer on a solid surface in He II produced by a large pulsewise heat input is necessary to evaluate whether a He II cooled superconducting magnet is quenched or not by an instantaneous thermal disturbance.

The steady state and transient heat transfer in He II has been mainly investigated for a He II channel with a uniform cross sectional area where the heat flow can be viewed as one-dimensional. Time delay before the onset of film boiling after the addition of stepwise heat input was observed. The results of the steady-state experiments[1,2] and transient experiments[3,4] have been analyzed in terms of two-fluid hydrodynamics and theory of mutual friction.

On the other hand, steady-state and transient heat transfer experiments for cylindrical test heaters were performed by several workers[5,-9]. Shiotsu et al.[7-9] reported that, in the heat transfer on a cylinder with the length far longer than its radius where the heat flow can be viewed as two-dimensional, the steady-state critical heat fluxes, q_{st}, were about 42 % lower than those predicted by the Gorter-Mellink equation for a cylinder. It is from this fact that the heat conductivity function of He II, $1/f(T)$, is effectively reduced to about 20 % (0.58^3) of that for the one-dimensional heat flow when the heat is expanded radially. It is assumed that this may be due to some turbulence in case of the heat flow expansion. The existence of a time delay before the onset of film boiling was also confirmed in the transient heat transfer on a cylindrical test heater produced by stepwise heat inputs.

Recently, the authors[10,11] made systematic experiments on the steady state and transient heat transfer from single flat plates of various widths. They reported that the values of q_{st} were higher for smaller widths of the test plates. The values of q_{st} for the test plate with the width w almost agreed with those for the cylinder with the diameter of w under the same conditions. Similar reduction of $1/f(T)$ seemed to exist for the heat flow to a pool of He II from the flat plates. However, the lifetime for a certain quasi-steady heat flux beyond q_{st} was almost independent of the plate width, although q_{st} depended on it. It is assumed from this results that the heat flow expansion from a flat plate and the resultant reduction of $1/f(T)$ would become more significant with the elapse of time after the addition of stepwise heat input. The transient heat transfer on a flat plate in a duct with expanding cross sectional area is experimentally studied in this work to clarify the effect of heat flow expansion and its dependence on time.

The purpose of this study is twofold. First is to obtain the experimental data of transient heat transfer produced by large stepwise heat inputs to a flat plate pasted on one end of a rectangular duct containing subcooled He II. Second is to clarify the effect of the ratio of the cross sectional duct area to the heater area, A_d / A_h, on the lifetime of quasi-steady state governed by Kapitza conductance and present a correlation of lifetime.

EXPERIMENTAL APPARATUS AND METHOD

The Claudet-type cryostat made of stainless steel was used. The inner He bath is 45 cm in diameter and 157 cm in height with liquid helium content of about 150 liters. The He II compartment (He II bath with no free surface) has a volume of about 74 liters. Maximum four test heaters can be installed in the He II compartment.

Four test heater plates made of Manganin with the same dimensions of 10 mm in width, 40 mm in length and 0.1 mm in thickness were used. Three of them were located at one end of rectangular ducts made of 4 mm thick fiber reinforced plastic (FRP) with the cross sections of 10×40 mm², 16.3 ×40 mm², and 20×40 mm², and all 100 mm in length. As shown in Fig. 1, one side of each test plate was thermally insulated by pasting it on an inner side of a FRP end-plate and the

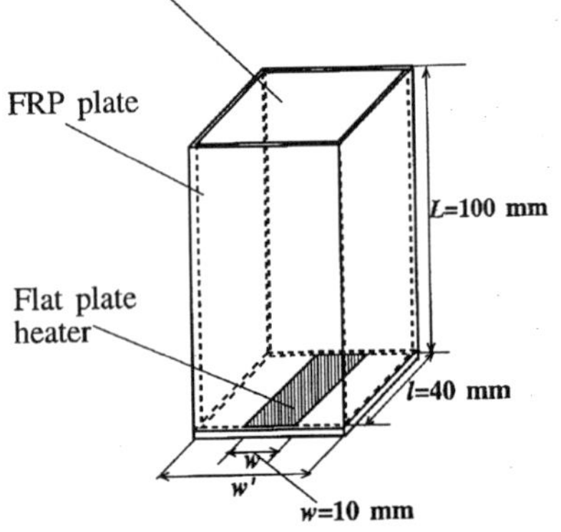

Fig. 1 Test heater in a duct.

other end of the duct was opened to a pool of pressurized He II. The fourth one was pasted on a FRP plate and supported in a pool of pressurized He II. Two fine 50-μm diameter platinum wires were spot welded as potential taps at around 5 mm from each end of the test plate. The thermal conductivity of the FRP is about 0.08 W/m · K. The leakage heat from the test plate to He II through the FRP end plate is estimated to be within 0.2 % of the heat flux beyond q_{st}. The leakage heat from other walls of the duct is negligible.

The heating current to the test plate was supplied by a power amplifier which can supply a direct current up to 300 A at a power level of 6 kW. The input signal of the power amplifier was controlled by an analog computer so that the heat generation rate in the test plate agrees with a desired heat input waveform.

The average temperature of the test heater was measured by resistance thermometry. A double bridge circuit including the test heater as a branch of the bridge was used to measure the electrical resistance of the test heater during its heating. The output voltages of the bridge circuit were amplified and passed to a digital memory system. These voltages are simultaneously sampled at the shortest available time interval, 0.5 μs, and stored in a computer. The average temperature of the test heater was calculated from its electrical resistance with the aid of the previously calibrated temperature resistance relation. The heat generation rate of the test heater was determined from the voltage drops across the potential taps of the test heater and across the standard resistance. The heat flux from the heater surface to the liquid is the difference between the heat generation rate per unit surface area and the rate of change of energy storage in the heater obtained from the temperature versus time curve.

The surface temperature of the test plate was obtained from the measured average temperature and heat generation rate by solving the conduction equation for the test heater with the heat flux set as a boundary condition at the surface. The thermal conductivity of the test heater that becomes lower with the decrease in temperature was approximately taken to be that for the average temperature in calculating the surface temperature.

STEADY-STATE CRITICAL HEAT FLUX

The steady-state critical heat fluxes obtained are highest for the test plate without duct and higher for a duct with larger cross sectional area. The steady-state critical heat for the test plate without duct, $q_{st,0}$, and those for the test plates with ducts, $q_{st,d}$, are expressed well by the following correlation given by the authors [11,12].
For a flat plate of length l and width w,

$$q_{st,0} = 0.58\left[\frac{2}{lw/\{2(l+w)\}}\int_{T_B}^{T_\lambda}\frac{1}{f(T)}dT\right]^{\frac{1}{3}} \tag{1}$$

and for the test plate in a rectangular duct

$$q_{st,d} = \left[k\int_{T_B}^{T_\lambda}\frac{1}{f(T)}dT\right]^{\frac{1}{3}} \tag{2}$$

where

$$\frac{1}{k} = \left\{\left(\frac{1}{M}\right)^{3/2} + \left(\frac{1}{N}\right)^{3/2}\right\}^{2/3}, \quad M = A_d^{\,3}/(A_h^{\,3}L), \quad N = 0.78/w,$$

A_d and A_h,are the cross sectional area of the duct and surface area of the test heater, $f(T)^{-1} = g(T_\lambda)[T_R^{6.8}(1-T_R^{6.8})]^3$, $g(T_\lambda) = \rho^2 s_\lambda^4 T_\lambda^3 / A_\lambda$, $T_R = T/T_\lambda$, $s_\lambda = 1559$ J/(kg K), $A_\lambda \cong 1150$ m s/kg .

TRANSIENT HEAT TRANSFER FOR STEPWISE HEAT INPUTS

Transient heat transfer coefficients for stepwise heat inputs with the heights larger than the values corresponding to q_{st} were measured for the bulk liquid temperatures of 1.8, 1.9, 2.0 and 2.1 K at atmospheric pressure.

Typical time traces of the heat input (heat generation rate per unit volume) Q, heat flux q, and surface temperature difference, ΔT, are shown in Fig. 2. Initially, the heat input rapidly increases in time and then it takes a constant value, Q_s, after $t = t_A$. As shown in the figure, the surface temperature difference and the heat flux remain constant at ΔT_s and q_s, respectively, for a certain duration ($t_B - t_A$), then they begin to increase and decrease, respectively, at $t = t_B$. The duration $t_L = t_B - t_A$ is defined as the lifetime of the quasi-steady-state heat flux q_s. The transient heat flux for the stepwise heat input increases along the Kapitza conductance curve and the curve's extrapolation, and then reaches a quasi-steady-state (ΔT_s, q_s) on the extrapolated curve.

The measured values of lifetime, t_L, for the plates in the ducts of $A_d / A_h = 1.0$, 1.6, and 2.0, and those for the plate in a pool of He II are shown versus q_s with bulk liquid temperature as a parameter in Figs. 3 to 6, respectively. In each figure, the lifetime is infinite for a step height corresponding to a heat flux lower than q_{st}, and it rapidly decreases with increasing the q_s from the q_{st} at each bulk liquid temperature. The decreasing rates of. t_L with the increments of q_s from q_{st} are lower for lower value of bulk liquid temperature.

For the transient heat transfer in a He II channel with $A_d / A_h = 1$ due to a sudden addition of the heat flux q at t=0 at one end, the following theoretical solution was given based on the two fluid model[13].

$$\Delta t^* = a^{-4} \overline{\rho c}\, \overline{f(T)^{-1}} (T_\lambda - T_B)^2 q^{-4} \tag{3}$$

Where Δt^* is the time to film boiling, a is the numerical coefficient of order unity, and the properties, $\bar{\rho}$, $\bar{c}$, and $\overline{f(T)^{-1}}$ are those averaged over from T_B to T_λ.

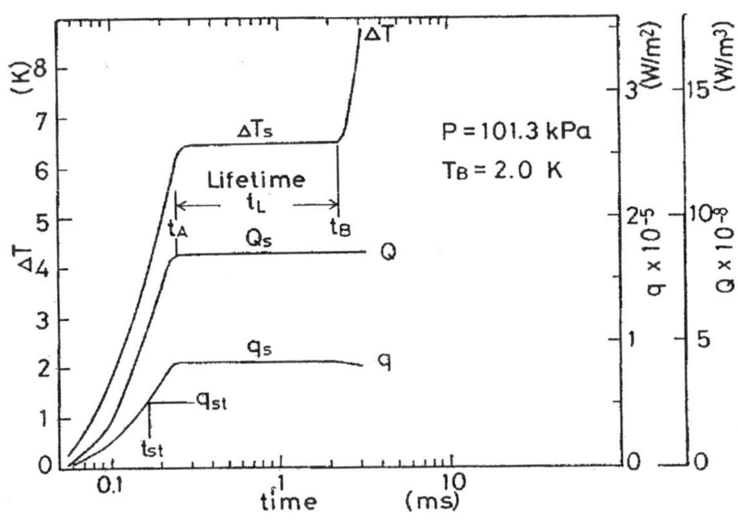

Fig. 2 Typical time traces of heat input, Q, heat flux, q, and surface temperature rise, ΔT

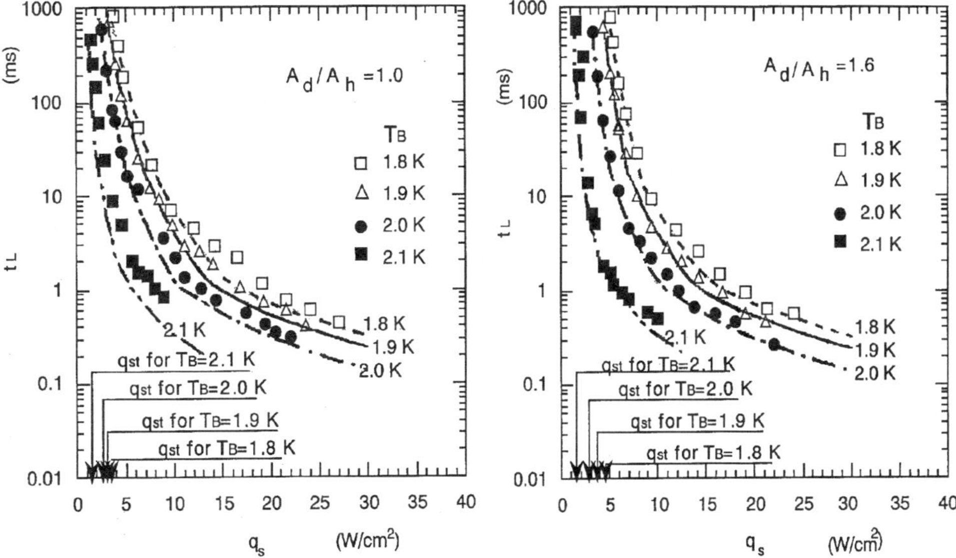

Fig. 3 Lifetime versus quasi-steady heat flux for the duct with Ad/Ah=1.0 .

Fig.4 Lifeteime versus quasi-steady heat flux for the duct with Ad/Ah=1.6

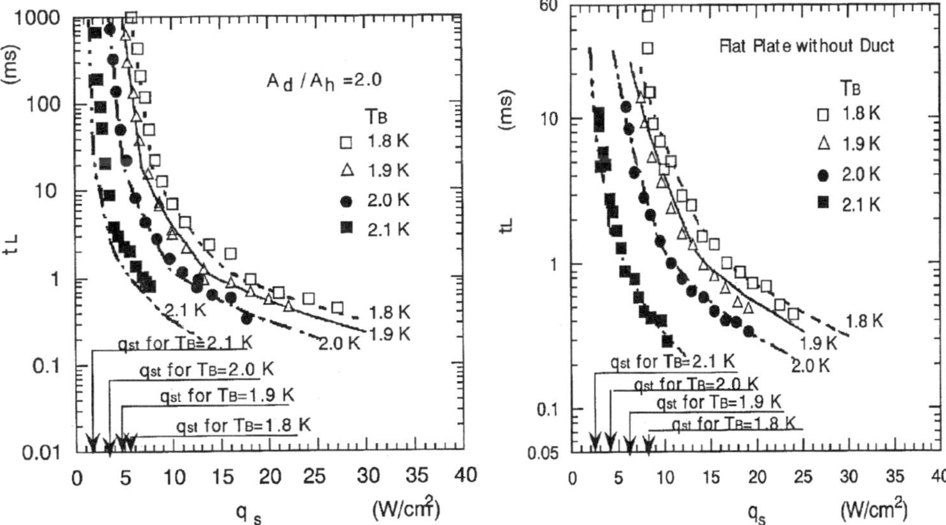

Fig.5 Lifetime versus quasi-steady heat flux for The duct with Ad/Ah=2.0 .

Fig.6 Lifetime versus quasi-steady heat flux for the test plate without duct.

Figure 7 shows the $\log(t_L)$ versus $\log(q_s)$ plot of the experimental data for the ducts with $A_d/A_h = 1.0$, at bulk liquid temperatures of 1.8 K. Data of Van Sciver[3] for 2-m-long tube are also shown. As shown in these figures, our data for t_L longer than around 1.2 ms and their data seem to be on a single line with the gradient of –4 on this graph. However, the value of t_L becomes longer than that given by the line for further increase of q_s (t_L <1.2 ms). The values of t_L on a flat plate in a pool of He II are also

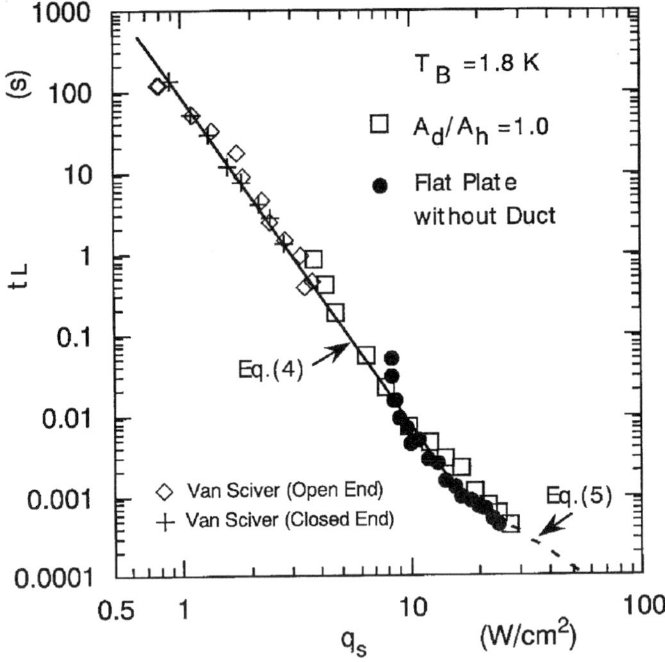

Fig. 7 The $\log(t_L)$ versus $\log(q_s)$ plot of the experimental data for the ducts with $A_d / A_h = 1.0$ and for a test plate without duct for bulk liquid temperature of 1.8 K.

shown in the figure for comparison. They agree well with those for the duct with $A_d / A_h = 1.0$ throughout the experimental range including the high q_s range (t_L <1.2 ms), although the q_{st} is higher than that for the test plate with a duct. It is from this fact that the two or three dimensional heat flow from a flat plate in a pool of He II occurs only near the q_{st}, and the transient heat transfer can be regarded as one-dimensional except this range near the q_{st}.

These data of t_L for the duct with $A_d / A_h = 1.0$ and for a test plate in a pool of He II are well described by the following equations already derived by the authors[10] based on Eq. (3) and the experimental data for the test plates with various widths in a pool of pressurized He II .

$$t_L = a^{-4}\overline{\rho c}\, f(T)^{-1}(T_\lambda - T_B)^2 q_s^{-4} \qquad \text{for} \quad t_L \geq 1.2 \text{ ms} \tag{4}$$

$$t_L = \overline{\rho c}\, B(T)^{-1}(T_\lambda - T_B)^2 q_s^{-2} \qquad \text{for} \quad t_L < 1.2 \text{ ms} \tag{5}$$

where $a = 1.16$, $B(T)^{-1} = s^2 T / A^*$, and $A^* = 8000$ m^3/(kg s).

Figure 8 shows the log(t_L) versus log(q_s) plot of the data on the flat plates in the ducts with the ratio of A_d / A_h higher than unity for the bulk liquid temperatures of 1.8 K. The curves given by Eqs. (4) and (5) are also shown in the figure for comparison. As shown in the figure, the data of t_L agree well with the curves for the q_s higher than a certain value. With the decrease of q_s from the value, they become longer than those given by Eq. (4) more significantly for the ducts with higher values of A_d / A_h. The data of t_L for the test plate without a duct are also shown in the figure. It should be noted that the threshold value of q_s (lower limit of q_s) for the one-dimensional heat flow regime is almost the same for the flat plates with and without ducts. For the q_s lower than the value, two- or three-dimensional heat flow expansion would occur most significantly for

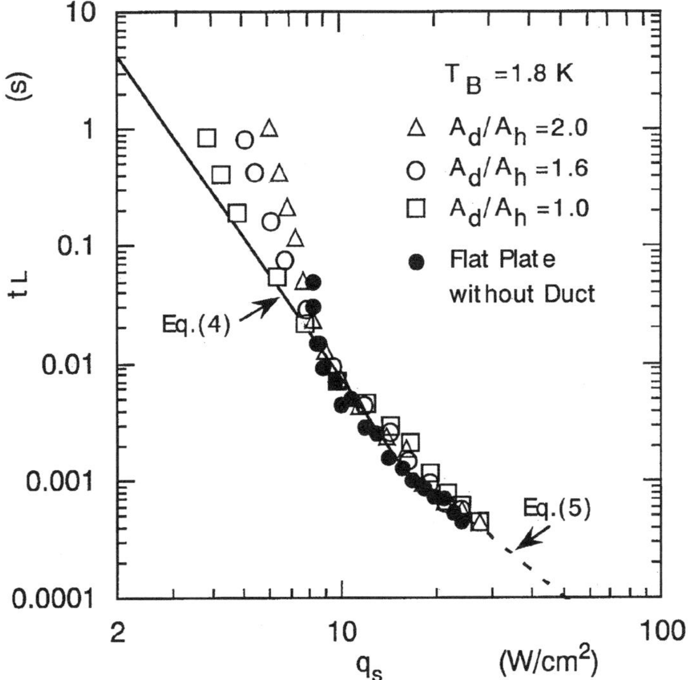

Fig. 8 The log(t_L) versus log(q_s) plot of the data on the flat plates in the ducts with the ratio of A_d / A_h higher than unity for the bulk liquid temperatures of 1.8 K.

the bare test plate in a pool of He II, less significantly for the ducts with smaller values of A_d / A_h.

The data of t_L on the flat plates in the ducts with the A_d / A_h higher than unity are expressed by the following equation:

$$t_L = t_L^* \exp[0.67(A_d / A_h)^2 R_p] \qquad \text{for}\, q_s \leq 1.1 \times q_{st0}, \qquad (6)$$

and Eqs. (4) and (5) for $q_s > 1.1 \times q_{st0}$, where t_L^* is the lifetime given by Eq. (4), and $R_p = (1.1 \times q_{st,0} - q_s)/(1.1 \times q_{st,0} - q_{st,d})$.

The values of t_L for the test plates in the ducts derived from Eqs. (4), (5) and (6), and those for the test plate without duct derived from Eqs. (4) and (5) are shown in Figs. 3 to 6. It can be seen from these figures that they area in good agreement with the experimental data.

CONCLUSIONS

Transient heat transfer caused by a stepwise heat input with the height, Q_s, larger than that corresponding to the steady-state critical heat flux, q_{st}, was measured for the flat plate heaters with and without ducts in subcooled He II at atmospheric pressure for the bulk liquid temperatures ranging from 1.8 to 2.1 K. Experimental results lead to the following conclusions.

Quasi-steady state exists with a lifetime on the extrapolation of Kapitza conductance curve. The surface temperature rapidly increases to film boiling after the depletion of the lifetime. The lifetime is shorter with increased values of the quasi-steady heat flux, q_s, corresponding to the step height.

The lifetime for a quasi-steady heat flux, q_s, for the test plate without a duct and for the duct with $A_d / A_h = 1.0$ agree well with each other, although the q_{st} is higher than that for the test plate in the duct. It is from this fact that the two or three dimensional heat flow from a flat plate in a pool of He II would occur only near the q_{st}, and the transient heat transfer can be regarded as one-dimensional except this range near the q_{st}. The lifetime is proportional to q_s^{-4} for the q_s lower than that corresponding to $t_L \cong 1.2$ ms, and to q_s^{-2} for q_s larger than the value.

The data of lifetime on the test plates in the ducts with the A_d / A_h higher than unity agree well with those for the test plate without a duct and for the duct with $A_d / A_h = 1.0$ for the q_s higher than around 1.1 times of $q_{st,0}$. Here, $q_{st,0}$ is the steady-state critical heat flux for the same sized flat plate in a pool of He II. With the decrease of q_s from the value, they become longer than the values more significantly for the ducts with higher values of A_d / A_h. Correlation of lifetime for the test plates in the ducts with various values of A_d / A_h and without duct was given based on the two fluid model and experimental data.

REFERENCES

1. G. Bon Mardion, G. Claudet, P. Seyfert, Steady state heat transport in superfluid helium at 1bar, in: "International Cryogenic Engineering Conference-7," IPC Science and Technology Press, London (1978) p.441.
2. S.W. Van Sciver, "Kapitza Conductance of Aluminum and Heat Transport through Subcooled He II," Cryogenics, Vol.18, (1978) , p.521.
3. S.W. Van Sciver, "Transient Heat Transport in He II," Cryogenics, Vol.19, (1979), p.385.
4. P. Sayfert, J. Lafferanderrie, and G. Claudet, "Time Dependent Heat Transport in Subcooled Superfluid Helium," Cryogenics, Vol.22, (1982), p.401.
5. T. Gradt, R. Wang, U. Ruppert, and K. Luders, Transient heat transfer to superfluid liquid helium, in "Advances in Cryogenic Engineering," Vol.35A, (1989) p.117.
6. S.W. Van Sciver, and R.L.Lee., in: Heat tranfsser from circular cylinders in He II, "Cryogenic Process and Equipment in Energy Systems," ASME Publication No. H00164 (1981) p.147.
7. M. Shiotsu., K. Hata, and A. Sakurai, Transient heat transfer for large stepwise heat inputs to a horizontal wire in subcooled He II, in: "Advances in Cryogenic Engineering," Vol.37A, (1992), p.37.
8. M. Shiotsu, K. Hata, and A. Sakurai, Effect of test heater diameter on critical heat flux in He II, in: "Advances in Cryogenic Engineering," Vol.39, (1994) p.1797.
9. M. Shiotsu., K. Hata, and A. Sakurai, Transient heat transfer from a horizontal wire in subcooled He II at atmospheric pressure for a wide Range of wire diameter, in: "Advances in Cryogenic Engineering," Vol.41, (1996) p.241.
10. M. Shiotsu , K. Hata, Y. Takeuchi, K. Hama and Y. Shirai., Transient heat transfer caused by a stepwise heat input to a flat plate in pressurized He II, in: "ICEC 17," DS.Dew-Hughes et al. ed. , Institute of Physics Publishing, Bristol & Philadelphia, (1998), p.687.
11. H. Tatsumoto, K. Hata, Y. Takeuchi, K. Hama, Y. Shirai and M. Shiotsu., Critical heat flux on various sized flat plates in pressurized He II, in: "ICEC 17," DS.Dew-Hughes et al. ed. , Institute of Physics Publishing, Bristol & Philadelphia (1998), p.683.
12. H. Tatsumoto, K. Hata, K. Hama, Y. Shirai and M. Shiotsu., Critical heat flux on a flat plate pasted on one end of a rectangular duct containing pressurized helium II, CEC/ICMC99, CDC- 7 , (1999).
13. L. Dresner, Transient heat transfer in superfluid helium, in: "Advances in Cryogenic Engineering", Vol.27, (1982), p.411.

CRITICAL HEAT FLUXES ON A FLAT PLATE ATTACHED TO ONE END OF A RECTANGULAR DUCT CONTAINING PRESSURIZED HE II

H. Tatsumoto,[1] K. Hata,[2] K. Hama,[2] Y. Shirai,[1] and M. Shiotsu[1]

[1]Dept. of Energy Science & Technology, Kyoto University,
Uji, Kyoto 611-0011, Japan
[2]Institute of Advanced Energy, Kyoto University,
Uji, Kyoto 611-0011, Japan

ABSTRACT

Critical heat fluxes (CHFs) were measured on single same sized flat plate heaters located at one end of six rectangular ducts with different cross-sectional areas in subcooled He II for bulk liquid temperature range of 1.8 to 2.1 K at atmospheric pressure in order to clarify the heat flow in a He II contained duct with a sudden change in its cross-sectional area. One side of the test plate was attached to the end plate of the duct made of Fiber-Reinforced Plastics (FRP), and the other end of the duct was opened to pressurized He II. The ratio of the cross-sectional duct area to the heater area was varied from 1.0 to 3.5. The CHFs increased with the increase in the ratio and approached a constant value: the CHF on the same sized flat plate in a pool of He II without duct. A CHF correlation can express the experimental data within 20 % error.

INTRODUCTION

Superfluid liquid helium (He II), especially pressurized He II, has excellent cooling properties, that is high heat transport and cooling stability for a local thermal disturbance on a winding of a large scale superconducting magnet, in comparison with normal liquid helium, He I. The superconducting magnets cooled by He II can obtain higher magnetic field. Therefore He II is expected as a coolant for a large-scale superconducting magnet for nuclear fusion facilities, accelerators and magnetic energy storage systems.

The cooling paths of the practical superconducting magnets are very complicated. It is important for cooling design of superconducting magnets to clarify heat flow in a duct containing He II in order to resolve the instability of the superconducting magnets caused by large pulsewise heat input for quenching due to conductor motion etc. Pfotenhauer[1] suggested that the cooling paths would be composed of a connection of ducts with various length and cross-sectional area. He treated the heat flow in the path as a series or parallel connection of the thermal resistance determined by the geometry of the ducts. However, whether such a treatment can be made or not is unresolved as far as the authors know.

The purpose of this study is twofold. The first is to obtain the experimental data of steady-state heat transfer and critical heat flux (CHF) on rectangular ducts with a sudden change in cross-sectional area in pressurized He II for wide ranges of experimental conditions. The second is to present a CHF correlation which can describe the experimental data.

EXPERIMENTAL APPARATUS AND METHOD

The cryostat[2] is the Claudet type made of stainless steel. The inner bath of the cryostat is 45 cm in diameter and 157 cm in height with liquid He content of about 150 liters. There is a glass epoxy separator in the middle height of the inner bath. The upper part of the bath contains saturated He I of about 74 liters with a free surface. The lower part of the bath contains subcooled He II of about 76 liters with no free surface. Four horizontal test heaters are supported in the He II compartment.

The shape of the test heaters is shown in Fig. 1. Six FRP (Fiber-Reinforced Plastics) ducts of the same lengths L, 100 mm, and the same thickness, 4 mm, with different cross-sectional area were used. One side of the flat plate heater made of Manganin was attached to the end plate of each FRP duct and the other end of the duct was opened to a pool of pressurized He II as shown in Fig. 1. All flat plate heaters were 10 mm in width, 40 mm in length and 0.1 mm in thickness. The inner widths of the cross-sectional duct area, w', were 10.0 mm, 16.1mm, 19.6 mm, 22.1 mm, 26.3 mm and 35.0 mm, respectively. Therefore, the ratio of the cross-sectional area A_d $(=l \times w')$ to the heater area A_h $(=l \times w)$, A_d/A_h, of six duct heaters was varied from 1.0 to 3.5. Two 50 μm-diameter wires were spot welded as

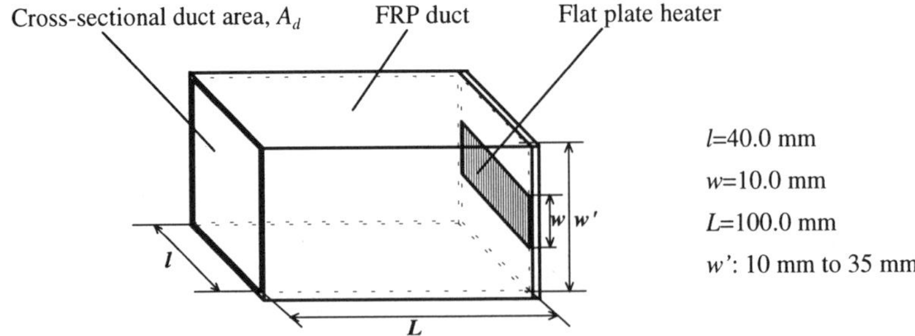

Figure 1. Shape of a duct heater

potential taps at about 5 mm from each end of the test heater. The thermal conductivity of the FPR for 1.8 K to 5.0 K is λ=0.082 W/m K. The leakage heat flow from the test heater to the He II through the FRP plate to which the test heater was attached was estimated to be within 0.2 % of the total heat flux.

The heating current to the test heater was supplied by a power amplifier. The power amplifier can supply a direct current of up to 300 A at a power level of 6 kW. The input signal of the power amplifier was controlled by an analog computer so that the heat generation rate in the test heater agreed with a desired value.

The average temperature of the test heater was measured by resistance thermometry using a double bridge circuit that included the test heater as a branch of the bridge. The output voltages of the bridge circuit together with the liquid temperature signal from a germanium diode in the He II bath were amplified and stored in a 64 k word/channel memory. The heater surface temperature was obtained from the measured average temperature and heat generation rate by solving the conduction equation for the heater. Experimental error is estimated to be 0.2 K in the heater surface temperature and 2 percent in the heat flux.

RESULTS AND DISCUSSION

Steady-State Heat Transfer

Figure 2 shows one of a typical heat transfer curve for a rectangular duct immersed in pressurized He II at atmospheric pressure for bulk liquid temperature of 2.0 K. The duct heater used in the figure has A_d/A_h=2.21. Vertical axis indicates the heat flux on the heater surface. Horizontal axis indicates ΔT that is the difference between heater surface temperature, T_s, and bulk liquid temperature, T_B. As the heat input increases quasi-steadily, the heat flux increases up to the critical heat flux, q_{cr}, where the heat transfer state continuously changes from Kapitza conductance regime to film boiling regime. The film

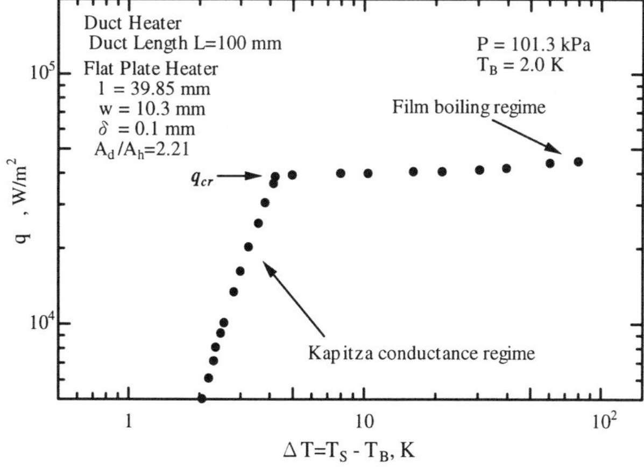

Figure 2. Typical heat transfer curve for a rectangular duct immersed in pressurized He II

boiling curve obtained by quasi-steadily increasing the heat input agrees well with that by decreasing it. No appreciable temperature jump was observed at CHF with each phase change from Kapitza conductance regime to film boiling regime.

CHF on Ducts with Various Cross Sectional Area

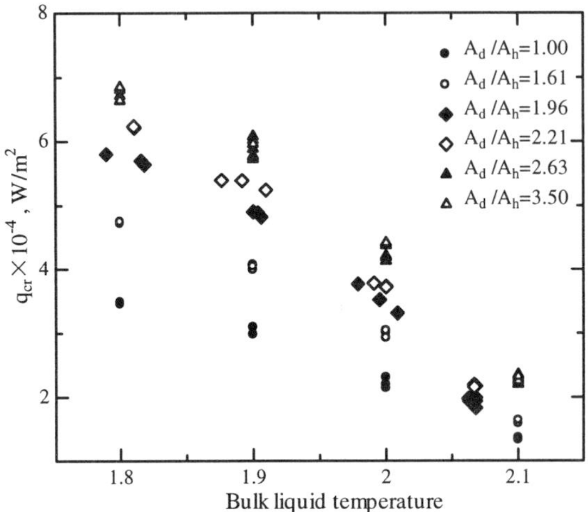

Figure 3. Relationship between critical heat flux, q_{cr}, and bulk liquid temperature, T_B,

Figure 3 shows the relationship between critical heat flux, q_{cr}, and bulk liquid temperature, T_B, at atmospheric pressure with A_d/A_h as a parameter. As shown in the figure, the CHF increases with the decrease in bulk liquid temperature, T_B. The values of CHF for a fixed T_B tend to increase and approach a constant value with increasing A_d/A_h. Suppose that the duct heater with very large A_d/A_h can be regarded as a part of infinitely long flat plate with the width of w as shown in Fig. 4. This is because the liquid flow along the length direction of the flat plate heater would be prevented by the vertical FRP plates attached to both sides of the flat plate. Therefore it is assumed that the CHF for the duct with large A_d/A_h would not exceed that on the infinitely long flat plate immersed in a pool of pressurized He II for the corresponding conditions. The CHF would approach the value for a flat plate

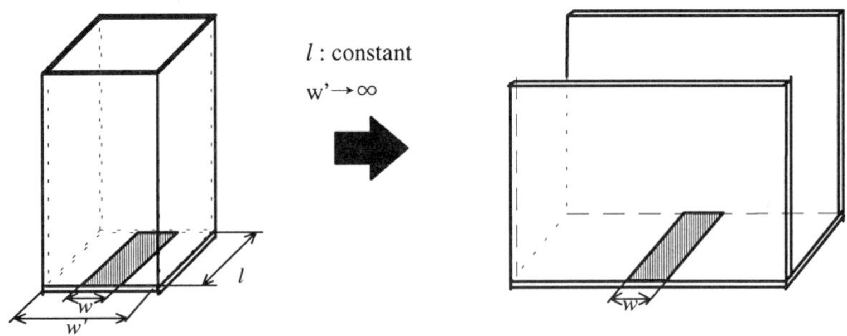

Figure 4. Schematic of a duct heater with large A_d/A_h

as the A_d/A_h increases. A CHF correlation for a duct with a sudden change in cross sectional duct area is derived in the next section based on the experimental data and the assumption.

CHF Correlation

The present authors[3] have already presented the following correlation for the critical heat flux on a flat plate with the length, l, and the width, w, in subcooled He II (for $T_{sat}(P_L) > T_\lambda$) by modifying a CHF correlation on a cylinder[4-7] obtained based on the Gorter-Mellink equations.

$$q_{cr} = K\left[\frac{2}{lw/\{2(l+w)\}}\int_{T_B}^{T_\lambda}\frac{1}{f(T)}dT\right]^{\frac{1}{3}} \tag{1}$$

where

$f(T)^{-1} = g(T_\lambda)[T_R^{6.8}(1- T_R^{6.8})]^3$, $g(T_\lambda) = \rho^2 s_\lambda^4 T_\lambda^3/A_\lambda$, $T_R = T/T_\lambda$, , $s_\lambda = 1559$ $J/(kg\ K)$, $A_\lambda = 1150\ m\ s/kg$

The modification coefficient K in Eq.(1) was determined to be 0.58 by using experimental data of CHF for cylinders with various diameters. The exponent of 6.8 in the equation of $f(T)^{-1}$ was changed from the original value of 5.7; the bulk liquid temperature at which the maximum of $f(T)^{-1}$ occurs was shifted from 1.923 K to 1.96 K with this modification. It has been confirmed that the correlation (Eq. (1)) can describe the experimental data well.

The CHF correlation for a duct with a sudden change in cross-sectional duct area was derived from a CHF model on a flat plate and on a Gorter-Mellink duct follows. First, if the heat flow rapidly expands from the flat plate attached to the end of the FRP duct to the duct containing He II and is uniform in the duct, the heat flux q averaged over the cross sectional area A_d can be expressed as

$$q = \frac{A_h}{A_d} q_0 \tag{2}$$

where q_0 is the heat flux on the heated surface. Substituting Eq. (2) into Gorter-Mellink equation yields the following form.

$$\frac{dT}{dx} = -f(T)\left(\frac{A_h}{A_d}\right)^3 q_0^{\ 3} \tag{3}$$

The CHF is determined by the condition that the liquid adjacent to the heated surface at x=0 reaches the λ-temperature and the boundary condition is that the liquid temperature at one end of the duct at $x = L$ is approximately equal to the bulk liquid temperature T_B, although this assumption does not hold for a very short duct.[8] Therefore integration of Eq. (3) under the conditions leads to the CHF for the duct with A_d / A_h .

$$q_{cr} = \left(\frac{A_d^{\ 3}}{A_h^{\ 3}} \frac{1}{L} \int_{T_B}^{T_\lambda} f(T)^{-1} dT \right)^{1/3} = \left(M \int_{T_B}^{T_\lambda} f(T)^{-1} dT \right)^{1/3} \tag{4}$$

where $M = A_d^{\ 3} / (A_h^{\ 3} L)$.

On the other hand, it is assumed from the experimental results that the values of the CHF for the duct with large A_d/A_h approaches the value for the flat plate with infinitely long length and width, w, under the corresponding conditions. Therefore the CHF for the infinitely long flat plate can be described by setting, l, equal to infinity in Eq. (1).

$$q_{cr} = \left(\frac{4 \times 0.58^3}{w} \int_{T_B}^{T_\lambda} f(T)^{-1} dT \right)^{1/3} = \left(N \int_{T_B}^{T_\lambda} f(T)^{-1} dT \right)^{1/3} \tag{5}$$

where $N = 0.78/w$. It can be seen from Eq. (4) and Eq. (5) that the value of the coefficients, M and N, of integrated value of $f(T)^{-1}$ in Eq.(4) and Eq.(5) are dependent only on the heater geometry. Therefore it is expected that the CHF for the duct with a sudden change in cross sectional area can also be expressed in terms of geometry-dependent coefficient k.

$$q_{cr} = \left(k \int_{T_B}^{T_\lambda} f(T)^{-1} dT \right)^{1/3} \tag{6}$$

The coefficient k may be expressed as the following approximate form by using the geometry coefficients, M and N.

$$\frac{1}{k} = \left\{ \left(\frac{1}{M} \right)^n + \left(\frac{1}{N} \right)^n \right\}^{1/n} \tag{7}$$

where the exponent n will be determined by the experimental data. Whether the above assumption is adequate or not is examined by comparison with the experimental data in the next section.

Comparison with Experimental Data

Figure 5 shows the relationship between the values of the coefficient k in Eq. (6) obtained from the experimental data and the ratio of the cross sectional duct area to the heater area, A_d/A_h with bulk liquid temperature as a parameter. The data of the coefficient for each bulk liquid temperature were obtained by substituting the measured values of CHF and integrated values of $f(T)^{-1}$ under each condition into Eq. (6). In this figure the values predicted by N in Eq.(4) and M in Eq.(5) are expressed in term of a curved line and a straight line, respectively. The values of M increase proportional to $(A_d/A_h)^3$, and the values of N hold constant independent of the value of A_d/A_h. When $A_d/A_h = 1$, the experimental data agree well with those by Eq. (4). The experimental data become lower than the values

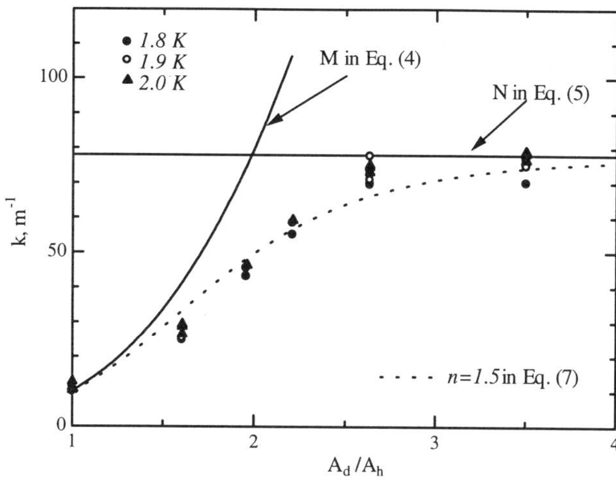

Figure 5. Relationship between coefficient and ratio of cross sectional duct area to heated surface area

predicted by Eq. (4) with increasing A_d/A_h and approach the constant value on the infinitely long flat plate in pool of pressurized He II. For instance, the data for A_d/A_h =1.63 are about 37% lower than the values predicted by Eq. (4) and the data for A_d/A_h =1.93 are 40% lower than the values predicted by Eq. (4). When A_d/A_h =2.63, the data are about 66% lower than the values predicted by Eq. (4) and then nearly agree with the values predicted by Eq. (5). Thus it is shown from the experimental results that the CHFs for the duct become far lower than the values by Eq. (4) in very small A_d/A_h region and agree with those on the flat plate. It is assumed that the heat flow for the duct with a sudden change in cross-sectional area would not rapidly expand from the flat plate heater to the duct because of the existence of the vortex near the heater unlike Photenhauer's treatment[1]. The exponent n in Eq. (7) was determined to be 1.5 based on the experimental data. The values of the coefficient predicted by Eq. (7) with n=1.5 were drawn as a broken line in the figure. It was confirmed that the CHF correlation can describe the data of the coefficient within 20 % difference.

CONCLUSIONS

The CHFs of steady-state heat transfer were measured on six ducts with the ratios of the cross sectional duct area to heater area, A_d/A_h, ranging from 1.0 to 3.5 in subcooled He II at atmospheric pressure for the bulk liquid temperatures ranging from 1.8 to 2.1 K. Experimental results lead to the following conclusions.

The CHFs increase with the increase in the ratio of A_d/A_h and approach a constant value that agrees with the CHF on the same-sized flat plate in a pool of He II. They are almost constant for the A_d/A_h>2.63. The CHFs for the A_d/A_h=1.93 and A_d/A_h=2.63 are about 40% and 66 % lower than the values obtained by assuming that the heat flow rapidly expands from the test heater and becomes uniform over the cross sectional area of the duct.

The CHF correlation for the duct with a sudden change of cross sectional duct area was obtained by modifying that on the flat plate and that on one-dimensional duct. It was confirmed that the correlation can describe the author's data on the ducts with the ratio, A_d/A_h, ranging from 1.0 to 3.5 within 20 % difference.

REFERENCES

1. J.M. Pfotenhauer, Geometry Dependent of Steady State Heat Transfer in He II, *Cryogenics.* 32:466(1992)
2. M. Shiotsu, K. Hata and A. Sakurai, Effect of Test Heater Diameter on Critical Heat Flux in He II, "Advances in Cryogenic Engineering," Vol.39, Plemum Press, New York (1994), p. 1797.
3. H. Tatsumoto, K. Hata, K. Hama Y. Shirai and M. Shiotsu, Critical Heat Fluxes on Flat Plates in Pressurized He II, "Proc. of 17th International Cryogenic Engineering Conference", IPC Science and Technology Press, Guildford (1998), p. 683
4. M. Shiotsu, K. Hata, and A. Sakurai, Transient Heat Transfer from a Horizontal wire in Superfluid Helium Caused by Exponential and Step Heat Inputs. "Superfluid Helium Heat Transfer," J.P. Kelly and W.J. Schneider, ed., ASME Publication No. HTD-134(1990), p. 9.
5. A. Sakurai, M. Shiotsu and K. Hata, Transient Heat Transfer for Large stepwise Heat Inputs to a Horizontal Wire in Saturated He II, "Advances in Cryogenic Engineering," Vol. 37A, Plenum Press, New York (1992), p. 25.
6. M. Shiotsu, K. Hata and A. Sakurai, Steady and Unsteady Heat Transfer from a Horizontal Wire in a Pool of Subcooled He II, "Heat Transfer and Superconducting Magnet Energy Storage," ASME Publication No. HTD-221(1992), p. 19.
7. M. Shiotsu, K. Hata and A. Sakurai, Transient Heat Transfer from a Horizontal Wire in Subcooled He II at Atmospheric Pressure for a Wide Range of Wire Diameter, "Advances in Cryogenic Engineering," Vol.41A(1992), p. 241.
8. K. Hata, M. Shiotsu, Y. Takeuchi and K. Hama, A Model for the CHF of One-wall Heated Rectangular He II Channel, "Proc. of 17th International Cryogenic Engineering Conference", IPC Science and Technology Press, Guildford (1998), p. 789

INFERRING THE STRUCTURE OF POROUS MATERIALS USING SUPERFLUID HELIUM

H. A. Snyder

University of Colorado at Boulder and
Ball Aerospace & Technologies Corp.
Boulder, Colorado, 80309

ABSTRACT

The operation of a phase separator for superfluid helium depends on the internal structure of the porous plug. We show how a phase separator apparatus can be used to study the flow through porous materials. This method appears to be better than conventional methods currently in use. Superfluid helium is a preferable test substance over ordinary fluids for two reasons. First, it has nearly zero contact angle for most substrates. Second, the superfluid film insulates the meniscus from surface contamination that causes variations of the contact angle. We present the theory of operation based on the governing equations. The largest term of the pressure balance equations is the jump across the meniscus. The jump is inversely proportional to the hydraulic diameter. We exploit this dependence to find the statistical properties of the hydraulic diameter that are helpful in predicting flows. We find the power spectrum of the hydraulic diameter from the measured data. All the useful probability distributions can be found from the power spectrum. We show how hysteresis loci arise due to the porous structure and discuss the probability distributions necessary to analyze hysteresis loci.

INTRODUCTION

Phase separators were used successfully on several space missions to vent superfluid helium: IRAS[1,2] COBE[3] and SHOOT[4]. A phase separator system consists of a storage dewar for a He II connected to a vent line by a porous plug. The vent line is pumped by space vacuum. A schematic of the system is shown below. Several investigators have published models to explain the operation of the device[5,6,7,8,9]. The author described a systems level model[10] that incorporates many of the features of previous theories, introduces some new phenomena and integrates the various effects.

Advances in Cryogenic Engineering, Volume 45.
Edited by Shu *et al.*, Kluwer Academic / Plenum Publishers, 2000.

Figure 1. A typical flow path through a granular porous material. The heavy dotted line is the central stream line.

The internal structure of the porous plug plays an important part in determining the operational characteristics of the phase separator in the author's model. Other models use an average pore diameter as the only structural parameter. Yuan et al.[8] pointed out that the plots of mass flow versus driving temperature gradient differ for uniform tubes with a distribution of diameters from that of the same number of tubes all with the average diameter. The present model assumes a distribution of flow paths with variable pore size along the path. A flow path might look like Fig. 1. This is a slice through a three dimensional structure and does not show possible multiple connections. The structural element is a suitably defined hydraulic diameter perpendicular to the central flow line plotted versus distance along the flow line. Statistical averages are applied to this element. Averages of the structural element are generally not available. However, we will show below how they can be deduced from a few measurements with a the phase separator apparatus. When these parameters are known, the operational curves of the device can be calculated.

The parameters needed to set up a model for the phase separator are also the most important for analyzing flows in porous media. The theory to follow shows how to find the probability distributions of: the hydraulic diameter, the convergence and divergence of the hydraulic diameter and the number and distance between "throats". Throats are the local minima of the hydraulic diameter. These statistical distributions are the basis for predicting: the spreading of pollutants in ground water, oil flows in petroleum engineering, the flows in aquifers, the operation of water purification plants, filtration processes and flows in related fields where flow in porous materials is important. This is an inverse problem in that properties of a system are deduced from measuring the output for a known input.

The purpose of this research is to show how the superfluid phase separator can be used to find important parameters used to calculate flows in porous media.

Advantages. The equations for the flow of superfluid helium through porous materials have been tested experimentally and are found to agree with calculations.[11] Atkins and Narahara[12] show that superfluidity does not modify the structure of a liquid-gas interface. The pressure differential is determined by the same formula as for an ordinary fluid. They also find a contact angle of zero. In their work the contact angle and the shape of the meniscus are measured with a film covering all surfaces as we assume. Reynolds and Satterlee[13] in their review article state that pure liquids on clean surfaces do not show contact angle hysteresis.

The largest term in the pressure balance equation is usually the jump across the gas-liquid interface. This is due to surface tension: $4\sigma/(d \cos \theta)$ where σ is the surface tension,

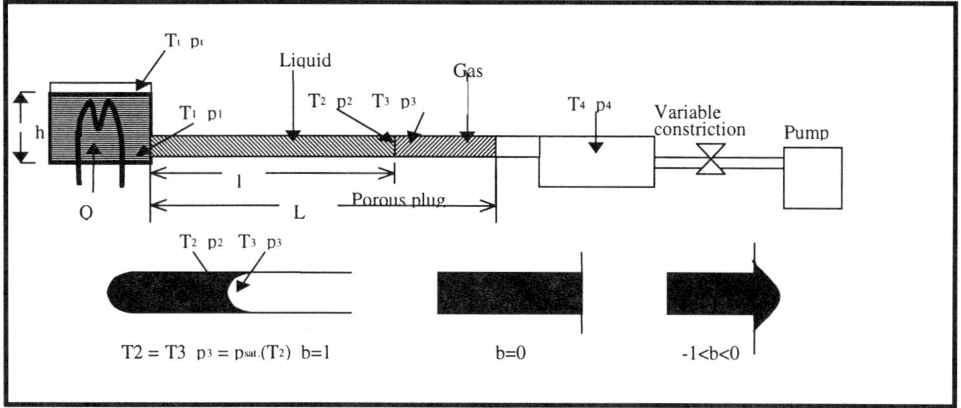

Figure 2. Schematic of phase separator apparatus with definition of symbols.

d the hydraulic diameter perpendicular to the interface and θ the contact angle. The discontinuity due to surface tension occurs in all fluids and is the physical mechanism usually used by conventional methods to find the distribution of hydraulic diameter. For superfluid helium θ is very close to zero for nearly all substrates and this is a substantial advantage.

The standard method of finding the distribution of hydraulic diameter is the capillary pressure measurement. The pressure needed to inject a nonwetting liquid, like mercury, into a sample is measured versus the fraction of void fraction filled. There are inherent uncertainties in this method due to unpredictable variations of θ. The contact angle for most liquids including mercury has a directional hysteresis. An advancing interface for the initial measurement has a well defined θ. When the interface recedes some of the liquid adheres to the solid surface so that the receding value of θ differs from the advancing value. A second measurement has a different advancing θ than the initial test.

Another difficulty with the capillary pressure measurement is that θ depends on adhered gas. The adhered gas may be that of the test liquid or a foreign gas. It takes a long time for the adhered layer to reach equilibrium and θ may vary significantly and randomly with the amount adhered and with time.

These problems do not occur with He II because the film covers all solid surfaces. The film insulates θ from the contamination of the surface and there is no measurable hysteresis. Also, the film forms very rapidly. The capillary pressure method does not find the distribution of throats. An analysis of hysteresis loci that occur in the phase separator measurements leads directly to those distributions. The He II method can define the hydraulic diameter more precisely than the capillary pressure method.

MODEL EQUATIONS

A simplified schematic of a phase separator is shown as Fig. 2. This diagram defines the relevant temperatures, *T*, and pressures, *p*, of the model. The vent line used in space applications is replaced by a vacuum pump and a variable constriction. The heat input rate to the system is *Q* and this includes the parasitic heat leak. When *Q* is specified, the mass flow rate *M* is determined by $M = Q/\Lambda$ where Λ is the latent heat. The pressure p_4 is set and controlled by the variable constriction at the inlet of the pump. The apparatus may be used in two ways: (a) to simulate a phase separator or (b) to find the properties of the porous plug. In the former case M together with the parameters of the vent line and the case p_4 can be used to vary the operating conditions over a range.

The porous plug may have a gas-liquid interface inside the plug at location l or the interface may be on the surface at $l = L$. Here, b is a dimensionless constant multiplier of the pressure jump formula. When the meniscus recedes into the plug as in the left sketch $b = 1$. When the interface is on the surface, the meniscus has the shape shown by the right sketch. This occurs at low values of Q. As Q increases, the curvature of the meniscus increases to infinity as shown in the middle sketch with $b = 0$. A further increase in Q causes the meniscus to recede into the plug as shown in the left sketch and b increases discontinuously to one.

Consider first the case when there are no transverse cross connections between the channels. It will be assumed henceforth that a steady state has been reached. Several relations may be used to reduce the number of unknowns. There are two liquid-gas interfaces: one in the storage tank and the other in the porous plug. The equation of the saturation line connects the temperature and pressure at the interfaces: $p_t = p_{sat}(T_t)$ and $p_3 = p_{sat}(T_3)$. There are no significant temperature gradients in the tank because of the high thermal conductivity of He II so that $T_1 = T_t$. If the down stream interface is at the surface of the plug, $p_3 = p_4$ and $T_3 = T_4$. In this case $-1 < b < 0$ and is unknown. Otherwise, $b = 1$ and since this is an isenthalpic flow and helium gas is ideal at these pressures, $T_3 = T_4$. There is no temperature jump across an interface. Thus, there are only two temperatures in the model, $T_1 = T_t$ and $T_2 = T_3 = T_4$. Also, $T_2 = T_{sat}(p_3)$. p_1 differs from p_3 due to the gravity head: $p_1 = p_{sat}(T_1) + \rho gh$. The variable p_2 differs from p_3 due to the surface tension jump: $p_2 = p_3 - 4\sigma b/d(l)$ and the pressure drop through the gas filled section of the porous plug follows Darcy's equation: $p_3 - p_4 = M\eta_g(T_2)(L-l)/A\varepsilon k_g$ where A is the area of the plug, ε the porosity, η_g the viscosity of the gas and k_g is the Darcy permeability for gas flow. This provides four algebraic equations connecting the unknowns. The usual four equations for the flow of the superfluid collapse into two because Q is given and $M = Q/\Lambda$. The superfluid equations now are differential equations, one for the gradient of T and the other for the gradient of p. In a previous paper[10] we showed that the property variations should be included in the solution of the superfluid equations for the simulation of phase separators. The author has described methods for solving these equations using the SUPERFLOW program[14].

Note that these equations apply to individual flow paths such as that shown in Fig. 1. Measurements represent averages over an ensemble of paths. The only variables that differ from an individual path to an average flow are $d(l)$ and l. It will be clear from context which interpretation is to be applied.

For the simulation of a phase separator the six governing equations can be used to find the unknowns: l, T_1, T_2, p_1, p_2, and p_3, when the variables Q, A, ε, h, p_4, k_g, k_l, and $d(l)$ are known. More details on the model equations and methods of solution are found in the previous paper[10].

The same equations are used to find the properties of a porous plug. However, now $T_t = T_1$ and $T_4 = T_2$ are measured. The variables l, p_1, p_2, and p_3 and are still unknown and there are six equations. It appears at first consideration that only two parameters of the structure can be treated as unknown. However, k_g, k_l and ε are likely to be independent of the operating conditions while $d(l)$ has a different value for each set of Q and p_4. If we add $d(l)$ to the unknown variables and consider k_g, k_l and ε as unknown constants, three measurements at different operating conditions are sufficient to find all the parameters. The 18 equations, six from each measurement, are solved simultaneously. If more measurements are made at different conditions, statistical methods can be used to reduce

Figure 3. An interface in a granular porous material. The heavy lines are the menisci, all with the same curvature.

the error bars. One of the results of this exercise is a plot of d versus l. This relation is the starting point for an analysis of the internal structure.

POROUS GEOMETRY

A typical granular porous material with interface is shown as Fig. 3. It must be kept in mind that this is a cut through a three dimensional matrix. There are transverse cross connections and there may be some multiply connected surfaces. But observations with a microscope shows that a hydralic diameter can be defined for most pores for nearly all materials. This is an assumption of the mercury injection method, that has been the standard way to find pore size distributions for over 50 years. Apparently it is a good assumption. At a liquid-gas interface a meniscus will cover a surface even if it is multiply connected and the curvature is inversely proportional to the hydraulic diameter of the path.

Our interest is focused on the statistical properties of the hydraulic diameter d as a function of the distance along a flow line. This distance includes the tortuosity. The pressure jump at the gas-liquid interface provides a mechanism to measure d. The location of the interface is determined by the pressure balance equations presented above. The value of l for an individual stream line is determined by two effects. As Q increases the interface recedes and the pressure across the porous plug changes due to the difference between the pressure drop through the liquid and the gas. The pressure drop through the gas ranges from two to four times larger than through the liquid depending on the temperature. Thus, the pressure drop from p_1 to p_4 increases when l decreases. As l decreases d may increase or decrease and the pressure jump across the interface changes inversely. The value of l must adjust for changes in Q so
the two effects taken together satisfy the balance equations. For a "soda straw" material only the former effect occurs. In a previous paper[10] it was shown that l is a solution of an equation of the form $d(l)/4\sigma b = 1/(c_1 l - c_2)$ where c_1 and c_2 are expressions of properties of He II and Q. Both are strongly dependent on the temperature. This equation represents the interaction of the two effects.

Figures 1 and 3 show that d is a rapidly varying function of l for an individual path in most materials. If the grains are of the order of 1-10 μm, the pressure jump changes from maximum to minimum in 3-30 μm. For this change in l, the pressure change due to the differential gradient between the gas and liquid is negligible compared to that caused by the change in d. The transverse cross connections of the void space prevents appreciable pressure gradients transverse to the interface. These two facts lead to the conclusion that the surface tension term dominates in the determination of l. Thus, the interface in porous

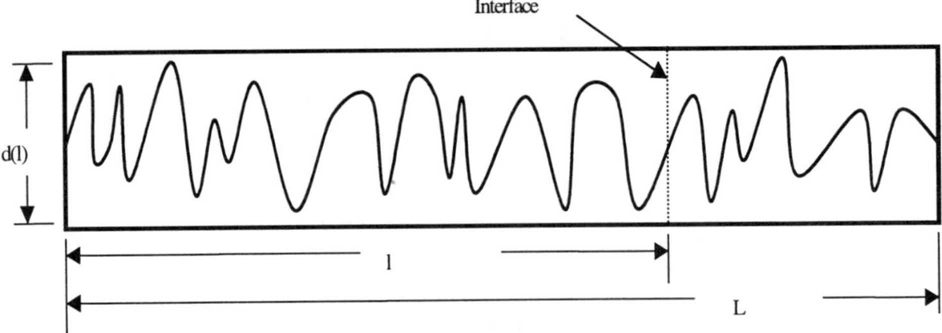

Figure 4. Plot of the hydraulic diameter perpendicular to a central stream line for an individual path, *d(l)* versus distance along a stream line *l*. The lengths *l* and *L* include the tortuosity.

material of the type shown in Fig. 3 consists of menisci all with nearly the same value of *d*. As *Q* increases each meniscus moves farther into the throat until a local minimum of *d* occurs. Then, the interface becomes unstable and reforms with different throats that have a smaller local minimum. The distance between local minima is of the order of the grain size and increases rapidly as *d* gets smaller than the grain size. It is rare that the interface recedes into the porous material more than 10 to 20 grain lengths or less than 100μ unless the standard deviation of *d* is very large. The transverse cross connections ensure that the equations for the flow of superfluid helium apply up to this point and the gradients of temperature and pressure are negligible from here to the interface.

STATISTICS

The plot of *d(l)* for the structural element is a random function with some limits. A typical plot is shown in Fig. 4. This curve extends throughout the length of the plug *L*. The interface is a distance *l* from the left edge of the diagram. There are a large number of paths such as this from inlet to exit. Some paths share the same throats. The equations above may be applied to individual paths or to averages over an ensemble of paths. For steady state conditions each path in the liquid portion ends at an interface with nearly the same curvature of the meniscus d. However, if the plots like Fig. 4 are random, each has a different length *l*. The interface is not plane but very uneven on the scale of the grain size as shown in Fig. 3. The variable *l* will vary from path to path for the same *d*. When experimental data is used as an input to the above equations, the predicted curve of *d(l)* is a statistical ensemble average over the paths.

The averaged frequency content of an ensemble of curves like Fig. 4 is the power spectrum. All the desired probability distributions can be derived from the spectrum. Those working in the field recommend a simple rectangular model: a flat spectrum with high and low frequency limits. The low frequency cut off f_a is related to the length of the largest pore space and the high frequency cut off f_b is set by the minimum throat. There are three frequency parameters: two cut off frequencies and the height of the rectangle w_0. The next task is to relate the frequency parameters to the experimental curve *d(l)*.

Many probability distributions associated with statistical elements such as Fig. 4 are available in the literature[15]. For a plane cut through a porous material the probability density for *d* is normal: $(2\pi\psi)^{-1/2}\ exp[-(d-d_m)^2/2\psi]$ where d_m is the mean value of *d* and ψ is determined by the power spectrum. For a rectangular spectrum $\psi = w_0\,(f_b - f_b)$.

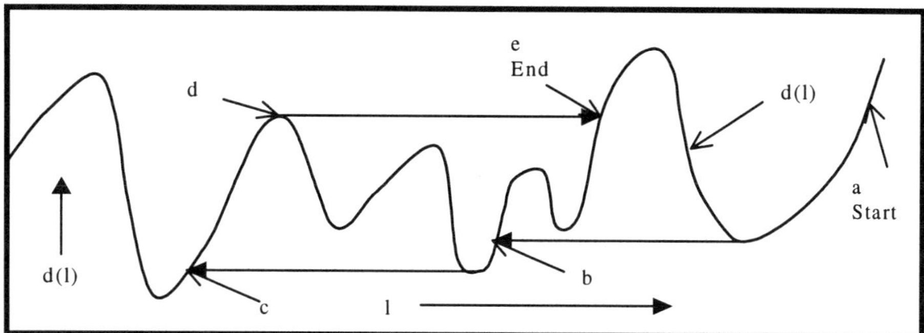

Figure 5. Hysteresis locus for an individual flow path.

This distribution applies to the face of the porous plug. If an experiment begins with Q very small so that the interface is at the downstream face of the porous plug, this is the distribution of d at $l = L$. As Q increases and the interface recedes into the plug, each path length l adjusts so that it ends at a constant value of $d = d_0$. The average value of l and d_0 are set by the balance equations discussed above. A result of an experimental run is a plot of the average value of l versus d_0, the $d(l)$ mentioned above. We want to relate $d(l)$ to ψ in order to find w_0, f_a and f_b,

For a receding meniscus only locations with positive slope of $d(l)$ are stable. Thus, we want the average distance from the interface to the location where $d = d_0$ and the slope of $d(l)$ is positive. This problem was first solved by Kac[15] in the context of electrical noise. For a power spectrum $w(f)$ the average distance to the interface τ $(\tau = L\text{-}l)$ is:

$$\tau = \frac{1}{2}\left[\frac{\int_0^\infty w(f)df}{\int_0^\infty f^2 w(f)df}\right]^{1/2} \exp\left[\frac{(d_0 - d_m)^2}{2\psi}\right] \qquad \psi = \int_0^\infty w(f)df \tag{1}$$

The average distance between two recurring values of d_0 is 2τ. For a rectangular spectrum the first bracket in Eq. 1 is $(\sqrt{3}/2)[(f_b - f_a)/(f_b^3 - f_a^3)]^{1/2}$ and $\psi = w_0\,[f_b\text{-}f_a]$. The three parameters of the assumed power spectrum can be found using Eq. 1 and three points from the experimental curve $d_0(l)$. If more points are available, the accuracy of the assumed $w(f)$ can be tested. It is important in applying these methods that the experiment start at sufficiently low values of Q so that the interface is on the surface of the plug. This is assumed in deriving Eq. 1.

Hysteresis. Hysteresis loci are observed in the operation of phase separators[16]. Each individual path has frequent jumps in $d(l)$ and the observed curves are averages over the ensemble. Figure 5 shows a hysteresis locus of an individual path. Starting at (a) as Q increases the meniscus moves to smaller and smaller values of d_0. When the first minima is reached, there is a jump to (b). Increasing Q moves the meniscus to another minimum and there is a jump to (c). If Q decreases at (c) d_0 must increase and the locus is to the point (d). At the maxima (d) there is a jump to (e). Jumps occur at a series of deeper minima when Q is increasing and a series of increasing maxima when Q is decreasing. Thus l is multivalued in d_0 and this causes jumps and hysteresis.

The averaging over the ensemble requires probability distributions for the maxima and minima. The average distance between maxima and minima is:

$$\left[\frac{\int_0^\infty f^2 w(f)df}{\int_0^\infty f^4 w(f)df}\right]^{1/2} \tag{2}$$

which is $(5/3)^{1/2}[(f_b^3 - f_a^3)/(f_b^5 - f_a^5)]^{1/2}$ for a rectangular spectrum. The probability that a maximum or minimum selected at random has a particular value is known but is complicated[15]. It is very nearly normal for large values of d_0. We need the probability of a decreasing series of minima. This is a conditional probability that the next minima/maxima is bigger than the previous. It appears that this distribution is not reported in standard texts. Much can be learned about the structure by analyzing hysteresis loci. There is need for further work in probability theory before an adequate procedure for analyzing the hysteresis data can be described.

REFERENCES

1. D. Petrac and P. Mason, Infrared Astronomical Satellite superfluid helium tank temperature control, *Proc. 1983 Space Helium Dewar Conference 163-169,* University of Alabama (1984).
2. A. R. Urbach and P. V. Mason, IRAS cryogenic flight performance report, *Adv. Cryo. Eng.* 29:651-667 (1984).
3. S. M. Volz, M. J. DiPirro, S. H. Castles, M. G. Rychkowitsch and R. Hopkins, *Adv. Cryo. Eng.* 37:1183-1192 (1992).
4. J. G. Tuttle, M. J. DiPirro and P. J. Shirron, *Adv. Cryo. Eng.* 39:121- (1994).
5. U. Schotte, He II phase separation with slits and porous plugs for space cryogenics, *Cryogenics*, 24: 536-547 (1984).
6. M. J. DiPirro and J. Zahniser, The liquid/vapor phase boundary in a porous plug, *Adv. Cryo. Eng.* 35:173-179 (1990).
7. A. Nakano, D. Petrac and C. Paine, He II liquid/vapor phase separator for large dynamic range operation, *Cryogenics* 36:823-828 (1996).
8. S. W. K. Yuan, D. J. Frank and C. Lagas, The dependence of choked flow and break through on pore size distribution in vapor-liquid phase separators of He II using porous media, *Adv. Cryo. Eng.* 41:1189-1194 (1996).
9. S. W. K. Yuan, A. R. Urbach, S. M. Volz and J. H. Lee, Vapor-liquid separation of He II, *Cryogenics* 38:921-925 (1998).
10. H. A. Snyder and A. J. Mord, Modeling superfluid phase separator systems, *Adv. Cryo. Eng.* 43:1377-1384 (1998).
11. G. L. Mills, A. J. Mord and H. A. Snyder, Pressure maxima in the flow of superfluid ^{4}helium, *Phys Rev. B* 49:666-669 (1994).
12. K. R. Atkins and Y. Narahara, Surface tension of superfluid helium, *Phys. Rev,* 138:437-441 (1965).
13. W. C. Reynolds and H. M. Satterlee, Liquid propellant behavior at low and zero g, *NASA SP 106:* 387-439 (1966).
14. H. A. Snyder and A. J. Mord, Calculation of He II flow in tubes, *J. Low Temp. Physics* 86:177-209 (1992).
15. S. O. Rice, Mathematical analysis of random noise, *Bell Tel. J.* 23:282 (1944); 25:46 (1945).
16. D. Elliott, JPL report, JPLD-11412 (1994).

HEAT TRANSFER PECULIARITIES AND THEIR APPLICATION TO NON-DRAINED LIQUIFIED GAS STORAGE

Zh. A.Suprunova, V.E.Seriogin and A.V.Latchenko

SR&DB in CT of the B.Verkin ILTPh&E
of the National Acad. Sci. of Ukraine,
47, Lenin Ave., Kharkov, 310164,UKRAINE

ABSTRACT

The features of heat-mass transfer are studied at different stages of the process and different states of the two-phase medium. A method of processing experimental results is proposed and adjusted to the goal of describing the pressure growth in the vessel containing the liquified gas. Recommendations are worked out for practical application of the results obtained.

INTRODUCTION

Although the heat and mass transfer inside the tank were studied in numerous works [1,2,3], some questions are still unanswered. These include the following two: What factors of the process are responsible for the features of the phase transitions at the phase boundary? How are the phase transition features related to the pressure rise rate in the vessel? The theoretical description of heat and mass transfer through the phase boundary is based on some simplifying assumptions [2], which in particular disregard the factors related to the vessel volume limits and to the liquid level in the vessel. Therefore, theoretical results sometimes appear to be inadequate in practice.

THE PROBLEM AND SOLUTION METHODS

A limited volume is filled with a liquid and its vapor. There is a free surface between the phases. In accordance with the problem set a preassigned heat flow to the outer surface of the vessel is time- invariable and uniformly distributed over the surface. Before the vent valve is closed, the system is at equilibrium, the initial pressure being P_0. The moment of closing the vent valve is taken as an onset of the process. The task is to describe the basic parameters of the process, in particular the pressure change in the vessel caused by the external heat fluxes. It is known that the pressure inside the vessel does not correspond to the temperature of the mean mass heating of the contents. According to the saturation curve, the pressure in the vessel reflects the temperature of the liquid surface under the condition of the temperature- induced stratification of the medium. This process can be described by differential equations of motion, energy continuum jointly with the edge conditions. The set of equations is supplemented with the thermodynamic equation of state for the system as a whole. Since the external heat fluxes essentially influence the medium properties, it is difficult to obtain a complete analytical solution. The decision is therefore to investigate the process experimentally involving the methods of the similarity theory.

RESULTS

Experimental studies were made on liquid oxygen, nitrogen and hexane in two spherical (d = 0.300 m, 0.625 m) and two toroidal (d_1=0.144 m, d_2=0.250 m and d_1=0.192 m, d_2=0.350 m) vessels. The experimental equipment and the measuring system are described in Ref. [4]. The analysis was made for the results taken in liquid H_2 vessels [4-7].

The process as a whole was studied using the energy equation in terms of thermodynamics :

$$dq_d = dU - v_e dP + dq_{d*} \tag{1}$$

or

$$1 = \frac{dU}{dq_d} - v_e \frac{dP}{dq_d} + \frac{dq_{d*}}{dq_d} \tag{2}$$

Here dq_{d*} is convective heat tranfer, $v_e dP$ describes the external work (this description of the work, as in [8], is only possible at a small degree of the process irreversibility), q_d is the heat quantity transferred to the system; $v_e = (1-x)v' + xv''$ is the equivalent specific volume of the system; u - specific internal energy.

As follows from Eq. (2), $v_e \frac{dP}{dq_d}$ can be written as a function of the following arguments (with according values of $\frac{P_0}{P_{cr}}$, x_0) :

$$v_e \frac{dP}{dq_d} = f\left(\frac{dU}{dq_d}, \frac{dq_{d*}}{dq_d}\right) \tag{3}$$

The index * in Eq. (1-3) shows the characteristic time τ_* of the process corresponding to the time during which a closed cycle of internal flows is formed in the bulk of the liquid [3]. Therefore the notions in Eq. (1) after its transformation into Eq. (3) represent the basic factors responsible for the pressure rise in the system: a change in the internal energy of the medium with the corresponding flux distribution in the system, the thermodynamic state of the medium including the equivalent system volume, the character of heat and mass transfer at different stages of the process.

The process was investigated in toroidal and spherical vessels with liquid oxygen, nitrogen and hexane. The initial system of similarity numbers was

$$Ho_* = \frac{q\tau_*}{ml\rho' r}, \text{ where } l = \frac{V}{S}; \quad m = \frac{V_l}{V}; \quad Fo = \frac{a\tau}{l^2};$$
$$Pr = \frac{\nu}{a}; \quad Ra_l^* = \frac{ql^4 g\beta}{\nu ak}; \quad x_0 = \frac{v_e - v'}{v'' - v'}; \quad \frac{P_0}{P_{cr}}; \quad \frac{S_l S_g d^2}{4V^2} \tag{4}$$

Here a - thermal diffusivity, C_p - specific heat, d - diameter, g - acceleration due to gravity, k - thermal conductivity, l - length, r - latent heat of evaporation, m - volume filling, P - pressure, q - heating rate per unit area, S - vessel surface area, T - temperature, V - volume, x - vapor mass fraction, β - coefficient of thermal expansion, ν - kinematic viscosity, ρ - density, τ - time, Pr - Prandtl number, Ra* - modifiied Raleigh number, Ho* - dimensionless time parameter, Fo - Fourier number.

Within the system (4), the following relation was obtained for the saturation temperature corresponding to the pressure in the vessel:

$$c_p \frac{(T_s - T_0)}{q\tau_* S} m\rho' V = 0.65(Ho_*)^{-0.24} \left(\frac{1-m}{m}\right)^{-0.30} \left(\frac{\tau}{\tau_*}\right)^{n_x}, \tag{5}$$

where the function:

$$n_x = f\left(x_0, \frac{P_0}{P_{cr}}, \frac{q\tau S}{q\tau_* S}\right) \tag{6}$$

is plotted in Figs. 1a,b for the transient and steady-state conditions of the process, respectively.

The characteristic time of the process in the dimensionless form is

$$Fo_* = \frac{a\tau_*}{l^2} = 4.16 Ra_l^{*-0.41} Pr^{-1.58} \left(\frac{S_l S_g d^2}{4V^2}\right)^{0.60} \tag{7}$$

Eq. (5) contain the integral characteristic of the process, such as the parameter n_x, which relates the rate of the pressure growth in the system and character of phase transitions at the interface (see Fig. 1).

INTERPRETATION OF RESULTS

The results presented are helpful in detecting and describing the features of phase transitions under the conditions of the process. According to the data in Fig.1, the pressure growth rate is dependent on the vapor contents: it can enhance when evaporation is predominant over other processes of mass transfer through the phase boundary; it can attenuate when condensation is in progress. It can also attain the value corresponding to equilibrium at the phase boundary. The data such as in Fig.1 permit us to assess the real conditions of phase transitions in the system.

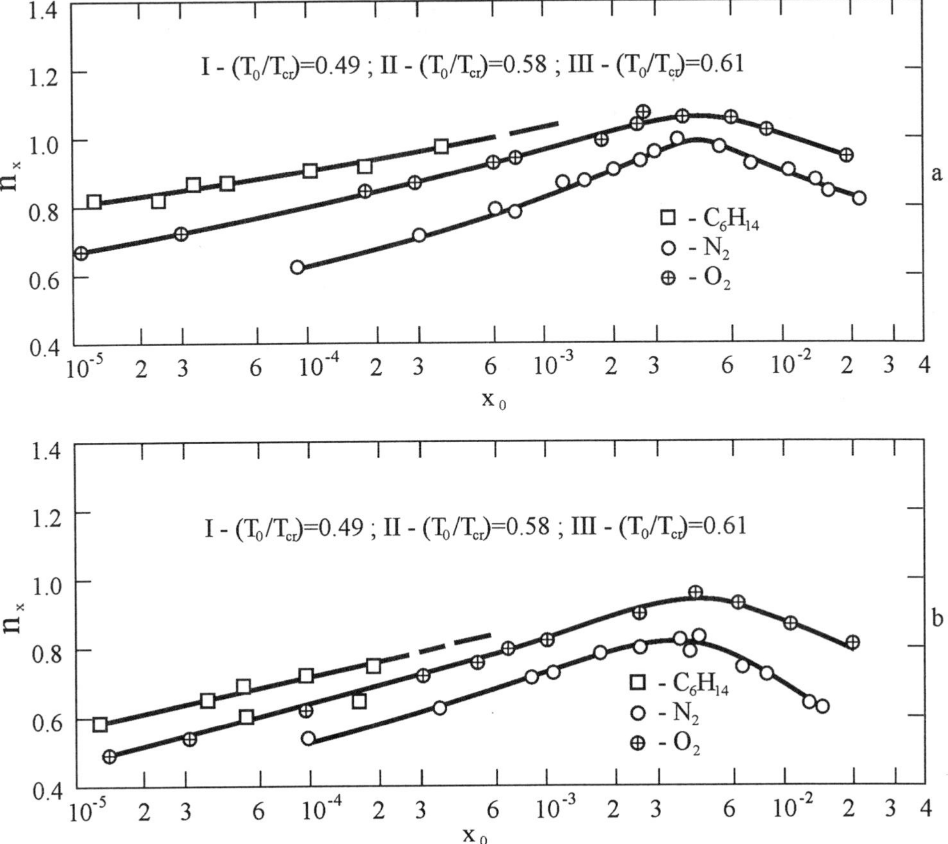

Figure1. Generalized dependencies of saturation temperature growth rate in toroidal vessels partially filled with liquid O_2, N_2, C_6H_{14}.
(a)- transient stage, $\tau < \tau_*$
(b)- steady-state stage, $\tau > \tau_*$.

The dependencies $n_x = f(x_0)$, similar to those in Fig.1 for hexane, oxygen and nitrogen, can be different for other liquids. As an example, Fig.3 shows the dependence for liquid hydrogen. According to Fig.3, the process of phase transitions due to changes in the vapor contents is characterized by repeatedly changing directions. Besides, at the initial stage the process has another effect described by the parameter $\frac{q\tau_*}{P_0 l}$, which may be induced by vapor condensation at the initial stage of the process [1].

Since the dependencies in Figs.1,3 are very complicated, for some $\frac{T_0}{T_{cr}}$=const the influence of the parameter $\frac{P_0}{P_{cr}}$ on the pressure growth rate can be detected only within a limited interval of the vapor content variations. To estimate qualitatively the influence of this parameter, the data in Fig. 1b for $x_0 = 10^{-5} - 10^{-3}$ were described by the relation $n_x = cx_0^{-0.36}$

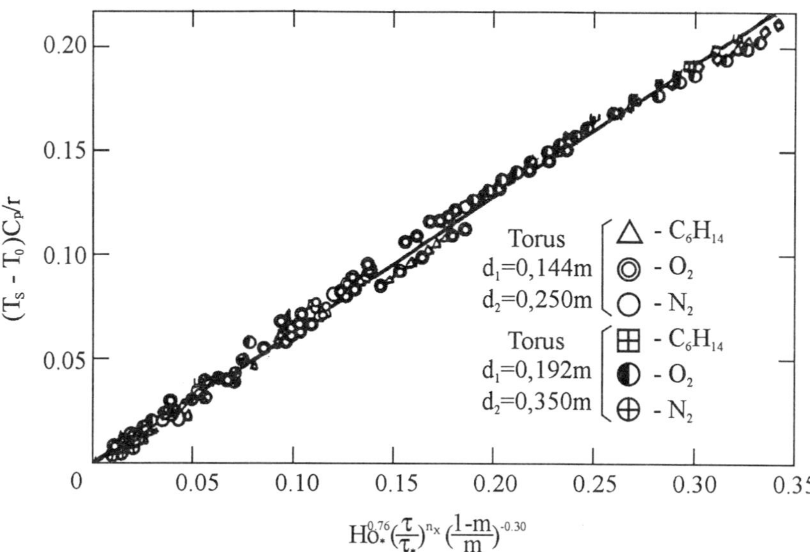

Figure 2. Generalized dependence of saturation temperature.

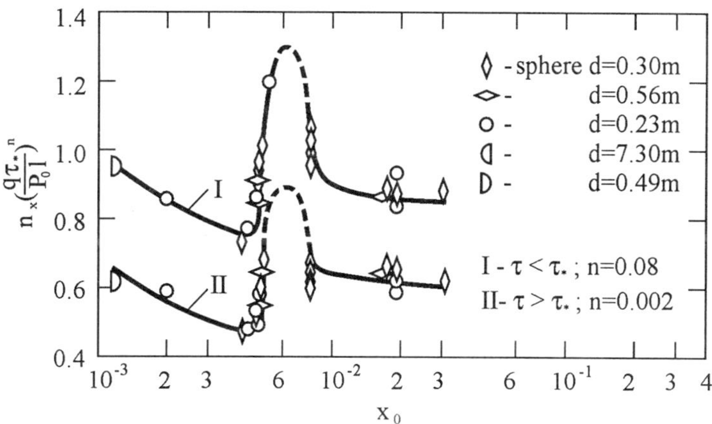

Figure3. Generalized dependence of saturation temperature growth rate in vessels filled with liquid H_2.

This permits estimation of the influence of the parameter $\frac{P_0}{P_{cr}}$ on the expression $n_x x_0^{0.36}$. This kind of data corresponding to m=0.85 for hexane, oxygen and nitrogen are shown in Fig.4

It is seen that the expression $n_x x_0^{0.36}$ decreases as the parameter $\frac{P_0}{P_{cr}}$ increases. Each liquid is described by its individual curve dependent on the molecular nature of the substance. To assess this factor, Fig.4 shows the dimensionless parameter $\Pi = R_\mu \frac{\rho_{cr} T_{cr}}{\mu P_{cr}}$ including the critical parameters of the liquids ρ_{cr}, T_{cr}, P_{cr} , the universal gas constant R_μ and molecule mass μ.

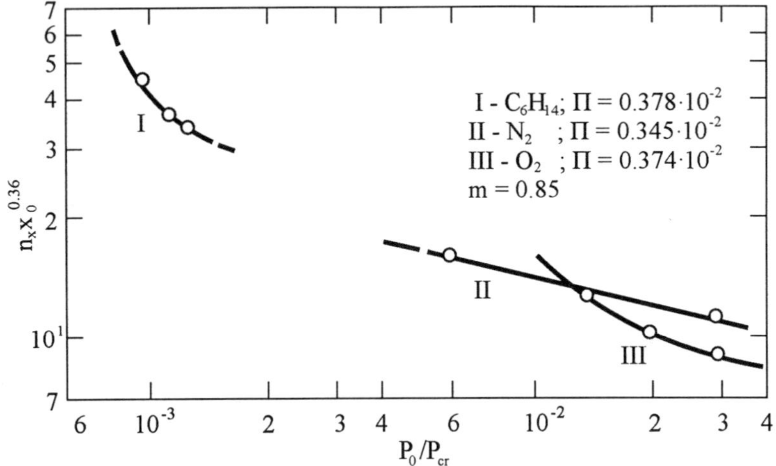

Figure 4. Dependence of saturation temperature growth rate in vessels on the initial pressure of the process and the molecular structure of the substance.

CONCLUSIONS

The data processing method described here permits working out techniques of calculation of the pressure growth for experimental and industrial objects - tanks containing liquefied gases of hydrogen, oxygen, nitrogen, methane, argon, helium and so on. Some of such techniques for calculating storage of hydrogen, nitrogen and oxygen have been developed . Data similar to those in Figs. 1, 3, 4 can be used to vary the storage conditions, such as the pressure and liquid levels in the vessels within a particular stage of the process.

The results presented were obtained for the following process conditions: $Ra_1^* = 10^{10} - 10^{16}$; $m = 0.40 - 0.90$; $\frac{P_0}{P_{cr}} = 10^{-3} - 10^{-1}$.

REFERENCES

1. J.A.Clark. A Review of pressurisation stratification and interfacial phenomena.,in Int. Adv. Cryog Engng. 10:200 (1965).
2. P.D.Thomas, F.H.Morthe. Analytic solution for the phase change in a suddenly pressurized liquid vapor system, in Int. Adv. Cryog. Engng., 8: 550, (1963).
3. B.O. Neff and C.W.Chiang. Free convection in a container of cryogenic liquid, in. Int. Adv. Cryog. Engng, 10: 112 (1965).
4. Yu.A. Kirichenko, J.A.Vasis, V.Ph. Solyanko et al. Test feasibility for study of heat exchange in liquefied gases, in Liquid and vacuum technologies, Issue 2., p.65. (IltPh&E Acad. Sci. Of Ukr SSR, Kharkov, (1972).
5. J.C.Aydelott, Normal gravity self-pressurization of 9-inch (23 cm) diameter spherical liquid hydrogen tankage, NASA Tech.Note. D4171 (1967). P.30.
6. J.C.Aydelott and C.M. Spucler. Effect of size of normal Gravity self pressurisation of spherical liquid hydrogen tankage, NASA Tech. Note. D5196 (1968), p.10.
7. D.H. Liebenberg and F.J. Edescuty . Pressurisation analysis of a large scale liquid hydrogen dewar, in Int. Adv. Cryog. Engng., 13: 284 (1968).
8. M.P. Vukalovich, I.I. Novikov. Thermodynamics, Moscow, Machionostroyenie, 1972.

THERMODYNAMIC ANALYSIS OF TEMPERATURE INCREASE AND ENERGY CONSUMPTION IN A CRYOGENIC CRUSHER

Y. Liang, W. Tang, P. R. Yang, and Y. Y. Guo

School of Chemical Engineering, Xi'an Jiaotong University
Xi'an, Shaanxi, 710049, P.R. China

ABSTRACT

Cryogenic comminution technology is widely regarded as a good method of recycling wastes such as used tires, plastics and other high polymers. Since the temperature of the material will unavoidably increase during the comminution processing, the temperature increase of crushed material becomes the most important characteristic of the cryogenic crusher to be investigated. In order to improve the performance of the cryogenic crusher, the energy consumption and temperature increase of the cryogenic crusher were investigated experimentally, and a thermodynamic analysis was adopted to study the energy consumption in the comminution process. Upon evaluating the classical energy consumption laws, a model of energy consumption was developed without consideration of the size distribution function. The variation in comminution temperature was introduced into this model. Furthermore, a series of experiments was carried out to verify the model. The crusher tested is a new type of high-speed cryogenic impact crusher used in several practical cryogenic recycling plants each of which can produce 3000 tons fine rubber powder per year. The various state parameters of this set were tested under different operating conditions. Results show that the comminution temperature could significantly affect the size distribution of the finished product. The experimental data are in accordance with the prediction of the model, and the thermodynamic analysis is deemed suitable.

Advances in Cryogenic Engineering, Volume 45.
Edited by Shu *et al.*, Kluwer Academic / Plenum Publishers, 2000.

INTRODUCTION

In recent years, the technology of cryogenic comminution has been widely applied in the field of chemical engineering, food making, medicine production, and particularly in recycling of waste materials. Because of the increasing pollution of waste tires and the shortage of raw rubber resource, the recycling process for waste rubber products has become important and commercially viable. This technology has shown a great number of advantages such as causing no environmental pollution, requiring low energy consumption and producing high quality products. Hence, the normal crusher which was used to reclaim materials, such as waste tires, nylon, plastic and many polymer materials at atmospheric temperature is being replaced by a cryogenic crusher. [1,2]

In the cryogenic crusher, the property of the milled material is usually very sensitive to temperature change. When a crusher is in operation, it will generate a great deal of heat that causes the material temperature increased. Once the temperature increases over the vitrification temperature, the material property will change and lose the brittle behavior causing the energy consumption to rise sharply. Consequently, the comminution process cannot be continued. Therefore, it is believed that the cryogenic crusher is the most critical component in the cryogenic comminution system. The research on the temperature increase and energy consumption in the cryogenic crusher is not only to reduce the energy consumption of the crasher, but also to reduce the energy consumption of the cryogenic system. In this paper, the energy consumption and temperature increase of a cryogenic crusher is studied experimentally. Results obtained are analyzed thermodynamically to determine the relationships among the temperature increase, heat of comminution, and energy consumption. Based on these parameters and the size distribution of the product, a mathematics model has been developed to describe the relationship of comminution energy and material size distribution.

EXPERIMENT

Initially, the rubber tire scraps were broken into particles with diameters of about 3mm. The particles were carried sequentially into the fluidized bed and then into the fixed bed to be frozen by the cryogenic air. The temperature of the rubber particles was reduced below the vitrification temperature of the material. Finally, the frozen particles were loaded into crusher to be milled. At the same time, the cryogenic air continued to flow into the crusher to absorb the comminution heat. The cryogenic comminution system is shown in Fig. 1.

This system is part of a practical industrial plant that can produce 3000 tons of 60-mesh fine rubber powder per year. In this plant, cryogenic air is available from a compression-expansion refrigeration system and used as the coolant instead of liquid nitrogen to chill the solid material. The cost of producing cryogenic air is far lower than liquid nitrogen and makes this system more economical.

A high-speed hammer crusher was employed in this system. Its structure is shown in Fig. 2. The rotor consists of a rotary shaft, three wheels and some hammers. The hammers are homogeneously attached to wheels. There are many triangular tines inside the gear ring.

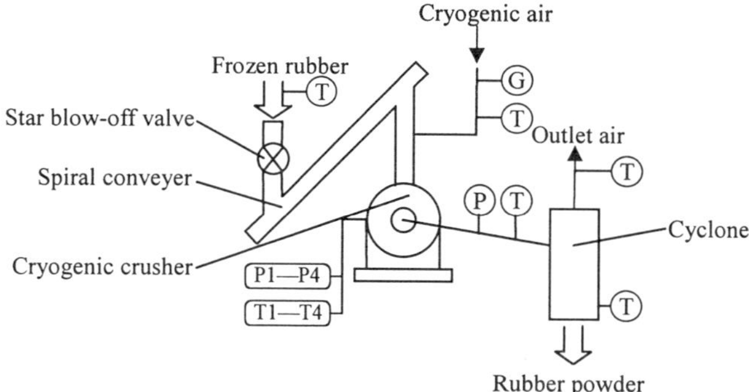

Figure 1. Schematic diagram of cryogenic comminution system

Initially, when the crusher is working, the solid particles are fed through feed inlet. Then, they are brought into the comminution area by centrifugal force. They are comminuted into small particles in the area between the gear ring and the hammers. Finally, the mixture of rubber powder and air is discharged from the back of the crusher through the discharge gate.

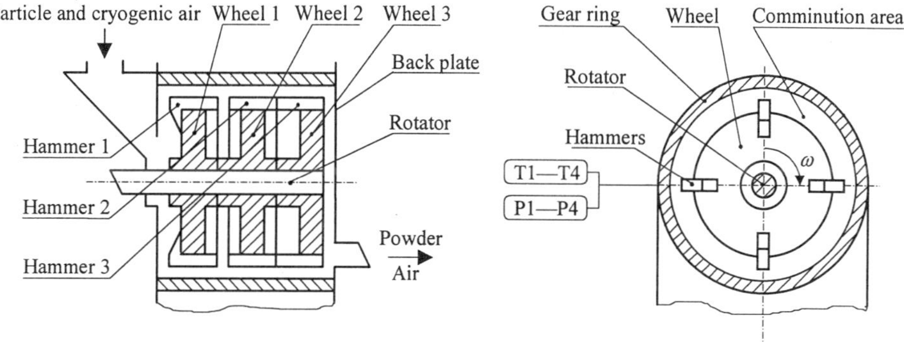

Figure 2. Structure of cryogenic hammer crusher

The parameters that were measured in the experiments include:

1. Temperature and pressure of the cryogenic air at the inlet and outlet of the crusher.
2. Variations of temperature and pressure inside the crusher.
3. Power consumption of the crusher.
4. Size distribution of input and output materials.

The electric current of the electromotor of the crusher was tested with a hand ampere meter to measure the power consumption. Comparing the power-current performance diagram of the electromotor, the input-power of the crusher can be determined. The size distribution is obtained by using a sieve method at room temperature.

The experiments were carried out in the following sequence:

1. Freeze the crusher itself by using cryogenic air.

2. Start-up the crusher.
3. Turn on the spiral conveyer and the star blow-off valve.
4. Add the cooled material into the crusher to be smashed.
5. Collect the rubber powder for analysis.

ANALYSIS OF THE COMMINUTION ENERGY USED IN THE CRYOGENIC CRUSHER

During the comminution process, the cryogenic crusher consumes a part of input energy to crush solid. This part of the energy is used to reduce the size of the solid and increase the surface area of the particles. The remaining part of the input energy is changed to the heat energy of comminution that elevates the temperature of the solid and air. There are many parameters, including property of materials, performance of crusher and size distribution of input and output of the solid that affect the energy consumption. When all other conditions remain constant, the energy consumption can be related directly to one parameter: size distribution of the input and output material.

It is well known that there are three well-recognized laws that were developed to estimate the comminution energy. There are the Rittinger law, the Kick law and the Bond law. These laws can be expressed uniformly by the Walker equation as

$$dE = -c\frac{dx}{x^n} \tag{1}$$

where E is the energy input, x is the particle diameter and c and n are constants. According to the value of n, the three laws are known as: Rittinger's for n=2, Kick's for n=1 and Bond's for n=1.5.

It has been suggested that each of these three laws for predicting the energy requirement are more applicable in certain areas of product size: Rittinger's being applicable for very small particle size, Kick's for large particle size, and Bond's for intermediate particle size — the most common range for comminution processes. In practice, however, there are few materials matching these laws precisely, so it is generally advisable to rely on experiment in order to predict energy requirements for the crushing of a particular material.[3~5]

Many researchers have analyzed the experimental data in an assumed size distribution function. In this study, we will analyze the size distribution directly to obtain the energy consumption equation for the cryogenic comminution crusher.[6~11]

The size of the crushed solid can be divided into N+1 levels. The diameters of each level are defined as $x_1, x_2, \ldots, x_{N+1}$ $(x_i > x_{i+1})$. The weight fractions of each level are defined as $m_1, m_2, \ldots, m_{N+1}$, respectively. The diagram of size distribution is shown in Fig. 3.

The energy factor e_i is defined as the energy consumption of crushing that the particle of unit weight requires to decrease from x_i to x_{i+1}; similarly, e_{ij} for x_i to x_j. It is assumed that the energy requirements do not relate to the comminution procedure. This means that the energy consumption e_{ij} is equal, no matter how the particles are to be crushed from x_i to x_j, or from x_i to an intermediate size x_k and then to x_j $(i<k<j)$. So e_{ij} can be expressed as

$$e_{ij} = \sum_{k=i}^{j-1} e_k \qquad (2)$$

It is assumed that all input particles are broken. All of the i level particles are broken into smaller particles, whose size is from x_{i+1} to x_{N+1}. The weight fraction of these smaller particles is m_{i+1} to m_{N+1}, respectively. Energy consumption (E_i) of the i level is given by

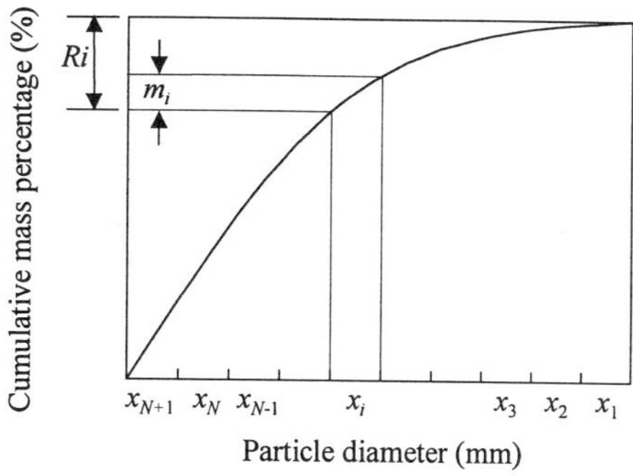

Figure 3. Diagram of particle size distribution

$$E_i = \sum_{j=i+1}^{N+1} (\sum_{k=i}^{j-1} e_k) m_{j,i} \qquad (i = 1,2,...N) \qquad (3)$$

Accumulating every E_i, the energy consumption of the complete input material E can be written as

$$E = \sum_{i=1}^{N} E_i = \sum_{i=1}^{N} \sum_{j=i+1}^{N+1} (\sum_{k=i}^{j-1} e_k) m_{j,i} \qquad (4)$$

Through an appropriate transformation, Eq. (4) can be changed to the following form

$$\begin{aligned} E &= \sum_{i=1}^{N} e_i (\sum_{k=1}^{i} \sum_{j=i+1}^{N+1} m_{j,k}) \\ &= \sum_{i=1}^{N} e_i [R_i(0) - R_i(t)] \end{aligned} \qquad (5)$$

where $R_i(0)$ and $R_i(t)$ are cumulative mass percentages of size x_i for the input and output material, respectively. Energy factor e_i should yield a form of the Walker's equation given earlier as

$$e_i = A(x_{i+1}^{1-n} - x_i^{n}) \qquad (1')$$

Combining Eq. (5) and (1'), the general form of the relation between the energy consumption and the particle size distribution can be shown as

$$E = \sum_{k=1}^{N+1} A(x_{k+1}^{1-n} - x_k^{1-n})[R_i(0) - R_i(t)] \qquad (6)$$

where A, n are undetermined coefficients. Because the energy consumption is related to the comminution temperature, coefficient n is also a function of the temperature as $n{=}C_1T^{C_2}$, where C_1 and C_2 are unknown constants. It is assumed that A is also a constant, so Eq. (6) is changed to the following form

$$E = \sum_{k=1}^{N+1} A(x_{k+1}^{1-C_1T^{C_2}} - x_k^{1-C_1T^{C_2}})[R_i(0) - R_i(t)] \qquad (7)$$

where the reference temperature T is defined as the average temperature between the inlet and outlet of the crusher. All of constants can be computed by a nonlinear method of least squares.

RESULTS AND DISCUSSION

Experiment Results

Experiments were carried out in following range: producing capacity of rubber is from 360kg/h to 466kg/h, and flow of air is from $370Nm^3/h$ to $600Nm^3/h$. Fig. 4 provides the size distribution of the product.

It is easy to determine from the figure that these distributions are susceptible to the outlet temperature. When the temperature increase, the mesh size of the rubber powder increases sharply, and the rate of finished product decreases significantly. It shoud be noted that the glass transition temperature of the rubber is not a single temperature but cover a wide temperature range. In this range, the property of the rubber changes continuously from an elastic to a brittle condition. The variation in the rate of the 60-mesh rubber powder as a function of different outlet temperature is diagramed in Fig. 5.

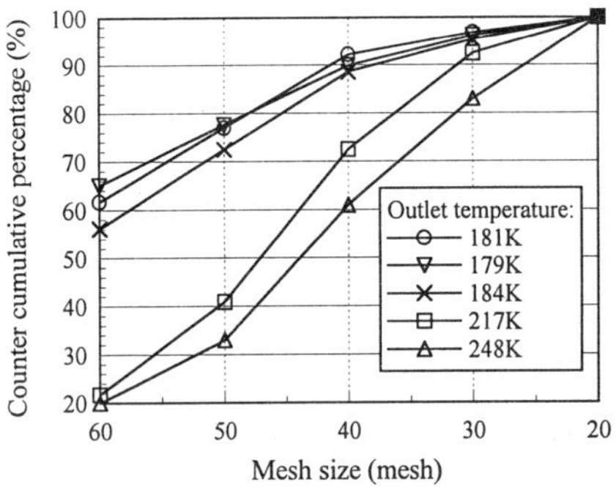

Figure 4. Size distribution of finished rubber powder

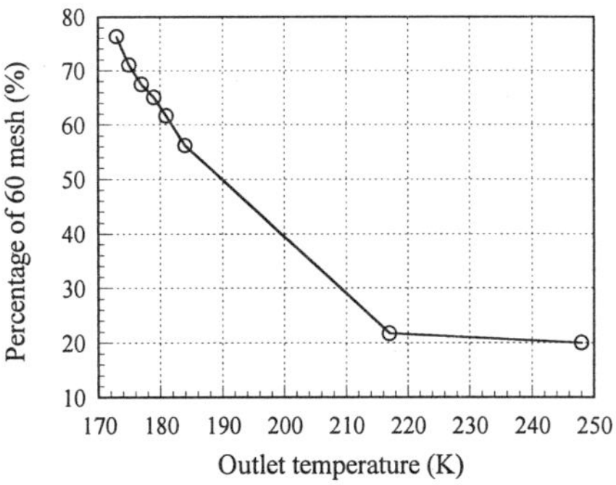

Figure 5. Variation of rate of 60-mesh rubber powder with different temperature

Analysis of Comminution Energy Consumption

Comminution energy consumption is usually difficult to be evaluated because it is difficult to obtain accurate data in a continuous crushing process. In this study, it is assumed that the input energy is essentially changed into two forms of energy: comminution heat and comminution work. The mechanical energy loss is ignored.

Comminution heat is a synthetically thermal effect. There are many reasons leading to this form of heat:

— The heat from high-speed rotation of the wheels which agitate air and solid in the crushing chamber.

— The heat released from solid breakage.
— The heat of friction between solid and solid, solid and wheels, and solid and gear ring.

The quantity of comminution heat can be obtained by measuring the state parameters using a black-box model of the crusher. The black-box model for the experimental cryogenic crusher is given in Fig. 6.

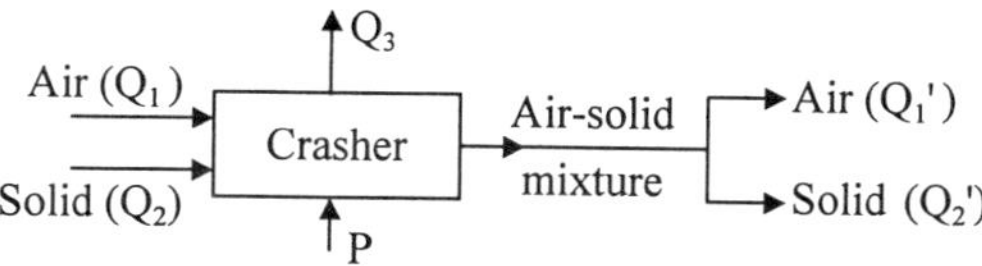

Figure 6. Black-box model for cryogenic comminution

The energy balance for the cryogenic crusher can be given as

$$P = P_0 + Q \tag{8}$$

where P is the input power of the crusher, P_0 is the power consumption of the comminution and Q is the heat of comminution. According the previous analysis,

$$\begin{aligned} Q &= \Delta Q_1 + \Delta Q_2 + Q_3 \\ &= (Q'_1 - Q_1) + (Q'_2 - Q_2) + Q_3 \end{aligned} \tag{9}$$

where Q_1 is the energy associated with the air, Q_2 is the energy with the solid and Q_3 is the loss of heat. Table 1 provides the calculated result of the comminution heat and power consumption of the comminution process. In the table, *Gs* and *Ga* represent the feed rate of solid and air, respectively. Here, the efficiency of the comminution is defined as $\eta = P_0/P$. This indicates the rate of usable power to the input power. In the range of the experimental parameter, η changes from 3.8% to 7.1%. The low efficiency of the comminution means that almost input energy has been transformed from mechanical energy to thermal energy. Compared with atmospheric experimental results of this kind of crusher whose efficiency is about 40~60%, the efficiency of the cryogenic crusher is much lower. Therefore, there are large opportunities for improving the performance of the cryogenic crusher.

Table 1. Heat and power consumption of cryogenic comminution

Gs(kg/s)	*Ga*(kg/s)	*P*(kw)	*Q*(kw)	$P_0 = P - Q$(kw)	η(%)
0.10	0.174	18.1	16.840	1.260	7.1
0.11	0.174	19.15	17.831	1.319	6.9
0.11	0.193	19.53	18.211	1.319	6.8
0.11	0.216	20.0	19.067	0.933	4.8
0.129	0.174	22.7	21.828	0.872	3.8

Relation Between Comminution Energy and Size Distribution

By a nonlinear method of least squares, the constants of Eq. (7) namely A, C_1, C_2 are computed as A=0.50112, C_1=0.078948, C_2=0.573313. In this paper, the value of n in Eq.(1) is shown a value between 1.6 and 1.9. These value indicated that the energy requirement for cryogenic comminution is between the values provided by Bond and Rittinger.

Although the amount of comminution energy is less than the heat produced, it must have a certain relation with the heat of comminution, since a major portion of the heat is unavoidable as it is associated with the breakage phenomenon. In order to reduce the energy requirement of the crusher and the cryogenic system, it is necessary in future work to determine the generation of this part of the heat.

CONCLUSION

The equations relating the operating of a cryogenic crusher have been developed to predict the comminution energy with product size distribution and the temperature increase. Cryogenic comminution experiments were carried out to determine the value of the constants in the equation. Several of the state parameters of the crusher have been measured in the experiments. A thermodynamic black-box model was used to analyze the energy equilibrium for the crusher. The results show that the efficiency of cryogenic crusher is lower, and there are many things can be done in future to improve the efficiency of the crusher.

REFERENCES

1. N.R. Braton, "Cryogenic Recycling and Processing", CRC Press Inc., 1980
2. S. Vivatpanachart, H. Nomura and Y. Mitahara, *Comminution* 5:473-480(1983)
3. Th. Folgner, K. Meltke and A. Schmandra, *Aufbereitungs-Technik* 37:357-366(1996)
4. R. Drogemeire and K. Leschonski, *Int. J. Miner. Process* 44-45:485-495(1996)
5. L. Blecher and J. Schwedes, *Int. J. Miner. Process* 44-45:617-627(1996)
6. L.G. Austin, V.K. Jindal, and C. Gotsis, *Powder Technology* 22:199-204(1979)
7. C. Gotsis, L.G. Austin, P.T. Luckic and K. Shoji, *Powder Technology* 42:209-216(1984)
8. I. Tschorbadjiski, H. Schallnus, and J.Schwedes, *Aufbereitungs-Technik* 10:555-562(1987)
9. D.W. Fuerstenau and M.R. Moharam, *Comminution* 18:388-395(1988)
10. N. Magdalinovic, M. Grbovic, I. Budic, Z. Markovic and Z. Mitrovic, *Aufbereitungs-Technik* 31:277-279(1990)
11. G. Timmel, K. Husemann and D. Esping, *Aufbereitungs-Technik* 37:249-258(1996)

STABILITY AND CONTROL OF SUPERCRITICAL HELIUM FLOW IN THE LHC CIRCUITS

E. Hatchadourian

LHC Division, CERN
1211 Geneva 23, Switzerland

ABSTRACT

The circulating particle beams of the Large Hadron Collider (LHC) will induce dynamic heat loads into the cryogenic system. Beam screens, maintained at a temperature between 5 K and 20 K by weakly supercritical helium -in order to avoid-two phase flow- are inserted inside the magnet cold bore to intercept most of these heat loads. Evidence has been presented in experimental and theoretical work that the main type of dynamic instability in long channels is that caused by the propagation of density waves due to multiple regenerative feedback. Oscillations are typically observed in circuits operating with low flow rate and/or high energy input. The study of the system behaviour under different operating cases permits assessment of the time constant of the system as well as its temperature-control parameters. A part of this work also concerns the study of flow stability in the other LHC cryogenic circuits working with supercritical helium.

INTRODUCTION

The design of the beam screen (figure 1) is based on a 1mm-thick stainless steel tube coated on its inner surface with 50 μm copper[1]. Weakly supercritical helium, flowing in two stainless steel tubes of 3.7 mm inner diameter running on each side of the beam screen, maintain its temperature below 20 K, while intercepting the heat loads deposited by synchrotron radiation, resistive dissipation of beam image currents, and impact of photo-electrons resonantly accelerated by the circulating proton bunches[2]. These beam-induced heat loads, which strongly depend on the intensity, amount to 0.46 W/m and 1.13 W/m respectively in so-called “nominal” and “ultimate” operating conditions[3]. After discussing the flow stability, we will analyse the control of the exit temperature in order to determine the parameters that are significant to the design of automatic control systems. The results of theoretical work has been validated on a full-scale test section which reproduces one of the cooling tubes.

HYDRAULIC STABILITY

Analytical approach. In order to study fluid behaviour through different LHC cryogenic lines, a mathematical investigation and experimental tests have been performed. Oscillations of a hydrodynamic system require a mechanism for temporary storage and release of some form of energy.

Advances in Cryogenic Engineering, Volume 45.
Edited by Shu *et al.*, Kluwer Academic / Plenum Publishers, 2000.

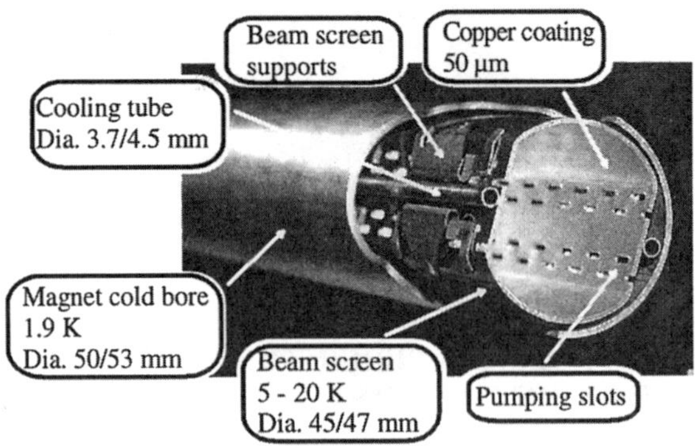

Figure 1. A view of the LHC beam screen.

In the case of constant heat load, at high flow rates little energy is stored and its amount increases with decreasing flow. The physical phenomenon can be described as follows : an inlet flow fluctuation creates an enthalpy perturbation in the high-density region, which travels with the flow along the channel. When this perturbation reaches the "change-of-density" boundary it is partly transmitted to the lower-density region. With correct timing, the perturbation can acquire appropriate phase and become self-sustained. The analysis is based on the assumption of an oscillatory inlet flow[4], and consists in writing and linearising the fundamental equations -the conservation of mass, energy and momentum- describing the fluid state. Other assumptions are made in order to produce a simple stability map that can be used for different cases :

- the equation of state is approximated by two regions ; the first (quasi-liquid) representing the subcooled liquid fluid has a constant density, the second (quasi-gas) is modelled as ideal gas. This equation of state is plotted in figure 2 for 3 bar, the normal operating pressure of the cooling tube. The dependence of the friction factor[4] on enthalpy (given by the Poiseuille and the Moody equations) is also shown in figure 2 at the same pressure. Exact properties for different operating pressures -between 2.4 and 3.6 bar- are represented by the cross-hatched area,
- the incompressible flow is assumed to be one-dimensional with uniform section along the heater line. The effect of diverse geometrical parameters (upstream and downstream adiabatic lengths, singularities) has been considered too,
- the thermal properties of the wall are not taken into account. The characteristic time due to thermal diffusivity in the tube wall is negligible in comparison with the period of the oscillations.

Stability criterion. The effect of a small perturbation of inlet velocity on the different regions (upstream length, quasi-liquid, quasi-gas, downstream length, singularities) can be found out from the linearised relations for conservation of mass and energy. Finally, the variation of the overall pressure drop (friction, acceleration and gravity terms in each region) is determined by integration of the linearised momentum relation.

The behaviour of an inlet velocity perturbation δu is determined by a dynamic system analysis in which the input signal is the perturbation in pressure drop over the different regions, and the output signal is the inlet velocity perturbation. The arbitrary choice of the forward transfer function (the response of inlet velocity to upstream pressure drop) and the feedback transfer function (the response of the heated tube and downstream pressure drop to inlet conditions) determines the hydrodynamic closed loop system, which is analysed using the standard Nyquist criterion[5]. When the real part of the denominator of the closed loop transfer function F (1) is negative and its imaginary part is zero, then the system becomes oscillatory.

$$F = \left(\frac{\delta \Delta P_{upstream}}{\delta u} + \frac{\delta \Delta P_{quasi-liquid}}{\delta u} + \frac{\delta \Delta P_{quasi-gas}}{\delta u} + \frac{\delta \Delta P_{downstream}}{\delta u} \right)^{-1} \tag{1}$$

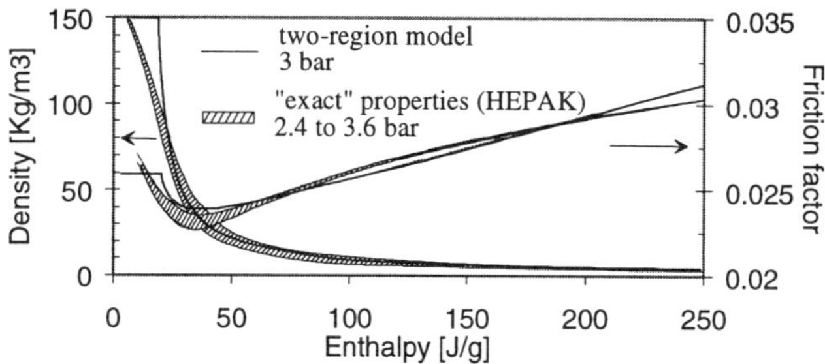

Figure 2. Density and frictional pressure drop coefficient versus enthalpy.

To account for the specific hydrodynamic characteristics of the different regions, it is necessary to treat separately the perturbation in each region.

Application to beam screen cooling tubes. From the above analysis, we expected oscillations for a wide range of operating conditions : inlet subcooling, low pressure, heat input, low flow rate and exit restriction. The inlet subcooling is defined by the enthalpy difference in the quasi-liquid region from inlet to the transition point. Figure 3 is a typical plot of the periods measured at the threshold of the stability versus heat input. The measured period of the oscillations ranged from 50 to 400 seconds, increasing continuously with subcooling.

A stability map -reduced inlet velocity (2) versus inlet subcooling enthalpy- for different upstream and downstream tube length and diameter is shown in figure 4. The notation in equation (2) is as follows : u is the dimensional inlet fluid velocity, L is the heated length, Q is the total heat input ,v is the ideal gas specific volume, h is the enthalpy and α is the expansion coefficient at constant pressure.

$$u_{reduced} = \frac{u}{LQ}\frac{1}{\alpha}, \quad \text{where } \alpha = \frac{1}{v}\frac{dv}{dh}\bigg|_P \tag{2}$$

Above the limit line, the flow is stable ; below its the flow is unstable. In figure 4a we can see the effect of unheated additional lengths (destabilising effect when downstream length increases). The length ratio is respectively 0.5 for downstream and 5 for upstream. Figure 4b shows the result of varying the unheated tube diameters (with 0.2 fractional unheated length and with a same 0.5 diameter ratio for the two limit lines). When the diameter of the

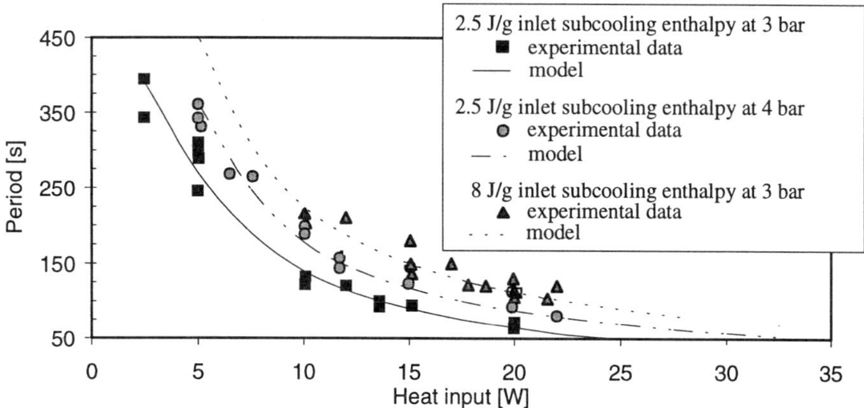

Figure 3. Period of oscillations versus heat input at the stability boundary.

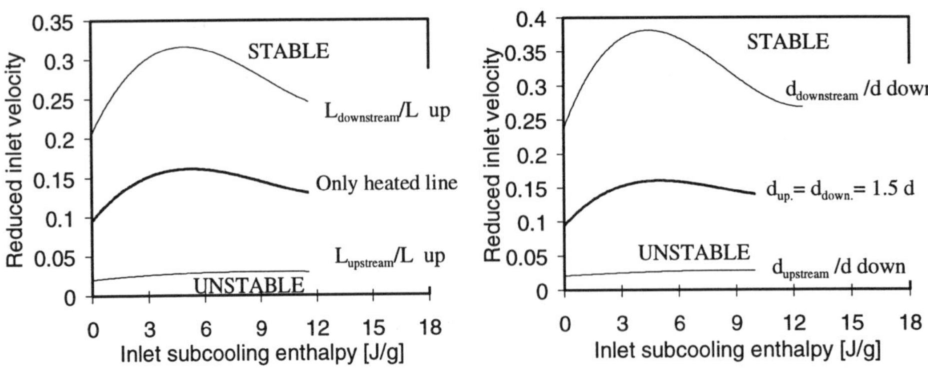

Figure 4. Stability boundaries for (a) different unheated tube length L and (b) unheated tube diameter d.

upstream or downstream lengths exceed 1.5 times that of the heated length, there is no effect on the stability diagram. At low subcooling an increase of subcooling renders the system less stable and at high subcooling, an increase of subcooling renders the system more stable.

Application to other LHC lines. While line C' (15 mm diameter) runs over the 53 m of a half cell, where it intercepts heat on the magnet supports before feeding the beam screen cooling tubes, line C (100 mm diameter) supplies supercritical helium over the 3.3 km length of an LHC sector. The cooling scheme for these 3 lines is shown in figure 5. No density oscillations are expected in the supply header (line C) because of the small heat loads deposited (principally radiation between 77K and 4.6 K) and the high mass flow rate (varying from 180 g/s at the beginning of the sector down to 20 g/s at the end).

Concerning line C', each of the 11 supports deposits 1.36 W locally, over a length of 30 cm (there are 2 supports on the quadrupole cryostat and 3 supports on each dipole cryostat). As long as the flow in the line remains in the pseudo-liquid domain, no oscillation is expected over the short length of the supports. If the mass flow rate decreases to less then 0.15 g/s with an operating pressure close to the critical pressure, the system could become unstable : this case occurs if the control valve at the end of the line is maintained at a set point of 20 K with very small heating deposited on the beam screen line. Table 1 summarises the different supercritical lines. It should be noted that the heat input brought by the supports increases the inlet beam screen temperature and thus reduces the possibility of density oscillations.

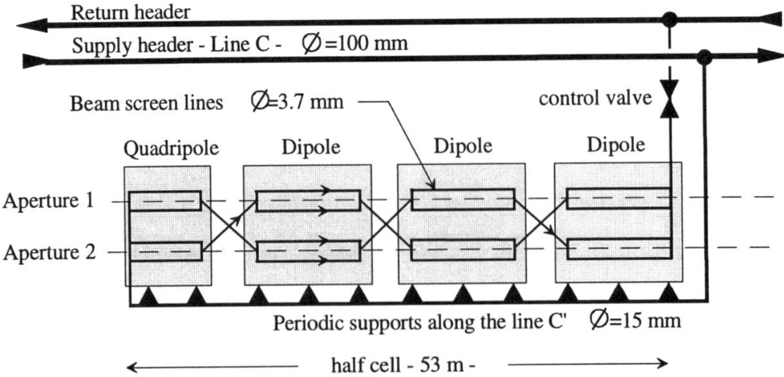

Figure 5. Supercritical helium circuits in the LHC.

Table 1 Characteristics of the supercritical helium lines in the LHC.

Lines	Length (m)	Diameter (mm)	Heat load		Density oscillations
C	3300	100	0.018 W/m	distributed	None
C'	53	15	1.360 W/support	local	Possible
Beam Screen	53	3.7	nominal : 0.458 W/m ultimate : 1.133 W/m	distributed	Possible

OUTLET TEMPERATURE CONTROL

The objective is to maintain the outlet temperature of the process fluid at its desired temperature set point (20 K) in the presence of variations of the process fluid flow or applied heating.

Transfer function. The feedback control scheme works as follows : the outlet temperature or controlled variable is measured with a sensor and transmitter that generates a signal T(mA) proportional to the temperature. The measurement is sent to a controller where it is compared to the set point. The signal difference between the measurement and the set point generates a controller output signal which is then converted and applied to the actuator of the control valve. The flow is then a function of the valve position. Variations in outlet temperature are measured by the sensor-transmitter and sent to the controller causing the controller output signal to vary. This in turn causes the control valve position and then the flow to vary. The block diagram for the feedback control loop is shown in figure 6. The symbols in this figure are :

E is the error signal, mA
F_c is the controller transfer function, mA/mA
F_v is the control valve transfer function, (kg/s)/mA
F_p is the process transfer function, K/(kg/s)
F_d is the disturbance transfer function, K/(kg/s)
H is the sensor-transmitter transfer function, mA/K
K is the scale factor for the temperature set point, mA/K

We can determine the closed-loop transfer function (3) :

$$T_{out} = \frac{KF_cF_vF_p}{1+HF_cF_vF_p} \times T_{control} + \frac{F_d}{1+HF_cF_vF_p} \times \text{Disturbance} \tag{3}$$

In order to adjust the controller parameter (Proportional-Integral type) to the characteristics of the rest of the components in the loop, we need to have access to the process steady-gain F_{gain}, the effective process dead time t_0 and the effective process time constants τ of the open-loop transfer function F of the equivalent block diagram ($F= F_v.F_p.H$). The procedure used is to work in open loop and apply a step change in the controller output signal. The process can then be characterised using a first-order dead-time model (4) :

$$F = \frac{F_{gain}\, e^{-t_0 s}}{1+\tau s} \tag{4}$$

The PI controller settings are then calculated using the Ziegler-Nichols tuning relations[6]. A typical test plot is sketched in Figure 7. The experimental data are compared with the first order model and with the transfer function of the density model described above (using inverse Laplace transform for a step input condition).

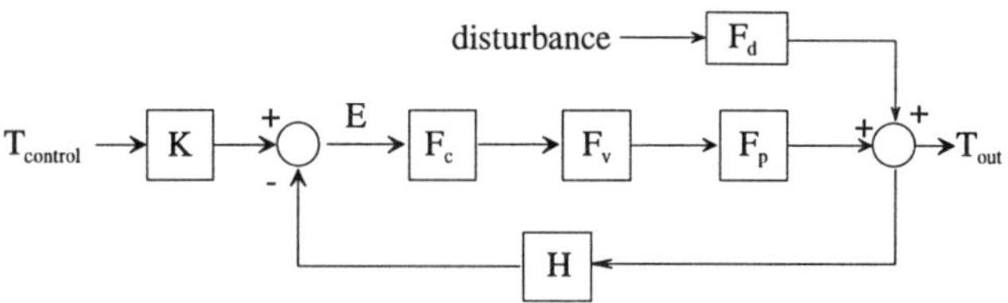

Figure 6. Block diagram for control loop of beam screen outlet temperature

Experimental results. The experiments for investigating the exit control temperature were of two classes : a heat load gradually increasing with time -0.25W per minute- up to "ultimate" LHC conditions, and heat load steps to simulate the "nominal" and "no beam" conditions (figure 8a). The controller settings have been obtained with the method described above. The operating range in these experiments was :

- inlet pressure between 2.5 and 4 bar with pressure drop less than 100 mbar,
- mass flow rate ranging from 0.1 to 0.3 g/s,
- a 3.75 W inlet heat load simulated the entry heat input for one cooling tube, which permitted an inlet temperature of the beam screen tube between 5.3 and 6.4 K (depending on pressure),
- the automatic control starts only when the temperature reaches the set point condition (20 K) because of lower limitation in control valve position.

In accordance with the analysis, the destabilising effect of the pressure decrease could be observed (figure 8). The amplitude of temperature oscillations decrease with increasing pressure, whereas the period of oscillation increases. The period ranges from 80 s to 120 s close to "ultimate" heat input and between 120 s to 210 s for "nominal" heat input. The result also shows that the controller exit valve reacts correctly in case of oscillations.

The smoothing effect of azimuthal thermal conduction in the stainless-steel wall of the beam screen can be characterised by a time constant τ_{screen} (5). In view of symmetry -figure 9a- only a quarter of the beam screen perimeter is considered. The effect of the conduction is weakest at the furthest points from the cooling tubes.

$$\tau_{screen} = \frac{\rho c l^2}{k} \tag{5}$$

In this formula k is the conductivity, ρ the density and c the specific heat of the beam screen stainless steel. In the range of temperatures encountered in the beam screen, this

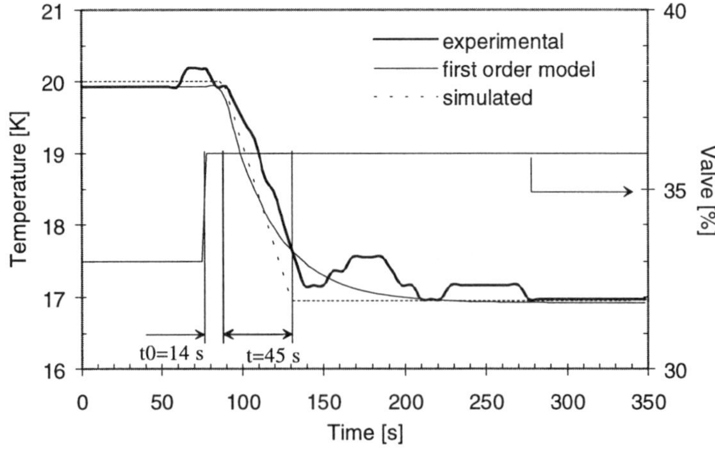

Figure 7. Open loop step response

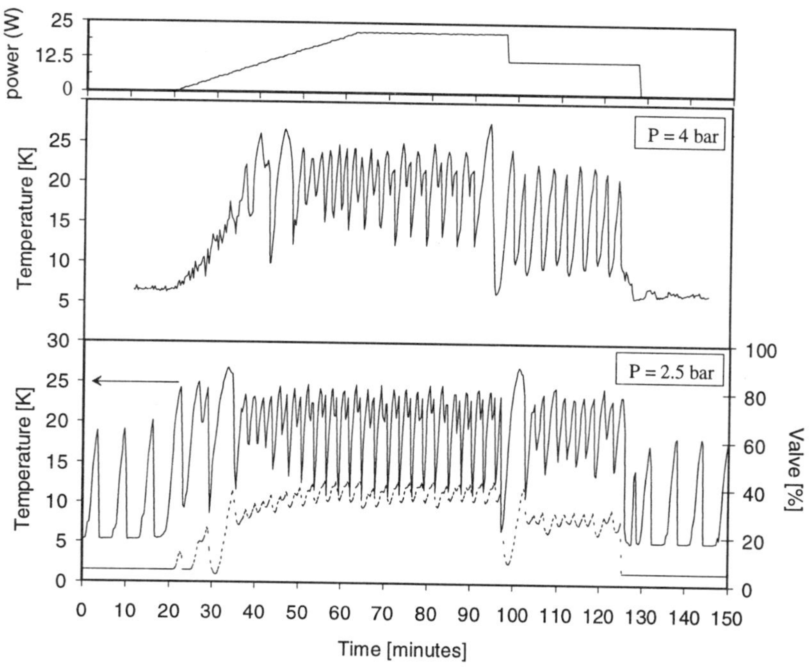

Figure 8a, b, c. Experimental exit temperature control

time constant (figure 9b) -for l=36 mm- shows little variation around 80 to 90 seconds. Above 20 K, τ_{screen} increases due to the rise in specific heat of stainless steel[7]. We can conclude that the beam screen has a smoothing effect for the control of the exit temperature under oscillating flow conditions, for periods of oscillation below τ_{screen} such as encountered in "ultimate" condition. For periods higher than τ_{screen}, a numerical implicit investigation of the azimuthal conduction through the beam screen has led us to conclude that only less then 30 % of the amplitude is propagated. A typical map showing the propagation of a temperature oscillation over length is represented in figure 10. Iso-temperature lines are shown.

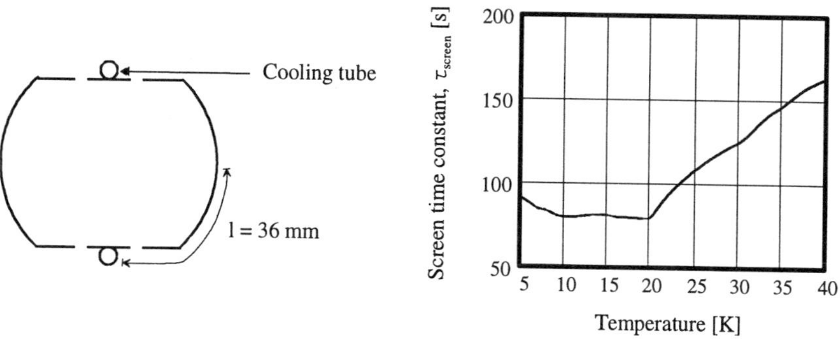

Figure 9. Transverse cross-section of beam screen (a) and screen time constant time versus temperature (b).

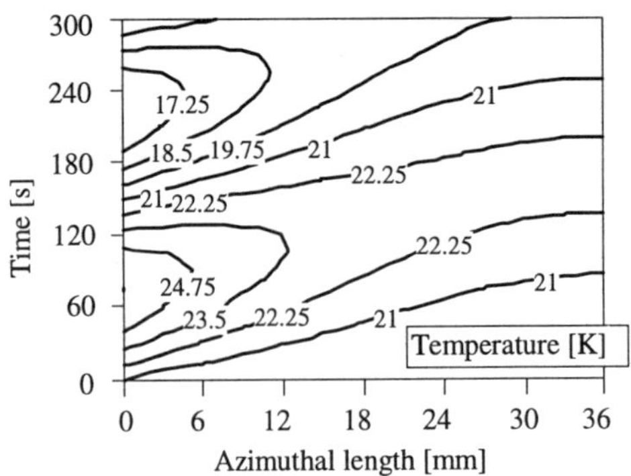

Figure 10. Azimuthal temperature propagation of a temperature oscillation with a period of 300 seconds and an amplitude of 10 K around 20 K.

CONCLUSION

Theory and experiment have both established the risk of instability in supercritical helium flow. Some effects are complex like an increase of subcooling which may have a positive or negative effect depending on the inlet conditions. The influence of geometry on the driving and damping terms for oscillations has also been shown. Series of experiments made at different pressures have permitted testing the exit temperature control under stable and oscillatory conditions. A simple study of the beam screen azimuthal conduction has permitted determination of the smoothing effect.

ACKNOWLEDGMENTS

The author wishes to acknowledge the guidance of P. Lebrun, L. Tavian (CERN), and L. Puech (CNRS). Thanks also go to A. Bézaguet and his team (CERN) for their help with the experimental setup.

REFERENCES

1. P. Cruikshank and al. "Mechanical Design Aspects of the LHC Beam Screen", CERN LHC Project Report 128, (1997).
2. V. Baglin and al. "Beam-Induced Electron Cloud in the LHC and possible Remedies", CERN LHC Project Report 188, (1998).
3. E. Hatchadourian, Ph. Lebrun, L. Tavian. "Supercritical Helium Cooling of the LHC Beam Screens", Proc. ICEC17, IOP, *793* (1998).
4. R.C. Martinelli, D.B. Nelson. "Pressure Drop during Forced-Circulation Boiling of Water", *Transactions of the ASME.* 8:695 (1948).
5. N. Zuber. "An Analysis of Thermally Induced Flow Oscillations", Report NAS8-11422 (1966).
6. K. Ogata. "Modern Control Engineering", Prentice Hall, New Jersey, (1997).
7. H.M. Rosenberg. "Low Temperature Solid State Physics", Oxford, (1963).

HYDRAULIC CHARACTERISTICS OF CABLE-IN-CONDUIT CONDUCTORS FOR LARGE HELICAL DEVICE

K. Takahata, A. Iwamoto, R. Maekawa, T. Mito, T. Satow, S. Satoh and O. Motojima

National Institute for Fusion Science
Oroshi, Toki, Gifu 509-5292, JAPAN

ABSTRACT

Poloidal field coils for the Large Helical Device (LHD) consist of Nb-Ti cable-in-conduit conductors. The first cool-down of LHD started in February of 1998, and the LHD experienced the cool-down three times up to the present time. Pressure drops of the conductors and coils have been measured during inspections at room temperature, cool-down, steady state and warm-up. Friction factors transformed from the pressure losses were compared with those of short samples at room temperature to confirm a normal operation of the coils. Correlations between the friction factor and the Reynolds number could be expressed as laminar flow in non-circular pipes when the Reynolds number was less than 20. In this region, the laminar friction constant, which is defined as the product of the friction factor and the Reynolds number, was found to depend on geometries of flow channels. For the LHD poloidal coils, the laminar friction constants were measured for short samples of the conductors, long conductors before winding and pancake coils, and the data were compared with each other. The Reynolds number varied between 10 and 3000 in cooling operation. The friction factors of the short samples were then measured over a wide range of Reynolds numbers at room temperature by the use of argon gas as a fluid. The friction factors of the coils agreed well with those of the short samples.

INTRODUCTION

LHD is a superconducting toroidal fusion device.[1] The plasma experiments of the LHD started on 31 March of 1998 after a successful eight-year construction. The superconducting coil system consists of a set of continuous helical coils and three pairs of poloidal coils.[2] Two different cooling methods are adopted for the coils: the poloidal coils are cooled with forced-flow supercritical helium of 4.4-4.7 K, 0.9-0.6 MPa in contrast to the pool-boiling helical coils. The poloidal coils consist of outer vertical (OV), inner shaping (IS) and inner vertical (IV) field coils. The main specifications of the coils are listed in Table 1. Conductors are Nb-Ti cable-in-conduit types of which cross-sections and specifications are shown in Fig. 1 and Table 2.

When using cable-in-conduit conductors, one should give special attention not only to stability but also to hydraulic characteristics because an unexpected choke inside the conductor causes degradation of the stability. In the case of the LHD conductors which are not equipped with a sub-channel, the conductor is easily clogged. In the experiment of a demonstration coil[3], the coil choked with frozen gases such as nitrogen, oxygen, carbon-dioxide and hydrocarbon. Since the experiment, regenerable filters have been installed to the coil inlet pipes. In addition, it was recognized that continuous monitoring of a choke was important.

Table 1. Specifications of the coils

	OV	IS	IV
Center diameter (m)	11.1	5.64	3.6
Height (m)	0.54	0.47	0.47
Number of turns	9×16=144	13×16=208	15×16=240
Number of flow paths	16	16	16
Length of a flow path (m)	314	230	170
Total conductor length (km)	5.0	3.7	2.7

Table 2. Specifications of the conductors

	Sub-scale	OV	IS, IV
Conduit inner dimensions (mm)	Circle 4.35	Rectangular 20.5×24.8	Rectangular 17.0×21.6
Strand diameter (mm)	0.35	0.89	0.76
Number of strands	0-90	3^4×6=486	3^4×6=486
Coolant fraction	>0.42	0.38	0.38
Hydraulic diameter (mm)	4.35-0.22	0.50	0.43

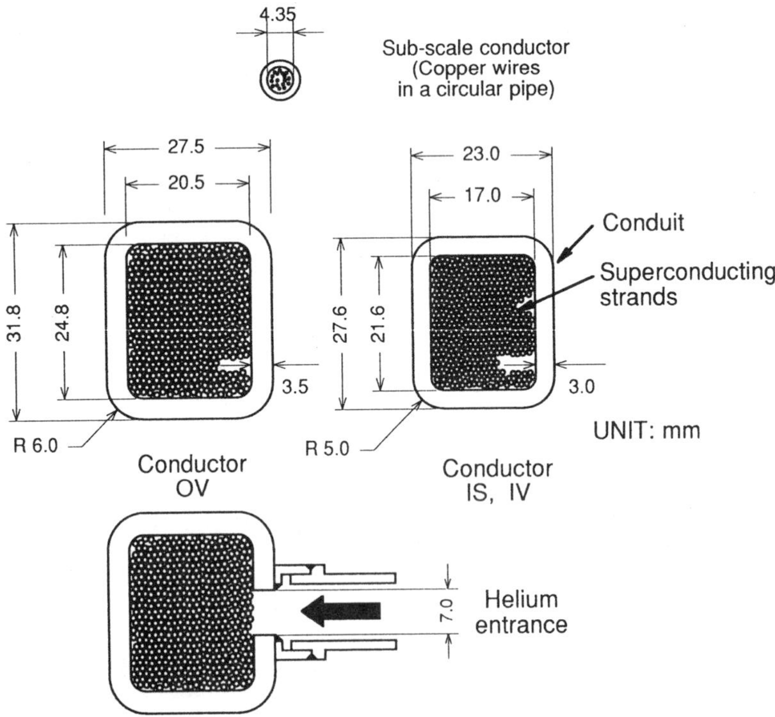

Figure 1. Cross-sectional drawings of the cable-in-conduit conductors examined in this study.

Previous studies show that different conductors have slightly different correlations between the friction factor and the Reynolds number.[4] In these situations, measured pressure drops in operation are invalid for detection of abnormalities. In this study, argon gas was passed through short conductors of about 1 m long at room temperature, and the correlation was obtained over a wide range of Reynolds numbers. The friction factors of the coil were investigated by referring to the correlation for the short samples.

The coil has 16 pancake coils, which is equivalent to 16 parallel paths with the same length. Because each path does not have a flow controlling valve, the flow distribution depends on the hydraulic characteristics of conductors. In the preliminary tests of the IV coil, unequal mass flow rates were observed between the pancake coils.[5] The importance of the uniformity of conductors was also recognized. The conductors were, therefore, inspected on the hydraulic characteristics under conditions of laminar flow at low Reynolds numbers in the manufacturing process. The results are described in detail in this paper.

TEST PROCEDURE

Pressure Drop Measurement for Coils

A continuous conductor was wound to a double-pancake coil. Therefore, the length of each conductor is equal to double the length of the flow path which is shown in Table 1. After fabrication of the conductor, one end was pressurized up to 1.1 MPa with a helium gas. The gas was bled to the atmosphere through the other end. The volume flow rate was measured at the outlet. The maximum flow rate was approximately 1 L/s.

Eight double-pancake coils were stacked in a coil. After stacking them, the pressure drops were similarly measured for each pancake coil. The helium entrance is then located at the middle of the conductor. Figure 1 also shows the structure of the helium entrance. A hole 7 mm in diameter was tapped into the conduit, and a pipe was welded to it. The maximum inlet pressure and flow rate were approximately 1.1 MPa and 2 L/s.

Finally the pressure drop of the coils was measured by pressure taps as the difference between inlet and outlet during the cool-down, steady-state and warm-up. A total mass flow rate of the coil was measured by an orifice flow meter. Inlet and outlet temperatures were observed by Pt-Co and carbon-glass-resistor (CGR) thermometers.

Pressure Drop Measurement for Short Conductors

As a preliminary test, sub-scale conductors were examined at room temperature to grasp the basic characteristics. The conductor was copper wires in a circular pipe as shown in Fig. 1. The inner diameter of the pipe was 4.35 mm, and the wire diameter 0.35 mm. The wires were put in a pipe without cabling. The number of wires was changed to examine the effect of the coolant fraction. To obtain data over a wide range of Reynolds numbers, argon gas was used as a working fluid in addition to helium. The density of argon is 1.6 kg/m^3 at atmospheric pressure and 300 K in contrast to 0.16 kg/m^3 for helium. Therefore, the Reynolds number can be increased by an order of magnitude under the same volume flow rate.

The cable-in-conduit conductors of LHD were cut to short samples of about 1 m. One end was pressurized with a helium or argon gas, and the pressure drops were measured by pressure gauges at room temperature. The flow rates were measured by a thermal-type flow meter for high flow rates and a float-type for low flow rates. The maximum flow rate was 10 L/s. In the next step, the helium entrance, which has the same configuration as the coils, was installed at the middle of the short conductor. Both ends were opened, and the inlet pipe was also pressurized with argon gas to estimate the pressure loss in the entrance structure.

Conversion to the Friction Factor

The hydraulic characteristics have been investigated by means of the Darcy friction factor. The relationship between the pressure loss (ΔP) and the friction factor (λ) can be expressed as

$$\Delta P = \lambda(L / D_h)(\rho V^2 / 2) \tag{1}$$

where L is the length of the flow path, D_h the hydraulic diameter, ρ the density of fluid and V the average velocity. The hydraulic diameter (D_h) is defined as

$$D_h = 4A / P \tag{2}$$

where A is the channel cross-sectional area and P the wetted perimeter. The perimeter is the sum of the circumferences of strands and the inside of a conduit in the case of cable-in-conduit conductors. The friction factor depends on roughness and the Reynolds number (Re_h) defined as

$$Re_h = \rho V D_h / \mu \tag{3}$$

in which μ is the coefficient of viscosity. The density in Eq. (1) and (3) and the viscosity in Eq. (3) depend on the pressure and temperature of the fluid. The mean value of the inlet and outlet pressures was used to evaluate ρ and μ. The inlet and outlet had different temperatures during the cool-down and the warm-up. The outlet temperature was used instead of the mean value to evaluate ρ and μ because the average temperature obtained by the conversion of the coil resistance was almost equal to the outlet temperature. This implies that the helium nearly reaches equilibrium with a short length from the entrance.

The pressure loss in the helium entrance can be written as

$$\Delta P = K(\rho V^2 / 2) \tag{4}$$

where K is the loss coefficient. Although the loss coefficient is a function of component geometry, it can be easily measured by the use of samples with the same configuration. The pressure loss in the exit component can be ignored because the conductor is connected to a joint box with large coolant channels. Therefore, the pressure loss of only the conductor was obtained by subtracting the loss of the entrance component from the pressure drop of the coil.

RESULTS AND DISCUSSIONS

Friction Factor of Sub-scale Conductors

Figure 2 shows the friction factors for sub-scale conductors as a function of the Reynolds number. The data with helium and argon, indicated by circles and triangles, were continuous. The argon gas enabled a flow with higher Reynolds number. For a pipe with no internal conductors, laminar and turbulent flows clearly appeared with a sharp transition at the Reynolds number of 3700. The friction factors then coincided with the well-known Hagen-Poiseuille and Blasius equations plotted on the figure. In contrast to the bare pipe, the sharp transition in the friction factor with the Reynolds number disappeared when eighty copper wires were put in the pipe. The complicated distribution of the fluid velocity was considered to contribute to the characteristics.[6] One explanation for this behavior is that one portion of the flow becomes turbulent at small Reynolds numbers, and the area of turbulence gradually widens with increasing Reynolds number. This behavior has been observed in non-circular conduits.[7] Although the transition from laminar to turbulent flow is not well-defined, complete laminar flow was observed for Reynolds numbers less than 30. The feature of laminar flow is

$$\lambda Re_h = \text{Constant}, \tag{5}$$

and the constant, called a laminar friction constant, is different for different geometries.[8] For circular pipes, the constant is 64 according to the Hagen-Poiseuille equation. For the sample with eighty copper wires, the constant decreased to 38. To examine the effect of geometry, the constant was measured for different numbers of wires corresponding to different coolant fractions. The measured constants are shown in Fig. 3 as a function of the coolant fraction. When the coolant fraction is unity, the constant is 64. For the closest packed wires, the coolant fraction corresponds to 0.093 ideally, and the constant should decrease to 26 because the geometry of channels becomes a three-point-star shape.[9] The measured constants, indicated by closed circles, varied between two theoretical values, indicated by open triangles. The laminar friction constant was also confirmed to depend on the geometries for the cable-in-conduit conductors.

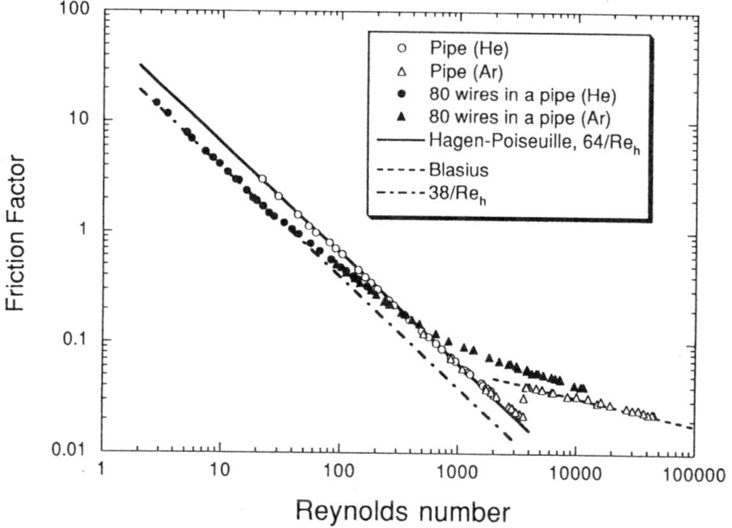

Figure 2. Friction factors of the sub-scale conductors as a function of the Reynolds number.

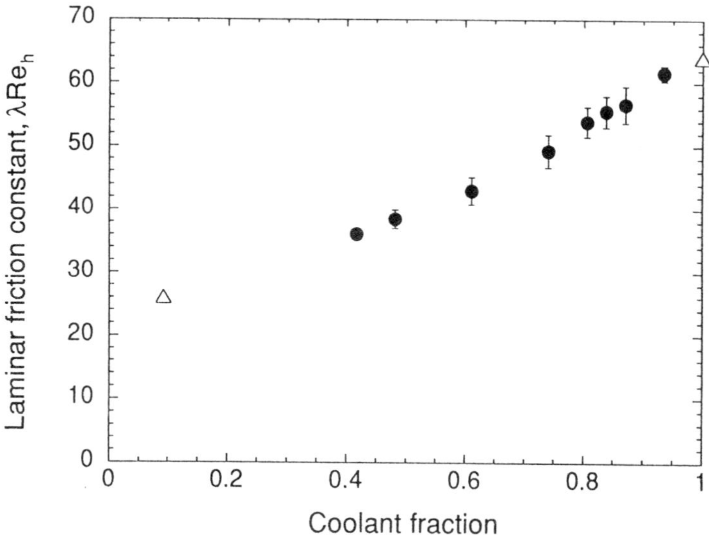

Figure 3. Laminar friction constants as a function of the coolant fraction for the sub-scale conductors.

Laminar Friction Constant of Cable-in-conduit Conductors

A change of laminar friction constants suggests an abnormality such as inclusion of contaminants and change of the coolant fraction. For the LHD poloidal coils, the laminar friction constants of the conductors were measured in the manufacturing process. The number of samples was 5 for the short samples, 16 for the long conductors and 32 for the pancake coils. Figure 4 shows the friction factors for the OV conductors for Reynolds numbers between 1 and 20. The factors were found to lie below the line of the Hagen-Poiseuille equations similar to the sub-scale conductors. In this Reynolds number range, the correlation had the feature of laminar flow, and the laminar friction constants could be obtained by

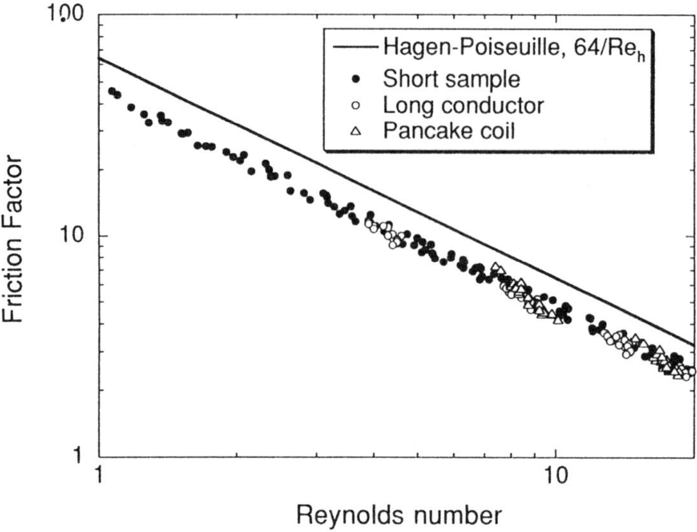

Figure 4. Friction factors of the OV coils and the short samples at low Reynolds numbers. The friction factors were measured in the manufacturing process at room temperature.

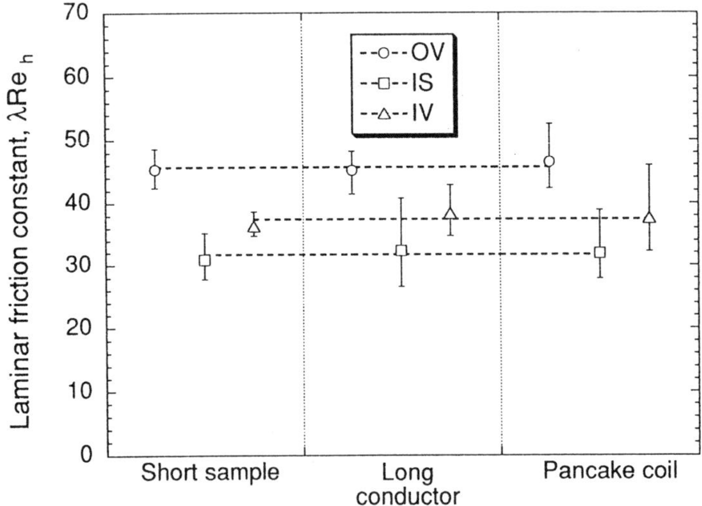

Figure 5. Laminar friction constants of the LHD coils in the manufacturing process. The marks are the mean values, and the error bars indicate the maximum and minimum for all samples.

approximate calculations. Figure 5 shows the laminar friction constants for three types of coils, OV, IS and IV. The marks are the mean values for all samples, and the error bars indicate the maximum and minimum. The constants were different for the different coils. It should be noted that the constants were different between IV and IS although the conductors have almost the same configurations. The only difference is Nb-Ti to copper ratio between the conductors. The ratios are 1:2.7 for IV and 1:3.4 for IS. The difference in hardness of strands might cause a slight difference in geometries because of differences in contact deformation of strands. This suggests that the constant is strongly influenced by geometries. For each coil, the mean values were confirmed to be constant. Furthermore, the variations showed the same levels, and an abnormality did not appear. Such an inspection is considered to be valid for manufacturing process of cable-in-conduit conductors.

Friction Factor of Coils in Operation

The coils were cooled down, while controlling the temperature difference between inlet and outlet within 50 K. Table 3 shows the cooling conditions in the third cool-down for the OV coil, as an example. In the cool-down and warm-up, the Reynolds number varied between 10 and 300 depending on the mass flow rate and the viscosity. For the steady

Table 3. Cooling conditions in the third cool-down for the OV coil.

Outlet temperature (K)	Inlet temperature (K)	Inlet pressure (MPa)	Pressure drop (MPa)	Mass flow rate (g/s)	Reynolds number
280	234	0.82	0.69	2.8	26
250	203	0.84	0.72	3.4	34
200	156	0.84	0.72	4.5	52
150	107	0.92	0.80	7.6	106
100	60	0.78	0.14	4.3	76
50	26	0.96	0.09	7.7	207
4.7	4.4	0.89	0.13	67	2685

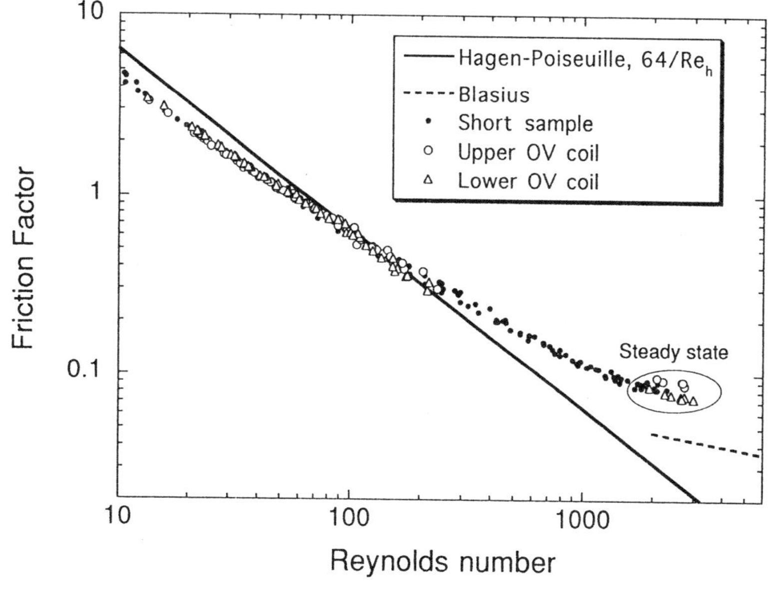

Figure 6. Friction factors of the OV coils in cooling operation and the short samples at room temperature.

state, the Reynolds number increased to 2000-3000 because the helium became supercritical and its density became comparable to that of liquid. It should be noted that the pressure loss in the entrance structure accounted for about 20 % of the total pressure drop in the steady-state, which was found by means of the component test at room temperature. For the OV coil, the estimated loss of the entrance was 0.023 MPa. The friction factor of the conductors should be estimated taking into account all components on pressure losses. The friction factors of the OV coils in operation were compared with those of the short samples. The results are shown in Fig. 6. The open circles and triangles indicate the factors for the upper and lower coils respectively. All data in three operational periods are plotted on the figure. The experimental results showed that the characteristics did not change for all operational periods and agreed well with those of the short samples indicated by closed circles. This suggested that no obstruction existed in the coils. The comparison with short samples is considered to be valid for detection of chokes.

The correlation between the friction factor and the Reynolds number was similar to that of the sub-scale sample indicated in Fig. 2. From the fact that the wires were not twisted for the sub-scale sample, it was considered that the correlation mainly depended on the configuration at the cross section instead of cabling. For Reynolds numbers more than 3000, additional losses due to cabling might become noticeable.

CONCLUSIONS

The hydraulic characteristics of the cable-in-conduit conductors for the Large Helical Device have been investigated to establish an efficient way to detect unexpected chokes. The measurements of laminar friction constants were available for inspection in the manufacturing process because the constant depends on geometries of flow channels. The pressure drops of the conductors could be estimated through a wide range of the Reynolds number by the use of short samples at room temperature. The friction factors of the coils in operation were compared with those of the short samples, and the smooth helium flow was confirmed. Before the coil fabrication, the short sample tests are recommended for accurate evaluation and later examination on the hydraulic characteristics.

REFERENCES

1. O. Motojima, et al., Initial physics achievements of Large Helical Device experiments, Physics of plasmas, 6:1843 (1999).
2. O. Motojima, et al., Superconducting magnet design and construction of Large Helical Device, "Fusion Energy 1996," International Atomic Energy Agency, Vienna (1997), p. 467.
3. K. Takahata, et al., Stability tests of the Nb-Ti cable-in-conduit superconductor with bare strands for demonstration of the Large Helical Device poloidal field coils, IEEE Trans. Magn., 30:1705 (1994).
4. H. Katheder, Optimum thermohydraulic operation regime for cable-in-conduit superconductors (CICS), Cryogenics, 34:595 (1994).
5. K. Takahata, et al., Cooldown performance of an inner vertical field coil for the Large Helical Device, IEEE Trans. Magn., 32:2252 (1996).
6. T. Amamo, et al., Flow visualization of coolant in cable-in-conduit conductor, IEEE Trans. Magn., 27:2112 (1991)
7. D. J. Gunn and C. W. W. Darling, Fluid flow and energy losses in non-circular conduits, Trans. Instn Chem. Engrs., 41:163 (1963).
8. P. M. Gerhart and R. J. Gross, "Fundamentals of Fluid Mechanics," Addison-Wesley, Massachusetts (1985).
9. F. S. Shih, Laminar flow in axisymmetric conduits by a rational approach, Can. J. Chem. Eng., 45:285 (1967).

FINAL THERMAL ANALYSIS OF THE G-ZERO MAGNETIC SPECTROMETER

T. A. Brandsberg,[1] T. A. Antaya, [1] P. D. Brindza[2]

[1]BWX Technologies, Inc.
Lynchburg, VA 24505-0785
[2]Thomas Jefferson National Accelerator Facility
Newport News, VA 23606

ABSTRACT

BWX Technologies, Inc. (BWXT) has completed the final design on a superconducting eight-coil toroidal magnetic spectrometer system for the G-Zero experiment. Conceptually designed by the University of Illinois at Urbana-Champaign, this magnet will be installed in Hall C at the Thomas Jefferson National Accelerator Facility.

This paper describes the thermal analysis performed as a part of the final design. Specific issues to be discussed include the finite element analysis of the magnet structure for evaluating cooldown rates, as well as steady state temperature distributions, and thermal siphon heat removal capabilities. The cooldown studies demonstrated that the magnet cold mass could be cooled to operating temperatures within the timeframe specified despite some limitations on the availability of coolant from the refrigeration system. The thermal siphon cooling system was analyzed with the ANSYS finite element analysis software package to assure that the thermal siphon cooling would function with the non-uniform heat-leak distribution into the cold mass structure. The finite element analysis modeling methods used to represent two phase liquid helium flow pressure drop provided very good agreement with previously published experimental data.

INTRODUCTION

The final design of the G^0 Superconducting Magnetic Spectrometer (SMS) has been completed by BWX Technologies, Inc. (BWXT) and is now in the manufacturing phase. This system is to be used for the G^0 parity experiment to be performed at the Thomas Jefferson National Accelerator Facility (TJNAF). The final design analysis included evaluation of the ability to cool the magnet with the cooling resources allocated.

This magnet consists of 8 superconducting coils that are assembled in a toroidal arrangement as shown in figure 1. Each coil is wound on a solid aluminum bobbin as two double pancakes with 36 turns per pancake. The bobbin (approximately a rectangular shape with rounded corners as shown in figure 2), is gun-drilled with three intersecting

Advances in Cryogenic Engineering, Volume 45.
Edited by Shu *et al.*, Kluwer Academic / Plenum Publishers, 2000.

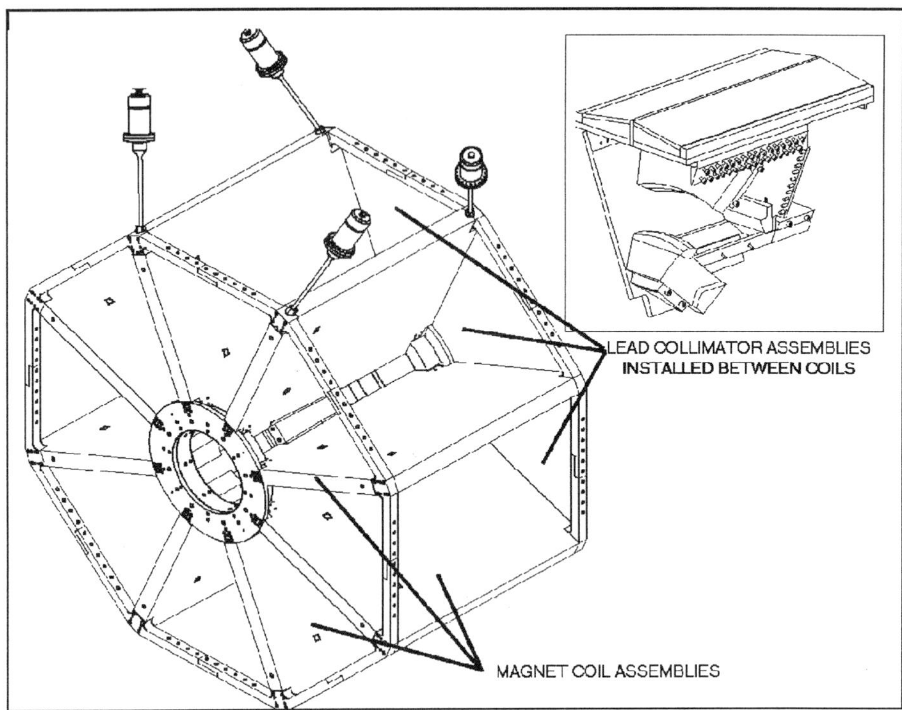

Fig.1. G^0 Superconducting Magnet Cold Mass & Support Rods

holes to form a U-shaped coolant channel. These coolant channels are connected to a liquid helium reservoir with stainless steel tubing in a fashion that promotes natural circulation cooling as shown in figure 3.

Note that the 8 coils are cooled in 4 parallel paths, and that the heat load in each of these paths will not be equal since only two coils have direct connection with the cold mass support rods. A fifth coolant channel is utilized to cool the splice joints and the lead bus. The driving force for the coolant flow is the difference in fluid density between the pipe supplying LHe to the bottom of the magnet, and the two-phase flow in the lines returning to the reservoir.

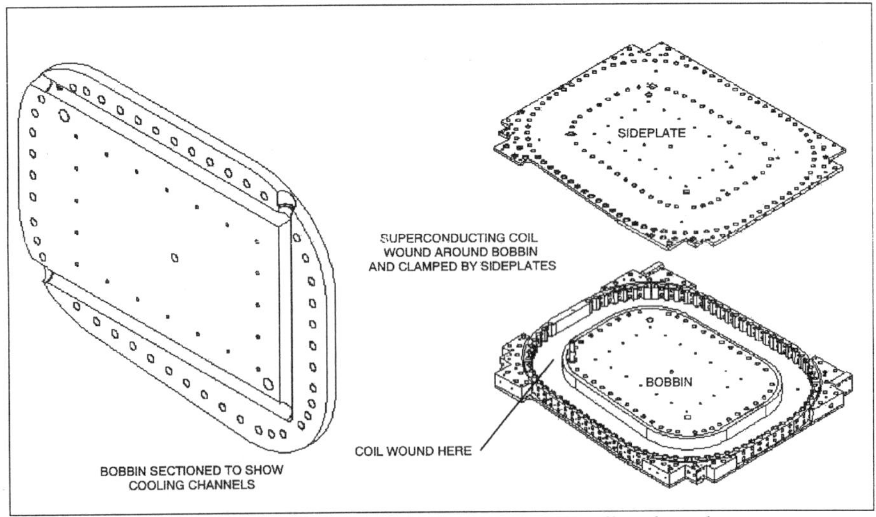

Fig.2. Coil Bobbin showing gun-drilled cooling channels

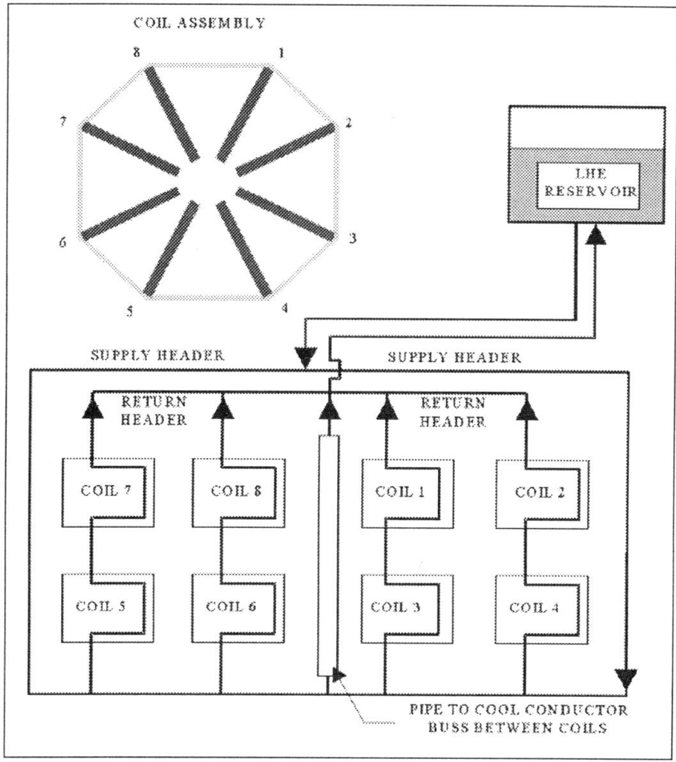

Fig.3. Thermal-Siphon Cooling System Schematic

The superconductor strand is made into a Rutherford cable and soldered into a channel formed in a copper strip. Electrical insulation is provided by 0.001" thick Kapton tape wound around the conductor with 50% overlap. Further electrical insulation is provided by sheets of G-10 which are placed against the bobbin and coil case hardware. The conductor is cooled by thermal conduction through these layers of insulation to the bobbin. Other components such as the lead collimators placed between the magnets will be cooled by conduction through their supports, to the coil sideplates and then to the bobbin.

ANALYSIS OBJECTIVES

The objectives of the thermal analysis included:

- Verify heat loads to the LHe and LN2 coolant did not exceed the specified values;
- Verify the steady state conductor temperatures for superconductor stability;
- Verify steady state and transient cooling capacity;
- Verify cooling system stability with non-symmetric heat loads;
- Verify cooldown capability with available cooling resources.

ANALYSIS METHODS

The heat loads required to be removed by the LHe coolant were calculated as shown in table 1 and used as input to the thermal analysis calculations. The G0 magnet cooling system was modeled using the FLUID66 element that is a part of the ANSYS[3] finite element software package. This element simulates the flow of a fluid through a pipe, including the pressure drop due to fluid friction, pressure changes due to bouyancy/elevation, and the variation in heat transfer due to the Reynolds number and physical properties of the fliud flow.

Although this element is intended for simulating single phase fluids, the two-phase behavior of the liquid helium coolant was simulated by transitioning from liquid properties to vapor properties over a very small temperature range. This was a very reasonable approximation since the pressure changes around the coolant system were small.

One key factor to keep in mind when using this element is that the fluid flowrate and specific heat must be input on a weight basis, whereas the specific heat of the solid structure is input on a mass basis. Additionally, the friction factor, f, should be based on the Moody friction factor rather than the Fanning friction factor. These two factors differ by 4:1 since they are associated with the following formulation for pressure drop:

Fanning form: $$\Delta p = \frac{2 f \rho u^2 L}{D_h} \qquad (1)$$

Moody form: $$\Delta p = \frac{f \rho u^2 L}{2 D_h} \qquad (2)$$

To verify the accuracy of the modeling of this element, two tests were performed to verify the friction pressure drop calculations, one for single phase and the other for two phase friction pressure drop calculations. Using an example found on page 245 of *Helium Cryogenics*[1], the ANSYS model calculated a single phase pressure drop of 2.478 kPa vs the textbook value of 2.5 kPa. The ANSYS input file for this test case is listed in table 3.

To allow this element to simulate two-phase flow, the fluid density and specific heat

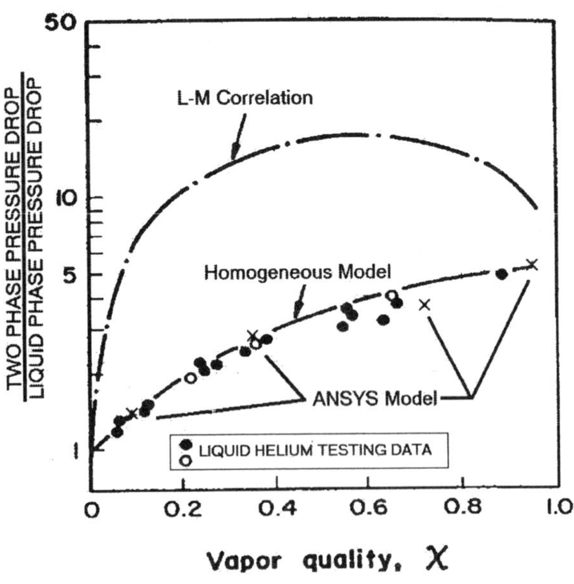

Fig.4. ANSYS FLUID66 element performance in simulating two phase pressure drop (X) matches the Homogeneous Model.

TABLE 1. CALCULATED HEAT LOADS TO LHe

Heat Source	Magnitude
Cold Mass Support Rods -	2.0 W
Lower Support Pin -	1.88 W
Radiation from shield -	0.75 W
Radiation from windows -	6.0 W
Radiation from target -	0.88 W
Residual Gas -	<1.0 W
Conductor Splices -	1.2 W
Nuclear Radiation -	<1.0 W
Summation (average per Coil)	**<1.85 W**

were input as a function of the fluid temperature, with the change from liquid to gas occurring over a range of 0.1 K. The specific heat was adjusted to include the latent heat of vaporization. Using this method, the pressure drop through a section of pipe was calculated for several values of vapor quality. This modeling method presumes that the vapor and fluid in a two-phase stream are well mixed and move at the same velocity (as assumed in the Homogeneous Model of two-phase flow). Thus, it is not surprising that the results correlate well to the data plotted in *Helium Cryogenics*[1] and shown in figure 4. Although some fluids and flow environments are better represented by the Lockhart-Martinelli (L-M) Correlation, the plotted test data indicates that liquid helium flowing in a pipe is not one.

ANALYSIS RESULTS

A. Steady State Cooling

Using the model shown in Figure 5, the coolant mass flow rate was calculated for a range of heat loads ranging from significantly smaller to significantly larger than that expected in operation. The calculated fluid flow increased as the heat load increased until the vapor quality reached about 25%, and then remained relatively constant as the vapor

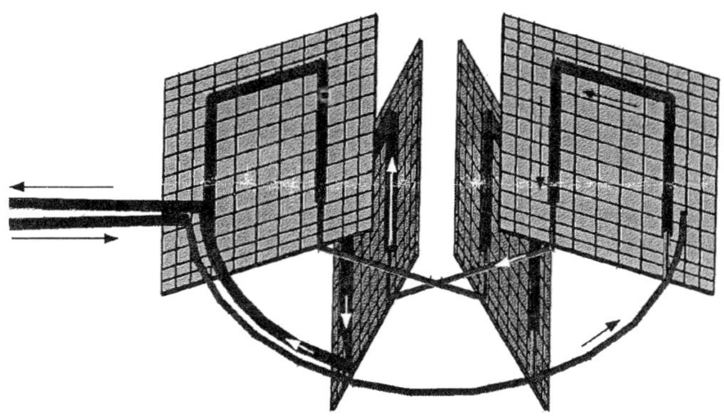

Fig.5. ANSYS model of the thermal-siphon cooling circuit (half symmetry)

TABLE 2. THERMO-SIPHON SENSITIVITY TO HOT SPOT HEATING

HOT SPOT FACTOR	TOTAL FLOW, GM/SEC	HOT LOOP FLOW, GM/SEC	COOL LOOP FLOW, GM/SEC
1	75	19.8	17.7
2	84	24.4	18.0
4	94.8	29.4	18.1
8	110.6	36.2	18.4

quality increased with greater heat load, as shown in figure 6. These results are qualitatively consistent with natural circulation behavior reported by Huang and Van Sciver[2].

Further calculations investigated the system behavior if one of the cooling legs received more heat than the others (perhaps due to quench or insulation defect). Non-symmetric heating was evaluated with one leg heated by factors 2, 4 and 8 times the baseline value. As reported in table 2, the total flow rate increased along with that of the hot leg, but the "cool" legs did not experience a reduction in flow rate. Therefore, the magnet will be adequately cooled despite individual coils being heated more than others.

For the expected steady state cooling scenario, coils 2, 3, 6 & 7 will each receive about 1.35 watts. Due to conduction heating through the supports, coils 1 and 8 will each receive 2.4 watts, while coils 4 and 5, 2.3 watts. Saturated LHe coolant was assumed to besupplied from the cryo-reservoir at 1.2 atmosphere with an initial temperature of 4.424 K.

Under these conditions, the total LHe flow through the magnet system was calculated to be 79.6 gm/sec with the warmest spot in the coil calculated to be 4.85 K at the outboard corners of coils 1 or 8 (adjacent to the cold mass support rods). This temperature is significantly below the current sharing temperature of the NbTi superconductor (5.9 K at 3.63 T max, 8 K at the outboard corners, 1.6 T).

The steady state cooling capability has been demonstrated to be stable for heat loads

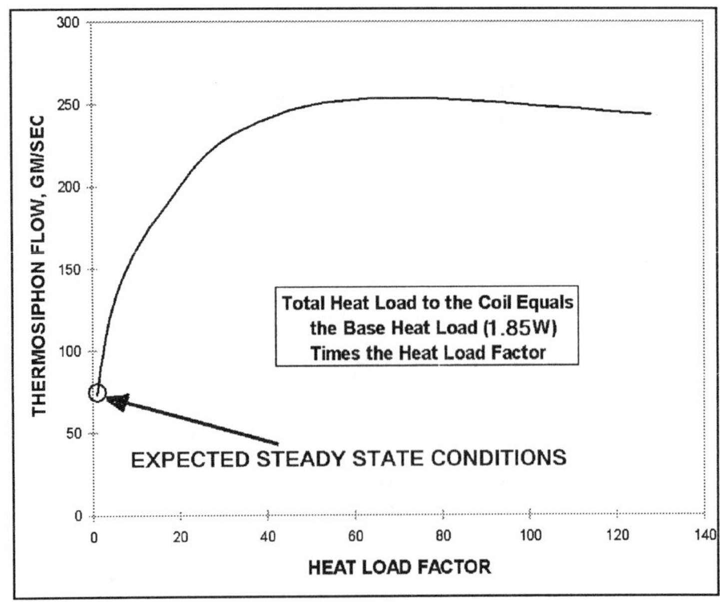

Fig.6. Thermal-siphon cooling mass flow rate as a function heat load. The heat load factor is the multiple of the expected heat load which can be handled by the system.

well over 100 times the normal expected values. Thus, it is obvious that this system design will allow the magnet to recover from a quench (should one ever occur). Further, it will be possible to use the natural circulation mode in the final stages of cooldown.

B. Cooldown To Operating Conditions

The magnet design specification required that the magnet be designed for cooldown to operating temperatures within 10 days. The initial phase of this cooldown was to utilize 2.5 g/s helium gas chilled by liquid nitrogen to chill the magnet down to about 80 to 100 K. The remainder of the cooldown would utilize LHe. Although the TJNAF refrigeration system could supply 4 g/s of LHe, this was not sufficient to finish the cooldown in 10 days. However, it was determined that this flow could be supplemented with LHe previously stored in a 10,000 liter dewar.

The final baseline cooldown scenario started with the chilled helium gas entering the magnet at 50K below the magnet hardware temperature. The gas temperature is then reduced at a rate of 1 degree K per hour until the gas is down to 100K. The coolant is changed over to LHe flowing at 10 g/s. Over the next 4 days, the coolant flow is gradually reduced to the steady state value of 4 g/s. At this time the valving can be adjusted to allow natural circulation flow, with the coolant flow provided at such a rate to maintain the LHe reservoir level.

In addition to the capability to transfer the heat into the coolant, this study also investigated the ability to transfer heat through the structure to the coolant channel walls. One of the key issues relating to the cooldown calculations is how to model the transfer of heat through mechanical joints in a vacuum. Many references were reviewed to determine a reasonable basis for modeling this phenomenon. They generally represent the contact heat transfer as a flow of heat per unit area per degree. In essence this is the same as a convection film coefficient for transfer of heat between a solid and a fluid. Typical values for contact thermal resistance in a vacuum were found to be about 1 to 19 $kW/m^2/K$ for contact stresses of about 5 to 7 MPa. A value at the middle of this range was selected for these calculations.

The final model represented the thermal mass of key components within the assembly as point masses (MASS71) thermally connected with various types of link elements from ANSYS, including LINK33 for conduction, LINK31 for thermal radiation and LINK34 for thermal convection.

The radiant heating was used to model the heat flow between components which did not have a direct mechanical connection. One area modeled with radiant heat transfer is between the cold mass and the LN2 shield. During the initial phase of the cooldown, the cold mass would be cooled by the LN2 shield, but later it would be part of the heat leaking into the cold mass. It was also used to evaluate cooling of the collimator assemblies located adjacent to the coil case side plates. It was determined that a direct conduction path should be established between the collimator and the side plates to assure that their temperatures did not lag too far behind the rest of the cold mass.

CONCLUSIONS

These calculations have demonstrated that the G0 magnet cooling system is robust enough to handle the heat loads which could enter the magnet cold mass structure. There is sufficient extra capability to respond to quench transients as well as assist in the completion of the cooldown transient. The thermal-siphon cooling system has been demonstrated to remain stable over a wide range of heating asymmetries.

TABLE 3
ANSYS Input File for Pressure Drop Test

```
!cryo1.mac    test single phase delta-p LHe example from Helium Cryogenics, Van Sciver, page 245

/prep7          ! DEFINE VALUES FOR PARAMETRIC VARIABLES
   flodia= .010                                         ! pipe ID,m
   lgth=  30.00                                         ! pipe length,m
   flovel=1.0                                           ! flow vel, m/s
! liq He at 4.0 k, 1.0 atm
   densityl=130.1                                             ! kg/m**3
   viscosl=3.34e-6                                      ! N*sec/m**2
   gacc=9.81                                            ! grav acc, m/s^2
   floarea=0.785*flodia**2                              ! flow area, m^2
   flor8=flovel*densityl*floarea*gacc                   ! flow weight/sec
   Re=densityl*flovel*flodia/viscosl                    ! Reynolds Number
   mffi=4*(0.0791/Re**.25)                              ! Moody frict fctr
   ET,2,FLUID66
   R,2,flodia,floarea,0,gacc,0,0,
   RMORE,0,0, ,1,0,
   mp,dens,  2, densityl                                ! kg/m**3
   mp,kxx,   2, thcond                                  ! w/m-deg k
   mp,c,     2, thcap                                   ! j/kg-deg k
   mp,visc,  2, viscosl                                 ! kg/m-s
   mp,mu,    2, mffi                                    ! Moody frict fctr

! CREATE SINGLE ELEMENT MODEL, APPLY B/C & SOLVE
   type,2  $ mat,2  $ real,2
   n,1,0,0,0  $  n,2,lgth,0,0
   e,1,2
/solu
   antype,static
   d,2,press,0  $  f,1,flow,flor8
   solve

! OBTAIN PRESSURE DROP RESULTS
/post1
  *GET,pres1,NODE,1,PRES,
/title,pressure drop is %pres1/1000% KPa vs 2.5 KPa from reference
```

The ANSYS FLUID66 element has been demonstrated to effectively model natural circulation cooling using two phase helium. However, care must be taken when supplying fluid properties, which are to be supplied on a weight basis. The input listing of table 2 offers an example for benchmarking these calculations.

The cooldown calculations demonstrated that the coolant resources available at the TJNAF could chill the coil below the current sharing temperature within the specified time frame. Additionally, due to the considerable steady state capability of the natural circulation cooling, the final stages of the cooldown will be significantly improved when the transition from forced flow is implemented.

REFERENCES

1. S.W. Van Sciver, *Helium Cryogenics*, New York, Plenum Press, 1986
2. X. Huang and S. Van Sciver, Performance of a venturi flow meter in two-phase helium flow, *Cryogenics*, 36:303 (1996).
3. ANSYS Finite element Software, ANSYS, Inc.

CRYOGENIC TARGET SYSTEM FOR Z-PINCH MACHINES

C.R. Gibson, N.B. Alexander, J.V. Del Bene,
D.T. Goodin, and G.E. Besenbruch

General Atomics
San Diego, California 92121

ABSTRACT

Recent advances in the technology of fast pulsed electrical power and load design have led to growing interest in the use of z-pinch machines as x-ray sources for fusion applications.[1] To achieve high yield, z-pinch machines require the use of cryogenic targets. These targets contain a spherical polymer capsule with a layer of solid deuterium-tritium (DT) on the inside wall. This DT layer is obtained by first permeation filling the capsule with high pressure DT gas at room temperature. The filled capsule is then cooled to approximately 19 K to solidify the DT. The uniform layer is created through a process called "beta layering," which involves placing the capsule in a very uniform temperature environment and allowing the natural heating due to beta decay of the tritium to provide the thermal energy to redistribute the DT into a uniform layer. General Atomics has performed preconceptual design studies of a system designed to fill a target with DT, cool it down to cryogenic temperatures, and then insert it into a z-pinch chamber. The baseline system design uses a supercritical helium stream to cool the target. The helium is compressed by a room temperature compressor and cooled using both liquid nitrogen and liquid helium reservoirs. The cold supercritical helium then travels to the target through an umbilical tube several meters long. This umbilical allows the target to be at the center of the chamber while the cryogenic equipment is safely outside. This paper will describe the preconceptual design of a z-pinch cryogenic target system and present the supporting thermal analysis.

INTRODUCTION

The z-pinch was one of the first plasma confinement concepts to be studied in the early days of controlled fusion energy research. In such a device, a large current flows axially down a cylindrical array of wires, as shown in Fig. 1. The current creates a large circumferential magnetic field producing a Lorentz force which implodes the wire array towards the center of the cylinder. The wires vaporize from the ohmic heating into a plasma which continues to carry the current. The Lorentz force accelerates the plasma inward to the center of the wire array where it is highly compressed and heated. This, in turn, causes the plasma to emit x-rays.

In the last three years, the x-ray output energy from Sandia National Laboratory's "Z-Machine" has increased from 0.1 to almost 2.0 MJ. This is the culmination of many decades of experimenting with z-pinches and pulsed power devices. The Z-Machine now produces the world's most powerful and energetic x-rays and can be used as a driver to study inertial confinement fusion (ICF). The x-rays can be used to compress capsules filled with a mixture of deuterium and tritium (DT). For an ICF device to reach ignition, the DT

Advances in Cryogenic Engineering, Volume 45.
Edited by Shu *et al.*, Kluwer Academic / Plenum Publishers, 2000.

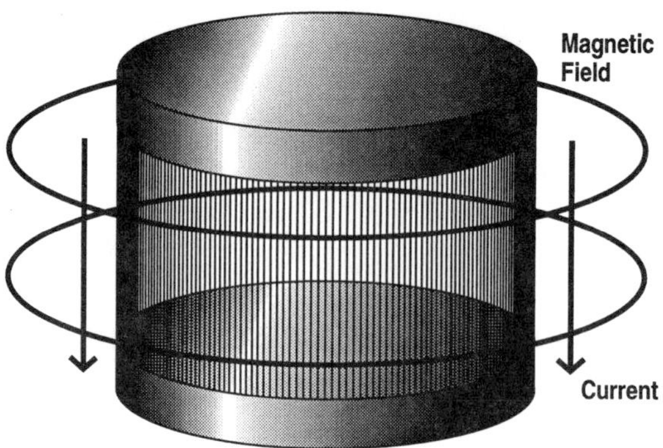

Figure 1. Sketch of z-pinch wire array showing the axial current and circumferential magnetic field.

must be a solid (~19 K) and must form a smooth, symmetrical layer on the inside of the capsule wall.

Sandia National Laboratory has proposed two follow-on z-pinch devices for the Z-Machine. The X-1 Advanced Radiation Source is designed to have a x-ray output of 16 MJ. This is enough energy to explore ignition and even high gain fusion. The ZX machine is an intermediate step between Z and X-1 and has a design output of 7 MJ.

Z-Pinch Cryogenic Target Systems

Both X-1 and ZX require cryogenic targets to meet their design goals. General Atomics in cooperation with the University of Rochester and Los Alamos National Laboratory, has designed a cryogenic target delivery system for the OMEGA laser at the University of Rochester's Laboratory for Laser Energetics.[2] This system has been fabricated and is in final testing in Rochester. The first laser shot of a cryogenic target is scheduled for September 1999.

Using the experience gained on the OMEGA system, we have produced design concepts for the proposed z-pinch machines. One major difference between OMEGA and the proposed z-pinch machines is the amount of damage caused by the shot. For OMEGA, everything within about one centimeter of the target is vaporized or severely damaged. For X-1, this radius will be about 50 cm. All components inside this spherical volume must be replaced after each shot. In addition, all unshielded regions inside the experiment chamber will be subject to shrapnel damage. This requirement forces all non-expendable equipment to be located safely outside of the chamber.

Z-PINCH CRYOGENIC TARGET SYSTEM

A pre-conceptual point design for a z-pinch cryogenic target system has been produced, as shown in Fig. 2. The system consists of the z-pinch target, the target cooling cryostat, the fill station, the insertion robot, and the auxiliary cooler.

Z-Pinch Target Design

Target designers working at Sandia and Lawrence Livermore National Laboratories have developed several z-pinch target designs.[3] A one-sided dynamic hohlraum design is shown in Fig. 3. The dynamic hohlraum design has the target centered inside of the wire array. During the shot, the high atomic number plasma is accelerated inward forming, in effect, a dynamic hohlraum wall. This wall implodes and stagnates on the foam cylinder emitting x-rays onto the capsule for its implosion.

At the center of the target is a spherical capsule holding the DT fuel. The target assembly is designed to cool the capsule to just below the triple point of DT (~19 K) in an isothermal environment (±0.1 K). Under these conditions, the DT will form a very symmetrical solid layer inside the capsule by a process called beta layering.[4] Heat energy from the

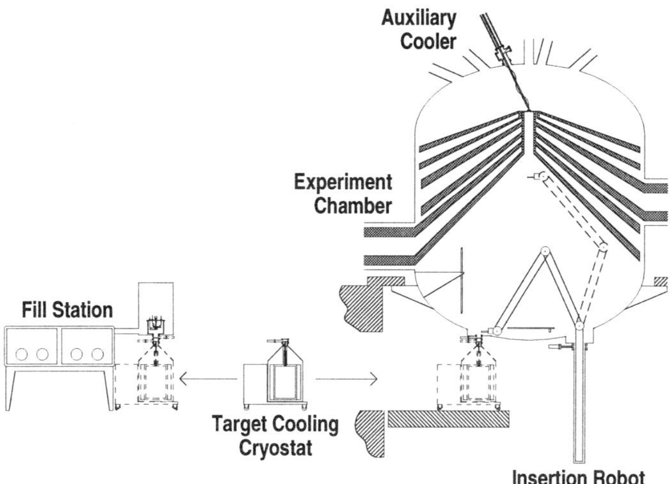

Figure 2. The z-pinch cryogenic target system includes a DT fill station, a target cooling cryostat, an experiment chamber, an insertion robot, and an auxiliary cooler.

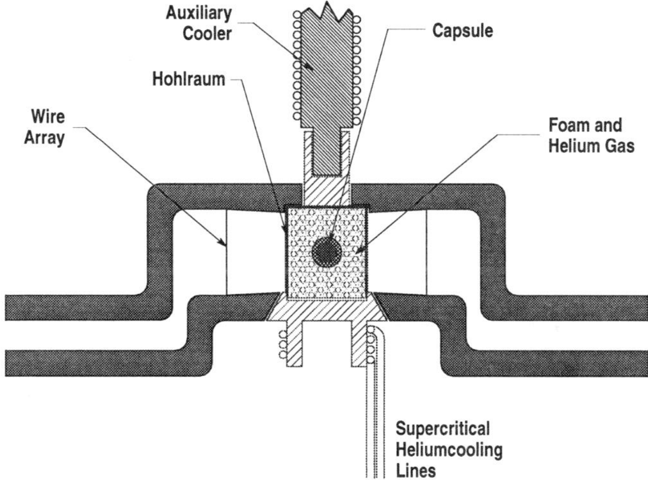

Figure 3. One-Sided dynamic hohlraum design.

beta decay of the tritium results in a redistribution of the DT into a uniform layer. This occurs due to the high vapor pressure of solid DT near its triple point. Non-uniformities in layer thickness cause local temperature variations, which translate into differences in local saturated vapor pressures. This results in net material movement from warm to cold spots until a smooth isothermal surface is created.

Immediately outside the capsule are two layers of foam of densities between 3 and 50 mg/cc. The first layer is spherical in shape while the second layer is cylindrical. These open cell foams serve to stagnate the wire plasma implosion and must include some helium gas to conduct the beta decay heat away from the capsule.

Surrounding the cylindrical foam is the cylindrical hohlraum wall. This wall is made of gold and is 2 to 4 μm thick. From a cryogenic point of view, the hohlraum wall acts like a thermal shield to keep the capsule cold and to conduct away heat. Windows may be placed in the hohlraum wall to allow access to the capsule for characterization.

Cooling tubes are attached to the bottom of the hohlraum wall to provide cooling. Supercritical helium at a pressure of approximately 15 atm and a temperature of about 13 K

Cooling tubes are attached to the bottom of the hohlraum wall to provide cooling. Supercritical helium at a pressure of approximately 15 atm and a temperature of about 13 K flows through the tubes. The helium is cooled to cryogenic temperatures in the target cooling cryostat as described in the following section. An auxiliary cooler is attached to the top of the target to provide more uniform cooling.

The target is designed to be inserted into the wire array from the bottom. To prevent a thermal short, the target is spaced away from the surrounding structure by a nominal 0.5 mm gap. A low thermal conductivity locking mechanism accurately positions the target assembly.

The pre-conceptual design of a cryogenic two-sided static hohlraum is shown in Fig. 4. The static hohlraum design has two sets of wire arrays located above and below the hohlraum and its capsule. The two wire arrays are connected in parallel to one power feed. During the shot, the wires are accelerated inward onto the foam cylinders heating them to ~100 eV. The x-ray radiation produced by the heating of this static cylinder then flows axially into the hohlraum and capsule.

The cryogenic design is similar to the dynamic hohlraum except that the axial length has been greatly increased. The interior spaces of the hohlraum are filled with CH foam and/or helium gas. The long aspect ratio of this target makes cooling this target more challenging than the dynamic hohlraum.

Z-Pinch Target Thermal Analysis

A finite-element thermal analysis has been performed on the one-sided dynamic hohlraum design shown in Fig. 3. There are two major thermal issues that must be addressed in the design of a z-pinch target. One is how the heat from radioactive beta decay of tritium inside the capsule is removed. The second issue is how to provide a sufficiently uniform thermal environment (±0.1 K) around the capsule to produce a uniform solid DT layer.

The thermal analysis results show that the foam surrounding the capsule must be filled with helium gas. The foam alone does not have sufficient thermal conductivity to remove the heat from the tritium beta decay and still allow the target to be at 19 K. Since theexperimental chamber must be under a high vacuum, the hohlraum walls surrounding the capsule must therefore be gas tight.

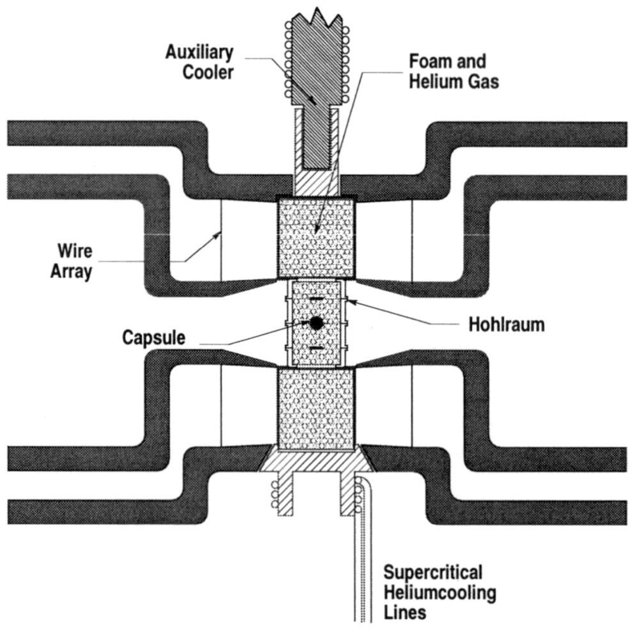

Figure 4. Two-sided static hohlraum design.

The thermal analysis also shows that cooling only from the bottom of the target cannot create the isothermal environment required for producing a uniform layer by beta layering. An auxiliary cooler has, therefore, been added to the design. This device attaches to the top of the target after the target has been placed in the center of the wire array by the insertion robot. The thermal analysis has shown that with the auxiliary cooler, a temperature uniformity of ±0.1 K can be achieved at the capsule wall.

Target Cooling Cryostat

The target cooling cryostat, shown in Fig. 5, is used during the fill, transport and insertion of z-pinch targets. It consists of three main parts: the extendible target holder, the liquid cryogen dewars and the umbilical.

The target is assembled separately at room temperature and then placed by hand on the target cooling cryostat's extendible target holder. This device elevates the target from its storage spot inside the cryostat to an elevation above the top of the cryostat. It is used to place the target into the fill station for target filling and partially into the experiment chamber. The extendible target holder consists of a welded stainless steel bellows which is extended by pressurizing the inside with helium gas. Precision rails guide the target holder and allow only vertical motion.

Three possible target cooling techniques, shown in Fig. 6, have been examined. Conduction cooling the target from a liquid helium reservoir [Fig. 6(a)] is conceptually the simplest approach, however there are several disadvantages. The reservoir must either be designed to be replaced after each shot or it must be located far away from the target. In addition, since the target must be kept cold at all times after filling, the reservoir must be inserted into the evacuated experiment chamber at the same time as the target. The other two concepts studied use flowing supercritical helium to transfer the heat from the target to the cooling source. Supercritical helium at approximately 1.5 MPa (15 atm) is first passed through a heat exchanger submerged in liquid nitrogen to cool the gas to about 77 K. The gas is then cooled to about 13 K by either liquid helium [Fig. 6(b)] or a mechanical cryocooler [Fig. 6(c)]. Several counterflow heat exchangers can be added to precool the gas with the return stream thereby increasing the efficiency of the system. Supercritical helium is used because of its low viscosity and excellent heat transfer characteristics. In addition, the flow oscillation and excessive pressure drop problems encountered with two-phase helium flow are avoided.

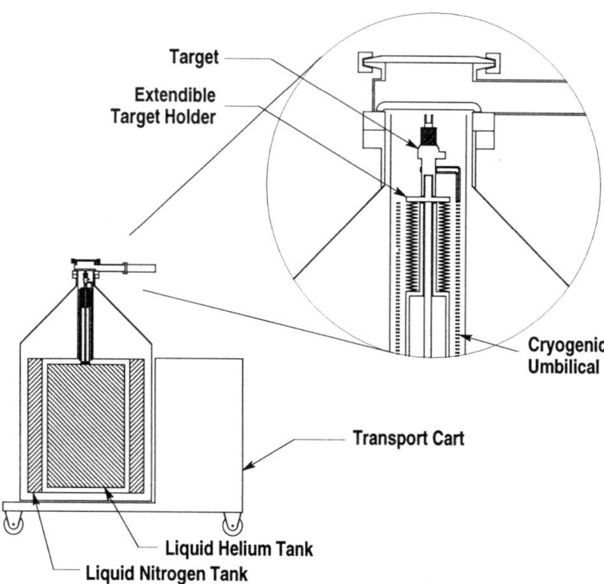

Figure 5. The target cooling cryostat moves the filled target from fill station to experiment chamber.

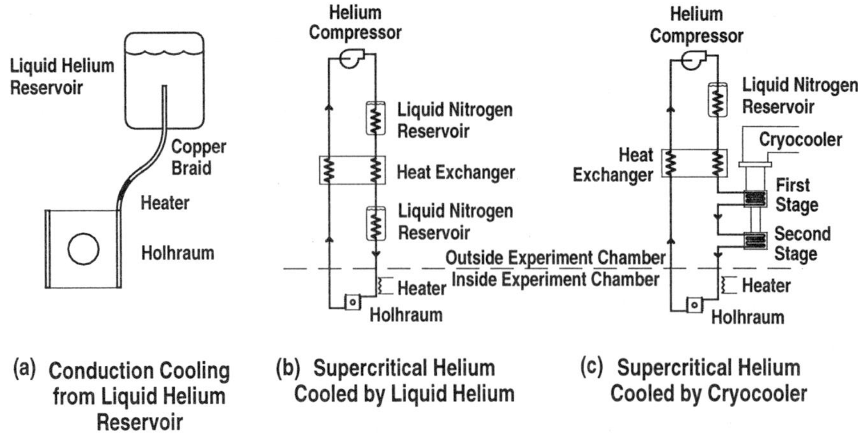

Figure 6. Several cooling schemes for target cooling were investigated.

The target cooling cryostat is located on a cart which moves between the fill station and the experiment chamber. The cart contains the helium compressor, support utilities, and instrumentation for the target cooling cryostat.

The baseline target cooling cryostat design presented here uses a supercritical helium stream cooled by liquid nitrogen and liquid helium as shown in Fig. 6(c). A preliminary thermal analysis predicts this system will reach steady state equilibrium in about 4 h. With a helium flow rate of 0.1 gm/s, the pressure drop through the system is about 0.4 MPa (4 atm) and the consumption rates are 0.3 L/h of liquid nitrogen and 6 L/h of liquid helium.

The umbilical carries supercritical helium between the liquid cryogen dewars and the target. This design allows the target cooling cryostat to remain safely outside the experiment chamberduring the shot. It consists of a helium supply tube, a helium return tube, and leads for temperature sensors and heaters on the hohlraum. The umbilical must extend from a span of about 0.5 m when inside the target cooling cryostat to approximately 5 m when the target is inside the wire array. This is accomplished by using a tightly coiled umbilical similar to a spring. Proof-of-concept tests were performed on 2 m long by 3.18 mm (0.125 in.) diameter tubing of various cryogenically compatible materials wound into a 9 cm (3.5 in.) diameter spiral. The sample umbilical with the best performance was made from a 304 stainless steel tube with a 0.71 mm (0.028 in.) wall. This sample umbilical showed the characteristics required for a full length umbilical: elastic deformations for small axial deflections and no kinking for large deflections.

Fill Station

The fill station includes several pieces of equipment, as shown in Fig. 7, which are designed to fill a polymer capsule by permeation of high pressure [up to 150 MPa (1500 atm)] DT gas through the capsule wall. The target cooling cryostat, with a target inside, is mated with the fill station. The extendible target holder is raised up to insert the target inside a permeation cell. The cell is then rotated 60° to engage a breech lock mechanism for containment of the high pressure DT. The cell is slowly pressurized with DT at a rate calculated to equal the rate of permeation through the capsule wall. The maximum pressure is determined by the required solid DT layer thickness. Once that pressure is reached, the permeation cell is gradually cooled to condense the DT. Both the permeation cell and target base include cooling tubes. The DT not trapped inside the capsule can now be pumped out of the cell. The breech lock mechanism is then disengaged and the target is lowered back down inside the target cooling cryostat. After the gate valve is closed, the target cooling cryostat can be disconnected from the Fill Station and transported to the experiment chamber.

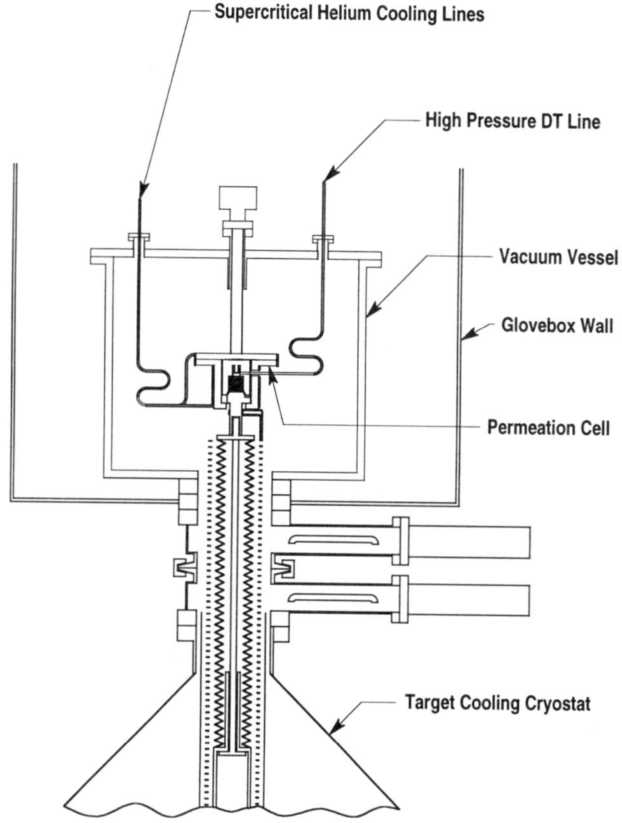

Figure 7. Target cooling cryostat extends target up into fill station for permeation filling.

Insertion Robot

The insertion robot is used to transfer targets from the target cooling cryostat to the center of the wire array, as shown in Fig. 2. Since the polymer capsule wall is not strong enough to withstand the room temperature pressure of the DT gas, the target must be kept at cryogenic temperatures until the shot. In order to maintain the cryogenic temperatures, the target must also be kept in a vacuum environment. Thus, any manipulation of the target must be done remotely.

After the target cooling cryostat is mated with the experiment chamber, the extendible target holder raises the target into the experiment chamber. The insertion robot then grasps the bottom of the target, lifts it into the experiment chamber, and places it on a long vertical end effector for placement within the wire array. The umbilical cooling lines spool out as the target is lifted, maintaining thermal contact between the target cooling cryostat and the target. The robot can also be used to position blast shields. Before the shot, the robot is removed from the experiment chamber to prevent neutron activation and damage from shrapnel.

Auxiliary Cooler

After the target is placed within the wire array, an auxiliary cooler extends from the top of the experiment chamberand makes thermal contact with the target. The arm which places the cooling tip in place is withdrawn to outside the experiment chamber. The auxiliary cooler allows both the bottom and top of the target to be actively cooled in order to create a uniform temperature environment required for beta layering the DT. In this baseline design, the cooling is provided by a supercritical helium loop similar to the one cooling the target base.

SUMMARY

The concepts described should allow the successful fielding of cryogenic targets on z-pinch machines, such as the existing z-machine or the proposed zx or x-1 facilities.

ACKNOWLEDGMENT

This is a report of work supported by General Atomics Internal Research and Development funds. The authors wish to acknowledge the many discussions with target designers at Sandia and Lawrence Livermore National Laboratories.

REFERENCES

1. G. Yonas, Fusion and the z-pinch, *Scientific American* **279**, 40 (1998).
2. C.R. Gibson, et al., Design of the fill/transfer station cryostat for the OMEGA cryogenic target system, *Advances in Cryogenic Engineering* **43**, 605 (1998).
3. R.E. Olson, et al., Indirect-drive ICF target concepts for the X-1 z-pinch facility, *Fusion Technology* **35**, 260 (1999).
4. J.K. Hoffer, et al., Forming a perfectly uniform shell of solid DT fusion fuel by the beta layering process, *Proc 14th International Conf. on Plasma Physics and Controlled Nuclear Fusion*, September (1992) p. 443.

A THREE-DIMENSIONAL NUMERICAL MODEL FOR HELIUM FLOW AND HEAT TRANSFER IN THE JOINT BETWEEN TWO CABLE-IN-CONDUIT CONDUCTORS

A. Martinez, J.L. Duchateau

Association Euratom-CEA
Département de Recherche sur la Fusion Contrôlée
CE Cadarache, F-13108 St. Paul Lez Durance, France

ABSTRACT

A Stainless Steel Full Size Joint Sample (SS-FSJS), representative of the European (EU) joint design proposal for both ITER Central Solenoid (CS) and Toroidal Field (TF) coils, has been tested in the SULTAN facility (CRPP, Switzerland). The design of this full size joint sample is based on the so-called twin box concept (separation of the cooling flows of the connected conductors). The purpose of the present work is to develop a three-dimensional simulation. This includes :

- the Joule effect associated with the resistance of the joint
- the heat exchange with the adjacent half joint through the copper sole
- the massflow repartition between the central region and the annular region as a function of the inlet conditions
- the transverse and longitudinal temperature gradient.

Using the experimental data of the thermohydraulic tests of the SS-FSJS, numerical results are obtained by this model in order to study the thermal response of the joint during the time-varying inlet helium temperature. This model will be used to prepare the ITER TF Model Coil experiments, which will take place in 2000. The exploration of the limits of this coil will also be carried out by propagating a helium heat slug in a similar way.

INTRODUCTION

During the TFMC (Toroidal Field Model Coil) quench experiments, a helium heat slug is injected through a pancake to heat the conductor. A part of the pipe supplying helium to

Advances in Cryogenic Engineering, Volume 45.
Edited by Shu *et al.*, Kluwer Academic / Plenum Publishers, 2000.

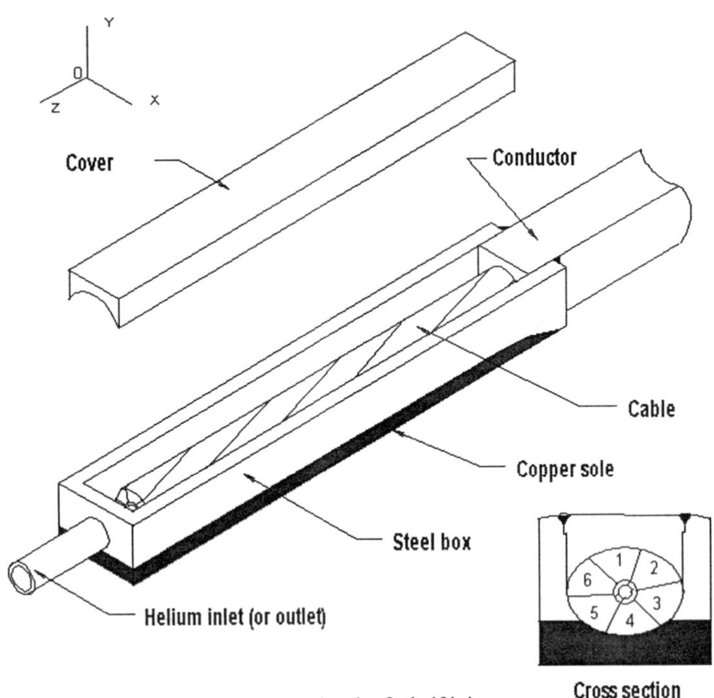

Figure 1. A sketch of a half joint

the TFMC is heated in order to reach, by forced circulation, the current sharing temperature on a point located on the conductor. During these experiments, the TFMC conductor is supplied with a current of 70 kA or 80 kA and the first double pancake is positioned in a maximum magnetic field of around 9 T. These pancakes, belonging to the same double pancake, are connected together by a joint, linking the two innermost turns of the pancakes. Although the joint is positioned in a magnetic field lower than the magnetic field around the conductor, it is not obvious that the joint will not quench before quenching the conductor. Actually, the high value of the Nb_3Sn uniaxial strain (-0.75% in the joint, -0.60% in the conductor) is a factor which could lead to a current sharing temperature in the joint lower than the current sharing temperature in the conductor. Moreover, the Joule losses dissipated in the joint (1.5 $n\Omega$) could also quench the joint before quenching the conductor. A joint is an important component that could restrict the performance of the superconducting coils. The advantage of three-dimensional numerical simulations is that they provide detailed information on the heat transfer taking place perpendicular to the flow direction combined with the axial fluid convection in the tube and in the cable bundle.

THE STAINLESS STEEL FULL SIZE JOINT SAMPLE

The Stainless Steel Full Size Joint Sample, as shown in Figure 1, was originally designed by CEA to be the EU Full Joint Sample for ITER[1]. In this respect, the joint design has followed the twin box concept (separation of the cooling flows of the connected conductors) proposed by the EU Home Team for the joints of the ITER Toroidal Field Model Coil. Each conductor end of a pancake is compacted inside a half-joint box machined in an explosively-bonded copper-steel plate. The two half-joint copper soles are soldered together with tin to ensure a good and reliable electrical connection. The cooling of a half-joint is in series with the associated conductor. The thick central tube (6 mm ID x 12 mm OD) is needed in order to ensure good compaction of the cable during the joint manufacturing.

EVALUATION OF THE CURRENT SHARING TEMPERATURE ALONG THE JOINT AND ALONG THE CONDUCTOR

In order to conclude whether the quench occurs in the joint or in the conductor, it is necessary to determine the current sharing temperature Tcs along the joint and along the conductor. The Summer's relation[2] with the parameters (B_{c20m}, T_{c0m}, C_0) deduced from experiment[3], is used to determine the current sharing temperature of the conductor as a function of the current, the field B and the Nb_3Sn uniaxial strain ε. An original calculation[4] has been performed in order to determine the field map along the joint and along the first few meters of the TFMC conductor (where the field is maximum). This variation is plotted in Figure 2, when the current in the TFMC is equal to 70 kA and when the current in the coil creating the background field is maximum. Also is plotted in Figure 2 the corresponding current sharing temperature along the joint and along the first meters of the TFMC conductor. As we can see in this figure, the difference between the minimum value of Tcs in the joint and the minimum value of Tcs in the conductor is only a few tenths of a Kelvin. So, an accurate knowledge of the thermal response of the joint during the time varying inlet helium temperature is very important for effective control of the location of the quench. As the minimum value of Tcs in the conductor (~7.4 K) is lower than the minimum value of Tcs in the joint (7.7 K), a possible method, in order to quench the conductor before quenching the joint, is to supply helium at 7.5 K to one half-joint and to absorb the Joule heat dissipated in the joint by supplying the second half-joint with helium at a temperature below 7.5 K.

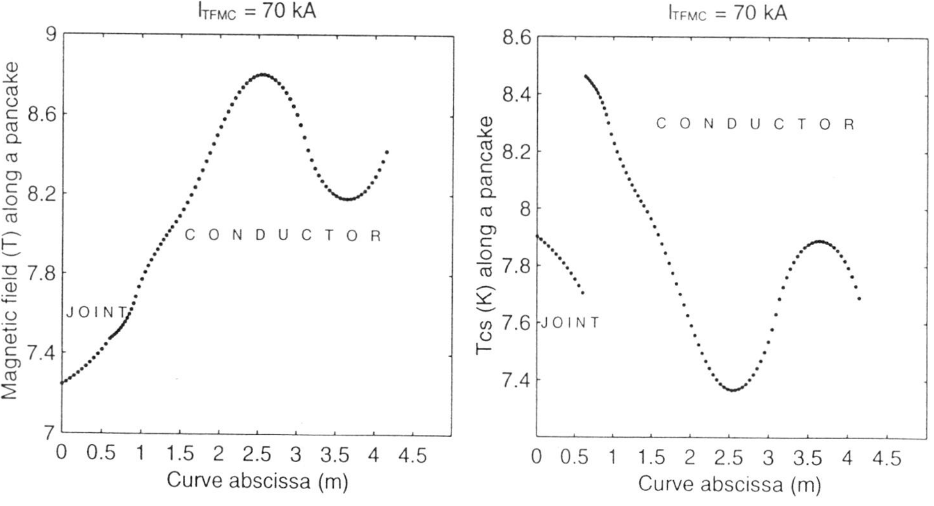

Figure 2. The magnetic field and the current sharing temperature alond the first meters of the pancake

THERMAL-CONVECTION ANALYSIS INSIDE A HALF JOINT BOX

The thermal process in the half joint, cooled with two parallel flows of supercritical helium, has been simulated in three-dimensions. In this code, the Joule effect associated with the resistance of the joint, the heat exchange with the adjacent half-joint through the copper sole, and the massflow repartition between the central tube and the cable bundle section as a function of the inlet conditions are taken into account. This investigation is restricted to 1-D flow and 3-D heat transfer.

The conservation equations in the cable bundle

The strands and the helium in the cable bundle are in thermal equilibrium and the entire cable bundle domain can be treated as a single region governed by the helium velocity w, a mean conductivity λ_{cab} and a mean specific heat $(\rho C_p)_{cab}$ defined respectively by :

$$\overline{\lambda}_{cab} = \lambda_{he}\alpha + \lambda_{str}(1-\alpha) \qquad \left(\rho C_p\right)_{cab} = \alpha\left(\rho C_p\right)_{he} + (1-\alpha)\left(\rho C_p\right)_{str} \tag{1}$$

In the above equations, α, λ_{he}, λ_{str}, $(\rho C_p)_{he}$ and $(\rho C_p)_{str}$ stand for the void fraction, the helium conductivity, the strand conductivity, the helium specific heat capacity and the specific heat capacity of strands respectively. Due to the small hydraulic diameter d_h of the the cable bundle (typically $d_h = 3.410^{-4}$m), it is assumed that the friction force in the momentum conservation equation is balanced by the pressure gradient force[5]. Neglecting helium inertia and assuming uniaxial flow in the z-direction, we obtain the following system of nonlinear, partial differential equations describing details of the flow and heat transfer in the cable bundle. The mass and momentum conservation equations are respectively (where ρ is the helium density and p the helium pressure) :

$$\frac{\partial \rho}{\partial t} + \frac{\partial(\rho w)}{\partial z} = 0 \qquad \text{and} \qquad \frac{\partial p}{\partial z} = -f_l \frac{\rho w|w|}{2d_h} \tag{2}$$

The friction factor f_l is given according to the correlation[6]:

$$f_l = \frac{1}{\alpha^{0.72}}\left[\frac{19.5}{Re_l^{0.7953}} + 0.0231\right] \quad \text{with } Re_l = \frac{\rho w d_h}{\mu} \tag{3}$$

Re_l is the Reynolds number of the flow in the cable bundle and μ is the dynamic viscosity. The energy conservation equation (which includes heat transfer by convection and diffusion, and source terms) is :

$$\frac{\partial}{\partial t}\left[\left(\rho C_p\right)_{cab} T\right] + \alpha\frac{\partial}{\partial z}\left[w\left(\rho C_p\right)_{he} T\right] = \nabla\left[\lambda_{cab}\nabla T\right] + T\beta\frac{dp}{dt} + Q_j \tag{4}$$

In the above equations, ρ, β and Q_j stand for the helium density, the helium thermal expansion coefficient and the Joule heating rate, respectively. The helium state equation $p(\rho,T)$ permits the number of conservation equations to match the number of unknowns : the pressure p, the temperature T and the axial helium velocity w.

The energy conservation equation in the different solids of the half joint

The governing energy equation in the half joint is written in the steel box, in the copper sole and in the steel tube as :

$$\int_V \frac{\partial(\rho C_p.T)}{\partial t} dV = -\int_S \vec{\varphi}.\vec{n}.ds \qquad (5)$$

where S and V are the surface and the volume of the elements, respectively, T is the local temperature, $\vec{n}$ is the normal to the surface and $\vec{\varphi}$ is the local heat flux density ($\varphi = -\lambda(T).\nabla(T)$). The material properties, the thermal conductivity λ, the density ρ and the specific heat C_P depend on temperature.

The conservation equations in the helium in the central tube

The helium flow in the tube can be described by the Navier-Stokes equation and the continuity equation. Because the helium properties depend on temperature (and pressure) these equations are coupled to the energy conservation equation. In Cartesian coordinates these equations read :

- The mass conservation equation, the Navier-Stokes equation and the helium equation-of-state are respectively :

$$\frac{\partial \rho_2}{\partial t} + \frac{\partial \rho_2 w_2}{\partial z} = 0 \qquad \frac{\partial \rho_2 w_2}{\partial t} + \frac{\partial \rho_2 w_2^2}{\partial z} + \frac{\partial p_2}{\partial z} = -f_2 \frac{\rho_2 w_2 |w_2|}{2 d_{h2}} \qquad p_2(\rho_2, T_2) \qquad (6)$$

The friction factor f_2 is given by the usual correlation for a turbulent, fully developed flow in a round smooth tube (of hydraulic diameter d_{h2}) where Re_2 is the Reynolds number:

$$f_2 = 0.046\, Re_2^{-0.2} \quad \text{with } Re_2 = \frac{\rho_2 w_2 d_{h2}}{\mu_2} \qquad (7)$$

- Energy conservation equation

$$\frac{\partial}{\partial t}\left[(\rho C_p)_{he2} T_2\right] + \frac{\partial}{\partial z}\left[w_2 (\rho C_p)_{he2} T_2\right] = \nabla\left[\lambda_{he2}(1+Nu)\nabla T_2\right] + T_2 \beta_2 \frac{dp_2}{dt} \qquad (8)$$

The presence of forced convection augments the heat transfer rate compared to that obtained by conduction alone. The heat transfer augmentation factor Nu, defined by the ratio of the heat transfer by convection to the heat transfer by conduction, is the Nusselt number as given by the standard Dittus-Bölter correlation. In addition to these equations, boundary conditions at the inlet, the oulet and the walls must be included. At the helium/solid interface, temperature and heat flux continuity must be imposed. It is assumed that the steel box walls are thermally insulated. The heat exchange with the adjacent half-joint from the copper sole has been calculated from the overall heat transfer coefficient H determined with a relation deduced from experiments[7]. This overall heat transfer coefficient is obtained as the inverse of the resultant of three thermal resistances : the convective resistance on the cable bundle side of the two half joints (labeled 1 and 2), depending on the Reynolds number Re, the Prandtl number Pr and the conductivity λ_{he} of the helium flow, and the conductive resistance of the copper sole of the joint, depending on the thickness e and the conductivity λ_{cop} of the copper sole. Then we have, where C is constant depending on the geometry of the joint:

$$\frac{1}{H} = \frac{e}{\lambda_{cop}} + C\left[\frac{1}{(\lambda \operatorname{Re} \operatorname{Pr})_1} + \frac{1}{(\lambda \operatorname{Re} \operatorname{Pr})_2}\right] \tag{9}$$

NUMERICAL METHOD

Due to symmetry about the vertical Y-axis, only one half of a half-joint needs to be modeled. To solve this set of equations in Cartesian coordinates, a fully implicit finite difference scheme is used. The advantage of this method is that they are unconditionally stable and hence impose no restrictions on the maximum allowable time step. This method, however, requires the solution of a set of nonlinear algebraic equations, at each time step, by using a matrix inversion algorithm. The governing equations are discretized based on the control volume formulation described by Patankar[8]. The power law scheme was used in order to evaluate the heat flux across the boundaries of each control volume. The calculations are carried out on an equidistant, rectilinear and rectangular grid. Moreover, a staggered grid was used. This means that the scalar quantities such as temperature and pressure are located in the centre of a computational cell, while the velocity vector is located at the boundary of a cell. The use of such a grid prevents unphysical oscillations of the computed pressure as is possible on a regular grid. In typical calculations, in the geometry of the parallelepiped, a 23 x 35 x 9 (x,y and z) grid was used with the grid spacings Δx = 0.15mm, Δy = 0.15mm and Δz = 5.6 cm. The time derivatives were discretized in accordance with the fully implicit scheme.

PROBLEM DESCRIPTION

Numerical solutions were developed for the transient problem in which (at $t > 0$) a change in temperature is suddenly imposed in the inlet plane of both half-joints or in one of the two half-joints. It is obvious that the helium in the central tube and the helium in the cable bundle represent two distinct hydraulic parallel flow paths. Subjected to a common pressure gradient, the total mass flow rate is divided into two flow paths according to some ratio of the local friction factors. So, the flow velocity in the central tube is several times (~20) greater than that in the cable bundle. Moreover, the convective heat transfer produced by the turbulent flow in the central tube has a negligible impact on the overall transverse heat transfer between the helium in the cable bundle and the helium in the central tube. In fact, the large thickness of the central tube (3 mm) decreases drastically this transverse heat transfer between these two circuits. Therefore, the Joule heat load remains homogeneously distributed within the cable bundle, and the large enthalpy rate supplied by the helium flow in the central tube is used only to minimize the helium temperature supplying the conductor, at the joint outlet. Several calculations were performed in order to determine the maximum cable bundle oulet temperature. Prior to the start of the time varying inlet temperature, the joint is in a thermal steady-state as a result of the Joule heating process. For all cases an initial temperature of 7 K, a pressure of 3.5 bar and an operating current of 70 kA were assumed. In table 1, we have summarized all the input parameters which

Table 1. Half joint input data for the analyses presented

Width x Height x Length	0.069m x 0.051m x 0.45 m
Cable bundle cross section	~1000 mm^2
Copper sole cross section	9mm x 0.064mm
i. o. central tube diameter	6mm x 12 mm

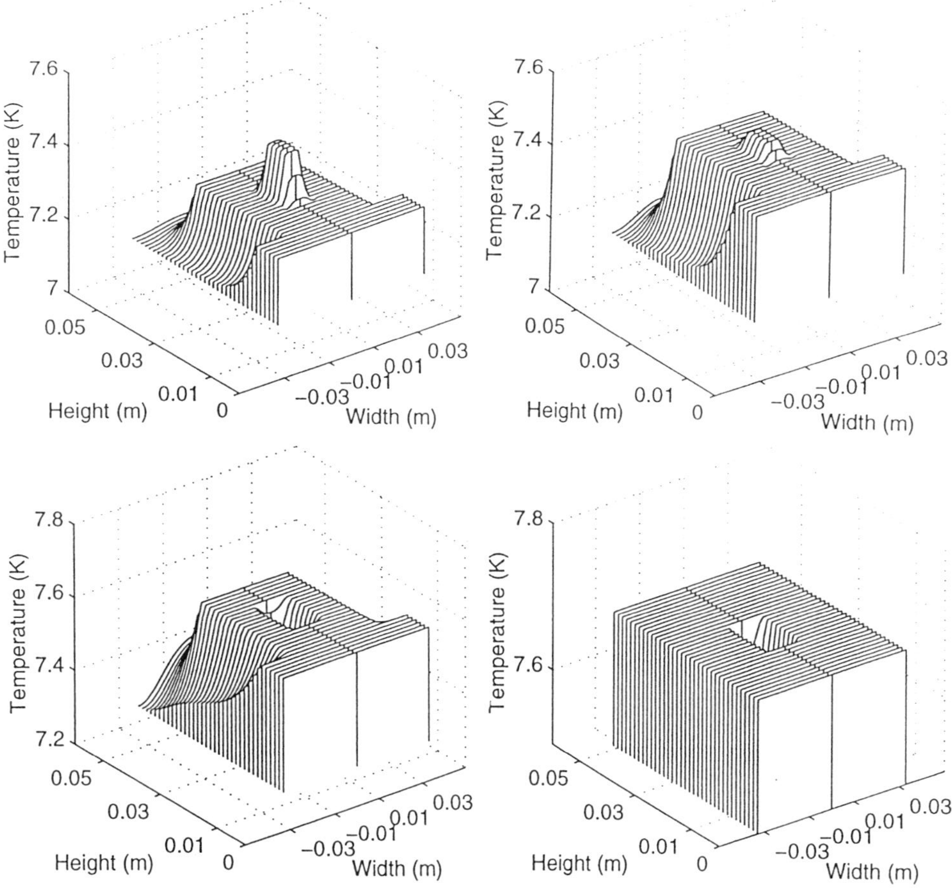

Figure 3. Temperature profile in a cross section of the half joint (xy plane), for different times.

characterize the physical problem.

In these simulations the circular steel tube has been replaced by a square tube with approximately the same cross section. For a total mass flow rate of 2 g/s, an inlet temperature equal to 7.5 K is suddenly imposed in the inlet plane of both half-joints (no heat transfer between the two half-joints). In order to minimize the delay time necessary to quench the conductor, the value of the inlet temperature is greater than the minimum value of the current sharing temperature in the conductor (7.36 K). The results of these calculations are reported in Figure 3. This figure shows the temperature profile in a cross section (xy plane) of the half joint at z = 0.17 m, for different times (1s, 2s, 7s and ∞). It is clear from these results that the large transverse temperature gradient in the cross section of the central tube has a very strong effect on the maximum cable bundle oulet temperature. In this case we can find that the maximum cable bundle oulet temperature (8.02 K) is greater than the minimum value of the current sharing temperature in the cable bundle of the half-joint (~7.7 K). The first way to reduce the maximum cable bundle oulet temperature in the half-joint is to supply helium at 7.5 K to one of the half-joints and to absorb the Joule heat dissipated in the joint by supplying the adjacent half-joint with helium at a temperature lower than 7.5 K. Figure 4 shows the steady state temperature profile in the oulet cross section of the half joint (xy plane) for two different coolings of the

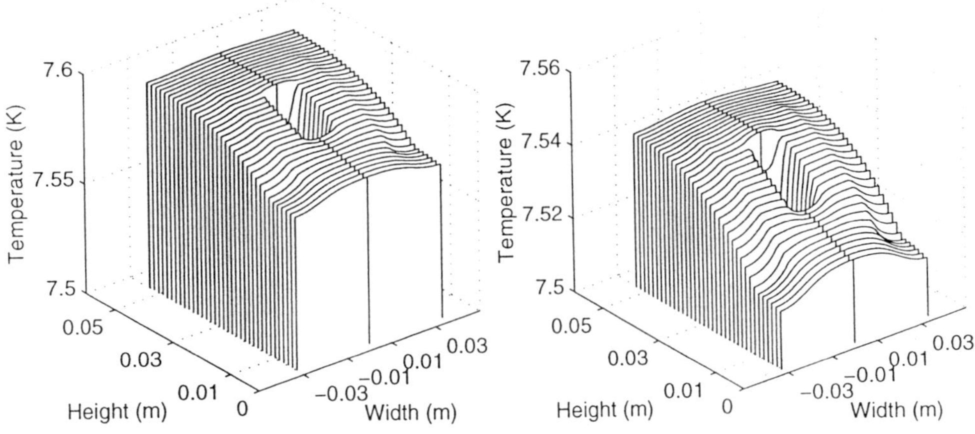

Figure 4. Steady state temperature profile in the oulet cross section of the hafl joint (xy plane)

adjacent half-joint. The first case is when the mean helium temperature in the adjacent half-joint is 7 K and a mass flow rate equal to 2 g/s (H = 200 W/mK), and the second case is when the mean helium temperature in the adjacent half-joint is 7.2 K and the mass flow rate equals 5 g/s (H = 400W/mK). In the two cases, the maximum cable bundle oulet temperature in the half-joint is smaller than the minimum current sharing temperature Tcs of the joint (7.7 K).

CONCLUSIONS

The thermal process in the half-joint, cooled with two parallel flows, has been simulated with a three-dimensional model. Considering the TFMC quench experiments, we have proposed a method, in order to quench the conductor before quenching the joint. This method consists in supplying one of the half-joints with helium at 7.5 K and in absorbing a part of the Joule power dissipated in the joint by supplying the adjacent half joint with helium at a temperature below 7.5 K.

REFERENCES

1. P. Libeyre et. al., "Development of joints in Europe for the ITER TFMC", Proc. of MT15, Beijing, (1997).
2. Summers et. al., "A Model for Prediction of Nb_3Sn Critical Current as a Function of Field, Temperature, Strain and Radiation damage", IEEE trans. on Mag. (1991), p.2041.
3. A. Martinez et. al., " Field and Temperature Dependencies of Critical Current Industrial Nb_3Sn Strands", Cryogenics, (1997), vol.37, p. 865.
4. European Home Team, "Minutes of the 6th ITER TF Model Coil Test Group Meeting", (1998).
5. A. Shajii, " Theory and Modeling of Quench in Cable in Conduit Superconducting Magnets", MIT PhD Thesis, PFC/RR-94-5, (1994).
6. S. Nicollet et. al., 'Hydraulics of the ITER Toroidal Field Model Coil Cable in Conduit Conductor", Proc. of the 20th SOFT, Marseille, (1998).
7. European Home Team, "Minutes of the 8th ITER TF Model Coil Test Group Meeting", (1999).
8. S.V. Patankar, "Numerical Heat Transfer and Fluid Flow", Hemisphere Publishing Corporation, Washington, (1980).

AN EXPERIMENTAL STUDY OF THE NON-LINEAR BEHAVIOR OF TEMPERATURE OSCILLATIONS IN HE II NOISY FILM BOILING

Jing WANG and Peng ZHANG

Institute of Thermophysics Engineering
Shanghai Jiao Tong University
200030, Shanghai, China

ABSTRACT

Analysis of the behavior of temperature and pressure fluctuation during noisy film boiling in He II is very important to study the heat transfer instability of cryogenic cooling systems.

According to the principles of deterministic chaotic dynamics, results from experimental data , such as the time series, power spectrum and phase-space trajectories can reveal the dynamic characteristics of the system in time domain and frequency domain. In particular, the Lyapunov Exponents provide a convenient parameter for mapping overall dynamic trends of cryogenic cooling system. In this paper, nonlinear behavior of temperature and pressure fluctuations during noisy film boiling in He II has been studied using the method of chaos time series analysis (CTSA) .

INTRODUCTION

Investigation of the dynamic characteristics of helium liquid-vapor two-phase boiling system is a very important research for safe and economical operation of space cryogenic applications, for example, IRTS (Infrared Telescope in Space), temperature must be cooled down to around 2K or even lower to provide it the maximum sensitivity and the capability of fulfiling cosmic task in far infrared sky. This kind of observition plan requires long life time of the whole system and high reliability. In several decades the dynamic characteristics of cryogenic boiling system and its heat transfer have been studied extensively, which are close to many important subjects such as propagation of the fluctuations of flow and heat transfer, unsteady of heat exchanger, on-line diagnose for cooling system and its save processes. In recent years, a lot of researchers focused their attention on pressure and temperature wave propagation in boiling or cooling system for that dynamic characteristics of the system are controlled by wave propagation in it.

The wave theory is a very useful method for studying unsteady boiling systems. Based on the Wallis wave theory[1], the primary research focus has been on nonlinear behavior of the bubble motion including, in particular, the existence of chaotic oscillations of bubble volume and shape. Feng and Leal[2] have shown when volume oscillations are forced by oscillations of the external pressure, the volume mode may lose stability for sufficiently large amplitudes of oscillation, and this instability may lead to chaotic oscillations of both the volume and the shape modes.

In fact, a boiling system is one of many nonlinear spatio-temporal systems where even very small system exhibits very complex behavior. Masahiro Shoji[3] suggested that the nonlinear behavior is one of the reasons that mechanistic predictive capabilities for the boiling process have remained elusive.

This paper, basing mainly on above recent research results and the chaos analysis method[4] used dynamic experimental data of noisy film boiling in a open He II bath has characterized the nonlinear behavior. Make use of probability density, FFT power spectrum, maximum entropy spectrum[5], strange attractor and Lyapunov exponents the nonlinear behavior of noisy film boiling in open He II are described more profoundly.

BOILING CHAOS AND CHAOS TIME SERIES ANALYSIS METHOD

Principle of nonlinear bubble dynamics of gas bubble

Consider a bubble of radius a in the weak viscous effects liquid. The surface tension is Γ, the density of the liquid is ρ, and the characteristic scales for length is $l_c = a$, for time is $t_c = (a^3\rho/\Gamma)^{1/2}$, for pressure is $p_c = \Gamma / a$. If the parameters $S = t_c / (a^3 / \nu)$ is small, the viscosity of the liquid is said to be weak. In general, when the air bubbles in water with radius is between 0.01-1.00 cm, their S = 0.015-0.00015, so we can assume S 《1. With the assumption that the deformation of the bubble from its equilibrium spherical shape is small, allows us to approximate the motion as potential flow with a thin boundary layer near the surface of the bubble. Thus the Laplace equation governing the liquid motion of non-dimensionalized velocity potential Φ for the liquid in spherical coordinates (r, n, θ) is:

$$\nabla^2\Phi = 0 \qquad \text{for } r\rangle 1 + f(\theta,t) \tag{1}$$

The kinematic boundary condition on the bubble surface is:

$$\frac{\partial F}{\partial t} + \nabla\Phi \cdot \nabla F = 0 \tag{2}$$

and the normal stress condition with the pressure correction and viscous stress term included is

$$-p_v + (\tilde{p} - p_0 + \frac{\partial\Phi}{\partial t} + \frac{1}{2}\nabla\Phi \cdot \nabla\Phi) + 2S\frac{\partial^2\Phi}{\partial n^{'2}} = A(\theta,t) + \nabla \cdot n \tag{3}$$

Where P_0 denotes the non-dimensionalized pressure at infinity; $\tilde{P}$ is the pressure inside the bubble; P_v is the pressure correction; $A(\theta,t)$ represents an imposed time-spatially dependent pressure at the bubble surface. In order to facilitate the problem there is:

$$A(\theta,t) = A_0 \cos 2\Omega t + A_n \cos\Omega t P_n(\cos\theta) \tag{4}$$

Where $P_n(\cos\theta)$ is the Legendre polynomial of $\cos\theta$. And Ω is the characteristic frequency. Introduce an additional scaling of Φ, t, $\tilde{t} = \Omega t, \tilde{\Phi} = \Phi/\Omega$, to obtain following Laplace equation:

$$\nabla^2 \tilde{\Phi} = 0 \tag{5}$$

Using an asymptotic approximation of the solution for a small parameter ε, study the bubble response to either an isotropic or an anisotropic pressure oscillation, with the boundary conditions at r=1 and the normal stress conditions, the general solution of the governing equation at $0(\varepsilon)$ is:

$$\begin{aligned} f_1 &= \frac{1}{2}\sum_{k=0}^{\infty} a_k \exp\left(\frac{i\omega_k \tilde{t}}{\omega_n}\right) P_k(\eta) + c.c., \\ \Phi_1 &= -\frac{i}{2}\sum_{k=0}^{\infty} \frac{a_k \omega_k}{(k+1)\omega_n r^{k+1}} \exp\left(\frac{i\omega_k \tilde{t}}{\omega_n}\right) P_k(\eta) + c.c. \end{aligned} \tag{6}$$

Then, using the method of superposition at $0(\varepsilon^2)$ can be obtained amplitude functions $\alpha_{1,0}$ and $\alpha_{1,n}$:

$$\frac{d\alpha_{1,0}}{d\tau} = -2\tilde{S}\alpha_{1,0} + (\beta_0 - 2\sigma_n) i\alpha_{1,0} + i\frac{1}{4\omega_n}\tilde{A}_0 + iH_5\alpha_{1,0}^2 \tag{7}$$

$$\frac{d\alpha_{1,n}}{d\tau} = -N\tilde{S}\alpha_{1,n} - i\sigma_n\alpha_{1,n} + i\frac{n+1}{2\omega_n}\tilde{A}_n + iH_6\alpha_{1,0}\alpha^*_{1,n} \tag{8}$$

Where N=(n+2)(2n+1), $H_5 = \dfrac{(4n-1)\omega_n}{16(n+1)(2n+1)}, H_6 = \dfrac{(4n-1)\omega_n}{4}$

Consequently, for fixed n, the dynamical system depends on four parameters, i.e. the dimensionless magnitude of viscosity $\tilde{S}$; the external frequency detuning σ_n; the internal frequency detuning β_0; the magnitude of the periodic forcing Δ_0, Δ_n.

According to analysis of Feng and Leal, the parameters σ_n, β_0 and Δ_0, Δ_n can be expressed as the function of S^2, A_0, A_n, Ω and ω_n. This means that a spherical gas bubble in a inviscid fluid will undergo periodic oscillation for any initial perturbation of the bubble radius.

Based on the analysis and simulation of above amplitude equations starting from the fixed points the bifurcation sets and bifurcation diagrams had been obtained by Miles[6]. His numerical results have shown that any periodic pressure perturbation would have led to chaotic oscillation of an ideal spherical gas bubble. Moreover, the occurrence of chaos is dependent on the forcing frequency, i.e. the external frequency detuning σ_n; the internal frequency detuning β_0. This important conclusion can be verified by the bifurcation sets in (σ, Δ_n)- space for fixed $\beta = 2$. These bifurcation sets illustrate the sequences of transition between periodic, quasi-periodic and chaotic bubble oscillations, and it can be used to guide the experiments to explain the mechanism of occurrence of chaos in cyogenic two-phase boiling system.

Chaos time series analysis (CTSA) method

According to the nonlinear theory Takens[7] developed the reconstruction parameters method by reconstructing an attractor and its characteristic numbers. The general approach to dimension and entropy calculation in nonlinear time series analysis starts from a measured experimental time series of n points $\{x_1, x_2, \cdots, x_n\}$. Then from this time series the attractor is composed in the reconstructed state space using the delay time coordinates, so the method is called the method of delays.

Based on the above researches a new method for nonlinear characteristics of pressure and temperature fluctuations in two-phase flow which is called the Chaos Time Series Analysis (CTSA) has been proposed by author[8]. The CTSA method is composed of two parts: Part 1 is a data acquisition system; Part 2 is a data treatment system which is mainly made of computer software package. The CTSA method is used as following steps: (a) Acquisition of the time series in two-phase flow system. In this step, it is important to determine the number of points, delay time, maximum scaling distance. (b) Calculation of Lyapunov exponents. According to the deterministic chaos theory the Lyapunov exponents are the average exponential rates of divergence or convergence of nearby orbits in phase space. And it is defined as following:

$$\lambda(\mu) = \lim_{k \to \infty} \frac{1}{k} \sum_{i=0}^{i=k} \log\left|f'(\mu, x_i)\right| \tag{9}$$

Clearly, Lyapunov exponents mean the space probability characteristics of the strange attractors. For the finite number of data, determine Lyapunov exponents from time series can be calculated by Equ.(10):

$$\lambda = \frac{1}{t_m - t_0} \sum_{k=1}^{m} \log \frac{L'(t_k)}{L'(t_{k-1})} \tag{10}$$

(c) Choice a phase space and construct stranger attractor from experimental data.

EXPERIMENTAL INVESTIGATION OF TEMPERATURE OSCILLATIONS DURING NOISY FILM BOILING OF LIQUID He II

In the present study, the temperature and pressure oscillations are measured simultaneously to investigate the nonlinear behavior of the noisy film boiling. In the saturated boiling case, pressure oscillations are measured with the pressure sensors, temperature oscillations are measured with the superconductor temperature sensors. The experimental set up and the analysis of pressure oscillations can be seen the Ph.D. thesis of Zhang[9] in detail. It is found that temperature oscillation is caused by the passage of thermal boundary and oscillation of vapor bubble on the planar heater in noisy film boiling case. As the bath temperature is far below the λ-temperature, the state of liquid He II near heater becomes nearly saturated for the case of not so large immersion depth. The saturated boiling state may be divided into the initial precursory boiling and film boiling according to both the magnitude of heating time t and the heat flux q.

Basically, there are two kinds of film boiling modes in He II: noisy film boiling and silent film boiling[10]. The former is characterized by very loud acoustic noise, and at the same time, the pressure oscillation induced by the contact of the liquid He II with the planar heater surface is very remarkable comparing to that of silent film boiling.

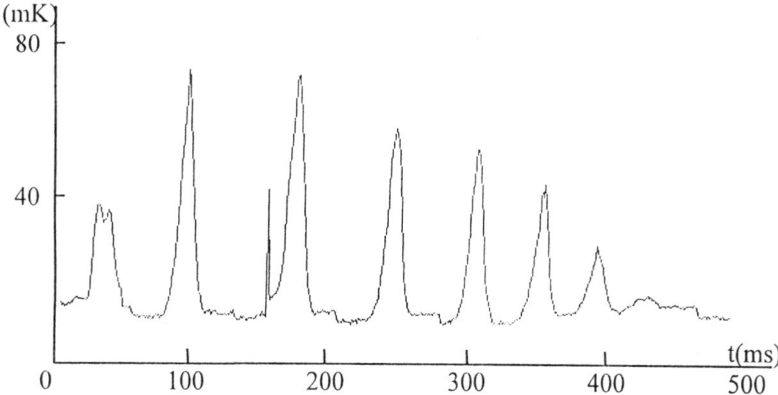

Figure 1. Temperature oscillations of a vapor bubble during slight noisy film boiling (T_b=1.8 K, q=30 W/cm^2, t_h=20ms).

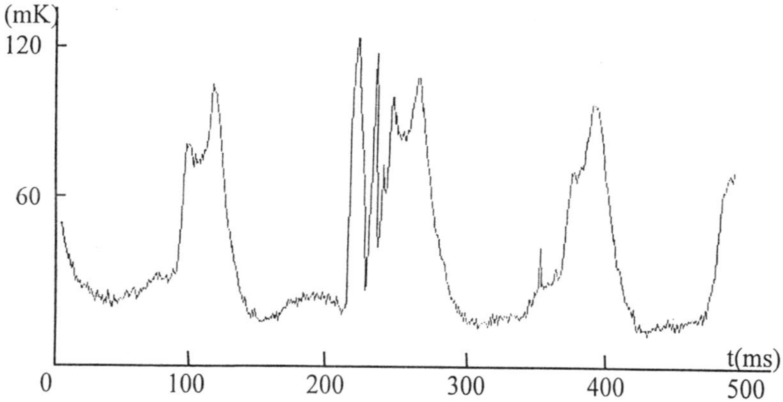

Figure 2. Temperature oscillations of a vapor bubble during violent noisy film boiling (T_b=1.6 K, q=30 W/cm^2, t_h=10ms).

During experiments, when the heating time t_h is long enough (t_h>5ms) for the heat flux of 5W/cm^2 or heat flux q is quite large (q> 20W/cm^2) for the heating time of 1 ms, noisy boiling happens with slight noise at the moment of onset of heating. Thus, the temperature rise during noisy film boiling can be detected by the superconductor temperature sensor (Fig. 1). Fig. 1 shows that the amplitude of temperature oscillations is about 0-80 mK. When the heating time is further prolonged, e.g. 1 s for the heat flux of 10W/cm^2, violent noisy film boiling appears, a big vapor bubble is found to grow and erupt repeatedly on the heater accompanied by large acoustic noise. It can be seen from Fig. 2 that the output of the superconductor temperature sensor is rather large. Fig. 2 shows that the amplitude of temperature oscillations is about 0-120 mK.

RESULTS BY CTSA METHOD AND DISCUSSIO

With above experimental output the chaos time series analysis have been done by our WLChaos program. In Fig. 3-4 (a) the phase space of phase attractor consists of value of the time series (x-coordinate) and value of the autocorrelation function of same time series (y-coordinate). The Lyapunov exponents are calculated from experimental data by WLChaos program.

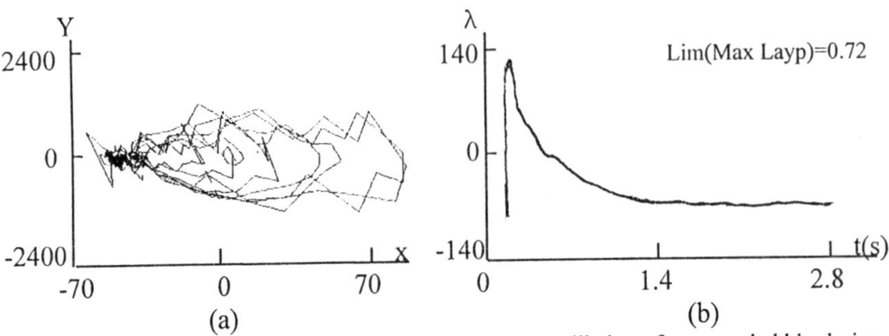

Figure 3. Chaos Time Series Analysis of Temperature oscillation of a vapor bubble during violent noisy film boiling (T_b=1.8 K, q=30 W/cm^2, t_h=20ms). (a) phase attractor (b) Lyapunov exponents

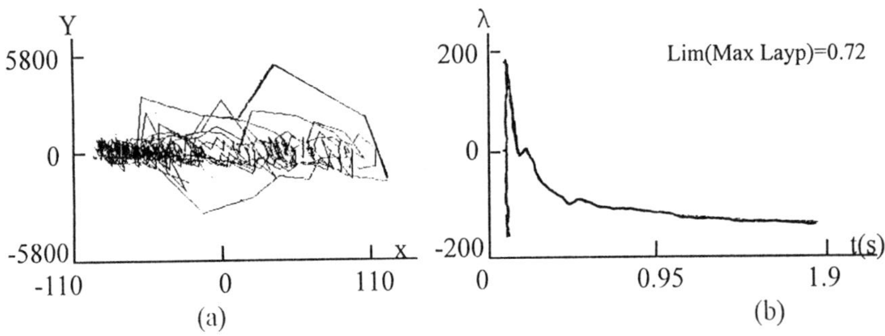

Figure 4. Chaos Time Series Analysis of Temperature oscillation of a vapor bubble during violent noisy film boiling (T_b=1.6 K, q=30 W/cm^2, t_h=10ms). (a) phase attractor (b) Lyapunov exponents

About temperature oscillations and Fig. 3-4 there are following discussion. First, experimental observations have shown that noise is generated by direct contact of He II and high temperature heater during noisy film boiling. In the experiments visualization pictures show that a density fluctuation happens in the bulk of He II during noisy film boiling, which is not found during silent film boiling. A series of pictures show that vapor bubble change its shape greatly in noisy film boiling, and the noisy film boiling is rarely affected by heater orientation, which indicates that the buoyant force has almost no influence on bubble in He II. What is the mechanism of the noise in He II during noisy film boiling? There would be several mechanisms which influence site interaction and nucleation. Secondly, it is clear that Fig.3 and Fig.4 mean during noisy film boiling the temperature fluctuations occur as a result of numerical results by Miles[6], i.e. any pressure perturbation would have led to chaotic oscillation of a gas or vapor bubble. Consequently, the occurrence of temperature oscillations is dependent on the frequency of vapor bubbles which grow and erupt repeatedly on the heater. So, as same as pressure oscillation data temperature oscillation data can be used to characterize dynamic behavior of cryogenic system. According to chaos time series analysis method Fig. 3-4 show that He II noisy film boiling came into unsteady nonperiodic fluctuations.

CONCLUSIONS

According to the results of above experiments and the CTSA method it has been shown that based on the deterministic chaos theory if a dynamic system have the positive Lyapunov exponents, it must be a chaos system[11]. So as shown in Fig.3 (b) and Fig.4 (b), the Lyapunov exponents is positive. Observably, these quantities might characterize that when boiling regime from silent film boiling to noisy film boiling the pressure oscillation and the temperature oscillation occure at same time which led boiling system to chaos. What is more, this conclusion has demonstrated the conclusion of above discussions by bubble dynamics.

On the other hand, the tools or techniques for the nonlinear analysis from the short-time-range data becomes very important, but the techniques have not yet validated for cryogenic noisy boiling time series. So development of experimental methods to measure local and spatial instantaneous characteristics must continue and the nonlinear behaviors such as of bubble generation, growth and of liqued rich layer(micro and macro layers) in cryogenic boiling system must be investigated more profoundly.

ACKNOWLEDGMENT

The authors would like to gratefully acknowledge the finance support from National Natural Science Foundation of China (No. 59676041) and Meng Minwei Foundation of Shanghai Jiao Tong University.

REFERENCES

1 G. B. Wallis, One - Dimensional Two-Phase Flow, McGraw - Hill, Inc. (1969).
2. Z. C. Feng and L. G. Leal, Bifurcation and chaos in shape and volume oscillations of a periodically driven bubble with two-to-one internal resonance, J. Fluid Mech., vol.266, pp.209-242, (1994).
3. Masahiro Shoji, Boiling Chaos and Modeling, Heat Transfer 1998, proc. The 11th International Heat Transfer Conference, vol.1, pp3-22, (1998).
4. A. Wolf, J. B. Swift et al, Determining Lyapunov Exponent from a Time Series, Physica D, vol. 16, pp. 285-317(1989).
5. Jing Wang, The Maximum Entropy Spectral Analysis on Void Fraction Fluctuation of Helium Liquid - Vapor Two-Phase Flow, Cryogenics, vol.34, pp. 349-352, (1994).
6. Miles, J., Resonant motion of a spherical pendulum, Physica 11 D; 309-323 (1984).
7. Takens, F., Detecting Strange Attractors in Turbulence, Lecture Notes in Mathematics. D.A.Rand and L.S.Young eds, Springier, Berlin, pp.366, (1981)
8. Jing Wang, The Exploration for the Dynamic Characteristics of Liquid-Vapor(Gas) Two-Phase Flow and Its Investigation Methods--Fluctuation, Nonlinear And Chaos Behaviour, Postdoctoral Works Report (in Chinese), HUST Wuhan (1995).
9. Peng Zhang, Study of Physical Mechanism of Film Boiling in He II , Ph.D Thesis , Shanghai Jiao Tong University , Shanghai (1999).
10. S.W.Van Sciver, Helium Cryogenics, Plenum Press, New York, (1986).
11. U. Parlitz, et al, Bifurcation structure of bubble oscillators, J. Acoust. Soc.Am. (88):1061-1077(1990)

R. Scurlock receives the 1999 Samuel C. Collins Award from K. Timmerhaus (University of Colorado).

TRANSIENT HEAT TRANSFER OF SUPERCRITICAL HELIUM IN HIGH CENTRIFUGAL ACCELERATION FIELD

M. Furuse,[1] O. Tsukamoto,[1] T. Takao,[2]
S. Fuchino,[3] I. Ishii,[3] M. Okano,[3] and N. Tamada[3]

[1]Yokohama National University
Yokohama, Kanagawa, 240-8501, Japan
[2]Sophia University
Chiyoda, Tokyo, 102-8554, Japan
[3]Electrotechnical Laboratory
Tsukuba, Ibaraki, 305-8568, Japan

ABSTRACT

Transient heat transfer of supercritical helium in centrifugal acceleration field has been measured for studying the stability characteristics of superconducting rotor windings of a superconducting generator. A temperature resistive platinum-0.5% cobalt (Pt-Co) thin wire, having small heat mass, was used as a temperature sensor and heater to measure the heat transfer characteristics. It was observed that the steady state heat transfer of supercritical helium in 3600G (G; gravity acceleration) was established in 200μsec. In the paper, data of transient heat transfer of supercritical helium in centrifugal acceleration field are shown and their specific characteristics are studied comparing with those of liquid helium.

INTRODUCTION

Premature quenches of superconducting rotor windings of superconducting generators are mainly caused by abrupt conductor motions. Duration times of such disturbances are very short, several 100μsec. The helium coolant for the rotor windings is in the supercritical state and subject to high centrifugal acceleration. Therefore, for stability analysis of the rotor windings, transient heat transfer data of supercritical helium in high centrifugal field are necessary.

Previously, we measured transient heat transfer characteristics of helium in centrifugal acceleration field up to 1300G where the pressure of the helium was up to 1.5×10^5Pa and the helium stayed in the liquid state. We found some specific characteristics of the transient heat transfer in the centrifugal field as follows,[1]

1. Take-off time, which is duration of quasi-nucleate boiling, was not dependent on the value of centrifugal acceleration field in the range of 1G to 1300G and relation between the take-off time τ and the heat flux of step heating q_0 was given by $q_0=2.6\times10^4\tau^{-0.25}$.
2. Steady state film boiling was established in the high centrifugal field much faster than in the case of 1G.
3. In the steady state film boiling region, the Pt-Co wire detected cyclic temperature fluctuations due to cyclic departure of bubbles from the wire surface. Average cycle time is only weakly dependent on the heat flux but proportional to the centrifugal acceleration (proportional to $a^{-1/3}$; a is the centrifugal acceleration).

In the previous experiment, due to the limitation in the rotating speed of the rotating cryostat, the centrifugal acceleration was limited up to 1300G. We conducted the heat transfer experiment by increasing the rotating speed of the rotating cryostat and obtained the transient heat transfer data of the cryogenic helium in the centrifugal acceleration up to 3600G where the pressure of the helium was 2.4×10^5Pa and the helium was in the supercritical state. We observed the heat transfer of the supercritical helium had some specific characteristics different from those of the liquid helium.

EXPERIMENTAL ARRANGEMENT

We used the same experimental arrangement used in the previous experiment. Details of the arrangement are described in the reference [1]. Here, outline of the experimental arrangement is described. The experimental arrangement is schematically shown in Figure 1. A heat transfer measurement sample made of a Pt-Co wire of 20μm diameter and 40mm length shown in Figure 1(b) was placed in the rotating cryostat illustrated in Figure 1(a). The Pt-Co wire was used as the heater and temperature sensor. Thermal time constant of Pt-Co wire is less than 10μsec, the thermal response of the Pt-Co wire is sufficiently fast to measure transient heat transfer characteristics in the range of several tens μsec. The temperature was determined from the resistance of the wire measured by the four terminal method. A step current of $I(t)$ was applied to the Pt-Co wire and the transient wave form of the voltage $V(t)$ between the voltage tap on the wire (length 40mm) was recorded by a transient recorder together with the current wave form. The surface heat flux $q(t)$ was calculated as $q(t)=IV/S$ where S was the surface area of the wire between the voltage taps. The Pt-Co wire has low thermal mass and we can assume that the wire surface temperature agreed to the wire temperature. During the step heating, the current was kept constant but $V(t)$ changed due to the temperature change of the wire, therefore, $q(t)$ changed. Our main concern is on the transient heat transfer characteristics for a few msec

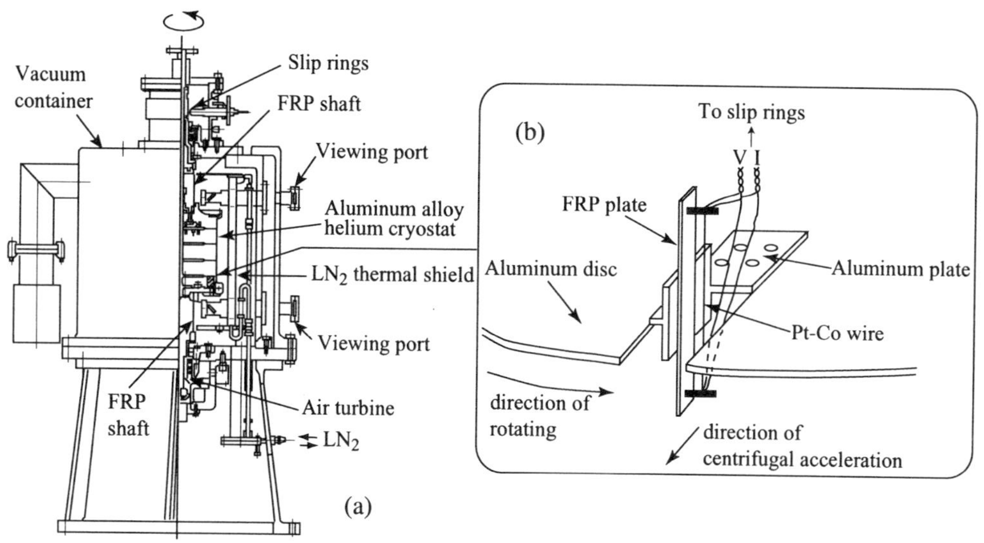

Figure 1. Cut-away vies of (a) rotational cryostat and (b) schematic drawing of the measurement sample.

after the step heating and in this time range, we assumed $q(t)$ did not change. We call the value of $q(t)$ at the beginning of the heating the initial heat flux. Data were taken by changing the rotational speed of the cryostat from 0rpm (1G) to 6000rpm (3600G). By monitoring the liquid helium level, the pressure of the helium around the Pt-Co wire was calculated. The helium stayed in liquid state below 4000rpm (1600G) and became supercritical state at 6000rpm (3600G). The bath temperature and pressure of the helium in the rotating cryostat varied 4.2K~4.7K and 1.0~2.4×10^5 Pa depending to the rotating speed of the cryostat.

EXPERIMENTAL RESULTS

Figure 2 shows temperature traces of the Pt-Co wire for step heating of 23kW/m^2~47kW/m^2 in various centrifugal acceleration field together with the case of 1G. It was found in our previous work that the steady state film boiling was established in early stage of the step heating for liquid helium in high centrifugal acceleration field and that cyclic temperature fluctuations caused by bubbling of the liquid helium were observed. In the supercritical helium of 3600G, the cyclic temperature fluctuations were also observed and the heat transfer reached steady state earlier for the step heating than in the liquid state helium. The average cycle of the temperature fluctuations T_b in the supercritical state are plotted against acceleration field a together with those in the liquid state in Figure 3. T_b's of the supercritical helium for a=3600G are shorter than those of the liquid helium of lower a and on the line of $T_b \propto a^{1/3}$ same as of T_b's of the liquid helium. Pressure of the supercritical helium at 3600G was about 2.4×10^5 Pa where the density of the supercritical helium is strongly dependent on the temperature. Therefore, it is considered that quasi-bubbling occurred in the supercritical helium and that large centrifugal acceleration aids departure of the bubble to shorten T_b.

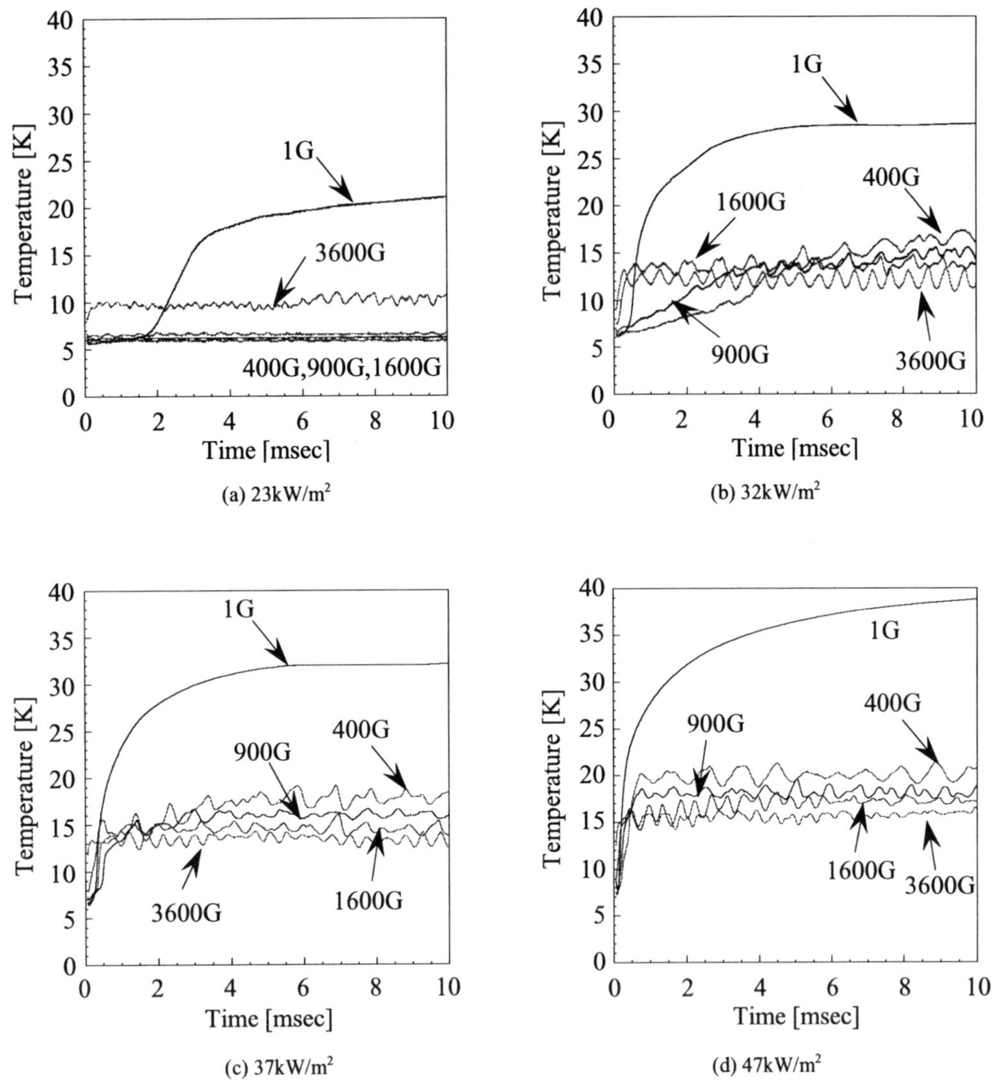

Figure 2. Transient time traces of temperature of Pt-Co wire for step heating in centrifugal field.

Obviously from Figure 2(a), the nucleate boiling region with good heat transfer appearing in the liquid state does not appear in the supercritical state because the supercritical helium is mono phase. Figure 4 shows the heat transfer characteristics at the early stage of the step heating for the heat flux 47kW/m^2. As seen in Figure 4, the heat transfer of supercritical helium reaches the steady state in 200μsec and the quasi-nucleate boiling which is observed in the liquid helium is not observed for the supercritical helium of 3600G and those characteristics were commonly observed in the range of the heat flux 23kW/m^2~47kW/m^2 where we conducted the experiment.

The temperature rise of the wire ΔT in the steady state is plotted against the heat flux for various centrifugal fields in Figure 5. The heat transfer in the supercritical state of 3600G is

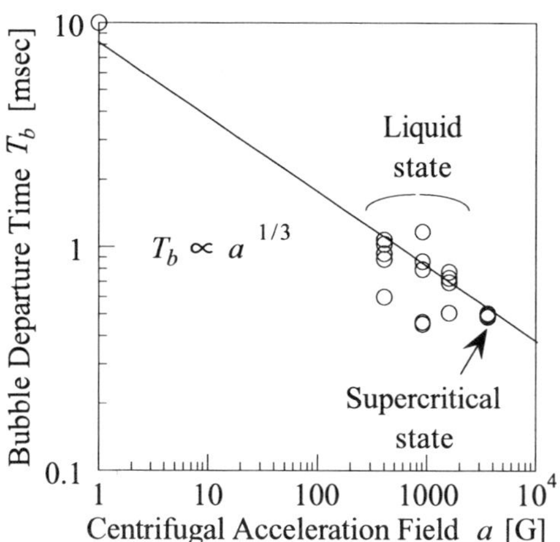

Figure 3. Plots of bubble departure time vs. centrifugal acceleration.

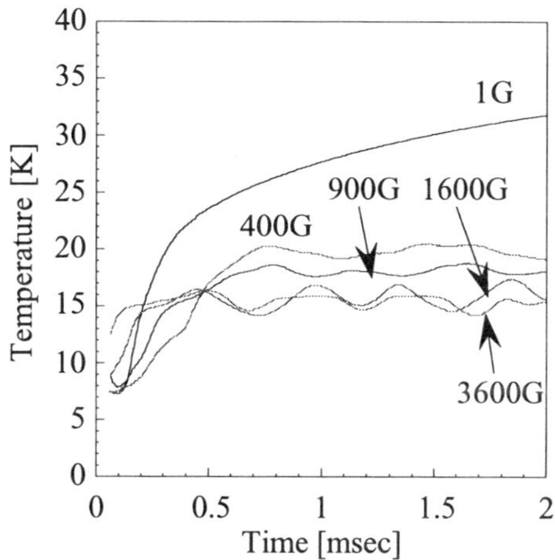

Figure 4. Enlargement of Figure 2. (d) $47kW/m^2$.

worse in the low heat flux region below $25kW/m^2$ and better in the high heat flux region above $26kW/m^2$ compared with that in the liquid state of 400G~1600G. The nucleate boiling occurs in the liquid helium even when the helium is subject to the centrifugal acceleration and good heat transfer is obtained. On the other hand, the nucleate boiling does not occur in the supercritical state and the heat is removed from the wire surface by free convection of the helium. The heat transfer in the nucleate boiling region is better than that by the free convection even in the high centrifugal field. Therefore in the low heat flux range, the heat transfer of the supercritical state is lower than that of the liquid state. When the heat flux in the liquid state exceeds a certain threshold, the maximum nucleate boiling heat flux, the boiling moves to the film boiling where the heat transfer by the boiling much decreases and the free convection cooling becomes dominant. The heat transfer by the free convection increases as the centrifugal acceleration increases. Therefore, the heat transfer in the supercritical state created by higher centrifugal field is higher than that in the liquid state subject to lower centrifugal field.

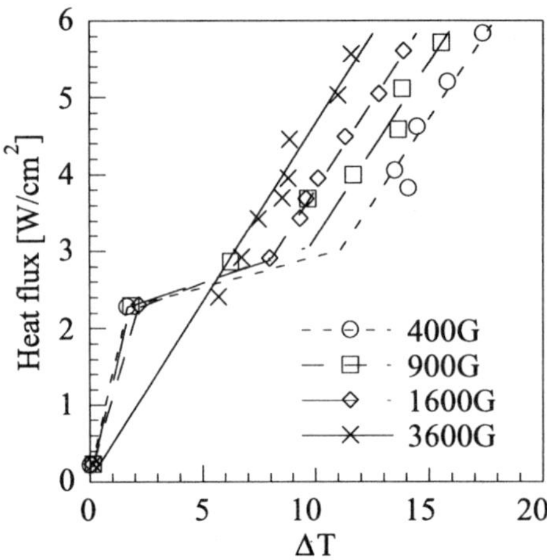

Figure 5. Steady state heat transfer characteristics.

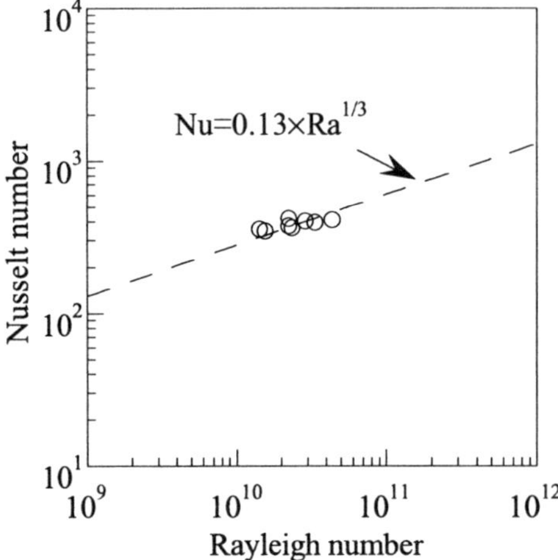

Figure 6. Steady state heat transfer data of supercritical helium of 3600G plotted in Rayleigh vs. Nusselt numbers.

Our steady state heat transfer data for the supercritical helium of 3600G are plotted in a graph of Nusselt vs. Rayleigh numbers in Figure 6. Our data are on the line $Nu=0.13\times Ra^{1/3}$ which is derived from the steady state heat transfer data for a flat heating surface in the supercritical helium in the centrifugal field.[2] The line in Figure 6 follows the well-known equation which is for the convection cooling.[2]

Based on the above study, we conclude that the heat transfer of the supercritical helium in high centrifugal field is dominated by the steady state free convection cooling and that the steady state heat transfer is established in about 200μsec after the start of the heating. For the stability analysis of the superconducting rotor windings subject to disturbances caused by conductor motions, we consider that the steady state heat transfer data can be used.

CONCLUSIONS

Transient heat transfer of supercritical helium in high centrifugal acceleration field was measured. The results of the study based on the experiment are summarized as follows.

1. The heat transfer reached the steady state in about 200μsec.
2. The nucleate boiling and quasi-nucleate boiling were not observed in the supercritical state while they were observed in the liquid state.
3. The steady state heat transfer of the supercritical helium was lower in the lower heat flux region and higher in the higher heat flux region than that of the liquid state.
4. The heat transfer is considered to be dominated by the free convection cooling because the data expressed in terms of Nusselt and Rayleigh numbers follow the line for the free convection cooling.
5. Based on the experimental results, we consider that the steady state heat transfer data in the supercritical helium can be used to the stability analysis of the superconducting rotor winding where duration of the disturbances causing premature quenches are several 100μsec.

REFERENCES

1. O. Tsukamoto, M. Furuse, T. Takao, N. Tamada, S. Fuchino, I. Ishii and N. Higuchi, Transient Heat Transfer Characteristics of Liquid Helium in Centrifugal Acceleration Field, *Advances in Cryogenic Engineering*, Vol.43B, Plenum Press, New York(1998), pp.1489-1496.
2. R. Nakajima, K. Sato, K. Miyake, M. Kumagai and Y. Kobayashi, Liquid and Supercritical Helium Heat Transfer of Horizontal Upward Facing Surfaces under High Centrifugal Acceleration Fields, ASME HTD-Vol.229, Heat Transfer in Superconducting Equipment, Book No. G00721 (1992), pp.39-44.

Directors from the ICMC and CEC Boards.

PREDICTION OF TRANSPORT PROPERTIES FOR MULTICOMPONENT CRYOGENIC MIXTURES USED IN J-T CRYOCOOLERS

M.Q. Gong, E.C. Luo, Y. Zhou, J.T. Liang, and L. Zhang

Cryogenic Laboratory, Chinese Academy of Sciences
Beijing, 100080, China

ABSTRACT

The Enskog dense-gas theory combined with an equation of state is a convenient and efficient method for predicting the transport properties of multicomponent mixtures. In this paper, an empirical modification of the Enskog equation is proposed to improve the prediction accuracy of the viscosity coefficient and thermal conductivity for mixtures. Extensive comparisons with experimental data are made for a wide range of fluid states for two mixtures. The total average absolute deviations are 2.88% and 1.63% for viscosity, and 0.81% and 3.69% for thermal conductivity, respectively. The proposed procedure for predicting the viscosity and thermal conductivity of mixtures is simple and straightforward.

INTRODUCTION

In recent years there has been an increasing interest in the usage of gas mixtures as refrigerants in the study of Joule-Thomson (J-T) cycle refrigerators. The thermodynamic performance of the J-T cycle refrigerator can be greatly improved by using multicomponent cryogenic mixtures as refrigerants. The use of the appropriate mixture as refrigerant in the J-T cryocooler makes it possible to use an air-conditioning compressor for driving the refrigerator. This broadens its commercial applications.

It is more and more important to utilize accurate property predictions of multicomponent cryogenic mixtures with the development of Joule-Thomson cycle

Advances in Cryogenic Engineering, Volume 45.
Edited by Shu *et al.*, Kluwer Academic / Plenum Publishers, 2000.

refrigerators. The properties of a fluid can be divided into two types, equilibrium properties and transport properties. The equilibrium properties such as density, enthalpy, entropy, vapor liquid equilibrium, and various excess properties can be predicted accurately and conveniently with an equation of state. The important advantage of using an equation of state is that it can provide a unique and consistent model for all equilibrium properties. However, our understanding of transport properties such as viscosity and thermal conductivity is far behind that of equilibrium properties. There is no satisfactory theory of transport properties for real dense gases, liquids, and their mixtures. Therefore, the generally used formulations of transport properties are either empirical or based on some theoretical foundation. Enskog dense-gas theory [1] is one of the very few theoretical approaches for predicting the transport properties of dense gas based on the distribution function. Many calculations of the transport properties at elevated densities are based on the theory of Enskog for a dense gas of rigid spherical molecules. Therefore, some modification is needed for real dense gases and their mixtures. Enskog proposed some modifications himself to extend the application for real dense gases and their mixture. W. Sheng et al [2] developed a method in which the Enskog theory is combined with an equation of state for predicting the transport properties of real fluids. This method is convenient; however, the accuracy of this method is not satisfactory.

In this work, another empirical modification of the Enskog theory is proposed to improve the prediction accuracy of the transport properties for multicomponent mixtures used in J-T refrigerators.

THEORY

Cubic Equation of State [3]

Many of the common two-parameter cubic equations of state (EOS) can be expressed by the equation

$$P = \frac{RT}{V-b} - \frac{a}{V^2 + \mu b' V + \delta b'^2} \tag{1}$$

Where, R is the gas constant. P, V and T are the pressure, volume, and temperature, respectively. The parameters a and b' are coefficients of the equation, and μ, δ are constants of the equation. An equivalent form of Eq. (1) is expressed as an equation for the compressibility factor:

$$Z^3 - (1 + B - \mu B)Z^2 + \left(A + \delta B^2 - \mu B - \mu B^2\right)Z - AB - \delta B^2 - \delta B^3 = 0 \tag{2}$$

where $A = \frac{aP}{R^2T^2}$

and $B = \frac{b'P}{RT}$

Table 1. Constants for four common cubic equations of state

No.	μ	δ	b	a
vdW	0	0	$\frac{RT_c}{8P_c}$	$\frac{27}{64}\cdot\frac{R^2T_c^2}{P_c}$
RK	1	0	$\frac{0.8664RT_c}{P_c}$	$\frac{0.42748R^2T_c^{2.5}}{P_cT^{0.5}}$
SRK	1	0	$\frac{0.08664RT_c}{P_c}$	$\frac{0.42748R^2T_c^2}{P_c}\left[1+k_{SRK}\left(1-T_r^{0.5}\right)\right]^2$ *1
PR	2	-1	$\frac{0.077796RT_c}{P_c}$	$\frac{0.457235R^2T_c^2}{P_c}\left[1+k_{PR}\left(1-T_r^{0.5}\right)\right]^2$ *2

*1 $k_{SRK} = 0.48 + 1.574 - 0.176\omega^2$

*2 $k_{PR} = 0.37464 + 1.54226 - 0.26992\omega^2$

Where: T_c, P_c, ω are the critical temperature, critical pressure, and eccentric factor, respectively. $T_r = T/T_c$ is the reduced temperature.

There are four well-known cubic equations of state: van der Waal (vdW), Redlich-Kwong (RK), Soave, and Peng-Robinson (PR) equations. For these four equations, μ and δ take on the integer values listed in Table 1. The expressions of a and b for the different equations are also listed in Table 1.

Enskog Theory[1]

According to the Enskog theory, the viscosity coefficient η and thermal conductivity λ for a gas of rigid spheres are represented by the following expressions:

$$\eta = \eta_0 b\rho(\frac{1}{b\rho\chi} + 0.800 + 0.7614b\rho\chi) \tag{3}$$

$$\lambda = \lambda_0 b\rho(\frac{1}{b\rho\chi} + 1.200 + 0.755b\rho\chi) \tag{4}$$

Where η_0 and λ_0 are the dilute-gas viscosity coefficient and thermal conductivity, respectively; ρ is the molecular density, $b = 2/3\pi\sigma^3$ is the co-volume, where σ is the molecular diameter, and χ is the value of the equilibrium radial distribution function at a distance σ. $b\rho\chi$ is determined by the "thermal pressure":

$$b\rho\chi = \frac{v}{R}(\frac{\partial P}{\partial T})_v - 1 \tag{5}$$

At low densities, it is necessary that $\chi \to 1$ as $\rho \to 0$:

$$\lim_{\rho \to 0} \chi = 1 \tag{6}$$

In this work, the cubic equation of state is used to evaluate the coefficients of Eq.(3) and Eq.(4). Substituting Eq.(1) into Eqs.(5) and (6):

$$\mathrm{b}\rho\chi = \frac{\mathrm{V}}{\mathrm{R}}[\frac{\mathrm{R}}{\mathrm{V} - \mathrm{b}'} - \frac{\beta}{\mathrm{V}^2 + \mu \mathrm{b}'\mathrm{V} + \delta \mathrm{b}'^2}] - 1 \tag{7}$$

$$b = \lim_{\rho \to 0} b\chi = b' - \frac{\beta}{R} \tag{8}$$

$$b\rho = \frac{1}{v}(b' - \frac{\beta}{R}) \tag{9}$$

$$\beta = \frac{\partial a(T)}{\partial T} = -\frac{ma(T)}{\sqrt{T_c T \alpha}} \tag{10}$$

For calculating mixture properties, the following mixing rules are adopted:

$$a = \sum_i \sum_j x_i x_j (1 - k_{ij}) \sqrt{a_i a_j} \tag{11}$$

$$b = \sum_i x_i b_i \tag{12}$$

$$\beta = -\frac{1}{2} \sum_i \sum_j x_i x_j a_{ij} (\frac{\beta_i}{a_i} + \frac{\beta_j}{a_j}) \tag{13}$$

Where x_i and x_j are mole fractions of components i and j in the mixture under consideration, k_{ij} is the interaction coefficient between component i and j.

With Eqs. (7) and (9), Eqs.(3) and (4) can be calculated directly. In order to make the results reliable, the three well-known cubic equations of state (RK, Soave, and PR EOS) are used to calculate the transport properties respectively. However, we established that the results calculated with Eqs. (3) and (4) are not accurate enough for realistic application in the design of the heat exchanger for the J-T refrigerator no matter which equation of state is used. So here two modified coefficients C_λ and C_η are used. C_λ and C_η are empirical parameters for the prediction of the thermal conductivity and viscosity, respectively. Then Eqs. (3) and (4) can be written as follows:

$$\eta = C_\eta \eta_0 b\rho(\frac{1}{b\rho\chi} + 0.800 + 0.7614 b\rho\chi) \tag{14}$$

$$\lambda = C_\lambda \lambda_0 b\rho(\frac{1}{b\rho\chi} + 1.200 + 0.755 b\rho\chi) \tag{15}$$

where C_λ and C_η are functions of the density, and can be obtained from the experimental data.

RESULTS AND DISCUSSION

Equations (14) and (15) are the basis of the present calculation when $b\rho$ and $b\rho\chi$ are calculated with Eqs.(7) and (9) (Here, the Peng-Robinson equation of state is used). In this paper, the method proposed above is used to calculate the viscosity and thermal conductivity of two different mixtures. One is N_2(36.7%), CH_4 (24.6%), C_2H_6 (12%), and C_3H_8 (26.7%) (referred to as No.1 mixture), the other is R22 (62%), Ar (6%), and CH_4 (32%) (mole fraction) (referred to as No.2 mixture).

The comparisons of calculated results and experimental data [4] of the two mixtures are shown in Table 2 and Table 3 for viscosity and Table 4 and Table 5 for thermal conductivity, respectively. From Table 2 and Table 3, the overall absolute average deviations (AAD) in the calculated viscosity are 2.88% and 1.63% for the No.1 and No.2 mixtures, respectively. The AADs in thermal conductivity are 0.81% and 3.69% for the No.1 mixture and No.2 mixture, respectively. The AAD is expressed as:

$$\text{AAD} = (1/NDP)\sum(|calculated - \exp erimental|)/\exp erimental \times 100\% \qquad (16)$$

where NDP is the number of data points.

The prediction procedure proposed here provides a convenient predictive method for transport properties of multicomponent mixtures. The calculation procedure is simple and straightforward. The agreement between calculated results and experimental data is quite good. However, the accuracy of the predicted results is worse when the density of the mixture is elevated. This is because that the Enskog theory is based on the rigid spherical molecule model. In addition, the proposed method must be successful for predicting nitrogen-hydrocarbon multicomponent mixtures. However, since the experimental evidence is rather limited for other mixtures, much effort is required. It is obvious that the effort is focused on how to get the empirical coefficients C_λ and C_η for multicomponent mixtures. Two formulations are presented in the Appendix for the evaluation of the two coefficients C_λ and C_η. They need to be verified by additional experimental data.

ACKNOWLEDGMENT

This work is financially supported by the National Natural Sciences Foundation of China under contract Number 59706002.

Table 2 Comparison of calculated results of viscosity for the No.1 mixture

No.	T (K)	P (MPa)	NDP	AAD (%) by Eq.(14)	AAD (%) by Eq.(3)
1	96.31-107.80	0.830-5.471	4	12.64	68.55
2	120.11-136.65	0.863-5.494	5	9.21	56.40
3	139.22-159.03	0.715-5.545	6	2.08	38.43
4	166.85-185.66	1.538-5.470	5	3.88	13.43
5	172.21-340.21	0.073-1.760	10	0.81	14.79
6	269.25-333.44	1.613-2.127	6	0.48	24.26
7	288.36-331.88	3.228-3.976	5	0.41	35.51
8	292.31-330.50	3.826-4.637	5	0.54	39.85
9	298.10-332.16	4.745-5.656	5	0.54	46.03
Total	96.31-333.44	0.073-5.656	51	2.88	34.40

Table 3 Comparison of calculated results of viscosity for the No.2 mixture

No.	T (K)	P (MPa)	NDP	AAD (%) by Eq.(14)	AAD (%) by Eq.(3)
1	110.22-122.11	1.257-5.450	5	6.01	80.35
2	117.89-140.28	0.471-5.303	6	2.37	73.51
3	131.09-157.33	0.851-5.003	5	3.30	68.74
4	152.17-180.28	1.765-5.192	4	0.65	61.87
5	178.33-197.08	3.137-5.275	4	1.36	53.70
6	223.03-332.83	0.090-0.144	8	1.13	4.75
7	287.42-334.17	1.265-1.517	4	0.41	3.58
8	308.24-345.78	2.248-2.615	6	0.74	8.80
9	318.31-352.41	3.111-3.574	6	0.27	16.75
10	327.62-353.30	3.946-4.396	5	0.61	24.35
11	328.95-353.71	4.250-4.723	5	1.04	28.39
12	332.54-354.12	4.832-5.307	5	1.98	37.28
Total	110.22-354.12	0.090-5.450	63	1.63	36.58

Table 4 Comparison of calculated results of thermal conductivity for the No.1 mixture

No.	T (K)	P (MPa)	NDP	AAD (%) by Eq.(15)	AAD (%) by Eq.(4)
1	94.1-106.2	0.787-5.493	4	0.63	20.69
2	122.1-136.9	1.420-5.564	4	1.37	14.82
3	140.3-158.7	0.979-5.464	5	1.07	20.46
4	165.3-185.7	1.215-5.478	5	0.33	29.36
5	170.4-211.8	0.072-0.097	6	4.18	10.25
6	233.0-332.1	0.110-0.171	10	0.53	0.73
7	257.2-335.5	0.980-1.352	10	0.36	5.94
8	281.7-331.4	2.516-3.163	8	0.32	17.00
9	290.1-331.2	3.555-4.364	8	0.32	25.27
10	296.6-332.0	4.457-5.335	7	0.34	32.7
11	299.1-329.5	4.881-5.715	6	0.42	35.85
Total	94.1-335.5	0.072-5.715	73	0.81	17.83

Table 5 Comparison of calculated results of thermal conductivity for the No.2 mixture

No.	T(K)	P(MPa)	NDP	AAD(%) by Eq.(15)	AAD(%) by Eq.(4)
1	107.5-120.9	0.297-5.024	4	1.63	18.58
2	117.2-141.8	0.375-5.685	6	2.37	15.11
3	131.4-158.4	0.900-5.173	5	1.07	10.56
4	153.3-198.2	1.758-5.403	8	1.33	6.21
5	222.1-331.9	0.089-0.143	14	3.21	38.31
6	256.7-332.4	0.381-0.510	10	4.53	39.16
7	282.3-330.2	0.960-1.157	9	3.36	37.21
8	299.8-331.7	1.580-1.789	8	3.81	35.10
9	306.8-333.1	2.201-2.451	7	3.32	32.41
10	307.1-334.4	2.735-3.075	9	3.34	29.72
11	323.1-343.3	3.665-3.992	7	2.34	23.63
12	326.6-344.1	4.049-4.368	8	3.56	21.56
13	330.7-343.7	4.490-4.757	7	2.78	19.38
Total	107.5-343.7	0.089-5.685	102	3.69	27.24

APPENDIX

When experimental data are not available, the accuracy of the prediction for mixtures without polar substances may be improved by using the following equations to formulate the coefficients:

For the vapor phase:

$$C_\eta = C_\lambda = 1 - b\rho$$

For the liquid phase:

$$C_\eta = \frac{V}{V-b} - \frac{aV}{(V^2 + 2bV + b^2)RT}$$

Where a and b are the coefficients of the PR equation of state.

REFERENCES

1. S.Chapman and T.G.Cowing, "The Mathematical Theory of Non-uniform Gases". Cambridge Univ. Press, London 3rd ed., chapter 16, (1970)
2. W.Sheng, G.J.Chen and H.C.Lu Prediction of transport properties of dense gases and liquids by Peng-Robinson (PR) equation of state, *Int. J. Thermophysics*, 10:133-144, (1989)
3. R. C. Reid, et al, "The Properties of Gases and Liquids", 4th edition pp.42-47, McGraw-Hill Book Company, New York, (1987)
4. The work report on Joule-Thomson cryocooler with gas mixture (052/6046), Odessa Institute of Refrigeration Engineering, Odessa, (1986), P.54 (in Russian)

"Best Paper" authors receive their awards.

VAPOR - LIQUID EQUILIBRIUM IN MULTICOMPONENT MIXTURES OF HYDROGEN'S ISOTOPES

Ioana Cristescu, I.Cristescu

Institute of Cryogenics & Isotope Separations
PO Box 10, 1000 Rm. Valcea, Romania

ABSTRACT

An important issue in the separation of hydrogen's isotopes using cryogenic distillation is the precise knowledge of the vapor-liquid equilibrium. To computes the equilibrium, we considered the non-idealities both in the vapor and in the liquid phase. We computed the specific volumes of hydrogen's isotopes with the aid of the virial equation of state and the Redlich-Kwong equation of state. The enthalpy of evaporation was computed with the aid of the Clausius-Clapeyron equation, where the saturation pressure for hydrogen's isotopes was expressed with the Mittelhauser equation. For the activity coefficients on the liquid phase, we made a comparative study of Hildebrand and Margules theories applied to mixtures of hydrogen's isotopes. The model is in good agreement with experimental data obtained for mixtures: H_2-D_2 and H_2-HD. Relevant data obtained on ternary mixtures are presented (as Gibbs diagrams).

INTRODUCTION

The models for simulation of cryogenic distillation use the concept of "theoretical plate" that simultaneously brings to thermodynamic equilibrium all the components in a multicomponent mixture. Owing to error propagation, it is very important to know with great accuracy the vapor-liquid equilibrium of the mixture. It must also correctly predict the heat balance for each stage considering the significant difference in heats of vaporization for different isotopic species.

BASIC ASSUMPTIONS

The conditions for coexistence of both phases result from the general thermodynamic conditions of equilibrium. Thermal and mechanical equilibrium requires equality of temperature and pressure in both phases. The material balance is valid for each individual component in the mixture and imposes the equality of chemical potentials in the boiling liquid and in the saturated vapor, equivalent to the equality of fugacities in both phases.

The fugacity of a component j in a mixture in the vapor phase can be expressed as:

$$RT\ln\frac{f_i}{y_i p} = \int_V^{\infty}\left[\frac{\partial p}{\partial n_i} - RT\right]dv - RT\ln z \tag{1}$$

The derivatives of pressure can be obtained by using an equation of state. For mixtures of hydrogen's isotopes, we investigate the virial equation of state and the Redlich-Kwong equation. The virial equation of state is [4]:

$$\frac{pv}{RT} = 1 + \frac{B(T)}{v} + \frac{C(T)}{v^2} + \ldots \tag{2}$$

The second virial coefficient B (T) and third virial coefficient C (T) were computed considering the Lennard-Jones potential function [3,4] and compared with some available data [2]. For mixtures, the virial coefficients were given by:

$$B(T)_{mixture} = \sum_i \sum_j B_{ij}(T) y_i y_j \qquad and \qquad C(T)_{mixture} = \sum_i \sum_j \sum_k y_i y_j y_k C_{ijk} \tag{3}$$

We also considered the Redlich-Kwong equation, successfully applied for various cryogenic fluids [7]:

$$p = \frac{RT}{v - b} - \frac{a}{\sqrt{T}\, v\,(v + b)} \tag{4}$$

The coefficients a and b for different isotopics form of hydrogen are summarized in Table 1.

Table 1. Redlich-Kwong Constants, Constants in Mittelhauser Equation for Hydrogen's Isotopes

	Redlich-Kwong Constants		Constants in Mittelhauser Equation			
	a[N· (m^2/mol)^2· K $^{1/2}$]	b[m^3/mol]	A_M	B_M	C_M	D_M
n-H_2	1.4250 ·10^4	1.8117	4.73642	-47.041	0.31609	0.02079
HD	1.5134·10^4	1.7115	5.69001	-59.9076	-0.10155	0.02123
HT	1.5850·10^4	1.6269	6.3826	-67.5537	-0.43028	0.02146
n-D_2	1.5821·10^4	1.6302	6.40296	-70.8737	-0.40539	0.02158
DT	1.6139·10^4	1.5875	7.0256	-78.4069	-0.68944	0.02186
n-T_2	1.6546·10^4	1.5619	7.62502	-85.8015	-0.96115	0.0216

For mixtures the coefficients a and b are related to those for pure components as:

$$b = \sum_i y_i b_i \qquad a = \sum_i \sum_j a_{ij} y_i y_j \qquad a_{ij} = \sqrt{a_{ii} a_{jj}} \tag{5}$$

After a comparative study of the two equations with available experimental data [10], we chose the Redlich-Kwong equation of state for computing the molar volumes and for computing the fugacity.

As a remark, binary mixture data are not need for computing the fugacities of components in multicomponent mixtures, only the critical points for pure components are needed.

For the liquid phase, the fugacity of a pure component is:

$$f_i^L = p_i^s \cdot \varphi_i^s \; exp \int_{p_i^s}^{P} \frac{v_i^L dp}{RT} \tag{6}$$

It is well known that the behavior of real mixtures greatly deviates from ideality. The molar Gibbs excess function is:

$$g^E = RT \sum_i x_i \, ln \, \gamma_i \tag{7}$$

where γ_i is the activity coefficient and can be computed by various approximations. For hydrogen and its isotopes, the best approximations are the theories of Hildebrand [3] and Margules [1].

The Hildebrand theory of regular solutions assumes that the components are mixing without changing the excess entropy or volume. The activity coefficient of component j in a mixture of liquid phases [3,5] is:

$$RT \ln \gamma_j = v_j(\delta_j - \sum_i \frac{x_i v_i}{\sum_j^m x_j v_j} \delta_i) \tag{8}$$

where δ_i are the solubility parameters defined as:

$$\delta_i = \sqrt{\left(\frac{\Delta u^v}{v} \right)_i} \tag{9}$$

and Δu^v is the vaporization energy.

An alternative way of computing the activity coefficients in the liquid phase is to consider the two-suffix Margules relation for the excess Gibbs potential:

$$g^E = \frac{1}{2} \sum_i \sum_j A_{ij} x_i x_j \tag{10}$$

Activity coefficients are then obtained by differentiating equation (10):

$$RT \, ln \, \gamma_k = \sum_i \sum_j \left(A_{ik} - \frac{1}{2} A_{ij} \right) x_i x_j \tag{11}$$

The most common way of evaluating the A_{ij} coefficients is from the Hildebrand semi-empirical method that relates them with the energy of vaporization [3]:

$$A_{ij} = \left(\sqrt{U_i^v} - \sqrt{U_j^v} \right)^2 \tag{12}$$

VAPORIZATION ENTHALPY FOR HYDROGEN AND ITS ISOTOPES

The vaporization energy was computed starting from the Clausius-Clapeyron equation. The pressure on the saturation curve varies with the temperature after the relation proposed by Mittelhauser [8]:

$$\log p = A_M + \frac{B_M}{T} + C_M \log T + D_M \frac{p}{T^2} \tag{13}$$

where coefficients A_M, B_M, C_M, D_M for hydrogen and its isotopes are given in Table 1.

By introducing equation (13) in the Clausius-Clapeyron equation for the vaporization enthalpy, we finally obtained:

$$\Delta h_v = \Delta v_v \left[\frac{C_M T^2 - B_M T - 2 D_M p}{(T^2 / p - D_M)} \right] \tag{14}$$

where Δv_v is the difference between the specific volume of the vapor and liquid phase on the saturation curve.

Due to the computational complexity of the models for the liquid state, it is preferably to use some semi-empirical evaluations for the specific volume. Most of these evaluations are based on the theory of corresponding state and therefore are applicable in the present case. We studied several models including [1] Yen-Woods, Chueh-Prausnitz, the compressibility factor, and Gunn - Yamada. An comparative analysis shows that the best results were given by the Gunn-Yamada method, where the specific volume of the saturated-liquid is:

$$v_{lic} = v_{sc} \cdot v_r^{(0)} (1 - \omega \, \Theta) \tag{15}$$

and where v_{sc} - scaling parameter, is a function of the critical temperature and pressure, and $v_r^{(0)}$ and Θ are functions of the reduced temperature [1].

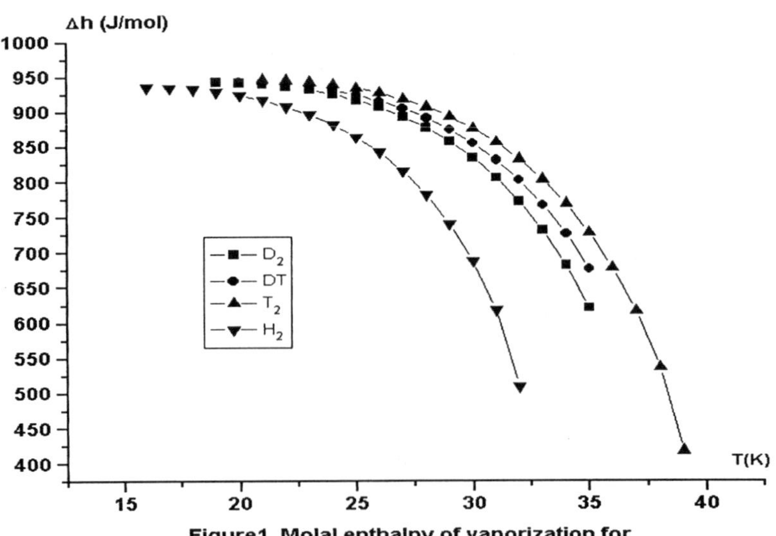

Figure1 Molal enthalpy of vaporization for hydrogen's isotopes as a function of temperature

Using these assumptions, we computed the molar enthalpy of vaporization for hydrogen's isotopes as a function of temperature, plotted in figure 1.

Unfortunately there is a lack of experimental data regarding the behavior of the liquid phase for multicomponent mixtures of hydrogen's' isotopes. We tried to investigate the accuracy of modeling with Hildebrand and Margules theory comparing the results with experimental data on the binary mixtures H_2-HD and H_2-D_2 [10]. Figure 2 presents the variations of the activity coefficient γ as function of the hydrogen concentration for the mixture H_2-HD.

As seen from figure 2 for this system, we obtained a better correlation with the experimental results using the Margules theory for modeling the activity coefficient. We obtained a similar conclusion for binary mixture H_2-D_2, therefore we decided to use the Margules theory in modeling mixtures of hydrogen's isotopes.

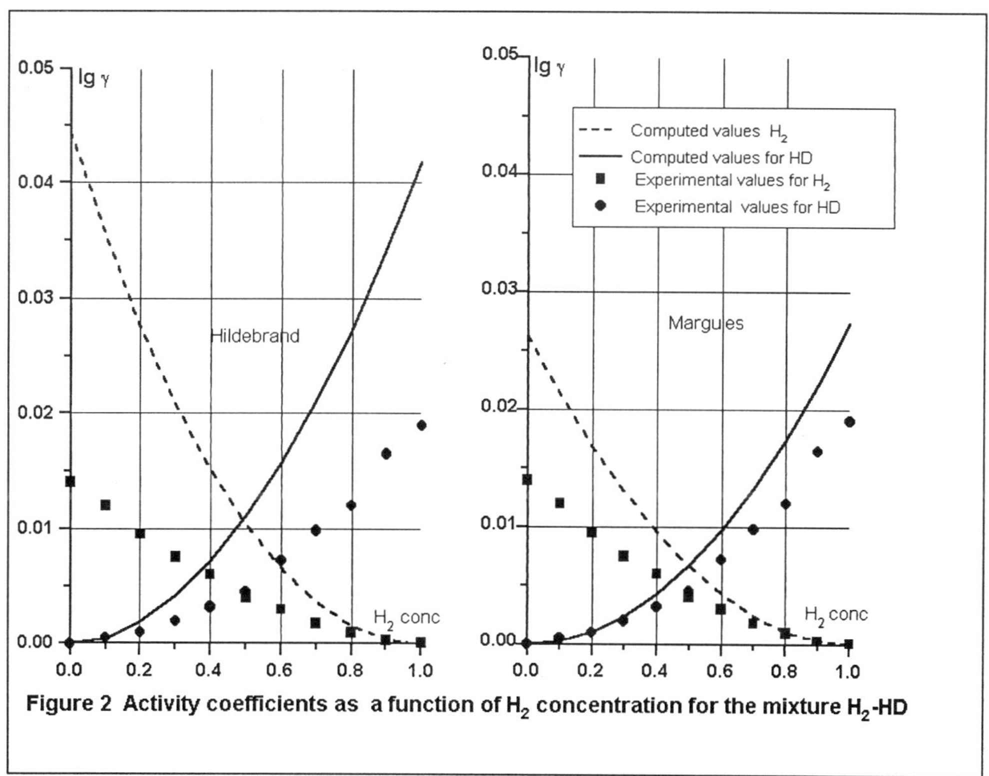

Figure 2 Activity coefficients as a function of H_2 concentration for the mixture H_2-HD

MODEL FOR COMPUTING VAPOR-LIQUID EQUILIBRIA FOR MIXTURES OF HYDROGEN'S ISOTOPES

The most common problem in the study of the phenomena appearing in distillation processes (boiling, separation, condensation) is the calculation of boiling dew-point temperature and vapor composition at a given pressure and known composition of the boiling liquid. Figure 3 shows the model for computing this type of equilibrium.

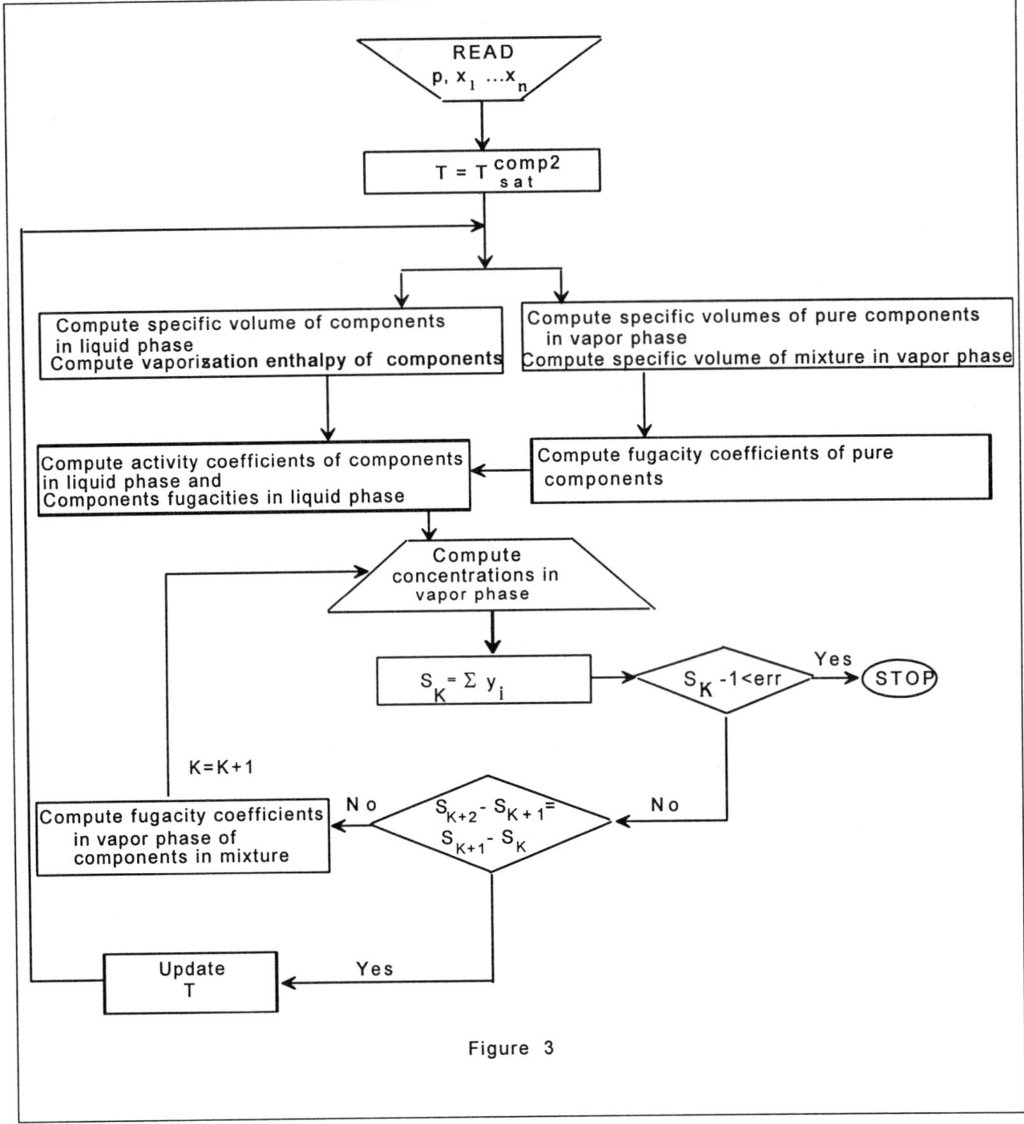

Figure 3

RESULTS

The Margules theory was used to compute the vapor-liquid equilibrium for multicomponent mixture of hydrogen's isotopes. For ternary mixtures, the equilibrium can be plotted with the aid of Gibbs diagrams. Figures 4 and 5 present such diagrams for the mixtures H_2-HD-D_2 and H_2-HD-HT for different values of pressure.

The model is designed for equilibrium of all 6 species of hydrogen's isotopes and their diatomic combinations: H_2, HD, D_2, HT, DT, and T_2.

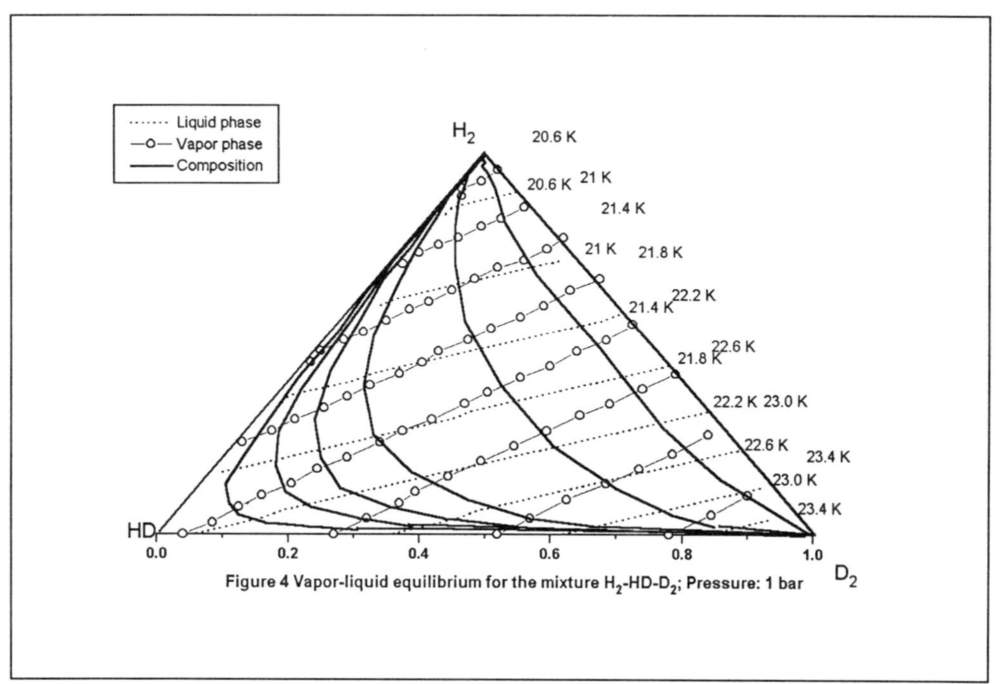

Figure 4 Vapor-liquid equilibrium for the mixture H_2-HD-D_2; Pressure: 1 bar

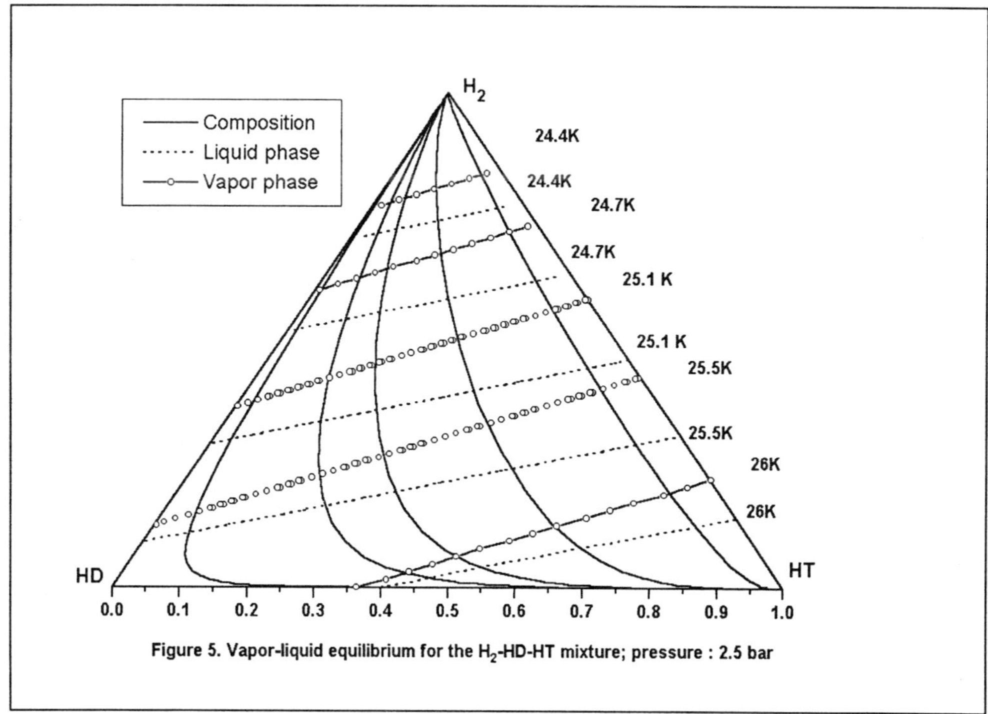

Figure 5. Vapor-liquid equilibrium for the H_2-HD-HT mixture; pressure : 2.5 bar

The model proposed in this paper for computing the vapor-liquid equilibrium is based only on data that characterized the pure components, and did not require any knowledge about the binary mixture behavior of hydrogen's isotopes. This is due to the applicability of the intermolecular forces and coresponding theory to cryogenic gases.

NOMENCLATURE

a,b - Redlich-Kwong coefficients
B, C - second order, third order virial coefficients
A_M, B_M, C_M, D_M - coefficients from Mittelhauser equation
f - fugacity
g - molar Gibbs potential
H - enthalpy
p - pressure
R - universal constant of gases
T - temperature
u - molar internal energy
v - specific volume
x - concentrations on liquid phase
y - concentrations on vapor phase
z - compressibility factor
ω - acentric factor
φ - fugacity coefficient
γ- activity coefficient

Superscripts

c - critical point
E - excess function
i,j,k - species of a multicomponent mixture
L - liquid
s- saturation
v - vapor

REFERENCES

1. R. Reid, J. Prausnitz, T. Sherwood - The Proprieties of Gases and Liquids, McGraw-Hill, New-York, 1977.
2. J. Dymond, E. Smith - The Virial Coefficients of Pure Gases and Mixtures, Clarendon Press, Oxford, 1980.
3. J. Prausnitz - Molecular Thermodynamics of Fluid - Phase Equilibria, Prentice Hall, New Jersey, 1969.
4. J. Hirschfelder, C. Curtiss, R. Bird - Molecular Theory of Gases and Liquids, John Wiley & Sons, New York, 1954.
5. I. Prigogine - The Molecular Theory of Solutions, North-Holland, Amsterdam, 1957.
6. Landolt-Bornstein - Zahlenwerte und Funktionen, aus Physik -Chemie -Astronomie - Geophysik und Technik, Band IV 4.Teil Warmetechnik, Springer Verlag, Berlin, 1967.
7. E. Bazua, J. Prausnitz - Cryogenics, **11**, 114-119 (1971).
8. H.M. Mittelhauser, G. Thodos - Cryogenics, **4,** 368-373 (1964).
9. C. Tsonopoulos, J.M. Prausnitz - Cryogenics, **9**, 315-327 (1969).
10. A. Rosen - Theory of Isotopes Separations in Columns, Goshimizdat, Moscow, 1960.

A PARTICLE SEEDING APPARATUS FOR CRYOGENIC FLOW VISUALIZATION

D. Çelik, M. R. Smith, and S. W. Van Sciver

National High Magnetic Field Laboratory, Tallahassee, FL

ABSTRACT

A particle seeding apparatus to create neutrally buoyant solid particles in liquid helium (He I and He II) has been developed and successfully tested as part of a flow visualization experiment. The apparatus consists of a liquid cavity to hold the hydrogen-deuterium mixture, a vacuum jacket, a plain orifice atomizer, a needle valve, a solenoid valve, heaters and temperature sensors. During the condensation process, the vacuum jacket around the hydrogen cavity was filled temporarily with helium gas. The needle valve, which is coated with indium to prevent any leaks, closes the gate between liquid mixture and helium, and operates via a solenoid. Heaters in the can control the temperature of the liquid mixture. Temperature is monitored by silicon diode temperature sensors placed on the can. Tests have been performed inside a glass cryostat for visual inspection, and have been recorded using a black and white video camera and a camera equipped with a macro lens. Depending on the flow velocity inside the nozzle, atomization with different particle sizes has been observed.

INTRODUCTION

Flow visualization is an important tool for fluid dynamicist, and there are well developed techniques that are applicable to flows at moderate or high temperatures. However, it becomes very complex and increasingly difficult to apply these techniques to the flows at liquid helium temperature. This is mainly due to the fact that most of these techniques require neutrally buoyant tracer particles fed into the flow field. From this point of view, the flows involving liquid helium have two major difficulties: finding tracer particles with the correct density for neutral buoyancy particles; and removal of these particles from the flow field to prepare the apparatus for the next run for continuous flow experiments.

Bielert and Stamm successfully used hollow glass spheres that were available commercially to visualize Taylor-Couette flow in liquid helium.[1] In general, however, solid tracer particles have two major drawbacks: first, finding the neutrally buoyant particles for all the flow conditions is difficult; second, these particles are not suitable for closed-loop or continuous experiments for it is difficult to remove the particles from the flow field. Murakami and Ichikawa created solid hydrogen-deuterium particles by condensing gas into a He II bath.[2] This technique is suitable for creating neutrally buoyant particles inside liquid helium at any temperature, as well as having the advantage of being used in closed loop experimental rigs. However, their particular technique required the presence of a free liquid surface. Due to these advantages of using species with higher freezing point than liquid helium temperature, a particle seeding apparatus that utilizes a mixture of hydrogen and deuterium has been developed and is described below.

APPARATUS

A picture of the apparatus, which has been used to crate particles, together with the atomizer, which is at the bottom of the apparatus, are shown in Figure 1. The innermost component is the cavity for the liquid mixture to be atomized. At the exit of this reservoir is a plane orifice atomizer. A wire heater wrapped around the cavity maintains the temperature of the liquid mixture. Temperature is monitored via two Lake Shore silicon diode temperature sensors placed on the liquid reservoir and the injector. Temperature of the atomizer is also maintained by a wire heater. Temperature is controlled by a Lake Shore Model 340 temperature controller unit. The liquid reservoir is surrounded by a vacuum jacket. The passage between the liquid reservoir and helium bath is controlled by a needle valve, which is coated with indium to attain as much sealing as possible. This valve is operated by a room temperature solenoid for fast opening and closing of the gate so that the exit pressure of the liquid to be dispersed resembles a square pulse as much as possible. Pressure of liquid helium is monitored by a Baratron pressure head connected to a MKS pressure controller. Calibrated pressure cylinders are used to measure the amount of gas to be condensed.

The apparatus is placed inside a liquid nitrogen cooled glass cryostat. Experimental runs are recorded using a black and white video camera. A camera with a macro lens is used to analyze particle sizes.

EXPERIMENTAL PROCEDURE

After pre-cooling the apparatus and the cryostat with liquid nitrogen, liquid helium is transferred into the cryostat. A small amount of helium gas is transferred into the vacuum jacket surrounding the liquid cavity during the cool down process. Following the reaching of the helium to the desired temperature, measured amounts of hydrogen and deuterium gas are transferred slowly into the cavity where the gas mixture is liquefied by removing heat via the heat exchange gas in the vacuum jacket to the helium bath, as well as the conduction through the atomizer. After this transfer process is completed, the temperature controller is turned on to maintain the liquid to be dispersed at the desired temperature. Helium gas is used to raise the pressure over the liquid mixture when a pressure above the vapor pressure is desired.

After the correct settings are reached, the solenoid is turned on to open the gate and inject the liquid into the helium bath. The solenoid remains on for a short time. This time interval is long enough for the jet to reach to the steady state and operate on that regime for at least three seconds. This transition is monitored visually. The entire process is recorded for later investigation.

Due to the small size of the atomizer, great care must be taken in order to prevent any contaminant from entering into the apparatus during the entire process.

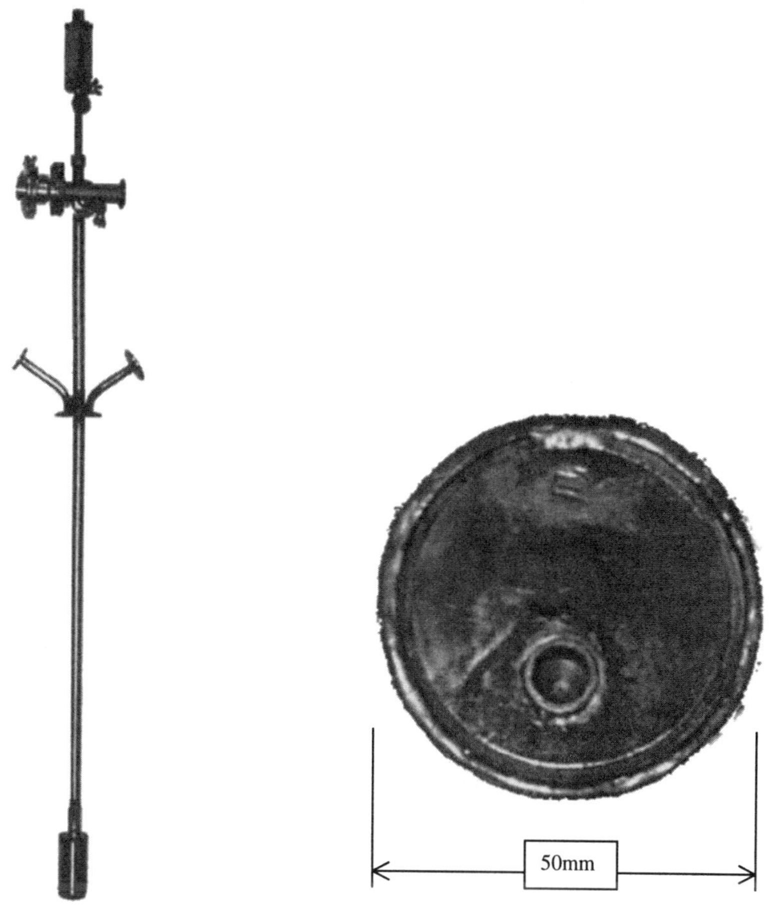

Figure 1. Particle seeding apparatus and atomizer (at the bottom of the apparatus).

DENSITY FOR NEUTRALLY BOUYANT PARTICLES

Since liquid hydrogen and deuterium are obtained by condensing gases of these species, pressure in the measuring cylinders that contain the gases directly controls the final mixture density inside the liquid cavity. Therefore, obtaining a relation between the desired liquid mixture density and the initial gas pressures is important. This relation can easily be found by assuming that the total volume of the liquid mixture is equal to the sum of individual volumes, and by considering the fact that mass is a preserved quantity. Therefore, the mass of the mixture is equal to the sum of the masses of liquid hydrogen and deuterium inside the liquid cavity:

$$V_{mix} = V_H + V_D \tag{1}$$

$$\rho_{mix} V_{mix} = \rho_H V_H + \rho_D V_D \tag{2}$$

where, V represents volume, ρ represents density, the indices H, D and mix represent hydrogen, deuterium and the mixture, respectively. For neutrally buoyant particles, neglecting the volume contraction during the solidification, mixture density is the same as the density of liquid helium at a given temperature. From these two equations, individual species volumes are found, and then these liquid quantities related to the gaseous state quantities through mass conservation, resulting in

$$P_H = \frac{\rho_H RT(\rho_D - \rho_{mix})}{W_H(\rho_D - \rho_H)}\left(\frac{V_{mix}}{V_C}\right) \quad \text{and} \quad P_D = \frac{\rho_D RT(\rho_{mix} - \rho_H)}{W_D(\rho_D - \rho_H)}\left(\frac{V_{mix}}{V_C}\right) \tag{3}$$

where, P represents gas pressure, T is temperature, R is the universal gas constant, W represents molecular weight, and the term V_C is the volume of the gas measuring cylinder. Here, compressibility effects are ignored and the gases are assumed to obey the ideal gas law. This is a valid assumption as long as the gases inside measuring cylinders are not over pressurized. Thus, Eq. (3) enables one to determine the amount of gases required based on the desired amount of liquid mixture.

For instance, when liquid hydrogen-deuterium mixture is at 19.5K and liquid helium is at 1.8K, volume ratios of liquid deuterium and hydrogen in the mixture becomes $V_D/V_H = 2.67$; while gas pressures inside calibrated measuring cylinders to obtain this liquid volume ratio becomes $P_D/P_H = 3.01$. Similarly, these ratios become $V_D/V_H = 1.13$ and $P_D/P_H = 1.30$ when liquid mixture is at 19.5K and liquid helium is at 4.2K.

PARTICLE SIZE CONTROL

The parameters influencing atomization through a plane orifice atomizer are: nozzle diameter, the velocity of the liquid at the nozzle exit, viscosities and densities of both the liquid to be dispersed and continuous liquid (the liquid into which injection occurs), and surface tension of the dispersed liquid.[3]

The velocity of the liquid at the nozzle exit is the only changeable parameter for given liquid helium and liquid mixture temperature, and for a particular nozzle. Therefore, a relation for the nozzle exit velocity is derived based on the energy equation along a streamline.[4] Since the diameter of the cavity is much greater than the nozzle diameter and the amount of liquid injected out is small, one can assume that the liquid level inside the cavity remains almost the same, and therefore the velocity at the liquid surface can be neglected. Then the nozzle exit velocity becomes:

$$U_E = \sqrt{2g\left[\frac{P_{mix} - P_E}{\gamma} - (h_f + h_s)\right]} \tag{4}$$

where g is the gravitational acceleration, γ is the specific gravity, P_{mix} is the pressure over the liquid mixture. The exit pressure P_E is assumed to be equal to the saturated vapor pressure of liquid helium, this is a direct consequence of assuming that the pressure at the nozzle exit due to the liquid helium column is small compared to the saturated vapor pressure, and therefore, can be neglected. Also pressure due to the liquid mixture column inside the cavity is ignored. In the above equation, h_f and h_s are, respectively, the total head loss terms due to friction and the cross section changes that the liquid encounters along the atomizer. h_s depends on the atomizer geometry only; while h_f is a function of both flow conditions and surface quality of the atomizer. These terms are given in the following forms[5]:

$$h_f = \sum_i f_i \frac{l_i}{D_i}\left(\frac{U_i^2}{2g}\right) \quad \text{and} \quad h_s = \sum_i K_{Li}\left(\frac{U_i^2}{2g}\right) \tag{5}$$

where, l is the length, D is diameter, f is friction coefficient, K_L is shape factor for the particular section of the atomizer. The subscript i (and j) represent different sections of the atomizer. For the calculation of the loss terms the velocities at different sections of the atomizer are related through the continuity equation, fluid is assumed to be incompressible:

$$U_i A_i = U_j A_j \tag{6}$$

where, A is the area.

For the particular atomizer used in the experiments, under fully turbulent flow assumption, and approximate surface roughness of 0.0015mm for the atomizer material, Eq. (4) takes the following form:

$$U_E = 0.88\sqrt{\frac{P_{mix} - P_E}{\rho_{mix}}} \tag{7}$$

RESULTS AND DISCUSSION

Successful and repeatable atomization has been obtained, and solid particles introduced into liquid helium using the above described apparatus and technique. Figure 2 shows the liquid hydrogen jet issuing into liquid helium; elapsed time is increasing from left to right. Particle size growth due to the collisions of the particles has been observed. However, adhesion of particles to one another seems to go away as the temperatures of these particles reach to liquid helium temperature, where the solidification process ends. On the other hand, particle growth can be reduced by creating a less dense initial spray. This can easily be accomplished by changing the atomizer configuration.

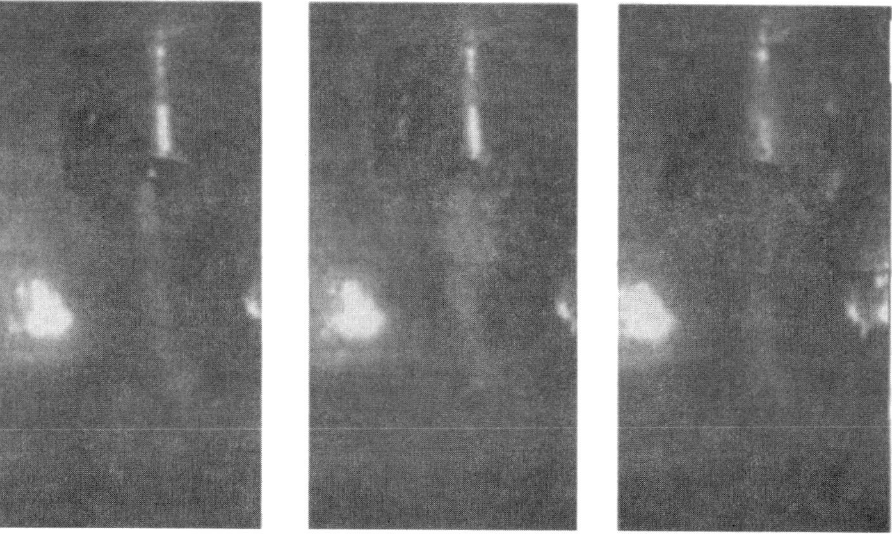

Figure 2. Liquid jet and solidification of droplets in liquid helium.

The adhesive nature of particles during solidification process suggests that the distance between the atomizer and the object around which the flow is to be visualized should be so chosen that the particles complete solidification process before reaching to the object.

ACKNOWLEDGMENT

This work is supported by the National Science Foundation-Thermal Transport Systems Division under grant CTS-9806725.

REFERENCES

1. F. Bielert and G. Stamm, Visualization of Taylor-Couette flow in superfluid helium, Cryogenics, 33: 938 (1993).
2. N. Ichikawa and M. Murakami, Application of flow visualization technique to superflow experiment, in: "High Reynolds Number Flows Using Liquid and Gaseous Helium", R. J. Donnelly, ed., Springer-Verlag, New York (1991), p.209.
3. A. H. Lefebvre. "Atomization and Sprays", Hemisphere Publishing Corporation, New York (1989).
4. D. N. Roy. "Applied Fluid Mechanics", Ellis Horwood Limited, West Sussex (1988).
5. I. E. Idelchik. "Handbook of Hydraulic Resistance", 2nd Ed., Hemisphere Publishing Corporation, New York (1986).

CRYOGENIC TECHNOLOGY INFORMATION DATABASE PROGRAM*

R. A. Mohling[1], W. L. Hufferd[2], E. D. Marquardt[3]

[1]Technology Applications, Inc.
Boulder, Colorado 80303

[2]Johns Hopkins University/CPIA
Columbia, Maryland 21044

[3]National Institute of Standards and Technology
Boulder, Colorado 80303

ABSTRACT

The Chemical Propulsion Information Agency (CPIA) and Technology Applications, Inc. (TAI), in collaboration with the National Institute of Standards and Technology (NIST), has reconstructed and updated the database of the former National Bureau of Standards (NBS) Cryogenic Data Center. NBS maintained a database of cryogenic technical documents that served the national need until the early 1980s. The electronic database, maintained on a mainframe computer, was a highly specific bibliography of cryogenic literature and thermophysical-property data that covered 100 years of data. However, since then, the database has not been maintained for use. We have undertaken a project to convert the NBS database to a personal-computer (PC) platform and to backfill the database with citations of cryogenic literature dating from 1980 up to the present.

Our first CD-ROM release of the reconstituted and updated retrieval system contained about 126,000 citations; 115,000 of them represent bibliographic entries from the former NBS database, and over 11,000 represent new citations with abstracts from recent literature. The retrieval system contains a new search engine with formatted printing capabilities. Semiannual releases of currently updated CD-ROMs are planned.

An additional on-going activity involves development of a database consisting of computer codes for generating thermophysical properties of cryogens and materials. Our objective for this task is to develop "user-friendly" standard properties of fluids and materials of databases traceable to NIST so that the government and industry can conduct and compare analyses using the same properties.

INTRODUCTION

One of NASA's greatest accomplishments is its development of cryogenic technologies that included a trained cadre of engineers; technical know-how through design, analyses, procedures, processes, and tests; and a database of cryogenic engineering. Much of this cryogenic technology was generated for the advanced launch vehicles, upper stages, and power systems needed in our race with the former Soviet Union to put man on the moon. These cryogenic technologies are a national asset and must be retained in order for U.S. industry to remain internationally competitive and for commercial opportunities to exist. Until 1980, much of this information was stored on mainframe computer tapes and libraries at the Cryogenic Data Center at the former National Bureau of Standards (NBS), now known as the National Institute of Standards and Technology (NIST), in Boulder, CO.

Since the early 1980s, however, there has been no centralized U.S. source of cryogenic technical data. Existing data are fragmented and dispersed, and coverage is incomplete for today's engineers and scientists. Over the past decade, there has been a national trend in which many experienced cryogenic engineers were either retiring or being relocated to other job specialties. During these moves, the departing engineers have taken their working tools and documents with them, leaving an information void for those remaining in the cryogenic group. This requires our engineers and scientists in many cases "to reinvent the wheel" to perform their assignments.

There is an obvious need in government and industry for a highly specific cryogenic database that can provide:

- a uniform, consistent source of cryogenic data,
- minimal duplication of effort, and
- access to previous analyses, design, materials and processes, and test data.

FORMER NBS CRYOGENIC DATA CENTER

Up to the early 1980s, the Cryogenic Data Center's responsibilities included data compilation, documentation, and technical services. A dedicated group of professionals supported these activities at NBS for several decades. A brief description of the three major database activities at the Cryogenic Data Center, taken from a 1967 NBS reference, is included below to give a "snapshot in time" and describe the thoroughness and extent of NBS professional staff involvement in maintaining the electronic and hardcopy databases.[1]

Data Compilation

Data compilation included the evaluation and compilation of data on the thermophysical properties (thermodynamic, transport, and other properties) for the principal fluids, and common mixtures of these fluids, used at low temperatures. These fluids were helium, hydrogen, neon, nitrogen, oxygen, air, carbon monoxide, fluorine, argon, methane, xenon, and krypton.

The literature for fluids was monitored on a continuing basis. As specific tasks were undertaken, comprehensive bibliographies were prepared and sometimes published. Task notebooks were made for preliminary selection of data and, where feasible, preliminary data sheets were issued. The senior staff, consisting of two physicists, one engineer (thermodynamic), one chemist, and one physical chemist, did critical evaluation.

Documentation

The documentation activities at the Cryogenic Data Center are summarized in Table 1.

Table 1. Summary of documentation activities

Activity	Description
Literature searching	Regular review of nearly two hundred periodicals and some fifteen abstract journals, and noting references in cryogenic documents; 150-200 items were noted weekly.
Literature procurement	Published literature was obtained from local, national, and foreign libraries; report literature was procured from the large national centers (NASA, DDC, and Clearinghouse); many reports were obtained directly from corporate sources as part of an information exchange program.
Cataloging, coding & processing	Data were coded into nine main subject categories such as properties of solids and fluids, cryogenic processes and equipment, instrumentation and laboratory apparatus, and cryogenic techniques; further characteristic coding was assigned to the type of document, temperature range, type, and range of data; comprehensive subject coding followed, based on a thesaurus of terms.
Bibliographic storage & retrieval	All cataloging and coding was converted to machine-readable form for automated processing on the NBS' Control Data Corporation 3600 computer; principal programs were used for searching, dictionary term identification, and for catalog tape output; custom bibliographies were prepared for specific subjects or for broad subject areas.
Distribution of literature & data	Announcements and abstract cards of new literature evolving from the Laboratory's Program were sent to more than 4,000 persons and institutions periodically; nearly 500 separate items of literature were available; fifteen to twenty thousand documents a year were distributed in response to some 2,000 orders.

Technical Services

The technical-service activities of the Cryogenic Data Center are summarized in Table 2.

Table 2. Description of technical services at the Cryogenic Data Center

Activity	Description
Current awareness	Weekly lists of new literature of cryogenic interest were prepared and distributed to subscribers; subscription price was $10 per year ($15 for foreign with airmail delivery).
Custom bibliographies	Over 42,000 accessions of cryogenic literature were entered into the Data Center's system; approximately 20,000 of these on properties of materials (for both fluids and solids) were processed for machine searching; detailed and/or extensive bibliographies could be prepared with computer facilities; likewise, some 2,000 patents, 2,000 articles on processes and equipment, and 1,000 articles on instrumentation were processed for machine retrieval; cost of custom searches was based on rate of $12 per minute of computer time plus 15 cents per reference for listing and indexing.
Preliminary data and advice	Data and advice on the thermodynamic and transport properties of cryogenic fluids and selected solids could be obtained from the Project Leader for the Data Compilation Group.
Announcements of reports	Available to anyone wishing to be placed on the mailing list; requests could be made for inclusion on the mailing list by completing and returning a postal reply card.
Local use of Data Center	Visitors were invited to use the Center's library, world literature file, catalog and abstract files, and microfilm facilities; the staff of both the Data Compilation and Documentation units helped with answers to questions, and with hard-to-find type of literature, or offered advice as to best sources of information.

Maintenance by NBS of the cryogenic database, which included both the electronic bibliography and hardcopy and microfilm documents, was discontinued in the early 1980s when the Boulder group's mission and scope were redefined by the Department of Commerce. The bibliographic tape was placed in storage; the fluid-property documents were retained in NBS libraries, but the remaining documents were either relocated or placed in archival storage at NBS.

CRYOGENIC TECHNICAL DATABASE PROJECT

Survey of Interest

In early 1997, a direct-mail survey was sent to 201 potential users of cryogenic information to determine whether a need existed for a revived Cryogenic Technology Database by industry and government personnel. One hundred responses were completed and returned; the results of the survey are given in Table 3.

Based upon the positive response of the survey, it was decided to proceed with an effort to reconstitute the former cryogenic database. For the initial phase, we were advised to prepare a proposal for the reconstitution effort that could be funded under a single government contract. It was agreed that funding from government sponsors would support the three-year startup effort to revive the NBS database. Once the database is established and in a maintenance mode, subscription and technical service fees would be required to support the upkeep and routine activities of the Cryogenic Technical Database.

Reconstruction Approach/Plan

A contract is in place at the NASA/Marshall Space Flight Center (MSFC) for the Chemical Propulsion Information Agency (CPIA) and Technology Applications, Inc. (TAI) to reconstruct the Cryogenic Technical Database Program. Mr. Garry Lyles of MSFC has the responsibility of the contract; he is totally supportive of the effort, expressing the opinion that the effort was several years overdue and that he was very willing to have his organization provide the contract administration services. The contract has been in place since April 8, 1998.

A three-year, three-phase program was proposed to reconstitute and update the cryogenic database. It was estimated that the startup phases of the reconstitution effort would take clerical, database, and staff personnel at the level of approximately 4-6 persons per year for three years. The contractor team consists of the CPIA and TAI. The phases and major tasks of the contract are:

1. *Phase I - Document Acquisition*
   - Locate NBS documents
   - Convert NBS database to CD-ROM format
2. *Phase II - Construction of Archive and Database*
   - Acquire documents and update cryogenics database
   - Annual updates of CD-ROM
3. *Phase III – Maintenance and Routine Operations*
   - Establish current awareness program
   - Transition document storage to permanent location
   - Establish technical and bibliographic inquiry services
   - Implement subscription funding

Table 3. Result of the 1997 survey indicating the need for cryogenic database

Number	Survey Question	% Affirmative Response
1	Use, develop or produce cryogenic technology, materials or processes frequently/most of the time?	68
2	Conduct research on cryogenic technology, materials or processes frequently/most of the time?	46
3	Require access to cryogenic technical literature or data?	93
4	Indicated willingness to contribute funds for the renewal/reestablishment of the cryogenics technology database (dependent upon level of support required)?	75

The immediate contract deliverable to each government sponsor is a CD-ROM of cryogenic data with a new search engine for key-word searches of the former NBS data. The CPIA/TAI team updates and releases the CD-ROM on a regular basis. The planned program deliverables for the cryogenic industry include: a centralized document archive and record base, bibliographies, technical analysis, state-of-the-art reviews, computer programs, expert referrals, specialized databases, current awareness service, and online internet service.

The reconstruction approach is to (1) build on the existing though-not-maintained NBS database, (2) use the Defense Technical Information Center (DTIC)[2] Information Analysis Center as a model, and (3) use NIST support in the reconstruction effort.

DTIC administratively manages and funds contractor-oriented Department of Defense (DoD) centers for analysis of scientific and technical information. There are 15 Information Analysis Centers (IACs) that provide DTIC users with access to specialized reference services and subject-matter experts for topics ranging from ceramics to high-temperature materials.

The IAC having the closest fit and significant overlap of cryogenic activity is the CPIA, located at Johns Hopkins University at Columbia, MD. CPIA is the U.S. national clearinghouse for worldwide information, data, and analysis on chemical, electrical, and nuclear propulsion. Many of the nation's chemical launch and upper-stage vehicles use cryogenic propellants including oxygen, hydrogen and methane. The Propulsion Information Retrieval System (PIRS), maintained by CPIA, is a combined database and retrieval system that consists of over 53,000 document citations that relate to missile, space, and gun technology.

The reconstruction plan includes a collaboration of CPIA, TAI, and NIST. CPIA provides the administrative management services and document/data-handling expertise using the DTIC Information Analysis Center model. The DTIC model includes the following: establish a clearinghouse of cryogenic information; acquire hardcopies of publications, papers, articles, and texts; store computerized, searchable records of bibliographic data for retrieval; update the record database continuously; train or retrain a cadre of specialized technical staff; and develop specialized databases as required.

TAI, a cryogenic systems engineering firm, is located in Boulder, CO. TAI provides the central staff to obtain, index, abstract, and process previous NBS documents and update the database by starting with current technical documents and working back to the 1980s to complete the database. TAI uses former NBS cryogenic specialists and retired cryogenic engineers located in the metropolitan Denver area, to acquire and index documents at TAI offices. As resources permit, the original NBS bibliography will be back-filled with abstract, index, and availability information.

NIST has agreed to support the CPIA/TAI team in reestablishing the cryogenic technology information database; all three parties believe a collaborative effort will enhance the technical product for use by U.S. engineers and scientists. NIST has exceptional capability and expertise in many areas of cryogenic technology including the custody of the bibliography tapes and many cryogenic documents from the Cryogenic Data Center. An agreement is in place with NIST/Boulder to make their files and library documents available to TAI for updating the database.

Program Status

A significant portion of the original NBS documents, data, and reports still remain at NIST; the data and documents for thermophysical properties of fluids were retained in Boulder. A larger portion of the original hardcopy and microfilm documents was sent for storage to the Naval Research Laboratory (NRL). NRL stored approximately 130 boxes of documents and microfilm at the National Archives in Washington, DC. These boxes have been recently returned to NIST and will soon be inventoried to determine the exact contents of the boxes. Other agencies have been queried regarding receipt of NBS hardcopy documents; none of these agencies report having received any NBS documents.

The original NBS bibliography of 115,000 entries has been transferred to CD-ROM format and released as the Cryogenics Information Retrieval System (CIRS). The original database contained bibliographic data (i.e., author and title) and index terms used for searches by subject; no abstract information was included in this bibliography. CIRS incorporates Dataware's BRS Windows® 95/98/NT-based search engine with formatted print output options that include abstracts, author indexes, corporate source, and subject terms. Numerous cryogenic citations have been added to the database. Table 4 indicates a summary of added capability for each released CD-ROM.

Figure 1 shows the time period for the journals and proceedings that have been added to the latest CD-ROM. Note in the figure that the former NBS bibliography has entries that cover a period of 100 years, with documents published from 1880 to 1980. Updates of both electronic-transfer data and manual citations from current literature back to 1980 have taken place in the past year. The electronic transfer data have come from the Propulsion Information Retrieval System (PIRS), DTIC, and NASA/RECON electronic databases. The NBS bibliography is updated by incorporating abstracts into the bibliography prior to 1980.

Planned Activities

We plan to continue updating the electronic database with recent cryogenic citations back to 1980 and to backfill abstract and availability information to the original NBS bibliography data beyond 1980. We recognize that some documents are more important and have therefore established a priority of adding documents to the database and to the CD-ROM. Our priority for adding documents is given in Table 5. Note that the dates in the table refer to the publication date of the document rather than the conference date.

Table 4. The additional CD-ROM capability added to each release

Version	Date	Entries/Citations	Summary of additional capability
0	7/98	115,000	The NIST-provided bibliography data of the former NBS Cryogenic Data Center converted from floppy disks to CD-ROM.
1	10/98	123,000	NBS bibliographic data converted to PC/DOS-based search engine; DTIC, NASA/RECON and PIRS data electronically merged into new database.
2	5/99	125,000	Windows version of search engine incorporated; entries corrected and duplicate entries removed; current citation entries added.
3	7/99	133,000	Enhanced search engine routines Incorporated; new data entries added.

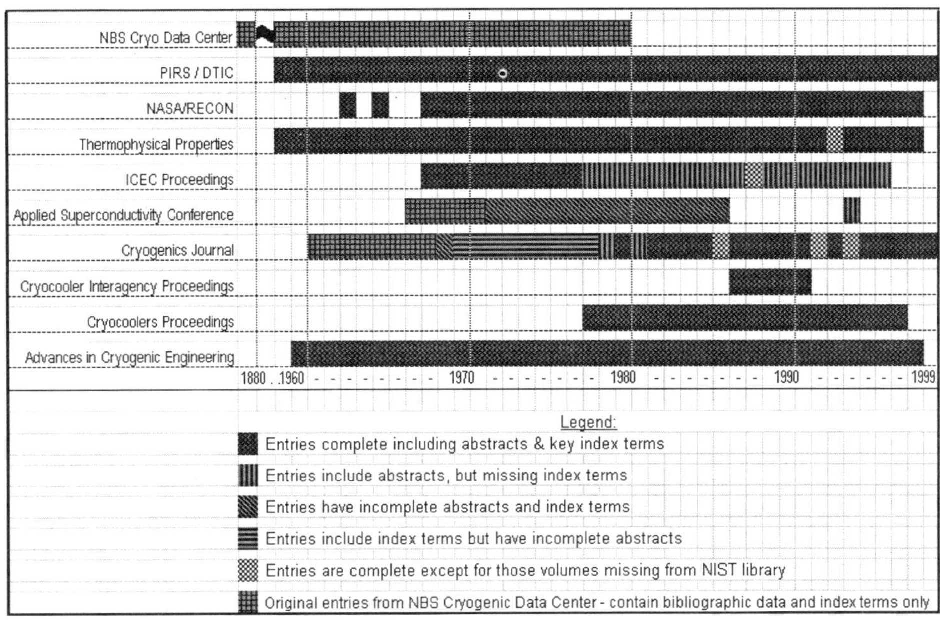

Figure 1. Journal and proceeding coverage in the database by year.

Table 5. Priority of document citations added to the database

Priority	Publication	Volumes	Dates	Status
1	Advances in Cryogenic Engineering	1-44	1960-99	Complete
2	Cryocooler Proceedings	0-9	1978-97	Complete
3	Interagency Cryocooler Meetings	2-4	1987-91	Complete
4	Thermophysical Properties Proceedings	1-10, 12-13	1959-98	Complete
5	Cryogenics Journal	9-58	1982-98	Complete
		1-8	1969-81	In work
6	International Cryogenic Engineering Conference	1-6	1968-76	Complete
		7-16	1978-96	In work
7	Applied Superconductivity		1967-95	In work
8	Patents			Pending
9	Government internal/contractor reports			Pending
10	Dissertations			Pending
11	International Institute of Refrigeration			Pending
12	Cryogenics and Refrigeration Conference			Pending
13	Low Temperature Physics Conference			Pending

Several groups within NASA have indicated the need for both improved material and fluid property software routines. Regarding fluid properties, there is interest in a user-friendly "standard" set of routines that would remove properties as an issue in contractor analyses and allow evaluators to focus on the real technical issues. An enhanced graphical user interface (GUI) for the NIST fluid-property routines is planned for (1) the NIST12 fluid properties including pure cryogens and hydrocarbons and (2) REFPROP6 for the common refrigerants. The desired capabilities of the GUI are:

- Load fluid properties directly into Excel spreadsheets,
- Access fluid properties in LabVIEW® for real-time experimental analysis,

- Cut-and-paste properties of interest into other Windows® applications,
- Ease of getting tabulated and plotted data in presentation format, and
- A well-documented user manual.

User-friendly "standard" material-property routines are needed for cryogenic applications. Comprehensive routines for many materials used in cryogenic applications either do not exist or are widely scattered. It is planned that hyperlinked electronic data sheets will be provided with:

- Tables and graphs,
- Both metals and non-metals (composites, insulators, etc.),
- English and metric units,
- Room-temperature properties,
- Temperature-dependent and integrated properties (yield strength, ultimate strength, thermal expansion, thermal conductivity, and specific heat), and
- Additional reference properties (manufacturer's data sheets, environmental, specifications, fluid compatibility).

CONCLUSIONS

Significant progress has been made in completing the Cryogenic Technical Database with the release of the Cryogenic Information Retrieval System (CIRS). CIRS is available on CD-ROM. Starting with the former bibliography of the NBS Cryogenic Data Center, there are over 133,000 cryogenic document citations now accessible on CD-ROM; the CD-ROM is planned for semi-annual release with additional citations from current literature. Over 130 boxes of original hardcopy and microfilm documents have been returned to Boulder to document support requests from electronic database searches.

A cadre of specialized technical staff have been trained to establish a clearinghouse of cryogenic information, acquire hardcopies of documents, store computerized records of bibliographic data for retrieval, and update the record base continuously. Much work is still required before we accomplish our goal of establishing an information center capable of providing technical analyses, state-of-the-art reviews, enhanced fluid and material property software routines, current-awareness service, and online internet service. We have made a significant step toward our ultimate goal of becoming a one-stop resource for cryogenic documents and technical data.

ACKNOWLEDGEMENT

Support for this project is being provided by the Air Force Research Laboratory, Propulsion Directorate, EAFB, CA and the Space Vehicles Directorate, Kirtland AFB, NM; NASA Ames Research Center, Moffett Field, CA, and the NASA Kennedy Space Center, Cape Kennedy, FL; the Naval Research Laboratory; and the Chemical Science and Technology Laboratory of the National Institute of Standards and Technology, Boulder, CO.

REFERENCES

1. "Cryogenic Data Center – Resume of Activities and Services," NBS, Boulder, CO (1967).
2. "Defense Technical Information Center," Directorate of User Services, Alexandria, VA (1994).

COMBINED THERMO-HYDRAULIC ANALYSIS OF A CRYOGENIC JET

M. Chorowski

Wroclaw University of Technology
Institute of Power Engineering and Fluid Mechanics
50-370 Wroclaw, Poland

ABSTRACT

A cryogenic jet is a phenomenon encountered in different fields like some technological processes and cryosurgery. It may also be a result of cryogenic equipment rupture or a cryogen discharge from the cryostats following resistive transition in superconducting magnets. Heat exchange between a cold jet and a warm steel element (e.g. a buffer tank wall or a transfer line vacuum vessel wall) may result in an excessive localisation of thermal strains and stresses. The objective of the analysis is to get a combined (analytical and experimental) one-dimensional model of a cryogenic jet that will enable estimation of heat transfer intensity between the jet and steel plate with a suitable accuracy for engineering applications. The jet diameter can only be determined experimentally. The mean velocity profile can be calculated from the fact that the total flux of momentum along the jet axis is conserved. The proposed model allows deriving the jet crown area with respect to the distance from the vent and the mean velocity profile along the jet axis. A simple formula to assess convective heat exchange between the jet and a solid obstacle has been proposed and experimentally verified.

INTRODUCTION

The paper describes a turbulent jet of a cryogenic fluid issuing from a nozzle into a space filled with the same or different fluid. The phenomenon may be encountered in some technological processes like grinding, where cryogenic cooling seems to have the edge on other coolants.[1] It is used in cryosurgery. The jet may be also a result of cryogenic equipment break or malfunctioning and create danger to people or equipment in its neighbourhood. It may also be observed inside buffer tanks as a result of a rapid cryostat discharge following resistive transitions in superconducting magnets.[2, 3] In all cases it is important to assess the jet basic hydrodynamic features as well as a convective heat exchange intensity between the jet and a solid object.

The form of the turbulent region in the jet results from similarity considerations and is conical. The diameter of the jet at a given distance from the nozzle cannot be defined theoretically but must be determined from experiments.[4] The velocity in the jet results from the fact that the total steady state flux of momentum through a spherical surface centred at the nozzle must be independent of the sphere radius.

Test Facilities

Figure 1 shows the conceptual scheme of the test rig used during the experiment. The vent was designed to allow changes of the orifice diameter (1, 3 and 5 mm). Gaseous nitrogen was supplied from a 0.15 m^3 dewar vessel of maximum pressure 2 MPa, while helium was vented from a 4.2 m^3 container of 0.33 MPa . The initial temperature of the gases was 115 K for nitrogen and 35 K for helium. The velocity at the outlet was equal to the speed of sound for both nitrogen and helium. Both fluids were vented to air at a temperature of 290 K and 275 K for nitrogen and helium respectively.

The jet temperature was measured by type T thermocouples (Cu-CuNi) placed on a thin wire along the jet axis, the velocity was measured by means of a vane anemometer of a 30 mm diameter, both at different distances from the vent. To measure the jet diameter, a simple device was used in form of a rod with light flaps located every 0.025 m. A flowing jet caused the flaps to move enabling an estimation of the jet diameter.

RESULTS

Diameter of the jet

The summary of the jet diameter measurements is shown in Figure 2. The cone angle does not depend on the orifice diameter or the gas vented but is constant and of about 13 degrees.

Temperature profile along the jet axis

During venting of nitrogen and helium, the temperature along the jet axis was measured. The results are shown in Figure 3. The orifice diameter was 3 mm and 5 mm for nitrogen, and 3 mm for helium. The results show that the biggest intake of air occurs in the first region of the jet. During the first 0.2 m the jet temperature rises approximately by 160 K for nitrogen and 180 K for helium.

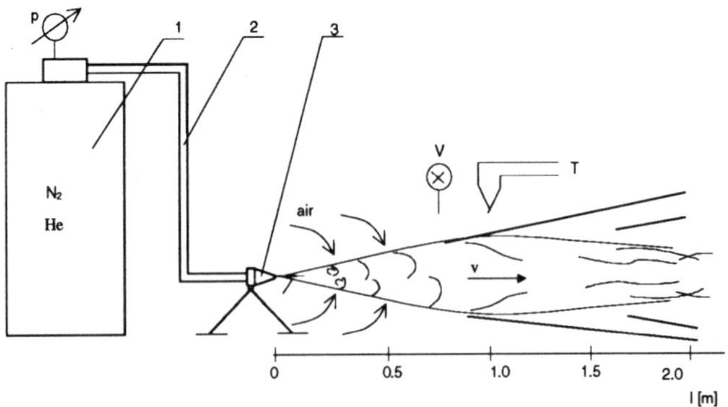

Figure 1. Test rig, 1 – LN2 / LHe vessel, 2 – transfer line, 3 – orifice, p – pressure gauge, v – velocity meter (vane anemometer), T – temperature sensor (thermocouple).

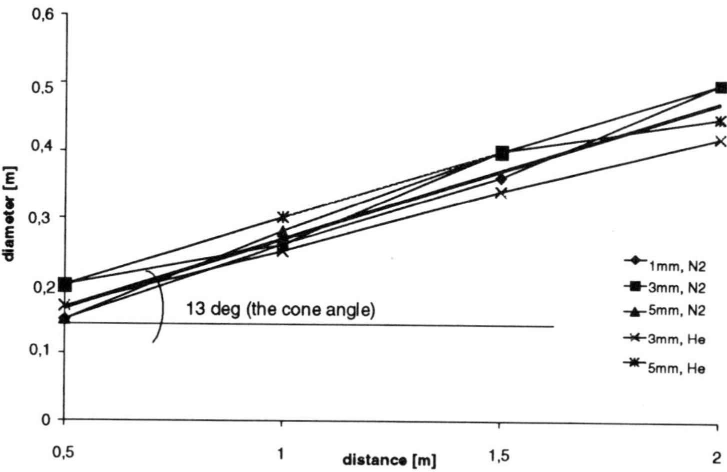

Figure 2. Jet diameter measured along the axis.

In the following 1.5 m the temperature rises only by about 15 K and 40 K, respectively. The jet temperature increase results from the mixing process and it approaches the temperature of the environment.

Velocity profile along the jet axis

The jet mean velocity decreases along the jet axis what directly results from the conservation of momentum principle, with the basic assumption that the momentum of the air entrained in the jet is negligible and the control surface can be defined as the actual cross section of the jet:

$$\frac{d(\rho \cdot v^2 \cdot A)}{dx} = 0 \tag{1}$$

where: ρ - cryogen density; v - cryogen mean velocity, A- jet cross section area, x – distance from the vent,

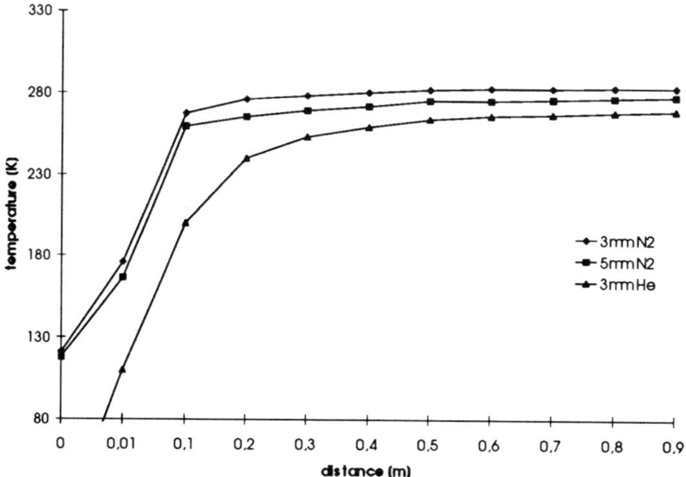

Figure 3. Temperature profile measured along the jet axis.

The relation defining velocity $v(x)$ variation along the jet axis is as follows:

$$v(x) = \sqrt{\frac{\rho_1 \cdot v_1^2 \cdot A_1}{\rho(x) \cdot A(x)}} \tag{2}$$

where subscript 1 denotes the jet outlet cross section.

$A(x)$ is a jet cross section area dependent on the distance from the orifice outlet and:

$$A(x) = \frac{\pi \cdot (D(x))^2}{4} \quad , \qquad D(x) = 0.23 \cdot x \tag{3}$$

where: $D(x)$ is the jet diameter and the number 0.23 results from the fact that a jet cone equals 13 degrees- compare Figure 2.

In the zone close to the orifice the cryogen undergoes a non-equilibrium free expansion until its pressure is equalised with the environment. The zone ends at x' and the distance depends on the pressure ratio. For the purpose of this analysis we assume that for x less than x' the jet mass flux is constant and no air is entrained – see Figure 4. The assumption enables to calculate the distance x' and then at larger distances the jet mean velocity from the gas conditions at the nozzle and the entrained gas parameters. The distance x' is of about 10 times the orifice diameter.

The jet density at the distance from the vent $x > x'$ is given by

$$\tilde{n}(x) = g_{air}(x) \cdot \tilde{n}_{air}(x) + g_{cr}(x) \cdot \tilde{n}(x)_{cr} \tag{4}$$

where: ρ_{cr} , ρ_{air} are the densities of the cryogen and air for $T = T(x)$ and g_{air} , g_{cr} are mass concentrations of air and cryogen in the jet, x denotes the distance from the vent.

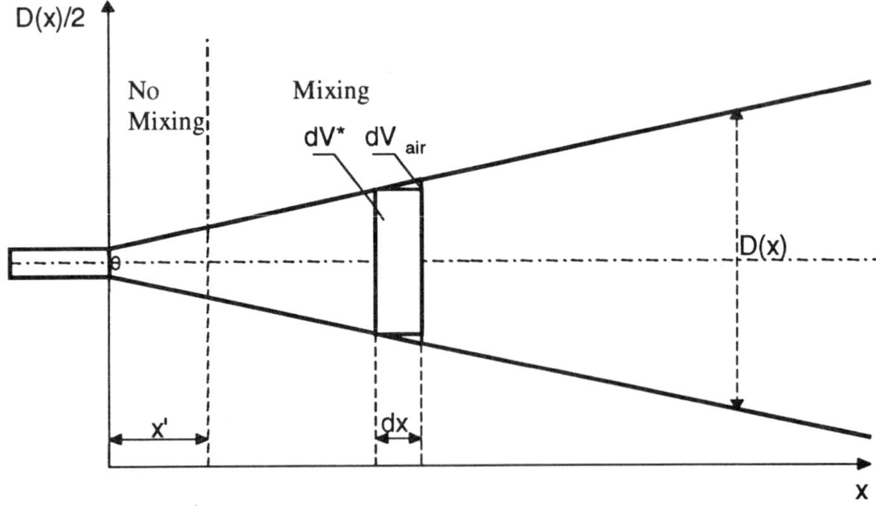

Figure 4. Schematic representation of a turbulent cryogenic jet development.

The average mass concentration of air in the jet at the distance $x > x'$ is given by Eq. (5, 6) resulting from the conical geometry of the jet.

$$g_{air}(x+dx) = \frac{dM_{air} + M(x)\cdot g_{air}(x)}{M(x+dx)} \tag{5}$$

where: dM_{air} is the mass of the air taken in the distance interval dx, $M(x)$ is the total mass of the jet at the distance x, and:

$$dM_{air} = \frac{dV_{air}}{dx}dx\cdot \rho_{air} \quad , \quad M(x+dx) = \frac{dM_{air}}{dx}dx + M(x) \quad , \quad \frac{dV_{air}(x)}{dx} = \frac{dV(x)}{dx} - \frac{dV^{*}(x)}{dx} \tag{6}$$

and $dV = dV_{air} + dV^{*}$ (compare Figure 4).

The calculated and measured nitrogen and helium velocity distribution along the jet axis is presented in Figure 5 and 6.

Heat transfer

The objective of this experiment was the estimation of the heat flux between the cryogenic jet and a solid obstacle represented here by a carbon steel plate, simulating a wall of a quench buffer vessel.[2] The conceptual scheme of the test rig is shown in Fig. 7. The diameter of the plate was 0.116 m and its thickness 0.02 m. The mass of the plate was 1.5 kg. The sides and back part of the plate were insulated. The temperature of the plate was

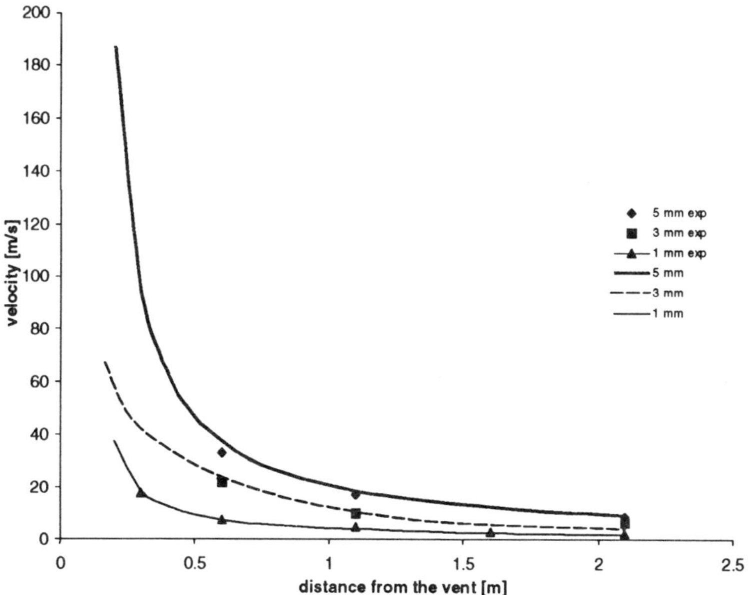

Figure 5. Nitrogen velocity along the jet axis (measured - dots and calculated - lines).

measured in three points. Due to the good thermal diffusivity of carbon steel, temperature differences across the plate cross-section were small and the average value was taken for further considerations.

The heat flux between the steel plate and the jet was calculated according to the formula:

$$\dot{Q} = M \cdot c_p \cdot \frac{\Delta T}{\Delta t} \tag{7}$$

where M – denotes mass of the plate, ΔT is an average temperature drop during the time interval Δt, c_p stands for heat capacity of the steel plate.

A heat transfer coefficient h results from Eq. (7) and is equal to:

$$h = \frac{\dot{Q}}{A_p \cdot (T_p - T_J) \cdot \Delta t} \tag{8}$$

where: T_p is the average steel plate temperature and T_J is the jet temperature close to the plate, A_p stands for the front area of the plate.

To describe the process of convective heat transfer between steel plate and jet the following formula has been proposed:

$$Nu = c \cdot Re^n \tag{9}$$

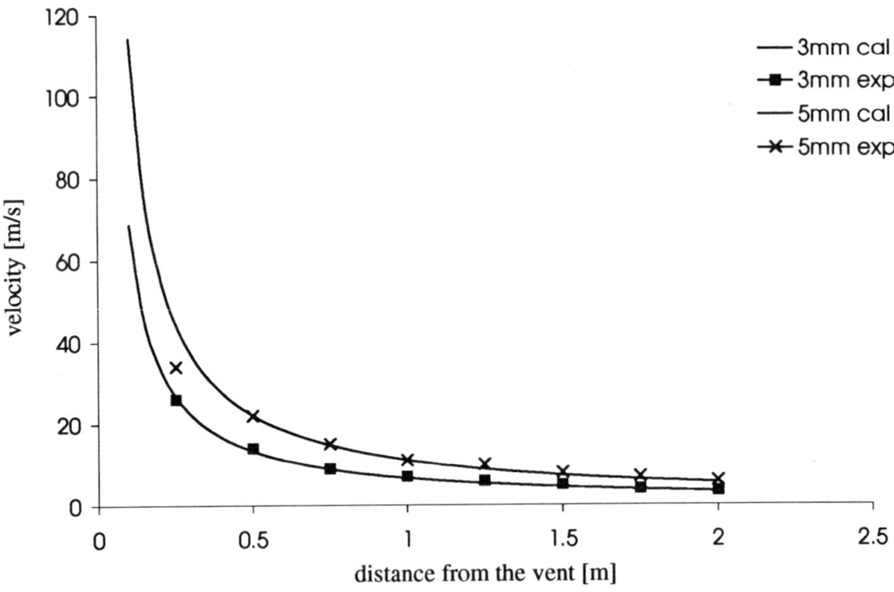

Figure 6. Helium velocity along the jet axis (measured - dots, calculated - lines).

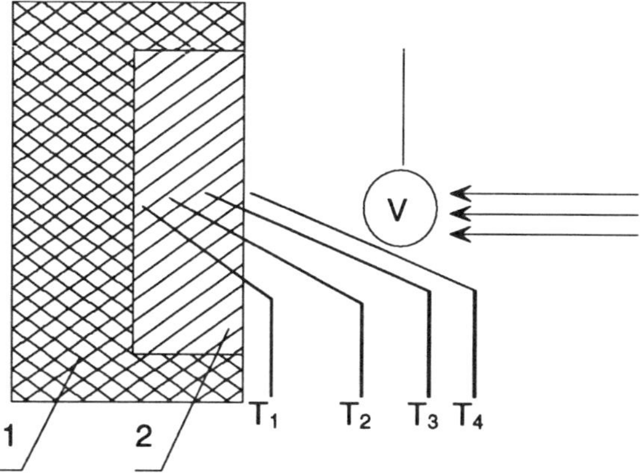

Figure 7. Conceptual scheme for heat transfer measurements between the jet and steel plate, 1 – insulation, 2 – steel plate, T – thermocouples (T_1-T_3 – steel plate, T_4 – boundary layer), V – vane anemometer.

The experimentally determined coefficients *c* and *n* are 0.107 and 0.77, respectively.

$$Re = \frac{v \cdot \rho \cdot \sqrt{A}}{\eta} \quad , \quad Nu = \frac{h \cdot \sqrt{A}}{\lambda} \tag{10}$$

where: ρ- jet density, *A*- jet cross section area, η- jet viscosity, λ- jet conductivity, *Nu*-Nusselt number, *Re*- Reynolds number, *h*- heat transfer coefficient.

Figure 8 shows the calculated heat transfer coefficient for nitrogen and helium jets as a function of the distance from the vent.

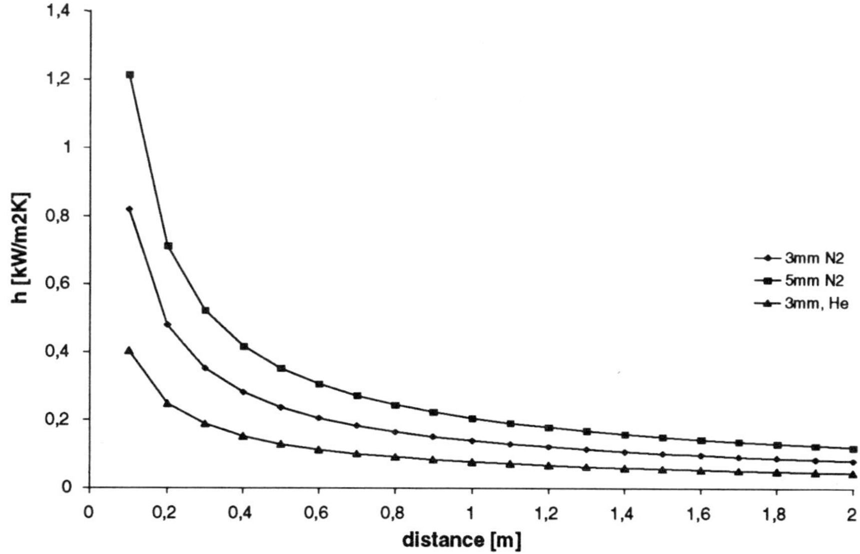

Figure 8. Calculated (Eq. 9) convective heat transfer coefficient versus distance from the vent (venting to air at the temperature of 290 K), initial conditions were: for nitrogen – pressure 1 MPa, saturated temperature, for helium – pressure 0.2 MPa, temperature 60 K.

CONCLUSIONS

It follows from literature data and the performed experiments that the cone angle of a cryogenic jet can be considered as constant and of about 13 degrees. It does not depend either on the orifice diameter or on the gas vented.

At distances large compared with the nozzle diameter (about 0.2 m in the studied case) the mean velocity can be calculated from the model based on the fact that the total flux of momentum along the jet axis is conserved.

At distances large compared with the nozzle diameter heat transfer between jet and any solid object can be calculated using a dimensionless Eq. (9).

A turbulent cryogenic jet is an efficient mixing mechanism ensuring fast equalising of the jet temperature and the environment.

ACKNOWLEDGEMENTS

The work reported here was conducted under contract with CERN. The author would like to thank W. Erdt, Ph. Lebrun and L. Tavian for fruitful discussions and support of the study.

REFERENCES

1. S. Paul and A.B. Chattopadhyay, Effects of cryogenic cooling by liquid nitrogen jet on forces, temperature and residual stresses in grinding steels, *Cryogenics*, 35,8 (1995).
2. M. Chorowski and B. Skoczen, Thermo-mechanical analysis of cold helium injection into gas storage tanks made of carbon steel following resistive transitions of LHC magnets, Proc. Of the XVII ICEC, Inst. of Phys., Bristol and Philadelphia (1998), p. 755.
3. M. Chorowski, P. Grzegory, G. Konopka, Thermo-hydraulic analysis of a cryogenic jet: application to helium recovery following resistive transitions in the LHC, LHC Project Note 177, CERN, Geneva, 1999.
4. L.D. Landau and E.M. Lifshitz, "Fluid Mechanics," Pergamon Press, Oxford (1966).

ULTRA-HIGH RAYLEIGH NUMBER CONVECTION RESEARCH CRYOSTAT

G.E. McIntosh,[1] J.H. Cloyd,[1] K. Pyziak,[1]
J.J. Niemela,[2] L. Skrbek,[2] R.J. Donnelly,[2]
and M. McAshan[3]

[1]Cryogenic Technical Services, Inc.
Longmont, Colorado, 80501

[2]University of Oregon
Eugene, Oregon, 97403-1274

[3]Fermi National Accelerator Laboratory
Batavia, Illinois, 60544

ABSTRACT

A fundamental problem in thermal turbulence is to understand heat transfer at very large Rayleigh numbers. For a fluid layer having a given imposed vertical temperature gradient, large Rayleigh numbers result from a combination of large isobaric thermal expansion and small kinematic viscosity and thermal diffusivity of the fluid. Furthermore, the Rayleigh number increases as the cube of the fluid layer height. Taking all these considerations into account, a convection apparatus has been designed and fabricated for the University of Oregon Turbulence Laboratory that uses cryogenic helium gas as the test fluid at temperatures between 4 and 6 K at pressures up to 5 bar. The working section consists of a volume of gas contained by a vertical cylinder, heated lower surface, and a cooler, temperature-controlled upper plate. For accurate results, the convection cell must be thermally stable and overall heat leak must be low with minimal solid conduction between the upper and lower plates. This 0.5 m diameter, 1.0 m high cell accesses stable Rayleigh numbers ranging from 10^9 to greater than 10^{16}, i.e., more than two orders of magnitude higher than have been previously attained under controlled conditions. Construction details, cooldown and operating experience, and typical experimental results are reported.

INTRODUCTION

This cryostat is designed for high Rayleigh number convection experiments and related turbulence research in helium vapor at temperatures of 4 to 6 K and pressures up to 5 bar. Operational drivers include thermal stability, a high degree of temperature uniformity on the heated and cooled surfaces, and ease of operation. Low heat leak combined with

Advances in Cryogenic Engineering, Volume 45.
Edited by Shu *et al.*, Kluwer Academic / Plenum Publishers, 2000.

cooldown speed and economy are also desired features. The 0.5 m diameter and 1.0 m height of the test cell are determined by the experimental design aimed at reaching Rayleigh numbers from 10^9 to the range of 10^{16} to 10^{17}. The cryostat is designed as a take-apart apparatus with several large indium seals in order to provide access to the experimental volume, to vary the test cell height, and, with some modifications, to accommodate other low temperature turbulence experiments. The combination of experimental and thermal requirements listed above dictate a large (1.07 m diameter X 2 m high) cryostat with a mass of some 500 kg as described below.

CRYOSTAT DESIGN

An assembly drawing of the cryostat is shown in Figure 1. Major elements include:

1. The 0.5 m diameter, 1.0 m high test cell cylinder. The test cell actually is made of two identical 0.5 m high flanged cylinders which can be configured for operation with either a 1:1 or 1:2 diameter to height ratio.
2. 0.0381 m (1.5 inch) OFHC (C101 copper) heater and refrigeration plates. A Minco brand Kapton™ surface heater is bonded to the bottom of the lower copper plate and is sandwiched between a 1.59 mm (1/16 inch) copper clamping plate to assure uniform heating.
3. The top surface of the upper copper plate is exposed to the so-called "soft vacuum" space between it and the bottom of the liquid helium reservoir. Varying helium gas pressure in the soft vacuum space from a few Pa up to nearly atmospheric provides very sensitive refrigeration control for the full range of convection experiments.
4. The 50 liter helium reservoir provides cooling for the convection cell and an attached conduction shield creates a nearly isothermal environment around the cell.
5. Helium vapor venting from the reservoir passes over a heat exchange surface attached to a secondary "20 K" shield which serves as an intermediate temperature heat intercept.
6. A conventional liquid nitrogen reservoir and shield is the major heat intercept for the cryostat and helps cool down the helium reservoir initially by means of a static heat pipe. The nitrogen-filled, thin wall stainless steel heat pipe provides one-way heat transfer via LN_2 for the initial cooldown, then shuts off heat transfer by freezing the nitrogen during the second cooldown phase.
7. Glass paper and aluminized Mylar™ MLI is used to insulate the nitrogen reservoir from the outer stainless steel vacuum jacket. Only the combination of high vacuum plus low emissivity surfaces is used for insulation below 77K.
8. Cryostat sub-systems include the nitrogen-filled heat pipe for partial cool down of the helium reservoir, a Gifford-McMahan cooler and related OFHC copper thermal link to cool the helium reservoir from approximately 80 to 30 K, a bleed valve from the helium reservoir to the convection cell volume used for pressure adjustment without the need of adding warm helium gas, and the 20 K vent gas heat exchanger mentioned above.

FABRICATION EXPERIENCE

While most aspects of the design are conventional, there are several sources of potential problems requiring special attention. First, there are four indium flange seals which are slightly over 0.5 m in diameter and must seal vacuum tight to 5 bar at 4 to 6 K. Second, two additional indium seals were eliminated by furnace brazing only to raise concerns about the feasibility of this approach. Third, multiple temperature level furnace brazing

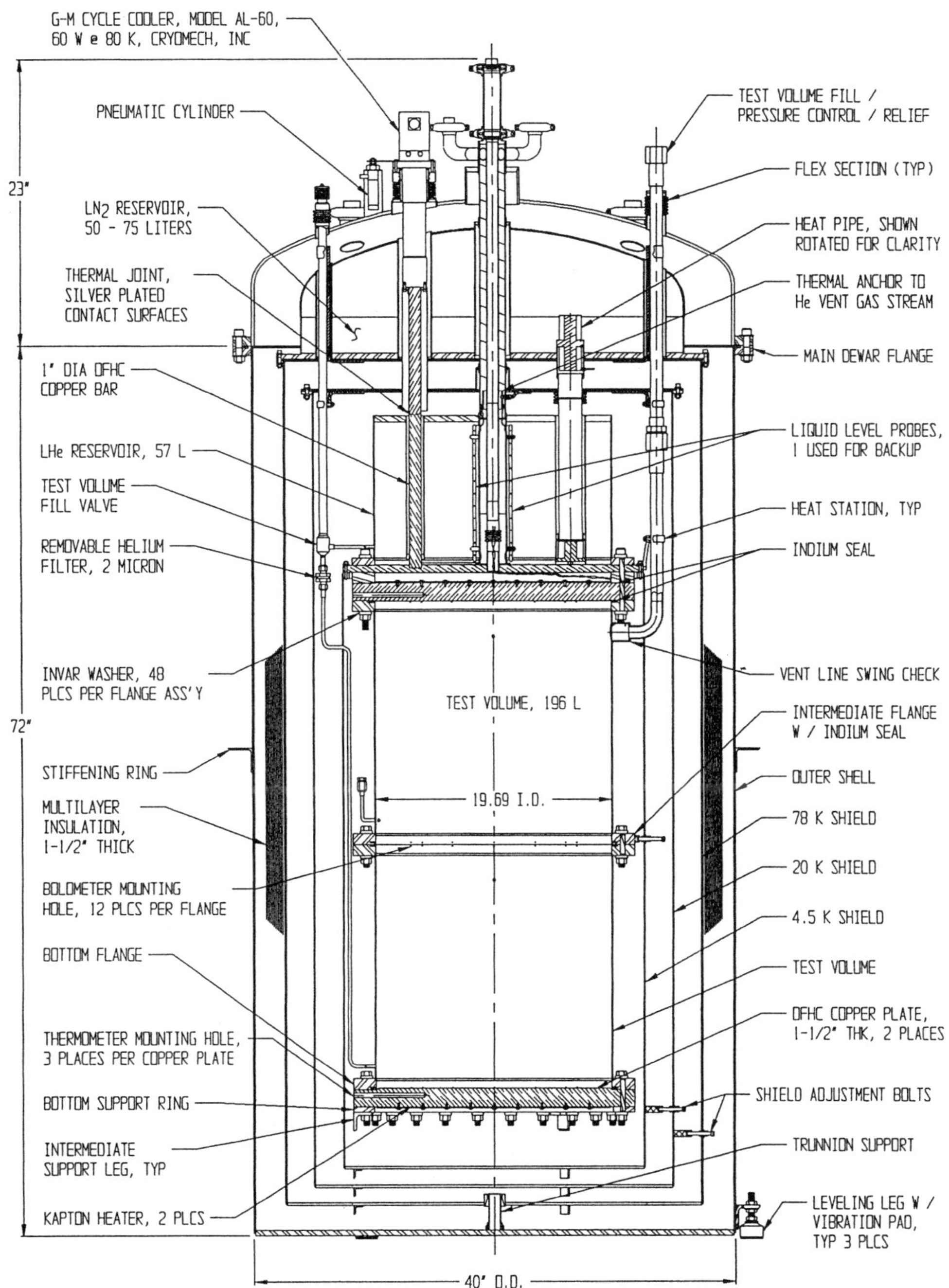

Figure 1. Cryostat assemby.

was employed for internal attachments in the helium and nitrogen reservoirs. Finally, there were questions about whether or not the thermal link between the G-M cooler and the helium reservoir would make good contact for cooldown and then separate cleanly when the helium reservoir reached 30 K.

The indium seal problem was solved by careful machining to get flat surfaces and by proper, sequential bolt tightening procedures. The difficulty of making up these joints was eased by sizing the clamp bolts for a maximum of 0.0127 mm (0.0005 inch) elongation at maximum pressure and by installing Invar washers which cause the joints to tighten at low temperature. A photograph of the cell flange assembly is shown as Figure 2.

Concerns about furnace brazing were valid and several costly re-heats were required. It appeared that the experience of the vendor was inadequate for the large, heavy components involved. Brazing copper to 304 stainless steel without flux in a hydrogen atmosphere was particularly unsuccessful. Good brazed joints were only achieved when the vendor used conventional brazing flux. The concept of using sequential melting temperature brazing materials is a recognized technique which is highly dependent on the time and temperature control of the furnace. Aside from the flux problem, it appeared that most of our brazing difficulties were caused by not holding the proper brazing temperature for long enough to achieve a thorough thermal "soak." This seemed to be particularly true for the low thermal diffusivity 304 stainless steel. A vendor qualification program would have been helpful in this case.

The auxiliary liquid helium reservoir cooldown system consists of an external G-M cooler, an OFHC copper bar to conduct heat down to the reservoir, a conical thermal contact joint attached to the reservoir copper bottom plate, and a spring-loaded air cylinder to drive the thermal link into contact. Heat transfer across the contact joint was a major concern. Material was needed which would be a thermal improvement over copper to copper and would separate cleanly when the thermal link was withdrawn. A conservative approach to this dilemma consisted of plating both surfaces of the conical joint with silver and then lapping them together for maximum contact. This assured less than optimum heat transfer but reduced the risk of a thermal short. Actual performance reflected this compromise design in that cooldown from 80 K was slow but 30 K was reached and no thermal short was observed.

OPERATIONAL EXPERIENCE

Initial Cooldown Procedures

Approximately 2 weeks were required for cooling down the apparatus. This relatively long time was primarily due to our attempts to avoid unnecessary stressing of the indium O-ring joints sealing the top plate to both the cell wall and the soft vacuum space above it. It is premature to comment on an optimized and regular cooldown method because the apparatus has only been cooled down twice. The following account gives a general idea of the steps involved. During the first few days, the LN2 reservoir was filled and the apparatus allowed to cool gradually by radiation and via the heat pipe connection between the LN2 and helium reservoirs. During this time, the common vacuum space was pumped using a small turbo pump (Varian) attached directly to a flange on top of the cryostat. We would subsequently allow liquid nitrogen into the "soft vacuum" space to initiate faster cooldown of the inner cell, which initially contained about an atmosphere of helium at room temperature. The liquid nitrogen would make direct contact with the upper copper plate of the cell and when the temperature of that plate dropped to about 80 K the liquid was pumped away. After the liquid was completely gone, the G-M cycle cryocooler was started in order to reduce the temperatures measured in the cell down to the range of 30 to 50K depending on location of the sensors. This method of cooling required from several days to a week but resulted in a considerable saving in liquid helium for each cooldown. The long cryocooler operating period was, as mentioned above, partly because of the quality of contact between the cryocooler and the helium reservoir, which was by a 1-inch copper rod tapered on its end to match a mating receptacle on the reservoir. The force of contact was manually re-adjusted as the apparatus cooled to account for contraction. Finally, the cryocooler was stopped and the copper rod retracted from the reservoir. Then, liquid helium was transferred to the apparatus very slowly. Nearly eight hours elapsed in cooling the reservoir from about 30 to 4 K. The transfer rate was then increased to accumulate liquid in the reservoir, which typically would require another 0.5-1 hour. Proper data were not obtained for another several days as temperatures stabilized in the apparatus.

Some Problems with the Cooldown

The first cooldown of the apparatus was entirely successful and it remained cold and running for several months. However, when very high heat fluxes were applied in an attempt to obtain the highest Rayleigh numbers, an O-ring at the top of the helium vent line froze up and caused air to leak into the soft vacuum space. The resultant ice ball which formed at the top of the cryostat strained and broke the top plate heater leads which made it necessary to terminate the experimental run. This problem was eliminated by replacing the Buna-N O-ring with an indium seal.

During the second cooldown, liquid nitrogen was admitted directly into the helium reservoir. This appeared to cause a large thermal gradient vertically across the upper copper plate. This plate is sealed by indium O-rings on both the top and bottom. The upper O-ring seals the soft vacuum space and the lower one seals the plate to the convection cell. Because of limited circumferential flange space, only a single set of bolts is available to apply the necessary clamping pressure. The resultant thermal stress caused a leak to open in the top indium O-ring seal during the cooldown and it was necessary to stop and warm up the cryostat again. After warm-up, the indium seals were remade with a slight difference: this time no vacuum grease was applied to the top O-ring (a practice designed to facilitate removal and replacement of the indium).

During the subsequent cooldown, LN_2 was allowed into the soft vacuum, but not into the helium reservoir, and no vacuum leaks developed. However, the cryostat was not without problems, as the cryogenic valve connecting the helium reservoir and the cell did not fully close, presumably due to frozen air or water on the valve seat. This eliminated low density operations, because the leak rate was equivalent to about 0.25 liters per hour. Plans are to install a heater on this valve in the future to enable de-icing the valve without a full cryostat warm-up. Installing a bellows sealed actuator at the warm end is also being considered because part of the problem with this valve may be due to leakage of the Buna-N O-rings used to seal the extended shaft at room temperature.

Operation of the Apparatus

With no applied heat load, the helium consumption of the apparatus was very small-only about 10 liters per day. With modest heating of the sample, this did not change significantly, and the bulk of the data were taken in this economical mode. At low heat loads, temperature stability of $\pm$ 0.0004 K could be maintained for 24 hours or more without adjustments. On the other hand, at very high heat loads consumption rates exceeded 100 liters per day. These rates corresponded to operating temperatures well above the critical temperature of helium, up to a maximum of 6K, and bottom plate heating of 1 Watt or more.

Sufficient helium gas was allowed into the soft vacuum space to provide a variable and uniform heat exchange between the reservoir and the top plate. This uniformity was especially important at high heat flux, where severe horizontal thermal gradients could have been established in the top copper plate without it.

The thermometry consists of semi-conductor temperature sensors, both standard encapsulated germanium thermometers as well as very small (side: 0.2mm) cubes of neutron-transmutation- doped (NTD) germanium used for looking at fluctuation spectra of the temperature field. The latter were placed in the gas at various locations and were chosen to have a high sensitivity near 5K. The neutron-transmutation method of doping provided extremely uniform temperature sensitivity compared with standard diffusion doping techniques. Their placement in the cell was as follows: four pairs spaced a few mm apart vertically and 90 degrees apart azimuthally, were placed near the sidewall. These were used to measure any large scale mean velocity in the cell by the time-correlation of two adjacent bolometers. An additional 5 bolometers were placed in the cell center along orthogonal axes.

The standard germanium thermometers were embedded in both the upper and lower copper plates. Nondimensional heat transport (Nusselt number) was calculated based on temperature measurements across the cell plus the power dissipated by the bottom plate heater. Some corrections had to be made before calculating the heat transfer, but were generally small. These include the calculated sidewall conductance which was subtracted from the measured conductance and the adiabatic temperature gradient resulting from the large absolute height of the cell. This gradient was subtracted from the measured temperature difference before Nusselt or Rayleigh numbers were calculated. Also, a minor correction for the temperature drop due to finite conductivity of the copper plates was required at the highest Rayleigh numbers.

Some Results

The main goal of the experiment is to investigate the nature of turbulent heat transfer under well-controlled conditions. The initial experiment has an aspect ratio of 0.5 as defined by the ratio of diameter to height of the working region. This was chosen to duplicate conditions used by other investigators to maximize the Rayleigh number Ra. So far, heat transfer data at constant Prandtl number (≈ 0.7) up to $Ra \sim 10^{15}$ has been obtained. This is an order of magnitude higher than any similar attempts. Properties of turbulent convection at even higher Ra values are being investigated currently but not at constant Prandtl number. Preliminary data analysis indicates that a single scaling law applies to the measured heat transfer in turbulent convection over the entire range studied. However, the measured power-law scaling exponent is between 5-10% less than the value of 1/3 predicted by simple models.

The temperature spectra obtained with the small NTD bolometers show also a power law scaling and are the subject of our current investigation. These same bolometers have been used to investigate the occurrence of a mean flow in the cell. This is done by monitoring the temperature fluctuations of two nearby bolometers and shifting the time axis of one or the other signals to minimize the difference. The delay time so calculated can be combined with the known separation of the bolometers to produce the apparent velocity. These experiments have shown that there are many "large" events in the temperature signal which can probably be identified as thermal plumes. It has been preliminarily observed that the "sense" of these events, whether they are cold or hot spikes, determines the sign of the calculated velocity. Steady mean flow has not been observed. Instead, the experimental flows are probably detached plumes which have velocities of approximately 10 cm/s.

One issue in turbulent heat transfer relates to the effect of surface roughness. It is generally observed that turbulent heat transport is enhanced when the roughness elements

Fig. 2 Convection Cell Assembly

approximate the thickness of the thermal boundary layers. These normally diffusive boundary layers form on the upper and lower boundaries of the cell and nearly all the temperature drop across the cell occurs across them. The boundary layers are typically on the order of 0.1 mm in thickness compared to the overall cell height of 1 m. The surface roughness of the heat transfer plates has been checked and they have a standard machine finish that is better than 10 microns.

Future plans include systematic investigation of the effects of surface roughness on turbulent convection. Experiments with higher aspect ratios of the convection cell will also be performed in order to obtain data more characteristic of existing theoretical treatments. These experiments will not allow higher Rayleigh numbers to be obtained.

ACKNOWLEDGEMENT

This work was supported by National Science Foundation Grant DMR9529609.

C. Heiden, (University of Giessen), receives the Honorable Mention Best Application Paper award from K. Timmerhaus (University of Colorado).

DETERMINATION OF THERMAL PROPERTIES OF SOLID POROUS MEDIA ON A SINGLE-BLOW TESTING APPARATUS

A. Frydman and J. A. Barclay

Cryofuel Systems Group
University of Victoria
Victoria, BC V8W 3P6, Canada

ABSTRACT

The efficiency of regenerative cycle refrigerators greatly depends upon the high performance of the regenerator. The design of high performance cryogenic regenerators is a challenge because several variables must be competitively balanced. Using published correlations, it is relatively easy to design extremely high performance regenerators but there are numerous manufacturing realities that limit observed regenerator performance. A methodology is presented to evaluate solid porous matrices that differ in nature and geometry, by balancing the effective thermal conductivity, heat transfer coefficient and friction factor. We have designed and built a single-blow testing apparatus to measure the design parameters for different porous solid matrices, by flowing a heated fluid through it and recording the temperature profiles as a function of time and location, for different values of Reynolds Number.[1] The determination of these parameters is obtained by fitting the experimental profiles with theoretical profiles from the solution of the energy balance equations in the fluid-solid system. Here, we present the results obtained on Dysprosium sphere particles with average particle size of 288 μm, as an initial demonstration of a characterization and testing methodology for high performance regenerator matrices. Important considerations with respect to the longitudinal conductance are also made.

INTRODUCTION

A regenerator is a compact periodic heat exchanger in which the fluid is in direct contact with the solid heat transfer area. It transfers heat to and from the same fluid in a periodic fashion. This is accomplished by an intermediate heat transfer step from the fluid to a solid matrix material. By reversing the flow direction of the fluid, heat is transferred from the matrix back to the fluid. Therefore, thermal energy is temporarily stored and used to heat or cool a fluid flowing through the matrix. Excellent reviews on regenerator theory development and analysis can be found in recent literature.[2,3]

Advances in Cryogenic Engineering, Volume 45.
Edited by Shu *et al.*, Kluwer Academic / Plenum Publishers, 2000.

The effectiveness of a high performance cryogenic regenerator depends upon several factors. An ideal regenerator has: infinite heat transfer area, no pressure drop in the fluid, no axial heat conduction, no void volume in the matrix, infinite thermal mass, infinite radial heat conduction, and outstanding mechanical integrity. Real regenerators have: finite heat transfer, appreciable pressure drop in the fluid flow, heat conduction in the axial direction, significant porosity, temperature dependent thermal mass, as well as the other features that are not ideal. By quantifying these properties, the effectiveness of the regenerator can be determined, as compared to an ideal regenerator. From a design point of view, it is important to determine the heat transfer coefficient, the effective thermal conductivity of the regenerator matrix, and the friction factor of the fluid flowing through the regenerator matrix (a function of the pressure drop).

Based on local entropy generation analysis[4], we have developed a methodology of regenerator design applicable to the practical use of regenerative cycles such as Stirling, Gifford-McMahon, or Regenerative Magnetic Cycles.[5,6] This methodology produces optimized designs for overall high thermal effectiveness, but it requires precise input parameters for the values of thermal properties and friction factor for a specific manufactured regenerator matrix.

Several experimental techniques have been reported to evaluate regenerators.[7] Among those, the single blow technique is known for its simple operation and fast response. The design of such an apparatus becomes a challenging project when the various sources of systematic error are meticulously tracked down.[1] The experimental validation of manufactured regenerators with extremely high performance is highly recommended in the initial stages of regenerator development to evaluate the performance of promising new geometries.

In the present work, we report results in the development of a procedure to determine heat transfer coefficient, effective thermal conductivity and friction factor of porous solid matrices based on modeling and experimental measurement of temperature profiles and pressure drop, directly and simultaneously, on a single-blow regenerator testing apparatus. We used Dy sphere particles in average particle size = 288 μm, an important refrigerant material, to evaluate the modeling technique, and identify sources of systematic error.

MODEL

The traditional methods for regenerator testing fall into two general types; integral and differential. The integral methods test a regenerator under realistic refrigerator conditions[7] and provide integral effectiveness data. They do not provide information about specific parameters of the regenerator. The differential methods are further subdivided into periodic and single-blow.[9,10] The periodic might be considered a variation of the single blow mode. In the differential method, the regenerator is first brought to thermal equilibrium with a fluid flowing at constant temperature. Once equilibrium is accomplished, the inlet temperature of the fluid is instantaneously raised in a step function mode. Loehrke emphasizes the importance of the inlet temperature history, because the effectiveness of the regenerator matrix might be underestimated if the generation of the initial temperature step is inefficient.[11] The method consists in measuring the temperature at various locations along the solid matrix with a fluid flowing through. The resultant sudden change in fluid temperature is spread out by finite heat transfer and effective longitudinal thermal conduction. Measurement of the temperature wave as a function of time provides information on thermal conductance and effective longitudinal conduction.

Analysis of differential method results has been done by maximum slope[10], average slope[12], or full curve matching[13] methods. Today's computer power allows the full curve matching method to be utilized without the need of obtaining analytical solutions for the energy balance equations governing the heat transfer between the solid and the fluid. The curves created by the energy balance should fit the experimental curves by varying the parameters in the equations. The best curve fit will determine the parameters that yield the best representation of the experimental curves. At this point, we must define the energy balance, the assumptions and the boundary conditions involved in the problem.

- Energy balance for the solid:

$$\begin{bmatrix}\text{Heat absorbed by}\\ \text{the solid or internal}\\ \text{energy change}\end{bmatrix} + \begin{bmatrix}\text{Heat transferred}\\ \text{through the solid}\\ \text{by conduction}\end{bmatrix} = \begin{bmatrix}\text{Heat transferred}\\ \text{to the solid by}\\ \text{convection}\end{bmatrix}$$

$$\left[\frac{m_S}{L}\cdot c_S\cdot\left(\frac{\partial T_S}{\partial t}\right)\cdot dx\right]-\left[k_S\cdot(1-\varepsilon)\cdot A_{REG}\cdot\left(\frac{\partial^2 T_S}{\partial x^2}\right)\cdot dx\right]=\left[h\cdot\frac{A_S}{L}\cdot(T_F-T_S)\cdot dx\right], \quad (1)$$

where: m_S is the mass of the solid;
L is the length of the solid;
c_S is the specific heat of the solid;
k_S is the effective thermal conductivity of the solid;
ε is the porosity of the solid matrix;
A_{REG} is the cross sectional area of the regenerator matrix;
h is the heat transfer coefficient between the solid and the fluid;
A_S is the heat transfer surface area of the solid; and
T_F and T_S are the temperatures of the fluid and the solid, respectively.

- Energy balance for the fluid:

$$\begin{bmatrix}\text{Heat transferred}\\ \text{from the fluid by}\\ \text{convection}\end{bmatrix} + \begin{bmatrix}\text{Heat transferred}\\ \text{through the fluid}\\ \text{by conduction}\end{bmatrix} = \begin{bmatrix}\text{Heat transferred}\\ \text{from the solid by}\\ \text{convection}\end{bmatrix}$$

$$\left[\dot{m}_F\cdot c_P\cdot\left(\frac{\partial T_F}{\partial x}\right)\cdot dx\right]-\left[k_F\cdot\varepsilon\cdot A_{REG}\cdot\left(\frac{\partial^2 T_F}{\partial x^2}\right)\cdot dx\right]=\left[h\cdot\frac{A_S}{L}\cdot(T_S-T_F)\cdot dx\right], \quad (2)$$

where: $\dot{m}_F$ is the mass flow rate of the fluid;
c_P is the specific heat of the fluid; and
k_F is the effective thermal conductivity of the fluid.

The assumptions and boundary conditions are the same as in Pucci *et al*[10], except that here we consider non-zero thermal conductivity of the fluid. Because we operate at near room temperature and the change is small (~20 K), the properties of the solid and the fluid are considered to remain constant with temperature. Both thermal conductivities, of the solid and the fluid, are assumed to be infinite in the radial direction (perpendicular to fluid flow direction). Initially the solid matrix is at constant temperature. At time zero, the inlet temperature of the fluid changes in a step function mode to another constant value. Experimentally, this was accomplished by means of a four-way switching valve introduced to the apparatus.[1] Finally, in the best case, the apparatus is adiabatic but in reality, there are both thermal mass effects and heat leaks due to conduction. We have tried to reduce the systematic errors from such effects thoroughly in the design of our regenerator testing apparatus.[1]

A numerical method was used to solve unsteady state one-dimensional differential equations. The technique is based on discretization using finite volume for the solid equation, and the upwind scheme for the fluid equation.[14] A Fortran code was written to solve the resultant equations and predict the progress of the temperature wave through the regenerator for different value of the parameters. The problem consists on dividing the regenerator length in "n" nodes separated by the same distance increment Δx. The temperature of the solid and the fluid are calculated at each node as a function of time, i.e., at every time step Δt. The computational time becomes increasingly large depending on the magnitude of the time step, the mass flow of fluid, and the precision set for the convergence. The accuracy of the model was examined against other models for various values of h and k_S.[8,13] The values predicted from our model falls below the 3% error range in comparison to those models from the literature.[15]

We considered non-zero longitudinal conduction for the fluid and the solid, which is not the case in other reports. Assuming the fluid longitudinal conduction to be zero is only acceptable at low Reynolds numbers where the molecular thermal conductivity is dominant. In regimes of high Reynolds numbers, eddy diffusivity caused by fluid mixing will increase the effective longitudinal conduction, and therefore, cannot be neglected. This effect also depends on the geometry of the solid matrix. For instance, spherical particles will promote fluid mixing to a higher degree than parallel horizontal plates. In our model, the effective thermal conductivity of the bed, k_{bed}, is the sum of two terms: the effective thermal conductivity of the solid (or static thermal conductivity of the solid), k_S, and the effective thermal conductivity of the fluid, k_F. The latter comprises the static thermal conductivity of the fluid, $k_{F,static}$, and the eddy diffusivity term, k_{eddy}. Eq. (3) gives total rate of longitudinal heat conduction, approximately. One should not take the effective thermal conductivity of the solid, k_S, for the bulk thermal conductivity of the solid, k_{bulk}. The bulk thermal conductivity is much larger than the effective thermal conductivity because of the point contacts within the solid porous matrix. K.J. Caudle[15] presents a review on literature correlations for k_S which and is measured with the solid matrix stagnant (Re=0).

$$\dot{q} \approx \left(\underbrace{k_S + \overbrace{k_{F,static} + \underbrace{\rho_F \cdot c_P \cdot D_L}_{k_{eddy}}}^{k_F}}_{k_{bed}} \right) \cdot \frac{\partial T}{\partial x}, \tag{3}$$

where: ρ_F is the density of the fluid; and
D_L is the longitudinal dispersion coefficient.[16]

According to regenerator theory, the influence of the effective longitudinal conduction is to reduce the performance of the regenerator for a given number of heat transfer units (NTU).[3] Some of the requirements for a high effectiveness regenerators dictate small aspect ratio (length of matrix/diameter), and utilization of high thermal conductivity materials. This is another example of a "trade-off" situation in regenerator design.[5,6]

EXPERIMENTAL

As stated before, the testing apparatus was specifically designed to reduce the systematic errors due to intrinsic and parasitic heat leaks. The mechanical design details are presented in another paper in this conference.[1] Among the techniques used, we have applied a simple procedure to correct the measured temperature profiles to compensate for inherent heat leaks in the temperature sensor. This procedure involves a discussion on the thermal resistances present in the system which must be accounted for.[1] A general schematic of the testing apparatus is presented in Figure 1.

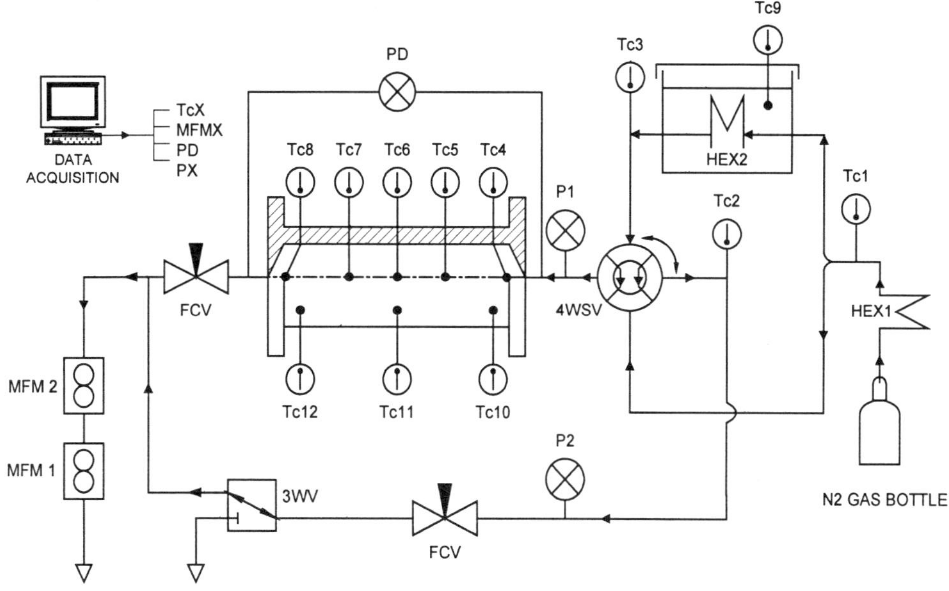

Figure 1. Schematic of the apparatus. TcX: thermocouples; HEX1: copper tube heat exchanger to ambient air; HEX2: copper tube heat exchanger to water bath; P1, P2: pressure transducers; PD: differential pressure transducer; 4WSV: four-way fluid switching valve; FCV: flow control valves; 3WV: three-way valve.

After making numerous improvements to the apparatus, the initial experiments reported were taken on Dy spheres. Table 1 shows some properties of Dysprosium (other materials are included for reference). The regenerator bed is 0.0254 m (1") I.D. and 0.0762 m (3") long. The bed was uniformly packed to 37% porosity. Testing was done by applying a temperature step of ~20 K above room temperature to a fluid, N_2, at ~0.69 MPa (100 psia). The range of N_2 mass flow rates was 0.4-4.0 g/s (Re = 13-133, calculated based on the hydraulic radius). The temperature step was generated by heating the fluid to a given temperature (in this case, ~317 K), bypassing the housing by means of a four-way, cross-flow valve. Simultaneously, this four-way valve allows a different stream of fluid at room temperature to flow through the housing. Once the hot stream of fluid has reached the desired temperature, it is simultaneously and suddenly switched into the housing as the cold fluid is switched from the housing. Five temperature probes (thermocouples type E, 36 AWG wires with PFA Teflon insulation) recorded the temperature at five specific locations along the housing as a function of time (location 0 - beginning, ¼ of L, ½ of L, ¾ of L, where L represents the total housing length). Signal from temperature probes, pressure transducers and mass flow meters was collected to NI-DAQ LabView 4.1 (Macintosh version).

Table 1. Bulk properties of some known regenerator materials

Material	Bulk density ρ_S (kg/m^3)	Bulk thermal conductivity k_S (W/m.K)	Specific heat c_S (J/kg.K)
Dysprosium	8550	10.7	167.4
Stainless steel 304	8030	16.3	502.1
Copper	8900	392.6	384.9

RESULTS AND DISCUSSION

According to our model, the temperature of the fluid and the solid will be calculated as a function of time and location. The experimental value of the temperature measured at a certain time and location inside the regenerator matrix is within the actual temperature of the fluid and the actual temperature of the solid. Therefore, the criteria we used to interpret the experimental temperature is an arithmetic average between the temperatures of the solid and fluid since we cannot determine their values with this experimental setup. This assumption is acceptable because our temperature span is narrow, ranging from near room temperature to ~317 K, and the range of number of heat transfer units is large (~200-400), so the difference between T_F and T_S is very small in these experiments.

An examination of Eqs (1) and (2) reveals three degrees of freedom and two equations, making this system of equations underdefined. In this case, we require a recursive methodology to solve the system. For instance, here, we assumed initial values for the unknowns, and used an objective-function to evaluate the values assumed for the unknowns. The objective-function ψ used is defined in Eq. (4) below. This function is evaluated at each temperature reading location along the solid matrix, and between the time interval of measurement from t_0 (initial time) to t_f (final time).

$$\psi = \sum_{t_0}^{t_f} \left(\frac{T_{exp} - T_{calc}}{T_{exp}} \right)^2 , \qquad (4)$$

where: T_{exp} is the experimental temperature as described above; and
T_{calc} is the temperature calculated by the model as an arithmetic average between the calculated temperatures of the fluid and solid;

The unknowns in this system are the heat transfer coefficient h, the effective thermal conductivity of the fluid k_F, and the effective thermal conductivity of the solid k_S. Because our model is not yet optimized to automatically search the best values of three unknowns simultaneously, for the sake of simplicity, we used literature correlations to determine the values of two unknowns: h and k_F. For spherical particles geometry, there are several correlations for h, and one of the most recommended is Handley and Heggs correlation.[17] As discussed above, k_F is ($k_{F,static}+k_{eddy}$). The value of $k_{F,static}$ is tabulated, and k_{eddy} is calculated using the correlation for D_L given by Edwards and Richardson.[16] Therefore, the problem becomes of one unknown only, i.e., k_S.

Figures 2 and 3 show examples of curve matching obtained for two different Re numbers. The curves represent temperature readings located according to the distribution mentioned in the Experimental section. Blank dots are calculated values, and filled dots are the corrected experimental values. The criteria for curve matching considered the temperature reading located inside the bed only. The topmost part of the calculated curves does not match the experimental curves in all cases. This is the region where heat transfer with the walls of the housing would influence (note that the model does not include the energy balance equation for the wall, and assumes an adiabatic system).

Figure 4 presents the values of k_S that resulted in the best curve fit. The average value of 3.26 ± 0.31 W/m.K is well below the solid Dy value, but well above the values predicted for some literature correlations for k_S.[18,19] However, values of stagnant thermal conductivity of porous solids might vary by a large degree depending on the nature of the stagnant fluid[20], nature of the solid, irregular particle shapes, and compaction of the bed. Figure 5 demonstrates that flow channeling did not occur in our measurements because our experimental friction factors closely match the values predicted from literature.[21]

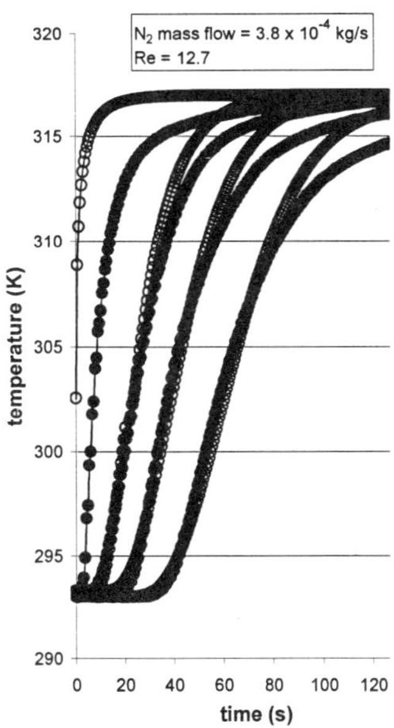

Figure 2. Example of the curve matching procedure at Re = 12.7 (based on hydraulic radius). Filled dots are the corrected experimental data points, and blank dots are the calculated. The first two profiles (blank and filled dots) were taken at the entrance of the bed, and were not used in the curve matching procedure.

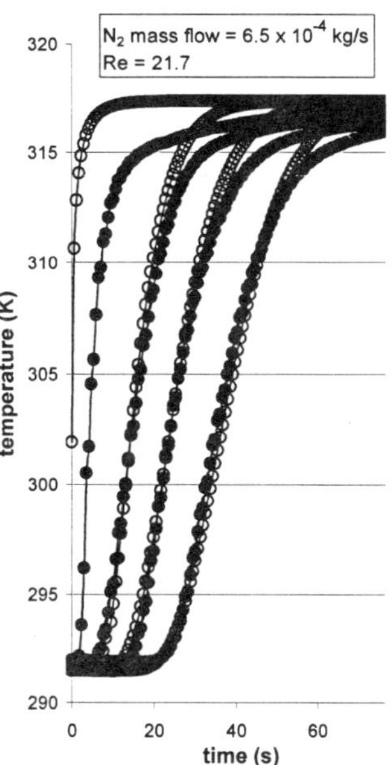

Figure 3. Example of the curve matching procedure at Re = 21.7 (based on hydraulic radius). Filled dots are the corrected experimental data points, and blank dots are the calculated. The first two profiles (blank and filled dots) were taken at the entrance of the bed, and were not used in the curve matching procedure.

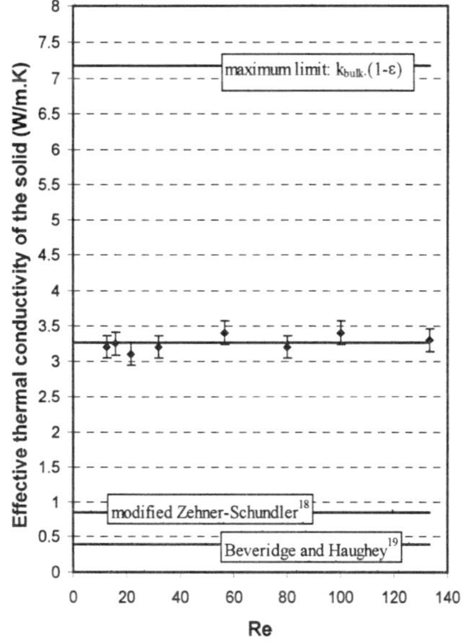

Figure 4. Effective thermal conductivity of the solid as fitted by the model (data points) compared to literature correlations, as a function of Re (based on hydraulic radius). Data points represent the values of k_S obtained from the best curve fit of three profiles of calculated-experimental temperatures (e.g.,Figs. 2/3).

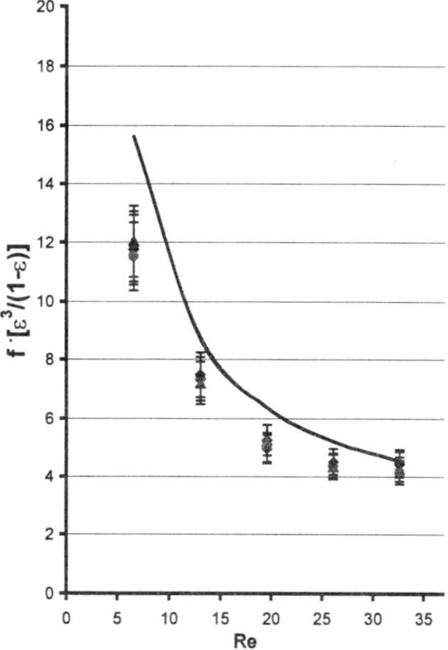

Figure 5. Friction factor measured for the packed bed of 288 μm Dy spheres (data points taken at 0.35, 0.69, 1.05, and 1.38 MPa, or 50, 100, 150 and 200 psia, respectively) as a function of Re (based on hydraulic radius). The line represents the modified Ergun correlation.[21]

The model and the curve matching procedure proved consistent because the value of k_S obtained was practically constant as we expected. However, other means to verify the validity of this value of k_S determined for a packed bed of Dy particles must be utilized. An independent experimental determination of the static effective thermal conductivity of these particles seems the most appropriate basis for comparison, rather than literature correlations. The matching procedure can be extended to fit three unknowns. The value of k_S determined from an independent technique would eliminate one unknown from the problem, thus remaining two to curve fit. An energy balance that considers the housing wall will improve the quality of the curve fit. Regardless, an automatic least squares procedure to evaluate the objective-function must be included as well. The initial results here presented were extremely useful in identifying advantages and problems with the proposed technique to determine thermal properties of porous solid media.

ACKNOWLEDGEMENT

The authors gratefully thank the financial support by Natural Resources Canada, Canadian Natural Sciences and Engineering Research Council, and Centra Gas BC, Inc., a subsidiary of WestCoast Energy, Inc.

REFERENCES

1. P. G. Reedeker, A. Frydman, and J. A. Barclay. A Laboratory Apparatus to Determine Fluid-Solid Heat Transfer Coefficient and Effective Thermal Conductivity of Solid Porous Media, "Proc. of the 1999 Cryogenic Eng. Conference", Montreal (1999).
2. A.J. Willmott. The Development of Thermal Regenerator Theory from 1931 to the Present, *J. of the Institute of Energy*, 66:54 (1993).
3. R.A. Ackerman. "Cryogenic Regenerative Heat Exchangers", Plenum Press, N.Y. (1997).
4. A. Bejan. "Entropy Generation Minimization", CRC Press, Boca Raton, FL (1996).
5. J.A. Barclay, and S. Sarangi. Selection of Regenerator Geometry for Magnetic Refrigerator Applications, *Cryogenic Processes and Equipment*, p. 51 (1984).
6. A. Frydman, and J.A. Barclay. Using Entropy Analysis to Design Regenerators for Active Magnetic Regenerative Cycle Refrigerators, "Proc. of 9th Canadian Hydrogen Conference", Vancouver, BC (1999).
7. K. Kratschmar. "The Development of a Cryogenic Regenerator Test Apparatus", M.Sc. Thesis, University of Victoria, Victoria, BC (1995).
8. P.J. Heggs, Experimental Techniques and Correlations for Heat Exchanger Surfaces: Packed Beds, "Low Reynolds Number Flow Heat Exchangers", Hemisphere Publishing Corp, p. 341 (1983).
9. J.H. Stang. "The Periodic Technique for Testing Compact Heat Exchanger Surfaces", M.Sc. Thesis, Marquette University, Milwaukee, WI (1968).
10. P.F. Pucci, C.P. Howard, and C.H. Piersall, Jr., The Single Blow Transient Testing Technique for Compact Heat Exchanger Surfaces, *J. of Engineering Power*, p. 29 (1967).
11. R.I Loehrke. Evaluating the Results of the Single Blow Transient Heat Exchanger Test, *Experimental Thermal and Fluid Science*, 3:574 (1990).
12. J.A. Barclay, W.C. Overton, Jr., W.F. Stewart, and S. Sarangi. Experiment to Determine Properties of Packed Particle Beds and Regenerators at Cryogenic Temperatures, "Advances in Cryogenic Engineering"., Plenum Publishing Co., 29:605 (1984).
13. M.P. Elliott, and C.W. Rapley. The Effect of Solid Conduction on the Single Blow Experimental Method, II UK National Conference on Heat Transfer, p.1623 (1988).
14. S.V. Patankar. "Numerical Heat Transfer and Fluid Flow", Hemisphere Publishing. Corp., N.Y. (1980).
15. K.J. Caudle. "Characterization of Regenerator Geometries Through Modeling and Experimentation", M.Sc. Thesis, University of Victoria, Victoria, BC (1997).
16. M.F. Edwards, and J.F. Richardson. Gas Dispersion in Packed Beds, *Chem.Eng.Sci.*, 23:109 (1968).
17. D. Handley, and P.J. Heggs. Momentum and Heat Transfer Mechanisms in Regular Shaped Packing, *Trans.IchemE*, 46:T251 (1968).
18. C.T. Hsu, P. Cheng, and K.W. Wong. *Int.Journal of Heat and Mass Transfer*, 37(17):2751 (1987).
19. G.S.G. Beveridge, and D.P. Hauguey. *Int.Journal of Heat and Mass Transfer*, 14:2751 (1971).
20. A.B. Duncan, G.P. Peterson, and L.S. Fletcher. *ASME Publication HTD*, 104(3):77 (1988).
21. I.F. Macdonald, M.S. El-Sayed, K. Mow, and F.A.L. Dullien. *Ind.Eng.Chem.Fundam.*, 18(3):199 (1979).

CRYOGENIC FUEL TANK DRAINING ANALYSIS MODEL

Donald Greer

Research Engineer, Fluid Dynamics
NASA Dryden Flight Research Center
Edwards, California 93523

ABSTRACT

One of the technological challenges in designing advanced hypersonic aircraft and the next generation of spacecraft is developing reusable flight-weight cryogenic fuel tanks. As an aid in the design and analysis of these cryogenic tanks, a computational fluid dynamics (CFD) model has been developed specifically for the analysis of flow in a cryogenic fuel tank. This model employs the full set of Navier-Stokes equations, except that viscous dissipation is neglected in the energy equation. An explicit finite difference technique in two-dimensional generalized coordinates, approximated to second-order accuracy in both space and time is used. The stiffness resulting from the low Mach number is resolved by using artificial compressibility. The model simulates the transient, two-dimensional draining of a fuel tank cross section. To calculate the slosh wave dynamics the interface between the ullage gas and liquid fuel is modeled as a free surface. Then, experimental data for free convection inside a horizontal cylinder are compared with model results. Finally, cryogenic tank draining calculations are performed with three different wall heat fluxes to demonstrate the effect of wall heat flux on the internal tank flow field.

NOMENCLATURE

CFD	computational fluid dynamics
He/LH2	helium and liquid hydrogen
K	kelvin
J/m^2-s	joules per meter squared per second
m	meter
m/s	meters per second
q	speed, meters per second

q_i	initial speed
T	temperature, K
Tavg	average temperature, K
T_i	initial temperature, K
u, v	velocity, meters per second
ρ	density, kilograms per meter cubed
ρ_i	initial density, kilograms per meter cubed

INTRODUCTION

Developing analytical models for thermodynamic predictions of cryogenic fuel tanks has not received a large amount of attention. However, a number of people have done work in this field. Grayson and Navickas[1] performed an axisymmetric Navier-Stokes analysis of a liquid hydrogen propellant tank. Their analysis treated the cryogenic liquid-pressurization gas interface as a free boundary and therefore the ullage region of the tank was not modeled. Their results show that sloshing is an important factor in predicting tank thermal gradients. Zhou and Graebel[2] developed an axisymmetric model for a cylindrical tank draining process. They used a boundary integral method and assumed the fluid was incompressible and inviscid. Heat transfer was not considered, as they were interested in the free surface motion near the tank drain hole. Several thermodynamic models[3,4,5] have been developed for analyzing cryogenic tanks. These models are based upon quasi-steady-state solutions of the first law of thermodynamics.

The computational fluid dynamic (CFD) model presented herein simulates the two-dimensional cross section of a cryogenic tank draining process. The CFD model is developed in generalized curvilinear body-fitted coordinates to allow for a variety of cross-sectional shapes. The time-dependent Navier-Stokes equations are modeled for both the cryogenic liquid and the pressurization gas. The interface between the pressurization gas and cryogenic liquid is modeled as a free surface to enable the prediction of slosh wave dynamics. The energy transfer across the interface is calculated, but mass transfer is neglected. To reduce stiffness and decrease the computational time the method of artificial compressibility developed by Chorin[6] is used. The development of the finite difference equations utilized a volume integral procedure as discussed by Lick[7]. The mathematical details of the algorithms development are presented in the Greer reference[8]. Use of trade names or names of manufacturers in this document does not constitute an official endorsement of such products or manufacturers, either expressed or implied, by the National Aeronautics and Space Administration.

MODEL COMPARISON TO EXPERIMENTAL DATA

A series of calculations were made to model the free convection process inside a horizontal cylinder. Experimental data from Martini and Churchill[9,10] is available for comparison. Briefly, the experiment of Martini and Churchill was for a horizontal cylinder which was 1 meter long and had an inside diameter of 0.1 meter. Each hemispherical side of the cylinder was held at constant but separate temperatures using a dual-sided water bath. After steady state was reached, temperature and velocity measurements were made at the midspan of the cylinder. The fluid inside the cylinder was air.

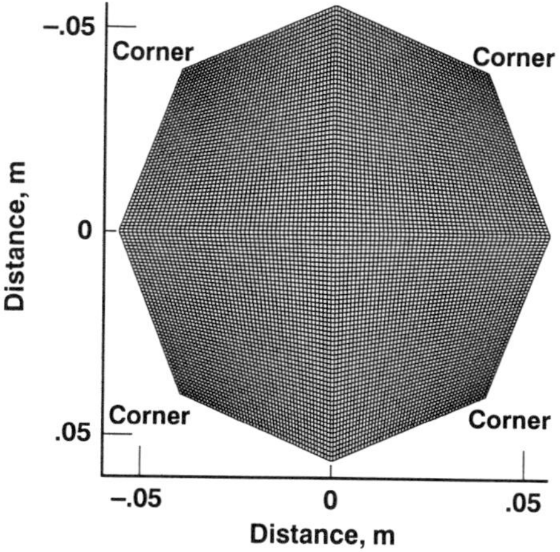

Figure 1. An 8-sided polygon for free convection inside a horizontal cylinder, 100 by 100 node grid.

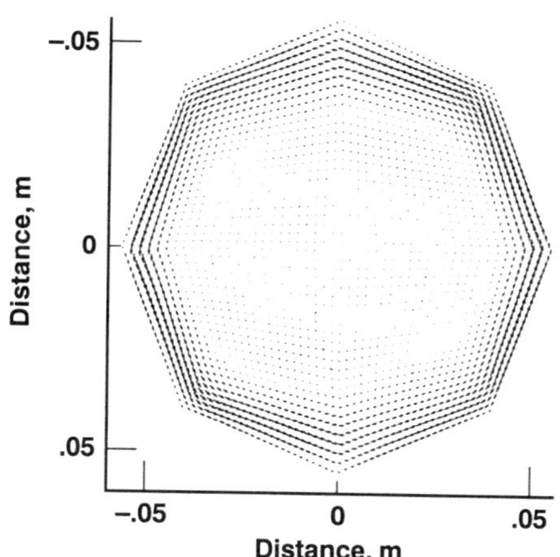

Figure 2. Velocity vectors at steady state for free convection inside a horizontal cylinder, vector skip index of 2.

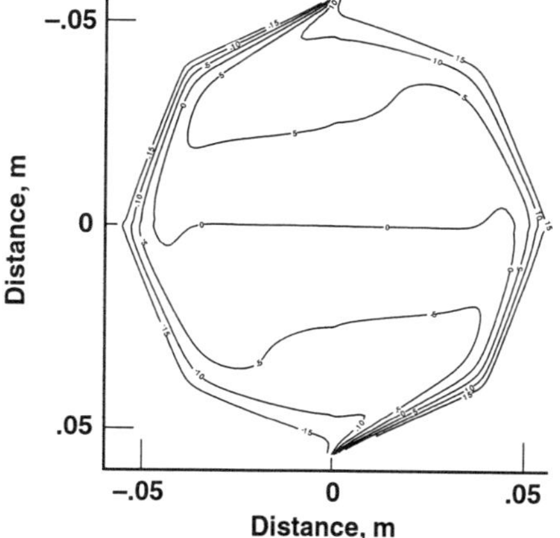

Figure 3. Temperature contours, T-Tavg, at steady state for free convection inside a horizontal cylinder.

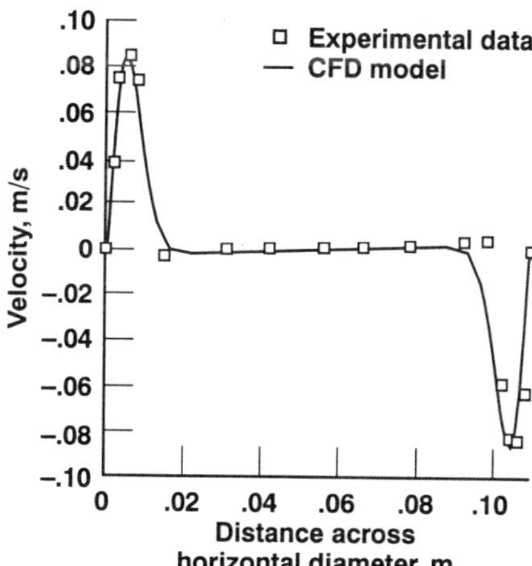

Figure 4. Comparison of velocity profiles across the horizontal diameter to experimental data of Martini and Churchill. Steady state free convection inside a horizontal cylinder.

An 8-sided polygon was used to simulate the experiment as shown in Figure 1. Calculations were performed for a Reynolds cell number of 5 (100 by 100 node grid), Courant-Friedrich-Levy number of 0.2, and Mach scaling factor of 250 (0.05 Mach number). Figures 2 and 3 present the velocity vectors and temperature contours. These figures show that the free convective flow travels in a boundary layer at the cylinder wall. The interior of the cylinder is relatively motionless. This is exactly what Martini and Churchill observed in their experiments. They concluded that the buoyancy forces were stronger than the viscous forces in the interior region. The temperature contours show that the fluid stratifies in the interior, which was experimentally observed. Figures 4 and 5 compare the velocity and temperature boundary layer profiles to the experimental data. The characteristics of the flow field between the model and the experimental data are in very good agreement.

FUEL TANK ANALYSIS

An 8-sided polygon was chosen as representative of a circular tank. (A circle can not be transformed into generalized coordinates because of the singular points at the corners.) The boundary and initial conditions are shown in Figure 6. The geometry for the 8-sided polygon is the same for all calculations. The drain calculations begin with the tank 70-percent full and the calculations are terminated when the tank is 30-percent full as was done for the rectangular tank. The initial and final grid geometry are shown in Figures 7 and 8. The fluids used are helium and liquid hydrogen. Calculations were performed for a Reynolds cell number of 8.2 (40 by 40 node grid), Courant-Friedrich-Levy number of 0.1, and Mach scaling factor of 100 (0.01 Mach number). The CPU time was 16 hours for each drain calculation performed on a Sun Ultra computer with a clock speed of 300MHz.

Figures 9 through 14 present velocity vectors of drain calculations for wall heat fluxes of 0, 1, and 2 J/m^2-s. The addition of heat flux at the wall causes a pair of convection vortices to form in the gaseous helium. As shown in the figures, the heat flux increases the helium flow velocities. The maximum flow velocities at the vertical side wall is 0.0017 m/s for q=0 J/m^2-s, 0.0035 m/s for q=1 J/m^2-s, and 0.0051 m/s for q=2 J/m^2-s. Figures 15 and 16 show temperature contours for the two cases with heat flux. The temperatures are symmetric and increase with higher heat flux. A secondary flow pattern develops in the gas region at the interface (near the center) for the q=2 J/m^2-s case (fig 14). This secondary flow pattern is a result of increased flow velocity inside the tank. It is caused by the wall heat flux and the geometry of decreasing tank width in the vicinity of the interface. Calculations with a rectangular geometry revealed no secondary flow patterns.

CONCLUDING REMARKS

A computational fluid dynamics model was developed to support the design, test, and analysis of cryogenic fuel tanks. This model uses a time-dependent finite difference technique in generalized coordinates. The finite difference algorithms were developed utilizing a volume integral method. To allow for the prediction of slosh wave dynamics, the interface between the liquid and the gas was modeled as a free surface. Artificial compressibility was used to decrease computational times. The model data compared well to experimental data for free convection inside a horizontal cylinder. A tank draining analysis was performed on an 8-sided polygon. Wall heat flux was found to be significant in the ullage gas, while it was insignificant in the liquid. This result is a function of the tank analysis parameters (drain rate, wall heat flux, etc.) and is not a general result for cryogenic fuel tanks. A secondary flow pattern was found to develop in the ullage gas for a large wall heat flux.

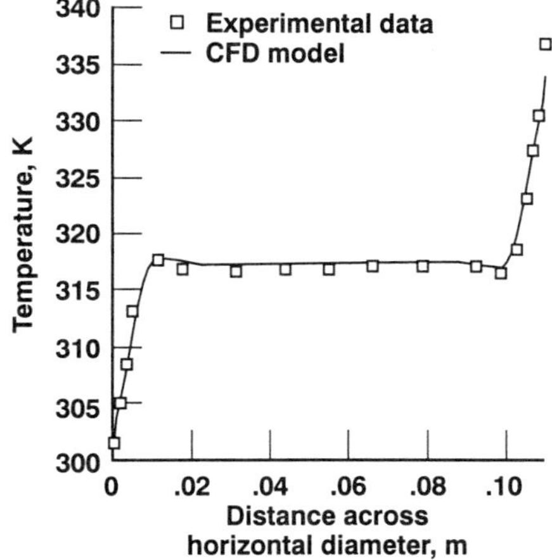

Figure 5. Comparison of temperature profiles across the horizontal diameter to experimental data of Martini and Churchill. Steady state free convection inside a horizontal cylinder.

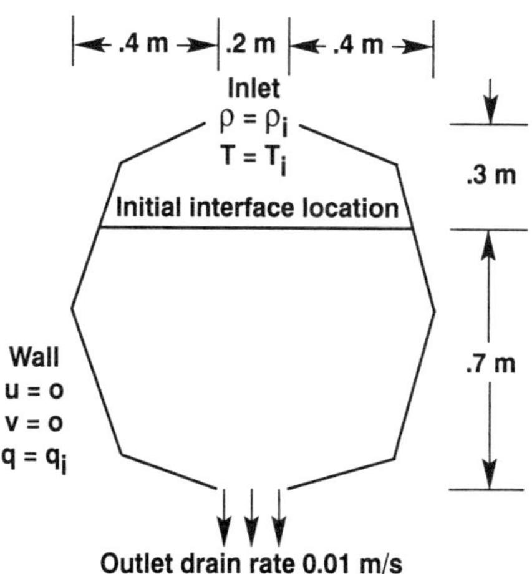

Figure 6. Geometry, boundary, and initial conditions for 8-sided polygon tank analysis. Tank is symmetric and 1 m by 1 m.

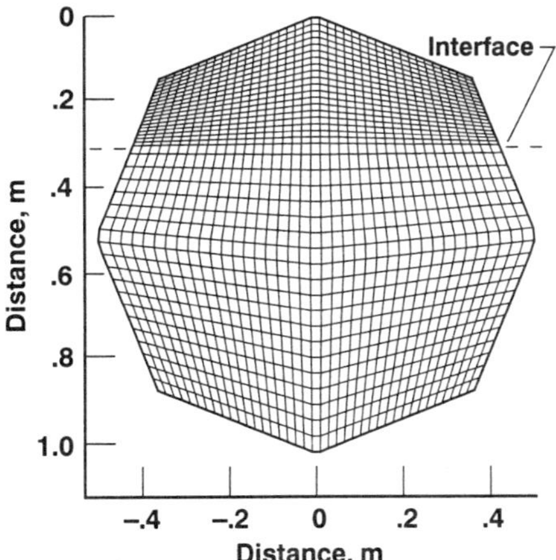

Figure 7. Initial 8-sided polygon grid. Drain time = 0 seconds.

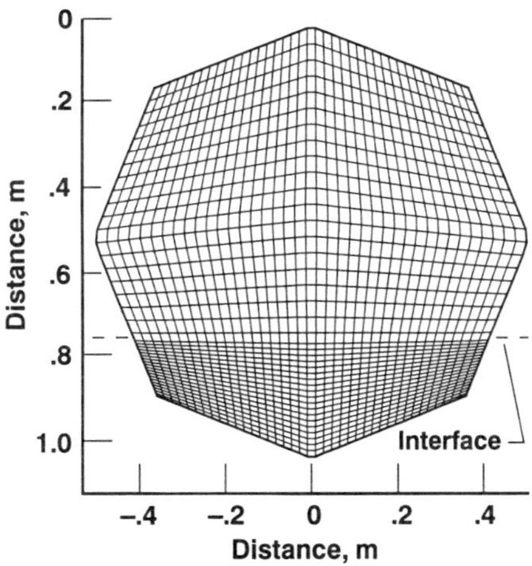

Figure 8. Final 8-sided polygon grid. Drain time = 300 seconds.

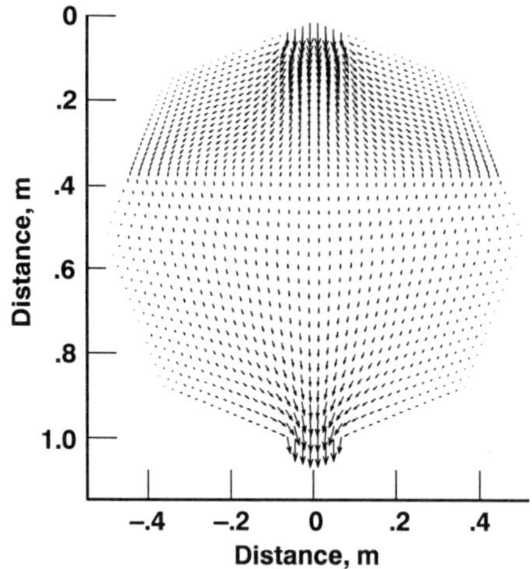

Figure 9. Velocity vectors at 50 seconds. He/LH2, q=0 J/m2-s.

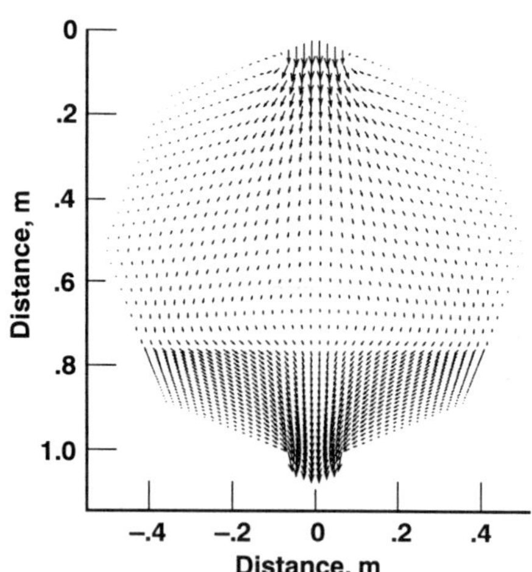

Figure 10. Velocity vectors at 300 seconds. He/LH2, q=0 J/m2-s.

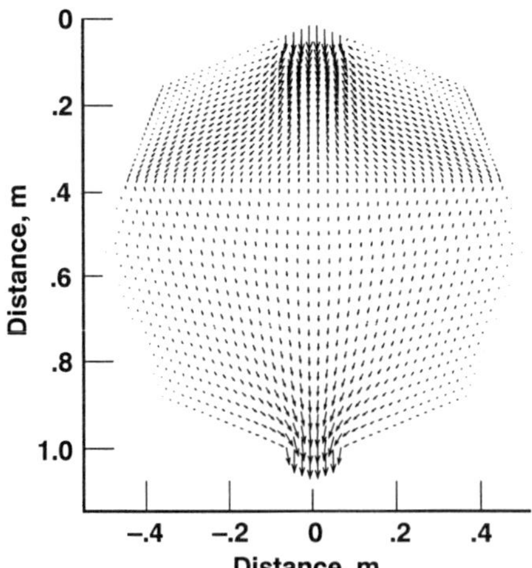

Figure 11. Velocity vectors at 50 seconds. He/LH2, q=1 J/m2-s.

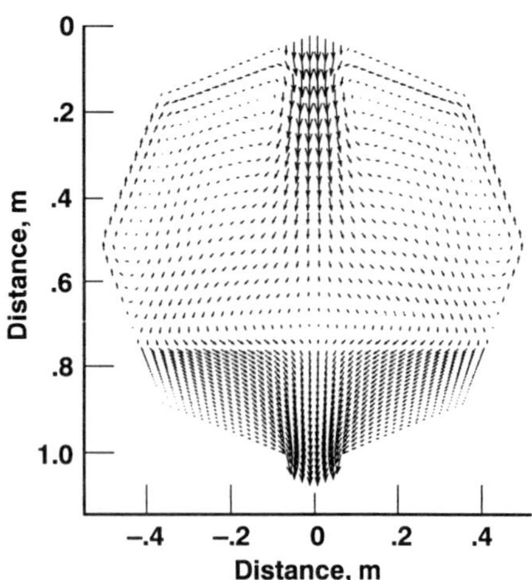

Figure 12. Velocity vectors at 300 seconds. He/LH2, q=1 J/m2-s.

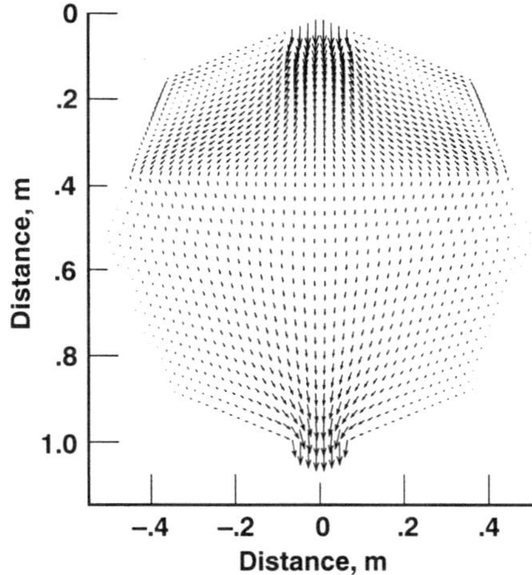

Figure 13. Velocity vectors at 50 seconds. He/LH2, q=2 J/m2-s.

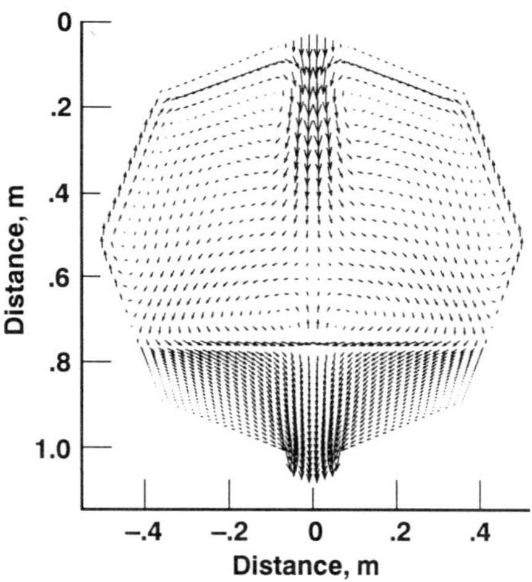

Figure 14. Velocity vectors at 300 seconds. He/LH2, q=2 J/m2-s.

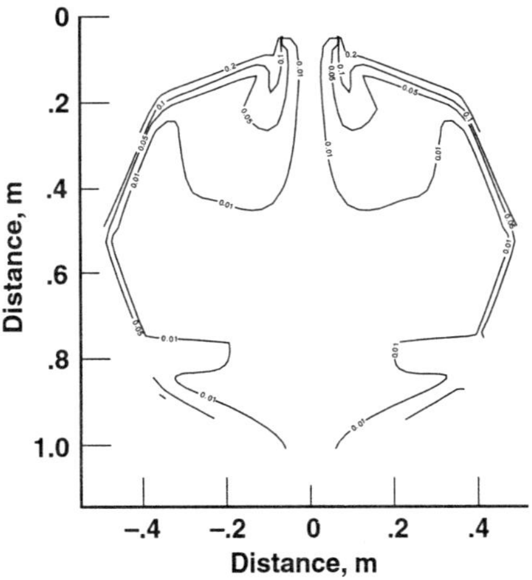

Figure 15. Temperature contours, $T - T_i$, at 300 seconds. He/LH2, q=1 J/m2-s.

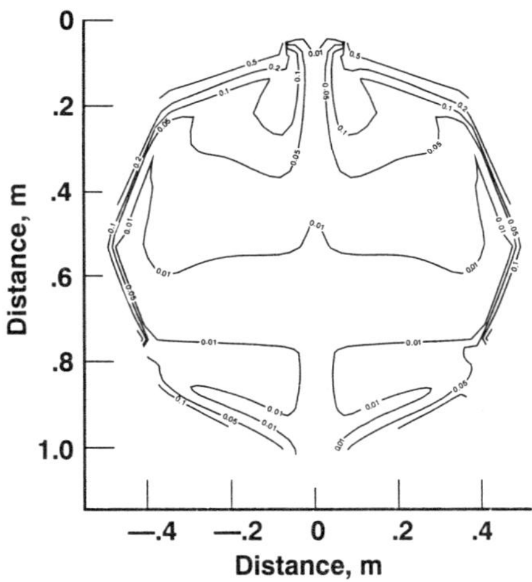

Figure 16. Temperature contours, $T - T_i$, at 300 seconds. He/LH2, q=2 J/m2-s.

REFERENCES

1. Grayson, G. D. and Navickas, J., Interaction Between Fluid-Dynamic and Thermodynamic Phenomena in a Cryogenic Upper Stage, AIAA-93-2753, AIAA 28th Thermophysics Conference, 1993.
2. Zhou, Qiao-Nain and Graebel, W. P., "Axisymmetric Draining of a Cylindrical Tank With a Free Surface," *J. Fluid Mech.*, vol. 221, 1990, pp. 511–532.
3. Willen, Scott, Hanna, Gregory J., and Anderson, Kevin R., "Cryogenic Tank Analysis Program," *Proceedings of the Eighth Annual Thermal and Fluid Analysis Workshop,* 1997.
4. Van Dresar, Neil T. and Haberbusch, Mark S., "Thermodynamic Models for Bounding Pressurant Mass Requirements of Cryogenic Tanks," *Cryogenic Engineering Conference*, Albuquerque, New Mexico, 1993.
5. Danilowicz, R. L., Temperature Profiles in a Propellant Tank, computer program, LEW-11034, NASA Lewis Research Center, COSMIC, 1994.
6. Chorin, Alexandre Joel, "A Numerical Method for Solving Incompressible Viscous Flow Problems," *Journal of Computational Physics*, vol. 2, 1967, pp. 12–26.
7. Lick, Wilbert J., *Difference Equations from Differential Equations*, Springer-Verlag, New York, New York, ISBN 0-387-50739-6, 1989.
8. Greer, Donald Stephen, *Numerical Modeling of the Flow in a Cryogenic Fuel Tank*, Ph. D. Thesis, Mechanical and Environmental Engineering Department, University of California, Santa Barbara, 1998.
9. Martini, William R. and Churchill, Stuart W., "Natural Convection Inside a Horizontal Cylinder," *AIChE Journal*, vol. 6, No. 2, 1960, pp. 251–257.
10. Martini, William R., *Natural Convection Inside a Horizontal Cylinder*, Ph. D. Thesis, University of Michigan, Ann Arbor, Michigan, 1957.

MODELING OF CRYOGENIC TRANSFER LINE COOL DOWN

A. R. Hasan[1], K. A. Haque[1], A. S. M. Rokanuzzaman[1], and M. M. Hasan[2]

[1]Univ. of North Dakota, Grand Forks, North Dakota, 58202

[2]NASA-Glenn Research Center, Cleveland, Ohio, 44135

ABSTRACT

The Cryogenic Storage and Propellant Feed System (CSPFS) of the space propulsion systems needs to provide propellant in controlled amounts during engine burns of variable duration. Thus, lines connecting the CSPFS to the engine/thruster will be subjected to cyclic cooling and heating. Here we present a model for cryogen transfer line chill-down rate. The model is based on an energy balance involving heat transfer from the outside to the pipe wall, from the wall to the flowing fluid, and energy depletion in the wall. The Morgan correlation is used for the outside heat transfer coefficient, h_a. For the inside heat transfer coefficient, h_{fb}, correlations of Giarrantano *et al.*, Hendricks *et al.*, and Ellerbrock *et al.* are examined.

Our model fit the literature data best when Giarrantano *et al.*'s superposition approach is used for h_{fb}. Even better agreement is obtained when this correlation is modified to account for convective heat transfer overestimation and pipe surface effect on heat transfer.

INTRODUCTION

Future HEDS (Human Exploration and Development of Space) vehicles and space propulsion systems, such as solar thermal upper stage, are likely to utilize cryogenic liquids as propellants. The lines connecting the CSPFS and the engine/thruster will be uninsulated, and may be heated, to allow the liquid from the CSPFS to evaporate, providing gaseous fuel to the thruster. The propellant flow rate, burn duration, and lockup period (duration between burns) may vary from orbit to orbit. Thus, these transfer lines will be subjected to cyclic cooling and heating of variable duration.

The transport of mass, momentum, and energy are often coupled. Thermohydraulic oscillations due to flashing and evaporation of a cryogenic propellant, aided by transient heat and momentum transport, may significantly affect flow rates, especially for short duration burns. Such transients may cause navigational difficulties and lead to instability of the fluid structure system. Consequently, modeling of transfer line transport processes is essential.

LITERATURE SURVEY

The potential use of cryogenic fuels for burns during orbit change and the economics of uninsulated transfer lines have led to many investigations into chill-down characteristics of cryogenic transfer lines. Understanding the mechanics of cryogenic fluid boiling is important in modeling chill-down. In the following, we present a survey of research work on chill-down. A brief discussion on boiling heat transfer will be presented thereafter.

Chill-down. Burke *et al.*[1] investigated liquid N_2 flow in a long horizontal line. They used an average heat transfer rate to avoid a transient analysis. Drake *et al.*[2] carried out similar experiments in vertical lines. Bronson *et al.*[3] proposed a transient analysis of chill-down; however they assumed a constant temperature difference between the pipe wall and the fluid.

Chi[4] presented a chill-down analysis neglecting heat transfer between the ambient and pipe wall. Thus its applicability to uninsulated lines is doubtful. Even the partially insulated transfer line data of Krishnamurthy *et al.*[5] show significant disagreement with his predictions. Chi[4] also related wall and fluid temperatures to the axial distance. However, it is difficult to reconcile his assumption of saturated two-phase flow in the line with the extent of variation in the axial fluid temperature. He considers only sensible heat, ignoring heat of vaporization, perhaps indicating the appropriateness of the analysis to somewhat different systems.

Srinivasan et al[6] chose a numerical approach to solving the governing equations. For film boiling, they used the Breen and Westwater[7] correlation for steady film pool boiling. Thus, their solution shows no effect of flow rate on the chill-down time.

Increasing flow rate is anticipated to increase the convective component of heat transfer mechanism operating in the vapor film blanketing the pipe wall. Higher flow rates should also cause the thicker liquid core to compress the vapor into a thinner film, reducing its conductive resistance. The independence of chill-down time from flow rate in Srinivasan *et al.*[6] model partly explains the significant differences between the predictions and some of their own data. Indeed, their steel tube data clearly show a dependence of the wall cooling rate on fluid flow rate. The later work of Krishnamurthy *et al.*[5] support a similar conclusion of faster chill-down at higher rates of the flowing fluid, which was also pointed out by Chi[4].

Boiling Heat Transfer Rate. One difficulty in establishing the effect of flow rate on chill-down is our incomplete knowledge of flow boiling in cryogenic fluids. Initial large temperature difference between the fluid and the wall causes rapid boiling of the liquid, forming a vapor blanket at the wall. This film boiling is expected to last for a very large portion of the chill-down process. While the phenomenon of pool boiling has been well researched, few works describe the effects of various parameters on film boiling of fluids flowing inside tubes. Studies involving internal flow of cryogens are even more scarce.

Most correlations express the Nusselt number for film boiling, $Nu_{fb} \equiv h_{fb}D/k$, as a function of the single-phase Nusselt number for the fluid, Nu_{sp}, the Lockhart-Martinelli parameter, X_{tt}, and the boiling number, as follows[8-10],

$$Nu_{fb} / Nu_{sp} = f(X_{tt}) Bo^{n} \quad (1)$$

The Lockhart-Martinelli parameter, X_{tt}, in Eq. 1 is calculated from quality, x, density, and viscosity of the fluid. Most researchers have used the following expression for X_{tt},

$$X_{tt} = \left(\frac{1-x}{x}\right)^{0.9} \left(\frac{\rho_v}{\rho_l}\right)^{0.5} \left(\frac{\mu_l}{\mu_v}\right)^{0.1} \quad (2)$$

Eq. 2 is valid only for turbulent flows of both the phases (hence the subscript, tt). The expressions for *X* for other cases are different from that shown in Eq. 2[10].

The single-phase Nusselt number, Nu_{sp}, is estimated from a Dittus-Boelter (or Sieder-Tate) type equation, $Nu_{sp} = a\, Re^{b}\, Pr^{c}$, sometimes with a viscosity correction of $(\mu/\mu_w)^{0.14}$. The boiling number, $Bo \equiv (q/A) / (\lambda G)$ – which involves heat flux per, q/A, and latent heat of vaporization, λ – can be interpreted as a ratio of the rate of vapor generation to the mixture mass flux, G. As such, Bo may be used as an index of the boiling induced flow normal to the channel axis. Differences among the various film boiling correlations lie in the values of the constants - a, b, c, and n - and in the way the Reynolds Number, Re, and the Prandtl Number, Pr, are evaluated.

Thus, Hendricks *et al.*[11] use a = 0.023, b = 0.8, c = 0.4, and n = 0 in Eq. 1. They use a two-phase homogeneous density, a velocity, u_m, calculated assuming the entire flow to be liquid ($u_m = m/A\rho_b$), and the vapor phase viscosity evaluated at the average film temperature to calculate the Reynolds Number. Thus, for the Hendricks *et al.*[11] correlation,

$$Nu_{sp} = 0.023(\mathrm{Re}_{fm})^{0.8}(\mathrm{Pr}_f)^{0.4} \quad (3)$$

$$\mathrm{Re}_{fm} = \frac{\rho_{fm} u_m D}{\mu_f} \qquad \mathrm{Pr}_f = \frac{Cp_f \mu_f}{k_f} \quad (4)$$

$$\rho_{fm} = \frac{1}{(x/\rho_f) + ((1-x)/\rho_l)} \qquad \rho_b = \frac{1}{(x/\rho_v) + ((1-x)/\rho_l)} \quad (5)$$

$$f(X_{tt}) = e^{\left(0.0527 - 0.416 \ln X_{tt} - 0.008(\ln X_{tt})^2\right)} \quad (6)$$

The subscript "f" represents vapor film property values at the film temperature $[=(T_w+T_l)/2]$, "v" is for vapor properties at the saturation temperature, while subscript "l" is for liquid.

Ellerbrock *et al.*[12] adopts Hendricks *et al.* approach (Eq. 1- 5) except with a value of 0.4 for the exponent, n, of Bo,

$$Nu_{fb} \equiv (h_{fb} D)/k_f = 0.023(\mathrm{Re}_{fm})^{0.8}(\mathrm{Pr}_f)^{0.4} f(X_{tt}) Bo^{0.4} \quad (1)$$

$$f(X_{tt}) = e^{\left(2.35-0.266\ln X_{tt}-0.025(\ln X_{tt})^2\right)} \quad \textbf{(7)}$$

Giarrantano and Smith[8], arguing that X_{tt} primarily represents quality effects, proposed a Nu_{fb} correlation in terms of quality alone,

$$Nu_{fb} = 0.026\,(\mathrm{Re}_v)^{0.8}(\mathrm{Pr}_v)^{0.33}(\mu_v/\mu_w)^{0.14} f(x) \quad \textbf{(1)}$$

$$f(x) = e^{\left(-0.185-0.25\ln x-0.00767(\ln x)^2\right)} \quad \textbf{(8)}$$

using vapor saturation properties to estimate dimensionless numbers. They[8] also proposed a superposition approach whereby h_{fb} is estimated by adding the pool boiling heat transfer coefficient, h_{film} (of ref[7]), to that due to convective boiling. Thus, $h_{fb} = h_{sp} + h_{film}$, where[7],

$$h_{film} = \left[4.94(\rho_l-\rho_f)^{0.375}/\sigma^{0.125} + 0.115\sigma^{0.375}/\left(D(\rho_l-\rho_f)^{0.125}\right)\right]F \quad \textbf{(9)}$$

$$h_{sp} = 0.026(\mathrm{Re}_v)^{0.8}(\mathrm{Pr}_v)^{0.33}(\mu_v/\mu_w)^{0.14}k_v/D \quad \textbf{(10)}$$

the property values must be in cgs units (g, cm, s) and F is defined by Eq. 18. An important objective of this work is to evaluate the effectiveness of these correlations in representing flowing cryogen boiling as indicated by their ability to predict transfer line chill-down data.

MODEL DEVELOPMENT

In the following we present an analysis which assumes that the fluid enters as a saturated liquid at the bulk temperature, T_b, and fills the transfer line in a negligibly short time. Thus, this analysis only examines pipe wall chill-down, neglecting flow and pressure transients. In the vast majority of the cases, the results of this approximate decoupled analysis are relevant.

The energy balance for the pipe wall involves heat transfer by convection and radiation from the ambient to the wall, heat transfer from the wall to the flowing fluid, and energy depletion in the wall. The heat transfer from the ambient to the pipe wall per unit length of the pipe, q_a, may be represented as follows, using an ambient heat transfer coefficient, h_a,

$$q_a = h_a(T_a - T_w)\pi D_o \quad \textbf{(11)}$$

Written thus, both radiative and natural convective heat transfer coefficients depend on T_a and T_w. However, we assume these to be constant. Similarly, heat transfer to the fluid, q_i, is

$$q_i = h_{fb}(T_w - T_b)\pi D_i \quad \textbf{(12)}$$

In Eqs. 11 and 12, T_a and T_w are the ambient and wall temperatures respectively, D_o, and D_i are tube outside and inside diameters, and h_{fb} is the inside (film boiling) heat transfer coefficient. Further, we neglect any axial and radial variation in wall temperature, which is quite appropriate for most lines made of thin metals. Using a pipe material density of ρ_w and specific heat of c_w, we write the energy balance equation for the pipe wall as,

$$\frac{\pi}{4}\left(D_o^2 - D_i^2\right)\rho_w c_w \frac{dT_w}{dt} = \pi\left[h_a D_o (T_a - T_w) - h_{fb} D_i (T_w - T_b)\right] \tag{13}$$

Rearranging, we obtain,

$$\frac{dT_w}{dt} + BT_w = A \tag{14}$$

where

$$A = 4\frac{h_a D_o T_a + h_{fb} D_i T_b}{\left(D_o^2 - D_i^2\right)\rho_w c_w} \quad and \quad B = 4\frac{h_a D_o + h_{fb} D_i}{\left(D_o^2 - D_i^2\right)\rho_w c_w} \tag{15}$$

With only minor dependence of property values and heat transfer coefficients (h_a and h_{fb}) on temperature, B may be taken as a constant. If the fluid in the line is saturated, then T_b, and hence A may also be assumed to remain constant. These assumptions make Eq. 14 a first order linear differential equation with the initial condition $T_w = T_a$ at $t = 0$. The solution is,

$$T_w = \frac{A}{B} + \left(T_a - \frac{A}{B}\right)e^{-Bt} \quad or \quad \frac{T_w - A/B}{T_a - A/B} = e^{-Bt} \tag{16}$$

We have assumed, in addition to constant h_a and h_{fb}, the fluid temperature to be invariant with time. For most cryogenic applications, even subcooled liquids quickly become saturated upon contact with the pipe walls. Thus, the applicability of Eq. 16 is not seriously limited.

RESULTS

Srinivasan *et al.*[6] studied chill-down of steel, copper, aluminum, and glass pipes for saturated liquid nitrogen (N_2) flow ranging from 105 to 182 kg/h. We examine the validity of our model and the applicability of various inside heat transfer coefficient correlations, h_{fb}, in light of these data. The Morgan[13] correlation is used in all cases for calculating h_a.

Fig.1 shows T_w predictions of Eq. 16 using the correlations of Hendricks *et al.*[11] (Eq. 6,

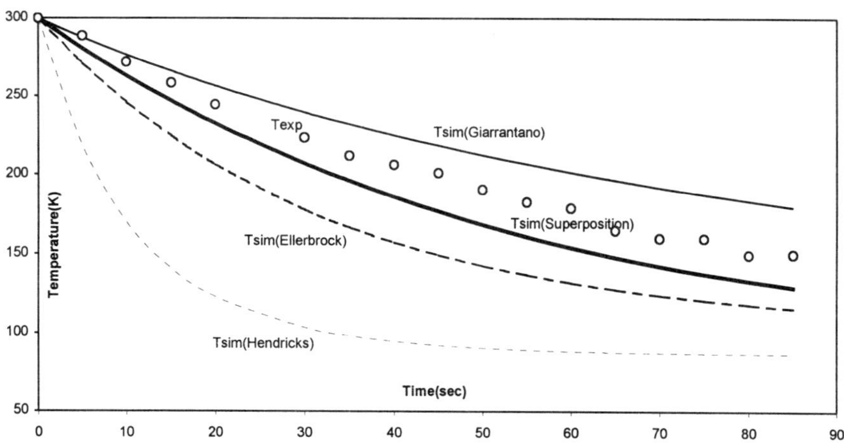

Fig. 1 Wall Temperature vs. Time for fluid flowing through a steel tube at 122 kg/hr

dotted line), Ellerbrock *et al.*[12] (Eq. 7, dashed line), Giarrantano *et al.*[8] (Eq. 8, light solid line) and their superposition approach[8] (Eqs. 9-10, heavy solid line). The circles represent data for 122 kg/h of saturated N_2 flow through a steel pipe. The figure shows that our model well represents the general pattern of wall temperature drop with time. It also shows that the superposition approach[8] (Eqs. 9-10) clearly represent the data better than the other h_{fb} correlations. Similar conclusions about better applicability of Eq. 9 than the other h_{fb} correlations may be drawn from Fig. 2 which shows data for 171 kg/h of saturated N_2 flowing through the glass tube. This general trend - better T_w predictions using Eqs. 9-10 than Eqs. 6, 7 or 8 - is observed for all T_w data reported by Srinivasan *et al.*[6]

Although the predictions were good, we noted general underestimation of Srinivasan *et al.*[6] data when Eqs. 9-10 are used, i.e., the superposition approach calculates too high values of h_{fb}. To obtain better agreement with the data we modified Eqs. 9-10 in two ways. First, we used a lower value of 0.019 for the parameter "a", instead of 0.023, in Eq. 10 for Nu_{sp}. Then, noting that the film heat transfer coefficient may be influenced by fluid-surface interaction, we used a parameter C_s in the factor F in Eq. 9 to reflect this effect. Thus,

$$h_{sp} = 0.019(\mathrm{Re}_v)^{0.8}(\mathrm{Pr}_v)^{0.33}(\mu_v/\mu_w)^{0.14}k_v/D \quad \textbf{(17)}$$

$$F = \left[\frac{C_s k_v^3 \rho_f}{\mu_f \Delta T}\right]^{1/4}\sqrt{\frac{\lambda + 0.34 Cp_f \Delta T}{\sqrt{\lambda}}} \quad \textbf{(18)}$$

Giarrantano *et al.*[8] didn't use C_s, i.e., $C_s = 1.0$ in their correlation. We used a value of 0.30 for all metal tubes and 0.25 for the glass tube for C_s in Eq. 18 for F. Fig. 3 show T_w predictions using the modified approach (Eqs. 9, 17, and 18) for the steel pipe data. Comparing Figs. 3 and 1, we see significant improvement in T_w predictions using the proposed method. Other Srinivasan *et al.*[6] data were also much better represented by Eqs. 9, 17-18 with appropriate C_s values (0.30 for metals, 0.25 for glass) than by the unmodified superposition approach.

Table 1 reports two statistical measures of goodness of fit obtained for the various data sets of Srinivasan *et al.*[6] and the predictions of Eqs.17-18 and the other h_{fb} correlations. The

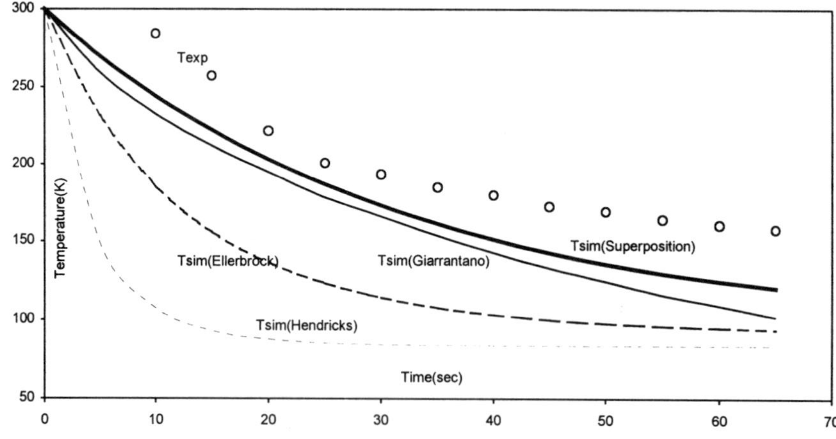

Fig. 2 Wall Temperature vs. Time for fluid flowing through a glass tube at 171 kg/hr

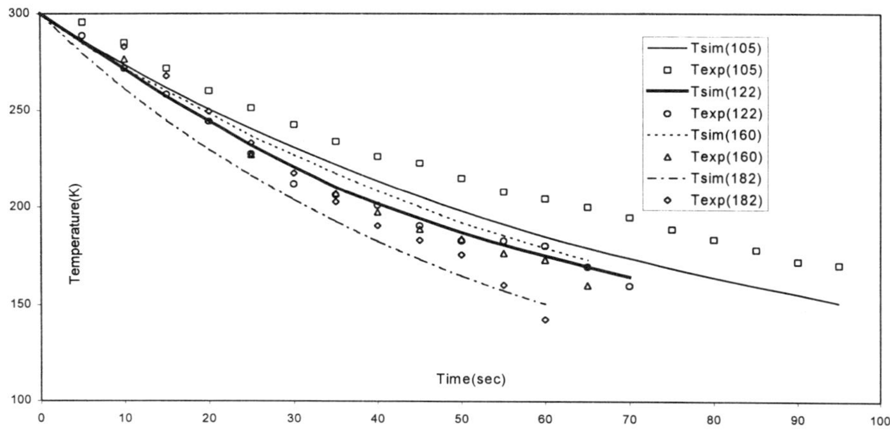

Fig. 3. Wall Temp. vs. Time for fluid flowing through a steel tube at different rates

pipe dimensions and flow rate conditions for the data sets can be found in the paper by Srinivasan *et al.*[6]. The average error, ε, for a given correlation and data set was calculated from the average of the differences between the experimental data and the predicted values. Thus, a positive ε value indicates a tendency by the correlation to generally underestimate the data. For example, Table 1 shows that Ellerbrock and Hendricks correlations always underestimate. The standard deviation, σ, is defined the usual way, $\sigma^2 = (1/(n-1))(\Sigma \text{error}^2)$. Standard deviation is probably the best overall indicator of goodness of fit. By that measure Table 1 clearly shows the modified superposition approach as the best correlation presently available for flow boiling of cryogenic fluids.

DISCUSSION AND CONCLUSIONS

We have presented a simple analytical model for estimating transfer line chill-down rate. We neglected wall temperature variation in the radial and axial directions, which may lead

Table 1. Statistical measures of goodness of fit for different correlations

Tube material	Flow rate (kg/hr)	Modified Superposition		Superposition		Ellerbrock		Giarrantano		Hendricks	
		ε	σ	ε	σ	ε	σ	ε	σ	ε	σ
Steel	105	14.68	15.84	38.96	42.8	57.89	60.55	-1.88	4.58	115.2	120.1
	122	-1.14	4.38	24.41	26.24	43.87	44.93	-14.04	15.80	96.84	98.82
	160	-11.11	13.10	15.74	17.06	45.12	47.38	-13.43	16.03	102.2	109.7
	182	-9.92	16.38	14.08	16.11	49.36	52.75	-6.53	15.06	115.7	124.3
Glass	148	15.75	18.97	39.90	44.30	91.51	101.02	31.83	35.55	148.1	161.4
	160	-6.61	35.44	24.30	40.91	77.45	92.71	14.61	38.23	111.5	132.4
	171	-1.36	10.11	28.25	30.62	70.48	74.43	20.92	23.15	111.7	121.7

to inaccurate T_w estimates in thick non-metallic pipes. Variation of h_{fb} and h_a with time (i.e., fluid temperature) are also neglected.

We note that the superposition approach of Giarrantano *et al.*[8] best predict the chill-down data for most cases, especially if the modifications represented by Eqs. 17-18 are used. The validity of this modification for other fluid-surface combinations needs to be verified.

ACKNOWLEDGMENT

We gratefully acknowledge the financial and other support received from NASA Glenn Research Center for this work.

REFERENCES

1. J. C. Burke, W. R. Byrnes, A. H. Post, and F. E. Ruccia, "Advances in Cryogenic Engineering," **4** (1960), p. 378.
2. E. M. Drake, F. E. Ruccia, and J.M. Ruder, "Advances in Cryogenic Engineering," **6** (1961), p. 323.
3. J. C. Bronson, *et al.*, "Advances in Cryogenic Engineering," **7** (1961), p. 198.
4. J. W. H. Chi, "Advances in Cryogenic Engineering," **10** (1965), p. 350.
5. M. V. Krishnamurthy, R. Chandra, S. Jackob, S. Kasthurirengan, and Karunanithi, R. Experimental studies on cool-down and mass flow characteristics of a demountable liquid nitrogen transfer line, *Cryogenics* **36** (1996), p. 453.
6. K. Srinivasan, S. Rao, and M. V. Krishnamurthy, Analytical and experimental investigation on cool-down short cryogenic transfer lines, *Cryogenics*, Sept. (1974), p. 459.
7. B. P. Breen, and J. W. Westwater, "Effect of diameters of horizontal tubes on film boiling heat transfer," *Chem. Eng. Prog.*, **58**, No. 7, p. 67.
8. P. J. Giarrantano and R. V. Smith, "Comparative study of forced convection boiling heat transfer correlations for cryogenic fluids," "Advances in Cryogenic Engineering," **11** (1965), p. 492.
   P. J. Brentari, P. J. Giarrantano, and Smith, R. V. "Boiling heat transfer for oxygen, nitrogen, hydrogen, and helium," NBS Tech. Note No. 317, Colo (1965).
9. A. R. Hasan and E. Rhodes, "Two-phase flow heat transfer in a horizontal steam-water system," *Chem. Eng. Comm.* **22** (1983), p. 205.
10. A. R. Hasan, "Low pressure boiling in straight tubes and a bend," Ph. D. Thesis, University of Waterloo (1979), p. 76.
11. R. C. Hendricks, R. W. Graham, Y. Y. Hsu, and R. Friedman, "Experimental heat transfer and pressure drop of liquid hydrogen flowing through a heated tube," NASA TND-765, 1961.
12. H. H. Ellerbrock, J. N. B. Livingood, and D. M. Straights, "Fluid flow and heat transfer problems in nuclear rockets, NASA SP-20, Dec. 1962.
13. F. P. Incropera and D .P. Dewitt "Introduction to Heat Transfer," John Wiley & Sons, New York, (1985), pp 397-398.

A GENERALIZED MODEL FOR THE TRANSPORT PROPERTIES OF AIR COMPONENTS AND MIXTURES

Z. Shan, R. T Jacobsen, and S. G. Penoncello

Center for Applied Thermodynamic Studies
College of Engineering
University of Idaho
Moscow, Idaho 83844-0905

ABSTRACT

A generalized correlation for the transport properties (viscosity and thermal conductivity) of pure fluids is presented. The new formulation, which is applicable to the whole fluid region, is based on Eyring's significant liquid structure theory. We have incorporated a modification of Eyring's reaction rate theory for liquid viscosity and gas kinetic theory for the dilute gas viscosity and thermal conductivity. An empirical hyperbolic secant function is used for the critical region transport property enhancements. In addition to transport properties for the pure cryogenic fluids, the formulation has been applied to the viscosity and thermal conductivity for mixtures of nitrogen, oxygen and argon, including air, in both gas and liquid phases. The accuracy of the formulation is assessed based upon comparisons to experimental data for both pure fluids and mixtures. The wide-range formulation is explicit in viscosity and thermal conductivity as functions of temperature and density. Wide-range equations of state developed earlier for each pure fluid and for air were used to calculate the densities of the fluids at specific temperatures and pressures in the development of the generalized correlation.

INTRODUCTION

The study described in this paper has been motivated by a need for a model for transport properties to serve as a companion to the generalized mixture model for equilibrium thermodynamic properties reported by Lemmon and Jacobsen[1] for a wide variety of pure fluids and mixtures. The thermodynamic property model was originally reported in the doctoral dissertation of Lemmon[2], and has been successfully applied to mixtures of refrigerants, hydrocarbons, and cryogens. The ultimate goal is a versatile package of computer programs capable of calculating all properties of interest to the design engineer over a wide range of temperatures and pressures. The application to cryogenic

Advances in Cryogenic Engineering, Volume 45.
Edited by Shu *et al.*, Kluwer Academic / Plenum Publishers, 2000.

fluid mixtures is motivated by the significant body of accurate data for pure argon, nitrogen, and oxygen and for air. In this work, air has been treated as both a pseudo-pure fluid and as a ternary mixture with an approximate composition from previous work[3].

This paper is not an exhaustive review of the literature on the correlation of transport properties, nor are all the available data for each fluid included. The authors have reviewed the literature extensively and have selected both correlation methods and data sets, which represent the thermal conductivity and viscosity for each of the fluids within the uncertainty of the available measured values.

There is considerable theoretical background for the representation of the transport properties of pure fluids and mixtures. The general approach of this work has been the selection of theories that appear to take into account the important aspects of molecular behavior, and the augmentation of the selected theories with empirical functions to assure accurate representation of the selected data by the models for each property. Two widely accepted molecular theories for fluid transport properties are kinetic theory for dilute gas transport properties (Hirschfelder *et al.*[4]) and Eyring's reaction rate theory for liquid viscosity (Glasstone *et al.*[5]). For other dense fluid states, the significant liquid structure theory of Eyring and Jhon[6] has been successfully applied to a variety of fluid systems (Hsu and Eyring[7], Leu *et al.*[8], Ree *et al.*[9], Jhon *et al.*[10], and Hildwein and Stephan[11]). The significant liquid structure theory, Eyring's reaction rate theory, and gas kinetic theory are the basis for the formulation used in this work for both viscosity and thermal conductivity. For the critical region, it is generally understood that there is an enhancement to both the viscosity and thermal conductivity. While the effects for thermal conductivity are well established by experiment, the effects for viscosity occur only in a very narrow range near the critical point and can generally be neglected in analyses for fluid states observed in practical systems.

The generalized thermodynamic property model of Lemmon and Jacobsen[1] can be used to determine the density of the fluid state from the experimentally measured values of pressure and temperature. The density and temperature specified or calculated from that thermodynamic property model are used as input information for the transport property model.

The proposed viscosity and thermal conductivity formulation is applicable to mixtures as well as to pure fluids. Air may be considered as a ternary mixture of nitrogen, oxygen, and argon. By applying appropriate mixing rules, the viscosity and thermal conductivity of air and nitrogen-oxygen-argon mixtures can be predicted directly from those of the pure components.

A summary of comparisons of values calculated using our model to selected data sets for viscosity and thermal conductivity is included. The analysis of the experimental data has been performed by other authors in previous research and is not repeated here. Representative analyses are given by Stephan *et al.*[12] for nitrogen, and by Laesecke *et al.*[13] for oxygen.

VISCOSITY AND THERMAL CONDUCTIVITY FORMULATION FOR PURE FLUIDS

The formulation used for both viscosity and thermal conductivity is an extension and modification of the significant liquid structure (SLS) theory of Eyring and Jhon[6] for liquid properties. The SLS theory treats the liquid as being composed of solid-like molecules and gas-like vacancies. The properties of the liquid are modeled as volume fraction averages of the properties of the solid-like molecules and the gas-like vacancies.

In this work, the SLS theory developed originally for liquid properties has been extended to all fluid states for both viscosity and thermal conductivity by some modifications. The proposed transport formulations for viscosity and thermal conductivity for the pure fluids are functions of temperature and density.

Generalized Equation for Transport Properties

The generalized model for both viscosity and thermal conductivity is

$$X = [(\rho_L - \rho)/\rho_L]^{c_1} X_{DG} + (\rho/\rho_L)^{c_1} X_L + \Delta X_c, \tag{1}$$

where X is the viscosity [μPa-s] or thermal conductivity [mW/m-K], ρ_L and c_1 are adjustable parameters, ρ is the density [mol/L], X_{DG} is the viscosity or thermal conductivity of the dilute gas, X_L is the viscosity or thermal conductivity of a liquid reference state, and ΔX_c is the critical region viscosity or thermal conductivity enhancement.

Critical Region Viscosity and Thermal Conductivity Enhancement

The transport properties of pure fluids exhibit enhancements in the neighborhood of the critical point. An equation based upon the expression for the hyperbolic secant has been used for the critical region viscosity and thermal conductivity enhancement as functions of temperature and density

$$\Delta X_c = \frac{4\Delta X_{\max}}{[\exp(\rho - \rho_c) + \exp(\rho_c - \rho)][\exp(T - T_c) + \exp(T_c - T)]}, \tag{2}$$

where ΔX_c is the critical region enhancement for the viscosity or thermal conductivity, $\Delta X_{\max}$ is an adjustable parameter, ρ_c and T_c are the critical density [mol/L] and critical temperature [K] and were taken from Jacobsen *et al.*[14], and T is the temperature [K].

Dilute Gas and Liquid Transport Property Relations

Dilute gas viscosity. The gas kinetic theory reported by Hirschfelder *et al.*[4] has been used here to describe the dilute gas viscosity and thermal conductivity. The theory gives the following viscosity expression for gases at low densities:

$$\eta_{DG}(T) = \frac{5}{16}\left(\frac{MkT}{1000\pi N_A}\right)^{0.5} \frac{10^{24}}{\sigma^2 \Omega(T^*)}, \tag{3}$$

where $T^* = (kT/\varepsilon)$, η_{DG} is the viscosity for dilute gas [μPa-s], M is the molecular weight [g/mol], k is the Boltzmann constant ($1.380658\text{x}10^{-23}$ J/K), N_A is the Avogadro number ($6.0221367\text{x}10^{23}$ mol^{-1}), σ is the molecular collision diameter [nm], and ε is the molecular interaction parameter, usually given as (ε/k) in [K]. The molecular collision integral, $\Omega(T^*)$, is often expressed by an empirical polynomial function of T^* as is given by Krauss *et al.*[15],

$$\Omega(T^*) = \exp \sum_{i=0}^{4} A_i (\ln T^*)^i . \tag{4}$$

The constants A_i in Eq. (4) determined in this work are listed in Table 1.

Table 1. Constants for intermolecular collision integral $\Omega(T^*)$

A_I	Value
A_0	0.490169763439
A_1	-0.468401398108
A_2	0.0443955660482
A_3	0.0186114605103
A_4	-0.00412543553120

Dilute gas thermal conductivity. A correction factor, λ_{CORR}, is applied to the kinetic theory for the dilute gas thermal conductivity. The correction factor accounts for energy transfer between internal and translational modes. The dilute gas thermal conductivity [mW/m-K] is

$$\lambda_{DG}(T) = \frac{75}{64}\left(\frac{N_A T}{\pi M}\right)^{0.5} k^{1.5} \frac{10^{21}}{\sigma^2 \Omega(T^*)}(1+\lambda_{CORR}), \text{ where} \tag{5}$$

$$\lambda_{CORR} = \sum_{i=1}^{2} B_i T^{k_i}(C_p^o / R - 2.5), \tag{6}$$

where M, k, N_A, σ, T, and $\Omega(T^*)$ are defined above, B_i and k_i are adjustable parameters, C_p^0 is the ideal gas heat capacity [J/mol-K] and R is the universal gas constant (8.31451 J/mol-K).

Liquid viscosity. Eyring's reaction rate theory (Glasstone *et al.*[5]) is an accepted model for the liquid viscosity. There have been a variety of modifications to Erying's theory. In this work, the modified reaction rate theory by Ree *et al.*[9] is further corrected and used for both viscosity and thermal conductivity. The liquid viscosity [μPa-s] is given by

$$\eta_L = c_2 \frac{\rho_L^2}{\rho_L - \rho} T^{1/2} \exp\left(\frac{\rho}{\rho_L - \rho}\frac{\Delta G^*}{RT}\right) + \Delta\eta_L, \text{ where} \tag{7}$$

$$\Delta\eta_L = N_1 \tau^{j_1} + N_2 \delta^{i_2} \tau^{j_2}, \tag{8}$$

where $\delta = \rho/\rho_c$, $\tau = T_c/T$, ΔG^*, c_2, N_1, N_2, j_1, i_2, and j_2 are adjustable parameters.

Liquid thermal conductivity. The functional form of Eq. (7) is also used for the liquid thermal conductivity. The liquid thermal conductivity [mW/m-K] is given by

$$\lambda_L = c_2 \frac{\rho_L^2}{\rho_L - \rho} T^{1/2} \exp\left(\frac{\rho}{\rho_L - \rho}\frac{\Delta G^*}{RT}\right) + \Delta\lambda_L, \text{ where} \tag{9}$$

$$\Delta\lambda_L = \sum_{k=1}^{2} N_k \delta^{i_k} \tau^{j_k} + \sum_{k=3}^{4} N_k \delta^{i_k} \tau^{j_k} \exp(-\delta^{l_k}), \tag{10}$$

where ΔG^*, c_2, N_k, i_k, j_k, and l_k are adjustable parameters, and δ and τ are defined above.

Correlation Results for Viscosity and Thermal Conductivity of Pure Fluids

The adjustable parameters in the proposed formulation for viscosity and thermal conductivity have been obtained by fitting the equations to the selected experimental data from the literature. The values of the constants and adjustable parameters for the viscosity model are listed in Table 2 for each pure fluid. Table 3 lists the values for the thermal conductivity model.

Table 2. Coefficients for the general formulation for viscosity

Parameter	Air (pseudo-pure)	Nitrogen	Oxygen	Argon
σ	0.374	0.3798	0.3553	0.3478
ε/k	74.9	71.4	88.2	105
c_1	1.17230418464	0.990295699197	1.17616516263	1.17493396028
c_2	0.0480476579028	0.0835799868325	0.0901494853476	0.0455136402476
ΔG^*	614	805	238	524.117757892
ρ_L	50.4	56.2	51.3	55.1
$\Delta\eta_{max}$	10000	10000	10000	9971
N_1	26.8284690051	-20.7452728895	5.67067418273	44.4888400425
j_1	0.23	-0.87	-0.25	-0.23
N_2	2.20290337467	0.0622058245025	0.231936367928	6.83100382857
i_2	3	5	3	3
j_2	0.32	-0.13	-0.43	0.75

Table 3. Coefficients for the general formulation for thermal conductivity

Parameter	Air (pseudo-pure)	Nitrogen	Oxygen	Argon
σ	0.374	0.3798	0.3553	0.3478
ε/k	74.9	71.4	88.2	105
c_1	1.08172780373	1	1.00303321452	0.953590251679
c_2	-2.76575058646	-4.57938632038	0.0477642690848	0.049245159582
ΔG^*	614	805	237.855851336	524.117757892
ρ_L	3083.92680815	2272.14509133	1618.96500652	129.905626537
$\Delta\lambda_{max}$	2387.72390865	521.819076559	451.044563754	2041.31344627
B_1	1.2058101727	0.0986018278832	0.3	0
k_1	-0.18	0.41	0	0
B_2	-10.5892263617	-0.0430800398678	0.000315859666333	0
k_2	-0.76	0.5	0	0
N_1	1979.13538798	1773.14671078	26.6784769263	59.1009134231
i_1	2	2	5	2
j_1	0.05	0.02	-0.9	0.23
N_2	103408.262822	118748.488681	1233.57838209	-0.202964105376
i_2	0	0	1	5
j_2	-0.49	-0.50	0.1	1.8
N_3	8474.73943744	6533.91660587	1469.99065934	291.785143838
i_3	1	1	2	1
j_3	2.29	1.9	6	3
l_3	2	2	2	2
N_4	13772.4498296	1139.3594058	2764.00050819	-0.271246121087
i_4	3	3	1	6
j_4	8	0.82	2	1.22
l_4	3	2	2	2

MIXTURE FORMULATION

The viscosity and thermal conductivity of air mixtures can be predicted from the pure component equations using the following mixing rules,

$$\ln X = \sum_i \sum_j x_i x_j \ln X_{ij}, \tag{11}$$

where X represents either viscosity or thermal conductivity, $X_{ij} = 0.5(X_i + X_j)$, $X_i = X_i(TT_{ci}/T_{cm}, \rho\rho_{ci}/\rho_{cm})$, x_i, T_{ci} and ρ_{ci} are the mole fraction, critical temperature, and critical density of component i. The mixing rules used for T_{cm} and ρ_{cm}, are given by Eqs. (12) and (13), and the binary parameters ξ_{ij} and ζ_{ij} are from Lemmon and Jacobsen[3].

$$\rho_{cm} = \left[\sum_{i=1}^{n} \frac{x_i}{\rho_{c_i}} + \sum_{i=1}^{n-1} \sum_{j=i+1}^{n} x_i x_j \xi_{ij} \right]^{-1} \tag{12}$$

$$T_{cm} = \sum_{i=1}^{n} x_i T_{c_i} + \sum_{i=1}^{n-1} \sum_{j=i+1}^{n} x_i x_j \zeta_{ij} \tag{13}$$

COMPARISONS OF CALCULATED PROPERTIES TO DATA

The calculated viscosity and thermal conductivity were compared with selected experimental data from the literature. The deviations between the calculated and the experimental data were analyzed using the relation, $\%\Delta X = 100(X_{EXP} - X_{CALC})/X_{EXP}$. The absolute average deviation is $AAD = \frac{1}{n}\sum_{i=1}^{n} |\%\Delta X_i|$, where X is any property and the subscripts $_{EXP}$ and $_{CALC}$ refer to measured and calculated properties, respectively. Data points with deviations greater than 10% were not included in the statistical analysis. Table 4 summarizes the accuracies and ranges of validity of the formulations for pure fluids and mixtures. Tables 5 and 6 summarize the comparisons for pure fluids and mixtures of air components, respectively.

Table 4. Accuracies and ranges of validity of the formulations

Fluid	P Range (MPa)	T Range (K)	No. of Points	No. of Data Sources	Estimated Accuracy, %	No. of Points $\Delta X_i > 10\%$
Viscosity						
Air (pseudo-pure)	0.01-325	70-1700	417	17	1.1	10
Air (ternary mixture)	0.01-325	70-1700	417	17	1.5	19
Argon	0.00-606	60-2103	689	21	1.3	26
Nitrogen	0.01-401	67-3273	677	19	1.2	11
Oxygen	0.01-78	70-3273	357	16	2.1	7
Binary mixtures	0.00-2.3	79-770	135	4	1.9	1
Thermal Conductivity						
Air (pseudo-pure)	0.00-70	70-1374	1436	13	1.0	12
Air (ternary mixture)	0.00-70	70-1374	1436	13	1.9	28
Argon	0.03-246	87-3306	1675	25	1.8	31
Nitrogen	0.00-1000	69-2473	1133	20	1.7	13
Oxygen	0.01-68	77-1502	1321	10	1.2	15
Binary mixtures	0.1-70	65-324	2013	3	1.6	19

Table 5. Comparisons of calculated properties of pure fluids with selected data

Fluid	Author	Year	No. of Points	AAD	P Range (MPa)	T Range (K)	No. of Points $\Delta X_i > 10\%$
Viscosity							
Air (psuedo-pure)	Diller *et al.*[16]	1991	64	1.33	3.4-32	71-130	1
	Kestin and Whitelaw[17]	1964	42	1.46	0.1-15	298-524	0
	Kurin and Golubev[18]	1974	36	1.28	9-325	293-323	0
	Matthews *et al.*[19]	1976	15	0.47	0.1	120-1700	0
	Timrot *et al.*[20]	1974	46	0.83	0.1-12	296-566	0
Argon	Gracki and Ross[21]	1969	47	1.02	0.46-17	173-298	0
	Kurin and Golubev[18]	1974	99	0.88	9.8-380	273-423	0
	Maitland and Smith[22]	1972	30	0.21	0.1	60-2000	1
	Timrot *et al.*[20]	1974	39	0.76	0.1-14	292-575	0
	Van Itterbeek *et al.*[23]	1966	21	1.54	0.1-9	90-234	0
Nitrogen	Diller[24]	1983	65	1.62	0.3-34	90-300	1
	Golubev and Kurin[25]	1974	70	0.58	9-401	273-423	0
	Matthews *et al.*[19]	1976	15	0.83	0.1	120-1700	0
Oxygen	Kestin and Leidenfrost[26]	1959	15	0.76	0.01-6	293-298	0
	Matthews *et al.*[19]	1976	15	0.57	0.1	120-1700	0
	Van Itterbeek *et al.*[23]	1966	32	2.93	0.01-9.8	70-90	0
Thermal Conductivity							
Air (psuedo-pure)	Geier and Schafer[27]	1961	12	0.72	0.1	273-1374	0
	Perkins and Cieszkiewicz[28]	1991	1066	0.67	0.1-70	70-304	0
Argon	Amirkhonov *et al.*[29]	1972	140	1.18	9-99	113-253	0
	Michels *et al.*[30]	1963	110	0.50	0.1-246	274-348	0
	Roder *et al.*[31]	1989	287	0.95	0.2-68	100-322	2
Nitrogen	Johns *et al.*[32]	1986	14	0.75	1-30	426-428	0
	Perkins *et al.*[33]	1991	377	1.10	0.3-71	81-303	6
	Roder[34]	1981	93	0.42	1-70	297-309	0
Oxygen	Geier and Schafer[27]	1961	12	0.75	0.1	273-1375	0
	Roder[35]	1982	1126	1.08	0.03-69	77-313	12

Table 6. Comparisons of calculated properties of binary mixtures with selected data

Author	Year	No. of Points	AAD	P Range (MPa)	T Range (K)	No. of Points $\Delta X_i > 10\%$
Viscosity						
DiPippo *et al.*[36]	1968	33	0.57	0.1-2.3	293-303	0
Hellemans *et al.*[37]	1973	46	1.45	0.0	298-770	1
Saji and Okuda[38]	1965	37	4.14	0.03-0.1	79-88	0
Trautz and Melster[39]	1930	19	1.12	0.1	295-551	0
Thermal Conductivity						
Perkins and Cieszkiewicz[28]	1991	1922	1.60	0.2-70	65-304	19
Yorizane *et al.*[40]	1983	67	2.38	0.1-9.8	301	0
Zheng *et al.*[41]	1984	24	2.49	0.1-9.8	324	0

ACKNOWLEDGEMENTS

The authors acknowledge the assistance of Dr. Eric W. Lemmon of the Physical and Chemical Properties Division (PCPD), Chemical Science and Technology Laboratory (CSTL), National Institute of Standards and Technology (NIST) in Boulder, Colorado, in the development of the functional forms and fitting techniques used in this research. The many helpful discussions have set the stage for continuing collaboration in research on thermophysical properties.

Richard Jacobsen acknowledges, with sincere appreciation, the financial support of the Chemical Science and Technology Laboratory, National Institute of Standards and Technology during his sabbatical leave for the period January 1 through June 30, 1999 during which the work in this paper was finished.

REFERENCES

1. E.W. Lemmon and R.T Jacobsen, *Int. J. Thermophys.* 20(3):825(1999).
2. E.W. Lemmon, A Generalized Model for the Prediction of the Thermodynamic Properties of Mixtures Including Vapor-Liquid Equilibrium, Ph.D. Dissertation, University of Idaho, Moscow (1996).
3. E.W. Lemmon, R.T Jacobsen, S.G. Penoncello, and D.G. Friend, to be submitted to *J. Phys. Chem. Ref. Data* (1999).
4. J.O. Hirschfelder, C.F. Curtiss, and R.B. Bird. "Molecular Theory of Gases and Liquids," Wiley, New York (1954).
5. S. Glasstone, K.J. Laidler, and H. Eyring. "The Theory of Rate Processes," McGraw-Hill, New York (1941).
6. H. Eyring and M.S. Jhon. "Significant Liquid Structures," Wiley, New York (1969).
7. C.C. Hsu and H. Eyring, *Proc. Nat. Acad. Sci. USA* 69(6):1342-1345 (1972).
8. A.L. Leu, S.-M. Ma, and H. Eyring, *Proc. Nat. Acad. Sci. USA* 72(3):1026-1030 (1975).
9. T.S. Ree, T. Ree, and H. Eyring, *J. Chem. Phys.* 68(11):3262-3267 (1964).
10. M.S. Jhon, T. Ree, and H. Eyring. *Chemistry* 54:1419-1426 (1965).
11. H. Hildwein and K. Stephan, *Chem. Eng. Sci.* 48(11):2005-2023 (1993).
12. K. Stephan, R. Krauss, and A. Laesecke, *J. Phys. Chem. Ref. Data* 16(4):993 (1987).
13. A. Laesecke, R. Krauss, K. Stephan, and W. Wagner, *J. Phys. Chem. Ref. Data* 19(5):1089 (1990).
14. R.T Jacobsen, S.G. Penoncello, and E.W. Lemmon, Thermodynamic properties of cryogenic fluids, in: "International Cryogenic Monograph Series," K. Timmerhaus, A. Clark, and C. Rizzuto, eds., Plenum Press (1997).
15. R. Krauss, V.C. Weiss, T.A. Edison, J.V. Sengers, and K. Stephan, *Int. J. Thermophys.* 17(4):731 (1996).
16. D.E. Diller, A.S. Aragon, and A. Laesecke, *Cryogenics* 31(12):1070 (1991).
17. J. Kestin and J.H. Whitelaw, Int. J. *Heat Mass Transfer* 7:1245 (1964).
18. I. Kurin and I. Golubev, *Teploenergetika* 21(11):84 (1974).
19. G.P. Matthews, C.M.S.R. Thomas, A.N. Dufty, and E.B. Smith, *J. Chem. Soc., Faraday Trans.* 172:238 (1976).
20. D.L. Timrot, M.A. Serednitskaya, and S.A. Traktueva, *Teploenergetika* 22(3):84 (1974).
21. J.A. Gracki and J. Ross, *J. Chem. Phys.* 51(9):3856 (1969).
22. G.C. Maitland and E.B. Smith, *J. Chem. Eng. Data* 17(2):150 (1972).
23. A. Van Itterbeek, J. Hellemans, H. Zink, and M. Van Cauteren, *Physica* 32:2171 (1966).
24. D.E. Diller, *Physica A* 119:92 (1983).
25. J. Golubev and I. Kurin, *Teploenergetika* 8:83 (1974).
26. J. Kestin and W. Leidenfrost, *Physica* 25:1033 (1959).
27. H. Geier and K. Schafer, *Allg. Warmetech.* 10(4):70 (1961).
28. R.A. Perkins and M.T. Cieszkiewicz, NISTIR 3961 (1991).
29. Kh.I. Amirkhonov, A.P. Adamov, and G.D. Gasanov, *Inz.-Fiz. Zh.* 22(5):835 (1972).
30. A. Michels, J.V. Sengers, and L.J.M. Van de Klundert, *Physica* 29:149 (1963).
31. H.M. Roder, R.A. Perkins, and C.A. Nieto de Castro, *Int. J. Thermophys.* 10(6):1141 (1989).
32. A.I. Johns, S. Rashid, T.R. Watson, and A.A. Clifford, *J. Chem. Soc., Faraday Trans.* 182:2235 (1986).
33. R.A. Perkins, H.M. Roder, and D.G. Friend, *Physica A* 173:332 (1991).
34. H.M. Roder, *J. Res. Natl. Bur. Stand.* 86(5):457 (1981).
35. H.M. Roder, *J. Res. Natl. Bur. Stand.* 87(4):279 (1982).
36. R. DiPippo, J. Kestin, and K. Oguchi, *J. Chem. Phys.* 46(12):4758 (1968).
37. J.M. Hellemans, J. Kestin, and S.T. Ro, *Physica* 65:362 (1973).
38. Y. Saji and T. Okuda, Adv. Cryo. Eng. 10:209 (1965).
39. M. Trautz and A. Melster, *Eingegangen* 27:409 (1930).
40. M. Yorizane, S. Yoshimura, H. Masuoka, and H. Yoshida, *Ind. Eng. Chem. Fundam.* 22:458 (1983).
41. X.Y. Zheng, S. Yamamoto, H. Yoshida, H. Masuoka, and M. Yorizane, *J. Chem. Eng. Japan* 17(3):237 (1984).

TRANSIENT BOILING ON THIN WIRES IN LIQUID NITROGEN

M.-C. Duluc, G. Defresne, and M.X. François

LIMSI-CNRS, UPR 3251
B.P. 133, 91403 Orsay Cedex, FRANCE

ABSTRACT

Transient boiling experiments are carried out on a small horizontal brass wire (∅=25μm) immersed in a liquid nitrogen bath under atmospheric pressure. Heat pulses of variable intensity ranging between 60% of the peak heat flux value and the peak heat flux are supplied to the wire in order to investigate the onset of a premature transition to film boiling. Various configurations are observed depending on the power level. At low heat flux, steady nucleate boiling is achieved in a few hundreds of milliseconds preceeded by a large temperature overshoot. Steady nucleate boiling is still observed on the wire at high heat flux but the transient period is drastically extended and more than ten seconds are sometimes required to achieve steady state. This feature is peculiar to the pulse heating procedure, which induces large wire temperature overshoots. Since the geometry is a thin wire, these conditions are favorable for the occurrence of mixed boiling states. Due to the important lifetime of these particular states, revealed by this transient heating procedure, this phenomenon has frequently been confused with a premature transition to film boiling.

INTRODUCTION

Transient boiling phenomena are involved in many cooling applications. The comprehension of these phenomena is of great importance in order to prevent any damage in cooling processes, in particular when a disturbance occurs in the system. Another point to be considered, concerns the necessity to predict the conditions in which a steady-state boiling configuration is settled and, in particular, to predict the possibility for an undesirable overheating of the solid surface to arise. In most practical applications, the study of the transient response of the cooling system is basically complicated by the large thermal inertia of the heating element leading the peculiar time constants of the fluid difficult to evaluate. Transient boiling on thin wires is then concurrently investigated with massive bodies. Data available in the literature concerning very thin wires (∅<100 μm) are obtained with fluids like FC-72, R113, methanol or nitrogen under atmospheric pressure. They show the possibility of a premature transition to film boiling for a heat flux smaller than the steady-state peak heat flux. This feature has been observed for platinum and nickel horizontal wires with FC-72, R113 and methanol with a linear increasing heating rate[1] and also in liquid nitrogen for horizontal and vertical platinum wires with a stepwise heating rate[2-4]. This premature transition may be initiated by the coalescence of neighboring bubbles, as nucleate boiling is present on the wire[2,4] but also directly from natural convection. In this case, the so-called "local dryout" phenomenon, a vapor sheath spreads all along the wire leading to a film boiling regime. The aim of the present paper is to investigate the conditions of transition using a brass wire (∅=25 μm) immersed in liquid nitrogen under atmospheric pressure conditions.

EXPERIMENTAL APPARATUS

Experimental cell

The experimental cell is immersed in a liquid nitrogen bath under saturation conditions at atmospheric pressure (T_{sat}=77.35 K). The liquid height above the cell is about 30 cm. The brass wire has a diameter D=25 μm and is of length L=5 cm. It has a thermoresistive nature, which makes it both a heater and a thermometer. The heat flux density dissipated by Joule effect $q_J(t)$ and the wire overheating $\Delta T(t)$ may then be calculated using the following relationships :

$$q_J(t) = \frac{U(t)I(t)}{\pi DL} \tag{1}$$

$$\Delta T(t) = T_w(t) - T_{sat} = \left(\frac{U(t)}{I(t)} - R_{sat} \right) \frac{1}{\gamma} \tag{2}$$

In order to measure the current intensity I(t) and the voltage U(t), the wire is then connected on four supports as depicted in Fig. 1. It is first soldered on the two external pins, for current supply. The voltage measurement is achieved by fixing the wire with silver paste on two internal contacts. Pasting is retained instead of soldering in order to reduce the thermomechanical effects which may lead to the wire breaking. These two internal contacts are designed large enough to set their temperature to the bath temperature. The knowledge of these boundary conditions is of practical interest when calculating the heat losses by the supports. Considering steady state conditions, these losses have been estimated being less than 3% as nucleate boiling regime is present on the wire (10% in the film boiling regime).

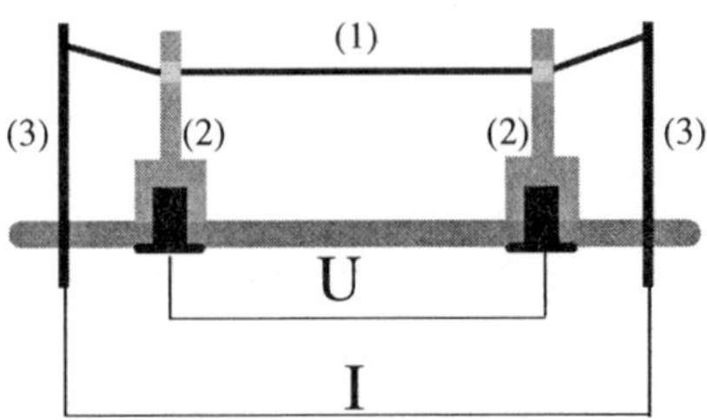

Figure. 1 Experimental cell: wire (1), internal contacts (2), pins (3)

R_{sat} is the wire electrical resistance at the bath temperature T_{sat} and is nearly equal to 5 Ω. γ is the wire thermal coefficient of constant value 0.01 Ω/K.

The present work reports investigations on transient phenomena induced by a heat pulse. Due to the very small value of the brass thermal coefficient γ, the heat pulse is achieved by the means of a current pulse. A schematic view of the whole electric circuit is displayed in Fig. 2. It includes two current supplies, a relay and two resistors. One of these resistors is the wire itself. The other one, which is a 1 Ω/8 watts type, is dedicated to the current intensity measurement. Its value is first very carefully controlled (1.015 Ω) as well as its ability to remain unchanged whatever the current intensity is ($I<1$ A and $\Delta R<\pm 0.0005$ Ω). The right part of the circuit (e_0, r_o, R_1 and R_w) allows the determination of the electrical resistance of the wire, R_{sat}, even without any heat pulse. The current value I_m delivered in this case is about 10 mA, which is low enough to ensure no selfheating effect ($q \leq 0.015$ W/cm²). The heat pulse is generated in the main branch of the circuit (E_0 , R_0 , R_1, R_w) once the relay is switched on. In order to achieve these two goals, the characteristics E_0, R_0, e_0 and r_0 are compliant with :

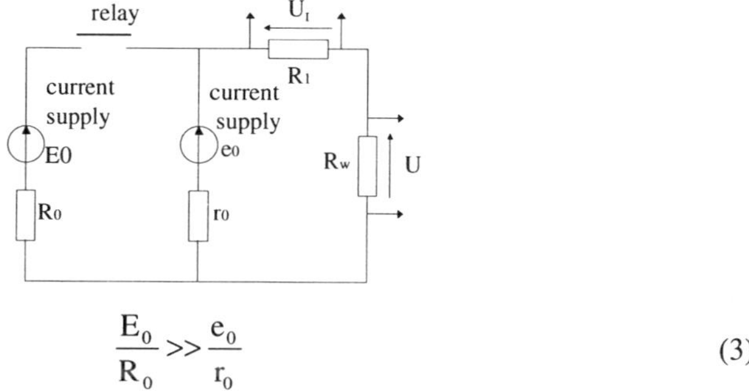

$$\frac{E_0}{R_0} >> \frac{e_0}{r_0} \quad (3)$$

Figure. 2 Electric circuit

Besides, the electrical resistor R_0 is also a 8 W type. The relay is a mercury type one. This technology provides closing time constants less than 50 μs and eliminates any overshoot phenomenon.

The tensions U_I and U are then connected to the high impedance inputs of two identical instrumentation differential amplifiers designed specifically for this experiment. As depicted in Fig 3, each amplifier includes two stages. The first one, whose gain value is either $G_1=1$ or $G_1=10$, delivers an output voltage U_{AO}. In order to achieve a zoom effect on the upper part of the signal, the mean value of U_{AO} is then suppressed by applying an opposite tension whose value is manually selected (offset). Finally, these variations are amplified once again by the second amplification stage ($G_2=10$) leading to U_{DO}. Using this procedure, the upper part of the signals U_{DO}, the only one to be relevant regarding of the physical phenomena, can be extracted then amplified without any saturation of the data acquisition board.

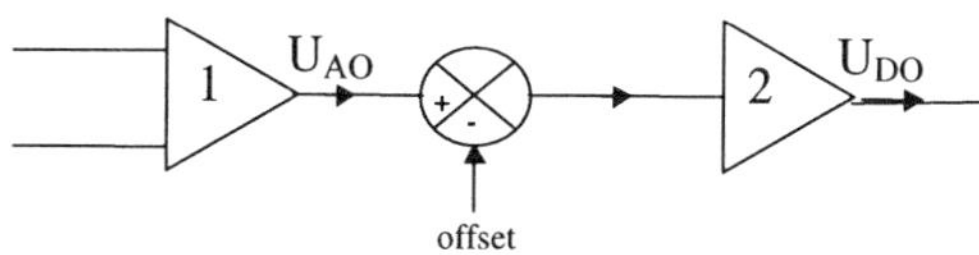

Figure. 3 Schematic view of the amplification device

The whole set of tensions U_{AO}, U_{DO}, $U_{offsetU}$, U_{IAO}, U_{IDO} and $U_{offsetI}$ is then connected to a 6-channel, simultaneously sampling, differential amplifier accessory which ensures a time skew between sampled channels less than 50 ns. The signals are finally recorded using a data acquisition board (0-10 V, 12 bits, 500 kHz). The sampling frequency ranges between 800 and 4000 Hz depending on the pulse width. The exact value of U(t) and I(t), from the occurence to the end of the heating pulse, is calculated by an automatic process, using the simple relashionships :

$$U(t) = \frac{U_{U_{DO}}(t)}{G_2} + Uoffset_U \tag{4}$$

$$I(t) = \frac{1}{R_1}\left(\frac{U_{I_{DO}}(t)}{G_2} + Uoffset_I\right) \tag{5}$$

Once these two parameters are known, it is possible to determine the mean wire overheating using equation (2). The heat flux density transmitted to the fluid $q_b(t)$ is given by the heat balance equation :

$$q_b(t) = q_J(t) - (\rho C)_{wire}\frac{D}{4}\frac{\partial T}{\partial t} - \frac{Q_s(t)}{\pi DL} \tag{6}$$

Where ρ and C are respectively the density and the specific heat of brass. Q_s is the conductive heat flux lost by the wire supports.

Due to the very small diameter of the wire, the last two terms of the right side are generally negligible and the heat flux generated by Joule effect can be regarded as the heat flux transmitted to the bath. This assumption is no longer true at earlier times when the rate of temperature increase of the wire is high and when large temperature gradients are present in the wire, i.e. for mixed boiling[5-6]. The whole experimental procedure leads to an uncertainty in the heat flux measurement of about 0.2 W/cm² and of about 1 K in the temperature measurement.

When investigating a transient process, the initial condition is of primary importance. In order to ensure a reproductive initial state, the wire is first degassed in air at 120°C for 30 minutes. Then, it is immersed in the nitrogen bath. Finally, before each heat pulse, a heat flux value equal to 90% of the critical heat flux is systematically supplied during 30 seconds to the wire in order to activate the whole set of nucleation sites. The heat pulse is then applied to the wire a few minutes later, once the wire temperature has returned to the bath temperature value.

RESULTS AND DISCUSSION

Stationary boiling curve

A classical stationary boiling curve obtained in liquid nitrogen under saturation conditions and atmospheric pressure with the 25 μm brass wire as a heating element is presented in Fig. 4. As is now commonly known, a thin wire leads to a particular feature. Some mixed boiling states, located in the transition region between the nucleate and the film boiling regimes are observed. The occurrence of these states, identified as the steady coexistence of nucleate and film boiling regimes on the same heating element[5-6] is connected to the very small diameter of a thin wire: although the wire displays areas of very different temperature, the resulting conductive heat flux remains however low enough to avoid the spreading of the film boiling part. And this particular feature would naturally not be observed with a massive heater.

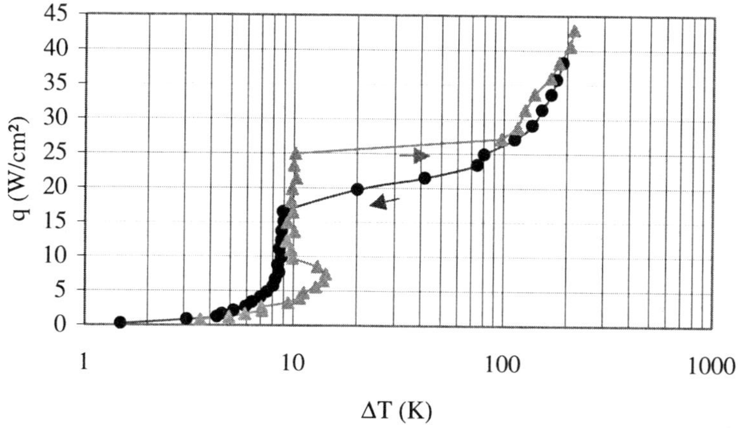

Figure 4. Stationary boiling curve. ▲ increasing heating rate, ● decreasing heating rate

Transient results

Several heat pulses increasing from 15.7 W/cm² to 24.4 W/cm² are supplied to the wire. As can be seen on the steady boiling curve, the critical heat flux value is 25 W/cm². The stationary state is then expected to be nucleate boiling.

Various kinds of response are observed depending on the heat flux level. The results obtained at relatively low heat flux values, i.e. less than 19 W/cm², are displayed in Fig. 5. In this case, the stationary state, characterized by a wire overheating of about 10 K, is nucleate boiling. The transient regime is characterized by a large temperature overshoot. Its value remains however always less than the homogeneous nucleation temperature, equal to 32 K in liquid nitrogen at atmospheric pressure[7]. The very fast temperature increase observed at earlier stages is due to conduction in the liquid phase as demonstrated by Giarratano using an analytical method[8].

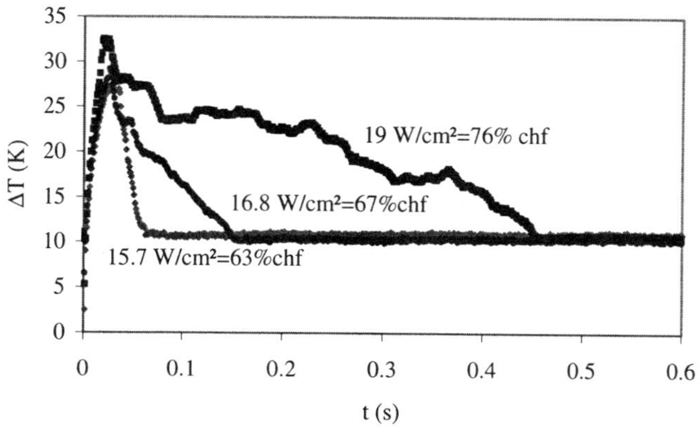

Figure 5. Wire temperature response as a function of time.
The step power is initiated at t=0. The corresponding heat flux intensity is indicated on each curve.

The decrease from the maximum wire overheating to the steady state may be achieved in a very short time (less than 0.03 s) as depicted in Fig. 5 by the 15.7 W/cm² curve but it may also require longer times as displayed by the 16.8 and 19 W/cm² curves. In order to explain such a feature, the stationary boiling curve, reported in Fig. 4, has to be considered. It shows that for a heat flux value less than 16.5 W/cm², nucleate boiling is the only possible regime. Beyond this value, a mixed boiling state may also be obtained provided that the wire temperature is large enough. Due to the very high overshoot temperatures induced by the pulse heating, some of these states are "caught". Thus, this transient procedure reveals their metastable nature and allows to determine their lifetime.

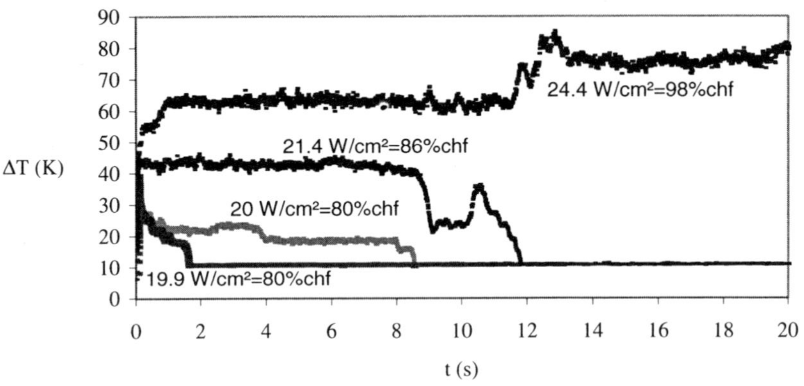

Figure 6. Wire temperature response as a function of time.
The step power is initiated at t=0. The corresponding heat flux intensity is indicated on each curve.

For larger heat fluxes (Fig. 6), the rate of wire temperature increase during conduction in the liquid phase is so fast that a 32 K temperature is achieved before nucleation could occur. The fluid in contact with the solid can no longer remain in the liquid phase and a vapor sheath develops along the wire, leading to the 'local dryout' phenomena. Next, a maximum is achieved. Its value is about 40 K for heat flux values less than 21.4 W/cm² and 44 K for 24.4 W/cm². The stationary boiling curve, displayed in Fig. 4, shows that the only two possible wire overheatings for a heat flux value of about 20 W/cm² are 20 K and 10 K (nucleate boiling). This explains why in a case (19.9 W/cm²), the solid-fluid system achieves the stationary state in a direct decrease and why in another case (20W/cm²), a flat part corresponding to a 20 K wire overheating is being observed for a few seconds. An analog scenario is displayed by the 21.4 W/cm² curve. After the first maximum, the wire temperature increases slightly, then is stabilized during almost 8 seconds at about 43 K. This boiling state (21.4 W/cm²; 43 K) is compliant with the stationary boiling curve. A nucleate boiling steady state is finally obtained, preceded by an oscillating process between two temperatures. These results show that nucleate boiling may sometimes require more than ten seconds to be established, which is quite unexpected on such a thin wire. To our knowledge, no other investigation has ever been carried out with heat pulses of such large duration. On the contrary, data reported in the literature correspond to very short time periods, i.e. less than 0.5 s [2-4]. In these conditions, a sharp temperature increase may easily be mistaken with a premature transition to film boiling.

Finally, a heat flux value close to the critical heat flux has been tested (24.4 W/cm²). The curve displayed in Fig. 6 shows that the wire overheating is stabilized at an average temperature of about 80 K. A transition to film boiling has happened and no steady nucleate boiling could be achieved.

CONCLUSION

The experiments carried out on a thin wire submerged in liquid nitrogen, using a pulse as a heating procedure, show that the process by which nucleate boiling is established on the wire may be quite complex. A heat pulse induces, at earlier stages, a very high wire temperature increasing rate, corresponding to conduction heat transfer in the liquid phase. Then, the wire may achieve a temperature high enough to allow a mixed boiling state to occur. However, these states are metastable and their lifetime, controlled by the heat flux intensity, may sometimes attain several seconds, accordingly delaying the establishment of the steady nucleate boiling regime. Furthermore, these results show that the stationary boiling curve is absolutely necessary in order to understand the transient characteristics.

REFERENCES

1. S.M. You, Y.S. Hong, J.P. O'Connor, The onset of film boiling on small cylinders : local dryout and hydrodynamic critical heat flux mechanisms, *Int. J. Heat Mass Transfer*, 37:2561-2569 (1994).
2. K. Okuyama, Y. Iida, Transient boiling heat transfer characteristics of nitrogen (bubble behavior and heat transfer rate at stepwise heat generation, *Int. J. Heat Mass Transfer*, 33:2065-2071 (1990).
3. E.W. Roth, E. Bogedom, L.C. Brodie, J.S. Semura, Transient heat transfer in liquid nitrogen, *Advances in Cryogenic Engineering*, 35:447-452 (1990).
4. O. Tsukamoto, T. Uyemura, T. Uyemura, Observation of bubble formation mechanism of liquid nitrogen subjected to transient heating, *Advances in Cryogenic Engineering*, 25:476-482 (1981).
5. S.A. Zhukov, V.V. Barelko, Non uniform steady states of the boiling process in the transition region between the nucleate and film boiling regimes, *Int. J. Heat Mass Transfer*, 26:1121-1130 (1983).
6. M.C. Duluc, M.X. François, Steady state transition boiling on thin wires in liquid nitrogen. The role of Taylor wavelength, *Cryogenics*, 38:631-638 (1998).
7. D.N. Sinha, L.C. Brodie, J.S. Semura, Liquid to vapor homogeneous nucleation in liquid nitrogen, P*hys. Rev. B*, 36:4082-4085 (1987).
8. P.J. Giarratano, Transient boiling heat transfer from two different heat sources : small diameter wire and thin film flat surface on a quartz substrate, *Int. J. Heat Mass Transfer*, 27:1311-1318 (1984).

From left to right: K. Timmerhaus (University of Colorado), S. Breon (Goddard Space Flight Center), Q. S. Shu (AMAC International), R.Witt (University of Wisconsin), and V. Bardos (Synchrony) at the Best Paper Awards event.

VISUALIZATION STUDY OF HEAT TRANSPORT MECHANISM IN SUPERCRITICAL NITROGEN

A. Nakano[1], M. Shiraishi[1], M. Nishio[1], S. Someya[1], T. Iida[2] and M. Murakami[3]

[1] Mechanical Engineering Laboratory, AIST, MITI
Namiki 1-2, Tsukuba, Ibaraki, 305-8564, Japan
[2] National Space Development Agency of Japan
Sengen 2-1-1, Tsukuba, Ibaraki, 305-0047, Japan
[3] Institute of Engineering Mechanics and Systems, University of Tsukuba
Tennodai 1-1-1, Tsukuba, Ibaraki, 305-8573, Japan

ABSTRACT

Heat transport mechanisms in supercritical nitrogen are examined by using a laser holographic interferometer. The experimental apparatus for the visualization study is designed to observe the process of thermal energy propagation under the supercritical state. The experimental cell is a rectangular parallelepiped surrounded with two optical windows, a ceiling, two side-walls made of Bakelite and a planar heater at the bottom. Heat is applied in step functions from the planar heater. Three cernox resister temperature detectors (RTDs) are set vertically in the cell to obtain the temperature distribution. The measured temperatures are compared with the experimental results obtained from the visualization study.

INTRODUCTION

Supercritical fluids have mainly been used in the chemical and the environmental engineering fields. Recently, the utilization of the supercritical fluids is considered in the energy-engineering field. An energy storage system in the form of liquid air is one of application technologies utilizing the thermophysical properties near the critical point of air.[1] The system is combined with a gas turbine system, which uses liquid natural gas (LNG) as fuel. The investigation of the thermophysical properties and the heat transfer characteristics of supercritical fluids are also expected for the future technologies of a geothermal heat application and nuclear power plants.

Fluids are in the supercritical state when their temperature and pressure are above their critical temperature and pressure. The vapor/liquid interface no longer exists and the density takes an intermediate state between the liquid and vapor. Near the critical point, the thermophysical properties exhibit very strange behavior. The specific heat and the isothermal compressibility at the critical point show a strong divergence, which leads to an unusual characteristic of heat transfer. That is called the " piston effect " which occurs near the critical point since the thermal energy propagates as an acoustic wave.[2~5] This effect originates from the high compressibility and the low thermal diffusivity of supercritical fluids. Heat transfer in dense fluids is normally achieved due to the basic mechanisms of convection, diffusion and radiation. The piston effect is considered to be a fourth mechanism of heat transfer.[2] It is also reported that heat transfer coefficient becomes large near the critical point because quasi-boiling phenomena enhances the heat transfer.[6] To investigate these strange behaviors in supercritical fluids, the visualization study is a powerful and useful method.

In this study, we try to apply a laser holographic interferometer to supercritical nitrogen. It is observed that a variety of thermodynamic phenomena occur when heat is applied in a step function from a planar heater. And we also report the phase transition process of nitrogen from the vapor/liquid equilibrium state to the supercritical state with the aid of shadowgraph visualization technique.

EXPERIMENTAL SET-UP AND PROCEDURE

The schematic illustration of the arrangement of optical elements for the laser holographic interferometer is shown in Figure 1. The infinite-fringe method with the double exposure technique is applied in the present study.[7] The wavelength of 632.8 nm from a He-Ne laser is used to record the hologram. The density variation appears in the form of interference fringes in a holographic image. The first exposure is made for recording an initial quiescent state, which is an initial temperature or density field, and then following the onset of heating the second exposure is done to record a disturbed state due to heating from a planar heater that is located at the bottom portion of the experimental cell. The variation

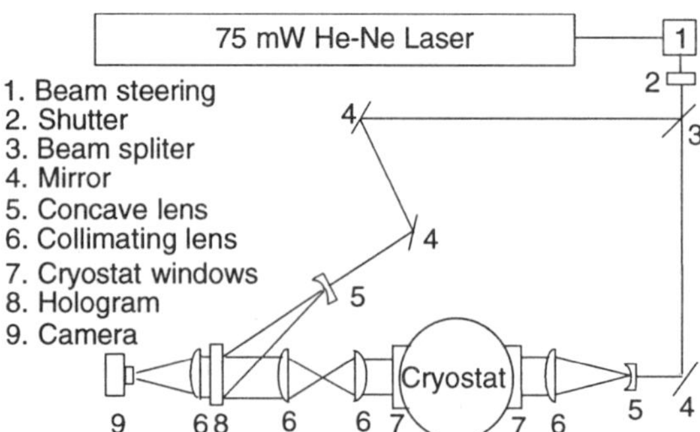

Figure 1. Arrangement of optical elements for laser holographic interferometer.

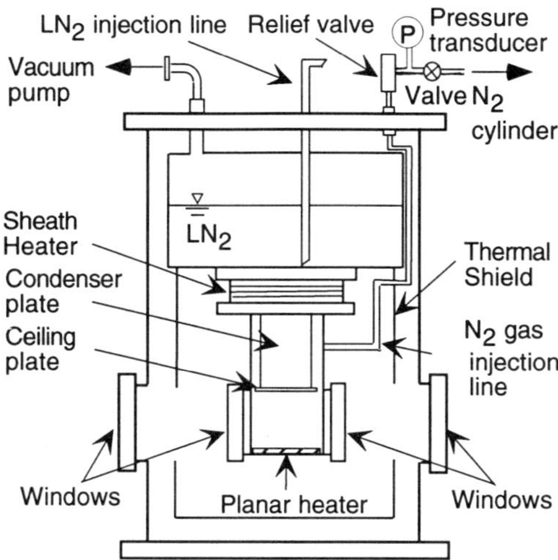

Figure 2. The schematic illustration of the cryostat

in refractive index due to the density variation gives the variation in optical path length along which the laser light passes. Figure 2 shows the schematic illustration of the cryostat. The experimental cell is a rectangular parallelepiped surrounded with two optical glass plates, a ceiling, two side-walls made of Bakelite with the thickness of 2 mm and a planar heater at the bottom. It is 16 mm wide, 65 mm deep between the window glass plates and 40 mm high. The planar heater, which is 20 mm wide and 46 mm deep, consists of a manganin wire and a thin copper plate with a thickness of 100 μm on a Bakelite plate. We made grooves with 1 mm interval and 0.2 mm deep on the Bakelite plate. The manganin wire with 0.2 mm in diameter is embedded in the grooves. The copper plate is glued by using Stycast 2850GT on the Bakelite plate. The electric resistance of the heater is 11.82 Ω. The ceiling plate is made of copper, 20 mm wide and 40 mm deep. Three cernox-1080 RTDs are set in the experimental cell to measure the inside temperature. They are set above the planar heater (RTD1), halfway on the side-wall (RTD2) and on the ceiling plate (RTD3). The pressure in the experimental cell is measured by a pressure transducer.

After sufficiently evacuating the inside of the experimental cell, the gaseous nitrogen is charged into the cell through the injection line. The gaseous nitrogen is condensed into liquid on the cooled condenser plate. Then the liquid flows down to the bottom of the cell and forms a pool. When the vapor/liquid interface exceeds half the experimental cell, the valve at the injection line is closed. After this procedure, the liquid nitrogen in the precooling vessel is extracted and the experimental cell is heated up by using a sheath (mineral insulated) heater mounted on the upper portion of the cell. When the temperature of the cell reaches about 100 K, the sheath heater is switched off. After that, the inside vapor pressure and the temperature in the cell gradually increase by thermal conduction and radiation from the surroundings and approach the critical point. Figure 3 shows the visualization condition by shadowgraph technique and laser holographic interferometry. The visualization using shadowgraph technique is started before the inside pressure and the temperature reach the critical point. The process of phase transition is recorded by a video camera. On the other hand, we take holograms after reaching the supercritical state by using a laser holographic interferometer. Hologram information is recorded on a holographic plate, Agfa 10E75. The reconstructed holographic image showing a fringe field is photographed by a still camera.

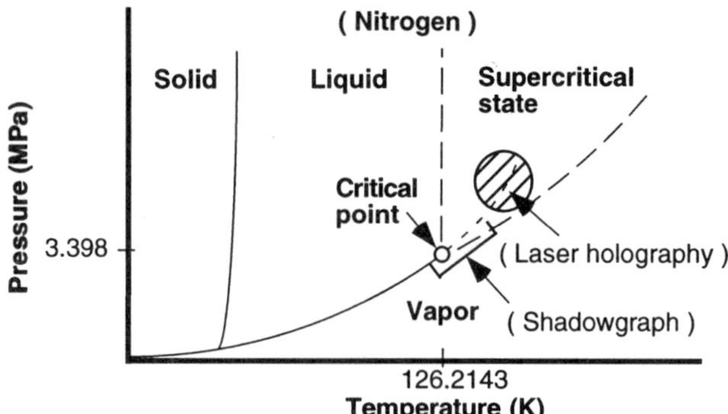

Figure 3. The visualization condition by shadowgraph technique and laser holographic interferometry.

EXPERIMENTAL RESULTS AND DISCUSSION

Shadowgraph visualization

We observe the phase transition from the vapor/liquid equilibrium state to the supercritical state by the use of a shadowgraph technique. The inside pressure and the temperature in the cell gradually increase by thermal radiation from the surroundings. The planar heater is not used during this experiment. Figure 4 shows the shadowgraph pictures, which indicate the phase transition of nitrogen. The inside pressure varies from 3.246 MPa to 3.530 MPa. The critical pressure and the critical temperature of nitrogen are 3.398 MPa and 126.2143 K, respectively. When the inside pressure approaches to the critical pressure, the thickness of the vapor/liquid interface becomes large. The strong scatter of light occurs

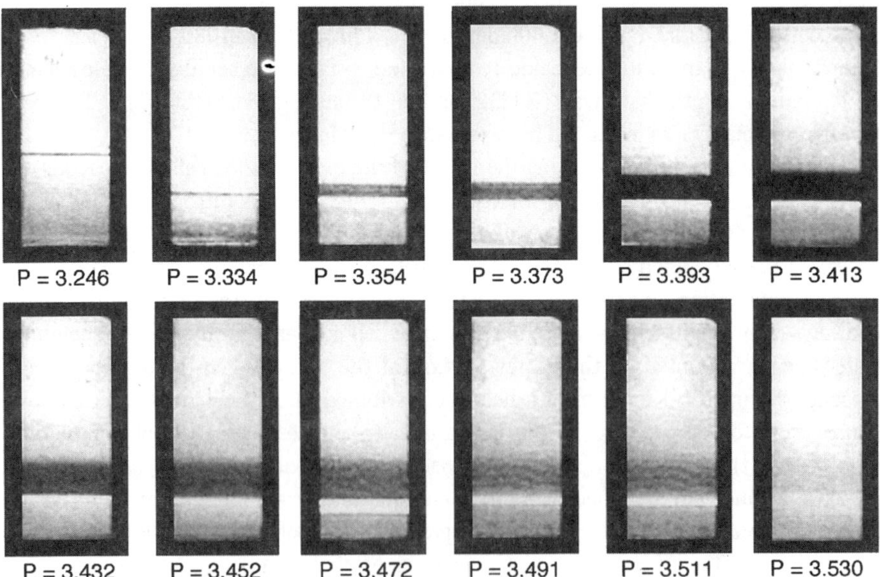

Figure 4. Phase transition of nitrogen from the vapor/liquid equilibrium state to the supercritical state. (Unit : MPa) The critical pressure is 3.398 MPa.

at the interface. As the inside pressure becomes large, the outline of the expanded interface has faded and a bright line appears at the bottom of the expanded interface. Further increasing the inside pressure, the outline of the expanded bright interface has also faded and the nitrogen in the cell turns to a single phase. It is found from this visualization study that the phase transition occurs first around the vapor/liquid interface.

Visualization study by using a laser holographic interferometer

The laser holographic interferometer method has not been applied to the observation of the thermodynamics phenomena in supercritical fluids. Therefore, to demonstrate the validity of the method in experiments of thermodynamics in supercritical fluids is one of objective of this study. We add heat from the planar heater that is located at the bottom of the experimental cell and observe the thermodynamic phenomena in the supercritical nitrogen.

Laser holographic interferometry

In the case of a laser holographic interferometry, the density, ρ, can be approximately related linearly to the refractive index, n, by

$$(n-1)/(n_0-1)=\rho/\rho_0 \tag{1}$$

Here, the subscript 0 shows the reference value. For an isobaric process, this results in

$$\Delta n=\left[(n_0-1)/\rho_0\right]\Delta\rho \tag{2}$$

The Δn shows λ/L, where λ is the wavelength of the laser light and L is the optical pass length across the cell, respectively. The density variation corresponding to an interval between two interference fringes can be written by

$$\Delta\rho=\lambda\rho_0/\left[(n_0-1)L\right] \tag{3}$$

On the other hand, the temperature variation corresponding to the interval of interference fringes is

$$\Delta \mathrm{T}=\lambda/\left[(1-\mathrm{n}_0)\mathrm{BL}\right] \tag{4}$$

where, the B shows the coefficient of thermal expansion, which is described by

$$B=-\frac{1}{\rho_0}\left(\frac{\partial\rho}{\partial T}\right)_P \tag{5}$$

However, it is well known that the B has an infinite value at the critical point. Therefore, the present experiment is carried out after the inside pressure reaches about 3.53 MPa, where the inside of the cell is a single phase as shown in Figure 4.

Refractive index of supercritical nitrogen

The refractive index of supercritical nitrogen, n_{sup}, is required for the quantitative measurement from the images of interference fringes produced by temperature or density variation. Therefore, we estimated the n_{sup} from a refractive index of gaseous nitrogen, n_g. The Clausius-Mossotti's equation gives a relationship between a relative dielectric constant

κ and a molecular constant α as,[8]

$$\frac{\kappa-1}{\kappa+2}=\frac{N\cdot\alpha}{3} \quad (6)$$

Here, the N shows the number of molecules per unit volume. The left side in Eq. (6) is proportional to the density ρ. From the Lorentz-Lorenz formula, the κ can be represented by a refractive index n, like $\kappa=n^2$.[9] Then, we can estimate the n_{sup} using a following relation,

$$\frac{n_{sup}^2-1}{n_{sup}^2+2}:\frac{n_g^2-1}{n_g^2+2}=\rho_{sup}:\rho_g \quad (7)$$

Here, the ρ_{sup} and the ρ_g are the densities of the supercritical nitrogen and the gaseous nitrogen, respectively. From Eq. (7), we found that the n_{sup} is about 1.075 near the critical point. The value is consistent with other researchers result.[10] It is noted from Eq. (4) that the ΔT becomes small as the condition approaches to the critical point, because the B diverges at the critical point.[11] In this experimental condition, the ΔT varies from about 0.1 mK to 3 mK.

Visualization of heat transfer in supercritical nitrogen

The experimental set-up does not have a control system of the inside temperature and the pressure. Therefore, the temperature and the pressure in the cell gradually increase by thermal radiation and conduction from the surroundings. It is important to grasp the background noise like intrusion heat from the surroundings before the experiment. Figure 5a ~ d show the interferograms in the case of q = 0 W/m^2. The T_{RTD1} and T_{RTD3} show the temperature of the bottom and the top surfaces in the experimental cell. The inside pressure and the mean temperature in the cell are referred to obtain the ΔT. The Δt shows the time interval between the first exposure and the second. The interference fringes are observed in Figures 5a, 5b and 5c. It is considered that the surface of the planar heater is heated by

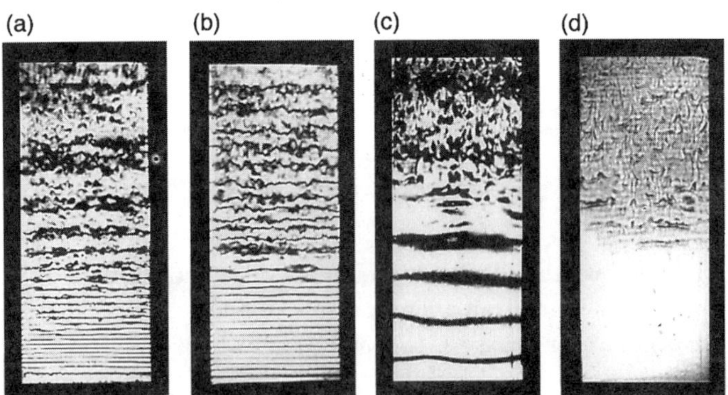

Figure 5. Infinite interferogram in the case of q=0 W/m^2.
(a) P = 3.530 MPa, T_{RTD1}= 126.45 K, T_{RTD3}= 126.75 K, ΔT = 0.50 mK, Δt = 5 sec
(b) P = 3.658 MPa, T_{RTD1}= 126.88 K, T_{RTD3}= 127.64 K, ΔT = 0.50 mK, Δt = 10 sec
(c) P = 3.756 MPa, T_{RTD1}= 128.18 K, T_{RTD3}= 129.19 K, ΔT = 0.42 mK, Δt = 20 sec
(d) P = 3.825 MPa, T_{RTD1}= 129.51 K, T_{RTD3}= 129.98 K, ΔT = 0.76 mK, Δt = 30 sec

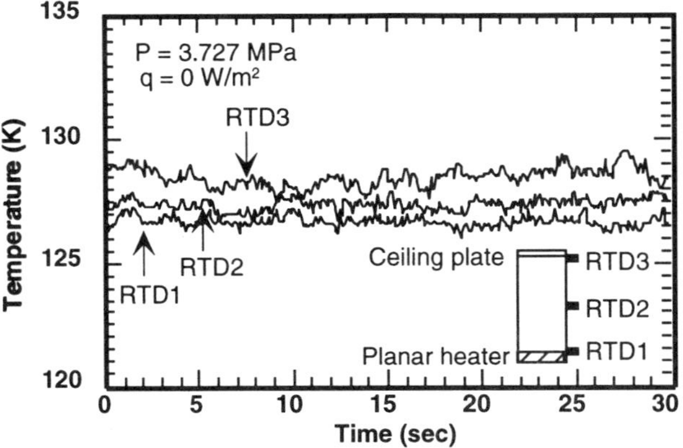

Figure 6. The time variations of temperature in the case of $q = 0$ W/m^2.

thermal radiation but the heat flux must be very small. As increasing the inside temperature and pressure, the interval of fringes become wide even though the Δt increases. And there is no interference fringe in Figure 5d. It suggests that the sensitivity becomes large as the system approaches to the critical point. Figure 6 shows the time variation of temperature when no heat flux is added from the planar heater. It is confirmed from the figure that the initial temperature and the final temperature of each position are almost the same in the time range. The figure suggests that the temperature of the bottom portion is lower than the upper part. The formation of the temperature gradient is always observed and that is due to a characteristic of this experimental set-up. It is considered that the temperature gradient suppresses the generation of convection. Figures 7a ~ d show the interferogram when the heat flux q of 237.3 W/m^2 is added from the planar heater. At glance, there are no interference fringes near the planar heater in the photos of the interferograms. The fringe free region grows as the Δt increases. It is observed by shadowgraph technique that convection occurs in the area. However, the rest of area is not affected by the disturbance due to the convection. It is noted from these figures that the temperature fields in the

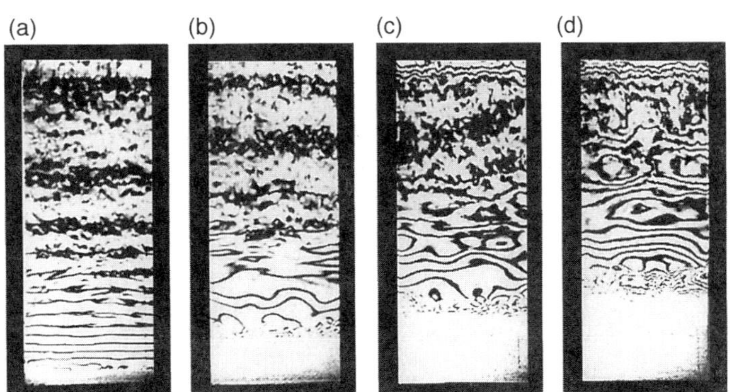

Figure 7. Infinite interferogram in the case of q=237.3 W/m^2.

(a) P = 3.579 MPa, T_{RTD1}= 126.60 K, T_{RTD3}=127.13 K, ΔT = 0.51 mK, Δt = 5 sec

(b) P = 3.668 MPa, T_{RTD1}= 127.60 K, T_{RTD3}=128.15 K, ΔT = 0.31 mK, Δt = 10 sec

(c) P = 3.756 MPa, T_{RTD1}= 128.62 K, T_{RTD3}=128.54 K, ΔT = 0.41 mK, Δt = 20 sec

(d) P = 3.834 MPa, T_{RTD1}= 129.07 K, T_{RTD3}=129.98 K, ΔT = 0.60 mK, Δt = 30 sec

middle portion in the photos are hardly affected by the disturbance and they are relatively stable. It suggests that a strong stability boundary exists between the fringe free region and the rest of area. It is considered that the initial temperature gradient discussed above is related to the formation of the stability boundary. It is also noted from figures 7c and 7d that some interference fringes are formed near the ceiling plate. The interference fringes in the photo suggests a typical temperature profile due to the piston effect. So we think that the typical phenomenon caused by the piston effect might be observed through the experiments although it has been reported that the effect could be observed only under the micro gravity condition. However, further investigation is required to confirm the piston effect.

CONCLUSIONS

We have investigated thermodynamic phenomena in supercritical nitrogen by means of visualization methods. The following conclusions are obtained:

1. It is observed that the phase transition from the vapor/liquid equilibrium state to the supercritical state occurs on the vapor/liquid interface.
2. We succeed in detecting the thermodynamic phenomena in supercritical nitrogen by using a laser holographic interferometer.
3. The temperature profile due to the piston effect might be observed in supercritical nitrogen although further investigation is required.

REFERENCES

1. H. Araki, K. Chino and M. Tsukamoto, Liquid air energy storage system, Proc. of 75th JSME (3): 425 (1998). (in Japanese)
2. P.Guenoun, B. Khalil, D. Beysens, Y. Garrabos, F. Kammoun, B. Le Neindre, and B. Zappoli, Thermal cycle around the critical point of carbon dioxide under reduced gravity, Phys. Rev. E. 47 (3): 1531 (1993).
3. A. Onuki, R. A. Ferrell, Adiabatic heating effect near the gas-liquid critical point, Physica A 164: 245 (1990).
4. T. Maekawa, K. Ishii, Temperature propagation in a single component critical fluid, Thermal Science and Eng. (The Heat Transfer Soc. of Japan) 5 (1): 15 (1997).
5. A. Nakano, M. Shiraishi, M. Nishio, F. Takemura and M. Murakami, Numerical analysis of heat transport mechanism in nitrogen near the critical point, Adv. in Cryog. Eng. 43: 1297 (1998).
6. K. Yamagata, K. Nishikawa, O. Hasegawa, S. Fujii and K. Miyabe, Heat transmission of natural convection in supercritical fluids, Transaction of JSME 32 (233): 98 (1946). (in Japanese)
7. T. Iida, M. Murakami, T. Shimazaki and H. Nagai, Visualization study on the thermo-hydrodynamic phenomena induced by pulsative heating in He II by the use of a laser holographic interferometer, Cryogenics 36: 943 (1996).
8. A. Sommerfeld "Electrodynamics" Academic Press, New York San Francisco London (1952).
9. A. Sommerfeld "Optics" Academic Press, New York and London (1959).
10. M. R. Moldover, J. V. Sengers, R. W. Gammon and R. J. Hocken, Gravity effects in fluids near the gas-liquid critical point, Rev. Mod. Phys. 51 (1): 79 (1979).
11. R. T. Jacobsen, R. B. Stewart and M. Jahangiri, Thermodynamic properties of nitrogen from the freezing line to 2000 K at pressures to 1000 MPa, J. Phys. Chem. Ref. Data 15 (2): 735 (1986).

RELIABLE LONG-TERM OPERATION OF THE CRYOGENIC SYSTEM FOR THE LARGE HELICAL DEVICE

T. Mito, R. Maekawa, T. Baba, S. Moriuchi, A. Iwamoto, A. Nishimura, S. Yamada, K. Takahata, S. Imagawa, N. Yanagi, H. Tamura, S. Hamaguchi, K. Oba, H. Sekiguchi, T. Satow, S. Satoh, and O. Motojima

National Institute for Fusion Science
322-6 Oroshi, Toki, Gifu 509-5292, JAPAN

ABSTRACT

Reliability is discussed during long-term operations of the cryogenic system for the Large Helical Device (LHD). The cooled objects of LHD are the helical coils, the poloidal coils, the supporting structure and the superconducting bus lines. The cryogenic system, which permits complicated cooling schemes for each cooled object, was designed to focus on the reliable long-term operation. Impurities in the helium gas should be controlled carefully to prevent a blocking of filters, malfunction of control valves and deterioration of heat exchanger performance. The first cycle operation of LHD started on February and ended on June in 1998, and the second cycle operation started on August and continued to the end of 1998. The performance during the first and second cycle long-term operations and the precise impurity measurement technique are reported.

INTRODUCTION

The LHD cryogenic system completed 6400-hours of operation during the first year, and proved its high reliability.[1-3] For a reliable long-term operation, it is essential to control contamination in the cryogenic system to as low a level as possible. The Large Helical Device (LHD) is a superconducting, heliotron type, experimental fusion device, which has the features of current-less and steady-state plasma confinement. The construction of LHD was completed by the end of 1997 as an eight-year project for the phase I experiment. The plasma experiment was started on March 31, 1998 as planned. Two experimental campaigns were successfully completed in 1998. The numbers of coil excitations and plasma discharges were 252 and 7132, respectively. The LHD has extended the envelope of confinement study on currentless helical plasmas by more than one-order of magnitude in plasma size. High-performance plasmas with temperature as high as 2 keV and energy confinement times as long as 0.23 seconds have been achieved by neutral beam heating. Long pulse operation has also been explored up to 22 seconds for neutral beam heated plasmas and 2 minutes for electron cyclotron heated plasmas.

Advances in Cryogenic Engineering, Volume 45.
Edited by Shu *et al.*, Kluwer Academic / Plenum Publishers, 2000.

MEASURE OF CRYOGENIC SYSTEM AGAINST IMPURITIES

The flow diagram for the LHD cryogenic system is shown in Fig. 1, and the specifications of the main components are listed in Table 1. The cold box of the helium refrigerator/liquefier simultaneously has a cooling capacity of 5.65 kW at 4.4 K, 20.6 kW from 40 K to 80 K and 650 L/h liquefaction. The helical coils are cooled with pool boiling liquid helium. The poloidal coils are cooled with forced flow supercritical helium. The supporting structure and the superconducting bus lines are cooled with forced flow two-phase helium and the liquid helium is used for the current leads. Three different cryogen systems (supercritical helium, liquid helium, 40 K - 80 K helium gas for radiation shields) are supplied from the cold box to the cooled objects. They are distributed and controlled by the helical valve box, the poloidal valve box and the current leads cryostats.

Purification System

Various contrivances for eliminating impurities and preventing contamination to the superconducting coils are made in the LHD cryogenic system. Before cool down, all the helium gas in the cryogenic system is purified with a helium purification system whose outlet impurity is less than 1 ppm. The helium purification system consists of the drier (molecular sieve adsorber) and the cryogenic purifier (liquid-nitrogen cooled activated-carbon adsorber). Both drier and purifier have twin units with an automatic-switching method every 12 hours. The helium purification system with a throughput of 50 g/s provides partial purification of the circulation gas, which is about 1/20 of the total flow of 960 g/s supplied by eight sets of oil injected screw compressors (main compressors). The latter helium purification system is also combined with a recovery compressor with a flow rate of 50 g/s, and purifies impure helium gas obtained from external gas containers. The resulting purified helium gas is then transferred to the helium buffer tanks. The amount of helium gas in the system before cool down is 21,000 Nm^3, when two spherical buffer tanks (a maximum-pressure of 2.0 MPa and a volume of 700 m^3) are filled to 1.5 MPa. After cool down, 26,000 Nm^3 of helium gas is required to maintain the steady state in which coil excitation is possible. Therefore, helium gas is added from an external source during the cool down operation. Very small quantities of impurities, which remain in the system after the cool down are removed by three stages of activated-carbon adsorbers (ADS-1, -2, -3), installed in the cold box of the helium refrigerator/liquefier. ADS-1 and ADS-2 have twin units, and these are regenerated every seven days.

Coil Inlet Filters

Filters are installed in the inlet of each superconducting coil and superconducting bus line for the purpose of coil protection and impurity elimination. The locations of the filters are shown in Fig. 2, and their specifications are listed in Table 2.

Since the helical coils are pool boiling coils, the filters used for inhibiting electrical breakdown by removing conductive foreign matter to the coils are attached in the inlets of the helical valve box and the helical coil. The filters at the upstream location are 5 μm slit type filters installed in a riser at the inlet of the helical valve-box, which essentially prevents the contamination when any activated-carbon powder migrate from the internal adsorber of the helium refrigerator. At the down-stream location, the 100 μm slit type filters are attached at the coil inlet as a protection for metal contamination. The 5-μm filter at the valve-box inlet also prevents solidification of impurities. This filter serves as a twin unit which enables filter regeneration on a continuous basis.

The poloidal coils are forced-flow-cooled with supercritical helium using a cable-in-conduit conductor without a cooling subchannel. For this reason, special caution needs to be paid to choking of the helium path by impurities, etc. The parallel 5-μm sintered-metal mesh filters are installed in the poloidal valve-box inlet. Furthermore, 20-μm slit type filters for inhibiting the contamination of solid impurities are installed in the inlet of the upper and lower poloidal-coil groups.

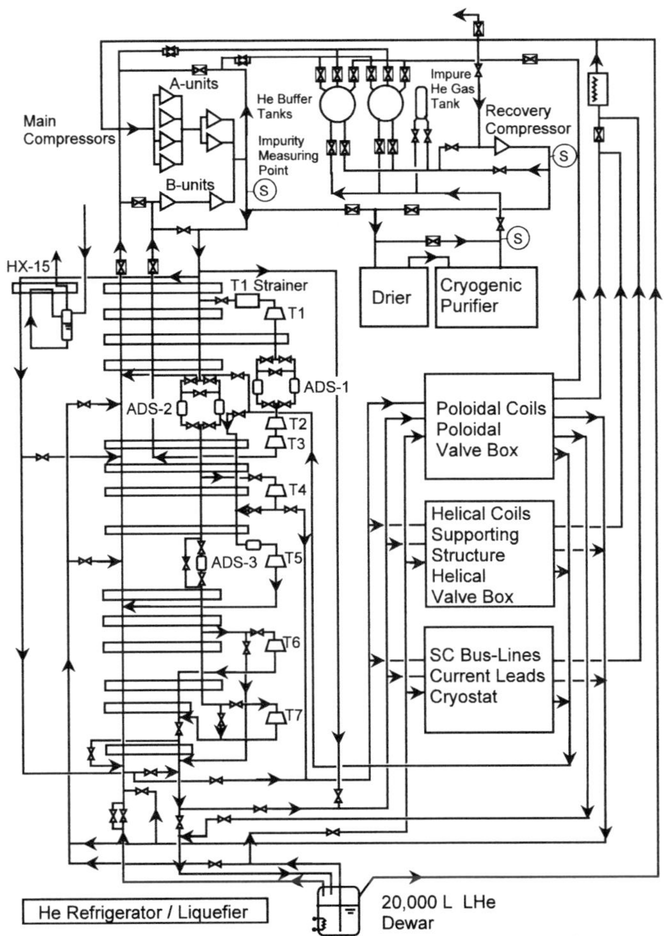

Figure 1. Flow diagram of the LHD cryogenic system.

Table 1. Specifications of the LHD cryogenic system main components

He Refrigerator/Liquefier		He Purification System	
Refrigeration capacity at 4.4 K	5670 W	Drier	Twin molecular sieve adsorbers
Liquefaction rate	650 L/h	Purifier	Twin LN_2-cooled charcoal adsorbers
Refrigeration capacity of 40 - 80 K	20.6 kW	Throughput	50 g/s
Main Compressors		Automatic-switching time	12 h
Oil injected screw compressors	8 units	Liquid He Dewar	
Inlet / outlet pressure of A-units	0.10 / 1.94 MPa	Capacity	20,000 L
Inlet / outlet pressure of B-units	0.20 / 1.94 MPa	Gas He Storage	
Mass flow rate of A-units	750 g/s	Design pressure	2.0 MPa
Mass flow rate of B-units	210 g/s	Buffer tanks	700 $m^3 \times 2$
Number of compressors	A-units: 6, B-unit: 2	Impure gas tank	100 m^3
Rated power	3,172 kW	Required He gas inventory	26,000 Nm^3

RESULTS OF LONG-TERM CONTINUOUS OPERATION

Progression of the operating modes for the 1st and the 2nd operating cycles of the LHD cryogenic system are shown in Fig. 3. Each cycle consists of a purification, cool down, steady state, and warm-up operation.

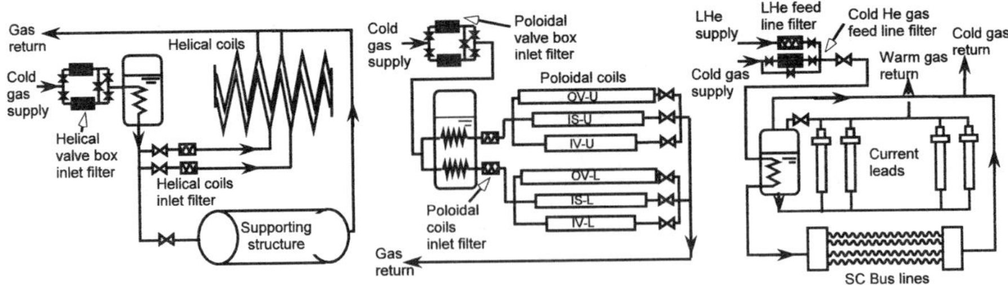

Figure 2. Position of the inlet filters for helical coils, poloidal coils and SC bus lines.

Table 2. Specifications of the inlet filter for helical coils, poloidal coils and SC bus lines

Position	Type	Mesh size (μm)	Element diameter (mm)	Element length (mm)	Number of elements	Effective element area (m^2)
Helical Coils						
Valve box inlet	Slit of winding	5	30	330	19	0.031
Coil inlet	Slit of winding	100	30	330	7	0.031
Poloidal Coils						
Valve box inlet	Sintered metal	5	120	300	2	0.072
Coil inlet	Slit of winding	20	30	460	7	0.014
SC Bus Lines						
Cold gas feed line	Pleated mesh	50	60	400	1	0.2
LHe feed line	Sintered metal	106	50	200	1	0.012
LHe feed line	Sintered metal	106	20	235	1	0.004

First Cycle Operation

During the 1st operating cycle, the compressor was started on February 9, 1998 and the system purification operation was performed for 2 weeks (325 hours). A criterion of 2 ppm or less for the impurity concentration was set for the completion of the purification process. The cool down was started on February 23 and finished on March 23. The steady-state operation was performed from March 23 to May 18. Impurity levels greater than the detection sensitivity of the analyzer were not observed during the cool down and steady-state operations. The inlet strainer of the 1st turbine (T1) was blocked with impurities twice during cool down. However, no other trouble from impurities occurred and the system was operated stably for over 1350 hours under steady-state conditions.

The warm-up operation for the entire system was initiated on May 18, and the warm up to room temperature was achieved on June 15. Impurity concentrations measured at the main compressor outlet and the coil return gas temperature are shown in Fig. 4. From May 27 to June 1, the compressor was shut down for system maintenance and the warm-up operation was interrupted. Accordingly, the sum of the actual warm-up operation time was 550 hours. Since the impurity concentration in the system was over 50 ppm (exceeding the upper limit of the analyzer), a partial purification of the circulation gas was conducted between May 26 and June 3 as noted by an arrowhead in Fig. 4. Nitrogen concentrations of over 50 ppm were reduced to around 10 ppm by the operation of the purification system on May 26. Nitrogen concentrations again increased to over 10 ppm just before the warm-up completion on June 8. This increased concentration was caused by connecting new lines currently not used for the warm-up operation, such as a liquid-helium feed line. Water concentration variations are similar to those reported for the nitrogen; however the maximum value was around 2 ppm.

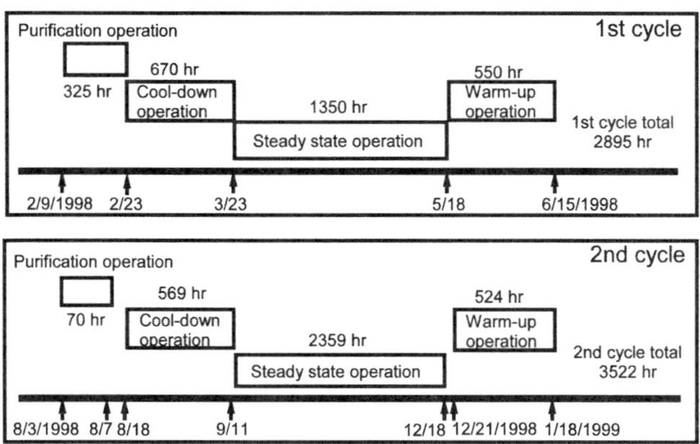

Figure 3. Operation of the 1st and the 2nd cycle of the LHD cryogenic system.

Second Cycle Operation

During the 2nd operating cycle, a purification operation of 70 hours was performed from August 3 to August 7, and the impurity concentrations in the system were reduced to 2 ppm or less. The main compressors were then shut down and restarted on August 18 to initiate the cool down. The impurity concentrations, measured at the outlets of the main compressors and the recovery compressor, are shown in Fig. 5. Nitrogen and water concentrations were maintained below the detection sensitivity at the main compressor outlet. Nitrogen concentration, measured at the recovery compressor outlet, increased to about 1 ppm, and water contamination increased to about 10 ppm when helium gas was added from an external container. After further gas purification to decrease the water concentration to 3 ppm or less, gas was transferred to the buffer tanks from the impure gas tanks. The purity of the buffer tanks (main compressor outlet) after the gas transfer was also maintained below the detection level of the analyzer.

The cool down was completed on September 11 and initiated a long-term continuous operation of 2359 hours until December 18. Although the T1 inlet strainer blocked seven times, no other trouble was experienced by impurities.

After performing the liquid helium recovery operation from December 18 to 21, warm up was started on December 21. The compressors were shut down from December 28, 1998 to January 5, 1999. The compressors were restarted again on January 5, and the warm-up was continued. Warm up was completed on January 18 and required 524 hours. Impurity concentrations and coil return gas temperatures during this time period are shown in Fig. 6. Although the coil outlet temperature increased to 230 K on December 28, the large release of accumulated impurities, as seen in the 1st operating cycle, was not observed. This is only mentioned because the cryogenic system was not disturbed during the period between the 1st and 2nd operating cycle. The system was filled with pure helium gas to prevent contamination of impurities. However, in the warm-up resumption on January 5 after the compressors were shut down, the nitrogen concentration increased to 4 ppm due to the admission of some air, which leaked into the system from a crack of the ceramic insulation joint in the room-temperature recovery piping.

Influence of Impurities

Records for the blockage of the 1st turbine (T1) inlet strainer (mesh size of 3 μm and surface-area of 0.03 m^2) and the T1 inlet temperature at that time are listed in Table 3. In the 2nd operating cycle, blockage of the T1 inlet strainer occurred seven times, four of which occurred during steady-state operation. The T1 inlet temperature during steady-state operation is usually 185 K to 200 K. Except for the event on October 5, it appears that

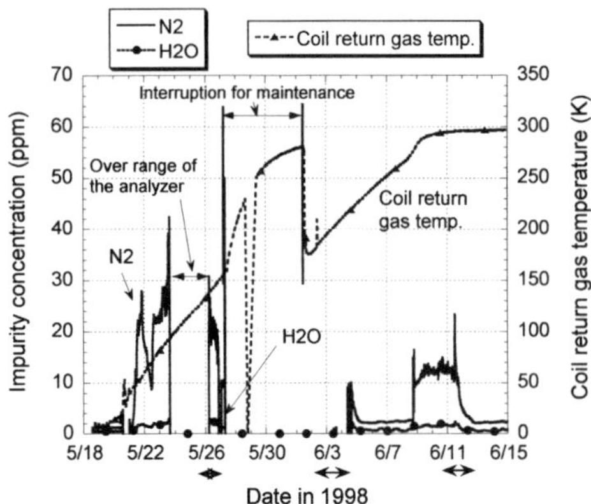

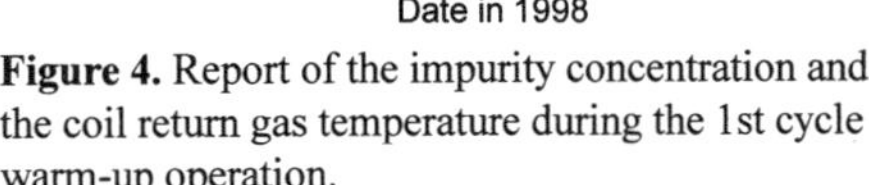

Figure 4. Report of the impurity concentration and the coil return gas temperature during the 1st cycle warm-up operation.

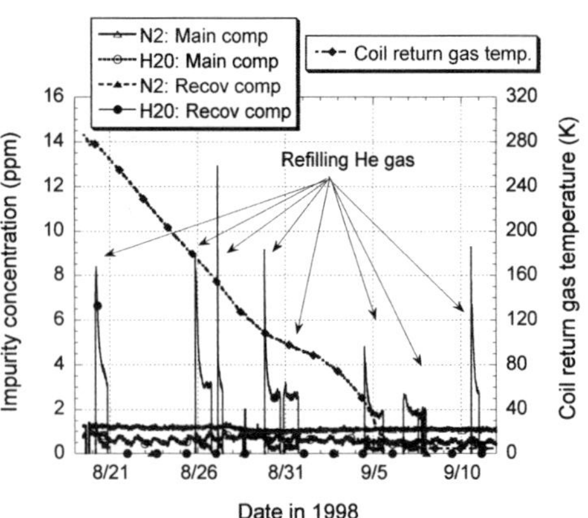

Figure 5. Impurity concentration measured at the outlet of the main and recovery compressors during the 2nd cool down operation.

when the T1 inlet temperature dropped below 180 K that the T1 inlet strainer blocked. When an operation was performed so that the T1 inlet temperature was maintained above 180 K, blockage of the T1 inlet strainer by the impurities stopped. Regeneration of the T1 inlet strainer could be performed while continuing the operation of the helium refrigerator/liquefier by utilizing the liquid-nitrogen heat exchanger HX-15 instead of turbines T1 - T3. The impurity concentration of the warming gas used for regeneration of the T1 inlet strainer was measured by a gas chromatograph and a dew point meter. Impurities, such as hydrocarbons, were not detected but a deterioration of the dew point was observed as shown in Fig. 7. From this, it was evident that solidification of the water caused the blockage of the T1 inlet strainer. It is believed that this water entered the system mainly during refilling with helium gas.

The time, date and temperature at the time of blockage of the coil inlet filters are listed in Table 4. The blockage of a coil inlet filter occurred only four times in both the cool down and steady-state operation. Therefore, it was demonstrated that the effective areas of the coil inlet filters have sufficient margin to maintain stable operation based on current observation of impurity concentration levels.

High-Sensitivity Measurement of Very-Small Quantity Impurities

Impurity concentrations during the steady state operation are below the detection sensitivity of the optical-emission-spectroscopy type analyzer made by Linde AG, whose measuring range is from 1 ppm to 50 ppm for N_2, O_2, H_2O, C_xH_y. It is very difficult to make the direct measurement of impurities at such small contamination levels. Therefore, we developed the low-temperature condensation type high-sensitive impurity measuring apparatus.

The block diagram of the apparatus is shown in Fig. 8. The high-precision measurement is carried out in the following manner. First, the gas to be analyzed is introduced into the filter, which is cooled with a small refrigerator, for a fixed time period and with a fixed flow rate, and impurities in this flow are allowed to condense. The inlet valve for the gas being analyzed is closed, and the filter is warmed, causing a circulation of the gas in the closed cycle of the filter and the small tank. Part of the circulation gas is sampled by the gas chromatograph and the dew point meter, and the condensed impurity concentrations are measured. The gas chromatograph made by Shimazu Inc. has measuring minimum sensitivity of 0.5 ppm for O_2, N_2, and 0.05 ppm for CO, CO_2, C_xH_y. The dew point meter made by Panametrics has a measuring range from –110 ℃ to 20 ℃. The impurity concentration is calculated from the amount of the impurity accumulated in the filter, and the quantity of the gas that was introduced into the filter.

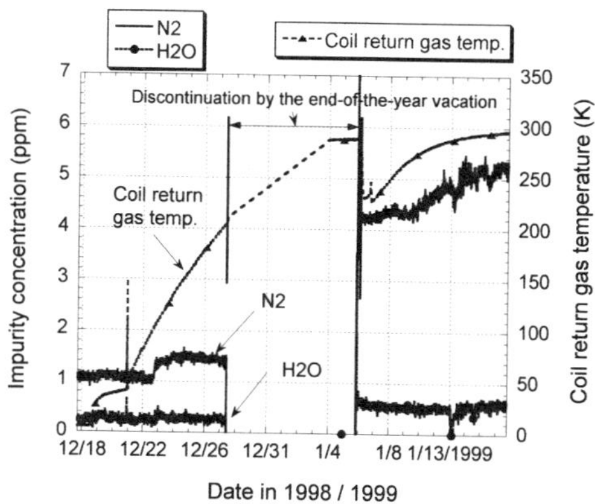

Figure 6. Time report of the impurity concentration and the coil return gas temperature during the 2nd cycle warm-up operation.

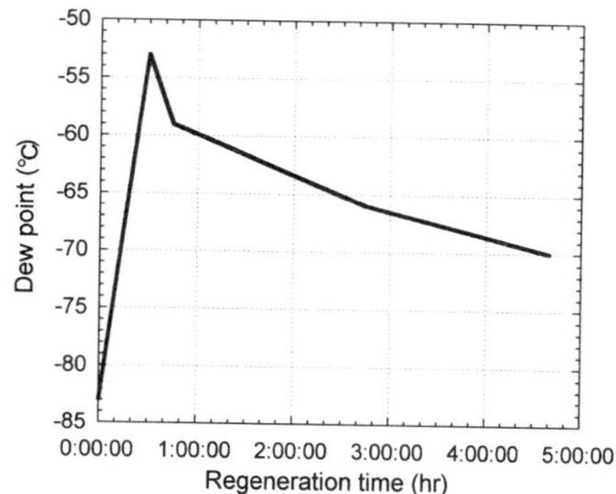

Figure 7. Deterioration of the dew point of the warming gas during the regeneration of the T1 inlet strainer.

Table 3. Records of the T1 inlet strainer blockage and T1 inlet temperature

Date (M/D/Y)	T1 Inlet Temp (K)	Regeneration Time (h)	Mode
1st Cycle			
3/17/1998	206	7	C. D.
3/23/1998	220	7	C. D.
2nd Cycle			
8/31/1998	173	10	C. D.
9/5/1998	176	8	C. D.
9/8/1998	174	8	C. D.
9/20/1998	173	8	S. S.
10/5/1998	188	8	S. S.
10/17/1998	175	8	S. S.
10/18/1998	174	10	S. S.
3rd Cycle			
6/261999	173	10	C. D.
7/9/1999	188	10	S. S.

C. D.: Cool down, S. S.: Steady State

Table 4. Records of blockage for the coil inlet filters

Date (M/D/Y)	Position	Temp. (K)	Mode
1st Cycle			
3/3/1998	Poloidal V. B.	136	C. D.
2nd Cycle			
9/4/1998	Helical V. B.	35	C. D.
10/24/1998	Poloidal V. B.	16	C. D.
3rd Cycle			
6/17/1999	Poloidal V. B.	150	C. D.

C. D.: Cool down, S. S.: Steady State

The impurity concentrations of the circulating helium gas in the system, which was measured by the above procedure during the 3rd cycle steady-state operation, are listed in Table 5. Impurity concentrations of hydrocarbons were less than 10 ppb. The concentration of oxygen and nitrogen varied from several tens to several hundreds ppb. It is believed that the accumulation in the filter of solidified O_2, N_2 and CO may vary considerably depending on the operating conditions of the impurity measuring apparatus, for example, the condensation time during which the measuring gas is introduced to the filter, the amount of the impurities accumulated up to that time, the surface condition, the temperature of the filter, etc. Therefore, there was large deviation for the measuring results of O_2, N_2 and CO levels in Table 5. It is a future subject to clarify the relation of the impurity accumulated with the operating condition of the impurity measuring apparatus, and to improve the accuracy of measurement. Water concentration during steady-state operation is around 15 ppb (-100 ℃ dew point). This value is in agreement with the operation results of the 2nd cycle, in which the T1 inlet strainer became blocked at a temperature level of 175 K (-98 ℃). Therefore, it appears that a very plausible explanation for the blockage of the T1 inlet strainer is the accumulation of solidified water.

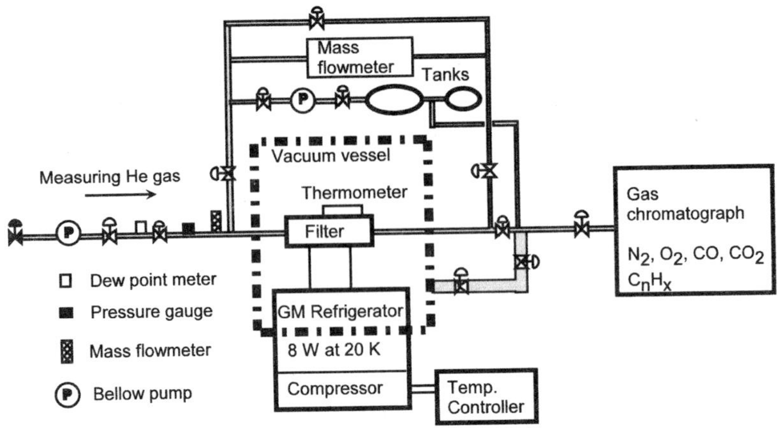

Figure 8. Block diagram of the low-temperature condensation type impurity measuring apparatus.

Table 5. Impurity concentration during the steady state operation measured by the low-temperature condensation type impurity measuring apparatus

Date (M/D/Y)	C. T.* (h)	CO (ppb)	CH_4 (ppb)	CO_2 (ppb)	C_2H_4 (ppb)	C_2H_6 (ppb)	O_2 (ppb)	N_2 (ppb)	H_2O (ppb)	H_2O (℃)
7/5/1999	84	6.0	0.1	3.1	-	-	255.0	801.8	17	-99
	8	0.6	0.2	1.0	-	-	40.9	97.5	14	-100
	8	0.7	0.1	0.9	-	-	34.2	103.7	11	-101
7/8/1999	72	2.1	-	3.0	-	-	-	255.1	17	-99

*C. T. means the condensation time during which the measuring gas introduced to the filter.

SUMMARY

The LHD cryogenic system operated successfully with high degree of reliability during long-term continuous operation in its 1st year. In order for the LHD cryogenic system to obtain this high reliability under continuous operation over a long period of time, careful attention was directed to the high-precision control of impurities during the design and construction phase. In addition, the advanced impurity management is performed during operation of the system. To measure the very low impurity concentrations during the period of steady-state operation, we developed a low-temperature condensation type impurity measuring apparatus. Concentrations of hydrocarbons during the steady-state operation were measured to be less than 10 ppb using this apparatus. Concentrations of oxygen and nitrogen were measured as several 10s to several 100s ppb. These results and actual continuous operation demonstrated that these impurity components with these concentration levels do not adversely influence continuous operation. However, it is easy to obtain an accumulation of water in the system. Even water with a low concentration level of 15 ppb (-100 ℃ dew point) can block a strainer, filter, etc. Management of the water contamination is critical for continuous operation of the cryogenic system.

REFERENCES

1. T. Mito, et al., Development of a cryogenic system for the Large Helical Device, in "Advances in Cryogenic Engineering," Vol. 43, Plenum Press, New York (1998), p. 589.
2. T. Mito, et al., First cool down performance of the LHD, IEEE Trans. Appl. Superconductivity (Proceedings of ASC'98), to be published.
3. R. Maekawa, et al., LHD cryogenic-control system performance under various operating conditions, presented at CEC/ICMC 1999 meeting, Montreal, Canada.

JEFFERSON LAB CEBAF ENERGY UPGRADE PLANS *

C. H. Rode, J. Benesch, Y. Chao, J. R. Delayen, J. Karn, J. Mammosser, W. Oren, J. Preble, W. J. Schneider, R. Wines, M. Wiseman

Thomas Jefferson National Accelerator Facility
12000 Jefferson Avenue
Newport News, VA 23606-1909

ABSTRACT

This paper presents interim conceptual plans for upgrading the Continuous Electron Beam Accelerator Facility (CEBAF) at the Thomas Jefferson National Accelerator Facility to extend Jefferson Lab's world leadership in nuclear physics research. The CEBAF accelerator was designed in the mid-1980s to provide beams of electrons at an energy of 4 GeV (billion electron volts) for use as probes of the atom's nucleus in CEBAF's three experiment halls. As of early 1999, the accelerator exceeds its design energy by routinely operating above 5.5 GeV. When upgraded, it will provide 11 GeV electron beams for studies in existing experimental halls and 12 GeV electrons to generate photon beams for related but qualitatively different nuclear studies in a new Hall D.[1]

PROJECT BACKGROUND AND CONTEXT

Jefferson Lab's CEBAF experimental program fulfills a major, two-decade-old priority of U.S. nuclear science: the construction and scientific use of a 4 GeV, continuous-beam electron accelerator capable of supporting a broad range of innovative research in nuclear physics.

The CEBAF facility is unique. It allows detailed exploration of the energy region where the transition occurs between two basic ways of viewing the nucleus. In the standard picture as seen at lower accelerator energies, the nucleus appears as a group of interacting nucleons—protons and neutrons. At higher energies, the quarks and gluons inside the nucleons must be explicitly included. Applicable at the high-energy end of the transition region between these two views are the essentially exact calculations of a field theory, perturbative quantum chromodynamics (QCD). In the lower-energy non-perturbative region characteristic of normal nuclear matter, an important new "strong QCD" framework is required. Elucidating the nature of this transition is one of the last

* Supported by U.S. DOE contract DE-AC05-84ER40150

frontiers in humankind's quest to understand ordinary matter. CEBAF is the key research tool for the task.

Therefore an intense scientific interest exists, not only nationwide but worldwide, for conducting experiments at CEBAF. During the mid-1990s, CEBAF's earliest operations for physics began to generate new contributions to understanding of the nucleus. As of spring 1999, more than 1400 nuclear physicists make up CEBAF's user group, and the backlog of planned experiments stretches to more than six years. Reference 2 presents a user-focused assessment of the prospects for upgrading CEBAF.

Jefferson Lab's Institutional Plan[3] expresses the vision for extending this success into the future. The lab's top two goals of "enabling and conducting a physics research program of the highest scientific priority at the nuclear/particle physics interface" and "continued world leadership in underlying core competencies" provide the basis for the envisioned upgrade. Increasing the maximum beam energy toward 12 GeV (intermediate-term goal) and the 24 GeV (long-range goal) will extend the scientific reach of the advancing experimental program. Jefferson Lab's main core competency in superconducting radio-frequency (SRF) electron acceleration is the enabling technology at the heart of the upgrade.

Within the nuclear physics community it is a given that in the long run a strong physics case exists for a CEBAF-like machine operating in the energy range beyond 24 GeV to perhaps 30 GeV. It is Jefferson Lab's intent to continue serving U.S. nuclear science by becoming the obvious cost-effective site for this machine. Two factors make upgrading CEBAF to these higher energies attractive:

- The needed accelerator tunnel already exists at Jefferson Lab. Since the existing racetrack-shaped, 1.3 km CEBAF tunnel was constructed with recirculation arcs of large radius, a 24 GeV machine could operate there without significant synchrotron-radiation-caused degradation of beam quality.
- The necessary expertise and infrastructure will exist at Jefferson Lab. The CEBAF SRF cryomodule development program described here will lead in the next decade to the capability for building the 24 GeV machine.

While the plan presented here does not directly address the 24 GeV upgrade, the future project does constitute an important part of the context, rationale, and planning for 12 GeV. Also, a clear consensus already exists amongst CEBAF's users in support of the currently envisioned 12 GeV upgrade. Thus, the next logical step toward 24 GeV is fully consistent with the nearer-term needs of nuclear physics researchers.

Though integrated with the longer-term vision, the 12 GeV upgrade also stands alone as a project that is both physics-driven and achievable with modest modifications and additions to the existing machine. Figures 1 through 3 represent what is to be upgraded. Figure 1 is the entire CEBAF accelerator complex as seen from the air; one of the twin linear accelerators (linacs) made up of 20 cryomodules consisting of four pairs of superconducting cavities each is shown in Figure 2. Figure 3 is a photograph of key components within the Central Helium Liquifier (CHL), which supplies liquid helium at 2 K for the superconducting operation of the accelerator.

Figure 1. Aerial view of the CEBAF accelerator complex. CEBAF's nuclear physics users conduct their experiments in the earthen-topped, domed, semi-underground structures in the foreground—from left to right, Experimental Halls A, B, and C. The underground accelerator's racetrack shape can be traced by the service buildings and road above ground. RF power reaches the accelerating components of each linac from within the long service buildings above the "straightaways." Service buildings above the recirculation arcs that interconnect the two linacs house DC power supplies for the recirculation-arc beam-transport magnets.

Figure 2. (on the left) Cryomodules in CEBAF's South Linac, with RF power waveguides descending from service buildings

Figure 3. (on the right) The 2 K and 4 K cold boxes in CEBAF's 4.8 kW Central Helium Liquifier (CHL), which represents the world's largest 2 K cryogenics capacity.

The following requirements were established to support the envisioned physics program and keep modifications and additions to the present machine modest:

- The highest-energy beam at 12 GeV must only be delivered for 8 GeV photon beam production to the new experimental hall, Hall D. Electron beam quality requirements are somewhat relaxed from original machine specifications.
- CW operation must be preserved.
- The maximum circulating beam current will be 430 μA (corresponding to an 85 μA beam for a five-pass design).
- The maximum installed refrigeration capacity will be 10.1 kW at 2 K.
- Technical choices should be made with the ultimate goal of 24 GeV in mind.
- Cost and impact on accelerator operation must be kept to a minimum.

Table 1 and Figure 4 summarize key elements of a 12 GeV upgrade plan that meets the above requirements.

Table 1. Selected key parameters of the CEBAF 12 GeV Upgrade

Parameter	Specification
Number of passes for Hall D	5.5 (add a tenth arc)
Energy to Hall D	12.113 GeV (for 8 GeV photons)
Number of passes for Halls A,B,C	5
Energy to Halls A,B,C	11.023 GeV
Duty factor	CW
Max. current to Halls A,C*	85 μA
Max current to Halls B,D	5 μA
New cryomodules	10 (5 per linac)
Replacement cryomodules	8 (3 per linac, 2 in injector)
Central Helium Liquifier upgrade	10.1 kW (from present 4.8 kW)

*Max. North Linac current is 430 μA; max. South Linac current is 425 μA.

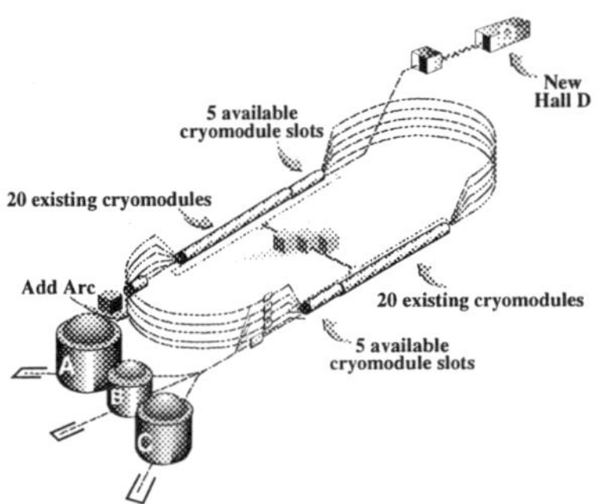

Figure 4. (on the left) Cryomodule slots available for upgrading existing CEBAF accelerator to a 12 GeV machine.

Figure 5. (on the right) Prototype seven-cell, 1497 MHz CEBAF SRF accelerating cavity.

SRF CRYOMODULES

The higher energy will be obtained by adding new, higher voltage cryomodules (five per linac) and replacing old units with new ones. The new cryomodules are being developed based on SRF technology improvements achieved at Jefferson Lab and elsewhere in the last decade. To increase the voltage provided by a cryomodule of given slot length requires increasing the cavity accelerating gradient, or increasing the effective accelerating length within the cryomodule, or both. However, maximizing the accelerating length involves less technical risk and, for CW operation, has the added advantage of lowering the dynamic refrigeration load. So the upgrade cryomodule will contain eight seven-cell cavities (Figure 5), as opposed to the four distinct pairs of five-cell cavities in the existing configuration. Since the cell design itself has not been changed for the upgrade, this means 8×70 cm of active length per cryomodule, rather than the previous 8×50 cm. To accommodate the increased cavity length within an unchanged cryomodule slot length, the upgrade has been redesigned to exclude the original pair-to-pair bridging sections. With gradients exceeding 12 MV/m routinely available from CEBAF SRF cavity technology in the late 1990s, 68 MV can be conservatively specified for the upgrade cryomodule. A Q of 6.5×10^9 at 12.5 MV/m has been specified for the seven-cell cavities, and has been met in tests of a seven-cell prototype.[4]

The key to the upgrade—and the majority of the project cost—lies in doubling the accelerating voltage of each superconducting radio-frequency (SRF) linac. Doubling the energy means:

- adding in available slots five new, improved-performance cryomodules to each linac,
- replacing three more in each linac, and
- as an option, replacing two more in the injector.

Table 2 details the 25 cryomodules per upgraded linac according to accelerating gradient, cryomodule energy, and total linac energy.

Table 2. Cryomodule (CM) and Linac Voltage (per Linac)in the 12 GeV Upgrade

Number of CMs	CM Active Length (meters)	CM gradient (MV/m)	CM energy (MV)	Linac energy MV)
(upgrade) 8	5.6	12.19	68.25	546.0
(existing) 17	4.0	8.00	32.00	544.0
Total: 25	112.8			1090.0

CHL UPGRADE

Overview

Besides the SRF effort summarized above, the upgrade project involves a number of much smaller but still significant tasks. The most costly of these is doubling the capacity of the Central Helium Liquifier (CHL), to support the increased SRF loads. For 6 GeV one refrigerator (CHL #1) cools both linacs, while for the upgrade each linac will have its own refrigerator.

The current setup has the existing CHL#1 feeding two linacs with the injector as an extension of the North Linac and Jefferson Lab's FEL plugged into a tee on the South Linac U-tube.[5] The 12 GeV upgrade has CHL#1 feeding only the North Linac, while the new CHL#2 is feeding the South Linac and the FEL.

CHL#2 consists of a compressor system very similar to CHL#1, a 300 to 80 K cold box, an 80 to 4.5 K cold box (the existing cold box from MFTF-B),[6] and the new "redundant 2 K cold box". CHL#2 is being designed to have 10% larger capacity than CHL#1 in order to handle a FEL upgrade.

Cycle Design

The fifth-stage cold compressor (CC) was installed in the "redundant 2 K cold box" in FY98-99 and will be installed in the original 2 K cold box next year. This permits much greater flexibility of operation. It also will permit operation at capacities of 70 to 120% and 140 to 220% of the original 4800 W. As the Q of the cavities improves, the BCS heat load fraction increases, which in turn changes the temperature optimization. For the 12 GeV upgrade we plan to operate at 1.95 K vs. the current 2.08 K.

Hardware

The new fifth-stage CC increases the total potential compression ratio by a factor of two, tremendously increasing the flexibility of the system. It provides:

- Higher discharge pressure: the system is easier to start and less sensitive to warm compressor upsets.
- Higher maximum capacity: the fifth stage can be used to reduce the load on the third and fourth stage.
- Lower minimum capacity: one can use the smaller stages 2 to 5 vs. 1 to 4.
- Lower operating temperature: one can reduce the operating pressure and return at the 35 K tap.

Distribution System

With the exception of eight bayonet valves (four zones in each linac) and the related controls, the distribution system is ready to accept the ten new cryomodules. The zone 26 "dummy cryomodules" required to keep the ends of the transfer lines cold will be removed.

Doubling the refrigeration capacity has two impacts. First due to higher flow rates, the return pressure drops will approximately double. Second with the ratio of dynamic vs. static heat loads increasing, the return temperature will decrease. This will permit the CCs to operate closer to their design point.

SUBSYSTEM UPGRADES

Relatively minor changes must also be made to other accelerator subsystems to accommodate the higher energy. A tenth arc beamline will be added so that the Hall D beam can be accelerated in a sixth pass through the north linac to Hall D, for a total of 5.5 passes through the entire machine. This yields a 12 GeV beam for 8 GeV photon production in Hall D utilizing hardware and processes addressed in Reference 1. Changes in DC power supplies and spreader/recombiner magnets represent the upgrade's other main impacts on the recirculation arcs. The project also necessitates ten new RF stations, plus minor modifications to other RF systems. Table 3 is an itemized list of major upgrade items.

Table 3. Major Upgrade Items

Number	Item
10	Additional upgrade CM & supporting RF
8	Replacement upgrade CM & RF upgrades
-	Double CHL capacity
-	CHL building addition
-	New Arc 10
-	Move injector beamline
17	New spreader/recombiner dipoles
55	Modified spreader/recombiner dipoles
-	Box PS upgrade (16 regulators, 25 rectifier modules)
5	Modified arcs (C to H style dipoles)
57	New quads (two new styles)
85	New 17 A, 55 V trim cards
130	New 60 A shunt modules
2	New extraction Lambertson magnets
5	New (higher-power) RF separator cavities
3	5 kW RF separator tubes
-	New Hall D transport line
2	Service building additions

Increased Bending and Focusing Strengths of Beam Transport Magnets

The increase in design energy from 6 GeV to 12 GeV demands stronger bending and focusing power from the magnets. In the recirculation arcs this can be achieved with modifications to the dipoles to reduce saturation. In five of the nine existing arcs the "C" dipoles with the highest fields saturate the backleg iron. We will replace the aluminum coil supports on the open face of the existing dipoles with a steel U-channels (Figure 6). This simple but time-consuming (160 dipoles) upgrade will be accomplished during bi-weekly maintenance shifts and dramatically improves the magnets saturation at higher currents (Figure 7).

Figure 6. Prototype BB with added H-Steel.

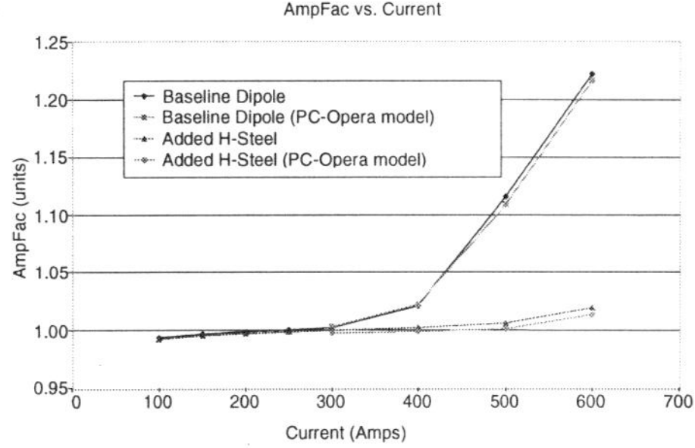

Figure 7. Saturation versus current for prototype BB dipole.

The most difficult of the non-SRF tasks will be the modification of the spreader/recombiner magnets, which will represent the largest contribution to upgrade driven accelerator down time. This is a very densely packed area with five or six vertically separated beams. Due to this limitation, the dipoles have a cross section aspect ratio reversed from normal dipoles, which leads to a severe saturation of the pole tips. In order to fix this and maintain good field quality, we are adding U-channel steel and four or six turns to the pole tip gap (see Figure 8). These coils will be in series with the main coils.

Quadrupole upgrades are scattered throughout the machine. The majority of these focusing issues are resolved by rearranging existing magnets and manufacturing two new styles which accommodate the more energetic beam.

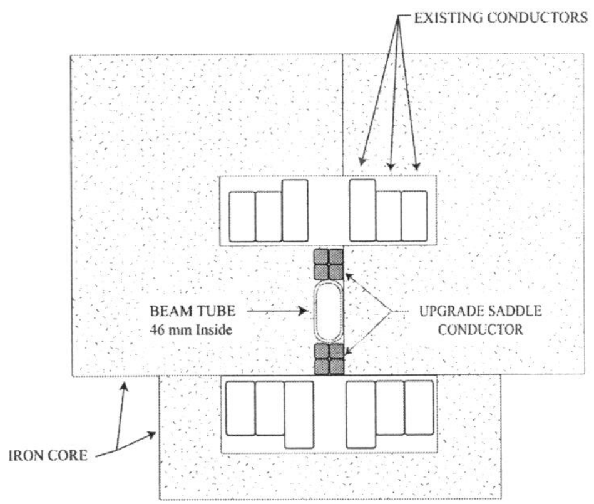

Figure 8. Upgrade 1 m spreader/recombiner dipole (at center).

Delivery of 5.5-Pass Beam to Experiment

Transporting the beam back to a sixth North Linac acceleration pass en route to Hall D, will require installation of a fifth arc at the west end of the accelerator (Arc 10). This will use longer 4 m "H" dipoles installed under the existing 3 m dipoles of Arc 8.

A new beam separation scheme allowing a six-way separation out of the North Linac is a major feature of the 12 GeV upgrade. This is motivated by an additional line transporting the highest-energy (12.113 GeV) beam to Hall D. The sixth-pass beamline will branch out of the northeast spreader at an elevation 0.5 m below the linac axis. This implies staged separation of the last three passes (versus the current two) using septum magnets while attending to tolerances on trajectory clearance inside septa and dispersion suppression within the existing longitudinal space.

FUTURE PLANS

Next Steps for FY 1999

- Continue upgrade CM design
- Test RF 6.5 kW upgrade
- Continue on 3D field quality calculations

Steps Planned for FY 2000

- Finalize CM design and fabricate cavities for 1/2 CM test
- Test 1 m Spreader/Recombiner dipole with six-turn gap coil
- Detailed cryo design and cost estimate
- Finalize Hall A, B, C extraction design and cost estimate
- High power test of spare RF separator cavity

Steps Planned for FY 2001

- Assemble and test 1/2 CM
- RF control module upgrade design and prototype

MAJOR OPEN ISSUES

1. RF control for the New Cryomodules. At 68 MV/CM (as opposed to the present nominal 30 MV/CM), there is a need to increase external Q leading to a bandwidth that is a factor of 4 or 5 narrower than at present. At 12.5 MV/m and 400 μA circulating current the maximum allowable amount of detuning (including static and microphonics) is 25 Hz. With the Lorentz detuning being much larger than the loaded bandwidth and the completely new tuning system hardware, a new more agile RF control interface will be required.

2. Injector recirculation. If the 12 GeV upgrade incorporated the added complexity of recirculation in the injector to increase injection energy, the two replacement cryomodules slated for this area could be dispensed with. For 24 GeV, injector recirculation is the baseline concept.

ACKNOWLEDGMENTS

This paper represents the efforts of the Jefferson Lab Accelerator Division.

REFERENCES

1. Jefferson Lab, *Photoproduction of Unusual Mesons and Gluonic Excitations: Hall D Preliminary Design Report*, January 1999.
2. L. S. Cardman, "Jefferson Lab at Higher Energies: Options and Opportunities," Jefferson Lab internal document, 15 June 1998.
3. The present report reflects the 2000-2004 update of Jefferson Lab's Institutional Plan; to be released in the fall of 1999.
4. I. E. Campisi et al., "Superconducting cavity development for the CEBAF upgrade," *Proceedings of the 1999 Particle Accelerator Conference*, Vol. 2, New York (1999) pp. 937–939.
5. C. H. Rode, "CEBAF Cryogenic System," *Proceedings of the 1995 Particle Accelerator Conference*, Vol. 3, (1996) p. 1994–1998.
6. D. S. Slack and W. C. Chronis, "The mirror fusion test facility cryogenic system – performance, management approach, and present equipment status," in: *Advances in Cryogenic Engineering*, Vol. 33, Plenum Press, New York (1988) pp. 585–589.

SPECIFICATION OF FOUR NEW LARGE 4.5 K HELIUM REFRIGERATORS FOR THE LHC

S. Claudet, P. Gayet and U. Wagner

LHC Division, CERN
1211 Geneva 23, Switzerland

ABSTRACT

The cooling capacity for the superconducting magnets in the Large Hadron Collider (LHC) at the European Laboratory for Particle Physics, CERN will be provided by eight helium refrigerators serving the eight 3.3 km long machine sectors. Of these eight refrigerators, four are already existing and are currently used for the Large Electron Positron Collider (LEP) project. These existing refrigerators have to be modified to fulfil the requirements for the LHC. Four new refrigerators providing cooling capacity down to 4.5 K will be added. All eight 4.5 K refrigerators will be connected to 1.8 K cooling stages. This presentation recalls the cryogenic architecture of the LHC, the constraints in process design resulting from it and from the desired capacity for steady state and transient operation. It then describes how these requirements were expressed in the technical specification for the four new 4.5 K refrigerators to be delivered between the years 2000 and 2002.

INTRODUCTION

A detailed description of the total cryogenic system for the LHC has already been published[1]. Figure 1 shows a simplified typical cryogenic block diagram for an even point of the LHC. It includes the existing[2] and the new 4.5 K refrigerator, the two 1.8 K refrigeration units, the gas storage tanks, the cryogenic interconnection box, the cryogenic distribution line and the magnet cryostats.

The refrigerators are divided into two units. A 4.5 K refrigerator covers the capacity for thermal shield cooling, current lead cooling, isothermal refrigeration at 4.5 K and cooling between 4.5 K and 20 K. Connected to this unit will be a 1.8 K refrigeration unit which cannot operate on its own but requires cooling capacity between 4.5 K and 20 K in order to supply the necessary refrigeration at 1.9 K for the LHC magnets.

Following the time schedule for the installation of the cryogenic equipment, the new 4.5 K refrigerators had to be purchased and delivered in advance of any other items, mainly to allow the testing of the 1.8 K refrigeration units. In the following we will describe the basic constraints that defined the specification for the new 4.5 K refrigerators of the LHC machine.

Advances in Cryogenic Engineering, Volume 45.
Edited by Shu *et al.*, Kluwer Academic / Plenum Publishers, 2000.

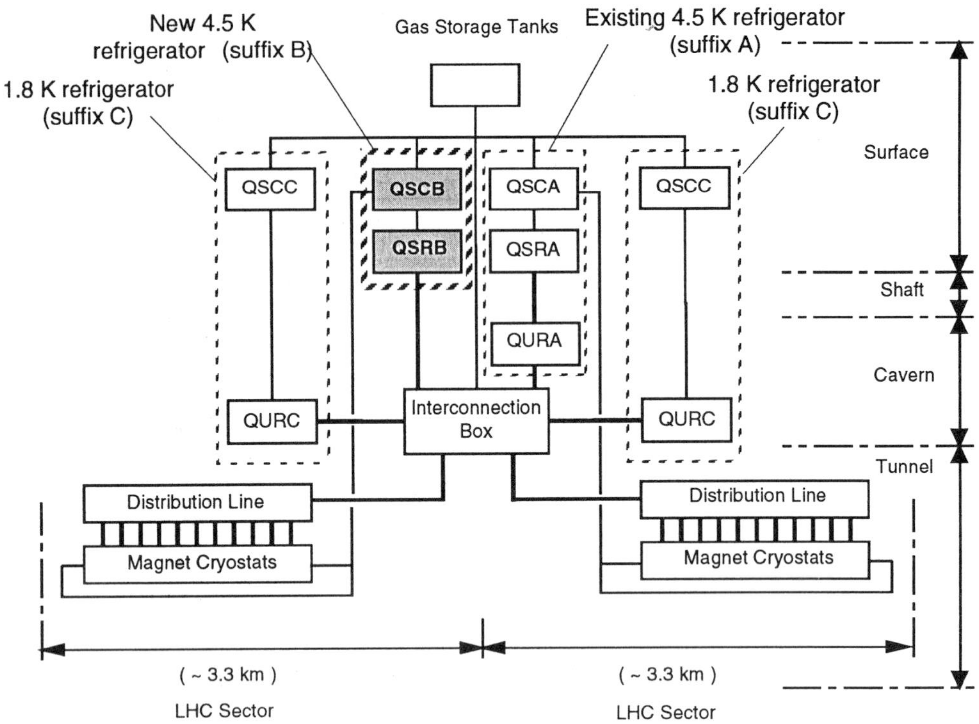

Figure 1. Simplified block diagram of the cryogenic installation in a typical even point of the LHC

DESIGN BASIS

The design for the new 4.5 K refrigerators is based on several constraints, which are defined partly by proper technical requirements, and partly by interface considerations. As concerns the latter, the new cold boxes must be comparable to the existing ones in order to allow for both types to be integrated equally into the LHC cryogenic system.

Division between 1.8 K and 4.5 K Refrigeration

In order to allow the identical 1.8 K refrigeration units to combine equally well with the new 4.5 K refrigerators for LHC and the existing LEP refrigerators, it was decided, at an early stage of the conceptual design studies, to separate the refrigerators for all LHC points into two systems: a 4.5 K unit and a 1.8 K unit[3]. As a consequence the new 4.5 K refrigerators need to interface with the rest of the cryogenic system at the temperature levels of 75 K, 50 K, 20 K and 4.5 K, like the already existing LEP cold boxes.

Given Interfaces

A principle block diagram of a new 4.5 K refrigerator together with the 1.8 K refrigeration unit, the cryogenic interconnection box and the cryogenic distribution system of the LHC machine is given in Figure 2. Since for the LHC no high-density vapour at 4.5 K flows back from the magnet ring to the cold boxes, it was decided to install the cold boxes completely at ground surface. This allows saving precious space at the underground level compared to the existing LEP cold boxes, which are of a split design[4].

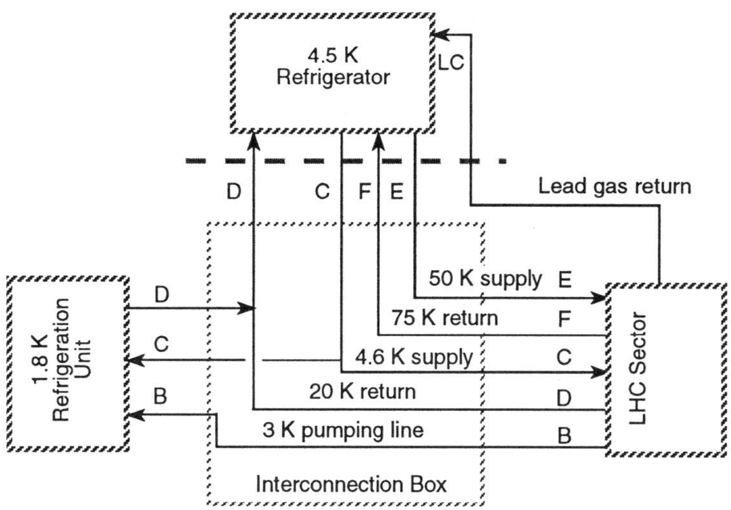

Figure 2. Simplified cryogenic block diagram for a new 4.5 K refrigerator including cryogenic interconnection box, 1.8 K refrigeration unit and the cryogenic distribution system of the machine.

The interface from the new 4.5 K refrigerator cold box to the LHC machine consists of five lines: 4.6 K supply (line C), 20 K return (line D), 50 K supply (line E), 75 K return (line F) and lead gas return at 280 K (line LC).

Cooling Requirements

Following the design of the different elements in the LHC machine, the required cooling capacities including contingency for overcapacity and uncertainty[5] are listed in Table 1.

As shown in Table 1, the cooling requirements are not identical for the different sectors. The sectors identified as "Low-load" sectors will be supplied by the existing LEP refrigerators that will undergo a final upgrade during the years 2001 to 2003 and be completed with a 1.8 K refrigeration unit. The new 4.5 K refrigerators consequently had to be specified in order to cover the load of the "High-load" sectors. The values for mass flows and helium properties which are to be supplied at the interface shown in Figure 2 are listed in Table 2.

Table 1. Required cooling capacity for the LHC machine sectors

Temperature level	50-75 K [W]	4.6-20 K [W]	4.5 K [W]	1.9 K [W]	3-4 K [W]	20-280 K [g/s]
High-load sector	33000	7700	300	2400	430	41
Low-load sector	31000	7600	150	2100	380	27

Table 2. Property and mass flow data at the interface of the new 4.5 K refrigerator

Line		C	D	E	F	LC
Temperature	[K]	4.6	20	50	75	280
Pressure	[bar]	3.0	1.3	18.5	16.0	1.1
Flow	[g/s]	235	194	251	251	41

Operation Modes

Due to the daily operation cycle of the LHC machine, four steady state operation modes which differ by the required capacity are defined [6]. To this adds a "75 K standby" mode at which only the shield cooling circuit between 50 and 75 K is operated, in order to keep the cold masses below 80 K during extended periods in which no physics is possible. Additionally the refrigerator has to cover the transient operation modes of cool-down and warm-up. At the end of the cool-down the magnets will accumulate about 50000 L of liquid helium, which results in the quasi steady-state mode of "liquid fill".

Liquid Nitrogen Pre-cooling

Large capacity of liquid nitrogen pre-cooling was never used at cryogenic installations of CERN. The magnets of the LHC machine will have a total mass to be cooled of 4500 tons per sector [7]. In order to limit the time for a machine cool-down to reasonable values, a liquid nitrogen pre-cooler for a capacity of 600 kW will be installed in each 4.5 K refrigerator. This unit will be used for the cool-down of the machine and to boost the liquefaction rate during the "liquid fill" mode at the end of the cool down. During all operation cycles with the magnets in cold conditions no pre-cooling with nitrogen is desired in order to eliminate the environmental impact by nitrogen deliveries during long-term operation.

SPECIFIED REQUIREMENTS

Capacity Requirements

Figure 3 shows how the cooling loops for the LHC machine are seen from a 4.5 K refrigerator. The refrigerator must be able to cope with the simultaneous cooling loads for each of the steady-state operation modes resulting from the operation modes of the machine described above and listed in Table 3 as "Installed", "Normal", "Low-intensity", "Injection standby" and "75 K standby".

The identification numbers of the heat loads listed in Table 3 refer to Figure 3.

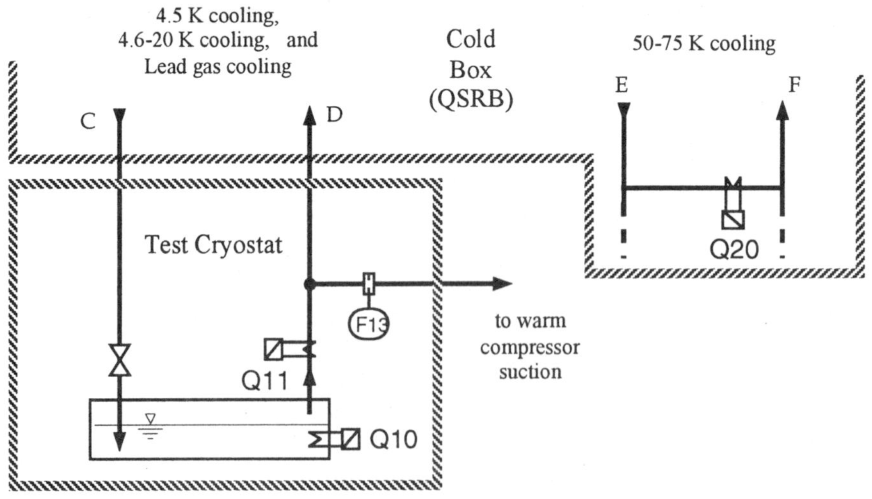

Figure 3. The LHC cooling capacities as seen from the 4.5 K refrigerator

Table 3. Cryogenic capacity requirements for a 4.5 K refrigerator in all steady-state LHC operation modes

Operation mode	Q10 4.5 K isothermal [W]	Q11 4.5 - 20 K non isothermal [W]	Q20 50 - 75 K non isothermal [W]	F13 20 - 280 K lead cooling flow [g/s]
Installed	4400	20700	33000	41
Normal	2600	12400	22000	27
Low-intensity	1600	7700	22000	27
Injection standby	1200	5700	22000	11
75 K standby	0	0	22000	0

As the purchasing rules of CERN, based on the concept of "conforming bid" do not allow to account for high safety margins in the process design of the different bids, the specification stipulates that the T-s diagram of the "Installed" mode must show 5% over–capacity on each load in order to compensate for manufacturing uncertainties, plus an additional 500 W capacity for regulation purpose in the phase separator of the cold box. The ambient-temperature compressor station was specified to satisfy the flow and pressure requirements resulting from this process calculation.

Test Cryostat

In order to limit the number of interfaces between CERN and the supplier that could influence the measured capacity during the reception test, and in view of tedious discussions in the past concerning such measurements, we decided to include in the specified supply a test cryostat dedicated only to the reception test. The supplier has to bear all static heat loads which are generated by this test cryostat. CERN accepts only the electrical heating capacity and the measured flow rate in the test cryostat as achieved capacity of the refrigerator. A simplified flow scheme of this test cryostat is included in Figure 3.

Redundancies and Over-capacities

No redundancy was specified for the refrigerators or any of their components except for the oil pumps of the compressor station. In view of the rather strong dependence of cooling requirements on LHC beam energy and intensity, it was accepted that in case of component failure, e.g. turbine or compressor, the supplied capacity would be reduced thus allowing operation of the collider at reduced performance. In fact the operation mode "Normal" which supplies the capacity to cover the nominal operation conditions of the machine represents only 60% of the cryogenic capacity installed. Moreover neighbouring refrigerators may be coupled at the level of the cryogenic interconnection box and a lack of capacity of one installation may be compensated by the capacity margin of the other one. As concerns the compressor flow, it is envisaged to install by-pass lines for the LP, MP and HP level and thus share compressor flow between the refrigerators installed on the same LHC point, if necessary.

Gas Purification Equipment

The level of impurities that the refrigerators will have to cope with is very difficult to assess. One has nevertheless to prepare for a high contamination by water during the first phase of each cool-down and the final phase of each warm-up. Besides this, contamination of oxygen, nitrogen and to a lesser extent, hydrogen and neon must be accepted. In order to cope with the gaseous contamination we specified in each cold box two parallel, switchable adsorbers operating between 80 K and 70 K for each stream being cooled down to temperatures below 70K. Each of these adsorbers has to be sized to retain the impurities of the maximum possible helium flow in the relevant line contaminated by up to 50 ppm by volume of air, for a duration of 50 h. In addition one single adsorber has to be installed operating at about 20 K for each stream being cooled down to temperatures below 20 K. This adsorber has to be sized to retain the impurities of the maximum possible helium flow in the relevant line contaminated by up to 1 ppm by volume of hydrogen or neon, for a duration of at least 200 h. The necessary equipment for fully automatic switching, regeneration and cool-down of the adsorber beds has to be provided.

Original plans to provide each refrigerator with an external full-flow purifier operating at 80 K for water and air were abandoned. Instead of this it is envisaged now to provide dryer beds for continuous operation, handling the full flow upstream of each cold box. Therefore it is specified that under maximum operation condition a pressure drop of one bar must be respected in the interconnecting HP piping between the compressor station and the cold box to allow for the pressure drop in the future dryers.

Liquid Nitrogen Pre-cooler

The liquid nitrogen pre-cooler was specified to be fully de-coupled from the main process heat exchangers in order to avoid problems in the case of return of large amounts of cold gas from the LHC machine, as well as to allow for operation even with a defective nitrogen heat exchanger. As the requirements are two-fold, for long-term cool-down operation and steady-state liquefaction boost, the connections to the main process lines are specified as shown in Figure 4. For cool-down operation the heat exchanger is used as indicated in Figure 4 A, for liquefaction boost as in Figure 4 B.

The heat exchanger is specified to have a nitrogen phase separator with included gas-liquid heat exchanger and a separate gas-gas heat exchanger.

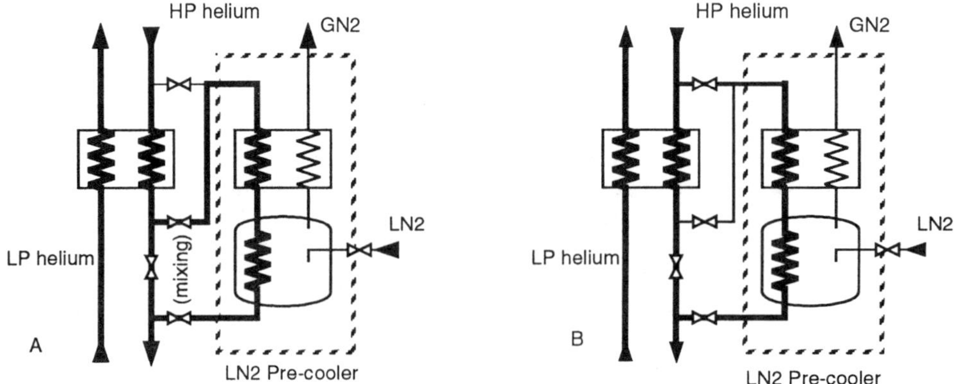

Figure 4. The specified process connections for the liquid nitrogen pre-cooler. A: Cool-down operation; B: Liquefaction boost

Liquid Helium Phase Separator

As no liquid helium storage dewar is foreseen, the integrated helium phase separator is specified to have sufficient volume in order to store 50% of the total helium hold-up in the cold box and the compressor station. The subcooler must be operational with the phase separator filled at only 30%. Thus it is possible to react on capacity changes by storing liquid helium due to excess capacity in the phase separator.

Process Control System

The process control system is not part of the supply for the refrigerators but will be purchased separately by CERN in order to allow a common process control system for the cryogenic equipment of the LHC machine. The supplier of the refrigerators has to deliver the necessary control logic in a pre-defined form compatible to international standards. The architecture of this control logic is based on the object-oriented philosophy validated already at CERN for the existing cryogenic installations of the LEP machine. The hardware safety as e.g. for turbo expanders and compressors is de-coupled from the main control system and must be provided with the supply. The necessary safety system may be based on logical relay arrangements or on the use of local process control units. The use of field bus connections is encouraged.

Integrated Cost Estimate for Adjudication

According to the CERN purchasing rules a contract is awarded to the bidder with the lowest price for the equipment specified. For a cryogenic refrigerator the cost for electrical consumption are an important factor and shadow all other operation cost[8]. Thus an installation with low investment cost but low coefficient of performance might prove more costly over the envisaged lifetime than an installation with better efficiency and higher investment cost. As for CERN only the total cost of investment and operation is relevant, we decided to leave the choice between efficiency and operation cost to the bidders.

For the specification of the new 4.5 K refrigerators for LHC, CERN therefore decided to use a formula for adjudication that includes the cost for investment and the expected cost for power consumption over 10 years of operation. The bidders were required to provide guaranteed data for the electrical input power of the 3.3 kV supply in the operation modes “Low-intensity” and “Installed”. The total cost of the supply was then calculated according to the formula below.

$$C = I + 4 * (1320 * P_{low} + 2640 * P_{inst}) \quad (1)$$

In this formula C represents the total cost, for adjudication purposes, in CHF for the four refrigerators, I the price in CHF as quoted by the bidder for the supply of the four refrigerators, P_{low} the guaranteed 3.3 kV power consumption in kW for one refrigerator in the operation mode "Low-intensity” and P_{inst} the guaranteed 3.3 kV power consumption in kW for one refrigerator in the operation mode "Installed". The factors 1320 and 2640 result from the expected 22000and 44000 hours respectively, of operation during ten years at a cost of 0.06 CHF per kWh.

The 3.3 kV power consumption measured during the reception tests will be compared to the guaranteed values of the bids and the difference will result in a compensation payment by CERN in case of lower power consumption than guaranteed, or by the supplier if the power consumption exceeds the guaranteed value.

Time Schedule

The time schedule specified for the four refrigerators is as given in Table 4.

Table 4. Time schedule for the 4.5 K refrigerators for LHC

	First refrigerator	Second refrigerator	Third refrigerator	Fourth refrigerator
Location	LHC point 18	LHC point 4	LHC point 8	LHC point 6
Earliest delivery date	15-04-2000	05-01-2001	01-04-2001	05-01-2002
Final reception test date	15-11-2000	30-06-2001	15-12-2001	01-07-2002

CONCLUSION

Following the specification for the new 4.5 K refrigerators for the LHC machine, CERN received two offers in February 1998. It was finally decided to split the supply in two and consequently two contracts for the supply of two 4.5 K refrigerators each were signed with Linde Kryotechnik AG, Switzerland and Air Liquide, France.

REFERENCES

1. V. Benda et al. "Conceptual design of the cryogenic system for the Large Hadron Collider (LHC)" 5th European Particle Accelerator Conference - EPAC '96 Sitges Barcelona, Spain ; 10 - 14 Jun 1996
2. S. Claudet, W.K. Erdt, P.K. Frandsen, Ph. Gayet, N.O. Solheim, Ch. Titcomb and G. Winkler, Four 12 kW / 4.5 K cryoplants at CERN, *Cryogenics* (1994) 34 ICEC Supplement, p. 83-86
3. F. Millet, P. Roussel, l. Tavian and U. Wagner, "A Possible 1.8 K Refrigeration Cycle for the Large Hadron
   Collider" Adv. Cryog. Eng., A : 43 (1998) 387-393
4. U. Wagner, "The LHC refrigerators with surface-located cold boxes for the temperature range 300-4.5 K," LHC Project Note 70 (1996)
5 L. Tavian and U. Wagner, "LHC sector heat loads and their conversion to LHC refrigerator capacities," LHC Project Note 140 (1998)
6. Ph. Lebrun, G. Riddone, L. Tavian and U. Wagner, Demands in refrigeration capacity for the Large Hadron Collider, in: "Proceedings of ICEC 16," T. Haruyama, T. Mitsui, K. Yamafuji, ed., Elsevier Science (1997), p. 95-98
7. Ph. Lebrun, G. Riddone, L. Tavian and U. Wagner, "Cooldown and Warmup Studies for the Large Hadron Collider" *Cryogenics* (1998) 17 ICEC Supplement, p. 813-816
8. Ph. Lebrun, "Estimation du Coût Marginal de L'Energie Consommée par la Cryogénie du LHC" LHC-Project-Note-92 (1997)

TWO LARGE 18 KW (EQUIVALENT POWER AT 4.5 K) HELIUM REFRIGERATORS FOR CERN'S LHC PROJECT, SUPPLIED BY AIR LIQUIDE

P. Dauguet, G.M. Gistau-Baguer, and P. Briend

Air Liquide, Advanced Technology Division
B.P N° 15, 38360 SASSENAGE, France

ABSTRACT

CERN in Switzerland has decided to build a new accelerator project called LHC (Large Hadron Collider). This 27 km long accelerator will, for the first time at a such large scale, operate superconducting magnets and radiofrequency cavities. For that purpose Air Liquide is now building two custom designed refrigerators of cryogenic power 18 kW equivalent at 4.5 K. The thermodynamical cycle, chosen to fit the LHC cryogenic loads with a very high efficiency, is discussed. A special emphasis is put on the cold end that makes use of a cryogenic expansion turbine discharging into the double phase domain. As these refrigerators are designed to be able to be operated at reduced cryogenic power with reduced electrical power consumption, chosen solutions to adapt the refrigerator operation to reduced load are described.

INTRODUCTION

The cryogenic duties to be provided by each refrigerator are presented in Table 1. The refrigerators have to fit very different operating loads. A large flexibility of operation of the machine is thus requested.

As the CERN call for tender was giving a very high importance to the electrical consumption of the plant as adjudication criteria, the choice of a very efficient cycle is as well mandatory[1]. The cycle design has to correspond to a refrigerator minimizing for CERN the sum of investment cost plus operating cost over 10 years of operation.

The third request to be taken into account in the design of the machine is the high availability of the cryogenic system needed by the LHC project. This is insured by the choice of reliable and technically proven components.

The cycle design presented hereafter is a compromise between efficiency, flexibility and reliability of operation.

Advances in Cryogenic Engineering, Volume 45.
Edited by Shu *et al.*, Kluwer Academic / Plenum Publishers, 2000.

Table 1. Cryogenic loads of the LHC refrigerators ("Intalled mode")

Loads Operation mode	4.5 K – 20 K (W)	20 K – 280 K (W)	50 K – 75 K (W)
Installed	25100	55400	33000
Normal	15000	36500	22000
Low Intensity	9300	36500	22000
Injection Standby	6900	14900	22000
75 K standby	0	0	22000

THE CYCLE DESIGN

The mass flow rates, pressures and temperatures given in the following description corresponds to the "Installed" operating mode.

The refrigerator cycle is presented in Figures 1 and 2. It is a three pressure cycle. Two stages of oil lubrificated screw compressors compress the cycle gas. The first stage compresses the low pressure mass flow rate (780 g/s) from 0.1 to 0.4 MPa. Then, this mass flow rate is mixed with the one coming from the medium pressure of the cold box (1600 g/s total) and is compressed by the second stage to the high pressure of the cycle : 2 MPa.

Two turbines in series (T1 and T2) are expending part of the gas from the high pressure to the medium pressure of the cycle, cooling the cycle gas from ambient temperature down to 75 K. Two other turbines (T3 and T4) which are arranged in parallel between the cycle high and medium pressures are cooling the helium gas down to 50 and 20 K respectively. At these temperature levels, the cold box provides 33 kW between 50 K and 75 K to the thermal shields of the LHC accelerator. This duty is fulfilled by part of the T3 turbine operation.

The mass flow rate to be delivered to the LHC accelerator at 4.5 K is then cooled through T5 and T6 turbines. Two turbines in parallel to the JT stream help the cooling of this mass flow rate : T7 and T8.

The nitrogen precooler is only used for cooling down of the accelerator and to fill it with liquid helium.

The flexibility of the refrigerator is insured by the choice of a cycle with most of the turbines in parallel. (Never more than two turbines are indeed mounted in series). The consequence is that the power extracted at each temperature level of the refrigerator can be controlled independently.

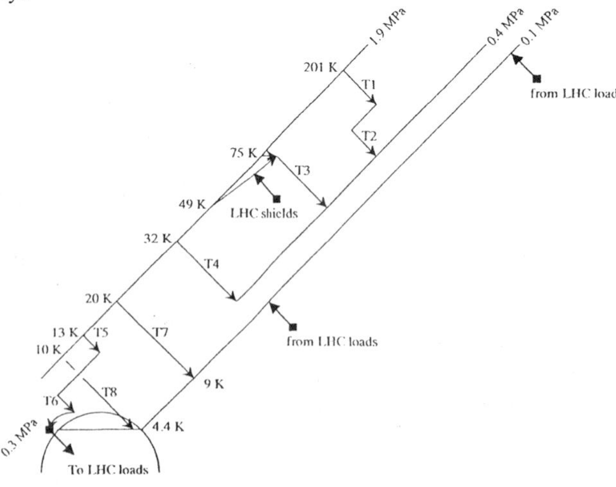

Figure 1. Schematic T-s diagram for "Installed mode"

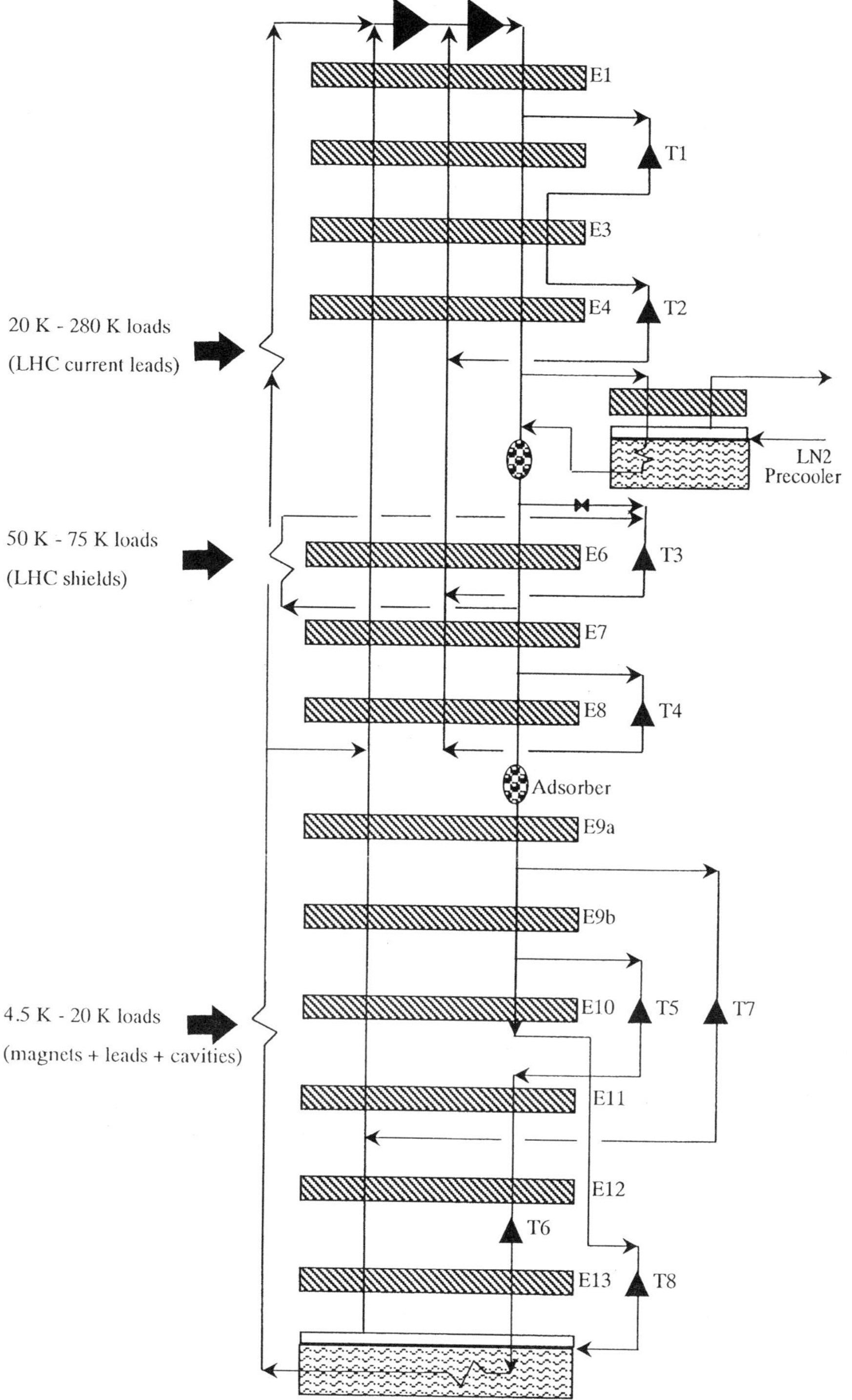

Figure 2. LHC refrigerators cycle design

The Cold End of the Cycle

Due to the fact that most of the duty of the refrigerator (more than 70 %) is to provide helium at 4.5 K to be returned to the cold box at 20 K (non-isothermal refrigeration), the downward and upward mass flow rates of helium in the cold exchangers (temperatures under 20 K) are not balanced : the high pressure flow is higher than the low pressure flow. This induces a bad operation of the exchangers, as the warm end temperature differences of the exchangers are increased by the mass flow imbalance. That is why a cold end design different from the one of a pure refrigerator (4.5 K isothermal refrigeration) has been chosen. An example of a "pure refrigerator" cold end (the one of the Air Liquide LEP refrigerators[2] at CERN) is presented in Figure 3.

In order to reduce the difference of the mass flow rates and to provide the cryogenic power induced by the remaining mass flow rate difference, turbines T7 and T8 are connected in parallel to the turbines T5 and T6 which are delivering the mass flow to the customer loads.(see Figure 2).

In the "Installed" operating conditions of the refrigerator defined in table 1, the inlet temperature of T7 is 20 K, which is the return temperature of the cold gas coming back form the LHC. For this reason the heat exchanger E9a is of no use in the "Installed" operating conditions. However, when the cryogenic load of the refrigerator is reduced, one consequence is that the temperature differences across the turbines are reduced, due to the reduction of the high pressure of the cycle (see for more details the chapter "operation at reduced cryogenic loads" hereafter). As the cold end temperature of the cycle remains constant (4.5 K), the consequence is that the inlet temperatures of the turbines are decreased. As the inlet temperature of T07 is reduced under 20 K, the heat exchanger block 9a begins to have some exchange duty at reduced cryogenic load. It contributes to the reduction of the electrical consumption of the plant with the reduction of the cryogenic load.

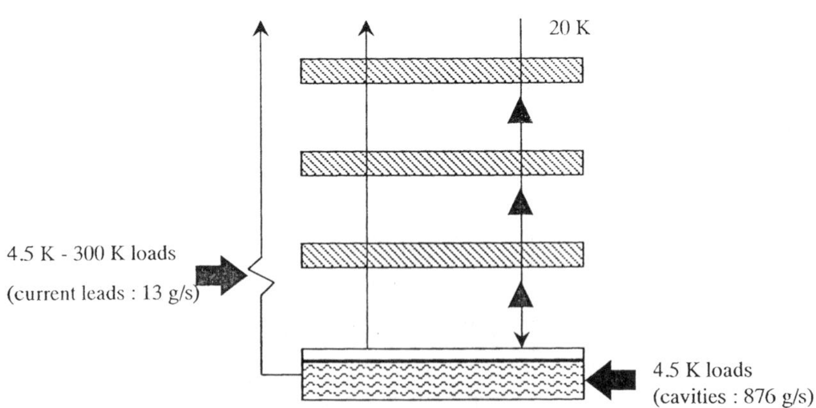

Figure 3. Cold end of a "pure refrigerator". Example of the Air Liquide plants LEP 12/18 kW for CERN[2].

THE COMPRESSOR STATION

A diagram of the compressor station is presented in Figure 4. As said before, the helium of the cycle is compressed in two stages.

The first stage is composed of three Aerzen machines (Germany) arranged in parallel, two of the 536 M type each processing 300 g/s (swept volume of each : 7460 m^3/h) and one of the 536 H type processing 230 g/s (5700 m^3/h).

The second stage is composed of two 536 H machines in parallel each processing 860 g/s.

At the discharge side of each compressor stage, the oil of the lubricated screw compressors is separated from helium by gravity in a bulk oil separator vessel. Then the helium and oil are cooled in water coolers.

The small amount of oil remaining in the helium after the oil separator vessel of the second compressor stage (less than 100 ppm by weight) is then separated with 3 coalescing cartridges in series (for aerosols) and an activated charcoal adsorber (for oil vapors).

At the outlet of this final oil removal system, the high pressure helium contains less than 10 ppb of oil by weight.

THE COLD BOX

The cold box vacuum enclosure is constituted with two main cylinders, a vertical one and an horizontal one, connected together, forming a L shape (see Figure 5).

In the vertical cylinder, one can find the process heat exchangers and 2 of the 8 turbines of the cycle.

In the horizontal cylinder, one can find, the cycle adsorbers, the phase separator, the nitrogen precooler, all cryogenic valves needed for the process control, and the 6 other turbines. All turbines are horizontal except for the "liquid discharging turbine" T8 which is vertical.

The heat exchangers used are aluminum alloy plate fin brazed type for the cycle and stainless steel plate welded type for the nitrogen precooler. Adsorbers at 80 K and 20 K, filled with activated charcoal, trap the impurities. The phase separator of the cold box allows subcooling of the supercritical helium, and regulation of the cold box load.

The dimensions of the cold box are the following. The vertical cold box has a diameter of 4 m and a length of 9 m. The horizontal cold box has a diameter of 3.5 m and a length of 12 m. The total weight of the L shape cold box constituted with the two cylinders is approximately 70 tons.

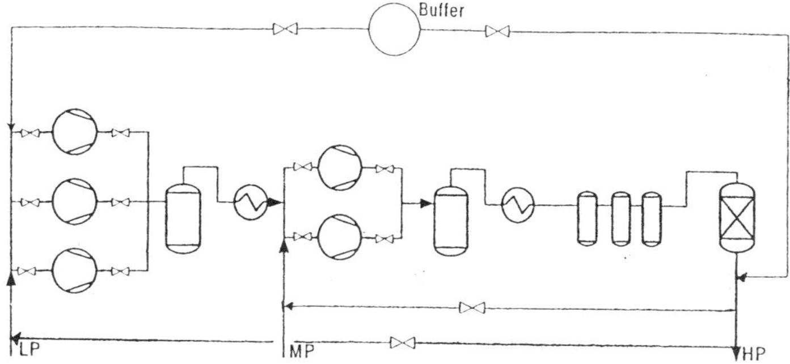

Figure 4. Diagram of the compressor stations of the LHC refrigerators

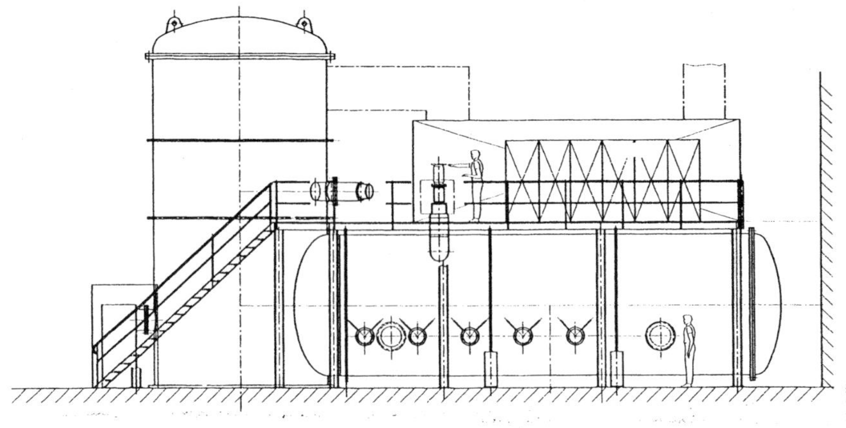

Figure 5. Sketch of the cold box

THE EXPANSION TURBINES

The turbines are of the industrial static gas bearing type[3]. They are designed and manufactured at the Advanced Technology Division of Air Liquide.

Most of them are radial-axial "3D" wheel turbines. With such cryogenic turbines, extracted power up to 125 kW on a single turbine has already been obtained[4].

The operating conditions of the LHC refrigerator turbines are presented in Table 2. Several helium turbines are already operated with success at CERN in similar operating conditions (12 kW Air Liquide refrigerators[2] upgraded to 18 kW for the operation of the CERN LEP 2 accelerator).

The Liquid Discharging Turbine

The coldest turbine in the LHC cycle is designed to produce helium in double phase. Its inlet is feed with helium gas at 1.84 MPa , 10 K. At its outlet (0.12 MPa , 4.4 K), part of the gas is transformed to liquid phase. The amount of liquid is 11 % by mass which correspond to only 2 % by volume.

Liquid discharging turbines are already operated on Air Liquide helium refrigerators[2,5], even with a higher liquid proportion at the outlet. For example, the LEP 12/18 kW refrigerators[2] liquid discharging turbine T8 has been operated during more than 8000 hours at CERN.

Table 2. Operating conditions of the LHC refrigerators' turbines.

Turbine number	1	2	3	4	5	6	7	8
Flow rate (g/s)	237	237	328	262	234	234	275	226
Extracted power (kW)	37	34	45	16	2.2	1.0	15	4
Speed (Hz)	2100	2100	2000	2200	800	700	2000	1400

OPERATION AT REDUCED CRYOGENIC LOADS

The refrigerator cycle has been designed : first, to fit the maximum load of the LHC accelerator with a very high efficiency ; secondly, to adapt to any operation mode of the LHC refrigerator (part cryogenic load) with a corresponding reduction of the electrical power consumption of the refrigerator.

The way to adapt to the LHC load is the following. When the cryogenic load of the refrigerator is decreased, the high pressure and medium pressure of the cycle are reduced. Consequently, as the mass flow processed by a turbine is proportional to its inlet pressure, the mass flow processed by each turbine, and consequently its power, is reduced.
Proceeding so, as the mass flow and the compression rate of the cycle is reduced, the global electrical consumption of the compression plant is reduced.

In order to optimize this electrical power reduction, the second stage compressors are operated at 100 % volumetric flow so that their isothermal efficiency is maximum. For the same reason, the rotation speed of the turbines are individually adjusted to the operation condition by regulation of the turbine brake pressure, in order to maximize their isentropic efficiency.

Operating so, a very high factor of merit of the plant (ratio of the electrical consumption versus the equivalent cryogenic power at 4.5 K) can be obtained even at reduced operating loads (See Table 3).

The adaptation of the operating parameters of the refrigerators to the variation of the cryogenic load is fully automatic as described hereafter.

PROCESS CONTROL

The process control system drives the plant automatically, according to the control program, for all transient and steady state regimes : cool-down, adaptation of the cryogenic power to the customer needs, either on a stand-by situation or during full power operation and warming up. Prior to start up, the control system checks all parameters, and makes sure that all conditions are correct to reach nominal operation. If some abnormal situation occurs during operation, the control system can react in different ways : if it is possible, it will correct the situation (regulation). If not, it will keep the plant operating at reduced power. And if the situation becomes critical, it will bring the plant to a stop (example : if air pressure becomes too low, the emergency stop procedure is launched).

An electrical power, cooling water or air breakdown brings the refrigerator to a temporary stop. As soon as the defective utility is back, the refrigerator restarts automatically.

Operated in this way the plant fully automatically brings reliability, flexibility and efficiency to the refrigerator.

Table 3. Expected electrical power consumptions of each LHC refrigerator versus the cryogenic load.

Operating mode	Cryogenic load (Exergetic equivalent at 4.5 K) kW and (% of max value)	Electrical consumption kW and (% of max value)	Factor of merit (W/W) (electrical consumption divided by cryogenic load)
Installed	17.5 (100)	4200 (100)	240
Normal	10.7 (61)	2800 (67)	260
Low Intensity	7.6 (43)	2300 (55)	300

CONCLUSION

A very high efficiency cycle has been custom designed for the LHC accelerator. The factor of merit of this cycle at maximum cryogenic power is 240 W/W, which corresponds to an efficiency of approximately 30 % compared to Carnot ratio.

The liquid discharging turbine is an interesting solution to get this high efficiency, as well as a special arrangement of the turbines operated below 20 K.

The way the plant is operated for reduced cryogenic duty enable it to keep a high cycle efficiency even at part loads down to 40 % of the installed power.

ACKNOWLEDGMENTS

The authors wish to thank their colleagues, members of the "LHC Refrigerators' Project Team" at AIR LIQUIDE, for their contributions to the design study of these refrigerators and especially P. Hirel, B. Hilbert, G. Marot, C. Guerin, J.P. Bacca, Y. Bonnet, U. Bassi, and P. Cuviller.

REFERENCES

1. U. Wagner et al., Specification of four new large 4.5 K refrigerators for the LHC, *Same volume*.
2. G.M. Gistau and J. Veaux, A 12/18 kW at 4.5 K Helium refrigerator for CERN's LEP superconducting Accelerating Cavities, *ICEC Proceedings, Cryogenics* 34:103 (1994).
3. G.M. Gistau and J.C. Villard, High power industrial gas bearing cryogenic expansion turbines, *ICEC Proceedings, Cryogenics* 32:87 (1992).
4. G. Marot and J.C. Villard, Recent developments of Air Liquide Cryogenics Expanders, *Same volume*.
5. G.M. Gistau and M. Bonneton, A 6 kW at 4.5 K helium refrigerator for CERN's cryogenic test station, *Advances in Cryogenic Engineering* 38 (1993).

TWO LARGE 18 KW (EQUIVALENT POWER AT 4.5 K) REFRIGERATORS FOR CERN'S LHC PROJECT SUPPLIED BY LINDE KRYOTECHNIK AG

J. Bösel, B. Chromec, A. Meier

Linde Kryotechnik AG, CH – 8422 Pfungen, Switzerland

ABSTRACT

Linde Kryotechnik AG has developed an efficient and reliable helium refrigeration system for the LHC project. The process cycle uses two stages of compression with five oil-lubricated screw compressors and ten expansion turbines. The refrigeration plants were optimized for reaching high efficiencies, also at part load, by using adapted compressor sizes for the 1st stage compression and making optimum use of the turbines. The exergetic efficiency of the coldbox for the design case reaches approx. 54%. Losses in turbines (31%) and heat exchangers (13%) are kept reasonably small. The plants will be assembled with well proven components and are scheduled to be commissioned in December 2001 and June 2002.

INTRODUCTION

For the Large Hadron Collider (LHC) CERN needs – in addition to the upgraded LEP-12 kW-plants[1)] – four new helium refrigerators, which are able to provide an exergetic equivalent of 18 kW at 4.5 K cooling power. Two of those plants will be supplied by Linde Kryotechnik AG. Design, construction and performance of the refrigeration plants are based on the following criteria:

- The operational and load requirements must meet CERN's Technical Specification[2)3)].
- Power input during main operation modes and transient conditions shall be economized.
- High reliability and easy maintainability must be achieved by using well proven components.

This paper describes the design of the process, the plant, the control concept and the main components of the refrigerator.

PLANT AND PROCESS DESCRIPTION

Outline of the Process

The helium refrigeration system chosen for the process consists of two major sub systems, the *compression system with oil removal and gas management* and *the refrigerator coldbox*. The selected process is a *Brayton cycle* with three pressure levels and five stages of expansion with a total of 10 turbines. Liquid nitrogen is used for cool-down purposes and liquid filling of the magnets of LHC only. Fig. 1 and 2 show the Process Flow Diagrams (PFDs) with the major process lines, compressors, turboexpanders, heat exchangers and main control loops.

Compression System

The compression system consists of three 1st stage compressors and two 2nd stage compressors (Fig. 1). Two different sizes of machines for the 1st stage are used in order to assure a good partload behavior.

The compressors provide the pressure levels for three different major helium streams in the coldbox:

- The *high pressure helium (HP) stream* which is discharged from the 2nd stage compressors 6C120 and 7C120.
- The *medium pressure (MP) return stream* which consists of the combined outlet streams of turboexpanders 3TU210, 5TU210 and 7TU210 (cp. Fig. 2). The MP stream returns to the suction header of compressors 6C120 and 7C120 and is combined with the discharge stream of the 1st stage compressors.
- The *low pressure (LP) return stream* which consists of helium gas from the phase separator, the discharge gas of turbine 9TU210 and the gas from the LHC experiment (cp. Fig. 2). It is returned to the combined suction header of compressors 1C120, 2C120 and 3C120.

The designed pressure levels for LP/MP/HP at full capacity are 1.01/4.1/21.6 bar.

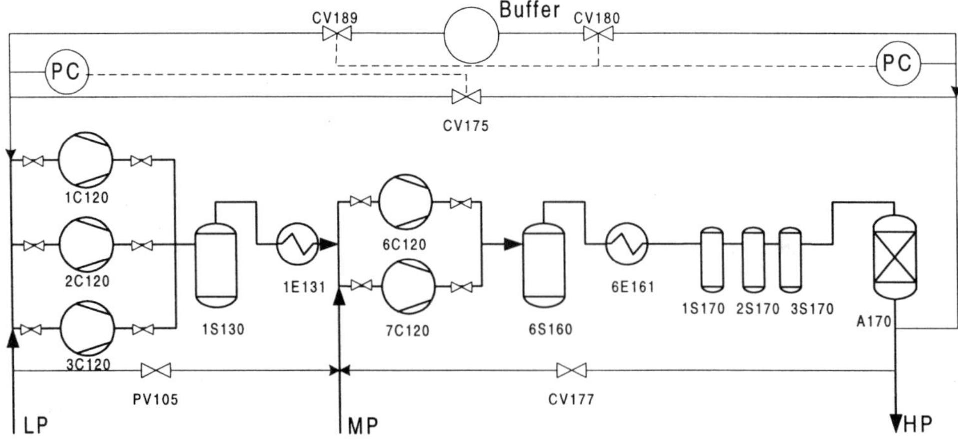

Figure 1. Process Flow Diagram (PFD) of the compressor station.

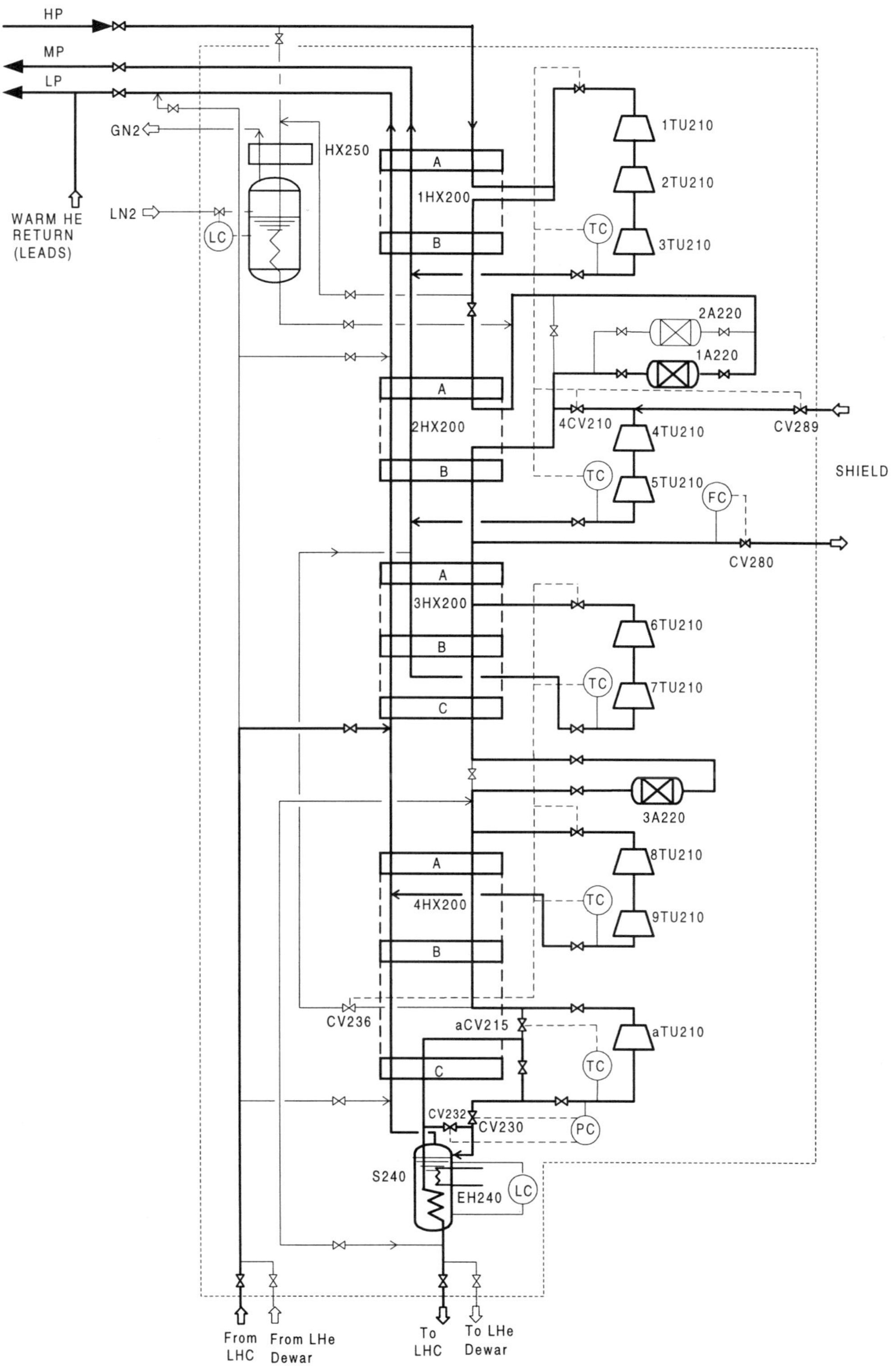

Figure 2. Process Flow Diagram (PFD) of the coldbox.

Coldbox

In the coldbox (Fig. 2) high pressure gas is cooled in counter current against medium pressure and low pressure gas in a total of four plate fin heat exchangers (in heat exchanger 4HX200 no MP path is present). The refrigeration is produced by a total of ten turbines which are arranged in five parallel side streams at different temperature levels. In all turbine streams gas is taken from the HP line and expanded either to the MP or to the LP level. Within the side streams the turbines are arranged in series without inter-cooling.

The upper turbine arrangement includes three turbines in series, the following lower three arrangements consist of two turbines each and the remaining stream has one single turbine.

Gas for shield cooling is taken from the HP stream downstream of heat exchanger 2HX200B. After being charged with the heat load from the shield, the gas is returned to the coldbox and joins the slightly throttled HP stream upstream of turbine 4TU210.

Downstream of the last turbine aTU210 a pressure of 3 bar is maintained during almost all operation modes. The main part is first cooled in the plate fin heat exchanger 4HX200C and then to 4.5 K by a heat exchanger submerged in the liquid helium of separator S240. The smaller part of the flow is throttled into the phase separator at LP level becoming partially liquefied. The liquid fraction serves as refrigerant for the supercritical (3-bar) helium to be cooled and is returned – together with the flash gas - to the 1st stage compressors through the heat exchangers. To optimize the heat transfer of heat exchanger 4HX200C the helium which is throttled into the phase separator can use two different paths: either through CV230 (more efficient during full capacity and cool down operations), or through heat exchanger 4HX200C and CV232 (for part load operations).

Within the LHC the subcooled helium is split again in two fractions. The smaller stream is used for leads cooling and returns at ambient temperature to the suction header of the 1st stage compressors. The larger stream returns at a temperature of approx. 20 K to the coldbox where it is combined with the low pressure gas between heat exchangers 4HX200 and 3HX200.

EXERGY ANALYSIS

Cooling power of the LHC plants is needed in the 4.5 to 20 K temperature range, in the 50 to 75 K temperature range (shield) and for leads-cooling (20 – 280 K). Because most of the cold gas returns to the plant at the 20 K level, the low temperature end of the plant represents a large liquefier. Fig. 3 displays a schematic T-s-diagram for *Installed Mode*[2),3)], indicating the different flow paths with pressures and temperatures. Turbines are present at nearly any temperature level: while the 1st and 3rd turbine strings mainly compensate heat exchanger losses and (together with the other strings) 'liquefaction' power for the gas stream cooling the leads and returning at ambient temperature, the 2nd turbine string provides cooling for the shield and the last two strings deliver the refrigeration needed in the 4.5 – 20 K temperature range. The size of the turbines (and such the massflow through them and cooling power of them) was selected according to these demands. For achieving large pressure ratios in helium turboexpanders high rotational speeds are required, especially at higher temperatures. For expanders with dynamic gas bearings two or even three turbines per string become necessary. Only at the cold end it is possible to expand gas from HP to 3 bar in one single stage (turbine aTU210). The maximum speed of the turbines is usually also the optimum speed.

The pressure ratio HP/MP stays nearly constant during most operation modes which keeps the efficiency of the 2nd stage compressors high.

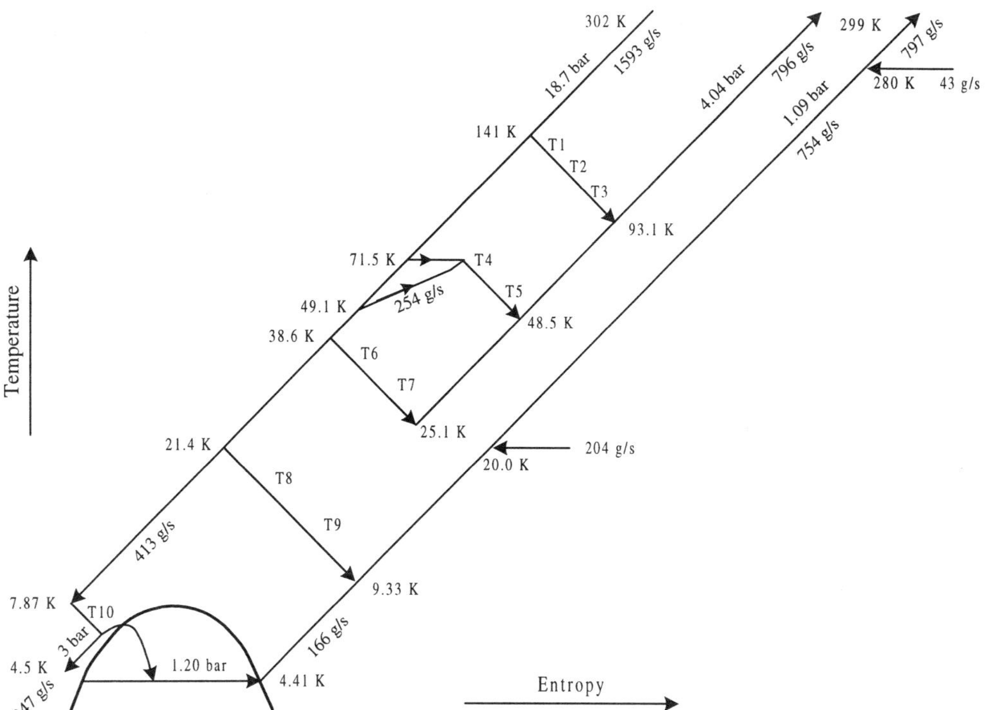

Figure 3. Schematic T-s diagram for *Installed Mode*.

The plant is very flexible: it covers a large range of refrigeration demands with good efficiencies also at part load and can even be used as a powerful liquefier. An exergetic analysis shows where the major losses in the coldbox occur (Tables 1 and 2).

For the design case (*Installed Mode*) the coldbox has an efficiency of approx. 54%. All heat exchanger losses sum up to about 13%, losses in turbines to about 31% of the total exergy supplied to the coldbox. Heat exchanger losses arise from temperature differences between the different streams. The warmest heat exchanger covers a large temperature range thus having the largest losses. Turbine power is not recovered and losses there are approx. proportional to massflow, except for the coldest turbine. Even though the pressure ratio is large and the isentropic efficiency only moderate, losses in this turbine are small due to real gas effects at the given pressures and temperatures (specific volumes at turbine inlet and outlet are similar). The same is valid for the JT-valve (CV230, CV232) whose losses are smaller than 2%.

When including the pressure losses in the interconnecting pipes between coldbox and compressors and compressor-losses (the isothermal efficiency of the compressors is larger than 50%) the overall efficiency of the plant reduces to approx. 27%.

Table 1. Exergy analysis of the LHC refrigeration plant (*Installed Mode*)

LHC-Ring	Leads	Shield	HX1	HX2	HX3	HX4	Turbine String 1	Turbine String 2	Turbine String 3	Turbine String 4	Turbine String 5	JT-Valve	Other Losses
40.7%	6.2%	7.0%	5.9%	1.5%	2.1%	3.1 %	6.6%	7.1%	2.8%	10.0%	4.2%	1.7%	1.0%
			12.6%				30.8%					2.7%	
53.9%			46.1%										

Table 2. Exergy analysis of the LHC refrigeration plant (*Low Intensity Mode*)

LHC-Ring	Leads	Shield	HX1	HX2	HX3	HX4	Turbine String 1	Turbine String 2	Turbine String 3	Turbine String 4	Turbine String 5	JT-Valve	Other Losses
30.8%	6.5%	8.1%	5.0%	3.4%	4.4%	6.2%	8.2%	7.5%	3.1%	8.5%	3.8%	3.1%	2.2%
			19.0%				31.0%					5.3%	
44.7%			55.3%										

Even at part load (e.g., *Low Intensity Mode*[2),3)] where approx. half of the refrigeration (exergetic equivalent) of *Installed Mode* is needed) the coldbox reaches an efficiency of 45%.

PROCESS CONTROL

Linde Kryotechnik AG will supply the control logic to establish the control software for the main Process Control System (PCS). PCS and control software will be provided by CERN. Process control is based on automatic operation during cool-down, warm-up and all operation modes and for the transition from one mode to the other. The refrigeration capacity will automatically be adjusted to the actual load and it will be possible to react quickly to an increase or decrease in required refrigeration capacity.

Pressure Control

Low Pressure (LP): This pressure is dictated by the helium equilibrium temperature of 4.5 K in the phase separator S240. Short term control of LP is achieved with the bypass valves CV175 from the high pressure line to low pressure line. For long term LP control the 1st stage slide valves are operated. During part load operation one of the 1st stage compressors, either the smaller machine or one of the larger machines, is switched off.

Medium Pressure (MP): As a function of the compressor size, MP adjusts automatically to the flow delivered from the 1st stage compressors and cold box return. A bypass valve from HP to MP (CV177) avoids that MP drops too low and limits the pressure ratio on the 2nd stage compressors. Usually the slide valves of the 2nd compressors can be kept closed.

High Pressure(HP): HP is controlled by the loading valve CV189 and the unloading valve CV180. The number of compressors in operation as well as the setpoint for HP is depending on the actual refrigeration requirements.

Temperature Control of Turbines

Each of the turbine strings has an individual low temperature limit. The turbines shall never be stopped for reasons of a low temperature and each of the turbine inlet valves always maintains a minimum position in order to run the turbines safely.

Because of the nearly constant shield load short-term disturbances at the cold end of the plant will not affect the 1st and 2nd turbine string. If the outlet temperature of the 1st or 2nd string falls below its preset limit, the inlet valve of the 1st string is throttled first followed by throttling the inlet valve of the 2nd string (respectively the shield return valve CV289), if the outlet temperature of the 2nd string still falls. Because the cooling power of the 3rd turbine string compared with the cooling power of string 4 is low, both inlet valves of string 3 and 4 are throttled simultaneously, no matter which of the turbine outlet temperatures sinks below its limit. Additionally cold gas can be bypassed via CV236 from the HP line downstream of 4HX200B into the MP line downstream of 3HX200A. The inlet valve of the last turbine aTU210 is only closed, if the turbine is directly endangered. If necessary, a bypass valve (aCV215) can be opened to reduce the flow through turbine aTU210, thus reducing the power, as well.

Temperature Control of Shield

The shield represents a very large thermal impedance. Therefore a flow control is applied. The shield flow return valve CV289 is usually completely open; by adjusting the opening of valve CV280 the required flow is established. If for any reason it is not possible to return the shield flow into the HP-line upstream of 4TU210, it may also be sent through a bypass directly into the MP line downstream of 2HX200A.

Liquid helium level of Phase Separator S240

The phase separator is designed to store at least 50% of the HP inventory of the plant. It serves as an effective buffer to react quickly in case of over- or under-capacity. At full capacity it is only filled about 40%. For decreased refrigeration capacity, liquid helium can be 'dumped' in the separator, lowering HP and thus reducing the cooling power of the turbines. If a quick increase in refrigeration is required, the control heaters are turned on and a large amount of helium can be evaporated, which in turn increases HP and the cooling power of the turbines.

PLANT COMPONENTS

Compressor Station and Oil Removal System

A system of oil-lubricated screw compressors made by Aerzen was chosen. The 1st stage units consist of two identical compressors type 536 M and one compressor type 536 H. The 2nd stage units consist of two identical compressors type 536 H. The compressors incorporate controlled slide valves by which the capacity can be varied with related power savings. The compressor skids include compressor block, main motor drive, isolation and non return valves and are mounted on a rigid base frame. Each compressor station has one common lube oil skid including oil retention vessel with bulk oil separator (1S130/ 6S160), dual lube oil pumps, dual oil filters and oil- and after-cooler (1E131/ 6E161).

The skid-mounted oil removal system consists of three stages of coalescers (1/2/3S170), a single, stand alone charcoal adsorber vessel (A170) and a dust filter. The first two coalescer stages reduce the oil content to less than 0.1 ppm and the third stage coalescer acts as a guard. The flow through the adsorber is downward, i.e. from top to bottom. The adsorber removes both vapor phase and liquid impurities from the helium to under 10 ppb by weight.

Coldbox

The refrigerator coldbox contains heat exchangers, turboexpanders, cryogenic valves, phase separator and all the necessary additional equipment which must be insulated against heat inleaks. It is a vacuum vessel of horizontal design with the valve plate on top for the penetrations and installation of the valves, turbines, instrument tubing and wiring.

Heat Exchangers: A total of four plate fin heat exchanger blocks made of aluminum are provided within the coldbox. Three of the heat exchangers (each consisting of two respectively three sections) are arranged horizontally while the last one, operating below 20 K, is in a vertical position. Linde Kryotechnik AG uses plate-fin exchangers made by its parent company, Linde AG.

The heat exchanger for precooling during cool-down is realized in a stainless steel plate heat exchanger submerged in a vessel containing liquid nitrogen (evaporator) and a stainless steel helium-to-gaseous-nitrogen plate heat exchanger.

Turboexpanders: Linde Kryotechnik AG uses its turboexpanders with self-acting dynamic gas bearings which do not require a separate bearing gas system during operation. Every turboexpander used is a single-stage centripetal turbine, braked by a directly-coupled, single-stage centrifugal compressor.

With the different turbine models used in the plant rotational speeds of 3600 rps (TGL22), 2500 rps (TGL32) and 1750 rps (TGL45) are possible. The speed is controlled through a throttle valve in the compressor gas loop.

Cryoadsorbers: Dual adsorbers are located at the 80 K level for removing traces of air. A single guard adsorber is located at the 20 K level to retain traces of hydrogen and neon. All adsorbers can be by-passed. Activated charcoal is fixed between rigid upper and lower filter packages, each consisting of wire mesh, a perforated plate, two layers of mineral wool filter and another perforated plate. Replacement of the adsorbent is possible. Regeneration with following cool-down of the adsorbers is possible at maximum refrigerator capacity. It is performed by the circulation of warm gas through the adsorbent followed by evacuation of the adsorber. The gas is heated by a heater located on the automatic regeneration skid outside the coldbox.

Phase Separator and Subcooler: The phase separator consists of a vessel of sufficiently large volume to store more than 50% of the HP compressor hold-up (in liquid form) and houses a subcooler and electric heaters. The subcooler is a plate-fin heat exchanger with the channels for the evaporating fluid open on both sides. The liquid level is measured by differential pressure.

OUTLOOK

Currently (July 1999) the process design (selection and specification of components) has reached its final stage. The mechanical design of the plant is under way and is scheduled to be finished in May 2000. The first plant will be installed in April 2001 and commissioned in December 2001. The installation and commissioning of the second plant is scheduled from January to June 2002.

REFERENCES

1. B. Chromec et al., A high efficient 12 kW helium refrigerator for the LEP 200 project at CERN, in: "Supercollider 5," P. Hale, ed., Plenum Press, New York (1994), p. 95.
2. European Organization for Nuclear Research (CERN), "Revised technical specification for four cryogenic helium refrigerators for the large hadron collider (LHC)", IT-2555/LHC/LHC, CERN, Geneva (1997).
3. S. Claudet et al., Specification of four new large 4.5 K helium refrigerators for the LHC, to be published in: "Advances in Cryogenic Engineering Vol. 45," Plenum Press, New York (2000).

OPERATION OF THE FOUR 12 KW AT 4.5K REFRIGERATORS FOR LEP

N.Bangert,[1] Ph. Gayet[2]

[1] Air-Product-Thomson
St Genis Pouilly, 01630, France

[2] CERN, European Organization for Nuclear Research
1211 Geneva 23, Switzerland

ABSTRACT

In 1998 the first energy upgrade of the LEP Electron/Positron collider, LEP2, was completed at CERN. Sixty-eight superconducting modules supplied by four 12 kW @ 4.5 K equivalent power refrigerators have been operated allowing a colliding beam energy of 94.5 GeV. Meanwhile, the operation and maintenance responsibilities were transferred to an industrial firm on the basis of a result-oriented contract. After a short description of the operational organization, we report on the operation of the LEP2 cryogenic system over the past three years. Particular attention is given to power availability, failure statistics and recovery time after interruptions. The most relevant problems and their solutions are exposed. Finally, we review the interactions between the cryogenic system and the particle beams, which are limiting the ultimate performance of the LEP collider.

INTRODUCTION

Operated since 1995 by a joint team of CERN/Air-Products-Thomson, the four cryoplants have accumulated a total of 71,000 hours of operation. During this period the upgrade of the LEP e^+/e^- collider from 45 GeV to 96 GeV per beam was achieved by gradually installing 68 modules on both sides of each of the four interaction points (IP) in the ring. In 1998 the 68 accelerating modules installed allowed a colliding energy of 94.5 GeV per beam in physics with a record luminosity of 200 pb^{-1}.

Within the same period the responsibility of the operation and maintenance was transferred to an industrial contractor allowing an important redeployment of CERN resources to the LHC activities without noticeable effect on the cryoplant performance.

During the LEP winter shutdown 1998/99, all four 12 kW@4.5 K refrigerators were upgraded to achieve the equivalent power of 18kW@4.5K necessary to reach the 100 GeV per beam ultimate potential of LEP.

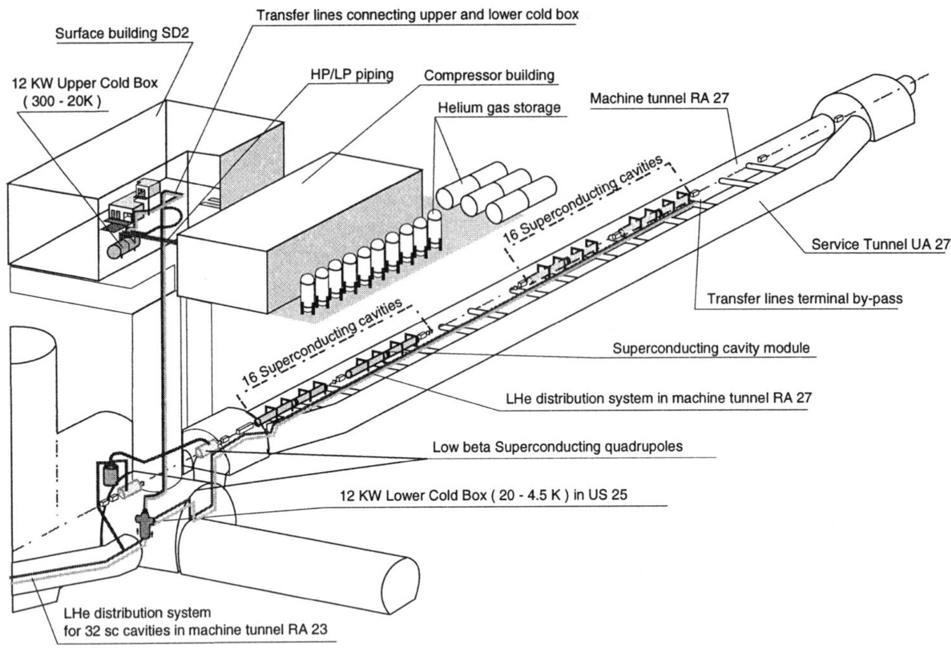

Figure 1. Typical layout of a LEP2 cryogenic system.

THE LEP2 CRYOGENIC SYSTEM

The LEP2 cryogenic system described in earlier publications [1,2,3,4] located at each of the four interaction points of LEP, of a cryoplant and its associated liquid helium distribution system, i.e., 200 m-long supply-and-return transfer lines to feed 8 to 10 superconducting modules on both sides of the point. A description of the liquid and gaseous helium circuits inside the 11 m modules can be found in the references [5]. In addition to the modules, the 12 kW plants are also cooling the low-beta quadrupoles at two LEP points. A typical layout of the cryogenic system of LEP2 is shown in Figure 1.

OPERATION

Operational Organization

In 1995 an operation and maintenance contract was established between CERN and the Air Products – Thomson Consortium with the initial objectives to enlarge the industrial support already present at CERN on the LEP2 operation.

This contract was structured in two phases. In a first contractual phase from 1995 to 1997, the consortium had to provide resources and infrastructure to operate and maintain the cryogenic installations under CERN's supervision and responsibility. In parallel, the contracting consortium had the duty to prepare, write and update operational documentation (operation procedures, control logic description, electrical and P&I Diagrams) and to establish a maintenance plan based on their industrial knowledge.

After three years the responsibility has been transferred to Air Products –Thomson with a new, result oriented, phase of the contract.

The tasks of the consortium for this new phase are:

- to provide and supervise qualified personnel ensuring efficient and uninterrupted operation of the installations with a minimum number of operators fixed within the contract.
- to organize and execute preventive maintenance as defined in the maintenance plan and to perform corrective interventions in case of failure.

In the new phase contract, an important fraction of the payment is based on the evaluation of the cryoplants' availability. This is evaluated by the recovery delay after utility failures and a tolerance related to cryogenic failures depending on accumulated running hours. Figure 2 presents the contract tolerance for the LEP2 cryoplants in case of utility failure.

The staff involved in the consortium consists of 37 persons, of which 30% are related to LEP2 operation and maintenance. For the LEP cryogenic system, the operation team consists of 5 operators and 1 engineer.

A small group of CERN-trained staff is still active and remain in close touch with the installations. The staff, involved in consolidation and upgrade work, operation and maintenance monitoring, as well as interface with the LEP machine, was reduced from 22 man-years in 1995 to 5 man-years in 1998.

Table 1. Overall statistics over three years

	NB of Utility failures	LEP hours lost	NB of Cryo failures	LEP hours lost Cryo origin
1996	2	32	2	6
1997	5	53	1	7
1998	5	113	4	20,5

Table 2. LEP2 cryogenics operation statistics in 1998

	Running Hours	Main Power cut 15/5/1998		Main Power cut 6/8/1998		Other Utilities failures				Cryo failures	total of hours lost for RF(h)	% total	% cryo origin
		Utility failure (h)	Recov. time (h)	Utility failure (h)	Recov. time (h)	Utility failure (h)	Recov. time (h)	Utility failure(h)	Recov. time (h)	Recov. time (h)			
IP2	5196:00	4:32	32:00	1:23	24:00					12:30	68:30	1.32	0.24
IP4	5043:00	1:47	28:00	1:23	12:00	0:08	5:00			3:00	48:00	0.95	0.06
IP6	5139:00	2:00	49:30	1:23	7:00					1:30	58:00	1.13	0.03
IP8	4995:00	3:13	33:00	1:23	27:00					3:30	63:30	1.27	0.07
SPS1	6010:00	3:16	16:20	2:12	7:53	0:02	2:58	1:17	8:15		35:26	0.59	0.00
SPS3	6010:00	3:16	12:50	2:12	4:53	0:02	1:22	1:17	6:15		25:20	0.42	0.00
total	32393:00		171:40		82:46		9:20		14:30	20:30	298:46	0.92	0.06
LEP	4224:00		49:30		27:00		7:58		8:15	20:30	113:13	2.68	0.49

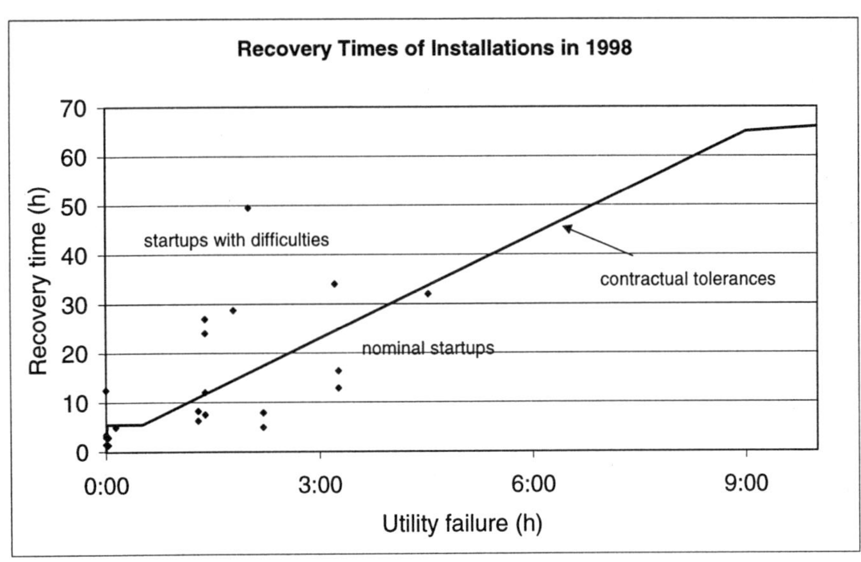

Figure 2. Time for recovery as function of the utility downtime in 1998

Downtime Statistics and Duration of Recovery

It turns out that interruption of operation and LEP downtime are mostly related to utility failures (water, electricity), especially in 1998 with 2 major power cuts resulting in 76 hours of LEP downtime. The time lost due to cryogenic failures, such as defective equipment, is low with a non-relevant increase in 1998 (Table 1). This result is due to the recent careful preventative maintenance on the cryogenic system.

As the installation of the superconducting modules was completed in 1998, the detailed statistics of the cryogenic and utility failures are given for that year in Table 2.

The recovery time for LEP is to be considered as the maximum cryogenic recovery time for each utility failure.

Most of the start-ups after a utility failure in 1998 were nominal as shown in Figure 2. If any difficulties, such as instrument or valve failures, appeared during start-up the recovery time went over the fixed tolerance limits.

For one of the cryoplants (IP8), longer recovery times, caused concerns about its efficiency. The lack of power was due to a leak in a valve between 20K and 4.5K, which was repaired during the last shutdown.

Helium Consumption

Each installation needs a minimum of 17,000 Nm3 of helium. To ensure a refill of the modules after accidental loss additional large storage tanks have been installed. Thus, a total of 24,000 Nm3 of helium can be stored. Over the past three years the annual consumption for the four installations was about 40,000 Nm3. Half of the losses are due to accidental stops of the installation, as the recovery of the 100 g/s of helium outgased by the SC modules at low pressure to the medium pressure storage cannot always be done (lack of utility,..). The other half is lost gradually over the year, mostly during the start-up phase, including purging and regeneration, of the cryoplants after the annual maintenance.

OBSERVED PROBLEMS

Turbine Filter Clogging

In 1996 it was observed in two cryoplants that the helium filter of a 90K turbine was clogging thereby reducing the power capacity of the cryoplant. Based on investigations done in 1998, it appears that the observed clogging was due to traces of water (<10 ppm) coming from the medium pressure storage or from the first heat exchanger.

The clogging induces a power reduction and may cause collapse of the filter. Periodic cleanings are required during the year. Each intervention takes 12-15 hours and was combined with other maintenance or access activities in order to reduce the impact on the running schedule of LEP.

Helium Leak in Inter Cold Box Transfer Lines

In 1996 a helium leak was detected on one of the Interconnecting Transfer Lines (ITL) linking the Upper and Lower Cold Box of one cryoplant. The leak appeared to be on a compensation bellows and induced by a design error existing on all ITL of two cryoplants. In view of the large amount of repair work, the run took place with the leaky part partially repaired, the insulating vacuum under active pumping and monitoring, plus a fallback solution based on spare flexible transfer lines. The repair of the ITL with a corrected design took place during the following shutdown. The work, which required the removal of all the lines from the access shaft to replace all the faulty parts and re-install the lines, was executed within a very tight schedule.

Thermomechanical Oscillations

At the end of the 1994 run a 100Hz oscillation of the accelerating field of two modules was observed [6]. Studies during the 1995 run showed a correlation between this oscillation and a mechanical vibration on the helium outlet of the modules. After elimination of other possible origins such as valve vibration and LHe level influence, a thermo-acoustical origin was identified. This oscillation was inducing the mechanical vibration of the outlet connector to the transfer lines, and as the frequency of this vibration corresponded to a mechanical resonance of the cavities, the accelerating field was affected. During the technical stop in October 1996, Kapton™ foils were placed between the male and female part of the outlet connector in order to limit the thermo-acoustical oscillation building up in the helium space along the connector. Lack of space on the female coupling prevented this installation for several modules. The results obtained are satisfying as most of the 100 Hz oscillations are now damped to an acceptable level for the RF operation. The installation was completed successfully during the 96/97 Shut-Down.

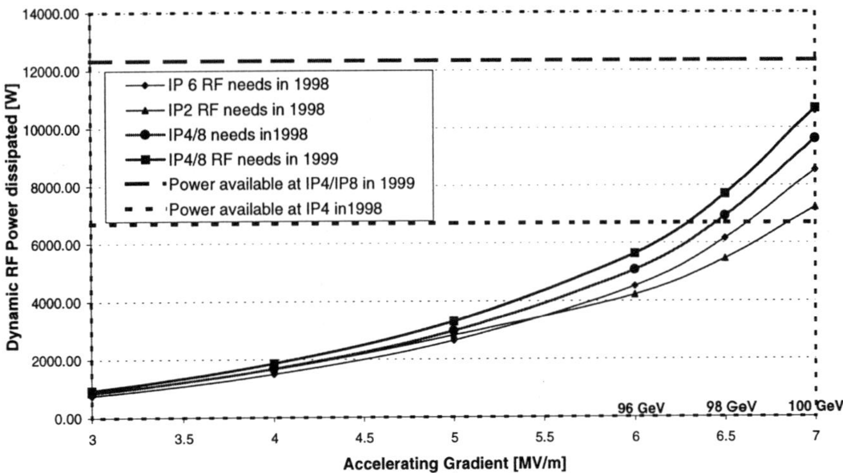

Figure 3. Dynamic power dissipated by RF into the cryogenic system

CRYOGENIC LIMITATION AND BEAM INDUCED THERMAL LOSSES

Maximum Energy

The 12 kW equivalent power at 4.5K in the cryoplant design is split into different parts : 1.6 kW is used for the cooling gas flow through the RF accessories in the modules, 0.4 kW for the shield cooling, and the remaining 10 kW of cooling power at 4.5K is available to compensate the static losses and the dynamic RF load. With all the static loads deducted from the real measured capacity of the cryoplant about 6.7kW are available for dynamic RF use [7].

The RF dynamic load is a quadratic function of the acceleration field divided by a quality factor Q. Figure 3 presents the predictions of the expected RF dynamic power based on measurements of the Quality factor performed during operation.

Influence of Circulating Beam

Since 1996 an influence of the beam on the cryogenic system has been noticed. This influence is correlated to the beam intensity and the bunch length. In order to compare different bunch configurations, an impedance Z has been defined. From the evolution of the power recovery in the phase separator as a function of the beam intensity in stable physics one can compute an average bunch impedance Z_b= 16 MΩ for a bunch length longer than 10mm.

An important sensitivity of losses on the bunch length exists. Bunches of 7 mm length induce twice the losses of 10 mm bunches. To limit this effect it was decided in 1998 to operate the LEP Machine with a bunch length longer than 10 mm.

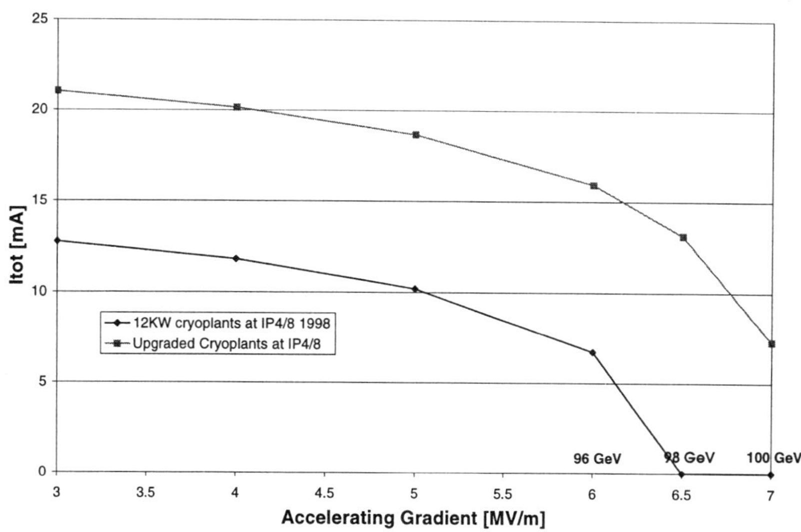

Figure 4. Beam intensity limits as function of the increasing LEP colliding energy

This beam effect reduces the operating margin and prevents physics runs with high bunch intensity affecting the luminosity of the LEP machine. Figure 4 presents the expected limitation in beam intensity as a function of the accelerating gradient [8].

Origin and Cure of the Beam Effect

The origin of the anomalous beam induced heat load on the LEP cryogenic system has been located in the cables of the RF antennas measuring the accelerating field in the cavities. In November 1998, eight cables of a module were successfully replaced by cables of larger cross sections. During the winter shutdown 1998/99 all the faulty cables were replaced. The observation completed during the first part of the 1999 run confirms this result.

HIGHER CURRENT, HIGHER GRADIENT, HIGH LUMINOSITY?

The cooling power available from the cryoplant was sufficient to cover the RF needs over the past years with an increased demand in power related to the rising objectives of the LEP2 energy up to 94.5 GeV. The limits observed in current intensity were coherent with constraints in other LEP systems and did not lower the performance of the LEP collider.

However, to reach the new colliding energy of 100 GeV, required for physics with enough luminosity, it was necessary to upgrade the cryoplant as presented in Figures 3 and 4.

The upgrade of the existing cryoplants provides four times 18kW @ 4.5 K and was installed during the last winter shutdown. Now the cryoplants are in operation and the generated cooling power is sufficient for the present LEP energy of 98 GeV and its future increase to 100 GeV.

The challenge is now to maintain the same level of availability of the cryoplant to keep the highest possible luminosity, required for the last two years of LEP operation.

REFERENCES

1. D. Güsewell, M. Barranco-Luque, S. Claudet, W.K. Erdt, P.K. Frandsen, Ph. Gayet, J. Schmid, N.O. Solheim, C. Titcomb and G. Winkler, Cryogenics for the LEP200 Superconducting Cavities at CERN, Proc. of *the Particle Accel. Conf.* 1993, 2956-2958
2. B. Chromec, W.K. Erdt, D. Güsewell, K. Löhlein, A. Meier, .A.E. Senn, T. Saugy, N.O. Solheim, U. Wagner, G. Winkler, B. Ziegler et. al, A High Efficient 12 kW Refrigerator for the LEP200 Project at CERN, Proc. of the IISSC 1993, Supercollider 5, Plenum Press (1994) 95-100
3. G. Gistau and J. Veaux, A 12/18 kW at 4.5 K Helium refrigerator for CERN's LEP superconducting accelerating cavities, *Cryogenics* (1994) 34 ICEC Supplement 103-106
4. S. Claudet, W.K. Erdt, P.K. Frandsen, Ph. Gayet, N.O. Solheim, C. Titcomb and G. Winkler, Four 12 kW/4.5 K Cryoplants at CERN, *Cryogenics* (1994) 34 ICEC Supplement 99-102
5. G. Winkler, Ph. Gayet, D. Güsewell and C. Titcomb, Cryogenics Operation and On Line Measurements of RF Losses in the SC Cavities of LEP2, *Particle Accel. Conf. 1995* Dallas
6. Ph. Gayet, S. Claudet, D. Kaiser, G. Winkler, *Cryogenic issues*, VII LEP Chamonix workshop, 1997
7. Ph. Gayet, S.Claudet, D. Kaiser, G. Riddone, LEP2 in 1998 and beyond, *Cryogenic issues,* VIII LEP Chamonix workshop, 1998
8. Ph. Gayet, S. Claudet, D. Kaiser, M. Sanmarti, *Cryogenic issues* for LEP2 in 1999, IX LEP Chamonix workshop, 1999

ECONOMICS OF LARGE HELIUM CRYOGENIC SYSTEMS: EXPERIENCE FROM RECENT PROJECTS AT CERN

S. Claudet, Ph. Gayet, Ph. Lebrun, L. Tavian and U. Wagner

LHC Division, CERN
1211 Geneva 23, Switzerland

ABSTRACT

Large projects based on applied superconductivity, such as particle accelerators, tokamaks or SMES, require powerful and complex helium cryogenic systems, the cost of which represents a significant, if not dominant fraction of the total capital and operational expenditure. It is therefore important to establish guidelines and scaling laws for costing such systems, based on synthetic estimators of their size and performance. Although such data has already been published for many years, the experience recently gathered at CERN with the LEP and LHC projects, which have *de facto* turned the laboratory into a major world cryogenic center, can be exploited to update this information and broaden the range of application of the scaling laws. We report on the economics of 4.5 K and 1.8 K refrigeration, cryogen distribution and storage systems, and indicate paths towards their cost-to-performance optimisation.

LARGE-SCALE CRYOGENICS AT CERN

On the spur of increasing demands from forefront accelerator and detector projects for high-energy physics, which all make intensive use of superconducting devices - electromagnets and acceleration cavities - CERN has become over the last decade a major center in helium cryogenics (Figure 1), with 19 refrigerators in operation all over the site, out of which 9 with capacities ranging from 0.8 to 12 kW @ 4.5 K[1].

LEP, the 26.7-km circumference electron-positron collider in operation since 1989, features four underground experimental areas equipped with high-luminosity insertions using superconducting quadrupoles, as well as large spectrometer solenoids, two of which also are superconducting. All these devices are cooled at 4.5 K by dedicated helium refrigerators and cryogenic systems[2], now totalising ten years of current operation. Over the nineties, the beam energy of LEP has gradually been increased by the addition of 288 superconducting acceleration cavities[3], installed in four strings, each 500-m long and cooled by a dedicated 12 kW @ 4.5 K cryogenic plant[4], through some 2 km of gas-shielded

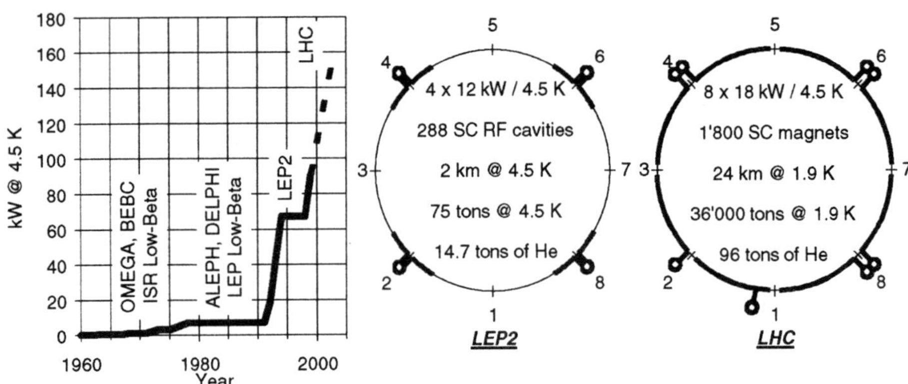

Figure 1. Installed capacity at CERN

Figure 2. LEP2 and LHC cryogenic systems

cryogenic distribution lines[5]. This constitutes the largest distributed cryogenic system in the world, and is being routinely operated as an integral part of the LEP2 collider[6,7].

The trend towards larger, more powerful systems will continue in the coming years with the procurement, installation and commissioning of the cryogenic system for the LHC (Figure 2), a high-energy, high-luminosity proton and ion collider using twin-aperture, high-field superconducting magnets operating in superfluid helium below 1.9 K[8]. Due to start operation in 2005, the LHC will reuse the LEP tunnel and infrastructure, among which the four large refrigerators, which have been upgraded in preparation for this purpose. It will also require four new 18 kW @ 4.5 K refrigerators, presently under construction in European industry[9], as well as eight 2.4 kW @ 1.8 K refrigeration units, partially fed from the 4.5 K cryogenic plants. The LHC cryogenic system will therefore add, to its sheer size and installed power, the qualitative novelty of large-scale cryogenics at 1.8 K, which has required the conduction of a dedicated development programme over the past years[10]. The LHC will also be characterised by its huge cold mass and large helium inventory, both of which strongly impact on the technical and economical choices of the cryogenic system.

The demanding projects briefly mentioned above have prompted us to devote significant effort to the technical and economical optimisation of their cryogenic systems, at the level of design, technical specification and industrial procurement of equipment, as well as methods for reliable operation and preventive maintenance. Conversely, they have provided us with detailed, up-to-date information on the economic aspects of large helium cryogenic systems, which we try and report in the following, confronting it when relevant with previously published material in the field.

ECONOMICS OF HELIUM REFRIGERATORS

Technical Scope

Since the now classical studies by Strobridge[11,12], the capital cost of helium refrigerators has been surveyed and tentatively correlated with a single performance estimator, originally the installed compressor power, and more recently the refrigeration capacity produced at 4.5 K[13,14]. This approach has the merit of extreme simplicity, but – as a consequence - requires careful application for large, complex refrigerators such as those of CERN, which for reason of site layout and underground implantation, often feature several cold boxes with interconnecting lines, and in most cases, provide cooling duties – isothermal and non-isothermal - at a variety of temperature levels[15].

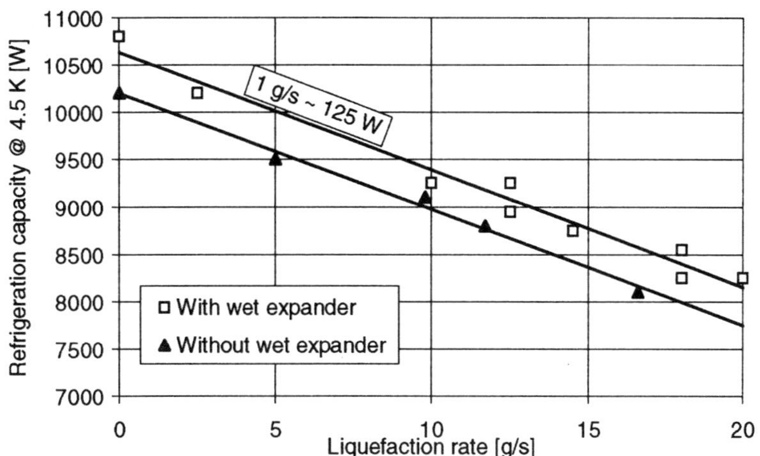

Figure 3. Measured liquefaction rate versus refrigeration capacity of 12 kW @ 4.5 K refrigerators

In the following, we shall therefore only consider multi-kW @ 4.5 K cryogenic plants of modern design, based on modified Claude cycles, using two stages of oil-injected screw compressors with intercooling, an oil-removal system composed of three stages of coalescing filters and an adsorber, coldboxes based on brazed aluminium plate-fin heat exchangers and turbo-expanders with helium-gas bearings. In line with widespread practice in Europe, the helium refrigeration cycles do not use liquid nitrogen precooling on a permanent basis, and, in cases when the cryogenic plants considered do feature a large-capacity liquid nitrogen precooler used for cooldown of the load, its specific cost has been substracted from the quoted numbers. A peculiarity of the large LEP refrigerators is their "split-coldbox" design. As a result, the "upper" and "lower" coldboxes are connected by compound cryogenic lines spanning their vertical separation, and installed in the technical shafts; the specific cost of these lines must also clearly be removed from significant cost estimates for "normal" cryogenic plants. Instrumentation and actuators, fault diagnosis and safety interlocks are included; however the process control hardware and software, which were purchased separately for the sake of uniformity among refrigerators of different origins, do not form part of the following estimates.

In order to approach overall exergetic efficiency, the LEP and LHC cryogenic systems make use of several levels of temperature for heat interception, each of which requiring its own cooling duty. For the sake of comparison, the multiple cooling duties produced by the cryogenic plants have been converted, by isoexergetic equivalence, into isothermal refrigeration capacity at 4.5 K. We are fully aware of the weakness of this *modus operandi*, particularly when trading liquefaction and refrigeration loads (Figure 3), but nevertheless find it an acceptable compromise, at least for cryogenic plants supplying the largest fraction of their exergetic refrigeration capacity at 4.5 K.

Capital Cost of 4.5 K Refrigerators

The capital cost of the different CERN refrigerators was actualized to 1998 CHF following[16]. Figure 4 also shows equation (1) given in[14] that was indexed from 1997 USD to 1998 USD[17] and converted into 1998 CHF.

$$\text{Cost[1998 MCHF]} = 2.6 * (\text{Capacity[kW@4.5K]})^{0.7} \quad (1)$$

Our best practical fit of the data is represented by equation (2)

$$\text{Cost[1998 MCHF]} = 2.2 * (\text{Capacity[kW@4.5K]})^{0.6} \quad (2)$$

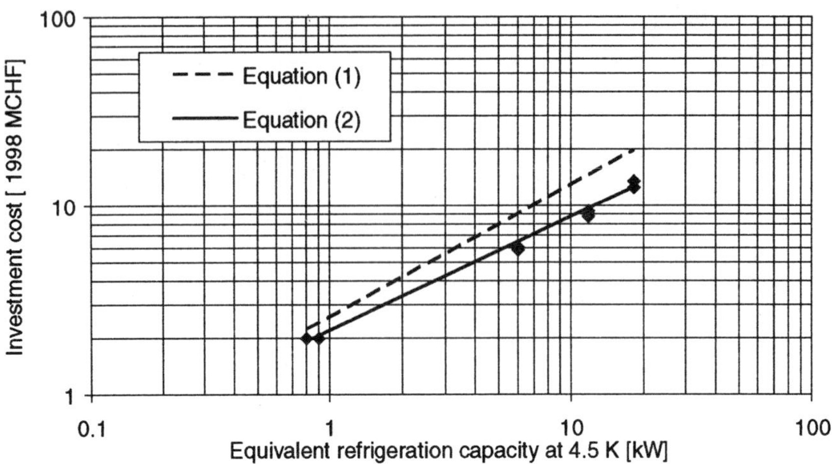

Figure 4. Investment cost of 4.5 K refrigerators at CERN versus equivalent capacity

Efficiency, Operation and Maintenance

As the CERN particle accelerators and experiments operate round-the-clock on a yearly basis – except for a 3- to 4-month long winter shut-down scheduled in the period of peaking electricity demand – it is legitimate to take operating expenses into consideration, in view of a global cost optimisation[9]. The key ingredients of such an optimisation are:

- the thermodynamic efficiency (C.O.P.) of the refrigerators,

- the annual number of hours of operation of the refrigerators, which has been taken at 6600 to account for the fact that cryogenics is usually required to start well in advance of physics (for equipment tests), and to terminate later (for managing and storing the helium inventory before shut-down),

- the number of years over which the integrated costs are calculated. For the purpose of equipment design, the technical lifetime of the LHC has been set at 20 years. However, in view of the fact that the operating costs over this period have not been actualised, the operation lifetime for the purpose of economic optimisation has been taken at 10 years,

- the mix of operational modes – each with different cooling duties and corresponding variable electrical power consumption – encountered in a typical year of operation,

- the marginal cost of electricity, based on the unit price – averaged over the annual period of operation to smear out strong seasonal price variations – and including externalities such as the marginal cost contributions of transport and conversion from 400 kV to 3.3 kV, as well as of rejection of the dissipated heat to the environment, globally amounting to 60 CHF/MWh[18],

- the cost of preventive maintenance, including spare parts and manpower. The cost of spare parts is mostly related to compressor maintenance and can be estimated for LEP2 at 2.5% per year of the capital cost of the compressor system. The manpower expenditure is equally distributed between the mechanical maintenance of the compressors (3% per year of the compressor capital cost) and the maintenance of instrumentation (80 CHF per channel per year),

- the cost of helium, which depends on the nature of the load. For a closed circuit such as LEP2 with a global helium inventory of 14700 kg, the annual losses amount to about 3400 kg (excluding accidental events),

- the cost of manpower for operation, which strongly depends on the organisation of the operating team and the level of automation[7].

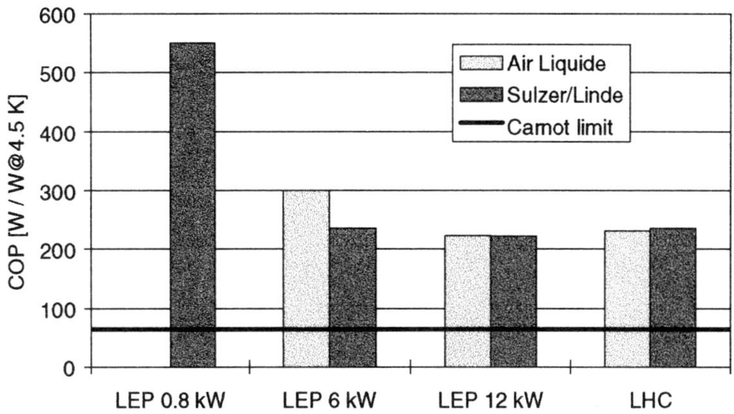

Figure 5. Coefficient of performance of 4.5 K refrigerators at CERN

As the operation costs are dominated by that of electricity, the integrated price estimates used for the purpose of comparing offers and adjudicating contracts included only capital and electricity costs[9]. An expected result of the large relative weight given to electricity costs, was to drive the refrigerator designs to higher efficiency, now amounting to about 220 W/W @ 4.5 K or 30% of the Carnot cycle (Figure 5). This was achieved not so much by improvement of efficiency of single components, such as heat exchangers or turbo-expanders, than by the arrangement of these components in carefully optimised cycles[19]. A less expected consequence of this technical-economical incentive policy has been to reduce the capital investment of the refrigerators. Contrary to the previously established belief that more efficient refrigerators were more expensive to build – although cheaper to operate - , this approach demonstrated that increasing the efficiency for the same, specified refrigeration output, enables to reduce the process flow-rate, and hence to down-size the heat exchangers, to gain on the size of the coldbox and compressors, and even in some cases to make the economy of one compressor unit, all of which result in overall savings in capital as well as in running costs.

1.8 K REFRIGERATION

Producing large capacity refrigeration at 1.8 K[20] requires the use of cold compressors[21] to compress helium up to a pressure at which warm compressors become feasible. Figure 6 shows a generic 1.8 K refrigeration scheme, which can be analysed in two parts. The first part concerns the 1.8 K refrigeration unit containing the cold compressors, the warm compressor station and a counter-flow heat exchanger. The second part is a standard 4.5 K refrigerator, producing non-isothermal refrigeration between 4.5 K and a temperature Tr, which directly depends on the cold compressor pressure ratio (CPR). Figure 7 shows the corresponding equivalent exergetic load at 4.5 K, as a function of the CPR. The investment and operation cost of this second part can be assessed as for other 4.5 K refrigerators. Figure 8 shows the relative cost of 1.8 K refrigeration as a function of the CPR. The investment and operation costs of the 1.8 K part decrease with increasing CPR. This reduction is however compensated by the investment and operation costs of the 4.5 K part. For CPR above 25, the total cost is about constant and other factors, such as compactness or operational flexibility determine the technical choice. Figure 9 shows the investment costs as a function of the 1.8 K refrigeration capacity. The total cost of 1.8 K refrigeration for the LHC is lower than given by this model, due to the use of larger, more cost effective 4.5 K cryogenic plants, which also provide other cooling duties.

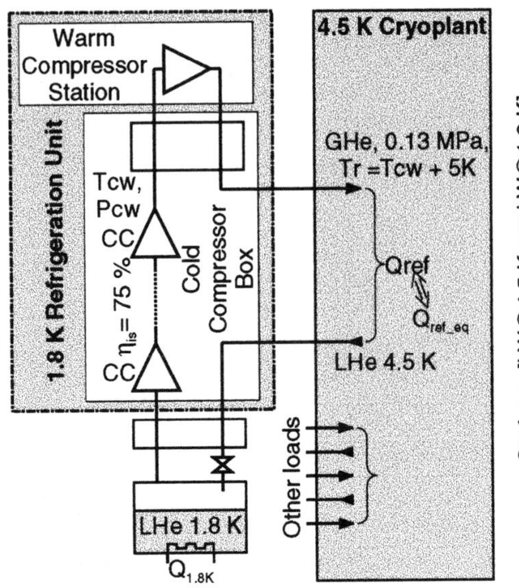

Figure 6. Generic 1.8 K refrigeration scheme

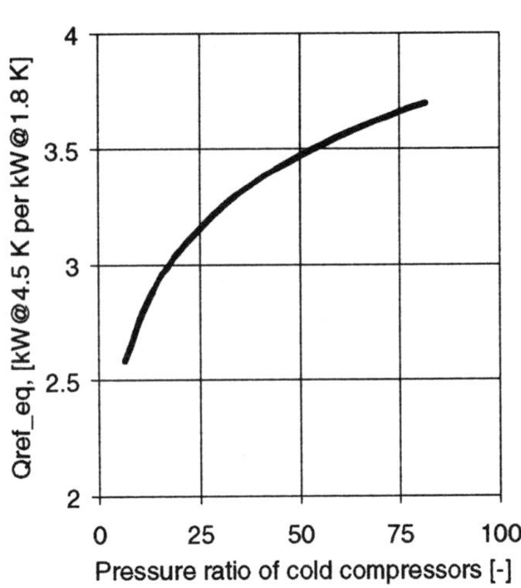

Figure 7. Equivalent 4.5 K load induced by 1.8 K refrigeration unit

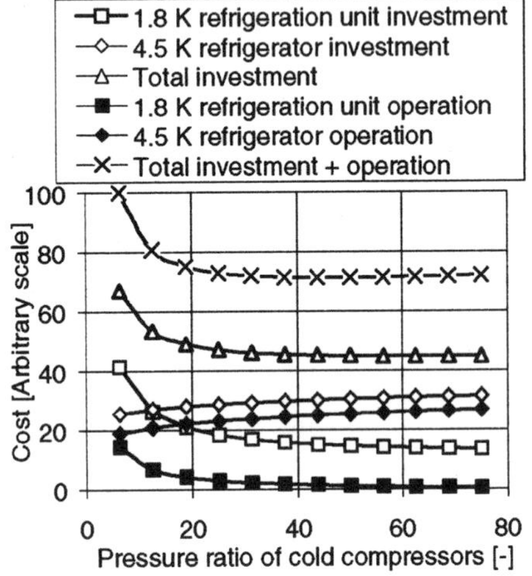

Figure 8. Cost structure of 1.8 K refrigeration

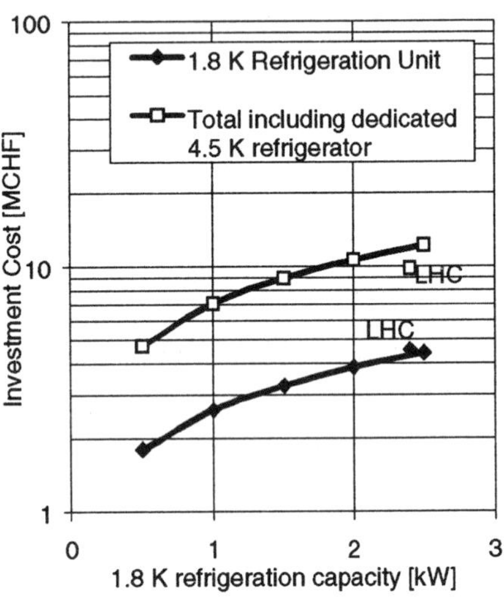

Figure 9. Investment cost of 1.8 K refrigeration

HELIUM STORAGE AND INVENTORY MANAGEMENT

For closed-circuit cryogenic systems which encounter relatively long shutdown periods without possibility of re-liquefaction, and fast helium discharge (magnet quenches), medium-pressure (MP) storage at 2 MPa and ambient temperature is commonly used. Figure 10 shows the cost of MP storage as a function of unit geometrical volume for carbon-steel storage vessels. The cost, driven by that of the material, was estimated for cylindrical and spherical geometries, including transportation and site erection. For large vessels, it matches the data in[22]. Due to the huge helium inventory of the LHC, the MP storage capacity is deliberately limited to 50% of the total, in order to minimise investment. Consequently, for cooldown and warmup of the machine, it is foreseen to shuffle helium from the LHC to the market, in the framework of a "virtual storage" type contract.

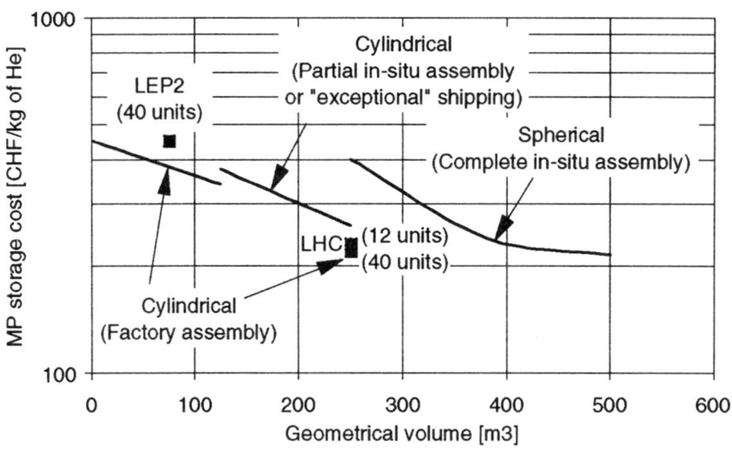

Figure 10. Cost of MP storage (2MPa, ambient temperature)

CRYOGENIC DISTRIBUTION LINES AND PIPEWORK

The cryogenic lines in the kilometre range used in CERN accelerators exhibit a variety of types, sizes, design choices and site layouts. While seeking a general estimator of their cost, we propose to use, as a basis, the cost of stainless steel piping (7.5 CHF/kg), multiplied by specific "cost multipliers" such as shown in Table 1. The relative cost of engineering is about constant for pipework at ambient temperature, whereas it strongly depends on overall tonnage and siting considerations for cryogenic distribution lines. The cost of singularities, such as valves or bayonet connections, should be added to the estimate.

Table 1. Cost structure of cryogenic lines and pipework: material cost multipliers

	Ambient pipework	Cryogenic lines
Material cost	1.0	1.5
Engineering	} 1.2	6 – 8
Installation		2.5 – 3.0
Tests		0.5 – 0.8
Total	2.2	10.5 – 13.3

CONCLUSION

With the advent of industrial-size helium cryogenic systems, ancillary to large research or industrial projects using applied superconductivity, a significant fraction of the total expenditure resides in the cost of cryogenics alone. In view of CERN's long experience in designing, procuring and operating several such large-capacity cryogenic systems, we are now able to assess their main economic features, and estimate their cost with engineering precision. As a result, the main procurement contracts for the cryogenics of the LHC, which have been adjudicated to specialised industry on a competitive basis over the last two years, generally fall within budget estimates, an important asset for the construction and funding on schedule of the project.

ACKNOWLEDGMENTS

The results presented here were obtained over more than a decade of hard work by our colleagues from the Accelerator Cryogenics and Experiment & Test Areas Cryogenics Groups at CERN. The competence, competitiveness and interest of specialised industry in CERN's projects constitutes an essential ingredient of success, which we also wish to acknowledge.

REFERENCES

1. J.P. Dauvergne, D. Delikaris, W. Erdt, D. Güsewell, F. Haug, Ph. Lebrun, G. Passardi, J.M. Rieubland, J. Schmid & G. Winkler, Application of liquid-helium cryoplants at CERN, *Adv. Cryo. Eng.* 39A:539 (1994)
2. J.P. Dauvergne, M. Firth, A. Juillerat, Ph. Lebrun & J.M. Rieubland, Helium cryogenics at the LEP experimental areas, *Adv. Cryo. Eng.* 35B:901 (1990)
3. M. Barranco-Luque, J.P. Dauvergne, W. Erdt, P. Frandsen, D. Güsewell, F. Haug, A. Juillerat, G. Passardi, J. Schmid & G. Winkler, The refrigeration system for the LEP energy upgrade, *Cryogenics* 30 ICEC Supplement:136 (1990)
4. S. Claudet, W. Erdt, P. Frandsen, Ph. Gayet, N. Solheim, Ch. Titcomb & G. Winkler, Four 12 kW / 4.5 K cryoplants at CERN, *Cryogenics* 34 ICEC Supplement:99 (1994)
5. D. Güsewell, M. Barranco-Luque, S. Claudet, W. Erdt, P. Frandsen, Ph. Gayet, J. Schmid, N. Solheim, Ch. Titcomb & G. Winkler, Cryogenics for the LEP200 superconducting cavities at CERN, in "Proc. PAC'93 Washington" 4:2956 (1993)
6. M. Barranco-Luque, S. Claudet, Ph. Gayet, N. Solheim & G. Winkler, Operation of the cryogenic system for the superconducting cavities of LEP, in: "Proc. ICEC16 Kitakyushu", Vol. 1, T. Haruyama, T. Mitsui & K. Yamafuji, eds., Elsevier Science, Oxford, New York & Tokyo (1997), p. 103
7. N. Bangert & Ph. Gayet, Operation of the four 12 kW at 4.5 K refrigerators for LEP, paper presented at this conference
8. Ph. Lebrun, Superfluid helium cryogenics for the Large Hadron Collider project at CERN, *Cryogenics* 34 ICEC Supplement:1 (1994)
9. S. Claudet, Ph. Gayet & U. Wagner, Specification of four new large 4.5 K refrigerators for the LHC, paper presented at this conference
10. Ph. Lebrun, L. Tavian & G. Claudet, Development of large-capacity refrigeration at 1.8 K for the Large Hadron Collider at CERN, in: "Proc. Kryogenika'96 Praha", IIR-IIF, Prague (1996), p.54
11. T.R. Strobridge, *IEEE Trans. Nucl. Sc. NS-16* 3:1:1104 (1969)
12. T.R. Strobridge, "Cryogenic Refrigerators: an Updated Survey", NBS Technical Note 655 (1974)
13. M.A. Green, R.A. Byrns & S.J. St. Lorant, Estimating the cost of superconducting magnets and the refrigerators needed to keep them cold, *Adv. Cryo. Eng.* 37A:637 (1992)
14. R.A. Byrns & M.A. Green, An update on estimating the cost of cryogenic refrigeration, *Adv. Cryo. Eng.* 43B:1661 (1998)
15. Ph. Lebrun, G. Riddone, L. Tavian & U. Wagner, Demands in refrigeration capacity for the Large Hadron Collider, in: "Proc. ICEC16 Kitakyushu", Vol. 1, , T. Haruyama, T. Mitsui & K. Yamafuji, eds., Elsevier Science, Oxford, New York & Tokyo (1997), p. 95
16. G. Lindecker, CERN materials index (excluding electricity, fluids and telecom), private communication
17. Gross Domestic Product Deflator Inflation Calculator, NASA Johnson Space Center, **http://www.jsc.nasa.gov/bu2/inflateGDP.html**
18. Ph. Lebrun, "Estimation du Coût Marginal de l'Energie Consommée par la Cryogénie du LHC", LHC Project Note 92 (1997)
19. M. Barranco-Luque, S. Claudet, W. Erdt, P. Frandsen, Ph. Gayet, D. Güsewell, Ph. Lebrun, J. Schmid, N. Solheim, Ch. Titcomb, U. Wagner & G. Winkler, Conclusions from procuring, installing and commissioning six large-scale helium refrigerators at CERN, *Adv. Cryo. Eng.* 41A:761 (1996)
20. F. Millet, P. Roussel, L. Tavian & U. Wagner, A possible 1.8 K refrigeration cycle for the Large Hadron Collider, *Adv. Cryo. Eng.* 43A:387 (1998)
21. A. Bézaguet, Ph. Lebrun & L. Tavian, Performance assessment of industrial prototype cryogenic helium compressors for the Large Hadron Collider, in: "Proc. ICEC17 Bournemouth", D. Dew-Hugues, R.G. Scurlock & J.H.P. Watson, eds., IoP, Bristol & Philadelphia (1998), p. 145
22. G.Y. Robinson Jr., Economics of cryogenic systems for superconducting magnets, *Adv. Cryo. Eng.* 25: 342 (1980)

PRELIMINARY RISK ANALYSIS OF THE LHC CRYOGENIC SYSTEM

M. Chorowski,[1] Ph. Lebrun,[2] G. Riddone[2]

[1]Wroclaw University of Technology
50-370 Wroclaw, Poland

[2]CERN, LHC Division
1211 Geneva, Switzerland

ABSTRACT

The Large Hadron Collider (LHC), presently under construction at CERN, will require a helium cryogenic system unprecedented in size and capacity, with more than 1600 superconducting magnets operating in superfluid helium and a total inventory of almost 100 tonnes of helium. The objective of the Preliminary Risk Analysis (PRA) is to identify all risks to personnel, equipment or environment resulting from failures that may accidentally occur within the cryogenic system of LHC in any phase of the machine operation, and that could not be eliminated by design. Assigning a gravity coefficient and one analyzing physical processes that will follow any of the recognised failure modes allows to single out worst case scenarios. Recommendations concerning lines of preventive and corrective defence, as well as for further detailed studies, are formulated.

INTRODUCTION

The safety philosophy of the LHC cryogenic system[1] rests on specific design features that differ from the options chosen for the cryogenic systems of other large superconducting accelerators,[2] and contribute to make the LHC cryogenic system inherently safe, with a limited number of possible risks and failures. These features are:

- the presence of a high-capacity, cold recovery header D (see Figures 1 and 2) which has a nominal working pressure of 1.3 bar (design pressure of 20 bar), a nominal working temperature of 20 K and a volume of about 60 m^3 per 3.3 km sector of the collider. Helium expelled from the magnet cold mass after a loss of insulation vacuum or magnet resistive transitions will discharge into header D. The header will also accommodate helium relieved from other cryogenic distribution line (QRL) headers after degradation of insulation vacuum. The only header protected with safety valves venting directly to underground areas is header B because of its low design pressure.

- the absence of nitrogen (liquid or gaseous) in the tunnel, in any operating mode; liquid nitrogen is only used for helium pre-cooling at ground surface during cool-down.
- the absence of forced helium flow by mechanical devices (pumps) in cooling loops, resulting in limited discharge of fluid in case of header rupture.

The cryogenic flow-scheme considered in the PRA is that described in reference [3]. The LHC operation modes have been defined on the basis of a tentative yearly operation schedule [4]: machine warm, drifting, 75 K stand-by, cool-down, 1.9 K stand-by, system test or ramp down, normal operation, warm up. Additionally, operation modes that will not happen on a regular basis have also been considered: short intervention on a cold sector and limited quench of a maximum of 8 cells of the LHC machine.

ELEMENTS OF PRELIMINARY RISK ANALYSIS

The cryogenic system of LHC is treated as being composed of separate helium enclosures, called nodes in the following. Each node is characterised by the amount and thermodynamic parameters of the helium enclosed. For the enclosures that are vacuum insulated, the volume of the vacuum space is also relevant. This conceptual scheme based on nodes is shown in Figure 3.

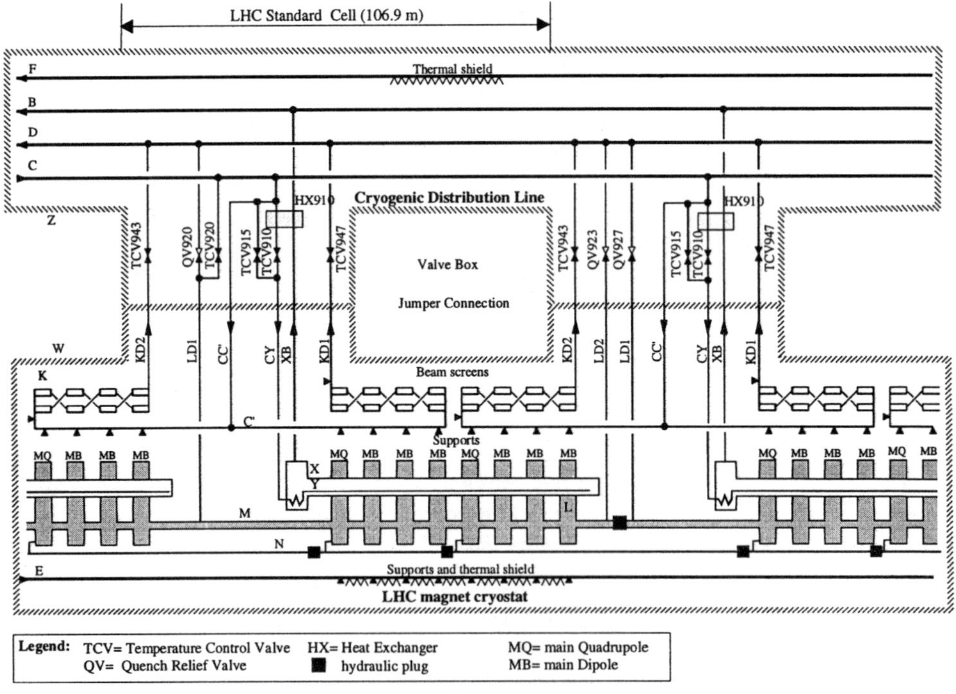

Figure 1. Flow-scheme of LHC elementary cooling loop

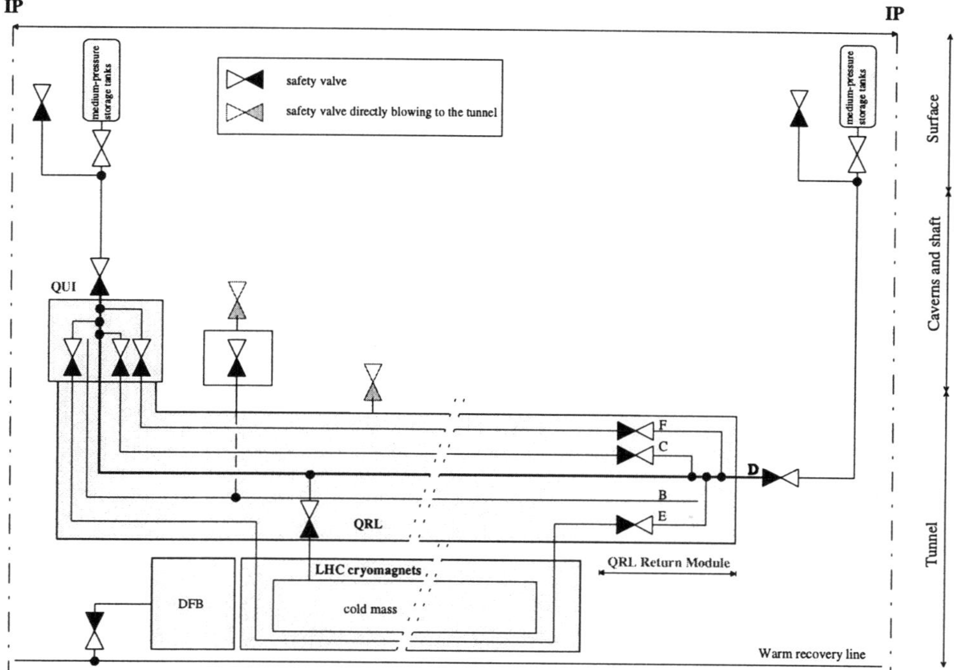

Figure 2. Simplified flow-scheme of a LHC sector (3.3 km) showing safety valve locations.

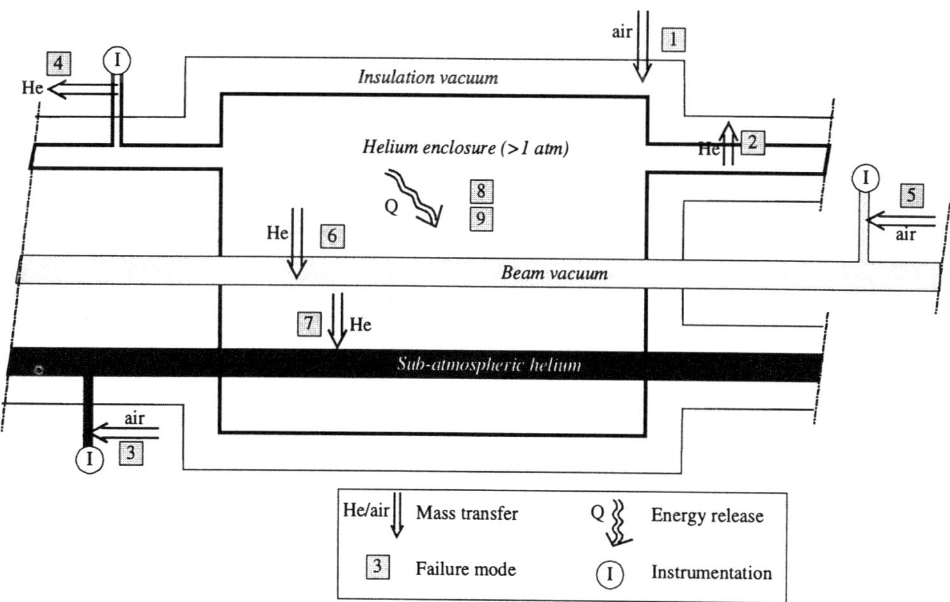

Figure 3. A generic view of the nodes in the LHC cryogenic system.

A cryogenic-related failure mode is defined as an accidental event (see Table 1) that may involve mass transfer of helium or air between helium enclosure, insulation or beam vacuum space and environment, a result of any constructional element break or malfunction (eg. broken bellows or leaking valve). A cryogenic failure mode may also be an unexpected energy release to helium in the magnet cold mass as a result of an extended resistive transition or an electrical arc. It is then possible to split a complex cryogenic system into a moderate number of nodes and perform a PRA even in the early stage of the system design. From their consequences, the LHC cryogenic related failures have been classified into three groups of increasing gravity, numbered from 1 to 3:

- Gravity 1: failure does not involve helium relief from the machine,
- Gravity 2: helium is blown out of the machine directly to the environment at ground surface,
- Gravity 3: helium is blown out of the machine to the confined area (e.g. tunnel).

For each cryogenic related failure mode, an attempt to assign a probability of mishap has been made and with following descriptive scale of probability used [5]:

- A (Frequent): failure likely to occur repeatedly during the life-cycle of the system.
- B (Occasional): failure likely to occur several times in the life-cycle of the system.
- C (Occasional): failure likely to occur sometimes in the life-cycle of the system
- D (Remote): failure not likely to occur in the life-cycle, but nevertheless possible.
- E (Improbable): probability of failure occurrence cannot be distinguished from zero.

The Preliminary Risk Analysis of the LHC cryogenic system has been performed in three steps:

- The first step (Identification) has consisted of listing all the machine nodes, cryogenics related failures and operation modes. Data have been collected through dedicated hearings, discussions and experience gathered at CERN.
- The objective of the second step (Combination) has been to list all potential failures for every node in any mode of the collider operation.
- The third step (Analysis) has been conducted in two parts. First, for each potential failure a YES/NO decision has been made, with the aim of retaining credible failures only. Then a descriptive analysis of the causes and consequences of each credible failure has been made and to each event a gravity level, as defined above, has been assigned. For the failures of gravity 2 and 3, the associated risks have been described and recommendations formulated. Finally worst case scenarios for the nodes located in the tunnel, caverns, shafts and surface have been identified and a descriptive probability for each of the mishaps estimated.

The technique used for performing the PRA of the LHC cryogenic system is based on FEMECA (Failure Mode and Effect Criticality Analysis) [6]. It fulfils the general requirements of the Preliminary Risk Analysis of a complex technical system, and in particular of a superconducting cryogenic system [7, 8].

Table 1. Cryogenic failure modes

No.	Cryogenic failure	No.	Cryogenic failure
1	Air flow to insulation vacuum	6	Helium flow to beam vacuum
2	Helium flow to insulation vacuum	7	Pressurised helium flow to sub-atmospheric helium
3	Air flow to sub-atmospheric helium	8	Energy release to cold mass helium due to a sector quench
4	Helium flow to environment	9	Energy release to cold mass helium due to electrical arc
5	Air flow to beam vacuum	10	Oil flow to environment (in case of helium compressors only)

RESULTS OF THE PRELIMINARY RISK ANALYSIS

The PRA of the cryogenic system for the whole LHC machine was performed [9], however this paper focuses on the detailed analysis concerning the nodes located in the tunnel; for the whole machine, we give only summary results.

A typical cross-section of the LHC tunnel is shown in Figure 4. The accelerator will be composed of the LHC cryomagnets paralleled by the cryogenic distribution line. Both will be linked by jumper connections every 107 m. The nodes of LHC cryogenic system located in the tunnel, together with the amount of helium enclosed and vacuum insulation volume are listed in Table 2 and schematically represented in Figure 5.

All potential failures at the cryogenic nodes located in the tunnel have been identified, which may happen in any phase of the collider operation. In this way, over five hundred (namely 540) of the failures have been defined. Subsequently, for each potential failure a YES/NO decision has been made to eliminate events that are not credible and 183 credible failures have been retained. Finally, some of the credible failures have been qualified as worst case scenarios on the criterion of mass of helium discharged and access to the tunnel. Twelve events of gravity level 3 have been identified. Table 3 lists these events, gives the amounts of helium that might be relieved to the tunnel together with approximate peak mass flow rates.

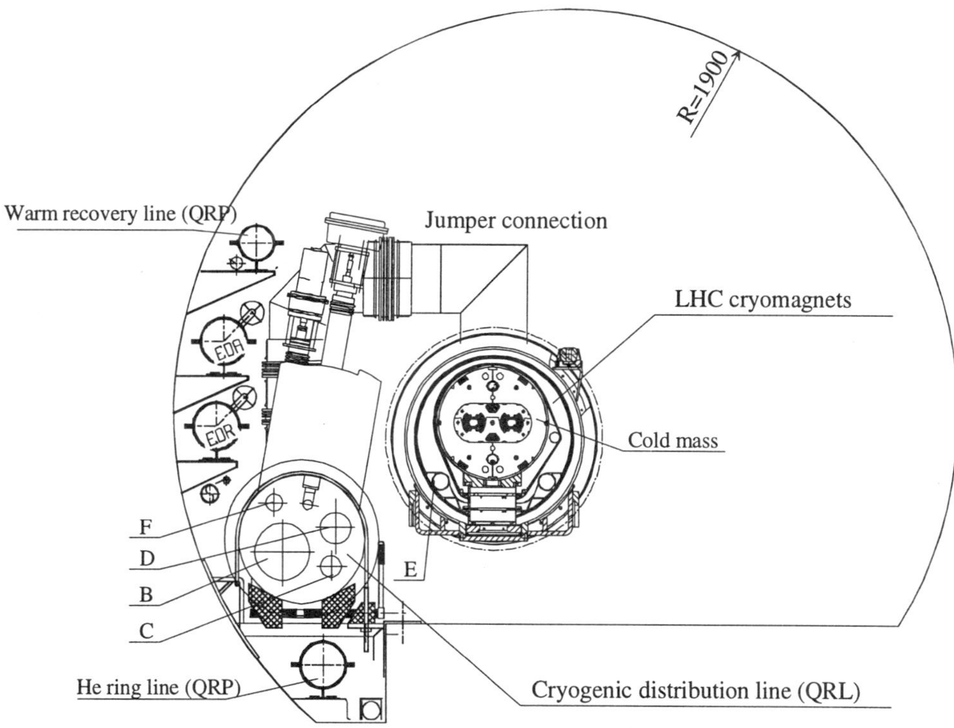

Figure 4. LHC machine components located in the tunnel.

Insulation Vacuum sectorization:

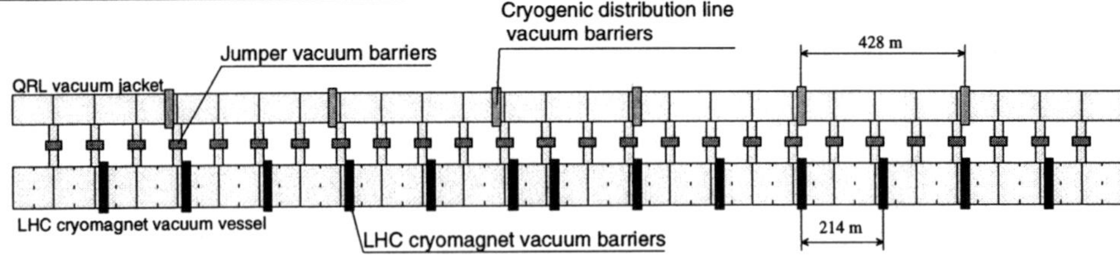

Cold-mass sectorization:

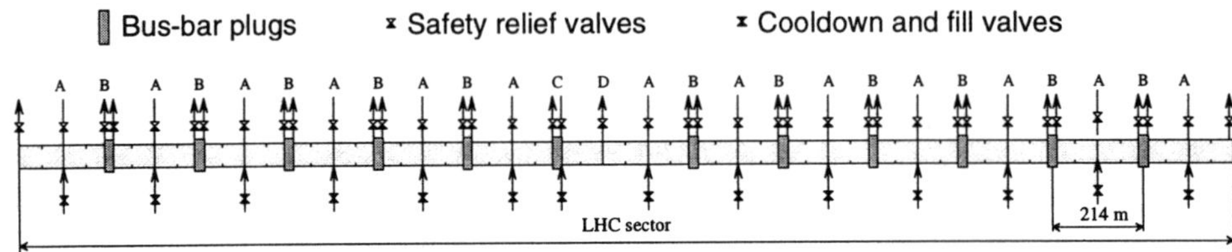

Figure 5. Vacuum and cold mass sub-sectorisation in LHC sector.

Table 2. Maximum He masses and volumes in the tunnel (see also Figure 5)

	P	T	Mass-flow	Mass per sub-sector	Residence time (1)	Volume per sub-sector	He sub-sectorisation	Vacuum sub-sectorisation
	[bar]	[K]	[kg/s]	[kg]	[s]	[m3]	[m]	[m]
QRL:								
Header B	0.016	4	0.125	37	296	195	3300	428
Header C	3	4.6	0.141	3300	23400	26	3300	428
Header D(2)	3	10	0.114	917(2)	8043	60	3300	428
Header F	19	75	0.234	195	833	17	3300	428
Cryomagnets:								
Cold mass	1.3	1.9	0	475	∞	3.21	214	214
Header E	19.5	75	0.234	195	833	17	3300	214
DFB(3)	1.3	4.4	0	79	∞	0.65(3)	15	300
RF cavities	1.3	4.4	0	34	∞	0.28(4)	13	13
He ring line	20	300	0.15	321	2140	104	26400	=
Warm recovery line	1.05	300	0.05	9	180	58	3300	=

(1) residence time is calculated as element length divided by the fluid velocity.
(2) helium parameters in header D – as after a limited quench (maximum 8 cells).
(3) DFB: Electrical feedbox. The longest DFB is situated at Point 2 left side.
(4) volume per module.

Table 3. Gravity 3 failures at the nodes located in the tunnel.

No.	Node and failure description , remarks	Probability	Maximum amount of helium relieved to the tunnel [kg]/ approximate peak flow rate [kg/s]
1	LHC cryomagnets: He flow to cryostat insulation vacuum	D	475 / < 2
2	LHC cryomagnets: He flow to air	D	475 / He leakage only
3	LHC cryomagnets He flow to beam vacuum	D	475 / < 2
4	LHC cryomagnets energy release – electrical arc Not directly related to cryogenics	D	475 kg / < 2
5	He flow to QRL insulation vacuum, assumption of header C break Worst case scenario of the probability higher than E	D	3300 kg / <2
6	He flow to air, assumption of jumper connection break Worst case scenario	E	4250 / < 20
7	He flow to air, He ring line break	D	321 kg / < 7
8	He flow to air, warm He recovery line break	D	9 kg / He leakage only
9	He flow to DFB insulation vacuum	D	79 kg / < 2
10	DFB He flow to air	D	79 kg / He leakage only
11	He flow to RF cavities insulation vacuum	D	34 kg / < 2
12	RF cavities cooling He flow to air	D	34 kg / He leakage only

As it follows from Table 3, two worst case scenarios can be identified based on the criterion of mass of helium involved and peak flow rate to the tunnel: helium flow to QRL insulation vacuum (assumption of a full break of header C) and a break of a jumper connection.

More generally, table 4 summarises the PRA of the LHC cryogenic system giving worst case scenarios following the failures of the modes located in the tunnel, caverns, shaft and surface buildings.

Table 4. Worst case scenario failures for the LHC cryogenic system nodes located in the tunnel, caverns, access shafts and surface buildings.

Location	Failure	Probability	Max. amount of helium relieved [kg] / peak flow rate [kg/s]
Tunnel	He flow to QRL insulation vacuum (break of header C)	D	3300 / < 2
Tunnel	He flow to air (jumper connection break)	E	4250 / < 20
Cavern & shaft	He flow to QURA[1] insulation vacuum, LP circuit break	D	176 / < 2
Surface building	He flow to QSRB[2] insulation vacuum	D	190 / < 2
Surface building	Break of LN2 storage vessel. Nitrogen flow to environment	D	40000 / may be of the order of hundreds

(1) QURA: Underground lower cold box. The given values correspond to the low pressure circuit.
(2) QSRB: Surface integrated cold box. The given values correspond to the subcooler.

CONCLUSIONS AND RECOMMENDATIONS

- Due to its specific design features (cold recovery header, absence of nitrogen in the tunnel, no pump driven forced flow in cooling loops) the LHC cryogenic system is inherently safe with a limited number of possible risks and failures, especially those resulting in helium relief to a confined space.
- Out of almost 1000 analysed failure modes there are only 29 events that may be followed by helium discharge to a confined space.
- Worst-case failure modes have been identified and potential amount of helium that might be vented into the machine tunnel, caverns, shafts and surface building as well as peak discharge flow rate have been estimated. More detailed studies are being conducted to assess the development over time of such events and the resulting propagation and diffusion of helium in confined spaces (tunnel, caverns, shafts buildings).

ACKNOWLEDGEMENTS

The authors would like to thank S. Claudet. L. Tavian, U. Wagner and M. Barranco, W. Erdt, G. Rau, V. Sergo, R. Trant and R. van Weelderen for their help during preparation of this study.

REFERENCES

1. Ph. Lebrun, Superfluid helium cryogenics for the Large Hadron Collider project at CERN, *Cryogenics* 34, ICEC Suppl. (1994).
2. G. Horlitz, Refrigeration of large scale superconducting systems for high energy accelerators, *Cryogenics* 32, ICEC Suppl., 44:49 (1992).
3. M. Chorowski, W. Erdt, Ph. Lebrun, G. Riddone, L. Serio, L. Tavian, U. Wagner and R. van Weldeeren, A simplified cryogenic distribution scheme for the Large Hadron collider, in: "Adv. Cryo, Eng." 43, Plenum Press, New York, (1998), pp. 395-402.
4. L. Bottura, P. Burla, R. Wolf, LHC main dipoles proposed baseline current ramping, LHC Project Report-172, CERN, Geneva (1998).
5. B. L. Hendrix, Application of system safety engineering techniques for hazard prevention at the superconducting super collider, in: "Supercollider 3", Plenum Press, New York (1991), pp 1005-1015.
6. H. Garin, "AMDEC, MADE, AEEL, L'essentiel de la méthode", AFNOR (1994).
7. S. W. Malasky, "System Safety", Hayden Book Company, Inc., ISBN 0-87671-559-5 (1974).
8. B. G. Blyukher, L.E. Reznikov, Design phase safety and risk assessment for superconducting cryogenic system, *ASME International*, 74:80 (1997).
9. M. Chorowski, Ph. Lebrun, G. Riddone, Preliminary risk analysis of the LHC cryogenic system, LHC Project Note 177, CERN, Geneva (1999).

APT CRYOGENIC SYSTEM

G.J. Laughon,[1] C.H. Rode, R. Ganni, W.C. Chronis, D.M. Arenius, B.S. Bevins[2]

[1]General Atomics
San Diego, CA

[2]Thomas Jefferson National Accelerator Facility*
Newport News, VA

ABSTRACT

In the Accelerator Production of Tritium (APT) project, a one-kilometer-long linear accelerator (linac) is used as part of a plant that will provide tritium for national defense purposes. The accelerator consists of a low-energy (LE) normally conductive, radiofrequency (rf) linac and a high-energy (HE) superconducting rf linac. The APT cryogenic system will supply cryogenic helium fluids to maintain the HE linac superconducting rf cavities at the required operating temperatures and pressures. Presently under continuing development, the cryogenic system was originally intended to provide 27 kW of refrigeration at 2.15 K and 0.046 mbar to cool the superconducting niobium rf cavities, and 82 kW of 4.5 K refrigeration for thermal shielding. Program redirection has led to a system design that supports 15 kW at 2.15 K and 66 kW at 4 K to 50 K, with the potential for upgrading to higher linac energies and refrigeration capacity. The cryogenic system will consist of multiple interconnected 2.15 K refrigerators (split into 4 K and 2 K coldboxes) with required compressors, gas and liquid storage, and a vacuum-jacketed cryogen distribution system. Each 2 K cold box will employ 4 stages of cold compressors in series. This will be the first industrial application of large, multiple cold compressor systems operated in parallel. System requirements, current design direction, and potential design options are presented.

INTRODUCTION

The original APT cryogenic system was modeled after the cryogenic systems at HERA (DESY research facility in Hamburg, Germany) and CEBAF (Continuous Electron Beam Accelerator Facility at the Thomas Jefferson National Accelerator Facility [TJNAF]). The

*Supported by U.S. DOE contract DE-AC05-84ER40150.

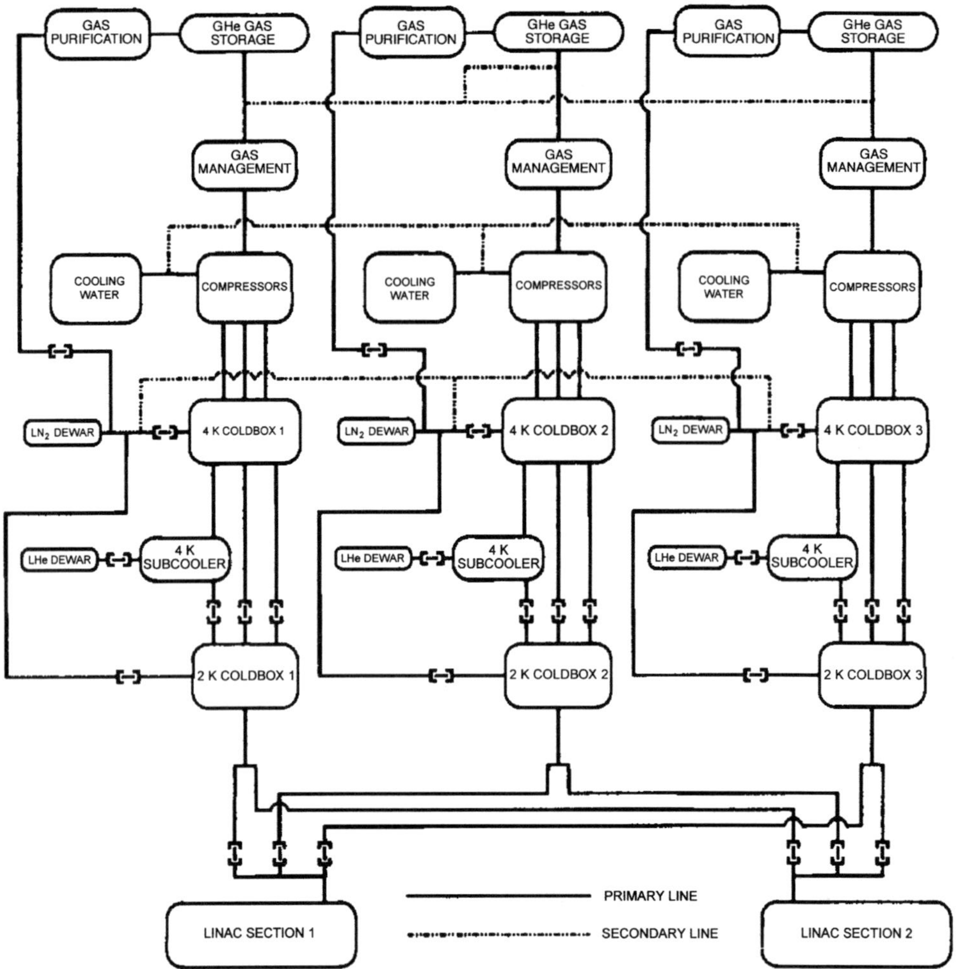

Figure 1. Cryogenic system.

system design was to support the cryogenic requirements of a 1700-MeV linac. Three large cryogenic plants (each with its own 4 K and 2 K coldboxes, two-stage warm helium compressor systems with gas purification and oil removal units, gas and liquid storage/management systems, and controls), would operate in parallel, semi-independently, with interconnection at the 2 K level. The cryoplant interconnection will provide sufficient redundancy to satisfy the system availability requirement.[1] A functional schematic of the cryogenic system is illustrated in Figure 1.

In late 1998, the Department of Energy redefined APT as a backup program for tritium production, resulting in a change in program direction and scope. The design emphasis was altered to optimize the cryogenic system for a 1030-MeV linac with the capability of being upgraded to 1700 MeV.

As these programmatic changes were occurring, the design of the superconducting cryomodules was significantly revised. An exhaustive study of the cryomodule heat loads was conducted, revealing appreciable differences in the system heat loads and fluid stream temperatures and pressures when compared to the original design. The number of required cryomodules and specific internal components (primarily the rf power coupler cooling circuits) were also changed. Results of these changes on the performance requirements are presented in Table 1.

Table 1. Cryogenic System Heat Loads and Pressures

	2 K Circuit Load (kW)	2 K Circuit Pressure (bar)	Shield Circuit Load (kW)	Shield Circuit Pressure (bar)
Original Design	27	3 (supply) .046 (return)	82 (at 4.5 K)	3 (supply) 2.8 (return)
1,030-MeV Optimized	15	3 (supply) .046 (return)	66 (at 4 K to 50 K)	18 (supply) 17.5 (return)
1,700-MeV Upgrade	10 (add'l) 25 (total)	3 (supply) .046 (return)	30 (at 4 K to 50 K)	18 (supply) 17.5 (return)

SYSTEM PERFORMANCE AND FUCTIONAL REQUIREMENTS

The performance and functional requirements that the cryogenic system must satisfy are summarized as follows:

- Provide sufficient cryogenic refrigeration to support the heat loads for the high-energy linac superconducting cryomodules (heat load requirements for the original and new designs are described in Table 1[2]);
- Efficient and flexible operation over a range of operating scenarios;
- Reliable and redundant design to achieve an overall system availability of >98%[3]; and
- Each cryoplant must be of identical process and physical design to minimize spares and maintenance costs and to maximize operational flexibility.

ORIGINAL DESIGN

Configuration. The cryogenic refrigerators, compressors, gas storage, and controls would be located in a single facility at the mass flow symmetry point of the linac. Cryogenic fluids would be transferred to and from the linac via the distribution system. Sufficient redundancy among all cryoplant components and system would be centralized at this facility.

Refrigerators. The 4 K coldbox design is based on the Central Helium Liquefier at TJNAF. This is a large cryogenic refrigerator utilizing three pressure streams, gas-bearing turbine expanders, and liquid nitrogen precooling. Process design changes that have been previously implemented or recognized as prudent to enhance performance will be included in the APT design. The coldbox provides 4.5 K helium gas at 3 bar to the 2 K coldbox and produces 4 K liquid for inventory storage. Each 4 K coldbox will be sufficiently sized so that any one cryoplant can maintain the high-energy linac at 4.5 K with the rf power off.

The 2 K coldbox will be patterned after the two existing units at TJNAF and will use four or five stages of cold compressors capable of processing up to 500 grams/sec of helium vapor. Recent developments at TJNAF regarding cold compressor operation and control will also be implemented in the APT units. An additional component, called the restart dewar, is integral to the 2 K coldbox and will provide an artificial load for the cold compressors after a disruption in the 2 K load. This is intended to significantly reduce the pumpdown and restart time for the cryogenic system and linac.

Distribution System. The three cryoplants are connected to the linac via a vacuum-jacketed distribution system, which comprises the interconnection system located at the cryo-

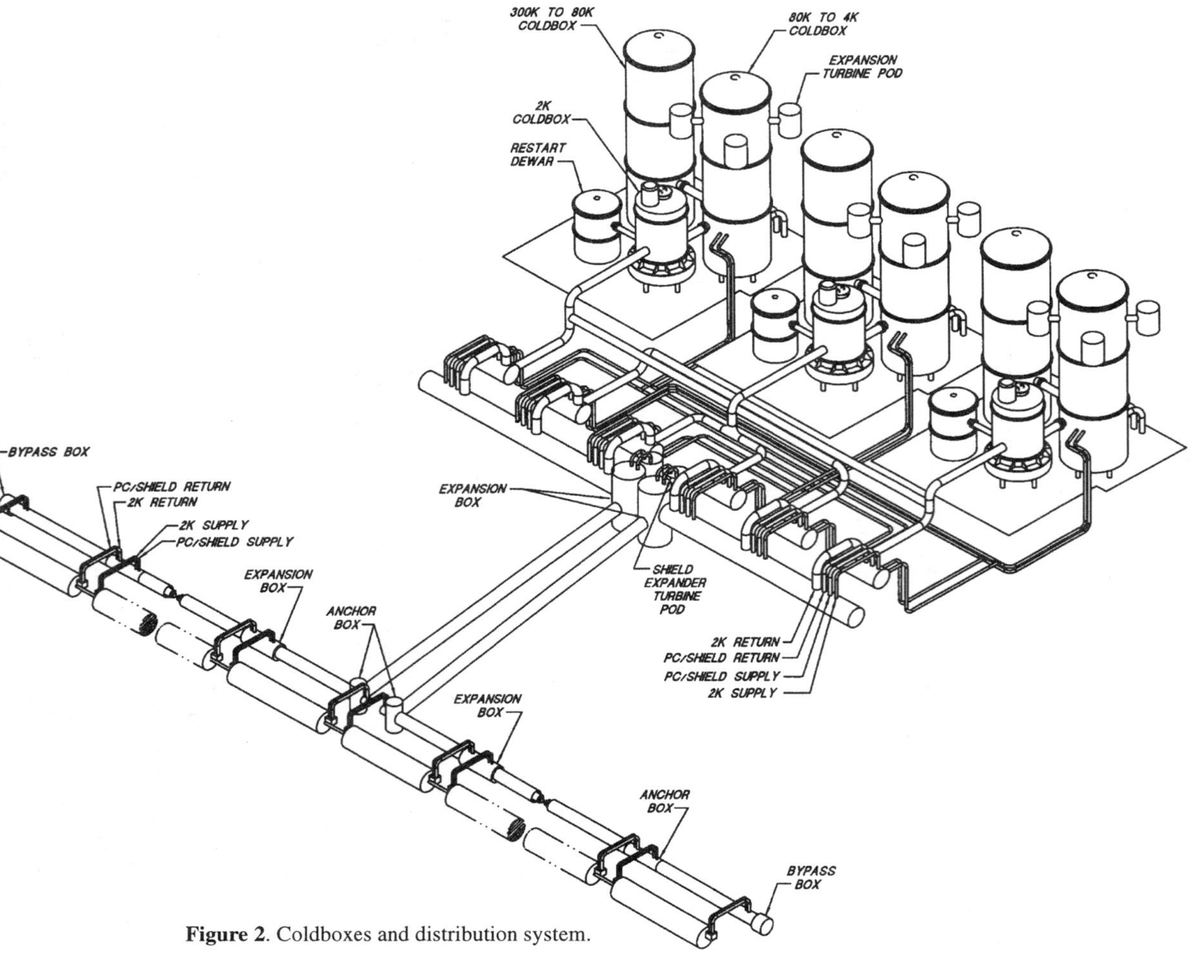

Figure 2. Coldboxes and distribution system.

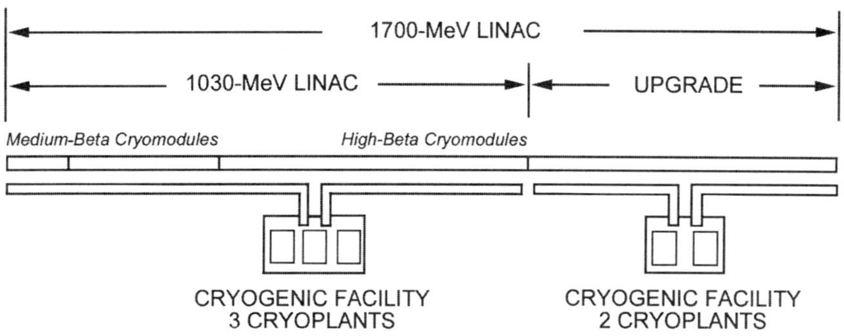

Figure 3. System configuration.

genic facility, two branch transfer lines connecting the facility to the two tunnel header transfer lines, and jumpers or "U-tubes" that connect the tunnel headers to the cryomodules. The branch lines and tunnel headers are of similar cross sections and are similar to the HERA transfer line design.[4] Both lines have expansion and anchor joints to mitigate stresses caused by thermal contractions. Figure 2 illustrates the general arrangement of the coldboxes and the distribution system, as well as relative locations of specific components.

A NEW DIRECTION

The changes resulting from the Energy Department's 1998 decision had a significant effect on the cryogenic system design. The original system was optimized for a 1700-MeV linac and, even though installing fewer large cryoplants could attain lower beam energies, this was not an ideal design. A trade study was conducted to determine the optimum system configuration. Several configuration options were investigated and evaluated for individual cryoplant capacity, overall system availability, and cost.

The recommended configuration is illustrated in Figure 3. Two independent cryogenic facilities, each with its own cryoplants and separate distribution system, will provide refrigeration for the cryomodules. The 1030-MeV facility contains three cryoplants, and the 1700-MeV upgrade facility has two. The general arrangement of multiple, semi-independent cryoplants within the facilities has not changed from the original design configuration. The distribution systems are completely independent, allowing construction of the 1700-MeV upgrade without impacting the 1030-MeV system. This configuration had the best combination of high availability and low cost differential. The five cryoplants are all the same capacity (approximately 5 kW at 2 K, which is similar to the TJNAF Central Helium Liquefier's capacity). This results in minimal spares and maintenance costs and represents the least developmental risk.[5]

FUTURE DESIGN PLANS

The cryogenic system is presently under continuing development, and significant work remains to be completed. Several design tasks are planned for the next few years. The system interconnection, which links the individual cryoplants, will be studied and concepts further developed. The refrigeration cycle analysis will continue through June of 2000, culminating in a report that will define the optimum candidate cycle. The distribution system heat loads, pressure drop, and structural analysis will follow. Detailed process and instrumentation drawings will also be developed. These and other tasks are to be completed, and associated results will be included in the APT preliminary design report due in 2002.

REFERENCES

1. *APT Conceptual Design Report*, LAUR 97-1329, Los Alamos National Laboratory, Los Alamos, New Mexico, April 15, 1997.
2. Waynert, J., *Preliminary Estimate of APT Cryomodule Heat Loads*, ESA-EPE: 99-029, Los Alamos National Laboratory, November 6, 1998.
3. Kern, K., "Cryogenics System Availability Allocation," APT-PPO-MEM-01994, Los Alamos National Laboratory, March 25, 1999.
4. Lierl, H., Personal Memo, DESY, June 2, 1997.
5. Laughon, G., "APT Cryogenic System Configuration Trade Study," to be published.

RHIC ACCELERATOR COMMISSIONING, CRYOGENIC TESTS AND INITIAL OPERATING EXPERIENCE OF THE 25 kW REFRIGERATOR AND DISTRIBUTION SYSTEM*

M. Iarocci, J. Sondericker, K.C. Wu, Y. Farah, C. Lac, A. Morgillo, A. Nicoletti, E. Quimby, M. Rehak, and A. Werner

Cryogenic Systems Group
Relativistic Heavy Ion Collider
Brookhaven National Laboratory
Upton, New York 11973

ABSTRACT

The installation, initial cooldown, and cryogenic testing of the RHIC accelerator magnets and cryogenic system are complete. This paper covers the final phase of cryogenic equipment installation, performance of the system during the first accelerator cooldown, and difficulties encountered with cooling and commissioning. The 25 kW refrigerator, although operated for limited time periods in the past, has been running continuously for months. Operational experiences of the refrigerator and warm compressor systems will be discussed.

INTRODUCTION

The cooling scheme used for RHIC, shown in Figure 1, uses a single "Barber Nichols" centrifugal pump per ring to circulate supercritical helium through the magnets at a nominal flow rate of 100 g/s. The temperature of the circulating gas rises as heat is absorbed from the magnets. This heat is then removed from the circulating helium as it passes through heat exchangers located periodically around the 3.8 kilometers in circumference rings. These heat exchangers, called recoolers, are fed from the 25 kW refrigerator and represent the distributed cooling system for the RHIC magnets.

The Collider has been under construction and installation since 1992 and in the spring of 1997 a test of some of the integrated components, including a single sextant of magnets, was performed. Until the present time, that two month test[1, 2] represented the longest single run period for the refrigerator since its acceptance in 1985.

* Work performed under contract with the US Department of Energy

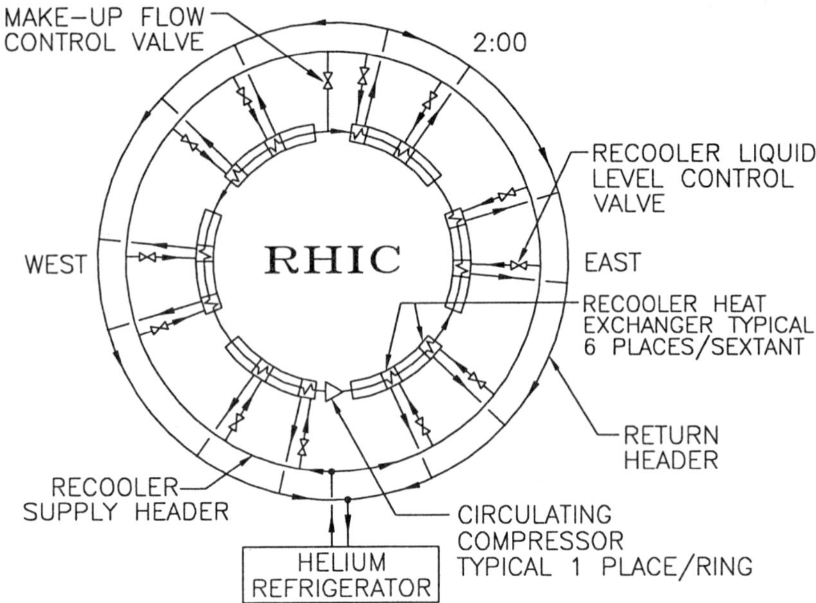

Figure 1. RHIC cooling scheme

The information derived from the sextant test[3] increased the confidence level in the basic design concepts and paved the way toward the completion of the project in the spring of this year.

During the last two years the major installation accomplishments include the completion of the process connections for the accelerator magnets, the installation of all magnet insulating vacuum components, completion of four kilometers of vacuum jacketed piping, installation of the balance of the twelve large valve boxes with cold crossing buss (CCB) and power leads, and installation of miscellaneous warm piping. At the completion of the project installation, the insulating vacuum system was nominally verified using a sniffing technique as the first leakcheck stage. Thirty large leaks were found and repaired. Upon establishing the insulating vacuum the process system was tested to "code" requirements, which resulted in a bellows squirm in a few areas. The first indication of the problem was detected as electrical shorts to ground found during the ongoing routine electrical certification of the magnet bus. Voltage taps were used to help locate the general areas of faults to guide us to the suspected bellows. Cryostats were opened for the replacement of magnet bus and bellows, and installation of corrective anti-squirm cans. Despite the squirm, no leaks were found in those areas. In addition to the mechanical failure, approximately fifteen more process leaks, not found in the first phase of leak checking, were located using helium gradient techniques. It's important to note that during the installation of the magnets and interconnects required approximately 25,000 discrete welded joints.

With the construction project complete, including the installation of all magnets, vacuum components, cryogenic distribution components, and controls, we cooled the rings during the period of April to July of this year. Since the last run, in early 1997, focus upon the construction project had once again left the cryogenic equipment idle. During the two year idle period the systems held up much better than previously documented[1] and the startup for the Collider cooldown allowed us to focus upon new construction issues. Nevertheless, we had numerous minor problems with legacy items including, motor controls and instrumentation. The root cause of these problems was found to be excessive electrical contact resistance and circuit component failures. These items are scheduled for capital improvement soon.

TESTS AND FINDINGS BEFORE COOLDOWN

Naturally with installation complete and all preliminary pre-start items covered, including vacuum, magnet electrical, cryogenic systems, and vacuum systems complete, we setup for the first integrated accelerator cooldown. The commissioning schedule was beginning to become condensed, with respect to delivering the demonstrated machine on time, so only three days were allotted for system cleanup and scrub. The entire machine was built to the intent of the ASME code and pneumatically tested to 110% of design pressure. For our case, after the process system was cleaned, the machine was pressurized to 20.5 atm. Prior to this step process connections were pressurized to three atmospheres for leak discovery. The low pressure leak test procedure was developed to conserve helium and time as this task was quite extensive. The intent of this high pressure test was twofold, in that we were looking for leaks missed during the initial leak check and verifying the mechanical integrity of the entire machine. During the low pressure phase of the leak testing, approximately thirty large leaks, >1E-4 std. cc/sec, had been identified and repaired. Left to do before cooldown was the final leak check with all process lines pressurized.

With the insulating vacuum established in all of the 28 discrete cryostats, the process lines were pressurized and leak detectors were setup to measure helium background during the pressure test. Fifteen more leaks were discovered in the range >1E-5 std. cc/sec. To find them, helium gradient techniques were used to isolate each leak to a single interconnect, followed by interconnect opening, leak pinpoint, and repair. Our intent was to focus upon finding and to fix all leaks >1E-5 std. cc/sec. It was during this process that unexpected magnet bus electrical shorts were detected during routine electrical high pot testing. Further investigation into the problem using magnet voltage taps to isolate the shorted wiring, revealed that we had some interconnects with squirmed bellows. Ironically, these squirmed bellows with offsets many times the pipe diameter did not produce any leaks, see Figure 2. The pressure test that was apparently successful had in fact disclosed a minor design error. Further investigation by analysis of the geometry did show this area prone to squirm. The design of these few interconnects was based upon an extension of the standard analyzed and tested interconnect design. The extension had naturally reduced the lateral pipe stiffness which led in this case to failure. The repair of each area entailed removal of each bellows assembly and damaged magnet superconducting bus and replacement of the same with new pieces. The prevention of recurring squirm was accomplished by installing a "clamshell brace" over each bellows that was attached to the pipe at each end of the bellows. See Figure 5. The moment of inertia in bending of the clamshell is much higher than the existing pipe therefore eliminating the problem. The clamshell was easily attached to the existing pipe system and clamped on one end of the bellows to prevent movement while guiding the pipe on the other end. This simple modification proved reliable during the subsequent pressure certification retest.

Figure 2. An example of bellows squirm found before the cooldown.

INITIAL COOLDOWN AND FINDINGS

The cooling of both rings was started in mid-April and was immediately interrupted by warm end water contamination, turbine faults, and turbine control skid faults. Verification of Oxygen Deficiency Hazard (ODH) safety was a requirement for continuance of our cooldown so we focused upon cooling sectors 4-5 to demonstrate that adequate oxygen concentration remains in the event of a maximum helium release at 50 K. Having successfully demonstrated that the oxygen levels were acceptable the cooldown continued. See Table 1 for the cooldown chronology.

With the 100 K wave through both rings the insulating vacuum at the 2:00 yellow valve box was spoiling. It was determined that the pressure would scale unfavorably from 100 K to 4 K, so the box and sextants on both sides were warmed in preparation for repair of the valve box. Cooling of the blue ring continued. The yellow valve box was opened for repair and it was discovered that the leaks were coming from two four-inch diameter flex lines in the magnet circuit. It was decided to repair this without disturbing the superconducting buss that runs through the magnet process as this approach would eliminate a few man-weeks of buss related rework. A piping analysis was performed and we determined that the flex lines were only used for piping misalignment and their elimination would not produce excessive pipe stress. Our repair entailed removal of the flex and installing pipe in a clamshell arrangement around the bus. This approach solved the problem. Later on, analysis of the leaking flex by scanning electron microscope and chemical identification in the affected area revealed cracks, see Figure 3, and high concentrations of zinc and chloride. Chemical analysis in non-affected regions showed normal makeup expected for stainless steel. The contaminants found in the affected zones were not just surface contaminants, but had etched the parent metal to produce what seemed to be a classic case of stress corrosion cracking (SCC). SCC is known to be a problem with stainless steel in the presence of a chloride rich environment. Further investigation shows zinc chloride is common to most of the flux used to soft solder stainless steel. Conclusions are that the flux was inadvertently deposited during the process of attaching the temperature diode mounts to the piping.

Table 1. Cooldown mid-April to mid-July

Weeks into cooldown	Summary of Events
1	Magnet cooling, water contamination of the warm turbine screens on numerous occasions, turbine contamination, turbine failures, and turbine oil skid failures caused numerous interruptions in cooling.
2	50 K spill test performed.
3	Approximately half of the magnets at 100 K, cold end heat exchangers show contamination. A local warmup to 200 K clears the problem, assume air contamination to blame. Continue 100 K wave.
4	Cooldown rate slows considerably toward the end of the 100 K wave although the refrigerator delivery conditions seem the same. Freezing water?
5	Partial yellow ring warmup, focus on blue ring cooling only. Blue ring at 4 K via liquid assist as makeup on two separate occasions, operators unable to balance the load once cold. Liquefying into dewars to reduce gas inventory.
6	After the third blue ring cooldown to 4 K we experienced severe refrigerator contamination. Decision to warm and clean the entire refrigerator, some gas lost due to lack of inventory space. Yellow 2:00 valve box repair started.
7	Refrigerator back on line. Blue ring cooled to ~8 K.
8	Refrigerator tested and healthy. Blue ring cooled to 6 K and in circulator mode. Experienced excessive heatload, due to valve leak-by from heat shield to 4 K return. Corrected same. Blue ring at 4.8 K. Yellow valve box repaired.
9-11	Yellow ring at 50 K and holding, blue ring sill stable at 4.8 K
12	Yellow ring cooled to 6 K, in circulator mode, and cooling.
13	Yellow ring at ~5.5 K. All systems are stable.

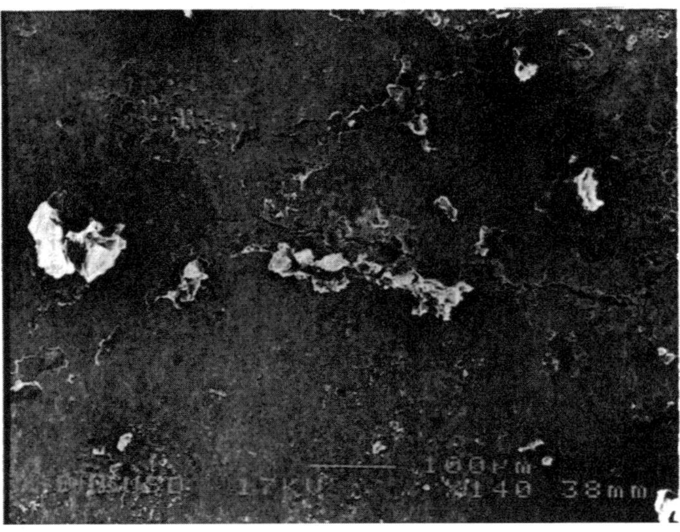

Figure 3. Scanningelectron microscope image along the peak of a convolution where stress corrosion cracking is observed.

COOLDOWN TO 4 K

During the refrigerator warmup, in the sixth week of cooldown, the 80 K external cryogenic purifier used in the loop for cleaning and drying the refrigerator plugged up three times. One of the clogs affected not only the drier section but the entire adsorber bed. One outstanding data point observed during one regeneration cycle of the drier was that gallons of water were observed draining from the warmed section of the drier.

We proceeded in a more cautious manner during the second phase of the cooldown. Instrumentation monitoring water and oxygen concentration in the helium supply to the refrigerator was upgraded prior to the second phase cooldown, and the refrigerator was monitored more closely with respect to maintaining temperatures at normal process levels. The refrigerator was also tested with its calorimeter to assure high efficiency prior to cooling the blue ring to 4 K. With most of the faults and contamination removed by the end of the sixth week, we finished the cooldown of the blue ring to 4 K and started cooling the yellow ring again. The yellow ring was at ~150 K on the west side and for the most part at 300 K on the east side. Once again we observed water in the warm-end heat exchangers and warm turbine inlet screens. The blue ring was in steady state operation, so the duty cycle for yellow cooling was reduced to ~20% for the first 3 days. The water returning from the yellow ring seemed to stop at that point.

The blue ring remained very stable for a few weeks and we took advantage of this mode to make a few 4 K heat load measurements. This measurement method was simple and the result represents only a heat load snapshot. The measurements were taken by logging the simultaneous liquid decay of each recooler with the supply valves closed over a time period of a few minutes. Naturally the rate of liquid decay translates to a heat input to the recoolers and is shown in Figure 4. In this figure notice the increased heat load at the left of each bank of 50W recoolers, located between the valve boxes, followed by the gradual decrease in load as you more to the right of each group. This characteristic is due to the extra heat load at the ends of each sextant, including the vacuum jacketed piping, valve boxes and especially leads mounted in the valve boxes. The heat load of the blue and yellow shields were also measured. Table 2 summarizes the results of the initial system heat load measurements.

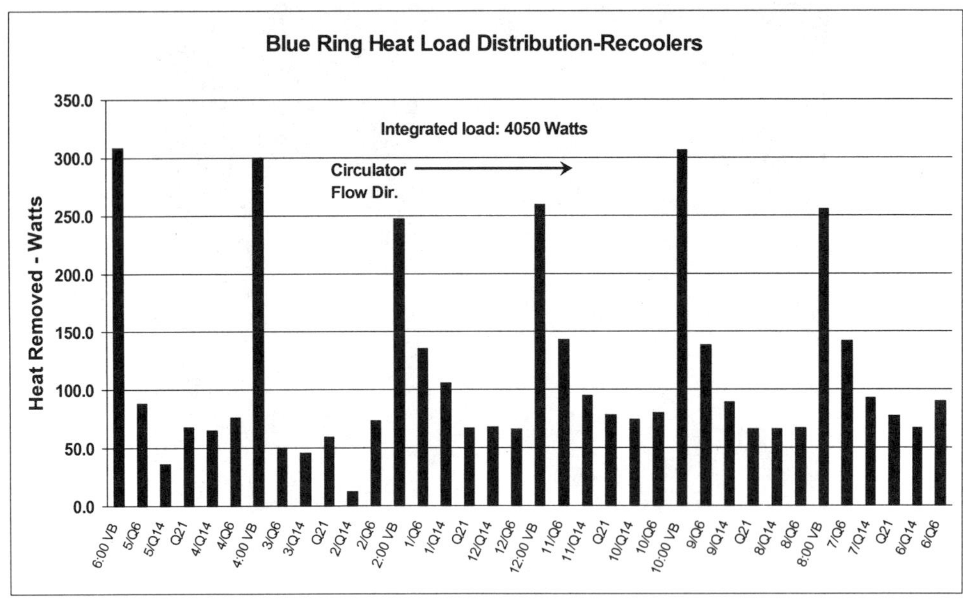

Figure 4. Heat load profile to 4 K recooler circuit during stable conditions, blue ring.

SUMMARY, LESSONS LEARNED AND PRESENT COLLIDER STATUS

Heat load and pressure drop measurements are close to projected for this system and the refrigerator has been running well since the cleanup period, showing only modest contamination in ten weeks of running with cold rings. The warm compressor system has been running well and handles most of the hot weather conditions without failures. It is not the source of trips, like it was in the past, as the number of events has fallen from many per day to 1 or 2 per week. The process control system is running reliability but still has a few evasive I/O problems that cause occasional trips. We are still working on some details with respect to inventory tracking but believe we are close to understanding gas usage and the magnitude of expected errors.

The lessons learned thus far are for the most part associated with contamination, refrigerator valve issues, and instrumentation redundancy. We did not anticipate and were surprised by the excessive amount of water in this system.

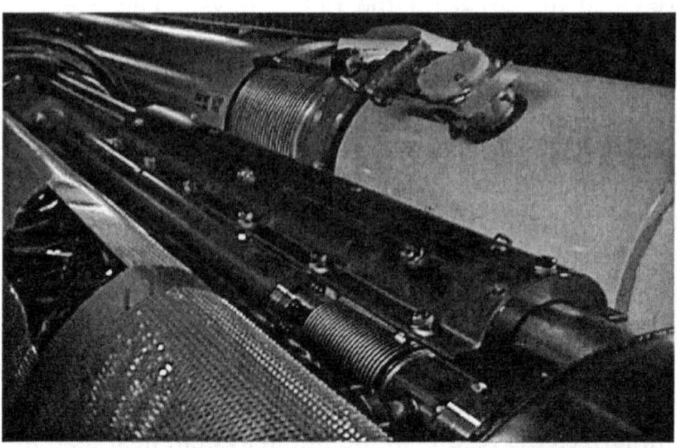

Figure 5. Clamshell brace installed over a helium process bellows is shown at the center of the figure.

Table 2. Summary of latest full ring heat load measurements

Process	Design Value	Measurement	Comments
Primary 4 K Heat Load Blue Ring	**4.3 kW**	**4.05 kW****	**Snapshot Measurement**
Primary 4 K Heat Load contribution from leads, one ring	**2.15 kW**	**2.3 kW***	**Individual lead verification**
Secondary 50 K Both Rings	**31 kW**	**36 kW*****	**Blue at 4 K, Yellow 4-30 K**

*** Projection from earlier single lead measurements.**
**** Projects to a total load of 12.8 kW with leads to 4 K cooling.**
***** 36 kW is slightly higher than projected, but well with the refrigerator's 50 kW shield cooling capacity.**

The subsequent refrigerator cooldown was accomplished with great respect for contamination and measurement of the same. During the post regeneration running period of the refrigerator, a decision to cool only one ring resulted from the non-availability of both rings. At first glance it seemed a prudent and logical approach to the present system readiness. We cooled the blue ring to its nominal operating point and at some time later also cooled the yellow ring, with most sections at room temperature, to its normal operating point. The difficulty in cooling the second ring was again water related. In order to maintain the blue ring at normal operating condition while cooling the yellow ring took much care and patients. After a few weeks of cooling the yellow ring is was clear that if the rings are contaminated with water it is better to cool them together to avoid the interactions associated with the two at radically different state points. This exercise affirmed the original goal of cooling both rings in parallel.

Besides water we discovered the lack of redundancy in some critical feedback control loops that almost upset the warm compressor system operation. Operator awareness and additional sensors solved the problem.

Finally the accelerator physics group has managed to achieve slightly greater than single turn beam operation at injection energy over the last couple of weeks.

ACKNOWLEDGMENT

The authors thank members of the cryogenic section engineering and technical staff for their excellence in installing and tenacity in operating the system thus far.

REFERENCES

1. J. Sondericker, et al., Performance and Cryogenic Operating Experience of the First RHIC Sextant Test, "Advances in Cryogenic Engineering," Vol. 43, P. Kittle, Ed., Plenum Press, New York (1997), p 237.
2. M. Iarocci, et al., RHIC 25 kW Refrigerator and Distribution System, Construction, Testing, and Initial Operating Experience, "Advances in Cryogenic Engineering," Vol. 43, P. Kittle, Ed., Plenum Press, New York (1997), p 499.
3. K.C. Wu, et. al., Thermal Hydraulic Performance for a Sextant of RHIC Magnets, to be published in these proceedings, July 1999 CEC.

Conference Plenary Speakers, from left to right: J. Gerhold (Technical University, Austria), R. Richardson (Nobel laureate, Cornell University), K. Johnson (Jet Propulsion Lab).

CRYOGENIC SYSTEM OF A SUPERCONDUCTING SOLENOID MAGNET FOR THE BELLE DETECTOR SYSTEM IN KEK

K. Aoki, Y. Doi, T. Haruyama, M. Kawai, T. Kobayashi, Y. Kondo and Y. Makida

High Energy Accelerator Research Organization, KEK
Tsukuba, Ibaraki, 305-0801, Japan

ABSTRACT

The cryogenic system of the large superconducting solenoid magnet for the BELLE detector system was constructed at KEK for B-meson physics. This system consists of a refrigerator, compressor and buffer tank, which were previously used and reassembled, and a recovery tank, subcooler and superconducting solenoid magnet, which were newly constructed. The cryogenic system is automatically controlled and operated from cool down to warm up according to programmed steps. It is monitored on the World Wide Web. The magnet is indirectly cooled by forced, two-phase, helium flow during steady state operation. In the case of an emergency stop of the refrigerator, the cooling method is switched to a thermo-syphon mode by an automatic process control. In the cooling test, the refrigeration power was 240 W at 4.4 K, which was almost the same as the previously measured value before reassembling, and the heat leakage into the magnet was deduced to be 31.8 W. After performance tests and two operations to tune the detectors, regular operations with the colliding beam started for the first time in May 1999.

INTRODUCTION

KEKB is an asymmetric-energy, two-ring, electron-positron collider constructed at KEK for B-meson physics. Electron (8 GeV) and positron (3.5 GeV) beams are accelerated in opposite directions and collide at one interaction point in the Tsukuba experimental hall. The BELLE detector system[1,2] is installed at this interaction point. The main purpose of this detector system is to measure CP violation in the decays of B-mesons. The detector system has the barrel and octagonal end yoke, a superconducting solenoid magnet[3,4,5] and inner detectors. The yoke has a structure composed of layers of flux-return iron plates and K_L/μ detector modules.

The cryogenic system for the BELLE magnet is required to cool the magnet smoothly, to recover from emergencies safely and quickly, and to be a reliable system for long continuous operation. It was designed with following features:

1) For safety and ease of operations, all the cryogenic system is controlled and operated automatically from the start of cool down to the end of warm up, and in an emergency.
2) For safety and reliability, a thermo-syphon cooling mode is selected automatically during magnet discharge, which occurs when the refrigerator stops in an emergency.
3) For convenience, all cryogenic system data are monitored on the World Wide Web. It contributes to safe operation.

All components of the cryogenic system and magnet were under construction at the Tsukuba experimental hall until September 1997. After performance tests, magnetic field measurements were performed for three months, starting in January 1998. A second period of operation continued for three months from January 1999 in order to calibrate the detectors. A third period of operation started in May 1999 for the first physics data taking with colliding beam.

Four other superconducting magnets, which are called IR magnets,[6,7] are aligned near the interaction point inside the BELLE solenoid bore as accelerator elements. They have their own cryogenic system. As for the control systems of these IR magnets and the BELLE cryogenic systems, the main processor and control program are completely independent, but peripheral devices are common to all.

OVERVIEW OF THE CRYOGENIC SYSTEM

Figure 1 shows the flow diagram of the cryogenic system. The system consists of a helium compressor, a cold box, a subcooler, transfer lines, a superconducting solenoid magnet, two helium gas tanks and a liquid nitrogen storage tank. Most of the system was removed from previous superconducting quadrupole magnet (QCS) systems[8] in other experimental halls and reassembled. Many components were constructed in 1990 and used for the TRISTAN experiment. Total operation time until 1997 was 28,000 hours. The subcooler, transfer lines, a recovery helium gas tank, a fifth oil separator after the compressor, and the magnet were newly constructed.

Compressed helium gas at 1.67 MPa from the compressor flows into the cold box and comes out at 6 K and 0.3 MPa after the first J-T valve. Passing through the subcooler where it is further cooled, the helium gas flows into the control dewar at the top of the magnet to be subcooled to 4.52 K at a pressure of 0.288 MPa. At this point, the helium condition is supercritical. After the second J-T valve, the helium is in the two-phase state at 4.46 K and 0.125 MPa and flows into the magnet through the chimney port. In the steady state, the magnet is indirectly cooled by forced, two-phase flow at a mass flow rate of 10 g/s.

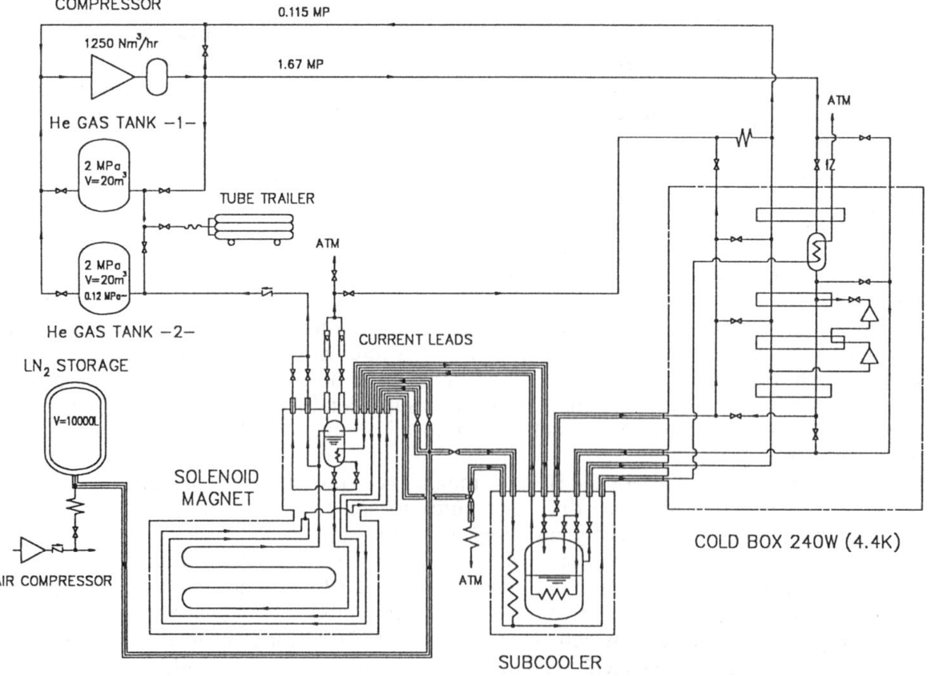

Figure 1. Flow diagram of the cryogenic system.

After cooling the magnet, two-phase helium returns to the subcooler by way of the control dewar. The boil-off helium gas from the reservoir inside the subcooler flows into the cold box. The returning cold gas comes back to the compressor, cooling the input helium line. In the case of a quench, the magnet is separated automatically from the subcooler and the boil-off helium gas from the magnet is recovered in the helium buffer tank 2.

Part of the helium returning from the magnet is used to cool the current leads. The returning route of the helium can take one of two possible paths. One path is the usual route, to compressor suction, and the other is to the atmosphere. These paths are selected and controlled automatically.

The main parameters of the cryogenic system are listed in Table 1.

SUBCOOLER

The main purpose of the subcooler is to control the total volume of liquid helium and ensure the regular circulation of the helium in the system. It houses a LHe reservoir containing a coil-shaped heat exchanger and a 400 W heater. This heater is used to control the total liquid volume of the system. The helium gas from the cold box is divided into two paths. One is cooled further in the heat exchanger inside the reservoir, and then flows into the control dewar on the magnet to cool the magnet. The other flows into the reservoir in the subcooler through a controlled J-T valve. This line supplies the return path of the circulation when a quench occurs.

SUPERCONDUCTING MAGNET

The BELLE superconducting solenoid magnet is indirectly cooled by two-phase helium flow through a serpentine tube welded on the outside of the support cylinder. The outer and inner radiation shields are cooled by liquid nitrogen.

Table 1. Main parameters of the cryogenic system

Refrigerator:			
	Type of refrigerator		Claude cycle
	Cold box:	Refrigeration power	240 W at 4.4 K
	Compressor:	Type	Two-stage screw compressor
		Flow rate	1,250 Nm^3/h
	Subcooler:	Volume of LHe reservoir	154.5 L
		Thermal load	10 W
	Transfer line:	Thermal load	
		To the magnet	20 W
		From the magnet	24 W
Superconducting solenoid magnet:			
	Cooling method:	In steady state	Indirect cooling with forced two-phase helium flow
		In refrigerator trouble	Thermo-syphon cooling
	LHe capacity:	LHe reservoir in control dewar	200 L
		Coil	65 L
	LN_2 capacity:	Shield	25 L
	Cold mass:		7.9 ton (Coil) +1.0 ton (Control dewar)
	Thermal load:	Coil	22.4 W at 4.4 K
		Current lead	15 L/h
		Shield	150.0 W at 80 K
	Precooling time:		180 h
	Central magnetic field:		1.5 T
	Nominal current:		4160 A
	Stored energy:		35 MJ

The magnet has the control dewar outside the barrel yoke on the chimney port. The control dewar houses a LHe reservoir of 200 L and two valves, which are the second J-T valve and the switching valve for the thermo-syphon cooling mode. The LHe reservoir contains a coil-shaped heat exchanger and a 200 Ω resistance to be used as a heater.

PROCESS CONTROL SYSTEM AND MONITOR SYSTEMS

Figure 2 shows the process control system and the monitor systems. The main processor and control program are independent from the control of the IR magnet's cryogenic system. The peripheral devices are used in common.

Hardware for the process control system. The process control computer system, a HITACHI EX1000A, was previously used and has been re-installed in the control room of Tsukuba experimental hall. The EX1000A monitors and controls most of the BELLE cryogenic system data.

There are two multi-controller (MLC) cubicles for the BELLE cryogenic system, and two pair of master and slave process operator's consoles (POC) for the BELLE and IR magnet's cryogenic systems. MLC cubicles have a function unit, common parts of which are duplicated, and process input and output units. The main processor of the function unit is a 16-bit micro-processor with 512 kbytes of memory. Usually a pair of the master and slave POCs is assigned for the BELLE cryogenic system, and the other pair is for the IR magnet's cryogenic system. Operation is possible equally with either pair. The system is used in 1 second scan mode and has 121 Direct Digital Control loops, 13 graphic pages, 19 cyclic trend pages (sampling interval: 60 sec, keeping period: 2 days) and 16 real time trend pages (sampling interval: 10 sec, keeping period: 4 h). These loops and pages were developed for the BELLE cryogenic system. The number of process input/output signals are listed in Table 2.

The system is connected to the duplex optical fiber data link and it is possible to communicate with the other pair of POCs in the Nikko experimental hall, and the gateway unit in the Accelerator control building.

Software for the process control system. A software program on the EX1000A was developed for the automatic and safe control of the BELLE cryogenic system for all operations: cool down, steady state, warming up and emergency. The program is composed of the following three sequence groups:

- Main modes: constituting the main part of the control flow.
- Basic sequences: having continuous actions and being called from the main modes.
- Emergency modes: being monitored in parallel with the main modes. But once an emergency occurs, the control flow is transferred completely to this mode.

A schematic flow diagram of the control is shown in Figure 3.

Three accidents are assumed as the emergency modes, and the recovery sequences are prepared as follows:

- Magnet quench.
- Sudden stop of the compressor.
- Turbine trip.

If the refrigerator stops due to the sudden stop of the compressor or a turbine trip, the cooling mode will be switched to the thermo-syphon mode as shown in Figure 4.

Table 2. Input/output signals of the cryogenic system

To or from EX1000A		
Analog input:	95 points	(pressure, 30; temperature, 42; flow rate, 5; others, 18)
Analog output:	33 points	(control valve, 31; heater, 2)
Digital input:	37 points	(status, 16; alarm, 21)
Digital output:	45 points	(on/off valve, 12; others, 33)
To BELLE magnet monitor		
Analog input:	152 points	(temperature, 63+10*; strain, 52; others, 19+8*)

* common signals with the EX1000A. A common signal is distributed to two from a sensor.

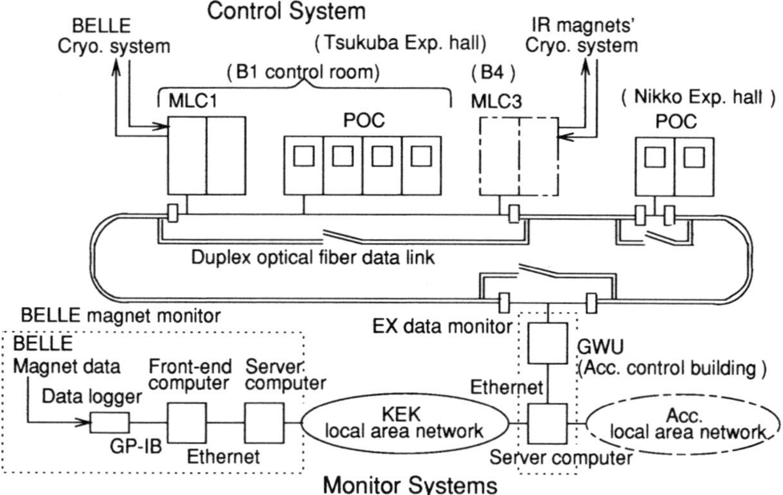

Figure 2. Control system and monitor systems. (POC: process operator's console. MLC: multi-controller, GWU: gateway unit)

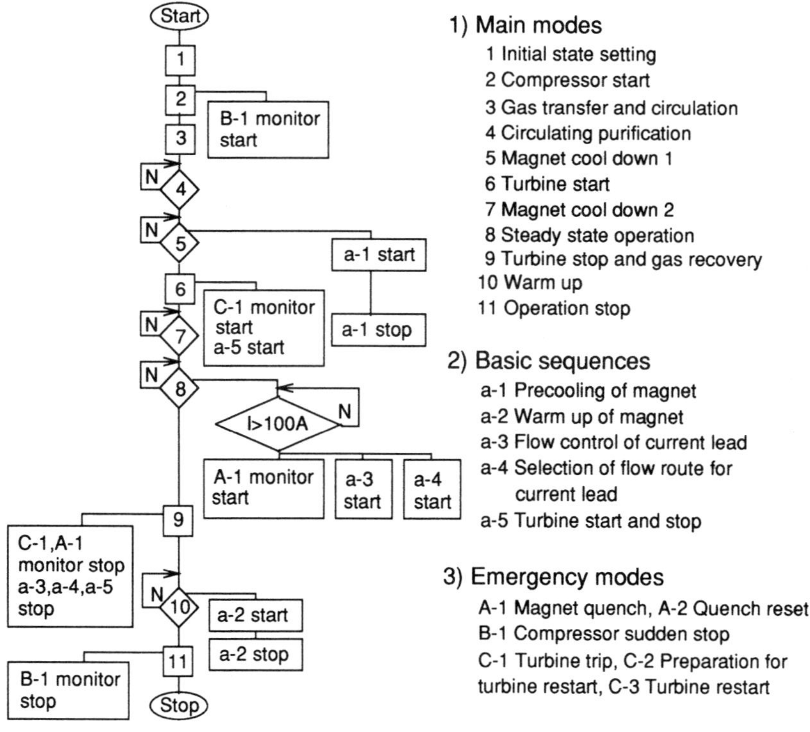

Figure 3. Flow chart of the automatic process control.

Monitor systems. There are two monitor systems. One is the BELLE magnet monitor for the magnet. This monitor is composed of a data logger (HP3497A), a front-end computer (ADVANTECH INDUSTRIAL COMPUTER 610), and a server computer (FIC BASIC COMPUTER 500). The data logger and the front-end computer are located on the same movable stand as the detectors. The data logger takes the data shown in Table 2 from the magnet and sends the data to the server computer through the front-end computer. The server computer, which links the KEK local area network, accumulates data every ten minutes and sends the current data of the magnet to the World Wide Web, updating every one minute. These time intervals are changeable.

The other monitor system is the EX monitor on the internet for the cryogenic system. The EX monitor aims to provide useful and convenient total monitoring information to the persons concerned. For this purpose, the following points were taken into account:

1) Easy access to the all data of the BELLE cryogenic system (anywhere, regardless of the type of computers used, and without complicated settings). Furthermore, the monitor requires comprehensible presentation with the ability to modify it.
2) Easy reference to all previous data in the database.
3) Easy graphing to see the time dependant data transitions.

For 1), we decided to use the World Wide Web on the internet. For 2) and 3), the World Wide Web also offers many tools. The monitor is composed of a gateway unit of the EX1000A series, and a server computer, HP9000 D210, which are located in the Accelerator control building. The gateway unit and the server computer are linked by a dedicated Ethernet line. The server computer also links to the KEK local area network. The server computer communicates with the EX1000A through the gateway unit to take all data of the BELLE cryogenic system at intervals of thirty seconds. Then it accumulates the data into a reliable mass-storage unit, a mirror type 18 Gbytes harddisk. The EX monitor has its own homepage on the internet and displays current data, updating every thirty seconds. Furthermore, two tools for referring to previous data were prepared on the same homepage. Realtime-data refers to the data at thirty second intervals for the most recent two hours. Cyclic-data refers to the data at ten minute intervals for the entire operation period. The homepage of the EX monitor also allows for easy graphs of the data, and refers via a software-link to the current data of the magnet from the BELLE magnet monitor. These time parameters are all changeable. The EX monitor has the following advantages:

- It is useful for watching the cryogenic system daily and during sudden troubles.
- Easy reference and graphing provide strong tools to analyze the system.

As the result, the EX monitor makes quick correspondence possible. It represents a contribution to safe operation.

PERFORMANCE

The performance test of the system was carried out and reasonable results were obtained. These are listed in Table 3. Figure 5 shows the relationship between the liquefaction rate and the refrigeration power compared with the data measured in the previous TRISTAN experiment before the reassembling. The refrigeration power was confirmed to be 240 W at 4.4 K. It was almost same as the previous value.

Figure 6 shows the relationship between the flow rate and the heater input power in the reservoir of the control dewar. It was measured under the condition of maintaining a constant liquid level. The minimum required helium flow rate, 1.7 g/s, was extrapolated

Table 3. Performance of the cryogenic system

Cold box of the refrigerator:	
Refrigeration	240 W at 4.4 K
Refrigeration and liquefaction	170 W + 25 L/h at 4.4 K
Liquefaction	92 L/h
Magnet:	
Minimum necessary flow rate	1.7 g/s
Thermal load	31.8 W

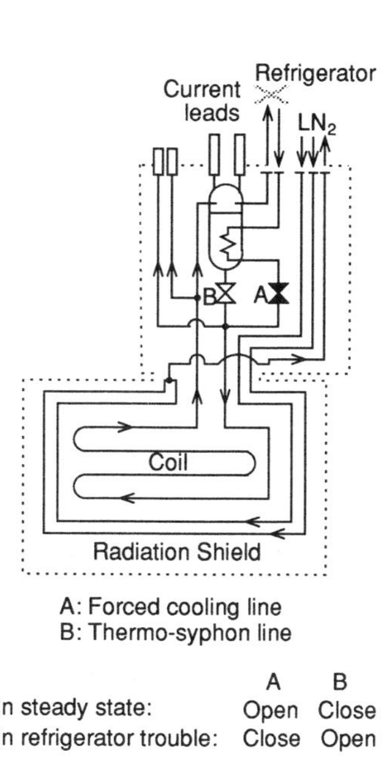

Figure 4. Flow diagram of the magnet. Two valves are switched in the emergency mode to change the cooling method.

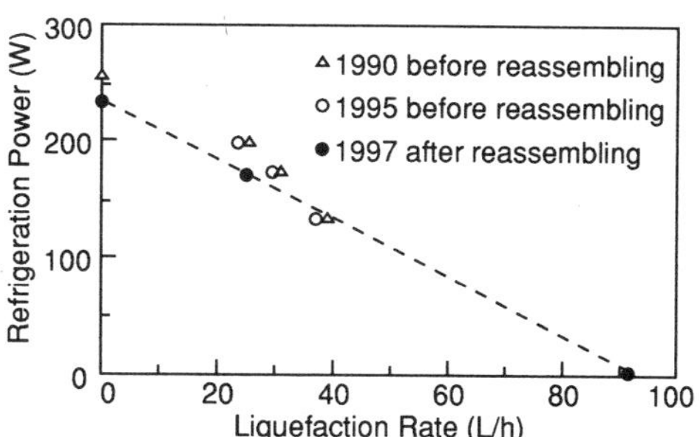

Figure 5. The relation between the liquefaction rate and the refrigeration power compared with the previous data.

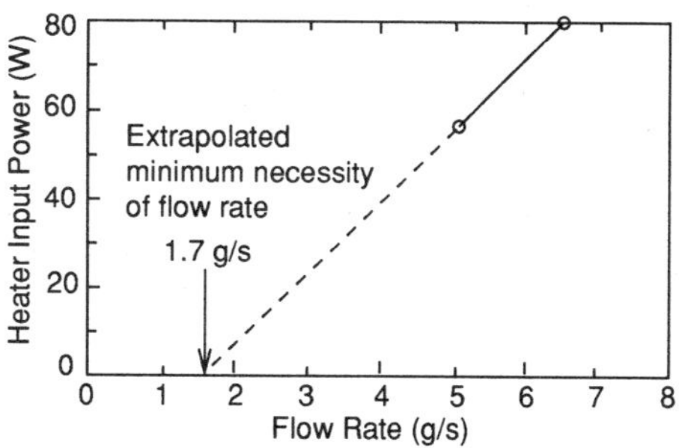

Figure 6. The relation of the flow rate and the heater input power in the reservoir used to keep the liquid level constant.

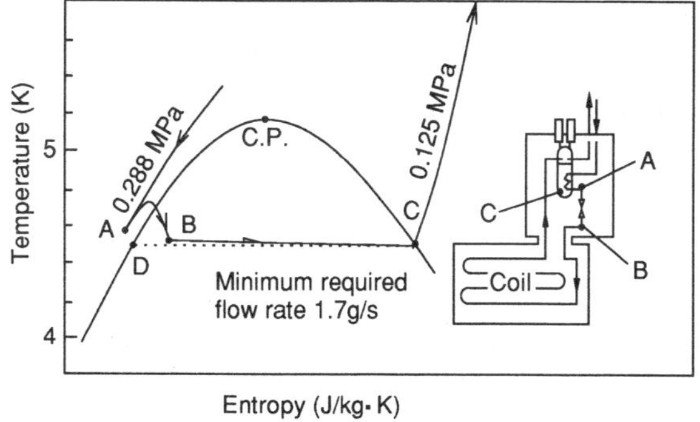

	Temperature (K)	Pressure (MPa)	Enthalpy (J/g)	Quality
A	4.52	0.228	11.84	
B	4.46	0.125	11.84	0.024
C	4.46	0.125	30.53	1
D	4.46	0.125	11.39	0

A: before 2nd J-T, B: after 2nd J-T, C: reservoir

Heat leakage = (30.53-11.84)J/g x 1.7g/s = 31.8W

Figure 7. T-S diagram along the cooling pipe and the calculation of the heat leakage to the magnet.

from the measured results, because it was difficult to control the mass flow rate at less than 5 g/s. The heat load of the magnet was calculated to be 31.8 W from this minimum required helium flow rate, as shown in Figure 7. This value is larger than the estimated value. But it is low enough for the refrigerator capacity.

Figure 8 shows the typical precooling curve of the magnet. It took 183 hours to cool the magnet from the room temperature to 4.4 K.

Switching to the thermo-syphon mode was tested. The circulation lasted during discharging from 1.5 T to zero at the normal ramp rate of -2 A/s. But a decrease of the pressure difference between inlet and outlet was observed. It is supposed that many

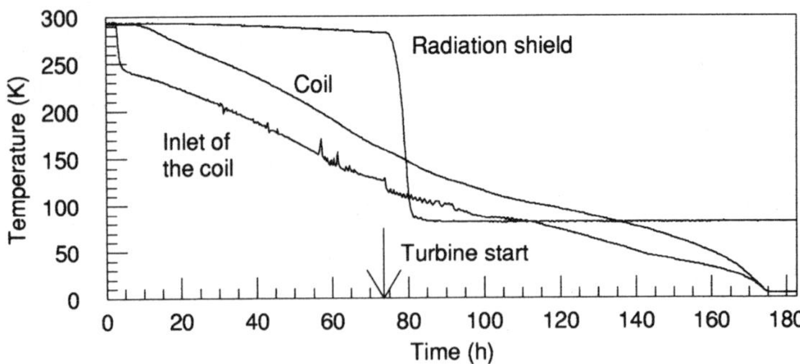

Figure 8. Precooling curve of the magnet. The turbines were started at a coil temperature of 160 K. The cooling of the LN_2 shield was started ten minutes later after the turbines reached their nominal power.

bubbles generated in the cooling pipe disturbed the gravitational liquid helium flow.

The automatic control process was checked and examined in detail. The control process was confirmed to work safely during all processes.

The EX monitor was confirmed to work as designed. The monitor proved its effectiveness when problems occurred. It was useful to see the data from a distant place and to compare the present data with the previous data.

SUMMARY

The cryogenic system for the BELLE superconducting solenoid magnet was constructed and tested. The refrigeration power of the reassembled refrigerator was confirmed to be 240 W at 4.4 K. It was almost the same as the previous value. The minimum required helium flow rate, 1.7 g/s, was extrapolated from the measured results. The heat load of the magnet was then estimated to be 31.8 W from this minimum required flow rate. Switching to the thermo-syphon mode and keeping it on during the discharge in the emergency process was tested and proved effective. Through all the tests, the automatic control system was confirmed and proved to work well and to be sufficiently safe. The EX monitor proved the monitoring on the World Wide Web is useful and convenient, especially when trouble occurs. The system experienced two other operation cycles, each for three months, after the tests and it has proven to be stable for long continuous operation.

ACKNOWLEDGMENT

The authors would like to express our gratitude to the BELLE collaborators and to the members of IR magnets group for their continuous support. We would also like to thank the engineers and technicians of Hitachi Ltd. for their skillful work.

REFERENCES

1. J. Haba, *Nuclear Instrument and Method in Physics Research*, A 368:74 (1995)
2. A Study of CP Violation in B Meson Decays, KEK Report 95-1, (1995)
3. Y. Makida et al., in: "Advances in Cryogenic Engineering, Vol. 43A", (1998), p. 221
4. H. Mukai et al., in: "Proc. of 15th Int'l Conf. on Magnet Technology (MT-15)", (1998), p. 1044
5. Y. Makida et al., Performance of a Superconducting Solenoid Magnet for BELLE Detector in KEKB B-factory, *IEEE Transaction on Applied Superconductivity*, Vol. 9, in press.
6. T. Ogitsu et al., in: "Proc. of the 6th European Accelerator Conference (EPAC 98)", (1998), p. 2038
7. K. Tsuchiya et al., Superconducting Magnets for the Interaction Region of KEKB, *IEEE Transaction on Applied Superconductivity*, Vol. 9, in press.
8. K. Tsuchiya et al., in: "Advances in Cryogenic Engineering, Vol. 37A", (1992), p. 667

LHD CRYOGENIC-CONTROL SYSTEM PERFORMANCE UNDER VARIOUS OPERATING CONDITIONS

R. Maekawa,[1] T. Mito,[1] K. Takahata,[1] S. Yamada,[1] A. Iwamoto,[1]
K. Ooba,[1] K. Watanabe,[1] T. Baba,[1] H. Yoshikawa,[2]
T. Nakashima,[2] F. Kishida,[2] K. Nakamura,[2] M. Nobutoki,[2]
K. Iimura,[2] T. Fukano,[2] S. Satoh,[1] and O. Motojima[1]

[1]National Institute for Fusion Science
322-6 Oroshi, Toki, Gifu 509-5292, JAPAN
[2]Nippon Sanso Corporation
6-2 Kojima, Kawasaki, Kanagawa 210-0861, JAPAN

ABSTRACT

The Large Helical Device (LHD), an experimental fusion apparatus, has conducted a second experiment cycle. The infrastructure of the LHD cryogenic control system (TESS) is a combination of a distributed control system with an integrated controller linked by multiple LANs. The integrated controller shares information with four subsystems: the helium refrigerator/liquefier, the helical coil, the poloidal coil and the superconducting bus-line, via reflective memories. Each subsystem has a VME bus and a workstation for process control and programming. Cooldown processes of each subsystem are implemented independently by the temperature levels. Sequence programs are developed to change cooling modes for subsystems corresponding to their temperatures. Program Control Units (PCUs) proceed cooldown and warmup process in terms of gaseous-helium supply temperature and mass-flow rates. Temperature distributions in the subsystems are monitored and process will be interrupted if temperature gradients exceed 50 K. In the case of coil quench, emergency sequence programs are executed to prevent any damage to the apparatus. The paper describes the performance of LHD-TESS during cooldown, steady-state operation, and the case of emergencies.

INTRODUCTION

Figure 1 shows an illustration of the LHD cryogenic system, which consists of a pair of helical coils and three sets of poloidal coils assembled in the supporting structure. Nine superconducting bus-lines connect superconducting coils with current leads. The system utilizes four cooling schemes: pool boiling for helical coils, forced-flow supercritical helium for poloidal coils, two-phase helium flow for superconducting bus-lines as well as supporting structure and forced-flow low-temperature helium gas (40 K ~ 80 K) for thermal

Advances in Cryogenic Engineering, Volume 45.
Edited by Shu *et al.*, Kluwer Academic / Plenum Publishers, 2000.

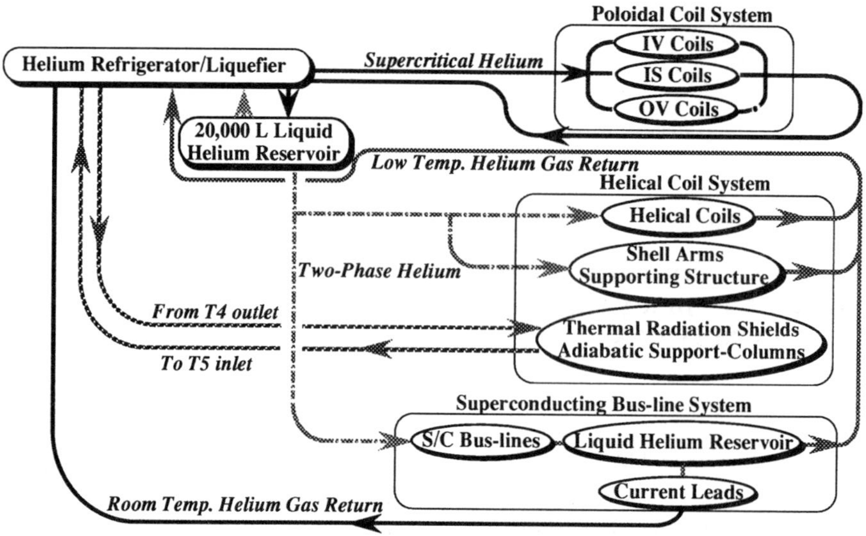

Figure 1. Illustration of the LHD Cryogenic System, which does not include the helical valve box and the poloidal valve box. Details of the LHD cryogenic system are described in reference 1.

radiation shields. The cold mass at 4.4 K is approximately 822 t, and for 40 K to 80 K is estimated to be 34 t. The 10kW-class helium refrigerator/liquefier produces coolants for the LHD.[1] There are approximately 205 PID control loops for the LHD cryogenic system, which control mass-flow rates, liquid helium levels and pressures. The complexity of the system demands a highly reliable control system.

LHD-TESS

Figure 2 shows the schematic for LHD-TESS, which consists of four subsystems and an integrated controller. The fundamental design concept is based on the open system that ensures expandability for future developments. The hardware architectures are workstations with UNIX™ and Versa Module Europe (VME™bus) with VxWorks™. Workstations are used to develop sequence programs and PID loop control, which are down-loaded to the VME bus for the process control.[2] The system is designed to build up advanced control algorithms. Thus, the programming itself requires knowledge of control dynamics rather than profound programming technique. The flexibility of the system allows for adjusting process control during operations.

Data from each subsystem are transferred to the integrated controller via reflective memories with 20Mb/s. The integrated controller also communicates with the LHD main control room and receives crucial information such as a quench signal, an emergency shut down signal and operating status of other LHD components. Therefore, the integrated controller is responsible for the overall operating condition. The operating graphic console units, DP1000™, are man-machine interfaces that provide real-time process monitoring and perform PID-control parameter setting. Graphic representations of the system greatly reduce the training time for operators and make it relatively easy to understand the process status on a real time basis.

LHD-TESS is required to have high reliability and availability since a control system failure could lead to significant damage and/or loss of the whole system. A fault-tolerant design was considered and modular redundancy was chosen. Each subsystem has identical replicas of hardware modules. Two fault diagnostics are used; a self-diagnostic for analog I/O and a voter diagnostic for digital I/O with Programmable a Logic Controller (PLC). Once the diagnostic system identifies failed component(s) within VME bus in service, the backup VME bus takes over the operation in terms of automatic switchover. The system operates without a back-up system until the failed components are repaired.

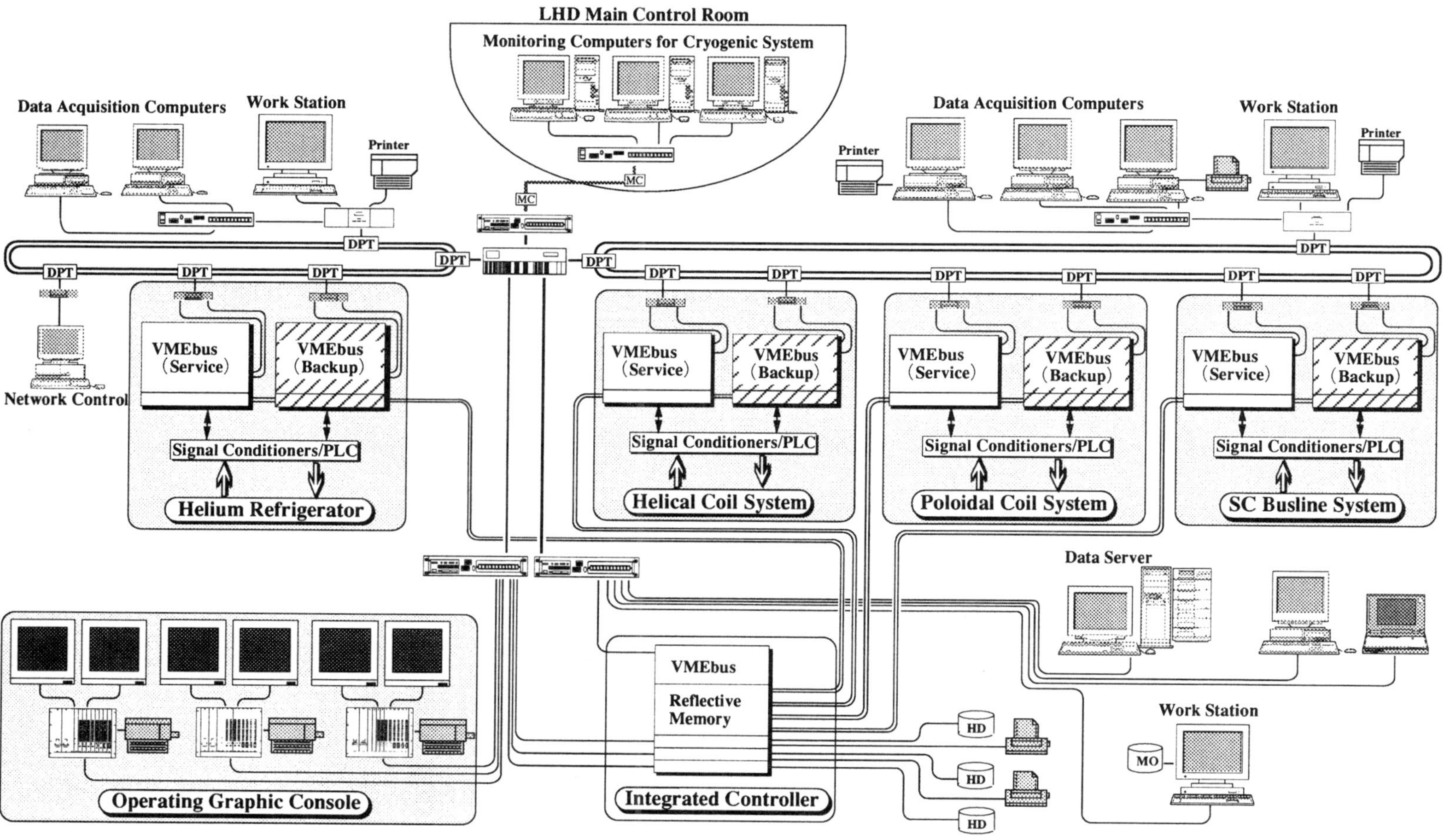

Figure 2. LHD-TESS (Cryogenic Control System)

Table 1. LHD Cryogenic System I/O

I/O	Helium Refrigerator	Helical Coil System	Poloidal Coil System	SC Bus-line System	Integrated Controller	Total
AI	160	315	122	448		1045
AO	96	32	24	64		216
DI	320	128	128	256	64	896
DO	192	128	128	256	64	768
IA	496	368	388	768	160	2180
ID	944	720	816	880	320	3680
Memory AI					1045	1045
Memory AO					216	216
Memory DI					896	896
Memory DO					768	768
Total	2208	1691	1606	2672	3533	11710

Table 1 shows the cryogenic system I/O list. Internal digital (ID) signals are used for the memory flags to identify the sequence program control and the process control status.

PERFORMANCE OF LHD-TESS

Cooldown Scenario

The LHD cooldown scenario, approximately a one-month operation, has been explicitly considered. The difficulty is that the subsystems have to cooldown simultaneously to maintain the temperature gradients within 50 K. Both Program Control Units (PCU) and the Temperature Monitoring System (TMS) are responsible for maintaining a semi-automatic cooldown process. PCUs are used to change the temperature and pressure of supply helium gas as well as mass flow rates for each subsystems, while TMS observes temperatures of four-groups; supporting structure (37pts), helical coils (17pts), poloidal coils (30pts) and thermal radiation shields (32pts). Cooldown by the PCU is interrupted if temperature gradients exceed 50 K. The program is designed to identify a high temperature region so that the mass-flow rates can be increased for that particular region. Cooldown processes of each subsystem are implemented independently by the temperature levels: gaseous-helium flow passes are changed with supply and discharge temperatures. Figure 3 shows an illustration of the cooldown process, which is primarily used before starting turbine operations.

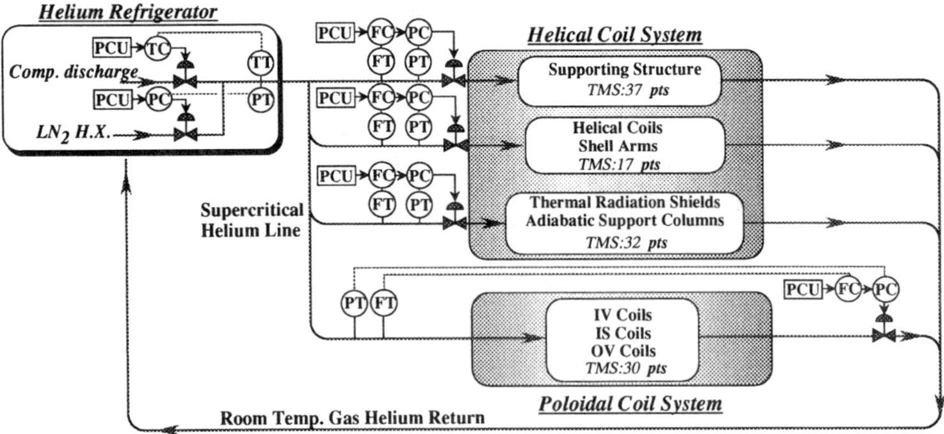

Figure 3. Illustration of hydraulic passes for the cooldown process. The figure only shows one control valve for each sub-group. All control valves were operated with cascade control to adjust mass-flow rates, depending on their cold mass. 24 cascade control valves were used for the cooldown. The same procedures were also used for the warm up process. Performance of first cooldown process is described in reference 3.

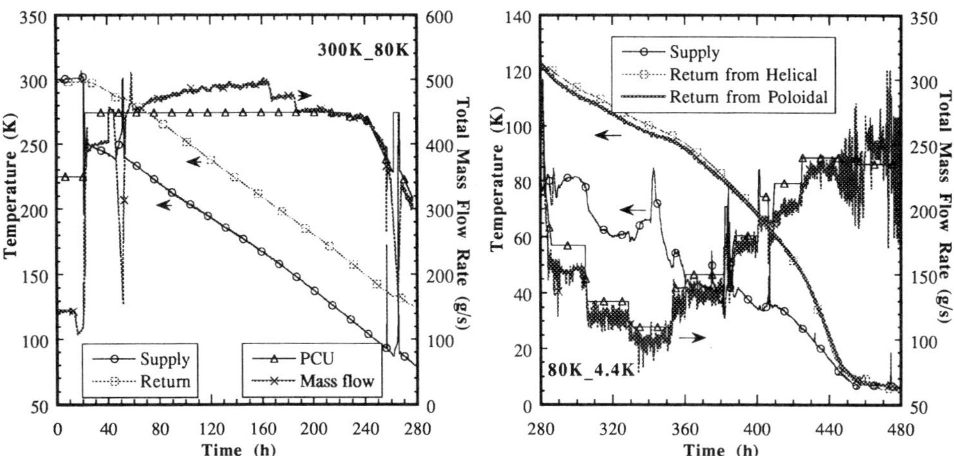

Figures 4 & 5. Cooldown curves for the second operation cycle. Operation from room temperature to 80 K was established with PCUs and TMS. Disturbances at t~50 h & 260 h were caused by power failures because of thunder- storm. The supply temperature increase at t~340 h was associated with a filter regeneration at the inlet of turbine 1.

Room Temperature to 80 K. Cooldown proceeded in terms of mixing the room temperature helium gas with 80 K helium gas, which was produced by the LN_2 heat exchanger. The cooldown speed was set at –1.0 K/h. A cooldown curve for the second operation cycle is shown in Figure 4 & 5.

80 K to Steady-State Operation. Cooldown from 80 K was started with turbine operation. Production of supply helium gas was achieved with turbine #1 to #5 operation. Mass flow rates of supply helium gas were reduced to cooldown heat exchangers in the cold boxes. Thermal radiation shields were cooled by turbine #4 and #5 so that the supply and return lines were changed from supercritical helium line to 80 K shield line by sequence program. As return gas temperature from the helical and the poloidal coils become less than 120 K, the return lines were changed from room temperature to low temperature return lines. Turbine #6 & #7 were started as the supply temperature reaches ~35 K. The 20,000 L liquid helium dewar was also cooled down, at that point. Sequence programs were used to change the supercritical helium line to two-phase helium line for the helical system as well as superconducting bus-line system. The upper and lower poloidal coils were connected in series as return gas temperature from poloidal coils reached 7 K.

Steady State Operation

The cryogenic system was under automatic control as the system reached a steady-state condition. The operating status was changed to energize the superconducting coils for plasma experiments. The Excitation Condition Monitoring (ECM) system was used to confirm a stand-by state for coil-excitation. The ECM was responsible for four groups, a helical coil (25 pts), a poloidal coil (53 pts), supporting structure & radiation shields (26 pts) and superconducting bus-line & current leads (112 pts). The parenthesis after each group represents the number of criteria, such as temperatures, liquid helium level, mass-flow rates and pressures. All criteria have to be satisfied to energize the superconducting coils. Each criterion has secondary condition to maintain a coil excitation-state. Even if some criteria were to deviate from set values, the excitation status could be sustained for 60 s, allowing the value to be restore. The system was designed to identify deviated conditions so that the operators can restore the criteria. If the restoration process fails, the integrated controller sends a discharge request to the main control room. However, the cryogenic system has operated successful without having a discharge request made by the ECM system.

Emergency Sequence Program

Even with a control system with high reliability, it is inevitable some failures will occur. Two predictable failures are a superconducting coil quench and a power failure. The cryogenic system has to take corrective actions for these failures to avoid significant damage and/or losses. Sequence programs have been developed to account for these failures.

Coil Quench. Although the superconducting coils are designed to have a high stability margin or a fully cryostable condition, a coil quench might be initiated with an unpredictable transient disturbance. The cryogenic system has to be protected in the case of quench, which results in high-pressure generation due to Joule heating and/or AC losses.

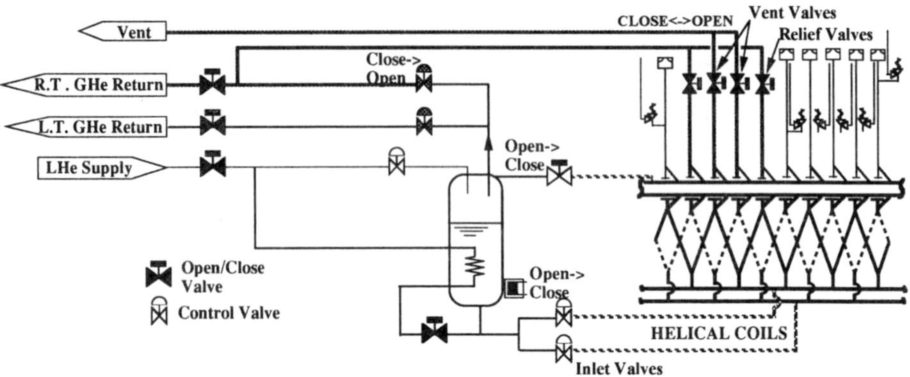

Figure 6. Quench sequence program for the helical coils. Inlet valves were closed with a discharge signal from integrated controller. Once the pressure within coils exceeds 0.137 MPa, pressure relief valves to the R.T. line opened. Vent valves were opened at 0.152 MPa to prevent a rupture disk operation.

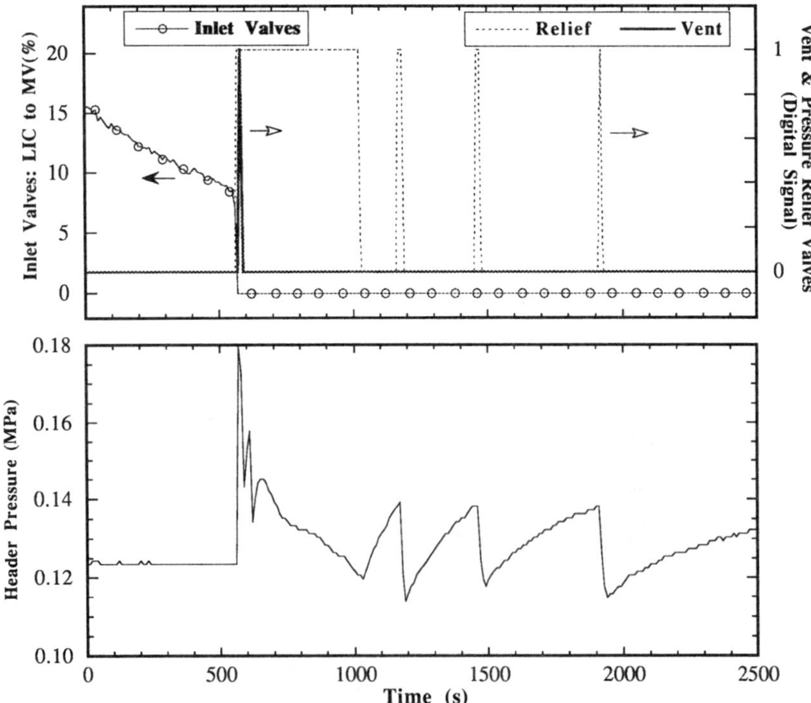

Figure 7. The result of quench sequence program execution during a second cycle experiment. Helical coil quench was initiated while energizing the coil up to a central magnetic field of 2.75 T.

Heat generation rates depend on the transport current to the superconducting coils. Sequence programs were considered based upon the transport current to each superconducting coil, which could distinguish a malfunction of the quench detector. Three kinds of discharge sequence programs were developed for the helical coils, poloidal coils and the superconducting bus-lines.

After receiving a quench signal from the LHD main control room, the integrated controller send discharge sequence signals to the subsystem. Superconducting coils are isolated from helium refrigerator/liquefier and any pressures increase due to discharge is relieved through vent valves and/or pressure relief valves. Otherwise, the helium refrigerat-

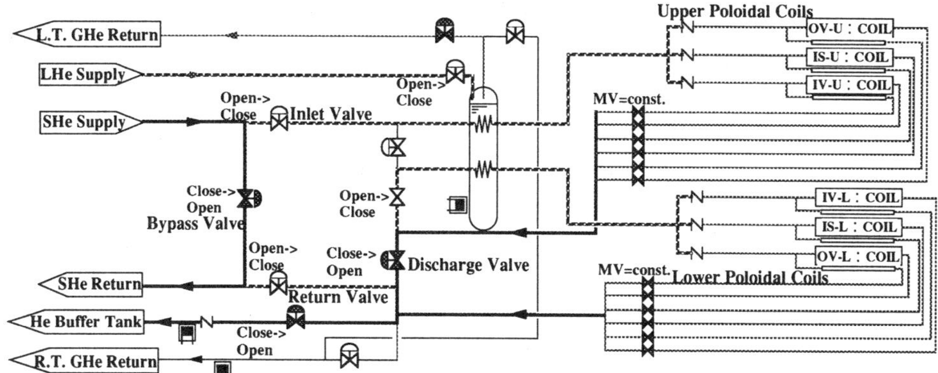

Figure 8. Sequence program for poloidal coils. Since the discharge of turbine #6 and #7 were directly connected to the poloidal coils, a bypass line was used to change the hydraulic flow pass. Therefore, inlet and return valves for the poloidal coil system were closed with a 60 s ramp speed, while a bypass valve is opened simultaneously. Outlets of the upper and lower poloidal coils were connected to the buffer for pressure relief.

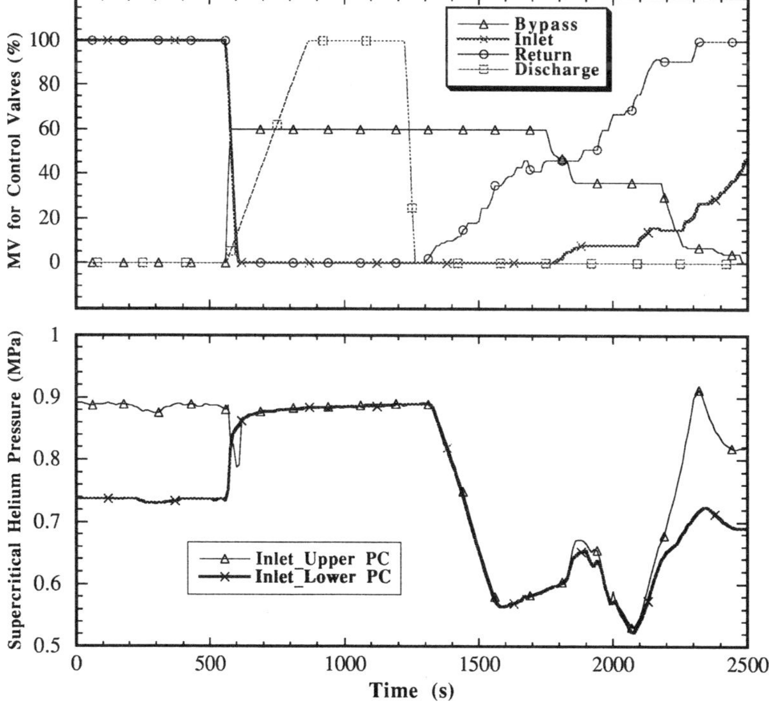

Figure 9. Results of the quench sequence program. Discharge for the poloidal coil did not show a large pressure increase due to AC losses. Turbines #6 & #7 successfully maintained their operating status. Some pressure adjustments was required by the operators.

or operation will fail due to a compressor trip and/or turbine trip. The former situation is primarily caused by the large quantity of liquid helium evaporation from the helical coil. The later situation is associated with supercritical helium pressures increase within the poloidal coils. Figures 6 & 7 show illustrations of the quench-sequence programs and the results for the helical coils, respectively. Similarly, Figures 8 & 9 show the case of poloidal coils.

Power Failure. In the case of power failure, the main compressors of the helium refrigerator fail and a recovery-compressor (20 g/s) takes over the gaseous helium recovery operation. The integrated controller sends a compressor-failure signal to the subsystems and the sequence programs are executed to isolate each subsystem. The subsystem is divided into two pressure regions: low pressure (~0.15 MPa) and high pressure (~1.0 MPa). Any pressure increases within the system are relieved by vent valves, which are controlled with setting pressure values. As a result, the subsystems are now in the self-sustaining condition. Helium gas recovery operation starts as the suction pressure of recovery compressor reaches 0.15 MPa. The SC bus-line system is connected to a room-temperature return line. The poloidal coil and the helical coil system follow as the suction pressure reaches 0.12 MPa. These operations are implemented independently for the each system. These sequence programs were developed after the system experienced several power failures caused by thunder-storm.

CONCLUSION

LHD-TESS demonstrated reliable operating conditions for the cryogenic system. The cooldown process with PCUs and TMS showed satisfactory results with 24 PID loop control (cascade control) valves. However, process optimization with the refrigerator is required to improve the efficiency. Sequence programs for the cooldown process were successfully executed without any disturbance to the refrigerator, which showed the potential for the fully automatic operation. The ECM system ensured secure superconducting coil excitation states, which greatly reduced overall operating loads. The quench sequence programs demonstrated smooth transitions in the hydraulic passes to isolate superconducting coils and superconducting bus-lines with a small transient disturbance to the refrigerator. The "Open System" has a potential to be used for a large scale process control system.

REFERENCES

1. T. Mito, S. Satoh, R. Maekawa, et. al., Development of a cryogenic system for the Large Helical Device, in: "Advances in Cryogenic Engineering, " Vol. 43A, Plenum Press, NY (1998), p. 589.
2. T. Mito, S. Satoh, R. Maekawa, et. al., "Cryogenic Control System for the Large Helical Device," Proceedings of the ICEC16/ ICMC, Vol. 1, Elsevier Science, Tokyo (1997), p. 79.
3. T. Mito, R. Maekawa, S. Yamada, et al., "First Cool-down Performance of the LHD," Presented in Applied Superconductivity Conference 1998 held at Palm Desert CA, USA.

CRYOGENIC SYSTEM FOR KEKB SUPER-CONDUCTING RF CAVITIES

A. Kabe,[1] K. Hara,[1] K. Hosoyama,[1] Y. Kojima,[1] Y. Morita,[1] H. Nakai,[1] T. Fujita,[2] K. Matsumoto,[2] and T. Kanekiyo[2]

[1] High Energy Accelerator Research Organization (KEK), Tsukuba-shi, Ibaraki-ken, 305-0801, Japan

[2] Hitachi Ltd., Kudamatsu-shi, Yamaguchi-ken, 744-0002, Japan

ABSTRACT

KEKB (KEK B-Factory) is a double-ring electron-positron collider used for studies of CP-violation and other topics on decays of B mesons. Because of the high beam current of KEKB, installation of superconducting cavities has been proposed for the KEKB, and 4 niobium single-cell superconducting cavities were installed in the KEKB ring in the summer of 1998. For cooling of these cavities, an existing cryogenic system with a capacity of 8 kW at 4.4 K and helium transfer lines, which were constructed and used for TRISTAN superconducting cavities, were reused. In this system newly developed, high performance, sub-transfer lines cooled with 80 K liquid nitrogen thermal shields were chosen and constructed to connect the cavity cryostats with the main transfer line. The operation of the complete system for the commissioning of KEKB was initiated at the end of 1998. This paper describes the cryogenic system and the operating experience gained during the commissioning.

INTRODUCTION

Construction of KEKB, a high luminosity, double-ring 8 x 3.5 GeV asymmetric electron-positron, high beam current collider in the existing TRISTAN tunnel was started in 1994 as a five year project. Figure 1 shows the layout of the KEKB accelerator ring. Installation of 10 single cell, superconducting, higher-order-modes (HOMs) damped acceleration cavities in the KEKB have been proposed. After intensive R&D studies of these superconducting cavities, 4 single-cell 509 MHz superconducting cavities with a design acceleration field gradient, Eacc, of 6 MV/m (acceleration voltage Vc=1.5 MV) have been constructed and installed in the straight section of the Nikko area of the high energy electron ring during the summer of 1998.

Following the installation of the 4 cryostats, each with a single cell superconducting cavity (SCC), in the KEKB tunnel, these cryostats and the connection boxes of the main transfer line were connected by the sub-transfer lines which were developed for the KEKB cryogenic system. Since these cryostats and connection boxes have valves, thermometers, pressure sensors, liquid helium level meters, etc., programs to control and monitor these devices were added to the control computer. The performance test for the complete system was completed by the end of October 1998 and cooldown of the superconducting cavities was initiated on November 17, 1998.

Advances in Cryogenic Engineering, Volume 45.
Edited by Shu *et al.*, Kluwer Academic / Plenum Publishers, 2000.

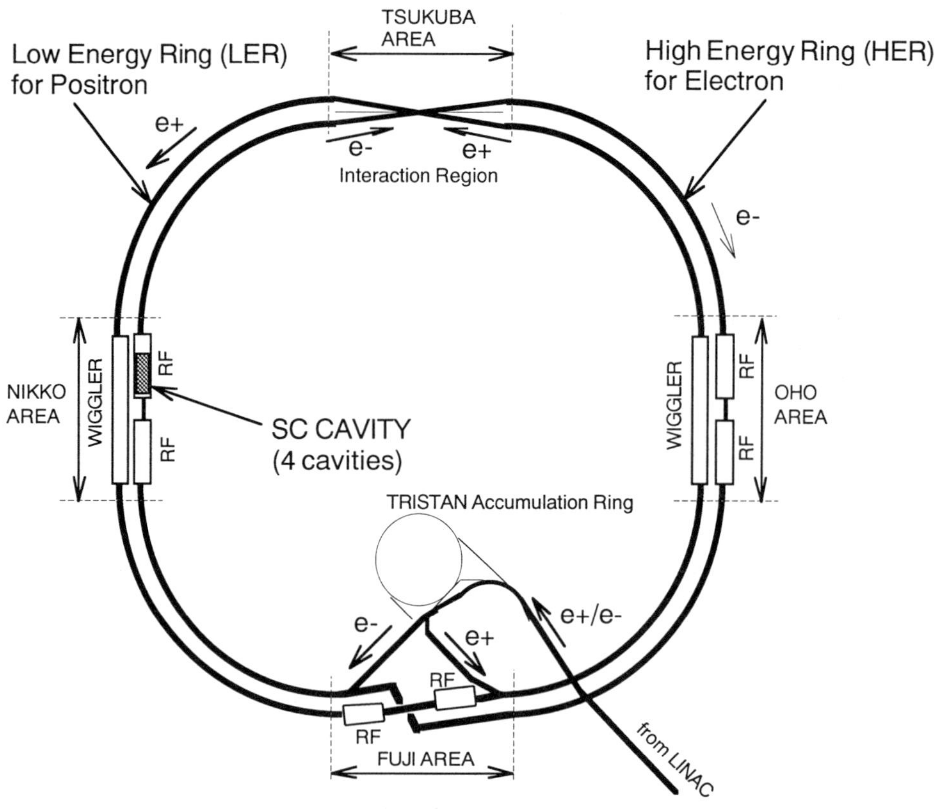

Figure 1. Schematic layout of the KEKB ring.

SUPERCONDUCTING CAVITY

Figure 2 shows the superconducting HOMs damped cavity installed in a cryostat. The spherical cell shape niobium cavity has large aperture beam pipes and RF absorbers fabricated of ferrite attached at the room temperature ends of the beam pipes. An electric heater installed in the cryostat is used to compensate the change of heat load caused by RF loss. The input coupler is cooled by vaporized helium gas. The liquefaction load of the input coupler is about 1 L/hour of liquid helium. The total liquefaction load of one cryomodule with heater is about 150 W.

CRYOGENIC SYSTEM

The flow diagram of the cryogenic system for the KEKB superconducting cavities is shown in Figure 3. This system was designed and constructed for the superconducting cavities for TRISTAN and operated for about 7 years[1]. We have reused this existing cryogenic system for the KEKB superconducting cavities. The cryogenic system consists of helium compressors, refrigerator cold box, liquid helium dewar, transfer lines, four cryostats for the SCC, liquid nitrogen circulation system, helium gas recovery system and helium gas tanks.

Compressor

An oil flooded two-stage screw type compressor system was selected to assure

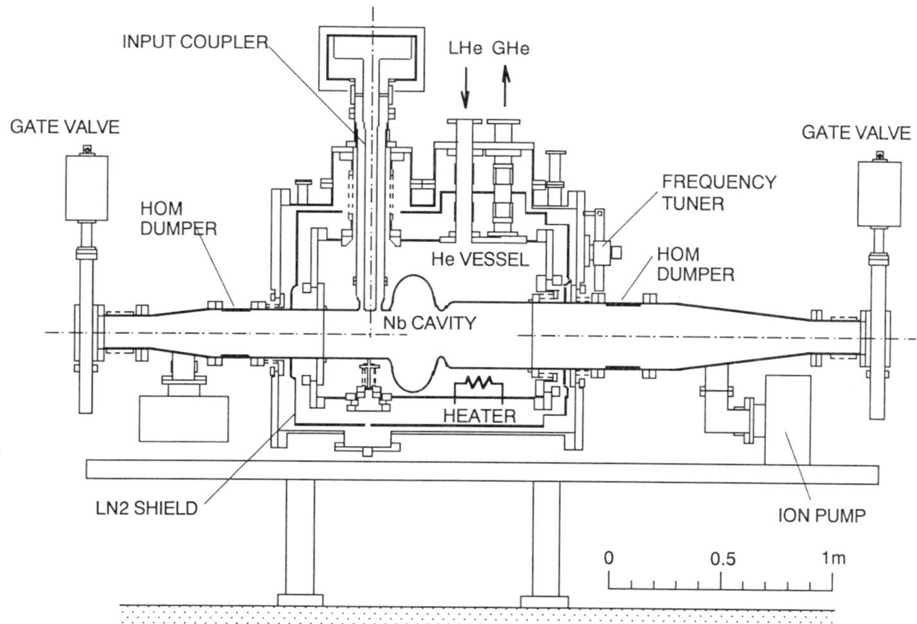

Figure 2. Superconducting cavity and associated cryostat.

reliability for long-term operation. Selected parameters of the compressors for the KEKB cryogenic system are listed in Table 1. The compression system consists of two 2-stage compressor units. One unit has three first stage compressors with a rotor diameter of 320 mm (MYCOM 320L) and one second stage compressor (MYCOM 320S). The second unit has one first stage compressor with a rotor diameter of 250 mm (MYCOM 250L) and one second stage compressor (MYCOM 250S). This compression system has a maximum flow rate of 12000 Nm^3/h. Total power of electric motors is 2.58 MW.

Refrigerator Cold Box

The refrigerator cold box consists of a main cold box and a sub cold box. Both are vertical-type vessels. The main cold box is 4 m in diameter and 6 m in height. The sub cold box is 2 m in diameter and 3 m in height. The refrigerator cold box has 6 heat exchangers, 10 control valves and 5 gas-bearing turbo-expanders. An 80 K charcoal filter is in the system to remove air impurities.

Table 1. Parameters for the compressors

	1st stage C1,C2,C3	2nd stage C4	1st stage C5	2nd stage C6
Number of compressor	3	1	1	1
Flow rate (Nm^3/h)	10,820	10,820	1397	3738
Inlet pressure (kPa)	100	517	100	400
Outlet pressure (MPa)	0.517	1.90	0.4	1.97
Rotor diameter (mm)	320	320	250	250
Rotor length (mm)	530	355	352	280
Rotor speed (rpm)	2950	2950	2950	2950
Isothermal efficiency(%)	49.9	51.0	47.7	46.5
Input power (kW)	373x3	913	127	423
Type	oil flooded screw compressor (Mayekawa)			
	320L	320S	250L	250S

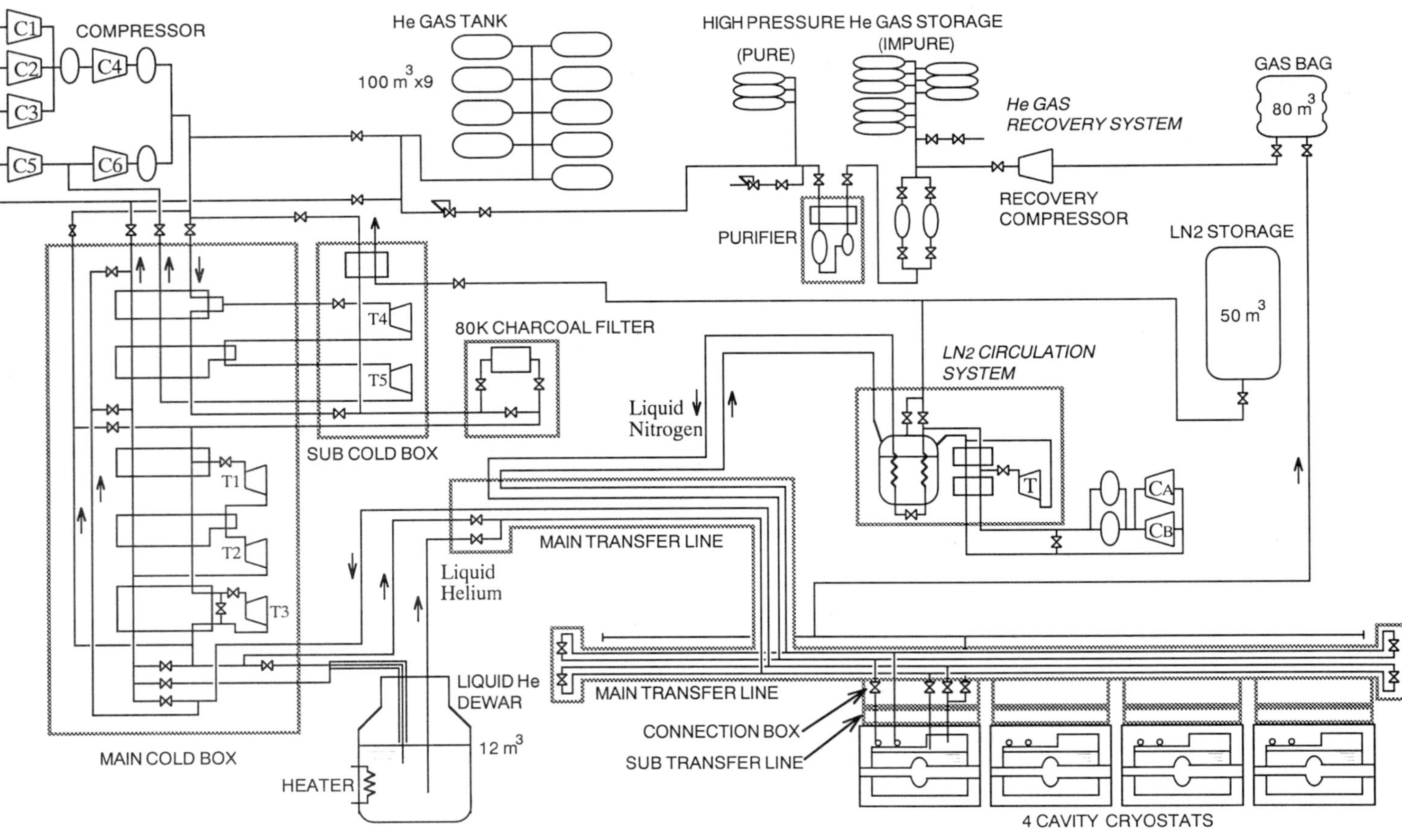

Figure 3. Flow diagram of the cryogenic system.

Table 2. Parameters of the turbo-expanders

	T1	T2	T3	T4	T5
Flow rate (g/s)	135.7	135.7	135.7	89.6	89.6
Inlet pressure (MPa)	1.58	0.57	1.60	1.60	1.0
Outlet pressure (kPa)	570	120	400	1000	400
Inlet temperature (K)	50.0	25.0	8.0	150.0	95.0
Isentropic efficiency (%)	75	75	60	60	55
Output power (W)	8,774	6,099	3,020	7,270	7,571
Rotor diameter (mm)	35	55	35	35	35
Rotor speed (rpm)	140,000	74,000	50,000	150,000	150,000
Fabricator			Hitachi		

In the main cold box, there are five temperature stages of heat exchangers (aluminum-alloy, plate-fin type). The sub cold box has one heat exchanger operating between 300 and 80 K. The five turbo-expanders in the cold box have static gas-bearings. Parameters for these turbo-expanders are summarized in Table 2. Helium liquefied by the cold box is stored in a dewar with a volume of 12 m^3, and is transferred to the cavity cryostats through the transfer lines.

Transfer Line

The existing transfer line and connection boxes which were designed for the TRISTAN SCC were used in the cryogenic system. The location of the connection boxes for the main transfer line were not optimized for the KEKB SCC. Long sub-transfer lines were required to connect the cavity cryostats. New high performance sub-transfer lines were developed to reduce the heat loss. Figure 4 shows the cross section of the sub-transfer lines. The helium lines of the sub-transfer line are protected by a 80 K liquid nitrogen shield to reduce the heat loss. The sub-transfer line has a bayonet type joint at its mid-point to provide easier handling capabilities. The estimated heat losses of the sub-transfer flow and return lines, including bayonet joint parts, are 3.5 W and 4.2 W respectively. Details of the sub-transfer lines are described elsewhere[2]. The connection box contains 4 control valves, and these valves control the cool-down speed and liquid helium level of the cryostat.

Liquid Nitrogen Circulation System

A liquid nitrogen circulation system was installed to supply liquid nitrogen (LN_2) to

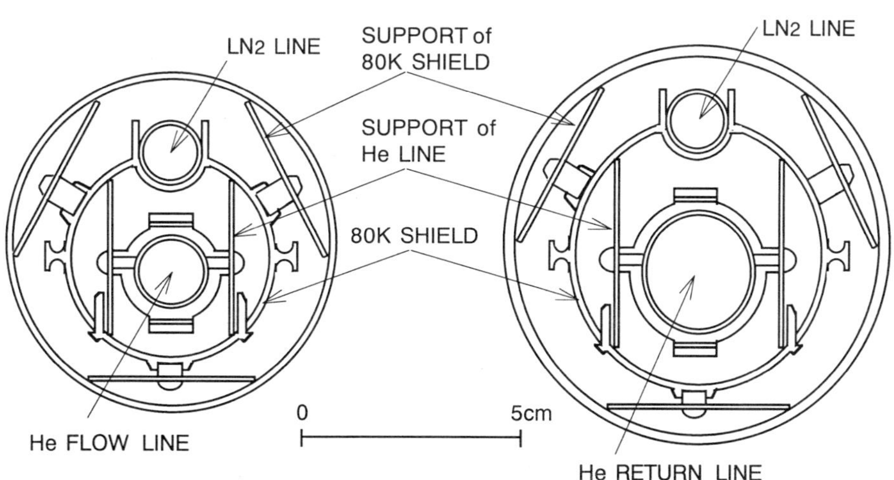

Figure 4. Cross section of the sub-transfer lines

the transfer lines and the cavity cryostats. This system has two modes of operation. One is "direct mode," the other is "circulation mode." In the direct mode, liquid nitrogen is sent directly to the cryogenic system from LN_2 storage through the LN_2 dewar which contains heat exchangers. In the circulation mode, vaporized nitrogen gas is compressed by a screw compressor and sent to the LN_2 cold box. The nitrogen gas is cooled by the cold return gas and is partially liquefied in a JT expansion valve and sent to the cavity cryostats as subcooled nitrogen. Liquid nitrogen is supplied from the LN_2 storage tank and a turbo expander is used to reduce LN_2 consumption.

CONTROL SYSTEM

The control system consists of 6 multi-controllers (MLC), 2 process operator consoles (POC), 2 printers and 2 color hard-copy units, illustrated in Figure 5. These control units are connected via a duplex twisted-pair data link named "data highway (DHW)". Each MLC controls a component of the cryogenic system. For example, MLC1 controls the refrigerator, MLC2 controls the compressors and helium recovery system, etc. MLC has a 16-bit micro-processor and a 256 kbyte memory which is protected by a battery. The MLC performs loop control, sequence control, etc.

The total number of input/output signals is 1120. I/O numbers of each MLC are shown in Table 3. Data of the analog signals are stored on a hard disk of the POC every 1 minute or 4 minutes, and are kept for 2 days. In order to keep these data, we transfer them to floppy disks. Temperature, pressure and level are controlled by a closed loop PID control. There are 63 PID control loops. For the safety of the cryogenic system, interlock controls are performed by the MLCs. The causes for interlock are many, including upper limit of the pressure and the level in the liquid helium vessel of the cavity cryostat, temperature difference in the cryostat, etc.

POC is a user interface for operating the cryogenic system. The control system has two POCs, one has 3 displays and 3 operation keyboards as shown in Figure 6, the other has 2 displays. An operator can monitor and operate the cryogenic system using the performances of POC which are trend graphs, instrument display, and graphical display of the cryogenic system, etc.

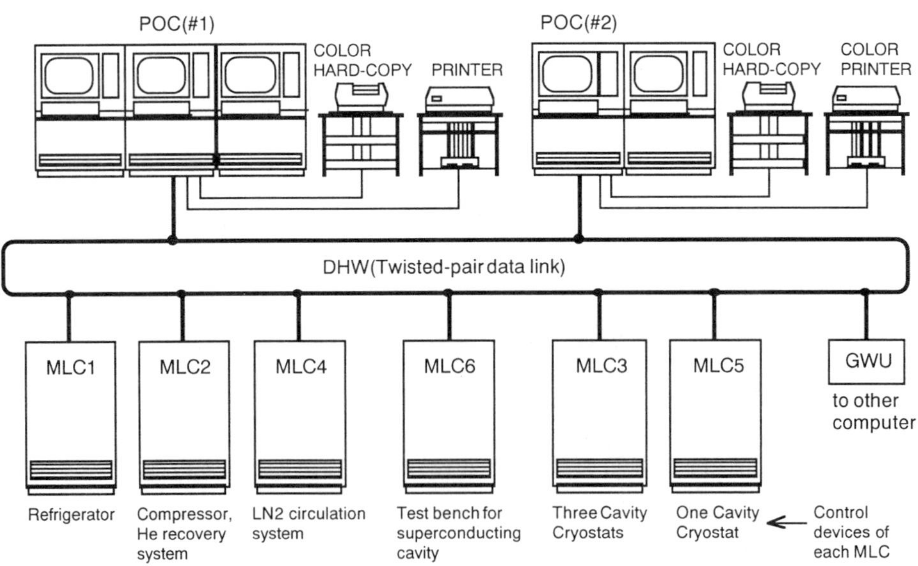

Figure 5. Configuration of the control system.

Table 3. Input/Output signals of the cryogenic system

Analog input	:	462 points
Analog output	:	99 points
Digital input	:	428 points
Digital output	:	131 points

Figure 6. View of the console of a process operator.

COOLDOWN OF THE SUPERCONDUCTING CAVITIES

The helium refrigerator liquefies about 7 m^3 of liquid helium in the 12 m^3 dewar before beginning of cooldown of the SCC. During the first stage of cooldown, cold helium gas, pre-cooled by liquid nitrogen, is used for the cooling. The cool-down rate is limited to about 10 K/hour. After the temperature of the SCC attains a temperature of about 120 K, the turbo-expanders are started and the SCC is cooled down to about 40 K. Then the liquid helium in the dewar is transferred to the SCC cryostats.

Figure 7 shows an example of the cool-down path for the cryogenic system and the superconducting cavity. It requires about 65 hours to cool down the cryogenic system and the 4 cavities from room temperature to filling the 4 cavity cryostats with liquid helium.

The output power of turbo-expander, heater in the dewar, and heaters in the cryostats are controlled suitably, to handle a small load of 4 cryomodules.

SUMMARY

The cryogenic system for the four single cell superconducting cavities for the KEKB ring has been fabricated. Before the commissioning of the KEKB, a performance test of the

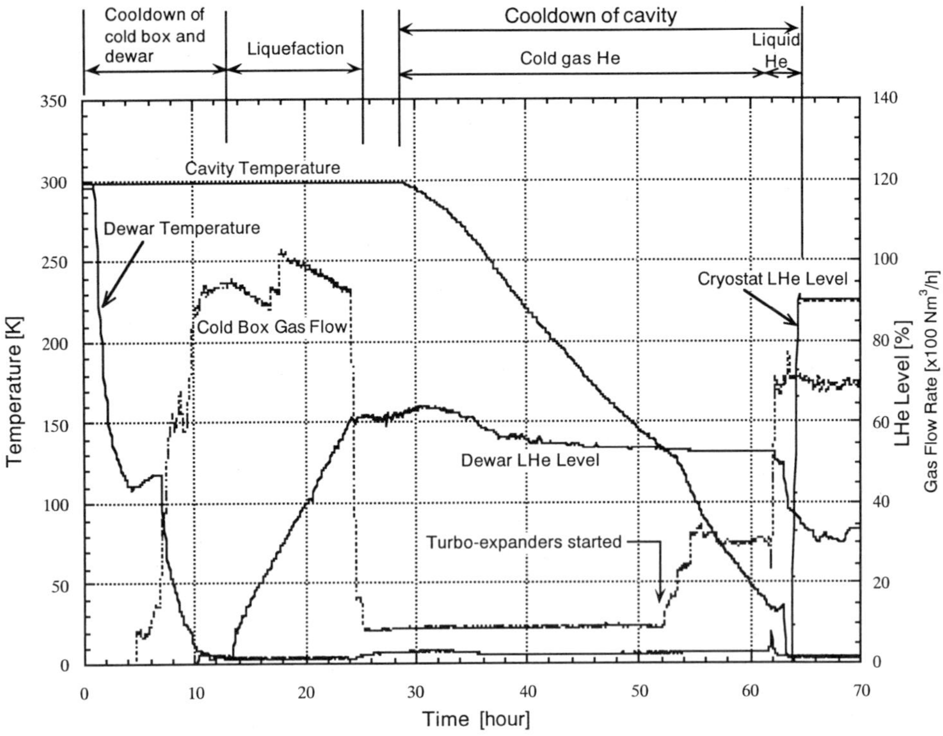

Figure 7. A typical example of the cooldown for the cryogenic system and superconducting cavity.

cryogenic system was held in October of 1998. In this test, the hardware of the cryogenic system and the software of the control computer were successfully tested with the cooldown of the four superconducting cavities.

Cooldown for commissioning of the KEKB began in November 1998, and the superconducting cavities have been in service from December 1998. Three warm-up and cool-down cycles were experienced because of required maintenance of the KEKB ring. The cryogenic system is now in full operation with a total operating time of the system at present of about 160 days.

ACKNOWLEDGMENTS

We would like to express our gratitude to Prof. S. Kurokawa for his continuous support and encouragement in this work and to S. Mitsunobu, T. Furuya and T. Tajima for their contributions as the superconducting cavity group. We would also like to thank the employees of Hitachi Techno Engineering Co., Ltd. for their operation of the cryogenic system.

REFERENCES

1. K. Hosoyama et al., Cryogenic system for TRISTAN superconducting RF cavities: Upgrading and present status, in: "Advances in Cryogenic Engineering", Vol.37, Plenum Press, New York (1992), p.683.
2. K. Hosoyama et al., Development of high performance transfer line system, presented at the CEC/ICMC 1999 meeting, Montreal, Canada.

TARGET COOLING SYSTEM FOR THE LASER BEAM MEGAJOULE FACILITY

D.Chatain[1], J.P. Perin[1], and D.Desenne[2]

[1]Département de Recherche Fondamentale sur la Matière Condensée
CEA/Grenoble. 17 rue des martyrs 38054 Grenoble cedex 9, France
[2]Département des Lasers de Puissance
CEA/Limeil-Valenton 94195 Villeneuve-Saint-Georges cedex, France

ABSTRACT

The cryogenic targets of the Laser Megajoule facility (LMJ) are hollow spheres. Their internal walls are covered with a solid layer of frozen deuterium-tritium (D-T). One issue of inertial confinement fusion experiments is to guarantee the quality of the geometry of fuel layer. Cryogenic targets must be cooled at a temperature near the triple point (19K) with a very good stability (0.2mK) for many hours. This period is used to position the target with an accuracy of ±5μm at the center of the experimental vacuum vessel where the 240 laser beams are focalized. A complex cryogenic infrastructure has been conceived to insure the continuity of the cryogenic chain from the filling station located at CEA/Valduc in Burgundy to the LMJ experimental chamber installed in the vicinity of Bordeaux. The design of the target and a detailed description of the infrastructure are presented. A first prototype of cryogenic grip has been fabricated and characterized. Some experimental results are given.

THE CRYOGENIC LMJ TARGET DESIGN

The cryogenic targets of the Laser Megajoule facility are two millimeter diameter hollow spheres in which the internal wall is covered with a solid layer of frozen deuterium and tritium (Figure 1). The thickness of this cryogenic layer is 0.2mm. Efficient energy production by laser D-T fusion depends on the quality of the geometry of the fuel layer. This layer must be uniform to a few percent both in sphericity and shell thickness to avoid the development of hydro-dynamical instabilities during the implosion. The sphere is held in a gold cylinder (the hohlraum) by a polymer membrane. It contains a mixture of hydrogen and helium which allows the cooling by conduction of the sphere. The 240 laser beams penetrate the hohlraum through the polymer windows.

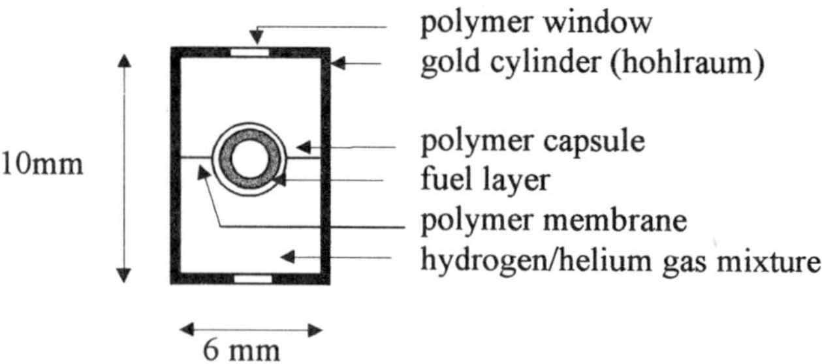

Figure 1. Design of the cryogenic LMJ target

As shown in Figure 2, the target is connected to the cold source by four pure aluminum or sapphire rods. These rods provide cooling of the target by conduction. With the aid of these rods the target can be cooled in a vacuum.

THE CRYOGENIC INFRASTRUCTURE (Figure 3)

The targets will be filled at the CEA/Valduc in Burgundy. They will then be transported in a cryogenic carrier to Bordeaux where they will be positioned in the center of the vacuum chamber of the 240 beams Laser Megajoule facility. Manipulation of the target must be performed at a cryogenic temperature of 19K. This requirement was met by the development of a complex cryogenic infrastructure that is described below.

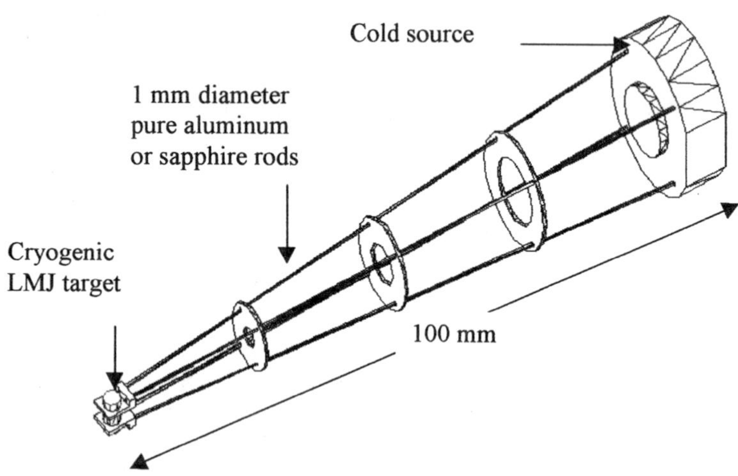

Figure 2 . Schematic showing the connection between the LMJ ignition target and the cold source.

Filling Station

To obtain a solid layer 0.2 mm in thickness, a plastic capsule is filled by permeation at about 350K and a pressure of about 1500 bars. At room temperature the microballoon can not support a pressure of 1500 bars. Thus, when this pressure is reached, the temperature of the filling box is decreased to just above the triple point of the D-T (19K) which decreases the pressure in the target to approximately 1 bar.

Cryogenic Carrier

The cryogenic carrier must permit the transport of one or several targets at cryogenic conditions from Valduc to Bordeaux a distance of about 700 km. A cryocooler or a helium bath with a 48 hour hold time provides the necessary cooling.

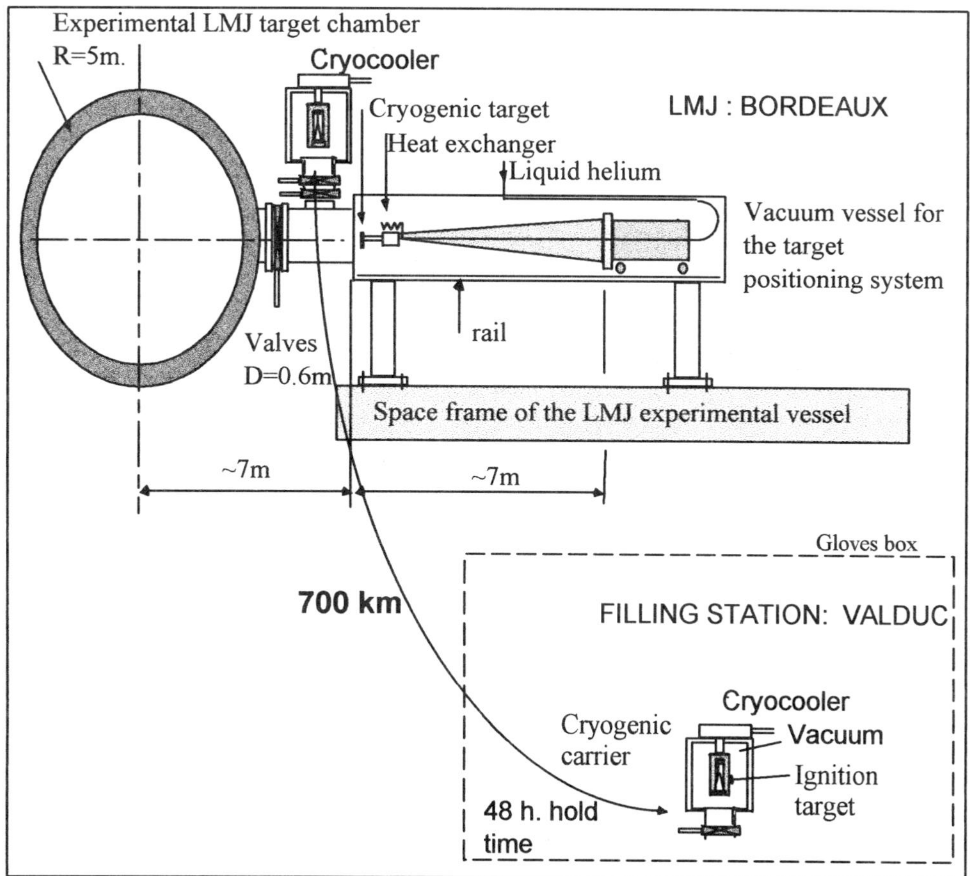

Figure 3 . General scheme of the cryogenic infrastructure.

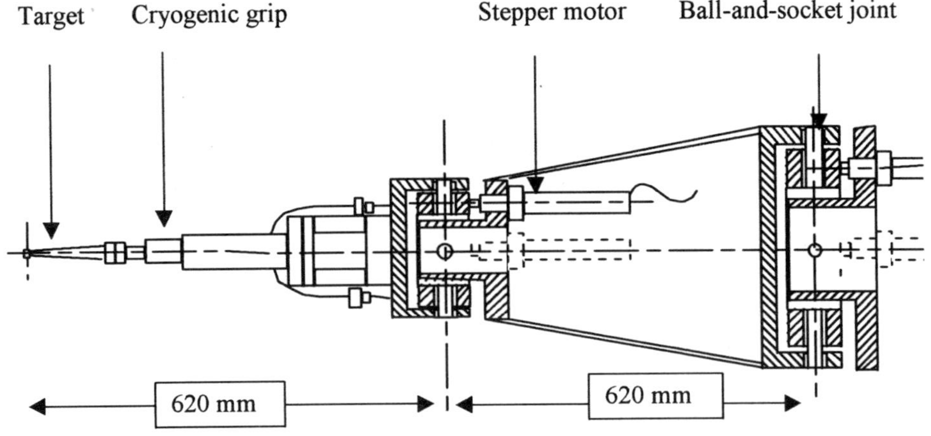

Figure 4. Schematic of the target positioning system extremity.

Target Positioning System

The cryogenic carrier from Valduc is connected to the vacuum vessel of the target positioning system. One target will be extracted with a cryogenic device and then connected to the cryogenic grip of the target positioning system. The target positioning system includes several parts. The major item is a conical carbon fiber tube, 7m in length, that is moved on rails in the vacuum vessel with a cog belt driven by a stepper motor.

As shown on Figure 4, the end of this tube is equipped with a cryogenic grip held on a pantograph driven by four stepper motors. This system must provide a positioning accuracy better than ± 5μm at the focus of the laser beam. The target must be maintained at a steady temperature during the period preceding the shot.

CRYOGENIC GRIP

The target must be moved from the filling box to the cryogenic carrier and then to the cryogenic positioning system. During each transfer, although the target can receive some thermal radiance from the 300K vessel, the temperature of the target must stay at about 19K. To verify that this can be done, tests on a cryogenic grip have been performed. This grip must have several functions; namely, 1) seize the target, 2) move the target, 3) keep the target cool. Figure 5 shows the design of the grip. It consists of a split cylinder that can expand because of a moving cone and a heat exchanger cooled by a flow of cold helium gas. By heating the target mock-up and by measuring the temperature difference between the target and the expansible grip, it is possible to evaluate the thermal contact resistance between the target and the grip.

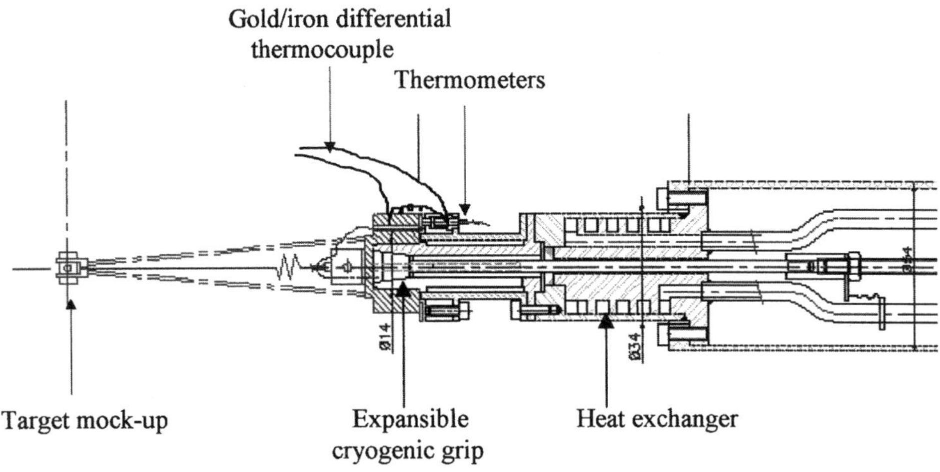

Figure 5 . Details of the expansible cryogenic grip.

EXPERIMENTAL RESULTS

The cryogenic grip was tested in a specially designed cryostat. The different parts of the cryostat are shown in Figure 6. In the same cryostat we have simulated the cryogenic carrier by a cold cylinder in which the target can be locked by a finger activated by a spring.

Case of the Target Cooled by the Grip

As shown in Figure 7 the thermal resistance of the contact between the grip and the target is large (about 100K/W). This is because we have used a copper/nickel alloy for the grip. This material was chosen for its elasticity properties. However, the thermal contact resistance must be improved because in some cases the target will be subjected to a power of about 150 mW. Another design based on an aluminum/copper contact will soon be tested. We expect to decrease the thermal contact resistance of the contact by a factor of ten.

Case of the Target Cooled by the Cryogenic Carrier

When the target was cooled by the cryogenic carrier, the thermal contact resistance measured was very low. The temperature difference between the target and the cryogenic carrier has been less than 1K when a power of 100 mW were dissipated in the target. This is because the cryogenic carrier is made from gold-plated copper and the contact force applied by the spring of the finger is 25 daN (Figure 7).

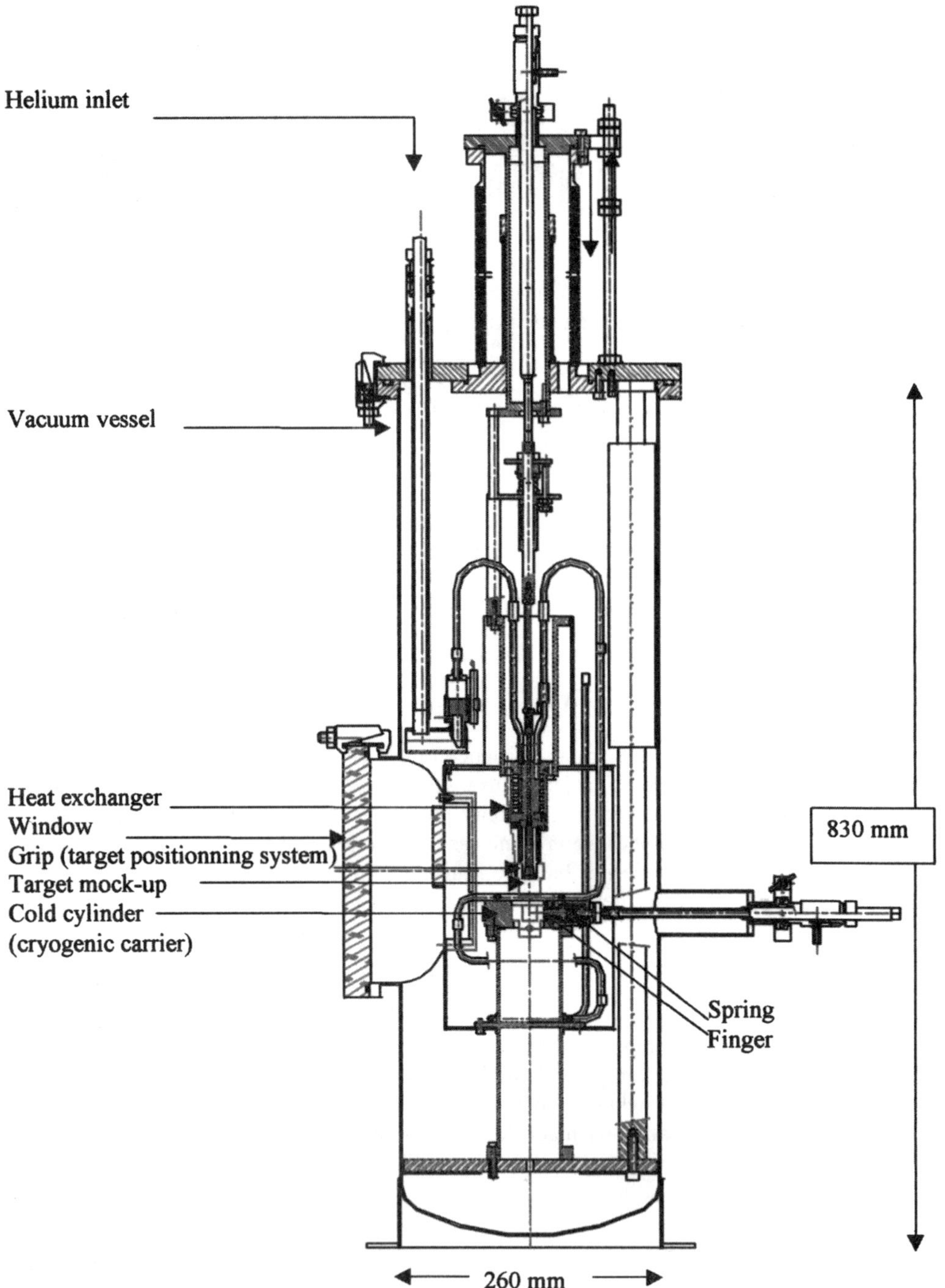

Figure 6. Schematic of the experimental cryostat.

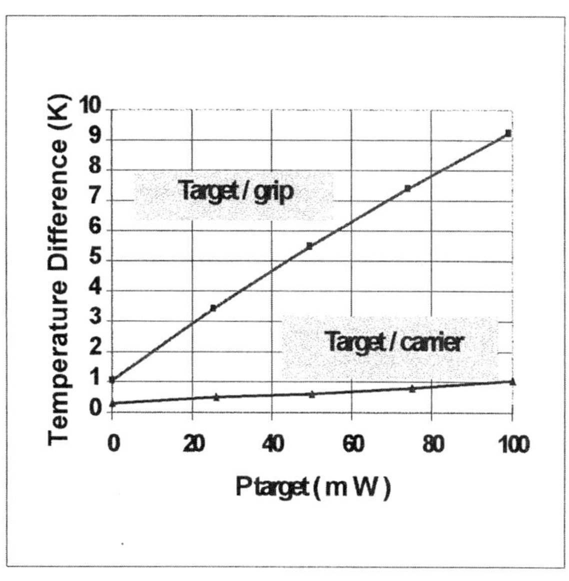

Figure 7. Experimental results: temperature difference between the target and the grip and the target and carrier versus the power dissipated in the target.

Temperature Stability

Extensive work has been done to obtain a good temperature stability for the target. By using germanium sensors, a phase sensitive detector, and a special helium transfer device, a temperature stability with a ±1 mK margin has been obtained on the target (Figure 8).

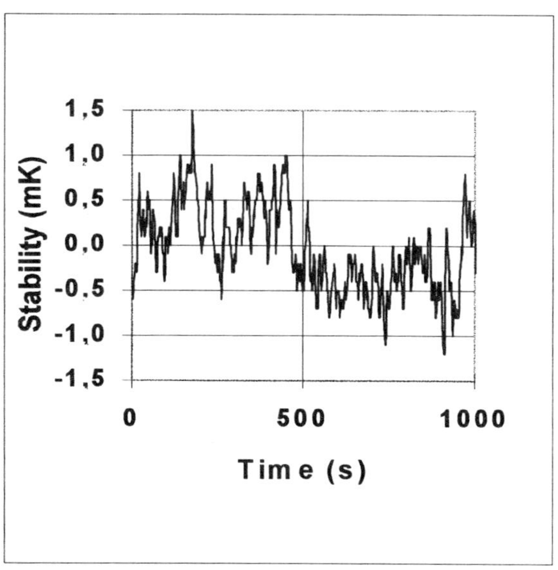

Figure 8. Temperature stability of the target mock-up at 19K.

CONCLUSIONS

The Laser Megajoule cryogenic facilities have been designed with care for detail. The design covers the filling station which will be installed at CEA/Valduc in Burgundy, the cryogenic carrier which will allow transporting of one or several frozen D-T targets from CEA/Valduc to CEA/Bordeaux, and the target positioning system which will position the target at the center of the vacuum chamber. A prototype of the cryogenic grip has been built and characterized in terms of thermal contact resistance and temperature stability. The results obtained provide confidence that the facilities will perform as required.

ACKNOWLEDGMENT

The authors are grateful to L.Guillemet and G. Rey Giraud for their technical support and the support by CEA/DAM's LMJ program.

REFRIGERATION OPTIONS FOR THE ADVANCED LIGHT SOURCE SUPERBEND DIPOLE MAGNETS

M. A. Green[1], E. H. Hoyer[1], R. D. Schlueter[1], C. E. Taylor[1], S. T. Wang[2] and J. Zbasnik[1]

1. Lawrence Berkeley National Laboratory
University of California
Berkeley, CA 94720

2. Wang NMR Incorporated
Livermore CA 94550

ABSTRACT

The 1.9 GeV Advance Light Source (ALS) at the Lawrence Berkeley National Laboratory (LBNL) produces photons with a critical energy of about 3.1 keV at each of its thirty-six 1.3 T gradient bending magnets. It is proposed that at three locations around the ring the conventional gradient bending magnets be replaced with superconducting bending magnets with a maximum field of 5.6 T. At the point where the photons are extracted, their critical energy will be about 12 keV. In the beam lines where the SuperBend superconducting magnets are installed, the X ray brightness at 20 keV will be increased over two orders of magnitude. This report describes three different refrigeration options for cooling the three SuperBend dipoles. The cooling options include: 1) liquid helium and liquid nitrogen cryogen cooling using stored liquids, 2) a central helium refrigerator (capacity 70 to 100 W) cooling all of the SuperBend magnets, 3) a Gifford McMahon (GM) cryocooler on each of the dipoles. This paper describes the technical and economic reasons for selecting a small GM cryocooler as the method for cooling the SuperBend dipoles on the LBNL Advanced Light Source.

INTRODUCTION

The ALS at LBNL is a national user facility for vacuum ultraviolet (6 eV to 6 keV) and soft x rays (>6 keV) synchrotron radiation. The 1.9 GeV ALS electron storage ring has a circumference of about 200 meters. The ALS ring consists of twelve cells each with three combined function C shaped 1.3 T bending magnets (with the return flux leg pointing to the inside of the ring) about one meter long. Each dipole generates photons with a critical energy of 3.1 keV when the electron beam energy is 1.9 GeV. These photons can be delivered to users through forty-eight ports around the ring to users outside of the synchrotron shielding. Each ring cell contains two defocusing quadrupoles, a long straight section with two pairs of quadrupoles (for insertion devices and RF), and four correction sextupoles. The conventional ALS electron storage ring has a periodicity of twelve. The twelve cell lattice configuration and the relatively low energy of the electron beam both contribute to the small beam size in the machine. The small electron beam size results in the creation of bright photon beams for current ALS users in the energy range from 1 keV to 5 keV.

Some years ago, a study was commissioned on ways to increase the photon energy from the ALS[1]. One approach that was suggested was to increase the energy of the photons from the ring by increasing the bending field in the ring dipoles. (The critical energy of the synchrotron radiation photons generated by a storage ring dipole is proportional to the dipole induction and the electron energy squared.) The photon energy in selected ports can be increased by increasing the dipole induction in selected places around the ring. A portion of the ring dipoles can be replaced with superconducting dipoles with minimum disturbance to the ring lattice and the other magnets in the ring. A number of test SuperBend test coils were built and tested. The fourth SuperBend coil was successful in that it did not train and could be charged to the its full design field in less than 100 seconds[2,3]. Once a successful test magnet with a suitable magnetic field was made, the SuperBend project could move forward.

Adding superconducting dipoles to the existing ALS lattice will change its periodicity (at least to second order). From a beam dynamics standpoint, the number of dipoles that can be changed is three, four, six or twelve. All of the studies concluded that the center dipole of the three dipoles in the cell was the one that should be changed. The addition of superconducting dipoles to the ALS lattice will increase the beam emittance due to the rebound effect of more energetic photons being generated by the higher field dipole. Since ALS has a dedicated user community in the 1 keV to 5 keV range, it was essential that the increase in emmittance be minimized. As a result, the higher field superconducting dipoles are put in only three locations around the ring.

It has been clear from the beginning that the superconducting dipoles must fit in the existing ALS ring with minimum disturbance to the users and existing ring components. The SuperBend superconducting dipoles must have the following characteristics 1) The dipole must provide 1.106 T m of bending when the stored electron energy is 1.9 GeV. (The beam must be bent 10 degrees.) 2) The dipole must be a C shaped dipole with the opening of the C facing away from the ring center. The magnetic field must be generated in a warm bore that fits around the existing ALS vacuum chamber. 3) The magnetic induction in the region where the extracted x rays are produced should be greater than 5 T so that the photons will have a critical energy near 12 keV 4) The superconducting dipole must increase its magnetic field as the ALS main ring beam energy increases from 1.5 to 1.9 GeV. The charge time from 1.5 GeV to 1.9 GeV must be the same as for the conventional magnet system, 100 seconds. 5) The superconducting dipoles must be powered at all times, so that their field can be adjusted along with the rest of the ring magnets. Persistent operation is feasible but not desirable. 6) Superconducting dipole quenches must be avoided during normal machine operation, in order to retain high levels of ALS availability. A power failure must not cause the superconducting magnets to quench.

REFRIGERATION REQUIREMENTS FOR THE SUPERBEND DIPOLES

The superconducting dipole has a C shaped iron yoke that is 380 mm long, in the direction of the beam. The height of the iron yoke is 820 mm; the width of the iron yoke is 700 mm. The gap between the poles on the yoke is 235 mm. Within this gap is a pair of race track shaped superconducting coils that are 67 mm thick. Between the race track shaped superconducting coils, there is a 100 mm gap that contains superinsulation, gap shields, the cryostat vacuum chamber walls, the ALS vacuum chamber and clearance for assembly and adjustment. The magnet system cold mass is about 1680 kg.

The cryostat is 1474 mm high, 902 mm wide and 648 mm thick in the beam direction. The center line of the magnet gap is 506 mm from the bottom of the magnet cryostat vessel. The warm gap in the SuperBend dipole is 54.9 mm. The ALS vacuum chamber thickness is 51.6 mm, which gives a total clearance of $\pm$1.65 mm between the SuperBend magnet cryostat opening and the ALS vacuum chamber. Within the cryostat envelope the magnet cold mass, a liquid helium vessel, a liquid nitrogen vessel, and the cold parts of a 4 K refrigeration system must reside.

The SuperBend dipole magnet cooling requirements are driven by the following considerations: 1) The maximum operating temperature for the magnets is 4.5 K. 2) The heat load in the 4 K part of the cryostat is less than 1 watt. 3) The magnet cold mass is about 1680 kg. Liquid cryogens can be used to cool the magnet system cold mass. 4) The superconducting coils are indirectly cooled. They are cooled by conduction from the iron yoke and through a copper strap from the helium vessel. The yoke is in turn cooled by conduction from a tank filled with liquid helium. 5) The superconducting magnet coils will be powered at all times. The electrical leads from 300 K to 4 K will be a factor in defining the refrigeration needed for the SuperBend magnet system. 6) The maximum magnet charge rate is 3 A per second, which means that the magnet is charged to its full design field in 100 seconds. 7) The magnet is cycled between 4.4 T and 5.6 T at about 0.06 T per second for injection and acceleration. Then the dipole field stays constant at 5.6 T for nearly four hours. 8) The cooling system must be very reliable, since the SuperBend magnets are integral an part of the ALS storage ring. 9) Cryogenic liquid cooling as a back up for mechanical refrigeration is acceptable. 10) Magnet cooling must not be affected by power shutdowns of an hour or less.

Table 1 Estimated Parasitic Heat Loads for a SuperBend Magnet

Heat Load Source	Heat Load Magnitude
Parasitic Heat Loads at 50 to 80 K from 300 K	
St. St. Tubes for He Vent, LN Tank Support & Cool Down	3.6 W
Radiation and Conduction through the MLI	6.7 W
Eight Cold Mass Suspension Bands	1.1 W
Instrumentation Wires	0.1 W
Total 50 to 80 K Heat Load from 300 K	11.5 W
Parasitic Heat loads at 4.0 K from a 50 K shield and Intercepts	
Thin Wall St. St. Neck Tubes and Cool Down Tubes	100.0 mW
Eight Cold Mass Suspension Bands	60.0 mW
Radiation and Conduction Through the MLI	30.0 mW
Twenty 0.13 mm Dia. Instrumentation Wires from 300 K	3.1 mW
Total 4.0 K Heat Load from 50 K	193.1 mW
Parasitic Heat loads at 4.0 K from a 80 K shield and Intercepts	
Thin Wall St. St. Neck Tubes and Cool Down Tubes	190.0 mW
Eight Cold Mass Suspension Bands	120.0 mW
Radiation and Conduction Through the MLI	50.0 mW
Twenty 0.13 mm Dia. Instrumentation Wires from 300 K	6.1 mW
Total 4.0 K Heat Load from 80 K	366.1 mW

The estimated parasitic heat loads at 4 K and at an intermediate shield and intercept temperature between 50 and 80 K is presented in Table 1. The parasitic heat leak to the shields and intercepts is of the order of 8 to 10 W. Heat leak due to magnet current leads may be added to the parasitic heat leak at 50 to 80 K. The heat leak into the 4 K region is strongly dependent on the temperature of the shields and intercepts. It is quite reasonable to expect this heat will be smaller than 0.5 W even when the shields operate with liquid nitrogen as a coolant. Care must be taken to ensure that the heat leak at 4 K is below 0.3 to 0.5 W. In all cases, there will be additional heat leaks at 4 K due to the 320 A magnet electrical leads. The type of leads used to carry current into the magnet has a strong effect on the 4 K refrigeration needed to cool the SuperBend magnet system.

COOLING OPTIONS FOR THE SUPERBEND DIPOLES

Three cooling options were investigated for the SuperBend magnet system. In all cases, the liquid helium reservoir is assumed to hold 91 liters and the liquid nitrogen reservoir is assumed to hold 60 liters. The three cooling options studied are as follows: 1) Liquid nitrogen and liquid helium cooling from dewars was considered. The LBNL team looked at cases with high Tc Superconductor (HTS) leads and conventional gas cooled copper leads cooled with the boil off gas from the helium dewar. 2) A 4.4 K central refrigerator was looked at to cool the three magnets. HTS leads and standard gas cooled leads were looked at for this option as well. 3) Individual cryocoolers for each magnet was studied. The type of cryocooler used depends strongly on whether conduction cooled or gas cooled leads are used on the magnets.

Stored liquid cryogen cooled systems were studied with and without a 30 K shield between the liquid nitrogen (80 K) shield and the 4.3 K region. This shield was cooled with a small amount of boil off helium from the 4.3 K region. The addition of a 30 K shield reduces the heat load at 4.3 K by a factor of three, if the magnet is in persistent mode with the leads retracted. Since this is not an option for SuperBend the effect of the shield is not so dramatic. With gas cooled leads, the boil-off is not changed when a 30 K shield is used. When 320 A gas cooled leads are used, the helium boil off is 1.16 liters per hour and the liquid nitrogen boil off is 0.19 liters per hour. HTS leads will reduce the helium boil off to 0.70 liters per hour without a 30 K shield and 0.42 liters per hour with a 30 K shield, but they will increase the nitrogen boil off to 0.82 liters per hour because the heat leak at 80 K increases due to heat leak down the 320 A solid copper electrical leads from 300 K.

The helium dewars must be filled every three to eight days depending on the type of leads and whether or not a gas cooled shield is used between the 80 K shield and the 4.3 K. The nitrogen dewar can be put on an automatic fill system that keeps the nitrogen dewar full. With liquid helium priced at $3.00 per liter and liquid nitrogen priced at $0.25 per liter and a transfer efficiency of 80 percent, the cryogen cost for ten years would be from 500 k$ to 1160 k$ depending on the type of leads used. Added to the ten year operating cost is the cost of the transfer line system, the cost of HTS leads and the ten years cost of the technician that keeps the dewars full. The estimated ten year refrigeration cost for three SuperBend magnets using liquid cryogen ranges from 630 k$ to 1280 k$ depending on the type of leads used. A schematic of a liquid cryogen cooled SuperBend magnet is shown in Figure 1a.

The central refrigeration system studied assumed that a single 70W Claude cycle refrigerator is used to cool all three SuperBend magnets. Cooling at 80 K is assumed to be provided by liquid nitrogen. The refrigeration required per each SuperBend magnet is dependent on the type of leads used on the magnet. For 320 A gas cooled leads, 4.7 W of refrigeration at 4.4 K is required. If HTS leads are used, the per magnet refrigeration requirement goes down to 0.4 to 0.6 W. The cooling needed to cool the three SuperBend magnets using a central refrigerator is dominated by the transfer line heat leak from the refrigerator to the magnets. The minimum length of transfer line to the magnets to and from the refrigerator is about 360 meters (from and back to the refrigerator). Shielded transfer lines in a common vacuum can achieve heat leaks of 0.1 W per running meter at 4.4 K. A single 70 W refrigerator can provide cooling to up to six SuperBend magnets operating with gas cooled leads or more than a dozen SuperBend magnets operating with HTS leads.

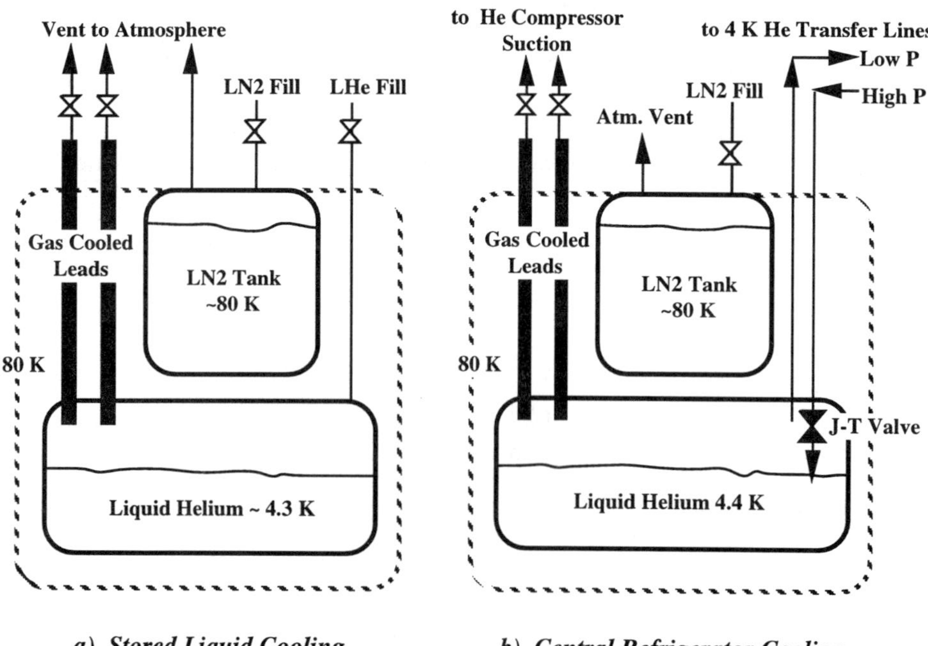

Figure 1 Schematic Representations of Stored Cryogen Cooling and Cooling with a Central Refrigerator for SuperBend Magnets with Gas Cooled Leads

In the event of a refrigerator failure, the magnets can be operated using stored liquid cryogens until the refrigerator is repaired. The transfer line system has to include a provision for precooling the lines before they can be reconnected to the magnets. The capital cost of the refrigerator (based on a PSI Model 1400) plus the around the ring transfer lines is about 450 k$ plus an additional 20 k$ per magnet[4]. The ten year operating cost for liquid nitrogen (at $0.25 per liter) plus electric power at $0.12 per kWh is about $670 k$ plus 20 k$ per magnet. The capital cost plus ten year operating cost for three SuperBend magnets is about 1240 k$. For twelve SuperBend magnets with HTS leads, the capital plus ten year operating cost goes up to about 1600 k$. The capital cost of the refrigeration system plus the ten year cost of cooling cost for six SuperBend magnet with gas cooled leads would be about 1460 k$. A schematic representation of cooling a SuperBend magnet with gas cooled leads using a central refrigerator is shown in Figure 1b.

Small cryocoolers can provide refrigeration to cool the shields (at 50 K) along with refrigeration at 4 K to cool the superconducting magnet. The SuperBend group looked at pure Gifford McMahon (GM) cryocoolers and GM cryocooler that use an additional heat exchanger and a J-T valve to achieve temperatures in 4 K range. The advantages of a GM cooler with a J-T circuit are the cooler can provide cold gas for conventional gas cooled leads and the temperature at the 4 K end of the cryocooler is relatively constant with the applied heat load. Pure GM cryocoolers can not provide cold helium to gas cooled leads and the low end temperature changes with the load. The primary advantages of the pure GM cryocoolers are greater reliability and lower cost. The technology that makes the use of pure GM cryocoolers useful for continuously powered superconducting magnets such as SuperBend is low heat leak HTS current leads between the first and second stages of the cryocooler. Without HTS current leads, continuous current excitation of the SuperBend magnet is not possible when it is cooled using a pure GM cryocooler. When a pure GM cryocooler is used with HTS leads, the 4 K heat leak into the magnets is between 0.3 W and 0.5 W. Because one is using HTS leads, the heat load at 50 K is about 40 W. It is clear that for SuperBend, the cryocooler must be sized for refrigeration at 50 K as well as refrigeration at 4 K. As a result, the SuperBend design team selected a Sumitomo GM 1.5 W (at 4.2 K) cryocooler for the SuperBend test magnet.

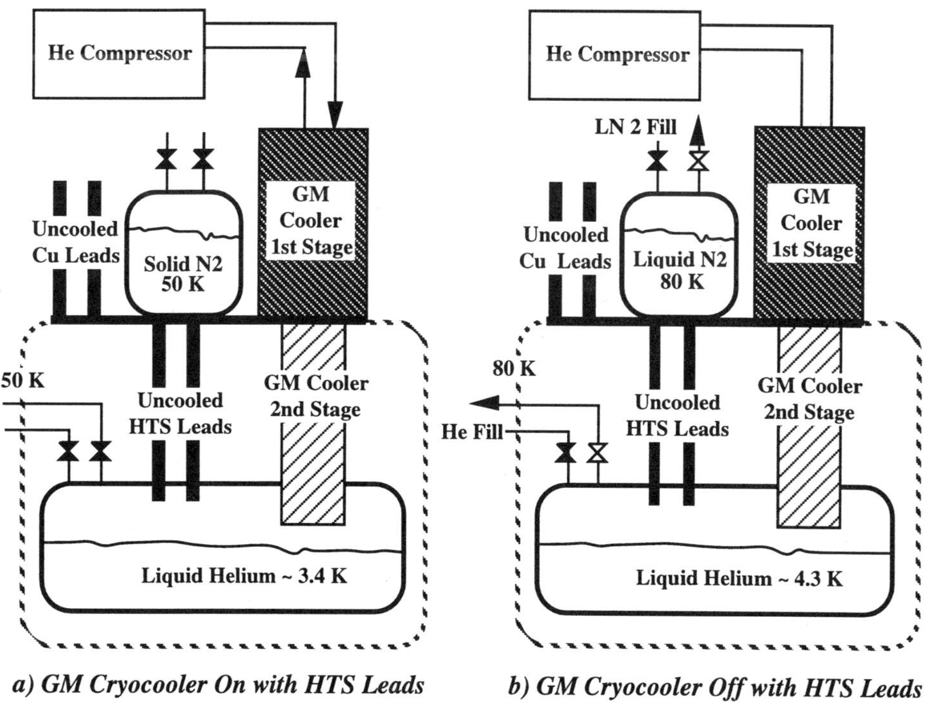

Figure 2 Schematic Representation of Cooling with a Small GM Cryocooler with the Cryocooler On and Off

The cost of HTS current leads and a pure GM cryocooler suitable for cooling the SuperBend magnet is about 50 k$ per magnet. The ten year cost of operation is about 75 k$ per magnet. The projected capital cost plus the cost of operation for ten years for three SuperBend magnets is 375 k$. The use of small cryocoolers to cool SuperBend magnets is clearly justified economically. Figure 2a shows a schematic representation of SuperBend cooling with a small GM cryocooler. The magnet system shown in Figure 2 is powered through HTS leads.

A GM CRYOCOOLER SYSTEM FOR THE SUPERBEND DIPOLES

A GM cryocooler can provide cooling at 4 K to the SuperBend superconducting magnet and 50 K cooling to the shields and intercepts without any stored cryogens. Since reliability is a key issue for the ALS, back-up cooling for SuperBend using stored liquid cryogens has been selected. This means that the second stage of the cryocooler must provide cooling to a tank of liquid helium as well as the cold mass of the magnet. In addition, the first stage of the cryocooler must cool a tank for liquid nitrogen as well as the shields. Figure 2 illustrates cooling with a cryocooler and back-up cooling with liquid cryogens. The additional cryogen storage tanks do not contribute to the heat leaks into the SuperBend magnet, but the stored cryogen tanks do add to the cost and create safety hazards associated with stored cryogens. The added reliability of having stored cryogens as a back-up for the cryocooler is considered to be worth the extra expense. The heat leaks and estimated boil off rates for stored cryogens for the SuperBend magnet cooled with a 1.5 W Sumitomo Cryocooler are shown in Table 2. Figure 3 shows a cross-section view of the SuperBend dipole. Shown in Figure 3 are the superconducting coils, the magnet iron, the two cryogen tanks and the cryocooler.

Table 2 SuperBend Dipole Heat Loads with the Cryocooler On and Heat Loads and Liquid Cryogen Boil Off Rates with the Cryocooler Off

SuperBend Heat Loads with a 1.5 W Sumitomo Cryocooler On	
Parasitic Heat Load at 40 K	11.5 W
Conduction Down the Copper Leads Operating at 300 A	27.0 W
Total Heat Load at 40 K	38.5 W
Available Cooling from the Cryocooler at 50 K	>60.0 W
Parasitic Heat Loads at 3.4 K	193. mW
Conduction Down HTS Leads Operating with 300 A	~250. mW
Total Heat Load at 3.4 K	423. mW
Available Cryocooler Cooling at 3.4 K	~420. mW
Available Cryocooler Cooling at 4.3 K	~1500. mW
SuperBend Heat Loads and Cryogen Boil Off with the Cryocooler Off	
Parasitic Heat Load at 80 K	11.5 W
Conduction Down the Copper Leads Operating at 300 A	27.0 W
Conduction Down Cryocooler First Stage	~50.0 W*
Total Heat Load at 80 K	~87.5 W
Liquid Nitrogen Boil Off	~2.0 l per hr
Parasitic Heat Leak at 4.3 K	366. mW
Conduction Down HTS Leads Operating at 300 A	~400. mW
Conduction Down the Cryocooler Second Stage	~1960. mW*
Total Heat Load at 4.3 K	~2726. mW
Liquid Helium Boil Off	~4.0 l per hr

* Sumitomo estimate for the second stage; LBNL estimate for the first stage.

From Table 2, it appears that the use of liquid cryogens is a viable back-up cooling method for a SuperBend dipole that is cooled with a pure GM cryocooler. A test cryostat has been set up by the SuperBend magnet group to measure the performance of the Sumitomo 1.5 W cryocooler operating at various heat loads[5]. The test cryostat also permits one to test the operation of HTS leads between the first and second stages of the cryocooler. The test cryostat will provide an accurate estimate of the heat load into the magnet vessels when the cryocooler has been shut off.

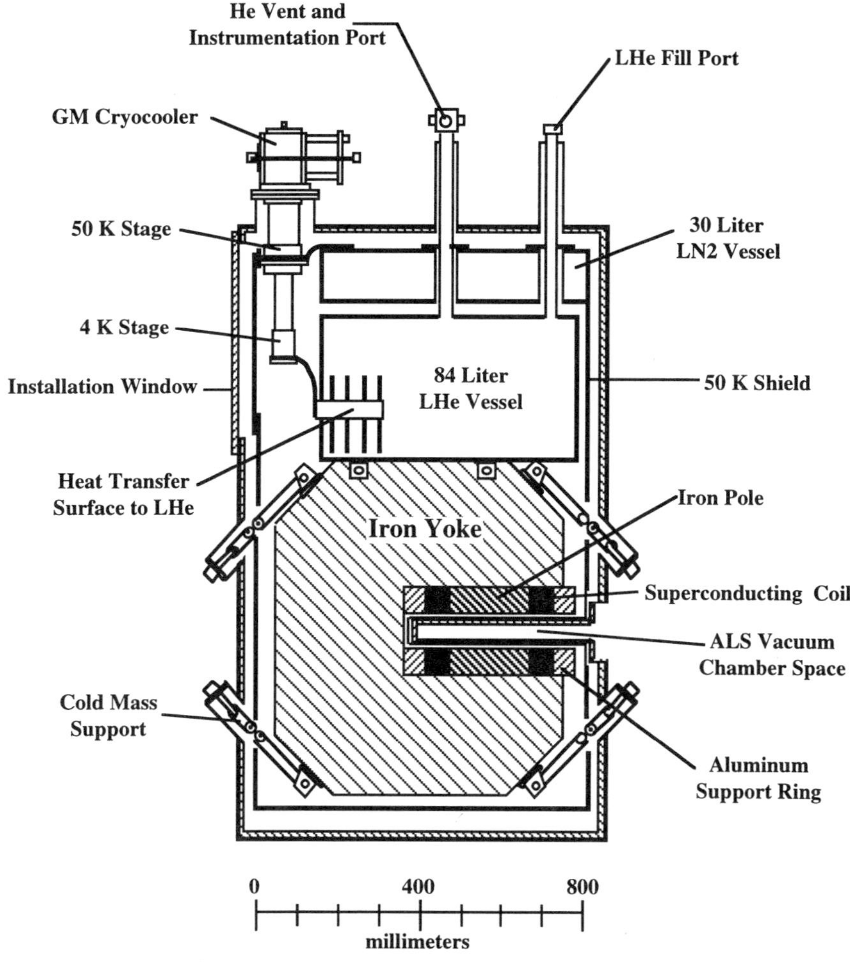

Figure 3 A Cross-section view of the SuperBend Dipole in a Plane Perpendicular to the ALS Beam

Figure 3 does not show the HTS leads. The leads will be mounted on either side of the cryocooler. The lower end of the HTS leads will be electrically insulated from but thermally connected to the liquid helium tank. The upper end of the HTS leads will be electrically insulated from but thermally connected to the liquid nitrogen tank. Niobium titanium leads will connect the lower end of the HTS leads to the superconducting coils. The niobium titanium leads will be cooled by conduction (through an electrical insulator) to the magnet case. The temperature margin for the niobium titanium leads should be at least 3 K

Figure 3 does not show the liquid cryogen cooling tubes for cooling the magnet down from room temperature (300 K) to its normal operating temperature 3.4 K. The Sumitomo cryocooler is capable of cooling down the SuperBend magnet from room 300 K to 3.4 K. The time needed to cool down the magnet to 80 K using the cryocooler alone is about 500 hours. An additional 100 hours is needed to bring the magnet temperature down to 4 K. It is clear that the magnet cool down should be done using liquid nitrogen and liquid helium. A liquid cryogen cooling circuit mounted on copper plates that will encase the iron yoke will allow liquid nitrogen and liquid helium to be used to cool down the SuperBend dipole. The use of liquid cryogens flowing through these tubes should speed up the magnet cool down by at least a factor of twenty.

A number of fault modes were studied. The results of the fault mode study are as follows: 1) The failure of a cryocooler compressor does not affect the operation of the SuperBend dipole. The compressor can be replaced quickly since the plumbing between the compressor and the cryocooler is designed so that the compressor can be disconnected from the cryocooler while the cryocooler heads are cold. 2) Cryocooler head failure is a rare occurrence, liquid cryogen cooling will keep the SuperBend magnets operating until the cryocooler head can be changed during a scheduled shutdown. 3) An electrical power failure will shut off the cryocooler. The magnet is kept cold using stored cryogens. Once power has been restored the cryocooler can be restarted. 4) A magnet quench should not affect the cryocooler, but the magnet will have to be cooled back down using liquid cryogens. The cryocooler can run while this cool down is taking place. 5) A high Tc superconducting lead turning normal should be a rare event that is indicative of some other fault in the system. A lead turning normal requires that the magnet be discharged to protect the HTS lead. Depending on the discharge time required to protect the lead, the magnet may or may not quench during the discharge. Once the magnet is cold, one should be able to repower the magnet.

CONCLUSION

Small GM cryocoolers appear to be the most cost effective cooling system for the three SuperBend dipoles for the ALS. A centralized refrigeration system with transfer lines around the ALS ring does not become competitive until the number of SuperBend dipoles exceeds twelve. GM cryocoolers can provide refrigeration at 50 K and 4 K simultaneously. Extra transfer lines are required for a centralized refrigeration system to do this. The use of high HTS current leads between 50 K and 4 K permits the SuperBend magnets to be powered continuously. The size of the GM cryocooler is dictated by the cooling requirements at 50 K not the cooling requirements at 4 K. The primary heat load at 50 K is the heat leak down the copper 300 A lead from 300 K to 50 K.

In an emergency, the SuperBend magnets can be kept cold using stored liquid helium and liquid nitrogen. The HTS leads must be capable of carrying up to 320 A while their upper end is at 80 K. The use of small GM cryocoolers is justified in terms of reliability as well as cost.

ACKNOWLEDGMENTS

This work was supported by the Director of the Office of Science, United States Department of Energy under contract number DE-AC03-76SF00098.

REFERENCES

1. C. E. Taylor and S. Caspi, "A. 6.3 T Bend Magnet for the Advanced Light Source," IEEE Trans. Magnetics **32**, No 4, (1996)
2. C. E. Taylor, et al, "Test of a High-Field Bend Magnet for the ALS," to be published in IEEE Trans. Appl. Superconductivity **9**, No. 2 (1999)
3. A. Lietzke, "SuperBend 4 Magnetic Measurements," LBNL Superconducting Magnet Group Internal Report SC-MAG 633, September 1998
4. M. A. green and R. Byrns, "An Update on Estimating the Cost of Cryogenic Refrigeration," Advances in Cryogenic Engineering **43**, p 1661, Plenum Press, (1997)
5. J. Zbasnik, et al, "Tests of a GM Cryocooler and High Tc Leads for Use on the ALS SuperBend Magnets," presented at the 1999 Cryogenic Engineering Conference, Montreal Quebec, Canada, 13 to 16 July 1999, this Proceedings Advances in Cryogenic Enginnering **45**, Plenum Press, New York (1999)

A LOW HEAT INLEAK CRYOGENIC STATION FOR TESTING HTS CURRENT LEADS FOR THE LARGE HADRON COLLIDER

A. Ballarino, A. Bézaguet, P. Gomes, L. Metral,
L. Serio and A. Suraci

LHC Division, CERN
1211 Geneva 23, Switzerland

ABSTRACT

The LHC will be equipped with about 8000 superconducting magnets of all types. The total current to be transported into the cryogenic enclosure amounts to some 3360 kA. In order to reduce the heat load into the liquid helium, CERN intends to use High Temperature Superconducting (HTS) material for leads having current ratings up to 13 kA. The resistive part of the leads is cooled by forced flow of gaseous helium between 20 K and 300 K. The HTS part of the lead is immersed in a 4.5 K liquid helium bath, operates in self cooling conditions and is hydraulically separated from the resistive part.

A cryogenic test station has been designed and built in order to assess the thermal and electrical performances of 13 kA prototype current leads.

We report on the design, commissioning and operation of the cryogenic test station and illustrate its performance by typical test results of HTS current leads.

INTRODUCTION

The reference design for the Large Hadron Collider[1] (LHC) at CERN is based on the generalised use of High Temperature Superconductor (HTS) current leads[2,3] so as to reduce the total liquefaction rate distributed over the eight cryogenic plants of the 27 km circumference future LHC machine.[4]

Cooling of conventional self-cooled leads would require a liquefaction rate of about 300 g/s, while the use of HTS leads would reduce the heat load into the liquid helium bath at nominal current by a factor of up to 19,[5] making use of the 20 K/0.13 MPa return gas to the refrigerators from the beam screen and magnet support cooling circuits.

The 8000 superconducting magnets of the LHC machine are cooled in a pressurized bath of superfluid helium at 1.9 K. Two electrical feed boxes at the end of each of the 8 sectors of the machine provide the warm-to-cold transition via the HTS leads down to 4.5 K, and via a lambda plate to feed the magnets at 1.9 K. The leads are integrated in the electrical feed boxes and connected to the magnets via Low Temperature Superconducting bus bars. A cryogenic distribution line, running along the magnets, provides the 4.5 K LHe

Figure 1. Prototype current leads prior to their installation in the test cryostat.

feeding of the leads bath and the 20 K GHe to ensure cooling of the resistive part. Warm gas from the leads and the LHe bath is recovered in a gas collector connected to the LP line of the refrigerator.

Prototype 13 kA HTS current leads have been specified by CERN and produced by several manufacturers in Europe, Japan and the US.

The specification defines the required thermo-electric performance and the geometric limitation imposed by the underground constraints of the accelerator. The proposed leads are composed of a resistive heat exchanger in which a flow of 20 K GHe is circulated and a HTS part that dips into the 4.5 K saturated helium bath. The temperature of the warm end of the HTS is controlled at or below 50 K by the flow of 20 K helium in the resistive heat exchanger. The vapor generated in the liquid helium bath is guided in a "skirt" surrounding the HTS element that is therefore operated in a self-cooling condition. The leads are optimised for operating at 13 kA with the warm end of the HTS below 50 K and the top part of the resistive heat exchanger at 290 K. Without current, the temperature of the gas outlet of the leads is maintained at 290 K by a thermostatically controlled heater at the warm terminal of the heat exchanger.

SYSTEM LAYOUT AND PRINCIPLES OF OPERATION

A low leat inleak cryogenic station for testing HTS current leads was designed in order to assess the thermal and electrical performance of the prototype leads ordered from industry. The cryostat provides the required working conditions: 4.5 K helium bath to cool the bottom part of the HTS leads, up to 2 x 1.2 g/s of 20 K helium gas to cool the resistive part of the leads, warm helium gas to quench the HTS, recovery of helium gas and the necessary instrumentation and valve actuators for control and diagnostic measurements. Furthermore, the design allows flexibility in the operation so as to optimize the control law for the optimum performance of the HTS current leads. The control and data acquisition system allows to monitor and automatically control transients such as cooldown, warm-up and current ramping/de-ramping, as well as steady-state operation.

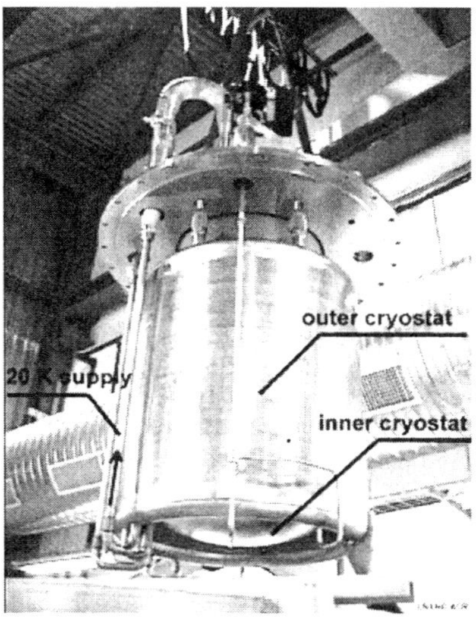

Figure 2. Inner and outer cryostat view.

System layout

As shown in figure 2, the test cryostat is composed of an annular (outer) cryostat providing radiation shielding at 4.5 K, neck thermal interception of the inner cryostat, and an inner cryostat in which the current leads are housed and cooled to maintain their nominal operating conditions. The two cryostats share a common insulation vacuum and are wrapped in multilayer thermal insulation.

The connection of the 20 K lines to the current leads in the inner cryostat is made via vacuum-insulated lines entering at and exiting from the top flange of the cryostats. This was necessary to allow easy demounting of the top flange of the inner cryostat to exchange current leads. Six radiation baffles in the inner cryostat provide thermal intercept of radiation from the top flange at ambient temperature. Gas from the inner and outer cryostats is warmed up via electrical heaters and recovered in a main manifold connected to the low pressure side of the refrigerator. The system makes use of the cryogenic infrastracture available, in particular a 6 kW @ 4.5 K cryogenic plant. Liquid helium is distributed via a vacuum-insulated and radiation-shielded transfer line, providing cooling and gas recovery for several test station. The outer cryostat is filled with liquid helium and acts as a supply of liquid for the inner cryostat via a cryogenic valve, and for cold gaseous helium for cooling the resistive part of the leads. An electrical heater can be used to increase the boil-off from the outer cryostat during powering of the leads when a higher mass flow is required. The inlet temperature of the gas supplied to the resistive part of the current leads is adjusted by mixing with 300 K gas.

Principles of operation

The test station has been designed for a maximum liquefaction requirement of about 4 g/s as shown in detail in Table 1.

The total nominal inventory of liquid helium in the two cryostats is about 150 l.

Two mass flow controllers adjust the necessary flow-rate to obtain the required gas temperature of 20 K. Additional electrical valves are used to add room temperature gas thereby increasing the temperature (up to 200 K) of the cooling flow in order to quench the

Table 1. Detailed design liquefaction requirements

Type of heat load	Equivalent load in g/s
1 pair of 13 kA leads in bath – 3 W	0.15
20 K cooling of leads	2.0
Anular cryostat – 5 W	0.25
Inner cryostat – 0.5 W	0.025
Transfer line and bayonet – 13 W	0.65
TOTAL	**3.075**
Design margin – 1.25	3.8

HTS. The gas is then recovered at the top part of each lead via an electro-pneumatic control valve, which regulates the mass flow in order to maintain the HTS warm temperature at or below 50 K. These valves can also control mass flow-rate depending on the current in the leads.

The pressure in the inner and outer cryostat is controlled via electro-pneumatic valves. The other electro-pneumatic control valve feeding LHe from the outer cryostat controls the level of liquid helium to cover the cold end of the HTS and the LTS short-circuit between the leads in the inner cryostat.

INSTRUMENTATION AND PROCESS CONTROL

Instrumentation

All the circuits' temperatures are monitored with Platinum 100 sensors (300 to 25 K) and Carbon or CERNOX™ sensors (25 to 4.2 K). The current leads are equipped with a high-voltage insulated Platinum 100 sensor for control of the HTS warm-end temperature and several uninsulated Platinum 100 sensors distributed over the resistive part of the lead to measure its temperature profile. Two CERNOX™ sensors monitor the HTS middle and cold end temperatures.

Liquid levels are measured with superconducting wire gauges that can be pulsed to decrease their heat input to the liquid helium bath. Two gauges of different dimensions (500 mm and 100 mm length) are used in the inner cryostat for control and diagnostic respectively, in order to achieve higher accuracy. The inner cryostat has been calibrated to assess heat loads by rate of level decrease with time.

Helium flow is measured at ambient temperature using thermal mass (0.5 g/s of full scale reading with 1 % accuracy) and V-cone™ flowmeters (1.2 g/s of full scale reading with 1 % accuracy). In view of its accuracy over the full range, the first type is used to measure the boil-off from the leads bath; the second type is employed at the warm outlet of the current leads to measure the flow of 20 K helium gas, thanks to its precision and low pressure drop. The flowmeters have all been calibrated with GHe in operating conditions and re-calibrated in-situ at the lower part of their ranges.

Differential pressure sensors check the pressure drop in the resistive part of the lead.

Process control

The process control is performed using an industrial Programmable Logic Controller (PLC) in which the number of I/O channels is of the order of 30 and there are about 7 closed control loops. The automatic operation modes include cooldown, normal operation, quench and warm-up. This equipment is interfaced with an industrial supervision system based on PCVUE32™.

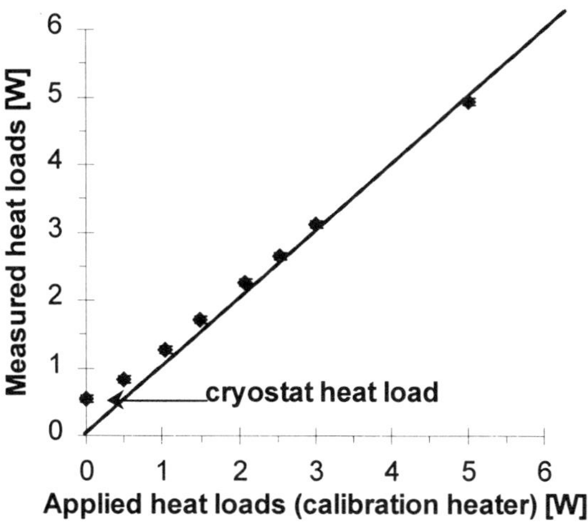

Figure 3. Test cryostat measured heat loads at various heater power (no leads).

The powering system of the leads is interlocked to the liquid level signal, the temperature of the warm part of the HTS and the cryogenic operator authorization.

The process control system is designed to run in fully automated mode 24h/24h. A shift operator is automatically called if a parameter of importance deviates from its expected range. The system can also be remotely accessed via Internet. Most of the experimental tests not involving powering of the leads are therefore performed overnight.

STAND-BY AND TRANSIENT OPERATION

Commissioning of the test cryostat

Prior to testing any of the prototype current leads, the cryostat underwent an extensive campaign of measurements to assess its thermal performance without leads. Furthermore, the cryostat itself, the instrumentation, the control and supervision system were fully commissioned and the instrumentation, where appropriate, re-calibrated in situ.

Cryostat characterisation

The HTS test cryostat needs about 24 h (with a flow of 0.2 g/s in the inner cryostat and 0.5 g/s in the outer cryostat) to reach stable operating conditions; this is mainly due to the masses involved and to the very good thermal insulation.

Once the cryostat is thermalized, the background heat loads (thermal conduction, radiation and superconducting liquid level gauge) in the inner cryostat can vary depending on the applied heat loads (current leads), which change the thermal characteristics of the cryostat (Figure 3).

During transient heat loads (changing the flow of evaporating helium from the bath), the neck and the radiative screen in the neck change their temperatures, thus varying the density of the helium in and around the radiative screen and modifying the steady-state flow measurements.

These two phenomena are taken into consideration to calculate the heat loads of the prototype current leads. A complete mapping of the thermal characteristics of the test cryostat at different applied heat loads prior to the insertion of the current leads allows one to determine the heat conducted to the 4.5 K liquid helium bath through the current leads with +/- 50 mW accuracy.

CRYOGENIC TESTS RESULTS

Each pair of leads requires at least one week to perform the specified performance tests. One week is also necessary to warm-up the cryostat, remove the previous leads, mount the new leads in the cryostat, check the instrumentation, condition the circuits and cool-down. A test program is established to study the behaviour of each lead in transient and stand-by operation.

Stand-by operation. The stand-by operation defines the thermal performance (Figure 4) of the leads with no current (heat loads in the liquid helium bath, flow-rate and pressure drop in the resistive part) at various HTS temperatures (30, 40 and 50 K).

Transient and steady operation. In transient operation we study the current ramping/de-ramping behaviour of the leads with 13 kA current (Figure 5), the behaviour of the lead during a loss of the 20 K helium flow cooling the resistive part with the current made to decay exponentially with a 120 seconds time constant, and the resistive transition of the superconducting part of the lead. In steady operation the thermal performance of the leads with 13 kA was observed.

The heat load in the liquid helium bath increases due to contact resistance when the leads are powered to 13 kA.

The self-cooling due to the skirt surrounding the HTS decreases the heat loads by pure conduction in the bath, thus reducing the absolute increase in heat loads when the leads are powered. This behaviour depends on the efficiency of the heat exchange in the HTS part.

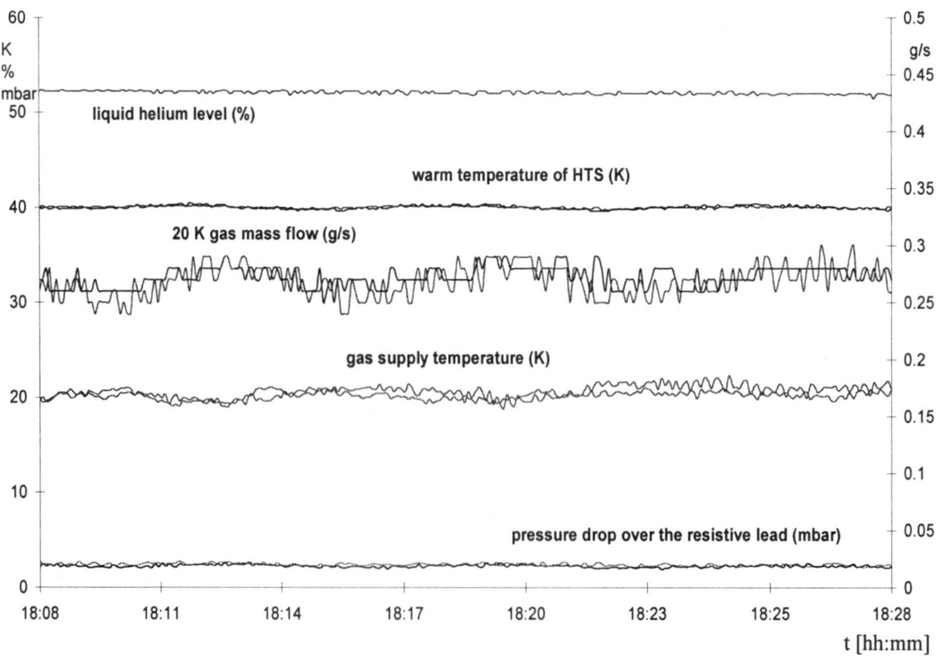

Figure 4. Typical steady-state thermal measurements on current leads (zero current).

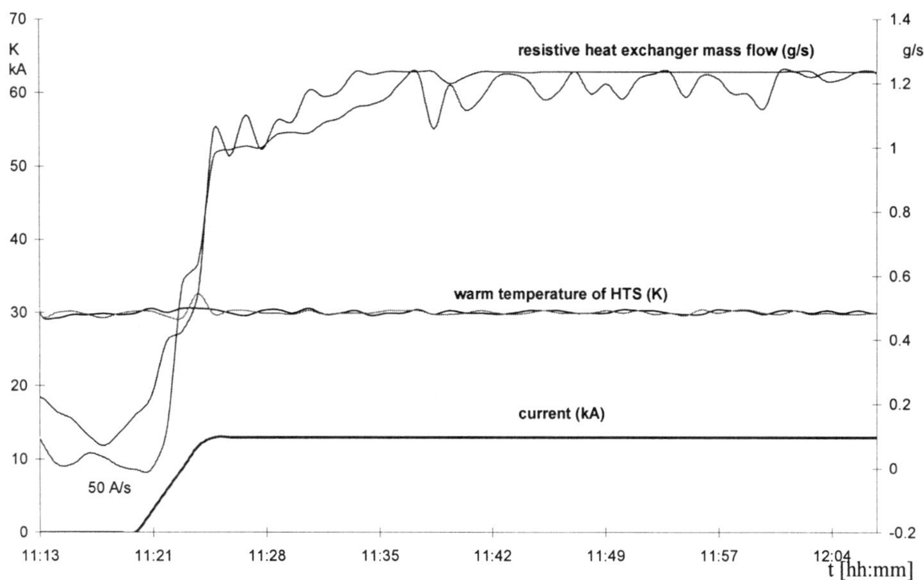

Figure 5. Typical response of cryogenic parameters during current ramping of leads.

Table 2 shows the typical cryogenic results obtained from a pair of HTS leads under test, including precision and stability. During stand-by, transient and steady operation the liquid helium level is maintained at its nominal value within +/- 0.25 cm, the precision of the liquid level gauge being +/- 0.1 cm.

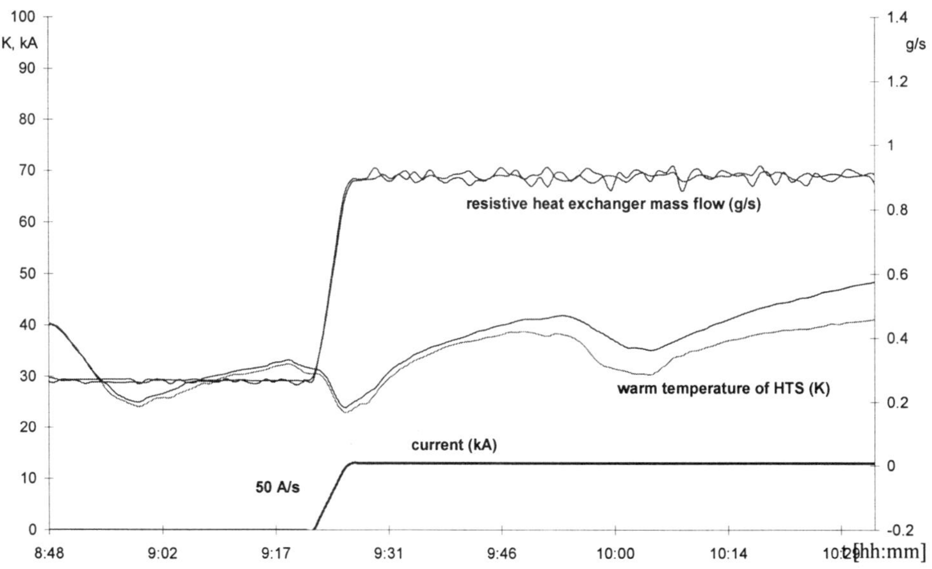

Figure 6. Typical current ramping of leads with control of the mass flow through the resistive heat part.

Table 2. Typical test results of a HTS lead pair

lead A lead B	HTS warm end [K]	20 K mass flow [g/s] 0 kA	Pressure drop [mbar] 0 kA	H.L. 4.5 K bath [W/lead] 0 kA	20 K mass flow [g/s] 13 kA	Pressure drop [mbar] 13 kA	H.L. 4.5 K bath [W/lead] 13 kA
CERN Specification	**< 50**	**< 0.6**	**< 50**	**< 1**	**< 1**	**< 50**	**< 1.5**
Measurement of a typical lead pair	**40** **40**	**0.42** **0.45**	**2.6** **1.9**	**0.73** **0.73**	**0.82** **0.86**	**3.3** **4**	**0.85** **0.85**
Precision	**+/- 0.1**	**+/- 0.01**	**+/- 1**	**+/- 0.05**	**+/- 0.01**	**+/- 1**	**+/- 0.05**
Stability	**+/- 0.5**	**+/- 0.03**	**+/- 1.5**	**+/- 0.08**	**+/- 0.1**	**+/- 3**	**+/- 0.08**

After having completely characterized the current leads in terms of their thermal performances (20 K mass flow at 0 and 13 kA, pressure drop over the heat exchanger, stability of the heat exchanger at various HTS temperatures and heat loads on the 4.5 K liquid helium bath), the control system can be set-up to control a constant mass-flow of 20 K GHe in the heat exchanger for each current value (Figure 6). This mass flow varies linearly with the current passing through the leads between its value measured at 0 A and 13 kA. The advantage of this control method is the stability and fast response achieved, independent from the temperature of the warm end of the HTS. Therefore the HTS remains below its maximum operating temperature value, otherwise an interlock prevents or aborts powering of the leads. The disadvantages are that the leads needs to be tested to set the control parameters (mass flow at 0 and 13 kA) and additional instrumentation is needed (flowmeter).

CONCLUSIONS

A low heat inleak test cryostat has been successfully designed, commissioned and operated at CERN to test prototype 13 kA HTS current leads.

Several pairs of HTS current leads have been tested and have shown that the cryogenic requirements set in the technical specification can be met. These leads have demonstrated that the heat load into the liquid helium bath can be reduce by a factor of up to 19 with respect to conventional self-cooled leads, however at the cost of GHe cooling at 20 K.

Additional tests to demonstrate the long term reliability (thermal and electrical cycles) will be performed.

Another test cryostat for 600 A HTS current leads of a similar design has been built and will be commissioned shortly for the test of prototype 600 A HTS leads.

REFERENCES

1. L. R. Evans, The Large Hadron Collider Project, in "Proceedings ICEC16", Editors Elsevier Science, Oxford UK (1997), p. 45.
2. T. Taylor, HTS Current Leads for the LHC, paper presented at ASC'98, Palm Desert, California (1998).
3. A. Ballarino, High Temperature Superconducting Current Leads for the Large Hadron Collider, paper presented at ASC'98, Palm Desert, California (1998).
4. A. Ballarino, A. Ijspeert, U. Wagner, Potential of High-Temperature Super Conductor Current Leads for LHC Cryogenics, in "Proceedings ICEC16", Editors Elsevier Science, Oxford UK (1997), p. 1139.
5. A. Ballarino, L. Serio, Test Results on the First 13 kA Prototype HTS Leads for the LHC, paper presented at PAC'99, New York (1999).

CHARACTERIZATION OF PROTOTYPE SUPERFLUID HELIUM SAFETY RELIEF VALVES FOR THE LHC MAGNETS

L. Dufay, A. Perin, and R. van Weelderen

LHC Division
CERN, European Organisation For Particle Physics
1211 Geneva 23, Switzerland

ABSTRACT

The Large Hadron Collider (LHC) at CERN will use high field superconducting magnets operating in pressurized superfluid helium (He II) at 1.9 K. Cold safety valves, with their inlet in direct contact with the He II bath, will be required to protect the cold masses in case of a magnet resistive transition. In addition to the safety function, the valves must limit their conduction heat load to the He II to below 0.3 W and limit their mass leakage when closed to below 0.01 g/s at 1.9 K with 100 mbar differential pressure. The valves must also have a high tolerance to contaminating particles in the liquid helium. The compliance with the specified performance is of crucial importance for the LHC cryogenic operation. An extensive test program is therefore being carried out on prototype industrial valves produced by four different manufacturers. The behavior of these valves has been investigated at room temperature and at 77 K. Precise heat load and mass leak measurements have been performed on a dedicated test facility at superfluid helium temperature. Results of cold and warm tests performed on as-delivered valves are presented.

INTRODUCTION

The Large Hadron Collider (LHC), presently under construction at the European Laboratory for Particle Physics (CERN), will use superconducting magnets operating at 1.9 K in pressurized superfluid He. In the eventuality of a resistive transition of the superconducting cables, a part of the energy stored in the magnetic field will be transferred to the refrigerant, resulting in a fast pressure rise in the magnet cold masses. To protect the cold masses against excessive pressure, safety relief valves, directly connected to the superfluid helium container are required. Although the main characteristic of these valves must be their reliable opening in case of a magnet quench, they must also not add an excessive heat load to the He II bath and must remain tight in presence of small debris coming from the cold masses. Approximately 400 valves, distributed every 107 m (the

Table 1. Main specified requirements of the safety relief valves

Parameter	Specified value
Valve type	valve lift must be proportional to inlet pressure and independent of outlet pressure
Valve size	DN40, proportional
Discharge capacity	$Kv^* > 30\ m^3/h$
Set pressure	12 ± 0.3 bar
Test mode set pressure	20 ± 0.6 bar
Blowdown⁺	≤ 1 bar
Operating temperature	inlet: 1.9 K ; outlet : 4.5 K
Operating pressures	inlet: 1.4 bar ; outlet 1.3 bar
Thermal anchoring temperature	100 K
Conduction heat load to 1.9K	< 0.3 W
Leak tightness over the seat	< 0.01 g/s at 1.9 K, ΔP =100 mbar

[*] Kv represents the valve's volumetric flow rate of water at NTP with a pressure drop 1 bar and is expressed in m^3/h.
[+] A valve blowdown is the difference between its set and its re-seating pressure.

length of a cooling cell) along the 27 km of the machine circumference[1,2] will provide the protection of the magnets against overpressure.

In a preliminary stage CERN developed a safety relief valve[3] that was connected to the first LHC prototype magnet string. For the LHC, the valves will be produced by industry. The compliance of these valves with the specified performance[4] is of crucial importance for the operation of the LHC. An extensive test program is therefore being carried out to investigate the characteristics of prototypes produced by four different manufacturers.

REQUIRED CHARACTERISTICS OF THE SAFETY RELIEF VALVES

The valve specification is based on the experience gained during the first stages of operation of the first LHC prototype magnet string.[4] The set pressure was fixed at 12 bar and, in addition, for pressure test purposes, the possibility to switch the set pressure to 20 bar was required.

The operating characteristics of the valves and the valve sizing have been determined by a numerical extrapolation of the results obtained on the LHC prototype magnet string[5] and on tests performed on the prototypes of the LHC magnets, resulting in a required Kv of 30 m^3/h.

The safety valves will be in direct contact with superfluid helium at 1.9 K and 1.3bar. When operating at such a low temperature, the minimization of the heat loads is essential. The contribution of the valves must not exceed 0.3 W per valve at 1.9 K. As any leak of superfluid helium translates directly into a heat load, the mass leakage across the seat must be lower than 0.01 g/s in operating conditions with a pressure difference of 100 mbar. Other requirements include the compatibility with the size of the LHC underground tunnel, easy exchangeability in cold conditions and tolerance to small sized contamination. A summary of the main requirements for the safety relief valves can be found in table 1.

INDUSTRIAL PROTOTYPES

Four manufacturers have been selected for the production of prototypes of the safety relief valves. Every manufacturer has produced two prototypes according to specified requirements.[4] The valve main dimensions and seat configurations are given in table 2 (The letter designating the valves have been randomly attributed and do not correspond to the

Table 2. Main characteristics of industrial safety relief valves

Valve	L1* [mm]	L2** [mm]	Seat configuration	Stem material
A	588	262	direct pressure	Fiberglass epoxy
B	660	239	sliding seat	Stainless steel
C	550	334	direct pressure	Fiberglass epoxy
D	660	239	direct pressure	Stainless steel
E	626	174	direct pressure	Stainless steel

* L1: length between thermal anchoring at 100 K and the valve outlet axis (see figure 2).
** L2: length between thermal anchoring at 100 K and 293 K (see figure 2).

alphabetical list given in the acknowledgements).

All manufacturers use a non-metallic material for the seal, the leak tightness being realized by either pressing the seal against a metallic seat (direct pressure in table 2) or by a specific seat configuration for the sliding seat of valve *B*. To reduce the heat conduction along the valve stem valves A and C use a fiberglass epoxy tube while the other manufacturers rely on optimized stainless steel stems.

ROOM TEMPERATURE AND 77 K TESTS

Test bench for room temperature and 77 K tests

A dedicated test bench has been constructed for set pressure and leakage testing. The valve properties can be measured either at room temperature or, with the valve lower part immersed in liquid nitrogen at 77 K.

The maximal gas flow of this test bench is 150 m^3/h (NTP) when operating at room temperature. At low temperature the flow must be kept at a much lower level to ensure a good thermalisation of the gas in the LN2/He heat exchanger integrated in the test bench. Leakage can be measured from 10^{-7} cm^3/s (NTP) up to 3 dm^3/s (NTP).

Set pressures at room temperature and 77 K

The maximal flow of the test bench is not sufficient to reach the full opening of the valves; therefore only the set pressure of the valves can be measured when the valve outlet is left open to atmosphere. For the investigation of the mechanical characteristics of the valves, advantage is taken of the independence of the set pressures against outlet pressure variations: the properties can be measured in quasi-static conditions by closing the valve outlet. Table 3 presents the results of measurements performed at room temperature and at LN2 temperature. It is to be noted that valve C was mistakenly designed for pressures 1 bar higher than the specified values.

Set pressure. In case of a magnet quench, the valves will experience a fast pressure increase[3,4] of about 100 bar/s, the first two lines of the table show the behavior of the valves for a pressure increase rate higher than 20 bar/s. It can be seen that all valves, except *A*, open within 0.8 bar of the specified set pressures, this both for the 12 bar and the 19 bar operating modes. In the case of valve *A*, the measured set pressure is strongly dependent on the pressure increase rate. The actuator of this valve is pilot operated, the pilot being connected to the valve inlet by a capillary tube that delays the pressure equilibration required to keep the valve closed.

To further investigate the mechanical characteristics of the valves, tests have been performed with closed outlet. This configuration allows one to reach the full opening of the

Table 3. Set pressures of the safety relief valves at 293 K and 77 K for two rates of pressure increase

Set pressure \ Valve		*A* 293 K/ 77 K	*B* 293 K/77 K	*C*✧ 293 K/77 K	*D* 293 K/77 K	*E* 293 K/77 K
12 bar mode, > 20 bar/s	[bar]	3.7 */ -	11.6/-	13.0/-	11.8/-	11.2 / -
20 bar mode, > 20 bar/s	[bar]	4.0 */ -	19.5/-	21.3/-	19.2/	18.3 / -
12 bar < 1bar/sec,CO⁺	[bar]	11.5 / 11.6	11.4 / 13.8	13.2 / 13.3	12.0 / 11.5	12.4 / 12.2
20 bar < 1bar/sec,CO⁺	[bar]	19.6 / 19.2	19.6 / 21.5	21.0 / 21.1	19.7 / 19.2	20.3 / 20.0
Blowdown, CO⁺	[bar]	0.3 / 0.3	1.1 / 2.8	0.5 / 1.2	0.6 / 0.7	0.4 / 0.4

✧ valve C was designed for set pressures 1 bar higher than the specified values.
* the opening pressure is strongly dependent on the pressurisation speed, values are for approx. 40 bar/s
⁺ CO: Closed Outlet

valves in quasi-static conditions. The set pressures in this case depend directly on the accuracy of the pressure compensation. At room temperature the set pressures are very well respected (most manufacturers used this configuration to adjust the set pressures). At LN2 temperature the shift in the set pressure is, with the exception of valve *B*, smaller than 0.4 bar. Valve *B* shows on the other hand an increase of more than 2 bar of the set pressure when operating at 77 K. This shift is apparently caused by an increased friction in the sliding seat.

The blowdown (difference between the set pressure and the re-seating pressure) values are directly related to the friction in the valves; at room temperature all valves, with again valve *B* being the exception, have a blowdown smaller than 0.6 bar, indicating a smooth movement with low friction forces. When operating at 77 K, valves *A*, *D* and *E* do not show an increase of the friction while for valve *C* the blowdown is slightly increased. In the case of valve *B* the strong increase in the blowdown indicates that temperature change is directly causing an increase of the friction forces in the valve seat.

Leakage. Figure 1 presents the low flow part of the measurements performed with gaseous helium on the valves at LN2 temperature. As can be seen on figure 1, there is a large difference between valves. For all valves, with the exception of valve *B* equipped with a sliding seat, the leakage increases faster than the square of the pressure difference, indicating that the size and/or number of the leak paths increases with increasing pressure.

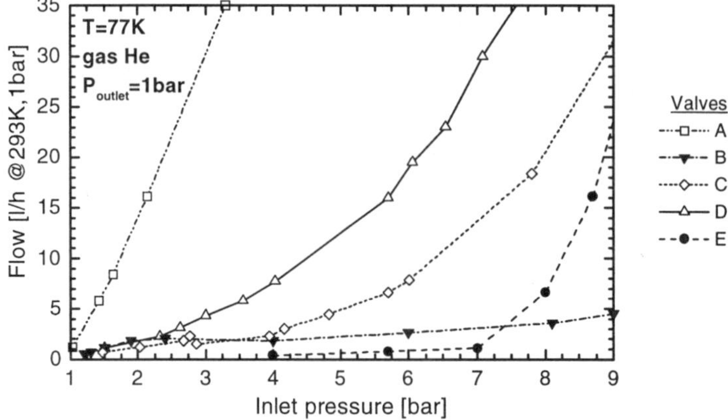

Figure 1. Valve leakage at 77K

THERMAL LOADS AND MASS LEAKAGE AT 1.9 K

Test Facility

A dedicated test bench was developed to reproduce the LHC temperature and pressure conditions. This test bench is described in detail in ref. 6. A schematic view of the test facility is given in figure 2. The valve inlet is connected to a volume (1) filled with pressurized superfluid helium at a temperature T_{press} between 1.9 K and 2 K, the pressure in the valve inlet can be varied from 1 bar to 3.5 bar. This volume is thermally connected to a pumped bath of saturated He II (3) via a Kapitza resistance heatmeter[6] (2). The temperature difference $T_{press} - T_{sat}$ is used to determine the heat flow Q_{kap} to the superfluid helium. The precision of the heatmeter is ±8 mW at $T_{press} = 1.9$ K and ±10 mW at $T_{press} = 2$ K. In addition, the heat flow measurement can be cross checked with a measurement of the heat flow Q_{vap} based on the vaporization of saturated helium. The outlet of the valve is connected to a discharge pipe which end temperature T_{disch} can be varied from 5 K to 20 K to study convection heat transfer phenomena.

The leaking helium is collected in the discharge pipe, warmed up to room temperature and measured through a mass flowmeter with a precision of 0.2 mg/s. Two thermometers, placed close to the valve seat, T_{seat}, and at the valve outlet, T_{outlet}, provide information on the temperature levels in the valve. A thermal anchoring at 100 K is provided and the heat flow to this heat sink is also measured.

Thermal loads

The heat flow to the superfluid helium has been measured for every valve. The measurements have been performed with the discharge pipe closed and zero leakage. During these tests, the temperatures close to the valve seat were, for all valves, well above the λ temperature (see table 4). We could not observe any measurable pressure difference

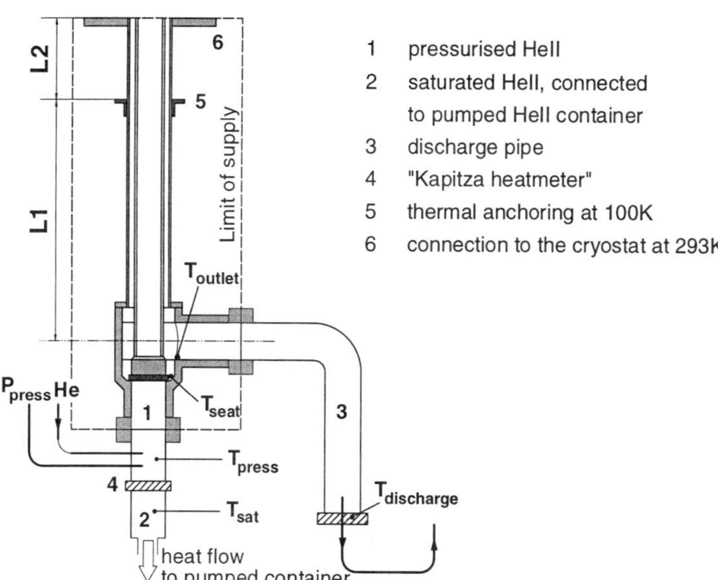

Figure 2. Schematic representation of the experimental configuration for the thermal characterisation of the safety relief valves

Table 4. Temperatures close to valve seat and valve outlet, T_{press}=2K

Valve		A	C	D (B)	E
T_{seat}	[K]	11.95	4.72	2.45	8.78
T_{outlet}	[K]	13.01	5.47	6.31	10.43

related to the fountain effect in contrast to what was observed on an early CERN developed safety valve.[3]

Table 5 presents a summary of the main thermal conduction properties of the safety valves. Valves *B* and *D*, produced by the same manufacturer, have the same thermal properties.

Heat flow to 100 K thermal anchoring. The measurements of the thermal flow to the 100 K heat sink, presented in the first row of table 5, show a strong variation with values ranging from less than 3 W for valves *B* and *D* up to more than 8 W for valve *E*. Although this was not specified for the prototype valves, a maximum heat load of about 3 W will be required for the LHC valves.

Heat flow to superfluid helium. The conduction heat load to the He II has been measured for two temperatures T_{press} = 1.9 K and T_{press} = 2 K. For these measurements a temperature T_{disch} = 5 K was used to suppress the convection heat flow in the discharge pipe. Under these operating conditions, a small part of the heat flow Q_{disch} flows to the discharge pipe end and must be added to the measurements. The results have been cross-checked with the helium vaporization measured heat flow Q_{vap}. Of all valves, only valves *B* and *D* comply with the specified requirement of 0.3 W, all other valves having a conduction heat load higher than the specified value.

Convection in the discharge pipe. To investigate the convection heat transport in the discharge pipe the temperature T_{disch} has been varied between 5 K and 20 K As it can be observed in figure 3, the convection heat load becomes a significant, and for some valves even the major part of the heat load when T_{disch} = 20 K. The measurements show that the convection is strongly dependent on the temperature of the valve outlet region, with the valves having the lowest temperatures (see table 4) being more sensitive to a variation of T_{disch}. This result, in good agreement with the previous experience[3], confirms the necessity of a convection limiting design of the discharge pipe for the LHC.

Table 5. Results of the thermal measurements performed on the safety relief valves

Heat load \ Valve		A	C	D (B)	E
100K heat load	[W]	3.1	3.93	(2.5) [1]	8.5
He II: $Q_{kap}(T_{press}$ =2K)	[mW]	404	505	209	508
He II: $Q_{kap}(T_{press}$ =1.9K)	[mW]	409	509	203	503
He II: $Q_{evap}(T_{press}$ =2K)	[mW]	412 ± 40	498 ± 40	200 ± 40	497 ± 40
Q_{disch} [2]	[mW]	+10	+6	+2	+2
Total heat flow	**[mW]**	**416 ± 10**	**513 ± 10**	**208 ± 10**	**508 ± 10**

[1] estimation, measurement not performed in nominal conditions
[2] heat flow to the discharge pipe end at 5K
All values with closed discharge pipe and zero leakage

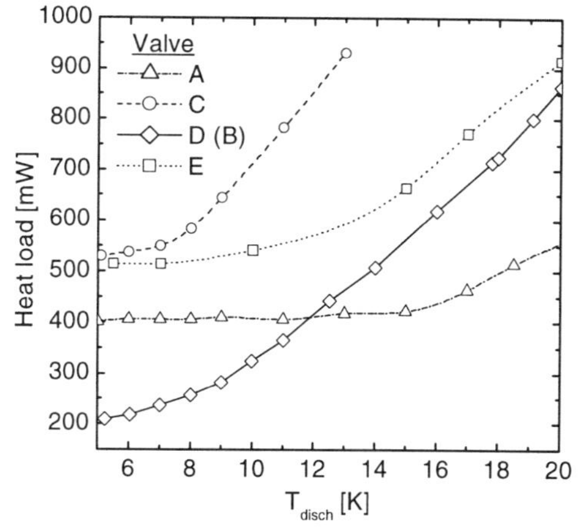

Figure 3. Heat flow to He II vs T_{disch}

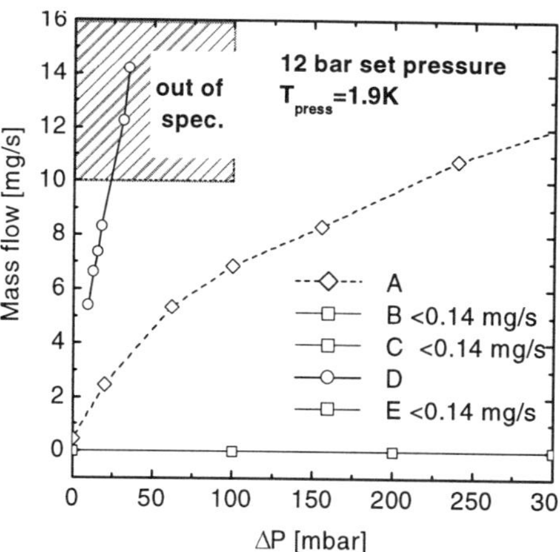

Figure 4. Leakage vs. ΔP for 12 bar set mode.

Leakage at 1.9K

For the mass leakage measurements, a pressure difference was established between the valve inlet and outlet. The first result of a helium leak is a strong reduction of the measured heat flow through the Kapitza heatmeter. This effect is due to the huge amount of heat removed by the helium flow and it completely hides the supplementary heat load required to cool the helium from 4.2 K to 1.9 K. The results are given in figure 4 and are summarized in table 6. For three valves, *A*, *B* and *E* the helium leakage was below the measuring resolution up to a ΔP of 300 mbar. These valves have also the lowest leakage at 77 K. Two valves have higher leaks: valve *A* has a high leak rate, although it remains within the specified limit of 10 mg/s while valve *D* exhibits a very high leakage, exceeding the specified flow with a ΔP well below 100 mbar.

CONCLUSIONS

The measurements performed at room temperature and at 77 K show that the required set pressures and functional characteristics can be obtained by industrially produced valves. The prototypes produced by three of the firms comply with the requirements, the pilot operated valve complies with most requirements, but its actuator design results in a set pressure that is strongly dependent upon the pressure increase rate. The sliding seat valve shows a high friction in the seat and a strong variation of its set pressures with decreasing temperature, which results in non-compliance with the specifications.

The conduction thermal loads and mass leakage have been measured in LHC operating conditions. Only one of the four valve designs complies with the required value of 0.3 W heat conduction to the He II. A global optimization of the thermal design will definitely be required for the final LHC safety valves. The measurements of the mass leakage show that the required leak tightness can be attained by four out of five prototypes. Most valves

Table 6. Leakage of the valves at 1.9 K for ΔP = 100mbar

Valve		***A***	***B***	***C***	***D***	***E***
Leak, 12 bar mode	[mg/s]	6.9	< 0.1	< 0.1	>15	< 0.1

showed leakage two orders of magnitude lower than the required value. These promising results are obtained with very clean and new valves and further investigation is required for determining the influence of the contamination found in the helium circuit.

The measurements of the high convection heat transport in the discharge pipe have confirmed the necessity of a convection limiting design of the discharge line to eliminate this heat load to the superfluid helium.

ACKNOWLEDGEMENTS

The prototype valves were developed and supplied by Adareg SA (France), Testfuchs GmbH (Austria), Techspace Aero SA (Belgium) and Weka AG (Switzerland), whom we gratefully thank for their collaboration.

REFERENCES

1. W. Erdt, G. Riddone, R. Trant, The Cryogenic Distribution Line for LHC: functional specification and conceptual design, paper presented at this conference.
2. M. Chorowski, W.Erdt, Ph. Lebrun, G. Riddone, L. Serio, L. tavian, U. Wagner, r. van Weelderen, A simplified cryogenic distribution scheme for the Large Hadron Collider, in "Advances in Cryogenic Engineering vol. 43", Plenum Press, New York (1998), p. 395.
3. H. Danielsson, G. Ferlin, B. Jenninger, C. Luguet, S.-E. Millner, J.-M. Rieubland, Peformance of a superfluid helium safety relief valve for the LHC superconducting magnets, in "Advances in Cryogenic Engineering vol. 41", Plenum Press, New York (1995), p. 805.
4 R. van Weelderen, Technical specification for the fabrication and supply of prototypes superfluid helium safety relief valves for the Large Hadron Collider, Technical Specification IT16786/LHC/LHC, CERN, Switzerland, (1996).
5. M. Chorowski, B. Hilbert, L.Serio, R. van Weelderen, Thermohydraulics of resistive transitions of the LHC prototype magnet string: Theoretical modeling and experimental results, in. "Advances in Cryogenic Engineering vol. 43", Plenum Press, New York (1998), p. 459.
6. A. Bézaguet, L. Dufay, G. Ferlin, A. Perin, G. Vandoni, R. van Weelderen, "A facility for accurate heat load and mass leak measurements on superfluid helium valve", paper presented at this conference.

THE CRYOGENIC DISTRIBUTION LINE FOR THE LHC: FUNCTIONAL SPECIFICATION AND CONCEPTUAL DESIGN

W. Erdt, G. Riddone, R. Trant

CERN, LHC Division
CH-1211 Geneva 23, Switzerland

ABSTRACT

The Large Hadron Collider (LHC)[1] currently under construction at CERN will make use of superconducting magnets operating in superfluid helium below 2 K. The cryogenic distribution scheme for each of the eight sectors, individually served by a refrigeration plant, is based on a separate Cryogenic Distribution Line (QRL) feeding helium at different temperatures and pressures to the elementary cooling loops. The QRL comprises two supply headers and three return headers including a sub-atmospheric one. Low heat inleak to all temperature levels is essential for the overall LHC cryogenic performance. With an overall length of 25.6 km the QRL has a very critical cost-to-performance ratio. Therefore, following an in-house feasibility study, CERN adjudicated in autumn 1998 three industrial contracts in parallel for the supply of Pre-Series Test Cells (~ 112 m) of the QRL, which will be tested at CERN in 2000. Installation of the QRL for LHC is scheduled from 2002 to mid 2004. This paper will present the general layout, the functional requirements as well as some aspects of the in-house conceptual design.

INTRODUCTION

The LHC cryogenic system[2], shown in Figure 1, is based on a five-point feed scheme with eight refrigeration plants serving the eight sectors of the LHC machine. Each of the eight sectors (~3.3 km) has its own self-standing cryogenic system. The cryogen provided by a refrigeration plant of 18 kW at 4.5 K equivalent power[3], is distributed at different temperatures and pressures via the Cryogenic Distribution Line (QRL) to the LHC magnet cryostats and other cryogenic consumers of the relevant sector. A 1.8 K refrigeration unit[4] maintains, via a large pumping line inside the QRL, the low pressure (1.6 kPa) for the local superfluid helium heat exchangers[5] inside the cold mass. The LHC cryogenic system will for the first time make use of a cold recovery line, also integrated into the QRL, to collect the helium discharged from the machine cryostats following a quench. The central

Advances in Cryogenic Engineering, Volume 45.
Edited by Shu *et al.*, Kluwer Academic / Plenum Publishers, 2000.

connecting element between all these parts of the cryogenic distribution system of a LHC feeding point is the Cryogenic Interconnection Box.

GENERAL LAYOUT

Each LHC sector, as shown in Figure 1, is composed of different sections, namely Long Straight Sections (LSS), Dispersion Suppressors (DS) and arc, which constitutes about ¾ of the total sector length. The QRL structure follows the LHC lattice, which is for the arc made of 23 identical cells, each of 106.9 m length. The DS is made up of 4 cells of length, which vary from 80 m to 91.6 m. The LHC lattice in the LSS is very particular in layout for each of these regions with QRL cell lengths from about 10 m to 170 m.

Each QRL cell consists of a Service Module and a Pipe Module as shown in Figure 2. Different cell lengths are accommodated by adjusting the Pipe Module length. The Service Module comprises two Valve Boxes and the so-called Jumper Connection, by which the QRL is linked to the magnet cryostat. The service module houses a subcooling helium heat exchanger, up to seven cryogenic control valves, one or two superfluid helium safety relief valves[6], a cryogenic mass flowmeter[7] as well as monitoring instrumentation. The Pipe Module is made of several Straight Pipe Elements and Compensation Units.

The cooling loop of a LHC standard cell, as shown in Figure 3, contains Service Modules of Type A and B, the most frequent ones. The QRL headers will supply helium to the LHC magnet cryostat at different temperatures and pressures. Supercritical helium at 4.6 K and 0.3 MPa tapped from header C will be used to supply the cold mass HeII heat exchanger and the beam screen circuit as well as for lower temperature heat interceptions and initial filling. The return line is header D that has also the function of cold recovery line. Helium at 4 K and very low pressure (1.6 kPa) will circulate in header B for maintaining the saturation pressure of the superfluid helium circuit. Finally, gaseous helium at 50 to 75 K will cool the thermal shields of the QRL and the LHC magnet cryostats. Inside the arc and DS the supply header E will be housed inside the LHC magnet cryostats while the return header F will always be integral part of the QRL.

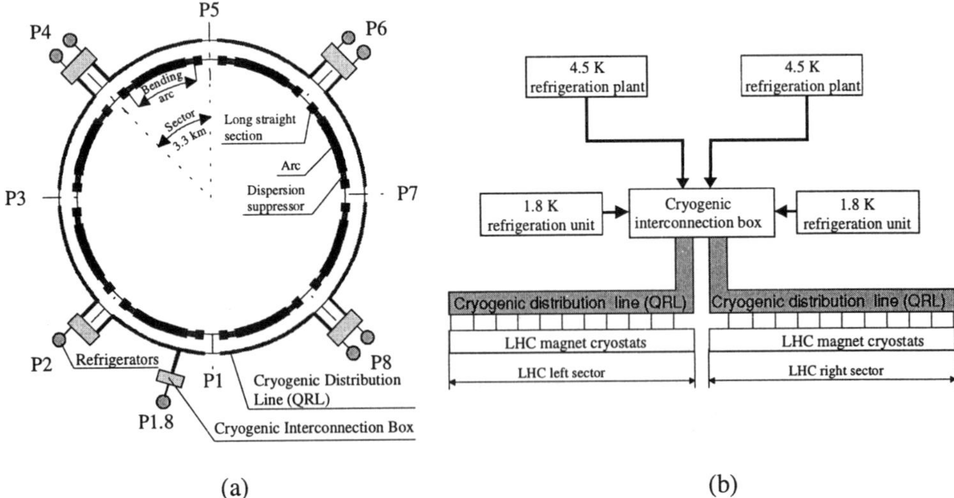

Figure 1. LHC cryogenic system: (a) layout and (b) architecture of a typical feeding point

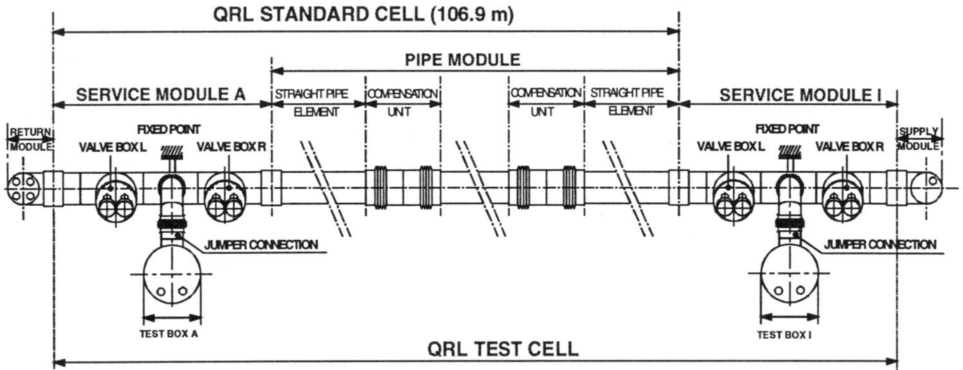

Figure 2. QRL standard cell and test cell (based on CERN feasibility study)

According to the overall LHC cryogenic flow scheme, all cooling loops are similar but not always identical. This leads to 26 different Service Module types supplying the cryogen needed to magnets operating at 1.9 K or 4.5 K, electrical feed-boxes and superconducting cavities. By minor modifications most of the types can be derived from a basic type.

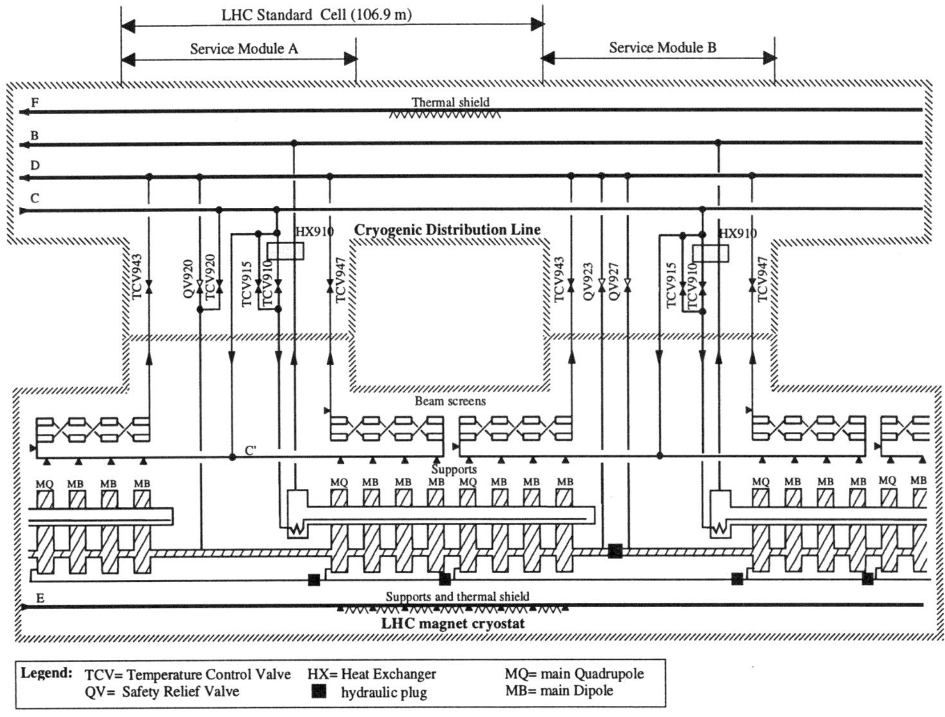

Figure 3. Typical LHC cryogenic flow scheme

BASIC REQUIREMENTS AND DESIGN CHALLENGES

The lifetime of the LHC is expected to be 20 years with a yearly operating time of 6600 hours for the cryogenic distribution system[8]. Cryogenic supply is needed all around the ring and tunnel access for maintenance purposes will be limited. Thereby reliability of the cryogenic system becomes very challenging. Furthermore all tunnel equipment must withstand the expected radiation environment[8].

The QRL design must be highly reliable, which is of first importance for the active components such as bellows and control valves. The cryogenic compensation units for the headers (~4000) and the metal hoses inside the Jumper Connections (~4000, see machine interface below) must be designed in accordance with well established international standards and will have to undergo extensive testing and qualification programs. Reliability as well as control and monitoring of the about 1500 cryogenic valves around the tunnel become important issues. CERN considers the use of new generation valve positioners, the so-called "intelligent" positioners, presently under evaluation[9]. They allow for automatic start-up, including self-calibration, remote configuration as well as preventive maintenance planning using diagnostic data.

The QRL design should be independent of the cooling rate, the cool-down sequence of the various headers and the pressurisation rate (e.g. header D in case of a helium discharge from a magnet cryostat following a quench). The nominal QRL operating conditions and required header diameters are listed in Table 1.

With a total length of about 25.5 km the QRL is a largely modular set-up of Pipe and Service Modules allowing a high degree of standardisation and consequently series production of the various elements. The industrial series production of highly reliable cryogenic modules with unprecedented low heat inleak for the lowest price per meter ever achieved is the real challenge of this project.

THERMAL REQUIREMENTS

The required heat inleak for a QRL Standard Cell (header E is not included) and the QRL sectors (headers are grouped with reference to the reception test scheme) are given in Table 2. The values are based on a CERN design and feasibility study. Considering the overall LHC thermal budget[10] on the different temperature levels, the heat inleaks of the QRL contribute for about 60 % to 50-75 K (headers E and F), 5 % to 4.6-20 K (headers C and D) and 95 % to 4 K (header B).

To limit the heat inleak by radiation from room temperature, the QRL cryostat will house a thermal shield and radiative insulation. The thermal shield, cooled by the return header F at 75 K, will surround the components which are at temperatures below 50 K. The outer surface of the thermal shield will be wrapped with multilayer insulation (30 layers). The cryogenic headers below 20 K will be individually or commonly wrapped by a "floating" insulation system (e.g. 10 layers). The choice of multilayer insulation is of crucial importance for meeting the total QRL budget, as at 4-20 K its contribution represents about ¾ of the total estimated heat inleaks. Indicative values, measured at CERN[11], of heat flux through multilayer insulation at 10^{-4} Pa (nitrogen equivalent measured with a room temperature gauge on the vacuum jacket) are 1 W/m^2 from 300 K to 75 K for 30 layers and 0.05 W/m^2 from 75 K to 5 K for 10 layers.

Table 1. QRL headers: diameter & nominal operating conditions

Header	Description	Inner Diameter [mm]	Nominal Temperature [K]	Nominal Pressure [MPa]	Design Pressure [MPa]	Nominal Mass flow rate [g/s]
B	pumping return	267	3.8 - 4.2	0.0016	0.4	125
C	4.6 K supply	100	4.6	0.36	2	215
D	20 K return	150	20	0.13	2	90
E	50 K supply	80	50 - 65	1.95	2.2	250
F	75 K return	80	65 - 75	1.9	2.2	250

The thermal shield and the header assembly will be supported by about 20 sliding and 4 fixed points per QRL Standard Cell. Low thermal conductivity components will have to be used together with heat interceptions to intermediate temperatures.

At working temperature without active pumping the required insulation for each insulation space (428 m) is 10^{-4} Pa. During nominal LHC operation the insulation vacuum of each of the vacuum spaces will be maintained without continuous mechanical pumping.

INTERFACE TO THE MACHINE CRYOSTATS (JUMPER CONNECTION)

The high alignment precision of the LHC machine has two implications for the QRL. Firstly the QRL must not introduce significant forces to the machine elements via the Jumper Connection every 106.9 m, which might lead to a misalignment. Therefore the interconnecting process pipes as well as the vacuum jacket must be equipped with flexible elements. Secondly the LEP (Large Electron Positron collider) operation has shown that the long-term stability of the tunnel plane due to geological movements is significantly less than the required alignment precision for the LHC. In order to realign machine elements without moving the QRL, the Jumper Connections must be able to compensate for movements of the beam tubes relative to the QRL of ± 10 mm in the horizontal plane and ± 20 mm in the vertical direction. For accumulated displacements exceeding these values the QRL external supports should give additional adjustment capability.

Table 2. Required heat inleak for (a) QRL Standard Cell (106.9 m) and (b) QRL sectors

(a)

Heat inleaks	Header B [W]	Header C [W]	Header D [W]	Header F [W]
Pipe Module (104.3 m)	6.4	2.5	4.0	270
Service Module type A (2.6 m)	0.7	1.6	1.2	30
Total	7.1	4.1	5.2	300

(b)

Heat inleaks	Headers E and F [W]	Headers C, B and D [W]
Low load sectors: 2-3, 3-4, 6-7, 7-8	9000	520
High load sectors: 1-2, 4-5, 5-6, 8-1	9850	580

TUNNEL INSTALLATION

The QRL will be placed inside the LHC tunnel between the machine cryostat and the outer tunnel wall, where the available space is very limited. Figure 4 shows standard tunnel cross-sections with a QRL design based on a CERN in-house feasibility study.

For the QRL with an expected outer diameter of about 610 mm a maximum outer envelope of 750 mm is foreseen, which includes available space for sleeves and expansion joints, as well as any kind of tooling like welding/cutting devices. In order to fit the valve boxes to the space constraints and to allow for exchanging stems of the cryogenic control valves, the CERN study foresees a tilt of the valve boxes by 10 degrees towards the machine cryostat.

PRE-SERIES TEST CELL

The aim of a Pre-series Test Cell is to qualify the design chosen and to verify its thermal and mechanical performance. The engineering of the Pre-series Test Cell fully takes into account the subsequent series production of the basic QRL elements and is as close as possible to the final concept, such as compensation units, multilayer insulation wrapping, installation procedures or polygonal geometry.

The QRL Test Cell will first be used for qualifying the manufacturing process and the validation of installation and quality control procedures. Through extensive testing at CERN, it will then be used to investigate heat inleaks and to perform thermal cycles and pressure tests in order to verify the design with respect to the assumed load breakdown. The test results may give input for simplification and subsequent reduction in cost for the modular QRL elements to be produced in series.

The QRL Test Cell (see Figure 2), approximately 112 m long, will consist of a QRL Standard Cell and an additional Service Module needed for test purposes (namely Service Module type I). To perform the reception tests as close as possible to the LHC working conditions, each QRL Test Cell will be individually fitted with the corresponding ancillary equipment such as a test supply module, a test return module and two test boxes. This test equipment will be connected to the Test Cell in the same way as the QRL sub-assembly interfaces and the future Jumper Connection to the LHC machine cryostats.

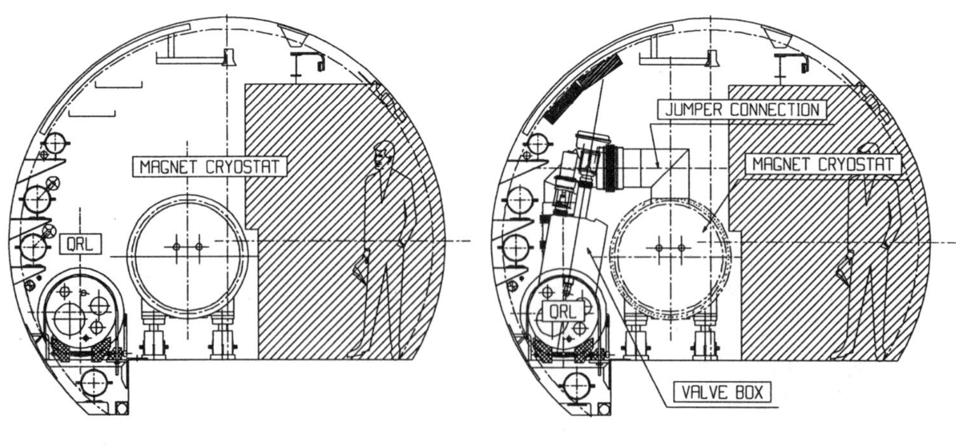

(a) with QRL service module

(b) with QRL pipe module

Figure 4. Standard LHC tunnel cross-sections

The ancillary equipment has been designed at CERN and three complete sets are under construction at CERN. Up to two QRL test cells will be tested in parallel, as the very tight time schedule will not allow for testing them sequentially.

The QRL test set-up will allow for separate measurements of the heat inleaks to each header and to the Service Modules. As foreseen for the LHC machine, header C will be used as a supply line and headers B, D and F as return lines. The QRL Test Cell will not operate exactly under the same nominal conditions as the QRL supplying the LHC magnet cryostats. The main difference concerns the header pressures which for the test cell will be around 0.1 MPa for all the headers. For the operational and thermal performance assessment tests, the QRL Test Cells will be equipped with instrumentation for precision measurement of temperatures, pressures and flowrates as well as forces.

To test the three QRL test cells a test area at CERN is being built. It consists of a concrete floor (130x13 m^2), on which concrete blocks will be placed to simulate the tunnel conditions. The whole structure will be closed at each extremity and will be covered by a metallic roof.

Once the QRL test cell will be installed in the dedicated test area, the pressure test will be performed. After this test has been passed successfully, the CERN ancillary equipment will be mounted to the QRL Test Cell. After the cool-down of a Test Cell, heat inleak measurements, mechanical and alignment tests and thermal cycles will be carried out. The heat inleak measurement will then be repeated. To verify the chosen safety concept, loss of insulation vacuum will be tested as well. Finally a visual inspection of the QRL Test Cell will be done. The vacuum jacket sleeves at the assembly interfaces will be opened and the main components such as MLI, thermal shield, inner supports and headers will be inspected. The total time foreseen for testing each QRL Test Cell is about 3 months.

QRL PROJECT PLANNING

In autumn 1998 CERN adjudicated three contracts in parallel for the supply of a Pre-series Test Cell in order to assess the very critical cost to performance ratio for the QRL project. The contractors are Air Liquide/France, the consortia Linde-Babcock/Germany and a group of five companies under the leadership of ABB Enertech/Switzerland. A formal Preliminary Design Review concluded this spring the detailed engineering design phase. The technical validation of the three Pre-series Test Cells at CERN is foreseen for the first half of 2000. The formal performance assessment with respect to the functional technical specification will conclude the present contracts by mid 2000. The suppliers having successfully passed this phase will then be invited to bid for the QRL series of the LHC. Installation will start in February 2002 and will end with the last commissioned sector mid 2004.

ACKNOWLEDGEMENT

We would like to acknowledge the contributions of our colleagues G. Mouron and N. Veillet of CERN. We would also like to thank our colleagues from DESY and BNL for helpful discussions.

REFERENCES

1. L. Evans, The Large Hadron Collider Project, in: "Proc. ICEC 16", Elsevier Science, Oxford, UK (1997) 45:52.
2. M. Chorowski, W. Erdt, Ph. Lebrun, G. Riddone, L. Serio, L. Tavian, U. Wagner and R. van Weldeeren, A simplified cryogenic distribution scheme for the Large Hadron collider, in: "Adv. Cryo, Eng." 43, Plenum Press, New York, (1998), 395:402.
3. S. Claudet, Ph. Gayet, U. Wagner, Specification of four new large 4.5 K refrigerators for LHC, paper presented at this conference.
4. A. Bézaguet, Ph. Lebrun, L. Tavian, Performance assessment of industrial prototype cryogenic helium compressors for the Large Hadron Collider, in: "Proc. ICEC 17", Institute of Physics Publishing, Bristol, UK (1999) 145:148.
5. Ph. Lebrun, L. Serio, L. Tavian and R. van Weelderen, Cooling strings of superconducting devices below 2 K: the helium II bayonet heat exchanger, in: "Adv. Cryo. Eng.", 43A (1998), 419:426.
6. L. Dufay, A. Perin, A. and R. van Weelderen, Characterisation of prototype superfluid helium safety relief valves for the LHC magnets, paper presented at this conference.
7. R. Losserand,-Madoux and L. Serio, Development of supercritical helium mass flow-meters for the Large Hadron Collider, paper presented at this conference.
8. P. Cruikshank *et. al.* , Design parameters for equipment installed in the LHC, LHC Engineering Specification LHC-PM-ES-0002.00, CERN, Geneva (1999).
9. W. Hees, R. Trant, Evaluation of electro pneumatic valve positioners for LHC cryogenics, LHC Project Note 190, CERN, Geneva (1999).
10. Ph. Lebrun, G Riddone, L. Tavian, L. and U. Wagner, Demands in refrigeration capacity for the Large Hadron Collider, in: "Proc. ICEC 16", Elsevier Science, Oxford, UK (1997), 95:98.
11. Ph. Lebrun, L. Mazzone, V. Sergo and B. Vullierme, Investigation and qualification of thermal insulation systems between 80 K and 4.2 K, *Cryogenics* 32, ICMC Supplement, 44:47 (1992).

DEVELOPMENT OF A HIGH PERFORMANCE TRANSFER LINE SYSTEM

K. Hosoyama,[1] K. Hara,[1] A. Kabe,[1] Y. Kojima,[1] Y. Morita,[1] H. Nakai,[1] T. Fujita,[2] T. Kanekiyo,[2] and K. Matsumoto[2]

[1] KEK, High Energy Accelerator Research Organization
Tsukuba, Ibaraki, 305 Japan

[2] Hitachi Ltd.
Kudamatsu, Yamaguchi, Japan

ABSTRACT

High performance transfer lines are key components for design and construction of cryogenic systems for superconducting magnets and cavities. We have developed two types of high performance transfer lines; main transfer lines with liquid helium flow and cold gas return dual path and sub-transfer line with single path, all helium lines in these are guarded by 80K liquid nitrogen cooled thermal shields. We have adopted the aluminum molding made by extrusion for the 80K thermal shields. The structure of the transfer lines was designed to be easy to assemble. The performance of these transfer lines were tested. The heat loss of the main transfer lines at straight section are 0.04W/m and 0.06W/m for supply and return line, respectively.

INTRODUCTION

The 80K thermal shield cooled by liquid nitrogen is very important to attain a very low heat loss transfer line. Especially critical is good thermal contact between the shielding component and liquid nitrogen piping. At the early stage of the R&D[1] we used thin copper pipe for the 80K thermal shielding. The thermal contact between liquid nitrogen pipe and the shielding pipe is attained by soldering. The thermal performance of these transfer lines were good but it required tedious work and took long time to assemble it.

In our R&D study of the transfer line we have concentrated our effort to improve not only the thermal performance but also to reduce the construction cost. We started the design study of the 80K thermal shield which has a structure that is easy to assemble and handle.

The cross sections of the newly developed multi-channel transfer lines are shown in Figure 1. These transfer lines have been designed and constructed for the cryogenic system of KEKB superconducting RF cavities[2].

We summarize the main features and selected parameters of these transfer lines in Table 1.

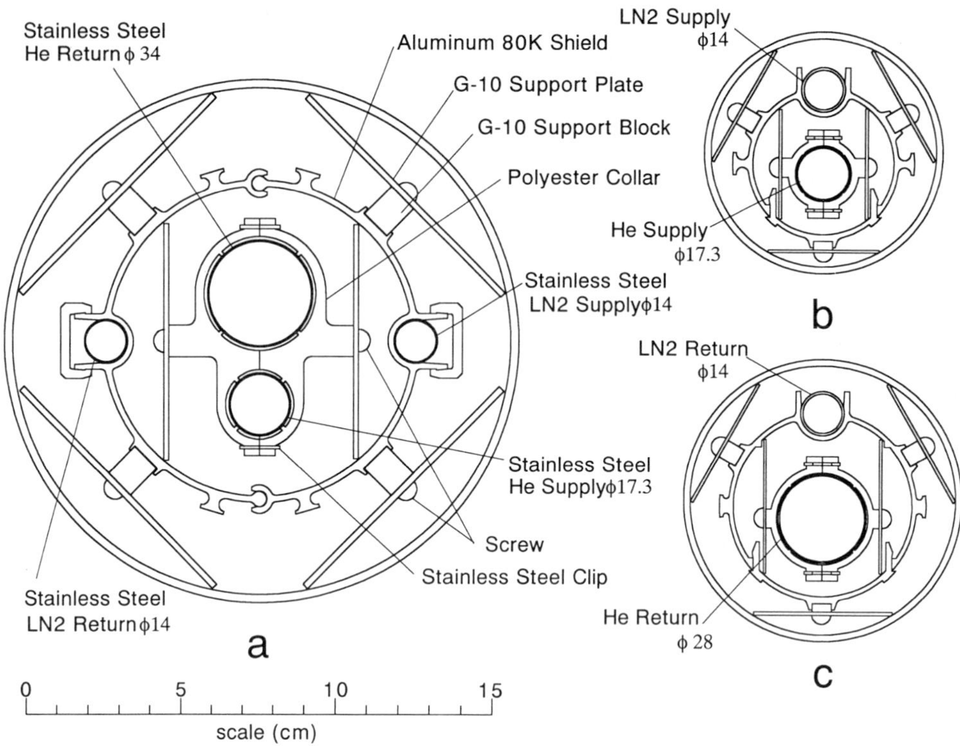

Figure 1. Cross sections of multi-channel transfer lines (a) two way main transfer line (b) one way sub-transfer line for liquid helium supply line (c) one way sub-transfer line for helium gas return line

Main Transfer Line

The multi-channel transfer line about 15m long, as shown in Figure 1-a, was constructed in a test stand of the KEKB superconducting cavity. A prototype KEKB superconducting higher order modes (HOMs) damped accelerator cavity in a horizontal cryostat has been cold tested at this test stand. The heat loss measurement of this newly designed transfer line was carried out and got very good result of about 0.05W/m at the straight section.

We have decided to use this type of multi-channel transfer line for a cryogenic system of high current beam test of the KEKB prototype superconducting HOMs damped cavity at the TRISTAN accumulation ring (AR). The multi-channel transfer line about 17m long which connects a helium refrigerator (TCF200 cooling capacity of 600W at 4.4K) on the ground and the superconducting cavity (maximum heat load of about 200W) in the underground AR tunnel has been constructed in 1996 for its feasibility study.

The whole system including the cryogenic system worked very well for about three months operation and a maximum beam current of 573mA was accelerated very stably by the KEKB superconducting HOMs damped cavity.[3]

Sub-transfer Line

A one way muti-channel transfer line shown in Figure 1-b has been designed and

Table 1 Selected parameters of newly developed transfer lines for KEKB superconducting cavities

Test stand for SC cavity	Main transfer line (He flow / return ϕ17.3/34 mm, Liq N2 ϕ14 mm) (15 m) (with old type sub-transfer lines)
Cryogenic system for SC cavity at AR	Main transfer line (He flow / return ϕ17.3/34 mm, Liq N2 ϕ14 mm) (17 m) (with old type sub-transfer lines)
Test stand for Crab cavity	Sub-transfer line (He flow ϕ17.3, Liq N2 ϕ14 mm) (8 m) (flow line with mid- joint))
Test stand for SC cavity	Sub-transfer line (He flow ϕ17.3, He return ϕ28, Liq N2 ϕ14 mm) (4 m x 2) (flow and return with mid-joint) 1 set
Cryogenic system for KEKB SC cavity	Sub-transfer line (He flow ϕ17.3, He return ϕ28, Liq N2 ϕ14 mm) (flow and return with mid-joint) 4 set

fabricated for liquid helium supply line to the test stand of KEKB superconducting crab cavity[4,5] and it worked satisfactory. We have decided to adopt these transfer lines for the sub-transfer lines of the KEKB superconducting HOMs damped cavity. Four cryo-modules and its sub-transfer lines have been installed in KEKB high energy ring (HER) in summer of 1998. The commissioning has been started in autumn 1998. The whole system worked very stably during about 5 months operation.[6]

DESIGN CONCEPT OF NEW TRANSFER LINE

Structure

A two-way multi-channel transfer line with liquid helium supply and gas helium return lines and liquid nitrogen supply and return lines for 80K thermal shield is shown in Figure1-a.

Two one-way multi-channel transfer lines with helium line and liquid nitrogen line for the 80K thermal shield are shown in Figure 1- b and 1-c, respectively. The characteristics and the design points of the newly developed transfer lines are listed in Table 2.

80K Thermal Shield

The heat leak from 300K vacuum jacket is intercepted by a 80K thermal shield and

Table 2 Characteristics of newly developed high performance transfer line

High performance	
low heat-leak	aluminum extrusion 80K thermal shield
small cold mass	good thermal contact between shield and piping
ease of operation	build-up structure of 80K shield
Low cost	simple mechanical support
easy to fabricate	thin-wall pipe
easy to install	interconnecting bayonet joint

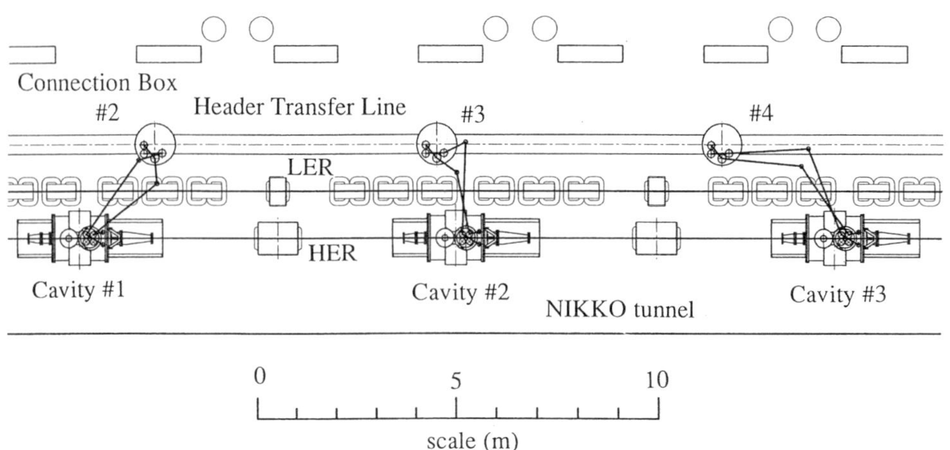

Figure 2. Layout of KEKB HOMs dumped superconducting cavity, header transfer line and sub-transfer lines in the KEKB tunnel.

conducted into liquid nitrogen in stainless steel piping. The liquid nitrogen, about 0.5MPa, is supplied to this pipe by liquid nitrogen circulation system. The 80K thermal shielding forms are made from aluminum alloy (A6063) by extrusion. The liquid nitrogen pipes of 14mm outer diameter are inserted into the grooves of the thermal shielding forms.

The thermal contact between the stainless steel pipe and the shielding form is very important to get good performance. In our design this is assured by good fitting due to very precise dimension and fine surface of the groove made by extrusion and slight deformation of the thin wall piping (about 0.5mm in thickness) and its surrounding part of the thermal shielding form. The clamping force are increased at operating temperature of about 80K because the thermal shrinkage of the aluminum alloy from 300K to 80K is larger than that of stainless steel pipe.

The good performance of the thermal contact between the nitrogen pipe and the extruded aluminum shielding was confirmed by the experiments at liquid nitrogen temperature.

Piping

We used thin-wall pipes to make the cold mass of the transfer lines small. This is very important to improve the thermal performance of the transfer lines and the stability of the cryogenic system for transient conditions such as when flow is restarted after long duration shutdown of the flow and warm up of the line. We asked industry to make the thin-wall, precise low carbon stainless steel (SUS316L; type 316 stainless steel) pipes for helium and nitrogen lines and for insulation pipes of bayonet joint.

The bending of the thin-wall piping were done by using specially designed bender. Before the bending process, inside of the pipes were filled with water and it was frozen to ice by liquid nitrogen.

All the connections of the pipes were TIG welded. We used special "welding flange" made by press forming from stainless steel (316L) thin sheet to solve the difficulty of thin-wall tubes welding.

The change of the pipe length by the thermal contraction during cool down is absorbed by expansion bellows attached at intervals of about every 2m.

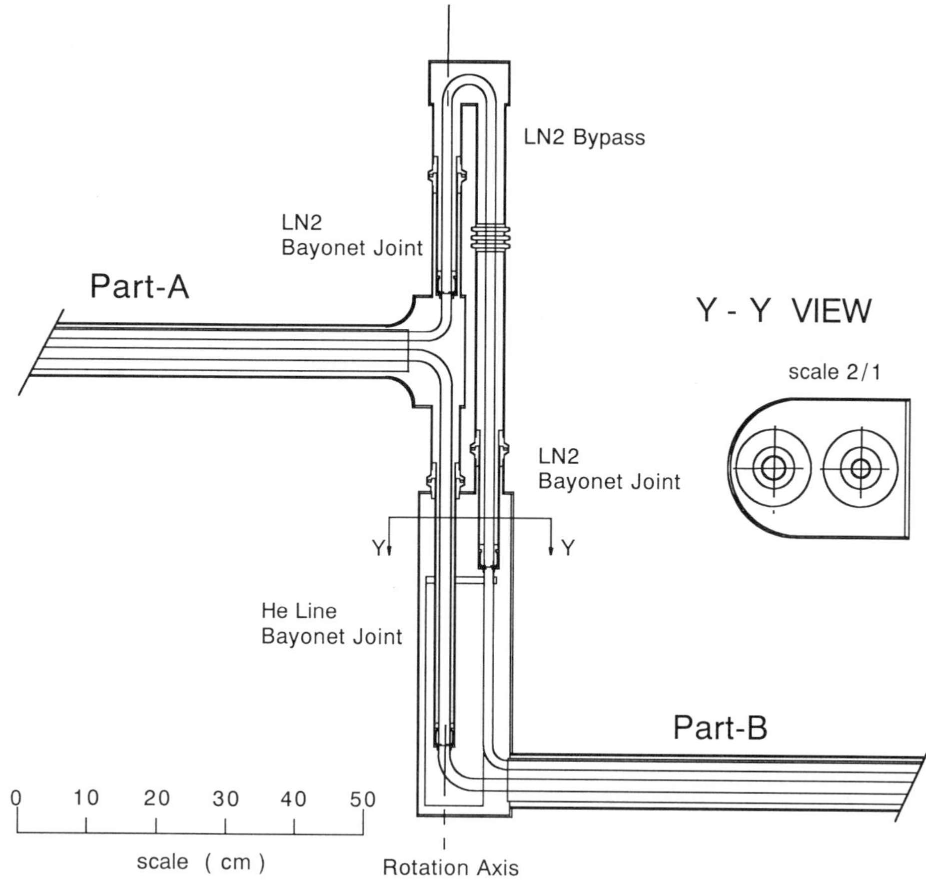

Figure 3. Mid-bayonet joint of sub-tansfer line (Flow line).

Supporting Structure

Mechanical supports of the helium piping are consisted of G-10 plates and collar rings. The collars are made by glass fiber reinforced polyester plate. Two piece of the parts are assembled into the collar by stainless steel clamping clips as shown in Figure 1.

Two G-10 supporting plates are screwed to collar to suspend helium pipes in the 80K shielding.

Four G-10 plates are screwed to the aluminum shielding parts through G-10 support blocks for supporting it in vacuum pipe. By this supporting design we could increase the heat path between 80K shield and 300K vacuum jacket pipe and at the same time, spring action of the G-10 plates make the insertion of the inner part into vacuum jacket very easy.

Inspection and Assembling

Each welded part of helium gas and liquid nitrogen lines is thoroughly inspected and mass spectrometer leak tested at room temperature and under liquid nitrogen temperature.

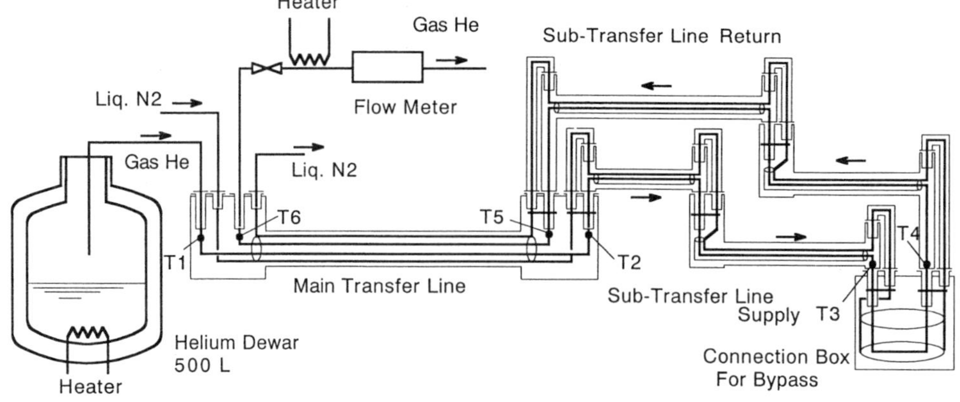

Figure 4. Test setup of the heat loss measurement.

The helium pipes and the 80K shield are wrapped by 10 and 30 layers of double sided aluminized 12μm thick polyester film with polyester inter layer net, respectively.

Bayonet Mid-joint Connection

Figure 2 shows the layout of KEKB superconducting HOMs damped cavities, header transfer line and sub-transfer lines in the straight section of KEKB tunnel. The sub-transfer lines which connect the existing header transfer line and KEKB superconducting cavities have the bayonet joints for helium and liquid nitrogen lines at the middle of it.

Figure 3 shows the details of the joint part of the sub-transfer line. The helium line of part-A is connected downward through the bayonet joint to that of part-B and the liquid nitrogen line goes upward and is connected to the other single transfer line on the same axis by a bayonet joint. The connected liquid nitrogen line is then bypassed to transfer line of part-B through one other bayonet joint. All bayonet joints utilize clamped connection.

By this configuration, the distance between the connection box of the header transfer line and cryostat can be easily adjusted by changing the angle of the joint part, reducing the installation cost. But unfortunately heat leak and fabrication cost of this transfer line increase because of the additional bayonet joint. We used thin-wall (0.5mm wall thickness) pipes for the insulation pipes of bayonet joints and a 80K thermal anchor is placed at the neck of the female bayonet joint to reduced the heat inleak.

The vacuum vessel has a U-shape cover as shown in Figure 3. By this cover the assembling of a very complicated end part of the transfer line becomes very easy. The cover is put in and welded at final stage of fabrication.

HEAT LOSS MEASUREMENTS

Test Setup

A test setup of the heat loss measurement is shown schematically in Figure 4. The test section consists of a 15m long main transfer line (two-way multi-channel transfer line) and about 4m long sub-transfer (one-way multi-transfer line) flow and return lines. These sub-transfer lines have bayonet type mid-joints to adjust its length. The vaporized cold helium gas from 500L dewar is supplied to the test cryostat through the flow line of the multi-

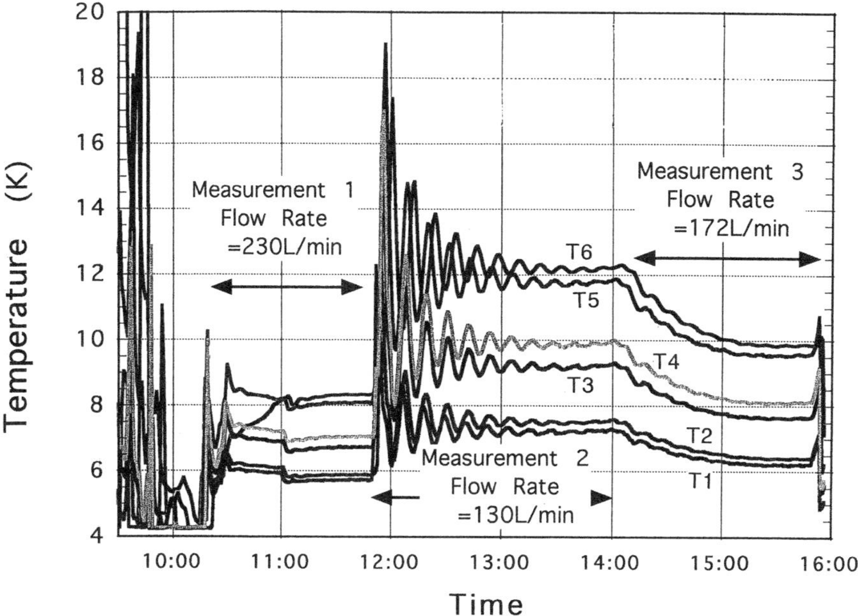

Figure 5. Measured temperatures of helium gas in the transfer line.

channel transfer line and the sub-transfer line. The cold helium gas is bypassed at the test cryostat and then return back through the return line.

The gas flow rate was measured by flow meter after warm up to room temperature by heater. Six carbon glass resistance temperature sensors T1-T6 were set inside the helium pipe to measure the helium gas temperature directly. The heat loss to the helium lines could be calculated by calorimetric method from the temperature rise and mass flow rate of helium gas. The mass flow rate of helium gas was controlled by the heater in the 500L dewar.

Test Result

We have carried out several measurements. At the first stage of the measurement it took long time to attain steady state condition due to the thermal oscillation caused by sudden warm up of the cold helium gas at exit narrow channel. We could overcome this by increasing the exit pipe size.

The measured temperatures of T1-T6 are shown in Figure 5. During this measurement mass flow rate of helium gas was controlled to be 230L/min, 130L/min and 172L/min. There was no significant mass flow dependence. The calculated heat losses are listed in Table 3.

The heat losses Q1-2 to flow line and Q5-6 to return line of main transfer line are 0.41W/m and 0.62W/m, respectively. Heat losses Q2-3 and Q4-5 to about 4m long flow and return lines of the sub-transfer lines are 3.5W and 4.2W, respectively. These measured values include the heat losses from two bayonet joints, i.e. end box of multi-transfer line and mid-joint of sub-transfer line.

SUMMARY

We have developed two types of the high performance helium transfer lines i.e. two-way type for main transfer line and one-way type for sub-transfer lines. All these transfer

Table 3 Measured heat loss to Main transfer line and sub- transfer lines

Q 1-2 /	Main Transfer Line Flow	0.61W	/ 0.041W/m
Q 5-6 /	Main Transfer Line Return	0.93W	/ 0.062W/m
Q 2-3 /	Sub-transfer Line Flow	3.48W	/ include 2 bayonet joints
Q 4-5 /	Sub-transfer Line Return	4.17W	/ include 2 bayonet joints
Q 3-4 /	Connection Box for Bypass	1.16W	/ include 2 bayonet joints

lines have 80K aluminum thermal shield cooled by liquid nitrogen line. Many new ideas were taken into the design not only for improving the thermal performance but also to make assembling and handling easy. These transfer lines were installed in the cryogenic systems for superconducting cavities and have been operated stably for long time without any trouble. The reliability and ease of handling for newly developed transfer lines are confirmed.

The performance test of the main transfer line was carried out at 15m long test section. The heat loss to the helium line at straight section was very small, about 0.05W/m. This will establish the technical basis for the design and construction of a high performance long transfer which will be used for satellite refrigeration system of KEKB crab cavity.[2]

The heat loss at sub-transfer lines were not so small, presumably due to the large heat leak at bayonet joints. We need further investigation to find the heat loss mechanism to reduce the heat loss at sub-transfer line.

ACKNOWLEDGMENT

The authors wish to thanks Professors S. Kurokawa and A. Ezura for their continuous support and encouragement and the staff of the superconducting cavity groups for their continuous support and many useful discussions.

REFERENCES

1. K. Hosoyama et al., A liquid helium transfer line system for super conducting RF cavity, in: "Advances in Cryogenic Engineering", Vol.31, Plenum Press, New York(1986),p.1027
2. K. Hosoyama, S. Mitsunobu, and T. Furuya, Design and performance of KEKB super conducting cavities and its cryogenic system, in: "Advances in Cryogenic Engineering", Vol.43, Plenum Press, New York(1998),p.123
3. T. Furuya et al., Beam test of a superconducting damped cavity for KEKB, in: "Proceedings of Particle Accelerator Conferencce", May 1997 Vancouver
4. K. Hosoyama et. al., Fabrication of a full scale crab cavity for KEKB, in: "Proceedings of The 8th workshop on RF superconductivity", Oct. 1997 Abano Terme (Padova), Italy, p.547
5. H. Nakai et al., Development of superconducting crab cavity for KEKB,: paper presented at this conference.
6. A. Kabe et al., Cryogenic system for KEKB super conducting RF cavity,: paper presented at this conference.

DESIGN AND USE OF A LARGE-SCALE LIQUID HELIUM CONVERSION SYSTEM

P. N. Knudsen

NASA Kennedy Space Center, MM-J2
Kennedy Space Center, Florida 32899

ABSTRACT

A large-scale liquid helium (LHe) to high-pressure (HP) gas conversion system has been implemented at the John F. Kennedy Space Center (KSC). Helium is used by the Space Shuttle, Titan, Atlas, and Delta programs for prelaunch processing, during launch count-down, and for postlaunch securing. The first phase of modifications to the Compressor Converter Facility (CCF), operational in April 1998, allowed the facility to accept bulk liquid helium from tanker containers and to off-load the helium at super-critical pressures. The second phase of modifications, planned to be operational by January 2001, will implement a 227-cubic-meter (m^3) on-site liquid helium storage system. This paper describes the design and operation of the current system and discusses the design and implementation for the second phase system.

INTRODUCTION

The design for a modification to the CCF, initiated in September 1995, was motivated by the impending closure of the United States Bureau of Land Management (BLM) Helium Field Operations (HFO) in Amarillo, Texas. Until the enactment of Public Law 104-273, the Helium Privatization Act of 1996,[1,2,3,4] on October 9, 1996, the BLM HFO [formerly under the supervision of the United States Bureau of Mines (BOM)] had provided KSC with helium by means of high-pressure gas railcars. KSC typically uses approximately 2.0 to 2.8 million standard cubic meters (std Mm^3) of helium per year[5] in support of both the Space Transportation System (STS) (namely, the Space Shuttle) and the Cape Canaveral Air Station (CCAS) expendable launch vehicle (ELV) programs (such as the Titan, Atlas, and Delta). At present, the CCF remains the focal point at KSC that provides helium support to the STS and CCAS ELV programs.

Advances in Cryogenic Engineering, Volume 45.
Edited by Shu *et al.*, Kluwer Academic / Plenum Publishers, 2000.

REQUIREMENT FOR HELIUM USAGE

The major requirement for helium usage at KSC is to support prelaunch and launch countdown (LCD) STS ground operations performed on the Space Shuttle's main engines (SSME's), orbiter main propulsion system (MPS), and external tank (ET).

Prior to launch, the ET is purged with helium in the vehicle assembly building (VAB) for approximately 1 hour until it meets the proper purity and moisture specifications. This is necessary because it is shipped by barge from the Michoud Assembly Facility (near New Orleans, Louisiana) with a nitrogen blanket pressure. Due to the shell structure of the ET, it is pressurized to approximately 48.3 kilopascals (kPa) (note: all pressures are in gauge unless otherwise noted) prior to leaving the VAB as an assembled vehicle integrated with the orbiter and solid rocket boosters.

The majority of the helium used for the MPS and SSME's is during LCD and begins 8 hours prior to launch, at the 6-hour mark before engine start (T-6 hours) when cryogenic propellant loading commences. Beginning at this time, until engine start, helium is required for SSME cavity/seal purges and a heated anti-icing purge of the volume between the ET ogive and liquid oxygen (LO_2) tank. After T-6 hours, helium is also used for MPS bottle pressurization [from 13.8 to 31.0 megapascals (MPa), absolute], MPS/SSME valve actuation, and the LO_2 POGO suppression system. Prior to pressurizing the ET LO_2/liquid hydrogen (LH_2) tanks and while loading the propellants, helium is bubbled from the bottom of the LO_2 tank to prevent geysering. Helium from the MPS bottles is used (1) during ascent to orbit to inert/pressurize portions of the SSME's, (2) upon and after engine shutdown for valve actuation and purging the MPS of any residual propellants, and (3) just prior to entry to inert/pressurize the MPS/SSME. During ascent the MPS helium bottles can also be used as a pneumatic backup, locking the SSME actuators in place, in the event of a hydraulic/auxiliary power unit system failure. Most of the helium required during LCD for the orbiter MPS/SSME's is delivered at a vehicle interface (T-0) pressure of 5.17 MPa (absolute, or more). After a successful launch, helium purges are initiated to secure the ground systems and cross-country lines. For an STS abort/recycle, the helium would also be used to secure the orbiter MPS/SSME and ET.

The CCAS ELV programs are provided helium from the CCF by means of railcars, compressed gas trailers (CGT's), and cylinders. The Titan, Atlas, and Delta III vehicles may have a Centaur-type stage (e.g., LO_2/LH_2) requiring a significant amount of helium for processing of the hydrogen systems. Helium for the Titan program is supplied by 25 railcars which are pressurized to 25.86 MPa prior to leaving the CCF. Only approximately a third of the usable helium is off-loaded since the Titan ground system requires a minimum pressure of 16.5 MPa. A significant off-site use of helium, provided by the CCF, is the NASA Shuttle Logistics Depot in Cape Canaveral. This facility uses 41.4 MPa helium CGT's for flow testing of mobile launcher platform hydrogen relief valves. There are many other requirements and uses for helium at KSC, including laboratories and shops.

CCF AND LAUNCH COMPLEX 39 (LC-39) HELIUM DISTRIBUTION SYSTEM

The CCF was installed and became operational between 1965 and 1966. During this same time approximately 9.7 kilometers (km) of high-pressure helium cross-country piping (the pipeline) interconnecting the CCF, launch pads A and B, and the VAB were installed by Catalytic Dow, Inc. Except for the expansion loop piping, the pipeline consisted of DN40 schedule XX carbon steel spools 16-m (minimum) long, connected by Grayloc hub/clamps, with a hydraulically formed, mechanically bonded, 304 stainless steel 1.27-mm thick liner due to cleanliness concerns. Forty-three 5.66 m^3, 41.37 MPa helium vertical-standing vessels at each launch pad and twenty-six 5.66 m^3, 41.37 MPa helium horizontal-standing vessels at the VAB were installed between 1965 and 1966. These vessels are connected to the helium system at only one port, serving as accumulator/storage devices. The pipeline forms the means for the LC-39 helium distribution system, by connecting the launch pads and VAB to the CCF. The vessels have since been de-rated to a maximum allowable working pressure (MAWP) of 40.46 MPa to meet section VIII, division 2 of the ASME Boiler and Pressure Vessel Code.

The CCF, originally called the High Pressure Gas and Converter Compressor Facility, was designed by an architect and engineering company in Detroit, Michigan, and the Army Corps of Engineers in 1963 and built by Morrison-Knudsen, Inc. During the Apollo era the CCF housed a high and low pressure liquid nitrogen (LN_2) conversion system that produced gaseous nitrogen (GN_2), supplied by a 1893 m^3 LN_2 storage sphere. This system operated until 1968, when it was deactivated shortly after the "Big-Three" nitrogen plant (off-site of KSC) came on-line. The inactive LN_2 storage sphere was not removed until 1990. Five 74.6 brake kilowatt (kW) (hereafter referred to as just kW) 5-stage Joy helium compressors, housed in the east end of the CCF, have been used for the Apollo, Skylab, and Space Shuttle programs and are still in service. Each Joy compressor is capable of delivering 12.5 grams per second (g/s) of helium, at an induction pressure above 1.38 MPa, discharging at 41.4 MPa. Prior to the activation of LHe conversion system at the CCF (refer to figure 1), railcars from the BLM supplied the five Joy compressors (and later, three more Henderson compressors). The compressors either pressurized the helium to the pipeline pressure (presently 39.0 MPa, maximum) and/or pressurized (filled-up) CGT's (to 16.5 or 41.4 MPa) and/or railcars (to 25.17 MPa). The first two 149 kW 4-stage, high-pressure Henderson compressors were installed in an addition to the CCF east end in 1991 and 1992, respectively, due to the de-rating of the launch pad and VAB storage vessels in 1987. These provided additional capacity to the STS and CCAS ELV programs. The storage vessels were pressurized to 41.4 MPa prior to de-rating the vessels. As a result of the de-rating, approximately 10,200 std m^3 of usable stored helium was forfeited. A third 149-kW high-pressure Henderson, made possible by an under-budget installation of the other two Henderson compressors, was installed in the CCF east end addition in 1994. Each Henderson compressor is capable of delivering 25.0 to 28.5 g/s of helium, at a minimum induction pressure of 1.72 MPa, discharging at 39.0 MPa. These compressors are configured tc automatically start-up and shut-down as controlled by a relay control logic system. Both the Joy and Henderson compressors have a "scavenger" compressor (i.e., a separate unit on the Joy compressors and 2 of the 6 cylinders on the Henderson compressors), which removes crankcase blow-by and feeds it back to the first stage. CCF personnel have increased the reliability of the Joy and east end Henderson compressors and have improved their output capacity by 0.8 to 1.6 g/s and 1.2 g/s, respectively. The Henderson compressors are only used to supply helium to the pipeline and are not used to pressurize 41.4 MPa CGT's. Several other modifications and enhancements have been added in past years.

The CCF has a total combined helium output capacity of 140 g/s at 39.0 MPa (maximum), filtered and desiccated to meet grade A, MIL-P-27407, as verified and continuously monitored by oxygen, moisture, and hydrocarbon analyzers. The total helium required for a successful Space Shuttle launch can range from 12,700 to 51,000 std m^3, with each abort/recycle consuming an additional 20,000 to 22,700 std m^3 (approximately). STS operations requires enough helium physically located at KSC 4 days prior to LCD ("call-to-stations," the start of LCD, begins at T-43 hours) to allow for three launch attempts (i.e., three abort/recycles). Prior to activation of the LHe Conversion System at KSC, this was accomplished by having at least 21 railcars, or approximately 0.15 std Mm^3 of helium on-site. At present, this requirement is met by having two (full) LHe tankers and nine railcars, filled to 25.17 MPa, on-site, with a third LHe tanker scheduled to arrive at KSC 24 hours after the scheduled launch, in the event of an abort/recycle. Typically the STS program will use about 60 percent of the helium delivered to the CCF, with the balance used by the CCAS ELV programs.

LHe CONVERSION SYSTEM

A study performed by a consulting and design engineering company in 1991 for KSC (known as the Ralph Hahn study) indicated that the most inexpensive method of providing helium to the CCF (considering only the commodity cost) was by commercial over-the-road, bulk LHe tanker delivery. However, it was not until 1995, faced with the impending closure of the BLM HFO in April 1998, that a preliminary engineering design and market survey was conducted to determine if industry could supply the quantities of gaseous helium (GHe) required at KSC. It was concluded that the only economical method was to deliver helium in bulk LHe tankers. This formed the basis for KSC's decision to modify the CCF to accept delivery of bulk LHe by converting the liquid to low-pressure gas, then pressurize the helium to supply the induction of the CCF east end Joy and Henderson compressors. The project was divided into two phases. The first phase would install the equipment necessary to the accept the bulk delivery of LHe. The second phase would install a storage capability to meet the STS prelaunch on-site helium requirements, without relying on railcars.

The facility modification construction and procurement of major equipment began in February 1997 and was completed in April 1998. A contract to supply the LHe had been awarded earlier in September 1997 to Air Products & Chemicals, Inc. (APCI), who would supply all of KSC's helium for at least 3 years. The first operation using LHe occurred in late April 1998, pressurizing a railcar to be used to support a Titan launch. The conversion system was not used to pressurize the pipeline until June 1998, at which time operations decided that there was sufficient confidence in the equipment and the purity of the helium output by the Conversion System. The first STS launch supported by the LHe Conversion System was STS-95 in late October 1998, the John Glenn launch. At present, over 73 LHe tanker deliveries have been off-loaded successfully using the LHe Conversion System.

The LHe Conversion System (see Figures 1 and 2) consists of a manifold allowing three tankers to simultaneously off-load LHe, though normally only two are connected at a given time. To off-load the LHe, the ullage of the tanker is pressurized to 276 kPa from a separate pressurization line. LHe is then off-loaded (at a super-critical pressure) through a 51-mm inside diameter, 4.6-m-long vacuum-jacketed flexhose into an uninsulated DN100 manifold. The (off-load) manifold feeds into a vertical, dual unit, electric defrostable fan-ambient vaporizer (manufactured by Thermax, Inc., model no. CD70-ED-HF-DC). A diverter valve on the vaporizer outlet switches the flow through each unit at a preset time interval defrosting the inactive unit. From the vaporizer outlet, there is a 40-kW electric trim heater (Thermax model TT40). However, in operation there is considerable heat added to the helium in the (uninsulated) off-load manifold, and consequently, the helium leaving the vaporizer is warm enough to not activate the trim heater. From the trim heater the helium flows into a low-pressure (LP) 8.33 m^3 (690 kPa MAWP) surge vessel (no. 1) and then to an induction manifold into three 149 kW each reciprocating compressors. Even at the highest flow rates, the pressure drop from the tanker to the compressor induction is no more than 14 kPa. The LP compressors, manufactured by Henderson International, Inc., are two-stage machines and have a capacity of approximately 45 g/s each, given an induction pressure of 241 kPa, discharging at up to 3.10 MPa. Typically, these compressors will be operated at a discharge pressure of only 2.60 to 2.93 MPa in order to run the second stage discharge temperature at not more than 204 degrees Celsius (°C). From the discharge of the LP Henderson compressors, the helium goes into a 5.87 m^3 (3.27 MPa MAWP) surge vessel (no. 2). A recycle loop renders the three west end LP Henderson compressors as a constant-pressure supply source, providing the flow rate of helium demanded by a given configuration of east end high-pressure Henderson and Joy compressors. This is accomplished using a control valve placed between the outlet of the second surge vessel and the inlet of the first surge vessel that regulates the second surge vessel pressure. The three LP Henderson compressors are housed in the west end of the CCF

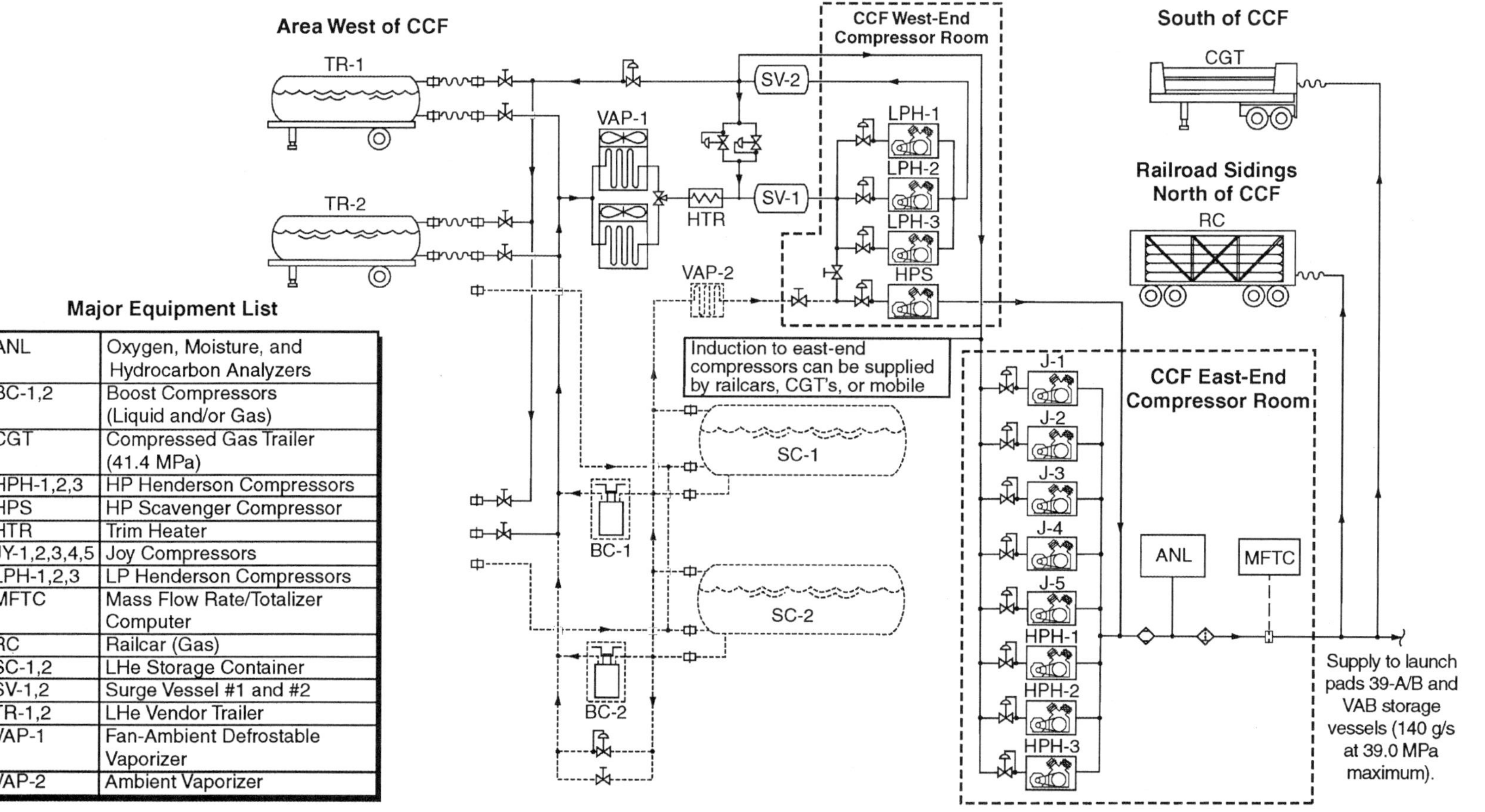

Major Equipment List

ANL	Oxygen, Moisture, and Hydrocarbon Analyzers
BC-1,2	Boost Compressors (Liquid and/or Gas)
CGT	Compressed Gas Trailer (41.4 MPa)
HPH-1,2,3	HP Henderson Compressors
HPS	HP Scavenger Compressor
HTR	Trim Heater
JY-1,2,3,4,5	Joy Compressors
LPH-1,2,3	LP Henderson Compressors
MFTC	Mass Flow Rate/Totalizer Computer
RC	Railcar (Gas)
SC-1,2	LHe Storage Container
SV-1,2	Surge Vessel #1 and #2
TR-1,2	LHe Vendor Trailer
VAP-1	Fan-Ambient Defrostable Vaporizer
VAP-2	Ambient Vaporizer

Figure 1. CCF Helium System Schematic

Figure 2. Aerial view of CCF liquid helium system (looking south).

with a fourth 149 kW Henderson compressor. This compressor is a five-stage HP compressor, with a capacity of 17.6 g/s, at an induction tied into the same induction manifold connected to the other three LP Henderson compressors. The intended use for this compressor was to depressurize the LHe tankers prior to returning to the vendor and to depressurize the LHe storage containers to be installed in the second phase of the project. All of the west end compressors are controlled by an Allen-Bradley programmable logic controller (PLC), which is tied into a master PLC. This provides future capability for automated start up, operation, and shut down of the system.

Operations found that the LP Henderson compressors can also be used to depressurize the LHe tankers to 83 kPa before the return trip. In preparing the LHe tanker for a return, the ullage pressurization supply is shut off, then helium from the tanker liquid withdrawal connection is removed for approximately 30 minutes using one east end Henderson and Joy compressor. After the tanker pressure has dropped to approximately 138 kPa, the east end HP Henderson is shut down, and helium is removed through the tanker vapor withdrawal connection for approximately 1/2 to 1 hour to reach 83 kPa in the ullage. When the tanker pressure reaches 83 kPa, the observed pressure drop between the tankers and compressor induction is 14 to 21 kPa. It is usually necessary to depressurize the tanker down to 83 kPa again the following day depending on how much cold helium is left in the tanker. The tanker to system connections are not usually disconnected until the third day, after initially depressurizing the tanker, to prevent unnecessary air-moisture intrusion. Just prior to completing a partial off-load, the pressurization supply to the tanker ullage is shut off, and the tanker is depressurized to 207 kPa. The west end to east end compressor capacities are matched as close as possible during operation to reduce the amount of energy wasted recompressing the west end to east end difference in helium flow rates. By operating one east end HP Henderson and Joy for every west end LP Henderson running, the west end compressors can be operated cooler. Important parameters to be monitored during operation are (1) the vaporizer discharge temperature, (2) the compressor induction pressure, (3) the compressor final stage discharge temperature, and (4) the second surge vessel pressure.

The pressurization source to the LHe tankers is presently supplied from the outlet of the second surge vessel, which then goes into the tanker ullage through a single stage of pressure regulation. This is a change to the original installation, which used a three-stage regulation panel to reduce the pressure from the distribution piping (at 39.0 MPa) to supply the tanker ullage. The change was made by CCF operations because of a concern that the purity of the helium in the tanker could be degraded. This concern arises for two reasons. First, MIL-P-27407, grade A, allows more neon [23 milligram per liter (mg/L), maximum] than some of the vendor's customers will accept. Also, the helium from the distribution piping is only required to meet an STS specification that is less stringent than grade A, MIL-P-27407.

CCF operations also observed a warm-gas collapse in the tankers. When the tankers first arrive at KSC, they are pressurized with warm helium to 276 kPa in order to sample and verify the purity. A pressure drop of approximately 70 kPa occurs over a quiescent period of a couple hours. Also, when the LHe tankers are depressurized to 83 kPa, the pressure will occasionally drop to approximately 14 kPa overnight, while quiescent. The later occurrence is not as severe since there is not as much liquid and cold gas in the tanker.

Several enhancements to the first phase are planned, including (1) a thermal mass flow meter installed upstream of the "recycle" loop, (2) a Coriolis mass flowmeter on the tanker pressurization line, and (3) a larger (DN40) tanker ullage to LHe off-load line bypass valve. The first two enhancements will simplify keeping track of how much helium has been off-loaded from the tankers. This will allow adequate residual helium to maintain a sufficiently cold tanker temperature during the return trip. The third enhancement will allow the tankers to be depressurized below 83 kPa, which is limited by the present bypass line size.

LHe STORAGE SYSTEM

The second phase of the project, presently in design, will install a 227-m^3 LHe storage system. Several other options of helium storage had been considered and evaluated in a life cycle cost analysis. Next to operating without LHe storage, as currently done, the analysis indicated that LHe storage was $373,000 less expensive than the next best option over a 5-year period. Presently, KSC depends on the "on-time" delivery of two LHe tankers, with a third ready at the vendor plant in the event of an STS launch abort/recycle. Also, railcars are pressurized in preparation for an STS launch and are available if needed.

Concurrently to the procurement of the equipment needed for the first phase of the modification, procurement was initiated for two 113.5-m^3 LHe storage containers manufactured by Gardner Cryogenics, Inc.[3] Delivered in May and June of 1998, respectively, these are the seventh and eighth LHe containers of this size built by the manufacturer. Each LHe storage container has an MAWP of 483 kPa and an empty weight (including the piping skid) of 34,000 kg. The container is 18.9 m long, 3.7 m wide, and 3.8 m high. The heat leak is guaranteed to be no more than 13.1 watts (i.e., 0.4 percent normal evaporation rate) when helium shielded, assuming a 29.4 °C ambient temperature. The storage containers are currently configured to be helium shielded but can be reconfigured to be LN_2 shielded. Each storage container is equipped with load cells and a weight-measuring system, allowing the amount of helium in the container to be monitored.

There are three routine operations (planned) involving the LHe storage containers. The first is off-loading LHe from the storage containers to supply the first phase LHe conversion system. It is uncertain the degree of usage that these containers will serve as the primary supply (rather than LHe tankers), since their original intent was for contingency supply for an STS launch abort/recycle. Instead of pressurizing the storage containers to 276 kPa as done on the LHe tanker containers, a pressure of around 7 kPa will attempt to be maintained as a target pressure.

In order to supply the existing system, a compressor/pump is necessary to boost the pressure to at least 241 kPa (termed a booster compressor). The second operation involving the LHe storage containers is refilling, by transferring the liquid inventory from a tanker in order to replenish the stored LHe. It is critical for an efficient transfer/refill that the storage (i.e., receiving) containers be at as low a pressure as practical. So as to not vent the GHe displaced by the incoming liquid (filling the storage container) and the GHe generated from the pressure drop and heat leak into the liquid transfer line (between the tanker and storage containers), a compressor is also necessary to boost the GHe pressure from the storage container to 241 kPa. Finally, the storage containers must be depressurized on a regular/intermittent basis to remove boiled-off liquid, rather than vent the vapor. For this operation also, a compressor is necessary to supply the existing system. Presently, KSC plans to use several cold gas compressors capable of accepting LHe and/or GHe to serve as a booster compressor. This is an attractive option because it is (1) significantly less expensive than an "ambient temperature" compressor (e.g., most likely a rotary-screw type), (2) is expected to be reliable, and (3) does not require a large area to install or a large power supply. The second phase is planned to begin implementation next spring.

CONCLUSIONS

The LHe Conversion System at the CCF is believed to be the largest facility of its type in the US, perhaps the world. Future programs, such as the Evolved Expendable Launch Vehicle and the Reusable Launch Vehicle, are anticipated to each require an amount of helium comparable to that used by STS operations. A proven reliability would make the CCF the focal point for the helium supply for those future programs.[6]

REFERENCES

1. Helium Annual Review – 1995, *CryoGas International*, Oct. 1996, pp. 19-23.
2. Helium Privatization Act of 1996, *CryoGas International*, Jan. 1997, pp. 15-16.
3. Helium and Cryogenics, *Cold Facts*, Vol. 14, No. 1, 1998, pp. 1, 8-10, 12, 15-18, 21, and 23.
4. Sale of Federal Helium Reserves Will Impact U.S. and International Helium Markets, *CryoGas International*, May 1999, pp. 30-33.
5. Helium, *Compressed Air Magazine*, July/Aug. 1995, pp. 6-12.
6. J.G. Jorgensen and H.T. Everett, Helium Pipeline for Evolved Expendable Launch Vehicle Program, paper presented at Space Congress '99, Cocoa Beach, Florida (1999).

CRYOGENIC SYSTEM FOR A HIGH-TEMPERATURE SUPERCONDUCTING POWER TRANSMISSION CABLE

J. A. Demko[1], J. W. Lue[1], M. J. Gouge[1], J. P. Stovall[1], R. Martin[2], U. Sinha[2], and R. L. Hughey[2]

[1]Oak Ridge National Laboratory
P.O. Box 2009, Oak Ridge, TN 37831-8071
[2]Southwire Company
One Southwire Drive, Carrollton, GA 30119

ABSTRACT

High-temperature superconducting (HTS) cable systems for power transmission are under development that will use pressurized liquid nitrogen to provide cooling of the cable and termination hardware. Southwire Company and Oak Ridge National Laboratory have been operating a prototype HTS cable system that contains many of the typical components needed for a commercial power transmission application. It is being used to conduct research in the development of components and systems for eventual commercial deployment. The cryogenic system was built by Air Products and Chemicals, Allentown, Pennsylvania, and can circulate up to 0.35 kg/s of liquid nitrogen at temperatures as low as 67 K at pressures of 1 to 10 bars. Sufficient cooling is provided for testing a 5-m-long HTS transmission cable system that includes the terminations required for room temperature electrical connections. Testing of the 5-m HTS transmission cable has been conducted at the design ac conditions of 1250 A and 7.5 kV line to ground. This paper contains a description of the essential features of the HTS cable cryogenic system and performance results obtained during operation of the system. The salient features of the operation that are important in large commercial HTS cable applications will be discussed.

INTRODUCTION

A series of tests of the Southwire Company's first 5-m high-temperature superconducting (HTS) transmission cable were conducted at Oak Ridge National Laboratory (ORNL). A simplified diagram for the HTS cable cryogenic system is provided in Fig. 1 that shows the main flow loop and instrumentation locations used in the pressure drop and calorimetric measurements of the HTS cable system. Figure 2 illustrates how the liquid nitrogen flows through the cable in a counterflow cooling mode.

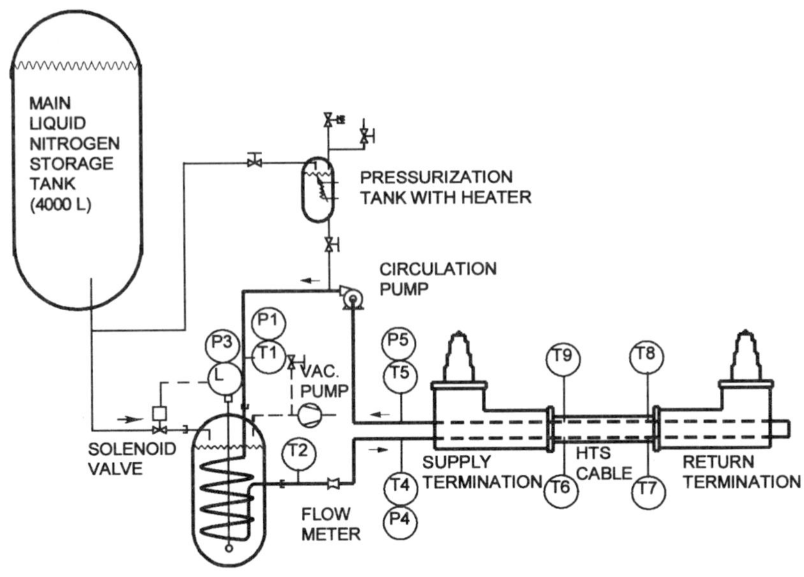

Figure 1. Simplified schematic of the cryogenically cooled HTS cable system.

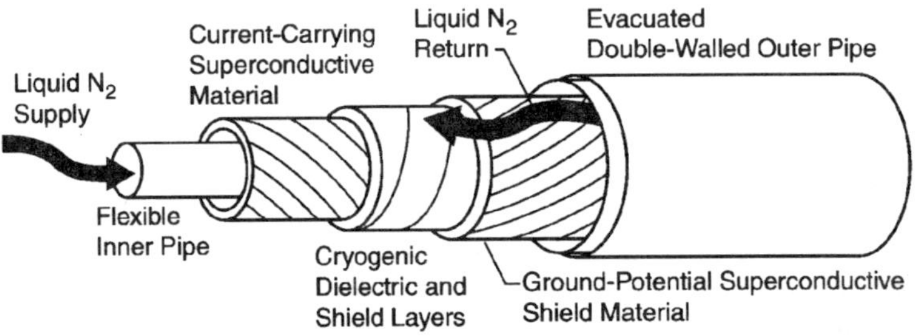

Figure 2. Simplified schematic of the cryogenically cooled HTS cable.

This paper presents the results of five runs of the system. The main operating parameters for the runs are listed in Table 1. The electrical characteristics and ac loss of the cable are discussed in another paper by Lue et al.[1] The first run was conducted to establish a baseline of the system performance without any current applied to the cable. This permitted measurement of the background heat loads and pressure drop dependency on flow rate. The second and third runs were conducted to extend the operating time of the cable. Steady state operation was not confirmed, so the fourth run was undertaken to

Table 1. Description of main operating parameters for each test run

Run	Duration (h)	Flow (kg/s)	Current (A rms)	Voltage (kV)
1	8	Varied	0	0
2	10	0.187	1250	7.5
3	12	0.210	1250	7.5
4	26	0.206	1250 to 0	7.5 to 0
5	80	Varied	1250 to 0	7.5 to 0

operate the cable at voltage and current for 26 h to achieve steady state operation for an extended period. A fifth run, conducted for 72 h, demonstrates the capability of operating the cable for extended periods at current and voltage. These runs simulated the normal operation of the cable expected in the field.

INSTRUMENTATION

This section is a brief discussion of the placement and accuracy of the instrumentation used in the HTS cable system. Table 2 lists the sensor types and accuracies as listed by the manufacturers. Thermometers T4 and T5 are factory-calibrated platinum resistance temperature devices (RTDs) mounted in stainless steel sheaths. They are installed in the liquid manifolds on the terminations as shown in Fig. 1. Thermometers T6 and T7 are factory-calibrated platinum RTDs mounted on specially designed Teflon™ holders to measure the temperature of the liquid nitrogen going into and out from the HTS cable through the HTS cable former. The Teflon™ holders were required to provide electrical isolation from the HTS cable when it is at high voltage. Thermometers T8 and T9 are factory-calibrated platinum RTDs that are mounted on the cable centering spiders and measure the liquid nitrogen temperature of the return flow over the outside of the cable. The flow rate of liquid nitrogen is measured with a turbine flowmeter given in Table 2 and has a factory calibration. The inlet and outlet pressures are measured with strain gage pressure transducers.

HTS CABLE CRYOGENIC SYSTEM TESTS

The pressure drops in the HTS cable system were measured as a function of flow without carrying current in Run 1. The HTS cable pressure drop is measured across the supply and return lines to the HTS cable (P_4 - P_5). The data fit well with the following expression for the pressure drops in the system:

$$\Delta P = \Delta P_g + K_L \dot{m}^2 \quad , \tag{1}$$

where the pressure drop ΔP is a combination of the contribution due to changes in elevation within the system ΔP_g and the mass flow rate through a modified loss coefficient K_L for the HTS cable and terminations [$K_L = 2.8 \times 10^{-6}$ bar/(g/s)^2].

Table 2. Instrument list and accuracies

Instrument	Type	Range	Accuracy	Vendor
T-4				
T-5				
T-6	Platinum RTD	14 K–325 K	±20 mK at 100 K	Lakeshore
T-7	(PT-103)		±35 mK at 300 K	Cryotronics
T-8				
T-9				
PS-1	Strain gage	0–10 bar gage	±1% FS	Omega
PS-4				
PS-5				
PS-3	Strain gage	0–1 bar absolute	±1% FS	Omega
T1	RTD	32.6 K–1128.6 K	±0.1 K	Doric
T2				
Flow	Turbine meter	3.7–38. L/min	±0.15 %	Sponsler Company

Temperatures are monitored at various locations throughout the system in order to perform calorimetry. During installation of the cable into the cryostat, one wire on the RTD T7 broke. Because this was a four-wire sensor, only one lead from the side with the broken lead was used to supply the measurement current and read the voltage across the sensor. This method gives the wrong resistance from the temperature. The value of T7 was corrected using either of two techniques. The first was to use the average heat load from the terminations and solve for the value of T7 based on calorimetry. For Run 1 the average termination heat load was around 291 W. A second method was to use an in-situ calibration. Both methods provide similar results, but are not accurate. Corrected values for T7 are not used in any calorimetry, but only for displaying the temperature distribution through the cable.

A 26-h test, Run 4, was conducted. The temperatures for this test are shown in Fig. 3. During the first hour, there is some cool down of the system. Temperature cycling was observed due to the periodic filling of the subcooler. During the subcooler filling process, the bath-side pressure changed either raising or lowering the saturation temperature of the bath, which in turn, changes the supply temperature of the liquid nitrogen to the HTS cable. After 14 h, the temperatures of the liquid nitrogen external to the cable (T8 and T9) reached a maximum and even decreased afterward. This demonstrates that the HTS cable system will operate stably over long time periods. Near the end of Run 4, the current and voltage were turned off, and the system was allowed to cool down. This is shown by the decrease in temperatures during the last 2 h of the run. This run provides some data for comparison of the system temperatures with and without the cable carrying full current.

The inlet and outlet temperatures for Run 5 are shown in Fig. 4 along with the corresponding flow of liquid nitrogen. At about 35 to 42 h into the run, there was an electrical problem in the power supply, and the current was turned off. This is the cause for the drop in the return temperatures during this period. This shows that the HTS cable system responds quickly to changes in current. From 47 h until the end of the run, the flow was changed to measure the cable response. The cable was operated with flows down to 9.5 L/min (0.126 kg/s) without any difficulty. Calorimetry is not reported for these flow variations because the time at each flow setting was not sufficient to ensure that steady state had been reached.

The temperatures throughout the system for Run 1 are shown in Fig. 5 for three different flow rates. For this graph, T7 was corrected using the calorimetric method using an average termination heat load of 291 W. Following the path taken by the liquid nitrogen,

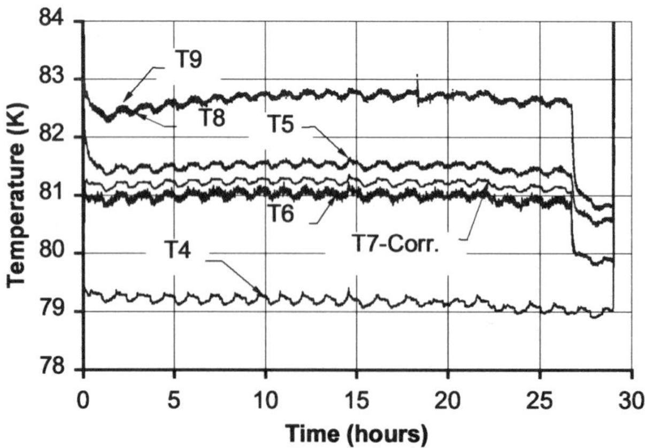

Figure 3. Measured HTS cable system temperatures for Run 4.

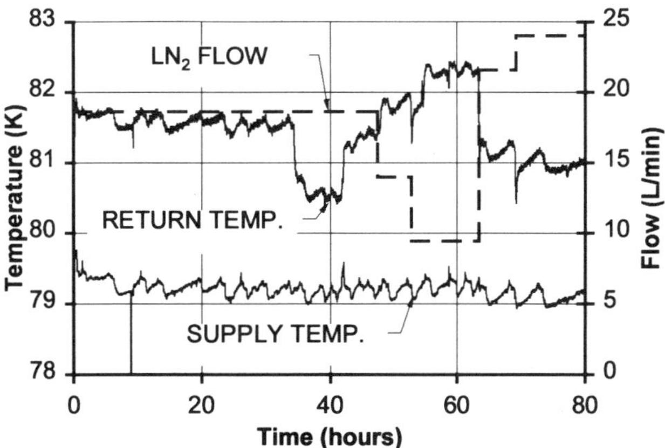

Figure 4. Measured HTS cable nitrogen supply, and return temperatures, and flow rate for Run 5.

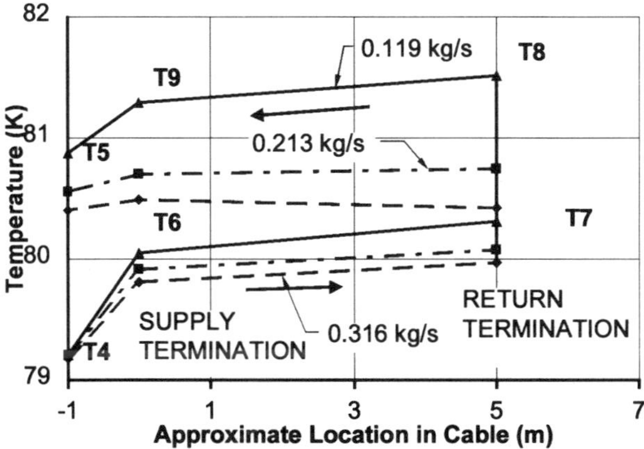

Figure 5. Measured temperature distribution in the HTS cable system for Run 1 at different flow rates using an average termination heat load of 291 W to estimate T7.

it enters the cable at T4, then leaves the termination at T6, enters the far end termination after T7, leaves the far end termination at T8, continues to flow over the outside of the HTS cable to T9, passes through the supply termination, and leaves the system at T5. The highest flow rate measured, 0.316 kg/s, produces the lowest system temperatures. In this case, the temperature of the liquid nitrogen increases all the way through the HTS cable. These temperature profiles can be expected in the non-ideal heat exchanger with heat transfer from an external source.[2] For this case, the liquid nitrogen gradually increases in temperature until it enters the supply termination where heat transfer occurs to the supply stream. The highest temperature in the system is just prior to entering the supply termination. For the medium flow case of 0.213 kg/s, the temperatures on the outside of the cable remained nearly the same. The former stream picks up most of the cable heat load. Finally for the lowest flow, 0.119 kg/s, the highest system temperature occurs at the exit of the return termination. This is what was determined earlier through analysis, because the HTS cable behaves similarly to a counterflow heat exchanger.

The portion of the flow path from T9 to T5 is the final section of the flow path for the liquid nitrogen before returning to the cryogenic skid. For the most efficient operation, it would be better if the temperature increased during this leg. This is not the case here. Instead, the stream leaving the cable heats up the stream supplied to the cable for cooling. This suggests that there may be a better way to design the cooling flow path through the termination.

The temperature distribution through the HTS cable system for Run 4 is shown in Fig. 6 for cases with and without current and voltage applied at a constant flow rate. In this figure the in-situ approximate calibration was applied to correct T7. The maximum system temperature is about 1.7 K lower for the zero current and voltage case. The overall system temperature drop is about 0.9 K.

Cryogenic system heat loads have been measured on large-scale systems such as in a prior study [3]. This reference covers many issues surrounding the accuracy of these measurements, and similar procedures are followed for the HTS cable system.

For the HTS cable the application of the first law of thermodynamics results in several energy balances for the heat loads on different components of interest in the HTS cable system. For the entire system, including the cryogenics, the heat load is given by:

$$Q_{TOT} = Q_{HTS,SYS} + Q_{CRYO} = \dot{m}(h_1 - h_2) = \dot{m}c_p(T_1 - T_2) \ . \tag{2}$$

The overall energy balance for the cable is obtained from measurements of the temperature rise and pressure drop across the entire system. From the piping and instrumentation diagram in Fig. 1, the appropriate measurements are the mass flow, P4, T4, P5, and T5. The resulting energy balance is given by Eq. (3).

$$Q_{HTS,SYS} = Q_{Terms} + Q_{HTS} + Q_{C-STAT} = \dot{m}(h_5 - h_4) = \dot{m}c_p(T_5 - T_4) \ . \tag{3}$$

In the energy balance, Q_{Terms} is the heat load for both terminations, Q_{HTS} is the cable heat generation, which is mainly the ac loss, and Q_{C-STAT} is the heat transfer through the walls of the 5-m cable cryostat.

One difficulty is encountered with calorimetric measurements because of the large uncertainties present when small temperature differences are present, such as normally occur in low-temperature systems. In general, the total uncertainty is calculated by a square root of the sum of the squared uncertainty terms. The temperature difference across the HTS cable system is small so that the specific heat of the liquid nitrogen is approximately constant. The heat load uncertainty can be estimated from the following equation in terms

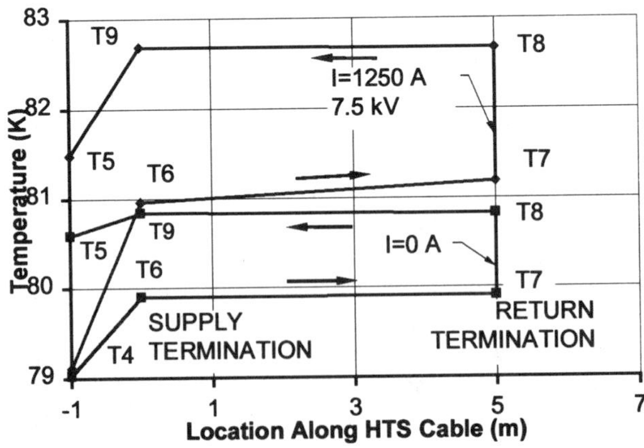

Figure 6. Temperature distributions throughout HTS cable system at the same flow rate with and without current applied. Data are from Run 4.

of the specific heat of liquid nitrogen, the measured quantities, and their respective errors as

$$\delta Q_{HTS,SYS} = \sqrt{\left(c_p (T_5 - T_4)\delta\dot{m}\right)^2 + \left(\dot{m}c_p \delta T_5\right)^2 + \left(\dot{m}c_p \delta T_4\right)^2} \tag{4}$$

The measured heat loads and their respective experimental uncertainties, for a 95% confidence limit, are provided in Tables 3 to 6 for Runs 1 to 4. In each case the uncertainty has been broken down into the contributions due to mass flow errors and temperature errors. It is clearly shown that the measurement error is far more sensitive to the accuracy of the temperatures than to the mass flow. Therefore, for accurate heat load measurements using a calorimetric approach, precision thermometers are required.

An estimate of the termination heat loads can be obtained using the available data as well, if it can be assumed that the heat loads from the pipe are on the order of 1 W/m, similar to values given elsewhere [4], or 5 W for the total 5-m cryostat. Also assuming the ac loss to be near 1 W/m or 5 W total at 1250 A rms, the load from the HTS cable section is estimated as 10 W. This value can be subtracted from the total measured HTS cable system heat load. This can be halved to estimate the average termination heat load. Using these measured data, the average termination heat loads with and without current are 491 W and 305 W, respectively. This is an increase of 186 W per termination due to joule heating and corresponds to an apparent termination resistance of 0.00031 Ω.

Table 3. Experimental heat load measurement uncertainties from Run 1

	Q at 0 A rms (W)	δM term (W)	δT term (W)	δQ (W)
Total system	1503	130.9	331.1	356.1
HTS cable and terminations	586	51.0	40.8	65.3
Terminations by average	291	-	-	-

Table 4. Experimental heat load measurement uncertainties from Run 2

	Q at 1250 A rms (W)	δM term (W)	δT term (W)	δQ (W)
Total system	1665	11.9	338.3	338.6
HTS cable and terminations	873	6.2	106.1	106.3
Terminations by average	432	-	-	-

Table 5. Experimental heat load measurement uncertainties from Run 3

	Q at 1250 A rms (W)	δM term (W)	δT term (W)	δQ (W)
Total system	1862	96.9	170.8	196.4
HTS cable and terminations	1116	58.1	81.6	100.2
Terminations by average	553	-	-	-

Table 6. Experimental heat load measurement uncertainties from Run 4

	Q at 1250 A rms (0 A rms) (W)	δM term (W)	δT term (W)	δQ (W)
Total system	1709 (1400)	27.5	174.0	176.2
HTS cable and terminations	985 (652)	24.0	92.3	92.4
Terminations by average	487 (320)	-	-	-

IMPLICATIONS TO LARGE-SCALE INSTALLATIONS

Using the measured heat load data for the terminations and existing commercially available vendor data for vacuum-jacketed flex hose, the refrigeration load can be extrapolated for large-scale installations. Normally, power transmission systems have three electrical phases. An estimate of the required refrigeration load can be performed assuming that each phase requires two terminations making the transition to room temperature and a separate, evacuated, multilayer, superinsulated cryostat is used for each phase of the cable. On a 1000-m basis, the heat loads are shown in Table 7. The ac loss was taken to be 0.5 W/m at the operating current. The cable cryostat heat loads become dominant at longer lengths. A study by Longsworth and Schoch[5] for a liquid-nitrogen-cooled aluminum conductor cable with a 2000-MVA rating shows that using a closed cell urethane foam insulation instead of evacuated multilayer superinsulation as used in this study resulted in cryostat losses an order of magnitude higher. The development of low heat leak, flexible cryostats for use in HTS cable applications is an important task to increase the benefits of these systems.

Table 7. Heat loads for a 1000-m-long, three-phase HTS cable installation at ~80 K

	I = 0 A	I = 1250 A
Number of HTS phases	3	3
Length, m	1000	1000
Total three-phase heat load cable only, kW	6.9	8.4
Total termination heat load, kW	1.8	2.9
Total heat load, kW	8.7	11.3

ACKNOWLEDGMENTS

This research was performed under a cooperative agreement between Southwire Company and ORNL, the latter is sponsored by the DOE Office of EE-RE, under contract DE-AC05-960R22464 with Lockheed Martin Energy Research Corp.

REFERENCES

1. J. W. Lue et al., 5-m single-phase HTS transmission cable tests, Invited paper presented at CEC/ICMC 1999 Meeting, Montreal Canada.
2. T. Ameel and L. Hewavitharana, Performance of Counter-Current Heat Exchangers Subjected to External Heating, in "Proc.Eighth International Energy Week Conference and Exhibition", Book VI - Intersociety Cryogenics, Penn Well Conferences and Exhibitions, Houston, (1997), p. 43.
3. K. Stifle and J. Demko, Evaluation of the accelerator systems string test (ASST) heat leak measurements and their uncertainties, Rept. SSCL-N-871, Superconducting Super Collider Laboratory, Waxahachie, Texas, July 1994.
4. CVI Catalog, Columbus, Ohio, 1998.
5. R. C. Longsworth and K. F. Schoch, A study of refrigeration for liquid nitrogen cooled power transmission cables, in "Advances in Cryogenic Engineering," Vol. 25, Plenum Press, New York (1980), p. 585.

DESIGN, CONSTRUCTION AND OPERATION OF A NEON REFRIGERATION SYSTEM FOR A HTS POWER TRANSMISSION CABLE

A. Anghel,[1] B.Jakob,[1] G. Pasztor,[1] R. Wesche,[1] A. M. Fuchs,[1] G.Vecsey,[1] H.Bieri,[2] H.Bauman,[3] F.X. Hansen[3]

[1]Centre de Recherches en Physique des Plasmas (CRPP)
Association EURATOM, CH-5232 Villigen, Switzerland
[2]Romabau AG, CH-8570 Weinfelden, Switzerland
[3]Sulzer-Burckhardt AG, CH-4002 Basel, Switzerland

ABSTRACT

A prototype of a gas-cooled high-temperature superconducting (HTS) power transmission cable has been manufactured and tested at variable cryogenic temperature in the range 30-70K. For this temperature range a Ne refrigeration system based on the Joule-Thomson cycle has been selected. The Ne refrigerator consists of two Ne-Ne counterflow heat-exchangers and a Ne-LN2 heat exchanger as pre-cooler. The main Joule-Thomson throttle valve reduces the pressure of the precooled Ne from 200 bar to the operating pressure of the cable rated at 15bar. The throttle effect lowers the temperature to about 45K. Lower temperatures are possible at a lower rated operating pressure whereas higher temperatures can be obtained by activating a heater. In order to improve the refrigeration power and to match the cable pressure to the suction pressure of the Ne compressor a second valve was used on the low pressure side of the refrigerator. The design and construction of the cryogenic system is presented, focusing on the practical aspects of heat-exchanger manufacture, cooling of the current leads, gas management system and upgrade of a commercial high-pressure, oil-free compressor for work with Ne.

INTRODUCTION

Upgrading power systems by retrofitting existing ducts is a quite promising application of high temperature superconductors. Investment and operating cost optimization studies[1] performed at CRPP indicated that the operating temperature of such cables should be in the range 30-60K, suggesting that neon gas cooling may be used. In a collaborative task between CRPP and Brugg Kabel AG a 5m long HTS prototype cable cooled by a Ne refrigeration system has been designed, manufactured and tested.

THE CRYOGENIC SYSTEM

The refrigeration system consists basically of the Ne refrigerator, two transfer lines, the two terminals provided with gas cooled current leads and gas distribution manifolds, the power cable in its cryogenic enclosure and the gas management system. The system is also equipped with a LN2 storage dewar and the high vacuum equipment necessary to achieve a vacuum better than 10^{-6} mbar in the superinsulated cable, refrigerator cold-box and terminals. A simplified flow scheme of the cryogenic system is illustrated in Fig. 1.

The Ne Refrigerator

The Ne refrigerator is based on a simple Joule-Thomson cycle with LN2 precooling. As shown in Fig.2, it consists of two stainless-steel Ne-Ne heat exchangers of tube-in-tube type, HX1 and HX2, a Ne-LN2 heat exchanger as pre-cooler and two cold valves: HCV601, the main throttle valve and HCV604, the secondary throttle valve. The secondary valve was used to bridge between the pressure in the cable which was operated between 5 and 10 bar and the suction pressure of the Ne compressor which is limited to about 1.1-1.2 bar.

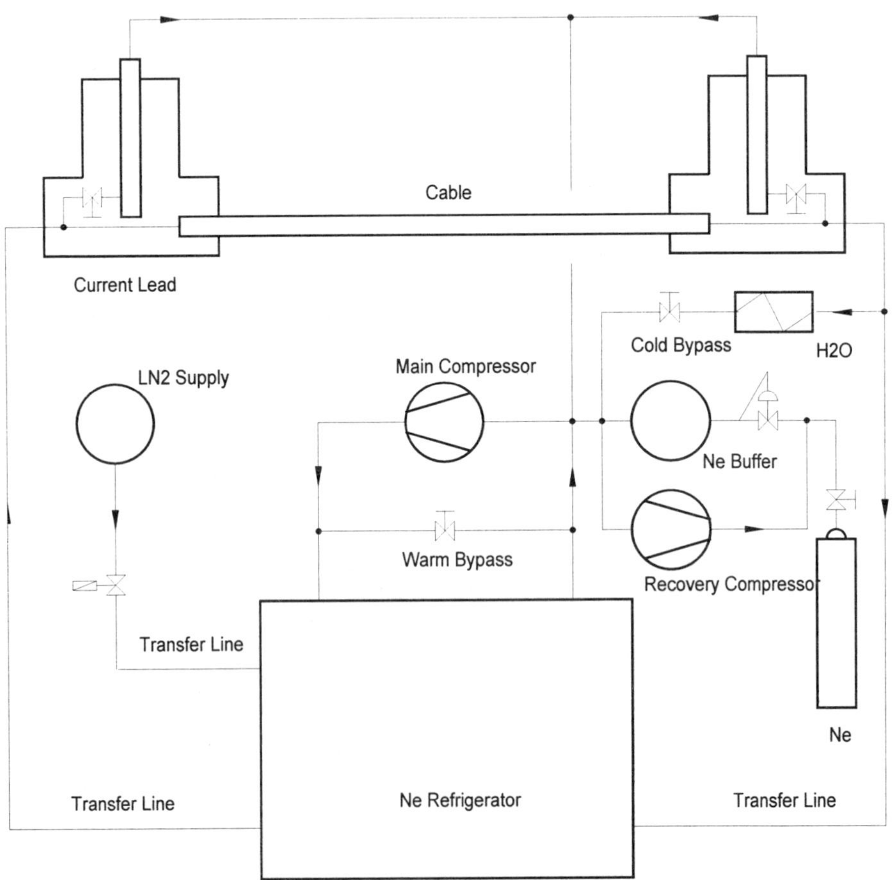

Figure 1. Flow scheme of the refrigeration system for the HTS Power Cable Demonstrator.

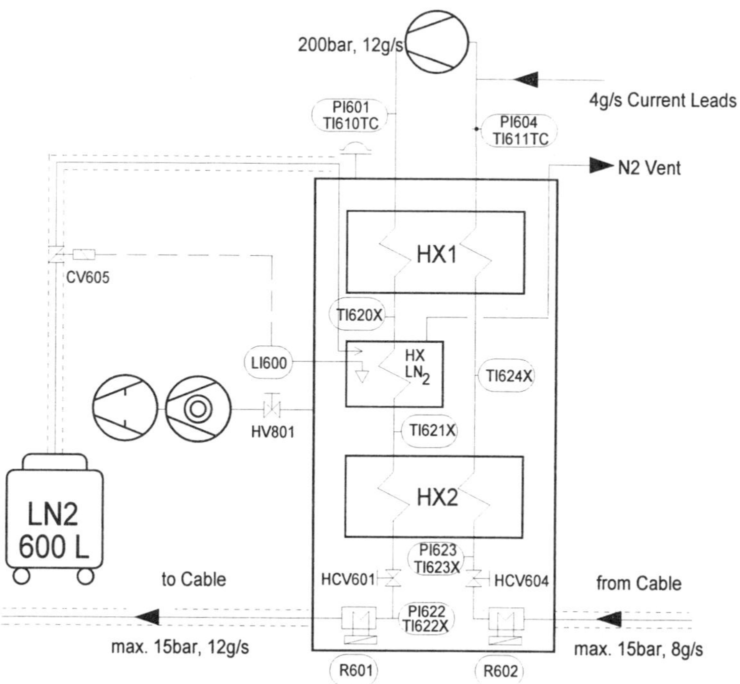

Figure 2. Flow scheme of the Ne refrigerator.

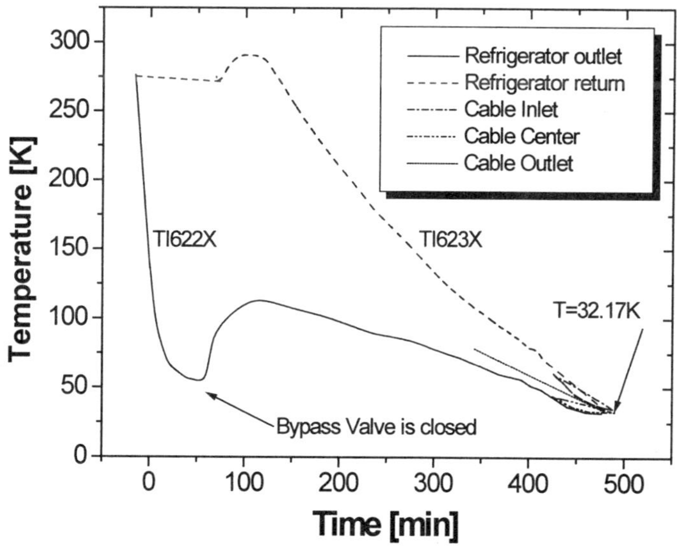

Figure 3. Cool-down history curves for the Ne refrigerator and cable after filling the LN2 precooling bath. Due to the excellent thermal insulation, subsequent runs began at lower temperatures even after a break of 2-3 days.

A typical cool-down curve of the refrigerator is illustrated in Fig.3. In order to shorten the cool-down time, a cold bypass valve was installed between the cable outlet and suction side of the Ne compressor. Closing this valve leads to the temperature increase shown in

Figure 4. Photograph of the Ne refrigerator showing the heat-exchanger HX1, the two heaters and the LN2 precooling bath.

Figure 5. One of the two power cable terminals (cable outlet), which enclose the Ne forced-flow gas cooled current lead and the throttle valve for thermal anchoring the copper joint.

Figure 6. A complete view of the Ne refrigeration system and test set-up of the 5 m long high temperature superconducting prototype cable.

Fig.3 on both refrigerator outlet and return lines. In this run the performance of the refrigerator to reach the lowest possible temperature was investigated. With a pressure of 5 bar in the cable a temperature of 32 K was reached leading to Ne liquefaction.

A view of the Ne refrigerator before being wrapped with superinsulation is shown in Fig.4. The heat-exchangers are made of stainless-steel and are mounted to the flange where the cold temperature valves are also fixed. At the bottom are shown two 340W heaters. The outlet and return connections of the refrigerator are of bayonet type. The LN2 bath is a double-walled hollow cylinder having a capacity of 60L. It works also as a thermal shield for the colder parts of the refrigerator i.e. second heat-exchanger and the two throttle valves.

The refrigerator is connected to the power cable by two superinsulated transfer lines, each about 6 m long with male bayonet couplings at both sides. Both ends of the 5 m long power cable are provided with cryostats each incorporating a 6 kA rated Ne cooled current lead, a throttle valve to produce additional cold at the bottom of the current lead and a small Ne chamber enclosing the copper-to-superconductor connections. A photograph of the outlet terminal is shown in Fig.5 and a general view of the whole system in Fig.6. The rubber-foam insulated conduit (black in figure) is connected to the cold bypass valve which accelerates the cool-down of the cable. Cold Ne gas leaving the cable through the bypass valve is warmed-up in a water-bath heat exchanger and then returned to the suction-side of the Ne-compressor.

System Design

The process in the Ne refrigerator is based on a LN2-precooled Joule-Thomson cycle with uneven flow distribution between the high- and low-pressure streams, and a second cold throttle valve (HCV604) on the low pressure side. The uneven flowrate is a consequence of using a fraction of the cold gas (about 3 g/s in normal operation, max. 4 g/s) after the main Joule-Thomson valve (HCV601) to cool the current leads. Cold Ne is throttled at the bottom of the current leads from the operating pressure of the cable (5-10 bar) to the suction pressure of the main Ne compressor producing an additional temperature drop and therefore an improved cooling. The two small cold valves attached to the bottom of the current leads (see Fig.1) accomplish this function.

The calculation of the cycle is standard. For the calculation we assumed real heat exchangers, expressed as an assumed temperature difference at the warm side of the heat exchangers: $\Delta T_1 = 8$ K for HX1 and $\Delta T_2 = 3$ K for HX2. For the precooler heat-exchanger a ΔT_3=1 K was assumed. Calculations have been done for the temperature range under which the cable was supposed to be investigated (45-75K). The results are presented in Table 1.

Table 1. Heat exchanger parameters

Heat Exchanger	**HX1**	**LN2**	**HX2**
Type: stainless-steel	concentric tubes	spiral coil	concentric tubes
High-pressure side, 200bar			
Inlet temperature [K]	300	154	78
Outlet temperature [K]	154	78	61.52
Heat transfer coefficient [W/m2-K]	991	1090	3674
Pressure drop [bar]	0.12	0.028	0.32
Low pressure side, 1bar			
Inlet temperature [K]	75	77.2	34.26
Outlet temperature [K]	292	77.2	75
Heat transfer coefficient [W/m2-K]	128.7	2000	171
Pressure drop [bar]	0.3	-	0.12
General properties			
NTU	7	8.936	3.706
Tube length [m]	21	10	13.5
Power transferred [W]	1791	1177	342
LN2 [Liter/hr]	-	26.5	-

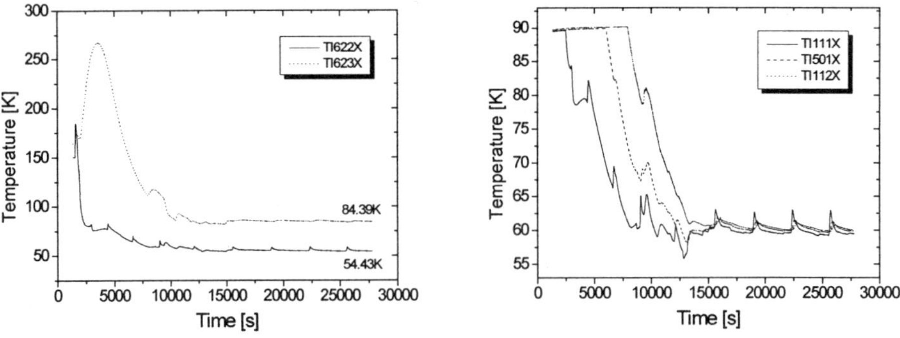

Figure 7. Temperature history during a cool-down in the refrigerator (left) and cable (right). The propagation of the cold slug from inlet to outlet takes about 1.5 hr.

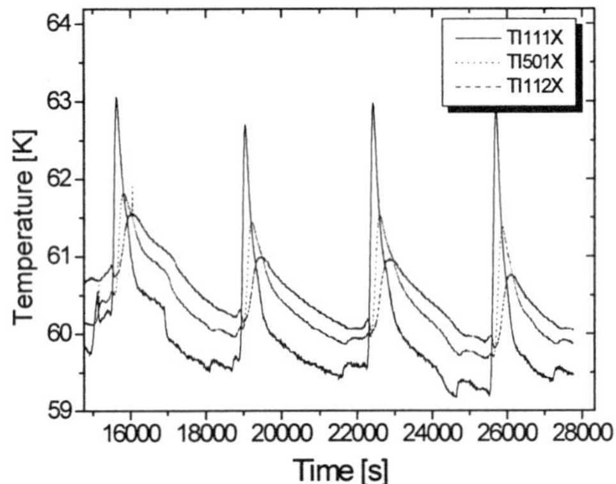

Figure 8. Disturbances in the cable temperature at three different locations: inlet (TI111), middle (TI501) and outlet (TI112) during the filling phase of the LN2 precooler.

Operational Experience and Performance

The operation of the Ne refrigerator was simple. After filling the precooler with liquid nitrogen and starting the compressor, the warm bypass valve is closed progressively in order to set the operating pressure. We worked usually at pressures in the range 150-190 bar. If it is a warm start, the cold bypass valve is open in the first 2-3 hours until the temperature at the outlet from the cable has reached 90 K. Then it is closed progressively. The main throttle valve and the return valve are adjusted in order to keep the compressor and cable pressures at specified values. The heaters are activated, first the return stream heater, R602, then, if necessary, also R601 and the desired temperature is set on. In case of a cold start the procedure is similar, the only difference is that the cold bypass valve is only slightly open or even not used depending on the temperature level at which the system was restarted. In the first 1 or 2 hours after the start, the temperature after the main throttle valve (HCV601) revealed a rapid decrease and reached easily 50 K or lower. At this level we got sometime problems with the valve clogging. Usually, by quick closing one, two turns and then reopening the valve again, the clog was washed-out. At temperatures above 55 K there was no clogging. Therefore in practice we applied another method. With this method, which could be used only when the desired temperature was not lower that 50 K, we slightly closed the cold bypass valve such that the temperature at TI622 never goes bellow say 55 K. When the bypass valve is closed, the heater R602 is used for the same purpose. In order to illustrate the operation of the Ne refrigerator, some history curves are presented in the Figs. 7 and 8.

The design cooling power was 340 W for 12 g/s at 200bar. With the current leads in operation at 4 g/s the cooling power was reduced to about 230W due to the warm gas return. The liquid nitrogen usage was measured as 40 L/hr. The compressor power was not directly measured, but assuming the installed power of the compressor which amounts to 30kW, figures of merit in the range 88 W/W - 130 W/W have been achieved.

NEON COMPRESSOR AND GAS MANAGEMENT

A critical component of the refrigeration system is the Ne compressor. A 5 stages, 200bar, air-cooled, oil-free unit, Fig.9, developed by SULZER-Burckhard for Natural Gas Vehicle (NGV) refueling was selected in view of the pressure and flowrate requirements of our system. This compressor was tested with other gases, such as nitrogen and helium, but there was no practical experience with neon. Factory tests with Ne revealed that, as expected, there is much more heating produced in the second stage of compression but this can be kept under control by working at lower suction pressure. The optimum range of suction pressure was found to be 50-100 mbar. At higher suction pressures the temperature of the second stage increases rapidly and the valve plates of this stage can be damaged. Therefore a rigorous control of the suction pressure is requested.

There is one situation where the control of the suction pressure is tricky, namely the change of the operating temperature of the cable. Suppose the refrigerator works at a very low temperature, say 45 K and a new temperature of say 65 K is wanted. The Ne mass in the system at 45 K is higher as that at 65 K since the Ne density at 45 K is higher than at 65 K. Therefore, by increasing the temperature in the cable the suction pressure increases due to the excess Ne mass. We solved this problem by using a second small Ne compressor, shown in Fig.10, attached on the suction side of the main neon compressor which pumps the Ne excess back into the gas storage bundle whenever it is requested. A pressure sensor on the suction side of the main compressor detects any increase in suction pressure and sends a signal to the small "recovery" compressor, which starts pumping neon back into the gas storage system. Another situation where the small neon recovery compressor is requested is the warm-up of the system. If a cold system is stopped for some reason, the system warms-up and the pressure in the system could increase to a dangerous level. In this situation the recovery compressor starts automatically and recovers the clean Ne back into the gas-storage cylinders. The specifications of the main and recovery Ne compressors are presented in Table 2. The purity grade of the Ne used was 4.0 (<1 ppm O_2, <5 ppm N_2, <2 ppm H_2O, <100 ppm He). The system was vacuum-pumped for 48 hr and swept 3 times with clean Ne before final filling.

CONCLUSIONS

A neon refrigeration system for a HTS cable was designed, built and successfully operated in the temperature range 32-75 K. The refrigerator was manufactured using standard cryogenic components and techniques. A commercially available, oil-free, gas compressor was adapted to work with Ne. The temperature stability of the cable was 0.56 K/hr, excepting the short time periods of filling the precooler with liquid nitrogen when temperature spikes of about 3 K could be observed.

Figure 9. The main Ne compressor.

Figure 10. The Ne recovery compressor

Table 2. Compressor specifications.

	Main Compressor	Recovery Compressor
Type	DM	DF
Discharge pressure	max. 220 bar	max. 200 bar
Suction pressure	50-100 mbar with Ne	max. 5.5 bar
Flowrate	50 Nm^3/hr	3 Nm^3/hr
No. of stages	5	4
Cooling	air	air
Power	30 kW	1.4 kW
Dimensions	2000×1100×2070	500×280×500
Weight	1000 kg	35 kg

The cool-down time of the 5 m long cable was about 8 hr for a warm start and 3 hr for a cold start. The estimated thermal load on the cable through superinsulation was 0.6 W/m. The heat input of one current lead was on the average 8.5 W. The cold bypass circuit was found as an important mean to improve the cool-down time of the system. Due to the high quality of the superinsulation, the facility could withstand shut-downs of a couple of days after which it can be restarted with a drastically reduced cool-down time i.e. from 8 hr to 3 hr.

ACKNOWLEDGMENT

This work was financially supported by the Swiss Federal Office of Energy, Brugg Kabel AG and PSEL.

REFERENCES

1. A.Anghel, G.Pasztor, R. Wesche, B.Jakob, and G.Vecsey, Neon Refrigeration for High Temperature Superconducting Power Transmission Cables, Proc. of 17th International Cryogenic Engineering Conference, Bournemouth, ICEC17, UK, 14-17 July 1998, Ed. D Dew-Hugues, RG Scurlock, JHP Watson, Institute of Physics Publishing, Bristol and Philadelphia, p. 337.

EXPERIENCE FROM THE TORE SUPRA CRYOGENIC SYSTEM, ASSOCIATED WITH THE STATUTORY TEN YEARS TESTS OF THE PRESSURE VESSELS

B. Gravil, F. Minot, D. Henry, J.-P. Serries, P. Prochet, A. Bocquillon
and the Tore Supra Cryogenic Team

Centre d'Etude de Cadarache Association EURATOM-CEA sur la fusion
Département de Recherches sur la Fusion Contrôlée
13108 Saint Paul-Lez-Durance France

ABSTRACT

After twelve years of operation of the TORE SUPRA cryogenic system, the installation was switched off for the statutory ten year tests of the pressure vessels. The vessels concerned are related to the hot circuits, compressor bank, oil removal system and buffer volume. These circuits are brought to atmospheric pressure, subjected to hydraulic tests under water or oil and then reconditioned before reuse. This paper present the main contamination problems encountered during the restart of the installation. It analyse the evolutions of the gaseous impurities and water contents of the various circuits involved in the tests, and report on the overall experience gained during this operation.

INTRODUCTION

The TORE SUPRA Tokamak has been operating since 1988 within EURATOM-CEA association at the Cadarache site in Provence, France. It is a medium sized machine (major radius, 2.4 m; plasma radius, 0.8 m), operating with a superconducting toroidal NbTi magnet providing a maximum field of 4.5 T on the plasma axis. The superconducting toroidal magnet, total mass 185,000 kg, is cooled by a cryogenic system extensively described elsewhere,[1,2] which provides the refrigerating power to the 3 levels of temperature, 80 K, 4.5 K and 1.8 K respectively, required by the magnet. The refrigerating powers are 300 W at 1.8 K, 970 W at 4.0 K and 10 to 32 kW at 80 K. The conductor is a filament of NbTi composite cooled in contact with pressurized superfluid helium at 1.8 K and $1.25{*}10^5$ Pa.

Advances in Cryogenic Engineering, Volume 45.
Edited by Shu *et al.*, Kluwer Academic / Plenum Publishers, 2000.

STATUTORY TEN YEAR TESTS OF THE PRESSURE VESSELS

French regulations impose routine checks of any device under gas pressure. For the fixed containers whose effective pressure of the gas phase is higher than $4.\ 10^5$ Pa and the product pressure, volume (PV)> $80\ 10^2$ Pa-m^3, two checks must be carried out by an approved organisation: internal and external inspection, every 3 years; and a hydraulic test with a pressure of 1.5 times the maximum pressure of service every 10 years.
In 1997, after ten years of operation of the cryogenic system, a postponement of hydraulic tests was requested and granted. The request to delay the hydraulic tests was justified by:

- The fact that the fluid, used in the various circuits concerned, is helium of very high purity, low water content <0.5 ppmv, and low oxygen content <1 ppmv.
- The fear that water contamination may occur as a result of these hydraulic tests
- The requirement to fit this major outage into the long term physics programme.

A two years delay was obtained for the machine room elements associated with the refrigeration cycle. A complete exemption was granted for the bank of storage cylinders of $2*10^7$ Pa helium; the hydraulic test being replaced by acoustic emission testing of the entire bank of cylinders. This non-destructive method, under development at the national level by the CETIM (Centre Technique des Industries Mécaniques), consists of measuring the sound emission due to the propagation of structural defects in the walls as incremental pressurisation steps, are made up to the operating pressure.

TESTS CONDITIONS

The conditions required for the hydraulic pressure tests are completely specified. The various pipes, other than the valves, safety valves and connecting flanges, are regarded as forming part of the device. The components concerned with the hydraulic tests relating to the hot circuits are presented in Table 1 and in Figure 1. For the devices, using very hygroscopic oil,[3] it was not appropriate to use water for the test, because a solvent cleaning of these tanks after the water test was forbidden. Hence, we had the choice between fuel oil or compressor synthetic oil. The gas oil presents a risk of deteriorating the joints and parts made from polymeric materials, and also runs the risk of producing a flammable fog in the event of leak under pressure. We chose to test these devices with compressor oil (BREOX B35). The tests proceeded in several phases as described hereafter.

Preparation Phase: Isolate the apparatus from the circuit. Remove and cap the ports of pressure and temperature sensors and valve elements that might be susceptible to leak or

Table 1. List of tested components

devices	Locate Fig.(1)	Capacity m^3	Max. pressure Operation Pa	Test pressure Pa	Material	Oil vapour
Oil separator	B	2.350	$15*10^5$	$22.5*10^5$	Steel	yes
Heat exchanger MP	C	0.110	$12*10^5$	$18*10^5$	Stainless steel	yes
Oil separator	D	2.350	$22*10^5$	$33*10^5$	Steel	yes
HP Heat exchanger	E	0.168	$19*10^5$	$28.5*10^5$	Stainless steel	yes
Coalescers	F	0.550	$20*10^5$	$30*10^5$	Steel	no
Coalescer outlet filter	G	0.040	$20*10^5$	$30*10^5$	Stainless steel	no
Charcoal filter	H	1.390	$20*10^5$	$30*10^5$	Steel	no
Dryers	I	0.126	$20*10^5$	$30*10^5$	Steel	no
Dryer filter	J	0.040	$20*10^5$	$30*10^5$	Steel	no
Oil filling system	K	0.282	$26*10^5$	$39*10^5$	Stainless steel	yes
Level gauge device	L	0.0048	$26*10^5$	$39*10^5$	Stainless steel	yes
Helium buffer vessel	M	20.242	$20*10^5$	$30*10^5$	Steel	no

accumulate test liquid, or make the vessel difficult to dry (e.g. seats of valves). Remove insulation covering and internal parts, which would also hinder drying. Establish a purge high up in the vessel to prevent gas pockets.

Pressure Test Phase: A hydraulic pump is used to produce a pressure equal to 1.5 times the operating pressure in order to confirm the tightness of the unit.

Drying Phase: Those vessels tested with water are dried by blowing through compressed air under $6*10^5$ Pa and then mechanical drying with clean lint free cloth. They are finally subjected to circulation with gaseous nitrogen at 360 K, while monitoring residual moisture at the exit. For vessels smaller than a cubic meter, a residual content of 10 ppmv of water vapour is obtained after 24 h of circulation. Up to 72 h are needed for larger volumes.

Re-assembly and Start-up Phase: This includes the leak-test and reconditioning of the various circuits. The vessels tested with oil were simply conditioned by cycles ofcompression and relaxation with pure helium before being put back in operation. Once the whole conditioning operation was ended, the compressors were put to use to sweep all the volumes tested and to circulate helium at room temperature through the dryers and active charcoal bed. The gas impurities are analysed by chromatography, the presence of water is measured by a hygrometer.

Impurities Analyse: During the purging and cleaning operation, it was noted that there was no contamination by CH_4 and other light hydrocarbons. Similarly no CO, CO_2 and H_2 were observed. During the first tests of hot circulation, N_2 and O_2 contamination were noted in proportions which do not correspond to the air. N_2 contamination probably originated from the insufficient helium conditioning of the activated charcoal bed after regeneration with ultra pure gaseous nitrogen. Oxygen contamination seems to come from the new molecular dryer sieves, which were conditioned but not regenerated before start-up. This contamination was reduced by progressive replacement of the He load by pure helium coming from an 80K purifier.

Water contamination: At the start of the drying phase, the sweep gas water content at the entry of the drier was about 35 ppmv. This content decreased gradually, reaching < 1 ppmv after 200 hours as shown in Figure 2. Water is, in fact, absorbed by the oil, a very hygroscopic polyalkyleneglycol, and re-released when it comes in contact with dry gas. A non-negligible quantity of water was obtained from the oil of the compressors, which were contaminated during the various preparations for tests. After having obtained a compressor water vapour content of < 0.5 ppmv, the restart of the oil ring pump, P2, exhibited a value of water vapour of 12 ppmv, which then decreased progressively while running in a closed loop as shown in Figure 3. However the analyser placed at the exit of the 1st stage of compression did not exhibit any significant of water content (< 0.5 ppmv). Quantity of oil present at the level of the 1st stage of compression retains the water, releasing it slowly over at least a week in closed cycle with the dryers.

CONCLUSION

The substantial and complex task of carrying out the 10 year statutory pressure tests of the many cryogenic pressure vessels at Tore-Supra was carried out smoothly and efficiently due to meticulous preparation and detailed planning. This substantial installation had to be warmed-up, vented, and much of it dismantled for these tests, allowing confirmation of:

- The satisfying state of the components,

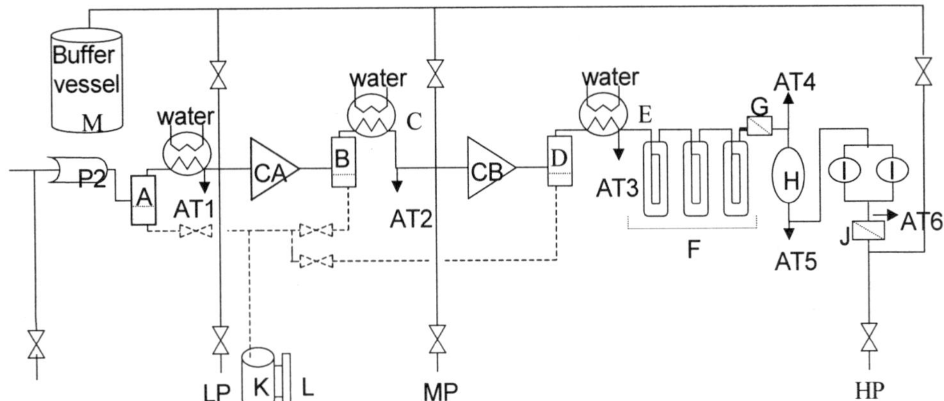

Figure 1. Devices involved in the hydraulic tests in the machine room

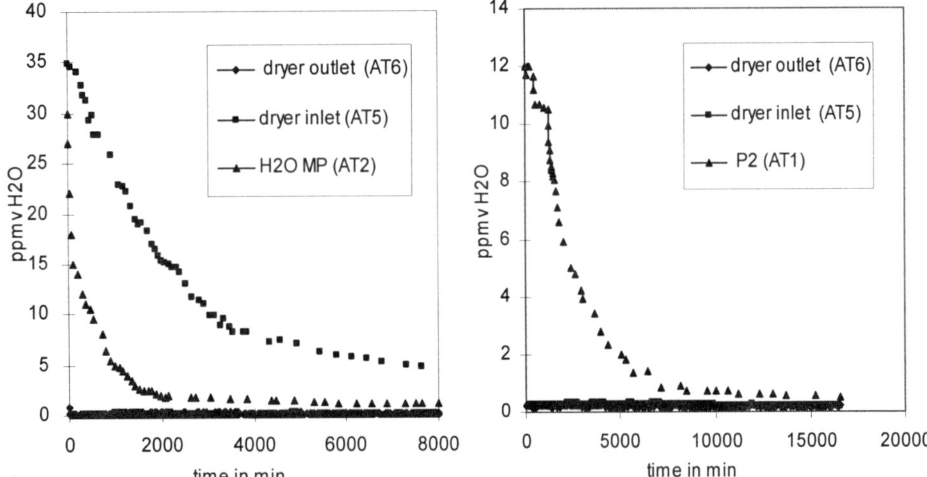

Figure 2. Water content during the re-starting phase.

Figure 3. Water content after starting P2 pump.

- The major role of the reconditioning of the circuits before start-up,
- The importance of checking of the emission rate of impurities during the reconditioning (gaseous impurities N_2, O_2 and H_2O),
- The absence of light hydrocarbons in the installation after restart using the original oil,
- The important part played by oil in trapping of moisture,
- The need to run the compressors in a closed cycle for at least a week in order to extract this trapped water present in the oil.

REFERENCES

1. G. Claudet et al., Design of the cryogenic system for the TORE SUPRA Tokamak, *Cryogenics*, 26:443 (1986).
2. G. Bon Mardion et al, Normal Operating of the 1.8K TORE SUPRA Cryogenic system, *Cryogenics,* 30S: 566 (1990).
3. F. Minot et al, Problems encountered with Sub-Atmospheric Circuits on the Tore Supra Cryogenic System, *ICEC 17, UK* (1998), p. 609.

THEORETICAL STUDY ON THE PERFORMANCE OF A WET TURBINE FOR HELIUM REFRIGERATORS

M. Obata[1], N. Saji[2], H. Asakura[2], S. Yoshinaga[2], T. Ishizawa[2]

[1]Kanazawa Institute of Technology,
7-1 Ohgigaoka, Nonoichi, Ishikawa, JP 921-8501

[2]Ishikawajima-Harima Heavy Industries Co.,Ltd,
2-16, Toyosu 3-chome, Koto-ku, Tokyo, JP 135-8733

ABSTRACT

In a large helium refrigerator, the power input can be reduced by using the expansion process of a wet turbine in place of a J-T valve (Joule Thomson valve). The cooled helium gas with a temperature between 6 K and 9 K and a pressure of 12 Bar is liquefied between 15 % and 80 % in the wet turbine, depending on the operating condition of the liquefying mode and the refrigerating mode. The turbine efficiency or performance is considered to be largely affected by the growth of the liquid phase in the expansion process of the turbine. In this study, a theoretical method is proposed to calculate the wetness effect of the helium on the turbine performance, and it is shown for a radial inflow wet turbine that the turbine output at 4.5 K is 5 % lower at 20 % of wetness and 20 % lower at 80 % wetness.

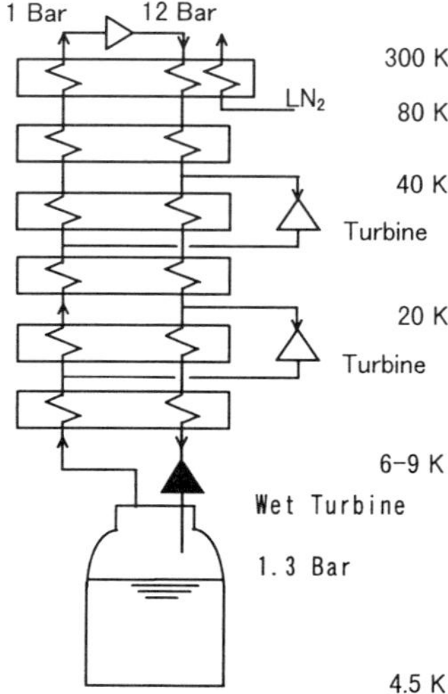

Figure 1. Flow chart of a typical helium refrigerator.

Advances in Cryogenic Engineering, Volume 45.
Edited by Shu *et al.*, Kluwer Academic / Plenum Publishers, 2000.

INTRODUCTION

In the helium refrigeration system for cooling large superconducting magnets, the system efficiency and refrigeration system capacity will be improved by 20 % or more by installing a wet-turbine and by changing the expansion process of approximately 6 K helium from isenthalpic expansion using a J-T valve to isentropic expansion. With a wet turbine, 15 % to 80 % of the helium passing through the turbine passage is liquefied depending on the refrigeration or liquefaction operation mode. The formation of two-phase conditions at the turbine outlet influences the efficiency by the portion of liquid phase at the outlet.

In the water steam turbine fields, the influence of steam wetness on the performance of turbines has been studied theoretically and experimentally in the range of wetness up to approximately 20 %[1]. However, in the development of helium wet turbines, the range of helium wetness will be much higher than those of steam turbines so that for the higher wetness range, the wetness effect on wet turbine should be clarified and the performance prediction method should be confirmed in advance. Performance confirmation tests for wet helium turbines will require some large test equipment and test costs when an actual wet turbine is tested under actual operating conditions.

This paper describes theoretical results about the growth time of helium droplet in the turbine rotor passage and the influence of wetness on turbine output performance.

WET TURBINE WORKING REGION AND GROWTH OF HELIUM DROPLETS IN THE TURBINE ROTOR PASSAGE

Working Region

Figure 1 shows the flow chart of a typical helium liquefaction refrigeration system. The wet turbine, instead of a J-T valve, is connected with the high pressure line exiting the coldest heat exchanger, and helium at the turbine outlet is directly introduced into the liquefaction tank. Figure 2 shows a T-S diagram at the inlet and outlet of the helium wet turbine rotor. In the liquefaction mode operation, the turbine inlet temperature is cooled down 8K to 9 K, the helium is expanded to reach point A of 1.3 Bar and 4.5 K, and the liquefaction rate becomes approximately 15 %. Meanwhile, in the refrigeration mode operation where flows at the high and low pressure side of the heat exchanger are equal, the turbine inlet temperature is cooled down to approximate 6 K, helium is expanded heading for point C of the same pressure and temperature and the liquefaction rate reaches near 80 %.

Growth Point of Helium Droplet in the Turbine Rotor Passage

We consider where in the turbine passage helium droplets occur and assume that the turbine inlet conditions are 12 Bar for pressure, 6K to 9 K for temperature and 1.3 Bar for turbine outlet pressure.

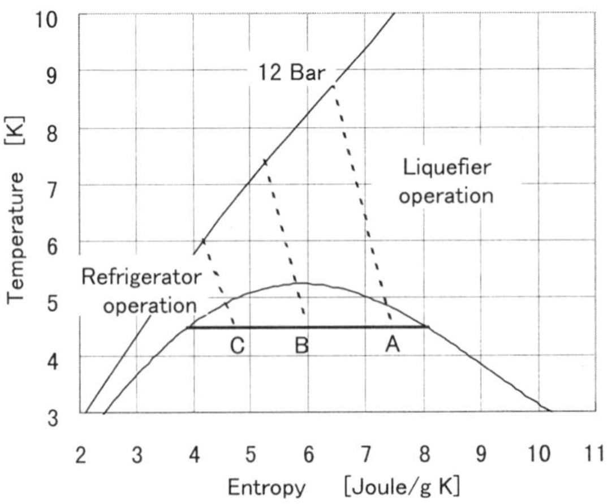

Figure 2. Wet turbine operation line on T-S diagram.

Table 1. Working region and the characteristics of helium expansion flow in the wet turbine passage

Item		Line A	Line B	Line C
Turbine inlet pressure	Bar	12	12	12
Inlet temperature	K	8.7	7.4	6.0
Outlet pressure	Bar	1.3	1.3	1.3
Total enthalpy drop	J/kg	10.6	7.6	5.9
Enthalpy drop of two phase part	%	9.5	13.7	6.9
Inlet density	kg/m^3	0.093	0.116	0.136
Average density at turbine outlet	kg/m^3	0.025	0.037	0.063
Density ratio of two phase part		0.71	0.54	0.62
Density ratio of SHE part		0.38	0.59	0.74
Density ratio between inlet and outlet		0.27	0.32	0.46
Quality of helium gas at turbine outlet		0.85	0.5	0.2

Then, we estimate at what point in the helium expansion in term of percent enthalpy change the liquefaction starts and how long the expansion process of the gas and liquid phase continues from the turbine inlet to the turbine outlet. From these, we can assume that the expansion process of the gas and liquid phase is most prolonged in the case where the line passes the critical point. It is where the line reaches point B in Figure 2 at the turbine outlet.

In the refrigeration operation mode with the lowest turbine inlet temperature, the density change from turbine inlet to outlet is small because the expansion line passes mostly through the super critical area, and the average helium density sharply changes with the start of liquefaction, passing over the saturated liquid curve. Generally the liquefaction starts near the turbine rotor outlet with a small vaporization ratio. It is where the line reaches point C in Figure 2.

On the contrary, in the liquefaction mode where the turbine inlet temperature is highest, reaching point A, the density decreases with the expansion line approaching the saturated vapor curve. The average density changing rate is comparatively small even with the expansion line passing over the saturated vapor curve. The liquefaction starts near the turbine outlet similar to the case of the refrigeration operation mode with a small liquefaction volume.

Table 1 shows a calculation example of the helium expansion process in the turbine passage for the typical three cases shown in Figure 1. The ratio of the enthalpy difference in the gas and liquid mixture region to the entire enthalpy difference in the turbine is not so large and helium droplets appear near the turbine rotor outlet region.

Growth Time of Droplet

Flow velocity in the wet turbine passage is on the order of 150 m/s. If the typical passage length is assumed to be approximately 20 mm, the time to pass through the turbine rotor will be on the order of 1.3×10^{-4}s. It is necessary to check whether helium is liquefied within this time. Assuming that the theoretical equation for growing droplets derived by Stdora [2] in the steam condensing process can be applied here, the time required for condensation from the growth speed of helium droplets is calculated. The droplet growth speed may be expressed as

$$\frac{d\xi}{dt}=\frac{1}{8}\frac{P}{\rho'}\sqrt{\frac{3}{RT_{S\varkappa}}}\left[\frac{\Delta T_C}{T_{G\xi}}-\left(\frac{2a}{P}+\frac{1}{T_{S\xi}}\right)dT'\right] \tag{1}$$

with

$$a=\left(\frac{dp_{S\xi}}{dT}\right)_{T=T'}\cong\frac{\Delta p}{dT'}$$

in which ξ is the droplet diameter, t the growth time, p the saturation pressure, ρ' the droplet density, R the gas constant, $T_{S\varkappa}$ and $p_{S\xi}$ the saturation temperature and pressure of the droplet surface, T' the droplet temperature, ΔT_C the droplet overcooled temperature, and $\delta T'$ the temperature difference between T' and $T_{S\xi}$. The relation between ΔT_c and $\delta T'$ is determined from equation (1) by assuming the droplet sizes to be the same. That is,

$$\delta T' \equiv \phi \Delta T_c , \tag{2}$$

where

$$\phi = \frac{1}{(2a / P)T_{S\xi} + 1}.$$

If an average value $a / P = 0.92$ is assumed from the working conditions, $\phi = 1/9$ is gained at $T_{S\xi} = 4.4$K, the value of $\delta T'$ obtained from equation (2) becomes one-order smaller than ΔT_c obtained from equation (2).

Therefore, when $\delta T'$ in equation (1) is neglected with $\rho' = 2.0$ kg/m^3 and R = 1350m^2/s^{2}K under the conditions of $T_{S\xi} = 4.4$ K and $p = 0.12$ MPa, the following equation is obtained:

$$d\xi/dt = 3.80\ \Delta T_c . \tag{3}$$

Then, if ΔT_c is assumed 0.5 K, the time Δ t until droplet diameter ξ grows up to 10^{-8}m is 5×10^{-8} s.

It is found that the time required until the helium changes from gas phase to liquid phase depends on ΔT_c, and it is a very short time compared with 10^{-4} s which is the time taken for the helium to pass through the turbine passage. It can be said that liquefaction phenomenon occurs while the helium passes through the passage.

DERIVATION OF THE EQUATIONS

Figure 3 shows the velocity diagram of a radial inflow turbine where the turbine rotor is operating at a constant angular velocity ω. The helium enters the rotor from the nozzle in gas phase with a relative velocity V_2 in the radial direction (an incidence angle of 0°) and flows out in two phases of gas and liquid produced in the expansion process in the rotor passage. In the present theory, the helium in gas phase is assumed to flow out from the rotor with a relative velocity V_{3G}, an outlet angle β_{3G} and a longitudinal absolute speed $C_{3G} = C_{3aG}$. Meanwhile, the helium in liquid phase is assumed to flow out with a relative velocity V_{3L} and an outlet angle β_{3L}. The following relative equations are formulated.

Calculation of Relative Output Ratio L/L_0

The torque T obtaining from the change of momentum at the rotor inlet and outlet is calculated from the following equation:

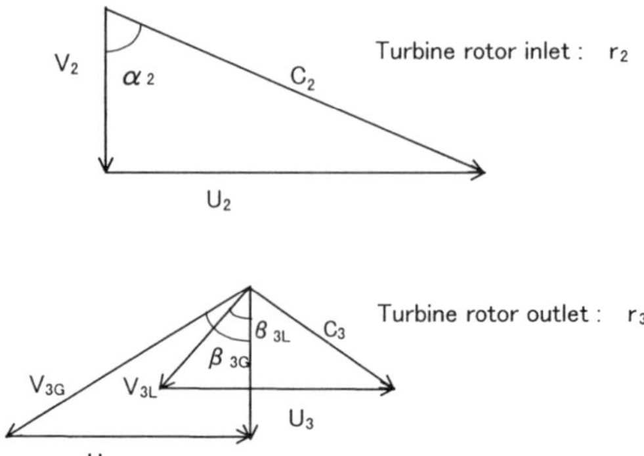

Figure 3. Velocity diagram of radial inflow turbine.

$$T = m_{G2} r_2 C_{2w} - (m_{G3} r_3 C_{3wG} + m_{L3} r_3 C_{3wL}) \tag{4}$$

in which m_{G2} and m_{G3} are the gas-phase mass flow rate at the rotor inlet and outlet and m_{L3} is the liquid-phase mass flow rate at the outlet. r_2 is the tip radius at the rotor inlet and r_3 is the average flow passage radius at the outlet. C_{2w}, C_{3wG} and C_{3wL} are the peripheral components of C_2 and C_3 and these are related with the peripheral velocity U_2 and U_3 as

$$\left.\begin{aligned} C_{2w} &= C_2 \sin\alpha_2 = U_2 \\ C_{3wG} &= 0 \\ C_{3wG} &= U_3 - V_{3L}\sin\beta_{3L} \\ U_3 &= V_{3G}\sin\beta_{3G} \end{aligned}\right\} \tag{5}$$

From the equations (4) and (5), the rotor output $L=T\,\omega$ is obtained as

$$\begin{aligned} L &= m_{G2} r_2 U_2 \frac{U_2}{r_2} - \left\{ m_{L3} r_3 \left(U_3 - V_{3L}\sin\beta_{3L} \right) \right\} \frac{U_3}{r_3} \\ &= m_{G2} U_2^2 - m_{L3}\left(U_3 - V_{3L}\sin\beta_{3L} \right) U_3 \end{aligned} \tag{6}$$

Then, the output L_0 without wetness (m_{L3}=0) is given by:

$$L_0 = m_{G2} U_2^2 \tag{7}$$

Therefore, the relative output ratio L/L_0 with and without wetness is calculated from equations (6) and (7) as

$$\begin{aligned} \frac{L_0}{L} &= 1 - \frac{m_{L3}}{m_{G2}} \left(\frac{U_3}{U_2} \right)^2 \left\{ 1 - \frac{V_{3L}\sin\beta_{3L}}{U_3} \right\} \\ &= 1 - (1-x)\left(\frac{U_3}{U_2} \right)^2 \left\{ 1 - f\,\frac{\sin\beta_{3l}}{\sin\beta_{3G}} \right\} \end{aligned} \tag{8}$$

in which x is the quality of the helium at the outlet and f is the ratio of the average helium flow velocity in the gas phase and the liquid phase. These are expressed as the following equations,

$$x = m_{G3} / m_{G2} = 1 - m_{L3} / m_{G2}, \tag{9}$$

$$f = V_{3L} / V_{3G} = \frac{1}{S}, \tag{10}$$

where S is the slip ratio.

Calculation of Average Helium Flow Velocity in Gas Phase and Liquid Phase

The helium mass velocities in the gas phase and the liquid phase are given by the definition of the two-phase flow theory[3]:

$$G_{G3} = m_{G3}/A\ , \qquad G_{L3} = m_{L3}/A\ , \tag{11}$$

where A is the throat area. The apparent helium flow velocities in the gas phase and the liquid phase are expressed with each density of ρ_G and ρ_L as

$$V'_{3G} = G_{G3}/\rho_G\ , \qquad V'_{3L} = G_{L3}/\rho_L\ . \tag{12}$$

Then, the actual helium flow velocities in the gas phase and the liquid phase are calculated with the void ratio ϕ, that is

$$V_{3G} = \frac{V'_{3G}}{\phi} = \frac{Gx}{\rho_G \phi} \quad , \qquad V_{3L} = \frac{V'_{3L}}{1-\phi} = \frac{G(1-x)}{\rho_L(1-\phi)} \quad , \tag{13}$$

where

$$G = G_{G3} + G_{L3} = G_{G2}.$$

Therefore, the ratio of the average helium flow velocities f in both phases is given from the equation (13) as

$$f = \frac{V_{3L}}{V_{3G}} = \frac{\rho_G}{\rho_L} \cdot \frac{(1-x)}{x} \cdot \frac{\phi}{(1-\phi)} \ . \tag{14}$$

Calculation of Void Ratio ϕ

In the two-phase flow in a pipe the average flow velocity can be expressed with the apparent helium flow velocities in the gas phase and the liquid phase, and the following relation may be established in the throat section:

$$V_o = B(V'_{3G} + V'_{3L}) \tag{15}$$

in which V_o is the average helium flow velocity at the throat section and B is an empirical constant. From equation (13) and (15), the calculation form of the Void ratio ϕ is given by the following equation:

$$\phi = \frac{1}{B(1+K)} \quad , \tag{16}$$

with

$$K = (\rho_G / \rho_L) \cdot (1-x)/x \ .$$

Average Velocity Ratio Obtained from Critical Flow Conditions at the Turbine Outlet

On the other hand, the flow in the turbine may be in the critical flow conditions at near the outlet region of the rotor passage. There are a few available theories for critical two-phase flows and in the present study, Moody flow model[4] of the two-phase mixture on a single component is adopted as an applicable model.

According to the Moody model, in which the same temperatures for gas and liquid phases and a slip ratio of each velocity are assumed, the slip ratio S at the critical flow condition is approximated by a simple power relationship as

$$S = \frac{V_G}{V_L} = \left(\frac{v_G}{v_L}\right)^{1/3} = \left(\frac{\rho_L}{\rho_G}\right)^{1/3} = \frac{1}{f} \tag{17}$$

in which v_G and v_L are the specific volumes in the gas and liquid phases. S is also related to the density ratio and the average velocity ratio f .

RESULTS AND DISCUSSION

Effect of Wetness Fraction on Output Work

Figure 4 compares calculated work output ratios with the wetness fraction of helium, 1- x, for a typical radial turbine. The operating conditions of the turbine are assumed $\beta_{3L} = \beta_{3G}$, $\rho_G / \rho_L = 0.6$,

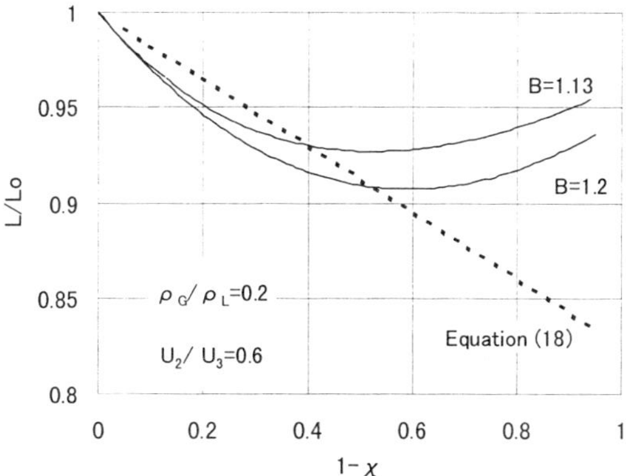

Figure 4. Typical case of effect of wetness fraction on work output. The calculated conditions are $\beta_{3L} = \beta_{3G}$, $\rho_G / \rho_L = 0.2$, $U_3/U_2 = 0.6$, and $B = 1.2$ and1.13.

$U_3/U_2 = 0.6$, and B = 1.2 and 1.13. The comparison is also made for an empirical correction[1] which is used to predict the dryness effect on the efficiency on wet steam turbine. That is

$$\eta_{wet}/\eta_{ss} = 1 - \{(1-x)/4\} log(v''/v') \, , \tag{18}$$

where η_{wet}/η_{ss} is the efficiency ratio in operating with wet and superheat steam, and v''/v' is the specific volume ratio of stream in liquid and vapor states.

The results show that the turbine output decreases with increasing wetness and after reaching the minimum value of nearly 50% wetness, it increases slowly as the wetness approaches unity. The calculated values from equations (8) and (18) are in agreement until approximately 50% wetness depending on the value of B, but quantitative differences exist.

Influence of the Turbine Outlet Temperature.

Figure 5 shows the influence of the turbine outlet temperatures on work output of the turbine, and the calculation conditions are $\beta_{3L} = \beta_{3G}$, $U_3/U_2 = 0.6$ and B = 1.16 in which T_e is the turbine outlet temperature. The result indicates that the work output decreases as the wetness increases and T_e

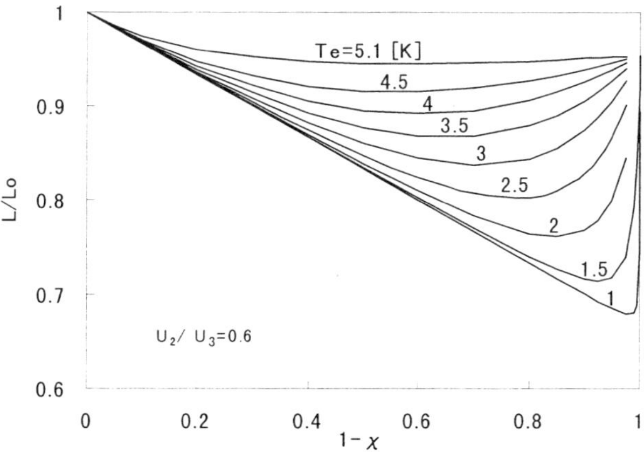

Figure 5. Calculation results of influence of the turbine outlet temperature. The conditions for calculation are B = 1.16. T_e values are the turbine outlet temperatures.

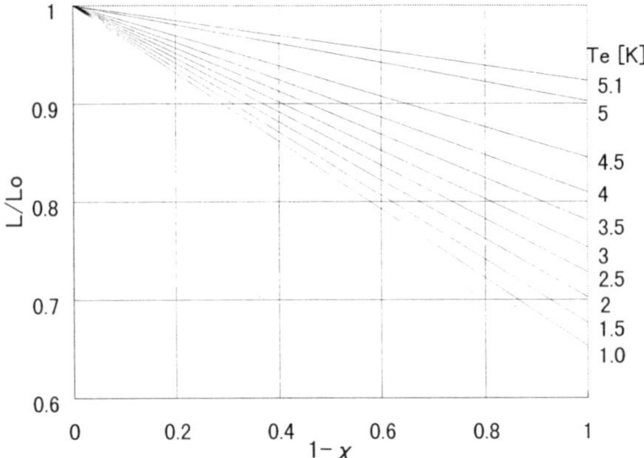

Figure 6. Effect of wetness fraction on work output. Calculation results of applied critical flow condition at turbine outlet.

decreases, but the output increases as the wetness approaches unity. This is due to the increase of helium density ratio of gas to liquid phase as the turbine outlet temperature decreases.

Result Applied Critical Flow Condition

The calculated values of work output applied at critical flow condition at the turbine outlet are shown as a function of wetness and T_e in Figure 6. Results indicate that the output decreases linearly as the wetness increases and T_e decreases, and that no turning point appears with increasing wetness. In the expansion process of an actual turbine operating with a high liquefaction rate, the density difference between gas and liquid phase is very small and the influence of wetness on the work output should be small. Therefore, a large decrease of turbine output cannot be considered and the critical flow condition at the turbine output gives an unrealistic result at the region where wetness approaches unity

CONCLUSION

The present study yielded the following results :

(1) For the working region of helium wet turbine, the liquefaction phenomenon in the rotor passage occurs in a sufficiently short time on the order of 10^{-8} s.
(2) A theoretical method for calculating the effect of wetness on the performance of a helium radial inflow wet turbine has been formulated and has shown that the turbine output at 4.5 K of the turbine outlet temperature is approximately 5 % lower at 20 % wetness and 20 % lower at 80 % Wetness.
(3) The performance of a wet turbine is largely affected by the turbine outlet temperature and slightly affected by the handling of two-phase flow in the turbine rotor.

REFERENCES

1. Б. М. Трояновский original, "Atomic Energy Turbine" (1997 S. Nagasima translating to Japanese).
2. I. I. Kirillov, R. M. Yablonik "Fundamentals of Theory of Turbines Operating on Wet Steam" Translation , NASA TTF-661(1970) 104.
3. JSME Data Book "Heat Transfer"4th Edition. (1986) 136,144.
4. F. J. Moody "Maximum Flow Rate of a Single Component, Two-phase Mixture" Trans. ASME, Ser. C, 87-1(1965)

PERFORMANCE OF 80K TURBO COMPRESSOR SYSTEM WITHOUT LN2 COOLING FOR HIGH RELIABLE AND EFFICIENT HELIUM REFRIGERATOR

H.Asakura[1], N.Saji[1], S.Yoshinaga[1], T.Ishizawa[1], M.Mori[1], A.Miyake[1], N.Ijichi[1], and K.Yamada[2]

[1]General Machinery Div. Ishikawajima-Harima Heavy Industries Co.,Ltd., 3-2-16 Toyosu Koto-ku Tokyo 135-8733 Japan
[2]Super-GM, 5-14-10 Nishitenma Kita-ku Osaka 530 Japan

ABSTRACT

80K cold compressor driven by coaxial helium turbine for high reliable and efficient refrigerator has been developed in the “Super-GM” project which is a national project in Japan. Since the inlet of the compressor is cooled by expansion gas through the driving turbine, the cold compressor system has been realized without using liquid nitrogen. The expansion turbine employs a variable geometry nozzle blade to keep high efficiency over a wide flow rate range. The performance test of this machine, especially for expansion turbine, was carried out by incorporating to the Brayton cycle test facility. In this paper, a construction of the new conceptional cold compressor prototype and the performance test results are reported.

INTRODUCTION

In the “Super-GM” project, which is a national project in Japan, we have been developing an advanced refrigeration system for a superconducting generator. The goals of this system are, achieving high reliability, high efficiency, and space saving. The key component is a compressor system which is usually a lubricated machine occupying a big space in the compressor house.

We have developed a new conceptional cold compressor, completely oil-free, and driven by an expansion turbine. This cold compressor does not require liquid nitrogen (LN2) because the inlet is cooled by the turbine-expanded gas. In addition to the oil-free characteristic, a highly efficient and space-saving refrigerator can be expected (see Fig.1).

Three turbines in parallel, a warm compressor, and a recuperator make up a Brayton cycle between 300K and 80K. In this system, the rotating machine becomes compact and the purifier can be excluded, so that the space we expect will be saved is up to approximately 40 % compared with that of the conventional refrigerator. This paper describes the construction and performance test results of the variable geometry turbine that is the key of this rotating machine.

Advances in Cryogenic Engineering, Volume 45.
Edited by Shu *et al.*, Kluwer Academic / Plenum Publishers, 2000.

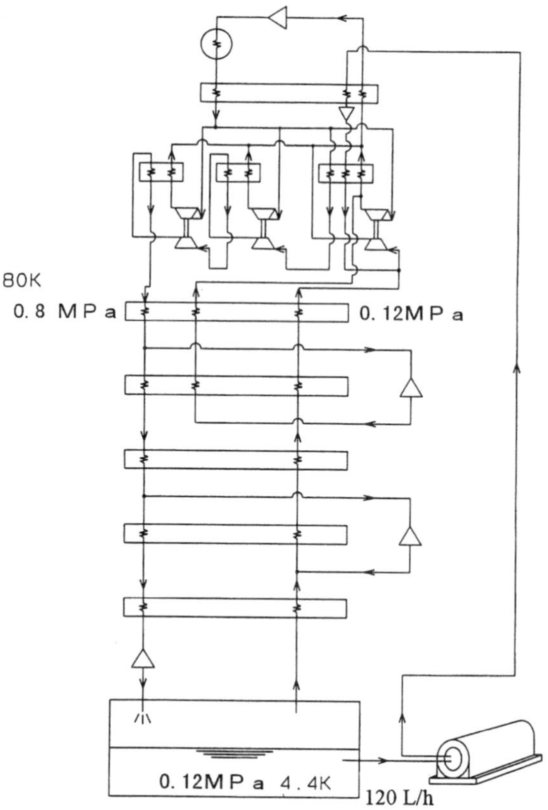

Figure 1. Advanced refrigeration system using new conceptional cold compressor.

COLD COMPRESSOR DRIVEN BY EXPANSION TURBINE

Construction Features

This machine is a high-speed rotating machine with a compressor impeller on one end, a turbine impeller on the other end, and a generator in the middle section. Figure 2 shows the cross-sectional view; Figure 3 shows the general view; Figure 4 shows the rotor shaft. The compressor and turbine are centrifugal type, and the generator is synchronous type with a permanent magnet on the rotor side. Besides the role of driving the rotor, the turbine generates low temperature helium of approximately 80K with expanded gas and cools the compressor inlet. Since the turbine output power, which absorbs helium gas enthalpy to make the temperature 80K, is bigger than the power to drive the compressor, the differential power is absorbed by the coaxial generator.

The bearing supporting the rotor is a foil-type dynamic gas bearing and is only moderately stiff. The rotor, therefore, might seize due to the large pull force into the stator. To cope with this problem, a toothless type of stator is employed for the generator.

Variable Geometry Turbine

The great characteristic of this machine is installing the variable geometry nozzle mechanism in the expansion turbine. The variable geometry turbine has the advantage of maintaining high efficiency over the wide flow range; the disadvantage lies in its lower overall efficiency due to leaks from the moving sections. The nozzle head is made so as to be obtuse, and the blade thickness is enlarged to make leakage as small as possible.

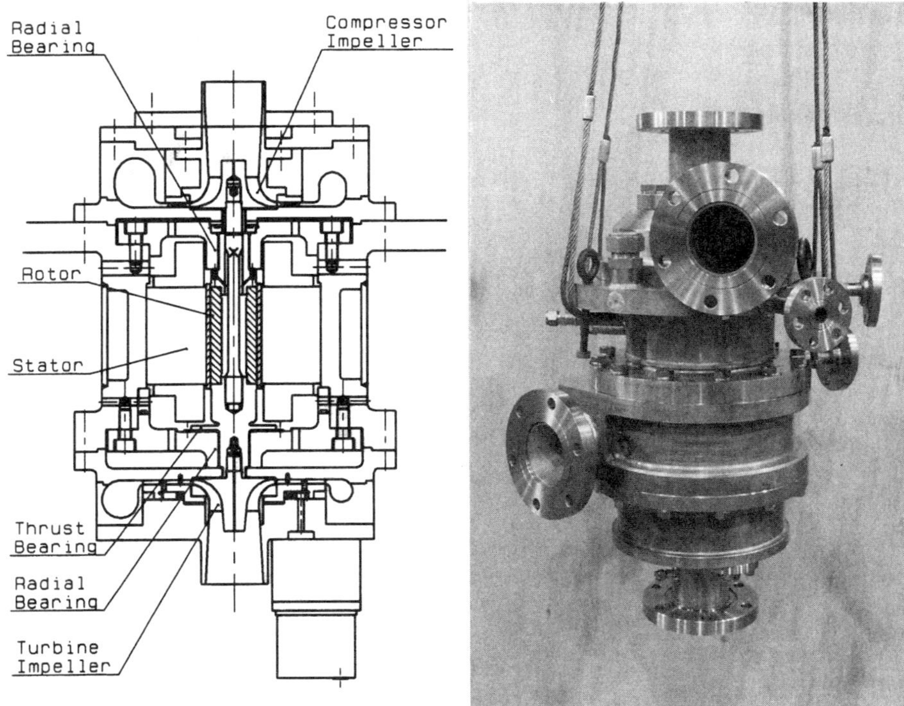

Figure 2. Cross-section of cold compressor driven by expansion turbine.

Figure 3. Appearance of cold compressor driven by expansion turbine.

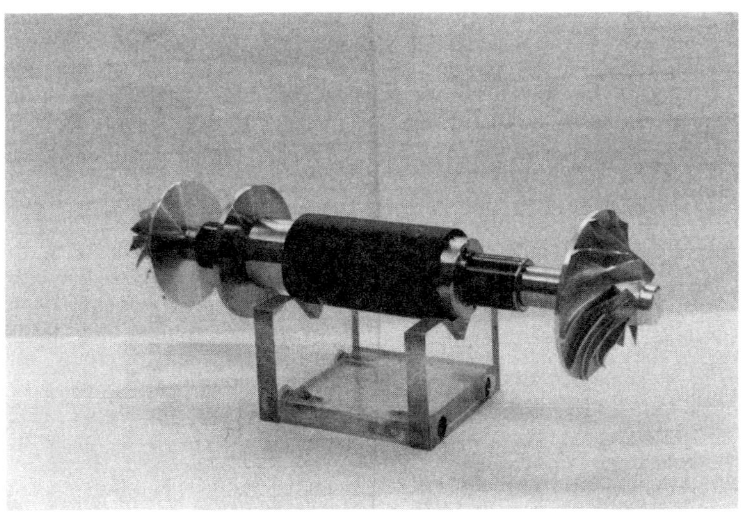

Figure 4. Rotor shaft. The right side is a compressor impeller, the left side is a turbine impeller, and the central part is a generator.

Figure 5. Variable geometry turbine nozzle at full-closed (left side) and at full-open (right side)

Besides, all of the variable nozzle blades are pushed to the passage wall by springs through a plate; this makes the clearance between the nozzle blades and the wall zero. The nozzle throat area is changed by altering the blade-installing angle with a pulse motor, which can work even at low temperature. Figure 5 shows the nozzle blades at full-open and full-closed conditions.

PERFORMANCE OF EXPANSION TURBINE

Test Facility

The turbine section of this machine operates in the Brayton cycle together with a warm compressor. Meanwhile, the compressor section liquefies helium in the Claude cycle together with the low-temperature turbine and heat exchanger. Therefore, in the tests of the present machine it is necessary to confirm that the turbine and compressor operate in good matching conditions in the two cycles. We confirmed performance of each component such as the turbine, compressor and generator, making the test facility as shown in Figure 6. Table 1 shows the specifications of the rotating machine and test facility. Figure 7 shows the appearance of the test facility.

Performance Test Results

Mechanical Performance. Figure 8 shows a FFT analysis result of the shaft vibration from start to the rated 120,000 rpm. Concerning the vibration, only a component synchronized with rotation speed was found and no harmful low-frequency components were observed. We confirmed that the amplitude was small enough and that the rotor system operated stably. We also measured generator performance for energy matching, and confirmed that the power is 11 kW at 120,000 rpm which covers the rated value sufficiently.

Fluid Dynamic Performance of Turbine. The performance to be confirmed for the developed turbine are flow rate and efficiency characteristics when the nozzle angle is changed. By incorporating the turbine in the facility shown in Figure 6, we measured the relation between the turbine pressure ratio and flow rate and efficiency, adjusting compressor side flow rate and generator speed.

Table 1. Specification of rotating machine and test facility

Warm compressor	Inlet pressure (MPa)	0.185
	Outlet pressure (MPa)	0.388
	Flow rate (g/s)	100
Expansion turbine	Inlet pressure (MPa)	0.388
	Inlet temperature (K)	99
	Outlet pressure (MPa)	0.185
	Flow rate (g/s)	87
Cold compressor	Inlet pressure (MPa)	0.12
	Inlet temperature (K)	80
	Outlet pressure (MPa)	0.185
	Flow rate (g/s)	50
Generator	Rated power (kW)	10 at 120,000 rpm
Heat exchanger	Plate-fin type	110 kW (High temperature side) 5.7 kW (Low temperature side)

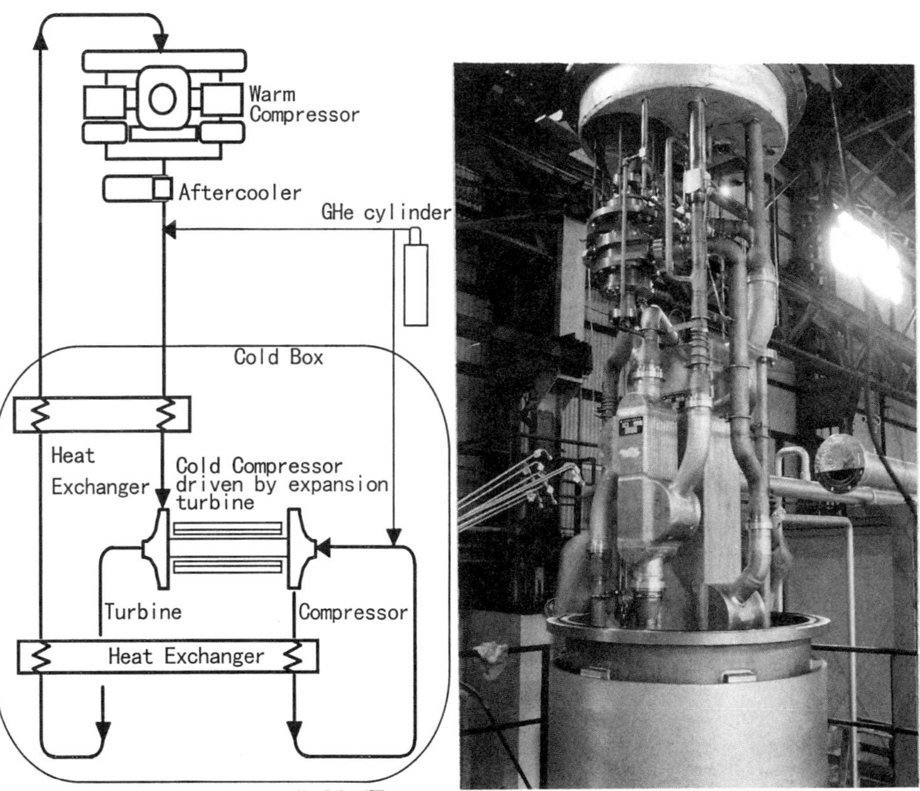

Figure 6. Flow diagram of the performance test facility. Expansion turbine is a component of the Brayton cycle.

Figure 7. Appearance of the performance test facility.

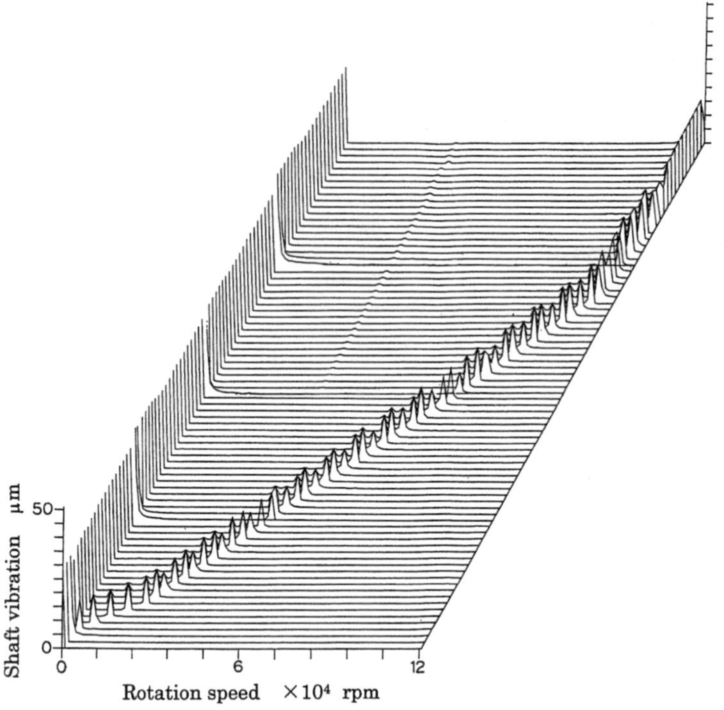

Figure 8. FFT analysis of the rotor vibration. From start to 120,000 rpm no harmful vibration component can be observed.

Figure 9 shows a flow rate characteristic when the nozzle angle was changed. Because it is difficult to set the inlet temperature at the constant condition in the turbine performance measurement, the following reduced flow rate is used for the flow rate.

$$m_r = \frac{m\sqrt{T}}{P} \quad (1)$$

where m: Mass flow rate [kg/s] T: Inlet temperature [K]
P: Inlet pressure [MPa]

Likewise, for rotation speed the following reduced rotation speed is used, which is proportional to Mach number.

$$N_r = \frac{N}{\sqrt{T}} \quad (2)$$

where N: rotation speed [rpm].

Figure 9 shows that the designed flow rate is obtained at a nozzle opening of 100 % and that the flow rate can be changed by altering the angle, maintaining the pressure ratio. Figure 10 shows turbine efficiency characteristics. Generally turbine efficiency is indicated by the ratio of impeller external peripheral speed U and isentropic velocity Co.

$$C_0 = \sqrt{2\Delta Had} \quad (3)$$

where △Had: Adiabatic head [J/kg].

The efficiency generally shows maximum value at the speed ratio around 0.7.

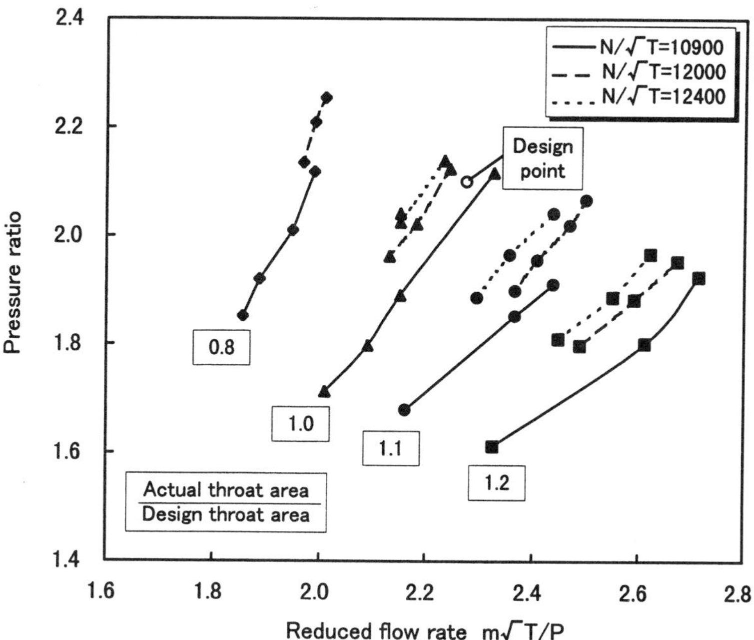

Figure 9. Flow rate characteristic of variable geometry turbine. The flow rate can be changed by altering the nozzle blade angle.

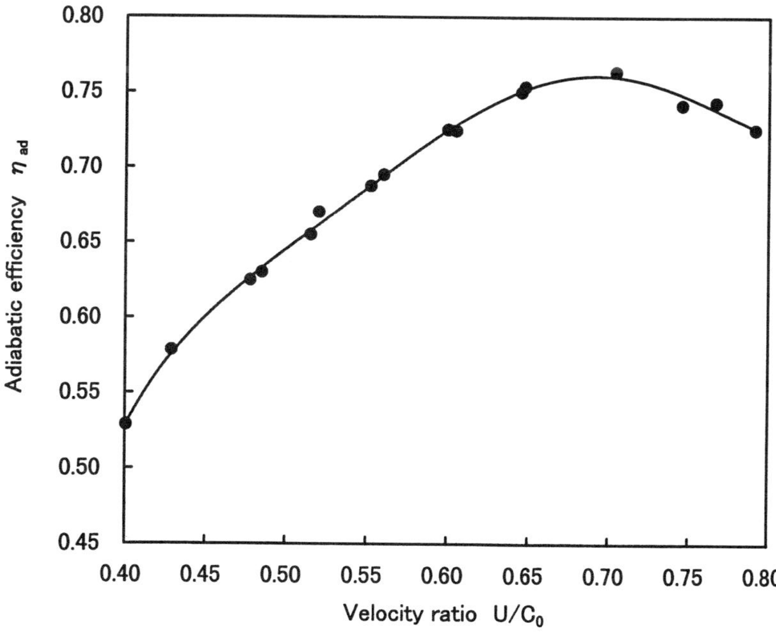

Figure 10. Efficiency characteristic of variable geometry turbine. The maximum efficiency of 76 % was measured at the speed ratio of 0.7.

As shown in the figure, the maximum turbine efficiency was 76 %. By analyzing the refrigeration system performance shown in Figure 1 with a turbine efficiency of 76 %, we confirmed that the power consumption per unit liquefaction rate becomes 2.0 kW/(L/h) which nearly reaches the target value.

CONCLUSION

We made a prototype of a cold compressor driven by expansion turbine which becomes a key component of the advanced refrigeration system for a superconducting generator, incorporated it in the test facility and carried out performance tests at low temperature.

The turbine is a variable geometry type which can change flow rate by altering the nozzle angle, having characteristics to maintain high efficiency over a wide operation range. We confirmed in the performance tests that the machine can operate over the 120 % flow range to the designed point by altering the nozzle angle. Besides, the highest turbine efficiency of 76 % was recorded.

From the present tests, the new conceptional cold compressor was confirmed to be practical, attaining 2.0 kW/(L/h) of performance when applied to the helium refrigerator.

ACKNOWLEDGMENT

This research has been carried out as a part of R&D on superconducting technology for electric power apparatuses under the New Sunshine Project of AIST, MITI, being consigned by NEDO.

REFERENCES

1. H.Asakura, N.Saji, Y.Kaneko, S.Yoshinaga, M.Mori, H.Yamaguchi, T.Nogaku, and T.Umeda, 80K Turbo Compressor System without LN2 Cooling for High Reliable and Efficient Helium Refrigerator, in: "Advances in Cryogenic Engineering ," Vol.43A, Plenum Press, New York (1998), P.667
2. N.Saji, S.Nagai, H.Asakura, Y.Kaneko, Design of Oilfree All Turbo-type Helium Refrigerator, *Cryogenics* 34 ICEC Supplement, P.115 (1994)

FERMILAB COLD COMPRESSOR BEARING LIFETIME IMPROVEMENTS

W. M. Soyars and J. D. Fuerst

Fermi National Accelerator Laboratory*
Batavia, IL, 60510

ABSTRACT

The Fermilab Tevatron liquid helium cryogenic system has high speed, cold turbocompressors, made by IHI Co. Ltd., that allow for subatmospheric, lower temperature operation (3.5K). The present cold compressor design utilizes spiral wound dynamic gas foil bearings for the radial journal bearings. Fermilab has experienced a variation in journal bearing lifetimes leading to more frequent maintenance. Bearing lifetime appears to be satisfactory when the cold compressor is operated in a stable environment, such as in a stand-alone cryostat. However, the operating configuration calls for the cold compressor units to be mounted on other equipment. The degradation in bearing lifetime is attributed to the cold compressor being exposed to an externally induced vibratory environment. The vibratory environment will be characterized. Possible solutions to improve the bearing life will be presented, including provisions for additional support structure to minimize vibrations, and redesign of the foil bearing to improve tolerance to vibrations.

COMPRESSOR AND BEARING DESCRIPTION

The IHI compressor, which consists of a single stage centrifugal impeller powered by a variable frequency induction motor, has been previously described.[1] The radial journal and thrust bearings are foil-type self acting (dynamic) gas bearings. The foils do not establish a supporting film until the shaft reaches an appropriate minimum lift off speed.

**Work supported by the U. S. Department of Energy under contract No. DE AC02 76CH03000.*

There is contact and therefore rubbing occurring during startup and shutdown. This contact is handled by a Teflon coating on the inner surface of the stainless steel foil. The maximum shaft speed is 90 krpm, with typical operating conditions of 50 krpm. Each machine has two spiral wrapped journal bearings on a shaft of approximately 1.6 cm.

OPERATION AND MAINTENANCE EXPERIENCE WITH IHI BEARINGS

The initial operating experience with the production units has been described in previous papers.[2,3] Over 130,000 total hours of operation have been accrued on these machines, at an average of over 4800 hours per machine. The IHI bearings have proven themselves operationally, even showing the capacity to survive off-design conditions of liquid inhalation onto the impeller. Yet, long-term operation has shown that journal bearings require more frequent replacement than expected. No problems have been experience with thrust bearings, so these will not be further addressed in this paper.

For reasons of saving space, the cold compressor cryostat mounts directly on top of the satellite refrigerator's valve box. The valve box houses a phase separator and piping to join the refrigerator to the magnets below ground. The valve box is fixed at the bottom and braced at the top, with internal piping runs which are not stiffly supported to limit heat leak. These mounting and flow restrictions place the turbocompressor in an environment where it is susceptible to external vibrations. The original prototype IHI cold compressor was tested in a free standing cryostat mounted directly to the floor. This prototype unit was run more than 10,000 hours at a variety of conditions, with numerous starts and stops, with no sign of bearing degradation. These tests, and prior vendor experience, led to expectations of 100,000 hours of journal bearing life. The reduction in production unit bearing life is attributed to their being mounted in a way that exposes them to an externally induced vibratory environment.

The initial indication of bearing wear occurred during planned bearing replacement. An upgraded version of the bearing, one that is designed to achieve operating stiffness at

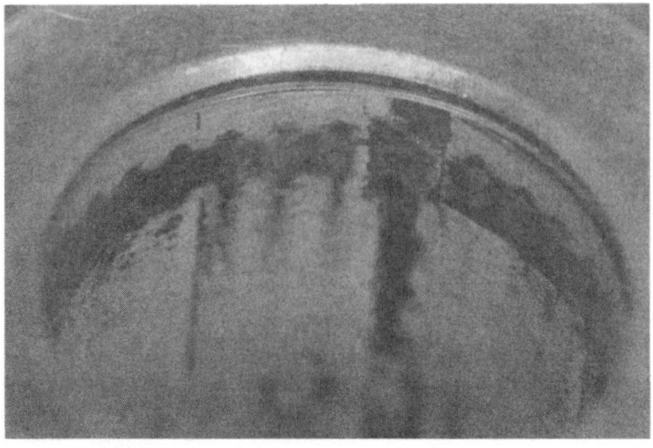

Figure 1. Representative journal bearing showing signs of severe wear.

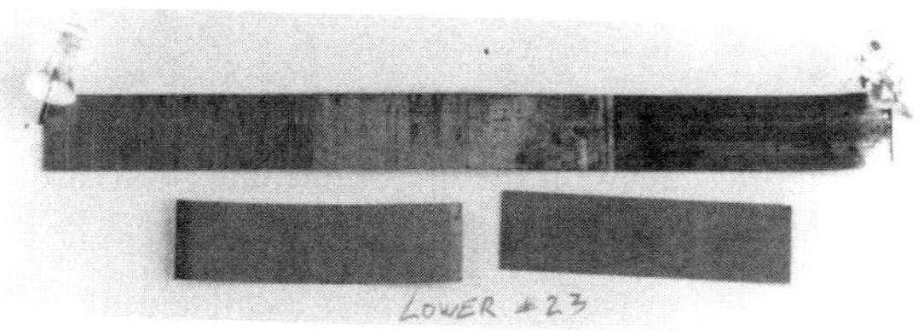

Figure 2. Failed journal bearing, unwrapped for display.

lower shaft speed (20 krpm vs. the original bearing's 40 krpm), was purchased and installed in August 1994. The original bearings had been operated for an average of 1060 hours on each of the 27 turbocompressors. Upon bearing disassembly, it was noted that the original bearings showed signs of more wear than was expected. Representative wear on a foil as it sits in its cartridge is shown in Figure 1.

During the next six months of upgraded bearing operations, three bearing failures occurred. The earliest one was after 2566 hours. The unwrapped foil from this high speed failure is shown in Figure 2. Two more failures occurred after approximately 3000 hours in units that only ran at minimum speed. Failures were characterized by the inverter reaching the maximum current when trying to start the motor drive. At least one of these units was operating in an environment where noticeable vibrations were present. In response, the minimum speed was elevated from 20 to 30 krpm to ensure bearing lift off, and no bearing failures have subsequently occurred.

This second set of bearings was removed and replaced in July 1998. This replacement was motivated by the desire to begin an impending Tevatron operation period with brand new bearings. Each machine had run for an average of at least 3800 hours. (The precise hour information is unavailable due to sporadic hour meter maintenance after this time.) Inspection revealed similar patterns and frequency of wear as seen during the 1994 bearing removal. Out of 24 units in service, 5 or 6 were deemed to have badly worn journal bearings.

In summary, the journal bearings showed wear that degraded their lifetime and in some cases led to premature failures of the rotating system. The wear was thought to be caused by excessive vibration, exceeding what this particular bearing can tolerate. With insufficient dampening, the vibration induced rubbing leads to wear and ultimately failure.

SOLUTIONS CONSIDERED

Ideally, the cause of the vibrations should be eliminated. Unfortunately in this case, this cannot be done because the cause is not well understood. It is speculated that turbulent, or possibly two-phase, liquid helium flow in the loosely supported valve box piping is the cause. Pulsation from the reciprocating wet expander could also contribute to excitation of the bearing system. Given the difficulty caused by modifying the valve box, it is not practical to alter flow paths, reduce noise source, or stiffen up piping. Therefore, we are limited to modifying the compressor system itself, as discussed below.

Enhanced support

Stiffening the cold compressor support with additional structure can reduce the amplitude of the vibration. A 3.35 m long, 10.2 cm square by 0.48 cm wall structural steel tubing column fixed at the floor and at the roof has been designed to provide additional stiffness to the IHI cold compressor structure. A photograph of this structure in a satellite refrigerator building is shown in Figure 3.

Isolation

One could further reduce vibration by decoupling the compressor from the valve box. This could be achieved with additional bellows on the intake and exhaust helium and vacuum circuits while the turbocompressor is held by some other support system. This would involve modifying the flow passages to the impeller, which could have an effect on performance.

A stand alone cryostat for the IHI cold compressor would be an extreme way to isolate it from the valve box. This was not pursued due to the previously mentioned space limitations. Furthermore, there would be additional heat leak and pressure drop.

In some cases, a dynamic absorber can reduce vibrations for frequency specific resonance. However, an operating deflection shape analysis, done by an outside consultant, showed that the apparently flow-induced vibrations cannot be matched to a specific frequency, thereby reducing the viability of this option.[4]

Improved bearing design

There are several styles of foil bearings. The style used on our machine, the spiral wound foil bearing, is an older concept. More modern designs often use either "tension" or "bending" style leaves.[5] Of particular interest is a style known as a "bump" foil, which has a spiral wound foil backed by a corrugated foil. For examples, see references.[6-9]

Figure 3. Support column installed for additional stiffening of cold compressor on valve box.

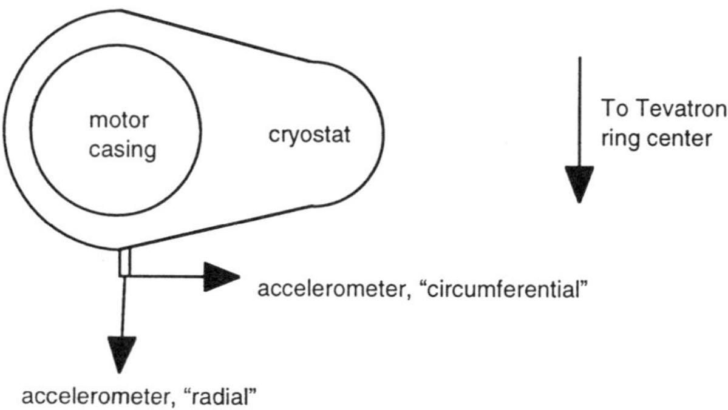

Figure 4. Schematic top view of IHI cold compressor, showing external accelerometer positions.

For our application, improved life could be gained by redesign of the bearings to a more state-of-the-art configuration. Bump foil bearings are an option, but a bearing expert would be required to predict the frequency dependent rotordynamic stiffness and dampening characteristics needed to develop a long life bearing system.

INSTRUMENTATION

External vibration measurements were taken on cold compressor units installed in operating refrigerators. The standard method used vibration transmitters to integrate accelerometer inputs to produce a DC signal for peak velocity. Peak velocity is a good measure for assessing the severity of mechanical roughness in our frequency range.[10] The transmitters used have a peak velocity range of 0-2.54 cm/s. Measurements were made in the horizontal plane circumferential to and radial to the accelerator ring. The mounting positions of the accelerometers is shown in Figure 4.

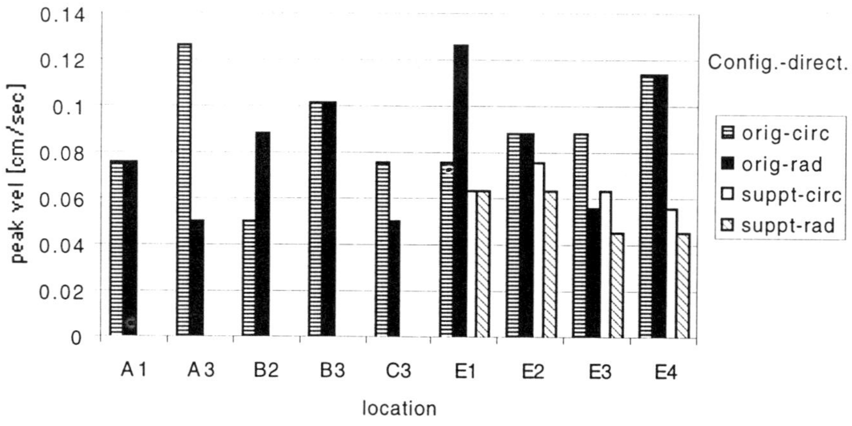

Figure 5. Cold compressor vibrations at various locations around Tevatron ring.

Table 1. High frequency accelerometer data at E3

Add'l Suppt?	direct	freq[hz]	pk accel[g]	pk vel[cm/s]	p-p δ [μm]
None	"radial"	27	0.0081	0.048	5.6
Yes	"radial"	29	0.0061	0.033	3.6
None	"circum."	31	0.013	0.066	6.6
Yes	"circum."	32.5	0.010	0.048	4.8

At one location, additional vibration measurements were taken at a high frequency. Raw accelerometer readings at 1000 hz were plotted to determine peak accelerations and frequencies. Assuming a sine wave profile, peak velocity and displacement can then be calculated.

RESULTS

Peak velocity measurements were taken at nine different satellite refrigerator buildings. The average readings are shown in Figure 5. The cryogenic system was functioning nominally, producing 4.5 K refrigeration without the cold compressors operating. The unsupported configuration characterizes the current vibration environment. The supported configuration shows the effect of the box beam column, which was installed at four locations. In general, we see about a 30% decrease in vibration peak velocity with the support.

Comparative information was gained from the high frequency accelerometer readings taken at one location. These results, given in Table 1, agreed well with simultaneous readings from the velocity transmitters .

Some additional observations were made to judge off-nominal conditions. First, a range of potential maintenance activities on the valve box was simulated. A gentle case was modeled by applying pressure on the valve box with finger tips while a severe case was simulated by a man climbing on the valve box. For the unsupported cold compressor, the gentle case increased the vibration peak velocity from 0.076 to 0.15 cm/s, while the severe case gave 0.38 cm/s. With the support installed, these readings were reduced to 0.076 cm/s for the gentle case and 0.10 cm/s for the severe case. Next, vibration measurements were taken while the cold compressor was started, sped up, and stopped. No change in exterior vibration level was seen with or without the additional support.

CONCLUSIONS

The lifetime of the cold compressor foil journal bearings needs to be increased for greater system reliability. The bearings suffer from exposure to the external environment that may generate about 0.09 cm/s peak velocity at the compressor. Adding a support column to some existing installations reduced these normal vibrations by about 30%. While this method offers some improvement, there may not be a significant increase in bearing life, and the additional structure would be a nuisance for certain maintenance activities.

The recommended action is to redesign the foil journal bearings. State-of-the-art foil bearings may provide longer life. A design study contract has been competitively awarded to a consulting firm, Turbotechnology Services Corporation of Scotia, New York, with expertise in foil bearing design.

REFERENCES

1. J. D. Fuerst, Selection of cold compressors for the Fermilab Tevatron, in: "Advances in Cryogenic Engineering," Vol. 39a, Plenum Press, New York (1994), p. 863.
2. J. C. Theilacker, Tevatron cold compressor operating experience, *Cryogenics* 34:107 (1994), ICEC Supplement.
3. B. L. Norris, Initial performance of upgraded Tevatron cryogenic system, in: "Advances in Cryogenic Engineering," Vol. 41a, Plenum Press, New York (1996), p. 693.
4. M.R. Drake, private correspondence "Valve box vibrations" (Nov. 3, 1997), Mohler Technology, Boonville, IN.
5. N. K. Arakere and H. D. Nelson, An analysis of gas-lubricated foil-journal bearings, *Tribology Transactions* 35:1 (1992).
6. G. L. Agarawal, Foil air bearings cleared to land, *Mechanical Engineering* 120(no. 7):78 (July 1998)
7. L. San Andres, Turbulent flow foil bearings for cryogenic applications, *Journal of Tribology* 117:185 (January 1995)
8. I. A. Davydenkov et. al., Development of cryogenic turboexpanders with gas dynamic foil bearings, , *Cryogenics* 32:80 (1992), ICEC Supplement.
9. H. Heshmet and C.P. R. Ku, Structural damping of self-acting compliant foil journal bearings, *Journal of Tribology* 116:76 (January 1994)
10. M. P. Blake, Vibration-part I: its effects, classification, and measurement, *National Safety News* (November 1975).

Centennial Conference staff who assisted in organizing the 1999 Conference. From the left: P. Pair, S. Pair, M. Park, K. Bass.

GAS LEAK INFLUENCE IN A CRYOGENIC SCROLL PUMP FOR SUPERFLUID HELIUM REFRIGERATION

F. Millet, P. D'Harboullé, R. Vallcorba, G. Claudet

CEA.Grenoble. Département de Recherche Fondamentale sur la Matière Condensée. Service des Basses Températures.
17 rue des martyrs, F38054 Grenoble cedex 9, France

ABSTRACT

The 1.8 K large capacity helium refrigerators (TORE SUPRA, CEBAF, LHC) require the use of cold pumps to reduce the size of the pumping system and to ease the design of the heat exchangers. In comparison to centrifugal pumps, scroll pumps feature at least two main advantages : a larger range of compression ratio and stable operation for changing mass flow leading to more flexibility for this volumetric machine.

CEA/SBT has developed a scroll pump designed to treat a 20 g/s helium mass flow at 6.2 K-4200 Pa. This scroll pump prototype should be installed in CEA/Grenoble within the 1.8 K helium test facilities which will provide a 400 W isothermal cooling capacity.

A numerical global model has been developed to predict the thermodynamic performances of the scroll machines. Efficiencies depend significantly upon leaks between the scrolls. A specific analysis of this aspect, based upon experimental leakage measurement in representative geometry and relevant correlation formulas determination, has been performed and is reported. The numerical model allows to quantify the contribution of gas leakage on the efficiency losses as discussed in the paper.

INTRODUCTION

The saturation curve of helium shows that the pressure on the helium bath is 1638 Pa at 1.8 K. This means that the compression ratio of a pumping system able to reach atmospheric pressure is closed to 100 / 1 when taking into account the pressure drops. In order to achieve the required high compression ratios, the pumping system on the liquid helium bath can be operated at room temperature for 1.8 K low capacity helium cryogenic systems, at cryogenic temperatures or a mixture of both for the large capacity cryogenic systems for 1.8 K helium refrigeration[1]. The 4.8 kW at 2 K for CEBAF[2] are produced exclusively by cryogenic pumping. The 300 W at 1.8 K for TORE SUPRA[3] or the 2.4 kW at 1.8 K for the LHC project[4] are supplied by cryogenic and then room temperature systems.

Cryogenic centrifugal (or dynamic) pumps are the most commonly used machines for large magnets and resonant cavities cooling at temperature lower than T(λ) at the present time. The use of cold pumps is required in order to reduce the size of the pumping system and to ease the design of the heat exchangers because their operating pressures become higher. Several stages are needed in order to reach the overall compression ratio. The cryogenic centrifugal pumps present the great disadvantage of having unstable operation especially when used in trains[2].

The cryogenic scroll pump will be able to pump on the 1.8 K liquid helium bath with a larger range of compression ratio and better operational stability due to volumetric characteristics. Reciprocating cold compressors or expanders with volumetric characteristics have been developed and tested by Fermilab[5] but the mechanical contacts inside cylinders decrease their reliability and impose regular internal maintenance. The scroll pump opens a way for a volumetric machine operated without any mechanical contact. These two points must allow to reach large stable operations and a high reliability level for a cryogenic pumping system. We carried out a first prototype of a scroll pump designed to pump a 20 g/s helium mass flow at 6.2 K-4200 Pa.

The prototype configuration, shown in Figure 1, consists essentially of :

- In the bottom of the pump, a pair of scrolls (one fixed, one orbiting) operating at low temperature (3 K < T < 8 K).
- In the central part of the prototype, cold orbiting scroll connected to the warm orbiting mechanism by a shaft. A rotating 80 K thermal shield is used to reduce the heat input on the pumped helium gas.
- In the top of the pump, a one-shaft orbiting mechanism operating at room temperature and developed in partnership with the French industrial firm NORMETEX[6].

NORMETEX is specialised in scroll pumps operating at room temperature for vacuum, we adapted this technology for a cryogenic use. The pumping part of the pump is done by two identical scrolls (see Figure 1). One of them is fixed and the other one described an orbital motion inside the previous one. Due to the orbiting movement, the gas volume is taken at the periphery of the scrolls where the suction port of the pump is located. This volume is pushed progressively with compression to the centre of the scrolls where the discharge port of the pump is located.

The scrolls are made of aluminium alloy in order to decrease the weight of the rotating parts of the pump. The main characteristics of the prototype are done in Table 1.

Table 1 . *Main characteristics of the scroll pump prototype*

Rotation speed	0 to 25 Hz
Scroll pitch	43 mm
Scroll outer diameter	440 mm
Number of scroll turns	4.25
Scroll tip height	90 mm
Scroll wall thickness	5 to 10 mm
Geometrical volume ratio	2.46
Maximum swept volume at 25 Hz	90 L/s

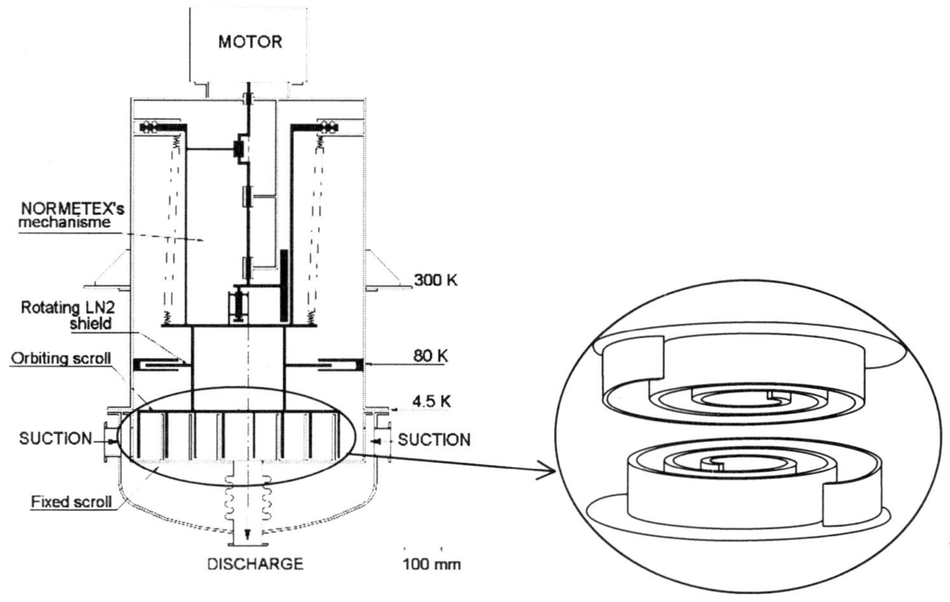

Figure 1 . *Sketch diagram of the scroll pump*

LEAKAGE SPECIFIC ANALYSIS

The necessary clearances between the two contact-less scrolls generate leaks, which degrade the performance of the pump[7]. Mass flow rate calculations[7,8] have been done to quantify the leakage influence for particular designs (low clearances and small scroll sizes). For the cryogenic scroll pump with so specific geometries (larger clearances and sizes), it is necessary to perform experimental measurements to determine the relevant correlation formulas.

Two kinds of leaks can be considered from a high pressure (HP) chamber to a low pressure (LP) chamber : the axial and radial clearances as shown on Figure 2 and Figure 3. We defined δa and δr as follows :

- δa : axial clearance between the two scrolls
- δr : radial clearance between the two scrolls

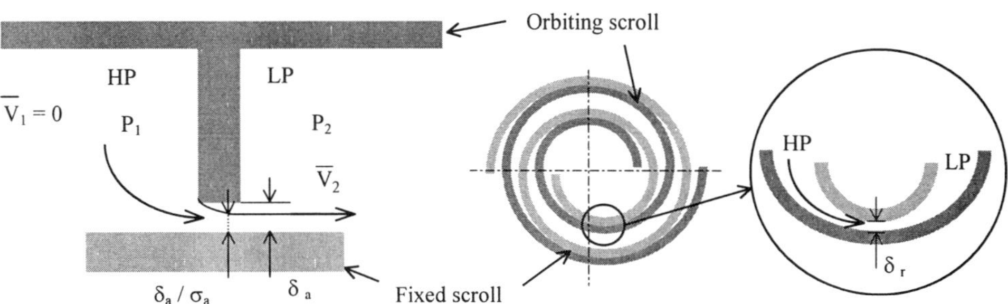

Figure 2. *Axial clearance for the wrap tip leaks*

Figure 3 .*Radial clearance for the flank wrap leaks*

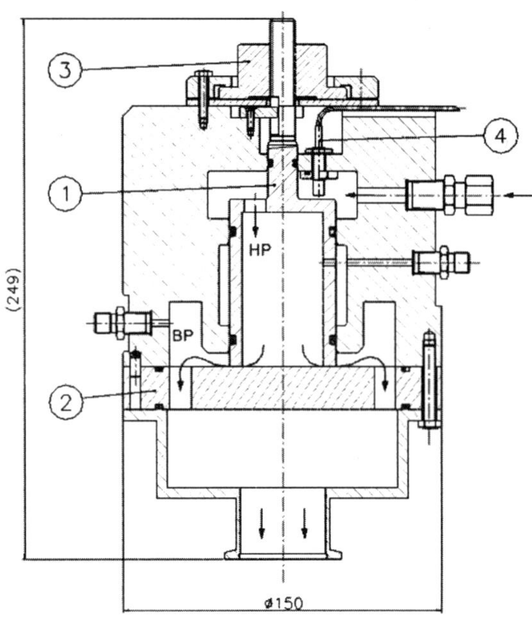

Figure 4. *Experimental device for measuring the leak flow through axial clearance*

Experimental devices

An experimental test bed was built in order to quantify the pressure drop, the wall thickness effect and the groove influence between HP and LP gas chambers. The Figure 4 presents the experimental system taking into account representative geometry for axial clearance. The cylinder ($\varnothing_{ext}$ = 50 mm) represents the small diameter of the scroll. The clearance between the extremity of the cylinder ① and the support ② is adjustable by a screw-nut system ③. Three position sensors ④ controlled the axial position of the cylinder. The precision of the sensor measurements is estimated at ± 3 μm. For a good tightness between the cylinder and the support, the two sides of these components were run in.

Due to the small tangential speed of the orbiting scroll (< 2.35 m/s) compared to the speed of the gas leak flow through the clearance, we assumed that a static experimental approach with a stationary cylinder was adequate for a realistic simulation of the axial phenomena. The fluid used for the measurements was air at room temperature. The pressure drop between HP chamber and LP chamber was chosen to simulate the Reynolds numbers existing inside the scroll pump prototype during nominal operations.

Leak rate correlation

For flow through the tip gap of a compressor blade, Rains[9] developed a model based on incompressible potential flow theory (Bernoulli's theory). For scroll pump, the Bernoulli's equation based flow model is applied between the upstream and downstream of the axial clearance to obtain the expression of the maximal mass flow rate.

As illustrated on Figure 2, the real mass flow rate through the axial scroll clearance is a function of a contraction ratio σ_a which is defined by Rains as follows :

$$\sigma_a = \frac{\dot{m}_{Bernoulli}}{\dot{m}_{real}} \tag{1}$$

with $\dot{m}_{real}$: measured mass flow kg/s

$$\dot{m}_{Bernoulli} = \rho_2 \overline{V}_2 S_2 = S_2 \sqrt{2\rho_2 (p_1 - p_2)} \quad \text{kg/s} \tag{2}$$

ρ_i : density of the gas chamber i with $p_1 > p_2$ kg/m^3

p_i : pressure inside gas chamber i Pa

$S_i = f(\delta a)$: section of leakage m^2

$$\overline{V}_2 = \sqrt{\frac{2(p_1 - p_2)}{\rho_2}} \quad \text{: gas speed} \quad \text{m/s} \tag{3}$$

The coefficient σ_a quantifies the pressure drop losses through the axial clearance and depends of the outlet Reynolds number Re_2 which is given by :

$$Re_2 = \frac{\rho_2 \overline{V}_2 D_{H2}}{\mu_2} = \frac{4\dot{m}_{real}}{\mu_2 L_2} \tag{4}$$

with D_{H2} is the outlet hydraulic diameter and is equivalent to $2\delta_a$. m

L_2 is the wet perimeter m

μ_i is the dynamic viscosity of the gas chamber i Pa.s

The values of the axial contraction ratio σ_a coming from experimental tests are presented in the Figure 5. This figure shows that :

- for Reynolds number lower than 3500 (laminar flow), the following correlation can be establish :

$$\sigma_a = \frac{70.9}{Re_2^{0.47}} \tag{5}$$

- for Reynolds number higher than 3500 (turbulent flow), the coefficient σ_a can be considered as a constant value equal to 1.53.

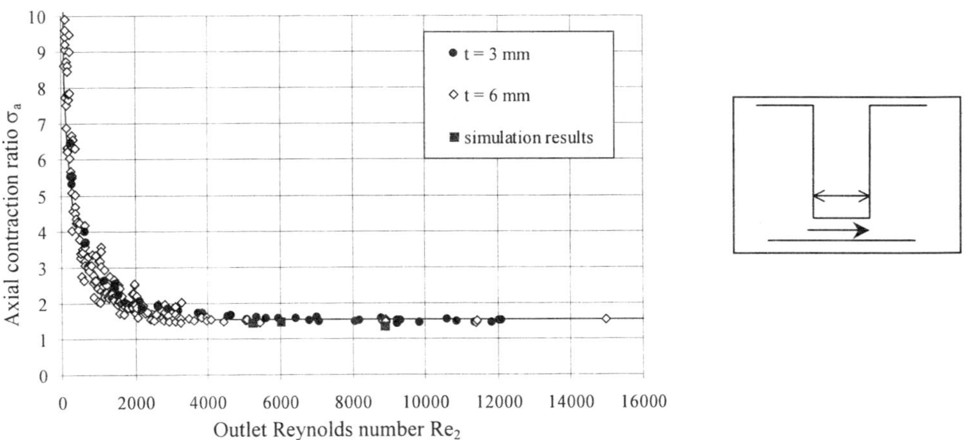

Figure 5. *Measured axial contraction ratio σ_a versus Reynolds number*

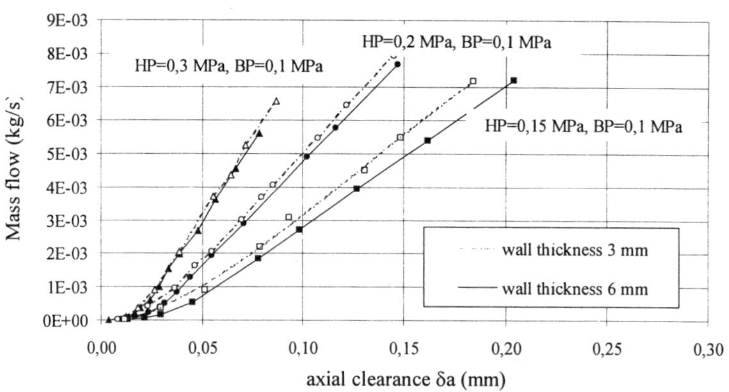

Figure 6 . *Measured mass flow versus axial clearance*

Experimental wall thickness and groove studies

After measurement and determination of the mass flow through the axial scroll clearances, we investigated how it was possible to reduce this leak flow by increasing the pressure drop losses between gas chambers. The first study was to increase the wall thickness. Figure 6 presents the data obtained for the clearance measurements with two different wall thickness, the influence on the axial leak flow is less than 15 %. Furthermore the differences between 3 mm and 6 mm wall thickness are inversely proportional with the pressure difference between HP and LP chambers.

With the same goal to reduce the leak flow by increasing the singularity along the wall, a 6 mm wall thickness cylinder with three 1 mm grooves at the tip was tested in order to experiment the labyrinth theory for scroll leakage. The decrease of the axial leak mass flow rate during this test is not significant with this type of grooved cylinder.

Numerical simulation

For the axial clearance flow, ANSYS/FLOTRAN[10] software calculations were compared against measurements and showed good agreement. The software simulates a viscous compressible two-dimensional helium flow through the relevant geometry. The results are closed to the experimental data with differences less than 10 % (see figure 5). The numerical simulation calculates the leak mass flow through the axial scroll clearance and allows observation of the pressure evolution along the wall thickness as shown in Figure 7. In this configuration, we can see that the losses due to the contraction in the cross section are most important than the friction losses of the wall.

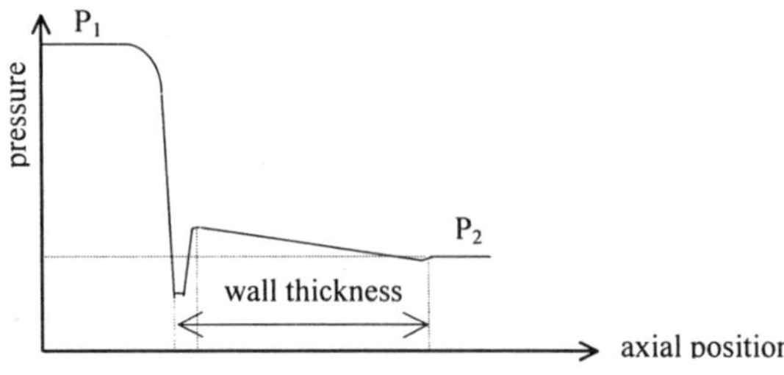

Figure 7. Pressure evolution inside wrap tip clearance

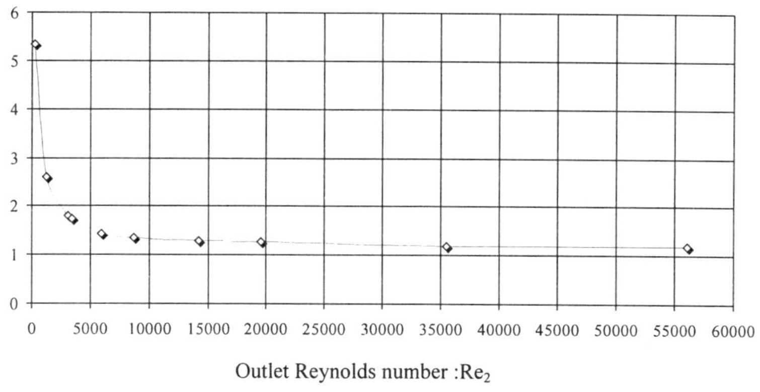

Figure 8. *Radial contraction ratio σ_r versus Reynolds*

After qualification of the model for axial leaks, this numerical simulation was applied for calculation of the radial contraction ratio for scroll pump leakages. The values of this radial contraction ratio σ_r versus outlet Reynolds number are presented in Figure 8. The figure shows :

- for Reynolds number lower than 14000, the following correlation can be establish :

$$\sigma_r = \frac{34.3}{Re_2^{\ 0.36}} \tag{6}$$

- for Reynolds number higher than 14000, the coefficient σ_r can be considered as a constant value equal to 1.19.

The axial and radial correlations obtained during this specific analysis were used in a global numerical model for the design of scroll pump or expander. We decided to take 1.53 as axial contraction ratio and 1.19 as radial one in order to keep some safety margins for the efficiency calculations of the scroll pump.

GLOBAL MODEL

A global model has been developed in the laboratory in order to predict the thermodynamic performances of the scroll machines. The global model is based on the conservation mass equation and on the thermodynamic first law. This tool takes into account geometric (diameter, scroll depth, thickness, number of scroll turns), dynamic (inter-chamber leaks as described before), thermodynamic (helium properties[11]) and heat transfer (conductivity, convection) characteristics of scroll pumps.

The Figure 9 shows the results of the calculation provided by the global model including leakage influence reported before and without heat inputs. For a pressure ratio of 3, the volumetric efficiency of the scroll pump is equal to 57 % for axial and radial clearances equal to 0.35 and 0.25 mm (fig 9a) and 79 % for axial and radial clearances equal to 0.15 and 0.35 mm (fig 9b).

The influence of the axial and radial clearances are not the same especially for high pressure ratios. As shown on Figure 9, the axial clearance is more important than the radial one because of the larger gas section offered for the axial mass leakage.

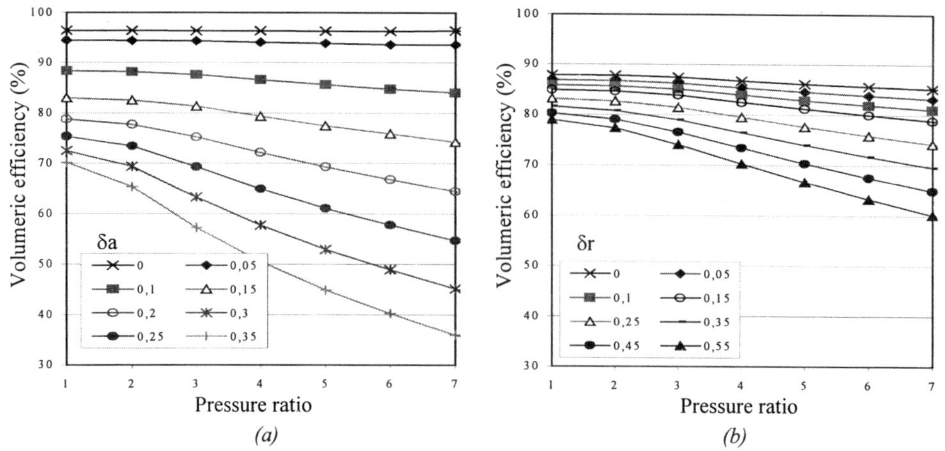

Figure 9 . *Volumetric efficiency versus compression ratio for inlet scroll conditions :* T_{asp}= *6.2K /* p_{asp} *=4200Pa and* $Q_{heat\ input}$ = *0W. (a) clearance δr = 0.25 mm and δa variable ; (b) clearance δa = 0.15 mm and δr variable.*

CONCLUSION

This paper reports the specific analysis performed for leakage influence determination inside cryogenic scroll pump dedicated to superfluid helium refrigeration system. The numerical simulations for axial leak rates are in good agreement with the experimental data with differences smaller than 10%. The leakage model leads to take the axial compression ratio σ_a equal to 1.53 and the radial compression ratio σ_r equal to 1.19. These relevant correlations are introduced inside a global model taking into account mass conservation and energy equilibrium. This model is used to design the scroll machines, the calculated performance will be compared with the results obtained with the scroll pump prototype developed for the helium test facilities designing by CEA/SBT to provide 400 W at 1.8 K.

REFERENCES

1. G.Gistau-Baguer, ''High Power Refrigeration at Temperatures Around 2.0 K'', Proceeding of 16th International Cryogenic Engineering Conference, Part 1, 1996
2. Bevins et al., ''Automatic Pumpdown of the 2K Cold Compressors For the CEBAF Central Helium Liquefier'', Advances in Cryogenics Engineering, Vol 40
3. B.Gravil et al., ''Experience with a Large Scale He II Refrigeration System at TORE SUPRA, Advances in Cryogenics Engineering, Vol 43, 1998
4. F.Millet et al., ''A Possible 1.8 K Refrigeration Cycle for the Large Hadron Collider'' Advances in Cryogenics Engineering, Vol 43, 1998
5. J.D.Fuerst, ''Design, Construction, and Operation of a Two Cylinder Reciprocating Cold Compressor'', Advances in Cryogenic Engineering, Vol. 37B, page 795
6. Brevet S.B.P.V (Société des Brevets P. VULLIEZ), Pompe à vide à cycle de translation circulaire, n° FR 96/00289, 1996
7. K.Suefuji et al., ''Performance Analysis of Hermetic scroll compressors'', Proceeding of 1992 International. Compressor Engineering Conference at Purdue, US, July, 1992
8. N.Ishii et al., ''Refrigerant Leakage Flow Evaluation for Scroll Compressors'', Proceeding of 1996 International. Compressor Engineering Conference at Purdue, US, July, 1996
9. Rains, ''Tip clearance in axial flow compressor and pumps'', California institute of technology, hydrodynamics and mechanical engineering laboratory, D.A., , Rep. n°5 ,1954
10. ANSYS/FLOTRAN, ANSYS INC.
11. HEPAK, Helium Properties Computer Program, CRYODATA INC.

DESIGN, DEVELOPMENT, AND TESTING OF A CRYOGENIC RECIPROCATING EXPANSION DEVICE

A. Sidi-Yekhlef, J. Maguire, P. Winn, B. Gamble, C. Gold, D. Bushko, D. Hannus

American Superconductor Corporation Westborough, MA

ABSTRACT

A lightweight, easily maintained helium expander system has been developed to provide reliable, low cost refrigeration in the 20 K to 40 K temperature range. The expander consists of a 2 inch diameter piston controlled by a combined hydraulic/pneumatic system. During the expansion process some of the energy released from the helium gas is stored in the hydraulic system. The stored energy is used to return the piston to its lower position. The expander assembly is built as a self- sealed removable pod that can be replaced with a new pod without warming or contaminating the main system. The expansion engine is controlled by a processor, which reads an LVDT signal and opens and closes valves in a timed sequence. The LVDT rod moves up and down with the piston giving a signal to the processor which controls the valve timing logic. This design is unique in the sense that the pressure of the gas after expansion can be easily adjusted by the valve timing. The hydraulic nature of the drive mechanism also allows us to compensate for the high back-pressure during the resetting of the piston. A typical reverse Brayton cycle operates with a compressor discharge of 1.7 MPa and a suction pressure of .1 to .138 MPa. This system is designed with the option of a higher suction pressure, allowing for higher throughput at similar power input. An adiabatic efficiency of 60% was demonstrated during testing.

INTRODUCTION

Progress has been made in the development of high efficiency high temperature superconducting (HTS) motors with the aid of Department of Energy funding under the Superconductivity Partnership Initiative [1]. The overall goal of this program is the commercialization of high efficiency, large scale, HTS motors. This effort includes the development, fabrication, and testing of a baseline refrigeration system as part of a 1000 hp motor system. The selected design requirement of this system is a cooling capacity of 60 watts at or below 30 K with a power input of no more than 10 kW. The system is based on a reverse Brayton cycle which includes a compressor skid built around an oil-lubricated scroll compressor pump, a cold box which houses a series of heat exchangers, and pod that contains a reciprocating expander. The expander pod is attached to the side of the cold box (Figure 1). The refrigerator system is connected to the superconducting motor by a way of a rotating transfer coupling. The development of the refrigeration system has been completed and it is now being packaged for shipment to Rockwell Automation for testing with the motor system. This paper summarizes the development and testing of the reciprocating expander.

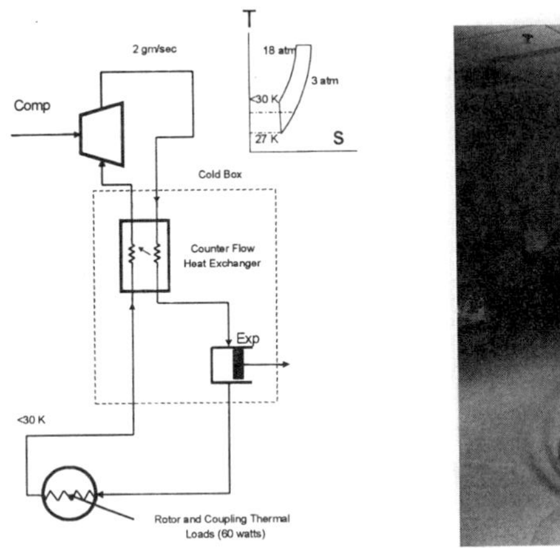

Figure 1. Baseline refrigerator system

EXPANDER DESIGN

Interchangeable Expansion Engine Assembly

Existing piston expanders typically have permanently installed cylinders with a work removal system that incorporates a large flywheel, crank rods, crosshead guide, and break motor. This assembly must be taken apart for maintenance every year. To achieve this the machine must be stopped and warmed, the connecting rods removed, and the flywheel/break assembly pulled back to get at the inner piston. This work requires experienced personnel on site and a minimum of a day of down time. One of the major goals for the design of this new system is to allow for scheduled maintenance that requires little or no down time and simple maintenance procedures. To accomplish this the engine assembly is constructed as a modular, removable pod that can be taken off the main system and replaced quickly with a new pod. The old pod can then be sent back to the manufacturer for routine maintenance. The key to this design is a set of self sealing couplings that allow an operator to quickly remove the pod without contaminating the system, and a light weight work extraction device that allows a single operator to lift the pod for removal and re-assembly.

To reduce the weight and size of the expansion engine assembly a hydraulic work extraction device was designed to mount the top of the helium expander, connected to the piston rod. The idea of a hydraulically operated expansion engine originated with S.C. Collins in his development of a prototype helium liquifier [2]. This concept was further developed by J.L Smith Jr. for use as a two-phase expander to enhance the performance of a helium liquifier system, [3]. The expansion engine (Figure 2) has a 51-mm diameter piston with an adjustable stroke of up to 102 mm. A phenolic piston is sealed inside a stainless steel cylinder by a set of redundant lubricated o-rings. This lower piston is connected to a hydraulic piston and two parallel pneumatic pistons via a connecting cross. The pneumatic pistons act as constant force springs that return the phenolic piston to the lowest portion of the cylinder after expansion of the helium gas. The phenolic piston and the hydraulic piston are mounted on the same axis and separated by a universal coupling that takes up misalignments. An LVDT rod is attached to the cross that runs from the hydraulic piston rod to the pneumatic pistons. This LVDT rod moves up and down with the piston giving a signal to a processor that controls the valve timing logic. The expansion engine valves consist of spring loaded flanges with plastic tips for sealing. The flanges are connected to rods that run up to the valve actuators. These rods are o-ring sealed at the top. The main cylinder, valve cylinders and an inlet filter are all housed inside a self contained vacuum jacket.

Expander Control

The expansion engine is controlled via a processor that reads the LVDT and opens and closes valves in a timed sequence. With the helium piston at the bottom of its stroke the inlet valve is opened via a linear solenoid on the top of the inlet valve rod. When the inlet valve opens compressed helium gas enters the cylinder. The helium gas expands against the phenolic piston forcing the piston upward. The attached hydraulic piston forces oil from the top of the hydraulic piston and through the control valve to the bottom of the hydraulic piston. As the pistons advance, the inlet valve is closed at a preset cut-off point. After the inlet valve closes, the helium pressure in the lower piston falls as the gas expands and the piston assembly slows. To maintain speed a solenoid valve (v-back in Figure 2) is opened allowing more oil to flow around the hydraulic piston loop. The work extracted from the gas is transferred as heat of friction through the control valve and v-back. This heat is transferred to the atmosphere via a fan that blows over the hydraulic assembly. As the piston reaches the top of the stroke the exhaust valve is opened allowing the helium gas to further de-pressurize and flow into the low side of the system. The exhaust valve is actuated at the top of the valve rod via a pneumatic cylinder that is controlled by a small solenoid valve. As the helium gas de-pressurizes the piston assembly is forced back down by the pneumatic spring cylinders. The oil from the bottom of the hydraulic piston then runs back up to the top of the hydraulic cylinder via v-back and a check valve to reduce the flow restriction. The exhaust valve is closed slightly before the bottom of the stroke allowing for

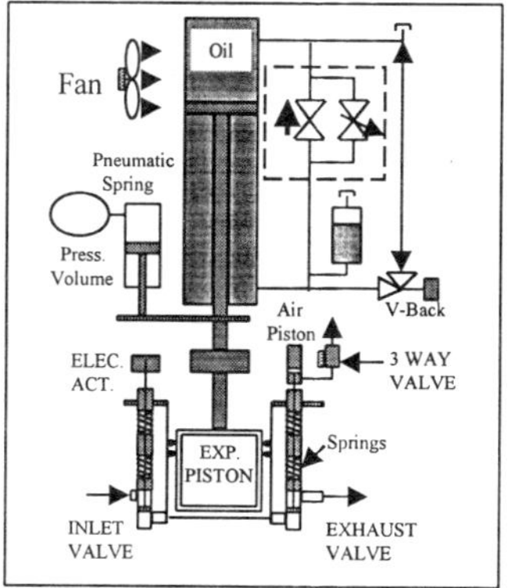

Figure 2. Schematic of the expander assembly

some recompression before the cycle begins again. The control system includes potentiometers to allow for the easy changing of cut-off, stroke, and re-compression.

Self-Sealing Couplings

The expansion engine pod is attached to the main cold box via a set of self-sealing bayonets. The bayonets have spring-loaded seals that open as the male and female bayonets are pushed together via preinstalled hand clamps. Figure 3 shows a sketch oft this concept. When the engine is removed the couplings seal both the expansion engine pod and the cold box, thereby keeping both systems clean. When a new expander is installed it is pre-cleaned and pressurized, ready for use.

TEST RESULTS

The development of the expander system has been completed and its performance and efficiency measured. The instrumentation during testing is shown in figure 4. It included pressure sensors installed at the inlet and out of the expander to monitor the helium pressure and a pressure transducer mounted at the bottom of the cylinder. The signals from the pressure transducer and the LVDT were used to monitor the P-V diagram on an oscilloscope. Diode temperature sensors were mounted at the inlet and outlet of the expander (s1 and s2 in Figure 4) for recording gas temperatures before and after the expander. In order to measure the capacity of the system; a vacuum insulated jumper was installed between the outlet of the expander and the return side of the cold box as shown in Figure 4. The jumper includes an electrical heater and two diodes (S3 and S4), one before the heater and the second one after the heater. The diode sensors were mounted directly on copper tubes to accurately measure the helium temperature. Another sensor (S5) was placed on the return side of the heat exchanger. This sensor allowed measurement of the efficiency of the heat exchanger and the heat loss across the jumper.

The system achieved a temperature of less than 15 K under no load conditions (with jumper losses but without any power to the heater). The capacity of the system was measured both under normal ambient condition as well as at 38 C temperature. The capacity as a function of the return temperature to the heat exchanger from the load is plotted in Figure 5.

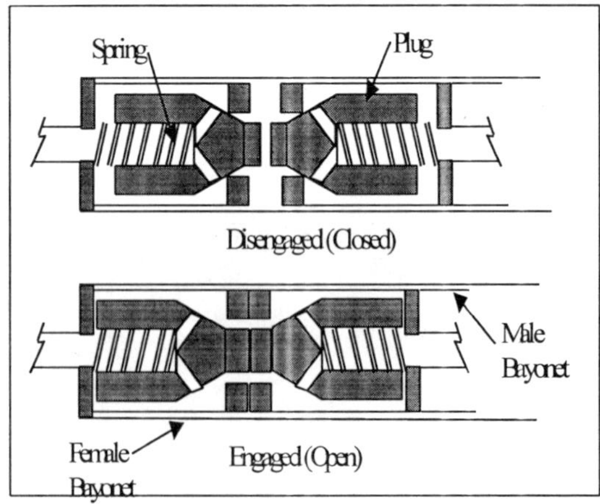

Figure 3. Self-sealing couplings

Without the N2 precooler (shown in Figure 4), the system can cool itself down to 20 K within six hours. A cooldown curve from 300 K down to 20 K is presented in figure 6. The N2 precooler consists of a liquid nitrogen boiler with an auto refill. It is designed with a coiled finned tube that circulates helium gas. Its purpose is to reduce the motor cooldown time.

To assure high reliability, the expander is designed to run at speeds between 1 and 1.5 Hz. At 1.0 Hz speed, an adiabatic efficiency of 60 % was demonstrated in testing. This efficiency was calculated at the following conditions, before and after the expander:

Inlet temperature = 35.8 K — Inlet pressure = 1.47 MPa

Outlet Temperature = 25 K — Outlet pressure = .40 MPa

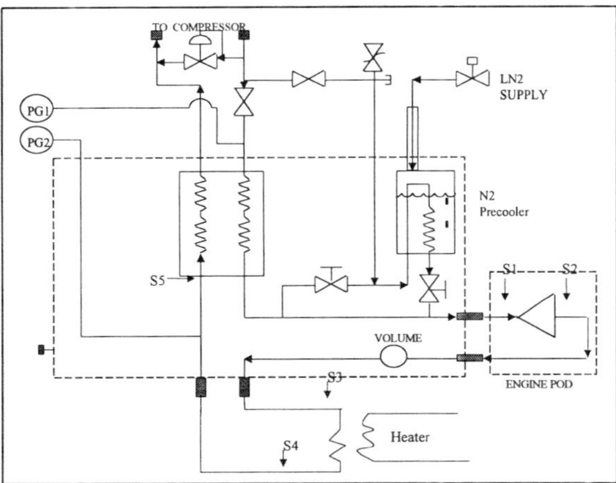

Figure 4. Flow schematics showing the instrumentation used during testing.

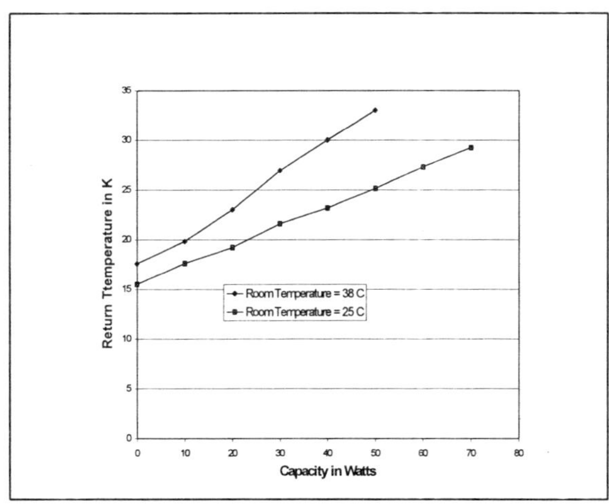

Figure 5. Capacity curve as a function of return temperature to the heat exchanger

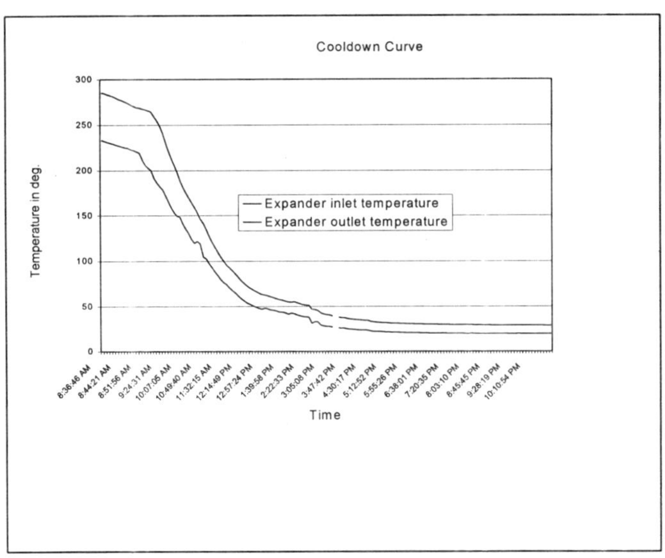

Figure 6. Cooldown curve: Inlet and outlet temperatures of the expander vs. time.

CONCLUSION

A 2 in diameter reciprocating expander has been developed and tested. This expander is part of a refrigeration system developed at American Superconductor Corporation for the 1000 hp HTS motor. A temperature of less 15 K was achieved with zero load. The system is capable of achieving 70 watts cooling capacity with a return temperature to the heat exchanger of less than 33 K. The main characteristics of this expander are its light weight, the ability to change speed, and to vary the timing easily. It is constructed as a removable pod, which can be interchanged without warming the system. This feature allows for redundancy in the system, which then allows for continuous performance.

ACKNOWLEDGEMENT

The work presented herein was supported in part under the Department of Energy Superconductivity Partnership Initiative Award DE-PC36-93CH10580.

REFERENCES

1. Development of Ultra Efficient Electric Motors Using High Temperature Superconducting Materials, DoE SPI Program Proposal, June 1995.
2. S. C. Collins " Proceedings of the 1956 Cryogenic Engineering Conference, Vol. 2 Paper A2 pp. 8-11.
3. R. W. Johnson, S.C. Collins, and J.L. Smith, Jr., "Hydraulically Operated Two-Phase Helium Expansion Engine", Advances in Cryogenic engineering, Vol. 12, pp. 595-601.

PERFORMANCE OF A HYBRID VACUUM PUMP AT LOW TEMPERATURE

J.P. Périn[1], J. Manzagol[1], J.J. Cordier[2], P. Garin[2], F. Samaille[2]

[1]C.E.A. Grenoble / Département de Recherche Fondamentale sur la Matière Condensée/SBT, 17 rue des Martyrs, 38054 Grenoble Cédex 9, France
[2]C.E.A. Cadarache / Département de Recherche sur la Fusion Contrôlée/STEP,
13108 Saint Paul lez Durance, France

ABSTRACT

In this paper a novel concept of mechanical vacuum pump is presented. This pump is able to run at low temperature (80K to 25K). A temperature gradient between the different pumping stages allows optimisation of the exhaust pressure. Since gas density varies inversely with temperature, a pump delivers much higher mass flow rate at low temperature than at room temperature for a given size. Advantages of this concept are reduction in size and weight of the pump compared to a conventional one scaled for the same mass flow rate at room temperature. This pump would be a solution for continuous tritium extraction and minimization of the mass inventory for fusion tokamaks fuel cycle control.

INTRODUCTION

News design of tokamaks will require the development of large tritium compatible high vacuum pumps. In large tokamaks, such as the International Tokamak Experimental Reactor (ITER), the plasma exhaust containing deuterium, tritium and a fraction of 5% helium must be pumped off in the burn and dwell regime (permanent regime after ignition). In this machine, the mass flow is 200 Pa-m^3/s (NPT) at a pressure of 50 Pa. Meeting these requirements, with present technology, necessitate the use of cryopumps[1] and turbomolecular pumps working at room temperature[2].

This paper provides the technical description of a vacuum pump prototype developed from the calculations and experimental results obtained previously[3,4] at low temperatures (80K and 25K) with a molecular drag pump (MDP) (Holweck type of 100 mm diameter) and with few stages of a turbomolecular pump (TMP) running at the same temperatures. This pump will be used for the future Toroidal Pumped Limiter of next tokamak generation, where the particle control needs high pumping speeds. It would be an engineering solution for ITER, which would minimize the scale-up and the tritium inventory.

LOW TEMPERATURE PUMPING CHARACTERISTICS

The main results obtained from the calculations and experimental set-up are the following:

(i) The proportionality between flow rate and gas density is verified within the performed experiments in the $1x10^{-1}$ Pa to 5 Pa range for TMP and MDP pumps and for different pure light gases like hydrogen, deuterium, helium and mixtures of hydrogen or deuterium with 5% of helium.

(ii) The maximum flow rate and the pressure ratio for MDP pump are higher if the gas is more viscous. Helium gas is more viscous than hydrogen or deuterium, so for mixtures the apparent viscosity is higher than for a pure gas as reported by the Mann law.

(iii) In MDP the compression ratio depends on the temperature and inlet pressure, so an optimum temperature level can be determined. The numerical model developed predicts this effect. At low inlet pressure ($1x10^{-1}$ Pa) the pump is working near the molecular flow range where the compression ratio (τ) is in a first approximation inversely proportional to the average velocity of the molecules ($\tau \sim \exp(1/T^{1/2})$). In the viscous range at high pressure (5 Pa) the pumping performance is determined by the variation in the dynamic viscosity of the gas with temperature ($\tau \sim T^a$, a is a dimensionless coefficient and its value ranges from 0.5 to 0.7). This temperature can be used as a parameter to optimize the outlet pressure.

(iv) For TMP, working at low temperature is particularly useful because the mass flow rate and the pressure ratio are both increasing[3,4].

At low temperature the TMP is working in the transition regime, so the thermal losses due to friction are higher than at room temperature for the same working pressure. They are indeed inversely proportional to the Knudsen number (ratio of the mean free path to the clearance between stages). They have to be taken into account in the design of the rotor blades. This point is very important for the cooling system capacity sizing.

Mechanical pumping of high specific weight gases, due to the decrease of the temperature, permits extraction of the light gas mixture at a very high flow rate while at a convenient pressure.

PROTOTYPE DESCRIPTION

Mechanical structure

The rotating speed was set as 21000 rpm to limit the mechanical stresses due to the centrifugal forces.

This pump is driven by an electrical motor provided with ball bearing at 300 K and a cold passive magnetic bearing at 25K. This magnetic bearing is a non contacting bearing essential for running at low temperature.

The materials used in constructing this prototype are titanium and aluminium alloys. To prevent differential expansion between the shaft and the pump body, both are made of titanium alloy. Titanium also reduces the axial thermal flux from ambient to the pumping stages. For the MDP stage or the TMP stage, aluminium alloys were used for the moving and static parts; thus the radial thermal resistance is low and the pumping stages can be assumed isothermal (fig.1).

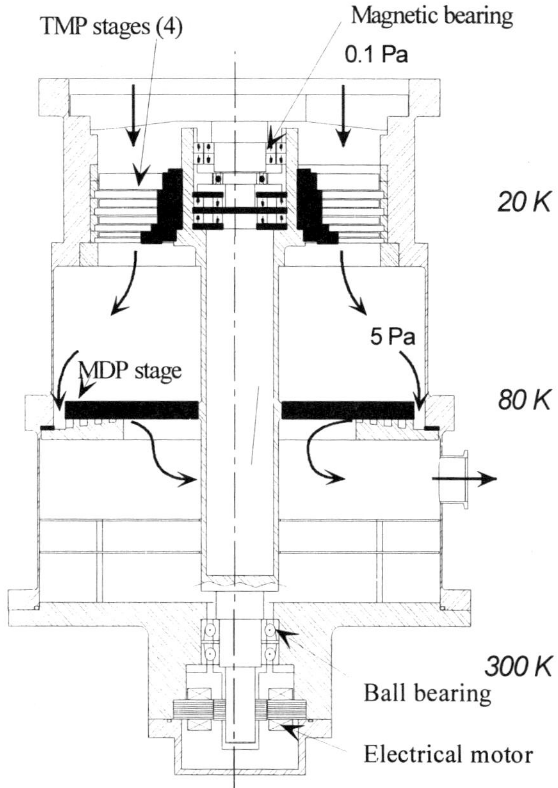

Figure 1 . Schematic layout of the pump.

Pumping System

To reach the desired requirement (mass flow 0.7 Pa-m^3/s NPT, inlet pressure 10^{-1} Pa outlet pressure 100 Pa), four turbomolecular stages and one molecular stage are needed. This design permits a pressure ratio of 10^3 (50 for TMP and 20 for MDP). The prototype is designed from industrial size TMP of Alcatel (290 mm diameter) and is shown in figure 2 and 3.

The TMP is made of 4 stages (3 compression stages 30° and 22° blade angle and one impeller stage 45° blade angle). The axial clearance between the stator and rotor blades is 4 mm, and the radial clearance for the same parts is 1.5 mm. The axial gap was defined to reduce thermal losses due to the friction torque between the gas and rotating blades.

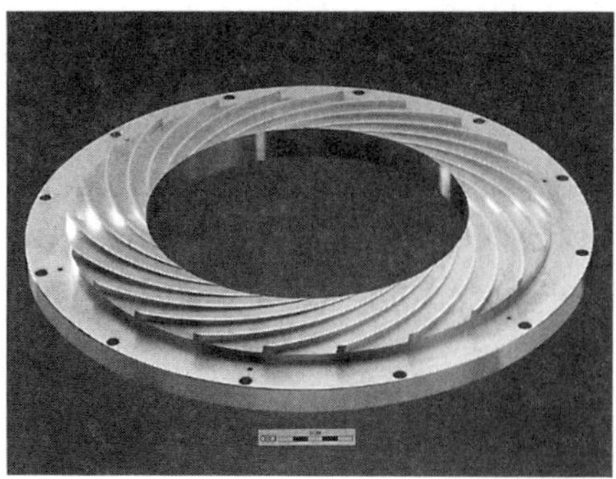

Figure 2 . MDP stator with logarithmic grooves.

Figure 3 . Pump rotor with TMP stages and 1 MDP stage

The MDP stage is a disc type pump (Seigbahn type) to avoid the problems due to thermal expansion and mechanical stresses, and also to allow an increase in pumping speed. The channels are grooved in the stator as a logarithmic spiral. With this design, the pumping channels follow more closely the direction of the pumped gas speed vector. The geometric parameters of the MDP stage are summarized as follows :

Numbers of channel = 22	Rotor diameter = 180 mm to 120 mm
Inlet cross section = 18 mm × 6 mm	Outlet cross section = 10 mm × 0.5 mm
Developed length = 168 mm	Gap rotor /stator = 0.2 mm

Cooling System

Two cooling systems are used for the prototype :

(i) Liquid nitrogen is used to cool down the MDP stage and the thermal shields at 80 K.

(ii) A closed hydrogen loop is used to cool down the TMP stage. For the hydrogen loop, the cold source is the second stage of a Gifford-Mc Mahon (GM) refrigerator fitted with a condenser. The condenser temperature is monitored at 15 K (H_2 triple point is 13.8 K). The heat transfer is obtained by evaporation of liquid hydrogen in the evaporator, which is attached to the TMP body. This technique reduces the thermal gradient to 1.5 K between the cold source and the TMP. The GM cryorefrigerator extracts 35 W @ 20 K with a ΔT of 1.5K. The thermal losses of the TMP stage are due to radiation and conduction (10 W) and to friction torque, between gas and blades, and Eddy courant torque of the magnetic bearing (25W).

PROTOTYPE TESTS

The prototype was tested on a full-scale model of the Tore Supra vacuum vessel. The pumped gas was first cooled down through the inlet 500 mm cylindrical pipe (300 K at its upper part), provided with two successive baffled heat exchangers. The first one is cooled to 80 K. The second one, cooled to 25 K, and is close to the entrance of the pump.

The figure 4 confirms increase for the pumping speed when the working temperature is low. For our pump we achieve a factor of four to ten increase in pumping speed in comparison with the same pump running at room temperature.

The prototype was tested at three temperatures (7K, 20K, 80K) using different cooling systems (helium flow, liquid hydrogen and liquid nitrogen). The main remarks are :

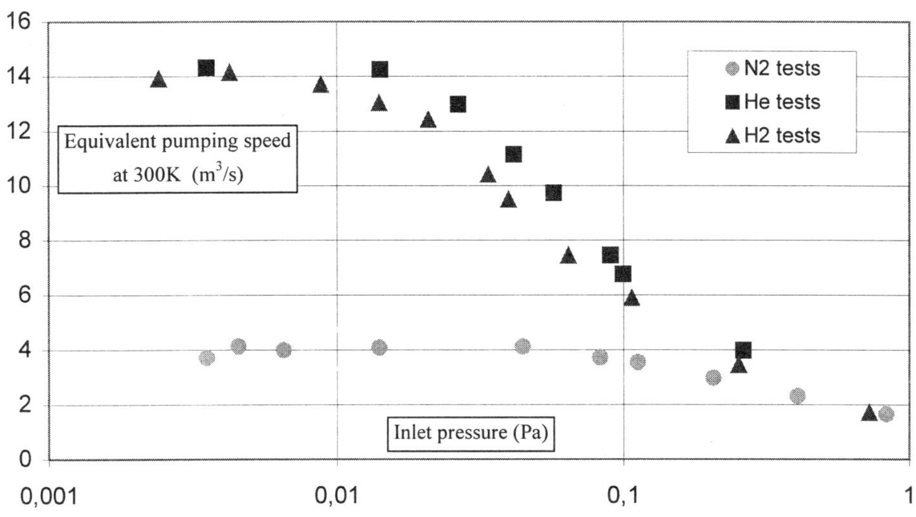

figure 4: Pumping speed for different temperatures versus inlet pressure measured above TMP blades

(i) The pumping speed are in the temperature ratio as shown in figure 4.

(ii) The pumped gas stay nearly at the same temperature whenever the pump body is at 7K (helium cooling) or at 20K (liquid hydrogen cooling). The explanation of this phenomena is that the heat is removed through the aluminium alloy of the pump body. Note that the thermal conductivity of this alloy is reduced by a factor two between 7K and 20K (see table 1).

(iii) This result shows that the real gain depends on the temperature ratio but also on the thermal properties of materials used for the mechanical parts of the pump.

Table 1.

Cold source	Thermal resistance of Al alloy	Power to be removed	Gas temperature
7 K	1 K/W	20 W	27 K
20 K	0.5K/W	20 W	30 K

When the flow is in the transition range, the heat losses due to the gas friction increase. The pinch point leaves the TMP stage at a point where pressure and temperature are the lowest and so the heat exchange is the worst.

A simplified model based on the following hypotheses was made :

(i) Principle of the 'slip flow' in the TMP stage so the shearing stress (τ) is described by a law like :

$$\tau = {dV}/{dz} \times 1/(1+2Kn), \qquad (1)$$

where ${dV}/{dz}$ is the gradient velocity between a static and a moving stage and Kn is the Knudsen number.

(ii) The pressure ratio is given by :

$$P_{i+1} = P_i^a \tag{2}$$

where $\boldsymbol{a}$ is the compression level of each stage.

(iii) The gas is considered isothermal

This model gives a good approximation of the heat losses generated in the TMP stage as described in the figure 5.

A new design of the TMP blades is necessary to improve the overall performance of the pump by decreasing the thermal losses (25W). The cost of the cooling system is directly link to these thermal losses.

The same analytical work has been done for the MDP stage which works at 80K. The channel section is decreasing from inlet to outlet and the flow is described by a Couette flow taking into account the 'slip flow' at the walls. The flow depends on the ratio depth / width of the channel

$$Q_v = \frac{VS}{2} \times \frac{b}{a} \times f(a/b) \tag{3}$$

where V is the velocity of the moving plate, S the channel section and a, b the channel dimensions.

$$F_c = \frac{V}{\mu} \times f(a/b) \tag{4}$$

F_c is the shearing strain for length unit of channel, μ the dynamic viscosity

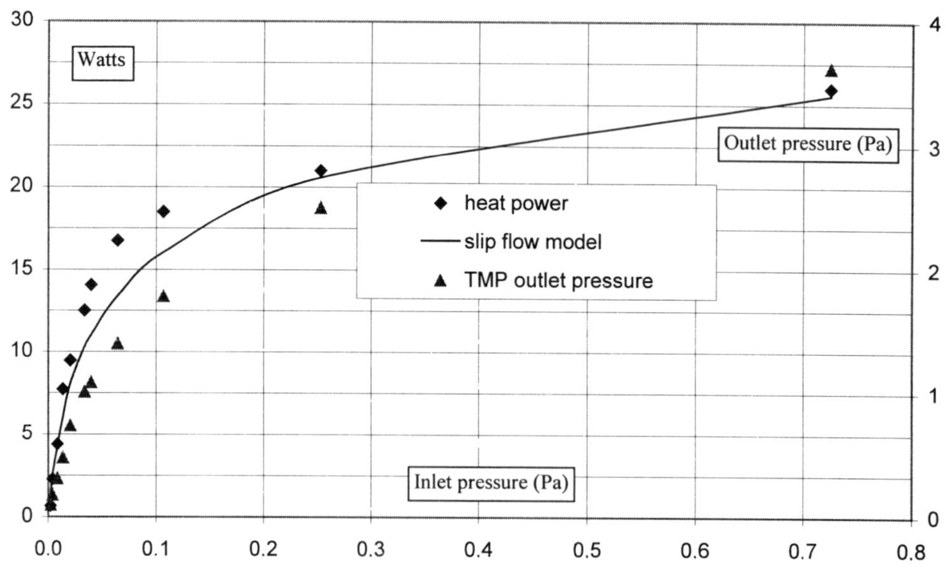

Figure 5 . Heat power due to the friction gas and outlet pressure of TMP stage (working temperature : 25 K)

The figure 6 describes the evolution of the ratio $Q_v/F_c = f(a/b)$ and the mass flow for different values of aspect ratio *(a/b)* of the channel. The continuous line is the region where the MDP is working. The aim of this calculation was the optimization, for a given compression ratio and mass flow rate, of the friction torque between the two parts of the MDP stage and the gas. The power dissipated in the MDP stage is evaluated at 20 to 30 watts depending on the inlet pressure (10^{-2} Pa to 10^{-1} Pa).

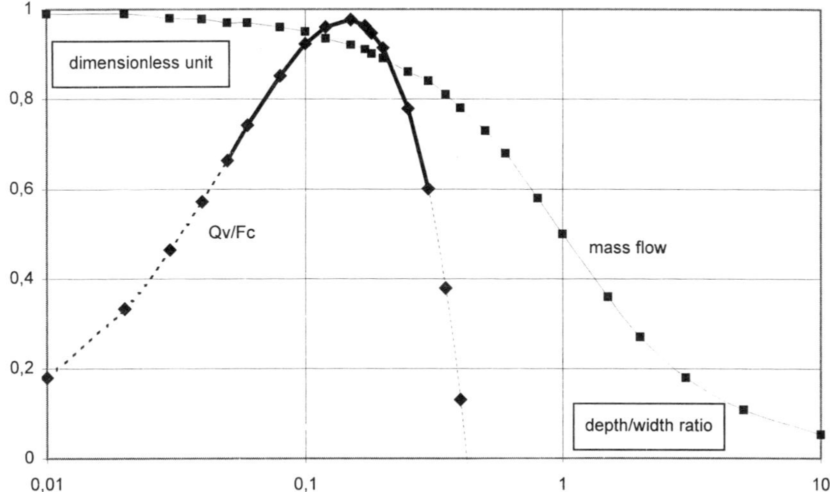

Figure 6 . Shearing stress and mass flow versus aspect ratio for a rectangular channel in MDP stage

From the deceleration curve (rotating speed vs time) at ultimate low pressure (10^{-6} Pa), we can deduce the heat losses due to the ball bearing torque and the Eddy current torque in the passive magnetic bearing and so a thermal balance of the pump can be made.

Table 2

F (Hz)	Measured ball bearing losses (W)	Calculated MDP losses (W)	Measured TMP losses (W)	Total (W)	Measured electrical motor power (W)	Error
0	0	0	0	0	0	0
150	10	5.5	5.9	21.4	23	-8.5%
200	17	8.9	8.6	34.5	35.5	-3%
250	23	13.9	12.5	49.4	51.5	-4%
300	31	19.8	16.5	67.3	70	-4%
350	48	27	21.5	96.5	100	-3.5%

The table 2 summarizes the heat balance of the pump for different rotation frequencies and for an inlet pressure of 10^{-1} Pa. The model shown good agreement between the different measurements and the overall evaluation made with the driving power of the motor.

The last issue to be demonstrated is the capability of operating rotating parts at high speed in close proximity to the stray magnetic field influence around a tokamak machine[5,6]. Some calculations were made for the turbomolecular stages where cooling conditions are worse, and under different magnetic field conditions ; time transient (1T/s) or spatial gradient (10^{-2} T/m). The results show that the time transient does not degrade the pump capacities because the energy dissipated is in the range of 1 to 35 Joules. This heat can be easily removed with a small temperature drift. The main problem is the spatial gradient; a dB(v)/dR of 10^{-2} T/m is equivalent to a time variable magnetic field of 2 T/s. Under this condition 10W has to be removed from the TMP stages in a permanent regime.

PROSPECT

All these results show that cold gas pumping is a good alternative for high gas flux but all the process conditions must be calculated for such a pump. The pump temperature can be adjusted to have a balanced system between the pumping speed and the pump environment. This parameter should be used to reduce the cost of the cooling system which could reduced the gain of the cold gas pumping.

CONCLUSION

This prototype has reaches a 0.7 Pa-m^3/s (NPT) flow. This technique seems very attractive for pumping high flux of light gases like deuterium, tritium and helium The achievement of the prototype test campaign implementation will set the start point for the fabrication of 10 standard cryomechanical pumps working at 80 K, for use on the Toroidal Pumped Limiter project. These pumps are scheduled to be running in 2001.

REFERENCES

1. J.C. Boissin et al., AIEA Vienna , 1991, p 732.
2. H. Yoshida et al., ITER-IL-FC-5.2-0-21 (1990).
3. J.P. Perin, G Claudet, F Disdier, 18th SOFT, Karlsruhe (Germany) 22/26 Aug. 1994.
4. J.P. Perin, R. Mathes, 13th Inter. Vacuum Congress, Yokohama (Japan) 25/29 Sept. 1995.
5. A. Nishide et al, J. Vac. Sci. Technol., 20(4), April 1982.
6. W. Becker et al, J. Vac. Sci. Technol., 15(2), March April 1978.

LIQUID CRYOGEN PUMPS INTEGRATED WITH SUPERCONDUCTING MOTORS

D. Dew-Hughes,[1] M. D. McCulloch,[1] K. Jim,[1] J. Aldwinckle,[1]
G. J. Barnes,[1] G. Jones,[1] J. R. Gaines Jr.[2] and S. Sengupta [2]

[1]Oxford University, Department of Engineering Science
Parks Rd, Oxford OX1 3PJ, UK
[2]Superconductive Components Inc.
1145 Chesapeake Ave., Columbus, OH 43212

ABSTRACT

Bulk superconducting material has an advantage over superconducting wires in that a significantly higher engineering current density is achievable. There are many different topologies in which bulk materials can be exploited. The most commercially viable application areas will be that where there is an existing cryogenic environment, such as the gas separation industry. To this end two designs of simple liquid nitrogen pumps driven by superconducting motors have been demonstrated. Existing liquid nitrogen pumps suffer from the problems of having the motor separate from the pump. Usually the motor is kept 'warm' and is connected to the pump via a long shaft. Often problems arise with the bearings of this assembly. These prototype pumps address the problem in a novel manner, by integrating the pump mechanism and the rotor of the motor, thus alleviating the need of a long shaft and complex bearings. The absence of external mechanical connections to the pump is thought to be of particular advantage for handling liquid hydrogen, where safety considerations are paramount.

INTRODUCTION

Electrical machines which use high temperature superconductors (HTS) suffer from the major drawback in that they are required to be kept cold. It is therefore envisaged that one near term application of such motors will be in systems which are inherently cold. A particular application is for pumping cryogenic liquids. This paper sets out the development of an integrated HTS pump-motor.

The practical application of high temperature superconductors in electrical machines has been impeded by the low critical current densities and sensitivity to magnetic field which is typical of these materials, allied to the difficulty of producing them in the form of flexible and robust conductor. Critical current densities approaching those typical of low temperature superconductors, i.e., in excess of 10^{10} Am^{-2}, are achieved in HTS only in epitaxial thin films. Single crystals and bulk materials in general have critical current densities that are several orders of magnitude below this value. The one remarkable exception is the seeded, melt-processed, bulk samples of the rare-earth bismuth copper oxide (REBCO) 123 phase. These samples consist of a single domain; strong flux pinning being provided by the presence of second phases, in particular the 211 phase, which may be present as a very fine dispersion occupying about 20% of the volume. Critical current densities of 10^8 Am^{-2}, at 77K and in externally applied fields up to 1 Tesla, are typical of the best samples.[1] These materials are able to trap inductions of several Tesla at 77K and may be thought of as rivals to permanent magnets. Their growth, structure, properties and applications have formed the subject of a recent conference.[2] Proposed applications for these remarkable materials include levitating systems, electromagnetic buffers for space docking, magnetic bearings for *inter alia* energy storage flywheels,[3] elements of both shielded core and resistive fault-current limiters (FCLs), and as the rotors in brushless ac motors[4] and generators.[5] It is with their application in motors, and in particular motors to drive liquid cryogen pumps, that this paper is concerned.

All electric motors work on the same basic principle: electric coils produce a magnetic field, which moves with respect to these coils, and the rotor adjusts its position to minimize the energy stored in the magnetic field. These coils may be found on either the stator or the rotor. The other element may be passive, i.e., contain no other source of magnetic field, or active, i.e., contain a source of magnetic field. For the passive type, the energy difference is caused by a salient magnetic structure. This structure may be singly salient, as in the synchronous reluctance machine, or doubly salient, as in the switched reluctance machine. For active machines, the field may be fixed with respect to the magnetic source, as in synchronous machines, or the field may move with respect to the magnetic source, as in the induction machine and the hysteresis machine. Sometimes real machines may exploit more than one principle, e.g., a synchronous machine may have a salient magnetic structure, as well as damper bars which act as an induction motor when starting.

The materials which are commonly used to make machines can be categorized either as good conductor, e.g., copper, as a hard magnetic material, e.g., NdFe, or as a soft magnetic material, e.g., magnetic steel. The former two are used as sources of magnetomotive force, while the latter is used to define the magnetic path and any saliences, if required. Like all materials, they have their limitations. For conductors, the current density is usually limited to at most 10 A/mm^2. For hard magnetic materials, the highest flux density that can be generated is about 1.2 T. Soft magnetic materials can usually carry magnetic fields up to a magnetic field density of about 1.5 T. HTS materials, which have a current density several orders of magnitude greater than copper and which can trap fields much larger than conventional permanent magnets, are an attractive choice as a material for motors.

SUPERCONDUCTING MOTORS

HTS wires can be wound as the main coils but AC losses, and the requirement to operate at low temperatures, 20 K, have delayed their use. Strong flux-pinning bulk HTS materials can be employed in most of the above types.

The difference in permeability of the *direct* and *quadrature* paths in synchronous reluctance machines can be enhanced by the use of ferromagnetic/superconducting composite rotors. The group at the Moscow Aviation Institute in collaboration with

Oswald Motoren GmbH have also explored this type of rotor.[6] Modeling indicates that the maximum output power can be increased by some thirty percent over that of a simple reluctance machine of the same size. The maximum power achieved with the superconducting reluctance motor to date is 9 kW.

Superconducting versions of induction motors have been constructed at Oxford.[7] These have either a squirrel cage rotor fabricated from silver dip-coated with Bi-2212 superconductor or a centrifugal melt-spun cylinder of Bi-2212. The torque versus speed characteristics measured at 77 K show the torque falling from a maximum value to zero well below synchronous speed. This is presumed to be due to flux flow in the superconductor giving rise to resistive behaviour. Any serious development of this type of machine will call for superconducting material with a much higher critical current than that exhibited by Bi-2212 at 77 K.

A type II superconductor, like a ferromagnetic material, exhibits magnetic hysteresis in a changing magnetic field. A solid superconducting rotor can act in exactly the same way as a ferromagnetic rotor. Relative motion of driving field and rotor will cause the superconductor to traverse its magnetization curve. The interaction between the magnetization of the rotor and the driving field will produce a torque. In the case of the ferromagnetic rotor, the field drags the rotor after it. Because the superconductor tries to exclude magnetic flux, and has a negative susceptibility, the superconducting rotor is pushed ahead by the stator field. A collaboration between the Moscow Aviation Institute and the IPHT-Jena has pioneered the development of superconducting hysteresis motors. The rotors are made up from blocks of melt-processed YBCO, and machines with output power up to 1kW have been tested.[8] The performance characteristics of these machines confirm their mode of operation as that of hysteresis machines. Their performance has been successfully modelled analytically on the assumption of hysteretic behaviour, treating the superconducting grains as a medium with an effective permeability.

If a superconducting version of either an induction or an hysteresis machine is made with a rotor of high current density and strong flux pinning, the induced currents or magnetic flux may become fixed in the rotor. On starting, the machine operates as either an hysteresis or an induction motor, but as the rotor speeds up and approaches synchronous speed the mode of operation changes as the flux becomes pinned and remains stationary with respect to the rotor. The machine then operates in a similar way as a permanent magnet motor, the rotor rotating in synchronism with the driving field. The flux, trapped in the rotor, locks on to the driving field.

This type of behaviour is expected of materials with supercurrent paths whose dimensions are of the same order as the dimensions of the rotor. Experiments at Oxford have confirmed that rotors made up from seeded, melt-processed single domain YBCO can act as both hysteresis and permanent magnet machines. The output torque is constant with rotational velocity up to the synchronous speed.[9,10] Finite element methods, based on the assumption of the field independent J_C Bean model for the superconductor, have been used to model flux and current distribution within the YBCO rotors, and to predict machine performance.[11] The model predicts that torque should vary as the cube of the driving field amplitude until the magnitude of the latter is such that flux has penetrated to the centre of the rotor. Thereafter the torque should become directly proportional to the field amplitude. This is in agreement with the results of the analytical modelling of the hysteresis motors mentioned above. The maximum field amplitude available in these experiments was 0.1 Tesla, a level at which the simple Bean model is expected to yield accurate results, and also insufficient to cause full flux penetration. The experimental results are in close agreement to the predictions. The model also predicts that a possibility for improvement of performance would be to preflux the rotor, by pulsing a DC current through the field coils.

This would strengthen the permanent magnet characteristic of the machine and could increase the stalling torque by fifty percent.

It was noticed during the course of the experiments on the melt-processed YBCO cylinders that the driving field not only caused the cylinders to rotate but could also provide a levitation force. The use of magnetic/superconducting bearings is a natural extension of the application of superconductors in rotating machines. By the appropriate shaping of the stator field, levitation and reaction to thrust can be incorporated in the machine operation.

BiSCCO LIQUID CRYOGEN PUMP

The particular advantage of machines with superconducting rotors would appear to be that they have a much greater power density in the rotor. Such machines are therefore likely to find applications in which the power to size or mass ratio is of prime importance. These applications are to be found in aircraft and space vehicles. These machines may also find application in already existing cryogenic environments, such as the gas separation industry. To this end, two configurations of simple liquid nitrogen pumps driven by superconducting rotors have been demonstrated at Oxford. Existing liquid nitrogen pumps suffer from the problems of having the motor separate from the pump. Usually the motor is at room temperature and is connected to the pump via a long shaft down which there is a considerable heat leak. Often problems arise with the bearings of this assembly. A conceptual design, illustrated in Figure 1, addresses the problem in a novel manner by integrating the pump mechanism into the rotor of the motor, thus alleviating the need of a long shaft and complex bearings. Circulation of the liquid cryogen is achieved by an Archimedean screw placed inside a hollow cylinder of superconductor that acts as the rotor. Recent modelling results indicate that a hollow cylinder should produce greater torque than a solid cylinder of the same external diameter.[12] The rotor is housed inside an insulated pipe; the stator coils are wound on the outside of this pipe. The stator coils are backed by soft iron to contain any stray field. Thrust reaction can be provided by additional coils, or by shaping the soft iron, so as to concentrate the field at either end of the stator. The absence of external mechanical connections to the pump is thought to be of particular advantage for handling liquid hydrogen, where safety considerations are paramount.

To demonstrate the principle of this type of pump an Archimedean screw machined from Delrin was fitted into the central bore of a centrifugal melt-spun Bi-2212 cylinder.[13] The cylinder had an internal diameter of 15 mm, external diameter of 25 mm and was 15 mm in length. The rotor just fitted inside the stator coils, which were free-standing three-phase copper coils supplied by Portescap. These coils were in turn placed inside an iron tube. It was felt that levitation of the rotor would prove an unnecessary complication at this

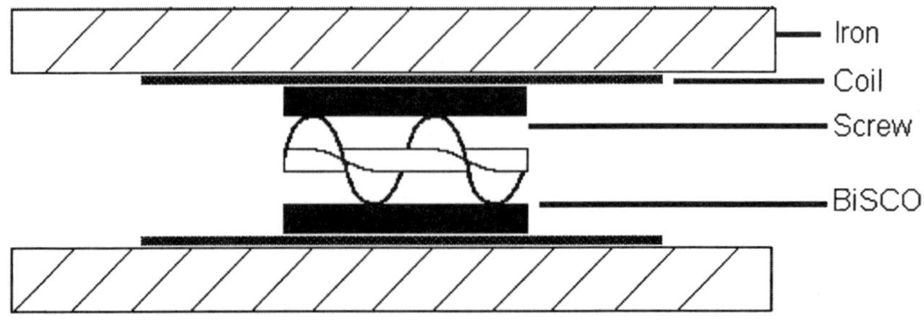

Figure 1. Schematic diagram of integrated superconducting motor and Archimedian pump.

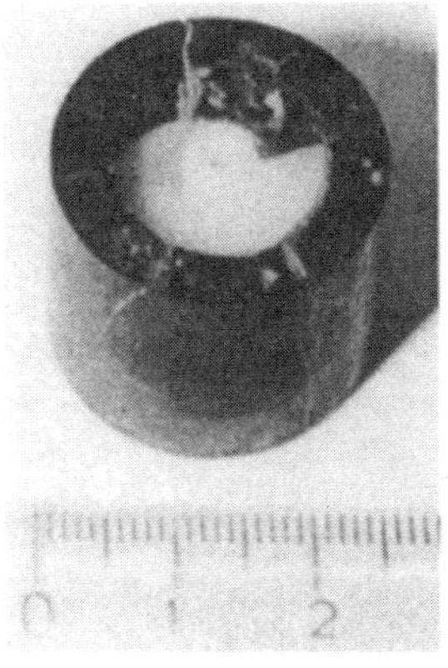

Figure 2. BiSCCO rotor with Archimedian screw.

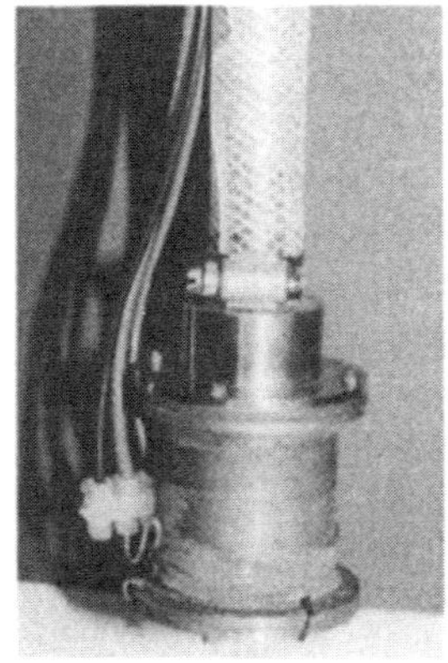

Figure 3. Pump assembly with delivery tube.

stage, and therefore the rotor was supported by simple needle bearings. The entire assembly is shown in Figures 2 and 3. A reinforced plastic delivery tube was attached to one end of the pump and operation is achieved by immersing the pump in a liquid nitrogen flask.

The best performance of this crude prototype is a rate of delivery of liquid nitrogen of 1 l/min. with a 10 cm head for a power input of about 45 W.[13] The problem of low efficiency results from the low rotational speed of the rotor which is limited by the inferior superconducting properties of the Bi-2212 at 77 K. Operation at 20 K, for example in pumping liquid hydrogen, should give much more efficient performance. An improved design, that would include the use of YBCO in the rotor and levitation of the rotor to obviate the need for needle bearings, is expected to lead to an increase in performance by several orders of magnitude. Despite its lack of performance, this project has demonstrated the feasibility of an integral superconducting liquid cryogen pump.

FABRICATION OF YBCO ROTOR

A subsequent project has been based on a YBCO rotor supplied by Superconductive Components, Inc.(SCI). The rotor was fabricated from seeded melt-processed single domainYBCO cylinders grown by the process developed by SCI for the levitator™ material.[14] The process involves synthesis of phase pure powder, single crystal growth, melt processing, and oxygenation. All of these steps are critical for successful fabrication of levitators™ with desired properties. Phase pure $YBa_2Cu_3O_x$ and Y_2BaCuO_5 powders were prepared by using a low-pressure calcination method. Typically, in a low pressure calcination process, the required oxides and carbonates are mixed and ball milled in a proper molar ratio. The resulting powders are then calcined in a low-pressure furnace to produce the desired phase. Calcination is carried out in 2-5 torr of oxygen. A partial vacuum is used to increase the efficiency of removal of CO_2. Once the reaction is completed, the vacuum is discontinued and the powders are cooled in an oxygen atmosphere at ambient pressure to fully oxygenate the superconducting powders. The low pressure calcination method is beneficial in reducing the carbon content and processing time while maintaining small particle size.

Phase pure powders of $YBa_2Cu_3O_x$, Y_2BaCuO_5 (~20-25 wt %) and PtO_2 (0-0.5 wt %) were mixed homogeneously by ball milling overnight (~ 8-10 hr.). The mixed powder was then pressed in to a 1.5 inch diameter pellet and seeded with a $Nd_{1+x}Ba_{2-x}Cu_3O_y$ single

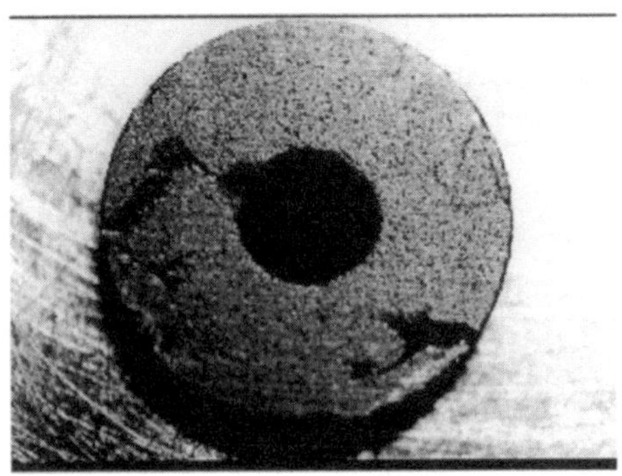

Figure 4. Photograph of a sliced YBCO singe domain.

crystal and melt-processed in a temperature gradient.[15] The pellets are heated to~1050 °C, slowly cooled through the peritectic temperature to ~950 °C and then finally cooled to room temperature at a rate of 100 °C/hr. The pellets are then oxygenated between 450 - 700 °C for 8-10 days in a flowing oxygen atmosphere.

The pellets were cut into 0.25"slices using a low speed diamond saw (Fig 4). Seven slices were then bonded, by using epoxy suitable for cryogenic application, to form a cylinder. The cylinder was then encased in a G10 cylinder to support the brittle ceramic. A central bore was drilled by water jet cutting. The G10 support was subsequently machined away and the cylinder then used as a rotor for the superconducting motor/pump.

YBCO LIQUID CRYOGEN PUMP

The central bore of this rotor, 11 mm dia., was insufficient to allow of the insertion of an Archimedian screw. It was therefore decided to use this rotor in a separate motor driving a centrifugal impeller.[16] The dimensions of the rotor, 25 mm o.d. and 45 mm long, were found to be almost identical to those of the ferrite rotor in a Heidolph 203.50 motor. The coils from such a motor, wound on a circular laminated iron core enclosed in a steel casing, are run from a 50 Hz, 230 V variable power supply. The YBCO rotor was mounted on a Delrin shaft with PTFE axial bearings and a needle thrust bearing and married to the Heidolph coils. The components and assembled motor and impeller are shown in Figures 5 and 6. The output torque at 77 K was found to vary as the cube of the current in, and hence the field generated by, the coils. This is exactly as predicted by theory up to fields below those sufficient to cause full penetration of the flux into the rotor.[11] The maximum efficiency, defined in terms of output power divided by input electrical power, was measured as 64.5%; this compares favourably with the rated efficiency of the Heidolph motor at 61.5%.

A plastic centrifugal impeller, sourced from a commercial garden feature water pump (Laguna 7000 Powerjet) was attached to the shaft of the superconducting motor, see Figure 6. The impeller had a plastic cover, not shown in Figure 6. By immersing the motor and impeller assembly in liquid nitrogen, pumping action was demonstrated. Best performance figures were a delivery rate of 12 l/min at a head of 0.6 m, reducing to 1.2 l/min at a head of 2 m. These values were obtained with 104 volts across the coil; the maximum voltage was limited by an acceptable rate of nitrogen boil-off due to heating of the coils. Clearly superconducting coils would provide an extra advantage, both in nitrogen economy and in attainable power levels and hence efficiency.

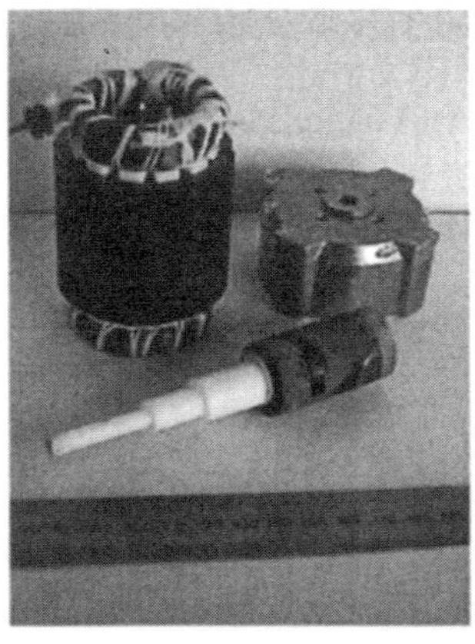

Figure 5. Components of YBCO motor

Figure 6. Assembled motor and impeller

It must be stressed that in neither of these demonstrators was any attempt made to optimise the fluid dynamics aspects of the pumps. The Archimedian screw was not fitted with guide vanes at either entry or exit. The centrifugal impeller was designed for water, and no alteration was made to take account of the different hydrodynamic properties of liquid nitrogen.

CONCLUSIONS

A near term application for HTS machines for pumping cryogenic liquids has been identified. This paper proposes that an integrated pump-motor design offers significant advantages over more conventional approaches. Two prototype designs have been built and tested to demonstrate proof of principle. In the first design, a Bi-2212 tube formed the basis of the rotor. An Archemedian screw was inserted into the middle of the tube to provide the pumping action. This protoype was capable of pumping up to 1 l/min. However this prototype suffered from two serious deficiences: the Bi-2212 operates in the flux-flow regime, and the Archemedian screw is not an efficient pump. The second prototype addressed these two issues by using a YBCO cylinder driving a centrifical impeller. This second pump was capable of delivering a flow of 12 l/min at a head of 0.6 m. These prototypes have demonstrated the proof of principle of an integrated HTS pump-motor.

ACKNOWLEDGEMENTS

The authors wish to thank Portescap and Heidolf AG for their generous contribution of coils and stators which enabled the completion of this work. This work was carried out as final year undergraduate engineering projects, in the Department of Engineering Science, Oxford University, by J Aldwinckle (1997-1998) and G Jones (1998-1999), working to a limited budget.

REFERENCES

1. M. Murakami, S. Goth, N. Koshizuka, S. Tanaka, T. Matsushita, S. Kambe and K.

Kitazawa, Critical currents and flux creep in melt processed high T_C oxide superconductors, *Cryogenics* **30**:390 (1990).

2. International Workshop on the Processing and Applications of Large Grain (RE)BCO High Temperature Superconductors, proceedings published in *Materials Science and Engineering* **53:**1 (1998).
3. T. S. Luhman, M. Strasik, A. C. Day, D. F. Garrigus, T. D. Martin, K. E. MacCrary, and H. G. Ahlstrom, H. G., Superconducting bearings and flywheel batteries for power quality applications, in: "Applied Superconductivity 1995," D. Dew-Hughes, ed., IOP Publishing, Bristol (1995), p. 35.
4. M. D. McCulloch, and D. Dew-Hughes, Brushless ac machines with high temperature superconducting rotors, Materials Science and Engineering **53**:211 (1998).
5. L. K. Kovalev et al, Alternators which use HTSC wire coils and bulk YBCO materials, in:"ICEC17," D. Dew-Hughes, R. G. Scurlock and J. H. P. Watson, eds., IOP Publishing, Bristol (1998), p. 379.
6. B. Oswald et al., Electric motors with HTS bulk material, in: "ICEC17," D. Dew-Hughes, R. G. Scurlock and J. H. P. Watson, eds., IOP Publishing, Bristol (1998), p. 547; L. K. Kovalev et al., HTS electric motors with compound HTS-ferromagnetic rotor, in: "ICEC17," D. Dew-Hughes, R. G. Scurlock and J. H. P. Watson, eds., IOP Publishing, Bristol (1998), p. 527.
7. M. D. McCulloch, D. Dew-Hughes, K. Jim and C. Morgan, The measurement and numerical calculation of the torque-speed curve of high temperature superconducting hysteresis motor, IEE Conference Publication No 444: Institution of Electrical Engineers, London (1987), p. 268.
8. L. K. Kovalev et al., Hysteresis electrical motors with bulk melt-textured YBCO, Materials Science and Engineering **B 53**: 216 (1998).
9. M. D. McCulloch, K. Jim, Y. Kawai and D. Dew-Hughes, Prospects for brushless ac machines with HTS rotors, in: "Applied Superconductivity 97," H. Rogalla and D. A. H. Blank, eds., IOP Publishing, Bristol (1997), p. 1519.
10. M. D. McCulloch and D. Dew-Hughes, Brushless ac machines with high temperature superconducting rotors, Materials Science and Engineering **B 53:**211 (1998).
11. G. J. Barnes, M. D. McCulloch and D. Dew-Hughes, The computer modelling of type II superconductors in applications, Superconductor Science and Technology **12**: 518 (1999).
12. G. J. Barnes, M. D. McCulloch and D. Dew-Hughes, Finite difference modelling of HTS hysteresis machines, presented at the 9th International Workshop on Critical Currents, Madison, Wisconsin, July 1999, and submitted for publication in Superconductor Science and Technology.
13. J. Aldwinckle, A superconducting liquid nitrogen pump, Final Year Project Report, Department of Engineering Science, Oxford University (1998); D. Dew-Hughes, M. D. McCulloch, J Aldwinckle and K. Jim, Specification for a pump, UK Patent Application No. 9810361.7, 14th May 1998.
14. S. Sengupta, J. Corpus, M. Agarwal, and J. R. Gaines, Jr., Feasibility of manufacturing large domain YBCO levitators by using melt processing techniques, in: "Impact of Recent Advances in Processing of Ceramic Superconductors," W. Wong-Ng, U. Balachandran and A.S. Bhalla, eds., The American Ceramic Society (1998), p.13.
15. V. R. Todt, S. Sengupta, D. Shi, J. R. Hull, P. R. Sahm, P. J. McGinn, R. B. Poeppel, Processing of large $YBa_2Cu_3O_x$ domains for levitation applications by $NdBa_2Cu_3O_x$ seeded melt-growth technique, J. of Electronic Materials, **23**:1127 (1994).
16. G. Jones, A superconducting pump for liquid nitrogen, Final Year Project Report, Department of Engineering Science, Oxford University (1999).

DESIGN AND TESTING OF EXPERIMENTAL FREE-PISTON CRYOGENIC EXPANDER

R. E. Jones and J. L. Smith Jr.

Massachusetts Institute of Technology
Cambridge, MA, 02139

ABSTRACT

An experimental free piston cryogenic expander was designed and constructed as a part of a feasibility test. Unlike most cryogenic expanders, there were no physical connections for controlling or monitoring the position of the piston. Rather, a linear variable differential transformer (LVDT) was designed and constructed on the outside of the cylinder wall to non-intrusively monitor the piston's position. Signals from the LVDT, pressure transducers and thermocouples were collected by a data acquisition system and integrated into a control routine. The control routine digitally switched valves on the warm and cold ends of the expander to control the pressure-volume relation in the cold end of the expander. Results from experimental testing were compared with theoretical thermodynamic and dynamic analysis of the single-stage apparatus to determine the expander's performance. Comparisons showed that the expander's behavior closely resembled the theoretical models, and identified design considerations and improvements for a second-generation expander. Although the expander was not insulated for thermal performance, significant cooling was evident under testing conditions.

INTRODUCTION

There are three processes commonly used to obtain cryogenic temperatures,[1,2,11] evaporation, isenthalpic expansion and isentropic expansion. Depending on the application and requirements of the system, a cryogenic system may incorporate any combination of these processes. The research in this paper dealt with the design and construction of a free-piston expander that isentropically cooled the working fluid. This work is a part of a larger effort to achieve the high efficiency of a Collins cycle helium liquefier without the complexity required for a mechanical cross head and cam operated valve. The objective is the simplicity of a pulse tube and the efficiency and pressure ratio of a Collins expander.[3]

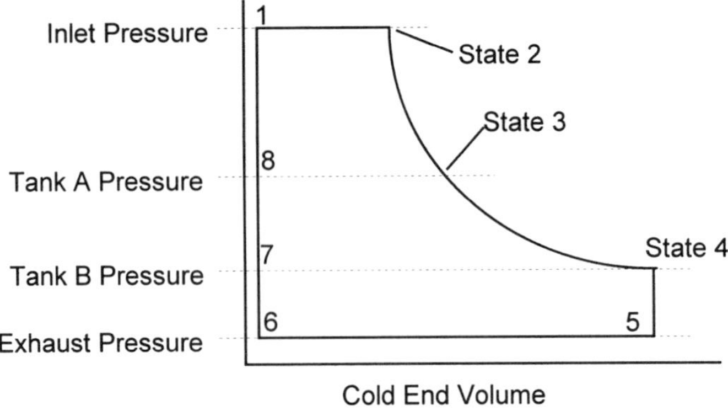

Figure 1. Ideal pressure vs. volume plot for the cold region of the expander.

The project was divided into three major areas: modeling and optimization, instrumentation and control, and experimental testing. To understand the process of the expander, consider Figure 1 and Figure 2. Figure 1 illustrated the ideal pressure versus volume relationship for the cold region of the expander, which was shown in the system configuration of Figure 2. At state 8 in Figure 1, the piston was at rest on the bottom of the cylinder, and the pressure in the cylinder was equal to that of tank A. At state 8 the inlet valve was opened and the pressure in the cold end increased to equal that of the inlet line before there was any significant displacement of the piston. During the process from state 1 through state 2, the charging process, the inlet valve remained opened to the cold end and the warm region was connected to tank A through the throttle valve. The spool valve only served to toggle the connection to the warm end between tank A and tank B. At state 2, which was determined by an adjustable valve for the cold-end volume at state 2, the inlet valve was closed and the compressed air in the warm region of the expander was throttled into tank A until the warm

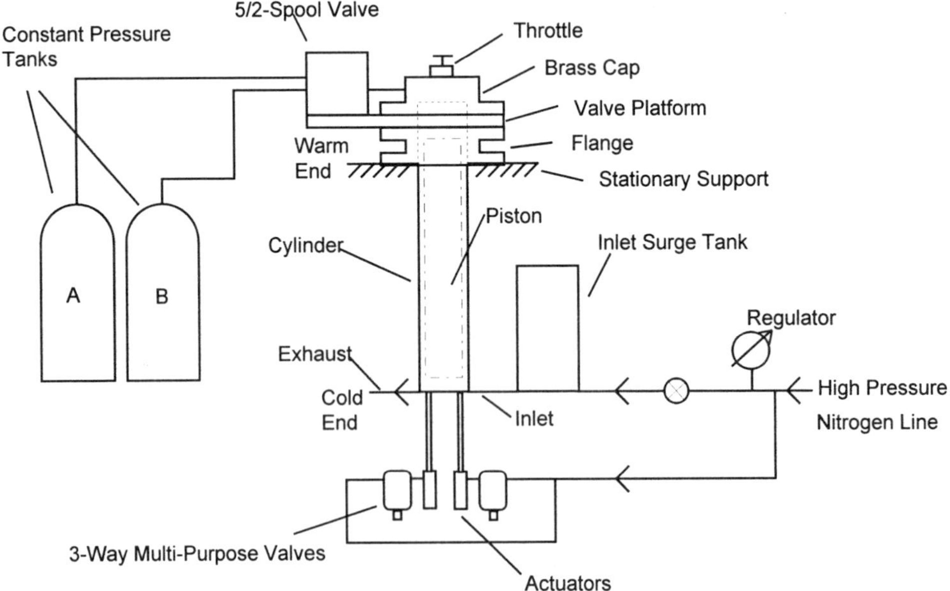

Figure 2. Sketch of system configuration used for testing.

end pressure equaled that of tank A, state 3. At this point, the spool valve was toggled, connecting the warm region to tank B, and the same throttling process occurred until the system reached state 4. It was during the process from state 2 to state 4 that the working fluid in the cold region was isentropically expanded and cooled. At state 4, the exhaust valve was opened and the pressure in the cold region dropped rapidly to that of the exhaust line, state 5. As the piston dropped in the cylinder, state 5 to state 6, air was throttled back into the warm end of the expander from tank B. When the piston reached the bottom of the cylinder, state 6, the exhaust valve was closed and the pressure in the warm end was charged to equal that of tank B, state 7. The spool valve was then toggled back to tank A and the system was charged to state 8.

MODELING AND OPTIMIZATION[1,2,5,9]

To model the process described above, three separate thermodynamic models needed to be developed. In Figure 1, it was assumed that the void space between the piston and the throttle at state 4 was equal to zero. Consequently, there was no volume change from state 4 to state 5. However, if the value for the cold end volume at state 2, V^*_2, was not set correctly, the piston would either never have reached the top of the cylinder or it may have reached the top of the stroke before the gas in the cold region had expanded to equal the pressure of tank B. Depending on the relationship between the two pressure tanks and the user-defined value for the V^*_2, the system behaved differently between state 3 and state 6. The three models developed were for a low stroke, matched stroke and high stroke scenario.

These three models only predicted the properties of the system at each of the states, and did not represent the behavior of the system between states. Also, these models were based on the assumption that the intermediate pressure tanks remained at a constant pressure. This assumption was verified by obtaining an iterative solution to the above models, and showed that the tanks would return to their respective pressures over several cycles.

With the models in place, the next stage was to optimize the system to obtain the maximum temperature decrease of the gas processed from inlet to exhaust during one cycle of the expander. After non-dimensionalizing the variables, the only independent variables to adjust were the inlet line pressure, P^*_{in}, and V^*_2. Therefore, for a given value of P^*_{in}, the system was optimized as only a function of V^*_2. The results of the optimization showed that the maximum temperature drop of the gas over one cycle was obtained when the system matched at state 4, as illustrated in Figure 1.

In order to build a dependable control system[4,8] for this expander, the more rapid processes of the cycle were dynamically modeled. Of particular interest was the blow-in process from state 8 to state 1. During this time there was a sudden pressure differential across the piston, which could induce a violent response that could damage either the piston or the cylinder. There was also a concern that the oscillation of the piston caused by the gas spring in the warm end would result in the control routine making an incorrect switch and damaging the system as well. This dynamic analysis of the system also provided insight into the viscous leak around the cylinder, and restricted our operating speeds to 0.6 Hz or faster to minimize its affect on the cycle.

DESIGN OF EXPERIMENTAL APPARATUS

A critical aspect in the design was the warm end assembly, Figure 3. This assembly was responsible for both the work dissipation and dynamic control of the expander. The flange, which bolted to the support shown in Figure 2, held a split ring in place around the cylinder. The other components, including the valve platform and the brass cap, were then bolted to the flange. The brass cap had to be machined in such a manner as to minimize the void space in the cylinder at top dead center and provide a static seal for the warm end of the expander. Figure 3, shows the relative position of the two throttles to the top of the cylinder. The throttles were soft soldered to the brass cap and a support plate was screwed down on top of the valves to provide extra stability. As illustrated, the seats on the throttles rested just above the top dead center position of the piston. This minimized the void space above the piston at state 4 of the cycle.

The two throttles were connected to a 5/2-spool valve,[10] which allowed for two configurations of testing. Either one throttle could be used to restrict the flow into both tank A and tank B, or both throttles could be used to restrict the flow into the tanks separately. Through testing, it was shown that although this affected the speed of various portions of the cycle, it did not affect the thermodynamic behavior of the system or the steady-state pressures of the tanks.

The most critical part of the system was the instrumentation and control routine. Pressure and temperature measurements were collected in both the warm and cold regions

Table 1. List of parts for Figure 3

Part #	Name	Notes
1	Brass Cap	Brass
2	Cylinder	Stainless Steel
3	Flange	Stainless Steel
4	O-Ring	Rubber
5	Split-Ring	Stainless Steel
6	Throttle	Brass
7	Valve Platform	Carbon Steel

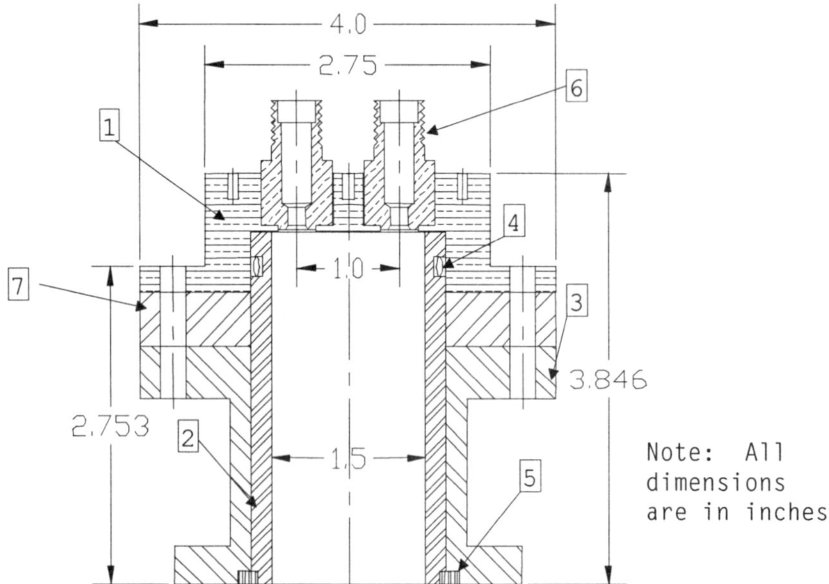

Figure 3. Warm end assembly cross-sectional view

of the expander to provide real-time feedback to the control routine. In addition to the thermodynamic properties, the exact location and velocity of the piston was required for the control routine to perform properly. To monitor the piston inside the cylinder, a LVDT was designed and integrated into the warm end of the system. Pick-up coils were wound around the top portion of the cylinder, and a 0.051mm ferro-magnetic shim was epoxied into a 0.152mm recess, which was machined around the circumference of the warm end of the piston. The coils were incorporated into a bridge circuit[6], which received a 6 KHz excitation voltage. The LVDT was designed in such a manner that the response of the system provided a linear output that was proportional to the position of the piston.

RESULTS

The plots illustrated in Figure 4 and Figure 5 show the cycle as predicted by the thermodynamic model and the actual data as collected by the control routine, respectively. There are several differences worth noting between these two plots and the one shown in Figure 1. First, there is clearly a difference in the blow-in process of Figure 4. Recall that the thermodynamic model only predicted the system's properties at each of the states. Consequently, the model predicted the volume for state 1 that corresponded with the cold region volume expanding and moving the piston up until the pressure in the warm region volume reaches the inlet pressure. The blow-in process of Figure 5 corresponds better with the ideal plot of Figure 1. However, the blow-in and blow-down processes of Figure 5 do not represent the actual behavior of the system accurately. Notice the piece-wise nature of these plots during the two processes. Due to the speed of these processes with respect to the rest of the cycle, the data acquisition was not able to obtain enough data point to produce a smooth plot. Also note the evidence of piston oscillation in Figure 5 between state 1 and state 2. Although evident, it was not significant enough to have an adverse effect on the control routine.

Figure 5 represented a cycle that was considered a matched scenario even though the plot does not match exactly at state 4. In reality, the pressure in the cylinder could never become equal to that of tank B before a switch was signaled. If the pressures matched exactly, the velocity of the piston would have become too slow and the viscous leak around the piston would have had a significant effect on the system.

To quantify the performance of the expander, the work represented by one complete cycle was computed and normalized to that of the ideal plot in Figure 1. For the experimental test shown in Figure 5, the work computed was 94% of the ideal work for the same operating parameters.

CONCLUSION

The goal of this project was to show that a free-piston expander could be constructed and controlled reliably. Results from the experimental tests were compared with the theoretical thermodynamic and dynamic analysis to determine the expander's performance. Work values computed from the P^*_{cold}-V^*_{cold} diagrams showed that the expander's behavior closely resembled the ideal models and identified design considerations for a second-generation expander. In particular, the experiments showed that the data acquisition and control system needed to be improved to obtain adequate data resolution and better control throughout the cycle. Despite the limitations, the single stage expander performed well and provided significant cooling under testing conditions.

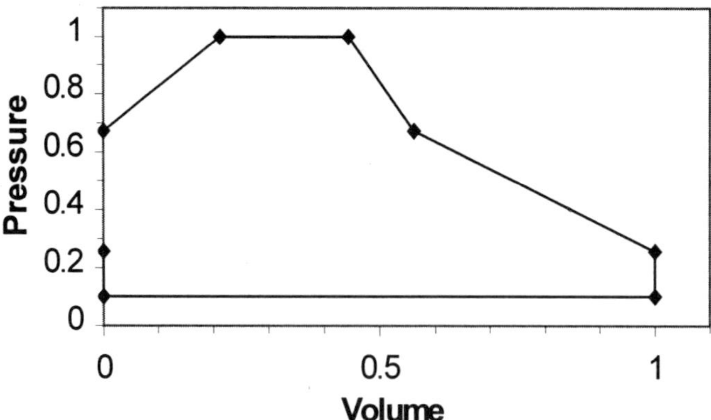

Figure 4. Pressure versus volume plot for the cold region of the expander. This plot was generated by the thermodynamic model for the matched stroke scenario.

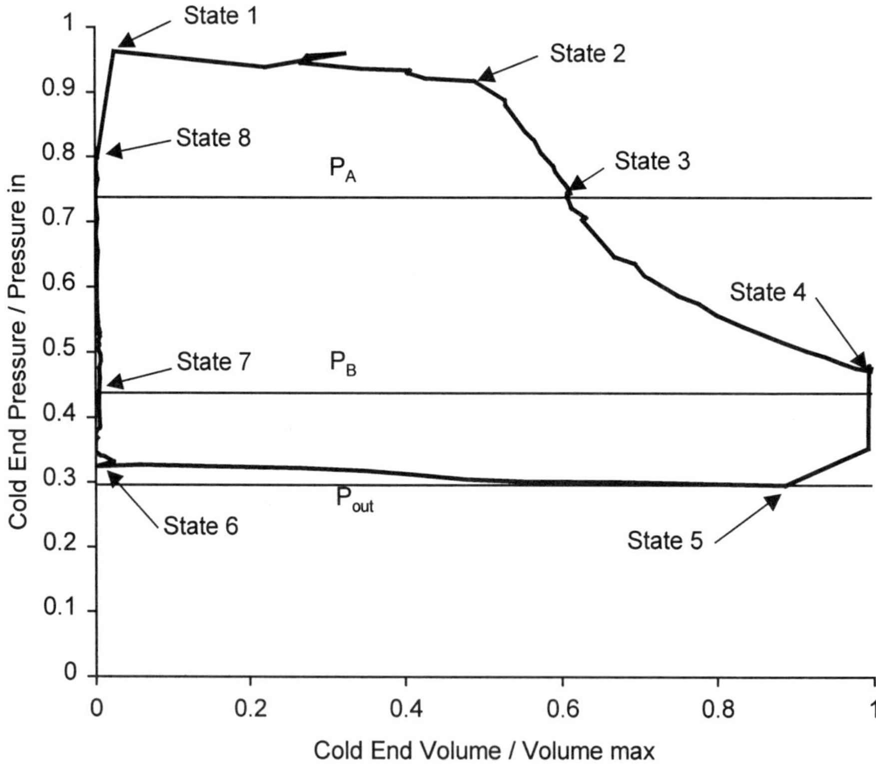

Figure 5. Pressure versus volume plot for the cold region of the expander. This plot was generated by the data points collected from an actual cycle for the matched stroke scenario.

REFERENCES

1. Bejan, Adrian, "Advanced Engineering Thermodynamics," John Wiley & Sons, New York (1988)
2. Boles, Micheal A., and □Cengel, Yunus A., "Thermodynamics, an Engineering Approach," McGraw-Hill, New York (1994).
3. Collins, S.C., and Cannaday, R.L., "Expansion Machines for Low Temperature Processes," Oxford University Press, New York (1958).
4. Hildebrand, Francis B., "Advanced Calculus for Applications," Prentice-Hall, Inc., New Jersey (1976).
5. Incropera, Frank P., and DeWitt, David P., "Fundamentals of Heat and Mass Transfer," John Wiley & Sons, New York (1990).
6. Johnson, David E., Johnson, Johnny R., Hilburn, John L., and Scott, Peter D., "Electric Circuit Analysis," Prentice-Hall, Inc., New Jersey (1997).
7. Ludwigsen, Jill L., "Configuration and Testing of a Saturated Vapor Helium Compressor," M.S. Thesis, Dept of Mechanical Eng., MIT (1985).
8. Marion, Jerry B., "Classical Dynamics of Particles and Systems," Academic Press, Inc., New York (1970).
9. Mills, A.F., "Heat and Mass Transfer," Irwin, Chicago (1995).
10. Minuteman Controls Catalog # 197 p. 540, Massachusetts (1997),.
11. Scott, R.B., "Cryogenic Engineering," Met-Chem Research, Inc., Colorado (1988).
12. Urieli, I., and Berchowitz, D.M., "Stirling Cycle Engine Analysis," Adam Hilger Ltd., Bristol (1984).

1999 CEC/ICMC Awards Luncheon.

RECENT DEVELOPMENTS OF AIR LIQUIDE CRYOGENIC EXPANDERS

G. Marot, J.C. Villard

Air Liquide, Advanced Technology Division
38360 Sassenage, France

ABSTRACT

The cryogenic expansion turbines supplied by the Advanced Technology Division of Air Liquide are presented, with a focus on a recent development: the integration of 3D wheels in order to increase the mass flow rate and the refrigeration power.

The comparison of 2D and 3D wheels is first described, with respect to mass flow rate, brake tuning capabilities and efficiency.

Mechanical aspects, linking the low vibration frequencies of the blades of 3D wheels as compared to 2D wheels, are also investigated, and solutions presented and discussed.

INTRODUCTION

Air Liquide has been manufacturing for 30 years cryogenics expansion turbines,[1] and more than 350 turbo-expanders are operating worldwide, installed on both industrial plants and refrigeration units for research institutes.[10] These turbines are characterized by the use of hydrostatic gas bearings, providing a unique reliability with a measured Mean Time Between Failure of 45000 hours.

The Air Liquide cryogenic turbines are provided in five casing sizes (TC3 to TC7). Among other features, each casing determines the maximum size of the turbine wheel. For many years, only so-called 2D wheels have been used (figure 1); in this technology, the blades are limited to the outer part of the wheel, and are therefore extremely robust. 2D wheels provide both excellent isentropic efficiencies (up to 80 %), and a total insensitivity to vibrations.

Recently, in order to satisfy the trend towards higher flow rates, the 3D wheel technology has been developed and successfully implemented on several projects. 3D wheels (figure 2) are characterized by the extension of the blades to the outlet section of the wheel.

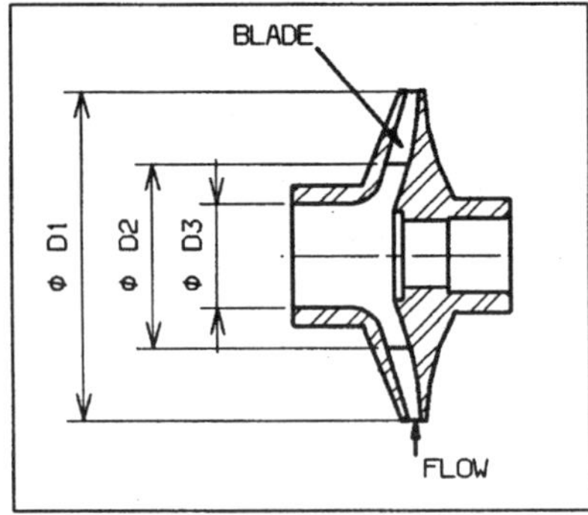

Figure 1. Cross section of a 2D wheel

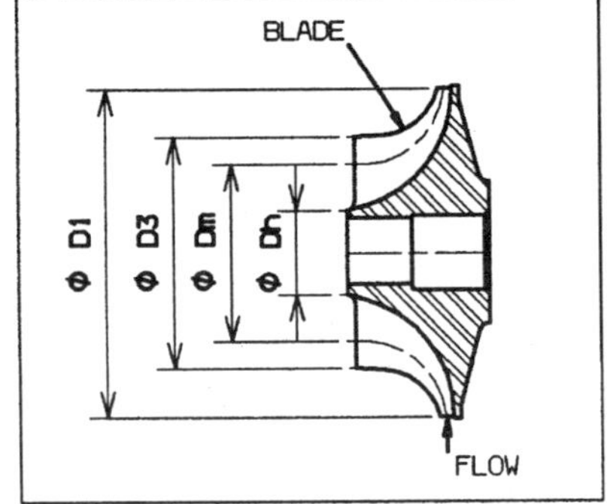

Figure 2. Cross section of a 3D wheel

This paper describes several aspects of this development: the first section is a comparison between 2D and 3D wheels. The second section is an analysis of the vibration aspects of 3D wheels, and adopted solutions.

COMPARISON OF 2D AND 3D WHEELS

In the present section, the motivation for using 3D wheels is first described in term of mass flow rate and rotation speed, followed by a discussion on efficiency.

Increase of flow rate

Analysis. The increase of mass flow rate is the most important reason for using a 3D wheel in place of a 2D wheel, for a given outer diameter, D_1. The order of magnitude of this increase can be deduced from the comparison of wheels with identical inlet and outlet thermodynamics conditions.

In a turbine, the mass flow rate, Q_M, is dictated by the outlet conditions of the wheel.[2] For given thermodynamics conditions, it is therefore proportional to the outlet cross-sectional area of the wheel, S_3. Ignoring for simplification the hub diameter, D_h, in the case of the 3D wheel, and referring to figures 1 and 2, the outlet cross-sectional area is itself proportional to the square of the external outlet diameter, D_3.

3D wheels are first considered: In order to avoid the boundary layer effect due to the change of direction of the fluid, the ratio of the external outlet, D_3, and of the wheel diameter, D_1, must be kept lower or equal to a limit value,[3] K_e. The flow rate of the 3D wheel, Q_M(3D), is therefore proportional to $K_e^2D_1^2$.

2D wheels are characterized by the outlet diameter of the blades: D_2. The change of direction of the fluid occurs between the diameter D_2 and the diameter D_3, therefore, as for the 3D wheel, Q_M(2D) is proportional to $K_e^2D_2^2$. Defining the blade ratio K_1 by the ratio D_2/D_1, the flow rate of the 2D wheel, Q_M(2D), is proportional to $K_1^2K_e^2D_1^2$, and:

$$\frac{Q_M(2D)}{Q_M(3D)} = K_1^2$$

K_1 must be low enough to produce a significant and efficient guidance of the wheel; in practice, it is kept lower than 0.7. K_1^2 will therefore be of the order of 0.5.

Conclusion. For a given wheel outlet diameter, and for similar inlet and outlet fluid conditions, the flow rate expanded by a 3D wheel is therefore at least twice larger than the flow rate expanded by a 2D wheel.

From a technical point of view, in a given casing, a 2D wheel can be replaced by a 3D wheel. For upgrade operations, it is therefore possible to increase significantly the flow rate without changing the casing. Furthermore, the maximum refrigeration power of each type expander increases by a factor of two.

Example. This last feature is illustrated by the following case, corresponding to an industrial unit presently under operation.

The working conditions are: inlet pressure: 32.9 10^5 Pa; inlet temperature: 250 K; outlet pressure: 2.6 10^5 Pa; Fluid: CO-N_2; flow rate: 1123 g/s.

By using the largest available casing, it would have been impossible to fit a 2D wheel working in good conditions, i.e. providing a good efficiency. A 3D wheel, with a diameter D_1 of 78 mm has been selected. The shaft nominal speed is 1340 RPS, with a guaranteed isentropic efficiency of 80 %. The refrigeration power of this expander is 130 kW.

Adaptation of the brake

Analysis. The mechanical power produced by the turbine is dissipated in the brake, which is a compressor on the same shaft as the turbine wheel, coupled with a heat exchanger. For a given compressor wheel, the dissipated power is proportional to P_bN^3, where P_b is the fluid pressure at the inlet of the compressor wheel, and N is the shaft rotation speed in RPS.[4] P_b is physically limited by the cycle compressor characteristics; consequently, it may be necessary to increase the shaft speed N to fit the brake capacity to the turbine capacity. The shaft rotation speed N depends on the tip speed of the turbine wheel U_1, and of the turbine wheel diameter D_1:

$$N = \frac{U_1}{\pi D_1}$$

The tip speed itself is closely dependent on the enthalpy drop of the expanded fluid, ΔH, and on the shape of the speed triangle (see figure 3, for example) at the inlet of the turbine wheel,[2] though the velocity ratio κ:

$$U_1 = \kappa\sqrt{2\Delta H}$$

κ has a maximum value of 0.707, corresponding to radial inlet wheels. Once κ is given the highest value, the tip speed U_1 is set at its maximum value, and the only way to increase the shaft rotation speed is to use a turbine wheel with a lower diameter; as explained in the previous section, the use of a 3D wheel may be of great interest for that purpose.

Example. This advantage is illustrated by the following case, corresponding to the T7 expander at CERN-LEP, redesigned for the upgrade operation[10] of January '99.

The working conditions are: inlet pressure: 7.1 10^5 Pa; inlet temperature: 5.77 K; outlet pressure: 2.43 10^5 Pa; Fluid: helium; flow rate: 935 g/s.

The first step of the upgrade design assumed a 2D wheel, outer diameter 55 mm; the nominal shaft speed is very low: 396 Hz. The required brake pressure is 17 10^5 Pa, too high with respect to the tuning requirements. The second step of the design assumed a 3D wheel, outer diameter 39.5 mm; the nominal shaft speed is now 455 Hz, with a brake pressure of 11.2 10^5 Pa, compatible with the refrigeration unit constraints

Comparison of the isentropic efficiencies

2D wheels operate with very high isentropic efficiencies (up to 80 %), and it is of great interest to compare their efficiency with 3D wheels.

The isentropic efficiency, η, is defined by the ratio of the achieved enthalpy drop of the expanded fluid, to the enthalpy drop, ΔH_s, which would be achieved with a perfect expander. The isentropic efficiency is therefore of direct interest to the designer of the refrigeration unit. η can be derived in the following simplified way, as a direct consequence of the first principle of thermodynamics:

$$\eta = \frac{W_s - W_l - W_f - W_c}{Q_M \Delta H_s}$$

In this equation, W_s, W_l, W_f, and W_c are respectively, the mechanical power on the shaft, the leak losses, the friction losses and the conduction losses. Theses four terms are now discussed, and an example given at the end of this section.

Shaft power, W_s. According to Euler law, and for a 2D wheel (figure 3):

$$W_s = (U_1 V_1 \cos(\alpha_1) - U_2 V_2 \cos(\alpha_2)) Q_M$$

Where subscript 1 refers to the wheel inlet, and subscript 2 to the blade outlet. U, V and α are respectively the wheel speed, the fluid speed and the angle between U and V. For a 3D wheel, the subscript 2 must be replaced by 3, as the blade outlet corresponds to the wheel outlet. Note that the hydraulic efficiency,[2] η_H, is defined by the ratio of W_s and $Q_M \Delta H_s$. Introducing the blade ratio K_1, defined by the ratio D_2/D_1 for a 2D wheel, and D_M/D_1 for a 3D wheel, the previous equation can be rewritten:

$$W_s = U_1 (V_1 \cos(\alpha_1) - K_1 V_2 \cos(\alpha_2)) Q_M$$

Comparing 2D and 3D wheels with identical diameters and working conditions, U_1 and $V_1 \cos(\alpha_1)$ will be identical for both wheels. The term $V_2 \cos(\alpha_2)$ is zero for a perfect wheel, and generally degrades the hydraulic efficiency for real wheels; as it is weighted by K_1, and as K_1(3D) is lower than K_1(2D) (compare figures 1 and 2), better hydraulic efficiency is expected for 3D wheels.

Conduction losses, W_c. The conduction losses depend mainly on the size of the casing. The wheel diameter has therefore very little effect on them, expect when the capability of 3D wheels to work with lower diameter leads to a reduction of the size of the casing.

Friction losses, W_f. The friction losses of the wheel, W_f, varies with D_1^n for given a tip speed,[5] with $n > 1$. The decrease of the wheel diameter has therefore a direct favorable effect on W_f.

Leak losses, W_l. W_l are the losses due to the leaking of fluid between the wheel and the shroud; by a good management of the functional parts, W_l for a 3D wheel will be intermediate between the losses of 2D shrouded wheel, and a 2D unshrouded wheel.

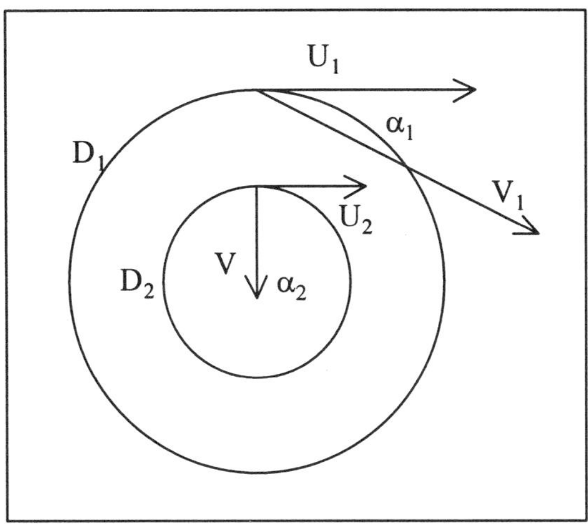

Figure 3. Speed triangles at the inlet (diameter D_1) and at the outlet (diameter D_2) of a 2D wheel. U is the wheel speed, and V is the fluid absolute speed.

Example. The following example is a comparison between the efficiency of a 3D wheel, and the efficiency of a 2D shrouded wheel.

The working conditions are: inlet pressure: 37.0 10^5 Pa; inlet temperature: 79 K; outlet pressure: 2.2 10^5 Pa; fluid: hydrogen; flow rate: 350 g/s.

A 2D wheel with a diameter of 55 mm leads to a guaranteed isentropic efficiency of 79.9 %; the hydraulic efficiency is 86.4 %. The casing corresponds to the TC5 size. A 3D wheel with the same diameter of 55 mm (i.e. same casing size) leads to a slightly higher isentropic efficiency (80.9 %), mainly due to the increase of the hydraulic efficiency (88 %), a consequence of a lower blade ratio, K_1. A 3D wheel with a lower diameter of 37.5 mm allows the use of a smaller casing size (TC4); the isentropic efficiency is the same as the 3D wheel of larger diameter, 80.8 %. This is the consequence of 3 compensating effects:

- the hydraulic efficiency is 86.3 %, i.e. lower than the 3D wheel of 55 mm, and at the level of the 2D wheel: for a given outlet wheel section and wheel geometry, the blade ratio, K_1, and therefore η_H, decreases with the wheel diameter.
- the conduction losses, W_c, are decreased by the use of a smaller casing.
- The friction losses, W_f, are decreased by the use of a smaller wheel diameter.

In this specific case, the use of a 3D wheel has a favorable but limited effect on the efficiency.

Conclusion

As explained in the present section, the use of a 3D wheel is to be considered when the flow rate is too large for a given casing case, or more generally, when a smaller wheel diameter is helpful. The consequence on the efficiency, although favorable, is limited, and is far from being the main motivation.

However, 3D wheels have a severe drawback compared to 2D wheels, as described in the next section: the frequency of the vibration modes of the blades are much smaller, and tend to interfere with both design and operation.[6] This is not the case for 2D wheel blades, which are much stiffer. 2D wheels will therefore be preferred if there is no firm advantage to the 3D wheel.

VIBRATION ANALYSIS OF 3D WHEEL BLADES

Before detailing the vibration analysis of the blades, it is of major importance to identify the excitation frequencies, in order to avoid them. In the present section, Z_w is the number of blades of the wheel, Z_n is the number of blades of the nozzles, and N_n is the nominal shaft rotation speed (figure 4).

Excitation frequencies

Rotation speed. Any single defect on the shaft, the pivoting system, the wheels or the nozzles will shake the wheel blade at the rotor frequency. As the shaft is general tested at a speed 20% higher than nominal (so-called overspeed), the excitation frequencies will range between the startup frequency (a few 10 Hz) and 1.2 N_n. The two no-dwell zones, corresponding to the two rigid modes of the shaft, are generally within the working range of the turbine. The quality of the balancing of the shaft is rigorous enough to allow some operation on these modes, and each turbine is tested accordingly before delivery. However, the time of operation on these modes is limited to a few minutes for safety. Due to this limited time, the two no-dwell zones are not considered as potentially dangerous excitation zones.

Nozzle blades. Due to the presence of nozzle blades, the fluid speed at the inlet of the wheel is not uniform: it is close to zero at the tip of each nozzle blade, and maximum between two nozzle blades. Consequently, each time an impeller blade passes in front of a nozzle blade, it suffers from a shock created by the fluid speed non-uniformity. At a given shaft speed N, the excitation frequency is therefore $N.Z_n$.

Impeller blades. Any single flow irregularity at the inlet of the wheel - other than the presence of nozzle blades - will shake the shaft Z_w times per turn. The excitation frequency is therefore N. Z_w. The amplitude of this excitation is probably much weaker than the previous one, as the flow discontinuity can reasonably be assumed to be smaller.

Nozzle and impeller blades. The flow irregularity at the exit of each nozzle blade will also shake the shaft each time an impeller blade passes in front of it; the excitation frequency is here N.($Z_w Z_n$).

Possible strategies

The possible strategies to prevent failure are as follows:

- To reject the vibration modes of the blade outside the working range of shaft speed , i.e.: lower than the startup speed, inside the no-dwell zones, or higher than the over-speed. This rejection is obtained by designing stiff blades, and minimizing the number of blades. This is by far the best solution, as it is fully transparent for the plant manager. An example is given in the next section.
- When the previous strategy is not possible, to add no-dwell zones; the position of these zones, in term of speed, must be closely discussed with the plant manager, and must integrate the different possible modes of operation, including the startup.

Example. In this test case, the wheel diameter D_1 is 78 mm; the outlet external diameter is 40 mm. The wheel has 9 long blades, and 9 short blades (figure 5).

The nominal speed is 1350 Hz, and the over-speed is 1620 Hz. In a first design run, the number of nozzle blades is set to 11, and the impeller blade thickness to 1.1 mm.

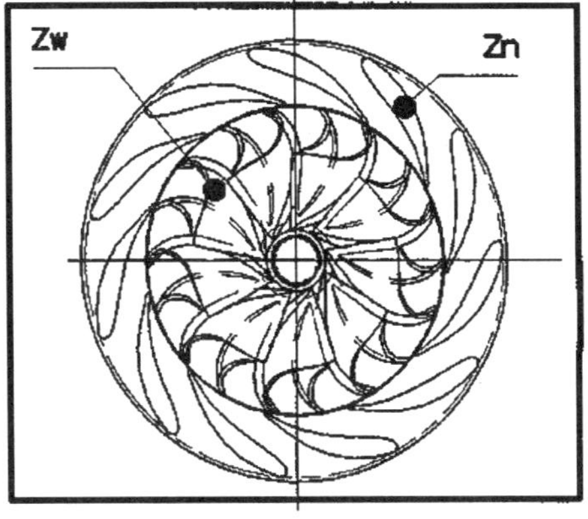

Figure 4. Front view of the nozzle blades and wheel blades assembly

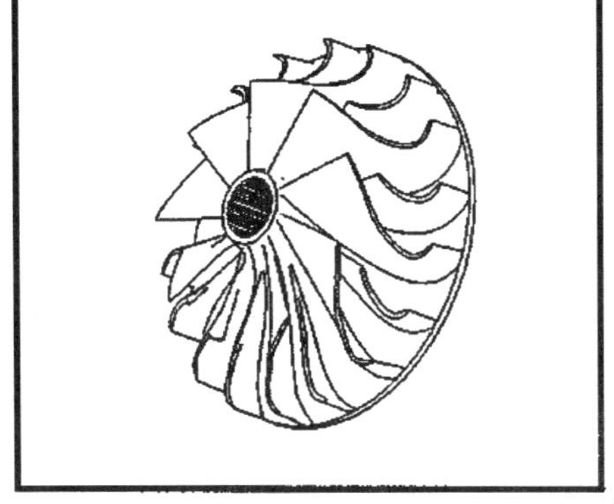

Figure 5. Isometric view of a 3D wheel

The frequency f of the first vibration modes, calculated by FEA,[8] is:

- for the short blades: 8760 Hz, 10720 Hz, 15187 Hz, 22456 Hz, etc ...
- for the long blades: 4770 Hz, 21050 Hz, 26 943 Hz, etc ....

These modes can be excited by the 11 nozzles at a speed of $N=f/Z_n$, i.e.:

- for the short blades: 796 Hz, 975 Hz, 1381 Hz, 2041 Hz
- for the long blades: 451 Hz, 1914 Hz, 2449 Hz

This configuration is not satisfactory, as it has at least five vibration modes which can be excited in the working range. In order to improve the situation, the number of nozzle blades is reduced to 5, and the thickness of the impeller blade increased to 1.6 mm. The frequency f of the first vibration modes, are now:

- for the short blades: 13480 Hz, 34410 Hz
- for the long blades: 6052 Hz, 15810 Hz, 23190Hz

These modes can be excited by the 5 nozzles at a speed of $N=f/Z_n$, i.e.:

- for the short blade: 2696 Hz, 6882 Hz
- for the long blade: 1210 Hz, 3162 Hz

The situation is much better than previously, but the excitation at 1210 Hz is still a problem, as it is close to the nominal speed. At this stage, the blade thickness can again be increased, but the mass of the wheel would increase as well, with negative effects on the shaft flexible modes. The chosen alternative is to use trapezoidal blades, with an average thickness of 1.6 mm, i.e. without any added mass. The number of nozzles is maintained at 5. A trapezoidal blade has two advantages over a rectangular blade, as far as vibrations are concerned : [9] the moving mass is smaller, and the stiffness in flexion in higher.

In our test case, the base of the blade is 2.4 mm thick, and the top free edge is 0.8 mm thick. The calculated frequency of the first vibration mode is 8480 Hz, rejecting the excitation frequency to outside the working zone.

This wheel is presently working in an industrial plant.

Conclusion

The use of 3D wheels implies severe constraints on the design of the blades, which constraints depend on the full range of working conditions of the turbine. Solutions can be found by changing the excitation frequencies either by changing the number of blades, or by stiffening the blades by using trapezoidal shapes. This latter solution has been successfully tested.

CONCLUSION

The use of 3D wheels in Air Liquide cryogenic expanders leads to new capabilities:
- for a given casing, the mass flow rate increases by at least a factor of two, and,
- for a given mass flow rate, the wheel diameter can be reduced, providing an extended range of design possibilities.

The efficiency of 2D and 3D wheels have been compared, and show, in the described case, a small advantage for the 3D wheels.

Mechanical aspects, linking the low vibration frequencies of the blades of 3D wheels as compared to 2D wheels, have also been investigated, and solutions have been presented and discussed.

The integration of 3D wheels in our product range is the first step in increasing the capacity of the Air Liquide expanders. Further steps are the increase in the size of the casings, the integration of variable nozzles, and the use of the brake as a booster for the cycle

ACKNOWLEDGMENTS

The authors are grateful to the Air Liquide staff members who actively participates in the turbomachinery activities at Air Liquid.

REFERENCES

1. J.C. Villard, Gas Bearing Expansion Turbines for Industry, Installation of the 300th Machine Worldwide, *CryoGas International*, (Aug./Sept. 1996).
2. S.L. Dixon. Fluid Mechanics, Thermodynamics of Turbomachinery Pergamon Press, Oxford (1975).
3. D. Japikse. Centrifugal Compressor Design and Performance Concepts ETI, Inc. (1996)
4. M. Sédille. Ventilateurs et Compresseurs Centrifuges et Axiaux, Masson, Paris (1973)
5. J. Poulain. Pompes rotodynamiques, Techniques de l'Ingénieur, Paris (1992).
6. J.S. Swearingen, S. Mafi, Experimental Investigation of Vibrations in High-Speed Rotating Machinery, *Proceedings of the Vibrations Conference*, Philadelphia, (1969).
7. R.R. Conte. Eléménts de Cryogénie, Masson, Paris (1970).
8. F. Janin, Internal communication.
9. C.M. Harris. Shock and Vibration Handbook, McGraw-Hill, New York (1987).
10. G.M. Gistau-Baguer, P. Briend and P. Dauguet, Two large 18 kW (equivalent power at 4.5 K) helium refrigerators for CERN's LHC project supplied by Air Liquide, *present conference.*

NOMINAL CURRENT TEST PERFORMANCE OF 2 KA-CLASS HIGH-T_C SUPERCONDUCTING CABLE CONDUCTORS AND ITS IMPLICATIONS FOR COOLING SYSTEMS FOR UTILITY CABLES

D.W.A. Willén[1], M. Däumling[1], C.N. Rasmussen[1], C. Træholt[2], S.K. Olsen[2], C. Rasmussen[2], K.H. Jensen[2], J. Østergaard[3], A. Kühle[2], O. Tønnesen[2]

[1]NKT Research Center, Priorparken 878
DK-2605 Brøndby, Denmark

[2]Technical University of Denmark, Electrical Eng. Dep., Bldg. 329
DK-2800 Lyngby, Denmark

[3]DEFU, Bldg. 325
DK-2800 Lyngby, Denmark

ABSTRACT

The current carrying performance of 3-10 m long superconducting cable conductor models has been evaluated. A reduced energy loss compared to conventional cables can be obtained using high-T_c superconducting materials due to the limited resistive and AC hysteresis losses in some conductor configurations. The conductors are characterised under DC and AC conditions. The current and voltage are recorded during the tests in order to determine the impedances and the losses of the cable models. Using a phase-sensitive measurement with two lock-in amplifiers, small losses can be accurately measured at high currents. The critical currents of these conductors are in the range of 1-3 kA, and AC losses smaller than 1 W/m are measured at 2 kArms. AC currents with peak values exceeding the DC critical currents are applied. Increased losses, in excess of the expected magnetisation losses are observed when individual layers in the cables saturate. The loss contributions from other components of the cable system are discussed, and the implications for the cooling apparatus for superconducting utility cables are determined.

INTRODUCTION

Superconducting (SC) cables are presently being developed and evaluated for use in utility networks. These cables can offer the advantages of lower loss, lighter weight, and more compact dimensions, as compared to conventional cables. In addition to the improvement of the energy efficiency of the utility grid, this can lead to easier and faster installation of the cable system, and reduced use of land. Also, advantages relating to the internal cooling of these cables can be expected, such as reduced requirements on the quality of the soil around the cable, as well as on the location of the cable. For example, the cable may intersect with heat sources such as central heating pipes and conventional cables, something that is undesirable for a conventional cable due to the sensitivity of the electrical insulation to high temperatures.

A central issue for utility applications is the reliability of the application. In a superconducting cable system of the room-temperature dielectric design, a conventional electrical insulation is used and it is therefore considered to be a low-risk factor. The reliability of such a cable system will therefore strongly depend on new components such as the cooling apparatus.

In this article, measurements of the different losses in a superconducting cable of the room-temperature dielectric design (RTD) will be presented. The results will be interpreted in terms of the requirements for the cryogenic cooling systems to be used in future utility applications.

LOSSES IN A SC-RTD CABLE

DC losses are small in a SC cable. The DC loss at $I_c(1\ \mu V/cm)$ is $P = 100\ \mu V/m \times I_c$. If I_c is 2 kA then the loss at these 2 kA is in the order of 0.2 W/m. The loss at higher and lower currents is determined by the detailed shape of the VI curve of the cable. There are also resistive loss-contributions from the current leads and contacts at each end of the cable. These losses can be in the range of 85-150 W per termination at 2 kArms, if the thermal heat inleak is included.[1, 2]

The AC loss in a layered superconducting cable is determined by two factors [3]: a) the magnetisation loss of the layers that carry the current and are exposed to the magnetic field it creates, and b) saturation loss determined by the overall layer current distribution, which is governed by the self and mutual layer inductances.[4] Unless special measures are taken, the current tends to concentrate in the outer layers of the cable. This leads to a strong dominance of the saturation loss, and a limiting case is the monoblock model.[5] In this limiting case, the saturation loss can be one or two orders of magnitude larger than the magnetisation loss.

If the current in the cable is distributed equally, then the magnetisation loss in cables of the present construction (8 layers) [6] is about 0.1 – 0.2 W/m at 50 Hz and 2 kA. Lower losses can be obtained if the filaments of the superconducting tapes are decoupled with respect to induced currents. This can be achieved by twisting the filaments with the appropriate twist pitch. An alternative but not so practical route is to decrease the layer (or tape) thickness while at the same time increasing the number of layers in the cable.

The heat leak through the cryostat wall is a result of thermal conduction through the gas and through the mechanical support, and by thermal radiation. From the Brookhaven demonstration [7] it is seen to be possible to maintain a vacuum at 10-20 Pa over 8 years without pumping. This resulted in a heat leak of 2 W/m, decreasing with time to 0.5 W/m as the vacuum space was cryopumped by the He coolant. In the present application, a thermal leak significantly below 1 W/m, will require a multi-layered vacuum insulation with a low pressure of 0.01-0.1 Pa which must be maintained over 30-40 years.

MEASUREMENTS ON 3-10 M CONDUCTOR MODELS

In order to evaluate different construction principles, several functional cable conductor models have been stranded both manually and by machine. These conductor models are 3-10 meters long and have ampere-meter ratings of 1-30 kAm (Fig. 1). The conductors were characterised in both DC and AC conditions, while being immersed in a liquid nitrogen bath and cooled by a closed-loop circulating cooling system. The circulating cooling system for this test setup is described elsewhere.[8] In these measurements the voltage probes are placed on the copper blocks that contact the superconducting tapes. The linear resistive part due to the copper resistance is deducted from the measured values. Two sample curves from a 10 m long experimental conductor are shown in Figure 2, where the curve a) is measured with the conductor immersed in liquid nitrogen at 77 K, and curve b) is measured using the pressurised circulation cooling system at a temperature 2-3 degrees higher. The reduction in the critical current, I_c, due to the increased temperature is clearly seen.

The AC loss was measured electrically. In these measurements, the large inductive signal that is superimposed onto the loss signal demands a very accurate measurement of the phase angle. This is accomplished with two lock-in amplifiers and an appropriate correction scheme for the phase angle. This method has previously been used for single tapes,[9] and has now been adapted to high-amperage conductors by using Rogowski coils as captors of current and phase angle.[6]

The AC loss from these measurements is shown in Figure 3. Here, four different conductors are shown. The samples "NRC 3", "NRC 4", and "Eltek 7" are 3 m long conductors where "NRC 3" has a lower I_c than "NRC 4". The conductor "Eltek 7" has a forced even current distribution through the introduction of increased contact resistances in each of the individually connected layers. The conductor "NRC 5" is a 10 m long conductor. All measurements are at 77 K. Conductors "NRC 3" and "NRC 4" illustrate the

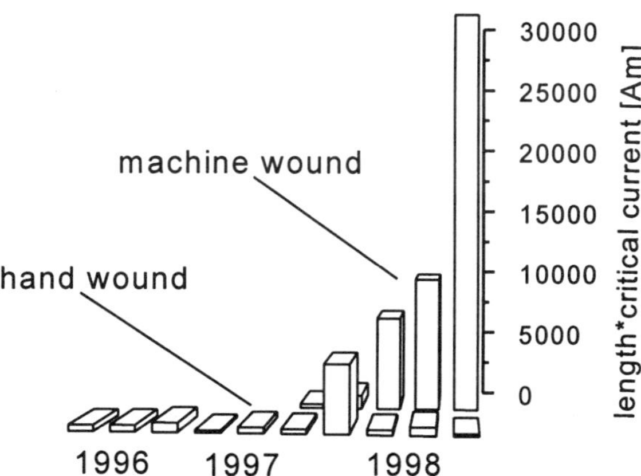

Fig. 1. Diagram of the ampere rating of the cable conductors tested 1996-1999.

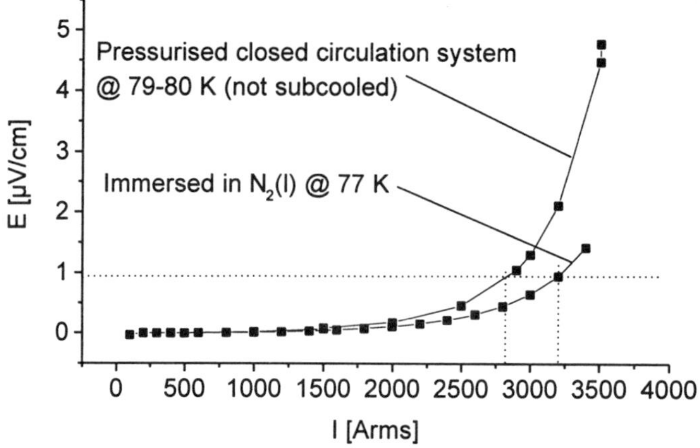

Fig. 2. DC measurements with a 10 m long cable conductor a) immersed in liquid nitrogen at 77 K, b) cooled by a closed-loop circulation system at 2-3 degrees higher temperature.

increased loss that occurs when certain layers saturate before others due to an uneven current distribution.[3] This loss can approach the monoblock loss. The loss curves of conductors "Eltek 7" and "NRC 5" show the possibility of reducing the AC loss by improving the current distribution between and within the layers. They also show that this can be achieved without significantly over-dimensioning the critical current, i.e. the cable can be operated with AC peak values close to its critical DC current.

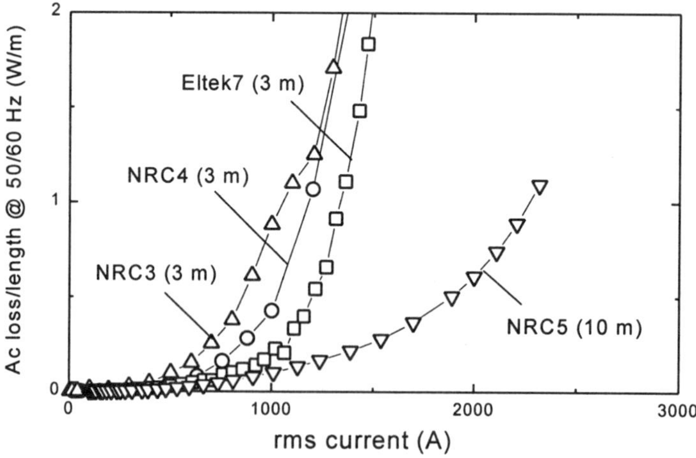

Fig. 3. AC loss of four different cable conductor models: "NRC 3" is a 3 m model, I_c=2300 A, "NRC 4" is a 3 m model, I_c=2800 A, "Eltek 7" is a 3 m model, I_c=1500 A, with forced equal current distribution, and "NRC 5" is a 10 m model, I_c=3200 A.

IMPLICATIONS FOR COOLING SYSTEMS FOR UTILITY CABLES

From the above measurements, it is seen that AC hysteresis losses as low as 0.6 W/m can be obtained at 2.0 kArms using todays commercially available superconducting tapes, and without significantly over-dimensioning the cable (I_{peak}/I_c=0.88). However, this requires an adequate construction of the cable conductor.

Numerical calculations [10] show that in the future, SC cables made from tapes with twisted and insulated filaments, or from significantly thinner tapes, can have a loss induced by the AC current in the order of 0.1-0.2 W/kAm, i.e. a factor 2-3 lower than at present. This can lead to an AC loss of 0.2-0.3 W/m per phase at 2 kArms. Only a slight over-dimensioning of the I_c of the cable will be required. At this level, the thermal losses from the cryostat will be dominant and in order to significantly improve the performance, a better cryostat will be needed. Improvements in the thermal insulation, can lead to thermal losses as low as 0.5-1.0 W/m. In addition, there may be with parasitic losses such as eddy current loss, in the order of 0.01-0.2 W/m at 2 kArms. These values imply that the total loss in a cable system rated for 2 kArms will amount to at least 0.7-1.5 W/m per phase. If coolers are placed at a distance of 4-10 km apart, these coolers have to remove a heating power of 6-7 kW/phase (assuming that the more efficient cable would have cooling stations further apart and vice versa). In addition, the cooling stations close to a termination will have an extra heat load of 100-150 W/phase. Also, splices will occur approximately 1 km apart and will be expected to have an increased heat loss, perhaps 20-30 W, adding another 100-200 W of loss per phase. These contributions are illustrated in Figure 4.

The efficiency of the cooling machine that cools and circulates the liquid coolant in the superconducting cable depends on the operating temperature of the cable. It can be expressed as the inverse of a penalty factor. Total system energy savings in the order of 50% compared to conventional cables can be obtained for cables operating at around 77 K if the penalty factor of the cooler can be as low as a factor of 9 to 10 at this temperature.[11] The power rating of the individual cooling machine will depend not only on the loss of the cable, but also on which redundancy scheme is acceptable to the utility. It is possible that a future HTS cable system would be engineered and commissioned in four steps:

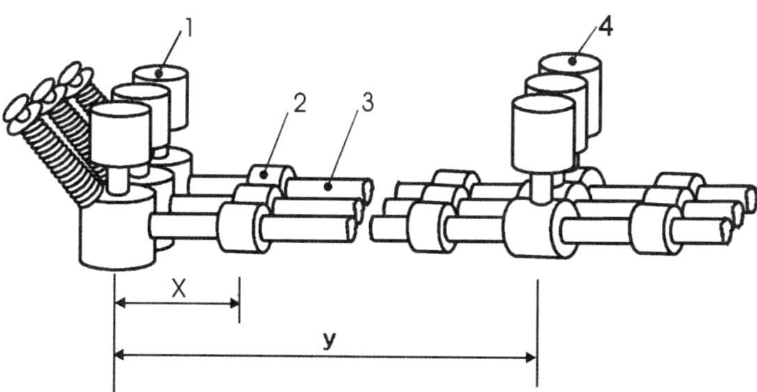

Fig. 4. Schematic of the sources of loss in a SC cable system and their approximate magnitude. The cable length x=1 km, the distance between cooling stations y=4-10 km. Losses per phase: 1) cable termination, 85-150 W; 2) cable splice, 20-30 W; 3) cable conductor and thermal insulation, 0.7-1.5 W/m; 4) splice with cooling station, 20-30 W.

1) The cable system is designed and over-dimensioned for future needs;
2) The cooling system is designed with three coolers/station for redundancy;
3) The cable system is installed with two coolers for the present rating;
4) The system capacity is upgraded with a third cooler when needed.

This translates into a cooler unit capacity not larger than 6.5-7.5 kW with a compressor rating not larger than 55-75 kW. The desired reliability that has been indicated by Electricité de France [12] is a maximum of 1-2 faults per 100 km per year. In the present scheme, this translates to 1-2 faults in 50 units per year. The desired service interval has been indicated by some utilities to be 2-3 yrs.

CONCLUSION

Superconducting cable models 3-10 m long have been manufactured for a rated current of 2 kArms, with AC losses as small as 0.6 W/m at 77K, 2 kArms. A future reduction of a factor 2-3 of these values and improvements in the thermal insulation, can lead to total losses per length of 0.7-1.5 W/m per phase at 2 kArms. Superconducting transmission cables will require cooling stations spaced 4-10 km apart. Taking into account the additional loss from terminations and splices, and the requirement for redundancy in the cooling supply, these cooling stations will consist of a minimum of 2-3 cooling units with a capacity at 77 K not larger than 6.5-7.5 kW each. This can lead to energy savings on the order of 50 % compared to conventional cables, as well as advantages related both to the reduced weight and dimensions of these cable systems.

ACKNOWLEDGEMENTS

This work was supported in part by the Danish Department of Energy (Energistyrelsen), I/S Eltra, and Elkraft, which is gratefully acknowledged.

REFERENCES

1. C. Rasmussen, A. Kühle, O. Tønnesen, C.N. Rasmussen, Design of a termination for a high temperature superconducting cable, *IEEE Trans. Appl. Supercond.* 9:1273 (1999)
2. T. Shimonosono, S. Nagaya, T. Masuda, S Isojima, "Development of a Termination for the 77 kV-class High Tc Superconducting Power Cable", *IEEE Trans. Power Delivery*, 12:33 (1997).
3. M.Däumling, Cryogenics, accepted for publication.
4. S.Krüger Olsen, C.Træholt, A.Kühle, O.Tønnesen, M.Däumling, J.Østergaard, Loss and inductance investigation in a 4-layer superconducting prototype cable, *IEEE Trans. Appl. Supercond.*, 9:833 (1999).
5. G.Vellego, P.Metra *Supercond. Sci. Technol.* 8:476 (1995).
6. S Krüger Olsen, A Kühle, C Træholt, C Rasmussen, O Tønnesen M Däumling, C N Rasmussen, D W A Willén, *Supercond. Sci. Techn.* 12:360 (1999).
7. EB Forsyth and R.A. thomas, Performance summary of the Brookhaven superconducting power transmission system, *Cryogenics*, Vol. 26 (1986).
8. C. Traeholt, C. Rasmussen, S. Krüger Olsen, K.H. Jensen, A. Kuhle, O. Tønnesen, D.W.A. Willén, M. Däumling, C.N. Rasmussen, and J. Østergaard, Cryogenic set-up for a 10 meter long nitrogen cooled superconducting power cable, *these proceedings*
9. K.-H. Müller and K.E. Leslie, Self-field ac loss of Bi-2223 superconducting tapes, *IEEE Trans. Appl. Supercond.*, 7:306 (1997).
10. M. Däumling, Unpublished.
11. Jacob Østergaard, Superconducting Power Cables in Denmark – A Case Study, *IEEE Trans. Appl. Supercond.*, 7:719 (1997).
12. Electricite de France presentation of HTS cable project, J-Cable conference (1999).

5-M SINGLE-PHASE HTS TRANSMISSION CABLE TESTS

J. W. Lue[1], G. C. Barber[1], J. A. Demko[1], M. J. Gouge[1], S. W. Schwenterly[1], J. P. Stovall[1], R. Martin[2], R. L. Hughey[2], U. Sinha[2], and J. C. Tolbert[2]

[1] Oak Ridge National Laboratory
Oak Ridge, TN 37831-8071
[2] Southwire Company
Carrollton, GA 30119

ABSTRACT

Two 5-m, single-phase, High-Temperature Superconducting (HTS) transmission cables have been built and tested at Oak Ridge National Laboratory (ORNL) in a joint program between Southwire Company and ORNL. The active conductor of the cable consists of layers of Bi-2223/Ag tapes helically wound on a flexible former, followed by layers of cryogenic dielectric tapes, and then HTS shield layers. Subcooled liquid nitrogen at temperatures of 70-80 K and pressures of 3-7 bar flows down the inner pipe, turns around at a termination, and returns through the annulus between the cable and the vacuum-jacketed outer pipe to cool the cable. The DC V-I curves of the cables have been mapped as a function of temperature and the critical currents measured. The cables have exceeded the performance goal of 1250-A AC at 7.2 kV line-ground voltage. The AC loss measurements showed a loss of about 1 W/m for one and 0.7 W/m for the second cable at 76 K. High voltage measurements showed partial discharge at only background level for voltage up to 18 kV AC (2.5 times design value). Stable cable temperature was maintained at 1250-A rms at temperatures up to 80.5 K, where the operating current is almost twice the cable's critical current. These test results provide confidence in the design approach and the extrapolation to the next phase: development of a 30-m, three-phase cable.

INTRODUCTION

The development of High Temperature Superconducting (HTS) transmission cables, headed by Southwire Company, went into the phase of building and testing cables with dielectrics, high voltage, and long-duration tests. This followed the construction and testing of several 1-m-long prototypes of core conductor for electrical and thermal evaluations.[1] Two 5-m, single-phase, HTS transmission cables were built with Bi-2223/Ag tapes, plus cold-dielectric, superconducting shield, and completed with vacuum-jacketed enclosures.

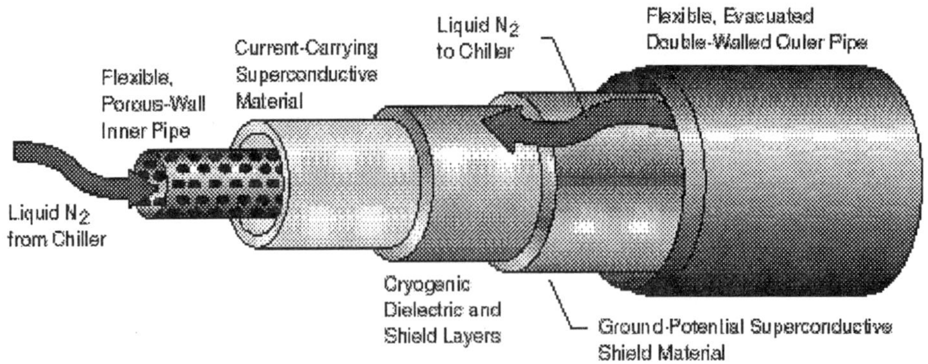

Figure 1. A cut-away view of the Southwire HTS 5-m cable.

A versatile, HTS cable test facility has been designed and built at Oak Ridge National Laboratory (ORNL) to evaluate the performance of prototype HTS power transmission cables at lengths on the order of 10 m.[2] The two 5-m cables were tested for their DC characteristics, AC losses, high voltage withstand, and long-duration performances.

THE 5-M HTS CABLES

The single-phase, 5-m, HTS cables were built based on a cold dielectric design. The core conductor consists of four layers of helically wound, Bi-2223/Ag tapes supplied by Intermagnetics General Corporation (IGC). Four layers of tapes were chosen to provide enough capability for the design current of 1250 A. The tapes were machine wound on a flexible stainless steel former. Cryoflex™ cold dielectric tapes were wrapped between the inner conductor and the outer HTS shield. The outer HTS conductor provides shielding of the field generated by the currents flowing on the inner conductor and thus eliminates magnetic fields and eddy currents in the external structure.

Figure 1 shows a cut-away view of the cable. The liquid nitrogen coolant flows down the center of the former and returns through the annulus between the HTS shield and the vacuum-jacketed outer pipe. The second cable was wound with an improved winding head, and a different technique was used for the shield terminal connections. Both cables were designed for 1250-A rms at a phase-to-ground voltage of 7.2 kV. These cables serve as a prototype of the longer 30-m, three-phase cable to be installed in late 1999 at the Southwire Company cable-manufacturing complex in Carrollton, Ga.

THE HTS CABLE TEST FACILITY

The HTS cable test facility consists of a pair of terminations, a high vacuum pumping station, a cryogenic cooling system, DC and AC high-current power supplies, AC high-voltage power supplies, and a data acquisition system. A sketch of the test facility is provided in Fig. 2. The 5-m HTS cable is located between two terminations — one on the east end of the facility, the other on the west end. The termination provides a transition from sub-cooled (70-80 K) liquid nitrogen at 3-7 bar in the cable space to a vacuum space, and a second transition from vacuum to atmosphere at room temperature. Bushings for the

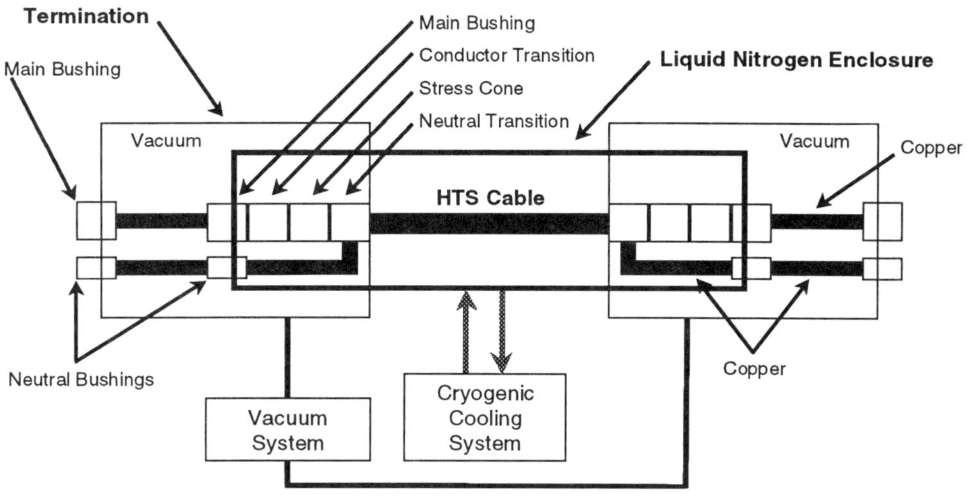

Figure 2. ORNL/Southwire HTS Cable Test Facility layout.

transitions are rated for full cable current and voltage. The two terminations are pumped through a common vacuum header by a pumping station to the 10^{-6} torr range to provide sufficient thermal and electrical insulation.

The cryogenic cooling system is built into a liquid-nitrogen skid that can provide about 1 kW of cooling at 70-80 K. It consists of a pressure/surge dewar, a subcooler, a circulation pump, and the associated piping. A vacuum pump is also included to pump the sub-cooler bath to 67-77 K. Not shown in the figure are a 3-kA DC power supply, a 2-kA rms AC power supply, a 25-kA DC power supply that can supply a pulse loading to the cable, and a 30-kV high-voltage power supply. The data acquisition system contains three ten-channel multimeters with a fiber optic link for the high-voltage signals and a personal computer using the LabVIEW program. Further details of the cryogenic system, performance of the skid, and the heat load of the terminations are reported elswhere.[3]

DC MEASUREMENTS

Cooldown of the HTS cable to the testing temperatures was achieved initially by pre-cooling the cable with nitrogen gas and liquid supplied from the storage dewar. The nitrogen skid was then started, and subcooled liquid nitrogen was flowed to the cable via one termination, through the center of the cable former, turned around at the other termination, through the annulus of the cable and the thermal pipe, and back to the skid for a closed loop cooling

A series of DC voltage-current scans was performed to characterize the 5-m cables. The observed DC V-I curves showed broad resistive transitions - characteristic of the HTS tapes supplied by IGC. The measured curves of resistivity (proportional to voltage drop) of the cable, ρ, *vs*. current density, J, can be reasonably well fitted by a power law: $\rho = k \cdot J^n$. Around the critical current, the curve fitted n-values averaged only about 2.2. The critical

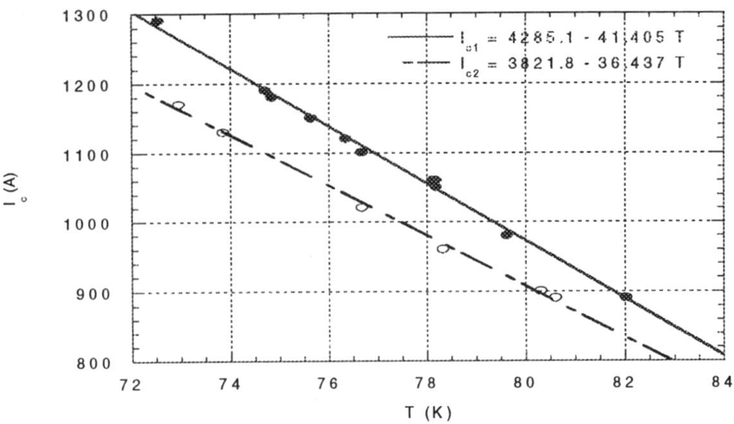

Figure 3. Critical currents of the two 5-m cables as a function of temperature.

currents of the two cables (as defined by the voltage drop criterion of 1 μV/cm) as a function of the average cable temperature are shown in Fig. 3. In the temperature range of 72 – 82 K, the measured critical currents increase nearly linearly as the cable temperature decreases.

AC LOSS MEASUREMENTS

AC loss of the cables was measured by an electrical technique similar to that used for the 1-m core conductor prototypes.[1] Voltage taps were placed on the two ends of the core conductor, and one lead was laid between the HTS and the dielectric tapes to follow the cable to the other end. A digital lock-in amplifier measured the voltage across the cable and the phase angle between the voltage signal and the current signal. The product of the current and the voltage in phase with the current gives the AC loss ($P = I \cdot V \cos\theta$) of the main conductor. A signal from a Rogowski coil with an integrator was used as the reference signal, and a second Rogowski coil corrected its phase error. Curve fitting of the loss voltage was also made to take out the linear term due to the resistive terminal joints.

Because of the existence of the shield layer, its effect on the measured loss voltage was a concern. Figure 4 shows the measured AC loss per unit length of the cable as a function of the AC rms current taken on two different runs. On the first test, the shield layer was grounded at one end only and open at the other end. On the second test, the shield was also grounded at one end, but an external copper cable was used to tie the two ends together to form a closed loop. The induced current in the shield layer should be quite different in these two tests. However, the same loss data were obtained. Thus, the main conductor AC loss measured by the present technique is independent of the shield current.

The AC losses of the cables were measured at several different cable temperatures. Figure 5 shows the AC loss of the first cable as a function of the AC rms current at four different temperatures. There are small but notable difference in the measured losses. Loss rates of 0.8 – 1.3 W/m were measured at the rated current of 1250-A rms.

The main effect of changing temperature to the cable is an increase in critical current as the temperature lowers. Thus, a good comparison in AC losses at different temperatures could be in terms of the peak AC current to the critical current ratio, I_p/I_c. But, as noted

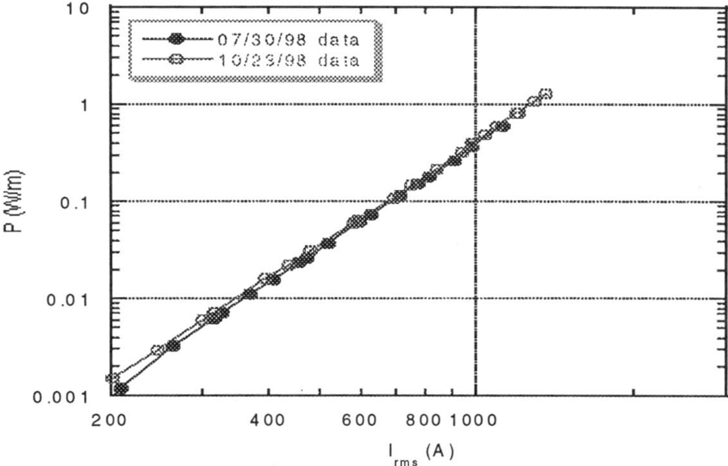

Figure 4. AC loss of the first 5-m cable measured with two different shield connections.

earlier, the broad resistive transition of the HTS tapes can be reasonably well fitted by a power law: $\rho = k \cdot J^{\,n}$. Under this assumption, Dresner has studied the AC loss of HTS cables.[4,5] In the limit of incomplete penetration (at lower currents), the dependence of AC loss of the cable P_{cable} on I_p/I_c was found to take the form:

$$P_{cable} \propto I_c^{\,\alpha} \cdot (I_p / I_c)^{\kappa},$$

$$\alpha = \frac{2n+3}{n+2}, \qquad \kappa = \frac{3n+4}{n+2}$$

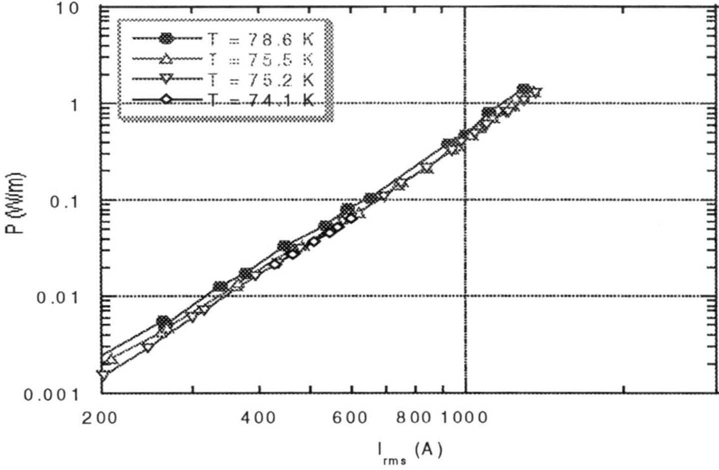

Figure 5. AC loss of the first 5-m cable measured at four different temperatures.

The proportional constant in the above equation depends mainly on the geometric factor of the cable. For conductors with a sharp resistive transition, *i.e.* with high n-values, $\alpha = 2$ and $\kappa = 3$, the same as predicted by the critical state model.[6] Also note that both α and κ are not too far from the critical state model values, even for the present very low n-value of 2.2.

The above equation indicated that the proper way to compare AC losses is not to simply plot P_{cable} *vs.* I_p/I_c, but to plot the normalized cable loss P_{cable} / I_c^{α} *vs.* I_p/I_c. For the present low n-value of 2.2, $\alpha = 1.76$. Thus, in Fig. 6 we plotted $P_{cable} / I_c^{1.76}$ *vs.* I_p/I_c for the data taken at the four different temperatures. The normalized AC losses at different temperatures converged onto a single curve. Thus, the present loss data agree well with the scaling law predicted by Dresner.

Similar results were obtained for the second 5-m cable. Figure 7 compares the AC losses of the two cables at temperatures of about 76 K. Although cable #2 had lower I_c than cable #1 as shown in Fig. 3, it had lower AC losses. At 1250-A rms, the AC loss of cable #2 is 0.7 W/m as compared to 1 W/m for cable #1. This may be the result of the tighter and more uniform winding achieved in cable #2.

INDUCED CURRENT IN THE SHIELD CONDUCTOR

Both ends of the HTS shield layer were brought out to room temperature by long lengths of Cu-braids through a cold liquid-nitrogen to vacuum and a vacuum to atmosphere feedthrough. During most of the cable test, the shield was grounded at one end, and an external copper cable was used to tie the two ends together to form a closed current loop. The external cable consists of two copper cables rated at 500 A. The induced current through the shield was measured for core conductor currents up to 1400 A. Figure 8 shows that the shield current increases linearly with the core current as expected, but its magnitude is only about 46% of the main current. In a later test, the shield external cable was re-routed to follow more closely the supply cable of the main conductor. The second curve in Fig.8 showed that the shield current increased to about 65% of the main current. Thus, the magnitude of the shield current in a single-phase test depends strongly on the relative loops of the primary and the secondary circuit.

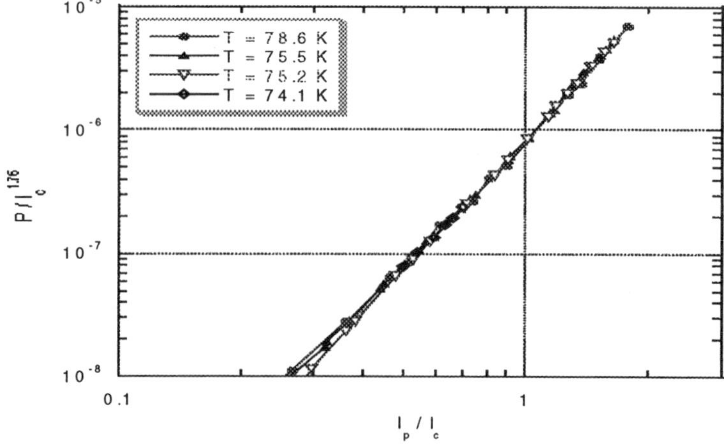

Figure 6. Normalized AC loss of the first 5-m cable measured at four different temperatures.

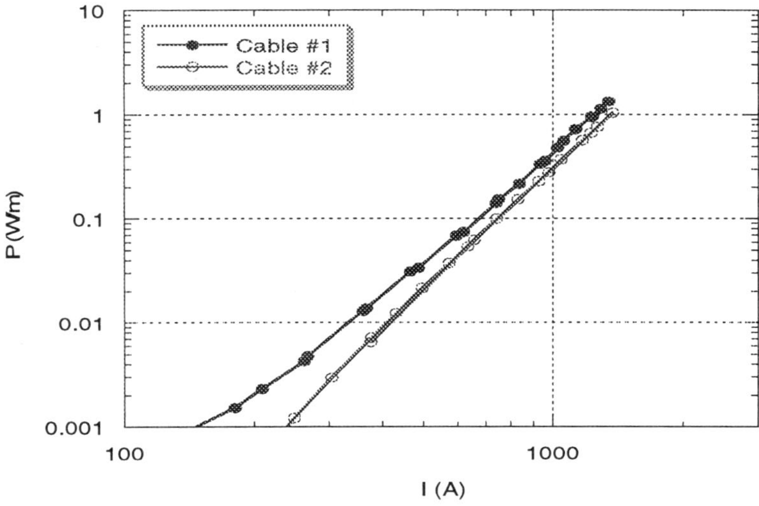

Figure 7. Comparison of the AC losses of the two 5-m cables taken at about 76 K.

A measurement of the shield current was also made after a long-duration (more than 3 days) test of the cable. It was found that the shield current had lowered from 46 to 43% of the core current. When one of the two external interconnecting copper cables was disconnected, the shield current further reduced to 37% of the core current. Apparently, the magnitude of the shield current also depends on the impedance in its loop. The reduction of shield current after a long-duration test can be explained by the warming up and the resultant increase in the resistance of the copper braids connecting the HTS shield to the room temperature feedthroughs.

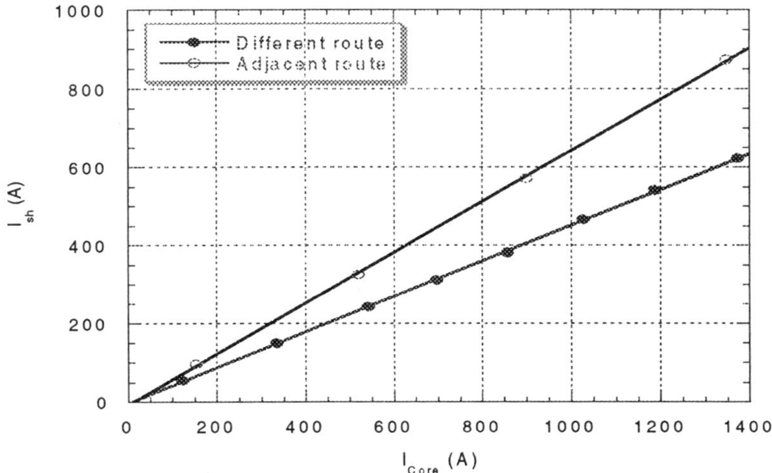

Figure 8. Measured shield currents as a function of the core conductor current with two different cable routing.

HIGH VOLTAGE TESTING

A high-voltage withstands test of the cable–termination system was made at a termination vacuum of 4.2 x 10^{-6} torr. Voltage was stepped up to 18 kV AC – 2.5 times the design operating voltage without a breakdown. Testing at this level was performed to assure that the cable and terminations would meet an industry voltage standard. The high voltage was held for 30 minutes at 18 kV. Figure 9 shows the measured partial discharge (in pico-Coulomb) and the leakage current as a function of the applied voltage. The partial discharge was independent of the applied voltage and stayed at the background value of about 23 pC. The leakage current increased linearly with the applied voltage. The measured leakage currents were consistent with the capacitances estimated for the HTS cable and the external supply cables.

A test was also made with the voltage held at the design operating value of 7.2-kV AC and the current increased from zero to 1420-A rms. The leakage current was steady at about 44 mA.

LONG DURATION TEST

Another test of the integrity of the HTS cable and the test facility is a long-duration run of the cable at its design operating conditions. The first 5-m cable was run at 7.5-kV AC line-to-ground voltage and 1250-A rms current for 27 hours. The inlet liquid nitrogen temperature remained constant at 79 K, and the outlet from the cable remained at 82 K during the entire test.

The second 5-m cable was run continuously for 80 hours. Figure 10 shows the inlet and outlet temperatures of the liquid nitrogen to and from the cable. Again the inlet temperature remained at about 79 K. For the first 36 hours, the flow rate was set at 250 g/s (5 gpm), and the outlet temperature remained constant at about 82 K. Then a fault in the power supply transformer circuit resulted in a no current for about 8 hours. During that time the outlet temperature of cable/termination lowered to about 80.5 K. After the current

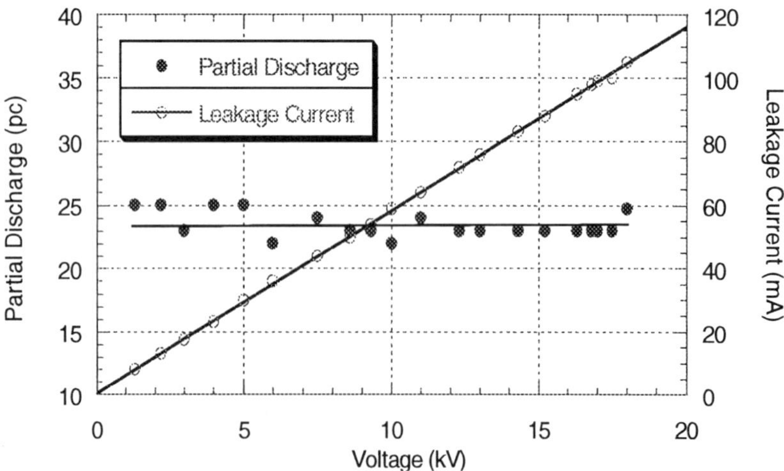

Figure 9. Partial discharge and leakage current of the cable-termination system as a function of the applied voltage.

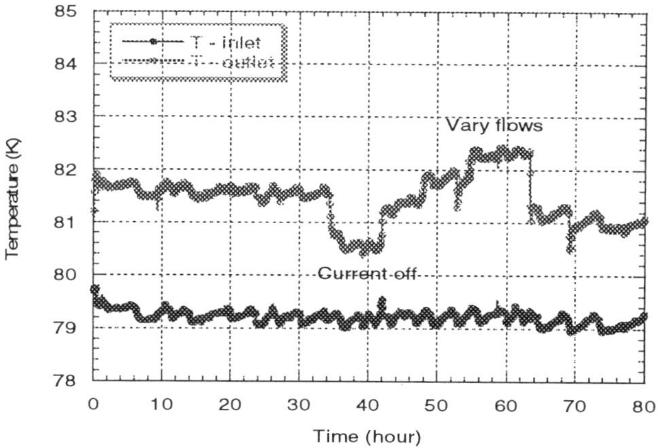

Figure 10. Inlet and outlet temperature of the coolant during the long duration test of the second 5-m cable.

was restored the outlet temperature went back to 82 K. The circulation rate of the coolant was next varied in several steps. This resulted in the corresponding changes in the outlet temperature shown in Fig. 10.

During both design-condition long-duration runs, the sub-cooler bath on the cryogenic skid was not pumped, and the cable main conductor temperature was stable at about 80.5 K. At this temperature the operating current has a peak value of 1.8 – 1.9 times the DC critical currents. Thus, the broad resistive transition of the HTS tapes made it possible to operate the cable stably at currents significantly above its critical current.

CONCLUSION

Two 5-m long, single-phase, HTS cables have been built and tested successfully in this phase of the development of HTS transmission cables by the Southwire/ORNL team. The cables were designed and built with cold dielectrics and a coaxial HTS shield. Test results showed that both cables met or exceeded the design parameters.

The DC characteristics of both of the cables have been measured at temperatures between 72 and 82 K. The critical currents were found to increase linearly with decreasing temperatures in this range. The AC loss of the cables was measured as a function of current at line frequency for several different temperatures. AC losses of 1 W/m for the first cable and of 0.7 W/m for the second cable were measured at the design current of 1250-A rms. The AC losses were also found to scale as the ratio of AC current to the critical current ratio as predicted by a theory by Dresner.

High-voltage testing of the cable-termination system proved the integrity of the cold-dielectric and the termination design. Both cables and the terminations surpassed the criteria of 2.5 times operating voltage – 18-kV AC without a breakdown. Both cables have been operated stably for extended periods of time (up to 72 hours) at the rated 7.2-kV line-to-ground voltage and 1250-A current continuously.

System heats loads, LN_2 flows, and pressure drops have also been measured and reported elsewhere.[3] This and the above test results provide confidence in the design approach. Valuable information from this 5-m cable test is being used in the design and construction of the 30-m, three-phase cables being installed at the Southwire Company site.

ACKNOWLEDGMENT

This work was sponsored by the Office of Energy Efficiency and Renewable Energy, U. S. Department of Energy under contract DE-AC05-960R22464 with Lockheed Martin Energy Systems, Inc. The installation work performed jointly by the Southwire and ORNL mechanics is gratefully acknowledged.

REFERENCES

1. J. W. Lue, *et al.*, AC losses of prototype HTS transmission cables, *IEEE Trans. Applied Superconductivity, **9-2**: 416-419 (1999).*
2. M. J. Gouge, *et al.*, HTS cable test facility: design and initial results, *IEEE Trans. Applied Superconductivity, **9-2**: 134-137 (1999).*
3. J. A. Demko, *et al.*, Cryogenic system for a high temperature superconducting power transmission cable, paper CAF-8 presented at this conference.
4. L. Dresner, Hysteresis losses in power-law cryoconductors, II, *Applied Supercoductivity, **4(7/8)**: 337-341 (1996).*
5. L. Dresner, AC losses in transmission line cables, paper MM2 presented at ICMC98, U. Twente, Netherlands, May 10-13, 1998.
6. W. T. Norris, Calculation of hysteresis losses in hard superconductors carrying ac: isolated conductors and edges of thin sheets, J. Phys. D. **3**: 489-507 (1970).

OPERATING A CRYOGENIC TEST RIG FOR A 10 METER LONG LIQUID NITROGEN COOLED SUPERCONDUCTING POWER CABLE

C. Træholt,[1] C. Rasmussen,[1] A. Kühle,[1,3] S. Krüger Olsen,[1] K. Høj Jensen,[1] O. Tønnesen,[1] D.W.A. Willén,[2] M. Däumling,[2] C.N. Rasmussen[2]

[1]Technical University of Denmark, Department of Electric Power Engineering, Bldg. 325, DK-2800 Lyngby, Denmark

[2]NKT Research Center, Priorparken 878, DK-2605 Brøndby, Denmark

[3]Danish Institute of Fundamental Metrology, Building 307, Anker Engelunds Vej 1, DK-2800 Lyngby, Denmark

ABSTRACT

One way to cool a high temperature superconducting cable is to circulate liquid nitrogen (LN2) by means of a mechanical pump through a sub-cooler and through the core of the cable.

We report on our experimental test rig for testing a 10-meter-long high temperature superconducting cable with a critical current of 3.2 kA (@77K). The test rig consists of custom designed cable end termination, current leads, coolant feed-throughs, electrical separation, liquid nitrogen closed loop circulation system and a 10-meter-long modified commercial vacuum insulated cryostat. The system is designed to yield both electrical and thermal data. With zero electrical load on the cable the complete system consumed about 400 W. The contributions could be identified as; 30-35 W from the cryostat with cable, 70 W from each termination and about 100 W from the feed lines and the electrical separation. The remaining 100 W could be attributed to the closed loop circulation station.

INTRODUCTION

Since the discovery of superconductivity[1] in 1911 people have been thinking of how to exploit the zero resistance characteristic of superconductivity in low loss electrical energy distribution. During the years 1972-1986 several projects,[2,3] of which the so-called Brookhaven project[3] is the most well known, addressed the use of low temperature superconductors in a superconducting cable. The aim was to construct a superconducting cable that ultimately would be commercially feasible to produce. These cable conductors were cooled by liquid helium and were encaged in a liquid nitrogen cryo-shield. Although technically feasible, the complexity and cost of the cable could not be justified commercially (i.e., not sufficiently attractive to the electrical utilities).

After the discovery of high temperature superconductors[4] (HTS) in 1986 the idea of applying superconductors in power transmission resurfaced. Among several HTS application possibilities, a superconducting cable based on HTS tapes was believed to present a relatively short prototype-to-market time. Thus, similar types of projects like the Brookhaven study were initiated on high temperature superconducting power transmission systems.[5-9]

Presently, there are several projects on-going with a commercial horizon including the project, *Superconductors in the Danish Energy Sector,* described in this paper. HTS tapes, the "raw material" of a superconducting cable, have already been produced on a commercial scale for several years.[10] Recently (1999), a pioneering project was announced on retrofitting HTS power cables for conventional cables in a utility grid,[11] paving the way for a possible commercial market within 5 years.

Besides the attractive electrical aspect of a superconducting cable, a means of cooling the cable has to be provided. A superconducting cable based on high temperature superconductors is at present expected to operate in the temperature range 50-90 K. Liquid nitrogen (LN2) is an obvious choice for a cooling media for two reasons. Its 1 atm boiling point at 77 K is in the desired temperature range and the cost is relatively low. Any heat leak into the system or produced within the system (flow friction, Joule heating and other ac-losses) needs to be removed. Because of the thermodynamic cost (12-14 W_{300K}/W_{70K}) in removing heat at 77 K it is paramount to be able to monitor and control the thermal performance of the cooling system. Such a cooling system needs to be tested and all individual components thermally evaluated.

A cooling system was purchased/built with the plan to drive a single phase superconducting cable up to 30 m long at 77 K. A 65 K extension for the system is under construction.

The system was tested on a 10-m-long test bed (set-up). First, it was operated with a dummy sample (tube without any conductors) but eventually a 10-m-long high temperature superconducting cable was installed into the cryostat. The set-up included electrical and thermal monitoring during operation. It is the scope of this paper to focus on the latter subject.

EXPERIMENTAL

The experimental set-up needs to fulfil several purposes:

1 Provide thermal performance data.
2 Provide electrical performance data.
3 Demonstrate a working system.
4 Provide operating experience with a superconducting cable in a closed loop LN2 cooling system.

The present experimental cooling system includes a cooling station, two custom designed cable terminations, a 10-m-long cryostat and LN2 feed lines. The system is outlined in Figure 1. With the cable installed in the cryostat it is possible to circulate LN2 in a closed loop through the cable core thereby keeping the cable cooled. With this set-up it is possible to do experiments with variable temperature, variable pressure and variable flow rate (cooling power). Each component of the closed loop system is described in detail.

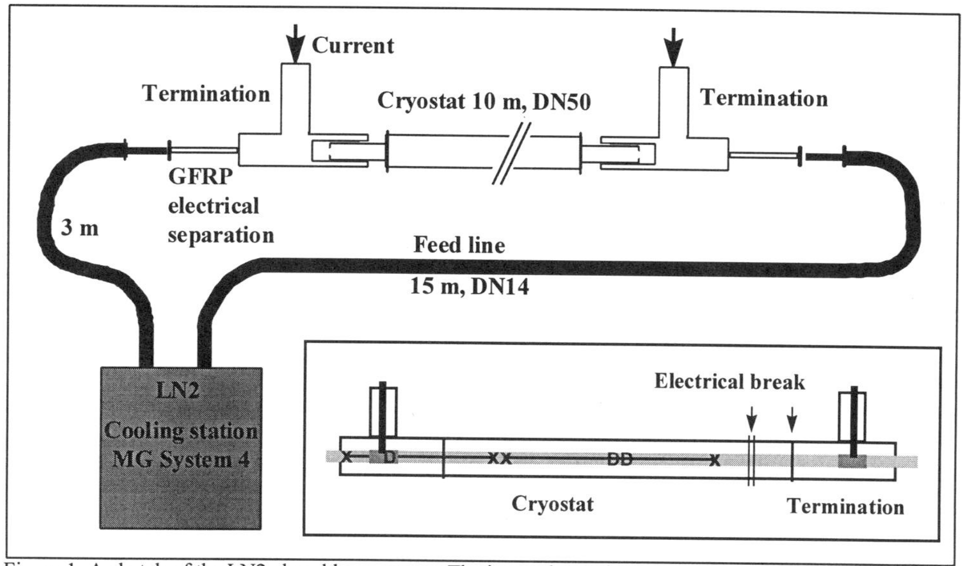

Figure 1. A sketch of the LN2 closed loop system. The insert shows the position of the thermocouples (**X**) and the diodes (**D**). The arrows indicate electrical breaks.

Two LN2 vacuum super insulated lines are used to connect (Johnston couplings) the cooling station with the cryostat/cable. One is 3 m long and the other is 15 m long. The heat leak is specified at 1.1 W/m. The heat leak of each Johnston coupling is specified to 3.7 W/m.

The 10 m cryostat was purchased as a custom modified commercial product.[12] In the outer corrugated cryostat wall an electrical break (30 cm long piece) is incorporated in order to reduce eddy current losses during ac operation of the cable. The cryostat is in effect a flexible vacuum insulated hose with a nominal inner bore diameter of 50 mm (DN50) and a heat leak of the cryostat specified at 2.5 W/m. A DN40 pump protrusion enables maintenance of the cryostat vacuum. The cryostat is connected to each cable end termination in a Johnston coupling (DN50).

The flexible cryostat at present is being used in a straight configuration, however, both the cryostat and the (superconducting) cable conductor[13] can operate in a bent configuration if necessary.

The cooling station consists of a commercial closed loop liquid nitrogen circulation system (Messer Griesheim Closed Loop System 4).[14] The figures of merit of the system are shown in Table 1. The closed loop circulation system is integrated with a sub-cooler via a heat exchanger. The cooling power is transferred to the closed loop by circulating LN2 through a heat exchanger that is immersed into the sub-cooler. In the present set-up the sub-cooler is a LN2 bath with a 150 l retention volume, which equals 6600 W·h, i.e. about 12 hours of continuous operation if the experiment consumes at a rate of about 500 W.

Table 1. Data for the closed loop circulation system

Cooling station (MG System 4)	Flow rate [kg/h]	Pressure range [bar]	Temperature range [K]	Max cooling power [W]
77 K	200 – 1500	-1 - +10	≈ 78	2200
65 K	200 – 1500	-1 - +10	65 – 80	1000

A control board displays the inlet and outlet temperature, the pressure at the inlet and outlet, the pump speed and flow rate.

The cooling station is capable of operating with an over pressure of up to 10 bar. The flow range is from 200 kg/h to 1500 kg/h. When the system is modified to a 65 K version (pumping on LN2 gas pressure) it can operate in the temperature range 65-80 K. The cooling station is capable of removing up to 2200 W (@ 78 K). An upgrading will enable the system to remove up to 1000 W at 65 K.

Each cable end there is:

1 Conductor termination: Joint between the superconducting tapes and the normal conductor (here a Cu cylinder).
2 Current lead.
3 Vacuum (thermal) insulated cryo-jacket.
4 LN2 feed through.
5 High voltage to ground transition (electrical insulator)

For simplicity, separate cooling of the current lead, such as gas cooling, has been avoided. This experiment operates with in-line forced LN2 cooling of the Cu cylinder. The Cu cylinder has been designed to anchor the temperature of the electrical transition area (normal conductor to superconductor) to the cable temperature. The thermal anchoring is accomplished by reducing the flow cross sectional area to 5 separate holes.

The cable end termination is integrated with a Johnston coupling (DN14) through a glass fibre reinforced tube (GFRP). The connection functions as both LN2 feed through and electrical separation between cable and cooling system. It is thermally insulated using insulating foams similar to what is used in district heating lines.

The stainless steel housing of the cable end and the current lead is vacuum insulated and integrated with a female Johnston coupling that fits on the 10 m cryostat.

In order to measure the heat transport of the LN2 in the cable, an E-type thermocouple stack is placed in the LN2 stream. The length between the pairs is 7 m in order to obtain the largest possible signal but at the same time to avoid end effects. Similarly, another thermocouple pair is placed in the LN2 stream spanning one cable end termination. Two diodes (type DT 471 SD from Lakeshore) are mounted in the middle of the cable conductor on the outside of the HTS tapes. One diode is thermally attached to the outer HTS tapes, the other one faces outwards and measures the temperature in the space between the cable conductor and the inner cryostat wall. A third diode is placed on the Cu cylinder immediately next to the current lead. The positions of the thermocouples and the diode are shown in Figure 1.

The signal from the thermocouples is displayed on a nanovoltmeter (HP 34420A). The three diodes are sequentially (via HP 34970A switch) displayed on a Lakeshore 330 temperature controller. Further, the data from the thermocouples and the diodes are automatically logged versus time using a HP-VEE program, developed for the purpose.

RESULTS/DISCUSSION

Initially the closed loop system was operated with a 10 m long plain tube mock-up installed in the 10 m flexible cryostat. This experiment was conducted in order to test the closed loop system, to gain operation experience with the system and to get a rough indication of the pressure loss and the heat leak contributed by each element of the system.

A pressure difference of 0.6-0.7 bar was measured between the inlet and outlet of the system with a constant LN2 flow of 480 kg/h. This value was obtained regardless of whether or not the cable (+ termination), the tube mock-up or nothing was inserted in the circulation loop.

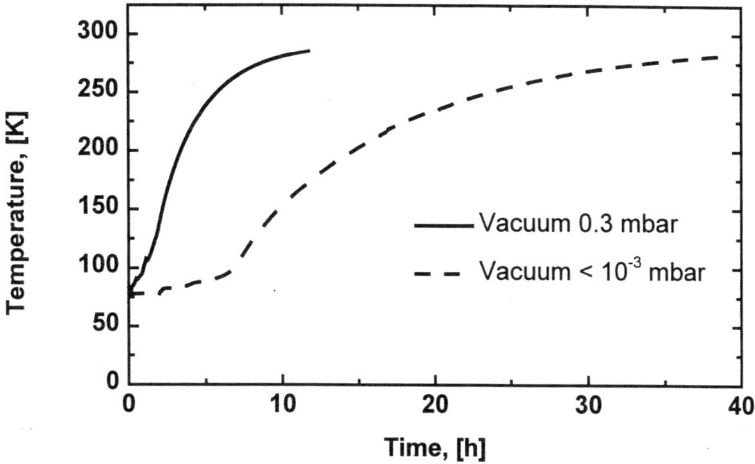

Figure 2. Typical transient response curve for the cryostat. The diode temperature is plotted versus time for a vacuum level of 0.3 (solid) and 10^{-3} mbar (dashed) respectively.

Thus, the pressure drop contribution of cable end terminations and the cable were below 0.1 bar. The major pressure loss occurred in the feed line and/or the GFRP tubes. If only half the pressure loss was attributed to the feed line, it follows that a 600 m line would have a pressure loss of 10 bar (18 m x 10 bar /0.3 bar = 600 m).

A typical warm up event is shown in Figure 2 (solid line), this corresponds to a heat leak of the order 10-14 W/m which can be associated with a measured vacuum of 0.3 mbar. The heat leak is estimated by relating the slope (@ 100 K) of the warm up curve with physical property data for the cryostat. When the vacuum level is brought down to less than 10^{-3} mbar the heat leak is reduced to less than 5 W/m as deduced from the warming-up curve in Figure 2 (dashed line).

From the curve in Figure 2 it is possible to estimate the thermal time constant of the cryostat to be 150-200 K/24 hours (< 10^{-3} mbar). Starting up the circulation system proved to be relatively fast. When the cooling station and the cable was charged with LN2, the circulation system reached a steady state within 15 minutes.

When a change occurred (e.g. in flow rate or heat input), it was almost immediately apparent in the temperature data. Data analysis showed a time constant for the cable-cryostat system of the order 0.5 – 1 hour.

A typical thermal measurement with the thermocouples is shown in Figure 3. The thermocouple voltage is displayed versus time. The data represents a time period of about 5 hours. In this time the flow rate has been varied from 235 kg/h to 550 kg/h. A small drift in the thermo-voltage from the cable has been eliminated. The drift is of the order 1 μV in 5 hours. The drift can probably be associated with a degradation of the cryostat vacuum. Such a drift is not observed (i.e. no vacuum degradation) in the data from the cable end termination. The observed voltage deviation (zero arbitrarily chosen) corresponds to each change in cooling power of the LN2, i.e. the flow rate. A spread of $\pm$1-2 μV and $\pm$0.5 μV is observed in the cable and in the termination data respectively. This uncertainty could be

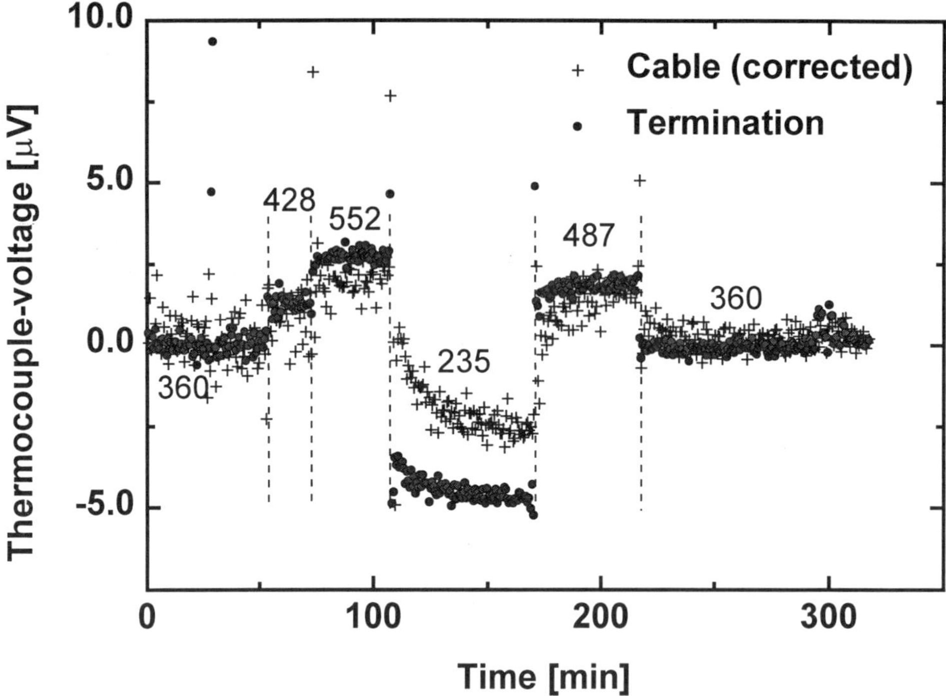

Figure 3. A typical curve showing the thermo-voltage versus time, taken from the LN2 in-stream mounted thermocouples. Cable dat is marked +, termination data is marked with dots.

reduced by averaging over a large number of data points. The data scattering can be associated with switching transients (electronic noise). Using the last 10 to 20 data points before each change in the flow rate the thermo-volttage was correlated with capacity rate ($m \cdot C_P$). The expression is:

$$\Delta U = \frac{1}{C_P} \cdot \frac{1}{m} \cdot A + B \tag{1}$$

where ΔU is the measured thermo-voltage, C_P is the heat capacity of LN2 (2050 J/kg·K), m is the flow rate in kg/s, A is the unknown variable connected to the heat leak and, B, is an off set connected to the arbitrarily chosen zero voltage.

The thermo-voltage is plotted against the flow rate in Figure 4, and fitted with a 1/m type of function (m = flow rate). Results are shown in Table 2. Also shown are temperature differences and heat leaks.

For a scaled down test set-up, the cable end termination represents a major heat leak to the system. The total heat contributed to the closed loop system by the cable end termination originates partly from the heat leak and partly from the heat generation in this component.

Table 2. Fitting parameters, temperature change and the heat leak

	A [W·V/K]	B [V]	ΔT [K]	Heat leak [W]
Cable	0.00199	-5.77E-6	0.108	22.5 (3.2 W/m)
Termination	0.00317	-8.67E-6	0.33	70

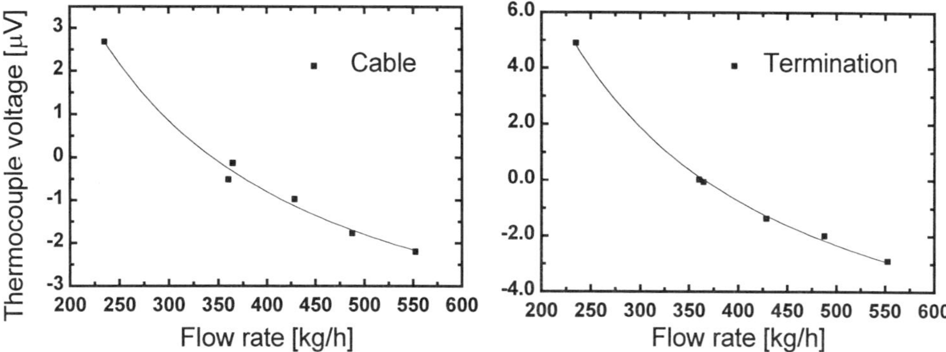

Figure 4. Thermocouple voltage plotted versus flow rate for the in-line mounted thermocouples. Left; Cable conductor. Right; Cable end termination. The solid line is in both cases a 1/m fit.

The heat generation (ohmic loss) in the Cu cylinder and in the joint between this and the superconductor is estimated to be a maximum of 1-2 W based on electrical measurements.

The thermal conductivity of the foam is of the order 0.03 W/m·K which leads to a heat leak of the order 2 @ 30 W. The heat leak along the stainless steel flanges can be estimated to 2 @ 5 W. The heat leak through the vacuum super insulated walls and the current lead (no current) is estimated at 5 W and 50 W for each termination respectively.[15-17] For a rated current of 2 kA the latter number is 85-90 W.[16] Using the numbers given for the feed line, Johnston couplings and the cryostat a total heat leak (excluding the cooling station) can be estimated at about 220 W.

The calorimetric measurements are summarized in Table 3: the cryostat 32 W, cable end termination 2 @ 70 W, feed lines and Johnston couplings 30 W (number from supplier), the electrical separation (GFRP connection) between terminations and feed lines 2 @ 30 W (estimated) which add up to 265 W.

The total loss can also be estimated independently to be 300 W from the volumetric flow and the difference between the inlet and outlet temperature of the cooling station. The discrepancy between the two experimental values may be related to the following circumstances. An uncertainty of 0.1 K on the inlet and outlet temperature corresponds to 20 W. In terms of loss the measured pressure drop of 0.7 bar results in 13 W. Also the data given by the suppliers on the cryostat and feed lines may deviate from actual values.

CONCLUSION

A closed loop circulation system for operating a 10 m long superconducting cable has been developed. The system has performed according to expectation and operating parameters have been obtained. The system has been operated at various LN2 flow rates and different vacuum levels.

Table 3. Heat leak of each constituent of the closed loop

	Heat leak/generation
Termination (x 2)	70 W + 30 W (el. separation)
LN2 feed line	(3 + 15) x 1.1 W
Johnston coupling (x 2)	3.7 W
Cooling station	100 W (@ idle)
Cable conductor	6 W (@ 2 kA, 77 K)
Cryostat	32 W

Thermal time constants of the experimental set-up have been measured. The warm up time is of the order of 1 day. Within 5-10 hours the whole set-up can be cooled down with circulating LN2. The time scale of cable-cryostat system for changes in heat load to be equilibrated is of the order 0.5-1 hour.

The no load heat leak has been extrapolated from thermal data. The heat leak for the cable-cryostat is 3.2 W/m; the heat leak for each cable end termination is 70 W.

The power consumption of the total system (minimum LN2 flow) is about 400 W (cooling station 100 W, remaining circuit 300 W) leading to a standby time of 12 hours.

The sensitivity of the thermocouple measurement in the cable and the termination is 0.02 K (flow rate 360 kg/h). For the cable this corresponds to 0.57 W/m.

The friction losses of the 10 m cable conductor is negligible (cable + termination < 0.1 bar).

ACKNOWLEDGEMENT

The work was carried out within the program *Superconductors in the Danish Energy Sector* and was funded in part by the Danish Energy Agency (Energistyrelsen) and the Electric Utilities of Zealand (ELKRAFT) and ELTRA, Denmark.

REFERENCE

1. H. K. Onnes, *Leiden Comm.* 120b, 122b, 124c (1911).
2. P. A. Klaudy, J. Gerhold, Practical conclusions from field trials of a superconducting cable, *IEEE Trans. Magnetics* 9:556 (1983).
3. E. B. Forsyth, R. A. Thomas, Performance summary of the Brookhaven superconducting power transmission system, *Cryogenics* 26:599 (1986).
4. J. G. Bednorz, K. A. Müller, *Z- Phys.* B64:189 (1986).
5. M. M. Rahman, M. Nassi, L. Gherardi, D. V. Dollen, Design, development and testing of the first factory-made high temperature superconducting cable for 115 kV – 400 MVA, CIGRÉ (1998).
6. J. A. Demko, J. W. Lue, U. Sinha, R.L. Hughey, L. Dresner, S.K. Olsen, Testing of the dependence of the number of layers on the performance of a HTS transmission cable prototype, *IEEE Transactions on Applied Superconductivity*, 9:126 (1999).
7. M. Leghissa, J. Rieger, J. Wiezoreck, H.-W. Neumuller , Advances in Solid State Physics, vol. 38, Vieweg Verlag, 1999.
8. S. Krüger Olsen, O. Tønnesen, J. Østergaard, Power applications for superconductingg cables in Denmark, *IEEE Trans. Appl. Supercond.* 9:1285 (1999).
9. S. Krüger Olsen, A. Kühle, C. Træholt, C. Rasmussen, O. Tønnesen, M. Däumling, C. N. Rasmussen, D. W. A. Willén, Alternating current losses of a 10 meter long low loss superconducting cable conductor determined from phase sensitive measurements, *Supercond. Sci. Technol.* 12:360 (1999).
10. Nordic Superconductor Technologies A/S, Priorparken 685, DK-2605 Brøndby, Denmark.
11. T. Moore, Superconducting Cable Packs a Power Punch, *EPRI Journal*, spring (1999) p8-15.
12. DeMaCo Holland BV, P. O. Box 1067, NL-1700 Heerhugowaard, The Netherlands.
13. C. N. Rasmussen, S. Hutchinson, M. Däumling, D. W. A. Willén, J. Rytter, M. Andersen, K. Høj Jensen, C. Træholt, O. Tønnesen, Bending tests of HTS-cable conductor models, to be published at EUCAS 99.
14. Messer Griesheim GmbH, Industriegase Deutschland, P. O. Box 140140, D-23516 Lübeck, Germany.
15. Yu. L. Buyanov, Current leads for use in cryogenic devices. Principle of design and formulae for design calculations, *Cryogenics* 25:94 (1985).
16. C. N. Rasmussen, C. Rasmussen, Optimization of Termination for a High Temperature Superconducting Cable with a Room Temperature Dielectric Design, *IEEE Trans. Appl. Supercond.* 9:45 (1999).
17. C. Rasmussen, A. Kühle, O. Tønnesen, C. N. Rasmussen, Design of a termination for a high temperature superconducting power cable, *IEEE Trans. Appl. Supercond.* 9:1273 (1999).

OVERALL OPERATING CHARACTERISTICS OF SUPERCONDUCTING CURRENT-FEEDER SYSTEM FOR THE LHD

[1]S. Yamada, [1]T. Mito, [1]R. Maekawa, [1]H. Chikaraishi, [1]A. Nishimura, [1]A. Iwamoto, [1]S. Moriuchi, [1]K. Ohba, [1]T. Baba, [2]T. Uede, [2]H. Hiue, [2]Y. Yasukawa, [2]I. Itoh, [1]T. Satow, [1]S. Satoh, [1] O. Motojima

[1]National Institute for Fusion Science
Toki-shi, Gifu-ken, 509-5292, Japan
[2]Fuji Electric Co., Ltd.
Kawasaki-shi, Kanagawa-ken, 210-8530, Japan

ABSTRACT

A superconducting (SC) current-feeder system is used as the current transmission lines for the LHD. It consists of nine SC bus lines with total length of 497 m and nine pairs of gas-cooled current leads. The cooling process is automatically switched from forced-flow of two-phase helium to pool boiling, when faults happen. Liquid helium consumption of the innermost tube of the SC bus line was negligible, because the liquid helium in the second inner channel absorbed the heat load of 0.34 - 0.42 W/m. Mass flow rates of the current leads could be controlled in constant values, whenever the pressure in the recovery line reached up from 0.106 MPa to 0.133 MPa. The allowable current carrying time on the fault protection mode is expected to be longer than 30 minutes of the design value. We successfully demonstrated that the SC current-feeder system with high current capacities was reliable and safe, and was useful for the SC experimental fusion device.

INTRODUCTION

The LHD is a large-scale SC toroidal fusion device of heliotron type.[1] The main objective of the LHD is to demonstrate the high potentiality of the helical-type device producing current-less steady-state plasma with a sufficiently large Lawson parameter without large current disruption. The body of the LHD consists of a pair of helical SC coils, three sets of poloidal SC coils and their supporting structure with a total cold mass of 820 tons. Each helical coil is divided into three block coils that can be excited independently. The upper and lower poloidal SC coils are connected in series. Therefore, nine SC bus lines with average length of 55 m are used for the current transmission lines between the SC coils and their power supplies.

The application of the SC bus lines leads to a decrease of the electrical power consumption for the power supplies and a reduction for installation work on site. Space for the heating devices and/or diagnostics apparatus around the LHD enable to expand, by locating the current leads apart from it. To develop such an SC current carrying system, a full scale model of an SC bus line with 20 m long was made. Experiments were conducted to study the characteristics of a full-scale model. The SC cables for the bus line were cryogenically stable at the rated current of 30 kA, and excited up to 40 kA without quenching.[2]

The construction of the LHD, including the SC current-feeder system, finished in February of 1998, and the experiments for the plasma confinements started on 31 March. The improved

Table 1. Parameters of the SC bus lines for the LHD.

Items	Specifications
SC Bus line	
Number of SC bus lines	Nine
Total length of nine SC bus lines	497 m
Rated current	dc 32 kA
Withstand Voltage	dc 5 kV @ 77 K helium gas
Minimum bending radius	1.5 m
Cryogenic transfer-line	five corrugated SUS tubes
Inserted SC cable	
Al : Cu : NbTi	8.4 : 0.5 : 0.5
Cross-section of aluminum	500 mm^2
RRR of aluminum	>2000
Number of SC strand wire	Nine
Numbers of SC filaments in a strand wire	>4000

Table 2. Descriptions of cryogenic transfer line for the SC bus line

Tube	Size of corrugate tube	Temperature	Purpose
Innermost tube	62 / 68 mm ϕ	4.4 K	2-phase He (S), SC cables
Second inner tube	77 / 85 mm ϕ	4.4 K	2-phase He (R)
Third tube	100 / 110 mm ϕ	50 – 80 K	Vacuum insulation space
Forth tube	130 / 143 mm ϕ	50 – 80 K	Shielding gas (S or R)
Outermost tube	198 / 220 mm ϕ	Room temp.	Vacuum insulation space

Remarks: (S); supply channel, (R); return channel

plasma confinement properties could be obtained through the first and second experimental cycles in the first year.[3] During the cooling-down of the second cycle, power interruptions caused by thunder storms occurred two times, and the main compressors and the turbines of a helium refrigerator/liquefier system were immediately tripped. On the other hand, a partial normal transition was observed in the SC conductor of the H1-I block coil, when engineering coil excitation tests under the high magnetic field regime were conducted. The fault protection programs in the power supplies, the cryogenic system and the center control system were successfully executed.

This paper describes the overall operating characteristics of the SC current-feeder system. At first, a helium flow circuit, its control scheme, and the operational properties in the steady state are presented. Then, the characteristics of the fault protection mode on the SC current-feeder system are described. Finally, the allowable current carrying time on the fault mode is also discussed by considering the heat load into the SC bus lines.

SC CURRENT-FEEDER SYSTEM

The SC current-feeder system has a sufficient safety margin exceeding that of the SC coils of the LHD, because the stored magnetic energy of the coils must be extracted safely through the SC bus lines when the halts happen. The system is designed to maintain its rated current carrying capacities for 30 minutes, even if the coolants supplied to the SC current-feeder system are accidentally stopped. The SC bus lines are designed to be flexible since the route from the SC coils to their power supplies have many corners and bends. The minimum bending radius is determined to be 1.5 m, because of the restrictions on height and width.[4]

An aluminum-stabilized, SC-compacted stranded cable that is the same size and same structure of the full-scale model, was applied to the SC bus lines for the LHD. A pair of SC cables was insulated electrically, and inserted into the cryogenic transfer line. To decrease the heat load into the SC bus lines, five-corrugated stainless-steel tubes with a thermal shielding channel were used. The specifications of the SC bus lines are listed in Table 1. Main parameters of the cryogenic transfer line are listed in Table 2.

Helium Flow Circuit and Its Control Scheme

Figure 1 shows a helium flow circuit of an SC current-feeder system for the LHD. The system is composed of two subcooler tanks, nine pairs of current leads and nine sets of SC bus lines,

Table 3. Specifications of the current leads

Target coil	Helical coils	IV, IS coils	OV Coil
Number of current leads	6 pairs	2 pairs	1 pair
Rated current	18 kA	22 kA	32 kA
Rated mass flow	1.1 g/s	1.2 g/s	1.9 g/s
(without current)	(0.8 g/s)	(0.9 g/s)	(1.4 g/s)
Length	1.5 m	1.5 m	1.5 m

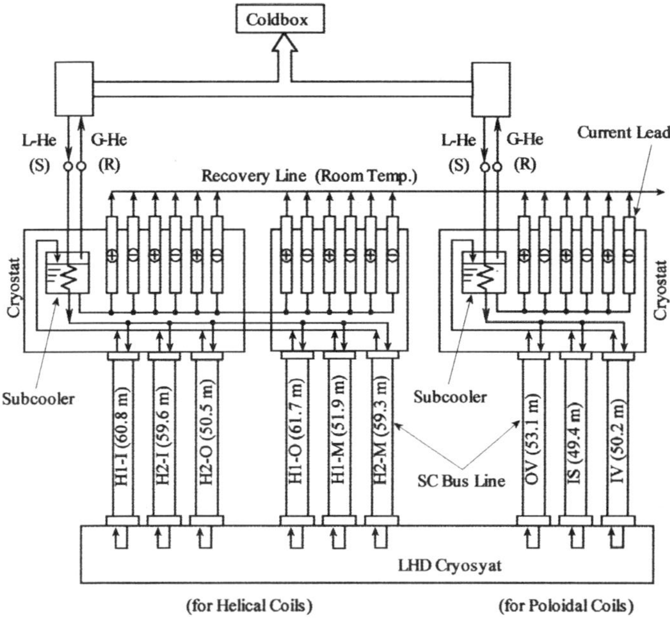

Figure 1. Helium flow circuit of the SC current-feeder system for the LHD.

including the cryogenic valves and their controller. The coolants of the SC current-feeder system are separated from that of the SC coil system of the LHD. The conventional gas-cooled current leads with the helium storage are applied to attain the stable current carrying properties at a rated current. Main parameters of the current leads are listed in Table 3.

Each process of a cooling-down, steady state and warming-up are controlled by the sequential program that is linked together with the program for the helium refrigerator/liquefier system. Four control programs are prepared for the cooling-down process; 1) initial cooling-down by the temperature controlled gas of 50 degrees lower than that of the SC bus lines, 2) cooling-down of the thermal shielding lines, 3) switching over the recovery line from room temperature to low temperature one, and 4) change the coolants from helium gas to liquid helium.

Figure 2 illustrates a control scheme of a helium flow for the SC current-feeder system. In steady-state, two-phase helium from the coldbox is cooled to 4.4 K by the heat exchanger in the subcooler tank, and is supplied to the SC bus lines. Returned two-phase helium from the SC bus line flows into the subcooler, and separates into liquid and gas phase. A part of the liquid helium cools the current leads. Most of the liquid helium evaporates by the heaters, and returns to the coldbox. The PID compensators for the cryogenic valves and the heaters adjust automatically for the manipulated values of the pressures in the subcoolers, and mass flow rates of the SC bus lines and current leads, and the liquid helium levels in the subcoolers. In steady-state, these heaters automatically adjust the overall heat balance between the SC current-feeder system and the refrigerator/liquefier system.

Steady State Operation

The experiments for the plasma confinements under 1.5 T, and the wall conditioning for the plasma vacuum vessel under 0.0875 T had been conducted alternately during the first experimental

Table 4. Main parameters of the steady state operations

Items	Specifications
SC Bus line	
Mass flow rate of two-phase helium	8.0 – 12.0 g/s
Mass flow rate of thermal shielding helium gas	1.8 – 2.2 g/s
Temperature of two-phase helium	4.4 – 4.5 K
Temperature of shielding helium gas	51 (inlet) – 69 (outlet) K
Subcoolers	
Pressures	0.115 – 0.118 MPa
Liquid helium levels	60 %

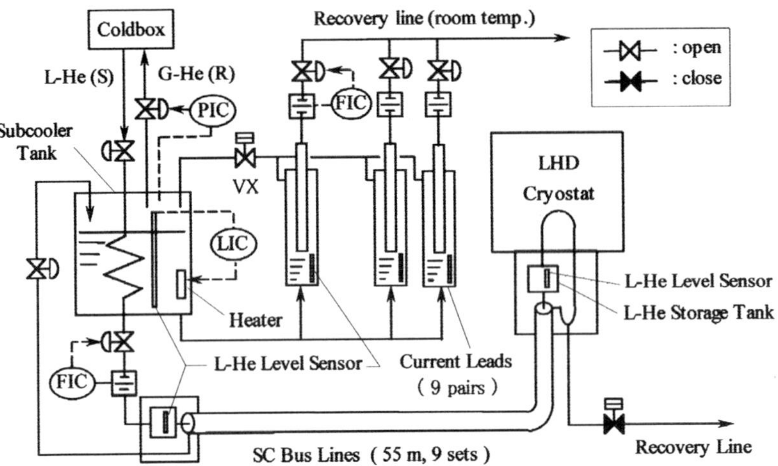

Figure 2. Control scheme of the helium flow in the SC current-feeder system.

cycle. On the second cycle, wide variations of the coil excitations, i.e., magnetic field from 1.5 T to 2.7 T, the changes of magnetic axis from center to inner suitable positions and/or changes of the elongation ratio for the plasma surface were performed. Before the plasma experiments under the high magnetic field regime up to 2.5 T, engineering excitation tests using magnetic field higher than the operational magnetic fields had been conducted.

Figure 3 gives a typical example of the steady-state operation that is the final plasma experiments on the second stage. The liquid helium level in the subcooler had been kept in constant value of 60 % by the control of the heater power. The mass flow rates in the current leads, the pressure difference between the subcooler and recovery line, and the current carrying rates influence the cooling conditions of the current leads. However, both liquid helium levels in the subcooler and current leads will automatically equalize by a siphon method without the pressure adjustment in the subcooler, when the valve VX shown in Fig. 2 is open. Even if the mass-flow rates of the current leads are less than that of the self-cooling condition, the evaporated helium gas from the current leads flows into the subcooler through the valve VX, and returns to the coldbox. To avoid the propagation of ice on the current leads, the mass-flow rate in the current leads had been decreased to 70 % of the rated value during the night. Main parameters of steady-state operations in the second cycle are summarized in Table 4.

FAULT PROTECTION MODE

The SC current-feeder system should be reliable and safe, even if faults such as power interruption during storms, compressor trip by momentary voltage drop, and other accidents in the cryogenic system occur. The coil-supporting structure and the helical-coil vessels do not have one-turn breaks, so the large volume of liquid helium will evaporate by ac losses, when the rapid current shutdown (called "1Q") occurs.

Receiving a "1Q" signal, a helium refrigerator/liquefier system is separated immediately from

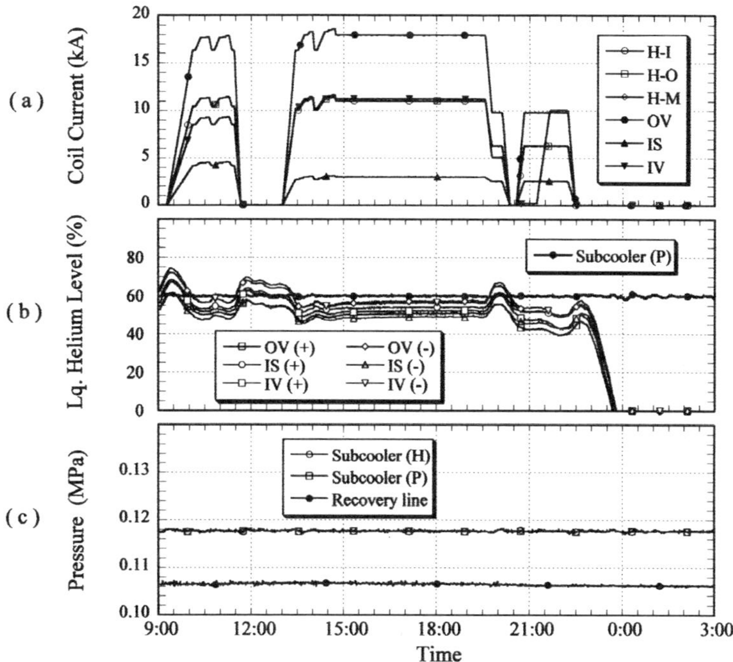

Figure 3. Typical examples of the steady state operation: (a) coil current, (b) liquid helium levels of subcooler and current leads, and (c) pressures of subcooler and recovery line.

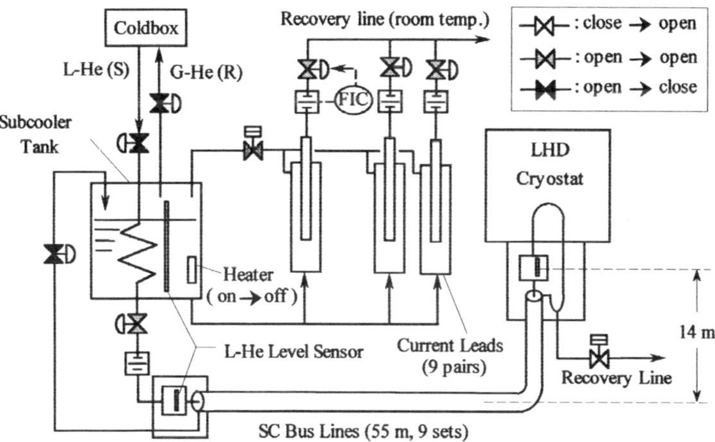

Figure 4. Control scheme of the fault protection mode.

the SC coils and SC current-feeder system, to avoid the compressor trip caused by the pressure increase by the evaporated helium gas return. On the contrary, the cooling process of the SC current-feeder system is automatically switched to the fault protection mode, as shown in Fig. 4.

Example of Rapid Current Shutdown

Figure 5 (a) shows a typical example of the rapid current discharge from 2.75 T to zero with a time constant of 20 sec. The cooling process of the SC bus lines was switched from forced flow of two-phase helium to pool boiling as shown in Fig 5 (b). The heater powers in both subcoolers stopped in force to eliminate the liquid helium consumption as shown in Fig. 5 (c). It was confirmed

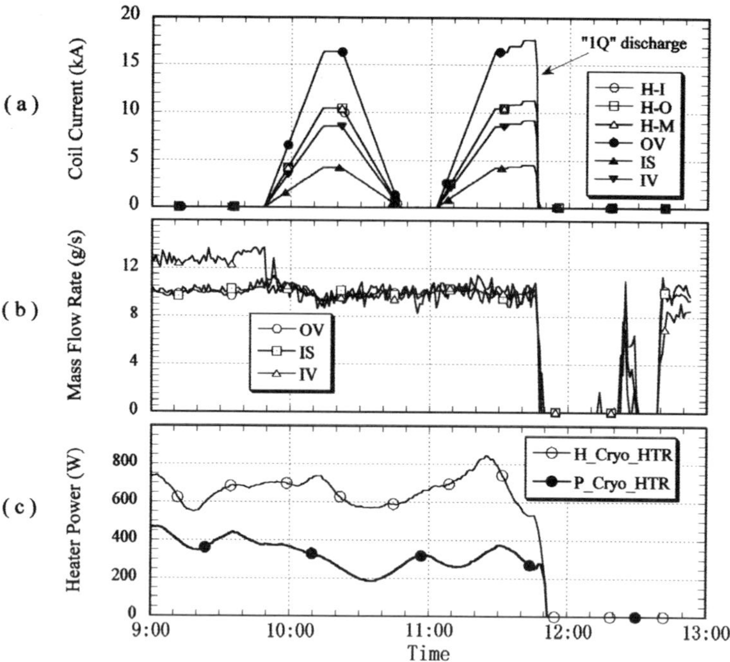

Figure 5. Typical examples of the fault protection mode: (a) the coil excitation currents, (b) mass flow rates of the SC bus lines, and (c) heater powers in the subcoolers.

that the cooling process in steady state was safely switched to that of the fault protection mode.

Figure 6 shows a typical example of the "1Q" influence; (a) the pressures in the subcooler tanks and recovery line, (b) the mass flow rates in the current leads, and (c) the liquid helium levels of the coil-side terminals of the SC bus lines, when the "1Q" protection as shown in Fig. 5(a) occurred. The pressure in the recovery line increased immediately from 0.106 MPa to 0.133 MPa by the ac losses of the supporting structure and helical coil vessels, and Joule heat loss of the normal transitioned H1-I block coil. On the contrary, the pressure in the subcoolers also increased from 0.118 MPa to 0.138 MPa through the valve VX, so that the cooling of the currents leads could be sustained by the self-cooling conditions. The most parts of liquid-helium levels in the coil-side terminals of the SC bus lines could be kept in 100 %. The liquid-helium level of the H1-I terminal was slightly decreased by thermal conduction from the H1-I block coil. However, the SC current-feeder system could be recovered to the steady-state condition within 30 minutes.

Heat-Load Measurements

Heat loads into the shielding gas channel (50-70 K) of the SC bus lines were measured simultaneously during the steady-state operation of the first cycle. The average heat load was 2.8 W/m. This value is less than that of the design value.[5] Heat loads from the shielding gas channel to the two-phase helium channel (4.5 K) were measured by the calorimetric method just before the warming-up of the LHD in the first experimental cycle. The test was conducted on the same condition as that shown in Fig. 4. The test results are shown in Fig. 7; (a) the mass flow rates of two-phase helium in the poloidal SC bus lines, (b) the liquid-helium levels of the coil-side terminals, and (c) the mass-flow rate of a evaporated helium gas in the recovery line from the coil-side terminals.

The liquid helium in the coil-side terminals could be sustained in 100 % during one half hours, when the coolant supplied to the SC bus lines was stopped. The liquid helium consumption of the innermost tube was negligible, because the liquid helium in the second inner channel absorbed the heat load from the shielding gas channel. In the early phase, the system switched to pool boiling (region (I) in Fig.7(c)), and the mass-flow rate of the evaporated helium gas increased transiently because of the thermal conduction from the recovery line. The measured mass flow rates were 2.5 -

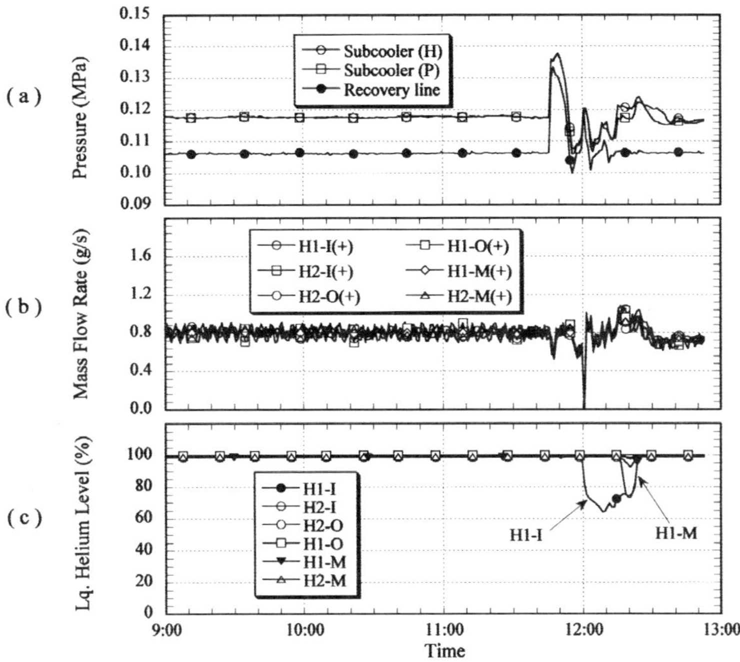

Figure 6. The influences for the fault protection: (a) pressures in the subcoolers and recovery line, (b) mass-flow rates of the current leads, and (c) liquid-helium levels in the coil-side terminals.

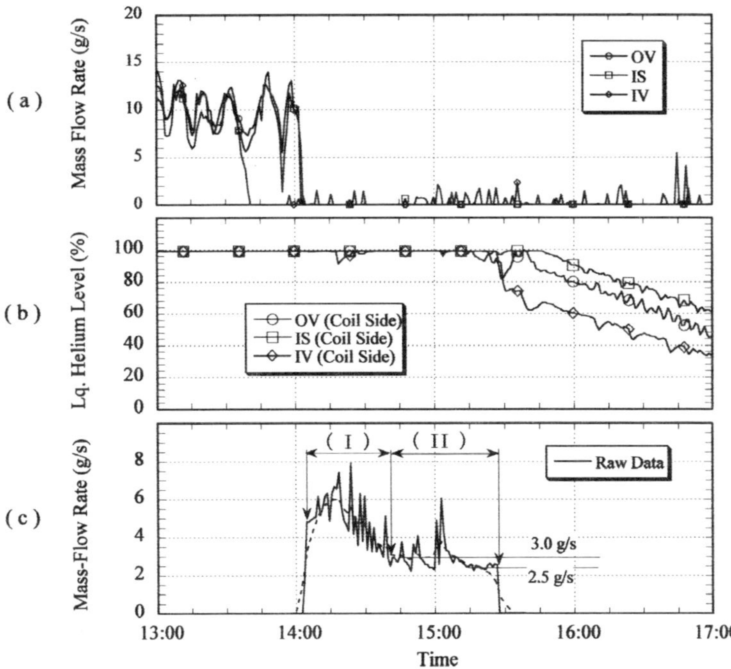

Figure 7. Test results: (a) the mass-flow rates of the SC bus lines , (b) liquid-helium levels of the coil-side terminal , and (c) mass-flow rate of the evaporated helium gas.

Table 5. Summary of the first and second experimental cycles

Items	Values
First Experimental Cycle	
Total shot numbers of plasma productions	1888 shots
Total times of coil excitations	692 hours
Second Experimental Cycle	
Total shot numbers of plasma productions	5244 shots
Total times of coil excitations	442 hours

3.0 g/s in the quiescent period (region (II) in Fig.7(c)). The total length of the tested SC bus lines is 152 m, so that the averaged heat load into the second inner channel was estimated to be 0.34 – 0.42 W/m. These values are nearly equal to that of the design value of 0.3 W/m.

The SC current-feeder system should be reliable, so that the fully cryostable superconductor at a rated current of 32 kA with a large cross section of aluminum (500 mm^2) was applied to the SC bus lines. The SC cables are cryostable, while they are soaked in liquid helium.[6] Therefore, the allowable current-carrying time with a rated current on the fault-protection mode is expected to be longer than 30 minutes of design value.

Table 5 shows a summary of the operations of the first and second experimental cycles. Wide variations of the coil excitations for the plasma-confinement studies, the engineering excitation tests in the high-magnetic regime, and/or long-term excitation for the wall conditioning of the plasma vacuum vessel (five days) were conducted in the first year after construction. The total time of the coil excitations exceeds 1100 hours in a year. The SC current-feeder system has been successfully operated with no trouble.

CONCLUSION

The overall operational characteristics of the SC current-feeder system for the LHD were investigated. The results are concluded as follows: (1) Mass-flow rates, pressures and liquid-helium levels of the SC bus lines and current leads are well controlled automatically in steady-state operation. (2) The cooling process is automatically switched from forced-flow of two-phase helium to pool boiling, when faults happen. (3) The mass-flow rates of the current leads could be controlled in constant value, even if the pressure of the recovery line was immediately increased. (4) The liquid helium consumption of innermost tube of the SC bus line was negligible, because the liquid helium in the second inner channel absorbed the heat load from the shielding helium gas channel. (5) We have demonstrated successfully that the SC current-feeder system with high-current capacities was reliable and safe, and was useful for the SC experimental fusion device.

ACKNOWLEDGEMENTS

The authors would like to express their appreciation to Professors S. Tanahashi, S. Imagawa, K. Takahata, S. Kitagawa and K. Murai, to Drs. N. Yanagi and H. Tamura, and to the other members of Engineering Division in NIFS for their kind advice and encouragement.

REFERENCES

1. M. Fujiwara, et al., "Large Helical Device (LHD) Program", *J. Fusion Energy*, 15: 7 (1996).
2. T. Mito, et al., "Development and Tests of a Flexible Superconducting Bus-line for Large Helical Device", *IEEE Trans. Magn.*, 30: 2090 (1994).
3. 3. O. Motojima, et al., "Initial Physics Achievements of Large Helical Device", *Physics of Plasmas*, 6: 1843 (1999).
4. S. Yamada, et al., "Superconducting Current feeder System for the Large Helical Device", *IEEE Trans. Magn.*, 32: 2422 (1996).
5. S. Yamada et al., "First cycle operation of SC current Feeder system for the LHD", *Proc. 17th ICEC*, Bournemouth, UK (1998) pp.443-446.
6. T. Mito et al., "Stability and Safety Estimates and Tests of a Superconducting Bus-line for Large Scale Superconducting Coils", *IEEE Trans. Applied Superconductivity*, 5: 917 (1995).

TESTING OF AN IMPROVED DESIGN NITROGEN HEAT PIPE THERMAL INTERCEPT FOR CURRENT LEADS

E. Roth, F.C. Prenger, J. Stewart, and P. Blumenfeld

Los Alamos National Laboratory
Los Alamos, New Mexico, 87545

ABSTRACT

With the increasing use of cryocoolers to cool superconducting magnets, attention is being focused on the problem of providing a thermally efficient intermediate heat sink for the current leads which also provides electrical isolation. The thermal intercept is a critical component for low heat leak HTS current leads because it significantly reduces the refrigeration requirement. Usually these two requirements are mutually exclusive for solid dielectric materials and one must accept a poorer thermal interface if electrical isolation is required. Further, the traditional methods of vapor cooled leads are not an option in a system cooled by a cryocooler because cryogens are usually absent. A novel method, developed at Los Alamos National Laboratory (LANL), utilizing gravity assisted heat pipes has been shown to circumvent this problem. We describe the fabrication and testing of a nitrogen-filled heat pipe for cooling of a current lead while providing electrical isolation. Previous efforts by our group at LANL had successfully produced a similar device but there were problems in creating and maintaining a reliable epoxy seal under the high pressure of the nitrogen gas and the stress of thermal cycling. A modified design and assembly method has helped but not completely solved the problem. Measurements of the thermal performance along with results from our heat pipe thermal model are presented. Peak heat transfer rates exceeding 150 watts were observed with good agreement with the thermal model. Testing of the electrical isolation between the heat pipe and the current lead was also performed and exceeded 1500 volts.

INTRODUCTION

Thermal intercepts are key components of low-heat-leak current leads because they enable significant reduction in overall refrigeration requirements.[1,2] The current leads for this project were conductively cooled and operated in a vacuum. The principal challenge in designing and building thermal intercepts for current leads is determining how to combine good electrical insulation with high thermal conductivity. Work at Argonne National Laboratory and Fermi National Accelerator Laboratory on low thermal resistance, high electrical isolation heat intercept connections has addressed this problem.[3,4] Niemann, Gonczy, and Nicol have developed a heat intercept design, which involves clamping, by thermal interference fit of an electrically insulating cylinder between an outer metallic ring and an inner metallic disc. The connections use the high clamping pressures made possible by the thermal interference fit to reduce the thermal resistance of joints between conducting and insulating materials.

We have been experimenting with composite heat pipes constructed of conducting and non-conducting materials for several years in an effort to develop an alternative approach to high thermal conductivity intercepts with good electrical isolation. We believe that our present heat pipe thermal intercept offers significant performance advantages in many cases.

HEAT PIPE OPERATION

Though this design is different from conventional heat pipe designs in that the condenser and evaporator are not at opposite ends of a sealed tube, the operating principles are similar. This device is more properly termed a gravity-assisted heat pipe or thermosyphon because it uses gravity rather than capillary action of a wick structure to return the liquid to the evaporator section. We use the generic term heat pipe as interchangeable with gravity-assisted heat pipe for all references to our design. A hollow copper cylinder with disc-shaped end caps serves as the heat pipe condenser. A solid copper center-rod (part of the current lead) penetrates the end caps and serves as the evaporator. A non conducting epoxy seal between the center-rod and the end caps forms a sealed chamber for the heat pipe working fluid. During operation, heat is transferred either by conduction, convection or boiling from the center rod to the liquid nitrogen in the bottom of the heat pipe cavity. The liquid then evaporates and the vapor flows upward where the heat is removed from the nitrogen vapor in the upper cavity by condensation on the outer wall. The liquid then flows back down to the evaporator reservoir. Because the heat pipe relies on two-phase heat transfer the operating range is between the triple point and critical point of the working fluid. In the case of nitrogen, that means an effective operating range of about 64 to 125 K. Below 64 the nitrogen freezes and heat transfer is though conduction only, above 125 K there is no longer any liquid remaining in the evaporator to vaporize. Based on the thermophysical properties of nitrogen, the best performance occurs at about 100 K. The evaporation and condensation mechanisms combine to allow high heat transfer rates with very low temperature differences. However, because these mechanisms are dependent on the heat flux temperature differences in the heat pipe can be quite variable. Also, there are upper limits to the heat load than can be carried that are also temperature dependent. For cryogenic heat pipes the limit is usually due to flooding of the heat pipe condenser wall by the upward flowing vapor stream. In severe cases this can result in entrainment of the liquid by the vapor. In all cases the result is a dryout of the evaporator and interruption of the fluid circulation. This limit is referred to as the flooding limit.

HEAT PIPE DESIGN AND ASSEMBLY

Because our heat pipe design has been reported in detail elsewhere[5], this section will contain a design summary and will point out differences from earlier designs. The most critical aspect of the design of the heat pipe is to ensure the mechanical and electrical integrity of the interface between the conducting and non-conducting portions of the assembly. We have chosen an epoxy to serve as both the nonconducting portion of the heat pipe and as the interface. This joint must be able to contain up to 138 bar (2,000 psi) of nitrogen gas, which is the room temperature filling pressure of the heat pipe. In addition, it must also remain leak tight during the cool down from room temperature to its operating temperature of 65-125 K. Figure 1 shows a schematic of the heat pipe. The high-pressure nitrogen charge is contained by the solid heat pipe body and the solder cap. These two ETP copper components are assembled with a silver-solder joint. The thickness of the epoxy joint between the heat pipe body and the center rod is minimized to reduce the forces exerted on it by the high pressure

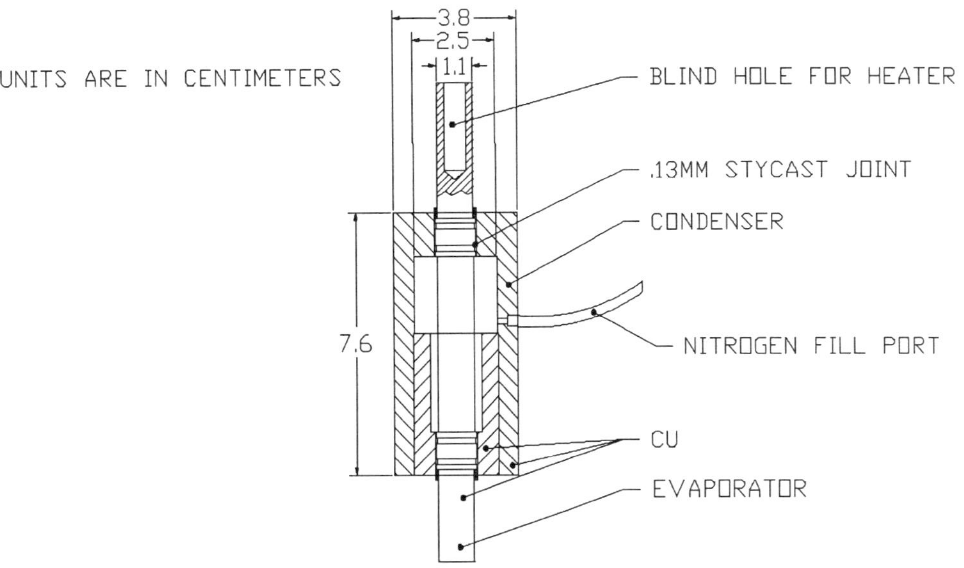

Figure 1. Heat pipe schematic

nitrogen gas and to insure that the joint does not fail due to differential thermal contraction on cool down. The joint, on the other hand, also has the opposing constraint that it must be thick enough to maintain electrical isolation The joint in this design is 0.127 mm (0.005") as opposed to our earlier design of 0.508 mm (0.020"). With this seal thickness and a 15,100 V/mm (385 V/mil) voltage rating for the Stycast 2850 FT epoxy at room temperature, the calculated electrical isolation is 1925 V. The joint is made using epoxy that has been degassed in a vacuum chamber to remove any bubbles that could compromise electrical isolation or the pressure seal. Cleaning of the copper surfaces exposed to the epoxy with an acid bath was also performed. Any soldering done on the heat pipe was completed before the epoxy joints were made. No soldering could be done on the heat pipe after filling because of excess pressure resulting from heating the nitrogen gas. Finite element modeling has shown that the maximum fill pressure (with a safety factor of 4) for this geometry is 179 bar (2600 psia). The joint was helium leak checked after the epoxy had cured at a slightly elevated temperature. In checking for leak tightness, this joint was not cold shocked by sudden immersion in liquid nitrogen as was done for the previous design.

Filling the heat pipe with 138 bar (2,000 psig) nitrogen gas at room temperature was the final step in the assembly. The shape of the cavity between the center rod and the heat pipe body was determined by thermal design requirements. When the heat pipe is cooled to its operating temperature the nitrogen will exist in the cavity as a two-phase fluid. The liquid nitrogen will just fill the well in the bottom of the cavity. This well was designed to optimize the area required for boiling heat transfer on the center rod with the upper area required for condensation on the heat pipe body. By making the heat pipe body in two pieces the center cavity with the required geometry, as discussed above, could be easily machined.

THERMAL TESTING

The heat pipe outer cylinder (condenser section) was clamped in a copper fixture with indium foil and heat sunk to a liquid nitrogen or argon bath. The cryogen bath could be pressurized or pumped on to control the temperature. The overall temperature could also be adjusted with a heater that was placed on the mounting block clamp. A schematic of the heat pipe test apparatus is shown in Figure 2. Though the rig relied on boiling cryogens as the heat sink, the system's thermal resistance was large enough that it limited our ability to test at the lower temperatures under moderate heat loads.

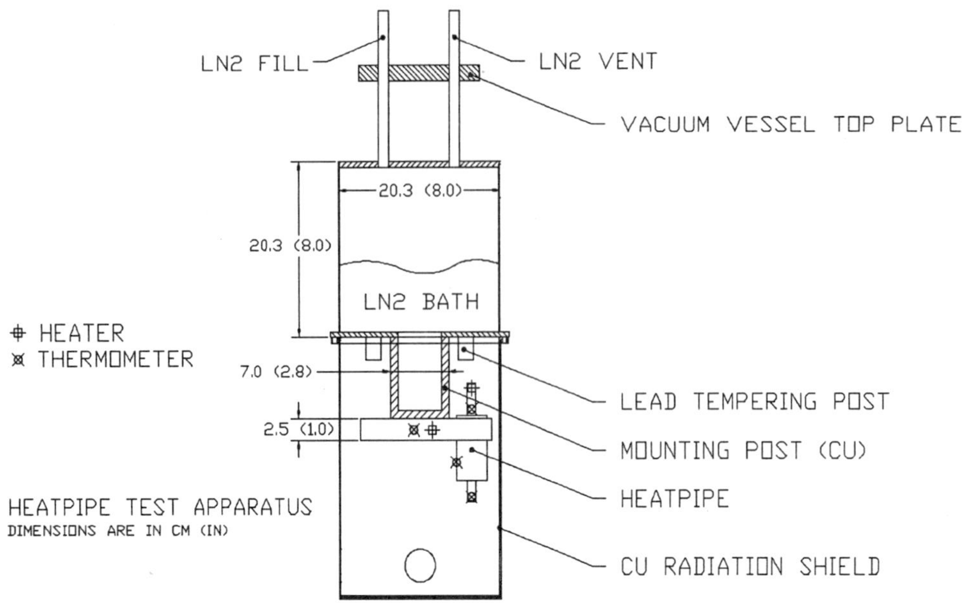

Figure 2. Heat Pipe Test Rig.

The thermal performance of the heat pipe was determined by measuring the heat input to the center conductor and the resulting temperature difference between the center conductor and the outer cylinder. An electrical heater with voltage taps was attached to the top end of the heat pipe center rod as this best simulated where a copper current lead and its associated heat load would be located. The heater power was determined by measuring the instantaneous voltage and current of the heater. Platinum resistance thermometers were used to measure the temperature difference across the heat pipe. They were mounted on the bottom end of the heat pipe center rod and on the heat pipe condenser near where the indium-filled clamp joint was made.

The steady state temperature difference across the heat pipe was measured as a function of the heater power. Conductances were calculated by dividing the heater output by the temperature difference between the evaporator and condenser section of the heat pipe. Data were taken under two conditions, 1) the condenser temperature was fixed and the heat pipe temperature was determined as a function of evaporator heater power and 2) the maximum heater power (before evaporator dryout) was found for a fixed condenser temperature. These performance limits for the heat pipe. Figure 3 shows the temperature difference between the evaporator and condenser as a function of heater power. In general, the differential change in the delta T's were largest at the low heat fluxes and decreased as heater power was increased. Our earlier work showed this effect more clearly; whereas, the present testing did not obtain much data at the lower power settings where the heat pipe turned on. Also, the present heat pipe configuration had a higher joint thermal conductance because of the thinner epoxy gap, which tended to mask the nonlinear turn-on effect. The data from Figure 3 were replotted as thermal conductance versus heat transfer rate in Figure 4. The thermal conductance increased from of 4 W/K at low heat transfer rates to 28 W/K at 170 W. The high residual value even when the heat pipe was "off" can be attributed to the two thin 0.127 mm, [0.005"] Stycast epoxy joints in the heat pipe. The results show that as the condenser temperature rises, the overall conductance increases as well. This effect is also predicted by the model, though at certain temperatures the turn-on effect occurs faster than the model predicts. This model is based on a conventional heat pipe layout where the evaporator and condenser are distinctly separated by an adiabatic section. This heat pipe does not have an adiabatic section. These values can be compared to a thermal conductance of 3.75 W/K (12 K temperature difference at 45 W heater power) reported by Argonne National Laboratory and Superconductivity, Inc. for a thermal interference fit current lead heat intercept. [4]

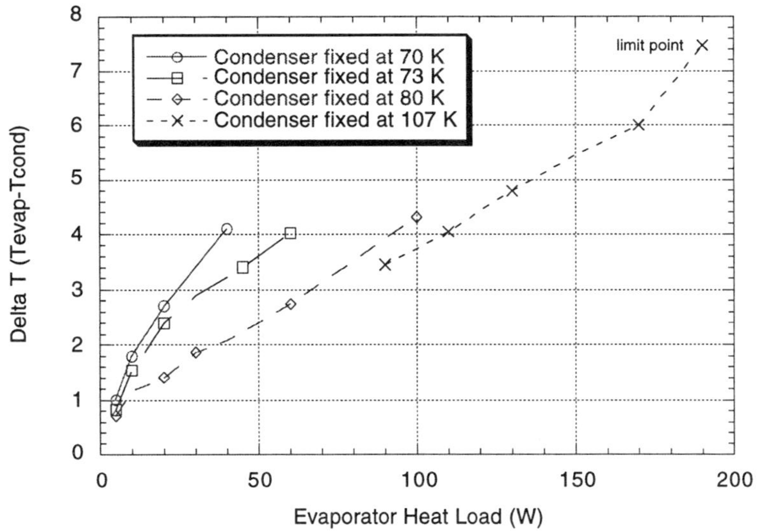

Figure 3. Experimental data of heat pipe temperature rise at various heat loads.

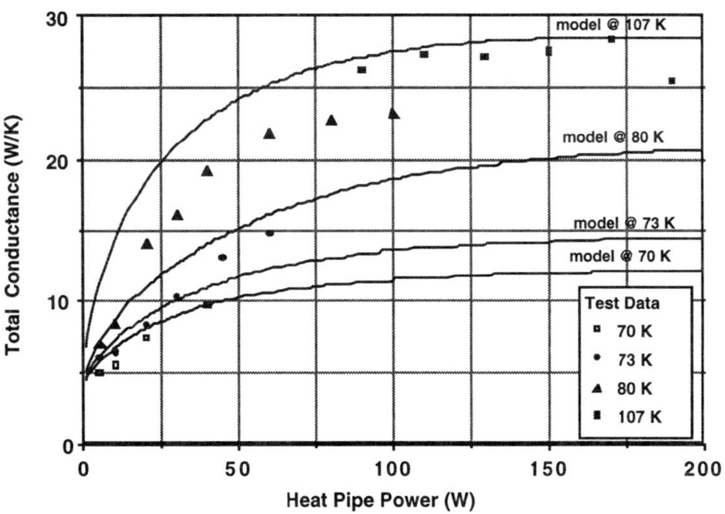

Figure 4. Experimental data and model results of heat pipe thermal conductance.

Figure 5 shows the liquid fraction in the heat pipe cavity as a function of the room temperature fill pressure and temperature. Liquid volume is important because it determines the liquid height in the evaporator. Increased wetted surface area on the evaporator reduces the boiling thermal resistance between the evaporator and the liquid. Further, if the fill pressure is too low and the heat pipe temperature too high, the evaporator will dry out. Dry out causes the heat pipe to cease operation. Because of the concentric design of this heat pipe, as opposed to traditionally designed heat pipes, the overall conductance is a trade off between the evaporator area and the condenser area. That is, as the evaporator liquid level decreases there is a corresponding increase in condenser area.

Figure 6 shows the conductances calculated by the model. It is a more detailed look at the "107 K" curve in Figure 4. The significant parameters in the heat pipe model include the heat transfer coefficients for filmwise condensation and nucleate boiling, as well as, the conduction through the thin epoxy gap separating the evaporator and condenser sections at the ends of the heat pipe. The film thickness was taken to be the average of the thickness predicted along the length of the condenser based on a falling film model. This assumes that the film gradually increases from zero thickness at the top of the condenser to some nonzero value at the bottom. The calculations indicate that at 107 K and above 80 watts the overall conductance remains constant at 24 W/K. At lower temperatures the conductance decreases.

Figure 7 shows performance limits for the heat pipe at higher temperatures. These points are maximum allowable heat loads for a given temperature and they correspond to where the evaporator dries out because of the flooding limit discussed previously. Performance limit data were only obtained at the higher temperatures because the heat pipe performance exceeded that of the test rig at lower temperatures. The solid curve is a heat pipe model prediction of the flooding limit. It agrees with the data fairly well.

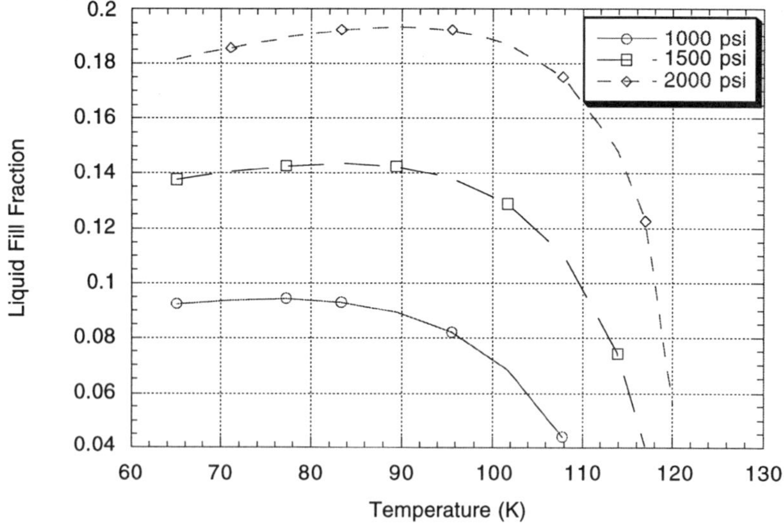

Figure 5. Calculated nitrogen liquid fraction at various temperatures and room temperature fill pressures.

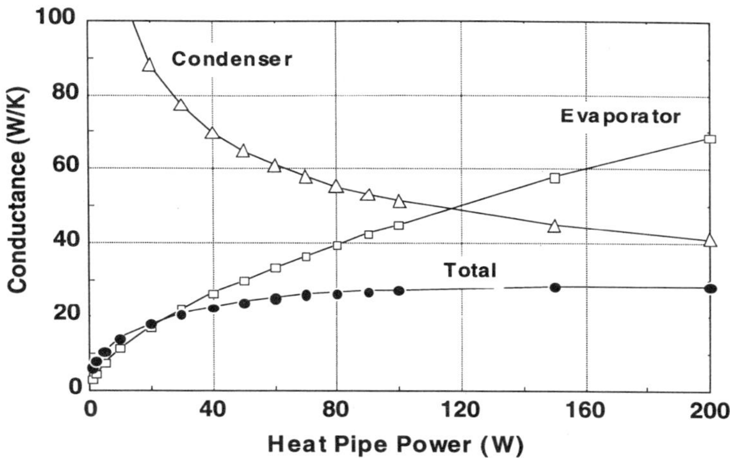

Figure 6. Heat pipe model results of heat pipe thermal conductance.

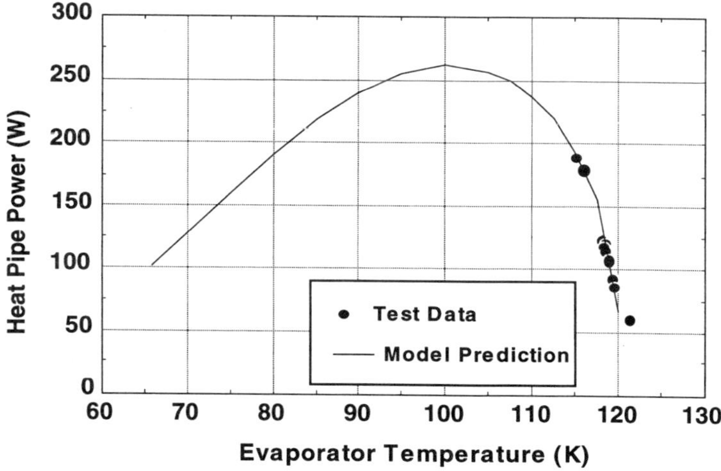

Figure 7. Experimental and calculated values of heat pipe performance limits at various temperatures.

ELECTRICAL TESTING

Electrical performance measurements were obtained by measuring the leakage current across the heat pipe as a function of the applied voltage. A positive voltage was applied to the heat pipe evaporator (center rod) with the heat pipe condenser (body) connected to electrical ground. Concerns that the bubbles produced by boiling at the heat pipe evaporator or center rod would significantly degrade electrical performance proved to be unfounded. Previously, measurements were made at room temperature with no heat load, however, no electrical breakdown was observed with applied voltages as high as 1,500 volts.

CONCLUSIONS

We have characterized the ability of a gravity-assisted heat pipe to function as a current lead thermal intercept. Near 100 K the device can transfer 100 W with a 4.3 K temperature difference, which gives a thermal conductance of 23 W/K, significantly higher than designs which rely on thermal conduction through solids. At lower temperatures (< 75 K) more pertinent to HTS materials, conductances exceed 10 W/K. The heat pipe thermal intercept successfully withstood 1,500 volts without electrical breakdown. Further work is being pursued to characterize the thermal conductance over the full range of the device. Also, a new glass pressure/dielectric joint is being designed so as to improve the joints reliability and dielectric strength. The rugged and compact heat pipe design can be applied to a variety of current lead configurations. The ability to vary the shape and size of the evaporator and condenser heat transfer surfaces allows the device to be customized to meet a wide range of design requirements. This flexibility is important because of the heat pipe's variable thermal conductance in certain temperature and heat load regimes. It should be possible to cool very large current leads using this design while maintaining a very compact geometry. Finally, the heat pipe is built of inexpensive materials and does not require precision tolerances or expensive machining for its components.

ACKNOWLEDGMENTS

This work was performed under the auspices of the United States Department of Energy, Office of Energy Management.

REFERENCES

1. K.Ueda, et al. Thermal Performance of a Pair of Current Leads Incorporating Bismuth Compound Superconductors, *Fifth International Symposium on Superconductivity*, Kobo, Japan, November 16-19, 1992.
2. R. Wesche and Am.M. Fuchs, Design of Superconducting Current Leads, *Cryogenics*, Volume 34, Number 2, p. 145, 1992
3. R.C. Nieman, J.D. Gonczy and T.H. Nicol, Low Thermal Resistance High Electrical Isolation Heat Intercept Connect, *Advances in Cryogenic Engineering*, Vol. 39, p.1665, 1994.
4. R.C. Nieman, J.D. Gonczy P.E. Phelan and T.H. Nicol, Design and performance of Low-Thermal-Resistance, High-Electrical-Isolation Heat Intercept Connect, *Cryogenics*, Volume 35, Number 2, p. 829, 1995
5. M.A. Daugherty et al, Assembly and Testing of a Composite Heat Pipe Thermal Intercept for HTS Current Leads, *Advances in Cryogenic Engineering*, Vol. 41A, p. 579, 1996.

OPTIMIZATION OF TWO-STAGE CONDUCTION COOLING FOR HTS CURRENT LEADS

H.-M. Chang

Hong Ik University, Department of Mechanical Engineering
Seoul, 121-791, Korea

ABSTRACT

An optimal two-stage cooling method for HTS current leads is developed to minimize the required refrigerator power input. The lead is a series combination of a normal metal conductor at the warmer part and an HTS at the colder part. The cold end of the HTS lead is conduction-cooled by a cryocooler and heat is also intercepted at an intermediate axial location. Mathematical expressions are derived for the required refrigerator power per unit current for typical cryocoolers. It is demonstrated that both the current density and the intercept temperature should be optimized simultaneously to minimize the refrigerator power per unit current and maintain the HTS in a superconducting state. In addition, it is verified that heat should be intercepted at the joint of the two parts or at an optimal axial position of the metal lead for the minimum refrigerator power. The design conditions for the two-stage cooling are quantitatively presented with the effects of the magnetic field and the stability margins of the HTS lead, when the cold end temperature is 4 K or 30 K.

INTRODUCTION

High Tc superconductors(HTS) are gradually replacing conventional metallic current leads in high field superconducting magnets, because they can reduce considerably the heat leak to the cryogenic temperatures and the required refrigerator power. The so-called binary lead is a series combination of a normal metal conductor at the warmer part and an HTS at the colder part. The cooling method for the binary lead may be quite different from the standard helium-vapor cooling of metal leads,[1-3] and a number of studies have been performed during past several years for an effective cooling technology of the bulk or the composite type of HTS leads.[4-11]

Conduction cooling of current leads has received close attention from many researchers since the successful development of 4 K Gifford-McMahon cryocoolers and liquid-free superconducting systems.[12] There is no liquid cryogen in the systems, so the lead should be cooled in vacuum by direct contact with cryocoolers. In addition, conduction cooling of

Advances in Cryogenic Engineering, Volume 45.
Edited by Shu *et al.*, Kluwer Academic / Plenum Publishers, 2000.

HTS leads is also applied to HTS magnet systems[13] where a high field is required and the operation temperature is lower than about 30 K.

The optimization techniques of conduction cooling for metal leads were studied earlier,[1,2] including single- or multi-stage (even distributed) methods. In these methods, the geometric conditions of the lead were designed such that the heat conduction and the heat generation may be balanced or the total refrigerator power required for the cooling loads may be minimized. Cooling of HTS leads without boil-off helium gas has been partly considered in some publications.[4-7] Most of these research works are, however, related with the analysis or the experiment for HTS leads whose ends were cooled by liquid nitrogen or liquid helium. Two recent papers[8,9] have tried to find the optimal intercept temperature for a given geometric condition of the binary lead.

The present author has lately published a new and generalized theory[10] to reveal that both the operating current density and the intercept temperature of binary lead should be optimized simultaneously for the most efficient cooling. The theory also provided the absolute minimum in refrigerator power for the binary lead as a thermodynamic limit for distributed or two-stage conduction cooling. He has also extended his original theory to a practical design tool by incorporating the performance of some actual cryocoolers and the realistic stability margins of the HTS[11].

This paper aims at a complement of the optimization technique for the two-stage conduction cooling by considering three additional factors that have not been included in the previous studies. The first is a theoretical consideration in association with the axial location of the heat intercept. In every previous study about two-stage cooling of binary leads, heat is intercepted at the joint of the metal and the HTS leads. A question has been raised how the refrigerator power would vary if the intercept location is not the joint. The second is a more practical and quantitative examination for the effect of the magnetic field on the optimal cooling conditions of the leads. Finally, the optimization theory is applied to binary leads for HTS magnet systems whose operating temperature is as high as 30 K, as the previous studies treated only the leads for the standard low Tc systems at 4 K.

OPTIMIZATION METHOD

Figure 1 shows schematically the two-stage conduction cooling of a binary current lead. The HTS part and the normal metal part in the binary lead are denoted by subscripts 1 and 2, respectively. Heat, Q_L, is removed from the cold end whose temperature is T_L, and Q_I is intercepted at an intermediate axial location where the temperature is T_I. T_J is the temperature at the joint of the two parts and identical to T_I in case of the intercept at joint as shown in (a). If heat is intercepted at a certain point on the metal or on the HTS, T_J would be smaller than T_I or greater than T_I as in (b) or (c), respectively. The warm end temperature, T_H, is the room temperature, and L is the length of the leads.

Required Refrigerator Power

The basic concept of the thermodynamic optimization for the two-stage cooling is to minimize the total required refrigerator power per unit current in steady-state. Since there exist two cooling loads at two different temperatures, the total refrigerator power per unit current can be expressed in general as

$$\frac{W_{ref}}{I} = \frac{Q_L}{I} \cdot \frac{1}{COP_L} + \frac{Q_I}{I} \cdot \frac{1}{COP_I} \tag{1}$$

where COP's are the coefficients of performance of the cryocooler at each stage.

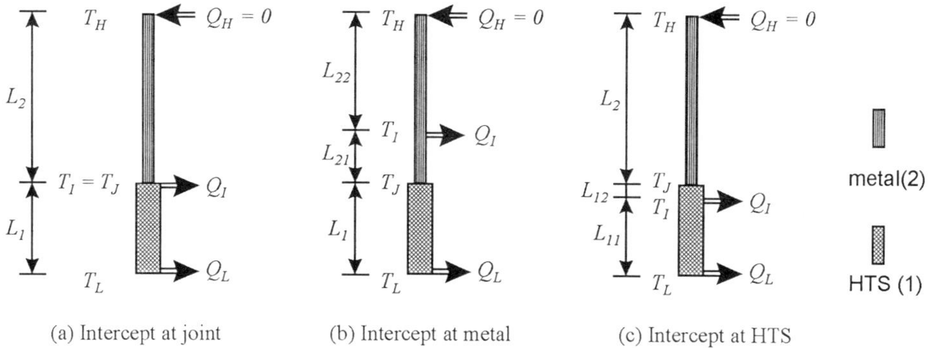

Figure 1. Schematic representation of two-stage conduction cooling method of a binary (metal + HTS) current lead with three different intercept locations.

Since the HTS lead should be always superconducting in normal operation, the current density of the HTS, J_1, cannot exceed its critical value, J_C. The critical current density of a superconductor is a strong function of the temperature and the magnetic field, B. Thus Eq. (1) should be minimized with a constraint,

$$J_1 \le J_C(T_J, B) \tag{2}$$

because the highest temperature in the HTS lead is T_J.

Cryocooler Performance

In order to calculate the required refrigerator power in Eq. (1), the coefficient of performance should be known for the specific cryocoolers to be used. In general, the *COP* of a cryocooler depends on the refrigeration temperature, the temperature at which heat is rejected, the type of refrigeration cycle, the cooling capacity, the performance of its components, and so on. For the purpose of a more quantitative demonstration of the optimization in this paper, simple, yet reasonable, *COP* models are selected from the previous study[11]. Figure 2 shows the *COP's* as a function of refrigeration temperature for medium or small sizes of various commercial cryocoolers, from which two simple mathematical functions have been fitted as the dashed curves or

$$COP_L = \frac{T_L}{T_H - T_L} \cdot \frac{0.1202 \cdot T_L + 1.316}{T_L + 84.81} \qquad 4\,\mathrm{K} \le T_L \le 40\,\mathrm{K} \tag{3}$$

$$COP_I = \frac{T_I}{T_H - T_I} \cdot \frac{0.4198 \cdot T_I + 2.740}{T_I + 1115} \qquad 20\,\mathrm{K} \le T_I \le 100\,\mathrm{K} \tag{4}$$

where $T_H = 300$ K.

Intercept Location and Cooling Loads

Two cooling loads per unit current in Eq. (1) can be expressed in terms of the temperature-dependent properties of the lead materials and the dimensional factors in each case of the intercept location. If the heat generation in the HTS is assumed to be zero and heat is intercepted at the joint,

$$\frac{Q_L}{I} = \frac{1}{J_1 L_1} \int_{T_L}^{T_I} k_1 \cdot dT \quad (5)$$

$$\frac{Q_I}{I} = \sqrt{2 \int_{T_I}^{T_H} k_2 \rho_2 \cdot dT} - \frac{Q_L}{I} \quad (6)$$

where k and ρ are the thermal conductivity and the electrical resistivity, respectively. In Eq. (6), the first term of the right-hand side is the minimum heat transfer rate per unit current from the metal part to the joint, which can be obtained when the dimension of the lead is optimized so that the conduction and the heat generation may be balanced or $Q_H = 0$.[1,10]

In case of heat intercept at a certain axial location of the metal lead, the metal lead is divided into two segments. Each segment can be optimized in the same manner as in Eq. (6), but with a different temperature span. Then, since the heat flow at the joint should be continuous, the two cooling loads per unit current can be expressed as

$$\frac{Q_L}{I} = \frac{1}{J_1 L_1} \int_{T_L}^{T_J} k_1 \cdot dT = \sqrt{2 \int_{T_J}^{T_I} k_2 \rho_2 \cdot dT} \quad (7)$$

$$\frac{Q_I}{I} = \sqrt{2 \int_{T_I}^{T_H} k_2 \rho_2 \cdot dT} \quad (8)$$

It should be noted that the length of each segment of the metal lead is not shown in Eq. (7) or (8), because it has already been optimized.

Even if the intercept at the HTS may not seem to be a practical option, the two cooling loads in case of Figure 1(c) can be theoretically obtained on an assumption that the HTS is superconducting and the dimension of the metal lead is optimized.

$$\frac{Q_L}{I} = \frac{1}{J_1 L_{11}} \int_{T_L}^{T_I} k_1 \cdot dT \quad (9)$$

$$\frac{Q_I}{I} = \frac{1}{J_1 L_{12}} \int_{T_I}^{T_J} k_1 \cdot dT - \frac{Q_L}{I} = \sqrt{2 \int_{T_J}^{T_H} k_2 \rho_2 \cdot dT} - \frac{Q_L}{I} \quad (10)$$

where L_{11} and L_{12} are the lengths of each segment of the HTS lead as shown in Figure 1(c).

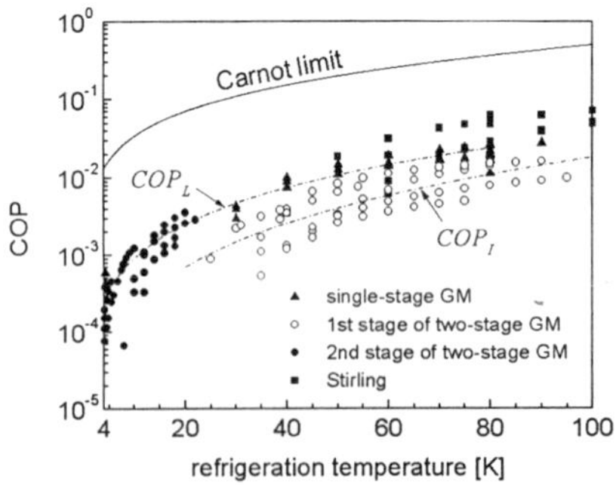

Figure 2. Coefficient of performance of various cryocoolers as a function of refrigeration temperature with the Carnot limit. COP_L and COP_I from Eq. (3) and (4) are shown by dashed curves.

Critical Current Density of HTS

The critical property of the HTS lead plays an important role as a constraint, Eq. (2), in the optimization. Again for the purpose of a more quantitative demonstration of the method, a reasonable behavior is assumed for the critical current density of HTS. For a bulk Bi-2223, it is reasonably accurate that J_C is a simple function[4,7] of temperature and magnetic field,

$$J_C(T_J, B) = 4 \cdot J_C(77\,\mathrm{K}, 0\,\mathrm{T}) \cdot \frac{1+B}{1+5.5B} \cdot \left[1 - \frac{T_J}{104}\right] \quad (T_J \text{ in K}, B \text{ in Tesla}) \tag{11}$$

where $J_C(77\ K,\ 0\ T)$ is the characteristic critical current density at 77 K and 0 T, depending upon the size, the shape and the fabrication method.

Stability Margin of HTS

Because of the nature of the superconductivity, the required refrigerator power has its minimum always at a marginally superconducting state[10,11] by the constraint, Eq. (2). In practice, the HTS current lead should be designed such that the current density is well below the critical value in order to be stable from a certain level of thermal disturbance. One of the design schemes to determine the operational current density as a fraction, ϕ, of the critical value.

$$J_1 = \phi \cdot J_C(T_J, B) \tag{12}$$

RESULTS AND DISCUSSION

Optimal Design

The refrigerator power per unit current, W_{ref}/I, is calculated by substituting Eq. (3) through (6) into Eq. (1), in case of the intercept at joint. When the materials of the lead and the two end temperatures are given, it is noted that W_{ref}/I is a function of J_l, L_l and T_J only. Figure 3(a) shows some contours of constant W_{ref}/I on a J_l-T_J diagram for Cu + Bi-2223 lead when L_l = 20 cm and T_L = 4 K. In the calculation, the properties of the materials are taken from two recent publications.[3,6] It is a general trend that W_{ref}/I decreases as J_l or T_J increases. However, if T_J is low or J_l is high, W_{ref}/I is almost independent of J_l, since the cooling load for the HTS is relatively small. On the contrary, if T_J is high or J_l is low, the cooling load for the HTS is relatively significant and J_l is a more important factor than T_J.

As mentioned above, the minimization of W_{ref}/I is subject to a constraint, Eq. (2). Figure 3(a) shows critical current density, J_C, as a function of T_J, Eq. (11), assuming that J_C(77 K, 0 T) = 1,000 A/cm^2 and B = 0 T. Clearly, there exists unique optimal values for J_l and T_J to minimize W_{ref}/I as indicated by a dot in a marginally superconducting state. It is also noted that if J_l is given as a constant, the optimal T_J to minimize W_{ref}/I may be significantly lower than T_C because of the concave shape of the contour curves, as discussed in other publications.[8,9] Therefore the present method in this paper can be considered as a two-dimensional optimization involving the previous one-dimensional method as a special case. The theoretical minimum of W_{ref}/I in this case is approximately 3.09 W/A, which is about 51 times the absolute minimum as a thermodynamic limit[10] that can be achieved by the most efficiently distributed Carnot refrigerators. The corresponding optimal values of J_l and T_J are approximately 378 A/cm^2 and 94.2 K, respectively. The stability margin of the HTS can be included as in Eq. (12) for a practical design. Two examples for ϕ = 0.5 and 0.2 are indicated in Figure 3 and the required minimum for W_{ref}/I is increased to about 3.54 W/A and 4.46 W/A, respectively. It is obvious that J_I and T_J should be smaller than the respective theoretical optima.

When heat is intercepted at an optimal location of the metal lead, the required W_{ref}/I is calculated by substituting Eq. (7) and (8) into Eq. (1). In this case, W_{ref}/I is explicitly a function of J_I, L_I, T_J and T_I for the given materials and the end temperatures. However, T_J and T_I are not independent because of Eq. (7), thus W_{ref}/I can be expressed as an implicit function of J_I, L_I, and T_J again. Figure 3(b) shows contours of W_{ref}/I on a J_I-T_J diagram in this case with the same conditions as (a). The values of W_{ref}/I are indistinguishable from those in case (a), except for the very low T_J region. It can be concluded that the minimum W_{ref}/I and the optimal conditions for J_I and T_J would be almost the same as the case of the intercept at joint, as far as each segment of the metal lead is optimally dimensioned.

The intercept at a certain location of the HTS lead may not be suitable both in its hardware construction and in its cooling, because of the huge difference in the thermal conductivity between the metal and the HTS. Just for a thermodynamic analysis, the required W_{ref}/I can be calculated by substituting Eq. (9) and (10) into Eq. (1). In this case, W_{ref}/I is an implicit function of L_{I1}, L_{I2}, J_I and T_J, since T_I are connected with T_J by Eq. (10). For a fair comparison, W_{ref}/I should calculated when the total length of the HTS lead is given as the same value or $L_I = L_{I1} + L_{I2}$. It can be verified, however, that the W_{ref}/I is a monotonously increasing function of L_{I2} and therefore has its minimum at $L_{I2} = 0$ for a given L_I. Obviously, there is no advantage for the intercept at any location of the HTS.

Effects of Magnetic Field and Cold End Temperature

As demonstrated in the previous section, the unique minimum in W_{ref}/I is obtained from the unique optima of J_I and T_J, for given L_I, J_C (77 K, 0 T), T_L, B and ϕ, if the cryocooler and the lead materials are given. It has been proven from a dimensional analysis[11] that the effect of L_I on the minimum W_{ref}/I is exactly the same as that of J_C (77 K, 0 T), so the product of L_I and J_C (77 K, 0 T) can be a single parameter in the optimization.

$$\left(\frac{W_{ref}}{I}\right)_{\min} = f\left[L_1 \cdot J_C(77\,K, 0\,T),\ T_L, B, \phi\,\right] \tag{13}$$

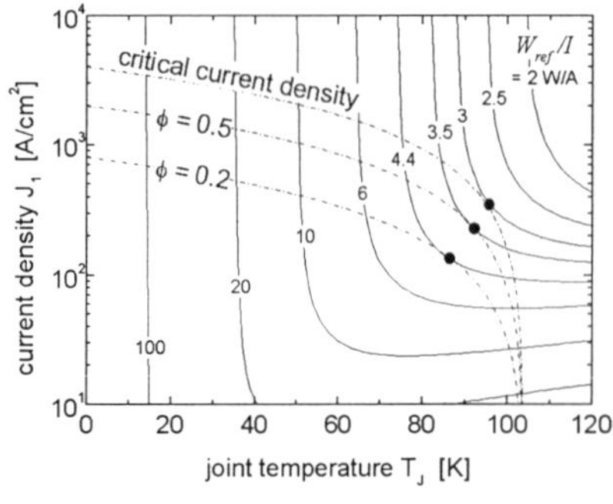

(a) Intercept at joint

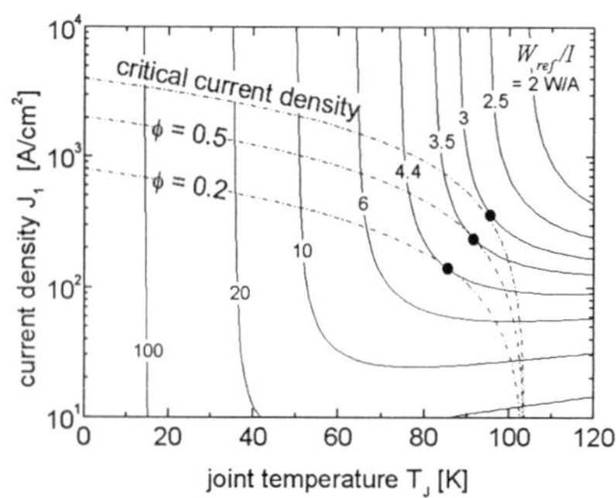

(b) Intercept at metal

Figure 3. Contours of refrigerator power per unit current on current density of HTS vs. joint temperature (J_I - T_J) diagram of Cu+Bi2223 lead when L_I = 20 cm and T_L = 4 K. Dots indicate the optimal conditions to minimize the refrigerator power per unit current.

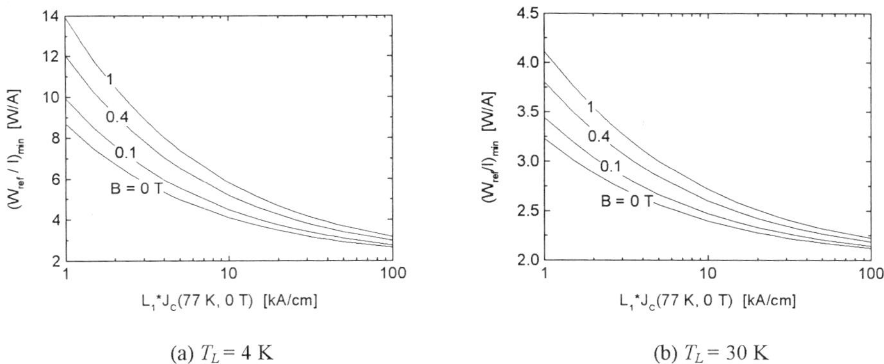

(a) $T_L = 4$ K (b) $T_L = 30$ K

Figure 4. Minimum refrigerator power per unit current as a function of the product of the critical current density and the length of HTS for Cu+Bi2223 lead when $\phi = 0.5$.

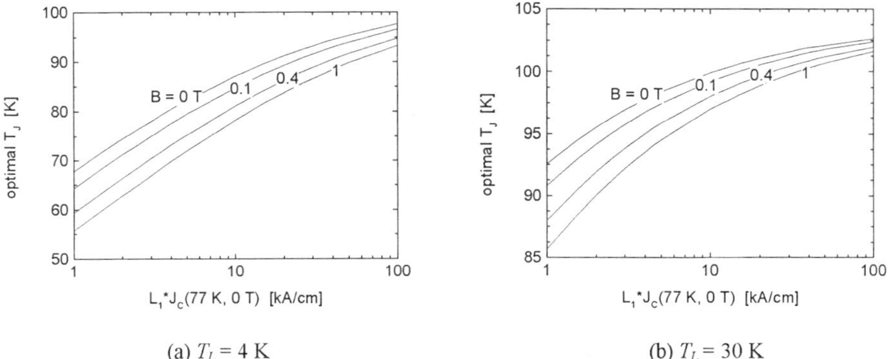

(a) $T_L = 4$ K (b) $T_L = 30$ K

Figure 5. Optimal intercept temperature as a function of the product of the critical current density and the length of the HTS for Cu+Bi2223 lead when $\phi = 0.5$.

The above optimization procedure has been repeated for various values of $L_1 \cdot J_C(77\,\mathrm{K}, 0\,\mathrm{T})$, T_L and B, and the some results for $\phi = 0.5$ have been plotted in Figures 4 through 6. The minimum of W_{ref}/I decreases as $L_1 \cdot J_C(77\,\mathrm{K}, 0\,\mathrm{T})$ increases or B decreases. However, W_{ref}/I does not vary significantly when $L_1 \cdot J_C(77\,\mathrm{K}, 0\,\mathrm{T})$ is greater than 20 kA/cm, which means for example that $L_l > 20$ cm if J_C (77 K, 0 T) = 1,000 A/cm^2 or $L_l > 10$ cm if J_C (77 K, 0 T) = 2,000 A/cm^2. It is noted that the length and the critical current density play exactly the same role in the cooling from the thermodynamic point of view, while the magnetic field is related in a more complicated manner with the critical current density as in Eq. (11). If $L_1 \cdot J_C(77\,\mathrm{K}, 0\,\mathrm{T})$ decreases to zero, W_{ref}/I for $B = 0$ approaches to its limit of the single-stage cooling of an optimized metal lead at the cold end.[11] On the contrary, if $L_1 \cdot J_C(77\,\mathrm{K}, 0\,\mathrm{T})$ is infinitely large, Q_L vanishes and W_{ref}/I would have the value

$$\frac{W_{ref}}{I} = \frac{1}{COP_I(T_C)} \sqrt{2 \int_{T_C}^{T_H} \rho_2 k_2 \cdot dT} \tag{14}$$

where T_C is the critical temperature of HTS. The minimum W_{ref}/I for $T_L = 30$ K is expected to be about 30 or 70% of that for $T_L = 4$ K, depending on B. As $L_1 \cdot J_C(77\,\mathrm{K}, 0\,\mathrm{T})$ increases or B decreases, the optimum for J_l increases and the optimum for T_J decreases. The optimal J_l for $T_L = 4$ K is higher by 5 to 30 K than that for $T_L = 4$ K in these conditions.

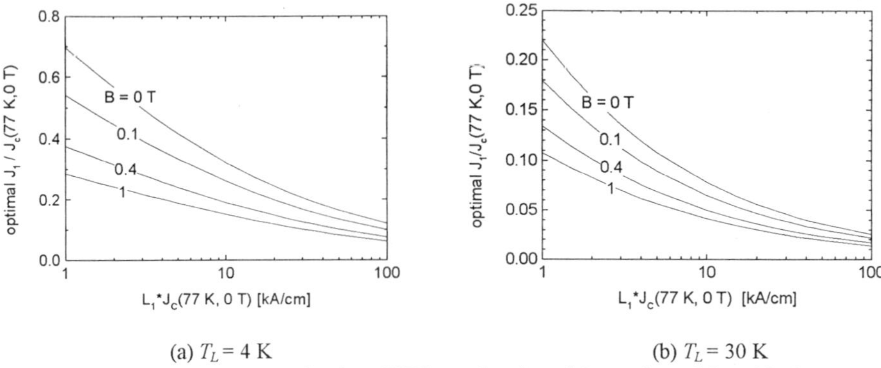

(a) $T_L = 4$ K (b) $T_L = 30$ K

Figure 6. Optimal operating current density of HTS as a function of the product of the critical current density and the length of HTS for Cu+Bi2223 lead when $\phi = 0.5$.

CONCLUSIONS

A comprehensive optimization technique of two-stage conduction cooling is presented for the metal-HTS current lead with some quantitative design data. The required refrigerator power per unit current for two-stage cooling is minimized by optimizing the dimensions of the lead and the heat intercept stage if the end temperatures and the properties of the lead materials are given. The procedure is demonstrated with every factor to be included in practical design, such as the refrigeration performance of the actual cryocoolers, the temperature dependent properties of the materials, the effect of magnetic field on the critical current density of HTS, and the stability margin of the HTS.

REFERENCES

1. Y.L. Buyanov et al., A review of current leads for cryogenic devices, *Cryogenics*, 15:193 (1975).
2. M.A. Hilal, Optimization of current leads for superconducting systems, *IEEE Trans. Magnetics*, MAG-13:690 (1977).
3. K. Maehata, K. Ishibashi, and Y. Wakuta, Design chart of gas-cooled current leads made of copper of different RRR values, *Cryogenics*, 34:935 (1994).
4. R. Wesche, and A.M. Fuchs, Design of superconducting current leads, *Cryogenics*, 34:145 (1994).
5. P.F. Herrmann et al., Cryogenic load calculation of high Tc current lead, *Cryogenics*, 33:555, (1993).
6. P.F. Herrmann et al., European project for the development of high Tc current leads, *IEEE Trans. App. Superconductivity*, 3:876, (1993).
7. U. Balachandran et al., Application of sinter-forged Bi-2223 bars to 1500 A AC power utility services as high frequency current leads in a 77-4 K temperature gradient, *App. Superconductivity*, 3:313 (1996).
8. S. Yang and J.M. Pfotenhauer, Optimization of intercept temperature for high temperature superconducting current lead, "Advances in Cryogenic Engineering," Vol.41, Plenum Press, New York (1996), pp.567-572.
9. C.M. Rey, Re-evaluation of the intercept temperature for HTS current leads based on practical and economical considerations, *Cryogenics*, 38:1217 (1998).
10. H.-M. Chang and S.W. Van Sciver, Thermodynamic optimization of conduction-cooled HTS current leads, *Cryogenics*, 38:729 (1998).
11. H.-M. Chang and S.W. Van Sciver, Optimal integration of binary current lead and cryocooler, 10th International Cryocooler Conference, Monterey, CA (1998).
12. T. Hasebe et al., Cryocooler cooled superconducting magnets and their applications, "Advances in Cryogenic Engineering," Vol.43, Plenum Press, New York (1998), pp.291-297.
13. T. Hase et al., Generation of 1 T, 0.5 Hz alternating magnetic field in room temperature bore of cryocooler-cooled Bi-2212 superconducting magnet, *Cryogenics*, 36:971(1996).

THERMAL TESTS OF 6 KA HTS CURRENT LEADS FOR THE TEVATRON

G. Citver[1], S. Feher[2], T. J. Peterson[2], and C. D. Sylvester[2]

[1]On leave from Jack and Pear Resnick Institute of Advanced Technology, Physics Department, Bar-Ilan University, Israel

[2]Fermi National Accelerator Laboratory*
Batavia, Illinois 60510

ABSTRACT

Prototype current leads incorporating High Temperature Superconductor (HTS) elements have been tested at Fermilab. Fermilab's Tevatron includes about 50 pair of 5 to 6 kA current leads, and Fermilab is investigating the feasibility of replacing some of these conventional leads with HTS leads. The prototype HTS current leads are cooled primarily by a countercurrent flow of liquid nitrogen from the 80 K intercept to the warm end of the leads, but also a small flow of helium gas cools the HTS section from the 4 K level. The HTS current leads carried the design current of 5 kA with good thermal and electrical stability. LN2 flow without current was 0.24 g/sec per lead and with 5 kA was 0.53 g/sec per lead, corresponding to heat inflows to the 80 K intercept of 46 Watts and 101 Watts, respectively. The heat input to the 4 K level was 0.6 W with no current and 0.7 W +/- 0.1 W per lead with 5 kA current, about 1/8 of the heat load via copper, vapor-cooled leads.

INTRODUCTION

Conventional, vapor-cooled power leads carry electric current from room temperature to 4.5 K in order to power the superconducting magnets in the Tevatron at Fermi National Accelerator Laboratory. Over 50 pair of leads each carry 4500 to 6000 amps of current and result in substantial heat loads for the cryogenic system. About 25% to 30% of the total Central Helium Liquefier capacity cools power leads. Reducing the total heat load to the liquid helium temperature level would allow either reduced operational costs or make more refrigeration available for lower temperature and higher energy operation of the Tevatron.

Conventional power leads have a well-known thermal performance limit of about 1.2 W/kA per lead.[1,2] However, with the use of High Temperature Superconductor (HTS), one

---

* Work supported by the U. S. Department of Energy under contract No. DE-AC02-76CH03000

can reduce the heat transport to the liquid helium level significantly below the limit for conventional, copper leads. Using HTS elements in a combined liquid nitrogen and liquid helium cooled power lead design, one can reduce the heat load to LHe through the lead by a factor of ten.[3] The replacement of some of the conventional power leads at Fermilab with more efficient HTS leads is under consideration.

As a first step toward realizing this plan, American Superconductor Corporation (ASC) and Intermagnetics General Corporation (IGC) each developed and built a pair of 5000A HTS current leads. The leads were similar in consisting of parallel tapes of BSCCO-2223 multifilamentary powder-in-tube conductor in a silver alloy matrix soldered to a conventional copper upper section. These leads went through extensive tests at Fermilab. The focus of this paper is on the thermal studies of the IGC current leads.

TEST APPARATUS

The HTS power lead test apparatus is located in the Magnet Test Facility (MTF) at Fermilab. The mechanical system consists of a liquid nitrogen shielded helium cryostat (Figure 1) with a baffle system that includes an 80 K intercept, and the instrumentation

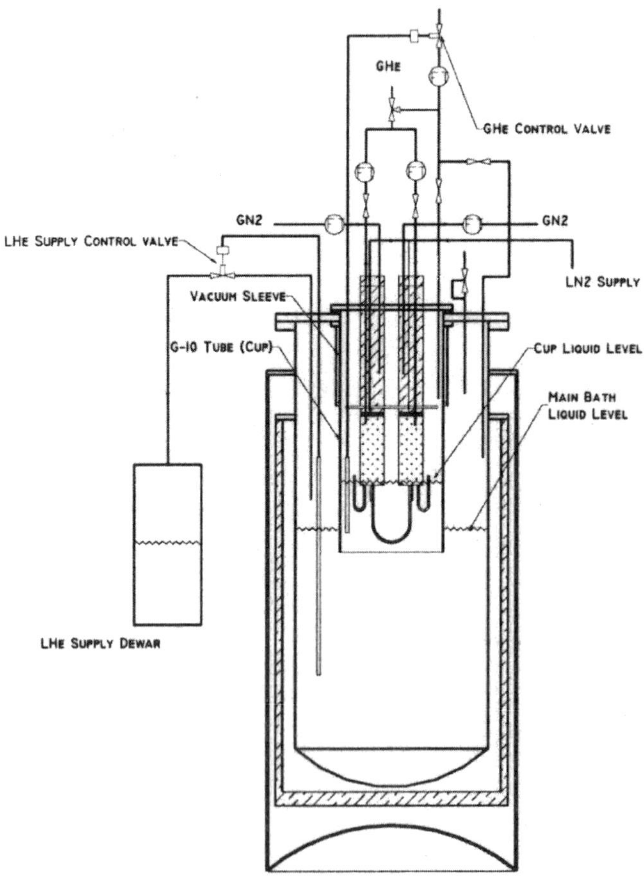

Figure 1. The current lead test schematic.

necessary to monitor mass flow rates, measure and regulate system pressures and liquid level and record system temperatures. The 51 cm diameter by 107 cm long helium vessel is sized to accommodate power lead pairs, which are spaced on a 10 cm center-to-center distance, and are up to 76 cm long. The leads are mounted on a plate that is separate from the main vessel cover plate. This feature allows for power lead removal without complete disassembly of the vessel cover plate. The remaining top surface area of the vessel accommodates fill and vent lines, valves, liquid level sensors, and other instrumentation.

To thermally separate the main dewar helium volume (referred to as the "main bath") from the volume immediately around the power leads, the leads are housed inside a glass/epoxy (G-10) tube (referred to as the "cup"). The cup extends from the cover plate to several centimeters below the minimum operating helium level in the main dewar, and thus the cup contains a separate liquid level around the leads. To further thermally isolate the two volumes, the power lead pair and G-10 tube assembly is mounted inside a vacuum-jacketed sleeve that extends to the depth of the liquid nitrogen cooled intercept. Anticipating differences in pressure due to the different thermal environment in each, the system provides for independent or simultaneous venting of the two volumes. A backpressure regulator, sensitive to 0.3 mbar pressure changes, is installed to help maintain a constant pressure in the main bath, while pressure changes may be occurring inside the G-10 tube.

The calibrated mass flow meters (FT in Figure 1) used in the system were sized to operate effectively over the expected range of flows. The HTS data acquisition (DAQ), and the quench detection and management systems used in this test are adapted and extended from those systems developed at MTF to conduct tests on superconducting R&D magnets.[4] Temperature and voltage measurements from the DAQ scans are monitored by a new software quench detection system[5] that triggers the quench management system to protect the leads from (relatively slow) quenches. An independent hardware backup system protects against ground faults as well as fast resistive voltage growth across the leads. A quench is detected when one of the analog signals or software process variables exceeds a (configurable) threshold or when a scan malfunctions. When triggered, the management system initiates fast quench data logging and slow power supply ramp down.

The quench thresholds were set low. The system tripped for any temperature rise of 5 K above the zero-current baseline temperature profile, HTS voltage greater than one millivolt, or copper section voltage greater than 32 millivolts. Temperature and voltage process variables were monitored and logged by two independent scan systems, which used complementary instrumentation schemes. The carefully wired and isolated sensors delivered typical noise levels of less than 1 K for temperatures, and less than 3 μV for voltage taps at 5000 A.

The Current Lead Test Modes

Thermal tests of the helium-cooled HTS section included three different modes of cooling. In "conduction mode", the HTS section was cooled by only heat conduction from 80 K to 4.5 K (Figure 2). Boil-off from the cup surrounding the leads was measured. In "regular mode", He cooling flow through the leads was adjusted to be 0.006, 0.011 and 0.025 g/sec. The balance of the boil-off from the cup surrounding the leads was vented and measured. In "self-sufficient mode", all boil-off helium inside the exits via the leads, and the flow through the leads was measured.

During the thermal tests, we tried to maintain the following parameters constant: pressure, liquid helium level inside the cup (directly under the leads), helium gas flow through the current leads, and LN2 flow to the 80 K intercept. Automatic control of the bath pressure and two liquid levels were implemented in order to help maintain steady conditions.

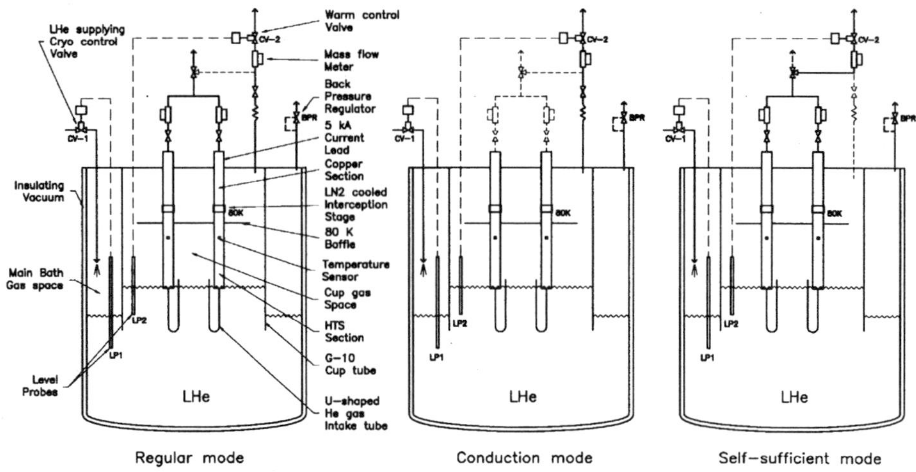

Figure 2. The current lead test modes.

POWER TEST RESULTS

Electrical test results and some preliminary thermal test results for these HTS current leads have been previously reported.[6] The IGC current leads, for which this paper describes the thermal test results, carried over 5000 Amps with stable voltages and temperatures.

THERMAL TESTS OF THE LN2-COOLED COPPER UPPER SECTION

To determine heat inflow to the 80 K intercept, the liquid nitrogen cooling flow was reduced stepwise until the temperature of the nitrogen intercept in the lead started to rise, indicating that all the LN2 supplied was boiled-off. Helium cooling flow was maintained as recommended by the vendor: 0.026 g/sec without current and 0.033 g/sec with 5 kA. It was found that the minimum required LN2 flow without current was 0.24 g/sec and with 5 kA was 0.53 g/sec, corresponding to heat inflows to the 80 K intercept of 46 Watts and 101 Watts, respectively.

THERMAL TESTS OF THE HELIUM-COOLED HTS SECTION

The results of the boil-off rate measurements for the HTS leads are shown in Table 1. The boil-off rate measurements were done both when liquid helium was transferred to maintain a constant main bath level and when no helium was transferred, so the main bath level dropped slowly. One can see in Table 1 that the boil-off rate was lower during helium transfer. As we will discuss later, the presence or absence of helium transfer and the

Table 1. Liquid helium boil-off rates under different modes (flows are per lead)

Mode	Conduction			Regular					Self-sufficient		
Helium transfer	yes	yes	no	yes	yes	yes	yes	no	yes	yes	no
Current (kA)	0	0	0	0	0	0	7	0	0	5	0
Lead flow (g/s)	0	0	0	0.006	0.01	0.024	0.024	0.025	0.025	0.029	0.059
Total flow (g/s)	0.05	0.058	0.086	0.044	0.035	0.029	0.037	0.035	0.025	0.029	0.059

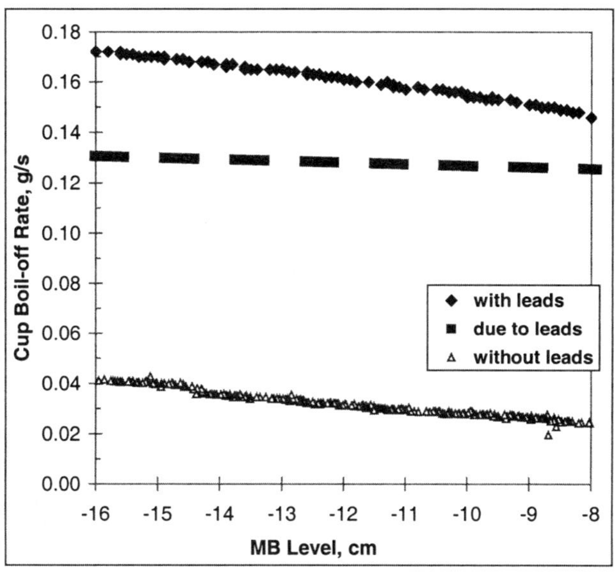

Figure 3. Subtraction of background heat load from conduction mode cooling, no helium transfer, as a function of main bath (MB) level.

venting location (e.g., self-sufficient mode or conduction mode) resulted in different vapor temperature profiles and convection rates on each side of the G-10 cup wall and different rates of heat transfer into or out of the current leads environment. This led to inconsistent results unless one accounts for these background effects.

In order to check the background heat load, boil-off rate measurements were made without current leads and as a function of main bath (MB) level. The vaporization rate without current leads versus main bath level is shown in Figure 3 in comparison to a current lead heat load measurement versus main bath level. For both this background run and the current lead run, helium exited via the cup vent (conduction mode) and the measurement was made without helium transfer to the main bath. One can see the dependence of boil-off rate on main bath liquid level. A higher main bath level provides more cooling of the cup through the epoxy-fiberglass tube wall. Subtracting the vent rate without current leads from the rate measured with current leads reveals a constant difference, 0.13 g/s, attributable to the pair of current leads in conduction mode.

Conduction mode had the problem that, due to the onset of thermo-acoustic oscillations when the insulated vent pipe was lowered to near the liquid level, the vent pipe entrance had to be raised to nearly the level of the 80 K intercept. The helium vapor then provided some external cooling of the current leads; some heat was still removed from the leads by convection. Thus, conduction mode results cannot be taken to indicate the heat that would be added to a passing forced-flow helium stream by a conductively cooled HTS section.

In order to get a more direct measure of the relationship between heat added to the bath within the cup and cup boil-off rate during helium transfer, measurements using a small electric heater were done while liquid helium was transferred into the main bath to maintain a constant main bath liquid level. The dependence of the liquid helium boil-off rate versus power dissipated in liquid helium in the cup is shown as the lower curve in Figure 4. Unlike the situation without helium transfer described in Figure 3 above, no correlation of boil-off rate with main bath level was noticed. Variation of main bath level and helium transfer rate resulted in some scatter of the results, with the resultant uncertainty shown in Figure 4. It

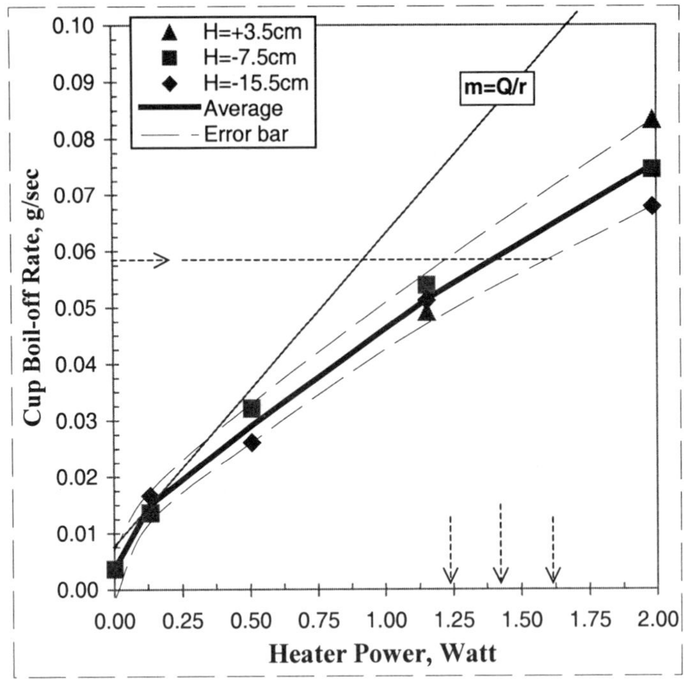

Figure 4. Evaporation from under the cup (both leads) versus heater power with liquid helium transfer

was expected that the dependence of LHe boil-off on the dissipated heater power should be linear (m=Q/r) starting from a background value at 0 Watts. However, the amount of increase in measured mass flow from the cup was less than that corresponding to the additional heat from the heater. The difference is probably due to the extra cooling of the cup volume through the G-10 cup wall.

One can see by comparing heater results with current lead boil-off measurements that in the self-sufficient cooling mode, with liquid helium transfer, the total current lead boiloff for the pair of leads of 0.050 g/s with no current corresponds to 1.2 W of heat input via the current leads. At 5 kA the vent rate of 0.058 g/s corresponds to 1.4 W +/- 0.2 W of heat input via the current leads. In a well-insulated environment, this would result in a self-cooling flow of 0.035 g/s +/- 0.005 g/s per lead.

DISCUSSION

A problem during these tests was that the G-10 tube, the "cup", was not a good thermal insulator, and heat flowed across the tube walls depending on the amount of turbulence-induced convection and the temperature difference between the vapor spaces above the main bath and in the cup. Helium transfer into the main bath affected the vapor flow rates and temperature distribution in the vapor space around the outside of the cup. With helium transfer, the vapor outside the cup over the main bath was colder and more turbulent, so some cooling of the cup volume occurred and resulted in a lower boil-off rate from the leads. Without helium transfer, the vapor temperatures increased to levels above the nominal lead temperature at the corresponding height, and significant heat was transferred to the current lead. Hence the higher lead flows with no helium transfer in Table 1.

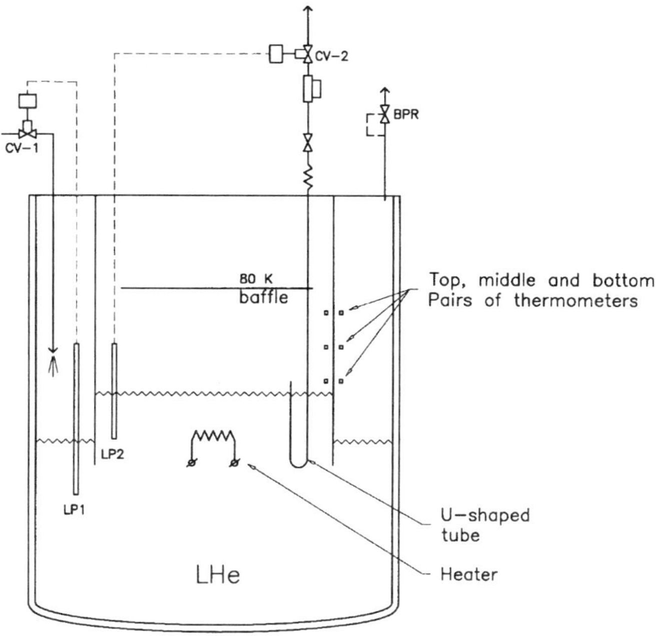

Figure 5. Background heat load measurement schematic.

After we understood the impact of the background heat leak on our head load measurements, we added some thermometry to the dewar (Figure 5). We measured temperature profiles with three pairs of thermometers installed at the same elevations inside and outside cup. The lowest pairs of sensors are at the elevation of the vent tube edge in order to measure the temperature at the vent gas level. The highest sensors are at the elevation of a thermometer embedded in the middle of the HTS section in order to measure the temperature difference between the lead and surrounding gas.

The additional thermometry confirmed that temperature distribution in the space inside the cup was affected by the mode of lead cooling. In conduction mode, cold vapor entered a vent above the HTS lead section, at about the 80 K intercept level (at this level to avoid oscillations), so the cup vapor was colder and was not stagnant. This vapor both cooled the leads and cooled the cup walls, affecting boil-off rate measurements. Our conduction mode results, therefore, are of questionable value. Self-sufficient mode, with all cooling through the leads, left the vapor above the leads largely undisturbed and provided probably the most reliable results.

To make the background boil-off measurements more like a current lead measurement in terms of vapor stratification, vapor temperatures, and background heat load, we inserted a vent tube at the current lead level with features to prevent oscillations (shown in Figure 5). This setup was used in making the heater tests described in Figure 4.

CONCLUSIONS

The focus of this paper is on the thermal studies of current leads supplied by Intermagnetics General Corporation consisting of parallel tapes of BSCCO-2223 multifilamentary powder-in-tube conductor in a silver alloy matrix soldered to a conventional copper upper section. The HTS current leads carried the design current of 5 kA with good thermal and electrical stability. LN2 flow without current was 0.24 g/sec per lead and with 5 kA was 0.53 g/sec per lead, corresponding to heat inflows to the 80 K intercept of 46 Watts and 101 Watts, respectively. Due to the effects of helium transfer and vapor flow on the current lead environment and background heating, the best 4 K heat load results were obtained in self-sufficient mode with a continuous transfer of liquid helium to the test dewar. In self-sufficient mode, the helium boiloff rate was about 0.03 g/s, corresponding to a heat input to the 4 K level of 0.6 W with no current and 0.7 W +/- 0.1 W per lead with 5 kA current, about 1/8 of the heat load via conventional copper leads.

ACKNOWLEDGEMENTS

The authors wish to thank C. Hess, W. Mumper, A. Rusy, and D. Parisi for their skill in assembling and operating the current lead test system. M. Tartaglia, D. Orris and D. Krause provided instrumentation, data readouts, and associated software and databases. J. Nogiec and J. Pachnik provided the software for data acquisition and display. The expertise and experience of these and other people in the Development and Test Department of the Technical Division at Fermilab were vital to the success of this project.

REFERENCES

1. J. M. Lock, Optimization of current leads into a cryostat, *Cryogenics*, 9 (1969), 438-442.
2. G. Citver, E. Barzi, A. Burov, S. Feher, P. Limon, T. Peterson, Steady state and transient current lead analysis, *ASC'98*, Palm Desert, CA (1998).
3. R.C. Niemann, Y.S. Cha, and J.R. Hull, Perfomance measurements of superconducting current leads with low helium boil-off rates, *IEEE Trans. Appl. Superconductivity* 3(1):392 (1993).
4. M. Lamm, et. al., A new facility to test superconducting accelerator magnets, *PAC'97*, p. 3395, Vancouver, B.C., Canada (1997).
5. J. Nogiec, et. al., Architecture of HTS leads software protection system, *PAC'99*, New York, NY, (1999).
6. G.Citver, S. Feher, P.J. Limon, D. Orris, T. Peterson, C. Sylvester, M.A. Tartaglia, J.C. Tompkins, HTS power lead test results, *PAC'99*, New York, NY, (1999).

PRACTICAL DESIGN AND MANUFACTURING OF CRYOGENIC HIGH CURRENT LEADS

V. I. Datskov,[1] V. Bartenev,[1] and J. A. Demko[2]

[1]Laboratory of High Energies
Joint Institute of Nuclear Research
Dubna, Moscow Region, Russia
[2]Oak Ridge National Laboratory
P.O. Box 2009, Oak Ridge, TN 37831

ABSTRACT

High current leads are required in many large superconducting magnet facilities and will be required for high-temperature superconductor power applications. Several important factors must be selected in order to design high current leads that include the choice of conductor properties and lead geometry (length, cross section, cooling surface area). The application of a mathematical model and optimization for a 13-kA lead design between liquid helium and room temperatures are discussed. A design and manufacturing technology for current leads with specific heat leak near 1 W/kA in forced-cooled mode are also described. The principles of selection of the current-carrying element material, geometric sizes, and their influence on the heat leak from the current leads are explained. Measurements of the typical copper residual resistivity ratio found at different locations of welded samples taken from prototype current leads are presented. These measurements provide direction in choice of material properties used in modeling a current lead design. Different modes of cooling the current leads, such as vapor cooled and forced-flow cooled, are considered. A comparison between some existing current lead designs is discussed.

INTRODUCTION

Cryogenic electrical devices may require that electric current, ranging to several thousand amperes, be brought into the cold region of the cryostat. Heat losses through the current leads are then of great importance. It is therefore necessary to find the optimum design for the leads, which would provide a minimum heat flow into the cold region or minimize the refrigeration, required to operate the leads. There are two main contributions of heat load from leads: thermal conduction and the dissipation of joule heat. For vapor or

Advances in Cryogenic Engineering, Volume 45.
Edited by Shu *et al.*, Kluwer Academic / Plenum Publishers, 2000.

forced-flow-cooled current leads, the first component depends mostly on the quality of the heat exchanger of the current-carrying conductor and on the ratio of the current-carrying conductor length to its cross section. The second component depends also on the quality of the current-carrying conductor material. The actual heat load of a current lead can be determined in the vapor-cooled mode only by measuring the flow of helium gas.

ANALYSIS OF THE EXISTING DESIGNS OF CURRENT LEADS

Analysis of current leads can be accomplished with a one-dimensional energy balance for the conductor and the cooling flow as given in Eqs. (1) and (2). In these equations, γ is the density, λ is the thermal conductivity, C is the specific heat, A is the cross-sectional area, h is the convective heat transfer coefficient, P is the heat transfer surface area per unit length, e_v is the vapor enthalpy, and ρ is the resistivity of the conductor. The subscripts c and v stand for the conductor and vapor, respectively.

$$\frac{\partial(\gamma_c A_c C_c T_c)}{\partial\tau} = \frac{\partial}{\partial x}(\lambda_c A_c \frac{\partial T_c}{\partial x}) + \frac{\rho I^2}{A_c} + hP(T_c - T_v) \cdot \tag{1}$$

$$\frac{\partial(\gamma_v C_v T_v)}{\partial\tau} = \dot{m}_v \frac{\partial e_v}{\partial x} + hP(T_v - T_c) \ . \tag{2}$$

The current-carrying conductors of the current leads have been made from ribbed pivots, strips, pipes and small-diameter wires, woven to bunches. The current leads with current-carrying conductors from ribbed pivots are described in Refs. 1–3. Defining a specific cooling surface as the ratio of the cooling surface area per unit length, or perimeter, to the current-carrying cross section, these lead designs have a specific cooling surface of 1 cm^2/cm^3 to 2 cm^2/cm^3. It is practically impossible to increase this value because the increase of the number of ribs leads to a fast growth of the current lead hydraulic resistance. The cross-sectional area of the current-carrying conductor of this current lead design is high enough for the selected nominal currents, and so they have a high heat influx.

Current leads with current-carrying conductors made from finned foils and small-diameter wires are described in Refs. 4–7. The current-carrying conductors of these designs can have a large specific cooling surface,[8] but this condition is not yet enough. The current flows by parallel strips or wires, and it is necessary to provide an identical removal of heat from each of them. When the current passes through parallel conductors, there arise electro-dynamic forces, which press the conductors against each other. The increased size of the cooling gas channels worsens considerably the performance of the heat exchanger.[9] The cross-sectional area of the current-carrying conductor of this design is large enough so they have a high heat influx. The performance of the current leads of these designs may degrade in time because there is an increase of the channels of passage of the cooling gas. This process has its limit and depends on the elasticity of strips and wires.

The box current lead design described in Ref. 10 does not have these disadvantages. It is formed using a stack of several copper braided conductors between two copper plates. The edges of the plates are welded closed, forming a box filled with copper strands. The current passes through the box walls, and the hard-secured shields are used as the heat exchanger surface. The performance of these current leads does not change for time periods over 10 years.

At the present time, there are many articles devoted to combined current leads, that is, current leads with the low-temperature section made of a high-temperature superconductor.[11–13] High-temperature superconductors are commonly made into tapes or in bulk form as tubes or rods. A current conductor from metal, copper, or stainless steel may be connected in parallel across superconductor sections to assure mechanical strength and help protect the HTS sections during fault currents. This decreases the benefits realized from the HTS section during operation at design conditions.

The refrigeration power consumed for cooling the heat influx of the conventional design[10] and that for a design using a silver alloy sheathed tape HTS section[12] is practically the same per kiloampere. Since most HTS materials are ceramic, combined current leads do not withstand current overloads or cooling loss as well as all copper constructions and may fracture during this kind of event. The advantages and disadvantages of these current leads are described in detail in Ref. 13.

PRACTICAL DESIGN AND MANUFACTURE OF CURRENT LEADS

The design of current leads is in practice reduced to the selection and optimization of geometrical features. The construction of the current lead with a current-carrying conductor from welded copper strips and shields, described in Ref. 10, is taken as a basis. For many situations, the minimum perimeter may be taken as 50 cm^2/cm^3 of current-carrying cross section. The effective heat exchanger of the current-carrying conductor with a specific cooling surface based on the first criterion permits a current density of 80–100 A/mm^2 in the cold region.[8] The cross section of the current-carrying conductor in the cold region has to be minimum because it determines the thermal conductance, that is, the product of the thermal conductivity and cross-sectional area. Figure 1 presents the dependence of the thermal conductance on the cross section of the current-carrying conductor for the heat exchanger design of Ref. 10 and a current-carrying conductor length of 600–800 mm in the force-cooled mode.

The distribution of temperatures, shown in Fig. 2, along a current lead of this design is taken from Ref. 14. The temperatures for the forced-cooled mode are lower than those for the vapor-cooled mode. In the vapor-cooled mode, the mass flow is determined by the heat leak into the liquid vaporizing the helium gas that is used to cool the lead. In the forced-cooled mode, the flow can be set by throttling the gas with a valve at the discharge from the

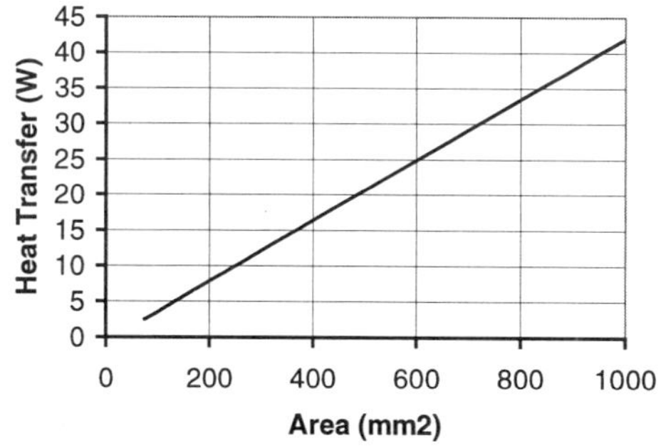

Figure 1. The dependence of the heat influx by conduction heat transfer along the current lead conductor cross section for the heat exchanger design[10] in the force-cooled mode.

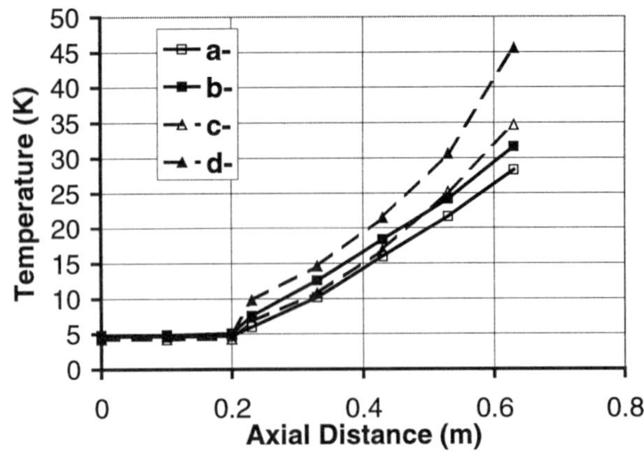

Figure 2. The temperature distribution along the current lead in different kinds of operations: (1) Force-cooled mode, a—temperatures at zero current and b—temperatures at 5 kA; (2) Vapor-cooled mode, c—temperatures at zero current and d—temperatures at 5 kA.

current lead. The current changes the temperature field in the cold region of the current-carrying conductor less than in the warm end, as shown in Fig. 2. In the force-cooled mode, the current lead operates at a nominal current at the mass flow rate through the lead of 0.049 g/s/kA, corresponding to a specific heat leak of 1 W/kA (Ref. 10). The length of the current-carrying conductor has a small influence on the current lead heat leak because the temperature field is practically the same in the cold region of the current-carrying conductor. To a large extent, the length of the current lead operating in the nominal mode determines the temperature of the warm end of the current lead. The length of the current lead, operating at a constant current, can be chosen assuming the warm end temperature is the same as the surroundings as an initial assumption. Increasing the current lead length directly increases to the cost of the current lead cryostat.

The copper used in the current-carrying conductor can be obtained from commercial sources with a residual resistivity ratio (RRR) = 30–50. The thermal conductivity of commercial and high-purity copper with RRR = 1800–2000 is practically the same in a temperature region of 50–295 K. The RRR was measured on a sample cut from a 6-kA lead of the design in Ref. 10 using a four-wire method. The measured RRRs of the plate and the weld region were 30 and 114, respectively. This demonstrates the effect of the weld on the material properties. Table 1 shows results from an analytical optimization of a

Table 1. Comparison of shape factor and heat load for conduction-cooled lead

RRR	IL/A (A/cm)	QI (W/A)
20	40200	0.044
40	46300	0.044
100	56100	0.044
1000	100400	0.043

conduction-cooled lead as discussed in McFee.[15] The trend is assumed valid for cooled leads and shows that the heat load per kiloampere is not very sensitive to the RRR but that the shape factor, hence the length-to-area ratio, varies significantly. The use of high-purity copper leads to very long, thin leads and to a higher cost of the current lead.

Currents leads, rated at a short-time operation to 8 kA, were manufactured as safety current leads based on the design described in Ref. 10, with a low heat load to the cryostat. The upper part of the current lead in a temperature region of 300–50 K is made of copper, and its cross-sectional area is 160 mm. In the lower part, located in a temperature region of 30 to 4.5 K, the copper strips are replaced with stainless steel, and the copper cross section is 40–45 mm^2. The static heat leak in the cryostat was 0.8 W per lead in the force-cooled mode and 0.45 W per lead in the vapor-cooled mode. The operating conditions were to apply a current of up to 8 kA for 4–5 s, then the current was turned off for 1–2 s. The current must be only off for a short time from the magnet system. A small jump in temperature was observed approximately in the middle of the insertion from low thermal conductivity material. Before applying current, a mass flow rate lead of 0.039 g/s, corresponding to a static heat leak of 0.8 W through the lead, was set. The operation of the current leads was stable in the vapor-cooled mode with a specific heat leak of 0.8 W/kA at 2.4 kA, which corresponds to the amount of copper (40–45 mm^2) in the lower part of the current lead.

At the present time, the magnets of superconducting accelerators[16] and large magnet systems,[17] placed in vacuum, are cooled by the force-cooled mode. Forced-cooled operation simplifies significantly the design of the magnet system and does not depend on orientation of the current lead (horizontal or vertical). Current lead operation in the force-cooled mode is described in Ref. 10. The helium supplied to the current lead is single phase by evaporating it just before entering the current lead. On cooling the current lead with single-phase helium (4.2 K), the static heat leak of the current lead may be determined by the flow rate of helium using the latent heat of liquid helium. When operating in the immersed mode, that is, when the cold end of the current lead is in liquid helium, one can operate with a lower helium flow through the lead. For a 6-kA current lead, the helium flow through the lead was decreased up to 50%, and the current lead operated in a stable mode with an apparent specific heat leak of 0.53 W/kA at 6 kA. A significant portion of the heat leak through the lead is taken out by evaporating the liquid helium.[14]

A typical superconducting magnet system consists of a superconducting magnet, current leads, and superconducting connecting cables. Figure 3 is a simplified schematic diagram of the electrical connections of the current lead with a magnet and helium cooling flow to the current lead. A more detailed schematic diagram of the electric and helium connections of the Nuclotron magnet-cryostat units is shown in Ref. 16.

The superconducting cable length, which connects the magnet and the current lead in the Nuclotron, is 1.8 m to 2 m long. In this connection, the long superconducting cable and two solder joints produce an extra heat influx to the current lead. This heat influx depends on the quality of the superconducting cable, the quality of solder joints. The superconducting cable, connecting the magnet and the current lead, is the weakest link in the magnet system. When the superconducting system transitions to the normal state, the current must be turned off for a short time, for example, in the Nuclotron for 50 ms, or the connecting superconducting cable must be protected from burning. The overheating stability of the current lead under cooling loss can be prevented only with a copper bus in parallel with the superconductor between the current leads at the cold end.

Without the parallel copper bus, the connecting superconducting cable burns up on cooling loss through the current lead when operating in the vapor-cooled mode[1,14] or in the

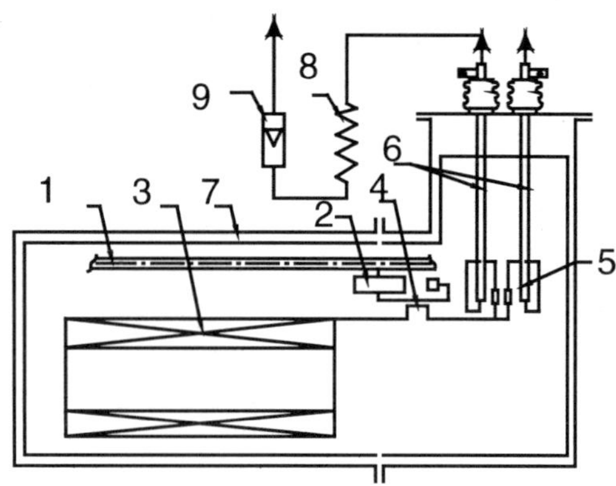

Figure 3. A schematic diagram of the electric and helium communications of the current lead with a magnet: 1—supply header, 2—insulating bushing, 3—superconducting magnet, 4—solder joint between superconductor cables, 5—solder joint between the superconductor cable and the current lead, 6—current lead, 7—cryostat, 8—heater, 9—flowmeter.

force-cooled mode. In large magnet systems with a slow current load-off, the cold ends of the current leads are connected to a copper stabilized superconductor cold bus.[18] The current lead allows a 40–50% overload at a corresponding addition of the mass flow rate through the lead. The Nuclotron current leads, designed for steady operation at 6.2 kA, have been operated for magnet testing at currents of up to 9.8 kA. Because the current leads operated at higher than optimum currents, the mass flow rate through the leads was increased from 1.6 g/s to 1.8 g/s so that the current leads might not cause the magnet to transition to the normal state.

The current lead-design parameters that meet the above-described conditions are shown in Table 2 for a 13-kA current lead. An analysis of this lead configuration following the procedures in Ref. 19 was conducted. The optimum operating flow, based on minimizing the Carnot refrigeration power, was at the minimum flow of 0.625 g/s for the 13-kA operating condition and around 0.1 g/s for the 0-kA operating condition. The temperature profiles for these two operating conditions are given in Fig. 4. The estimated heat load and top end temperatures are given in Fig 5. Using the experimentally determined overall heat transfer coefficient from Ref. 19, it is seen that the predicted operating temperatures of the lead should be at or above 250 K.

Table 2. Design parameters for force-cooled mode, current-carrying, copper-welded conductor

Description	Size
Length of the current-carrying conductor	1.0 m
Current carrying cross section, upper part	0.00075 m^2
Current carrying cross section, lower part	0.00017 m^2
Mass flow rate through the lead	0.65 g/s
Specific heat leak per lead	<1 W/kA
Hydraulic resistance	10–12 kPa
Warm end temperature	>250 K

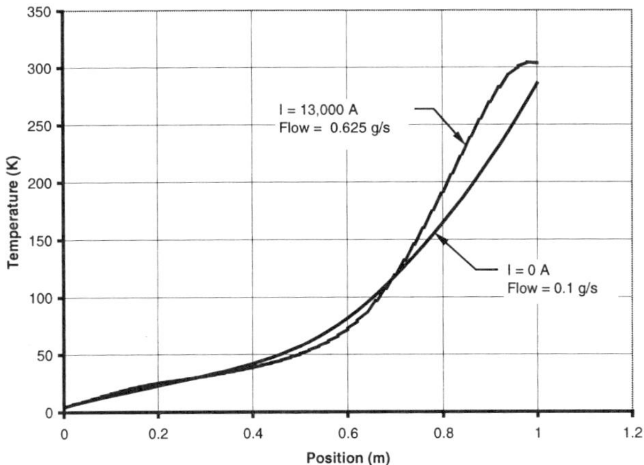

Figure 4. Calculated temperature distribution for 13,000-A current lead at different operating currents.

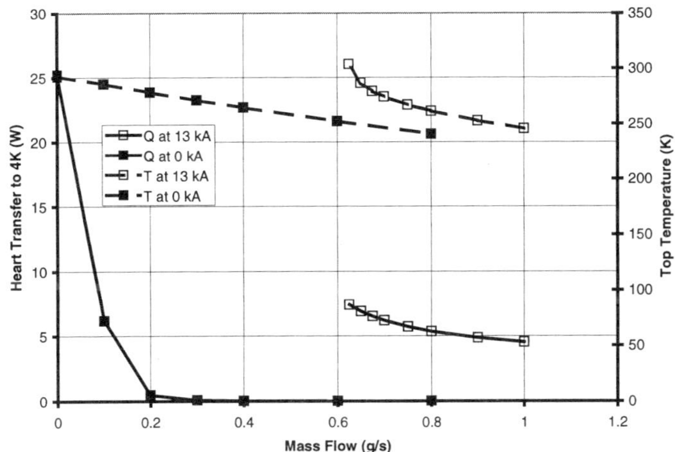

Figure 5. Calculated heat load and top end temperature for a two-stage, 13-kA, forced-cooled current lead at different cooling flow rates.

CONCLUSIONS

According to Ref. 10, manufactured current leads of the box conductor design operate with a specific heat leak of 1 W/kA in the vapor-cooled mode. They are compact and reliable in operation. The actual current lead heat influx is best determined in the vapor-cooled mode. For the design of current leads, the proper selection of the main geometric features (length, cross section, and heat transfer surface area) must be accomplished. Some useful guidelines have been discussed based on previous design experiences.

ACKNOWLEDGMENTS

The author thanks Yu. A. Shishov for useful consultations and discussions regarding the problems considered in the review. Research is partially sponsored by the Office of Energy Efficiency and Renewable Energy, U.S. Department of Energy under contract DE-AC05-96OR22464 with Lockheed Martin Energy Research Corp.

REFERENCES

1. V. E. Keilin and I. A. Kovalev, Electrical leads up to 10 kA for the superconductor magnet system, *Prib. Tekh. Eksp.* 6:178 (1974).
2. H. Katheder and L. Shappals, Design and test of a 10-kA gas cooled current lead for superconducting magnets, *IEEE Transactions on Magnetics*, V.MAG-17, 15:2071–2074 (1981).
3. K. Maehata, S. Kawasaki, K. Ishibashi, Y. Wakuta, H. Kawamata, and T. Shintomi, Operational performance of spiral-fin current leads, *Cryogenics* 33:680 (1993).
4. D. Leroy and D. Oberli, "18 kA vapor-cooled current leads using a stack of copper finned foils," Proc. Twelfth Int. Cryo. Eng. Conf., Southampton, U.K. (1988), p. 237.
5. J. W. Lue, W. A. Fietz, R. E. Stamps, G. R. Zahn, and J. R. Miller, "Test results of the vapor-cooled leads for the JFSMTF," *Adv. Cryo. Eng.* 31:217 (1986).
6. D. Hagedorn et al., "18 kA vapor-cooled current leads to test superconducting magnet models for the proposed large hadron collider at CERN using wire mesh heat exchangers," Proc. Twelfth Int. Cryo. Eng. Conf., Southampton, U.K. (1988), p. 242.
7. E. Tada et al., "Development of 30 kA vapor-cooled current leads for fusion devices," Proc. Eleventh Int. Cryo. Eng. Conf., Berlin-West, Germany (1986), p. 528.
8. E. I. Mikulin and Yu. A. Shevich, Matrix heat exchange devices, Moscow, Engineering Industry, 1983.
9. V. D. Bartenev, The force-cooled 2.5 kA current leads for force-cooled superconducting magnets of the Nuclotron, Preprint *JINR*, Dubna (1990).
10. V. D. Bartenev and Y. A. Shishov, Force-cooled current leads for the force-cooled superconducting magnets of the Nuclotron, *Cryogenics* 31:985 (1991).
11. R. C. Niemann, Y. S. Cha, J. R. Hull, C. M. Rey, and K. D. Dixon, "Design of a high-temperature superconductor current lead for electric utility SMES," *IEEE Trans. Appl. Supercond.* 5:789 (1995).
12. R. Heller and J. R. Hull, "Conceptual design of a 20 kA current leads using force-flow cooling and Ag-alloy-sheathed Bi-2233 high-temperature superconductors," *IEEE Trans. Appl. Supercond.* 5:797 (1995).
13. A. Ballarino and A. Ijspeert, "Expected advantages and disadvantages of high-Tc current leads for the large hadron collider orbit correctors," *Adv. Cryo. Eng.* 39:1959 (1994).
14. V. D. Bartenev, Operation of current leads at different kinds of cooling, Preprint *JINR*, Dubna (1995).
15. R. McFee, "Optimum Input Leads for Cryogenic Apparatus," *Rev. Sci. Instr.* 30(2) (1959).
16. H. G. Khodzhibagiyan and A. A. Smirnov, "The concept of a superconducting magnet system for the Nuclotron," Proc. Twelfth Int. Cryo. Eng. Conf., Southampton, U.K. (1988), p. 841.
17. H. Desportes, I. Le Bars, and G. Mayaux, Construction and test of the CELLO thin-wall solenoid, *Adv. Cryo. Eng.* 25:175 (1980).
18. M. Heiberger, D. G. Morris, D. Tallorin, A SMES shuntable vapor-cooled current lead array reduces liquid helium usage and improves reliability, *Adv. Cryo. Eng.* 39:1967 (1984).
19. Q. S. Shu, J. A. Demko, R. Domam, D. Finan, T. Peterson, I. Syromyatnikov, and A. Zolotov, "The thermal optimum analyses and mechanical design of 10 kA vapor cooled power leads for the SSC superconducting magnet tests at MTL," *IEEE Trans. Appl. Supercond.* 3(1) (1993), p. 408.

YBCO COATED CONDUCTOR CURRENT LEAD TESTING

P.E. Blumenfeld, J.Y. Coulter, S.R. Foltyn, D.E. Peterson, F.C. Prenger, E.W. Roth, J.A. Stewart

Los Alamos National Laboratory
Los Alamos, NM, 87545

ABSTRACT

Current leads formed from YBCO coated conductors have been tested at the Superconductivity Technology Center (STC) at Los Alamos National Laboratory. The test facility at STC provides conductive cooling for binary current leads, i.e. a warmer conventional lead section joined to a colder high temperature superconducting (HTS) section. Cryogenic heat pipes filled with nitrogen are used as electrically isolated thermal intercepts at the junction of the conventional and the HTS sections. Temperatures are controlled at the warm and cold ends of the HTS sections. Refrigeration is provided by a mechanical cryocooler. Self-field critical current measurements ranging from 0 to 200 A are reported for conduction cooled leads in a vacuum with warm and cold end temperatures varied as parameters. These critical current measurements are compared with critical current measurements made on the same components in a 75 K liquid nitrogen bath. It is found that the critical current of the conduction-cooled leads is nearly twice that measured in liquid nitrogen when the warm end of the leads is held at 70 K and the cold end is held at 60 K.

INTRODUCTION

Superconducting devices operating at low temperature are not as simple to configure and operate as their conventional counterparts. Economic viability, however, requires that benefits outweigh costs when new technologies compete with old. There is continuing motivation, then, to improve the quality, efficiency and reliability of not just superconducting appliances such as magnets and transformers and motors, but of the peripheral systems that connect these things with the world at large. The research presented here is immediately concerned with one such peripheral system: current leads which carry electricity to superconducting devices, and less directly with another: refrigeration systems which maintain cryogenic environments.

In fact these systems are intimately connected. Well-designed current leads reduce refrigeration loads while refrigeration system design impacts the performance of current leads. A brief survey of current lead systems will demonstrate this point. The simplest current lead configuration is a metallic conductor running from room temperature directly to a cryogenic environment. This design's geometry can be optimized to reduce the cold end heat load.[1] If the refrigeration system uses a liquid cryogen, the cold vapors boiled off can be forced back up the lead, with a consequent reduction in total refrigeration power.[2] Thermally anchoring a metallic lead at an intermediate temperature also reduces its ultimate cold end heat load and the total refrigerator work rate.[1] but the refrigeration system must then provide heat sinks at multiple temperatures. Reductions in cold end heat load can be achieved by use of high temperature superconducting (HTS) lead materials which divorce electrical properties from thermal properties in a way that no other material can, but these also require an anchored, or "binary" configuration with an intermediate temperature thermal intercept.[3] Finally, vapor cooling of HTS leads can decrease their cold end heat loads as well.[4]

Although liquid cryogens can increase current lead performance and are convenient for laboratory use, they are not always a good choice. Widespread acceptance of commercial superconducting technology may, in fact, be impeded by continued reliance on liquid cryogens. Mechanical refrigerators, running on ubiquitous AC electricity are increasingly sought as the refrigeration system of choice for small, medium, and even large scale superconducting appliances. With this in mind, the present research is focused on development of current lead components for use in conduction-cooled, mechanically refrigerated environments.

Next considering the choice of lead materials, it is clear that HTS leads can be a great benefit in a mechanically refrigerated system, reducing the cold end heat load and thus reducing the cost of the refrigeration system. Further, the thermodynamics of such refrigerators often results in a staged design convenient for thermal anchoring of the warm end of a superconducting lead section. HTS leads present their own design challenges: leads formed from bulk HTS materials are brittle or stiff or both, and may require careful strain relief at their terminations. Newer HTS materials in a flexible tape form are simpler to install and more mechanically reliable, but are susceptible to damage from overheating during system faults. Burnout protection in the form of metallic conductors can be added to either bulk or tape HTS leads but is always traded against increased cold end heat load.

The present research considers a relatively new kind of HTS tape material, YBCO coated conductors[5], for use as current leads in a conduction-cooled, mechanically refrigerated environment.

COATED CONDUCTOR PROPERTIES

YBCO coated conductors have properties that are desirable in HTS current leads. They are thin, strong, and flexible; they have a low thermal conductance and a high critical current density.

Configuration

Figure 1 presents a comparison of the layer thicknesses in a typical coated conductor. The alloy substrate constitutes the bulk of the material. The silver layer is quite thin and is of variable thickness as it serves mainly to protect the YBCO. It is possible to manufacture these conductors with no silver layer at all, if desired, except at the ends where the silver is used for electrical contacts. The alloy substrate renders the conductor strong and flexible. It has been shown that these materials can be bent to a radius of 1 cm with only a small and reversible derating of their critical current[6]. This flexibility is desirable in an HTS current lead: it allows the system designer to bend the leads to suit the application. Coated conductor current leads may thus make HTS systems simpler, more reliable, and less expensive by eliminating mechanical constraints imposed by bulk HTS leads.

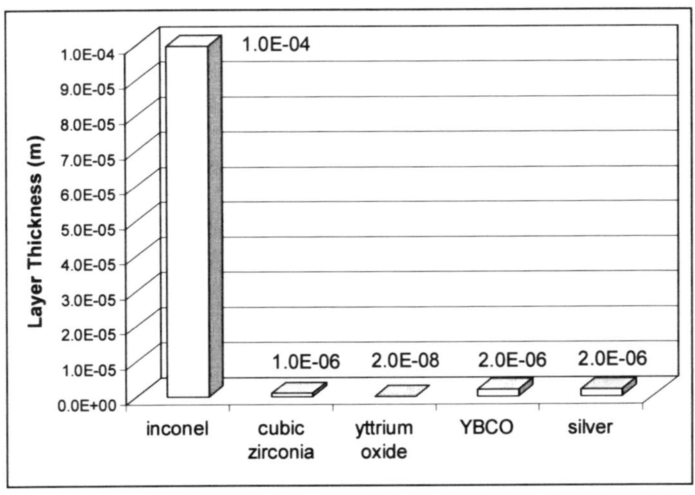

Figure 1. Layer thicknesses in typical YBCO coated conductor

Thermal Properties

HTS current leads are used to reduce the heat load to a cryogenic electrical system's cold end. Coated conductors' thermal properties suggest that they may perform as well in this regard as bulk HTS materials have previously done. Figure 2 presents the heat transfer rates predicted for each layer in a 10 cm length of the conductor represented by Figure 1, assuming end temperatures typical for a mechanically-refrigerated HTS application. Temperature-dependent thermal properties[7] were integrated between the boundary temperatures to make these predictions. The cubic zirconia and yttrium oxide layers are very thin and have low thermal conductivities. The silver layer is the greatest contributor to cold end heat load.

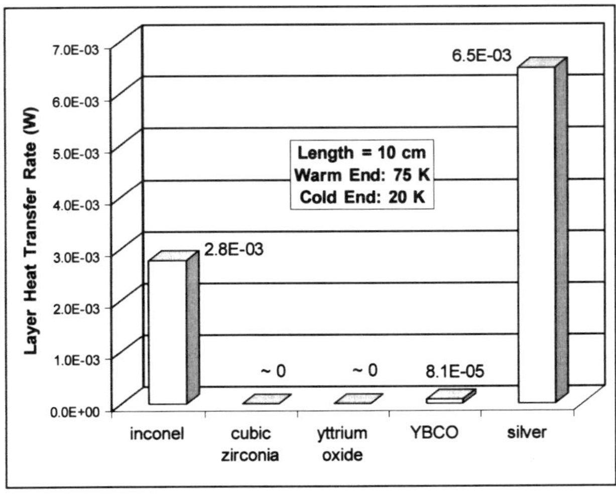

Figure 2. Conduction heat transfer rates in typical YBCO coated conductor layers

The total heat load predicted for the three contributing layers is 9.4×10^{-3} watts. A coated conductor of this design may carry 100 A, thus a normalized heat load of 9.4×10^{-5} W/A is predicted.

Thermal Performace with Protection

This prediction is made for a bare conductor with no protection against burnout. Optimization studies are in progress to design a passive protection shunt system appropriate for use with coated conductor materials. Preliminary results indicate that normalized cold end heat loads below 1.0×10^{-3} W/A may be achieved in combination with full passive protection. This heat load is comparable to that achieved with bulk HTS materials and is further incentive to test coated conductors configured for use as current leads.

TEST CONFIGURATION AND PROCEDURE

YBCO coated conductors are configured as conduction-cooled current leads and the critical current of the current lead system thus formed is measured in self-field while varying the HTS leads' warm and cold end temperatures as parameters.

Thermal/Mechanical Configuration

The thermal and mechanical configuration of the test equipment has been previously described[8]; an overview is presented here. A cryogen-free, conduction-cooled thermal environment is created to simulate the conditions under which current leads must operate in a mechanically refrigerated HTS system. A binary current lead configuration is used which includes coated conductor HTS sections, conventional lead sections, an electrically isolated thermal intercept and low resistance thermal/electrical junctions.

Mechanical Refrigeration. A Gifford-McMahon cycle 2-stage cryocooler provides refrigeration. This is an APD brand model 204 cryocooler which develops approximately 17 watts of cooling capacity at 77 K on its first stage. Its second stage operates at 8 to 10 K at the low heat loads carried by the coated conductor lead sections.

Conventional Current Lead Sections. Current is carried from the room-temperature top plate of the experiment's vacuum vessel to the HTS sections through conventional current leads. These are formed from flexible copper braid, and are optimized in their length and cross-sectional area for minimal heat transfer (about 9 watts per lead) while carrying 200 amperes.

Electrically Isolated Thermal Intercepts. An intercept with both high thermal conductance and high electrical resistance is required at the junction of the conventional and HTS lead sections. A composite structure is typically used, a combination of an electrical insulator with a thermal conductor. The thermal conductance of the structure is determined in part by the interface of the two materials and can be increased by increasing the surface area and/or the pressure at the interface.

A different method is used in the present work to increase the thermal conductance of the interface. An intercept is built around a nitrogen-filled cryogenic heat pipe. This device is described in detail elsewhere.[9] It consists of a heavy-walled copper cylinder (condenser) with copper end plugs. A copper rod (evaporator) passes through bores in the plug with some annular clearance and protrudes from either end. The clearance between the rod and the plug bores is sealed with epoxy, thus forming a cavity which is filled with nitrogen gas to a nominal gage pressure of 13.8 MPa (2000 psi) at room temperature.

The conventional lead section is attached to one end of the evaporator rod and the HTS section is attached to the other end. The condenser body is mounted to the first stage of the cryocooler. When the condenser temperature is between nitrogen's triple and critical point temperatures a two-phase convective cell is formed in the cavity. The high heat flux associated with two-phase heat transfer gives the heat pipe a high thermal conductance. The epoxy seal provides voltage isolation.

Junctions. Thermal/mechanical junctions between the coated conductors and conventional conductors are made with clamped joints. Flat surfaces are machined onto the conventional conductors and then polished. Silver-bearing grease is then applied to the polished surfaces and the silver side of the coated conductor is then clamped firmly to this prepared surface. This junction design allows for installation and removal of the coated conductor sections without soldering and desoldering, while maintaining good thermal and electrical conductance.

Electrical Configuration

The coated conductor current leads are electrically shunted by a busbar at the cold end. This busbar is thermally anchored to the cryocooler's second stage. Current is passed through the top plate of the vacuum vessel to the warm ends of the conventional lead sections by electrical feedthroughs rated at 1000 A. Two power supplies are used during the tests: a Power Ten brand model 3030 which can source up to 110 A, and a Hewlett-Packard model 6464 which can source up to 1200 A.

Instrumentation and Control

Temperature Sensors and Heaters. Silicon diode sensors on a Lakeshore brand model 805 temperature controller monitor the cryocooler's first and second stage temperatures. The controller also powers heaters located inside of the heat pipe evaporator rods. The heaters are used to control first stage temperature and to measure thermal conductance in the heat pipes. The second stage temperature is measured on the cold end busbar and is controlled there by a heater. The coated conductors' cold end temperatures are indicated by the busbar temperature. Silicon diode sensors are also located on the evaporator side of the warm end junctions, and on the warm ends of the coated conductors. Diode sensors on the coated conductors are calibrated to an accuracy of +\-0.25 K in the temperature range of interest; all other diode sensors are calibrated to an accuracy of +\-0.5 K.

Current Lead Voltage and Current. The voltage drop across each of the coated conductor leads is measured by means of spring-loaded clips that contact the conductors at the warm and cold ends near the mounting junctions. Thus voltage is measured over a conductor length of slightly less than 8 cm. Resolution in voltage measurement is +\- 0.8 μV as compared with an 8 μV criterion for critical current. Current through the leads is indicated by voltage drop across a calibrated external shunt and is measured with a resolution better than +\- 0.5 A.

Data Acquisition and Control. Voltage signals from diode sensors not monitored by the temperature controller are read by a Hewlett Packard model 34970A scanning digital voltmeter. The voltmeter also monitors lead and current shunt voltages. The temperature controller and the voltmeter communicate with LabVIEW® software running on a personal computer through an IEEE-488.2 interface. Data from the voltmeter are converted in software to temperature and current readings, then displayed and written to disk files along with data from the temperature controller. The voltmeter also contains a multifunction card which can source voltage under software control.

Critical Current Measurement. LabVIEW® software is also used to control lead current during critical current measurements. Maximum current and current ramp rate are programmed and a corresponding voltage ramp generated by the digital voltmeter's multifunction card in turn controls the current supplied to the leads. Ramps proceed in 0.5 second steps with the current incremented slightly higher each step and lead voltages immediately checked for indications of resistance in the leads. As soon as a programmed threshold voltage is observed across either of the leads the current is set to zero. It is possible to use this active lead protection system because there is no inductive load in the circuit. To measure critical current the voltage threshold is set to a value greater than the 8 μV critical current criterion, usually 50 μV. After allowing the system to settle on the desired set of warm and cold end temperatures a series of current ramps is made, increasing the maximum current each time until the voltage threshold is crossed. The lead end temperatures are rechecked to be sure that they are still near the desired temperatures after the ramps. The critical current for that set of temperatures is then inferred from the voltage vs. current curve generated during the last ramp.

RESULTS AND DISCUSSION

Bath Cooled Critical Currents

Figure 3 presents the results of previous critical current tests performed in a 75 K liquid nitrogen bath. Note that the first and tenth centimeter of each tape is clamped into the junctions for the conduction cooled tests; the middle eight centimeters are in vacuum. The section showing the lowest critical current on lead 1 is at 6 cm which has a critical current of 100 A, and the lowest on lead 2 is at 10 cm with a critical current of 95 A.

Conduction-Cooled Critical Currents

Compare Figure 3 with Figure 4 which presents the results of the conduction-cooled tests. Here critical currents are not measured locally as in Figure 3, but are measured over the

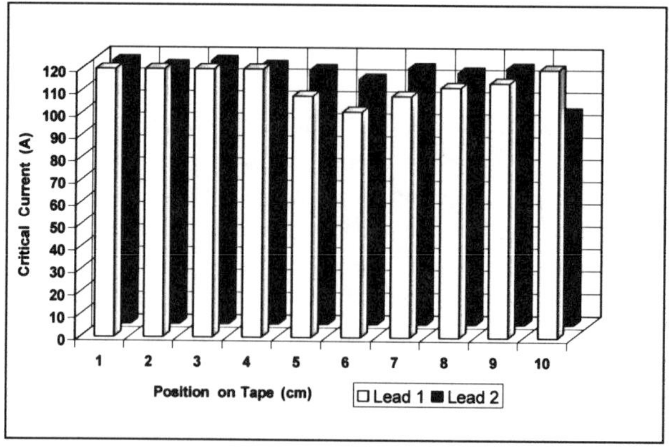

Figure 3. Critical currents measured locally on each lead in a 75 K liquid nitrogen bath

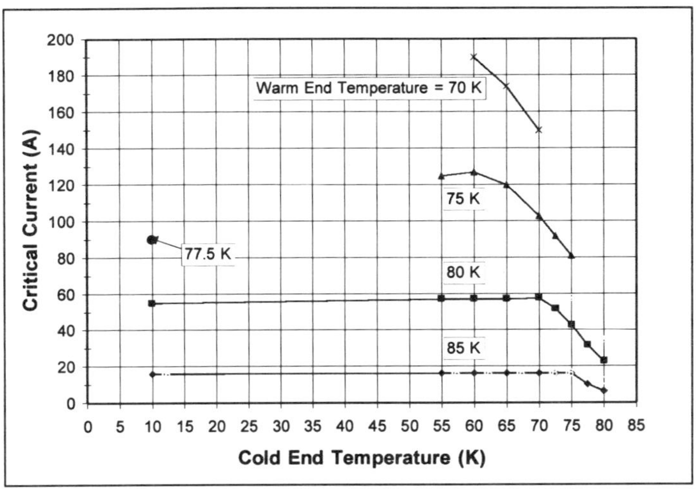

Figure 4. Conduction-cooled critical currents with end temperatures varied as parameters

8 cm length between the junctions. A strong dependence of critical current on warm end temperature is observed. A weaker dependence of critical current on cold end temperature is also observed. Note the data set representing the 75 K warm end temperature. With the cold end of the lead also at 75 K the critical current is somewhat less than the lowest value observed for any section in bath cooling. For end temperature differences greater than 10 K, however, the critical current is equal to or greater than the highest value observed anywhere on the lead in bath cooling. The 75 K warm end curve appears to level off for end temperature differences greater than 10 K, as do the higher temperature curves. Currents greater than 190 A were not measured, but it may be inferred that the 70 K curve levels off near 200 A for end temperature differences greater than 10 or 15 K.

Heat Pipe Thermal Conductance

The thermal conductance of the cryogenic heat pipes is measured over a range of temperatures and heat transfer rates. Table 1 presents these results for a 75 K evaporator temperature along with conservative estimates for lead currents that would put these heat loads onto a heat pipe in this application.

Table 1. Cryogenic Heat Pipe Thermal Conductance

Heat Transfer Rate (W)	Equivalent Lead Current (A)	Thermal Conductance (W/K)
12	250	3.4 (measured)
25	500	5.0 (extrapolated)
50	1000	5.8 (extrapolated)

Thus at 1000 A a temperature rise of 8.6 K is predicted across the heat pipe thermal intercept. The breakdown voltage of the heat pipes is found to be approximately 1000 V.

CONCLUSIONS AND FUTURE WORK

Self-field critical current measurements made on conduction-cooled YBCO coated conductor samples indicate that these materials can be put to good use as HTS current leads in mechanically refrigerated applications. It is found that the nominal critical current of a 10 cm length of coated conductor as measured in a liquid nitrogen bath is achieved in conduction cooling by holding the warm end of the conductor at 75 K and cooling the cold end to 65 K or below. Further, the critical current in conduction-cooled leads is nearly doubled over the nominal bath-cooled value by establishing warm and cold end temperatures of 70 and 60 K, respectively.

These results are encouraging for development of cryocooler-based superconducting technologies. The compact size and high current carrying capacity of these coated conductor materials, combined with their strength and flexibility have potential to greatly ease system design constraints. The results reported here indicate that a safety factor of two in current capacity over nominal critical currents measured in isothermal baths may be achieved while efficiently operating a two-stage cryocooler with its first stage near 70 K.

Results from the heat pipe conductance tests are also encouraging. The increased thermal conductance with increased heat load is a desirable feature which will help to keep a cryocooler's first stage temperature high, thus maintaining refrigeration cycle efficiency. Recent measurements[10] of an improved heat pipe design show thermal conductances more than twice those reported here. New designs are also being evaluated to improve the dielectric strength of the device.

Future work on YBCO coated conductor current leads will include development of a passive burnout protection system, construction of kiloamp-scale leads formed from multiple coated conductors and lead testing in magnetic field. Improvements to electrically isolated cryogenic heat pipe technology will be pursued as a promising adjunct to current lead development.

ACKNOWLEDGMENT

This work was supported by the U.S. Department of Energy Office of Energy Efficiency and Renewable Energy.

REFERENCES

1 R. McFee, Optimum input leads for cryogenic apparatus, *Rev. Sci. Instrum.* 30,2:98 (1959).

2 Y.L Buyanov, Current leads for use in cryogenic devices: principle of design and formulas for design calculations, *Cryogenics* 25,2:94 (1985).

3 C.M. Rey, Re-evaluation of the intercept temperature for HTS current leads based on practical and economical considerations, *Cryogenics* 38,12:1217 (1998).

4 J.R Hull, High temperature superconducting current leads for cryogenic apparatus, *Cryogenics* 29,12:1116 (1989).

5 P.N. Arendt, et al., YBCO/YSZ coated conductors on flexible Ni alloy substrates, *Applied Superconductivity* 4,10-11:429 (1996)

6 S.R. Foltyn, et al., High-Tc coated conductors – performance of meter-long YBCO/IBAD flexible tapes, Applied Superconductivity Conference, Palm Desert (1998)

7 V. Arp, Cryocomp properties module V3.0, Cryodata, Inc, Louisville (1996)

8 P.E. Blumenfeld, et al., High temperature superconducting current lead test facility with heat pipe intercepts, Applied Superconductivity Conference, Palm Desert (1998)

9 M.A. Daugherty, F.C. Prenger, D.D. Hill, D.E. Daney, K.A. Woloshun, HTS current lead using a composite heat pipe, *IEEE Trans. App. Superconductivity* 5,2:773 (1995)

10 E.W. Roth, F.C. Prenger, J.A. Stewart, P.E. Blumenfeld, Testing of an improved nitrogen heat pipe thermal intercept for current leads, 1999 Cryogenic Engineering Conference, Montreal (1999).

PROTOTYPE 13 kA HIGH TEMPERATURE SUPERCONDUCTING CURRENT LEADS FOR THE LARGE HADRON COLLIDER

Y. Yasukawa[1], S. Nose[1], M. Nozawa[1], M. Konno[2], K. Sakaki[2], T. Uede[2], A. Ballarino[3] and T. Taylor[3]

[1]Fuji Electric Corporate Research and Development, Ltd.
7 Yawata-kaigandori, Ichihara-shi, Chiba, 290-8511 Japan

[2]Fuji Electric Co.,Ltd.
1-1 Tanabeshinden, Kawasaki-ku, Kawasaki-shi, Kanagawa, 210-8530 Japan

[3]CERN
1211 Geneva 23, Switzerland

ABSTRACT

Fuji Electric has designed and manufactured prototype 13 kA high temperature superconducting (HTS) current leads for the LHC project. The current leads are required to have high thermal performance and high reliability in compact dimensions. The design was optimized in terms of material requirements, dimensions and cooling scheme to enhance thermal performance. The resistive part of the current lead is composed of a large number of phosphorus deoxidized copper strands and the HTS part of the current lead is composed of Bi-2223 silver-sheathed tapes. Connections between current carrying parts are made by welding or soldering methods to minimize contact resistance. The performance tests have been carried out by European Organization for Nuclear Research (CERN). The current leads showed the thermo-electric performance which satisfies the technical specifications.

INTRODUCTION

Fuji Electric has developed conventional current leads, especially large capacity ones since 1983. Fuji Electric began a feasibility study of a current lead using high-temperature-superconductor (HTS) in 1989. 1 kA HTS current leads cooled with boiled-off helium gas have already been developed and other 1 kA HTS current leads of the intermediate cooling type have also been developed[1,2]. For those leads, the heat load at the low temperature end is reduced to 10 % of that of the low temperature end of an optimal conventional lead. For ac use, 2 kA HTS current leads were designed and tested. The current leads demonstrated the possibility in ac use[3].

HTS current leads, however, are the most suitable for large cryogenic systems because heat loads from current leads are a major part of total loss in the system. Fuji Electric began the study of HTS applicability to large capacity current leads in 1995. We demonstrated the applicability of 10 kA-class HTS current leads to fusion use in collaboration with Japan Atomic Energy Research Instituite[4, 5].

On the other hand, CERN has been developing the 13 kA HTS current leads for the Large Hadron Collider (LHC)[6, 7]. Fuji Electric has designed and manufactured prototype 13 kA HTS current leads based on the technical specifications of CERN. This report describes the concept of design and the results of tests carried out by CERN on these leads.

DESIGN OF THE CURRENT LEAD

Technical Specifications

The 13 kA HTS current leads required by CERN are to be of high thermo-electric performance . The nominal operating current and stand-off voltage are 13 kA and 3.5 kV. The mass flow rates required during 13 kA operation and zero current are less than 1.0 g/s/lead and 0.6 g/s/lead, and the heat loads into liquid helium during 13 kA operation and zero current are less than 1.5 W/lead and 1.0 W/lead, respectively. The pressure drop thorough the resistive part of the lead is less than 50 mbar. In addition, the overall length of the current lead is limited to a maximum of 1500 mm. A design which satisfies these required values is described below.

Resistive part of the lead

The thermal performance of the HTS current lead depends greatly on the performance of the resistive part of the lead. As the cooling gas for the resistive part occupies main portion of the actual electric power of the refrigerator for the current lead cooling. The resistive part is limited in a adequate length because the overall length of the lead must be less than 1500 mm. In addition, there should be no frost and condensation at the room temperature terminal. Therefore, the conductor of the 13 kA current lead is required to have high heat transfer effectiveness from 50 K to 300 K in a short length.

A parameter survey for thermo-electric performance was carried out in order to optimize the main parameters of the resistive part of the lead. Parameters that should be considered are structure, material, density of conductor, length of heat exchanger and flow rate of the gas helium.

The bundle structure which has a cable-in-conduit with copper strands was adopted for the resistive part on the basis of our experience. The parameter survey was carried out by the thermal analysis program that was developed for the bundle type current leads.

Oxygen free copper (OFC) is sometimes suitable for the current lead material for a long length current lead. When the overall length is allowed to be more than 1 m, leads made of OFC show high thermal performance. However, as the length of the resistive part of the lead is limited as above mentioned, phosphorus deoxidized copper (PDC) is an appropriate material to avoid this problem. The resistive part of the lead made of PDC has another advantage for decreasing the required flow rate during zero current operation. We chose PDC as the material of the resistive part of the current lead. The current density and length of the heat exchanger has been optimized using by the thermal analysis program in the aspects of heat load at the warm end of the HTS, pressure drop, outlet gas temperature and quenching behavior. The main parameters of the resistive part of the current lead are given in Table 1.

Design of the HTS part of the lead

The design concept on the HTS part of the lead was based on the 10 kA-class HTS current lead[4]. The HTS part of the lead has been optimized taking account the measured thermo-electric performance of the 10kA leads in order to satisfy both the steady state and the quenching state requirement.

The HTS part of the lead is constructed from 20 parallel HTS elements in a cylindrical array. The HTS elements are attached to a stainless steel tube to insure good thermal contact. In order to decrease the heat load into the cryostat, the stacked number of the HTS elements in the lower temperature region is tempered. The HTS element is 6 stacked HTS tapes in the higher temperature region and 4 stacked HTS tapes in the lower temperature region. The heat exchanging length of an HTS element is 30 mm in the higher temperature region and 270 mm in the lower temperature region. In addition, the cooling gas penetrates the HTS elements from inside of the tube as well as from the outside of it.

For the current lead use, the silver-gold alloy is used to utilize as the HTS tape matrix. The HTS matrix is silver alloy of Ag-Au 1.0 at %. The HTS material is BSCCO-2223. The cross-sectional dimensions of the HTS tape are 6.5 mm$\times$0.45 mm.

The critical currents of HTS tape are more than 123 A at 77 K, 250 A at 60 K and 387 A at 50 K without external field. Figure 1 shows the magnetic field dependence of the normalize critical current at 60 K. A magnetic field analysis of HTS elements arranged in a cylindrical array was carried out with the ANSYS® program. The load lines of HTS elements in standing alone and in a cylindrical array are shown in Figure 1. The critical current in a cylindrical array enhances the potential ability compared with that in standing alone position. This is because the critical current of an HTS is more sensitive to a field that is perpendicular to the tape surface than it is to a field that is parallel to the tape surface. The total critical current per lead is predicted to be 17.2 kA at 60 K according to Figure 1. The critical current margin at 60 K is more than 30 %.

The HTS elements are produced by Sumitomo Electric Industries.

The design parameter of the HTS part of the lead are also given in Table 1. Figure 2 shows the photograph of the prototype of 13 kA HTS current lead.

Table 1. Design parameters of the 13 kA HTS current lead

Parameter	Design value
Nominal current/voltage	13 kA/ 3.5 kV
Overall length	1434 mm
Resistive part of the lead	
Heat exchanging length	600 mm
Material	Phosphorus deoxidized copper
Structure	Bundle type
Flow rate required during 13 kA operation	0.95 g/s
Flow rate required during zero current	0.5 g/s
Temperature of the low temperature end	50 K
Cooling for the resistive part	Helium gas of 20 K
HTS part of the lead	
Heat exchanging length	300 mm
HTS material	BSCCO-2223
Material of matrix	Ag-Au 1.0 at %
HTS element	Stacked HTS tapes
Structure	Parallel HTS elements in a cylindrical array
Temperature of the warm end of HTS	50 K
Critical current at 60 K	17.2 kA
Heat load at 4.5 K during 13 kA operation	less than 1.5 W
Heat load at 4.5 K during zero current	less than 1.0 W

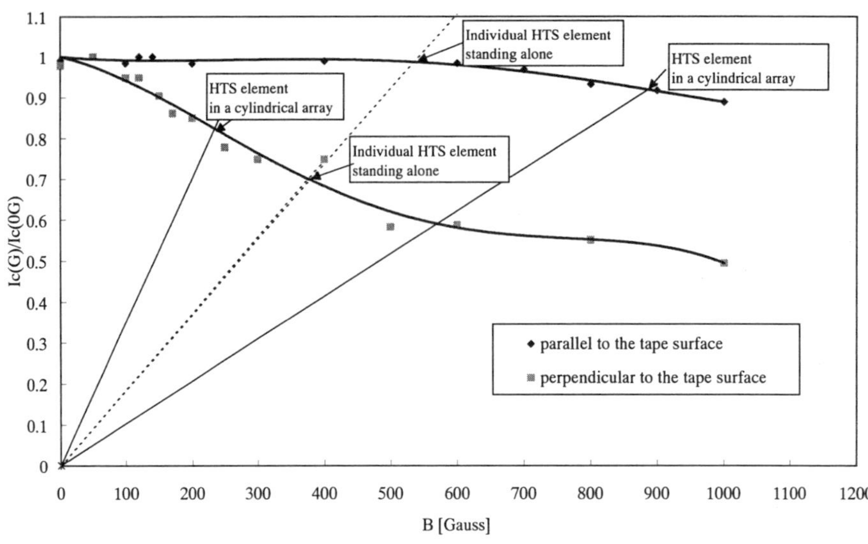

Figure 1. Magnetic field dependence of the normalized critical current at 60 K. The two real lines are indicate the load lines of HTS elements in a standing alone position. The two dotted lines are indicate the load lines of HTS elements in a cylindrical array.

Figure 2. Photograph of prototype 13 kA HTS current lead. This photograph was taken at the factory of Fuji Electric before sending to CERN.

Design of the transition part of the lead

The transition part of the lead means the connection between the resistive part and the HTS part. In order to minimize contact resistance, the copper strands of the resistive part of the lead are connected to a copper block by welding. The copper block is connected to the HTS elements with silver solder.

TEST RESULTS

The test set-up has been built at CERN to test the prototype 13 kA HTS current leads[8]. The tests were carried out by CERN. The current leads are shorted at the low temperature terminal with superconductors (NbTi). Instrumentation, consisting of temperature sensors and voltage taps, are installed on the current leads.

Thermo-electric performance in the steady state

The heat load into liquid helium at 4.5 K was measured from the boil-off rate of liquid helium in the cryostat. The measurement was kept for at least 12 hours. During zero current operation, the measured heat load at 4.5 K is 0.64 W per a lead when the temperature of the warm HTS end ($T_{HTS\ upper}$) is 40 K. The calculated value in the same conditions is 0.74 W per a lead. During the nominal current (13 kA) operation, the measured heat load at 4.5 K is 0.715 W per a lead at $T_{HTS\ upper}$ of 40 K. The calculated value at the nominal current, which is assumed to be the same contact resistance as the measured resistance between the HTS and the lower temperature terminal of 0.7 nΩ, is 0.80 W per a lead. The measured results are fairly in good agreement with the calculated values. These results indicate that the HTS part of the lead is cooled enough with penetrating helium gas. According to the test result, the heat load at 4.5 K during 13 kA operation is equivalent to 0.055 W/kA which is the value by a factor of 20 compared with that of the optimized conventional current lead .

Figure 3 shows the minimum required flow rate for the resistive part of the lead at various current operations. The dots show the test results for lead B and the real line shows the calculation result. The measured values agree with the calculated values up to 10 kA operation. The required flow rate during zero current operation is 0.27 g/s for both lead A and lead B at $T_{HTS\ upper}$ of 40 K. At 13 kA operation, lead A required the minimum flow rate of 1.12 g/s in order to cool the resistive heat exchanger and lead B required that of 0.98 g/s at $T_{HTS\ upper}$ of 40 K. The measured voltage drops through the resistive part of the lead are then 99.5 mV for lead A and 84.0 mV for lead B. The calculated voltage drop for lead B is 88.7 mV. The measured pressure drops are 16 mbar for lead A and 19 mbar for lead B while the calculated pressure drop is 25 mbar at 13 kA operation.

Figure 4 shows the temperature profiles. The real lines show the calculated results which correspond to the temperatures on the terminals and the flow rate of the resistive part of the lead under the same conditions as the test results. The measured and calculated values are fairly in good agreement with each other during zero current operation and 13 kA operation. The contact resistance between the copper block and the HTS at the higher temperature side at 40 K is 1.15 nΩ for both leads. The contact resistances between the HTS in the lower temperature side and the copper terminal are 0.5 nΩ for lead A and 0.7 nΩ for lead B.

When the coefficients of the actual refrigerator are assumed to be 23.5 % at 4.5 K and 26 % at 20 K (relevant to the LHC project), the actual electric power of the refrigerator for current lead cooling is 35 % of that of required a optimized conventional current lead .

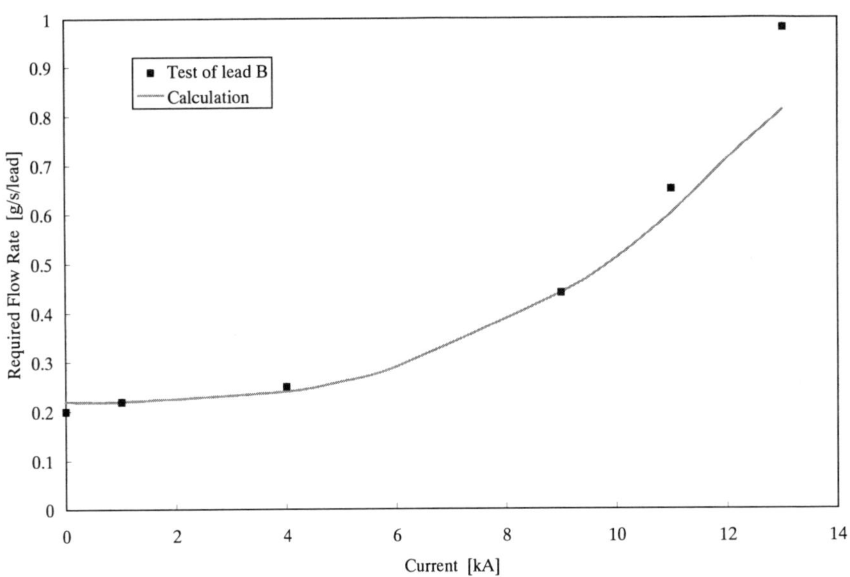

Figure 3. The flow rate minimum required for the resistive part of the lead at various current operations. The test results during 0, 1, 4, 9, 11 and 13 kA operations are plotted .

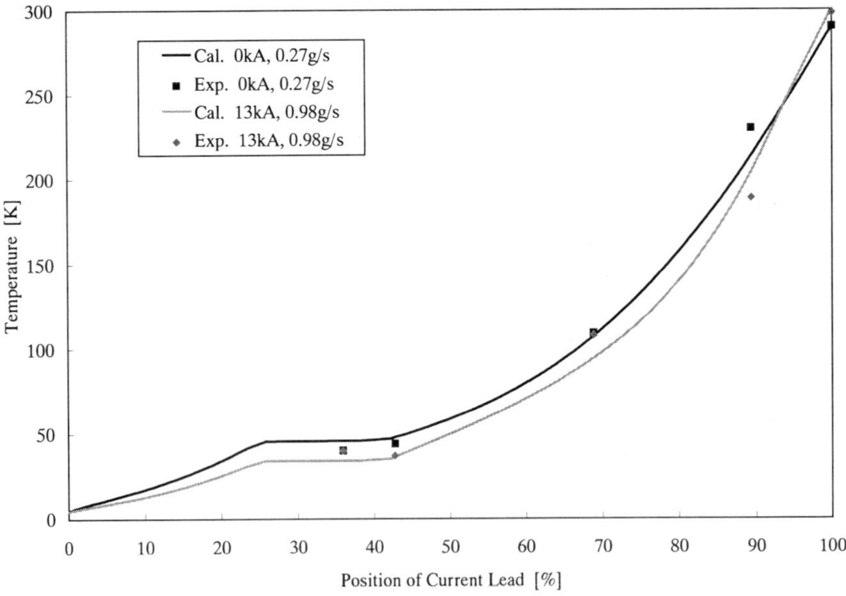

Figure 4. Temperature profiles during zero current and the nominal current operations. The real lines show the calculated results which are corresponded with the temperatures on the terminals and the flow rate of the resistive part of the lead under the same conditions of the test results.

Dielectric Test

The dielectric test of the current lead was carried out by applying 3.5 kV (d.c.) between the current carrying part and the ground in helium atmosphere at the room temperature. The current lead withstood the stipulated voltage. The leakage current measured is less than 3.5 μA corresponding to an insulation resistance of better than 1 GΩ.

Quench Test

The quench test was carried out. The quench detection voltage was increased in steps of 1 mV from 1 mV to 5 mV. Assuming a critical current is achieved when the HTS voltage reaches 5 mV (0.167 mV/cm), the critical temperatures are 86.9 K for lead A and 86.3 K for lead B. For a quench detection voltage of 5 mV, the resistive zone propagates and the voltage increases up to a peak of 0.1 V for lead A and 0.11 V for lead B. For this test, on detection of a quench, the current is made to decay exponentially with a time constant of 120 seconds.

The lead A and lead B presented a comparable behavior in the quench test. After quench tests, the current leads were able to conduct the nominal current and showed the same thermo-electric performance as before the quench test.

Summary of the test

The measured values of lead A and lead B in the comparison with calculated values are given in Table 2.

Table 2. The thermo-electric performance of the 13 kA HTS current leads

Parameter	Measured values lead A	Measured values lead B	Calculated values
zero current			
Temperature of the warm end of HTS	40 K	40 K	45.6 K
Flow rate required	0.27±0.01 g/s	0.27±0.01 g/s	0.2 g/s
Pressure drop	2.5±1 mbar	2.5±1 mbar	0.88 mbar
Heat load at 4.5 K	0.64±0.04 W	0.64±0.04 W	0.74 W
13 kA operation			
Temperature of the warm end of HTS	40 K	40 K	34.0 K
Flow rate required	1.12 ±0.01 g/s	0.98±0.01 g/s	0.81 g/s
Pressure drop	16±1 mbar	19±1 mbar	25 mbar
Voltage drop	99.5 mV	84.0 mV	88.7 mV
Heat load at 4.5 K	0.715±0.04 W	0.715±0.04 W	0.80 W
Critical temperature @0.167 mV/cm	86.9 K	86.3 K	-
Contact resistance at warm end of HTS	1.15 nΩ	1.15 nΩ	-
Contact resistance at cold end of HTS	0.5 nΩ	0.7 nΩ	-

CONCLUSIONS

Fuji Electric has designed and manufactured prototype 13 kA HTS current leads for the LHC project. Performance tests have been carried out by CERN. The heat load into liquid helium was 0.055 W/kA during the nominal current operation which is lower by a factor of 20 compared with optimized conventional current lead. The lead B required a minimum flow rate of 0.98 g/s/lead. The actual electric power of the refrigerator for current lead cooling was reduced to 35 % of that for the optimized conventional lead. In addition, quench tests of the HTS part of the lead were carried out. The current leads were evaluated on a quench detection voltage up to 5 mV. After quench test, the current leads were able to conduct the nominal current and showed the same performance as before the quench test.

Fuji Electric continues to study more reliable and more high performance HTS current leads such as those required for the LHC.

REFERENCES

1. K. Ueda, T. Bohno, K. Takita, K. Mukae, T. Uede, I. Itoh, M. Mimura, N. Uno and T.Tanaka, "Design and testing of a pair of current leads using bismuth compound superconductor ", *IEEE Trans. Appl. Superconductivity*, **3**, (1993), 400-403.
2. K. Ueda, K. Takita, T. Uede, M. Mimura, T. Kinoshita and Y. Tanaka, "Design study and model tests of high Tc superconductor current leads", *Supercollider* 5, (1994), 575-578.
3. Y. Yasukawa, K. Takita, H. Hiue, I. Itoh, M. Mimura, K. Iwashita, Y. Tanaka, M. Iwakuma, K. Funaki, M. Takeo, K. Yamafuji, "Development of 2 kA High-Temperature-Superconductor Current Lead System for AC application", *IEEE Trans. Mag.*, **32**, (1996), 2671-2674.
4. T. Ando, T. Isono, K. Hamada, H. Tsuji, Y. Yasukawa, H. Yamada, M. Konno, K. Sakaki, T. Kato, K. Hayashi, K. Sato and S. Shimamoto, "Design and testing of 10 kA current leads for fusion magnet using high temperature superconductors", *Proceedings of 15th International Conference on Magnet Technology*, **2**, Beijing, (1998), 847-850.
5. T. Isono, K. Hamada, T. Ando, H. Tsuji, Y. Yasukawa, A. Tomioka, M. Nozawa, M. Konno and K. Sakaki, "Test results of high temperature superconductor current lead at 14.5 kA operation", *ASC '98*, Palm Springs, to be published , 1999.
6. A. Ballarino, "High temperature superconducting current leads for the large hadron collider", *ASC '98*, Palm Springs, to be published , 1999.
7. A. Ballarino and L. Serio, "Test results on the first 13 kA prototype HTS leads for the LHC", *PAC '99*, to be published, 1999.
8. A. Ballarino, A. Bezaguet, P. Gomes, L. Metral, L. Serio and A.Suraci, "A low heat inleak cryogenic station for testing HTS current leads for the large hadron collider", to be presented at *CEC/ICMC '99*, Montreal, 1999.

THE SUPERCONDUCTING BUSBAR CABLE POWERING THE AUXILIARY MAGNETS OF THE LHC

R. Herzog, K. Dahlerup-Petersen, C. Parente,
R. Schmidt, F. Sonnemann, M. Teng

CERN, LHC Division, 1211 Geneva 23, Switzerland

ABSTRACT

The tuning of LHC beam parameters and the correction of field errors in the main magnets require several thousand small superconducting magnets operating at currents of up to 600 A, powered in about 40 electrical circuits per octant. All corrector magnets installed close to the main quadrupole magnets will be connected by 53 m long segments of cable with 36 superconducting wires. The cable will be installed in a tube which is welded to the helium pressure vessel of the main magnets and filled with 1.9 K helium. This paper describes electrical measurements on a 32 m long prototype cable installed in the LHC Magnet String. The quench current was determined as a function of applied transverse magnetic field. Quenches initiated by a spot heater were used to measure the quench propagation velocity as a function of the current in the cable. A separate experiment on a single wire helped to understand the quench behavior at thermally insulated wire sections.

INTRODUCTION

In the LHC about 6000 corrector magnets with currents up to 600 A will be installed along a large part of the 27 km circumference of the accelerator. Because power converters can only be located in caverns at both ends of the eight octants, about 1200 km of wire are required to connect the magnets to the converters. To minimize resistive losses, superconducting NbTi multifilamentary wires are used, which are routed within the cryostat system of the main dipoles and the main quadrupoles. During operation they will be immersed in 1.9 K superfluid helium at a pressure of about 1 bar.

The regular arc of the LHC is built up of half-cells, each consisting of three 15 m long dipoles and one short straight section containing a quadrupole (see Fig. 1). Corrector magnets which are powered in series in one circuit are members of one family. Families of the so called spool pieces are located next to the main dipoles. The magnets of all other families, like the quadrupole magnets for the correction of the betatron tunes, the sextupole

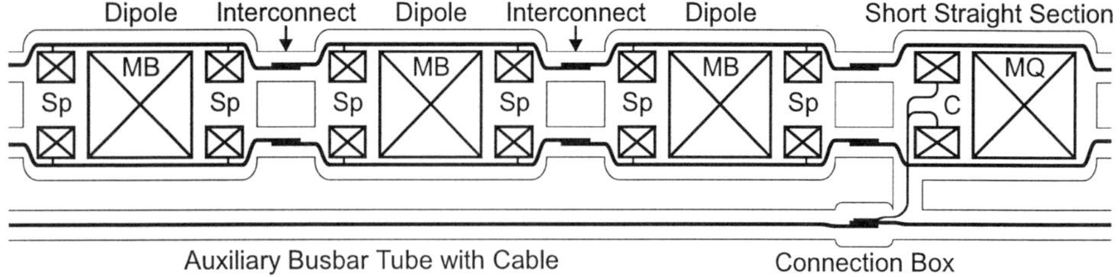

Figure 1. Block diagram of the location and powering of the corrector magnets in one LHC half-cell. Sp: spool pieces at both ends of the main dipoles (MB), independently powered for both apertures. MQ: main quadrupoles in the short straight sections. C: Correctors in the short straight sections (Usually there are more than one per aperture.). The hair line shows the cryostat, which contains 1.9 K helium.

magnets for the correction of the chromaticity and the octupole magnets providing Landau damping are located in the short straight section, close to the main quadrupole magnets.

The busbars for the spool piece magnets are routed inside the cold mass of the dipoles and the short straight sections. Consequently a splice is required at every magnet interconnect (about every 15 m). The total number of splices in the circuits for spool piece magnets is thus about 32000. For the correctors in the short straight sections an even larger number of circuits is required, but they only need to be connected to magnets every 53 m, in the short straight sections. By pulling 53 m long superconducting cable segments into a duct attached to the cold masses (auxiliary busbar tube) it is possible to reduce the number of connections required for these circuits by a factor of 4 compared to the routing of the spool piece circuits. It is considered important to minimize the number of splices in the LHC in order to reduce resistive losses caused by the contact resistance[1] and to minimize the number of faults caused by erroneous or sub-quality connections. The development and testing of a suitable multiconductor cable are the subject of this paper.

The electrical tests conducted on a first 32 m long prototype cable at 1.9 K were intended to prove the satisfactory performance of the cable at nominal conditions and to explore the limits and available margins of the design at higher than nominal currents and applied fields. Quenches triggered by a spot heater allowed the determination of the quench propagation velocity, an important parameter for the design of a quench protection scheme.

CABLE DESIGN

Requirements

The recently studied corrector magnet powering scheme[2] of the LHC requires at least 32 wires for the short straight section correctors in all half cells of the machine and a maximum of 64 wires towards the ends of the arcs. To compensate for thermal contraction during the cooling of the machine from 300 to 1.9 K, the cable will be installed with expansion loops. Geometrical constraints and the installation procedure require that the cable is flexible enough to be bent to a radius of curvature of 200 mm without any damage. On the other hand the conductors of the cable should be firmly held together to prevent quenches caused by their movement within the cable.

The critical current in the superconducting busbar has to largely exceed the critical current of the corrector magnets so that the busbars do not limit the magnet performance. The largest magnetic field at the location of the busbar tube occurs at a section of 0.3 m at the ends of the dipole magnets where their iron yoke is replaced by non-magnetic material and hence the stray field increased to 0.3 T. At other locations the stray field reaches only 40 mT.

Table 1. Required cable properties

Parameter	Value	Parameter	Value
Min / max number of conductors	32 / 64	Insulation withstand voltage	3 kV
Minimum bending radius of cable	200 mm	Radiation dose the cable has to withstand	60 kGy
Nominal current per conductor at 1.9 K	600 A	Copper cross section of one wire	1.8 mm^2
Min. critical current at 1.9 K and 0.3 T	1 kA	Minimum RRR of copper	100

In case of a quench[3] the copper cross section must be large enough to carry the current during the discharge of the magnets without heating the busbar to more than 300 K. The maximum temperature after a quench depends on the RRR value of the copper, the time to detect a quench in a magnet (less than 100 ms), the current I, the inductance L of the circuit and the extraction resistance R, which is switched into the circuit at the time of detection. The time constant of the current decay is $\tau = L/R$. Assuming adiabatic conditions during the magnet discharge it is possible to calculate[4] the peak temperature with the parameters listed above and the quench load $\int I^2\,dt$.

The circuit with the largest inductance consists of 10 sextupole magnets powered in series, each with an inductance of 42 mH. With a resistor for energy extraction of 1 Ω the time constant for discharge is about 0.42 s. The quench load is 111000 A^2s corresponding to a hot spot temperature of less than 50 K. In the unlikely event that a quench starts in a busbar, for example because of a bad connection, the detection takes longer due to the low quench propagation velocity. However, even for a detection time of 1 s, the hot spot temperature still does not rise above 300 K.

The operation of the machine and required testing procedures prior to commissioning demand a withstand voltage of the insulation of 3 kV at room temperature and 600 V at 1.9 K. The insulation has to maintain its properties after an exposure to a radiation dose of 60 kGy (expected dose at the location of the cable over the lifetime of the LHC times a safety factor of three). The conductors should be protected by a suitable outer layer of the cable to prevent any damage, particularly during the installation procedure. The required cable properties are summarized in Table 1.

First Prototype

The LHC layout suggests the use of one cable with about 40 wires (including some spares) in the two middle quarters of the arcs and two such cables in the outer two quarters. 100 m of prototype cable were produced by industry with the cross section shown in Figure 2. Two layers of 18 and 24 conductors, respectively, were wound in opposite directions onto the core, formed by three twisted polyethylene strands of 5 mm diameter. For mechanical protection a copper braid was added as the outermost layer. The 1.6 mm diameter conductors consist of 18 NbTi filaments embedded in a copper matrix with a Cu:NbTi ratio of 9:1. The insulation is made of two layers of FEP (fluorinated propylene ethylene) coated polyimide tape, 35 μm thick, which were wrapped around the wire in opposite directions, each with an overlap of 66 % to obtain an insulation thickness of 0.2 mm. The insulated conductors have thus an overall diameter of 2 mm. The FEP was cured for a short time at 200 °C to seal the insulation.

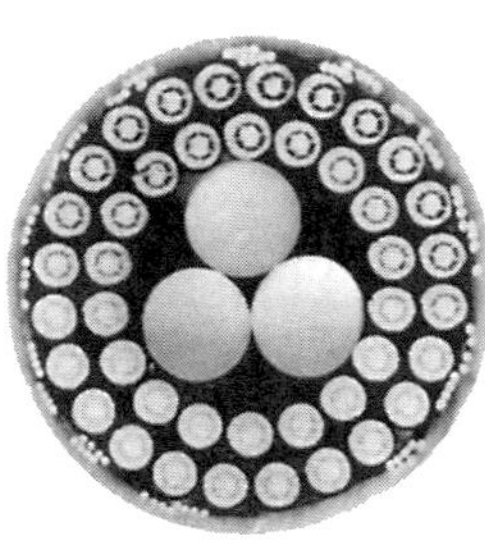

Figure 2. Cross section of the first prototype cable. Centre: 3 polyethylene strands; two layers of superconducting wires with the NbTi filaments clearly visible; protective copper braid. Overall diameter: 15.5 mm

TEST INSTALLATION

The prototype cable was installed in the LHC magnet string. The string[5], a half cell consisting of three prototype dipoles and one prototype quadrupole, was installed in 1994 to validate the simultaneous operation of LHC component systems like magnets, vacuum and cryogenics. A 32 m stainless steel tube with an inner diameter of 30 mm was installed in one of the beam tubes to accommodate the cable to be tested. A transverse magnetic field up to several tesla could be applied. The prototype cable was connected to existing busbars and current leads according to the circuit diagram shown in Figure 3. A small spot heater could be used to initiate a quench at a well defined location. Voltage taps were installed at one end of the cable, thereby monitoring the voltage drop across a pair of wires. Depending on the connection between the power converter and the warm end of the current leads, three configurations could be realized. In configuration 1 the current in neighboring wires of the inner layer was flowing in opposite directions (↑↓↑↓), in configuration 2 it was flowing parallel in one neighbor and antiparallel in the other (↑↓↓↑). With the third configuration the wire pair 1&2 with the spot heater installed could be powered.

PERFORMANCE AT NOMINAL CONDITIONS

During about 40 current ramps to 600 A, the nominal current of the corrector circuits, no quenches were observed with applied fields up to 0.3 T, in configuration 1 (↑↓↑↓) as well as in configuration 2 (↑↓↓↑). However, voltage spikes with a duration of a few ms appeared in magnetic fields above 0.2 T, indicating the movement of conductors within the cable. Only at applied fields larger than 0.3 T or at currents above 600 A some of the spikes were followed by a quench (see below).

QUENCHES INDUCED BY CONDUCTOR MOVEMENT OR BY EXCEEDING I_C

In a series of measurements the current in the cable was increased until a quench was detected by observing a voltage across the taps shown in Figure 3. At the start of a quench two types of voltage signals have been observed. With applied magnetic fields larger than 0.5 T, quenches were usually preceded by a voltage spike (Figure 4) indicating a movement

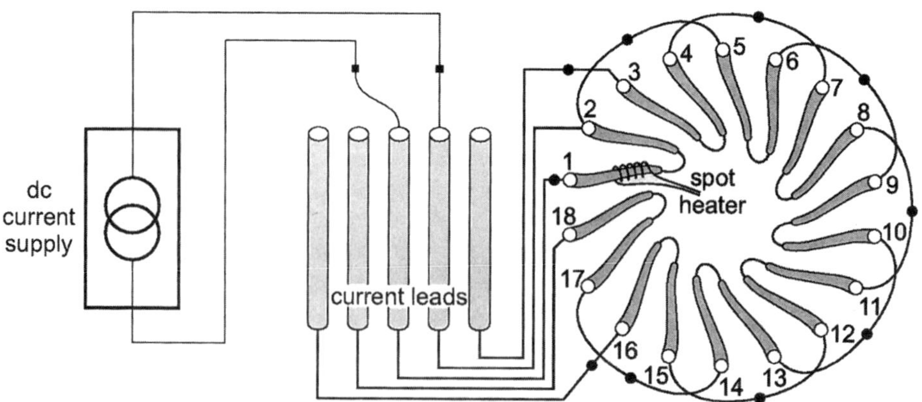

Figure 3. Block diagram of the test circuit. The 18 conductors of the inner layer of the cable are drawn at the right. The connection end is drawn at the periphery of the circle (with numbers), the 'far' end at the center. The conductors were connected to each other and to the current leads as shown. Components in the gray area were cooled to 1.9 K during the tests. At the dots (•) voltage taps were connected.

of the conductors that releases energy and causes the quench. For other quenches the voltage started to rise continuously without an observable initial spike. This was the case if the current in the cable exceeded I_c, or if the movement of the conductor was too small to generate a detectable voltage (at small applied fields). Similar observations were made with configurations 1 and 2.

To study the influence of the quench history on the quench behavior, a measurement series was carried out powering wire pair 1&2 only. With no applied field (triangles in Figure 5, branch 0) a training behavior similar to that of a superconducting magnet was observed. The first quench occurred at a current of 1245 A, the eighth quench at 1675 A, which is close to the critical current of short samples of the conductor.

The quench current dropped substantially with increasing field (Figure 5, branch 1). When the magnetic field was lowered (branch 2), the quench current increased again and reached the initial value. However, hysteretic behavior can be clearly seen: for the same applied field the quench current is higher for branch 2 compared to branch 1, which can be explained by training during the increase of the applied field along branch 1. The Lorentz forces acting on the conductors increase with the applied field, thus forcing them to move to increasingly more stable configurations. When the magnetic field was applied in the reverse direction, the direction of the Lorentz force changed and the conductors moved to a configuration more stable under these new (reversed force) conditions. The movements lead to quenches at currents much smaller than at zero field (branch 3). The observed voltage spikes and the training-like behavior indicate that movements of the conductors are the cause for quenches below the critical current in the prototype cable, like they are during the training of superconducting magnets. The retraining necessary after a reversal of the Lorentz force supports this conclusion.

All observed quench currents lie clearly outside the operational region of currents in the auxiliary busbar cable which is marked by the gray areas in Figure 5 (600 A maximum current and with 40 to 300 mT stray fields, depending on location).

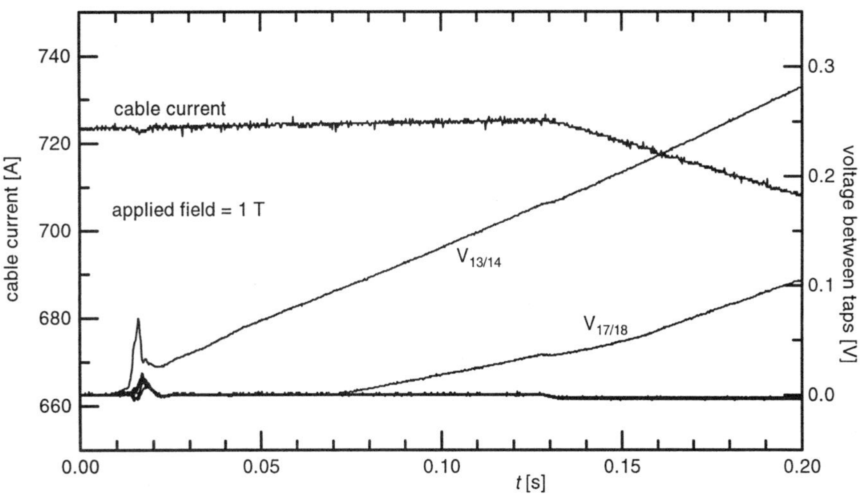

Figure 4. Quench of cable induced by the movement of a conductor in an applied field of 1 T. Conductor 13 or 14 moves causing a voltage spike. The movement also triggers a quench and the voltage starts to rise. About 60 ms after the initial spike the quench has spread to an adjacent conductor, seen by the voltage rise across wires 17 and 18. Another 55 ms later the protection system detects the quench and reduces the current. (Note that conductors 14 and 17 are directly connected. See Fig. 3)

QUENCHES INDUCED BY A LOCAL HEATER PULSE

The quench propagation velocity is an important parameter for the protection scheme of any superconducting installation. If it is low, the voltage across the quenched zone increases slowly, and it takes a long time to detect a quench. During this time the temperature of the resistive zone might become unacceptably high. In order to design a reliable protection system, it is useful to simulate possible quench scenarios. The quench propagation velocity measured as a function of the current in the conductor is used as one input parameter for such simulations.

To set off a quench at a specific location, a heater wire wrapped around conductor 1 at the far end of the cable (Figure 3) generated a short heat pulse. This pulse caused a resistive transition of the conductor. From the rate of the increase of resistance R and the resistivity ρ of the conductor just above T_c, the quench propagation velocity v can be calculated: $v = (dR/dt)\cdot(A/2\rho)$ (A is the conductor cross section). Up to about 20 K the resistance of the normal zone is independent of temperature. At the beginning of the quench the resistance increase thus results from the expansion of the normal zone only.

The measured quench propagation velocities are shown in Figure 6, together with the results of one-dimensional finite element simulations. In these simulations the following heat balance equation was integrated over time to reproduce the development of a propagating quench.

$$\pi r^2 C(T)\frac{dT(x,t)}{dt}\Delta x=\pi r^2\lambda(T)\frac{d^2T(x,t)}{dx^2}\Delta x+\frac{\rho(T)}{\pi r^2}I^2\Delta x-2\pi rk(T)\Delta x \qquad (1)$$

The four terms are: the increase of thermal energy in element Δx per second, the heat inflow per second from neighboring elements, the ohmic heating rate and the heat flow rate into the helium bath. C(T), λ(T) and ρ(T) are the heat capacity, thermal conductivity and electrical resistivity of the conductor as a function of temperature respectively. The radius of the conductor is r. The unknown function is k(T), the heat transfer rate into the helium per m^2 of conductor surface. Since the conductor is insulated it was assumed that $k(T) = K\cdot(T - T_{bath})$ with the helium bath temperature $T_{bath} = 1.9$ K. The results shown in Figure 6 were obtained by setting K to 30 $Wm^{-2}K^{-1}$.

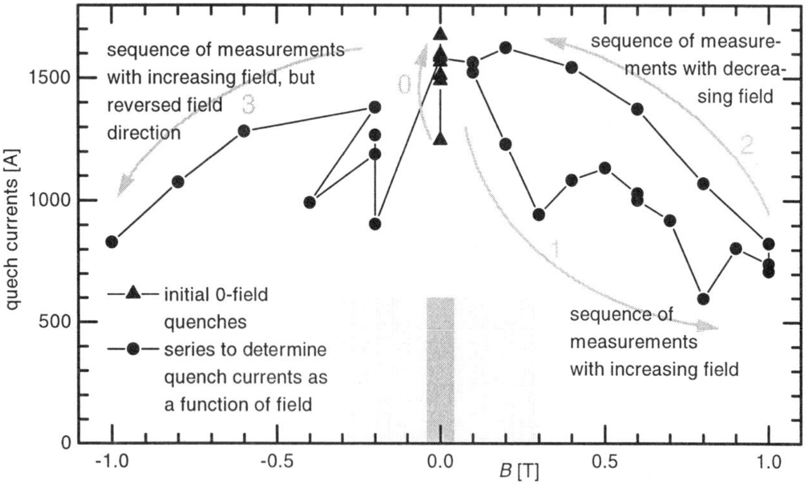

Figure 5. Quench currents as a function of applied magnetic fields. After initial quenches with no applied magnetic field (0), the field was increased (1), decreased (2) and increased in opposite direction (3). The gray area marks the operational region of the auxiliary busbar cable in the LHC.

The simulation reproduces the experimental data well, in particular a collapsing normal zone below a current of 240 A. At these low currents the cooling of the conductor is stronger than the ohmic heat generation in the normal zone, which consequently shrinks and vanishes. The conductor is thus cryostable[6] below 240 A.

This behavior might cause a problem at locations where the heat transfer into surrounding material is strongly reduced, for example in a leak tight feedthrough. If a quench starts at such a location it might be trapped, because the well cooled adjacent sections of the conductor prevent the expansion of the normal zone. The hot spot temperature could exceed acceptable values before the voltage across this non-expanding normal zone reaches the quench detection threshold.

In order to investigate this situation, an experiment was conducted at 4.2 K on a short piece of superconducting wire with a copper cross section of 1.58 mm^2, slightly smaller than the wires in the busbar cable mentioned so far. A heater covered a 10 cm long section of the wire, which was inserted into a stainless steel tube subsequently filled with Stycast. The temperature was estimated from the resistance of the heated and insulated section, assuming the temperature profile along the section to be flat. A current of 600 A was applied and the wire quenched by the heater. The current was switched off when the voltage across the heated section exceeded 1.5 V, much more than the 0.2 V typically foreseen for the quench detection threshold of the corrector circuits. During the 1.9 s until the 1.5 V were reached the measured temperature of the wire increased to about 600 K (Figure 7). From the quench load during this experiment and assuming adiabatic conditions, the wire temperature should have exceeded 1000 K. The discrepancy between experiment and adiabatic calculation shows that over the typical time scale of the experiments the adiabatic assumption is not justified, even at rather well insulated sections of the wire. Since the busbar copper cross section was dimensioned assuming adiabatic conditions (worst case) the actual cooling of the wire provides a large margin for the quench protection system.

CONCLUSIONS

The first prototype of the superconducting auxiliary busbar cable for the LHC could fulfill all technical requirements during extensive tests in the string. An exploration of the performance limits of the cable showed, however, that further development would lead to

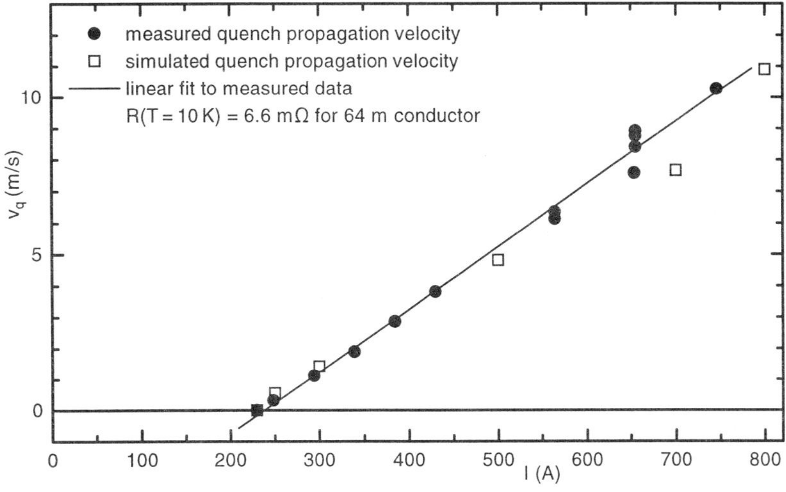

Figure 6. Quench propagation velocity as a function of the dc current in the conductor.

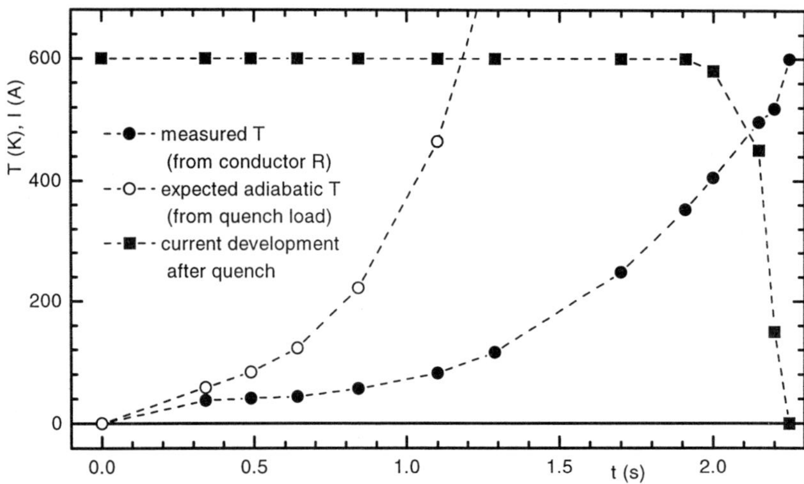

Figure 7. Time development of measured temperature and temperature calculated from quench load with adiabatic assumptions after a quench of an insulated conductor section (1.58 mm^2 wire). After 1.9 s the voltage limit of the power converter was reached and the current switched off.

improved operational margins. In particular it now seems desirable to design a cable with the conductors held more firmly in their positions to better prevent their movement. Without compromising the necessary longitudinal flexibility, this can be done by replacing the three polyethylene strands with a smaller metallic former and winding the conductors with high prestress and a shorter pitch length.

ACKNOWLEDGEMENT

The experiments could not have been performed without the help of R. Saban, A. Jacquemod, R. Moiroux and other members of the CRI group. J. Kragh kindly assisted with the quench detection electronics. F. Rodriguez-Mateos and P. Proudlock contributed many valuable ideas. F.Bordry and colleagues kindly supplied the dc power converter.

REFERENCES

1. R. Herzog and D. Hagedorn, Inductive method to measure very small joint resistances of superconducting wire, in: "International Cryogenic Engineering Conference 1998", D. Dew-Hughes et al., eds., Institute of Physics Publishing, Bristol (1998), p. 495.
2. P. Proudlock, Update on the powering strategy of LHC, Particle Accelerator Conference 1999, 29 April-2 May, New York, USA.
3. K. Dahlerup-Petersen et al., The Protection System for the Superconducting Elements of the Large Hadron Collider at CERN, Particle Accelerator Conference 1999, 29 April-2 May, New York, USA.
4. M.N. Wilson, "Superconducting Magnets", Oxford University Press, Oxford (1983).
5. F. Bordry, R. Saban et al., The LHC Magnet String Programme, Particle Accelerator Conference 1999, 29 April-2 May, New York, USA.
6. Z.J.J. Stekly and J.L. Zar, Stable Superconducting Coils *IEEE Trans. on Nuclear Science 12* p. 367 (1965).

DEVELOPMENT OF A 1000A CLASS BISMUTH-BASED HTS MODEL CABLE

Y. B. Lin, L. Z. Lin, S. P. Li, F. Y. Zhang, S. F. Chen,
Y. S. Wang, L. Xu, N. H. Song, H. M. Wen, and J. Li

Institute of Electrical Engineering, Chinese Academy of Sciences
Beijing, Beijing 100080, China

ABSTRACT

The first Bismuth-based HTS model cable in China has been developed and tested successfully at the Institute of Electrical Engineering, Chinese Academy of Sciences. The cable consists of a flexible former, conductor layer, inter-layer insulation and an outer insulation-binding layer. The conductor layer of is wound from 171 Bi-2223/Ag multifilamentary tapes which are divided into six layers. The Bi-2223/Ag tapes have been developed by the Northwest Institute for Nonferrous Metal Research and Beijing General Research Institute for Nonferrous Metals. The cable has a 45.2 mm outer diameter and 1.1m length. The tests at 77 K show that the critical current of the cable is more than 1180 A (at 1 μV/cm criterion) and the total joint resistance is less than 0.06 $\mu\Omega$. The critical current of the cable was decreased from 1221 A to 1188 A after 2 thermal cycles – a 2.7% degradation. The HTS cable operated with 1000 A steadily and reliably.

The main properties of the Bi-2223/Ag tape and the design, fabrication and test results of the HTS cable are presented in this paper.

INTRODUCTION

The HTS cables enable to transmit the large power capability with a considerable lower loss level and a smaller diameter in comparison with conventional cables. HTS cables can be installed in existing conduits to at least double the amount of power being transmitted , so the cost of cable system construction and operation could be reduced greatly. HTS power cables are the most promising of the various potential applications of high temperature superconductivity in power systems.

We have initiated a project to develop a 2000 A / 6 m Bismuth-based HTS model DC cable. The first step of the project is the development of a 1000 A / 1 m class Bismuth-based HTS model cable. The Bi-2223/Ag tapes were developed by the Northwest Institute for Nonferrous Metal Research and Beijing General Research Institute for Nonferrous Metals.

Making measurement on properties of the Bi-2223/Ag tapes and experimental study on joint technique between SC and normal conductor of the cable, the 1000 A HTS cable was designed, fabricated and tested successfully at 77 K. The development of the 2000A class HTS cable is being conducted at present.

*This work is supported by the National Center for R & D on Superconductivity of P.R. China.

PROPERTIES OF THE BISMUTH-BASED TAPES

The Bi-2223/Ag multifilamentary tapes produced by the powder-in-tube (PIT) technique were used in our model cable. The tapes are 0.25 mm thick by 4.0 mm wide and contain 27, 37 and 61 Bi-2223 filaments, respectively. The Ag /SC ratio is 4. The properties of the HTS tapes form the basis for the design of the HTS cable. We made an investigation on the properties of the HTS tapes before development of the HTS cable, such as *I-V* characteristics, the effect of external magnetic field, the mechanical properties and the effect of thermal cycling.

I-V Characteristics

The *I-V* characteristic of Bi-2223/Ag tape can usually be approximated by the empirical power-law equation $V=V_0(I/I_0)^n$, where V_0 and I_0 is a reference voltage and current respectively. The n value as a important parameter reflects the general shape of *I-V* curve. The n value is calculated as the slope of the plot log V versus log I in the range 0.1 μV/cm to 1 μV/cm.[1]

The *I-V* characteristics of Bi-2223/Ag tapes are measured with a four-terminal technique. The critical current of the HTS tapes are determined from *I-V* curve at criterion of 1 μV/cm. The average critical current of the used tapes at 77K is 11.6 A, and the average n value is 15.3. They were measured from 171 Bi-2223/Ag multifilamentary tapes for the HTS cable.

Effect of Applied Magnetic Field

The critical current of HTS tape degrades obviously with the increase of external magnetic field at 77 K. The effects have the well-known anisotropy. The magnetic field produced by transmission current can cause the degradation of critical current of the tapes in the cable. The effect must be taken into account in designing HTS cable. Fig. 1 shows the applied magnetic field dependence of the normalized critical current for our Bi-2223/Ag tape at 77K.

Mechanical Properties

The mechanical properties of Bi-2223/Ag tapes are essential to the design and fabrication of the HTS cable. The effects of the tensile strain and bending strain of the tapes on the critical current at 77K were investigated. Fig.2 shows the dependence of normalized I_c on the tensile strain ε_t. Up to a critical tensile strain of ε_{ct}=0.16% at 77K, there is no degradation of the critical current. The results of the bending strain ε_b tests are shown in Fig.3. The results show that up to a critical bending strain of ε_{cb}=0.39% at 77K, there is almost no influence on the critical current.

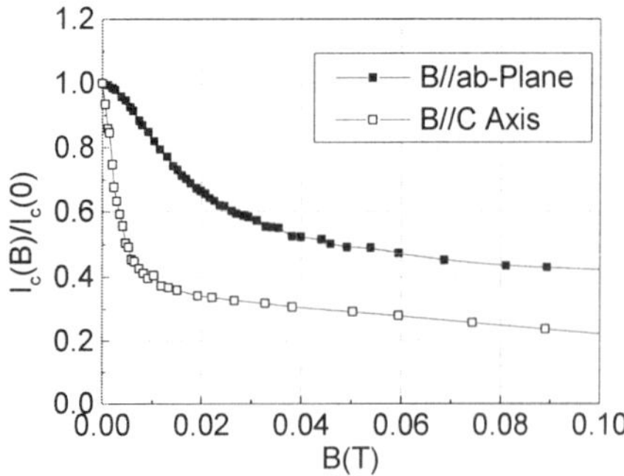

Figure 1. Magnetic field dependence of normalized I_c for Bi-2223/Ag tape at 77K

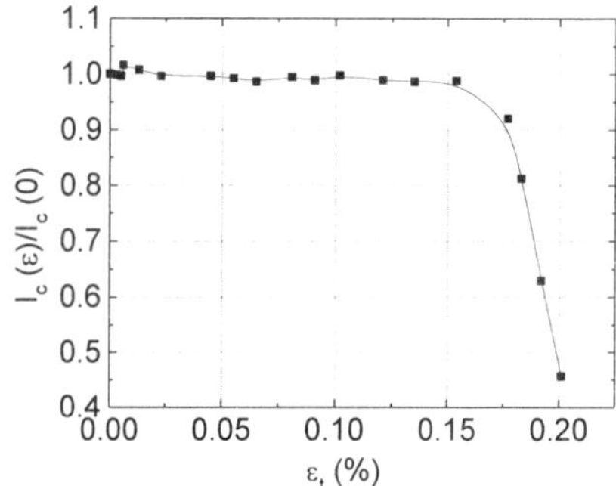

Figure 2. Relationship between ε_t and normalized I_c for tape at 77K

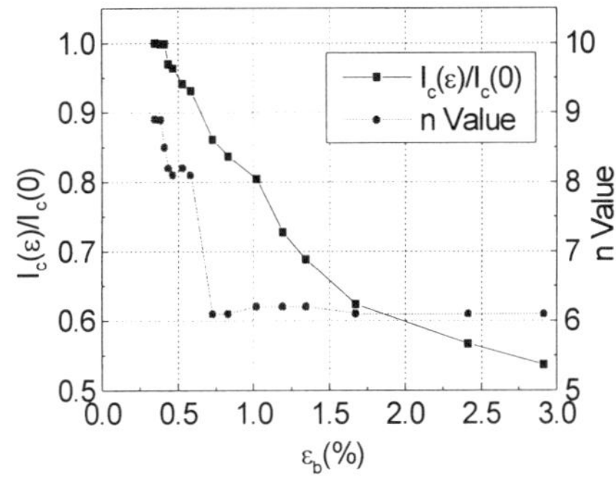

Figure 3. Relationship between ε_b and both normalized I_c and n value for Bi-2223/Ag tape

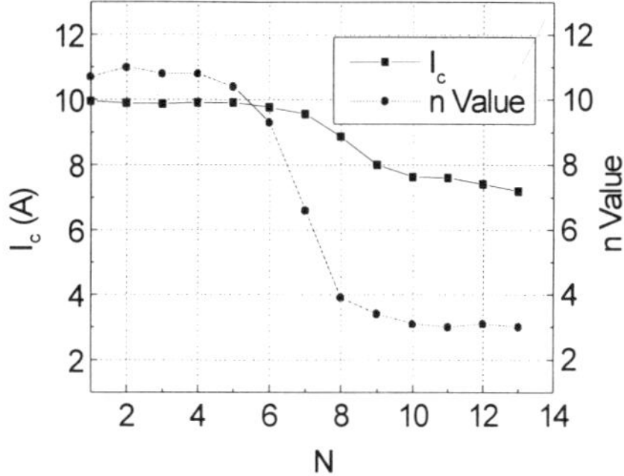

Figure 4. The effects of thermal cycling on I_c and n value of Bi-2223/Ag tape

Effect of Thermal Cycling

A major problem of the Bi-2223/Ag tapes produced by PIT process is the degradation in I_c with thermal cycling. The cable has to undergo many times thermal cycling from room temperature to 77K during measurement and operation. The effects of the thermal cycling on I_c and n value of the tapes must be considered in designing HTS cable. Figure 4 shows the typical variation of I_c and n value for our HTS tapes with number of thermal cycles N between LN_2 and room temperature.

DESIGN OF CABLE

Requirements for Cable Design

A. The tightness of laminated structure of the cable should be ensured and the degradation of mechanical properties caused by the thermal strain in the tapes should be prevented on cooldown of the cable.

B. The axial magnetic field in the cable should be reduced as far as possible.

C. The rating of the cable is 1000A/1000V.

D. The total joint resistance of the cable should be less than 1 $\mu\Omega$.

Structure of Cable

Compared with rigid SC cable, the flexible SC cable has the advantages of longer fabrication length, accommodation of thermal contraction and easier transportation and installation. The flexible structure was adopted in the design for our HTS cable. The cable consists of a flexible former, the conductor layer, the inter-layer insulation and an outer insulation-binding layer.

The former is a stainless steel corrugated tube having a 40.3 mm outer diameter and 1m length. Two copper terminals of the cable were welded to both ends of the corrugated tube. The terminals have the turning steps for soldering both ends of the HTS tapes. The conductor layer with multi-layer structure was formed from a large number of Bi-2223/Ag tapes. The tapes were wound spirally on the former side by side in order to prevent the damage of tapes from the thermal strain during cooldown from room temperature to LN_2 temperature. Each double layer of conjugate tapes in the conductor layer was wound in helices of opposite sense and at equal pitches to eliminate the axial magnetic field. The 0.1mm thick Mylar insulating tapes were wound on the former and between layers of the laminated tapes in the conductor layer to reduce AC losses due to the electromagnetic coupling of each layer during ramping current of the cable. The outer insulation-binding layer having 0.1mm thickness was wound on the conductor layer using Kapton insulating tape with a helix angle of 80^0. The structure of our cable is depicted in Fig. 5.

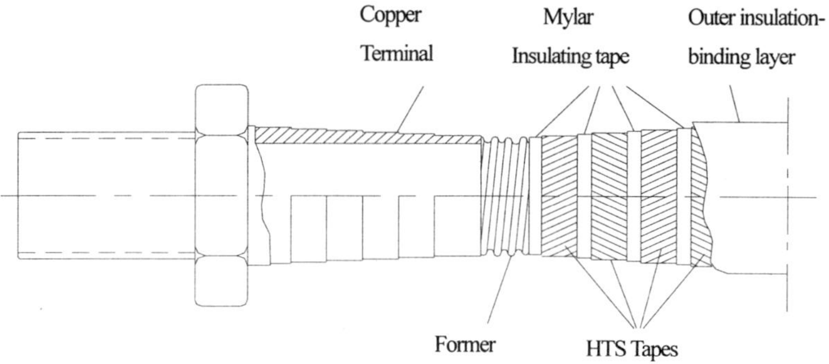

Figure 5. The structure of 1000A class HTS model cable

Determination of Parameters for Conductor Layer

Helix Angle. The fractional radial thermal contraction ε_r of the tape helices in the HTS cable on cooldown is given by [2]

$$\varepsilon_r = \frac{\varepsilon_l - \varepsilon_p}{\sin^2 \alpha} + \varepsilon_p \tag{1}$$

where ε_l and ε_p are the fractional changes in the tape length and pitch length respectively, and α is the helix angle. The fractional change in tape length is given by

$$\varepsilon_l = \varepsilon_t - \varepsilon_s \tag{2}$$

where ε_t is the free thermal contraction of the tape and ε_s is the tensile strain generated on cooldown.

From Eqs.(1) and (2) the dependence of helix angle on the tensile strain is obtained by

$$\alpha = \sin^{-1}\left(\sqrt{\frac{\varepsilon_t - \varepsilon_s - \varepsilon_p}{\varepsilon_r - \varepsilon_p}}\right) \tag{3}$$

The bending strain of a bent HTS tape is given by [3]

$$\varepsilon_b = \frac{t}{2R} \tag{4}$$

where t and R are the thickness and the curvature radius of the bent tape respectively. For a helix tape wound with helix angle α on a cylinder having a radius r, the curvature radius of the bent tape is given by [4]

$$R = \frac{r}{\sin^2 \alpha} \tag{5}$$

From Eqs.(4) and (5) the dependence of helix angle on the bending strain is obtained by

$$\alpha = \sin^{-1}\left(\sqrt{\frac{2r\varepsilon_b}{t}}\right) \tag{6}$$

To ensure the tightness of the cable on cooldown the radial thermal contraction ε_r of the tape helices should be slightly more than 0.28%, which is that of the stainless steel corrugated former. We can take ε_r=0.3%. To avoid the degradation of critical current of the tapes due to the tensile strain and bending strain on cooldown of the cable it is necessary to ensure $\varepsilon_s < \varepsilon_{ct}$ and $\varepsilon_b < \varepsilon_{cb}$. We can take $\varepsilon_s \leqslant 0.15\%$ and $\varepsilon_b \leqslant 0.3\%$. Taking ε_p=0.05% and ε_t=0.25%, for inner radius of the conductor layer r_1=20.25 mm, from Eqs.(3) and (6) the helix angle of the tapes must be laid by $26.6° \leqslant \alpha \leqslant 44.2°$.

Number of Tapes. Supposing the average critical current of tapes having total number N in the cable is $I_c(0)$, the transport current I_t of the cable is determined by

$$I_t = N I_c(0) k_1 k_2 k_3 k_4 \tag{7}$$

where k_1 is the degradation rate due to the self-field of the cable, k_2 is the degradation rate due to strain during the cable fabrication and k_2=0.8 was taken, k_3 is the degradation rate caused by the thermal cycling of the tapes, k_4 is a safety factor and k_4=0.9 was generally taken. The number of tapes in the inner layer of conductor layer is given by

$$N_1 = \frac{2\pi r_1 \cos\alpha}{w+g} \tag{8}$$

where w is width of the tape and g is small gap between the tapes.

The coefficient k_1 is determined from the dependence of the normalized I_c on applied magnetic field in Fig. 1. Assuming a uniform current distribution within the conductor layer, there is only the circular field within the conductor layer because the net axial field is zero in the design of cable. The circular field distribution is given by

$$B(r) = \frac{\mu_0 I_t (r^2 - r_1^2)}{2\pi r (r_2^2 - r_1^2)} \tag{9}$$

where r_1 and r_2 are inner and outer radius of the conductor layer respectively. There is a maximum circular field $B_m = B(r_2) = \mu_0 I_t/(2\pi r_2)$ at the outer radius of the conductor layer. Taking r_2=22.25 mm corresponding to 6 layers of the tapes, a maximum circular field B_m=89.9 Gs is generated when I_t=1000 A. The direction of the circular field runs parallel with the surface of the tapes. From Fig. 1, $I_c(B_m)/I_c(0)$=0.84 was obtained and k_1=0.8 can be taken.

Concerning I_c degradation of Bi-2223/Ag tape with thermal cycling, we have adopted some measures to improve the effects and k_3=0.9 can be taken. k_4=0.9 can be used also. Taking $I_c(0)$=11.6A and I_t=1000A, N=166 was obtained from Eq.(7).

Taking w=4.0mm, g=0.1 mm and $\alpha=27^0$, we have N_1=27 from Eq.(8), hence the laminated layer number in the conductor layer is 6.

Pitch and Helix Angle. The pitch of the tape helix with helix angle α can be expressed as follows.

$$p = \frac{2\pi r}{\tan\alpha} \tag{10}$$

where r is the radius of the tape helix. The net axial field is zero when the pitches of each double layer of conjugate tapes in the conductor layer are equal. Hence the zero axial field condition is

$$r_1 \tan\alpha_2 = r_2 \tan\alpha_1 \tag{11}$$

where r_1,r_2 and α_1, α_2 are the radii and helix angles of the two layers of conjugate tapes respectively.

The pitches and helix angles of each double layer in the conductor layer can be obtained by means of Eqs.(3),(6),(10) and (11).

FABRICATION OF CABLE

The HTS model cable was wound manually according to the designed parameters. The tapes with better electric properties should be laid on outer laminated layers in conductor layer. Two total I_c of each conjugate layer in the condutor layer should be approaching. The pitches of each double layer in conductor layer were ensured to be equal, adjusting the helix angles to around 27^0 during the fabrication of cable. The tapes of conductor layer were soldered to the copper terminals using low-melting-point In-Ag alloy solder.

The main parameters of the model cable are given in Table 1. The appearance of the cable is shown in Fig 6.

Table 1. Main parameters of the model cable

Rating	1000A/1000V
Length of cable	1.1 m
Outer diameter of cable	45.2 mm
Inner and outer diameter of conductor layer	40.5/45.0 mm
Layer number of conductor layer	6
Total length of the tapes	211 m
Number of tapes	171
Pitches and helix angles	
First and second layer	251.3mm, $27.1^0/27.5^0$
Third and fourth layer	261.3mm, $27.0^0/27.4^0$
Fifth and sixth layer	271.3mm, $26.9^0/27.0^0$

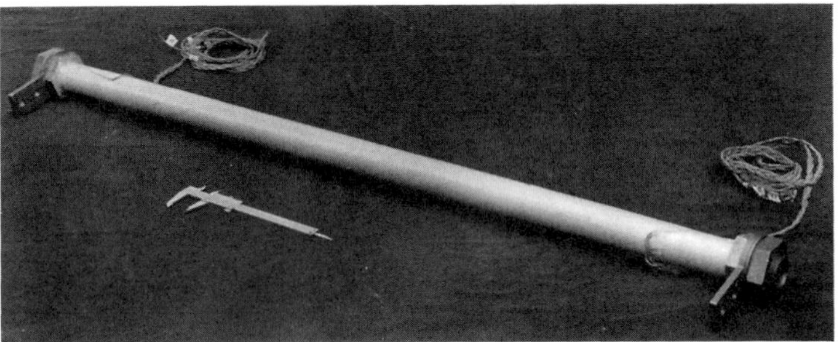

Figure 6. Appearance of 1000A HTS model cable

TESTS OF CABLE

The current carrying test of the cable was conducted in the self-field at 77 K. The DC power supply used for test had a maximum output current of 1180A. A data acquisition and an auto-control systems were used for measurement. The distance between voltage taps from the outer layer tapes was 98 cm. A Keithley model 182 sensitive digital voltmeter was used for test.

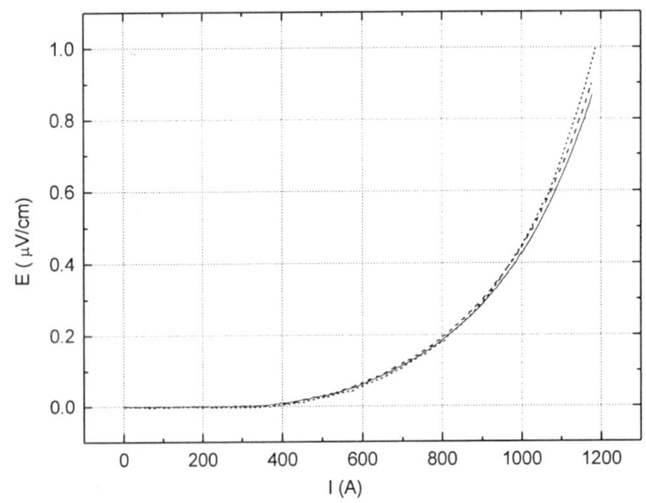

Figure 7. E-I curves of the HTS cable

The *E-I* curves of the cable is shown in Fig.7. The lower curve on Fig.7 corresponds to the first test result in LN_2. The middle and upper curves correspond to the test results at 77K after first and second thermal cycling respectively. The transport currents in the cable achieved the maximum output current of 1180A corresponding to a measured electric field of 0.88 μV/cm which is less than 1 μV/cm criteria. Using power law fitting the critical current for the cable is 1221A and the n value is 3.9 before thermal cycles. The critical current of the cable is 61.6% of the sum of the used tapes I_c. The total joint resistance is less than 0.06 $\mu\Omega$. Both get the better of the design specifications. The cable operated in 30 minutes with 1000A steadily and reliably.

The critical currents of the cable after the thermal cycles were obtained from Figure 7 using power law fitting: I_c=1220A after first thermal cycle and I_c=1188A after second one. The thermal cycle tests show that the critical current of the cable was only degraded 2.7% and the n value was same after 2 thermal cycles.

CONCLUSION

The first HTS model cable has been developed and tested successfully in China. The tests show that the critical current of the model cable is more than 1180 A and the total joint resistance is less than 0.06 $\mu\Omega$ at 77 K. Both get the better of the design specifications. The model cable operated with 1000A steadily and reliably.

The tests of the cable show that the design method and the fabrication technique for the model cable are correct and reasonable. The development of the model cable lays a foundation of the industrial application for HTS cable in china.

ACKNOWLEDGMENT

The authors gratefully acknowledge the Northwest Institute for Nonferrous Metal Research and Beijing General Research Institute for Nonferrous Metals for their cooperation and support the HTS tapes.

REFERENCES

1. International Electrotechnical Commission (IEC) 1788-1-2,1996.
2. D. A. Swift, A flexible SC ac cable: radial thermal contraction and the x-ray examination of a sample length of cold core. Cryogenics. 18:33 (1978).
3. Z. Han, et al., Mechanical properties of Bi-2223/Ag composite tapes for large scale applications, in:MT-15 Proceedings, L. Z. Lin, et al., ed., Science Press, Beijing (1998), p.956.
4. J. Sutton and D. A. Ward, Design of flexible coaxial cores for ac superconducting cable, Cryogenics.17:495 (1977).

L. Huget and S. Augustynowicz were representing the Cryogenics Society.

CURRENT LEADS FOR TESLA

M. Kauschke, Ch. Haberstroh, H. Quack

Technische Universität Dresden
Institut für Energiemaschinen und Maschinenlabor
Lehrstuhl für Kälte- und Kryotechnik
D-01062 Dresden, Germany

ABSTRACT

For TESLA, a linear collider project, HTS current leads are planned. These current leads have to fulfill the electrical demands as well as the economic expectations. Therefore a design study and different prototypes will be introduced. Numerical calculations of the transient behaviour were done to propose design changes after the working regime is settled concerning the current level. A test rig for the prototype current leads is presented.

INTRODUCTION

For the next millennium a new linear $e^+ e^-$-collider, TESLA, is planned at DESY, Hamburg[1]. For the planned accelerator a large number of superconducting correction magnets is needed. Altogether 812 groups of correction magnets are to be installed, in each case consisting of one quadrupole and two dipole magnets (inductance < 1H). These have to be operated independently in each case. Therefore, six current leads are needed per unit.

Due to the large number and the corresponding total thermal load the use of hybrid current leads is intended. For the required cooling of these current leads two methods are possible. They may be gas or conduction cooled. From the thermodynamic view gas cooled current leads should be preferred. The amount of energy saving depends on the cryoplant used. For cryoplants with discrete temperature levels the energy consumption will not increase by using conduction cooled current leads. But for large cryoplants, using several expansion turbines and therefore providing helium on several temperature levels, the losses will be noticeable.

On the other hand for complex systems like TESLA other details have to be taken in account. As gas cooling requires a lot of control elements for the mass flow and an additional gas return line, the conduction cooled alternative is favored for TESLA. In the

Advances in Cryogenic Engineering, Volume 45.
Edited by Shu *et al.*, Kluwer Academic / Plenum Publishers, 2000.

Table 1. Technical requirements of a current lead for TESLA

The distance between the coupling at 65 K and the helium cryostat:	0.1 m
The distance between that 65 K-level and the external wall:	0.5 m
The maximum current:	110 A
The average current:	50 A
Flashover stability:	1000 V
Magnetic field:	neglectable
radiation:	low

cryogenic layout of TESLA a shield cooling at a temperature level between 40 K and 80 K is foreseen. This shield is cooled by the helium returning from the magnets. This shield or the supply line will be used as a heat sink at an intermediate temperature for cooling for the current leads. Therefore, the current lead will be thermally connected to the cryogenic system at the 4.4K liquid helium bath and at the cooling shield.

TECHNICAL REQUIREMENTS

For this study some data as available today from TESLA given in Table 1. These numbers may still change in an further step of the TESLA project. Especially the figures concerning the average and maximum current are subject to changes. The geometric dimensions for the current lead are sketched in Figure 1. The decision for a new particle accelerator in dimensions like TESLA will be influenced by the estimated investment and operating costs. Therefore in respect to the current leads three characteristics have to be optimized:

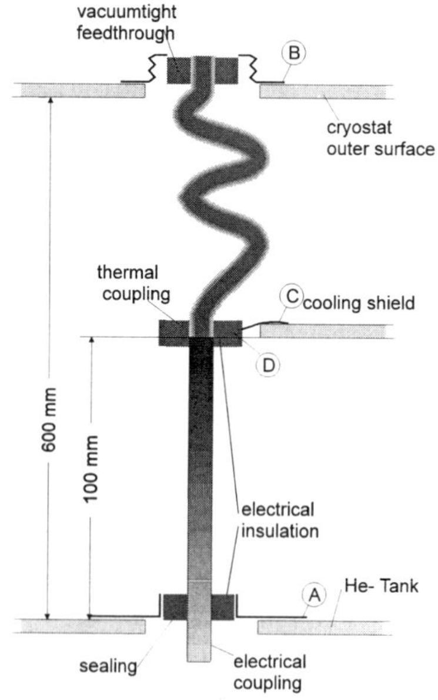

Figure 1. Geometry of current leads for TESLA

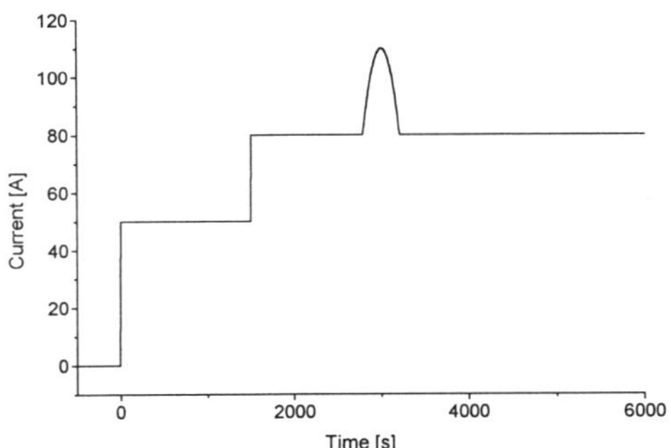

Figure 2. Current through the leads versus time

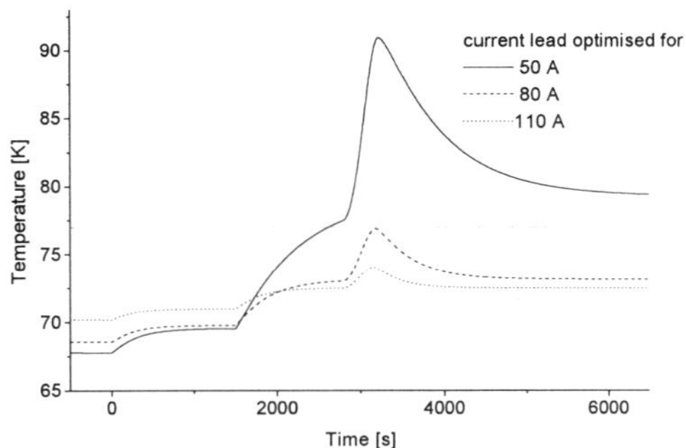

Figure 3. Transient behaviour of the temperature in the middle of the current lead, (point D in Figure 1).

- minimization of the heat leak through the current leads, and investigation of the heat load for different operating modes as the cryoplant may be optimized due to these heat loads.
- easy handling and mounting of the current leads
- use of standard 'catalogue'- products as often as possible

To make the assembly as simple as possible it is required to design the normal conducting "warm" end of the current lead flexible. The construction should permit a bending radius of approximately 50 mm without large forces. Special attention is given to the 'cold' electrical feed through into the helium vessel. Both easy mounting and leak tightness must be realized at the same time.

In co-operation with DESY current leads will be specified and prototypes will be tested at the Technical University of Dresden for later application in the accelerator.

CALCULATIONS

For the theoretical investigation upon the current leads a numerical simulation of the transient thermal behavior of the leads was done. The task of the transient simulation is to analyse different operating modes which the lead has to resist. As no electrical heat is produced in HTS as long as the temperature is kept low, the temperature of the coupling between the HTS and the copper has to be monitored. Due to the character of correction magnets, which get supplied by the current leads, it might be advantageous to optimize the current leads for the average current instead of the maximum current.

The transient behaviour of the different components i of the lead can be described in an energy balance as follows[2]:

$$\frac{\partial c_i T_i}{\partial \tau} = \left(\frac{\partial}{\partial x} \left(\lambda_i \, A_i \frac{\partial T_i}{\partial x} \right) + \left(\frac{e}{A} \right)_i I^2 \right) \left(\rho_i \, A_i \right)^{-1},$$

where the heat capacity c, the heat conductivity λ and the density ρ are functions of the temperature and therefore from time. For the simulation some assumptions have to be made.

Due to a lack of information on heat resistances of electrical insulation materials, a fixed temperature difference between the LN_2 bath and the copper binding of 1.5 K is assumed. Also the heat resistance of the soldering is neglected.

In Figure 2 the transient of the current through the lead is given for these simulation. This is an exemplary curve, which is characterized by an average current of 80 A, a

Table 2. Heat loads for different working conditions of the current leads onto the cooling levels

d_{con} [m]		0.004			0.0033			0.0029		
electrical power [W]		0	.97	2.78	0	1.53	5.25	0	2.2	11.6
Heat load on [W]	outer flange	- 2.43	-1.79	-.64	-1.67	-.68	1.57	-1.3	0.1	5.38
	LN_2	1.71	2.05	2.72	.97	1.53	3.	.6	1.41	5.56
	LHe	.72	.73	.74	.70	.71	.74	.7	.72	.81

current pulse of 425 sec and a maximum current of 110 A. For this current the temperature at the connection between the HTS and the normal conductor was calculated for three different diameters of the normal-conducting part of the lead. The behaviour of the temperature at the soldering between the normal conductor and the HTS is shown in Figure 3.

For steady state conditions the heat loads for three different diameters d_{con} of the normal conductor on the three different temperature levels (ambient, coolling shield, LHe) are listed in Table 2. The electrical power varies between stand-by (0 A), the average current (80 A) and the maximum current (110 A).Heat is given as a load, therefore a negative load marks the case of heat being brought into the system by conduction. During the transient change of the current the heat loads do not exceed the given numbers. As the heat load onto the intercooling is dropping down by 25 percent by optimizing the diameter for the most frequent current there should be a careful examination about the occuring current. On the other hand we see from these calculations that a short overloading of the current does not harm the lead. If a larger safety margin in this respect is required, the cold mass at the connecting point might be intresting, especially if the heat resistance between the HTS and the shield cooling increases due to the electrical insulation. As TESLA will run a long time between cool downs, the addional cold mass will not influence the overall heat balance.

REALIZATION

For testing and examination of different proposed designs of current leads and components of the system industry was asked to offer prototypes of leads. These current leads will be tested at a test rig at TU Dresden. The test rig is shown in Figure 4. The shield cooling will be simulated by a liquid nitrogen bath pumped down to 65 K.

In Table 4 the main differences between the different design proposals are given. After testing the different solutions for the feed throughs and electrical insulation will be combined and completed with some new develpoments for materials and working regimes.

Table 3. Comparison of the various prototypes of current leads

Suppliers	Eurus	Accel	ASC
Feed through and electrical insulation at LHe	HTS is led into the liquid helium bath where the superconductor can be can be directly connected. The helium tightness is reached by using CF- flanges The electrical insulation is realized in the flange.	HTS is led into the liquid helium bath where the superconductor can be can be directly connected. The helium tightness is reached by using CF- flanges The electrical insulation is realized in the flange.	A normal conducting feed through is proposed, as the helium tightness of the combination of copper and insulation materials is investigated in more detail.
HTS material	a) YBCO fiber b) Ag/ BSSCO tape c) Ag/Au/BSCCO tape all assembled with epoxy resin into stainless steel tubes.	BSCCO 2212 bulk material (supplied by Aventis) wrapped with glass fiber	tapes silver matrix assembled to form a bar, than covered with GFK
Coupling at the LN_2 level	copper braided wires screw fixing on both sides; at current lead and at the LN_2	copper braided wires, screw fixing on both sides of the connecting element	copper rope soldered onto the current lead screw fixing at LN_2 vessel
Electrical insulation at the LN_2 – level	electrical insulation realized at the connecting between HTS and copper	electrical insulating foil at the cold fix point of the connecting element	sapphire or synthetic foil at the cold end of the connector suggested.

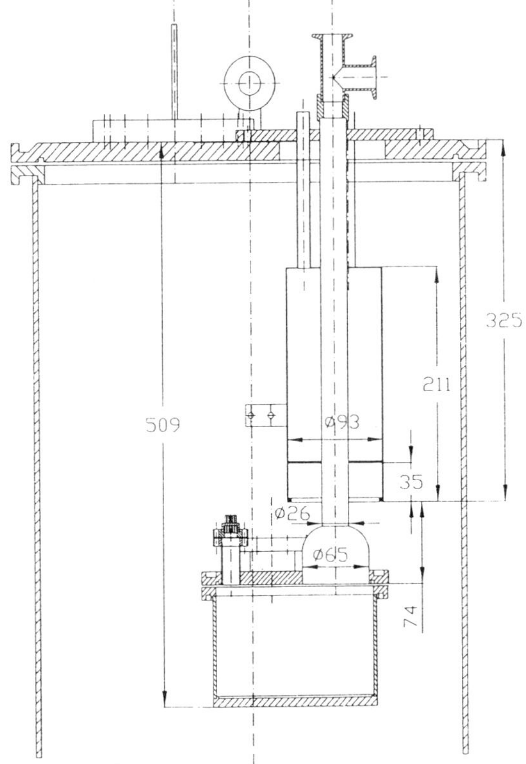

Figure 4. Test rig for current leads

ACKNOWLEDGMENTS

We would like to thank DESY (Deutsches Elektronen-Synchrotron) for the support of this research project and the Gesellschaft von Freunden und Förderen der TU Dresden eV for supporting our attendance at the CEC '99 at Montreal.

REFERENCES

1. R. Brinkmann, G. Materlik, J. Rossbach, A. Wagner, Conceptual Design of a 500 GeV e+e- Linear Collider with Integrated X-ray Laser Facilty, DESY 1997-048/ ECFA 1997-182
2. H. D. Baehr, K. Stephan: Wärme- und Stoffübertragung, Springer-Verlag, Heidelberg, 1994
3. J.G. Weisend II;:Handbook of Cryogenic Engineering, Taylor & Francis, Philadelphia; 1998

CRYOGENIC POWER INVERTERS FOR MAGNETIC RESONANCE IMAGING (MRI) SYSTEMS

O.M. Mueller, E.K. Mueller

LTE
Ballston Lake, NY, 12019

ABSTRACT

The new concept of Cryogenic Power Conversion (CPC) can provide a drastic size, weight, and therefore cost reduction in electronic high-power systems for certain suitable applications such as magnetic resonance imaging (MRI) and adjustable speed motor drives (ASDs). It enables the miniaturization of power electronics, based on the fact that the conduction losses of power MOSFETs are reduced by one to two orders of magnitude through cryo-cooling. Using these Cryo-MOSFETs and new soft-switching techniques, one can achieve a very high conversion efficiency (>99.7 %, without the cooling penalty) and therefore reduce the size of heatsinks and equipment drastically. The various required components (switches, diodes, capacitors, HTS- and Cu-inductors, drivers) and two topologies for a DC/AC pulse-width-modulated (PWM) amplifier/inverter circuit have been evaluated immersed in liquid nitrogen (77 K). Switch assemblies of 4 x 4 MOSFETs (4x50 A, 500 V) were used. Test results for the switching frequency range of 100 to 300 kHz for voltage and pulse current levels of 350 V and 220 A, respectively, will be reported.

INTRODUCTION: MRI

Magnetic resonance imaging (MRI) systems have revolutionized the field of medical diagnostics since 1982. A seldom considered key feature of these systems is that they also represent the first large-scale, commercial, profitable and successful application of cryogenics and low-temperature superconductivity (LTS) used in their high-field (0.3 - 4 Tesla) magnets. More then 10,000 whole-body MRI systems have been installed in hospitals worldwide (1997), providing a power base for Cryogenics which has not yet been exploited. Their price tag is still very high: $ 1 - 2 million. Therefore, most people do not yet have access to the benefits of this superb new diagnostic tool. An MRI scan still costs around $ 300-800. Drastic cost reduction of such MRI systems is therefore desirable. It can be achieved by using the new concept of Cryogenic Power Conversion (CPC).[1,5,8,10]

The key (potentially cryogenic) components of an MRI machine (excluding the digital computer) are in the sequence of costs:

◇ The superconducting whole-body magnet (0.3, 0.5, 1.0, 1.5, 3, 4 Tesla).
◇ Three gradient power amplifiers for the generation of the gradient magnetic fields.
◇ The RF pulse power amplifier (1-20 kW, 13, 21, 42, 64, 126, 168 MHz).
◇ RF receive coils using high-temperature superconductors plus cryo-preamplifiers.

◇ Add-on: Cryo-cooled laser diodes for laser surgery [2] and hyperpolarized inert gas MR imaging as well as: Cryo-cooled high-speed computers.

There is room and opportunity for the implementation of size and cost reduction in all of these components by the application of the new concept of CPC. Thus, one can envision a complete Cryo-MRI system which would be much smaller and lower cost than the existing ones because all the components can be integrated into the main magnet assembly. This concept is illustrated in Figure 1.

A start has been made with researching the feasibility of a CPC implementation of the three MRI gradient amplifiers which require high-speed PWM switching. They have to provide 100-200 A current pulses at high voltage into the 3 gradient coils (300 V in general, 30-60 kVA, 1kV - 2 kV for high-speed imaging [15] . Commercial applications include cost- and size-reduced Magnetic Resonance Imaging systems (MRI) providing improved performance for low-cost health care, MRI mass screening, etc., especially in rural areas and for export. Miniaturized, small size, high-efficiency cryogenic power conversion circuits will also find applications in airplanes, ships, buses, trains, etc., especially in combination with high-temperature superconducting components.

In summary, it is believed that Magnetic Resonance Imaging (MRI) machines using superconducting magnets are well suited for a first implementation and demonstration of this CPC concept because cryogenics is already available, the required power levels are high, the duty cycles are low and various other system components can also be miniaturized and optimized by cryo-cooling.

WHY CROGENIC POWER CONVERSION?

A look at the history of the past half century clearly shows that there is no better size, weight and, therefore, cost reduction technology available than that provided by semiconductors. For example, a computer with the calculating power available in today's lap-top computer required an entire office room thirty years ago. Unfortunately, power electronics has not yet benefited to the full possible extent from the potential provided by this silicon integrated circuit technology as have the fields of computers and communications considering size and cost reduction. The reason is that the power density in power semiconductor devices is still too high. The key problem in the field of power electronics is to remove the high dissipation power generated by the semiconductor switching devices. A better solution is to reduce and minimize that dissipation at the source. This can be done by cryogenically cooling certain suitable power semiconductor devices such as MOSFETs: Cryo-MOSFET.

A system which requires the cooling of several high-power components is now the proposed Cryo-MRI machine. It requires the cooling of, first, the superconducting magnet, second, third and fourth the 3 gradient amplifiers (30 - 60 kVA each) required for the NMR signal resolution of the 3 axes x, y, and z in the 3-dimensional space [4], and fifth, the RF pulse power amplifier (1 - 10 kW pulsed, low duty cycle) [13]. In addition, one can add more cryo-cooled components such as the gradient coils themselves [12], RF receive coils using thin-film HTS inductors and cryo-preamplifiers [9, 14], circuits using laser diodes for laser surgery [2], diode-lasers for hyperpolarized inert gas generation used in MR air ways imaging, and cryo-surgery equipment. A conventional gas laser transmitter for laser surgery could cost as much as $ 20k -50k while a cryo-laser diode may be much cheaper. The light output of semiconductor light-emitting diodes (LED) and lasers also increases by an order of magnitude or more through cryo-cooling. GaAs laser diodes for surgery as replacement for expensive gas lasers are under development [2]. In summary, one can envision 5 or more components of an MRI system which could benefit from cryo-cooling. Then a centralized cooling system becomes economical. In the case of superconducting (HTS) gradient coils [12] one would have very low losses and the high reactive gradient energy can be recovered at each pulse cycle.

An assumption for this project is that the MOSFET switching losses can be drastically reduced by suitable "soft-switching" techniques and the conduction losses by cryo-cooling.

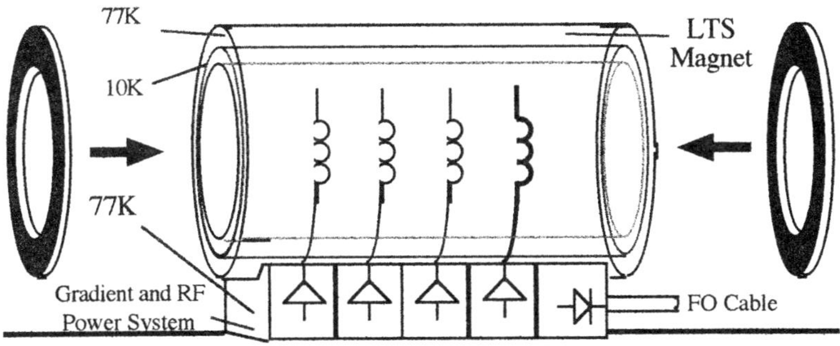

Figure 1. The Vision: MRI system integrating gradient and RF power cryo-amplifiers as well as cryo-laser diodes, HTS receive coils, etc. , with the LTS (or HTS) magnet.

COOLING SYSTEM and POWER DISSIPATION REDUCTION

An important question arises: Are the benefits of cryo-cooling higher than its cost? Two cooling system approaches exist for CPC: Cryocoolers or liquid nitrogen cooling. At the present time, the former are in general not yet suitable for the implementation of the concept of Cryogenic Power Conversion (CPC) because of their high price ($10 k - $ 20 k) and low efficiency (30-40 W/W at 77 K). But this may change in the near future. Using a cryo-cooler for each circuit or system to be cooled is analogous to having a separate electrical generator for every house, a scheme which was found to be uneconomical a century ago when large generator stations in the multi-gigawatt range where built. Insistence on this approach will kill the CPC concept. The second approach is to search for, to develop and to implement systems where a centralized cooling agent such as LN2 is employed in order to provide the cooling for several CPC circuits. One such system is the Cryo-MRI machine.

Since billions of dollars have already been invested in the development of high-temperature superconductors (HTS) for cable applications, one can expect that earlier or later HTS cables should be available. Then, they will supply power as well as liquid nitrogen as cooling fluid. This will open up the way for a completely new scenario of power / energy distribution and conversion systems. In other words, the proposed MRI system assumes the availability of low-cost liquid nitrogen supplied by HTS cables.

The ideal power required to produce refrigeration at the normal boiling point of liquid nitrogen (LN2, 77K) is 2.88 Watt/Watt (ideal Carnot efficiency [17]). In large quantities, LN2 can be generated [17] with 40 - 50 % of this ideal Carnot efficiency: (2.88/0.4) W/W=7.2 W/W. Taking into account some transportation and other losses one can use for LN2 an inefficiency number of 8-10 W/W. The ideal liquefaction energy for LN2 is 173 Wh/liter [17]. Assuming the inefficiency of 10W/W or 10/2.88 = 3.47 one can generate one liter of LN2 with an energy of E = 3.47 x 173 Wh/l = 600 Wh/liter. Assuming a cost of $ 0.10 for 1 kWh, a liter LN2 would then cost 6 cent. With overhead, transportation, etc. a price of $ 0.10 per liter LN2 can be used.

Switchmode circuits have two loss mechanisms: Conduction and switching losses. For a "back of the envelope" estimate one can assume that with the use of soft-switching techniques the latter can be made negligible. (R. DeDoncker: "The switching losses can be virtually eliminated" [7].) Then the reduction in the (conduction) losses due to cryo-cooling corresponds to that of the loss producing on-resistance of the power MOSFET switches employed in the circuitry. This means that as soon as the on-resistance reduction factor F is larger than the LN2 cooling inefficiency of 10 W/W, energy can be saved even taking into account the cooling penalty. As shown later the loss reduction factor for 600 V MOSFETs is F(300K/77K) ~ 10 for room temperature comparison, and F(400K/77K) ~ 20 since normally a power device operates at a junction temperature of about 400 K. For 1000 V devices is F(400 K/77K) ~ 35. These factors are even higher for the newly developed Cool-MOS devices [16]. Conclusion: Since F(400K/77K) is larger than 10 cryocooling saves energy even with the cooling penalty.

Let's assume for comparison 3 conventional switchmode gradient amplifiers with a rated total power of 200 kVA, a duty cycle of 10 % and a typical efficiency of 95 %. The peak and average (10 %) power dissipation is then 10 kW and 1 kW respectively for room temperature operation requiring heavy heatsinks for the power equipment at RT. For cryo-cooled (77K) operation with F(400K/77K) = 20 these numbers would be reduced to 500 Watt peak and 50 Watt (average) respectively. These power levels can easily be handled by a few liters of liquid nitrogen. One liter of LN2 is evaporated by 45 Wh of dissipation energy. The 50 W loss corresponds to an LN2 consumption of ~1.2 liters per hour. The other advantages of drastic size, weight and, therefore, cost reduction must here, of course, also be considered. The "load shedding" capability of LN2 which can be generated in off-peak hours is a useful feature to be mentioned. The cooling penalty can also be reduced by operating at cryogenic temperatures above 77 K in the range of 80 - 200 K. Cooling fluids for this range are available. A typical 2-stage cryocooler for a LT superconducting MRI magnet provides 1 Watt at 4 K and 50 - 60 W at 77 K and costs $ 20,000.

OBJECTIVES and DESIGN PHILOSOPHY

The key technical research objective of this project was to investigate the feasibility of the concept of 'Cryogenic Power Conversion' for MRI switch-mode gradient amplifiers with high-speed PWM switching (100 - 500 kHz). Gradient amplifiers should provide:

- Higher gradient pulse currents for smaller field-of-view imaging (magnification).
- Faster gradient rise times for increased imaging speed.
- Dramatically reduced size, weight, and space requirements for the power system.
- Higher energy efficiency, and thus reduced air-conditioning load in the hospital.
- Considerably reduced cost of equipment and operation.

The key component for these gradient amplifiers are the power switches best implemented with MOSFETs.

THE CRYO-MOSFET

High-efficiency power conversion circuits require fast, low-loss switches. From the series of available switches (GTOs, MCTs, SCRs, MTOs, IGBTs, Bipolar transistors, BITs) the MOSFET is the fastest, one of its great advantages. Gradient amplifiers have to exhibit a large power bandwidth (>20 kHz) requiring a high PWM switching frequency (100 - 300 kHz). This means that today the Cryo-MOSFET has no competitor for the application in MRI gradient amplifiers.

It has been found [5, 10] that commercially available, non-FREDFET, power MOSFETs in low-cost plastic packages (TO-247, TO-264 etc.) work well when immersed in liquid nitrogen at a temperature of 77 K, even if they were in no way designed for such a "cool" operation. Figure 2 shows the measured on-resistance of the new device to be used as a function of the drain current at 300 K and 77 K: R_{ON}: 100 μs pulse, APT5010LVR (APT-MOS-V technology, ratings: 500 V, 47 A, 0.1 Ω, TO-264). The on-resistance improvement factor is about F = 9 in this pulsed condition at low currents and then increasing. It is higher (~ 20) compared to actual operation at RT with internal junction heating up to about 400 K. In 1000 V MOSFETs the improvement factor F(300K/77K) is about 15. One can see the drastic reduction of the loss-producing on-resistance by cryo-cooling in addition to a remarkable increase in the current handling capability up to 100 A - 125 A. The on-resistance can now be made as small as desirable by paralleling more MOSFETs, a feature not possible with minority carrier devices such as IGBTs, MCTs, GTOs, SCRs, etc. which exhibit an on-state threshold voltage of 1-2 V. ("Silicon is cheap!").

The new COOL-MOS devices from Siemens and IRC have even lower on-resistance both at 300 K and 77 K [16]. In a 500 V device the improvement factor comparing conventional MOSFET at 300 K with a COOL-MOS at 77 K is about F=25.

These measurements demonstrate that the conduction losses of MOSFETs can be reduced by an order of magnitude through cryo-cooling. The reduction of dissipation power permits the elimination of large heat sinks and provides the possibility to place the

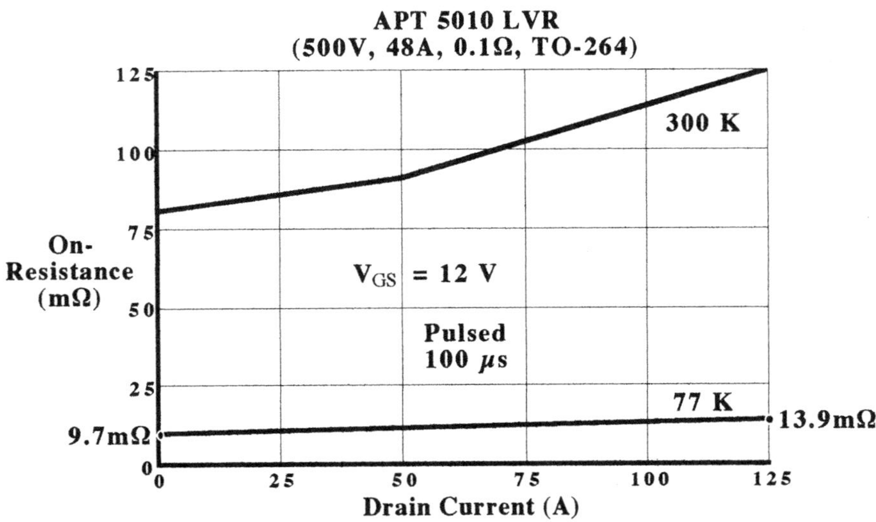

Figure 2. Measured on-resistance ($R_{DS\text{-}ON}$) of APT5010LVR power MOSFET as a function of drain current (I_D). Note: High current (125 A) capability at 77K.

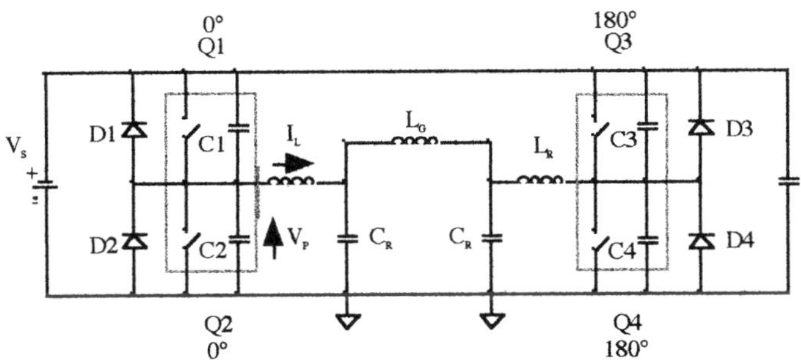

Figure 3. The Full-Bridge Circuit: Two Half-Bridges

MOSFET chips close together. As a result power circuits can be miniaturized. The switching losses of MOSFETs are lower than those of other devices since they are fast switching and do not exhibit the so-called "tail current".

CIRCUIT TOPOLOGIES AND TEST RESULTS

The Half- and Full-Bridge:

The key element of most high power electronics[3] is the so-called half- or full-bridge (=2 half-bridges) circuit shown in Figure 3. For successful cryogenic power conversion one has to minimize all the losses in order to compensate for the cooling cost. The CPC concept for MRI is based on the assumption that even at higher frequencies the switching losses can be more or less eliminated by suitable soft-switching techniques to be found and implemented.

The following losses have to be considered:

- The conduction losses which are reduced by a factor ~20 at low currents and more at high currents by cryo-cooling and paralleling.
- The voltage / current overlap losses to be reduced by fast- and soft-switching.
- The losses due to MOSFET drain-source output capacity discharge during turn-on.
- The MOSFET source-drain diode reverse recovery losses eliminated or reduced by

using the bi-directionality of the MOSFET (MOSFET commutation) or through the use of fast recovery diodes (FREDs).

The overlap and capacity discharge losses can be minimized in the half-bridge topology by a technique called zero-voltage-switching / clamped voltage (ZVS-CV), also known as quasi-resonant, pseudo-resonant or transition-resonant ZVS techniques. (Ref. 3, page 186-189). It is based on the assumption that the active devices can be switched fast compared to the voltage transition time. This condition can best be implemented with the fast-switching MOSFET. All other devices, IGBT, MCTs, GTO, etc. exhibit longer turn-off, turn-on times. In addition, the MOSFET has the required commutating diodes (Figure 3) already built-in. Thus, at least at lower frequencies, no external diodes are required. The on-state source-drain diode losses can be reduced in the Cryo-MOSFET by turning it on (MOSFET commutation). This feature is due to the MOSFET bi-directionality (positive / negative current flow) not available in any other power device: A great advantage. The ZVS-CV soft- switching technique[3] can be explained with Figure 3. Lets assume the upper MOSFET Q1 is turned on carrying a filter coil current I_L. Q2 is off. At time t_1 Q1 is turned off fast compared to the total transition switching time, ideally in zero nanoseconds. Now, for a dead time Δt (= t_2 - t_1) both switches are off. The inductive coil current I_L cannot change instantly, therefore the current flows now through the capacitors C1 and C2 charging C1 and discharging C2. At the same time the center pole voltage V_P changes from ~V_B down to -0.7 V, turning on the lower diode D2. It now carries the current until after the dead time Δt at time t2 the lower switch Q2 is turned on. The voltage transition is free of loss since no current flows through the MOSFETs or diodes: No overlap losses. With no load the filter capacity C_R now acts as the voltage source and reverses the current direction. I_L has a triangular shape at 50% duty cycle (See Figure 5). Ideally, the dead time Δt should be made equal to the capacitor discharge time Δt_C which is given by:

$$\Delta t_C = \frac{(C_1 + C_2)\, V_S}{I_L}$$

At the turn-on of switch Q2 the inductor current I_L changes its direction and flows first through D2 and then through Q2. After half a cycle time the same soft-switching transition repeats itself at time t_3 for the current flowing through Q2.

A problem occurs at large (or small) duty cycles if the current I_L is large and does not change its polarity. In this case the (C_1+C_2) capacity charge and discharge mechanism does not work and soft-switching occurs only at one and not at both transitions of a full cycle. At the turn-off of switch Q2 the positive current transfers now from the switch to the commutating diode and the reverse recovery problems of that diode enter the picture. Unfortunately, the fast recovery diode MOSFETs (FREDFETs) do not work at 77 K due to some gate channel related "carrier freeze-out" effect[12]. But the reverse recovery charge/time of a normal MOSFET decreases with cryo-cooling, even if not as much as one would have hoped. Paralleling fast FRED-diodes does not help since the on-state voltage of the MOSFET source-drain diode is lower than that of the fast recovery diodes. Using the MOSFET source-drain diode for the commutation function works fine at low frequencies (< 10 kHz) if one slows down the switching speed of the transistor with large gate resistors (10 - 100 Ω). This is necessary because the fast recovery of the diode produces dangerous dV/dt slopes which could retrigger the MOSFET to be turned off. In other words, while the MOSFET itself can handle fast voltage transients (20 V/ns - 100 V/ns), its source-drain diode, if used, limits that permissible dV/dt value to much lower numbers: 2-5 V/ns. Since its reverse recovery time is relatively large (500 - 700 ns), and therefore the related reverse recovery loss high, the diode limits efficient high frequency switching.

The OCIA or S-Topology (Stanley-Topology): A very elegant solution to this MOSFET / commutating diode dilemma has just recently (PESC-97) been proposed by Gerald Stanley and Ken Bradshaw from Techron, a division of Crown International, Inc.[6]. It separates the MOSFET switch and the diode commutation functions thus permitting the use of Cryo-MOSFETs as well as high-speed FRED-diodes together. This Stanley-topology or S-topology is shown in Figure 4. "The interleaved four-quadrant PWM power stage results in a doubling of the output ripple frequency while greatly reducing the ripple

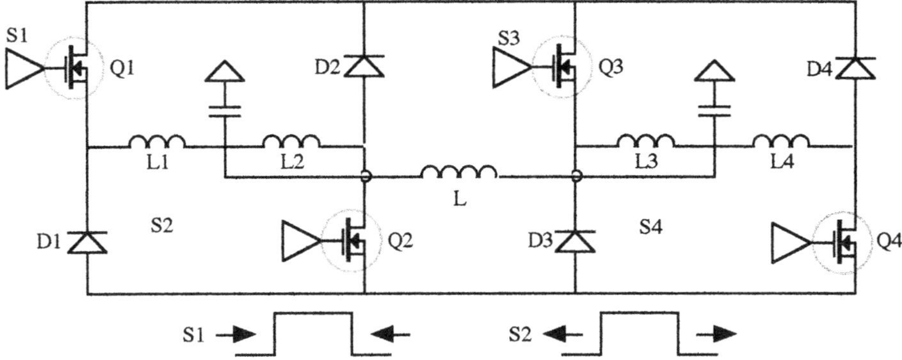

Figure 4. Opposed Current or Stanley-Topolygy (S-Topology): A solution to the diode problem

current amplitude. Shoot-through currents are reduced to low di/dt events that are readily controlled allowing zero deadtime operation" [6]. Here the MOSFET does not work as commutator. A key feature is that the pulse width of control signal S1 of an OC (opposed current) half-bridge is changed in the opposite direction to that of S2. A cryo-germanium diode would be best for this topology.

Test Results for the Full-Bridge: A full-bridge circuit (Figure 3) using 4 x 4 MOSFETs APT5010LVR (FB-1) has been tested. The dead time was 400 ns, the current capability at 77 K is 4 x 50 A = 200 A. The 2-filter ferrite or air inductors were 18 μH each (stranded AWG10 wire). Figure 5 shows waveforms of pole voltage (half-bridge centerpoint) and quasi-resonant filter inductor current I_L for 400 kHz at a supply voltage of 350 V measured with the power circuit immersed in LN2. The triangular coil current was measured with a Pearson (415-494-6444) current monitor No. 2879 where 100 mV=10 A. One can see that smooth soft-switching waveforms were obtained up to 400 - 500 kHz with the drive circuitry developed. The main limitation at higher frequencies is the increased power dissipation in the driver ICs (TC4422). The peak-to-peak coil currents were 25.2 A at 250 kHz, 15.6 A at 400 kHz and 8.3 A at 500 kHz. At higher frequencies the coil current is decreased because of the reduced coil charging time. Then the current is no longer sufficient to provide a clean charge / discharge of the MOSFET output capacitors as required for ZVS-CV soft-switching. The waveforms start to show distortions and ringing (at 500 kHz). This is not shown due to space limitations. An MRI gradient coil was simulated with a 580 μH inductor having a DC resistance of 100 mΩ.

Test Results for the S-Topology at 77 K: The half-bridge S-topology (Stanley-topology) of Figure 4 with 4 paralleled APT5012LNR MOSFETs (500 V, 47 A, 0.12 Ω), 2 APT60D60B FRED-diodes (600 V, 60 A, <70 ns) and with 2 air coils of 18.6 μH each has been tested. The dead-time was 360 ns. As expected the MOSFET turn-off transition of the pole voltage is soft-switching and the FRED-diode turn-off transition is fast- and hard-switching. Clean 220 A current pulses at 310 V with a PWM switching frequency of 150 kHz have been generated at 77 K with this circuit into a 580 μH gradient coil or a 1.3 Ω load resistor respectively (Figure 6 for 1.3 Ω load).

SUMMARY AND CONCLUSIONS

It has been shown that switchmode gradient amplifiers can be designed operating at high frequencies (100 - 600 kHz) and at cryogenic temperatures (77 K). A high-speed fiber-optic cryo-cooled MOSFET driver system was developed providing good (no-load) soft-switching performance of 4 MOSFET (4x50 A) switch assemblies operating at 77 K up to 600 kHz at a 350 V supply. The classical full-bridge circuit with 16 MOSFETs as well as the new OCA- or S-Topology (16 MOSFETs and 8 FREDs) have been built and tested. Since cryo-cooling speeds up diodes and MOSFETs very good RF (or even

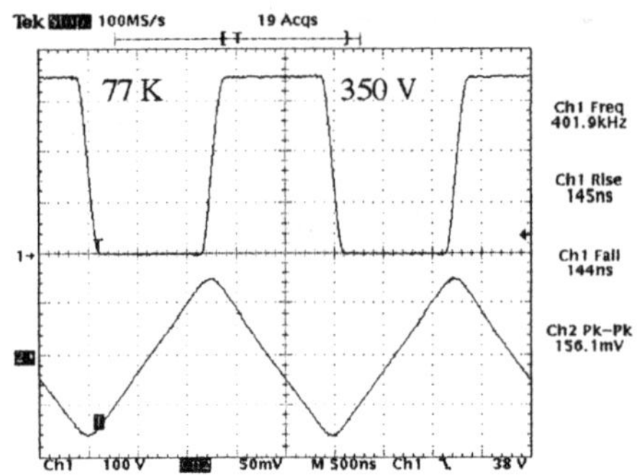

Figure 5. FB-1: Full-bridge waveforms. 100V/div, 5A/div

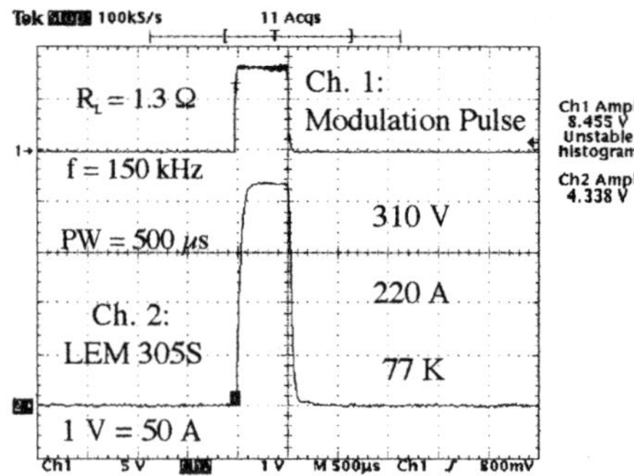

Figure 6. S-Topology: Current pulse into resisitive load. Ch.2: 1V = 50 A

microwave) layout techniques must be employed for reliable cryo-power circuits. In summary, the results are promising and justify continued research and design work on the CPC concept for a full Cryo-MRI system and many other applications.

ACKNOWLEDGEMENT

Thanks are due to the National Science Foundation for SBIR funds (Grant DMI-9796142) and to reviewers for comments. LTE is also grateful to Gerald Stanley and Ken Bradshaw at Techron for helpful discussions on gradient amplifiers and the S-topology.

REFERENCES

1. R. Singh, B.J. Baliga: Cryogenic Operation of Silicon Power Devices, Kluwer, Boston, 1998
2. T.E. Bell: Innovations: "GaAs versus Gas Lasers for Surgery." IEEE Spectrum, February 1997, p.16.
3. N. Mohan, T. Undeland, and W. Robbins, Power Electronics: John Wiley & Sons, pp. 154-203
4. O. Mueller, W.A. Edelstein, and P.B. Roemer, "The cryogenic NMR gradient amplifier," 8th Annual Meeting, Society of Mag. Res. in Medicine, Book of Abstracts, Part 2, p. 970, Amsterdam, NL, 1989
5. O. Mueller, "On-resistance, thermal resistance, and reverse recovery time of power MOSFETs at 77 K," Cryogenics, vol. 29, pp. 1006-1014, October 1989
6. G. Stanley and K. Bradshaw, "Precision DC-to-AC Power Conversion by Optimization of the Output Current Waveform - the Half-Bridge Revisited," PESC-97, pp. 993-999.
7. R.W. DeDoncker, J.P. Lyons: "The Auxiliary Commutated Pole Converter", IEEE-IAS Conf. Record, 1990, pp. 1228-1235. and: "The Auxiliary Quasi-Resonant DC Link Inverter", PESC-91, pp. 248-253.
8. O. Mueller and K. Herd, "Ultra-High Efficiency Power Conversion Using Cryogenic MOSFETs and HT Superconductors," IEEE PESC 93, pp. 772-778, 1993
9. R.D. Black, T.A. Early, P.B. Roemer, O.M. Mueller, et al.,"A high-temperature superconducting receiver for Nuclear Magnetic Resonance Microscopy," Science, vol. 259, pp. 793-795, Feb. 5, 1993
10. O. Mueller, "Properties of High-Power Cryo-MOSFETs", Conf. Record of the 1996 Annual IEEE IAS Meeting, vol. 3, pp. 1443-1448 (IEEE 96CH35977), San Diego, CA, October 1996
11. A.I. Gardiner, S.A. Johnson, and E. Schempp, "Operation of power electronic converters at cryogenic temperatures for utility energy conditioning applications," IECEC, vol. 14, pp. 2209-2214, 1996
12. E. Laskaris, B. Dorri, M. Vermilya, and O. Mueller, "Refrigerated Superconducting MR Magnet with Integrated Cryogenic Gradient Coils," U.S. Patent No. 5,278,502, January 11, 1994
13. O. Mueller and W.A. Edelstein, "The cryogenic NMR RF power amplifier," Society of Magnetic Res. in Medicine (SMRM), 9th Annual Meeting, NY, Book of Abstract, p. 205, August 18-24, 1990
14. R.D. Black, P.B. Roemer, and O. Mueller, "Electronics for a HTS receiver system for Magnetic Resonance Microimaging," IEEE Trans. on Biomed. Eng., vol. 41, no. 2, pp. 195-197, Feb 1994
15. O. Mueller et al., "A new 'quasi-linear,' high-efficiency, non-resonant, high-power MRI gradient system," Proceedings of the Society of Magnetic Resonance in Medicine, Vol. 1, p. 312, Aug. 1993.
16. A. Schlögl, et.al.: Properties of COOL-MOS between 420 K and 80 K - The ideal device for cryogencis applications, ISPSD-99, May 1999, Toronto.
17. E.C.Guyer, D.L. Brownell: "Handbook of Applied Thermal Design", MacGraw-Hill, 1989, p. 11.12

APPENDIX: HIGH-SPEED CRYO-CMOS DRIVER CIRCUITS

Introduction: High-speed switching, and thus a good RF / microwave layout, are required for the minimization of switching losses in Cryogenic Power Conversion (CPC) systems. This means that special attention has to be given to an optimized design of the gate driver circuits for the Cryo-MOSFET assemblies (200 A, 500 V) of a cryogenic inverter (10 - 100 kVA). Various CMOS integrated circuits were evaluated for operation at 77 K. A high-speed, optically coupled, cryo-cooled MOSFET driver system as well as a pulse-width modulator (PWM) enabling good switching performance of the power devices up to frequencies of 600 kHz at 350 V has been designed and built. In addition, the minimization of the gate-source-driver-output loop lead inductance was achieved. The voltage and frequency dependence of the selected driver IC current both at room temperature (300 K) and after immersion in liquid nitrogen (77 K) has been tested. The temperature and voltage dependence of rise and fall times as well as the propagation delay of the cryo-driver will be reported. It was found that cryo-cooling also decreases the driver power dissipation.

The driver design goals are (a) to provide high-speed switching, (b) to avoid the use of series gate resistance, (c) to minimize the lead inductances where possible, and (d) to eliminate the need for a negative gate-source voltage pulse. The minimization of MOSFET gate-source-driver-output loop inductances was done by giving each of the 4 paralleled MOSFETs its own driver intergrated circuit (IC). This helps to solve two problems at once: High output currents of several drivers (4x9 A = 36 A) make high-speed switching possible and the driver output is placed physically as close to the transistor gate as possible.

Selecting Driver ICs and 77 K Testing: After some searching a suitable 9 A driver made by Telcom, Inc. was found: TC4422. It is built with the so-called ‘tough CMOS’ technology and works nicely immersed in LN2 whereas the MC4422 (BiCMOS) does not. Figures A1 and A2 show the measured current consumption of a TC4422 as a function of supply voltage and frequency at 300 K and 77 K for a load capacity of 0 nF and 7.5 nF (MOSFET input capacity). A nice feature is that cryo-cooling reduces the power dissipation considerably and that the driver works even at high frequencies (500 - 600 kHz). The measurements of Figure A3 and A4 show that the rise and fall times as well as the propagation delay are also reduced by cooling.

Conclusion: Driver ICs which work well at cryogenic temperatures have been found and are available on the market despite the fact that none of these have been specifically designed for cryogenic operation. Their speed increases, and their power dissipation decreases by cryo-cooling. These drivers permit high-speed switching to 400 - 600 kHz in 200 A switches at 77 K.

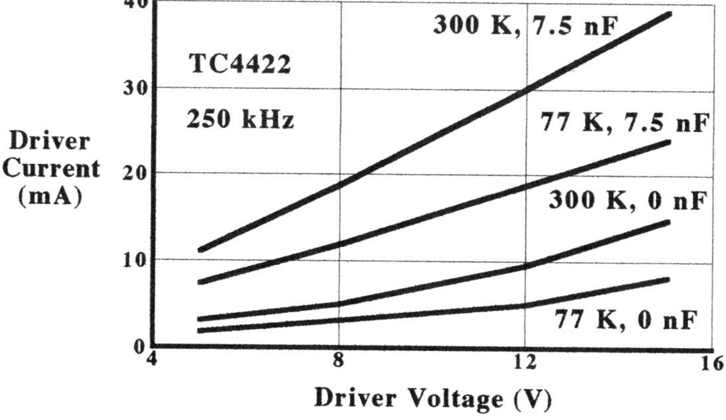

Figure A1. Driver current (I_{DD}) as a function of driver voltage (V_{DD}) for load capacitances of 7.5 and 0 nF at a frequency of 250 kHz.

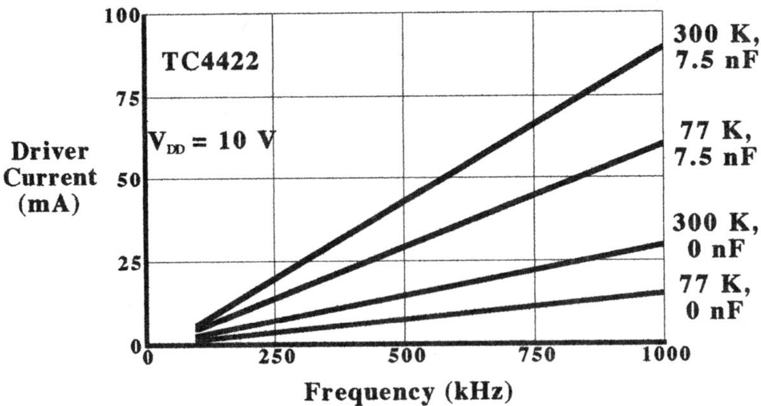

Figure A2. Driver current (I_{DD}) as a function of Frequency (f) for load capacitances of 7.5 nF and 0 nF.

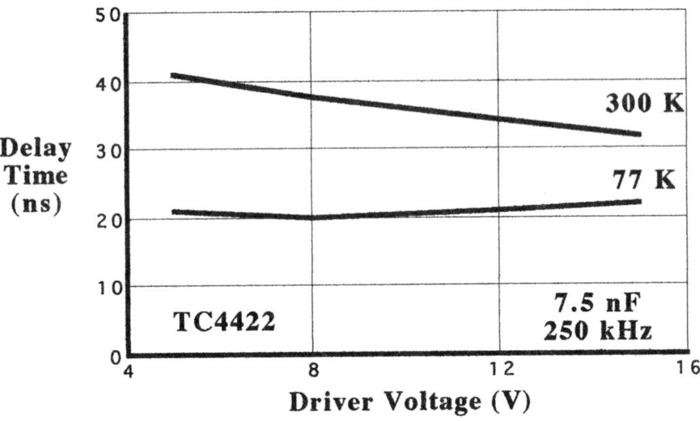

Figure A3. Propagation delay time as a function of driver voltage (V_{DD}).

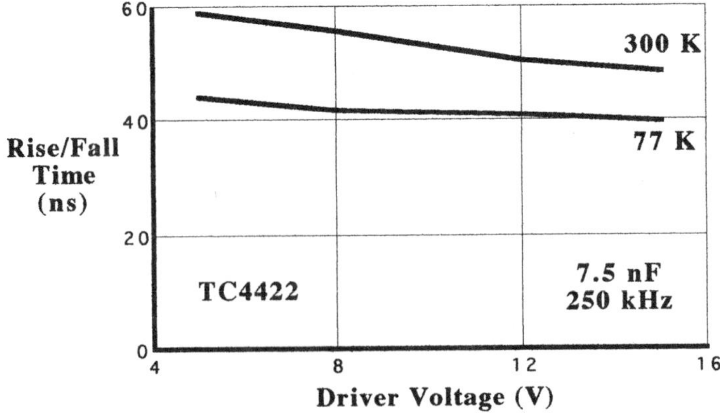

Figure A4. Rise and fall time (the average of the two was taken) as a function of driver voltage (V_{DD}).

PERFORMANCE OF A MIXED-REFRIGERANT SYSTEM DESIGNED FOR COMPUTER COOLING

Michael J. Ellsworth, Jr.,[1] Eric K. Moser,[2] Ajay Khatri,[3] and Mikhail Boiarski[3]

[1] IBM Corporation, Poughkeepsie, NY
[2] IGC-Technology Development, Latham, NY
[3] IGC-APD Cryogenics, Allentown, PA

ABSTRACT

CMOS-based computer processors will operate at increased frequencies when cooled to sub-ambient temperatures by a refrigeration system. This benefit must be weighed against many refrigerator considerations, the most noteworthy being thermal performance, efficiency, reliability, and cost. With this in mind a throttle-cycle, mixed-refrigerant cryocooler using a single-stage, oil-lubricated compressor was modified for increased capacity at 173K (–100°C) and its thermal performance was characterized. This mixed-refrigerant system cooled 119 Watts (with an average heat flux of better than 5 Watts/cm^2) at an interface temperature of 173K, with a Coefficient of Performance (COP) of more than 22%. The maximum in cooling capacity was 140 Watts at 183K (–90°C). The ratio of the COP relative to Carnot (ideal) efficiency held very constant at 16% over the temperature range of 153 to 183K (–120 to –90°C). This performance (relative to Carnot) is comparable to that of refrigeration systems with similar capacity operating at much higher temperatures. These results establish the good efficiency obtainable using mixed-refrigerant technology, making cooled CMOS applications attractive. Based on experimental data and computer simulations, we anticipate that better than 20% of Carnot efficiency can be achieved with a more optimal design.

INTRODUCTION

Computing performance levels continue to rise at a frenzied pace. This trend is expected to continue in the future, despite all technological obstacles. In this context, thermal management becomes an ever more challenging problem, and active refrigeration is an increasingly attractive technique for enhancing performance.

The paper reviews the benefits of computer cooling and presents an introduction on mixed-refrigerant systems, with a brief discussion of the challenges to be addressed in refrigerators specifically made for computer cooling applications, and a historical perspective on computer cooling. The mixed refrigerant system designed for 173K (–100°C) operation is presented, and its performance described and discussed.

Advances in Cryogenic Engineering, Volume 45.
Edited by Shu *et al.*, Kluwer Academic / Plenum Publishers, 2000.

BENEFITS OF COOLING COMPUTERS

Cooling a CMOS processor does not automatically increase performance, but there are benefits associated with cooling that enable performance increases. These benefits can be subdivided into device level improvements (improvements in the characteristics of the transistor devices themselves) and interconnect improvements (improvements in the characteristics of the on-chip "wiring" connecting the devices). Carrier mobility is increased, and devices feature reduced threshold voltages, steeper transitions, and lower leakage currents. Lower interconnect resistances reduce the interconnect delay times. The thermal conductivity of silicon also improves, aiding in heat transfer.

The Semiconductor Industry Association (SIA) provides a well-know "roadmap" projecting state-of-the-art processor technology into the future. The roadmap is exemplary of the rapid rate of advancement in computer performance.

Table 1: 1997 SIA Roadmap Projections.

Year	1997	1999	2001	2003	2006	2009	2012
Minimum Feature Size (μm)	0.25	0.18	0.15	0.13	0.10	0.07	0.05
Processor Speed (MHz)	400	600	800	1000	1200	1400	1500
Chip Size (mm^2)	480	800	850	900	1000	1100	1300
Device Voltage (Vdd, Volts)	1.8-2.5	1.5-1.8	1.2-1.5	1.2-1.5	0.9-1.2	0.6-0.9	0.5-0.6
Number of Interconnect Levels	6	6-7	7	7	7-8	8-9	9
Transistors Per Chip	11M	21M	40M	76M	200M	520M	>1G

The traditional method of increasing CMOS performance is "scaling," or the reduction of all device dimensions, shrinking the circuit and providing higher circuit density. Scaling approaches are the basis for Moore's law (the well-known rule for doubling of transistors per processor every 18 months), as well as the aggressive SIA roadmap projections. However, as feature sizes decrease below the 0.1μm level, scaling becomes much more difficult, as the dimensions become comparable to characteristic lengths for diffusion layers and device field zones. At reduced temperatures, CMOS can be operated at reduced supply voltages, thus providing some relief. A unique feature of low-temperature operation is that scaling *can* be applied in circuit designs with sub-0.1μm feature sizes.

Radical advances are implicit in the SIA roadmap. Recent years have seen advances in lithography and interconnect material. Reduced temperature has not been fully exploited as an avenue to higher performance, though it is now mentioned in the roadmap[1] as a "potential solution."

As a practical matter, overall processor speed is limited by either the devices *or* the interconnects. For a given processor, there is a maximum allowable clock frequency, F_{max}, and decreasing the processor temperature results in an increase in F_{max}. Increasing the clock frequency above "designed for" levels is termed "overclocking," especially as it applies to personal computing. While overclocking is a specialized operating option of limited use, it provides an indication of the speed enhancements possible with actively cooled computers.

The percentage increase in F_{max} rises monotonically with decreasing temperature, a bit faster than linearly, the slope depending on the extent to which the circuit is designed to operate at reduced temperatures[2]. For instance, generic CMOS circuits cooled to 173K might feature a 50% increase in F_{max}, while CMOS with optimized doping could be operated at 80% higher clock frequency when cooled to 173K.

HISTORICAL PERSPECTIVE ON COMPUTER COOLING

Low-temperature operation has historically been a desirable and yet unattained goal for high-performance computing[3]. Many attempts have been made in the past, with no sustained commercial success. IBM and Carrier Corp. collaborated on a computer cooling system using Stirling technology in the early 1980s[4], but the refrigerator proved too cumbersome. The ETA-10 was a true cryogenic computer, utilizing liquid nitrogen (77 K) for cooling[5]. However, liquid nitrogen-cooled computers are no longer commercial. RSFQ superconducting circuits operated at liquid helium temperature (4 to 5 K) show promise and are presently at the early development stage[6].

Kryotech, Inc., Columbia, SC, originated as a unique company dedicated to the commercialization of cooled computing[7]. Competitors have also arisen. Kryotech's current cooling systems utilize vapor-phase refrigeration technology based on Joule-Thomson expansion to actively cool processors to -40°C (233K).

The most appropriate temperature range for cooled computers is a matter of opinion, since a number of complicated factors apply. The potential computing benefits are enhanced by operating at very low temperatures, but the refrigerators are limited by fundamental performance limits. Carnot's law[8] places an upper bound on refrigerator efficiency. Coefficient of performance is often expressed as a fraction of Carnot efficiency. Our rationale for building a refrigerator in the 173K range is that there may be a "sweet spot" where performance increases are fairly balanced with the increased input power and added complexity associated with active cooling.

MIXED-GAS REFRIGERATION SYSTEMS

Refrigerators based on the principle of Joule-Thomson (JT) expansion provide a proven, reliable method for localized cooling. Two common examples of "vapor phase" JT systems are air conditioners and kitchen refrigerators. Similar systems using mixed gas refrigerants[9] enable JT refrigerators to operate at lower, including cryogenic[10], temperatures. A generic diagram for a JT refrigerator for cryogenic cooling is shown in Figure 1.

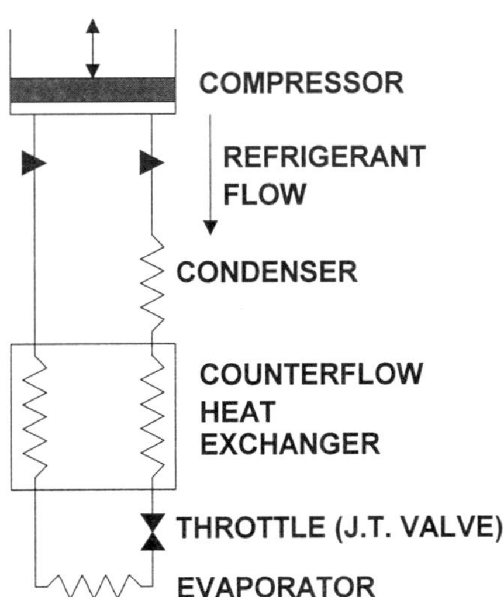

Figure 1. The Joule-Thomson Cycle.

The system is recuperative, with refrigerant flowing continuously in one direction. The particular components required and the configuration vary somewhat with factors including the desired temperature of operation, capacity, and efficiency. For example, the counterflow heat exchanger and oil separator (not pictured) would not be needed at relatively high temperatures.

The IGC-APD CRYOTIGER® system is a commercial low-temperature JT system with blended refrigerant, designed to provide temperatures in the range of 70 to 120K. The CRYOTIGER® cools to these temperatures using an innovative, patented blend of refrigerants. The compressor uses 450 Watts of input power and it can provide over 2 Watts of cooling at 70K, over 20 Watts at 125K, or over 80 Watts at 200K by charging the system with properly selected refrigerant blends. The CRYOTIGER® provides a valuable benchmark for computer cooling, though it is designed for lower-temperature operation.

REFRIGERATOR ISSUES

Generally speaking, refrigerator issues include evaporator temperature and heat capacity, efficiency, reliability, size, and cost. When considering the use of refrigeration in a computer cooling application, evaporator temperature distribution, temperature stability, and initial cool down time also come into play. Furthermore, overall system availability is critical. To assure continuous system operation, IBM has incorporated a redundant design where two independent refrigerant loops pass through a single evaporator[11].

Packaging the refrigeration unit in the computer system is also important. The refrigeration system must be transparent to the end user (i.e., the end user shouldn't have to know or care that refrigeration is taking place.) Refrigeration system heat must therefore be transferred to the ambient air in the same manor as do conventional computer cooling systems today. Condensation must also be prevented from forming on components within the computer as condensation on electronics components is detrimental to reliability and functionality. This is a major issue[12], but further consideration is beyond the scope of this paper.

A wide variety of different approaches to building a refrigerator for computer cooling are possible. Active cooling systems for computers must compete with passive alternatives such as air cooling and heat pipes as well as an array of refrigeration technologies including thermoelectrics, Gifford-McMahon systems, pulse tubes, and Stirling Engines. Commercially successful refrigerators and cryocoolers are generally highly adapted to meet specific applications. Computer coolers have context in the broader classification of spot-cooling electronic components such as filters, detectors, lasers, and superconducting devices. No "perfect" refrigeration technology exists, so expansion of refrigerators into new markets depends on successful compromise between the benefits of cooling and the hassles of incorporating the refrigerator. This having been said, we return our focus to mixed-gas systems as computer coolers.

THE 173K (–100°C) PROTOTYPE MIXED-GAS REFRIGERATOR

Several modifications were made to an "off-the-shelf" CRYOTIGER® system to achieve 173K cooling with good efficiency. A custom blended refrigerant was used, the oil separator within the compressor unit was modified, and both the heat exchanger and evaporator area were increased in size to enable cooling capacity above 100 Watts. A larger diameter capillary tube (throttle) was also used to accommodate the higher refrigerant mass flow rate.

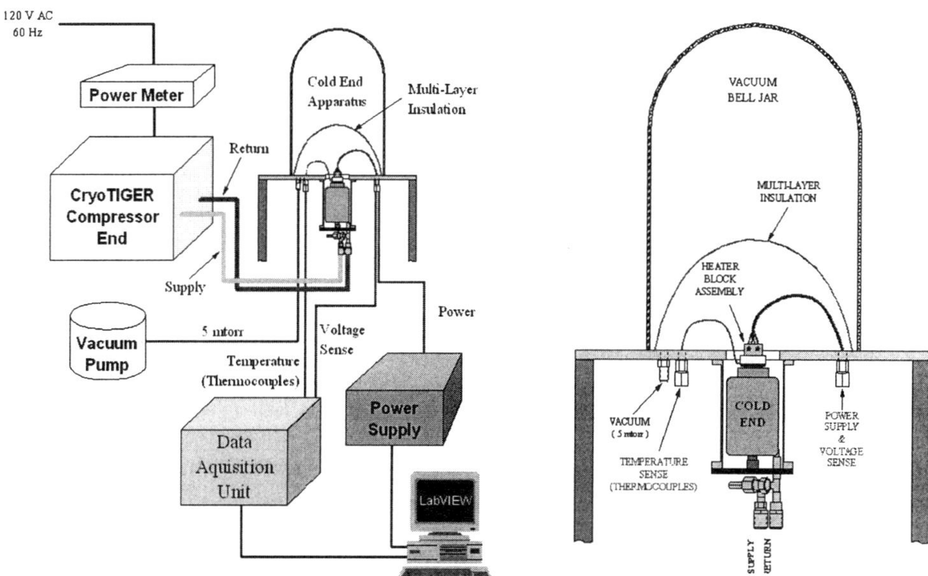

Figure 2. Experimental apparatus.

The prototype unit was tested and characterized in the IBM Advanced Thermal Laboratory. The experimental setup, with some detail on the test apparatus, is depicted in Figure 2.

Care was taken in the design of the experimental apparatus to minimize the infiltration of heat into the system by virtue of the temperature difference that exists between the cold end and the environment during operation. This parasitic heat load is minimized so that the unit could be accurately characterized in terms of its heat capacity-evaporator surface temperature behavior. The cold end, comprised of the heat exchanger, capillary tube, and evaporator, are maintained at a vacuum environment of approximately 5 mtorr to minimize the infiltration of heat by convection. Multilayer insulation is placed over the evaporator end to minimize radiative heat infiltration. The majority of parasitic heat load, therefore, comes from conduction through the mechanical support of the components. By characterizing the thermal capacitance of the system (referenced to the evaporator surface temperature) and measuring the rate at which the system increased in temperature with no heating or cooling load applied, it was estimated that the parasitic heat load at 173K was approximately 2.5 Watts, which constitutes roughly 2.1% of the applied heat load at said temperature.

Electric resistance heaters were used to apply the heat load. The voltage drop across the heaters (measured as close to the heater as practicable) and the voltage drop across a precision resistor (i.e. a known resistance) was used to determine the power dissipated by the heaters. It can be assumed that 100% of the heat dissipated within the heaters is absorbed by the evaporator. Temperature was measured using type E (Nickel-Chromium vs. Copper Nickel) thermocouples. Type E thermocouples are ideally suited for low temperature measurements because of their high Seebeck coefficient (58 μV/K). Five thermocouples are placed in contact with the evaporator surface: three located within the center; one located towards one corner, and the last located in the corner diagonally opposed from the previous thermocouple. Voltage and thermocouple measurements were made using an HP3852 data acquisition system. The data acquisition and the application of electrical heater power were controlled using a personal computer. Finally, the electrical power supplied to the compressor end was measured so that coefficient of performance could be determined.

A number of performance characteristics were measured. First was the time required to bring the system down from room temperature to operating temperature. The system was not able to absorb the requisite heat load until it was at operating temperature. This translates in a computer application to not being able to turn on the computer electronics until the cooling system is down to operating temperature. For large system servers with expectations of 99.9999% or better availability this time delay is critical. It took the prototype unit over 50 minutes on average to get down to operating temperature (see Figure 3).

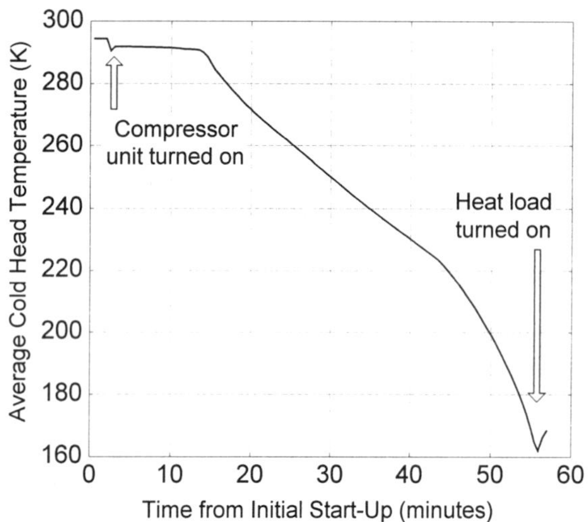

Figure 3. Cooldown (initial configuration).

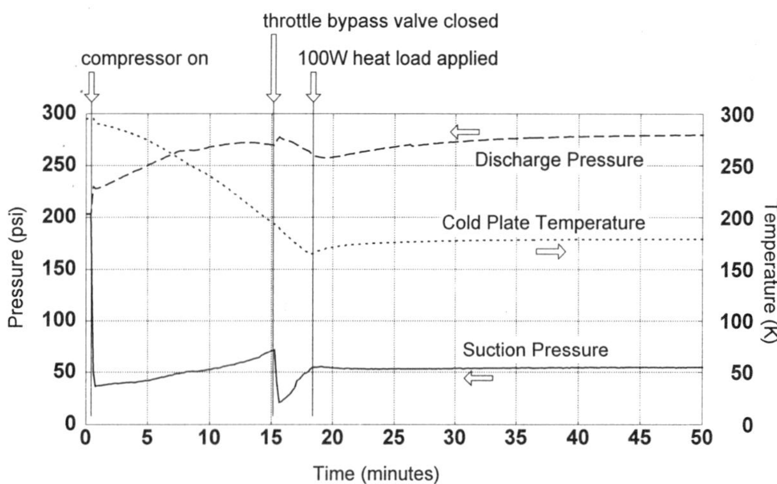

Figure 4. Improved cooldown.

Adjusting this result to take into account the thermal capacitance of a large system server, the cool down time would approach 75 minutes. This turned out to be an obvious concern. To address this concern, a second system was built utilizing a special throttle designed to enable high flow rate during the cool down period, then switch over to a continuous operation setting. Figure 4 shows that the cool down period could be improved to less than 20 minutes using this throttling scheme.

The next characteristic to evaluate was the heat load - temperature behavior of the system. Figure 5 shows how heat load varies as a function of average evaporator (cold head) temperature. The unit could absorb 119 Watts (approximately 5.3 Watts/cm^2) at 173K. The maximum load the system would sustain a stable temperature at was found to be 140.3 Watts at 182.3K. Power above this level would result in temperature runaway. The data in Figure 5 also suggests that the sensitivity in change to heat load is roughly 0.5K/Watt. The transient temperature response for a step change in heat load was observed to behave similarly to the step response of an electrical RC circuit.

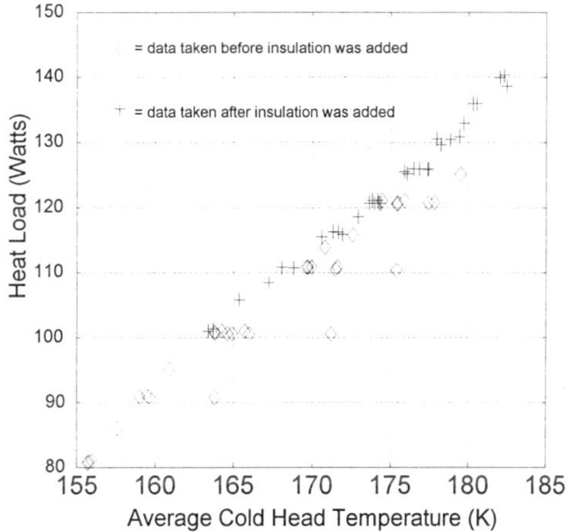

Figure 5. Capacity vs. temperature.

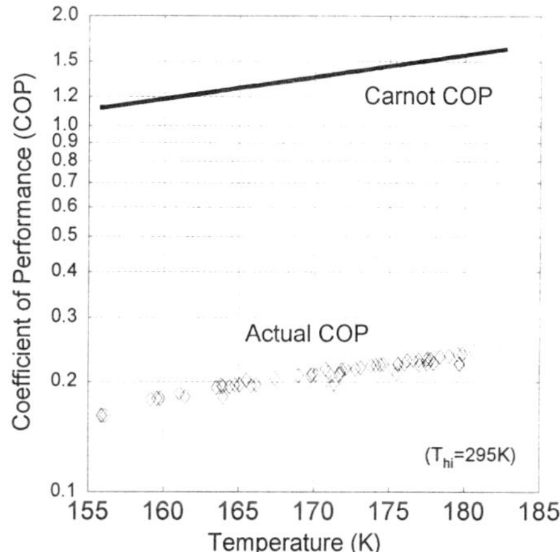

Figure 6. COP vs. temperature.

Finally, the coefficient of performance (COP) was evaluated as a function of average evaporator surface temperature (Figure 6). The unit was found to operate at a COP of 0.22 W/W at 173K. Furthermore, the unit appears to operate consistently at 16% of Carnot COP over the range of 153 to 183K (-120°C to -90°C). This compares very favorably with other refrigeration technologies.

This system described in this paper demonstrated cooling capacity, efficiency, and cooldown time that are adequate to cool a high performance CMOS chip for operation in the 150K to 180K temperature range. Further obstacles remain before the refrigerator can truly be regarded as ready for computer cooling.

FUTURE IMPROVEMENTS

In order to attain maximal refrigerator efficiency, careful designs are needed matching refrigerant, compressor, heat exchanger, and cold plate attributes. Computer simulations have proven useful for exploring these design issues. A detailed discussion on mixed refrigeration systems operating in the temperature range of 120K to 200K is given in an accompanying paper by Boiarski, et al.[13]

SUMMARY

The benefits of computer cooling are well known. There are many challenges associated with the marriage of refrigeration technology with computer technology. Mixed-gas systems are an attractive refrigeration technology for computer cooling.

ACKNOWLEDGEMENTS

We acknowledge support from others involved in this work including George Manelski, Robert Simons, and Richard Chu at IBM, Terry Rufer, Noel Fasnacht, and Stephen Harold at APD, and Boris Yudin of Intercool. We acknowledge useful discussions with John Peeples of Kryotech and Kenneth Rose of Rensselaer Polytechnic Institute.

REFERENCES

[1] Further SIA roadmap information is available at the Sematech website, www.sematech.org/public.

[2] Discussed by Y. Taur and R. Dennard of the IBM Watson Research Center at the *IEEE Cool Electronics Workshop*, Washington DC, 15-16 October, 1998.

[3] R. Solomon, Opportunities and problems for refrigeration in small computers, *1991 IIR Proceedings*, p. 202. See also J. Sloan, Cold computing: the future of high-speed processing, *Superconductor Industry* Fall 1996, p. 30, and R. K. Kirschman, Low-temperature electronics, *IEEE Circuits and Devices Magazine* 6:12 (1990).

[4] R. E. Schwall and W. S. Harris, Packaging and cooling low temperature electronics, *Advances in Cryogenic Engineering* 37-A:587 (1991).

[5] D. M. Carlson, D. C. Sullivan, R. E. Bach, and D. R. Resnick, The ETA10 LN cooled supercomputer, *IEEE Trans on Electron Devices* ED-36:1404 (1989).

[6] Aaron Bitterman, Superconductors & cryoelectronics in the petaflops-scale computer project, *Superconductor & Cryoelectronics*, Spring 1999 Issue, p. 17.

[7] Further information is available on the Kryotech, Inc. website, www.kryotech.com.

[8] Carnot's law of refrigeration states that the efficiency of an ideal refrigerator is given by $T/\Delta T$ where T is the refrigeration temperature and ΔT is the difference between the refrigeration temperature and the ambient temperature, in Kelvin.

[9] M. Boiarski, V. Brodianski, R. Longsworth, Retrospective of mixed-refrigerant technology and modern status of cryocoolers based on one-stage, oil-lubricated compressors, *Advances in Cryogenic Engineering* 43:1701 (1998).

[10] Strictly speaking, "Cryogenic" refers to temperatures below 120K (or -153C).

[11] P. Singh, D. Becker, V. Cozzolino, M. Ellsworth, R. Schmidt, and E. Seminaro, System packaging of a CMOS mainframe, *Advanced Microelectronics*, 25:12 (1998).

[12] This topic was discussed by J. Peeples of Kryotech at the *Cryoelectronics Technologies in Microelectronics* workshop, sponsored by the Knowledge Foundation, San Francisco, 18-19 February, 1999. See also J. W. Peeples, The deskside supercomputer, *Advances in Cryogenic Engineering* 43:849 (1998).

[13] M. Boiarski, A. Khatri, and O. Podcherniaev, Enhanced refrigeration performance of throttle-cycle coolers operating with mixed refrigerants, *1999 CEC/ICMC*, Montreal, Canada.

CRYOGENIC ASPECTS OF A FETAL HEART MONITOR BASED ON HIGH-Tc SQUIDS

A.P. Rijpma, H.J.M. ter Brake, M.J. Peters, M.R. Bangma and H. Rogalla

University of Twente, Department of Applied Physics
PO Box 217, Enschede, The Netherlands

ABSTRACT

A high-T_c SQUID-based magnetometer system for measuring fetal heart activity in standard clinical environments is the goal of the so-called FHARMON project. To lower the threshold for the application of this fetal heart monitor, it should be low-cost and, moreover, simple to operate. It is, therefore, advantageous to replace the liquid cryogen bath by a closed-cycle refrigerator. In this paper the requirements with respect to the cryogenic system are discussed for this specific application. These include operating temperature (55-80 K), temperature stability (<0.5 K drift in 10 hours), cooling power (>0.5 W), mechanical vibrations, electromagnetic interference and costs (<$US 5000). Next to the requirements, the paper also reviews the most relevant options for the realization of the cryogenic system. In this respect two categories are distinguished: systems incorporating coolers with 'noisy' cold heads (e.g. GM and Stirling) versus those using coolers with a low-noise cold head (e.g. pulse tube and JT) requiring respectively long and short thermal interfaces.

INTRODUCTION

The electrocardiogram (ECG) is a well-known and inexpensive tool for monitoring the condition of the adult heart. It is based on measuring the differences in electrical potential on the body caused by the flow of currents in the heart muscle. As currents generate a magnetic field as well, it is also possible to measure the so-called magnetocardiogram (MCG). In the past decades several groups showed that MCG-recordings were not only feasible, but on occasion even displayed a clearer indication of heart disorder than the ECG-measurements.[1] Nevertheless, the additional information that is obtained does not outweigh the higher cost and complexity of an MCG system. For fetal heart monitoring, however, the situation is different. In this case, the measurement of the ECG often fails or becomes distorted, whereas the MCG-recordings are mostly successful.[2,3]

In order to develop fetal MCG as a standard clinical tool, the medical value of these recordings has to be demonstrated. For that purpose, we perform measurements in a

Advances in Cryogenic Engineering, Volume 45.
Edited by Shu *et al.*, Kluwer Academic / Plenum Publishers, 2000.

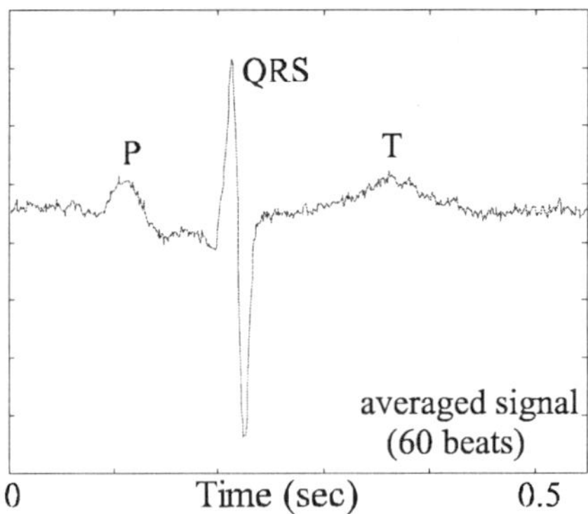

Figure 1. Fetal magnetocardiogram recorded with low-T_c system in shielded room.

shielded environment with a helium-cooled 19-channel low-T_c system.[4] An example of such a fetal MCG-recording is shown in Figure 1. The P, QRS and T-waves correspond to respectively the contraction of the atria, the contraction of the ventricles and the relaxation of the ventricles. The wave corresponding with the relaxation of the atria is obscured by the QRS-complex. Our experiments indicate two important advantages of fMCG when compared to alternative techniques as ultrasound and fECG.[5] Firstly, the fetal heart rate (FHR) can be measured on a beat to beat basis. An example is shown in Figure 2, in which the heart occasionally skips a beat. Also shown is the fetal heart rate derived using standard ultrasound data processing. Due to the poor signal-to-noise-ratio, data processing for ultrasound measurements usually makes use of an autocorrelation step over several seconds. This tends to average out the single missing beat. Secondly, information can be obtained by examining the time-intervals and amplitudes within a single (averaged) heart beat. Examples are the time between QRS-peak and T-wave occurrence and the ratio of the amplitudes. Both van Leeuwen[6] and Quinn[7] et al. observed a statistical significance for such parameters in fMCG-recordings.

Parallel to the low-T_c measurements, a fetal heart monitor is under development for clinical use (FHARMON). The major differences with the system in use are: it should be able to operate without the use of a shielded room, it is based on high-T_c SQUID sensors and the cooling should not require technical/cryogenic knowledge. Operation outside a shielded environment eliminates the need for an expensive shielded room and, thereby, also makes it possible to move the system around on occasion. The main reason for using high-T_c SQUIDs is the relaxation of the cooling requirements, leading to both smaller and less expensive coolers and cryostats. In general, it also makes operation of the cooling system less complicated.

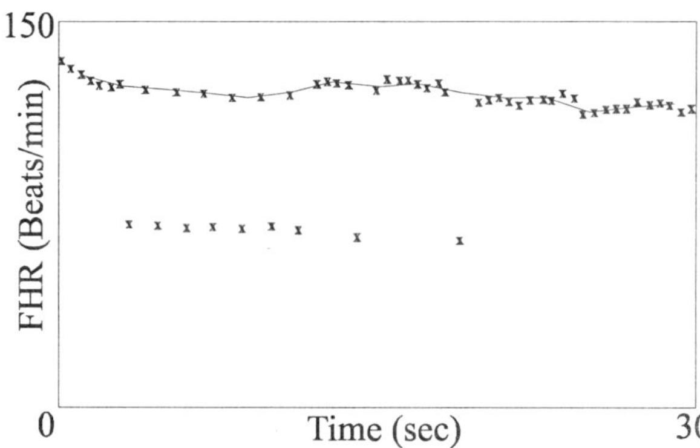

Figure 2. Fetal heart rate (FHR) determined from fMCG data (**x**) and as determined by ultrasound data processing (—) of these same data.

FHARMON SYSTEM

Signal and Noise fields. The FHARMON system is planned to make use of SQUID sensors with a white noise level below 50 fT/√Hz. This is required, as the fetal heart signal has an amplitude of only 1-10 pT. As the system should be used outside a shielded room, it has to deal with the environmental magnetic fields. The largest field is the earth's field, which is about 50 μT strong. A bigger problem, however, is caused by the environmental disturbances overlapping the signal frequency bandwidth of 1-100 Hz. Although Vrba[8] reports noise levels of 1 nT/√Hz for frequencies above 10 Hz rising to 10 nT/√Hz at 1 Hz, it is expected that, in general, a site can be found where the noise level is about a factor 10 lower. This is supported by our measurements in the local hospital, where in one of three available rooms, the noise level was in the order of 100 pT/√Hz. An exception is the (50 Hz) power line frequency, for which the amplitude can easily reach 100 nT.

Gradiometer Noise Suppression. The discrimination between the magnetic field of the heart and the field of the environmental disturbances can to a small extent be realized by using a band-pass filter tuned to the frequency content of the fetal heart: 1-100 Hz. However, the main suppression of the environmental field comes from spatial separation by using a 2nd order axial gradiometer.[8] In high-T_c materials no superconducting wire is available to construct a flux transformer. Therefore, three independent magnetometers are used. As a result, the gradiometer noise level is about 2.4 times the noise of a single sensor. We aim at a level of 100 fT/√Hz. The sensitivity of the gradiometer for magnetic fields depends on the distance between the magnetometers: the baseline. For very small baselines the intrinsic sensor noise limits the signal-to-noise-ratio, whereas for larger baselines the environmental disturbances set a limit. To find a proper value, a series of measurements was performed with a low-T_c system with variable baseline inside a shielded room. On the basis of these measurements and a 2nd order noise field gradient of 10 $pT/m^2\sqrt{Hz}$, a baseline of 6 cm was calculated to give a good compromise between signal sensitivity and noise suppression. Because of small errors in the construction, the output signal of a gradiometer is expressed as:

$$B_{sensed} \approx G^{(2)} + \mathbf{C}_B \bullet \mathbf{B} + \mathbf{C}_G : \mathbf{G}^{(1)} \qquad (1)$$

In this equation, $G^{(2)}$ is the output of an ideal 2nd order gradiometer and $\mathbf{C}_B$ is the remaining sensitivity for homogeneous fields **B.** The last term is the scalar product of the gradient imbalance $\mathbf{C}_G$ with the environmental 1st order gradient matrix $\mathbf{G}^{(1)}$. For proper operation C_B should be below 10^{-4}: the ratio of system noise over environmental field noise. The magnitude of C_B is primarily determined by tilt of the sensors relative to one another, causing sensitivity for orthogonal field components. A tilt of merely 10^{-4} radians results in an imbalance of $C_B=10^{-4}$. As this corresponds to 1 μm height difference over the sensor area of roughly 1×1 cm^2, it is in practice impossible to place the sensors with the required accuracy. Three methods are available to reduce C_B after the initial construction: the orientation of (one of) the SQUIDs can be adjusted by mechanical means[9], the magnetic field can be distorted locally by superconducting vanes[10] or a correction can be made electronically by measuring the orthogonal fields with reference sensors[11]. The last option was chosen, as the other two involve mechanical constructions at cryogenic temperatures. Although it is possible to improve the balance electronically, the aim is still to align the principle sensors to within 0.1 degree. This has two reasons. Firstly, the reference sensors may have a relatively high noise level and, therefore, the electronic correction should be as small as possible. Secondly, if the hardware imbalance is too high, the term C_G may become relevant.[8]

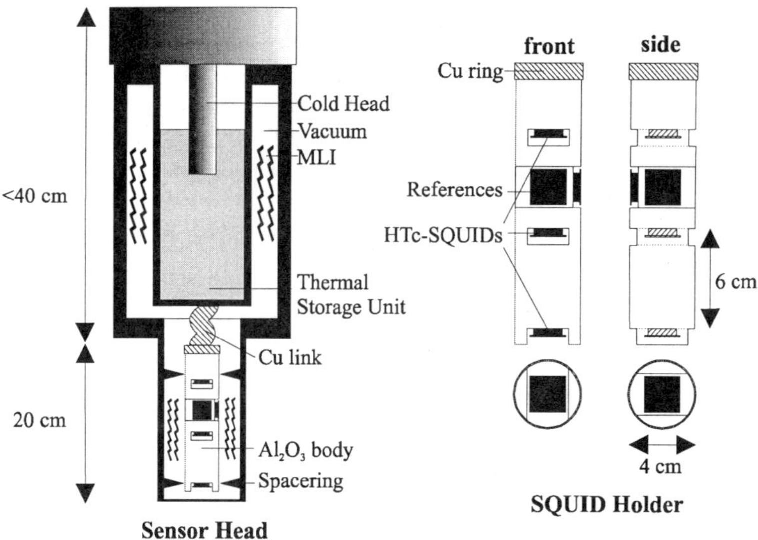

Figure 3. Preliminary design of the FHARMON sensor head.

Sensor head design. Figure 3 shows the sensor head of the system with an Al_2O_3 holder, which contains the 3 SQUIDs needed to form the 2nd order gradiometer. The Al_2O_3 SQUID holder will be placed inside a vacuum housing. The connection to the cold tip will be made with a flexible copper link, to suppress mechanical vibrations as well as temperature fluctuations from the cooler. To keep the insert in position some means of spacering will be utilized. Optionally, an rf-shield may be placed next to the outer wall of the vacuum space. This is a thin metallic sheet that shields the sensors from high-frequency (>100 kHz) magnetic noise fields that negatively influence SQUID operation. Finally, the whole head will need some support structure. This structure has two main functions. Before an actual measurement is done, it should aid the medical staff in changing the position and orientation of the sensor head while a good measurement position is sought. During the recording, it should keep the sensor head in place as rigidly as possible.

Active Compensation Coil. In an unshielded hospital environment large low-frequency disturbances are present, for instance due to moving magnetic objects. Also a displacement of the measuring head in the earth's field will cause a large flux change in the SQUID sensors. In order to keep the separate SQUIDs in locked-loop operation, these low-frequency fluctuations will be compensated in an active manner. The axial field is measured with a reference sensor (e.g. a fluxgate magnetometer) and the output of this sensor is used to create a compensatory field by means of a coil set in the direct vicinity of the SQUIDs.[12] In the FHARMON system we aim at a 40 dB suppression of the earth's field.

Cooling options. The need to refill a system daily or weekly with cryogenic liquids is a disadvantage for use in a clinical environment. Therefore, a cryocooler is used to extract the heat. For the interfacing between the cooler and the SQUID holder three options are conceivable:

1. continuous operation of a cooler with the cold tip placed in the sensor head
2. non-continuous operation of a cooler with the cold tip placed in the sensor head
3. continuous operation of a cooler with the cold tip spatially separated from the sensor head

Option 1 can only be applied for coolers, with a 'low-noise' cold head. If the cold head creates mechanical or magnetic interference during operation only, then option 2 becomes sound. In this case, however, a thermal storage unit (TSU) is required in order to keep the temperature stable while the cooler is turned off. This can be a small thermal buffer, if the cooler is only briefly turned off during a recording. It might also be a larger buffer, as shown in Figure 3, which is charged overnight. If the cooler is too large, heavy or (magnetically) noisy, the cooler needs to be spatially separated from the SQUID holder. This requires a cryogenic connection of about 2 meters, which can be achieved by circulating a cryogen[13] or using a copper strap[14].

COOLER REQUIREMENTS

The cooler requirements are listed in Table 1. Here, three types of requirements can be distinguished. The first type covers the commercial conditions, such as price, lifetime and maintenance. The second includes the elementary conditions for SQUID operation, such as operating temperature and cooling power. The third type deals with noise from the cooler: thermal, mechanical and magnetic disturbances. It is important to note that these interference requirements may be relaxed if appropriate measures are taken. For instance, temperature control can be applied to reduce cold-tip temperature drift.

Commercial requirements

The commercial requirements are not specified for prototype coolers but are envisioned for fetal heart monitors that are sold in relatively large quantities. The cost of the cooler is set at $US 5,000. This would be 10% of the cost of the total FHARMON system, which we aim to keep below $US 50,000. The lifetime is chosen at 10 years on the basis of 10 hours runtime per day (25,000 hrs). Maintenance is acceptable, provided the costs do not exceed $US 500 and the downtime is not significantly more than 1 week, both values referring to a yearly basis. The cool-down time is chosen such that it will not take more than one day to cool the system and thus make it ready for use.

Basic system requirements

The three high-T_c SQUID magnetometer sensors form the basis of the FHARMON system. As the sensors are made from superconducting material (YBaCuO), they have to be cooled to below the critical temperature T_c at which the material becomes superconducting. This is typically around 90 K. For proper operation it is necessary to stay well below this temperature. Therefore, the cooler should at least reach 80 K with an expected heat load

Table 1. Requirements for FHARMON cooler.

Requirement	Value	Remark
Cost	$US 5,000	(not for prototype)
Lifetime	10 years, 10 hours/day	(not for prototype)
Cool-down time with 0.3 kg Cu	< 8 hour	
Operating temperature	55 - 80 K	
Cooling power	>0.5 W >2.0 W	 (with TSU)
Operation + flexibility	Up to 90 degr. tilt Transportable	
Temperature drift	< 0.1 K in 10 hours < 0.5 K in 10 hours	(77 K) (55 K)
Temperature fluctuations (periodic; e.g. drive frequency f)	< 10f mK_{RMS}/Hz < 35f mK_{RMS}/Hz	(77 K,1-100Hz) (55 K,1-100Hz)
Mechanical interference	$< 2.10^{-5}$ rad/√Hz	
EMI at 10 cm from cold tip	< 10 nT_{RMS} < 2 pT_{RMS}	(50 Hz) (1-100Hz)

from the environment of 0.5 Watt. However, as the SQUID properties improve at lower temperatures (i.e $0.6T_c$)[15], it would be advantageous to be able to cool to 55 K. In case a thermal storage unit (TSU) is incorporated in the sensor head, it is assumed that the cooler is only operated during the night. It, therefore, has to cope with the 24 hour a day heat load in half the time raising the required cooling power to 1 Watt. As the TSU also makes it necessary to increase the cold volume, the required power rises further to 2 Watt. For the cooling option incorporating a thermal link the cooling power should also be about 2 Watt minimum as the cryogenic connection gives rise to an additional heat load. The cryogenic connection becomes relatively complex, as it should be flexible enough to allow the search for a good measurement position. This search also requires that the SQUID holder can be tilted to a maximum of 90°. Preferably, those coolers that are placed in the sensor head should be able to follow this tilt without violating the requirements.

Cooler interference considerations

Temperature effects. The effective measuring area of a SQUID is temperature dependent because of the London penetration depth.[16] As a result, the gradiometer balance can be affected by temperature drift. This effect may clearly not exceed the acceptable imbalance of $C_B=10^{-4}$. The temperature dependence of the SQUID effective area is given by:

$$\frac{1}{A_{eff}}\frac{d}{dT}A_{eff}=\frac{1}{T_c}\frac{2\tau}{1-\tau^2}=\frac{1}{T_c}\alpha(\tau)\ , \qquad (2)$$

where τ is the dimensionless temperature T/T_c.[16] In closed loop operation a magnetic signal is fed back to the SQUID by means of a feedback coil. This feedback coupling exhibits similar temperature dependence. However, the temperature dependence is expected to be different by up to 1% due to the difference in field shape of the environmental and feedback fields. Therefore, the temperature drift is required to stay below:

$$T_{drift}<T_cC_B\,/\,0.01\alpha(\tau) \qquad (3)$$

With the appropriate substitutions this results in the values presented in Table 1. Allowing for daily calibrations of the system, we require the temperature to be stable for a full workday (10 hrs). A less obvious temperature effect is caused by temperature fluctuations in the measuring band. Now, the noise contribution is determined by the background dc-field times the fluctuations in effective area, leading to much stronger temperature requirements:

$$T_{fluctuation}<T_c(B_{noise}\sqrt{BW}\,/\,0.01B_{earth})\,/\,\alpha(\tau) \qquad (4)$$

The term between brackets can be considered as the acceptable fluctuation in the imbalance factor, where the factor 0.01 represents the axial field suppression of the compensation coil. The bandwidth BW is included to make the step from noise density to amplitude. Here, the relaxation in α is not allowed, because the effect is caused by a dc-field that is not compensated by the feedback circuit of the sensors (field present at cool-down of SQUIDs or at reset of electronics). Substituting the appropriate values for the elements in Eq. (4), an acceptable temperature fluctuation of 0.1 mK_{rms} is found (55 K). Fortunately, temperature fluctuations at higher frequencies are strongly damped by thermal masses and resistance's. So, if we include a thermal resistance of the link of 1Watt/K and add the mass of the insert, the fluctuation requirement for the cold tip temperature relaxes to those stated in Table 1. The frequency dependence is a result of the low-pass filter behavior of the above mentioned thermal resistance and mass.

Vibrations. As the SQUID holder is assumed to be a rigid unit, vibrations have the effect of translating and rotating the total insert within the environmental field. Referring to Eq. (1), the derivatives in the various directions give the sensitivity for translations. The most significant term is expected to be due to the gradient of the earth's field. After reducing the elements of the equation to scalars for estimating the acceptable translations x, the following requirement is obtained:

$$x < \mathrm{B}_{noise} / \mathrm{C}_{\mathrm{B}} \mathrm{G}^{(1)}_{\boldsymbol{earth}} \tag{5}$$

As the gradient of the earth's field measured inside a building is typically 200 nT/m[8], this leads to translations in the order of millimeters. This should not be a problem. Unfortunately, rotations are a much bigger problem:

$$\varphi < \mathrm{B}_{noise} / \mathrm{C}_{\mathrm{B}} \mathrm{B}_{earth} \tag{6}$$

With a imbalance C_B of 10^{-4}, an acceptable noise B_{noise} of 100 fT/√Hz and an earth's field of 50 μT, this gives an acceptable rotation of $\varphi=2*10^{-5}$ rad/√Hz in the frequency band of interest. An exception is a rotation with a frequency corresponding to the power line frequency, where much higher noise levels are present in the environment. The rotational requirements can be reduced by applying three-axis field compensation. However, this will only work if the homogeneity of the compensation system in the sensor volume is better than 10^{-4}. This can, unfortunately, only be reliably obtained with room-size coil systems. This is a disadvantage with regard to transportability of the system and will, therefore, not be applied.

Electromagnetic interference. The strongest requirement on the electromagnetic disturbances is found by assuming that the cooler is located close to the sensor and thus exhibits a large magnetic field gradient. To calculate the acceptable magnetic noise from the cooler, the cooler was imagined at 10 cm above the top SQUID. Next, the field of the cooler was assumed to decrease with r^{-3} as the field of a magnetic dipole. That strength of the equivalent dipole was calculated that gave a system response equal to the system noise level of 100 fT/√Hz. By assuming a bandwidth of interest of 100 Hz, this led to 2 pT_{rms} as the acceptable noise level at 10 cm from the cooler. An exception is the power-line frequency, where the environment generates so much noise that a cooler contribution of 10 nT_{pp} is acceptable.

DISCUSSION OF COOLING OPTIONS

The available coolers can be split in two categories. The low-noise coolers (JT-coolers, GM-type pulse tube coolers) can be operated continuously, even when the cold tip is located in the sensor head. For other coolers (Stirling, GM) either time or space separation is required, as operation during a measurement would disturb the fetal heart measurement. Space separation will require a thermal interface of about 2 m in length. As a result, additional cooling power is required, which is a drawback. An even bigger drawback is seen in the practical realization of the thermal interface. It will need a cryogenic fluid circulation or a thermal strap. This significantly adds to the complexity of the system and is likely to be a costly and life-limiting component. For this reason, low-noise coolers are the preferred option. Also, the use of noisy coolers with the added complexity of a TSU is still an option. As the system should facilitate the search for a good measurement position, it should be possible to tilt the system without affecting its properties. Therefore, a disadvantage is seen in the orientation dependence of the cooling power, which is a well-known side effect in pulse tube coolers.[17] However, this could be dealt with by means of temperature control, provided adequate power is available even with the cooler tilted 90 degrees with respect to the vertical axis. At present, coolers are selected and characterized on thermodynamic and interference aspects. A first demonstrator version of the fetal heart monitor is planned to be constructed in the first half of the year 2000.

ACKNOWLEDGEMENT

The project is supported by the Dutch Technology Foundation (STW), the Institute for Biomedical Technology (BMTI), Philips Medical Systems and Signaal USFA.

REFERENCES

1. K. Ziegert, D. Selbig, H. Soltner, L. Trahms, P. Hanrath and Ch. Stellbrink, Evaluation of a new fragmentation score of the QRS complex for risk stratification using a high-Tc-SQUID-magnetocardiography-system, The Official Journal of the Int. Society for Holter and Noninvasive Electrocardiology, Inc.; A.N.E. 3 (3), Part 2, (1998) 61.
2. J.A. Crowe, M.S. Woolfson, et al., Antenatal assessment using the fECG obtained via abdominal electrodes, Journ. of Perinatal Medicine 24 (1) 43:53 (1996).
3. H.W.P. Quartero, M.J. Peters, J.G. Stinstra, J.A.F. Huirne, Clinical applications for fetal magnetocardiography. Present and future., Fetal Biomagnetism (Satellite Symposium of the 4th Int. Hans Berger Conf.), Jena (Sept. 1999).
4. H.J.M. ter Brake, J. Flokstra, W. Jaszczuk, R. Stammis, G.K. van Ancum and H. Rogalla, The UT 19-channel dc SQUID based neuromagnetometer, J. Clin. Phys. Physiol. Meas. 12B 45:50 (1991).
5. M.J. Peters, J.A. Crowe, J-F. Pieri, H.W.P. Quartero, B.R. Hayes-Gill, D.K. James, J.G. Stinstra and S.A. Shakespeare, Monitoring the foetal heart non-invasively: A review of methods, submitted to: American Journ. of Obstetr. and Gynaecol.
6. P. van Leeuwen, S. Lange, M. Schüßler, L. Lajoie-Junge, Determination of changes in fetal cardiac time intervals and T/QRS ratio during pregnancy using magnetocardiography, Biomag96, Santa Fe, 1996.
7. A. Quinn, A. Weir, U. Shahani, R. Bain, P. Maas and G. Donaldson, Antenatal fetal magnetocardiography: a new method for fetal surveillance?, British Journ. of Obst. And Gyn. 101 866:870 (1994).
8. J. Vrba, SQUID gradiometers in real environments, in: "SQUID sensors: fundamentals, fabrication and applications," H. Weinstock, ed., Kluwer, Dordrecht, (1995) 117.
9. Y.A. Tavrin, Y. Xu, A.I. Braginski, C. Heiden, A.S. Linnik, V.A. Puvliuk and V.V. Schapovalenko, Two-hole bulk high Tc SQUID electronic gradiometer for magnetocardiography in unshielded space, Cryogenics 33 (7) 719:723 (1993).
10. J.A. Overweg and M.J. Peters, The design of a system of adjustable superconducting plates for balancing a gradiometer, Cryogenics 18 529:534 (1978).
11. H.J.M. ter Brake, Z. Dunajski, W.A.G. van der Mheen and J. Flokstra, electronic balancing of multichannel SQUID magnetometers, Journal of Physics E: Scientific Instr. 22 560:564 (1989).
12. A.P. Rijpma, Y. Seppenwoolde, H.J.M. ter Brake, M.J. Peters and H. Rogalla, Application of SQUID magnetometers in fetal magnetocardiography, in: " Proc. EUCAS '97, Applied Superconductivity 2," H. Rogalla and D.H.A. Blank, eds., IOP Publishing, Bristol (1997), p. 771:774.
13. P.J. van den Bosch, H.J. Holland, H.J.M. ter Brake and H. Rogalla, Closed-cycle gas flow system for cooling of highTc dc SQUID magnetometers, Cryogenics 35 (2) 109:116 (1995).
14. P.J. van den Bosch, H.J.M. ter Brake, H.J. Holland, H.A. de Boer, J.F.C. Verberne and H. Rogalla, Cryogenic design of a high-Tc SQUID-based heart scanner cooled by small Stirling cryocoolers, Cryogenics 37 139:151 (1997).
15. H. Rogalla, Superconducting electronics, Cryogenics 34 (ICEC suppl.) 25:30 (1994).
16. H.J.M. ter Brake, W.A.M. Aarnink, P.J. van den Bosch, J.W.M. Hilgenkamp and H. Rogala, Temperature dependence of the effective sensing area of high-Tc dc SQUIDs, Superc. Sci. Techn. 10 512:515 (1997).
17. G. Thummes, M. Schreiber, R. Landgraf and C. Heiden, Convective heat losses in pulse tube coolers: effect of pulse tube inclination, in: "Cryocoolers 9," R.G. Ross, ed., Plenum Press, New York (1997), p. 393.

A MULTI-STAGE CONTINUOUS-DUTY ADIABATIC DEMAGNETIZATION REFRIGERATOR

P.J. Shirron, E.R. Canavan, M.J. DiPirro, J.G. Tuttle, and C.J. Yeager

Code 552
NASA/Goddard Space Flight Center
Greenbelt, MD 20771

ABSTRACT

The design of a multi-stage adiabatic demagnetization refrigerator (ADR) that can provide continuous cooling at very low temperatures is presented. The ADR is being developed for use in x-ray, IR and sub-millimeter space astronomy missions which will employ large format detector arrays operating at 50 mK and lower and which may dissipate up to 10 μW. It is also being designed to reject heat slowly to a relatively warm heat sink (in the 6-10 K range), so that future missions may use mechanical cryocoolers instead of liquid helium for pre-cooling. The continuous nature of the device gives it a much higher cooling power per unit mass, allowing it to be much smaller and lighter than existing ADRs with comparable performance. Design details are discussed.

INTRODUCTION

High-resolution detectors used in space astronomy and in some laboratory and industrial applications must be cooled to extremely low temperature (< 0.1 K) in order to obtain the required sensitivity. These detectors include infrared (IR) bolometers using superconducting transition-edge sensors[1] and silicon x-ray microcalorimeters[2], as well as superconducting tunnel junction detectors using rf-SETs[3] (Single Electron Transistors that use a radio frequency excitation). For space applications, adiabatic demagnetization refrigerators (ADRs) have an inherent advantage over other techniques (He-3/He-4 dilution refrigerators and He-3 refrigerators) because they don't rely on gravity to function, and because they have the highest thermodynamic efficiency. ADRs, though, are single-shot devices which provide cooling for a period of time, then they must be taken off line and recycled. This periodic interruption in cooling leads to a number of significant constraints on the design of ADRs for space missions. The two most important are that the hold time must be long, typically on the order of 1 day, and the time spent recycling the ADR must be short, typically 1 hour or less.

These requirements have been manageable up until now, but two trends for future missions are proving to be quite challenging. The first is the need for progressively larger cooling power and lower temperature operation. With recent improvements in micromachining of silicon, it is now possible to produce large format detector arrays containing thousands, even millions of pixels. The heat load from these arrays will be one to two orders of magnitude larger (10 μW at 50 mK) than, for example, the 32-pixel array used on the Astro-E XRS mission[2] (0.3 μW at 60 mK).

The second trend is toward the use of mechanical cryocoolers instead of liquid helium. Mechanical coolers are less desirable because of their higher base temperature (in the 4-10 K

Advances in Cryogenic Engineering, Volume 45.
Edited by Shu *et al.*, Kluwer Academic / Plenum Publishers, 2000.

range depending on the type of cooler used) and their limited cooling power. The higher base temperature increases the parasitic heat loads on the ADR and increases the magnetic fields needed to recycle it. The limited cooling power makes the situation worse because the cryocooler's base temperature is a function of the heat load. As the ADR is recycled, the large momentary heat load drives the temperature up, further increasing the magnetic field requirements and at some point making it impossible to use low temperature superconducting magnets.

To counter these problems, we have developed a design for a multi-stage, continuous ADR. The concept can be compared to the quasi-continuous designs of Heer, Barnes and Daunt[4], and Zimmerman, McNutt and Bohm[5], which used a high heat capacity reservoir (a paramagnetic salt) connected to a single-stage ADR through a superconducting heat switch. The reservoir reduced temperature excursions as the ADR was recycled. In our design we actively regulate the reservoir's temperature to achieve true continuous cooling. Additional warmer stages are also included to allow the device to operate from a relatively warm heat sink. The continuous operation allows the ADR to achieve a much higher cooling power per unit mass, and the particular arrangement of stages allows heat to be rejected to the heat sink over an extended period of time. This significantly reduces the peak heat load on the mechanical cooler, and therefore maintains the lowest possible base temperature. A preliminary design has been generated for a device that can continuously absorb 10 μW at 50 mK, operating from a 6-10 K heat sink. The projected mass is a factor of 2 less than the single-stage ADR used on XRS which had a 4 μW capacity (including parasitics) at 60 mK with a 1 day hold time.

CONCEPT DESIGN AND OPERATING MODES

The continuous ADR concept is shown schematically in Figure 1. In this version, containing three stages arranged serially, S1, S2, and S3 are heat switches and M1, M2, and M3 are the magnetic refrigerants. Each refrigerant is surrounded by a superconducting magnet and some thickness of magnetic shielding. The following discussion refers to M1 as the "continuous" stage, and to M2 and M3 collectively as the "main" ADR. The intent is to describe the general operation of the ADR, without reference to component characteristics which may impose significant constraints on the detailed design. These will be discussed in the following sections.

We first describe the steady state operation of the ADR, then how this state is reached from a warm start.

Steady State Operation

The concept of the continuous ADR is quite simple. In steady state, the continuous stage spends most of its time thermally isolated from the main ADR (i.e. S1 is open). As it absorbs heat, the coil surrounding M1 is slowly demagnetized to maintain constant temperature. During this time, the main ADR is regulated at an intermediate temperature to

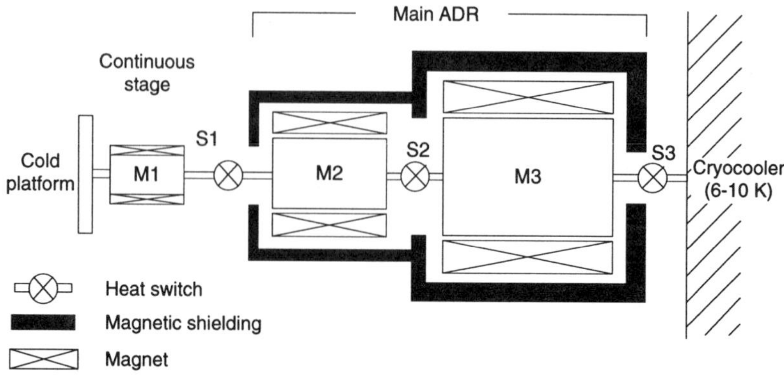

Figure 1. Schematic of a 3-stage continuous ADR.

minimize the parasitic heat load on M1. Before M1 is completely demagnetized, it is recycled by cycling M2 to a lower temperature and closing S1. As heat is transferred from M1 to M2, the temperature control system magnetizes M1 to maintain constant temperature. Once M1 is fully magnetized, S1 is opened and the slow demagnetization of M1 resumes.

M2 is restored to its normal operating temperature and the main ADR is then recycled to cascade the heat up to the heat sink. M2 is recycled first, dumping its heat to M3, then M3 is recycled to dump its heat to the cryocooler. M2 is recycled in the same manner as the continuous stage: M3 is demagnetized to a lower temperature than M2 and heat switch S2 is closed. After the heat transfer is complete, S2 is opened and M3 is returned to its normal temperature. M3 is then recycled using the usual procedure for single-stage ADRs[6].

Recycling the main ADR sequentially has two significant advantages. First, it allows M2 to be recycled at a much lower temperature than the cryocooler. This greatly reduces the magnetic field required and minimizes the parasitic heat conducted back to the continuous stage. Second, M3 can be recycled over an extended time period to reduce the peak heat load on the cryocooler. This in turn allows the cryocooler to maintain a low base temperature, and minimizes the size of the magnet needed to recycle M3. The only constraint on recycling M3 is that it must be demagnetized and cooled down in time to accept heat from M2 on the next cycle. The maximum amount of time is therefore on the same order as the hold time of the continuous stage.

One disadvantage to recycling the stages sequentially is a need to isolate each refrigerant from magnetic fields generated by magnets on other stages. The level of shielding, however, may turn out to be very modest since each stage is actively temperature controlled and stray fields can be cancelled out, on average, by reducing the dc field. For this, we can use combinations of active shield coils and passive ferromagnetic shielding.

Warm Start

The continuous ADR can be brought on line in a relatively short time period. After cooling the ADR to the heat sink temperature with all heat switches closed, the first step is to fully magnetize all three stages. This will transfer the entropy of the spin systems to the heat sink. Once the system relaxes back to the sink temperature, S3 is opened and M3 is demagnetized fully. The effect will be to cool the entire ADR. If M3 consists of a material such as Gadolinium Gallium Garnet (GGG) or Dysprosium Gallium Garnet (DGG), the lower temperature stages will cool to as low as about 0.5 K depending on their heat capacity. From this temperature, M2 can be demagnetized fully (with S2 open) to bring the continuous stage to near its operating temperature. If not, the warmer stages can be repetitively isolated and recycled to achieve a lower initial temperature. From that point, the continuous stage can be isolated and operated as described above.

Benefits of the Continuous ADR Concept

The main benefit of this technique is that the hold time of each stage can be very short - there is no need to absorb heat over long periods of time because there is no interruption in cooling when the ADR is recycled. The optimal hold time will depend on the characteristics of the heat switches, especially on/off transition times and the on-state conductance of the low temperature switch, but a realistic estimate is 2-3 hours. Compared to the 24-hour hold time that is typical for single-shot devices, each stage in the continuous ADR will be an order of magnitude smaller. This allows us to achieve cooling powers of 10 μW or more in a relatively low mass ADR.

PRELIMINARY DESIGN

Our preliminary design is based on a set of performance goals which envelop the requirements for a number of future missions. The ADR is to have a detector cooling power of 10 μW at 50 mK, a mass of less than 15 kg, and be able to reject heat to a cryocooler capable of sinking 10-100 mW at 6-10 K. We are using the characteristics of Creare's miniature turbo-Brayton cooler[7] instead of other coolers capable of lower temperature operation for two reasons. First, its vibrationless operation makes it very attractive, if not the only practical choice, for many space applications. Second, the relatively high thermodynamic efficiency of ADRs suggests that the optimal system uses an ADR up to the highest possible temperature. The need for superconducting magnets in the ADR, however, puts a practical limit at 8-10 K.

Heat Switches

The success of the continuous ADR concept depends on having heat switches that can efficiently transfer heat between adjacent stages with small temperature drops while providing adequate isolation with large temperature gradients. In particular, the switch that services the continuous stage must efficiently transfer heat at very low temperature with a temperature difference of only ΔT~5-10 mK.

Table 1 lists the types of heat switch that have either been proposed or developed for the temperature ranges of interest here. The table identifies constraints on the warm and cold end temperatures in the on- and off-states, and other factors to consider in selecting a switch. In most cases, (e.g., the superconducting switch) the constraints represent the limits beyond which the performance is too poor to be of use. In others, such as switches using superfluid 4He, the heat transfer mechanism fails beyond certain temperature thresholds. The list is roughly in ascending order of operating temperature.

The ADR was designed by selecting the heat switch for the coldest stage, establishing the operating parameters for the next warmer stage, then selecting its heat switch, and so on up to the warmest stage. The first step was to determine the on-state conductance of the continuous stage's heat switch, which depends on the fraction of time, f, that we spend recycling the stage. The values of f for each stage depend on a large number of factors, including heat switch conductances and on/off transitions times. Because these parameters are not yet known, we have made initial assignments and will adjust them as the design matures.

The on conductance is obtained by dividing the total heat load on the stage (including parasitics) by f. The parasitics will depend on the on-state conduction of the switch, so in general we need to iterate the design for each heat switch type. However, the parasitics should be small, and a reasonable estimate of the conductance can be obtained by neglecting them. As an example, if 25% of the time is spent recycling the continuous stage, the on-state conductance must be $\dot{Q}_{ON}$=40 μW. The required conductivity is $K_{ON}=\dot{Q}_{ON}/\Delta T$, where ΔT is the temperature difference across the switch. Any ΔT, though, degrades the thermodynamic efficiency by the ratio of ΔT to the absolute temperature, so we are forced to keep it very small. Our initial design assumes a value of ΔT=5 mK, giving K_{ON}~8 mW/K.

Table 1 identifies superconducting, mechanical, and magnetoresistive switches as viable candidates at 50 mK. Mechanical switches would be very attractive for their nearly zero off-state conductance, but it is doubtful that the necessary on conductance can be achieved. The clamping force needed to reach 8 mW/K at 50 mK for gold or gold-plated copper contacts is on the order of 3600 N[10]. While such a requirement is not impossible to meet - in fact, low-stroke, high-force magnetostrictive actuators have been developed with performance near this level - the long-term stability of such a joint undergoing 10 or so mechanical cycles per day is unknown. On the other hand, both superconducting and magnetoresistive switches use a normal metal for on-state conduction and will have little difficulty in meeting the required on conductance. Superconducting switches have a large advantage, though, in having much

Table 1. Heat Switch Operating Temperatures

Heat Switch Type	OFF state	ON state	Comments	Refs.
Superconducting	$T_H \lesssim 0.4$ K	Any temp.		8,9
3He liquid gap	Any temp.	$0.1 \lesssim T_H < 4.2$ K	concept only; requires large area to overcome Kapitza resistance at low T	
Superfluid diode	$T_H < 1.5$ K	$0.6 \lesssim T_H < 2.1$ K	completely passive	11
Superfluid film	Any temp.	$0.8 \lesssim T_C < 2.1$ K		12
3He gas-gap	Any temp.	$T_C \gtrsim 0.6$ K		13
4He gas-gap	Any temp.	$T_C \gtrsim 1$ K		13
Mechanical	Any temp.	Any temp.	essentially zero OFF conductance	14
Magnetoresistive	$T_H \lesssim 20$ K	Any temp.	high magnetic field required in OFF state	15

higher on/off conductance ratios (~6×10^5 at 50 mK[8], compared to at most ~1000 for magnetoresistive switches). Even so, the parasitic (off-state) conduction provides a significant constraint on the ADR's design. Using data from Mueller[8] for an Al switch adjusted to have 8 mW/K at 50 mK, the off-state conduction is 1 μW for a warm end temperature of 0.2 K, and 5 μW at 0.25 K. Using data from Konter[9] for a similarly sized Sn switch, the temperatures would be 0.3 and 0.44 K, respectively. The conductivities of Sn and In are very similar, so we have baselined the use of an In switch.

The upshot is that M2 must be maintained at temperatures below about 0.3 K in order to keep the parasitic load on the continuous stage to an acceptable level. This means M3 must be capable of reaching temperatures less than 0.3 K. It must also be able to dump its own heat at temperatures between 6-10 K. There is no magnetocaloric material which can provide useful cooling between these temperature extremes for any reasonable magnetic field strength. We therefore have added another stage to bridge the temperature gap. The arrangement is shown in Figure 2. A thermal model was developed to optimize the mass of the ADR as a function of the intermediate temperature. The best performance occurs when the warmest stage is cycled between the cryocooler heat sink and about 1 K, and the intermediate stage between 1 K and 0.2 K.

For the two warmest stages, mechanical and gas-gap switches will work well, especially considering the slow rate at which these stages can be recycled. The 0.3 K to 1 K range is more difficult. However, we can make certain compromises to make the problem tractable. For instance, the higher temperature at which stage M2 is recycled allows us to use a much larger ΔT to transfer heat through the switch, greatly reducing the required on-state conductance. This can be estimated the same way as for S1, taking the average heat load on M2 to be $Q_1 T_2/T_1$, where Q_1 is the average heat load on M1, and T_1=45 mK and T_2=200 mK are the low and high temperature extremes for M2's cycle. We have previously estimated Q_1 to be 10 μW of detector load plus 1 μW of parasitics. Assuming that we spend 25% of the time recycling M2, we find Q_{ON}~200 μW, and if we allow the ΔT across the heat switch to be as large as 50 mK, the required conductance is K_{ON}=4 mW/K at 0.2 K. A mechanical switch can achieve this using a force of ~300 N. A magnetoresistive switch could also achieve this conductance, with an off-state conduction between 0.2 and 1 K of about 13 μW.

Magnetocaloric Materials

Selecting refrigerants for each stage is fairly straightforward. Since the ADR works by cascading heat from the continuous stage up to the heat sink, the entropy capacity of each stage must be roughly the same. Therefore the choice comes down to which material has the

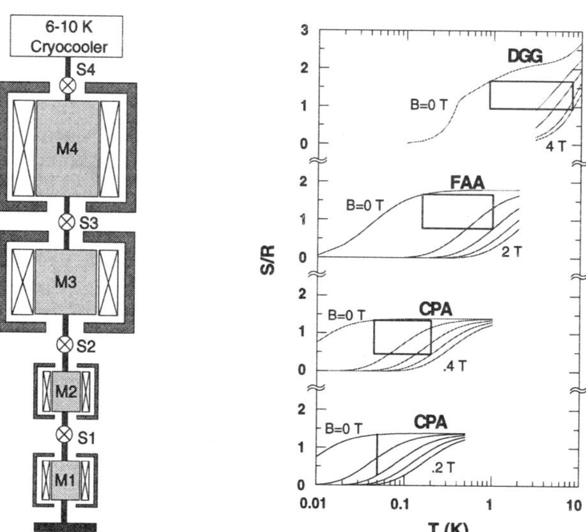

Figure 2. 4-stage continuous ADR is shown at left. At right are entropy versus temperature curves showing the cycle of all four stages. Entropies are given for zero field and four equally spaced field values up to the indicated maximum.

Table 2. Magnetocaloric Materials

Material	J	g	b (T)	T_C (K)	Ref.
Cerium Magnesium Nitrate, CMN	1/2	2	.002	.0015	
Chromic Potassium Alum, CPA	3/2	2	.009	.009	17
Chromic Cerium Alum, CCA	3/2	2		.01	
Ferric Ammonium Alum, FAA	5/2	2	.038	.026	17
Gadolinium Gallium Garnet, GGG	7/2	2	.48	~.38	18
Dysprosium Gallium Garnet, DGG	15/2*	8		~.4	19

* Below ~1 K, the behavior of DGG resembles that of a J=1/2 spin material

largest entropy capacity per unit mass (of refrigerant and magnet) for each operating temperature range. Table 2 lists the relevant parameters for some of the most common materials used in ADRs. A large number of other materials can be used (for example, certain rare-earth compounds called elpasolites[16] have very high entropy capacity, low ordering temperature and low field requirements), but very few have been subjected to careful measurements of entropy or heat capacity at the temperatures and fields of interest, and many would be difficult to process into a usable form. For the present, we will only use materials for which thermodynamic properties and fabrication techniques are well-established.

We can judge the relative merits of various magnetocaloric materials by examining how their entropy varies with magnetic field and temperature. The entropy of an ideal, non-interacting magnetic spin system as a function of field, B, and temperature is described by

$$S/R = x\coth(x) - (2J+1)x\coth((2J+1)x) + \ln\left(\frac{\sinh((2J+1)x)}{\sinh(x)}\right) \quad (1)$$
$$x \equiv \mu_B gB/2k_BT = (0.336\ \mathrm{K/T})\ gB/T$$

where μ_B is the Bohr magneton, g is the Landé g-factor and J is the magnetic spin for the particular material, and k_B is Boltzmann's constant. R is the gas constant. For real materials, the interaction of neighboring spins causes them to self-align at low temperature. This can be taken into account in Eq. 1 by defining an effective field such that $B^2 = B_{ext}^2 + b^2$, where B_{ext} is the external field and b is the self-field arising from the interaction. The significance of b is that it gives rise to an "ordering" temperature, T_c, below which the material cannot produce refrigeration. Values listed in Table 2 were determined through least-squares fits to zero-field entropy data.

Noting that entropy depends only on J and the ratio of gB/T, we can see that the ideal material would have high J (for the greatest entropy capacity), low T_c (for low temperature operation) and high g (for low magnetic fields required to drive the cycle). However, the correspondence between T_c and J that can be seen in Table 2 means that entropy capacity must be sacrificed for the low temperature stages to get sufficiently low T_c. At the highest temperature, T_c is less of a concern, so the materials with highest J and g are favored.

Since M1 and M2 must be capable of reaching 50 mK, only the hydrated salts with T_c<50 mK are suitable. Within this group, CPA and CCA have the largest entropy capacity at 50 mK even though they do not have the largest J value. This can be seen from Figure 3 in which ΔS=S(0 T, 50 mK)-S(B_i, 50 mK) for CPA and FAA are plotted against temperature

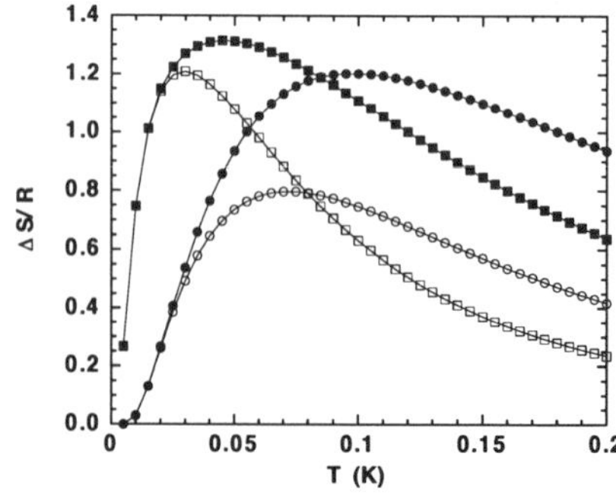

Figure 3 . Entropy capacity per mole of FAA (circles) and CPA (squares) versus temperature for initial magnetic fields of 0.1 (open symbols) and 0.2 T (solid symbols).

for field strengths of B_i=0.1 and 0.2 T. CPA will be used instead of CCA because its higher solubility in water makes it easier to fabricate a salt pill.

The coil surrounding M1 will generate fields up to 0.1 T. This sacrifices about 20% of the entropy capacity, but allows the coil to be made with Nb wire, instead of NbTi. Using a Type I superconductor avoids hysteresis losses which could severely degrade thermodynamic efficiency at such low temperatures. For stage M2, a field of at least 0.3 T is needed for it to have an entropy capacity per unit mass comparable to M1.

For the next warmer stage, M3, which operates between 0.3 and 1 K, we choose FAA because it has the highest *J* of the hydrated salts. A field of 1 T gives the FAA a higher entropy change per mole than the colder CPA stages. The drawback to FAA is its relatively low melting point of 38°C, compared to 89°C for CPA. Anecdotal evidence suggests that FAA is not at risk when stored and handled at room temperature, but it is not known how much warmer it can be taken without damage. Therefore care must be taken to ensure an FAA salt pill is never exposed to temperatures above about 25°C.

The choice of refrigerant for the warmest stage is driven almost exclusively by magnetic field considerations. The high fields that will be necessary for it to dump its heat at high temperature (6-10 K) mean that the magnet will be the dominant mass. The refrigerant, therefore, should have the highest possible *g* value to minimize the magnetic fields required. The best choice is DGG which would require fields of only 3 T for a 6 K heat sink, and 5 T for a 10 K sink. For these field strengths, standard superconducting NbTi magnets can be used at sink temperatures up to about 6 K. Beyond that temperature we must resort to higher temperature materials such as Nb_3Sn or Nb_3Al.

DETAILED DESIGN

Since the configuration of the continuous ADR depends strongly on the performance of the superconducting heat switch, we are focusing our efforts on developing that switch and the continuous stage's salt pill. These components, and the procedure of recycling the continuous stage, can be tested with an existing single-stage ADR that contains a large FAA salt pill. Its cooling capacity at 50 mK is an order of magnitude larger than necessary, allowing us to simulate the performance of stage M2 over a wide temperature range, from as low as 40 mK to well above 0.3 K. We will measure the thermal conductance of the heat switch and thermodynamic efficiency of the continuous stage, and feed the results back into the full ADR design to continue development of the remaining stages.

Superconducting Heat Switch

The prototype heat switch was made from high purity indium (99.99+%) with OFHC copper end pieces. The modular design, shown in Figure 4, is a departure from most previous designs[8,9] in that it uses a bolted interface at each end. This was done so that the switch could be built and tested independent of the salt pill. However there is only a modest performance penalty since the required on-state conductance is relatively low and the extra thermal impedances of the copper and the bolted joints contribute less than 25% of the total.

The In switch is expected to have an on/off ratio of ~2000 at 50 mK and an on-state conductance of 8 mW/K at 50 mK. It will conduct less than 3.5 μW back to the continuous

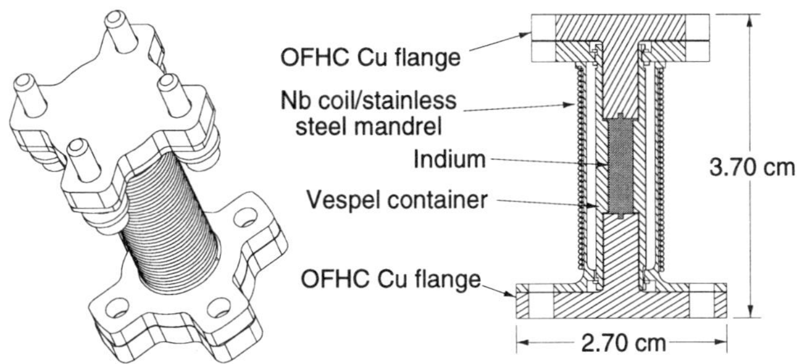

Figure 4. Cross-section of the superconducting indium heat switch.

stage when the warm end is held at 0.2 K. This switch is less capable than the Al and Sn switches discussed in references 8 and 9, but it will allow us to conduct a proof-of-principle demonstration. The switch has been built and is currently under test.

CPA Salt Pill

One of the main advantages of the continuous ADR is an increase in the cooling power per mass of refrigerant. This also introduces certain difficulties. Most magnetocaloric materials are polycrystalline and have poor thermal conductivity. An internal thermal bus is needed to distribute heat through the material so that the temperature drop from the external heat source and the refrigerant is minimized. This is particularly important for very low temperature operation where a temperature difference of just a few milliKelvin can degrade the thermodynamic efficiency by 10-20%. The standard practice[20] is to use high purity gold wires, and either pack the refrigerant in among the wires (if it is in powder form) or grow a crystal within the wire matrix. A typical ADR uses enough wire to provide on the order of 1 cm^2 of contact area per cm^3 of material, displacing only a few percent of the total volume. Gold wire is used because it has high thermal conductivity and is stable in the presence of all of the usual salts. Salts like FAA are extremely corrosive to copper.

In our case, we are attempting to increase the cooling power over existing ADRs by an order of magnitude using an order of magnitude less refrigerant. This requires a much more substantial thermal bus than described above. The thermal bottleneck is the boundary conductance between the salt and the wires, so it is necessary to greatly increase the contact area between the two. This suggests using a larger number of finer wires, but in such a small volume it will be difficult to maintain a uniform distribution, particularly at the end where the wires must be attached to the external thermal anchor. Instead, we have designed a thermal bus that is cut from a single piece of metal using the wire EDM (electric discharge machine) technique. In our case we can use copper since it is not attacked by CPA.

The bus is formed by two series of equally spaced cuts along the axis of a right circular cylinder of OFHC copper. Each cut is made from the same end, and ends about 1 cm from the other end. The cuts are made first in one orientation then after a 90° rotation about the cylinder's axis. The result resembles a bristle brush. Ideally, the wire diameter is equal to the desired spacing of the conductors, but it is possible to make the spacing wider by cutting into the copper, then moving the wire laterally and making a second cut as the wire is retracted. Figure 5 is an exploded view of the entire salt pill, showing the thermal bus with the stainless steel hermetic enclosure.

The main benefits of this automated process are an ability to produce a very uniform distribution of conductors with a high packing density. Unfortunately this last feature is proving more difficult than expected. Internal strains cause the copper to bend as the cuts are made. If the copper bends in the direction of the slot being cut, it is sheared off when the wire is retracted. Annealing the copper should have relieved these strains, but it actually made the problem worse. This has forced us to space the cuts farther apart and increase the size of the conductors to limit the bending. The minimum size that can be reliably machined is t=0.075 cm.

Given this constraint, we can calculate the spacing of the conductors which maximizes

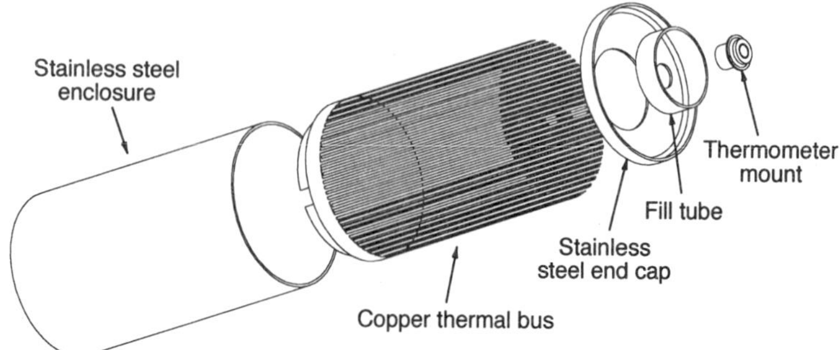

Figure 5. Exploded view of the copper thermal bus and stainless steel hermetic enclosure. The thermal bus is fabricated by wire EDM to define a square array of copper "fingers" measuring 0.075 cm square by 4.5 cm length.

the hold time of the salt pill for a given heat load. If the center-to-center spacing is d, then the surface area of copper per unit volume is $4t/d^2$, and the fill fraction of salt is $f=1-t^2/d^2$. The entropy capacity of the salt pill is $\Delta S=(S(B=0,\ T_{salt})-S(B=0.1\ \mathrm{T},\ T_{salt}))*f\rho V/M$, where T_{salt} is the temperature at which the salt must be regulated to keep the interface to the salt at 50 mK, ρ is the salt density of 2.06 g/cm^3, V is the volume of the salt pill and M is the molecular weight of CPA. In the presence of a heat load $\dot{Q}$, $T_{salt}\approx T_{copper}-\dot{Q}(4t/d^2)V/K(T_{salt})$, where the boundary conductance per unit area between copper and CPA[20], $K(T)$, is approximately 0.059 T^3 W/cm^2•K^4. For $\dot{Q}$=10 μW and T_{copper}=50 mK, we find that ΔS is a maximum for d=0.10 cm. At this spacing the thermal bus occupies 19% of the salt pill's volume. Our salt pill measures 5 cm in length and 3.1 cm diameter and will contain 65 grams of CPA. Its hold time will be about 1.5 hours at 10 μW.

The thermal bus is currently being machined. Upon its completion, the thin-wall stainless steel enclosure will be welded in place. The top end of the enclosure includes a copper button for mounting a thermometer, and a 1.5 cm diameter tube of about 2 cm length. The salt will be grown so that it will be in intimate thermal contact with the copper button. The tube provides access for growing the salt and will be sealed after the crystal is formed by welding on a cap.

We are currently experimenting with salt pill growing techniques. The usual procedure is to prepare a saturated solution at or slightly above room temperature. Because of CPA's relatively low solubility in water (244 g/liter[20]), a reservoir volume is often attached to the top end of the salt container, and both it and the salt container are filled with solution. Upon cooling in a refrigerator to ~5°C, approximately half of the dissolved CPA will crystallize onto the thermal bus. To minimize the likelihood that pockets of dilute solution will become trapped by crystals growing from multiple directions, it is customary to anchor the thermal bus in a way that it cools more rapidly than the walls of the salt pill and reservoir. This imposes a heat flow through the thermal bus and biases the crystal growth toward the bottom of the container, so that the crystal grows steadily up from the bottom.

This technique can work for our salt pill, but it is complicated by the fact that we have such a large thermal bus. Even when the heat flow is exaggerated by heating the reservoir to ~40°C, crystals begin to nucleate throughout the copper fingers within a day or so. We have switched to using a small peristaltic circulating pump with two pumping lines. One delivers saturated solution at room temperature from a remote reservoir into the salt container and the other withdraws depleted solution and returns it to the reservoir. The return line's inlet is initially positioned close to the bottom of the salt container, and we use larger tubing than the supply line to give it a 4 times higher flow rate. With this setup, the solution is maintained at a level just below the return line's inlet. By discretely raising the return line as the salt grows, we have produced salt pills with no voids or pockets of spent solution.

Fabrication of the salt pill is expected to be complete by August 1999. It will then be integrated with the superconducting heat switch to measure its thermodynamic efficiency and to test its ability to produce continuous cooling.

SUMMARY

A 4-stage ADR is proposed that can produce continuous cooling at temperature of 50 mK and below with a cooling power of 10 μW. Its design and operating modes have been tailored to allow it to dump its waste heat to a high temperature heat sink, in the 6-10 K range. This will allow it to operate coupled to a mechanical cryocooler instead of a liquid helium bath. The device is being developed for x-ray, IR and submillimeter space missions that will use large format detector arrays that must be cooled to such low temperature. The continuous nature allows the ADR to be made much smaller than existing single-stage refrigerators that have far lower cooling power. We have developed a preliminary design for the entire device, and a detailed design for the continuous stage and its heat switch. These components are presently being fabricated and tested. Testing of the complete continuous stage will be done with an existing single-stage ADR from the XRS program to simulate the warmer stages. Results of these tests will be used to complete the detailed design of the ADR, leading to the fabrication and demonstration of a full-scale device.

ACKNOWLEDGMENTS

This work was sponsored by the Director's Discretionary Fund at NASA's Goddard Space Flight Center.

REFERENCES

1. K.D. Irwin, G.C. Hilton, D.A. Wollman, and J.M. Martinis, X-ray detection using a superconducting transition-edge sensor microcalorimeter with electrothermal feedback, *Appl. Phys. Lett.* 69:1945 (1996).
2. R.L. Kelley, et al., The Astro-E high resolution X-Ray Spectrometer, Proceedings of the SPIE Conference on EUV, X-Ray, and Gamma-Ray Instrumentation for Astronomy X, Vol. 3765 (1999).
3. R.J. Schoelkopf, S.H. Moseley, C.M. Stahle, P. Wahlgren and P. Delsing, A concept for a submillimeter-wave single-photon counter, submitted for publication in *IEEE Transactions on Applied Superconductivity*, vol. 9 (1999).
4. C.V. Heer, C.B. Barnes and J.G. Daunt, *Rev. Sci. Instrum.* 25:11 (1954).
5. J.E. Zimmerman, J.D. McNutt and H.V. Bohm, A magnetic refrigerator employing superconducting solenoids, *Cryogenics* 2:153 (1962).
6. A.T. Serlemitsos, M. SanSebastian and E. Kunes, Design of a spaceworthy adiabatic demagnetization refrigerator, *Cryogenics* 32:117 (1992).
7. W.L. Swift, M.V. Zagarola, G.F. Nellis, J.A. McCormick and J.A. Gibbon, Developments in turbo–Brayton technology for low temperature applications, submitted for publication in *Adv. Cryo. Eng.* 45 (2000).
8. R.M. Mueller, C. Buchal, T. Oversluizen and F. Pobell, Superconducting aluminum heat switch and plated press-contacts for use at ultralow temperatures, *Rev. Sci. Instrum.* 49:515 (1978).
9. J.A. Konter, R. Hunik and W.J. Huiskamp, Nuclear demagnetization experiments on copper, *Cryogenics* 17:145 (1977).
10. Lionel Duband, private communication.
11. C.D. Fulton, C.F. Hwang, W.M. Fairbank, and M.M. Vilas, Helium thermal rectifiers for magnetic cooling, University of Wisconsin Conference Proceedings, p. 221 (1958).
12. M.J. DiPirro, P.J. Shirron and D.C. McHugh, A liquid helium film heat pipe/heat switch, *Adv. Cryo. Eng.* 43:1497 (1998).
13. E.R. Canavan, J.G. Tuttle, P.J. Shirron, M.J. DiPirro, Performance of the XRS ADR heat switch, submitted for publication in *Adv. Cryo. Eng.* 45 (2000).
14. C. Hagmann and P.L. Richards, Two-stage magnetic refrigerator for astronomical applications with reservoir temperatures above 4K, *Cryogenics* 34:221 (1994).
15. J.M.L. Engels, et al., *Cryogenics* 12:141 (1972); R.J. Radebaugh, *J. Low Temp. Phys.* 27:91 (1977).
16. M.R. Roser, J. Xu, S.J. White and L.R. Corruccini, Magnetic order in three rare-earth elapsolite compunds Cs_2NaRCl_6, *Phys. Rev. B* 45:12337 (1992).
17. O.E. Vilches and J.C. Wheatley, Measurements of the specific heat of three magnetic salts at low temperatures, *Phys. Rev.* 148:509 (1966).
18. E.W. Hornung, R.A. Fisher, G.E. Brodale and W.F. Giauque, Magnetothermodynamics of gadolinium gallium garnet, *J. Chem. Phys.* 61:282 (1974).
19. A. Tomokiyo, H. Yayama, T. Hashimoto, T. Aomine, M. Nishida and S. Sakaguchi, Specific heat and entropy of dysprosium gallium garnet in magnetic fields, *Cryogenics* 25:271 (1985); D.P. Goshorn, D.G. Onn and J.P. Remeika, Specific heat of europium, praseodymium, and dysprosium gallium garnets between 0.4 and 90 K, *Phys. Rev. B* 15:3527 (1977).
20. C. Hagmann, D.J. Benford and P.L. Richards, Paramagnetic salt pill design for magnetic refrigerators used in space applications, *Cryogenics* 34:213 (1994).

A LONG LIFETIME BALLOON-BORNE CRYOSTAT AND MAGNETIC REFRIGERATOR

J. O. Gundersen[1,3], M. Jirmanus[2], P. Timbie[1], Z. Zhao[2], S. Cordone[1], and Y. Lin[2]

[1]Department of Physics, University of Wisconsin
Madison, WI, 53706

[2]Janis Research Company, Inc.
Wilmington, MA 01887

[3]Current Address: Department of Physics, Princeton University
Princeton, NJ 08544

ABSTRACT

A long lifetime, low cost, vapor-shielded liquid neon and liquid helium cryostat and an adiabatic demagnetization refrigerator have been developed for cooling a bolometric radiometer to 0.1 K for mm and sub-mm observations of the cosmic microwave background. This is the first phase of a development effort for a very long hold time system for use in NASA's 100 Day Ultra Long Duration Balloon Project. This type of system has additional applications in sub-orbital and ground-based research in millimeter, sub-millimeter, infrared, and x-ray astrophysics. A detailed description of the design, analysis, and performance of the system is presented and plans for follow-up design improvements are discussed.

INTRODUCTION

A prototype long lifetime cryostat and two-stage adiabatic demagnetization refrigerator (ADR) have been constructed to support the National Aeronautics and Space Administration's (NASA) Ultra-Long Duration Balloon (ULDB) Project.[1] The primary design goals for this prototype instrument were a 10 day lifetime for the liquid helium (LHe) and liquid neon (LNe) cryogens and a 48 hour lifetime between recycling at 0.1 K for the two-stage ADR. This instrument is suitable for cooling far-infrared bolometers[2] for sub-orbital and ground-based research in millimeter, sub-millimeter, and infrared astrophysics, as well as laboratory testing of x-ray microcalorimeters for sub-orbital research in x-ray astrophysics.[3,4] Many of the design features from this prototype can be implemented in an instrument that supports the 100 day lifetime requirements of NASA's ULDB Project. In addition to astrophysics, there are several commercial applications that this type of cryogenic system can support. These applications include the cooling of x-ray microcalorimeters used in microanalysis[5] for the semiconductor industry and for cooling superconducting tunnel junctions[6,7] and transition edge sensors used in heavy particle detection[8] for studying biological samples and DNA sequencing. The project objectives are summarized, the cryostat and two-stage ADR design and performance are reviewed, and future plans for this type of system are discussed.

PROJECT OBJECTIVES

The principal objective was to design and construct a lightweight, low cost, long lifetime LHe-LNe cryostat and a two-stage ADR capable of cooling bolometric detectors to 0.1 K. Since this project was funded by the first phase of a NASA Small Business Innovative Research[9] (SBIR) grant, the design, fabrication, and testing of this cryogenic system had to be completed in six months. In the process of modeling and building this prototype we have gained the experience that is necessary to construct a 100 day hold time LHe-LNe ADR system for NASA's ULDB program. The technical objectives are divided into those that relate to the LHe-LNe cryostat, those that relate to the two-stage ADR, and those that relate to the integration and testing of the combined ADR and the LHe-LNe cryostat.

As with spacecraft instrumentation, one of the primary constraints on balloon-borne instrumentation is weight. The maximum weight of the science instrumentation for a 10 day, long duration balloon[10] (LDB) flight around Antarctica is 1360 kg, while the science payload limit for the demonstration 100 day ULDB payload is 910 kg.[1] For this project, a target weight of 140 kg was set and achieved for the filled cyrostat and two-stage ADR. The remaining science weight is typically used for the gondola superstructure, the optical system, the electronics, and the power, guidance, and telemetry systems. For comparison, this cryostat is a factor of three lighter than the LHe/liquid nitrogen (LN) cooled ^{3}He cryostat flown in the BOOMERANG LDB project.[11] Unlike spacecraft instrumentation, balloon-borne projects are funded at considerably lower levels – typically an order of magnitude below a Small Explorer class satellite. Hence the relative cost of a balloon-borne cryostat compared to a space-borne system has to reflect this funding difference. This cryogenic system was constructed with the aid of a SBIR Phase I grant of $70,000 and an additional $17,000 of supplemental internal funding. This cryostat is designed to support a LDB flight, which typically has a duration of 10 days. The cryostat incorporates a rugged design to withstand the parachute shock that occurs at the termination of a balloon flight and causes as much as 10 g vertical deceleration and 5 g at 45° from the vertical.[12]

For the above applications, the ADR is required to cool detectors below 0.1 K. Single-stage ADRs can reach these temperatures when operated from a bath temperature of ~2 K. The two-stage ADR is a natural extension of the single stage ADR that was developed[13] for NASA's Space Infrared Telescope Facility (SIRTF) and is based on the concept forwarded by Hagmann and Richards.[2] The detailed design for the two-stage ADR was obtained from J. Martinis and collaborators[5] at the National Institute of Standards and Technology (NIST). In order to fulfil the project goals within the stringent time constraints of a Phase I SBIR grant, we chose to replicate the NIST design with minor modifications. The primary motivation for building a two-stage ADR (versus a single stage ADR) is that it can operate either from an unpumped, 4 K LHe bath or a 4 K cryocooler. A LHe bath was

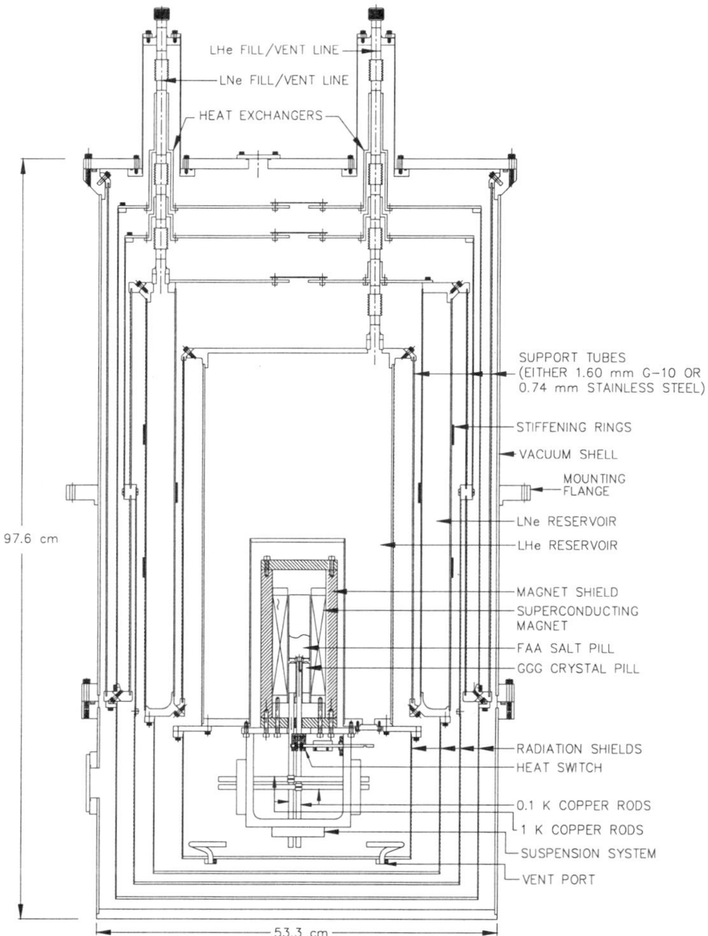

Figure 1. Schematic of LNe/LHe cryostat and two-stage ADR.

used rather than a 4 K cryocooler because of the vibration and power consumption constraints of balloon-borne, bolometric observations. The capability of operating from a 4 K bath simplifies the ground based operations immensely and alleviates the need for a superfluid bath at balloon altitudes. Balloon-borne LHe cryostats typically allow the atmosphere to pump on the LHe bath as the payload ascends to 40 km where the ambient pressure is 4-5 Torr. Roughly half the LHe is lost during this process. In a balloon flight that uses a two-stage ADR, the pressure above the LHe bath could be regulated to one atmosphere, doubling the LHe hold time.

The cryostat and two-stage ADR were integrated for testing at the University of Wisconsin, Madison (UWM). The primary objectives of the testing procedure were to measure the temperatures of the vapor cooled shields, measure the evaporation rates of the cryogens to establish hold times, and verify the operation of the two-stage ADR. All but the last of these objectives were met within the time constraint of the SBIR. Due to vendor-related delays associated with the delivery of the cryostat support tubes, the testing of the two-stage ADR didn't occur until several weeks after the SBIR deadline. The results from these tests are discussed in more detail below.

LIQUID HELIUM-LIQUID NEON CRYOSTAT

Cryostat Design

The cryostat design, shown in Figure 1, incorporates two superinsulated, vapor-cooled shields, a LHe reservoir and a LNe reservoir. The interwoven design of radiation shields and support tubes was inspired by the design of a rocket-borne ADR system.[3,4] This modular design allows full access to all reservoirs and radiation shields and includes a 22 L LNe reservoir and a 19.5 L LHe reservoir. The escaping cryogen vapors pass through heat exchangers[14] at each warmer stage on the fill/vent lines. The LHe reservoir has a re-entrant space to accommodate the two-stage ADR. The LNe volume is toroidal in shape and is used as the secondary cryogen for three reasons:

1. LNe was chosen over LN because its lower temperature significantly reduces the conductive and radiative loads on the LHe.
2. LNe was chosen over liquid hydrogen for safety reasons.
3. LNe was chosen over a pure LHe cryostat because a liter of LNe can absorb approximately seven times more heat than a liter of LHe used to vapor cool a shield at 27 K, assuming 100% vapor cooling efficiency (VCE).[15] VCE is defined as the ratio of the enthalpy used to the total enthalpy available at each heat exchanger.

The additional cost of using LNe (~$120/L) versus LN (~$1/L) for a balloon flight is trivial compared to the overall costs of ballooning. For preflight and laboratory use, the LNe is easily replaced with LN and a several day hold time is obtained.

The primary compromise in the design of the cryostat occurred when the 1.6 mm thick G-10 tubes were delivered late and out of specification by an outside vendor. In order to finish the project on a six month schedule, a decision was made to replace the G-10 with 0.74 mm thick stainless tubes. This reduced the ultimate hold time significantly, but the 10 day target hold time was still met by minor adjustments in the design and the LNe and LHe reservoir capacities. This modification was accompanied by a detailed finite element analysis[16] (FEA) thermal model of the cryostat in three different configurations:

1. Stainless steel tubing and a 300 K vacuum shell
2. G-10 tubing and a 300 K vacuum shell
3. G-10 tubing and a 240 K vacuum shell

The third configuration simulates balloon-borne temperatures and assumes one atmosphere pressure regulation of the liquid cryogens. Results of this analysis and the cryostat performance are given in the next section.

Cryostat Performance

The first extended cold test of the cryostat used LHe and LN (instead of LNe) and did not include the ADR and its associated wiring. This test resulted in a static evaporation rate of 250 mL/hr of LHe, 19 mL/hr of LN and temperatures of 210 K and 150 K for the outer and inner vapor cooled shields (OVCS/IVCS). A FEA model was constructed using these experimental results. The model predicts the final VCS temperatures and the heat load into the two reservoirs as a function of the VCE at the neon reservoir and the two VCS. The result of this analysis indicated that the data is best fit with a VCE of 50% at the neon reservoir, 32% at the IVCS and 43% at the OVCS. The refined FEA performed after the first test was subsequently used to predict the performance of the cryostat under various conditions. The results are shown in Table 1. Due to time constraints, no further testing and effort was expended to improve the heat exchange between the escaping gas and the various stages.

Table 1. Predicted cryostat performance

Case Number	Support Tube	Vacuum Shell (K)	OVCS (K)	IVCS (K)	LHe/LNe Evaporation Rate (mL/hour)	LHe/LNe Hold Time (Days)
1*	S/S	300	205	121	50/93	16/10
2	G-10	300	253	150	20/30	41/31
3	G-10	240	195	115	20/15.5	41/59

*Measured performance for Case 1 was within 5% of predicted performance.

After installing the ADR, the heat switch, and the associated temperature sensors into the cryostat, additional measurements with LHe and LNe were performed to establish the system performance. Both the temperatures and evaporation rates agreed within 5% of the FEA model for Case 1. The FEA model also calculates the total radiative and conductive loads on each of the cryogenic stages. The chart in Figure 2 shows these results for the Case 1 configuration of the cryostat. The FEA model was then used to predict two other configurations given above. The estimated temperatures, evaporation rates, and hold times for each of these configurations are given in Table 1.

This cryostat met the goal of a 10 day hold time with the LNe being the limiting cryogen. If this cryostat were instrumented for a LDB flight using infrared bolometers, a vacuum window would be required along with openings down to the LHe level. This, combined with extra loading from instrumentation wiring, would make this cryostat unusable for a LDB flight in its current configuration. However, if this cryostat were refitted with G-10 supports, then it would be more than adequate for a LDB flight. Furthermore, based on the Case 3 analysis, a scaled up version of this cryostat, with more sophisticated heat exchangers, would be well suited for a ULDB flight. In the last section, additional improvements are discussed which could reduce the reservoir sizes.

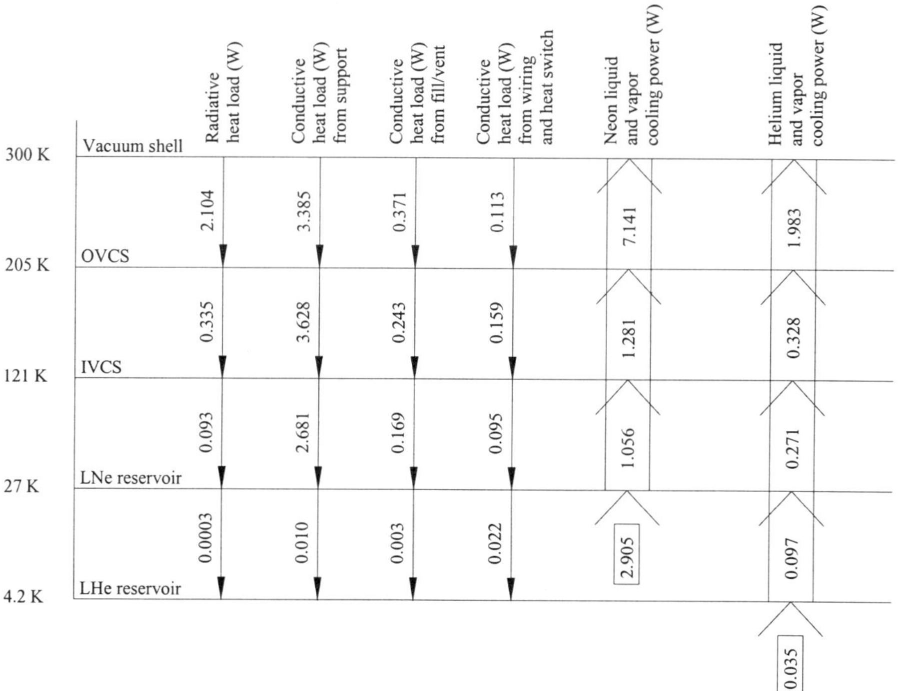

Figure 2. Heat map for the LHe/LNe cryostat with stainless support tubes (Case 1) with 50 and 100 mL/hour evaporation rates for LHe and LNe, respectively.

TWO-STAGE ADIABATIC DEMAGNETIZATION REFRIGERATOR

ADR Design

The two-stage ADR was fabricated based on a design developed at NIST[5] and is shown inside the cryostat in Figure 1. A brief review of the design is given here, while the interested reader should contact J. Martinis[17] for specific details. The two-stage ADR consists of two paramagnetic pills, a Kevlar®[18]-69 suspension system, a magnet, a magnet shield, and a heat switch. The two paramagnetic pills are composed of a ferric ammonium alum (FAA) salt and a gadolinium-gallium garnet (GGG) crystal. The FAA pill was grown at the UWM and is based on the SIRTF design.[13] Annealed gold wires of 99.999% purity permeate the FAA and couple heat from a gold-plated copper rod that is attached to the suspension system. The GGG pill was purchased commercially[19] and is similarly coupled to a gold-plated copper rod that is attached to the suspension system. During operation the GGG pill reaches a temperature of ~1 K and intercepts the heat load from the 4 K stage. The pills reside within the bore of a 4 T, superconducting magnet.[20] The magnet and pills are surrounded by a 1.51 cm thick magnet shield of heat treated Hiperco[21] (a Fe-Co-V alloy generally known as vanadium permendur) which reduces the field outside the shield to less than 1 mT[13] (10 gauss). Commercially available high T_c superconducting leads[22] are used between the LNe and LHe stage to minimize the Joule heating of the LHe during the magnetization process and to minimize conductive loading. A manually actuated heat switch[23] simultaneously couples the FAA and GGG pills to the 4 K bath during magnetization. A segmented tube (not shown in Figure 1) is coupled to a rotary feedthrough at 300 K and a right angle gear box at 4 K, and the tube is rotated to actuate the heat switch. The segmented tube is composed of thin wall stainless sections that run between the radiation shields and small copper sections that are used at the rotary feedthroughs of each radiation shield to assist in heat sinking.

ADR Performance

The ADR was tested in a series of measurements at UWM. Since the primary objective in these first tests was to confirm device operation, a very simple experimental set up was used: A stable power supply was used to directly set the magnet current, while the salt pill temperatures were monitored using germanium resistance thermometers. Other than the Kevlar® suspension system, the only connection to the cold stages was the thermometry wiring. The wiring to the 0.1 K stage was buffered by heat sinking on the 1 K stage.

A general discussion of magnetic refrigeration may be found in many references.[24] The operation of this ADR involves the following steps:

1. The magnet is ramped up to full current (~8.5 A) with the heat switch closed so that the heat of magnetization generated in the pills is rejected to the 4K bath.
2. The heat switch is opened, thermally isolating the pills from the bath, and the magnet is demagnetized.

Roughly, since the entropy $S=S(B,T)$ and the system is adiabatically isolated ($\Delta S=0$), controlling the magnetic field B applied to the pills controls the salt pill temperatures. In practice, this provides a convenient means for temperature regulation in the sub-Kelvin temperature range. The magnet current can be reduced until the desired operating temperature is reached, and then an electrical feedback circuit may be used to further reduce the magnet current to compensate for parasitic loading, such that the cold stage is thermostatically regulated. During the initial ADR test, the magnet was ramped to zero field and a minimum temperature of 0.06 K was obtained. For our 83 g FAA pill this can be translated to an ability to sink 91 mJ while regulating at 0.1 K. The magnetic field was then increased until the FAA stabilized at a temperature of 0.1 K. The warm up rate at this temperature was measured to be 650 μK/hour, which corresponds to a parasitic load of 160 nW. Given this amount of loading, we estimate the hold time we would achieve regulating at 0.1 K to be six days. This represents an upper limit, since other demands on the cooling capacity, such as additional wiring for detector signals, would reduce the hold time. This estimate is also complicated by the fact that the GGG pill does not remain at constant temperature when the FAA pill temperature is regulated, since they share the bore of the same magnet. The GGG reached 0.9 K with 55 mT applied field (the field at which the FAA pill reached 0.1 K), and warmed 2.2 mK/hour at this temperature. The actual temperature drifts in the GGG that would be induced due to changes in the field regulating the 0.1 K stage and the resulting changes in conductive loading on the FAA are difficult to estimate. Nevertheless, it is clear that even a pessimistic estimate of the hold time, when compared to the cycling time of ~45 minutes, yields a 0.1 K duty cycle of better than 99%.

FUTURE DIRECTIONS

There are several modifications that would improve this cryogenic system's performance in either ground-based or balloon-borne operations.

1) The stainless support system needs to be replaced with a G-10 support system. Janis has recently manufactured a similar cryostat with this type of G-10 support system.[25] This will significantly decrease the conductive heat loads at all stages in the cryostat.
2) A remotely operable heat switch is required for balloon-borne operations, and this would also make ground based operations more convenient. One solution is to place a computer controlled stepper motor on the externally actuated heat switch.
3) The high T_c leads are the largest contributors to the 4 K conductive loading assigned to the wiring and heat switch in Figure 2. The length of these leads can easily be doubled reducing this conductive heat load by almost a factor of two.
4) A magnet controller is required to stabilize the ADR temperature for bolometric measurements. This is currently being built based on an existing rocket-borne ADR system.[3,4]

A list of other improvements that would be suitable for a future ULDB cryogenic system is given here.

1) A more efficient heat exchanger system would be required for an ULDB cryostat. This has been realized in a cryogenic system that has recently been built by Janis.[25]
2) A system to solidify the LNe would further reduce the conductive loads on the LHe. A solid Ne reservoir has worked well in a cryogenic system for the ASTRO-E satellite.[26]
3) A more compact ADR suspension system would allow for a larger instrument volume and reduce the area of the radiation shields yielding a reduction in the radiative loads on all stages.

Finally, we are investigating a low vibration, low power consumption 4 K cryocooler that could be coupled to the two-stage ADR. This would require a complete cryostat redesign; however, if the cryocooler induced microphonics can be minimized, a cryocooled two-stage ADR is an ideal solution for an ULDB cryogenic system.

ACKNOWLEDGEMENTS

The authors wish to acknowledge Dan McCammon, Scott Porter and Peter Shirron for helpful discussions regarding cryostat design, John Martinis for enabling the fabrication of the two-stage ADR, and Karen Lewis and Chris O'Dell for assistance in salt pill fabrication. This work was performed at Janis Research Company and the University of Wisconsin-Madison and was supported by NASA Phase I SBIR contract # NAS5-98119.

REFERENCES

1. http://www.wff.nasa.gov/~uldb/.
2. C. Hagmann and P.L. Richards, Two-stage magnetic refrigerator for astronomical applications with reservoir temperatures above 4 K, *Cryogenics*, 34:221(1994).
3. W. Cui, R. Almy, S. Deiker, D. McCammon, J. Morgenthaler, W.T. Sanders, R.L. Kelley, F. J. Marshall, S.H. Moseley, C.K. Stahle, and A.E. Szymkowiak, A sounding rocket experiment employing microcalorimeter detectors to obtain a high-resolution spectrum of the diffuse X-ray background, *SPIE Proc.* 2280:362 (1994).
4. D. McCammon, R. Almy, S. Deiker, J. Morgenthaler, R.L. Kelley, F.J. Marshall, S.H. Moseley, C.K. Stahle, and A.E. Szymkowiak, A sounding rocket payload employing high-resolution microcalorimeters, *Nucl. Instr. Meth. Phys. Res. A* 370:266 (1996).
5. D.A. Wollman, K.D. Irwin, G.C. Hilton, L.L. Dulcie, D.E. Newbury, and J.M. Martinis, High-resolution energy-dispersive microcalorimeter spectrometer for X-ray microanalysis, *J. Of Microscopy,* 188, Part 3, p. 196 (1997).
6. C.A. Mears, S.E. Labov, M. Frank, M.A. Lindeman, L.J. Hiller, H. Netel, and A.T. Barfknecht, Analysis of pulse shape from a high resolution superconducting tunnel junction and x-ray spectrometer, *Nucl. Inst. Meth. Phys. Res. A* 370:53 (1996).
7. T. Twerenbold, J.L. Vuilleumier, D. Gerber, A. Tadsen, B.v.d. Brandt, and P.M. Gillevet, Detection of single macromolecules using a cryogenic particle detector coupled to a biopolymer mass spectrometer. *Appl. Phys. Lett.* 68 (24), 3503 (1996).
8. G.C. Hilton, J.M. Martinis, D. Twerenbold, P.M. Gillever, D.A. Wollman, K.D. Irwin, L.L. Dulcie, and D. Gerber, Impact energy measurement in time-of-flight mass spectrometry with cryogenic microcalorimeters, *Nature,* 391:672 (1998).
9. http://sbir.nasa.gov/.
10. http://master.nsbf.nasa.gov/.
11. http://astro.caltech.edu/~lgg/boom/.
12. National Scientific Balloon Facility Handbook, Palestine, TX.
13. P.T. Timbie, G.M. Bernstein, and P.L. Richards, Development of an adiabatic demagnetization refrigerator for SIRTF, *Cryogenics,* 30:271 (1990).
14. D. Steffensrud, J. Bahls, E. Christenson, K. Fassnacht, E. Guckel, D. Harriman, S. Nahn, P. Vu, and J. Xhang, Efficiency of short heat exchangers for helium vapor cooling, *Rev. Sci. Instr.* 62, 1, 214 (1991).
15. J.E. Jensen, W.A. Tuttle, R.B. Stewart, H. Brechna, and A. Prodell, Eds. "Brookhaven National Laboratory Selected Cryogenic Data Notebook," National Technical Information Service, Springfield, VA (1980), II-F-1, V-E-1.
16. ANSYS, Inc., Cannonsburg, PA, release 5.3, thermal and structural finite element analysis software.
17. J. Martinis, National Institute of Standards and Technology, Boulder, CO 80303.
18. Kevlar® is a registered DuPont trademark.
19. Litton Industries, Inc., Airtron Div., SYNOPTICS, Charlottesville, NC.
20. American Magnetics, Inc., Oak Ridge, TN.
21. Arnold Engineering Co., Marengo, IL.
22. BICC Cables Limited, Wrexham, England.
23. Infrared Laboratories, Tuscon, AZ.
24. Pobell, F., "Matter and Methods at Low Temperature," Springer (1996), p. 169.
25. A. Nash, P. Shields, M. Jirmanus, and Z. Zhao, Development of the FACET cryostat, preprint submitted for publication in *Adv Cryog Eng* (1999).
26. S.M. Volz, K. Mitsuda, H. Inoue, Y. Ogawara, M. Hirabayashi, and M. Kyoya, The X-ray Spectrometer (XRS): a multi-stage cryogenic instrument of the Astro-E x-ray astrophysics mission, *Cryogenics,* 36:763 (1996).

PROGRESS IN TWO-STAGE SUPERFLUID STIRLING REFRIGERATION

A.B. Patel and J.G. Brisson

Cryogenic Engineering Laboratory
Massachusetts Institute of Technology
Cambridge, Massachusetts 02139

ABSTRACT

The superfluid Stirling refrigerator (SSR) uses a ^{3}He-^{4}He mixture to cool below 1 K. Using a two-stage SSR we have achieved a temperature of 248 mK while rejecting waste heat above 1 K; under similar operating conditions, single-stage SSR's have been limited to ultimate temperatures near 300 mK. This new record in the ultimate temperature was achieved despite a heat leak due to exchange gas in the vacuum insulation around the SSR.

INTRODUCTION

The superfluid Stirling refrigerator (SSR) is an example of a new class of refrigerators that use low temperature moving parts to provide efficient sub-Kelvin refrigeration. The SSR's working fluid is a superfluid ^{3}He-^{4}He mixture. By using pistons bypassed with superleaks, the ^{3}He component of the ^{3}He-^{4}He mixture, which behaves as an ideal Boltzmann gas, is forced to execute the Stirling cycle to provide cooling below 1 K.

The first proof-of-principle SSR was built by Kotsubo and Swift in 1990.[1] This machine achieved temperatures of 590 mK and had typical cooling powers of 5 μW at 700 mK. The primary weakness of this machine was the use of a liquid reservoir of pure ^{3}He as the high heat capacity material for the regenerator. The low thermal diffusivity of ^{3}He limited the refrigerator to low operating frequency (typically 4 mHz) and hence to low cooling power. Brisson and Swift[2] recognized this limitation and developed a recuperative superfluid Stirling refrigerator where two Stirling refrigerators operate 180 degrees out of phase with each other. In this way, the fluid flow from one Stirling refrigerator could be passed through a recuperator to regenerate the flow in the opposing

Stirling machine. This change increased the operating speed by almost two orders of magnitude (0.1 Hz) and, with this machine, Brisson and Swift achieved temperatures of 296 mK.

All these SSR's had been operated with the high temperature reservoir held at approximately 1 Kelvin. Watanabe et al.,[3] modified an SSR designed for high temperature operation and pre-cooled the SSR compressor to 383 mK with a ^{3}He evaporation refrigerator. They achieved an ultimate temperature of 168 mK.

Patel and Brisson realized that further progress in lowering the ultimate temperature of the SSR depended on developing plastic heat exchangers to lower the deleterious effect of Kapitza conductance[4] in the recuperator. They demonstrated these heat exchangers in a single-stage SSR that delivered typical cooling powers of 2000 μW at 700 mK,[5] an improvement by a factor of 400 over the Kotsubo and Swift machine and a factor of 12.5 over the Brisson and Swift machine. Their ultimate temperature of 291 mK is comparable to the Brisson and Swift value of 296 mK. Patel has numerically modeled recuperative superfluid Stirling refrigerators and has shown theoretically that the flow imbalances in the recuperators limit single-stage SSR's to ultimate temperatures of 200 mK when exhausting waste heat above 1 K.[6]

From the experimental results, it became apparent that to obtain significant cooling at temperatures below 300 mK, a two-stage SSR would be necessary. In 1998, Patel and Brisson successfully demonstrated a proof of the two-stage principle achieving an ultimate temperature of 282 mK.[7] They also reported that the upper stage recuperator was inadequately sized to achieve low ultimate temperatures.

This article presents experimental results on a two-stage SSR with a larger upper stage recuperator. In this work, a new record of 248 mK has been set for the ultimate temperature of an SSR while discharging its waste heat at 1 Kelvin.

APPARATUS AND PROCEDURE

A schematic diagram of the two-stage SSR is shown in Figure 1. The refrigerator consists of three piston platforms: a compressor platform, an upper expander platform and a lower expander platform. Each of these platforms was machined from blocks of OFHC copper to ensure good thermal contact within the platform. The compressor platform consists of two bellows, which act as piston/cylinders. These bellows are rigidly connected so that one bellows compresses the superfluid mixture as the other expands the mixture. The enclosed volumes of these bellows are connected by a Vycor glass[8] superleak which allows the superfluid ^{4}He component of the mixture to flow freely from one piston to the other. The normal (viscous) ^{3}He component of the mixture cannot flow through the 4 nm pores of the Vycor; and consequently is confined to one side or the other of the superleak. The compressor platform is maintained at a temperature of approximately 1.05 K by a ^{4}He evaporation refrigerator.

The upper expander and lower expander platforms are similar to the compressor platform. The piston pairs on the upper and lower expander platforms are also mechanically linked to ensure that both sets of expanders operate in phase with each other. The compressor pistons and expander pistons are driven sinusoidally by drive

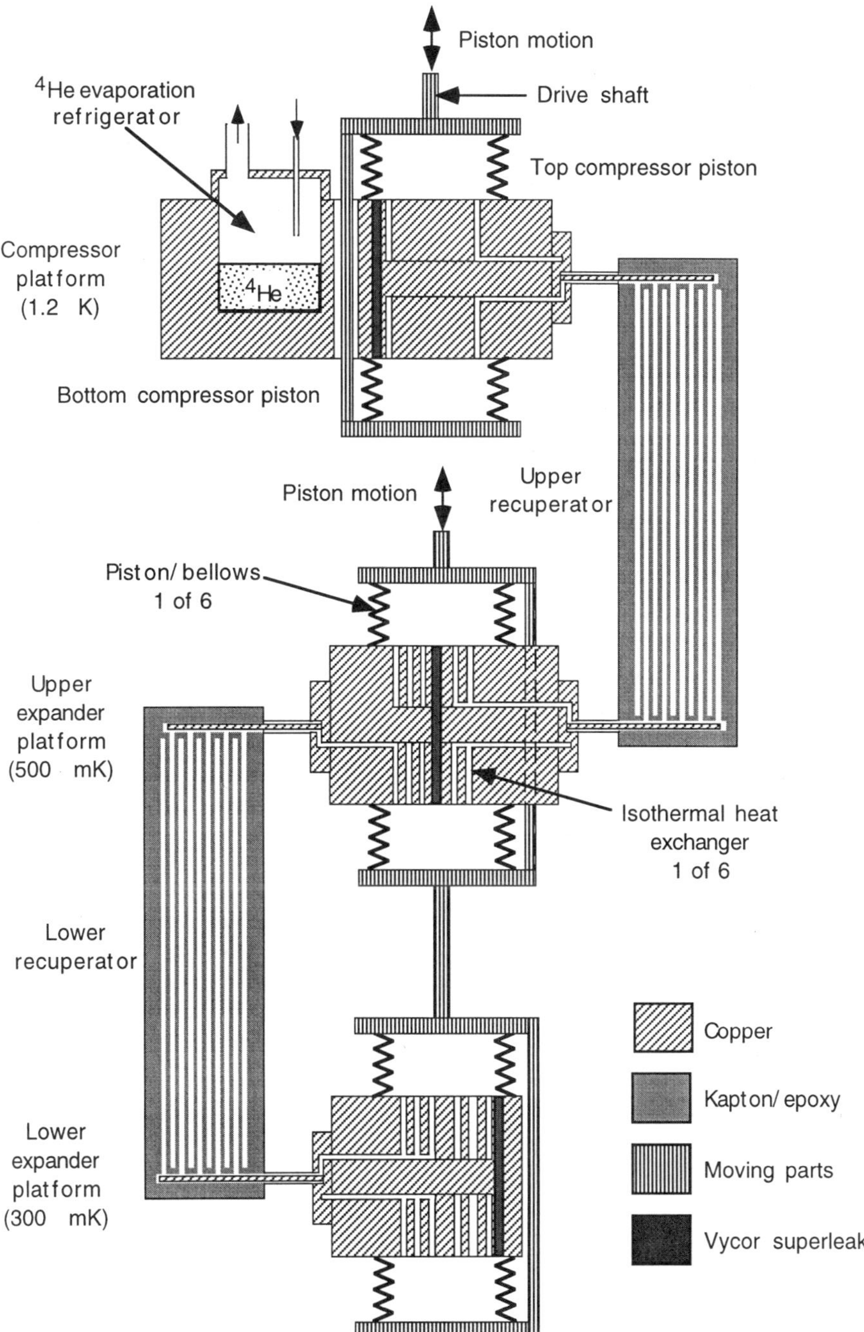

Figure 1. A schematic diagram of the two-stage superfluid Stirling refrigerator. The compressor platform is maintained at a temperature of order 1.2 K by a ^{4}He evaporation refrigerator. The design temperatures for the upper and lower expansion platforms are shown. To avoid confusion, only part of the fluid passages are shown. For example, the connection between the top compressor piston and the appropriate upper recuperator passages are not shown but can be reconstructed by symmetry from the bottom compressor piston.

shafts connected to cams at room temperature. There is a 90 degree phase shift between the compressor and expander drives.

The recuperators are built using a Kapton-Stycast 1266 epoxy composite.[4] These recuperators have 127 μm x 2.38 mm x 20 cm long fluid channels. Each channel is separated from neighboring counterflow channels by 25 μm Kapton septum. The upper (high temperature) recuperator has 200 such channels (100 per side) for a total volume of 12.1 cm^3 devoted to recuperative heat transfer. The lower (low temperature) recuperator has a total of 50 channels (25 per side) for a total volume of 4.8 cm^3 devoted to recuperative heat transfer.

The compressor, upper expander and lower expander pistons are made from 60055, 60050 and 60035-2 Senior Flexonics stainless steel welded bellows,[9] respectively. These bellows were chosen because they have nested convolutions that minimize the piston clearance volume. The compressor piston stroke was 1 cm corresponding to a swept volume of 17.74 cm^3. The upper and lower expander pistons were swept through a 0.69 cm stroke corresponding to swept volumes of 9.38 cm^3 and 5.30 cm^3, respectively.

Each piston bellows has associated with it a uniform temperature heat exchanger made of nested copper cylinders press fit into the piston platform. In the center of these nested copper cylinders resides a press fit Invar plug (not shown in Figure 1). As the refrigerator is cooled from room temperature to 1 K differential thermal contraction between the copper platform, copper cylinders and the Invar plug swages the copper cylinders and the copper platform together, ensuring intimate thermal contact between the platform and the cylinders. These large area uniform temperature heat exchangers guarantee good thermal contact between the ^3He-^4He mixture and their respective platforms.

The heat exchanger associated with a single compressor piston consists of a 3.80 cm diameter copper cylinder with a 76 μm annular gap between the cylinder and the compressor platform. During the operation of the SSR, the mixture flows through this gap to remove the heat of compression from the mixture. To provide an even distribution of flow through this annulus, 0.64 mm deep by 0.32 cm wide flow distribution channels are machined into the top and bottom of the cylinder. The total area available for heat transfer in a single compressor heat exchanger is 51 cm^2. The upper-expander-piston heat exchanger is similar to the compressor piston heat exchanger except that two cylinders are used which increases the heat transfer area of this heat exchanger to 211 cm^2. Finally, the lower-expander-piston heat exchanger is made with four cylinders and has a total surface area of 617 cm^2.

It may seem contradictory that the heat exchanger that transfers the most heat (the compressor piston heat exchanger) has the smallest surface area. The overall heat transfer coefficient in all three of these heat exchangers is strongly influenced by the Kapitza conductance between the helium mixture and the copper heat exchanger, which scales as temperature cubed.[10] It is this quantum effect that determines the surface area of the SSR heat exchangers.

The SSR was cooled to 1.05 K and 3% ^3He-^4He mixture was condensed into the SSR from room temperature storage tanks. The total internal volume of the SSR is 129 cm^3 (64.5 cm^3 per side), excluding the 15 cm^3 in the Vycor superleaks. When the SSR was filled with liquid, low temperature valves mounted on the compressor platform (not shown in Figure 1) were closed to seal the mixture into the SSR.

Low temperature electrical switches provided known fixed positions to calibrate low-temperature linear position transducers.[11] These linear transducers were used, in turn, to center the SSR pistons prior to loading the mixture and to ensure symmetric displacements in the pistons during SSR operation.

Temperatures were determined using calibrated ruthenium oxide[12] and germanium thermometers[13] mounted on each of the piston platforms. The precision of our measurements is $\pm$0.67 mK at 1.0 K and $\pm$1.02 mK at 300 mK. The temperature of each of the platforms oscillates with an amplitude of approximately 10 mK when the SSR is operated in steady state. The temperatures that we report here are mean temperatures determined by averaging the maximum and the minimum temperatures observed in one SSR cycle.

RESULTS AND DISCUSSION

The two-stage SSR achieved an ultimate lower-expander temperature of 248 mK. This temperature was obtained when the upper expander temperature was 432 mK, the compressor temperature was 1.034 K, and the cycle period was 23 seconds. The SSR was also operated with a cycle period of 15 seconds. It then achieved an ultimate temperature of 253 mK when the upper expander temperature was 470 mK and the compressor temperature was 1.041 K.

During these experiments, there was a small leak into the SSR's vacuum space that limited the time over which data could be taken. We believe that the ultimate temperatures achieved by this SSR were limited by a parasitic heat load due to the presence of helium in the vacuum space.

The earlier two-stage SSR reported by Patel and Brisson[7] achieved an ultimate temperature of 282 mK with a drive period of 25 seconds. Throughout their measurements, the upper expander temperature remained within 60 mK of 0.8 K regardless of the 0 μW to 2 mW variation of the heat load to the lower expander. The high operating temperature of the upper expander combined with this stiffness in the temperature was attributed to a small upper recuperator between the compressor and the upper expander. In the two-stage SSR reported here, the upper recuperator area is four times larger than in the earlier case which now allows the upper expander to cool to almost half the temperature of the previous SSR.

In this experiment, the recuperator between the upper and lower expanders is still smaller than optimum. In addition, any parasitic heating effects due to helium gas in the vacuum space affect the lower-expander-platform temperature more than the upper-expander-platform temperature. Design calculations suggest that the low temperature performance of this refrigerator will improve significantly with a larger low temperature recuperator.

CONCLUSIONS AND FUTURE WORK

The two-stage superfluid Stirling refrigerator shows promise of cooling to 200 mK. Efforts are now underway to build a two-stage SSR with adequately sized heat exchangers to demonstrate this. In addition, a three-stage machine, now under development at MIT, could potentially cool to 100 mK. Cold-cycle dilution refrigerators, cooled with a two-stage SSR have been proposed[14] and could achieve temperatures of a few milliKelvin.

ACKNOWLEDGMENT

The authors gratefully acknowledge the support of the National Science Foundation for this work.

REFERENCES

1. V. Kotsubo and G.W. Swift, “Superfluid Stirling-cycle refrigeration below 1 Kelvin”, J. Low Temp. Phys. **83**, 217 (1991).
2. J.G. Brisson and G.W. Swift, “Superfluid Stirling refrigerator with a counterflow regenerator”, in Proceedings of the Seventh Cryocooler Conference, (Phillips Laboratory, Kirtland AFB, NM 1993) p 460.
3. A. Watanabe, G.W. Swift and J.G. Brisson, "Measurements with a recuperative superfluid Stirling refrigerator", in Advances in Cryogenic Engineering, **41**, (Plenum, New York, 1996) pp. 1527-1533.
4. A.B. Patel and J.G. Brisson, "Design and construction of plastic heat exchangers for sub-Kelvin use", To be published.
5. A.B. Patel and J.G. Brisson, "Experimental performance of a single stage superfluid Stirling refrigerator using a small plastic recuperator", J. Low Temp. Phys. **111**, 217 (1998).
6. A.B. Patel, Doctoral Thesis, Massachusetts Institute of Technology (Expected in July 1999).
7. A.B. Patel and J.G. Brisson, "Preliminary experimental results using a two stage superfluid Stirling refrigerator", accepted for publication in Cryocoolers 10, (Plenum, New York, 1999).
8. Corning Glass Works, Corning, NY, USA.
9. Metal Bellows Division, Senior Flexonics Inc., Sharon, MA, USA.
10. See for example: F. Pobell, Matter and Methods at Low Temperatures, (Springer-Verlag, Berlin, 1996) p. 84.
11. K.A. Backes and J.G. Brisson, "An inductive position sensor for the measurement of large displacements at low temperature", Rev. Sci. Instrum. **69**, 599, (1998).
12. Oxford Instruments Ltd., Oxford, England.
13. Lakeshore Cryotronics, Inc., Westerville, OH, USA.
14. J.G. Brisson, "Cold-cycle dilution refrigeration", J. Low Temp. Phys. **111**, 181 (1998).

OPTIMISATION OF SUPERCONDUCTING MOTORS WITH YBCO BULK MATERIAL

B. OSWALD,[1] M. KRONE,[1] M. SÖLL,[2] T. STRAßER,[1] J. OSWALD,[1] K.-J. BEST[3]

[1] OSWALD Elektromotoren GmbH, 63897 Miltenberg, Germany
[2] Fachhochschule Landshut, 84036 Landshut, Germany
[3] Best Engineering, 63452 Hanau, Germany

ABSTRACT

In order to optimize superconducting motors with YBCO bulk material operating at 77 K, several motors have been constructed and tested up to more than 20 kW. These reluctance motors have been operated at different speeds in synchronous mode. The rotors consist of a combination of iron poles and HTS-segments. Special attention has been paid to internal loss mechanisms to improve the efficiency. With this respect the rotor geometry and the stator design have been varied. Different motor types "PILZ" and "ZEBRA" types have been constructed and tested. The test results are reported and compared. A computer program was installed and is the basis of theoretical considerations. The experimental results are in good accordance with the theory. The small size and low weight, as well as the high dynamics of these motors, can be considered as important advantages for various applications. The specific requirements of potential applications are discussed.

INTRODUCTION

Former results on superconducting reluctance motors of mushroom (PILZ)-type have been encouraging with respect to the power obtained. Electric motors with bulk $YBa_2Cu_3O_{7-x}$ melt-textured high-temperature superconductors (YBCO) promise high output power at low size. These motors are operated in liquid nitrogen (77 K). The rotors have been equipped with melt-textured YBCO bulk material from IPHT, Jena, and IFW, Dresden. Motor SRS 40-240 showed an increase of the specific power by a factor 1.5 to 2 by use of this kind of superconducting material. In case of ZEBRA-type motors, however, this increase was lower. Therefore it was necessary to study the theory of both motor types in detail, taking into account the complete set of parameters. The aim of these considerations was to find out to what extent these motors could be improved by use of YBCO bulk material. This is regarded in view of the dominant criteria of power density and power factor. To facilitate optimization a computer program has been developed that includes the magnetic circuit, where the most important parameters can be varied. With the curves obtained, parameters can be derived to obtain optimum results.

DESIGN OF "PILZ" AND "ZEBRA" TYPES

The cross-section of the "PILZ" type is shown in Fig. 1. The stator consists of a stack of iron lamination. The slots contain a three-phase two-pole winding of normal conductive copper producing a rotating field that depends on the applied frequency. The rotor consists of solid or laminated iron poles and

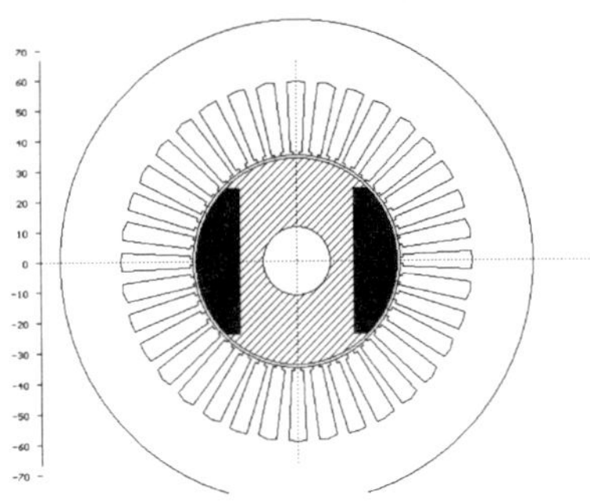

Figure 1. Cross section of a "PILZ" type reluctance motor.

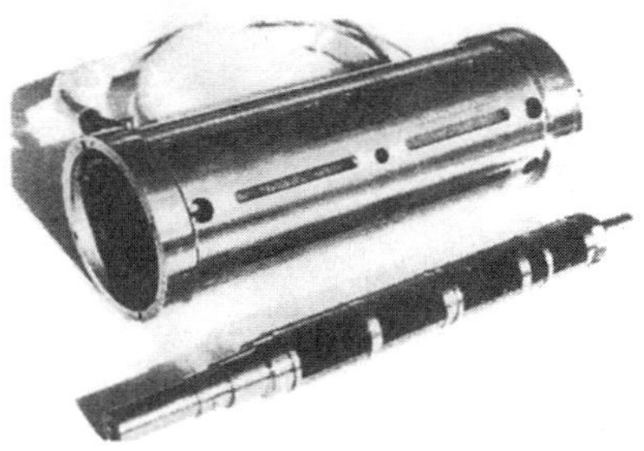

Figure 2. Superconducting 2 pole "PILZ" type reluctance motor.

YBCO bulk superconductors filling the pole gaps. The field repelling current in the bulk superconductor produces a high field gradient at the pole edges. The quality of the superconductor and the size of the grains therefore are important for achieving high forces. The cross-section of a 2-pole ZEBRA type motor is shown in Fig. 3. The stator is comparable to the same design and type as described before.

The rotor consists of a stack of iron and superconducting YBCO-bulk plates. The task of the superconducting segments is to expel the horizontal field in order to force the field into the vertical direction (d-axis). The construction of ZEBRA type motors is more complicated than in the case of PILZ motors. Also the manufacturing of bulk plates with a small thickness and very low geometric tolerances turns out to be very expensive. In order to transfer the forces across the stack and to avoid cracks in the superconductor, the total rotor was impregnated by a special method.

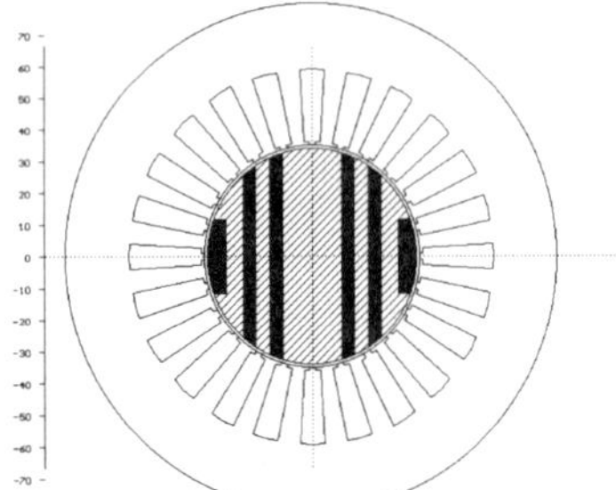

Figure 3. Cross section of a superconducting 2 pole "ZEBRA" type reluctance motor.

EXPERIMENTAL RESULTS

In order to evaluate the efficiency of the superconductor, tests with and without superconductors were performed. The difference of the load line in case of the PILZ motor SRS 40-240 is shown in Fig. 4. The torque values represent the maximum torque at the corresponding currents.

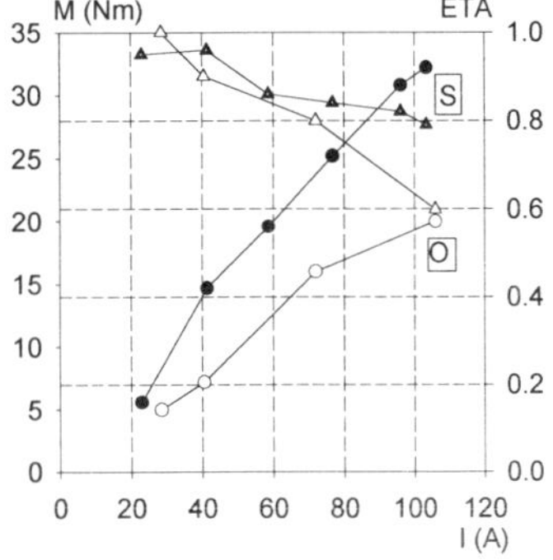

Figure 4. Reluctance motor SRS 40-240 ("PILZ" type) with (S) and without (O) superconductor. Maximum torque (M) and efficiency (ETA) vs current.

If there is no superconductor placed in the pole gaps we also get a torque corresponding to the field gradient between the poles and the gaps. However, if we use superconducting material which expels the field in the gap, much higher torque could be achieved. The efficiency of the superconductor in this case can be derived from the theoretical calculation. The motor type SRS 40/240 was not optimized with respect to losses and efficiency. Therefore the eta values must not be considered as the optimum. There are still pulsation losses caused by various harmonics and eddy current losses in the solid iron poles. Typical load characteristics at constant voltage are shown in Fig. 5.

THEORETICAL APPROACH

In order to compare the test results with the theory a computer program was developed that is based on the equations described in Ref. #1. The maximum torque is approximately given by

$$M = \frac{3p}{2\pi f_1} \cdot U_1^{\,2} \cdot \frac{X_d - X_q}{2X_d \cdot X_q}$$

where X_d and X_q is the reactance in longitudinal and orthogonal direction, p is the number of pole pairs, f_1 is the applied frequency and U_1 the primary voltage. The reactances strongly depend on the gap geometry and on the effective permeability of the gap and must be exactly calculated by FEM. The leakage reactance of the stator is included in X_d and X_q. The program also includes the stator resistance, which actually is very low.

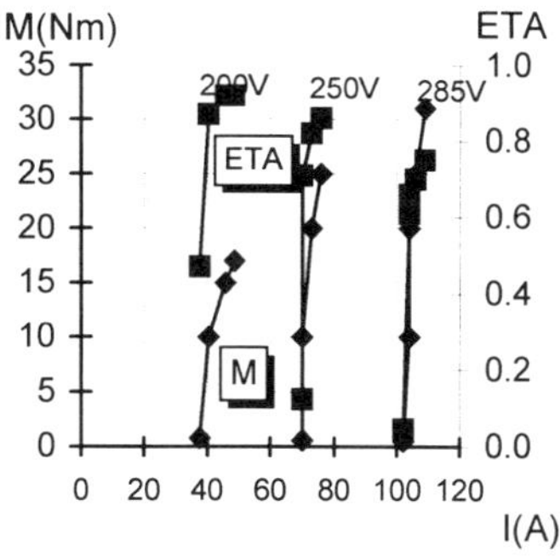

Figure 5. SRS 40-240 ("PILZ" type) load characteristics for 3 different voltages: torque M and efficiency ETA vs current.

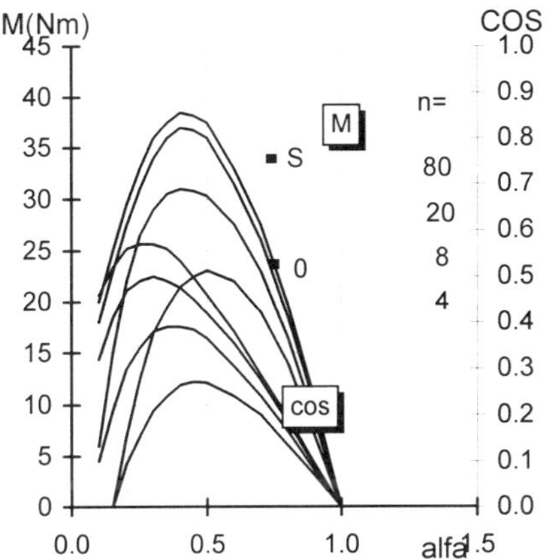

Figure 6. Reluctance motor SRS 40-240 2 poles ("PILZ" type). Maximum torque (M) and power factor (cosφ) vs alfa for different parameter values of n at constant current 100 A. Experimental data with (S) and without (O) superconductor. Alfa = pole width / pole pitch. In this diagram the n-values assumerd were 4, 8, 20, and 80.

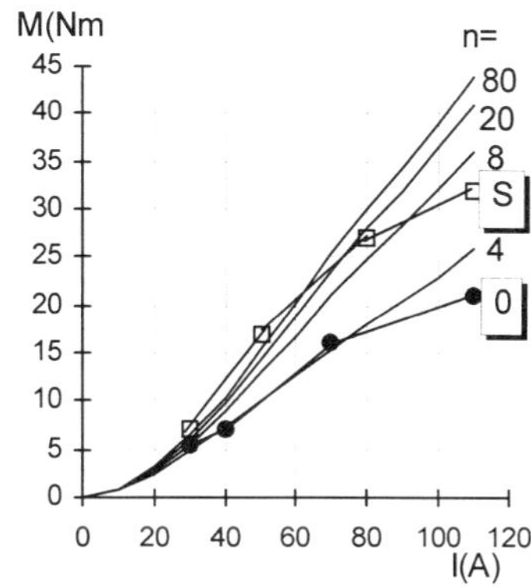

Figure 7. Reluctance motor SRS 40-240 2 poles ("PILZ" type). Calculated maximum torque vs current for different n-values. Experimental data with (S) and without (O) superconductor.

The effectiveness of the superconducting material can be described by the factor n: $n = B_o / B_g$

where B_o is the (fictive) induction if there exists no gap and B_g is the real effective inductance in the gap. The smaller B_g, the larger is the number n. If we assume a pole gap geometry as shown in Fig. 1 the value of n would be approximately 4 in case no superconductor is placed in the gap. If the gap is filled by a superconducting segment the n value could be much higher depending on the current density in the superconductor.

The following diagrams show the computer results of the program used for various relationships. It was of special interest to evaluate the optimum pole width related to the pole pitch indicated by the factor alfa. Fig.6 shows the torque and the power factor depending on the factor alfa. The better the superconductor (n>4) the higher the torque and power factor. From this diagram the optimum value of alfa could be assumed to be between 0.4 and 0.45, if only the maximum torque is considered.

For the same motor type the calculated maximum torque curves related to the current are shown in Fig. 7. The calculation of the data in Fig. 7 presumes constant temperature and constant permeability of the pole gap over the current. In fact the behavior of the superconducting material strongly depends on the transversal field and on the temperature. The local temperature certainly rises with increasing current. Both, field and temperature lead to load lines, changing from high to lower n-values. The difference of the measured load characteristic and the calculated data could be explained in such a way.

From our point of view it is of special interest to compare PILZ-type reluctance motors to ZEBRA-types, shown in Fig. 3. Therefore, we constructed two ZEBRA-motors with rotor diameters of 68 mm and active rotor length of 100 mm and respectively 200 mm. Some tests were performed up to more than 20 kW (in case of the bigger motor). The calculation of these motors shows that we could expect a definite improvement of the motors by using superconducting segments. In fact, the experiments didn't show much influence of the superconductors, which could be due to temperature problems inside the motor. Therefore, only the theoretical results shall be shown in Fig. 8. The calculated plot: torque versus apparent power demonstrates for this specific motor design the benefit of the incorporated superconducting segments, whereas the experimental data show saturation effects in the upper power range.

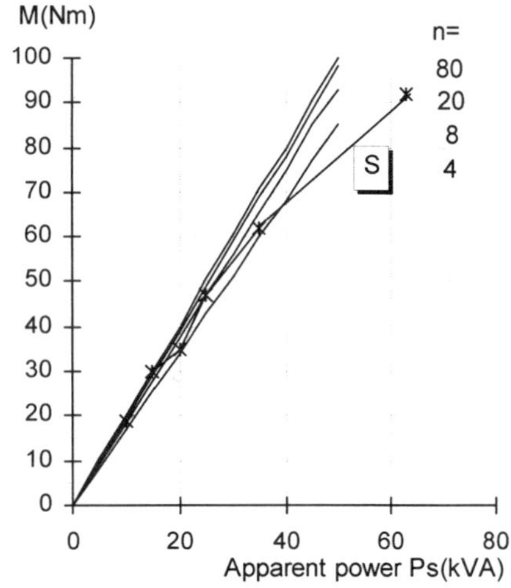

Figure 8. Reluctance ZEBRA motor SRC 70-200 2 poles. Rotor dia 68 mm, length 200 mm. Calculated torque vs apparent power for different n-values compared to experimental data (S).

DISCUSSION

The specific advantage of superconducting reluctance motors is characterized by a relative high specific force related to the rotor volume, respective to the rotor surface. This means high specific power of the total motor and also extremely high dynamics. We already achieved maximum specific forces of 5 to 9 N/cm^2 with different motors of this type tested so far, which is 4 to 5 times higher compared to conventional motors working at room temperature. This topic has to be taken into account especially in case of applications where the weight or the volume are important parameters. The significant advantage of using superconducting bulk material has been verified in the case of PILZ-type reluctance motors. The rather simple rotor design and the geometry of the superconducting segments of these motor types are certainly less complicated and less expensive than those of ZEBRA-type motors. This must be taken into account if both types are compared.

All tests so far were performed at a temperature of 77 K. In order to achieve improved behavior of the superconductors, we also are interested in lower working temperatures down to LH_2 applications. In this case, significantly higher n-values could be obviously expected.

It could be shown that the efficiency of these motor types is rather high if special attention is paid to an optimized design of the rotor and the stator. However, reluctance motors of this types are limited in power according to the iron magnetization curves and the maximum current density in the stator windings. The next step of improvement could be done by using superconducting wires or tapes for the AC windings producing the rotating field of the stator.

SUMMARY

The comparison of different types of reluctance motors with superconducting bulk material was performed experimentally and theoretically in order to optimize the design and construction. In multiple tests high specific power could be achieved and verified by computer simulation. The effectiveness of the superconducting material depends on the motor design but also on the geometry of the segments used. In future developments more attention will be paid on the behavior at lower working temperatures and on improved designs.

ACKNOWLEDGEMENT

The present investigation is part of a cooperation with the following institutions:

IPHT / Jena, MAI / Moscow, IEMA / Stuttgart, IFW / Dresden

The project is supported by the BMBF with the notification 13N 6856/5

REFERENCES

1. G. R. Slemon: „Electrical Machines and Drives“ Addison-Wesley Publishing Company, (1992).
2. W. Düchting: „Ermittlung der günstigsten Läuferabmessungen bei Drehstrom-Reluktanzmotoren“ ETZ-A 84 (1963).
3. B. Oswald, M. Krone, M. Söll, J. Oswald, K.-J. Best: „Superconducting Motors with YBCO-bulk Material“ 4th Meeting 9. 4. (1998) Moscow
4. L.K. Kovalev, K.V. Ilushin, S.M.-A. Koneev, V.T. Penkin, V.T. Poltavets, W. Gawalek, T. Habisreuther, B. Oswald, and K.-J. Best, “Hysteresis and Reluctance Electric Machines with bulk HTS rotor elements” IEEE Trans. Appl. Supercon. 9:1261 (1999).

HYSTERESIS AND RELUCTANCE ELECTRIC MACHINES WITH BULK HTS ELEMENTS

Recent Results and Future Development

L.K. Kovalev,[1] W. Gawalek,[2] B. Oswald,[3]
K.V. Ilushin,[1] S.M.-A. Koneev,[1] K.L. Kovalev,[1]
V.T. Penkin,[1] K.A. Modestov,[1] S.A. Larionoff,[1]
T. Habisreuther,[2] K.-J. Best[3]

[1]Moscow State Aviation Institute (Technical University) (MAI),
Moscow, Volokolamskoe shosse 4, Russia
[2]Institut fur Physikalische Hochtechnologie (IPHT),
Jena, Helmholzweg 4, Germany
[3]Oswald Elektromotoren GmbH, Miltenberg, Benzstrasse 12, Germany

ABSTRACT

Two new types of HTS electric machines are considered. The first type is hysteresis motors and generators with cylindrical and disk rotors containing bulk HTS elements. The second type is reluctance motors with compound HTS-ferromagnetic rotors. The compound HTS-ferromagnetic rotors, consisting of joined alternating bulk HTS (YBCO) and ferromagnetic (iron) plates, provide a new active material for electromechanical purposes. Such rotors have anisotropic properties (ferromagnetic in one direction and diamagnetic in perpendicular one). Theoretical and experimental results for HTS hysteresis and reluctance motors are presented. A series of hysteresis HTS motors with output power rating from 1 kW (at 50 Hz) up to 4 kW (at 400 Hz) and a series of reluctance HTS motors with output power 2-18.5 kW (at 50 Hz) were constructed and successfully tested. It was shown that HTS reluctance motors could reach 2 to 5 times better overall dimensions and specific power than conventional asynchronous motors of the same size and will have higher values of power factor ($\cos\varphi \geq 0.7$ to 0.8).

INTRODUCTION

HTS Hysteresis Motors

A construction scheme of HTS hysteresis machines (HM) is given in Fig. 1.a. The rotor consists of HTS cylindrical elements. Each element can be manufactured as a whole cylinder (left part) or can be glued from circular sectors in such a way that the axis *C* has the radial orientation (right part).

It was shown that, contrary to conventional electric motors (e.g. synchronous, asynchronous, hysteresis), the mechanical torque of the HTS HM results from the repulsion of the magnetic poles, induced into HTS rotor by the rotating field of the stator winding.[1-4] The useful torque magnitude is defined by the interaction of the domain and transport currents into the bulk HTS elements with the rotating magnetic field. It was shown that the torque is in linear proportion to the total hysteresis loss in the HTS rotor and does not depend on the rotor angular velocity.[1] At the temperature of liquid nitrogen the specific mass-dimension parameters of the hysteresis HTS machines are 3 to 5 times greater than for similar conventional hysteresis electrical machines.[3]

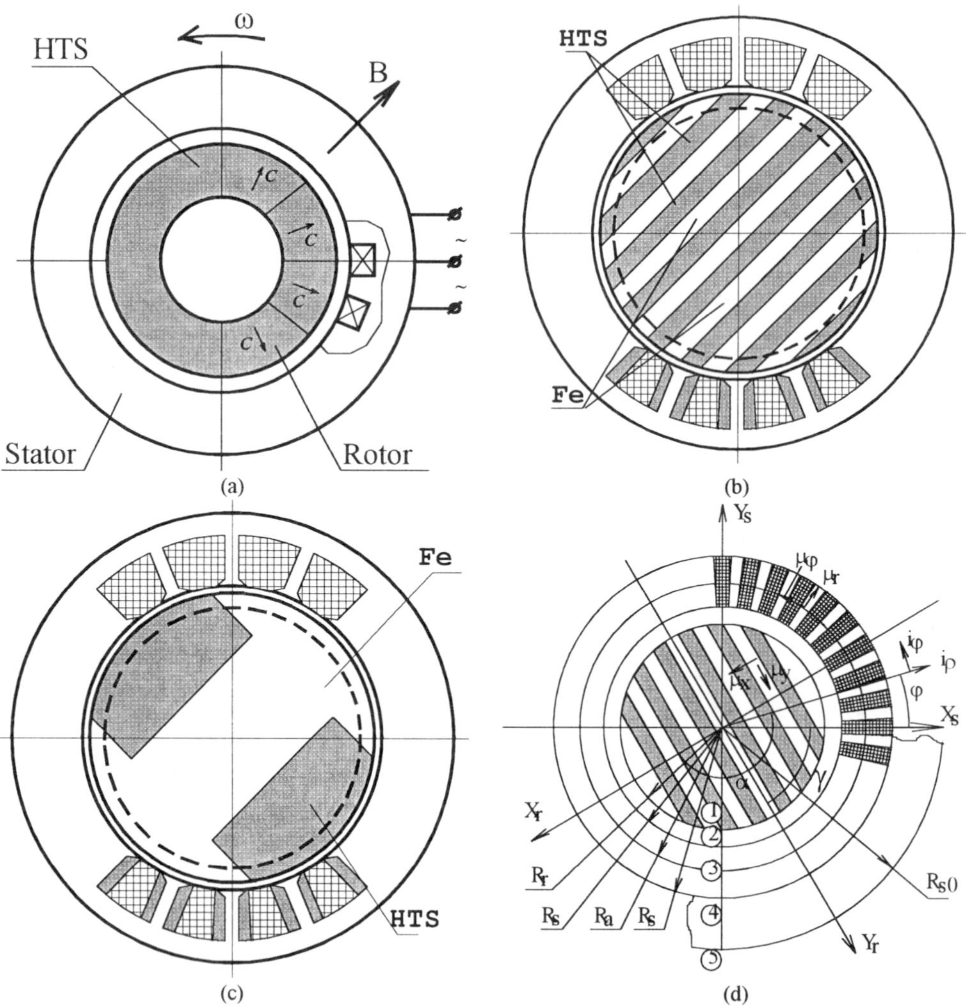

Fig. 1. Schematics of HTS motors: (a) - hysteresis HTS motor, (b) - reluctance HTS motor with compound rotor, (c) - reluctance HTS motor with rotor containing bulk HTS blocks; (d) - calculation scheme for HTS reluctance motor.

HTS Reluctance Motors

The maximum output power and power characteristics of traditional reluctance motors (RM) are basically determined by the relation between the magnetic permeabilities on longitudinal (y) and cross (x) axes of the compound rotor. In the conventional RM this is achieved by use of a compound rotor of magnetic and non-magnetic materials. In HTS RM non-magnetic materials are replaced by HTS elements which possesses diamagnetic properties. It allows to realize much larger ratios between the magnetic permeability on axes y and x and therefore improves the power characteristics of the motors. Simulation and experimental data show that the cryogenic HTS reluctance motors operating in liquid nitrogen have overall dimensions and specific power 3 to 5 times better in comparison with traditional reluctance and asynchronous motors and have at the output power of 5 to 10 kW values of power factor $\cos\varphi = 0.7$ to 0.8.[5]

The main schematics of the HTS RM are shown in Figs. 1.b, c. In Figs. 1.b and 1.d a two-pole motor with layered HTS compound rotor is presented. Along the longitudinal axis y of the motor the ferromagnetic properties with relative magnetic permeability $\mu_y \cong \mu_{Fe} \cdot k_{Fe} >> 1$ and along the cross axis x diamagnetic properties with $\mu_x \cong \mu_{HTS}/(1-k_{Fe}) \leq 1$ are realized. (Here "Fe" and "HTS" accordingly correspond to the relative magnetic permeability of the iron and HTS elements; k_{Fe} - volume fraction of iron in the compound rotor).

In Fig. 1.c a HTS reluctance motor has a compound rotor which consists of a continuous iron core with finite HTS blocks. In this scheme HTS blocks work as the concentrators of the magnetic flux along the longitudinal axis y. Such motors are practically feasible and have good starting characteristics, however, have lower efficiency and power factor in comparison with HTS motors with compound layered rotor.

PROBLEM FORMULATION AND SOLUTION

HTS Hysteresis Motors

A theoretical method for calculation of hysteretic losses, as well as for electrical, magnetic and current fields, was developed for bulk HTS elements of different shapes (spheroids, cylinders, plates) under influence of external magnetic fields of alternative polarization which takes place in the various models of electrical machines.[1-4]

The analytical dependencies were obtained from the solution of a set of two-dimensional and three-dimensional tasks.[1,2] The Bean-London model was used for the critical current description when HTS elements have high current carrying ability $\xi = J_t \cdot \Delta / J_s \cdot a >> 1$ (where J_t and J_s are transport and domain current densities, Δ and a - characterizing dimensions of HTS element and HTS domain respectively). The spherical grain model (average sphere approach) was used for the elements of HTS materials possessing relatively low current carrying ability $\xi << 1$ (e.g. melt-textured YBCO containing large domains). The mutual influence of magnetic moments of the HTS grains was realized with Lorenz approach.[3]

Using the above mentioned theoretical models for electromagnetic and hysteresis phenomena in bulk HTS elements, a method was developed for the calculation and development of hysteresis electrical machines with a range of output power from 0.1 kW to 10 kW. The applied theory for hysteresis machines with bulk HTS patterns from melt-textured YBCO ceramics was adjusted in accordance to the test results and used for development of a series of experimental hysteresis machines of 100 W, 300 W, 500 W, 1 kW and 4 kW output power rating.[3]

HTS Reluctance Motors

The magnetic field distribution, main inductive resistances and mechanical torque of the HTS RM of various constructions with long compound rotors (L/D>3-4) were determined from the solution of the two-dimensional electrodynamic problems described by Maxwell equations:

$$rot\vec{H} = \vec{J} \quad div\vec{B} = 0 \quad \vec{B} = \hat{\mu}\vec{H} \tag{1}$$

with the boundary conditions for the regions with different magnetic permeabilities:

$$H_{\tau+} - H_{\tau-} = I_p \quad B_{n+} = B_{n-} \tag{2}$$

here the indexes "+" and "–" concern to the parameters at the different sides of the boundaries, I_p - is a surface current for the problems in which the stator windings were presented as a current layer.

The solutions were done for the main harmonic of the current I of m-phase stator winding:

$$I = I_0 \cos(\varpi t - p\varphi) \tag{3}$$

As a main variant for calculation, we assumed a two-pole HTS RM with layered compound rotor shown in Fig. 1.d (the solution for multi-pole motors can be obtained via complex function transformation $z=w^p$, where p is a pole number).

The models of anisotropic magnetic permeability were used for the problem solution:

$$\hat{\mu} = \begin{pmatrix} \overline{\mu}_x & 0 \\ 0 & \overline{\mu}_y \end{pmatrix}; \quad \overline{B} = \hat{\mu}\overline{H} \tag{4}$$

Here $\overline{\mu}_x$ and $\overline{\mu}_y$ - accordingly relative magnetic permeability along the axes x and y; $\overline{B}$ and $\overline{H}$ - average magnetic induction and tension, $\overline{\mu}_x$ and $\overline{\mu}_y$ can be determined as:

$$\overline{\mu}_x = \frac{\overline{\mu}_{Fe}\mu_s}{k_{Fe}\mu_s + (1-k_{Fe})\mu_{Fe}} \qquad \overline{\mu}_y = k_{Fe}\overline{\mu}_{Fe} + (1-k_{Fe})\mu_s \tag{5}$$

In the theoretical model the teeth zone (3 in Fig. 1.d) of the stator winding was presented as an anisotropic layer with effective current density J_c and anisotropic magnetic permeability $\hat{\mu}(\overline{\mu}_r, \overline{\mu}_\varphi)$ on the radial and azimuth axes (r, φ).[5] The $\overline{\mu}_r$ and $\overline{\mu}_\varphi$ are determined by the geometry of the slots and teeth with the formulas similar to Eq (5). The stator core has constant magnetic permeability μ_{Fe}. The calculation scheme in this case consists of five zones (see Fig. 1.d): compound rotor (1), "air-gap" (2), teeth zone (3), stator core (4) and outward part of the motor (5). The analytical solutions are too long to be presented here.

The analytical solution of the problem was obtained by taking into account the dependence of anisotropic magnetic permeability of the compound rotor ($\overline{\mu}_r = \mu(H)$) and stator $\overline{\mu}_{st} = \mu_{st}(H)$. These solutions were found for each zone (see Fig. 1.d) in the form of magnetic flux function $A(0,0,A_z)$ (B=rot$A$) ($A_z = C_{1i}f_1(r)\cos(\omega t - p\varphi) + C_{2i}f_2(r)\sin(\omega t - p\varphi)$) and then the unknown coefficients C_{1i} and C_{2i} were determined from the boundary conditions (2).[5]

The saturation of the stator core and teeth was approximately taken into account by introducing the real dependencies $\mu(H)$ in the final solutions. The influence of the frontal parts of the stator windings, inductive dissipation factors of the teeth zone and losses in steel and windings of the stator were also taken into account.[5]

THEORETICAL AND EXPERIMENTAL RESULTS

Another theoretical approach for the simulation of the processes in HTS reluctance motors was developed on the basis of the finite element modeling. The examples of the calculations of the magnetic flux density distributions for the different HTS reluctance motors design are shown in Fig. 2.

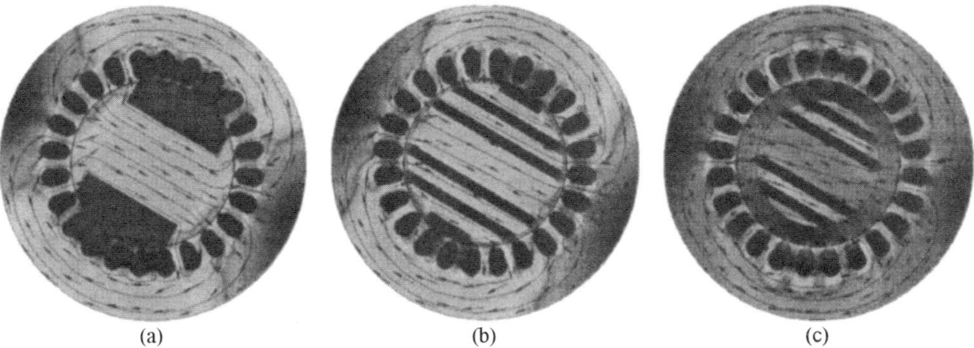

(a) (b) (c)

Fig. 2. Distributions of the magnetic flux density in the active zone of the different HTS reluctance motors: (a) - rotor with the finite HTS blocks, (b) - compound HTS-ferromagnetic layered rotor, (c) - HTS motor with the YBCO cylinder with "trapped field".

In these simulations the real dependencies of the magnetic permeabilities of the stator and rotor steels and HTS critical parameters were taken into account. Analysis of the calculation results showed that the HTS motor construction shown in Fig. 2.b has a 1.5 to 2.0 higher output power than the motor construction presented in Fig. 2.a. If the trapped field in the cylindrical YBCO shell (Fig. 2.c) is more than 0.5 to 0.75 T, then the power factor can achieve values up to ~ 0.9 to 0.95.

The series of HTS hysteresis motors was designed, manufactured and tested. The HTS hysteresis motor HTSHM-4000 with operating frequency 400 Hz is presented in Fig. 3. The motor parameters are the following: stator length and rotor diameter - $L \times \varnothing = 70 \times 40$ mm, synchronous rotating velocity 24000 rpm, output power 4 kW. The HTS for the rotor's plates was manufactured by IPHT (Jena, Germany). The optimal stator windings construction was found. It was shown that the output power of HTS HM is in 3 to 5 times higher than for conventional hysteresis motors. The test results for this motor are shown in Fig. 3.b.

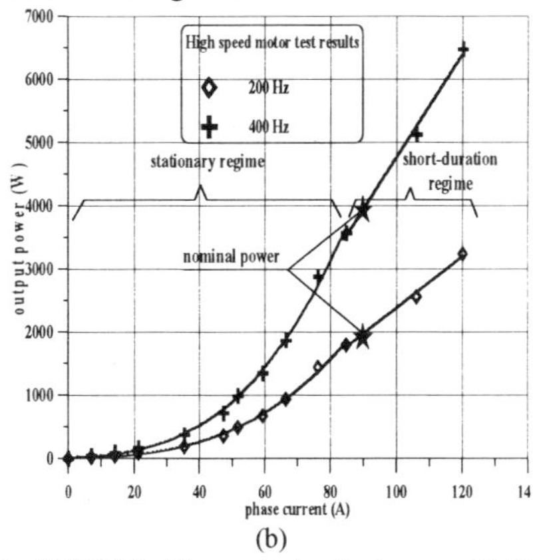

(a) (b)

Fig. 3. General view (a) and experimental parameters (b) of the HTS HM with nominal output power 4 kW and operating frequency 400 Hz (24000 rpm).

The series of HTS reluctance motors with the compound layered rotors (see Fig. 1.b) and iron rotor with finite YBCO blocks (Fig. 1.c) was designed, manufactured and tested. The output power of these experimental HTS motors was from 0.5 kW up to 10 kW and 18.5 kW. It was shown that the HTS reluctance motors with compound YBCO-iron layered rotors have specific power output 1.5 to 2 times higher than the HTS reluctance motors with the finite YBCO blocks. [5]

A general view, experimental and theoretical results for 10 kW HTS reluctance motor with compound rotor (L×∅=164×62 mm) are given in Fig. 4.

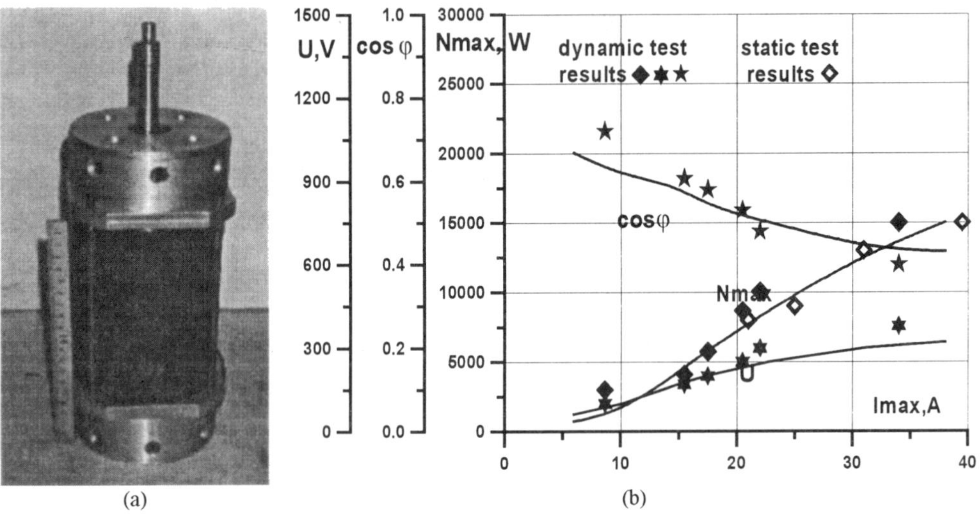

(a) (b)

Fig. 4. General view (a); experimental (dots) and theoretical (lines) parameters (b) of the HTS RM with nominal output power 10 kW at the temperature 77 K.

The experiments were carried out at the temperature of liquid nitrogen (77 K). The first experiments for the HTS reluctance motors for the temperature of 66 K (liquid nitrogen at the lower pressure) showed that the output power increases ~ 1.5 to 2 times in comparison to the test results at 77 K (see Fig. 5).

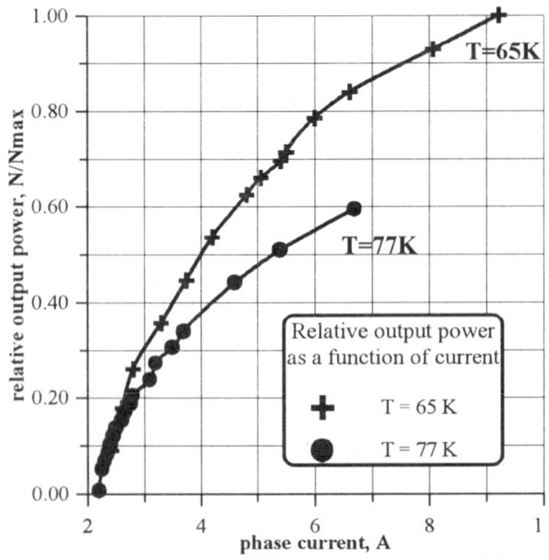

Fig. 5. Comparison of the output power for the different temperatures of the coolant (77 K and 66 K).

FUTURE DEVELOPMENT

According to the developed theoretical models [1-6] for hysteresis and reluctance HTS motors the specific output power of the motor is determined as a function of the domain current load $J_s \cdot a$ (here J_s is domain current and a is a domain size). The comparison of the specific output power parameters of HTS hysteresis and reluctance motors with the conventional hysteresis and asynchronous motors is given in Fig. 6. The experimental points for the achieved level of HTS motors are plotted in Fig. 6 as well. It is clearly seen, that even now the specific output power of the new types of liquid nitrogen cooled HTS motors with bulk YBCO elements can be higher than for electric machines of traditional design. [6]

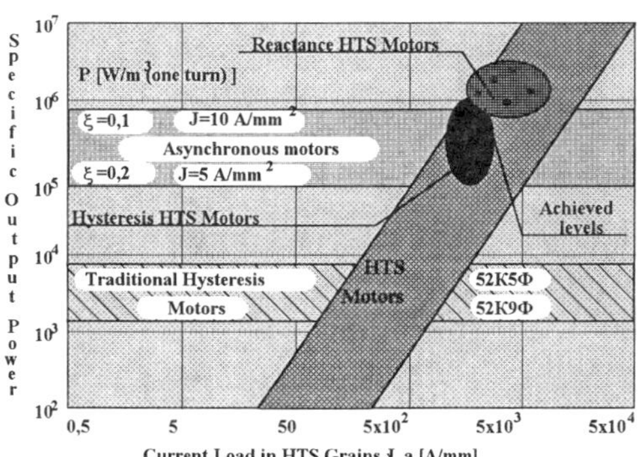

Fig. 6. Comparison of HTS motors with the motors of traditional design.

Theoretical and experimental results show that a new class of HTS motors can successfully operate at the temperature of liquid nitrogen and produce an output power of 10-20 kW (and more) with rather high efficiency (0.95) and power factor (0.7 to 0.8).

Future investigations of the HTS motors with bulk HTS blocks will be aimed at the following aspects:

- experimental study of the HTS reluctance motors operating in the medium of liquid nitrogen with the higher output power (up to 250 kW and more) and higher power factor ($\cos \varphi \geq 0.75$);
- theoretical and experimental study of new schematics of HTS motors with bulk materials (such as "trapped field" HTS motors, HTS motors with the permanent magnets installed on the compound HTS-ferromagnetic layered rotor core etc.).

REFERENCES

1. L.K.Kovalev, K.V.Ilushin, V.T.Penkin, K.L.Kovalev, Hysteresis electrical machines with high-temperature superconducting rotors, *Electrical Technology*, 2, p 145 (1994).
2. L.K.Kovalev, K.V.Ilushin, W.Gawalek, K.L.Kovalev et al., Hysteresis electrical machines with bulk melt-textured YBCO, *Materials Science & Engineering*, B53, p 216 (1998).
3. L.K.Kovalev, K.V.Ilushin, K.L.Kovalev, W.Gawalek et al., Hysteresis electrical machines with bulk YBCO and high output power, Proc. of the 3d Eur. Conf. on Applied Superconductivity, The Netherlands, 1997.
4. L.K.Kovalev, K.V.Ilushin, K.L.Kovalev, W.Gawalek, et al., Electrical hysteresis motors and generators with bulk melt textured YBCO, *Applied Physics*, Elsevier Science Ltd., UK, p 24 (1996).
5. L.K.Kovalev, W.Gawalek et al., HTS Electric machines with compound HTS-ferromagnetic rotor. Proc. of the 17th International Cryogenic Engineering Conference, 14-17 July, 1998, Bournemouth, UK
6. L.K.Kovalev, K.V.Ilushin, K.L.Kovalev, W.Gawalek et al., HTS motors design. Recent results and future development, *Superconductivity: Research and Development*, No. 9–10, p 69 (1998).

A BEARING SYSTEM FOR AN ENERGY STORAGE FLYWHEEL

T.A. Coombs, A.M. Campbell

IRC in Superconductivity, Cambridge University, Madingley Road, Cambridge CB3 0HE, United Kingdom

ABSTRACT

We have constructed an experimental rig which is designed to test superconducting bearing systems for energy storage flywheels. The rig is built in two main sections, a demountable "top hat" cylindrical section, and this is supported on an aluminium base plate. The resulting experimental chamber has a volume of 0.4 m^3. The flywheel that is currently being tested was originally supplied by British Nuclear Fuels. It weighs 43.6 kg and is representative of the Pirouette flywheel system. The experimental rig incorporates a loading mechanism for inducing and measuring out of balance forces, together with optical measurement systems for measuring rotation speeds and displacements. Evacuation is performed by a rotary pump and a turbomolecular pump. Cooling is provided by a CTI 350CP cryocooler. Control and datalogging is by computer program written under Visual Basic 6.0. The flywheel rig is supported by a vibration analysis rig, which is used to test the effect of vibrations by providing biaxial stimulation in a range of known waveforms to individual YBCO pucks. The information provided by this rig, together with that from computer modelling, is used in the design of the flywheel bearing system.

INTRODUCTION

We are investigating the use of superconducting bearings for levitating energy storage flywheels. We are using a Pirouette flywheel [1] (supplied by URENCO UK) which is a carbon and glass fibre cylinder weighing approximately 43.6 kg. The system under test uses a thrust bearing based around an Evershed system in which the levitation force is provided by a pair of rare earth magnets and the superconductors are used to stabilise the levitation magnets. Within this paper are given details of the vacuum chamber and the thermal considerations taken into account when designing the cryogenics and the magnet configuration. In addition, data are provided which show the rates of cooling and the characteristics of the bearing itself.

Advances in Cryogenic Engineering, Volume 45.
Edited by Shu *et al.*, Kluwer Academic / Plenum Publishers, 2000.

DESIGN

The design of the bearing system itself is described in detail elsewhere.[2] However, Fig. 1 gives a schematic arrangement of the energy storage rig. The superconductors may be either field-cooled, in which case the rotor is cooled in its operating position, or zero-field cooled with the rotor lifted 10 cm from its operating point. The bearing system is based around an Evershed system. The levitation force is provided by a pair of magnets (B in Fig. 1) acting in attraction. These magnets are stabilised by a ring magnet (C) interacting with a ring of twenty sections of bulk YBCO (E) each of OD 35 mm.

Problems have been encountered with the uniformity of the magnetic field provided by the ring magnet (c). Because of the large diameter of the ring it is not possible to fabricate it from a single section of NdFeB. The magnet was therefore made in sections glued together and held within an aluminium retaining ring. This led to discontinuities in the magnetic field. In order to ameliorate this it was decided to wind a magnet from flexible NdFeB. This is NdFeB powder in a rubber matrix and is available in metre long strips. This produced a much more uniform field pattern, however the remnance of the strips is much lower than that of the bonded NdFeB blocks. In order to produce a surface field of 0.18 Tesla both magnets (glued section & flexible strip) were used stacked on top of each other. This produced an acceptably uniform field.

To test the characteristics of the bearing, the flywheel (D) can be displaced from its equilibrium position by a loading plate (A). This applies an out of balance force to the rotor via an arrangement of three load cells. This mechanism is also used to centre the bearing and thus to set its initial position.

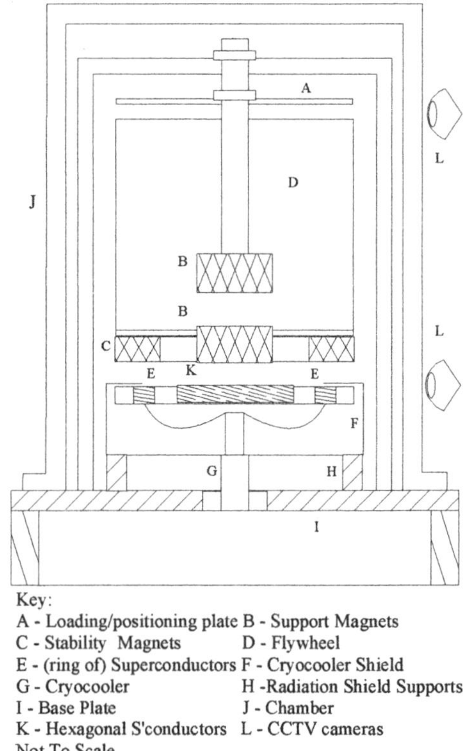

Figure 1. Schematic representation of energy storage flywheel rig.

Two motors are used, one drives the loading plate up and down, and the other raises and lowers the whole assembly, taking away or bringing it towards the superconductors which are contained in the cryostat. Absolute position is measured using an optical switch which switches on and off once per revolution of the motors. The motors are connected via a 500:1 gearbox and a drive belt to the lead screw which has a pitch of 1 mm. Hence, each revolution of the motor will move either the whole assembly or the loading plate by 2 microns.

The rotor is contained inside an evacuated chamber (J). The chamber is constructed from rolled steel sheet and is made in three sections: a top plate with lifting attachment, the walls of the cylinder itself, and a flange around the base. The chamber is sealed by an O-ring which is clamped between the flange and the base plate (I). All services are routed via the base plate (I). In addition, the rotor and bearing are supported on an independent structure which is mounted on the base plate and enclosed by the vacuum cylinder. Thus, the cylinder itself is a simple capped tube whose wall thickness is calculated by checking for radial buckling under one atmosphere pressure.[3] Access to the bearing system is gained by raising the cylinder using a dedicated hoist.

During levitated operation, viewing is by CCTV camera located within the vacuum chamber. Two CCTV cameras are used. One is at the bottom end of the rotor, and the other is at the top end. These are connected via a capture board to the data-logging and control computer, and either camera can be selected. The image from the camera is used to check the position of the rotor when it is levitating. The image is processed using a standard Sobel filter to detect edges, and the software then searches for the (horizontal) bottom or top edge of the rotor. The approximate resolution of this measurement is 0.09 mm (11 TV lines per mm).

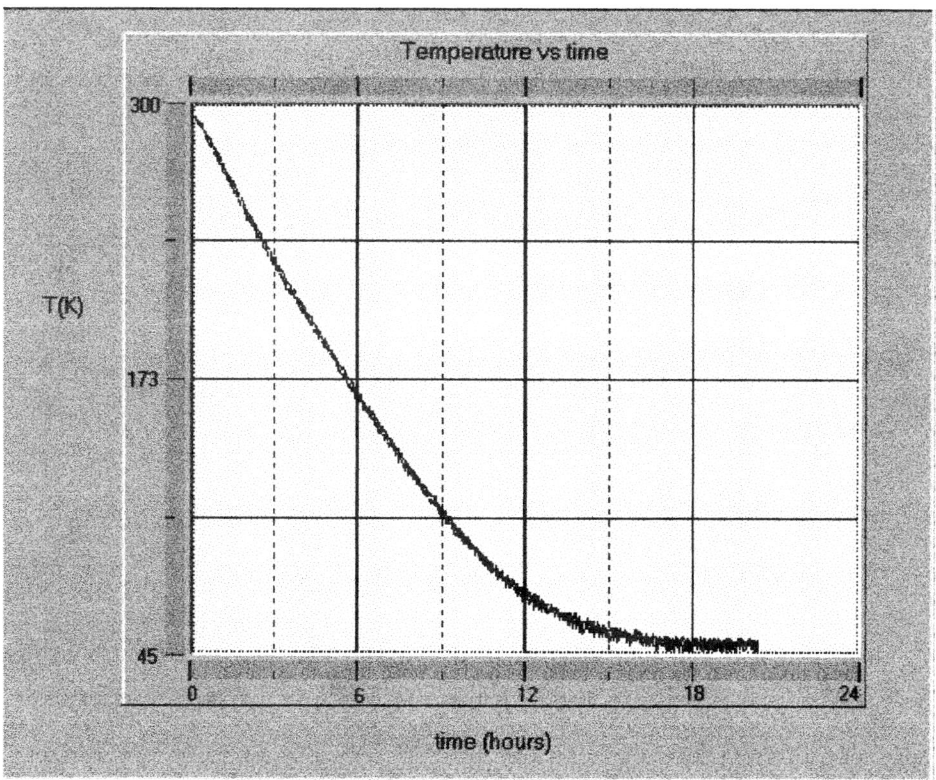

Figure 2. Sample cool down curve.

Type E thermocouples are used for temperature measurement. These are fixed to the cryostat and show that at the very perimeter of the cryostat the temperature reached is 52 kelvin. A sample cool down curve is shown in Fig. 2.

The device originally used twenty circular sections of bulk YBCO (35 mm OD) held in brass canisters, screwed into an aluminium body. The canisters, together with the body, make up the main cryostat.

There is an additional weight overhead (3.6 kg) represented by the flexible NdFeB. Although this is not supported directly by the YBCO, lifting force is provided by the pair of rare earth magnets (B), additional YBCO is required. This is because the lifting magnets provide greater force, hence they are closer together, so the stiffness of this interface is increased. Thus the superconductor magnet interface stiffness must be correspondingly increased.

For this reason an additional cryostat has been added at position (k) in figure 1. This cryostat holds 18 sections of hexagonal YBCO (20 mm across). These samples were supplied by Superconductivity Inc. These hexagonal samples interact with the lower of the two levitation magnets (B), providing the additional stiffness. A Cryogenics 350 CP cryocooler is used. This is a 2 stage device. The first stage cools a copper radiation shield and the second stage is connected, via a copper plate, (not shown in Fig. 3) to the cryostat. The overall arrangement used can be seen on the right hand side of Fig. 3. Data showing the output of the first and second stage are shown in Figure 4.

Figure 3. Energy storage flywheel rig.

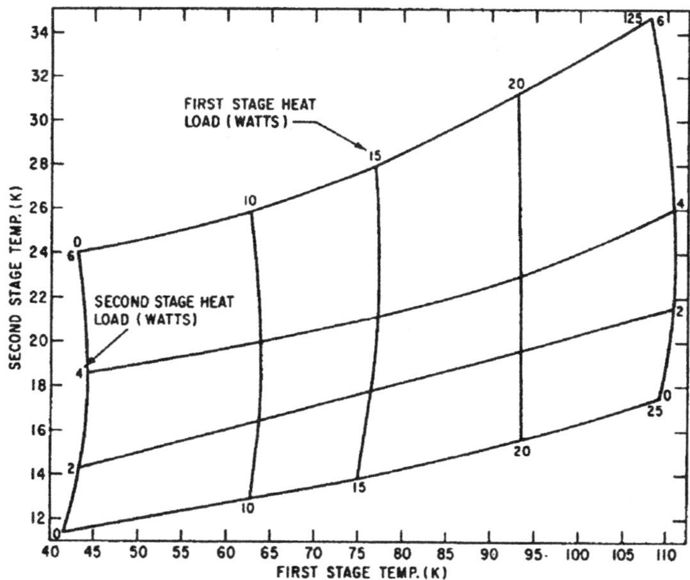

Figure 4. Typical refrigeration capacity of the Model 350CP Cryodyne Refrigerator .

The major sources of loss will be radiation and conduction through the cryostat supports. Both windage and conduction can be effectively ignored if a good enough vacuum is achieved. The radiation losses are given by the following standard equation :

$$q = \sigma \varepsilon k A(T_1^4 - T_2^4) \tag{1}$$

Where σ = Stefan-Boltzmann constant, ε= emissivity of the surface, k = shape constant, and A is the area of the surface under consideration. Since the surface of the cryostat is in close proximity to the rotor, we take the shape factor as 1. The average emissivity is 0.06. At 77 kelvin this gives a total heat loss of 1.5 watt.

The cryostat is supported on five tufnol legs, and the conduction down these is again approximately 1.5 watts. Both the radiation loss and the conduction loss may be reduced by adding additional insulation in the final production system. However, for the system as built the loss is between 1.5 watts and 3 watts, and this agrees well with Fig. 5.

In order to evacuate the rig, two pumps are used, a rotary pump for roughing, and a turbomolecular pump for finishing. Evacuation time is approximately half an hour (down to a vacuum of $2*10^{-5}$ mbar), and the best vacuum which has been obtained so far (during cooling) is $1*10^{-6}$ mbar.

RESULTS

The bearing characteristic may be altered by one of two methods. The first is to vary the temperature of the superconductors. Reducing the temperature increases the critical current, reducing the depth to which current will penetrate into the superconductors. This has the effect of increasing the stiffness of the superconductor to magnet interface. The second method is to vary the gap between the magnet to magnet pair (B in Fig. 1).

Increasing the magnet to magnet gap decreases the stiffness of this interface and thus reduces the overall stiffness of the bearing. If the gap is increased sufficiently, then the stiffness will first reduce to zero and then change sign. This is the condition which is required for stable levitation. Figure 5 shows a typical set of force displacement graphs for the bearing.

In the uppermost trace the initial position of the levitation magnets was a gap of 1.5 mm and the magnet to superconductor gap was 4.0 mm. There are five traces in all (each consisting of two lines – applying load using the loading plate (A) and then removing it again). Between each trace the initial levitation magnet gap has been increased by 0.5 mm while keeping the superconductor magnet gap constant at 4 mm.

In the initial position of each trace there is an out of balance force because the rotor is being held up by the levitation magnets. As the loading plate is lowered, the net force from the levitation magnets gradually reduces. At the same time, there is a force being developed between the superconductors and the rotor magnets. In the final trace (initial levitation magnet gap 3.5 mm) the net upward force has been reduced to zero between a superconductor magnet gap of between 2.8 and 1.6 mm and then increases again as the superconductor magnet gap further decreases. When the loading plate is raised again, the superconductor is left levitating. The peak stiffness on this final trace is circa 100 N/mm

The behaviour of the bearing as described here has been predicted by considering the magnet to magnet interface and the superconductor to magnet interface separately and summing the force displacement graphs.[4,5] Two methods were used for predicting the behaviour of the superconductors. One is to assume perfect diamagnetism and use a standard finite element package such as Quickfield. The second is a discrete method for evaluating the Bean model which predicts the distribution of currents within the superconductor in a varying magnetic field.

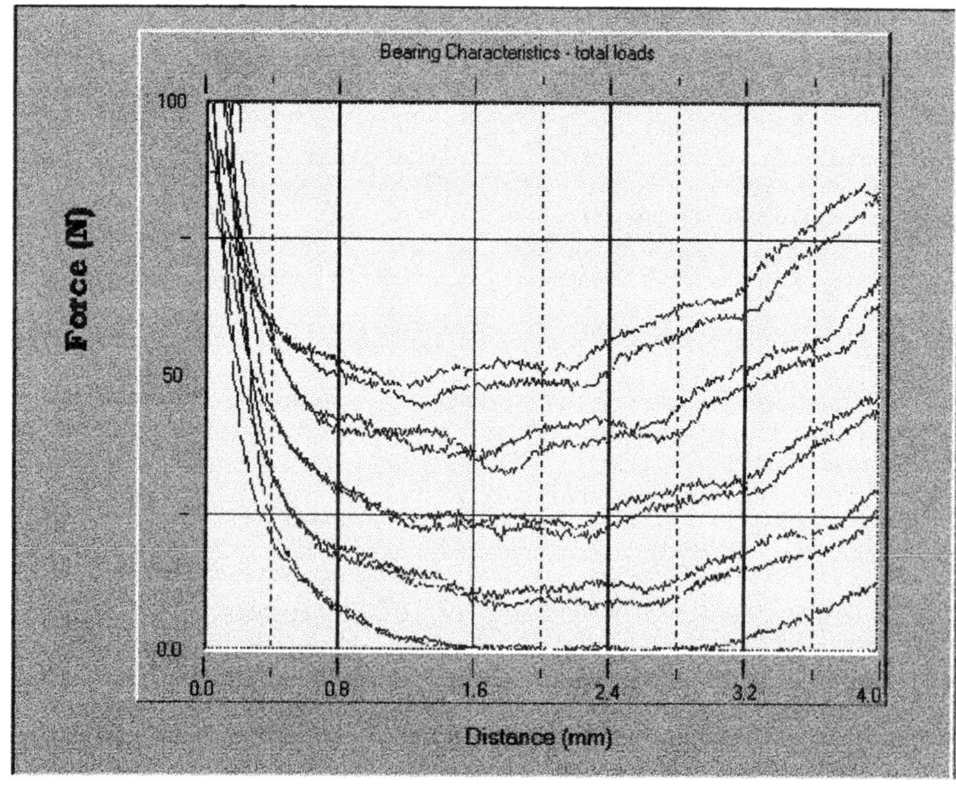

Figure 5. Typical force displacement curve.

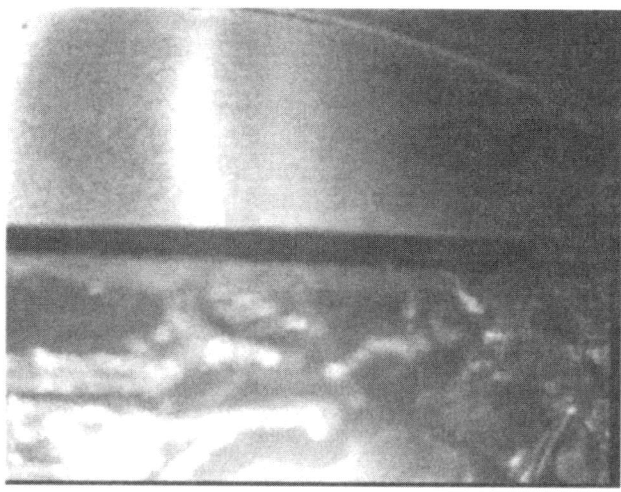

Figure 6. Picture from lower CCTV camera.

A picture from the lower CCTV camera, showing the bottom of the rotor above the cryostat is given in Figure 6.

CONCLUSIONS

The system as described within this paper is designed so as to minimise creep and travel. The majority of the levitation force is provided by permanent magnets which are stabilised by the action of superconductors. This is termed an Evershed bearing. The bearing is shown levitated at a gap of 3.5 mm. We have designed a low loss cryogenic system which enables the behaviour of the bearing system to be examined over a range of temperatures. The bearing and the cryogenics are housed in a demountable vacuum chamber which enables easy inspection and setting up.

ACKNOWLEDGEMENT

This project is supported by the EPSRC. In addition, thanks are due to Superconductive Components Inc. who have supplied superconductors for the project and to CryoPhysics who have supplied the cryocooler.

REFERENCES

1. R. Dettmer, Spinning reserve, *IEE Review* (January 1997) p. 36.
2. T.A..Coombs, A.M. Campbell , I. Ganney, W. Lo, T. Twardowski, B. Dawson, Superconducting bearings in high-speed rotating machinery, *IOP Conference Series* 158:1531 (1997).
3. S. Timoshenko " Theory of Elastic Stability", McGraw Hill, New York.
4. T. A. Coombs, D.A. Cardwell, A.M. Campbell, Dynamic properties of Superconducting Magnetic Bearings, *IEEE Transactions on Applied Superconductivity* 7: 924 (1997).
5. T.A. Coombs, A. M. Campbell, R. Storey, R Weller, Superconducting Magnetic Bearings for Energy Storage Flywheels, *IEEE Transactions on Applied Superconductivity* 9: 968 (1999).

THERMAL PERFORMANCE OF MULTILAYER INSULATION DOWN TO 4.2 K

A. Senthil Kumar[1], M.V Krishna Murthy [1], S. Jacob[2] and S. Kasthurirengan[2]

[1] Department of Mechanical Engineering, Indian Institute of Science
Bangalore 560 012, India
[2] Centre for Cryogenic Technology, Indian Institute of Science
Bangalore 560 012, India

ABSTRACT

The thermal performance of different types of multilayer insulation (MLI) down to 4.2 K is presented. A double guarded cylindrical boil-off calorimeter has been designed and fabricated to evaluate thermal performance of multilayer insulations between 300-77 K and 77-4.2 K. Studies on steady state heat flux between 300-77 K and 77-4.2 K of multilayer insulations as a function of number of layers and that of aluminum foil adhesive taped cold boundary (with no MLI) surfaces have been carried out. Use of aluminum foil adhesive tape over the bare copper surface has reduced the heat flux by a factor of ≅ 3 for both temperature ranges. Application of a few layers of MLI on the aluminum foil adhesive taped surface resulted in the decrease of heat flux for the 300 – 77 K range. On the contrary, it increased the heat flux for the 77 – 4.2 K range. The comparative performance of different types of insulations for both temperature ranges are discussed.

INTRODUCTION

The most efficient cryogenic insulation is the multilayer insulation also known as superinsulation. The heat transfer characteristics of MLI have been studied in detail by several authors[1, 2, 3, 4] between room temperature and liquid nitrogen temperature (300-77 K). However only a few experimental data have been reported[5, 6, 7, 8] on the evaluation of MLI down to 4.2 K. The limited experimental investigations of the latter have not given a clear picture whether the heat flux for the 77-4.2 K temperature range can be reduced by the use of specific types of MLI at an optimum number of layers as compared to aluminum foil adhesive taped surface. Presently, there is lack of information on the thermal performance of a few layers of MLI on the 4.2 K surface, as compared to the application of many layers.

It is also important to evaluate whether a type of MLI having good thermal performance in the 300-77 K range maintains the same level of performance in the 77-4.2 K range. The present work is aimed to study the thermal performance of different types of MLI systems (NRC-2, dimpled aluminised Mylar, aluminised Mylar with thermal spacer and aluminum foil with thermal spacer) and aluminum foil adhesive taped cold boundary surface with no MLI, between 300-77 K and 77-4.2 K.

EXPERIMENTAL SET-UP

A double guarded cylindrical boil-off calorimeter shown in Figure 1 has been designed and fabricated for conducting the above studies in the 300-77 K and 77-4.2 K temperature ranges. The general design features of the calorimeter follow that of the IHI [9] calorimeter except that the test surface area is larger. The calorimeter consists of a test chamber in the middle, an upper and a lower guard chamber, which can be filled with liquid nitrogen / liquid helium. The test chamber is made up of copper to ensure uniform surface temperature irrespective of liquid level. These chambers are surrounded by two shield tanks which are filled with LN_2 providing a 77 K boundary for conducting 77-4.2 K experiments.

The above structures are enclosed in a common vacuum chamber. The test chamber has a surface area of 0.1414 m^2 and has a volume of 3 litres. The total guard chamber volume is about 4.2 litres. The test and guard chambers are each provided with a single fill/vent line designed to avoid thermoacoustic oscillations observed in the case of closed fill lines [10]. The bigger LN_2 shield tank has a volume of 50 litres, and is suspended by glass fibre strips from the top flange. The smaller LN_2 shield tank of 2.2 litres capacity is attached to the fill / vent lines of the test and guard chambers at its top to minimise the conductive heat flux to them.

The vacuum chamber and the insulation structure are evacuated using a high vacuum pumping system. The pressure in the vacuum chamber is measured using a cold cathode Penning gauge.

The evaporation rate of the cryogen in the test chamber is measured using a Hastings (low pressure drop) mass flow meter (Model ENALL) with an accuracy $\pm$ 3 %. To avoid fluctuations in the boil-off rate due to variations in atmospheric pressure during measurements a pressure controller is used in the down stream side of the mass flow meter[2].

For measuring the temperature profiles of the MLI layers, Chromel-CuFe (0.15%) thermocouples (36 AWG, Lakeshore Cryotronics, USA) are used in the temperature range 77-4.2 K and Chromel-Constantan thermocouples (36 SWG, TC England) are used for the 300-77 K range. The output of the thermocouples and that of the mass flowmeter are acquired by a PC using a Keithley Scanner (Model 701) and Keithley DMM (Model 192).

EXPERIMENTAL PROGRAMME

Aluminum foil adhesive tape and the different types of MLI have been used for the experimental investigations and their specifications are given in Table 1.

Bare Copper and Aluminum Foil Adhesive taped test surface

For experiments in the 300-77 K temperature range, the chemically cleaned stainless steel surface of the unfilled LN_2 shield tank constitutes the warm boundary and the test and

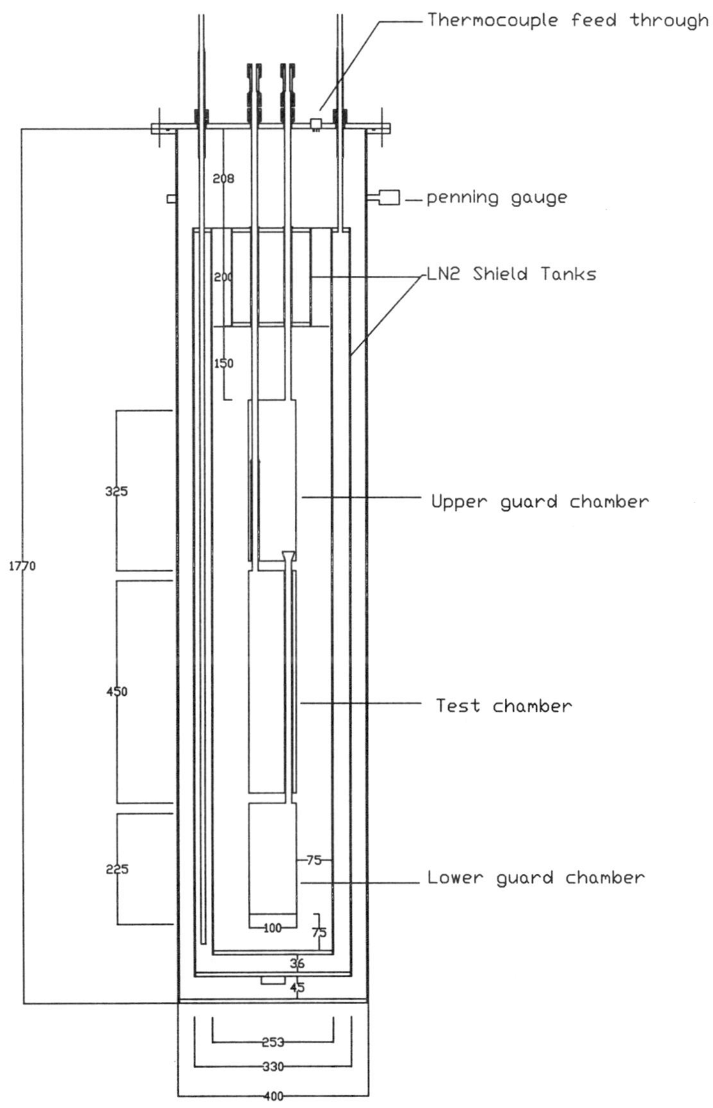

Figure1. Schematic of the double guarded cylindrical boil-off calorimeter

guard chambers are filled with LN_2. For conducting 77-4.2 K experiments, both the LN_2 shield tanks are filled with LN_2 and the test as well as guard chambers are filled with liquid helium.

Table 1. Specifications of the MLI and aluminum foil adhesive tape

Sl.No.	Nomenclature	Manufacturer`	Specifications	Estimated Weight/unit area kg/m^2
1.	Aluminum foil adhesive tape	Advance tapes, India	Commercial grade aluminum foil thickness: 70 μm	158×10^{-3}
2.	Lydall CRS wrap	Lydall Manning USA	Aluminum foil 1145-0, 7μm thickness + Cryotherm 1303 (microfiber glass non-woven binder free thermal spacer)	40×10^{-3}
3.	Jehier Insulray IR 305	Jehier, France	Double aluminised Mylar 6 μm thickness , 400A° coating thickness on each side, perforation diameter 2mm at a pitch of 56mm + polyester tulle with weight of 5×10^{-3} kg/m^2	14.3×10^{-3}
4.	NRC-2	Oxford Instruments, UK	Crinkled single side aluminised Mylar with thickness 6 μm	8.8×10^{-3}
5.	IHI dimpled MLI	IHI, Japan	12μm dimpled, double aluminised Mylar, dimple diameter 2mm at a pitch of 6mm, dimple height 200μm	9.24×10^{-3}

Initially experiments were conducted on the bare copper test surface for both the 300-77 K range and 77-4.2 K ranges. After the above experiments, the bare copper test chamber surface was covered with aluminum foil adhesive tape and the experiments were conducted for the above two temperature ranges. A dynamic chamber pressure of 3.5×10^{-6} mbar was reached for the 300-77K experiments, while a static chamber pressure of 2.0×10^{-7} mbar for the 77-4.2 K experiments. The steady state heat flux is estimated from the boil-off measurements of the test chamber, with corrections for the retained fraction of the boil-off in the calorimeter.

Different types of MLI

In all the experiments with MLI, the cold boundary is an aluminum foil adhesive taped surface and the warm boundary is bare stainless steel. The MLI is wound over the test chamber and the guard chambers (separately, to prevent lateral heat transfer) such that it neither is compressed too much nor it sags down. Care is taken to maintain a uniform layer density by measuring and controlling growth in the circumference of the insulation winding at regular intervals. The layer density for the experiments were in the range of 17 to 20 layers / cm for aluminised Mylar MLI systems and about 8 layers / cm for the Lydall aluminum foil insulation. Initially the maximum number of layers (equal to 50, except in the case of IHI insulation) is wound over the test and the guard chambers ensuring that there is a clear gap between the outermost layer of the insulation and the warm boundary. Experiments with fewer layers are conducted by unwinding the insulation until the required number is reached. The MLI is baked to ≅ 350K and evacuated until a dynamic pressure level of 1×10^{-4} mbar is obtained. The baking is then stopped, however continuing the evacuation to reach the dynamic pressure level of ≅ 5×10^{-5} mbar. For a given MLI at different number of layers, experiments are initially conducted by filling the test and guard chambers with LN_2. After cool-down the dynamic pressure level reached is nearly the same as that for aluminum foil adhesive taped surface experiments. Heat flux and layer temperature profile measurements are made at regular intervals till steady state conditions are reached.

After completion of the 300-77 K range experiment for a particular number of layers of MLI, the 77-4.2 K experiment is carried out. For this, the shield tanks are filled with LN_2 and the test and guard chambers are filled with LHe, after removal of LN_2. Heat flux and layer temperature profile measurements are carried out.

The vacuum levels reached after cool-down for MLI experiments, were nearly the same values as for the aluminum adhesive tape experiments, for both temperature ranges.

RESULTS AND DISCUSSION

Bare Copper and Aluminum Foil Adhesive Taped Surfaces between 300-77 K and 77-4.2 K

In the present experiments, the heat transfer due to solid conduction is absent in view of the absence of MLI and the residual gas conduction heat transfer is also negligible. Hence, the heat transfer is mainly by radiation. With the bare copper cold surface, a heat flux of 43 W/m^2 was measured for the 300-77 K range. A steady state heat flux of 12.5 W/m^2 has been measured for the 300-77 K range for aluminum adhesive taped surface. Thus the application of aluminum foil adhesive tape over the bare copper surface has reduced the heat flux by a factor of 3.5. However the heat flux values using the 3M aluminum tape applied to both warm and cold surfaces have been reported[5,6] to be in the range of 5-6 W/m^2.

For the 4.2 K bare copper surface, heat flux measured is 0.12 W/m^2 in the 77-4.2 K range. For the aluminum adhesive taped cold surface for the same temperature range a steady state heat flux of 0.044 W/m^2 has been measured, which indicates that the application of the aluminum tape has reduced the heat flux by a factor of ≅ 2.7. In the literature, however, much lower heat flux values (≅ 0.012 W/m^2) have been reported [5, 6] when 3M aluminum tapes are used.

The above experiments have confirmed that the heat flux to the cold surfaces can be substantially reduced by the application of aluminum adhesive tapes over them. Using the 77-4.2 K experimental data, the emissivity of aluminum tape at 4.2 K can be estimated to be 0.02, (assuming an emissivity of 0.05 for the chemically cleaned stainless steel surface at 77 K). Since the pure aluminum in annealed condition can have an emissivity value of 0.011 at 4.2 K[11] the observed increase in emissivity could be due to the factors such as insufficient purity level of the aluminum, annealing, mechanical handling conditions, etc.

Different types of MLI (300-77 K)

Figure 2 shows the heat flux measured for different types of MLI as a function of the number of layers. It is observed that the heat flux decreases exponentially with increasing number of layers. On using a small number of layers of MLI over the aluminum foil adhesive taped surface, the heat flux reduces substantially. For instance, using 2 layers of Lydall MLI results in a reduction of heat flux by a factor of ≅ 3.5 on the heat flux for aluminum taped surface. Using other MLI systems based on aluminised Mylar, leads to higher heat fluxes compared to that of Lydall MLI. These measurements indicate that for restricted gaps between 300-77 K surfaces in cryogenic systems, a few layers of Lydall MLI should be preferred.

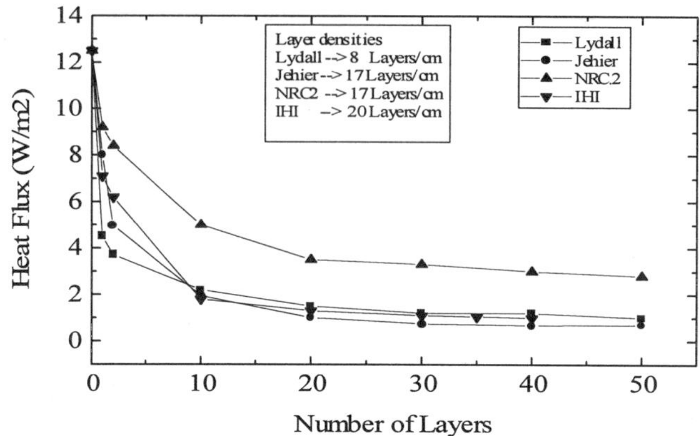

Figure 2. Heat flux as a function of number of layers in the 300 – 77 K range

However with more layers, MLI systems based on aluminised Mylar (except NRC-2) have shown lower heat flux, compared to Lydall MLI. The use of Jehier MLI resulted in the lowest heat flux of 0.68 W/m^2 at 40 layers.

Tests using NRC-2 have shown higher heat flux ($\cong 3 W/m^2$) compared to other MLI systems. This can be attributed to the higher effective emissivity values of this single side aluminised insulation and also due to the fact that in the 300-77 K range, the predominant heat transfer mode is radiation. The gas conduction heat transfer appears to be negligible since, even with smaller number of layers, where the insulation would have been pumped quite well without any interlayer gas pressure gradient, the observed heat flux is high. Hence, due to its poor radiative properties, a much larger number of layers would perhaps be needed for NRC-2 insulation.

For specific applications where low insulation weight (Table 1) combined with high performance is desired, MLI such as Jehier and IHI appear to be better suited.

Different Types of MLI (77-4.2 K)

Figure 3 shows the heat flux measured with MLI between 77 – 4.2 K. It can be seen from this figure that unlike the case of 300-77 K, the application of a few layers of MLI over the aluminum taped cold surface increases the heat flux. This effect is more predominant for spacerless insulation such as IHI and NRC-2. The temperature of the first layer of the insulation for experiments with different number of layers is shown in Figure 4.

Using the single layer temperature and the total heat flux, the emissivity of the single layer was calculated since in this case, the heat transfer from the 77 K warm boundary to this layer is by radiation alone. Using this data, the solid conduction and radiation heat fluxes were calculated and these results are shown in Table 2.

It is observed that in the case of Lydall insulation, more than 99% of the total heat flux is by solid conduction. This behaviour is not expected to be very different for larger number of layers, since the first layer temperature continues to be nearly the same, as can be seen from Figure 4.

For other insulations studied, there is a considerable decrease in the first layer temperature of the insulation, when larger number of layers are used. Since in the 77-4.2 K temperature region, the main heat transfer mode is solid conduction, this results in the decrease in heat flux through the insulation with increasing number of layers. For number of layers greater than 20, only marginal variations in the first layer temperature are observed resulting in the nearly constant value of the solid conduction heat transfer.

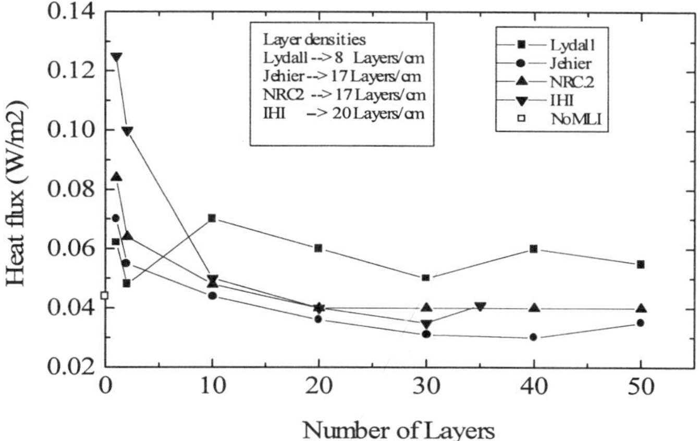

Figure 3. Heat flux as a function of number of layers in the 77 – 4.2 K range

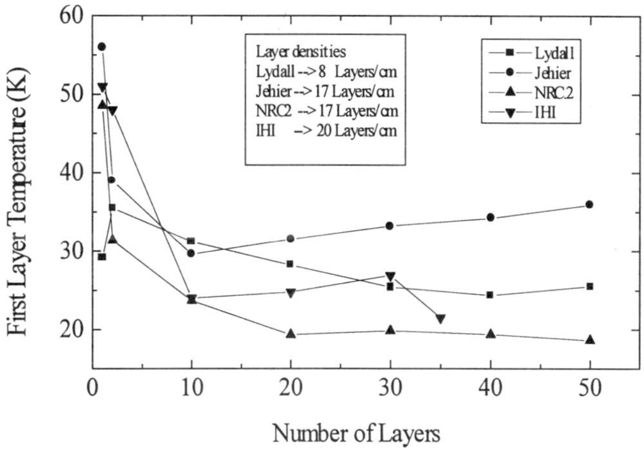

Figure 4. Temperature of the first layer as a function of number of layers in the 77 – 4.2 K range

Table 2. First layer temperatures, emissivities, solid conduction & radiation heat fluxes

Sl. No	MLI System	Single Layer Temperature (K)	Emissivity of first layer	Solid Conduction heat flux Q_s (W/m^2)	Radiation heat flux Q_r (W/m^2)	Total Heat flux Q_t (W/m^2)	First Layer Temperature(K) 30 layer case
1	Jehier	56.0	0.055	0.0617	0.0083	0.070	31.5
2	IHI	51.0	0.0948	0.1185	0.0064	0.125	24.7
3	Lydall	29.1	0.0331	0.0594	0.0005	0.060	28.2
4	NRC-2	48.5	0.0566	0.0793	0.0047	0.084	20.0

CONCLUSIONS

Studies on different types of MLI between 300-77 K and 77-4.2 K have been performed using a double guarded cylindrical boiloff calorimeter. To our knowledge, not much published experimental results are available, comparing the thermal performance of several MLI systems, both in the 300-77 K and 77-4.2 K ranges. Using aluminum foil adhesive tape on the bare copper test surface has resulted in the reduction of heat flux by a factor of $\cong$ 3.5 and $\cong$ 2.7 in the 300-77 K and 77-4.2 K ranges respectively. However, the measured values with the aluminum foil adhesive tape are larger than the reported values using 3M aluminum adhesive tape. The application of a few MLI layers on the aluminum foil adhesive taped test surface resulted in a decrease in the heat flux from 300-77 K, but an increase in the heat flux from 77-4.2 K. For number of layers greater than 20 for all MLI except Lydall, the heat flux values are either comparable or lower in some cases to that of aluminum foil adhesive tape for 77-4.2 K range. More experiments with better quality (lower emissivity) aluminum adhesive tapes are needed to estimate the lowest possible heat flux achievable in the 77-4.2 K range, as compared to using large number of MLI layers. Since MLI could prevent a sudden increase in heat flux due to variation in vacuum levels it will be useful to extend the work of Boroski et al.[12] down to the cold boundary temperature of 4.2 K at different chamber pressures.

ACKNOWLEDGEMENT

We would like to thank Dr. T. Ohmori, IHI, Japan for providing us the sample of dimpled insulation. We also thank staff of Centre for Cryogenic Technology, Indian Institute of Science for their help in the fabrication of calorimeter and conduction of experiments. The first author, who is on study leave from Vellore Engineering College, Vellore, India, thanks the management for deputing him for higher studies.

REFERENCES

1. R. G. Scurlock and B. Saull, Development of multilayer insulation with thermal conductivities below 0.1 μW Cm^{-1} K^{-1}, *Cryogenics* 16:303 (1976)
2. S. Jacob, S. Kasthurirengan and R. Karunanithi, Investigations into the thermal performance of multilayer insulations (300-77 K) Part 1. Calorimetric Studies, *Cryogenics,* 32:1137 (1992).
3. W. Barth and W. Lehmann, Experimental investigations of multilayer insulation model equipped with carbons paper *Cryogenics,* 28:317 (1988)
4. J.D. Gronczy, W.N. Boroski and R.D. Niemann, Thermal performance measurements of a 100 percent polyester MLI system for the superconducting super collider Part II: Laboratory tests (300K – 80K), in: "Adv. Cryo.. Eng.", 35 (1990), p. 497.
5. T.R. Gathright. and P.A. Reeve, Effect of multilayer insulation on radiation heat transfer at cryogenic temperatures, *IEEE Transactions on Magnetics*, 24:1105 (1988).
6. E.M.W. Leung, R.W. Fast, H.L. Hart and J. R. Heim, Techniques for reducing radiation heat transfer between 77 and 4.2 K, in: "Adv. Cryo. Eng.", 25 (1979) p. 489.
7. Q.S. Shu, R.W. Fast and H.L. Hart, An experimental study of heat transfer in multilayer insulation systems from room temperature to 77 K, in: "Adv. Cryo. Eng.", 31 (1986) p. 455.
8. E. Spradley, T. C. Nast and D. J. Frank, Experimental studies of MLI systems at very low boundary temperatures, in: "Adv. Cryo. Eng.", 35 (1989) p. 477.
9. T. Ohmori and K. Takahashi, Thermal Performance of Evacuated Multilayer Insulations, *Proc. Thermal Engineering Symp. (ISME)* 920-57: 48 (1992).
10. H. Luck and Ch. Trepp, Thermoacoustic oscillations in cryogenics. Part 1: Basic theory and experimental verification, *Cryogenics*, 32:690 (1992).
11. R.B. Scott, "Cryogenic Engineering", D. Van. Nostrand 153 (1959).
12. W. N. Boroski, T. H. Nicol and C. J. Schoo, Thermal performance of various multilayer insulation systems below 80 K, *Proc. 4th IISSC*, 235-242 (1992).

INSULATION TESTING USING CRYOSTAT APPARATUS WITH SLEEVE

J. E. Fesmire[1] and S. D. Augustynowicz[2]

[1]NASA Kennedy Space Center, MM-J2
Kennedy Space Center, FL 32899

[2]Dynacs Engineering Company, Inc., DNX-3
Kennedy Space Center, FL 32899

ABSTRACT

The method and equipment for testing continuously rolled insulation materials is presented in this paper. Testing of blanket and molded products is also facilitated. Materials are installed around a cylindrical copper sleeve using a wrapping machine. The sleeve is slid onto the vertical cold mass of the cryostat. The gap between the cold mass and the sleeve measures less than 1 mm. The cryostat apparatus is a liquid nitrogen boiloff calorimeter system that enables direct measurement of the apparent thermal conductivity (k-value) of the insulation system at any vacuum level between $5x10^{-5}$ and 760 torr. Sensors are placed between layers of the insulation to provide complete temperature-thickness profiles. The temperatures of the cold mass [maintained at 77.8 kelvin (K)], the sleeve [cold boundary temperature (CBT)], the insulation outer surface [warm boundary temperature (WBT)], and the vacuum can (maintained at 313 K by a thermal shroud) are measured. Plots of CBT, WBT, and layer temperature profiles as functions of vacuum level show the transitions between the three dominant heat transfer modes. For this cryostat apparatus, the measureable heat gain is from 0.2 to 20 watts. The steady-state measurement of k-value is made when all temperatures and the boiloff rate are stable.

INTRODUCTION

The advantages of using cryogen boiloff calorimeters for measuring the thermal performance of insulation systems assembled in a cylindrical manner have been previously described.[1,2,3] The main challenge in the execution of this technique is to obtain stability of the cryogen inside the three chambers coincident with stability of the boundary conditions in the vacuum space. That is, the liquid inside each of the three chambers of the cold mass must be brought to very near saturated conditions at the ambient pressure while the vacuum level and temperatures across the insulation thickness are maintained at constant values. The main problem with testing high performance materials such as multilayer insulation (MLI) is the extreme care that must be exercised in the fabrication and installation of such highly anisotropic test articles. As emphasized by De Haan[4], inconsistency in MLI wrapping techniques is the dominant source of error and poses a basic problem in the comparison of such insulating materials. Improper treatment of the ends or seams, for example, can render results that are several times worse than predicted. Localized compression effects, sensor installation, and outgassing are further complications. To eliminate the seam and minimize these other problems, a sleeve method of fabricating and testing insulation has been developed. The improved method and corresponding equipment for testing continuously rolled materials are the subject of this paper.

DESCRIPTION OF TEST EQUIPMENT

The cryostat apparatus, shown in Figure 1, is a sleeved liquid nitrogen boiloff calorimeter that enables direct measurement of the apparent thermal conductivity (k-value) of the insulation system at a vacuum level between 5×10^{-5} and 760 torr. The cryostat cold mass is a 167-mm-diameter by 900-mm-long cylinder constructed from heavy-wall stainless steel which provides high thermal stability. The cold mass consists of a 10-liter capacity test chamber with 2.5-liter (each) guard chambers on top and bottom over which the instrumented, insulated sleeve is installed. A removable vacuum can allows quick changeover of test articles. The apparatus is supplied with liquid nitrogen, subcooled to approximately 77.8 K, as depicted in the simplified system schematic of Figure 2. The measurable total heat transfer rate is from 0.2 to 20 watts, corresponding to a boiloff flow rate of from 50 to 5000 cubic centimeters per minute of gas at standard temperature and pressure. The vacuum pumping system consists of a combination of turbopumps and mechanical pumps plus a finely metered gaseous nitrogen supply to control pumping speed.

In testing, five operational sequences are performed: (1) heating and vacuum pumping, (2) liquid nitrogen cooling and filling, (3) cold soak, (4) replenish boiloff, and (5) steady-state boiloff. Steady-state conditions are typically achieved in less than 12 hours after an initial cooling and thermal stabilization period of at least 24 hours. During the boiloff replenish phase, the guard chambers are maintained at approximately 4.0 ±1 kPa. In the final phase, liquid nitrogen replenishment of the guard chambers is terminated as the cold mass approaches thermal equilibrium at near ambient pressure saturated conditions. The upper guard chamber is kept at a slightly higher pressure (1.0 ±0.3 kPa) than the test chamber to preclude the condensation of any boiloff gas as it exits through the center of the guard.

Materials are installed around a cylindrical copper sleeve using the custom-built 1-meter wide wrapping machine shown in Figure 3. Testing of blanket, loose fill, poured, and molded products is also facilitated by this sleeve method. Insulation test articles 170-mm inside diameter by 900 mm in length by up to 50 mm in thickness can be fabricated and tested. Using an effective length of the cold mass of 575 mm, the mean surface area for heat transfer through a typical 25-mm-thick insulation test article is 0.35 m^2.[5] After fabrication of the insulation system, the sleeve is simply slid onto the vertical cold mass of the cryostat. The gap between the cold mass and the sleeve measures less than 1 mm. Sensors are placed between layers of the insulation to obtain complete temperature-thickness profiles. Temperatures of the cold mass (maintained at near 77.8 K), the sleeve (CBT), the insulation outer surface (WBT), and the vacuum can (maintained at approximately 313 K by a thermal shroud) are measured as shown in Figure 4.

The steady-state measurement of insulation performance is made when all temperatures and the boiloff flow are stable. The k-value of the insulation is directly determined from the measured boiloff rate, temperature difference (WBT-CBT), latent heat of vaporization, and geometry of the insulation. All measurements are recorded on a custom built 32-channel data acquisition system using LabView software.

Figure 1. Cryostat test apparatus showing the insulation test article inside (extra sleeve shown for reference).

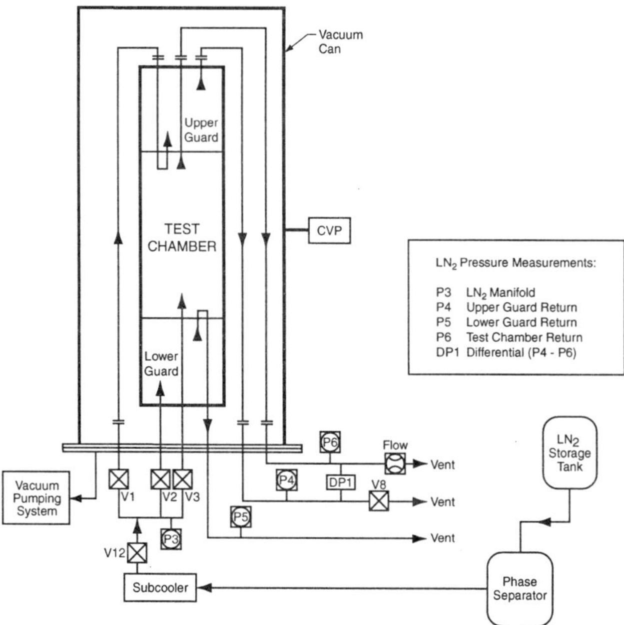

Figure 2. Simplified schematic of the cryostat test apparatus indicating key system measurements.

Figure 3. Multilayer insulation wrapping machine and copper sleeve assembly.

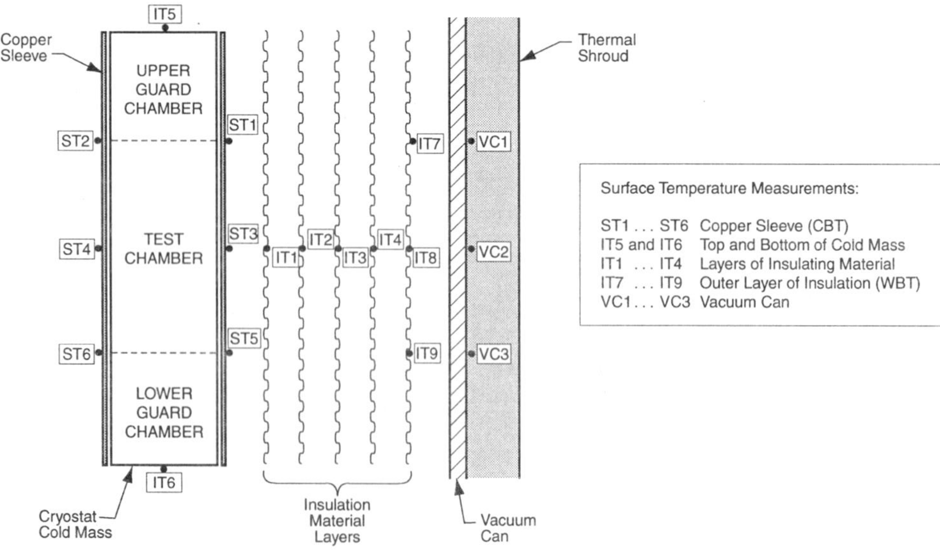

Figure 4. Diagram of temperature sensor locations for a typical insulation test configuration.

MEASUREMENT SUMMARY

Accurate measurements require repeatable and precise operation of each element of the total system. Boiloff measurements in the milliwatt range require long stabilization times and a carefully executed process to achieve thermal stability. This stability comes about through reaching a saturated liquid condition inside the test chamber that precisely matches the guard chambers. Tests at cold vacuum pressure (CVP) above 0.1 torr are further complicated by the influence of viscous gas conduction and convection which hinder maintaining constant boundary temperatures and a fixed vacuum level. The four charts of Figure 5 give a summary of the key system parameters for a typical test (including cold soak, replenish boiloff, and steady-state boiloff) of an evacuated insulation system.[6]

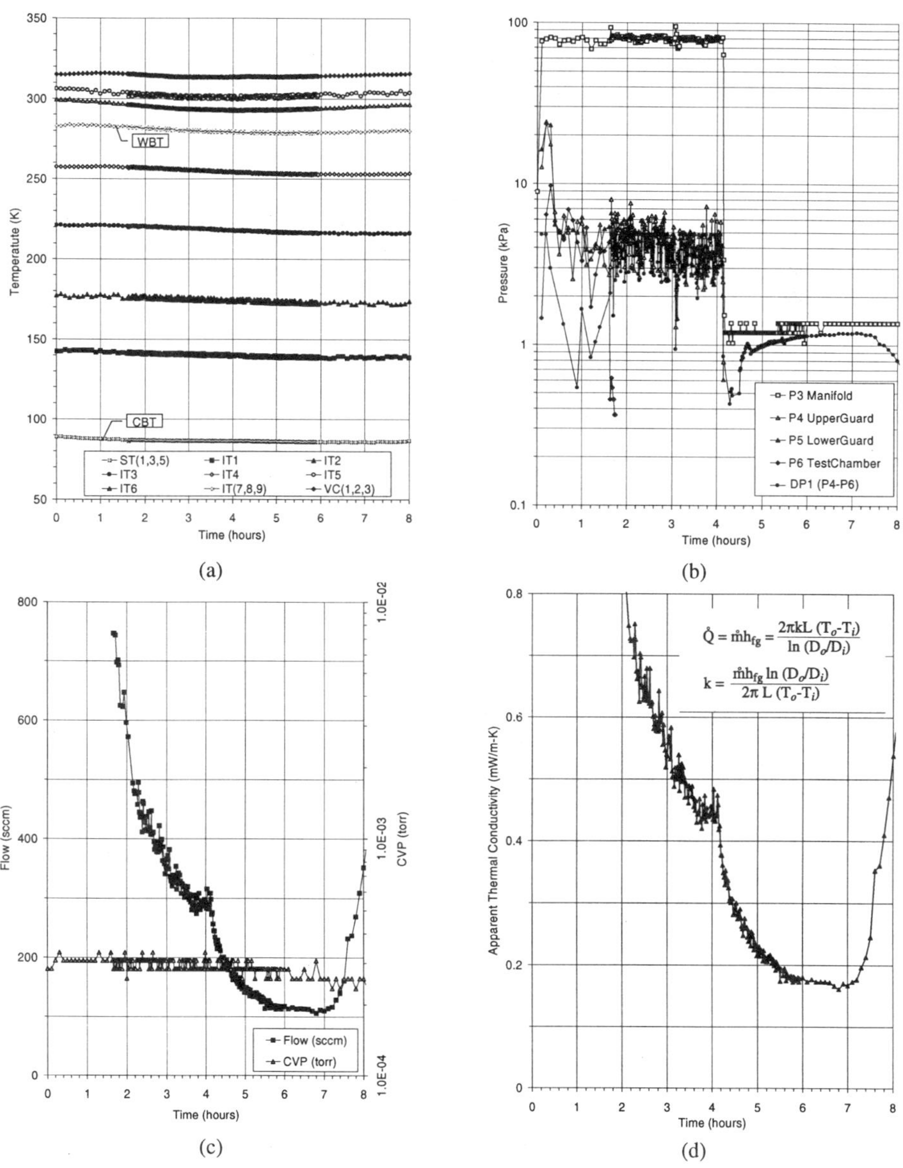

Figure 5. Test measurements for a typical evacuated insulation system: (a) layer temperature distribution, (b) chamber LN_2 pressures, (c) boiloff flow and CVP, and (d) calculated k-value.

SLEEVE GAP EFFECTS

A series of boiloff tests with only the sleeve installed was performed to understand the effects of the gap between the sleeve and cold mass on the thermal stability of the system. The temperature profiles in Figure 6(a) clearly show the transition from radiation to gas-conduction-dominated heat transfer. Observations showed another transition to convection-dominated heat transfer at about 40 torr. Apparent thermal conductivity values for the gap and the vacuum space are given in Figure 6(b). The measured k-values at the 10^{-4} torr level correspond to typical experimental values for unpolished surfaces. Reported values for a pure vacuum insulation system range from about 0.5 to 5 mW/m-K for boundary temperatures of 300 and 77 K, depending on the emissivities of the surfaces.[7]

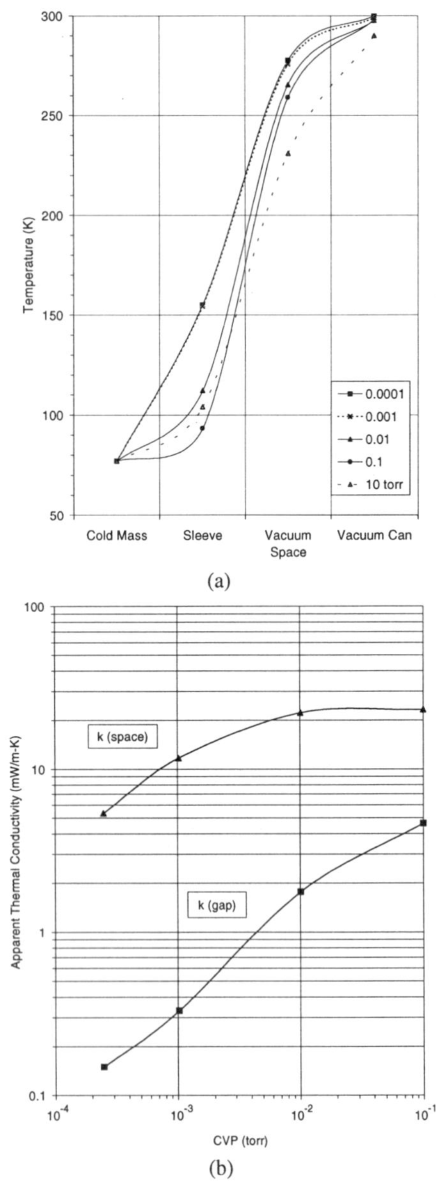

Figure 6. Tests with the sleeve only: (a) temperature profiles and (b) k-values as a function of CVP.

BOUNDARY TEMPERATURE EFFECTS

Variations of the warm and cold boundary temperatures with CVP are shown for selected insulation test articles in Figure 7.[6] The influence of gas conduction can be clearly seen starting at 10^{-3} torr with a minimum CBT reached at 10^{-1} torr. The temperature sensors used were thermocouples with standard reference junctions. Overall trends, not individual accuracy, are therefore emphasized. The vacuum can temperature (thermal shroud) was maintained at approximately 313 K in order to obtain a WBT of about 290 K for most insulations over the widest range of vacuum levels. In contrast, detailed studies of a single insulation system would be conducted by keeping the WBT at a prescribed value.

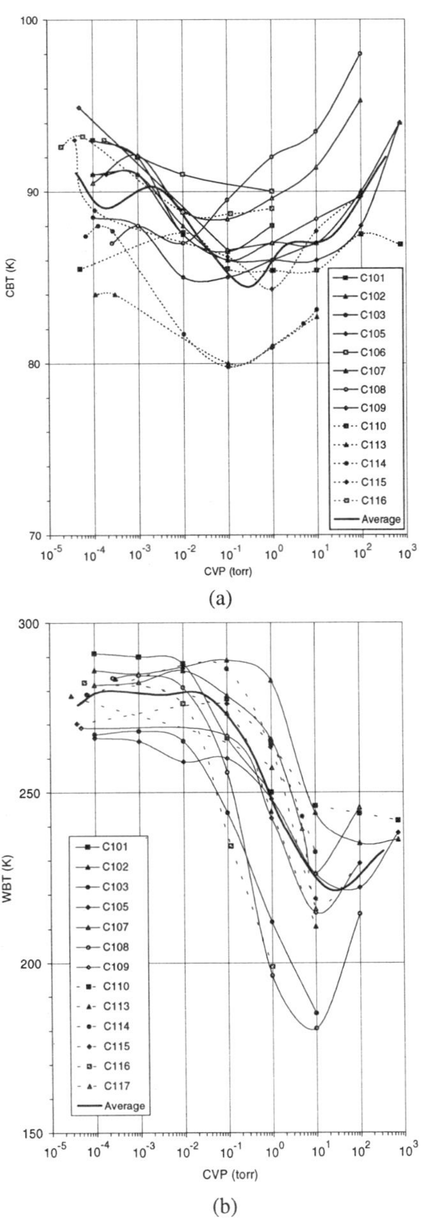

Figure 7. Variation of boundary temperatures with CVP for selected insulations: (a) CBT and (b) WBT.

CONCLUSIONS

An improved method of testing cryogenic insulation systems using a sleeve for insulation fabrication and test on a cylindrical boiloff calorimeter has been developed. The sleeve method is particularly useful for multilayer insulation test and evaluation by eliminating the error inherent with seams and reducing other problems with fabrication and handling. The sleeve allows test articles to be fabricated off-site and facilitates the testing of almost any product form. The corresponding wrapping method for continuously rolled products offers convenience in fabrication, installation, and instrumentation while providing a precise, repeatable process with no handling of the materials required. Over 150 tests of 18 different test articles have been performed that demonstrated the measurement range of the system.[6] For continuously rolled MLI products, the apparent thermal conductivity measurements were found to be essentially the same with or without the sleeve when compared to the data in the literature.[1,3] Boundary temperatures were in the range of 80 to 300 K. The sleeve method can be used for vacuum levels up to 760 torr for comparative results but should be limited to about 10 torr for accurate quantitative results.

A smaller boiloff apparatus with a fully removable cold mass has recently been built to address these limitations. This system, a calibrated-type cryostat, can support testing at a rate of up to five cylindrical test articles per week. The cold mass can also be configured for flat plate insulation geometries to test compression effects in MLI or to test smaller size samples. Optimization studies, for example, may be performed, followed by tests using the larger cryostat for the final performance measurements. Both cryostats are presently being used in thermal insulation studies and applications development in the Cryogenics Test Laboratory at NASA Kennedy Space Center.

REFERENCES

1. S.L. Bapat, K.G. Narayankhedkar, and T.P. LuKose, Experimental investigations of multilayer insulation, *Cryogenics*, Vol. 30, 1990, pp. 711-719.
2. R.M. Coston and C.A. Zierman, Cryogenic thermal-conductivity measurements of insulating materials, *Thermal-Conductivity Measurements of Insulating Materials at Cryogenic Temperatures*, ASTM STP 411, American Society for Testing Materials, 1967, pp. 25-42.
3. M.G. Kaganer, *Thermal Insulation in Cryogenic Engineering*, Israel Program for Scientific Translations Ltd., Israel, 1969.
4. J.R. De Haan, A simple multilayer insulation calorimeter, *Thermal Conductivity Measurements of Insulating Materials at Cryogenic Temperatures*, ASTM STP 411, American Society for Testing Materials, 1967, pp. 95-109.
5. ASTM C 745, Standard test method for heat flux through evacuated insulations using a guarded flat plate boiloff calorimeter, American Society for Testing Materials, 1992.
6. S.D. Augustynowicz and J.E. Fesmire, Cryogenic insulation system for soft vacuum, Cryogenic Engineering Conference, Montreal, 1999.
7. T.M. Flynn, *Cryogenic Engineering*, Marcel Dekker, New York, 1997, pp. 440-441.

CRYOGENIC INSULATION SYSTEM FOR SOFT VACUUM

S. D. Augustynowicz[1] and J. E. Fesmire[2]

[1]Dynacs Engineering Company, Inc., DNX-3
Kennedy Space Center, Florida 32899

[2]NASA Kennedy Space Center, MM-J2
Kennedy Space Center, Florida 32899

ABSTRACT

The development of a cryogenic insulation system for operation under soft vacuum is presented in this paper. Conventional insulation materials for cryogenic applications can be divided into three levels of thermal performance, in terms of apparent thermal conductivity [k-value in milliwatt per meter-kelvin (mW/m-K)]. System k-values below 0.1 can be achieved for multilayer insulation operating at a vacuum level below $1x10^{-4}$ torr. For fiberglass or powder operating below $1x10^{-3}$ torr, k-values of about 2 are obtained. For foam and other materials at ambient pressure, k-values around 30 are typical. New industry and aerospace applications require a versatile, robust, low-cost thermal insulation with performance in the intermediate range. The target for the new composite insulation system is a k-value below 4.8 mW/m-K (R-30) at a soft vacuum level (from 1 to 10 torr) and boundary temperatures of approximately 77 and 293 kelvin (K). Many combinations of radiation shields, spacers, and composite materials were tested from high vacuum to ambient pressure using cryostat boiloff methods. Significant improvement over conventional systems in the soft vacuum range was demonstrated. The new layered composite insulation system was also shown to provide key benefits for high vacuum applications as well.

INTRODUCTION

The technological developments of this century have led to insulation systems that approached the ultimate limit of performance. More technologies and markets forecast for rapid expansion into the 21st century will require, in many cases, not superinsulations but efficient, low-cost systems for these broader cryogenic markets. The cooperative research program *Comparative Study of Cryogenic Vacuum Insulation Systems* between NASA Kennedy Space Center and MVE, Inc., was conducted in 1997 through 1998. Details of this study, from an energy and economics viewpoint, can be found in a paper recently submitted to the 20th International Refrigeration Congress.[1] A main target of the study, whichis the subject of this paper, was to develop a new soft vacuum insulation system (that is, from about 1 to 10 torr) that provides an intermediate level of performance (k-value below 4.8 mW/m-K). The design should be robust and efficient, providing flexibility in the design, manufacturing, and implementation of a variety of thermal insulation systems.

OVERVIEW OF THERMAL INSULATION IN CRYOGENICS

Cryogenic insulation system development began around the turn of the century, coinciding with the first liquefaction of key industrial gases during the period 1877 to 1908. D'Arsonval first demonstrated the vacuum flask in 1887.[2] This design was significantly improved by Dewar in 1893 by silvering the walls of the flask. The concept of filling the vacuum space with powder was illustrated by Stanley in 1912.[3] Stanley notes the key advantage that "the same degree of heat insulation may be obtained at a much less reduced gaseous pressure in the vacuous space." Early insulation system designs for cryogenic tanks, shown in Figure 1, were advanced by Dana in 1939 and by Cornell in 1947.[4,5] Multilayer insulation (MLI), which could provide an order of magnitude improvement in performance, was first demonstrated by Peterson in 1951.[6] MLI systems were well developed by about 1960 through the work of Matsch, Kropschot, Hnilicka, and others.[7,8,9]

Conventional insulation materials for cryogenic applications can be divided into three categories of apparent thermal conductivity (k-value): around 30 mW/m-K for materials at ambient pressure, about 1.5 mW/m-K for bulk materials at good vacuum (below 10^{-3} torr), and below 0.1 mW/m-K for MLI at high vacuum (below 10^{-4} torr) (for boundary temperatures of about 300 and 77 K). Thermal performance of MLI degrades rapidly for vacuum levels above 10^{-3} torr. In addition to the high vacuum requirement, other drawbacks of MLI are its high inplane heat conduction, sensitivity to compressive loads and edge effects, the extreme care needed during installation, and its limitation to more simple structures. Furthermore, the steps of evacuation, heating, and vacuum retention are costly and time consuming. It is important to recognize that there are three levels of thermal performance for MLI: ideal, laboratory, and actual. Actual system performance is typically several times worse than the laboratory performance and often 10 times worse than the ideal.[10,11]

An insulation system that performs well in soft vacuum fills the performance gap between high vacuum systems (R-1500) and ambient pressure systems (R-5), representing a substantial new market area. The expansion of process applications at extreme temperatures combined with the need for higher energy efficiencies translates to an increasing demand for improved thermal insulation systems.

EXPERIMENTAL METHODS

The liquid nitrogen boiloff method using a cylindrical cryostat with sleeve was used for all tests. The system and corresponding methods are further described in a paper by Fesmire and Augustynowicz.[12] Continuously rolled materials were installed around a copper sleeve that was then slid onto the vertical cold mass of the cryostat. Sensors were placed between layers of the insulation to obtain temperature-thickness profiles. The temperatures of the cold mass (maintained at 77.8 K), the sleeve [cold boundary temperature (CBT)], the insulation outer surface [warm boundary temperature (WBT)], and the vacuum can were measured. The cold vacuum pressure (CVP) was adjusted for the desired vacuum level. Test articles were, in most cases, heated and pumped to below 10^{-4} torr to begin the

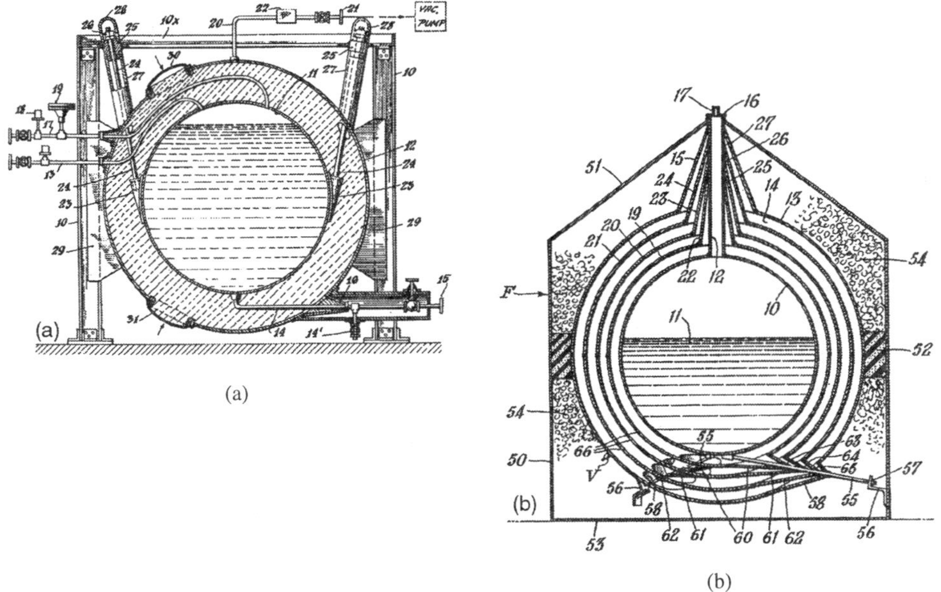

Figure 1. Early insulation system designs for cryogenic tanks: (a) double-walled tank with evacuated powder[5] and (b) vacuum-insulated container with multiple radiation shields[6].

test series. The residual gas was nitrogen for all tests. When the vacuum level, all temperatures, and the boiloff flow are stable, the k-value is determined from Fourier's law of heat conduction for a cylindrical wall. All tests were run with the same copper sleeve outfitted with six surface temperature sensors. The CBT was constant at around 80 K. The outer heat shroud was maintained at approximately 313 K, which gave WBT ranging from 290 K (at high vacuum levels) down to about 190 K (for near ambient pressure).

INSULATION MATERIALS

Conventional commercially available materials were tested for reference and calibration purposes. Novel materials and unique combinations of conventional materials were also tested. The test articles included combinations of aluminum foil, fiberglass paper, polyester fabric, silica aerogel composite blanket, fumed silica, silica aerogel powder, and syntactic foam. Characteristics of the syntactic foam test article are given by Tupper.[13] Further description of the aerogel materials is given by Fesmire, Rouanet, and Ryu.[14] Standard industrial MLI composed of a reflective shield (aluminum foil 0.00724 mm thick) and spacer (fiberglass paper 0.061 mm thick) at a density of 1.8 layers per mm served as the benchmark for comparison with other material systems. The installed thickness for most test articles was around 25 mm. This paper focuses on a layered composite insulation system that includes aluminum foil for reflective shields and novel materials for spacers.

TEST RESULTS

A total of 142 tests of 17 different insulation systems was performed. Table 1 reports the key measurements and conditions for select test articles. The key data points are those at high vacuum (10^{-4} torr) and soft vacuum (0.1, 1, and 10 torr). The boundary temperatures given in the table must be considered for accurate comparison with the measurements reported by others. Figure 2 gives a summary of the variation of apparent thermal conductivity with CVP for four groups of material systems: (a) MLI systems (and a syntactic foam system for reference), (b) layered composites with paper, (c) layered composites with fabric, and (d) aerogel composite blankets. The benchmark MLI system (C108) is in good agreement with the experimental data for similar systems as reported by Kaganer, Hnilicka, and Black.[2,9,15] The Kaganer line in Figure 1(a) is for the following system: aluminum foil and fiberglass spacer, 1.5 layers per mm, 293 and 90 K boundary temperatures, and air as residual gas. Summaries of the temperature profiles of representative materials, which can be used to estimate thermal conductivity as a function of temperature, are presented in Figure 3. Overall views of the apparent thermal conductivity measurements as a function of CVP are shown in Figure 4.

Table 1. Summary of measurements and conditions for select test articles

Test No.	Description of Insulation System	Installed Density (kg/m^3)	Total Thickness (mm)	CVP (torr)	k-value (mW/m-K)	CBT (K)	WBT (K)
C102	Aerogel composite blanket (silica aerogel / fiber matrix), 21 mm, plus 10 layers MLI	81	30.2	0.0001	0.19	91	286
				0.001	0.34	91	285
				0.1	2.9	86	289
C104	Syntactic foam composite	107	39.9	0.0003	7.10	80	127
				0.150	13.2	80	192
				1	12.8	80	192
				10	13.3	80	184
C105	Aerogel composite blanket with carbon dispersion, 20 mm, plus 10 layers MLI	79	27.9	0.0002	0.23	91	280
				0.1	3.06	86	260
				1	4.54	86	248
				10	6.63	86	226
C107	Layered composite insulation with fiberglass paper and fumed silica dispersion	52	24.8	0.0001	0.09	91	281
				0.1	1.23	88	279
				1	2.40	90	269
				10	6.07	91	225
C108	MLI (aluminum foil and fiberglass paper spacer), 40 layers at 1.8 layers/mm	58	22.3	0.0005	0.08	87	281
				0.1	2.68	90	256
				1	9.01	92	196
				10	13.6	94	181
C109	Layered composite insulation with polyester fabric and fumed silica dispersion	59	28.7	0.0001	0.24	95	269
				0.1	1.91	87	267
				1	4.81	87	242
				10	9.66	88	215
C110	Aerogel composite blanket (silica aerogel / fiber matrix), 2 layers, 16 mm each layer	125	32.0	0.0001	0.55	86	214
				0.1	1.16	86	278
				1	3.31	85	265
				10	5.80	85	246
C113	Layered composite insulation with fiberglass paper and fumed silica dispersion	51	20.9	0.0003	0.12	84	284
				0.1	1.34	80	273
				1	2.93	81	244
				10	7.71	83	211
C114	Layered composite insulation with fiberglass paper and fumed silica dispersion	53	24.2	0.0001	0.15	88	278
				0.1	1.11	80	286
				1	2.66	81	265
				10	6.82	83	232
C115	Layered composite insulation with polyester fabric and fumed silica dispersion	64	25.1	0.0001	0.28	93	271
				0.1	1.51	86	276
				1	3.85	84	263
				10	8.78	88	219
C116	Layered insulation with polyester fabric	51	18.7	0.0001	0.15	93	284
				0.1	4.98	89	234
				1	8.49	89	200

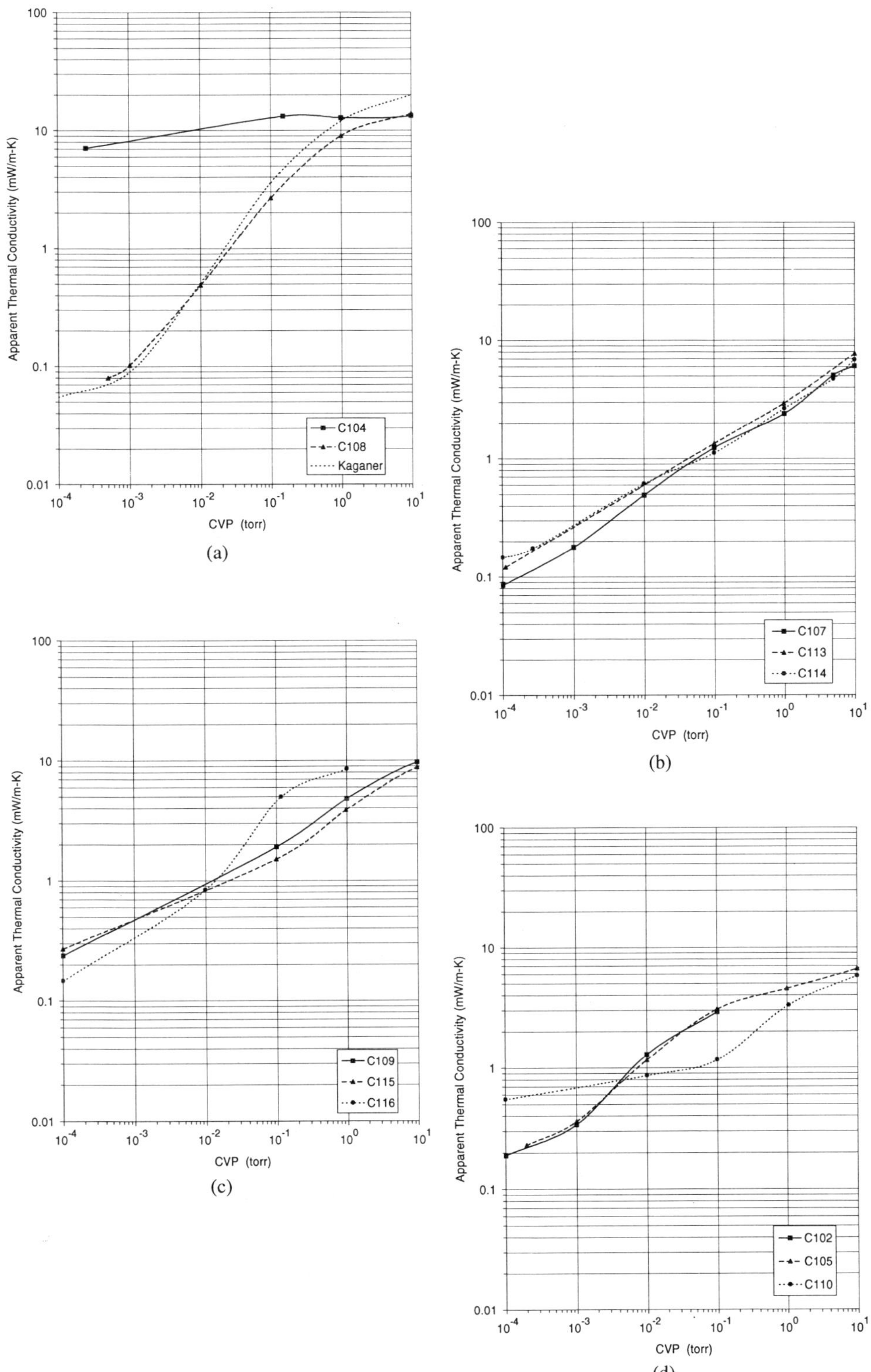

Figure 2. Variation of apparent thermal conductivity with CVP: (a) MLI and foam, (b) layered composite insulation with paper, (c) layered composite insulation with fabric, and (d) aerogel composite blankets.

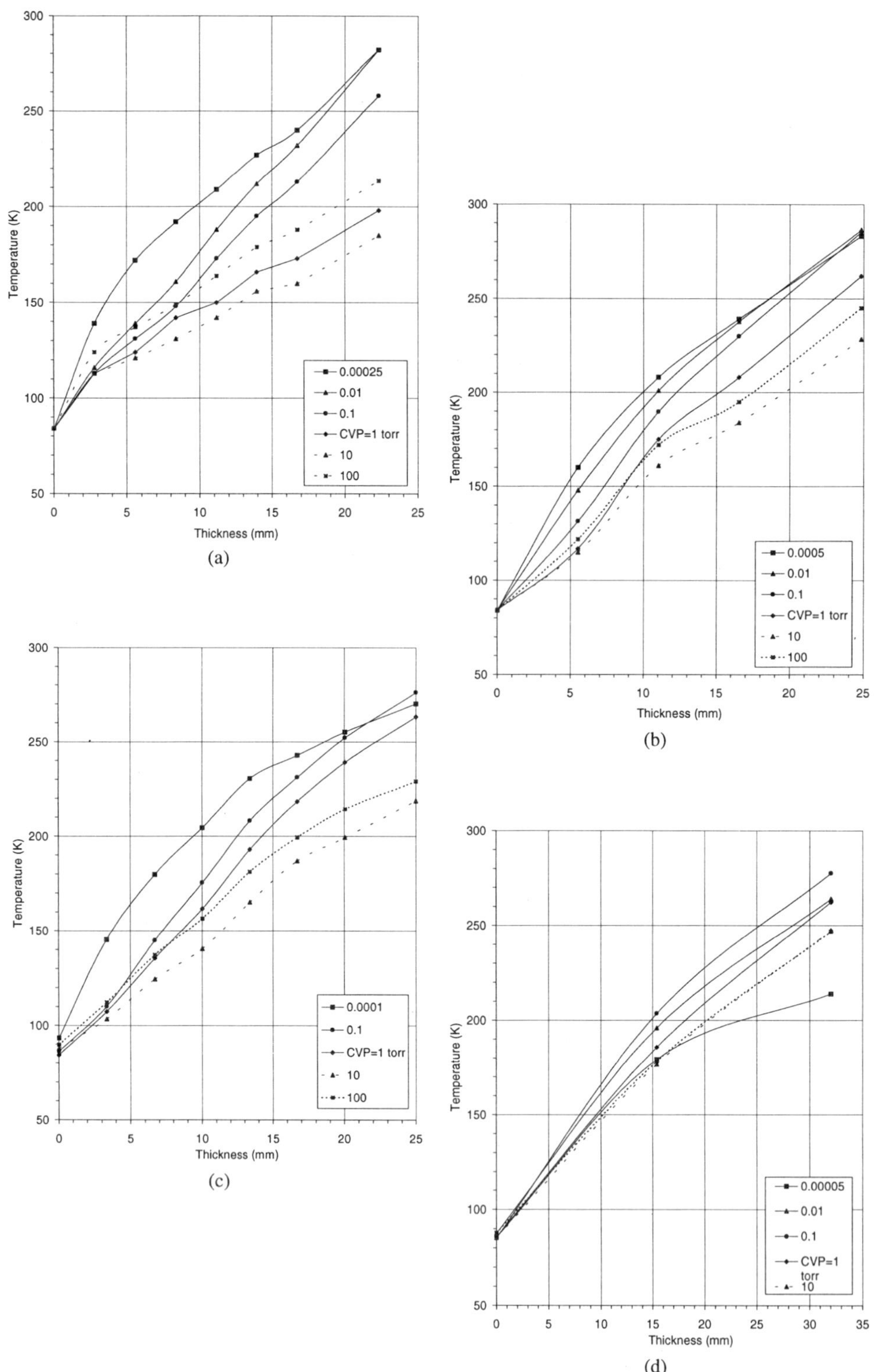

Figure 3. Temperature profiles through insulation at different vacuum levels: (a) MLI, (b) layered composite insulation with paper, (c) layered composite insulation with fabric, and (d) aerogel composite blankets.

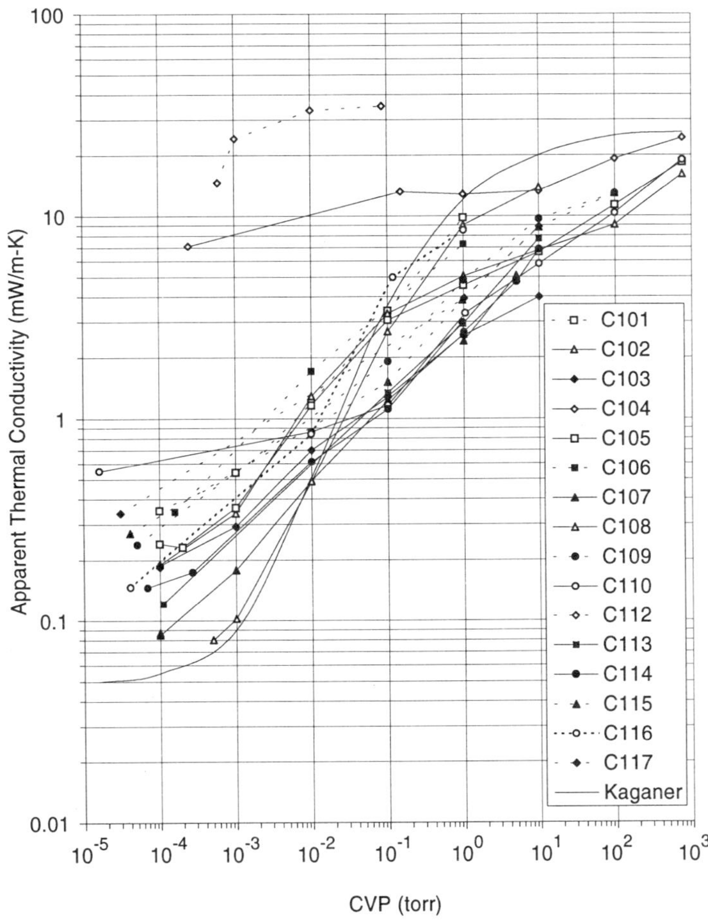

Figure 4. Overall view of liquid nitrogen boiloff measurements: apparent thermal conductivity as a function of CVP.

SYSTEM DESIGN CONSIDERATIONS

The appropriate choice of a thermal insulation system depends on matching the performance level with the overall cost. That is, the performance must justify the cost. Consideration of the actual operating conditions and an estimation of the total heat leak of the mechanical system is needed to determine the insulation requirements. The amount of heat flow through the insulation relative to the amount of heat flow from all the other sources must be understood in order to select an appropriate insulation system. The main factors to consider are: (1) operating conditions of the system, (2) total heat leak of the mechanical system, (3) material properties such as density and compatibility, and (4) method of testing and evaluation. Attention should also be given to offering advantages in easier fabrication, maintenance, and modification where possible. For the cost of a bulk storage container (with standard MLI), more than 25 percent is attributed to the insulation.[16] The materials are only a small fraction of this cost; but the heating, vacuum pumping, testing, material handling, and other steps necessary for manufacturing a high-vacuum vessel are costly. In summary, the overall effectiveness of the insulation system design depends on: (1) thermal performance, (2) versatility and durability, (3) ease of use in manufacturing and installation, and (4) costs of operations and maintenance.

CONCLUSIONS

A number of the presented systems have k-values better than the target 4.8 mW/m-K (R-30) at a 1-torr vacuum level and boundary temperatures of approximately 280 and 80 K. For example, the layered composite C107 gave superior performance of 2.4 mW/m-K (R-60) at 1 torr which is about four times better than the benchmark MLI C108. The layered composite insulation systems are targeted for low-cost, intermediate performance uses but were found to offer advantages in high-vacuum superinsulation applications as well. Thermal performance of the layered composite with paper was comparable to the benchmark MLI at high vacuum (0.09 versus 0.08 mW/m-K). The actual performance of the more robust composite could exceed that of highly evacuated MLI systems when the factors of edge effects and compression are considered. The "vacuum burden" of fabricating 1-torr systems versus 0.0001-torr systems is accordingly reduced. The new layered composite is being developed into a family of thermal insulation systems, targeting both soft vacuum and high vacuum cryogenic applications. Work on material optimization and application design is currently being performed at the Cryogenics Test Laboratory at NASA Kennedy Space Center.

REFERENCES

1. S.D. Augustynowicz, J.E. Fesmire, and J.P. Wikstrom, Cryogenic insulation systems, 20th International Refrigeration Congress, Sydney, 1999.
2. M.G. Kaganer, "Thermal Insulation in Cryogenic Engineering," Israel Program for Scientific Translations, Jerusalem, 1969.
3. W. Stanley, Heat-insulated receptacle, US Patent No. 1,071,817, 1912.
4. L.I. Dana, Insulated container for liquefied gases and the like, US Patent No. 2,396,459, 1939.
5. W.D. Cornell, Radiation shield supports in vacuum insulated containers, US Patent No. 2,643,022, 1947.
6. I.A. Black, et al, Development of high-efficiency insulation, *Advances in Cryogenic Engineering*, Vol. 5, Plenum Press, New York, 1960, pp. 181-188.
7. L.C. Matsch, Thermal insulation, US Patent No. 3,007,596, 1956.
8. R.H. Kropschot, et al, Multiple-layer insulation, *Advances in Cryogenic Engineering*, Vol. 5, Plenum Press, New York, 1960, pp. 189-197.
9. M.P. Hnilicka, Engineering aspects of heat transfer in multilayer reflective insulation and performance of NRC insulation, *Advances in Cryogenic Engineering*, Vol. 5, Plenum Press, New York, 1960, pp. 199-208.
10. American Society for Testing Materials, ASTM C740, Standard practice for evacuated reflective insulation in cryogenic service, 1996.
11. T. Nast, Multilayer insulation systems, *In:* Weisend, J., *Handbook of Cryogenic Engineering,* Taylor and Francis, Philadelphia, 1998, pp. 195-196.
12. J.E. Fesmire and S.D. Augustynowicz, Insulation testing using cryostat apparatus with sleeve, Cryogenic Engineering Conference, Montreal, 1999.
13. M. Tupper, Composite Technology Development, Inc., Insulation, adhesives, and coatings specially formulated for space cryogenic applications, Space Cryogenics Workshop, University of Oregon, 1997.
14. J. Fesmire, S. Rouanet, and J. Ryu, Aerogel-based cryogenic superinsulation, *Advances in Cryogenic Engineering*, Vol. 44, Plenum Press, New York, 1998, pp. 219-226.
15. I.A. Black and P.E. Glaser, Progress report on the development of high-efficiency insulation, *Advances in Cryogenic Engineering*, Vol. 6, Plenum Press, New York, 1960, pp. 32-41.
16. J. Wikstrom, personal communication, MVE, Inc., Burnsville, MN.

DESIGN AND FABRICATION OF A LARGE CRYOSTAT FOR TESTING SMES APPARATUS

R. D. Luther

Meyer Tool & Manufacturing, Inc.
Oak Lawn, IL 60453

ABSTRACT

The design and fabrication of a large liquid helium cryostat for testing of SMES components is described. The cryostat was built in 1996 and 1997 by Meyer Tool & Mfg., Inc. for the Westinghouse Science and Technology Center and the US Navy, for installation at the National High Magnetic Field Laboratory in Tallahassee, FL. The cryostat is a system of stainless steel pressure/vacuum vessels designed for containment of liquid helium at atmospheric pressure and 4.2 K. The outer containment is an eleven-foot (3.35 m) diameter vacuum vessel with a dished bottom head and a full-body flange at the top. The other components are supported from the flat annular top head of the vacuum vessel. These include an annular background magnet case, a seventy-nine inch (2.0 m) diameter test vessel, and a thirty-six inch (9.14m) diameter lead vessel. The overall height of the system is about twelve feet (3.66 m). Large lateral magnetic interaction loads are transferred within the cryostat by three 1.75-inch (44.45 mm) diameter Nitronic 50 rods and a system of brackets and gussets on the vessel heads. Heat leaks are controlled by multilayer vacuum insulation and a system of nitrogen-cooled aluminum shields. Design details will be described, as will the rationale behind various design decisions.

INTRODUCTION

This paper describes various aspects of the design and fabrication of a large liquid helium cryostat to be used for testing of components and apparatus for superconducting magnet energy storage (SMES) systems. The cryostat was built in 1996 and 1997 by Meyer Tool & Mfg., Inc. (MTM) under contract to the Westinghouse Science and Technology Center. The cryostat was to be installed at the National High Magnetic Field Laboratory in Tallahassee, Florida for use by the US Navy Weapons Research Laboratory. The cryostat was designed to provide structural support for the massive internal components under magnetic and gravity loads, and to efficiently contain the cryogenic fluids required to cool

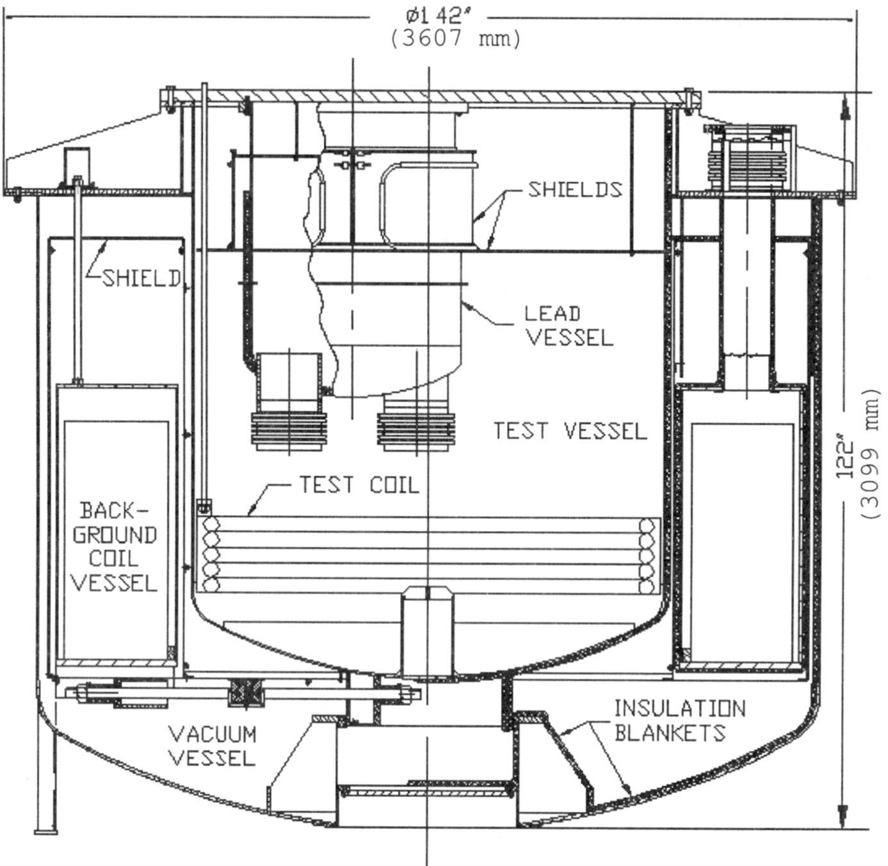

Figure 1. Cross-section of the SMES Cryostat system

the components to operating temperatures. The primary cryostat components are described and the rationale behind certain design decisions are delineated.

COMPONENT DESCRIPTIONS

The cryostat is a system of Type 304 stainless steel pressure and vacuum vessels designed for containment of liquid helium at 4.2 K. (See Figure 1.) Supercooled magnetic components are contained by three separate pressure vessels, all of which are enclosed by a vacuum-tight insulating outer vessel. The vacuum insulation system includes multilayer superinsulation blankets and a series of liquid nitrogen-cooled aluminum shields. The primary components of the cryostat are described below.

Vacuum Vessel

The outer containment, denoted as the vacuum vessel, is a vertical tank, approximately eleven feet (3.36 m) in diameter by nine feet (2.74 m) tall, with a full-body flange at the top and an ASME flanged and dished (F & D) head at the bottom. The vacuum vessel cover is a stiffened annular plate assembly to which the inner helium

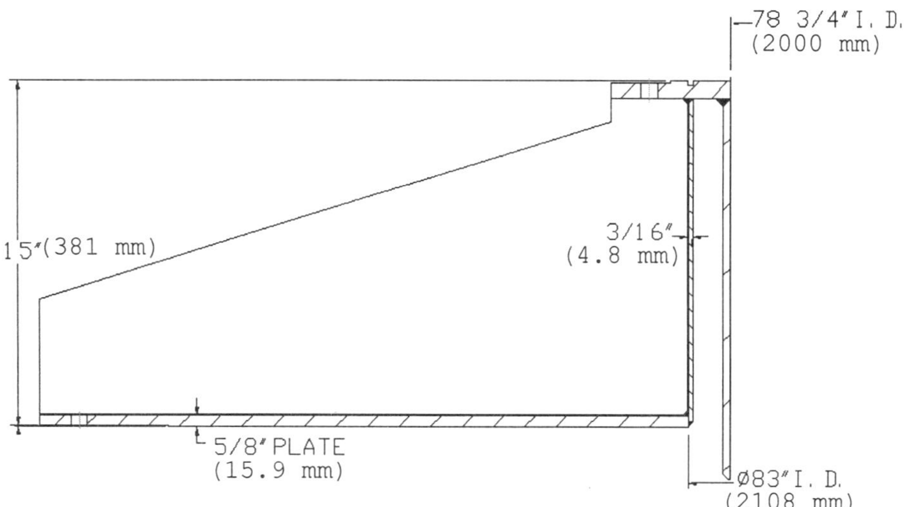

Figure 2. Cross section of the vacuum vessel cover plate.

container, called the Test Vessel, is welded, and from which the Background Magnet is hung. (See Figure 2.) This cover is the most important structural component of the cryostat. It must transfer both gravity loads and magnetic interaction loads, and at the same time must be stiff enough to maintain vacuum-tight o-ring seals at its inner and outer diameters and at several nozzle locations. Structurally the cover is a circular ring loaded by distributed moments and forces acting perpendicular to the plane of the ring.

In the center of the bottom head of the vacuum vessel, a re-entrant manway with an aluminum blind provides access to the bottom of the test vessel and its structural attachments. Temporary support legs were added to the vacuum vessel to support the cryostat during fabrication and shipping. In service the cryostat will be supported at the vacuum vessel top flange by components designed and supplied by others.

Test Vessel

The largest of the pressure vessels is called the test vessel because it contains the superconducting Test Coil magnet. The ASME Code-stamped vessel is a 78 inch (3.07 m) diameter vertical cylinder with a bolted top cover plate and an F & D bottom head. The full-body flange at the top of the test vessel forms the compression ring of the vacuum vessel cover plate described above. Large gusset plates on the vacuum vessel cover are welded to the outer surface of the test vessel and to the bottom surface of the test vessel flange. Consequently, the sealing surfaces of the flange were machined after welding.

The test vessel cover is a two inch thick, ninety inch (2.29 m) diameter flat plate containing openings for the two vapor-cooled leads for the Test Coil Magnet as well as a number of nozzles for utilities and services. The leads are relatively delicate electrical/cryogenic components that are subject to substantial loads due to magnetic interactions. The necessary structural support and electrical and thermal insulation is provided by a 1.5 inch (38 mm) thick G-10 plate suspended from the cover between two large brackets welded to the underside of the plate. The Test Coil Magnet is suspended from the cover plate by three rods that penetrate the cover plate to allow for positional adjustment from above. The lead vessel, described in detail below, is bolted directly to the bottom of the cover plate and is sealed by an elastomeric o-ring whose seal surface is machined into the bottom face of the cover plate.

Large structural attachments are welded to both the inside and outside surfaces of the test vessel bottom head. These are required to transfer the lateral interaction loads and to allow alignment of the test vessel and background magnet vessels after assembly. Inside the head is stiffened by large gussets extending from a short, centrally-located vertical pipe almost to the knuckle area of the head. Lateral loads from the Test Coil are transferred to the pipe by a wagon wheel or spider assembly attached to the coil (by others). A heavy circular ring is welded to the outside of the bottom head along with three heavy gusseted boxes spaced around the ID of the ring. The boxes are attachment points for three 1.75 inch (44.45 mm) diameter rods which are also attached to the bottom of the Background Coil vessel and serve to transfer lateral interaction loads between the two vessels. The material of the rods is Nitronic 50, a high-strength stainless steel. The circular ring is also machined to accept three G-10 bumpers which transfer lateral loads between the test vessel and the vacuum vessel. The bumpers were machined after the initial assembly to ensure a close fit, and will also allow some realignment of the vessels if necessary at a later date.

Lead Vessel

The Lead Vessel is a thirty-four inch (864 mm) diameter by thirty-nine inch (991 mm) long cylinder with an F & D bottom head and a compact full-body flange at the top. The vessel must be installed from below after the leads are installed in the test vessel cover plate and must be sealed at the bottom surface of the cover plate with a single buna-n o-ring and at the bottom of the leads with indium wire seals. Two nozzles in the bottom head are fitted with stainless steel formed bellows to provide compatible interfaces with the bottom of the leads. In some modes of operation, the lead vessel must be thermally insulated from the test vessel. To facilitate this, the bottom head and most of the vessel shell are covered by an evacuated jacket containing fifty layers of superinsulation. The lead vessel was designed, fabricated, tested, and stamped in accordance with the rules of the ASME Code.

Background Coil Vessel

The Background Magnet is housed in its own Code-stamped liquid helium containment pressure vessel, called the background coil vessel. The vessel is a torus with a rectangular cross section. The annular bottom plate of the vessel was machined flat to serve as the base of the magnet coil. The side walls and top plate of the vessel were fabricated and welded complete and this assembly was seal welded to the bottom plate in the MTM shop and completely leak tested. The seal welds were then removed and the vessel was shipped to Westinghouse for assembly of the coil on the bottom plate. Once the coil was complete and tested, MTM personnel re-installed the top part of the vessel, completed the welding and helium leak-checking, and then pressure-tested and Code-stamped the vessel in the Westinghouse shop. Upon assembly in Florida, the background coil vessel will be wrapped with superinsulation and suspended within the vacuum vessel annular space by three struts connected to the vacuum vessel cover plate. Electrical, cryogenic, and instrumentation connections to the magnet are made through three service stacks that exit the cover plate through formed bellows assemblies. The background coil vessel will also be connected to the bottom of the test vessel by three Nitronic 50 rods as described above.

Thermal Shields

Heat entering the cryostat is intercepted by two actively-cooled (LN2 at ~80 K) thermal shield assemblies, one in the vacuum vessel and one in the test vessel. Each of the shields is an assembly of aluminum sheeting with aluminum angles and bars used as

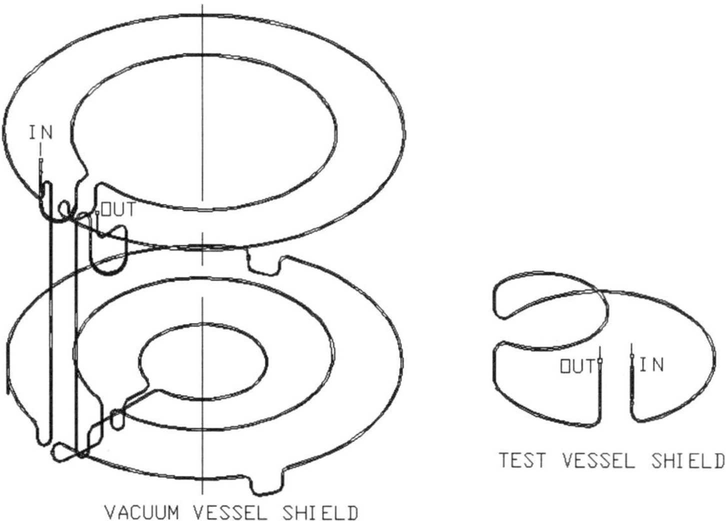

Figure 3. Thermal shield flow diagrams.

corners and stiffeners where necessary. Each also includes an independent series circuit of aluminum trace tubing routed throughout the shield assembly. (See Figure 3.) The trace tubing is extruded "D" tubing welded to the sheeting. Connections are made with lengths of circular tubing fitted to the ID of the D tube and socket welded. The inlet and outlet for each circuit are 1/2 inch (13 mm) CVI bayonets mounted on the respective vessel cover plate. Differential thermal expansion is accommodated by lengths of flexible stainless steel hose and by loops of aluminum tubing. The shields are suspended from the vessel cover plates by threaded rods with nuts and G-10 washers.

The vacuum vessel shield completely surrounds the background coil vessel and also shields the cylinder and bottom head of the test vessel from radiation. The shield system comprises an outer cylinder, an inner cylinder, a top plate spanning between the two cylinders, and a bottom plate that stretches from the outer cylinder, beyond the inner cylinder toward the center to shield the bottom of the test vessel. Corners are formed by angles welded to the sheets, and vertical seams between cylindrical sheet sections are intermittently welded to provide electrical breaks in the cylinders. In the MTM shop the bottom plate and the bottom outer angle ring were attached to the cylinders by rivets to facilitate the removal of the bottom shield plate, which is required in order to install the Background Coil Vessel in the field. All tubing joints were welded in the shop to ensure vacuum integrity, and aluminum couplings were provided to facilitate reassembly of the circuit in the field after installation of the magnet.

The Nitronic 50 lateral support rods are located outside the shield assembly and will be wrapped with insulation. However, they are attached at both ends (through G-10 bushings) to brackets that are welded to cold helium vessels. Because of their large diameter (1.75 inch, 44.45 mm), the rods are potential conduits for substantial heat leak into both the test vessel and the background coil vessel. To preclude this, a copper heat station will be clamped to each rod with the other end riveted to the bottom plate of the vacuum vessel shield assembly after assembly and tensioning of the rods.

The test vessel shield system protects the test vessel shell and its helium pool, and the lead vessel and its helium pool. The system consists of a flat plate suspended in the test vessel at the same elevation as the upper liquid circuit tube located outside the test vessel. A short vertical cylinder extends upward from the flat shield plate to surround the non-jacketed portion of the lead vessel. At the top of this cylinder a two-piece aluminum collar is inserted between the shield cylinder and the lead vessel and clamped to form a heat station to protect the lead vessel o-ring seal from cold temperatures. A passive shield plate is also located within the lead vessel at the elevation of the collar clamp. To facilitate attachment of the lead vessel to the test vessel cover plate, a portion of the flat shield plate can be removed to improve access to the vessel attachment studs.

HEAT LEAKS

Heat leaks were estimated for the 4 K and 80 K systems for three operating modes – Standby, Cable in Conduit, and Bath. Standby is the non-test mode in which the Backgound Coil is maintained at 4 K while the test vessel is open to atmosphere (300 K).

Table 1. Heat leaks to the 4 K and 80 K systems in Watts

Mode	Standby		Cable in conduit		Bath	
System	**4 K**	**80 K**	**4 K**	**80 K**	**4 K**	**80 K**
Source						
Lead Vessel						
Conduction						
Vessel wall to top flange	0.00	0.00	42.30	0.00	42.30	0.00
Internal radiation shield supports	0.00	0.00	0.00	2.00	0.00	2.00
Radiation						
Internal radiation shield	0.00	0.00	0.43	0.00	0.43	0.00
Radiation to test vessel	0.00	0.00	3.70	0.00	3.70	0.00
Test Vessel						
Conduction						
Vessel wall to top flange	0.00	0.00	0.00	81.33	92.40	0.00
Vessel G10 bumper supports	0.00	0.00	0.00	9.10	1.20	0.00
Lateral support rods	0.00	7.20	0.00	0.00	14.70	0.00
Internal shield support rods	0.00	0.00	0.00	0.20	0.00	0.20
External shield support rods	0.00	0.30	0.00	0.30	0.00	0.30
External shield lateral G-10 supports	0.00	8.75	0.00	0.00	1.20	0.00
External shield bayonets	0.00	5.00	0.00	5.00	0.00	5.00
Internal shield bayonets	0.00	0.00	0.00	5.00	0.00	5.00
Radiation						
Internal shield	0.00	0.00	0.00	2.20	0.75	2.20
External shield	0.00	38.40	0.00	27.90	4.61	27.90
Background Vessel						
Conduction						
Support rods to top flange	9.10	0.00	9.10	0.00	9.10	0.00
Lateral support rods	0.94	0.00	0.94	0.00	0.94	0.00
Service piping to top flange (3)	11.50	0.00	11.50	0.00	11.50	0.00
Radiation						
External shields	10.60	0.00	10.60	0.00	10.60	0.00
Totals	**32.14**	**59.65**	**78.57**	**133.03**	**193.43**	**42.60**

Cable in Conduit and Bath modes refer to the type of cooling of the superconducting Test Coil magnet during a test. In the Cable in Conduit mode the liquid helium flows through the coil itself; the test vessel is evacuated, and the lead vessel and background coil vessel both contain liquid helium at 4 K. In the Bath Mode, all three vessels contain liquid helium at 4 K. Heat leaks for various components are given for all three modes in Table 1.

CONCLUSION

The design and fabrication of a large cryostat for testing SMES apparatus in either pool boiling or cable-in-conduit configuration has been described. This unit was built by Meyer Tool & Mfg., Inc. in Oak Lawn, IL, and is currently being integrated with magnets, cryogenics, and other systems at the National High Magnetic Field Laboratory in Tallahassee, FL. Details regarding the performance of the cryostat, the test apparatus, and the tested components were unavailable to the author at the time of publication.

USE OF HIGH-TEMPERATURE SUPERCONDUCTING FILMS IN SUPERCONDUCTING BEARINGS

A. Cansiz,[1] J. R. Hull,[1] T. M. Mulcahy,[1] and B. Lehndorff[2]

[1]Energy Technology Division, Argonne National Laboratory
Argonne, IL 60439

[2]Dept. of Physics, University of Wuppertal
D-42097 Wuppertal, Germany

ABSTRACT

We have investigated the effect of high-temperature superconductor (HTS) films deposited on substrates that are placed above bulk HTSs in an attempt to reduce rotational drag in superconducting bearings composed of a permanent magnet levitated above the film/bulk HTS combination. According to the critical state model, hysteresis energy loss is inversely proportional to critical current density J_c, and because HTS films typically have much higher J_c values than those of bulk HTSs, the film/bulk combination was expected to reduce rotational losses by at least one order of magnitude in the coefficient of friction, which in turn is a measure of the hysteresis losses. We measured rotational losses of a superconducting bearing in a vacuum chamber and compared the losses with and without a film present. The experimental results showed that contrary to expectation, the rotational losses are increased by the film. These results are discussed in terms of flux drag through the film, as well as of the critical state model.

INTRODUCTION

The use of high-temperature superconductors (HTSs) in levitation applications such as magnetic bearings has seen considerable development during the past decade.[1-3] The most common configuration for a HTS bearing incorporates a rotatable permanent magnet (PM) stably levitated in close proximity to a stationary bulk HTS. The bulk HTS is usually melt-textured Y-Ba-Cu-O in the form of single domains of up to ≈10 cm in linear dimension and with a critical current density $J_c > 10^4$ A/cm^2. The domains are typically oriented with the c-axis perpendicular to the HTS surface that faces the PM. These configurations are capable of providing levitation pressures > 20 N/cm^2.

More recently, interest in the use of thin-film HTSs in superconducting bearings has increased.[4-8] Although YBCO thin films often have J_c values of >1 MA/cm^2 at 77 K, the thickness of these films is only ≈1 μm and they do not provide much levitation force. The magnetization quickly saturates, and very large hysteresis loops can be produced if this occurs. However, when unsaturated, the stiffness of such films is often similar to that produced by bulk material of greater thickness.[5,7]

The use of high-J_c thin-films in HTS bearings has long been thought to offer the potential for reduced rotational drag.[9,10] This conjecture is based on the generally accepted hysteretic loss mechanism for HTS bearings that is based on the critical-state model. Hysteretic loss is produced whenever there is a cyclic change in applied magnetic field, and the energy loss per cycle is[11]

$$E_h = K\mu_o(\Delta H)^3/J_c \qquad (1)$$

where E_h is the hysteretic energy loss per unit area per cycle, K is a geometric coefficient of order unity, $\mu_o = 4\pi \times 10^{-7}$ N/A^2 is the magnetic permeability of vacuum, $\Delta H = \Delta B/\mu_o$ is the peak-to-peak amplitude of the varying magnetic field, and J_c is the critical current density in the HTS. When a levitated PM spins over a HTS, rotational loss occurs because of azimuthal inhomogeneities of the magnetic field of the PM, which causes a ΔB at a surface location on the HTS over a complete rotation of the PM. The amplitude of the ΔB increases if there is any appreciable whirl amplitude of the PM, in which case the radial gradient of the magnetic field also contributes to the ΔB. One may conceptualize the magnetics of the bearing system as the PM providing a constant magnetic field B that interacts with the HTS to provide levitation force and a circumferentially varying magnetic field ΔB that produces the rotational loss. In general, the ΔB associated with inhomogeneity is much less than the constant magnetic fields B that provide the levitation force.

A further consequence of the critical-state model, beyond the energy loss described in Eq. (1), is that the interior of the HTS is shielded from the varying magnetic field, so that the hysteresis losses occur only on the surface of the HTS. Thus, it was conjectured that if a thin-film HTS were interposed between the PM and bulk HTS, the levitation force would remain approximately the same and most of the hysteretic loss would occur in the thin-film HTS. The thin-film HTS would shield the bulk HTS from the effects of the applied ΔB, so that the bulk HTS experiences only a constant applied magnetic field. Because J_c is so much higher in the thin film, the total hysteretic loss (and therefore the rotational drag on the levitated PM) should be much lower with the HTS film installed.

The above considerations motivated us to create and test the experimental arrangement, shown schematically in Fig. 1, to measure the rotational drag of a PM levitated over a combination of bulk HTS with a thin-film HTS covering the bulk. This article reports the detailed arrangement and results of these tests. A theoretical interpretation of the results is also provided.

ROTATIONAL DRAG MEASUREMENTS

The experimental apparatus for rotational drag measurements is shown in Fig. 1. A PM disk rotor was levitated over a HTS in a vacuum chamber, with an oil-diffusion pump reducing the pressure to <100 µPa.[12] The HTS was inside a room-pressure cryochamber, through which liquid nitrogen flowed from a gravity feed at ≈ 3 kPa. The cryochamber walls and top were electrically insulating fiberglass to avoid eddy-current losses while the PM was rotating. The cryochamber was covered on the top and sides with three thin layers of aluminized Mylar to reduce radiation heat input. Before cooling the HTS, the PM was held by a thin flat plate that was movable in the z (vertical) direction and rotatable around its support post.

Following HTS cooldown and positioning of the PM rotor, the support plate was moved away from the rotor to prevent interference during the spin tests. The PM rotation

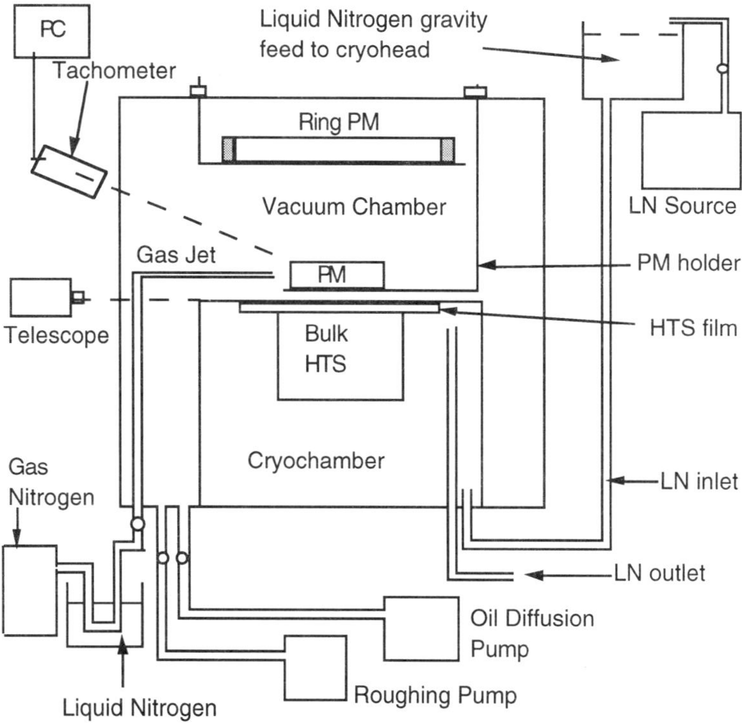

Figure 1. Schematic diagram of apparatus for rotational drag measurements.

was accelerated and decelerated by a cold gaseous nitrogen jet impinging approximately tangentially on the PM perimeter from a small-diameter copper tube. The nitrogen gas, supplied from a pressurized bottle through a regulator, was cooled by passing it through a heat exchanger submerged in liquid nitrogen before it entered the vacuum chamber; this averted a possible temperature rise in the HTS that would reduce the J_c and possibly decrease the levitation height. Rotation of the PM was continuously monitored by a tachometer. Once the PM reached the desired speed, the gas jet was shut off. To obtain good statistics, the PM was allowed to freely spin down for 3-5 min after acceleration, and speeds were recorded at 5-s intervals. When we were interested in details of resonance behavior, we recorded the data every second. Linear regression was used on the data from the free-spin period to calculate the slope of the rotational frequency versus time function (df/dt), which in turn was used to calculate the coefficient of friction.[12]

The height of the levitated PM was measured with a traveling telescope at intervals throughout the experiment, and no change in height from the initial levitation value was observed to within 10 μm. The levitation height in these experiments includes the thickness of the cryochamber (3.5 mm) and the gap between the PM and the top of the cryochamber. The height was recorded after the system had cooled, so as to avoid obfuscation by any changes due to thermal contraction.

Three configurations of HTS were used: (a) bulk HTS, (b) thin-film HTS, and (c) thin-film HTS over bulk HTS.

The bulk HTS was a melt-textured Y-Ba-Cu-O cylinder with its c-axis aligned along the vertical, J_c = 20 kA/cm^2, critical temperature T_c = 92 K, diameter of 32 mm, and thickness of 22 mm. We chose a thin-film HTS with a diameter significantly larger than that of the bulk HTS, hoping that the film would significantly shield the bulk. The disk-shaped thin-film Y-Ba-Cu-O HTS had J_c = 3.7-4.1 MA/cm^2, critical temperature

T_c = 89.2-89.6 K, diameter of 51 mm, and thickness of 350 nm. The film was deposited on a La-Al-O_3 substrate with a thickness of 0.5 mm. When the film was used together with the bulk, they were coaxial and the film was immediately above the bulk.

Degradation of the thin-film HTS by contact with moisture was a concern. The thin film and substrate were therefore encased in a thin, sealed polyethylene bag. In addition, before cooling with liquid nitrogen, the cryochamber was evacuated with a mechanical pump and then backfilled with gaseous nitrogen at ambient temperature. We also flowed liquid nitrogen through the inlet line for some time before directing it to the cryochamber, so that most of the trapped air in this line would be removed.

The same PM rotor was used in all the tests. The PM rotor is an axially polarized NdFeB disk with magnetization of $\mu_0 M \approx 1.1$ T, diameter of 25.4 mm, height of 6.35 mm, and mass of 35.59 gm. The levitation force provided by the thin-film HTS was too low to levitate this PM. The levitational force was augmented by placing a stationary PM above the rotor in an Evershed configuration to produce an attractive force between the two PMs.[13] The distance between the two PMs was adjusted to that the upward force due to magnetic attraction was slightly lower than the gravitational force downward. Because the additional PM in the Evershed configuration does not move, it does not contribute to hysteresis loss in the thin-film HTS. The velocities of the experiment are low enough that loss due to eddy currents induced by relative velocity of the PMs will be negligible compared to the hysteresis loss.

RESULTS

Rotational loss of an HTS bearing is evidenced by the decay rate of the rotational frequency f and is characterized by the coefficient of friction (COF),

$$\mathrm{COF} = -[2\pi R_\gamma^2/(g R_D)]\ df/dt \tag{2}$$

where R_γ is the radius of gyration of the rotor, g is the acceleration of gravity, t is time, and R_D is the weighted mean radius at which the drag force acts and is given by[14]

$$R_D = \Sigma\, r_i\, (\Delta B_i)^3 \,/\, \Sigma\, (\Delta B_i)^3 \tag{3}$$

where r_i is the radius for the ith term and ΔB_i is the peak-to-peak value of the variation of the vertical component of magnetic field measured in the circumferential path at r_i at the surface of the HTS. For the PM used in our experiments, R_D = 10.78 mm.

COF versus f at different heights for the three HTS configurations, bulk only, film only, and bulk plus film, are shown in Figs. 2-4, respectively. In all cases, COF increases with decreasing levitation height. Each data set is characterized by a peak in COF that corresponds to a resonance in the radial vibration amplitude that is readily observed visually. The behavior of the rotational losses is thus divided into three regions: below the resonance, resonance region, and above the resonance. For rotation rates away from the resonance, the loss is mostly independent of frequency, which is consistent with a hysteretic energy loss behavior. Below the resonance, rotational loss is caused only by circumferential inhomogeneity in the magnetic field of the PM. The COF is higher above the resonance than below the resonance due to rotation about the PM center of mass, rather than its center of magnetism. Here, there is an additional angular variation in magnetic field seen by the HTS, due to the radial gradient of the PM magnetic field combined with the whirl amplitude about the center of mass.

Figures 2-4 also show that the resonant frequency shifts to higher values as the levitation height decreases. This is expected, because more magnetic flux lines are trapped in the HTS at lower field-cooling heights, and thus the magnetomechanical stiffness is

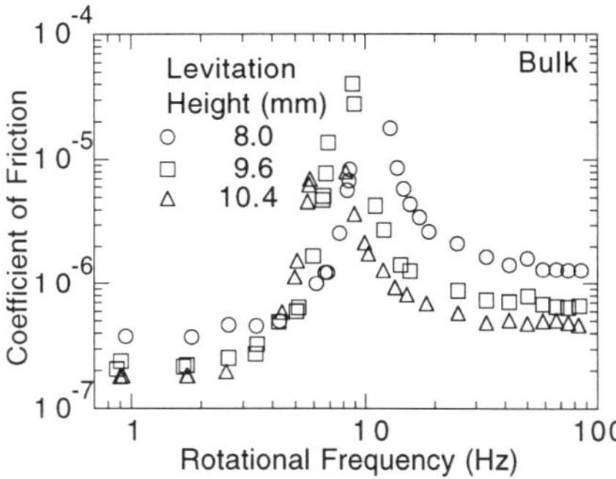

Figure 2. COF vs. frequency at different heights for PM levitated over bulk HTS.

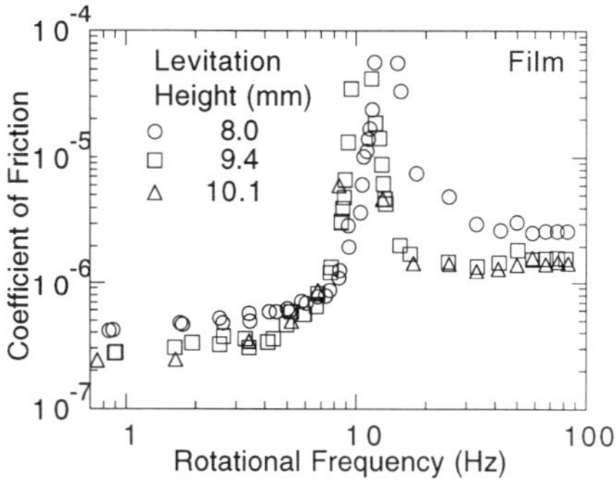

Figure 3. COF vs. frequency at different heights for PM levitated over thin-film HTS.

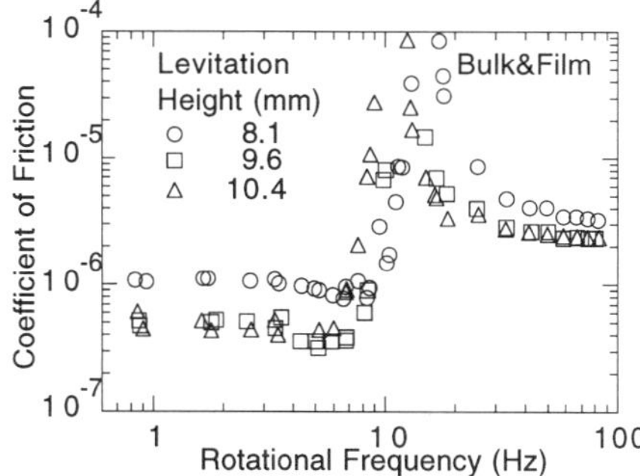

Figure 4. COF vs. frequency at different heights for PM levitated over combination of bulk HTS and thin-film HTS.

higher. Figure 4 shows a slight dip in the COF immediately before the resonance region for all three heights. Figures 2 and 3 do not show this behavior.

In Figs. 5-7, the data sets are rearranged to show the behavior for each HTS configuration at approximately the same levitation height. For all the levitation heights, the resonant frequency is consistently lowest for the bulk alone and highest for the combination of bulk and film. For all the levitation heights, both above and below the resonance, the minimum loss occurred when bulk HTS was used alone, and the loss was maximum when the bulk and film were used together. In these regions, the COF of the

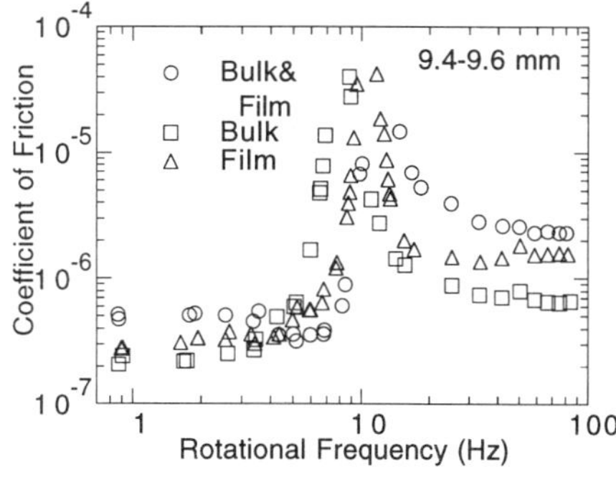

Figure 5. COF vs. frequency at 9.4-9.6 mm height for PM levitated over HTS in various HTS configurations.

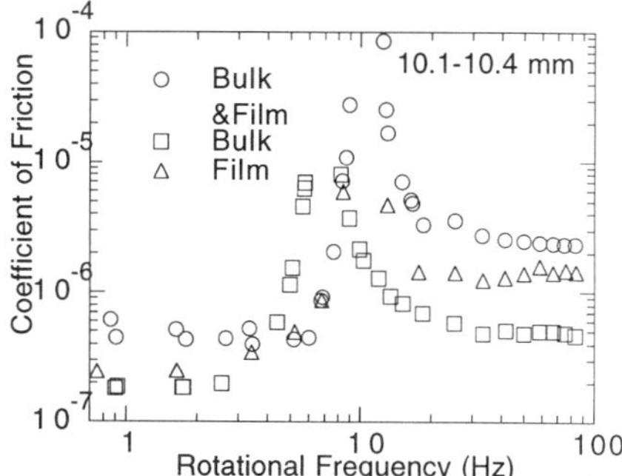

Figure 6. COF vs. frequency at 10.1-10.4 mm height for PM levitated over HTS in various HTS configurations.

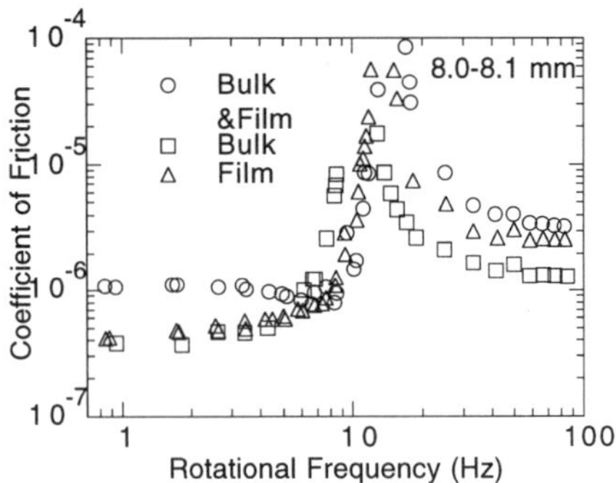

Figure 7. COF vs. frequency at 8.0-8.1 mm height for PM levitated over HTS in various HTS configurations.

bulk and film is approximately equal to the sum of the COFs for the bulk alone and the film alone.

We assumed that the substrate on which the thin film was deposited was not contributing to the rotational loss. An experiment was performed in which a bare substrate (without an HTS film) was placed above the bulk HTS. The rotational loss in this experiment was essentially identical to that of the bulk HTS alone, consistent with our assumption.

DISCUSSION

The results of the rotational loss measurements are opposite to our original expectations. At this time, we are unable to present a complete theory of the rotational loss when HTS films are used. However, we discuss here the germane physical concepts that we believe govern the behavior and are consistent with the results obtained.

The critical-state calculation of Eq. (1) is generally valid when the magnetic field is parallel to the HTS surface. When the magnetic field is perpendicular to the surface of a film, the aspect ratio plays an important part in determining the behavior of the penetration of the field.[15-17] Experimentally, the hysteresis loss of an HTS film can be almost three orders of magnitude higher with the field perpendicular to the film surface than when the field is parallel to the surface.[16] The case of a thin disk in a uniform field has undergone considerable analysis, but the configuration of our experiment is considerably more complicated and seemingly not susceptible to elementary analysis. The experimental configuration can be conceptualized as a spot of flux being dragged in a circle over the surface of the HTS film. The diameter of the spot is considerably smaller than the diameter of the film, the flux has components parallel and perpendicular to the film surface, and part of the flux may be shielded out while part of it may be dragged through.

To illustrate the situation, we can perform several simplified example calculations. First, consider the bulk HTS with J_c = 20 kA/cm^2 and calculate the ΔB required to produce a COF of 2 x 10^{-7}. We assume an HTS surface area of 270 mm^2, equivalent to a radial swath of ≈4 mm around the perimeter of the HTS upper surface. The weight of the PM is F_L = 0.36 N, so the drag force is F_D = 72 nN. The hysteretic energy loss per revolution is then 4.9 nJ, and according to Eq. (1), the associated ΔB is 146 μT. This value is consistent with measured values of ΔB_z for the PM.

Next, as an upper bound, consider application of the 146 μT field over the entire top surface of the film. The surface area is about 8 times larger than that of the hypothetical flux ring discussed above, but J_c is 200 times higher. Based only on these relations, one would expect the COF to be lower by at least a factor of 25. Because the COF is not lower for the film, plus the COF when bulk and film are used together are essentially the sum of the COFs when they are used separately, we believe that there is probably a nearly complete lack of shielding of the bulk HTS by the film. This result leads us to investigate the three-dimensional nature of the currents in the film. If a magnetic dipole were oriented perpendicular to the film surface, we should be able to shield the field out of the film interior with currents running only in the plane of the film. However, if the dipole moment is oriented parallel to the film surface, then currents must flow perpendicular to the film surface to shield the interior of the film. This creates a counterflowing current on the back side of the film, which would account for the lack of shielding of the bulk HTS, since from the bulk's point of view, the currents flowing in the film would essentially cancel the magnetic field produced. If current flows perpendicular to the film surface, it will do so at a greatly reduced J_c, resulting in a greater hysteresis loss in the film.

The difference in shielding of a perpendicular dipole versus a parallel dipole by a thin-film HTS is an experimentally testable hypothesis, and we hope to perform such experiments in the future.

One may show that dragging of flux through the film is probably not occurring. Consider a spot of perpendicular flux of radius R_f and mean magnetic induction B that is dragged in a circle of radius R_D through the HTS film. The force per unit volume is[18]

$$F = B\,J_c, \tag{4}$$

which yields

$$COF = \pi R_f^2 hB\,J_c/F_L \tag{5}$$

where h = 350 nm is the film thickness. Consider a 10 μT field and a spot size of 1 mm. The calculated COF in this case is 1.2×10^{-6}. Even with the low values of field considered, this COF is much higher than the experimental results.

A further consideration may be important in understanding our results. The thickness of the film is of the order of the penetration depth, which might explain the incomplete shielding. Furthermore, because the film is always completely penetrated by flux, the flux lines may be oriented parallel to the a-b planes, which implies that they are pinned intrinsically.

CONCLUSIONS

We have experimentally measured the rotational drag of a permanent magnet (PM) levitated above a high-temperature superconductor (HTS). We examined HTS configurations of bulk, thin film, and a combination of thin film over a bulk. The loss in the thin-film HTS was significantly higher than predicted by a simplified critical-state model, and the film did not shield the bulk HTS from magnetic-field inhomogeneities of the PM. A comprehensive theory of these phenomena is not available, but we believe that the three-dimensional nature of the currents in the film provide a plausible physical explanation for the increased losses.

ACKNOWLEDGMENTS

This work was partially supported by the U.S. Dept. of Energy, Energy Efficiency and Renewable Energy, as part of a program to develop electric power technology, under Contract W-31-109-Eng-38. Author Cansiz gratefully acknowledges the support of the Turkish government. The authors are grateful to K. Salama for providing the bulk HTS used in the experiments and to Markus Getta for providing the thin-film HTS.

REFERENCES

1. J. Hull, Using high-temperature superconductors for levitation applications, *JOM* 51(11):13 (July 1999).
2. J.R. Hull, Flywheels on a roll, *Spectrum* 34(7):20 (July 1997).
3. J.R. Hull, Superconducting levitation, in "Encyclopedia of Electrical and Electronics Engineering," J.G. Webster, ed., John Wiley and Sons, New York (1999), vol. 20, p.729.
4. B. Lehndorff, H.-G. Kurschner, and B. Lucke, Determination of pinning strength of $YBa_2Cu_3O_{7-\delta}$ from magnetic stiffness measurements, *Appl. Phys. Lett.* 67:1932 (1995).
5. B.R. Weinberger, Further studies of magnetic levitation using high temperature superconductors, *Appl. Supercond.* 2:511 (1994).
6. P. Schonhuber and F.C. Moon, Levitation forces, stiffness and force-creep in YBCO high-T_c superconducting thin films, *Appl. Supercond.* 2:523 (1994).
7. B. Lucke, H.-G. Kurschner, B. Lehndorff, M. Lenkens, and H. Piel, Levitation force of a stack of epitaxial $YBa_2Cu_3O_{7-\delta}$ films, *Physica C* 259:151 (1996).
8. T.H. Johansen, A.B. Riise, H. Bratsberg, and Y.Q. Shen, Magnetic levitation with high-T_c superconducting thin films, *J. Supercond.* 11:519 (1998).
9. B.R. Weinberger, L. Lynds, Jr., and J.R. Hull, Flywheel energy storage with superconductor magnetic bearings, U.S. Patent No. 5,086,034 (1993).
10. B.R. Weinberger and L. Lynds, Jr., Magnetically levitated superconducting bearing, U.S. Patent #5,256,638 (1993).
11. J.R. Clem, AC losses in high-T_c materials, *Adv. Supercond. X* 1:23 (1997).
12. J.R. Hull, T.M. Mulcahy, K.L. Uherka, R.A. Erck, and R.G. Abboud, Flywheel energy storage using superconducting magnetic bearings, *Appl. Supercond.* 2:449 (1994).
13. J.R. Hull, T.M. Mulcahy, and J.F. Labataille, Velocity dependence of rotational loss in Evershed-type superconducting bearings, *Appl. Phys. Lett.* 70:655 (1997).
14. J.R. Hull, J.F. Labataille, T.M. Mulcahy, and J.A. Lockwood, Reduced hysteresis loss in superconducting bearings, *J. Appl. Supercond.* 4:1 (1996).
15. E.H. Brandt, Superconductor disks and cylinders in an axial magnetic field. I. flux penetration and magnetization curves, *Phys. Rev. B* 58:6506 (1998).
16. A. Rastogi, H Yamasaki, and A. Sawa, Surface quality effects in hysteretic magnetization of $YBa_2Cu_3O_{7-\delta}$ films, *Adv. Supercond. X* 1:561 (1997).
17. X.N. Xu, A.M. Sun, M.J. Qin, S.Y. Ding, X. Jin, X.X. Yao, and S.L. Yan, Current and magnetic field distribution of disk-shaped superconducting film, *Physica C* 291:315 (1997).
18. E.W. Collings, "Applied Superconductivity Vol. 1," Plenum, New York (1986), p. 570.

DEVELOPMENT AND RESEARCH OF A CRYOGENIC PLANT FOR RECYCLING AND PROCESSING WASTED TIRES

Y. Y. Guo,[1] Y. Liang,[1] P. R. Yang,[1] Q. H. Xiao[2]

[1] School of Chemical Engineering, Xi'an Jiaotong University, Xi'an, Shaanxi, 710049, P. R. China
[2] Xi'an Yutong New Technology Development Co. Ltd.

ABSTRACT

In order to process and recycle worn-out tires of automobiles and trucks a novel cold air freezing system is presented in this paper. This system produces refrigerated air at temperature below -100?C by means of a stage-by-stage refrigeration system. The cold air goes through a moving bed to chill the tire chips, then returns to the first stage compressor of the refrigeration system. Several tests, such as precooling and chilling the tire, comminuting them into powder, rewarming the rubber powder, recovering the cooling energy and so on, were conducted. The overall processing efficiency of the whole system is about 1.2 kW/kg tire.

INTRODUCTION

Since 1970's, scientists and engineers from developed countries have been developing technology for recycling and processing of wasted tires.[1] This not only eliminates the environmental pollution due to wasted tires that are abandoned, but also supplements the shortage of natural rubber in the commercial market. In the last decade, however, it is noted that rapid progress of the technologies for recovering wasted tires in China has been made in several ways. They can be categorized generally into three types as follows.

Ambient temperature pulverization

By various means such as cutting, grinding, shearing, impacting etc., wasted rubber tires can be processed, and the resultant product is in the form of particles that predominantly fall in the 20 to 40 mesh size. Then, those particles are desulfurized to become reclaimed rubber, or they are activated to become activated rubber powders. The processing of rubber tires is not very difficult technologically. However, the production line of ambient temperature comminution for reclaimed rubber or activated rubber powders requires considerable energy consumption and causes environmental pollution too. In addition, the final products fail to maintain the original performance of material by the generated heat during processing. In order to achieve the finer powders of 60 to 80 mesh, a new mill has developed. Nevertheless, its output of fine rubber is small and some relevant technologies are still not mature.

Cryogenic pulverization using liquid nitrogen

Rubber is fragile and easily pulverized at less than about -70°C. Applying this principle to obtain rubber particles of 60 to 80 mesh is known as cryogenic pulverization in which liquid nitrogen is used as cryogen. Research in this method has been carried out since 1970. The predominant advantages of this production method are simplifying the equipment, keeping the original performance of the attainable rubber powder, and better safety of operation because of the very low temperature of the cryogen. However the consumption of liquid nitrogen per ton of produced rubber powder is high. And it would make cryogenic processing economically unfeasible.

Cryogenic pulverization using cooling air

Going through compressor and expansion, air as a medium becomes refrigerated air at a temperature of less than -70°C. Chilling the processing tire rubber to its embrittlement temperature can be accomplished by using the refrigerated air directly. These systems, however, appear to offer attractive economic advantage over the liquid cryogen system where high tonnage processing is to be carried out, and have no more environmental pollution. But the following problems have to be considered carefully: how to build a cold air freezing system with economic advantage, how to chill the processing materials to their embrittlement temperature by using the refrigerated air in a short cooling down time, and how to make the warm-up of cryogenically processed materials to avoid forming frost on powder surfaces exposed to the atmosphere.

In this paper a production line for processing wasted tires of 3000 tons per year has been developed by using an air-cooling system. The mentioned problems were answered finely.

AN AIR FREEZING SYSTEM

To work out the air freezing system, low energy consumption is required in association with benefits to rubber pulverization. When the refrigeration system is planned, three primary questions need to be specified: (1) air temperature, (2) refrigeration capacity, and (3) reheat shortage loss.

In general, the brittle temperature of worn-out tires of automobiles and trucks is about -70°C to -80°C. Therefore, less than -80°C cryogenic gas is enough to obtain the lowest energy consumption. These lead one to reduce the heat transfer temperature difference between the rubber and cryogenic gas as well as increase the heat transfer equipment size. Meanwhile, removal of the heat generated in the pulverizer needs a lot of cryogenic gas. And as the cryogen is charged into the pulverizer much more, it can cause thicker particles of pulverization.

Refrigeration capacity is applied to freeze worn-out tire blocks to reach the temperature below their brittle temperature and to make up for the requirement of pulverization heat. The former is easily determined; the latter is determined with difficulty. It is also the heart of the problem if the cooling energy of processed rubber powder is recovered effectively.

Tire rubber has a low thermal conductivity, so does the air. The heat transfer between both is gas-solid unsteady heat transfer. The magnitude of the reheat shortage loss seriously affects refrigeration capacity and energy consumption of refrigeration system.

The core of the above mentioned is: firstly what effective pattern and equipment of heat transfer between tire rubber and cryogen is employed, and how to determine the coefficient of heat transfer of the employed heat exchanger; secondly, what pattern of pulverization machine should be adopted so as to assure the required mesh of products and the required production. With view of these two items, many experimental and analytical works have been done. It is shown that: (1) The smaller the initial rubber particle is, the stronger the heat transfer between gas and solid becomes, the shorter the cooling down times for all rubber particles to embrittlement temperature become, and the smaller the heat transfer temperature difference in heat end; (2) Heat generated in the pulverizer is more than 90% of the pulverization power applied. Therefore, the combination of room temperature pulverization and cryogenic pulverization should be adopted primarily. This means that the wasted tires are cut into pieces up to about $3\times3\times3mm^3$ particles with not much power consumed at ambient temperature. And then, after cooling them to the brittle temperature by cryogen, the rubber particles are comminuted into powders. Because the temperature of the milled products is usually still at a brittle temperature after milling, the gas that is used to remove generated heat of pulverization with the same temperature as the milled products can be used to precool the rubber blocks to be cooled. A cooled air from recovering the cold of cryogenically processed rubber powders may also precool the initial rubber particles to save the cooling energy consumption greatly. Furthermore, as far as the cooling cycle is concerned, the following measures have been taken: (1) The various cooling sources maintain the different equivalent temperature levels required in the cold air freezing system. This makes the thermodynamic economy as good as possible. (2) Use of a plate-fin heat exchanger for heat exchanging of air is to reduce the irreversibility loss of heat exchange. (3) As an optimally designed boost-turboexpander is used, it not only has its efficiency of operation up to 84%, but also recovers the expansion work, so that it save the energy in the refrigeration cycle.

Hence, for an established air cryogenic pulverization cycle with the annual output of 3000 tons of fine rubber powders, the energy cost per 1kg 60 mesh tire rubber powders can save 40% of that of one at ambient temperature, and 65% of that of one using liquid nitrogen as the cryogen. The overall processing efficiency of the whole pulverization plant is about 1.2 kW per 1kg rubber powder. The process diagram is shown in Figure 1, and the diagram of the refrigeration system is shown in Figure 2.

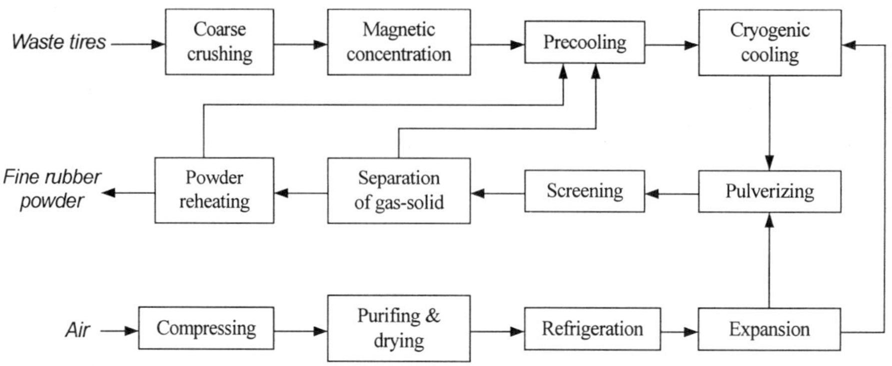

Figure 1. Process flow diagram.

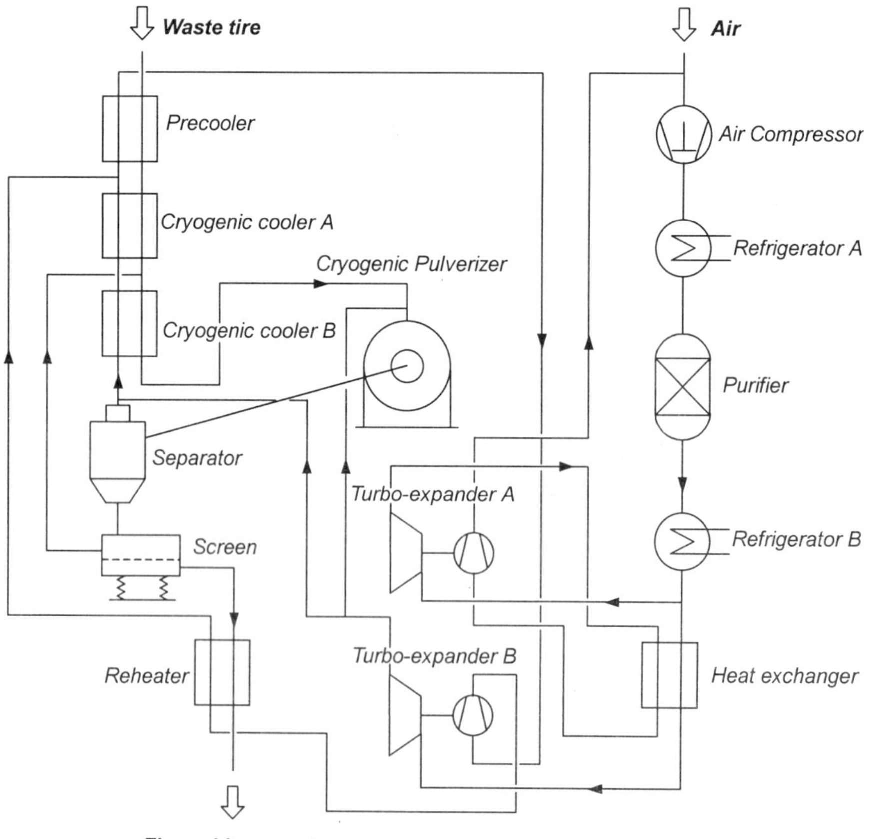

Figure 2. Cooling system diagram.

RESEARCH AND DEVELOPMENT OF KEY TECHNOLOGIES

Heat Transfer Equipment

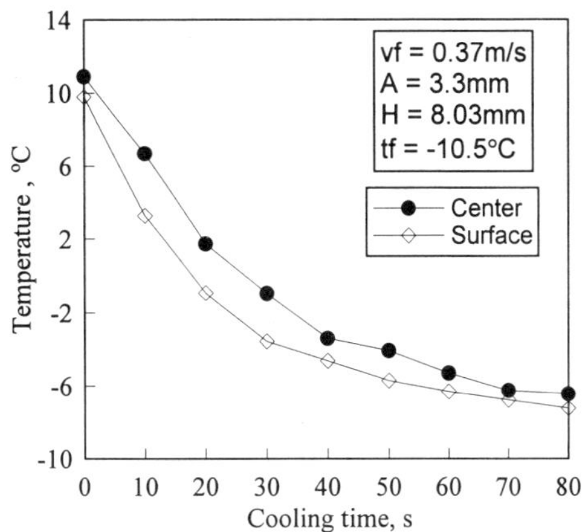

Figure 3. Cooling curve of rubber particle.

Precooling To precool the tire particles, a vibrated fluidized bed as a new cooling mean has been proposed in which the tire rubber particles are in contact with the cooling air and are cooled down. A vibrated fluidized bed can be regarded as a composition of a fluidized bed and its vibration. It is used to improve the fluidizing state including heat transfer between the particles and gas, between the particles. The temperature distribution and the particle distribution in the bed are more uniform than that in the fixed bed. The surface of heat exchange in the bed is big; the rate of heat exchange of gas-solid is high. Because rubber particles flow poorly and are difficult to fluidize in nature, the rubber particles are forced to be fluidizing by using vibration. The materials can be tossed and traveled by vibration force instead of fan pressure. The traveling speed of material is monitored in a stepless control. It is also easy to change the operation condition. The machine is simple in its structure and easy to maintain. The temperature decline curve on the surface of particles and the center of particles has measured as shown in Figure 3 with the bed amplitude of 3.3 mm, the material thickness of 8.03 mm, and cooling gas temperature of -10.5°C. At the outlet of the fluidized bed, the temperature on the surface of the particles is constantly about -6°C after 80 seconds. At this moment the temperature difference between the surface and center of the rubber particles is about 1°C.[2] Although the vibrated fluidized bed has the advantage of heat transfer performance, its heat preservation at the cryogenic condition is a bit difficult due to the usage of vibrated fluidized motor. Therefore, this is only good equipment for precooling rubber particles at present.[3]

Cryogenic cooling After precooling, the rubber particles need to be further cooled. The continuum of particles at the specified initial temperature flows uniformly upstream to downstream in a container, while the cryogenic gas flows uniformly downwards to upwards through the particle groups. At the outlet of the container, the temperature at the center of particles can be cooled down below the brittle temperature of the material. This equipment, including the heat exchange of gas-solid direct contacting, is known as a moving bed. In an adiabatic moving bed, the flow temperature at any cross-section area is generally uniform, and it goes higher and higher gradually downstream to upstream. The interior temperature of the particles is changed unsteadily since the thermal conductivity of the particles themselves varies with the temperature and there exists collision and turnover of the particle groups in the flowing. Experimental research and computing simulation have been conducted on this type of heat transfer in cryogenic condition with respect to the size of various particles, the flow speed of the particles, and the flow speed of gas. Meanwhile, the declined rate of temperature of the particles has been studied. The heat transfer formula for the design has been regressed.[4] This provides a reliable base for designing gas-solid moving bed.

Pulverization Technology

The type of selected pulverization machine, the mesh of the pulverized particles, production, energy consumption, and temperature to be controlled in the pulverization chamber are able to affect the reasonable usage and distribution of cooling energy. This is also the key point to determine that machine works economically and beneficially. However, in this respect, the research is limited to the initial theoretical analysis and partial experimental studies. It is noteworthy that the design theory of the pulverization machine structure is also not mature and not quantified. After comprehensive analysis of various pulverization machines and combination of the knowledge that cooled rubber particles

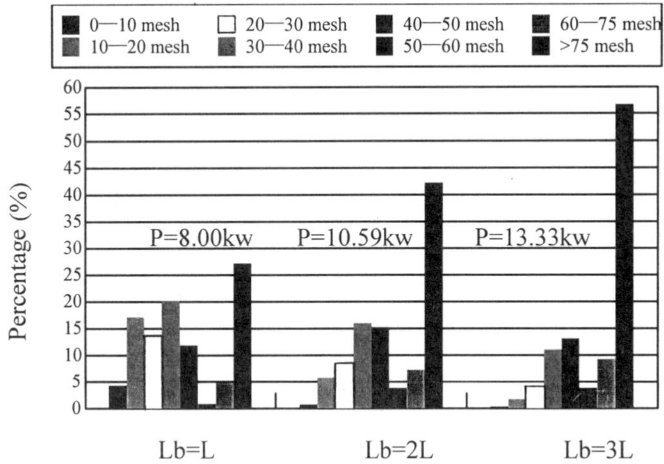

Figure 4. The effect of different lengths of pulverization chamber.

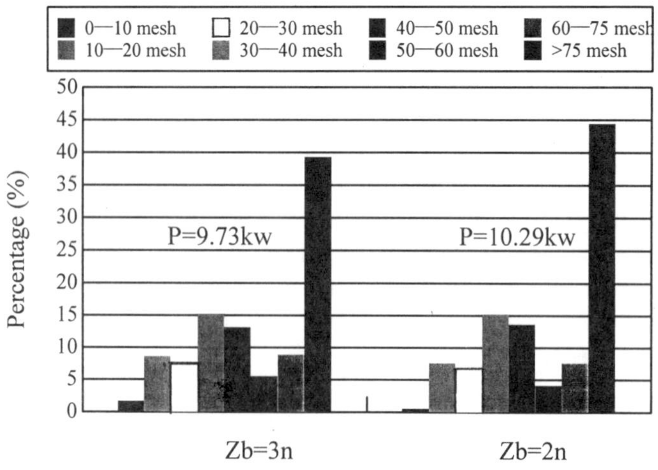

Figure 5. The effect of two different hammers number.

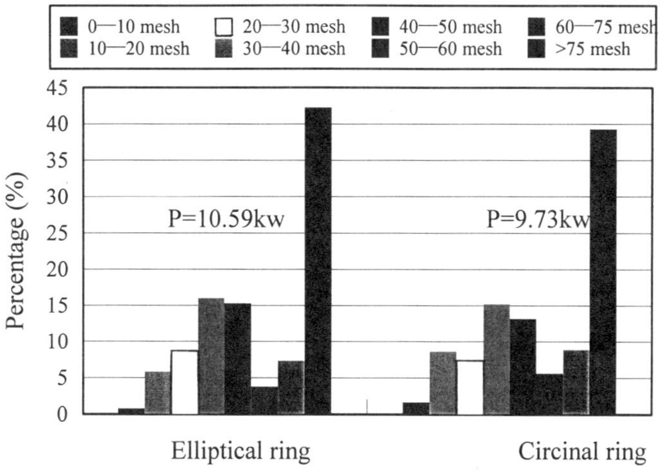

Figure 6. The effect of elliptical and circinal ring.

have the same fatigue as glass, the impact pulverization machine was recognized. Some key factors of the pulverization machine structure such as the number of hammer plates, the axial length of pulverization chamber, structure type of gear ring, the gap between the wheel plate and gear ring, the front clearance of wheel plate and so on have been studied initially.

In order to minimize the research cost, some fatigue materials at room temperature were studied. The prototype is similar to the industrial one on size. During the experiments, the criteria can be regarded as the distribution of the meshed particles and the energy consumed. The experimental results are shown in Figures 4, 5 and 6 respectively, in terms of the length of pulverization chamber, the number of hammer plates and the type of the pulverization chamber.

From these diagrams, it is shown that increasing the length of the pulverization chamber results in significantly finer particles. There exists an optimum number of hammer plates to minimize the energy consumed in the pulverization machine and to produce the finest particles. This specific reasonable type of pulverization chamber effectively destroys gas-particles recycling layers and produces the fined particles. In conclusion, the structure parameters of the impact pulverization machine have a great effect on its performance. The main structure parameters can be determined on the basis of the experiments. Then the influences of outlet temperature on meshed particles, the relationships between

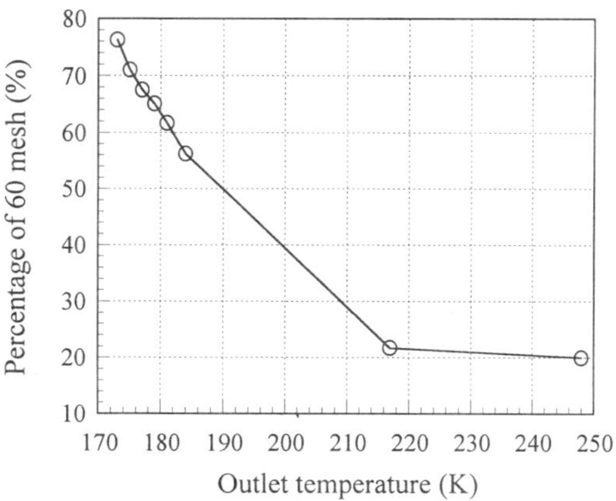

Figure 7. Relation between pulverization temperature and the rate of finished products.

the production and pulverization heat, and the relationships between the blowing capacity and pulverization heat have been studied in the cryogenic condition. These experimental results in the cryogenic conditions concluded that the temperature inside the pulverization machine is the main constraint parameter, as shown in Figure 7. The great amount of pulverization heat accounts for more than 90% of the motor input work.

CONCLUSIONS

(1) A production line for processing worn-out tires by using a cooling air system has been developed.

(2) The overall processing efficiency of the whole pulverization plant is about 1.2 kW per 1 kg rubber powder. Its annual output is 3000 tons of fine rubber powder with size of 60 mesh. Its cost of processing wasted tires is cheaper than that of the one using liquid nitrogen.

(3) The effective equipment for precooling and chilling the tire, as well as rewarming the processed rubber powders were proposed.

(4) The temperature inside the impact pulverizer is the main parameter that affects the mesh of processed material.

REFERENCES

1. N. R. Braton. "Cryogenic Recycling and Processing", CRC Press Inc., (1980).
2. Y. Liang, Y. Y. Guo et al, Experimental investigation of heat transfer between gas and solid in cryogenic vibrated fluidized bed, *ICEC 17 Proceedings*: 667-670 (1998).
3. X. X. Wang, Y. Liang, Y. Y. Guo et al, Numerical analysis of heat transfer between gas and solid in vibrated fluidized bed, *ICCR'98 Proceedings*: 317-320 (1998).
4. Y. Y. Guo et al, Characteristic investigation on unsteady heat transfer between gas and solid, *Cryogenics*, 34: 369-372(1994).

A LABORATORY APPARATUS TO DETERMINE FLUID-SOLID HEAT TRANSFER COEFFICIENT AND EFFECTIVE THERMAL CONDUCTIVITY OF SOLID POROUS MEDIA

P. G. Reedeker, A. Frydman, and J. A. Barclay

Cryofuel Systems Group, University of Victoria
Victoria, B.C., V8W 3P6, CANADA

ABSTRACT

A single-blow apparatus was designed and built to specifically eliminate most systematic errors in characterizing thermal properties of solid porous media. By matching temperature profiles calculated from the solid-fluid transient energy balance in a porous bed to experimental profiles, the heat transfer coefficient between solid and fluid and the effective longitudinal thermal conductivity can be determined. The specifications of such an apparatus require strict accounting of sources of heat leak and other systematic errors. Here, we present a detailed study in identifying and minimizing these systematic errors. Details on the design, construction and testing of this apparatus are shown, including utilization of low thermal conductivity materials for components, and meticulous design and construction of temperature probes. We also present for the first time a methodology to correct for parasitic heat leaks inherently present in the system. This correction procedure to correct the temperature profiles eliminates the effect of the major systematic errors. This apparatus is being used as a screening-design tool to characterize overall effectiveness of high performance regenerators for regenerative cycle refrigerators and liquefiers.

INTRODUCTION

Fluid dynamics and heat transfer in solid porous media are dominant mechanisms in diverse commercial devices such as cryogenic regenerators and high temperature catalytic converters. It is of critical importance to the design of such devices to use accurate heat transfer coefficients, effective thermal conductivities, and friction factors. These parameters are specific to the contact between the fluid and the solid, and the nature and geometry of the porous media in question. It is also relatively easy to design extremely high performance regenerators, but it is not easy to attain this performance during fabrication. Experimental validation of performance becomes critically important for highly efficient regenerative liquefier development.

A regenerator is a compact type of heat exchanger in which the fluid is in direct contact with the heat transfer surface. Heat is transferred to and from the fluid in a periodic fashion. This is accomplished by an intermediate heat transfer step from the fluid to a solid matrix material. By reversing the flow direction of the fluid, heat is transferred from the matrix back to the fluid. Therefore, thermal energy is temporarily stored and used to heat or cool a fluid flowing through the matrix. A recent comprehensive work on regenerator analysis and applications is found in R.A. Ackerman.[1]

Traditional methods for regenerator testing are basically two: integral and differential. The integral method is applicable to testing a regenerator under realistic refrigerator conditions.[2] The differential method presents two major variations: periodic and single-blow, however, the periodic might be considered a variation of the single blow mode.[3,4] In the differential method, the regenerator is brought to thermal equilibrium with a fluid flowing at constant temperature. The inlet temperature of the fluid is instantaneously raised in a step function mode. This technique is recommended in the initial stage of regenerator matrix development, and is the subject of the present paper.

Experimental measurement of the thermal properties of a fluid flowing through a solid porous media is not a trivial task. The assumptions usually consider that the experimental system is adiabatic, i.e., it does not exchange energy with the external environment and, as the single-blow technique requires, the fluid inlet temperature must change according to a step function. Throughout our preliminary study of regenerator testing methods, we have identified three major sources of systematic errors involving experimental units that used the single-blow technique for testing regenerators:[4] (a) utilization of high thermal conductivity and high thermal mass components, (b) inefficient method of generation of the temperature step; and (c) inability to take measurements in the range of low Re numbers. We have thoroughly examined these sources of systematic errors to develop an apparatus used as design tool to select high performance solid porous media for use as thermal regenerators to be incorporated into efficient regenerative cycle refrigerators or liquefiers. Description of a procedure to characterize regenerator matrices using this apparatus and energy balance modeling is published elsewhere.[5]

In the present work, the process of designing, assembling and testing the apparatus for characterization and testing of regenerator matrices is described. Several heat leak sources in the design of the experimental apparatus were reduced, eliminated or corrected to improve the accuracy of the measurements. We also present for the first time a procedure to correct the measured temperature profiles for parasitic heat leaks.

BACKGROUND THEORY

When a fluid at a certain temperature T flows through a regenerator material at a temperature lower than T, the fluid will exchange heat with the material and lower its initial temperature. However, one has to be concerned with heat transfer to/from the container of the regenerator material under test, namely, the container that holds it (polymer, stainless steel, titanium alloys, etc.) and other bodies (fittings, sensors, etc.) introduced to the system. In our apparatus, the following mechanisms of heat leak could be identified: (a) longitudinal (or axial) heat transfer along the walls of the regenerator housing; (b) radial heat transfer through the walls of the regenerator housing; and (c) heat transfer between the container and the temperature probes. The latter is relevant to the temperature correction procedure described here.

To determine the heat transfer coefficient and the effective thermal conductivity of a fluid flowing through a solid porous matrix, the temperature profile at certain locations along the regenerator matrix as a function of time must be recorded. This set of temperature profiles is used to match the profiles obtained from modeling the solid-fluid energy balance.[4] The temperature probes used to obtain temperature measurements within the bed have several implications with respect to the design of the apparatus. The temperature probes have heat transfer along their length from both the container wall where they are mounted as well as from the fluid. This energy transfer to/from the fluid must be properly sensed by the probe to register the temperature of the fluid. Because each probe possesses a measurable mass and finite thermal conductivity, it's transient response to a sudden temperature change in the fluid is not instantaneous. Reducing the mass (or size) of the temperature probe is advantageous, however, an almost "massless" probe is difficult to obtain. A good compromise for the transient temperature probe is a thermocouple of fine wire gauge. If the magnitude of the thermal resistance between the container and the tip of the thermocouple is large enough, the thermocouple readings will gradually asymptote to the real value of T_f (Figure 1, curve b). As a consequence of finite thermal resistance along the thermocouples, parasitic heat leaks will cause the thermocouple readings to stabilize at a value $T_{tc} < T_f$ (Figure 1, curve c). In our system, the thermocouple probe, the fittings, and the housing are the dominant sources of finite heat leak into the tip of the thermocouple. Figure 1, curve a, illustrates the ideal behavior represented by a step function in temperature as a sudden temperature change in the fluid is introduced into the bed. A simple schematic of the thermocouple arrangement is shown in Figure 2.

The total sum of these thermal resistances, or R_{total}, can be schematically represented by an analogy to electrical circuits, in this case, between the thermocouple bead at T_{tc}, and the container wall at T_{wall}. The rate of heat transferred between these two points:

$$\dot{Q}_{tc} = \frac{\left(T_{tc} - T_{wall}\right)}{R_{total}} \tag{1}$$

The thermocouple senses the temperature of the fluid by an energy transfer with the fluid at temperature T_f that can be calculated by:

$$\dot{Q}_{fluid} = h \cdot A_{bead} \cdot \left(T_f - T_{tc}\right) , \tag{2}$$

where A_{bead} is the surface area of the thermocouple bead, and h is the heat transfer coefficient between the fluid and the bead estimated using correlations for fluid flowing over spheres.[6,7]

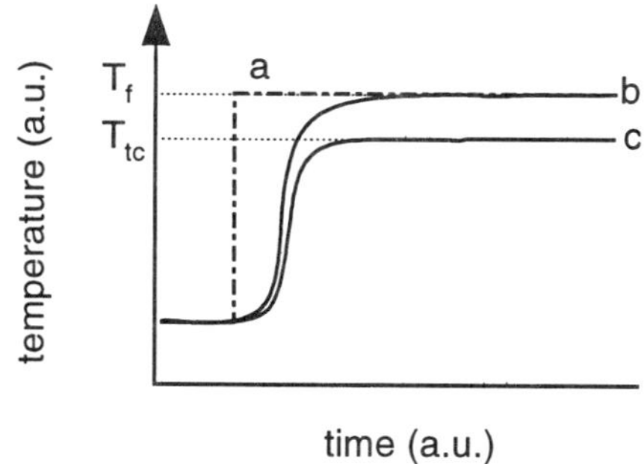

Figure 1. Graphic representation of the temperature response curve. Curve (a) shows the ideal behavior. Curves (b) and (c) show the corrected and the experimental responses, respectively.

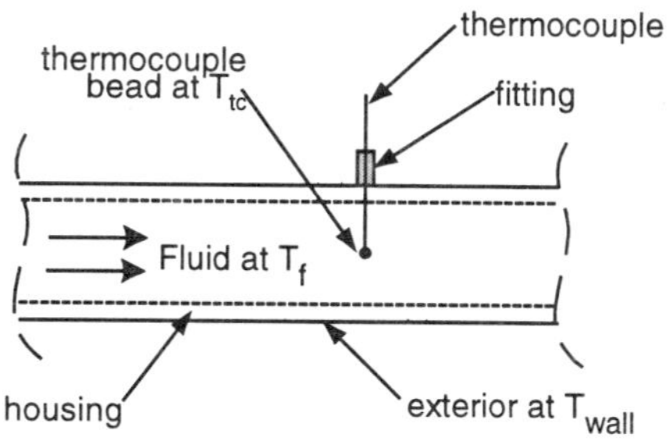

Figure 2. Simplified representation of the fluid path to the temperature probe.

At thermal equilibrium, Eqs. (1) and (2) can be equated, and the corrected temperature of the fluid can be calculated by:

$$T_f = \left[\frac{T_{tc} - T_{wall}}{R_{total}}\right] \cdot \left[\frac{1}{h \cdot A_{bead}}\right] + T_{tc} \qquad (3)$$

Eq. (3) can be used to compensate for the thermal resistances present in the system, to correct the thermocouple reading. At the lower limit, when R_{total} goes to zero, the equilibrium thermocouple reading will greatly diverge from the real temperature of the fluid. At the other limit, when R_{total} goes to infinity, the equilibrium thermocouple reading will precisely match the real temperature of the fluid. The value of R_{total} is given by:

$$R_{total} = \frac{1}{k_{total} \cdot S_{total}} , \qquad (4)$$

where k_{total} is the total thermal conductivity of the set of thermal resistance sources in the mounted thermocouple sensor, and S_{total} is a shape factor.

The value of R_{total} can be determined experimentally by measuring T_{tc} and T_{wall} as a function of time for an empty regenerator housing. Assuming that T_f is known immediately before entering the regenerator housing, a curve of thermal resistance as a function of time can be generated from Eq. (3). Eventually, the thermocouple will reach thermal equilibrium with the fluid, and the thermal resistance curve will saturate at R_{total} (Figure 3).

The value of R_{total} will be specific to the operating conditions of the testing apparatus. However, the corrected T_{tc} obtained from this procedure is a very good estimate of the temperature of the fluid. If the apparatus is altered in any way, or other elements are introduced in the system, a new value of R_{total} must be determined. This value of R_{total} represents the total sum of thermal resistances present, which means there might be other sources of heat leak, inherently included in R_{total}, that we might not even recognize.

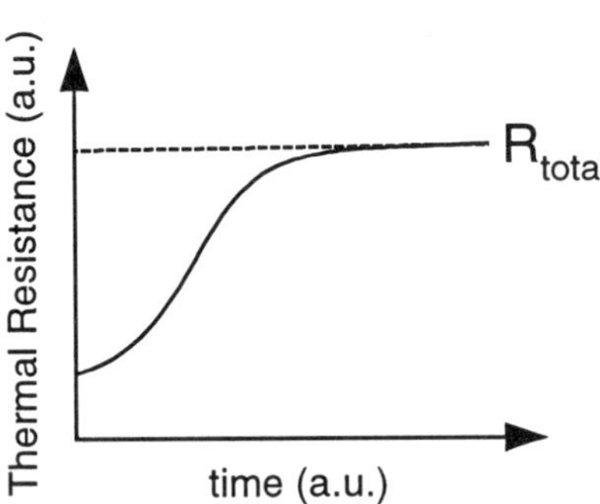

Figure 3. Graphic representation thermal resistance variation as a function of time.

Table 1. Material properties of the main components of the test apparatus [8,9]

Component	Material	Tensile Strength MPa (ksi)	Specific Heat kJ/kg.K	Density kg/m^3	Thermal Conductivity W/m.K
Housing	Unfilled Acetal	65 (9.5)	1.47	1410	0.231
Fittings	Nylon 6/6	79 (11.5)	1.63	1150	0.245
Thermocouple tube	PEEK	103 (15)	1.34	1320	0.252
Ref. material	Stainless steel 304	552 (80)	0.50	8030	16.3
Ref. material	Copper (ETP)	455 (66)	0.38	8360	393

DESIGN AND MANUFACTURING

Many factors were included in the design of the regenerator housing (Figure 4). Design parameters considered for functionality included media containment and replacement, working pressure to 1.28 MPa (200 psig), robust sealing, and minimizing length to reduce heat leak to the environment. Design parameters for material selection include minimized thermal mass and thermal conductivity, strength for required pressure, ease of machining, and cost. Table 1 lists relevant properties of main components.

Acetal, 100% unfilled, was selected as the primary structural material due to its low thermal conductivity, ease of machining, availability and cost. Fittings were chosen of nylon material for much the same reasons. PEEK (Polyetheretherketone)[8] was the material of choice for the thermocouple feed-throughs due to its low thermal conductivity, hardness, strength, and availability in very small tube sizes.

Media containment is achieved with two layers of nylon mesh screen at each end of the housing. The screen is retained as a sandwich between two washers – one with a groove and one with a lip. The washers were then bonded together with the groove/lip interface producing a knife edge clamp on the mesh. Each washer assembly is held in place with bushings retained in the housing. To replace the regenerator bed media, only one end (flange/bushing/screen) needs to be removed.

The design wall thickness of the housing body was calculated using working pressure, tensile strength of acetal material, and hoop stress calculations with a safety factor of two. This wall thickness was considered a minimum in the trade-off between thermal mass and thermal conductivity to the environment. Sealing of the fluid is ensured by standard flange-type static seal for the end flanges and triangular groove seals at the washer to prevent fluid from blowing past the bed at the walls. Due to the geometric nature of the media/housing interface, there is a possibility of fluid channeling along the walls, so keeping the fluid inlet and outlet within the perimeter of the media is very important.

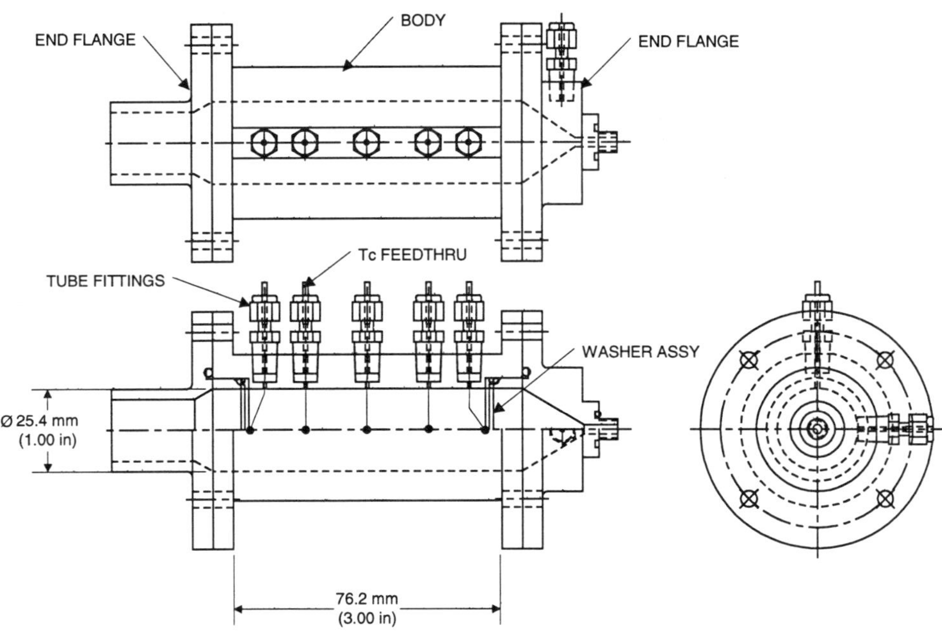

Figure 4. Drawing of regenerator housing assembly.

Keeping a minimum length of fluid flow path between the HEX2 water bath and the regenerator bed is crucial to reduce intrinsic heat leaks to the suddenly changed fluid associated with the thermal mass of the entrance section of the apparatus. Bath and bed must also be separated by a four-way bypass switching valve for switching of hot to cold fluid instantaneously. Because of this, a short acetal tube was used with very small threaded fittings to connect the four-way valve. The valve was of acetal construction and purchased off-the-shelf with a rating of 0.64 MPa (100 psig). Reverse engineering and redesign modifications enabled the valve to safely seal to 1.28 MPa (200 psig). This four-way bypass valve is a critical element to provide rapid heating of the fluid in a mode that closely approximates the step function needed for the single-blow regenerator testing technique.

A requirement of the thermocouple feed-throughs was that they be removable yet sealed to the apparatus using a proven sealing technique. This introduced an innovative design element of potted thermocouple wires inside a ∅1.58 mm (0.063") plastic tube. For this reason, the thermocouple feed-throughs were fabricated by potting the thermocouple wire inside a PEEK tube, which was then sealed to the housing using a Swagelok® nylon tube fitting. The potting compound was Stycast 2850 epoxy. Complete potting was accomplished by vacuum pumping on one end of the tube to pull the epoxy inward. The yield using this process was about 50%. Defective fittings were all successfully repaired employing the same technique using Loctite® 414 cyanoacrylate adhesive.

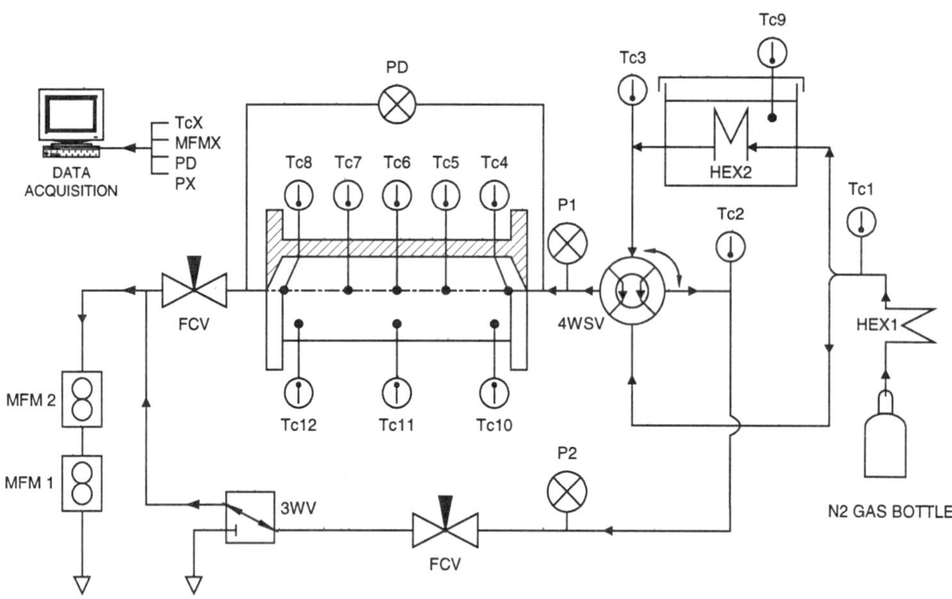

Figure 5. Schematic of the apparatus. TcX: thermocouples; HEX1: copper tube heat exchanger to ambient air; HEX2: copper tube heat exchanger to water bath; P1, P2: pressure transducers; PD: differential pressure transducer; 4WSV: four-way fluid switching valve; FCV: flow control valves; 3WV: three-way valve; MFM1: mass flow meter > 1g/s; MFM2: mass flow meter < 1 g/s.

TESTING

Testing and evaluation of the empty housing was done by applying a temperature step of ~20 K above room temperature to a fluid, N_2 gas, flowing at 0.1-1.0 g/s (Re = 320-3,200), ~0.69 MPa (100 psia). This step is generated by heating the fluid to a given temperature (in this case, ~315 K), bypassing the housing by means of a four-way, cross-flow valve. Simultaneously, this four-way valve allows a different stream of fluid at room temperature to flow through the housing. Once the hot stream of fluid has reached the desired temperature, it is switched into the housing simultaneously with the cold fluid being switched into bypass with the housing. Five temperature probes (thermocouples type E, 36 AWG wires with PFA Teflon insulation) recorded the temperature at five specific locations along the housing as a function of time (location 0, ¼ of L, ½ of L, ¾ of L, and L, where L represents the total housing length). Signal from temperature probes, pressure transducers and mass flow meters is collected to NI-DAQ LabView 4.1 (Macintosh version). A general schematic of the entire apparatus is shown in Figure 5. A detailed photograph of the housing, showing the four-way bypass valve, is presented in Figure 6.

The value of the thermal resistances was determined by measuring external wall temperatures as a function of time at various positions along the housing. Given the temperature change of the fluid, the value of R_{total} could be determined by taking the value of the thermal resistance at which the curve saturates (Figure 7). This value was taken for each temperature profile to generate a corrected temperature curve as illustrated in Figure 8. It is clearly noticed that the correction procedure elevated the equilibrium temperature to match the temperature of the fluid initially set by the step function. This corrected profile still does not match the step function profile of the fluid because of the inevitable thermal mass of the system, but now both curves, ideal and corrected, equilibrate at the same level. The difference between the steady state temperatures, before and after correction, represents, qualitatively, the magnitude of parasitic heat leaks that caused the experimental temperature profiles to stabilize to a value lower than the expected. One should be aware that such correction procedure should only be applied after most sources of heat leak have been minimized. A previous design of the test housing fabricated in stainless steel resulted in a temperature profile that saturated at ~8 K below the experimental curve in Figure 8.

The evaluation of the regenerator testing apparatus here reported, as well as the procedure to correct for parasitic heat leaks were extremely positive, demonstrating its effectiveness and potential to characterize high performance regenerator matrices.

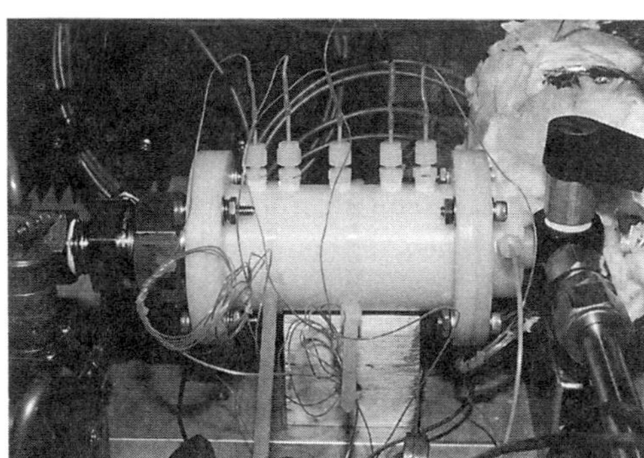

Figure 6. Photograph showing the acetal housing assembled in the testing apparatus. The miniature four-way bypass valve appears on the right hand side.

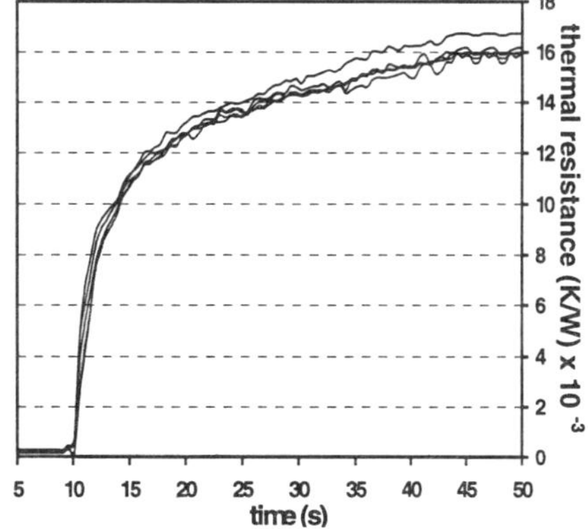

Figure 7. Thermal resistance profile as a function of time until saturation (each line represents a probe location along the housing). Re number = 1,600 (N_2 mass flow = 0.5 g/s, ~0.69 MPa).

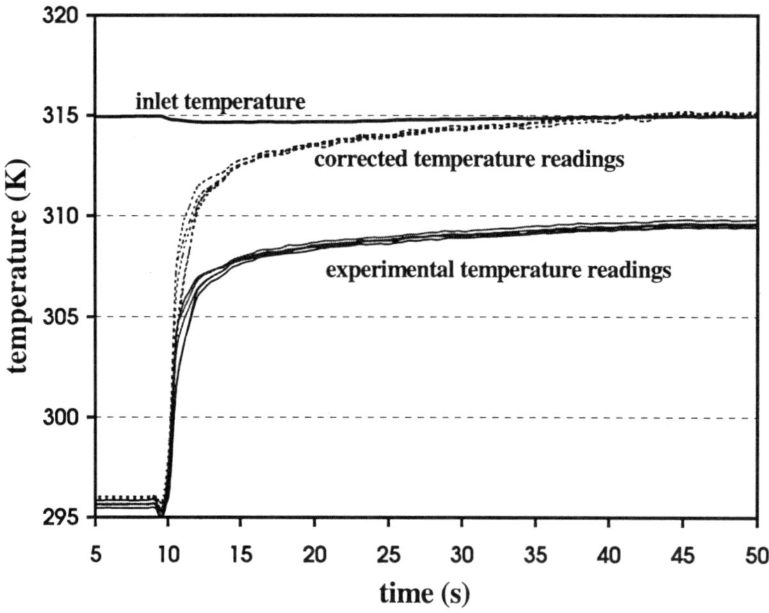

Figure 8. Correction of the experimental temperature profiles following the methodology described in the text (each line represents a temperature reading location along the housing). Re number = 1,600 (N_2 mass flow = 0.5 g/s, ~0.69 MPa).

ACKNOWLEDGEMENT

The authors gratefully thank the financial support by Natural Resources Canada, Canadian Natural Sciences and Engineering Research Council, and Centra Gas B.C., Inc., a subsidiary of WestCoast Energy, Inc.

REFERENCES

1. R.A. Ackerman. "Cryogenic Regenerative Heat Exchangers", Plenum Press, N.Y. (1997).
2. K. Kratschmar. "The Development of a Cryogenic Regenerator Test Apparatus", M.Sc. Thesis, University of Victoria, Victoria, BC (1995).
3. J.H. Stang. "The Periodic Technique for Testing Compact Heat Exchanger Surfaces", M.Sc. Thesis, Marquette University, Milwaukee, WI (1968).
4. K.J. Caudle. "Characterization of Regenerator Geometries Through Modeling and Experimentation", M.Sc. Thesis, University of Victoria, Victoria, BC (1997).
5. A. Frydman, and J.A. Barclay, *Proceedings of the Cryogenic Eng. Conference*, Montreal (1999). To be published.
6. W.H. McAdams. "Heat Transmission", 3rd ed., McGraw-Hill Book Co., N.Y. (1954).
7. S. Whitacker. Forced convection heat transfer correlations for flow in pipes, past flat plates, single cylinders, single spheres, and flow in packed beds and tube bundles, *AIChe Journal*, 18:361 (1972).
8. DSM Engineering Plastics Product, Inc. (company's catalog), printed in USA (1997).
9. E. Oberg, F.D. Jones, H.L. Horton, and H.H. Ryffel. "Machinery's Handbook", 25th ed., Industrial Press Inc., N.Y., 24 (1996).

DEVELOPMENT OF PILOT PLANT FOR CRYOGRINDING OF SPICES: A METHOD FOR QUALITY IMPROVEMENT

S.Jacob, S.Kasthurirengan, R.Karunanithi and Upendra Behera

Centre for Cryogenic Technology, Indian Institute of Science
Bangalore-560 012, India

ABSTRACT

The development of a pilot plant for cryogrinding of spices is discussed in this paper. The unit has an indigenously designed and developed cryogenic screw cooler for optimal utilization of liquid nitrogen along with a digital signal processing system (DSP) for liquid nitrogen spray and automatic temperature control. In the first part, the advantages of cryogrinding over conventional grinding are described. The latter part discusses the details of the pilot plant, along with preliminary experimental studies using pepper as the sample. Experimental results clearly indicate the quality improvement of the cryoground spice in terms of increased volatile oil contents and flavour components.

INTRODUCTION

India is the largest grower and consumer of spices with an estimated annual output of more than two million tons of spices. However it exports only on the average 200,000 tons of various spices, mostly in the unprocessed form. The emphasis is fast shifting to exporting value added products, i.e., processed spices in place of whole spices. However, for the acceptance of processed spices by the importers, the quality of processed spices using conventional spice grinding mills is causing grave concern. This is primarily because of the substantial loss of etheric oils by conventional grinding and hence poor quality, as well as the deterioration of the processed spices due to an environment conducive to bacterial growth. Cryogrinding technology overcomes many of these problems and provides a product of superior quality.

The paper discusses the technology of cryogrinding and the pilot plant studies carried out at Indian Institute of Science, Bangalore.

FRACTURE MECHANICS

The disintegration of materials is a process of subjecting solid substances to mechanical forces. A particle fractures when the internal molecular binding forces are overcome to an extent that a minimum depth of cleavage is produced. A major breakthrough occurs as an extension of this cleavage rather easily. For an ideal continuous crystal structure, the theoretical force required to achieve the minimum cleave is very high. However, in practice the force required is only a fraction of this theoretically estimated value. This is due to the defects in the material structure, which can initiate the process of disintegration with a lower applied force. The energy required to achieve lower particle size in a size reduction process is higher as the particle size gets smaller. This can be attributed to the lesser number of defects in the already fragmented particles. A wide variety of milling techniques can be employed to achieve the desired particle size distribution.

The substance to be ground can be hard (rocks, ores etc.,) or soft (plastics, foodstuffs etc). While hard materials (up to No. 6 in the Mohs hardness scale) disintegrate fairly readily and operating temperature is not a major problem, for softer materials the situation is very different. Plastics and elastomers (not highly cross linked) respond to stress by uncoiling their long chain molecules and sliding over one another, the thermal energy of the material at room temperature aiding this process. This makes milling of these products very difficult. Foodstuffs are difficult to grind because of their varying degree of moisture or fat and often fibrous structures.

The process of cooling the material to be ground results in contraction, which aids in the propagation of cracks from defects or flaws in the molecular bonds. Also the thermal energy of the material, which allows certain ductility of the material, is reduced so that brittle fracture can be initiated.

PROBLEMS OF CONVENTIONAL GRINDING OF SPICES

Loss of etheric oils

In most of the grinding operations the applied energy (>99%) gets dissipated in the form of heat. This is primarily due to the non-uniformity of the feed material in size, structure, defects, dislocation etc., so that an accurately judged application of force to all sample elements is not possible. Also, the fracture process results in particles of varying size, needing unnecessarily large forces to get the product milled to a particular particle size.

Spices owe their characteristic aroma, and therefore their value as a spice, to the enclosed etheric oils contained in the oil cells or intercellular containers of spices[1] from which, in most cases, the etheric oils can be obtained by crushing. This is the primary reason, why most of the spices are sold and used as a ground product. However during conventional milling of spices, with no cooling of the mill and the product, the temperature increases to over 363 K in most of the fast rotating mills.[2] Since etheric oils have boiling points ranging from 323 K to 593 K,[3] some of the precious components are lost due to conventional milling, resulting in an inferior product.

Clogging and gumming of the product

When the material to be ground contains high levels of fat or oils (nutmeg, clove, etc.), they cause clogging of the mill, affecting throughput and quality of the product.[4] A high moisture content also gives problems in milling. In some cases it is possible to grind only after drying (e.g. chilli). High moisture content materials often tend to stick to the parts of the mill. The acceptable moisture content in practice varies from material to material.

Oxidation and related degradation

In the conventional grinding of spices, aromatic substances oxidize and become rancid as a result of the intimate cyclone effect of air in the vicinity of the grinding zone. This process is intensified by the fact that fresh surfaces available from grinding are readily reactant. In certain cases, conventional milling degrades the natural colour of the product (e.g. capsicum).

ADVANTAGES OF CRYOGRINDING OF SPICES WITH LIQUID NITROGEN

Most of the difficulties and degradations experienced in conventional grinding are overcome by grinding in plants refrigerated with liquid nitrogen. Some of the significant advantages of cryogrinding are listed below.

Higher retention of etheric oils

The lower operating temperature of the mill prevents excessive temperature of the grinding zone and results in higher retention of etheric oils.[4] The cryoground product retains nearly the original oil content.

Prevention of Oxidation and rancidity

Use of liquid nitrogen cuts down the oxidation of the spice oils and resulting rancidity by way of the evaporating nitrogen expelling any air in the mill and producing an inert atmosphere from the start to the end of milling. It also provides inexpensive, dry and inert atmosphere of nitrogen for storage and package of the ground product.

Increase in throughput and reduction in grinding power consumption

As discussed earlier, chilling of the spices makes it brittle. It also keeps oil and moisture in the product in the crystallized conditions during milling and avoids clogging.[4] These aspects result in substantial increase in throughput[4] and decrease in specific energy for grinding compared to conventional grinding.

Possibility to grind difficult spices

Low temperature causes fibers to shatter and thus enables grinding of fibrous spices such as ginger in one pass through the cryomill. Cryogrinding enables easy grinding of high oil content spices (e.g. nutmeg).[4] It enables grinding of chilli without pre-drying and retention of its original colour.[5] In fact, fine grinding of green spice is a potential application.

Finer particle sizes

Cryogrinding results in finer particle sizes of the ground spices.[6] This in turn eliminates speckling problems and settling rate of spice powders in liquid preparations is considerably reduced.

Reduction in microbial load

The very low temperature environment in crygrinding can reduce specific microbial load of the spices.

TECHNOLOGY OF CRYOGRINDING

Three cryogenic cooling methods can be adopted for cryogrinding.

(i) The feed is cooled in a LN_2 bath and then transported from the bath to the mill by means of a screw conveyor. The mill is cooled by the feed itself and also by the gaseous nitrogen released during the cooling process. The process is grossly inefficient and the grinding temperature cannot be controlled optimally.

(ii) LN_2 is injected to cool-down the feed stock while it is transported through a screw conveyer. Here again, the chilled feed as well as evaporated nitrogen cool the mill . The available enthalpy of the evaporated nitrogen is not fully utilized due to the absence of pre-cooling of the feed stock. In spite of the process inefficiency, all the advantages of cryogrinding can be achieved by this process and the plant is not complex.

(iii) The stock is initially precooled by the enthalpy of the evaporated nitrogen and the operating temperature of the mill can be controlled suitably for a specific application. Utilization of the cold gas circuit makes the process efficient. However the system becomes more complex.

In many cases it is possible to adapt the existing grinding systems for cryogenic operation. However, the material of construction of the mill should be compatible for low temperature operation. AISI 316 or AISI 304 stainless steels are the preferred materials for cryogenically exposed grinding elements.

The liquid nitrogen supply for a cryomill is done from a stationary tank, the capacity of which depends on the production requirement. The tank should be situated reasonably close to the mill to reduce loses from the pipelines. This tank consists of an inner vessel containing LN_2 and an outer jacket surrounding it. The interspace in high vacuum with multilayer insulation reduces the heat flux to liquid nitrogen to a minimum. The tank should be installed accessible for a LN_2 road tanker to refill the tank as and when needed.

The consumption of liquid nitrogen is of importance in optimizing the cost of cryogrinding. For continuous operation of the mill, the consumption of LN_2 comprises the following components:

(i) Cooling the raw material and mill (initially) to operating temperatures.
(ii) Compensating the heat liberated in the grinding process.
(iii) Thermal insulation loss of the grinding installation and cryogen systems.
(iv) Residual coldness of the ground product that leaves the mill.
(v) Residual refrigeration of the gaseous nitrogen that is vented out.

Optimization of the above five parameters can result in the economical use of liquid nitrogen for cryogrinding.

DEVELOPMENT OF PILOT PLANT FOR CRYOGRINDING

The work had the following objectives:

(i) Indigenous development of technology for cryogrinding of spices.
(ii) Design of the cryocooling and control systems for cryogrinding process.
(iii) Development of a pilot plant for cryogrinding and evaluation studies of spices.

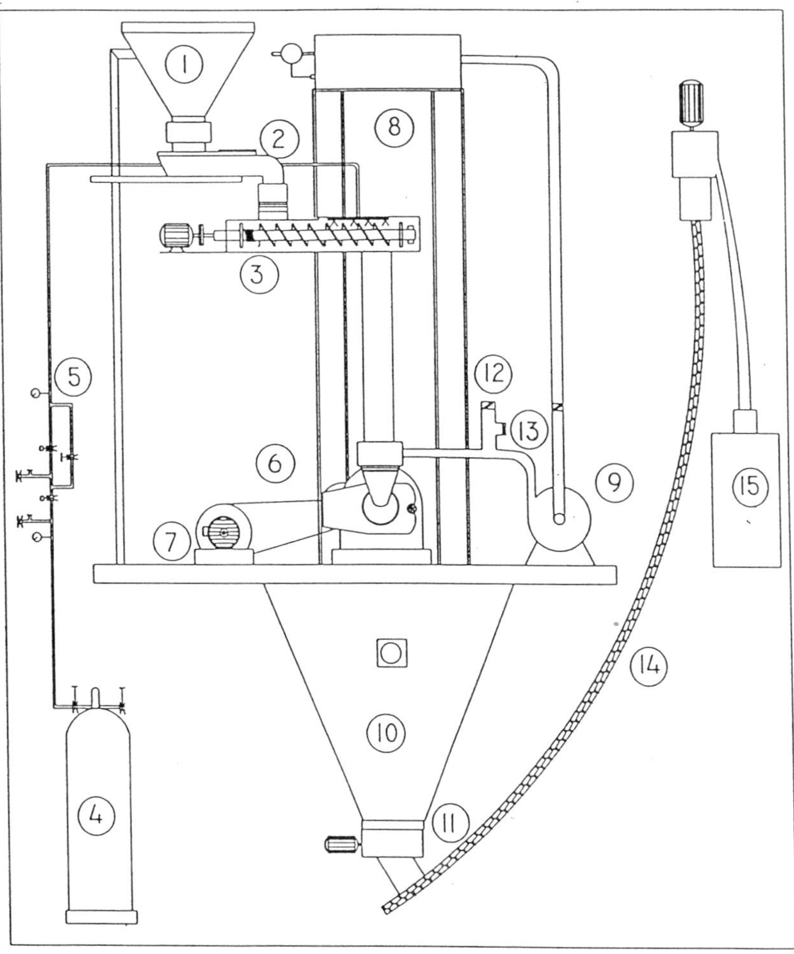

Figure 1. Schematic of pilot plant.

The schematic of the pilot plant designed and developed at the Indian Institute of Science is shown in Figure 1.

From the filling hopper(1), the spice to be ground enters the vibratory feeder(2), which has provision to control feed rate. The material falls into screw conveyor(3) driven by motor, reduction gear and inverter control. Liquid nitrogen from a storage container(4) is sprayed into the screw conveyor, and the time of stay of spice in screw conveyor is adjusted by the speed of the drive. A sensor monitors the temperature at the grinding zone and the liquid nitrogen spray is optimized using automatic feed back control by a digital signal processing system and associated electronics. The ground product is collected in collecting bin(10) and through a rotary a valve(11) and a Hapman Helix Conveyor(14) it is transferred to the storage receiver(15). The vapourised nitrogen from the mill is sucked by a blower(9) through the filter assembly(8) and fed back to the mill.

The pilot plant is totally designed and fabricated using stainless steel and other food quality materials.

SPECIFICATION OF MAJOR COMPONENTS

Cryogenic Screw Conveyor

- Designed at Indian Institute of Science and fabricated by M/s. Maschinen Fabrik, Calcutta.
- Maximum throughput 120 kg/hour for pepper.
- Suitable for operation between room temperature and 77 K.
- Liquid nitrogen consumption is 0.5 kg/kg of pepper.
- Instrumentation control software through DSP system developed at Indian Institute of Science.

Grinding mill

- Fine impact UPZ 160, Alpine AG, Germany.
- Grinding elements - stud disc (pin mill), fan beater and swing beater.

Temperature measurement and control circuits for Cryogrinding Mill

Platinum resistor temperature sensors (PT100) are used for monitoring temperatures at various locations. Temperatures are monitored at the following places:
(i) Inlet to the screw conveyor, (ii) Intermediate position in screw conveyor, (iii) Mid point of screw conveyor, (iv) End of screw conveyor (Mill inlet), (v) Mill bearing rinsing point, (vi) Product outlet and (vii) Blower suction side.

A personal computer with IEEE 488 interface card along with a scanner, digital multimeter (DMM), current source, etc., are used for this measurement. The product outlet temperature from the mill is used for controlling the liquid nitrogen supply rate in the screw conveyor. For this purpose, a DSP system and associated electronic circuit are used with the same PC for controlling the opening and closing periods of the solenoid valve. Keeping the frequency constant, the duty cycles of 'on' and 'off' periods were varied according to the temperature. This way the LN_2 consumption can be optimized. When the product outlet temperature reaches 233 K, the liquid nitrogen supply should be stopped. For this purpose, the main solenoid valve will be closed by a programmable power supply. It is also possible to control the screw conveyor speed using the computer.

EXPERIMENTAL STUDIES WITH PILOT PLANT

Experimental studies have been conducted with pepper and the results have been analysed for volatile oil and flavour components, particle size distribution and throughput of the mill.

Volatile oil and flavour components

Table 1. Comparison of ground samples of pepper

Sl No	Components Percentage	Cryoground	Conventional Ground
1	Moisture %	13.00	11.00
2	Volatile oil %	2.61	1.15
3	Flavour Compounds (Relative Concentration)		
	α-Pinene %	1.40	0.29
	Limonene %	8.30	1.18

Table 2. Particle size analysis of ground pepper

Particle size range (μm)	Cryoground (%)	Conventional ground (%)
<106	46.0	35.0
106-125	21.2	14.0
125-150	9.0	13.0
150-180	11.0	13.0
180-212	3.0	4.0
212-250	3.3	6.0
250-300	4.0	6.0
300-425	1.8	4.0
425-500	0.4	1.0
>500	0.3	4.0

Table 1 compares the results for cryoground sample with that of conventional ground sample of pepper and shows that, the cryoground sample has higher percentage of volatile oils and flavour components than the conventional ground samples.

Particle size distribution

Cryogrinding leads to finer particle sizes compared to conventional grinding and large particle size residue is negligible in cryogrinding, as shown in Table 2.

Throughput of grinding mill

- Conventional grinding ≅ 100 kg/hr (pepper).
- Cryogrinding (268 K) ≅ 225 kg/hr (pepper).

FUTURE TECHNOLOGY PROSPECTS

It is proposed to extend the studies to other spices such as coriander, ginger, cardamom etc. Studies on Cryogrinding of industrial plastic wastes as a sequence to the present project, in view of the economic advantage and the possibility to process non-biodegradable plastic wastes can also be taken up.

CONCLUSIONS

Cryogrinding using liquid nitrogen as a coolant has emerged as a viable and economically competitive technology for grinding of spices. This process enables much greater retention of etheric oils and aromatic substances of the whole spices and prevents oxidation and rancidity in the processed spices. Cryogrinding enables preservation of natural properties of the spices and provides a means of grinding specific spices which cause clogging in conventional grinding.

The experimental results on cryogrinding of pepper shows that 67% of the cryoground pepper has particle size less than 125 μm, compared to 49% obtained by conventional room temperature grinding from the same mill. The retention of volatile oils content increased by 125% under cryogrinding, which is significantly higher than that of conventional ground pepper. Another advantage of cryogrinding is that the throughput of the mill increases by a factor of 2.25, since the specific energy consumption is substantially lower as compared to conventional grinding. An integrated packaging unit will provide an inert nitrogen gas environment for packaging, which will prevent bacterial growth and retain spice quality during storage and transport.

ACKNOWLEDGMENT

The above Technology Development Mission project was funded by Ministry of Human Resources Development, through Planning Commission, Government of India and this valuable financial support is gratefully acknowledged. We are thankful to M/s Bhavani Foods Pvt. Ltd., for providing support in various aspects as industry partner. We also thank the staff of Centre for Cryogenic Technology, Indian Institute of Science, Bangalore for their help in setting up and operation of the pilot plant.

REFERENCES

1. V.B. Singh and Kirti Singh "Spices", New Age International (P) Limited Publisher, New Delhi (1996).
2. C.A. Pesek , L.A. Wilson and E.G. Hammond, Spice quality: effect of cryogenic and ambient grinding on volatiles, *Journal of Food Science*, 50:559 (1985).
3. Wilfried Lissack, Cold grinding of spices, *Lebensmittel Technik, 7 (1975).*
4. Advanced Techniques in Spice Plant, Cryogenic grinding, "Carousel" *Material Handling, Food Engineering, 33* (1976).
5. C.A. Pesek and L.A. Wilson, Spice quality: effect of cryogenic and ambient grinding on colour, *Journal of Food Science*, 51(5):1386 (1986).
6. Cryogenics Aims for Perfect Powder Particles, *Food Process Engineering,* (1973).

HIGHLY SENSITIVE MAGNETIC SENSOR CONSTRUCTED WITH A YBCO THICK FILM

Mineo Itoh,[1] Kazuto Yamagata,[1] Atsushi Omura,[1]
Masahiro Ishidoh,[2] Takumi Minemoto,[2] and Kazuya Mori[3]

[1]Dept. of Electronic Engi., Kinki University
Higashi-Osaka, Osaka 577-8502, Japan
[2]Div. of System Science, Kobe University
Kobe, Hyogo 657-8501, Japan
[3]R&D Unit, Tokin Corporation
Sendai, Miyagi 982-8510, Japan

ABSTRACT

In the present research, use was made of a Y-Ba-Cu-O (YBCO) thick film to construct a highly sensitive magnetic sensor using the spray technique. The YBCO magnetic sensor was fabricated to retain a residual resistance R_{res} at the boiling point of liquid nitrogen (77.4 K). The value of resistance R_{ms} for the magnetic sensor, in the transition region ΔT of the thick film, was highly sensitive to the excitation magnetic field B_{ex}. Here, ΔT is the region between the transitional temperature T_c^{onset} and residual resistance temperature T_c^{res}. The value of R_{ms} for an applied magnetic field of 50×10^{-4} T (= 50 gauss) exhibited an increase to a value of about 680% greater than the resistance found in a zero magnetic field, and did not display any evidence of hysteresis. The present paper examines the fabrication conditions and magnetic responses of a highly sensitive sensor constructed with a YBCO thick film. In addition, a study was done on the characteristics of a magnetic measuring system that uses this sensor.

INTRODUCTION

There is an increasing need for a magnetic sensor that is capable of application to a wide range of fields. Examples of such needs are the magnetic head for magnetic recording systems, a force sensor for automobiles and industrial robots, medical devices used for magnetocardiogram and magnetoencephalogram, urban traffic control systems, magnetic environmental sensors, among others. In addition, the yearly production output of magnetic sensors, such as the Hall, giant magnetoresistance (GMR), and magneto-impedance (MI) sensors, to name a few, has greatly increased.[1]

Advances in Cryogenic Engineering, Volume 45.
Edited by Shu *et al.*, Kluwer Academic / Plenum Publishers, 2000.

For use in constructing highly sensitive magnetic sensors, the present authors have been studying the preparation of Y-Ba-Cu-O (YBCO) thick films. The YBCO thick film as the magnetic sensor was fabricated by use of the spray technique, constructed to retain a residual resistance R_{res} at the boiling point of liquid nitrogen (77.4 K). The resistance R_{ms} for the YBCO magnetic sensor in the transition temperature region ΔT, was found to be highly sensitive to the excitation magnetic field B_{ex}. Here, ΔT is the range of temperature in the transition region between the resistance drop T_c^{onset} and residual resistance T_c^{res} temperatures.

The paste, consisting of YBCO powder mixed with ethylene glycol ($HOCH_2CH_2OH$), was sprayed on a MgO substrate and dried on a hot plate. This procedure was repeated about 130 times. The resulting YBCO thick film was then sintered in an electric furnace. When the film was cooled to 77.4 K, the resistance R_{ms} of the sensor for a B_{ex} of 50×10^{-4} T (= 50 gauss) exhibited an increase to a value 680% greater than the resistance in a zero magnetic field. In addition, the sensor displayed no evidence of hysteresis between 0 gauss and ±50 gauss. The change ratio of resistance for this sensor matches that of the MI sensor, which has the highest sensitivity of sensors in commercial use.

Experimental results revealed the characteristics that include the dependence of the R_{ms} on temperature T, and the dependence of the resistance ratio of R_{ms} at 50 gauss to R_{ms} at 0 gauss on the residual resistivity ρ_{res}. Also revealed are the fabrication process of the YBCO magnetic sensor and the characteristics of a magnetic measuring system that uses this sensor.

EXPERIMENTAL PROCEDURE

The YBCO powder, a commercial powder (Dowa, DSC-010002), was finely milled by use of a planetary type agate grinder. The grain size of the YBCO particles after milling was about 2 μm or smaller. The paste[2] for constructing the magnetic sensor was prepared by homogeneously mixing the fine YBCO powder with ethylene glycol using an agate mortar and pestle. The weight ratio of YBCO powder to ethylene glycol was 1:3. The paste was sprayed on a vicinal MgO(100) single-crystal substrate (20 mm long, 10 mm wide, and 0.5 mm thick) by use of an air brush, and then dried on a hot plate at 220 °C for 10 minutes. This process was repeated about 130 times to fabricate the magnetic sensor, such as illustrated in Fig. 1. Prior to the spraying, the MgO substrate was masked, using masking tape in order to fabricate the magnetic sensor with a geometry such as illustrated in Fig. 2. The masking tape was removed after completion of the spraying process, and prior to sintering.

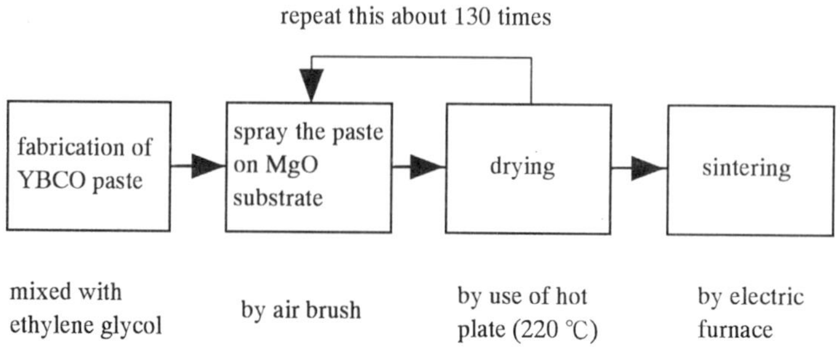

Figure 1. Schematic diagram illustrating the fabrication process of the magnetic sensor.

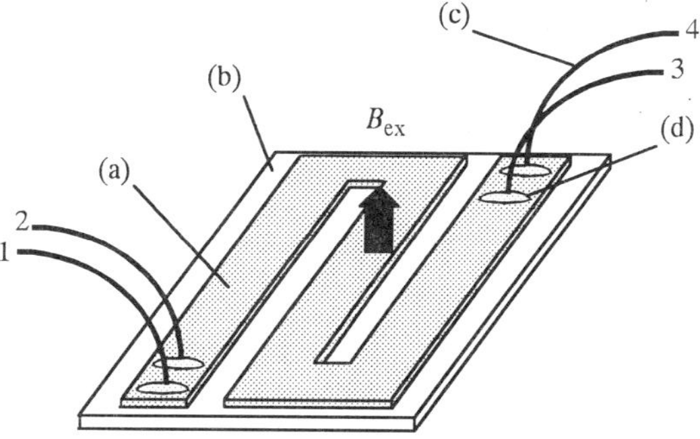

Figure 2. Schematic illustration of a typical YBCO magnetic sensor.

The specimens were heated at the rate of 4.5 °C/min in a flow of dry air to a temperature of 933 °C, using a programmable electric furnace, maintained at 933 °C for 30 minutes, and then cooled to room temperature at the rate of 4.5 °C/min. During the heating and cooling processes, two plateaus were maintained, one at 200 °C for one hour, the other at 570 °C for one hour. These plateaus were necessary for drying the thick film and the growth of crystallized (perovskite structure) grain in the film, respectively. The flow rate of the dry air was 4 l/min, which continued from a temperature of 300 °C during the heating process to 300 °C in the cooling process, a total 6 hours. The dry air flowed into the electric furnace through three filters, consisting of coconut charcoal, silica gel, and soda lime filters. Figure 2 illustrates the geometry of the YBCO magnetic sensor. In this experiment, four copper leads (diameter 0.1 mm) were attached to the sensor by use of a silver paste, such as described in Refs. 3-5.

In the present research, four magnetic sensors were fabricated (Sensor #1 through Sensor #4). These sensors differed in length, width, thickness, and number of spray applications. See Table 1 for the characteristics of the magnetic sensors.

Using a calibrated Pt thermometer at temperatures ranging between 77.4 K and 110 K, the temperature dependence of the resistance of the sensors was measured at a cooling rate of 0.5 K/min. The sensors and Pt thermometer were fixed to a block composed of oxygen free copper (OFC) with an OFC spiral for controlling a temperature. Magnetic characteristics were evaluated with the magnetic sensors placed in a homogeneous DC excitation of magnetic flux density B_{ex}, applied perpendicular to the surface of sensor. The values of B_{ex} were measured with a calibrated GaAs Hall sensor.

Table 1. Properties of the YBCO thick film as magnetic sensors

Sensor Number	#1	#2	#3	#4
T_c^{onset} (K)	96.04	96.80	97.54	101.7
ΔT (K)	16.0	15.7	15.5	16.7
ρ_{res} (77.4 K) ($\times 10^{-6}$ Ω m)	6.31	34.97	52.57	575.4
$\{R_{ms}$ (50 gauss) $- R_{ms}$ (0 gauss)$\}/R_{ms}$ (0 gauss)	6.81	1.52	1.66	0.51
Width (mm)	1.90	1.82	2.30	1.83
Length (mm)	15.0	47.0	50.6	39.2
Thickness (μm)	422	273	222	296
Number of spray applications	123	140	70	30

RESULTS AND DISCUSSION

Basic Characteristics of the YBCO Magnetic Sensor

Table 1 lists the results for Sensors #1, #2, #3, and #4, including the temperature of the onset of the resistance drop T_c^{onset}, the transition temperature region ΔT, resistivity of residual resistance ρ_{res} (77.4 K) without a magnetic field, and the ratio of the resistance R_{ms} (50 gauss) to the resistance R_{ms} (0 gauss) at a temperature of 77.4 K. Sensor #1 employed MgO substrates measuring 10 mm in length, 10 mm in width, and 0.5 mm thickness.

Figure 3 demonstrates the typical temperature T dependence of the resistance R_{ms}, in this case for Sensor #1, in the absence of an excitation magnetic field (B_{ex} = 0 gauss), with a current I_{ms} application of 100 μA (current density 125 A/m^2).

Figure 4 (a) shows the typical dependence of the resistance R_{ms} on the excitation magnetic field B_{ex} for Sensor #1. It is found that no hysteresis characteristics are observed over the range of B_{ex} from 0 gauss to ±50 gauss. Furthermore, no hysteresis characteristics were observed for Sensors #2, #3, and #4 for B_{ex} values of less than 50 gauss, respectively. In addition, the characteristics as shown in Fig. 4 (a) exhibit a symmetry with respect to the direction of the applied B_{ex}. The value of R_{ms} at 50 gauss was 0.91 Ω, about 680% greater than the resistance found in a zero magnetic field. The ratio of the resistance R_{ms} for a B_{ex} of 50 gauss to the resistance of a zero magnetic field (B_{ex} = 0 gauss) is termed the change ratio of the resistance. The change ratio of this resistance to the magnetic field is the same as that for a magnetoimpedance (MI) sensor,[1] which has the highest sensitivity of all sensors used commercially. However, the change ratio of the resistance for the MI sensor, about 330% in the range between 0 gauss and 20 gauss, decreases for B_{ex} values of more than 20 gauss. Similar characteristics such as those shown in Fig. 4 (a) were obtained in the frequency region from 10 Hz to 200 kHz, with a constant AC current of 70 μA rms (not shown). For B_{ex} values greater than 50 gauss, hysteresis characteristics are observed, such as shown in Fig. 4 (b). The value of R_{res} for a B_{ex} of ±200 gauss, however, exhibited an increase of about 2,000% over the resistance in a zero magnetic field.

Figure 5 displays the change ratio, namely, the ratio of the resistance R_{ms} (B_{ex} = 50 gauss) to the resistance R_{ms} (B_{ex} = 0 gauss) for Sensor #1 as function of the residual resistivity ρ_{res} (77.4 K). It can be seen that the ratio increases as the value of ρ_{res} (77.4 K)

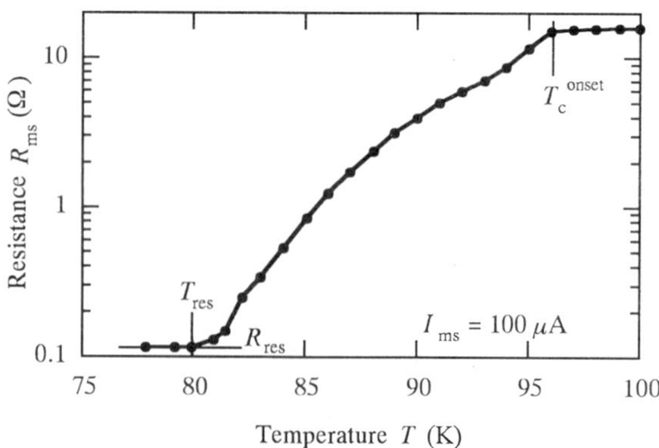

Figure 3. Typical temperature T dependence of the resistance R_{ms} for Sensor #1 in the absence of an excitation magnetic field (B_{ex} = 0 gauss), with a current I_{ms} application of 100 μA.

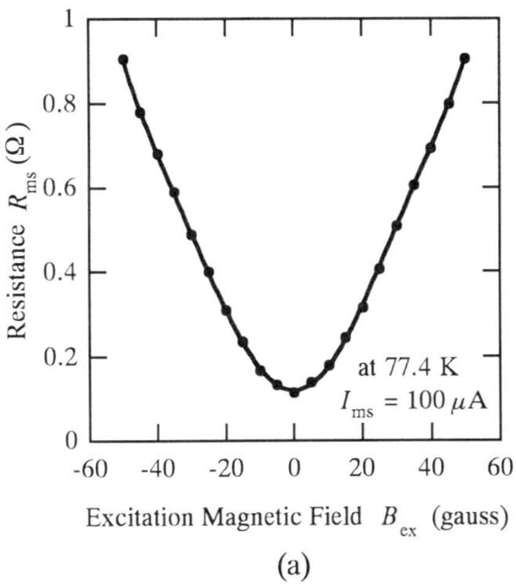

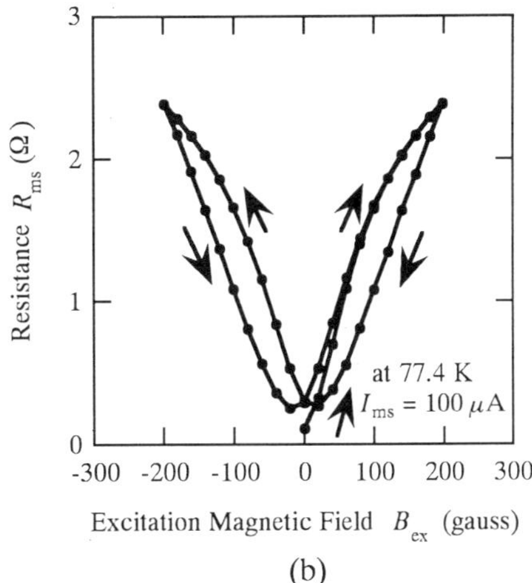

Figure 4. Dependence of the resistance R_{ms} on the excitation magnetic field B_{ex} for Sensor #1 for (a) the characteristics in the range of B_{ex} from -50 gauss to 50 gauss, and (b) the characteristics in the range of B_{ex} from -200 gauss to 200 gauss.

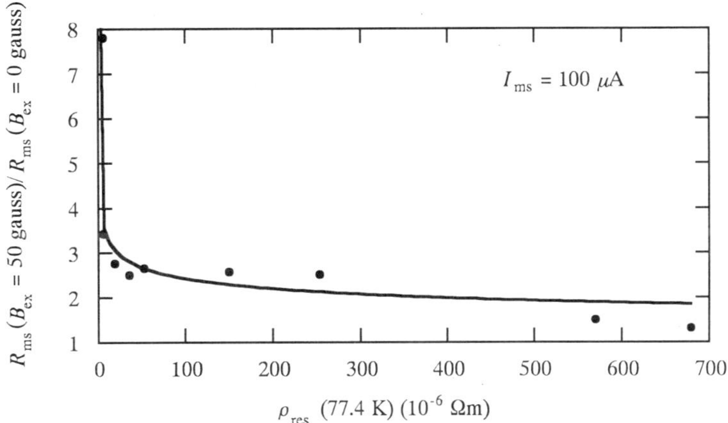

Figure 5. The ratio of the resistance R_{ms} (B_{ex} = 50 gauss) to the resistance R_{ms} (B_{ex} = 0 gauss) as function of the residual resistivity ρ_{res} (77.4 K).

decreases. This means that the ratio, i.e., the magnetic sensitivity of the sensor, can be estimated from values of ρ_{res} (77.4 K), which are more easily measured than the ratio of R_{ms} (B_{ex} = 50 gauss)/R_{ms} (B_{ex} = 0 gauss).

The X-ray diffraction measurements were carried out on Sensor #1 to analyze the crystallographic structure, The X-ray source was copper K_α, and an example of a diffraction pattern is shown in Fig. 6. This diffraction pattern was compared with that of an $Y_1Ba_2Cu_3O_7$ pellet specimen used as a standard, such as shown Ref. 3. Although the diffraction pattern of Sensor #1 exhibits the influence of the MgO(100) substrate, the dominant characteristics of the pattern are similar to those of the pellet specimen, which exhibits an orthorhombic structure. Similar X-ray diffraction patterns such were obtained for Sensors #2, #3, and #4 (not shown).

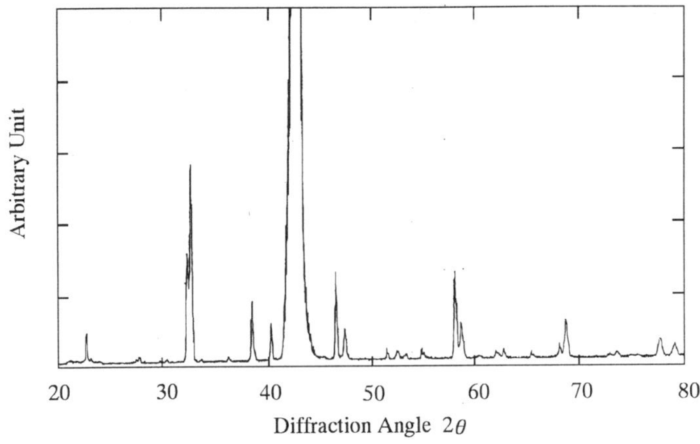

Figure 6. Typical X-ray diffraction pattern for Sensor #1.

Results of a Magnetic Measuring System Using a YBCO Magnetic Sensor

The magnetic measuring system was biased for a homogeneous DC magnetic flux density B_{ex} of 30 gauss, applied perpendicular to the surface of the YBCO magnetic sensor. The measuring magnetic flux density B_{meas} was then applied perpendicular to the surface of sensor. The output of a magnetic sensor, the resistance change due to the applied B_{meas}, was converted to a voltage signal by a Wheatstone bridge.

Figure 7 displays the ratio of the resistance change ΔR to the resistance R_{ms} (B_{ex} = 30 gauss) for Sensor #1, $100\Delta R/R_{ms}$ (B_{ex} = 30 gauss) (%), as function of the B_{meas} from 0 gauss to ±10 gauss. Here, the values of ΔR are determined by

$$\Delta R = R_{ms}(B_{meas}) - R_{ms}(B_{ex} = 30 \text{ gauss}). \qquad (1)$$

The sensitivity of the measuring system is about 4 %/gauss, with directional and linear

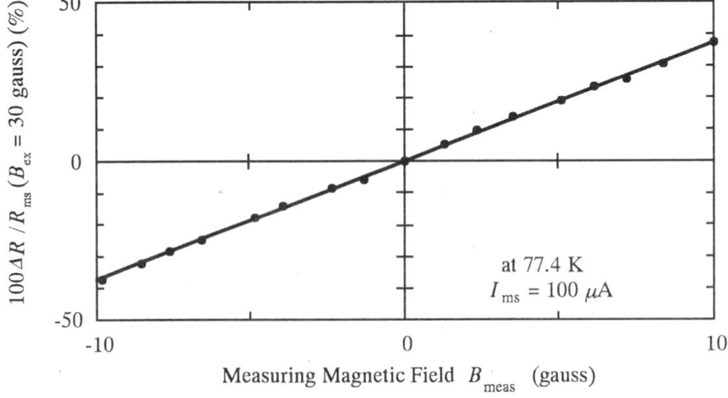

Figure 7. Typical dependence of the ratio of resistance change ΔR to the resistance R_{ms} (B_{ex} = 30 gauss) on the measuring magnetic field B_{meas}, for the magnetic measuring system.

characteristics. The sensitivity of this system is about 4 times that of the giant magnetoresistance (GMR) sensor.[1] In addition, this sensing system displayed no evidence of hysteresis in the magnetic region between 0 gauss and ±10 gauss.

CONCLUSIONS

As one of the basic areas of research for the development of a highly sensitive, high-critical temperature superconductor (HTS) magnetic sensor, the present paper has examined a highly sensitive magnetic sensor fabricated with a Y-Ba-Cu-O (YBCO) thick film. The resistance R_{ms} of the YBCO magnetic sensor for the excitation magnetic field B_{ex} of 50×10^{-4} T (= 50 gauss) exhibited to a value about 680% greater than the resistance in a zero magnetic field, such as shown in Fig. 4 (a). In addition, the sensor displayed no evidence of hysteresis in the magnetic region between 0 gauss and ±50 gauss. The change ratio of resistance for this sensor matches that of the magnetoimpedance (MI) sensor which has the highest magnetic sensitivity in commercial use. However, the change ratio of resistance for the MI sensor decreases in the range above 20 gauss.[1]

It was found that the ratio of the sensor resistance R_{ms} (B_{ex} = 50 gauss) to the sensor resistance R_{ms} (B_{ex} = 0 gauss) increased as the value of ρ_{res} (77.4 K), the resistivity of the YBCO magnetic sensor at a temperature of 77.4 K, decreased such as was shown in Fig. 5. This means that the ratio, i.e., the magnetic sensitivity of the sensor, can be estimated from the values of ρ_{res} (77.4 K), which are more easily measured than the ratio of R_{ms} (B_{ex} = 50 gauss)/R_{ms} (B_{ex} = 0 gauss).

The magnetic measuring system that was fabricated in the present research, was biased for a B_{ex} of 30 gauss, applied perpendicular with a homogeneous DC magnetic flux density, to the surface of the YBCO magnetic sensor. The sensitivity of this system, $100\Delta R/R_{ms}$ (B_{ex} = 30 gauss) (%), was determined as about 4 %/gauss, in the magnetic range of 0 gauss to ±10 gauss, with directional and linear characteristics. The sensitivity of this system is about 4 times that of the giant magnetoresistance (GMR) sensor. The measuring system also displayed no evidence of hysteresis in the magnetic region between 0 gauss and ±10 gauss.

ACKNOWLEDGEMENT

The authors would like to thank Mr. Yukio Hotta of Tokin EMC Engi. Co., Ltd. for cooperation in performing the experiments.

REFERENCES

1. For example, M. Mohri, "Magnetic Sensor", Corona, Tokyo (1998).
2. M. Itoh and H. Ishigaki, Preparation of superconducting Y-Ba-Cu-O thick film, *Japanese J. Appl. Physics* 27: L420 (1986).
3. M. Itoh and H. Ishigaki, and K. Demizu, Influence of carbon on critical temperatures of superconducting Y-Ba-Cu-O, *Japanese J. Appl. Physics* 28: L1527 (1988).
4. M. Itoh, H. Ishigaki, and T. Ohyama, Effect of Ag on magnetic shield of superconducting Y-Ba-Cu-O cylinder, *Cryogenic* 30: 863 (1990).
5. M. Itoh, H. Ishigaki, T. Ohyama, T. Minemoto, H. Nojiri, and M. Motokawa, Influence of silver on critical current of the Y-Ba-Cu-O superconductor, *J. Mater. Res.* 6: 2272 (1991).

EXPERIMENTAL STUDY ON MAGNETIC REFRIGERATION FOR THE LIQUEFACTION OF HYDROGEN

Katsuhide Ohira, Kenji Nakamichi and Hitoshi Furumoto

Mitsubishi Heavy Industries, Ltd. Nagasaki R&D Center,
5-717-1 Fukahori-machi Nagasaki 851-0392, Japan

ABSTRACT

An experimental study of magnetic refrigeration has been conducted for the liquefaction of hydrogen. The magnetic material selected is GGG ($Gd_3Ga_5O_{12}$). The experimental magnetic refrigeration system principally consists of a superconducting pulsed magnet, a Gifford-McMahon type refrigerator as heat expelling portion and a thermal siphon type heat pipe as a thermal switch on the low temperature side. In the GGG heat-pipe system, gaseous hydrogen coming up from liquid hydrogen reservoir condenses on the GGG surface, then falls down to the reservoir. Liquefaction rate is measured from the input power of the electrical heater in the hydrogen bath. The authors try to find the suitable gap width of the GGG heat-pipe system considering heat capacity around GGG and interaction between liquefied and gaseous hydrogen. Subsequently we conducted a hydrogen liquefaction test and obtained a liquefaction rate of 3.55 g/h (50 cc/h) or refrigeration power of 0.4 W at 20.3 K, and these experimental results agreed with the calculated results of cycle simulation.

INTRODUCTION

Recently, liquid hydrogen is being used not only as fuel for rockets but as clean energy. When demands for hydrogen increase, a major issue will be reduction of its liquefaction cost. Magnetic refrigeration can be expected to be more efficient than gas cycle refrigeration for the liquefaction of hydrogen. The authors had previously conducted a first step in an experimental series to present an experimentally made magnetic refrigeration test apparatus using GGG for its magnetic material and to evaluate the temperature-entropy characteristics of GGG and its refrigeration power using an electrical heater attached to GGG, in order to obtain the fundamental characteristics of magnetic refrigeration cycle which operates at a temperature level of 20 K.[1] In the present study, we improved the magnetic refrigeration test apparatus as described above by adding a newly designed heat pipe for the liquefaction of hydrogen and conducted a hydrogen liquefaction cycle test for the purpose of hydrogen liquefaction using the magnetic refrigeration cycle.

Advances in Cryogenic Engineering, Volume 45.
Edited by Shu *et al.*, Kluwer Academic / Plenum Publishers, 2000.

EXPERIMENTAL APPARATUS

Overall Structure and Magnetic Refrigerant

Figure 1 shows the overall structure of the magnetic refrigeration test apparatus. This apparatus is a static magnetic refrigerator which operates in Carnot cycle, consisting of a superconducting pulsed magnet, a heat-pipe type absorption heat switch, a magnetic material and a Gifford-McMahon(G-M) type refrigerator as heat expelling portion. For the magnetic material, cylindrical shape of GGG (0.035 m in diameter, 0.1 m high, 0.7 mol) was used.

Magnet

The magnetic field is added to the magnetic material by changing the energizing current of the solenoidal superconducting pulsed magnet, which had an external diameter of 0.116 m, a bore diameter of 0.085 m, a height of 0.2 m, maximum magnetic field of 5 T (energizing current of 150 A) and a maximum magnetization and demagnetization sweep rate of 0.36 T/s.

Rejection Heat Switch

Figure 2 shows a detailed configuration of the magnetic material and the heat switches. Above the magnetic material, the rejection heat switch consisting of SiO_2 and oxygen free high purity copper (OFHC) is provided and also, connected to the G-M refrigerator as heat expelling portion. The heat rejection switch moves with this switch contacting the OFHC block retained to the cold head of the G-M refrigerator, expelling heat of the magnetic material to the G-M refrigerator when contacting the magnetic material. Although the cold head of the G-M refrigerator can reach a temperature approximately 10 K, the temperature was controlled above 20.3 K using an electrical heater attached to the cold head. Furthermore, the contact surface between the magnetic material and the rejection heat switch was polished to 0.5 μm to improve the heat transfer performance of the rejection heat switch. In addition, bellows and springs were used in such a manner that surfaces could contact each other through spring force. The chamber for the rejection heat switch was sealed with helium gas (60 torr) to improve its heat transfer performance. Carbon glass resistance thermometers, which were not influenced by the magnetic field, were used for the temperature measurement of GGG and heat switch, and a Hall probe was used for the measurement of the magnetic field.

Thermal Design of Absorption Heat Switch

As shown in Fig. 2, the absorption heat switch used for this experiment was a heat pipe that condenses hydrogen vapor from the liquid hydrogen bath directly on the surface of the magnetic material. The key point in the heat pipe design relates to the gap width between the magnetic material wall and its container wall at the condensation portion of the heat pipe. The optimal gap width of the heat pipe for helium liquefaction was studied with experiment and analysis.[2] In the hydrogen liquefaction process, when the temperature of the magnetic material drops below the boiling point of gaseous hydrogen, gaseous hydrogen condenses over the magnetic material wall and hydrogen vapor equivalent to this condensation rate is supplied from the liquid hydrogen bath. If a fast vapor flows between the container wall and a liquid film developing over the surface of the magnetic material during this process, the liquid film would ripple and obstruct the smooth supply of hydrogen vapor to the gap in the condenser portion and thereby decrease the liquefaction rate.

The Carnot type magnetic refrigeration cycle consists of the magnetization and the

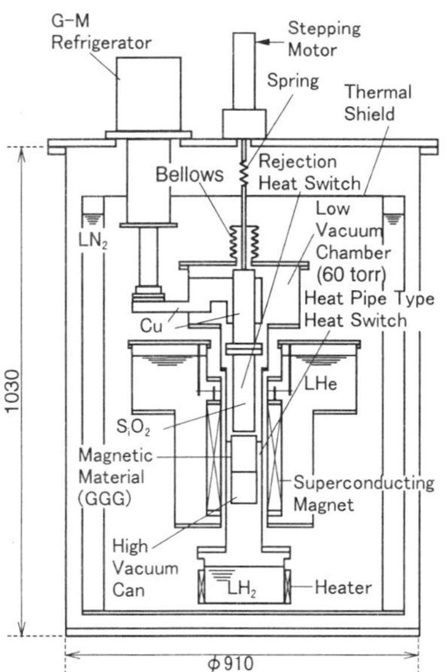

Figure 1. Cross sectional view of magnetic refrigeration test apparatus.

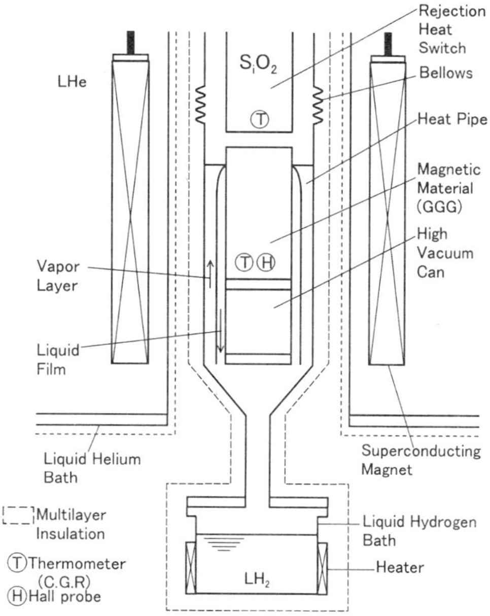

Figure 2. Schematic details of magnetic material and heat switches.

demagnetization process. There exists hydrogen vapor, around the magnetic material, in three processes of adiabatic magnetization, isothermal magnetization and adiabatic demagnetization in the gap of condensation portion. During the adiabatic magnetization process, heat from the magnetic material flows to heat capacities of the gaseous hydrogen in the gap and the container of magnctic material; whereas during the adiabatic demagnetization process, heat flows to the magnetic material. As a result, its cycle would deviate from the ideal Carnot cycle, decreasing the refrigeration power. Therefore, there should be an optimal gap width in the heat pipe condensation portion when the following considerations are taken into account:

(1) Interaction between hydrogen vapor supplied from the liquid hydrogen bath to the gap in the condensation portion and the falling liquid film, and

(2) Heat capacity losses during adiabatic magnetization and demagnetization process.

Design conditions of the heat pipe for hydrogen liquefaction are listed in Table 1.

Table 1. Parameters used for design of heat pipe

Parameter	Value
Size of the GGG	D = 0.035 m in diameter, L = 0.1 m in height
Heat transfer area	$A = 0.01143\ m^2$
Heat transfer coefficient[3]	$hm = 1757\ W/m^2K$
Temperature difference	$\Delta T = 1.3$ K
(Temperature of GGG 19 K)	
(Temperature of GH_2 20.3 K)	
Operating temperature	20.3 K−25 K

Interaction between the Liquefied and Gaseous Hydrogen

In the two-phase opposite flow, where a liquid film falls down along the surface of magnetic material and vapor comes up, the liquid film shows rippling and a portion of the liquid begins to come up when the vapor velocity reaches a certain limit. The vapor velocity at this point is called the flooding velocity. If the velocity of the gaseous hydrogen supplied from the liquid hydrogen bath and passing through the gap in the condensation portion increases, this flooding phenomenon is more likely to occur. If flooding occurs, the gap in the condensation portion would be blocked, obstructing smooth supply of hydrogen vapor to the surface of the magnetic material and decreasing the liquefaction rate. Therefore, we studied the gap width in the condensation portion to determine factors that could inhibit this flooding.

For a stable liquid-vapor interface to exist in liquid-vapor opposite flow, the wavelength of a standing wave must satisfy the relationship determined from the Helmholtz's instability theory.[4] (Symbols in equations (1)~(5) are listed in nomenclature table at end of the paper.)

$$\frac{\lambda}{2\pi(1+\rho_g/\rho_\ell)} \le \frac{\sigma}{\rho_g(u_g/u_\ell)^2} \tag{1}$$

Here, there is a relationship shown below between the wave length of the standing wave λ and the film thickness δ. The constant value k is determined from experiment in air-water system.[4] We choose three as value of k which shows a low limit of flooding velocity.

$$\lambda = 4\pi(1+\rho_g/\rho_\ell)k\delta \tag{2}$$

Because it was considered that flooding is most likely to occur in the lowest portion of the condensation area showing the fastest liquid velocity, the film velocity u_δ was used for u_ℓ. Film thickness and velocity of liquid and vapor were calculated by the following equations. (Equations (3) and (4) are Nusselt's laminar condensation theory)

$$\delta = \left(\frac{4K_\ell \mu_\ell L \Delta T}{g\rho_\ell(\rho_\ell-\rho_g)h_{fg}}\right)^{1/4} \tag{3}$$

$$u_\delta = \frac{g(\rho_\ell-\rho_g)\delta^2}{2\mu_\ell} \tag{4}$$

$$u_g = \frac{h_m A \Delta T}{\rho_g A_{gap} h_{fg}} \tag{5}$$

At the parameters of Table 1., $\delta = 0.075$ mm, $u_\delta = 0.15$ m/s and $u_g = 0.33$ m/s. Figure 3 shows relationship of the gaseous hydrogen velocity relative to the hydrogen liquid film versus the gap width in the condensation portion. If the gap width in the condensation portion is increased to 0.3 mm or more in the heat pipe, flooding can not occur.

Heat Capacity Losses

Figure 4 shows the relationship between gap widths and heat capacity losses in the condensation portion. Heat capacity losses were calculated from enthalpy changes of the heat load around the magnetic material (gaseous hydrogen in the gap of the condensation portion and the magnetic material container made of stainless steel SUS304L), assuming that the temperature of the heat load is between 20 K and 25 K which was equal to the operating temperature.

Based on the foregoing study, 1.5 mm at room temperature (1.4 mm at 20 K) was chosen for the gap width in the condensation portion. This gap is wide enough to inhibit flooding, and the ratio of heat capacity losses to refrigeration heat capacity in the magnetic refrigeration cycle at this gap width can be inhibited at about 25%.

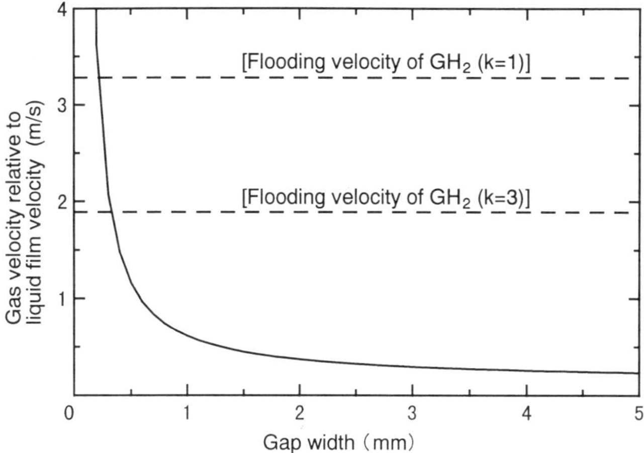

Figure 3. Relationship between gap width and gas velocity relative to liquid film velocity.

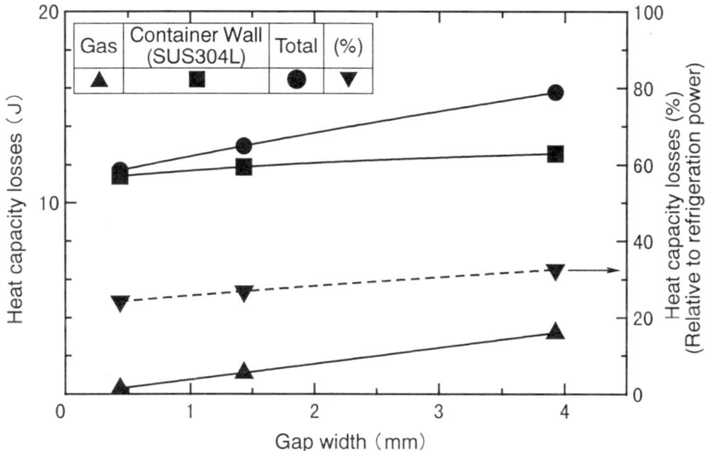

Figure 4. Heat capacity losses of the test apparatus.

EXPERIMENTAL RESULTS AND DISCUSSION

Experimental Procedure and Cycle Operation

First, the rejection heat switch is precooled to a temperature level of 20 K using the G-M refrigerator and sealed with gaseous helium after the precooling process. Then the liquid helium and liquid para hydrogen is stored in the liquid helium and liquid hydrogen baths respectively. After precooling, the magnetic refrigeration cycle is started. Figure 5 shows changes with time of the GGG temperature, the magnetic field and the pressure of the liquid hydrogen bath in the magnetic refrigeration cycle, which operates between T_l (20.3 K) and T_h (25 K). In the magnetization process, heat is rejected by bringing the rejection heat switch in contact with the GGG, whereas in the demagnetization process, gaseous hydrogen condenses and is liquefied over the GGG surface if the GGG temperature drops below the temperature for hydrogen liquefaction (20.3 K). Repeating this cycle, the electrical heater input is adjusted such that pressure and temperature of the liquid hydrogen bath can be stabilized, and then, the hydrogen liquefaction rate is measured.

Characteristics of Thermal Switch

Figure 5 also shows temperature changes of the rejection heat switch. Because the contact between the GGG and the rejection heat switch was insufficient in the heat switch experimentally made earlier by the authors,[1] its heat transfer coefficient was 0.07 W/K; however, in the present study, the heat transfer coefficient has been improved to 0.34 W/K by improving its contact method using bellows and springs.

From the experimental results, the heat absorption switch coefficient can be predicted to be 6 W/K. The maximum liquefaction efficiency (ratio of the refrigeration power of the cycle in the experiment to the refrigeration power of the ideal cycle) was 78% obtained at the sweep rate of 0.36 T/s. This indicated that heat capacity losses were inhibited well and no flooding occurred in the heat pipe, so design of the heat pipe for hydrogen liquefaction was appropriate.

Magnetic Refrigeration Test

The authors conducted a hydrogen liquefaction test using the sweep rate as a parameter. For the test conditions, 5 Tesla was used as the maximum magnetic field, 0.08 T/s to 0.36 T/s as the sweep rate, and 25 K as the temperature of rejection heat. The hydrogen liquefaction cycle was calculated,[5] using input data of coefficients of absorption and rejection heat switches, etc., as obtained from the present test. Figure 6 compares Carnot cycles as obtained from the experiment and the calculated results of cycle simulation shown on the temperature-entropy diagram of the GGG. It has been verified that the experimental results agreed well with the calculated results of cycle simulation. Figure 7 shows the effect of the sweep rate on the magnetic refrigeration cycle. If the sweep rate is too slow, distortion of this cycle would be enhanced, decreasing the temperature of rejection heat and also the change of enthalpy in the isothermal heat absorb process, resulting in a decreased refrigeration power. Figure 8 shows the effect of the sweep rate on the refrigeration power (liquefaction rate) using varied sweep rates when the peak magnetic field is fixed at 5 Tesla and the temperature of rejection heat at 25 K. The experimental results approximately agree with the simulated results and this shows that the results experimentally obtained were appropriate.

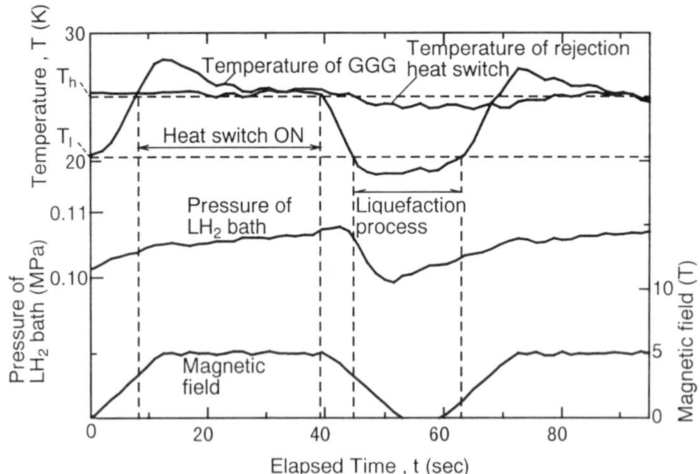

Figure 5. Magnetic refrigeration process.

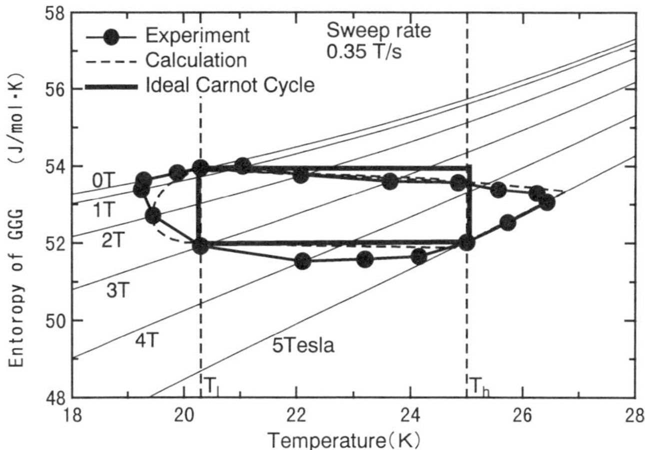

Figure 6. Experimental and calculated results of magnetic Carnot cycle on T-S diagram of GGG.

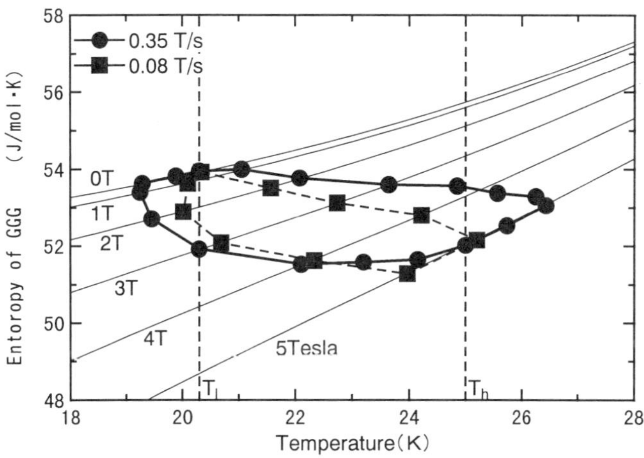

Figure 7. Experimental results of magnetic Carnot cycle at various ramp rates on T-S diagram of GGG.

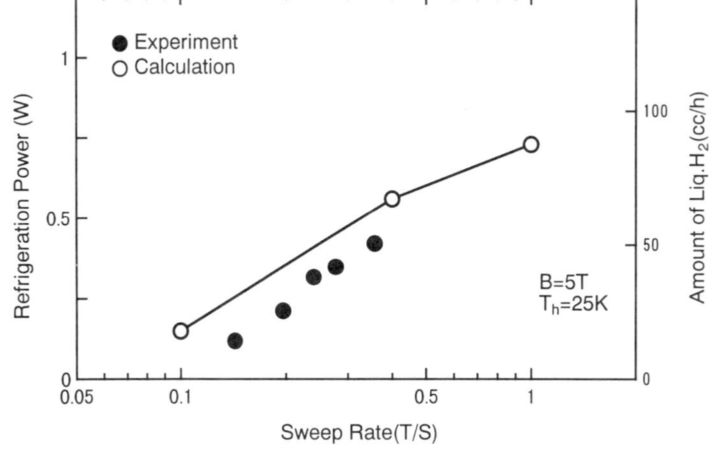

Figure 8. Experimental and calculated results of refrigeration power at various ramp rates. (Maximum field of 5 T and heat rejection temperature of 25 K were constant on the cycle.)

The maximum refrigeration power of 0.4 W was obtained at the fastest sweep rate of 0.36 T/s. (This rate can be converted to a liquefaction rate of 3.55 g/h (50 cc/h).)

CONCLUSIONS

The gap width of a heat pipe for hydrogen liquefaction was studied in consideration of flooding and heat capacity losses at the condensation portion of the heat pipe and 1.5 mm at the room temperature gap width was chosen. Hydrogen liquefaction tests by magnetic refrigeration cycles were conducted. The experimental results were a refrigeration power of 0.4 W with liquefaction efficiency of 78% and Carnot efficiency (FOM : Figure of Merit) of 37% at the fastest sweep rate of 0.36 T/s which indicated that the design of the gap width for the heat pipe was appropriate.

NOMENCLATURE

A	: area of condensing surface
A_{gap}	: area of gap
D	: diameter
g	: acceleration of gravity
h_{fg}	: latent heat of vaporization
h_m	: mean condensing heat transfer coefficient
k	: constant value ($1 \leqq k \leqq 3$)
K_ℓ	: thermal conductivity of liquid
u_g	: velocity of vapor
u_ℓ	: velocity of liquid
ΔT	: temperature difference (between GGG wall and vapor)
δ	: film thickness
λ	: wavelength of standing wave
μ_ℓ	: viscosity of liquid
ρ_g	: density of vapor
ρ_ℓ	: density of liquid
σ	: surface tension

REFERENCES

1. Ohira, K., Matsuo, S., Furumoto, H., "The characteristics of magnetic refrigeration operating at the temperature of 20 K", *Proceedings of 16th ICEC* 403-406 (1996).
2. Numazawa, T., Hashimoto, T., Nakagome, H., "Improvement of liquefaction efficiency of the heat pipe type magnetic refrigerator", *Adv.Cryog.Eng.* Vol.31: 771-777 (1986).
3. Ohira, K., Furumoto, H., "An experimental investigation of film-condensation heat transfer of hydrogen in a vertical tube", *Adv.Cryog.Eng.* Vol.35: 421-428 (1990).
4. Monde, M., Mitsutake, Y., Kurihara, A., "Analysis of critical heat flux in two-phase thermosyphon", *JSME* Vol.61(591): 303-310 (1995).
5. Numazawa, T., Kimura, H., Sato, M. and Maeda, H., "Carnot magnetic refrigerator operating between 1.4 and 10 K", *6th International Cryocooler Conference Vol. II* 199-213 (1990).

CRYOGENIC POWER / ENERGY DISTRIBUTION SYSTEM

O.M. Mueller, E.K. Mueller

LTE
Ballston Lake, NY 12019

ABSTRACT

An innovative power and energy distribution system for high-rise structures and large buildings is proposed and described. It is based on the integration of two emerging technologies, namely those of High-Temperature Superconductivity (HTS) and the new concept of Cryogenic Power Conversion (CPC) using Low-Temperature Operated Semiconductor Devices (LOTOS). In the proposed system HTS cables supply DC power and cooling fluid to a set of distributed ultra-high efficiency, low-cost, miniature Cryo-Silicon Transformers (CST) acting as DC/AC inverters. The CSTs implemented with Cryo-MOSFETs or Cryo-IGBTs/MTOs and HTS filter inductors will enable drastic reductions in size and weight in comparison to conventional copper/iron transformers. They will probably also be less expensive than HTS transformers. Integrated into the complete energy distribution system the CST will save due to its much smaller footprint expensive floor space in any high-rise building. In addition considerable electrical energy savings of 5 % to 10 % in future building complexes should be achievable, especially in hot areas. Similar thermal energy savings are possible in the power station. The use of DC distribution solves also any problem with AC losses in the HTS cables. The proposed system could be a first profitable application of HTS cables as well as CPC. It can also be integrated with the air-conditioning and LN2 fire extinguishing systems.

INTRODUCTION

This paper describes an innovative energy distribution and conversion system suitable for high-rise structures, large building complexes, factories, ships, airplanes, city blocks, etc., based on two new and emerging technologies: High-Temperature Superconductors (HTS) coupled with the new concept of Cryogenic Power and Energy Conversion (CPEC) using Low-Temperature Operated Semiconductor Devices (LOTOS). A key element of this system is the high efficiency, low-cost Cryo-Silicon Transformer / Inverter (CST) implemented using Cryo-MOSFETs or Cryo-Cool-MOS devices. The CST will demonstrate drastic reductions in size and weight in comparison to copper/iron transformers or conventional inverters. Integrated into the complete energy distribution system, it will save expensive floor space and help to achieve electrical energy savings of 5 to 10% in the distribution system. This corresponds to a total energy saving of 10% to 20% assuming a 50% generator efficiency. In addition, load shedding, power quality, and fire protection are provided. In hot areas the new system supports also the air conditioning / cooling system or it can be integrated with it. The basic circuitry developed for the CST can also be used for high power cryo-motor drives.

The silicon technology that led the computer industry to revolutionize the world has not

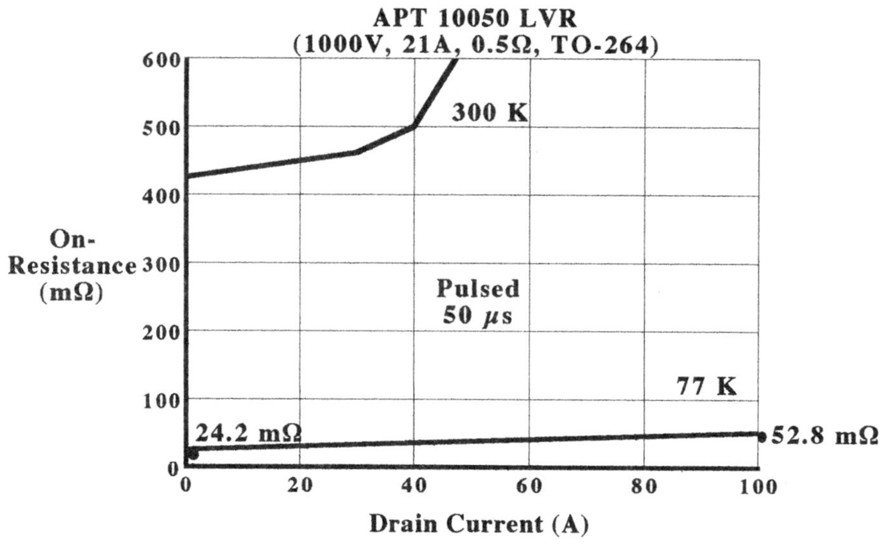

Figure 1. On-resistance (R_{DS-ON}) as a function of drain current (I_D) of the MOSFET APT10050LVR for room temperature (300 K) and 77 K. Pulsed measurement.

yet been seriously applied to the copper-and-iron realm of power engineering. However, this is now a possibility through the application of the Cryogenic Power Conversion (CPC) concept. The basic research for CPC has been done and awaits implementation[2-18].

The concept of Cryogenic Power Conversion is based on the fact that the conduction losses of power MOSFET switches and the switching losses of IGBTs, MTOs, Thyristors, GTOs, etc. are drastically reduced by cooling the devices to the temperature range of 77 K (liquid nitrogen) to 200 K. This can be seen in the on-resistance measurements of Figure 1 for the APT10050LVR (1000 V, 21 A, 0.5 Ω, TO-264). The area between the 300 K and 77 K curves represent the Cryo-Gain which is even larger taking into account the higher junction temperatures (400-425 K) in actual operation.

The physics behind this effect is the drastic increase at low temperatures of the majority carrier electron mobility in the drain-drift region of a high-voltage n-channel power MOSFET. MOSFETs are the fastest switching power devices available. Even further improvements, i.e. further reductions in on-resistance, are possible using the new Cool-MOS power devices developed by Siemens [2-3] and recently also by International Rectifier Corporation. Cryo-MOSFETs are best for applications below 800 V.

In minority carrier devices such as IGBTs, MTOs, MCTs, etc. for the higher voltage range (1-6 kV) [4] charge storage limits the switching speed. These charges are proportional to the minority carrier lifetime which in turn is drastically reduced by cryo-cooling thus reducing switching times, and herewith also the switching losses considerably. Figure 2 shows how the resistive (100 Ω load) turn-off time of a 1700 V IGBT is reduced from 288 ns at 300 K to 39 ns at 77 K when operated at 1200 V.

Figure 3 shows the concept of Cryogenic Power Conversion. The power dissipation of a high power circuit is reduced at the source by cryo-cooling thus permitting a drastic size, weight and cost reduction by eliminating big heatsinks, etc.

The U.S. Department of Energy (DOE) is funding several efforts to develop high-temperature superconducting cables. An added benefit of the proposed system is that only relatively short cable lengths are required, and these do not have to be designed for high-voltage (>100 kV) applications. Thus, the HTS industry finds a much needed new market for their cables, and the entire system can be realized in a period of a few years for any new high-rise building to be constructed. One can assume that sooner or later, such lossless cables will be available for applications in Cryogenic Energy Distribution. HTS cables will solve the key problem of CPC and the Cryogenic Energy Distribution System, the cryo-cooling, by providing a cryogenic fluid such as liquid nitrogen.

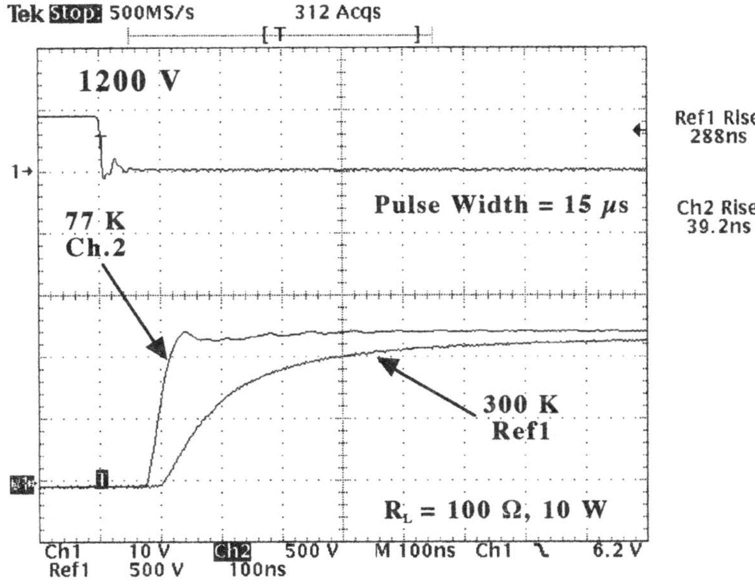

Figure 2. The resistive turn-off waveforms of a 1700 V IGBT (APT15GF170BRX) at 1200 V and 12 A. A reduction of the turn-off time from 288 ns to 39 ns at 77 K is demonstrated.

Even if it should turn out that HTS cables are too expensive [5], the CED system could be implemented using liquid-nitrogen-cooled cooper or aluminum cables, whose DC resistance drops by about the same factor (~ x 7-8) that applies to LN2 generation (7-10 W/W). Another advantage of cryogenic operation is the drastic reduction of the thermal conductivity of silicon and the substrate materials such as copper, beryllium-oxide, etc.

BRIEF SUMMARY OF THE CED SYSTEM

The existing electric distribution infrastructure has remained unchanged for almost a century. LTE proposes a system change which will bring the benefits of higher efficiency, reduced energy consumption, power load shedding, voltage regulation, power quality, power factor correction and voltage control. Additional benefits will be described.

The proposed new Cryogenic Energy and Power Distribution (CED) system (patent pending) is shown in Figure 4 based on the application of the two new technologies of High-Temperature Superconductivity (HTS) and Cryogenic Power Electronics (CPE). The CED system combines the following concepts, features and configurations:

- Installation of a central high-temperature superconducting DC cable (650 VDC or higher) cooled by liquid nitrogen (LN2, 77 K) and extending from the top floor to the basement of the high-rise (or any other) building through its center (Figure4)
- HTS Cable Power Distribution: High-Temperature Superconducting (HTS) DC cables supply electrical power and the cooling medium - liquid nitrogen - to the main power loads. The use of direct current permits a doubling of the transmitted energy per wire and the elimination of AC losses in the HTS cables, thereby allowing for reduced cable weight. The DC voltage is 650 VDC (480 VAC rectified) or any other suitable voltage (1.2 kV, 2.4 kV). Above 800 VDC one may use cryo-IGBTs for the CSTs.
- Cryogenic Power Conversion (CPC): The conventional heavy and bulky copper / iron core transformers (for example 460 VAC/120VAC, 75 kVA: 500 pound, 1.5 square meter footprint) are replaced by ultra-small, lightweight DC/AC solid-state inverters or DC/DC converters made with cryo-cooled power MOSFETs.

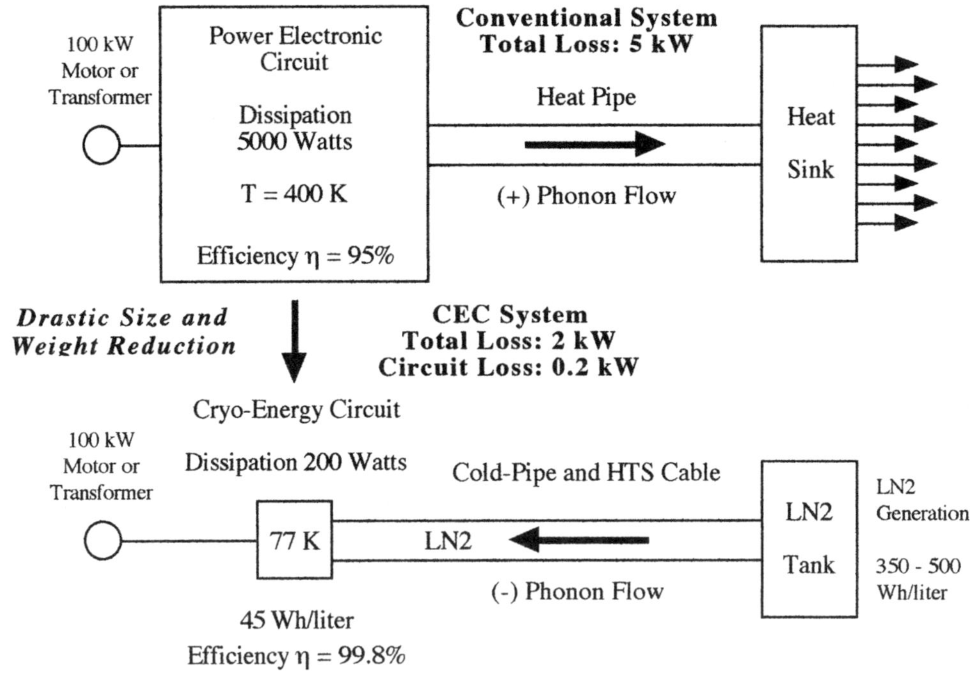

FIGURE 3. The Cryo-Energy Conversion (CEC) Concept: Loss reduction permits miniaturization.

- These cryo-silicon transformers (650 VDC to 120/240 VAC) may be integrated directly with the HTS cable structure due to their small size, exhibit an electrical efficiency of >99.8 % (without cooling penalty), and save a lot of valuable, expensive rental space.
- With some minor modifications, the cryo-transformer / inverter circuits can also be used as motor drives and adjustable speed drives (ASDs). They have a high overdrive capability.
- Cryogenic Power Electronics (CPE) provides drastically increased reliability and lifetime. In addition, it permits power quality, power factor correction, voltage regulation, HTS current limiting, load shedding and so on.
- The thermal insulation of the circuit dewar as well as the HTS cables can be implemented using polymer thin-film multi-layer isolation (MLI) in a vacuum, which can also be used to increase the capacity of a capacitive energy storage system (patent pending). The thermal losses of HTS cables or Cryo-Inverters can be dimensioned in such a way that they add to air-conditioning and space cooling in hot areas if desired. In this case the cables can be made at a lower cost.
- The ubiquitous availability of liquid nitrogen in a CED system provides an excellent new means for fire protection and suppression. LN2 was used to extinguish the burning oil wells in Kuwait!

The proposed CED system can be used for commercial energy and power distribution in all new large buildings, especially high-rise buildings, factories, large-scale offices, ships, etc. A high-rise building in Boston, for example, has 62 floors and 124 large and bulky cooper-iron distribution transformers. The implementation of this system also provides the feature of electrical power load shedding of 7-15%, depending on the air-conditioning load (outside temperature), and an electrical energy saving of 5-12% in high rise city blocks. Additional savings are possible in these high-rise buildings due to the small footprint of the silicon transformer compared to conventional copper/iron transformers. Rental costs range from $150 to $600 per square meter per year. Thus, the rental value of one square meter of space for a 50 year time span is about $7,500 to

$30,000. This price is much higher than the cost of a power transformer itself. The cryo-silicon transformer technology can also be applied to motor drives, especially in combination with HTS motors.

The complete CED system is shown in Figure 4. The Cryo-Silicon Transformers are cooled by LN2, which is supplied by the HTS cable. It is assumed that each floor requires at least 1 - 2 inverters / transformers. At the top of the building, a liquid nitrogen tank is placed, as shown in Figure 5. The HTS cable supplies the electrical power via 650 VDC obtained after rectification of the usual 3-phase, 480 VAC power source delivered by a conventional or HTS transformer from a high voltage source such as 13.8 kV or a fuel cell. This transformer provides DC isolation to the outside power grid.

Synchronous Cryo-MOSFET rectification can also be used in order to reduce the losses of the rectifier circuitry. Note that DC cables permit the transmission of twice the power for the same voltage ratings and wire cross-sections.

EPRI, the Electrical Power Research Institute, Palo Alto, CA, issued a technical report in 1995 entitled "Proof of Principle of the Solid-State Transformer" [1]. It is believed that, if the silicon transformer is feasible at room temperature as shown by EPRI, it will be even better at cryogenic temperatures (77 K) in combination with HTS cables.

For a 60 Hz silicon transformer the PWM (pulse-width-modulated) switching frequency of a switch-mode inverter can be low (1 - 10 kHz) so that switching losses are also small or negligible if in addition soft-switching techniques are applied. The efficiency is determined by the ratio of on-state voltage and voltage swing. The APT MOSFET (Advanced Power Technology's Metal-Oxide Silicon Field-Effect Transistor) APT 10050 LVR, rated 1000 V, 21 A, and 0.5 Ω (at 300 K), has an on-resistance of 24.2 mΩ at 77 K, i.e. immersed in liquid nitrogen (LN2). For a supply of 650 V, a current of 10 A, and 2 MOSFETs in series (bridge circuit), the on-state voltage to voltage swing ratio is:

$$L = 2\,\frac{0.242\ \text{V}}{650\ \text{V}} = 0.00075$$

This corresponds to an efficiency of >99.9 %. Assuming a cooling penalty of a factor 10 (10 W/W for LN2 production) and negligible switching losses at these low frequencies an overall transformer efficiency of >99.0% should be possible. By paralleling more MOSFETs one can further reduce the losses to any desirable level: "Silicon is cheap"!

Considering "Load Shedding" one can use the 99.9 % figure, since the cooling LN2 can be produced at a time of low load power consumption.

The main advantage of the cryo-silicon transformer is its potential for greatly reduced size, weight and cost. Nothing beats semiconductors as far as cost and size reduction are concerned! A cryo-silicon transformer which does not require a large LN2 tank or an expensive cryo-cooler can be quite small. It can be put into a container of a few liters. Where a 75 kVA conventional copper / iron transformer weighs about 500 lbs, a cryo-silicon transformer may have a weight of a few pounds given the availability of a central LN2 source. This is supplied by the HTS cable. It is believed that a cryo-silicon transformer can be made even so small that it can be mounted inside the wall or floor of a high-rise building, thus saving a lot of expensive rental space. Figure 6 compares a conventional copper/iron transformer with a cryo-silicon transformer inverter.

DETAILED DESCRIPTION OF A CED SYSTEM

The efficiency of conventional transformers (75 kVA) at full load is about 95 - 96%, and their weight around 500 pounds. The core losses are 400 - 600 W, the (reactive) magnetization or excitation power 750 - 2400 VA. The efficiency decreases rapidly at lower load levels due to the required magnetization power. This means a conventional transformer uses considerable power even in open circuit operation under no-load conditions. If one multiplies these numbers with the millions of transformers installed worldwide, one obtains quite substantial loss figures. A silicon transformer does not require magnetization power and has zero core losses. Its own losses can be made quite small. A problem remains: The filter inductors.

A cryo-silicon transformer can operate with an efficiency of >99.8%, not counting the cooling penalty. On a hot summer day in a large southern city the air conditioning system

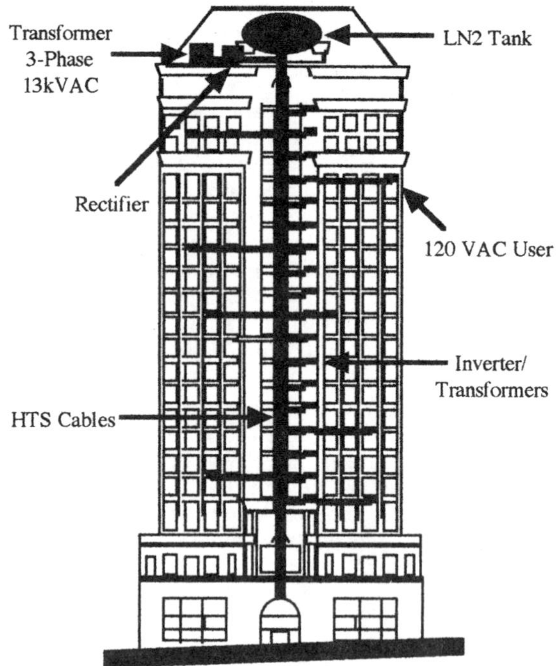

Figure 4: Cryogenic energy distribution system applied to a high-rise building

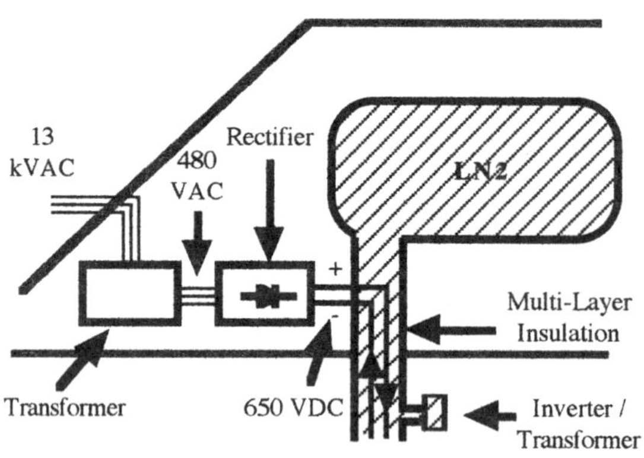

Figure 5: The heart of the cryo energy distribution system.

of any high-rise building has to take out all the heat losses generated by the transformers, cables and hundreds of lights, computers, etc. Assuming a COP (Coefficient of Performance) of 2 for the air-conditioning (AC) system, the removal of 2 kW of heat requires 1 kW of AC power. One can prepare the following loss table:

- Conventional Transformer Loss: 5 %
- Conventional Cable Loss: 3 %
- Air-conditioning COP=2: 4 % [(5% + 3% = 8%) / 2 = 4%]
- Sum: 12 %

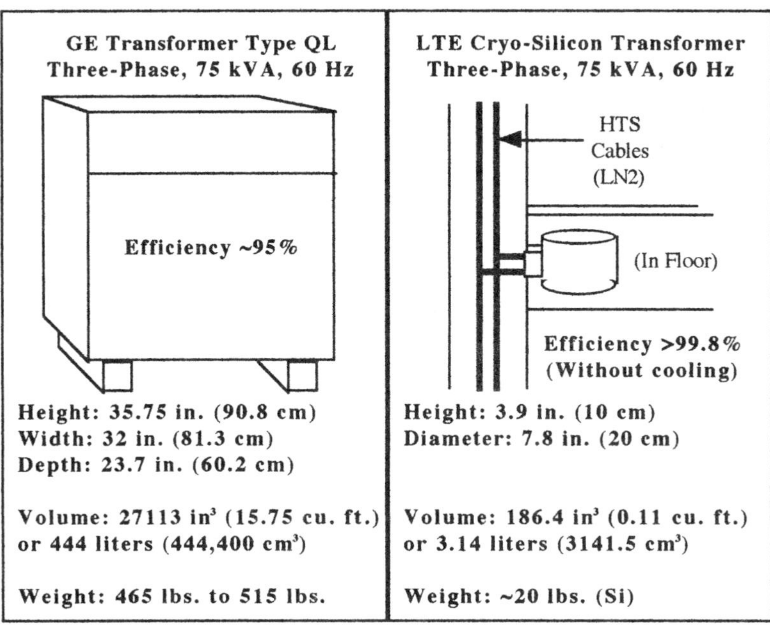

Figure 6. Comparison of a conventional Cu/Fe transformer with the CST

In a 1-megawatt building, this corresponds to a loss of 120 kW, which, at peak load, can basically be eliminated by using the proposed Cryo-Energy Distribution (CED) system. Thus, the CED system uniquely provides 'load shedding' as well as energy savings. The on-site generation of liquid nitrogen called for by the CED system can be carried out in low-load, off-peak hours such as during the night or on weekends. There is no reason why one cannot eventually put an air liquefication plant into a dense city district filled with many high-rise buildings using the CED system. Hospitals in that city require large amounts of liquid oxygen and nitrogen. Liquid nitrogen may also be used for fire protection and extinguishing. Liquid oxygen which can be dangerous may be used for sewage treatment in that same district: It should be obvious that the proposed system could be very beneficial for big cities.

PROBLEMS OF A CED SYSTEM

The most serious problem for a speedy implementation of the proposed CED system is the cryo-phobia of many designers, engineers, architects, etc. But this is a general attitude which has to be overcome anyhow by all potential users of high-temperature superconductor technologies. Up to the present time cryogenics was used mostly by physicists and only in relatively few commercial applications such as magnetic resonance imaging (MRI). The CEC community has worked in the past in such exotic fields as high-energy particle physics and aerospace applications, etc. The proposed Cryogenic Energy Distribution (CED) system could be a further step in the direction of widespread, commercial application of cryogenic engineering bringing its advantages "down to earth".

It is believed that other problems such as harmonics content due to the inverters, etc. can be solved by good engineering design. Repairs of highly integrated circuits such as the cryo-silicon transformer will be done by replacement as is the case generally with integrated circuits. DC isolation between the outside high-voltage grid is established by the input HTS transformer shown in Figure 5.

SUMMARY AND CONCLUSIONS

- The CED system combines Cryo-Silicon Transformers with HTS cables.
- 460 VAC, 3-Phase system is replaced by 650 VDC: Doubling of power handling capability.
- DC Operation: No HTS AC-losses! (Very important! AC losses in HTS cables are a real problem!)
- Low-weight silicon transformer replaces heavy iron-core transformer.
- At 75 kVA: Weight reduction: 250 kg to about 5-10 kg (factor of 25-50)
- Improvement: 0.3 kVA/kg to 15 kVA/kg
- At 75 kVA: Size reduction: 0.17 kVA/liter to 10-20 kVA/liter.
- Cooling problem solved by HTS cryo-pipe distribution cable
- Single LN2 tank needed for multiple transformer units (5-100)
- Additional uses: Air-conditioning (cooling) support, fire protection/extinction
- DC-HTS 650 VDC cable: Only short lengths needed
- CED system suitable for all large buildings, for ships, airplanes, etc.
- HTS wires with reduced specifications for CED.
- No magnetization power required: Energy savings
- Excellent power factor correction due to DC operation:
- Solution to power quality problems. Voltage regulation possible
- UPS system using capacitive energy storage can be added.
- Built-in power load shedding capability via LN2.

REFERENCES

1. EPRI: "Proof of Principle of the Solid-State Transformer". Report TR-105067, August 1995.
2. L. Lorenz, M. Maerz, G. Deboy: "Improved MOSFET - An important milestone toward a new power MOSFET generation", PCIM, Sept. 1998, pp. 14-22.
3. A.E. Schlögl, et.al: "Properties of Cool-MOS between 420 K and 80 K - The ideal device for cryogenic applications," 11th International Symposium of Power Semiconductor Devices, May 1999.
4. F. Auerbach, et.al.: "6.5 kV IBGT Modules," PCIM-99 Conference, Nürnberg, June 1999
5. P.M. Grant, T.OP. Sheahen: "Cost projections for high-temperature superconductors," Applied Superconductivity Conference, ASC-98, Palm Desert, Ca. Sept. 1998.
6. R. Singh, B.J. Baliga: Cryogenic Operation of Silicon Power Devices, Kluwer, Boston, 1998
7. O. Mueller, "On-resistance, thermal resistance, and reverse recovery time of power MOSFETs at 77 K," Cryogenics, vol. 29, pp. 1006-1014, October 1989
8. G. Stanley and K. Bradshaw, "Precision DC-to-AC power conversion by optimization of the output current waveform - the half-bridge revisited," 28th Annual IEEE PESC-97, June 1997, pp. 993-999.
9. J.L. Hudgins, S. Menhart, W.M. Portnoy, V.A. Sankaran: "Temperature variation effects on the switching characteristics of MOS-gate devices." EPE Firenze, 1991, pp. O-261/O-266.
10. O. Mueller and K. Herd, "Ultra-high efficiency power conversion using cryogenic MOSFETs and HT superconductors," IEEE PESC 93, pp. 772-778, 1993
11. W. Jackson, O. Mazzoni, and E. Schempp, "Characteristics of semiconductor devices at cryogenic temperatures," IECEC, Vol. 2, pp. 676-681, 1996
12. O. Mueller, "Properties of high-power Cryo-MOSFETs," 1996 Annual IEEE Industrial Applications Society Meeting (IAS-96), vol. 3, pp. 1443-1448 (IEEE 96CH35977), October 1996
13. E. Schempp and W. Jackson, "Capacitive energy storage with hypercapacitors and cryogenic power electronics," Power Systems World, vol. 9, pp. 381-395, 1996
14. A.I. Gardiner, S.A. Johnson, and E. Schempp, "Operation of power electronic converters at cryogenic temperatures for utility energy conditioning applications," IECEC, vol. 14, pp. 2209-2214, 1996
15. R. Severns: "Superconductivity and low-temperature power converters". Powertechnics Magazine, vol. 4, no. 4, April 1988, pp. 32-34.
16. O.M. Mueller, E.K. Mueller: "Cryogenic power inverters for MRI". CEC-ICMC-99.
17. O. Mueller, "Cryogenic MOSFET power conversion," Proceedings of the IEEE Workshop on Low Temperature Semiconductor Electronics, pp. 94-98, Univ. of Vermont, 1989 (IEEE 89TH0252-7)
18. O. Mueller, "Cryogenic power conversion: Combining superconductors and semiconducting devices," American Institute of Physics, "Superconductivity and its Applications," vol. 251, pp. 746-760, 1992

PRODUCTION AND INNOVATIVE APPLICATIONS OF CRYOGENIC SOLID PELLETS

M. J. Gouge, L. R. Baylor, S. K. Combs, P. W. Fisher,
C. A. Foster, C. R. Foust, and S. L. Milora

Oak Ridge National Laboratory
P.O. Box 2009, Oak Ridge, TN 37831-8071

ABSTRACT

For over two decades Oak Ridge National Laboratory has been developing cryogenic pellet injectors for fueling hot, magnetic fusion plasmas. Cryogenic solid pellets of all three hydrogen isotopes have been produced in a size range of 1- to 10-mm diameter and accelerated to speeds from <100 to ~3000 m/s. The pellets have been formed discretely by cryocondensation in gun barrels and also by extrusion of cryogenic solids at mass flow rates up to ~0.26 g/s and production rates up to ten pellets per second. The pellets traverse the hot plasma in a fraction of a millisecond and continuously ablate, providing fresh hydrogenic fuel to the interior of the plasma. From this initial application, uses of this technology have expanded to include (1) cryogenic xenon drops or solids for use as a debris-less target in a laser plasma source of X-rays for advanced lithography systems, (2) solid argon and carbon dioxide pellets for surface cleaning or decontamination, and (3) methane pellets in a liquid hydrogen bath for use as an innovative moderator of cold neutrons. Methods of production and acceleration/transport of these cryogenic solids will be described, and examples will be given of their use in prototype systems.

INTRODUCTION

Oak Ridge National Laboratory (ORNL) has been developing cryogenic pellet systems for over two decades with a primary emphasis on production and acceleration of solid pellets of the three hydrogen isotopes for fueling the interior of hot plasmas. Building on this technology, other applications of cryogenic pellets have been developed to the prototype stage, including (1) cryogenic xenon drops or solids for use as a debris-less target in a laser plasma X-ray source, (2) solid argon and carbon dioxide pellets for surface cleaning or decontamination, and (3) methane pellets in a liquid hydrogen bath for use as a cold neutron moderator. Each of these innovative applications of cryogenic pellets will be described in more detail.

Advances in Cryogenic Engineering, Volume 45.
Edited by Shu *et al.*, Kluwer Academic / Plenum Publishers, 2000.

HYDROGENIC PELLETS FOR PLASMA FUELING

It was recognized early in the magnetic fusion development program that fueling of hot plasmas by the traditional method of puffing hydrogenic (H_2, D_2, T_2) gas into the vacuum vessel would be problematic because the gas molecules do not penetrate far into the plasma core where they are needed to replenish hydrogenic ions consumed by the deuterium-tritium (D-T) or deuterium-deuterium (D-D) fusion reactions. An alternate method of fueling was proposed where solid pellets of all three hydrogen isotopes are injected into the plasma at high speeds and continuously ablate, providing fuel to the plasma interior. For reference, relevant properties of the hydrogen isotopes are provided in Table 1. Due to the low boiling points, typically liquid helium or cold gaseous helium has been used as the cryogenic fluid to cool and solidify the hydrogenic feed gas, although more recently Gifford-McMahon (GM) cryocoolers have been used for discrete pellet production. For radioactive tritium pellets, the cryogenic cooling also has to accommodate the volume heat source caused by the beta decay of about 1 W/mole.

There has been much progress in the physics and technology of pellet fueling since this technique was first applied to magnetic fusion experiments over 20 years ago.[1,2] From first experiments with a single, small hydrogen pellet at hundreds of meters per second, the technology has advanced to hydrogen/deuterium/tritium (H/D/T) pellets in longer pulse length systems, with inventories up to about 1000 pellets and speeds in the kilometer-per-second range. Table 2 below provides a summary of the progress of this technology.

Pellets are typically accelerated by pneumatic (light gas gun) injectors using hydrogen or helium propellant gas or via centrifuge acceleration.[2] The low density of the hydrogenic solids as noted in Table 1 allows respectable speeds of 1–2 km/s with single-stage light gas guns with propellant gas pressures on the order of 10–100 bar. Two-stage light gas guns operate repetitively at 1 Hz with 3-mm pellets and speeds in the range of 2.5 to 3 km/s. Single 4-mm-diam pellets have been accelerated to speeds above 4 km/s with two-stage guns. Centrifuges have accelerated 1- to 4-mm pellets to speeds over 1 km/s.

Hydrogenic pellets have been produced discretely in-situ[2] by admitting feed gas into the interior of a barrel in which a short section (a few millimeters) is cooled to ~6–12 K by a helium-cooled, copper block attached to the outer diameter of the barrel. Alternatively, a GM cryocooler could cool the copper block to 8–12 K. The feed gas forms a cylindrical solid pellet inside the gun barrel; typical dimensions are a few to 10 mm diameter and aspect ratios (length to diameter) from 1–2. Another method of hydrogen ice production is

Table 1. Properties of hydrogen isotopes

Property	Hydrogen	Deuterium	Tritium
Density, g/cm^3	0.09	0.2	0.32
Boiling point at 1 atm, K	20.4	23.7	25
Triple point, K	13.8	18.7	20.6
Triple point pressure, torr	54	129	162
Shear strength, MPa	0.1 (10 K)	0.5 (9 K)	1.2 (9 K)

Table 2. Progress in pellet technology 1978–1999

Parameter	Status 1978	Status 1999
Pellet isotope	Hydrogen	H/D/T
Pellet size	One to a few millimeters	0.5–10 mm
Pellet inventory	One to few	1–1200
Pellet feed rate	N/A	0.26 g/s
Pellet speed	Hundreds of meters per second	Thousands of meters per second
Reliability	Unknown	Sufficient

by extrusion.[2] Extruders are generally capable of moving solid material at a sufficiently high rate of speed to provide a "real time" pellet feed system. The extruder section (see Fig. 1) consists of a liquid reservoir positioned above a cylindrical freezing chamber, which is fitted at its outlet with a copper extrusion assembly including the nozzle. A motor-driven (electric or pneumatic) screw press activates a polyimide piston that moves vertically inside the cylindrical bore of the freezing chamber where the solid hydrogenic charge is located. The extruder assembly has two cryostats, each made of a block of oxygen-free, high-conductivity (OFHC) copper that has square cooling channels with a spiral pitch machined into it. A ~25- to 50-W nichrome heater is wrapped around and epoxied to the perimeter of each cryostat for cryogenic temperature control. The extrusion nozzle assembly, which terminates just above the pellet feed mechanism position, provides a smooth transition from the cylindrical extruder bore to a smaller cross section nozzle; this area determines the size of the accelerated pellet. Present deuterium extruders typically operate at ~14 K where the shear strength of the ice is sufficiently low to prevent excessive piston forces at high extrusion rates. In recent experiments at the Tritium Systems Test Assembly facility at Los Alamos National Laboratory, the Tritium Proof-of-Principle Phase II (TPOP-II) pellet injector has produced good quality tritium and D-T extrusions at temperatures in the range 13–14 K (Ref. 3).

The TPOP-II injector will be described in more detail because it has produced tritium and D-T pellets of typical dimensions (8–10 mm) required for refueling of a fusion reactor.[4] Figure 2 is a sketch of the TPOP-II repeating pneumatic injector. An extruder assembly similar to that shown in Fig. 1 is housed in a guard vacuum chamber, which provides vacuum insulation of the cold extruder components. Principal components include the extruder mechanism with two copper cryostats, the pellet cutting and chambering mechanism with adjacent copper cryostat, the ~8-mm-ID gun barrel, and the ^3He separator that removes the ^3He by-product from the tritium radioactive decay from the incoming tritium supply gas stream. The extruder mechanism freezes input gas into a column of hydrogenic solid in a batch process for use at a later time. As the extruder feeds a continuous supply of solid deuterium (or tritium) to the gun assembly, a reciprocating, electromagnetically driven, breech-side cutting mechanism (punch tube) forms and chambers cylindrical pellets from the extrusion. The chambering mechanism contains a single cryostat, the main body of which is also a block of OFHC.

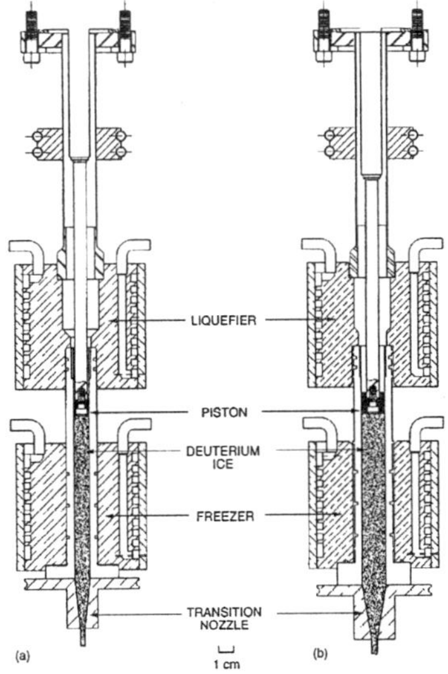

Figure 1. ORNL hydrogenic extruders.

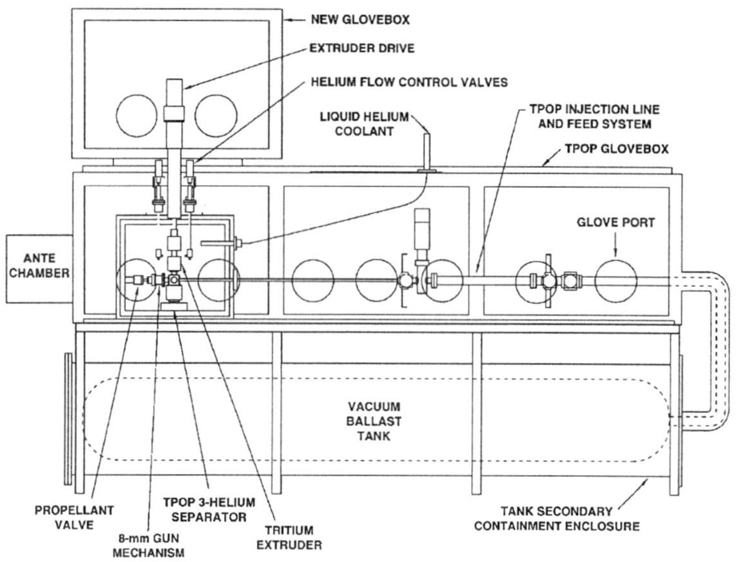

Figure 2. TPOP-II pellet injector.

Tritium and D-T pellets are accelerated in the gun barrel to high speeds with controlled amounts of compressed deuterium gas delivered by a fast electromagnetic propellant valve. This valve is a fast solenoid valve (<1-ms opening time) developed by ORNL. The gun barrel is a continuous 8-mm-ID, 1.25-m-long, stainless steel tube with a wall thickness of 0.89 mm. Liquid helium enters and exits the guard vacuum chamber through two vacuum-insulated, O-ring sealed, bayonet fittings (Linde-type) on the front plate. The liquid helium supply is split into four parallel cooling paths inside the guard vacuum box (the two extruder cryostats, the gun cryostat, and the ^{3}He separator), each with a flow control valve. These four parallel paths finally combine and are then discharged through the exhaust bayonet fitting. The ^{3}He separator is basically a small copper counterflow heat exchanger. Helium coolant passes through the central cooling channel, and tritium is cryopumped into the external finned surfaces. D-T pellets up to 10 mm in size (see Fig. 3) have been extruded at rates up to 0.26 g/s for short pulses only; this pellet size and feed rate is sufficient for fueling fusion reactors at the gigawatt power level.[4]

The cryogenic pellet injector technology development described previously has resulted in a pneumatic or centrifuge pellet fueling system on virtually every major magnetic fusion experiment in the past two decades.[1]

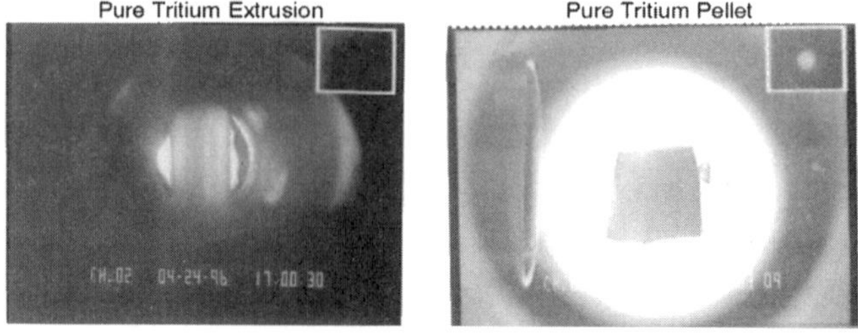

Figure 3. Pure tritium extrusion and pellet.

XENON DROPLET GENERATOR FOR USE IN A COMPACT LASER PLASMA X-RAY SOURCE

Soft X-ray projection and proximity lithography systems are being developed for the advanced semiconductor industry for high-resolution feature size printing. A key element in a commercial lithography system would be a compact plasma X-ray source. Compact systems have been developed using a solid state or excimer laser beam impacting a rotating metallic (typically gold or copper) target as a plasma source.[5] Debris formation and subsequent transport result in impurity deposition or impact damage to process optical components. To mitigate fragmentation damage in a continuous (repetitive) plasma source, discrete droplets delivered at low-to-moderate repetition rates (~10–100 Hz) are desired. The droplets should be small (mass limited, about 10–50 μm in diameter) to ensure complete vaporization by the high-power, short-pulse laser beam. The resulting plasma would generate soft X-rays for use in an advanced lithography system.

To achieve still further improvements in debris reduction, a cryogenic droplet target system for use in an advanced laser plasma source for X-ray lithography has been designed, fabricated, and tested at ORNL.[6] The advantage of the cryogenic droplets is that small fragments and vapor from the laser-droplet interaction can be collected preferentially on a cold surface and pumped away later at room temperature. It was determined that drop-on-demand droplet generators similar to those used for ink jet printing have potential performance requirements meeting the lithography application. The drop-on-demand feature allows control of the lithography process on a drop-to-drop basis, which is attractive due to the inherent interruptions of the semiconductor wafer etching process for indexing and wafer replacement. This also results in the most efficient use of the feed gas and minimizes the recycled gas stream.

A fundamental decision is the choice of fluid for the cryogenic droplet. This choice is largely driven by the potential to produce intense X rays from the droplet-laser interaction in the wavelength region of interest. For X-ray lithography systems, solid xenon has excellent X-ray conversion efficiencies for wavelengths in the range of 1–14 nm. Xenon was chosen as the initial cryogenic fluid to use in this system due to the X-ray production efficiencies discussed above. An obvious benefit of xenon is that the boiling and triple point temperatures are high enough that liquid nitrogen is a feasible coolant to maintain the xenon liquid at the specified temperature. The fluid properties of liquid xenon present some unique challenges to droplet production. Water is the most common fluid used in droplet production systems. Liquid xenon has about one-third the surface tension, one-fourth the viscosity, and about three times the density of water. In fact, liquid xenon is denser than several light metals such as aluminum. The Weber number is the ratio of the inertial force to the surface tension force; this dimensionless number is a key parameter in drop-on-demand systems as surface tension forces hold the meniscus in the orifice in between drop-on-demand cycles. The Weber number for liquid xenon is about 40 times higher than a water droplet of the same size. A schematic of the xenon drop production device with support components is shown in Fig. 4.

The xenon drops are produced on demand by pulsing a piezoelectric disc attached to a rectangular copper block that has a supply reservoir of liquid xenon. The copper block is maintained at 160–175 K by conduction cooling via copper straps from a copper cold finger of a liquid nitrogen dewar. A signal generator and radio frequency (RF) amplifier provide the ~20- to 100-V square wave drive pulse to produce each drop. The droplet generator is housed in a vacuum chamber that is a six-way cross assembly with 10-in. conflat flanges. Windows are installed on four of the six flanges. An MKS pressure monitoring and differential pressure control system (range 0–10 torr) maintains the helium background pressure in the chamber at typical values in the range 900–1100 torr with a

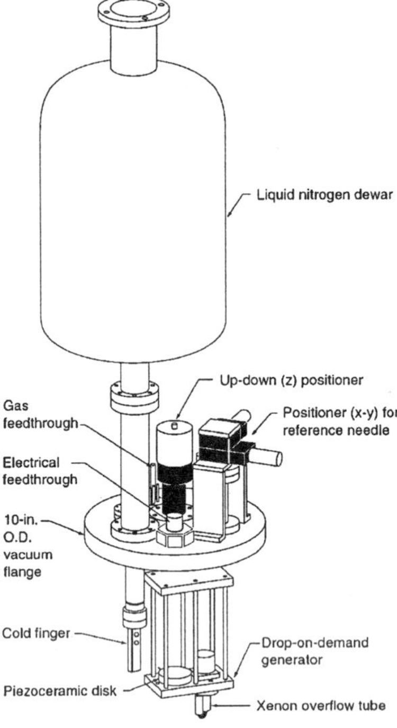

Figure 4. Piezoelectric drop-on-demand generator and support systems.

near-zero differential pressure across the orifice, which has liquid xenon pressure on the interior side and vacuum chamber pressure on the exterior side. Diagnostics include vacuum chamber, helium feed gas and xenon supply gas pressure, differential pressure across the orifice, silicon diode temperature sensors on the copper block, and a CCD camera with microscope lens to view the ~25- to 100-μm droplets through one of the windows. Parameters affecting system operation include the temperature of the droplet generator, the background helium pressure in the vacuum chamber, and the orifice differential pressure. An acceptable operating point for drops-on-demand is at a temperature of 172–173 K on the droplet generator, 1000–1025 torr of helium in the vacuum tank, and close to zero pressure differential across the droplet generator orifice. The operating point once achieved is stable; a recent run was made producing ~50-μm-diam drops at 30 Hz for ~42 min (about 75,000 drops). Drop terminal velocities are in all cases <1 m/s, and Reynolds numbers are <1.

CARBON DIOXIDE PELLETS FOR SURFACE CONDITIONING

ORNL has developed a technology that uses cryogenic pellets for surface conditioning to replace solvent-intensive and sandblast technology. An ORNL centrifuge driver, which accelerates frozen CO_2 (or argon) pellets to high speeds, has potential as a high-performance, energy efficient, and environmentally friendly means for surface cleaning and paint removal. In comparison to the more conventional compressed air "sandblast" pellet accelerators, the ORNL centrifugal accelerator system can achieve higher pellet speeds, has precise speed control, and is more energy efficient. Furthermore, the use of frozen CO_2 pellets instead of conventional metal, plastic, sand, or other abrasive materials that remain solid at room temperature, minimizes the waste stream. This apparatus has been used to demonstrate cleaning of various surfaces, including removal of paint, oxide coatings, metal coatings, organic coatings, and oil and grease coatings from a variety of surfaces. In the centrifuge acceleration technique, pellets are loaded onto a spinning arbor (accelerator wheel) that accelerates them and then slings them off its outer edge at high speeds. The centrifuge accelerator concept is illustrated in Fig. 5. The arbor with a diameter of 0.35 m is spun using a brushless, 11 kW dc motor. The motor and arbor are designed to be mounted on a robotic apparatus that can move the device over the desired surface region at a controlled scan rate. Experimental results have shown that the unit will deliver CO_2 pellets at feed rates as high as 450 kg/h with pellet velocities of 350 m/s. The pellets are distributed over a swath about 0.13 m wide as it is scanned across a surface. In an initial test, epoxy-based paint was removed from aluminum panels at a rate of about 6.7 m^2/h with the panels at room temperature. By raising the panel temperature to about 100°C, paint stripping rates greater than 22 m^2/h are possible.

METHANE DROPS FOR USE IN AN ADVANCED COLD NEUTRON MODERATOR

A variety of cold moderators have been developed and installed in pulsed and continuous, reactor and accelerator-based, neutron sources for the production of cold neutrons. For cold moderators, the principal material choices have been liquid H_2, liquid D_2, solid D_2O, liquid CH_4, and solid CH_4. The technology of liquid H_2 and D_2 is well established. Methane has a higher density of protons than all moderators listed except ammonia; solid methane has an almost 70% larger proton density than liquid hydrogen. This accounts, in part, for its superior performance as a moderator; the other key factor being a high density of internal rotational modes, which are nearly free even in the low-temperature solid. Liquid methane can be produced with liquid nitrogen cooling and solid methane by cooling with liquid or supercritical hydrogen or cold gaseous helium.

During this decade, this potential has resulted in installation of liquid and/or solid methane moderators at several neutron science facilities. Radiation-induced effects, however, have limited the ability of solid methane moderators to be used in high-power neutron sources. The practical effects of irradiation of methane are the production of hydrogen and the formation of solid hydrocarbon deposits in the moderator vessel. Solid methane moderators used to date are monolithic blocks that can only be operated for limited periods before the radiation-induced reactions described above cause swelling of the containment vessel and production of oily and tar-like polymerization products. The amount of beam power a solid moderator of this type can handle is also limited due to the low thermal conductivity of solid methane that limits the ability to remove the heat generated.

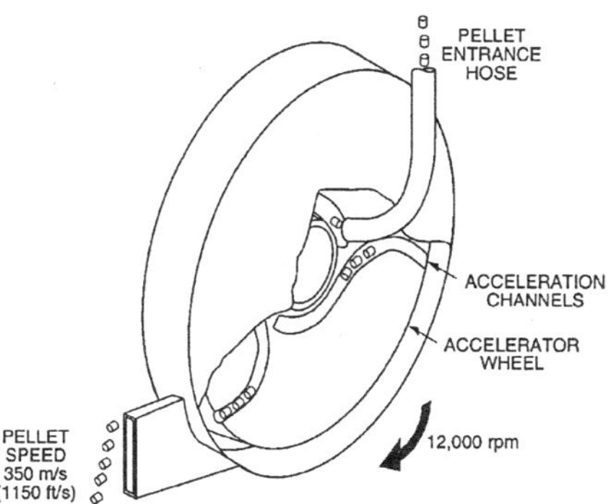

Figure 5. ORNL centrifuge accelerator used for surface conditioning.

ORNL has proposed an innovative solid methane moderator system capable of operating at power levels up to 1 MW on target. This would yield a flux of low-energy (1- to 10-meV) neutrons three to four times as intense as that from liquid hydrogen moderators. The basic concept is to use millimeter-size, solid methane spheres surrounded by liquid or supercritical hydrogen at 20 K (Ref. 7). The hydrogen maintains the methane spheres at 20 K and is also used to transport the spheres to the active moderator region.

In 1996 an initial scoping experiment was conducted. The goal was to evaluate the feasibility of creating and maintaining a fluidized bed of discrete methane spheres for a cold neutron source. This brief experiment involved the production of ~1- to 2-mm-diam methane droplets and the subsequent study of their collective behavior in a research dewar containing liquid hydrogen. Methane gas was fed into a copper droplet generator where liquid methane was produced in the 90–100 K reservoir. The liquid was then formed into 1- to 2-mm droplets via a magnetostrictive plunger action on a diaphragm. The drops fell through the transfer line to the first floor station, where the partially solid drops were collected in the bottom of the research cryostat. Windows on the cryostat allowed observation of the drops and provided insight on their collective behavior in a background bath of liquid hydrogen (~50 cm^3). Most of the experiment goals were achieved:

- The methane droplet generator, operating at 92 K with liquid nitrogen cooling, produced consistent, spherical, methane droplets of diameter ~1 mm at repetition rates from 1–100 Hz. In the final run, over 300,000 drops were generated at 30 Hz for 3 h.
- The research cryostat located some 9 m below the droplet generator produced reliable pools of liquid hydrogen (at 18–19 K), which were used as a reservoir for accumulation of the vertically falling methane droplets.
- The solubility of methane in liquid hydrogen was measured at <1 ppm, which was the limit of sensitivity of the measurement.

This brief test campaign showed that continuous production of consistent drops was indeed feasible. However, it proved difficult to fully freeze the drops during their freefall to the liquid hydrogen in the research dewar, which limited the quality of the fluid-solid bed. A more direct solid pellet production method is probably required, such as extrusion or molding solids via batch processing.

ACKNOWLEDGMENT

Research at Oak Ridge National Laboratory was sponsored by the U.S. Department of Energy under contract DE-AC05-96OR22464 with Lockheed Martin Energy Research Corp.

REFERENCES

1. S. L. Milora et al., Review paper: pellet fuelling, *Nucl. Fusion* 35:657 (1995).
2. S. K. Combs, Pellet injection technology, *Rev. Sci. Instrum.* 67:1679 (1993).
3. P. W. Fisher and M. J. Gouge, Extrusion of tritium and DT pellets for ITER fueling, *Fusion Technology* 30:845 (1996).
4. M. J. Gouge, Fuelling of ITER-scale fusion plasmas, *Fusion Technology* 34:435 (1998).
5. D. J. Nagel, in: "VLSI Electronics," M. Brown, ed., Academic, New York (1984), Vol. 8.
6. M. J. Gouge and P. W. Fisher, A cryogenic xenon droplet generator for use in a compact laser plasma X-ray source, *Rev. Sci. Instrum.* 68:2158 (1997).
7. A. T. Lucas, A combined H_2/CH_4 moderator for a short pulsed neutron source, "Proceedings of ICANS-X Conference, Los Alamos, New Mexico, October 3–7, 1988," D. K. Dyer, ed., Institute of Physics Conference Series No. 97, Institute of Physics, Bristol and New York (1989).

OXYGEN SEPARATION USING CRYOGENIC VORTEX TUBE

S. Jacob, S. Kasthurirengan, R. Karunanithi and T. Jagadish

Centre for Cryogenic Technology, Indian Institute of Science
Bangalore 560012, India

ABSTRACT

An experimental system has been designed and fabricated to study the two phase flow behaviour through a vortex tube (Ranque-Hilsch tube). Using counter flow cryogenic heat exchangers utilizing the enthalpy of liquid nitrogen and the cold output of the vortex tube, a dry and oil free air stream at 0.7 MPa is cooled down to an optimum temperature around 95 K. This partially condensed air is injected to the vortex chamber, where the liquid phase enriched in oxygen gets thrown to the periphery and flows out through the peripheral discharge end of the vortex tube. The vapour phase predominantly enriched in nitrogen flows out through the central discharge opening of the vortex chamber. The oxygen purity for the peripheral discharge is measured to be about 90%. The paper describes the dependence on various parameters for cryogen phase separation using vortex tube.

INTRODUCTION

Vortex tube known as Ranque-Hilsch tube[1,2] is a device with no moving parts in which, compressed gas injected through tangential nozzles into a vortex chamber results in the separation of the inlet flow into two streams, one of which is warmer than the inlet gas while the other is cooler. The peripheral warm end stream is a free vortex for which the swirl velocity increases with decreasing radius and the inner cold stream is a forced vortex for which the swirl velocity is proportional to radial distance. The emerging flow patterns in a vortex tube are shown schematically in Figure 1. The energy separation initiated in the radial plane continues as the flow progresses axially but with decreasing swirl towards the warm end of the vortex tube. For a few times diameter length, the flow somewhat follows Taylor-Proudman profile of velocity field parallel to the axis of rotation, except at the inlet flow zone and at the zone where the flow reversal occurs for the central core of flow.

The mechanisms of energy separation have been studied by several investigators[3,4,5,6] but so far no conclusive theory has emerged. However to a large extent, the energy separation can be attributed to the effect of forced vortex to cause a velocity gradient, resulting in kinetic energy separation and a radial pressure gradient enabling turbulent

Advances in Cryogenic Engineering, Volume 45.
Edited by Shu *et al.*, Kluwer Academic / Plenum Publishers, 2000.

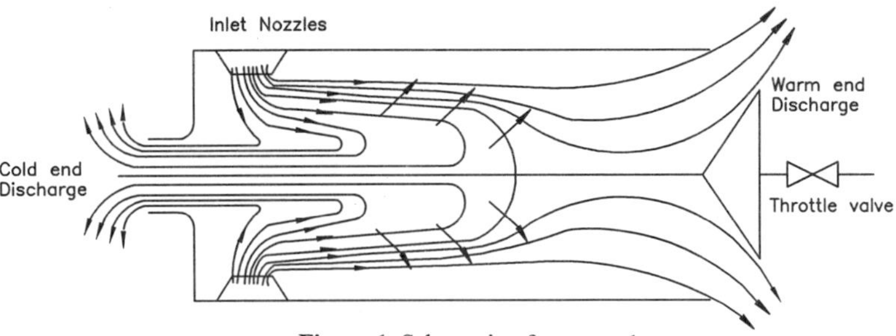

Figure 1. Schematic of vortex tube

fluctuations to enhance the kinetic energy separation by transporting thermal energy radially outwards. These processes are further modified by two thermal effects. The axial convection extends the turbulent thermal transport to an appreciable distance to the warm end discharge of the vortex tube, providing an axial field for energy separation. However it is the viscous dissipation of flow, which converts the kinetic energy separation into a thermal separation. Practical applications of the thermal separation provided by vortex tubes have been now commercially well established where moderate heating and cooling capacities are required.[7]

Apart from thermal separation, it has been observed that vortex tubes can be used for mixture separation.[8,9] However, there are few published papers on operating the vortex tubes at cryogenic temperatures and using them as cryogenic phase separators, except for some patent literature.[10]

The work reported here is an attempt to provide thermal data on the performance of a vortex tube at cryogenic temperatures and its application to separate out a high purity oxygen stream from a partially condensed air stream.

MIXTURE SEPARATION IN VORTEX TUBES

Cockerill[6] has provided a simple theory for separation of binary mixtures of gases. It has been shown that any changes in the mole fraction composition across the vortex radius are primarily influenced by the swirl velocity and the molar mass difference of the constituents. Therefore centrifuge force is the key for mixture separation in the vortex tubes. The axial flow patterns in the vortex tube increase the resident time period of the flow, increasing the higher molar mass concentration distribution at the warm end of the vortex tube.

The above concepts can be very effectively used for gas-liquid mixture separation as well in vortex tubes. If the inlet to discharge pressure ratio is larger than a critical value, then for an optimally designed vortex tube, the inlet velocity is nearly sonic and the resultant rotational velocity of the forced vortex generates a centrifugal acceleration which is in the range of million times the earth's gravity.

Unlike the separation process based on solution equilibrium at earth's gravity, as for cryogenic distillation towers, separation in a vortex tube occurs at the vapour-liquid interface, in a high centrifugal field where solubility is less of a force and specific gravity values of the liquid and vapour phases are more important. Therefore, the vortex separation process is unaffected by gravity, acceleration and orientation changes.

To achieve appreciable separation effect, the difference in the specific gravities of gas species or phases must be high. There was insignificant separation of methane from hydrogen gas with a specific gravity ratio of 8:1 in a vortex tube.[11] However, for cryogenic two phase fluids, the specific gravity ratio at the boiling points can be fairly high. For example, at atmospheric pressure the liquid to vapour density ratio for oxygen at 90 K is about 255, while that of nitrogen at 77 K is about 175. Therefore if a two phase flow can be generated in a vortex tube, separation of the liquid-vapour phases is feasible. Further, from the temperature-composition diagrams of oxygen-nitrogen mixtures, it can be observed that for inlet temperatures closer to the dew point, the oxygen concentration will be highest in the liquid phase. Though under a high centrifugal field, the dew point may be different than at earth's gravity and closer to the boiling point, still there could be an optimum inlet temperature for the partially liquefied air to achieve this separation. The time duration of the axial flow towards the warm end of the vortex tube and the heat transfer coefficient of tube wall will also influence the oxygen purity and separation efficiency.

EXPERIMENTAL PROGRAMME

Figure 2 shows the schematic of the experimental system designed and built to study the two-phase flow behaviour of air through a vortex tube.

Dry and oil free air supply at 0.7 MPa and throughput of 0.0194 kg/s is split into two major streams. One of them passes through two counter flow heat exchangers HE1 and HE2 in series and the other through a submerged heat exchanger coil (HE3) in a liquid nitrogen cryostat. The air passing through the submerged heat exchanger gets partially condensed, which causes the evaporation of liquid nitrogen. The enthalpy of the nitrogen vapour is utilized in HE2 to further cool down the air stream pre-cooled in HE1 by the cold end flow from the vortex tube. The partially condensed air from HE3 and the cooled air from HE2 get mixed in a flow mixing chamber to generate air at optimum temperature condition for inlet to the vortex tube (ITW vortec: Model 208-35L-SS). The fraction of airflow through the warm and cold end of the vortex tube is optimized by the adjustment of the throttling screw at the warm end. Experiments reported here have been conducted for the vortex tube throttling valve opening of ¼ turns. The experimental system is instrumented to measure the temperature, pressure and flow rate of different flow passages. The oxygen purities of the cold end and warm end discharges from the vortex tube are measured using a Teledyne analytical oxygen analyser (Model 3000 PA). Before conducting experiments, it is calibrated for span and zero setting so that the percentage of oxygen in air is read as 20.9 % by volume.

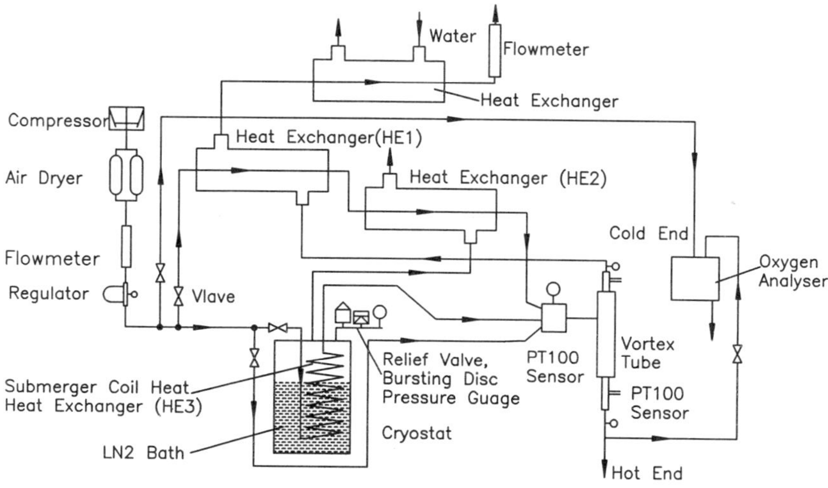

Figure 2. Schematic of vortex tube experimental system

RESULTS AND DISCUSSION

Room Temperature Experiments

The warm and cold end temperatures and pressures of the vortex tube for air at 300 K as a function of vortex inlet pressures are shown in Figure 3 and Figure 4 respectively. It can be noted that the maximum warm end temperature attained was 345 K where as the minimum cold end temperature reached was 272 K, for the throttle valve opening of ¼ turns. As shown by Ahlborn et. al.[5] the normalised pressure drop at the entrance plane (Pi-Pc)/Pi determines the kinetic energy of the inlet flow, where Pi is the inlet pressure and Pc is the pressure of the cold end discharge. For the maximum observed temperature difference, the value of normalised pressure drop has been about 0.85, which indicates the flow inside the vortex is crossing over the sonic limit.

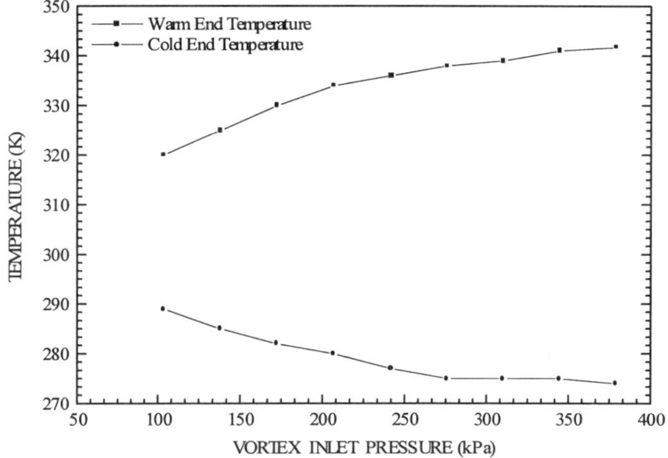

Figure 3. Vortex inlet pressure vs. warm and cold end temperature

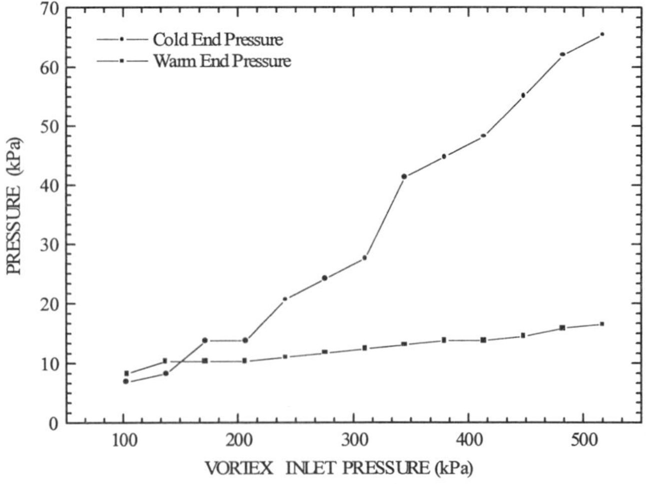

Figure 4. Vortex inlet pressure vs. warm and cold end pressure

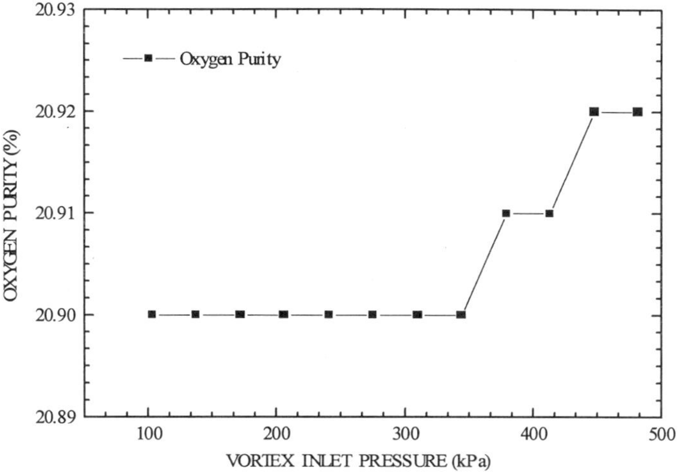

Figure 5. Oxygen purity vs. vortex inlet pressure

Figure 5 shows the oxygen purity measured for the warm end discharge as a function of inlet pressure. It is seen that for ambient air, the increase in oxygen concentration in the warm end discharge has been rather minimal.

Low Temperature Experiments

Cool down characteristics. Figure 6 shows the typical temperature profile of the inlet stream to the vortex as well as that of the hot and cold end discharges as a function of time, while the air is progressively cooled down in the heat exchangers. A significant observation has been the conventional warm end discharge stream which maintains a higher temperature profile for room temperature inlet experiment, drops its temperature below that of the inlet stream and even that of the conventional cold end stream after some period of cool down time and reaches a steady value close to 97 K. The other two streams reach almost this temperature value after some delay.

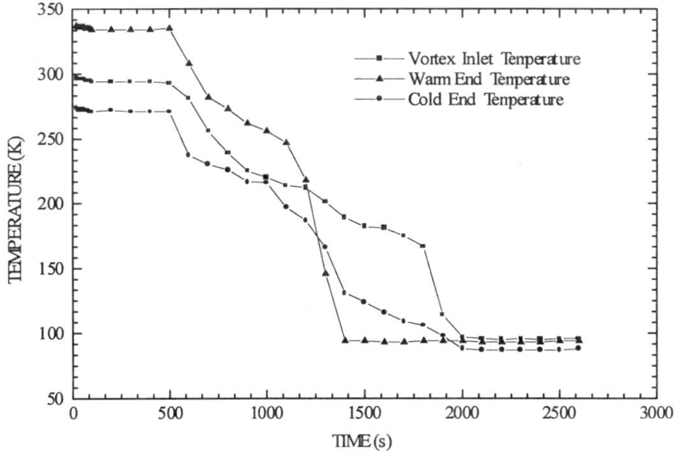

Figure 6. Temperature profile of vortex inlet, warm end, cold end vs. time

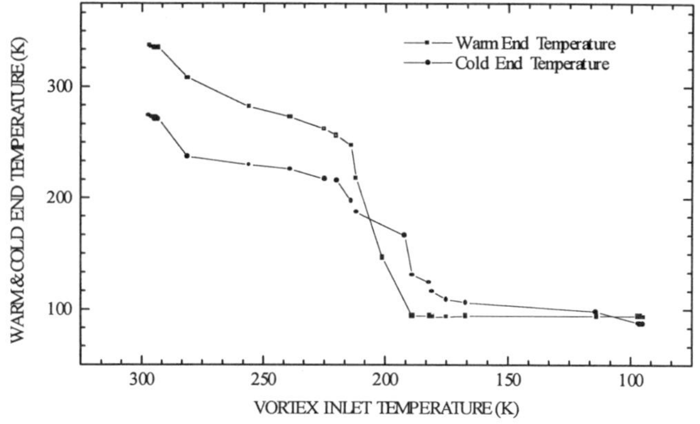

Figure 7. Temperature profile of warm & cold end vs. vortex inlet temperature

The temperature profiles of the conventional warm and cold end streams as the vortex inlet temperature is progressively lowered is shown in Figure 7. It is observed that when the vortex inlet temperature reduces to nearly 215 K, the temperature of the conventional warm end stream goes down below that of the conventional cold end stream and at around an inlet temperature of 190 K, the warm end temperature reaches a steady value of 97 K. The crossing over of the temperatures of the inlet and conventional cold end stream by the conventional warm end stream indicates the formation of liquid phase within the vortex tube.

Oxygen Purity Level of the Warm End Stream

Figure 8 shows the purity of the conventional warm end stream as a function of vortex inlet temperature. It is observed that there is no increase in oxygen purity until the inlet temperature reaches about 215 K, a temperature close to which the conventional warm end stream temperature equals that of the vortex inlet temperature. From then on till about 190 K, there is a sharp increase in oxygen purity to a value near to 70%. As discussed earlier, 190 K is roughly the temperature of the vortex inlet, at which the conventional warm end temperature reaches its minimum value of 97 K. As the inlet temperature decreases further to 95 K, the oxygen purity reaches a value close to 90%.

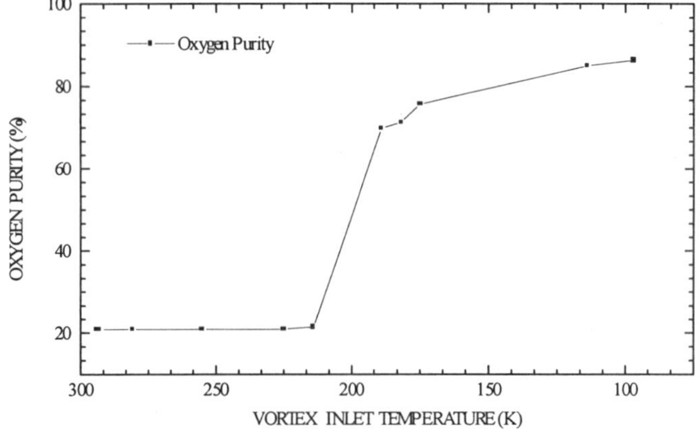

Figure 8. Oxygen purity of conventional warm end vs. vortex inlet temperature

It was observed that the conventional warm end stream carried a two phase flow enriched in oxygen. The oxygen purity level is shown in Figure 8 for the liquid fraction and the oxygen purity of the vapour fraction was about 24%. This indicates the advantage of selectively withdrawing the liquid fraction with higher oxygen purity from a location in the vortex tube warm end ahead of the stagnation point of flow. It may further enhance the purity of the oxygen in the liquid fraction.

CONCLUSION

1. A vortex tube is capable of achieving separation of a pre-cooled air stream into an enriched oxygen stream through its conventional warm end and thereby a depleted oxygen stream through its conventional cold end.
2. A vortex inlet air temperature of $\cong$ 215 K was required to initiate phase separation and the highest purity level was reached when it reached a temperature $\cong$ 95 K.
3. The conventional warm end discharge is a two phase flow, with the liquid fraction having a purity of $\cong$ 90% and the vapour fraction of $\cong$ 24%. The liquid fraction could be withdrawn at an optimum distance from the inlet plane, which is likely to enhance its oxygen purity level.

ACKNOWLEDGMENTS

The work was supported by the Grant-in-aid project sanctioned by DTSR, New Delhi. The authors wish to acknowledge the help received from the staff of the Centre for Cryogenic Technology in the fabrication and installation of the experimental set up.

REFERENCES

1. G.J. Ranque, Experiences sur la datente giratoire avec productions simultanes d'un chand et d'un échappement d'air froid, *J. Physics, Radium* IV 112-4 (1993).
2. R. Hilsh, The use of the expansion of gases in a centrifugal field as cooling process, *Rev. Sci. Instrum*, 18(2): 108 (1951).
3. F. Schultz-Grunow, Turbulenter warmedurchgang im zentrifugalfeld, Forsch. Ing, 17:65(1951).
4. R.G. Deissler and M. Perlmutter, Analysis of the flow and energy separation in a turbulent vortex, *Int. J. Heat Mass Transfer* 1:173 (1960).
5. B. Ahlborn, J.U. Keller, R. Staudt, G. Treitz and E. Rebhan, Limits of temperature separation in a vortex tube, *Phys.D: Appl. Phys*., 27:480 (1994).
6. T.T. Cockerill, "Thermodynamics and Fluid Mechanics of a Ranque Hilsch Vortex Tube," Masters Thesis, University of Cambridge, England (1995).
7. ITW Vortec, "Product Applications," Vortec Corporation, Cincinnati, USA (1994).
8. C. U. Linderstrom-Lang, Gas Separation in the Ranque-Hilsch vortex tube, *Int. J. Heat Mass Transfer*, 7:1195 (1964).
9. J. Marshal, Effect of operating conditions physical size and flow characteristics in gas separation performance of a Linderstrom-Lang vortex tube, *Int. J. Heat Mass Transfer*, 20:227 (1977).
10. D.L. Bennet, K.A. Ludwig, R.F. Weimer, "Process and Apparatus for Producing Nitrogen and Oxygen", U.S. Patent No.5 305610, (1994).
11. L. A. Fekete, Vortex tube separator may solve weight/space limitations, World Oil,7:40(1986).

CRYOGENIC AND FROSTPROOF CONCRETE MADE FROM CONVENTIONAL AGGREGATES AND PORTLAND CEMENT

R.G. Scurlock and K.B. Mohd Yusof *

Kryos Technology
22, Brookvale Road,
Southampton, SO17 1QP
England

*University of Kuala Lumpur
Dept. Civil Engineering
Selangor
Malaysia

ABSTRACT

Using proton (hydrogen) nuclear magnetic resonance (NMR), heat capacity and thermal strain measurements over temperatures from 273 K down to 77 K, on many concrete samples, a formula has been developed for the manufacture of cryogenic concrete and frostproof concrete. The same formula may apply to the retro-processing of existing concrete to make it frostproof against temperatures down to 77 K, including Arctic and Antarctic temperatures down to 193 K (-80 degrees C).

Studies at Southampton have revealed that beta-water, part of the non-chemically bound water in the capillaries of the cement paste, may be responsible for frost damage, including break-up of the concrete composite matrix, when it freezes to ice between 273 K and 213 K. (Beta-water is characterised by a strongly temperature dependent peak at the side of the temperature-independent alpha peak from tightly bound chemical water in the NMR spectrum.)

Removal of beta-water, followed by application of a water-repellant coating to prevent re-wetting by ingress from water or high humidity air, appears to reduce frost damage significantly when the beta-water free concrete is cycled to temperatures below 273 K. Further steps appear to be needed to make cryogenic concrete, and these steps will also be discussed in the paper. Air entrainment is not necessary.

Cube samples of 10 cm side of cryogenic concrete, made according to the formula, suffered no frost damage under extreme thermal cycling by repeated dunking in liquid nitrogen.

Advances in Cryogenic Engineering, Volume 45.
Edited by Shu *et al.*, Kluwer Academic / Plenum Publishers, 2000.

INTRODUCTION

In the 1970s and '80s, a great deal of research work was carried out on the mechanical properties of concretes at temperatures below 273 K toward the development of cryogenic concretes. This work demonstrated the variable behaviour of concretes when frozen and the parameters which could be varied (water/cement ratio, aggregate type, degree of air entrainment etc.) to achieve optimum strength (See for example Marshall (1)). All this work depended on standard tests, using Portland cement in 10 x 10 x 10 cm cubes which had been moist-cured for 28 days, in line with the conventional procedure for testing concrete. The purpose of the moist-curing , in which the concrete surfaces were maintained saturated with liquid water, was to ensure that the chemical hydration of the cement paste within the concrete matrix proceeded to approximately 90% completion in the presence of excess water.

Following this work, an EPRSC funded research contract at the Institute of Cryogenics, University of Southampton, set out to understand more clearly the behaviour of cryogenic concretes from a solid-state physics point of view, rather than as a concrete engineering project. The results of this contract show that water in concrete can be divided between weakly and strongly bound water, and that removal of weakly bound water is the way forward for making cryogenic and frost-proof concretes without any need for air entrainment techniques.

EARLY FINDINGS

If a 10-cm cube of 28-day moist-cured concrete at ambient temperature is "dunked" into liquid nitrogen, with cooling rates ranging up to 100 K per minute, the cube will crumble and break into pieces. If it is cooled slowly by gaseous heat transfer, the cube will survive cooling and its compression strength at 77 K will probably be higher than at 300 K, in a complicated way depending on several factors, including cooling rate, water / cement ratio etc. If, on the other hand, the cube is dried at ambient temperature in a low humidity environment or by heating until there is no further weight loss by the removal of " evaporable " water, then the cube will survive " dunking" in liquid nitrogen without crumbling or spalling. These early observations suggested that first, some of the non-chemically bonded water in concrete is responsible for the reduction in strength and structural damage (frost damage) of concrete cooled below 273 K, since this water expands on freezing to ice, and second,studying saturated moist-cured concrete cubes below 273 K is not the most fruitful way to develop cryogenic or frostproof concretes.

TERMINOLOGY

Before describing the experimental results and conclusions, some definitions are provided for use in this paper (2).

Concrete. Starting with a mixture of aggregate and Portland cement powder (a dehydrated mixture of largely calcium aluminates and calcium alumino-silicates), the addition of water causes chemical hydration of the cement paste at a relatively slow rate. After 28 days in the presence of excess water, the hydration or hardening process or curing process has proceeded to the point where the concrete composite of aggregates and hydrated cement paste has approximately 90% of its final strength.

Hydrated water is chemically bonded water within the cement paste with a binding energy of approximately 13500 kJ/kg.
Free water is water with zero binding energy.
Capillary water is physically bound water which is retained in the capillaries and pores of the cement paste by surface tension forces, with binding energies of less than 100 kJ/kg.
Adsorbed water is physically bound water which is adsorbed on the walls of the porous structure of the cement paste with several degrees of binding, with binding energies comparable to the latent heat of freezing of ice in the range 1000 - 2000 kJ/kg.
Aggregate water is physically bound water which is contained in the structure of the aggregate particles of sand, gravel or ballast.
Evaporable water is the physically bound water and free water which is removed by drying to no further weight loss, in a low humidity environment at ambient or specified elevated temperatures, e.g. 100 degrees C.
Moist -cured concrete - has been kept in water at 20 degrees C for 28 days.
Rewetting water - water that goes into moist-cured concrete which has been dried to constant weight at ambient temperature (in a drying chamber) or under accelerated conditions at higher specified temperatures such as 100 degrees C.
Alpha water is hydrated water plus adsorbed water.
Beta water is free water plus capillary water plus some of the more weakly bound adsorbed water and includes evaporable water and /or rewetting water.

EXPERIMENTAL RESULTS

To resolve between the various water states, and their effect on concrete properties during freezing from 273 K to 200 K, a three year programme of complementary studies was made, using NMR techniques, enthalpy-temperature measurements and thermal strain - temperature measurements (3,4). The following are examples of the extensive number of studies made:

Nuclear magnetic resonance NMR studies

These showed immediately two well-separated NMR peaks, the alpha peak associated with chemically bound water, hence alpha-water, and the beta peak associated with some of the weakly, physically bound liquid water - hence beta-water.

On lowering the temperature of a saturated moist-cured concrete sample below 0 degreesC a change in the relative size of the alpha and beta peaks is apparent. (Fig.1) with the beta peak diminishing to zero at -53 degrees C , as the beta-water freezes to ice.

On raising the temperature of a dried and re-wetted concrete sample, in a dry atmosphere, it can be seen (Fig.2) that the beta peak diminishes with time. From this, it is concluded that beta-water,corresponding to the beta peak, is at least part of the evaporable water which is more rapidly removed by heating to 120 degrees C.
It should also be noted that on re-wetting the dried sample, the beta peak reappears.

On lowering the temperature of a dried and re-wetted concrete sample, it can be seen (Fig.3) that the beta peak diminishes with decreasing temperature to zero at -53 degrees C, again as the beta-water freezes to ice. It should be noted that capillary water has been predicted to freeze at lower temperatures in smaller diameter pores down to a limit of -70 degrees C, or so. However, ice and unfrozen water in the smaller pores will tend to be absorbed by frozen ice in the larger pores as a result of differential capillary forces and slow cooling rates.

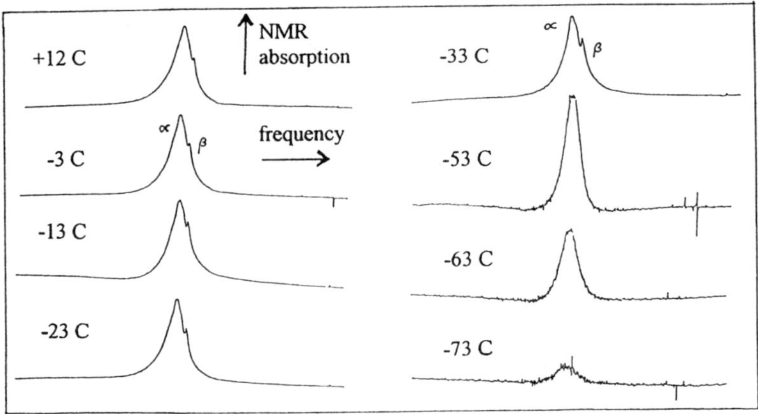

Figure 1. Proton NMR in a saturated moist-cured concrete sample W/C =0.37.
Effect of reducing temperature.

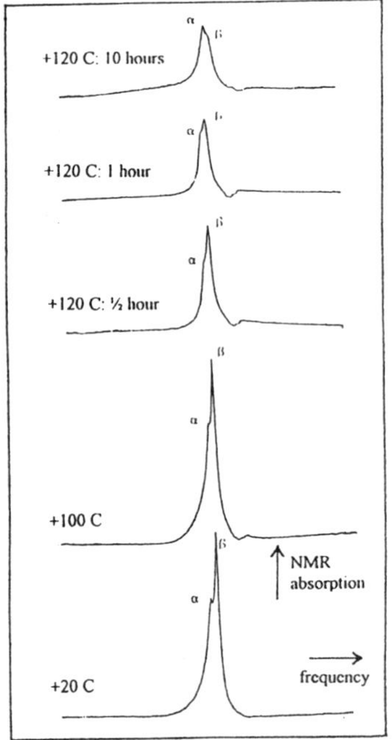

Figure 2. NMR in dried and rewetted concrete sample.
Effect of heating to 120 degrees C.

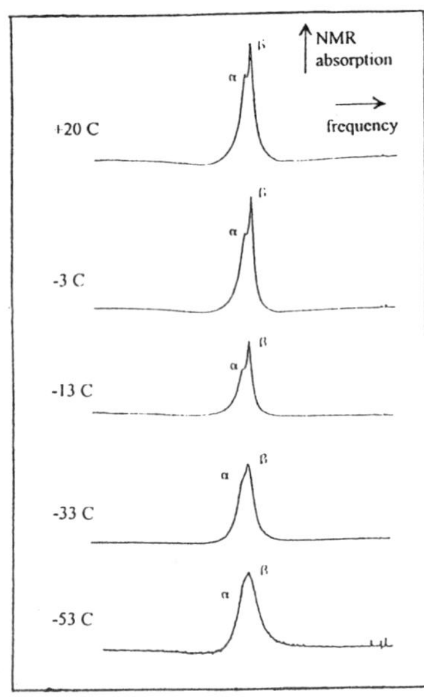

Figure 3. NMR in dried and rewetted concrete sample.
Effect of cooling to -53 degrees C.

These examples indicate how NMR spectra provide a qualitative means for observing the behaviour of different types of water in concrete. With appropriate calibration, a quantitative picture could easily be developed.
The strength of the beta peak during re-wetting of the concrete sample is particularly noteworthy. In particular, a continuous increase in beta peak is observed in a high relative humidity atmosphere - illustrating the hygroscopic behaviour of concrete.

Enthalpy - temperature measurements

A large calorimeter, capable of taking cylindrical samples of concrete 96 mm diameter and 200 mm long, was used to produce enthalpy-temperature plots down to -80 degrees C. The normalised H-T curves for dry, saturated and re-wetted concrete samples and water/ice are shown in Fig. 4 . The effective heat capacity plots for saturated, moist-cured samples are shown in Fig. 5 and illustrate the distributed excess heat capacity associated with physically bound water freezing over the temperature range 0 to -55 degrees C.
The effective heat capacity for a re-wetted sample is shown in Fig.6. In this case, the excess heat capacity associated with the re-wetting is confined to a narrower temperature range from 0 to -10 degrees C.

Strain - temperature measurements

Strain measurements were made on cylindrical samples, 96 mm diameter and 200 mm long, made from ordinary Portland cement and a marine flint aggregate of about 5-mm size. The aggregate had been dried beforehand. Fig.7 shows the thermal strain behaviour of a moist-cured concrete sample with water/cement ratio of 0.4, and after drying for three weeks in a drying chamber. The dilation on freezing can be seen to disappear after drying.
Fig.8 shows the effect of re-wetting the same sample. The dilation and hysteretic behaviour has returned, and there is a permanent increase in volume after each cycle - i.e., there is an indication of frost damage to the sample after each freeze-thaw cycle.

CONCLUSIONS

1. There is a good correlation between the three sets of studies in relation to the effect of evaporable water and re-wetting water on the properties of concrete at low temperatures betwen 273 K (0 degrees C) and 193 K (-80 degrees C).
2. The NMR spectra provide an observable quantity, the beta peak, which can be used to monitor the behaviour of that part of the physically bound water which gives rise to the beta peak, namely the beta-water.
3. When beta-water is removed by drying, the indications leading to frost damage are also removed.
4. When concrete is re-wetted after drying, the re-wetted beta-water is particularly severe in creating permanent dilation after freeze/thaw cycles.
5. From these observations it follows that the beta-water, whether in saturated moist-cured concrete or in dried and re-wetted concrete, is a mixture of free water and capillary water which has small binding energies and constitutes a major proportion of evaporable water. As re-wetting water, beta-water occupies the largest diameter capillaries, so that on cooling below 273 K (0 degrees C), the majority freezes to ice by 263 K (-10 degrees C) with expansion of the cement paste matrix. Cooling is accompanied by preferential ice crystal growth in the larger capillaries, with further expansion of the cement paste.

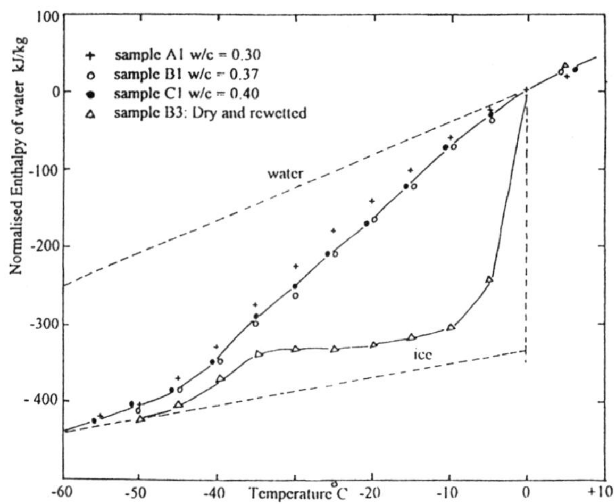

Figure 4. Normalised enthalpy of water in concrete samples.

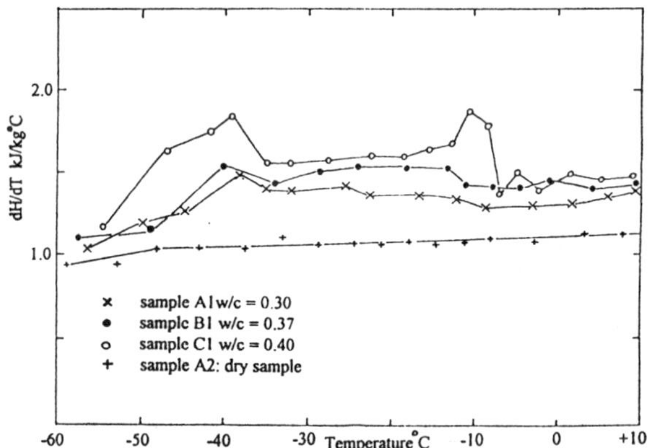

Figure 5. Specific heat of concrete samples.

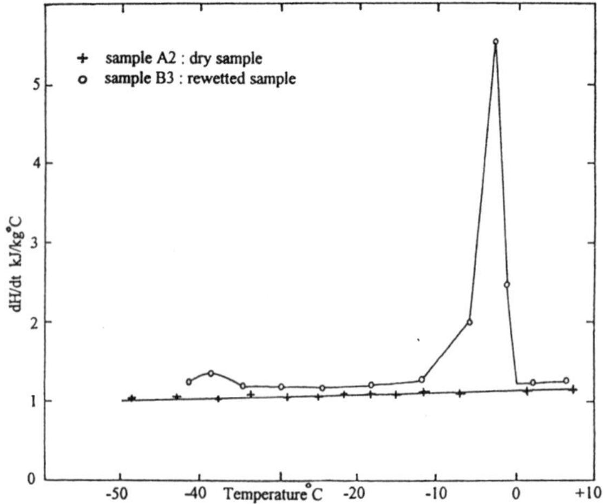

Figure 6. Specific heat of rewetted concrete sample.

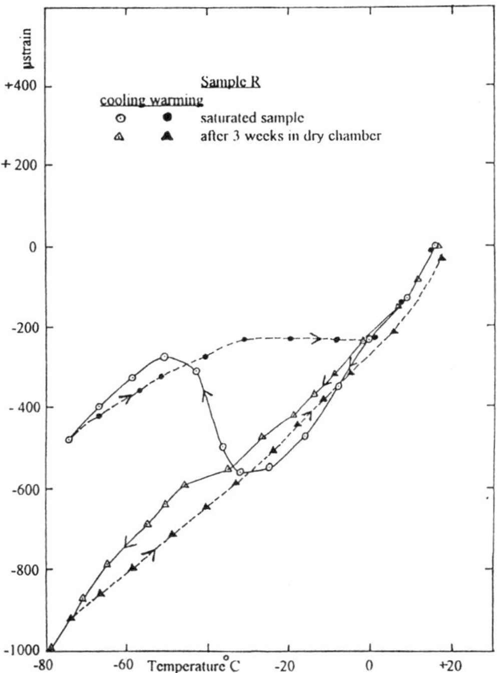

Figure 7. Thermal strain of moist-cured concrete samples W/C =0.40

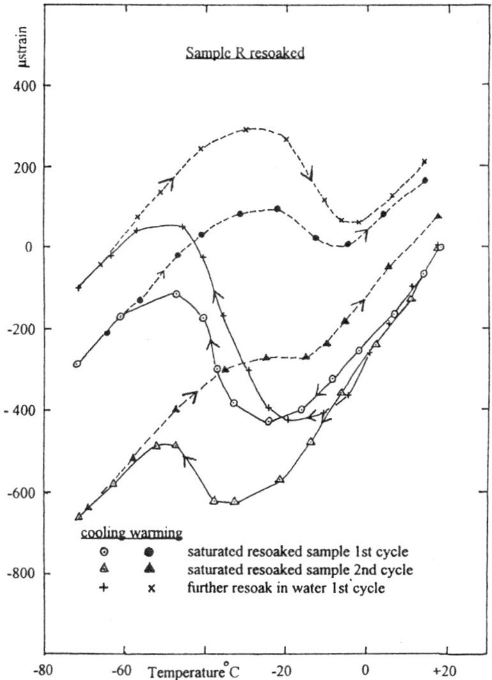

Figure 8. Thermal strain of rewetted concrete samples W/C =0.40 after repeated cycles.

6. In the absence of beta-water, there is no thermal strain, hysteresis or evidence of frost damage during thermal cycling below 273 K (0 degrees C).
7. It therefore follows that the manufacture of frost-proof concrete requires the removal of beta-water by adequate air or vacuum drying after moist-curing is complete, followed by the application of a water repellant or sealant coating, to prevent re-wetting by ingress of water in contact with the surface of the concrete, or from air with high relative humidity.
8. This simple formula is an extrapolation of our experimental observations and has been found to work for freshly prepared moist-cured concrete and retrospectively by drying and sealing existing concrete structures exposed to temperatures down to 203 K (-70 degrees C).
9. The same formula applies to the manufacture of cryogenic concretes for use at temperatures down to 77 K (-196 degrees C), but attention must also be paid to the thermal strains of the aggregate and the cement paste. These should be identical to ensure adhesion is maintained between the two at low temperatures and under thermal cycling.
10. When this formula is applied, 10-cm cube samples survive without damage after dunking in liquid nitrogen, with cooling rates of 100 K per minute, the severest test of proof against frost damage down to 77 K (-196 degrees C).
11. These findings are based on three years of measurements on a wide range of samples with different water/cement ratios. The results described in this paper are a small fraction of those obtained.
12. While further work is needed to substantiate these findings , the applications of the formula for making frostproof concrete could be put in hand immediately. After moist curing and adequate drying, steps should be taken to keep out the real menace of re-wetting water, and then no frost damage will occur.

ACKNOWLEDGEMENTS

This work was carried out with the help of a research grant funded by EPSRC.

REFERENCES

1. A.L.Marshall Cryogenic concrete. *Cryogenics* 32: 555 (1982).
2. V.S.Ramachandran, R.F.Feldman and J.J.Beaudoir "Concrete Science, "Heyden, London (1981).
3. R.G.Scurlock and Kamarudin M.Yusof Thermal behaviour of concretes between -80 degrees C and +20 degrees C. Proc. 2nd Int.Conf.Cryogenic Concrete, Amsterdam, (1983).
4. Kamarudin bin Mohd Yusof "The Effects of Moisture on the Thermal Properties of Concrete between -80 degrees and 0 degrees C." University of Southampton, Ph.D. thesis (1984).

DEVELOPMENT OF SUPERCONDUCTIVE HOT-WIRE ANEMOMETER FOR USE AROUND 2 K

H. Okubo, T. Furukawa and M. Murakami

Institute of Engineering Mechanics and Systems,
University of Tsukuba
Tennoudai 1-1-1, Tsukuba, Ibaraki 305-8573, Japan

ABSTRACT

A superconductive hot-wire anemometer has been developed to be used in the measurement of low speed helium vapor flow around 2 K for the study of evaporation from a HeII free surface. The hot-wire element is made of gold-tin superconductive thin film vacuum deposited on the outer surface of fine quartz fiber with a diameter of about 40 μm. It is a modified version of the superconductive temperature sensor in the structure and material which we had developed, but is operated in the constant temperature mode excited by larger bias electric current for sufficiently large temperature ratio to enhance the sensitivity to the gas velocity. We also developed a miniature cryogenic helium gas wind tunnel for the purpose of calibration, in which evaporated helium vapor from He II is used as the working gas. It is confirmed that the hot-wire anemometer can be used for the velocity measurement of gaseous helium at temperatures from 1.65 K to 2.15 K in the velocity range up to 4.5 m/s at 2.0 K.

INTRODUCTION

A superconductive hot-wire anemometer has been requested for velocity measurements in a cryogenic environment and for potential high performance. At liquid-helium temperatures, the temperature dependent property of the electric resistance of normal hot-wires disappears and consequently they cannot be used under such temperature conditions. The attempt to develop a superconductive hot-wire would be quite natural for these conditions. In fact, a new type of superconductive hot wire, composed of thin films of resistive and superconducting metals vacuum deposited on a glassy micron-size substrate, has been developed to be utilized in gas flow measurement around 4.2 K.[3,4] It was designed to be applied to experiments of flows at ultra-high Reynolds and Rayleigh numbers, where helium gas at cryogenic temperatures is used as a test fluid.

The primary objective of the present study is also to develop a superconductive hot-wire which can be used for the measurement of the velocity of gaseous ^{4}He around 2 K in the study of evaporation from a HeII free surface.[1] The hot-wire anemometer operates at a temperature 2K lower than previous designs, which used lead-indium. The lower temperature requires use of tin-gold, which has been used for superconductive temperature sensors.[1,5]

The technical specification of the present design is decided considering the application to the evaporation experiment. In the experiment, evaporation is caused by stepwise heating of a HeII free surface by the impingement of a second sound thermal pulse on the free surface. The flow velocity of evaporated helium vapor is as large as 5 m/s at the temperature of around 2 K. Accordingly, some special attention is required in the present hot-wire application. The temperature and the velocity may vary in the evaporation process, which may cause a difficulty in separating the velocity variation from the total variation in the output signal. It should be noted that the pressure, which also varies in the evaporation process, is decided almost uniquely by the temperature along the saturated vapor pressure relation.

Furthermore, a miniature cryogenic helium gas wind tunnel is also developed, in which the hot-wire anemometer developed in the present study was calibrated and tested. It is found that the sensitivity to the temperature variation can be diminished by selecting a higher temperature ratio of the element than that in conventional hot-wire applications, and the variation in the velocity is satisfactorily separated from the temperature variation with the aid of computer software.

SUPERCONDUCTIVE HOT-WIRE ANEMOMETER

Figure 1 shows an illustration of the superconductive hot-wire anemometer. Thin quartz fiber with a diameter of about 40 μm, on the surface of which the thin film sensor element is formed, is fixed at the tips of the inconel needles with gold paste. The portion is fired in an electric furnace to treat the gold paste for tight connection. After the firing, it is cooled down in a room temperature environment. The sensor element thin film is vacuum-deposited on the surface of the quartz fiber. The film is composed of gold with a thickness of 230 Å and tin with a thickness of 990 Å. To enhance the long-term

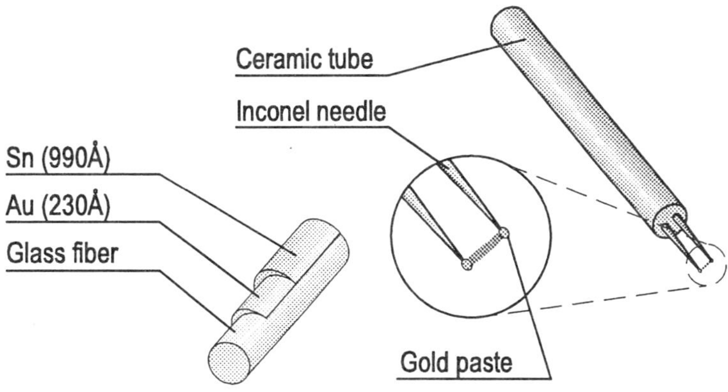

Figure 1. Illustration of the superconductive hot-wire anemometer.

stability of the sensor element, the sensor is annealed at low temperature for a short time.

The connection of the quartz fiber with the inconel needle with gold paste is found to be a key for the long term stability of the hot-wire. The aging degradation problem found in the original design in which silver paste was used is almost solved by using gold paste. It is considered that the firing treatment at higher temperature in the case of gold paste than that for silver paste results in less residual impurity in the connecting portion, which led to considerable variation in the electric resistance after thermal cycles between the 2 K and the room temperature environments in the case of silver paste. The sensor is operated in the constant temperature mode with the commercially available hot-wire signal conditioner (KANOMAX MODEL1010). The resistance of the sensor is about 30 Ω in the cryogenic helium gas environment.

CRYOGENIC MINIATURE HELIUM GAS WIND TUNNEL FOR THE CALIBRATION OF THE ANEMOMETER

Figure 2-(a) is a schematic overview of the wind tunnel system. Figure 2-(b) is an enlarged illustration of the main parts of the wind tunnel section. It consists of two main parts, the measuring section made of brass where the hot-wire is mounted and the evaporator made of acrylic acid resin to generate helium vapor as the working fluid. The evaporator is immersed in liquid helium. The evaporator supplies helium vapor and compensates for the flow-out of the vapor to keep the pressure in the helium dewar constant. The carbon resistance thermometer measures the temperature in the evaporator. The pressure transducer, the Cernox temperature sensor and the hot-wire are fixed in the measuring section to measure the pressure, the temperature and the velocity of the helium vapor in the wind tunnel. The pressure transducer is installed at the end of the measuring pipe to measure the static pressure of the vapor. The measuring section is thermally connected to the liquid helium to keep the temperature of the section as close as possible to that of the liquid helium. The evaporator and the

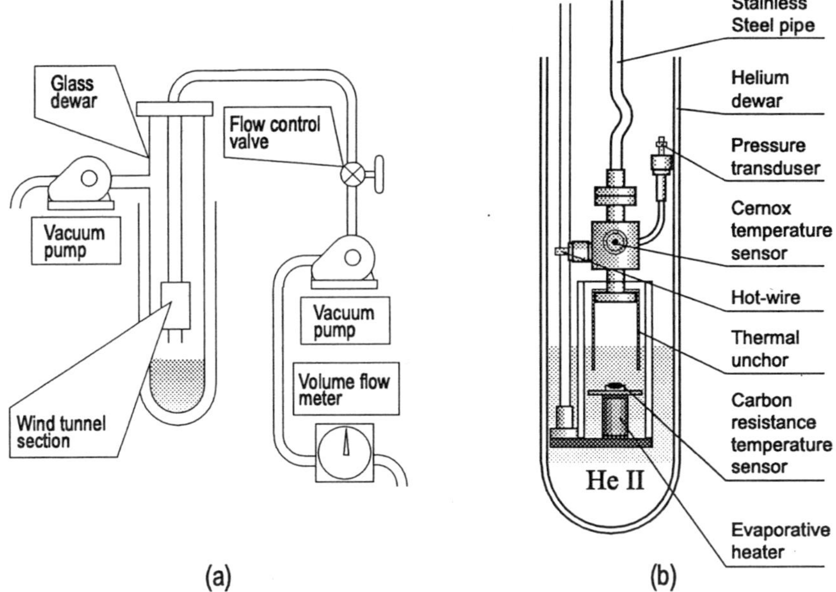

Fig. 2. Schematic illustration of the wind tunnel for the hot-wire calibration.
(a). Schematic overview of the wind tunnel system.
(b). Enlarged illustration of the main parts of the wind tunnel section.

measuring section can be moved up and down and thus the relative distance between these sections and the liquid surface can be adjusted arbitrarily. The flow rate control valve is installed between the wind tunnel section and the vacuum pump outside of the glass dewar. The flow meter connected to the outlet of the vacuum pump measures the evacuated vapor flow rate to be converted into the flow velocity in the wind tunnel section.

RESULTS AND DISCUSSION

Flow Performance of the Miniature Wind Tunnel for Calibration

The flow performance of the miniature wind tunnel is shown in Fig. 3. The experiments are conducted at three vapor pressures by varying the valve opening to control the flow velocity. The flow velocity, which is the average velocity of the vapor flow through the test section, is converted from the volume flow rate of the evacuated vapor measured with the gas flow meter connected to the exhaust port of the vacuum pump. It is demonstrated that the fundamental flow performance satisfies the requirement for the calibration wind tunnel, that is, the maximum velocity of about 4.5

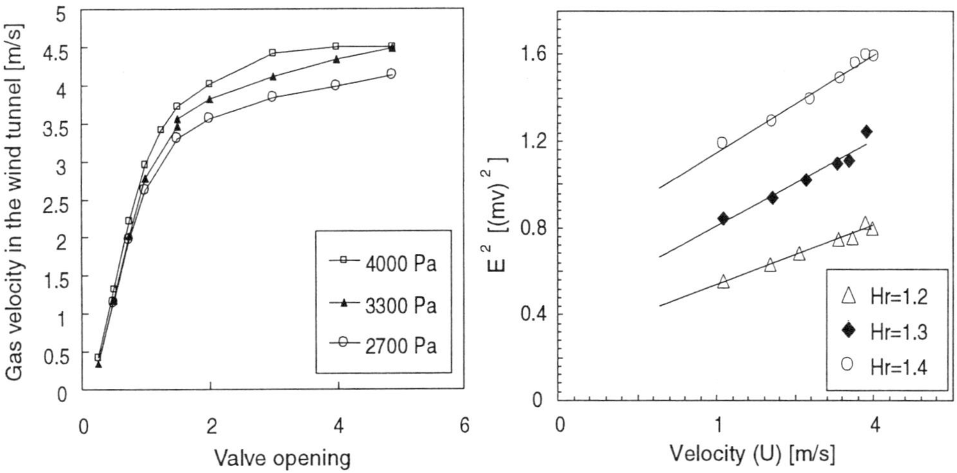

Fig. 3. Flow performance of the wind tunnel; flow velocity vs valve opening at three pressures.

Fig. 4. Hot-wire measurements for three heating ratios. The solid lines are least square approximations. T = 1.94 K.

m/s, which is anticipated to arise in the HeII evaporation experiment.[1] It is well known that the temperature of the working fluid in the wind tunnel has a significant effect on the conventional hot-wire measurement, because the electric resistance of a hot-wire element is sensitive to the temperature. This is also quite true for the superconductive hot-wire. It is, therefore, important to control the gas temperature in the wind tunnel to construct a reliable calibration formula. In the present wind tunnel system the control of the gas temperature is rather easily attained using saturated vapor, the temperature of which is automatically controlled through the control of the vapor pressure with an accuracy of 3 mK.

Calibration Result of the Hot-wire Anemometer

The hot-wire is operated in the constant temperature mode, and, consequently, the heating ratio is very important for optimal operation. The measurement data is shown in Fig. 4 for three values of the heating ratio at a vapor pressure of 2700 Pa. In the figure the square of the output voltage from the anemometer signal conditioner is plotted against the flow velocity, which is King's law plot, and the solid lines are the least square approximation for the measurement data. It is fair to conclude that each data set can be well correlated by a straight line for the flow velocity larger than 1 m/s. This fact implies that King's law holds for the present type of the superconductive hot-wire measurement, as in the case of conventional hot-wire anemometers, as long as the gas is kept at a constant temperature. The sensitivity of the superconductive hot-wire to the velocity is found to be large for larger heating ratio, as found in conventional hot-wire applications. However, excessive bias current to increase the hot-wire heating ratio should be avoided because of the possibility of break down of superconductivity of the hot-wire element. So, we should select the appropriate heating ratio for optimal performance.

In the application to the HeII evaporation experiment the temperature will change with the vapor velocity. The measurement is conducted at different temperatures of the working fluid. The result is shown in Fig. 5. No reliable data is available below 1 m/s because of the influence of natural convection. The solid lines are least square approximations to the measurement data, of which formulae are given in the figure. It is seen that the inclination of the formulae becomes larger as the temperature rises. This character is also true for conventional hot-wire anemometers. The constant term, that is the A coefficient of King's law, is also derived as a function of temperature from the data plot. As the final result, we predict that the calibration formula is established in terms of A(T) and B(T) as follows:

$$E^2 = A(T) + B(T)\sqrt{U} \ , \tag{1}$$

where E and U are the output voltage and the flow velocity, respectively. In actual application, reliable interpolation formulae for A(T) and B(T) should be prepared for the computer aided anemometry to be applied to the velocity field where the temperature also varies, though the detailed investigation of A(T) is still required.

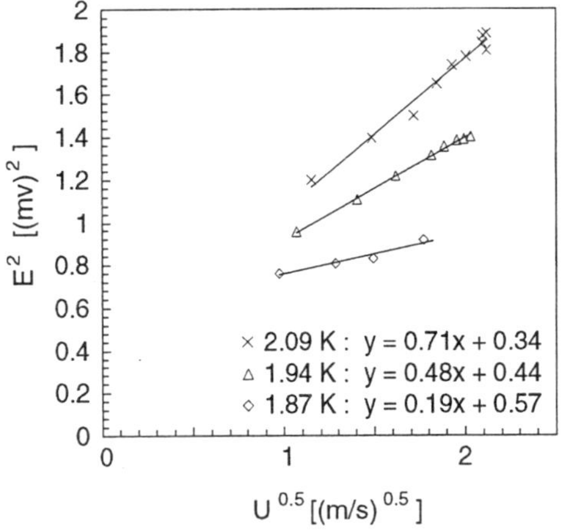

Fig. 5. Hot-wire measurements for three temperatures. The solid lines are the least square approximation to the measurement data. Hr = 1.2.

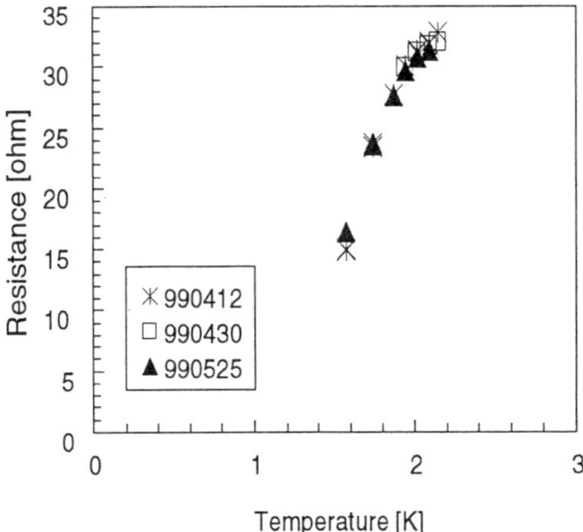

Fig. 6. Thermal cycle test results of long term stability.

Long Term Stability Data of the Superconductive Hot-wire Anemometer Element

The most serious concern with the superconductive temperature sensor, which is the original form of the present superconductive hot wire, is the long term stability after thermal cycles. It had been recognized that the aging of elements was mostly caused by the degradation of the silver paste connecting the quartz fiber with the support needle leading to considerable change in the normal electrical resistance of the element. In the present design, gold paste is adopted instead of silver paste, as mentioned previously. The result of a long-term thermal cycling test is given in Fig. 6, in which the variation of the electrical resistance with the temperature is measured for several thermal cycles between the room temperature and 2 K. The test result indicates that no appreciable change of the resistance resulted after several thermal cycles. One can conclude that the new design using gold paste achieves the long term stability required for practical applications.

CONCLUDING REMARKS

1. King's law holds for the superconductive hot-wire anemometers, as long as the gas is kept at a constant temperature, as is the case of conventional hot-wire anemometers.
2. The superconductive hot-wire has similar qualitative characteristics with respect to the heating ratio and the calibration formula as conventional hot-wires.
3. We confirmed that the present superconductive hot-wire anemometer element has long-term stability after several thermal cycles.
4. It is confirmed that the miniature wind tunnel is useful for calibration of the superconductive hot-wire anemometer.

ACNOWLEDGEMENT

This study is financially supported by Grant-in-Aid for Scientific Research (B) from The Ministry Education, Science, Sports and Culture.

REFERENCES

1. T. Furukawa, M. Murakami, and H. Nagai, Transient evaporation phenomena in a pure vapor-liquid system, in: "Heat Pipe Technology", J. Andrews, A. Akbarzadeh, and I. Sauciuc, ed., Elsevier Science (1997), p. 255-260.
2. Y. Sone, and K. Aoki, "Molecule Gas Dynamics", Asakura Publishing company inc. (1994), p. 112-113.
3. B. Castaing, B.Chabaud, F. Chillia, and B. Hebral, Hot wire anemometer operating at cryogenic temperatures, *Rev. Sci. Instrum.* 63(9):4167-4173 (1992).
4. B. Castaing, B.Chabaud, F. Chillia, B. Hebral, A Naert and J. Peinke, Anemometry in gaseous 4He around 4K, *J. Phys.III France 4* :671-674 (1994).
5. T. Shimazaki, M. Murakami, T. Iida, Second sound wave heat transfer, thermal boundary layer formation and boiling: highly transient heat transport phenomena in HeII, *Cryogenics*, 35:645-651 (1995).

CONTROLS UPGRADE OF THE FERMILAB MAGNET TEST FACILITY 1500 W HELIUM REFRIGERATOR

R. H. Carcagno and R. J. Rabehl

Fermi National Accelerator Laboratory*
Batavia, Illinois 60510

ABSTRACT

The control system of the Fermilab Magnet Test Facility 1500 W helium refrigerator has been successfully upgraded using modern industrial control system components and standards. This upgrade was performed to provide automation for unattended operation of the system and to increase system availability by simplifying and replacing old and unreliable electrical wiring and control devices, improving system troubleshooting through trending capabilities, and providing an alarm management system. A noteworthy aspect of this project was the constraint of continued operation of the Magnet Test Facility during the upgrade. This required a novel approach to the project that avoided impacting the magnet testing schedule. A core team of in-house engineers and technicians formed to work on the upgrade produced extensive documentation, off-line simulations, and standards. The field work and system integration were completed within short windows of opportunity. The resulting operational system uses Siemens 545 programmable logic controllers running Siemens SoftShop software communicating over an Ethernet network. Windows NT stations run human-machine interface software produced using Intellution FIX32 products. The project approach and metrics will be presented, as well as examples illustrating the achievement of the controls upgrade objectives.

INTRODUCTION

The Fermilab Magnet Test Facility 1500 W Helium Refrigerator[1,2] was commissioned in October 1977, and has since operated to provide cooling for the development and testing of superconducting components such as Tevatron magnets and spool pieces, SSC magnets, and LHC High-Gradient Quadrupole (HGQ) magnets.

Unanticipated refrigerator shutdowns due to an aging control system have been an increasing source of problems for this facility. An increasing rate of costly turbine failures

---

* Work supported by the U. S. Department of Energy under contract No. DE-AC02-76CH03000

seemed to indicate that the control system was not providing adequate protection for this critical equipment. Sensor failures made it difficult to obtain the necessary information for troubleshooting and optimization of operating strategies. The lack of accurate documentation and troubleshooting tools to quickly locate and repair control system failures resulted in longer downtimes than necessary.

In 3QFY01 the operating duty of this refrigerator is expected to increase to near 100% to support testing of production LHC HGQ magnets. This provided a strong incentive to initiate in 3QFY98 a controls upgrade project for the refrigerator. The major goals of the project were to (1) increase the control system availability and (2) increase the automation of the machine to allow extended periods of unattended operation. The controls upgrade was to take place without impacting the testing schedule of the Magnet Test Facility, and had to be completed within short windows of scheduled refrigerator downtime. This was successfully accomplished by dividing the project into phases.

After analyzing the state of the original control system, the project team decided that in order to meet the availability goals of this project it was necessary to replace the entire control system. This included the electrical system, controllers, and operator interfaces as well as the replacement of some sensors such as temperature diodes in the cold box. Sensors and valve actuators were also added at new locations to support new control strategies.

This work has been successfully completed on schedule and within budget. The plant has been running with a new control system that has all the necessary features to implement the second project goal of enhanced automation.

PROJECT APPROACH

The controls upgrade project followed an established approach.[3] A project core team of five engineers and technicians representing the various disciplines involved (e.g., Project, Process, Controls, Instrumentation, Electrical, and Operations) was formed to concentrate effort on this project. The team was expanded during fieldwork activities to a total of approximately ten people. All team members were staff members of the Development and Test Department of the Fermilab Technical Division. The project was divided into three phases. Phase I was the compressor plant controls upgrade. Phase II was the cold box system controls upgrade. Phase III is the automation and optimization upgrade. The first two phases are the most capital- and labor-intensive phases. Phase III is mostly a software upgrade and is expected to be completed by 3QFY00.

During design and development periods of each phase, the core team met weekly. After the design for each phase was completed and before starting fieldwork activities, a safety review took place. In addition, Siemens Westinghouse engineers reviewed safety aspects of the compressor 480 V Motor Control Centers and verified that the upgrade complied with the National Electrical Code (NEC).

The following items were available before starting fieldwork activities for each phase: a complete and well-documented design, purchases of all major and long-lead items, and a detailed schedule to plan and keep track of the many concurrent implementation tasks. During fieldwork activities, the team members met daily to monitor progress and resolve conflicts.

The first two phases have been successfully completed on schedule and within budget. The first phase was completed in mid-December 1998. For four months, the compressor plant operated with the new controls and the cold box with the old controls until fieldwork activities for phase II started in mid-April 1999. At the completion of phase II in mid-June 1999, the entire refrigerator was operated with the new control system. The phased implementation approach was successful in minimizing the impact of the controls upgrade project on the testing schedule of the Magnet Test Facility. Table 1 lists some additional information for each phase.

Table 1. Metrics of the refrigerator controls upgrade project

	Phase I Compressor Plant	Phase II Cold Box	Total
Physical I/O Points	Total: 77 (22 AI, 5 AO, 28 DI, 22 DO)	Total: 109 (59 AI, 11 AO, 25 DI, 14 DO)	Total: 186 (81 AI, 16 AO, 53 DI, 36 DO)
PID Loops	4	10	14
Database Tagnames	197	311	578
Operator Displays	11	18	29
Fieldwork	Mid-October to mid-December 1998	Mid-April to mid-June 1999	All fieldwork was completed in two two-month windows.

THE ORIGINAL CONTROL SYSTEM

The original control system was based on 1970's technology. The operator interface consisted of a control console with pushbuttons, LEDs, single-loop PID controllers, and indication lamps. Temperatures and pressures were monitored one at a time, and there was no real-time or historical trending. All electrical wiring had to be brought to the control console, resulting in a large amount of cables routed from field devices to the control console in the control room. Many control devices required 110 VAC to operate, creating a safety hazard during troubleshooting.

Control functions were provided by several systems, namely: (a) An old-style PLC (5TI model) for basic interlock and discrete logic functions; (b) Hardwired electrical circuits with relays, timers, and math blocks to supplement the PLC discrete logic; (c) Single-loop PID controllers for analog valve control; (d) Pushbuttons, indicator lamps, and LEDs indicators for operator interface. The limitations of these systems resulted in extensive, complicated wiring with a large amount of field devices and redundant instrumentation to service the separate systems. For example, in some cases up to three pressure transducers were installed in the same location: one pressure transmitter for single-loop controller input, another pressure transmitter for pressure monitoring LED input, and a pressure switch for discrete PLC logic input.

The original control system had gone through several modifications over the many years of operation of this refrigerator. Unfortunately these modifications were not well-documented, and accurate as-built documentation did not exist.

THE NEW CONTROL SYSTEM

Using modern control systems technology it is possible to overcome many of the limitations and complexities of the original control system. In the new control system, the operator interface consists of CRT-based terminals. All process variables can be monitored simultaneously. Real-time and historical trending is available. An alarm management system and first-fault diagnostics logic helps to quickly identify and troubleshoot problems.

A single network cable is brought into the control room. Field devices are wired to remote Input/Output cards that are geographically distributed in three strategic locations to minimize cable length. Discrete control signals were changed from 110 VAC to 24 VDC, eliminating electrical safety hazards during troubleshooting. Analog control signals were standardized to 4 to 20 mA or 0 to 5 VDC signals.

Table 2. Hardware, software, and documentation tools for the new control system

System	Choice
Hardware and Software	
Programmable Logic Controller (PLC)	Two Siemens 505-454 PLCs (one for the compressor plant and another for the Cold Box)
Input/Output (I/O) Cards	Siemens 505 series
Control Network	Ethernet with TCP/IP in a routed network
PLC Programming Language	Siemens 505 SoftShop, v2.20
Operator Interface Software	FIX32 v6.15 from Intellution running on two Windows NT stations.
Electrical Loops	24 VDC for discrete I/O, 4-20 mA or 0-5 VDC for analog I/O. 24 VDC control relays in a separate enclosure was used to switch 110 VAC for MCC components such as breakers and contactors, pumps, heaters, etc.
Temperature sensors inside cold box	Lakeshore DT-470 diodes with Lakeshore Model 231 temperature transmitter card. Redundant sensors in critical locations.
Liquid Helium Level Probes	American Magnetics superconducting level sensors and Model 135 liquid helium level monitor
External Interface	ODBC interface to FIX32 real-time database
Documentation Tools	
Piping and Instrumentation Diagrams (P&IDs)	AutoCAD
Electrical Schematics	Electrical CAD system EPLAN21
Instrumentation Database	MS Access 97
Technical Notes	MS Word and MS Excel
Schedule	MS Project

All control functions were integrated into the same system. Interlocks and discrete logic, PID controllers, timers, math functions, and operator interfaces are now all software functions. Because of this integration, redundant instrumentation and control field devices were eliminated. This reduction of field devices, plus remote location of distributed I/O modules and standardization of signals, resulted in a much simpler and safer electrical wiring.

Extensive documentation was produced for the new control system. As-built P&IDs, electrical schematics, and an instrumentation database are available in addition to numerous engineering notes describing the design and choices made. Engineering standards and recommended engineering practices[3] were followed during the design and implementation of the electrical system.

Table 2 summarizes the choices made for the new control system.

Siemens series 505 PLCs and I/O cards from the SSC project were used. The decision was made to use the newer Ethernet control network instead of the traditional proprietary TIWAY control network for this system. An Ethernet TCP/IP card is available from Siemens (model 505-CP2572), and an Ethernet I/O driver for the 505-CP2572 card (TIE) is available from Intellution . For additional security and protection, the Ethernet control network was implemented as a routed network. The Ethernet control network configuration worked very well and has been very reliable.

An advantage of using the Ethernet network is that the PLC program can be accessed over the existing computer network, avoiding the need for having a PC connected to the PLC via a serial cable interface.

The Intellution FIX32 software for operator interfaces was configured as one SCADA node (or station) and one VIEW node. The only difference is that the real-time database and I/O drivers reside only in the SCADA node. The system can be monitored and controlled from either node, and more VIEW nodes can be easily added to the system. The real-time database has an SQL (Standard Query Language) interface that uses ODBC (Open

Database Connectivity) to provide two-way access between the refrigerator control system and other databases. This capability is important to provide tight integration between the refrigerator control system and the Magnet Test Facility Data Acquisition System, a feature that will be implemented during phase III of this project. An interface between the FIX32 real-time database and a FORTRAN program was developed to allow accurate on-line turbine efficiency calculations using a commercial helium thermodynamics package written in FORTRAN.

The control system was designed so that in the event of a loss of the SCADA operator station the refrigerator will keep running under the control and protection of the PLC software alone. Operators will lose the ability to monitor and control the system, trends will stop, and the alarm system will cease to function. However, the equipment will continue operating protected by the PLC logic, and the PLC will initiate a system shutdown if an unsafe condition is reached. Any PC can be quickly configured to boot as a SCADA node, so operators will be able to quickly resume normal operation without the need for system shutdown.

Another feature of the system is that the PLCs and I/O cards provide self-diagnosis information. If a failure occurs, the information is passed to the operator interface and an alarm condition is set. This allows operators to quickly locate a control system hardware failure. A PLC "Heartbeat" signal was programmed into the system, and if this signal is lost an alarm condition is set. "First-Fault" logic was programmed into the PLC. This logic allows the operator to quickly identify which event initiated a shutdown sequence for the compressors or turbines.

A significant effort of this upgrade was the replacement of all 21 original diode temperature sensors inside the cold box. Seven of these sensors were no longer working, and we determined that it would be beneficial to have sensors in additional locations. Redundant sensors were installed at critical locations. A total of 33 new Lakeshore DT-470 sensors were installed, 30 inside the cold box and 3 external. Each diode has a corresponding Model 231 Lakeshore Temperature Transmitter card that provides the diode excitation current, reads the temperature-dependent voltage, and outputs a linear 4-20 mA output signal that was wired directly to a standard Analog Input card. Depending on the accuracy required, temperature transmitter cards were loaded with either a standard curve 10 EPROM, or with an EPROM containing calibration information for a specific diode sensor. Recommended Lakeshore practices were followed during installation and wiring of these sensors. Adequate packaging, pre-installation testing at liquid nitrogen temperature and careful routing of wires inside the cold box resulted in no diode sensor failures.

PROCESS ISSUES

The controls upgrade project addresses two important issues related to the helium liquefaction process. The first issue is process troubleshooting. With the old controls system, the combination of limited process condition information and a lack of historical trending gave an incomplete picture of system performance. Diagnosis of poor system performance often relied on process data hand-recorded by refrigerator operators at eight-hour intervals, but the snapshot nature of this information made pinpointing the initiation of a problem even more difficult. The second issue is process improvement toward the goal of system automation. Successful automation of the helium refrigerator will require that each system component behaves in a regular, well-understood manner. The first two phases of the controls upgrade project have addressed several aspects of refrigerator operation that required improvement before automation can be realized. Examples of each issue and their solutions are included.

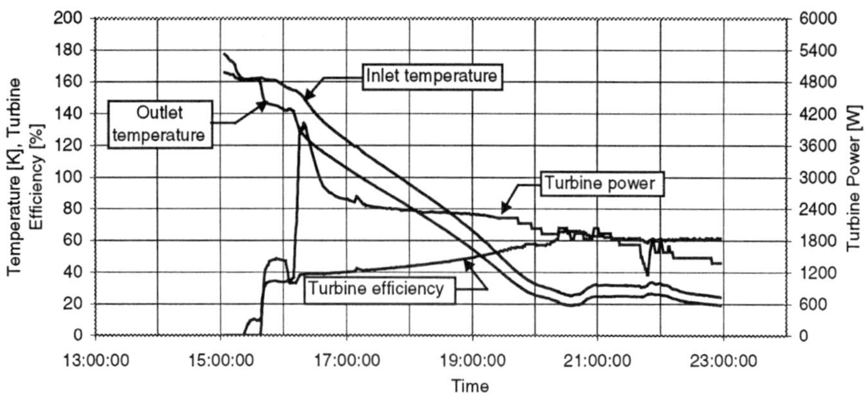

Figure 1. Historical trend plot of a T1 turbine startup. Inlet temperature, outlet temperature, turbine power and turbine efficiency are plotted over an eight hour period.

Turbine Performance

The MTF 1500 W refrigerator uses two turboexpanders. The T1 turbine takes a portion of the high pressure flow, expands it, and returns it as a medium pressure flow. The T2 turbine expands the remainder of the high pressure flow before it goes through a Joule-Thomson valve to yield a two-phase mixture. The performance of these turbines is critical to the liquefaction rate.

Turbine performance could not be studied prior to completion of phase II of this project. Although pressure and temperature sensors were in place at both the inlet and outlet of each turbine, neither turbine exit temperature sensor was operational. The phase II fieldwork not only replaced the temperature sensors at the inlet and outlet of both turbines, but redundant sensors were also placed at each of these locations.

A FORTRAN program carries out the turbine performance calculations. The calculated figures of merit are the turbine efficiencies of both the T1 and T2 turbines and the turbine power of the T1 turbine only. The turbine performance parameters are calculated continuously.

Figure 1 shows a historical trend plot of T1 turbine performance at the start of a cooldown from liquid nitrogen temperatures toward liquid helium temperatures. Turbine efficiency, turbine power, and inlet and outlet temperatures are plotted over time.

Similar information is expected to aid in system automation. Turbine performance studies at various operating conditions and speeds will lead to system optimization through performance at peak efficiency.

Turbine Shutdown Logic

Over a twelve month period beginning in March 1998, the Fermilab Magnet Test Facility experienced five turbine failures. In at least one case, the control logic did not shut down the turbine as expected. Turbine failures are costly in terms of system operations scheduling, requiring a thermal cycle of the plant to remove the damaged turbine and install an operationally ready turbine. Implementation of the new control system has restored fully functional turbine control logic. The new turbine control logic is expected to be effective in preventing future turbine failures.

On each of the turbines, there are several interlocks on process conditions that will close the turbine inlet valve and stop the turbine. Interlock conditions retained from the original control system include turbine speed, brake circuit cooling water temperature, and brake circuit cooling water flow. Two important turbine interlocks were added. A turbine will shut down on low inlet temperature. The trip points are 20 K for the T1 turbine and 6.5 K for the T2 turbine. The T2 turbine also has a shutdown interlock condition on the combination of outlet pressure and outlet temperature. The T2 turbine must operate in the supercritical region to eliminate the possibility of producing liquid in the turbine.

T1 Turbine Inlet Valve

Understanding the operational characteristics of the turbine inlet valves is necessary for future automation of the MTF refrigerator. Smooth, predictable operation of each inlet valve is required before turbine operation can be automated. The controls upgrade project has included a detailed study of the T1 turbine inlet valve.

Two problems existed with the T1 turbine inlet valve. The first problem was unsteady operation. As the T1 turbine inlet valve was opened to start the turbine, flow was suddenly choked off when the valve was 20-25% open. This behavior resulted in a momentary slow down of the turbine until the inlet valve was more than 25% open. These unsteady startups may have contributed to turbine failures. The second problem was leakage through the inlet valve. The resulting pressurization of the T1 turbine circuit necessitated a special startup procedure to allow this pressurization to be bled off quickly enough so that the turbine exit valve could be opened without causing a rush of gas through the turbine, possibly upsetting the turbine and damaging its bearings.

The T1 turbine inlet valve body was pulled out of the cold box, and the valve plug was examined. The brass valve plug is approximately 3.78 cm (1.5 in) in diameter and has four 1.52 cm (0.6 in) wide triangular notches through which the helium flows. It was also found that the valve plug had a small step in its diameter. The top one-quarter of the plug had a diameter 0.051 mm (0.002 in) less than that of the remainder of the plug. This step was choking off the helium flow, resulting in unsteady turbine starts. A new valve plug was machined. This new valve plug is different from the old plug in two ways. First, it does not have a step in its diameter. Second, the triangular notches were replaced with notches designed for equal percentage operation of the valve.

Figures 2a and 2b compare the performance of the two valve plugs. Figure 2a plots the differential pressure across the T1 turbine inlet valve against the valve position. Operation at the low end of the range is more gentle with the new plug. Figure 2b plots the valve coefficient Cv of the T1 turbine inlet valve against the valve position. The new plug displays equal percentage performance.

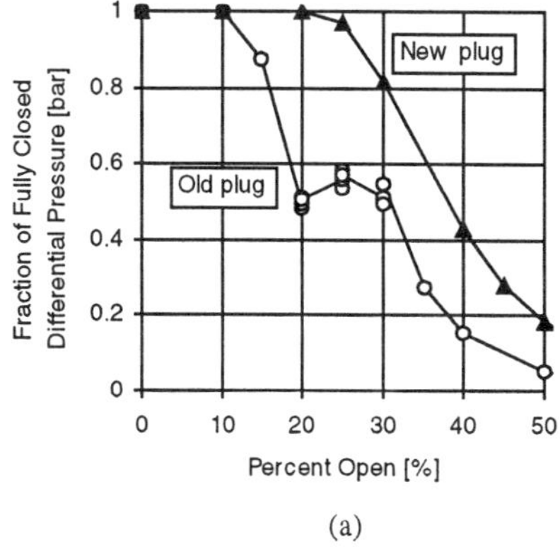

(a)

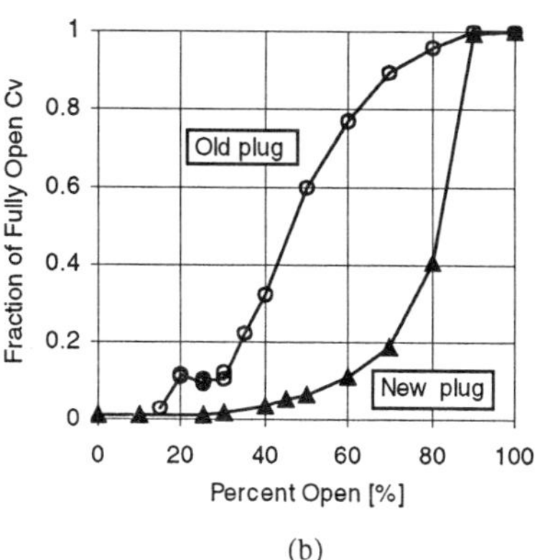

(b)

Figure 2. (a) Pressure differential across the T1 turbine inlet valve, and (b) valve coefficient Cv of the T1 turbine inlet valve as a function of valve position.

In addressing the leakage problem, the seat seal was found to be slightly warped so that the valve did not seal around its entire circumference. After a slight machining of the seat seal and reinstalling the valve, its leakage rate was found to be greatly improved. Before machining, a differential pressure of 8.6 bar (125 psi) or higher across the valve caused an unacceptable leakage rate. Any warm-up flow to the T1 turbine circuit was overpowered by the leakage through the inlet valve. With the newly machined seat seal, atmospheric pressure was maintained downstream of the valve with 13.8 bar (200 psia), 110 K helium upstream of the valve. The remaining leakage produced a downstream pressure rise of only 0.1 bar/hr (1.5 psi/hr).

CONCLUSIONS

The original control system of the 22-year old Fermilab Magnet Test Facility 1500 W helium refrigerator has been successfully replaced by a modern control system. The replacement was implemented in phases to minimize the impact on the testing schedule of the facility.

Immediate benefits expected of this replacement are an increase of control system availability by an increase of the mean time between failures and a reduction of the mean time to repair. The new system also includes sensors and historical trending to better understand the system performance and design control strategies to enhance its automation. The next project phase is the implementation of this enhanced automation to extend periods of unattended operation.

ACKNOWLEDGMENTS

An important aspect of this controls upgrade was the replacement of the electrical system. Jim Garvey and Jeff Arthur provided the electrical expertise and the design for the new electrical system, and they coordinated and helped implement the fieldwork associated with this system.

George Kirschbaum's experience as an operator of the system was invaluable during the design efforts of this upgrade, and his willingness to help and tireless dedication during all aspects of fieldwork are greatly appreciated.

The authors wish to thanks the numerous other Fermilab personnel that contributed to this project: T. Peterson, B. Bianchi, D. Orris, G. Citver, D. Massengill, D. Parisi, A. Rusy, and F. Wilson.

Dana Arenius of the Thomas Jefferson National Accelerator Facility was important during efforts to understand this system and the changes made over the years.

REFERENCES

1. J. E. Cooper et al., "Cryogenic System for Production Testing and Measurement of Fermilab Energy Saver Superconducting Magnets," 1983 Particle Accelerator Conference, Santa Fe.
2. A. J. Bianchi et al., "Operational History of Fermilab's 1500 W Refrigerator Used for Energy Saver Magnet Production Testing," 1985 Cryogenic Engineering Conference, Massachusetts Institute of Technology, Cambridge, Massachusetts, August 12-16, 1985.
3. R. Carcagno, "Process Control," Handbook of Accelerator Physics and Engineering, Edited by A. W. Chao and M. Tigner, World Scientific Publishing, 1999.

DEVELOPMENT OF DISCRETE ARRAY TYPE LIQUID LEVEL INDICATOR FOR CRYOGENIC FLUIDS

R. Karunanithi, S. Jacob and S. Kasthurirengan

Centre for Cryogenic Technology, Indian Institute of Science
Bangalore 560012, India

ABSTRACT

In large cryogenic storage systems, the level of cryogenic fluids such as liquid nitrogen, liquid oxygen, etc., are usually monitored using capacitance type indicators. They are both heavy and inaccurate. We report here the development of discrete array type liquid level indicators. A special class of analog multiplexer IC, which operates at cryogenic temperatures, is used in the unit. Operating it at liquid helium temperature and switching the channels one million times without any failure tested the reliable operation of the IC at cryogenic temperature. The system is capable of detecting the liquid level of any cryogen. Only a software tuning is required for operation with each fluid. The developed unit has an active length of 400 mm and six regions with sensitivities ranging from ± 5 mm to ± 0.5 mm. Redundancy of level measurement is achieved through software. The interaction of the system that has 138 sensors to ambient electronics is through only 15 wires. The integrated unit has been cycled several times between room temperature and 77 K.

INTRODUCTION

Liquid level measuring devices in cryogenics are usually of two types - point liquid level sensors and continuous liquid level sensors. The former type of sensors will detect the liquid vapour interface at a given point. The latter type of sensors is a continuous liquid level sensor. Under the point level sensors, hot wire, capacitance, optical, acoustic and vibrating paddle type sensors are used. The continuous liquid-level sensors are of the following types:

i. Direct weighing scheme,
ii. Differential pressure,
iii. Capacitance,
iv. Acoustic and
v. Nuclear radiation attenuation type.

Advances in Cryogenic Engineering, Volume 45.
Edited by Shu *et al.*, Kluwer Academic / Plenum Publishers, 2000.

Direct weighing schemes have been used for liquid hydrogen but are handicapped by the dead weight of the container compared to that of the contents. Differential pressure measurements are simple to visualize but the signal to noise ratio is low. Capacitance sensors are widely used. They do not truly follow the interface since the output is that of total content. Hence, the higher dielectric constant of cold vapour causes error in measurement. In order to increase the accuracy of measurement, the area of the capacitor plates must be increased and the distance between them reduced. Due to the practical difficulties of maintaining uniform distance between the plates when it is reduced beyond a limit, the option of increasing the area of the capacitor plates would usually be preferred. This results in unacceptable weight for space applications.

Acoustic type units use liquid as the transmitting medium. By measuring the time taken for the pulse to travel and by knowing the velocity in the medium, the liquid level can be calculated. Some times, an acoustic racetrack of known path length and variable time of flight is included to provide a measure of the density of the fluid. The product of the two measurements tends to give an indication of the mass of the tank.

Nuclear radiation attenuation (NRA) is used in LOX and LN_2 systems. But this method is not useful for LH_2 as it is transparent to commonly available sources of nuclear radiation.[1] In general, capacitance type of liquid level sensors are used for very large cryogenic containers.

The present investigation makes use of point type detectors arranged in the form of an array with fixed inter-space between them. They are multiplexed with the help of an analog multiplexer system operating in the cryogenic environment and the particular sensor, which is at the liquid vapour interface, can be detected. This way, it is a hybrid type system. This is a novel type of liquid level measurement system and not reported elsewhere to our knowledge. Compared to the capacitance type sensors that are normally used with large cryogenic containers, the developed units are more accurate and have less weight.

PRINCIPLE OF OPERATION

Silicon and germanium diodes are widely used as temperature sensors at cryogenic temperatures. The operations of Si & Ge diodes at cryogenic temperatures are mainly influenced by the effect of carrier freeze out at low temperatures. In addition to the freeze out effect, the increase in the junction potential due to Fermi level shift also affects the operation of the diodes. The forward bias voltage drop V_0 can be a measure of the junction potential. V_0, which is about 0.2V for Ge diode at room temperature, becomes 0.6 V at 77 K and 0.75 V at 4.2 K. The temperature dependence of the junction barrier potential is useful when these diodes are used as temperature sensors for low temperatures. The I,V characteristics of silicon and germanium diodes at 300 K, 77 K and 4.2 K are shown in Figure 1.[2] From the figure, it is clear that for a given current, the potential drop across the junction (V_0) is a function of temperature. V_0 for silicon diode, which is ~0.58 V at 300 K goes to ~1.02 V at 90 K and ~1.04 V at 77 K for a given current. The above principle and the difference in heat transfer coefficient for liquid and vapour are used in the design of cryogenic liquid level units with silicon diodes as liquid-vapour interface detectors. The use of commercially available silicon diodes for cryogenic thermometer applications was extensively studied by Gmelin et. al.[3] For our application of liquid vapour interface detector, commercially available, glass encapsulated type switching diodes (1N4148) are selected. An array of detectors is switched with the help of a cryogenically operated analog multiplexer to locate the sensor in the interface. The position of the sensor indicates the liquid level.

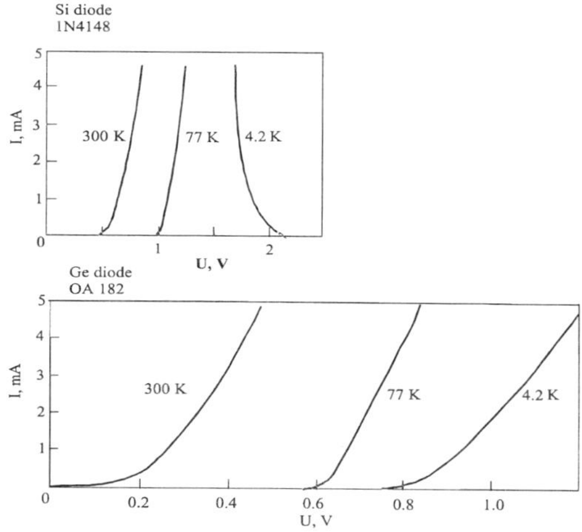

Figure 1. I,V Characteristics of a Silicon diode (1N4148) and a Germanium diode (OA 182).

DESCRIPTION OF THE SYSTEM

The liquid level detectors (silicon diodes) are mounted in a PCB as an array. The positioning of the sensors depends on the accuracy of measurement required in that region. The analog multiplexer ICs and passive components also are accommodated in the same PCB. The block diagram of the setup is shown in the Figure 2. The PCB is kept inside the cryogenic container in an upright position to measure the liquid level. The active length of the PCB is 400 mm. It is divided into six regions. The gap between the sensors and the number of sensors in each region are shown in the Table 1. This particular arrangement was adapted here as a special case in which the degree of accuracy required at different positions inside the container is different. Also, the total height to which the liquid level has to be monitored is 400 mm in this case. It is possible to place the sensors with uniform gap and also to change the total height of the PCB. In the present case, the data of Table 1 are fed to the software so that on identifying the sensor in the liquid vapour interface, the percentage of the liquid will be calculated by it. Hence, a liquid column of 400 mm from the bottom of the unit will be shown as 100%. A digital I/O card and an IEEE card are assembled in the motherboard of an IBM PC. Programmable voltmeter is interfaced to the PC through the IEEE488 bus.

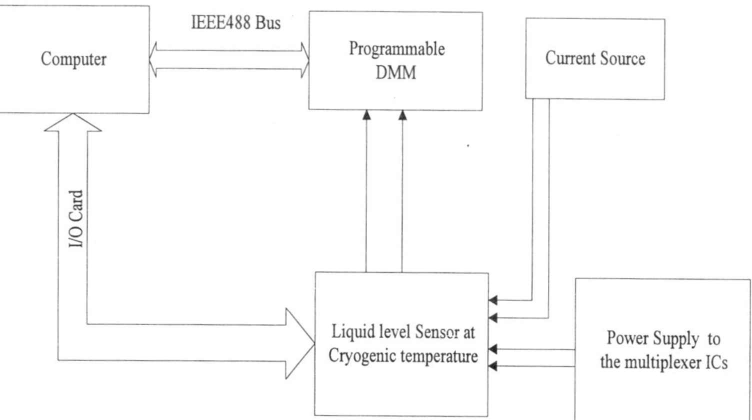

Figure 2. Block diagram of the cryogenic liquid level measurement setup.

Table 1. Details of the arrangement of sensors in the PCB

Region	No. of sensors	Height of this region in mm	Inter-space between consecutive sensors in mm	Sensor Nos.
6	10	90	10	137-128
5	19	90	5	127-109
4	20	20	1	108-89
3	09	40	5	88-80
2	66	65	1	79-14
1	14	65	5	13-0

The sensors used in the circuit are subjected to thermal cycling for more than 30 times between room temperature and 77 K and five times to 4.2 K. This ensures better reproducibility.[3] The survival rate of the sensors that have undergone this thermal cycling will also be better. The software first initializes the programmable voltmeter, IEEE card and the Digital I/O card. Totally 138 sensors with different spacing schemes of six regions are sampled by using the analog multiplexer circuit by sending the corresponding 8 bit address from the PC through the I/O card. By using a set of two multiplexer circuits in parallel, four probe measurement of connecting a constant current of 1 mA to the selected sensor and measuring the voltage drop across it can be accomplished. The software will diagnose the measured voltage to find out whether it is in vapour, liquid or not working. Accordingly, the selection scheme will select the other sensors and find out the one that is in the liquid vapour interface. Depending on the position of that particular sensor, the liquid level in the container will be shown as a graphical display in the computer monitor. If any sensor is found to be bad, that one will be skipped for the rest of the measurements.

The software works in such a way that only the sensors which are in the vicinity of the liquid level will be repeatedly sampled and it will follow the liquid level. Further, only one sensor in the circuit will be selected and the constant current will be supplied only to it. Hence, the power dissipation of the circuit will be much less. This aspect is very important for cryogenic systems, as otherwise, evaporation loss from the containers will be increased. As mentioned earlier, the percentage of the liquid level is calculated by the software on determining the sensor in the liquid vapour interface taking into account the different regions of the sensors, the gap between two adjacent regions, the spacing of the sensors within the regions etc. As per the algorithm, a set of five sensors around the liquid vapour interface will be repeatedly sampled and when the liquid level changes, the sampling will automatically follow the level and identify the current liquid level.

The main attraction of this unit is that it can be used to measure the liquid level of any cryogenic fluid by tuning the threshold voltage corresponding to that particular liquid temperature. There is a provision in the software to modify the threshold voltage.

MULTIPLEXER CIRCUIT

The sixteen-channel high-speed CMOS analog multiplexer/demutliplexer ICs are used in the circuit.[4] The power supply voltage used for these ICs is ~ 5 V D.C. Figure 3 shows the schematic block diagram of the multiplexing circuit used for switching the sensors. IC1 selects one of the nine sets of ICs, 2A, 2B; 3A, 3B, etc., which in turn select a set of sixteen sensors each. Out of eight digital address bits, the four most significant bits (D4 - D7) are used to select one of the nine IC sets and the remaining four least significant bits (D0 - D3) are used to select one of the sixteen sensors connected to the selected IC sets. This way, one of the 138 sensors can be selected one at a time and connected to the I in (+) and V out (+) terminals via the multiplexers. There is a delay between the selection and energization of the sensor with 1 mA current and measurement of the voltage drop across it. The delay time is user settable through the software. The threshold voltage also is settable by the user through the software and hence, the user can tune the unit to operate with any cryogenic liquid. The delay between the energization of the sensor and measurement of the voltage drop is needed for the following. It enables the sensor to make use of the self-heating effect due to the current flow and the difference in heat transfer coefficient for liquid and vapour to settle to a voltage either above or below the threshold value depending on its position. The four-probe measurement is accomplished here by way of using a set of two mulitiplexers for switching current and voltage separately. The PC through an I/O card supplies the digital address bits needed for operating the multiplexer circuit. From the PCB, only fifteen wires are taken out via feed-through. They include eight digital I/O lines, four wires for I-in and V-out, two wires for power supply (+5 V) and one for enabling the main multiplexer.

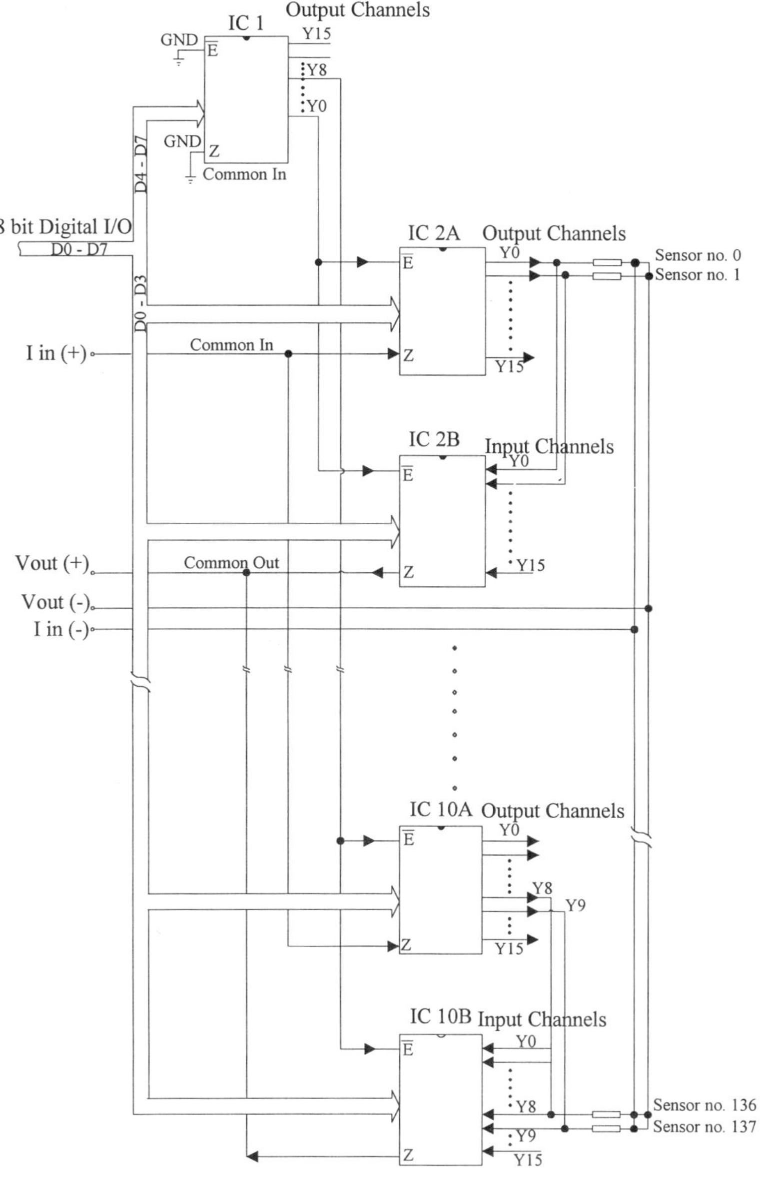

Figure 3. Schematic block diagram of the multiplexing circuit.

RESULTS AND DISCUSSIONS

The unit was tested with both liquid nitrogen and liquid oxygen inside a glass cryostat. The liquid level shown by the system as a percentage of total height of the unit was cross-checked by way of visual examination. They match fairly accurately. Also, by keeping the liquid level constant, the unit was moved up and down. The liquid level shown in the computer monitor followed the movement very well.

As the software checks only the sensors in the vicinity of the liquid-vapour interface repeatedly, the measurement gets fine tuned automatically and over a period of time, if the liquid level change is not much, the accuracy will be more. This unit at present can take care of a filling rate of 0.5 mm/sec at the most sensitive region. In other regions, it can be yet faster.

CONCLUSIONS

The discrete array type liquid level indicator is a good alternative to the conventional capacitance type liquid level measuring unit - particularly for very large cryogenic containers and in applications where the weight should be as low as possible. The accuracy of measurement also is at least five times better than that of the capacitance type unit. In the case of cryostage of space launch vehicles, the improved accuracy of the unit may result in optimum fluid management and the substantial reduction in weight of the unit, as compared to that of commonly used capacitance type level sensors, will lead to an increased payload fraction.

ACKNOWLEDGMENT

The work was carried out as a part of the consultancy project with the Liquid Propulsion Systems Centre (LPSC) of ISRO. Discussions with O. Vijayan and M.S. Bhat of LPSC were very helpful in the execution of the work. The project staff S.V. Chavan and the staff of the Centre for Cryogenic Technology are gratefully acknowledged for their association in the work.

REFERENCES

1. Flynn, M. Thomas. "Cryogenic Engineering," Marcel Dekker, Inc., New York 10016 (1996).
2. B. Lengeler, Semiconductor devices suitable for use in cryogenic environments, *Cryogenics* 14:439 (1974).
3. E. Gmelin, K.-H. Greuble, Martina Knoop and M. Their, Cryogenic thermometry with common electronic diodes, *Cryogenics* 30:442 (1990).
4. Z. Szücs, R. Karunanithi and U. Ruppert, High-speed CMOS multiplexer for cryogenic application, *Proc. Symp. Low Temp Electronics and High Temp. Superconductors* 537 (1988).

APPROXIMATING THE RESISTANCE-TEMPERATURE RELATIONSHIP OF PLATINUM RESISTANCE THERMOMETERS FROM 20 K TO 273 K

R. W. Phillips

BFGoodrich Aircraft Sensors Division
Eagan, Minnesota, 55121

ABSTRACT

A platinum resistance thermometer (PRT) must be calibrated to accurately define the resistance-temperature relationship R(T). For the temperature range of 20 K to 273 K, this can be problematic due to the non-linearity of the R(T) and lack of convenient calibration media near 20 K. A method of approximating the R(T) has been developed based on a simple 3-point calibration using the boiling point of helium, boiling point of nitrogen, and ice point. This method has been investigated with laboratory grade and rugged aerospace sensor designs, including wire diameter between 8 and 80 μm. A correlation between resistance ratio values allows use of a standard ITS-90 interpolation formula with uncertainty less than 0.1 K over the range 20 K to 273 K.

INTRODUCTION

The platinum resistance thermometer (PRT) is widely used in industry and can be configured to meet the needs of numerous applications. Whether designed as a precision laboratory standard or a rugged aerospace sensor, the performance of the PRT depends on the resistance-temperature relationship R(T). The temperature range 20 K to 273 K is of particular interest for launch vehicle applications that require accurate measurement of liquid fuel and oxygen temperatures.

PRTs of the same design are often used with a nominal R(T) curve, allowing multiple thermometers to be used interchangeably within a specified tolerance. The use of a single curve simplifies data acquisition but the accuracy is limited by the manufacturer's tolerance. As temperatures approach 20 K, impurities and residual strain in the platinum dominate unit-to-unit variation, making it difficult to achieve tolerances better than ±1 K. When a nominal R(T) is not available or manufacturing tolerances are too large, the individual PRT must be calibrated to establish a unique R(T).

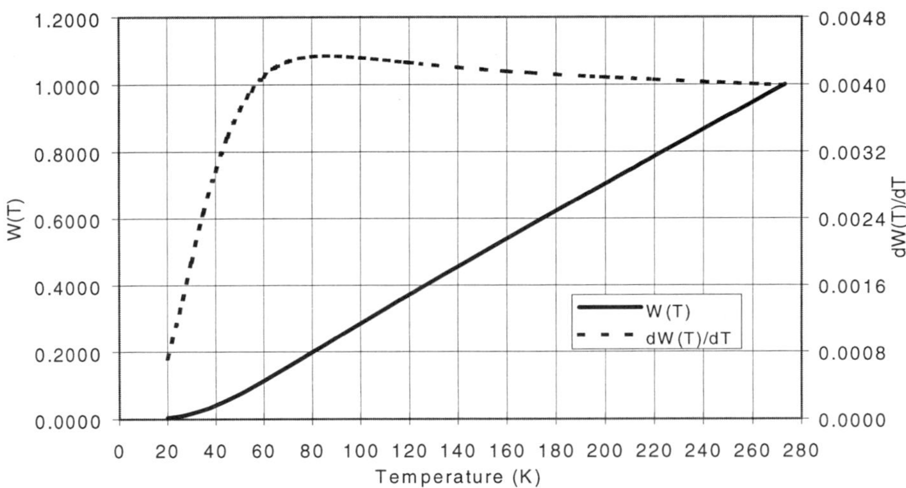

Figure 1. Nominal resistance ratio R(T)/R(273.16 K) versus temperature for platinum.

Calibration consists of measuring the PRT resistance at known temperatures and fitting the results with an interpolation equation. The calibration temperatures can be fixed points with assigned temperature values such as reproducible freezing points, boiling points, or triple points of pure materials.[1,2] Alternatively, the PRT is calibrated by comparison against a temperature standard with known calibration. The latter approach offers greater flexibility provided a suitable calibration medium is available. Unfortunately, between the boiling points of liquid nitrogen (77 K) and helium (4.2 K), calibration generally requires use of a temperature controlled cryostat apparatus that adds considerable time and expense to the calibration. As shown in Figure 1, the PRT response to temperature is complicated below 90 K. Several tradeoffs must be considered to determine the optimum level of calibration, including accuracy, cost, and thermometer stability.

The International Temperature Scale of 1990 (ITS-90) prescribes calibration points, interpolation formula, and acceptance criteria for the standard platinum resistance thermometer (SPRT) from 13.8033 K to 1234.93 K.[1] The defined procedures represent the lowest possible uncertainty but are not appropriate for all PRT applications. However, by using fewer and different calibration points, the ITS-90 methods can be simplified for a broad class of thermometers including SPRTs and rugged PRT designs. These approximations have been demonstrated using calibrations performed by the National Institute of Standards and Technology (NIST) and BFGoodrich Aircraft Sensors Division (BFG).

THERMOMETERS

BFG designed and manufactured the thermometers used in this study. The PRTs were fabricated using reference grade platinum wire drawn from multiple bars having initial purity not less than 99.999%. Wire diameters between 8 μm and 80 μm were used to construct elements with ice point resistance values between 25 ohms and 8100 ohms. The wire was mounted and annealed as part of the element assembly.

A diverse sample of configurations was investigated, including the strain-free SPRT and more rugged aerospace PRTs.[3] Although the wire is fully constrained, the rugged PRTs had temperature coefficient of resistance values greater than 0.391 %/K and thermal hysteresis less than 0.01% of span.[4]

INTERPOLATION EQUATIONS

The proposed interpolation equations are identical to those defined as part of the ITS-90 for the subrange 83.8058 K to 273.16 K.[1] To summarize, the PRT is calibrated in terms of the ratio of resistance at temperature T and the resistance at the triple point of water:

$$W(T) = R(T) / R(273.16\text{ K}). \tag{1}$$

The W(T) characteristics are described relative to a reference function $W_r(T)$ using the deviation equation:

$$\Delta W(T) = W(T) - W_r(T) = a[W(T) - 1] + b[W(T) - 1]\ell n[W(T)]. \tag{2}$$

The reference function accurately describes the resistance ratio of a real SPRT as a function of temperature T (K) from 13.8033 K to 273.16 K:

$$\ell n[W_r(T)] = A_0 + \sum_{i=1}^{12} A_i \{[\ell n(T / 273.16) + 1.5] / 1.5\}^i, \tag{3}$$

where constants A_i are given in Table 1.

The coefficients a and b in Eq. (2) are determined by calibrating the PRT at three points including the triple point of water. For the expanded range 20 K to 273 K, comparison calibration near 77 K (LN_2) and 20 K are preferred. The R(273.16 K) may also be determined by comparison method or corrected from the ice point using the equation:

$$R(273.16\text{ K}) = R(273.15\text{ K}) / 0.99996. \tag{4}$$

APPROXIMATING W(20 K)

The equations described above require a calibration point near 20 K to cover the range of interest. To eliminate the need for cryostat calibration, the W(20 K) can be estimated from calibration at approximately 4.2 K using liquid helium (LHe). This is accomplished by using a correlation developed between ΔW(T) and W(LHe):

$$\Delta W(T) = W(T) - W_r(T) = x_0 + x_1 W(\text{LHe}). \tag{5}$$

The specific correlation temperature is arbitrary. For T=20.141 K, the correlation is shown in Figure 2 and constants x_0 and x_1 were determined to be -4.186×10^{-4} and 1.2057 respectively. Given $W_r(T)$ from Eq. (3), the W(T) is determined directly from W(LHe).

The thermometers used to develop the correlation satisfy the following criteria:

$$\Delta W(20\text{ K}) < 0.00195 \text{ and } \Delta W(77\text{ K})/\Delta W(20\text{ K}) < 1.2. \tag{6}$$

Table 1. ITS-90 reference function constants

A_0=	-2.13534729	A_5=	-0.61899395	A_{10}=	0.04459872
A_1=	3.18324720	A_6=	-0.05332322	A_{11}=	0.11868632
A_2=	-1.80143597	A_7=	0.28021362	A_{12}=	-0.05248134
A_3=	0.71727204	A_8=	0.10715224		
A_4=	0.50344027	A_9=	-0.29302865		

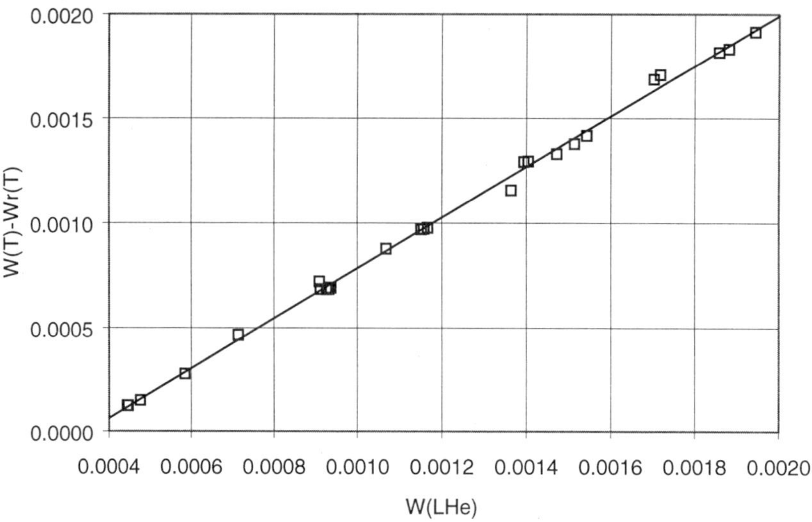

Figure 2. Correlation between W(LHe) and $\Delta W(T)$ at T=20.141 K. Based on data from 25 thermometers representing 9 different designs.

The correlation of Eq. (5) was established as a local definition without correcting for actual LHe temperature. Barometric pressure variation has minimal influence on W(LHe). Using the ITS-90 ^{4}He temperature-pressure relation, a local boiling point temperature of 4.189 K was calculated for nominal pressure of 98.2 kPa with variation between 4.162 K and 4.206 K for extreme conditions. The effect on W(LHe) is less than 5.7×10^{-7} assuming resistance ratio sensitivity of 1.3×10^{-8} per mK.[5]

The ITS-90 does not define the $W_r(T)$ below 13.8033 K. However, it is interesting to note that Eq. (5) predicts a W_r(LHe) of 0.000347 for $\Delta W=0$.

DISCUSSION

Comparing interpolated results with actual calibration data tested the equations. A capsule SPRT, calibrated by the NIST over the range 13.8033 K to 273.16 K, provides a baseline with uncertainty (k=2) less than 1 mK.[6] Using this calibration, the W(20 K) and W(77 K) values were calculated and then refit using Eq. (2). The resulting W(T) was compared to the original 8-point NIST calibration over the range 20 K to 273 K. Figure 3 includes the results of several model 162D capsule SPRTs using available data.[7,8]

The same group of SPRTs was also calibrated by BFG at 273.16 K, 77 K (LN_2) and 4.2 K (LHe). Calibration at 77 K was performed by comparison method using a separate laboratory SPRT also calibrated by the NIST. The W(LHe) values for the SPRTs measured between 0.00044 and 0.00058 and were used to approximate the W(20 K) with Eq. (5). The estimated W(20 K) and measured W(77 K) were fitted using Eq. (2). The resulting W(T) was again compared to the original NIST 8-point calibration as shown in Figure 4. As expected, the BFG results exhibited greater deviation from the NIST due to additional uncertainty of the 77 K calibration and W(20 K) approximation. The SPRT may also have experienced some drift between the time BFG and the NIST performed the calibrations. This is likely for SPRTs 5 through 7, which were tested more than 6 years after the original NIST calibration.

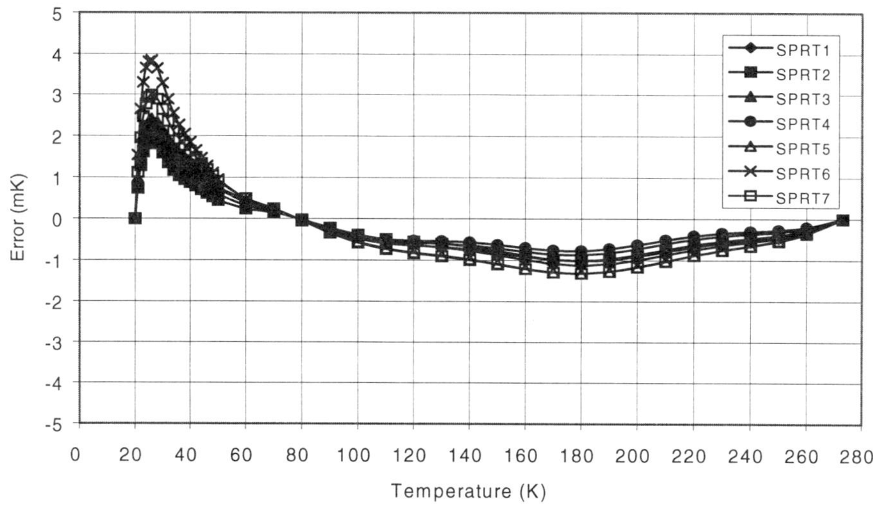

Figure 3. Difference between temperatures indicated from NIST 3-point and NIST 8-point calibrations for measured value of W(T).

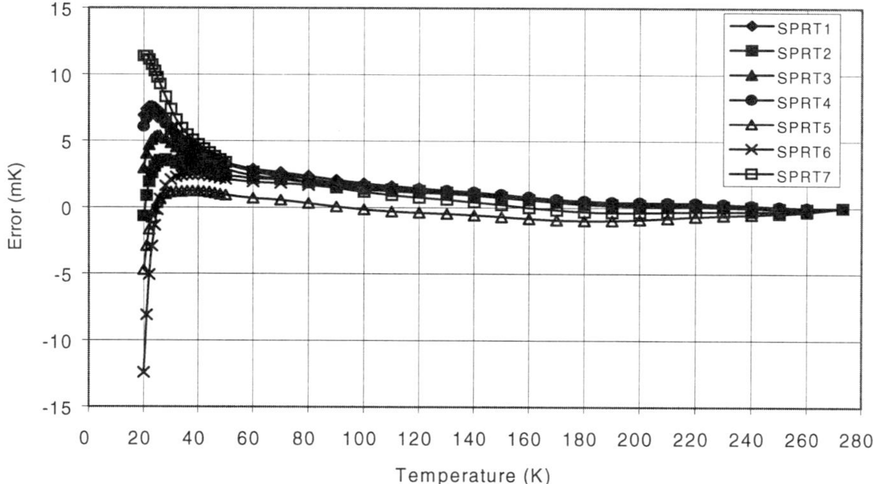

Figure 4. Difference between temperatures indicated from BFG 3-point and NIST 8-point calibrations for measured value of W(T).

The equations are more difficult to evaluate with more rugged PRTs due to the relatively large calibration uncertainty. Repeatability errors as large as 0.03 K are common and have a significant influence on any comparison made between interpolated and actual calibration data. Figure 5 shows typical deviations observed with rugged PRT designs representing W(LHe) values of approximately 0.0010 and 0.0019. The thermometers were calibrated at LHe, LN_2, and ice point as previously described to establish the W(T). The resistance was also measured at random checkpoints to compare the temperature indicated from the 3-point calibration with the actual checkpoint temperature. Some checkpoints were repeated to evaluate PRT instability. The checkpoints were performed using a NIST calibrated SPRT, a temperature controlled cryostat, and stirred liquid baths.

Using the 3-point calibration with LHe, the largest uncertainty occurs near 20 K. A distribution of errors associated with the W(20 K) approximation is shown in Figure 6. The sample includes routine production data from 93 rugged PRTs of identical design manufactured over 5 years. For this sample, 95% of units showed errors less than 0.1 K. Although this design exhibited the greatest variation of the PRTs studied, the errors reflect a factor of 2 improvement compared with other 3-point methods.[9]

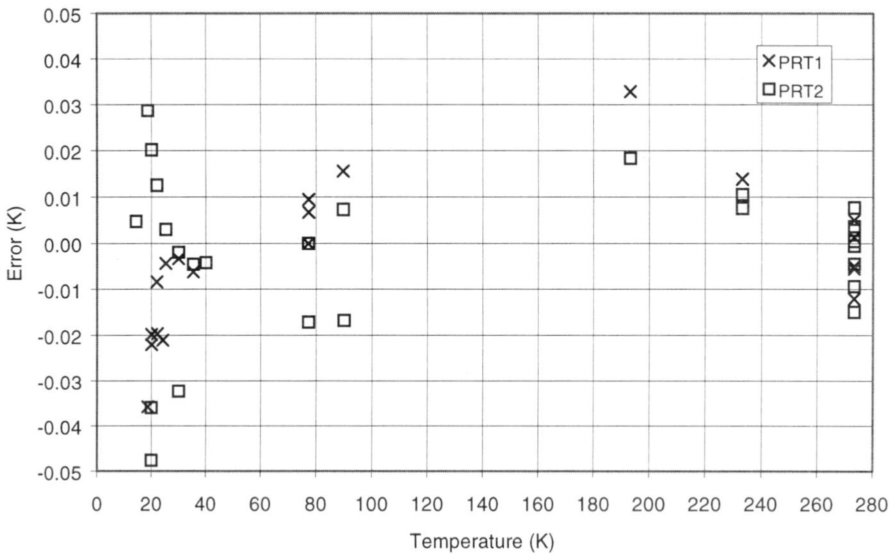

Figure 5. Difference between indicated temperature using BFG 3-point calibration and checkpoint temperature for measured value of R(T). Typical variation for rugged PRT.

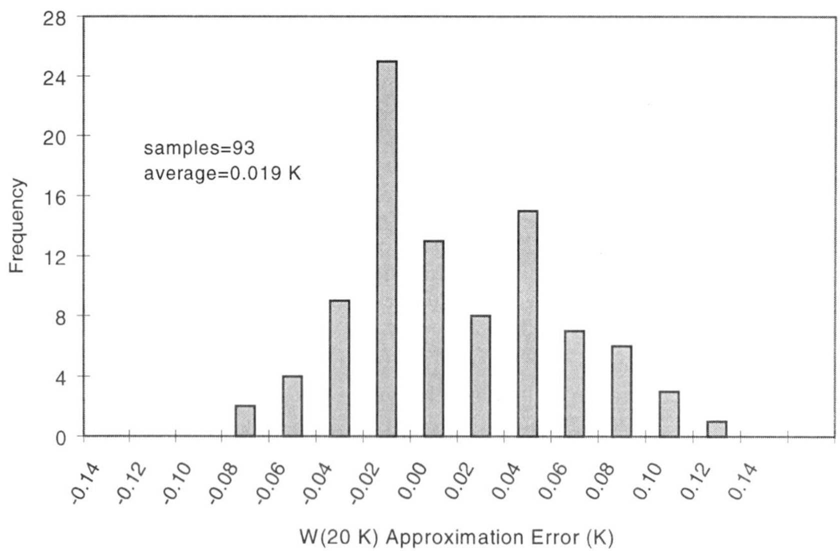

Figure 6. Difference between approximated and actual W(20 K) for a rugged PRT design.

CONCLUSIONS

The ITS-90 methods simplify calibration by using a detailed reference function and simple deviation equation. The standard 2-term deviation equation requires only three calibration points and can accommodate the expanded range 20 K to 273 K for most applications. These equations are applicable for a diverse class of PRTs provided the element is constructed with high purity platinum and mounted essentially strain-free. International recognition and compatibility with many commercial readout devices are also beneficial.

The W(LHe) can be used to accurately estimate the resistance ratio at higher temperatures. When combined with the ITS-90 equations, PRTs can be calibrated at considerable cost savings relative to conventional cryostat methods. The approximation introduces errors typically less than 0.1 K for rugged PRTs and less than 0.015 K for SPRTs.

ACKNOWLEDGMENTS

Rosemount Aerospace Inc. was acquired by BFGoodrich in 1993 and operates as BFGoodrich Aircraft Sensors Division. The author would like to acknowledge M. T. Smith of BFG for performing the PRT calibrations. Appreciation is also expressed to W. L. Tew of the NIST for several helpful discussions.

REFERENCES

1. H. Preston-Thomas, "The International Temperature Scale of 1990 (ITS-90)," *Metrologia*, Vol. 27, pp. 3-10, (1990). For errata see *ibid*, Vol. 27, p. 107, (1990).
2. R. E. Bedford, G. Bonnier, H. Maas, and F. Pavese, "Recommended values of temperature on the International Temperature Scale of 1990 for selected set of secondary reference points," *Metrologia*, Vol. 33, pp. 133-154, (1996).
3. Rosemount Aerospace Inc. Product Data Sheets: 2086, 2267, 2268, 2271, 2337, and 2445.
4. D. J. Curtis, "Thermal hysteresis and stress effects in platinum resistance thermometers," *Temperature Its Measurement and Control in Science and Industry*, J. F. Schooley, ed., Vol. 5, pp. 803-812, American Institute of Physics, New York, (1982).
5. R. J. Berry, "Platinum Resistance Thermometry Below 10 K," *Metrologia*, Vol. 3, pp. 53-57, (1967).
6. G. F. Strouse and W. L. Tew, "Assessment of Uncertainties of Calibration of Resistance Thermometers at the National Institute of Standards and Technology," NISTIR 5319, (1994).
7. W. L. Tew and G. F. Strouse, "Standard Reference Material 1750, Standard Platinum Resistance Thermometer, 13.8033 K to 429.7485 K," NIST Special Publication 260, (in press).
8. NIST Report of Calibration, Test Numbers 224143, 241690, and 245640.
9. D. J. Curtis, "Temperature Calibration and Interpolation Methods for Platinum Resistance Thermometers," Rosemount Report 68023F, (1980).

INFLUENCE OF THERMAL CYCLING ON CRYOGENIC THERMOMETERS

Ch. Balle, J. Casas, J.M. Rieubland,
A. Suraci, F. Togny and N. Vauthier

CERN, European Laboratory for Particle Physics
1211 Geneva 23, Switzerland

ABSTRACT

The stringent requirements on temperature control of the superconducting magnets for the Large Hadron Collider (LHC), impose that the cryogenic temperature sensors meet compelling demands such as long-term stability, radiation hardness, readout accuracy better than 5 mK at 1.8 K and compatibility with industrial control equipment.

This paper presents the results concerning long-term stability of resistance temperature sensors submitted to cryogenic thermal cycles. For this task a simple test facility has been designed, constructed and put into operation for cycling simultaneously 115 cryogenic thermometers between 300 K and 4.2 K. A thermal cycle is set to last $7^1/_4$ hours: 3 hours for either cooling down or warming up the sensors and 1 respectively 1/4 hour at steady temperature conditions at each end of the temperature cycle. A Programmable Logic Controller (PLC) drives automatically this operation by reading 2 thermometers and actuating on 3 valves and 1 heater. The first thermal cycle was accomplished in a temperature calibration facility and all the thermometers were recalibrated again after 10, 25 and 50 cycles. Care is taken in order not to expose the sensing elements to moisture that can reputedly affect the performance of some of the sensors under investigation. The temperature sensors included Allen-Bradley® and TVO® carbon resistors, Cernox™, thin-film germanium, thin-film and wire-wound Rh-Fe sensors.

INTRODUCTION

Within the framework of various tests and developments of a cryogenic thermometer[1-4] for the LHC, the long-term stability of the temperature characteristics in correlation with thermal cycles had to be evaluated for the various LHC prototype temperature sensors (RhFe thin film and wire-wound, Allen-Bradley®, TVO® and Cernox™). In the past much work has been published concerning either the long-term or thermal cycling stability of cryogenic temperature sensors[5-10]. Data is missing for some of the sensors under investigation and also it is often unreliable or at least difficult to interpret. For instance many contradictory reports concerning the performance of Allen-Bradley® temperature sensors exist, it is possible to find many recipes[9-10] to improve their characteristics by encapsulation techniques, sudden thermal cycles, exposure to high temperature, etc. To further complicate this situation it is worth pointing out that Allen-Bradley® resistors were produced for the electronic industry and their characteristics from a cryogenic point of view may change from year to year. Allen-Bradley® thermometer is

Advances in Cryogenic Engineering, Volume 45.
Edited by Shu *et al.*, Kluwer Academic / Plenum Publishers, 2000.

Table 1. LHC-conditions

Items	Value
Expected LHC-lifetime	20 year
LHC-Temperature Range	1.8...300 K
Cooling-speed	60 K/day
Warming-speed	40 K/day
Cool-down and Warm-up	1/year
300 – 1.8 K thermal cycles	20/lifetime

Table 2. LHC cryogenic thermometer needs

Temperature Range/K	Accuracy [mK]	Quantity
1.6...2.2	10	2418
2.2...4.0	20	2550
4.0...6.0	30	4061
6.0...25	1000	3603
25...300	5000	3603

still a LHC prototype even though the fabrication of these sensors was ceased in 1997; CERN still holds a stock large enough for satisfying the LHC requirements.

The measurement of temperature is critical for the proper operation of the LHC and it was then decided to investigate the stability of LHC prototype temperature sensors when subjected to thermal cycles between 4.2 and 300 K. It is important to note that such cycles are relatively slow and the temperature sensors are never exposed to humidity. The maximum cooling rate is given by the test of individual magnets. The main parameters for such a test are dictated by the relevant LHC-specifications that are shown in Tables 1 and 2. The accuracy budget is defined by the LHC process constraints and it is arbitrarily divided in equal parts between the uncertainties related to the sensing element and the conditioning electronics.

TEST

A total number of 100 thermal cycles between 300 K and 4.2 K will be attained by the end of summer 1999. The number of cycles is higher than the specifications of the LHC machine, which represents a safety margin for taking into account uncertainties due to the relatively small number of sensors under test and to the variability of sensor fabrication. For simplicity no monitoring of the thermometers is done when using the cycling cryostat. The thermometers are calibrated at CERN's main calibration facility during the first thermal cycle and then on cycles 10, 25, 50 and 100. This calibration facility has been described elsewhere[4].

Thermal cycling setup

To simulate the above-mentioned conditions, a simple cycling facility has been built. Apart from thermometers many other cryogenic devices can be installed for investigating ageing processes provoked by thermal cycling without exposure to humidity or air.

Cryostat. The main part is the cylindrical helium chamber made of copper (Figure 2). which is located inside a cryostat under vacuum (Figure 1) and equipped with:

- three baskets (ø 5 cm x 11 cm) to carry up to 120 thermometers,
- an ISO-KF flange to open and close the chamber for manipulating the baskets. Leak tightness is provided by an indium gasket,
- a tube inside the chamber fixed on the centre of the ISO-KF flange, used to guide the baskets,
- two control thermometers TT4 and TT5 measuring two different locations inside the chamber,
- an electrical connector on the ISO-KF flange to connect the two control thermometers,
- the exchange gas flow controlled by inlet and outlet tubes,
- an external cooling loop to cool down the chamber,
- an external electrical heater EH6 to warm up the chamber,
- an external thermal radiation screen which is cooled down by the outlet from the liquid helium transfer line,
- liquid helium from a 500 l storage dewar, forced through the transfer line.

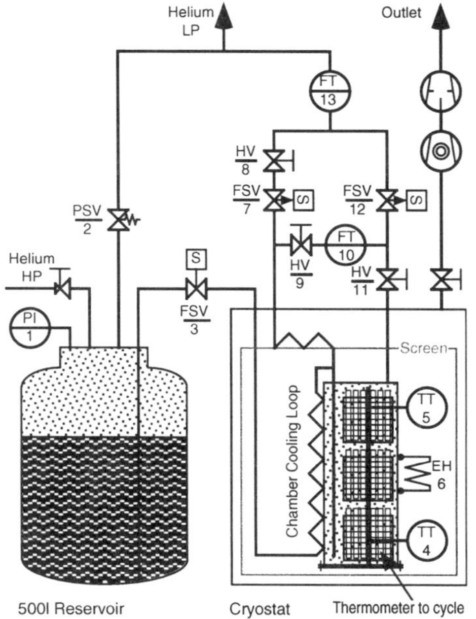

Figure 1. Piping & instrumentation diagram

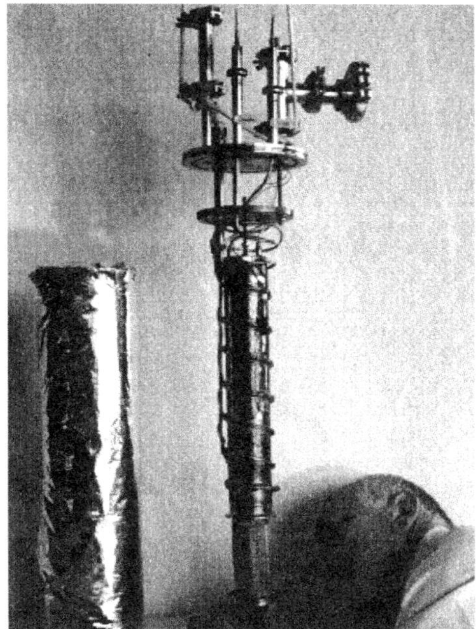

Figure 2. Loading of the helium chamber

Instrumentation. This cycling facility is intended to run without operator intervention. Thus instrumentation and the relevant control should be simple and robust. A Programmable Logic Controller (PLC) takes care of all automatic procedures and stops the cycling after a given number of cycles. The instrumentation controlled by the PLC is:

- FSV3, a switching valve controlling the transfer of LHe into the cryostat,
- Control thermometers TT4 and TT5 (Allen-Bradley® type). The supply of a proper excitation current and the 4-20 mA output signal is ensured by signal conditioners,
- EH6, an electrical heater from Minco® with a power of 30 Watt,
- FSV7 and FSV12, two solenoid valves.

Before starting the automatic procedure it is necessary to adapt the manual valves to the thermal load of the devices under test. The following equipment is used:

- HV7, HV9, HV11, three precision needle hand-valves from Hoke® for adjusting in combination with FSV7 and FSV12 the cooling speed,
- FT10 and FT13, two flow meters 0 to 2.2 m^3/h and 0 to 15 m^3/h as a help for the operator when adjusting the precision needle hand-valves,
- the pumping group, to establish a vacuum better then 10^{-6} mbar.

For diagnostics a 32-channel paper recorder monitors all electrical signals from thermometers, heaters and valves positioners.

Functional principle. Via an operator interface the duration of the 4.2 and 300 K plateaus and the total number of thermal cycles to perform is entered into the PLC. One thermal cycle is composed of four successive phases (Figure 4). Table 3 shows the setup for this experiment when cycling 120 thermometers. The main heat capacity is stored in the copper used for lodging the sensors, its approximate total weight is about 0.95 kg.

Table 3. Setup for cycling 120 thermometers

Actuator/Sensor	300 K to 20 K	20 K to 4.2 K	4.2 K to 300 K
FSV3	open	open	closed
FSV7	open	closed	open
FSV12	closed	open	open
EH6	0	0	30 W
FT10	0.7 m^3/h	3.0 m^3/h	0
FT13	0	0.8 m^3/h	0

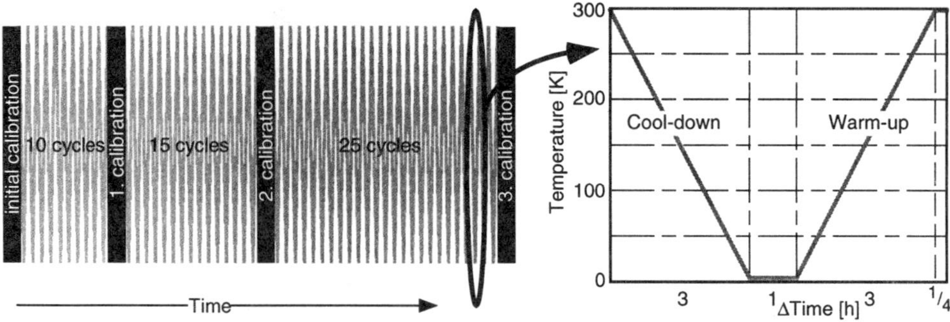

Figure 3. Operation

Figure 4. Definition of one thermal cycle

Procedure

Load. The facility was loaded with the following resistance thermometers:

15 x CRT (Carbon Resistance Thermometer) from Allen-Bradley®, USA, Model 100 Ohm, 1/8 Watt

4 x CRT from Allen-Bradley®, USA, Model 100 Ohm, 1/8 Watt, placed inside a copper canister and sealed with Stycast®, by CNRS-IPN, France

16 x CRT from TVO®, Russia

15 x CX (Cernox™) from Lake Shore®, USA, Model CX-1050-SD

Adjustment. The hand-valves were adjusted to obtain a cooling and warming speed of approximately 100 K/h. The plateaus at 4.2 K and 300 K were set to 75 minutes (Figure 4). Those parameters are imposed to some extent by the aim of completing 100 thermal cycles plus 5 long calibrations, within a time span of one year.

Operation. Figure 3 shows how the calibration procedures are interleaved within the thermal cycles. Arbitrarily it was assumed that the most significant changes in the sensor characteristics would occur during the first cycles, which explains why 0, 10, 25, and 50 cycles are chosen as calibration cycles. Common sense and previous experience with other sensing elements led us to tentatively perform the tests in this manner. In principle when an element drifts, it either drifts indefinitely or it may reach a stable status after a certain number of thermal cycles.

Results

The drift ΔT of the T-R-characteristic of each thermometer was calculated after 10, 25 and 50 cycles as follows:

- generation of an individual calibration function T(R) for each thermometer by fitting the initial calibration points obtained during the first thermal cycle,
- application of this calibration function T(R) on the later calibrations,
- elimination of the mathematical approximation error by subtracting the difference of the temperature reference between the cycle under investigation and the initial calibration:

$$\Delta T_n = (T(R_n) - T(R_0)) - (Tref_0 - Tref_n) \qquad (1)$$

Table 4 shows the numerical results of the variation of the sensors characteristics after 10, 25 and 50 thermal cycles. Due to the higher sensitivity at low temperatures all the sensors show good results at liquid helium temperatures and most of the population complies with the LHC specifications listed in Table 2. It should be noted however that some Allen-Bradley® sensors below 2.2 K exceed the uncertainty of 5 mK as shown in Figure 5. At 20 K it is possible to see that the spread in the uncertainty is lower for sensors with higher sensitivity, this explains the better performance of Cernox™ when compared with TVO®, and TVO® when compared with Allen-Bradley®. This difference is the more obvious the higher the temperature.

Concerning the Allen-Bradley® it is worth to note that no difference in quality can be observed between the off-the-shelf and the copper-canister encapsulated sensors. Also one of the encapsulated samples was damaged by the thermal cycles after about 25 cycles. Their sensitivity above 100 K is very poor and this type of sensor is typically used with a Platinum sensor for measuring the higher temperatures. According to the published literature we would have expected a much worse degradation for Allen-Bradley®. Actually with some selection procedure, Allen-Bradley® temperature sensors are from the thermal cycling point of view within the LHC specifications. At present, selection rules by performing thermal cycles and measuring data at room temperature only are being investigated but so far without success. The mean average temperature difference above 20 K seems to reach a stable value for Cernox™. This could be an indication that this sensor benefits from being exposed to 25 thermal cycles before use. When delivered Cernox™ are accompanied by a low temperature test sheet implying that they have been cycled at least once before delivery. For TVO® and Allen-Bradley® no such stabilization behaviour is observed.

Table 4. Numeric results

		Calibration 1 after 10 cycles		Calibration 2 after 10+15 cycles		Calibration 3 after 10+15+25 cycles	
T [K]	Sensor Name	$\overline{\Delta T_1}$ [mK]	σT_1 [mK]	$\overline{\Delta T_2}$ [mK]	σT_2 [mK]	$\overline{\Delta T_3}$ [mK]	σT_3 [mK]
1.8	CX	-0.7882	0.5535	0.5037	0.6802	-0.6513	0.4330
	TVO	-0.0455	0.6606	0.3939	1.0351	-0.3835	1.2027
	AB	2.7427	2.0045	3.3095	2.8231	-1.9148	4.2703
2	CX	-0.8491	0.4570	0.9300	0.5250	-0.7064	0.3422
	TVO	0.0978	0.6431	0.6417	1.1550	-0.5706	1.3620
	AB	3.0962	2.2987	3.9094	3.2699	-2.5211	4.9080
3	CX	-1.0763	0.3334	0.2618	0.3449	-0.8783	0.2863
	TVO	-0.4328	1.2796	-0.4419	1.9982	-1.1056	2.3854
	AB	7.6528	4.1628	7.2454	5.9068	-3.4715	8.6093
4.7	CX	1.7374	2.0156	-2.9928	2.1274	-5.2280	2.0487
	TVO	0.7607	2.1165	-1.5941	3.9934	-2.2640	4.3788
	AB	17.3037	8.3450	13.5193	12.7921	-8.3744	17.5300
20	CX	3.0921	2.4428	-2.2611	2.5180	-1.3179	2.5545
	TVO	-1.3065	18.2370	-14.8668	31.5075	-20.3284	36.2329
	AB	207.4665	110.8418	139.2465	206.5669	-246.6199	421.5318
50	CX	4.7892	5.0310	-14.1263	5.1998	-14.9530	5.2845
	TVO	-13.7263	57.6131	-74.9515	93.2377	-93.4202	108.2195
	AB	811.1598	499.0036	483.8471	951.0693	-1327.8682	2471.7225
80	CX	9.1676	6.9240	-20.7874	7.3440	-25.6557	7.3030
	AB	1388.3302	969.4480	798.4592	1825.3065	-2768.8821	5457.5423
	TVO	-12.7281	83.8031	-116.6825	140.3333	-150.2212	154.0933
180	CX	13.0871	13.5885	-43.7546	13.4965	-53.3914	13.2617
	TVO	96.4536	215.0169	-149.6401	216.1717	-191.2922	262.8603
270	CX	-28.2666	18.5580	-105.6869	16.2912	-120.5796	18.4223
	TVO	-110.2638	157.1709	-485.0942	133.4017	-635.3769	124.2295

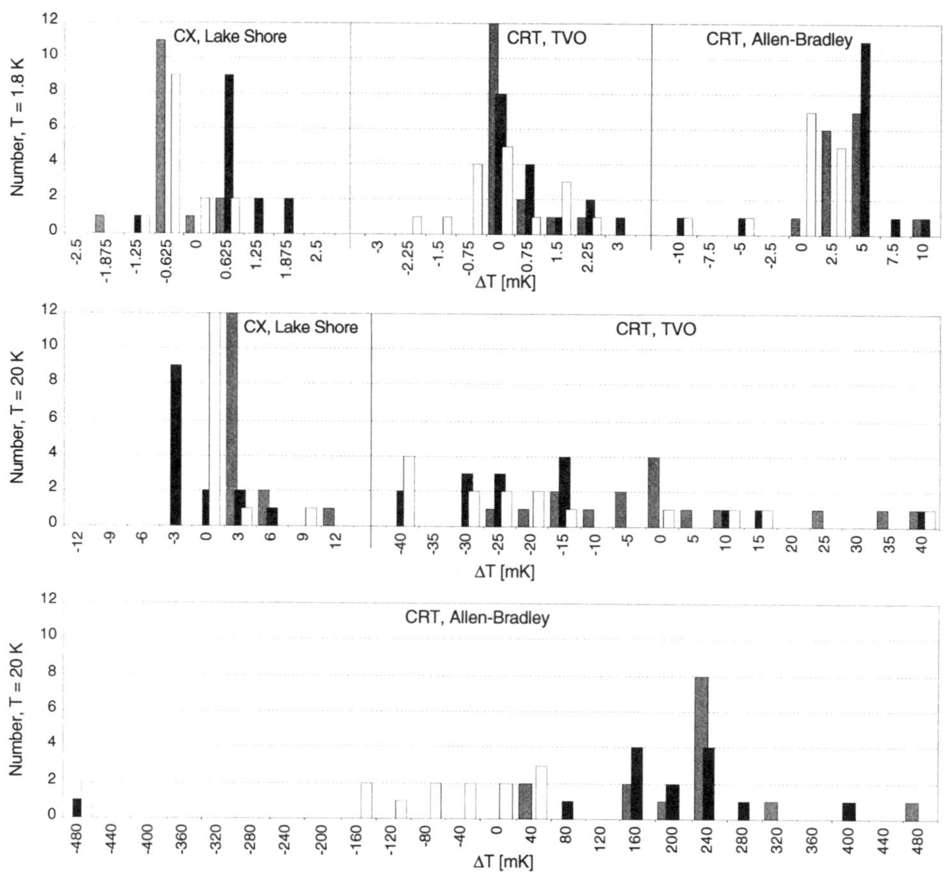

Figure 5. Distribution curves at 1.8 K and 20 K for different temperature sensors after 10 (black), 25 (grey) and 50 (white) thermal cycles (keep in mind the different ΔT-scales).

CONCLUSION

A simple and robust, automatic cycling facility has been designed, commissioned and operated. The first results of the thermal cycled cryogenic thermometers clearly show that sensors with higher sensitivity across the temperature operating range have a much better reproducibility of their characteristics. Below 4 K all sensors presented behave similarly but for temperatures above 20 K, Cernox™ has much better stability against thermal cycles. However, Allen-Bradley®, Cernox™ and TVO® satisfy the LHC specifications.

One of the aims of this test was to identify selection rules for the Allen-Bradley®. No correlation in performance has been found between stability at low temperatures and variations of resistance at room temperature, encapsulation, etc. From our results it is clear that it is extremely difficult to improve the performance of off-the-shelf Allen-Bradley® resistors by some simple recipe.

The experimental data suggest that Cernox™ characteristics are improved if they are previously exposed to at least 25 thermal cycles.

RhFe wire-wound and thin-film temperature sensors have also been investigated although more work is still required for analysing their results. The main difficulty is the poor sensitivity of this type of temperature sensors below 20 K, which is up to one order of magnitude lower than Allen-Bradley®.

We intend to complete 100 thermal cycles by the third quarter of 1999 and later we will also investigate other types of temperature sensors like germanium (bulk and thin-film), diodes, etc. Although not considered for equipping the LHC ring, these other sensors may find potential applications in other components of the complex cryogenic system for CERN's accelerators and detectors.

REFERENCES

1. Ch. Balle and J. Casas, Industrial-type Cryogenic Thermometer with Built-in Heat Interception, in: "Advances in Cryogenic Engineering", Vol. 41B, Plenum, NY (1996), p. 1715.
2. E. Chanzy et al., Cryogenic Calibration System using a Helium Cooling Loop and a Temperature Controller, in: "International Cryogenic Engineering Conference 1998", Institute of Physics Publishing, Bristol (1998), p. 751.
3. J-F. Amand et al., Neutron Irradiation Tests in Superfluid Helium of LHC Cryogenic Thermometers, in: "International Cryogenic Engineering Conference 1998", Institute of Physics Publishing, Bristol (1998), p. 751.
4. Ch. Balle, J. Casas and J-P. Thermeau, Cryogenic Thermometer Calibration Facility at CERN, in:"Advances in Cryogenic Engineering", Vol. 43A, Plenum, NY (1997), p. 741.
5. L.M. Besley, The Long-term Stability of Resistance Thermometers for Temperatures below 30 K, in: "International Cryogenic Engineering Conference 1980", p. 807.
6. A.L. Zahariev, D.A. Dimitrov and J.K. Georgiev, Characteristic Stabilization of Thin Film Platinum Thermometers during Thermal Cycling between 5 K and 300 K, in: "Cryogenics", Volume 36, Number 8, Elsevier (1996), p. 631.
7. L.M. Besley, Stability of some Cryogenic Carbon Resistance Thermometers, in: "The Review of Scientific Instruments", Vol. 54, Number 9, (1983), p. 1213.
8. D. Giraudi, P.P.M. Steur, D. Ferri, F. Pavese, Resistance-temperature characteristics of the new Cernox cryogenic thermometers in the range 1.5-350 K, in: "Proceeding of TEMPMEKO-96", (1996), p 155
9. F.J. Kopp and T. Ashworth, Carbon Resistors as Low Temperature Thermometers, in: "The Review of Scientific Instruments", Vol. 43, Number 2, (1972), p. 327.
10. W.L. Johnson and A.C. Anderson, The Stability of Carbon Resistance Thermometers, in: "The Review of Scientific Instruments", Vol. 42, Number 9, (1971), p. 1296.

CRYOGENIC THERMOMETER CALIBRATION FACILITY FOR THE LHC

J.P. Thermeau[1], C. Balle[2], P. Blache[1], S. Buhler[1], J. Casas-Cubillos[2], E. Chanzy[1], F. Chatelet[1], C. Joly[1], T. Junquera[1]

[1] Institut de Physique Nucleaire, CNRS/IN2P3
Orsay, F 91406, France

[2] CERN-LHC Division,
Geneve 23, CH 1211, Switzerland

ABSTRACT

The "Institut de Physique Nucléaire" (IPN) and the European organization for Particle Research (CERN) collaborate on the design and the construction of a cryogenic thermometer calibration facility for the Large Hadron Collider (LHC). The thermometers are calibrated within the temperature range 1.6 K to 300 K, with an accuracy of ± 5 mK from 1.6 K to 4.2 K and about 1 % for higher temperatures. The calibration system has a capacity of 94 thermometers including 4 RhFe working temperature references. All thermometers are mounted on an isothermal copper block hung in a vacuum vessel and thermally insulated. A circulating helium heat exchanger cools down the copper block. The temperature is stabilized by means of a Proportional Integral regulator which controls a resistive heater mounted near the heat exchanger. For each step of temperature, calibration data are obtained by comparison of the thermometer resistance with the temperature deduced from the working references. The thermometers can be calibrated either in vacuum or immersed in saturated liquid helium. The particularities of this apparatus are its active cooling loop and controlled heating which enable to stabilize the temperature at previously fixed values with good dynamic characteristics. This paper presents the cryogenic thermometer calibration facility and the results of the first calibrations.

INTRODUCTION

The LHC accelerator is a ring of 27 km, which is composed of 1800 superconducting magnets operating in superfluid helium. Cryogenic thermometers are needed to control the maximum temperature at which the magnets can operate. The superconducting magnets are immersed inside static baths of pressurized He II at 1.9 K, 1 bar. The pressurized baths are cooled by conduction by means of a heat exchanger tube and by vaporization of the helium II liquid inside of this heat exchanger tube. The temperature regulation allows to optimize the liquid helium quantity necessary to maintain the magnets at the operating temperature.

About 5000 thermometers are needed for the operation process of the LHC, 1800 thermometers for the magnets and the others for the cryogenic lines and equipment. Since the superconducting coils require severe conditions, the thermometers must be calibrated with an accuracy of ± 5 mK from 1.6 K to 4.2 K. During cool-down or warm-up, the thermometers are used as diagnostic, therefore an accuracy of 1 % is sufficient.

The IPN and CERN collaborate on a cryogenic thermometry program which includes sensor irradiation damage at low temperature[1-2], design of a calibration facility and storing and analysis of the sensor calibration data with an Oracle® database.

A cryogenic thermometer calibration facility is in operation at CERN with a maximum output of 60 thermometers per week[3]. This station calibrates all thermometers necessary for the LHC cryogenic prototypes. For the thermometer calibration in series, the IPN has developed a fully automated cryogenic calibration facility[4], which will enable the processing of 90 thermometers per week in the 1.6 K to 300 K temperature range.

The calibration is made by comparison with 4 working standard thermometers (RhFe, Rhodium Iron thermometer). To maintain the working standard thermometers, the CERN uses the secondary standards associated with a calibration apparatus supplied by the Institute for Physical Research (VNIIFTRI, Moscow, Russia). The "Istituto di Metrologia, G. Colonnetti" (Torino, Italy) provides the secondary standard thermometers (RhFe) which are traceable in the range 1.6 K to 30 K and have an uncertainty of ± 2 mK to ± 5 mK. For the temperature above 30 K, the calibration of these standards is better than ± 10 mK and CERN can control it with a Platinum standard thermometer (14 K – 300 K) provided by the National Physical Laboratory (Teddington, England).

CALIBRATION FACILITY

In the LHC cryogenic equipment, the thermometers will be mounted either under vacuum conditions or in liquid helium to get the best accuracy. The calibration facility has been studied in order to realize the thermometer calibrations in both conditions. Furthermore, a thermometric block[5] has been developed by CERN to thermalize the leads, to reduce the radiation heat transfer and to evacuate the heat produced by the Joule self heating. This thermometric block allows to simplify the thermometer mounting and to guarantee a good reproducibility of the temperature measurement.

The calibration facility uses a vertical cryostat, a calibration insert (see Figure 1) and ancillary equipment which are necessary for the automation process.

Calibration apparatus

The vertical cryostat is composed of a cylindrical cold vessel with a large neck (∅300 mm) in order to house the calibration insert. The radiation heat shield is hung on the cold vessel neck and is cooled down to 80 K with liquid nitrogen circulation. The cold vessel is enclosed in the shield and the vacuum vessel. The different phases of cryostat conditioning are realized automatically using ancillary and cryogenic equipment. The cryogenic valves are used for transferring or pumping of the liquid helium. They particularly fill with liquid helium the cold vessel by intermittent transfer to keep the liquid helium level constant.

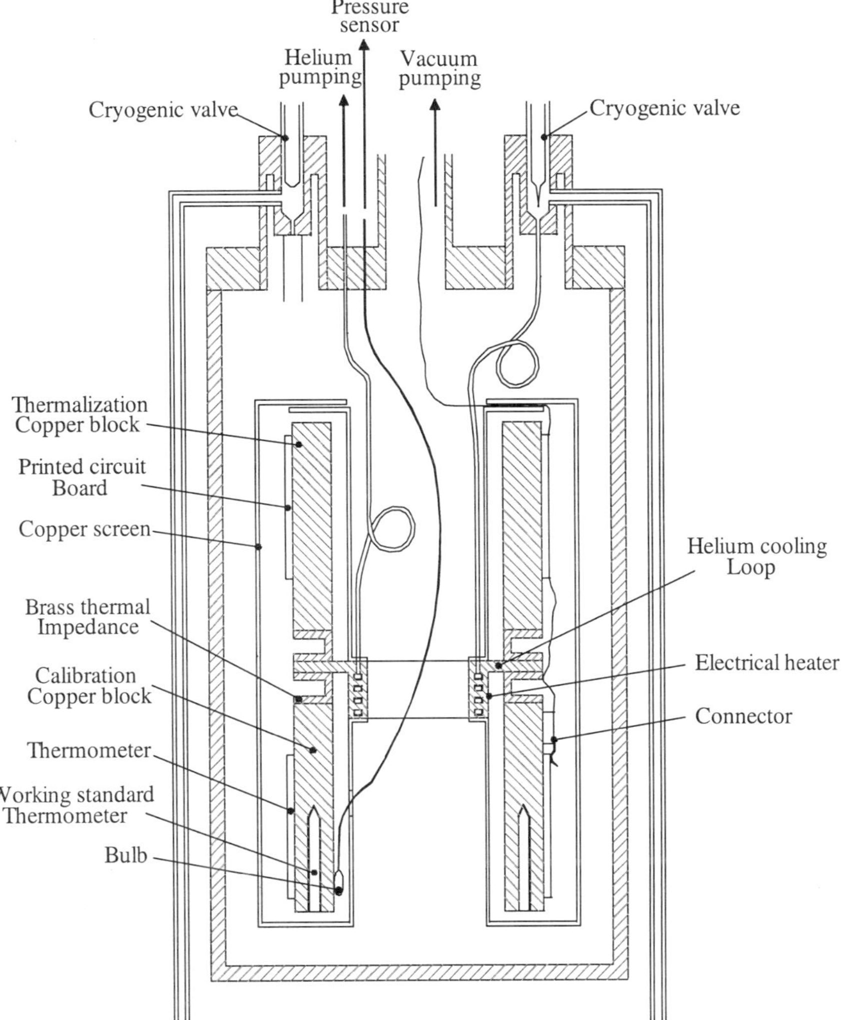

Figure 1. Calibration insert.

The thermalization block, the calibration block and the helium cooling loop are surrounded by a radiation copper shield and a vacuum enclosure; they form the calibration insert. The thermalization and the calibration copper blocks are screwed to the helium cooling loop with thermal impedances. These impedances reduce the high frequency perturbation transmission due to helium flow instabilities in the helium cooling loop and to the very low specific heat of copper at low temperature. An electrical heater, glued on the helium cooling loop, is used to adjust the calibration block temperature. Printed circuit boards are anchored on the thermalization cooper block and thermalize the thermometer wires at the helium cooling loop temperature. The thermometers are fed with manganin wires, 20×24 pins vacuum-tight connectors at room temperature are used as pressure-vacuum feedthrough. The thermometers are mounted on the calibration block; the working standards are introduced inside holes while the thermometers to be calibrated are screwed. Apiezon® grease is used to improve thermal contact between the thermometer and the calibration block. The thermometers can be easily disconnected without soldering using special slide connectors. The copper shield thermally anchored to the cooling loop helium allows the calibration under adiabatic conditions. The copper blocks (thermalization, calibration, heat exchanger, ...) are made of O.F.H.C. copper and their total mass is about 60 kg. They are supported by three fiberglass rods.

Process

The program automatically performs the thermometer calibration. It communicates with the measurement system and with the Programmable Logic Controller (PLC) which controls the cryogenic and ancillary equipment. Before a calibration run, the temperature set points and their regulation parameters are loaded in a program. This program runs with two calibration modes, one under vacuum condition and the other in liquid/gas helium condition. In both modes, it sends set points for each temperature step and controls the working standard temperature. When the temperature is reached, it controls the temperature stability. The calibration measurement cycle starts when the temperature is stable (stability better than 1 mK/min at 4 K). The program realizes four acquisition cycles per temperature step before getting to the next set point.

Under vacuum condition (about 10^{-6} mbar). The calibration insert is cooled by a liquid helium loop. For each temperature set point, the program sets the insert cryogenic valve position. When the temperature is approached, the valve is adjusted to cool slightly the calibration block and the Proportional Integral regulator (PI) maintains the temperature using the heater.

In liquid/gas helium condition. For temperature below 4.5 K, the calibration insert is filled with liquid helium through a cryogenic valve, which sucks the cryostat liquid helium. The temperature is stabilized by control of the insert bath pressure, which is regulated with a room temperature set: pump, valve and pressure sensor.

For temperature above 4.5 K, the insert pressure is about 1 bar, the temperature is measured during warm-up of the calibration copper block which temperature is controlled by the heater.

INSTRUMENTATION

A LabVIEW® program running on a PC controls the acquisition equipment through an IEEE 488 bus (GPIB). The measurement system is composed of a low-noise channel multiplexer (contact potential < 500 nV), a programmable current source and a nano-voltmeter with a measurement uncertainty of $\pm$ (6.5 10^{-5} of reading + 8 10^{-6} of range). The resistance of thermometers is measured by the 4-wire method, successively with positive and negative polarity. RhFe thermometers have a poor sensitivity and require an instrumentation with a very high accuracy. To increase the measurement system accuracy, a standard resistor reference is used (accuracy of $\pm$ 2 10^{-6}) which allows to control the current value. The current uncertainty becomes $\pm$ (9 10^{-5} of reading + 9 10^{-6} of range). The resistance measurement system has an uncertainty of 1 10^{-4} which gives an temperature uncertainty for the working standard of $\pm$ 3 mK at 1.6 K and $\pm$ 40 mK at 300 K. This uncertainty includes the random and the systematic uncertainties associated with the measurements, at the level of 95 % of confidence. Two independent measurement systems are used, one for the working standard thermometers and the other for the thermometers to be calibrated. That allows to measure continuously the 4 working standards during the acquisition cycle of thermometers to be calibrated. The scanning cycle is performed in about 7 minutes for 90 thermometers.

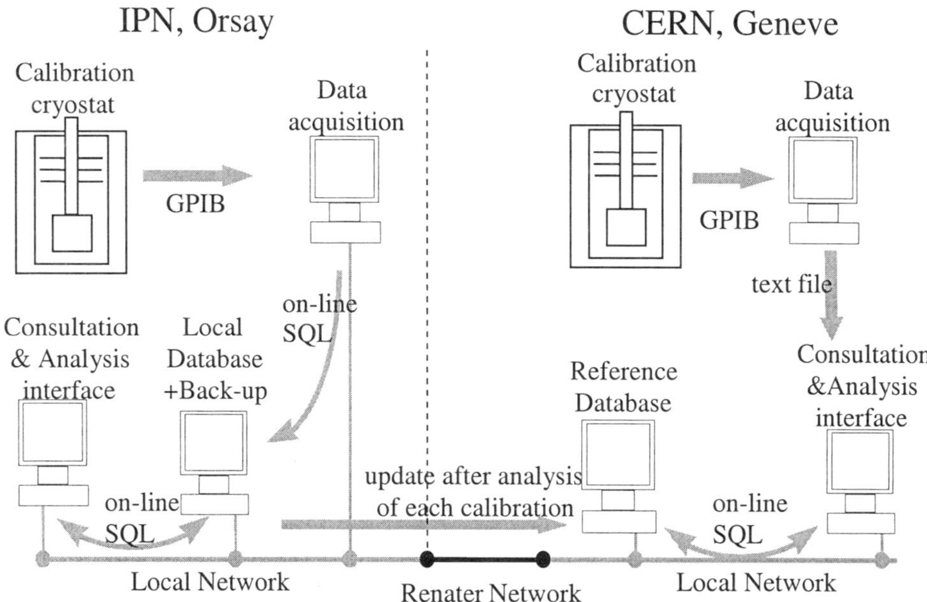

Figure 2. Two calibration facilities layout.

DATABASE AND DATA FLOW

5000 thermometers will be individually calibrated. A large quantity of data has to be recorded and organized: calibration conditions, calibration points, and history of each thermometer. In order to simplify data manipulation, two databases have been created with Oracle®: a reference database at CERN, and a local database at IPN. Calibrations are transferred from IPN database to CERN database once they are completed (see Figure 2).

The database structure allows to record thermometer description and identification, raw calibration data, and mathematical data analysis results (such as mathematical calibration functions).

Using the consultation and analysis interface, one can create thermometers and define the calibration run. Before processing a calibration run, the data acquisition program loads all needed information from the database: thermometers' name, position on the calibration block, scanner channel, and room-temperature resistance. A validation test is then performed.

During a calibration run, data acquisition program records the measures in the database (through the local network). After a calibration run, one can use the database interface in order to check the validity of the calibration, to compute the best calibration mathematical function for each thermometer and to generate a calibration sheet.

EXPERIMENTAL RESULTS

The calibration insert can be cooled down from 300 K to 4.2 K in about 6 hours. The good efficiency of the heat exchanger allows to cool down the calibration insert (60 kg of copper) with 100 L of liquid helium. To realize a full thermometer calibration, 250 L are necessary (cryostat and insert consumption).

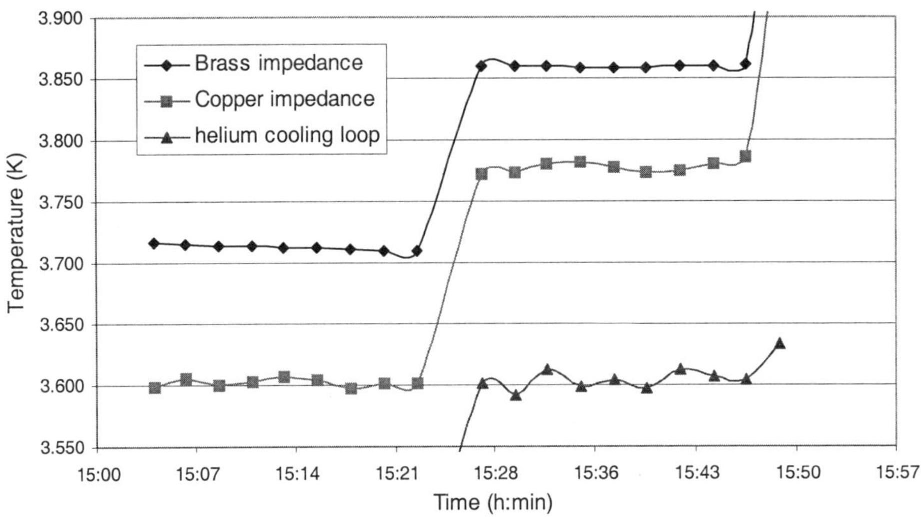

Figure 3. Temperature stability measured on three insert components: i) the temperature helium cooling loop which is regulated, ii) the thermalization block with a copper impedance, iii) the calibration blocks with a brass impedance.

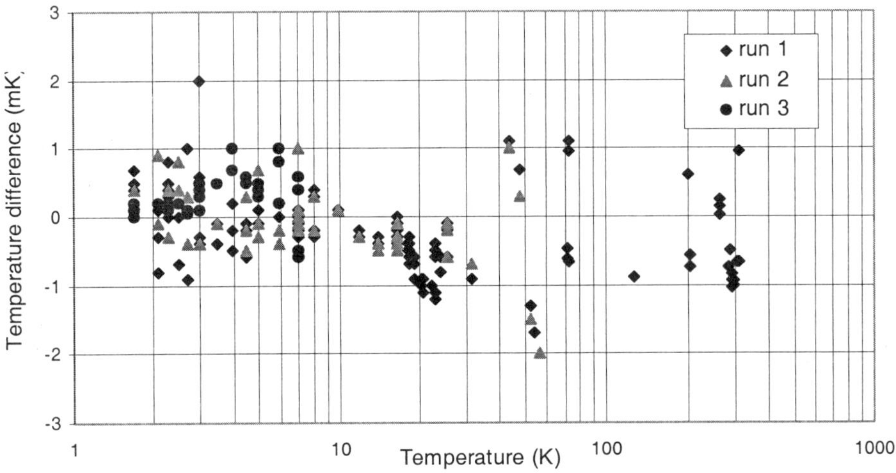

Figure 4. Temperature difference between two working standards.

For temperature below 10 K, 15 min are necessary to reach the temperature stability better than 1mK/min on the calibration block. At higher temperature and for the identical stability parameter, the stabilization time of the calibration block is 1 hour at 200 K. The brass thermal impedances reduce the temperature oscillations at low temperature (see Figure 3) but they increase the stabilization time at higher temperature. The LHC project needs a calibration accuracy about 1 % for the temperature above 4.2 K, therefore, a very high temperature stability is not necessary and the calibration time can be reduced. A full calibration, 15 calibration points in the temperature range 1.6 K to 4.2 K and 10 calibration points with a drift of 50 mK/min for higher temperatures can be performed in less than three days (continuously running).

Four RhFe thermometers are used as temperature references, which are very well correlated in the full temperature range. Their dispersion is smaller than ± 2 mK, which is within the thermometer measurement error (see Figure 4). A vapor pressure bulb which can operate with different fluids (He, H2, ...), is soldered on the calibration block. This bulb connected to a pressure sensor forms a primary temperature reference which allows to control the working standard thermometer calibration. In our configuration, the thermomolecular effect and the correction for aerostatic pressure are 0.5 mK at 1.6 K. The estimated uncertainty of vapor pressure bulb is about ± 1.5 mK from 1.6 K to 4.2 K which is due essentially to pressure sensor uncertainty.

CONCLUSION

The Orsay calibration facility can calibrate a large number of thermometers with a high accuracy, ± 5 mK in the temperature range 1.6 K to 30 K. For higher temperatures, the accuracy depends on the temperature reference accuracy and the number of calibration points realized during a calibration run. Today, we can calibrate thermometers with an accuracy of 0.25 % from 1.6 K to 300 K.

This calibration facility can produce 1500 calibrated thermometers per year, with two weeks shifts: a week to mount and to test thermometers, the other week to calibrate. To optimize the occupation rate of the calibration facility and to increase the calibration throughput, a second calibration insert may be built in year 2000.

In order, to avoid data manipulation errors and to address a large quantity of data, CERN experts have recommended to use the Oracle® database. This database contains the calibration points and the history for each thermometer. Before calibration, each thermometer is created in the database (sensor, model, manufacturer, mounting, ...). Afterwards, the database is used to initialize the process parameters of the calibration system and to store the calibration points. Furthermore, the analysis and the calculation of the thermometer calibration function are realized with a database consultation tool, which runs with LabVIEW®.

ACKNOWLEDGEMENTS

We would like to thank the members of both laboratories involved in this project and specially D. Grolet, D. Ruffier, J.P. Leclercq, F. Carrey of the cryogenics group at IPN and finally, Ph. Lebrun and the whole LHC Cryogenics group at CERN.

REFERENCES

1. T. Junquera, J.F. Amand, J.P. Thermeau, J. Casas-Cubillos, Neutron Irradiation Tests of Calibrated Cryogenic Sensors at Low Temperatures, in: "Advances in Cryogenic Engineering", Vol. 43A, Plenum, NY (1997), p. 765

2. J.F. Amand, J. Casas-Cubillos, T. Junquera, J.P. Thermeau, Neutron Irradiation Tests in Superfluid Helium of LHC Cryogenic Thermometers, in: "International Cryogenic Engineering Conference", Vol. 17, IOP, Bristol (1998), p. 727

3. C. Balle, J. Casas-Cubillos, J.P. Thermeau, Cryogenic Thermometer Calibration Facility at CERN, in: "Advances in Cryogenic Engineering", Vol. 43A, Plenum, NY (1997), p. 741

4. E. Chanzy and al., Cryogenic Thermometer Calibration Facility Using a Helium Cooling Loop and a Temperature Controller, in: "International Cryogenic Engineering Conference", Vol. 17, IOP, Bristol (1998), p. 751

5. C. Balle and J. Casas-Cubillos, Industrial-type cryogenic thermometer with built heat interception, in: "Advances in Cryogenic Engineering", Vol. 41B, Plenum, NY (1996), p. 1715

TRANSIENT CALIBRATION OF CRYOGENIC THERMOMETERS

J.M. Pfotenhauer and W.P. Cole

Applied Superconductivity Center
University of Wisconsin – Madison
Madison, WI 53706

ABSTRACT

A novel method for rapidly and accurately calibrating cryogenic temperature sensors has been developed and demonstrated using a set of Cernox™ thermometers. The calibration approach is based on a thermal configuration for which the Biot number (defined as the ratio of conductive resistance to convective resistance) is much less than 0.1. This condition is met by mounting uncalibrated and calibrated thermometers together inside a sufficiently large copper block. The calibration is carried out in a transient mode by slowly lifting the copper block out of a bath of liquid cryogen (helium in this case), and allowing it to warm to room temperature, while simultaneously recording voltages of both the calibrated and uncalibrated thermometers. A complete calibration from 4.2 K to room temperature may be accomplished in approximately 30 minutes. Temperature uniformity within the copper block is theoretically determined by Biot number, and practically measured through the use of more than one calibrated thermometer. Calibration results, error analysis, and details regarding a unique data acquisition requirement are presented.

INTRODUCTION

Calibrating cryogenic thermometers typically requires a significant investment of time, very stable temperature control, and depending on the final accuracy required, a significant effort to obtain primary temperature references. All of these factors contribute to the noticeable expense associated with calibrated cryogenic temperature sensors. In this report, we present an alternative method to the usual approach involving numerous 'equilibrium' temperature points for calibrating cryogenic sensors[1]. The method is transient in nature, simple to implement, and inexpensive. The issues influencing the final calibration accuracy, such as the thermometer mounting, temperature uniformity during the transient process, the influence of thermal emf's, and the rate of temperature change are discussed below. The results from a sample set of Lakeshore Cernox™ sensors presented do not represent the best accuracy one can obtain using this method, but do illustrate the possibility of doing so.

The transient calibration method is founded on the observation that when a convectively cooled (or warmed) solid satisfies the 'lumped capacitance' condition given by

Advances in Cryogenic Engineering, Volume 45.
Edited by Shu *et al.*, Kluwer Academic / Plenum Publishers, 2000.

$$Bi = \frac{hL}{k} << 0.1 \tag{1}$$

the solid may be approximated as an isothermal structure. Here Bi is the Biot number, h the convection coefficient, k the thermal conductivity, and L a length defined by the ratio of the volume to the surface area of the solid. The Biot number represents the ratio of the conductive thermal resistance (L / kA_x) to the convective thermal resistance ($1 / hA_s$). Here A_x is the cross sectional area through which the heat flows in the solid, while A_s is the surface area. The smaller the Biot number, the more isothermal the structure will be.

Using this observation, a transient calibration has been performed on a set of LakeShore Cernox™ model 1070 SD thermometers mounted within a copper block and warmed from 4.2 K to room temperature through a combination of natural and forced convection.

The speed with which the copper block is warmed to room temperature is governed by the thermal time constant of the solid

$$\tau = \frac{\rho V c}{h A_s} = \frac{k}{h L} \cdot \frac{L^2}{\alpha} = \frac{t}{Fo \cdot Bi}, \tag{2}$$

and the time dependence of the surrounding temperature. In equation (2), ρ is the solid density, V is the solid volume, c is the heat capacity, α is the thermal diffusivity and Fo is the Fourier number. In cases where the lumped capacitance condition is satisfied, the temperature uniformity within the copper block at any instant in time is determined by the Biot number alone. For example, if $Bi < 0.01$, 99% of the temperature difference between the convective fluid temperature T_f and the temperature T_L at a distance L within the solid from the convective surface occurs within the convective boundary layer. The remaining 1% of T_f - T_L occurs within the solid in accordance with Fourier's law. Temperature sensors mounted anywhere within the solid all experience the same temperature at any given time within the 1% variation of $T_f - T_L$. Mounting the sensors all at the same distance from the convective surface ensures temperature uniformity to an even greater degree. A one-to-one correspondence between the resistance of calibrated and un-calibrated sensors mounted in this way is then possible at any given time during the convective warming process.

This transient approach affords a rapid calibration of thermometers, but in practice its accuracy is dependent on four primary factors: (1) the inherent accuracy of primary, or previously calibrated sensor(s), (2) the temperature uniformity as determined by the Biot number, (3) the method used to mount the thermometers, and (4) the degree to which thermal emf's generated in the measuring circuits can be canceled during the warming process. The following sections detail the approach we have used to ensure optimal accuracy as dependent on the latter three factors. The accuracy of the primary sensors is limited by its own calibration. For example, the absolute temperature accuracy of the Cernox™ 1070 SD sensors used as primary temperature references in this report typically vary from 4 mK at 1 K to 140 mK at 300 K[1]. In addition to the efforts used to optimize the calibration accuracy, we have also used the finite element heat transfer (FEHT[2]) software to model the temperature uniformity within the copper block during the transient process. The results from this model provide quantitative guidelines for future use of this technique.

EQUIPMENT

The calibration rig, as shown in figures 1 and 2, accommodates ten thermometers. The resistances of each thermometer are measured using a four-wire technique. Nine of the thermometers are electrically connected in series with the tenth in an independent electric circuit. In keeping with standard cryogenic techniques (see for example Hust[3]), instrumentation wiring extends from room temperature through a single stainless steel tube and is heat-sunk to the copper mounting block through the use of non-inductive windings

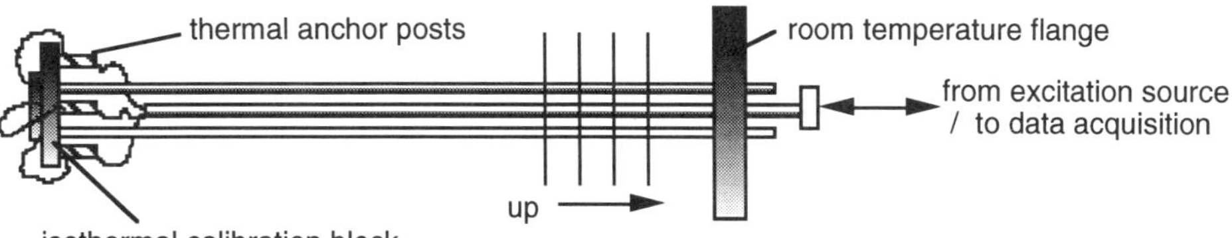

Figure 1. Simple rig for transient calibration procedure. Note the lack of any vacuum requirements.

on posts extending from the copper block. A forty-pin connector block permits plug-in connections between the thermometers and the heat-sunk instrumentation wiring. The 4-wire circuit for each thermometer typically involves 14 solder joints, and two mechanical joints, all of which have the potential for introducing thermal emf's during the warm-up process. Subsequent calibration rig designs would be improved by reducing the number of these joints.

Figure 2 displays the configuration of the thermometers as mounted in the copper block. The type 1070 SD Cernox™ thermometers with dimensions 3.2 mm x 1.9 mm x 1.0 mm fit within machined grooves at a single radius in the lower copper block. The disk shaped geometry includes significant symmetry as far as the thermal conditions seen by the ten thermometers during the calibration process. Prior to mounting the thermometers, a single layer of 25 micron-thick cigarette paper is applied to the copper with IMI 7031 varnish to ensure electrical isolation between the thermometers and the copper block. Thermometers are mounted on the lower copper block using a thin layer of Apiezon-M grease to enhance the thermal contact between them and the copper. The grease thickness between the upper and lower thermometer surfaces and the adjacent copper surfaces is estimated to be less than 0.1 mm. With this type of thermometer, the dominant heat transfer to the sensors occurs through its upper and lower surfaces, and the geometry of the copper block has been configured to provide optimal thermal contact between those surfaces and the copper. For many other thermometers, alternate mounting methods would be required to ensure optimum thermal contact between the lead-wires and the copper block. After mounting the thermometers on the lower copper disk, the assembly is mated to the upper copper block and secured in place with a stainless steel bolt extending up through the radial center of the two copper disks. Two calibrated thermometers are positioned opposite each other within the copper block to allow measurement of the actual temporal temperature uniformity both at steady state and in transient conditions.

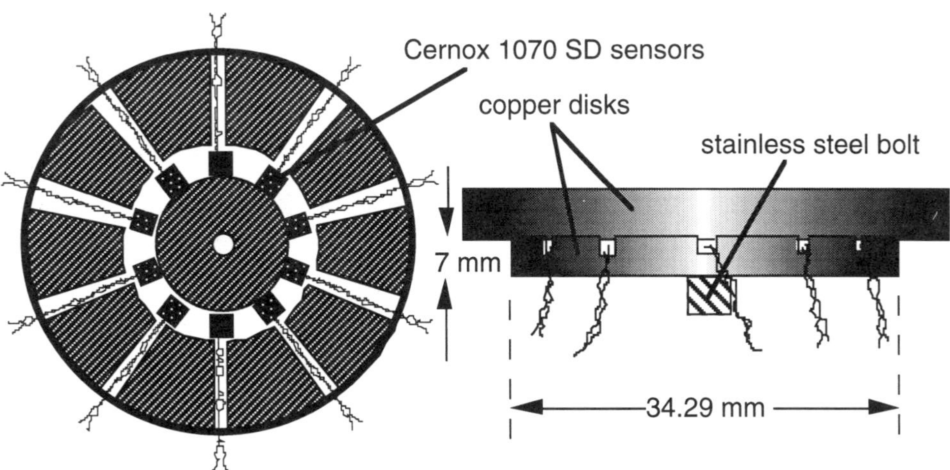

Figure 2. Copper 'sandwich' mounting block for Cernox™ 1070 SD sensors.

PROCEDURE

The calibration procedure includes 1) room temperature preparations, 2) cool-down to steady state conditions in liquid helium, and 3) extraction and warm up to room temperature. After securing the thermometers to the calibration rig, standard procedures are carried out at room temperature to ensure continuity of all circuits, the absence of any shorts between the thermometers and the calibration rig, and the ability to measure room temperature resistance values. The entire calibration rig is then inserted into a dewar and liquid helium is transferred into the dewar sufficient to cover both copper disks and the thermal anchor posts for the instrumentation wiring. Resistance values at 4.2 K are gathered using a DC measurement, after an equilibrium condition has been established.

The transient calibration is achieved by smoothly lifting the calibration rig out of the dewar with a hoist. Typically this is accomplished over a 15 – 30 minute time span. To speed up the warming process once the rig is completely out of the dewar, a warm breeze is applied using a hot-air gun from an approximate distance of 0.5 meters directly below the rig. However, before the extraction process begins, a square wave excitation signal is applied to the thermometers. As will be shown below, the square wave excitation signal can eliminate the effects of parasitic voltages in the measurement circuit. These voltages arise primarily from material inhomogeneities and joints in the wiring circuit, which experience strong temperature gradients during the warming, process (see for example, Walstrom[4]). From an investigation of the self-heating of the two calibrated sensors it was determined that a current of 10 μ-amps introduces less than a 1 mK offset at 4.2 K. The 0 – 10 μ-amp square-wave excitation signal is supplied by an HP 3245A waveform generator. Voltage signals from the four-wire circuit of the thermometers are amplified by an Ectron model 750 differential amplifier, and then converted to digital signals via the A/D converter in a National Instruments MIO-16X A/D board before being stored on the computer. Signal uncertainty introduced by the excitation signal and the voltage measurement add less than 0.1% to the relative uncertainty in a given temperature measurement.

The frequency of the square wave excitation signal is set to 0.1 Hz. The impact of this frequency can only be appreciated by first discussing the purpose for the square wave signal. The voltage (V) measured by the computer is related to the thermometer resistance (R) at some time t1 through the relation:

$$V_1 = G \cdot (I_1 R + \delta V_1) \tag{3}$$

where G is the circuit gain, I is the current, and δV represents parasitic voltages caused by thermal emf's in the circuits, ground offsets, etc. If at a time t2 sufficiently close to t1 such that $\delta V_2 = \delta V_1$ the current is changed and a second voltage is recorded:

$$V_2 = G \cdot (I_2 R + \delta V_2), \tag{4}$$

then the resistance can be determined from the relationship

$$R = \frac{V_1 - V_2}{G(I_1 - I_2)} \tag{5}$$

and is independent of the parasitic voltages. Independent measurements of the parasitic voltages during the calibrations revealed them to be significant and strongly dependent on warm-up rates and electrical isolation (i.e. handling) of the calibration rig during extraction. Nevertheless, by using the square wave excitation current and measuring the signal voltage in both the high and low states of the square wave allowed the parasitic voltages to be canceled. Further measurements are required to quantitatively define the degree to which the parasitic voltages are completely eliminated by the 0.1 Hz square wave, however the

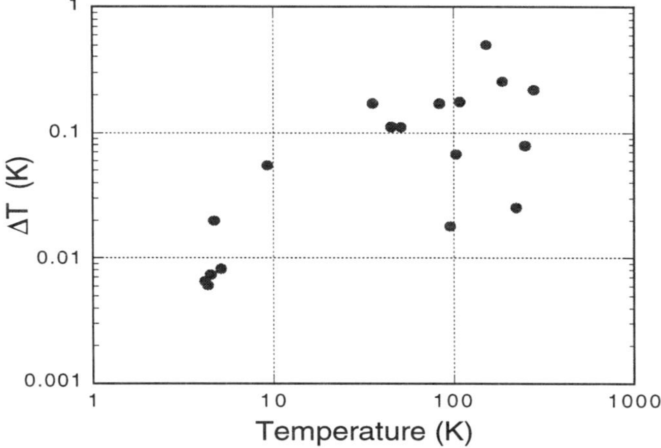

Figure 3. Measured temperature variation between calibrated sensors during transient calibration.

sample rate was set after consideration of the time dependence of the measured parasitic voltages.

The temperature agreement between the two calibrated sensors measured during the transient calibration process is shown in figure 3. The level of agreement is similar to the inherent uncertainty of the sensors' original calibration and suggests that little, if any temperature uncertainty has been added by the transient calibration process. The typical expenditure of liquid helium for such a calibration procedure is less than 30 liters.

THERMAL ANALYSIS

Model definition

A finite element heat transfer model has been developed to quantify the expected temperature variations during the transient calibration procedure. The heat transfer model is defined by its geometry, material properties, heat generation, and surface heat transfer coefficient. Because of the inherent symmetry of the calibration block shown in figure 2, the model incorporates a 2 dimensional axially symmetric geometry as shown in figure 4

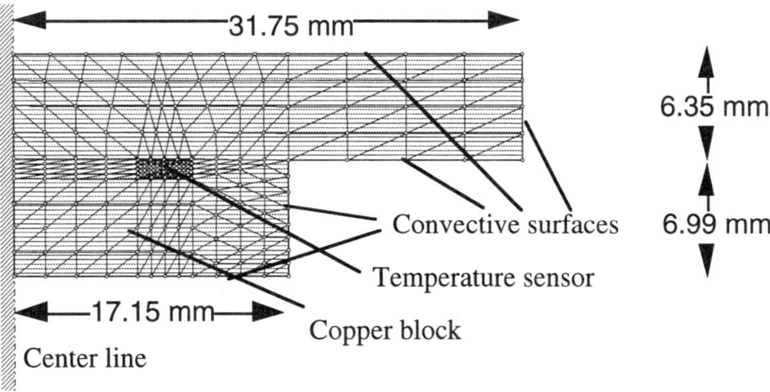

Figure 4. Side view of axially symmetric copper block including nodal arrangement of thermal model.

and is made up of 464 nodes. The computer software FEHT[2] permits definition of thermal properties both in terms of time and temperature. Therefore, polynomial and power fits are used to define the temperature-dependent thermal conductivity and heat capacity of the copper block, and the Apiezon N-grease. The sensor has also been modeled as a piece of copper, although because of its size, its own thermal properties do not contribute significantly to the overall time dependent temperatures. Heat generation (self heating) in the thermometer was also found to have an undetectable influence on the time dependent temperatures even within the sensor itself.

As shown in figure 4, the sensor is separated from the copper block by a 0.1 mm layer of grease. As a worst case assumption, one could imagine the grease introducing sufficient thermal isolation so that the sensor would assume the temperature of one random contact point with the surrounding copper. In this case, the temperature uncertainty would be equal to the temperature variation realized in the copper over the spatial extent of the sensor. This temperature variation is defined by the variable ΔT_sensor.

The heat transfer coefficient at the surface of the copper block is calculated using a natural convection correlation

$$h_{conv} = 0.27\frac{k_{helium}}{d}Ra^{0.25} \approx 40.38 * T_{cu}^{-0.487} \tag{6}$$

where the approximation assumes a largest possible heat transfer coefficient and is based on the assumption that the temperature difference between the copper surface and the surrounding vapor is less than 150 K. In addition to the natural convection coefficient, thermal radiation from the surrounding environment is characterized as a convection coefficient according to the relationship:

$$h_{rad} = \varepsilon\,\sigma\,(T_{cu} + T_{vapor})\,(T_{cu}^{2} + T_{vapor}^{2}), \tag{7}$$

and the total convective heat transfer is governed by the sum of these two heat transfer coefficients. The process of extracting the calibration rig from the helium dewar is modeled by using a linearly increasing vapor temperature, from 4.2 K to 300 K. The convective heating by the hot-air gun is modeled by a constant heat transfer coefficient of 100 W/m^2K and a vapor temperature of 350 K.

Model results

The time dependent results generated by the thermal model for both the copper and vapor temperatures, as well as the Biot number are displayed in figure 5. Note the abrupt increase in the Biot number when the hot-air gun is applied. The corresponding values of ΔT_sensor, the temperature variation in the copper in the vicinity of the temperature sensor, and the Biot number are given in figure 6. A significant correlation is displayed between the Biot number and the temperature variation around the temperature sensor. Quantitatively one can observe that a Biot number less than 0.001 is required to ensure temperature uniformity below 0.1 K. For temperature uniformity 10 mK or less, the Biot number should be on the order of 10^{-5}. Finally, it should be noted that the temperature variations displayed in figure 6 qualitatively match those experimentally measured and displayed in figure 3.

CALIBRATION REPORT

Following a well established commercial convention[1], a complete calibration report for the eight sensors calibrated in a single run can be easily provided in terms of Chebychev polynomials for separate temperature ranges, (typically 4.2 K – 25 K, 25 K – 100 K, and 100 K – 300 K), graphical presentation of resistance vs. temperature data, and interpolation tables. For a 6th order polynomial definition, one must supply at least 7 data points in a

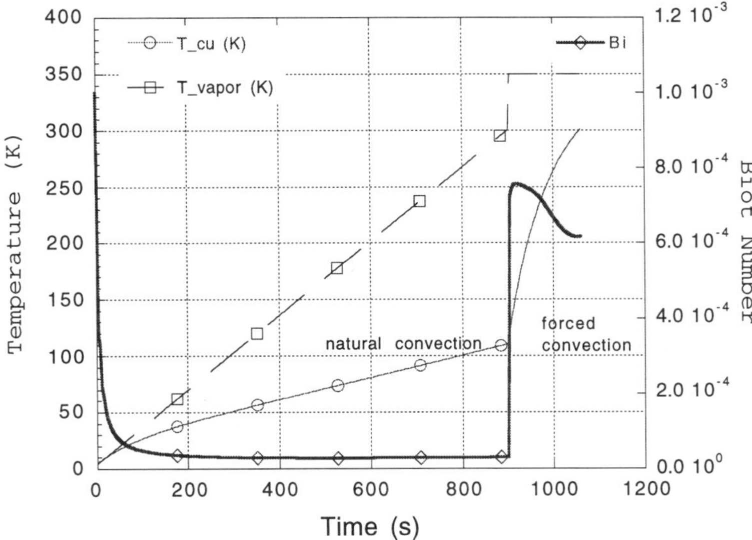

Figure 5. Thermal model results for vapor and copper temperatures, and Biot number during transient calibration process.

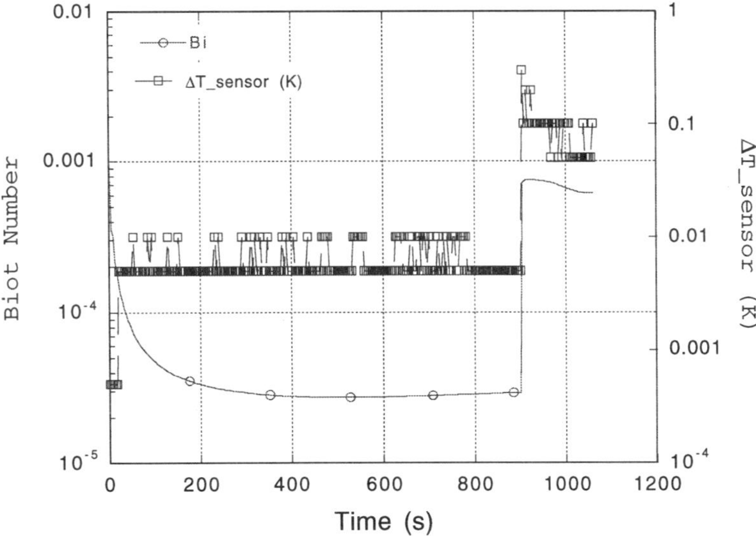

Figure 6. Thermal model results: Biot number and temperature variation in the vicinity of the sensor.

given temperature range. This consideration must also be taken into account when determining the extraction rate and the excitation square-wave frequency.

CONCLUSIONS

An inexpensive and relatively simple method for calibrating cryogenic thermometers has been demonstrated. The success and accuracy of the transient calibration method is determined by four primary factors: 1) the inherent accuracy of the primary (calibrated) thermometer, 2) an isothermal environment for the batch of calibrated and un-calibrated thermometers, 3) mounting technique that provides good thermal contact between the thermometers and the isothermal environment, and 4) elimination of parasitic voltages, especially as caused by thermal emf's.

Results from this test demonstrate that the second factor is determined by the Biot number of the calibration block, and that the temperature variation within the calibration block can be limited to 0.1 K or 10 mK when Bi < 0.001 or < 10^{-4} respectively. Additionally these results demonstrate that the use of two calibrated sensors provides a valuable measurement of the actual temperature uniformity during the transient calibration. The modeling results show that a 0.1 mm layer of Apiezon N grease provides an excellent thermal contact between the sensor and copper block. Finally, we have shown that a square wave excitation signal effectively eliminates the problems associated with signal noise due to the strong thermal emf's which develop during the extraction process. The frequency of the square wave excitation signal should be matched with the extraction rate so that parasitic voltages do not change within the time associated with one period of the square wave signal. In addition, the data sampling rate must be chosen to provide sufficient number of data points within a given temperature range to adequately define the temperature dependence in that range. In practice we find that 6th order Chebychev polynomials adequately describe the temperature variation of model 1070 SD Cernox™ thermometers in the ranges of 4 – 25 K, 25 – 100 K and 100 – 300 K. A minimum of seven data points in each temperature range is sufficient to completely define these polynomials.

ACKNOWLEDGEMENT

The authors wish to thank Sara Fowler for her assistance in the execution of the measurements and in the data analysis. This work was funded in part by the Department of Energy through Lawrence Livermore National Laboratory, subcontract number B340832.

REFERENCES

1. "Temperature Measurement and Control, 1995 Part 1 of 2," LakeShore Measurement and Control Technologies, Westerville, OH. (1995), pp. 1-22, 1-23, and A-1 through A-54.
2. "Finite Element Heat Transfer, version 7.101" F-Chart Software, 4406 Fox Bluff Rd., Middleton, WI (1996-1999).
3. J.G. Hust, "Thermal Anchoring of Wires in Cryogenic Apparatus," *Rev. Sci. Instr.* 41:622 (1970).
4. P.L. Walstrom, "Spatial Dependence of Thermoelectric Voltages and Reversible Heats," *Am. J. Phys.* 56(10):890 (1988).

STABILITY OF CERNOX™ RESISTANCE TEMPERATURE SENSORS

S.S. Courts and P.R. Swinehart

Lake Shore Cryotronics, Inc.
Westerville, OH, 43082

ABSTRACT

Cryogenic temperature sensing devices fabricated from sputtered zirconium oxy-nitride thin films have been commercially available from Lake Shore Cryotronics, Inc. under the tradename Cernox™ Resistance Thermometers (CXRTs) since 1992. These sensors possess many qualities desired in a cryogenic thermometer and are presently widely used. To date, no long term stability data has been available in the literature. Over the past six years, thirty-nine temperature sensors from six of the initial wafer production lots have been calibrated periodically and monitored for long term calibration stability and aging effects. These devices were stored at room temperature between calibration. Stability comparisons are given in terms of elapsed time, package type, wafer lot, and sensitivity level. Analysis of calibration data over the six year period show that these devices have repeated their initial calibration to within an average of ±0.08% of temperature over the 1.4 K to 325 K temperature range. Small qualitative and quantitative differences in stability were measured with respect to wafer lot, sensor sensitivity and package.

INTRODUCTION

Resistors fabricated from sputtered zirconium nitride thin films were shown to be possible cryogenic thermometers by Yotsuya, *et al.*[1] Further work performed at Lake Shore Cryotronics, Inc. resulted in the development of commercially available devices under the tradename Cernox™ resistance temperature sensors (CXRTs).[2,3] Sputter deposition occurs in an atmosphere consisting of argon, nitrogen, and oxygen. Judicious selection of the deposition parameters allows incorporation of oxygen into the ZrN lattice. The lattice enlarges as more oxygen is incorporated into the ZrN lattice, steadily changing the electrical conductivity from metallic behavior to percolating behavior.[3] Resistance temperature sensors can thus be produced with sensitivities designed to cover desired temperature ranges in a manner similar to germanium resistance temperature sensors. These devices have many useful properties including a wide temperature range, small physical size, fast thermal response time, low magnetoresistance, and excellent resistance to ionizing radiation. Data detailing these characteristics are available either from the literature or from the manufacturer.[2,5,6,7,8,9,10] To date, however, the long term stability of these types of devices have not been quantified. The present work was undertaken to provide long term stability data for Cernox™ resistance temperature sensors.

EXPERIMENTAL PROCEDURE

Zirconium oxy-nitride films were reactively sputtered in an argon/nitrogen/oxygen atmosphere. The deposition parameters were varied to produce a series of wafers whose room temperature dimensionless sensitivity $S_T \equiv (T/R)(dR/dT)$ varied from about -0.5 to -1.5. The dimensionless sensitivity is independent of geometry and allows for quick comparison between films independent of the final sensor pattern. Test sensors were fabricated from each wafer using standard photolithography techniques to define the active sensor area and subsequent contacts. Contacts were formed in the shape of inter-digitated fingers to allow trimming sensor resistance values upward to a desirable level. Figure 1 shows a finished sensor chip layout and scale. The lower temperature limits of these sensors are designed to cover three temperature ranges of commercial cryogenic importance as outlined in Table 1. For each model designation, the resistance is limited to 100 kΩ at the minimum temperature. Typical room temperature resistances are on the order of 30 to 60 Ω. Figure 2 shows a typical resistance versus temperature curve for a sample of each wafer tested. Figure 3 shows the typical sensitivity, dR/dT, for each sample plotted in Figure 2. Finally, Figure 4 shows the dimensionless temperature sensitivity, (T/R)(dR/dT) for each of the samples shown in Figure 2.

To prevent film surface damage/contamination and facilitate handling, these bare chips were packaged in one of two ways. The first method utilizes a 3 mm diameter by 8 mm long gold-plated copper canister identical to those used for other common cryogenic thermometers such as germanium and carbon glass resistance thermometers. Two 50.8 μm diameter gold wires are ball bonded to the Cernox chip contact pads. The chip is then epoxied to a beryllium oxide header. The gold leads are spot welded to phosphor bronze lead wires that pass through the beryllium oxide header. The gold plated copper can and phosphor bronze lead wires are then Stycast epoxied to the header creating a hermetic seal. The can is filled with He-4 gas prior to sealing to enhance thermal connection. In the second

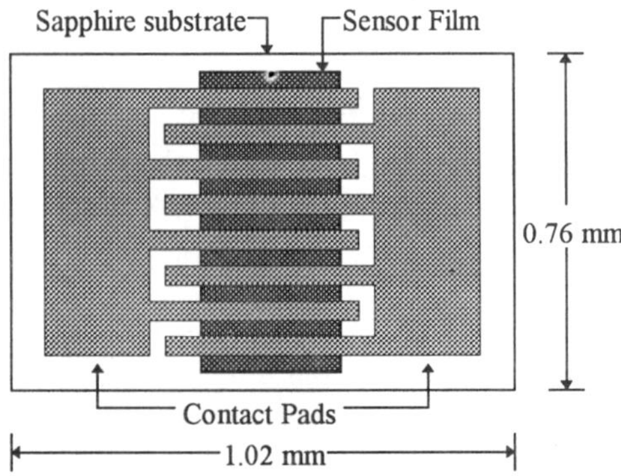

Figure 1. Chip layout and scale.

Table 1. Cernox™ RTD Temperature range and model designations

Wafer ID	Temperature range	Model Designation	Cryogenic Purpose	Number of samples
2Z184A	0.3 K - 325 K	CX-1030	Pumped He-3 system	4
2Z243B	1.4 K - 325 K	CX-1050	Pumped He-4 system	13
2Z245A	4 K - 325 K	CX-1070	Unpumped He-4 system	16
2Z251A	4 K - 325 K	CX-1070	Unpumped He-4 system	3
2Z255A	0.3 K - 325 K	CX-1030	Pumped He-3 system	1
3Z392A	1.4 K - 325 K	CX-1050	Pumped He-4 system	2

method, the Cernox chip is soldered into the cavity of a 3 mm long by 1.84 mm wide by 0.98 mm tall ceramic package. The base of the package is sapphire while the sides and top are alumina. Electrical connections are made by ball bonding one 25.4 μm diameter gold wire from each chip contact pad to the package electrical feedthrough bond pad. After packaging, the devices are thermally cycled from room temperature to liquid nitrogen temperature 200 times for conditioning.

Temperature calibrations from 1.4 K to 325 K were performed in Lake Shore Cryotronics' commercial Temperature Calibration Facility. Temperature and resistance was measured using standards grade platinum and germanium thermometers in conjunction with a Keithley Model 224 current source, Hewlett Packard Model 3456A DVM, and Guildline Model 9330 standard resistors (resistance values from 10 Ω to 100 kΩ in decade steps). The device under test was placed in series with a standard resistor of comparable value. The current was varied to a minimum of 0.1 μA to maintain a nominal 2 mV signal level across each sensor during calibration. The voltmeter was used in a ratiometric form with readings taken across both the standard resistor and device under test. Current reversal was performed to eliminate thermal EMFs. Approximately 84 temperature - resistance data points were taken for each sensor over the 1.4 K to 325 K temperature range. Two points beyond each extreme were taken for curve fitting purposes. After each calibration, the data for each sensor was smoothed using a combination of a least squares fit and a Chebychev polynomial fit. After smoothing the data, a cubic spline fit was used to generate a table of resistance, temperature sensitivity, and dimensionless temperature sensitivity at a chosen set

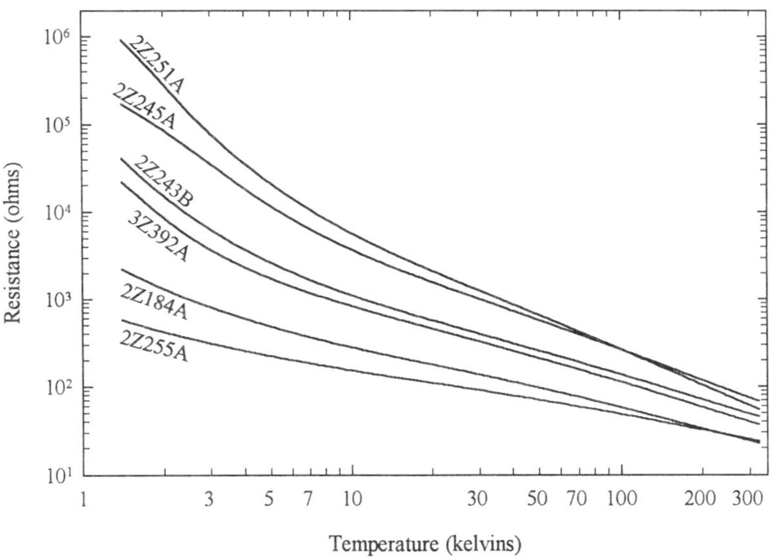

Figure 2. Typical resistance versus temperature characteristic for each wafer tested for stability.

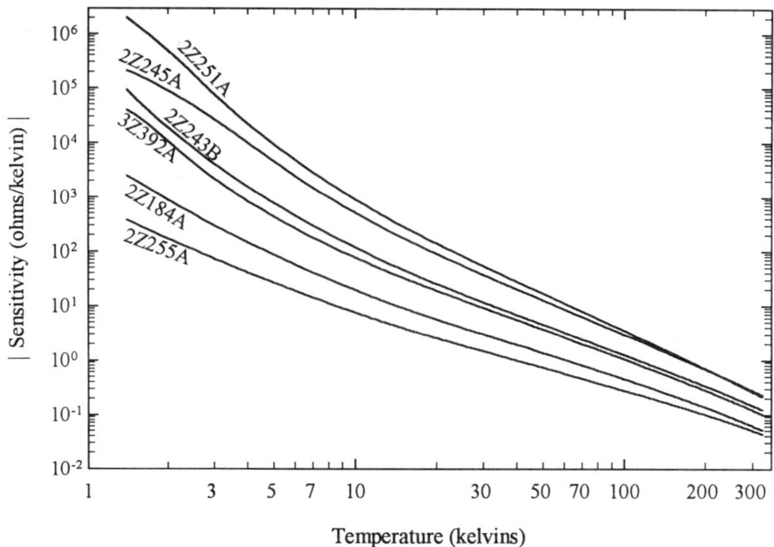

Figure 3. Typical temperature sensitivity versus temperature characteristic for each wafer tested for stability.

of comparison temperatures. The resulting calibration shifts were calculated from these tables as $\Delta T=(R_{final} - R_{initial})/(dR/dT)$. The calibration shift at each temperature is considered to be indicative of the stability at each temperature.

The resulting calibration accuracy is temperature dependent. The contributions to the accuracy come from stability of the current source, stability of the resistance standards, stability of the voltmeter, and accuracy of the NIST traceable thermometers. Typical values of accuracy range from ±4 mK at 4.2, ±8 mK at 20 K, ±30 mK at 100 K, and ±180 mK at

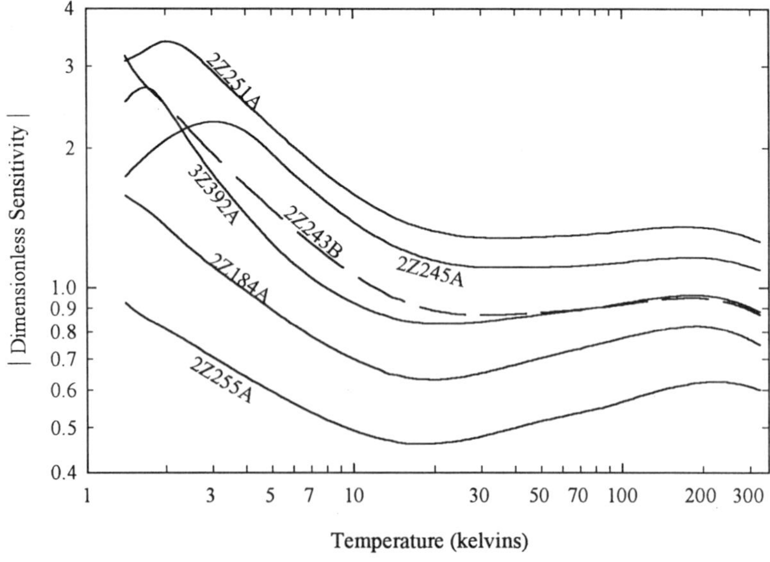

Figure 4. Typical dimensionless temperature sensitivity versus temperature characteristic for each wafer tested for stability.

300 K. Given a worst case scenario, it is possible for the recalibration inaccuracy to be twice the specification of a single calibration. It should be noted that the resolution is typically an order of magnitude or more better than the accuracy.

EXPERIMENTAL RESULTS

Calibration shift data was analyzed in four ways: 1) by time, 2) by mounting package after 4.4 years, 3) by wafer after 4.4 years, and 4) by sensitivity (model) after 4.4 years. Comparison at the 4.4 year time mark was chosen because all sensors in the study were recalibrated at that point. Subsets were calibrated at other times. In each comparison, the offset shown is relative to the initial calibration as ΔT/T (%) versus temperature. Data for the model CX-1070 sensors are considered only over that model's designed 4 K to 325 K temperature range. Below 4 K, self heating occurs in this model as a consequence of the large resistances, obscuring the stability measurement.

Stability dependence upon time. Various subsets of the 39 sensors were calibrated at seven different elapsed times over a 5.8 year period after construction and initial calibration. These elapsed times are 0.27 years (4 samples), 1.1 years (8 samples), 1.6 years (15 samples), 2.2 years (8 samples), 3.0 years (28 samples), 4.4 years (39 samples), and 5.8 years (25 samples). The sensors tested at each point in time were varied to ensure that the total cycling effects on each individual sensor was minimized. The average calibration shift as a function of temperature at each calibration time is shown in Figure 5. Notable is the general trend toward a negative temperature shift at temperatures starting around 50 K and going to room temperature. Since these are negative temperature coefficient devices, this negative temperature shift is indicative of a higher resistance either in the body of the film, in the contacts, or in both. Although there is a shift toward negative temperature, the shift does not grow at a constant rate (i.e. the shift does not correlate well with time). Over the 5.8 year test cycle, the average calibration offsets are less than ±0.02% of temperature below 40 K and ±0.08% of temperature above 50 K at all times tested.

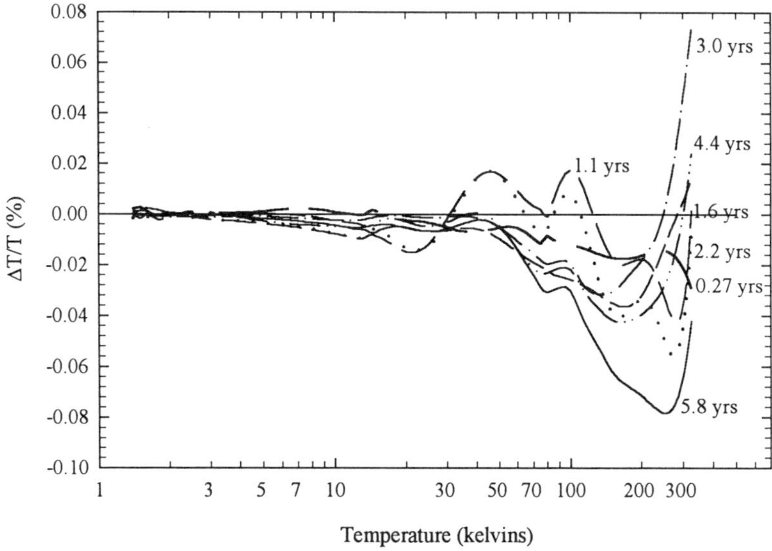

Figure 5. Average calibration shift as a function of temperature at each point in time for Cernox™ RTDs.

Stability dependence upon package. Sensors were packaged in either a copper can (8 samples) or in a flat, hermetically sealed package (31 samples). The average calibration shift as a function of temperature for these two groups of devices 4.4 years after construction and initial calibration is shown in Figure 6. There is a marked difference between long term stability indicated for the two packages. It would be expected that the copper can package would yield a more stable sensor due to decreased stresses transmitted from the package to the sensor. The data shows that sensors packaged in these copper cans have drifted to larger negative temperature shifts with ΔT going roughly as 0.10% of temperature. As mentioned earlier, these are negative temperature coefficient devices, so a negative temperature shift is indicative of a higher resistance either in the body/contacts system. On the other hand, Cernox™ devices packaged in the flat hermetic packages have small deviations around zero with the offset below ±0.05% of temperature over the 4.4 years. This difference might be attributed to better matching of thermal expansion coefficients between the Cernox chip and the flat hermetic package, yielding less stress on the sensor film upon thermal cycling. In the flat hermetic package, a solder attachment is formed between two pieces of metallized sapphire, one of which contains the Cernox™ thin film. An epoxied sapphire-to-BeO attachment is formed for the copper can. It is also possible that the flat hermetic package, with its solder ring seal, yields better hermeticity. This would keep contaminants out of the package and away from the sensor film better than the copper can with its Stycast epoxy seal.

Stability dependence upon wafer. Sensors were fabricated from six different wafers. Each wafer produces about 1300 sensors. The number of samples tested from each wafer (small in some cases) is given in Table 1. The average calibration shifts as a function of temperature are presented in Figure 7 for these six groups of devices 4.4 years after construction and initial calibration. Similar behavior is noted among all of the wafers from 40 K to 325 K, where small negative temperature drifts are typical. Below 40 K, five wafers show a tendency to drift negatively in temperature. The sixth wafer, 2Z255A, also drifted in the negative temperature direction, but to a level roughly three time larger than the other five. Overall, the long term stability for all wafers is better than ±0.25% of temperature below 40 K and better than ±0.10% of temperature above 40 K over the 4.4 year period.

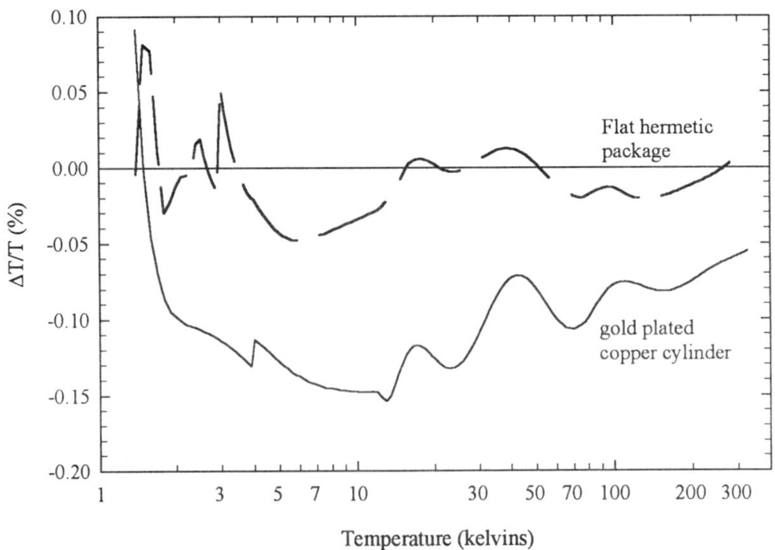

Figure 6. Calibration shift at 4.4 years as a function of temperature for two package types of CXRTs.

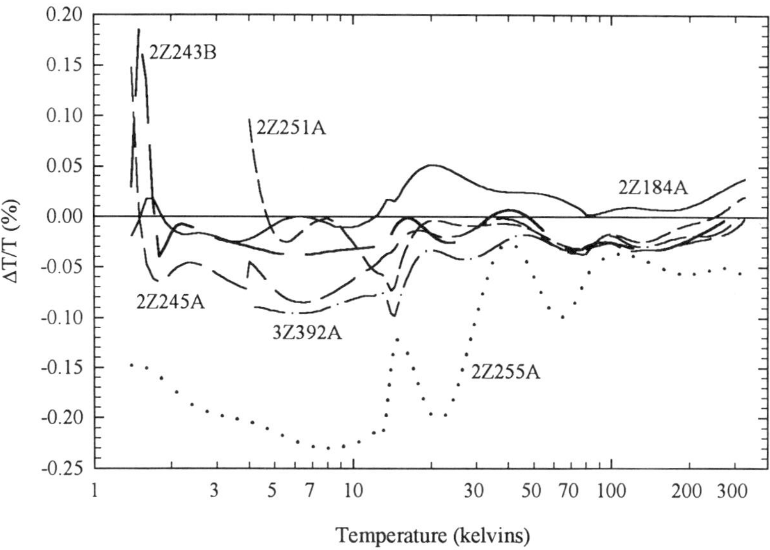

Figure 7. Calibration shift at 4.4 years as a function of temperature for each wafer used in this experiment.

Stability dependence upon sensitivity (model). The sensors were purposely designed with the best sensitivities to cover specific temperature ranges. The Lake Shore model numbers designated for these sensors are CX-1030 (least sensitive, widest range), CX-1050, and CX-1070 (most sensitive, narrowest range). The number of samples of each model type is listed in Table 1. The average calibration shifts as a function of temperature for these three models of Cernox devices 4.4 years after construction and initial calibration are shown in Figure 8. Note that the shape of the offset is very consistent between the three levels of sensitivity. As might be expected, the model CX-1030 with the lowest sensitivity has the highest offset. Again, the negative temperature shift is indicative of an increase in resistance. For all three models, the offset is less than ±0.07% of temperature.

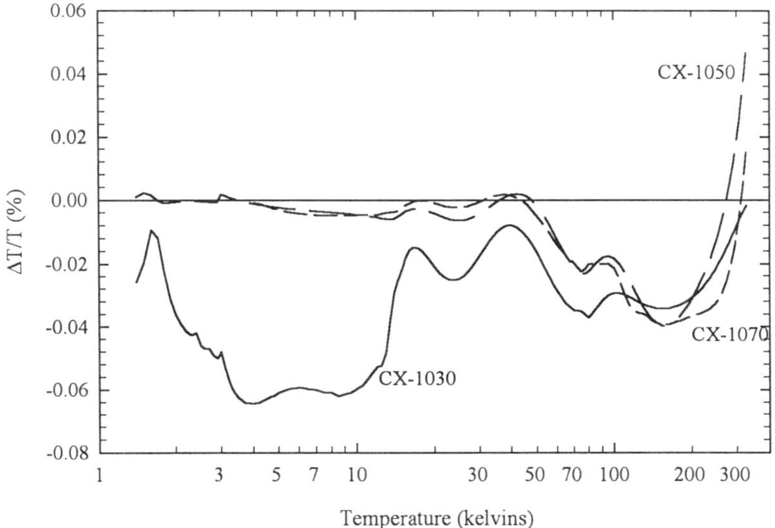

Figure 8. Calibration shift at 4.4 years as a function of temperature for three levels of sensitivity of CXRTs.

CONCLUSIONS

There does appear to be a tendency toward a higher resistance/negative temperature shift with the passage of time with the CXRT sensors stored at room temperature, but the size of the shift does not correlate well with time. Over the 5.8 year test cycle, the average calibration offsets are less than ±0.02% of temperature below 40 K and ±0.08% of temperature above 50 K at all times tested.

At the 4.4 year time mark, the CXRTs mounted in a flat hermetic package were shown to be much more stable (ΔT less than ± 0.05% of temperature) than those mounted in the gold plated can (ΔT less than ±0.1% of temperature). This might be attributed to the better matching of thermal expansion coefficients in mounting the Cernox chip in the flat hermetic package and the better hermiticity of the flat hermetic package compared to the copper can package.

At the 4.4 year time mark, similar long term stability was found among all wafers over the 40 K to 100 K temperature range. Below that temperature, one wafer showed a quantitatively different stability. Overall, the long term stability for all wafers is better than ±0.25% of temperature below 40 K and ±0.10% of temperature above 40 K.

The long term stability was measured to be very similar among all three levels of sensitivity (models) tested. The model with the lowest sensitivity showed the highest offset. For all three models, the offset is less than ±0.07% of temperature.

FUTURE STUDIES

The fabricated Cernox™ sensors for this experiment were inherently two-lead devices subject to calibration shift due to changes in contact resistance. A better design would incorporate a true four-lead measurement which would be independent of the changes in contact resistance. Future studies will incorporate appropriately designed four-lead CXRTs.

REFERENCES

1. T. Yotsuya, M. Yoshitake, and J. Yamamoto, New type cryogenic thermometer using sputtered Zr-N films, *Appl. Phys. Lett.* **51**:235 (1987).
2. Cernox temperature sensors available from Lake Shore Cryotronics, Inc. Westerville, OH, 43082.
3. P.R. Swinehart, S.S. Courts, and D.S. Holmes, Metal oxy-nitride resistance films and methods of making the same, U.S. Patent No. 5367285, 1994.
4. D.S. Holmes, S.S. Courts, and P.R. Swinehart, Electrical properties of metal-oxy-nitride nanocomposities, Proc. of Micro Materials, 1997 (Berlin, Germany), 16-18 April 1997.
5. H.D. Ramsbottom, S. Ali, and D.P. Hampshire, Response of a new ceramic-oxynitride (Cernox) resistance sensor in high magnetic fields, *Cryogenics*. **36**:61 (1996).
6. G. Heine, and W. Lang, Magnetoresistance of the new ceramic "Cernox" thermometer from 4.2 K to 300 K in magnetic fields up to 13 T, *Cryogenics* **38**:377-379 (1998).
7. B. L. Brandt, D.W. Liu, and L.G. Rubin, Low temperature thermometry in high magnetic fields. VII. Cernox™ sensors to 32 T, *Rev. Sci. Instr.* **70**:104 (1999).
8. S.S. Courts, D.S. Holmes, and P.R. Swinehart, Neutron and gamma radiation effects on cryogenic temperature sensors, in: "Temperature, It's Measurement and Control in Science and Industry, Volume 6," J.F. Schooley, ed., American Institute of Physics, New York (1992), p.1237.
9. S.S. Courts and D.S. Holmes, Effects of cryogenic irradiation on Temperature sensors, in: "Advances in Cryogenic Engineering, Volume 41B," P. Kittel, ed., Plenum Press, New York (1996), p.1707.
10. J.F. Amand, J. Casas-Cubillos, T. Junquera, and J.P. Thermeau, "Neutron irradiation tests in superfluid helium of LHC cryogenic thermometers," Presented at ICEC/ICMC 17, 1998 (Bournemouth, England), 14-17 July 1998.

THERMAL RESISTANCES OF CRYOGENIC TEMPERATURE SENSORS FROM 1 - 300 K

S. Scott Courts, W.E. Davenport and D. Scott Holmes

Lake Shore Cryotronics, Inc.
Westerville, Ohio 43082

ABSTRACT

Operation of most temperature sensors requires the dissipation of power in the sensor. The flow of the heat generated by the measurement creates a temperature difference and a potential temperature measurement error. The self-heating temperature difference is directly proportional to the thermal resistance. A procedure for measuring the thermal resistance of a mounted temperature sensor is described.

Thermal resistances were measured at cryogenic temperatures (1 to 300 K) on several commercially available temperature sensors. The sensors were mounted to a copper block in either a vacuum or liquid helium environment. The thermal resistance was found to depend on temperature, thermal environment and details of sensor mounting and packaging.

Minimization of the temperature measurement uncertainty requires a balance between the uncertainties due to self-heating and measurement of the output signal. Equations, corrected from those published previously, are provided for calculating the operating point for minimum combined temperature measurement uncertainty.

INTRODUCTION

Any difference between the temperature of the sensor and the environment the sensor is intended to measure produces a temperature measurement error or uncertainty. Dissipation of power in the temperature sensor will cause its temperature to rise above that of the surrounding environment. Power dissipation in the sensor is also necessary to make a measurement with most temperature sensors (exceptions include thermocouples and optical pyrometers). Minimization of the temperature measurement uncertainty thus requires balancing the uncertainties due to self heating and output signal measurement. The possibility that other experimental considerations might impose more stringent limitations on the power which can be dissipated in the temperature sensor should also be considered.

The self-heating can be affected by changes in the thermal resistance between the sensor and its surroundings due to remounting, temperature dependent properties, or environment. Sensor mounting techniques are reviewed by Sparks.[1] An example of how sensor mounting can affect performance is given by Kobel, et al.[2] Note that cooling history can also have an affect; Kreitman and Callahan determined that Apiezon N-grease surrounded by vacuum has a thermal conductivity almost 50 times higher if cooled in the presence of helium gas rather than vacuum.[3]

Calculation of an effective thermal resistance for a mounted temperature sensor is extremely difficult. Details of construction unknown to a user (e.g. thickness of epoxy used to mount a sensor element within a canister, or thickness and composition of a ruthenium oxide thick-film) can strongly affect the thermal performance. The thermal properties of many commonly used materials are frequently unknown at cryogenic temperatures. The Kapitza resistance becomes important below about 4 K and is difficult to estimate.

Following are two approaches to dealing with the problem of self heating:

1) Choose an excitation that allows acceptable instrumentation measurement uncertainty and check to make sure self heating is negligible at one or two points where it is likely to be most significant. An easy way to check for self heating is to increase the power dissipation and check for an indicated temperature rise. Unfortunately, this procedure will not work with non-linear devices such as semiconductor diodes. An indication of the self heating error can be made by reading the diode temperature in both a liquid bath and in a vacuum at the same temperature, as measured by a second thermometer not dissipating enough power to self heat significantly.

2) Measure the thermal resistance in the temperature range of interest and calculate the optimum operating point.

Designers of cryogenic instrumentation and experiments usually must predict performance before the thermal resistance can be measured. In this case, thermal resistances measured under similar conditions can be used to estimate optimal sensor excitation, self-heating effects, and overall measurement system performance.

Previous measurements of the thermal resistances of cryogenic temperature sensors include those of Rusby[4] on a standards grade rhodium-iron sensor from 0.4 to 280 K; Holmes and Courts[5] on most of the same sensors reported in this paper, but from 0.05 to 10 K; and Dötzer and Schoepe[6] on a thick film resistor from 15 to 460 mK.

THEORETICAL DEVELOPMENT

The self-heating depends on the excitation power according to the equation

$$\Delta T_{sh} = P_s R_t = I^2 R_e R_t = V^2 R_t / R_e \tag{1}$$

where ΔT_{sh} is the temperature rise due to self heating, P_s is the power dissipated in the sensor, I is the excitation current, R_e is the electrical resistance, and R_t is the effective thermal resistance between the sensor and its environment. The thermal resistance is usually temperature dependent, but a linear relationship between power dissipation and temperature rise usually holds for small values of ΔT.

The thermal resistance is extremely difficult to calculate for all but the simplest cases and is best determined experimentally using the following procedure:

1) mount the sensor as it will be used on a temperature controlled block or directly in liquid,

2) record the output voltage as a function of excitation current (*I-V* curve) until significant self heating is observed (when $R_e = V/I$ is no longer constant),

3) replot the data as sensor temperature reading versus power dissipated (*T* versus *P*),

4) fit the data with a linear equation of the form $T = To + R_t P_s$ to find the thermal resistance, R_t . If fitting with higher power polynomials appears helpful, determination of a linear thermal resistance is probably not justified.

The experimentally determined quantity R_t is termed the effective thermal resistance because it relates the measured power dissipation to the temperature rise. In some sensor designs (e.g. germanium or carbon-glass resistors) the actual power dissipated in the sensor is significantly higher than that calculated by multiplying the voltage output by the current through the sensor because the voltage taps only extend across a portion of the resistive element. This does not affect the usefulness of the calculated effective thermal resistance.

Temperature measurement uncertainties u_T arise from many causes. As an example, Equation 1 only applies to self-heating caused by power dissipated within the sensor, but heat conducted along the sensor leads or thermal radiation can produce similar effects. The combined uncertainty u_c resulting from several independent measurement uncertainties is usually estimated by assuming a statistical distribution of uncertainties, in which case the uncertainties are summed in quadrature using the formula

$$u_c = \sqrt{u_1^2 + u_2^2 + \cdots + u_i^2 + \cdots + u_n^2} \tag{2}$$

Information helpful to calculating uncertainties of cryogenic temperature measurements is given elsewhere.[7, 8, 9] Note that attempting to correct for self-heating errors by calculation or extrapolation is not considered good practice. An estimate of the self-heating error should be included in the total uncertainty calculation instead.

A temperature sensor produces some indirect temperature-dependent output such as a voltage, resistance or capacitance. Uncertainties in the measured quantity are not in units of temperature and must be converted before use in Equation 2 . In the case of an output voltage V, the temperature uncertainty u_T is related to the voltage uncertainty u_V by the dimensionless formula

$$\frac{u_T}{T} = \frac{u_V / V}{(T/V)(dV/dT)} \equiv \frac{u_V / V}{S_T} \tag{3}$$

The quantity in the denominator, $S_T = (T/R)(\mathrm{d}R/\mathrm{d}T)$ is defined as the dimensionless temperature sensitivity, where R is the sensor resistance at absolute temperature T. The dimensionless sensitivity is also equal to $\mathrm{d}(\ln R)/\mathrm{d}(\ln T)$, the slope of the voltage versus temperature on a log-log plot. Note that Equation 3 can be made to apply to thermometers based on other temperature-dependent properties (capacitance, voltage or pressure) by replacing R with C, V, or P as appropriate.

Examination of Equation 3 leads to the conclusion that an increase in the sensor output voltage will result in a decreasing temperature uncertainty, so long as the voltage uncertainty remains constant. This is possible with an ohmic sensor by increasing the excitation current. Unfortunately, a larger excitation will dissipate more power in the temperature sensor, raising its temperature above the surroundings.

The optimum operating point minimizes the combined temperature measurement uncertainty due to instrumentation and self-heating. The optimum is determined by a) writing the instrument uncertainty (Equation 3) and the self-heating (Equation 1) in terms of sensor power, b) including these two terms on the right hand side of Equation 2 under the square root as shown in Equation 4, c) differentiating with respect to sensor power - note that all measurement uncertainties proportional to the excitation will drop out, d) setting the result equal to zero, and e) solving for the sensor power. The resulting expression for the optimum sensor power is given in Equation 5.

$$u_c = \sqrt{(PR_t)^2 + \left(\frac{T\sqrt{\sum_i (u_{Vi})^2}}{S_T \sqrt{PR_e}} \right)^2} \tag{4}$$

$$P_{s\,opt} = I^2 R_e = \frac{V^2}{R_e} = \left[\frac{T^2}{2 R_e S_T^2 R_t^2} \sum_i \left(u_{Vi}\right)^2 \right]^{1/3} \tag{5}$$

where the voltage measurement uncertainties have been written as a sum of terms to allow for a variety of statistically independent effects. ONLY uncertainties independent of the excitation should be included in the summation as those terms proportional to the excitation have dropped out. Note that Equation 5 is different from expressions published earlier by Lake Shore [5, 7, 10] because terms proportional to the excitation had been erroneously retained. Also be aware that other experimental considerations might impose more stringent limitations on the power which can be dissipated in the temperature sensor.

EXPERIMENTAL PROCEDURE AND RESULTS

Several commercially available cryogenic temperature sensors were mounted in a standard Lake Shore calibration probe. In some cases, two sensors of the same model were tested, thus providing a check on sensor packaging and mounting variability. The mounting block and calibration procedure are described elsewhere. Apiezon N-grease was used to improve thermal contact between the sensor and the copper calibration block. The sensors were not remounted between tests. Self-heating measurements were made at five temperatures in one of two environmental conditions: either a) vacuum with no helium exchange gas used during cooling, or b) liquid helium or liquid nitrogen surrounding the mounting block. The lowest excitation current used was one tenth that required to develop a 2 mV output signal and was then increased by a factor of $10^{0.05} = 1.122$ until one of the following conditions was reached: a) I = 100 mA, b) P > 100 μW, c) $\Delta T/T > 0.5$, or d) $\Delta T > 10$ K. Roughly 30 to 100 data points were produced for each sensor and temperature, far more than were taken for our previously reported data.[5]

An example of the data taken and the fit to the data is given in **Figure 1**. As the self-heating temperature rise ΔT approaches the temperature of the environment, T, the relationship between the temperature rise and the power is seen to become nonlinear. Linear fits were performed only to data points with $\Delta T/T < 5\%$.

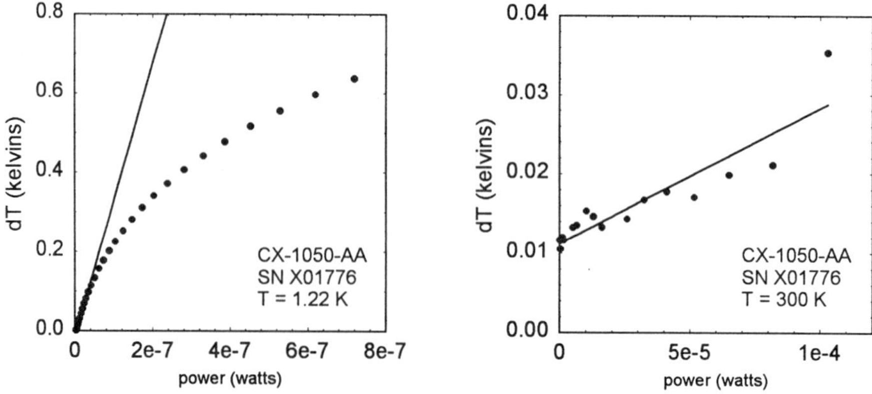

Figure 1. Self-heating temperature rise as a function of power for one sensor at low and high temperature extremes where the fits had the greatest errors. First order fits used to determine the thermal resistance, R_t, are also shown.

Thermal resistance values determined from the first order curve fits are given in **Table 1**. The thermal resistance is found to depend on temperature and the details of sensor mounting. The thermal resistances at 1.22 K differ by a factor of 1000 among the sensors tested.

The thermal resistance is shown as a function of temperature for the germanium resistors in **Figure 2**. The 200A models are mounted in standard gold-plated copper canisters of 3 mm diameter and the 200B models are in smaller, 2.36 mm diameter canisters. The two 200B models tested appear to be much more strongly affected by operation in the 2-3 K temperature range. One possible explanation is that these sensors have significant helium gas remaining inside the canister and the helium is liquefying and forming a highly conductive film on the sensor.

The thermal resistance is shown as a function of temperature for the Cernox resistors in **Figure 3**. The Cernox sensors with the bare chip (-BC) packages had the largest thermal resistances above 10 K. The bare chips are mounted with both grease and cigarette paper, and the paper might contribute significantly to the thermal resistance. The bare chip packages are most useful when immersed in liquid or when very low mass is required.

The two Model CX-1050-AA Cernox sensors had very different thermal resistances, reaching a maximum of about 18x at 2.2 K. These two sensors were measured previously and had nearly equal thermal resistances. The sensor with serial number X01776 appears to have greatly increased its thermal resistance over a period of several years. Additional measurements will have to be done after remounting the sensor to check if mounting was a problem.

Thermal resistances typically differ by less than a factor of two for sensors of the same model. Similar germanium (GR) and carbon-glass (CGR) resistors show more variability than the Cernox™ (CX), possibly because the difference between 2-wire and 4-wire resistances are greater, which would vary the actual power dissipated.

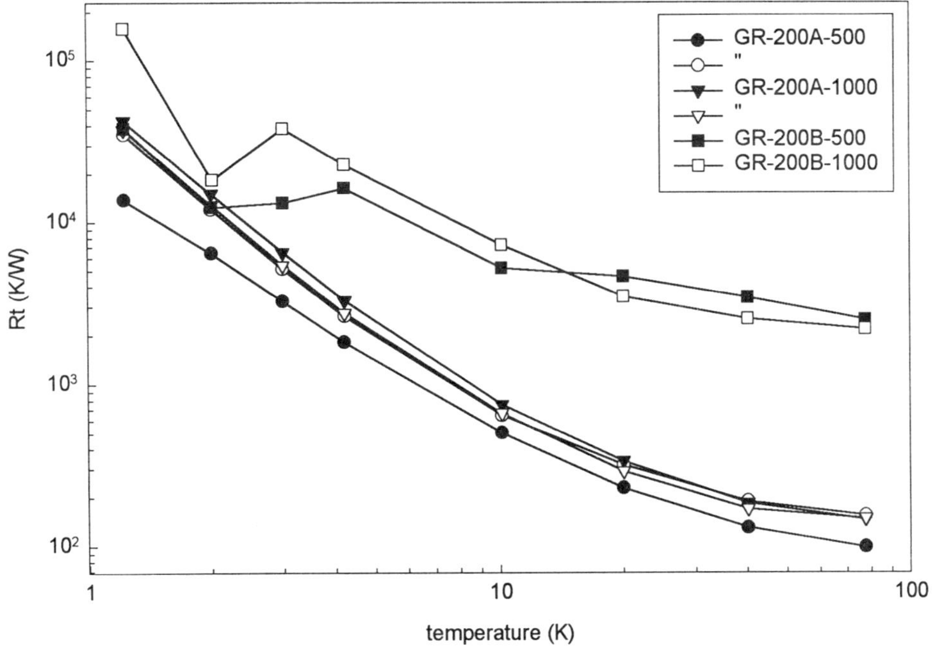

Figure 2. Thermal resistance, R_t, as a function of temperature for several germanium (GR) temperature sensors in vacuum.

Table 1. Thermal resistances calculated using first order polynomial fits to ΔT versus power data for sensors mounted in vacuum

Sensor	serial	thermal resistance, R_t (K/W)									
Model No.	number	1.22 K	2.01 K	2.99 K	4.21 K	10.1 K	20.1 K	40.6 K	77.5 K	150.1 K	300.1 K
CGR-1-1000	C15351	1.580E+05	9.574E+04	6.590E+04	4.985E+04	1.329E+04	6.573E+03	3.254E+03	1.795E+03	1.116E+03	8.068E+02
CGR-1-1000	C15352	1.916E+05	1.167E+05	8.360E+04		1.515E+04	7.604E+03	3.835E+03	2.333E+03	1.464E+03	7.779E+02
CGR-1-500	C15303	2.126E+05	1.358E+05	9.450E+04	7.106E+04	1.405E+04	6.867E+03	3.376E+03	1.894E+03	1.201E+03	9.421E+02
CX-1050-AA	X01770	2.430E+05	3.909E+04	1.170E+04	4.592E+03	8.632E+02	3.343E+02	1.762E+02	1.140E+02	9.231E+01	3.007E+02
CX-1050-AA	X01776	3.451E+06	7.052E+05	1.370E+05	3.741E+04	3.080E+03	7.938E+02	3.124E+02	1.534E+02	1.243E+02	1.705E+02
CX-1050-BC	X09743	2.344E+05	5.565E+04	2.309E+04	1.374E+04	4.688E+03	2.555E+03	2.255E+03	2.374E+03	1.461E+03	3.640E+02
CX-1050-SD	X01748	4.609E+05	7.428E+04	2.121E+04	8.232E+03	1.225E+03	3.634E+02	1.458E+02	6.350E+01	8.377E+01	2.468E+02
CX-1050-SD	X01749	5.731E+05	7.944E+04	2.259E+04	8.988E+03	1.644E+03	6.045E+02	2.601E+02	1.375E+02	1.135E+02	2.406E+02
CX-1070-AA	X01450	5.858E+05	7.314E+04	1.618E+04	5.681E+03	8.123E+02	2.724E+02	1.213E+02	5.815E+01	7.694E+01	1.839E+02
CX-1070-BC	X01466	1.540E+07	1.216E+06	2.605E+05	5.558E+04	3.659E+03	1.025E+03	4.635E+02	3.601E+02	2.407E+02	3.234E+02
CX-1070-SD	X01647	3.451E+06	7.052E+05	1.370E+05	3.741E+04	3.080E+03	7.912E+02	3.124E+02	1.469E+02	1.097E+02	1.451E+02
CX-1070-SD	X01648	3.498E+06	5.925E+05	1.127E+05	3.195E+04	2.939E+03	7.896E+02	3.208E+02	1.766E+02	7.427E+01	1.292E+02
GR-200A-1000	# 26392	4.291E+04	1.517E+04	6.540E+03	3.309E+03	7.561E+02	3.366E+02	1.854E+02	1.472E+02		
GR-200A-1000	# 26395	3.782E+04	1.290E+04	5.431E+03	2.763E+03	6.674E+02	2.902E+02	1.702E+02	1.492E+02		
GR-200A-500	# 26352	1.381E+04	6.474E+03	3.258E+03	1.827E+03	5.052E+02	2.297E+02	1.306E+02	9.950E+01		
GR-200A-500	# 26353	3.530E+04	1.214E+04	5.179E+03	2.649E+03	6.494E+02	3.178E+02	1.896E+02	1.554E+02		
GR-200B-1000	#25891	1.586E+05	1.857E+04	3.848E+04	2.308E+04	7.282E+03	3.486E+03	2.537E+03	2.204E+03		
GR-200B-500	# 25893	3.847E+04	1.250E+04	1.333E+04	1.650E+04	5.229E+03	4.640E+03	3.440E+03	2.514E+03		
RO104	# 854	3.440E+06	2.245E+06	1.468E+06	3.882E+05	3.043E+05	1.290E+05				
RO600	# 856		1.303E+04	6.368E+03	3.653E+03	1.295E+03	6.347E+02	3.255E+02	2.367E+02	3.827E+02	

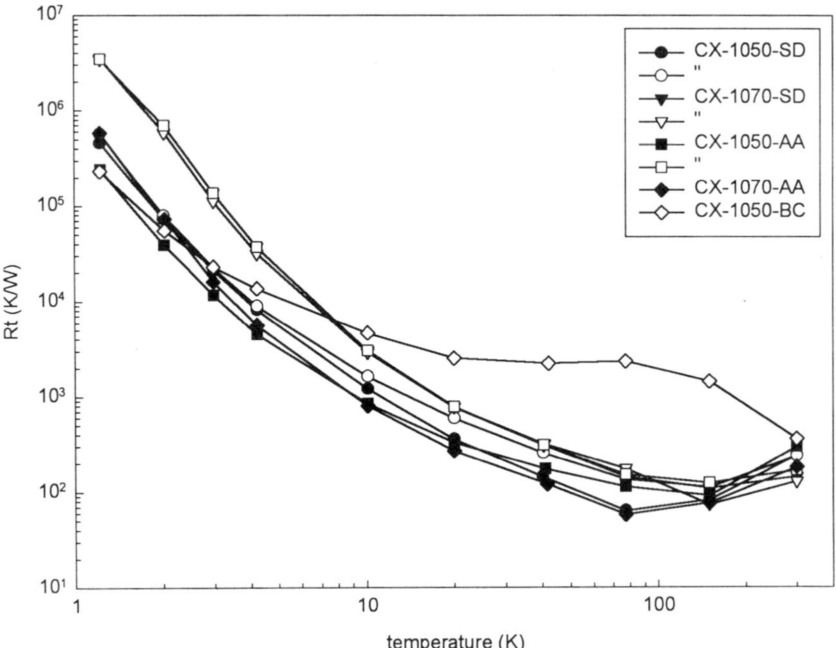

Figure 3. Thermal resistance, R_t, as a function of temperature for several Cernox (CX) temperature sensors in vacuum.

CONCLUSIONS

Measurement of the thermal resistance is not difficult and should be standard procedure when making critical cryogenic temperature measurements. The thermal resistance is sensitive to construction and mounting details in sometimes surprising ways. More data is needed on variations in thermal resistances for other sensors, other mounting configurations, and on remounting of sensors. The thermal resistance can be used to improve the accuracy of temperature measurements by allowing calculations of excitation conditions necessary to minimize the overall measurement uncertainty.

EXAMPLE CALCULATION

As an example of a thermal resistance calculation and excitation optimization, consider a Cernox model CX-1050-SD temperature sensor attached to a copper block operating at 1.6 K and surrounded by vacuum. The voltmeter to be used to make the measurement is a Keithley Instruments model 2000 with a one year dc voltage accuracy specification of ±(50 ppm + 3.5 μV). The sensor electrical resistance R_e = 46 240 Ω and dimensionless sensitivity S_T = 2.21 come from the Cernox data plots.[8] The thermal resistance R_t = 150 000 K/W from **Table 1** is about twice that of the CX-1050-AA model shown in **Figure 3**. From the voltmeter accuracy specification, $uVlin/V$ = 50 ppm = 5e-5 with an additional offset voltage uncertainty $uVoff$ = 3.5 μV. Assume an additional noise voltage $uVnoise$ = 1 μV. The following values can be calculated:

$$P_{s\,opt} = \left[\frac{(1.6\mathrm{K})^2}{2(46240\,\Omega)(2.21)^2(150000\,\mathrm{K/W})^2} \left[(3.5\mathrm{x}10^{-6}\ \mathrm{V})^2 + (1\mathrm{x}10^{-6}\ \mathrm{V})^2 \right] \right]^{1/3}$$

$$P_{s\,opt} = 1.5\text{x}10^{-9}\ \text{W} = 1.5\,\text{n}W$$

$$I = \sqrt{\frac{P_{s\,opt}}{R_e}} = \sqrt{\frac{1.5\text{x}10^{-9}\ \text{W}}{46240\,\Omega}} = 1.8\text{x}10^{-7}\ \text{A} = 0.18\,\mu\text{A}$$

$$V = IR_e = (1.8\text{x}10^{-7}\ \text{A})(46240\,\Omega) = 0.0083\ \text{V} = 8.3\,\text{mV}$$

$$u_{Tsh} = \Delta T_{sh} = P_s R_t = (1.5\text{x}10^{-9}\ \text{W})(150000\,\text{K / W}) = 224\,\mu\text{K}$$

$$u_{Tlin} = \left(\frac{T}{S_T}\right)\left(\frac{u_{Vlin}}{V}\right) = \left(\frac{1.6\,\text{K}}{2.21}\right)\left(5\text{x}10^{-5}\right) = 36.2\,\mu\text{K}$$

$$u_{Toff} = \left(\frac{T}{S_T}\right)\left(\frac{u_{Vi}}{V}\right) = \left(\frac{1.6\,\text{K}}{2.21}\right)\left(\frac{3.5\text{x}10^{-6}\ \text{V}}{0.0083\ \text{V}}\right) = 305\,\mu\text{K}$$

$$u_{Tnoise} = \left(\frac{T}{S_T}\right)\left(\frac{u_{Vi}}{V}\right) = \left(\frac{1.6\,\text{K}}{2.21}\right)\left(\frac{1\text{x}10^{-6}\ \text{V}}{0.0083\ \text{V}}\right) = 87.2\,\mu\text{K}$$

$$u_T = \sqrt{u_{Tsh}^2 + \sum_i \left(u_{Ti}^2\right)} = \sqrt{224^2 + 36^2 + 305^2 + 87^2} = 390\,\mu\text{K}$$

$$\frac{u_T}{T} = \frac{390\,\mu\text{K}}{1.6\,\text{K}} = 2.4\text{x}10^{-5} = 240\,\text{ppm}$$

Note that this calculated optimal sensor excitation of 8.3 mV is above the 1-3 mV range conservatively used in the Lake Shore calibration facility where sensors must be removable and thus can not be thermally sunk as well as in a permanent application. Note also that the uncertainty due to voltmeter nonlinearity, u_{Vlin}, is proportional to the excitation, so is not included in the calculation of the optimum power, but does affect the overall temperature measurement uncertainty.

REFERENCES

1. L.L. Sparks, Temperature, strain, and magnetic field measurements, in: "Materials at Low Temperatures," R.P. Reed and A.F. Clark, eds., American Society for Metals, Metals Park, Ohio (1983), pp. 515-521.
2. T.A. Kobel, M.A. Kozyrczak, S.W. Schwenterly and W.M. Bell, Effects of mounting methods on temperature sensor accuracy below 10 K, in "Supercollider 4," J. Nonte, ed., Plenum Press, New York (1992) 619-626.
3. M.M. Kreitman and J.T. Callahan, Thermal conductivity of Apiezon N grease at liquid helium temperatures, Cryogenics 10 (1971) 155-159.
4. R.L. Rusby, The rhodium-iron resistance thermometer: Ten years on, Temperature: Its Measurement and Control in Science and Industry, Vol. 5, part 2, J.F. Schooley, ed., American Inst. Phys., New York (1982) 829-833, Fig. 3.
5. D.S. Holmes, S.S. Courts, Thermal resistances of mounted cryogenic temperature sensors, "Advances in Cryogenic Engineering," Vol. 41B, Plenum Press, New York (1996), pp. 1699-1706.
6. R. Dötzer and W. Schoepe, Thermal impedance between a thick-film resistor and liquid helium below 1 K, Cryogenics 33 (1993) 936-937.
7. D.S. Holmes and S.S. Courts, Cryogenic instrumentation, in: "Handbook of Cryogenic Engineering," J.G. Weisend II, ed., Taylor & Francis, Philadelphia (1998), pp. 203-257.
8. Temperature measurement resolution and uncertainty, in: "Temperature Measurement and Control," Lake Shore Cryotronics catalog (1999) A-47.
9. Guide to the Expression of Uncertainty in Measurement, ISO/TAG 4/WG 3: June 1992, International Standards Organization, Genève, Switzerland.
10. Temperature measurement accuracy, in: "Temperature Measurement and Control," part 1, Lake Shore Cryotronics catalog (1995) A-50.

CAPACITIVE FORCE SENSORS FOR PRESTRESS IN ACCELERATOR MAGNET COILS

L. Vieillard, B. Gallet, C. Gourdin and P. Védrine

DSM/DAPNIA/STCM
CEA/Saclay
GIF SUR YVETTE, 91191, France

ABSTRACT

The capacitive force sensors are set up to determine the compressive stress in the collared coil assemblies of superconducting quadrupole magnets throughout assembly and cold testing.

The sensors are designed as printed circuits, consisting of an alternated stack of dielectric and electrodes. The dielectric is a polyimide film without adhesives, and the electrodes are made of copper. Compressive stresses bring about a variation in the thickness of the dielectric and thus, capacitance. The calibration curves determine the force as a function of the capacitance and the temperature. The first results are encouraging as the sensors exhibit near linear capacitance-force curves, a low hysteresis and a good accuracy. The sensors are insensitive to magnetic field. The paper describes the design of the sensors, reviews the results of characterization and shows an application.

INTRODUCTION

Two quadrupole magnet prototypes for the Large Hadron Collider (L.H.C.) are being developed and built at CEA/Saclay in collaboration with CERN[1]. To validate the main concepts of the mechanical design, the stresses inside the magnets must be measured during each step of the assembly and the tests of the prototypes.

This measurement is a difficult task. The main difficulty is to integrate a sensor inside a very complex mechanical assembly without modifying the mechanical behavior. The capacitive gauges offer a simple and reliable solution because of their small thickness and their ability to be easily shape-designed. Studies on capacitive gauges were first performed at the SSCL[2]. Further developments have been made at C.E.R.N.[3] and the results achieved were very promising. CERN has transferred its technology to the CEA in the framework of the LHC collaboration.

Advances in Cryogenic Engineering, Volume 45.
Edited by Shu *et al.*, Kluwer Academic / Plenum Publishers, 2000.

The principle of such sensors is to use the effect of the variation of the dielectric thickness on the capacitance of a gauge when submitted to a compressive force. The capacitance of a dielectric material of thickness δ and electric permittivity ε placed between two parallel plane electrodes of area S is given by (without fringe field effect) :

$$C = \frac{\varepsilon S}{\delta} \tag{1}$$

With a mechanical loading in compression, the thickness of the dielectric is reduced and the capacitance becomes :

$$C = \frac{\varepsilon S}{\delta - \Delta\delta} \tag{2}$$

If the gauge material behavior is linear elastic, we can rewrite the equation (2) as a function of applied stress σ and the modulus of elasticity (Young's modulus) E :

$$C = \frac{\varepsilon S}{\delta\left(1 - \frac{\sigma}{E}\right)} \tag{3}$$

The sensivity P of the gauge is then given approximately by (with $\varepsilon \ll 1$):

$$P = \frac{dC}{d\sigma} = \frac{\varepsilon S}{\delta E} \tag{4}$$

To manufacture reliable and reproducible capacitive gauges, a new technique of fabrication based on the printed-circuit technology has been studied and developed. More than 70 capacitive sensors have been produced with this technique.

They must work in a range from 1.8K to 300K, from 0 to 150 MPa and in a range of magnetic fields from 0 to 10T. A calibration process allows characterization of its valid domain. These capacitive gauges have been calibrated and characterized at 300K, 77K and 4.2K. Finally these sensors have been inserted into a short model of superconductiong magnet.

FABRICATION

The technology used before was essentially based on a manual technique of fabrication. This method can introduce the following potential risks : bad reproducibility and homogeneity, dielectric constant of the adhesive different from the dielectrics of the polyimide, differences between theoretical and measured capacitance, risk of short circuit.

As the gauges must work inside magnets, the material used for the electrodes shouldn't be ferromagnetic. The dielectric material must stay in the elastic region under the maximum stresses (150MPa in our application) applied to the transducer.

To increase the sensitivity P (Eq.(4)), the gauge must use materials with high dielectric constant ε and low modulus of elasticity E. A new design of capacitive gauges which allows an industrial method of fabrication is described hereafter.

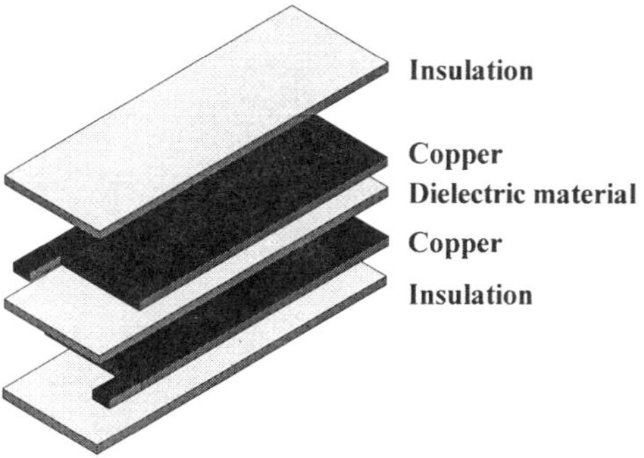

Figure 1. Layout of an elementary capacitive force sensor

The elementary gauge consists of a laminate (Pyralux ® AP [4]), two copper foils (electrodes) with, as dielectric, a polyimide film bonded between (Kapton® EKJ [4]) without any other adhesives (Fig.1). Copper was chosen because it is the most common material used by the printed-circuit manufacturer. The Kapton® EKJ is a self-Adhering Polyimide composite film with a polyimide core and a polyimide adhesive layer which can be bonded at temperature ranging from 300-350°C and under a 1.3 MPa pressure. The geometry is obtained by using a printed-circuit process. A chemical or ultraviolet attack through masks gives the shape of the electrodes. The width of the dielectric is slightly wider than the width of the copper to prevent contacts between electrodes. An advantage of this process is the ability to obtain any geometry such as rectangular, circular or more complex one. Several elementary layouts can be arranged on one larger printed-circuit laminate, chemically-attacked and individually cut.

The thickness of the different constituents are: 17 μm for the copper foil, 50 μm for the polyimide films (Kapton® EKJ), and 50 μm for the external insulation (Pyralux® LF Coverlay [4]). This sheet adhesive has a thickness of 50 μm and is used instead of external insulation between gauges. The thickness of an elementary sensor (Fig. 2) is 184 μm.

To increase the sensitivity, considering the available space in the magnet, a capacitive force gauge can be designed with several elementary sensors. In our case, four layers are used. An acrylic glue (Pyralux® LF Sheet adhesive [4]) bonds the different layers, but doesn't play any role in the dielectric constant. The curing of the glue is performed at 190°C under a pressure of 1.4-2.8 MPa during one or two hours. The glue has been chosen to have a curing temperature well below the bonding temperature of the Kapton® EKJ. The dimensions of the gauges are about 600μm thick and $3065 mm^2$ in surface.

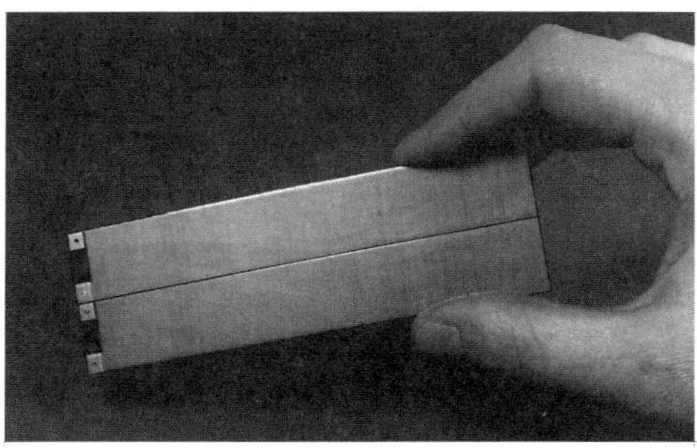

Figure 2. Layout of an elementary capacitive force sensor

The connections between the layers are made with plated-through holes. Once drilled the holes are cleaned with a process called active sputtering to avoid bad contacts. The leads are soldered with a SnPb solder. The external shape of the gauge is cut with a tool, leaving an extra-part of insulation.

At reception, the gauges have a capacitance (~6.95 nF) very close to the theoretical value (- 0.1%), a good reproducibility on a production of more than 70 gauges (capacitance ± 0.4%) and a good homogeneity (surface evenness ± 10μm i.e ~2% of the thickness). A specific calibration process is performed with each sensor.

Before calibration, each gauge is cycled 20 times up to 150 MPa, then kept at 80 MPa during 2 hours, and finally cycled again 20 times. After cycling the capacitance is slightly increased, as is the difference between gauges.

CALIBRATION

An hydraulic press (200 t) is used to calibrate and characterize the capacitive force sensor. A special apparatus was designed and is shown in Fig. 3. In order to acquire signals of capacitance and stress, a system based on a LCRmeter (Hewlett-packard 4284A), an analogue multiplexer and a computer with Labview® program, is used.

Typical calibration curves (capacitance versus stress) are shown Fig 4. At the beginning (stress less than 2 MPa), the curves have a small non-linearity. But after, the gauges work linearly with a low hysteresis (3 MPa at 300K, less than 1 MPa at 77K and 4K). The initial value of the capacitance depends also on the electrical environment (surrounding conducting materials, etc..) but the sensitivity does not.

The zero-shift of the gauge between 300K and 4K is only 70 pF. The sensitivity of the gauges (~1.36 pF/MPa with S = 3065 mm^2) at 300 K is close to the theoretical one (–5%). The variation of the average sensitivity for different gauges is within 0.5 % for 20 sensors and 1% for 35 other sensors. The reduction of the sensitivity between 300K and 77K is about 16% and about 22% between 300K and 4.2K (Fig. 4) (four sensors have been cold tested).

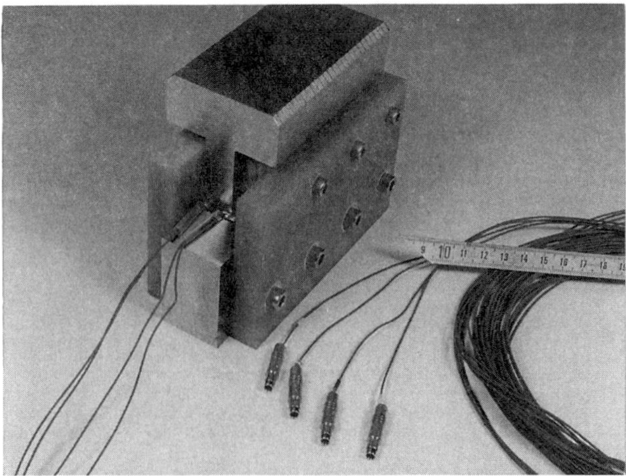

Figure 3. Special apparatus used to calibrate and characterize the capacitve force sensor (S = 3065 mm^2)

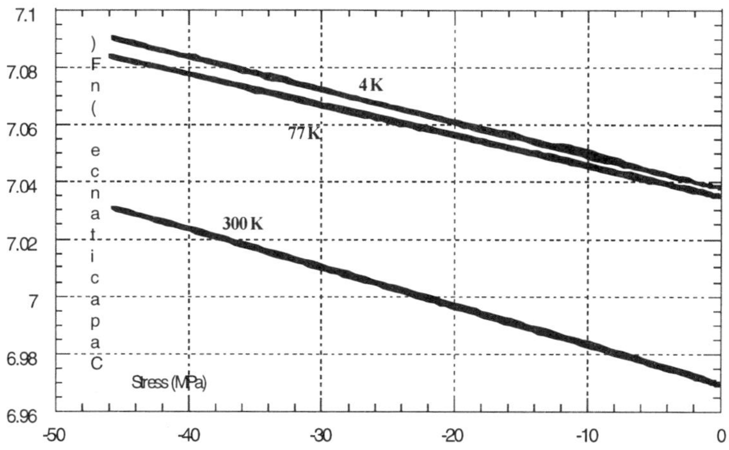

Figure 4. Typical calibration curve of a capacitive force sensor at different temperatures

Creep tests have shown a good stability of the measurements under constant stress for hours. The maximum of variation observed is about 3 MPa (during two hours). The gauges are not disturbed by the magnetic fields at 300K, 77K and 4.2K.

The repeatability of the measurement on one gauge is shown Fig. 5. For each stress level, the measurement has been repeated 10 times. Finally the gauges give an ultimate precision of stress better than 3 MPa at 300K, and better than 1 MPa at 77K and 4K.

APPLICATION

Capacitive gauges have been used at CEA/Saclay on one application related to the prototype for the arc LHC quadrupole magnet [5]. The aim of this application is to determine the pre-stress into the coils during the collaring process and the cool down at liquid nitrogen temperature. This study is made on a small scale model (Fig. 6) (200 mm length and full scale section).

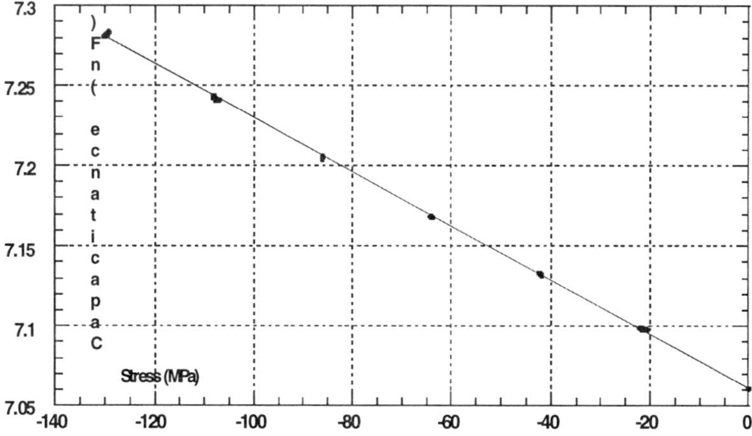

Figure 5. Repeatability curve of a capacitive force sensor at 300K (10 measurements for each step of stress)

Figure 6. The short model of prototype quadrupole magnet with capacitive force sensors

The model is instrumented with eight gauges. Each gauge can measure the pre-stress applied on the inner and the outer layer separately. Two tests have been made using coils with different azimuthal dimensions.

The collaring process is made in three stages. The first stage consists of pressing the side frame to push radially the collars. The second is the pressing on intermediate keys. And the last stage is pressing on the final keys. The evolution of the average pre-stress during these three stages is illustrated on Fig.7.

After the collaring process, the average pre-stress level in the first collared coil is equal to 96 MPa and to 89 MPa for the second collared coil. Fig. 8 shows that the variation of the pre-stress between coils in each octant of the model is less then 10%. The test results show that the differences between pre-stress measured on the inner and on the outer layer are negligible.

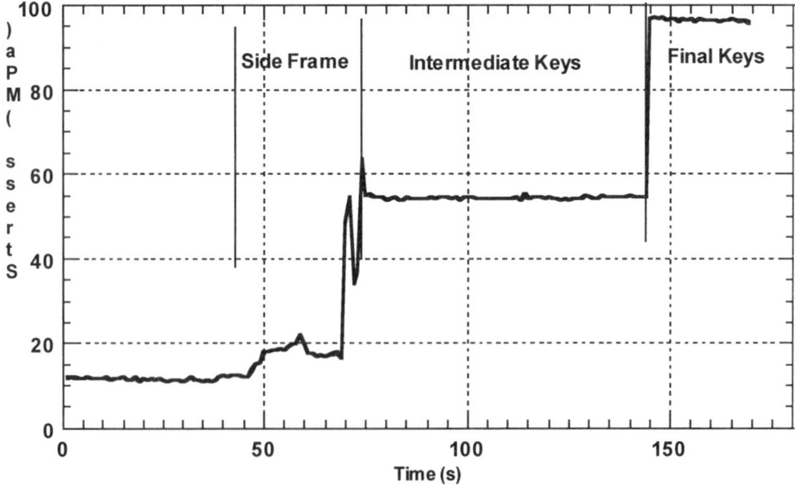

Figure 7. Evolution of the average pre-stress during the collaring process

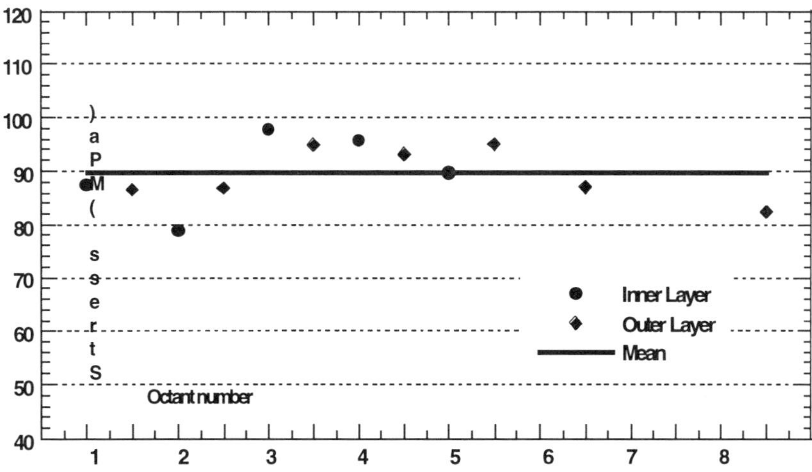

Figure 8. Variation of the coil pre-stress measured by capacitive gauges after the collaring process in each octant of the model.

After assembly, the short models were cooled down to liquid nitrogen temperature. The loss of pressure was 23 MPa for the first coil and 18 MPa for the second coil. The variation of the pressure was similar to the collaring prestress, i.e. less than 10%. The difference between inner and outer layer are also negligible.

CONCLUSION

A manufacturing process of capacitive force sensors using well-known printed circuit techniques has been developed and used. The quality and the reproducibility of the gauges have been established through the manufacturing and the calibration of more than 70 sensors. It is planned to use such gauges in the full length prototype quadrupole magnet to monitor the pre-stress all along the operations.

ACKNOWLEDGEMENT

The authors would like to thank all the technical staff of Saclay who have been involved in the tests and specially D. Thomas and also the company ATI, 6 rue Jean Mermoz, F-91080 Courcouronnes, France for its help in the manufacturing of the gauges.

REFERENCES

1. J.M. Rifflet, M. Peyrot, P. Vedrine, P. Rohmig, T. Tortschanoff., "Status of the fabrication and test of the prototype LHC lattice quadrupole magnets", IEEE Trans. Magn., in press.
2. C.E. Dickey, Bulk Modulus Capacitor Load Cells, SSC Laboratory, Plenum Press, New York (1990)
3. N. Siegel, D. Tommasini and I. Vanenkov, Design and Use of Capacitive Force Transducers for Superconducting Magnet Models for the LHC, LHC Project Report 173, Geneva, (1998)
4. DuPont de Nemours S.A, Contern, L-2984 Luxembourg
5. P. Védrine, J.M. Rifflet, J. Perot, B. Gallet, C. Lyraud, P. Giovannoni and F. LeCoz, Mechanical Tests on the Prototype LHC Lattice Quadrupole, IEEE Transactions on Magnetics, 30:2475 (1994)

NEUTRON IRRADIATION TESTS OF PRESSURE TRANSDUCERS IN LIQUID HELIUM

J.F. Amand,[1] T. Bager,[1] J. Casas-Cubillos,[1] and J.P. Thermeau[2]

[1] CERN-LHC Division
CH-1211, Geneva 23, Switzerland
[2] Institut de Physique Nucléaire d'Orsay, CNRS/IN2P3
Orsay, 91406 Cedex, France

ABSTRACT

The superconducting magnets of the future Large Hadron Collider (LHC) at CERN will operate in pressurised superfluid helium (1 bar, 1.9 K). About 500 pressure transducers will be placed in the liquid helium bath for monitoring the filling and the pressure transients after resistive transitions. Their precision must remain better than 100 mbar at pressures below 2 bar and better than 5% for higher pressures (up to 20 bar), with temperatures ranging from 1.8 K to 300 K. All the tested transducers are based on the same principle: the fluid or gas is separated from a sealed reference vacuum by an elastic membrane; its deformation indicates the pressure. The transducers will be exposed to high neutron fluence (2 kGy, 10^{14} n/cm^2 per year) during the 20 years of machine operation. This irradiation may induce changes both on the membranes characteristics (leakage, modification of elasticity) and on gauges which measure their deformations. To investigate these effects and select the transducer to be used in the LHC, a neutron irradiation program is being performed at the CERI cyclotron (CNRS Orléans, France): a cryostat is installed on a beam line, transducers are immersed in liquid helium and irradiated by neutrons (1-20 MeV, 10^{15} n/cm^2). The tested transducers measure the helium bath pressure, the true value of which is given by a warm, unirradiated sensor. Every readout is acquired on-line. This paper presents the results of the first experiments performed during spring, 1999.

INTRODUCTION

For controlling the cryogenic parameters of the LHC magnets, the inner pressure of the helium loop is an important measurement. It allows the control of vessel filling, and can help in magnet quench propagation studies. About 500 pressure transducers are needed to monitor the pressure of cryogenic vessels around 27 km of LHC ring. The aim of this study is to investigate the possibility of immersing the transducers in the liquid helium of the cold masses. They must work in liquid helium and gas, at a temperature ranging from 1.8 K to 300 K, and must keep their precision better than 100 mbar below 2 bar, and better than 5% for higher pressure (up to 20 bar). Two types of problems may be encountered with cold transducers. The first is their intrinsic behaviour in liquid helium at cryogenic temperature, because most of the transducers were not primarily designed to be used for such conditions; this is the object of another paper [2]. The second possible problem, which is the subject of this paper, is the effect of high neutron fluence: if placed in the cold mass, close to the quadrupoles, the transducers may receive about 10^{14} n/cm^2 per year (2 kGy neutrons) [1]. A neutron irradiation campaign is being performed to investigate the possible effects. To be as close as possible to the LHC conditions, the sensors are irradiated by neutrons at low temperature, immersed in liquid helium.

TESTED TRANSDUCERS

All the tested transducers are based on the same principle (Figure 1). The pressure is deduced from the flexion of an elastic membrane placed between the measured fluid and a reference vacuum. The displacement of the membrane is measured by resistive strain gauges placed on its surface on three different types. The transducer contains built-in passive electronic devices to supply the power to the strain gauges and deliver an analog signal. Several sensors, the signals of which are strongly temperature dependent, contain an integrated thermometer. The possible effects that may occur under irradiation are leakage of the reference vacuum, degradation of the strain gauges, modification of membranes elasticity and problems with built-in electronics or temperature sensors. Four types of commercial transducers have been tested in the irradiation facility (Table 1). However some modifications were necessary: the stainless steel bodies are incompatible with neutron irradiation because of the high activation of this material. The stainless steel of the tested samples has been replaced by low-activation metal (aluminium or copper-beryllium alloy).

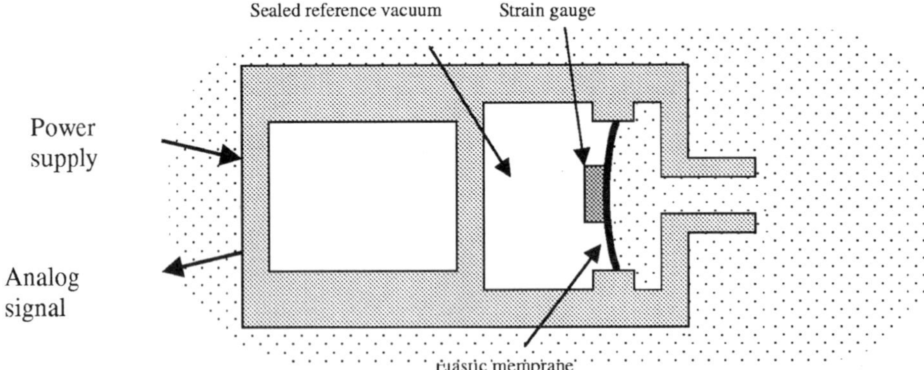

Figure 1. Scheme of a generic cold pressure transducer. On some sensors, the strain gauge may be replaced by an inductive sensing system.

Table 1. List of tested pressure transducers

	Sensor 1	Sensor 2	Sensor 3	Sensor 4
Membrane	Metal (W1.4542)	Metal (Inconel)	SiO_2	SiO_2
Body (modified)	Al	Cu-Be alloy	Al	Al
Built-in Thermometer	Pt resistor	-	Si diode	-
Dimensions D × l (mm)	20 mm × 50 mm	50 mm × 15 mm	20 mm × 40 mm	15 mm × 40 mm
Sensing principle	Metallic resistive strain gauge	Inductive	Piezoresistive strain gauge	Metallic resistive strain gauge
Number of tested samples	2	1	2	1

EXPERIMENTAL FACILITY

The irradiation facility is composed of a cryostat (Figure 2) installed on a beam line of the CERI cyclotron (Orléans, France). This set-up reproduces partially the LHC conditions: the samples are irradiated at low temperature (immersed in liquid helium, from 1.7 K to 4.2 K); the total neutron dose is equivalent to that expected in the machine. An on-line acquisition system takes the data during the whole experiment.

Irradiation Aspect of the Facility

The neutrons are generated by a deuteron beam (20 MeV, up to 35 μA) colliding a beryllium target. The $^9Be + d \rightarrow {}^{10}B + n$ reaction produces neutrons with energy between 1 and 20 MeV and low gamma dose (20%). The total dose of 10^{15} n/cm^2 (equivalent of 10 years of LHC operation at full intensity) can be attained in 20 hours of exposure. The dose is measured by small nickel foils (mass: 15 mg), the neutron activation of which gives the total fluence.

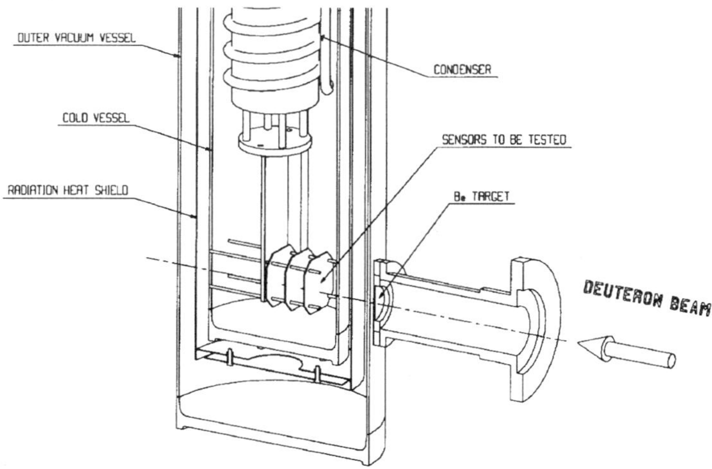

Figure 2. The irradiation cryostat.

Cryogenic Aspects

The cryostat is made of a liquid helium vessel (diameter 27 cm, height 60 cm), a radiation heat shield and an outer vacuum vessel. The set-up is fabricated of either aluminium or copper to avoid activation problems. The bath can be filled at 4.2 K and cooled to 1.8 K by pumping. A PID controlled valve regulates the bath pressure. Unirradiated sensors measure the bath pressure. The temperature is obtained both by a helium bulb and resistive thermometers. The irradiation cryostat, in its actual configuration, cannot operate in pressurised helium. The sensors are limited to measure only the bath saturation pressure. At 1.8 K, the pressure is then 60 times below LHC operating conditions.

EXPERIMENTAL PROCEDURE AND RESULTS

Irradiation at low Temperature

The effects during irradiation in saturated helium at 4.2 K (1 bar) and 1.8 K (16.4 mbar) are investigated. A preliminary in-situ calibration allows deduction of the correspondence between signal and measured pressure. The presented data concern only the latest irradiation, performed in June 1999. The first beam lasted eight hours for a total neutron dose of 3.4×10^{14} n/cm^2 (T=4.2 K); the second beam, performed at 1.8 K, induced a neutron dose of 1.6×10^{14} n/cm^2 in three hours of irradiation; and the last beam was performed at 4.2 K for 2.5×10^{14} n/cm^2 in five hours (Table 2).

Sensor 1. No visible effect could be observed during the irradiation of this sensor.

Sensor 2. We observe two phenomena (Figure 3): when the beam starts, the measured pressure increases by about 7 mbar with an exponential saturation curve ($y = y0 + A.(1-e^{-t/\tau})$); then drops continuously; this descent does not stop with the beam. It indicates a radiation-induced leak on the reference vacuum, the inner pressure of which tends to reach the outer pressure, resulting in a lower pressure difference and a decrease of the signal. This leak makes the analysis of the following runs difficult as the pressure varies from 1 bar to 16 mbar.

Table 2. The experiment performed in June 1999

Event	Temperature	Pressure	Duration	Accumulated neutron dose
Warm calibration	300 K	1 bar => 4 mbar	-	0
Cold calibration	4.2 K => 1.8 K	1 bar => 16 mbar	-	0
Irradiation 1	4.2 K	1 bar	8 hours	3.4×10^{14} n/cm^2
Irradiation 2	1.8 K	17 mbar	3 hours	5×10^{14} n/cm^2
Irradiation 3	4.2 K	1 bar	5 hours	7.5×10^{14} n/cm^2
Cold calibration	4.2 K => 1.8 K	1 bar => 16 mbar	-	
Warm calibration	300 K	1 bar => 4 mbar	-	

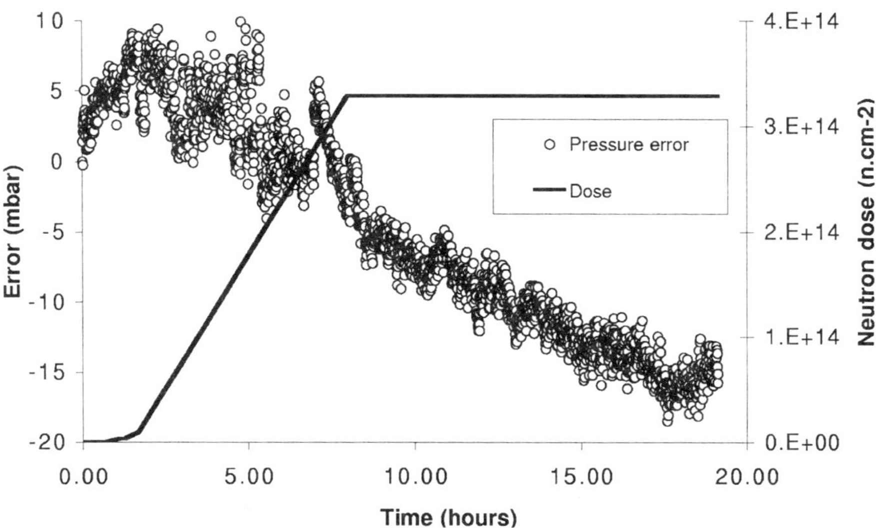

Figure 3. Influence of neutron irradiation on sensor 2 when irradiated at 4.2 K. The long descent of the measured pressure indicates a leak in the sealed reference vacuum.

Sensor 3. The signal of this sensor drops when the beam is on (Figure 4). The descent follows an exponential saturation curve. A high influence of neutron irradiation on the internal diode-type thermometer of this transducer is observed: its read-out decreases proportionally to the neutron dose. No leak appears on this sensor. After this first drop at 4.2 K, the sensor seems "saturated", and no further radiation effect occurs.

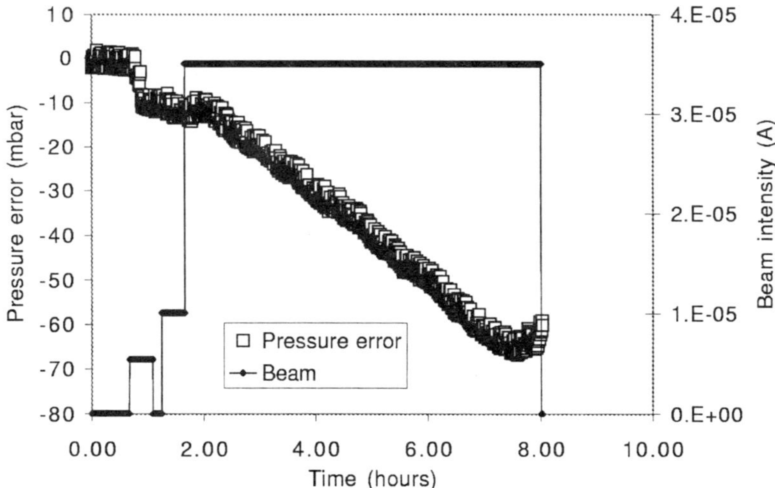

Figure 4. Neutron irradiation of sensor 3, first beam performed at 4.2K. The measured pressure drops when the beam is on

Sensor 4. A high noise level is measured on this sensor (σ_P = 10 mbar at atmospheric pressure), and its calibration was not possible in superfluid helium, because of the important nonlinearity of the transducer readout with the temperature. At 4.2 K a decrease (-30 mbar) of the measured pressure with a saturation curve is observed. After the irradiation, the sensor recovers its initial reading in less than one hour. The same kind of behaviour is observed at 1.8 K. No leak was observed at low temperature.

Variation of Calibration Curve at Low Temperature

The transducers were calibrated in saturated liquid helium before and after the irradiation (between 1.8 K and 4.2 K, corresponding to a bath pressure ranging from 19 mbar to 1 bar). As the transducer read-outs are temperature dependant, this experiment cannot be considered as a pressure calibration which would have to be performed in pressurised liquid helium. It is just a qualitative appreciation of the radiation stability of the transducer's characteristics. The total neutron dose received between these calibrations is 7.5×10^{14} n/cm^2 (15 kGy) in16 hours of irradiation. The curves of Figure 5 show the results of these calibrations. The effects are low on sensors 2 and 3, which just show a small offset on the output signal. For sensor 1, the signal at 1 bar is almost unchanged, but the curve is modified for lower pressures.

Behaviour of the Transducers after Final Warm-up

After irradiation, a warm calibration is performed (T=300 K). Three of the measured sensors are leaking, which makes them unusable (Figure 6). The shift of the sensors, the sensitivity (estimated with a quick 1-0-1 bar pressure cycle) and the time constant of the leaks are measured. All results are summarised in table 3.

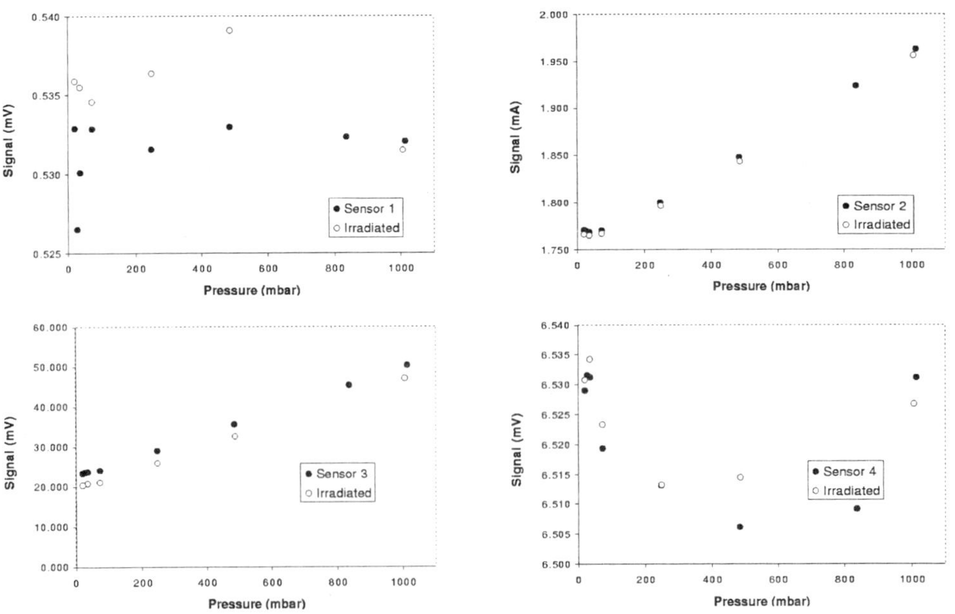

Figure 5. Comparison of cold calibrations performed before and after irradiation

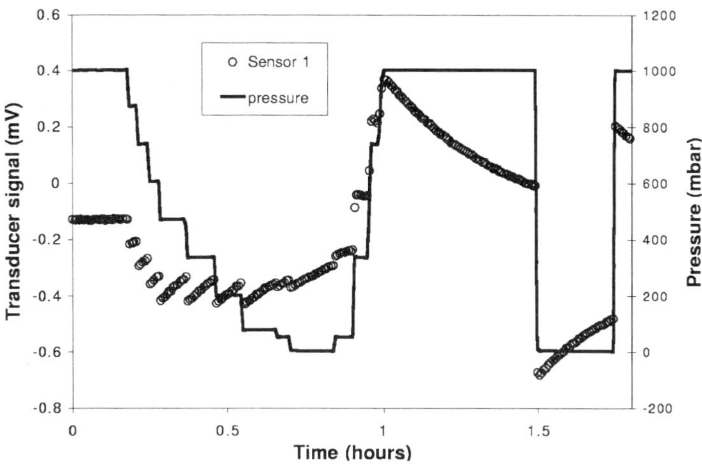

Figure 6. Evidence of a leak at room temperature after cold irradiation of sensor 1. The signal changes with the pressure, and then reaches (with a time constant of 15 minutes) the "vacuum" signal (-0.14 mV).

Table 3. Summary of the experimental campaign

	Sensor 1	Sensor 2	Sensor 3	Sensor 4
On-line observed drift at 4.2 K (3.4×10^{14} n/cm^2)	0	*	-60 mbar	-30 mbar
Observed radiation-induced leak in liquid helium	No	Yes	No	No
Radiation-induced pressure error at 1 bar, 4.2 K (7.5×10^{14} n/cm^2)	**	-100 mbar	-150 mbar	-50 mbar
Radiation-induced pressure shift at 1 bar, 300 K (warm calibration)	*	*	-300 mbar	*
Measured signal change for ΔP=1 bar before irradiation	0.69 mV	0.22 mA	10.7 mV	1.15 mV
Measured signal change for ΔP=1 bar after irradiation	0.63 mV	0.21 mA	10.6 mV	1.25 mV
Observed leak at room temperature after irradiation	Yes	Yes	No	Yes
Time constant of the warm leak	15 minutes	2 hours	-	30 minutes

* The leak on the sensor makes an estimation impossible

** The calibration curve is modified by irradiation

CONCLUSION

The irradiation campaign was difficult to conduct. To limit radiation hazards the manufacturers were requested to provide pressure sensors with a minimum possible amount of stainless steel as this material has a strong activation when exposed to neutrons. The observed leaks are probably due to a lack of expertise of the manufacturers to work with lighter materials such as aluminium or copper-beryllium. From the experiments, one can draw conclusions concerning the elastic membrane and the deformation gauge; some of the sensors exhibited a drift varying with the received radiation dose. To investigate whether leaks were produced on the membrane by irradiation it will be necessary to proceed to electron microscopic analysis.

ACKNOWLEDGEMENTS

We would like to thank F. Châtelet, C. Joly (IPN Orsay) and P. Romand (CERN) for their participation to the experiments. We also want to thank C. Balle, G. Bertrand, J.C. Grivat, R. Orfila, P. Portier and N. Vauthier (CERN); J. Collot, Ph. Martin and G. Mondin (ISN Grenoble, CNRS-IN2P3, France); and J.C. Bajard, G. Blondiaux and the cyclotron group (CERI, CNRS Orléans, France) for their help. Finally, we thank L. Thomé (CSNSM Orsay, CNRS-IN2P3, France) for fruitful discussions, W. Erdt and Ph. Lebrun for their support.

REFERENCES

1. K.M. Potter, H. Schonbacher, G.R. Stevenson, "Estimates of Dose to Components in the Arcs of the LHC due to Beam-Loss and Beam-Gas Interactions", CERN-LHC report #18 (1995)
2.T. Bager, J. Casas-Cubillos and L. Metral, "Cryogenic pressure calibration facility using a cold force reference", paper submitted to CEC-ICMC 1999 (Montréal)

CRYOGENIC PRESSURE CALIBRATION FACILITY USING A COLD FORCE REFERENCE

T. Bager, J. Casas and L. Metral

LHC Division
CERN, European Laboratory for Particle Physics
1211 Geneva 23, Switzerland

ABSTRACT

Presently various commercial cryogenic pressure sensors are being investigated for installation in the LHC collider, they will eventually be used to assess that the magnets are fully immersed in liquid and to monitor fast pressure transients.

In the framework of this selection procedure, a cryogenic pressure calibration facility has been designed and built. It is based on a cryogenic primary pressure reference made of a bellows that converts the pressure into a force measurement. For that, a shaft transfers this force to a precision force transducer at room temperature. Knowing the liquid bath pressure and the surface area of the bellows, the pressure applied to the transducers under calibration is calculated; corrections due to thermal contraction are introduced. To avoid loss of force in the bellows wall, its length is maintained constant; a cold capacitive displacement sensor measures this. The calibration temperature covers 1.5 K to 4.2 K and the pressure 0 to 20 bar. In contrast with more classical techniques that refer to a pressure reference at room temperature, the method presented in this paper avoid errors due to the uncertainty on the hydrostatic head calculation, to thermoacoustic oscillations and to pressure variation caused by temperature drift along the sensing capillary.

INTRODUCTION

The test setup is shown in Figure 1.

Cryostat

The cryostat has two liquid helium (LHe) reservoirs in the same vacuum vessel. The outer reservoir is shaped like a cylindrical shell, it is continuously filled with LHe at 4.2 K and acts as a thermal screen for the inner one. A cryogenic valve (18) is used for filling the inner reservoir.

The inner reservoir dimensions are ø30 cm x 1 m, a high capacity primary pump (8) regulates its saturated LHe temperature between 1.5 and 4.2 K. The saturation pressure is measured with a high accuracy pressure sensor (7).

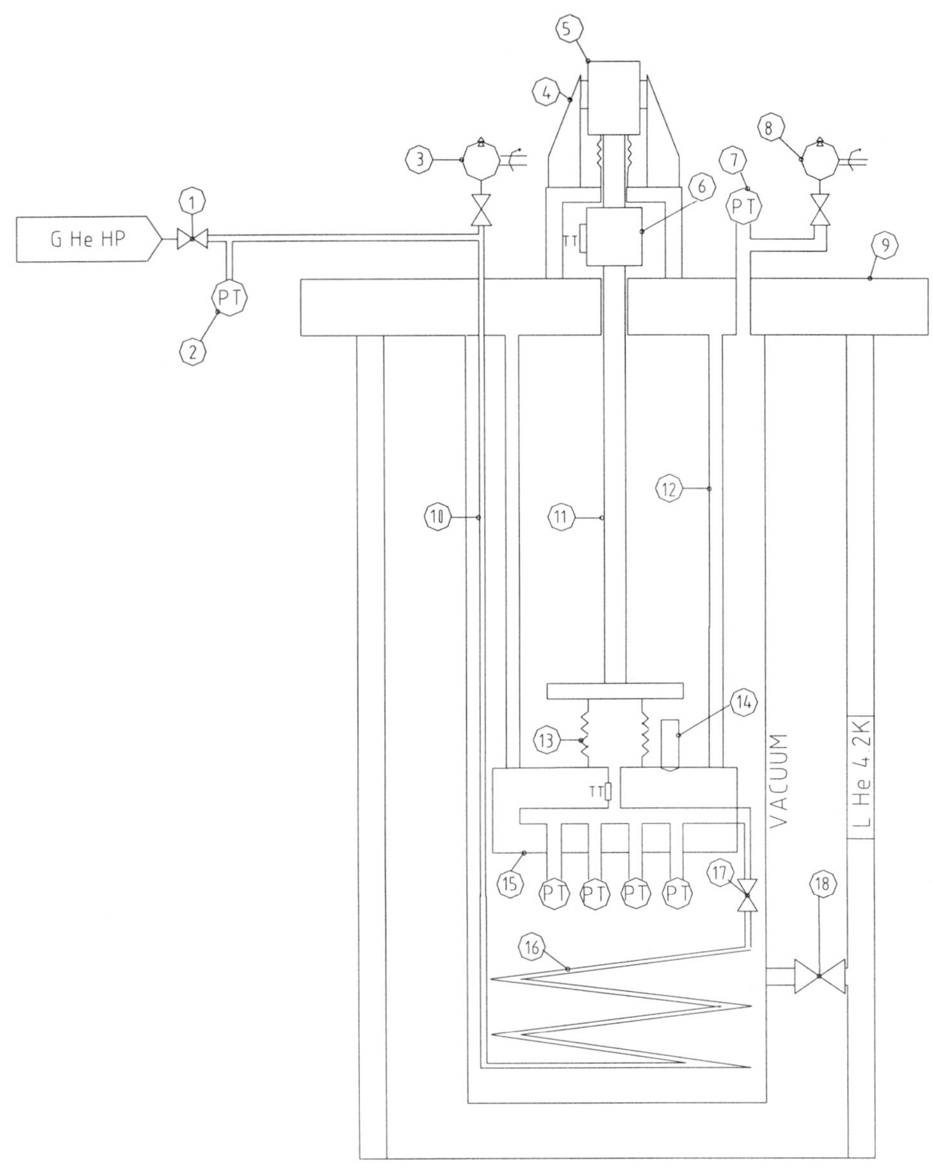

Figure 1. Schematic view of the cryostat

Pressurised test volume

A gaseous helium (GHe) cylinder at 20 MPa is connected to a pressure regulator (1) resulting in a variable test pressure of 0.1 – 2 MPa. Sub-atmospheric pressures are obtained by means of a primary vacuum pump (3). A capillary (10) is used as hydraulic connection between the warm and cold parts of the experiment and its pressure is always monitored by a high accuracy warm pressure reference (2). The capillary is wound into a helical heat exchanger (16) to cool the helium entering the test volume, which can be isolated by a cryogenic shut-off valve (17).

The distribution flange (15) is made of two 316L stainless steel (SS) disks welded together and it is attached to the main flange (9) by metallic tubes (12). Threaded holes are machined in the distribution flange allowing the installation of cryogenic pressure sensors. A bellows (13) is used to convert the inner pressure into a force that is measured with a warm force transducer (6). Risk of buckling imposes a minimum value to the bellows stiffness. The bellows length is measured with a capacitive sensor (14) and it is maintained constant by a linear actuator (5).

MECHANICAL CONSIDERATIONS

Pressure-to-force conversion

The bellows is made of electrodeposited nickel (Ni) and has two 316L SS end caps. The effective end cap surface area and the bellows inner and outer pressure determine the force transferred through the shaft (11) that by design is limited to about 1060 N.

The effective surface area is measured in warm conditions and thermal contraction effects are taken into account when operating in cryogenic conditions. With liquid helium the correction factor for effective surface area is:

$$\frac{A_{316L,\,4K}}{A_{316L,\,293K}} = (1 - \int_{4K}^{293K} dl/l)^2 = (1 - 3.31 \times 10^{-3})^2 \\ = 0.9934 \tag{1}$$

Alignment

A critical source of uncertainty is the alignment of shaft, bellows and force transducer. The bellows end-cap is aligned by design with the shaft (11), a transversal displacement of 0.35 mm of the shaft end-cap position is observed when varying the bellows inner pressure from vacuum to 2 MPa. This corresponds to an angular variation of about 0.021° that results in a negligible correction. At cryogenic temperatures it is assumed lower due to higher stiffness.

Torque is minimised since shaft (11) and bellows assembly (13) are not attached to each other by a fixed connection. At low force the bellows is almost free to rotate without transferring torque to the load cell. No shaft rotation is observed during pressurising at 293 K.

Elastic deformation

Ideally the bellows length should be maintained constant because its finite stiffness is a source of uncertainty. However elastic deformations are unavoidable and their effects should be investigated and compensated if necessary in the load cell, the shaft, the support tubes and the actuator support.

The shaft and support tubes are made of respectively one and three 316L SS tubes. The actuator support consists of three tubes, two are made of 316L SS and one of aluminium.

The elastic compression/expansion is calculated by:

$$\Delta L = \frac{L \times F}{A \times E} \tag{2}$$

where ΔL is the elastic expansion, A the effective cross section area, F the force and E the modulus of elasticity. Table 1 shows the results for each critical component, a load cell expansion of 108 μm is deduced from manufacturer data.

Table 1. Mechanical characteristics of force transmission components.

	Shaft	Support tubes	Actuator support 1	Actuator support 2	Actuator support 3
Force	1058 N	1058 N	1058 N	1058 N	1058 N
Tube length	970 mm	810 mm	100 mm	100 mm	100 mm
Tube cross section	Ø28x1	3 *(Ø16x0.5)	Ø70x2	Ø100x2	Ø70x3
Worst case modulus	195 GPa	195 GPa	195 GPa	195 GPa	70 GPa
Calculated max elastic expansion of tubes at full load	62 μm	60 μm	1.3 μm	0.9 μm	3.5 μm

Thermal deformations

Thermal contraction is also a source of uncertainty and the main contributors are the support tubes, the shaft (both made of SS) and the bellows. A first approximation is to calculate the thermal contraction difference of Ni and 316L over the bellows length of 13.9 mm.

$$\Delta L_{316L} = \int_{4K}^{293K} dl/l \times L = 3.31 \times 10^{-3} \times 13.9 = 46\ \mu m \tag{3}$$

$$\Delta L_{Ni} = \int_{4K}^{293K} dl/l \times L = 2.30 \text{ x } 10^{-3} \times 13.9 = 32\ \mu m \tag{4}$$

$$\Delta L_{thermal} = \Delta L_{316L} - \Delta L_{Ni} = 14\ \mu m \tag{5}$$

Overall deformation

The total deflection of the construction is due to the additive effects of elastic and thermal deformation:

$$\Delta L_{total} = \Delta L_{thermal} + \Delta L_{elastic} = 249\ \mu m \tag{6}$$

Without any compensation, such variation in the bellows length would result in part of the force being taken by the stiffness of the bellows wall. At room temperature this force can be estimated by using manufacturer data:

$$F_{bellows\ wall,293K} = springrate_{bellows,293K} \times \Delta L_{total} = 11\ N \tag{7}$$

This effect might result in an uncertainty of 1%. At cryogenic temperatures the bellows stiffness is higher and it is difficult to implement a compensation algorithm. In consequence a capacitive displacement sensor and a linear actuator are used to maintain the bellows length constant to within ± 5 μm. When operating in LHe, the displacement sensor output should be corrected by approximately 5% as shown in Table 2.

Table 2. Relative dielectric constant in various relevant media

Media	Relative dielectric constant
air, 300 K	1.0005
LHe, 4.2 K	1.049
LHe, 1.8 K	1.057

OPERATION PRINCIPLE

Setting the initial bellows position

Since the bellows is not attached mechanically to the shaft, it is operated with a slight pre-compression, and the applied force through the shaft is always oriented in the same direction independently of the operating temperature. The compression of the bellows is obtained by manipulating the linear displacement actuator while monitoring the force transducer. With a Δp of 0.1 MPa, a springrate of 45 N/mm^2 and an area of 529 mm^2, the minimal required pre-compression is as a first approximation:

$$Bellows\ compression_{\max} = \frac{\Delta p \times A_{bellows}}{springrate_{bellows}} = 1.2\ mm \tag{8}$$

By taking a safety margin of about 20 % the actual pre-compression is 1.4 mm, thus the resulting compression force is 9 N, thus reducing the force transducer dynamic range by 0.08 %.

Bellows effective area

The bellows effective surface area for converting pressure into force is determined at room temperature. In this case the warm pressure reference is used and the cryogenic shut-off valve is open. A set of measurements is taken and the best linear fit between measured pressure and force (Figure 2) is used for determining the effective surface area; the repeatability of these measurements is better than 0.15 %. Corrections are used to take into account thermal contraction.

Operation at cryogenic temperatures

When calibrating a set of cryogenic sensors the pressurized volume is gradually filled with gaseous helium until a given pressure setpoint (measured by the warm pressure reference) is reached. During the filling, the bellows length is monitored by the capacitive sensor and is maintained constant by means of the linear displacement actuator. The cryogenic shut-off valve is then closed and a combined measure of the force transducer, inner cryostat pressure and LHe level of the inner reservoir is done. The pressurized helium temperature is measured by using a resistance thermometer calibrated between 1.6 K and 300 K. Pressure sensors under test are energized according to manufacturers recommendations (usually 10 V_{DC}) and their outputs recorded.

The applied pressure is calculated by:

$$p_{sensor} = \frac{F_{LC}}{A_{bellows\ area}} + p_{inner\ cryostat} + p_{liquid\ head} \tag{9}$$

The liquid level head contributes with a pressure of (see Table 3):

$$p_{liquid\ head} = L_{LHe} \times g \times \rho_{LHe} \tag{10}$$

Table 3. Pressure formed by liquid head

$p_{liquid\ head}$	L = 10 cm	L = 40 cm
T = 4.2 K	0.12 kPa	0.49 kPa
T = 1.8 K	0.14 kPa	0.57 kPa

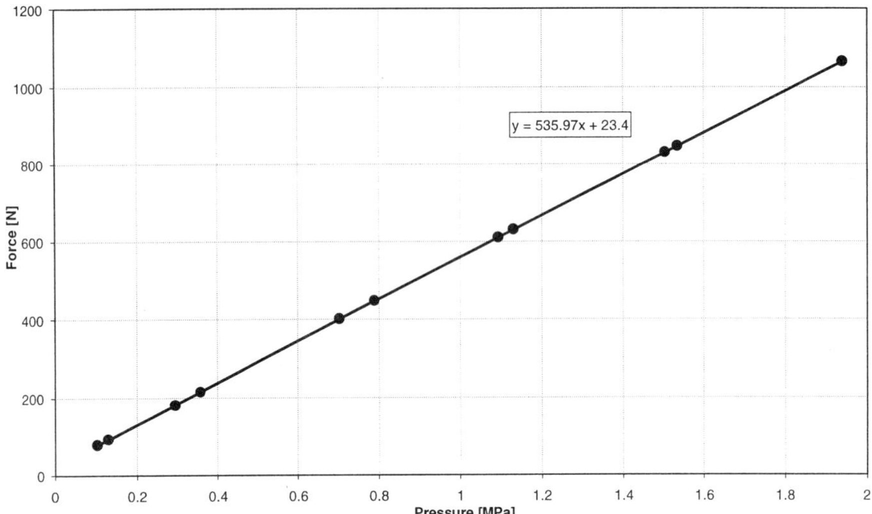

Figure 2. Calculation of bellows area

The pressure inside the test volume is varied from vacuum to the maximum range of the pressure sensors or 2 MPa, whatever is lower. A measurement in vacuum conditions permits elimination of bellows effects due to the pre-compression and to the dependence with temperature of bellows stiffness.

Sensors under test are located 10 cm lower than bellows, thus for pressures above saturation pressure the liquid head inside the pressurised chamber is 0.12/0.14 kPa, which must be added to sensor pressure. For pressures below saturation pressure the hydrostatic head consists of gas only and is below 16 Pa.

Apparatus

A MacIntosh™ computer running LabVIEW™ is used for control and data acquisition. IEEE-488 is used for communication between the computer and the instruments. A digital multi-meter with an integrated 20-channel scanner is used for monitoring the temperatures and other diagnostic instruments. The pressure reference is directly read through the bus.

All the LHC-prototype pressure sensors have a membrane for separating a reference vacuum chamber from the pressurized media. The displacement of the membrane is a function of the applied pressure. Depending on the sensor, this displacement is measured by strain gauges, reluctance coils or optical techniques.

For strain gauges an HBM™ ML10 amplifier was used. This instrument is used for measuring the force transducer and up to 5 pressure sensors simultaneously. The excitation voltage is typically 10 V_{DC} and the accuracy is 0.03 % of full scale.

For reluctance coils and optical type pressure sensors, a specific instrument for conditioning the output is supplied by the sensor manufacturer. In such cases it is not

Table 4. Combined errors of instruments.

Bellows area	Pressure inner cryostat	Force measurement	Liquid head	Bellows position	Conditioner for sensors under test
0.17 %	0.29 kPa (~145 ppm)	0.12 % FS	29 Pa (~15 ppm)	20 nm	300 ppm FS

possible to discriminate whether the uncertainties come from the electronic unit or the sensor itself.

The pressure inside the capillary is measured by a Mensor™ 2106 warm pressure reference with a range of 0 to 2.6 MPa and an uncertainty of 0.65 kPa. The liquid helium level is measured by using a superconducting probe with an uncertainty of about 2 cm. The inner reservoir pressure is measured by an industrial pressure transducer and the HBM™ amplifier, the uncertainty is 0.29 kPa.

The linear displacement actuator is programmed via RS232 by a computer and controlled during operation by a keypad. The maximal displacement is 20 mm, the resolution better than 1 μm and has backlash of about 12 μm. This actuator is used regularly during cool-down and warm-up requiring displacement of up to 200 μm.

Instrument errors are summarised in Table 4.

EXPERIMENTAL RESULTS

Operation at 4.2 K

Preliminary tests have been performed at 4.2 K and four different types of pressure sensors have been tested. A relatively good linearity between sensor output and applied pressure is observed. The deviation from best straight line is for all sensors less than 3 kPa (~ 0.15 % FS). See Figure 3. A short overview of sensor types tested can be found in Table 5.

The accuracy obtained has been limited by a leak in the cryogenic shut-off valve that result in oscillations and drift in the force transducer. The seal of this valve will be changed and we expect its leak-tightness not to be a limiting factor in the ultimate uncertainty that can be obtained with this calibration facility.

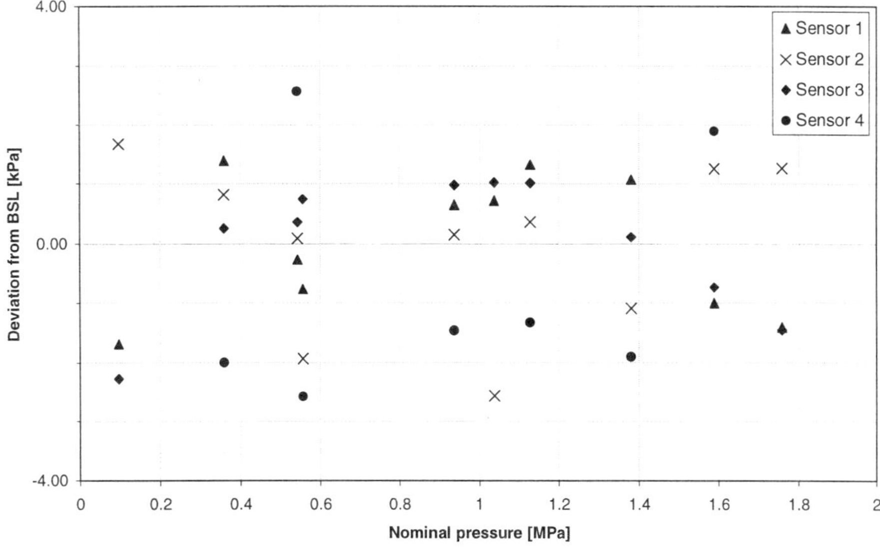

Figure 3. Sensors under test, deviation from best straight line

Table 5. Main characteristics of sensors tested

	Sensor 1	Sensor 2	Sensor 3	Sensor 4
Diaphragm	metal (w 1.4542)	metal (15.5PH)	SiO_2	Metal (Inconel)
Sensing principle	Strain gauge metallic resistance	Strain gauge metallic resistance	Strain gauge metallic resistance	Inductive, brass target
Excitation	10 V_{DC}, 5 kΩ	10 V_{DC}, 1.25 kΩ	10 V_{DC}, 0.7 kΩ	56 kHz AC
Nominal output @FS	20 mV	30 mV	28 mV	4-20 mA (1)

Note (1) : Output of dedicated signal conditioner

CONCLUSION

A facility for calibrating cryogenic pressure sensors has been fabricated and commissioned. It is based in a primary cryogenic pressure reference made of a bellows and a piston transmitting the resulting force into a high-accuracy force transducer, in order to eliminate uncertainties related to an unknown pressure head inside the sensing capillary. The pressurized volume can be isolated from the capillary strongly reducing the effects of pressure oscillations that can be observed with the warm pressure reference.

ACKNOWLEDGMENT

Thanks for good ideas to Ph. Lebrun, CERN, for manufacturing components to P. Portier and J. Beltron, CERN, for help on operating to P. Romand, CERN.

REFERENCES

H.E.Krex. "Maskinståbi", Teknisk Forlag, København (1986).
J.G. Weisend II. "Handbook of Cryogenic Enginering", Taylor & Francis, Philadelphia (1998).
Hepak Software, Cryodata Inc., Niwot, Co (1993).

SIGNAL CONDITIONING FOR CRYOGENIC THERMOMETRY IN THE LHC

J. Casas, P. Gomes, K.N. Henrichsen, U. Jordung, M.A. Rodriguez Ruiz

CERN, European Laboratory for Particle Physics
1211 Geneva 23, Switzerland

ABSTRACT

Temperature measurement is a key issue in the Large Hadron Collider (LHC), as it will be used to regulate the cooling of the superconducting magnets. The compromise between available cooling power and the coil superconducting characteristics leads to a restricted temperature control band, around 1.9 K. An absolute accuracy of 10 mK below 2.2 K, and 5 K above 25 K, is necessary. For resistive thermometers covering the full temperature range, and having a negative dR/dT sensitivity, this is typically equivalent to a relative accuracy $\Delta R/R$ of $3\ 10^{-3}$ over 3 resistance decades. Also, to limit the thermometer's self-heating, the sensing current must be limited to few μA. Furthermore, the radiation levels next to the accelerator are expected to significantly degrade the performance of conventional analog electronics.

As these stringent requirements are not met by commercial conditioners, three different architectures have been developed at CERN. The first compresses the input dynamic range using a logarithmic transfer function; the second partitions the input range into three linear regions; the third converts resistance linearly into the frequency of a square wave. They fulfill the above specifications and provide industrial robustness in terms of thermal drift, galvanic protection, and compact packaging, while optimizing cost-to-performance ratio. This paper describes the principles of their design, compares their characteristics and shows results of field tests. Future developments include Application Specific Integrated Circuit versions, Fieldbus interfacing, and radiation tolerant re-design.

INTRODUCTION

The various components of the LHC cryogenic system work at temperatures from ambient down to 1.6 K. Depending on the actual temperature value, different accuracies are required on its measurement. Between 300 K and 25 K, an uncertainty of 5 K can be tolerated to monitor the warmer components and the general cool-down. However, at the nominal operation of superconducting magnets (below 2.2 K) only 10 mK inaccuracy is allowed, to give enough room for the regulation band of the cryogenic controller, while avoiding magnet quench and minimizing the cooling effort of the cryogenic system.

Table 1 shows the allowed uncertainty on temperature measurement on the LHC machine, for the different temperature ranges. The accuracy budget is to be evenly shared between the sensor and the signal conditioning (Figure 1). The aimed resolution has to be ten times better than the overall accuracy (dT < 1mK, below 2.2 K).

Table 1. Required overall T accuracy and resolution

T span [K]	accuracy [mK]	resolution [mK]
1.6 ↔ 2.2	10	1
2.2 ↔ 4.0	20	2
4.0 ↔ 6.0	30	3
6.0 ↔ 25	1 000	100
25 ↔ 300	5 000	500

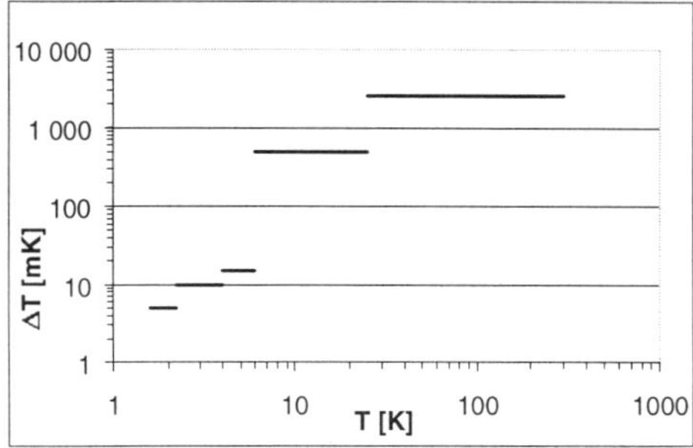

Figure 1. Required conditioning temperature accuracy

Sensors

Cryogenic temperature sensors currently used at CERN can be classified according to four main attributes, as shown in Table 2. CERNOX™ (CX), TVO® and RhFe cover the full temperature range with a single sensor. AllenBradley® (AB) and Pt100 can be combined to cover respectively low and high temperature scales, or used alone in applications not requiring full range measurements.

In terms of resistive values, CX's span is the largest among all sensors (3 decades), requiring wide dynamic range signal conditioning. Also covering the whole temperature range, RhFe spans over only 2 decades of resistance, with the advantage of less demanding dynamic range, but with the consequence of limited sensitivity.

Sensors with negative dR/dT (Figure 2), like CX, TVO® and AB, show high resistance and high sensitivity (dR/R / dT/T) (Figure 3) at low temperatures, where measurement accuracy has to be at its best. This semiconductor behavior relaxes the constraints on conditioner accuracy for low temperature measurement. On the other hand, at low temperature metallic sensors like RhFe exhibit a sensitivity one order of magnitude worse, demanding much more accurate signal conditioning.

Signal conditioning

The basic principle used to read resistive sensors consists in sending a known sensing current, over a pair of wires, and reading the voltage developed at the resistor leads, via another pair of wires. After amplification and correction, the read signal can be sent to the process controller under different formats, either analog or digital.

Table 2. Typical characteristics of cryogenic temperature sensors currently used at CERN

	T span [K]	R span [Ω]	dR/dT [Ω/K]	(dR/R) / (dT/T)
CX	1.6 ↔ 300	30 000 ↔ 30	-40 000 ↔ - 0.1	-2.7 ↔ - 1.0
TVO®	1.6 ↔ 300	9 000 ↔ 900	-7 000 ↔ - 0.7	-1.3 ↔ - 0.2
RhFe	1.6 ↔ 300	6 ↔ 110	+0.7 ↔ +0.4	+0.2 ↔ +1.0
AB	1.6 ↔ 100	10 000 ↔ 100	-12 000 ↔ - 0.3	-3.0 ↔ - 0.2
Pt100	73 ↔ 300	18 ↔ 110	+0.4 ↔ +0.4	+2.0 ↔ +1.0

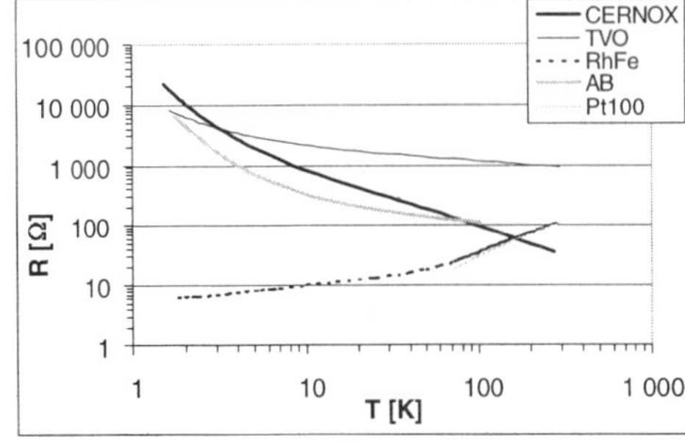

Figure 2. Typical thermometers transfer function R(T)

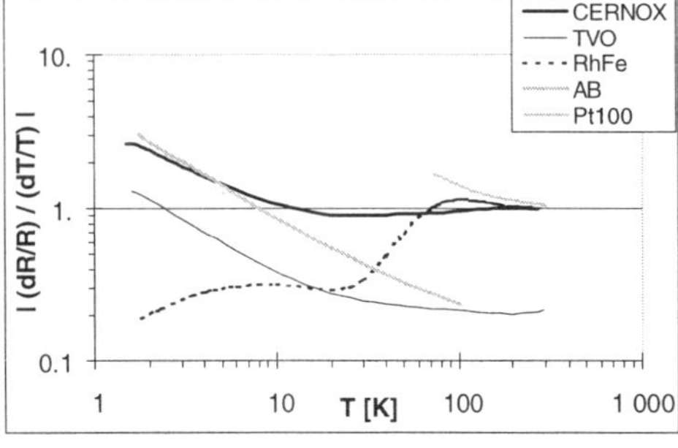

Figure 3. Dimensionless sensititvity vs temperature

Table 3. Conditioner's accuracy and resolution required by each type of sensor, below and above T = 6 K

	T < 6 K			T > 6 K		
sensor	R span [Ω]	ΔR/R [10^{-3}]	dR [mΩ]	R span [Ω]	ΔR/R [10^{-3}]	dR [mΩ]
CX	1 500 ↔ 30 000	3.3	900	30 ↔ 1 500	8.9	60
TVO®	2 600 ↔ 9 000	1.3	700	900 ↔ 2 600	1.8	300
RhFe	6 ↔ 8.5	0.5	0.6	8.5 ↔ 110	6.0	15
AB	500 ↔ 10 000	3.1	300	100 ↔ 500	5.8	100
Pt100				18 ↔ 110	9.8	200

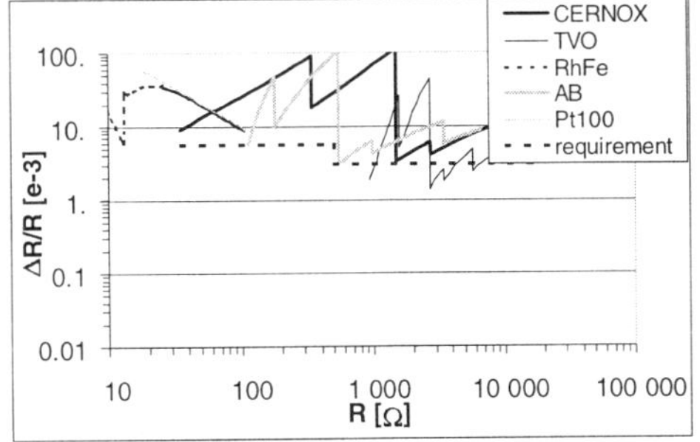

Figure 4. Required accuracy

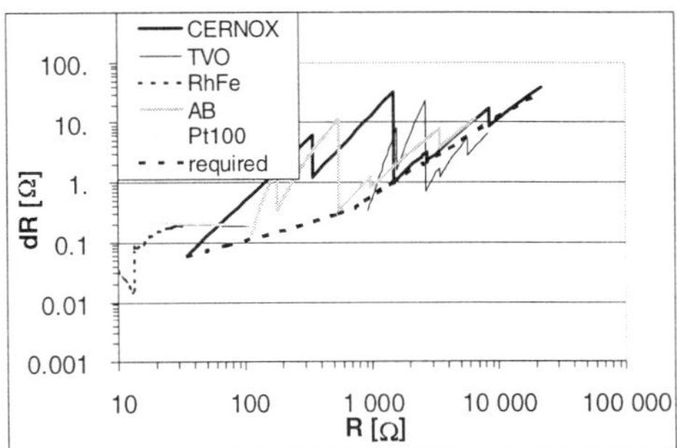

Figure 5. Required resolution

For each type of thermometer, the signal conditioning accuracy and resolution requirements are summarized in Table 3 and plotted in Figure 4 and Figure 5. The RhFe sensor imposes quite severe constraints on the conditioning electronics, with $\Delta R/R < 0.5\ 10^{-3}$ for $R < 8.5\Omega$, thus requiring techniques beyond the scope of the conditioners described in this paper.

A conditioner satisfying $\Delta R/R < 3\ 10^{-3}$ for $R > 500\Omega$ and $\Delta R/R < 6\ 10^{-3}$ for $R < 500\Omega$ is adequate for CX, AB and Pt100 thermometers. A TVO® sensor would impose a three times better accuracy on the resistance measurement.

As the sensor's resistance will be calculated linearly from the read voltage (Eq. 1), the relative accuracy is the addition of the voltage and the sensing current uncertainties.

$$R_{sensor} = U_{read} / I_{sense} \qquad \Delta R/R = \Delta U/U + \Delta I/I \tag{1}$$

The read signal has to be digitized in order to be usable by computerized control or diagnostics. Industrial control equipment typically employs 12-bit Analog to Digital Converters (ADC) and the targeted temperature accuracy cannot be satisfied unless some type of signal compression is implemented. This paper presents two compression architectures that are mixed logarithmic+linear (LOG+LIN) and linear piecewise (LIN pw).

The resistance measurement accuracy must be split between the accuracy of the analog circuitry and that of the ADC, which can be expressed in terms of the ADC Least Significant Bit (LSB). Figure 6 and Figure 7 show respectively the accuracy and resolution imposed by the resistance measurement, and also the accuracy and resolution attained by the combination of different compression algorithms and a minimum size ADC.

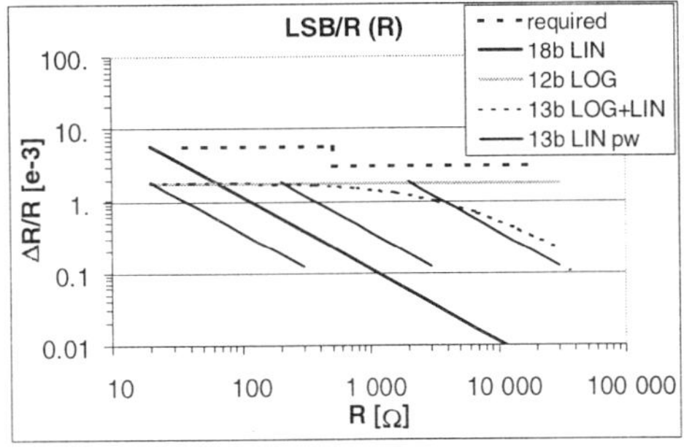

Figure 6. Required ADC accuracy

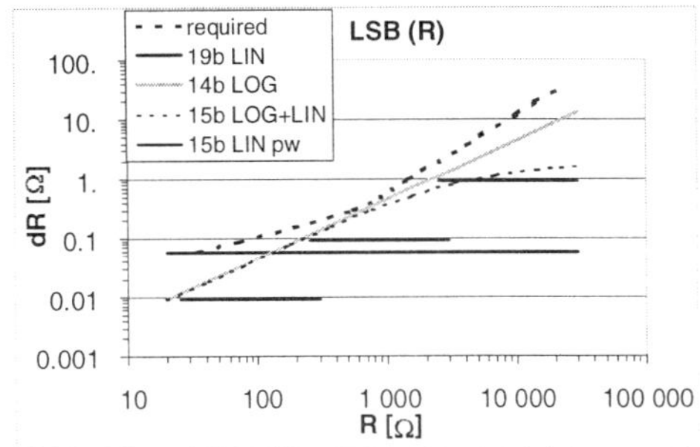

Figure 7. Required ADC resolution

Table 4. ADC accuracy and resolution required for each type of conditioner transfer function

	for CX, AB, Pt100 and TVO®		for CX, AB, and Pt100 only	
conditioner	**accuracy [bit]**	**resolution [bit]**	**accuracy [bit]**	**resolution [bit]**
LIN	18	19	18	19
LOG	13	15	12	14
LOG+LIN	14	16	13	15
LIN pw	14	16	13	15

Table 4 summarizes these requirements. Due to its lower sensitivity, the TVO® typically needs 1 more bit than other sensors. For all sensors, the resolution normally requires 2 more bits than the accuracy.

It is clear that a 12-bit resolution ADC is not sufficiently accurate (above 4.5 K) for all the mentioned conditioning architectures. We are thus considering the use of larger ADC, with a suitable network interface.

LOGARITHMIC CONDITIONER

The first conditioner (Figure 8) incorporates a logarithmic compression of the input dynamic range. A DC sensing current of 10 μA develops a voltage (0.2 mV ↔ 200 mV) across the thermometer, which is raised by an instrumentation amplifier. The offset and gain, before and after the logarithmic amplifier, have to be manually adjusted. Power supply, input and output are galvanically isolated to 750 V.

The overall performance against ambient temperature variation is limited by the Logarithmic IC. This explains the improvement in temperature drift at higher resistances, for LOG+LIN in comparison with LOG (Figure 9 and Figure 10). The relationship between input and output is given by the following equations for LOG and LOG+LIN respectively:

$$I_{loop} = 4 + 16/3 \cdot \text{Log}_{10}(R/20) \quad \text{[mA]} \qquad (2)$$

$$I_{loop} = 4 + 8/3 \cdot \text{Log}_{10}(R/20) + 8/(20k-20) \cdot (R-20) \quad \text{[mA]} \qquad (3)$$

The rejection of power supply drift is good for $R > 500\ \Omega$ ($|\Delta R/R\ /\Delta V| < 0.7\ 10^{-3}$ /V, Figure 11), but quite poor for small R.

Several tens of these conditioners have been installed in cryogenic experiments at CERN, yielding good results (Figure 12), provided the ambient temperature does not change more than a few degrees.

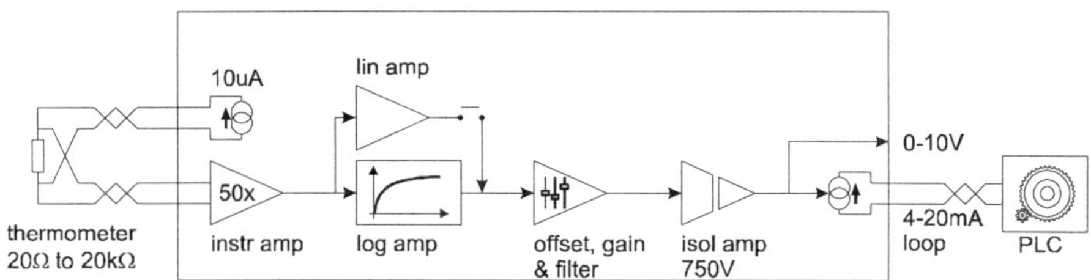

Figure 8. Logarithmic conditioner block diagram

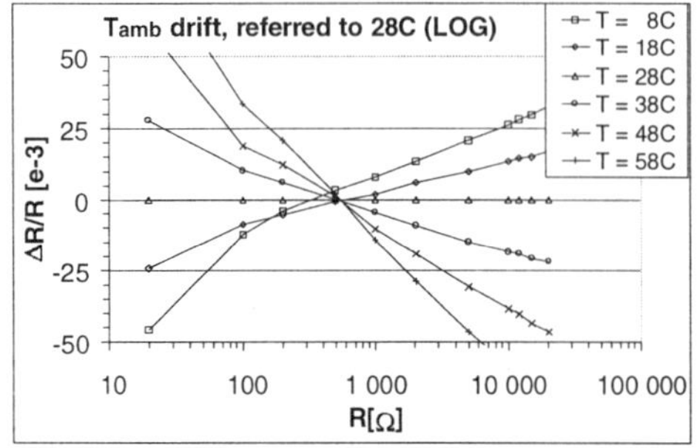

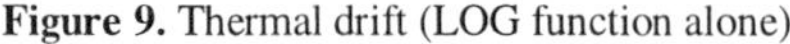

Figure 9. Thermal drift (LOG function alone)

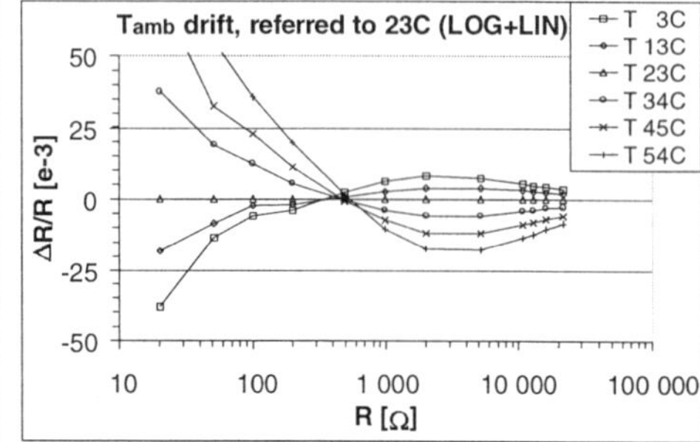

Figure 10. Thermal drift (combined LOG+LIN)

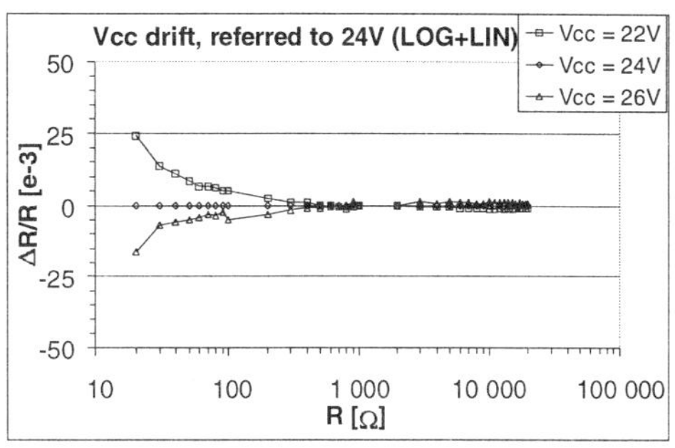

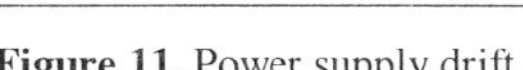

Figure 11. Power supply drift

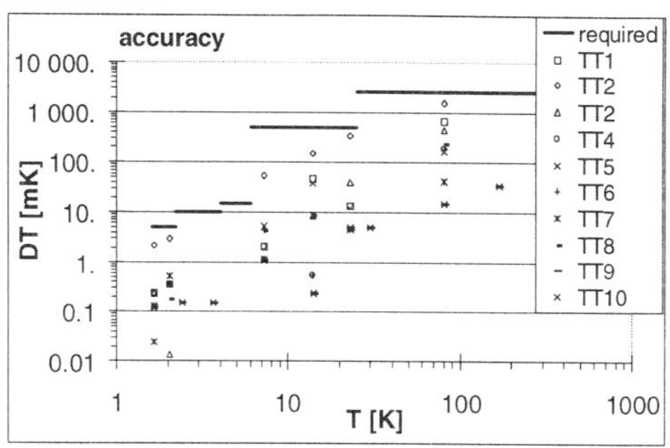

Figure 12. Field measurements

LINEAR MULTI-RANGE CONDITIONER

In order to reduce the ADC size, a linear multi-range conditioner has been developed. The input span is partitioned into three regions (Table 5), one decade wide each, with the gain proportional to the sensing current. The thermal drift due to the logarithmic IC is thus avoided, at the expense of more complex circuitry for controlling the gain (Figure 13).

Furthermore, the sensing current can be lower for high R values (low T), reducing the self-heating of the sensor, and higher for small R values, increasing the voltage developed at the sensor and its signal to noise ratio. The thermocouple voltages are cancelled by the use of a bipolar sensing current, oscillating at 4 Hz.

The sensing current creates a voltage (±3.5 mV ↔ ±40 mV) across the thermometer, which is increased, by an instrumentation amplifier, (to ±0.875 V ↔ ±10 V), and then rectified and smoothed. When this voltage raises above 10 V, the gain controller selects a sensing current 10 times lower, leading to the same reduction in voltage. If the voltage drops below 0.875 V, a 10 times higher current is selected. The gap between 0.875 V and 1 V corresponds to a hysteresis in the transfer function (Figure 14), that prevents oscillation between consecutive ranges when close to the switching point.

The selected range is indicated by two bit or a three level analog signal. Manual adjustments are necessary to correct general gain and offset and differential offset of the rectifier. Power supply, input and output are galvanically isolated to 750 V.

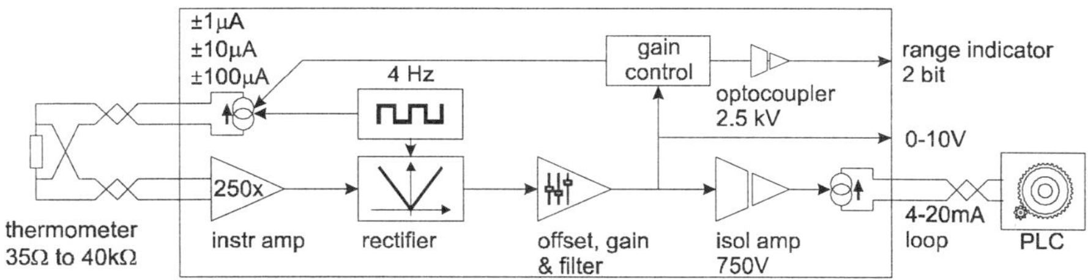

Figure 13. Linear multi-range conditioner block diagram

Table 5. Sensing current and conditioner transfer function, for each input range

R span [Ω]	I_{sense} [μA]	I_{loop} [mA]
35 ↔ 400	100	$4 + 16 \cdot R / 400$
350 ↔ 4 000	10	$4 + 16 \cdot R / 4\,000$
3 500 ↔ 40 000	1	$4 + 16 \cdot R / 40\,000$

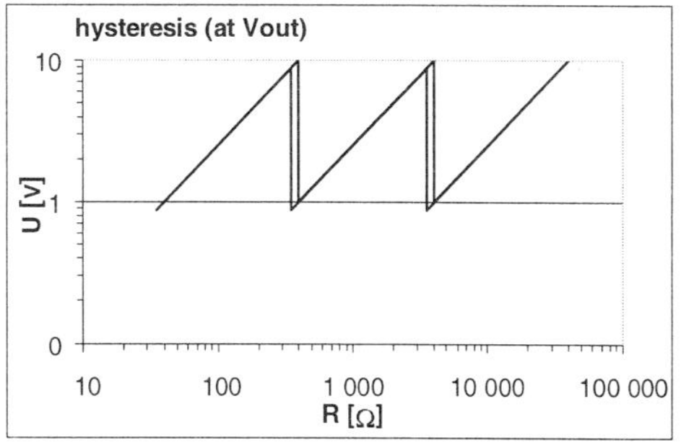

Figure 14. Hysteresis in the transfer function

The sensitivity to thermal drift ($|\Delta R/R \ / \Delta T_{amb}| < 0.11 \ 10^{-3}$ /°C, Figure 15) is very low compared to LOG conditioner. The thermal performance is further improved (Figure 16) at the voltage output, without galvanic isolation.

The insensitivity to power supply drift is good for the whole input span, both at the current output ($|\Delta R/R \ / \Delta V| < 1.0 \ 10^{-3}$ /V, Figure 17) and the voltage output ($|\Delta R/R \ / \Delta V| < 0.4 \ 10^{-3}$ /V, Figure 18).

Figure 19 shows the excellent reproducibility of the conditioners' characteristics between three samples.

Several tens of these conditioners have been installed in cryogenic experiments at CERN, behaving in accordance with the requirements (Figure 20).

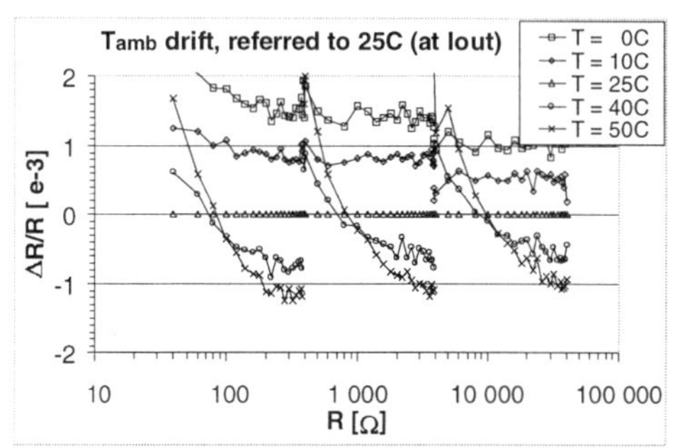

Figure 15. Thermal drift at the current output

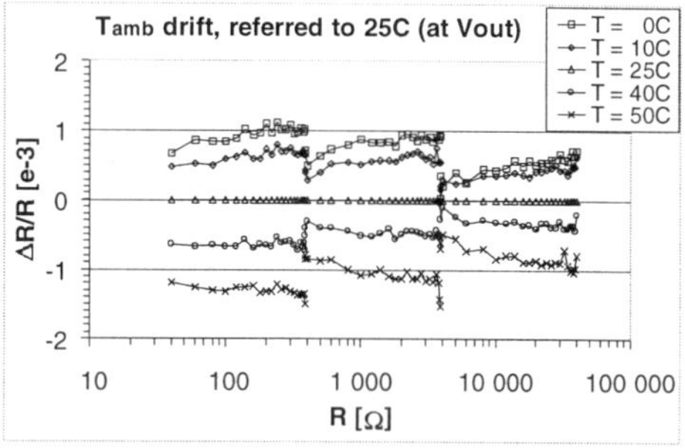

Figure 16. Thermal drift at the voltage output

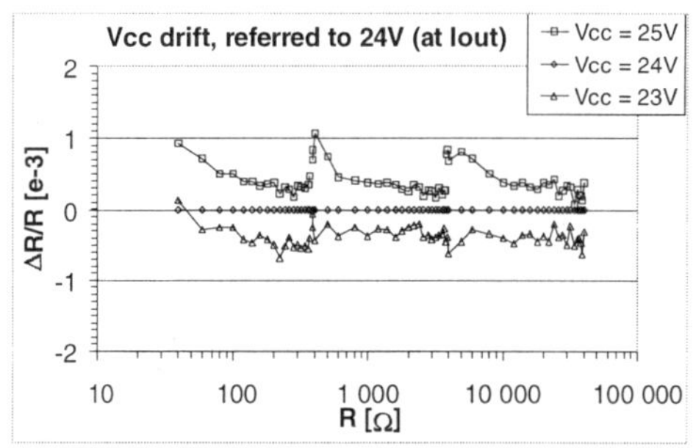

Figure 17. Power supply drift at current output

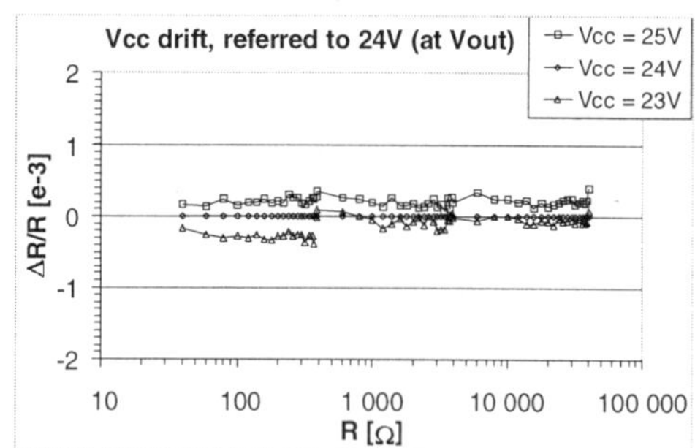

Figure 18. Power supply drift at voltage output

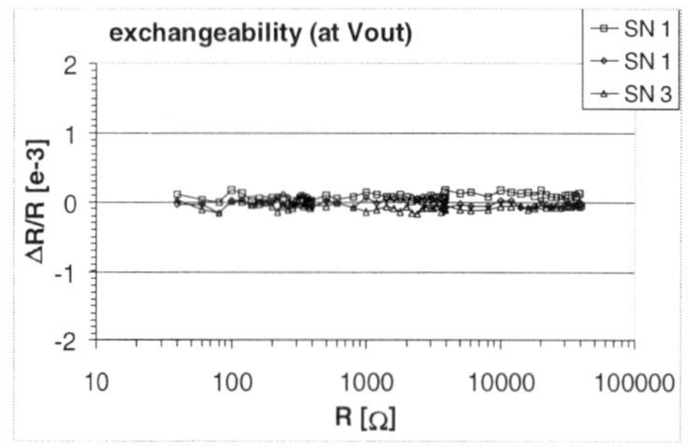

Figure 19. Exchangeability

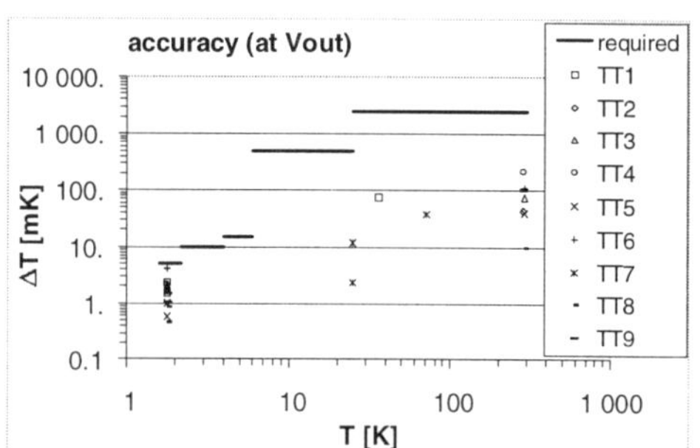

Figure 20. Field measurements

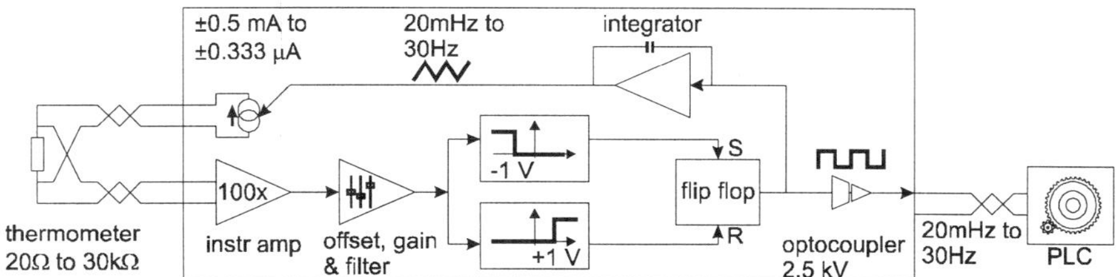

Figure 21. T2F conditioner block diagram

T2F CONDITIONER

This conditioner converts the voltage across the sensor resistance linearly into a frequency (Figure 21). The requirements on accuracy and resolution are no longer put on voltage amplitude measurement but rather on frequency measurement. The output frequency spans over three decades of dynamic range, like the measured resistance. The logarithmic compression nuisances or the complexity of a multi-range controller are traded by a simpler concept.

The sensing current follows a ramp of fixed slope (±10 μA/s). Once reached the upper (10 mV) or lower (-10 mV) threshold referred to the input, the ramp sign is inverted. This fixed voltage amplitude restricts the thermometer self-heating to a relatively low value.

Despite the fact that the working frequencies are below 50 Hz, the current control loop easily picks-up the mains noise. This effect is tackled by the use of a 50 Hz notch filter. In order to avoid spurious trigger of the flip-flop, a glitch detector is implemented, which introduces an extra delay on the oscillation period (Eq. 4).

$$f = 1 / (10^{-3} + 10^{3}/R) \qquad [\mathrm{Hz}] \qquad (4)$$

Manual adjustments are necessary for notch filter width and center frequency, for the glitch filter delay and for the current slope.

Figure 22 shows the deviation ($|\Delta R/R| < 2\ 10^{-3}$) between two conditioners, using the same transfer function. The accuracy obtained in field measurements (Figure 23) is well within the requirements. Further tests are under way to investigate temperature and power supply drifts.

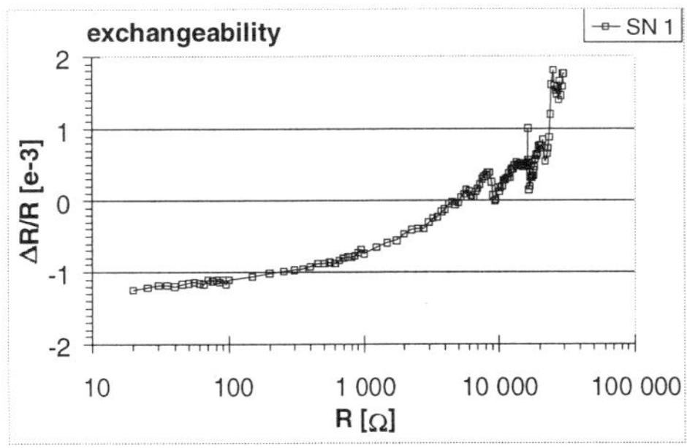

Figure 22. Exchangeability relative to SN2

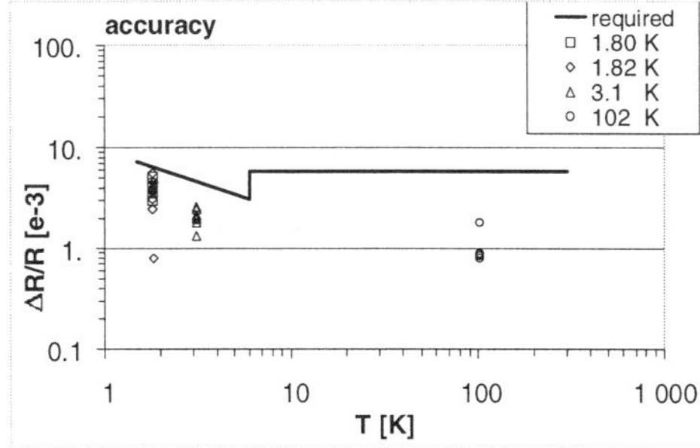

Figure 23. Field measurements

CONCLUSIONS

The cryogenic thermometer signal conditioners presented in this paper are used in a variety of applications. When fast response is required the LOG or LOG+LIN conditioner can be used. For higher accuracy at low temperature it would be necessary to reduce the sensing current. However, in this case the thermal drift of the input amplifier will be the dominant source of error for low values of the thermometric resistance.

The LINpw and T2F conditioners satisfy the LHC accuracy requirements. However their installation inside the LHC tunnel impose their redesign by using radiation tolerant integrated components, both passive and active. This problem is being solved by performing irradiation tests on commercial devices, and by porting the conditioners' design into a radiation hardened Application Specific Integrated Circuit (ASIC).

Once the design of a suitable signal conditioner for thermometers has been finalized, it will be necessary to adapt its characteristics in order to being able to use it as front end of other instruments like pressure sensors, liquid helium level gauges, etc.

The cryogenic control system requires temperature measurements distributed around the 27 km circumference LHC machine. The data exchange can be done with either point-to-point analog signal transmission or by using an industrial field network. Analog transmission implies a large investment in cabling and installation. In order to reduce this cost we are also investigating the performance of network components and ADCs, to integrate them together with the signal conditioners.

LINEAR MODEL-BASED PREDICTIVE CONTROL OF THE LHC 1.8 K CRYOGENIC LOOP

E. Blanco Viñuela[1], J. Casas Cubillos[1], C. de Prada Moraga[2]

[1]CERN, European Center for Particle Physics
1211 Geneva 23, Switzerland
[2]Ingeniería de Sistemas y Automática, UVA
47011 Valladolid, Spain

ABSTRACT

The LHC accelerator will employ 1800 superconducting magnets (for guidance and focusing of the particle beams) in a pressurized superfluid helium bath at 1.9 K. This temperature is a severely constrained control parameter in order to avoid the transition from the superconducting to the normal state. Cryogenic processes are difficult to regulate due to their highly non-linear physical parameters (heat capacity, thermal conductance, etc.) and undesirable peculiarities like non self-regulating process, inverse response and variable dead time. To reduce the requirements on either temperature sensor or cryogenic system performance, various control strategies have been investigated on a reduced-scale LHC prototype built at CERN (String Test). Model Based Predictive Control (MBPC) is a regulation algorithm based on the explicit use of a process model to forecast the plant output over a certain prediction horizon. This predicted controlled variable is used in an on-line optimization procedure that minimizes an appropriate cost function to determine the manipulated variable. One of the main characteristics of the MBPC is that it can easily incorporate process constraints; therefore the regulation band amplitude can be substantially reduced and optimally placed. An MBPC controller has completed a run where performance and robustness has been compared against a standard PI controller (Proportional and Integral).

INTRODUCTION

Due to the complexity of the LHC accelerator, it was decided to install and operate several full-length prototype magnets in a test String[1]. This is a fully working model of the future LHC apart from the absence of circulating particle beams. The goal of the String was to optimize operational aspects of all accelerator systems. In particular for the cryogenic system the two-phase superfluid helium flow in the heat exchanger tube, the temperature control of the magnets, and the thermohydraulic effect of resistive transitions, were investigated.

The String (Figure 1) consists of one quadrupole and three dipoles (total length of 50 meters), it is mounted with a slope of 1.4% to match the steepest inclination in the actual tunnel. The magnets operate below 1.9 K in a bath of pressurised helium. The heat

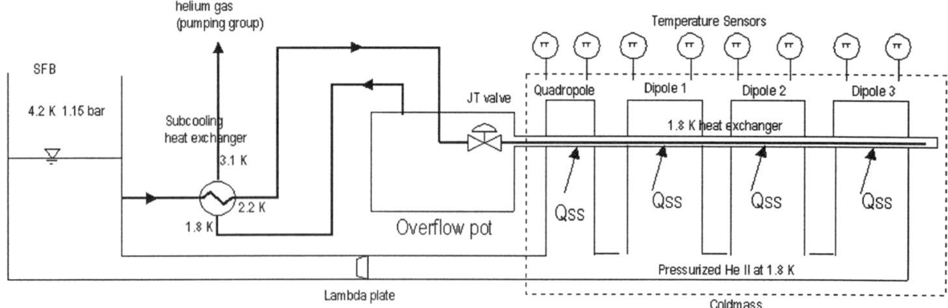

Figure 1. String facility: schematic diagram of the LHC 1.8 K Cooling Loop

deposited on the bath is extracted by gradual vaporization of saturated superfluid helium flowing along the wetted length of a heat exchanger (HX) tube.

The liquid helium is taken from the main reservoir (SFB). Subcooled helium is expanded to saturation in the Joule-Thomson valve and transported by a small-diameter feeder pipe to the end of the HX tube. The helium flows back towards the overflow pot that, in normal operation, remains empty. The helium vapor is taken out from the overflow pot and through the subcooling-heat exchanger, thus providing the subcooling for the incoming pressurised liquid.

PLANT MODELLING AND DYNAMICAL BEHAVIOUR

Modelling and simulation are central to the design of control systems. They result in a better knowledge of the process and of the quality of both the regulation and the overall process optimization.

A first principles model (based on basic physical laws) gives the possibility of implementing the strongest non-linearities of the process. On the other hand the main drawbacks are the time required for developing a sufficiently accurate model and its adaptation into the non-linear controller strategy.

A non-linear model based on basic physics has been developed and validated using real data[2]. It is used to get more knowledge about the process, generate a linear model for the MBPC controller, and to tune the MBPC controller before porting it to the real plant. This model has been simulated using a general-purpose process simulation language (ACSL®). The main non-linearities of the process are that it is non-self regulating (integrating response), it exhibits inverse response and it has a variable dead time, mainly due to the transportation lag in the HX pipe.

System identification theory allows the creation of linear mathematical models describing relationships between inputs and outputs taken from the real plant (measurements). Essentially this is done by adjusting parameters within a given model until its output coincides as well as possible with the measured output. A set of experiments has to be done in order to get the data set with adequate information contents. The design of these experiments includes selection and determination of input signals, sampling time, measuring time period, equipment and filtering. Evidently a rather sound knowledge of the process helps to plan these experiments. Again the drawback of using linear models is their limited operational range. Performance might be severely degraded outside the identification conditions for a strongly non-linear process.

A phase of identification was carried out at the String with the goal of collecting input-output data for designing linear models. A reasonably good performance is observed by using a transfer function (Eq. (1)) of second order that was identified with a sampling period of 20 seconds using Matlab® (Identification® and *Hiden*® toolboxes). The $G(q^{-1})$

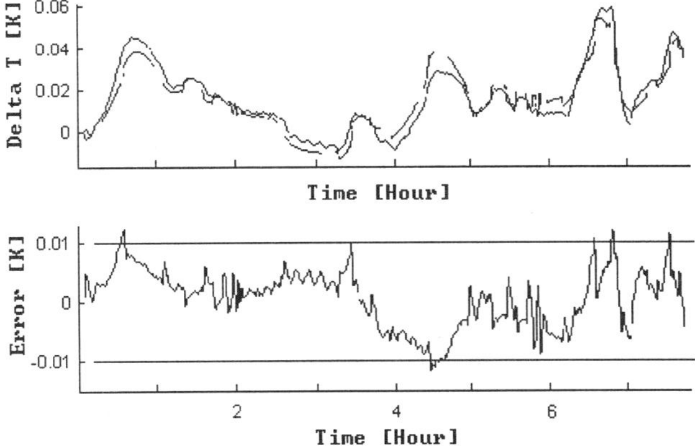

Figure 2. Δ(temperature vs. model ouput) @ 1.87 K

represents the transfer function model on the backward shift operator q^{-1}, u(t) is the input (valve), and y(t) is the output (temperature).

$$y(t) = G(q^{-1})u(t) \quad ; \quad G(q^{-1}) = \frac{2.532\text{e} - 005\, q^{-1} - 2.702\text{e} - 005\, q^{-2}}{1 - 1.918\, q^{-1} + 0.9179\, q^{-2}} \quad ; \quad q^{-1} \cdot u(t) = u(t-1) \tag{1}$$

The identified model performance is checked comparing measured and simulated data (Figure 2). Residuals are within a band narrower than approximately 10 mK, which is a good result.

CONTROL SYSTEM

Regulation goal

The regulation goal is to keep the temperature of the superconducting magnets as constant as possible within strict operating constraints imposed by: the maximum temperature at which the magnets can operate, the cooling capacity of the cryogenic system, the heat loads, and the accuracy of the instrumentation.

The Joule-Thomson valve is the manipulated variable, and the warmest temperature sensor located in the cold masses (two on each magnet) provides the controlled variable. Disturbances are of two different types: heat loads and variations in the flow through the Joule-Thomson valve. Heat loads are produced by heat inleaks from the higher temperature levels, magnet current ramping and particle beam losses (simulated in this case by electrical heaters). The set point is the saturation temperature of the liquid helium flowing through the HX plus a certain ΔT, typically 0.06 K.

Up to now, a PI regulator has controlled the pilot plant (String), but despite its ability to bring the controlled variable to the desired value in the presence of heat load variations, a substantial improvement is expected by using a more suitable control approach. Such controller should have a reduced regulation band that would allow relaxing the constraints on either the instrumentation accuracy or the cryogenic system capacity margin.

MBPC controller algorithm

Model Based Predictive Control (MBPC) methods[3] use a dynamic model of the process to predict the controlled variable ($y(t)$). This prediction is used in an on-line optimization procedure that minimizes an appropriate cost function to determine the manipulated variable ($u(t)$). Usually the cost function depends on the quadratic error between the future reference variable and the future controlled variable within a limited time horizon. This procedure is repeated every sampling time with actual process data (receding strategy)

MBPC presents a series of advantages like, among others, adaptability to a great variety of processes (i.e.: long time delays, non-minimum phase), it introduces feedforward in a natural way for compensating measurable disturbances, it considers delay times, it has a complete treatment of constraints and the multivariable case can be easily dealt with.

The methodology[4] is composed of:

(a) The future outputs, $y(t+j)$, for a determined time horizon, N, are predicted at each instant, t, using a process model (Eq. 2). This predicted output depends on the past history (input, output) and the future control signals. In the case of GPC (Generalized Predictive Control) the relationship between the control and manipulated variable is:

$$Ay(t) = Bu(t) + n(t) \tag{2}$$

where $A(q\text{-}1)$, $B(q\text{-}1)$ are polynomials on the $q\text{-}1$ shift operator, $u(t)$, $y(t)$ the input and output data respectively, and $n(t)$ a noise signal.

(b) A future reference trajectory $w(t+j)$ is defined (Figure 3), which describes how the process should be driven from the actual $y(t)$ to the desired set-point $r(t+j)$

(c) The set of future controls ($u(t)$, $u(t+1)$, $u(t+2)$, ..., $u(t+Nu)$) is usually calculated by minimizing a quadratic function (Eq. (3)) of the errors between the predicted output and the reference trajectory, usually also including penalties on the control moves.

$$I = \sum_{j=N1}^{N2}[w(t+j)-\hat{y}(t+j)]^2 + \beta\sum_{j=0}^{Nu}[\Delta u(t+j)]^2 \tag{3}$$

N1, N2 : prediction horizon
Nu : control horizon
β : tuning parameter

(d) The manipulated variable, $u(t)$, at time t, is sent to the process. This procedure is repeated every sampling time.

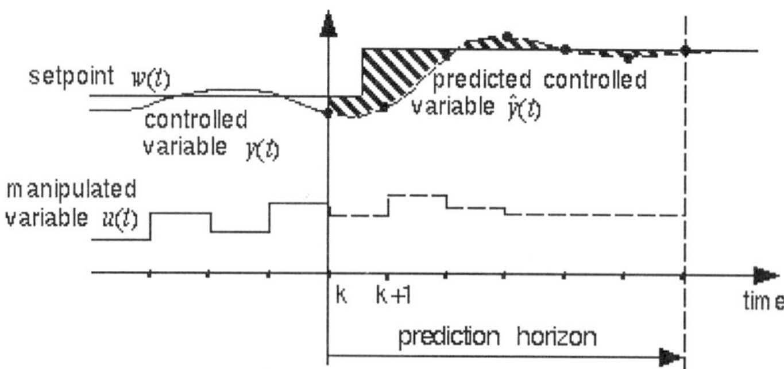

Figure 3. Basic principle of model predictive control

(e) This optimization procedure can include restrictions (Eq. (4)) in the future values of the process variables:

$$\begin{aligned} L &\leq u(t+j) \leq H \\ L' &\leq \Delta u(t+j) \leq H' \\ L'' &\leq y(t+j) \leq H'' \end{aligned} \tag{4}$$

Control Architecture

Figure 4 shows the regulatory control strategy implemented in the String test. Two Programmable Logic Controllers (PLC) are used concurrently. The first (S5) is a standard PLC where a PI control loop is implemented. The M7 represents an industrial PC using a real time operation system (RMOS®) where high level programs (C++) run over. The communication between them is done via serial port (3964R Siemens® protocol)

The S5 controller takes the temperature measurements from the magnets and acts on the Joule Thomson valve. During normal conditions the PI is idle, and the manipulated and controlled variable are exchanged with the MBPC (M7); once the M7 receives the controlled variable, a predictive control algorithm calculates and sends back the appropriate valve sequence to the S5 PLC. A watchdog is always checking that the M7 is working properly; if an anomaly is detected, the PI implemented on the S5 takes over the control yielding a high degree of reliability.

A desktop PC has been connected to the M7 via the serial port (RS-232-C) for controller and process monitoring, controller tuning and data archiving. For these purposes, a Windows application was developed using the Object Windows Library (OWL®). This application allows mainly:

- On-line visualization of the controller predictions, optimum valve sequence calculations and internal controller variables.
- On-line controller tuning (models, horizons, weights, constraints…)
- Data archiving of the measured and predicted variables, optimal sequence calculation and some internal variables of the controller.

EXPERIMENTAL RESULTS

The control loop should be tuned either for set-point changes, process disturbances, or reduced overshoot. Tuning is usually performed for one strategy, after which the acceptability of the tuning with regard to the other strategy is verified and compromises are made as necessary.

Stability is the keystone of the LHC 1.8 K cooling loop because, in principle, it does not require set point changes during normal operation. Control loop response can be checked by observing the process variable when process upsets are simulated. In this case, two heaters inside the dipoles 1 and 3 were used to perturb the process with a quantified heat load representative of the future LHC operational scenario. Disturbances of 0.2 watts/meter were used in typical experiments, and an exceptional 0.6 watts/meter heat load was used to check the behavior of the controller with higher temperature excursions.

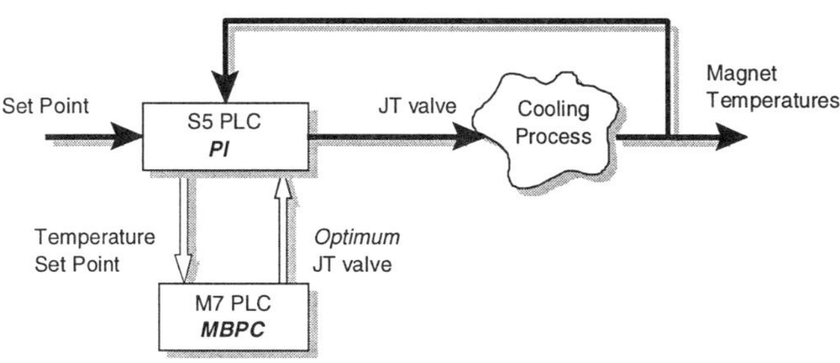

Figure 4. Control architecture

Control loop performance can be quantified by the overshoot percentage, time to return to set point, the integral of the square of the error, ISE, and the integral of time multiplied by the absolute value of the error, ITAE, Eq. (5).

$$ISE = \int_0^T e^2(t) \cdot dt \qquad ITAE = \int_0^T t \cdot |e(t)| \cdot dt \tag{5}$$

Results I: Controller Robustness

As explained before, the controller is designed to be robust against possible perturbations. To achieve this objective, extensive experiments on tuning were performed and robustness of the controller was checked producing perturbations (heat load between 0.2 and 0.6 W/m).

The MBPC controller performance is shown in comparison with a PI controller which tuning parameters have been experimentally optimized. Processes, which show delays and inverse response, oblige to slow down the controller action in order to preserve stability. PI parameters that seem to be the more appropriate for the majority of the different working ranges of the process may produce unstable actions if the heat loads change.

By applying a heat load of 0.2 W/m (Figure 5), it is possible to notice that using a MBPC with constraints on temperature excursions (allowed band of 6 mK around the set point), could improve largely the PI robustness. The indexes are improved in 60.5% for the IAE and a 38.5% for the ITAE one. Furthermore the temperature excursion and settling time are reduced by 36.4% and 26.4% respectively.

With larger heat loads of up to 0.6 W/m (Figure 6) both the PI and MBPC performances are degraded since their parameters where optimized in a different operational region. However the MBPC shows a far superior behavior on settling time and temperature excursion.

Clearly the recovery of the temperature excursion shows that the controller overreacts and a correction is required afterwards. This is a clear symptom of deteriorated predictions either provoked by the model-process mismatch or by using a linear model outside of its operational range. Anyway the robustness observed for MBPC is superior to that of the PI, which suffers from a strong oscillatory behavior when applying sudden heat loads. Corroborating this fact, several linear MBPC techniques tested on the String have also shown better behavior[5].

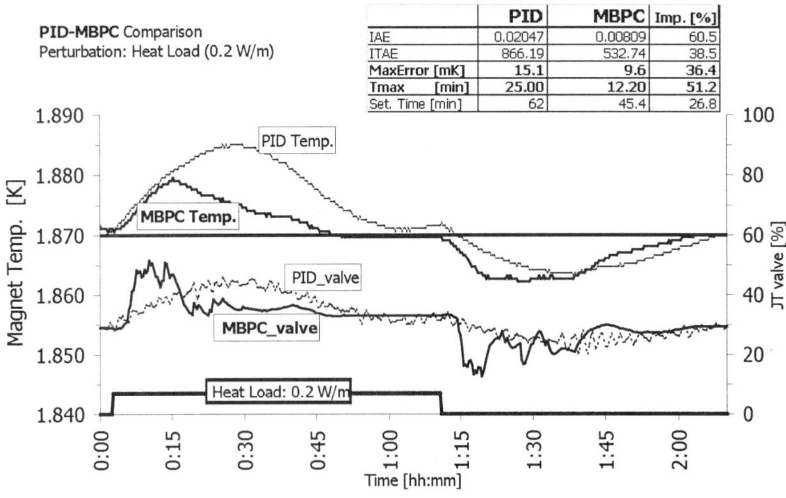

	PID	MBPC	Imp. [%]
IAE	0.02047	0.00809	60.5
ITAE	866.19	532.74	38.5
MaxError [mK]	15.1	9.6	36.4
Tmax [min]	25.00	12.20	51.2
Set. Time [min]	62	45.4	26.8

Figure 5. Compared performance of MBPC vs. PI (Disturbance: 0.2 W/m)

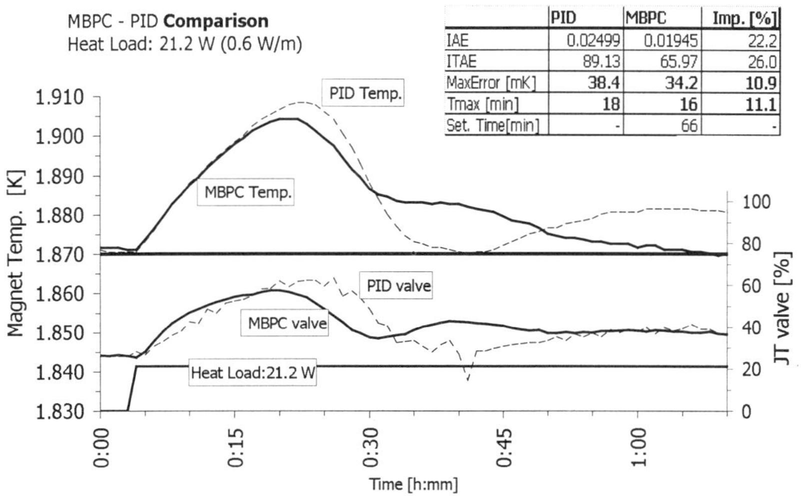

	PID	MBPC	Imp. [%]
IAE	0.02499	0.01945	22.2
ITAE	89.13	65.97	26.0
MaxError [mK]	38.4	34.2	10.9
Tmax [min]	18	16	11.1
Set. Time[min]	-	66	-

Figure 6. Compared performance of MBPC vs. PI (Disturbance: 0.6 W/m)

Results II: Controller Performance

Although the controller has been designed and tuned specifically for robustness against disturbances, its performance has been tested under changes on the set point, even though this is not representative of the LHC operation.

MBPC response, under a set-point change (15 mK), is compared with that of the PI (Figure 7), both, with aggressive parameters [K=100,T_i=60] and with the sub-optimal parameters [K=50, T_i=60]. In a classical PI controller, K represents the controller gain and T_i represents the integral time. With the first couple of parameters and the set-point step up, we clearly get an unstable situation with oscillatory behavior. In the step down, the parameters have been set to the conservative ones [K=50,T_i=60] getting a better response though still worse than the MBPC.

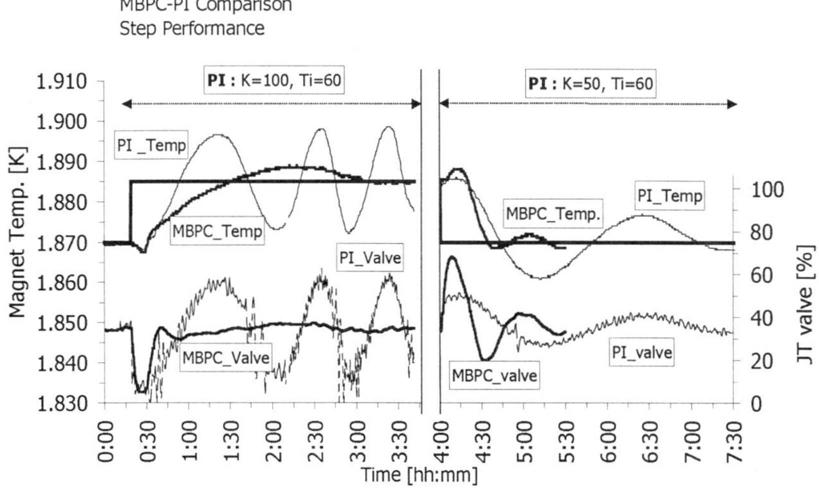

Figure 7. Compared performance of MBPC vs. PI (Step test)

CONCLUSIONS

After four years operation, the test-bed called LHC Test String[6] was switched off. During the experimental program the equipment installed in the String operated for almost 13000 hours below 2 K. This corresponds to a good simulation of the operating conditions expected at the LHC. The new Test String (String 2) will be commissioned in early 2001, and it will represent a full cell of the LHC accelerator consisting in two quadrupoles and six dipoles with a length of 106.9 meters.

Choosing an MBPC technology for a given application is a fairly complex question. In our particular case, MBPC has shown a substantial regulation improvement, and it did demonstrate the potential of using new advanced control techniques to the regulation of very complex processes with high non-linearities like the one exposed in this paper. Further improvements should still be expected when applying non-linear predictive control[7] algorithms.

The major challenge to the String 2 implementation will be the usage of non-linear models embedded on the controller to perform non-linear predictive control. The controller algorithm has to be reformulated and a new approach has to be programmed based on a first principles model. Extensive work has been done on the process modeling, but still much work is required to develop a non-linear controller.

The more important feature of the LHC 1.8 K Cooling Loop is the fact that the operating conditions could change abruptly and then the process characteristics vary largely as modifications in the dead time and inverse response amplitude. The String 2 is expected to exhibit longer delays, although the pressure drop will be largely reduced by using the new heat exchanger smooth tube instead of the former corrugated one. It is then essential that the controller cope with such disturbances. Improvement on robustness should be obtained for a model covering a larger working zone without having problems with the prediction calculations.

ACKNOWLEDGEMENTS

This work has been done under the framework of the European Training and Mobility of Researchers (TMR) Program (TMR contract: ERBFMBICT 972793). Special gratitude to the ISA group (University of Valladolid) and the String Team (CERN).

REFERENCES

1. J. Casas Cubillos, et. al., Operation, testing and long term behavior of the LHC Test String cryogenic system, Proceedings ICEC17, Institute of Physics Publishing, Bristol (1998), pp. 747-750.
2. B. Flemsaeter, E. Blanco Viñuela, J. Casas Cubillos, C. de Prada, S. Saelid, Applying advanced control techniques for temperature regulation of the LHC superconducting magnets, Proceedings ICEC17, Institute of Physics Publishing, Bristol (1998), pp. 631-634.
3. S. Joe Qin, T. A. Badgwell, An overview of industrial model predictive control technology, AIChE Symposium Series No. 315, 15, (1997), pp. 232-256.
4. D.W. Clarke, C. Mohtadi, P.S. Tuffs, Generalized predictive control: the basic algorithm I, Automatica, vol. 23. Nº 2, Great Britain (1987), pp. 137-148.
5. E. Blanco Viñuela, C. de Prada Moraga, J. Casas Cubillos, S. Cristea, Predictive temperature control of the LHC superconducting magnets, European Control Conference. Karlsruhe, Germany, (1999).
6. F. Bordry, J. Casas Cubillos, P. Cruikshank, et al., The LHC magnet String programme: status and future plans, Particle Accelerator Conference, New York, (1999).
7. S.J. Qin, T. A. Badgwell, An overview of nonlinear predictive control applications, Preprints international symposium on nonlinear model predictive control: assessment and future directions, Ascona, Switzerland, (1998), pp. 128-143

PELTIER HEAT FLUX SENSOR FOR CRYOGENIC USE

T. Haruyama

KEK, High Energy Accelerator Research Organization
1-1 Oho, Tsukuba, Ibaraki 305-0801, Japan

ABSTRACT

A thermoelectric element, known as a Peltier element, is evaluated as a heat flux sensor at cryogenic temperature. Several commercially available Peltier elements were tested at 77 K to detect heat flux generated by the heater. Preliminary experimental results show that an output voltage of the Peltier element is nearly 10 times larger than that of the same-size conventional heat flux sensor at 77 K. It implies a possible use of the Peltier element as a heat flux sensing device for cryogenic applications. This element can be used as a kind of *in situ* sensor for radiant heat flux measurement. Also, it will be a strong tool for investigation of insulation techniques, e.g. performance check at the overlapping part of multi-layer insulation etc.

A newly developed heat meter was used to calibrate the heat flux through the Peltier element. It is made of a thermal resistance material (stainless steel: SS), sandwiched by the oxygen free copper block by using the HIP (hot isostatic pressing) method.

INTRODUCTION

In order to design and fabricate cryogenic equipment properly, it is important to know the amount of radiant heat flux and where it enters the system. A boil-off method by using a heat meter[1,2] has been applied to measure the amount of radiant heat leak; however, this method can only be used to measure the integrated heat flux to the system. *In situ* heat flux measurement has not been done because there is no appropriate sensor applicable to cryogenic temperature. Of course, there are many conventional heat flux sensors for high temperature applications, for measuring heat flux from an iron furnace, heat leak from LNG tank to ground and so on. The amount of heat is around several hundred to tens of thousands of W/m^2, and the operating temperature is not as cold as 77 K. Most materials of these sensors can not be used at cryogenic temperature. One conventional heat flux sensor is claiming possible measurement of the radiant heat flux at cryogenic temperature. However, the output signal of the sensor seems to be too small for cryogenic applications.

In this study, three commercially available Peltier elements with the same surface area and different number of *p-n* pairs were tested at 77 K. Preliminary experimental results show that the Peltier element has nearly 10 times larger output than the conventional heat flux sensor. It implies a possible use of the Peltier element for heat flux measurement at cryogenic temperature.

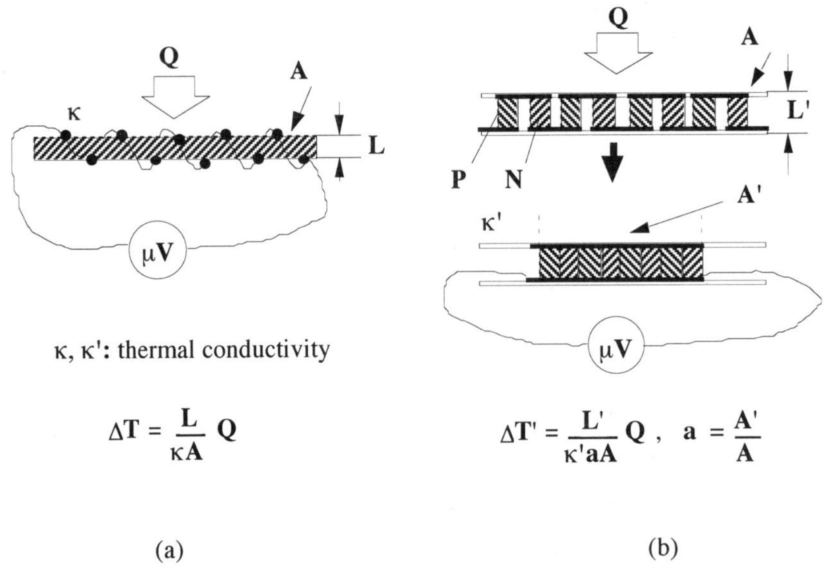

Figure 1. Principle of heat flux measurement by a conventional sensor (a), and the Peltier element (b).

PELTIER HEAT FLUX SENSOR

A principle of heat flux measurement by a conventional sensor is shown in Figure 1 (a). Thermocouples are connected in series on both sides of a thermal resistance material to measure a temperature difference across the material. When heat flux passes through the material, a temperature difference of both sides, ΔT, can be detected by the thermocouples. For low temperature applications, a thin polyimide sheet is used as the thermal resistance material utilizing a 40-thermopile construction. The performance specification of the company brochure[3] says that a normal sensitivity at 300 K is 1-2 μV/K.

The Peltier element consists of a number of small thermoelements connected electrically in series. Typical thermoelements are made by *n*-type and *p*-type bismuth telluride for commercial use. A few hundred of the 1x1x1.5 mm^3 bismuth telluride thermoelement are soldered to the bottom and top plates. The plates are subsequently slit to form separate links. Finally, the plates are sandwiched between two ceramic plates[4]. As shown in Figure 1 (b), these elements are placed loosely, so there is a geometrical merit to make a large temperature difference across the elements. Taking into account the thickness, surface areas and thermal conductivity of materials, ΔT of the Peltier element is nearly twice as large as the conventional sensor. The most effective reason to use the Peltier element is that the thermoelectric power of the semiconductor, Seebeck coefficient, is about 10 times larger than that of a metallic thermocouple[5].

Table 1 gives detailed information of the Peltier elements tested in this study. The surface area of each of the three elements is 30x30 mm^2. Two elements have 127 *p-n* pairs and one element has 71 *p-n* pairs. Pictures of the element are shown in Figure 2. Many small thermoelements of bismuth telluride can be seen in the side view picture. The hot and cold plates are made of aluminum oxide ceramic whose thermal conductivity is nearly 10 times larger than that of bismuth telluride at 300 K. There are many types of commercially available Peltier elements.

Table 1. Peltier elements tested in this study

Peltier element	Supplier	Number of *p-n* pairs	Dimension (mm^3)
(1) CP-1.0-127-05L	Melcor	127	30x30x3.27
(2) M010	--	127	30x30x4.00
(3) KSM-06071A	Komatsu	71	30x30x3.75

Figure 2. Peltier element, left: 127 *p-n* pairs, 40x40x3.5 mm^3(not tested in this study), 30x30x3.5 mm^3; right: side view of an element (30x30x3.27mm^3).

EXPERIMENTS

Set Up

A special sample holder has been prepared to measure the relationship between the heat flux and the output voltage of the Peltier element. The holder is shown in Figure 3. It consists of a top plate with a 20 Ω MINCO heater and a body with a newly developed heat meter section. The detailed schematic drawing of the sample holder is shown in Figure 4. The Peltier element is clamped between the top plate and the body by using four plastic screws. The heat leak through these plastic screws is estimated to be nearly 0.3 % of the total heat input. A sheet heater is glued tightly by using a special resin prepared by MINCO products, Inc. No serious problems such as peeling off or deformation of the heater occurred. The heater is used to generate a known amount of heat flux to the Peltier element and the heat meter.

The platinum thin film thermometers are used to measure temperatures across the Peltier element and the thermal resistance material (SS). The CRZ-2005 platinum resistance thermometer by Hayashi Electric, Japan, is as small as 2x5x1.5 mm^3. These thermometers were calibrated using a commercially calibrated platinum thermometer, PT-103, supplied by Lake Shore Cryotronics, Inc.

Figure 3. Top plate with heater (left) and a body with a heat meter (right). A SS block of 20 mm diameter and 10 mm long is used as a thermal resistance material for heat meter.

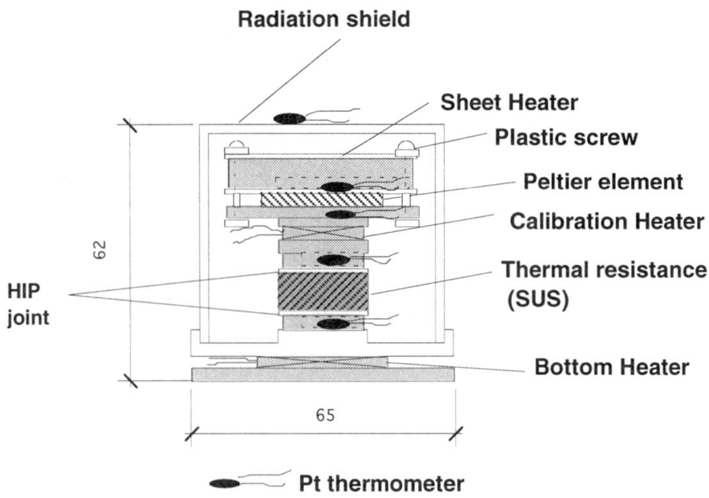

Figure 4. Sample holder with a calibration heat meter. Kapton sheet heater, MINCO 20 Ω, is used to generate heat flux to the Peltier element.

A heat meter was set in the middle of the body for calibration purpose. A stainless steel (SS) rod of 20 mm diameter and 10 mm long is joined to the oxygen free copper rod by the HIP (hot isostatic pressing) method. A manganin wire of 0.2 mm diameter was wound at the upper side of the heat meter to generate the calibrated heat flux. As the thermal conductivity of the thermal resistance material (SS) is well known at cryogenic temperature, ΔT across the SS material simply indicates the amount of heat flux through it.

An additional heater was set at the bottom of the sample holder to change the operation temperature. However, the heating capacity of the manganin wire was too small to heat up the holder to a high enough temperature.

Silicon vacuum grease was used to assure thermal contact between the sample and the sample holder at cryogenic temperature. This grease was also used to fix the sample holder to the bottom of the cryostat. No screw was used to fix the sample holder. The grease seemed to be a good thermal contact and also worked as a glue at cryogenic temperature.

A copper radiation shield was capped to prevent any radiant heat flow from the surroundings. The temperature of the shield was also monitored by a platinum thermometer.

Procedure

A schematic drawing of the final set up in the cryostat is shown in Figure 5. The cryostat was made by a GFRP pipe of 100 mm outer diameter and 5 mm thickness. The top flange and the bottom copper plate were glued by Stycast 2850 epoxy. The Peltier element was set between the top plate and the body of the sample holder and fixed by four plastic screws. Then, the sample holder was fixed at the bottom of the cryostat. A turbo molecular pump was started to evacuate the cryostat down to 10^{-3} Pa. Liquid nitrogen was used to cool the cryostat slowly to avoid excess heat shock to the equipment. Finally, the cryostat was fully immersed into the liquid nitrogen bath. Platinum thermometers were excited with a 1 mA constant D.C. current (TR6142, Advantest) and the output voltages were measured with 0.1 μV resolution (Keithley 2000). The sensitivity of the platinum thermometer is 0.43 mV/K at 1 mA near 77 K. The 0.1 μV resolution of the system results in a measurable temperature change of 0.23 mK. Cooling down characteristics and temperature profiles

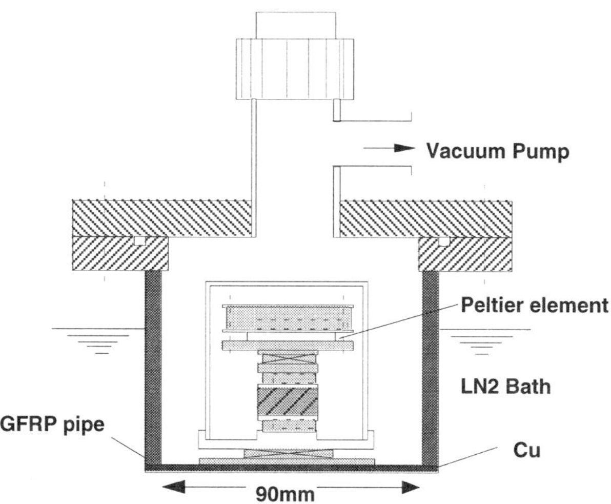

Figure 5. Experimental set up - sample holder is fixed on a copper plate of the cryostat, then the cryostat is evacuated and immersed in the liquid nitrogen bath.

were also recorded (LR8100, Yokogawa). It took approximately 30 minutes to cool down the sample holder from 300 K to 77 K. After the temperature of the sample holder stabilized, the sheet heater of the top plate was energized to generate a certain amount of heat flux to the Peltier element. The amount of power applied to the Peltier element was 0 to 0.057 W, corresponding to a heat flux of 0 to nearly 60 W/m^2 to the samples with a surface area of 30x30 mm^2. Assuming that the emissivity of the materials is 0.1, the amount of radiant heat flux from the 300 K surface to the 77 K surface is calculated to be about 40 W/m^2.

RESULTS

Calibration

At first, a calibration of the heat meter was carried out. A typical result is shown in Figure 6. The calibration heater was energized from 0 to 0.02 W, and the temperature differences between the thermal resistance material (SS) were measured and plotted as closed circles in Figure 6. A sheet heater at the top plate was also energized to approximately 0.045 W. Measured results in ΔT are plotted as open circles. These two calibration data are well-fit on a straight line. From the slope of the ΔT-Q line, the thermal conductivity of SS is calculated as 8.16 W/mK for 20 mm diameter and 10 mm long. This is close to the standard value of thermal conductivity of SS at 77 K (8.154 W/mK). There are small temperature offsets in each sensor, and so, ΔT does not indicate zero even at Q=0. The calibration results imply that there is no serious heat leak and the amount of heat flux added to the sample can be calculated by reading ΔT across the SS material, or reading an electric power added to the sheet heater.

Output voltage characteristics

Figure 7 shows the relationship between the output voltage and heat flux for two different elements with 127 and 71 *p-n* pairs. The thermoelectric power of each element per unit heat flux (W/m^2) is 27 μV for 127 *p-n* pairs and 15 μV for 71 *p-n* pairs. The sensitivities of these Peltier elements are nearly 10 times larger than that of a conventional

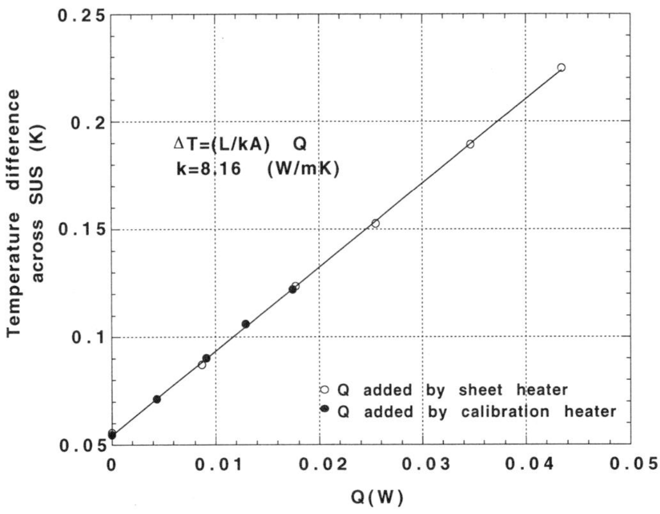

Figure 6. Heat flux calibration by the calibration heater and the sheet heater. Thermal conductivity of SUS in the heat meter is estimated from the slope of the ΔT-Q line.

heat flux sensor. The output voltage is large enough to use only a simple DC amplifier for practical applications. As the size of each bismuth telluride block is almost the same, it can be said that the thermoelectric power per *p-n* pair is 0.209 μV/(W/m^2). It is not shown in Figure 7 but a similar result was obtained for the M010 element as that of the CP1.0-127-05L element.

Temperature dependence of the output voltage was also tested preliminarily by energizing the bottom heater of the sample holder. As mentioned before, the heater wire

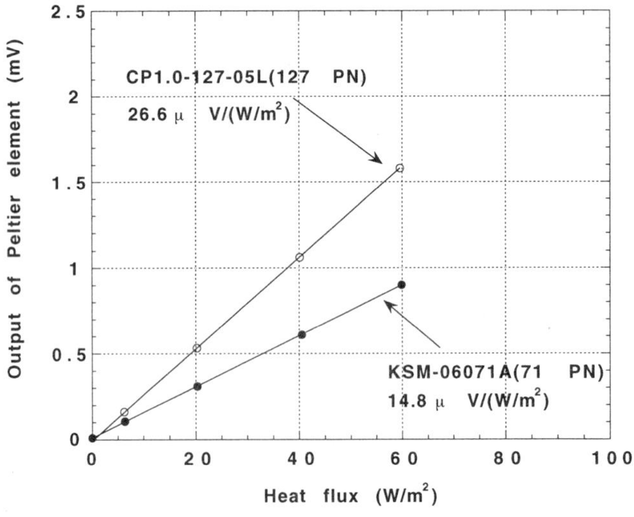

Figure 7. Relationship between the output voltage and heat flux for two different Peltier elements with 127 and 71 *p-n* pairs.

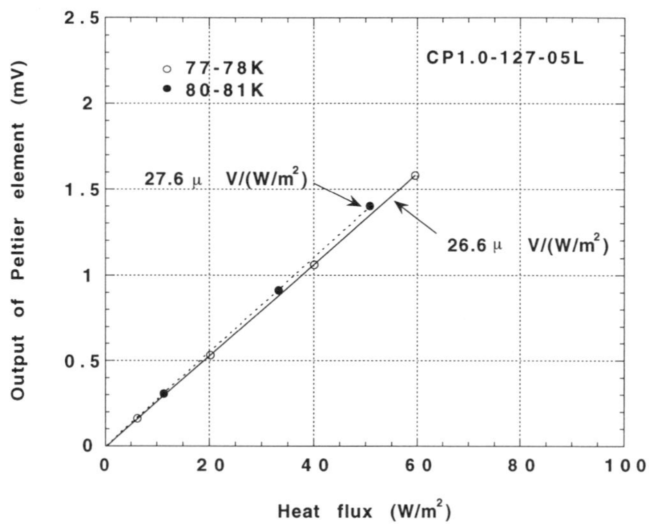

Figure 8. Output voltage at 77 K and 80 K for 127 *p-n* pairs element.

was not large enough to raise temperature of the sample holder significantly and only a few degrees change was obtained in this study. Figure 8 shows the typical output voltage at temperatures of 77 K and 80 K. For practical applications, more results of the temperature dependence in a wide temperature range should be obtained.

The thermal resistance of the Peltier element was estimated by measuring the temperature difference across the element when the heater power was supplied. The sheet heater was energized up to 0.045 W. Figure 9 shows the ΔT-Q relationship of the Peltier element with 127 *p-n* pairs at 77 K. The relationship between Q and ΔT can be written as $\Delta T = WQ$, where W is the thermal resistance defined as $W = (L/A\kappa)$. From the slope of the

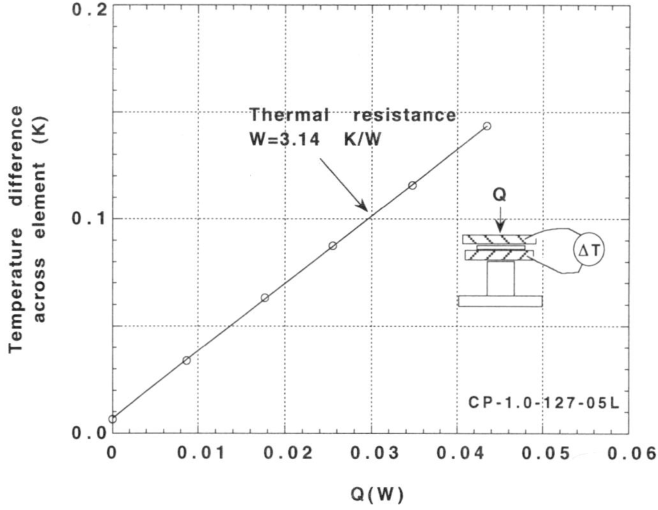

Figure 9. Thermal resistance of the Peltier element at 77 K.

line in Figure 9, the thermal resistance of the element is calculated as 3.14 K/W. This is close to the theoretical value of 2.98 K/W for the Peltier element with 256 small pieces of bismuth telluride thermoelement of 1x1x1.5 mm^3. The small discrepancy in the thermal resistance may be attributed to the additional thermal resistance of the silicon grease.

SUMMARY

Commercially available Peltier elements were tested near 77 K for possible application to radiant heat flux measurement. For the element with 127 *p-n* pairs and surface area of 30x30mm^2, the output voltage was 26 $\mu V/(W/m^2)$ which is nearly 10 times larger than that of a conventional heat flux sensor at 77 K. This implies a possible use of the Peltier element for *in situ* measurement of cryogenic heat flux. It can be concluded preliminarily that the temperature dependence of the thermoelectric power of the Peltier element is not large near 77 K. Also, it is confirmed that a simple heat meter by the HIP (hot isostatic pressing) method is useful for a heat flux calibration. More experimental research should be done, especially on the temperature dependence of the thermoelectric power.

ACKNOWLEDGEMENT

This research was partly supported by the Grant-in-Aid for Scientific Research (C) of the Ministry of Education, Science Sports and Culture (09650258).

REFERENCES

1. M. Kichnir, J.D. Gonczy and J.L. Tague, Measuring heat leak with a heat meter, “Advances in Cryogenic Engineering 31”, Plenum Press, New York (1988), p 1285.
2. J.D. Gonczy, M. Kuchnir, T.H. Nicol, R.C. Nieman and R.J. Powers, Heat leak measurement facility, “Advances in Cryogenic Engineering 31”, Plenum Press, New York (1988), p 1291.
3. Micro foil heat flux sensor Model 27060, RdF Corp., Hudson, NH 03051, USA
4. H.J. Goldsmid. “Thermoelectric refrigeration”, Plenum Press, New York (1964)
5. S. Yamaguchi, H. Nakamura, K. Ikeda, T. Sakurai, I. Yoshida, S. Tanuma, S. Tobise and K. Koumoto, Peltier current lead experiments with a thermoelectric semiconductor near 77 K, *Proceedings of MT-15*: 855-858 (1998)

MRI COMPATIBLE CRYOSURGICAL MACHINE

B-Z. Maytal

Rafael, Cryogenic Section
P.O.Box 2250 (39), Haifa 31021, Israel

ABSTRACT

The strong magnetic field of the interventional magnetic resonance imaging (IMRI) challenges the designer of surgical and particularly of cryosurgical devices. It strongly interferes with electronic, electromechanical and magnetic media. The proposed solution incorporates an argon based Joule-Thomson cryosurgical machine. The entire main frame, including control, monitoring and pressure vessels, is installed outside the MRI affected chamber, remaining totally out of the influence of the magnetic field. Gas supply tubing provides argon to a panel close to the site of surgery or on the magnet frame itself. It includes ports for five cryoprobes and their control enabling autonomous operation, sufficiently handy and close to the surgical site. In addition, the view of the outer main frame monitor that displays the parameters of cryosurgical treatment is transmitted to the MRI display monitor. So far, successful experience is accumulated on about ten IMRI medical center stations.

INTRODUCTION

The Interventional (or Intraoperative) Magnetic Resonance Imaging (IMRI), actually the MRI assisted surgery is recognized as a potential and promising therapeutic technique [1, 2, 3, 4, 5, 6]. Rubinsky[7, 8] reviewed the advantages of conducting a cryosurgical procedure supported by MRI. Each phase of the treatment may enjoy the support of MRI. Prior to an operation, the tumor is observed for planning the cryo ablative strategy. One may track the position of invasive devices [9, 10]. During the cryocooling phase it can be used to monitor the location and shape [11, 12] of frozen tissue. Measuring the temperature distribution through the MRI at the unfrozen adjacent tissues [13, 14] may be used to calculate in real time[15, 16] temperature distribution in the frozen bulk, velocity of the freezing front, and observe its overlap with the target tumor. After thawing, post surgical cryodamage may be evaluated, whether by following the development of the edema or by examining treated tissue viability by spectroscopic analysis [17, 18], generally of phosphor, carbon and sodium. More about MRI assisted cryosurgery is recommended in references [19, 20, 21, 22, 23, 24].

Advances in Cryogenic Engineering, Volume 45.
Edited by Shu *et al.*, Kluwer Academic / Plenum Publishers, 2000.

However, the high magnetic field environment of the MRI does challenge indeed, the engineering of such combinations. The state-of-the-art open magnetic loop MRI induces fields of about 0.5 T while the closed type interventional MRI maintains 1.5 to 2 T.

The objective of the present summary is to introduce an IMRI compatible cryosurgical machine (patent pending) with multiplicity of probes and variety of control options. Itzhak et al. [25] reported ten cryosurgical experiments on animals conducted with this machine under 0.5 T.

THE PROBLEM

Two levels of interference should be considered. The elementary level is of the probe itself, in direct contact with the undesirable target living tissues, designated to cooldown and perform the cryo-ablation. Tissues around the probe are the objects of the imaging process. Therefore, the probe should be compatible with the high magnetic field and should not induce artifacts, which interfere with the identified pattern [26]. This issue is handled by proper choice of materials with vanishing magnetic signature. Isoda [11, 12] employed gauze immersed in liquid nitrogen. Matsumoto [17] filled a non metallic cup with liquid nitrogen. Rubinski [7] applied non-metallic materials. The probe that contains liquid nitrogen is made of Pyrex glass. The stereotactic holder is constructed by plexiglass with nylon screws with silicon rubber tubing for inlet and exhaust lines. There are some metallic alloys that do not induce artifacts.

The second and much more complicated problem is related to using the cryosurgical machine in the vicinity of the MRI. The test for successful operation is simple, at least conceptually whether is it possible to operate the entire cryosurgical machine as if it would run in a regular surrounding, out of the MRI affected zone.

Locating a cryosurgical machine close to the surgeon means it is near the MRI magnet. That results in two harmful influences: (a) the bulk of the machine disturbs the uniformity of the magnetic filed, and (b) the machine itself is subjected to the undesirable influence of the magnetic field.

One should suggest constructing the whole machine solely by MRI compatible metals and non-metallic materials. That would become extremely expensive, demanding in some case developments and study activities. Furthermore, some elements must use ferromagnetic materials, like operating valves and solenoids. The electronic devices and control would demand remarkable efforts for being protected against the magnetic influence and still the success might remain questionable.

One should distinguish between two MRI environments: the closed magnetic loop IMRI and the open version. The former is more challenging since in general it induces larger magnetic fields (up to 2 T). Furthermore, the patient is treated out of the magnet and then inserted with the cryoprobes into the magnet for monitoring. Therefore, it demands more distant operating cryoprobes as the open MRI provides direct access.

THE SOLUTION

The following principles rest beyond the general concept, as schematically displayed in Figure 1:

1. The basic cryosurgical machine for the present application is driven by pressurized gas [27] at room temperature. It incorporates a Joule-Thomson cryocooler at the tip of each probe [28, 29], creating a bath of liquid cryogen upon demand. The noble gases are

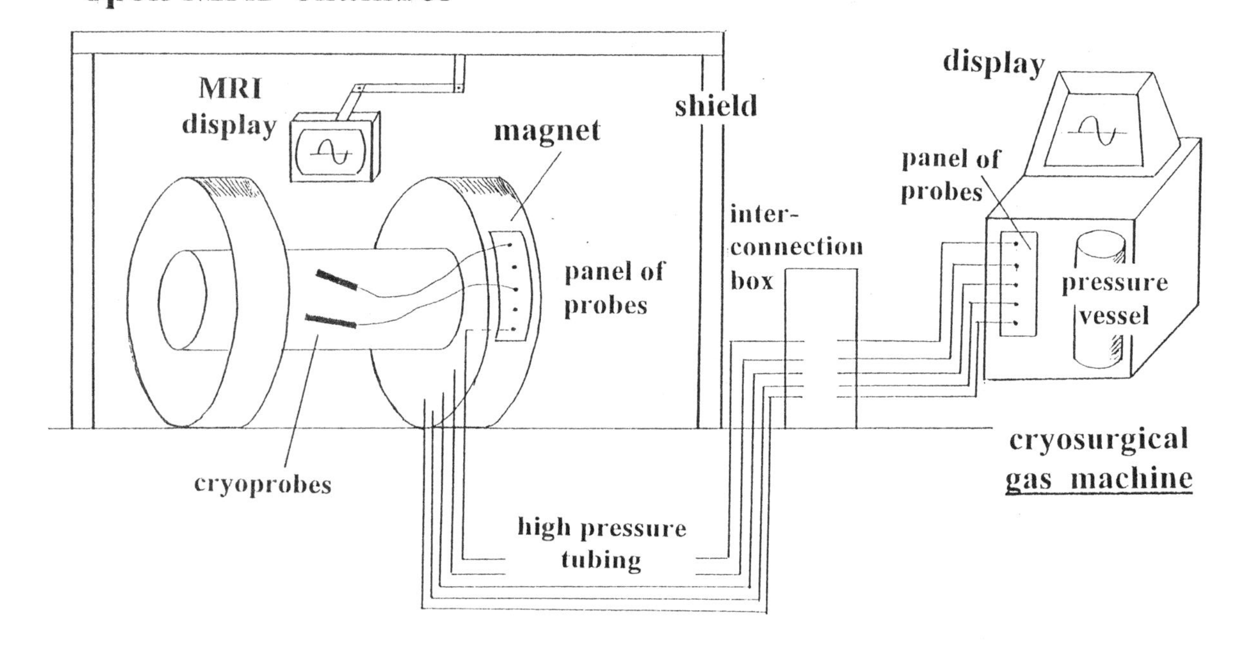

Figure 1. Schematic layout of MRI compatible cryosurgical machine.

the favorite candidates for Joule-Thomson cryocooling [30]. Hence, argon is the preferable coolant while its low boiling point enables the freezing of a substantial bulk (large iceball) of undesirable tissues. A multi-probe configuration [31] facilitates handling of irregular shapes of tumors and magnifies the ablative power. The main frame of the machine includes pressure vessels for storage of argon.

2. The entire main frame cryosurgical machine, excluding the probes themselves, is located out of the MRI affected room [32].

3. Multiplicity of probes is mounted in the close vicinity of the surgical site. The desirable probes are connected (and disconnected) at this panel. Put another way, probes are prolonged from the machine located outside the operating MRI chamber.

4. This extension includes a control device for operating the probes and selecting the desirable function: stick, freeze, thaw and stop. This way the has full control over the cryosurgical machine even though it is remotely located.

5. High-pressure tubing is installed into the MRI room, one tube for each probe, conducting pressurized gas from the vessels of the cryosurgical machine. Tubes below 2-mm diameter are tolerated and still keep reasonable pressure drop at the applied flow rates. A special isolator is used to avoid electrical current flow through this tubing out of the room.

6. Each pressure tube comprises an intermediate reservoir, which continues to feed the probe after turning off the gas supply. However, its volume is about 15 cm^3 per line of 10 m. Typical flow rates range from 15 to 25 standard liters per minute. Hence, the delay lasts only for few seconds.

7. This arrangement preserves the special thawing capability of the above mentioned cryosurgical gas machine [28, 29]: Helium instead of argon is pressurized into the probe and the inverse Joule-Thomson effect warms up the probe. The small capacity of the transfer lines enables trouble free switching from argon to helium and vice versa. Since the pressure tubes remain at room temperature and the total thermal inertia of the probe is very low, warming is obtained within few seconds.

8. Upon request the MRI monitor takes over the display of the monitor of the cryosurgical machine outside. The surgeon then observes the status and history of each probe or follows execution of new commands and protocol.

DISCUSSION

Currently, more operating machines are based on liquid nitrogen storage (in a dewar) than the gas machines. Hence, examining the potential of these machines to be combined in the MRI environment is unavoidable. It seems impractical to position the machine inside the MRI chamber, but the outside option is not trouble free either. It would require maintaining flow of liquid nitrogen in 5 (to 7) parallel and isolated channels, each longer than 10 meters. The degraded thermodynamic efficiency of the transfer lines will necessitate a larger flow rate, dewar and driving system. Instead of a cryosurgical machine (which is not too compact in its usual version) one will get a much larger station of liquid nitrogen supply. Furthermore, the lines will become intermediate reservoirs of liquid nitrogen and it would become impossible to flow warming gas for fast thawing, as possible with proposed devices. Warming by electricity at the tip of each probe is less attractive.

The gas technology of cryocooling seems to be superior for this purpose. The Cryo-Hit™ machine (33) of Galil Medical, Ltd. brand proved this capability. It does operate successfully with the General Electric brand 0.5 T open MRI in Boston MA, Stanford CA, Jackson MI, Quebec Canada, Oslo Norway, London England, Leipzig Germany, Tel-Aviv Israel, with the Siemens brand 0.5 T in Mainz Germany, and with the Phillips brand 1.5 T interventional MRI in Aachen Germany.

REFERENCES

1. Kaufman, L., Arakawa, M., et al., Accessible magnetic resonance imaging, *Magn Reson Qurat,* Vol, 5, p.283-297, (1989).
2. Jolesz, F.A. and Blumenfeld, S.M., Interventional use of magnetic resonance imaging, *Magn Reson Qurat,* Vol, 10, p. 85-96, (1994).
3. Jolesz, F.A. and Silverma, S.G., Interventional magnetic resonance therapy, *Sem Interven Radiol,* Vol. 12, p. 20-27, (1995).
4. Lufkin, R.B. et al., Interventional MRI: update, *Eur Radiol,* Vol. 7, (Suppl. 5), p. 187-200, (1997).
5. Alexander, E., Moriarty,T.M., et al., The present and future role of intraoperative MRI in neurosurgical procedures, *Stereotact Func Neurosurg,* Vol. 68, p. 7-10, (1997).
6. Jolesz, F.A., Interventional and intraoperational MRI:: a general overview of the field", *J Magn Reson Imaging,* Vol. 8, No. 1, p. 3-7, (1998).
7. Rubinsky, et al., "Magnetic resonance imaging assisted cryosurgery, US Patent No. 5,706,810, Filed June 2, 1995, Patented June 13, 1998, The Regents of the University of California, Oakland, CA.
8. Rubinsky, B. et al., "Monitoring cryosurgery in the brain and in the prostate with proton NMR", *Cryobiology,* Vol. 30, pp. 191-199, April 1993.
9. Dumoulin, C.L., Souza, S.P. and Darrow, R.D., "Real time position monitoring of interventional devices using magnetic resonance", *Magn Reson Med,* Vol. 31, p. 309-314, (1994).
10. Leung, D.A., Debatin, J.F. and Wildermouth, S., "Real time biplanar needle tracking foe interventional MR procedures", *Radiology,* Vol. 197, p. 485-488, (1995).
11. Isoda, H., "Sequential MRI and CT monitoring in cryosurgery – an experimental study in rats", *Nippon Acta Radiologica,* Vol. 49, No. 12, p. 17, December 25, 1989.
12. Isoda, H., "Sequential MRI and CT monitoring in cryosurgery – an experimental study in polivinil alcohol gel phantom", *Nippon Acta Radiologica,* Vol. 49, No. 12, p. 6, December 25, 1989.
13. Parker, D.L. et al., Temperature distribution measurements in two dimensional NMR imaging", *Med Phys*, Vol. 10, p. 321-325, (1983).
14. Dickinson, R.J. et al., Measurment of changes in tissue temperature using MR imaging", *Journal of Computer Assisted Topography,* Vol. 10, p. 468-472, (1986).
15. Hong, J.S., Wong, S.T., et al., "MR imaging assisted temperature calculations during cryosurgery", *Magn. Reson. Imaging,* Vol. 12, p. 1021-1031, (1994).
16. Gilbert, J.C., Rubinsky, B., et al., Temperature determination in the frozen Rregion during cryosurgry of rabbit liver using MR liver analysis", *Magnetic Resonance Imaging,* Vol. 15, p. 657-667, (1997).
17. Parivar, F., Hricak, H., et al., Detection of locally current prostate cancer after cryosurgery: evaluation by transrectal ultrasound, magnetic imaging and three-

dimensional proton magnetic resonance imaging, *Urology,* Vol. 48, p. 594596, (1996).

18. Kurhanewicz, J., Vigneron, D.B. et al., Prostate Cancer: Metabolic Response to Cryosurgery as Detected with 3D H-1 MR Spectroscopic Imaging, *Radiology,* Vol. 200, p. 489-496, (1996).
19. Gilbert, J.C., Rubinsky, B., Ross, M.S. et al., MRI-monitored cryosurgery in rabbit Brain, *Magn Reson Imag,* Vol. 11, p. 1155-1164, (1993).
20. Matsumoto, R., Selig, A.M. et al., MR monitoring during cryotherapy of liver: predictability of histological outcome, *JMRI,* Vol. 3, p. 770-776, (1993).
21. Pease, G.R., Wong, S.T., et al., MR image-guided control of cryosurgery, *JMRI,* Vol. 5, p. 753-760, (1995).
22. Gilbert, J., Current Advances in interventional MRI-guided cryosurgery, Abstract at the *Proceedings of the Second Interventional MRI Symposium,* Dusseldorf, Germany, October 17-18, *Eur Radiol,* Vol. 7, p. 1165, (1997).
23. Daniel, B.L., Butts, K. and Block, W.F., Magnetic resonance imaging of frozen tissues-temperature development MR signal characteristics and Relevance for MR Monitoring of Cryosurgery", Abstract for the *Proceedings of the International Society for Magnetic Resonance in Medicine,* (1998).
24. Tacke, J., Adam, G. et al., MR-guided interstitial cryotherapy of the liver with a novel nitrogen-cooled cryoprobe", *Magnetic Resonance in Medicine,* Vol. 39, p. 354-360, (1998).
25. Itzhak, Y, Ravid, M. and Schatzberger, S., New cryogenic gas system for interventional MRI procedures, Abstract at the *Proceedings of the Second Interventional MRI Symposium,* Dusseldorf, Germany, October 17-18, (1997).
26. Jolesz, F.A. and Morrison, P.R. et al., Compatble instrumentation for intraoperative MRI: expanding resources, *J Magn Reson,* Vol. 8, No. 1, p. 8-11, (1998).
27. Maytal, B-Z., Fast Joule-Thomson CryoCycling Apparatus for Cryosurgical Applications, *Advances in Cryogenic Engineering,* Plenum Press, New York, Vol. 43A, p. 911-917, (1998).
28. Maytal, B-Z., Fast Changing Heating Cooling Device and Method Particularly for Cryogenic and/or Surgical Use, US Patent No. 5,522,870, Filed January 7, 1994, Patented June 4, 1996, and US Patent 5,702,435, Filed February 9, 1996, Patented December 30, 1997, Priority (both patents) January 25, 1993. Rafael, Haifa, Israel.
29. Maytal, B-Z., Controlled Cryogenic Contact System, US Patent No. 5,5540,062, Filed February 14, 1994, Patented July 30, 1996, and US Patent 5,540,062, Filed February 14, 1994, Patented July 30, 1996, Priority (both) November 1, 1993, Rafael, Haifa, Israel.
30. Maytal, B-Z., Noble Gases as Favorite Coolants for Joule-Thomson Cryostats, *Advances in Cryogenic Engineering,* Vol. 39B, pp. 1935-1941, Plenum Press, New York, (1994).
31. Maytal, B-Z., Multi-probe cryosurgical apparatus, US Patent No. 5,603,221, Filed June 29, 1995, Patented February 18, 1997, Priority January 30, 1994, Rafael, Haifa, Israel..
32. Maytal, B-Z. et al., System and method for MRI guided cryosurgery, European Patent No. EP-927,542, Filed January 4, 1999, Patented July 7, 1999, Priority January 5, 1998, Galil Medical, Ltd., Yokneam, Israel.
33. Galil Medical, Ltd., P.O.Box 224, Yokneam, Israel.

IMPROVEMENT OF RF-SYSTEM TO MEASURE VOID FRACTION OF CRYOGENS

Yu.P. Filippov, A.M. Kovrizhnykh, and S.V. Romanov

Joint Institute for Nuclear Research, Dubna, Russia

ABSTRACT

This work continues the activity to determine the thermodynamic state of both vapor-liquid and single phase cryogens taking into account the pressure or temperature and dielectric permittivity ε. To find the value of ε depending on void fraction, φ, one can use the radio frequency (RF) method when an investigated medium fills a volume whose resonant frequency is a function of ε(φ). It is clear that a reading of the sensor depends on the ε-values of both the phases and flow pattern as well. The technique enabling to measure void fraction of the flow in vertical channels was presented at ICEC17. The sensor with a uniform electric field with the constant direction within the measuring volume was used in that case. This paper presents a new version of the RF sensor with a round cross-section for asymmetrical flows which can be observed in horizontal channels. The key feature of the sensor is that its sensitive part is fabricated by means of a twisted meander line formed on a dielectric tube, and this sensor can be used to determine an integral φ-characteristic of the flow. The feature of the RF measuring system is the use of a method based on a sweep generator and subsequent detecting the signal of the sensor, its digital transformation and mathematical processing. Such an approach allows one to provide a relatively small error of the measurement even at room temperature when RF-quality of the sensor is relatively low. The operation characteristics and test results for the developed sensor and measuring device are reported.

INTRODUCTION

Monitoring of multiphase forced flows is an important topic in both cryogenic and room temperature applications. Thus, in cryogenics where two-phase cooling flows are used, it is necessary to determine the vapor content and flow rate, and at ambient temperatures the oil content in compressed gas, for example, or the level of mechanical impurities in the gas derived from absorbents, etc. [3] must be established. The underlying principle, common for these examples, is that the control can be performed using the capacitance method, where measurements are carried out by taking into account the difference in the dielectric permittivity, ε, of the components. The direct measurement of the capacitance dependence on the ε value is rather complicated. So, to find $\varepsilon(\varphi)$, where φ is the void fraction of the flow, we have used a more convenient radio frequency method when the investigated medium fills a special cavity and the characteristics of this medium are defined from measurements of the corresponding resonant frequencies and qualities.[1,2,3] In this case the accuracy of the measurement of the capacitance characteristic is of principal importance.

The radio frequency sensor may be characterized by the value of the signal characteristic $H = (df_{tp}/d\varphi)/f_{tp} = \alpha(\varphi)\Delta\varepsilon/2$, where $\Delta\varepsilon$ is the difference in the dielectric permittivity of the liquid and vapor phases, and f_{tp} is the resonant frequency of the sensor at a given value of ε_{tp}. The geometric factor α [2] significantly depends on the configuration of the sensor flow part (round or annular cross-section) and it is always less than 1. As shown in Reference 2, the relation $\alpha_{annular}/\alpha_{round} \approx 3...4$ if the mass of the dielectric material around the meander line is minimized. The main characteristics of the sensors depend, in particular, on the value of H. For example, the so-called instrumental error [2] is $\delta\varphi_{inst} = 2(\delta f_{inst})/Hf$, where $\delta\varphi_{inst}$ is the accuracy of the measuring device and δf_{inst} corresponds to the measured resonant frequency f_{tp}. The value of f_{tp}, in turn, depends on the geometry of the sensor. For example, in the first approximation, f_{tp} is inversely proportional to the diameter of the channel for the sensor with a round cross-section.

The characteristics of the cryogenic sensors have already been discussed in greater details in References 1,2. One can note that in these papers the main attention was paid to the sensors for research purposes. This report analyses the design and test results of new sensors and measuring device for industrial applications. A similar sensor of a round cross-section with the inner diameter of 71 mm was described in Reference 5. It was intended for the vertical symmetrical two-phase hydrogen flow at a pressure of about 2 MPa. It is clear that a reading of the sensor depends not only on the ε-values for both phases of the cryogen but also on the flow pattern. Besides, the factor α for the industrial sensor is approximately two times smaller with respect to the sensors for research purposes due to its design features. So it requires a higher accuracy of measurement. In this work the main attention was paid to solving the technical problems for the sensor with asymmetrical flow which can be observed in horizontal channels, and to the accuracy of the measurements of the resonant frequency at different temperatures from 1.6 K to 300 K, which depends significantly on the characteristics of the resonant curve, RF-quality and amplitude.

RF-SENSOR FOR HORIZONTAL FLOW

For adequate measurements of the void fraction, φ, one needs a device where the electric field distribution in the sensitive element is close to uniform [1,2] that is valid for all orientations of the flows. As noted above, the way to measure the value of φ in the vertical pipe is presented in Reference 5. The sensitive part is fabricated by means of the parallel (with respect to the axis) straight meander line located on the outer surface of the dielectric tube. [1,2] When resonant oscillations on the first mode are excited in the meander line of the sensor, the potential distribution over the dielectric tube is close to that of the harmonic. If the sensor is comparatively long and the meander pitch is sufficiently small, high electric field uniformity is ensured in the measuring volume. This design allows one to avoid any distortion of the flow pattern within the sensitive part of the sensor and the corresponding influence on the velocity profile, slip ratio, and sensor reading. Even though, there are some difficulties caused by the fact that even the ideally uniform field is distorted by a non-uniform dielectric, and the resonant frequency, f, depends on the orientation of the phase boundaries with respect to the direction of the electric field. This is best manifested in a stratified flow or a flow close to the stratified one (wavy pattern, for example).

There are two ways to decrease the corresponding error. The first is to use several pairs of communication rods [1] shown in Figure 1. In this case one can apply the same meander line as in the design used earlier. [1,2,5] The resulting value of φ is calculated as a mean $\varphi = \varphi\ (\ \varphi_1,\ \ \varphi_2,\ \dots\ \varphi_n)$ of the φ_i values obtained for each pair of the rods. The advantage of this approach is that it allows one to find not only the mean value of void fraction but allows to develop a device to investigate the internal flow patterns in a round tube. Thus, analyzing the first harmonic frequencies f_{0n} of the sensor for different pairs of the rods and similar frequency sets, f_{in}, f_{mn}, of higher harmonics [1], one can reconstruct the two-dimensional flow pattern of the phase distribution in the flow by using mathematical methods of tomography. This design of a «dielectric tomographer» resembles the «flow image» capacitance device presented in Reference 4. The disadvantage of this approach is that it requires a relatively large amount of RF-cables entering the cold volume. The second approach is to use a twisted meander line that is illustrated in Figure 2. In the sensor with this design, most of the errors are automatically corrected. An additional practical advantage of this sensor is the use of only one pair of communication rods. However, the twisted meander sensor can be mainly used to determine integral characteristics.

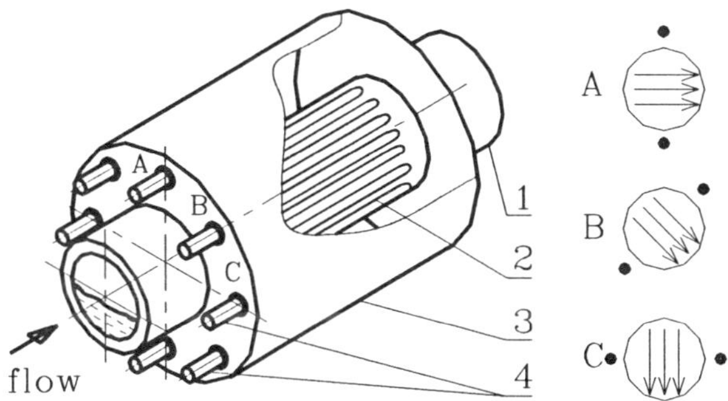

Figure 1. Sensitive part of sensor with several pairs of communication rods and directions of the electric field inside. 1 - dielectric tube, 2 - straight meander line, 3 - shield, 4 - communication rod.

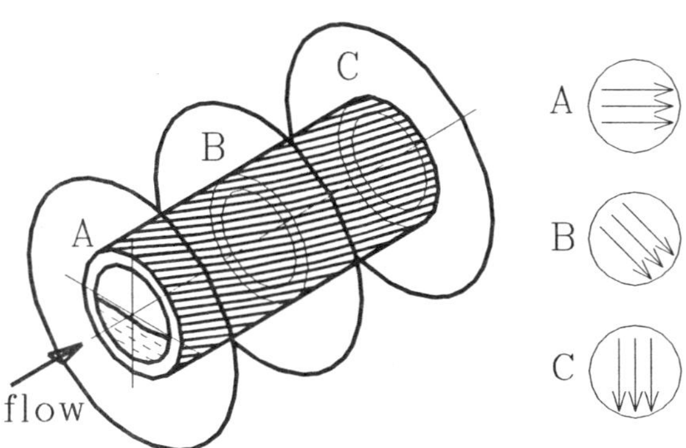

Figure 2. Sensitive part of sensor with a twisted meander line and directions of the electric field inside.

During fabrication the sensor models with the twisted and straight meander lines particular attention was paid to their reproducibility, which depends on the characteristics of the pair of materials for the metal body and dielectric tubes. To fabricate the dielectric tubes the same material was used as for the hydrogen sensor mentioned above.[5] The previous tests have shown that the difference in thermal contraction coefficients for the dielectric tubes and stainless steel pipe does not exceed 1 % in the range from 80 K to 300 K. It was revealed that this pair also works successfully down to 1.6 K. Two model sensitive parts were fabricated: their inner diameters are 57 mm and 70 mm. Each model consists of two parts having a common meander line. For the model with the inner diameter of 57 mm, for example, the length of the first part with the twisted meander line is 70 mm and the length of the part with the straight meander line is 100 mm. These sensor models are intended for the tests at room temperature. In addition, the sensor with the straight meander line, inner diameter of 70 mm and 240 mm length, was designed and fabricated for the tests down to 1.6 K.

MEASURING SYSTEM

In principle, one can arrange the measurement of the resonant frequency in two ways: first is a scanning method based on application of the sweep generator and subsequent signal detection from the sensor and its digital transformation and corresponding mathematical processing, and second, the well known tracking method when the frequency of the sweep generator is maintained within some range of the sensor resonant frequency by means of the tracking system with a feedback to the sensor. Details of both approaches are considered below.

Tracking method. The structural schematic of the radio frequency board with the usage of the tracking feedback was described elsewhere.[1,2] To determine the resonant frequency, the sweep generator sends a signal to the sensor. From the sensor the signal goes to the extreme regulator which controls the sweep generator and maintains the resonant frequency. The sensor is connected with the measuring board through the detector which includes the input preamplifier, detector, and buffer amplifier to provide its operation with a comparatively long cable if necessary. The modification with the removed detector is used in the system[5] to measure void fraction of the two-phase hydrogen flow. The distance between the RF-board and removed detector located near the cryostat, can be as large as 30 m. The advantage of this approach is that it allows one to measure the resonant frequency of the sensor, and the reliability of measurements depends mainly on the sensor errors. The disadvantage is the relative complexity and the necessity to retune the RF-board for a specific sensor.

Scanning method. One can realize the so-called scanning method of a «direct action» illustrated in Figure 3. According to this version, the frequency is stored in the registers of the direct digital synthesizer, DDS, generating the given frequency. After that, the amplitude corresponding to the stored frequency is measured. Thus, if several measurements are made at different frequencies, one can obtain the resonant curve and determine the value of the sought resonant frequency. The advantage of this method is that the electronics allows one to measure the amplitude of the signal with the resolution about $\delta f = 1/2^{16}$ and to adjust the frequency with the accuracy about $125/2^{24}$ MHz which is more than redundant for reasonable technical requirements. The disadvantage of this method is that its speed is comparatively low because of the need to make a relatively large number of measurements (about 10^5) and the subsequent mathematical processing to yield the result.

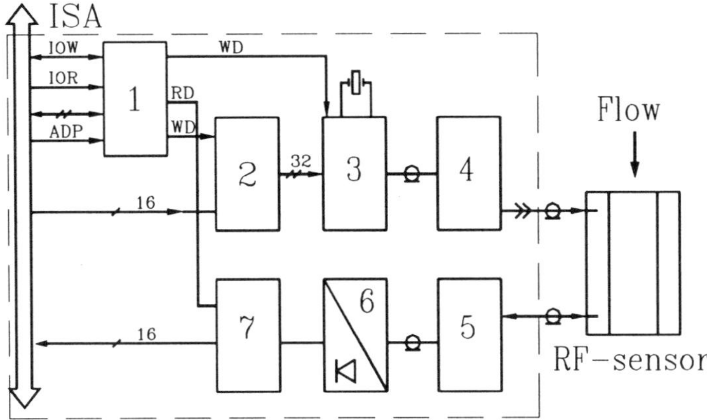

Figure 3. Structural schematic of the radio frequency board to realize the scanning method of measuring the resonant frequency of the RF-sensor. 1 - interface, 2 - register, 3 - DDS, 4 and 5 - amplifiers, 6 - linear RF-detector, 7 - ADC.

Structural schematic of the measuring device. The modular industrial computer MIC2000 based on the ISA-bus and micro PC, was used to construct the measuring device. Its structural schematic is shown in Figure 4. Usage of the cross-bus with 8 slots, allows one to enlarge the number of measuring channels and add complementary possibilities for other measurements, for example, temperature, pressure, etc. There are two operating regimes: stand-alone, and with the removed PC.

The developed software provides the following: [5] an internal self-test of the device; calculation of the dielectric permittivity and void fraction of the chosen cryogen, for example helium, using the measured resonant frequency; calculation of the necessary parameters of cryogen via dielectric permittivity and temperature (or pressure) using the cryogen state equation; measurement of the temperature; reading the user input from the keyboard, and translating the input data into internal commands whose validity is verified; reading the input from the interface (GPIB or Ethernet); and displaying the results of measurements and corresponding calculations.

RESULTS

First tests of the fabricated sensors and measuring RF-boards were carried out at room temperature. To simulate the reading of the sensor with inner diameter of 57 mm while filling it with gas, liquid helium and liquid methane, air and samples of foam-plastic with $\varepsilon_l \approx$ 1.1 and teflon ($\varepsilon_l \approx 2.2$) were used. The shapes of the plastic samples of 56 mm in diameter are: cylinders 35 and 70 mm long and half-cylinders of the same length to simulate the stratified helium flow. The results are as follows: the resonant frequency and RF quality of the sensor filled with air ($\varepsilon \approx 1$) are $f_g \approx 49.594$ MHz and 42, respectively; $f_l \approx 49.500$ MHz – with the 70 mm foam-plastic cylinder of round cross-section. The tests have shown that the values of f_l do not practically depend on the position of the foam-plastic samples within both sensitive volumes: the found maximum deviations of δf_l do not exceed 1 kHz. It is valid also for the half-cylinder (l = 70 mm) located in the horizontal and vertical positions with respect to the direction of the electric field. It allows one to come to the conclusion that there is no need to fabricate a twisted meander line for helium RF sensors. The results for the half-cylinder with $\varepsilon_l \approx 2.2$, l = 50 mm, located vertically and horizontally, have shown

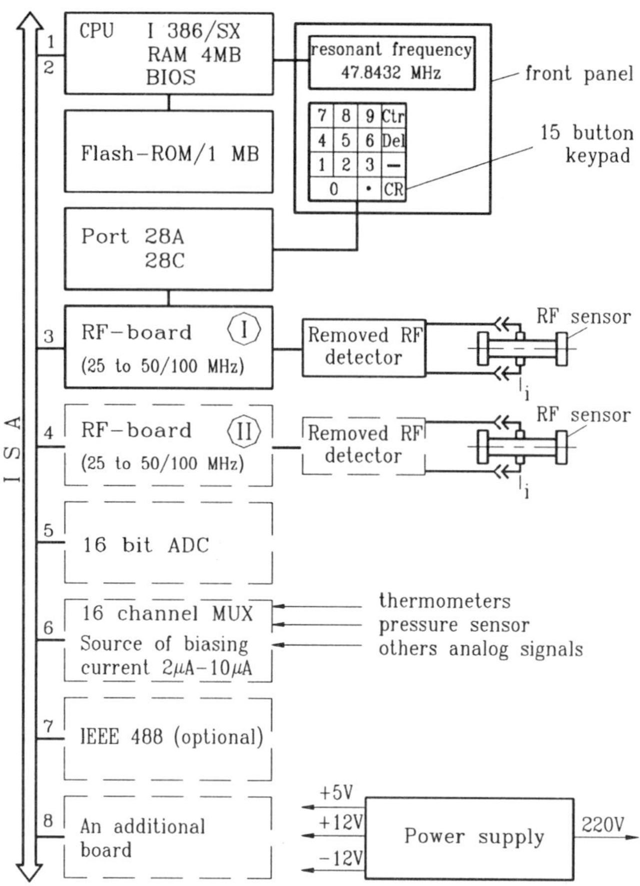

Figure 4. Structural schematic of the measuring device based on modular industrial computer MIC2000.

that the difference, δf_l, may be about 2-3 kHz in the twisted part, however this value can reach 10-12 kHz in the part with the straight meander. This confirms the usefulness of the twisted configuration of the sensitive part intended for cryogens with $\varepsilon_l \approx 1.4 - 1.6$.

Experiments with the sensor of 70 mm inner diameter were carried out at room and cryogenic temperatures using both methods to measure the resonant frequency. The main parameters of this sensor and characteristics of both RF-boards are presented in Table 1.One can see that at cryogenic temperatures both RF-boards operate successfully due to relatively high values of the RF-quality of the sensor. However, it is preferable to use the tracking method of measurement since its cost is lower, and its speed is higher which is important for practical applications. In this case even working with helium one can provide $\delta\varphi_{inst} = 2\delta f_{inst}100/(f_g-f_l) \approx 0.6$ %. At room temperature the use of the scanning method is preferred. Especially it is valid for the case when the sensor is filled with a medium having large dielectric permittivity. Thus, it was impossible to use the RF-board for the tracking method while working with water at low signal amplitude and RF-quality of the sensor. The scanning method could ensure $\delta\varphi_{inst} < 0.1\%$ for measuring the water content in the mixture consisted of oil and water. Finally, the scanning method allows one to reach excellent results in performing the calibration procedures.

Table 1. Characteristics of the RF-sensor, ID = 70 mm, filled with different media

Medium	Temperature, K	Frequency, MHz	Amplitude, unit-less	RF-quality	Accuracy, kHz Tracking/Scanning
Water, 50%	300	26.478	0.3	20	-- / 2.0
Oil, 50 %	300	29.702	0.7	39	1.0 / 0.2
Air	300	30.517	1	40	1.0 / 0.2
Nitrogen					
gas	77.3	30.900	4	140	0.5 / 0.1
liquid	77.3	30.210			
Helium					
gas	4.2	31.184	10	240	0.3 / 0.1
liquid	4.2	31.081			
Helium					
gas	1.6	31.190	10	240	0.3 /0.1

CONCLUSIONS

The design and technology developed allow one to fabricate industrial RF-sensors with a round cross-section of 30 - 100 mm in diameter to measure void fraction of different dielectric media in the range of temperatures from 1.6 K to 300 K, pressures up to 10 MPa and higher without distortion of the flow structure within sensitive parts.

The use of the twisted meander line for the sensitive part of the RF-sensor allows one to decrease the error due to influence of the flow pattern on the reading of the sensor and to determine an integral characteristic of the investigated flow which is especially important for the stratified flow in horizontal channels. In the meantime, there is no need to carried out the twisted meander line for helium RF-sensors because of its comparatively low dielectric permittivity even for the liquid phase ($\varepsilon_l \approx 1.05$ at T<1.8 K).

While measuring the signal of the RF-sensor, the use of the tracking method (based on the fact that the frequency of the sweep generator is maintained within certain range of the sensor resonant frequency using the tracking system with a feedback to the sensor) allows one to provide sufficiently high accuracy, $\delta f \leq 0.5$ kHz at $f \approx 30 - 50$ MHz, for all cryogens.

The use of the scanning way to measure the signal of the RF-sensor (based on application of the sweep generator and subsequently detecting the signal of the sensor and its digital transformation and corresponding mathematical processing) allows one to provide higher accuracy of measurements, $\delta f \leq 0.2$ kHz @ $f \approx 30 - 50$ MHz, at room temperatures even for medium with a relatively large value of dielectric permittivity, for example, oil ($\varepsilon_l \approx 2$). This way can be applied in the oil industry for measuring moisture content of the oil/water mixture and during calibration with cryogens.

REFERENCES

1. Yu.P. Filippov, A.I. Alexeyev, N.I. Lebedev, I.S. Mamedov, and S.V. Romanov, Monitoring of cryogenic flows: realization of the radio frequency method, in: «Advances in Cryogenic Engineering,» R.W. Fast, ed., Vol. 37 B, Plenum Press, New York (1991), p. 1461.
2. A.I. Alexeyev, Yu.P. Filippov, I.S. Mamedov and S.V. Romanov, Application of radio frequency method to measurements in cryogenics, *Cryogenics* 31:1020 (1991).

3. Yu.P. Filippov, A.I. Alexeyev, A.V. Ivanov, and B.V. Lunkin, Expanding application of cryogenic void fraction sensors, *Cryogenics* 33:828 (1993).
4. S.M. Huang, A.B. Plaskowski, C.C. Xie, and M.S. Beck, *J. Phys. E.: Sci. Instrum.* 3:173 (1989).
5. Yu.P. Filippov, A.M. Kovrizhnykh, and S.V. Romanov, RF Sytem to measure void fraction of two-phase hydrogen flow, in: Proceeding of the 17 International Cryogenic Engineering Conference, D. Dew-Hughes, R.G. Scurlock, and J.H.P. Watson, ed., Institute of Physics Publishing, Bristol and Philadelphia (1998), p. 767.

MEASURING THERMAL CONDUCTIVITY OF INSULATORS AT CRYOGENIC TEMPERATURES

B.P.M. Helvensteijn[1], L.J. Salerno[2], P.R. Roach[2], P. Kittel[2] and S.M. White[2]

[1] Atlas Scientific
Sunnyvale, CA 94086

[2] NASA Ames Research Center
Moffett Field, CA 94035-1000

ABSTRACT

Traditionally, launch vehicles have consisted of a structure containing the vehicle engines, propellant tanks, and electronics. NASA's next generation of space transportation vehicles is planned to utilize the innovative concept of conformal propellant tanks. This concept makes the tanks part of the vehicle structure, thereby, offering significant reduction in launch mass. Key to successful implementation of this approach will be the development of a new material that serves not only as cryogenic insulation but also as a heat shield, protecting the vehicle. This material must be precisely characterized at both elevated and cryogenic temperatures. The present paper discusses the apparatus set up to establish the thermal conductivity of candidate materials at low temperatures. The apparatus enables cryogenic heat flow measurements in the presence of a gas. A finite difference model, accounting for the system configuration, is employed to convert the heat flow measurements into thermal conductivity as a function of temperature.

INTRODUCTION

One of the most significant factors in the development of any spacecraft is the reduction of launch mass. Reusable vehicles are especially tied to this precept considering that with each flight fuel may be saved and other expenditures reduced. Future space exploration will rely increasingly on reusable launch vehicles making it the more relevant to design low launch mass as an embedded feature into the next generation launch vehicles.

The traditional launch vehicle consists of an overall external shell with an internal structure supporting the various modules. The propulsion system, propellant tanks and payload are all attached to the basic vehicle structure, each item adding mass with its individual container and support struts. Significant mass reduction becomes possible when each sizable item on board the spacecraft takes on multiple functions. The largest items on board are the propellant tanks, the basic vehicle structure and the envelope. This implies that the greatest launch mass reduction may be achieved by merging the functionalities of these major components of the spacecraft. The next generation spacecraft will apply this concept by employing conformal tanks in which the propellant tanks constitute the main structure of the vehicle.

Advances in Cryogenic Engineering, Volume 45.
Edited by Shu *et al.*, Kluwer Academic / Plenum Publishers, 2000.

The great promise of conformal design may be realized by researching the various aspects involved in the actualization of this concept. The aspect we intend to investigate is the functionality as a cryogenic insulator of the thermal protection system (TPS) that is to be attached to the outside of the cryogenic propellant tanks, in particular the slush hydrogen tank. Given the conformal design concept, the TPS performs the dual function of insulating the cryogenic tanks from the open-air environment, preventing excessive boil off, and of protecting the spacecraft as a thermal shield during re-entry of the earth's atmosphere. Presently, several candidate materials have been identified while other potentially suitable materials are yet in the development stages.

Characterizing the prospective materials as a cryogenic insulator involves two major issues: 1) thermal conduction from the open-air environment through the vast TPS insulating blanket to the tank wall; 2) adsorption of air, water and other substances from the environment. The present paper concentrates solely on the thermal conduction aspect of the TPS materials' research. The design of the apparatus built to conduct the thermal conduction measurements will be discussed. The method chosen to analyze the data will be illustrated. Then the results of initial tests on a sample insulator of plexiglass (acrylic) will be presented and compared with available data.

EXPERIMENTAL SETUP

The purpose of the TPS is to guard the cryogen tanks both against destruction, due to re-entry heating, and excessive loss of cryogen, due to heat exchange with the atmosphere. Clearly, this requires excellent insulation qualities. Any material with potential will exhibit large temperature drops even at low heating powers. Therefore, in order not to disturb the thermal conductivity experiments the apparatus has been designed to minimize parasitic heat leaks into the sample material. Considering that in actual use the TPS will be exposed to the open air, the setup is designed so that the samples may be engulfed in gas without introducing heat exchange that cannot be accounted for.

The configuration chosen in order to reduce parasitics including convection heat transfer is that of an undivided isothermal enclosure around the entire test sample. Having the sample completely surrounded by an isothermal shell of known temperature eliminates all concerns about potential heat exchange with any external substance or object. An isothermal hot plate with an electrical heater is placed in the center of the sample. Hence, all heat that is applied to the hot plate must flow through the sample once the temperature differential between the isothermal surfaces has stabilized. For this particular configuration, the presence of gas will not sway the thermal conductivity measurements but will show changes in the sample's thermal conductivity with gas pressure if they occur.

A sketch of the thermal conduction apparatus (TCA) is shown in Figure 1. The isothermal sample enclosure is made out of 6 mm (1/4") thick OFHC copper. The side and bottom plates are all bolted together forming an isothermal box around the sample while the top plate is simply resting on top of the sample. Flexible 40 mesh copper straps connect the top plate to the box ensure thermal equilibrium. The hot plate is equipped with a Si diode thermometer, an RTD (Pt thermometer) and an electrical heater. These items are each mounted inside an individual mounting disc that mounts wholly inside the hot plate. The box is equipped with a Si diode thermometer and an electrical heater. The hot plate heater is used to apply heat for the thermal conductivity measurements; the box heater assists in raising the overall temperature. The copper box is heat sunk to and suspended inside an isothermal (OFHC copper) vacuum tight sample chamber. This chamber is placed inside the vacuum can and equipped with a heater and a Si diode thermometer. This arrangement allows variation of the temperature of the sample chamber and its contents well above that of the cryogen bath surrounding the vacuum can (LN2 or LHe). Two different size square boxes plus two cylindrical boxes each with a corresponding hot plate have been constructed.

In order to connect the instrumentation on the hot plate to a terminal outside the box, 4 copper and 16 manganin wires are drawn, each 0.8 m long and 0.13 mm in diameter. These wires run in a spiral through the sample from the hot plate to a terminal just outside the box.

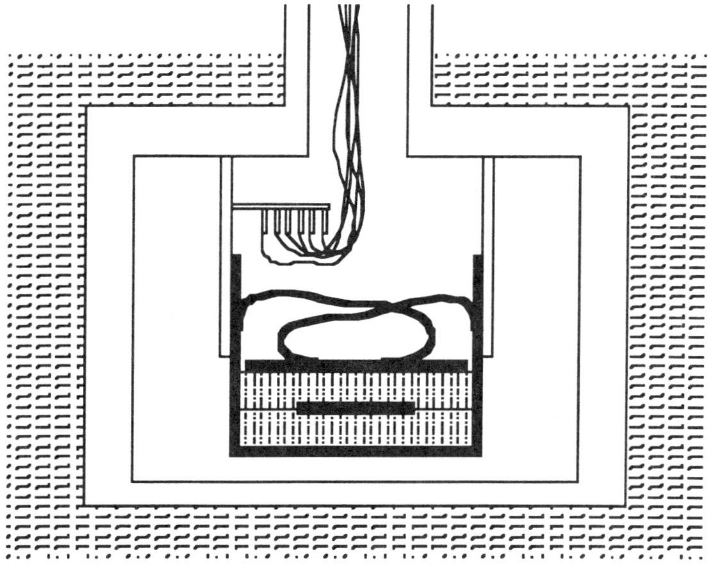

Figure 1. Schematic of the test setup.

The first test sample inserted into the TCA has been a set of 2 identical square plates of plexiglass (acrylic). The center of each plexiglass plate is machined to make room for half the thickness of the hot plate. The hot plate measures 0.9 x 7.6 x 7.6 cm^3; the inside of the box touching the sandwiched plexiglass plates is 3.3 x 15.2 x 15.2 cm^3.

THERMAL ANALYSIS

The experimental setup has the advantage of eliminating uncertainties regarding any heat loss to unintended heat sinks even in the presence of gas. However, the square box geometry does not lend itself to a direct conversion of the test data into thermal conductivity by means of a single closed formula. Our solution to this problem is illustrated in Figure 2. Shown is a quarter segment of a square two-dimensional geometry. The isothermal center disc has a hypothetical 1 W heat load and is assumed to be at 200K; the isothermal perimeter is at 100K. All internal nodes, such as the one labeled "0", are at some intermediate temperature. However, there is no exact formula by which the temperature of this node, or any other node, can be established. Instead, the temperatures at the internal nodes may be approximated iteratively. The following discusses the Relaxation or Gauss/Seidel method[1] (R/GS) used in this paper to solve the nodal temperatures for steady state heat flow.

Assumed are two isothermal boundaries (heavy lines in Fig. 2) to the medium that conducts the heat. The medium is represented by the square array of nodes. Initial values are assigned to the temperatures at all the nodes, i.e. all initial temperatures assigned to the internal nodes are merely educated guesses. The only error free nodes are on the isothermal boundaries. In steady state the heat flow into all internal nodes cancels, but for the assigned values the heat flow does not cancel at all nodes. For a square geometry of equally spaced nodes, we would find for instance for node "0":

$$\dot{q}_0 = (k A / L)(T_1+T_2+T_3+T_4 - 4T_0) \neq 0 \qquad (1)$$

Adjusting T_0 such that $\dot{q}_0 = 0$, we find a new value for T_0:

$$T_0' = T_0 + \Delta T_0 = (T_1+T_2+T_3+T_4) / 4 \qquad (2)$$

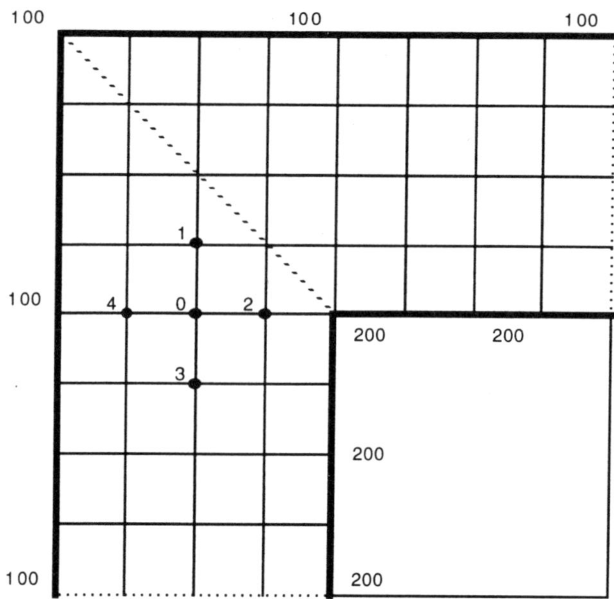

Figure 2. Arrangement of nodes for a two-dimensional square geometry. At convergence the net heat flow into each node internal to the insulator (e.g. node "0") cancels.

ΔT_0 is one adjustment out of the many needed for one full iteration in which all internal temperatures are adjusted (the boundary temperatures stay constant). Note that each iteration does not reflect a physical change but merely alters our guesses of the internal temperature profile. In each iteration errors in the initial assumptions are further divided out while error-free contributions from the isothermal boundaries permeate into the internal array of nodes.

A single iteration consists of applying Eqn. (2) to all internal nodes consecutively. The iterative application of Eqn. (2) results in an ever changing but converging temperature profile. After convergence a small error in all internal temperatures remains that is nearly the same at all the internal nodes. This implies an inherent difference in the thermal conductivity derived from Eqn. (1) when the converged temperature profile is applied to assess the heat flow out of, repectively into the insulator. This difference is a measure of the accuracy of the calculation. If the error is too large then one needs to refine the grid, i.e. add nodes.

The convergence for the R/GS method is slow when the node spacing becomes small, in particular for three-dimensional configurations as is the case here. Despite the slow convergence for small node spacing we believe this method yields a more accurate solution to the nodal equations than the direct inversion of a large matrix. A slight gain in speed is achieved by starting the calculations on the diagonals where the initial guesses have the greatest errors. Further, our method takes benefit from the square box geometry, implying mirror symmetry across various planes (x=0; y=0; z=0; x=y; dashed lines in Fig. 2). The symmetries eliminate the need to calculate the temperature profile throughout the whole sample and in fact reduces the field of computation to a triangular section $1/16^{th}$ the size of the insulator. In order to reduce the need for very fine node spacing, we have used a hybrid method of successive refinement of the nodal grid and extrapolation of the results to an infinite number of nodes. The graph in Fig. 3 shows the converged computational results with the extrapolation. Shown is the computed ratio of heat flow to thermal conductivity as well as the thermal conductivity versus the number of nodes in the positive x, y and z directions. A hypothetical 1 W heat load is assumed for a 100 K temperature differential.

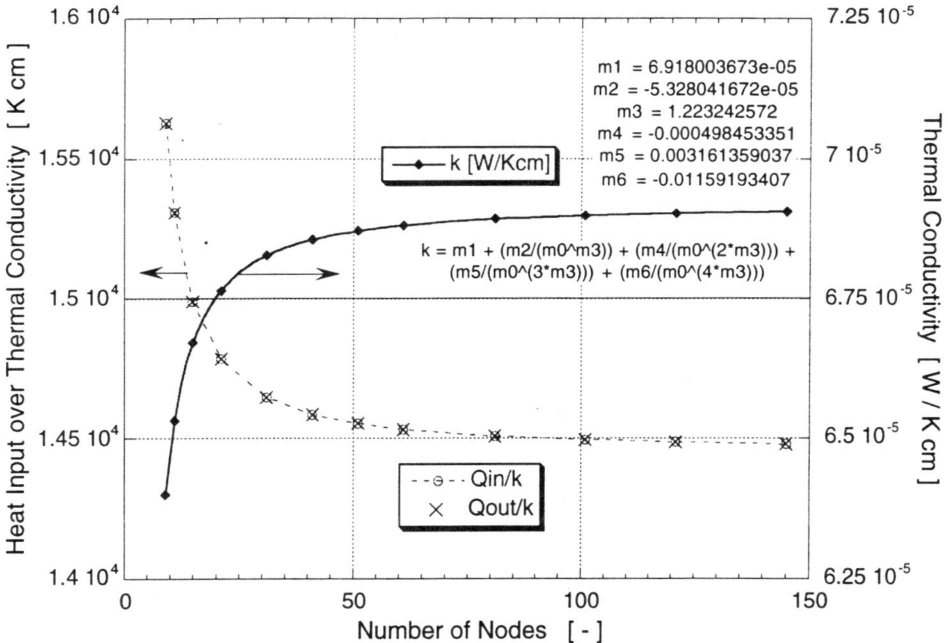

Figure 3. Asymptotic approach of nodal spacing refinement. Note that the heat flow (1 W) and temperature drop (100 K) used to calculate these data are hypothetical.

The R / GS method has been applied for the square box geometry defined above. The extrapolation of the computational results, applying a power series as shown in Fig. 3, gives for the thermal conductivity: $k_{R/GS} = 6.918*10^{-5}$ W/Km. The measured temperature drop and heat load are likely to be different from these values, however, the linear relation in Eqn. (1) allows factoring in the actual data as follows:

$$\frac{k_{actual}}{k_{R/GS}} = \frac{\left(\dot{q}_{actual}/\Delta T_{actual}\right)}{\left(\dot{q}_{R/GS}/\Delta T_{R/GS}\right)} \tag{3}$$

This equation is applied to the experimental data to arrive at the thermal conductivity of our test sample.

RESULTS AND DISCUSSION

The experimental measurements determine the actual ratio of heat flow to temperature drop for the plexiglass sample. Comparing this ratio with the values used in the computer model gives the actual thermal conductivity by virtue of Eqn. (3). The experiments have been conducted with the sample chamber at 290 K, 78 K and 4.2 K. The results are depicted in Figure 4 together with reference data[2]. Smooth lines are drawn between the data points to guide the eye. The reference data set has significant scatter. Our measurements fall within the same range.

The measured thermal conductivity data deviate from the reference values by up to 20%. This discrepancy is larger than expected but is on the same order as the scatter in the reference data. The discrepancy cannot be due to parasitic heat leaks. The only heat flow path in parallel with the sample is formed by the instrumentation wires. The wires can only account for about 0.4 mW out of the experimental 0.3 W heater power that corresponds to a

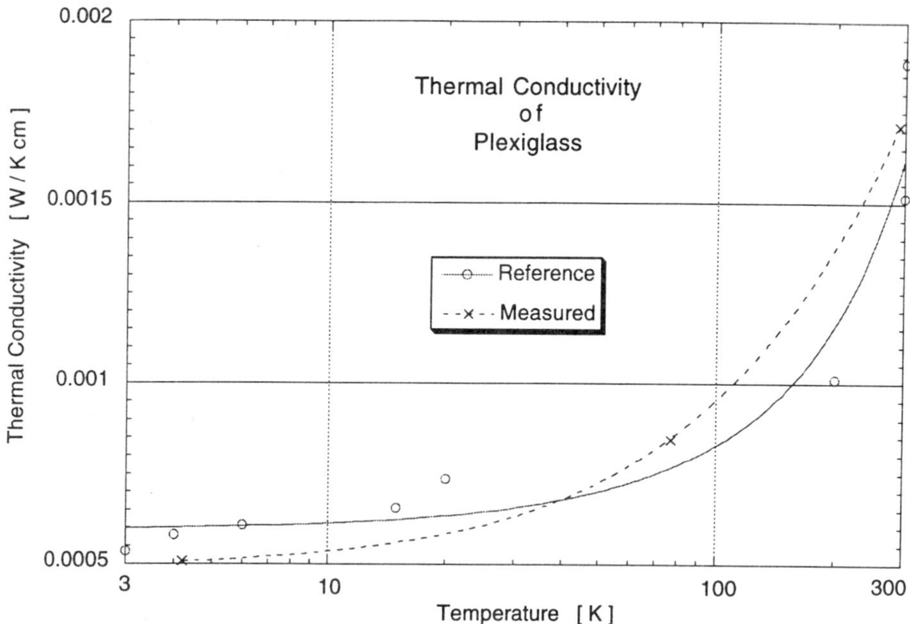

Figure 4. Comparison of experimental results with reference data.

10 K hot plate and a 5 K box. However, the calculations do not consider any potential temperature drop at the walls between the sample and the neighboring hardware. A calculation based on Chan's work[3] demonstrates that a temperature drop of about 0.5 Kcm^{-1} may occur at the inner boundary. The roughly 3 times larger surface area at the external boundary reduces the external boundary temperature differential to about 0.2 Kcm^{-1}. In these gas gap calculations it is assumed that there is sufficient helium gas in the sample chamber so that the gas in the gaps between the copper plates and the sample is in the continuum regime. To satisfy this condition sufficient helium gas has been supplied to the sample chamber in all tests. Such has been verified by observing the temperature drop between the hot plate and the box versus the helium gas pressure. Estimating the gaps to be on the order of 0.02 cm or less, the gas gap boundary resistances account in total for less than 0.02 K temperature drop, that is 0.4% of the measured differential. Thus, also this factor is inconsequential for explaining the discrepancies between the data sets. Still, in order to verify that the heat flow through the gas layer at the boundaries is indeed sufficient, thermometers will be embedded just inside the sample surface in future experiments.

CONCLUSIONS

The comparison between our experimental results and the reference data verifies that the TCA is suitable for measuring thermal conduction of highly insulating materials. The Relaxation / Gauss-Seidel method of data analysis has proven to be adequate for the task. The results on the test sample agree with the reference data.

REFERENCES

1. Holman, J.P., "Heat Transfer", McGraw-Hill Book Company, 4th Ed. (1976) p 65, 82.
2. "Thermophysical Properties of Matter".
3. Chan, C.K., "Gas Adsorption/Absorption Heat Switch", JPL Publication 87-7 (1987).

NOMENCLATURE

k thermal conductivity

A heat flow cross-sectional area

L length of heat flow path

ΔT temperature drop

q heat flow .

AUTHOR INDEX

SUBJECT INDEX